NFPA 70®

National Electrical Code®

2017 Edition

This edition of *NFPA 70, National Electrical Code,* was prepared by the National Electrical Code Committee and acted on by NFPA at its June Association Technical Meeting held June 13–16, 2016, in Las Vegas, NV. It was issued by the Standards Council on August 4, 2016, with an effective date of August 24, 2016, and supersedes all previous editions.

This document has been amended by one or more Tentative Interim Amendments (TIAs) and/or Errata. See "Codes & Standards" at www.nfpa.org for more information.

This edition of *NFPA 70* was approved as an American National Standard on August 24, 2016.

History and Development of the *National Electrical Code* ®

The National Fire Protection Association has acted as sponsor of the *National Electrical Code* since 1911. The original Code document was developed in 1897 as a result of the united efforts of various insurance, electrical, architectural, and allied interests.

In accordance with the Regulations Governing the Development of NFPA Standards, a National Electrical Code First Draft Report containing proposed amendments to the 2014 National Electrical Code was published by NFPA in July 2015. This report recorded the actions of the various Code-Making Panels and the Correlating Committee of the National Electrical Code Committee on each public input and first revision that had been made to revise the 2014 Code. The report was published at www.nfpa.org/70. Following the close of the public comment period, the Code-Making Panels met, acted on each comment, and created some second revisions, which were reported to the Correlating Committee. NFPA published the National Electrical Code Second Draft Report in April 2016, which recorded the actions of the Code-Making Panels and the Correlating Committee on each public comment on the National Electrical Code Committee First Draft Report. The National Electrical Code First Draft Report and the National Electrical Code Second Draft Report were presented to the 2016 June Association Technical Meeting for adoption.

NFPA has an Electrical Section that provides particular opportunity for NFPA members interested in electrical safety to become better informed and to contribute to the development of the National Electrical Code and other NFPA electrical standards. At the Electrical Section Codes and Standards Review Session held at the 2016 NFPA Conference and Expo, Section members had the opportunity to discuss and review the report of the National Electrical Code Committee prior to the adoption of this edition of the Code by the Association at its 2016 June Technical Session.

This 54th edition supersedes all other previous editions, supplements, and printings dated 1897, 1899, 1901, 1903, 1904, 1905, 1907, 1909, 1911, 1913, 1915, 1918, 1920, 1923, 1925, 1926, 1928, 1930, 1931, 1933, 1935, 1937, 1940, 1942, 1943, 1947, 1949, 1951, 1953, 1954, 1955, 1956, 1957, 1958, 1959, 1962, 1965, 1968, 1971, 1975, 1978, 1981, 1984, 1987, 1990, 1993, 1996, 1999, 2002, 2005, 2008, 2011, and 2014.

This Code is purely advisory as far as NFPA is concerned. It is made available for a wide variety of both public and private uses in the interest of life and property protection. These include both use in law and for regulatory purposes and use in private self-regulation and standardization activities such as insurance underwriting, building and facilities construction and management, and product testing and certification.

Contents

CONTENTS

CONTENTS

Chapter 7 Special Conditions

Chapter 8 Communications Systems

Chapter 9 Tables

NATIONAL ELECTRICAL CODE COMMITTEE

These lists represent the membership at the time the Committee was balloted on the final text of this edition. Since that time, changes in the membership may have occurred. A key to classifications is found at the back of this document.

Correlating Committee on National Electrical Code®

Michael J. Johnston, *Chair*
National Electrical Contractors Association, MD [IM]

Mark W. Earley, *Secretary (Nonvoting)*
National Fire Protection Association, MA

Kimberly L. Shea, *Recording Secretary (Nonvoting)*
National Fire Protection Association, MA

James E. Brunssen, Telcordia Technologies (Ericsson), NJ [UT]
    Rep. Alliance for Telecommunications Industry Solutions
Kevin L. Dressman, U.S. Department of Energy, MD [U]
Palmer L. Hickman, Electrical Training Alliance, MD [L]
    Rep. International Brotherhood of Electrical Workers
David L. Hittinger, Independent Electrical Contractors of Greater Cincinnati, OH [IM]
    Rep. Independent Electrical Contractors, Inc.
Richard A. Holub, The DuPont Company, Inc., DE [U]
    Rep. American Chemistry Council

John R. Kovacik, UL LLC, IL [RT]
Alan Manche, Schneider Electric, KY [M]
Richard P. Owen, Oakdale, MN [E]
    Rep. International Association of Electrical Inspectors
James F. Pierce, Intertek Testing Services, OR [RT]
Vincent J. Saporita, Eaton's Bussmann Business, MO [M]
    Rep. National Electrical Manufacturers Association

Alternates

Lawrence S. Ayer, Biz Com Electric, Inc., OH [IM]
    (Alt. to David L. Hittinger)
Roland E. Deike, Jr., CenterPoint Energy, Inc., TX [UT]
    (Voting Alt.)
James T. Dollard, Jr., IBEW Local Union 98, PA [L]
    (Alt. to Palmer L. Hickman)
Stanley J. Folz, Morse Electric Company, NV [IM]
    (Alt. to Michael J. Johnston)
Ernest J. Gallo, Telcordia Technologies (Ericsson), NJ [UT]
    (Alt. to James E. Brunssen)

Robert A. McCullough, Tuckerton, NJ [E]
    (Alt. to Richard P. Owen)
Mark C. Ode, UL LLC, AZ [RT]
    (Alt. to John R. Kovacik)
Christine T. Porter, Intertek Testing Services, WA [RT]
    (Alt. to James F. Pierce)
George A. Stranicro, AFC Cable Systems, Inc., NJ [M]
    (Alt. to Vincent J. Saporita)

Nonvoting

Timothy J. Pope, Canadian Standards Association, Canada [SE]
    Rep. CSA/Canadian Electrical Code Committee
William R. Drake, Fairfield, CA [M]
    (Member Emeritus)

D. Harold Ware, Libra Electric Company, OK [IM]
    (Member Emeritus)

Mark W. Earley, NFPA Staff Liaison

CODE-MAKING PANEL NO. 1

Articles 90, 100, 110, Chapter 9, Table 10, Annex A, Annex H, Annex I, and Annex J

Kenneth P. Boyce, *Chair*
UL LLC, IL [RT]

Michael A. Anthony, University of Michigan, MI [U]
Rep. Association of Higher Education Facilities Officers

Louis A. Barrios, Shell Global Solutions, TX [U]
Rep. American Chemistry Council

Roland E. Deike, Jr., CenterPoint Energy, Inc., TX [UT]
Rep. Electric Light & Power Group/EEI

Ernest J. Gallo, Telcordia Technologies (Ericsson), NJ [U]
Rep. Alliance for Telecommunications Industry Solutions

Palmer L. Hickman, Electrical Training Alliance, MD [L]
Rep. International Brotherhood of Electrical Workers

David L. Hittinger, Independent Electrical Contractors of Greater Cincinnati, OH [IM]
Rep. Independent Electrical Contractors, Inc.

Donald R. Iverson, National Electrical Manufacturers Association, MI [M]
Rep. National Electrical Manufacturers Association

James F. Pierce, Intertek Testing Services, OR [RT]

Harry J. Sassaman, Forest Electric Corporation, NJ [IM]
Rep. National Electrical Contractors Association

Kent A. Sayler, P2S Engineering, Inc., CA [U]
Rep. Institute of Electrical & Electronics Engineers, Inc.

Mohinder P. Sood, City of Alexandria, VA [E]
Rep. International Association of Electrical Inspectors

Alternates

James E. Brunssen, Telcordia Technologies (Ericsson), NJ [U]
(Alt. to Ernest J. Gallo)

Michael J. Johnston, National Electrical Contractors Association, MD [IM]
(Alt. to Harry J. Sassaman)

Gary W. Jones, City of Aledo, Texas, TX [E]
(Alt. to Mohinder P. Sood)

Joseph Marquardt, ExxonMobil Production Company, TX [U]
(Alt. to Louis A. Barrios)

Dirk R. F. Mueller, UL LLC, Germany [RT]
(Alt. to Kenneth P. Boyce)

Donald R. Offerdahl, Intertek Testing Services, ND [RT]
(Alt. to James F. Pierce)

Michael C. Stone, National Electrical Manufacturers Association, CA [M]
(Alt. to Donald R. Iverson)

Frank E. Tyler, The DuPont Company, Inc., DE [U]
(Alt. to Kent A. Sayler)

Nonvoting

Ark Tsisserev, Applied Engineering Solutions, Canada [SE]
Rep. CSA/Canadian Electrical Code Committee

William T. Fiske, Intertek Testing Services, NY [RT]
(Member Emeritus)

CODE-MAKING PANEL NO. 2

Articles 210, 215, 220, Annex D, Examples D1 through D6

Mark R. Hilbert, *Chair*
MR Hilbert Electrical Inspections & Training, NH [E]
Rep. International Association of Electrical Inspectors

Charles L. Boynton, The DuPont Company, Inc., TX [U]
Rep. American Chemistry Council

Daniel Buuck, National Association of Home Builders, DC [U]
Rep. National Association of Home Builders

Steve Campolo, Leviton Manufacturing Company, Inc., NY [M]

Frank Coluccio, New York City Department of Buildings, NY [E]

Thomas A. Domitrovich, Eaton Corporation, MD [M]
Rep. National Electrical Manufacturers Association

Ronald E. Duren, PacifiCorp, WA [UT]
Rep. Electric Light & Power Group/EEI

Thomas L. Harman, University of Houston-Clear Lake, TX [SE]

Donald M. King, IBEW Local Union 313, DE [L]
Rep. International Brotherhood of Electrical Workers

Alan Manche, Schneider Electric, KY [M]

James E. Mitchem, JEM Electrical Consulting Services, CO [U]
Rep. Institute of Electrical & Electronics Engineers, Inc.

Frederick P. Reyes, UL LLC, NY [RT]

Stephen J. Thorwegen, Jr., FSG Electric, TX [IM]
Rep. Independent Electrical Contractors, Inc.

Thomas H. Wood, Cecil B. Wood, Inc., IL [IM]
Rep. National Electrical Contractors Association

Alternates

David A. Dini, UL LLC, IL [RT]
(Alt. to Frederick P. Reyes)

James M. Imlah, City of Hillsboro, OR [E]
(Alt. to Mark R. Hilbert)

Andrew Kriegman, Leviton Manufacturing Company, Inc., NY [M]
(Alt. to Steve Campolo)

Ed Larsen, Schneider Electric USA, IA [M]
(Alt. to Alan Manche)

John McCamish, NECA IBEW Electrical Training Center, OR [L]
(Alt. to Donald M. King)

Roger D. McDaniel, Georgia Power Company, GA [UT]
(Alt. to Ronald E. Duren)

William J. McGovern, Intertek Testing Services, TX [RT]
(Voting Alt.)

Fernando E. Pacheco, Methanex Chile SA, TX [U]
(Alt. to Charles L. Boynton)

Brian E. Rock, Hubbell Incorporated, CT [M]
(Alt. to Thomas A. Domitrovich)

Edward E. Rodriguez, IEC Texas Gulf Coast, TX [IM]
(Alt. to Stephen J. Thorwegen, Jr.)

Michael Weaver, M&W Electric, OR [IM]
(Alt. to Thomas H. Wood)

Nonvoting

Douglas A. Lee, U.S. Consumer Product Safety Commission, MD [C]

Andrew M. Trotta, U.S. Consumer Product Safety Commission, MD [C]

CODE-MAKING PANEL NO. 3

Articles 300, 590, 720, 725, 727, 728, 760, Chapter 9, Tables 11(A) and (B), and Tables 12(A) and (B)

Paul J. Casparro, *Chair*
Scranton Electricians JATC, PA [L]
Rep. International Brotherhood of Electrical Workers

Douglas P. Bassett, XFinity Home, FL [IM]
   Rep. Electronic Security Association
   (VL to 720, 725, 727, 760)

Larry G. Brewer, Intertek Testing Services, NC [RT]

William A. Brunner, Main Electric Construction Inc., ND [IM]
   Rep. National Electrical Contractors Association

Steven D. Burlison, Progress Energy, FL [UT]
   Rep. Electric Light & Power Group/EEI

Shane M. Clary, Bay Alarm Company, CA [M]
   Rep. Automatic Fire Alarm Association, Inc.

Adam D. Corbin, Corbin Electrical Services, Inc., NJ [IM]
   Rep. Independent Electrical Contractors, Inc.

Les Easter, Atkore International, IL [M]
   Rep. National Electrical Manufacturers Association

Ray R. Keden, Pentair-ERICO, CA [M]
   Rep. Building Industry Consulting Services International

T. David Mills, T. David Mills Associates, LLC, GA [U]
   Rep. Institute of Electrical & Electronics Engineers, Inc.

Steven J. Owen, Steven J. Owen, Inc., AL [IM]
   Rep. Associated Builders & Contractors

David A. Pace, Olin Corporation, AL [U]
   Rep. American Chemistry Council

Susan Newman Scearce, City of Humboldt, TN, TN [E]
   Rep. International Association of Electrical Inspectors

John E. Sleights, Travelers Insurance Company, CT [I]

Susan L. Stene, UL LLC, CA [RT]

Alternates

Richard S. Anderson, RTKL Associates Inc., VA [M]
   (Alt. to Ray R. Keden)

Jorge L. Arocha, Florida Power & Light, FL [UT]
   (Alt. to Steven D. Burlison)

Sanford E. Egesdal, Egesdal Associates PLC, MN [M]
   (Alt. to Shane M. Clary)

Michael J. Farrell III, Lucas County Building Regulation, MI [L]
   (Alt. to Paul J. Casparro)

Danny Liggett, The DuPont Company, Inc., TX [U]
   (Alt. to David A. Pace)

Mark C. Ode, UL LLC, AZ [RT]
   (Alt. to Susan L. Stene)

Dmitriy V. Plotnikov, Intertek Testing Services, NJ [RT]
   (Alt. to Larry G. Brewer)

Rick D. Sheets, DIRECTV, TX [IM]
   (VL to 720, 725, 727, 760)
   (Alt. to Douglas P. Bassett)

George A. Straniero, AFC Cable Systems, Inc., NJ [M]
   (Alt. to Les Easter)

Joseph J. Wages, Jr., International Association of Electrical Inspectors, TX [E]
   (Alt. to Susan Newman Scearce)

Nonvoting

Edward C. Lawry, Oregon, WI [E]
   (Member Emeritus)

CODE-MAKING PANEL NO. 4

Articles 225, 230, 690, 691, 692, 694, 705, 710

Ronald J. Toomer, *Chair*
Toomer Electrical Company Inc., LA [IM]
Rep. National Electrical Contractors Association

Malcolm Allison, Mersen USA Newburyport-MA, LLC, NH [M]
Rep. National Electrical Manufacturers Association

Ward I. Bower, Solar Energy Industries Association, NM [U]
Rep. Solar Energy Industries Association
(VL to 690, 692, 705)

Bill F. Brooks, Brooks Engineering, CA [U]
Rep. Photovoltaic Industry Code Council
(VL to 690, 692, 705)

Thomas E. Buchal, Intertek Testing Services, NY [RT]

James G. Cialdea, Three-C Electrical Company Inc., MA [IM]
Rep. InterNational Electrical Testing Association

Todd Fries, HellermannTyton, WI [M]

Mark D. Gibbs, Consolidated Nuclear Security, TN [U]
Rep. Institute of Electrical & Electronics Engineers, Inc.

Roger D. McDaniel, Georgia Power Company, GA [UT]
Rep. Electric Light & Power Group/EEI

Matthew Paiss, San Jose Fire Department, CA [L]
Rep. International Association of Fire Fighters

David J. Picatti, Picatti Bros. Inc., DBA Industrial Service & Electric, WA [IM]
Rep. Independent Electrical Contractors, Inc.

James J. Rogers, Towns of Oak Bluffs, Tisbury, West Tisbury, MA [E]
Rep. International Association of Electrical Inspectors

Rebecca S. Templet, Shell Chemical, LA [M]
Rep. American Chemistry Council

Wendell R. Whistler, Alaska Joint Electrical Apprenticeship Training Trust, AK [L]
Rep. International Brotherhood of Electrical Workers

Robert H. Wills, Intergrid, LLC, NH [U]
Rep. American Wind Energy Association
(VL to 690, 692, 694, 705)

Stephen P. Wurmlinger, SunPower Corporation, TX [M]
Rep. Large-Scale Solar Association

Timothy P. Zgonena, UL LLC, IL [RT]

Alternates

Paul D. Barnhart, UL LLC, NC [RT]
(Alt. to Timothy P. Zgonena)

Alex Z. Bradley, The DuPont Company, Inc., DE [U]
(Alt. to Rebecca S. Templet)

Larry D. Cogburn, Cogburn Bros., Inc., FL [IM]
(Alt. to Ronald J. Toomer)

Lee M. Kraemer, First Solar, OH [M]
(Alt. to Stephen P. Wurmlinger)

Howard Liu, Intertek Testing Services, NY [RT]
(Alt. to Thomas E. Buchal)

Harold C. Ohde, IBEW 134/Electrical Joint Apprenticeship Training & Trust, IL [L]
(Alt. to Wendell R. Whistler)

Rhonda Parkhurst, City of Palo Alto, CA [E]
(Alt. to James J. Rogers)

Robert W. Preus, National Renewable Energy Lab, CO [U]
(VL to 690, 692, 694, 705)
(Alt. to Robert H. Wills)

Karl Reighard, Delmarva Power and Light, DE [UT]
(Alt. to Roger D. McDaniel)

Patrick G. Salas, General Electric Company, CT [M]
(Alt. to Malcolm Allison)

Nonvoting

Stephen W. Douglas, QPS Evaluation Services Inc., Canada [SE]
Rep. CSA/Canadian Electrical Code Committee

CODE-MAKING PANEL NO. 5

Articles 200, 250, 280, 285

Nathan Philips, *Chair*
Integrated Electronic Systems, OR [IM]
Rep. National Electrical Contractors Association

Paul W. Abernathy, Encore Wire Corporation, TX [M]
Rep. The Aluminum Association, Inc.

Gary A. Beckstrand, Utah Electrical JATC, UT [L]
Rep. International Brotherhood of Electrical Workers

Trevor N. Bowmer, Telcordia (Ericsson), NJ [U]
Rep. Alliance for Telecommunications Industry Solutions

David Brender, Copper Development Association, Inc., NY [M]
Rep. Copper Development Association Inc.

Martin J. Brett, Jr., Wheatland Tube Company, DE [M]
Rep. Steel Tube Institute of North America

Paul Dobrowsky, Innovative Technology Services, NY [U]
Rep. American Chemistry Council

G. Scott Harding, F. B. Harding, Inc., MD [IM]
Rep. Independent Electrical Contractors, Inc.

Joseph Harding, Power Tool Institute, OH [M]

William J. Helfrich, U.S. Department of Labor, PA [E]

Charles F. Mello, UL LLC, WA [RT]

Daleep C. Mohla, DCM Electrical Consulting Services, Inc., TX [U]
Rep. Institute of Electrical & Electronics Engineers, Inc.

Mike O'Meara, Arizona Public Service Company, AZ [UT]
Rep. Electric Light & Power Group/EEI

Charles J. Palmieri, Town of Norwell, MA [E]
Rep. International Association of Electrical Inspectors

Christine T. Porter, Intertek Testing Services, WA [RT]

Nick Sasso, State of Wyoming, WY [E]

Gregory J. Steinman, Thomas & Betts Corporation, TN [M]
Rep. National Electrical Manufacturers Association

Alternates

Joseph F. Andre, Steel Tube Institute, WA [M]
(Alt. to Martin J. Brett, Jr.)

Derrick L. Atkins, Minneapolis Electrical JATC, MN [L]
(Alt. to Gary A. Beckstrand)

Joseph P. DeGregoria, UL LLC, NY [RT]
(Alt. to Charles F. Mello)

Ernest J. Gallo, Telcordia Technologies (Ericsson), NJ [U]
(Alt. to Trevor N. Bowmer)

Bobby J. Gray, Hoydar/Buck, Inc., WA [IM]
(Alt. to Nathan Philips)

Buster Grissett, Mississippi Power Company, MS [UT]
(Alt. to Mike O'Meara)

Ronald Lai, Burndy LLC, NH [M]
(Alt. to Gregory J. Steinman)

William A. Pancake, III, North Naples Fire Control & Rescue District, FL [E]
(Alt. to Charles J. Palmieri)

Paul R. Picard, AFC Cable Systems, Inc., MA [M]
(Alt. to Paul W. Abernathy)

Phil Simmons, Simmons Electrical Services, WA [M]
(Alt. to David Brender)

Fred Song, Intertek Testing Services, China [RT]
(Alt. to Christine T. Porter)

David B. Stump, Independent Electrical Contractors, TX [IM]
(Alt. to G. Scott Harding)

Nonvoting

Robert A. Nelson, Canadian Standards Association, Canada [RT]

CODE-MAKING PANEL NO. 6

Articles 310, 400, 402, Chapter 9, Tables 5 through 9, and Annex B

Michael W. Smith, *Chair*
Schaeffer Electric Company, Inc., MO [IM]
Rep. National Electrical Contractors Association

Edwin F. Brush, BBF & Associates, ME [U]
Rep. Institute of Electrical & Electronics Engineers, Inc.

Samuel B. Friedman, General Cable Corporation, RI [M]
Rep. National Electrical Manufacturers Association

Robert L. Huddleston, Jr., Eastman Chemical Company, TN [U]
Rep. American Chemistry Council

Gerald W. Kent, Kent Electric & Plumbing Systems, TX [IM]
Rep. Independent Electrical Contractors, Inc.

William F. Laidler, IBEW Local 223 JATC, MA [L]
Rep. International Brotherhood of Electrical Workers

Paul R. Picard, AFC Cable Systems, Inc., MA [M]
Rep. The Aluminum Association, Inc.

Kenneth Riedl, Intertek Testing Services, NY [RT]

John Stacey, City of St. Louis, MO [E]
Rep. International Association of Electrical Inspectors

Carl Timothy Wall, Alabama Power Company, AL [UT]
Rep. Electric Light & Power Group/EEI

Mario Xerri, UL LLC, NY [RT]

Joseph S. Zimnoch, The Okonite Company, NJ [M]
Rep. Copper Development Association Inc.

Alternates

John J. Cangemi, UL LLC, NY [RT]
(Alt. to Mario Xerri)

Scott Cline, McMurtrey Electric, Inc., CA [IM]
(Alt. to Michael W. Smith)

Todd Crisman, IBEW Local 22 JATC, NE [L]
(Alt. to William F. Laidler)

Joseph W. Cross, Eastman Chemical Company, TN [U]
(Alt. to Robert L. Huddleston, Jr.)

Fred Echeverri, AFC Cable Systems, MA [M]
(Alt. to Paul R. Picard)

Christel K. Hunter, General Cable Corporation, NV [M]
(Alt. to Samuel B. Friedman)

Armando M. Lozano, MSF Electric, Inc., TX [IM]
(Alt. to Gerald W. Kent)

William Maxwell, National Grid, NY [UT]
(Alt. to Carl Timothy Wall)

Charles David Mercier, Southwire Company, GA [M]
(Alt. to Joseph S. Zimnoch)

Borgia Noel, State of Wyoming Fire Marshal's Office, WY [E]
(Alt. to John Stacey)

CODE-MAKING PANEL NO. 7

Articles 320, 322, 324, 326, 328, 330, 332, 334, 336, 338, 340, 382, 394, 396, 398, 399

David A. Williams, *Chair*
Delta Charter Township, MI [E]
Rep. International Association of Electrical Inspectors

Thomas H. Cybula, UL LLC, NY [RT]

Vincent Della Croce, eti Conformity Services, FL [RT]

Chris J. Fahrenthold, Facility Solutions Group, TX [IM]
Rep. Independent Electrical Contractors, Inc.

Herman J. Hall, Austin, TX [M]
Rep. The Vinyl Institute

Christel K. Hunter, General Cable Corporation, NV [M]
Rep. The Aluminum Association, Inc.

Samuel R. La Dart, City of Memphis, TN [L]
Rep. International Brotherhood of Electrical Workers

Charles David Mercier, Southwire Company, GA [M]
Rep. National Electrical Manufacturers Association

Ronald G. Nickson, National Multifamily Housing Council, NC [U]

Dennis A. Nielsen, Lawrence Berkeley National Laboratory, CA [U]
Rep. Institute of Electrical & Electronics Engineers, Inc.

John W. Ray, Duke Energy Corporation, NC [UT]
Rep. Electric Light & Power Group/EEI

Gregory L. Runyon, Eli Lilly and Company, IN [U]
Rep. American Chemistry Council

George A. Straniero, AFC Cable Systems, Inc., NJ [M]
Rep. Copper Development Association Inc.

Wesley L. Wheeler, National Electrical Contractors Association, MD [IM]
Rep. National Electrical Contractors Association

Alternates

J. Richard Barker, General Cable Corporation, CA [M]
(Alt. to Christel K. Hunter)

Richard C. Bennett, Cerro Wire LLC, AL [M]
(Alt. to Charles David Mercier)

Timothy Earl, GBH International, MI [M]
(Alt. to Herman J. Hall)

Rachel E. Krepps, Baltimore Gas & Electric Company, MD [UT]
(Alt. to John W. Ray)

Keith Owensby, Chattanooga Electrical JATC, TN [L]
(Alt. to Samuel R. La Dart)

Kevin T. Porter, Encore Wire Corporation, TX [M]
(Alt. to George A. Straniero)

Irozenell Pruitt, The DuPont Company, Inc., TX [U]
(Alt. to Gregory L. Runyon)

Michael W. Smith, Schaeffer Electric Company, Inc., MO [IM]
(Alt. to Wesley L. Wheeler)

Susan L. Stene, UL LLC, CA [RT]
(Alt. to Thomas H. Cybula)

Allen R. Turner, James City County, Virginia, VA [E]
(Alt. to David A. Williams)

CODE-MAKING PANEL NO. 8

Articles 342, 344, 348, 350, 352, 353, 354, 355, 356, 358, 360, 362, 366, 368, 370, 372, 374, 376, 378, 380, 384, 386, 388, 390, 392, Chapter 9, Tables 1 through 4, Example D13, and Annex C

Larry D. Cogburn, *Chair*
Cogburn Bros., Inc., FL [IM]
Rep. National Electrical Contractors Association

David F. Allen, National Grid, MA [UT]
Rep. Electric Light & Power Group/EEI

David M. Campbell, AFC Cable Systems, Inc., MA [M]
Rep. The Aluminum Association, Inc.

David A. Gerstetter, UL LLC, IL [RT]
Rep. Underwriters Laboratories Inc.

Kenneth W. Hengst, Walker Engineering, Inc., TX [IM]
Rep. Independent Electrical Contractors, Inc.

Pete Jackson, City of Bakersfield, California, CA [E]
Rep. International Association of Electrical Inspectors

David H. Kendall, Thomas & Betts Corporation, TN [M]
Rep. The Vinyl Institute

Richard E. Loyd, R & N Associates, AZ [M]
Rep. Steel Tube Institute of North America

Michael C. Martin, ExxonMobil Research & Engineering, TX [U]
Rep. American Chemistry Council

Paul W. Myers, PCS Nitrogen, OH [U]
Rep. Institute of Electrical & Electronics Engineers, Inc.

Donald R. Offerdahl, Intertek Testing Services, ND [RT]

Rhett A. Roe, IBEW Local Union 26 JATC, MD [L]
Rep. International Brotherhood of Electrical Workers

Rodney J. West, Schneider Electric, OH [M]
Rep. National Electrical Manufacturers Association

Alternates

Richard J. Berman, UL LLC, IL [RT]
(Alt. to David A. Gerstetter)

Rachel Guenther, Thomas & Betts Corporation, TN [M]
(Alt. to David H. Kendall)

J. Grant Hammett, Colorado State Electrical Board, CO [E]
(Alt. to Pete Jackson)

Raymond W. Horner, Allied Tube & Conduit, IL [M]
(Alt. to Richard E. Loyd)

Gary K. Johnson, The Dow Chemical Company, LA [U]
(Alt. to Michael C. Martin)

Stephen P. Poholski, Newkirk Electric Associates, Inc., MI [IM]
(Alt. to Larry D. Cogburn)

Dan Rodriguez, IBEW Local Union 332, CA [L]
(Alt. to Rhett A. Roe)

Frederic F. Small, Hubbell Incorporated, CT [M]
(Alt. to Rodney J. West)

Raul L. Vasquez, Independent Electrical Contractors, TX [IM]
(Alt. to Kenneth W. Hengst)

Dave Watson, Southwire, GA [M]
(Alt. to David M. Campbell)

Nonvoting

Stephen W. Douglas, QPS Evaluation Services Inc., Canada [SE]
Rep. CSA/Canadian Electrical Code Committee

CODE-MAKING PANEL NO. 10

Article 240

Julian R. Burns, *Chair*
Quality Power Solutions, Inc., NC [IM]
Rep. Independent Electrical Contractors, Inc.

Scott A. Blizard, American Electrical Testing Company, Inc., MA [IM]
    Rep. InterNational Electrical Testing Association

Dennis M. Darling, Stantec, Canada [U]
    Rep. Institute of Electrical & Electronics Engineers, Inc.

James T. Dollard, Jr., IBEW Local Union 98, PA [L]
    Rep. International Brotherhood of Electrical Workers

Carl Fredericks, The Dow Chemical Company, TX [U]
    Rep. American Chemistry Council

Jeffrey H. Hidaka, UL LLC, WA [RT]

Robert J. Kauer, Building Inspection Underwriters, Inc., PA [E]
    Rep. International Association of Electrical Inspectors

Kenneth J. Rempe, Siemens Industry Inc., GA [M]
    Rep. National Electrical Manufacturers Association

Vincent J. Saporita, Eaton's Bussmann Business, MO [M]

Richard Sobel, Quantum Electric Corporation, NY [IM]
    Rep. National Electrical Contractors Association

Christopher R. Vance, National Grid, NY [UT]
    Rep. Electric Light & Power Group/EEI

Alternates

Christopher M. Jensen, North Logan City, UT [E]
    (Alt. to Robert J. Kauer)

Kevin J. Lippert, Eaton Corporation, PA [M]
    (Alt. to Vincent J. Saporita)

Richard E. Lofton, II, IBEW Local Union 280, OR [L]
    (Alt. to James T. Dollard, Jr.)

Alan Manche, Schneider Electric, KY [M]
    (Alt. to Kenneth J. Rempe)

Kathleen McKitish, Baltimore Gas & Electric, MD [UT]
    (Alt. to Christopher R. Vance)

Bruce M. Rockwell, American Electrical Testing Company, Inc., NJ [IM]
    (Alt. to Scott A. Blizard)

Roy K. Sparks, III, Eli Lilly and Company, IN [U]
    (Alt. to Carl Fredericks)

Steve A. Struble, Freeman's Electric Service, Inc., SD [IM]
    (Alt. to Julian R. Burns)

Steven E. Townsend, General Motors Company, MI [U]
    (Alt. to Dennis M. Darling)

CODE-MAKING PANEL NO. 11

Articles 409, 430, 440, 460, 470, Annex D, Example D8

John M. Thompson, *Chair*
UL LLC, NC [RT]

Luis M. Bas, Intertek Testing Services, FL [RT]

Terry D. Cole, Hamer Electric, Inc., WA [IM]
Rep. Independent Electrical Contractors, Inc.

Zivorad Cosic, ABB Inc., WI [M]

Robert G. Fahey, City of Janesville, WI [E]
Rep. International Association of Electrical Inspectors

James M. Fahey, IBEW Local Union 103, MA [L]
Rep. International Brotherhood of Electrical Workers

Stanley J. Folz, Morse Electric Company, NV [IM]
Rep. National Electrical Contractors Association

Paul E. Guidry, Fluor Enterprises, Inc., TX [U]
Rep. Associated Builders & Contractors

Stephen M. Jackson, Southern Company, GA [UT]
Rep. Electric Light & Power Group/EEI

Arthur S. Neubauer, Arseal Technologies, GA [U]
Rep. American Petroleum Institute

George J. Ockuly, Technical Marketing Consultants, MO [M]

Charles L. Powell, Eastman Chemical Company, TN [U]
Rep. American Chemistry Council

Arthur J. Smith, III, Waldemar S. Nelson & Company, Inc., LA [U]
Rep. Institute of Electrical & Electronics Engineers, Inc.

Ron Widup, Shermco Industries, TX [IM]
Rep. InterNational Electrical Testing Association

James R. Wright, Siemens Industry, Inc., IL [M]
Rep. National Electrical Manufacturers Association

Alternates

John E. Cabaniss, Eastman Chemical Company, TN [U]
(Alt. to Charles L. Powell)

Seth J. Carlton, UL LLC, IL [RT]
(Alt. to John M. Thompson)

Gregory J. Clement, Fluor Enterprises, Inc., TX [U]
(Alt. to Paul E. Guidry)

Eric Gesualdi, Shell Oil Company, TX [U]
(Alt. to Arthur S. Neubauer)

Tim Hinson, Miller Electric Company, FL [IM]
(Alt. to Stanley J. Folz)

Rodney B. Jones, Clackamas County, Oregon, OR [E]
(Alt. to Robert G. Fahey)

Tim LaLonde, Haskin Electric, Inc., WA [IM]
(Alt. to Terry D. Cole)

Ed Larsen, Schneider Electric USA, IA [M]
(Alt. to James R. Wright)

Jebediah J. Novak, Cedar Rapids Electrical JATC, IA [L]
(Alt. to James M. Fahey)

Vincent J. Saporita, Eaton's Bussmann Business, MO [M]
(Alt. to George J. Ockuly)

Carl Timothy Wall, Alabama Power Company, AL [UT]
(Alt. to Stephen M. Jackson)

Bobby A. Walton, Intertek, TX [RT]
(Alt. to Luis M. Bas)

CODE-MAKING PANEL NO. 12

Articles 610, 620, 625, 626, 630, 640, 645, 647, 650, 660, 665, 668, 669, 670, 685, and Annex D, Examples D9 and D10

Scott Cline, *Chair*
McMurtrey Electric, Inc., CA [IM]
Rep. National Electrical Contractors Association

Frank Anthony Belio, International Union of Elevator Constructors, CA [L]

Jeffrey W. Blain, Schindler Elevator Corporation, NY [M]
Rep. National Elevator Industry Inc.
(VL to 610, 620, 630)

Thomas R. Brown, Intertek Testing Services, NY [RT]

James L. Brown, DTE Energy, MI [UT]
Rep. Electric Light & Power Group/EEI

Philip Clark, City of Southfield, MI [E]
Rep. International Association of Electrical Inspectors

Karl M. Cunningham, Alcoa, Inc., PA [M]
Rep. The Aluminum Association, Inc.
(VL to 610, 625, 630, 645, 660, 665, 668, 669, 685)

Joel Goergen, Cisco Systems, Inc., CA [M]

Jeffrey L. Holmes, IBEW Local Union 1 JATC, MO [L]
Rep. International Brotherhood of Electrical Workers

Angelo G. Horiates, Navy Crane Center, VA [U]
(VL to 610)

Robert E. Johnson, ITE Safety, MA [U]
Rep. Information Technology Industry Council
(VL to 640, 645, 647, 685)

Stanley Kaufman, CableSafe, Inc./OFS, GA [M]
Rep. Society of the Plastics Industry, Inc.
(VL to 640, 645, 646, 650)

John R. Kovacik, UL LLC, IL [RT]

Todd F. Lottmann, Easton's Bussmann Business, MO [M]
Rep. National Electrical Manufacturers Association

Jeffrey S. Menig, General Motors Company, MI [U]
Rep. SAE Hybrid/EV Technical Standards Committee

Duke W. Schamel, Electrical Service Solutions, Inc., CA [IM]
Rep. Independent Electrical Contractors, Inc.

Arthur E. Schlueter, Jr., A. E. Schlueter Pipe Organ Company, GA [M]
Rep. American Pipe Organ Builders
(VL to 640, 650)

Robert C. Turner, Inductotherm Corporation, PA [M]
(VL to 610, 630, 665, 668, 669)

Alternates

Joseph M. Bablo, UL LLC, IL [RT]
(Alt. to John R. Kovacik)

William B. Crist, Jr., IES Residential Inc., TX [IM]
(Alt. to Duke W. Schamel)

Vincent Della Croce, eti Conformity Services, FL [L]
(Alt. to Jeffrey L. Holmes)

Jody B. Greenwood, Navy Crane Center, VA [U]
(VL to 610)
(Alt. to Angelo G. Horiates)

Jacob Haney, General Cable Corporation, IN [M]
(VL to 610, 625, 630, 645, 660, 665, 668, 669, 685)
(Alt. to Karl M. Cunningham)

John D. (Doug) Henderson, ThyssenKrupp Elevator Manufacturing Inc., TN [M]
(VL to 610, 620, 630)
(Alt. to Jeffrey W. Blain)

Todd R. Konieczny, Intertek Testing Services, MA [RT]
(Alt. to Thomas R. Brown)

Michael Owen, White Electrical, TN [IM]
(Alt. to Scott Cline)

Joseph F. Prisco, IBM Corporation, MN [U]
(VL to 640, 645, 647, 685)
(Alt. to Robert E. Johnson)

Emad Tabatabaei, Inductotherm Corporation, NJ [M]
(VL to 610, 630, 665, 668, 669)
(Alt. to Robert C. Turner)

James E. Tarchinski, General Motors Company, MI [U]
(Alt. to Jeffrey S. Menig)

Frank Tse, Leviton Manufacturing Company, Inc., NY [M]
(Alt. to Todd F. Lottmann)

Phillip J. Yehl, City of Peoria, IL [E]
(Alt. to Philip Clark)

Nonvoting

Andre R. Cartal, Yardley, PA [E]
(Member Emeritus)

CODE-MAKING PANEL NO. 13

Articles 445, 455, 480, 695, 700, 701, 702, 706, 708, 712, 750, Annex F, and Annex G

Linda J. Little, *Chair*
IBEW Local 1 Electricians JATC, MO [L]

Martin D. Adams, Adams Electric, Inc., CO [IM]
Rep. National Electrical Contractors Association

George M. Brandon, One World Technologies, SC [M]
Rep. Portable Generator Manufacturers' Association

Daniel J. Caron, Bard, Rao + Athanas Consulting Engineers, LLC, MA [SE]

Timothy M. Croushore, FirstEnergy Technologies, PA [UT]
Rep. Electric Light & Power Group/EEI

Richard D. Currin, Jr., North Carolina State University, NC [U]
Rep. American Society of Agricultural & Biological Engineers

Neil A. Czarnecki, Reliance Controls Corporation, WI [M]
Rep. National Electrical Manufacturers Association

James E. Degnan, Stantec, WA [U]
Rep. American Society for Healthcare Engineering

Steven F. Froemming, City of Franklin, WI [E]
Rep. International Association of Electrical Inspectors

Ronald A. Keenan, M. C. Dean, Inc., VA [IM]
Rep. Independent Electrical Contractors, Inc.

Daniel R. Neeser, Eaton's Bussmann Division, MO [M]

Mark C. Ode, UL LLC, AZ [RT]

Shawn Paulsen, CSA Group, Canada [RT]

Arnoldo L. Rodriguez, LyondellBasell Industries, TX [U]
Rep. American Chemistry Council

Michael L. Savage, Sr., City of Rio Rancho, NM [E]

Mario C. Spina, Verizon Wireless, OH [U]
Rep. Institute of Electrical & Electronics Engineers, Inc.

David Tobias, Jr., Intertek Testing Services, OH [RT]

Kendall M. Waterman, Draka Cableteq, MA [M]
Rep. Copper Development Association Inc.

James R. White, Shermco Industries, Inc., TX [IM]
Rep. InterNational Electrical Testing Association

Herbert V. Whittall, Electrical Generating Systems Association, FL [M]
Rep. Electrical Generating Systems Association

Timothy P. Windey, Cummins Power Generation, MN [M]

Alternates

Lawrence S. Ayer, Biz Com Electric, Inc., OH [IM]
(Alt. to Ronald A. Keenan)

Barry S. Bauman, Alliant Energy, WI [U]
(Alt. to Richard D. Currin, Jr.)

Krista McDonald Biason, HGA Architects and Engineers, MN [U]
(Alt. to James E. Degnan)

William P. Cantor, TPI Corporation, PA [U]
(Alt. to Mario C. Spina)

James S. Conrad, RSCC Wire & Cable, CT [M]
(Alt. to Kendall M. Waterman)

Timothy Crnko, Eaton's Bussmann Business, MO [M]
(Alt. to Daniel R. Neeser)

Herbert H. Daugherty, Electric Generating Systems Association, FL [M]
(Alt. to Herbert V. Whittall)

James T. Dollard, Jr., IBEW Local Union 98, PA [L]
(Alt. to Linda J. Little)

Lawrence W. Forshner, Bard, Rao + Athanas Consulting Engineers, LLC, MA [SE]
(Alt. to Daniel J. Caron)

Travis Foster, Shell Oil Company, TX [U]
(Alt. to Arnoldo L. Rodriguez)

Robert E. Jordan, Alabama Power Company, AL [UT]
(Alt. to Timothy M. Croushore)

Chad Kennedy, Schneider Electric, SC [M]
(Alt. to Neil A. Czarnecki)

John R. Kovacik, UL LLC, IL [RT]
(Alt. to Mark C. Ode)

Greg Marchand, Briggs & Stratton, [M]
(Alt. to George M. Brandon)

Rich Scroggins, Cummins Power Generation, MN [M]
(Alt. to Timothy P. Windey)

Michael Wilson, CSA Group, Canada [RT]
(Alt. to Shawn Paulsen)

CODE-MAKING PANEL NO. 14

Articles 500, 501, 502, 503, 504, 505, 506, 510, 511, 513, 514, 515, and 516

Robert A. Jones, *Chair*
Independent Electrical Contractors, Inc., TX [IM]
Rep. Independent Electrical Contractors, Inc.

Harold G. Alexander, American Electric Power Company, OH [UT]
Rep. Electric Light & Power Group/EEI

Donald W. Ankele, UL LLC, IL [RT]

Marc J. Bernsen, National Electrical Contractors Association, ID [IM]
Rep. National Electrical Contractors Association

Steven J. Blais, Appleton Group, IL [M]
Rep. National Electrical Manufacturers Association

Corey Cahill, U.S. Coast Guard, DC [E]

Mark Goodman, Mark Goodman Electrical Consulting, CA [U]
Rep. American Petroleum Institute

Haywood Kines, Prince William County Building Development, VA [E]
Rep. International Association of Electrical Inspectors

William G. Lawrence, Jr., FM Global, MA [I]

L. Evans Massey, Baldor Electric Company, SC [M]
Rep. Instrumentation, Systems, & Automation Society

William E. McBride, Northern Electric Company, AK [U]
Rep. Institute of Electrical & Electronics Engineers, Inc.

Jeremy Neagle, U.S. Bureau of Alcohol, Tobacco, Firearms & Explosives, MD [U]

Ryan Parks, Intertek Testing Services, TX [RT]

John L. Simmons, Florida East Coast JATC, FL [L]
Rep. International Brotherhood of Electrical Workers

David B. Wechsler, Consultant, TX [U]
Rep. American Chemistry Council

Mark C. Wirfs, R & W Engineering, Inc., OR [U]
Rep. Grain Elevator and Processing Society

Alternates

Dave Burns, Shell P&T: Innovation/R&D, TX [U]
(Alt. to Mark Goodman)

Larry W. Burns, Burns Electric, Inc., TX [IM]
(Alt. to Robert A. Jones)

Thomas E. Dunne, Long Island Joint Apprenticeship & Training Committee, NY [L]
(Alt. to John L. Simmons)

Mitch Feininger, North Dakota State Electrical Board, ND [E]
(Alt. to Haywood Kines)

Andrew Hernandez, AstraZeneca Pharmaceuticals, DE [U]
(Alt. to William E. McBride)

Richard A. Holub, The DuPont Company, Inc., DE [U]
(Alt. to David B. Wechsler)

Paul T. Kelly, UL LLC, IL [RT]
(Alt. to Donald W. Ankele)

Edmund R. Leubner, Eaton's Crouse-Hinds Business, NY [M]
(Alt. to Steven J. Blais)

Arkady Levi, Exelon Power, MD [UT]
(Alt. to Harold G. Alexander)

Eddie Ramirez, FM Global, MA [I]
(Alt. to William G. Lawrence, Jr.)

Ted H. Schnaare, Rosemount Incorporated, MN [M]
(Alt. to L. Evans Massey)

Steven C. Trapp, Christenson Electric Inc., OR [IM]
(Alt. to Marc J. Bernsen)

Wesley Van Hill, Intertek Testing Services, AB [RT]
(Alt. to Ryan Parks)

Nonvoting

Michael E. Aaron, JENSEN HUGHES, IL [SE]
Rep. TC on Airport Facilities

Timothy J. Pope, Canadian Standards Association, Canada [RT]

Eduardo N. Solano, Estudio Ingeniero Solano S.A., Argentina [SE]

CODE-MAKING PANEL NO. 15

Articles 517, 518, 520, 522, 525, 530, 540

Lawrence E. Todd, *Chair*
Intertek Testing Services, KY [RT]

Chad E. Beebe, ASHE - AHA, WA [U]

David A. Dagenais, Wentworth-Douglass Hospital, NH [U]
Rep. NFPA Health Care Section

Matthew B. Dozier, IDesign Services, TN [U]
Rep. Institute of Electrical & Electronics Engineers, Inc.

Joe L. DuPriest, Orange County Public Schools, FL [E]
Rep. International Association of Electrical Inspectors

Kenneth J. Gilbert, Florida Power & Light Company, FL [UT]
Rep. Electric Light & Power Group/EEI

Mitchell K. Hefter, Philips Lighting, TX [IM]
Rep. Illuminating Engineering Society of North America
(VL to 518, 520, 525, 530, 540)

Kim Jones, Funtastic Shows, OR [U]
Rep. Outdoor Amusement Business Association, Inc.
(VL to 525)

Edwin S. Kramer, Radio City Music Hall, NY [L]
Rep. International Alliance of Theatrical Stage Employees
(VL to 518, 520, 525, 530, 540)

Gary J. Krupa, U.S. Department of Veterans Affairs, NE [U]

Stephen M. Lipster, The Electrical Trades Center, OH [L]
Rep. International Brotherhood of Electrical Workers

Hugh O. Nash, Jr., Nash-Consult, TN [SE]
Rep. TC on Electrical Systems

Kevin T. Porter, Encore Wire Corporation, TX [M]
Rep. The Aluminum Association, Inc.

Brian E. Rock, Hubbell Incorporated, CT [M]
Rep. National Electrical Manufacturers Association

James C. Seabury III, Enterprise Electric, LLC, TN [IM]
Rep. Independent Electrical Contractors, Inc.

Bruce D. Shelly, Shelly Electric Company, Inc., PA [IM]
Rep. National Electrical Contractors Association

Michael D. Skinner, CBS Studio Center, CA [U]
Rep. Alliance of Motion Picture and Television Producers
(VL to 518, 520, 525, 530, 540)

Donald J. Talka, UL LLC, NY [RT]

Kenneth E. Vannice, Portland, OR [M]
Rep. U.S. Institute for Theatre Technology, Inc.
(VL to 518, 520, 525, 530, 540)

Alternates

Gary A. Beckstrand, Utah Electrical JATC, UT [L]
(Alt. to Stephen M. Lipster)

David M. Campbell, AFC Cable Systems, Inc., MA [M]
(Alt. to Kevin T. Porter)

Carmon A. Colvin, Bright Future Electric, LLC, AL [IM]
(Alt. to James C. Seabury III)

Samuel B. Friedman, General Cable Corporation, RI [M]
(Alt. to Brian E. Rock)

Pamela Gwynn, UL LLC, NC [RT]
(Alt. to Donald J. Talka)

Don W. Jhonson, Interior Electric, Inc., FL [IM]
(Alt. to Bruce D. Shelly)

Jay Y. Kogoma, Intertek Testing Services, CA [RT]
(Alt. to Lawrence E. Todd)

Frank Novitzki, U.S. Department of Veterans Affairs, VA [U]
(Alt. to Gary J. Krupa)

Douglas Rheinheimer, Paramount Pictures, CA [U]
(VL to 518, 520, 525, 530, 540)
(Alt. to Michael D. Skinner)

Alan M. Rowe, International Alliance of Theatrical Stage
Employees, CA [L]
(VL to 518, 520, 525, 530, 540)
(Alt. to Edwin S. Kramer)

Clinton Bret Stoddard, City of Rexburg, ID [E]
(Alt. to Joe L. DuPriest)

Steven R. Terry, Electronic Theatre Controls Inc., NY [M]
(VL to 518, 520, 525, 530, 540)
(Alt. to Kenneth E. Vannice)

R. Duane Wilson, George C. Izenour Associates, Inc., NM [IM]
(VL to 518, 520, 525, 530, 540)
(Alt. to Mitchell K. Hefter)

CODE-MAKING PANEL NO. 16

Articles 770, 800, 810, 820, 830, 840

Thomas E. Moore, *Chair*
City of Beachwood, OH [E]
Rep. International Association of Electrical Inspectors

George Bish, MasTec, NC [IM]
Rep. Satellite Broadcasting & Communications Association

James E. Brunssen, Telcordia Technologies (Ericsson), NJ [U]
Rep. Alliance for Telecommunications Industry Solutions

Fred C. Dawson, Chemours, Canada [U]
Rep. American Chemistry Council

Gerald Lee Dorna, Belden Wire & Cable Co., IN [M]
Rep. Insulated Cable Engineers Association Inc

Randolph J. Ivans, UL LLC, NY [RT]

Robert W. Jensen, dbi-Telecommunication Infrastructure Design, TX [M]
Rep. Building Industry Consulting Services International

Steven C. Johnson, Johnson Telecom, LLC, CA [UT]
Rep. National Cable & Telecommunications Association

William J. McCoy, Telco Sales, Inc., TX [U]
Rep. Institute of Electrical & Electronics Engineers, Inc.

Jack McNamara, Bosch Security Systems, NY [M]
Rep. National Electrical Manufacturers Association

Michael F. Murphy, Intertek Testing Services, MA [RT]

Harold C. Ohde, IBEW 134/Electrical Joint Apprenticeship Training & Trust, IL [L]
Rep. International Brotherhood of Electrical Workers

Thomas J. Parrish, Telgian Corporation, MI [M]
Rep. Automatic Fire Alarm Association, Inc.

W. Douglas Pirkle, Pirkle Electric Company, Inc., GA [IM]
Rep. National Electrical Contractors Association

Luigi G. Prezioso, M. C. Dean, Inc., VA [IM]
Rep. Independent Electrical Contractors, Inc.

Leo Zieman, Florida Power & Light (Nextera Energy), FL [UT]
Rep. Electric Light & Power Group/EEI

Alternates

Rendell K. Bourg, National Fire Protection Company Inc., HI [M]
(Alt. to Thomas J. Parrish)

Trevor N. Bowmer, Telcordia (Ericsson), NJ [U]
(Alt. to James E. Brunssen)

Larry Chan, City of New Orleans, LA [E]
(Alt. to Thomas E. Moore)

Terry C. Coleman, Electrical Training Alliance, TN [L]
(Alt. to Harold C. Ohde)

Timothy D. Cooke, Times Fiber Communications, Inc., VA [UT]
(Alt. to Steven C. Johnson)

John A. Kacperski, Tele Design Services, CA [M]
(Alt. to Robert W. Jensen)

Stanley Kaufman, CableSafe, Inc./OFS, GA [M]
(Alt. to Gerald Lee Dorna)

Eric Lawrence, Berk-Tek, A Nexans Company, PA [M]
(Voting Alt. to TIA rep)

David M. Lettkeman, Dish Network Service, LLC, CO [IM]
(Alt. to George Bish)

Rodger Reiswig, Tyco/SimplexGrinnell, FL [M]
(Alt. to Jack McNamara)

David B. Schrembeck, DBS Communications, Inc., OH [IM]
(Alt. to Luigi G. Prezioso)

Anthony Tassone, UL LLC, NY [RT]
(Alt. to Randolph J. Ivans)

CODE-MAKING PANEL NO. 17

Articles 422, 424, 425, 426, 427, 680, 682

Donald R. Cook, *Chair*
Shelby County Department of Development Services, AL [E]
Rep. International Association of Electrical Inspectors

Thomas V. Blewitt, UL LLC, NY [RT]

E. P. Hamilton, III, E. P. Hamilton & Associates, Inc., TX [M]
Rep. Association of Pool & Spa Professionals
(VL to 680)

Randal Hunter, Eaton Bussmann, NV [M]
Rep. National Electrical Manufacturers Association

Don W. Jhonson, Interior Electric, Inc., FL [IM]
Rep. National Electrical Contractors Association

Rachel E. Krepps, Baltimore Gas & Electric Company, MD [UT]
Rep. Electric Light & Power Group/EEI

Dennis Michael Querry, Trinity River Authority, TX [IM]
Rep. Independent Electrical Contractors, Inc.

Chester L. Sandberg, Shell Exploration & Production Inc., CA [U]
Rep. Institute of Electrical & Electronics Engineers, Inc.

Ronald F. Schapp, Intertek Testing Services, OH [RT]

Kenneth M. Shell, Pentair Thermal Management, CA [M]
Rep. Copper Development Association Inc.
(VL to 426, 427)

Peter C. Swim, Whirlpool Corporation, MI [M]
Rep. Air-Conditioning, Heating, & Refrigeration Institute
(VL to 422, 424)

Matt B. Williams, Association of Home Appliance Manufacturers, DC [M]
Rep. Association of Home Appliance Manufacturers
(VL to 422, 424)

Randy J. Yasenchak, IBEW Local Union 607, PA [L]
Rep. International Brotherhood of Electrical Workers

Alternates

Paul W. Abernathy, Encore Wire Corporation, TX [M]
(VL to 426, 427)
(Alt. to Kenneth M. Shell)

Bernie Donnie Bell, Gulf Power Company, FL [UT]
(Alt. to Rachel E. Krepps)

Peter E. Bowers, Satellite Electric Company, Inc., MD [IM]
(Alt. to Dennis Michael Querry)

Ira "Lee" Douglas, Murfreesboro, TN [E]
(Alt. to Donald R. Cook)

Stephen Macey, Watkins Manufacturing Corporation, CA [M]
(VL to 680)
(Alt. to E. P. Hamilton, III)

Wayne E. Morris, Association of Home Appliance Manufacturers, DC [M]
(VL to 422, 424)
(Alt. to Matt B. Williams)

Brian Myers, IBEW Local Union 98, PA [L]
(Alt. to Randy J. Yasenchak)

Gary L. Siggins, UL LLC, CA [RT]
(Alt. to Thomas V. Blewitt)

Kam Fai Siu, Intertek Testing Services, Hong Kong [RT]
(Alt. to Ronald F. Schapp)

Marcelo E. Valdes, GE Energy Industrial Solutions, CT [M]
(Alt. to Randal Hunter)

Nonvoting

Andrew M. Trotta, U.S. Consumer Product Safety Commission, MD [C]

Douglas A. Lee, U.S. Consumer Product Safety Commission, MD [C]

CODE-MAKING PANEL NO. 18

Articles 393, 406, 411, 600, 605

Bobby J. Gray, *Chair*
Hoydar/Buck, Inc., WA [IM]
Rep. National Electrical Contractors Association

Ron D. Alley, Northern New Mexico IEC, NM [IM]
Rep. Independent Electrical Contractors, Inc.

Frederick L. Carpenter, Acuity Brands Lighting, GA [M]
Rep. National Electrical Manufacturers Association

Kurt J. Clemente, Clark Nexsen, Inc., VA [U]
Rep. Institute of Electrical & Electronics Engineers, Inc.

Paul Costello, NECA and IBEW Local 90 JATC, CT [L]
Rep. International Brotherhood of Electrical Workers

Hakim Hasan, Intertek, GA [RT]

Jack E. Jamison, Jr., Miller Engineering, Inc., WV [E]
Rep. International Association of Electrical Inspectors

Charles S. Kurten, UL LLC, NY [RT]

William Ross McCorcle, American Electric Power, OK [UT]
Rep. Electric Light & Power Group/EEI

Michael S. O'Boyle, Philips Lightolier, MA [M]
Rep. American Lighting Association
(VL to 410, 411)

Wesley J. Wilkens, Persona, Inc., SD [M]
Rep. International Sign Association
(VL to 600)

Randall K. Wright, RKW Consulting, PA [SE]

Alternates

Donald Berlin, Intermatic Inc., IL [M]
(VL to 410, 411)
(Alt. to Michael S. O'Boyle)

Steve Campolo, Leviton Manufacturing Company, Inc., NY [M]
(Alt. to Frederick L. Carpenter)

Joseph R. Chandler, Independent Electrical Contractors-Dallas, TX [IM]
(Alt. to Ron D. Alley)

Richard Hollander, City of Tucson, AZ [E]
(Alt. to Jack E. Jamison, Jr.)

Jesse Sprinkle, IBEW Local 461, IL [L]
(Alt. to Paul Costello)

Paul Yesbeck, Acu Sign Corporation, FL [IM]
(Alt. to Bobby J. Gray)

CODE-MAKING PANEL NO. 19

Articles 545, 547, 550, 551, 552, 553, 555, 604, 675, and Annex D, Examples D11 and D12

Ron B. Chilton, *Chair*
North Carolina Department of Insurance, NC [E]
Rep. International Association of Electrical Inspectors

Aisha Bajwa, General Cable Corporation, CA [M]
Rep. The Aluminum Association, Inc.

Barry S. Bauman, Alliant Energy, WI [U]
Rep. American Society of Agricultural & Biological Engineers

Wade Elliott, Utility Services Group, Inc., WA [U]
Rep. National Association of RV Parks & Campgrounds
(VL to 550, 551, 552)

Robert A. Garcia, Cavco Industries/Fleetwood Homes, Inc., AZ [M]

John P. Goodsell, Hubbell Incorporated, CT [M]
Rep. National Electrical Manufacturers Association

Bruce A. Hopkins, Recreation Vehicle Industry Association, VA [M]
(VL to 550, 551, 552)

Ryan Hyer, Testing Engineers International, UT [RT]

David W. Johnson, CenTex IEC, TX [IM]
Rep. Independent Electrical Contractors, Inc.

Thomas R. Lichtenstein, UL LLC, IL [RT]

Doug Mulvaney, Kampgrounds of America, Inc., MT [U]
(VL to 550, 551, 552, 555)

Richard A. Paredes, IBEW Local 164 JATC, NJ [L]
Rep. International Brotherhood of Electrical Workers

Darrell M. Sumbera, Centerpoint Energy, TX [UT]
Rep. Electric Light & Power Group/EEI

Wesley L. Wheeler, National Electrical Contractors Association, MD [IM]
Rep. National Electrical Contractors Association

Michael L. Zieman, RADCO, CA [M]
Rep. Manufactured Housing Institute
(VL to 545, 550, 551, 552)

Donald W. Zipse, Zipse Electrical Forensics, LLC, PA [U]
Rep. Institute of Electrical & Electronics Engineers, Inc.

Alternates

William Bruce Bowman, Fox Systems, Inc., GA [IM]
(Alt. to David W. Johnson)

Garry D. Cole, Shelby/Mansfield KOA, OH [U]
(VL to 550, 551, 552)
(Alt. to Wade Elliott)

Gerald D. Dix, Hampton Roads Joint Apprenticeship Training Committee, VA [L]
(Alt. to Richard A. Paredes)

Chris Fairlee, Kampgrounds of America, Inc., MT [U]
(VL to 550, 551, 552, 555)
(Alt. to Doug Mulvaney)

Robert J. Fick, Alliant Energy, WI [U]
(Alt. to Barry S. Bauman)

Dean C. Hunter, Minnesota Department of Labor & Industry, MN [E]
(Alt. to Ron B. Chilton)

Kent Perkins, Recreation Vehicle Industry Association, VA [M]
(VL to 550, 551, 552)
(Alt. to Bruce A. Hopkins)

Thomas L. Pottschmidt, Indianapolis Power & Light, IN [UT]
(Alt. to Darrell M. Sumbera)

Paul J. Reis, AFC Cable Systems, Inc., MA [M]
(Alt. to Aisha Bajwa)

Stephen G. Rood, Legrand North America, NY [M]
(Alt. to John P. Goodsell)

Eugene W. Wirth, UL LLC, WA [RT]
(Alt. to Thomas R. Lichtenstein)

Committee Scope: This Committee shall have primary responsibility for documents on minimizing the risk of electricity as a source of electric shock and as a potential ignition source of fires and explosions. It shall also be responsible for text to minimize the propagation of fire and explosions due to electrical installations.

NFPA Electrical Engineering Division Technical Staff

NFPA 70

National Electrical Code

2017 Edition

IMPORTANT NOTE: This NFPA document is made available for use subject to important notices and legal disclaimers. These notices and disclaimers appear in all publications containing this document and may be found under the heading "Important Notices and Disclaimers Concerning NFPA Standards." They can also be viewed at www.nfpa.org/disclaimers or obtained on request from NFPA.

UPDATES, ALERTS, AND FUTURE EDITIONS: New editions of NFPA codes, standards, recommended practices, and guides (i.e., NFPA Standards) are released on scheduled revision cycles. This edition may be superseded by a later one, or it may be amended outside of its scheduled revision cycle through the issuance of Tentative Interim Amendments (TIAs). An official NFPA Standard at any point in time consists of the current edition of the document, together with all TIAs and Errata in effect. To verify that this document is the current edition or to determine if it has been amended by TIAs or Errata, please consult the National Fire Codes® Subscription Service or the "List of NFPA Codes & Standards" at www.nfpa.org/docinfo. In addition to TIAs and Errata, the document information pages also include the option to sign up for alerts for individual documents and to be involved in the development of the next edition.

This 2017 edition includes the following usability features as aids to the user. Changes other than editorial are indicated with gray shading within sections. An entire figure caption with gray shading indicates a change to an existing figure. New sections, tables, and figures are indicated by a bold, italic **_N_** in a gray box to the left of the new material. An **_N_** next to an Article title indicates that the entire Article is new. Where one or more complete paragraphs have been deleted, the deletion is indicated by a bullet (•) between the paragraphs that remain.

ARTICLE 90
Introduction

90.1 Purpose.

(A) Practical Safeguarding. The purpose of this *Code* is the practical safeguarding of persons and property from hazards arising from the use of electricity. This *Code* is not intended as a design specification or an instruction manual for untrained persons.

(B) Adequacy. This *Code* contains provisions that are considered necessary for safety. Compliance therewith and proper maintenance result in an installation that is essentially free from hazard but not necessarily efficient, convenient, or adequate for good service or future expansion of electrical use.

Informational Note: Hazards often occur because of overloading of wiring systems by methods or usage not in conformity with this *Code*. This occurs because initial wiring did not provide for increases in the use of electricity. An initial adequate installation and reasonable provisions for system changes provide for future increases in the use of electricity.

(C) Relation to Other International Standards. The requirements in this *Code* address the fundamental principles of protection for safety contained in Section 131 of International Electrotechnical Commission Standard 60364-1, *Electrical Installations of Buildings.*

Informational Note: IEC 60364-1, Section 131, contains fundamental principles of protection for safety that encompass protection against electric shock, protection against thermal effects, protection against overcurrent, protection against fault currents, and protection against overvoltage. All of these potential hazards are addressed by the requirements in this *Code*.

90.2 Scope.

(A) Covered. This *Code* covers the installation and removal of electrical conductors, equipment, and raceways; signaling and communications conductors, equipment, and raceways; and optical fiber cables and raceways for the following:

(1) Public and private premises, including buildings, structures, mobile homes, recreational vehicles, and floating buildings

(2) Yards, lots, parking lots, carnivals, and industrial substations

(3) Installations of conductors and equipment that connect to the supply of electricity

(4) Installations used by the electric utility, such as office buildings, warehouses, garages, machine shops, and recreational buildings, that are not an integral part of a generating plant, substation, or control center

(B) Not Covered. This *Code* does not cover the following:

(1) Installations in ships, watercraft other than floating buildings, railway rolling stock, aircraft, or automotive vehicles other than mobile homes and recreational vehicles

Informational Note: Although the scope of this *Code* indicates that the *Code* does not cover installations in ships, portions of this *Code* are incorporated by reference into Title 46, Code of Federal Regulations, Parts 110–113.

(2) Installations underground in mines and self-propelled mobile surface mining machinery and its attendant electrical trailing cable

(3) Installations of railways for generation, transformation, transmission, energy storage, or distribution of power used exclusively for operation of rolling stock or installations used exclusively for signaling and communications purposes

(4) Installations of communications equipment under the exclusive control of communications utilities located outdoors or in building spaces used exclusively for such installations

(5) Installations under the exclusive control of an electric utility where such installations

a. Consist of service drops or service laterals, and associated metering, or

b. Are on property owned or leased by the electric utility for the purpose of communications, metering, generation, control, transformation, transmission, energy storage, or distribution of electric energy, or

c. Are located in legally established easements or rights-of-way, or

d. Are located by other written agreements either designated by or recognized by public service commissions, utility commissions, or other regulatory agencies having jurisdiction for such installations. These written agreements shall be limited to installations for the purpose of communications, metering, generation, control, transformation, transmission, energy storage, or distribution of electric energy where legally established easements or rights-of-way cannot be obtained. These installations shall be limited to federal lands, Native American reservations through the U.S. Department of the Interior Bureau of Indian Affairs, military bases, lands controlled by port authorities and state agencies and departments, and lands owned by railroads.

Informational Note to (4) and (5): Examples of utilities may include those entities that are typically designated or recognized by governmental law or regulation by public service/utility commissions and that install, operate, and maintain electric supply (such as generation, transmission, or distribution systems) or communications systems (such as telephone, CATV, Internet, satellite, or data services). Utilities may be subject to compliance with codes and standards covering their regulated activities as adopted under governmental law or regulation. Additional information can be found through consultation with the appropriate governmental bodies, such as state regulatory commissions, the Federal Energy Regulatory Commission, and the Federal Communications Commission.

(C) Special Permission. The authority having jurisdiction for enforcing this *Code* may grant exception for the installation of conductors and equipment that are not under the exclusive control of the electric utilities and are used to connect the electric utility supply system to the service conductors of the premises served, provided such installations are outside a building or structure, or terminate inside at a readily accessible location nearest the point of entrance of the service conductors.

90.3 Code Arrangement. This *Code* is divided into the introduction and nine chapters, as shown in Figure 90.3. Chapters 1, 2, 3, and 4 apply generally. Chapters 5, 6, and 7 apply to special occupancies, special equipment, or other special conditions and may supplement or modify the requirements in Chapters 1 through 7.

Chapter 8 covers communications systems and is not subject to the requirements of Chapters 1 through 7 except where the requirements are specifically referenced in Chapter 8.

Chapter 9 consists of tables that are applicable as referenced.

Informative annexes are not part of the requirements of this *Code* but are included for informational purposes only.

90.4 Enforcement. This *Code* is intended to be suitable for mandatory application by governmental bodies that exercise legal jurisdiction over electrical installations, including signaling and communications systems, and for use by insurance inspectors. The authority having jurisdiction for enforcement of the *Code* has the responsibility for making interpretations of the rules, for deciding on the approval of equipment and materials, and for granting the special permission contemplated in a number of the rules.

By special permission, the authority having jurisdiction may waive specific requirements in this *Code* or permit alternative methods where it is assured that equivalent objectives can be achieved by establishing and maintaining effective safety.

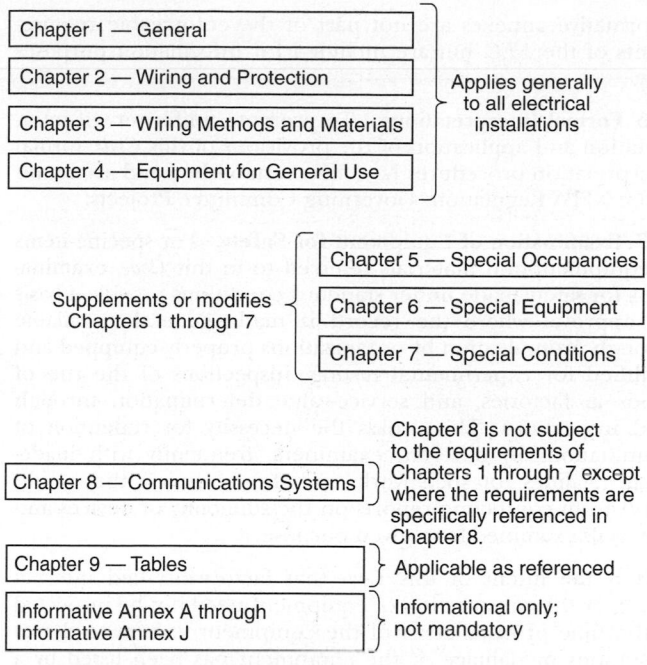

FIGURE 90.3   Code Arrangement.

This *Code* may require new products, constructions, or materials that may not yet be available at the time the *Code* is adopted. In such event, the authority having jurisdiction may permit the use of the products, constructions, or materials that comply with the most recent previous edition of this *Code* adopted by the jurisdiction.

90.5 Mandatory Rules, Permissive Rules, and Explanatory Material.

(A) Mandatory Rules. Mandatory rules of this *Code* are those that identify actions that are specifically required or prohibited and are characterized by the use of the terms *shall* or *shall not*.

(B) Permissive Rules. Permissive rules of this *Code* are those that identify actions that are allowed but not required, are normally used to describe options or alternative methods, and are characterized by the use of the terms *shall be permitted* or *shall not be required.*

(C) Explanatory Material. Explanatory material, such as references to other standards, references to related sections of this *Code,* or information related to a *Code* rule, is included in this *Code* in the form of informational notes. Such notes are informational only and are not enforceable as requirements of this *Code.*

Brackets containing section references to another NFPA document are for informational purposes only and are provided as a guide to indicate the source of the extracted text. These bracketed references immediately follow the extracted text.

Informational Note: The format and language used in this *Code* follows guidelines established by NFPA and published in the *NEC Style Manual.* Copies of this manual can be obtained from NFPA.

(D) Informative Annexes. Nonmandatory information relative to the use of the *NEC* is provided in informative annexes.

Informative annexes are not part of the enforceable requirements of the *NEC*, but are included for information purposes only.

90.6 Formal Interpretations. To promote uniformity of interpretation and application of the provisions of this *Code*, formal interpretation procedures have been established and are found in the NFPA Regulations Governing Committee Projects.

90.7 Examination of Equipment for Safety. For specific items of equipment and materials referred to in this *Code*, examinations for safety made under standard conditions provide a basis for approval where the record is made generally available through promulgation by organizations properly equipped and qualified for experimental testing, inspections of the run of goods at factories, and service-value determination through field inspections. This avoids the necessity for repetition of examinations by different examiners, frequently with inadequate facilities for such work, and the confusion that would result from conflicting reports on the suitability of devices and materials examined for a given purpose.

It is the intent of this *Code* that factory-installed internal wiring or the construction of equipment need not be inspected at the time of installation of the equipment, except to detect alterations or damage, if the equipment has been listed by a qualified electrical testing laboratory that is recognized as having the facilities described in the preceding paragraph and that requires suitability for installation in accordance with this *Code*. Suitability shall be determined by application of requirements that are compatible with this *Code*.

> Informational Note No. 1: See requirements in 110.3.
>
> Informational Note No. 2: *Listed* is defined in Article 100.
>
> Informational Note No. 3: Informative Annex A contains a list of product safety standards that are compatible with this *Code*.

90.8 Wiring Planning.

(A) Future Expansion and Convenience. Plans and specifications that provide ample space in raceways, spare raceways, and additional spaces allow for future increases in electric power and communications circuits. Distribution centers located in readily accessible locations provide convenience and safety of operation.

(B) Number of Circuits in Enclosures. It is elsewhere provided in this *Code* that the number of circuits confined in a single enclosure be varyingly restricted. Limiting the number of circuits in a single enclosure minimizes the effects from a short circuit or ground fault.

90.9 Units of Measurement.

(A) Measurement System of Preference. For the purpose of this *Code*, metric units of measurement are in accordance with the modernized metric system known as the International System of Units (SI).

(B) Dual System of Units. SI units shall appear first, and inch-pound units shall immediately follow in parentheses. Conversion from inch-pound units to SI units shall be based on hard conversion except as provided in 90.9(C).

(C) Permitted Uses of Soft Conversion. The cases given in 90.9(C)(1) through (C)(4) shall not be required to use hard conversion and shall be permitted to use soft conversion.

(1) Trade Sizes. Where the actual measured size of a product is not the same as the nominal size, trade size designators shall be used rather than dimensions. Trade practices shall be followed in all cases.

(2) Extracted Material. Where material is extracted from another standard, the context of the original material shall not be compromised or violated. Any editing of the extracted text shall be confined to making the style consistent with that of the *NEC*.

(3) Industry Practice. Where industry practice is to express units in inch-pound units, the inclusion of SI units shall not be required.

(4) Safety. Where a negative impact on safety would result, soft conversion shall be used.

(D) Compliance. Conversion from inch-pound units to SI units shall be permitted to be an approximate conversion. Compliance with the numbers shown in either the SI system or the inch-pound system shall constitute compliance with this *Code*.

> Informational Note No. 1: Hard conversion is considered a change in dimensions or properties of an item into new sizes that might or might not be interchangeable with the sizes used in the original measurement. Soft conversion is considered a direct mathematical conversion and involves a change in the description of an existing measurement but not in the actual dimension.
>
> Informational Note No. 2: SI conversions are based on IEEE/ASTM SI 10-1997, *Standard for the Use of the International System of Units (SI): The Modern Metric System.*

Chapter 1 General

ARTICLE 100
Definitions

Scope. This article contains only those definitions essential to the application of this *Code*. It is not intended to include commonly defined general terms or commonly defined technical terms from related codes and standards. In general, only those terms that are used in two or more articles are defined in Article 100. Other definitions are included in the article in which they are used but may be referenced in Article 100.

Part I of this article contains definitions intended to apply wherever the terms are used throughout this *Code*. Part II contains definitions applicable to installations and equipment operating at over 1000 volts, nominal.

Part I. General

Accessible (as applied to equipment). Admitting close approach; not guarded by locked doors, elevation, or other effective means. (CMP-1)

Accessible (as applied to wiring methods). Capable of being removed or exposed without damaging the building structure or finish or not permanently closed in by the structure or finish of the building (CMP-1)

Accessible, Readily (Readily Accessible). Capable of being reached quickly for operation, renewal, or inspections without requiring those to whom ready access is requisite to take actions such as to use tools (other than keys), to climb over or under, to remove obstacles, or to resort to portable ladders, and so forth. (CMP-1)

> Informational Note: Use of keys is a common practice under controlled or supervised conditions and a common alternative to the ready access requirements under such supervised conditions as provided elsewhere in the NEC.

Adjustable Speed Drive. Power conversion equipment that provides a means of adjusting the speed of an electric motor. (CMP-11)

> Informational Note: A variable frequency drive is one type of electronic adjustable speed drive that controls the rotational speed of an ac electric motor by controlling the frequency and voltage of the electrical power supplied to the motor.

Adjustable Speed Drive System. A combination of an adjustable speed drive, its associated motor(s), and auxiliary equipment. (CMP-11)

Ampacity. The maximum current, in amperes, that a conductor can carry continuously under the conditions of use without exceeding its temperature rating. (CMP-6)

Appliance. Utilization equipment, generally other than industrial, that is normally built in standardized sizes or types and is installed or connected as a unit to perform one or more functions such as clothes washing, air-conditioning, food mixing, deep frying, and so forth. (CMP-17)

Approved. Acceptable to the authority having jurisdiction. (CMP-1)

Arc-Fault Circuit Interrupter (AFCI). A device intended to provide protection from the effects of arc faults by recognizing characteristics unique to arcing and by functioning to de-energize the circuit when an arc fault is detected. (CMP-2)

Askarel. A generic term for a group of nonflammable synthetic chlorinated hydrocarbons used as electrical insulating media. (CMP-9)

> Informational Note: Askarels of various compositional types are used. Under arcing conditions, the gases produced, while consisting predominantly of noncombustible hydrogen chloride, can include varying amounts of combustible gases, depending on the askarel type.

Associated Apparatus [as applied to Hazardous (Classified) Locations]. Apparatus in which the circuits are not necessarily intrinsically safe themselves but that affects the energy in the intrinsically safe circuits and is relied on to maintain intrinsic safety. Such apparatus is one of the following:

(1) Electrical apparatus that has an alternative type of protection for use in the appropriate hazardous (classified) location

(2) Electrical apparatus not so protected that shall not be used within a hazardous (classified) location

(CMP-14)

> Informational Note No. 1: Associated apparatus has identified intrinsically safe connections for intrinsically safe apparatus and also may have connections for nonintrinsically safe apparatus.

> Informational Note No. 2: An example of associated apparatus is an intrinsic safety barrier, which is a network designed to limit the energy (voltage and current) available to the protected circuit in the hazardous (classified) location, under specified fault conditions.

Associated Nonincendive Field Wiring Apparatus [as applied to Hazardous (Classified) Locations]. Apparatus in which the circuits are not necessarily nonincendive themselves but that affect the energy in nonincendive field wiring circuits and are relied upon to maintain nonincendive energy levels. Such apparatus are one of the following:

(1) Electrical apparatus that has an alternative type of protection for use in the appropriate hazardous (classified) location

(2) Electrical apparatus not so protected that shall not be used in a hazardous (classified) location

(CMP-14)

> Informational Note: Associated nonincendive field wiring apparatus has designated associated nonincendive field wiring apparatus connections for nonincendive field wiring apparatus and may also have connections for other electrical apparatus.

Attachment Plug (Plug Cap) (Plug). A device that, by insertion in a receptacle, establishes a connection between the conductors of the attached flexible cord and the conductors connected permanently to the receptacle. (CMP-18)

Authority Having Jurisdiction (AHJ). An organization, office, or individual responsible for enforcing the requirements of a

100.
DEFINITIONS

code or standard, or for approving equipment, materials, an installation, or a procedure. (CMP-1)

> Informational Note: The phrase "authority having jurisdiction," or its acronym AHJ, is used in NFPA documents in a broad manner, since jurisdictions and approval agencies vary, as do their responsibilities. Where public safety is primary, the authority having jurisdiction may be a federal, state, local, or other regional department or individual such as a fire chief; fire marshal; chief of a fire prevention bureau, labor department, or health department; building official; electrical inspector; or others having statutory authority. For insurance purposes, an insurance inspection department, rating bureau, or other insurance company representative may be the authority having jurisdiction. In many circumstances, the property owner or his or her designated agent assumes the role of the authority having jurisdiction; at government installations, the commanding officer or departmental official may be the authority having jurisdiction.

Automatic. Performing a function without the necessity of human intervention. (CMP-1)

Bathroom. An area including a basin with one or more of the following: a toilet, a urinal, a tub, a shower, a bidet, or similar plumbing fixtures. (CMP-2)

Battery System. Interconnected battery subsystems consisting of one or more storage batteries and battery chargers, and can include inverters, converters, and associated electrical equipment. (CMP-13)

Bonded (Bonding). Connected to establish electrical continuity and conductivity. (CMP-5)

Bonding Conductor or Jumper. A reliable conductor to ensure the required electrical conductivity between metal parts required to be electrically connected. (CMP-5)

Bonding Jumper, Equipment. The connection between two or more portions of the equipment grounding conductor. (CMP-5)

Bonding Jumper, Main. The connection between the grounded circuit conductor and the equipment grounding conductor at the service. (CMP-5)

Bonding Jumper, System. The connection between the grounded circuit conductor and the supply-side bonding jumper, or the equipment grounding conductor, or both, at a separately derived system. (CMP-5)

Branch Circuit. The circuit conductors between the final overcurrent device protecting the circuit and the outlet(s). (CMP-2)

Branch Circuit, Appliance. A branch circuit that supplies energy to one or more outlets to which appliances are to be connected and that has no permanently connected luminaires that are not a part of an appliance. (CMP-2)

Branch Circuit, General-Purpose. A branch circuit that supplies two or more receptacles or outlets for lighting and appliances. (CMP-2)

Branch Circuit, Individual. A branch circuit that supplies only one utilization equipment. (CMP-2)

Branch Circuit, Multiwire. A branch circuit that consists of two or more ungrounded conductors that have a voltage between them, and a grounded conductor that has equal voltage between it and each ungrounded conductor of the circuit and

that is connected to the neutral or grounded conductor of the system. (CMP-2)

Building. A structure that stands alone or that is separated from adjoining structures by fire walls. (CMP-1)

Cabinet. An enclosure that is designed for either surface mounting or flush mounting and is provided with a frame, mat, or trim in which a swinging door or doors are or can be hung. (CMP-9)

Cable Routing Assembly. A single channel or connected multiple channels, as well as associated fittings, forming a structural system that is used to support and route communications wires and cables, optical fiber cables, data cables associated with information technology and communications equipment, Class 2, Class 3, and Type PLTC cables, and power-limited fire alarm cables in plenum, riser, and general-purpose applications. (CMP-16)

Charge Controller. Equipment that controls dc voltage or dc current, or both, and that is used to charge a battery or other energy storage device. (CMP-13)

Circuit Breaker. A device designed to open and close a circuit by nonautomatic means and to open the circuit automatically on a predetermined overcurrent without damage to itself when properly applied within its rating. (CMP-10)

> Informational Note: The automatic opening means can be integral, direct acting with the circuit breaker, or remote from the circuit breaker.

Adjustable (as applied to circuit breakers). A qualifying term indicating that the circuit breaker can be set to trip at various values of current, time, or both, within a predetermined range.

Instantaneous Trip (as applied to circuit breakers). A qualifying term indicating that no delay is purposely introduced in the tripping action of the circuit breaker.

Inverse Time (as applied to circuit breakers). A qualifying term indicating that there is purposely introduced a delay in the tripping action of the circuit breaker, which delay decreases as the magnitude of the current increases.

Nonadjustable (as applied to circuit breakers). A qualifying term indicating that the circuit breaker does not have any adjustment to alter the value of the current at which it will trip or the time required for its operation.

Setting (of circuit breakers). The value of current, time, or both, at which an adjustable circuit breaker is set to trip.

Clothes Closet. A nonhabitable room or space intended primarily for storage of garments and apparel. (CMP-1)

Coaxial Cable. A cylindrical assembly composed of a conductor centered inside a metallic tube or shield, separated by a dielectric material, and usually covered by an insulating jacket. (CMP-16)

Combustible Dust [as applied to Hazardous (Classified) Locations]. Dust particles that are 500 microns or smaller (i.e., material passing a U.S. No. 35 Standard Sieve as defined in ASTM E11-2015, *Standard Specification for Woven Wire Test Sieve Cloth and Test Sieves*), and present a fire or explosion hazard when dispersed and ignited in air. (CMP-14)

> Informational Note: See ASTM E1226-2012a, *Standard Test Method for Explosibility of Dust Clouds*, or ISO 6184-1, *Explosion*

protection systems — Part 1: Determination of explosion indices of combustible dusts in air, for procedures for determining the explosibility of dusts.

Combustible Gas Detection System [as applied to Hazardous (Classified) Locations]. A protection technique utilizing stationary gas detectors in industrial establishments. (CMP-14)

Communications Equipment. The electronic equipment that performs the telecommunications operations for the transmission of audio, video, and data, and includes power equipment (e.g., dc converters, inverters, and batteries), technical support equipment (e.g., computers), and conductors dedicated solely to the operation of the equipment. (CMP-16)

> Informational Note: As the telecommunications network transitions to a more data-centric network, computers, routers, servers, and their powering equipment, are becoming essential to the transmission of audio, video, and data and are finding increasing application in communications equipment installations.

Communications Raceway. An enclosed channel of nonmetallic materials designed expressly for holding communications wires and cables; optical fiber cables; data cables associated with information technology and communications equipment; Class 2, Class 3, and Type PLTC cables; and power-limited fire alarm cables in plenum, riser, and general-purpose applications. (CMP-16)

Composite Optical Fiber Cable. A cable containing optical fibers and current-carrying electrical conductors. (CMP-16)

Concealed. Rendered inaccessible by the structure or finish of the building. (CMP-1)

> Informational Note: Wires in concealed raceways are considered concealed, even though they may become accessible by withdrawing them.

Conductive Optical Fiber Cable. A factory assembly of one or more optical fibers having an overall covering and containing non–current-carrying conductive member(s) such as metallic strength member(s), metallic vapor barrier(s), metallic armor or metallic sheath. (CMP-16)

Conductor, Bare. A conductor having no covering or electrical insulation whatsoever. (CMP-6)

Conductor, Covered. A conductor encased within material of composition or thickness that is not recognized by this *Code* as electrical insulation. (CMP-6)

Conductor, Insulated. A conductor encased within material of composition and thickness that is recognized by this *Code* as electrical insulation. (CMP-6)

Conduit Body. A separate portion of a conduit or tubing system that provides access through a removable cover(s) to the interior of the system at a junction of two or more sections of the system or at a terminal point of the system.

Boxes such as FS and FD or larger cast or sheet metal boxes are not classified as conduit bodies. (CMP-9)

Connector, Pressure (Solderless). A device that establishes a connection between two or more conductors or between one or more conductors and a terminal by means of mechanical pressure and without the use of solder. (CMP-1)

Continuous Load. A load where the maximum current is expected to continue for 3 hours or more. (CMP-2)

Control Circuit. The circuit of a control apparatus or system that carries the electric signals directing the performance of the controller but does not carry the main power current. (CMP-11)

Control Drawing [as applied to Hazardous (Classified) Locations]. A drawing or other document provided by the manufacturer of the intrinsically safe or associated apparatus, or of the nonincendive field wiring apparatus or associated nonincendive field wiring apparatus, that details the allowed interconnections between the intrinsically safe and associated apparatus or between the nonincendive field wiring apparatus or associated nonincendive field wiring apparatus. (CMP-14)

Controller. A device or group of devices that serves to govern, in some predetermined manner, the electric power delivered to the apparatus to which it is connected. (CMP-1)

Cooking Unit, Counter-Mounted. A cooking appliance designed for mounting in or on a counter and consisting of one or more heating elements, internal wiring, and built-in or mountable controls. (CMP-2)

Coordination, Selective (Selective Coordination). Localization of an overcurrent condition to restrict outages to the circuit or equipment affected, accomplished by the selection and installation of overcurrent protective devices and their ratings or settings for the full range of available overcurrents, from overload to the maximum available fault current, and for the full range of overcurrent protective device opening times associated with those overcurrents. (CMP-10)

Copper-Clad Aluminum Conductors. Conductors drawn from a copper-clad aluminum rod, with the copper metallurgically bonded to an aluminum core, where the copper forms a minimum of 10 percent of the cross-sectional area of a solid conductor or each strand of a stranded conductor. (CMP-6)

N **Cord Connector [as applied to Hazardous (Classified) Locations].** A fitting intended to terminate a cord to a box or similar device and reduce the strain at points of termination and may include an explosionproof, a dust-ignitionproof, or a flameproof seal. (CMP-14)

Cutout Box. An enclosure designed for surface mounting that has swinging doors or covers secured directly to and telescoping with the walls of the enclosure. (CMP-9)

Dead Front. Without live parts exposed to a person on the operating side of the equipment. (CMP-9)

Demand Factor. The ratio of the maximum demand of a system, or part of a system, to the total connected load of a system or the part of the system under consideration. (CMP-2)

Device. A unit of an electrical system, other than a conductor, that carries or controls electric energy as its principal function. (CMP-1)

Disconnecting Means. A device, or group of devices, or other means by which the conductors of a circuit can be disconnected from their source of supply. (CMP-1)

Dust-Ignitionproof [as applied to Hazardous (Classified) Locations]. Equipment enclosed in a manner that excludes dusts and does not permit arcs, sparks, or heat otherwise generated or liberated inside of the enclosure to cause ignition of exterior accumulations or atmospheric suspensions of a specified dust on or in the vicinity of the enclosure. (CMP-14)

Informational Note: For further information on dust-ignitionproof enclosures, see ANSI/UL 1202-2013, *Enclosures for Electrical Equipment*, and ANSI/UL 1203-2013, *Explosionproof and Dust-Ignitionproof Electrical Equipment for Hazardous (Classified) Locations.*

Dusttight. Enclosures constructed so that dust will not enter under specified test conditions. (CMP-14)

Informational Note No. 1: Enclosure Types 3, 3S, 3SX, 4, 4X, 5, 6, 6P, 12, 12K, and 13, per ANSI/NEMA 250-2014, *Enclosures for Electrical Equipment*, are considered dusttight and suitable for use in unclassified locations and in Class II, Division 2; Class III; and Zone 22 hazardous (classified) locations.

Informational Note No. 2: For further information, see ANSI/ISA-12.12.01-2013, *Nonincendive Electrical Equipment for Use in Class I and II, Division 2, and Class III, Divisions 1 and 2 Hazardous (Classified) Locations.*

Duty, Continuous. Operation at a substantially constant load for an indefinitely long time. (CMP-1)

Duty, Intermittent. Operation for alternate intervals of (1) load and no load; or (2) load and rest; or (3) load, no load, and rest. (CMP-1)

Duty, Periodic. Intermittent operation in which the load conditions are regularly recurrent. (CMP-1)

Duty, Short-Time. Operation at a substantially constant load for a short and definite, specified time. (CMP-1)

Duty, Varying. Operation at loads, and for intervals of time, both of which may be subject to wide variation. (CMP-1)

Dwelling, One-Family. A building that consists solely of one dwelling unit. (CMP-1)

Dwelling, Two-Family. A building that consists solely of two dwelling units. (CMP-1)

Dwelling, Multifamily. A building that contains three or more dwelling units. (CMP-1)

Dwelling Unit. A single unit, providing complete and independent living facilities for one or more persons, including permanent provisions for living, sleeping, cooking, and sanitation. (CMP-2)

Effective Ground-Fault Current Path. An intentionally constructed, low-impedance electrically conductive path designed and intended to carry current under ground-fault conditions from the point of a ground fault on a wiring system to the electrical supply source and that facilitates the operation of the overcurrent protective device or ground-fault detectors. (CMP-5)

Electric Power Production and Distribution Network. Power production, distribution, and utilization equipment and facilities, such as electric utility systems that deliver electric power to the connected loads, that are external to and not controlled by an interactive system. (CMP-13)

Electric Sign. A fixed, stationary, or portable self-contained, electrically operated and/or electrically illuminated utilization equipment with words or symbols designed to convey information or attract attention. (CMP-18)

Electric-Discharge Lighting. Systems of illumination utilizing fluorescent lamps, high-intensity discharge (HID) lamps, or neon tubing. (CMP-18)

Electrical Circuit Protective System A system consisting of components and materials intended for installation as protection for specific electrical wiring systems with respect to the disruption of electrical circuit integrity upon exterior fire exposure. (CMP-16)

Electronically Actuated Fuse. An overcurrent protective device that generally consists of a control module that provides current-sensing, electronically derived time–current characteristics, energy to initiate tripping, and an interrupting module that interrupts current when an overcurrent occurs. Such fuses may or may not operate in a current-limiting fashion, depending on the type of control selected. (CMP-10)

Enclosed. Surrounded by a case, housing, fence, or wall(s) that prevents persons from accidentally contacting energized parts. (CMP-1)

Enclosure. The case or housing of apparatus, or the fence or walls surrounding an installation to prevent personnel from accidentally contacting energized parts or to protect the equipment from physical damage. (CMP-1)

Informational Note: See Table 110.28 for examples of enclosure types.

Energized. Electrically connected to, or is, a source of voltage. (CMP-1)

Equipment. A general term, including fittings, devices, appliances, luminaires, apparatus, machinery, and the like used as a part of, or in connection with, an electrical installation. (CMP-1)

Explosionproof Equipment. Equipment enclosed in a case that is capable of withstanding an explosion of a specified gas or vapor that may occur within it and of preventing the ignition of a specified gas or vapor surrounding the enclosure by sparks, flashes, or explosion of the gas or vapor within, and that operates at such an external temperature that a surrounding flammable atmosphere will not be ignited thereby. (CMP-14)

Informational Note: For further information, see ANSI/UL 1203-2009, *Explosion-Proof and Dust-Ignition-Proof Electrical Equipment for Use in Hazardous (Classified) Locations.*

Exposed (as applied to live parts). Capable of being inadvertently touched or approached nearer than a safe distance by a person. (CMP-1)

Informational Note: This term applies to parts that are not suitably guarded, isolated, or insulated.

Exposed (as applied to wiring methods). On or attached to the surface or behind panels designed to allow access. (CMP-1)

Externally Operable. Capable of being operated without exposing the operator to contact with live parts. (CMP-1)

Feeder. All circuit conductors between the service equipment, the source of a separately derived system, or other power supply source and the final branch-circuit overcurrent device. (CMP-2)

Festoon Lighting. A string of outdoor lights that is suspended between two points. (CMP-18)

N **Field Evaluation Body (FEB).** An organization or part of an organization that performs field evaluations of electrical or other equipment. [790, 2012] (CMP-1)

N **Field Labeled (as applied to evaluated products).** Equipment or materials to which has been attached a label, symbol, or other identifying mark of an FEB indicating the equipment or materials were evaluated and found to comply with requirements as described in an accompanying field evaluation report. (CMP-1)

Fitting. An accessory such as a locknut, bushing, or other part of a wiring system that is intended primarily to perform a mechanical rather than an electrical function. (CMP-1)

Garage. A building or portion of a building in which one or more self-propelled vehicles can be kept for use, sale, storage, rental, repair, exhibition, or demonstration purposes. (CMP-1)

> Informational Note: For commercial garages, repair and storage, see Article 511.

Ground. The earth. (CMP-5)

Ground Fault. An unintentional, electrically conductive connection between an ungrounded conductor of an electrical circuit and the normally non–current-carrying conductors, metallic enclosures, metallic raceways, metallic equipment, or earth. (CMP-5)

Grounded (Grounding). Connected (connecting) to ground or to a conductive body that extends the ground connection. (CMP-5)

Grounded, Solidly. Connected to ground without inserting any resistor or impedance device. (CMP-5)

Grounded Conductor. A system or circuit conductor that is intentionally grounded. (CMP-5)

Ground-Fault Circuit Interrupter (GFCI). A device intended for the protection of personnel that functions to de-energize a circuit or portion thereof within an established period of time when a current to ground exceeds the values established for a Class A device. (CMP-2)

> Informational Note: Class A ground-fault circuit interrupters trip when the current to ground is 6 mA or higher and do not trip when the current to ground is less than 4 mA. For further information, see UL 943, *Standard for Ground-Fault Circuit Interrupters.*

Ground-Fault Current Path. An electrically conductive path from the point of a ground fault on a wiring system through normally non–current-carrying conductors, equipment, or the earth to the electrical supply source. (CMP-5)

> Informational Note: Examples of ground-fault current paths are any combination of equipment grounding conductors, metallic raceways, metallic cable sheaths, electrical equipment, and any other electrically conductive material such as metal, water, and gas piping; steel framing members; stucco mesh; metal ducting; reinforcing steel; shields of communications cables; and the earth itself.

Ground-Fault Protection of Equipment. A system intended to provide protection of equipment from damaging line-to-ground fault currents by operating to cause a disconnecting means to open all ungrounded conductors of the faulted circuit. This protection is provided at current levels less than those required to protect conductors from damage through the operation of a supply circuit overcurrent device. (CMP-5)

Grounding Conductor, Equipment (EGC). The conductive path(s) that provides a ground-fault current path and connects normally non–current-carrying metal parts of equipment together and to the system grounded conductor or to the grounding electrode conductor, or both. (CMP-5)

> Informational Note No. 1: It is recognized that the equipment grounding conductor also performs bonding.

> Informational Note No. 2: See 250.118 for a list of acceptable equipment grounding conductors.

Grounding Electrode. A conducting object through which a direct connection to earth is established. (CMP-5)

Grounding Electrode Conductor. A conductor used to connect the system grounded conductor or the equipment to a grounding electrode or to a point on the grounding electrode system. (CMP-5)

Guarded. Covered, shielded, fenced, enclosed, or otherwise protected by means of suitable covers, casings, barriers, rails, screens, mats, or platforms to remove the likelihood of approach or contact by persons or objects to a point of danger. (CMP-1)

Guest Room. An accommodation combining living, sleeping, sanitary, and storage facilities within a compartment. (CMP-2)

Guest Suite. An accommodation with two or more contiguous rooms comprising a compartment, with or without doors between such rooms, that provides living, sleeping, sanitary, and storage facilities. (CMP-2)

Handhole Enclosure. An enclosure for use in underground systems, provided with an open or closed bottom, and sized to allow personnel to reach into, but not enter, for the purpose of installing, operating, or maintaining equipment or wiring or both. (CMP-9)

Hermetic Refrigerant Motor-Compressor. A combination consisting of a compressor and motor, both of which are enclosed in the same housing, with no external shaft or shaft seals, with the motor operating in the refrigerant. (CMP-11)

Hermetically Sealed [as applied to Hazardous (Classified) Locations]. Equipment sealed against the entrance of an external atmosphere where the seal is made by fusion, for example, soldering, brazing, welding, or the fusion of glass to metal. (CMP-14)

> Informational Note: For further information, see ANSI/ISA-12.12.01-2013, *Nonincendive Electrical Equipment for Use in Class I and II, Division 2, and Class III, Divisions 1 and 2 Hazardous (Classified) Locations.*

Hoistway. Any shaftway, hatchway, well hole, or other vertical opening or space in which an elevator or dumbwaiter is designed to operate. (CMP-12)

Hybrid System. A system comprised of multiple power sources. These power sources could include photovoltaic, wind, micro-hydro generators, engine-driven generators, and others, but do not include electric power production and distribution network systems. Energy storage systems such as batteries, flywheels, or superconducting magnetic storage equipment do not constitute a power source for the purpose of this definition. The energy regenerated by an overhauling (descending) elevator does not constitute a power source for the purpose of this definition. (CMP-4)

Identified (as applied to equipment). Recognizable as suitable for the specific purpose, function, use, environment, application, and so forth, where described in a particular *Code* requirement. (CMP-1)

> Informational Note: Some examples of ways to determine suitability of equipment for a specific purpose, environment, or application include investigations by a qualified testing laboratory (listing and labeling), an inspection agency, or other organizations concerned with product evaluation.

In Sight From (Within Sight From, Within Sight). Where this *Code* specifies that one equipment shall be "in sight from," "within sight from," or "within sight of," and so forth, another equipment, the specified equipment is to be visible and not more than 15 m (50 ft) distant from the other. (CMP-1)

Industrial Control Panel. An assembly of two or more components consisting of one of the following: (1) power circuit components only, such as motor controllers, overload relays, fused disconnect switches, and circuit breakers; (2) control circuit components only, such as push buttons, pilot lights, selector switches, timers, switches, and control relays; (3) a combination of power and control circuit components. These components, with associated wiring and terminals, are mounted on, or contained within, an enclosure or mounted on a subpanel.

The industrial control panel does not include the controlled equipment. (CMP-11)

Information Technology Equipment (ITE). Equipment and systems rated 1000 volts or less, normally found in offices or other business establishments and similar environments classified as ordinary locations, that are used for creation and manipulation of data, voice, video, and similar signals that are not communications equipment as defined in Part I of Article 100 and do not process communications circuits as defined in 800.2. (CMP-12)

> Informational Note: For information on listing requirements for both information technology equipment and communications equipment, see UL 60950-1-2014, *Information Technology Equipment — Safety — Part 1: General Requirements* or UL 62368-1-2014, *Audio/Video Information and Communication Technology Equipment Part 1: Safety Requirements.*

Innerduct. A nonmetallic raceway placed within a larger raceway. (CMP-16)

Interactive Inverter. An inverter intended for use in parallel with an electric utility to supply common loads that may deliver power to the utility. (CMP-13)

Interactive System. An electric power production system that is operating in parallel with and capable of delivering energy to an electric primary source supply system. (CMP-4)

Interrupting Rating. The highest current at rated voltage that a device is identified to interrupt under standard test conditions. (CMP-10)

> Informational Note: Equipment intended to interrupt current at other than fault levels may have its interrupting rating implied in other ratings, such as horsepower or locked rotor current.

Intersystem Bonding Termination. A device that provides a means for connecting intersystem bonding conductors for communications systems to the grounding electrode system. (CMP-16)

Intrinsically Safe Apparatus. Apparatus in which all the circuits are intrinsically safe. (CMP-14)

Intrinsically Safe System [as applied to Hazardous (Classified) Locations]. An assembly of interconnected intrinsically safe apparatus, associated apparatus, and interconnecting cables, in that those parts of the system that may be used in hazardous (classified) locations are intrinsically safe circuits. (CMP-14)

> Informational Note: An intrinsically safe system may include more than one intrinsically safe circuit.

Isolated (as applied to location). Not readily accessible to persons unless special means for access are used. (CMP-1)

Kitchen. An area with a sink and permanent provisions for food preparation and cooking. (CMP-2)

Labeled. Equipment or materials to which has been attached a label, symbol, or other identifying mark of an organization that is acceptable to the authority having jurisdiction and concerned with product evaluation, that maintains periodic inspection of production of labeled equipment or materials, and by whose labeling the manufacturer indicates compliance with appropriate standards or performance in a specified manner. (CMP-1)

Lighting Outlet. An outlet intended for the direct connection of a lampholder or luminaire. (CMP-18)

Lighting Track (Track Lighting). A manufactured assembly designed to support and energize luminaires that are capable of being readily repositioned on the track. Its length can be altered by the addition or subtraction of sections of track. (CMP-18)

Listed. Equipment, materials, or services included in a list published by an organization that is acceptable to the authority having jurisdiction and concerned with evaluation of products or services, that maintains periodic inspection of production of listed equipment or materials or periodic evaluation of services, and whose listing states that either the equipment, material, or service meets appropriate designated standards or has been tested and found suitable for a specified purpose. (CMP-1)

> Informational Note: The means for identifying listed equipment may vary for each organization concerned with product evaluation, some of which do not recognize equipment as listed unless it is also labeled. Use of the system employed by the listing organization allows the authority having jurisdiction to identify a listed product.

Live Parts. Energized conductive components. (CMP-1)

Location, Damp. Locations protected from weather and not subject to saturation with water or other liquids but subject to moderate degrees of moisture. (CMP-1)

> Informational Note: Examples of such locations include partially protected locations under canopies, marquees, roofed open porches, and like locations, and interior locations subject to moderate degrees of moisture, such as some basements, some barns, and some cold-storage warehouses.

Location, Dry. A location not normally subject to dampness or wetness. A location classified as dry may be temporarily subject to dampness or wetness, as in the case of a building under construction. (CMP-1)

Location, Wet. Installations underground or in concrete slabs or masonry in direct contact with the earth; in locations subject to saturation with water or other liquids, such as vehicle washing areas; and in unprotected locations exposed to weather. (CMP-1)

Luminaire. A complete lighting unit consisting of a light source such as a lamp or lamps, together with the parts designed to position the light source and connect it to the power supply. It may also include parts to protect the light source or the ballast or to distribute the light. A lampholder itself is not a luminaire. (CMP-18)

Mobile Equipment. Equipment with electrical components suitable to be moved only with mechanical aids or is provided with wheels for movement by person(s) or powered devices. (CMP-14)

Motor Control Center. An assembly of one or more enclosed sections having a common power bus and principally containing motor control units. (CMP-11)

Multioutlet Assembly. A type of surface, flush, or freestanding raceway designed to hold conductors and receptacles, assembled in the field or at the factory. (CMP-18)

Neutral Conductor. The conductor connected to the neutral point of a system that is intended to carry current under normal conditions. (CMP-5)

Neutral Point. The common point on a wye-connection in a polyphase system or midpoint on a single-phase, 3-wire system, or midpoint of a single-phase portion of a 3-phase delta system, or a midpoint of a 3-wire, direct-current system. (CMP-5)

Informational Note: At the neutral point of the system, the vectorial sum of the nominal voltages from all other phases within the system that utilize the neutral, with respect to the neutral point, is zero potential.

Nonautomatic. Requiring human intervention to perform a function. (CMP-1)

Nonconductive Optical Fiber Cable. A factory assembly of one or more optical fibers having an overall covering and containing no electrically conductive materials. (CMP-16)

Nonincendive Circuit [as applied to Hazardous (Classified) Locations]. A circuit, other than field wiring, in which any arc or thermal effect produced under intended operating conditions of the equipment, is not capable, under specified test conditions, of igniting the flammable gas–air, vapor–air, or dust–air mixture. (CMP-14)

Informational Note: Conditions are described in ANSI/ISA-12.12.01-2013, *Nonincendive Electrical Equipment for Use in Class I and II, Division 2, and Class III, Divisions 1 and 2 Hazardous (Classified) Locations.*

Nonincendive Component [as applied to Hazardous (Classified) Locations]. A component having contacts for making or breaking an incendive circuit and the contacting mechanism is constructed so that the component is incapable of igniting the specified flammable gas–air or vapor–air mixture. The housing of a nonincendive component is not intended to exclude the flammable atmosphere or contain an explosion. (CMP-14)

Informational Note: For further information, see ANSI/ISA-12.12.01-2013, *Nonincendive Electrical Equipment for Use in Class I and II, Division 2, and Class III, Divisions 1 and 2 Hazardous (Classified) Locations.*

Nonincendive Equipment [as applied to Hazardous (Classified) Locations]. Equipment having electrical/electronic circuitry that is incapable, under normal operating conditions, of causing ignition of a specified flammable gas–air, vapor–air, or dust–air mixture due to arcing or thermal means. (CMP-14)

Informational Note: For further information, see ANSI/ISA-12.12.01-2013, *Nonincendive Electrical Equipment for Use in Class I and II, Division 2, and Class III, Divisions 1 and 2 Hazardous (Classified) Locations.*

Nonincendive Field Wiring [as applied to Hazardous (Classified) Locations]. Wiring that enters or leaves an equipment enclosure and, under normal operating conditions of the equipment, is not capable, due to arcing or thermal effects, of igniting the flammable gas–air, vapor–air, or dust–air mixture. Normal operation includes opening, shorting, or grounding the field wiring. (CMP-14)

Nonincendive Field Wiring Apparatus [as applied to Hazardous (Classified) Locations]. Apparatus intended to be connected to nonincendive field wiring. (CMP-14)

Informational Note: For further information, see ANSI/ISA-12.12.01-2013, *Nonincendive Electrical Equipment for Use in Class I and II, Division 2, and Class III, Divisions 1 and 2 Hazardous (Classified) Locations.*

Nonlinear Load. A load where the wave shape of the steady-state current does not follow the wave shape of the applied voltage. (CMP-1)

Informational Note: Electronic equipment, electronic/electric-discharge lighting, adjustable-speed drive systems, and similar equipment may be nonlinear loads.

Oil Immersion [as applied to Hazardous (Classified) Locations]. Electrical equipment immersed in a protective liquid in such a way that an explosive atmosphere that may be above the liquid or outside the enclosure cannot be ignited. (CMP-14)

Optical Fiber Cable. A factory assembly or field assembly of one or more optical fibers having an overall covering. (CMP-16)

Informational Note: A field-assembled optical fiber cable is an assembly of one or more optical fibers within a jacket. The jacket, without optical fibers, is installed in a manner similar to conduit or raceway. Once the jacket is installed, the optical fibers are inserted into the jacket, completing the cable assembly.

Outlet. A point on the wiring system at which current is taken to supply utilization equipment. (CMP-1)

Outline Lighting. An arrangement of incandescent lamps, electric-discharge lighting, or other electrically powered light sources to outline or call attention to certain features such as the shape of a building or the decoration of a window. (CMP-18)

Overcurrent. Any current in excess of the rated current of equipment or the ampacity of a conductor. It may result from overload, short circuit, or ground fault. (CMP-10)

Informational Note: A current in excess of rating may be accommodated by certain equipment and conductors for a given set of conditions. Therefore, the rules for overcurrent protection are specific for particular situations.

Overcurrent Protective Device, Branch-Circuit. A device capable of providing protection for service, feeder, and branch circuits and equipment over the full range of overcurrents between its rated current and its interrupting rating. Such devices are provided with interrupting ratings appropriate for the intended use but no less than 5000 amperes. (CMP-10)

Overcurrent Protective Device, Supplementary. A device intended to provide limited overcurrent protection for specific applications and utilization equipment such as luminaires and appliances. This limited protection is in addition to the protection provided in the required branch circuit by the branch-circuit overcurrent protective device. (CMP-10)

Overload. Operation of equipment in excess of normal, full-load rating, or of a conductor in excess of rated ampacity that, when it persists for a sufficient length of time, would cause damage or dangerous overheating. A fault, such as a short circuit or ground fault, is not an overload. (CMP-10)

Panelboard. A single panel or group of panel units designed for assembly in the form of a single panel, including buses and automatic overcurrent devices, and equipped with or without switches for the control of light, heat, or power circuits; designed to be placed in a cabinet or cutout box placed in or against a wall, partition, or other support; and accessible only from the front. (CMP-9)

Photovoltaic (PV) System. The total components and subsystem that, in combination, convert solar energy into electric energy for connection to a utilization load. (CMP-4)

Plenum. A compartment or chamber to which one or more air ducts are connected and that forms part of the air distribution system. (CMP-3)

Portable Equipment. Equipment with electrical components suitable to be moved by a single person without mechanical aids. (CMP-14)

Power Outlet. An enclosed assembly that may include receptacles, circuit breakers, fuseholders, fused switches, buses, and watt-hour meter mounting means; intended to supply and control power to mobile homes, recreational vehicles, park trailers, or boats or to serve as a means for distributing power required to operate mobile or temporarily installed equipment. (CMP-19)

Premises Wiring (System). Interior and exterior wiring, including power, lighting, control, and signal circuit wiring together with all their associated hardware, fittings, and wiring devices, both permanently and temporarily installed. This includes (a) wiring from the service point or power source to the outlets or (b) wiring from and including the power source to the outlets where there is no service point.

Such wiring does not include wiring internal to appliances, luminaires, motors, controllers, motor control centers, and similar equipment. (CMP-1)

> Informational Note: Power sources include, but are not limited to, interconnected or stand-alone batteries, solar photovoltaic systems, other distributed generation systems, or generators.

Pressurized [as applied to Hazardous (Classified) Locations]. The process of supplying an enclosure with a protective gas with or without continuous flow, at sufficient pressure to prevent the entrance of combustible dust or ignitible fibers/flyings. (CMP-14)

N **Process Seal [as applied to Hazardous (Classified) Locations].** A seal between electrical systems and flammable or combustible process fluids where a failure could allow the migration of process fluids into the premises' wiring system. (CMP-14)

Purged and Pressurized [as applied to Hazardous (Classified) Locations]. The process of (1) purging, supplying an enclosure with a protective gas at a sufficient flow and positive pressure to reduce the concentration of any flammable gas or vapor initially present to an acceptable level; and (2) pressurization, supplying an enclosure with a protective gas with or without continuous flow at sufficient pressure to prevent the entrance of a flammable gas or vapor, a combustible dust, or an ignitible fiber. (CMP-14)

> Informational Note: For further information, see ANSI/NFPA 496-2013, *Purged and Pressurized Enclosures for Electrical Equipment.*

Qualified Person. One who has skills and knowledge related to the construction and operation of the electrical equipment and installations and has received safety training to recognize and avoid the hazards involved. (CMP-1)

> Informational Note: Refer to NFPA 70E-2012, *Standard for Electrical Safety in the Workplace,* for electrical safety training requirements.

Raceway. An enclosed channel designed expressly for holding wires, cables, or busbars, with additional functions as permitted in this *Code.* (CMP-8)

> Informational Note: A raceway is identified within specific article definitions.

Rainproof. Constructed, protected, or treated so as to prevent rain from interfering with the successful operation of the apparatus under specified test conditions. (CMP-1)

Raintight. Constructed or protected so that exposure to a beating rain will not result in the entrance of water under specified test conditions. (CMP-1)

Receptacle. A contact device installed at the outlet for the connection of an attachment plug, or for the direct connection of electrical utilization equipment designed to mate with the corresponding contact device. A single receptacle is a single contact device with no other contact device on the same yoke. A multiple receptacle is two or more contact devices on the same yoke. (CMP-18)

Receptacle Outlet. An outlet where one or more receptacles are installed. (CMP-18)

Remote-Control Circuit. Any electrical circuit that controls any other circuit through a relay or an equivalent device. (CMP-3)

Retrofit Kit. A general term for a complete subassembly of parts and devices for field conversion of utilization equipment. (CMP-18)

Sealable Equipment. Equipment enclosed in a case or cabinet that is provided with a means of sealing or locking so that live parts cannot be made accessible without opening the enclosure. (CMP-1)

> Informational Note: The equipment may or may not be operable without opening the enclosure.

Separately Derived System. An electrical source, other than a service, having no direct connection(s) to circuit conductors of any other electrical source other than those established by grounding and bonding connections. (CMP-5)

Service. The conductors and equipment for delivering electric energy from the serving utility to the wiring system of the premises served. (CMP-4)

Service Cable. Service conductors made up in the form of a cable. (CMP-4)

Service Conductors. The conductors from the service point to the service disconnecting means. (CMP-4)

Service Conductors, Overhead. The overhead conductors between the service point and the first point of connection to the service-entrance conductors at the building or other structure. (CMP-4)

Service Conductors, Underground. The underground conductors between the service point and the first point of connection to the service-entrance conductors in a terminal box, meter, or other enclosure, inside or outside the building wall. (CMP-4)

Informational Note: Where there is no terminal box, meter, or other enclosure, the point of connection is considered to be the point of entrance of the service conductors into the building.

Service Drop. The overhead conductors between the utility electric supply system and the service point. (CMP-4)

Service-Entrance Conductors, Overhead System. The service conductors between the terminals of the service equipment and a point usually outside the building, clear of building walls, where joined by tap or splice to the service drop or overhead service conductors. (CMP-4)

Service-Entrance Conductors, Underground System. The service conductors between the terminals of the service equipment and the point of connection to the service lateral or underground service conductors. (CMP-4)

Informational Note: Where service equipment is located outside the building walls, there may be no service-entrance conductors or they may be entirely outside the building.

Service Equipment. The necessary equipment, usually consisting of a circuit breaker(s) or switch(es) and fuse(s) and their accessories, connected to the load end of service conductors to a building or other structure, or an otherwise designated area, and intended to constitute the main control and cutoff of the supply. (CMP-4)

Service Lateral. The underground conductors between the utility electric supply system and the service point. (CMP-4)

Service Point. The point of connection between the facilities of the serving utility and the premises wiring. (CMP-4)

Informational Note: The service point can be described as the point of demarcation between where the serving utility ends and the premises wiring begins. The serving utility generally specifies the location of the service point based on the conditions of service.

Short-Circuit Current Rating. The prospective symmetrical fault current at a nominal voltage to which an apparatus or system is able to be connected without sustaining damage exceeding defined acceptance criteria. (CMP-10)

Show Window. Any window, including windows above doors, used or designed to be used for the display of goods or advertising material, whether it is fully or partly enclosed or entirely open at the rear and whether or not it has a platform raised higher than the street floor level. (CMP-2)

Signaling Circuit. Any electrical circuit that energizes signaling equipment. (CMP-3)

Simple Apparatus [as applied to Hazardous (Classified) Locations]. An electrical component or combination of components of simple construction with well-defined electrical parameters that does not generate more than 1.5 volts, 100 mA, and 25 mW, or a passive component that does not dissipate more than 1.3 watts and is compatible with the intrinsic safety of the circuit in which it is used. (CMP-14)

Informational Note: The following apparatus are examples of simple apparatus:

(1)   Passive components; for example, switches, junction boxes, resistance temperature devices, and simple semiconductor devices such as LEDs

(2)   Sources of stored energy consisting of single components in simple circuits with well-defined parameters; for example, capacitors or inductors, whose values are considered when determining the overall safety of the system

(3)   Sources of generated energy; for example, thermocouples and photocells, that do not generate more than 1.5 volts, 100 mA, and 25 mW

Special Permission. The written consent of the authority having jurisdiction. (CMP-1)

Stand-Alone System. A system that supplies power independently of an electrical production and distribution network. (CMP-4)

Structure. That which is built or constructed, other than equipment. (CMP-1)

Surge Arrester. A protective device for limiting surge voltages by discharging or bypassing surge current; it also prevents continued flow of follow current while remaining capable of repeating these functions. (CMP-5)

Surge-Protective Device (SPD). A protective device for limiting transient voltages by diverting or limiting surge current; it also prevents continued flow of follow current while remaining capable of repeating these functions and is designated as follows:

Type 1: Permanently connected SPDs intended for installation between the secondary of the service transformer and the line side of the service disconnect overcurrent device.

Type 2: Permanently connected SPDs intended for installation on the load side of the service disconnect overcurrent device, including SPDs located at the branch panel.

Type 3: Point of utilization SPDs.

Type 4: Component SPDs, including discrete components, as well as assemblies. (CMP-5)

Informational Note: For further information on Type 1, Type 2, Type 3, and Type 4 SPDs, see UL 1449, *Standard for Surge Protective Devices*.

Switch, Bypass Isolation. A manually operated device used in conjunction with a transfer switch to provide a means of directly connecting load conductors to a power source and of disconnecting the transfer switch. (CMP-13)

Switch, General-Use. A switch intended for use in general distribution and branch circuits. It is rated in amperes, and it is capable of interrupting its rated current at its rated voltage. (CMP-9)

Switch, General-Use Snap. A form of general-use switch constructed so that it can be installed in device boxes or on box covers, or otherwise used in conjunction with wiring systems recognized by this *Code*. (CMP-9)

Switch, Isolating. A switch intended for isolating an electrical circuit from the source of power. It has no interrupting rating, and it is intended to be operated only after the circuit has been opened by some other means. (CMP-9)

Switch, Motor-Circuit. A switch rated in horsepower that is capable of interrupting the maximum operating overload current of a motor of the same horsepower rating as the switch at the rated voltage. (CMP-11)

Switch, Transfer. An automatic or nonautomatic device for transferring one or more load conductor connections from one power source to another. (CMP-13)

Switchboard. A large single panel, frame, or assembly of panels on which are mounted on the face, back, or both, switches, overcurrent and other protective devices, buses, and usually instruments. These assemblies are generally accessible from the rear as well as from the front and are not intended to be installed in cabinets. (CMP-9)

Switchgear. An assembly completely enclosed on all sides and top with sheet metal (except for ventilating openings and inspection windows) and containing primary power circuit switching, interrupting devices, or both, with buses and connections. The assembly may include control and auxiliary devices. Access to the interior of the enclosure is provided by doors, removable covers, or both. (CMP-9)

Informational Note: All switchgear subject to *NEC* requirements is metal enclosed. Switchgear rated below 1000 V or less may be identified as "low-voltage power circuit breaker switchgear." Switchgear rated over 1000 V may be identified as "metal-enclosed switchgear" or "metal-clad switchgear." Switchgear is available in non–arc-resistant or arc-resistant constructions.

Thermal Protector (as applied to motors). A protective device for assembly as an integral part of a motor or motor-compressor that, when properly applied, protects the motor against dangerous overheating due to overload and failure to start. (CMP-11)

Informational Note: The thermal protector may consist of one or more sensing elements integral with the motor or motor-compressor and an external control device.

Thermally Protected (as applied to motors). The words *Thermally Protected* appearing on the nameplate of a motor or motor-compressor indicate that the motor is provided with a thermal protector. (CMP-11)

Unclassified Locations [as applied to Hazardous (Classified) Locations]. Locations determined to be neither Class I, Division 1; Class I, Division 2; Class I, Zone 0; Class I, Zone 1; Class I, Zone 2; Class II, Division 1; Class II, Division 2; Class III, Division 1; Class III, Division 2; Zone 20; Zone 21; Zone 22; nor any combination thereof. (CMP-14)

Ungrounded. Not connected to ground or to a conductive body that extends the ground connection. (CMP-5)

Uninterruptible Power Supply. A power supply used to provide alternating current power to a load for some period of time in the event of a power failure. (CMP-13)

Informational Note: In addition, it may provide a more constant voltage and frequency supply to the load, reducing the effects of voltage and frequency variations.

Utilization Equipment. Equipment that utilizes electric energy for electronic, electromechanical, chemical, heating, lighting, or similar purposes. (CMP-1)

Ventilated. Provided with a means to permit circulation of air sufficient to remove an excess of heat, fumes, or vapors. (CMP-14)

Volatile Flammable Liquid. A flammable liquid having a flash point below 38°C (100°F), or a flammable liquid whose temperature is above its flash point, or a Class II combustible liquid that has a vapor pressure not exceeding 276 kPa (40 psia) at 38°C (100°F) and whose temperature is above its flash point. (CMP-14)

Voltage (of a circuit). The greatest root-mean-square (rms) (effective) difference of potential between any two conductors of the circuit concerned. (CMP-1)

Informational Note: Some systems, such as 3-phase 4-wire, single-phase 3-wire, and 3-wire direct current, may have various circuits of various voltages.

Voltage, Nominal. A nominal value assigned to a circuit or system for the purpose of conveniently designating its voltage class (e.g., 120/240 volts, 480Y/277 volts, 600 volts). (CMP-1)

Informational Note No. 1: The actual voltage at which a circuit operates can vary from the nominal within a range that permits satisfactory operation of equipment.

Informational Note No. 2: See ANSI C84.1-2011, *Voltage Ratings for Electric Power Systems and Equipment (60 Hz).*

Informational Note No. 3: Certain battery units may be considered to be rated at nominal 48 volts dc, but may have a charging float voltage up to 58 volts. In dc applications, 60 volts is used to cover the entire range of float voltages.

Voltage to Ground. For grounded circuits, the voltage between the given conductor and that point or conductor of the circuit that is grounded; for ungrounded circuits, the greatest voltage between the given conductor and any other conductor of the circuit. (CMP-1)

Watertight. Constructed so that moisture will not enter the enclosure under specified test conditions. (CMP-1)

Weatherproof. Constructed or protected so that exposure to the weather will not interfere with successful operation. (CMP-1)

Informational Note: Rainproof, raintight, or watertight equipment can fulfill the requirements for weatherproof where varying weather conditions other than wetness, such as snow, ice, dust, or temperature extremes, are not a factor.

Part II. Over 1000 Volts, Nominal

Electronically Actuated Fuse. An overcurrent protective device that generally consists of a control module that provides current sensing, electronically derived time–current characteristics, energy to initiate tripping, and an interrupting module that interrupts current when an overcurrent occurs. Electronically actuated fuses may or may not operate in a current-limiting fashion, depending on the type of control selected. (CMP-10)

Fuse. An overcurrent protective device with a circuit-opening fusible part that is heated and severed by the passage of overcurrent through it. (CMP-10)

Informational Note: A fuse comprises all the parts that form a unit capable of performing the prescribed functions. It may or may not be the complete device necessary to connect it into an electrical circuit.

Controlled Vented Power Fuse. A fuse with provision for controlling discharge circuit interruption such that no solid material may be exhausted into the surrounding atmosphere.

> Informational Note: The fuse is designed so that discharged gases will not ignite or damage insulation in the path of the discharge or propagate a flashover to or between grounded members or conduction members in the path of the discharge where the distance between the vent and such insulation or conduction members conforms to manufacturer's recommendations.

Expulsion Fuse Unit (Expulsion Fuse). A vented fuse unit in which the expulsion effect of gases produced by the arc and lining of the fuseholder, either alone or aided by a spring, extinguishes the arc.

Nonvented Power Fuse. A fuse without intentional provision for the escape of arc gases, liquids, or solid particles to the atmosphere during circuit interruption.

Power Fuse Unit. A vented, nonvented, or controlled vented fuse unit in which the arc is extinguished by being drawn through solid material, granular material, or liquid, either alone or aided by a spring.

Vented Power Fuse. A fuse with provision for the escape of arc gases, liquids, or solid particles to the surrounding atmosphere during circuit interruption.

Multiple Fuse. An assembly of two or more single-pole fuses. (CMP-10)

Substation. An assemblage of equipment (e.g., switches, interrupting devices, circuit breakers, buses, and transformers) through which electric energy is passed for the purpose of distribution, switching, or modifying its characteristics. (CMP-9)

Switching Device. A device designed to close, open, or both, one or more electrical circuits. (CMP-1)

Circuit Breaker. A switching device capable of making, carrying, and interrupting currents under normal circuit conditions, and also of making, carrying for a specified time, and interrupting currents under specified abnormal circuit conditions, such as those of short circuit.

Cutout. An assembly of a fuse support with either a fuseholder, fuse carrier, or disconnecting blade. The fuseholder or fuse carrier may include a conducting element (fuse link) or may act as the disconnecting blade by the inclusion of a nonfusible member.

Disconnecting Means. A device, group of devices, or other means whereby the conductors of a circuit can be disconnected from their source of supply.

Disconnecting (or Isolating) Switch (Disconnector, Isolator). A mechanical switching device used for isolating a circuit or equipment from a source of power.

Interrupter Switch. A switch capable of making, carrying, and interrupting specified currents.

Oil Cutout (Oil-Filled Cutout). A cutout in which all or part of the fuse support and its fuse link or disconnecting blade is mounted in oil with complete immersion of the contacts and the fusible portion of the conducting element (fuse link) so that arc interruption by severing of the fuse link or by opening of the contacts will occur under oil.

Oil Switch. A switch having contacts that operate under oil (or askarel or other suitable liquid).

Regulator Bypass Switch. A specific device or combination of devices designed to bypass a regulator.

ARTICLE 110
Requirements for Electrical Installations

Part I. General

110.1 Scope. This article covers general requirements for the examination and approval, installation and use, access to and spaces about electrical conductors and equipment; enclosures intended for personnel entry; and tunnel installations.

> Informational Note: See Informative Annex J for information regarding ADA accessibility design.

110.2 Approval. The conductors and equipment required or permitted by this *Code* shall be acceptable only if approved.

> Informational Note: See 90.7, Examination of Equipment for Safety, and 110.3, Examination, Identification, Installation, and Use of Equipment. See definitions of *Approved, Identified, Labeled,* and *Listed.*

110.3 Examination, Identification, Installation, Use, and Listing (Product Certification) of Equipment.

(A) Examination. In judging equipment, considerations such as the following shall be evaluated:

(1) Suitability for installation and use in conformity with the provisions of this *Code*

> Informational Note No. 1: Equipment may be new, reconditioned, refurbished, or remanufactured.

> Informational Note No. 2: Suitability of equipment use may be identified by a description marked on or provided with a product to identify the suitability of the product for a specific purpose, environment, or application. Special conditions of use or other limitations and other pertinent information may be marked on the equipment, included in the product instructions, or included in the appropriate listing and labeling information. Suitability of equipment may be evidenced by listing or labeling.

(2) Mechanical strength and durability, including, for parts designed to enclose and protect other equipment, the adequacy of the protection thus provided
(3) Wire-bending and connection space
(4) Electrical insulation
(5) Heating effects under normal conditions of use and also under abnormal conditions likely to arise in service
(6) Arcing effects
(7) Classification by type, size, voltage, current capacity, and specific use
(8) Other factors that contribute to the practical safeguarding of persons using or likely to come in contact with the equipment

(B) Installation and Use. Listed or labeled equipment shall be installed and used in accordance with any instructions included in the listing or labeling.

N **(C) Listing.** Product testing, evaluation, and listing (product certification) shall be performed by recognized qualified electrical testing laboratories and shall be in accordance with applicable product standards recognized as achieving equivalent and effective safety for equipment installed to comply with this *Code.*

> Informational Note: The Occupational Safety and Health Administration (OSHA) recognizes qualified electrical testing laboratories that perform evaluations, testing, and certification of certain products to ensure that they meet the requirements of both the construction and general industry OSHA electrical standards. If the listing (product certification) is done under a qualified electrical testing laboratory program, this listing mark signifies that the tested and certified product complies with the requirements of one or more appropriate product safety test standards.

110.4 Voltages. Throughout this *Code,* the voltage considered shall be that at which the circuit operates. The voltage rating of electrical equipment shall not be less than the nominal voltage of a circuit to which it is connected.

110.5 Conductors. Conductors normally used to carry current shall be of copper or aluminum unless otherwise provided in this *Code.* Where the conductor material is not specified, the sizes given in this *Code* shall apply to copper conductors. Where other materials are used, the size shall be changed accordingly.

> Informational Note: For copper-clad aluminum conductors, see 310.15.

110.6 Conductor Sizes. Conductor sizes are expressed in American Wire Gage (AWG) or in circular mils.

110.7 Wiring Integrity. Completed wiring installations shall be free from short circuits, ground faults, or any connections to ground other than as required or permitted elsewhere in this *Code.*

110.8 Wiring Methods. Only wiring methods recognized as suitable are included in this *Code.* The recognized methods of wiring shall be permitted to be installed in any type of building or occupancy, except as otherwise provided in this *Code.*

110.9 Interrupting Rating. Equipment intended to interrupt current at fault levels shall have an interrupting rating at nominal circuit voltage at least equal to the current that is available at the line terminals of the equipment.

Equipment intended to interrupt current at other than fault levels shall have an interrupting rating at nominal circuit voltage at least equal to the current that must be interrupted.

110.10 Circuit Impedance, Short-Circuit Current Ratings, and Other Characteristics. The overcurrent protective devices, the total impedance, the equipment short-circuit current ratings, and other characteristics of the circuit to be protected shall be selected and coordinated to permit the circuit protective devices used to clear a fault to do so without extensive damage to the electrical equipment of the circuit. This fault shall be assumed to be either between two or more of the circuit conductors or between any circuit conductor and the equipment grounding conductor(s) permitted in 250.118. Listed equipment applied in accordance with their listing shall be considered to meet the requirements of this section.

110.11 Deteriorating Agents. Unless identified for use in the operating environment, no conductors or equipment shall be located in damp or wet locations; where exposed to gases, fumes, vapors, liquids, or other agents that have a deteriorating effect on the conductors or equipment; or where exposed to excessive temperatures.

> Informational Note No. 1: See 300.6 for protection against corrosion.

> Informational Note No. 2: Some cleaning and lubricating compounds can cause severe deterioration of many plastic materials used for insulating and structural applications in equipment.

Equipment not identified for outdoor use and equipment identified only for indoor use, such as "dry locations," "indoor use only," "damp locations," or enclosure Types 1, 2, 5, 12, 12K, and/or 13, shall be protected against damage from the weather during construction.

> Informational Note No. 3: See Table 110.28 for appropriate enclosure-type designations.

> Informational Note No. 4: Minimum flood provisions are provided in NFPA 5000-2015 *Building Construction and Safety Code,* the *International Building Code (IBC),* and the *International Residential Code for One- and Two-Family Dwellings (IRC).*

110.12 Mechanical Execution of Work. Electrical equipment shall be installed in a neat and workmanlike manner.

> Informational Note: Accepted industry practices are described in ANSI/NECA 1-2015, *Standard for Good Workmanship in Electrical Construction,* and other ANSI-approved installation standards.

(A) Unused Openings. Unused openings, other than those intended for the operation of equipment, those intended for mounting purposes, or those permitted as part of the design for listed equipment, shall be closed to afford protection substantially equivalent to the wall of the equipment. Where metallic plugs or plates are used with nonmetallic enclosures, they shall be recessed at least 6 mm (¼ in.) from the outer surface of the enclosure.

(B) Integrity of Electrical Equipment and Connections. Internal parts of electrical equipment, including busbars, wiring terminals, insulators, and other surfaces, shall not be damaged or contaminated by foreign materials such as paint, plaster, cleaners, abrasives, or corrosive residues. There shall be no damaged parts that may adversely affect safe operation or mechanical strength of the equipment such as parts that are broken; bent; cut; or deteriorated by corrosion, chemical action, or overheating.

110.13 Mounting and Cooling of Equipment.

(A) Mounting. Electrical equipment shall be firmly secured to the surface on which it is mounted. Wooden plugs driven into holes in masonry, concrete, plaster, or similar materials shall not be used.

(B) Cooling. Electrical equipment that depends on the natural circulation of air and convection principles for cooling of exposed surfaces shall be installed so that room airflow over such surfaces is not prevented by walls or by adjacent installed equipment. For equipment designed for floor mounting, clearance between top surfaces and adjacent surfaces shall be provided to dissipate rising warm air.

Electrical equipment provided with ventilating openings shall be installed so that walls or other obstructions do not prevent the free circulation of air through the equipment.

110.14 Electrical Connections. Because of different characteristics of dissimilar metals, devices such as pressure terminal or pressure splicing connectors and soldering lugs shall be identified for the material of the conductor and shall be properly installed and used. Conductors of dissimilar metals shall not be intermixed in a terminal or splicing connector where physical contact occurs between dissimilar conductors (such as copper and aluminum, copper and copper-clad aluminum, or aluminum and copper-clad aluminum), unless the device is identified for the purpose and conditions of use. Materials such as solder, fluxes, inhibitors, and compounds, where employed, shall be suitable for the use and shall be of a type that will not adversely affect the conductors, installation, or equipment.

Connectors and terminals for conductors more finely stranded than Class B and Class C stranding as shown in Chapter 9, Table 10, shall be identified for the specific conductor class or classes.

(A) Terminals. Connection of conductors to terminal parts shall ensure a thoroughly good connection without damaging the conductors and shall be made by means of pressure connectors (including set-screw type), solder lugs, or splices to flexible leads. Connection by means of wire-binding screws or studs and nuts that have upturned lugs or the equivalent shall be permitted for 10 AWG or smaller conductors.

Terminals for more than one conductor and terminals used to connect aluminum shall be so identified.

(B) Splices. Conductors shall be spliced or joined with splicing devices identified for the use or by brazing, welding, or soldering with a fusible metal or alloy. Soldered splices shall first be spliced or joined so as to be mechanically and electrically secure without solder and then be soldered. All splices and joints and the free ends of conductors shall be covered with an insulation equivalent to that of the conductors or with an identified insulating device.

Wire connectors or splicing means installed on conductors for direct burial shall be listed for such use.

(C) Temperature Limitations. The temperature rating associated with the ampacity of a conductor shall be selected and coordinated so as not to exceed the lowest temperature rating of any connected termination, conductor, or device. Conductors with temperature ratings higher than specified for terminations shall be permitted to be used for ampacity adjustment, correction, or both.

(1) Equipment Provisions. The determination of termination provisions of equipment shall be based on 110.14(C)(1)(a) or (C)(1)(b). Unless the equipment is listed and marked otherwise, conductor ampacities used in determining equipment termination provisions shall be based on Table 310.15(B)(16) as appropriately modified by 310.15(B)(7).

(a) Termination provisions of equipment for circuits rated 100 amperes or less, or marked for 14 AWG through 1 AWG conductors, shall be used only for one of the following:

(1) Conductors rated 60°C (140°F).

(2) Conductors with higher temperature ratings, provided the ampacity of such conductors is determined based on the 60°C (140°F) ampacity of the conductor size used.

(3) Conductors with higher temperature ratings if the equipment is listed and identified for use with such conductors.

(4) For motors marked with design letters B, C, or D, conductors having an insulation rating of 75°C (167°F) or higher shall be permitted to be used, provided the ampacity of such conductors does not exceed the 75°C (167°F) ampacity.

(b) Termination provisions of equipment for circuits rated over 100 amperes, or marked for conductors larger than 1 AWG, shall be used only for one of the following:

(1) Conductors rated 75°C (167°F)

(2) Conductors with higher temperature ratings, provided the ampacity of such conductors does not exceed the 75°C (167°F) ampacity of the conductor size used, or up to their ampacity if the equipment is listed and identified for use with such conductors

(2) Separate Connector Provisions. Separately installed pressure connectors shall be used with conductors at the ampacities not exceeding the ampacity at the listed and identified temperature rating of the connector.

Informational Note: With respect to 110.14(C)(1) and (C)(2), equipment markings or listing information may additionally restrict the sizing and temperature ratings of connected conductors.

N **(D) Installation.** Where a tightening torque is indicated as a numeric value on equipment or in installation instructions provided by the manufacturer, a calibrated torque tool shall be used to achieve the indicated torque value, unless the equipment manufacturer has provided installation instructions for an alternative method of achieving the required torque.

110.15 High-Leg Marking. On a 4-wire, delta-connected system where the midpoint of one phase winding is grounded, only the conductor or busbar having the higher phase voltage to ground shall be durably and permanently marked by an outer finish that is orange in color or by other effective means. Such identification shall be placed at each point on the system where a connection is made if the grounded conductor is also present.

110.16 Arc-Flash Hazard Warning.

(A) General. Electrical equipment, such as switchboards, switchgear, panelboards, industrial control panels, meter socket enclosures, and motor control centers, that is in other than dwelling units, and is likely to require examination, adjustment, servicing, or maintenance while energized, shall be field or factory marked to warn qualified persons of potential electric arc flash hazards. The marking shall meet the requirements in 110.21(B) and shall be located so as to be clearly visible to qualified persons before examination, adjustment, servicing, or maintenance of the equipment.

(B) Service Equipment. In other than dwelling units, in addition to the requirements in (A), a permanent label shall be field or factory applied to service equipment rated 1200 amps or more. The label shall meet the requirements of 110.21(B) and contain the following information:

(1) Nominal system voltage

(2) Available fault current at the service overcurrent protective devices

(3) The clearing time of service overcurrent protective devices based on the available fault current at the service equipment

(4)   The date the label was applied

Exception: Service equipment labeling shall not be required if an arc flash label is applied in accordance with acceptable industry practice.

Informational Note No. 1: *NFPA 70E -2015, Standard for Electrical Safety in the Workplace*, provides guidance, such as determining severity of potential exposure, planning safe work practices, arc flash labeling, and selecting personal protective equipment.

Informational Note No. 2: ANSI Z535.4-2011, *Product Safety Signs and Labels*, provides guidelines for the design of safety signs and labels for application to products.

Informational Note No. 3: Acceptable industry practices for equipment labeling are described in *NFPA 70E -2015 Standard for Electrical Safety in the Workplace*. This standard provides specific criteria for developing arc-flash labels for equipment that provides nominal system voltage, incident energy levels, arc-flash boundaries, minimum required levels of personal protective equipment, and so forth.

110.18 Arcing Parts.   Parts of electrical equipment that in ordinary operation produce arcs, sparks, flames, or molten metal shall be enclosed or separated and isolated from all combustible material.

Informational Note: For hazardous (classified) locations, see Articles 500 through 517. For motors, see 430.14.

110.19 Light and Power from Railway Conductors.   Circuits for lighting and power shall not be connected to any system that contains trolley wires with a ground return.

Exception: Such circuit connections shall be permitted in car houses, power houses, or passenger and freight stations operated in connection with electric railways.

110.21 Marking.

(A) Equipment Markings.

(1) General.   The manufacturer's name, trademark, or other descriptive marking by which the organization responsible for the product can be identified shall be placed on all electrical equipment. Other markings that indicate voltage, current, wattage, or other ratings shall be provided as specified elsewhere in this *Code*. The marking or label shall be of sufficient durability to withstand the environment involved.

N (2) Reconditioned Equipment.   Reconditioned equipment shall be marked with the name, trademark, or other descriptive marking by which the organization responsible for reconditioning the electrical equipment can be identified, along with the date of the reconditioning.

Reconditioned equipment shall be identified as "reconditioned" and approval of the reconditioned equipment shall not be based solely on the equipment's original listing.

Exception: In industrial occupancies, where conditions of maintenance and supervision ensure that only qualified persons service the equipment, the markings indicated in 110.21(A)(2) shall not be required.

Informational Note: Industry standards are available for application of reconditioned and refurbished equipment. Normal servicing of equipment that remains within a facility should not be considered reconditioning or refurbishing.

(B) Field-Applied Hazard Markings.   Where caution, warning, or danger signs or labels are required by this *Code*, the labels shall meet the following requirements:

(1)   The marking shall warn of the hazards using effective words, colors, symbols, or any combination thereof.

Informational Note: ANSI Z535.4-2011, *Product Safety Signs and Labels*, provides guidelines for suitable font sizes, words, colors, symbols, and location requirements for labels.

(2)   The label shall be permanently affixed to the equipment or wiring method and shall not be handwritten.

Exception to (2): Portions of labels or markings that are variable, or that could be subject to changes, shall be permitted to be handwritten and shall be legible.

(3)   The label shall be of sufficient durability to withstand the environment involved.

Informational Note: ANSI Z535.4-2011, *Product Safety Signs and Labels*, provides guidelines for the design and durability of safety signs and labels for application to electrical equipment.

110.22 Identification of Disconnecting Means.

(A) General.   Each disconnecting means shall be legibly marked to indicate its purpose unless located and arranged so the purpose is evident. The marking shall be of sufficient durability to withstand the environment involved.

(B) Engineered Series Combination Systems.   Equipment enclosures for circuit breakers or fuses applied in compliance with series combination ratings selected under engineering supervision in accordance with 240.86(A) shall be legibly marked in the field as directed by the engineer to indicate the equipment has been applied with a series combination rating. The marking shall meet the requirements in 110.21(B) and shall be readily visible and state the following:

CAUTION — ENGINEERED SERIES COMBINATION SYSTEM RATED _____ AMPERES. IDENTIFIED REPLACEMENT COMPONENTS REQUIRED.

(C) Tested Series Combination Systems.   Equipment enclosures for circuit breakers or fuses applied in compliance with the series combination ratings marked on the equipment by the manufacturer in accordance with 240.86(B) shall be legibly marked in the field to indicate the equipment has been applied with a series combination rating. The marking shall meet the requirements in 110.21(B) and shall be readily visible and state the following:

CAUTION — SERIES COMBINATION SYSTEM RATED ____ AMPERES. IDENTIFIED REPLACEMENT COMPONENTS REQUIRED.

Informational Note: See IEEE 3004.5-2014 *Recommended Practice for the Application of Low-Voltage Circuit Breakers in Industrial and Commercial Power Systems*, for further information on series tested systems.

110.23 Current Transformers.   Unused current transformers associated with potentially energized circuits shall be short-circuited.

110.24 Available Fault Current.

(A) Field Marking.   Service equipment at other than dwelling units shall be legibly marked in the field with the maximum available fault current. The field marking(s) shall include the date the fault-current calculation was performed and be of sufficient durability to withstand the environment involved. The calculation shall be documented and made available to those authorized to design, install, inspect, maintain, or operate the system.

Informational Note: The available fault-current marking(s) addressed in 110.24 is related to required short-circuit current ratings of equipment. *NFPA 70E -2015, Standard for Electrical Safety in the Workplace,* provides assistance in determining the severity of potential exposure, planning safe work practices, and selecting personal protective equipment.

(B) Modifications. When modifications to the electrical installation occur that affect the maximum available fault current at the service, the maximum available fault current shall be verified or recalculated as necessary to ensure the service equipment ratings are sufficient for the maximum available fault current at the line terminals of the equipment. The required field marking(s) in 110.24(A) shall be adjusted to reflect the new level of maximum available fault current.

Exception: The field marking requirements in 110.24(A) and 110.24(B) shall not be required in industrial installations where conditions of maintenance and supervision ensure that only qualified persons service the equipment.

110.25 Lockable Disconnecting Means. If a disconnecting means is required to be lockable open elsewhere in this *Code*, it shall be capable of being locked in the open position. The provisions for locking shall remain in place with or without the lock installed.

Exception: Locking provisions for a cord-and-plug connection shall not be required to remain in place without the lock installed.

Part II. 1000 Volts, Nominal, or Less

110.26 Spaces About Electrical Equipment. Access and working space shall be provided and maintained about all electrical equipment to permit ready and safe operation and maintenance of such equipment.

(A) Working Space. Working space for equipment operating at 1000 volts, nominal, or less to ground and likely to require examination, adjustment, servicing, or maintenance while energized shall comply with the dimensions of 110.26(A)(1), (A)(2), (A)(3), and (A)(4) or as required or permitted elsewhere in this *Code.*

Informational Note: NFPA 70E-2015, *Standard for Electrical Safety in the Workplace,* provides guidance, such as determining severity of potential exposure, planning safe work practices, arc flash labeling, and selecting personal protective equipment.

(1) Depth of Working Space. The depth of the working space in the direction of live parts shall not be less than that specified in Table 110.26(A)(1) unless the requirements of 110.26(A)(1)(a), (A)(1)(b), or (A)(1)(c) are met. Distances shall be measured from the exposed live parts or from the enclosure or opening if the live parts are enclosed.

(a) *Dead-Front Assemblies.* Working space shall not be required in the back or sides of assemblies, such as dead-front switchboards, switchgear, or motor control centers, where all connections and all renewable or adjustable parts, such as fuses or switches, are accessible from locations other than the back or sides. Where rear access is required to work on nonelectrical parts on the back of enclosed equipment, a minimum horizontal working space of 762 mm (30 in.) shall be provided.

(b) *Low Voltage.* By special permission, smaller working spaces shall be permitted where all exposed live parts operate at not greater than 30 volts rms, 42 volts peak, or 60 volts dc.

(c) *Existing Buildings.* In existing buildings where electrical equipment is being replaced, Condition 2 working clearance shall be permitted between dead-front switchboards, switchgear, panelboards, or motor control centers located across the aisle from each other where conditions of maintenance and supervision ensure that written procedures have been adopted to prohibit equipment on both sides of the aisle from being open at the same time and qualified persons who are authorized will service the installation.

(2) Width of Working Space. The width of the working space in front of the electrical equipment shall be the width of the equipment or 762 mm (30 in.), whichever is greater. In all cases, the work space shall permit at least a 90 degree opening of equipment doors or hinged panels.

(3) Height of Working Space. The work space shall be clear and extend from the grade, floor, or platform to a height of 2.0 m (6½ ft) or the height of the equipment, whichever is greater. Within the height requirements of this section, other equipment that is associated with the electrical installation and is located above or below the electrical equipment shall be permitted to extend not more than 150 mm (6 in.) beyond the front of the electrical equipment.

Exception No. 1: In existing dwelling units, service equipment or panelboards that do not exceed 200 amperes shall be permitted in spaces where the height of the working space is less than 2.0 m (6½ ft).

Exception No. 2: Meters that are installed in meter sockets shall be permitted to extend beyond the other equipment. The meter socket shall be required to follow the rules of this section.

Exception No. 3: On battery systems mounted on open racks, the top clearance shall comply with 480.10(D).

N **(4) Limited Access.** Where equipment operating at 1000 volts, nominal, or less to ground and likely to require examination, adjustment, servicing, or maintenance while energized is required by installation instructions or function to be located in a space with limited access, all of the following shall apply:

Table 110.26(A)(1) Working Spaces

Nominal Voltage to Ground	Minimum Clear Distance		
	Condition 1	Condition 2	Condition 3
0–150	900 mm (3 ft)	900 mm (3 ft)	900 mm (3 ft)
151–600	900 mm (3 ft)	1.0 m (3 ft 6 in.)	1.2 m (4 ft)
601–1000	900 mm (3 ft)	1.2 m (4 ft)	1.5 m (5 ft)

Note: Where the conditions are as follows:

Condition 1 — Exposed live parts on one side of the working space and no live or grounded parts on the other side of the working space, or exposed live parts on both sides of the working space that are effectively guarded by insulating materials.

Condition 2 — Exposed live parts on one side of the working space and grounded parts on the other side of the working space. Concrete, brick, or tile walls shall be considered as grounded.

Condition 3 — Exposed live parts on both sides of the working space.

(a)   Where equipment is installed above a lay-in ceiling, there shall be an opening not smaller than 559 mm × 559 mm (22 in. × 22 in.), or in a crawl space, there shall be an accessible opening not smaller than 559 mm × 762 mm (22 in. × 30 in.).

(b)   The width of the working space shall be the width of the equipment enclosure or a minimum of 762 mm (30 in.), whichever is greater.

(c)   All enclosure doors or hinged panels shall be capable of opening a minimum of 90 degrees.

(d)   The space in front of the enclosure shall comply with the depth requirements of Table 110.26(A)(1). The maximum height of the working space shall be the height necessary to install the equipment in the limited space. A horizontal ceiling structural member or access panel shall be permitted in this space.

N **(5) Separation from High-Voltage Equipment.**   Where switches, cutouts, or other equipment operating at 1000 volts, nominal, or less are installed in a vault, room, or enclosure where there are exposed live parts or exposed wiring operating over 1000 volts, nominal, the high-voltage equipment shall be effectively separated from the space occupied by the low-voltage equipment by a suitable partition, fence, or screen.

(B) Clear Spaces.   Working space required by this section shall not be used for storage. When normally enclosed live parts are exposed for inspection or servicing, the working space, if in a passageway or general open space, shall be suitably guarded.

(C) Entrance to and Egress from Working Space.

(1) Minimum Required.   At least one entrance of sufficient area shall be provided to give access to and egress from working space about electrical equipment.

(2) Large Equipment.   For equipment rated 1200 amperes or more and over 1.8 m (6 ft) wide that contains overcurrent devices, switching devices, or control devices, there shall be one entrance to and egress from the required working space not less than 610 mm (24 in.) wide and 2.0 m (6½ ft) high at each end of the working space.

A single entrance to and egress from the required working space shall be permitted where either of the conditions in 110.26(C)(2)(a) or (C)(2)(b) is met.

(a)   *Unobstructed Egress.* Where the location permits a continuous and unobstructed way of egress travel, a single entrance to the working space shall be permitted.

(b)   *Extra Working Space.* Where the depth of the working space is twice that required by 110.26(A)(1), a single entrance shall be permitted. It shall be located such that the distance from the equipment to the nearest edge of the entrance is not less than the minimum clear distance specified in Table 110.26(A)(1) for equipment operating at that voltage and in that condition.

(3) Personnel Doors.   Where equipment rated 800 A or more that contains overcurrent devices, switching devices, or control devices is installed and there is a personnel door(s) intended for entrance to and egress from the working space less than 7.6 m (25 ft) from the nearest edge of the working space, the door(s) shall open in the direction of egress and be equipped with listed panic hardware.

(D) Illumination.   Illumination shall be provided for all working spaces about service equipment, switchboards, switchgear, panelboards, or motor control centers installed indoors. Control by automatic means only shall not be permitted. Additional lighting outlets shall not be required where the work space is illuminated by an adjacent light source or as permitted by 210.70(A)(1), Exception No. 1, for switched receptacles.

(E) Dedicated Equipment Space.   All switchboards, switchgear, panelboards, and motor control centers shall be located in dedicated spaces and protected from damage.

Exception: Control equipment that by its very nature or because of other rules of the Code must be adjacent to or within sight of its operating machinery shall be permitted in those locations.

(1) Indoor.   Indoor installations shall comply with 110.26(E)(1)(a) through (E)(1)(d).

(a)   *Dedicated Electrical Space.* The space equal to the width and depth of the equipment and extending from the floor to a height of 1.8 m (6 ft) above the equipment or to the structural ceiling, whichever is lower, shall be dedicated to the electrical installation. No piping, ducts, leak protection apparatus, or other equipment foreign to the electrical installation shall be located in this zone.

Exception: Suspended ceilings with removable panels shall be permitted within the 1.8-m (6-ft) zone.

(b)   *Foreign Systems.* The area above the dedicated space required by 110.26(E)(1)(a) shall be permitted to contain foreign systems, provided protection is installed to avoid damage to the electrical equipment from condensation, leaks, or breaks in such foreign systems.

(c)   *Sprinkler Protection.* Sprinkler protection shall be permitted for the dedicated space where the piping complies with this section.

(d)   *Suspended Ceilings.* A dropped, suspended, or similar ceiling that does not add strength to the building structure shall not be considered a structural ceiling.

(2) Outdoor.   Outdoor installations shall comply with 110.26(E)(2)(a) through (c).

(a)   *Installation Requirements.* Outdoor electrical equipment shall be the following:

(1)   Installed in identified enclosures
(2)   Protected from accidental contact by unauthorized personnel or by vehicular traffic
(3)   Protected from accidental spillage or leakage from piping systems

(b)   *Work Space.* The working clearance space shall include the zone described in 110.26(A). No architectural appurtenance or other equipment shall be located in this zone.

Exception: Structural overhangs or roof extensions shall be permitted in this zone.

(c)   *Dedicated Equipment Space.* The space equal to the width and depth of the equipment, and extending from grade to a height of 1.8 m (6 ft) above the equipment, shall be dedicated to the electrical installation. No piping or other equipment foreign to the electrical installation shall be located in this zone.

(F) Locked Electrical Equipment Rooms or Enclosures. Electrical equipment rooms or enclosures housing electrical apparatus that are controlled by a lock(s) shall be considered accessible to qualified persons.

110.27 Guarding of Live Parts.

(A) Live Parts Guarded Against Accidental Contact. Except as elsewhere required or permitted by this *Code*, live parts of electrical equipment operating at 50 to 1000 volts, nominal shall be guarded against accidental contact by approved enclosures or by any of the following means:

(1) By location in a room, vault, or similar enclosure that is accessible only to qualified persons.

(2) By permanent, substantial partitions or screens arranged so that only qualified persons have access to the space within reach of the live parts. Any openings in such partitions or screens shall be sized and located so that persons are not likely to come into accidental contact with the live parts or to bring conducting objects into contact with them.

(3) By location on a balcony, gallery, or platform elevated and arranged so as to exclude unqualified persons.

(4) By elevation above the floor or other working surface as follows:

    a. A minimum of 2.5 m (8 ft) for 50 volts to 300 volts between ungrounded conductors

    b. A minimum of 2.6 m (8 ft 6 in.) for 301 volts to 600 volts between ungrounded conductors

    c. A minimum of 2.62 m (8 ft 7 in.) for 601 volts to 1000 volts between ungrounded conductors

(B) Prevent Physical Damage. In locations where electrical equipment is likely to be exposed to physical damage, enclosures or guards shall be so arranged and of such strength as to prevent such damage.

(C) Warning Signs. Entrances to rooms and other guarded locations that contain exposed live parts shall be marked with conspicuous warning signs forbidding unqualified persons to enter. The marking shall meet the requirements in 110.21(B).

Informational Note: For motors, see 430.232 and 430.233. For over 1000 volts, see 110.34.

110.28 Enclosure Types.
Enclosures (other than surrounding fences or walls covered in 110.31) of switchboards, switchgear, panelboards, industrial control panels, motor control centers, meter sockets, enclosed switches, transfer switches, power outlets, circuit breakers, adjustable-speed drive systems, pullout switches, portable power distribution equipment, termination boxes, general-purpose transformers, fire pump controllers, fire pump motors, and motor controllers, rated not over 1000 volts nominal and intended for such locations, shall be marked with an enclosure-type number as shown in Table 110.28.

Table 110.28 shall be used for selecting these enclosures for use in specific locations other than hazardous (classified) locations. The enclosures are not intended to protect against conditions such as condensation, icing, corrosion, or contamination that may occur within the enclosure or enter via the conduit or unsealed openings.

Part III. Over 1000 Volts, Nominal

110.30 General.
Conductors and equipment used on circuits over 1000 volts, nominal, shall comply with Part I of this article and with 110.30 through 110.41, which supplement or modify Part I. In no case shall the provisions of this part apply to equipment on the supply side of the service point.

110.31 Enclosure for Electrical Installations.
Electrical installations in a vault, room, or closet or in an area surrounded by a wall, screen, or fence, access to which is controlled by a lock(s) or other approved means, shall be considered to be accessible to qualified persons only. The type of enclosure used in a given case shall be designed and constructed according to the nature and degree of the hazard(s) associated with the installation.

For installations other than equipment as described in 110.31(D), a wall, screen, or fence shall be used to enclose an outdoor electrical installation to deter access by persons who are not qualified. A fence shall not be less than 2.1 m (7 ft) in height or a combination of 1.8 m (6 ft) or more of fence fabric and a 300 mm (1 ft) or more extension utilizing three or more strands of barbed wire or equivalent. The distance from the fence to live parts shall be not less than given in Table 110.31.

Informational Note: See Article 450 for construction requirements for transformer vaults.

(A) Electrical Vaults. Where an electrical vault is required or specified for conductors and equipment 110.31(A)(1) to (A)(5) shall apply.

(1) Walls and Roof. The walls and roof shall be constructed of materials that have adequate structural strength for the conditions, with a minimum fire rating of 3 hours. For the purpose of this section, studs and wallboard construction shall not be permitted.

(2) Floors. The floors of vaults in contact with the earth shall be of concrete that is not less than 102 mm (4 in.) thick, but where the vault is constructed with a vacant space or other stories below it, the floor shall have adequate structural strength for the load imposed on it and a minimum fire resistance of 3 hours.

(3) Doors. Each doorway leading into a vault from the building interior shall be provided with a tight-fitting door that has a minimum fire rating of 3 hours. The authority having jurisdiction shall be permitted to require such a door for an exterior wall opening where conditions warrant.

Exception to (1), (2), and (3): Where the vault is protected with automatic sprinkler, water spray, carbon dioxide, or halon, construction with a 1-hour rating shall be permitted.

(4) Locks. Doors shall be equipped with locks, and doors shall be kept locked, with access allowed only to qualified persons. Personnel doors shall swing out and be equipped with panic bars, pressure plates, or other devices that are normally latched but that open under simple pressure.

(5) Transformers. Where a transformer is installed in a vault as required by Article 450, the vault shall be constructed in accordance with the requirements of Part III of Article 450.

Informational Note No. 1: For additional information, see ANSI/ASTM E119-2015, *Method for Fire Tests of Building Construction and Materials*, and NFPA 80-2016, *Standard for Fire Doors and Other Opening Protectives.*

Table 110.28 Enclosure Selection

Provides a Degree of Protection Against the Following Environmental Conditions	For Outdoor Use									
	Enclosure Type Number									
	3	3R	3S	3X	3RX	3SX	4	4X	6	6P
Incidental contact with the enclosed equipment	X	X	X	X	X	X	X	X	X	X
Rain, snow, and sleet	X	X	X	X	X	X	X	X	X	X
Sleet*	—	—	X	—	—	X	—	—	—	—
Windblown dust	X	—	X	X	—	X	X	X	X	X
Hosedown	—	—	—	—	—	—	X	X	X	X
Corrosive agents	—	—	—	X	X	X	—	X	—	X
Temporary submersion	—	—	—	—	—	—	—	—	X	X
Prolonged submersion	—	—	—	—	—	—	—	—	—	X

Provides a Degree of Protection Against the Following Environmental Conditions	For Indoor Use									
	Enclosure Type Number									
	1	2	4	4X	5	6	6P	12	12K	13
Incidental contact with the enclosed equipment	X	X	X	X	X	X	X	X	X	X
Falling dirt	X	X	X	X	X	X	X	X	X	X
Falling liquids and light splashing	—	X	X	X	X	X	X	X	X	X
Circulating dust, lint, fibers, and flyings	—	—	X	X	—	X	X	X	X	X
Settling airborne dust, lint, fibers, and flyings	—	—	X	X	X	X	X	X	X	X
Hosedown and splashing water	—	—	X	X	—	X	X	—	—	—
Oil and coolant seepage	—	—	—	—	—	—	—	X	X	X
Oil or coolant spraying and splashing	—	—	—	—	—	—	—	—	—	X
Corrosive agents	—	—	—	X	—	—	X	—	—	—
Temporary submersion	—	—	—	—	—	X	X	—	—	—
Prolonged submersion	—	—	—	—	—	—	X	—	—	—

*Mechanism shall be operable when ice covered.

Informational Note No. 1: The term *raintight* is typically used in conjunction with Enclosure Types 3, 3S, 3SX, 3X, 4, 4X, 6, and 6P. The term *rainproof* is typically used in conjunction with Enclosure Types 3R and 3RX. The term *watertight* is typically used in conjunction with Enclosure Types 4, 4X, 6, and 6P. The term *driptight* is typically used in conjunction with Enclosure Types 2, 5, 12, 12K, and 13. The term *dusttight* is typically used in conjunction with Enclosure Types 3, 3S, 3SX, 3X, 5, 12, 12K, and 13.

Informational Note No. 2: Ingress protection (IP) ratings may be found in ANSI/IEC 60529, *Degrees of Protection Provided by Enclosures.* IP ratings are not a substitute for Enclosure Type ratings.

Informational Note No. 2: A typical 3-hour construction is 150 mm (6 in.) thick reinforced concrete.

(B) Indoor Installations.

(1) In Places Accessible to Unqualified Persons. Indoor electrical installations that are accessible to unqualified persons shall be made with metal-enclosed equipment. Switchgear, transformers, pull boxes, connection boxes, and other similar associated equipment shall be marked with appropriate caution signs. Openings in ventilated dry-type transformers or similar openings in other equipment shall be designed so that foreign objects inserted through these openings are deflected from energized parts.

(2) In Places Accessible to Qualified Persons Only. Indoor electrical installations considered accessible only to qualified persons in accordance with this section shall comply with 110.34, 110.36, and 490.24.

Table 110.31 Minimum Distance from Fence to Live Parts

Nominal Voltage	Minimum Distance to Live Parts	
	m	ft
1001–13,799	3.05	10
13,800–230,000	4.57	15
Over 230,000	5.49	18

Note: For clearances of conductors for specific system voltages and typical BIL ratings, see ANSI/IEEE C2-2012, *National Electrical Safety Code.*

(C) Outdoor Installations.

(1) In Places Accessible to Unqualified Persons. Outdoor electrical installations that are open to unqualified persons shall comply with Parts I, II, and III of Article 225.

(2) In Places Accessible to Qualified Persons Only. Outdoor electrical installations that have exposed live parts shall be accessible to qualified persons only in accordance with the first paragraph of this section and shall comply with 110.34, 110.36, and 490.24.

(D) Enclosed Equipment Accessible to Unqualified Persons. Ventilating or similar openings in equipment shall be designed such that foreign objects inserted through these openings are deflected from energized parts. Where exposed to physical damage from vehicular traffic, suitable guards shall be provided. Equipment located outdoors and accessible to unqualified persons shall be designed such that exposed nuts or bolts cannot be readily removed, permitting access to live parts. Where equipment is accessible to unqualified persons and the bottom of the enclosure is less than 2.5 m (8 ft) above the floor or grade level, the enclosure door or hinged cover shall be kept locked. Doors and covers of enclosures used solely as pull boxes, splice boxes, or junction boxes shall be locked, bolted, or screwed on. Underground box covers that weigh over 45.4 kg (100 lb) shall be considered as meeting this requirement.

110.32 Work Space About Equipment. Sufficient space shall be provided and maintained about electrical equipment to permit ready and safe operation and maintenance of such equipment. Where energized parts are exposed, the minimum clear work space shall be not less than 2.0 m (6½ ft) high (measured vertically from the floor or platform) or not less than 914 mm (3 ft) wide (measured parallel to the equipment). The depth shall be as required in 110.34(A). In all cases, the work space shall permit at least a 90 degree opening of doors or hinged panels.

110.33 Entrance to Enclosures and Access to Working Space.

(A) Entrance. At least one entrance to enclosures for electrical installations as described in 110.31 not less than 610 mm (24 in.) wide and 2.0 m (6½ ft) high shall be provided to give access to the working space about electrical equipment.

(1) Large Equipment. On switchgear and control panels exceeding 1.8 m (6 ft) in width, there shall be one entrance at each end of the equipment. A single entrance to the required working space shall be permitted where either of the conditions in 110.33(A)(1)(a) or (A)(1)(b) is met.

(a) *Unobstructed Exit.* Where the location permits a continuous and unobstructed way of exit travel, a single entrance to the working space shall be permitted.

(b) *Extra Working Space.* Where the depth of the working space is twice that required by 110.34(A), a single entrance shall be permitted. It shall be located so that the distance from the equipment to the nearest edge of the entrance is not less than the minimum clear distance specified in Table 110.34(A) for equipment operating at that voltage and in that condition.

(2) Guarding. Where bare energized parts at any voltage or insulated energized parts above 1000 volts, nominal, are located adjacent to such entrance, they shall be suitably guarded.

(3) Personnel Doors. Where there is a personnel door(s) intended for entrance to and egress from the working space less than 7.6 m (25 ft) from the nearest edge of the working space, the door(s) shall open in the direction of egress and be equipped with listed panic hardware.

(B) Access. Permanent ladders or stairways shall be provided to give safe access to the working space around electrical equipment installed on platforms, balconies, or mezzanine floors or in attic or roof rooms or spaces.

110.34 Work Space and Guarding.

(A) Working Space. Except as elsewhere required or permitted in this *Code*, equipment likely to require examination, adjustment, servicing, or maintenance while energized shall have clear working space in the direction of access to live parts of the electrical equipment and shall be not less than specified in Table 110.34(A). Distances shall be measured from the live parts, if such are exposed, or from the enclosure front or opening if such are enclosed.

Exception: Working space shall not be required in back of equipment such as switchgear or control assemblies where there are no renewable or adjustable parts (such as fuses or switches) on the back and where all connections are accessible from locations other than the back. Where rear access is required to work on nonelectrical parts on the back of enclosed equipment, a minimum working space of 762 mm (30 in.) horizontally shall be provided.

(B) Separation from Low-Voltage Equipment. Where switches, cutouts, or other equipment operating at 1000 volts, nominal, or less are installed in a vault, room, or enclosure where there are exposed live parts or exposed wiring operating at over 1000 volts, nominal, the high-voltage equipment shall be effectively separated from the space occupied by the low-voltage equipment by a suitable partition, fence, or screen.

Exception: Switches or other equipment operating at 1000 volts, nominal, or less and serving only equipment within the high-voltage vault, room, or enclosure shall be permitted to be installed in the high-voltage vault, room, or enclosure without a partition, fence, or screen if accessible to qualified persons only.

(C) Locked Rooms or Enclosures. The entrance to all buildings, vaults, rooms, or enclosures containing exposed live parts or exposed conductors operating at over 1000 volts, nominal, shall be kept locked unless such entrances are under the observation of a qualified person at all times.

Table 110.34(A) Minimum Depth of Clear Working Space at Electrical Equipment

Nominal Voltage to Ground	Minimum Clear Distance		
	Condition 1	Condition 2	Condition 3
1001–2500 V	900 mm (3 ft)	1.2 m (4 ft)	1.5 m (5 ft)
2501–9000 V	1.2 m (4 ft)	1.5 m (5 ft)	1.8 m (6 ft)
9001–25,000 V	1.5 m (5 ft)	1.8 m (6 ft)	2.8 m (9 ft)
25,001 V–75 kV	1.8 m (6 ft)	2.5 m (8 ft)	3.0 m (10 ft)
Above 75 kV	2.5 m (8 ft)	3.0 m (10 ft)	3.7 m (12 ft)

Note: Where the conditions are as follows:
(1) **Condition 1** — Exposed live parts on one side of the working space and no live or grounded parts on the other side of the working space, or exposed live parts on both sides of the working space that are effectively guarded by insulating materials.
(2) **Condition 2** — Exposed live parts on one side of the working space and grounded parts on the other side of the working space. Concrete, brick, or tile walls shall be considered as grounded.
(3) **Condition 3** — Exposed live parts on both sides of the working space.

Permanent and conspicuous danger signs shall be provided. The danger sign shall meet the requirements in 110.21(B) and shall read as follows:

DANGER — HIGH VOLTAGE — KEEP OUT

(D) Illumination. Illumination shall be provided for all working spaces about electrical equipment. Control by automatic means only shall not be permitted. The lighting outlets shall be arranged so that persons changing lamps or making repairs on the lighting system are not endangered by live parts or other equipment.

The points of control shall be located so that persons are not likely to come in contact with any live part or moving part of the equipment while turning on the lights.

(E) Elevation of Unguarded Live Parts. Unguarded live parts above working space shall be maintained at elevations not less than required by Table 110.34(E).

(F) Protection of Service Equipment, Switchgear, and Industrial Control Assemblies. Pipes or ducts foreign to the electrical installation and requiring periodic maintenance or whose malfunction would endanger the operation of the electrical system shall not be located in the vicinity of the service equipment, switchgear, or industrial control assemblies. Protection shall be provided where necessary to avoid damage from condensation leaks and breaks in such foreign systems. Piping and other facilities shall not be considered foreign if provided for fire protection of the electrical installation.

110.36 Circuit Conductors. Circuit conductors shall be permitted to be installed in raceways; in cable trays; as metal-clad cable Type MC; as bare wire, cable, and busbars; or as Type MV cables or conductors as provided in 300.37, 300.39, 300.40, and 300.50. Bare live conductors shall comply with 490.24.

Insulators, together with their mounting and conductor attachments, where used as supports for wires, single-conductor cables, or busbars, shall be capable of safely withstanding the maximum magnetic forces that would prevail if two or more conductors of a circuit were subjected to short-circuit current.

Exposed runs of insulated wires and cables that have a bare lead sheath or a braided outer covering shall be supported in a manner designed to prevent physical damage to the braid or sheath. Supports for lead-covered cables shall be designed to prevent electrolysis of the sheath.

110.40 Temperature Limitations at Terminations. Conductors shall be permitted to be terminated based on the 90°C (194°F) temperature rating and ampacity as given in Table 310.60(C)(67) through Table 310.60(C)(86), unless otherwise identified.

Table 110.34(E) Elevation of Unguarded Live Parts Above Working Space

Nominal Voltage Between Phases	Elevation	
	m	ft
1001–7500 V	2.7	9
7501–35,000 V	2.9	9 ft 6 in.
Over 35 kV	Add 9.5 mm per kV above 35 kV	Add 0.37 in. per kV above 35 kV

N 110.41 Inspections and Tests.

(A) Pre-energization and Operating Tests. Where required elsewhere in this *Code*, the complete electrical system design, including settings for protective, switching, and control circuits, shall be prepared in advance and made available on request to the authority having jurisdiction and shall be tested when first installed on-site.

(B) Test Report. A test report covering the results of the tests required in 110.41(A) shall be available to the authority having jurisdiction prior to energization and made available to those authorized to install, operate, test, and maintain the system.

Part IV. Tunnel Installations over 1000 Volts, Nominal

110.51 General.

(A) Covered. The provisions of this part shall apply to the installation and use of high-voltage power distribution and utilization equipment that is portable, mobile, or both, such as substations, trailers, cars, mobile shovels, draglines, hoists, drills, dredges, compressors, pumps, conveyors, underground excavators, and the like.

(B) Other Articles. The requirements of this part shall be additional to, or amendatory of, those prescribed in Articles 100 through 490 of this *Code*.

(C) Protection Against Physical Damage. Conductors and cables in tunnels shall be located above the tunnel floor and so placed or guarded to protect them from physical damage.

110.52 Overcurrent Protection. Motor-operated equipment shall be protected from overcurrent in accordance with Parts III, IV, and V of Article 430. Transformers shall be protected from overcurrent in accordance with 450.3.

110.53 Conductors. High-voltage conductors in tunnels shall be installed in metal conduit or other metal raceway, Type MC cable, or other approved multiconductor cable. Multiconductor portable cable shall be permitted to supply mobile equipment.

110.54 Bonding and Equipment Grounding Conductors.

(A) Grounded and Bonded. All non–current-carrying metal parts of electrical equipment and all metal raceways and cable sheaths shall be solidly grounded and bonded to all metal pipes and rails at the portal and at intervals not exceeding 300 m (1000 ft) throughout the tunnel.

(B) Equipment Grounding Conductors. An equipment grounding conductor shall be run with circuit conductors inside the metal raceway or inside the multiconductor cable jacket. The equipment grounding conductor shall be permitted to be insulated or bare.

110.55 Transformers, Switches, and Electrical Equipment. All transformers, switches, motor controllers, motors, rectifiers, and other equipment installed belowground shall be protected from physical damage by location or guarding.

110.56 Energized Parts. Bare terminals of transformers, switches, motor controllers, and other equipment shall be enclosed to prevent accidental contact with energized parts.

110.57 Ventilation System Controls. Electrical controls for the ventilation system shall be arranged so that the airflow can be reversed.

110.58 Disconnecting Means. A switch or circuit breaker that simultaneously opens all ungrounded conductors of the circuit shall be installed within sight of each transformer or motor location for disconnecting the transformer or motor. The switch or circuit breaker for a transformer shall have an ampere rating not less than the ampacity of the transformer supply conductors. The switch or circuit breaker for a motor shall comply with the applicable requirements of Article 430.

110.59 Enclosures. Enclosures for use in tunnels shall be dripproof, weatherproof, or submersible as required by the environmental conditions. Switch or contactor enclosures shall not be used as junction boxes or as raceways for conductors feeding through or tapping off to other switches, unless the enclosures comply with 312.8.

Part V. Manholes and Other Electrical Enclosures Intended for Personnel Entry

110.70 General. Electrical enclosures intended for personnel entry and specifically fabricated for this purpose shall be of sufficient size to provide safe work space about electrical equipment with live parts that is likely to require examination, adjustment, servicing, or maintenance while energized. Such enclosures shall have sufficient size to permit ready installation or withdrawal of the conductors employed without damage to the conductors or to their insulation. They shall comply with the provisions of this part.

Exception: Where electrical enclosures covered by Part V of this article are part of an industrial wiring system operating under conditions of maintenance and supervision that ensure that only qualified persons monitor and supervise the system, they shall be permitted to be designed and installed in accordance with appropriate engineering practice. If required by the authority having jurisdiction, design documentation shall be provided.

110.71 Strength. Manholes, vaults, and their means of access shall be designed under qualified engineering supervision and shall withstand all loads likely to be imposed on the structures.

Informational Note: See ANSI C2-2007, *National Electrical Safety Code,* for additional information on the loading that can be expected to bear on underground enclosures.

110.72 Cabling Work Space. A clear work space not less than 900 mm (3 ft) wide shall be provided where cables are located on both sides, and not less than 750 mm (2½ ft) where cables are only on one side. The vertical headroom shall be not less than 1.8 m (6 ft) unless the opening is within 300 mm (1 ft), measured horizontally, of the adjacent interior side wall of the enclosure.

Exception: A manhole containing only one or more of the following shall be permitted to have one of the horizontal work space dimensions reduced to 600 mm (2 ft) where the other horizontal clear work space is increased so the sum of the two dimensions is not less than 1.8 m (6 ft):

(1) Optical fiber cables as covered in Article 770
(2) Power-limited fire alarm circuits supplied in accordance with 760.121
(3) Class 2 or Class 3 remote-control and signaling circuits, or both, supplied in accordance with 725.121

110.73 Equipment Work Space. Where electrical equipment with live parts that is likely to require examination, adjustment, servicing, or maintenance while energized is installed in a manhole, vault, or other enclosure designed for personnel access, the work space and associated requirements in 110.26 shall be met for installations operating at 1000 volts or less. Where the installation is over 1000 volts, the work space and associated requirements in 110.34 shall be met. A manhole access cover that weighs over 45.4 kg (100 lb) shall be considered as meeting the requirements of 110.34(C).

110.74 Conductor Installation. Conductors installed in manholes and other enclosures intended for personnel entry shall be cabled, racked up, or arranged in an approved manner that provides ready and safe access for persons to enter for installation and maintenance. The installation shall comply with 110.74(A) or 110.74(B), as applicable.

(A) 1000 Volts, Nominal, or Less. Wire bending space for conductors operating at 1000 volts or less shall be provided in accordance with the requirements of 314.28.

(B) Over 1000 Volts, Nominal. Conductors operating at over 1000 volts shall be provided with bending space in accordance with 314.71(A) and (B), as applicable.

Exception: Where 314.71(B) applies, each row or column of ducts on one wall of the enclosure shall be calculated individually, and the single row or column that provides the maximum distance shall be used.

110.75 Access to Manholes.

(A) Dimensions. Rectangular access openings shall not be less than 650 mm × 550 mm (26 in. × 22 in.). Round access openings in a manhole shall be not less than 650 mm (26 in.) in diameter.

Exception: A manhole that has a fixed ladder that does not obstruct the opening or that contains only one or more of the following shall be permitted to reduce the minimum cover diameter to 600 mm (2 ft):

(1) Optical fiber cables as covered in Article 770
(2) Power-limited fire alarm circuits supplied in accordance with 760.121
(3) Class 2 or Class 3 remote-control and signaling circuits, or both, supplied in accordance with 725.121

(B) Obstructions. Manhole openings shall be free of protrusions that could injure personnel or prevent ready egress.

(C) Location. Manhole openings for personnel shall be located where they are not directly above electrical equipment or conductors in the enclosure. Where this is not practicable, either a protective barrier or a fixed ladder shall be provided.

(D) Covers. Covers shall be over 45 kg (100 lb) or otherwise designed to require the use of tools to open. They shall be designed or restrained so they cannot fall into the manhole or protrude sufficiently to contact electrical conductors or equipment within the manhole.

(E) Marking. Manhole covers shall have an identifying mark or logo that prominently indicates their function, such as "electric."

110.76 Access to Vaults and Tunnels.

(A) Location. Access openings for personnel shall be located where they are not directly above electrical equipment or conductors in the enclosure. Other openings shall be permitted over equipment to facilitate installation, maintenance, or replacement of equipment.

(B) Locks. In addition to compliance with the requirements of 110.34, if applicable, access openings for personnel shall be arranged such that a person on the inside can exit when the access door is locked from the outside, or in the case of normally locking by padlock, the locking arrangement shall be such that the padlock can be closed on the locking system to prevent locking from the outside.

110.77 Ventilation. Where manholes, tunnels, and vaults have communicating openings into enclosed areas used by the public, ventilation to open air shall be provided wherever practicable.

110.78 Guarding. Where conductors or equipment, or both, could be contacted by objects falling or being pushed through a ventilating grating, both conductors and live parts shall be protected in accordance with the requirements of 110.27(A)(2) or 110.31(B)(1), depending on the voltage.

110.79 Fixed Ladders. Fixed ladders shall be corrosion resistant.

Chapter 2   Wiring and Protection

ARTICLE 200
Use and Identification of Grounded Conductors

200.1 Scope. This article provides requirements for the following:

(1) Identification of terminals
(2) Grounded conductors in premises wiring systems
(3) Identification of grounded conductors

Informational Note: See Article 100 for definitions of *Grounded Conductor, Equipment Grounding Conductor,* and *Grounding Electrode Conductor.*

200.2 General. Grounded conductors shall comply with 200.2(A) and (B).

(A) Insulation. The grounded conductor, if insulated, shall have insulation that is (1) suitable, other than color, for any ungrounded conductor of the same circuit for systems of 1000 volts or less, or impedance grounded neutral systems of over 1000 volts, or (2) rated not less than 600 volts for solidly grounded neutral systems of over 1000 volts as described in 250.184(A).

(B) Continuity. The continuity of a grounded conductor shall not depend on a connection to a metallic enclosure, raceway, or cable armor.

Informational Note: See 300.13(B) for the continuity of grounded conductors used in multiwire branch circuits.

200.3 Connection to Grounded System. Premises wiring shall not be electrically connected to a supply system unless the latter contains, for any grounded conductor of the interior system, a corresponding conductor that is grounded. For the purpose of this section, *electrically connected* shall mean connected so as to be capable of carrying current, as distinguished from connection through electromagnetic induction.

Exception: Listed utility-interactive inverters identified for use in distributed resource generation systems such as photovoltaic and fuel cell power systems shall be permitted to be connected to premises wiring without a grounded conductor where the connected premises wiring or utility system includes a grounded conductor.

200.4 Neutral Conductors. Neutral conductors shall be installed in accordance with 200.4(A) and (B).

(A) Installation. Neutral conductors shall not be used for more than one branch circuit, for more than one multiwire branch circuit, or for more than one set of ungrounded feeder conductors unless specifically permitted elsewhere in this *Code.*

(B) Multiple Circuits. Where more than one neutral conductor associated with different circuits is in an enclosure, grounded circuit conductors of each circuit shall be identified or grouped to correspond with the ungrounded circuit conductor(s) by wire markers, cable ties, or similar means in at least one location within the enclosure.

Exception No. 1: The requirement for grouping or identifying shall not apply if the branch-circuit or feeder conductors enter from a cable or a raceway unique to the circuit that makes the grouping obvious.

Exception No. 2: The requirement for grouping or identifying shall not apply where branch-circuit conductors pass through a box or conduit body without a loop as described in 314.16(B)(1) or without a splice or termination.

200.6 Means of Identifying Grounded Conductors.

(A) Sizes 6 AWG or Smaller. An insulated grounded conductor of 6 AWG or smaller shall be identified by one of the following means:

(1) A continuous white outer finish.
(2) A continuous gray outer finish.
(3) Three continuous white or gray stripes along the conductor's entire length on other than green insulation.
(4) Wires that have their outer covering finished to show a white or gray color but have colored tracer threads in the braid identifying the source of manufacture shall be considered as meeting the provisions of this section.
(5) The grounded conductor of a mineral-insulated, metal-sheathed cable (Type MI) shall be identified at the time of installation by distinctive marking at its terminations.
(6) A single-conductor, sunlight-resistant, outdoor-rated cable used as a grounded conductor in photovoltaic power systems, as permitted by 690.31, shall be identified at the time of installation by distinctive white marking at all terminations.
(7) Fixture wire shall comply with the requirements for grounded conductor identification as specified in 402.8.
(8) For aerial cable, the identification shall be as above, or by means of a ridge located on the exterior of the cable so as to identify it.

(B) Sizes 4 AWG or Larger. An insulated grounded conductor 4 AWG or larger shall be identified by one of the following means:

(1) A continuous white outer finish.
(2) A continuous gray outer finish.
(3) Three continuous white or gray stripes along the conductor's entire length on other than green insulation.
(4) At the time of installation, by a distinctive white or gray marking at its terminations. This marking shall encircle the conductor or insulation.

(C) Flexible Cords. An insulated conductor that is intended for use as a grounded conductor, where contained within a flexible cord, shall be identified by a white or gray outer finish or by methods permitted by 400.22.

(D) Grounded Conductors of Different Systems. Where grounded conductors of different systems are installed in the same raceway, cable, box, auxiliary gutter, or other type of enclosure, each grounded conductor shall be identified by system. Identification that distinguishes each system grounded conductor shall be permitted by one of the following means:

(1) One system grounded conductor shall have an outer covering conforming to 200.6(A) or (B).
(2) The grounded conductor(s) of other systems shall have a different outer covering conforming to 200.6(A) or

200.6(B) or by an outer covering of white or gray with a readily distinguishable colored stripe other than green running along the insulation.

(3) Other and different means of identification allowed by 200.6(A) or (B) shall distinguish each system grounded conductor.

The means of identification shall be documented in a manner that is readily available or shall be permanently posted where the conductors of different systems originate.

(E) Grounded Conductors of Multiconductor Cables. The insulated grounded conductors in a multiconductor cable shall be identified by a continuous white or gray outer finish or by three continuous white or gray stripes on other than green insulation along its entire length. Multiconductor flat cable 4 AWG or larger shall be permitted to employ an external ridge on the grounded conductor.

Exception No. 1: Where the conditions of maintenance and supervision ensure that only qualified persons service the installation, grounded conductors in multiconductor cables shall be permitted to be permanently identified at their terminations at the time of installation by a distinctive white marking or other equally effective means.

Exception No. 2: The grounded conductor of a multiconductor varnished-cloth-insulated cable shall be permitted to be identified at its terminations at the time of installation by a distinctive white marking or other equally effective means.

Informational Note: The color gray may have been used in the past as an ungrounded conductor. Care should be taken when working on existing systems.

200.7 Use of Insulation of a White or Gray Color or with Three Continuous White or Gray Stripes.

(A) General. The following shall be used only for the grounded circuit conductor, unless otherwise permitted in 200.7(B) and (C):

(1) A conductor with continuous white or gray covering
(2) A conductor with three continuous white or gray stripes on other than green insulation
(3) A marking of white or gray color at the termination

(B) Circuits of Less Than 50 Volts. A conductor with white or gray color insulation or three continuous white stripes or having a marking of white or gray at the termination for circuits of less than 50 volts shall be required to be grounded only as required by 250.20(A).

(C) Circuits of 50 Volts or More. The use of insulation that is white or gray or that has three continuous white or gray stripes for other than a grounded conductor for circuits of 50 volts or more shall be permitted only as in (1) and (2).

(1) If part of a cable assembly that has the insulation permanently reidentified to indicate its use as an ungrounded conductor by marking tape, painting, or other effective means at its termination and at each location where the conductor is visible and accessible. Identification shall encircle the insulation and shall be a color other than white, gray, or green. If used for single-pole, 3-way or 4-way switch loops, the reidentified conductor with white or gray insulation or three continuous white or gray stripes shall be used only for the supply to the switch, but not as a return conductor from the switch to the outlet.
(2) A flexible cord having one conductor identified by a white or gray outer finish or three continuous white or

gray stripes, or by any other means permitted by 400.22, that is used for connecting an appliance or equipment permitted by 400.10. This shall apply to flexible cords connected to outlets whether or not the outlet is supplied by a circuit that has a grounded conductor.

Informational Note: The color gray may have been used in the past as an ungrounded conductor. Care should be taken when working on existing systems.

200.9 Means of Identification of Terminals. The identification of terminals to which a grounded conductor is to be connected shall be substantially white in color. The identification of other terminals shall be of a readily distinguishable different color.

Exception: Where the conditions of maintenance and supervision ensure that only qualified persons service the installations, terminals for grounded conductors shall be permitted to be permanently identified at the time of installation by a distinctive white marking or other equally effective means.

200.10 Identification of Terminals.

(A) Device Terminals. All devices, excluding panelboards, provided with terminals for the attachment of conductors and intended for connection to more than one side of the circuit shall have terminals properly marked for identification, unless the electrical connection of the terminal intended to be connected to the grounded conductor is clearly evident.

Exception: Terminal identification shall not be required for devices that have a normal current rating of over 30 amperes, other than polarized attachment plugs and polarized receptacles for attachment plugs as required in 200.10(B).

(B) Receptacles, Plugs, and Connectors. Receptacles, polarized attachment plugs, and cord connectors for plugs and polarized plugs shall have the terminal intended for connection to the grounded conductor identified as follows:

(1) Identification shall be by a metal or metal coating that is substantially white in color or by the word *white* or the letter *W* located adjacent to the identified terminal.
(2) If the terminal is not visible, the conductor entrance hole for the connection shall be colored white or marked with the word *white* or the letter *W*.

Informational Note: See 250.126 for identification of wiring device equipment grounding conductor terminals.

(C) Screw Shells. For devices with screw shells, the terminal for the grounded conductor shall be the one connected to the screw shell.

(D) Screw Shell Devices with Leads. For screw shell devices with attached leads, the conductor attached to the screw shell shall have a white or gray finish. The outer finish of the other conductor shall be of a solid color that will not be confused with the white or gray finish used to identify the grounded conductor.

Informational Note: The color gray may have been used in the past as an ungrounded conductor. Care should be taken when working on existing systems.

(E) Appliances. Appliances that have a single-pole switch or a single-pole overcurrent device in the line or any line-connected screw shell lampholders, and that are to be connected by (1) a permanent wiring method or (2) field-installed attachment plugs and cords with three or more wires (including the equipment grounding conductor), shall have means to identify the terminal for the grounded circuit conductor (if any).

200.11 Polarity of Connections. No grounded conductor shall be attached to any terminal or lead so as to reverse the designated polarity.

ARTICLE 210
Branch Circuits

Part I. General Provisions

210.1 Scope. This article provides the general requirements for branch circuits.

210.3 Other Articles for Specific-Purpose Branch Circuits. Table 210.3 lists references for specific equipment and applications not located in Chapters 5, 6, and 7 that amend or supplement the requirements of this article.

210.4 Multiwire Branch Circuits.

(A) General. Branch circuits recognized by this article shall be permitted as multiwire circuits. A multiwire circuit shall be permitted to be considered as multiple circuits. All conductors of a multiwire branch circuit shall originate from the same panelboard or similar distribution equipment.

Informational Note No. 1: A 3-phase, 4-wire, wye-connected power system used to supply power to nonlinear loads may necessitate that the power system design allow for the possibility of high harmonic currents on the neutral conductor.

Informational Note No. 2: See 300.13(B) for continuity of grounded conductors on multiwire circuits.

(B) Disconnecting Means. Each multiwire branch circuit shall be provided with a means that will simultaneously disconnect

Table 210.3 Specific-Purpose Branch Circuits

Equipment	Article	Section
Air-conditioning and refrigerating equipment		440.6, 440.31, 440.32
Busways		368.17
Central heating equipment other than fixed electric space-heating equipment		422.12
Fixed electric heating equipment for pipelines and vessels		427.4
Fixed electric space-heating equipment		424.3
Fixed outdoor electrical deicing and snow-melting equipment		426.4
Infrared lamp industrial heating equipment		422.48, 424.3
Motors, motor circuits, and controllers	430	
Switchboards and panelboards		408.52

all ungrounded conductors at the point where the branch circuit originates.

Informational Note: See 240.15(B) for information on the use of single-pole circuit breakers as the disconnecting means.

(C) Line-to-Neutral Loads. Multiwire branch circuits shall supply only line-to-neutral loads.

Exception No. 1: A multiwire branch circuit that supplies only one utilization equipment.

Exception No. 2: Where all ungrounded conductors of the multiwire branch circuit are opened simultaneously by the branch-circuit overcurrent device.

(D) Grouping. The ungrounded and grounded circuit conductors of each multiwire branch circuit shall be grouped in accordance with 200.4(B).

210.5 Identification for Branch Circuits.

(A) Grounded Conductor. The grounded conductor of a branch circuit shall be identified in accordance with 200.6.

(B) Equipment Grounding Conductor. The equipment grounding conductor shall be identified in accordance with 250.119.

(C) Identification of Ungrounded Conductors. Ungrounded conductors shall be identified in accordance with 210.5(C)(1) or (2), as applicable.

(1) Branch Circuits Supplied from More Than One Nominal Voltage System. Where the premises wiring system has branch circuits supplied from more than one nominal voltage system, each ungrounded conductor of a branch circuit shall be identified by phase or line and system at all termination, connection, and splice points in compliance with 210.5(C)(1)(a) and (b).

(a) *Means of Identification.* The means of identification shall be permitted to be by separate color coding, marking tape, tagging, or other approved means.

(b) *Posting of Identification Means.* The method utilized for conductors originating within each branch-circuit panelboard or similar branch-circuit distribution equipment shall be documented in a manner that is readily available or shall be permanently posted at each branch-circuit panelboard or similar branch-circuit distribution equipment. The label shall be of sufficient durability to withstand the environment involved and shall not be handwritten.

Exception: In existing installations where a voltage system(s) already exists and a different voltage system is being added, it shall be permissible to mark only the new system voltage. Existing unidentified systems shall not be required to be identified at each termination, connection, and splice point in compliance with 210.5(C)(1)(a) and (b). Labeling shall be required at each voltage system distribution equipment to identify that only one voltage system has been marked for a new system(s). The new system label(s) shall include the words "other unidentified systems exist on the premises."

(2) Branch Circuits Supplied from Direct-Current Systems. Where a branch circuit is supplied from a dc system operating at more than 60 volts, each ungrounded conductor of 4 AWG or larger shall be identified by polarity at all termination, connection, and splice points by marking tape, tagging, or other approved means; each ungrounded conductor of 6 AWG or smaller shall be identified by polarity at all termination, connection, and splice points in compliance with 210.5(C)(2)(a) and (b). The identification methods utilized for conductors originating within each branch-circuit panelboard or similar branch-circuit distribution equipment shall be documented in a manner that is readily available or shall be permanently posted at each branch-circuit panelboard or similar branch-circuit distribution equipment.

(a) *Positive Polarity, Sizes 6 AWG or Smaller.* Where the positive polarity of a dc system does not serve as the connection point for the grounded conductor, each positive ungrounded conductor shall be identified by one of the following means:

(1) A continuous red outer finish
(2) A continuous red stripe durably marked along the conductor's entire length on insulation of a color other than green, white, gray, or black
(3) Imprinted plus signs (+) or the word POSITIVE or POS durably marked on insulation of a color other than green, white, gray, or black and repeated at intervals not exceeding 610 mm (24 in.) in accordance with 310.120(B)
(4) An approved permanent marking means such as sleeving or shrink-tubing that is suitable for the conductor size, at all termination, connection, and splice points, with imprinted plus signs (+) or the word POSITIVE or POS durably marked on insulation of a color other than green, white, gray, or black

(b) *Negative Polarity, Sizes 6 AWG or Smaller.* Where the negative polarity of a dc system does not serve as the connection point for the grounded conductor, each negative ungrounded conductor shall be identified by one of the following means:

(1) A continuous black outer finish
(2) A continuous black stripe durably marked along the conductor's entire length on insulation of a color other than green, white, gray, or red
(3) Imprinted minus signs (−) or the word NEGATIVE or NEG durably marked on insulation of a color other than green, white, gray, or red and repeated at intervals not exceeding 610 mm (24 in.) in accordance with 310.120(B)
(4) An approved permanent marking means such as sleeving or shrink-tubing that is suitable for the conductor size, at all termination, connection, and splice points, with imprinted minus signs (−) or the word NEGATIVE or NEG durably marked on insulation of a color other than green, white, gray, or red

210.6 Branch-Circuit Voltage Limitations. The nominal voltage of branch circuits shall not exceed the values permitted by 210.6(A) through (E).

(A) Occupancy Limitation. In dwelling units and guest rooms or guest suites of hotels, motels, and similar occupancies, the voltage shall not exceed 120 volts, nominal, between conductors that supply the terminals of the following:

(1) Luminaires
(2) Cord-and-plug-connected loads 1440 volt-amperes, nominal, or less or less than ¼ hp

(B) 120 Volts Between Conductors. Circuits not exceeding 120 volts, nominal, between conductors shall be permitted to supply the following:

(1) The terminals of lampholders applied within their voltage ratings
(2) Auxiliary equipment of electric-discharge lamps

Informational Note: See 410.137 for auxiliary equipment limitations.

(3) Cord-and-plug-connected or permanently connected utilization equipment

(C) 277 Volts to Ground. Circuits exceeding 120 volts, nominal, between conductors and not exceeding 277 volts, nominal, to ground shall be permitted to supply the following:

(1) Listed electric-discharge or listed light-emitting diode-type luminaires
(2) Listed incandescent luminaires, where supplied at 120 volts or less from the output of a stepdown autotransformer that is an integral component of the luminaire and the outer shell terminal is electrically connected to a grounded conductor of the branch circuit
(3) Luminaires equipped with mogul-base screw shell lampholders
(4) Lampholders, other than the screw shell type, applied within their voltage ratings
(5) Auxiliary equipment of electric-discharge lamps

Informational Note: See 410.137 for auxiliary equipment limitations.

(6) Cord-and-plug-connected or permanently connected utilization equipment

(D) 600 Volts Between Conductors. Circuits exceeding 277 volts, nominal, to ground and not exceeding 600 volts, nominal, between conductors shall be permitted to supply the following:

(1) The auxiliary equipment of electric-discharge lamps mounted in permanently installed luminaires where the luminaires are mounted in accordance with one of the following:

a. Not less than a height of 6.7 m (22 ft) on poles or similar structures for the illumination of outdoor areas such as highways, roads, bridges, athletic fields, or parking lots
b. Not less than a height of 5.5 m (18 ft) on other structures such as tunnels

Informational Note: See 410.137 for auxiliary equipment limitations.

(2) Cord-and-plug-connected or permanently connected utilization equipment other than luminaires
(3) Luminaires powered from direct-current systems where either of the following apply:

a. The luminaire contains a listed, dc-rated ballast that provides isolation between the dc power source and the lamp circuit and protection from electric shock when changing lamps.
b. The luminaire contains a listed, dc-rated ballast and has no provision for changing lamps.

Exception No. 1 to (B), (C), and (D): For lampholders of infrared industrial heating appliances as provided in 425.14.

Exception No. 2 to (B), (C), and (D): For railway properties as described in 110.19.

(E) Over 600 Volts Between Conductors. Circuits exceeding 600 volts, nominal, between conductors shall be permitted to supply utilization equipment in installations where conditions of maintenance and supervision ensure that only qualified persons service the installation.

210.7 Multiple Branch Circuits. Where two or more branch circuits supply devices or equipment on the same yoke or mounting strap, a means to simultaneously disconnect the ungrounded supply conductors shall be provided at the point at which the branch circuits originate.

210.8 Ground-Fault Circuit-Interrupter Protection for Personnel. Ground-fault circuit-interrupter protection for personnel shall be provided as required in 210.8(A) through (E). The ground-fault circuit interrupter shall be installed in a readily accessible location.

> Informational Note No. 1: See 215.9 for ground-fault circuit-interrupter protection for personnel on feeders.

> Informational Note No. 2: See 422.5(A) for GFCI requirements for appliances.

For the purposes of this section, when determining distance from receptacles the distance shall be measured as the shortest path the cord of an appliance connected to the receptacle would follow without piercing a floor, wall, ceiling, or fixed barrier, or passing through a door, doorway, or window.

(A) Dwelling Units. All 125-volt, single-phase, 15- and 20-ampere receptacles installed in the locations specified in 210.8(A)(1) through (10) shall have ground-fault circuit-interrupter protection for personnel.

(1) Bathrooms
(2) Garages, and also accessory buildings that have a floor located at or below grade level not intended as habitable rooms and limited to storage areas, work areas, and areas of similar use
(3) Outdoors

Exception to (3): Receptacles that are not readily accessible and are supplied by a branch circuit dedicated to electric snow-melting, deicing, or pipeline and vessel heating equipment shall be permitted to be installed in accordance with 426.28 or 427.22, as applicable.

(4) Crawl spaces — at or below grade level
(5) Unfinished portions or areas of the basement not intended as habitable rooms

Exception to (5): A receptacle supplying only a permanently installed fire alarm or burglar alarm system shall not be required to have ground-fault circuit-interrupter protection.

> Informational Note: See 760.41(B) and 760.121(B) for power supply requirements for fire alarm systems.

Receptacles installed under the exception to 210.8(A)(5) shall not be considered as meeting the requirements of 210.52(G).

(6) Kitchens — where the receptacles are installed to serve the countertop surfaces
(7) Sinks — where receptacles are installed within 1.8 m (6 ft) from the top inside edge of the bowl of the sink
(8) Boathouses
(9) Bathtubs or shower stalls — where receptacles are installed within 1.8 m (6 ft) of the outside edge of the bathtub or shower stall
(10) Laundry areas

(B) Other Than Dwelling Units. All single-phase receptacles rated 150 volts to ground or less, 50 amperes or less and three-phase receptacles rated 150 volts to ground or less, 100 amperes or less installed in the following locations shall have ground-fault circuit-interrupter protection for personnel.

(1) Bathrooms
(2) Kitchens
(3) Rooftops

Exception: Receptacles on rooftops shall not be required to be readily accessible other than from the rooftop.

(4) Outdoors

Exception No. 1 to (3) and (4): Receptacles that are not readily accessible and are supplied by a branch circuit dedicated to electric snow-melting, deicing, or pipeline and vessel heating equipment shall be permitted to be installed in accordance with 426.28 or 427.22, as applicable.

Exception No. 2 to (4): In industrial establishments only, where the conditions of maintenance and supervision ensure that only qualified personnel are involved, an assured equipment grounding conductor program as specified in 590.6(B)(3) shall be permitted for only those receptacle outlets used to supply equipment that would create a greater hazard if power is interrupted or having a design that is not compatible with GFCI protection.

(5) Sinks — where receptacles are installed within 1.8 m (6 ft) from the top inside edge of the bowl of the sink

Exception No. 1 to (5): In industrial laboratories, receptacles used to supply equipment where removal of power would introduce a greater hazard shall be permitted to be installed without GFCI protection.

Exception No. 2 to (5): For receptacles located in patient bed locations of general care (Category 2) or critical care (Category 1) spaces of health care facilities other than those covered under 210.8(B)(1), GFCI protection shall not be required.

(6) Indoor wet locations
(7) Locker rooms with associated showering facilities
(8) Garages, service bays, and similar areas other than vehicle exhibition halls and showrooms
(9) Crawl spaces — at or below grade level
(10) Unfinished portions or areas of the basement not intended as habitable rooms

(C) Boat Hoists. GFCI protection shall be provided for outlets not exceeding 240 volts that supply boat hoists installed in dwelling unit locations.

(D) Kitchen Dishwasher Branch Circuit. GFCI protection shall be provided for outlets that supply dishwashers installed in dwelling unit locations.

N (E) Crawl Space Lighting Outlets. GFCI protection shall be provided for lighting outlets not exceeding 120 volts installed in crawl spaces.

210.9 Circuits Derived from Autotransformers. Branch circuits shall not be derived from autotransformers unless the circuit supplied has a grounded conductor that is electrically connected to a grounded conductor of the system supplying the autotransformer.

Exception No. 1: An autotransformer shall be permitted without the connection to a grounded conductor where transforming from a nominal 208 volts to a nominal 240-volt supply or similarly from 240 volts to 208 volts.

Exception No. 2: In industrial occupancies, where conditions of maintenance and supervision ensure that only qualified persons service the installation, autotransformers shall be permitted to supply nominal 600-volt loads from nominal 480-volt systems, and 480-volt loads from nominal 600-volt systems, without the connection to a similar grounded conductor.

210.10 Ungrounded Conductors Tapped from Grounded Systems. Two-wire dc circuits and ac circuits of two or more ungrounded conductors shall be permitted to be tapped from the ungrounded conductors of circuits that have a grounded neutral conductor. Switching devices in each tapped circuit shall have a pole in each ungrounded conductor. All poles of multipole switching devices shall manually switch together where such switching devices also serve as a disconnecting means as required by the following:

(1)   410.93 for double-pole switched lampholders
(2)   410.104(B) for electric-discharge lamp auxiliary equipment switching devices
(3)   422.31(B) for an appliance
(4)   424.20 for a fixed electric space-heating unit
(5)   426.51 for electric deicing and snow-melting equipment
(6)   430.85 for a motor controller
(7)   430.103 for a motor

210.11 Branch Circuits Required. Branch circuits for lighting and for appliances, including motor-operated appliances, shall be provided to supply the loads calculated in accordance with 220.10. In addition, branch circuits shall be provided for specific loads not covered by 220.10 where required elsewhere in this *Code* and for dwelling unit loads as specified in 210.11(C).

(A) Number of Branch Circuits. The minimum number of branch circuits shall be determined from the total calculated load and the size or rating of the circuits used. In all installations, the number of circuits shall be sufficient to supply the load served. In no case shall the load on any circuit exceed the maximum specified by 220.18.

(B) Load Evenly Proportioned Among Branch Circuits. Where the load is calculated on the basis of volt-amperes per square meter or per square foot, the wiring system up to and including the branch-circuit panelboard(s) shall be provided to serve not less than the calculated load. This load shall be evenly proportioned among multioutlet branch circuits within the panelboard(s). Branch-circuit overcurrent devices and circuits shall be required to be installed only to serve the connected load.

(C) Dwelling Units.

(1) Small-Appliance Branch Circuits. In addition to the number of branch circuits required by other parts of this section, two or more 20-ampere small-appliance branch circuits shall be provided for all receptacle outlets specified by 210.52(B).

(2) Laundry Branch Circuits. In addition to the number of branch circuits required by other parts of this section, at least one additional 20-ampere branch circuit shall be provided to supply the laundry receptacle outlet(s) required by 210.52(F). This circuit shall have no other outlets.

(3) Bathroom Branch Circuits. In addition to the number of branch circuits required by other parts of this section, at least one 120-volt, 20-ampere branch circuit shall be provided to supply the bathroom(s) receptacle outlet(s). Such circuits shall have no other outlets.

Exception: Where the 20-ampere circuit supplies a single bathroom, outlets for other equipment within the same bathroom shall be permitted to be supplied in accordance with 210.23(A)(1) and (A)(2).

N (4) Garage Branch Circuits. In addition to the number of branch circuits required by other parts of this section, at least one 120-volt, 20-ampere branch circuit shall be installed to supply receptacle outlets in attached garages and in detached garages with electric power. This circuit shall have no other outlets.

Exception: This circuit shall be permitted to supply readily accessible outdoor receptacle outlets.

210.12 Arc-Fault Circuit-Interrupter Protection. Arc-fault circuit-interrupter protection shall be provided as required in 210.12(A), (B), and (C). The arc-fault circuit interrupter shall be installed in a readily accessible location.

(A) Dwelling Units. All 120-volt, single-phase, 15- and 20-ampere branch circuits supplying outlets or devices installed in dwelling unit kitchens, family rooms, dining rooms, living rooms, parlors, libraries, dens, bedrooms, sunrooms, recreation rooms, closets, hallways, laundry areas, or similar rooms or areas shall be protected by any of the means described in 210.12(A)(1) through (6):

(1)   A listed combination-type arc-fault circuit interrupter, installed to provide protection of the entire branch circuit

(2)   A listed branch/feeder-type AFCI installed at the origin of the branch-circuit in combination with a listed outlet branch-circuit type arc-fault circuit interrupter installed at the first outlet box on the branch circuit. The first outlet box in the branch circuit shall be marked to indicate that it is the first outlet of the circuit.

(3)   A listed supplemental arc protection circuit breaker installed at the origin of the branch circuit in combination with a listed outlet branch-circuit type arc-fault circuit interrupter installed at the first outlet box on the branch circuit where all of the following conditions are met:

     a.   The branch-circuit wiring shall be continuous from the branch-circuit overcurrent device to the outlet branch-circuit arc-fault circuit interrupter.
     b.   The maximum length of the branch-circuit wiring from the branch-circuit overcurrent device to the first outlet shall not exceed 15.2 m (50 ft) for a 14 AWG conductor or 21.3 m (70 ft) for a 12 AWG conductor.
     c.   The first outlet box in the branch circuit shall be marked to indicate that it is the first outlet of the circuit.

(4)   A listed outlet branch-circuit type arc-fault circuit interrupter installed at the first outlet on the branch circuit in combination with a listed branch-circuit overcurrent protective device where all of the following conditions are met:

     a.   The branch-circuit wiring shall be continuous from the branch-circuit overcurrent device to the outlet branch-circuit arc-fault circuit interrupter.
     b.   The maximum length of the branch-circuit wiring from the branch-circuit overcurrent device to the first outlet shall not exceed 15.2 m (50 ft) for a 14 AWG conductor or 21.3 m (70 ft) for a 12 AWG conductor.

c. The first outlet box in the branch circuit shall be marked to indicate that it is the first outlet of the circuit.

d. The combination of the branch-circuit overcurrent device and outlet branch-circuit AFCI shall be identified as meeting the requirements for a system combination–type AFCI and shall be listed as such.

(5) If RMC, IMC, EMT, Type MC, or steel-armored Type AC cables meeting the requirements of 250.118, metal wireways, metal auxiliary gutters, and metal outlet and junction boxes are installed for the portion of the branch circuit between the branch-circuit overcurrent device and the first outlet, it shall be permitted to install a listed outlet branch-circuit type AFCI at the first outlet to provide protection for the remaining portion of the branch circuit.

(6) Where a listed metal or nonmetallic conduit or tubing or Type MC cable is encased in not less than 50 mm (2 in.) of concrete for the portion of the branch circuit between the branch-circuit overcurrent device and the first outlet, it shall be permitted to install a listed outlet branch-circuit type AFCI at the first outlet to provide protection for the remaining portion of the branch circuit.

Exception: Where an individual branch circuit to a fire alarm system installed in accordance with 760.41(B) or 760.121(B) is installed in RMC, IMC, EMT, or steel-sheathed cable, Type AC or Type MC, meeting the requirements of 250.118, with metal outlet and junction boxes, AFCI protection shall be permitted to be omitted.

Informational Note No. 1: For information on combination-type and branch/feeder-type arc-fault circuit interrupters, see UL 1699-2011, *Standard for Arc-Fault Circuit Interrupters.* For information on outlet branch-circuit type arc-fault circuit interrupters, see UL Subject 1699A, *Outline of Investigation for Outlet Branch Circuit Arc-Fault Circuit-Interrupters.* For information on system combination AFCIs, see UL Subject 1699C, *Outline of Investigation for System Combination Arc-Fault Circuit Interrupters.*

Informational Note No. 2: See 29.6.3(5) of *NFPA 72* -2013, *National Fire Alarm and Signaling Code,* for information related to secondary power-supply requirements for smoke alarms installed in dwelling units.

Informational Note No. 3: See 760.41(B) and 760.121(B) for power-supply requirements for fire alarm systems.

(B) Dormitory Units. All 120-volt, single-phase, 15- and 20-ampere branch circuits supplying outlets and devices installed in dormitory unit bedrooms, living rooms, hallways, closets, bathrooms, and similar rooms shall be protected by any of the means described in 210.12(A)(1) through (6).

N **(C) Guest Rooms and Guest Suites.** All 120-volt, single-phase, 15- and 20-ampere branch circuits supplying outlets and devices installed in guest rooms and guest suites of hotels and motels shall be protected by any of the means described in 210.12(A)(1) through (6).

(D) Branch Circuit Extensions or Modifications — Dwelling Units and Dormitory Units. In any of the areas specified in 210.12(A) or (B), where branch-circuit wiring is modified, replaced, or extended, the branch circuit shall be protected by one of the following:

(1) A listed combination-type AFCI located at the origin of the branch circuit

(2) A listed outlet branch-circuit-type AFCI located at the first receptacle outlet of the existing branch circuit

Exception: AFCI protection shall not be required where the extension of the existing conductors is not more than 1.8 m (6 ft) and does not include any additional outlets or devices.

210.13 Ground-Fault Protection of Equipment. Each branch-circuit disconnect rated 1000 A or more and installed on solidly grounded wye electrical systems of more than 150 volts to ground, but not exceeding 600 volts phase-to-phase, shall be provided with ground-fault protection of equipment in accordance with the provisions of 230.95.

Informational Note: For buildings that contain health care occupancies, see the requirements of 517.17.

Exception No. 1: The provisions of this section shall not apply to a disconnecting means for a continuous industrial process where a nonorderly shutdown will introduce additional or increased hazards.

Exception No. 2: The provisions of this section shall not apply if ground-fault protection of equipment is provided on the supply side of the branch circuit and on the load side of any transformer supplying the branch circuit.

210.17 Guest Rooms and Guest Suites. Guest rooms and guest suites that are provided with permanent provisions for cooking shall have branch circuits installed to meet the rules for dwelling units.

Part II. Branch-Circuit Ratings

N **210.18 Rating.** Branch circuits recognized by this article shall be rated in accordance with the maximum permitted ampere rating or setting of the overcurrent device. The rating for other than individual branch circuits shall be 15, 20, 30, 40, and 50 amperes. Where conductors of higher ampacity are used for any reason, the ampere rating or setting of the specified overcurrent device shall determine the circuit rating.

Exception: Multioutlet branch circuits greater than 50 amperes shall be permitted to supply nonlighting outlet loads on industrial premises where conditions of maintenance and supervision ensure that only qualified persons service the equipment.

210.19 Conductors — Minimum Ampacity and Size.

(A) Branch Circuits Not More Than 600 Volts.

Informational Note No. 1: See 310.15 for ampacity ratings of conductors.

Informational Note No. 2: See Part II of Article 430 for minimum rating of motor branch-circuit conductors.

Informational Note No. 3: See 310.15(A)(3) for temperature limitation of conductors.

Informational Note No. 4: Conductors for branch circuits as defined in Article 100, sized to prevent a voltage drop exceeding 3 percent at the farthest outlet of power, heating, and lighting loads, or combinations of such loads, and where the maximum total voltage drop on both feeders and branch circuits to the farthest outlet does not exceed 5 percent, provide reasonable efficiency of operation. See Informational Note No. 2 of 215.2(A)(1) for voltage drop on feeder conductors.

(1) General. Branch-circuit conductors shall have an ampacity not less than the maximum load to be served. Conductors shall be sized to carry not less than the larger of 210.19(A)(1)(a) or (b).

(a) Where a branch circuit supplies continuous loads or any combination of continuous and noncontinuous loads, the minimum branch-circuit conductor size shall have an allowable ampacity not less than the noncontinuous load plus 125 percent of the continuous load.

(b) The minimum branch-circuit conductor size shall have an allowable ampacity not less than the maximum load to be served after the application of any adjustment or correction factors.

Exception: If the assembly, including the overcurrent devices protecting the branch circuit(s), is listed for operation at 100 percent of its rating, the allowable ampacity of the branch-circuit conductors shall be permitted to be not less than the sum of the continuous load plus the noncontinuous load.

(2) Branch Circuits with More than One Receptacle. Conductors of branch circuits supplying more than one receptacle for cord-and-plug-connected portable loads shall have an ampacity of not less than the rating of the branch circuit.

(3) Household Ranges and Cooking Appliances. Branch-circuit conductors supplying household ranges, wall-mounted ovens, counter-mounted cooking units, and other household cooking appliances shall have an ampacity not less than the rating of the branch circuit and not less than the maximum load to be served. For ranges of 8¾ kW or more rating, the minimum branch-circuit rating shall be 40 amperes.

Exception No. 1: Conductors tapped from a 50-ampere branch circuit supplying electric ranges, wall-mounted electric ovens, and counter-mounted electric cooking units shall have an ampacity of not less than 20 amperes and shall be sufficient for the load to be served. These tap conductors include any conductors that are a part of the leads supplied with the appliance that are smaller than the branch-circuit conductors. The taps shall not be longer than necessary for servicing the appliance.

Exception No. 2: The neutral conductor of a 3-wire branch circuit supplying a household electric range, a wall-mounted oven, or a counter-mounted cooking unit shall be permitted to be smaller than the ungrounded conductors where the maximum demand of a range of 8¾-kW or more rating has been calculated according to Column C of Table 220.55, but such conductor shall have an ampacity of not less than 70 percent of the branch-circuit rating and shall not be smaller than 10 AWG.

(4) Other Loads. Branch-circuit conductors that supply loads other than those specified in 210.3 and other than cooking appliances as covered in 210.19(A)(3) shall have an ampacity sufficient for the loads served and shall not be smaller than 14 AWG.

Exception No. 1: Tap conductors shall have an ampacity sufficient for the load served. In addition, they shall have an ampacity of not less than 15 for circuits rated less than 40 amperes and not less than 20 for circuits rated at 40 or 50 amperes and only where these tap conductors supply any of the following loads:

(a) Individual lampholders or luminaires with taps extending not longer than 450 mm (18 in.) beyond any portion of the lampholder or luminaire

(b) A luminaire having tap conductors as provided in 410.117

(c) Individual outlets, other than receptacle outlets, with taps not over 450 mm (18 in.) long

(d) Infrared lamp industrial heating appliances

(e) Nonheating leads of deicing and snow-melting cables and mats

Exception No. 2: Fixture wires and flexible cords shall be permitted to be smaller than 14 AWG as permitted by 240.5.

(B) Branch Circuits Over 600 Volts. The ampacity of conductors shall be in accordance with 310.15 and 310.60, as applicable. Branch-circuit conductors over 600 volts shall be sized in accordance with 210.19(B)(1) or (B)(2).

(1) General. The ampacity of branch-circuit conductors shall not be less than 125 percent of the designed potential load of utilization equipment that will be operated simultaneously.

(2) Supervised Installations. For supervised installations, branch-circuit conductor sizing shall be permitted to be determined by qualified persons under engineering supervision. Supervised installations are defined as those portions of a facility where both of the following conditions are met:

(1) Conditions of design and installation are provided under engineering supervision.

(2) Qualified persons with documented training and experience in over 600-volt systems provide maintenance, monitoring, and servicing of the system.

210.20 Overcurrent Protection. Branch-circuit conductors and equipment shall be protected by overcurrent protective devices that have a rating or setting that complies with 210.20(A) through (D).

(A) Continuous and Noncontinuous Loads. Where a branch circuit supplies continuous loads or any combination of continuous and noncontinuous loads, the rating of the overcurrent device shall not be less than the noncontinuous load plus 125 percent of the continuous load.

Exception: Where the assembly, including the overcurrent devices protecting the branch circuit(s), is listed for operation at 100 percent of its rating, the ampere rating of the overcurrent device shall be permitted to be not less than the sum of the continuous load plus the noncontinuous load.

(B) Conductor Protection. Conductors shall be protected in accordance with 240.4. Flexible cords and fixture wires shall be protected in accordance with 240.5.

(C) Equipment. The rating or setting of the overcurrent protective device shall not exceed that specified in the applicable articles referenced in Table 240.3 for equipment.

(D) Outlet Devices. The rating or setting shall not exceed that specified in 210.21 for outlet devices.

210.21 Outlet Devices. Outlet devices shall have an ampere rating that is not less than the load to be served and shall comply with 210.21(A) and (B).

(A) Lampholders. Where connected to a branch circuit having a rating in excess of 20 amperes, lampholders shall be of the heavy-duty type. A heavy-duty lampholder shall have a rating of not less than 660 watts if of the admedium type, or not less than 750 watts if of any other type.

(B) Receptacles.

(1) Single Receptacle on an Individual Branch Circuit. A single receptacle installed on an individual branch circuit shall have an ampere rating not less than that of the branch circuit.

Exception No. 1: A receptacle installed in accordance with 430.81(B).

Exception No. 2: A receptacle installed exclusively for the use of a cord-and-plug-connected arc welder shall be permitted to have an ampere rating not less than the minimum branch-circuit conductor ampacity determined by 630.11(A) for arc welders.

Informational Note: See the definition of *receptacle* in Article 100.

(2) Total Cord-and-Plug-Connected Load. Where connected to a branch circuit supplying two or more receptacles or outlets, a receptacle shall not supply a total cord-and-plug-connected load in excess of the maximum specified in Table 210.21(B)(2).

(3) Receptacle Ratings. Where connected to a branch circuit supplying two or more receptacles or outlets, receptacle ratings shall conform to the values listed in Table 210.21(B)(3), or, where rated higher than 50 amperes, the receptacle rating shall not be less than the branch-circuit rating.

Exception No. 1: Receptacles installed exclusively for the use of one or more cord-and-plug-connected arc welders shall be permitted to have ampere ratings not less than the minimum branch-circuit conductor ampacity determined by 630.11(A) or (B) for arc welders.

Exception No. 2: The ampere rating of a receptacle installed for electric discharge lighting shall be permitted to be based on 410.62(C).

(4) Range Receptacle Rating. The ampere rating of a range receptacle shall be permitted to be based on a single range demand load as specified in Table 220.55.

210.22 Permissible Loads, Individual Branch Circuits. An individual branch circuit shall be permitted to supply any load for which it is rated, but in no case shall the load exceed the branch-circuit ampere rating.

210.23 Permissible Loads, Multiple-Outlet Branch Circuits. In no case shall the load exceed the branch-circuit ampere rating. A branch circuit supplying two or more outlets or receptacles shall supply only the loads specified according to its size as specified in 210.23(A) through (D) and as summarized in 210.24 and Table 210.24.

Table 210.21(B)(2) Maximum Cord-and-Plug-Connected Load to Receptacle

Circuit Rating (Amperes)	Receptacle Rating (Amperes)	Maximum Load (Amperes)
15 or 20	15	12
20	20	16
30	30	24

Table 210.21(B)(3) Receptacle Ratings for Various Size Circuits

Circuit Rating (Amperes)	Receptacle Rating (Amperes)
15	Not over 15
20	15 or 20
30	30
40	40 or 50
50	50

(A) 15- and 20-Ampere Branch Circuits. A 15- or 20-ampere branch circuit shall be permitted to supply lighting units or other utilization equipment, or a combination of both, and shall comply with 210.23(A)(1) and (A)(2).

Exception: The small-appliance branch circuits, laundry branch circuits, and bathroom branch circuits required in a dwelling unit(s) by 210.11(C)(1), (C)(2), and (C)(3) shall supply only the receptacle outlets specified in that section.

(1) Cord-and-Plug-Connected Equipment Not Fastened in Place. The rating of any one cord-and-plug-connected utilization equipment not fastened in place shall not exceed 80 percent of the branch-circuit ampere rating.

(2) Utilization Equipment Fastened in Place. The total rating of utilization equipment fastened in place, other than luminaires, shall not exceed 50 percent of the branch-circuit ampere rating where lighting units, cord-and-plug-connected utilization equipment not fastened in place, or both, are also supplied.

(B) 30-Ampere Branch Circuits. A 30-ampere branch circuit shall be permitted to supply fixed lighting units with heavy-duty lampholders in other than a dwelling unit(s) or utilization equipment in any occupancy. A rating of any one cord-and-plug-connected utilization equipment shall not exceed 80 percent of the branch-circuit ampere rating.

(C) 40- and 50-Ampere Branch Circuits. A 40- or 50-ampere branch circuit shall be permitted to supply cooking appliances that are fastened in place in any occupancy. In other than dwelling units, such circuits shall be permitted to supply fixed lighting units with heavy-duty lampholders, infrared heating units, or other utilization equipment.

(D) Branch Circuits Larger Than 50 Amperes. Branch circuits larger than 50 amperes shall supply only nonlighting outlet loads.

210.24 Branch-Circuit Requirements — Summary. The requirements for circuits that have two or more outlets or receptacles, other than the receptacle circuits of 210.11(C)(1), (C)(2), and (C)(3), are summarized in Table 210.24. This table provides only a summary of minimum requirements. See 210.19, 210.20, and 210.21 for the specific requirements applying to branch circuits.

210.25 Branch Circuits in Buildings with More Than One Occupancy.

(A) Dwelling Unit Branch Circuits. Branch circuits in each dwelling unit shall supply only loads within that dwelling unit or loads associated only with that dwelling unit.

(B) Common Area Branch Circuits. Branch circuits installed for the purpose of lighting, central alarm, signal, communications, or other purposes for public or common areas of a two-family dwelling, a multifamily dwelling, or a multi-occupancy building shall not be supplied from equipment that supplies an individual dwelling unit or tenant space.

Part III. Required Outlets

210.50 General. Receptacle outlets shall be installed as specified in 210.52 through 210.64.

Informational Note: See Informative Annex J for information regarding ADA accessibility design.

Table 210.24 Summary of Branch-Circuit Requirements

Circuit Rating	15 A	20 A	30 A	40 A	50 A
Conductors (min. size):					
Circuit wires[1]	14	12	10	8	6
Taps	14	14	14	12	12
Fixture wires and cords — see 240.5					
Overcurrent Protection	**15 A**	**20 A**	**30 A**	**40 A**	**50 A**
Outlet devices:					
Lampholders permitted	Any type	Any type	Heavy duty	Heavy duty	Heavy duty
Receptacle rating[2]	15 max. A	15 or 20 A	30 A	40 or 50 A	50 A
Maximum Load	**15 A**	**20 A**	**30 A**	**40 A**	**50 A**
Permissible load	See 210.23(A)	See 210.23(A)	See 210.23(B)	See 210.23(C)	See 210.23(C)

[1]These gauges are for copper conductors.

[2]For receptacle rating of cord-connected electric-discharge luminaires, see 410.62(C).

(A) Cord Pendants. A cord connector that is supplied by a permanently connected cord pendant shall be considered a receptacle outlet.

(B) Cord Connections. A receptacle outlet shall be installed wherever flexible cords with attachment plugs are used. Where flexible cords are permitted to be permanently connected, receptacles shall be permitted to be omitted for such cords.

(C) Appliance Receptacle Outlets. Appliance receptacle outlets installed in a dwelling unit for specific appliances, such as laundry equipment, shall be installed within 1.8 m (6 ft) of the intended location of the appliance.

210.52 Dwelling Unit Receptacle Outlets. This section provides requirements for 125-volt, 15- and 20-ampere receptacle outlets. The receptacles required by this section shall be in addition to any receptacle that is:

(1) Part of a luminaire or appliance, or
(2) Controlled by a wall switch in accordance with 210.70(A)(1), Exception No. 1, or
(3) Located within cabinets or cupboards, or
(4) Located more than 1.7 m (5½ ft) above the floor

Permanently installed electric baseboard heaters equipped with factory-installed receptacle outlets or outlets provided as a separate assembly by the manufacturer shall be permitted as the required outlet or outlets for the wall space utilized by such permanently installed heaters. Such receptacle outlets shall not be connected to the heater circuits.

> Informational Note: Listed baseboard heaters include instructions that may not permit their installation below receptacle outlets.

(A) General Provisions. In every kitchen, family room, dining room, living room, parlor, library, den, sunroom, bedroom, recreation room, or similar room or area of dwelling units, receptacle outlets shall be installed in accordance with the general provisions specified in 210.52(A)(1) through (A)(4).

(1) Spacing. Receptacles shall be installed such that no point measured horizontally along the floor line of any wall space is more than 1.8 m (6 ft) from a receptacle outlet.

(2) Wall Space. As used in this section, a wall space shall include the following:

(1) Any space 600 mm (2 ft) or more in width (including space measured around corners) and unbroken along the floor line by doorways and similar openings, fireplaces, and fixed cabinets that do not have countertops or similar work surfaces
(2) The space occupied by fixed panels in walls, excluding sliding panels
(3) The space afforded by fixed room dividers, such as free-standing bar-type counters or railings

(3) Floor Receptacles. Receptacle outlets in or on floors shall not be counted as part of the required number of receptacle outlets unless located within 450 mm (18 in.) of the wall.

(4) Countertop and Similar Work Surface Receptacle Outlets. Receptacles installed for countertop and similar work surfaces as specified in 210.52(C) shall not be considered as the receptacle outlets required by 210.52(A).

(B) Small Appliances.

(1) Receptacle Outlets Served. In the kitchen, pantry, breakfast room, dining room, or similar area of a dwelling unit, the two or more 20-ampere small-appliance branch circuits required by 210.11(C)(1) shall serve all wall and floor receptacle outlets covered by 210.52(A), all countertop outlets covered by 210.52(C), and receptacle outlets for refrigeration equipment.

Exception No. 1: In addition to the required receptacles specified by 210.52, switched receptacles supplied from a general-purpose branch circuit as defined in 210.70(A)(1), Exception No. 1, shall be permitted.

Exception No. 2: In addition to the required receptacles specified by 210.52, a receptacle outlet to serve a specific appliance shall be permitted to be supplied from an individual branch circuit rated 15 amperes or greater.

(2) No Other Outlets. The two or more small-appliance branch circuits specified in 210.52(B)(1) shall have no other outlets.

Exception No. 1: A receptacle installed solely for the electrical supply to and support of an electric clock in any of the rooms specified in 210.52(B)(1).

Exception No. 2: Receptacles installed to provide power for supplemental equipment and lighting on gas-fired ranges, ovens, or counter-mounted cooking units.

(3) Kitchen Receptacle Requirements. Receptacles installed in a kitchen to serve countertop surfaces shall be supplied by not fewer than two small-appliance branch circuits, either or both of which shall also be permitted to supply receptacle outlets in the same kitchen and in other rooms specified in 210.52(B)(1). Additional small-appliance branch circuits shall be permitted to supply receptacle outlets in the kitchen and other rooms specified in 210.52(B)(1). No small-appliance branch circuit shall serve more than one kitchen.

(C) Countertops and Work Surfaces. In kitchens, pantries, breakfast rooms, dining rooms, and similar areas of dwelling units, receptacle outlets for countertop and work surfaces shall be installed in accordance with 210.52(C)(1) through (C)(5).

(1) Wall Countertop and Work Surface. A receptacle outlet shall be installed at each wall countertop and work surface that is 300 mm (12 in.) or wider. Receptacle outlets shall be installed so that no point along the wall line is more than 600 mm (24 in.) measured horizontally from a receptacle outlet in that space.

Exception: Receptacle outlets shall not be required on a wall directly behind a range, counter-mounted cooking unit, or sink in the installation described in Figure 210.52(C)(1).

(2) Island Countertop Spaces. At least one receptacle shall be installed at each island countertop space with a long dimension of 600 mm (24 in.) or greater and a short dimension of 300 mm (12 in.) or greater.

(3) Peninsular Countertop Spaces. At least one receptacle outlet shall be installed at each peninsular countertop long dimension space with a long dimension of 600 mm (24 in.) or greater and a short dimension of 300 mm (12 in.) or greater. A peninsular countertop is measured from the connected perpendicular wall.

(4) Separate Spaces. Countertop spaces separated by range-tops, refrigerators, or sinks shall be considered as separate countertop spaces in applying the requirements of 210.52(C)(1). If a range, counter-mounted cooking unit, or sink is installed in an island or peninsular countertop and the depth of the countertop behind the range, counter-mounted cooking unit, or sink is less than 300 mm (12 in.), the range, counter-mounted cooking unit, or sink shall be considered to divide the countertop space into two separate countertop spaces. Each separate countertop space shall comply with the applicable requirements in 210.52(C).

(5) Receptacle Outlet Location. Receptacle outlets shall be located on or above, but not more than 500 mm (20 in.) above, the countertop or work surface. Receptacle outlet assemblies listed for use in countertops or work surfaces shall be permitted to be installed in countertops or work surfaces. Receptacle outlets rendered not readily accessible by appliances fastened in place, appliance garages, sinks, or rangetops as covered in 210.52(C)(1), Exception, or appliances occupying dedicated space shall not be considered as these required outlets.

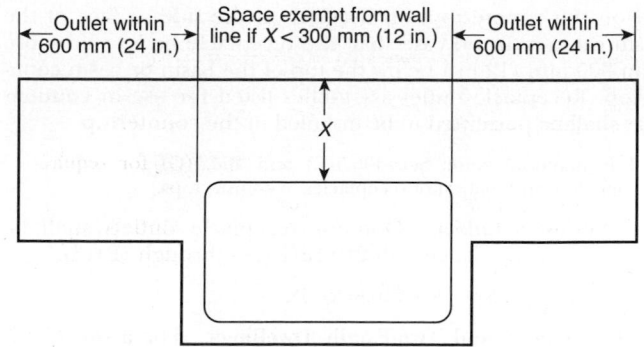

Range, counter-mounted cooking unit extending from face of counter

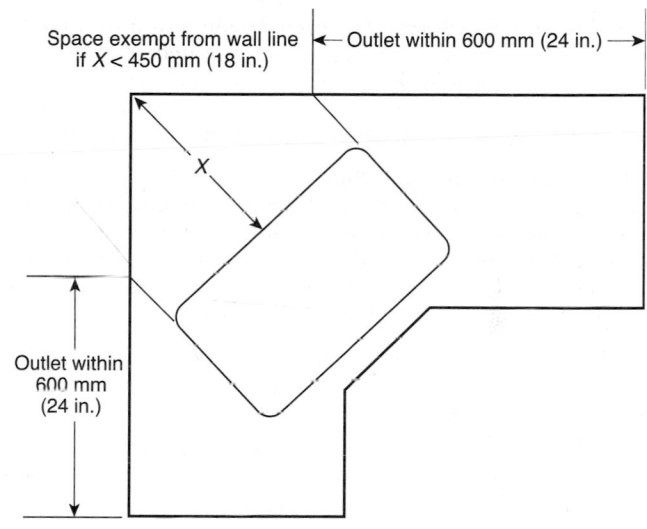

Range, counter-mounted cooking unit mounted in corner

FIGURE 210.52(C)(1)   Determination of Area Behind a Range, Counter-Mounted Cooking Unit, or Sink.

Informational Note: See 406.5(E) and 406.5(G) for requirements for installation of receptacles in countertops and 406.5(F) and 406.5(G) for requirements for installation of receptacles in work surfaces.

Exception to (5): To comply with the following conditions (1) and (2), receptacle outlets shall be permitted to be mounted not more than 300 mm (12 in.) below the countertop or work surface. Receptacles mounted below a countertop or work surface in accordance with this exception shall not be located where the countertop or work surface extends more than 150 mm (6 in.) beyond its support base.

(1)   Construction for the physically impaired
(2)   On island and peninsular countertops or work surface where the surface is flat across its entire surface (no backsplashes, dividers, etc.) and there are no means to mount a receptacle within 500 mm (20 in.) above the countertop or work surface, such as an overhead cabinet

(D) Bathrooms. At least one receptacle outlet shall be installed in bathrooms within 900 mm (3 ft) of the outside edge of each basin. The receptacle outlet shall be located on a wall or partition that is adjacent to the basin or basin countertop, located on the countertop, or installed on the side or face of the basin cabinet. In no case shall the receptacle be located more than 300 mm (12 in.) below the top of the basin or basin countertop. Receptacle outlet assemblies listed for use in countertops shall be permitted to be installed in the countertop.

> Informational Note: See 406.5(E) and 406.5(G) for requirements for installation of receptacles in countertops.

(E) Outdoor Outlets. Outdoor receptacle outlets shall be installed in accordance with 210.52(E)(1) through (E)(3).

> Informational Note: See 210.8(A)(3).

(1) One-Family and Two-Family Dwellings. For a one-family dwelling and each unit of a two-family dwelling that is at grade level, at least one receptacle outlet readily accessible from grade and not more than 2.0 m (6 ½ ft) above grade level shall be installed at the front and back of the dwelling.

(2) Multifamily Dwellings. For each dwelling unit of a multifamily dwelling where the dwelling unit is located at grade level and provided with individual exterior entrance/egress, at least one receptacle outlet readily accessible from grade and not more than 2.0 m (6½ ft) above grade level shall be installed.

(3) Balconies, Decks, and Porches. Balconies, decks, and porches that are attached to the dwelling unit and are accessible from inside the dwelling unit shall have at least one receptacle outlet accessible from the balcony, deck, or porch. The receptacle outlet shall not be located more than 2.0 m (6½ ft) above the balcony, deck, or porch walking surface.

(F) Laundry Areas. In dwelling units, at least one receptacle outlet shall be installed in areas designated for the installation of laundry equipment.

Exception No. 1: A receptacle for laundry equipment shall not be required in a dwelling unit of a multifamily building where laundry facilities are provided on the premises for use by all building occupants.

Exception No. 2: A receptacle for laundry equipment shall not be required in other than one-family dwellings where laundry facilities are not to be installed or permitted.

(G) Basements, Garages, and Accessory Buildings. For one- and two- family dwellings, at least one receptacle outlet shall be installed in the areas specified in 210.52(G)(1) through (3). These receptacles shall be in addition to receptacles required for specific equipment.

(1) Garages. In each attached garage and in each detached garage with electric power, at least one receptacle outlet shall be installed in each vehicle bay and not more than 1.7 m (5½ ft) above the floor.

(2) Accessory Buildings. In each accessory building with electric power.

(3) Basements. In each separate unfinished portion of a basement.

(H) Hallways. In dwelling units, hallways of 3.0 m (10 ft) or more in length shall have at least one receptacle outlet.

As used in this subsection, the hallway length shall be considered the length along the centerline of the hallway without passing through a doorway.

(I) Foyers. Foyers that are not part of a hallway in accordance with 210.52(H) and that have an area that is greater than

5.6 m² (60 ft²) shall have a receptacle(s) located in each wall space 900 mm (3 ft) or more in width. Doorways, door-side windows that extend to the floor, and similar openings shall not be considered wall space.

210.60 Guest Rooms, Guest Suites, Dormitories, and Similar Occupancies.

(A) General. Guest rooms or guest suites in hotels, motels, sleeping rooms in dormitories, and similar occupancies shall have receptacle outlets installed in accordance with 210.52(A) and (D). Guest rooms or guest suites provided with permanent provisions for cooking shall have receptacle outlets installed in accordance with all of the applicable rules in 210.52.

(B) Receptacle Placement. In applying the provisions of 210.52(A), the total number of receptacle outlets shall not be less than the minimum number that would comply with the provisions of that section. These receptacle outlets shall be permitted to be located conveniently for permanent furniture layout. At least two receptacle outlets shall be readily accessible. Where receptacles are installed behind the bed, the receptacle shall be located to prevent the bed from contacting any attachment plug that may be installed or the receptacle shall be provided with a suitable guard.

210.62 Show Windows. At least one 125-volt, single-phase, 15- or 20-ampere-rated receptacle outlet shall be installed within 450 mm (18 in.) of the top of a show window for each 3.7 linear m (12 linear ft) or major fraction thereof of show window area measured horizontally at its maximum width.

210.63 Heating, Air-Conditioning, and Refrigeration Equipment Outlet. A 125-volt, single-phase, 15- or 20-ampere-rated receptacle outlet shall be installed at an accessible location for the servicing of heating, air-conditioning, and refrigeration equipment. The receptacle shall be located on the same level and within 7.5 m (25 ft) of the heating, air-conditioning, and refrigeration equipment. The receptacle outlet shall not be connected to the load side of the equipment disconnecting means.

> Informational Note: See 210.8 for ground-fault circuit-interrupter requirements.

Exception: A receptacle outlet shall not be required at one- and two-family dwellings for the service of evaporative coolers.

210.64 Electrical Service Areas. At least one 125-volt, single-phase, 15- or 20-ampere-rated receptacle outlet shall be installed in an accessible location within 7.5 m (25 ft) of the indoor electrical service equipment. The required receptacle outlet shall be located within the same room or area as the service equipment.

Exception No. 1: The receptacle outlet shall not be required to be installed in one- and two-family dwellings.

Exception No. 2: Where the service voltage is greater than 120 volts to ground, a receptacle outlet shall not be required for services dedicated to equipment covered in Articles 675 and 682.

210.70 Lighting Outlets Required. Lighting outlets shall be installed where specified in 210.70(A), (B), and (C).

(A) Dwelling Units. In dwelling units, lighting outlets shall be installed in accordance with 210.70(A)(1), (A)(2), and (A)(3).

(1) Habitable Rooms. At least one wall switch–controlled lighting outlet shall be installed in every habitable room, kitchen, and bathroom.

Exception No. 1: In other than kitchens and bathrooms, one or more receptacles controlled by a wall switch shall be permitted in lieu of lighting outlets.

Exception No. 2: Lighting outlets shall be permitted to be controlled by occupancy sensors that are (1) in addition to wall switches or (2) located at a customary wall switch location and equipped with a manual override that will allow the sensor to function as a wall switch.

(2) Additional Locations. Additional lighting outlets shall be installed in accordance with the following:

(1) At least one wall switch–controlled lighting outlet shall be installed in hallways, stairways, attached garages, and detached garages with electric power.

(2) For dwelling units, attached garages, and detached garages with electric power, at least one wall switch–controlled lighting outlet shall be installed to provide illumination on the exterior side of outdoor entrances or exits with grade-level access. A vehicle door in a garage shall not be considered as an outdoor entrance or exit.

(3) Where one or more lighting outlet(s) are installed for interior stairways, there shall be a wall switch at each floor level, and landing level that includes an entryway, to control the lighting outlet(s) where the stairway between floor levels has six risers or more.

Exception to (A)(2)(1), (A)(2)(2), and (A)(2)(3): In hallways, in stairways, and at outdoor entrances, remote, central, or automatic control of lighting shall be permitted.

(4) Lighting outlets controlled in accordance with 210.70(A)(2)(3) shall not be controlled by use of dimmer switches unless they provide the full range of dimming control at each location.

(3) Storage or Equipment Spaces. For attics, underfloor spaces, utility rooms, and basements, at least one lighting outlet containing a switch or controlled by a wall switch shall be installed where these spaces are used for storage or contain equipment requiring servicing. At least one point of control shall be at the usual point of entry to these spaces. The lighting outlet shall be provided at or near the equipment requiring servicing.

(B) Guest Rooms or Guest Suites. In hotels, motels, or similar occupancies, guest rooms or guest suites shall have at least one wall switch–controlled lighting outlet installed in every habitable room and bathroom.

Exception No. 1: In other than bathrooms and kitchens where provided, one or more receptacles controlled by a wall switch shall be permitted in lieu of lighting outlets.

Exception No. 2: Lighting outlets shall be permitted to be controlled by occupancy sensors that are (1) in addition to wall switches or (2) located at a customary wall switch location and equipped with a manual override that allows the sensor to function as a wall switch.

(C) All Occupancies. For attics and underfloor spaces, utility rooms, and basements, at least one lighting outlet containing a switch or controlled by a wall switch shall be installed where these spaces are used for storage or contain equipment requiring servicing. At least one point of control shall be at the usual point of entry to these spaces. The lighting outlet shall be provided at or near the equipment requiring servicing.

N **210.71 Meeting Rooms.**

(A) General. Each meeting room of not more than 93 m² (1000 ft²) in other than dwelling units shall have outlets for nonlocking-type, 125-volt, 15- or 20-ampere receptacles. The outlets shall be installed in accordance with 210.71(B). Where a room or space is provided with movable partition(s), each room size shall be determined with the partition in the position that results in the smallest size meeting room.

Informational Note No. 1: For the purposes of this section, meeting rooms are typically designed or intended for the gathering of seated occupants for such purposes as conferences, deliberations, or similar purposes, where portable electronic equipment such as computers, projectors, or similar equipment is likely to be used.

Informational Note No. 2: Examples of rooms that are not meeting rooms include auditoriums, schoolrooms, and coffee shops.

(B) Receptacle Outlets Required. The total number of receptacle outlets, including floor outlets and receptacle outlets in fixed furniture, shall not be less than as determined in (1) and (2). These receptacle outlets shall be permitted to be located as determined by the designer or building owner.

(1) Receptacle Outlets in Fixed Walls. Receptacle outlets shall be installed in accordance with 210.52(A)(1) through (A)(4).

(2) Floor Receptacle Outlets. A meeting room that is at least 3.7 m (12 ft) wide and that has a floor area of at least 20 m² (215 ft²) shall have at least one receptacle outlet located in the floor at a distance not less than 1.8 m (6 ft) from any fixed wall for each 20 m² (215 ft²) or major portion of floor space.

Informational Note No. 1: See Section 314.27(B) for floor boxes used for receptacles located in the floor.

Informational Note No. 2: See Article 518 for assembly occupancies designed for 100 or more persons.

ARTICLE 215
Feeders

215.1 Scope. This article covers the installation requirements, overcurrent protection requirements, minimum size, and ampacity of conductors for feeders.

Exception: Feeders for electrolytic cells as covered in 668.3(C)(1) and (C)(4).

215.2 Minimum Rating and Size.

(A) Feeders Not More Than 600 Volts.

(1) General. Feeder conductors shall have an ampacity not less than required to supply the load as calculated in Parts III, IV, and V of Article 220. Conductors shall be sized to carry not less than the larger of 215.2(A)(1)(a) or (b).

(a) Where a feeder supplies continuous loads or any combination of continuous and noncontinuous loads, the minimum feeder conductor size shall have an allowable ampacity not less than the noncontinuous load plus 125 percent of the continuous load.

Exception No. 1: If the assembly, including the overcurrent devices protecting the feeder(s), is listed for operation at 100 percent of its rating, the allowable ampacity of the feeder conductors shall be permitted to be not less than the sum of the continuous load plus the noncontinuous load.

Exception No. 2: Where a portion of a feeder is connected at both its supply and load ends to separately installed pressure connections as covered in 110.14(C)(2), it shall be permitted to have an allowable ampacity not less than the sum of the continuous load plus the noncontinuous load. No portion of a feeder installed under the provisions of this exception shall extend into an enclosure containing either the feeder supply or the feeder load terminations, as covered in 110.14(C)(1).

Exception No. 3: Grounded conductors that are not connected to an overcurrent device shall be permitted to be sized at 100 percent of the continuous and noncontinuous load.

(b)   The minimum feeder conductor size shall have an allowable ampacity not less than the maximum load to be served after the application of any adjustment or correction factors.

> Informational Note No. 1: See Examples D1 through D11 in Informative Annex D.
>
> Informational Note No. 2: Conductors for feeders, as defined in Article 100, sized to prevent a voltage drop exceeding 3 percent at the farthest outlet of power, heating, and lighting loads, or combinations of such loads, and where the maximum total voltage drop on both feeders and branch circuits to the farthest outlet does not exceed 5 percent, will provide reasonable efficiency of operation.
>
> Informational Note No. 3: See 210.19(A), Informational Note No. 4, for voltage drop for branch circuits.

(2) Grounded Conductor. The size of the feeder circuit grounded conductor shall not be smaller than that required by 250.122, except that 250.122(F) shall not apply where grounded conductors are run in parallel.

Additional minimum sizes shall be as specified in 215.2(A)(3) under the conditions stipulated.

(3) Ampacity Relative to Service Conductors. The feeder conductor ampacity shall not be less than that of the service conductors where the feeder conductors carry the total load supplied by service conductors with an ampacity of 55 amperes or less.

(B) Feeders over 600 Volts. The ampacity of conductors shall be in accordance with 310.15 and 310.60 as applicable. Where installed, the size of the feeder-circuit grounded conductor shall not be smaller than that required by 250.122, except that 250.122(F) shall not apply where grounded conductors are run in parallel. Feeder conductors over 600 volts shall be sized in accordance with 215.2(B)(1), (B)(2), or (B)(3).

(1) Feeders Supplying Transformers. The ampacity of feeder conductors shall not be less than the sum of the nameplate ratings of the transformers supplied when only transformers are supplied.

(2) Feeders Supplying Transformers and Utilization Equipment. The ampacity of feeders supplying a combination of transformers and utilization equipment shall not be less than the sum of the nameplate ratings of the transformers and 125 percent of the designed potential load of the utilization equipment that will be operated simultaneously.

(3) Supervised Installations. For supervised installations, feeder conductor sizing shall be permitted to be determined by qualified persons under engineering supervision. Supervised installations are defined as those portions of a facility where all of the following conditions are met:

(1)   Conditions of design and installation are provided under engineering supervision.
(2)   Qualified persons with documented training and experience in over 600-volt systems provide maintenance, monitoring, and servicing of the system.

215.3 Overcurrent Protection. Feeders shall be protected against overcurrent in accordance with the provisions of Part I of Article 240. Where a feeder supplies continuous loads or any combination of continuous and noncontinuous loads, the rating of the overcurrent device shall not be less than the noncontinuous load plus 125 percent of the continuous load.

Exception No. 1: Where the assembly, including the overcurrent devices protecting the feeder(s), is listed for operation at 100 percent of its rating, the ampere rating of the overcurrent device shall be permitted to be not less than the sum of the continuous load plus the noncontinuous load.

Exception No. 2: Overcurrent protection for feeders between 600 and 1000 volts shall comply with Parts I through VIII of Article 240. Feeders over 1000 volts, nominal, shall comply with Part IX of Article 240.

215.4 Feeders with Common Neutral Conductor.

(A) Feeders with Common Neutral. Up to three sets of 3-wire feeders or two sets of 4-wire or 5-wire feeders shall be permitted to utilize a common neutral.

(B) In Metal Raceway or Enclosure. Where installed in a metal raceway or other metal enclosure, all conductors of all feeders using a common neutral conductor shall be enclosed within the same raceway or other enclosure as required in 300.20.

215.5 Diagrams of Feeders. If required by the authority having jurisdiction, a diagram showing feeder details shall be provided prior to the installation of the feeders. Such a diagram shall show the area in square feet of the building or other structure supplied by each feeder, the total calculated load before applying demand factors, the demand factors used, the calculated load after applying demand factors, and the size and type of conductors to be used.

215.6 Feeder Equipment Grounding Conductor. Where a feeder supplies branch circuits in which equipment grounding conductors are required, the feeder shall include or provide an equipment grounding conductor in accordance with the provisions of 250.134, to which the equipment grounding conductors of the branch circuits shall be connected. Where the feeder supplies a separate building or structure, the requirements of 250.32(B) shall apply.

215.7 Ungrounded Conductors Tapped from Grounded Systems. Two-wire dc circuits and ac circuits of two or more ungrounded conductors shall be permitted to be tapped from the ungrounded conductors of circuits having a grounded neutral conductor. Switching devices in each tapped circuit shall have a pole in each ungrounded conductor.

215.9 Ground-Fault Circuit-Interrupter Protection for Personnel. Feeders supplying 15- and 20-ampere receptacle branch circuits shall be permitted to be protected by a ground-fault circuit interrupter installed in a readily accessible location in lieu of the provisions for such interrupters as specified in 210.8 and 590.6(A).

215.10 Ground-Fault Protection of Equipment. Each feeder disconnect rated 1000 amperes or more and installed on solidly grounded wye electrical systems of more than 150 volts to ground, but not exceeding 600 volts phase-to-phase, shall be provided with ground-fault protection of equipment in accordance with the provisions of 230.95.

> Informational Note: For buildings that contain health care occupancies, see the requirements of 517.17.

Exception No. 1: The provisions of this section shall not apply to a disconnecting means for a continuous industrial process where a nonorderly shutdown will introduce additional or increased hazards.

Exception No. 2: The provisions of this section shall not apply if ground-fault protection of equipment is provided on the supply side of the feeder and on the load side of any transformer supplying the feeder.

215.11 Circuits Derived from Autotransformers. Feeders shall not be derived from autotransformers unless the system supplied has a grounded conductor that is electrically connected to a grounded conductor of the system supplying the autotransformer.

Exception No. 1: An autotransformer shall be permitted without the connection to a grounded conductor where transforming from a nominal 208 volts to a nominal 240-volt supply or similarly from 240 volts to 208 volts.

Exception No. 2: In industrial occupancies, where conditions of maintenance and supervision ensure that only qualified persons service the installation, autotransformers shall be permitted to supply nominal 600-volt loads from nominal 480-volt systems, and 480-volt loads from nominal 600-volt systems, without the connection to a similar grounded conductor.

215.12 Identification for Feeders.

(A) Grounded Conductor. The grounded conductor of a feeder, if insulated, shall be identified in accordance with 200.6.

(B) Equipment Grounding Conductor. The equipment grounding conductor shall be identified in accordance with 250.119.

(C) Identification of Ungrounded Conductors. Ungrounded conductors shall be identified in accordance with 215.12(C)(1) or (C)(2), as applicable.

(1) Feeders Supplied from More Than One Nominal Voltage System. Where the premises wiring system has feeders supplied from more than one nominal voltage system, each ungrounded conductor of a feeder shall be identified by phase or line and system at all termination, connection, and splice points in compliance with 215.12(C)(1)(a) and (b).

(a) *Means of Identification.* The means of identification shall be permitted to be by separate color coding, marking tape, tagging, or other approved means.

(b) *Posting of Identification Means.* The method utilized for conductors originating within each feeder panelboard or similar feeder distribution equipment shall be documented in a manner that is readily available or shall be permanently posted at each feeder panelboard or similar feeder distribution equipment.

(2) Feeders Supplied from Direct-Current Systems. Where a feeder is supplied from a dc system operating at more than 60 volts, each ungrounded conductor of 4 AWG or larger shall be identified by polarity at all termination, connection, and splice points by marking tape, tagging, or other approved means; each ungrounded conductor of 6 AWG or smaller shall be identified by polarity at all termination, connection, and splice points in compliance with 215.12(C)(2)(a) and (b). The identification methods utilized for conductors originating within each feeder panelboard or similar feeder distribution equipment shall be documented in a manner that is readily available or shall be permanently posted at each feeder panelboard or similar feeder distribution equipment.

(a) *Positive Polarity, Sizes 6 AWG or Smaller.* Where the positive polarity of a dc system does not serve as the connection for the grounded conductor, each positive ungrounded conductor shall be identified by one of the following means:

(1) A continuous red outer finish
(2) A continuous red stripe durably marked along the conductor's entire length on insulation of a color other than green, white, gray, or black
(3) Imprinted plus signs (+) or the word POSITIVE or POS durably marked on insulation of a color other than green, white, gray, or black, and repeated at intervals not exceeding 610 mm (24 in.) in accordance with 310.120(B)
(4) An approved permanent marking means such as sleeving or shrink-tubing that is suitable for the conductor size, at all termination, connection, and splice points, with imprinted plus signs (+) or the word POSITIVE or POS durably marked on insulation of a color other than green, white, gray, or black

(b) *Negative Polarity, Sizes 6 AWG or Smaller.* Where the negative polarity of a dc system does not serve as the connection for the grounded conductor, each negative ungrounded conductor shall be identified by one of the following means:

(1) A continuous black outer finish
(2) A continuous black stripe durably marked along the conductor's entire length on insulation of a color other than green, white, gray, or red
(3) Imprinted minus signs (−) or the word NEGATIVE or NEG durably marked on insulation of a color other than green, white, gray, or red, and repeated at intervals not exceeding 610 mm (24 in.) in accordance with 310.120(B)
(4) An approved permanent marking means such as sleeving or shrink-tubing that is suitable for the conductor size, at all termination, connection, and splice points, with imprinted minus signs (−) or the word NEGATIVE or NEG durably marked on insulation of a color other than green, white, gray, or red

ARTICLE 220
Branch-Circuit, Feeder, and Service Load Calculations

Part I. General

220.1 Scope. This article provides requirements for calculating branch-circuit, feeder, and service loads. Part I provides general requirements for calculation methods. Part II provides calculation methods for branch-circuit loads. Parts III and IV provide calculation methods for feeder and service loads. Part V provides calculation methods for farm loads.

> Informational Note No. 1: See examples in Informative Annex D.
>
> Informational Note No. 2: See Figure 220.1 for information on the organization of Article 220.

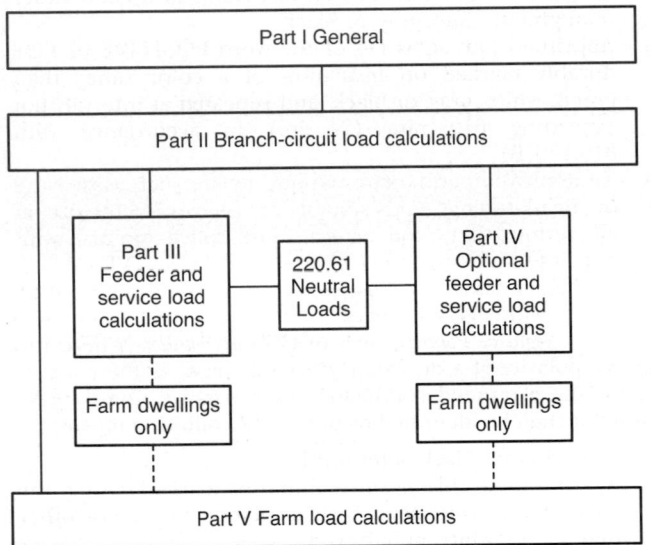

FIGURE 220.1 **Branch-Circuit, Feeder, and Service Load Calculation Methods.**

Table 220.3 **Specific-Purpose Calculation References**

Calculation	Article	Section (or Part)
Air-conditioning and refrigerating equipment, branch-circuit conductor sizing	440	Part IV
Fixed electric heating equipment for pipelines and vessels, branch-circuit sizing	427	427.4
Fixed electric space-heating equipment, branch-circuit sizing	424	424.3
Fixed outdoor electric deicing and snow-melting equipment, branch-circuit sizing	426	426.4
Motors, feeder demand factor	430	430.26
Motors, multimotor and combination-load equipment	430	430.25
Motors, several motors or a motor(s) and other load(s)	430	430.24
Over 600-volt branch-circuit calculations	210	210.19(B)
Over 600-volt feeder calculations	215	215.2(B)
Phase converters, conductors	455	455.6
Storage-type water heaters	422	422.11(E)

220.3 Other Articles for Specific-Purpose Calculations. Table 220.3 shall provide references for specific-purpose calculation requirements not located in Chapters 5, 6, or 7 that amend or supplement the requirements of this article.

220.5 Calculations.

(A) Voltages. Unless other voltages are specified, for purposes of calculating branch-circuit and feeder loads, nominal system voltages of 120, 120/240, 208Y/120, 240, 347, 480Y/277, 480, 600Y/347, and 600 volts shall be used.

(B) Fractions of an Ampere. Calculations shall be permitted to be rounded to the nearest whole ampere, with decimal fractions smaller than 0.5 dropped.

Part II. Branch-Circuit Load Calculations

220.10 General. Branch-circuit loads shall be calculated as shown in 220.12, 220.14, and 220.16.

220.12 Lighting Load for Specified Occupancies. A unit load of not less than that specified in Table 220.12 for occupancies specified shall constitute the minimum lighting load. The floor area for each floor shall be calculated from the outside dimensions of the building, dwelling unit, or other area involved. For dwelling units, the calculated floor area shall not include open porches, garages, or unused or unfinished spaces not adaptable for future use.

> Informational Note: The unit values are based on minimum load conditions and 100 percent power factor and may not provide sufficient capacity for the installation contemplated.

Exception No. 1: Where the building is designed and constructed to comply with an energy code adopted by the local authority, the lighting load shall be permitted to be calculated at the values specified in the energy code where the following conditions are met:

(1)   *A power monitoring system is installed that will provide continuous information regarding the total general lighting load of the building.*

(2)   *The power monitoring system will be set with alarm values to alert the building owner or manager if the lighting load exceeds the values set by the energy code.*

(3)   *The demand factors specified in 220.42 are not applied to the general lighting load.*

Exception No. 2: Where a building is designed and constructed to comply with an energy code adopted by the local authority and specifying an overall lighting density of less than 13.5 volt-amperes/13.5 m² (1.2 volt-amperes/1.2 ft²), the unit lighting loads in Table 220.12 for office and bank areas within the building shall be permitted to be reduced by 11 volt-amperes/11 m² (1 volt-amperes/1 ft²).

220.14 Other Loads — All Occupancies. In all occupancies, the minimum load for each outlet for general-use receptacles and outlets not used for general illumination shall not be less than that calculated in 220.14(A) through (L), the loads shown being based on nominal branch-circuit voltages.

Exception: The loads of outlets serving switchboards and switching frames in telephone exchanges shall be waived from the calculations.

(A) Specific Appliances or Loads. An outlet for a specific appliance or other load not covered in 220.14(B) through (L) shall be calculated based on the ampere rating of the appliance or load served.

Table 220.12 General Lighting Loads by Occupancy

Type of Occupancy	Unit Load	
	Volt-amperes/ m²	Volt-amperes/ ft²
Armories and auditoriums	11	1
Banks	39[b]	3½ [b]
Barber shops and beauty parlors	33	3
Churches	11	1
Clubs	22	2
Courtrooms	22	2
Dwelling units[a]	33	3
Garages — commercial (storage)	6	½
Hospitals	22	2
Hotels and motels, including apartment houses without provision for cooking by tenants[a]	22	2
Industrial commercial (loft) buildings	22	2
Lodge rooms	17	1½
Office buildings	39[b]	3½ [b]
Restaurants	22	2
Schools	33	3
Stores	33	3
Warehouses (storage)	3	¼
In any of the preceding occupancies except one-family dwellings and individual dwelling units of two-family and multifamily dwellings:		
Assembly halls and auditoriums	11	1
Halls, corridors, closets, stairways	6	½
Storage spaces	3	¼

[a]See 220.14(J).
[b]See 220.14(K).

(B) Electric Dryers and Electric Cooking Appliances in Dwellings and Household Cooking Appliances Used in Instructional Programs. Load calculations shall be permitted as specified in 220.54 for electric dryers and in 220.55 for electric ranges and other cooking appliances.

(C) Motor Outlets. Loads for motor outlets shall be calculated in accordance with the requirements in 430.22, 430.24, and 440.6.

(D) Luminaires. An outlet supplying luminaire(s) shall be calculated based on the maximum volt-ampere rating of the equipment and lamps for which the luminaire(s) is rated.

(E) Heavy-Duty Lampholders. Outlets for heavy-duty lampholders shall be calculated at a minimum of 600 volt-amperes.

(F) Sign and Outline Lighting. Sign and outline lighting outlets shall be calculated at a minimum of 1200 volt-amperes for each required branch circuit specified in 600.5(A).

(G) Show Windows. Show windows shall be calculated in accordance with either of the following:

(1) The unit load per outlet as required in other provisions of this section
(2) At 200 volt-amperes per linear 300 mm (1 ft) of show window

(H) Fixed Multioutlet Assemblies. Fixed multioutlet assemblies used in other than dwelling units or the guest rooms or guest suites of hotels or motels shall be calculated in accordance with (H)(1) or (H)(2). For the purposes of this section, the calculation shall be permitted to be based on the portion that contains receptacle outlets.

(1) Where appliances are unlikely to be used simultaneously, each 1.5 m (5 ft) or fraction thereof of each separate and continuous length shall be considered as one outlet of not less than 180 volt-amperes.
(2) Where appliances are likely to be used simultaneously, each 300 mm (1 ft) or fraction thereof shall be considered as an outlet of not less than 180 volt-amperes.

(I) Receptacle Outlets. Except as covered in 220.14(J) and (K), receptacle outlets shall be calculated at not less than 180 volt-amperes for each single or for each multiple receptacle on one yoke. A single piece of equipment consisting of a multiple receptacle comprised of four or more receptacles shall be calculated at not less than 90 volt-amperes per receptacle. This provision shall not be applicable to the receptacle outlets specified in 210.11(C)(1) and (C)(2).

(J) Dwelling Occupancies. In one-family, two-family, and multifamily dwellings and in guest rooms or guest suites of hotels and motels, the outlets specified in (J)(1), (J)(2), and (J)(3) are included in the general lighting load calculations of 220.12. No additional load calculations shall be required for such outlets.

(1) All general-use receptacle outlets of 20-ampere rating or less, including receptacles connected to the circuits in 210.11(C)(3)
(2) The receptacle outlets specified in 210.52(E) and (G)
(3) The lighting outlets specified in 210.70(A) and (B)

(K) Banks and Office Buildings. In banks or office buildings, the receptacle loads shall be calculated to be the larger of (1) or (2):

(1)   The calculated load from 220.14(I)
(2)   11 volt-amperes/m² or 1 volt-ampere/ft²

(L) Other Outlets. Other outlets not covered in 220.14(A) through (K) shall be calculated based on 180 volt-amperes per outlet.

220.16 Loads for Additions to Existing Installations.

(A) Dwelling Units. Loads added to an existing dwelling unit(s) shall comply with the following as applicable:

(1)   Loads for structural additions to an existing dwelling unit or for a previously unwired portion of an existing dwelling unit, either of which exceeds 46.5 m² (500 ft²), shall be calculated in accordance with 220.12 and 220.14.
(2)   Loads for new circuits or extended circuits in previously wired dwelling units shall be calculated in accordance with either 220.12 or 220.14, as applicable.

(B) Other Than Dwelling Units. Loads for new circuits or extended circuits in other than dwelling units shall be calculated in accordance with either 220.12 or 220.14, as applicable.

220.18 Maximum Loads. The total load shall not exceed the rating of the branch circuit, and it shall not exceed the maximum loads specified in 220.18(A) through (C) under the conditions specified therein.

(A) Motor-Operated and Combination Loads. Where a circuit supplies only motor-operated loads, Article 430 shall apply. Where a circuit supplies only air-conditioning equipment, refrigerating equipment, or both, Article 440 shall apply. For circuits supplying loads consisting of motor-operated utilization equipment that is fastened in place and has a motor larger than ⅛ hp in combination with other loads, the total calculated load shall be based on 125 percent of the largest motor load plus the sum of the other loads.

(B) Inductive and LED Lighting Loads. For circuits supplying lighting units that have ballasts, transformers, autotransformers, or LED drivers, the calculated load shall be based on the total ampere ratings of such units and not on the total watts of the lamps.

(C) Range Loads. It shall be permissible to apply demand factors for range loads in accordance with Table 220.55, including Note 4.

Part III. Feeder and Service Load Calculations

220.40 General. The calculated load of a feeder or service shall not be less than the sum of the loads on the branch circuits supplied, as determined by Part II of this article, after any applicable demand factors permitted by Part III or IV or required by Part V have been applied.

> Informational Note: See Examples D1(a) through D10 in Informative Annex D. See 220.18(B) for the maximum load in amperes permitted for lighting units operating at less than 100 percent power factor.

220.42 General Lighting. The demand factors specified in Table 220.42 shall apply to that portion of the total branch-circuit load calculated for general illumination. They shall not be applied in determining the number of branch circuits for general illumination.

220.43 Show-Window and Track Lighting.

(A) Show Windows. For show-window lighting, a load of not less than 660 volt-amperes/linear meter or 200 volt-amperes/linear foot shall be included for a show window, measured horizontally along its base.

> Informational Note: See 220.14(G) for branch circuits supplying show windows.

(B) Track Lighting. For track lighting in other than dwelling units or guest rooms or guest suites of hotels or motels, an additional load of 150 volt-amperes shall be included for every 600 mm (2 ft) of lighting track or fraction thereof. Where multicircuit track is installed, the load shall be considered to be divided equally between the track circuits.

Exception: If the track lighting is supplied through a device that limits the current to the track, the load shall be permitted to be calculated based on the rating of the device used to limit the current.

220.44 Receptacle Loads — Other Than Dwelling Units. Receptacle loads calculated in accordance with 220.14(H) and (I) shall be permitted to be made subject to the demand factors given in Table 220.42 or Table 220.44.

Table 220.42 Lighting Load Demand Factors

Type of Occupancy	Portion of Lighting Load to Which Demand Factor Applies (Volt-Amperes)	Demand Factor (%)
Dwelling units	First 3000 at	100
	From 3001 to 120,000 at	35
	Remainder over 120,000 at	25
Hospitals*	First 50,000 or less at	40
	Remainder over 50,000 at	20
Hotels and motels, including apartment houses without provision for cooking by tenants*	First 20,000 or less at	50
	From 20,001 to 100,000 at	40
	Remainder over 100,000 at	
		30
Warehouses (storage)	First 12,500 or less at	100
	Remainder over 12,500 at	50
All others	Total volt-amperes	100

*The demand factors of this table shall not apply to the calculated load of feeders or services supplying areas in hospitals, hotels, and motels where the entire lighting is likely to be used at one time, as in operating rooms, ballrooms, or dining rooms.

Table 220.44 Demand Factors for Non-Dwelling Receptacle Loads

Portion of Receptacle Load to Which Demand Factor Applies (Volt-Amperes)	Demand Factor (%)
First 10 kVA or less at	100
Remainder over 10 kVA at	50

220.50 Motors. Motor loads shall be calculated in accordance with 430.24, 430.25, and 430.26 and with 440.6 for hermetic refrigerant motor-compressors.

220.51 Fixed Electric Space Heating. Fixed electric space-heating loads shall be calculated at 100 percent of the total connected load. However, in no case shall a feeder or service load current rating be less than the rating of the largest branch circuit supplied.

Exception: Where reduced loading of the conductors results from units operating on duty-cycle, intermittently, or from all units not operating at the same time, the authority having jurisdiction may grant permission for feeder and service conductors to have an ampacity less than 100 percent, provided the conductors have an ampacity for the load so determined.

220.52 Small-Appliance and Laundry Loads — Dwelling Unit.

(A) Small-Appliance Circuit Load. In each dwelling unit, the load shall be calculated at 1500 volt-amperes for each 2-wire small-appliance branch circuit as covered by 210.11(C)(1). Where the load is subdivided through two or more feeders, the calculated load for each shall include not less than 1500 volt-amperes for each 2-wire small-appliance branch circuit. These loads shall be permitted to be included with the general lighting load and subjected to the demand factors provided in Table 220.42.

Exception: The individual branch circuit permitted by 210.52(B)(1), Exception No. 2, shall be permitted to be excluded from the calculation required by 220.52.

(B) Laundry Circuit Load. A load of not less than 1500 volt-amperes shall be included for each 2-wire laundry branch circuit installed as covered by 210.11(C)(2). This load shall be permitted to be included with the general lighting load and shall be subjected to the demand factors provided in Table 220.42.

220.53 Appliance Load — Dwelling Unit(s). It shall be permissible to apply a demand factor of 75 percent to the nameplate rating load of four or more appliances fastened in place, other than electric ranges, clothes dryers, space-heating equipment, or air-conditioning equipment, that are served by the same feeder or service in a one-family, two-family, or multi-family dwelling.

220.54 Electric Clothes Dryers — Dwelling Unit(s). The load for household electric clothes dryers in a dwelling unit(s) shall be either 5000 watts (volt-amperes) or the nameplate rating, whichever is larger, for each dryer served. The use of the demand factors in Table 220.54 shall be permitted. Where two or more single-phase dryers are supplied by a 3-phase, 4-wire feeder or service, the total load shall be calculated on the basis of twice the maximum number connected between any two phases. Kilovolt-amperes (kVA) shall be considered equivalent to kilowatts (kW) for loads calculated in this section.

220.55 Electric Cooking Appliances in Dwelling Units and Household Cooking Appliances Used in Instructional Programs. The load for household electric ranges, wall-mounted ovens, counter-mounted cooking units, and other household cooking appliances individually rated in excess of 1¾ kW shall be permitted to be calculated in accordance with Table 220.55. Kilovolt-amperes (kVA) shall be considered equivalent to kilowatts (kW) for loads calculated under this section.

Table 220.54 Demand Factors for Household Electric Clothes Dryers

Number of Dryers	Demand Factor (%)
1–4	100
5	85
6	75
7	65
8	60
9	55
10	50
11	47
12–23	47% minus 1% for each dryer exceeding 11
24–42	35% minus 0.5% for each dryer exceeding 23
43 and over	25%

Where two or more single-phase ranges are supplied by a 3-phase, 4-wire feeder or service, the total load shall be calculated on the basis of twice the maximum number connected between any two phases.

Informational Note No. 1: See the examples in Informative Annex D.

Informational Note No. 2: See Table 220.56 for commercial cooking equipment.

220.56 Kitchen Equipment — Other Than Dwelling Unit(s). It shall be permissible to calculate the load for commercial electric cooking equipment, dishwasher booster heaters, water heaters, and other kitchen equipment in accordance with Table 220.56. These demand factors shall be applied to all equipment that has either thermostatic control or intermittent use as kitchen equipment. These demand factors shall not apply to space-heating, ventilating, or air-conditioning equipment.

However, in no case shall the feeder or service calculated load be less than the sum of the largest two kitchen equipment loads.

220.60 Noncoincident Loads. Where it is unlikely that two or more noncoincident loads will be in use simultaneously, it shall be permissible to use only the largest load(s) that will be used at one time for calculating the total load of a feeder or service.

220.61 Feeder or Service Neutral Load.

(A) Basic Calculation. The feeder or service neutral load shall be the maximum unbalance of the load determined by this article. The maximum unbalanced load shall be the maximum net calculated load between the neutral conductor and any one ungrounded conductor.

Exception: For 3-wire, 2-phase or 5-wire, 2-phase systems, the maximum unbalanced load shall be the maximum net calculated load between the neutral conductor and any one ungrounded conductor multiplied by 140 percent.

Table 220.55 Demand Factors and Loads for Household Electric Ranges, Wall-Mounted Ovens, Counter-Mounted Cooking Units, and Other Household Cooking Appliances over 1¾ kW Rating (Column C to be used in all cases except as otherwise permitted in Note 3.)

| Number of Appliances | Demand Factor (%) (See Notes) | | Column C Maximum Demand (kW) (See Notes) (Not over 12 kW Rating) |
	Column A (Less than 3½ kW Rating)	Column B (3½ kW through 8¾ kW Rating)	
1	80	80	8
2	75	65	11
3	70	55	14
4	66	50	17
5	62	45	20
6	59	43	21
7	56	40	22
8	53	36	23
9	51	35	24
10	49	34	25
11	47	32	26
12	45	32	27
13	43	32	28
14	41	32	29
15	40	32	30
16	39	28	31
17	38	28	32
18	37	28	33
19	36	28	34
20	35	28	35
21	34	26	36
22	33	26	37
23	32	26	38
24	31	26	39
25	30	26	40
26–30	30	24	15 kW + 1 kW for each range
31–40	30	22	
41–50	30	20	25 kW + ¾ kW for each range
51–60	30	18	
61 and over	30	16	

Notes:

1. Over 12 kW through 27 kW ranges all of same rating. For ranges individually rated more than 12 kW but not more than 27 kW, the maximum demand in Column C shall be increased 5 percent for each additional kilowatt of rating or major fraction thereof by which the rating of individual ranges exceeds 12 kW.

2. Over 8¾ kW through 27 kW ranges of unequal ratings. For ranges individually rated more than 8¾ kW and of different ratings, but none exceeding 27 kW, an average value of rating shall be calculated by adding together the ratings of all ranges to obtain the total connected load (using 12 kW for any range rated less than 12 kW) and dividing by the total number of ranges. Then the maximum demand in Column C shall be increased 5 percent for each kilowatt or major fraction thereof by which this average value exceeds 12 kW.

3. Over 1¾ kW through 8¾ kW. In lieu of the method provided in Column C, it shall be permissible to add the nameplate ratings of all household cooking appliances rated more than 1¾ kW but not more than 8¾ kW and multiply the sum by the demand factors specified in Column A or Column B for the given number of appliances. Where the rating of cooking appliances falls under both Column A and Column B, the demand factors for each column shall be applied to the appliances for that column, and the results added together.

4. Branch-Circuit Load. It shall be permissible to calculate the branch-circuit load for one range in accordance with Table 220.55. The branch-circuit load for one wall-mounted oven or one counter-mounted cooking unit shall be the nameplate rating of the appliance. The branch-circuit load for a counter-mounted cooking unit and not more than two wall-mounted ovens, all supplied from a single branch circuit and located in the same room, shall be calculated by adding the nameplate rating of the individual appliances and treating this total as equivalent to one range.

5. This table shall also apply to household cooking appliances rated over 1¾ kW and used in instructional programs.

Table 220.56 Demand Factors for Kitchen Equipment — Other Than Dwelling Unit(s)

Number of Units of Equipment	Demand Factor (%)
1	100
2	100
3	90
4	80
5	70
6 and over	65

(B) Permitted Reductions. A service or feeder supplying the following loads shall be permitted to have an additional demand factor of 70 percent applied to the amount in 220.61(B)(1) or portion of the amount in 220.61(B)(2) determined by the following basic calculations:

(1) A feeder or service supplying household electric ranges, wall-mounted ovens, counter-mounted cooking units, and electric dryers, where the maximum unbalanced load has been determined in accordance with Table 220.55 for ranges and Table 220.54 for dryers

(2) That portion of the unbalanced load in excess of 200 amperes where the feeder or service is supplied from a 3-wire dc or single-phase ac system; or a 4-wire, 3-phase system; or a 3-wire, 2-phase system; or a 5-wire, 2-phase system

Informational Note: See Examples D1(a), D1(b), D2(b), D4(a), and D5(a) in Informative Annex D.

(C) Prohibited Reductions. There shall be no reduction of the neutral or grounded conductor capacity applied to the amount in 220.61(C)(1), or portion of the amount in (C)(2), from that determined by the basic calculation:

(1) Any portion of a 3-wire circuit consisting of 2 ungrounded conductors and the neutral conductor of a 4-wire, 3-phase, wye-connected system

(2) That portion consisting of nonlinear loads supplied from a 4-wire, wye-connected, 3-phase system

Informational Note: A 3-phase, 4-wire, wye-connected power system used to supply power to nonlinear loads may necessitate that the power system design allow for the possibility of high harmonic neutral conductor currents.

Part IV. Optional Feeder and Service Load Calculations

220.80 General. Optional feeder and service load calculations shall be permitted in accordance with Part IV.

220.82 Dwelling Unit.

(A) Feeder and Service Load. This section applies to a dwelling unit having the total connected load served by a single 120/240-volt or 208Y/120-volt set of 3-wire service or feeder conductors with an ampacity of 100 or greater. It shall be permissible to calculate the feeder and service loads in accordance with this section instead of the method specified in Part III of this article. The calculated load shall be the result of adding the loads from 220.82(B) and (C). Feeder and service-entrance conductors whose calculated load is determined by this optional calculation shall be permitted to have the neutral load determined by 220.61.

(B) General Loads. The general calculated load shall be not less than 100 percent of the first 10 kVA plus 40 percent of the remainder of the following loads:

(1) 33 volt-amperes/m^2 or 3 volt-amperes/ft^2 for general lighting and general-use receptacles. The floor area for each floor shall be calculated from the outside dimensions of the dwelling unit. The calculated floor area shall not include open porches, garages, or unused or unfinished spaces not adaptable for future use.

(2) 1500 volt-amperes for each 2-wire, 20-ampere small-appliance branch circuit and each laundry branch circuit covered in 210.11(C)(1) and (C)(2).

(3) The nameplate rating of the following:

   a. All appliances that are fastened in place, permanently connected, or located to be on a specific circuit

   b. Ranges, wall-mounted ovens, counter-mounted cooking units

   c. Clothes dryers that are not connected to the laundry branch circuit specified in item (2)

   d. Water heaters

(4) The nameplate ampere or kVA rating of all permanently connected motors not included in item (3).

(C) Heating and Air-Conditioning Load. The largest of the following six selections (load in kVA) shall be included:

(1) 100 percent of the nameplate rating(s) of the air conditioning and cooling.

(2) 100 percent of the nameplate rating(s) of the heat pump when the heat pump is used without any supplemental electric heating.

(3) 100 percent of the nameplate rating(s) of the heat pump compressor and 65 percent of the supplemental electric heating for central electric space-heating systems. If the heat pump compressor is prevented from operating at the same time as the supplementary heat, it does not need to be added to the supplementary heat for the total central space heating load.

(4) 65 percent of the nameplate rating(s) of electric space heating if less than four separately controlled units.

(5) 40 percent of the nameplate rating(s) of electric space heating if four or more separately controlled units.

(6) 100 percent of the nameplate ratings of electric thermal storage and other heating systems where the usual load is expected to be continuous at the full nameplate value. Systems qualifying under this selection shall not be calculated under any other selection in 220.82(C).

220.83 Existing Dwelling Unit. This section shall be permitted to be used to determine if the existing service or feeder is of sufficient capacity to serve additional loads. Where the dwelling unit is served by a 120/240-volt or 208Y/120-volt, 3-wire service, it shall be permissible to calculate the total load in accordance with 220.83(A) or (B).

(A) Where Additional Air-Conditioning Equipment or Electric Space-Heating Equipment Is Not to Be Installed. The following percentages shall be used for existing and additional new loads.

Load (kVA)	Percent of Load
First 8 kVA of load at	100
Remainder of load at	40

Load calculations shall include the following:

(1) General lighting and general-use receptacles at 33 volt-amperes/m² or 3 volt-amperes/ft² as determined by 220.12

(2) 1500 volt-amperes for each 2-wire, 20-ampere small-appliance branch circuit and each laundry branch circuit covered in 210.11(C)(1) and (C)(2)

(3) The nameplate rating of the following:

  a. All appliances that are fastened in place, permanently connected, or located to be on a specific circuit

  b. Ranges, wall-mounted ovens, counter-mounted cooking units

  c. Clothes dryers that are not connected to the laundry branch circuit specified in item (2)

  d. Water heaters

(B) Where Additional Air-Conditioning Equipment or Electric Space-Heating Equipment Is to Be Installed. The following percentages shall be used for existing and additional new loads. The larger connected load of air conditioning or space heating, but not both, shall be used.

Load	Percent of Load
Air-conditioning equipment	100
Central electric space heating	100
Less than four separately controlled space-heating units	100
First 8 kVA of all other loads	100
Remainder of all other loads	40

Other loads shall include the following:

(1) General lighting and general-use receptacles at 33 volt-amperes/m² or 3 volt-amperes/ft² as determined by 220.12

(2) 1500 volt-amperes for each 2-wire, 20-ampere small-appliance branch circuit and each laundry branch circuit covered in 210.11(C)(1) and (C)(2)

(3) The nameplate rating of the following:

  a. All appliances that are fastened in place, permanently connected, or located to be on a specific circuit

  b. Ranges, wall-mounted ovens, counter-mounted cooking units

  c. Clothes dryers that are not connected to the laundry branch circuit specified in item (2)

  d. Water heaters

220.84 Multifamily Dwelling.

(A) Feeder or Service Load. It shall be permissible to calculate the load of a feeder or service that supplies three or more dwelling units of a multifamily dwelling in accordance with Table 220.84 instead of Part III of this article if all the following conditions are met:

(1) No dwelling unit is supplied by more than one feeder.

(2) Each dwelling unit is equipped with electric cooking equipment.

Exception: When the calculated load for multifamily dwellings without electric cooking in Part III of this article exceeds that calculated under Part IV for the identical load plus electric cooking (based on 8 kW per unit), the lesser of the two loads shall be permitted to be used.

(3) Each dwelling unit is equipped with either electric space heating or air conditioning, or both. Feeders and service conductors whose calculated load is determined by this optional calculation shall be permitted to have the neutral load determined by 220.61.

(B) House Loads. House loads shall be calculated in accordance with Part III of this article and shall be in addition to the dwelling unit loads calculated in accordance with Table 220.84.

(C) Calculated Loads. The calculated load to which the demand factors of Table 220.84 apply shall include the following:

(1) 33 volt-amperes/m² or 3 volt-amperes/ft² for general lighting and general-use receptacles

(2) 1500 volt-amperes for each 2-wire, 20-ampere small-appliance branch circuit and each laundry branch circuit covered in 210.11(C)(1) and (C)(2)

(3) The nameplate rating of the following:

  a. All appliances that are fastened in place, permanently connected, or located to be on a specific circuit

  b. Ranges, wall-mounted ovens, counter-mounted cooking units

  c. Clothes dryers that are not connected to the laundry branch circuit specified in item (2)

  d. Water heaters

(4) The nameplate ampere or kVA rating of all permanently connected motors not included in item (3)

(5) The larger of the air-conditioning load or the fixed electric space-heating load

Table 220.84 Optional Calculations — Demand Factors for Three or More Multifamily Dwelling Units

Number of Dwelling Units	Demand Factor (%)
3–5	45
6–7	44
8–10	43
11	42
12–13	41
14–15	40
16–17	39
18–20	38
21	37
22–23	36
24–25	35
26–27	34
28–30	33
31	32
32–33	31
34–36	30
37–38	29
39–42	28
43–45	27
46–50	26
51–55	25
56–61	24
62 and over	23

220.85 Two Dwelling Units. Where two dwelling units are supplied by a single feeder and the calculated load under Part III of this article exceeds that for three identical units calculated under 220.84, the lesser of the two loads shall be permitted to be used.

220.86 Schools. The calculation of a feeder or service load for schools shall be permitted in accordance with Table 220.86 in lieu of Part III of this article where equipped with electric space heating, air conditioning, or both. The connected load to which the demand factors of Table 220.86 apply shall include all of the interior and exterior lighting, power, water heating, cooking, other loads, and the larger of the air-conditioning load or space-heating load within the building or structure.

Feeders and service conductors whose calculated load is determined by this optional calculation shall be permitted to have the neutral load determined by 220.61. Where the building or structure load is calculated by this optional method, feeders within the building or structure shall have ampacity as permitted in Part III of this article; however, the ampacity of an individual feeder shall not be required to be larger than the ampacity for the entire building.

This section shall not apply to portable classroom buildings.

220.87 Determining Existing Loads. The calculation of a feeder or service load for existing installations shall be permitted to use actual maximum demand to determine the existing load under all of the following conditions:

(1) The maximum demand data is available for a 1-year period.

Exception: If the maximum demand data for a 1-year period is not available, the calculated load shall be permitted to be based on the maximum demand (the highest average kilowatts reached and maintained for a 15-minute interval) continuously recorded over a minimum 30-day period using a recording ammeter or power meter connected to the highest loaded phase of the feeder or service, based on

the initial loading at the start of the recording. The recording shall reflect the maximum demand of the feeder or service by being taken when the building or space is occupied and shall include by measurement or calculation the larger of the heating or cooling equipment load, and other loads that may be periodic in nature due to seasonal or similar conditions.

(2) The maximum demand at 125 percent plus the new load does not exceed the ampacity of the feeder or rating of the service.

(3) The feeder has overcurrent protection in accordance with 240.4, and the service has overload protection in accordance with 230.90.

220.88 New Restaurants. Calculation of a service or feeder load, where the feeder serves the total load, for a new restaurant shall be permitted in accordance with Table 220.88 in lieu of Part III of this article.

The overload protection of the service conductors shall be in accordance with 230.90 and 240.4.

Feeder conductors shall not be required to be of greater ampacity than the service conductors.

Service or feeder conductors whose calculated load is determined by this optional calculation shall be permitted to have the neutral load determined by 220.61.

Part V. Farm Load Calculations

220.100 General. Farm loads shall be calculated in accordance with Part V.

220.102 Farm Loads — Buildings and Other Loads.

(A) Dwelling Unit. The feeder or service load of a farm dwelling unit shall be calculated in accordance with the provisions for dwellings in Part III or IV of this article. Where the dwelling has electric heat and the farm has electric grain-drying systems, Part IV of this article shall not be used to calculate the dwelling load where the dwelling and farm loads are supplied by a common service.

(B) Other Than Dwelling Unit. Where a feeder or service supplies a farm building or other load having two or more separate branch circuits, the load for feeders, service conductors, and service equipment shall be calculated in accordance with demand factors not less than indicated in Table 220.102.

Table 220.86 Optional Method — Demand Factors for Feeders and Service Conductors for Schools

Connected Load		Demand Factor (Percent)
First 33 VA/m² Plus,	(3 VA/ft²) at	100
Over 33 through 220 VA/m² Plus,	(3 through 20 VA/ft²) at	75
Remainder over 220 VA/m²	(20 VA/ft²) at	25

Table 220.88 Optional Method — Permitted Load Calculations for Service and Feeder Conductors for New Restaurants

Total Connected Load (kVA)	All Electric Restaurant Calculated Loads (kVA)	Not All Electric Restaurant Calculated Loads (kVA)
0–200	80%	100%
201–325	10% (amount over 200) + 160.0	50% (amount over 200) + 200.0
326–800	50% (amount over 325) + 172.5	45% (amount over 325) + 262.5
Over 800	50% (amount over 800) + 410.0	20% (amount over 800) + 476.3

Note: Add all electrical loads, including both heating and cooling loads, to calculate the total connected load. Select the one demand factor that applies from the table, then multiply the total connected load by this single demand factor.

Table 220.102 Method for Calculating Farm Loads for Other Than Dwelling Unit

Ampere Load at 240 Volts Maximum	Demand Factor (%)
The greater of the following:	
All loads that are expected to operate simultaneously, or	100
125 percent of the full load current of the largest motor, or	
First 60 amperes of the load	
Next 60 amperes of all other loads	50
Remainder of other loads	25

220.103 Farm Loads — Total. Where supplied by a common service, the total load of the farm for service conductors and service equipment shall be calculated in accordance with the farm dwelling unit load and demand factors specified in Table 220.103. Where there is equipment in two or more farm equipment buildings or for loads having the same function, such loads shall be calculated in accordance with Table 220.102 and shall be permitted to be combined as a single load in Table 220.103 for calculating the total load.

Table 220.103 Method for Calculating Total Farm Load

Individual Loads Calculated in Accordance with Table 220.102	Demand Factor (%)
Largest load	100
Second largest load	75
Third largest load	65
Remaining loads	50

Note: To this total load, add the load of the farm dwelling unit calculated in accordance with Part III or IV of this article. Where the dwelling has electric heat and the farm has electric grain-drying systems, Part IV of this article shall not be used to calculate the dwelling load.

ARTICLE 225
Outside Branch Circuits and Feeders

225.1 Scope. This article covers requirements for outside branch circuits and feeders run on or between buildings, structures, or poles on the premises; and electrical equipment and wiring for the supply of utilization equipment that is located on or attached to the outside of buildings, structures, or poles.

Informational Note: For additional information on wiring over 1000 volts, see ANSI/IEEE C2-2012, *National Electrical Safety Code.*

225.3 Other Articles. Application of other articles, including additional requirements to specific cases of equipment and conductors, is shown in Table 225.3.

Part I. General

225.4 Conductor Covering. Where within 3.0 m (10 ft) of any building or structure other than supporting poles or towers,

Table 225.3 Other Articles

Equipment/Conductors	Article
Branch circuits	210
Class 1, Class 2, and Class 3 remote-control, signaling, and power-limited circuits	725
Communications circuits	800
Community antenna television and radio distribution systems	820
Conductors for general wiring	310
Electrically driven or controlled irrigation machines	675
Electric signs and outline lighting	600
Feeders	215
Fire alarm systems	760
Fixed outdoor electric deicing and snow-melting equipment	426
Floating buildings	553
Grounding and bonding	250
Hazardous (classified) locations	500
Hazardous (classified) locations — specific	510
Marinas and boatyards	555
Messenger-supported wiring	396
Mobile homes, manufactured homes, and mobile home parks	550
Open wiring on insulators	398
Over 1000 volts, general	490
Overcurrent protection	240
Radio and television equipment	810
Services	230
Solar photovoltaic systems	690
Swimming pools, fountains, and similar installations	680
Use and identification of grounded conductors	200

open individual (aerial) overhead conductors shall be insulated for the nominal voltage. The insulation of conductors in cables or raceways, except Type MI cable, shall be of thermoset or thermoplastic type and, in wet locations, shall comply with 310.10(C). The insulation of conductors for festoon lighting shall be of the rubber-covered or thermoplastic type.

Exception: Equipment grounding conductors and grounded circuit conductors shall be permitted to be bare or covered as specifically permitted elsewhere in this Code.

225.5 Size of Conductors 600 Volts, Nominal, or Less. The ampacity of outdoor branch-circuit and feeder conductors shall be in accordance with 310.15 based on loads as determined under 220.10 and Part III of Article 220.

225.6 Conductor Size and Support.

(A) Overhead Spans. Open individual conductors shall not be smaller than the following:

(1) For 1000 volts, nominal, or less, 10 AWG copper or 8 AWG aluminum for spans up to 15 m (50 ft) in length, and 8 AWG copper or 6 AWG aluminum for a longer span unless supported by a messenger wire

(2) For over 1000 volts, nominal, 6 AWG copper or 4 AWG aluminum where open individual conductors, and 8 AWG copper or 6 AWG aluminum where in cable

(B) Festoon Lighting. Overhead conductors for festoon lighting shall not be smaller than 12 AWG unless the conductors are supported by messenger wires. In all spans exceeding 12 m (40 ft), the conductors shall be supported by messenger wire. The messenger wire shall be supported by strain insulators. Conductors or messenger wires shall not be attached to any fire escape, downspout, or plumbing equipment.

225.7 Lighting Equipment Installed Outdoors.

(A) General. For the supply of lighting equipment installed outdoors, the branch circuits shall comply with Article 210 and 225.7(B) through (D).

(B) Common Neutral. The ampacity of the neutral conductor shall not be less than the maximum net calculated load current between the neutral conductor and all ungrounded conductors connected to any one phase of the circuit.

(C) 277 Volts to Ground. Circuits exceeding 120 volts, nominal, between conductors and not exceeding 277 volts, nominal, to ground shall be permitted to supply luminaires for illumination of outdoor areas of industrial establishments, office buildings, schools, stores, and other commercial or public buildings.

(D) 1000 Volts Between Conductors. Circuits exceeding 277 volts, nominal, to ground and not exceeding 1000 volts, nominal, between conductors shall be permitted to supply the auxiliary equipment of electric-discharge lamps in accordance with 210.6(D)(1).

225.8 Calculation of Loads 1000 Volts, Nominal, or Less.

(A) Branch Circuits. The load on outdoor branch circuits shall be as determined by 220.10.

(B) Feeders. The load on outdoor feeders shall be as determined by Part III of Article 220.

225.10 Wiring on Buildings (or Other Structures). The installation of outside wiring on surfaces of buildings (or other structures) shall be permitted for circuits not exceeding 1000 volts, nominal, as the following:

(1)	Auxiliary gutters
(2)	Busways
(3)	Cable trays
(4)	Cablebus
(5)	Electrical metallic tubing (EMT)
(6)	Flexible metal conduit (FMC)
(7)	Intermediate metal conduit (IMC)
(8)	Liquidtight flexible metal conduit (LFMC)
(9)	Liquidtight flexible nonmetallic conduit (LFNC)
(10)	Messenger-supported wiring
(11)	Multiconductor cable
(12)	Open wiring on insulators
(13)	Reinforced thermosetting resin conduit (RTRC)
(14)	Rigid metal conduit (RMC)
(15)	Rigid polyvinyl chloride conduit (PVC)
(16)	Type MC cable
(17)	Type MI cable
(18)	Type UF cable
(19)	Wireways

Circuits of over 1000 volts, nominal, shall be installed as provided in 300.37.

225.11 Feeder and Branch-Circuit Conductors Entering, Exiting, or Attached to Buildings or Structures. Feeder and branch-circuit conductors entering or exiting buildings or structures shall be installed in accordance with the requirements of 230.52. Overhead branch circuits and feeders attached to buildings or structures shall be installed in accordance with the requirements of 230.54.

225.12 Open-Conductor Supports. Open conductors shall be supported on knobs, racks, brackets, or strain insulators, that are made of glass, porcelain, or other approved materials.

225.14 Open-Conductor Spacings.

(A) 1000 Volts, Nominal, or Less. Conductors of 1000 volts, nominal, or less, shall comply with the spacings provided in Table 230.51(C).

(B) Over 1000 Volts, Nominal. Conductors of over 1000 volts, nominal, shall comply with the spacings provided in 110.36 and 490.24.

(C) Separation from Other Circuits. Open conductors shall be separated from open conductors of other circuits or systems by not less than 100 mm (4 in.).

(D) Conductors on Poles. Conductors on poles shall have a separation of not less than 300 mm (1 ft) where not placed on racks or brackets. Conductors supported on poles shall provide a horizontal climbing space not less than the following:

(1) Power conductors below communications conductors — 750 mm (30 in.)
(2) Power conductors alone or above communications conductors:
    a. 300 volts or less — 600 mm (24 in.)
    b. Over 300 volts — 750 mm (30 in.)
(3) Communications conductors below power conductors — same as power conductors
(4) Communications conductors alone — no requirement

225.15 Supports over Buildings. Supports over a building shall be in accordance with 230.29.

225.16 Attachment to Buildings.

(A) Point of Attachment. The point of attachment to a building shall be in accordance with 230.26.

(B) Means of Attachment. The means of attachment to a building shall be in accordance with 230.27.

225.17 Masts as Supports. Only feeder or branch-circuit conductors specified within this section shall be permitted to be attached to the feeder and/or branch-circuit mast. Masts used for the support of final spans of feeders or branch circuits shall be installed in accordance with 225.17(A) and (B).

(A) Strength. The mast shall have adequate strength or be supported by braces or guys to safely withstand the strain imposed by the overhead feeder or branch-circuit conductors. Hubs intended for use with a conduit serving as a mast for support of feeder or branch-circuit conductors shall be identified for use with a mast.

(B) Attachment. Feeder and/or branch-circuit conductors shall not be attached to a mast where the connection is between a weatherhead or the end of the conduit and a coupling where the coupling is located above the last point of securement to the building or other structure, or where the coupling is located above the building or other structure.

225.18 Clearance for Overhead Conductors and Cables. Overhead spans of open conductors and open multiconductor cables of not over 1000 volts, nominal, shall have a clearance of not less than the following:

(1) 3.0 m (10 ft) — above finished grade, sidewalks, or from any platform or projection that will permit personal contact where the voltage does not exceed 150 volts to ground and accessible to pedestrians only

(2) 3.7 m (12 ft) — over residential property and driveways, and those commercial areas not subject to truck traffic where the voltage does not exceed 300 volts to ground

(3) 4.5 m (15 ft) — for those areas listed in the 3.7 m (12 ft) classification where the voltage exceeds 300 volts to ground

(4) 5.5 m (18 ft) — over public streets, alleys, roads, parking areas subject to truck traffic, driveways on other than residential property, and other land traversed by vehicles, such as cultivated, grazing, forest, and orchard

(5) 7.5 m (24 ½ ft) — over track rails of railroads

225.19 Clearances from Buildings for Conductors of Not over 1000 Volts, Nominal.

(A) Above Roofs. Overhead spans of open conductors and open multiconductor cables shall have a vertical clearance of not less than 2.7 m (8 ft 6 in.) above the roof surface. The vertical clearance above the roof level shall be maintained for a distance not less than 900 mm (3 ft) in all directions from the edge of the roof.

Exception No. 1: The area above a roof surface subject to pedestrian or vehicular traffic shall have a vertical clearance from the roof surface in accordance with the clearance requirements of 225.18.

Exception No. 2: Where the voltage between conductors does not exceed 300, and the roof has a slope of 100 mm in 300 mm (4 in. in 12 in.) or greater, a reduction in clearance to 900 mm (3 ft) shall be permitted.

Exception No. 3: Where the voltage between conductors does not exceed 300, a reduction in clearance above only the overhanging portion of the roof to not less than 450 mm (18 in.) shall be permitted if (1) not more than 1.8 m (6 ft) of the conductors, 1.2 m (4 ft) horizontally, pass above the roof overhang, and (2) they are terminated at a through-the-roof raceway or approved support.

Exception No. 4: The requirement for maintaining the vertical clearance 900 mm (3 ft) from the edge of the roof shall not apply to the final conductor span where the conductors are attached to the side of a building.

(B) From Nonbuilding or Nonbridge Structures. From signs, chimneys, radio and television antennas, tanks, and other nonbuilding or nonbridge structures, clearances — vertical, diagonal, and horizontal — shall not be less than 900 mm (3 ft).

(C) Horizontal Clearances. Clearances shall not be less than 900 mm (3 ft).

(D) Final Spans. Final spans of feeders or branch circuits shall comply with 225.19(D)(1), (D)(2), and (D)(3).

(1) Clearance from Windows. Final spans to the building they supply, or from which they are fed, shall be permitted to be attached to the building, but they shall be kept not less than 900 mm (3 ft) from windows that are designed to be opened and from doors, porches, balconies, ladders, stairs, fire escapes, or similar locations.

Exception: Conductors run above the top level of a window shall be permitted to be less than the 900 mm (3 ft) requirement.

(2) Vertical Clearance. The vertical clearance of final spans above or within 900 mm (3 ft) measured horizontally of platforms, projections, or surfaces that will permit personal contact shall be maintained in accordance with 225.18.

(3) Building Openings. The overhead branch-circuit and feeder conductors shall not be installed beneath openings through which materials may be moved, such as openings in farm and commercial buildings, and shall not be installed where they obstruct entrance to these openings.

(E) Zone for Fire Ladders. Where buildings exceed three stories or 15 m (50 ft) in height, overhead lines shall be arranged, where practicable, so that a clear space (or zone) at least 1.8 m (6 ft) wide will be left either adjacent to the buildings or beginning not over 2.5 m (8 ft) from them to facilitate the raising of ladders when necessary for fire fighting.

225.20 Protection Against Physical Damage. Conductors installed on buildings, structures, or poles shall be protected against physical damage as provided for services in 230.50.

225.21 Multiconductor Cables on Exterior Surfaces of Buildings (or Other Structures). Supports for multiconductor cables on exterior surfaces of buildings (or other structures) shall be as provided in 230.51.

225.22 Raceways on Exterior Surfaces of Buildings or Other Structures. Raceways on exteriors of buildings or other structures shall be arranged to drain and shall be listed or approved for use in wet locations.

225.24 Outdoor Lampholders. Where outdoor lampholders are attached as pendants, the connections to the circuit wires shall be staggered. Where such lampholders have terminals of a type that puncture the insulation and make contact with the conductors, they shall be attached only to conductors of the stranded type.

225.25 Location of Outdoor Lamps. Locations of lamps for outdoor lighting shall be below all energized conductors, transformers, or other electric utilization equipment, unless either of the following apply:

(1) Clearances or other safeguards are provided for relamping operations.

(2) Equipment is controlled by a disconnecting means that lockable in accordance with 110.25.

225.26 Vegetation as Support. Vegetation such as trees shall not be used for support of overhead conductor spans.

225.27 Raceway Seal. Where a raceway enters a building structure from outside, it shall be sealed. Spare or unused raceways shall also be sealed. Sealants shall be identified for use with cable insulation, conductor insulation, bare conductor, shield, or other components.

Part II. Buildings or Other Structures Supplied by a Feeder(s) or Branch Circuit(s)

225.30 Number of Supplies. A building or other structure that is served by a branch circuit or feeder on the load side of a service disconnecting means shall be supplied by only one feeder or branch circuit unless permitted in 225.30(A) through (E). For the purpose of this section, a multiwire branch circuit shall be considered a single circuit.

Where a branch circuit or feeder originates in these additional buildings or other structures, only one feeder or branch circuit shall be permitted to supply power back to the original building or structure, unless permitted in 225.30(A) through (E).

(A) Special Conditions. Additional feeders or branch circuits shall be permitted to supply the following:

(1) Fire pumps
(2) Emergency systems
(3) Legally required standby systems
(4) Optional standby systems
(5) Parallel power production systems
(6) Systems designed for connection to multiple sources of supply for the purpose of enhanced reliability
(7) Electric vehicle charging systems listed, labeled, and identified for more than a single branch circuit or feeder

(B) Special Occupancies. By special permission, additional feeders or branch circuits shall be permitted for either of the following:

(1) Multiple-occupancy buildings where there is no space available for supply equipment accessible to all occupants
(2) A single building or other structure sufficiently large to make two or more supplies necessary

(C) Capacity Requirements. Additional feeders or branch circuits shall be permitted where the capacity requirements are in excess of 2000 amperes at a supply voltage of 1000 volts or less.

(D) Different Characteristics. Additional feeders or branch circuits shall be permitted for different voltages, frequencies, or phases, or for different uses such as control of outside lighting from multiple locations.

(E) Documented Switching Procedures. Additional feeders or branch circuits shall be permitted to supply installations under single management where documented safe switching procedures are established and maintained for disconnection.

225.31 Disconnecting Means. Means shall be provided for disconnecting all ungrounded conductors that supply or pass through the building or structure.

225.32 Location. The disconnecting means shall be installed either inside or outside of the building or structure served or where the conductors pass through the building or structure. The disconnecting means shall be at a readily accessible location nearest the point of entrance of the conductors. For the purposes of this section, the requirements in 230.6 shall be utilized.

Exception No. 1: For installations under single management, where documented safe switching procedures are established and maintained for disconnection, and where the installation is monitored by qualified individuals, the disconnecting means shall be permitted to be located elsewhere on the premises.

Exception No. 2: For buildings or other structures qualifying under the provisions of Article 685, the disconnecting means shall be permitted to be located elsewhere on the premises.

Exception No. 3: For towers or poles used as lighting standards, the disconnecting means shall be permitted to be located elsewhere on the premises.

Exception No. 4: For poles or similar structures used only for support of signs installed in accordance with Article 600, the disconnecting means shall be permitted to be located elsewhere on the premises.

225.33 Maximum Number of Disconnects.

(A) General. The disconnecting means for each supply permitted by 225.30 shall consist of not more than six switches or six circuit breakers mounted in a single enclosure, in a group of separate enclosures, or in or on a switchboard or switchgear. There shall be no more than six disconnects per supply grouped in any one location.

Exception: For the purposes of this section, disconnecting means used solely for the control circuit of the ground-fault protection system, or the control circuit of the power-operated supply disconnecting means, installed as part of the listed equipment, shall not be considered a supply disconnecting means.

(B) Single-Pole Units. Two or three single-pole switches or breakers capable of individual operation shall be permitted on multiwire circuits, one pole for each ungrounded conductor, as one multipole disconnect, provided they are equipped with identified handle ties or a master handle to disconnect all ungrounded conductors with no more than six operations of the hand.

225.34 Grouping of Disconnects.

(A) General. The two to six disconnects as permitted in 225.33 shall be grouped. Each disconnect shall be marked to indicate the load served.

Exception: One of the two to six disconnecting means permitted in 225.33, where used only for a water pump also intended to provide fire protection, shall be permitted to be located remote from the other disconnecting means.

(B) Additional Disconnecting Means. The one or more additional disconnecting means for fire pumps or for emergency, legally required standby or optional standby system permitted by 225.30 shall be installed sufficiently remote from the one to six disconnecting means for normal supply to minimize the possibility of simultaneous interruption of supply.

225.35 Access to Occupants. In a multiple-occupancy building, each occupant shall have access to the occupant's supply disconnecting means.

Exception: In a multiple-occupancy building where electric supply and electrical maintenance are provided by the building management and where these are under continuous building management supervision, the supply disconnecting means supplying more than one occupancy shall be permitted to be accessible to authorized management personnel only.

225.36 Type of Disconnecting Means. The disconnecting means specified in 225.31 shall be comprised of a circuit breaker, molded case switch, general-use switch, snap switch, or other approved means. Where applied in accordance with 250.32(B), Exception No. 1, the disconnecting means shall be suitable for use as service equipment.

225.37 Identification. Where a building or structure has any combination of feeders, branch circuits, or services passing through it or supplying it, a permanent plaque or directory shall be installed at each feeder and branch-circuit disconnect location denoting all other services, feeders, or branch circuits supplying that building or structure or passing through that building or structure and the area served by each.

Exception No. 1: A plaque or directory shall not be required for large-capacity multibuilding industrial installations under single management, where it is ensured that disconnection can be accomplished by establishing and maintaining safe switching procedures.

Exception No. 2: This identification shall not be required for branch circuits installed from a dwelling unit to a second building or structure.

225.38 Disconnect Construction. Disconnecting means shall meet the requirements of 225.38(A) through (D).

(A) Manually or Power Operable. The disconnecting means shall consist of either (1) a manually operable switch or a circuit breaker equipped with a handle or other suitable operating means or (2) a power-operable switch or circuit breaker, provided the switch or circuit breaker can be opened by hand in the event of a power failure.

(B) Simultaneous Opening of Poles. Each building or structure disconnecting means shall simultaneously disconnect all ungrounded supply conductors that it controls from the building or structure wiring system.

(C) Disconnection of Grounded Conductor. Where the building or structure disconnecting means does not disconnect the grounded conductor from the grounded conductors in the building or structure wiring, other means shall be provided for this purpose at the location of the disconnecting means. A terminal or bus to which all grounded conductors can be attached by means of pressure connectors shall be permitted for this purpose.

In a multisection switchboard or switchgear, disconnects for the grounded conductor shall be permitted to be in any section of the switchboard or switchgear, if the switchboard section or switchgear section is marked to indicate a grounded conductor disconnect is contained within the equipment.

(D) Indicating. The building or structure disconnecting means shall plainly indicate whether it is in the open or closed position.

225.39 Rating of Disconnect. The feeder or branch-circuit disconnecting means shall have a rating of not less than the calculated load to be supplied, determined in accordance with Parts I and II of Article 220 for branch circuits, Part III or IV of Article 220 for feeders, or Part V of Article 220 for farm loads. Where the branch circuit or feeder disconnecting means consists of more than one switch or circuit breaker, as permitted by 225.33, combining the ratings of all the switches or circuit breakers for determining the rating of the disconnecting means shall be permitted. In no case shall the rating be lower than specified in 225.39(A), (B), (C), or (D).

(A) One-Circuit Installation. For installations to supply only limited loads of a single branch circuit, the branch circuit disconnecting means shall have a rating of not less than 15 amperes.

(B) Two-Circuit Installations. For installations consisting of not more than two 2-wire branch circuits, the feeder or branch-circuit disconnecting means shall have a rating of not less than 30 amperes.

(C) One-Family Dwelling. For a one-family dwelling, the feeder disconnecting means shall have a rating of not less than 100 amperes, 3-wire.

(D) All Others. For all other installations, the feeder or branch-circuit disconnecting means shall have a rating of not less than 60 amperes.

225.40 Access to Overcurrent Protective Devices. Where a feeder overcurrent device is not readily accessible, branch-circuit overcurrent devices shall be installed on the load side, shall be mounted in a readily accessible location, and shall be of a lower ampere rating than the feeder overcurrent device.

Part III. Over 1000 Volts.

225.50 Sizing of Conductors. The sizing of conductors over 1000 volts shall be in accordance with 210.19(B) for branch circuits and 215.2(B) for feeders.

225.51 Isolating Switches. Where oil switches or air, oil, vacuum, or sulfur hexafluoride circuit breakers constitute a building disconnecting means, an isolating switch with visible break contacts and meeting the requirements of 230.204(B), (C), and (D) shall be installed on the supply side of the disconnecting means and all associated equipment.

Exception: The isolating switch shall not be required where the disconnecting means is mounted on removable truck panels or switchgear units that cannot be opened unless the circuit is disconnected and that, when removed from the normal operating position, automatically disconnect the circuit breaker or switch from all energized parts.

225.52 Disconnecting Means.

(A) Location. A building or structure disconnecting means shall be located in accordance with 225.32, or, if not readily accessible, it shall be operable by mechanical linkage from a readily accessible point. For multibuilding industrial installations under single management, it shall be permitted to be electrically operated by a readily accessible, remote-control device in a separate building or structure.

(B) Type. Each building or structure disconnect shall simultaneously disconnect all ungrounded supply conductors it controls and shall have a fault-closing rating not less than the maximum available short-circuit current available at its supply terminals.

Exception: Where the individual disconnecting means consists of fused cutouts, the simultaneous disconnection of all ungrounded supply conductors shall not be required if there is a means to disconnect the load before opening the cutouts. A permanent legible sign shall be installed adjacent to the fused cutouts and shall read DISCONNECT LOAD BEFORE OPENING CUTOUTS.

Where fused switches or separately mounted fuses are installed, the fuse characteristics shall be permitted to contribute to the fault closing rating of the disconnecting means.

(C) Locking. Disconnecting means shall be lockable in accordance with 110.25.

Exception: Where an individual disconnecting means consists of fused cutouts, a suitable enclosure capable of being locked and sized to contain all cutout fuse holders shall be installed at a convenient location to the fused cutouts.

(D) Indicating. Disconnecting means shall clearly indicate whether they are in the open "off" or closed "on" position.

(E) Uniform Position. Where disconnecting means handles are operated vertically, the "up" position of the handle shall be the "on" position.

Exception: A switching device having more than one "on" position, such as a double throw switch, shall not be required to comply with this requirement.

(F) Identification. Where a building or structure has any combination of feeders, branch circuits, or services passing through or supplying it, a permanent plaque or directory shall be installed at each feeder and branch-circuit disconnect location that denotes all other services, feeders, or branch circuits supplying that building or structure or passing through that building or structure and the area served by each.

225.56 Inspections and Tests.

(A) Pre-Energization and Operating Tests. The complete electrical system design, including settings for protective, switching, and control circuits, shall be prepared in advance and made available on request to the authority having jurisdiction and shall be performance tested when first installed on-site. Each protective, switching, and control circuit shall be adjusted in accordance with the system design and tested by actual operation using current injection or equivalent methods as necessary to ensure that each and every such circuit operates correctly to the satisfaction of the authority having jurisdiction.

(1) Instrument Transformers. All instrument transformers shall be tested to verify correct polarity and burden.

(2) Protective Relays. Each protective relay shall be demonstrated to operate by injecting current or voltage, or both, at the associated instrument transformer output terminal and observing that the associated switching and signaling functions occur correctly and in proper time and sequence to accomplish the protective function intended.

(3) Switching Circuits. Each switching circuit shall be observed to operate the associated equipment being switched.

(4) Control and Signal Circuits. Each control or signal circuit shall be observed to perform its proper control function or produce a correct signal output.

(5) Metering Circuits. All metering circuits shall be verified to operate correctly from voltage and current sources in a similar manner to protective relay circuits.

(6) Acceptance Tests. Complete acceptance tests shall be performed, after the substation installation is completed, on all assemblies, equipment, conductors, and control and protective systems, as applicable, to verify the integrity of all the systems.

(7) Relays and Metering Utilizing Phase Differences. All relays and metering that use phase differences for operation shall be verified by measuring phase angles at the relay under actual load conditions after operation commences.

(B) Test Report. A test report covering the results of the tests required in 225.56(A) shall be delivered to the authority having jurisdiction prior to energization.

Informational Note: For an example of acceptance specifications, see ANSI/NETA ATS-2013, *Acceptance Testing Specifications for Electrical Power Distribution Equipment and Systems*, published by the InterNational Electrical Testing Association.

225.60 Clearances over Roadways, Walkways, Rail, Water, and Open Land.

(A) 22 kV, Nominal, to Ground or Less. The clearances over roadways, walkways, rail, water, and open land for conductors and live parts up to 22 kV, nominal, to ground or less shall be not less than the values shown in Table 225.60.

(B) Over 22 kV Nominal to Ground. Clearances for the categories shown in Table 225.60 shall be increased by 10 mm (0.4 in.) per kV above 22,000 volts.

(C) Special Cases. For special cases, such as where crossings will be made over lakes, rivers, or areas using large vehicles such as mining operations, specific designs shall be engineered considering the special circumstances and shall be approved by the authority having jurisdiction.

Informational Note: For additional information, see ANSI/IEEE C2-2012, *National Electrical Safety Code.*

Table 225.60 Clearances over Roadways, Walkways, Rail, Water, and Open Land

Location	Clearance	
	m	ft
Open land subject to vehicles, cultivation, or grazing	5.6	18.5
Roadways, driveways, parking lots, and alleys	5.6	18.5
Walkways	4.1	13.5
Rails	8.1	26.5
Spaces and ways for pedestrians and restricted traffic	4.4	14.5
Water areas not suitable for boating	5.2	17.0

225.61 Clearances over Buildings and Other Structures.

(A) 22 kV Nominal to Ground or Less. The clearances over buildings and other structures for conductors and live parts up to 22 kV, nominal, to ground or less shall be not less than the values shown in Table 225.61.

(B) Over 22 kV Nominal to Ground. Clearances for the categories shown in Table 225.61 shall be increased by 10 mm (0.4 in.) per kV above 22,000 volts.

Informational Note: For additional information, see ANSI/IEEE C2-2012, *National Electrical Safety Code.*

Table 225.61 Clearances over Buildings and Other Structures

Clearance from Conductors or Live Parts from:	Horizontal		Vertical	
	m	ft	m	ft
Building walls, projections, and windows	2.3	7.5	—	—
Balconies, catwalks, and similar areas accessible to people	2.3	7.5	4.1	13.5
Over or under roofs or projections not readily accessible to people	—	—	3.8	12.5
Over roofs accessible to vehicles but not trucks	—	—	4.1	13.5
Over roofs accessible to trucks	—	—	5.6	18.5
Other structures	2.3	7.5	—	—

ARTICLE 230
Services

230.1 Scope. This article covers service conductors and equipment for control and protection of services and their installation requirements.

Informational Note: See Figure 230.1.

Part I. General

230.2 Number of Services. A building or other structure served shall be supplied by only one service unless permitted in 230.2(A) through (D). For the purpose of 230.40, Exception No. 2 only, underground sets of conductors, 1/0 AWG and larger, running to the same location and connected together at their supply end but not connected together at their load end shall be considered to be supplying one service.

(A) Special Conditions. Additional services shall be permitted to supply the following:

(1) Fire pumps
(2) Emergency systems
(3) Legally required standby systems
(4) Optional standby systems
(5) Parallel power production systems
(6) Systems designed for connection to multiple sources of supply for the purpose of enhanced reliability

(B) Special Occupancies. By special permission, additional services shall be permitted for either of the following:

(1) Multiple-occupancy buildings where there is no available space for service equipment accessible to all occupants
(2) A single building or other structure sufficiently large to make two or more services necessary

(C) Capacity Requirements. Additional services shall be permitted under any of the following:

(1) Where the capacity requirements are in excess of 2000 amperes at a supply voltage of 1000 volts or less

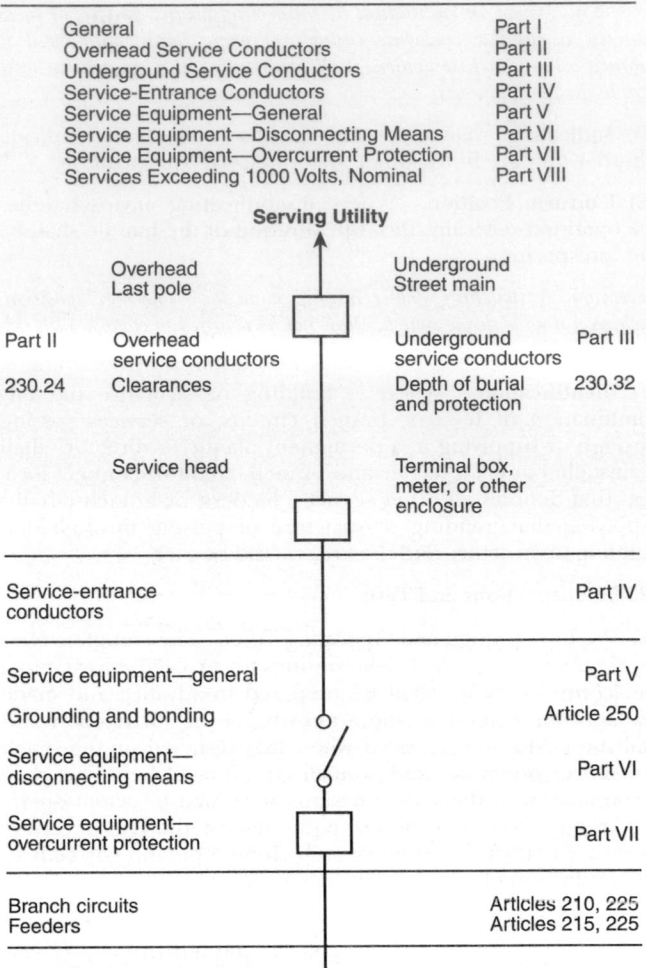

General	Part I	
Overhead Service Conductors	Part II	
Underground Service Conductors	Part III	
Service-Entrance Conductors	Part IV	
Service Equipment—General	Part V	
Service Equipment—Disconnecting Means	Part VI	
Service Equipment—Overcurrent Protection	Part VII	
Services Exceeding 1000 Volts, Nominal	Part VIII	

FIGURE 230.1 Services.

(2) Where the load requirements of a single-phase installation are greater than the serving agency normally supplies through one service
(3) By special permission

(D) Different Characteristics. Additional services shall be permitted for different voltages, frequencies, or phases, or for different uses, such as for different rate schedules.

(E) Identification. Where a building or structure is supplied by more than one service, or any combination of branch circuits, feeders, and services, a permanent plaque or directory shall be installed at each service disconnect location denoting all other services, feeders, and branch circuits supplying that building or structure and the area served by each. See 225.37.

230.3 One Building or Other Structure Not to Be Supplied Through Another. Service conductors supplying a building or other structure shall not pass through the interior of another building or other structure.

230.6 Conductors Considered Outside the Building. Conductors shall be considered outside of a building or other structure under any of the following conditions:

(1) Where installed under not less than 50 mm (2 in.) of concrete beneath a building or other structure

(2) Where installed within a building or other structure in a raceway that is encased in concrete or brick not less than 50 mm (2 in.) thick

(3) Where installed in any vault that meets the construction requirements of Article 450, Part III

(4) Where installed in conduit and under not less than 450 mm (18 in.) of earth beneath a building or other structure

(5) Where installed within rigid metal conduit (Type RMC) or intermediate metal conduit (Type IMC) used to accommodate the clearance requirements in 230.24 and routed directly through an eave but not a wall of a building

230.7 Other Conductors in Raceway or Cable. Conductors other than service conductors shall not be installed in the same service raceway or service cable in which the service conductors are installed.

Exception No. 1: Grounding electrode conductors or supply side bonding jumpers or conductors shall be permitted within service raceways.

Exception No. 2: Load management control conductors having overcurrent protection shall be permitted within service raceways.

230.8 Raceway Seal. Where a service raceway enters a building or structure from an underground distribution system, it shall be sealed in accordance with 300.5(G). Spare or unused raceways shall also be sealed. Sealants shall be identified for use with the cable insulation, shield, or other components.

230.9 Clearances on Buildings. Service conductors and final spans shall comply with 230.9(A), (B), and (C).

(A) Clearances. Service conductors installed as open conductors or multiconductor cable without an overall outer jacket shall have a clearance of not less than 900 mm (3 ft) from windows that are designed to be opened, doors, porches, balconies, ladders, stairs, fire escapes, or similar locations.

Exception: Conductors run above the top level of a window shall be permitted to be less than the 900 mm (3 ft) requirement.

(B) Vertical Clearance. The vertical clearance of final spans above, or within 900 mm (3 ft) measured horizontally of platforms, projections, or surfaces that will permit personal contact shall be maintained in accordance with 230.24(B).

(C) Building Openings. Overhead service conductors shall not be installed beneath openings through which materials may be moved, such as openings in farm and commercial buildings, and shall not be installed where they obstruct entrance to these building openings.

230.10 Vegetation as Support. Vegetation such as trees shall not be used for support of overhead service conductors or service equipment.

Part II. Overhead Service Conductors

230.22 Insulation or Covering. Individual conductors shall be insulated or covered.

Exception: The grounded conductor of a multiconductor cable shall be permitted to be bare.

230.23 Size and Rating.

(A) General. Conductors shall have sufficient ampacity to carry the current for the load as calculated in accordance with Article 220 and shall have adequate mechanical strength.

(B) Minimum Size. The conductors shall not be smaller than 8 AWG copper or 6 AWG aluminum or copper-clad aluminum.

Exception: Conductors supplying only limited loads of a single branch circuit — such as small polyphase power, controlled water heaters, and similar loads — shall not be smaller than 12 AWG hard-drawn copper or equivalent.

(C) Grounded Conductors. The grounded conductor shall not be less than the minimum size as required by 250.24(C).

230.24 Clearances. Overhead service conductors shall not be readily accessible and shall comply with 230.24(A) through (E) for services not over 1000 volts, nominal.

(A) Above Roofs. Conductors shall have a vertical clearance of not less than 2.5 m (8 ft) above the roof surface. The vertical clearance above the roof level shall be maintained for a distance of not less than 900 mm (3 ft) in all directions from the edge of the roof.

Exception No. 1: The area above a roof surface subject to pedestrian or vehicular traffic shall have a vertical clearance from the roof surface in accordance with the clearance requirements of 230.24(B).

Exception No. 2: Where the voltage between conductors does not exceed 300 and the roof has a slope of 100 mm in 300 mm (4 in. in 12 in.) or greater, a reduction in clearance to 900 mm (3 ft) shall be permitted.

Exception No. 3: Where the voltage between conductors does not exceed 300, a reduction in clearance above only the overhanging portion of the roof to not less than 450 mm (18 in.) shall be permitted if (1) not more than 1.8 m (6 ft) of overhead service conductors, 1.2 m (4 ft) horizontally, pass above the roof overhang, and (2) they are terminated at a through-the-roof raceway or approved support.

Informational Note: See 230.28 for mast supports.

Exception No. 4: The requirement for maintaining the vertical clearance 900 mm (3 ft) from the edge of the roof shall not apply to the final conductor span where the service drop or overhead service conductors are attached to the side of a building.

Exception No. 5: Where the voltage between conductors does not exceed 300 and the roof area is guarded or isolated, a reduction in clearance to 900 mm (3 ft) shall be permitted.

(B) Vertical Clearance for Overhead Service Conductors. Overhead service conductors, where not in excess of 600 volts, nominal, shall have the following minimum clearance from final grade:

(1) 3.0 m (10 ft) — at the electrical service entrance to buildings, also at the lowest point of the drip loop of the building electrical entrance, and above areas or sidewalks accessible only to pedestrians, measured from final grade or other accessible surface only for overhead service conductors supported on and cabled together with a grounded bare messenger where the voltage does not exceed 150 volts to ground

(2) 3.7 m (12 ft) — over residential property and driveways, and those commercial areas not subject to truck traffic where the voltage does not exceed 300 volts to ground

(3) 4.5 m (15 ft) — for those areas listed in the 3.7 m (12 ft) classification where the voltage exceeds 300 volts to ground

(4) 5.5 m (18 ft) — over public streets, alleys, roads, parking areas subject to truck traffic, driveways on other than residential property, and other land such as cultivated, grazing, forest, and orchard

(5) 7.5 m (24½) over tracks of railroads

(C) Clearance from Building Openings. See 230.9.

(D) Clearance from Swimming Pools. See 680.9.

(E) Clearance from Communication Wires and Cables. Clearance from communication wires and cables shall be in accordance with 830.44(A)(4).

230.26 Point of Attachment. The point of attachment of the overhead service conductors to a building or other structure shall provide the minimum clearances as specified in 230.9 and 230.24. In no case shall this point of attachment be less than 3.0 m (10 ft) above finished grade.

230.27 Means of Attachment. Multiconductor cables used for overhead service conductors shall be attached to buildings or other structures by fittings identified for use with service conductors. Open conductors shall be attached to fittings identified for use with service conductors or to noncombustible, nonabsorbent insulators securely attached to the building or other structure.

230.28 Service Masts as Supports. Only power service-drop or overhead service conductors shall be permitted to be attached to a service mast. Service masts used for the support of service-drop or overhead service conductors shall be installed in accordance with 230.28(A) and (B).

(A) Strength. The service mast shall be of adequate strength or be supported by braces or guys to withstand safely the strain imposed by the service-drop or overhead service conductors. Hubs intended for use with a conduit that serves as a service mast shall be identified for use with service-entrance equipment.

(B) Attachment. Service-drop or overhead service conductors shall not be attached to a service mast between a weatherhead or the end of the conduit and a coupling, where the coupling is located above the last point of securement to the building or other structure or is located above the building or other structure.

230.29 Supports over Buildings. Service conductors passing over a roof shall be securely supported by substantial structures. For a grounded system, where the substantial structure is metal, it shall be bonded by means of a bonding jumper and listed connector to the grounded overhead service conductor. Where practicable, such supports shall be independent of the building.

Part III. Underground Service Conductors

230.30 Installation.

(A) Insulation. Underground service conductors shall be insulated for the applied voltage.

Exception: A grounded conductor shall be permitted to be uninsulated as follows:

(1) Bare copper used in a raceway

(2) Bare copper for direct burial where bare copper is approved for the soil conditions

(3) Bare copper for direct burial without regard to soil conditions where part of a cable assembly identified for underground use

(4) Aluminum or copper-clad aluminum without individual insulation or covering where part of a cable assembly identified for underground use in a raceway or for direct burial

(B) Wiring Methods. Underground service conductors shall be installed in accordance with the applicable requirements of this *Code* covering the type of wiring method used and shall be limited to the following methods:

(1) Type RMC conduit
(2) Type IMC conduit
(3) Type NUCC conduit
(4) Type HDPE conduit
(5) Type PVC conduit
(6) Type RTRC conduit
(7) Type IGS cable
(8) Type USE conductors or cables
(9) Type MV or Type MC cable identified for direct burial applications
(10) Type MI cable, where suitably protected against physical damage and corrosive conditions

230.31 Size and Rating.

(A) General. Underground service conductors shall have sufficient ampacity to carry the current for the load as calculated in accordance with Article 220 and shall have adequate mechanical strength.

(B) Minimum Size. The conductors shall not be smaller than 8 AWG copper or 6 AWG aluminum or copper-clad aluminum.

Exception: Conductors supplying only limited loads of a single branch circuit — such as small polyphase power, controlled water heaters, and similar loads — shall not be smaller than 12 AWG copper or 10 AWG aluminum or copper-clad aluminum.

(C) Grounded Conductors. The grounded conductor shall not be less than the minimum size required by 250.24(C).

230.32 Protection Against Damage. Underground service conductors shall be protected against damage in accordance with 300.5. Service conductors entering a building or other structure shall be installed in accordance with 230.6 or protected by a raceway wiring method identified in 230.43.

230.33 Spliced Conductors. Service conductors shall be permitted to be spliced or tapped in accordance with 110.14, 300.5(E), 300.13, and 300.15.

Part IV. Service-Entrance Conductors

230.40 Number of Service-Entrance Conductor Sets. Each service drop, set of overhead service conductors, set of underground service conductors, or service lateral shall supply only one set of service-entrance conductors.

Exception No. 1: A building with more than one occupancy shall be permitted to have one set of service-entrance conductors for each service, as defined in 230.2, run to each occupancy or group of occupancies. If the number of service disconnect locations for any given classification of service does not exceed six, the requirements of 230.2(E) shall apply at each location. If the number of service disconnect locations exceeds six for any given supply classification, all service disconnect locations for all supply characteristics, together with any branch circuit or feeder supply sources, if applicable, shall be clearly described using suitable graphics or text, or both, on one or more plaques located in an approved, readily accessible location(s) on the building or structure served and as near as practicable to the point(s) of attachment or entry(ies) for each service drop or service lateral, and for each set of overhead or underground service conductors.

Exception No. 2: Where two to six service disconnecting means in separate enclosures are grouped at one location and supply separate loads from one service drop, set of overhead service conductors, set of underground service conductors, or service lateral, one set of service-entrance conductors shall be permitted to supply each or several such service equipment enclosures.

Exception No. 3: A one-family dwelling unit and its accessory structures shall be permitted to have one set of service-entrance conductors run to each from a single service drop, set of overhead service conductors, set of underground service conductors, or service lateral.

Exception No. 4: Two-family dwellings, multifamily dwellings, and multiple occupancy buildings shall be permitted to have one set of service-entrance conductors installed to supply the circuits covered in 210.25.

Exception No. 5: One set of service-entrance conductors connected to the supply side of the normal service disconnecting means shall be permitted to supply each or several systems covered by 230.82(5) or 230.82(6).

230.41 Insulation of Service-Entrance Conductors. Service-entrance conductors entering or on the exterior of buildings or other structures shall be insulated.

Exception: A grounded conductor shall be permitted to be uninsulated as follows:

(1) *Bare copper used in a raceway or part of a service cable assembly*
(2) *Bare copper for direct burial where bare copper is approved for the soil conditions*
(3) *Bare copper for direct burial without regard to soil conditions where part of a cable assembly identified for underground use*
(4) *Aluminum or copper-clad aluminum without individual insulation or covering where part of a cable assembly or identified for underground use in a raceway, or for direct burial*
(5) *Bare conductors used in an auxiliary gutter*

230.42 Minimum Size and Rating.

(A) General. Service-entrance conductors shall have an ampacity of not less than the maximum load to be served. Conductors shall be sized to carry not less than the largest of 230.42(A)(1) or (A)(2). Loads shall be determined in accordance with Part III, IV, or V of Article 220, as applicable. Ampacity shall be determined from 310.15. The maximum allowable current of busways shall be that value for which the busway has been listed or labeled.

(1) Where the service-entrance conductors supply continuous loads or any combination of noncontinuous and continuous loads, the minimum service-entrance conductor size shall have an allowable ampacity not less than the sum of the noncontinuous loads plus 125 percent of continuous loads.

Exception No. 1: Grounded conductors that are not connected to an overcurrent device shall be permitted to be sized at 100 percent of the sum of the continuous and noncontinuous load.

Exception No. 2: The sum of the noncontinuous load and the continuous load if the service-entrance conductors terminate in an overcurrent device where both the overcurrent device and its assembly are listed for operation at 100 percent of their rating shall be permitted.

(2) The minimum service-entrance conductor size shall have an ampacity not less than the maximum load to be served after the application of any adjustment or correction factors.

(B) Specific Installations. In addition to the requirements of 230.42(A), the minimum ampacity for ungrounded conductors for specific installations shall not be less than the rating of the service disconnecting means specified in 230.79(A) through (D).

(C) Grounded Conductors. The grounded conductor shall not be smaller than the minimum size as required by 250.24(C).

230.43 Wiring Methods for 1000 Volts, Nominal, or Less. Service-entrance conductors shall be installed in accordance with the applicable requirements of this *Code* covering the type of wiring method used and shall be limited to the following methods:

(1) Open wiring on insulators
(2) Type IGS cable
(3) Rigid metal conduit (RMC)
(4) Intermediate metal conduit (IMC)
(5) Electrical metallic tubing (EMT)
(6) Electrical nonmetallic tubing
(7) Service-entrance cables
(8) Wireways
(9) Busways
(10) Auxiliary gutters
(11) Rigid polyvinyl chloride conduit (PVC)
(12) Cablebus
(13) Type MC cable
(14) Mineral-insulated, metal-sheathed cable, Type MI
(15) Flexible metal conduit (FMC) not over 1.8 m (6 ft) long or liquidtight flexible metal conduit (LFMC) not over 1.8 m (6 ft) long between a raceway, or between a raceway and service equipment, with a supply-side bonding jumper routed with the flexible metal conduit (FMC) or the liquidtight flexible metal conduit (LFMC) according to the provisions of 250.102(A), (B), (C), and (E)
(16) Liquidtight flexible nonmetallic conduit (LFNC)
(17) High density polyethylene conduit (HDPE)
(18) Nonmetallic underground conduit with conductors (NUCC)
(19) Reinforced thermosetting resin conduit (RTRC)

230.44 Cable Trays. Cable tray systems shall be permitted to support service-entrance conductors. Cable trays used to support service-entrance conductors shall contain only service-entrance conductors and shall be limited to the following methods:

(1) Type SE cable
(2) Type MC cable
(3) Type MI cable
(4) Type IGS cable
(5) Single conductors 1/0 and larger that are listed for use in cable tray

Such cable trays shall be identified with permanently affixed labels with the wording "Service-Entrance Conductors." The labels shall be located so as to be visible after installation with a spacing not to exceed 3 m (10 ft) so that the service-entrance conductors are able to be readily traced through the entire length of the cable tray.

Exception: Conductors, other than service-entrance conductors, shall be permitted to be installed in a cable tray with service-entrance conductors, provided a solid fixed barrier of a material compatible with the cable tray is installed to separate the service-entrance conductors from other conductors installed in the cable tray.

230.46 Spliced Conductors. Service-entrance conductors shall be permitted to be spliced or tapped in accordance with 110.14, 300.5(E), 300.13, and 300.15.

230.50 Protection Against Physical Damage.

(A) Underground Service-Entrance Conductors. Underground service-entrance conductors shall be protected against physical damage in accordance with 300.5.

(B) All Other Service-Entrance Conductors. All other service-entrance conductors, other than underground service entrance conductors, shall be protected against physical damage as specified in 230.50(B)(1) or (B)(2).

(1) Service-Entrance Cables. Service-entrance cables, where subject to physical damage, shall be protected by any of the following:

(1)   Rigid metal conduit (RMC)
(2)   Intermediate metal conduit (IMC)
(3)   Schedule 80 PVC conduit
(4)   Electrical metallic tubing (EMT)
(5)   Reinforced thermosetting resin conduit (RTRC)
(6)   Other approved means

(2) Other Than Service-Entrance Cables. Individual open conductors and cables, other than service-entrance cables, shall not be installed within 3.0 m (10 ft) of grade level or where exposed to physical damage.

Exception: Type MI and Type MC cable shall be permitted within 3.0 m (10 ft) of grade level where not exposed to physical damage or where protected in accordance with 300.5(D).

230.51 Mounting Supports. Service-entrance cables or individual open service-entrance conductors shall be supported as specified in 230.51(A), (B), or (C).

(A) Service-Entrance Cables. Service-entrance cables shall be supported by straps or other approved means within 300 mm (12 in.) of every service head, gooseneck, or connection to a raceway or enclosure and at intervals not exceeding 750 mm (30 in.).

(B) Other Cables. Cables that are not approved for mounting in contact with a building or other structure shall be mounted on insulating supports installed at intervals not exceeding 4.5 m (15 ft) and in a manner that maintains a clearance of not less than 50 mm (2 in.) from the surface over which they pass.

(C) Individual Open Conductors. Individual open conductors shall be installed in accordance with Table 230.51(C). Where exposed to the weather, the conductors shall be mounted on insulators or on insulating supports attached to racks, brackets, or other approved means. Where not exposed to the weather, the conductors shall be mounted on glass or porcelain knobs.

230.52 Individual Conductors Entering Buildings or Other Structures. Where individual open conductors enter a building or other structure, they shall enter through roof bushings or through the wall in an upward slant through individual, noncombustible, nonabsorbent insulating tubes. Drip loops shall be formed on the conductors before they enter the tubes.

230.53 Raceways to Drain. Where exposed to the weather, raceways enclosing service-entrance conductors shall be listed or approved for use in wet locations and arranged to drain. Where embedded in masonry, raceways shall be arranged to drain.

230.54 Overhead Service Locations.

(A) Service Head. Service raceways shall be equipped with a service head at the point of connection to service-drop or overhead service conductors. The service head shall be listed for use in wet locations.

(B) Service-Entrance Cables Equipped with Service Head or Gooseneck. Service-entrance cables shall be equipped with a service head. The service head shall be listed for use in wet locations.

Exception: Type SE cable shall be permitted to be formed in a gooseneck and taped with a self-sealing weather-resistant thermoplastic.

(C) Service Heads and Goosenecks Above Service-Drop or Overhead Service Attachment. Service heads on raceways or service-entrance cables and goosenecks in service-entrance cables shall be located above the point of attachment of the service-drop or overhead service conductors to the building or other structure.

Exception: Where it is impracticable to locate the service head or gooseneck above the point of attachment, the service head or gooseneck location shall be permitted not farther than 600 mm (24 in.) from the point of attachment.

(D) Secured. Service-entrance cables shall be held securely in place.

Table 230.51(C) Supports

Maximum Volts	Maximum Distance Between Supports		Minimum Clearance			
			Between Conductors		From Surface	
	m	ft	mm	in.	mm	in.
1000	2.7	9	150	6	50	2
1000	4.5	15	300	12	50	2
300	1.4	4½	75	3	50	2
1000*	1.4*	4½*	65*	2½*	25*	1*

*Where not exposed to weather.

(E) Separately Bushed Openings. Service heads shall have conductors of different potential brought out through separately bushed openings.

Exception: For jacketed multiconductor service-entrance cable without splice.

(F) Drip Loops. Drip loops shall be formed on individual conductors. To prevent the entrance of moisture, service-entrance conductors shall be connected to the service-drop or overhead service conductors either (1) below the level of the service head or (2) below the level of the termination of the service-entrance cable sheath.

(G) Arranged That Water Will Not Enter Service Raceway or Equipment. Service-entrance and overhead service conductors shall be arranged so that water will not enter service raceway or equipment.

230.56 Service Conductor with the Higher Voltage to Ground. On a 4-wire, delta-connected service where the midpoint of one phase winding is grounded, the service conductor having the higher phase voltage to ground shall be durably and permanently marked by an outer finish that is orange in color, or by other effective means, at each termination or junction point.

Part V. Service Equipment — General

230.62 Service Equipment — Enclosed or Guarded. Energized parts of service equipment shall be enclosed as specified in 230.62(A) or guarded as specified in 230.62(B).

(A) Enclosed. Energized parts shall be enclosed so that they will not be exposed to accidental contact or shall be guarded as in 230.62(B).

(B) Guarded. Energized parts that are not enclosed shall be installed on a switchboard, panelboard, or control board and guarded in accordance with 110.18 and 110.27. Where energized parts are guarded as provided in 110.27(A)(1) and (A)(2), a means for locking or sealing doors providing access to energized parts shall be provided.

230.66 Marking. Service equipment rated at 1000 volts or less shall be marked to identify it as being suitable for use as service equipment. All service equipment shall be listed or field labeled. Individual meter socket enclosures shall not be considered service equipment but shall be listed and rated for the voltage and ampacity of the service.

Exception: Meter sockets supplied by and under the exclusive control of an electric utility shall not be required to be listed.

Part VI. Service Equipment — Disconnecting Means

230.70 General. Means shall be provided to disconnect all conductors in a building or other structure from the service-entrance conductors.

(A) Location. The service disconnecting means shall be installed in accordance with 230.70(A)(1), (A)(2), and (A)(3).

(1) Readily Accessible Location. The service disconnecting means shall be installed at a readily accessible location either outside of a building or structure or inside nearest the point of entrance of the service conductors.

(2) Bathrooms. Service disconnecting means shall not be installed in bathrooms.

(3) Remote Control. Where a remote control device(s) is used to actuate the service disconnecting means, the service disconnecting means shall be located in accordance with 230.70(A)(1).

(B) Marking. Each service disconnect shall be permanently marked to identify it as a service disconnect.

(C) Suitable for Use. Each service disconnecting means shall be suitable for the prevailing conditions. Service equipment installed in hazardous (classified) locations shall comply with the requirements of Articles 500 through 517.

230.71 Maximum Number of Disconnects.

(A) General. The service disconnecting means for each service permitted by 230.2, or for each set of service-entrance conductors permitted by 230.40, Exception No. 1, 3, 4, or 5, shall consist of not more than six switches or sets of circuit breakers, or a combination of not more than six switches and sets of circuit breakers, mounted in a single enclosure, in a group of separate enclosures, or in or on a switchboard or in switchgear. There shall be not more than six sets of disconnects per service grouped in any one location.

For the purpose of this section, disconnecting means installed as part of listed equipment and used solely for the following shall not be considered a service disconnecting means:

(1) Power monitoring equipment
(2) Surge-protective device(s)
(3) Control circuit of the ground-fault protection system
(4) Power-operable service disconnecting means

(B) Single-Pole Units. Two or three single-pole switches or breakers, capable of individual operation, shall be permitted on multiwire circuits, one pole for each ungrounded conductor, as one multipole disconnect, provided they are equipped with identified handle ties or a master handle to disconnect all conductors of the service with no more than six operations of the hand.

Informational Note: See 408.36, Exception No. 1 and Exception No. 3, for service equipment in certain panelboards, and see 430.95 for service equipment in motor control centers.

230.72 Grouping of Disconnects.

(A) General. The two to six disconnects as permitted in 230.71 shall be grouped. Each disconnect shall be marked to indicate the load served.

Exception: One of the two to six service disconnecting means permitted in 230.71, where used only for a water pump also intended to provide fire protection, shall be permitted to be located remote from the other disconnecting means. If remotely installed in accordance with this exception, a plaque shall be posted at the location of the remaining grouped disconnects denoting its location.

(B) Additional Service Disconnecting Means. The one or more additional service disconnecting means for fire pumps, emergency systems, legally required standby, or optional standby services permitted by 230.2 shall be installed remote from the one to six service disconnecting means for normal service to minimize the possibility of simultaneous interruption of supply.

(C) Access to Occupants. In a multiple-occupancy building, each occupant shall have access to the occupant's service disconnecting means.

Exception: In a multiple-occupancy building where electric service and electrical maintenance are provided by the building management and where these are under continuous building management supervision, the service disconnecting means supplying more than one occupancy shall be permitted to be accessible to authorized management personnel only.

230.74 Simultaneous Opening of Poles. Each service disconnect shall simultaneously disconnect all ungrounded service conductors that it controls from the premises wiring system.

230.75 Disconnection of Grounded Conductor. Where the service disconnecting means does not disconnect the grounded conductor from the premises wiring, other means shall be provided for this purpose in the service equipment. A terminal or bus to which all grounded conductors can be attached by means of pressure connectors shall be permitted for this purpose. In a multisection switchboard or switchgear, disconnects for the grounded conductor shall be permitted to be in any section of the switchboard or switchgear, if the switchboard or switchgear section is marked to indicate a grounded conductor disconnect is located within.

230.76 Manually or Power Operable. The service disconnecting means for ungrounded service conductors shall consist of one of the following:

(1)   A manually operable switch or circuit breaker equipped with a handle or other suitable operating means

(2)   A power-operated switch or circuit breaker, provided the switch or circuit breaker can be opened by hand in the event of a power supply failure

230.77 Indicating. The service disconnecting means shall plainly indicate whether it is in the open (off) or closed (on) position.

230.79 Rating of Service Disconnecting Means. The service disconnecting means shall have a rating not less than the calculated load to be carried, determined in accordance with Part III, IV, or V of Article 220, as applicable. In no case shall the rating be lower than specified in 230.79(A), (B), (C), or (D).

(A) One-Circuit Installations. For installations to supply only limited loads of a single branch circuit, the service disconnecting means shall have a rating of not less than 15 amperes.

(B) Two-Circuit Installations. For installations consisting of not more than two 2-wire branch circuits, the service disconnecting means shall have a rating of not less than 30 amperes.

(C) One-Family Dwellings. For a one-family dwelling, the service disconnecting means shall have a rating of not less than 100 amperes, 3-wire.

(D) All Others. For all other installations, the service disconnecting means shall have a rating of not less than 60 amperes.

230.80 Combined Rating of Disconnects. Where the service disconnecting means consists of more than one switch or circuit breaker, as permitted by 230.71, the combined ratings of all the switches or circuit breakers used shall not be less than the rating required by 230.79.

230.81 Connection to Terminals. The service conductors shall be connected to the service disconnecting means by pressure connectors, clamps, or other approved means. Connections that depend on solder shall not be used.

230.82 Equipment Connected to the Supply Side of Service Disconnect. Only the following equipment shall be permitted to be connected to the supply side of the service disconnecting means:

(1)   Cable limiters or other current-limiting devices.

(2)   Meters and meter sockets nominally rated not in excess of 1000 volts, if all metal housings and service enclosures are grounded in accordance with Part VII and bonded in accordance with Part V of Article 250.

(3)   Meter disconnect switches nominally rated not in excess of 1000 V that have a short-circuit current rating equal to or greater than the available short-circuit current, if all metal housings and service enclosures are grounded in accordance with Part VII and bonded in accordance with Part V of Article 250. A meter disconnect switch shall be capable of interrupting the load served. A meter disconnect shall be legibly field marked on its exterior in a manner suitable for the environment as follows:
METER DISCONNECT
NOT SERVICE EQUIPMENT

(4)   Instrument transformers (current and voltage), impedance shunts, load management devices, surge arresters, and Type 1 surge-protective devices.

(5)   Taps used only to supply load management devices, circuits for standby power systems, fire pump equipment, and fire and sprinkler alarms, if provided with service equipment and installed in accordance with requirements for service-entrance conductors.

(6)   Solar photovoltaic systems, fuel cell systems, wind electric systems, energy storage systems, or interconnected electric power production sources.

(7)   Control circuits for power-operable service disconnecting means, if suitable overcurrent protection and disconnecting means are provided.

(8)   Ground-fault protection systems or Type 2 surge-protective devices, where installed as part of listed equipment, if suitable overcurrent protection and disconnecting means are provided.

(9)   Connections used only to supply listed communications equipment under the exclusive control of the serving electric utility, if suitable overcurrent protection and disconnecting means are provided. For installations of equipment by the serving electric utility, a disconnecting means is not required if the supply is installed as part of a meter socket, such that access can only be gained with the meter removed.

Part VII. Service Equipment — Overcurrent Protection

230.90 Where Required. Each ungrounded service conductor shall have overload protection.

(A) Ungrounded Conductor. Such protection shall be provided by an overcurrent device in series with each ungrounded service conductor that has a rating or setting not higher than the allowable ampacity of the conductor. A set of fuses shall be considered all the fuses required to protect all the ungrounded conductors of a circuit. Single-pole circuit breakers, grouped in accordance with 230.71(B), shall be considered as one protective device.

Exception No. 1: For motor-starting currents, ratings that comply with 430.52, 430.62, and 430.63 shall be permitted.

Exception No. 2: Fuses and circuit breakers with a rating or setting that complies with 240.4(B) or (C) and 240.6 shall be permitted.

Exception No. 3: Two to six circuit breakers or sets of fuses shall be permitted as the overcurrent device to provide the overload protection. The sum of the ratings of the circuit breakers or fuses shall be permitted to exceed the ampacity of the service conductors, provided the calculated load does not exceed the ampacity of the service conductors.

Exception No. 4: Overload protection for fire pump supply conductors shall comply with 695.4(B)(2)(a).

Exception No. 5: Overload protection for 120/240-volt, 3-wire, single-phase dwelling services shall be permitted in accordance with the requirements of 310.15(B)(7).

(B) Not in Grounded Conductor. No overcurrent device shall be inserted in a grounded service conductor except a circuit breaker that simultaneously opens all conductors of the circuit.

230.91 Location. The service overcurrent device shall be an integral part of the service disconnecting means or shall be located immediately adjacent thereto. Where fuses are used as the service overcurrent device, the disconnecting means shall be located ahead of the supply side of the fuses.

230.92 Locked Service Overcurrent Devices. Where the service overcurrent devices are locked or sealed or are not readily accessible to the occupant, branch-circuit or feeder overcurrent devices shall be installed on the load side, shall be mounted in a readily accessible location, and shall be of lower ampere rating than the service overcurrent device.

230.93 Protection of Specific Circuits. Where necessary to prevent tampering, an automatic overcurrent device that protects service conductors supplying only a specific load, such as a water heater, shall be permitted to be locked or sealed where located so as to be accessible.

230.94 Relative Location of Overcurrent Device and Other Service Equipment. The overcurrent device shall protect all circuits and devices.

Exception No. 1: The service switch shall be permitted on the supply side.

Exception No. 2: High-impedance shunt circuits, surge arresters, Type 1 surge-protective devices, surge-protective capacitors, and instrument transformers (current and voltage) shall be permitted to be connected and installed on the supply side of the service disconnecting means as permitted by 230.82.

Exception No. 3: Circuits for load management devices shall be permitted to be connected on the supply side of the service overcurrent device where separately provided with overcurrent protection.

Exception No. 4: Circuits used only for the operation of fire alarm, other protective signaling systems, or the supply to fire pump equipment shall be permitted to be connected on the supply side of the service overcurrent device where separately provided with overcurrent protection.

Exception No. 5: Meters nominally rated not in excess of 600 volts shall be permitted, provided all metal housings and service enclosures are grounded.

Exception No. 6: Where service equipment is power operable, the control circuit shall be permitted to be connected ahead of the service equipment if suitable overcurrent protection and disconnecting means are provided.

230.95 Ground-Fault Protection of Equipment. Ground-fault protection of equipment shall be provided for solidly grounded wye electric services of more than 150 volts to ground but not exceeding 1000 volts phase-to-phase for each service disconnect rated 1000 amperes or more. The grounded conductor for the solidly grounded wye system shall be connected directly to ground through a grounding electrode system, as specified in 250.50, without inserting any resistor or impedance device.

The rating of the service disconnect shall be considered to be the rating of the largest fuse that can be installed or the highest continuous current trip setting for which the actual overcurrent device installed in a circuit breaker is rated or can be adjusted.

Exception: The ground-fault protection provisions of this section shall not apply to a service disconnect for a continuous industrial process where a nonorderly shutdown will introduce additional or increased hazards.

(A) Setting. The ground-fault protection system shall operate to cause the service disconnect to open all ungrounded conductors of the faulted circuit. The maximum setting of the ground-fault protection shall be 1200 amperes, and the maximum time delay shall be one second for ground-fault currents equal to or greater than 3000 amperes.

(B) Fuses. If a switch and fuse combination is used, the fuses employed shall be capable of interrupting any current higher than the interrupting capacity of the switch during a time that the ground-fault protective system will not cause the switch to open.

(C) Performance Testing. The ground-fault protection system shall be performance tested when first installed on site. This testing shall be conducted by a qualified person(s) using a test process of primary current injection, in accordance with instructions that shall be provided with the equipment. A written record of this testing shall be made and shall be available to the authority having jurisdiction.

Informational Note No. 1: Ground-fault protection that functions to open the service disconnect affords no protection from faults on the line side of the protective element. It serves only to limit damage to conductors and equipment on the load side in the event of an arcing ground fault on the load side of the protective element.

Informational Note No. 2: This added protective equipment at the service equipment may make it necessary to review the overall wiring system for proper selective overcurrent protection coordination. Additional installations of ground-fault protective equipment may be needed on feeders and branch circuits where maximum continuity of electric service is necessary.

Informational Note No. 3: Where ground-fault protection is provided for the service disconnect and interconnection is made with another supply system by a transfer device, means or devices may be needed to ensure proper ground-fault sensing by the ground-fault protection equipment.

Informational Note No. 4: See 517.17(A) for information on where an additional step of ground-fault protection is required for hospitals and other buildings with critical areas or life support equipment.

Part VIII. Services Exceeding 1000 Volts, Nominal

230.200 General. Service conductors and equipment used on circuits exceeding 1000 volts, nominal, shall comply with all the applicable provisions of the preceding sections of this article and with the following sections that supplement or modify the preceding sections. In no case shall the provisions of Part VIII apply to equipment on the supply side of the service point.

Informational Note: For clearances of conductors of over 1000 volts, nominal, see ANSI/IEEE C2-2012, *National Electrical Safety Code.*

230.202 Service-Entrance Conductors. Service-entrance conductors to buildings or enclosures shall be installed to conform to 230.202(A) and (B).

(A) Conductor Size. Service-entrance conductors shall not be smaller than 6 AWG unless in multiconductor cable. Multiconductor cable shall not be smaller than 8 AWG.

(B) Wiring Methods. Service-entrance conductors shall be installed by one of the wiring methods covered in 300.37 and 300.50.

230.204 Isolating Switches.

(A) Where Required. Where oil switches or air, oil, vacuum, or sulfur hexafluoride circuit breakers constitute the service disconnecting means, an isolating switch with visible break contacts shall be installed on the supply side of the disconnecting means and all associated service equipment.

Exception: An isolating switch shall not be required where the circuit breaker or switch is mounted on removable truck panels or switchgear units where both of the following conditions apply:

(1)   Cannot be opened unless the circuit is disconnected
(2)   Where all energized parts are automatically disconnected when the circuit breaker or switch is removed from the normal operating position

(B) Fuses as Isolating Switch. Where fuses are of the type that can be operated as a disconnecting switch, a set of such fuses shall be permitted as the isolating switch.

(C) Accessible to Qualified Persons Only. The isolating switch shall be accessible to qualified persons only.

(D) Connection to Ground. Isolating switches shall be provided with a means for readily connecting the load side conductors to a grounding electrode system, equipment ground busbar, or grounded steel structure when disconnected from the source of supply.

A means for grounding the load side conductors to a grounding electrode system, equipment grounding busbar, or grounded structural steel shall not be required for any duplicate isolating switch installed and maintained by the electric supply company.

230.205 Disconnecting Means.

(A) Location. The service disconnecting means shall be located in accordance with 230.70.

For either overhead or underground primary distribution systems on private property, the service disconnect shall be permitted to be located in a location that is not readily accessible, if the disconnecting means can be operated by mechanical linkage from a readily accessible point, or electronically in accordance with 230.205(C), where applicable.

(B) Type. Each service disconnect shall simultaneously disconnect all ungrounded service conductors that it controls and shall have a fault-closing rating that is not less than the maximum short-circuit current available at its supply terminals.

Where fused switches or separately mounted fuses are installed, the fuse characteristics shall be permitted to contribute to the fault-closing rating of the disconnecting means.

(C) Remote Control. For multibuilding, industrial installations under single management, the service disconnecting means shall be permitted to be located at a separate building or structure. In such cases, the service disconnecting means shall be permitted to be electrically operated by a readily accessible, remote-control device.

230.206 Overcurrent Devices as Disconnecting Means. Where the circuit breaker or alternative for it, as specified in 230.208 for service overcurrent devices, meets the requirements specified in 230.205, it shall constitute the service disconnecting means.

230.208 Protection Requirements. A short-circuit protective device shall be provided on the load side of, or as an integral part of, the service disconnect, and shall protect all ungrounded conductors that it supplies. The protective device shall be capable of detecting and interrupting all values of current, in excess of its trip setting or melting point, that can occur at its location. A fuse rated in continuous amperes not to exceed three times the ampacity of the conductor, or a circuit breaker with a trip setting of not more than six times the ampacity of the conductors, shall be considered as providing the required short-circuit protection.

Informational Note: See Table 310.60(C)(67) through Table 310.60(C)(86) for ampacities of conductors rated 2001 volts and above.

Overcurrent devices shall conform to 230.208(A) and (B).

(A) Equipment Type. Equipment used to protect service-entrance conductors shall meet the requirements of Article 490, Part II.

(B) Enclosed Overcurrent Devices. The restriction to 80 percent of the rating for an enclosed overcurrent device for continuous loads shall not apply to overcurrent devices installed in systems operating at over 1000 volts.

230.209 Surge Arresters. Surge arresters installed in accordance with the requirements of Article 280 shall be permitted on each ungrounded overhead service conductor.

Informational Note: Surge arresters may be referred to as lightning arresters in older documents.

230.210 Service Equipment — General Provisions. Service equipment, including instrument transformers, shall conform to Article 490, Part I.

230.211 Switchgear. Switchgear shall consist of a substantial metal structure and a sheet metal enclosure. Where installed over a combustible floor, suitable protection thereto shall be provided.

230.212 Over 35,000 Volts. Where the voltage exceeds 35,000 volts between conductors that enter a building, they shall terminate in a switchgear compartment or a vault conforming to the requirements of 450.41 through 450.48.

ARTICLE 240
Overcurrent Protection

Part I. General

240.1 Scope. Parts I through VII of this article provide the general requirements for overcurrent protection and overcurrent protective devices not more than 1000 volts, nominal. Part VIII covers overcurrent protection for those portions of supervised industrial installations operating at voltages of not more than 1000 volts, nominal. Part IX covers overcurrent protection over 1000 volts, nominal.

Informational Note: Overcurrent protection for conductors and equipment is provided to open the circuit if the current reaches a value that will cause an excessive or dangerous temperature in conductors or conductor insulation. See also 110.9 for requirements for interrupting ratings and 110.10 for requirements for protection against fault currents.

240.2 Definitions.

Current-Limiting Overcurrent Protective Device. A device that, when interrupting currents in its current-limiting range, reduces the current flowing in the faulted circuit to a magnitude substantially less than that obtainable in the same circuit if the device were replaced with a solid conductor having comparable impedance.

Supervised Industrial Installation. For the purposes of Part VIII, the industrial portions of a facility where all of the following conditions are met:

(1) Conditions of maintenance and engineering supervision ensure that only qualified persons monitor and service the system.
(2) The premises wiring system has 2500 kVA or greater of load used in industrial process(es), manufacturing activities, or both, as calculated in accordance with Article 220.
(3) The premises has at least one service or feeder that is more than 150 volts to ground and more than 300 volts phase-to-phase.

This definition excludes installations in buildings used by the industrial facility for offices, warehouses, garages, machine shops, and recreational facilities that are not an integral part of the industrial plant, substation, or control center.

Tap Conductor. A conductor, other than a service conductor, that has overcurrent protection ahead of its point of supply that exceeds the value permitted for similar conductors that are protected as described elsewhere in 240.4.

240.3 Other Articles. Equipment shall be protected against overcurrent in accordance with the article in this *Code* that covers the type of equipment specified in Table 240.3.

240.4 Protection of Conductors. Conductors, other than flexible cords, flexible cables, and fixture wires, shall be protected against overcurrent in accordance with their ampacities specified in 310.15, unless otherwise permitted or required in 240.4(A) through (G).

Informational Note: See ICEA P-32-382-2007 for information on allowable short-circuit currents for insulated copper and aluminum conductors.

Table 240.3 Other Articles

Equipment	Article
Air-conditioning and refrigerating equipment	440
Appliances	422
Assembly occupancies	518
Audio signal processing, amplification, and reproduction equipment	640
Branch circuits	210
Busways	368
Capacitors	460
Class 1, Class 2, and Class 3 remote-control, signaling, and power-limited circuits	725
Cranes and hoists	610
Electric signs and outline lighting	600
Electric welders	630
Electrolytic cells	668
Elevators, dumbwaiters, escalators, moving walks, wheelchair lifts, and stairway chairlifts	620
Emergency systems	700
Fire alarm systems	760
Fire pumps	695
Fixed electric heating equipment for pipelines and vessels	427
Fixed electric space-heating equipment	424
Fixed outdoor electric deicing and snow-melting equipment	426
Generators	445
Health care facilities	517
Induction and dielectric heating equipment	665
Industrial machinery	670
Luminaires, lampholders, and lamps	410
Motion picture and television studios and similar locations	530
Motors, motor circuits, and controllers	430
Phase converters	455
Pipe organs	650
Receptacles	406
Services	230
Solar photovoltaic systems	690
Switchboards and panelboards	408
Theaters, audience areas of motion picture and television studios, and similar locations	520
Transformers and transformer vaults	450
X-ray equipment	660

(A) Power Loss Hazard. Conductor overload protection shall not be required where the interruption of the circuit would create a hazard, such as in a material-handling magnet circuit or fire pump circuit. Short-circuit protection shall be provided.

Informational Note: See NFPA 20-2013, *Standard for the Installation of Stationary Pumps for Fire Protection.*

(B) Overcurrent Devices Rated 800 Amperes or Less. The next higher standard overcurrent device rating (above the ampacity of the conductors being protected) shall be permitted to be used, provided all of the following conditions are met:

(1) The conductors being protected are not part of a branch circuit supplying more than one receptacle for cord-and-plug-connected portable loads.

(2) The ampacity of the conductors does not correspond with the standard ampere rating of a fuse or a circuit breaker without overload trip adjustments above its rating (but that shall be permitted to have other trip or rating adjustments).

(3) The next higher standard rating selected does not exceed 800 amperes.

(C) Overcurrent Devices Rated over 800 Amperes. Where the overcurrent device is rated over 800 amperes, the ampacity of the conductors it protects shall be equal to or greater than the rating of the overcurrent device defined in 240.6.

(D) Small Conductors. Unless specifically permitted in 240.4(E) or (G), the overcurrent protection shall not exceed that required by (D)(1) through (D)(7) after any correction factors for ambient temperature and number of conductors have been applied.

(1) 18 AWG Copper. 7 amperes, provided all the following conditions are met:

(1) Continuous loads do not exceed 5.6 amperes.

(2) Overcurrent protection is provided by one of the following:

    a. Branch-circuit-rated circuit breakers listed and marked for use with 18 AWG copper wire

    b. Branch-circuit-rated fuses listed and marked for use with 18 AWG copper wire

    c. Class CC, Class J, or Class T fuses

(2) 16 AWG Copper. 10 amperes, provided all the following conditions are met:

(1) Continuous loads do not exceed 8 amperes.

(2) Overcurrent protection is provided by one of the following:

    a. Branch-circuit-rated circuit breakers listed and marked for use with 16 AWG copper wire

    b. Branch-circuit-rated fuses listed and marked for use with 16 AWG copper wire

    c. Class CC, Class J, or Class T fuses

(3) 14 AWG Copper. 15 amperes

(4) 12 AWG Aluminum and Copper-Clad Aluminum. 15 amperes

(5) 12 AWG Copper. 20 amperes

(6) 10 AWG Aluminum and Copper-Clad Aluminum. 25 amperes

(7) 10 AWG Copper. 30 amperes

(E) Tap Conductors. Tap conductors shall be permitted to be protected against overcurrent in accordance with the following:

(1) 210.19(A)(3) and (A)(4), Household Ranges and Cooking Appliances and Other Loads

(2) 240.5(B)(2), Fixture Wire

(3) 240.21, Location in Circuit

(4) 368.17(B), Reduction in Ampacity Size of Busway

(5) 368.17(C), Feeder or Branch Circuits (busway taps)

(6) 430.53(D), Single Motor Taps

(F) Transformer Secondary Conductors. Single-phase (other than 2-wire) and multiphase (other than delta-delta, 3-wire) transformer secondary conductors shall not be considered to be protected by the primary overcurrent protective device. Conductors supplied by the secondary side of a single-phase transformer having a 2-wire (single-voltage) secondary, or a three-phase, delta-delta connected transformer having a 3-wire (single-voltage) secondary, shall be permitted to be protected by overcurrent protection provided on the primary (supply) side of the transformer, provided this protection is in accordance with 450.3 and does not exceed the value determined by multiplying the secondary conductor ampacity by the secondary-to-primary transformer voltage ratio.

(G) Overcurrent Protection for Specific Conductor Applications. Overcurrent protection for the specific conductors shall be permitted to be provided as referenced in Table 240.4(G).

Table 240.4(G) Specific Conductor Applications

Conductor	Article	Section
Air-conditioning and refrigeration equipment circuit conductors	440, Parts III, VI	
Capacitor circuit conductors	460	460.8(B) and 460.25(A)–(D)
Control and instrumentation circuit conductors (Type ITC)	727	727.9
Electric welder circuit conductors	630	630.12 and 630.32
Fire alarm system circuit conductors	760	760.43, 760.45, 760.121, and Chapter 9, Tables 12(A) and 12(B)
Motor-operated appliance circuit conductors	422, Part II	
Motor and motor-control circuit conductors	430, Parts II, III, IV, V, VI, VII	
Phase converter supply conductors	455	455.7
Remote-control, signaling, and power-limited circuit conductors	725	725.43, 725.45, 725.121, and Chapter 9, Tables 11(A) and 11(B)
Secondary tie conductors	450	450.6

240.5 Protection of Flexible Cords, Flexible Cables, and Fixture Wires. Flexible cord and flexible cable, including tinsel cord and extension cords, and fixture wires shall be protected against overcurrent by either 240.5(A) or (B).

(A) Ampacities. Flexible cord and flexible cable shall be protected by an overcurrent device in accordance with their ampacity as specified in Table 400.5(A)(1) and Table 400.5(A)(2). Fixture wire shall be protected against overcurrent in accordance with its ampacity as specified in Table 402.5. Supplementary overcurrent protection, as covered in 240.10, shall be permitted to be an acceptable means for providing this protection.

(B) Branch-Circuit Overcurrent Device. Flexible cord shall be protected, where supplied by a branch circuit, in accordance with one of the methods described in 240.5(B)(1), (B)(3), or (B)(4). Fixture wire shall be protected, where supplied by a branch circuit, in accordance with 240.5(B)(2).

(1) Supply Cord of Listed Appliance or Luminaire. Where flexible cord or tinsel cord is approved for and used with a specific listed appliance or luminaire, it shall be considered to be protected when applied within the appliance or luminaire listing requirements. For the purposes of this section, a luminaire may be either portable or permanent.

(2) Fixture Wire. Fixture wire shall be permitted to be tapped to the branch-circuit conductor of a branch circuit in accordance with the following:

(1)　20-ampere circuits — 18 AWG, up to 15 m (50 ft) of run length
(2)　20-ampere circuits — 16 AWG, up to 30 m (100 ft) of run length
(3)　20-ampere circuits — 14 AWG and larger
(4)　30-ampere circuits — 14 AWG and larger
(5)　40-ampere circuits — 12 AWG and larger
(6)　50-ampere circuits — 12 AWG and larger

(3) Extension Cord Sets. Flexible cord used in listed extension cord sets shall be considered to be protected when applied within the extension cord listing requirements.

(4) Field Assembled Extension Cord Sets. Flexible cord used in extension cords made with separately listed and installed components shall be permitted to be supplied by a branch circuit in accordance with the following:

20-ampere circuits — 16 AWG and larger

240.6 Standard Ampere Ratings.

(A) Fuses and Fixed-Trip Circuit Breakers. The standard ampere ratings for fuses and inverse time circuit breakers shall be considered as shown in Table 240.6(A). Additional standard ampere ratings for fuses shall be 1, 3, 6, 10, and 601. The use of fuses and inverse time circuit breakers with nonstandard ampere ratings shall be permitted.

(B) Adjustable-Trip Circuit Breakers. The rating of adjustable-trip circuit breakers having external means for adjusting the current setting (long-time pickup setting), not meeting the requirements of 240.6(C), shall be the maximum setting possible.

(C) Restricted Access Adjustable-Trip Circuit Breakers. A circuit breaker(s) that has restricted access to the adjusting means shall be permitted to have an ampere rating(s) that is equal to the adjusted current setting (long-time pickup

N **Table 240.6(A) Standard Ampere Ratings for Fuses and Inverse Time Circuit Breakers**

Standard Ampere Ratings				
15	20	25	30	35
40	45	50	60	70
80	90	100	110	125
150	175	200	225	250
300	350	400	450	500
600	700	800	1000	1200
1600	2000	2500	3000	4000
5000	6000	—	—	—

setting). Restricted access shall be defined as located behind one of the following:

(1)　Removable and sealable covers over the adjusting means
(2)　Bolted equipment enclosure doors
(3)　Locked doors accessible only to qualified personnel

240.8 Fuses or Circuit Breakers in Parallel. Fuses and circuit breakers shall be permitted to be connected in parallel where they are factory assembled in parallel and listed as a unit. Individual fuses, circuit breakers, or combinations thereof shall not otherwise be connected in parallel.

240.9 Thermal Devices. Thermal relays and other devices not designed to open short circuits or ground faults shall not be used for the protection of conductors against overcurrent due to short circuits or ground faults, but the use of such devices shall be permitted to protect motor branch-circuit conductors from overload if protected in accordance with 430.40.

240.10 Supplementary Overcurrent Protection. Where supplementary overcurrent protection is used for luminaires, appliances, and other equipment or for internal circuits and components of equipment, it shall not be used as a substitute for required branch-circuit overcurrent devices or in place of the required branch-circuit protection. Supplementary overcurrent devices shall not be required to be readily accessible.

240.12 Electrical System Coordination. Where an orderly shutdown is required to minimize the hazard(s) to personnel and equipment, a system of coordination based on the following two conditions shall be permitted:

(1)　Coordinated short-circuit protection
(2)　Overload indication based on monitoring systems or devices

Informational Note: The monitoring system may cause the condition to go to alarm, allowing corrective action or an orderly shutdown, thereby minimizing personnel hazard and equipment damage.

240.13 Ground-Fault Protection of Equipment. Ground-fault protection of equipment shall be provided in accordance with the provisions of 230.95 for solidly grounded wye electrical systems of more than 150 volts to ground but not exceeding 1000 volts phase-to-phase for each individual device used as a building or structure main disconnecting means rated 1000 amperes or more.

The provisions of this section shall not apply to the disconnecting means for the following:

(1) Continuous industrial processes where a nonorderly shutdown will introduce additional or increased hazards

(2) Installations where ground-fault protection is provided by other requirements for services or feeders

(3) Fire pumps

240.15 Ungrounded Conductors.

(A) Overcurrent Device Required. A fuse or an overcurrent trip unit of a circuit breaker shall be connected in series with each ungrounded conductor. A combination of a current transformer and overcurrent relay shall be considered equivalent to an overcurrent trip unit.

> Informational Note: For motor circuits, see Parts III, IV, V, and XI of Article 430.

(B) Circuit Breaker as Overcurrent Device. Circuit breakers shall open all ungrounded conductors of the circuit both manually and automatically unless otherwise permitted in 240.15(B)(1), (B)(2), (B)(3), and (B)(4).

(1) Multiwire Branch Circuits. Individual single-pole circuit breakers, with identified handle ties, shall be permitted as the protection for each ungrounded conductor of multiwire branch circuits that serve only single-phase line-to-neutral loads.

(2) Grounded Single-Phase Alternating-Current Circuits. In grounded systems, individual single-pole circuit breakers rated 120/240 volts ac, with identified handle ties, shall be permitted as the protection for each ungrounded conductor for line-to-line connected loads for single-phase circuits.

(3) 3-Phase and 2-Phase Systems. For line-to-line loads in 4-wire, 3-phase systems or 5-wire, 2-phase systems, individual single-pole circuit breakers rated 120/240 volts ac with identified handle ties shall be permitted as the protection for each ungrounded conductor, if the systems have a grounded neutral point and the voltage to ground does not exceed 120 volts.

(4) 3-Wire Direct-Current Circuits. Individual single-pole circuit breakers rated 125/250 volts dc with identified handle ties shall be permitted as the protection for each ungrounded conductor for line-to-line connected loads for 3-wire, direct-current circuits supplied from a system with a grounded neutral where the voltage to ground does not exceed 125 volts.

Part II. Location

240.21 Location in Circuit. Overcurrent protection shall be provided in each ungrounded circuit conductor and shall be located at the point where the conductors receive their supply except as specified in 240.21(A) through (H). Conductors supplied under the provisions of 240.21(A) through (H) shall not supply another conductor except through an overcurrent protective device meeting the requirements of 240.4.

(A) Branch-Circuit Conductors. Branch-circuit tap conductors meeting the requirements specified in 210.19 shall be permitted to have overcurrent protection as specified in 210.20.

(B) Feeder Taps. Conductors shall be permitted to be tapped, without overcurrent protection at the tap, to a feeder as specified in 240.21(B)(1) through (B)(5). The provisions of 240.4(B) shall not be permitted for tap conductors.

(1) Taps Not over 3 m (10 ft) Long. If the length of the tap conductors does not exceed 3 m (10 ft) and the tap conductors comply with all of the following:

(1) The ampacity of the tap conductors is

    a. Not less than the combined calculated loads on the circuits supplied by the tap conductors, and

    b. Not less than the rating of the equipment containing an overcurrent device(s) supplied by the tap conductors or not less than the rating of the overcurrent protective device at the termination of the tap conductors.

> *Exception to b: Where listed equipment, such as a surge protective device(s) [SPD(s)], is provided with specific instructions on minimum conductor sizing, the ampacity of the tap conductors supplying that equipment shall be permitted to be determined based on the manufacturer's instructions.*

(2) The tap conductors do not extend beyond the switchboard, switchgear, panelboard, disconnecting means, or control devices they supply.

(3) Except at the point of connection to the feeder, the tap conductors are enclosed in a raceway, which extends from the tap to the enclosure of an enclosed switchboard, switchgear, a panelboard, or control devices, or to the back of an open switchboard.

(4) For field installations, if the tap conductors leave the enclosure or vault in which the tap is made, the ampacity of the tap conductors is not less than one-tenth of the rating of the overcurrent device protecting the feeder conductors.

> Informational Note: For overcurrent protection requirements for panelboards, see 408.36.

(2) Taps Not over 7.5 m (25 ft) Long. Where the length of the tap conductors does not exceed 7.5 m (25 ft) and the tap conductors comply with all the following:

(1) The ampacity of the tap conductors is not less than one-third of the rating of the overcurrent device protecting the feeder conductors.

(2) The tap conductors terminate in a single circuit breaker or a single set of fuses that limit the load to the ampacity of the tap conductors. This device shall be permitted to supply any number of additional overcurrent devices on its load side.

(3) The tap conductors are protected from physical damage by being enclosed in an approved raceway or by other approved means.

(3) Taps Supplying a Transformer [Primary Plus Secondary Not over 7.5 m (25 ft) Long]. Where the tap conductors supply a transformer and comply with all the following conditions:

(1) The conductors supplying the primary of a transformer have an ampacity at least one-third the rating of the overcurrent device protecting the feeder conductors.

(2) The conductors supplied by the secondary of the transformer shall have an ampacity that is not less than the value of the primary-to-secondary voltage ratio multiplied by one-third of the rating of the overcurrent device protecting the feeder conductors.

(3) The total length of one primary plus one secondary conductor, excluding any portion of the primary conductor that is protected at its ampacity, is not over 7.5 m (25 ft).

(4) The primary and secondary conductors are protected from physical damage by being enclosed in an approved raceway or by other approved means.

(5) The secondary conductors terminate in a single circuit breaker or set of fuses that limit the load current to not more than the conductor ampacity that is permitted by 310.15.

(4) Taps over 7.5 m (25 ft) Long. Where the feeder is in a high bay manufacturing building over 11 m (35 ft) high at walls and the installation complies with all the following conditions:

(1) Conditions of maintenance and supervision ensure that only qualified persons service the systems.

(2) The tap conductors are not over 7.5 m (25 ft) long horizontally and not over 30 m (100 ft) total length.

(3) The ampacity of the tap conductors is not less than one-third the rating of the overcurrent device protecting the feeder conductors.

(4) The tap conductors terminate at a single circuit breaker or a single set of fuses that limit the load to the ampacity of the tap conductors. This single overcurrent device shall be permitted to supply any number of additional overcurrent devices on its load side.

(5) The tap conductors are protected from physical damage by being enclosed in an approved raceway or by other approved means.

(6) The tap conductors are continuous from end-to-end and contain no splices.

(7) The tap conductors are sized 6 AWG copper or 4 AWG aluminum or larger.

(8) The tap conductors do not penetrate walls, floors, or ceilings.

(9) The tap is made no less than 9 m (30 ft) from the floor.

(5) Outside Taps of Unlimited Length. Where the conductors are located outside of a building or structure, except at the point of load termination, and comply with all of the following conditions:

(1) The tap conductors are protected from physical damage in an approved manner.

(2) The tap conductors terminate at a single circuit breaker or a single set of fuses that limits the load to the ampacity of the tap conductors. This single overcurrent device shall be permitted to supply any number of additional overcurrent devices on its load side.

(3) The overcurrent device for the tap conductors is an integral part of a disconnecting means or shall be located immediately adjacent thereto.

(4) The disconnecting means for the tap conductors is installed at a readily accessible location complying with one of the following:

a. Outside of a building or structure

b. Inside, nearest the point of entrance of the tap conductors

c. Where installed in accordance with 230.6, nearest the point of entrance of the tap conductors

(C) Transformer Secondary Conductors. A set of conductors feeding a single load, or each set of conductors feeding separate loads, shall be permitted to be connected to a transformer secondary, without overcurrent protection at the secondary, as specified in 240.21(C)(1) through (C)(6). The provisions of 240.4(B) shall not be permitted for transformer secondary conductors.

Informational Note: For overcurrent protection requirements for transformers, see 450.3.

(1) Protection by Primary Overcurrent Device. Conductors supplied by the secondary side of a single-phase transformer having a 2-wire (single-voltage) secondary, or a three-phase, delta-delta connected transformer having a 3-wire (single-voltage) secondary, shall be permitted to be protected by overcurrent protection provided on the primary (supply) side of the transformer, provided this protection is in accordance with 450.3 and does not exceed the value determined by multiplying the secondary conductor ampacity by the secondary-to-primary transformer voltage ratio.

Single-phase (other than 2-wire) and multiphase (other than delta-delta, 3-wire) transformer secondary conductors are not considered to be protected by the primary overcurrent protective device.

(2) Transformer Secondary Conductors Not over 3 m (10 ft) Long. If the length of secondary conductor does not exceed 3 m (10 ft) and complies with all of the following:

(1) The ampacity of the secondary conductors is

a. Not less than the combined calculated loads on the circuits supplied by the secondary conductors, and

b. Not less than the rating of the equipment containing an overcurrent device(s) supplied by the secondary conductors or not less than the rating of the overcurrent protective device at the termination of the secondary conductors.

Exception: Where listed equipment, such as a surge protective device(s) [SPD(s)], is provided with specific instructions on minimum conductor sizing, the ampacity of the tap conductors supplying that equipment shall be permitted to be determined based on the manufacturer's instructions.

(2) The secondary conductors do not extend beyond the switchboard, switchgear, panelboard, disconnecting means, or control devices they supply.

(3) The secondary conductors are enclosed in a raceway, which shall extend from the transformer to the enclosure of an enclosed switchboard, switchgear, a panelboard, or control devices or to the back of an open switchboard.

(4) For field installations where the secondary conductors leave the enclosure or vault in which the supply connection is made, the rating of the overcurrent device protecting the primary of the transformer, multiplied by the primary to secondary transformer voltage ratio, shall not exceed 10 times the ampacity of the secondary conductor.

Informational Note: For overcurrent protection requirements for panelboards, see 408.36.

(3) Industrial Installation Secondary Conductors Not over 7.5 m (25 ft) Long. For the supply of switchgear or switchboards in industrial installations only, where the length of the secondary conductors does not exceed 7.5 m (25 ft) and complies with all of the following:

(1) Conditions of maintenance and supervision ensure that only qualified persons service the systems.

(2) The ampacity of the secondary conductors is not less than the secondary current rating of the transformer, and the

sum of the ratings of the overcurrent devices does not exceed the ampacity of the secondary conductors.

(3) All overcurrent devices are grouped.

(4) The secondary conductors are protected from physical damage by being enclosed in an approved raceway or by other approved means.

(4) Outside Secondary Conductors. Where the conductors are located outside of a building or structure, except at the point of load termination, and comply with all of the following conditions:

(1) The conductors are protected from physical damage in an approved manner.

(2) The conductors terminate at a single circuit breaker or a single set of fuses that limit the load to the ampacity of the conductors. This single overcurrent device shall be permitted to supply any number of additional overcurrent devices on its load side.

(3) The overcurrent device for the conductors is an integral part of a disconnecting means or shall be located immediately adjacent thereto.

(4) The disconnecting means for the conductors is installed at a readily accessible location complying with one of the following:

   a. Outside of a building or structure

   b. Inside, nearest the point of entrance of the conductors

   c. Where installed in accordance with 230.6, nearest the point of entrance of the conductors

(5) Secondary Conductors from a Feeder Tapped Transformer. Transformer secondary conductors installed in accordance with 240.21(B)(3) shall be permitted to have overcurrent protection as specified in that section.

(6) Secondary Conductors Not over 7.5 m (25 ft) Long. Where the length of secondary conductor does not exceed 7.5 m (25 ft) and complies with all of the following:

(1) The secondary conductors shall have an ampacity that is not less than the value of the primary-to-secondary voltage ratio multiplied by one-third of the rating of the overcurrent device protecting the primary of the transformer.

(2) The secondary conductors terminate in a single circuit breaker or set of fuses that limit the load current to not more than the conductor ampacity that is permitted by 310.15.

(3) The secondary conductors are protected from physical damage by being enclosed in an approved raceway or by other approved means.

(D) Service Conductors. Service conductors shall be permitted to be protected by overcurrent devices in accordance with 230.91.

(E) Busway Taps. Busways and busway taps shall be permitted to be protected against overcurrent in accordance with 368.17.

(F) Motor Circuit Taps. Motor-feeder and branch-circuit conductors shall be permitted to be protected against overcurrent in accordance with 430.28 and 430.53, respectively.

(G) Conductors from Generator Terminals. Conductors from generator terminals that meet the size requirement in 445.13 shall be permitted to be protected against overload by the generator overload protective device(s) required by 445.12.

(H) Battery Conductors. Overcurrent protection shall be permitted to be installed as close as practicable to the storage battery terminals in an unclassified location. Installation of the overcurrent protection within a hazardous (classified) location shall also be permitted.

240.22 Grounded Conductor. No overcurrent device shall be connected in series with any conductor that is intentionally grounded, unless one of the following two conditions is met:

(1) The overcurrent device opens all conductors of the circuit, including the grounded conductor, and is designed so that no pole can operate independently.

(2) Where required by 430.36 or 430.37 for motor overload protection.

240.23 Change in Size of Grounded Conductor. Where a change occurs in the size of the ungrounded conductor, a similar change shall be permitted to be made in the size of the grounded conductor.

240.24 Location in or on Premises.

(A) Accessibility. Switches containing fuses and circuit breakers shall be readily accessible and installed so that the center of the grip of the operating handle of the switch or circuit breaker, when in its highest position, is not more than 2.0 m (6 ft 7 in.) above the floor or working platform, unless one of the following applies:

(1) For busways, as provided in 368.17(C).

(2) For supplementary overcurrent protection, as described in 240.10.

(3) For overcurrent devices, as described in 225.40 and 230.92.

(4) For overcurrent devices adjacent to utilization equipment that they supply, access shall be permitted to be by portable means.

Exception: The use of a tool shall be permitted to access overcurrent devices located within listed industrial control panels or similar enclosures.

(B) Occupancy. Each occupant shall have ready access to all overcurrent devices protecting the conductors supplying that occupancy, unless otherwise permitted in 240.24(B)(1) and (B)(2).

(1) Service and Feeder Overcurrent Devices. Where electric service and electrical maintenance are provided by the building management and where these are under continuous building management supervision, the service overcurrent devices and feeder overcurrent devices supplying more than one occupancy shall be permitted to be accessible only to authorized management personnel in the following:

(1) Multiple-occupancy buildings

(2) Guest rooms or guest suites

(2) Branch-Circuit Overcurrent Devices. Where electric service and electrical maintenance are provided by the building management and where these are under continuous building management supervision, the branch-circuit overcurrent devices supplying any guest rooms or guest suites without permanent provisions for cooking shall be permitted to be accessible only to authorized management personnel.

(C) Not Exposed to Physical Damage. Overcurrent devices shall be located where they will not be exposed to physical damage.

Informational Note: See 110.11, Deteriorating Agents.

(D) Not in Vicinity of Easily Ignitible Material. Overcurrent devices shall not be located in the vicinity of easily ignitible material, such as in clothes closets.

(E) Not Located in Bathrooms. In dwelling units, dormitories, and guest rooms or guest suites, overcurrent devices, other than supplementary overcurrent protection, shall not be located in bathrooms.

(F) Not Located over Steps. Overcurrent devices shall not be located over steps of a stairway.

Part III. Enclosures

240.30 General.

(A) Protection from Physical Damage. Overcurrent devices shall be protected from physical damage by one of the following:

(1) Installation in enclosures, cabinets, cutout boxes, or equipment assemblies
(2) Mounting on open-type switchboards, panelboards, or control boards that are in rooms or enclosures free from dampness and easily ignitible material and are accessible only to qualified personnel

(B) Operating Handle. The operating handle of a circuit breaker shall be permitted to be accessible without opening a door or cover.

240.32 Damp or Wet Locations. Enclosures for overcurrent devices in damp or wet locations shall comply with 312.2.

240.33 Vertical Position. Enclosures for overcurrent devices shall be mounted in a vertical position unless that is shown to be impracticable. Circuit breaker enclosures shall be permitted to be installed horizontally where the circuit breaker is installed in accordance with 240.81. Listed busway plug-in units shall be permitted to be mounted in orientations corresponding to the busway mounting position.

Part IV. Disconnecting and Guarding

240.40 Disconnecting Means for Fuses. Cartridge fuses in circuits of any voltage where accessible to other than qualified persons, and all fuses in circuits over 150 volts to ground, shall be provided with a disconnecting means on their supply side so that each circuit containing fuses can be independently disconnected from the source of power. A current-limiting device without a disconnecting means shall be permitted on the supply side of the service disconnecting means as permitted by 230.82. A single disconnecting means shall be permitted on the supply side of more than one set of fuses as permitted by 430.112, Exception, for group operation of motors and 424.22(C) for fixed electric space-heating equipment.

240.41 Arcing or Suddenly Moving Parts. Arcing or suddenly moving parts shall comply with 240.41(A) and (B).

(A) Location. Fuses and circuit breakers shall be located or shielded so that persons will not be burned or otherwise injured by their operation.

(B) Suddenly Moving Parts. Handles or levers of circuit breakers, and similar parts that may move suddenly in such a way that persons in the vicinity are likely to be injured by being struck by them, shall be guarded or isolated.

Part V. Plug Fuses, Fuseholders, and Adapters

240.50 General.

(A) Maximum Voltage. Plug fuses shall be permitted to be used in the following circuits:

(1) Circuits not exceeding 125 volts between conductors
(2) Circuits supplied by a system having a grounded neutral point where the line-to-neutral voltage does not exceed 150 volts

(B) Marking. Each fuse, fuseholder, and adapter shall be marked with its ampere rating.

(C) Hexagonal Configuration. Plug fuses of 15-ampere and lower rating shall be identified by a hexagonal configuration of the window, cap, or other prominent part to distinguish them from fuses of higher ampere ratings.

(D) No Energized Parts. Plug fuses, fuseholders, and adapters shall have no exposed energized parts after fuses or fuses and adapters have been installed.

(E) Screw Shell. The screw shell of a plug-type fuseholder shall be connected to the load side of the circuit.

240.51 Edison-Base Fuses.

(A) Classification. Plug fuses of the Edison-base type shall be classified at not over 125 volts and 30 amperes and below.

(B) Replacement Only. Plug fuses of the Edison-base type shall be used only for replacements in existing installations where there is no evidence of overfusing or tampering.

240.52 Edison-Base Fuseholders. Fuseholders of the Edison-base type shall be installed only where they are made to accept Type S fuses by the use of adapters.

240.53 Type S Fuses. Type S fuses shall be of the plug type and shall comply with 240.53(A) and (B).

(A) Classification. Type S fuses shall be classified at not over 125 volts and 0 to 15 amperes, 16 to 20 amperes, and 21 to 30 amperes.

(B) Noninterchangeable. Type S fuses of an ampere classification as specified in 240.53(A) shall not be interchangeable with a lower ampere classification. They shall be designed so that they cannot be used in any fuseholder other than a Type S fuseholder or a fuseholder with a Type S adapter inserted.

240.54 Type S Fuses, Adapters, and Fuseholders.

(A) To Fit Edison-Base Fuseholders. Type S adapters shall fit Edison-base fuseholders.

(B) To Fit Type S Fuses Only. Type S fuseholders and adapters shall be designed so that either the fuseholder itself or the fuseholder with a Type S adapter inserted cannot be used for any fuse other than a Type S fuse.

(C) Nonremovable. Type S adapters shall be designed so that once inserted in a fuseholder, they cannot be removed.

(D) Nontamperable. Type S fuses, fuseholders, and adapters shall be designed so that tampering or shunting (bridging) would be difficult.

(E) Interchangeability. Dimensions of Type S fuses, fuseholders, and adapters shall be standardized to permit interchangeability regardless of the manufacturer.

Part VI. Cartridge Fuses and Fuseholders

240.60 General.

(A) Maximum Voltage — 300-Volt Type. Cartridge fuses and fuseholders of the 300-volt type shall be permitted to be used in the following circuits:

(1) Circuits not exceeding 300 volts between conductors
(2) Single-phase line-to-neutral circuits supplied from a 3-phase, 4-wire, solidly grounded neutral source where the line-to-neutral voltage does not exceed 300 volts

(B) Noninterchangeable — 0–6000-Ampere Cartridge Fuseholders. Fuseholders shall be designed so that it will be difficult to put a fuse of any given class into a fuseholder that is designed for a current lower, or voltage higher, than that of the class to which the fuse belongs. Fuseholders for current-limiting fuses shall not permit insertion of fuses that are not current-limiting.

(C) Marking. Fuses shall be plainly marked, either by printing on the fuse barrel or by a label attached to the barrel showing the following:

(1) Ampere rating
(2) Voltage rating
(3) Interrupting rating where other than 10,000 amperes
(4) Current limiting where applicable
(5) The name or trademark of the manufacturer

The interrupting rating shall not be required to be marked on fuses used for supplementary protection.

(D) Renewable Fuses. Class H cartridge fuses of the renewable type shall be permitted to be used only for replacement in existing installations where there is no evidence of overfusing or tampering.

240.61 Classification. Cartridge fuses and fuseholders shall be classified according to voltage and amperage ranges. Fuses rated 1000 volts, nominal, or less shall be permitted to be used for voltages at or below their ratings.

N **240.67 Arc Energy Reduction.** Where fuses rated 1200 A or higher are installed, 240.67(A) and (B) shall apply. This requirement shall become effective January 1, 2020.

(A) Documentation. Documentation shall be available to those authorized to design, install, operate, or inspect the installation as to the location of the fuses.

(B) Method to Reduce Clearing Time. A fuse shall have a clearing time of 0.07 seconds or less at the available arcing current, or one of the following shall be provided:

(1) Differential relaying
(2) Energy-reducing maintenance switching with local status indicator

(3) Energy-reducing active arc flash mitigation system
(4) An approved equivalent means

Informational Note No. 1: An energy-reducing maintenance switch allows a worker to set a disconnect switch to reduce the clearing time while the worker is working within an arc-flash boundary as defined in *NFPA 70E* -2015, *Standard for Electrical Safety in the Workplace*, and then to set the disconnect switch back to a normal setting after the potentially hazardous work is complete.

Informational Note No. 2: An energy-reducing active arc flash mitigation system helps in reducing arcing duration in the electrical distribution system. No change in the disconnect switch or the settings of other devices is required during maintenance when a worker is working within an arc flash boundary as defined in *NFPA 70E* -2015, *Standard for Electrical Safety in the Workplace*.

Informational Note No. 3: IEEE 1584, *IEEE Guide for Performing Arc Flash Hazard Calculations*, is one of the available methods that provides guidance in determining arcing current.

Part VII. Circuit Breakers

240.80 Method of Operation. Circuit breakers shall be trip free and capable of being closed and opened by manual operation. Their normal method of operation by other than manual means, such as electrical or pneumatic, shall be permitted if means for manual operation are also provided.

240.81 Indicating. Circuit breakers shall clearly indicate whether they are in the open "off" or closed "on" position.

Where circuit breaker handles are operated vertically rather than rotationally or horizontally, the "up" position of the handle shall be the "on" position.

240.82 Nontamperable. A circuit breaker shall be of such design that any alteration of its trip point (calibration) or the time required for its operation requires dismantling of the device or breaking of a seal for other than intended adjustments.

240.83 Marking.

(A) Durable and Visible. Circuit breakers shall be marked with their ampere rating in a manner that will be durable and visible after installation. Such marking shall be permitted to be made visible by removal of a trim or cover.

(B) Location. Circuit breakers rated at 100 amperes or less and 1000 volts or less shall have the ampere rating molded, stamped, etched, or similarly marked into their handles or escutcheon areas.

(C) Interrupting Rating. Every circuit breaker having an interrupting rating other than 5000 amperes shall have its interrupting rating shown on the circuit breaker. The interrupting rating shall not be required to be marked on circuit breakers used for supplementary protection.

(D) Used as Switches. Circuit breakers used as switches in 120-volt and 277-volt fluorescent lighting circuits shall be listed and shall be marked SWD or HID. Circuit breakers used as switches in high-intensity discharge lighting circuits shall be listed and shall be marked as HID.

(E) Voltage Marking. Circuit breakers shall be marked with a voltage rating not less than the nominal system voltage that is indicative of their capability to interrupt fault currents between phases or phase to ground.

240.85 Applications. A circuit breaker with a straight voltage rating, such as 240V or 480V, shall be permitted to be applied in a circuit in which the nominal voltage between any two conductors does not exceed the circuit breaker's voltage rating. A two-pole circuit breaker shall not be used for protecting a 3-phase, corner-grounded delta circuit unless the circuit breaker is marked 1φ–3φ to indicate such suitability.

A circuit breaker with a slash rating, such as 120/240V or 480Y/277V, shall be permitted to be applied in a solidly grounded circuit where the nominal voltage of any conductor to ground does not exceed the lower of the two values of the circuit breaker's voltage rating and the nominal voltage between any two conductors does not exceed the higher value of the circuit breaker's voltage rating.

> Informational Note: Proper application of molded case circuit breakers on 3-phase systems, other than solidly grounded wye, particularly on corner grounded delta systems, considers the circuit breakers' individual pole-interrupting capability.

240.86 Series Ratings. Where a circuit breaker is used on a circuit having an available fault current higher than the marked interrupting rating by being connected on the load side of an acceptable overcurrent protective device having a higher rating, the circuit breaker shall meet the requirements specified in (A) or (B), and (C).

(A) Selected Under Engineering Supervision in Existing Installations. The series rated combination devices shall be selected by a licensed professional engineer engaged primarily in the design or maintenance of electrical installations. The selection shall be documented and stamped by the professional engineer. This documentation shall be available to those authorized to design, install, inspect, maintain, and operate the system. This series combination rating, including identification of the upstream device, shall be field marked on the end use equipment.

For calculated applications, the engineer shall ensure that the downstream circuit breaker(s) that are part of the series combination remain passive during the interruption period of the line side fully rated, current-limiting device.

(B) Tested Combinations. The combination of line-side overcurrent device and load-side circuit breaker(s) is tested and marked on the end use equipment, such as switchboards and panelboards.

> Informational Note to (A) and (B): See 110.22 for marking of series combination systems.

(C) Motor Contribution. Series ratings shall not be used where

(1) Motors are connected on the load side of the higher-rated overcurrent device and on the line side of the lower-rated overcurrent device, and

(2) The sum of the motor full-load currents exceeds 1 percent of the interrupting rating of the lower-rated circuit breaker.

240.87 Arc Energy Reduction. Where the highest continuous current trip setting for which the actual overcurrent device installed in a circuit breaker is rated or can be adjusted is 1200 A or higher, 240.87(A) and (B) shall apply.

(A) Documentation. Documentation shall be available to those authorized to design, install, operate, or inspect the installation as to the location of the circuit breaker(s).

(B) Method to Reduce Clearing Time. One of the following means shall be provided:

(1) Zone-selective interlocking
(2) Differential relaying
(3) Energy-reducing maintenance switching with local status indicator
(4) Energy-reducing active arc flash mitigation system
(5) An instantaneous trip setting that is less than the available arcing current
(6) An instantaneous override that is less than the available arcing current
(7) An approved equivalent means

> Informational Note No. 1: An energy-reducing maintenance switch allows a worker to set a circuit breaker trip unit to "no intentional delay" to reduce the clearing time while the worker is working within an arc-flash boundary as defined in *NFPA 70E -2015, Standard for Electrical Safety in the Workplace*, and then to set the trip unit back to a normal setting after the potentially hazardous work is complete.

> Informational Note No. 2: An energy-reducing active arc flash mitigation system helps in reducing arcing duration in the electrical distribution system. No change in the circuit breaker or the settings of other devices is required during maintenance when a worker is working within an arc flash boundary as defined in *NFPA 70E -2015, Standard for Electrical Safety in the Workplace*.

> Informational Note No. 3: An instantaneous trip is a function that causes a circuit breaker to trip with no intentional delay when currents exceed the instantaneous trip setting or current level. If arcing currents are above the instantaneous trip level, the circuit breaker will trip in the minimum possible time.

> Informational Note No. 4: IEEE 1584–2002, *IEEE Guide for Performing Arc Flash Hazard Calculations*, is one of the available methods that provide guidance in determining arcing current.

Part VIII. Supervised Industrial Installations

240.90 General. Overcurrent protection in areas of supervised industrial installations shall comply with all of the other applicable provisions of this article, except as provided in Part VIII. The provisions of Part VIII shall be permitted to apply only to those portions of the electrical system in the supervised industrial installation used exclusively for manufacturing or process control activities.

240.91 Protection of Conductors. Conductors shall be protected in accordance with 240.91(A) or 240.91(B)

(A) General. Conductors shall be protected in accordance with 240.4.

(B) Devices Rated Over 800 Amperes. Where the overcurrent device is rated over 800 amperes, the ampacity of the conductors it protects shall be equal to or greater than 95 percent of the rating of the overcurrent device specified in 240.6 in accordance with (B)(1) and (2).

(1) The conductors are protected within recognized time vs. current limits for short-circuit currents

(2) All equipment in which the conductors terminate is listed and marked for the application

240.92 Location in Circuit. An overcurrent device shall be connected in each ungrounded circuit conductor as required in 240.92(A) through (E).

(A) Feeder and Branch-Circuit Conductors. Feeder and branch-circuit conductors shall be protected at the point the conductors receive their supply as permitted in 240.21 or as otherwise permitted in 240.92(B), (C), (D), or (E).

(B) Feeder Taps. For feeder taps specified in 240.21(B)(2), (B)(3), and (B)(4), the tap conductors shall be permitted to be sized in accordance with Table 240.92(B).

(C) Transformer Secondary Conductors of Separately Derived Systems. Conductors shall be permitted to be connected to a transformer secondary of a separately derived system, without overcurrent protection at the connection, where the conditions of 240.92(C)(1), (C)(2), and (C)(3) are met.

(1) Short-Circuit and Ground-Fault Protection. The conductors shall be protected from short-circuit and ground-fault conditions by complying with one of the following conditions:

(1) The length of the secondary conductors does not exceed 30 m (100 ft), and the transformer primary overcurrent device has a rating or setting that does not exceed 150 percent of the value determined by multiplying the secondary conductor ampacity by the secondary-to-primary transformer voltage ratio.

(2) The conductors are protected by a differential relay with a trip setting equal to or less than the conductor ampacity.

Table 240.92(B) Tap Conductor Short-Circuit Current Ratings

Tap conductors are considered to be protected under short-circuit conditions when their short-circuit temperature limit is not exceeded. Conductor heating under short-circuit conditions is determined by (1) or (2):

(1) *Short-Circuit Formula for Copper Conductors*
$$(I^2/A^2)t = 0.0297 \log_{10}[(T_2 + 234)/(T_1 + 234)]$$
(2) *Short-Circuit Formula for Aluminum Conductors*
$$(I^2/A^2)t = 0.0125 \log_{10}[(T_2 + 228)/(T_1 + 228)]$$

where:

I = short-circuit current in amperes

A = conductor area in circular mils

t = time of short circuit in seconds (for times less than or equal to 10 seconds)

T_1 = initial conductor temperature in degrees Celsius

T_2 = final conductor temperature in degrees Celsius

Copper conductor with paper, rubber, varnished cloth insulation, $T_2 = 200$

Copper conductor with thermoplastic insulation, $T_2 = 150$

Copper conductor with cross-linked polyethylene insulation, $T_2 = 250$

Copper conductor with ethylene propylene rubber insulation, $T_2 = 250$

Aluminum conductor with paper, rubber, varnished cloth insulation, $T_2 = 200$

Aluminum conductor with thermoplastic insulation, $T_2 = 150$

Aluminum conductor with cross-linked polyethylene insulation, $T_2 = 250$

Aluminum conductor with ethylene propylene rubber insulation, $T_2 = 250$

Informational Note: A differential relay is connected to be sensitive only to short-circuit or fault currents within the protected zone and is normally set much lower than the conductor ampacity. The differential relay is connected to trip protective devices that de-energize the protected conductors if a short-circuit condition occurs.

(3) The conductors shall be considered to be protected if calculations, made under engineering supervision, determine that the system overcurrent devices will protect the conductors within recognized time vs. current limits for all short-circuit and ground-fault conditions.

(2) Overload Protection. The conductors shall be protected against overload conditions by complying with one of the following:

(1) The conductors terminate in a single overcurrent device that will limit the load to the conductor ampacity.

(2) The sum of the overcurrent devices at the conductor termination limits the load to the conductor ampacity. The overcurrent devices shall consist of not more than six circuit breakers or sets of fuses mounted in a single enclosure, in a group of separate enclosures, or in or on a switchboard or switchgear. There shall be no more than six overcurrent devices grouped in any one location.

(3) Overcurrent relaying is connected [with a current transformer(s), if needed] to sense all of the secondary conductor current and limit the load to the conductor ampacity by opening upstream or downstream devices.

(4) Conductors shall be considered to be protected if calculations, made under engineering supervision, determine that the system overcurrent devices will protect the conductors from overload conditions.

(3) Physical Protection. The secondary conductors are protected from physical damage by being enclosed in an approved raceway or by other approved means.

(D) Outside Feeder Taps. Outside conductors shall be permitted to be tapped to a feeder or to be connected at a transformer secondary, without overcurrent protection at the tap or connection, where all the following conditions are met:

(1) The conductors are protected from physical damage in an approved manner.

(2) The sum of the overcurrent devices at the conductor termination limits the load to the conductor ampacity. The overcurrent devices shall consist of not more than six circuit breakers or sets of fuses mounted in a single enclosure, in a group of separate enclosures, or in or on a switchboard or switchgear. There shall be no more than six overcurrent devices grouped in any one location.

(3) The tap conductors are installed outdoors of a building or structure except at the point of load termination.

(4) The overcurrent device for the conductors is an integral part of a disconnecting means or is located immediately adjacent thereto.

(5) The disconnecting means for the conductors are installed at a readily accessible location complying with one of the following:

a. Outside of a building or structure

b. Inside, nearest the point of entrance of the conductors

c. Where installed in accordance with 230.6, nearest the point of entrance of the conductors

(E) Protection by Primary Overcurrent Device. Conductors supplied by the secondary side of a transformer shall be permitted to be protected by overcurrent protection provided on the primary (supply) side of the transformer, provided the primary device time–current protection characteristic, multiplied by the maximum effective primary-to-secondary transformer voltage ratio, effectively protects the secondary conductors.

Part IX. Overcurrent Protection over 1000 Volts, Nominal

240.100 Feeders and Branch Circuits.

(A) Location and Type of Protection. Feeder and branch-circuit conductors shall have overcurrent protection in each ungrounded conductor located at the point where the conductor receives its supply or at an alternative location in the circuit when designed under engineering supervision that includes but is not limited to considering the appropriate fault studies and time–current coordination analysis of the protective devices and the conductor damage curves. The overcurrent protection shall be permitted to be provided by either 240.100(A)(1) or (A)(2).

(1) Overcurrent Relays and Current Transformers. Circuit breakers used for overcurrent protection of 3-phase circuits shall have a minimum of three overcurrent relay elements operated from three current transformers. The separate overcurrent relay elements (or protective functions) shall be permitted to be part of a single electronic protective relay unit.

On 3-phase, 3-wire circuits, an overcurrent relay element in the residual circuit of the current transformers shall be permitted to replace one of the phase relay elements.

An overcurrent relay element, operated from a current transformer that links all phases of a 3-phase, 3-wire circuit, shall be permitted to replace the residual relay element and one of the phase-conductor current transformers. Where the neutral conductor is not regrounded on the load side of the circuit as permitted in 250.184(B), the current transformer shall be permitted to link all 3-phase conductors and the grounded circuit conductor (neutral).

(2) Fuses. A fuse shall be connected in series with each ungrounded conductor.

(B) Protective Devices. The protective device(s) shall be capable of detecting and interrupting all values of current that can occur at their location in excess of their trip-setting or melting point.

(C) Conductor Protection. The operating time of the protective device, the available short-circuit current, and the conductor used shall be coordinated to prevent damaging or dangerous temperatures in conductors or conductor insulation under short-circuit conditions.

240.101 Additional Requirements for Feeders.

(A) Rating or Setting of Overcurrent Protective Devices. The continuous ampere rating of a fuse shall not exceed three times the ampacity of the conductors. The long-time trip element setting of a breaker or the minimum trip setting of an electronically actuated fuse shall not exceed six times the ampacity of the conductor. For fire pumps, conductors shall be permitted to be protected for overcurrent in accordance with 695.4(B)(2).

(B) Feeder Taps. Conductors tapped to a feeder shall be permitted to be protected by the feeder overcurrent device where that overcurrent device also protects the tap conductor.

ARTICLE 250
Grounding and Bonding

Part I. General

250.1 Scope. This article covers general requirements for grounding and bonding of electrical installations, and the specific requirements in (1) through (6).

(1) Systems, circuits, and equipment required, permitted, or not permitted to be grounded
(2) Circuit conductor to be grounded on grounded systems
(3) Location of grounding connections
(4) Types and sizes of grounding and bonding conductors and electrodes
(5) Methods of grounding and bonding
(6) Conditions under which guards, isolation, or insulation may be substituted for grounding

Informational Note: See Figure 250.1 for information on the organization of Article 250 covering grounding and bonding requirements.

250.2 Definition.

Bonding Jumper, Supply-Side. A conductor installed on the supply side of a service or within a service equipment enclosure(s), or for a separately derived system, that ensures the required electrical conductivity between metal parts required to be electrically connected.

250.3 Application of Other Articles. For other articles applying to particular cases of installation of conductors and equipment, grounding and bonding requirements are identified in Table 250.3 that are in addition to, or modifications of, those of this article.

250.4 General Requirements for Grounding and Bonding. The following general requirements identify what grounding and bonding of electrical systems are required to accomplish. The prescriptive methods contained in Article 250 shall be followed to comply with the performance requirements of this section.

(A) Grounded Systems.

(1) Electrical System Grounding. Electrical systems that are grounded shall be connected to earth in a manner that will limit the voltage imposed by lightning, line surges, or unintentional contact with higher-voltage lines and that will stabilize the voltage to earth during normal operation.

Informational Note No. 1: An important consideration for limiting the imposed voltage is the routing of bonding and grounding electrode conductors so that they are not any longer than necessary to complete the connection without disturbing the permanent parts of the installation and so that unnecessary bends and loops are avoided.

Informational Note No. 2: See NFPA 780-2014, *Standard for the Installation of Lightning Protection Systems,* for information on installation of grounding and bonding for lightning protection systems.

(2) Grounding of Electrical Equipment. Normally non–current-carrying conductive materials enclosing electrical conductors or equipment, or forming part of such equipment, shall be connected to earth so as to limit the voltage to ground on these materials.

(3) Bonding of Electrical Equipment. Normally non–current-carrying conductive materials enclosing electrical conductors or equipment, or forming part of such equipment, shall be connected together and to the electrical supply source in a manner that establishes an effective ground-fault current path.

(4) Bonding of Electrically Conductive Materials and Other Equipment. Normally non–current-carrying electrically conductive materials that are likely to become energized shall be connected together and to the electrical supply source in a manner that establishes an effective ground-fault current path.

(5) Effective Ground-Fault Current Path. Electrical equipment and wiring and other electrically conductive material likely to become energized shall be installed in a manner that creates a low-impedance circuit facilitating the operation of the overcurrent device or ground detector for high-impedance grounded systems. It shall be capable of safely carrying the maximum ground-fault current likely to be imposed on it from any point on the wiring system where a ground fault may occur

to the electrical supply source. The earth shall not be considered as an effective ground-fault current path.

(B) Ungrounded Systems.

(1) Grounding Electrical Equipment. Non–current-carrying conductive materials enclosing electrical conductors or equipment, or forming part of such equipment, shall be connected to earth in a manner that will limit the voltage imposed by lightning or unintentional contact with higher-voltage lines and limit the voltage to ground on these materials.

> Informational Note: See NFPA 780-2014, *Standard for the Installation of Lightning Protection Systems*, for information on installation of grounding and bonding for lightning protection systems.

(2) Bonding of Electrical Equipment. Non–current-carrying conductive materials enclosing electrical conductors or equipment, or forming part of such equipment, shall be connected together and to the supply system grounded equipment in a manner that creates a low-impedance path for ground-fault current that is capable of carrying the maximum fault current likely to be imposed on it.

(3) Bonding of Electrically Conductive Materials and Other Equipment. Electrically conductive materials that are likely to become energized shall be connected together and to the supply system grounded equipment in a manner that creates a low-impedance path for ground-fault current that is capable of carrying the maximum fault current likely to be imposed on it.

(4) Path for Fault Current. Electrical equipment, wiring, and other electrically conductive material likely to become energized shall be installed in a manner that creates a low-impedance circuit from any point on the wiring system to the electrical supply source to facilitate the operation of overcurrent devices should a second ground fault from a different phase occur on the wiring system. The earth shall not be considered as an effective fault-current path.

250.6 Objectionable Current.

(A) Arrangement to Prevent Objectionable Current. The grounding of electrical systems, circuit conductors, surge arresters, surge-protective devices, and conductive normally non–current-carrying metal parts of equipment shall be installed and arranged in a manner that will prevent objectionable current.

(B) Alterations to Stop Objectionable Current. If the use of multiple grounding connections results in objectionable current and the requirements of 250.4(A)(5) or (B)(4) are met, one or more of the following alterations shall be permitted:

(1) Discontinue one or more but not all of such grounding connections.
(2) Change the locations of the grounding connections.
(3) Interrupt the continuity of the conductor or conductive path causing the objectionable current.
(4) Take other suitable remedial and approved action.

(C) Temporary Currents Not Classified as Objectionable Currents. Temporary currents resulting from abnormal conditions, such as ground faults, shall not be classified as objectionable current for the purposes specified in 250.6(A) and (B).

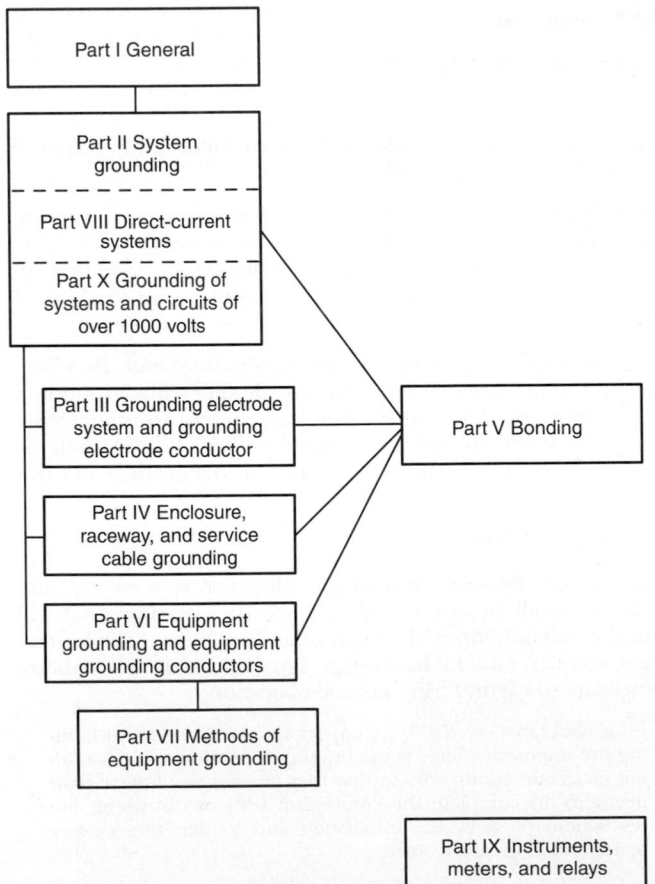

FIGURE 250.1   Grounding and Bonding.

Table 250.3 Additional Grounding and Bonding Requirements

Conductor/Equipment	Article	Section
Agricultural buildings		547.9 and 547.10
Audio signal processing, amplification, and reproduction equipment		640.7
Branch circuits		210.5, 210.6, 406.3
Cablebus		370.9
Cable trays	392	392.60
Capacitors		460.10, 460.27
Circuits and equipment operating at less than 50 volts	720	
Communications circuits	800	
Community antenna television and radio distribution systems		820.93, 820.100, 820.103, 820.106
Conductors for general wiring	310	
Cranes and hoists	610	
Electrically driven or controlled irrigation machines		675.11(C), 675.12, 675.13, 675.14, 675.15
Electric signs and outline lighting	600	
Electrolytic cells	668	
Elevators, dumbwaiters, escalators, moving walks, wheelchair lifts, and stairway chairlifts	620	
Fixed electric heating equipment for pipelines and vessels		427.29, 427.48
Fixed outdoor electric deicing and snow-melting equipment		426.27
Flexible cords and cables		400.22, 400.23
Floating buildings		553.8, 553.10, 553.11
Grounding-type receptacles, adapters, cord connectors, and attachment plugs		406.9
Hazardous (classified) locations	500–517	
Health care facilities	517	
Induction and dielectric heating equipment	665	
Industrial machinery	670	
Information technology equipment		645.15
Intrinsically safe systems		504.50
Luminaires and lighting equipment		410.40, 410.42, 410.46, 410.155(B)
Luminaires, lampholders, and lamps	410	
Marinas and boatyards		555.15
Mobile homes and mobile home park	550	
Motion picture and television studios and similar locations		530.20, 530.64(B)
Motors, motor circuits, and controllers	430	
Natural and artificially made bodies of water	682	682.30, 682.31, 682.32, 682.33
Network powered broadband communications circuits		830.93, 830.100, 830.106
Optical fiber cables		770.100
Outlet, device, pull, and junction boxes; conduit bodies; and fittings		314.4, 314.25
Over 600 volts, nominal, underground wiring methods		300.50(C)
Panelboards		408.40
Pipe organs	650	
Radio and television equipment	810	
Receptacles and cord connectors		406.3
Recreational vehicles and recreational vehicle parks	551	
Services	230	
Solar photovoltaic systems		690.41, 690.42, 690.43, 690.45, 690.47
Swimming pools, fountains, and similar installations	680	
Switchboards and panelboards		408.3(D)
Switches		404.12
Theaters, audience areas of motion picture and television studios, and similar locations		520.81
Transformers and transformer vaults		450.10
Use and identification of grounded conductors	200	
X-ray equipment	660	517.78

(D) Limitations to Permissible Alterations. The provisions of this section shall not be considered as permitting electronic equipment from being operated on ac systems or branch circuits that are not connected to an equipment grounding conductor as required by this article. Currents that introduce noise or data errors in electronic equipment shall not be considered the objectionable currents addressed in this section.

(E) Isolation of Objectionable Direct-Current Ground Currents. Where isolation of objectionable dc ground currents from cathodic protection systems is required, a listed ac coupling/dc isolating device shall be permitted in the equipment grounding conductor path to provide an effective return path for ac ground-fault current while blocking dc current.

250.8 Connection of Grounding and Bonding Equipment.

(A) Permitted Methods. Equipment grounding conductors, grounding electrode conductors, and bonding jumpers shall be connected by one or more of the following means:

(1) Listed pressure connectors
(2) Terminal bars
(3) Pressure connectors listed as grounding and bonding equipment
(4) Exothermic welding process
(5) Machine screw-type fasteners that engage not less than two threads or are secured with a nut
(6) Thread-forming machine screws that engage not less than two threads in the enclosure
(7) Connections that are part of a listed assembly
(8) Other listed means

(B) Methods Not Permitted. Connection devices or fittings that depend solely on solder shall not be used.

250.10 Protection of Ground Clamps and Fittings. Ground clamps or other fittings exposed to physical damage shall be enclosed in metal, wood, or equivalent protective covering.

250.12 Clean Surfaces. Nonconductive coatings (such as paint, lacquer, and enamel) on equipment to be grounded shall be removed from threads and other contact surfaces to ensure good electrical continuity or be connected by means of fittings designed so as to make such removal unnecessary.

Part II. System Grounding

250.20 Alternating-Current Systems to Be Grounded. Alternating-current systems shall be grounded as provided for in 250.20(A), (B), (C), or (D). Other systems shall be permitted to be grounded. If such systems are grounded, they shall comply with the applicable provisions of this article.

> Informational Note: An example of a system permitted to be grounded is a corner-grounded delta transformer connection. See 250.26(4) for conductor to be grounded.

(A) Alternating-Current Systems of Less Than 50 Volts. Alternating-current systems of less than 50 volts shall be grounded under any of the following conditions:

(1) Where supplied by transformers, if the transformer supply system exceeds 150 volts to ground
(2) Where supplied by transformers, if the transformer supply system is ungrounded
(3) Where installed outside as overhead conductors

(B) Alternating-Current Systems of 50 Volts to 1000 Volts. Alternating-current systems of 50 volts to 1000 volts that supply premises wiring and premises wiring systems shall be grounded under any of the following conditions:

(1) Where the system can be grounded so that the maximum voltage to ground on the ungrounded conductors does not exceed 150 volts
(2) Where the system is 3-phase, 4-wire, wye connected in which the neutral conductor is used as a circuit conductor
(3) Where the system is 3-phase, 4-wire, delta connected in which the midpoint of one phase winding is used as a circuit conductor

(C) Alternating-Current Systems of over 1000 Volts. Alternating-current systems supplying mobile or portable equipment shall be grounded as specified in 250.188. Where supplying other than mobile or portable equipment, such systems shall be permitted to be grounded.

(D) Impedance Grounded Neutral Systems. Impedance grounded neutral systems shall be grounded in accordance with 250.36 or 250.187.

250.21 Alternating-Current Systems of 50 Volts to 1000 Volts Not Required to Be Grounded.

(A) General. The following ac systems of 50 volts to 1000 volts shall be permitted to be grounded but shall not be required to be grounded:

(1) Electrical systems used exclusively to supply industrial electric furnaces for melting, refining, tempering, and the like
(2) Separately derived systems used exclusively for rectifiers that supply only adjustable-speed industrial drives
(3) Separately derived systems supplied by transformers that have a primary voltage rating of 1000 volts or less if all the following conditions are met:
   a. The system is used exclusively for control circuits.
   b. The conditions of maintenance and supervision ensure that only qualified persons service the installation.
   c. Continuity of control power is required.
(4) Other systems that are not required to be grounded in accordance with the requirements of 250.20(B)

(B) Ground Detectors. Ground detectors shall be installed in accordance with 250.21(B)(1) and (B)(2).

(1) Ungrounded ac systems as permitted in 250.21(A)(1) through (A)(4) operating at not less than 120 volts and at 1000 volts or less shall have ground detectors installed on the system.
(2) The ground detection sensing equipment shall be connected as close as practicable to where the system receives its supply.

(C) Marking. Ungrounded systems shall be legibly marked "Caution: Ungrounded System Operating — _____Volts Between Conductors" at the source or first disconnecting means of the system. The marking shall be of sufficient durability to withstand the environment involved.

250.22 Circuits Not to Be Grounded. The following circuits shall not be grounded:

(1) Circuits for electric cranes operating over combustible fibers in Class III locations, as provided in 503.155

(2) Circuits in health care facilities as provided in 517.61 and 517.160

(3) Circuits for equipment within electrolytic cell line working zones as provided in Article 668

(4) Secondary circuits of lighting systems as provided in 411.6(A)

(5) Secondary circuits of lighting systems as provided in 680.23(A)(2)

(6) Class 2 load side circuits for suspended ceiling low-voltage power grid distribution systems as provided in 393.60(B)

250.24 Grounding Service-Supplied Alternating-Current Systems.

(A) System Grounding Connections. A premises wiring system supplied by a grounded ac service shall have a grounding electrode conductor connected to the grounded service conductor, at each service, in accordance with 250.24(A)(1) through (A)(5).

(1) General. The grounding electrode conductor connection shall be made at any accessible point from the load end of the overhead service conductors, service drop, underground service conductors, or service lateral to, including the terminal or bus to which the grounded service conductor is connected at the service disconnecting means.

> Informational Note: See definitions of *Service Conductors, Overhead; Service Conductors, Underground; Service Drop;* and *Service Lateral* in Article 100.

(2) Outdoor Transformer. Where the transformer supplying the service is located outside the building, at least one additional grounding connection shall be made from the grounded service conductor to a grounding electrode, either at the transformer or elsewhere outside the building.

Exception: The additional grounding electrode conductor connection shall not be made on high-impedance grounded neutral systems. The system shall meet the requirements of 250.36.

(3) Dual-Fed Services. For services that are dual fed (double ended) in a common enclosure or grouped together in separate enclosures and employing a secondary tie, a single grounding electrode conductor connection to the tie point of the grounded conductor(s) from each power source shall be permitted.

(4) Main Bonding Jumper as Wire or Busbar. Where the main bonding jumper specified in 250.28 is a wire or busbar and is installed from the grounded conductor terminal bar or bus to the equipment grounding terminal bar or bus in the service equipment, the grounding electrode conductor shall be permitted to be connected to the equipment grounding terminal, bar, or bus to which the main bonding jumper is connected.

(5) Load-Side Grounding Connections. A grounded conductor shall not be connected to normally non–current-carrying metal parts of equipment, to equipment grounding conductor(s), or be reconnected to ground on the load side of the service disconnecting means except as otherwise permitted in this article.

> Informational Note: See 250.30 for separately derived systems, 250.32 for connections at separate buildings or structures, and

250.142 for use of the grounded circuit conductor for grounding equipment.

(B) Main Bonding Jumper. For a grounded system, an unspliced main bonding jumper shall be used to connect the equipment grounding conductor(s) and the service-disconnect enclosure to the grounded conductor within the enclosure for each service disconnect in accordance with 250.28.

Exception No. 1: Where more than one service disconnecting means is located in an assembly listed for use as service equipment, an unspliced main bonding jumper shall bond the grounded conductor(s) to the assembly enclosure.

Exception No. 2: Impedance grounded neutral systems shall be permitted to be connected as provided in 250.36 and 250.187.

(C) Grounded Conductor Brought to Service Equipment. Where an ac system operating at 1000 volts or less is grounded at any point, the grounded conductor(s) shall be routed with the ungrounded conductors to each service disconnecting means and shall be connected to each disconnecting means grounded conductor(s) terminal or bus. A main bonding jumper shall connect the grounded conductor(s) to each service disconnecting means enclosure. The grounded conductor(s) shall be installed in accordance with 250.24(C)(1) through 250.24(C)(4).

Exception: Where two or more service disconnecting means are located in a single assembly listed for use as service equipment, it shall be permitted to connect the grounded conductor(s) to the assembly common grounded conductor(s) terminal or bus. The assembly shall include a main bonding jumper for connecting the grounded conductor(s) to the assembly enclosure.

(1) Sizing for a Single Raceway or Cable. The grounded conductor shall not be smaller than specified in Table 250.102(C)(1).

(2) Parallel Conductors in Two or More Raceways or Cables. If the ungrounded service-entrance conductors are installed in parallel in two or more raceways or cables, the grounded conductor shall also be installed in parallel. The size of the grounded conductor in each raceway or cable shall be based on the total circular mil area of the parallel ungrounded conductors in the raceway or cable, as indicated in 250.24(C)(1), but not smaller than 1/0 AWG.

> Informational Note: See 310.10(H) for grounded conductors connected in parallel.

(3) Delta-Connected Service. The grounded conductor of a 3-phase, 3-wire delta service shall have an ampacity not less than that of the ungrounded conductors.

(4) High Impedance. The grounded conductor on a high-impedance grounded neutral system shall be grounded in accordance with 250.36.

(D) Grounding Electrode Conductor. A grounding electrode conductor shall be used to connect the equipment grounding conductors, the service-equipment enclosures, and, where the system is grounded, the grounded service conductor to the grounding electrode(s) required by Part III of this article. This conductor shall be sized in accordance with 250.66.

High-impedance grounded neutral system connections shall be made as covered in 250.36.

(E) Ungrounded System Grounding Connections. A premises wiring system that is supplied by an ac service that is ungrounded shall have, at each service, a grounding electrode conductor connected to the grounding electrode(s) required by Part III of this article. The grounding electrode conductor shall be connected to a metal enclosure of the service conductors at any accessible point from the load end of the overhead service conductors, service drop, underground service conductors, or service lateral to the service disconnecting means.

250.26 Conductor to Be Grounded — Alternating-Current Systems. For ac premises wiring systems, the conductor to be grounded shall be as specified in the following:

(1) Single-phase, 2-wire — one conductor
(2) Single-phase, 3-wire — the neutral conductor
(3) Multiphase systems having one wire common to all phases — the neutral conductor
(4) Multiphase systems where one phase is grounded — one phase conductor
(5) Multiphase systems in which one phase is used as in (2) — the neutral conductor

250.28 Main Bonding Jumper and System Bonding Jumper. For a grounded system, main bonding jumpers and system bonding jumpers shall be installed as follows:

(A) Material. Main bonding jumpers and system bonding jumpers shall be of copper or other corrosion-resistant material. A main bonding jumper and a system bonding jumper shall be a wire, bus, screw, or similar suitable conductor.

(B) Construction. Where a main bonding jumper or a system bonding jumper is a screw only, the screw shall be identified with a green finish that shall be visible with the screw installed.

(C) Attachment. Main bonding jumpers and system bonding jumpers shall be connected in the manner specified by the applicable provisions of 250.8.

(D) Size. Main bonding jumpers and system bonding jumpers shall be sized in accordance with 250.28(D)(1) through (D)(3).

(1) General. Main bonding jumpers and system bonding jumpers shall not be smaller than specified in Table 250.102(C)(1).

(2) Main Bonding Jumper for Service with More Than One Enclosure. Where a service consists of more than a single enclosure as permitted in 230.71(A), the main bonding jumper for each enclosure shall be sized in accordance with 250.28(D)(1) based on the largest ungrounded service conductor serving that enclosure.

(3) Separately Derived System with More Than One Enclosure. Where a separately derived system supplies more than a single enclosure, the system bonding jumper for each enclosure shall be sized in accordance with 250.28(D)(1) based on the largest ungrounded feeder conductor serving that enclosure, or a single system bonding jumper shall be installed at the source and sized in accordance with 250.28(D)(1) based on the equivalent size of the largest supply conductor determined by the largest sum of the areas of the corresponding conductors of each set.

250.30 Grounding Separately Derived Alternating-Current Systems. In addition to complying with 250.30(A) for grounded systems, or as provided in 250.30(B) for ungrounded

systems, separately derived systems shall comply with 250.20, 250.21, 250.22, or 250.26, as applicable. Multiple separately derived systems that are connected in parallel shall be installed in accordance with 250.30.

> Informational Note No. 1: An alternate ac power source, such as an on-site generator, is not a separately derived system if the grounded conductor is solidly interconnected to a service-supplied system grounded conductor. An example of such a situation is where alternate source transfer equipment does not include a switching action in the grounded conductor and allows it to remain solidly connected to the service-supplied grounded conductor when the alternate source is operational and supplying the load served.

> Informational Note No. 2: See 445.13 for the minimum size of conductors that carry fault current.

(A) Grounded Systems. A separately derived ac system that is grounded shall comply with 250.30(A)(1) through (A)(8). Except as otherwise permitted in this article, a grounded conductor shall not be connected to normally non–current-carrying metal parts of equipment, be connected to equipment grounding conductors, or be reconnected to ground on the load side of the system bonding jumper.

> Informational Note: See 250.32 for connections at separate buildings or structures and 250.142 for use of the grounded circuit conductor for grounding equipment.

Exception: Impedance grounded neutral system grounding connections shall be made as specified in 250.36 or 250.187, as applicable.

(1) System Bonding Jumper. An unspliced system bonding jumper shall comply with 250.28(A) through (D). This connection shall be made at any single point on the separately derived system from the source to the first system disconnecting means or overcurrent device, or it shall be made at the source of a separately derived system that has no disconnecting means or overcurrent devices, in accordance with 250.30(A)(1)(a) or (b). The system bonding jumper shall remain within the enclosure where it originates. If the source is located outside the building or structure supplied, a system bonding jumper shall be installed at the grounding electrode connection in compliance with 250.30(C).

Exception No. 1: For systems installed in accordance with 450.6, a single system bonding jumper connection to the tie point of the grounded circuit conductors from each power source shall be permitted.

Exception No. 2: If a building or structure is supplied by a feeder from an outdoor separately derived system, a system bonding jumper at both the source and the first disconnecting means shall be permitted if doing so does not establish a parallel path for the grounded conductor. If a grounded conductor is used in this manner, it shall not be smaller than the size specified for the system bonding jumper but shall not be required to be larger than the ungrounded conductor(s). For the purposes of this exception, connection through the earth shall not be considered as providing a parallel path.

Exception No. 3: The size of the system bonding jumper for a system that supplies a Class 1, Class 2, or Class 3 circuit, and is derived from a transformer rated not more than 1000 volt-amperes, shall not be smaller than the derived ungrounded conductors and shall not be smaller than 14 AWG copper or 12 AWG aluminum.

(a)   *Installed at the Source.* The system bonding jumper shall connect the grounded conductor to the supply-side bonding jumper and the normally non–current-carrying metal enclosure.

(b) *Installed at the First Disconnecting Means.* The system bonding jumper shall connect the grounded conductor to the supply-side bonding jumper, the disconnecting means enclosure, and the equipment grounding conductor(s).

(2) Supply-Side Bonding Jumper. If the source of a separately derived system and the first disconnecting means are located in separate enclosures, a supply-side bonding jumper shall be installed with the circuit conductors from the source enclosure to the first disconnecting means. A supply-side bonding jumper shall not be required to be larger than the derived ungrounded conductors. The supply-side bonding jumper shall be permitted to be of nonflexible metal raceway type or of the wire or bus type as follows:

(a) A supply-side bonding jumper of the wire type shall comply with 250.102(C), based on the size of the derived ungrounded conductors.

(b) A supply-side bonding jumper of the bus type shall have a cross-sectional area not smaller than a supply-side bonding jumper of the wire type as determined in 250.102(C).

Exception: A supply-side bonding jumper shall not be required between enclosures for installations made in compliance with 250.30(A)(1), Exception No. 2.

(3) Grounded Conductor. If a grounded conductor is installed and the system bonding jumper connection is not located at the source, 250.30(A)(3)(a) through (A)(3)(d) shall apply.

(a) *Sizing for a Single Raceway.* The grounded conductor shall not be smaller than specified in Table 250.102(C)(1).

(b) *Parallel Conductors in Two or More Raceways.* If the ungrounded conductors are installed in parallel in two or more raceways, the grounded conductor shall also be installed in parallel. The size of the grounded conductor in each raceway shall be based on the total circular mil area of the parallel derived ungrounded conductors in the raceway as indicated in 250.30(A)(3)(a), but not smaller than 1/0 AWG.

Informational Note: See 310.10(H) for grounded conductors connected in parallel.

(c) *Delta-Connected System.* The grounded conductor of a 3-phase, 3-wire delta system shall have an ampacity not less than that of the ungrounded conductors.

(d) *Impedance Grounded System.* The grounded conductor of an impedance grounded neutral system shall be installed in accordance with 250.36 or 250.187, as applicable.

(4) Grounding Electrode. The building or structure grounding electrode system shall be used as the grounding electrode for the separately derived system. If located outdoors, the grounding electrode shall be in accordance with 250.30(C).

Exception: If a separately derived system originates in equipment that is listed and identified as suitable for use as service equipment, the grounding electrode used for the service or feeder equipment shall be permitted to be used as the grounding electrode for the separately derived system.

Informational Note No. 1: See 250.104(D) for bonding requirements for interior metal water piping in the area served by separately derived systems.

Informational Note No. 2: See 250.50 and 250.58 for requirements for bonding all electrodes together if located at the same building or structure.

(5) Grounding Electrode Conductor, Single Separately Derived System. A grounding electrode conductor for a single separately derived system shall be sized in accordance with 250.66 for the derived ungrounded conductors. It shall be used to connect the grounded conductor of the derived system to the grounding electrode in accordance with 250.30(A)(4) , or as permitted in 250.68(C)(1) and (2). This connection shall be made at the same point on the separately derived system where the system bonding jumper is connected.

Exception No. 1: If the system bonding jumper specified in 250.30(A)(1) is a wire or busbar, it shall be permitted to connect the grounding electrode conductor to the equipment grounding terminal, bar, or bus if the equipment grounding terminal, bar, or bus is of sufficient size for the separately derived system.

Exception No. 2: If the source of a separately derived system is located within equipment listed and identified as suitable for use as service equipment, the grounding electrode conductor from the service or feeder equipment to the grounding electrode shall be permitted as the grounding electrode conductor for the separately derived system, if the grounding electrode conductor is of sufficient size for the separately derived system. If the equipment grounding bus internal to the equipment is not smaller than the required grounding electrode conductor for the separately derived system, the grounding electrode connection for the separately derived system shall be permitted to be made to the bus.

Exception No. 3: A grounding electrode conductor shall not be required for a system that supplies a Class 1, Class 2, or Class 3 circuit and is derived from a transformer rated not more than 1000 volt-amperes, provided the grounded conductor is bonded to the transformer frame or enclosure by a jumper sized in accordance with 250.30(A)(1), Exception No. 3, and the transformer frame or enclosure is grounded by one of the means specified in 250.134.

(6) Grounding Electrode Conductor, Multiple Separately Derived Systems. A common grounding electrode conductor for multiple separately derived systems shall be permitted. If installed, the common grounding electrode conductor shall be used to connect the grounded conductor of the separately derived systems to the grounding electrode as specified in 250.30(A)(4). A grounding electrode conductor tap shall then be installed from each separately derived system to the common grounding electrode conductor. Each tap conductor shall connect the grounded conductor of the separately derived system to the common grounding electrode conductor. This connection shall be made at the same point on the separately derived system where the system bonding jumper is connected.

Exception No. 1: If the system bonding jumper specified in 250.30(A)(1) is a wire or busbar, it shall be permitted to connect the grounding electrode conductor tap to the equipment grounding terminal, bar, or bus, provided the equipment grounding terminal, bar, or bus is of sufficient size for the separately derived system.

Exception No. 2: A grounding electrode conductor shall not be required for a system that supplies a Class 1, Class 2, or Class 3 circuit and is derived from a transformer rated not more than 1000 volt-amperes, provided the system grounded conductor is bonded to the transformer frame or enclosure by a jumper sized in accordance with 250.30(A)(1), Exception No. 3, and the transformer frame or enclosure is grounded by one of the means specified in 250.134.

(a) *Common Grounding Electrode Conductor.* The common grounding electrode conductor shall be permitted to be one of the following:

(1) A conductor of the wire type not smaller than 3/0 AWG copper or 250 kcmil aluminum
(2) A metal water pipe that complies with 250.68(C)(1)

(3)   The metal structural frame of the building or structure that complies with 250.68(C)(2) or is connected to the grounding electrode system by a conductor not smaller than 3/0 AWG copper or 250 kcmil aluminum

(b)   *Tap Conductor Size.* Each tap conductor shall be sized in accordance with 250.66 based on the derived ungrounded conductors of the separately derived system it serves.

Exception: If the source of a separately derived system is located within equipment listed and identified as suitable for use as service equipment, the grounding electrode conductor from the service or feeder equipment to the grounding electrode shall be permitted as the grounding electrode conductor for the separately derived system, if the grounding electrode conductor is of sufficient size for the separately derived system. If the equipment grounding bus internal to the equipment is not smaller than the required grounding electrode conductor for the separately derived system, the grounding electrode connection for the separately derived system shall be permitted to be made to the bus.

(c)   *Connections.* All tap connections to the common grounding electrode conductor shall be made at an accessible location by one of the following methods:

(1)   A connector listed as grounding and bonding equipment.
(2)   Listed connections to aluminum or copper busbars not smaller than 6 mm thick × 50 mm wide (¼ in. thick × 2 in. wide) and of sufficient length to accommodate the number of terminations necessary for the installation. If aluminum busbars are used, the installation shall also comply with 250.64(A).
(3)   The exothermic welding process.

Tap conductors shall be connected to the common grounding electrode conductor in such a manner that the common grounding electrode conductor remains without a splice or joint.

(7) Installation. The installation of all grounding electrode conductors shall comply with 250.64(A), (B), (C), and (E).

(8) Bonding. Structural steel and metal piping shall be connected to the grounded conductor of a separately derived system in accordance with 250.104(D).

(B) Ungrounded Systems. The equipment of an ungrounded separately derived system shall be grounded and bonded as specified in 250.30(B)(1) through (B)(3).

(1) Grounding Electrode Conductor. A grounding electrode conductor, sized in accordance with 250.66 for the largest derived ungrounded conductor(s) or set of derived ungrounded conductors, shall be used to connect the metal enclosures of the derived system to the grounding electrode as specified in 250.30(A)(5) or (6), as applicable. This connection shall be made at any point on the separately derived system from the source to the first system disconnecting means. If the source is located outside the building or structure supplied, a grounding electrode connection shall be made in compliance with 250.30(C).

(2) Grounding Electrode. Except as permitted by 250.34 for portable and vehicle-mounted generators, the grounding electrode shall comply with 250.30(A)(4).

(3) Bonding Path and Conductor. A supply-side bonding jumper shall be installed from the source of a separately derived system to the first disconnecting means in compliance with 250.30(A)(2).

(C) Outdoor Source. If the source of the separately derived system is located outside the building or structure supplied, a grounding electrode connection shall be made at the source location to one or more grounding electrodes in compliance with 250.50. In addition, the installation shall comply with 250.30(A) for grounded systems or with 250.30(B) for ungrounded systems.

Exception: The grounding electrode conductor connection for impedance grounded neutral systems shall comply with 250.36 or 250.187, as applicable.

250.32 Buildings or Structures Supplied by a Feeder(s) or Branch Circuit(s).

(A) Grounding Electrode. Building(s) or structure(s) supplied by feeder(s) or branch circuit(s) shall have a grounding electrode or grounding electrode system installed in accordance with Part III of Article 250. The grounding electrode conductor(s) shall be connected in accordance with 250.32(B) or (C). Where there is no existing grounding electrode, the grounding electrode(s) required in 250.50 shall be installed.

Exception: A grounding electrode shall not be required where only a single branch circuit, including a multiwire branch circuit, supplies the building or structure and the branch circuit includes an equipment grounding conductor for grounding the normally non–current-carrying metal parts of equipment.

(B) Grounded Systems.

(1) Supplied by a Feeder or Branch Circuit. An equipment grounding conductor, as described in 250.118, shall be run with the supply conductors and be connected to the building or structure disconnecting means and to the grounding electrode(s). The equipment grounding conductor shall be used for grounding or bonding of equipment, structures, or frames required to be grounded or bonded. The equipment grounding conductor shall be sized in accordance with 250.122. Any installed grounded conductor shall not be connected to the equipment grounding conductor or to the grounding electrode(s).

Exception No. 1: For installations made in compliance with previous editions of this Code that permitted such connection, the grounded conductor run with the supply to the building or structure shall be permitted to serve as the ground-fault return path if all of the following requirements continue to be met:

(1)   An equipment grounding conductor is not run with the supply to the building or structure.
(2)   There are no continuous metallic paths bonded to the grounding system in each building or structure involved.
(3)   Ground-fault protection of equipment has not been installed on the supply side of the feeder(s).

If the grounded conductor is used for grounding in accordance with the provision of this exception, the size of the grounded conductor shall not be smaller than the larger of either of the following:

(1)   That required by 220.61
(2)   That required by 250.122

Exception No. 2: If system bonding jumpers are installed in accordance with 250.30(A)(1), Exception No. 2, the feeder grounded circuit conductor at the building or structure served shall be connected to the equipment grounding conductors, grounding electrode conductor, and the enclosure for the first disconnecting means.

(2) Supplied by Separately Derived System.

(a) *With Overcurrent Protection.* If overcurrent protection is provided where the conductors originate, the installation shall comply with 250.32(B)(1).

(b) *Without Overcurrent Protection.* If overcurrent protection is not provided where the conductors originate, the installation shall comply with 250.30(A). If installed, the supply-side bonding jumper shall be connected to the building or structure disconnecting means and to the grounding electrode(s).

(C) Ungrounded Systems.

(1) Supplied by a Feeder or Branch Circuit. An equipment grounding conductor, as described in 250.118, shall be installed with the supply conductors and be connected to the building or structure disconnecting means and to the grounding electrode(s). The grounding electrode(s) shall also be connected to the building or structure disconnecting means.

(2) Supplied by a Separately Derived System.

(a) *With Overcurrent Protection.* If overcurrent protection is provided where the conductors originate, the installation shall comply with (C)(1).

(b) *Without Overcurrent Protection.* If overcurrent protection is not provided where the conductors originate, the installation shall comply with 250.30(B). If installed, the supply-side bonding jumper shall be connected to the building or structure disconnecting means and to the grounding electrode(s).

(D) Disconnecting Means Located in Separate Building or Structure on the Same Premises. Where one or more disconnecting means supply one or more additional buildings or structures under single management, and where these disconnecting means are located remote from those buildings or structures in accordance with the provisions of 225.32, Exception No. 1 and No. 2, 700.12(B)(6), 701.12(B)(5), or 702.12, all of the following conditions shall be met:

(1) The connection of the grounded conductor to the grounding electrode, to normally non–current-carrying metal parts of equipment, or to the equipment grounding conductor at a separate building or structure shall not be made.

(2) An equipment grounding conductor for grounding and bonding any normally non–current-carrying metal parts of equipment, interior metal piping systems, and building or structural metal frames is run with the circuit conductors to a separate building or structure and connected to existing grounding electrode(s) required in Part III of this article, or, where there are no existing electrodes, the grounding electrode(s) required in Part III of this article shall be installed where a separate building or structure is supplied by more than one branch circuit.

(3) The connection between the equipment grounding conductor and the grounding electrode at a separate building or structure shall be made in a junction box, panelboard, or similar enclosure located immediately inside or outside the separate building or structure.

(E) Grounding Electrode Conductor. The size of the grounding electrode conductor to the grounding electrode(s) shall not be smaller than given in 250.66, based on the largest ungrounded supply conductor. The installation shall comply with Part III of this article.

250.34 Portable and Vehicle-Mounted Generators.

(A) Portable Generators. The frame of a portable generator shall not be required to be connected to a grounding electrode as defined in 250.52 for a system supplied by the generator under the following conditions:

(1) The generator supplies only equipment mounted on the generator, cord-and-plug-connected equipment through receptacles mounted on the generator, or both, and

(2) The normally non–current-carrying metal parts of equipment and the equipment grounding conductor terminals of the receptacles are connected to the generator frame.

(B) Vehicle-Mounted Generators. The frame of a vehicle shall not be required to be connected to a grounding electrode as defined in 250.52 for a system supplied by a generator located on this vehicle under the following conditions:

(1) The frame of the generator is bonded to the vehicle frame, and

(2) The generator supplies only equipment located on the vehicle or cord-and-plug-connected equipment through receptacles mounted on the vehicle, or both equipment located on the vehicle and cord-and-plug-connected equipment through receptacles mounted on the vehicle or on the generator, and

(3) The normally non–current-carrying metal parts of equipment and the equipment grounding conductor terminals of the receptacles are connected to the generator frame.

(C) Grounded Conductor Bonding. A system conductor that is required to be grounded by 250.26 shall be connected to the generator frame where the generator is a component of a separately derived system.

Informational Note: For grounding portable generators supplying fixed wiring systems, see 250.30.

250.35 Permanently Installed Generators. A conductor that provides an effective ground-fault current path shall be installed with the supply conductors from a permanently installed generator(s) to the first disconnecting mean(s) in accordance with (A) or (B).

(A) Separately Derived System. If the generator is installed as a separately derived system, the requirements in 250.30 shall apply.

(B) Nonseparately Derived System. If the generator is installed as a nonseparately derived system, and overcurrent protection is not integral with the generator assembly, a supply-side bonding jumper shall be installed between the generator equipment grounding terminal and the equipment grounding terminal, bar, or bus of the disconnecting mean(s). It shall be sized in accordance with 250.102(C) based on the size of the conductors supplied by the generator.

250.36 High-Impedance Grounded Neutral Systems. High-impedance grounded neutral systems in which a grounding impedance, usually a resistor, limits the ground-fault current to a low value shall be permitted for 3-phase ac systems of 480 volts to 1000 volts if all the following conditions are met:

(1) The conditions of maintenance and supervision ensure that only qualified persons service the installation.

(2) Ground detectors are installed on the system.

(3) Line-to-neutral loads are not served.

High-impedance grounded neutral systems shall comply with the provisions of 250.36(A) through (G).

(A) Location. The grounding impedance shall be installed between the grounding electrode conductor and the system neutral point. If a neutral point is not available, the grounding impedance shall be installed between the grounding electrode conductor and the neutral point derived from a grounding transformer.

(B) Conductor Insulation and Ampacity. The grounded system conductor from the neutral point of the transformer or generator to its connection point to the grounding impedance shall be fully insulated.

The grounded system conductor shall have an ampacity of not less than the maximum current rating of the grounding impedance but in no case shall the grounded system conductor be smaller than 8 AWG copper or 6 AWG aluminum or copper-clad aluminum.

(C) System Grounding Connection. The system shall not be connected to ground except through the grounding impedance.

> Informational Note: The impedance is normally selected to limit the ground-fault current to a value slightly greater than or equal to the capacitive charging current of the system. This value of impedance will also limit transient overvoltages to safe values. For guidance, refer to criteria for limiting transient overvoltages in ANSI/IEEE 142-2007, *Recommended Practice for Grounding of Industrial and Commercial Power Systems.*

(D) Conductor Routing. The conductor connecting the neutral point of the transformer or generator to the grounding impedance shall be permitted to be installed in a separate raceway from the ungrounded conductors. It shall not be required to run this conductor with the phase conductors to the first system disconnecting means or overcurrent device.

(E) Equipment Bonding Jumper. The equipment bonding jumper (the connection between the equipment grounding conductors and the grounding impedance) shall be an unspliced conductor run from the first system disconnecting means or overcurrent device to the grounded side of the grounding impedance.

(F) Grounding Electrode Conductor Connection Location. For services or separately derived systems, the grounding electrode conductor shall be connected at any point from the grounded side of the grounding impedance to the equipment grounding connection at the service equipment or the first system disconnecting means of a separately derived system.

(G) Equipment Bonding Jumper Size. The equipment bonding jumper shall be sized in accordance with (1) or (2) as follows:

(1)   If the grounding electrode conductor connection is made at the grounding impedance, the equipment bonding jumper shall be sized in accordance with 250.66, based on the size of the service entrance conductors for a service or the derived phase conductors for a separately derived system.

(2)   If the grounding electrode conductor is connected at the first system disconnecting means or overcurrent device, the equipment bonding jumper shall be sized the same as the neutral conductor in 250.36(B).

Part III. Grounding Electrode System and Grounding Electrode Conductor

250.50 Grounding Electrode System. All grounding electrodes as described in 250.52(A)(1) through (A)(7) that are present at each building or structure served shall be bonded together to form the grounding electrode system. Where none of these grounding electrodes exist, one or more of the grounding electrodes specified in 250.52(A)(4) through (A)(8) shall be installed and used.

Exception: Concrete-encased electrodes of existing buildings or structures shall not be required to be part of the grounding electrode system where the steel reinforcing bars or rods are not accessible for use without disturbing the concrete.

250.52 Grounding Electrodes.

(A) Electrodes Permitted for Grounding.

(1) Metal Underground Water Pipe. A metal underground water pipe in direct contact with the earth for 3.0 m (10 ft) or more (including any metal well casing bonded to the pipe) and electrically continuous (or made electrically continuous by bonding around insulating joints or insulating pipe) to the points of connection of the grounding electrode conductor and the bonding conductor(s) or jumper(s), if installed.

N (2) Metal In-ground Support Structure(s). One or more metal in-ground support structure(s) in direct contact with the earth vertically for 3.0 m (10 ft) or more, with or without concrete encasement. If multiple metal in-ground support structures are present at a building or a structure, it shall be permissible to bond only one into the grounding electrode system.

> Informational Note: Metal in-ground support structures include, but are not limited to, pilings, casings, and other structural metal.

(3) Concrete-Encased Electrode. A concrete-encased electrode shall consist of at least 6.0 m (20 ft) of either (1) or (2):

(1)   One or more bare or zinc galvanized or other electrically conductive coated steel reinforcing bars or rods of not less than 13 mm (½ in.) in diameter, installed in one continuous 6.0 m (20 ft) length, or if in multiple pieces connected together by the usual steel tie wires, exothermic welding, welding, or other effective means to create a 6.0 m (20 ft) or greater length; or

(2)   Bare copper conductor not smaller than 4 AWG

Metallic components shall be encased by at least 50 mm (2 in.) of concrete and shall be located horizontally within that portion of a concrete foundation or footing that is in direct contact with the earth or within vertical foundations or structural components or members that are in direct contact with the earth. If multiple concrete-encased electrodes are present at a building or structure, it shall be permissible to bond only one into the grounding electrode system.

> Informational Note: Concrete installed with insulation, vapor barriers, films or similar items separating the concrete from the earth is not considered to be in "direct contact" with the earth.

(4) Ground Ring. A ground ring encircling the building or structure, in direct contact with the earth, consisting of at least 6.0 m (20 ft) of bare copper conductor not smaller than 2 AWG.

(5) Rod and Pipe Electrodes. Rod and pipe electrodes shall not be less than 2.44 m (8 ft) in length and shall consist of the following materials.

(a) Grounding electrodes of pipe or conduit shall not be smaller than metric designator 21 (trade size ¾) and, where of steel, shall have the outer surface galvanized or otherwise metal-coated for corrosion protection.

(b) Rod-type grounding electrodes of stainless steel and copper or zinc coated steel shall be at least 15.87 mm (⅝ in.) in diameter, unless listed.

(6) Other Listed Electrodes. Other listed grounding electrodes shall be permitted.

(7) Plate Electrodes. Each plate electrode shall expose not less than 0.186 m² (2 ft²) of surface to exterior soil. Electrodes of bare or electrically conductive coated iron or steel plates shall be at least 6.4 mm (¼ in.) in thickness. Solid, uncoated electrodes of nonferrous metal shall be at least 1.5 mm (0.06 in.) in thickness.

(8) Other Local Metal Underground Systems or Structures. Other local metal underground systems or structures such as piping systems, underground tanks, and underground metal well casings that are not bonded to a metal water pipe.

(B) Not Permitted for Use as Grounding Electrodes. The following systems and materials shall not be used as grounding electrodes:

(1) Metal underground gas piping systems
(2) Aluminum
(3) The structures and structural reinforcing steel described in 680.26(B)(1) and (B)(2)

Informational Note: See 250.104(B) for bonding requirements of gas piping.

250.53 Grounding Electrode System Installation.

(A) Rod, Pipe, and Plate Electrodes. Rod, pipe, and plate electrodes shall meet the requirements of 250.53(A)(1) through (A)(3).

(1) Below Permanent Moisture Level. If practicable, rod, pipe, and plate electrodes shall be embedded below permanent moisture level. Rod, pipe, and plate electrodes shall be free from nonconductive coatings such as paint or enamel.

(2) Supplemental Electrode Required. A single rod, pipe, or plate electrode shall be supplemented by an additional electrode of a type specified in 250.52(A)(2) through (A)(8). The supplemental electrode shall be permitted to be bonded to one of the following:

(1) Rod, pipe, or plate electrode
(2) Grounding electrode conductor
(3) Grounded service-entrance conductor
(4) Nonflexible grounded service raceway
(5) Any grounded service enclosure

Exception: If a single rod, pipe, or plate grounding electrode has a resistance to earth of 25 ohms or less, the supplemental electrode shall not be required.

(3) Supplemental Electrode. If multiple rod, pipe, or plate electrodes are installed to meet the requirements of this section, they shall not be less than 1.8 m (6 ft) apart.

Informational Note: The paralleling efficiency of rods is increased by spacing them twice the length of the longest rod.

(B) Electrode Spacing. Where more than one of the electrodes of the type specified in 250.52(A)(5) or (A)(7) are used, each electrode of one grounding system (including that used for strike termination devices) shall not be less than 1.83 m (6 ft) from any other electrode of another grounding system. Two or more grounding electrodes that are bonded together shall be considered a single grounding electrode system.

(C) Bonding Jumper. The bonding jumper(s) used to connect the grounding electrodes together to form the grounding electrode system shall be installed in accordance with 250.64(A), (B), and (E), shall be sized in accordance with 250.66, and shall be connected in the manner specified in 250.70.

(D) Metal Underground Water Pipe. If used as a grounding electrode, metal underground water pipe shall meet the requirements of 250.53(D)(1) and (D)(2).

(1) Continuity. Continuity of the grounding path or the bonding connection to interior piping shall not rely on water meters or filtering devices and similar equipment.

(2) Supplemental Electrode Required. A metal underground water pipe shall be supplemented by an additional electrode of a type specified in 250.52(A)(2) through (A)(8). If the supplemental electrode is of the rod, pipe, or plate type, it shall comply with 250.53(A). The supplemental electrode shall be bonded to one of the following:

(1) Grounding electrode conductor
(2) Grounded service-entrance conductor
(3) Nonflexible grounded service raceway
(4) Any grounded service enclosure
(5) As provided by 250.32(B)

Exception: The supplemental electrode shall be permitted to be bonded to the interior metal water piping at any convenient point as specified in 250.68(C)(1), Exception.

(E) Supplemental Electrode Bonding Connection Size. Where the supplemental electrode is a rod, pipe, or plate electrode, that portion of the bonding jumper that is the sole connection to the supplemental grounding electrode shall not be required to be larger than 6 AWG copper wire or 4 AWG aluminum wire.

(F) Ground Ring. The ground ring shall be installed not less than 750 mm (30 in.) below the surface of the earth.

(G) Rod and Pipe Electrodes. The electrode shall be installed such that at least 2.44 m (8 ft) of length is in contact with the soil. It shall be driven to a depth of not less than 2.44 m (8 ft) except that, where rock bottom is encountered, the electrode shall be driven at an oblique angle not to exceed 45 degrees from the vertical or, where rock bottom is encountered at an angle up to 45 degrees, the electrode shall be permitted to be buried in a trench that is at least 750 mm (30 in.) deep. The upper end of the electrode shall be flush with or below ground level unless the aboveground end and the grounding electrode conductor attachment are protected against physical damage as specified in 250.10.

(H) Plate Electrode. Plate electrodes shall be installed not less than 750 mm (30 in.) below the surface of the earth.

250.54 Auxiliary Grounding Electrodes. One or more grounding electrodes shall be permitted to be connected to the equipment grounding conductors specified in 250.118 and shall not be required to comply with the electrode bonding requirements of 250.50 or 250.53(C) or the resistance requirements of 250.53(A)(2) Exception, but the earth shall not be used as an effective ground-fault current path as specified in 250.4(A)(5) and 250.4(B)(4).

250.58 Common Grounding Electrode. Where an ac system is connected to a grounding electrode in or at a building or structure, the same electrode shall be used to ground conductor enclosures and equipment in or on that building or structure. Where separate services, feeders, or branch circuits supply a building and are required to be connected to a grounding electrode(s), the same grounding electrode(s) shall be used.

Two or more grounding electrodes that are bonded together shall be considered as a single grounding electrode system in this sense.

250.60 Use of Strike Termination Devices. Conductors and driven pipes, rods, or plate electrodes used for grounding strike termination devices shall not be used in lieu of the grounding electrodes required by 250.50 for grounding wiring systems and equipment. This provision shall not prohibit the required bonding together of grounding electrodes of different systems.

> Informational Note No. 1: See 250.106 for the bonding requirement of the lightning protection system components to the building or structure grounding electrode system.

> Informational Note No. 2: Bonding together of all separate grounding electrodes will limit voltage differences between them and between their associated wiring systems.

250.62 Grounding Electrode Conductor Material. The grounding electrode conductor shall be of copper, aluminum, copper-clad aluminum, or the items as permitted in 250.68(C). The material selected shall be resistant to any corrosive condition existing at the installation or shall be protected against corrosion. Conductors of the wire type shall be solid or stranded, insulated, covered, or bare.

250.64 Grounding Electrode Conductor Installation. Grounding electrode conductors at the service, at each building or structure where supplied by a feeder(s) or branch circuit(s), or at a separately derived system shall be installed as specified in 250.64(A) through (F).

(A) Aluminum or Copper-Clad Aluminum Conductors. Bare aluminum or copper-clad aluminum grounding electrode conductors shall not be used where in direct contact with masonry or the earth or where subject to corrosive conditions. Where used outside, aluminum or copper-clad aluminum grounding electrode conductors shall not be terminated within 450 mm (18 in.) of the earth.

(B) Securing and Protection Against Physical Damage. Where exposed, a grounding electrode conductor or its enclosure shall be securely fastened to the surface on which it is carried. Grounding electrode conductors shall be permitted to be installed on or through framing members.

(1) Not Exposed to Physical Damage. A 6 AWG or larger copper or aluminum grounding electrode conductor not exposed to physical damage shall be permitted to be run along the surface of the building construction without metal covering or protection.

(2) Exposed to Physical Damage. A 6 AWG or larger copper or aluminum grounding electrode conductor exposed to physical damage shall be protected in rigid metal conduit (RMC),

intermediate metal conduit (IMC), rigid polyvinyl chloride conduit (PVC), reinforced thermosetting resin conduit Type XW (RTRC-XW), electrical metallic tubing (EMT), or cable armor.

(3) Smaller Than 6 AWG. Grounding electrode conductors smaller than 6 AWG shall be protected in RMC, IMC, PVC, RTRC-XW, EMT, or cable armor.

(4) In Contact with the Earth. Grounding electrode conductors and grounding electrode bonding jumpers in contact with the earth shall not be required to comply with 300.5, but shall be buried or otherwise protected if subject to physical damage.

(C) Continuous. Except as provided in 250.30(A)(5) and (A)(6), 250.30(B)(1), and 250.68(C), grounding electrode conductor(s) shall be installed in one continuous length without a splice or joint. If necessary, splices or connections shall be made as permitted in (1) through (4):

(1) Splicing of the wire-type grounding electrode conductor shall be permitted only by irreversible compression-type connectors listed as grounding and bonding equipment or by the exothermic welding process.

(2) Sections of busbars shall be permitted to be connected together to form a grounding electrode conductor.

(3) Bolted, riveted, or welded connections of structural metal frames of buildings or structures.

(4) Threaded, welded, brazed, soldered or bolted-flange connections of metal water piping.

(D) Building or Structure with Multiple Disconnecting Means in Separate Enclosures. If a building or structure is supplied by a service or feeder with two or more disconnecting means in separate enclosures, the grounding electrode connections shall be made in accordance with 250.64(D)(1), 250.64(D)(2), or 250.64(D)(3).

(1) Common Grounding Electrode Conductor and Taps. A common grounding electrode conductor and grounding electrode conductor taps shall be installed. The common grounding electrode conductor shall be sized in accordance with 250.66, based on the sum of the circular mil area of the largest ungrounded conductor(s) of each set of conductors that supplies the disconnecting means. If the service-entrance conductors connect directly to the overhead service conductors, service drop, underground service conductors, or service lateral, the common grounding electrode conductor shall be sized in accordance with Table 250.66, note 1.

A grounding electrode conductor tap shall extend to the inside of each disconnecting means enclosure. The grounding electrode conductor taps shall be sized in accordance with 250.66 for the largest service-entrance or feeder conductor serving the individual enclosure. The tap conductors shall be connected to the common grounding electrode conductor by one of the following methods in such a manner that the common grounding electrode conductor remains without a splice or joint:

(1) Exothermic welding.
(2) Connectors listed as grounding and bonding equipment.
(3) Connections to an aluminum or copper busbar not less than 6 mm thick × 50 mm wide (¼ in. thick × 2 in. wide) and of sufficient length to accommodate the number of terminations necessary for the installation. The busbar shall be securely fastened and shall be installed in an accessible location. Connections shall be made by a listed

connector or by the exothermic welding process. If aluminum busbars are used, the installation shall comply with 250.64(A).

(2) Individual Grounding Electrode Conductors. A grounding electrode conductor shall be connected between the grounding electrode system and one or more of the following, as applicable:

(1) Grounded conductor in each service equipment disconnecting means enclosure
(2) Equipment grounding conductor installed with the feeder
(3) Supply-side bonding jumper

Each grounding electrode conductor shall be sized in accordance with 250.66 based on the service-entrance or feeder conductor(s) supplying the individual disconnecting means.

(3) Common Location. A grounding electrode conductor shall be connected in a wireway or other accessible enclosure on the supply side of the disconnecting means to one or more of the following, as applicable:

(1) Grounded service conductor(s)
(2) Equipment grounding conductor installed with the feeder
(3) Supply-side bonding jumper

The connection shall be made with exothermic welding or a connector listed as grounding and bonding equipment. The grounding electrode conductor shall be sized in accordance with 250.66 based on the service-entrance or feeder conductor(s) at the common location where the connection is made.

(E) Raceways and Enclosures for Grounding Electrode Conductors.

(1) General. Ferrous metal raceways and enclosures for grounding electrode conductors shall be electrically continuous from the point of attachment to cabinets or equipment to the grounding electrode and shall be securely fastened to the ground clamp or fitting. Ferrous metal raceways and enclosures shall be bonded at each end of the raceway or enclosure to the grounding electrode or grounding electrode conductor to create an electrically parallel path. Nonferrous metal raceways and enclosures shall not be required to be electrically continuous.

(2) Methods. Bonding shall be in compliance with 250.92(B) and ensured by one of the methods in 250.92(B)(2) through (B)(4).

(3) Size. The bonding jumper for a grounding electrode conductor raceway or cable armor shall be the same size as, or larger than, the enclosed grounding electrode conductor.

(4) Wiring Methods. If a raceway is used as protection for a grounding electrode conductor, the installation shall comply with the requirements of the appropriate raceway article.

(F) Installation to Electrode(s). Grounding electrode conductor(s) and bonding jumpers interconnecting grounding electrodes shall be installed in accordance with (1), (2), or (3). The grounding electrode conductor shall be sized for the largest grounding electrode conductor required among all the electrodes connected to it.

(1) The grounding electrode conductor shall be permitted to be run to any convenient grounding electrode available

in the grounding electrode system where the other electrode(s), if any, is connected by bonding jumpers that are installed in accordance with 250.53(C).

(2) Grounding electrode conductor(s) shall be permitted to be run to one or more grounding electrode(s) individually.

(3) Bonding jumper(s) from grounding electrode(s) shall be permitted to be connected to an aluminum or copper busbar not less than 6 mm thick × 50 mm wide (¼ in. thick × 2 in wide.) and of sufficient length to accommodate the number of terminations necessary for the installation. The busbar shall be securely fastened and shall be installed in an accessible location. Connections shall be made by a listed connector or by the exothermic welding process. The grounding electrode conductor shall be permitted to be run to the busbar. Where aluminum busbars are used, the installation shall comply with 250.64(A).

250.66 Size of Alternating-Current Grounding Electrode Conductor. The size of the grounding electrode conductor at the service, at each building or structure where supplied by a feeder(s) or branch circuit(s), or at a separately derived system of a grounded or ungrounded ac system shall not be less than given in Table 250.66, except as permitted in 250.66(A) through (C).

(A) Connections to a Rod, Pipe, or Plate Electrode(s). If the grounding electrode conductor or bonding jumper connected to a single or multiple rod, pipe, or plate electrode(s), or any combination thereof, as described in 250.52(A)(5) or (A)(7), does not extend on to other types of electrodes that require a larger size conductor, the grounding electrode conductor shall not be required to be larger than 6 AWG copper wire or 4 AWG aluminum wire.

(B) Connections to Concrete-Encased Electrodes. If the grounding electrode conductor or bonding jumper connected to a single or multiple concrete-encased electrode(s), as described in 250.52(A)(3), does not extend on to other types of electrodes that require a larger size of conductor, the grounding electrode conductor shall not be required to be larger than 4 AWG copper wire.

(C) Connections to Ground Rings. If the grounding electrode conductor or bonding jumper connected to a ground ring, as described in 250.52(A)(4), does not extend on to other types of electrodes that require a larger size of conductor, the grounding electrode conductor shall not be required to be larger than the conductor used for the ground ring.

250.68 Grounding Electrode Conductor and Bonding Jumper Connection to Grounding Electrodes. The connection of a grounding electrode conductor at the service, at each building or structure where supplied by a feeder(s) or branch circuit(s), or at a separately derived system and associated bonding jumper(s) shall be made as specified 250.68(A) through (C).

(A) Accessibility. All mechanical elements used to terminate a grounding electrode conductor or bonding jumper to a grounding electrode shall be accessible.

Exception No. 1: An encased or buried connection to a concrete-encased, driven, or buried grounding electrode shall not be required to be accessible.

Exception No. 2: Exothermic or irreversible compression connections used at terminations, together with the mechanical means used to attach such terminations to fireproofed structural metal whether or not the mechanical means is reversible, shall not be required to be accessible.

(B) Effective Grounding Path. The connection of a grounding electrode conductor or bonding jumper to a grounding electrode shall be made in a manner that will ensure an effective grounding path. Where necessary to ensure the grounding path for a metal piping system used as a grounding electrode, bonding shall be provided around insulated joints and around any equipment likely to be disconnected for repairs or replacement. Bonding jumpers shall be of sufficient length to permit removal of such equipment while retaining the integrity of the grounding path.

(C) Grounding Electrode Conductor Connections. Grounding electrode conductors and bonding jumpers shall be permitted to be connected at the following locations and used to extend the connection to an electrode(s):

(1) Interior metal water piping that is electrically continuous with a metal underground water pipe electrode and is located not more than 1.52 m (5 ft) from the point of entrance to the building shall be permitted to extend the connection to an electrode(s). Interior metal water piping located more than 1.52 m (5 ft) from the point of entrance to the building shall not be used as a conductor

Table 250.66 Grounding Electrode Conductor for Alternating-Current Systems

Size of Largest Ungrounded Service-Entrance Conductor or Equivalent Area for Parallel Conductors[a] (AWG/kcmil)		Size of Grounding Electrode Conductor (AWG/kcmil)	
Copper	Aluminum or Copper-Clad Aluminum	Copper	Aluminum or Copper-Clad Aluminum[b]
2 or smaller	1/0 or smaller	8	6
1 or 1/0	2/0 or 3/0	6	4
2/0 or 3/0	4/0 or 250	4	2
Over 3/0 through 350	Over 250 through 500	2	1/0
Over 350 through 600	Over 500 through 900	1/0	3/0
Over 600 through 1100	Over 900 through 1750	2/0	4/0
Over 1100	Over 1750	3/0	250

Notes:

1. If multiple sets of service-entrance conductors connect directly to a service drop, set of overhead service conductors, set of underground service conductors, or service lateral, the equivalent size of the largest service-entrance conductor shall be determined by the largest sum of the areas of the corresponding conductors of each set.

2. Where there are no service-entrance conductors, the grounding electrode conductor size shall be determined by the equivalent size of the largest service-entrance conductor required for the load to be served.

[a]This table also applies to the derived conductors of separately derived ac systems.

[b]See installation restrictions in 250.64(A).

to interconnect electrodes of the grounding electrode system.

Exception: In industrial, commercial, and institutional buildings or structures, if conditions of maintenance and supervision ensure that only qualified persons service the installation, interior metal water piping located more than 1.52 m (5 ft) from the point of entrance to the building shall be permitted as a bonding conductor to interconnect electrodes that are part of the grounding electrode system, or as a grounding electrode conductor, if the entire length, other than short sections passing perpendicularly through walls, floors, or ceilings, of the interior metal water pipe that is being used for the conductor is exposed.

(2) The metal structural frame of a building shall be permitted to be used as a conductor to interconnect electrodes that are part of the grounding electrode system, or as a grounding electrode conductor. Hold-down bolts securing the structural steel column that are connected to a concrete-encased electrode that complies with 250.52(A)(3) and is located in the support footing or foundation shall be permitted to connect the metal structural frame of a building or structure to the concrete encased grounding electrode. The hold-down bolts shall be connected to the concrete-encased electrode by welding, exothermic welding, the usual steel tie wires, or other approved means.

(3) A rebar-type concrete-encased electrode installed in accordance with 250.52(A)(3) with an additional rebar section extended from its location within the concrete to an accessible location that is not subject to corrosion shall be permitted for connection of grounding electrode conductors and bonding jumpers. The rebar extension shall not be exposed to contact with the earth without corrosion protection.

250.70 Methods of Grounding and Bonding Conductor Connection to Electrodes. The grounding or bonding conductor shall be connected to the grounding electrode by exothermic welding, listed lugs, listed pressure connectors, listed clamps, or other listed means. Connections depending on solder shall not be used. Ground clamps shall be listed for the materials of the grounding electrode and the grounding electrode conductor and, where used on pipe, rod, or other buried electrodes, shall also be listed for direct soil burial or concrete encasement. Not more than one conductor shall be connected to the grounding electrode by a single clamp or fitting unless the clamp or fitting is listed for multiple conductors. One of the following methods shall be used:

(1) A pipe fitting, pipe plug, or other approved device screwed into a pipe or pipe fitting

(2) A listed bolted clamp of cast bronze or brass, or plain or malleable iron

(3) For indoor communications purposes only, a listed sheet metal strap-type ground clamp having a rigid metal base that seats on the electrode and having a strap of such material and dimensions that it is not likely to stretch during or after installation

(4) An equally substantial approved means

Part IV. Enclosure, Raceway, and Service Cable Connections

250.80 Service Raceways and Enclosures. Metal enclosures and raceways for service conductors and equipment shall be connected to the grounded system conductor if the electrical system is grounded or to the grounding electrode conductor for electrical systems that are not grounded.

Exception: Metal components that are installed in a run of underground nonmetallic raceway(s) and are isolated from possible contact by a minimum cover of 450 mm (18 in.) to all parts of the metal components shall not be required to be connected to the grounded system conductor, supply side bonding jumper, or grounding electrode conductor.

250.84 Underground Service Cable or Raceway.

(A) Underground Service Cable. The sheath or armor of a continuous underground metal-sheathed or armored service cable system that is connected to the grounded system conductor on the supply side shall not be required to be connected to the grounded system conductor at the building or structure. The sheath or armor shall be permitted to be insulated from the interior metal raceway or piping.

(B) Underground Service Raceway Containing Cable. An underground metal service raceway that contains a metal-sheathed or armored cable connected to the grounded system conductor shall not be required to be connected to the grounded system conductor at the building or structure. The sheath or armor shall be permitted to be insulated from the interior metal raceway or piping.

250.86 Other Conductor Enclosures and Raceways.
Except as permitted by 250.112(I), metal enclosures and raceways for other than service conductors shall be connected to the equipment grounding conductor.

Exception No. 1. Metal enclosures and raceways for conductors added to existing installations of open wire, knob-and-tube wiring, and nonmetallic-sheathed cable shall not be required to be connected to the equipment grounding conductor where these enclosures or wiring methods comply with (1) through (4) as follows:

(1) Do not provide an equipment ground
(2) Are in runs of less than 7.5 m (25 ft)
(3) Are free from probable contact with ground, grounded metal, metal lath, or other conductive material
(4) Are guarded against contact by persons

Exception No. 2: Short sections of metal enclosures or raceways used to provide support or protection of cable assemblies from physical damage shall not be required to be connected to the equipment grounding conductor.

Exception No. 3: Metal components shall not be required to be connected to the equipment grounding conductor or supply-side bonding jumper where either of the following conditions exist:

1) The metal components are installed in a run of nonmetallic raceway(s) and isolated from possible contact by a minimum cover of 450 mm (18 in.) to any part of the metal components.

2) The metal components are part of an installation of nonmetallic raceway(s) and are isolated from possible contact to any part of the metal components by being encased in not less than 50 mm (2 in.) of concrete.

Part V. Bonding

250.90 General.
Bonding shall be provided where necessary to ensure electrical continuity and the capacity to conduct safely any fault current likely to be imposed.

250.92 Services.

(A) Bonding of Equipment for Services. The normally non–current-carrying metal parts of equipment indicated in 250.92(A)(1) and (A)(2) shall be bonded together.

(1) All raceways, cable trays, cablebus framework, auxiliary gutters, or service cable armor or sheath that enclose, contain, or support service conductors, except as permitted in 250.80

(2) All enclosures containing service conductors, including meter fittings, boxes, or the like, interposed in the service raceway or armor

(B) Method of Bonding at the Service. Bonding jumpers meeting the requirements of this article shall be used around impaired connections, such as reducing washers or oversized, concentric, or eccentric knockouts. Standard locknuts or bushings shall not be the only means for the bonding required by this section but shall be permitted to be installed to make a mechanical connection of the raceway(s).

Electrical continuity at service equipment, service raceways, and service conductor enclosures shall be ensured by one of the following methods:

(1) Bonding equipment to the grounded service conductor in a manner provided in 250.8

(2) Connections utilizing threaded couplings or threaded hubs on enclosures if made up wrenchtight

(3) Threadless couplings and connectors if made up tight for metal raceways and metal-clad cables

(4) Other listed devices, such as bonding-type locknuts, bushings, or bushings with bonding jumpers

250.94 Bonding for Communication Systems.
Communications system bonding terminations shall be connected in accordance with (A) or (B).

(A) The Intersystem Bonding Termination Device. An intersystem bonding termination (IBT) for connecting intersystem bonding conductors shall be provided external to enclosures at the service equipment or metering equipment enclosure and at the disconnecting means for any additional buildings or structures. If an IBT is used, it shall comply with the following:

(1) Be accessible for connection and inspection.

(2) Consist of a set of terminals with the capacity for connection of not less than three intersystem bonding conductors.

(3) Not interfere with opening the enclosure for a service, building or structure disconnecting means, or metering equipment.

(4) At the service equipment, be securely mounted and electrically connected to an enclosure for the service equipment, to the meter enclosure, or to an exposed nonflexible metallic service raceway, or be mounted at one of these enclosures and be connected to the enclosure or to the grounding electrode conductor with a minimum 6 AWG copper conductor.

(5) At the disconnecting means for a building or structure, be securely mounted and electrically connected to the metallic enclosure for the building or structure disconnecting means, or be mounted at the disconnecting means and be connected to the metallic enclosure or to the grounding electrode conductor with a minimum 6 AWG copper conductor.

(6)   The terminals shall be listed as grounding and bonding equipment.

Exception: In existing buildings or structures where any of the intersystem bonding and grounding electrode conductors required by 770.100(B)(2), 800.100(B)(2), 810.21(F)(2), 820.100(B)(2), and 830.100(B)(2) exist, installation of the intersystem bonding termination is not required. An accessible means external to enclosures for connecting intersystem bonding and grounding electrode conductors shall be permitted at the service equipment and at the disconnecting means for any additional buildings or structures by at least one of the following means:

(1)   *Exposed nonflexible metallic raceways*
(2)   *An exposed grounding electrode conductor*
(3)   *Approved means for the external connection of a copper or other corrosion-resistant bonding or grounding electrode conductor to the grounded raceway or equipment*

Informational Note No. 1: A 6 AWG copper conductor with one end bonded to the grounded nonflexible metallic raceway or equipment and with 150 mm (6 in.) or more of the other end made accessible on the outside wall is an example of the approved means covered in 250.94, Exception item (3).

Informational Note No. 2: See 770.100, 800.100, 810.21, 820.100, and 830.100 for intersystem bonding and grounding requirements for conductive optical fiber cables, communications circuits, radio and television equipment, CATV circuits and network-powered broadband communications systems, respectively.

N (B) Other Means.   Connections to an aluminum or copper busbar not less than 6 mm thick × 50 mm wide (¼ in. thick × 2 in. wide) and of sufficient length to accommodate at least three terminations for communication systems in addition to other connections. The busbar shall be securely fastened and shall be installed in an accessible location. Connections shall be made by a listed connector. If aluminum busbars are used, the installation shall also comply with 250.64(A).

Exception to (A) and (B): Means for connecting intersystem bonding conductors are not required where communications systems are not likely to be used.

Informational Note: The use of an IBT can reduce electrical noise on communication systems.

250.96 Bonding Other Enclosures.

(A) General.   Metal raceways, cable trays, cable armor, cable sheath, enclosures, frames, fittings, and other metal non–current-carrying parts that are to serve as equipment grounding conductors, with or without the use of supplementary equipment grounding conductors, shall be bonded where necessary to ensure electrical continuity and the capacity to conduct safely any fault current likely to be imposed on them. Any nonconductive paint, enamel, or similar coating shall be removed at threads, contact points, and contact surfaces or shall be connected by means of fittings designed so as to make such removal unnecessary.

(B) Isolated Grounding Circuits.   Where installed for the reduction of electrical noise (electromagnetic interference) on the grounding circuit, an equipment enclosure supplied by a branch circuit shall be permitted to be isolated from a raceway containing circuits supplying only that equipment by one or more listed nonmetallic raceway fittings located at the point of attachment of the raceway to the equipment enclosure. The metal raceway shall comply with provisions of this article and

shall be supplemented by an internal insulated equipment grounding conductor installed in accordance with 250.146(D) to ground the equipment enclosure.

Informational Note: Use of an isolated equipment grounding conductor does not relieve the requirement for grounding the raceway system.

250.97 Bonding for Over 250 Volts.   For circuits of over 250 volts to ground, the electrical continuity of metal raceways and cables with metal sheaths that contain any conductor other than service conductors shall be ensured by one or more of the methods specified for services in 250.92(B), except for (B)(1).

Exception: Where oversized, concentric, or eccentric knockouts are not encountered, or where a box or enclosure with concentric or eccentric knockouts is listed to provide a reliable bonding connection, the following methods shall be permitted:

(1)   *Threadless couplings and connectors for cables with metal sheaths*
(2)   *Two locknuts, on rigid metal conduit or intermediate metal conduit, one inside and one outside of boxes and cabinets*
(3)   *Fittings with shoulders that seat firmly against the box or cabinet, such as electrical metallic tubing connectors, flexible metal conduit connectors, and cable connectors, with one locknut on the inside of boxes and cabinets*
(4)   *Listed fittings*

250.98 Bonding Loosely Jointed Metal Raceways.   Expansion fittings and telescoping sections of metal raceways shall be made electrically continuous by equipment bonding jumpers or other means.

250.100 Bonding in Hazardous (Classified) Locations.   Regardless of the voltage of the electrical system, the electrical continuity of non–current-carrying metal parts of equipment, raceways, and other enclosures in any hazardous (classified) location, as defined in 500.5, 505.5, and 506.5, shall be ensured by any of the bonding methods specified in 250.92(B)(2) through (B)(4). One or more of these bonding methods shall be used whether or not equipment grounding conductors of the wire type are installed.

Informational Note: See 501.30, 502.30, 503.30, 505.25, or 506.25 for specific bonding requirements.

250.102 Grounded Conductor, Bonding Conductors, and Jumpers.

(A) Material.   Bonding jumpers shall be of copper, aluminum, copper-clad aluminum, or other corrosion-resistant material. A bonding jumper shall be a wire, bus, screw, or similar suitable conductor.

(B) Attachment.   Bonding jumpers shall be attached in the manner specified by the applicable provisions of 250.8 for circuits and equipment and by 250.70 for grounding electrodes.

(C) Size — Supply-Side Bonding Jumper.

(1) Size for Supply Conductors in a Single Raceway or Cable. The supply-side bonding jumper shall not be smaller than specified in Table 250.102(C)(1).

(2) Size for Parallel Conductor Installations in Two or More Raceways or Cables.   Where the ungrounded supply conductors are paralleled in two or more raceways or cables, and an individual supply-side bonding jumper is used for bonding these raceways or cables, the size of the supply-side bonding jumper for each raceway or cable shall be selected from Table 250.102(C)(1) based on the size of the ungrounded supply conductors in each raceway or cable. A single supply-side bonding jumper installed for bonding two or more raceways or cables shall be sized in accordance with 250.102(C)(1).

> Informational Note No. 1: The term *supply conductors* includes ungrounded conductors that do not have overcurrent protection on their supply side and terminate at service equipment or the first disconnecting means of a separately derived system.

> Informational Note No. 2: See Chapter 9, Table 8, for the circular mil area of conductors 18 AWG through 4/0 AWG.

(D) Size — Equipment Bonding Jumper on Load Side of an Overcurrent Device. The equipment bonding jumper on the load side of an overcurrent device(s) shall be sized in accordance with 250.122.

A single common continuous equipment bonding jumper shall be permitted to connect two or more raceways or cables if the bonding jumper is sized in accordance with 250.122 for the largest overcurrent device supplying circuits therein.

(E) Installation. Bonding jumpers or conductors and equipment bonding jumpers shall be permitted to be installed inside or outside of a raceway or an enclosure.

(1) Inside a Raceway or an Enclosure. If installed inside a raceway, equipment bonding jumpers and bonding jumpers or conductors shall comply with the requirements of 250.119 and 250.148.

(2) Outside a Raceway or an Enclosure. If installed on the outside, the length of the bonding jumper or conductor or equipment bonding jumper shall not exceed 1.8 m (6 ft) and shall be routed with the raceway or enclosure.

Exception: An equipment bonding jumper or supply-side bonding jumper longer than 1.8 m (6 ft) shall be permitted at outside pole locations for the purpose of bonding or grounding isolated sections of metal raceways or elbows installed in exposed risers of metal conduit or other metal raceway, and for bonding grounding electrodes, and shall not be required to be routed with a raceway or enclosure.

(3) Protection. Bonding jumpers or conductors and equipment bonding jumpers shall be installed in accordance with 250.64(A) and (B).

250.104 Bonding of Piping Systems and Exposed Structural Metal.

(A) Metal Water Piping. The metal water piping system shall be bonded as required in (A)(1), (A)(2), or (A)(3) of this section.

(1) General. Metal water piping system(s) installed in or attached to a building or structure shall be bonded to any of the following:

(1) Service equipment enclosure
(2) Grounded conductor at the service
(3) Grounding electrode conductor if of sufficient size
(4) One or more grounding electrodes used, if the grounding electrode conductor or bonding jumper to the grounding electrode is of sufficient size

The bonding jumper(s) shall be installed in accordance with 250.64(A), 250.64(B), and 250.64(E). The points of attachment of the bonding jumper(s) shall be accessible. The bonding jumper(s) shall be sized in accordance with Table

Table 250.102(C)(1) Grounded Conductor, Main Bonding Jumper, System Bonding Jumper, and Supply-Side Bonding Jumper for Alternating-Current Systems

Size of Largest Ungrounded Conductor or Equivalent Area for Parallel Conductors (AWG/kcmil)		Size of Grounded Conductor or Bonding Jumper* (AWG/kcmil)	
Copper	Aluminum or Copper-Clad Aluminum	Copper	Aluminum or Copper-Clad Aluminum
2 or smaller	1/0 or smaller	8	6
1 or 1/0	2/0 or 3/0	6	4
2/0 or 3/0	4/0 or 250	4	2
Over 3/0 through 350	Over 250 through 500	2	1/0
Over 350 through 600	Over 500 through 900	1/0	3/0
Over 600 through 1100	Over 900 through 1750	2/0	4/0
Over 1100	Over 1750	See Notes 1 and 2.	

Notes:
1. If the ungrounded supply conductors are larger than 1100 kcmil copper or 1750 kcmil aluminum, the grounded conductor or bonding jumper shall have an area not less than 12½ percent of the area of the largest ungrounded supply conductor or equivalent area for parallel supply conductors. The grounded conductor or bonding jumper shall not be required to be larger than the largest ungrounded conductor or set of ungrounded conductors.
2. If the ungrounded supply conductors are larger than 1100 kcmil copper or 1750 kcmil aluminum and if the ungrounded supply conductors and the bonding jumper are of different materials (copper, aluminum, or copper-clad aluminum), the minimum size of the grounded conductor or bonding jumper shall be based on the assumed use of ungrounded supply conductors of the same material as the grounded conductor or bonding jumper and will have an ampacity equivalent to that of the installed ungrounded supply conductors.
3. If multiple sets of service-entrance conductors are used as permitted in 230.40, Exception No. 2, or if multiple sets of ungrounded supply conductors are installed for a separately derived system, the equivalent size of the largest ungrounded supply conductor(s) shall be determined by the largest sum of the areas of the corresponding conductors of each set.
4. If there are no service-entrance conductors, the supply conductor size shall be determined by the equivalent size of the largest service-entrance conductor required for the load to be served.
*For the purposes of applying this table and its notes, the term *bonding jumper* refers to main bonding jumpers, system bonding jumpers, and supply-side bonding jumpers.

250.102(C)(1) except as permitted in 250.104(A)(2) and 250.104(A)(3).

(2) Buildings of Multiple Occupancy. In buildings of multiple occupancy where the metal water piping system(s) installed in or attached to a building or structure for the individual occupancies is metallically isolated from all other occupancies by use of nonmetallic water piping, the metal water piping system(s) for each occupancy shall be permitted to be bonded to the equipment grounding terminal of the switchgear, switchboard, or panelboard enclosure (other than service equipment) supplying that occupancy. The bonding jumper shall be sized in accordance with 250.102(D).

(3) Multiple Buildings or Structures Supplied by a Feeder(s) or Branch Circuit(s). The metal water piping system(s) installed in or attached to a building or structure shall be bonded to any of the following:

(1) Building or structure disconnecting means enclosure where located at the building or structure
(2) Equipment grounding conductor run with the supply conductors
(3) One or more grounding electrodes used

The bonding jumper(s) shall be sized in accordance with Table 250.102(C)(1), based on the size of the feeder or branch-circuit conductors that supply the building or structure. The bonding jumper shall not be required to be larger than the largest ungrounded feeder or branch-circuit conductor supplying the building or structure.

(B) Other Metal Piping. If installed in or attached to a building or structure, a metal piping system(s), including gas piping, that is likely to become energized shall be bonded to any of the following:

(1) Equipment grounding conductor for the circuit that is likely to energize the piping system
(2) Service equipment enclosure
(3) Grounded conductor at the service
(4) Grounding electrode conductor, if of sufficient size
(5) One or more grounding electrodes used, if the grounding electrode conductor or bonding jumper to the grounding electrode is of sufficient size

The bonding conductor(s) or jumper(s) shall be sized in accordance with Table 250.122, and equipment grounding conductors shall be sized in accordance with Table 250.122 using the rating of the circuit that is likely to energize the piping system(s). The points of attachment of the bonding jumper(s) shall be accessible.

> Informational Note No. 1: Bonding all piping and metal air ducts within the premises will provide additional safety.

> Informational Note No. 2: Additional information for gas piping systems can be found in Section 7.13 of NFPA 54 -2015, *National Fuel Gas Code.*

(C) Structural Metal. Exposed structural metal that is interconnected to form a metal building frame and is not intentionally grounded or bonded and is likely to become energized shall be bonded to any of the following:

(1) Service equipment enclosure
(2) Grounded conductor at the service
(3) Disconnecting means for buildings or structures supplied by a feeder or branch circuit
(4) Grounding electrode conductor, if of sufficient size
(5) One or more grounding electrodes used, if the grounding electrode conductor or bonding jumper to the grounding electrode is of sufficient size

The bonding conductor(s) or jumper(s) shall be sized in accordance with Table 250.102(C)(1) and installed in accordance with 250.64(A), 250.64(B), and 250.64(E). The points of attachment of the bonding jumper(s) shall be accessible unless installed in compliance with 250.68(A) Exception No. 2.

(D) Separately Derived Systems. Metal water piping systems and structural metal that is interconnected to form a building frame shall be bonded to separately derived systems in accordance with 250.104(D)(1) through 250.104(D)(3).

(1) Metal Water Piping System(s). The grounded conductor of each separately derived system shall be bonded to the nearest available point of the metal water piping system(s) in the area served by each separately derived system. This connection shall be made at the same point on the separately derived system where the grounding electrode conductor is connected. Each bonding jumper shall be sized in accordance with Table 250.102(C)(1) based on the largest ungrounded conductor of the separately derived system.

Exception No. 1: A separate bonding jumper to the metal water piping system shall not be required if the metal water piping system is used as the grounding electrode for the separately derived system and the water piping system is in the area served.

Exception No. 2: A separate water piping bonding jumper shall not be required if the metal frame of a building or structure is used as the grounding electrode for a separately derived system and is bonded to the metal water piping in the area served by the separately derived system.

(2) Structural Metal. If exposed structural metal that is interconnected to form the building frame exists in the area served by the separately derived system, it shall be bonded to the grounded conductor of each separately derived system. This connection shall be made at the same point on the separately derived system where the grounding electrode conductor is connected. Each bonding jumper shall be sized in accordance with Table 250.102(C)(1) based on the largest ungrounded conductor of the separately derived system.

Exception No. 1: A separate bonding jumper to the building structural metal shall not be required if the metal frame of a building or structure is used as the grounding electrode for the separately derived system.

Exception No. 2: A separate bonding jumper to the building structural metal shall not be required if the water piping of a building or structure is used as the grounding electrode for a separately derived system and is bonded to the building structural metal in the area served by the separately derived system.

(3) Common Grounding Electrode Conductor. If a common grounding electrode conductor is installed for multiple separately derived systems as permitted by 250.30(A)(6), and exposed structural metal that is interconnected to form the building frame or interior metal piping exists in the area served by the separately derived system, the metal piping and the structural metal member shall be bonded to the common grounding electrode conductor in the area served by the separately derived system.

Exception: A separate bonding jumper from each derived system to metal water piping and to structural metal members shall not be required if the metal water piping and the structural metal members in the area served by the separately derived system are bonded to the common grounding electrode conductor.

250.106 Lightning Protection Systems. The lightning protection system ground terminals shall be bonded to the building or structure grounding electrode system.

> Informational Note No. 1: See 250.60 for use of strike termination devices. For further information, see NFPA 780-2014, *Standard for the Installation of Lightning Protection Systems,* which

contains detailed information on grounding, bonding, and side-flash distance from lightning protection systems.

Informational Note No. 2: Metal raceways, enclosures, frames, and other non–current-carrying metal parts of electrical equipment installed on a building equipped with a lightning protection system may require bonding or spacing from the lightning protection conductors in accordance with NFPA 780-2014, *Standard for the Installation of Lightning Protection Systems.*

Part VI. Equipment Grounding and Equipment Grounding Conductors

250.110 Equipment Fastened in Place (Fixed) or Connected by Permanent Wiring Methods. Exposed, normally non–current-carrying metal parts of fixed equipment supplied by or enclosing conductors or components that are likely to become energized shall be connected to an equipment grounding conductor under any of the following conditions:

(1) Where within 2.5 m (8 ft) vertically or 1.5 m (5 ft) horizontally of ground or grounded metal objects and subject to contact by persons
(2) Where located in a wet or damp location and not isolated
(3) Where in electrical contact with metal
(4) Where in a hazardous (classified) location as covered by Articles 500 through 517
(5) Where supplied by a wiring method that provides an equipment grounding conductor, except as permitted by 250.86, Exception No. 2, for short sections of metal enclosures
(6) Where equipment operates with any terminal at over 150 volts to ground

Exception No. 1: If exempted by special permission, the metal frame of electrically heated appliances that have the frame permanently and effectively insulated from ground shall not be required to be grounded.

Exception No. 2: Distribution apparatus, such as transformer and capacitor cases, mounted on wooden poles at a height exceeding 2.5 m (8 ft) above ground or grade level shall not be required to be grounded.

Exception No. 3: Listed equipment protected by a system of double insulation, or its equivalent, shall not be required to be connected to the equipment grounding conductor. Where such a system is employed, the equipment shall be distinctively marked.

250.112 Specific Equipment Fastened in Place (Fixed) or Connected by Permanent Wiring Methods. Except as permitted in 250.112(F) and (I), exposed, normally non–current-carrying metal parts of equipment described in 250.112(A) through (K), and normally non–current-carrying metal parts of equipment and enclosures described in 250.112(L) and (M), shall be connected to an equipment grounding conductor, regardless of voltage.

(A) Switchgear and Switchboard Frames and Structures. Switchgear or switchboard frames and structures supporting switching equipment, except frames of 2-wire dc switchgear or switchboards where effectively insulated from ground.

(B) Pipe Organs. Generator and motor frames in an electrically operated pipe organ, unless effectively insulated from ground and the motor driving it.

(C) Motor Frames. Motor frames, as provided by 430.242.

(D) Enclosures for Motor Controllers. Enclosures for motor controllers unless attached to ungrounded portable equipment.

(E) Elevators and Cranes. Electrical equipment for elevators and cranes.

(F) Garages, Theaters, and Motion Picture Studios. Electrical equipment in commercial garages, theaters, and motion picture studios, except pendant lampholders supplied by circuits not over 150 volts to ground.

(G) Electric Signs. Electric signs, outline lighting, and associated equipment as provided in 600.7.

(H) Motion Picture Projection Equipment. Motion picture projection equipment.

(I) Remote-Control, Signaling, and Fire Alarm Circuits. Equipment supplied by Class 1 circuits shall be grounded unless operating at less than 50 volts. Equipment supplied by Class 1 power-limited circuits, by Class 2 and Class 3 remote-control and signaling circuits, and by fire alarm circuits shall be grounded where system grounding is required by Part II or Part VIII of this article.

(J) Luminaires. Luminaires as provided in Part V of Article 410.

(K) Skid-Mounted Equipment. Permanently mounted electrical equipment and skids shall be connected to the equipment grounding conductor sized as required by 250.122.

(L) Motor-Operated Water Pumps. Motor-operated water pumps, including the submersible type.

(M) Metal Well Casings. Where a submersible pump is used in a metal well casing, the well casing shall be connected to the pump circuit equipment grounding conductor.

250.114 Equipment Connected by Cord and Plug. Under any of the conditions described in 250.114(1) through (4), exposed, normally non–current-carrying metal parts of cord-and-plug-connected equipment shall be connected to the equipment grounding conductor.

Exception: Listed tools, listed appliances, and listed equipment covered in 250.114(2) through (4) shall not be required to be connected to an equipment grounding conductor where protected by a system of double insulation or its equivalent. Double insulated equipment shall be distinctively marked.

(1) In hazardous (classified) locations (see Articles 500 through 517)
(2) Where operated at over 150 volts to ground

Exception No. 1: Motors, where guarded, shall not be required to be connected to an equipment grounding conductor.

Exception No. 2: Metal frames of electrically heated appliances, exempted by special permission, shall not be required to be connected to an equipment grounding conductor, in which case the frames shall be permanently and effectively insulated from ground.

(3) In residential occupancies:
   a. Refrigerators, freezers, and air conditioners
   b. Clothes-washing, clothes-drying, dish-washing machines; ranges; kitchen waste disposers; information technology equipment; sump pumps and electrical aquarium equipment

    c. Hand-held motor-operated tools, stationary and fixed motor-operated tools, and light industrial motor-operated tools

    d. Motor-operated appliances of the following types: hedge clippers, lawn mowers, snow blowers, and wet scrubbers

    e. Portable handlamps

(4) In other than residential occupancies:

    a. Refrigerators, freezers, and air conditioners

    b. Clothes-washing, clothes-drying, dish-washing machines; information technology equipment; sump pumps and electrical aquarium equipment

    c. Hand-held motor-operated tools, stationary and fixed motor-operated tools, and light industrial motor-operated tools

    d. Motor-operated appliances of the following types: hedge clippers, lawn mowers, snow blowers, and wet scrubbers

    e. Portable handlamps

    f. Cord-and-plug-connected appliances used in damp or wet locations or by persons standing on the ground or on metal floors or working inside of metal tanks or boilers

    g. Tools likely to be used in wet or conductive locations

Exception: Tools and portable handlamps likely to be used in wet or conductive locations shall not be required to be connected to an equipment grounding conductor where supplied through an isolating transformer with an ungrounded secondary of not over 50 volts.

250.116 Nonelectrical Equipment. The metal parts of the following nonelectrical equipment described in this section shall be connected to the equipment grounding conductor:

(1) Frames and tracks of electrically operated cranes and hoists

(2) Frames of nonelectrically driven elevator cars to which electrical conductors are attached

(3) Hand-operated metal shifting ropes or cables of electric elevators

Informational Note: Where extensive metal in or on buildings or structures may become energized and is subject to personal contact, adequate bonding and grounding will provide additional safety.

250.118 Types of Equipment Grounding Conductors. The equipment grounding conductor run with or enclosing the circuit conductors shall be one or more or a combination of the following:

(1) A copper, aluminum, or copper-clad aluminum conductor. This conductor shall be solid or stranded; insulated, covered, or bare; and in the form of a wire or a busbar of any shape.

(2) Rigid metal conduit.

(3) Intermediate metal conduit.

(4) Electrical metallic tubing.

(5) Listed flexible metal conduit meeting all the following conditions:

    a. The conduit is terminated in listed fittings.

    b. The circuit conductors contained in the conduit are protected by overcurrent devices rated at 20 amperes or less.

    c. The size of the conduit does not exceed metric designator 35 (trade size 1¼).

    d. The combined length of flexible metal conduit and flexible metallic tubing and liquidtight flexible metal conduit in the same ground-fault current path does not exceed 1.8 m (6 ft).

    e. If used to connect equipment where flexibility is necessary to minimize the transmission of vibration from equipment or to provide flexibility for equipment that requires movement after installation, an equipment grounding conductor shall be installed.

(6) Listed liquidtight flexible metal conduit meeting all the following conditions:

    a. The conduit is terminated in listed fittings.

    b. For metric designators 12 through 16 (trade sizes ⅜ through ½), the circuit conductors contained in the conduit are protected by overcurrent devices rated at 20 amperes or less.

    c. For metric designators 21 through 35 (trade sizes ¾ through 1¼), the circuit conductors contained in the conduit are protected by overcurrent devices rated not more than 60 amperes and there is no flexible metal conduit, flexible metallic tubing, or liquidtight flexible metal conduit in trade sizes metric designators 12 through 16 (trade sizes ⅜ through ½) in the ground-fault current path.

    d. The combined length of flexible metal conduit and flexible metallic tubing and liquidtight flexible metal conduit in the same ground-fault current path does not exceed 1.8 m (6 ft).

    e. If used to connect equipment where flexibility is necessary to minimize the transmission of vibration from equipment or to provide flexibility for equipment that requires movement after installation, an equipment grounding conductor shall be installed.

(7) Flexible metallic tubing where the tubing is terminated in listed fittings and meeting the following conditions:

    a. The circuit conductors contained in the tubing are protected by overcurrent devices rated at 20 amperes or less.

    b. The combined length of flexible metal conduit and flexible metallic tubing and liquidtight flexible metal conduit in the same ground-fault current path does not exceed 1.8 m (6 ft).

(8) Armor of Type AC cable as provided in 320.108.

(9) The copper sheath of mineral-insulated, metal-sheathed cable Type MI.

(10) Type MC cable that provides an effective ground-fault current path in accordance with one or more of the following:

    a. It contains an insulated or uninsulated equipment grounding conductor in compliance with 250.118(1).

    b. The combined metallic sheath and uninsulated equipment grounding/bonding conductor of interlocked metal tape–type MC cable that is listed and identified as an equipment grounding conductor

    c. The metallic sheath or the combined metallic sheath and equipment grounding conductors of the smooth or corrugated tube-type MC cable that is listed and identified as an equipment grounding conductor

(11) Cable trays as permitted in 392.10 and 392.60.

(12) Cablebus framework as permitted in 370.60(1).

(13) Other listed electrically continuous metal raceways and listed auxiliary gutters.

(14) Surface metal raceways listed for grounding.

Informational Note: For a definition of *Effective Ground-Fault Current Path*, see Article 100.

250.119 Identification of Equipment Grounding Conductors. Unless required elsewhere in this *Code*, equipment grounding conductors shall be permitted to be bare, covered, or insulated. Individually covered or insulated equipment grounding conductors shall have a continuous outer finish that is either green or green with one or more yellow stripes except as permitted in this section. Conductors with insulation or individual covering that is green, green with one or more yellow stripes, or otherwise identified as permitted by this section shall not be used for ungrounded or grounded circuit conductors.

Exception No. 1: Power-limited Class 2 or Class 3 cables, power-limited fire alarm cables, or communications cables containing only circuits operating at less than 50 volts where connected to equipment not required to be grounded in accordance with 250.112(I) shall be permitted to use a conductor with green insulation or green with one or more yellow stripes for other than equipment grounding purposes.

Exception No. 2: Flexible cords having an integral insulation and jacket without an equipment grounding conductor shall be permitted to have a continuous outer finish that is green.

Informational Note: An example of a flexible cord with integral-type insulation is Type SPT-2, 2 conductor.

Exception No. 3: Conductors with green insulation shall be permitted to be used as ungrounded signal conductors where installed between the output terminations of traffic signal control and traffic signal indicating heads. Signaling circuits installed in accordance with this exception shall include an equipment grounding conductor in accordance with 250.118. Wire-type equipment grounding conductors shall be bare or have insulation or covering that is green with one or more yellow stripes.

(A) Conductors 4 AWG and Larger. Equipment grounding conductors 4 AWG and larger shall comply with 250.119(A)(1) and (A)(2).

(1) An insulated or covered conductor 4 AWG and larger shall be permitted, at the time of installation, to be permanently identified as an equipment grounding conductor at each end and at every point where the conductor is accessible.

Exception: Conductors 4 AWG and larger shall not be required to be marked in conduit bodies that contain no splices or unused hubs.

(2) Identification shall encircle the conductor and shall be accomplished by one of the following:

   a. Stripping the insulation or covering from the entire exposed length
   b. Coloring the insulation or covering green at the termination
   c. Marking the insulation or covering with green tape or green adhesive labels at the termination

(B) Multiconductor Cable. Where the conditions of maintenance and supervision ensure that only qualified persons service the installation, one or more insulated conductors in a multiconductor cable, at the time of installation, shall be permitted to be permanently identified as equipment grounding conductors at each end and at every point where the conductors are accessible by one of the following means:

(1) Stripping the insulation from the entire exposed length.

(2) Coloring the exposed insulation green.
(3) Marking the exposed insulation with green tape or green adhesive labels. Identification shall encircle the conductor.

(C) Flexible Cord. Equipment grounding conductors in flexible cords shall be insulated and shall have a continuous outer finish that is either green or green with one or more yellow stripes.

250.120 Equipment Grounding Conductor Installation. An equipment grounding conductor shall be installed in accordance with 250.120(A), (B), and (C).

(A) Raceway, Cable Trays, Cable Armor, Cablebus, or Cable Sheaths. Where it consists of a raceway, cable tray, cable armor, cablebus framework, or cable sheath or where it is a wire within a raceway or cable, it shall be installed in accordance with the applicable provisions in this *Code* using fittings for joints and terminations approved for use with the type raceway or cable used. All connections, joints, and fittings shall be made tight using suitable tools.

Informational Note: See the UL guide information on FHIT systems for equipment grounding conductors installed in a raceway that are part of an electrical circuit protective system or a fire-rated cable listed to maintain circuit integrity.

(B) Aluminum and Copper-Clad Aluminum Conductors. Equipment grounding conductors of bare or insulated aluminum or copper-clad aluminum shall be permitted. Bare conductors shall not come in direct contact with masonry or the earth or where subject to corrosive conditions. Aluminum or copper-clad aluminum conductors shall not be terminated within 450 mm (18 in.) of the earth.

(C) Equipment Grounding Conductors Smaller Than 6 AWG. Where not routed with circuit conductors as permitted in 250.130(C) and 250.134(B) Exception No. 2, equipment grounding conductors smaller than 6 AWG shall be protected from physical damage by an identified raceway or cable armor unless installed within hollow spaces of the framing members of buildings or structures and where not subject to physical damage.

250.121 Use of Equipment Grounding Conductors. An equipment grounding conductor shall not be used as a grounding electrode conductor.

Exception: A wire-type equipment grounding conductor installed in compliance with 250.6(A) and the applicable requirements for both the equipment grounding conductor and the grounding electrode conductor in Parts II, III, and VI of this article shall be permitted to serve as both an equipment grounding conductor and a grounding electrode conductor.

250.122 Size of Equipment Grounding Conductors.

(A) General. Copper, aluminum, or copper-clad aluminum equipment grounding conductors of the wire type shall not be smaller than shown in Table 250.122, but in no case shall they be required to be larger than the circuit conductors supplying the equipment. Where a cable tray, a raceway, or a cable armor or sheath is used as the equipment grounding conductor, as provided in 250.118 and 250.134(A), it shall comply with 250.4(A)(5) or (B)(4).

Equipment grounding conductors shall be permitted to be sectioned within a multiconductor cable, provided the combined circular mil area complies with Table 250.122.

(B) Increased in Size. Where ungrounded conductors are increased in size from the minimum size that has sufficient ampacity for the intended installation, wire-type equipment grounding conductors, where installed, shall be increased in size proportionately, according to the circular mil area of the ungrounded conductors.

(C) Multiple Circuits. Where a single equipment grounding conductor is run with multiple circuits in the same raceway, cable, or cable tray, it shall be sized for the largest overcurrent device protecting conductors in the raceway, cable, or cable tray. Equipment grounding conductors installed in cable trays shall meet the minimum requirements of 392.10(B)(1)(c).

(D) Motor Circuits. Equipment grounding conductors for motor circuits shall be sized in accordance with (D)(1) or (D)(2).

(1) General. The equipment grounding conductor size shall not be smaller than determined by 250.122(A) based on the rating of the branch-circuit short-circuit and ground-fault protective device.

(2) Instantaneous-Trip Circuit Breaker and Motor Short-Circuit Protector. Where the overcurrent device is an instantaneous-trip circuit breaker or a motor short-circuit protector, the equipment grounding conductor shall be sized not smaller than that given by 250.122(A) using the maximum permitted rating of a dual element time-delay fuse selected for branch-circuit short-circuit and ground-fault protection in accordance with 430.52(C)(1), Exception No. 1.

(E) Flexible Cord and Fixture Wire. The equipment grounding conductor in a flexible cord with the largest circuit conductor 10 AWG or smaller, and the equipment grounding conductor used with fixture wires of any size in accordance with 240.5, shall not be smaller than 18 AWG copper and shall not be smaller than the circuit conductors. The equipment grounding conductor in a flexible cord with a circuit conductor larger than 10 AWG shall be sized in accordance with Table 250.122.

(F) Conductors in Parallel. For circuits of parallel conductors as permitted in 310.10(H), the equipment grounding conductor shall be installed in accordance with (1) or (2).

N **(1) Conductor Installations in Raceways, Auxiliary Gutters, or Cable Trays.**

(a)　*Single Raceway or Cable Tray.* If conductors are installed in parallel in the same raceway or cable tray, a single wire-type conductor shall be permitted as the equipment grounding conductor. The wire-type equipment grounding conductor shall be sized in accordance with 250.122, based on the over-current protective device for the feeder or branch circuit. Wire-type equipment grounding conductors installed in cable trays shall meet the minimum requirements of 392.10(B)(1)(c). Metal raceways or auxiliary gutters in accordance with 250.118 or cable trays complying with 392.60(B) shall be permitted as the equipment grounding conductor.

(b)　*Multiple Raceways.* If conductors are installed in parallel in multiple raceways, wire-type equipment grounding conductors, where used, shall be installed in parallel in each raceway. The equipment grounding conductor installed in each raceway shall be sized in compliance with 250.122 based on the overcurrent protective device for the feeder or branch circuit. Metal raceways or auxiliary gutters in accordance with 250.118 or cable trays complying with 392.60(B) shall be permitted as the equipment grounding conductor.

N **(2) Multiconductor Cables.**

(a)　If multiconductor cables are installed in parallel, the equipment grounding conductor(s) in each cable shall be connected in parallel.

(b)　If multiconductor cables are installed in parallel in the same raceway, auxiliary gutter, or cable tray, a single equipment grounding conductor that is sized in accordance with 250.122 shall be permitted in combination with the equipment grounding conductors provided within the multiconductor cables and shall all be connected together.

(c)　Equipment grounding conductors installed in cable trays shall meet the minimum requirements of 392.10(B)(1)(c). Cable trays complying with 392.60(B), metal raceways in accordance with 250.118, or auxiliary gutters shall be permitted as the equipment grounding conductor.

(d)　Except as provided in 250.122(F)(2)(b) for raceway or cable tray installations, the equipment grounding conductor in each multiconductor cable shall be sized in accordance with 250.122 based on the overcurrent protective device for the feeder or branch circuit.

(G) Feeder Taps. Equipment grounding conductors run with feeder taps shall not be smaller than shown in Table 250.122 based on the rating of the overcurrent device ahead of the feeder but shall not be required to be larger than the tap conductors.

250.124 Equipment Grounding Conductor Continuity.

(A) Separable Connections. Separable connections such as those provided in drawout equipment or attachment plugs and mating connectors and receptacles shall provide for first-make, last-break of the equipment grounding conductor. First-make, last-break shall not be required where interlocked equipment, plugs, receptacles, and connectors preclude energization without grounding continuity.

(B) Switches. No automatic cutout or switch shall be placed in the equipment grounding conductor of a premises wiring system unless the opening of the cutout or switch disconnects all sources of energy.

250.126 Identification of Wiring Device Terminals. The terminal for the connection of the equipment grounding conductor shall be identified by one of the following:

(1)　A green, not readily removable terminal screw with a hexagonal head.
(2)　A green, hexagonal, not readily removable terminal nut.
(3)　A green pressure wire connector. If the terminal for the equipment grounding conductor is not visible, the conductor entrance hole shall be marked with the word *green* or *ground*, the letters *G* or *GR*, a grounding symbol, or otherwise identified by a distinctive green color. If the terminal for the equipment grounding conductor is readily removable, the area adjacent to the terminal shall be similarly marked.

Informational Note: See Informational Note Figure 250.126.

Table 250.122 Minimum Size Equipment Grounding Conductors for Grounding Raceway and Equipment

Rating or Setting of Automatic Overcurrent Device in Circuit Ahead of Equipment, Conduit, etc., Not Exceeding (Amperes)	Size (AWG or kcmil)	
	Copper	Aluminum or Copper-Clad Aluminum*
15	14	12
20	12	10
60	10	8
100	8	6
200	6	4
300	4	2
400	3	1
500	2	1/0
600	1	2/0
800	1/0	3/0
1000	2/0	4/0
1200	3/0	250
1600	4/0	350
2000	250	400
2500	350	600
3000	400	600
4000	500	750
5000	700	1200
6000	800	1200

Note: Where necessary to comply with 250.4(A)(5) or (B)(4), the equipment grounding conductor shall be sized larger than given in this table.

*See installation restrictions in 250.120.

Informational Note Figure 250.126   One Example of a Symbol Used to Identify the Grounding Termination Point for an Equipment Grounding Conductor.

Part VII. Methods of Equipment Grounding

250.130 Equipment Grounding Conductor Connections. Equipment grounding conductor connections at the source of separately derived systems shall be made in accordance with 250.30(A)(1). Equipment grounding conductor connections at service equipment shall be made as indicated in 250.130(A) or (B). For replacement of non–grounding-type receptacles with grounding-type receptacles and for branch-circuit extensions only in existing installations that do not have an equipment grounding conductor in the branch circuit, connections shall be permitted as indicated in 250.130(C).

(A) For Grounded Systems. The connection shall be made by bonding the equipment grounding conductor to the grounded service conductor and the grounding electrode conductor.

(B) For Ungrounded Systems. The connection shall be made by bonding the equipment grounding conductor to the grounding electrode conductor.

(C) Nongrounding Receptacle Replacement or Branch Circuit Extensions. The equipment grounding conductor of a grounding-type receptacle or a branch-circuit extension shall be permitted to be connected to any of the following:

(1)   Any accessible point on the grounding electrode system as described in 250.50

(2)   Any accessible point on the grounding electrode conductor

(3)   The equipment grounding terminal bar within the enclosure where the branch circuit for the receptacle or branch circuit originates

(4)   An equipment grounding conductor that is part of another branch circuit that originates from the enclosure where the branch circuit for the receptacle or branch circuit originates

(5)   For grounded systems, the grounded service conductor within the service equipment enclosure

(6)   For ungrounded systems, the grounding terminal bar within the service equipment enclosure

Informational Note: See 406.4(D) for the use of a ground-fault circuit-interrupting type of receptacle.

250.132 Short Sections of Raceway. Isolated sections of metal raceway or cable armor, where required to be grounded, shall be connected to an equipment grounding conductor in accordance with 250.134.

250.134 Equipment Fastened in Place or Connected by Permanent Wiring Methods (Fixed) — Grounding. Unless grounded by connection to the grounded circuit conductor as permitted by 250.32, 250.140, and 250.142, non–current-carrying metal parts of equipment, raceways, and other enclosures, if grounded, shall be connected to an equipment grounding conductor by one of the methods specified in 250.134(A) or (B).

(A) Equipment Grounding Conductor Types. By connecting to any of the equipment grounding conductors permitted by 250.118.

(B) With Circuit Conductors. By connecting to an equipment grounding conductor contained within the same raceway, cable, or otherwise run with the circuit conductors.

Exception No. 1: As provided in 250.130(C), the equipment grounding conductor shall be permitted to be run separately from the circuit conductors.

Exception No. 2: For dc circuits, the equipment grounding conductor shall be permitted to be run separately from the circuit conductors.

Informational Note No. 1: See 250.102 and 250.168 for equipment bonding jumper requirements.

Informational Note No. 2: See 400.10 for use of cords for fixed equipment.

250.136 Equipment Considered Grounded. Under the conditions specified in 250.136(A) and (B), the normally non–current-carrying metal parts of the equipment shall be considered grounded.

(A) Equipment Secured to Grounded Metal Supports. Electrical equipment secured to and in electrical contact with a metal rack or structure provided for its support and connected to an equipment grounding conductor by one of the means indicated in 250.134. The structural metal frame of a building shall not be used as the required equipment grounding conductor for ac equipment.

(B) Metal Car Frames. Metal car frames supported by metal hoisting cables attached to or running over metal sheaves or drums of elevator machines that are connected to an equipment grounding conductor by one of the methods indicated in 250.134.

250.138 Cord-and-Plug-Connected Equipment. Non–current-carrying metal parts of cord-and-plug-connected equipment, if grounded, shall be connected to an equipment grounding conductor by one of the methods in 250.138(A) or (B).

(A) By Means of an Equipment Grounding Conductor. By means of an equipment grounding conductor run with the power supply conductors in a cable assembly or flexible cord properly terminated in a grounding-type attachment plug with one fixed grounding contact.

Exception: The grounding contacting pole of grounding-type plug-in ground-fault circuit interrupters shall be permitted to be of the movable, self-restoring type on circuits operating at not over 150 volts between any two conductors or over 150 volts between any conductor and ground.

(B) By Means of a Separate Flexible Wire or Strap. By means of a separate flexible wire or strap, insulated or bare, connected to an equipment grounding conductor, and protected as well as practicable against physical damage, where part of equipment.

250.140 Frames of Ranges and Clothes Dryers. Frames of electric ranges, wall-mounted ovens, counter-mounted cooking units, clothes dryers, and outlet or junction boxes that are part of the circuit for these appliances shall be connected to the equipment grounding conductor in the manner specified by 250.134 or 250.138.

Exception: For existing branch-circuit installations only where an equipment grounding conductor is not present in the outlet or junction box, the frames of electric ranges, wall-mounted ovens, counter-mounted cooking units, clothes dryers, and outlet or junction boxes that are part of the circuit for these appliances shall be permitted to be connected to the grounded circuit conductor if all the following conditions are met.

(1) The supply circuit is 120/240-volt, single-phase, 3-wire; or 208Y/120-volt derived from a 3-phase, 4-wire, wye-connected system.

(2) The grounded conductor is not smaller than 10 AWG copper or 8 AWG aluminum.

(3) The grounded conductor is insulated, or the grounded conductor is uninsulated and part of a Type SE service-entrance cable and the branch circuit originates at the service equipment.

(4) Grounding contacts of receptacles furnished as part of the equipment are bonded to the equipment.

250.142 Use of Grounded Circuit Conductor for Grounding Equipment.

(A) Supply-Side Equipment. A grounded circuit conductor shall be permitted to ground non–current-carrying metal parts of equipment, raceways, and other enclosures at any of the following locations:

(1) On the supply side or within the enclosure of the ac service-disconnecting means

(2) On the supply side or within the enclosure of the main disconnecting means for separate buildings as provided in 250.32(B)

(3) On the supply side or within the enclosure of the main disconnecting means or overcurrent devices of a separately derived system where permitted by 250.30(A)(1)

(B) Load-Side Equipment. Except as permitted in 250.30(A)(1) and 250.32(B) Exception, a grounded circuit conductor shall not be used for grounding non–current-carrying metal parts of equipment on the load side of the service disconnecting means or on the load side of a separately derived system disconnecting means or the overcurrent devices for a separately derived system not having a main disconnecting means.

Exception No. 1: The frames of ranges, wall-mounted ovens, counter-mounted cooking units, and clothes dryers under the conditions permitted for existing installations by 250.140 shall be permitted to be connected to the grounded circuit conductor.

Exception No. 2: It shall be permissible to ground meter enclosures by connection to the grounded circuit conductor on the load side of the service disconnect where all of the following conditions apply:

(1) No service ground-fault protection is installed.

(2) All meter enclosures are located immediately adjacent to the service disconnecting means.

(3) The size of the grounded circuit conductor is not smaller than the size specified in Table 250.122 for equipment grounding conductors.

Exception No. 3: Direct-current systems shall be permitted to be grounded on the load side of the disconnecting means or overcurrent device in accordance with 250.164.

Exception No. 4: Electrode-type boilers operating at over 1000 volts shall be grounded as required in 490.72(E)(1) and 490.74.

250.144 Multiple Circuit Connections. Where equipment is grounded and is supplied by separate connection to more than one circuit or grounded premises wiring system, an equipment grounding conductor termination shall be provided for each such connection as specified in 250.134 and 250.138.

250.146 Connecting Receptacle Grounding Terminal to Box. An equipment bonding jumper shall be used to connect the grounding terminal of a grounding-type receptacle to a grounded box unless grounded as in 250.146(A) through (D). The equipment bonding jumper shall be sized in accordance with Table 250.122 based on the rating of the overcurrent device protecting the circuit conductors.

(A) Surface-Mounted Box. Where the box is mounted on the surface, direct metal-to-metal contact between the device yoke and the box or a contact yoke or device that complies with 250.146(B) shall be permitted to ground the receptacle to the box. At least one of the insulating washers shall be removed from receptacles that do not have a contact yoke or device that complies with 250.146(B) to ensure direct metal-to-metal contact. This provision shall not apply to cover-mounted receptacles unless the box and cover combination are listed as providing satisfactory ground continuity between the box and the receptacle. A listed exposed work cover shall be permitted to be the grounding and bonding means when (1) the device is attached to the cover with at least two fasteners that are permanent (such as a rivet) or have a thread locking or screw or nut locking means and (2) when the cover mounting holes are located on a flat non-raised portion of the cover.

(B) Contact Devices or Yokes. Contact devices or yokes designed and listed as self-grounding shall be permitted in conjunction with the supporting screws to establish equipment bonding between the device yoke and flush-type boxes.

(C) Floor Boxes. Floor boxes designed for and listed as providing satisfactory ground continuity between the box and the device shall be permitted.

(D) Isolated Ground Receptacles. Where installed for the reduction of electrical noise (electromagnetic interference) on the grounding circuit, a receptacle in which the grounding terminal is purposely insulated from the receptacle mounting means shall be permitted. The receptacle grounding terminal shall be connected to an insulated equipment grounding conductor run with the circuit conductors. This equipment grounding conductor shall be permitted to pass through one or more panelboards without a connection to the panelboard grounding terminal bar as permitted in 408.40, Exception, so as to terminate within the same building or structure directly at an equipment grounding conductor terminal of the applicable derived system or service. Where installed in accordance with the provisions of this section, this equipment grounding conductor shall also be permitted to pass through boxes, wireways, or other enclosures without being connected to such enclosures.

> Informational Note: Use of an isolated equipment grounding conductor does not relieve the requirement for grounding the raceway system and outlet box.

250.148 Continuity and Attachment of Equipment Grounding Conductors to Boxes. If circuit conductors are spliced within a box or terminated on equipment within or supported by a box, all equipment grounding conductor(s) associated with any of those circuit conductors shall be connected within the box or to the box with devices suitable for the use in accordance with 250.8 and 250.148(A) through (E).

Exception: The equipment grounding conductor permitted in 250.146(D) shall not be required to be connected to the other equipment grounding conductors or to the box.

(A) Connections. Connections and splices shall be made in accordance with 110.14(B) except that insulation shall not be required.

(B) Grounding Continuity. The arrangement of grounding connections shall be such that the disconnection or the removal of a receptacle, luminaire, or other device fed from the box does not interfere with or interrupt the grounding continuity.

(C) Metal Boxes. A connection shall be made between the one or more equipment grounding conductors and a metal box by means of a grounding screw that shall be used for no other purpose, equipment listed for grounding, or a listed grounding device.

(D) Nonmetallic Boxes. One or more equipment grounding conductors brought into a nonmetallic outlet box shall be arranged such that a connection can be made to any fitting or device in that box requiring grounding.

(E) Solder. Connections depending solely on solder shall not be used.

Part VIII. Direct-Current Systems

250.160 General. Direct-current systems shall comply with Part VIII and other sections of Article 250 not specifically intended for ac systems.

250.162 Direct-Current Circuits and Systems to Be Grounded. Direct-current circuits and systems shall be grounded as provided for in 250.162(A) and (B).

(A) Two-Wire, Direct-Current Systems. A 2-wire, dc system supplying premises wiring and operating at greater than 60 volts but not greater than 300 volts shall be grounded.

Exception No. 1: A system equipped with a ground detector and supplying only industrial equipment in limited areas shall not be required to be grounded where installed adjacent to or integral with the source of supply.

Exception No. 2: A rectifier-derived dc system supplied from an ac system complying with 250.20 shall not be required to be grounded.

Exception No. 3: Direct-current fire alarm circuits having a maximum current of 0.030 ampere as specified in Article 760, Part III, shall not be required to be grounded.

(B) Three-Wire, Direct-Current Systems. The neutral conductor of all 3-wire, dc systems supplying premises wiring shall be grounded.

250.164 Point of Connection for Direct-Current Systems.

(A) Off-Premises Source. Direct-current systems to be grounded and supplied from an off-premises source shall have the grounding connection made at one or more supply stations. A grounding connection shall not be made at individual services or at any point on the premises wiring.

(B) On-Premises Source. Where the dc system source is located on the premises, a grounding connection shall be made at one of the following:

(1) The source
(2) The first system disconnection means or overcurrent device
(3) By other means that accomplish equivalent system protection and that utilize equipment listed and identified for the use

250.166 Size of the Direct-Current Grounding Electrode Conductor. The size of the grounding electrode conductor for a dc system shall be as specified in 250.166(A) and (B), except as permitted by 250.166(C) through (E). The grounding electrode conductor for a dc system shall meet the sizing requirements in this section but shall not be required to be larger than 3/0 copper or 250 kcmil aluminum.

(A) Not Smaller Than the Neutral Conductor. Where the dc system consists of a 3-wire balancer set or a balancer winding with overcurrent protection as provided in 445.12(D), the grounding electrode conductor shall not be smaller than the neutral conductor and not smaller than 8 AWG copper or 6 AWG aluminum.

(B) Not Smaller Than the Largest Conductor. Where the dc system is other than as in 250.166(A), the grounding electrode conductor shall not be smaller than the largest conductor supplied by the system, and not smaller than 8 AWG copper or 6 AWG aluminum.

(C) Connected to Rod, Pipe, or Plate Electrodes. Where connected to rod, pipe, or plate electrodes as in 250.52(A)(5) or (A)(7), that portion of the grounding electrode conductor that is the sole connection to the grounding electrode shall not be required to be larger than 6 AWG copper wire or 4 AWG aluminum wire.

(D) Connected to a Concrete-Encased Electrode. Where connected to a concrete-encased electrode as in 250.52(A)(3), that portion of the grounding electrode conductor that is the sole connection to the grounding electrode shall not be required to be larger than 4 AWG copper wire.

(E) Connected to a Ground Ring. Where connected to a ground ring as in 250.52(A)(4), that portion of the grounding electrode conductor that is the sole connection to the grounding electrode shall not be required to be larger than the conductor used for the ground ring.

250.167 Direct-Current Ground-Fault Detection.

(A) Ungrounded Systems. Ground-fault detection systems shall be required for ungrounded systems.

(B) Grounded Systems. Ground-fault detection shall be permitted for grounded systems.

(C) Marking. Direct-current systems shall be legibly marked to indicate the grounding type at the dc source or the first disconnecting means of the system. The marking shall be of sufficient durability to withstand the environment involved.

> Informational Note: *NFPA 70E* -2015 identifies four dc grounding types in detail.

250.168 Direct-Current System Bonding Jumper. For direct-current systems that are to be grounded, an unspliced bonding jumper shall be used to connect the equipment grounding conductor(s) to the grounded conductor at the source or the first system disconnecting means where the system is grounded. The size of the bonding jumper shall not be smaller than the system grounding electrode conductor specified in 250.166 and shall comply with the provisions of 250.28(A), (B), and (C).

250.169 Ungrounded Direct-Current Separately Derived Systems. Except as otherwise permitted in 250.34 for portable and vehicle-mounted generators, an ungrounded dc separately derived system supplied from a stand-alone power source (such as an engine–generator set) shall have a grounding electrode conductor connected to an electrode that complies with Part III of this article to provide for grounding of metal enclosures, raceways, cables, and exposed non–current-carrying metal parts of equipment. The grounding electrode conductor connection shall be to the metal enclosure at any point on the separately derived system from the source to the first system disconnecting means or overcurrent device, or it shall be made at the source of a separately derived system that has no disconnecting means or overcurrent devices.

The size of the grounding electrode conductor shall be in accordance with 250.166.

Part IX. Instruments, Meters, and Relays

250.170 Instrument Transformer Circuits. Secondary circuits of current and potential instrument transformers shall be grounded where the primary windings are connected to circuits of 300 volts or more to ground and, where installed on or in switchgear and on switchboards, shall be grounded irrespective of voltage.

Exception No. 1: Circuits where the primary windings are connected to circuits of 1000 volts or less with no live parts or wiring exposed or accessible to other than qualified persons.

Exception No. 2: Current transformer secondaries connected in a three-phase delta configuration shall not be required to be grounded.

250.172 Instrument Transformer Cases. Cases or frames of instrument transformers shall be connected to the equipment grounding conductor where accessible to other than qualified persons.

Exception: Cases or frames of current transformers, the primaries of which are not over 150 volts to ground and that are used exclusively to supply current to meters.

250.174 Cases of Instruments, Meters, and Relays Operating at 1000 Volts or Less. Instruments, meters, and relays operating with windings or working parts at 1000 volts or less shall be connected to the equipment grounding conductor as specified in 250.174(A), (B), or (C).

(A) Not on Switchgear or Switchboards. Instruments, meters, and relays not located on switchgear or switchboards operating with windings or working parts at 300 volts or more to ground, and accessible to other than qualified persons, shall have the cases and other exposed metal parts connected to the equipment grounding conductor.

(B) On Switchgear or Dead-Front Switchboards. Instruments, meters, and relays (whether operated from current and potential transformers or connected directly in the circuit) on switchgear or switchboards having no live parts on the front of the panels shall have the cases connected to the equipment grounding conductor.

(C) On Live-Front Switchboards. Instruments, meters, and relays (whether operated from current and potential transformers or connected directly in the circuit) on switchboards having exposed live parts on the front of panels shall not have their cases connected to the equipment grounding conductor. Mats of insulating rubber or other suitable floor insulation shall be provided for the operator where the voltage to ground exceeds 150.

250.176 Cases of Instruments, Meters, and Relays — Operating at 1000 Volts and Over. Where instruments, meters, and relays have current-carrying parts of 1000 volts and over to ground, they shall be isolated by elevation or protected by suitable barriers, grounded metal, or insulating covers or guards. Their cases shall not be connected to the equipment grounding conductor.

Exception: Cases of electrostatic ground detectors where the internal ground segments of the instrument are connected to the instrument case and grounded and the ground detector is isolated by elevation.

250.178 Instrument Equipment Grounding Conductor. The equipment grounding conductor for secondary circuits of instrument transformers and for instrument cases shall not be smaller than 12 AWG copper or 10 AWG aluminum. Cases of instrument transformers, instruments, meters, and relays that are mounted directly on grounded metal surfaces of enclosures or grounded metal of switchgear or switchboard panels shall be considered to be grounded, and no additional equipment grounding conductor shall be required.

Part X. Grounding of Systems and Circuits of over 1000 Volts

250.180 General. Where systems over 1000 volts are grounded, they shall comply with all applicable provisions of the preceding sections of this article and with 250.182 through 250.194, which supplement and modify the preceding sections.

250.182 Derived Neutral Systems. A system neutral point derived from a grounding transformer shall be permitted to be used for grounding systems over 1 kV.

250.184 Solidly Grounded Neutral Systems. Solidly grounded neutral systems shall be permitted to be either single point grounded or multigrounded neutral.

(A) Neutral Conductor.

(1) Insulation Level. The minimum insulation level for neutral conductors of solidly grounded systems shall be 600 volts.

Exception No. 1: Bare copper conductors shall be permitted to be used for the neutral conductor of the following:

(1) Service-entrance conductors
(2) Service laterals or underground service conductors
(3) Direct-buried portions of feeders

Exception No. 2: Bare conductors shall be permitted for the neutral conductor of overhead portions installed outdoors.

Exception No. 3: The grounded neutral conductor shall be permitted to be a bare conductor if isolated from phase conductors and protected from physical damage.

Informational Note: See 225.4 for conductor covering where within 3.0 m (10 ft) of any building or other structure.

(2) Ampacity. The neutral conductor shall be of sufficient ampacity for the load imposed on the conductor but not less than 33⅓ percent of the ampacity of the phase conductors.

Exception: In industrial and commercial premises under engineering supervision, it shall be permissible to size the ampacity of the neutral conductor to not less than 20 percent of the ampacity of the phase conductor.

(B) Single-Point Grounded Neutral System. Where a single-point grounded neutral system is used, the following shall apply:

(1) A single-point grounded neutral system shall be permitted to be supplied from (a) or (b):

   a. A separately derived system
   b. A multigrounded neutral system with an equipment grounding conductor connected to the multigrounded neutral conductor at the source of the single-point grounded neutral system

(2) A grounding electrode shall be provided for the system.
(3) A grounding electrode conductor shall connect the grounding electrode to the system neutral conductor.
(4) A bonding jumper shall connect the equipment grounding conductor to the grounding electrode conductor.
(5) An equipment grounding conductor shall be provided to each building, structure, and equipment enclosure.

(6) A neutral conductor shall only be required where phase-to-neutral loads are supplied.
(7) The neutral conductor, where provided, shall be insulated and isolated from earth except at one location.
(8) An equipment grounding conductor shall be run with the phase conductors and shall comply with (a), (b), and (c):

   a. Shall not carry continuous load
   b. May be bare or insulated
   c. Shall have sufficient ampacity for fault current duty

(C) Multigrounded Neutral Systems. Where a multigrounded neutral system is used, the following shall apply:

(1) The neutral conductor of a solidly grounded neutral system shall be permitted to be grounded at more than one point. Grounding shall be permitted at one or more of the following locations:

   a. Transformers supplying conductors to a building or other structure
   b. Underground circuits where the neutral conductor is exposed
   c. Overhead circuits installed outdoors

(2) The multigrounded neutral conductor shall be grounded at each transformer and at other additional locations by connection to a grounding electrode.
(3) At least one grounding electrode shall be installed and connected to the multigrounded neutral conductor every 400 m (1300 ft).
(4) The maximum distance between any two adjacent electrodes shall not be more than 400 m (1300 ft).
(5) In a multigrounded shielded cable system, the shielding shall be grounded at each cable joint that is exposed to personnel contact.

250.186 Grounding Service-Supplied Alternating-Current Systems.

(A) Systems with a Grounded Conductor at the Service Point. Where an ac system is grounded at any point and is provided with a grounded conductor at the service point, a grounded conductor(s) shall be installed and routed with the ungrounded conductors to each service disconnecting means and shall be connected to each disconnecting means grounded conductor(s) terminal or bus. A main bonding jumper shall connect the grounded conductor(s) to each service disconnecting means's enclosure. The grounded conductor(s) shall be installed in accordance with 250.186(A)(1) through (A)(4). The size of the solidly grounded circuit conductor(s) shall be the larger of that determined by 250.184 or 250.186(A)(1) or (A)(2).

Exception: Where two or more service disconnecting means are located in a single assembly listed for use as service equipment, it shall be permitted to connect the grounded conductor(s) to the assembly common grounded conductor(s) terminal or bus. The assembly shall include a main bonding jumper for connecting the grounded conductor(s) to the assembly enclosure.

(1) Sizing for a Single Raceway or Overhead Conductor. The grounded conductor shall not be smaller than the required grounding electrode conductor specified in Table 250.102(C)(1) but shall not be required to be larger than the largest ungrounded service-entrance conductor(s).

(2) Parallel Conductors in Two or More Raceways or Overhead Conductors. If the ungrounded service-entrance conductors are installed in parallel in two or more raceways or as overhead parallel conductors, the grounded conductors shall also be installed in parallel. The size of the grounded conductor in each raceway or overhead shall be based on the total circular mil area of the parallel ungrounded conductors in the raceway or overhead, as indicated in 250.186(A)(1), but not smaller than 1/0 AWG.

> Informational Note: See 310.10(H) for grounded conductors connected in parallel.

(3) Delta-Connected Service. The grounded conductor of a 3-phase, 3-wire delta service shall have an ampacity not less than that of the ungrounded conductors.

(4) Impedance Grounded Neutral Systems. Impedance grounded neutral systems shall be installed in accordance with 250.187.

(B) Systems Without a Grounded Conductor at the Service Point. Where an ac system is grounded at any point and is not provided with a grounded conductor at the service point, a supply-side bonding jumper shall be installed and routed with the ungrounded conductors to each service disconnecting means and shall be connected to each disconnecting means equipment grounding conductor terminal or bus. The supply-side bonding jumper shall be installed in accordance with 250.186(B)(1) through (B)(3).

Exception: Where two or more service disconnecting means are located in a single assembly listed for use as service equipment, it shall be permitted to connect the supply-side bonding jumper to the assembly common equipment grounding terminal or bus.

(1) Sizing for a Single Raceway or Overhead Conductor. The supply-side bonding jumper shall not be smaller than the required grounding electrode conductor specified in Table 250.102(C)(1) but shall not be required to be larger than the largest ungrounded service-entrance conductor(s).

(2) Parallel Conductors in Two or More Raceways or Overhead Conductors. If the ungrounded service-entrance conductors are installed in parallel in two or more raceways or overhead conductors, the supply-side bonding jumper shall also be installed in parallel. The size of the supply-side bonding jumper in each raceway or overhead shall be based on the total circular mil area of the parallel ungrounded conductors in the raceway or overhead, as indicated in 250.186(A)(1), but not smaller than 1/0 AWG.

(3) Impedance Grounded Neutral Systems. Impedance grounded neutral systems shall be installed in accordance with 250.187.

250.187 Impedance Grounded Neutral Systems. Impedance grounded neutral systems in which a grounding impedance, usually a resistor, limits the ground-fault current shall be permitted where all of the following conditions are met:

(1) The conditions of maintenance and supervision ensure that only qualified persons service the installation.
(2) Ground detectors are installed on the system.
(3) Line-to-neutral loads are not served.

Impedance grounded neutral systems shall comply with the provisions of 250.187(A) through (D).

(A) Location. The grounding impedance shall be inserted in the grounding electrode conductor between the grounding electrode of the supply system and the neutral point of the supply transformer or generator.

(B) Identified and Insulated. The neutral conductor shall comply with both of the following:

(1) The neutral conductor shall be identified.
(2) The neutral conductor shall be insulated for the maximum neutral voltage.

> Informational Note: The maximum neutral voltage in a three-phase wye system is 57.7 percent of the phase-to-phase voltage.

(C) System Neutral Conductor Connection. The system neutral conductor shall not be connected to ground, except through the neutral grounding impedance.

(D) Equipment Grounding Conductors. Equipment grounding conductors shall be permitted to be bare and shall be electrically connected to the ground bus and grounding electrode conductor.

250.188 Grounding of Systems Supplying Portable or Mobile Equipment. Systems supplying portable or mobile equipment over 1000 volts, other than substations installed on a temporary basis, shall comply with 250.188(A) through (F).

(A) Portable or Mobile Equipment. Portable or mobile equipment over 1000 volts shall be supplied from a system having its neutral conductor grounded through an impedance. Where a delta-connected system over 1000 volts is used to supply portable or mobile equipment, a system neutral point and associated neutral conductor shall be derived.

(B) Exposed Non–Current-Carrying Metal Parts. Exposed non–current-carrying metal parts of portable or mobile equipment shall be connected by an equipment grounding conductor to the point at which the system neutral impedance is grounded.

(C) Ground-Fault Current. The voltage developed between the portable or mobile equipment frame and ground by the flow of maximum ground-fault current shall not exceed 100 volts.

(D) Ground-Fault Detection and Relaying. Ground-fault detection and relaying shall be provided to automatically de-energize any component of a system over 1000 volts that has developed a ground fault. The continuity of the equipment grounding conductor shall be continuously monitored so as to automatically de-energize the circuit of the system over 1000 volts to the portable or mobile equipment upon loss of continuity of the equipment grounding conductor.

(E) Isolation. The grounding electrode to which the portable or mobile equipment system neutral impedance is connected shall be isolated from and separated in the ground by at least 6.0 m (20 ft) from any other system or equipment grounding electrode, and there shall be no direct connection between the grounding electrodes, such as buried pipe and fence, and so forth.

(F) Trailing Cable and Couplers. Trailing cable and couplers of systems over 1000 volts for interconnection of portable or mobile equipment shall meet the requirements of Part III of Article 400 for cables and 490.55 for couplers.

250.190 Grounding of Equipment.

(A) Equipment Grounding. All non–current-carrying metal parts of fixed, portable, and mobile equipment and associated fences, housings, enclosures, and supporting structures shall be grounded.

Exception: Where isolated from ground and located such that any person in contact with ground cannot contact such metal parts when the equipment is energized, the metal parts shall not be required to be grounded.

Informational Note: See 250.110, Exception No. 2, for pole-mounted distribution apparatus.

(B) Grounding Electrode Conductor. If a grounding electrode conductor connects non–current-carrying metal parts to ground, the grounding electrode conductor shall be sized in accordance with Table 250.66, based on the size of the largest ungrounded service, feeder, or branch-circuit conductors supplying the equipment. The grounding electrode conductor shall not be smaller than 6 AWG copper or 4 AWG aluminum.

(C) Equipment Grounding Conductor. Equipment grounding conductors shall comply with 250.190(C)(1) through (C)(3).

(1) General. Equipment grounding conductors that are not an integral part of a cable assembly shall not be smaller than 6 AWG copper or 4 AWG aluminum.

(2) Shielded Cables. The metallic insulation shield encircling the current carrying conductors shall be permitted to be used as an equipment grounding conductor, if it is rated for clearing time of ground-fault current protective device operation without damaging the metallic shield. The metallic tape insulation shield and drain wire insulation shield shall not be used as an equipment grounding conductor for solidly grounded systems.

(3) Sizing. Equipment grounding conductors shall be sized in accordance with Table 250.122 based on the current rating of the fuse or the overcurrent setting of the protective relay.

Informational Note: The overcurrent rating for a circuit breaker is the combination of the current transformer ratio and the current pickup setting of the protective relay.

250.191 Grounding System at Alternating-Current Substations. For ac substations, the grounding system shall be in accordance with Part III of Article 250.

Informational Note: For further information on outdoor ac substation grounding, see IEEE 80-2013, *IEEE Guide for Safety in AC Substation Grounding.*

250.194 Grounding and Bonding of Fences and Other Metal Structures. Metallic fences enclosing, and other metal structures in or surrounding, a substation with exposed electrical conductors and equipment shall be grounded and bonded to limit step, touch, and transfer voltages.

(A) Metal Fences. Where metal fences are located within 5 m (16 ft) of the exposed electrical conductors or equipment, the fence shall be bonded to the grounding electrode system with wire-type bonding jumpers as follows:

(1) Bonding jumpers shall be installed at each fence corner and at maximum 50 m (160 ft) intervals along the fence.
(2) Where bare overhead conductors cross the fence, bonding jumpers shall be installed on each side of the crossing.
(3) Gates shall be bonded to the gate support post, and each gate support post shall be bonded to the grounding electrode system.
(4) Any gate or other opening in the fence shall be bonded across the opening by a buried bonding jumper.
(5) The grounding grid or grounding electrode systems shall be extended to cover the swing of all gates.

(6) The barbed wire strands above the fence shall be bonded to the grounding electrode system.

Alternate designs performed under engineering supervision shall be permitted for grounding or bonding of metal fences.

Informational Note No. 1: A nonconducting fence or section may provide isolation for transfer of voltage to other areas.

Informational Note No. 2: See IEEE 80-2013, *IEEE Guide for Safety In AC Substation Grounding,* for design and installation of fence grounding.

(B) Metal Structures. All exposed conductive metal structures, including guy wires within 2.5 m (8 ft) vertically or 5 m (16 ft) horizontally of exposed conductors or equipment and subject to contact by persons, shall be bonded to the grounding electrode systems in the area.

ARTICLE 280
Surge Arresters, Over 1000 Volts

Part I. General

280.1 Scope. This article covers general requirements, installation requirements, and connection requirements for surge arresters installed on premises wiring systems over 1000 volts.

280.3 Number Required. Where used at a point on a circuit, a surge arrester shall be connected to each ungrounded conductor. A single installation of such surge arresters shall be permitted to protect a number of interconnected circuits, if no circuit is exposed to surges while disconnected from the surge arresters.

280.4 Surge Arrester Selection. The surge arresters shall comply with 280.4(A) and (B).

(A) Rating. The rating of a surge arrester shall be equal to or greater than the maximum continuous operating voltage available at the point of application.

(1) Solidly Grounded Systems. The maximum continuous operating voltage shall be the phase-to-ground voltage of the system.

(2) Impedance or Ungrounded System. The maximum continuous operating voltage shall be the phase-to-phase voltage of the system.

(B) Silicon Carbide Types. The rating of a silicon carbide-type surge arrester shall be not less than 125 percent of the rating specified in 280.4(A).

Informational Note No. 1: For further information on surge arresters, see IEEE C62.11-2012, *Standard for Metal-Oxide Surge Arresters for Alternating-Current Power Circuits (>1 kV);* and IEEE C62.22-2009, *Guide for the Application of Metal-Oxide Surge Arresters for Alternating-Current Systems.*

Informational Note No. 2: The selection of a properly rated metal oxide arrester is based on considerations of maximum continuous operating voltage and the magnitude and duration of overvoltages at the arrester location as affected by phase-to-ground faults, system grounding techniques, switching surges, and other causes. See the manufacturer's application rules for selection of the specific arrester to be used at a particular location.

Part II. Installation

280.11 Location. Surge arresters shall be permitted to be located indoors or outdoors. Surge arresters shall be made inaccessible to unqualified persons, unless listed for installation in accessible locations.

N **280.12 Uses Not Permitted.** A surge arrester shall not be installed where the rating of the surge arrester is less than the maximum continuous phase-to-ground voltage at the power frequency available at the point of application.

280.14 Routing of Surge Arrester Grounding Conductors. The conductor used to connect the surge arrester to line, bus, or equipment and to a grounding conductor connection point as provided in 280.21 shall not be any longer than necessary and shall avoid unnecessary bends.

Part III. Connecting Surge Arresters

280.21 Connection. The arrester shall be connected to one of the following:

(1)   Grounded service conductor
(2)   Grounding electrode conductor
(3)   Grounding electrode for the service
(4)   Equipment grounding terminal in the service equipment

280.23 Surge-Arrester Conductors. The conductor between the surge arrester and the line and the surge arrester and the grounding connection shall not be smaller than 6 AWG copper or aluminum.

280.24 Interconnections. The surge arrester protecting a transformer that supplies a secondary distribution system shall be interconnected as specified in 280.24(A), (B), or (C).

(A) Metal Interconnections. A metal interconnection shall be made to the secondary grounded circuit conductor or the secondary circuit grounding electrode conductor, if, in addition to the direct grounding connection at the surge arrester, the following occurs:

(1) Additional Grounding Connection. The grounded conductor of the secondary has elsewhere a grounding connection to a continuous metal underground water piping system. In urban water-pipe areas where there are at least four water-pipe connections on the neutral conductor and not fewer than four such connections in each mile of neutral conductor, the metal interconnection shall be permitted to be made to the secondary neutral conductor with omission of the direct grounding connection at the surge arrester.

(2) Multigrounded Neutral System Connection. The grounded conductor of the secondary system is a part of a multigrounded neutral system or static wire of which the primary neutral conductor or static wire has at least four grounding connections in each 1.6 km (1 mile) of line in addition to a grounding connection at each service.

(B) Through Spark Gap or Device. Where the surge arrester grounding electrode conductor is not connected as in 280.24(A), or where the secondary is not grounded as in 280.24(A) but is otherwise grounded as in 250.52, an interconnection shall be made through a spark gap or listed device as required by 280.24(B)(1) or (B)(2).

(1) Ungrounded or Unigrounded Primary System. For ungrounded or unigrounded primary systems, the spark gap or listed device shall have a 60-Hz breakdown voltage of at least twice the primary circuit voltage but not necessarily more than 10 kV, and there shall be at least one other ground on the grounded conductor of the secondary that is not less than 6.0 m (20 ft) distant from the surge-arrester grounding electrode.

(2) Multigrounded Neutral Primary System. For multigrounded neutral primary systems, the spark gap or listed device shall have a 60-Hz breakdown of not more than 3 kV, and there shall be at least one other ground on the grounded conductor of the secondary that is not less than 6.0 m (20 ft) distant from the surge-arrester grounding electrode.

(C) By Special Permission. An interconnection of the surge-arrester ground and the secondary neutral conductor, other than as provided in 280.24(A) or (B), shall be permitted to be made only by special permission.

280.25 Grounding Electrode Conductor Connections and Enclosures. Except as indicated in this article, surge-arrester grounding electrode conductor connections shall be made as specified in Article 250, Parts III and X. Grounding electrode conductors installed in metal enclosures shall comply with 250.64(E).

ARTICLE 285
Surge-Protective Devices (SPDs), 1000 Volts or Less

Part I. General

285.1 Scope. This article covers general requirements, installation requirements, and connection requirements for surge-protective devices (SPDs) permanently installed on premises wiring systems of 1000 volts or less.

> Informational Note: Surge arresters 1000 volts or less are also known as Type 1 SPDs.

285.3 Uses Not Permitted. An SPD device shall not be installed in the following:

(1)   Circuits over 1000 volts
(2)   On ungrounded systems, impedance grounded systems, or corner grounded delta systems unless listed specifically for use on these systems
(3)   Where the rating of the SPD is less than the maximum continuous phase-to-ground voltage at the power frequency available at the point of application

285.4 Number Required. Where used at a point on a circuit, the SPD shall be connected to each ungrounded conductor.

285.6 Listing. An SPD shall be a listed device.

285.7 Short-Circuit Current Rating. The SPD shall be marked with a short-circuit current rating and shall not be installed at a point on the system where the available fault current is in excess of that rating. This marking requirement shall not apply to receptacles.

Part II. Installation

285.11 Location. SPDs shall be permitted to be located indoors or outdoors and shall be made inaccessible to unqualified persons, unless listed for installation in accessible locations.

285.12 Routing of Connections. The conductors used to connect the SPD to the line or bus and to ground shall not be any longer than necessary and shall avoid unnecessary bends.

285.13 Type 4 and Other Component Type SPDs. Type 4 component assemblies and other component type SPDs shall only be installed by the equipment manufacturer.

Part III. Connecting SPDs

285.21 Connection. Where an SPD device is installed, it shall comply with 285.23 through 285.28.

285.23 Type 1 SPDs. Type 1 SPDs shall be installed in accordance with 285.23(A) and (B).

(A) Installation. Type 1 SPDs shall be installed as follows:

(1) Type 1 SPDs shall be permitted to be connected to the supply side of the service disconnect as permitted in 230.82(4), or

(2) Type 1 SPDs shall be permitted to be connected as specified in 285.24.

(B) At the Service. When installed at services, Type 1 SPDs shall be connected to one of the following:

(1) Grounded service conductor
(2) Grounding electrode conductor
(3) Grounding electrode for the service

(4) Equipment grounding terminal in the service equipment

285.24 Type 2 SPDs. Type 2 SPDs shall be installed in accordance with 285.24(A) through (C).

(A) Service-Supplied Building or Structure. Type 2 SPDs shall be connected anywhere on the load side of a service disconnect overcurrent device required in 230.91, unless installed in accordance with 230.82(8).

(B) Feeder-Supplied Building or Structure. Type 2 SPDs shall be connected at the building or structure anywhere on the load side of the first overcurrent device at the building or structure.

(C) Separately Derived System. The SPD shall be connected on the load side of the first overcurrent device in a separately derived system.

285.25 Type 3 SPDs. Type 3 SPDs shall be permitted to be installed on the load side of branch-circuit overcurrent protection up to the equipment served. If included in the manufacturer's instructions, the Type 3 SPD connection shall be a minimum 10 m (30 ft) of conductor distance from the service or separately derived system disconnect.

285.26 Conductor Size. Line and grounding conductors shall not be smaller than 14 AWG copper or 12 AWG aluminum.

285.27 Connection Between Conductors. An SPD shall be permitted to be connected between any two conductors — ungrounded conductor(s), grounded conductor, equipment grounding conductor, or grounding electrode conductor. The grounded conductor and the equipment grounding conductor shall be interconnected only by the normal operation of the SPD during a surge.

285.28 Grounding Electrode Conductor Connections and Enclosures. Except as indicated in this article, SPD grounding connections shall be made as specified in Article 250, Part III. Grounding electrode conductors installed in metal enclosures shall comply with 250.64(E).

Chapter 3   Wiring Methods and Materials

ARTICLE 300
General Requirements for Wiring Methods and Materials

Part I. General Requirements

300.1 Scope.

(A) All Wiring Installations. This article covers general requirements for wiring methods and materials for all wiring installations unless modified by other articles in Chapter 3.

(B) Integral Parts of Equipment. The provisions of this article are not intended to apply to the conductors that form an integral part of equipment, such as motors, controllers, motor control centers, or factory-assembled control equipment or listed utilization equipment.

(C) Metric Designators and Trade Sizes. Metric designators and trade sizes for conduit, tubing, and associated fittings and accessories shall be as designated in Table 300.1(C).

300.2 Limitations.

(A) Voltage. Wiring methods specified in Chapter 3 shall be used for 1000 volts, nominal, or less where not specifically limited in some section of Chapter 3. They shall be permitted for over 1000 volts, nominal, where specifically permitted elsewhere in this *Code*.

(B) Temperature. Temperature limitation of conductors shall be in accordance with 310.15(A)(3).

300.3 Conductors.

(A) Single Conductors. Single conductors specified in Table 310.104(A) shall only be installed where part of a recognized wiring method of Chapter 3.

Table 300.1(C) Metric Designators and Trade Sizes

Metric Designator	Trade Size
12	⅜
16	½
21	¾
27	1
35	1¼
41	1½
53	2
63	2½
78	3
91	3½
103	4
129	5
155	6

Note: The metric designators and trade sizes are for identification purposes only and are not actual dimensions.

Exception: Individual conductors shall be permitted where installed as separate overhead conductors in accordance with 225.6.

(B) Conductors of the Same Circuit. All conductors of the same circuit and, where used, the grounded conductor and all equipment grounding conductors and bonding conductors shall be contained within the same raceway, auxiliary gutter, cable tray, cablebus assembly, trench, cable, or cord, unless otherwise permitted in accordance with 300.3(B)(1)through (B)(4).

(1) Paralleled Installations. Conductors shall be permitted to be run in parallel in accordance with the provisions of 310.10(H). The requirement to run all circuit conductors within the same raceway, auxiliary gutter, cable tray, trench, cable, or cord shall apply separately to each portion of the paralleled installation, and the equipment grounding conductors shall comply with the provisions of 250.122. Parallel runs in cable tray shall comply with the provisions of 392.20(C).

Exception: Conductors installed in nonmetallic raceways run underground shall be permitted to be arranged as isolated phase, neutral, and grounded conductor installations. The raceways shall be installed in close proximity, and the isolated phase, neutral, and grounded conductors shall comply with the provisions of 300.20(B).

(2) Grounding and Bonding Conductors. Equipment grounding conductors shall be permitted to be installed outside a raceway or cable assembly where in accordance with the provisions of 250.130(C) for certain existing installations or in accordance with 250.134(B), Exception No. 2, for dc circuits. Equipment bonding conductors shall be permitted to be installed on the outside of raceways in accordance with 250.102(E).

(3) Nonferrous Wiring Methods. Conductors in wiring methods with a nonmetallic or other nonmagnetic sheath, where run in different raceways, auxiliary gutters, cable trays, trenches, cables, or cords, shall comply with the provisions of 300.20(B). Conductors in single-conductor Type MI cable with a nonmagnetic sheath shall comply with the provisions of 332.31. Conductors of single-conductor Type MC cable with a nonmagnetic sheath shall comply with the provisions of 330.31, 330.116, and 300.20(B).

(4) Column-Width Panelboard Enclosures. Where an auxiliary gutter runs between a column-width panelboard and a pull box, and the pull box includes neutral terminations, the neutral conductors of circuits supplied from the panelboard shall be permitted to originate in the pull box.

(C) Conductors of Different Systems.

(1) 1000 Volts, Nominal, or Less. Conductors of ac and dc circuits, rated 1000 volts, nominal, or less, shall be permitted to occupy the same equipment wiring enclosure, cable, or raceway. All conductors shall have an insulation rating equal to at least the maximum circuit voltage applied to any conductor within the enclosure, cable, or raceway.

Secondary wiring to electric-discharge lamps of 1000 volts or less, if insulated for the secondary voltage involved, shall be permitted to occupy the same luminaire, sign, or outline lighting enclosure as the branch-circuit conductors.

Informational Note No. 1: See 725.136(A) for Class 2 and Class 3 circuit conductors.

Informational Note No. 2: See 690.4(B) for photovoltaic source and output circuits.

(2) Over 1000 Volts, Nominal. Conductors of circuits rated over 1000 volts, nominal, shall not occupy the same equipment wiring enclosure, cable, or raceway with conductors of circuits rated 1000 volts, nominal, or less unless otherwise permitted in 300.3(C)(2)(a) through 300.3(C)(2)(d).

(a) Primary leads of electric-discharge lamp ballasts insulated for the primary voltage of the ballast, where contained within the individual wiring enclosure, shall be permitted to occupy the same luminaire, sign, or outline lighting enclosure as the branch-circuit conductors.

(b) Excitation, control, relay, and ammeter conductors used in connection with any individual motor or starter shall be permitted to occupy the same enclosure as the motor-circuit conductors.

(c) In motors, transformers, switchgear, switchboards, control assemblies, and similar equipment, conductors of different voltage ratings shall be permitted.

(d) In manholes, if the conductors of each system are permanently and effectively separated from the conductors of the other systems and securely fastened to racks, insulators, or other approved supports, conductors of different voltage ratings shall be permitted.

Conductors having nonshielded insulation and operating at different voltage levels shall not occupy the same enclosure, cable, or raceway.

300.4 Protection Against Physical Damage. Where subject to physical damage, conductors, raceways, and cables shall be protected.

Informational Note: Minor damage to a raceway, cable armor, or cable insulation does not necessarily violate the integrity of either the contained conductors or the conductors' insulation.

(A) Cables and Raceways Through Wood Members.

(1) Bored Holes. In both exposed and concealed locations, where a cable- or raceway-type wiring method is installed through bored holes in joists, rafters, or wood members, holes shall be bored so that the edge of the hole is not less than 32 mm (1¼ in.) from the nearest edge of the wood member. Where this distance cannot be maintained, the cable or raceway shall be protected from penetration by screws or nails by a steel plate(s) or bushing(s), at least 1.6 mm (¹⁄₁₆ in.) thick, and of appropriate length and width installed to cover the area of the wiring.

Exception No. 1: Steel plates shall not be required to protect rigid metal conduit, intermediate metal conduit, rigid nonmetallic conduit, or electrical metallic tubing.

Exception No. 2: A listed and marked steel plate less than 1.6 mm (¹⁄₁₆ in.) thick that provides equal or better protection against nail or screw penetration shall be permitted.

(2) Notches in Wood. Where there is no objection because of weakening the building structure, in both exposed and concealed locations, cables or raceways shall be permitted to be laid in notches in wood studs, joists, rafters, or other wood members where the cable or raceway at those points is protected against nails or screws by a steel plate at least 1.6 mm (¹⁄₁₆ in.) thick, and of appropriate length and width, installed to cover the area of the wiring. The steel plate shall be installed before the building finish is applied.

Exception No. 1: Steel plates shall not be required to protect rigid metal conduit, intermediate metal conduit, rigid nonmetallic conduit, or electrical metallic tubing.

Exception No. 2: A listed and marked steel plate less than 1.6 mm (¹⁄₁₆ in.) thick that provides equal or better protection against nail or screw penetration shall be permitted.

(B) Nonmetallic-Sheathed Cables and Electrical Nonmetallic Tubing Through Metal Framing Members.

(1) Nonmetallic-Sheathed Cable. In both exposed and concealed locations where nonmetallic-sheathed cables pass through either factory- or field-punched, cut, or drilled slots or holes in metal members, the cable shall be protected by listed bushings or listed grommets covering all metal edges that are securely fastened in the opening prior to installation of the cable.

(2) Nonmetallic-Sheathed Cable and Electrical Nonmetallic Tubing. Where nails or screws are likely to penetrate nonmetallic-sheathed cable or electrical nonmetallic tubing, a steel sleeve, steel plate, or steel clip not less than 1.6 mm (¹⁄₁₆ in.) in thickness shall be used to protect the cable or tubing.

Exception: A listed and marked steel plate less than 1.6 mm (¹⁄₁₆ in.) thick that provides equal or better protection against nail or screw penetration shall be permitted.

(C) Cables Through Spaces Behind Panels Designed to Allow Access. Cables or raceway-type wiring methods, installed behind panels designed to allow access, shall be supported according to their applicable articles.

(D) Cables and Raceways Parallel to Framing Members and Furring Strips. In both exposed and concealed locations, where a cable- or raceway-type wiring method is installed parallel to framing members, such as joists, rafters, or studs, or is installed parallel to furring strips, the cable or raceway shall be installed and supported so that the nearest outside surface of the cable or raceway is not less than 32 mm (1¼ in.) from the nearest edge of the framing member or furring strips where nails or screws are likely to penetrate. Where this distance cannot be maintained, the cable or raceway shall be protected from penetration by nails or screws by a steel plate, sleeve, or equivalent at least 1.6 mm (¹⁄₁₆ in.) thick.

Exception No. 1: Steel plates, sleeves, or the equivalent shall not be required to protect rigid metal conduit, intermediate metal conduit, rigid nonmetallic conduit, or electrical metallic tubing.

Exception No. 2: For concealed work in finished buildings, or finished panels for prefabricated buildings where such supporting is impracticable, it shall be permissible to fish the cables between access points.

Exception No. 3: A listed and marked steel plate less than 1.6 mm (¹⁄₁₆ in.) thick that provides equal or better protection against nail or screw penetration shall be permitted.

(E) Cables, Raceways, or Boxes Installed in or Under Roof Decking. A cable, raceway, or box, installed in exposed or concealed locations under metal-corrugated sheet roof decking, shall be installed and supported so there is not less than 38 mm (1½ in.) measured from the lowest surface of the roof decking to the top of the cable, raceway, or box. A cable, raceway, or box shall not be installed in concealed locations in metal-corrugated, sheet decking–type roof.

Informational Note: Roof decking material is often repaired or replaced after the initial raceway or cabling and roofing installation and may be penetrated by the screws or other mechanical devices designed to provide "hold down" strength of the waterproof membrane or roof insulating material.

Exception: Rigid metal conduit and intermediate metal conduit shall not be required to comply with 300.4(E).

(F) Cables and Raceways Installed in Shallow Grooves. Cable- or raceway-type wiring methods installed in a groove, to be covered by wallboard, siding, paneling, carpeting, or similar finish, shall be protected by 1.6 mm (1/16 in.) thick steel plate, sleeve, or equivalent or by not less than 32-mm (1¼-in.) free space for the full length of the groove in which the cable or raceway is installed.

Exception No. 1: Steel plates, sleeves, or the equivalent shall not be required to protect rigid metal conduit, intermediate metal conduit, rigid nonmetallic conduit, or electrical metallic tubing.

Exception No. 2: A listed and marked steel plate less than 1.6 mm (1/16 in.) thick that provides equal or better protection against nail or screw penetration shall be permitted.

(G) Insulated Fittings. Where raceways contain 4 AWG or larger insulated circuit conductors, and these conductors enter a cabinet, a box, an enclosure, or a raceway, the conductors shall be protected by an identified fitting providing a smoothly rounded insulating surface, unless the conductors are separated from the fitting or raceway by identified insulating material that is securely fastened in place.

Exception: Where threaded hubs or bosses that are an integral part of a cabinet, box, enclosure, or raceway provide a smoothly rounded or flared entry for conductors.

Conduit bushings constructed wholly of insulating material shall not be used to secure a fitting or raceway. The insulating fitting or insulating material shall have a temperature rating not less than the insulation temperature rating of the installed conductors.

(H) Structural Joints. A listed expansion/deflection fitting or other approved means shall be used where a raceway crosses a structural joint intended for expansion, contraction or deflection, used in buildings, bridges, parking garages, or other structures.

300.5 Underground Installations.

(A) Minimum Cover Requirements. Direct-buried cable, conduit, or other raceways shall be installed to meet the minimum cover requirements of Table 300.5.

(B) Wet Locations. The interior of enclosures or raceways installed underground shall be considered to be a wet location. Insulated conductors and cables installed in these enclosures or raceways in underground installations shall comply with 310.10(C).

(C) Underground Cables and Conductors Under Buildings. Underground cable and conductors installed under a building shall be in a raceway.

Exception No. 1: Type MI cable shall be permitted under a building without installation in a raceway where embedded in concrete, fill, or other masonry in accordance with 332.10(6) or in underground runs

where suitably protected against physical damage and corrosive conditions in accordance with 332.10(10).

Exception No. 2: Type MC cable listed for direct burial or concrete encasement shall be permitted under a building without installation in a raceway in accordance with 330.10(A)(5) and in wet locations in accordance with 330.10(A)(11).

(D) Protection from Damage. Direct-buried conductors and cables shall be protected from damage in accordance with 300.5(D)(1) through (D)(4).

(1) Emerging from Grade. Direct-buried conductors and cables emerging from grade and specified in columns 1 and 4 of Table 300.5 shall be protected by enclosures or raceways extending from the minimum cover distance below grade required by 300.5(A) to a point at least 2.5 m (8 ft) above finished grade. In no case shall the protection be required to exceed 450 mm (18 in.) below finished grade.

(2) Conductors Entering Buildings. Conductors entering a building shall be protected to the point of entrance.

(3) Service Conductors. Underground service conductors that are not encased in concrete and that are buried 450 mm (18 in.) or more below grade shall have their location identified by a warning ribbon that is placed in the trench at least 300 mm (12 in.) above the underground installation.

(4) Enclosure or Raceway Damage. Where the enclosure or raceway is subject to physical damage, the conductors shall be installed in electrical metallic tubing, rigid metal conduit, intermediate metal conduit, RTRC-XW, Schedule 80 PVC conduit, or equivalent.

(E) Splices and Taps. Direct-buried conductors or cables shall be permitted to be spliced or tapped without the use of splice boxes. The splices or taps shall be made in accordance with 110.14(B).

(F) Backfill. Backfill that contains large rocks, paving materials, cinders, large or sharply angular substances, or corrosive material shall not be placed in an excavation where materials may damage raceways, cables, conductors, or other substructures or prevent adequate compaction of fill or contribute to corrosion of raceways, cables, or other substructures.

Where necessary to prevent physical damage to the raceway, cable, or conductor, protection shall be provided in the form of granular or selected material, suitable running boards, suitable sleeves, or other approved means.

(G) Raceway Seals. Conduits or raceways through which moisture may contact live parts shall be sealed or plugged at either or both ends. Spare or unused raceways shall also be sealed. Sealants shall be identified for use with the cable insulation, conductor insulation, bare conductor, shield, or other components.

Informational Note: Presence of hazardous gases or vapors may also necessitate sealing of underground conduits or raceways entering buildings.

(H) Bushing. A bushing, or terminal fitting, with an integral bushed opening shall be used at the end of a conduit or other raceway that terminates underground where the conductors or cables emerge as a direct burial wiring method. A seal incorporating the physical protection characteristics of a bushing shall be permitted to be used in lieu of a bushing.

Table 300.5 Minimum Cover Requirements, 0 to 1000 Volts, Nominal, Burial in Millimeters (Inches)

Location of Wiring Method or Circuit	Type of Wiring Method or Circuit									
	Column 1 Direct Burial Cables or Conductors		Column 2 Rigid Metal Conduit or Intermediate Metal Conduit		Column 3 Nonmetallic Raceways Listed for Direct Burial Without Concrete Encasement or Other Approved Raceways		Column 4 Residential Branch Circuits Rated 120 Volts or Less with GFCI Protection and Maximum Overcurrent Protection of 20 Amperes		Column 5 Circuits for Control of Irrigation and Landscape Lighting Limited to Not More Than 30 Volts and Installed with Type UF or in Other Identified Cable or Raceway	
	mm	in.	mm	in.	mm	in.	mm	in.	mm	in.
All locations not specified below	600	24	150	6	450	18	300	12	150[a, b]	6[a, b]
In trench below 50 mm (2 in.) thick concrete or equivalent	450	18	150	6	300	12	150	6	150	6
Under a building	0 (in raceway or Type MC or Type MI cable identified for direct burial)	0	0	0	0	0	0 (in raceway or Type MC or Type MI cable identified for direct burial)	0	0 (in raceway or Type MC or Type MI cable identified for direct burial)	0
Under minimum of 102 mm (4 in.) thick concrete exterior slab with no vehicular traffic and the slab extending not less than 152 mm (6 in.) beyond the underground installation	450	18	100	4	100	4	150 (direct burial) 100 (in raceway)	6 / 4	150 (direct burial) 100 (in raceway)	6 / 4
Under streets, highways, roads, alleys, driveways, and parking lots	600	24	600	24	600	24	600	24	600	24
One- and two-family dwelling driveways and outdoor parking areas, and used only for dwelling-related purposes	450	18	450	18	450	18	300	12	450	18
In or under airport runways, including adjacent areas where trespassing prohibited	450	18	450	18	450	18	450	18	450	18

[a] A lesser depth shall be permitted where specified in the installation instructions of a listed low-voltage lighting system.
[b] A depth of 150 mm (6 in.) shall be permitted for pool, spa, and fountain lighting, installed in a nonmetallic raceway, limited to not more than 30 volts where part of a listed low-voltage lighting system.

Notes:

1. Cover is defined as the shortest distance in mm (in.) measured between a point on the top surface of any direct-buried conductor, cable, conduit, or other raceway and the top surface of finished grade, concrete, or similar cover.

2. Raceways approved for burial only where concrete encased shall require concrete envelope not less than 50 mm (2 in.) thick.

3. Lesser depths shall be permitted where cables and conductors rise for terminations or splices or where access is otherwise required.

4. Where one of the wiring method types listed in Columns 1 through 3 is used for one of the circuit types in Columns 4 and 5, the shallowest depth of burial shall be permitted.

5. Where solid rock prevents compliance with the cover depths specified in this table, the wiring shall be installed in a metal raceway, or a nonmetallic raceway permitted for direct burial. The raceways shall be covered by a minimum of 50 mm (2 in.) of concrete extending down to rock.

(I) Conductors of the Same Circuit. All conductors of the same circuit and, where used, the grounded conductor and all equipment grounding conductors shall be installed in the same raceway or cable or shall be installed in close proximity in the same trench.

Exception No. 1: Conductors shall be permitted to be installed in parallel in raceways, multiconductor cables, or direct-buried single conductor cables. Each raceway or multiconductor cable shall contain all conductors of the same circuit, including equipment grounding conductors. Each direct-buried single conductor cable shall be located in close proximity in the trench to the other single conductor cables in the same parallel set of conductors in the circuit, including equipment grounding conductors.

Exception No. 2: Isolated phase, polarity, grounded conductor, and equipment grounding and bonding conductor installations shall be permitted in nonmetallic raceways or cables with a nonmetallic covering or nonmagnetic sheath in close proximity where conductors are paralleled as permitted in 310.10(H), and where the conditions of 300.20(B) are met.

(J) Earth Movement. Where direct-buried conductors, raceways, or cables are subject to movement by settlement or frost, direct-buried conductors, raceways, or cables shall be arranged so as to prevent damage to the enclosed conductors or to equipment connected to the raceways.

Informational Note: This section recognizes "S" loops in underground direct burial cables and conductors to raceway transitions, expansion fittings in raceway risers to fixed equipment, and, generally, the provision of flexible connections to equipment subject to settlement or frost heaves.

(K) Directional Boring. Cables or raceways installed using directional boring equipment shall be approved for the purpose.

300.6 Protection Against Corrosion and Deterioration. Raceways, cable trays, cablebus, auxiliary gutters, cable armor, boxes, cable sheathing, cabinets, elbows, couplings, fittings, supports, and support hardware shall be of materials suitable for the environment in which they are to be installed.

(A) Ferrous Metal Equipment. Ferrous metal raceways, cable trays, cablebus, auxiliary gutters, cable armor, boxes, cable sheathing, cabinets, metal elbows, couplings, nipples, fittings, supports, and support hardware shall be suitably protected against corrosion inside and outside (except threads at joints) by a coating of approved corrosion-resistant material. Where corrosion protection is necessary and the conduit is threaded in the field, the threads shall be coated with an approved electrically conductive, corrosion-resistant compound.

Informational Note: Field-cut threads are those threads that are cut in conduit, elbows, or nipples anywhere other than at the factory where the product is listed.

Exception: Stainless steel shall not be required to have protective coatings.

(1) Protected from Corrosion Solely by Enamel. Where protected from corrosion solely by enamel, ferrous metal raceways, cable trays, cablebus, auxiliary gutters, cable armor, boxes, cable sheathing, cabinets, metal elbows, couplings, nipples, fittings, supports, and support hardware shall not be used outdoors or in wet locations as described in 300.6(D).

(2) Organic Coatings on Boxes or Cabinets. Where boxes or cabinets have an approved system of organic coatings and are

marked "Raintight," "Rainproof," or "Outdoor Type," they shall be permitted outdoors.

(3) In Concrete or in Direct Contact with the Earth. Ferrous metal raceways, cable armor, boxes, cable sheathing, cabinets, elbows, couplings, nipples, fittings, supports, and support hardware shall be permitted to be installed in concrete or in direct contact with the earth, or in areas subject to severe corrosive influences where made of material approved for the condition, or where provided with corrosion protection approved for the condition.

(B) Aluminum Metal Equipment. Aluminum raceways, cable trays, cablebus, auxiliary gutters, cable armor, boxes, cable sheathing, cabinets, elbows, couplings, nipples, fittings, supports, and support hardware embedded or encased in concrete or in direct contact with the earth shall be provided with supplementary corrosion protection.

(C) Nonmetallic Equipment. Nonmetallic raceways, cable trays, cablebus, auxiliary gutters, boxes, cables with a nonmetallic outer jacket and internal metal armor or jacket, cable sheathing, cabinets, elbows, couplings, nipples, fittings, supports, and support hardware shall be made of material approved for the condition and shall comply with (C)(1) and (C)(2) as applicable to the specific installation.

(1) Exposed to Sunlight. Where exposed to sunlight, the materials shall be listed as sunlight resistant or shall be identified as sunlight resistant.

(2) Chemical Exposure. Where subject to exposure to chemical solvents, vapors, splashing, or immersion, materials or coatings shall either be inherently resistant to chemicals based on their listing or be identified for the specific chemical reagent.

(D) Indoor Wet Locations. In portions of dairy processing facilities, laundries, canneries, and other indoor wet locations, and in locations where walls are frequently washed or where there are surfaces of absorbent materials, such as damp paper or wood, the entire wiring system, where installed exposed, including all boxes, fittings, raceways, and cable used therewith, shall be mounted so that there is at least a 6-mm (¼-in.) airspace between it and the wall or supporting surface.

Exception: Nonmetallic raceways, boxes, and fittings shall be permitted to be installed without the airspace on a concrete, masonry, tile, or similar surface.

Informational Note: In general, areas where acids and alkali chemicals are handled and stored may present such corrosive conditions, particularly when wet or damp. Severe corrosive conditions may also be present in portions of meatpacking plants, tanneries, glue houses, and some stables; in installations immediately adjacent to a seashore and swimming pool areas; in areas where chemical deicers are used; and in storage cellars or rooms for hides, casings, fertilizer, salt, and bulk chemicals.

300.7 Raceways Exposed to Different Temperatures.

(A) Sealing. Where portions of a raceway or sleeve are known to be subjected to different temperatures, and where condensation is known to be a problem, as in cold storage areas of buildings or where passing from the interior to the exterior of a building, the raceway or sleeve shall be filled with an approved material to prevent the circulation of warm air to a colder section of the raceway or sleeve. An explosionproof seal shall not be required for this purpose.

(B) Expansion, Expansion-Deflection, and Deflection Fittings. Raceways shall be provided with expansion, expansion-deflection, or deflection fittings where necessary to compensate for thermal expansion, deflection, and contraction.

> Informational Note: Table 352.44 and Table 355.44 provide the expansion information for polyvinyl chloride (PVC) and for reinforced thermosetting resin conduit (RTRC), respectively. A nominal number for steel conduit can be determined by multiplying the expansion length in Table 352.44 by 0.20. The coefficient of expansion for steel electrical metallic tubing, intermediate metal conduit, and rigid metal conduit is 1.170×10^{-5} (0.0000117 mm per mm of conduit for each °C in temperature change) [0.650×10^{-5} (0.0000065 in. per inch of conduit for each °F in temperature change)].
>
> A nominal number for aluminum conduit and aluminum electrical metallic tubing can be determined by multiplying the expansion length in Table 352.44 by 0.40. The coefficient of expansion for aluminum electrical metallic tubing and aluminum rigid metal conduit is 2.34×10^{-5} (0.0000234 mm per mm of conduit for each °C in temperature change) [1.30×10^{-5} (0.000013 in. per inch of conduit for each °F in temperature change)].

300.8 Installation of Conductors with Other Systems. Raceways or cable trays containing electrical conductors shall not contain any pipe, tube, or equal for steam, water, air, gas, drainage, or any service other than electrical.

300.9 Raceways in Wet Locations Abovegrade. Where raceways are installed in wet locations abovegrade, the interior of these raceways shall be considered to be a wet location. Insulated conductors and cables installed in raceways in wet locations abovegrade shall comply with 310.10(C).

300.10 Electrical Continuity of Metal Raceways and Enclosures. Metal raceways, cable armor, and other metal enclosures for conductors shall be metallically joined together into a continuous electrical conductor and shall be connected to all boxes, fittings, and cabinets so as to provide effective electrical continuity. Unless specifically permitted elsewhere in this *Code*, raceways and cable assemblies shall be mechanically secured to boxes, fittings, cabinets, and other enclosures.

Exception No. 1: Short sections of raceways used to provide support or protection of cable assemblies from physical damage shall not be required to be made electrically continuous.

Exception No. 2: Equipment enclosures to be isolated, as permitted by 250.96(B), shall not be required to be metallically joined to the metal raceway.

300.11 Securing and Supporting.

(A) Secured in Place. Raceways, cable assemblies, boxes, cabinets, and fittings shall be securely fastened in place.

(B) Wiring Systems Installed Above Suspended Ceilings. Support wires that do not provide secure support shall not be permitted as the sole support. Support wires and associated fittings that provide secure support and that are installed in addition to the ceiling grid support wires shall be permitted as the sole support. Where independent support wires are used, they shall be secured at both ends. Cables and raceways shall not be supported by ceiling grids.

(1) Fire-Rated Assemblies. Wiring located within the cavity of a fire-rated floor–ceiling or roof–ceiling assembly shall not be secured to, or supported by, the ceiling assembly, including the ceiling support wires. An independent means of secure support shall be provided and shall be permitted to be attached to the assembly. Where independent support wires are used, they shall be distinguishable by color, tagging, or other effective means from those that are part of the fire-rated design.

Exception: The ceiling support system shall be permitted to support wiring and equipment that have been tested as part of the fire-rated assembly.

> Informational Note: One method of determining fire rating is testing in accordance with ANSI/ASTM E119-2015, *Method for Fire Tests of Building Construction and Materials.*

(2) Non–Fire-Rated Assemblies. Wiring located within the cavity of a non–fire-rated floor–ceiling or roof–ceiling assembly shall not be secured to, or supported by, the ceiling assembly, including the ceiling support wires. An independent means of secure support shall be provided and shall be permitted to be attached to the assembly. Where independent support wires are used, they shall be distinguishable by color, tagging, or other effective means.

Exception: The ceiling support system shall be permitted to support branch-circuit wiring and associated equipment where installed in accordance with the ceiling system manufacturer's instructions.

(C) Raceways Used as Means of Support. Raceways shall be used only as a means of support for other raceways, cables, or nonelectrical equipment under any of the following conditions:

(1) Where the raceway or means of support is identified as a means of support

(2) Where the raceway contains power supply conductors for electrically controlled equipment and is used to support Class 2 circuit conductors or cables that are solely for the purpose of connection to the equipment control circuits

(3) Where the raceway is used to support boxes or conduit bodies in accordance with 314.23 or to support luminaires in accordance with 410.36(E)

(D) Cables Not Used as Means of Support. Cable wiring methods shall not be used as a means of support for other cables, raceways, or nonelectrical equipment.

300.12 Mechanical Continuity — Raceways and Cables. Raceways, cable armors, and cable sheaths shall be continuous between cabinets, boxes, fittings, or other enclosures or outlets.

Exception No. 1: Short sections of raceways used to provide support or protection of cable assemblies from physical damage shall not be required to be mechanically continuous.

Exception No. 2: Raceways and cables installed into the bottom of open bottom equipment, such as switchboards, motor control centers, and floor or pad-mounted transformers, shall not be required to be mechanically secured to the equipment.

300.13 Mechanical and Electrical Continuity — Conductors.

(A) General. Conductors in raceways shall be continuous between outlets, boxes, devices, and so forth. There shall be no splice or tap within a raceway unless permitted by 300.15; 368.56(A); 376.56; 378.56; 384.56; 386.56; 388.56; or 390.7.

(B) Device Removal. In multiwire branch circuits, the continuity of a grounded conductor shall not depend on device connections such as lampholders, receptacles, and so forth, where the removal of such devices would interrupt the continuity.

300.14 Length of Free Conductors at Outlets, Junctions, and Switch Points. At least 150 mm (6 in.) of free conductor, measured from the point in the box where it emerges from its raceway or cable sheath, shall be left at each outlet, junction, and switch point for splices or the connection of luminaires or devices. Where the opening to an outlet, junction, or switch point is less than 200 mm (8 in.) in any dimension, each conductor shall be long enough to extend at least 75 mm (3 in.) outside the opening.

Exception: Conductors that are not spliced or terminated at the outlet, junction, or switch point shall not be required to comply with 300.14.

300.15 Boxes, Conduit Bodies, or Fittings — Where Required. A box shall be installed at each outlet and switch point for concealed knob-and-tube wiring.

Fittings and connectors shall be used only with the specific wiring methods for which they are designed and listed.

Where the wiring method is conduit, tubing, Type AC cable, Type MC cable, Type MI cable, nonmetallic-sheathed cable, or other cables, a box or conduit body shall be installed at each conductor splice point, outlet point, switch point, junction point, termination point, or pull point, unless otherwise permitted in 300.15(A) through (L).

(A) Wiring Methods with Interior Access. A box or conduit body shall not be required for each splice, junction, switch, pull, termination, or outlet points in wiring methods with removable covers, such as wireways, multioutlet assemblies, auxiliary gutters, and surface raceways. The covers shall be accessible after installation.

(B) Equipment. An integral junction box or wiring compartment as part of approved equipment shall be permitted in lieu of a box.

(C) Protection. A box or conduit body shall not be required where cables enter or exit from conduit or tubing that is used to provide cable support or protection against physical damage. A fitting shall be provided on the end(s) of the conduit or tubing to protect the cable from abrasion.

(D) Type MI Cable. A box or conduit body shall not be required where accessible fittings are used for straight-through splices in mineral-insulated metal-sheathed cable.

(E) Integral Enclosure. A wiring device with integral enclosure identified for the use, having brackets that securely fasten the device to walls or ceilings of conventional on-site frame construction, for use with nonmetallic-sheathed cable, shall be permitted in lieu of a box or conduit body.

Informational Note: See 334.30(C); 545.10; 550.15(I); 551.47(E), Exception No. 1; and 552.48(E), Exception No. 1.

(F) Fitting. A fitting identified for the use shall be permitted in lieu of a box or conduit body where conductors are not spliced or terminated within the fitting. The fitting shall be accessible after installation.

(G) Direct-Buried Conductors. As permitted in 300.5(E), a box or conduit body shall not be required for splices and taps in direct-buried conductors and cables.

(H) Insulated Devices. As permitted in 334.40(B), a box or conduit body shall not be required for insulated devices supplied by nonmetallic-sheathed cable.

(I) Enclosures. A box or conduit body shall not be required where a splice, switch, terminal, or pull point is in a cabinet or cutout box, in an enclosure for a switch or overcurrent device as permitted in 312.8, in a motor controller as permitted in 430.10(A), or in a motor control center.

(J) Luminaires. A box or conduit body shall not be required where a luminaire is used as a raceway as permitted in 410.64.

(K) Embedded. A box or conduit body shall not be required for splices where conductors are embedded as permitted in 424.40, 424.41(D), 426.22(B), 426.24(A), and 427.19(A).

(L) Manholes and Handhole Enclosures. A box or conduit body shall not be required for conductors in manholes or handhole enclosures, except where connecting to electrical equipment. The installation shall comply with the provisions of Part V of Article 110 for manholes, and 314.30 for handhole enclosures.

300.16 Raceway or Cable to Open or Concealed Wiring.

(A) Box, Conduit Body, or Fitting. A box, conduit body, or terminal fitting having a separately bushed hole for each conductor shall be used wherever a change is made from conduit, electrical metallic tubing, electrical nonmetallic tubing, nonmetallic-sheathed cable, Type AC cable, Type MC cable, or mineral-insulated, metal-sheathed cable and surface raceway wiring to open wiring or to concealed knob-and-tube wiring. A fitting used for this purpose shall contain no taps or splices and shall not be used at luminaire outlets. A conduit body used for this purpose shall contain no taps or splices, unless it complies with 314.16(C)(2).

(B) Bushing. A bushing shall be permitted in lieu of a box or terminal where the conductors emerge from a raceway and enter or terminate at equipment, such as open switchboards, unenclosed control equipment, or similar equipment. The bushing shall be of the insulating type for other than lead-sheathed conductors.

300.17 Number and Size of Conductors in Raceway. The number and size of conductors in any raceway shall not be more than will permit dissipation of the heat and ready installation or withdrawal of the conductors without damage to the conductors or to their insulation.

Informational Note: See the following sections of this *Code*: intermediate metal conduit, 342.22; rigid metal conduit, 344.22; flexible metal conduit, 348.22; liquidtight flexible metal conduit, 350.22; PVC conduit, 352.22; HDPE conduit, 353.22; RTRC, 355.22; liquidtight nonmetallic flexible conduit, 356.22; electrical metallic tubing, 358.22; flexible metallic tubing, 360.22; electrical nonmetallic tubing, 362.22; cellular concrete floor raceways, 372.22; cellular metal floor raceways, 374.22; metal wireways, 376.22; nonmetallic wireways, 378.22; surface metal raceways, 386.22; surface nonmetallic raceways, 388.22; underfloor raceways, 390.6; fixture wire, 402.7; theaters, 520.6; signs, 600.31(C); elevators, 620.33; audio signal processing, amplification, and reproduction equipment, 640.23(A) and 640.24; Class 1, Class 2, and Class 3 circuits, Article 725; fire alarm circuits, Article 760; and optical fiber cables and raceways, Article 770.

300.18 Raceway Installations.

(A) Complete Runs. Raceways, other than busways or exposed raceways having hinged or removable covers, shall be installed complete between outlet, junction, or splicing points prior to the installation of conductors. Where required to facilitate the installation of utilization equipment, the raceway shall be permitted to be initially installed without a terminating connection at the equipment. Prewired raceway assemblies shall be permitted only where specifically permitted in this *Code* for the applicable wiring method.

Exception: Short sections of raceways used to contain conductors or cable assemblies for protection from physical damage shall not be required to be installed complete between outlet, junction, or splicing points.

(B) Welding. Metal raceways shall not be supported, terminated, or connected by welding to the raceway unless specifically designed to be or otherwise specifically permitted to be in this *Code*.

300.19 Supporting Conductors in Vertical Raceways.

(A) Spacing Intervals — Maximum. Conductors in vertical raceways shall be supported if the vertical rise exceeds the values in Table 300.19(A). At least one support method shall be provided for each conductor at the top of the vertical raceway or as close to the top as practical. Intermediate supports shall be provided as necessary to limit supported conductor lengths to not greater than those values specified in Table 300.19(A).

Exception: Steel wire armor cable shall be supported at the top of the riser with a cable support that clamps the steel wire armor. A safety device shall be permitted at the lower end of the riser to hold the cable in the event there is slippage of the cable in the wire-armored cable support. Additional wedge-type supports shall be permitted to relieve the strain on the equipment terminals caused by expansion of the cable under load.

(B) Fire-Rated Cables and Conductors. Support methods and spacing intervals for fire-rated cables and conductors shall comply with any restrictions provided in the listing of the electrical circuit protective system used and in no case shall exceed the values in Table 300.19(A).

(C) Support Methods. One of the following methods of support shall be used:

(1) By clamping devices constructed of or employing insulating wedges inserted in the ends of the raceways. Where clamping of insulation does not adequately support the cable, the conductor also shall be clamped.

(2) By inserting boxes at the required intervals in which insulating supports are installed and secured in an approved manner to withstand the weight of the conductors attached thereto, the boxes being provided with covers.

(3) In junction boxes, by deflecting the cables not less than 90 degrees and carrying them horizontally to a distance not less than twice the diameter of the cable, the cables being carried on two or more insulating supports and additionally secured thereto by tie wires if desired. Where this method is used, cables shall be supported at intervals not greater than 20 percent of those mentioned in the preceding tabulation.

(4) By other approved means.

300.20 Induced Currents in Ferrous Metal Enclosures or Ferrous Metal Raceways.

(A) Conductors Grouped Together. Where conductors carrying alternating current are installed in ferrous metal enclosures or ferrous metal raceways, they shall be arranged so as to avoid heating the surrounding ferrous metal by induction. To accomplish this, all phase conductors and, where used, the grounded conductor and all equipment grounding conductors shall be grouped together.

Exception No. 1: Equipment grounding conductors for certain existing installations shall be permitted to be installed separate from their associated circuit conductors where run in accordance with the provisions of 250.130(C).

Exception No. 2: A single conductor shall be permitted to be installed in a ferromagnetic enclosure and used for skin-effect heating in accordance with the provisions of 426.42 and 427.17.

(B) Individual Conductors. Where a single conductor carrying alternating current passes through metal with magnetic properties, the inductive effect shall be minimized by (1) cutting slots in the metal between the individual holes through which the individual conductors pass or (2) passing all the conductors in the circuit through an insulating wall sufficiently large for all of the conductors of the circuit.

Exception: In the case of circuits supplying vacuum or electric-discharge lighting systems or signs or X-ray apparatus, the currents carried by the conductors are so small that the inductive heating effect can be ignored where these conductors are placed in metal enclosures or pass through metal.

Informational Note: Because aluminum is not a magnetic metal, there will be no heating due to hysteresis; however, induced currents will be present. They will not be of sufficient magnitude to require grouping of conductors or special treatment in passing conductors through aluminum wall sections.

Table 300.19(A) Spacings for Conductor Supports

		Conductors			
		Aluminum or Copper-Clad Aluminum		Copper	
Conductor Size	Support of Conductors in Vertical Raceways	m	ft	m	ft
18 AWG through 8 AWG	Not greater than	30	100	30	100
6 AWG through 1/0 AWG	Not greater than	60	200	30	100
2/0 AWG through 4/0 AWG	Not greater than	55	180	25	80
Over 4/0 AWG through 350 kcmil	Not greater than	41	135	18	60
Over 350 kcmil through 500 kcmil	Not greater than	36	120	15	50
Over 500 kcmil through 750 kcmil	Not greater than	28	95	12	40
Over 750 kcmil	Not greater than	26	85	11	35

300.21 Spread of Fire or Products of Combustion. Electrical installations in hollow spaces, vertical shafts, and ventilation or air-handling ducts shall be made so that the possible spread of fire or products of combustion will not be substantially increased. Openings around electrical penetrations into or through fire-resistant-rated walls, partitions, floors, or ceilings shall be firestopped using approved methods to maintain the fire resistance rating.

Informational Note: Directories of electrical construction materials published by qualified testing laboratories contain many listing installation restrictions necessary to maintain the fire-resistive rating of assemblies where penetrations or openings are made. Building codes also contain restrictions on membrane penetrations on opposite sides of a fire-resistance-rated wall assembly. An example is the 600-mm (24-in.) minimum horizontal separation that usually applies between boxes installed on opposite sides of the wall. Assistance in complying with 300.21 can be found in building codes, fire resistance directories, and product listings.

300.22 Wiring in Ducts Not Used for Air Handling, Fabricated Ducts for Environmental Air, and Other Spaces for Environmental Air (Plenums). The provisions of this section shall apply to the installation and uses of electrical wiring and equipment in ducts used for dust, loose stock, or vapor removal; ducts specifically fabricated for environmental air; and other spaces used for environmental air (plenums).

Informational Note: See Article 424, Part VI, for duct heaters.

(A) Ducts for Dust, Loose Stock, or Vapor Removal. No wiring systems of any type shall be installed in ducts used to transport dust, loose stock, or flammable vapors. No wiring system of any type shall be installed in any duct, or shaft containing only such ducts, used for vapor removal or for ventilation of commercial-type cooking equipment.

(B) Ducts Specifically Fabricated for Environmental Air. Equipment, devices, and the wiring methods specified in this section shall be permitted within such ducts only if necessary for the direct action upon, or sensing of, the contained air. Where equipment or devices are installed and illumination is necessary to facilitate maintenance and repair, enclosed gasketed-type luminaires shall be permitted.

Only wiring methods consisting of Type MI cable without an overall nonmetallic covering, Type MC cable employing a smooth or corrugated impervious metal sheath without an overall nonmetallic covering, electrical metallic tubing, flexible metallic tubing, intermediate metal conduit, or rigid metal conduit without an overall nonmetallic covering shall be installed in ducts specifically fabricated to transport environmental air. Flexible metal conduit shall be permitted, in lengths not to exceed 1.2 m (4 ft), to connect physically adjustable equipment and devices permitted to be in these fabricated ducts. The connectors used with flexible metal conduit shall effectively close any openings in the connection.

Exception: Wiring methods and cabling systems, listed for use in other spaces used for environmental air (plenums), shall be permitted to be installed in ducts specifically fabricated for environmental air-handling purposes under the following conditions:

(1) The wiring methods or cabling systems shall be permitted only if necessary to connect to equipment or devices associated with the direct action upon or sensing of the contained air, and

(2) The total length of such wiring methods or cabling systems shall not exceed 1.2 m (4 ft).

(C) Other Spaces Used for Environmental Air (Plenums). This section shall apply to spaces not specifically fabricated for environmental air-handling purposes but used for air-handling purposes as a plenum. This section shall not apply to habitable rooms or areas of buildings, the prime purpose of which is not air handling.

Informational Note No. 1: The space over a hung ceiling used for environmental air-handling purposes is an example of the type of other space to which this section applies.

Informational Note No. 2: The phrase "Other Spaces Used for Environmental Air (Plenum)" as used in this section correlates with the use of the term "plenum" in NFPA 90A-2015, *Standard for the Installation of Air-Conditioning and Ventilating Systems,* and other mechanical codes where the plenum is used for return air purposes, as well as some other air-handling spaces.

Exception: This section shall not apply to the joist or stud spaces of dwelling units where the wiring passes through such spaces perpendicular to the long dimension of such spaces.

(1) Wiring Methods. The wiring methods for such other space shall be limited to totally enclosed, nonventilated, insulated busway having no provisions for plug-in connections, Type MI cable without an overall nonmetallic covering, Type MC cable without an overall nonmetallic covering, Type AC cable, or other factory-assembled multiconductor control or power cable that is specifically listed for use within an air-handling space, or listed prefabricated cable assemblies of metallic manufactured wiring systems without nonmetallic sheath. Other types of cables, conductors, and raceways shall be permitted to be installed in electrical metallic tubing, flexible metallic tubing, intermediate metal conduit, rigid metal conduit without an overall nonmetallic covering, flexible metal conduit, or, where accessible, surface metal raceway or metal wireway with metal covers.

Nonmetallic cable ties and other nonmetallic cable accessories used to secure and support cables shall be listed as having low smoke and heat release properties.

Informational Note: One method to determine low smoke and heat release properties is that the nonmetallic cable ties and other nonmetallic cable accessories exhibit a maximum peak optical density of 0.50 or less, an average optical density of 0.15 or less, and a peak heat release rate of 100 kW or less when tested in accordance with ANSI/UL 2043-2008, *Fire Test for Heat and Visible Smoke Release for Discrete Products and Their Accessories Installed in Air-Handling Spaces.*

(2) Cable Tray Systems. The provisions in (a) or (b) shall apply to the use of metallic cable tray systems in other spaces used for environmental air (plenums), where accessible, as follows:

(a) *Metal Cable Tray Systems.* Metal cable tray systems shall be permitted to support the wiring methods in 300.22(C)(1).

(b) *Solid Side and Bottom Metal Cable Tray Systems.* Solid side and bottom metal cable tray systems with solid metal covers shall be permitted to enclose wiring methods and cables, not already covered in 300.22(C)(1), in accordance with 392.10(A) and (B).

(3) Equipment. Electrical equipment with a metal enclosure, or electrical equipment with a nonmetallic enclosure listed for use within an air-handling space and having low smoke and heat release properties, and associated wiring material suitable for the ambient temperature shall be permitted to be installed in such other space unless prohibited elsewhere in this *Code.*

Informational Note: One method to determine low smoke and heat release properties is that the equipment exhibits a maximum peak optical density of 0.50 or less, an average optical density of 0.15 or less, and a peak heat release rate of 100kW or less when tested in accordance with ANSI/UL 2043-2013, *Fire Test for Heat and Visible Smoke Release for Discrete Products and Their Accessories Installed in Air-Handling Spaces.*

Exception: Integral fan systems shall be permitted where specifically identified for use within an air-handling space.

(D) Information Technology Equipment. Electrical wiring in air-handling areas beneath raised floors for information technology equipment shall be permitted in accordance with Article 645.

300.23 Panels Designed to Allow Access. Cables, raceways, and equipment installed behind panels designed to allow access, including suspended ceiling panels, shall be arranged and secured so as to allow the removal of panels and access to the equipment.

Part II. Requirements for over 1000 Volts, Nominal

300.31 Covers Required. Suitable covers shall be installed on all boxes, fittings, and similar enclosures to prevent accidental contact with energized parts or physical damage to parts or insulation.

300.32 Conductors of Different Systems. See 300.3(C)(2).

300.34 Conductor Bending Radius. The conductor shall not be bent to a radius less than 8 times the overall diameter for nonshielded conductors or 12 times the overall diameter for shielded or lead-covered conductors during or after installation. For multiconductor or multiplexed single-conductor cables having individually shielded conductors, the minimum bending radius is 12 times the diameter of the individually shielded conductors or 7 times the overall diameter, whichever is greater.

300.35 Protection Against Induction Heating. Metallic raceways and associated conductors shall be arranged so as to avoid heating of the raceway in accordance with the applicable provisions of 300.20.

300.37 Aboveground Wiring Methods. Aboveground conductors shall be installed in rigid metal conduit, in intermediate metal conduit, in electrical metallic tubing, in RTRC and PVC conduit, in cable trays, in auxiliary gutters, as busways, as cablebus, in other identified raceways, or as exposed runs of metal-clad cable suitable for the use and purpose. In locations accessible to qualified persons only, exposed runs of Type MV cables, bare conductors, and bare busbars shall also be permitted. Busbars shall be permitted to be either copper or aluminum.

Exception: Airfield lighting cable used in series circuits that are powered by regulators and installed in restricted airport lighting vaults shall be permitted as exposed cable installations.

Informational Note: FAA L-824 cables installed as exposed runs within a restricted vault area are common applications.

300.38 Raceways in Wet Locations Above Grade. Where raceways are installed in wet locations above grade, the interior of these raceways shall be considered to be a wet location. Insulated conductors and cables installed in raceways in wet locations above grade shall comply with 310.10(C).

300.39 Braid-Covered Insulated Conductors — Exposed Installation. Exposed runs of braid-covered insulated conductors shall have a flame-retardant braid. If the conductors used do not have this protection, a flame-retardant saturant shall be applied to the braid covering after installation. This treated braid covering shall be stripped back a safe distance at conductor terminals, according to the operating voltage. Where practicable, this distance shall not be less than 25 mm (1 in.) for each kilovolt of the conductor-to-ground voltage of the circuit.

300.40 Insulation Shielding. Metallic and semiconducting insulation shielding components of shielded cables shall be removed for a distance dependent on the circuit voltage and insulation. Stress reduction means shall be provided at all terminations of factory-applied shielding.

Metallic shielding components such as tapes, wires, or braids, or combinations thereof, shall be connected to a grounding conductor, grounding busbar, or a grounding electrode.

300.42 Moisture or Mechanical Protection for Metal-Sheathed Cables. Where cable conductors emerge from a metal sheath and where protection against moisture or physical damage is necessary, the insulation of the conductors shall be protected by a cable sheath terminating device.

300.45 Warning Signs. Warning signs shall be conspicuously posted at points of access to conductors in all conduit systems and cable systems. The warning sign(s) shall be legible and permanent and shall carry the following wording:

DANGER—HIGH VOLTAGE—KEEP OUT

300.50 Underground Installations.

(A) General. Underground conductors shall be identified for the voltage and conditions under which they are installed. Direct-burial cables shall comply with the provisions of 310.10(F). Underground cables shall be installed in accordance with 300.50(A)(1), (A)(2), or (A)(3), and the installation shall meet the depth requirements of Table 300.50.

(1) Shielded Cables and Nonshielded Cables in Metal-Sheathed Cable Assemblies. Underground cables, including nonshielded, Type MC and moisture-impervious metal sheath cables, shall have those sheaths grounded through an effective grounding path meeting the requirements of 250.4(A)(5) or (B)(4). They shall be direct buried or installed in raceways identified for the use.

(2) Industrial Establishments. In industrial establishments, where conditions of maintenance and supervision ensure that only qualified persons service the installed cable, nonshielded single-conductor cables with insulation types up to 2000 volts that are listed for direct burial shall be permitted to be directly buried.

(3) Other Nonshielded Cables. Other nonshielded cables not covered in 300.50(A)(1) or (A)(2) shall be installed in rigid metal conduit, intermediate metal conduit, or rigid nonmetallic conduit encased in not less than 75 mm (3 in.) of concrete.

(B) Wet Locations. The interior of enclosures or raceways installed underground shall be considered to be a wet location. Insulated conductors and cables installed in these enclosures or raceways in underground installations shall be listed for use in wet locations and shall comply with 310.10(C). Any

Table 300.50 Minimum Cover[a] Requirements

	General Conditions (not otherwise specified)						Special Conditions (use if applicable)					
	Column 1		Column 2		Column 3		Column 4		Column 5		Column 6	
Circuit Voltage	Direct-Buried Cables[b]		RTRC, PVC, and HDPE Conduit[c]		Rigid Metal Conduit and Intermediate Metal Conduit		Raceways Under Buildings or Exterior Concrete Slabs, 100 mm (4 in.) Minimum Thickness[d]		Cables in Airport Runways or Adjacent Areas Where Trespass Is Prohibited		Areas Subject to Vehicular Traffic, Such as Thoroughfares and Commercial Parking Areas	
	mm	in.	mm	in.	mm	in.	mm	in.	mm	in.	mm	in.
Over 1000 V through 22 kV	750	30	450	18	150	6	100	4	450	18	600	24
Over 22 kV through 40 kV	900	36	600	24	150	6	100	4	450	18	600	24
Over 40 kV	1000	42	750	30	150	6	100	4	450	18	600	24

General Notes:
1. Lesser depths shall be permitted where cables and conductors rise for terminations or splices or where access is otherwise required.
2. Where solid rock prevents compliance with the cover depths specified in this table, the wiring shall be installed in a metal or nonmetallic raceway permitted for direct burial. The raceways shall be covered by a minimum of 50 mm (2 in.) of concrete extending down to rock.
3. In industrial establishments, where conditions of maintenance and supervision ensure that qualified persons will service the installation, the minimum cover requirements, for other than rigid metal conduit and intermediate metal conduit, shall be permitted to be reduced 150 mm (6 in.) for each 50 mm (2 in.) of concrete or equivalent placed entirely within the trench over the underground installation.

Specific Footnotes:
[a]Cover is defined as the shortest distance in millimeters (inches) measured between a point on the top surface of any direct-buried conductor, cable, conduit, or other raceway and the top surface of finished grade, concrete, or similar cover.
[b]Underground direct-buried cables that are not encased or protected by concrete and are buried 750 mm (30 in.) or more below grade shall have their location identified by a warning ribbon that is placed in the trench at least 300 mm (12 in.) above the cables.
[c]Listed by a qualified testing agency as suitable for direct burial without encasement. All other nonmetallic systems shall require 50 mm (2 in.) of concrete or equivalent above conduit in addition to the table depth.
[d]The slab shall extend a minimum of 150 mm (6 in.) beyond the underground installation, and a warning ribbon or other effective means suitable for the conditions shall be placed above the underground installation.

connections or splices in an underground installation shall be approved for wet locations.

(C) Protection from Damage. Conductors emerging from the ground shall be enclosed in listed raceways. Raceways installed on poles shall be of rigid metal conduit, intermediate metal conduit, RTRC-XW, Schedule 80 PVC conduit, or equivalent, extending from the minimum cover depth specified in Table 300.50 to a point 2.5 m (8 ft) above finished grade. Conductors entering a building shall be protected by an approved enclosure or raceway from the minimum cover depth to the point of entrance. Where direct-buried conductors, raceways, or cables are subject to movement by settlement or frost, they shall be installed to prevent damage to the enclosed conductors or to the equipment connected to the raceways. Metallic enclosures shall be grounded.

(D) Splices. Direct burial cables shall be permitted to be spliced or tapped without the use of splice boxes, provided they are installed using materials suitable for the application. The taps and splices shall be watertight and protected from mechanical damage. Where cables are shielded, the shielding shall be continuous across the splice or tap.

Exception: At splices of an engineered cabling system, metallic shields of direct-buried single-conductor cables with maintained spacing between phases shall be permitted to be interrupted and overlapped. Where shields are interrupted and overlapped, each shield section shall be grounded at one point.

(E) Backfill. Backfill containing large rocks, paving materials, cinders, large or sharply angular substances, or corrosive materials shall not be placed in an excavation where materials can damage or contribute to the corrosion of raceways, cables, or other substructures or where it may prevent adequate compaction of fill.

Protection in the form of granular or selected material or suitable sleeves shall be provided to prevent physical damage to the raceway or cable.

(F) Raceway Seal. Where a raceway enters from an underground system, the end within the building shall be sealed with an identified compound so as to prevent the entrance of moisture or gases, or it shall be so arranged to prevent moisture from contacting live parts.

ARTICLE 310
Conductors for General Wiring

Part I. General

310.1 Scope. This article covers general requirements for conductors and their type designations, insulations, markings, mechanical strengths, ampacity ratings, and uses. These requirements do not apply to conductors that form an integral part of equipment, such as motors, motor controllers, and similar equipment, or to conductors specifically provided for elsewhere in this *Code*.

Informational Note: For flexible cords and cables, see Article 400. For fixture wires, see Article 402.

310.2 Definitions.

Electrical Ducts. Electrical conduits, or other raceways round in cross section, that are suitable for use underground or embedded in concrete.

Thermal Resistivity. As used in this *Code*, the heat transfer capability through a substance by conduction.

Informational Note: Thermal resistivity is the reciprocal of thermal conductivity and is designated Rho, which is expressed in the units °C-cm/W.

Part II. Installation

310.10 Uses Permitted. The conductors described in 310.104 shall be permitted for use in any of the wiring methods covered in Chapter 3 and as specified in their respective tables or as permitted elsewhere in this *Code*.

(A) Dry Locations. Insulated conductors and cables used in dry locations shall be any of the types identified in this *Code*.

(B) Dry and Damp Locations. Insulated conductors and cables used in dry and damp locations shall be Types FEP, FEPB, MTW, PFA, RHH, RHW, RHW-2, SA, THHN, THW, THW-2, THHW, THWN, THWN-2, TW, XHH, XHHW, XHHW-2, Z, or ZW.

(C) Wet Locations. Insulated conductors and cables used in wet locations shall comply with one of the following:

(1) Be moisture-impervious metal-sheathed
(2) Be types MTW, RHW, RHW-2, TW, THW, THW-2, THHW, THWN, THWN-2, XHHW, XHHW-2, or ZW
(3) Be of a type listed for use in wet locations

(D) Locations Exposed to Direct Sunlight. Insulated conductors or cables used where exposed to direct rays of the sun shall comply with (D)(1) or (D)(2):

(1) Conductors and cables shall be listed, or listed and marked, as being sunlight resistant
(2) Conductors and cables shall be covered with insulating material, such as tape or sleeving, that is listed, or listed and marked, as being sunlight resistant

(E) Shielding. Nonshielded, ozone-resistant insulated conductors with a maximum phase-to-phase voltage of 5000 volts shall be permitted in Type MC cables in industrial establishments where the conditions of maintenance and supervision ensure that only qualified persons service the installation. For other establishments, solid dielectric insulated conductors operated above 2000 volts in permanent installations shall have ozone-resistant insulation and shall be shielded. All metallic insulation shields shall be connected to a grounding electrode conductor, a grounding busbar, an equipment grounding conductor, or a grounding electrode.

Informational Note: The primary purposes of shielding are to confine the voltage stresses to the insulation, dissipate insulation

leakage current, drain off the capacitive charging current, and carry ground-fault current to facilitate operation of ground-fault protective devices in the event of an electrical cable fault.

Exception No. 1: Nonshielded insulated conductors listed by a qualified testing laboratory shall be permitted for use up to 2400 volts under the following conditions:

(a) Conductors shall have insulation resistant to electric discharge and surface tracking, or the insulated conductor(s) shall be covered with a material resistant to ozone, electric discharge, and surface tracking.

(b) Where used in wet locations, the insulated conductor(s) shall have an overall nonmetallic jacket or a continuous metallic sheath.

(c) Insulation and jacket thicknesses shall be in accordance with Table 310.104(D).

Exception No. 2: Nonshielded insulated conductors listed by a qualified testing laboratory shall be permitted for use up to 5000 volts to replace existing nonshielded conductors, on existing equipment in industrial establishments only, under the following conditions:

(a) Where the condition of maintenance and supervision ensures that only qualified personnel install and service the installation.

(b) Conductors shall have insulation resistant to electric discharge and surface tracking, or the insulated conductor(s) shall be covered with a material resistant to ozone, electric discharge, and surface tracking.

(c) Where used in wet locations, the insulated conductor(s) shall have an overall nonmetallic jacket or a continuous metallic sheath.

(d) Insulation and jacket thicknesses shall be in accordance with Table 310.104(D).

Informational Note: Relocation or replacement of equipment may not comply with the term *existing* as related to this exception.

Exception No. 3: Where permitted in 310.10(F), Exception No. 2.

(F) Direct-Burial Conductors. Conductors used for direct-burial applications shall be of a type identified for such use.

Cables rated above 2000 volts shall be shielded.

Exception No. 1: Nonshielded multiconductor cables rated 2001–2400 volts shall be permitted if the cable has an overall metallic sheath or armor.

The metallic shield, sheath, or armor shall be connected to a grounding electrode conductor, grounding busbar, or a grounding electrode.

Exception No. 2: Airfield lighting cable used in series circuits that are rated up to 5000 volts and are powered by regulators shall be permitted to be nonshielded.

Informational Note to Exception No. 2: Federal Aviation Administration (FAA) Advisory Circulars (ACs) provide additional practices and methods for airport lighting.

Informational Note No. 1: See 300.5 for installation requirements for conductors rated 1000 volts or less.

Informational Note No. 2: See 300.50 for installation requirements for conductors rated over 1000 volts.

(G) Corrosive Conditions. Conductors exposed to oils, greases, vapors, gases, fumes, liquids, or other substances having a deleterious effect on the conductor or insulation shall be of a type suitable for the application.

(H) Conductors in Parallel.

(1) General. Aluminum, copper-clad aluminum, or copper conductors, for each phase, polarity, neutral, or grounded circuit shall be permitted to be connected in parallel (electrically joined at both ends) only in sizes 1/0 AWG and larger where installed in accordance with 310.10(H)(2) through (H)(6).

Exception No. 1: Conductors in sizes smaller than 1/0 AWG shall be permitted to be run in parallel to supply control power to indicating instruments, contactors, relays, solenoids, and similar control devices, or for frequencies of 360 Hz and higher, provided all of the following apply:

(a) They are contained within the same raceway or cable.

(b) The ampacity of each individual conductor is sufficient to carry the entire load current shared by the parallel conductors.

(c) The overcurrent protection is such that the ampacity of each individual conductor will not be exceeded if one or more of the parallel conductors become inadvertently disconnected.

Exception No. 2: Under engineering supervision, 2 AWG and 1 AWG grounded neutral conductors shall be permitted to be installed in parallel for existing installations.

> Informational Note to Exception No. 2: Exception No. 2 can be used to alleviate overheating of neutral conductors in existing installations due to high content of triplen harmonic currents.

(2) Conductor and Installation Characteristics. The paralleled conductors in each phase, polarity, neutral, grounded circuit conductor, equipment grounding conductor, or equipment bonding jumper shall comply with all of the following:

(1) Be the same length.
(2) Consist of the same conductor material.
(3) Be the same size in circular mil area.
(4) Have the same insulation type.
(5) Be terminated in the same manner.

(3) Separate Cables or Raceways. Where run in separate cables or raceways, the cables or raceways with conductors shall have the same number of conductors and shall have the same electrical characteristics. Conductors of one phase, polarity, neutral, grounded circuit conductor, or equipment grounding conductor shall not be required to have the same physical characteristics as those of another phase, polarity, neutral, grounded circuit conductor, or equipment grounding conductor.

(4) Ampacity Adjustment. Conductors installed in parallel shall comply with the provisions of 310.15(B)(3)(a).

(5) Equipment Grounding Conductors. Where parallel equipment grounding conductors are used, they shall be sized in accordance with 250.122. Sectioned equipment grounding conductors smaller than 1/0 AWG shall be permitted in multiconductor cables, if the combined circular mil area of the sectioned equipment grounding conductors in each cable complies with 250.122.

(6) Bonding Jumpers. Where parallel equipment bonding jumpers or supply-side bonding jumpers are installed in raceways, they shall be sized and installed in accordance with 250.102.

310.15 Ampacities for Conductors Rated 0–2000 Volts.

(A) General.

(1) Tables or Engineering Supervision. Ampacities for conductors shall be permitted to be determined by tables as provided in 310.15(B) or under engineering supervision, as provided in 310.15(C).

> Informational Note No. 1: Ampacities provided by this section do not take voltage drop into consideration. See 210.19(A), Informational Note No. 4, for branch circuits and 215.2(A), Informational Note No. 2, for feeders.

> Informational Note No. 2: For the allowable ampacities of Type MTW wire, see Table 13.5.1 in NFPA 79-2015, *Electrical Standard for Industrial Machinery.*

(2) Selection of Ampacity. Where more than one ampacity applies for a given circuit length, the lowest value shall be used.

Exception: Where different ampacities apply to portions of a circuit, the higher ampacity shall be permitted to be used if the total portion(s) of the circuit with lower ampacity does not exceed the lesser of 3.0 m (10 ft) or 10 percent of the total circuit.

> Informational Note: See 110.14(C) for conductor temperature limitations due to termination provisions.

(3) Temperature Limitation of Conductors. No conductor shall be used in such a manner that its operating temperature exceeds that designated for the type of insulated conductor involved. In no case shall conductors be associated together in such a way, with respect to type of circuit, the wiring method employed, or the number of conductors, that the limiting temperature of any conductor is exceeded.

> Informational Note No. 1: The temperature rating of a conductor [see Table 310.104(A) and Table 310.104(C)] is the maximum temperature, at any location along its length, that the conductor can withstand over a prolonged time period without serious degradation. The allowable ampacity tables, the ampacity tables of Article 310 and the ampacity tables of Informative Annex B, the ambient temperature correction factors in 310.15(B)(2), and the notes to the tables provide guidance for coordinating conductor sizes, types, allowable ampacities, ampacities, ambient temperatures, and number of associated conductors. The principal determinants of operating temperature are as follows:
>
> (1) Ambient temperature — ambient temperature may vary along the conductor length as well as from time to time.
> (2) Heat generated internally in the conductor as the result of load current flow, including fundamental and harmonic currents.
> (3) The rate at which generated heat dissipates into the ambient medium. Thermal insulation that covers or surrounds conductors affects the rate of heat dissipation.
> (4) Adjacent load-carrying conductors — adjacent conductors have the dual effect of raising the ambient temperature and impeding heat dissipation.

> Informational Note No. 2: Refer to 110.14(C) for the temperature limitation of terminations.

(B) Tables. Ampacities for conductors rated 0 to 2000 volts shall be as specified in the Allowable Ampacity Table 310.15(B)(16) through Table 310.15(B)(19), and Ampacity Table 310.15(B)(20) and Table 310.15(B)(21) as modified by 310.15(B)(1) through (B)(7).

The temperature correction and adjustment factors shall be permitted to be applied to the ampacity for the temperature rating of the conductor, if the corrected and adjusted ampacity does not exceed the ampacity for the temperature rating of the termination in accordance with the provisions of 110.14(C).

Informational Note: Table 310.15(B)(16) through Table 310.15(B)(19) are application tables for use in determining conductor sizes on loads calculated in accordance with Article 220. Allowable ampacities result from consideration of one or more of the following:

(1) Temperature compatibility with connected equipment, especially the connection points.
(2) Coordination with circuit and system overcurrent protection.
(3) Compliance with the requirements of product listings or certifications. See 110.3(B).
(4) Preservation of the safety benefits of established industry practices and standardized procedures.

(1) General. For explanation of type letters used in tables and for recognized sizes of conductors for the various conductor insulations, see Table 310.104(A) and Table 310.104(B). For installation requirements, see 310.1 through 310.15(A)(3) and the various articles of this *Code.* For flexible cords, see Table 400.4, Table 400.5(A)(1), and Table 400.5(A)(2).

(2) Ambient Temperature Correction Factors. Ampacities for ambient temperatures other than those shown in the ampacity tables shall be corrected in accordance with Table 310.15(B)(2)(a) or Table 310.15(B)(2)(b), or shall be permitted to be calculated using the following equation:

$$[310.15(B)(2)]$$

$$I' = I \sqrt{\frac{T_c - T_a'}{T_c - T_a}}$$

where:

I' – ampacity corrected for ambient temperature
I = ampacity shown in the tables
T_c – temperature rating of conductor (°C)
T_a' = new ambient temperature (°C)
T_a = ambient temperature used in the table (°C)

(3) Adjustment Factors.

(a) *More than Three Current-Carrying Conductors.* Where the number of current-carrying conductors in a raceway or cable exceeds three, or where single conductors or multiconductor cables are installed without maintaining spacing for a continuous length longer than 600 mm (24 in.) and are not installed in raceways, the allowable ampacity of each conductor shall be reduced as shown in Table 310.15(B)(3)(a). Each current-carrying conductor of a paralleled set of conductors shall be counted as a current-carrying conductor.

Where conductors of different systems, as provided in 300.3, are installed in a common raceway or cable, the adjustment factors shown in Table 310.15(B)(3)(a) shall apply only to the number of power and lighting conductors (Articles 210, 215, 220, and 230).

Informational Note No. 1: See Annex B for adjustment factors for more than three current-carrying conductors in a raceway or cable with load diversity.

Informational Note No. 2: See 366.23 for adjustment factors for conductors and ampacity for bare copper and aluminum bars in

Table 310.15(B)(2)(a) Ambient Temperature Correction Factors Based on 30°C (86°F)

For ambient temperatures other than 30°C (86°F), multiply the allowable ampacities specified in the ampacity tables by the appropriate correction factor shown below.

Ambient Temperature (°C)	Temperature Rating of Conductor			Ambient Temperature (°F)
	60°C	75°C	90°C	
10 or less	1.29	1.20	1.15	50 or less
11–15	1.22	1.15	1.12	51–59
16–20	1.15	1.11	1.08	60–68
21–25	1.08	1.05	1.04	69–77
26–30	1.00	1.00	1.00	78–86
31–35	0.91	0.94	0.96	87–95
36–40	0.82	0.88	0.91	96–104
41–45	0.71	0.82	0.87	105–113
46–50	0.58	0.75	0.82	114–122
51–55	0.41	0.67	0.76	123–131
56–60	—	0.58	0.71	132–140
61–65	—	0.47	0.65	141–149
66–70	—	0.33	0.58	150–158
71–75	—	—	0.50	159–167
76–80	—	—	0.41	168–176
81–85	—	—	0.29	177–185

auxiliary gutters and 376.22(B) for adjustment factors for conductors in metal wireways.

(1) Where conductors are installed in cable trays, the provisions of 392.80 shall apply.

(2) Adjustment factors shall not apply to conductors in raceways having a length not exceeding 600 mm (24 in.).

(3) Adjustment factors shall not apply to underground conductors entering or leaving an outdoor trench if those conductors have physical protection in the form of rigid metal conduit, intermediate metal conduit, rigid polyvinyl chloride conduit (PVC), or reinforced thermosetting resin conduit (RTRC) having a length not exceeding 3.05 m (10 ft), and if the number of conductors does not exceed four.

(4) Adjustment factors shall not apply to Type AC cable or to Type MC cable under the following conditions:

a. The cables do not have an overall outer jacket.

b. Each cable has not more than three current-carrying conductors.

c. The conductors are 12 AWG copper.

d. Not more than 20 current-carrying conductors are installed without maintaining spacing, are stacked, or are supported on "bridle rings."

Exception to (4): If cables meeting the requirements in 310.15(B)(3)(4)a through c with more than 20 current-carrying conductors are installed longer than 600 mm (24 in.) without maintaining spacing, are stacked, or are supported on bridle rings, a 60 percent adjustment factor shall be applied.

(b) *Raceway Spacing.* Spacing between raceways shall be maintained.

Table 310.15(B)(2)(b) Ambient Temperature Correction Factors Based on 40°C (104°F)

For ambient temperatures other than 40°C (104°F), multiply the allowable ampacities specified in the ampacity tables by the appropriate correction factor shown below.

Ambient Temperature (°C)	Temperature Rating of Conductor						Ambient Temperature (°F)
	60°C	75°C	90°C	150°C	200°C	250°C	
10 or less	1.58	1.36	1.26	1.13	1.09	1.07	50 or less
11–15	1.50	1.31	1.22	1.11	1.08	1.06	51–59
16–20	1.41	1.25	1.18	1.09	1.06	1.05	60–68
21–25	1.32	1.2	1.14	1.07	1.05	1.04	69–77
26–30	1.22	1.13	1.10	1.04	1.03	1.02	78–86
31–35	1.12	1.07	1.05	1.02	1.02	1.01	87–95
36–40	1.00	1.00	1.00	1.00	1.00	1.00	96–104
41–45	0.87	0.93	0.95	0.98	0.98	0.99	105–113
46–50	0.71	0.85	0.89	0.95	0.97	0.98	114–122
51–55	0.50	0.76	0.84	0.93	0.95	0.96	123–131
56–60	—	0.65	0.77	0.90	0.94	0.95	132–140
61–65	—	0.53	0.71	0.88	0.92	0.94	141–149
66–70	—	0.38	0.63	0.85	0.90	0.93	150–158
71–75	—	—	0.55	0.83	0.88	0.91	159–167
76–80	—	—	0.45	0.80	0.87	0.90	168–176
81–90	—	—	—	0.74	0.83	0.87	177–194
91–100	—	—	—	0.67	0.79	0.85	195–212
101–110	—	—	—	0.60	0.75	0.82	213–230
111–120	—	—	—	0.52	0.71	0.79	231–248
121–130	—	—	—	0.43	0.66	0.76	249–266
131–140	—	—	—	0.30	0.61	0.72	267–284
141–160	—	—	—	—	0.50	0.65	285–320
161–180	—	—	—	—	0.35	0.58	321–356
181–200	—	—	—	—	—	0.49	357–392
201–225	—	—	—	—	—	0.35	393–437

(c) *Raceways and Cables Exposed to Sunlight on Rooftops.* Where raceways or cables are exposed to direct sunlight on or above rooftops, raceways or cables shall be installed a minimum distance above the roof to the bottom of the raceway or cable of 23 mm (⅞ in.). Where the distance above the roof to the bottom of the raceway is less than 23 mm (⅞ in.), a temperature adder of 33°C (60°F) shall be added to the outdoor temperature to determine the applicable ambient temperature for application of the correction factors in Table 310.15(B)(2)(a) or Table 310.15(B)(2)(b).

Exception: Type XHHW-2 insulated conductors shall not be subject to this ampacity adjustment.

Informational Note: One source for the ambient temperatures in various locations is the ASHRAE *Handbook — Fundamentals.*

(4) Bare or Covered Conductors. Where bare or covered conductors are installed with insulated conductors, the temperature rating of the bare or covered conductor shall be equal to the lowest temperature rating of the insulated conductors for the purpose of determining ampacity.

(5) Neutral Conductor.

(a) A neutral conductor that carries only the unbalanced current from other conductors of the same circuit shall not be required to be counted when applying the provisions of 310.15(B)(3)(a).

(b) In a 3-wire circuit consisting of two phase conductors and the neutral conductor of a 4-wire, 3-phase, wye-connected system, a common conductor carries approximately the same current as the line-to-neutral load currents of the other

Table 310.15(B)(3)(a) Adjustment Factors for More Than Three Current-Carrying Conductors

Number of Conductors[1]	Percent of Values in Table 310.15(B)(16) Through Table 310.15(B)(19) as Adjusted for Ambient Temperature if Necessary
4–6	80
7–9	70
10–20	50
21–30	45
31–40	40
41 and above	35

[1]Number of conductors is the total number of conductors in the raceway or cable, including spare conductors. The count shall be adjusted in accordance with 310.15(B)(5) and (6). The count shall not include conductors that are connected to electrical components that cannot be simultaneously energized.

conductors and shall be counted when applying the provisions of 310.15(B)(3)(a).

(c) On a 4-wire, 3-phase wye circuit where the major portion of the load consists of nonlinear loads, harmonic currents are present in the neutral conductor; the neutral conductor shall therefore be considered a current-carrying conductor.

(6) Grounding or Bonding Conductor. A grounding or bonding conductor shall not be counted when applying the provisions of 310.15(B)(3)(a).

(7) Single-Phase Dwelling Services and Feeders. For one-family dwellings and the individual dwelling units of two-family and multifamily dwellings, service and feeder conductors supplied by a single-phase, 120/240-volt system shall be permitted to be sized in accordance with 310.15(B)(7)(1) through (4).

For one-family dwellings and the individual dwelling units of two-family and multifamily dwellings, single-phase feeder conductors consisting of 2 ungrounded conductors and the neutral conductor from a 208Y/120 volt system shall be permitted to be sized in accordance with 310.15(B)(7)(1) through (3).

(1) For a service rated 100 through 400 amperes, the service conductors supplying the entire load associated with a one-family dwelling, or the service conductors supplying the entire load associated with an individual dwelling unit in a two-family or multifamily dwelling, shall be permitted to have an ampacity not less than 83 percent of the service rating.

(2) For a feeder rated 100 through 400 amperes, the feeder conductors supplying the entire load associated with a one-family dwelling, or the feeder conductors supplying the entire load associated with an individual dwelling unit in a two-family or multifamily dwelling, shall be permitted to have an ampacity not less than 83 percent of the feeder rating.

(3) In no case shall a feeder for an individual dwelling unit be required to have an ampacity greater than that specified in 310.15(B)(7)(1) or (2).

(4) Grounded conductors shall be permitted to be sized smaller than the ungrounded conductors, if the requirements of 220.61 and 230.42 for service conductors or the requirements of 215.2 and 220.61 for feeder conductors are met.

Where correction or adjustment factors are required by 310.15(B)(2) or (3), they shall be permitted to be applied to the ampacity associated with the temperature rating of the conductor.

Informational Note No. 1: The service or feeder ratings addressed by this section are based on the standard ampacity ratings from 240.6(A).

Informational Note No. 2: See Example D7 in Annex D.

(C) Engineering Supervision. Under engineering supervision, conductor ampacities shall be permitted to be calculated by means of the following general equation:

[310.15(C)]

$$I = \sqrt{\frac{T_c - T_a}{R_{dc}(1 + Y_c)R_{ca}}} \times 10^3 \, \text{amperes}$$

where:

T_c = conductor temperature in degrees Celsius (°C)

T_a = ambient temperature in degrees Celsius (°C)

R_{dc} = dc resistance of 305 mm (1 ft) of conductor in micro-ohms at temperature, T_c

Y_c = component ac resistance resulting from skin effect and proximity effect

R_{ca} = effective thermal resistance between conductor and surrounding ambient

310.60 Conductors Rated 2001 to 35,000 Volts.

(A) Ampacities of Conductors Rated 2001 to 35,000 Volts. Ampacities for solid dielectric-insulated conductors shall be permitted to be determined by tables or under engineering supervision, as provided in 310.60(B) and (C).

(1) Selection of Ampacity. Where more than one calculated or tabulated ampacity could apply for a given circuit length, the lowest value shall be used.

Exception: Where two different ampacities apply to adjacent portions of a circuit, the higher ampacity shall be permitted to be used beyond the point of transition, a distance equal to 3.0 m (10 ft) or 10 percent of the circuit length calculated at the higher ampacity, whichever is less.

Informational Note: See 110.40 for conductor temperature limitations due to termination provisions.

(B) Engineering Supervision. Under engineering supervision, conductor ampacities shall be permitted to be calculated by using the following general equation:

[310.60(B)]

$$I = \sqrt{\frac{T_c - (T_a + \Delta T_d)}{R_{dc}(1 + Y_c)R_{ca}}} \times 10^3 \, \text{amperes}$$

where:

T_c = conductor temperature (°C)

T_a = ambient temperature (°C)

ΔT_d = dielectric loss temperature rise

R_{dc} = dc resistance of conductor at temperature T_c

Y_c = component ac resistance resulting from skin effect and proximity effect

R_{ca} = effective thermal resistance between conductor and surrounding ambient

Informational Note: The dielectric loss temperature rise (ΔT_d) is negligible for single circuit extruded dielectric cables rated below 46 kV.

Table 310.15(B)(16) (formerly Table 310.16) Allowable Ampacities of Insulated Conductors Rated Up to and Including 2000 Volts, 60°C Through 90°C (140°F Through 194°F), Not More Than Three Current-Carrying Conductors in Raceway, Cable, or Earth (Directly Buried), Based on Ambient Temperature of 30°C (86°F)*

	Temperature Rating of Conductor [See Table 310.104(A).]						
Size AWG or kcmil	60°C (140°F)	75°C (167°F)	90°C (194°F)	60°C (140°F)	75°C (167°F)	90°C (194°F)	Size AWG or kcmil
	Types TW, UF	Types RHW, THHW, THW, THWN, XHHW, USE, ZW	Types TBS, SA, SIS, FEP, FEPB, MI, RHH, RHW-2, THHN, THHW, THW-2, THWN-2, USE-2, XHH, XHHW, XHHW-2, ZW-2	Types TW, UF	Types RHW, THHW, THW, THWN, XHHW, USE	Types TBS, SA, SIS, THHN, THHW, THW-2, THWN-2, RHH, RHW-2, USE-2, XHH, XHHW, XHHW-2, ZW-2	
	COPPER			ALUMINUM OR COPPER-CLAD ALUMINUM			
18**	—	—	14	—	—	—	—
16**	—	—	18	—	—	—	—
14**	15	20	25	—	—	—	—
12**	20	25	30	15	20	25	12**
10**	30	35	40	25	30	35	10**
8	40	50	55	35	40	45	8
6	55	65	75	40	50	55	6
4	70	85	95	55	65	75	4
3	85	100	115	65	75	85	3
2	95	115	130	75	90	100	2
1	110	130	145	85	100	115	1
1/0	125	150	170	100	120	135	1/0
2/0	145	175	195	115	135	150	2/0
3/0	165	200	225	130	155	175	3/0
4/0	195	230	260	150	180	205	4/0
250	215	255	290	170	205	230	250
300	240	285	320	195	230	260	300
350	260	310	350	210	250	280	350
400	280	335	380	225	270	305	400
500	320	380	430	260	310	350	500
600	350	420	475	285	340	385	600
700	385	460	520	315	375	425	700
750	400	475	535	320	385	435	750
800	410	490	555	330	395	445	800
900	435	520	585	355	425	480	900
1000	455	545	615	375	445	500	1000
1250	495	590	665	405	485	545	1250
1500	525	625	705	435	520	585	1500
1750	545	650	735	455	545	615	1750
2000	555	665	750	470	560	630	2000

*Refer to 310.15(B)(2) for the ampacity correction factors where the ambient temperature is other than 30°C (86°F). Refer to 310.15(B)(3)(a) for more than three current-carrying conductors.

**Refer to 240.4(D) for conductor overcurrent protection limitations.

Table 310.15(B)(17) (formerly Table 310.17) Allowable Ampacities of Single-Insulated Conductors Rated Up to and Including 2000 Volts in Free Air, Based on Ambient Temperature of 30°C (86°F)*

Size AWG or kcmil	Temperature Rating of Conductor [See Table 310.104(A).]						Size AWG or kcmil
	60°C (140°F)	75°C (167°F)	90°C (194°F)	60°C (140°F)	75°C (167°F)	90°C (194°F)	
	Types TW, UF	Types RHW, THHW, THW, THWN, XHHW, ZW	Types TBS, SA, SIS, FEP, FEPB, MI, RHH, RHW-2, THHN, THHW, THW-2, THWN-2, USE-2, XHH, XHHW, XHHW-2, ZW-2	Types TW, UF	Types RHW, THHW, THW, THWN, XHHW	Types TBS, SA, SIS, THHN, THHW, THW-2, THWN-2, RHH, RHW-2, USE-2, XHH, XHHW, XHHW-2, ZW-2	
	COPPER			ALUMINUM OR COPPER-CLAD ALUMINUM			
18	—	—	18	—	—	—	—
16	—	—	24	—	—	—	—
14**	25	30	35	—	—	—	—
12**	30	35	40	25	30	35	12**
10**	40	50	55	35	40	45	10**
8	60	70	80	45	55	60	8
6	80	95	105	60	75	85	6
4	105	125	140	80	100	115	4
3	120	145	165	95	115	130	3
2	140	170	190	110	135	150	2
1	165	195	220	130	155	175	1
1/0	195	230	260	150	180	205	1/0
2/0	225	265	300	175	210	235	2/0
3/0	260	310	350	200	240	270	3/0
4/0	300	360	405	235	280	315	4/0
250	340	405	455	265	315	355	250
300	375	445	500	290	350	395	300
350	420	505	570	330	395	445	350
400	455	545	615	355	425	480	400
500	515	620	700	405	485	545	500
600	575	690	780	455	545	615	600
700	630	755	850	500	595	670	700
750	655	785	885	515	620	700	750
800	680	815	920	535	645	725	800
900	730	870	980	580	700	790	900
1000	780	935	1055	625	750	845	1000
1250	890	1065	1200	710	855	965	1250
1500	980	1175	1325	795	950	1070	1500
1750	1070	1280	1445	875	1050	1185	1750
2000	1155	1385	1560	960	1150	1295	2000

*Refer to 310.15(B)(2) for the ampacity correction factors where the ambient temperature is other than 30°C (86°F).
**Refer to 240.4(D) for conductor overcurrent protection limitations.

Table 310.15(B)(18) (formerly Table 310.18) Allowable Ampacities of Insulated Conductors Rated Up to and Including 2000 Volts, 150°C Through 250°C (302°F Through 482°F). Not More Than Three Current-Carrying Conductors in Raceway or Cable, Based on Ambient Air Temperature of 40°C (104°F)*

	Temperature Rating of Conductor [See Table 310.104(A).]				
	150°C (302°F)	200°C (392°F)	250°C (482°F)	150°C (302°F)	
	Type Z	Types FEP, FEPB, PFA, SA	Types PFAH, TFE	Type Z	
Size AWG or kcmil	COPPER		NICKEL OR NICKEL-COATED COPPER	ALUMINUM OR COPPER-CLAD ALUMINUM	Size AWG or kcmil
14	34	36	39	—	14
12	43	45	54	30	12
10	55	60	73	44	10
8	76	83	93	57	8
6	96	110	117	75	6
4	120	125	148	94	4
3	143	152	166	109	3
2	160	171	191	124	2
1	186	197	215	145	1
1/0	215	229	244	169	1/0
2/0	251	260	273	198	2/0
3/0	288	297	308	227	3/0
4/0	332	346	361	260	4/0

*Refer to 310.15(B)(2) for the ampacity correction factors where the ambient temperature is other than 40°C (104°F). Refer to 310.15(B)(3)(a) for more than three current-carrying conductors.

Table 310.15(B)(19) (formerly Table 310.19) Allowable Ampacities of Single-Insulated Conductors, Rated Up to and Including 2000 Volts, 150°C Through 250°C (302°F Through 482°F), in Free Air, Based on Ambient Air Temperature of 40°C (104°F)*

	Temperature Rating of Conductor [See Table 310.104(A).]				
	150°C (302°F)	200°C (392°F)	250°C (482°F)	150°C (302°F)	
	Type Z	Types FEP, FEPB, PFA, SA	Types PFAH, TFE	Type Z	
Size AWG or kcmil	COPPER		NICKEL, OR NICKEL-COATED COPPER	ALUMINUM OR COPPER-CLAD ALUMINUM	Size AWG or kcmil
14	46	54	59	—	14
12	60	68	78	47	12
10	80	90	107	63	10
8	106	124	142	83	8
6	155	165	205	112	6
4	190	220	278	148	4
3	214	252	327	170	3
2	255	293	381	198	2
1	293	344	440	228	1
1/0	339	399	532	263	1/0
2/0	390	467	591	305	2/0
3/0	451	546	708	351	3/0
4/0	529	629	830	411	4/0

*Refer to 310.15(B)(2) for the ampacity correction factors where the ambient temperature is other than 40°C (104°F).

Table 310.15(B)(20) (formerly Table 310.20) Ampacities of Not More Than Three Single Insulated Conductors, Rated Up to and Including 2000 Volts, Supported on a Messenger, Based on Ambient Air Temperature of 40°C (104°F)*

	Temperature Rating of Conductor [See Table 310.104(A).]				
	75°C (167°F)	90°C (194°F)	75°C (167°F)	90°C (194°F)	
Size AWG or kcmil	Types RHW, THHW, THW, THWN, XHHW, ZW	Types MI, THHN, THHW, THW-2, THWN-2, RHH, RHW-2, USE-2, XHHW, XHHW-2, ZW-2	Types RHW, THW, THWN, THHW, XHHW	Types THHN, THHW, RHH, XHHW, RHW-2, XHHW-2, THW-2, THWN-2, USE-2, ZW-2	Size AWG or kcmil
	COPPER		ALUMINUM OR COPPER-CLAD ALUMINUM		
8	57	66	44	51	8
6	76	89	59	69	6
4	101	117	78	91	4
3	118	138	92	107	3
2	135	158	106	123	2
1	158	185	123	144	1
1/0	183	214	143	167	1/0
2/0	212	247	165	193	2/0
3/0	245	287	192	224	3/0
4/0	287	335	224	262	4/0
250	320	374	251	292	250
300	359	419	282	328	300
350	397	464	312	364	350
400	430	503	339	395	400
500	496	580	392	458	500
600	553	647	440	514	600
700	610	714	488	570	700
750	638	747	512	598	750
800	660	773	532	622	800
900	704	826	572	669	900
1000	748	879	612	716	1000

*Refer to 310.15(B)(2) for the ampacity correction factors where the ambient temperature is other than 40°C (104°F). Refer to 310.15(B)(3)(a) for more than three current-carrying conductors.

(C) Tables. Ampacities for conductors rated 2001 to 35,000 volts shall be as specified in Table 310.60(C)(67) through Table 310.60(C)(86). Ampacities for ambient temperatures other than those specified in the ampacity tables shall be corrected in accordance with 310.60(C)(4).

Informational Note No. 1: For ampacities calculated in accordance with 310.60(A), reference IEEE 835-1994, *Standard Power Cable Ampacity Tables*, and the references therein for availability of all factors and constants.

Informational Note No. 2: Ampacities provided by this section do not take voltage drop into consideration. See 210.19(A), Informational Note No. 4, for branch circuits and 215.2(A), Informational Note No. 2, for feeders.

(1) Grounded Shields. Ampacities shown in Table 310.60(C)(69), Table 310.60(C)(70), Table 310.60(C)(81), and Table 310.60(C)(82) shall apply for cables with shields grounded at one point only. Where shields for these cables are grounded at more than one point, ampacities shall be adjusted to take into consideration the heating due to shield currents.

Informational Note: Tables other than those listed contain the ampacity of cables with shields grounded at multiple points.

(2) Burial Depth of Underground Circuits. Where the burial depth of direct burial or electrical duct bank circuits is modified from the values shown in a figure or table, ampacities shall be permitted to be modified as indicated in (B)(2)(a) and (B)(2)(b).

(a) Where burial depths are increased in part(s) of an electrical duct run, a decrease in ampacity of the conductors shall not be required, provided the total length of parts of the duct run increased in depth is less than 25 percent of the total run length.

(b) Where burial depths are deeper than shown in a specific underground ampacity table or figure, an ampacity derating factor of 6 percent per 300 mm (1 ft) increase in depth for all values of rho shall be permitted.

No ampacity adjustments shall be required where the burial depth is decreased.

(3) Electrical Ducts in Figure 310.60(C)(3). At locations where electrical ducts enter equipment enclosures from underground, spacing between such ducts, as shown in Figure 310.60(C)(3), shall be permitted to be reduced without requiring the ampacity of conductors therein to be reduced.

Table 310.15(B)(21) (formerly Table 310.21) Ampacities of Bare or Covered Conductors in Free Air, Based on 40°C (104°F) Ambient, 80°C (176°F) Total Conductor Temperature, 610 mm/sec (2 ft/sec) Wind Velocity

| Copper Conductors | | | | AAC Aluminum Conductors | | | |
| Bare | | Covered | | Bare | | Covered | |
AWG or kcmil	Amperes	AWG or kcmil	Amperes	AWG or kcmil	Amperes	AWG or kcmil	Amperes
8	98	8	103	8	76	8	80
6	124	6	130	6	96	6	101
4	155	4	163	4	121	4	127
2	209	2	219	2	163	2	171
1/0	282	1/0	297	1/0	220	1/0	231
2/0	329	2/0	344	2/0	255	2/0	268
3/0	382	3/0	401	3/0	297	3/0	312
4/0	444	4/0	466	4/0	346	4/0	364
250	494	250	519	266.8	403	266.8	423
300	556	300	584	336.4	468	336.4	492
500	773	500	812	397.5	522	397.5	548
750	1000	750	1050	477.0	588	477.0	617
1000	1193	1000	1253	556.5	650	556.5	682
—	—	—	—	636.0	709	636.0	744
—	—	—	—	795.0	819	795.0	860
—	—	—	—	954.0	920	—	—
—	—	—	—	1033.5	968	1033.5	1017
—	—	—	—	1272	1103	1272	1201
—	—	—	—	1590	1267	1590	1381
—	—	—	—	2000	1454	2000	1527

(4) Ambient Temperature Correction. Ampacities for ambient temperatures other than those specified in the ampacity tables shall be corrected in accordance with Table 310.60(C)(4) or shall be permitted to be calculated using the following equation:

[310.60(C)(4)]

$$I' = I\sqrt{\frac{T_c - T_a'}{T_c - T_a}}$$

where:
I' = ampacity corrected for ambient temperature
I = ampacity shown in the table for T_c and T_a
T_c = temperature rating of conductor (°C)
T_a' = new ambient temperature (°C)
T_a = ambient temperature used in the table (°C)

Part III. Construction Specifications

310.104 Conductor Constructions and Applications. Insulated conductors shall comply with the applicable provisions of Table 310.104(A) through Table 310.104(E).

Informational Note: Thermoplastic insulation may stiffen at temperatures lower than −10°C (+14°F). Thermoplastic insulation may also be deformed at normal temperatures where subjected to pressure, such as at points of support.

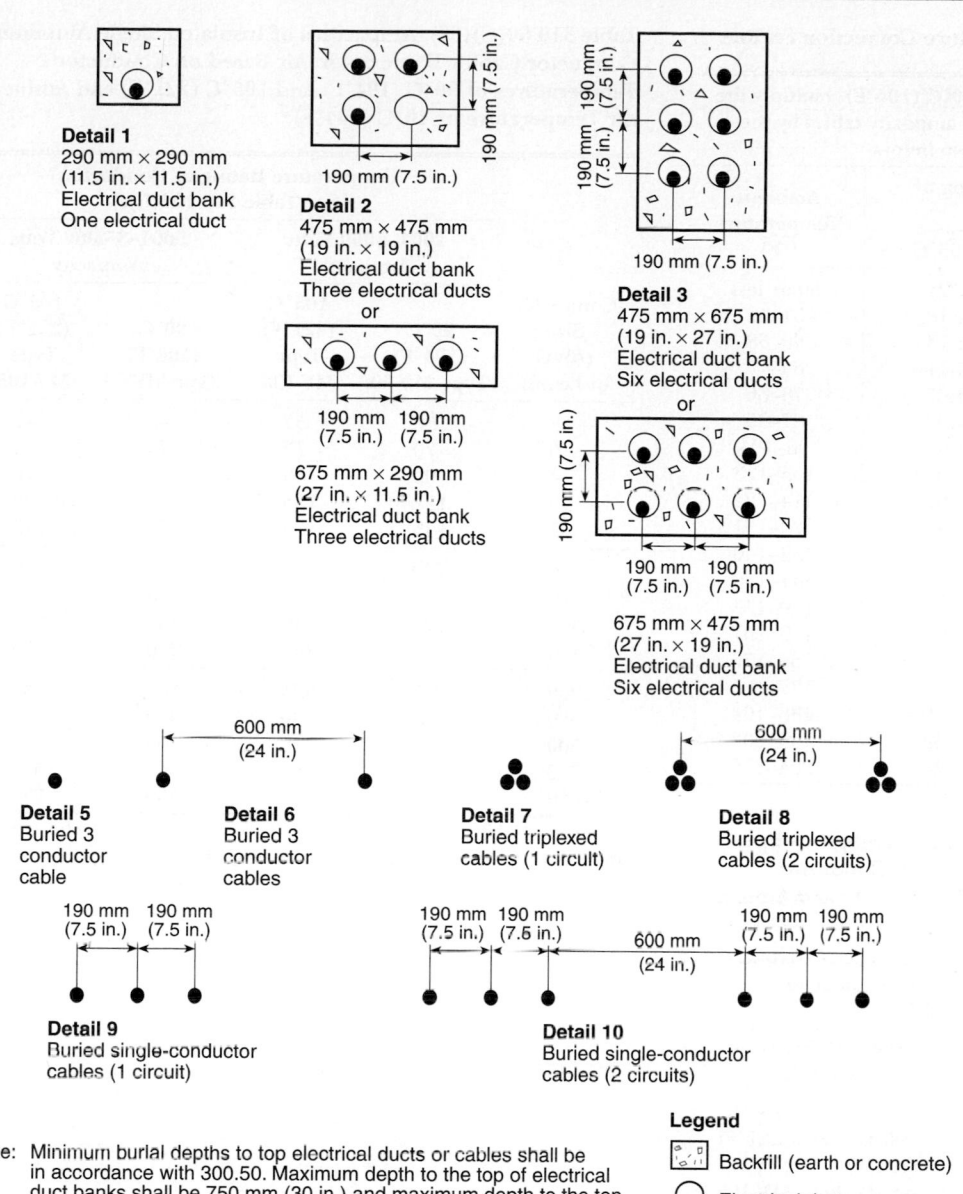

Detail 1
290 mm × 290 mm
(11.5 in. × 11.5 in.)
Electrical duct bank
One electrical duct

Detail 2
475 mm × 475 mm
(19 in. × 19 in.)
Electrical duct bank
Three electrical ducts
or
675 mm × 290 mm
(27 in. × 11.5 in.)
Electrical duct bank
Three electrical ducts

Detail 3
475 mm × 675 mm
(19 in. × 27 in.)
Electrical duct bank
Six electrical ducts
or
675 mm × 475 mm
(27 in. × 19 in.)
Electrical duct bank
Six electrical ducts

Detail 5
Buried 3
conductor
cable

Detail 6
Buried 3
conductor
cables

Detail 7
Buried triplexed
cables (1 circuit)

Detail 8
Buried triplexed
cables (2 circuits)

Detail 9
Buried single-conductor
cables (1 circuit)

Detail 10
Buried single-conductor
cables (2 circuits)

Legend

▢ Backfill (earth or concrete)

◯ Electrical duct

● Cable or cables

Note: Minimum burial depths to top electrical ducts or cables shall be in accordance with 300.50. Maximum depth to the top of electrical duct banks shall be 750 mm (30 in.) and maximum depth to the top of direct-buried cables shall be 900 mm (36 in.).

FIGURE 310.60(C)(3)   Cable Installation Dimensions for Use with Table 310.60(C)(77) Through Table 310.60(C)(86).

Table 310.60(C)(4) Ambient Temperature Correction Factors

For ambient temperatures other than 40°C (104°F), multiply the allowable ampacities specified in the ampacity tables by the appropriate factor shown below.

Ambient Temperature (°C)	Temperature Rating of Conductor		Ambient Temperature (°F)
	90°C	105°C	
10 or less	1.26	1.21	50 or less
11–15	1.22	1.18	51–59
16–20	1.18	1.14	60–68
21–25	1.14	1.11	69–77
26–30	1.10	1.07	78–86
31–35	1.05	1.04	87–95
36–40	1.00	1.00	96–104
41–45	0.95	0.96	105–113
46–50	0.89	0.92	114–122
51–55	0.84	0.88	123–131
56–60	0.77	0.83	132–140
61–65	0.71	0.78	141–149
66–70	0.63	0.73	150–158
71–75	0.55	0.68	159–167
76–80	0.45	0.62	168–176
81–85	0.32	0.55	177–185
86–90	—	0.48	186–194
91–95	—	0.39	195–203
96–100	—	0.28	204–212

Table 310.60(C)(67) Ampacities of Insulated Single Copper Conductor Cables Triplexed in Air Based on Conductor Temperatures of 90°C (194°F) and 105°C (221°F) and Ambient Air Temperature of 40°C (104°F)*

Conductor Size (AWG or kcmil)	Temperature Rating of Conductor [See Table 310.104(C).]			
	2001–5000 Volts Ampacity		5001–35,000 Volts Ampacity	
	90°C (194°F) Type MV-90	105°C (221°F) Type MV-105	90°C (194°F) Type MV-90	105°C (221°F) Type MV-105
8	65	74	—	—
6	90	99	100	110
4	120	130	130	140
2	160	175	170	195
1	185	205	195	225
1/0	215	240	225	255
2/0	250	275	260	295
3/0	290	320	300	340
4/0	335	375	345	390
250	375	415	380	430
350	465	515	470	525
500	580	645	580	650
750	750	835	730	820
1000	880	980	850	950

*Refer to 310.60(C)(4) for the ampacity correction factors where the ambient air temperature is other than 40°C (104°F).

Table 310.60(C)(68) Ampacities of Insulated Single Aluminum Conductor Cables Triplexed in Air Based on Conductor Temperatures of 90°C (194°F) and 105°C (221°F) and Ambient Air Temperature of 40°C (104°F)*

Conductor Size (AWG or kcmil)	Temperature Rating of Conductor [See Table 310.104(C).]			
	2001–5000 Volts Ampacity		5001–35,000 Volts Ampacity	
	90°C (194°F) Type MV-90	105°C (221°F) Type MV-105	90°C (194°F) Type MV-90	105°C (221°F) Type MV-105
8	50	57	—	—
6	70	77	75	84
4	90	100	100	110
2	125	135	130	150
1	145	160	150	175
1/0	170	185	175	200
2/0	195	215	200	230
3/0	225	250	230	265
4/0	265	290	270	305
250	295	325	300	335
350	365	405	370	415
500	460	510	460	515
750	600	665	590	660
1000	715	800	700	780

*Refer to 310.60(C)(4) for the ampacity correction factors where the ambient air temperature is other than 40°C (104°F).

Table 310.60(C)(69) Ampacities of Insulated Single Copper Conductor Isolated in Air Based on Conductor Temperatures of 90°C (194°F) and 105°C (221°F) and Ambient Air Temperature of 40°C (104°F)*

| | Temperature Rating of Conductor [See Table 310.104(C).] | | | | | |
| | 2001–5000 Volts Ampacity | | 5001–15,000 Volts Ampacity | | 15,001–35,000 Volts Ampacity | |
Conductor Size (AWG or kcmil)	90°C (194°F) Type MV-90	105°C (221°F) Type MV-105	90°C (194°F) Type MV-90	105°C (221°F) Type MV-105	90°C (194°F) Type MV-90	105°C (221°F) Type MV-105
8	83	93	—	—	—	—
6	110	120	110	125	—	—
4	145	160	150	165	—	—
2	190	215	195	215	—	—
1	225	250	225	250	225	250
1/0	260	290	260	290	260	290
2/0	300	330	300	335	300	330
3/0	345	385	345	385	345	380
4/0	400	445	400	445	395	445
250	445	495	445	495	440	490
350	550	615	550	610	545	605
500	695	775	685	765	680	755
750	900	1000	885	990	870	970
1000	1075	1200	1060	1185	1040	1160
1250	1230	1370	1210	1350	1185	1320
1500	1365	1525	1345	1500	1315	1465
1750	1495	1665	1470	1640	1430	1595
2000	1605	1790	1575	1755	1535	1710

*Refer to 310.60(C)(4) for the ampacity correction factors where the ambient air temperature is other than 40°C (104°F).

Table 310.60(C)(70) Ampacities of Insulated Single Aluminum Conductor Isolated in Air Based on Conductor Temperatures of 90°C (194°F) and 105°C (221°F) and Ambient Air Temperature of 40°C (104°F)*

| | Temperature Rating of Conductor [See Table 310.104(C).] | | | | | |
| | 2001–5000 Volts Ampacity | | 5001–15,000 Volts Ampacity | | 15,001–35,000 Volts Ampacity | |
Conductor Size (AWG or kcmil)	90°C (194°F) Type MV-90	105°C (221°F) Type MV-105	90°C (194°F) Type MV-90	105°C (221°F) Type MV-105	90°C (194°F) Type MV-90	105°C (221°F) Type MV-105
8	64	71	—	—	—	—
6	85	95	87	97	—	—
4	115	125	115	130	—	—
2	150	165	150	170	—	—
1	175	195	175	195	175	195
1/0	200	225	200	225	200	225
2/0	230	260	235	260	230	260
3/0	270	300	270	300	270	300
4/0	310	350	310	350	310	345
250	345	385	345	385	345	380
350	430	480	430	480	430	475
500	545	605	535	600	530	590
750	710	790	700	780	685	765
1000	855	950	840	940	825	920
1250	980	1095	970	1080	950	1055
1500	1105	1230	1085	1215	1060	1180
1750	1215	1355	1195	1335	1165	1300
2000	1320	1475	1295	1445	1265	1410

*Refer to 310.60(C)(4) for the ampacity correction factors where the ambient air temperature is other than 40°C (104°F).

Table 310.60(C)(71) Ampacities of an Insulated Three-Conductor Copper Cable Isolated in Air Based on Conductor Temperatures of 90°C (194°F) and 105°C (221°F) and Ambient Air Temperature of 40°C (104°F)*

| Conductor Size (AWG or kcmil) | Temperature Rating of Conductor [See Table 310.104(C).] | | | |
| | 2001–5000 Volts Ampacity | | 5001–35,000 Volts Ampacity | |
	90°C (194°F) Type MV-90	105°C (221°F) Type MV-105	90°C (194°F) Type MV-90	105°C (221°F) Type MV-105
8	59	66	—	—
6	79	88	93	105
4	105	115	120	135
2	140	154	165	185
1	160	180	185	210
1/0	185	205	215	240
2/0	215	240	245	275
3/0	250	280	285	315
4/0	285	320	325	360
250	320	355	360	400
350	395	440	435	490
500	485	545	535	600
750	615	685	670	745
1000	705	790	770	860

*Refer to 310.60(C)(4) for the ampacity correction factors where the ambient air temperature is other than 40°C (104°F).

Table 310.60(C)(72) Ampacities of an Insulated Three-Conductor Aluminum Cable Isolated in Air Based on Conductor Temperatures of 90°C (194°F) and 105°C (221°F) and Ambient Air Temperature of 40°C (104°F)*

| Conductor Size (AWG or kcmil) | Temperature Rating of Conductor [See Table 310.104(C).] | | | |
| | 2001–5000 Volts Ampacity | | 5001–35,000 Volts Ampacity | |
	90°C (194°F) Type MV-90	105°C (221°F) Type MV-105	90°C (194°F) Type MV-90	105°C (221°F) Type MV-105
8	46	51	—	—
6	61	68	72	80
4	81	90	95	105
2	110	120	125	145
1	125	140	145	165
1/0	145	160	170	185
2/0	170	185	190	215
3/0	195	215	220	245
4/0	225	250	255	285
250	250	280	280	315
350	310	345	345	385
500	385	430	425	475
750	495	550	540	600
1000	585	650	635	705

*Refer to 310.60(C)(4) for the ampacity correction factors where the ambient air temperature is other than 40°C (104°F).

Table 310.60(C)(73) Ampacities of an Insulated Triplexed or Three Single-Conductor Copper Cables in Isolated Conduit in Air Based on Conductor Temperatures of 90°C (194°F) and 105°C (221°F) and Ambient Air Temperature of 40°C (104°F)*

Conductor Size (AWG or kcmil)	2001–5000 Volts Ampacity		5001–35,000 Volts Ampacity	
	90°C (194°F) Type MV-90	105°C (221°F) Type MV-105	90°C (194°F) Type MV-90	105°C (221°F) Type MV-105
8	55	61	—	—
6	75	84	83	93
4	97	110	110	120
2	130	145	150	165
1	155	175	170	190
1/0	180	200	195	215
2/0	205	225	225	255
3/0	240	270	260	290
4/0	280	305	295	330
250	315	355	330	365
350	385	430	395	440
500	475	530	480	535
750	600	665	585	655
1000	690	770	675	755

*Refer to 310.60(C)(4) for the ampacity correction factors where the ambient air temperature is other than 40°C (104°F).

Table 310.60(C)(74) Ampacities of an Insulated Triplexed or Three Single-Conductor Aluminum Cables in Isolated Conduit in Air Based on Conductor Temperatures of 90°C (194°F) and 105°C (221°F) and Ambient Air Temperature of 40°C (104°F)*

Conductor Size (AWG or kcmil)	2001–5000 Volts Ampacity		5001–35,000 Volts Ampacity	
	90°C (194°F) Type MV-90	105°C (221°F) Type MV-105	90°C (194°F) Type MV-90	105°C (221°F) Type MV-105
8	43	48	—	—
6	58	65	65	72
4	76	85	84	94
2	100	115	115	130
1	120	135	130	150
1/0	140	155	150	170
2/0	160	175	175	200
3/0	190	210	200	225
4/0	215	240	230	260
250	250	280	255	290
350	305	340	310	350
500	380	425	385	430
750	490	545	485	540
1000	580	645	565	640

*Refer to 310.60(C)(4) for the ampacity correction factors where the ambient air temperature is other than 40°C (104°F).

Table 310.60(C)(75) Ampacities of an Insulated Three-Conductor Copper Cable in Isolated Conduit in Air Based on Conductor Temperatures of 90°C (194°F) and 105°C (221°F) and Ambient Air Temperature of 40°C (104°F)*

| Conductor Size (AWG or kcmil) | Temperature Rating of Conductor [See Table 310.104(C).] | | | |
| | 2001–5000 Volts Ampacity | | 5001–35,000 Volts Ampacity | |
	90°C (194°F) Type MV-90	105°C (221°F) Type MV-105	90°C (194°F) Type MV-90	105°C (221°F) Type MV-105
8	52	58	—	—
6	69	77	83	92
4	91	100	105	120
2	125	135	145	165
1	140	155	165	185
1/0	165	185	195	215
2/0	190	210	220	245
3/0	220	245	250	280
4/0	255	285	290	320
250	280	315	315	350
350	350	390	385	430
500	425	475	470	525
750	525	585	570	635
1000	590	660	650	725

*Refer to 310.60(C)(4) for the ampacity correction factors where the ambient air temperature is other than 40°C (104°F).

Table 310.60(C)(76) Ampacities of an Insulated Three-Conductor Aluminum Cable in Isolated Conduit in Air Based on Conductor Temperatures of 90°C (194°F) and 105°C (221°F) and Ambient Air Temperature of 40°C (104°F)*

| Conductor Size (AWG or kcmil) | Temperature Rating of Conductor [See Table 310.104(C).] | | | |
| | 2001–5000 Volts Ampacity | | 5001–35,000 Volts Ampacity | |
	90°C (194°F) Type MV-90	105°C (221°F) Type MV-105	90°C (194°F) Type MV-90	105°C (221°F) Type MV-105
8	41	46	—	—
6	53	59	64	71
4	71	79	84	94
2	96	105	115	125
1	110	125	130	145
1/0	130	145	150	170
2/0	150	165	170	190
3/0	170	190	195	220
4/0	200	225	225	255
250	220	245	250	280
350	275	305	305	340
500	340	380	380	425
750	430	480	470	520
1000	505	560	550	615

*Refer to 310.60(C)(4) for the ampacity correction factors where the ambient air temperature is other than 40°C (104°F).

Table 310.60(C)(77) Ampacities of Three Single-Insulated Copper Conductors in Underground Electrical Ducts (Three Conductors per Electrical Duct) Based on Ambient Earth Temperature of 20°C (68°F), Electrical Duct Arrangement in Accordance with Figure 310.60(C)(3), 100 Percent Load Factor, Thermal Resistance (RHO) of 90, Conductor Temperatures of 90°C (194°F) and 105°C (221°F)

Conductor Size (AWG or kcmil)	Temperature Rating of Conductor [See Table 310.104(C).]			
	2001–5000 Volts Ampacity		5001–35,000 Volts Ampacity	
	90°C (194°F) Type MV-90	105°C (221°F) Type MV-105	90°C (194°F) Type MV-90	105°C (221°F) Type MV-105
One Circuit [See Figure 310.60(C)(3), Detail 1.]				
8	64	69	—	—
6	85	92	90	97
4	110	120	115	125
2	145	155	155	165
1	170	180	175	185
1/0	195	210	200	215
2/0	220	235	230	245
3/0	250	270	260	275
4/0	290	310	295	315
250	320	345	325	345
350	385	415	390	415
500	470	505	465	500
750	585	630	565	610
1000	670	720	640	690
Three Circuits [See Figure 310.60(C)(3), Detail 2.]				
8	56	60	—	—
6	73	79	77	83
4	95	100	99	105
2	125	130	130	135
1	140	150	145	155
1/0	160	175	165	175
2/0	185	195	185	200
3/0	210	225	210	225
4/0	235	255	240	255
250	260	280	260	280
350	315	335	310	330
500	375	405	370	395
750	460	495	440	475
1000	525	565	495	535
Six Circuits [See Figure 310.60(C)(3), Detail 3.]				
8	48	52	—	—
6	62	67	64	68
4	80	86	82	88
2	105	110	105	115
1	115	125	120	125
1/0	135	145	135	145
2/0	150	160	150	165
3/0	170	185	170	185
4/0	195	210	190	205
250	210	225	210	225
350	250	270	245	265
500	300	325	290	310
750	365	395	350	375
1000	410	445	390	415

Table 310.60(C)(78) Ampacities of Three Single-Insulated Aluminum Conductors in Underground Electrical Ducts (Three Conductors per Electrical Duct) Based on Ambient Earth Temperature of 20°C (68°F), Electrical Duct Arrangement in Accordance with Figure 310.60(C)(3), 100 Percent Load Factor, Thermal Resistance (RHO) of 90, Conductor Temperatures of 90°C (194°F) and 105°C (221°F)

Conductor Size (AWG or kcmil)	Temperature Rating of Conductor [See Table 310.104(C).]			
	2001–5000 Volts Ampacity		5001–35,000 Volts Ampacity	
	90°C (194°F) Type MV-90	105°C (221°F) Type MV-105	90°C (194°F) Type MV-90	105°C (221°F) Type MV-105
One Circuit [See Figure 310.60(C)(3), Detail 1.]				
8	50	54	—	—
6	66	71	70	75
4	86	93	91	98
2	115	125	120	130
1	130	140	135	145
1/0	150	160	155	165
2/0	170	185	175	190
3/0	195	210	200	215
4/0	225	245	230	245
250	250	270	250	270
350	305	325	305	330
500	370	400	370	400
750	470	505	455	490
1000	545	590	525	565
Three Circuits [See Figure 310.60(C)(3), Detail 2.]				
8	44	47	—	—
6	57	61	60	65
4	71	80	77	83
2	96	105	100	105
1	110	120	110	120
1/0	125	135	125	140
2/0	145	155	145	155
3/0	160	175	165	175
4/0	185	200	185	200
250	205	220	200	220
350	245	265	245	260
500	295	320	290	315
750	370	395	355	385
1000	425	460	405	440
Six Circuits [See Figure 310.60(C)(3), Detail 3.]				
8	38	41	—	—
6	48	52	50	54
4	62	67	64	69
2	80	86	80	88
1	91	98	90	99
1/0	105	110	105	110
2/0	115	125	115	125
3/0	135	145	130	145
4/0	150	165	150	160
250	165	180	165	175
350	195	210	195	210
500	240	255	230	250
750	290	315	280	305
1000	335	360	320	345

Table 310.60(C)(79) Ampacities of Three Insulated Copper Conductors Cabled Within an Overall Covering (Three-Conductor Cable) in Underground Electrical Ducts (One Cable per Electrical Duct) Based on Ambient Earth Temperature of 20°C (68°F), Electrical Duct Arrangement in Accordance with Figure 310.60(C)(3), 100 Percent Load Factor, Thermal Resistance (RHO) of 90, Conductor Temperatures of 90°C (194°F) and 105°C (221°C)

Conductor Size (AWG or kcmil)	Temperature Rating of Conductor [See Table 310.104(C).]			
	2001–5000 Volts Ampacity		5001–35,000 Volts Ampacity	
	90°C (194°F) Type MV-90	105°C (221°F) Type MV-105	90°C (194°F) Type MV-90	105°C (221°F) Type MV-105
One Circuit [See Figure 310.60(C)(3), Detail 1.]				
8	59	64	—	—
6	78	84	88	95
4	100	110	115	125
2	135	145	150	160
1	155	165	170	185
1/0	175	190	195	210
2/0	200	220	220	235
3/0	230	250	250	270
4/0	265	285	285	305
250	290	315	310	335
350	355	380	375	400
500	430	460	450	485
750	530	570	545	585
1000	600	645	615	660
Three Circuits [See Figure 310.60(C)(3), Detail 2.]				
8	53	57	—	—
6	69	74	75	81
4	89	96	97	105
2	115	125	125	135
1	135	145	140	155
1/0	150	165	160	175
2/0	170	185	185	195
3/0	195	210	205	220
4/0	225	240	230	250
250	245	265	255	270
350	295	315	305	325
500	355	380	360	385
750	430	465	430	465
1000	485	520	485	515
Six Circuits [See Figure 310.60(C)(3), Detail 3.]				
8	46	50	—	—
6	60	65	63	68
4	77	83	81	87
2	98	105	105	110
1	110	120	115	125
1/0	125	135	130	145
2/0	145	155	150	160
3/0	165	175	170	180
4/0	185	200	190	200
250	200	220	205	220
350	240	270	245	275
500	290	310	290	305
750	350	375	340	365
1000	390	420	380	405

Table 310.60(C)(80) Ampacities of Three Insulated Aluminum Conductors Cabled Within an Overall Covering (Three-Conductor Cable) in Underground Electrical Ducts (One Cable per Electrical Duct) Based on Ambient Earth Temperature of 20°C (68°F), Electrical Duct Arrangement in Accordance with Figure 310.60(C)(3), 100 Percent Load Factor, Thermal Resistance (RHO) of 90, Conductor Temperatures of 90°C (194°F) and 105°C (221°C)

Conductor Size (AWG or kcmil)	Temperature Rating of Conductor [See Table 310.104(C).]			
	2001–5000 Volts Ampacity		5001–35,000 Volts Ampacity	
	90°C (194°F) Type MV-90	105°C (221°F) Type MV-105	90°C (194°F) Type MV-90	105°C (221°F) Type MV-105
One Circuit [See Figure 310.60(C)(3), Detail 1.]				
8	46	50	—	—
6	61	66	69	74
4	80	86	89	96
2	105	110	115	125
1	120	130	135	145
1/0	140	150	150	165
2/0	160	170	170	185
3/0	180	195	195	210
4/0	205	220	220	240
250	230	245	245	265
350	280	310	295	315
500	340	365	355	385
750	425	460	440	475
1000	495	535	510	545
Three Circuits [See Figure 310.60(C)(3), Detail 2.]				
8	41	44	—	—
6	54	58	59	64
4	70	75	75	81
2	90	97	100	105
1	105	110	110	120
1/0	120	125	125	135
2/0	135	145	140	155
3/0	155	165	160	175
4/0	175	185	180	195
250	190	205	200	215
350	230	250	240	255
500	280	300	285	305
750	345	375	350	375
1000	400	430	400	430
Six Circuits [See Figure 310.60(C)(3), Detail 3.]				
8	36	39	—	—
6	46	50	49	53
4	60	65	63	68
2	77	83	80	86
1	87	94	90	98
1/0	99	105	105	110
2/0	110	120	115	125
3/0	130	140	130	140
4/0	145	155	150	160
250	160	170	160	170
350	190	205	190	205
500	230	245	230	245
750	280	305	275	295
1000	320	345	315	335

Table 310.60(C)(81) Ampacities of Single Insulated Copper Conductors Directly Buried in Earth Based on Ambient Earth Temperature of 20°C (68°F), Arrangement per Figure 310.60(C)(3), 100 Percent Load Factor, Thermal Resistance (RHO) of 90, Conductor Temperatures of 90°C (194°F) and 105°C (221°C)

| Conductor Size (AWG or kcmil) | Temperature Rating of Conductor [See Table 310.104(C).] | | | |
| | 2001–5000 Volts Ampacity | | 5001–35,000 Volts Ampacity | |
	90°C (194°F) Type MV-90	105°C (221°F) Type MV-105	90°C (194°F) Type MV-90	105°C (221°F) Type MV-105
One Circuit, Three Conductors [See Figure 310.60(C)(3), Detail 9.]				
8	110	115	—	—
6	140	150	130	140
4	180	195	170	180
2	230	250	210	225
1	260	280	240	260
1/0	295	320	275	295
2/0	335	365	310	335
3/0	385	415	355	380
4/0	435	465	405	435
250	470	510	440	475
350	570	615	535	575
500	690	745	650	700
750	845	910	805	865
1000	980	1055	930	1005
Two Circuits, Six Conductors [See Figure 310.60(C)(3), Detail 10.]				
8	100	110	—	—
6	130	140	120	130
4	165	180	160	170
2	215	230	195	210
1	240	260	225	240
1/0	275	295	255	275
2/0	310	335	290	315
3/0	355	380	330	355
4/0	400	430	375	405
250	435	470	410	440
350	520	560	495	530
500	630	680	600	645
750	775	835	740	795
1000	890	960	855	920

Table 310.60(C)(82) Ampacities of Single Insulated Aluminum Conductors Directly Buried in Earth Based on Ambient Earth Temperature of 20°C (68°F), Arrangement per Figure 310.60(C)(3), 100 Percent Load Factor, Thermal Resistance (RHO) of 90, Conductor Temperatures of 90°C (194°F) and 105°C (221°F)

| Conductor Size (AWG or kcmil) | Temperature Rating of Conductor [See Table 310.104(C).] | | | |
| | 2001–5000 Volts Ampacity | | 5001–35,000 Volts Ampacity | |
	90°C (194°F) Type MV-90	105°C (221°F) Type MV-105	90°C (194°F) Type MV-90	105°C (221°F) Type MV-105
One Circuit, Three Conductors [See Figure 310.60(C)(3), Detail 9.]				
8	85	90	—	—
6	110	115	100	110
4	140	150	130	140
2	180	195	165	175
1	205	220	185	200
1/0	230	250	215	230
2/0	265	285	245	260
3/0	300	320	275	295
4/0	340	365	315	340
250	370	395	345	370
350	445	480	415	450
500	540	580	510	545
750	665	720	635	680
1000	780	840	740	795
Two Circuits, Six Conductors [See Figure 310.60(C)(3), Detail 10.]				
8	80	85	—	—
6	100	110	95	100
4	130	140	125	130
2	165	180	155	165
1	190	200	175	190
1/0	215	230	200	215
2/0	245	260	225	245
3/0	275	295	255	275
4/0	310	335	290	315
250	340	365	320	345
350	410	440	385	415
500	495	530	470	505
750	610	655	580	625
1000	710	765	680	730

Table 310.60(C)(83) Ampacities of Three Insulated Copper Conductors Cabled Within an Overall Covering (Three-Conductor Cable), Directly Buried in Earth Based on Ambient Earth Temperature of 20°C (68°F), Arrangement per Figure 310.60(C)(3), 100 Percent Load Factor, Thermal Resistance (RHO) of 90, Conductor Temperatures of 90°C (194°F) and 105°C (221°F)

| Conductor Size (AWG or kcmil) | Temperature Rating of Conductor [See Table 310.104(C).] | | | |
| | 2001–5000 Volts Ampacity | | 5001–35,000 Volts Ampacity | |
	90°C (194°F) Type MV-90	105°C (221°F) Type MV-105	90°C (194°F) Type MV-90	105°C (221°F) Type MV-105
One Circuit [See Figure 310.60(C)(3), Detail 5.]				
8	85	89	—	—
6	105	115	115	120
4	135	150	145	155
2	180	190	185	200
1	200	215	210	225
1/0	230	245	240	255
2/0	260	280	270	290
3/0	295	320	305	330
4/0	335	360	350	375
250	365	395	380	410
350	440	475	460	495
500	530	570	550	590
750	650	700	665	720
1000	730	785	750	810
Two Circuits [See Figure 310.60(C)(3), Detail 6.]				
8	80	84	—	—
6	100	105	105	115
4	130	140	135	145
2	165	180	170	185
1	185	200	195	210
1/0	215	230	220	235
2/0	240	260	250	270
3/0	275	295	280	305
4/0	310	335	320	345
250	340	365	350	375
350	410	440	420	450
500	490	525	500	535
750	595	640	605	650
1000	665	715	675	730

Table 310.60(C)(84) Ampacities of Three Insulated Aluminum Conductors Cabled Within an Overall Covering (Three-Conductor Cable), Directly Buried in Earth Based on Ambient Earth Temperature of 20°C (68°F), Arrangement per Figure 310.60(C)(3), 100 Percent Load Factor, Thermal Resistance (RHO) of 90, Conductor Temperatures of 90°C (194°F) and 105°C (221°F)

| Conductor Size (AWG or kcmil) | Temperature Rating of Conductor [See Table 310.104(C).] | | | |
| | 2001–5000 Volts Ampacity | | 5001–35,000 Volts Ampacity | |
	90°C (194°F) Type MV-90	105°C (221°F) Type MV-105	90°C (194°F) Type MV-90	105°C (221°F) Type MV-105
One Circuit [See Figure 310.60(C)(3), Detail 5.]				
8	65	70	—	—
6	80	88	90	95
4	105	115	115	125
2	140	150	145	155
1	155	170	165	175
1/0	180	190	185	200
2/0	205	220	210	225
3/0	230	250	240	260
4/0	260	280	270	295
250	285	310	300	320
350	345	375	360	390
500	420	450	435	470
750	520	560	540	580
1000	600	650	620	665
Two Circuits [See Figure 310.60(C)(3), Detail 6.]				
8	60	66	—	—
6	75	83	80	95
4	100	110	105	115
2	130	140	135	145
1	145	155	150	165
1/0	165	180	170	185
2/0	190	205	195	210
3/0	215	230	220	240
4/0	245	260	250	270
250	265	285	275	295
350	320	345	330	355
500	385	415	395	425
750	480	515	485	525
1000	550	590	560	600

Table 310.60(C)(85) Ampacities of Three Triplexed Single Insulated Copper Conductors Directly Buried in Earth Based on Ambient Earth Temperature of 20°C (68°F), Arrangement per Figure 310.60(C)(3), 100 Percent Load Factor, Thermal Resistance (RHO) of 90, Conductor Temperatures 90°C (194°F) and 105°C (221°F)

Conductor Size (AWG or kcmil)	Temperature Rating of Conductor [See Table 310.104(C).]			
	2001–5000 Volts Ampacity		5001–35,000 Volts Ampacity	
	90°C (194°F) Type MV-90	105°C (221°F) Type MV-105	90°C (194°F) Type MV-90	105°C (221°F) Type MV-105
One Circuit, Three Conductors [See Figure 310.60(C)(3), Detail 7.]				
8	90	95	—	—
6	120	130	115	120
4	150	165	150	160
2	195	205	190	205
1	225	240	215	230
1/0	255	270	245	260
2/0	290	310	275	295
3/0	330	360	315	340
4/0	375	405	360	385
250	410	445	390	410
350	490	580	470	505
500	590	635	565	605
750	725	780	685	740
1000	825	885	770	830
Two Circuits, Six Conductors [See Figure 310.60(C)(3), Detail 8.]				
8	85	90	—	—
6	110	115	105	115
4	140	150	140	150
2	180	195	175	190
1	205	220	200	215
1/0	235	250	225	240
2/0	265	285	255	275
3/0	300	320	290	315
4/0	340	365	325	350
250	370	395	355	380
350	445	480	425	455
500	535	575	510	545
750	650	700	615	660
1000	740	795	690	745

Table 310.60(C)(86) Ampacities of Three Triplexed Single Insulated Aluminum Conductors Directly Buried in Earth Based on Ambient Earth Temperature of 20°C (68°F), Arrangement per Figure 310.60(C)(3), 100 Percent Load Factor, Thermal Resistance (RHO) of 90, Conductor Temperatures 90°C (194°F) and 105°C (221°F)

Conductor Size (AWG or kcmil)	Temperature Rating of Conductor [See Table 310.104(C).]			
	2001–5000 Volts Ampacity		5001–35,000 Volts Ampacity	
	90°C (194°F) Type MV-90	105°C (221°F) Type MV-105	90°C (194°F) Type MV-90	105°C (221°F) Type MV-105
One Circuit, Three Conductors [See Figure 310.60(C)(3), Detail 7.]				
8	70	75	—	—
6	90	100	90	95
4	120	130	115	125
2	155	165	145	155
1	175	190	165	175
1/0	200	210	190	205
2/0	225	240	215	230
3/0	255	275	245	265
4/0	290	310	280	305
250	320	350	305	325
350	385	420	370	400
500	465	500	445	480
750	580	625	550	590
1000	670	725	635	680
Two Circuits, Six Conductors [See Figure 310.60(C)(3), Detail 8.]				
8	65	70	—	—
6	85	95	85	90
4	110	120	105	115
2	140	150	135	145
1	160	170	155	170
1/0	180	195	175	190
2/0	205	220	200	215
3/0	235	250	225	245
4/0	265	285	255	275
250	290	310	280	300
350	350	375	335	360
500	420	455	405	435
750	520	560	485	525
1000	600	645	565	605

Table 310.104(A) Conductor Applications and Insulations Rated 600 Volts[1]

Trade Name	Type Letter	Maximum Operating Temperature	Application Provisions	Insulation	Thickness of Insulation — AWG or kcmil	Thickness of Insulation — mm	Thickness of Insulation — mils	Outer Covering[2]
Fluorinated ethylene propylene	FEP or FEPB	90°C (194°F)	Dry and damp locations	Fluorinated ethylene propylene	14–10 / 8–2	0.51 / 0.76	20 / 30	None
		200°C (392°F)	Dry locations — special applications[3]	Fluorinated ethylene propylene	14–8	0.36	14	Glass braid
					6–2	0.36	14	Glass or other suitable braid material
Mineral insulation (metal sheathed)	MI	90°C (194°F)	Dry and wet locations	Magnesium oxide	18–16[4] / 16–10	0.58 / 0.91	23 / 36	Copper or alloy steel
		250°C (482°F)	For special applications[3]		9–4 / 3–500	1.27 / 1.40	50 / 55	
Moisture-, heat-, and oil-resistant thermoplastic	MTW	60°C (140°F)	Machine tool wiring in wet locations	Flame-retardant, moisture-, heat-, and oil-resistant thermoplastic		(A) / (B)	(A) / (B)	(A) None (B) Nylon jacket or equivalent
		90°C (194°F)	Machine tool wiring in dry locations. Informational Note: See NFPA 79.		22–12	0.76 / 0.38	30 / 15	
					10	0.76 / 0.51	30 / 20	
					8	1.14 / 0.76	45 / 30	
					6	1.52 / 0.76	60 / 30	
					4–2	1.52 / 1.02	60 / 40	
					1–4/0	2.03 / 1.27	80 / 50	
					213–500	2.41 / 1.52	95 / 60	
					501–1000	2.79 / 1.78	110 / 70	
Paper		85°C (185°F)	For underground service conductors, or by special permission	Paper				Lead sheath
Perfluoro-alkoxy	PFA	90°C (194°F)	Dry and damp locations	Perfluoro-alkoxy	14–10 / 8–2	0.51 / 0.76	20 / 30	None
		200°C (392°F)	Dry locations — special applications[3]		1–4/0	1.14	45	
Perfluoro-alkoxy	PFAH	250°C (482°F)	Dry locations only. Only for leads within apparatus or within raceways connected to apparatus (nickel or nickel-coated copper only)	Perfluoro-alkoxy	14–10 / 8–2 / 1–4/0	0.51 / 0.76 / 1.14	20 / 30 / 45	None
Thermoset	RHH	90°C (194°F)	Dry and damp locations		14–10	1.14	45	Moisture-resistant, flame-retardant, nonmetallic covering[2]
					8–2	1.52	60	
					1–4/0	2.03	80	
					213–500	2.41	95	
					501–1000	2.79	110	
					1001–2000	3.18	125	
Moisture-resistant thermoset	RHW	75°C (167°F)	Dry and wet locations	Flame-retardant, moisture-resistant thermoset	14–10	1.14	45	Moisture-resistant, flame-retardant, nonmetallic covering
					8–2	1.52	60	
	RHW-2	90°C (194°F)			1–4/0	2.03	80	
					213–500	2.41	95	
					501–1000	2.79	110	
					1001–2000	3.18	125	
Silicone	SA	90°C (194°F)	Dry and damp locations	Silicone rubber	14–10	1.14	45	Glass or other suitable braid material
					8–2	1.52	60	
		200°C (392°F)	For special application[3]		1–4/0	2.03	80	
					213–500	2.41	95	
					501–1000	2.79	110	
					1001–2000	3.18	125	

(continues)

Table 310.104(A) *Continued*

Trade Name	Type Letter	Maximum Operating Temperature	Application Provisions	Insulation	Thickness of Insulation			Outer Covering[2]
					AWG or kcmil	mm	mils	
Thermoset	SIS	90°C (194°F)	Switchboard and switchgear wiring only	Flame-retardant thermoset	14–10 8–2 1–4/0	0.76 1.14 2.41	30 45 55	None
Thermoplastic and fibrous outer braid	TBS	90°C (194°F)	Switchboard and switchgear wiring only	Thermoplastic	14–10 8 6–2 1–4/0	0.76 1.14 1.52 2.03	30 45 60 80	Flame-retardant, nonmetallic covering
Extended polytetra-fluoro-ethylene	TFE	250°C (482°F)	Dry locations only. Only for leads within apparatus or within raceways connected to apparatus, or as open wiring (nickel or nickel-coated copper only)	Extruded polytetra-fluoroethylene	14–10 8–2 1–4/0	0.51 0.76 1.14	20 30 45	None
Heat-resistant thermoplastic	THHN	90°C (194°F)	Dry and damp locations	Flame-retardant, heat-resistant thermoplastic	14–12 10 8–6 4–2 1–4/0 250–500 501–1000	0.38 0.51 0.76 1.02 1.27 1.52 1.78	15 20 30 40 50 60 70	Nylon jacket or equivalent
Moisture- and heat-resistant thermoplastic	THHW	75°C (167°F) 90°C (194°F)	Wet location Dry location	Flame-retardant, moisture- and heat-resistant thermoplastic	14–10 8 6–2 1–4/0 213–500 501–1000 1001–2000	0.76 1.14 1.52 2.03 2.41 2.79 3.18	30 45 60 80 95 110 125	None
Moisture- and heat-resistant thermoplastic	THW	75°C (167°F) 90°C (194°F)	Dry and wet locations Special applications within electric discharge lighting equipment. Limited to 1000 open circuit volts or less. (Size 14-8 only as permitted in 410.68.)	Flame-retardant, moisture- and heat-resistant thermoplastic	14–10 8 6–2 1–4/0 213–500 501–1000 1001–2000	0.76 1.14 1.52 2.03 2.41 2.79 3.18	30 45 60 80 95 110 125	None
	THW-2	90°C (194°F)	Dry and wet locations					
Moisture- and heat-resistant thermoplastic	THWN	75°C (167°F)	Dry and wet locations	Flame-retardant, moisture- and heat-resistant thermoplastic	14–12 10 8–6 4–2 1–4/0 250–500 501–1000	0.38 0.51 0.76 1.02 1.27 1.52 1.78	15 20 30 40 50 60 70	Nylon jacket or equivalent
	THWN-2	90°C (194°F)						
Moisture-resistant thermoplastic	TW	60°C (140°F)	Dry and wet locations	Flame-retardant, moisture-resistant thermoplastic	14–10 8 6–2 1–4/0 213–500 501–1000 1001–2000	0.76 1.14 1.52 2.03 2.41 2.79 3.18	3 45 60 80 95 110 125	None
Underground feeder and branch-circuit cable — single conductor (for Type UF cable employing more than one conductor, see Article 340).	UF	60°C (140°F) 75°C (167°F) [5]	See Article 340.	Moisture-resistant Moisture- and heat-resistant	14–10 8–2 1–4/0	1.52 2.03 2.41	60[6] 80[6] 95[6]	Integral with insulation

(continues)

Table 310.104(A)   *Continued*

Trade Name	Type Letter	Maximum Operating Temperature	Application Provisions	Insulation	Thickness of Insulation			Outer Covering[2]
					AWG or kcmil	mm	mils	
Underground service-entrance cable — single conductor (for Type USE cable employing more than one conductor, see Article 338).	USE	75°C (167°F) [5]	See Article 338.	Heat- and moisture-resistant	14–10	1.14	45	Moisture-resistant nonmetallic covering (See 338.2.)
	USE-2	90°C (194°F)	Dry and wet locations		8–2	1.52	60	
					1–4/0	2.03	80	
					213–500	2.41	95 [7]	
					501–1000	2.79	110	
					1001–2000	3.18	125	
Thermoset	XHH	90°C (194°F)	Dry and damp locations	Flame-retardant thermoset	14–10	0.76	30	None
					8–2	1.14	45	
					1–4/0	1.40	55	
					213–500	1.65	65	
					501–1000	2.03	80	
					1001–2000	2.41	95	
Thermoset	XHHN	90°C (194°F)	Dry and damp locations	Flame-retardant thermoset	14–12	0.38	15	Nylon jacket or equivalent
					10	0.51	20	
					8–6	0.76	30	
					4–2	1.02	40	
					1–4/0	1.27	50	
					250–500	1.52	60	
					501–1000	1.78	70	
Moisture-resistant thermoset	XHHW	90°C (194°F)	Dry and damp locations	Flame-retardant, moisture-resistant thermoset	14–10	0.76	30	None
		75°C (167°F)	Wet locations		8–2	1.14	45	
					1–4/0	1.40	55	
					213–500	1.65	65	
					501–1000	2.03	80	
					1001–2000	2.41	95	
Moisture-resistant thermoset	XHHW-2	90°C (194°F)	Dry and wet locations	Flame-retardant, moisture-resistant thermoset	14–10	0.76	30	None
					8–2	1.14	45	
					1–4/0	1.40	55	
					213–500	1.65	65	
					501–1000	2.03	80	
					1001–2000	2.41	95	
Moisture-resistant thermoset	XHWN	75°C (167°F)	Dry and wet locations	Flame-retardant, moisture-resistant thermoset	14–12	0.38	15	Nylon jacket or equivalent
					10	0.51	20	
					8–6	0.76	30	
	XHWN-2	90°C (194°F)			4–2	1.02	40	
					1–4/0	1.27	50	
					250–500	1.52	60	
					501–1000	1.78	70	
Modified ethylene tetrafluoro-ethylene	Z	90°C (194°F)	Dry and damp locations	Modified ethylene tetrafluoro-ethylene	14–12	0.38	15	None
					10	0.51	20	
		150°C (302°F)	Dry locations — special applications[3]		8–4	0.64	25	
					3–1	0.89	35	
					1/0–4/0	1.14	45	
Modified ethylene tetrafluoro-ethylene	ZW	75°C (167°F)	Wet locations	Modified ethylene tetrafluoro-ethylene	14–10	0.76	30	None
		90°C (194°F)	Dry and damp locations		8–2	1.14	45	
		150°C (302°F)	Dry locations — special applications[3]					
	ZW-2	90°C (194°F)	Dry and wet locations					

[1] Conductors can be rated up to 1000 V if listed and marked.

[2] Some insulations do not require an outer covering.

[3] Where design conditions require maximum conductor operating temperatures above 90°C (194°F).

[4] For signaling circuits permitting 300-volt insulation.

[5] For ampacity limitation, see 340.80.

[6] Includes integral jacket.

[7] Insulation thickness shall be permitted to be 2.03 mm (80 mils) for listed Type USE conductors that have been subjected to special investigations. The nonmetallic covering over individual rubber-covered conductors of aluminum-sheathed cable and of lead-sheathed or multiconductor cable shall not be required to be flame retardant. For Type MC cable, see 330.104. For nonmetallic-sheathed cable, see Article 334, Part III. For Type UF cable, see Article 340, Part III.

Table 310.104(B) Thickness of Insulation for Nonshielded Types RHH and RHW Solid Dielectric Insulated Conductors Rated 2000 Volts

Conductor Size (AWG or kcmil)	Column A[1]		Column B[2]	
	mm	mils	mm	mils
14–10	2.03	80	1.52	60
8	2.03	80	1.78	70
6–2	2.41	95	1.78	70
1–2/0	2.79	110	2.29	90
3/0–4/0	2.79	110	2.29	90
213–500	3.18	125	2.67	105
501–1000	3.56	140	3.05	120
1001–2000	3.56	140	3.56	140

[1]Column A insulations are limited to natural, SBR, and butyl rubbers.

[2]Column B insulations are materials such as cross-linked polyethylene, ethylene propylene rubber, and composites thereof.

Table 310.104(C) Conductor Application and Insulation Rated 2001 Volts and Higher

Trade Name	Type Letter	Maximum Operating Temperature	Application Provision	Insulation	Outer Covering
Medium voltage solid dielectric	MV-90 MV-105*	90°C 105°C	Dry or wet locations	Thermoplastic or thermosetting	Jacket, sheath, or armor

*Where design conditions require maximum conductor temperatures above 90°C.

Table 310.104(D) Thickness of Insulation and Jacket for Nonshielded Solid Dielectric Insulated Conductors Rated 2001 to 5000 Volts

Conductor Size (AWG or kcmil)	Dry Locations, Single Conductor						Wet or Dry Locations					
	Without Jacket Insulation		With Jacket				Single Conductor				Multiconductor Insulation*	
			Insulation		Jacket		Insulation		Jacket			
	mm	mils	mm	mils	mm	mils	mm	mils	mm	mils	mm	mils
8	2.79	110	2.29	90	0.76	30	3.18	125	2.03	80	2.29	90
6	2.79	110	2.29	90	0.76	30	3.18	125	2.03	80	2.29	90
4–2	2.79	110	2.29	90	1.14	45	3.18	125	2.03	80	2.29	90
1–2/0	2.79	110	2.29	90	1.14	45	3.18	125	2.03	80	2.29	90
3/0–4/0	2.79	110	2.29	90	1.65	65	3.18	125	2.41	95	2.29	90
213–500	3.05	120	2.29	90	1.65	65	3.56	140	2.79	110	2.29	90
501–750	3.30	130	2.29	90	1.65	65	3.94	155	3.18	125	2.29	90
751–1000	3.30	130	2.29	90	1.65	65	3.94	155	3.18	125	2.29	90
1001–1250	3.56	140	2.92	115	1.65	65	4.32	170	3.56	140	2.92	115
1251–1500	3.56	140	2.92	115	2.03	80	4.32	170	3.56	140	2.92	115
1501–2000	3.56	140	2.92	115	2.03	80	4.32	170	3.94	155	3.56	140

*Under a common overall covering such as a jacket, sheath, or armor.

Table 310.104(E) Thickness of Insulation for Shielded Solid Dielectric Insulated Conductors Rated 2001 to 35,000 Volts

Conductor Size (AWG or kcmil)	2001–5000 Volts 100 Percent Insulation Level[1]		5001–8000 Volts 100 Percent Insulation Level[1]		133 Percent Insulation Level[2]		173 Percent Insulation Level[3]		8001–15,000 Volts 100 Percent Insulation Level[1]		133 Percent Insulation Level[2]		173 Percent Insulation Level[3]		15,001–25,000 Volts 100 Percent Insulation Level[1]		133 Percent Insulation Level[2]		173 Percent Insulation Level[3]	
	mm	mils	mm	mils	mm	mils	mm	mils	mm	mils	mm	mils	mm	mils	mm	mils	mm	mils	mm	mils
8	2.29	90	—	—	—	—	—	—	—	—	—	—	—	—	—	—	—	—	—	—
6–4	2.29	90	2.92	115	3.56	140	4.45	175	—	—	—	—	—	—	—	—	—	—	—	—
2	2.29	90	2.92	115	3.56	140	4.45	175	4.45	175	5.59	220	6.60	260	—	—	—	—	—	—
1	2.29	90	2.92	115	3.56	140	4.45	175	4.45	175	5.59	220	6.60	260	6.60	260	8.13	320	10.67	420
1/0–2000	2.29	90	2.92	115	3.56	140	4.45	175	4.45	175	5.59	220	6.60	260	6.60	260	8.13	320	10.67	420

Conductor Size (AWG or kcmil)	25,001–28,000 Volts 100 Percent Insulation Level[1]		133 Percent Insulation Level[2]		173 Percent Insulation Level[3]		28,001–35,000 Volts 100 Percent Insulation Level[1]		133 Percent Insulation Level[2]		173 Percent Insulation Level[3]	
	mm	mils	mm	mils	mm	mils	mm	mils	mm	mils	mm	mils
1	7.11	280	8.76	345	11.30	445	—	—	—	—	—	—
1/0–2000	7.11	280	8.76	345	11.30	445	8.76	345	10.67	420	14.73	580

[1] **100 Percent Insulation Level.** Cables in this category shall be permitted to be applied where the system is provided with relay protection such that ground faults will be cleared as rapidly as possible but, in any case, within 1 minute. While these cables are applicable to the great majority of cable installations that are on grounded systems, they shall be permitted to be used also on other systems for which the application of cables is acceptable, provided the above clearing requirements are met in completely de-energizing the faulted section.

[2] **133 Percent Insulation Level.** This insulation level corresponds to that formerly designated for ungrounded systems. Cables in this category shall be permitted to be applied in situations where the clearing time requirements of the 100 percent level category cannot be met and yet there is adequate assurance that the faulted section will be de-energized in a time not exceeding 1 hour. Also, they shall be permitted to be used in 100 percent insulation level applications where additional insulation is desirable.

[3] **173 Percent Insulation Level.** Cables in this category shall be permitted to be applied under all of the following conditions:

(1) In industrial establishments where the conditions of maintenance and supervision ensure that only qualified persons service the installation

(2) Where the fault clearing time requirements of the 133 percent level category cannot be met

(3) Where an orderly shutdown is essential to protect equipment and personnel

(4) There is adequate assurance that the faulted section will be de-energized in an orderly shutdown

Also, cables with this insulation thickness shall be permitted to be used in 100 or 133 percent insulation level applications where additional insulation strength is desirable.

310.106 Conductors.

(A) Minimum Size of Conductors. The minimum size of conductors shall be as shown in Table 310.106(A), except as permitted elsewhere in this *Code*.

Table 310.106(A) Minimum Size of Conductors

Conductor Voltage Rating (Volts)	Minimum Conductor Size (AWG)	
	Copper	Aluminum or Copper-Clad Aluminum
0–2000	14	12
2001–5000	8	8
5001–8000	6	6
8001–15,000	2	2
15,001–28,000	1	1
28,001–35,000	1/0	1/0

(B) Conductor Material. Conductors in this article shall be of aluminum, copper-clad aluminum, or copper unless otherwise specified.

Solid aluminum conductors 8, 10, and 12 AWG shall be made of an AA-8000 series electrical grade aluminum alloy conductor material. Stranded aluminum conductors 8 AWG through 1000 kcmil marked as Type RHH, RHW, XHHW, THW, THHW, THWN, THHN, service-entrance Type SE Style U, and SE Style R shall be made of an AA-8000 series electrical grade aluminum alloy conductor material.

(C) Stranded Conductors. Where installed in raceways, conductors 8 AWG and larger, not specifically permitted or required elsewhere in this *Code* to be solid, shall be stranded.

(D) Insulated. Conductors, not specifically permitted elsewhere in this *Code* to be covered or bare, shall be insulated.

Informational Note: See 250.184 for insulation of neutral conductors of a solidly grounded high-voltage system.

310.110 Conductor Identification.

(A) Grounded Conductors. Insulated or covered grounded conductors shall be identified in accordance with 200.6.

(B) Equipment Grounding Conductors. Equipment grounding conductors shall be in accordance with 250.119.

(C) Ungrounded Conductors. Conductors that are intended for use as ungrounded conductors, whether used as a single conductor or in multiconductor cables, shall be finished to be clearly distinguishable from grounded and grounding conductors. Distinguishing markings shall not conflict in any manner with the surface markings required by 310.120(B)(1). Branch-circuit ungrounded conductors shall be identified in accordance with 210.5(C). Feeders shall be identified in accordance with 215.12.

Exception: Conductor identification shall be permitted in accordance with 200.7.

310.120 Marking.

(A) Required Information. All conductors and cables shall be marked to indicate the following information, using the applicable method described in 310.120(B):

(1) The maximum rated voltage.
(2) The proper type letter or letters for the type of wire or cable as specified elsewhere in this *Code*.
(3) The manufacturer's name, trademark, or other distinctive marking by which the organization responsible for the product can be readily identified.
(4) The AWG size or circular mil area.

> Informational Note: See Conductor Properties, Table 8 of Chapter 9, for conductor area expressed in SI units for conductor sizes specified in AWG or circular mil area.

(5) Cable assemblies where the neutral conductor is smaller than the ungrounded conductors shall be so marked.

(B) Method of Marking.

(1) Surface Marking. The following conductors and cables shall be durably marked on the surface. The AWG size or circular mil area shall be repeated at intervals not exceeding 610 mm (24 in.). All other markings shall be repeated at intervals not exceeding 1.0 m (40 in.).

(1) Single-conductor and multiconductor rubber- and thermoplastic-insulated wire and cable
(2) Nonmetallic-sheathed cable
(3) Service-entrance cable
(4) Underground feeder and branch-circuit cable
(5) Tray cable
(6) Irrigation cable
(7) Power-limited tray cable
(8) Instrumentation tray cable

(2) Marker Tape. Metal-covered multiconductor cables shall employ a marker tape located within the cable and running for its complete length.

Exception No. 1: Type MI cable.

Exception No. 2: Type AC cable.

Exception No. 3: The information required in 310.120(A) shall be permitted to be durably marked on the outer nonmetallic covering of Type MC, Type ITC, or Type PLTC cables at intervals not exceeding 1.0 m (40 in.).

Exception No. 4: The information required in 310.120(A) shall be permitted to be durably marked on a nonmetallic covering under the metallic sheath of Type ITC or Type PLTC cable at intervals not exceeding 1.0 m (40 in.).

> Informational Note: Included in the group of metal-covered cables are Type AC cable (Article 320), Type MC cable (Article 330), and lead-sheathed cable.

(3) Tag Marking. The following conductors and cables shall be marked by means of a printed tag attached to the coil, reel, or carton:

(1) Type MI cable
(2) Switchboard wires

(3) Metal-covered, single-conductor cables
(4) Type AC cable

(4) Optional Marking of Wire Size. The information required in 310.120(A)(4) shall be permitted to be marked on the surface of the individual insulated conductors for the following multiconductor cables:

(1) Type MC cable
(2) Tray cable
(3) Irrigation cable
(4) Power-limited tray cable
(5) Power-limited fire alarm cable
(6) Instrumentation tray cable

(C) Suffixes to Designate Number of Conductors. A type letter or letters used alone shall indicate a single insulated conductor. The letter suffixes shall be indicated as follows:

(1) D — For two insulated conductors laid parallel within an outer nonmetallic covering
(2) M — For an assembly of two or more insulated conductors twisted spirally within an outer nonmetallic covering

(D) Optional Markings. All conductors and cables contained in Chapter 3 shall be permitted to be surface marked to indicate special characteristics of the cable materials. These markings include, but are not limited to, markings for limited smoke, sunlight resistant, and so forth.

ARTICLE 312
Cabinets, Cutout Boxes, and Meter Socket Enclosures

Part I. Scope and Installation

312.1 Scope. This article covers the installation and construction specifications of cabinets, cutout boxes, and meter socket enclosures. It does not apply to equipment operating at over 1000 volts, except as specifically referenced elsewhere in the *Code*.

312.2 Damp and Wet Locations. In damp or wet locations, surface-type enclosures within the scope of this article shall be placed or equipped so as to prevent moisture or water from entering and accumulating within the cabinet or cutout box, and shall be mounted so there is at least 6-mm (¼-in.) airspace between the enclosure and the wall or other supporting surface. Enclosures installed in wet locations shall be weatherproof. For enclosures in wet locations, raceways or cables entering above the level of uninsulated live parts shall use fittings listed for wet locations.

Exception: Nonmetallic enclosures shall be permitted to be installed without the airspace on a concrete, masonry, tile, or similar surface.

> Informational Note: For protection against corrosion, see 300.6.

312.3 Position in Wall. In walls of concrete, tile, or other noncombustible material, cabinets shall be installed so that the front edge of the cabinet is not set back of the finished surface more than 6 mm (¼ in.). In walls constructed of wood or other combustible material, cabinets shall be flush with the finished surface or project therefrom.

312.4 Repairing Noncombustible Surfaces. Noncombustible surfaces that are broken or incomplete shall be repaired so there will be no gaps or open spaces greater than 3 mm (⅛ in.) at the edge of the cabinet or cutout box employing a flush-type cover.

312.5 Cabinets, Cutout Boxes, and Meter Socket Enclosures. Conductors entering enclosures within the scope of this article shall be protected from abrasion and shall comply with 312.5(A) through (C).

(A) Openings to Be Closed. Openings through which conductors enter shall be closed in an approved manner.

(B) Metal Cabinets, Cutout Boxes, and Meter Socket Enclosures. Where metal enclosures within the scope of this article are installed with messenger-supported wiring, open wiring on insulators, or concealed knob-and-tube wiring, conductors shall enter through insulating bushings or, in dry locations, through flexible tubing extending from the last insulating support and firmly secured to the enclosure.

(C) Cables. Where cable is used, each cable shall be secured to the cabinet, cutout box, or meter socket enclosure.

Exception: Cables with entirely nonmetallic sheaths shall be permitted to enter the top of a surface-mounted enclosure through one or more nonflexible raceways not less than 450 mm (18 in.) and not more than 3.0 m (10 ft) in length, provided all of the following conditions are met:

(1) Each cable is fastened within 300 mm (12 in.), measured along the sheath, of the outer end of the raceway.

(2) The raceway extends directly above the enclosure and does not penetrate a structural ceiling.

(3) A fitting is provided on each end of the raceway to protect the cable(s) from abrasion and the fittings remain accessible after installation.

(4) The raceway is sealed or plugged at the outer end using approved means so as to prevent access to the enclosure through the raceway.

(5) The cable sheath is continuous through the raceway and extends into the enclosure beyond the fitting not less than 6 mm (¼ in.).

(6) The raceway is fastened at its outer end and at other points in accordance with the applicable article.

(7) Where installed as conduit or tubing, the cable fill does not exceed the amount that would be permitted for complete conduit or tubing systems by Table 1 of Chapter 9 of this Code and all applicable notes thereto. Note 2 to the tables in Chapter 9 does not apply to this condition.

Informational Note: See Table 1 in Chapter 9, including Note 9, for allowable cable fill in circular raceways. See 310.15(B)(3)(a) for required ampacity reductions for multiple cables installed in a common raceway.

312.6 Deflection of Conductors. Conductors at terminals or conductors entering or leaving cabinets or cutout boxes and the like shall comply with 312.6(A) through (C).

Exception: Wire-bending space in enclosures for motor controllers with provisions for one or two wires per terminal shall comply with 430.10(B).

(A) Width of Wiring Gutters. Conductors shall not be deflected within a cabinet or cutout box unless a gutter having a width in accordance with Table 312.6(A) is provided. Conductors in parallel in accordance with 310.10(H) shall be judged on the basis of the number of conductors in parallel.

(B) Wire-Bending Space at Terminals. Wire-bending space at each terminal shall be provided in accordance with 312.6(B)(1) or (B)(2).

(1) Conductors Not Entering or Leaving Opposite Wall. Table 312.6(A) shall apply where the conductor does not enter or leave the enclosure through the wall opposite its terminal.

(2) Conductors Entering or Leaving Opposite Wall. Table 312.6(B) shall apply where the conductor does enter or leave the enclosure through the wall opposite its terminal.

Exception No. 1: Where the distance between the wall and its terminal is in accordance with Table 312.6(A), a conductor shall be permitted to enter or leave an enclosure through the wall opposite its terminal, provided the conductor enters or leaves the enclosure where the gutter joins an adjacent gutter that has a width that conforms to Table 312.6(B) for the conductor.

Exception No. 2: A conductor not larger than 350 kcmil shall be permitted to enter or leave an enclosure containing only a meter socket(s) through the wall opposite its terminal, provided the distance between the terminal and the opposite wall is not less than that specified in Table 312.6(A) and the terminal is a lay-in type, where the terminal is either of the following:

(a) Directed toward the opening in the enclosure and within a 45 degree angle of directly facing the enclosure wall

(b) Directly facing the enclosure wall and offset not greater than 50 percent of the bending space specified in Table 312.6(A)

Informational Note: *Offset* is the distance measured along the enclosure wall from the axis of the centerline of the terminal to a line passing through the center of the opening in the enclosure.

(C) Conductors 4 AWG or Larger. Installation shall comply with 300.4(G).

312.7 Space in Enclosures. Cabinets and cutout boxes shall have approved space to accommodate all conductors installed in them without crowding.

312.8 Switch and Overcurrent Device Enclosures. The wiring space within enclosures for switches and overcurrent devices shall be permitted for other wiring and equipment subject to limitations for specific equipment as provided in (A) and (B).

(A) Splices, Taps, and Feed-Through Conductors. The wiring space of enclosures for switches or overcurrent devices shall be permitted for conductors feeding through, spliced, or tapping off to other enclosures, switches, or overcurrent devices where all of the following conditions are met:

(1) The total of all conductors installed at any cross section of the wiring space does not exceed 40 percent of the cross-sectional area of that space.

(2) The total area of all conductors, splices, and taps installed at any cross section of the wiring space does not exceed 75 percent of the cross-sectional area of that space.

(3) A warning label complying with 110.21(B) is applied to the enclosure that identifies the closest disconnecting means for any feed-through conductors.

N (B) Power Monitoring Equipment. The wiring space of enclosures for switches or overcurrent devices shall be permitted to contain power monitoring equipment where all of the following conditions are met:

(1) The power monitoring equipment is identified as a field installable accessory as part of the listed equipment, or is

a listed kit evaluated for field installation in switch or overcurrent device enclosures.

(2) The total area of all conductors, splices, taps, and equipment at any cross section of the wiring space does not exceed 75 percent of the cross-sectional area of that space.

312.9 Side or Back Wiring Spaces or Gutters. Cabinets and cutout boxes shall be provided with back-wiring spaces, gutters, or wiring compartments as required by 312.11(C) and (D).

Part II. Construction Specifications

312.10 Material. Cabinets, cutout boxes, and meter socket enclosures shall comply with 312.10(A) through (C).

(A) Metal Cabinets and Cutout Boxes. Metal enclosures within the scope of this article shall be protected both inside and outside against corrosion.

(B) Strength. The design and construction of enclosures within the scope of this article shall be such as to secure ample strength and rigidity. If constructed of sheet steel, the metal thickness shall not be less than 1.35 mm (0.053 in.) uncoated.

(C) Nonmetallic Cabinets. Nonmetallic cabinets shall be listed, or they shall be submitted for approval prior to installation.

312.11 Spacing. The spacing within cabinets and cutout boxes shall comply with 312.11(A) through (D).

(A) General. Spacing within cabinets and cutout boxes shall provide approved spacing for the distribution of wires and cables placed in them and for a separation between metal parts of devices and apparatus mounted within them in accordance with 312.11(A)(1), (A)(2), and (A)(3).

(1) Base. Other than at points of support, there shall be an airspace of at least 1.59 mm (0.0625 in.) between the base of the device and the wall of any metal cabinet or cutout box in which the device is mounted.

(2) Doors. There shall be an airspace of at least 25.4 mm (1.00 in.) between any live metal part, including live metal parts of enclosed fuses, and the door.

Exception: Where the door is lined with an approved insulating material or is of a thickness of metal not less than 2.36 mm (0.093 in.) uncoated, the airspace shall be not less than 12.7 mm (0.500 in.).

(3) Live Parts. There shall be an airspace of at least 12.7 mm (0.500 in.) between the walls, back, gutter partition, if of metal, or door of any cabinet or cutout box and the nearest exposed current-carrying part of devices mounted within the cabinet where the voltage does not exceed 250. This spacing shall be increased to at least 25.4 mm (1.00 in.) for voltages of 251 to 1000, nominal.

Exception: Where the conditions in 312.11(A)(2), Exception, are met, the airspace for nominal voltages from 251 to 600 shall be permitted to be not less than 12.7 mm (0.500 in.).

(B) Switch Clearance. Cabinets and cutout boxes shall be deep enough to allow the closing of the doors when 30-ampere branch-circuit panelboard switches are in any position, when combination cutout switches are in any position, or when other single-throw switches are opened as far as their construction permits.

Table 312.6(A) Minimum Wire-Bending Space at Terminals and Minimum Width of Wiring Gutters

Wire Size (AWG or kcmil)		Wires per Terminal									
	Compact Stranded AA-8000 Aluminum Alloy Conductors (see Note 2)	1		2		3		4		5	
All Other Conductors		mm	in.	mm	in.	mm	in.	mm	in.	mm	in.
14–10	12–8	Not specified		—	—	—	—	—	—	—	—
8–6	6–4	38.1	1½	—	—	—	—	—	—	—	—
4–3	2–1	50.8	2	—	—	—	—	—	—	—	—
2	1/0	63.5	2½	—	—	—	—	—	—	—	—
1	2/0	76.2	3	—	—	—	—	—	—	—	—
1/0–2/0	3/0–4/0	88.9	3½	127	5	178	7	—	—	—	—
3/0–4/0	250–300	102	4	152	6	203	8	—	—	—	—
250	350	114	4½	152	6	203	8	254	10	—	—
300–350	400–500	127	5	203	8	254	10	305	12	—	—
400–500	600–750	152	6	203	8	254	10	305	12	356	14
600–700	800–1000	203	8	254	10	305	12	356	14	406	16
750–900	—	203	8	305	12	356	14	406	16	457	18
1000–1250	—	254	10	—	—	—	—	—	—	—	—
1500–2000	—	305	12	—	—	—	—	—	—	—	—

Notes:

1. Bending space at terminals shall be measured in a straight line from the end of the lug or wire connector (in the direction that the wire leaves the terminal) to the wall, barrier, or obstruction.

2. This column shall be permitted to be used to determine the minimum wire-bending space for compact stranded aluminum conductors in sizes up to 1000 kcmil and manufactured using AA-8000 series electrical grade aluminum alloy conductor material in accordance with 310.106(B). The minimum width of the wire gutter space shall be determined using the all other conductors value in this table.

Table 312.6(B) Minimum Wire-Bending Space at Terminals

Wire Size (AWG or kcmil)		Wires per Terminal							
		1		2		3		4 or More	
All Other Conductors	Compact Stranded AA-8000 Aluminum Alloy Conductors (See Note 3.)	mm	in.	mm	in.	mm	in.	mm	in.
14–10	12–8	Not specified		—	—	—	—	—	—
8	6	38.1	1½	—	—	—	—	—	—
6	4	50.8	2	—	—	—	—	—	—
4	2	76.2	3	—	—	—	—	—	—
3	1	76.2	3	—	—	—	—	—	—
2	1/0	88.9	3½	—	—	—	—	—	—
1	2/0	114	4½	—	—	—	—	—	—
1/0	3/0	140	5½	140	5½	178	7	—	—
2/0	4/0	152	6	152	6	190	7½	—	—
3/0	250	165[a]	6½ [a]	165[a]	6½ [a]	203	8	—	—
4/0	300	178[b]	7[b]	190[c]	7½ [c]	216[a]	8½ [a]	—	—
250	350	216[d]	8½ [d]	229[d]	8½ [d]	254[b]	9[b]	254	10
300	400	254[e]	10[e]	254[d]	10[d]	279[b]	11[b]	305	12
350	500	305[e]	12[e]	305[e]	12[e]	330[e]	13[e]	356[d]	14[d]
400	600	330[e]	13[e]	330[e]	13[e]	356[e]	14[e]	381[e]	15[e]
500	700–750	356[e]	14[e]	356[e]	14[e]	381[e]	15[e]	406[e]	16[e]
600	800–900	381[e]	15[e]	406[e]	16[e]	457[e]	18[e]	483[e]	19[e]
700	1000	406[e]	16[e]	457[e]	18[e]	508[e]	20[e]	559[e]	22[e]
750	—	432[e]	17[e]	483[e]	19[e]	559[e]	22[e]	610[e]	24[e]
800	—	457	18	508	20	559	22	610	24
900	—	483	19	559	22	610	24	610	24
1000	—	508	20	—	—	—		—	
1250	—	559	22	—	—	—		—	
1500	—	610	24	—	—	—		—	
1750	—	610	24	—	—	—		—	
2000	—	610	24	—	—	—		—	

Notes:

1. Bending space at terminals shall be measured in a straight line from the end of the lug or wire connector in a direction perpendicular to the enclosure wall.

2. For removable and lay-in wire terminals intended for only one wire, bending space shall be permitted to be reduced by the following number of millimeters (inches):

[a] 12.7 mm (½ in.)

[b] 25.4 mm (1 in.)

[c] 38.1 mm (1½ in.)

[d] 50.8 mm (2 in.)

[e] 76.2 mm (3 in.)

3. This column shall be permitted to determine the required wire-bending space for compact stranded aluminum conductors in sizes up to 1000 kcmil and manufactured using AA-8000 series electrical grade aluminum alloy conductor material in accordance with 310.106(B).

(C) Wiring Space. Cabinets and cutout boxes that contain devices or apparatus connected within the cabinet or box to more than eight conductors, including those of branch circuits, meter loops, feeder circuits, power circuits, and similar circuits, but not including the supply circuit or a continuation thereof, shall have back-wiring spaces or one or more side-wiring spaces, side gutters, or wiring compartments.

(D) Wiring Space — Enclosure. Side-wiring spaces, side gutters, or side-wiring compartments of cabinets and cutout boxes shall be made tight enclosures by means of covers, barriers, or partitions extending from the bases of the devices contained in the cabinet, to the door, frame, or sides of the cabinet.

Exception: Side-wiring spaces, side gutters, and side-wiring compartments of cabinets shall not be required to be made tight enclosures where those side spaces contain only conductors that enter the cabinet directly opposite to the devices where they terminate.

Partially enclosed back-wiring spaces shall be provided with covers to complete the enclosure. Wiring spaces that are required by 312.11(C) and are exposed when doors are open shall be provided with covers to complete the enclosure. Where space is provided for feed-through conductors and for splices as required in 312.8, additional barriers shall not be required.

ARTICLE 314
Outlet, Device, Pull, and Junction Boxes; Conduit Bodies; Fittings; and Handhole Enclosures

Part I. Scope and General

314.1 Scope. This article covers the installation and use of all boxes and conduit bodies used as outlet, device, junction, or pull boxes, depending on their use, and handhole enclosures. Cast metal, sheet metal, nonmetallic, and other boxes such as FS, FD, and larger boxes are not classified as conduit bodies. This article also includes installation requirements for fittings used to join raceways and to connect raceways and cables to boxes and conduit bodies.

314.2 Round Boxes. Round boxes shall not be used where conduits or connectors requiring the use of locknuts or bushings are to be connected to the side of the box.

314.3 Nonmetallic Boxes. Nonmetallic boxes shall be permitted only with open wiring on insulators, concealed knob-and-tube wiring, cabled wiring methods with entirely nonmetallic sheaths, flexible cords, and nonmetallic raceways.

Exception No. 1: Where internal bonding means are provided between all entries, nonmetallic boxes shall be permitted to be used with metal raceways or metal-armored cables.

Exception No. 2: Where integral bonding means with a provision for attaching an equipment bonding jumper inside the box are provided between all threaded entries in nonmetallic boxes listed for the purpose, nonmetallic boxes shall be permitted to be used with metal raceways or metal-armored cables.

314.4 Metal Boxes. Metal boxes shall be grounded and bonded in accordance with Parts I, IV, V, VI, VII, and X of Article 250 as applicable, except as permitted in 250.112(I).

Part II. Installation

314.15 Damp or Wet Locations. In damp or wet locations, boxes, conduit bodies, outlet box hoods, and fittings shall be placed or equipped so as to prevent moisture from entering or accumulating within the box, conduit body, or fitting. Boxes, conduit bodies, outlet box hoods, and fittings installed in wet locations shall be listed for use in wet locations. Approved drainage openings not smaller than 3 mm (⅛ in.) and not larger than 6 mm (¼ in.) in diameter shall be permitted to be installed in the field in boxes or conduit bodies listed for use in damp or wet locations. For installation of listed drain fittings, larger openings are permitted to be installed in the field in accordance with manufacturer's instructions.

Informational Note No. 1: For boxes in floors, see 314.27(B).

Informational Note No. 2: For protection against corrosion, see 300.6.

314.16 Number of Conductors in Outlet, Device, and Junction Boxes, and Conduit Bodies. Boxes and conduit bodies shall be of an approved size to provide free space for all enclosed conductors. In no case shall the volume of the box, as calculated in 314.16(A), be less than the fill calculation as calculated

in 314.16(B). The minimum volume for conduit bodies shall be as calculated in 314.16(C).

The provisions of this section shall not apply to terminal housings supplied with motors or generators.

Informational Note: For volume requirements of motor or generator terminal housings, see 430.12.

Boxes and conduit bodies enclosing conductors 4 AWG or larger shall also comply with the provisions of 314.28.

(A) Box Volume Calculations. The volume of a wiring enclosure (box) shall be the total volume of the assembled sections and, where used, the space provided by plaster rings, domed covers, extension rings, and so forth, that are marked with their volume or are made from boxes the dimensions of which are listed in Table 314.16(A). Where a box is provided with one or more securely installed barriers, the volume shall be apportioned to each of the resulting spaces. Each barrier, if not marked with its volume, shall be considered to take up 8.2 cm³ (½ in³) if metal, and 16.4 cm³ (1.0 in³) if nonmetallic.

(1) Standard Boxes. The volumes of standard boxes that are not marked with their volume shall be as given in Table 314.16(A).

(2) Other Boxes. Boxes 1650 cm³ (100 in.³) or less, other than those described in Table 314.16(A), and nonmetallic boxes shall be durably and legibly marked by the manufacturer with their volume(s). Boxes described in Table 314.16(A) that have a volume larger than is designated in the table shall be permitted to have their volume marked as required by this section.

(B) Box Fill Calculations. The volumes in paragraphs 314.16(B)(1) through (B)(5), as applicable, shall be added together. No allowance shall be required for small fittings such as locknuts and bushings. Each space within a box installed with a barrier shall be calculated separately.

(1) Conductor Fill. Each conductor that originates outside the box and terminates or is spliced within the box shall be counted once, and each conductor that passes through the box without splice or termination shall be counted once. Each loop or coil of unbroken conductor not less than twice the minimum length required for free conductors in 300.14 shall be counted twice. The conductor fill shall be calculated using Table 314.16(B). A conductor, no part of which leaves the box, shall not be counted.

Exception: An equipment grounding conductor or conductors or not over four fixture wires smaller than 14 AWG, or both, shall be permitted to be omitted from the calculations where they enter a box from a domed luminaire or similar canopy and terminate within that box.

(2) Clamp Fill. Where one or more internal cable clamps, whether factory or field supplied, are present in the box, a single volume allowance in accordance with Table 314.16(B) shall be made based on the largest conductor present in the box. No allowance shall be required for a cable connector with its clamping mechanism outside the box.

A clamp assembly that incorporates a cable termination for the cable conductors shall be listed and marked for use with specific nonmetallic boxes. Conductors that originate within the clamp assembly shall be included in conductor fill calculations covered in 314.16(B)(1) as though they entered from outside the box. The clamp assembly shall not require a fill

Table 314.16(A) Metal Boxes

Box Trade Size			Minimum Volume		Maximum Number of Conductors* (arranged by AWG size)						
mm	in.		cm³	in.³	18	16	14	12	10	8	6
100 × 32	(4 × 1¼)	round/octagonal	205	12.5	8	7	6	5	5	5	2
100 × 38	(4 × 1½)	round/octagonal	254	15.5	10	8	7	6	6	5	3
100 × 54	(4 × 2⅛)	round/octagonal	353	21.5	14	12	10	9	8	7	4
100 × 32	(4× 1¼)	square	295	18.0	12	10	9	8	7	6	3
100 × 38	(4 × 1½)	square	344	21.0	14	12	10	9	8	7	4
100 × 54	(4 × 2⅛)	square	497	30.3	20	17	15	13	12	10	6
120 × 32	(4¹¹⁄₁₆ × 1¼)	square	418	25.5	17	14	12	11	10	8	5
120 × 38	(4¹¹⁄₁₆ × 1½)	square	484	29.5	19	16	14	13	11	9	5
120 × 54	(4¹¹⁄₁₆ × 2⅛)	square	689	42.0	28	24	21	18	16	14	8
75 × 50 × 38	(3 × 2 × 1½)	device	123	7.5	5	4	3	3	3	2	1
75 × 50 × 50	(3 × 2 × 2)	device	164	10.0	6	5	5	4	4	3	2
75× 50 × 57	(3× 2 × 2¼)	device	172	10.5	7	6	5	4	4	3	2
75 × 50 × 65	(3 × 2 × 2½)	device	205	12.5	8	7	6	5	5	4	2
75 × 50 × 70	(3 × 2 × 2¾)	device	230	14.0	9	8	7	6	5	4	2
75 × 50 × 90	(3 × 2 × 3½)	device	295	18.0	12	10	9	8	7	6	3
100 × 54 × 38	(4 × 2⅛ × 1½)	device	169	10.3	6	5	5	4	4	3	2
100 × 54 × 48	(4 × 2⅛ × 1⅞)	device	213	13.0	8	7	6	5	5	4	2
100 × 54 × 54	(4 × 2⅛ × 2⅛)	device	238	14.5	9	8	7	6	5	4	2
95 × 50 × 65	(3¾ × 2 × 2½)	masonry box	230	14.0	9	8	7	6	5	4	2
95 × 50 × 90	(3¾ × 2 × 3½)	masonry box	344	21.0	14	12	10	9	8	7	4
min. 44.5 depth	FS — single cover (1¾)		221	13.5	9	7	6	6	5	4	2
min. 60.3 depth	FD — single cover (2⅜)		295	18.0	12	10	9	8	7	6	3
min. 44.5 depth	FS — multiple cover (1¾)		295	18.0	12	10	9	8	7	6	3
min. 60.3 depth	FD — multiple cover (2⅜)		395	24.0	16	13	12	10	9	8	4

*Where no volume allowances are required by 314.16(B)(2) through (B)(5).

allowance, but the volume of the portion of the assembly that remains within the box after installation shall be excluded from the box volume as marked in 314.16(A)(2).

(3) Support Fittings Fill. Where one or more luminaire studs or hickeys are present in the box, a single volume allowance in accordance with Table 314.16(B) shall be made for each type of fitting based on the largest conductor present in the box.

(4) Device or Equipment Fill. For each yoke or strap containing one or more devices or equipment, a double volume allowance in accordance with Table 314.16(B) shall be made for each yoke or strap based on the largest conductor connected to a device(s) or equipment supported by that yoke or strap. A device or utilization equipment wider than a single 50 mm (2 in.) device box as described in Table 314.16(A) shall have double volume allowances provided for each gang required for mounting.

(5) Equipment Grounding Conductor Fill. Where one or more equipment grounding conductors or equipment bonding jumpers enter a box, a single volume allowance in accordance with Table 314.16(B) shall be made based on the largest equipment grounding conductor or equipment bonding jumper present in the box. Where an additional set of equipment grounding conductors, as permitted by 250.146(D), is present in the box, an additional volume allowance shall be made based on the largest equipment grounding conductor in the additional set.

Table 314.16(B) Volume Allowance Required per Conductor

Size of Conductor (AWG)	Free Space Within Box for Each Conductor	
	cm³	in.³
18	24.6	1.50
16	28.7	1.75
14	32.8	2.00
12	36.9	2.25
10	41.0	2.50
8	49.2	3.00
6	81.9	5.00

(C) Conduit Bodies.

(1) General. Conduit bodies enclosing 6 AWG conductors or smaller, other than short-radius conduit bodies as described in 314.16(C)(3), shall have a cross-sectional area not less than twice the cross-sectional area of the largest conduit or tubing to which they can be attached. The maximum number of conductors permitted shall be the maximum number permitted by Table 1 of Chapter 9 for the conduit or tubing to which it is attached.

(2) With Splices, Taps, or Devices. Only those conduit bodies that are durably and legibly marked by the manufacturer with their volume shall be permitted to contain splices, taps, or devices. The maximum number of conductors shall be calculated in accordance with 314.16(B). Conduit bodies shall be supported in a rigid and secure manner.

(3) Short Radius Conduit Bodies. Conduit bodies such as capped elbows and service-entrance elbows that enclose conductors 6 AWG or smaller, and are only intended to enable the installation of the raceway and the contained conductors, shall not contain splices, taps, or devices and shall be of an approved size to provide free space for all conductors enclosed in the conduit body.

314.17 Conductors Entering Boxes, Conduit Bodies, or Fittings. Conductors entering boxes, conduit bodies, or fittings shall be protected from abrasion and shall comply with 314.17(A) through (D).

(A) Openings to Be Closed. Openings through which conductors enter shall be closed in an approved manner.

(B) Metal Boxes and Conduit Bodies. Where metal boxes or conduit bodies are installed with messenger-supported wiring, open wiring on insulators, or concealed knob-and-tube wiring, conductors shall enter through insulating bushings or, in dry locations, through flexible tubing extending from the last insulating support to not less than 6 mm ($\frac{1}{4}$ in.) inside the box and beyond any cable clamps. Where nonmetallic-sheathed cable or multiconductor Type UF cable is used, the sheath shall extend not less than 6 mm ($\frac{1}{4}$ in.) inside the box and beyond any cable clamp. Except as provided in 300.15(C), the wiring shall be firmly secured to the box or conduit body. Where raceway or cable is installed with metal boxes or conduit bodies, the raceway or cable shall be secured to such boxes and conduit bodies.

(C) Nonmetallic Boxes and Conduit Bodies. Nonmetallic boxes and conduit bodies shall be suitable for the lowest temperature-rated conductor entering the box. Where nonmetallic boxes and conduit bodies are used with messenger-supported wiring, open wiring on insulators, or concealed knob-and-tube wiring, the conductors shall enter the box through individual holes. Where flexible tubing is used to enclose the conductors, the tubing shall extend from the last insulating support to not less than 6 mm ($\frac{1}{4}$ in.) inside the box and beyond any cable clamp. Where nonmetallic-sheathed cable or multiconductor Type UF cable is used, the sheath shall extend not less than 6 mm ($\frac{1}{4}$ in.) inside the box and beyond any cable clamp. In all instances, all permitted wiring methods shall be secured to the boxes.

Exception: Where nonmetallic-sheathed cable or multiconductor Type UF cable is used with single gang boxes not larger than a nominal size 57 mm × 100 mm (2$\frac{1}{4}$ in. × 4 in.) mounted in walls or ceilings, and where the cable is fastened within 200 mm (8 in.) of the box measured along the sheath and where the sheath extends through a cable knockout not less than 6 mm ($\frac{1}{4}$ in.), securing the cable to the box shall not be required. Multiple cable entries shall be permitted in a single cable knockout opening.

(D) Conductors 4 AWG or Larger. Installation shall comply with 300.4(G).

Informational Note: See 110.12(A) for requirements on closing unused cable and raceway knockout openings.

314.19 Boxes Enclosing Flush Devices. Boxes used to enclose flush devices shall be of such design that the devices will be completely enclosed on back and sides and substantial support

for the devices will be provided. Screws for supporting the box shall not also be used to attach a device.

314.20 Flush-Mounted Installations. Installations within or behind a surface of concrete, tile, gypsum, plaster, or other noncombustible material, including boxes employing a flush-type cover or faceplate, shall be made so that the front edge of the box, plaster ring, extension ring, or listed extender will not be set back of the finished surface more than 6 mm ($\frac{1}{4}$ in.).

Installations within a surface of wood or other combustible surface material, boxes, plaster rings, extension rings, or listed extenders shall extend to the finished surface or project therefrom.

314.21 Repairing Noncombustible Surfaces. Noncombustible surfaces that are broken or incomplete around boxes employing a flush-type cover or faceplate shall be repaired so there will be no gaps or open spaces greater than 3 mm ($\frac{1}{8}$ in.) at the edge of the box.

314.22 Surface Extensions. Surface extensions shall be made by mounting and mechanically securing an extension ring over the box. Equipment grounding shall be in accordance with Part VI of Article 250.

Exception: A surface extension shall be permitted to be made from the cover of a box where the cover is designed so it is unlikely to fall off or be removed if its securing means becomes loose. The wiring method shall be flexible for an approved length that permits removal of the cover and provides access to the box interior and shall be arranged so that any grounding continuity is independent of the connection between the box and cover.

314.23 Supports. Enclosures within the scope of this article shall be supported in accordance with one or more of the provisions in 314.23(A) through (H).

(A) Surface Mounting. An enclosure mounted on a building or other surface shall be rigidly and securely fastened in place. If the surface does not provide rigid and secure support, additional support in accordance with other provisions of this section shall be provided.

(B) Structural Mounting. An enclosure supported from a structural member or from grade shall be rigidly supported either directly or by using a metal, polymeric, or wood brace.

(1) Nails and Screws. Nails and screws, where used as a fastening means, shall secure boxes by using brackets on the outside of the enclosure, or by using mounting holes in the back or in a single side of the enclosure, or they shall pass through the interior within 6 mm ($\frac{1}{4}$ in.) of the back or ends of the enclosure. Screws shall not be permitted to pass through the box unless exposed threads in the box are protected using approved means to avoid abrasion of conductor insulation. Mounting holes made in the field shall be approved.

(2) Braces. Metal braces shall be protected against corrosion and formed from metal that is not less than 0.51 mm (0.020 in.) thick uncoated. Wood braces shall have a cross section not less than nominal 25 mm × 50 mm (1 in. × 2 in.). Wood braces in wet locations shall be treated for the conditions. Polymeric braces shall be identified as being suitable for the use.

(C) Mounting in Finished Surfaces. An enclosure mounted in a finished surface shall be rigidly secured thereto by clamps, anchors, or fittings identified for the application.

(D) Suspended Ceilings. An enclosure mounted to structural or supporting elements of a suspended ceiling shall be not more than 1650 cm³ (100 in.³) in size and shall be securely fastened in place in accordance with either 314.23(D)(1) or (D)(2).

(1) Framing Members. An enclosure shall be fastened to the framing members by mechanical means such as bolts, screws, or rivets, or by the use of clips or other securing means identified for use with the type of ceiling framing member(s) and enclosure(s) employed. The framing members shall be supported in an approved manner and securely fastened to each other and to the building structure.

(2) Support Wires. The installation shall comply with the provisions of 300.11(A). The enclosure shall be secured, using identified methods, to ceiling support wire(s), including any additional support wire(s) installed for ceiling support. Support wire(s) used for enclosure support shall be fastened at each end so as to be taut within the ceiling cavity.

(E) Raceway-Supported Enclosure, Without Devices, Luminaires, or Lampholders. An enclosure that does not contain a device(s), other than splicing devices, or supports a luminaire(s), a lampholder, or other equipment and is supported by entering raceways shall not exceed 1650 cm³ (100 in.³) in size. It shall have threaded entries or identified hubs. It shall be supported by two or more conduits threaded wrenchtight into the enclosure or hubs. Each conduit shall be secured within 900 mm (3 ft) of the enclosure, or within 450 mm (18 in.) of the enclosure if all conduit entries are on the same side.

Exception: The following wiring methods shall be permitted to support a conduit body of any size, including a conduit body constructed with only one conduit entry, provided that the trade size of the conduit body is not larger than the largest trade size of the conduit or tubing:

(1) *Intermediate metal conduit, Type IMC*
(2) *Rigid metal conduit, Type RMC*
(3) *Rigid polyvinyl chloride conduit, Type PVC*
(4) *Reinforced thermosetting resin conduit, Type RTRC*
(5) *Electrical metallic tubing, Type EMT*

(F) Raceway-Supported Enclosures, with Devices, Luminaires, or Lampholders. An enclosure that contains a device(s), other than splicing devices, or supports a luminaire(s), a lampholder, or other equipment and is supported by entering raceways shall not exceed 1650 cm³ (100 in.³) in size. It shall have threaded entries or identified hubs. It shall be supported by two or more conduits threaded wrenchtight into the enclosure or hubs. Each conduit shall be secured within 450 mm (18 in.) of the enclosure.

Exception No. 1: Rigid metal or intermediate metal conduit shall be permitted to support a conduit body of any size, including a conduit body constructed with only one conduit entry, provided the trade size of the conduit body is not larger than the largest trade size of the conduit.

Exception No. 2: An unbroken length(s) of rigid or intermediate metal conduit shall be permitted to support a box used for luminaire or lampholder support, or to support a wiring enclosure that is an integral part of a luminaire and used in lieu of a box in accordance with 300.15(B), where all of the following conditions are met:

(1) *The conduit is securely fastened at a point so that the length of conduit beyond the last point of conduit support does not exceed 900 mm (3 ft).*

(2) *The unbroken conduit length before the last point of conduit support is 300 mm (12 in.) or greater, and that portion of the conduit is securely fastened at some point not less than 300 mm (12 in.) from its last point of support.*

(3) *Where accessible to unqualified persons, the luminaire or lampholder, measured to its lowest point, is at least 2.5 m (8 ft) above grade or standing area and at least 900 mm (3 ft) measured horizontally to the 2.5 m (8 ft) elevation from windows, doors, porches, fire escapes, or similar locations.*

(4) *A luminaire supported by a single conduit does not exceed 300 mm (12 in.) in any direction from the point of conduit entry.*

(5) *The weight supported by any single conduit does not exceed 9 kg (20 lb).*

(6) *At the luminaire or lampholder end, the conduit(s) is threaded wrenchtight into the box, conduit body, integral wiring enclosure, or identified hubs. Where a box or conduit body is used for support, the luminaire shall be secured directly to the box or conduit body, or through a threaded conduit nipple not over 75 mm (3 in.) long.*

(G) Enclosures in Concrete or Masonry. An enclosure supported by embedment shall be identified as suitably protected from corrosion and securely embedded in concrete or masonry.

(H) Pendant Boxes. An enclosure supported by a pendant shall comply with 314.23(H)(1) or (H)(2).

(1) Flexible Cord. A box shall be supported from a multiconductor cord or cable in an approved manner that protects the conductors against strain, such as a strain-relief connector threaded into a box with a hub.

(2) Conduit. A box supporting lampholders or luminaires, or wiring enclosures within luminaires used in lieu of boxes in accordance with 300.15(B), shall be supported by rigid or intermediate metal conduit stems. For stems longer than 450 mm (18 in.), the stems shall be connected to the wiring system with flexible fittings suitable for the location. At the luminaire end, the conduit(s) shall be threaded wrenchtight into the box, wiring enclosure, or identified hubs.

Where supported by only a single conduit, the threaded joints shall be prevented from loosening by the use of setscrews or other effective means, or the luminaire, at any point, shall be at least 2.5 m (8 ft) above grade or standing area and at least 900 mm (3 ft) measured horizontally to the 2.5 m (8 ft) elevation from windows, doors, porches, fire escapes, or similar locations. A luminaire supported by a single conduit shall not exceed 300 mm (12 in.) in any horizontal direction from the point of conduit entry.

314.24 Depth of Boxes. Outlet and device boxes shall have an approved depth to allow equipment installed within them to be mounted properly and without likelihood of damage to conductors within the box.

(A) Outlet Boxes Without Enclosed Devices or Utilization Equipment. Outlet boxes that do not enclose devices or utilization equipment shall have a minimum internal depth of 12.7 mm (½ in.).

(B) Outlet and Device Boxes with Enclosed Devices or Utilization Equipment. Outlet and device boxes that enclose devices or utilization equipment shall have a minimum internal depth that accommodates the rearward projection of the equipment and the size of the conductors that supply the equipment. The internal depth shall include, where used, that of any extension boxes, plaster rings, or raised covers. The internal depth shall comply with all applicable provisions of 314.24(B)(1) through (B)(5).

(1) Large Equipment. Boxes that enclose devices or utilization equipment that projects more than 48 mm (1⅞ in.) rearward from the mounting plane of the box shall have a depth that is not less than the depth of the equipment plus 6 mm (¼ in.).

(2) Conductors Larger Than 4 AWG. Boxes that enclose devices or utilization equipment supplied by conductors larger than 4 AWG shall be identified for their specific function.

Exception to (2): Devices or utilization equipment supplied by conductors larger than 4 AWG shall be permitted to be mounted on or in junction and pull boxes larger than 1650 cm³ (100 in.³) if the spacing at the terminals meets the requirements of 312.6.

(3) Conductors 8, 6, or 4 AWG. Boxes that enclose devices or utilization equipment supplied by 8, 6, or 4 AWG conductors shall have an internal depth that is not less than 52.4 mm (2¹⁄₁₆ in.).

(4) Conductors 12 or 10 AWG. Boxes that enclose devices or utilization equipment supplied by 12 or 10 AWG conductors shall have an internal depth that is not less than 30.2 mm (1³⁄₁₆ in.). Where the equipment projects rearward from the mounting plane of the box by more than 25 mm (1 in.), the box shall have a depth not less than that of the equipment plus 6 mm (¼ in.).

(5) Conductors 14 AWG and Smaller. Boxes that enclose devices or utilization equipment supplied by 14 AWG or smaller conductors shall have a depth that is not less than 23.8 mm (¹⁵⁄₁₆ in.).

Exception to (1) through (5): Devices or utilization equipment that is listed to be installed with specified boxes shall be permitted.

314.25 Covers and Canopies. In completed installations, each box shall have a cover, faceplate, lampholder, or luminaire canopy, except where the installation complies with 410.24(B). Screws used for the purpose of attaching covers, or other equipment, to the box shall be either machine screws matching the thread gauge or size that is integral to the box or shall be in accordance with the manufacturer's instructions.

(A) Nonmetallic or Metal Covers and Plates. Nonmetallic or metal covers and plates shall be permitted. Where metal covers or plates are used, they shall comply with the grounding requirements of 250.110.

> Informational Note: For additional grounding requirements, see 410.42 for metal luminaire canopies, and 404.12 and 406.6(B) for metal faceplates.

(B) Exposed Combustible Wall or Ceiling Finish. Where a luminaire canopy or pan is used, any combustible wall or ceiling finish exposed between the edge of the canopy or pan and the outlet box shall be covered with noncombustible material if required by 410.23.

(C) Flexible Cord Pendants. Covers of outlet boxes and conduit bodies having holes through which flexible cord pendants pass shall be provided with identified bushings or shall have smooth, well-rounded surfaces on which the cords may

bear. So-called hard rubber or composition bushings shall not be used.

314.27 Outlet Boxes.

(A) Boxes at Luminaire or Lampholder Outlets. Outlet boxes or fittings designed for the support of luminaires and lampholders, and installed as required by 314.23, shall be permitted to support a luminaire or lampholder.

(1) Vertical Surface Outlets. Boxes used at luminaire or lampholder outlets in or on a vertical surface shall be identified and marked on the interior of the box to indicate the maximum weight of the luminaire that is permitted to be supported by the box if other than 23 kg (50 lb).

Exception: A vertically mounted luminaire or lampholder weighing not more than 3 kg (6 lb) shall be permitted to be supported on other boxes or plaster rings that are secured to other boxes, provided that the luminaire or its supporting yoke, or the lampholder, is secured to the box with no fewer than two No. 6 or larger screws.

(2) Ceiling Outlets. At every outlet used exclusively for lighting, the box shall be designed or installed so that a luminaire or lampholder may be attached. Boxes shall be required to support a luminaire weighing a minimum of 23 kg (50 lb). A luminaire that weighs more than 23 kg (50 lb) shall be supported independently of the outlet box, unless the outlet box is listed for not less than the weight to be supported. The interior of the box shall be marked by the manufacturer to indicate the maximum weight the box shall be permitted to support.

(B) Floor Boxes. Boxes listed specifically for this application shall be used for receptacles located in the floor.

Exception: Where the authority having jurisdiction judges them free from likely exposure to physical damage, moisture, and dirt, boxes located in elevated floors of show windows and similar locations shall be permitted to be other than those listed for floor applications. Receptacles and covers shall be listed as an assembly for this type of location.

(C) Boxes at Ceiling-Suspended (Paddle) Fan Outlets. Outlet boxes or outlet box systems used as the sole support of a ceiling-suspended (paddle) fan shall be listed, shall be marked by their manufacturer as suitable for this purpose, and shall not support ceiling-suspended (paddle) fans that weigh more than 32 kg (70 lb). For outlet boxes or outlet box systems designed to support ceiling-suspended (paddle) fans that weigh more than 16 kg (35 lb), the required marking shall include the maximum weight to be supported.

Where spare, separately switched, ungrounded conductors are provided to a ceiling-mounted outlet box, in a location acceptable for a ceiling-suspended (paddle) fan in one-family, two-family, or multifamily dwellings, the outlet box or outlet box system shall be listed for sole support of a ceiling-suspended (paddle) fan.

(D) Utilization Equipment. Boxes used for the support of utilization equipment other than ceiling-suspended (paddle) fans shall meet the requirements of 314.27(A) for the support of a luminaire that is the same size and weight.

Exception: Utilization equipment weighing not more than 3 kg (6 lb) shall be permitted to be supported on other boxes or plaster rings that are secured to other boxes, provided the equipment or its supporting yoke is secured to the box with no fewer than two No. 6 or larger screws.

N **(E) Separable Attachment Fittings.** Outlet boxes required in 314.27 shall be permitted to support listed locking support and mounting receptacles used in combination with compatible attachment fittings. The combination shall be identified for the support of equipment within the weight and mounting orientation limits of the listing. Where the supporting receptacle is installed within a box, it shall be included in the fill calculation covered in 314.16(B)(4).

314.28 Pull and Junction Boxes and Conduit Bodies. Boxes and conduit bodies used as pull or junction boxes shall comply with 314.28(A) through (E).

Exception: Terminal housings supplied with motors shall comply with the provisions of 430.12.

(A) Minimum Size. For raceways containing conductors of 4 AWG or larger that are required to be insulated, and for cables containing conductors of 4 AWG or larger, the minimum dimensions of pull or junction boxes installed in a raceway or cable run shall comply with 314.28(A)(1) through (A)(3). Where an enclosure dimension is to be calculated based on the diameter of entering raceways, the diameter shall be the metric designator (trade size) expressed in the units of measurement employed.

(1) Straight Pulls. In straight pulls, the length of the box or conduit body shall not be less than eight times the metric designator (trade size) of the largest raceway.

(2) Angle or U Pulls, or Splices. Where splices or where angle or U pulls are made, the distance between each raceway entry inside the box or conduit body and the opposite wall of the box or conduit body shall not be less than six times the metric designator (trade size) of the largest raceway in a row. This distance shall be increased for additional entries by the amount of the sum of the diameters of all other raceway entries in the same row on the same wall of the box. Each row shall be calculated individually, and the single row that provides the maximum distance shall be used.

Exception: Where a raceway or cable entry is in the wall of a box or conduit body opposite a removable cover, the distance from that wall to the cover shall be permitted to comply with the distance required for one wire per terminal in Table 312.6(A).

The distance between raceway entries enclosing the same conductor shall not be less than six times the metric designator (trade size) of the larger raceway.

When transposing cable size into raceway size in 314.28(A)(1) and (A)(2), the minimum metric designator (trade size) raceway required for the number and size of conductors in the cable shall be used.

(3) Smaller Dimensions. Listed boxes or listed conduit bodies of dimensions less than those required in 314.28(A)(1) and (A)(2) shall be permitted for installations of combinations of conductors that are less than the maximum conduit or tubing fill (of conduits or tubing being used) permitted by Table 1 of Chapter 9.

Listed conduit bodies of dimensions less than those required in 314.28(A)(2), and having a radius of the curve to the centerline not less than that indicated in Table 2 of Chapter 9 for one-shot and full-shoe benders, shall be permitted for installations of combinations of conductors permitted by Table 1 of Chapter 9. These conduit bodies shall be marked to show they

have been specifically evaluated in accordance with this provision.

Where the permitted combinations of conductors for which the box or conduit body has been listed are less than the maximum conduit or tubing fill permitted by Table 1 of Chapter 9, the box or conduit body shall be permanently marked with the maximum number and maximum size of conductors permitted. For other conductor sizes and combinations, the total cross-sectional area of the fill shall not exceed the cross-sectional area of the conductors specified in the marking, based on the type of conductor identified as part of the product listing.

Informational Note: Unless otherwise specified, the applicable product standards evaluate the fill markings covered here based on conductors with Type XHHW insulation.

(B) Conductors in Pull or Junction Boxes. In pull boxes or junction boxes having any dimension over 1.8 m (6 ft), all conductors shall be cabled or racked up in an approved manner.

(C) Covers. All pull boxes, junction boxes, and conduit bodies shall be provided with covers compatible with the box or conduit body construction and suitable for the conditions of use. Where used, metal covers shall comply with the grounding requirements of 250.110.

(D) Permanent Barriers. Where permanent barriers are installed in a box, each section shall be considered as a separate box.

(E) Power Distribution Blocks. Power distribution blocks shall be permitted in pull and junction boxes over 1650 cm³ (100 in.³) for connections of conductors where installed in boxes and where the installation complies with 314.28(E)(1) through (5).

Exception: Equipment grounding terminal bars shall be permitted in smaller enclosures.

(1) Installation. Power distribution blocks installed in boxes shall be listed. Power distribution blocks installed on the line side of the service equipment shall be listed and marked "suitable for use on the line side of service equipment" or equivalent.

(2) Size. In addition to the overall size requirement in the first sentence of 314.28(A)(2), the power distribution block shall be installed in a box with dimensions not smaller than specified in the installation instructions of the power distribution block.

(3) Wire Bending Space. Wire bending space at the terminals of power distribution blocks shall comply with 312.6.

(4) Live Parts. Power distribution blocks shall not have uninsulated live parts exposed within a box, whether or not the box cover is installed.

(5) Through Conductors. Where the pull or junction boxes are used for conductors that do not terminate on the power distribution block(s), the through conductors shall be arranged so the power distribution block terminals are unobstructed following installation.

314.29 Boxes, Conduit Bodies, and Handhole Enclosures to Be Accessible. Boxes, conduit bodies, and handhole enclosures shall be installed so that the wiring contained in them can be rendered accessible without removing any part of the building or structure or, in underground circuits, without excavating sidewalks, paving, earth, or other substance that is to be used to establish the finished grade.

Exception: Listed boxes and handhole enclosures shall be permitted where covered by gravel, light aggregate, or noncohesive granulated soil if their location is effectively identified and accessible for excavation.

314.30 Handhole Enclosures. Handhole enclosures shall be designed and installed to withstand all loads likely to be imposed on them. They shall be identified for use in underground systems.

> Informational Note: See ANSI/SCTE 77-2002, *Specification for Underground Enclosure Integrity,* for additional information on deliberate and nondeliberate traffic loading that can be expected to bear on underground enclosures.

(A) Size. Handhole enclosures shall be sized in accordance with 314.28(A) for conductors operating at 1000 volts or below, and in accordance with 314.71 for conductors operating at over 1000 volts. For handhole enclosures without bottoms where the provisions of 314.28(A)(2), Exception, or 314.71(B)(1), Exception No. 1, apply, the measurement to the removable cover shall be taken from the end of the conduit or cable assembly.

(B) Wiring Entries. Underground raceways and cable assemblies entering a handhole enclosure shall extend into the enclosure, but they shall not be required to be mechanically connected to the enclosure.

(C) Enclosed Wiring. All enclosed conductors and any splices or terminations, if present, shall be listed as suitable for wet locations.

(D) Covers. Handhole enclosure covers shall have an identifying mark or logo that prominently identifies the function of the enclosure, such as "electric." Handhole enclosure covers shall require the use of tools to open, or they shall weigh over 45 kg (100 lb). Metal covers and other exposed conductive surfaces shall be bonded in accordance with 250.92 if the conductors in the handhole are service conductors, or in accordance with 250.96(A) if the conductors in the handhole are feeder or branch-circuit conductors.

Part III. Construction Specifications

314.40 Metal Boxes, Conduit Bodies, and Fittings.

(A) Corrosion Resistant. Metal boxes, conduit bodies, and fittings shall be corrosion resistant or shall be well-galvanized, enameled, or otherwise properly coated inside and out to prevent corrosion.

> Informational Note: See 300.6 for limitation in the use of boxes and fittings protected from corrosion solely by enamel.

(B) Thickness of Metal. Sheet steel boxes not over 1650 cm³ (100 in.³) in size shall be made from steel not less than 1.59 mm (0.0625 in.) thick. The wall of a malleable iron box or conduit body and a die-cast or permanent-mold cast aluminum, brass, bronze, or zinc box or conduit body shall not be less than 2.38 mm (³⁄₃₂ in.) thick. Other cast metal boxes or conduit bodies shall have a wall thickness not less than 3.17 mm (⅛ in.).

Exception No. 1: Listed boxes and conduit bodies shown to have equivalent strength and characteristics shall be permitted to be made of thinner or other metals.

Exception No. 2: The walls of listed short radius conduit bodies, as covered in 314.16(C)(2), shall be permitted to be made of thinner metal.

(C) Metal Boxes Over 1650 cm³ (100 in.³). Metal boxes over 1650 cm³ (100 in.³) in size shall be constructed so as to be of ample strength and rigidity. If of sheet steel, the metal thickness shall not be less than 1.35 mm (0.053 in.) uncoated.

(D) Grounding Provisions. A means shall be provided in each metal box for the connection of an equipment grounding conductor. The means shall be permitted to be a tapped hole or equivalent.

314.41 Covers. Metal covers shall be of the same material as the box or conduit body with which they are used, or they shall be lined with firmly attached insulating material that is not less than 0.79 mm (¹⁄₃₂ in.) thick, or they shall be listed for the purpose. Metal covers shall be the same thickness as the boxes or conduit body for which they are used, or they shall be listed for the purpose. Covers of porcelain or other approved insulating materials shall be permitted if of such form and thickness as to afford the required protection and strength.

314.42 Bushings. Covers of outlet boxes and conduit bodies having holes through which flexible cord pendants may pass shall be provided with approved bushings or shall have smooth, well-rounded surfaces on which the cord may bear. Where individual conductors pass through a metal cover, a separate hole equipped with a bushing of suitable insulating material shall be provided for each conductor. Such separate holes shall be connected by a slot as required by 300.20.

314.43 Nonmetallic Boxes. Provisions for supports or other mounting means for nonmetallic boxes shall be outside of the box, or the box shall be constructed so as to prevent contact between the conductors in the box and the supporting screws.

314.44 Marking. All boxes and conduit bodies, covers, extension rings, plaster rings, and the like shall be durably and legibly marked with the manufacturer's name or trademark.

Part IV. Pull and Junction Boxes, Conduit Bodies, and Handhole Enclosures for Use on Systems over 1000 Volts, Nominal

314.70 General.

(A) Pull and Junction Boxes. Where pull and junction boxes are used on systems over 1000 volts, the installation shall comply with the provisions of Part IV and with the following general provisions of this article:

(1) Part I, 314.2; 314.3; and 314.4
(2) Part II, 314.15; 314.17; 314.20; 314.23(A), (B), or (G); 314.28(B); and 314.29
(3) Part III, 314.40(A) and (C); and 314.41

(B) Conduit Bodies. Where conduit bodies are used on systems over 1000 volts, the installation shall comply with the provisions of Part IV and with the following general provisions of this article:

(1) Part I, 314.4
(2) Part II, 314.15; 314.17; 314.23(A), (E), or (G); 314.28(A)(3); and 314.29
(3) Part III, 314.40(A) and 314.41

(C) Handhole Enclosures. Where handhole enclosures are used on systems over 1000 volts, the installation shall comply with the provisions of Part IV and with the following general provisions of this article:

(1) Part I, 314.3 and 314.4
(2) Part II, 314.15; 314.17; 314.23(G); 314.28(B); 314.29; and 314.30

314.71 Size of Pull and Junction Boxes, Conduit Bodies, and Handhole Enclosures. Pull and junction boxes and handhole enclosures shall provide approved space and dimensions for the installation of conductors, and they shall comply with the specific requirements of this section. Conduit bodies shall be permitted if they meet the dimensional requirements for boxes.

(A) For Straight Pulls. The length of the box shall not be less than 48 times the outside diameter, over sheath, of the largest shielded or lead-covered conductor or cable entering the box. The length shall not be less than 32 times the outside diameter of the largest nonshielded conductor or cable.

(B) For Angle or U Pulls.

(1) Distance to Opposite Wall. The distance between each cable or conductor entry inside the box and the opposite wall of the box shall not be less than 36 times the outside diameter, over sheath, of the largest cable or conductor. This distance shall be increased for additional entries by the amount of the sum of the outside diameters, over sheath, of all other cables or conductor entries through the same wall of the box.

Exception No. 1: Where a conductor or cable entry is in the wall of a box opposite a removable cover, the distance from that wall to the cover shall be permitted to be not less than the bending radius for the conductors as provided in 300.34.

Exception No. 2: Where cables are nonshielded and not lead covered, the distance of 36 times the outside diameter shall be permitted to be reduced to 24 times the outside diameter.

(2) Distance Between Entry and Exit. The distance between a cable or conductor entry and its exit from the box shall not be less than 36 times the outside diameter, over sheath, of that cable or conductor.

Exception: Where cables are nonshielded and not lead covered, the distance of 36 times the outside diameter shall be permitted to be reduced to 24 times the outside diameter.

(C) Removable Sides. One or more sides of any pull box shall be removable.

314.72 Construction and Installation Requirements.

(A) Corrosion Protection. Boxes shall be made of material inherently resistant to corrosion or shall be suitably protected, both internally and externally, by enameling, galvanizing, plating, or other means.

(B) Passing Through Partitions. Suitable bushings, shields, or fittings having smooth, rounded edges shall be provided where conductors or cables pass through partitions and at other locations where necessary.

(C) Complete Enclosure. Boxes shall provide a complete enclosure for the contained conductors or cables.

(D) Wiring Is Accessible. Boxes and conduit bodies shall be installed so that the conductors are accessible without removing any fixed part of the building or structure. Working space shall be provided in accordance with 110.34.

(E) Suitable Covers. Boxes shall be closed by suitable covers securely fastened in place. Underground box covers that weigh over 45 kg (100 lb) shall be considered meeting this requirement. Covers for boxes shall be permanently marked "DANGER — HIGH VOLTAGE — KEEP OUT." The marking shall be on the outside of the box cover and shall be readily visible. Letters shall be block type and at least 13 mm (½ in.) in height.

(F) Suitable for Expected Handling. Boxes and their covers shall be capable of withstanding the handling to which they are likely to be subjected.

ARTICLE 320
Armored Cable: Type AC

Part I. General

320.1 Scope. This article covers the use, installation, and construction specifications for armored cable, Type AC.

320.2 Definition.

Armored Cable, Type AC. A fabricated assembly of insulated conductors in a flexible interlocked metallic armor. See 320.100.

N **320.6 Listing Requirements.** Type AC cable and associated fittings shall be listed.

Part II. Installation

320.10 Uses Permitted. Type AC cable shall be permitted as follows:

(1) For feeders and branch circuits in both exposed and concealed installations
(2) In cable trays
(3) In dry locations
(4) Embedded in plaster finish on brick or other masonry, except in damp or wet locations
(5) To be run or fished in the air voids of masonry block or tile walls where such walls are not exposed or subject to excessive moisture or dampness

Informational Note: The "Uses Permitted" is not an all-inclusive list.

320.12 Uses Not Permitted. Type AC cable shall not be used as follows:

(1) Where subject to physical damage
(2) In damp or wet locations
(3) In air voids of masonry block or tile walls where such walls are exposed or subject to excessive moisture or dampness
(4) Where exposed to corrosive conditions
(5) Embedded in plaster finish on brick or other masonry in damp or wet locations

320.15 Exposed Work. Exposed runs of cable, except as provided in 300.11(A), shall closely follow the surface of the building finish or of running boards. Exposed runs shall also be permitted to be installed on the underside of joists where supported at each joist and located so as not to be subject to physical damage.

320.17 Through or Parallel to Framing Members. Type AC cable shall be protected in accordance with 300.4(A), (C), and (D) where installed through or parallel to framing members.

320.23 In Accessible Attics. Type AC cables in accessible attics or roof spaces shall be installed as specified in 320.23(A) and (B).

(A) Cables Run Across the Top of Floor Joists. Where run across the top of floor joists, or within 2.1 m (7 ft) of the floor or floor joists across the face of rafters or studding, the cable shall be protected by guard strips that are at least as high as the cable. Where this space is not accessible by permanent stairs or ladders, protection shall only be required within 1.8 m (6 ft) of the nearest edge of the scuttle hole or attic entrance.

(B) Cable Installed Parallel to Framing Members. Where the cable is installed parallel to the sides of rafters, studs, or ceiling or floor joists, neither guard strips nor running boards shall be required, and the installation shall also comply with 300.4(D).

320.24 Bending Radius. Bends in Type AC cable shall be made such that the cable is not damaged. The radius of the curve of the inner edge of any bend shall not be less than five times the diameter of the Type AC cable.

320.30 Securing and Supporting.

(A) General. Type AC cable shall be supported and secured by staples; cable ties listed and identified for securement and support; straps, hangers, or similar fittings; or other approved means designed and installed so as not to damage the cable.

(B) Securing. Unless otherwise permitted, Type AC cable shall be secured within 300 mm (12 in.) of every outlet box, junction box, cabinet, or fitting and at intervals not exceeding 1.4 m (4½ ft).

(C) Supporting. Unless otherwise permitted, Type AC cable shall be supported at intervals not exceeding 1.4 m (4½ ft).

Horizontal runs of Type AC cable installed in wooden or metal framing members or similar supporting means shall be considered supported where such support does not exceed 1.4 m (4½ ft) intervals.

(D) Unsupported Cables. Type AC cable shall be permitted to be unsupported and unsecured where the cable complies with any of the following:

(1) Is fished between access points through concealed spaces in finished buildings or structures and supporting is impracticable

(2) Is not more than 600 mm (2 ft) in length at terminals where flexibility is necessary

(3) Is not more than 1.8 m (6 ft) in length from the last point of cable support to the point of connection to a luminaire(s) or other electrical equipment and the cable and point of connection are within an accessible ceiling

For the purposes of this section, Type AC cable fittings shall be permitted as a means of cable support.

320.40 Boxes and Fittings. At all points where the armor of AC cable terminates, a fitting shall be provided to protect wires from abrasion, unless the design of the outlet boxes or fittings is such as to afford equivalent protection, and, in addition, an insulating bushing or its equivalent protection shall be provided between the conductors and the armor. The connector or clamp by which the Type AC cable is fastened to boxes or cabinets shall be of such design that the insulating bushing or its equivalent will be visible for inspection. Where change is made from Type AC cable to other cable or raceway wiring methods, a box, fitting, or conduit body shall be installed at junction points as required in 300.15.

320.80 Ampacity. The ampacity shall be determined in accordance with 310.15.

(A) Thermal Insulation. Armored cable installed in thermal insulation shall have conductors rated at 90°C (194°F). The ampacity of cable installed in these applications shall not exceed that of a 60°C (140°F) rated conductor. The 90°C (194°F) rating shall be permitted to be used for ampacity adjustment and correction calculations; however, the ampacity shall not exceed that of a 60°C (140°F) rated conductor.

(B) Cable Tray. The ampacity of Type AC cable installed in cable tray shall be determined in accordance with 392.80(A).

Part III. Construction Specifications

320.100 Construction. Type AC cable shall have an armor of flexible metal tape and shall have an internal bonding strip of copper or aluminum in intimate contact with the armor for its entire length.

320.104 Conductors. Insulated conductors shall be of a type listed in Table 310.104(A) or those identified for use in this cable. In addition, the conductors shall have an overall moisture-resistant and fire-retardant fibrous covering. For Type ACT, a moisture-resistant fibrous covering shall be required only on the individual conductors.

320.108 Equipment Grounding Conductor. Type AC cable shall provide an adequate path for fault current as required by 250.4(A)(5) or (B)(4) to act as an equipment grounding conductor.

320.120 Marking. The cable shall be marked in accordance with 310.120, except that Type AC shall have ready identification of the manufacturer by distinctive external markings on the cable armor throughout its entire length.

ARTICLE 322
Flat Cable Assemblies: Type FC

Part I. General

322.1 Scope. This article covers the use, installation, and construction specifications for flat cable assemblies, Type FC.

322.2 Definition.

Flat Cable Assembly, Type FC. An assembly of parallel conductors formed integrally with an insulating material web specifically designed for field installation in surface metal raceway.

N 322.6 Listing Requirements. Type FC and associated fittings shall be listed.

Part II. Installation

322.10 Uses Permitted. Flat cable assemblies shall be permitted only as follows:

(1) As branch circuits to supply suitable tap devices for lighting, small appliances, or small power loads. The rating of the branch circuit shall not exceed 30 amperes.

(2) Where installed for exposed work.

(3) In locations where they will not be subjected to physical damage. Where a flat cable assembly is installed less than 2.5 m (8 ft) above the floor or fixed working platform, it shall be protected by a cover identified for the use.

(4) In surface metal raceways identified for the use. The channel portion of the surface metal raceway systems shall be installed as complete systems before the flat cable assemblies are pulled into the raceways.

322.12 Uses Not Permitted. Flat cable assemblies shall not be used as follows:

(1) Where exposed to corrosive conditions, unless suitable for the application

(2) In hoistways or on elevators or escalators

(3) In any hazardous (classified) location, except as specifically permitted by other articles in this *Code*

(4) Outdoors or in wet or damp locations unless identified for the use

322.30 Securing and Supporting. The flat cable assemblies shall be supported by means of their special design features, within the surface metal raceways.

The surface metal raceways shall be supported as required for the specific raceway to be installed.

322.40 Boxes and Fittings.

(A) Dead Ends. Each flat cable assembly dead end shall be terminated in an end-cap device identified for the use.

The dead-end fitting for the enclosing surface metal raceway shall be identified for the use.

(B) Luminaire Hangers. Luminaire hangers installed with the flat cable assemblies shall be identified for the use.

(C) Fittings. Fittings to be installed with flat cable assemblies shall be designed and installed to prevent physical damage to the cable assemblies.

(D) Extensions. All extensions from flat cable assemblies shall be made by approved wiring methods, within the junction boxes, installed at either end of the flat cable assembly runs.

322.56 Splices and Taps.

(A) Splices. Splices shall be made in listed junction boxes.

(B) Taps. Taps shall be made between any phase conductor and the grounded conductor or any other phase conductor by means of devices and fittings identified for the use. Tap devices shall be rated at not less than 15 amperes, or more than 300 volts to ground, and shall be color-coded in accordance with the requirements of 322.120(C).

Part III. Construction Specifications

322.100 Construction. Flat cable assemblies shall consist of two, three, four, or five conductors.

322.104 Conductors. Flat cable assemblies shall have conductors of 10 AWG special stranded copper wires.

322.112 Insulation. The entire flat cable assembly shall be formed to provide a suitable insulation covering all the conductors and using one of the materials recognized in Table 310.104(A) for general branch-circuit wiring.

322.120 Marking.

(A) Temperature Rating. In addition to the provisions of 310.120, Type FC cable shall have the temperature rating durably marked on the surface at intervals not exceeding 600 mm (24 in.).

(B) Identification of Grounded Conductor. The grounded conductor shall be identified throughout its length by means of a distinctive and durable white or gray marking.

Informational Note: The color gray may have been used in the past as an ungrounded conductor. Care should be taken when working on existing systems.

(C) Terminal Block Identification. Terminal blocks identified for the use shall have distinctive and durable markings for color or word coding. The grounded conductor section shall have a white marking or other suitable designation. The next adjacent section of the terminal block shall have a black marking or other suitable designation. The next section shall have a red marking or other suitable designation. The final or outer section, opposite the grounded conductor section of the terminal block, shall have a blue marking or other suitable designation.

ARTICLE 324
Flat Conductor Cable: Type FCC

Part I. General

324.1 Scope. This article covers a field-installed wiring system for branch circuits incorporating Type FCC cable and associated accessories as defined by the article. The wiring system is designed for installation under carpet squares.

324.2 Definitions.

Bottom Shield. A protective layer that is installed between the floor and Type FCC flat conductor cable to protect the cable from physical damage and may or may not be incorporated as an integral part of the cable.

Cable Connector. A connector designed to join Type FCC cables without using a junction box.

FCC System. A complete wiring system for branch circuits that is designed for installation under carpet squares.

Informational Note: The FCC system includes Type FCC cable and associated shielding, connectors, terminators, adapters, boxes, and receptacles.

Insulating End. An insulator designed to electrically insulate the end of a Type FCC cable.

Metal Shield Connections. Means of connection designed to electrically and mechanically connect a metal shield to another metal shield, to a receptacle housing or self-contained device, or to a transition assembly.

Top Shield. A grounded metal shield covering under-carpet components of the FCC system for the purposes of providing protection against physical damage.

Transition Assembly. An assembly to facilitate connection of the FCC system to other wiring systems, incorporating (1) a means of electrical interconnection and (2) a suitable box or covering for providing electrical safety and protection against physical damage.

Type FCC Cable. Three or more flat copper conductors placed edge-to-edge and separated and enclosed within an insulating assembly.

324.6 Listing Requirements. Type FCC cable and associated fittings shall be listed.

Part II. Installation

324.10 Uses Permitted.

(A) Branch Circuits. Use of FCC systems shall be permitted both for general-purpose and appliance branch circuits and for individual branch circuits.

(B) Branch-Circuit Ratings.

(1) Voltage. Voltage between ungrounded conductors shall not exceed 300 volts. Voltage between ungrounded conductors and the grounded conductor shall not exceed 150 volts.

(2) Current. General-purpose and appliance branch circuits shall have ratings not exceeding 20 amperes. Individual branch circuits shall have ratings not exceeding 30 amperes.

(C) Floors. Use of FCC systems shall be permitted on hard, sound, smooth, continuous floor surfaces made of concrete, ceramic, or composition flooring, wood, and similar materials.

(D) Walls. Use of FCC systems shall be permitted on wall surfaces in surface metal raceways.

(E) Damp Locations. Use of FCC systems in damp locations shall be permitted.

(F) Heated Floors. Materials used for floors heated in excess of 30°C (86°F) shall be identified as suitable for use at these temperatures.

(G) System Height. Any portion of an FCC system with a height above floor level exceeding 2.3 mm (0.090 in.) shall be tapered or feathered at the edges to floor level.

324.12 Uses Not Permitted. FCC systems shall not be used in the following locations:

(1) Outdoors or in wet locations
(2) Where subject to corrosive vapors
(3) In any hazardous (classified) location
(4) In residential buildings
(5) In school and hospital buildings, other than administrative office areas

324.18 Crossings. Crossings of more than two Type FCC cable runs shall not be permitted at any one point. Crossings of a Type FCC cable over or under a flat communications or signal cable shall be permitted. In each case, a grounded layer of metal shielding shall separate the two cables, and crossings of more than two flat cables shall not be permitted at any one point.

324.30 Securing and Supporting. All FCC system components shall be firmly anchored to the floor or wall using an adhesive or mechanical anchoring system identified for this use. Floors shall be prepared to ensure adherence of the FCC system to the floor until the carpet squares are placed.

324.40 Boxes and Fittings.

(A) Cable Connections and Insulating Ends. All Type FCC cable connections shall use connectors identified for their use, installed such that electrical continuity, insulation, and sealing against dampness and liquid spillage are provided. All bare cable ends shall be insulated and sealed against dampness and liquid spillage using listed insulating ends.

(B) Polarization of Connections. All receptacles and connections shall be constructed and installed so as to maintain proper polarization of the system.

(C) Shields.

(1) Top Shield. A metal top shield shall be installed over all floor-mounted Type FCC cable, connectors, and insulating ends. The top shield shall completely cover all cable runs, corners, connectors, and ends.

(2) Bottom Shield. A bottom shield shall be installed beneath all Type FCC cable, connectors, and insulating ends.

(D) Connection to Other Systems. Power feed, grounding connection, and shield system connection between the FCC system and other wiring systems shall be accomplished in a transition assembly identified for this use.

(E) Metal-Shield Connectors. Metal shields shall be connected to each other and to boxes, receptacle housings, self-contained devices, and transition assemblies using metal-shield connectors.

324.41 Floor Coverings. Floor-mounted Type FCC cable, cable connectors, and insulating ends shall be covered with carpet squares not larger than 1.0 m (39.37 in.) square. Carpet squares that are adhered to the floor shall be attached with release-type adhesives.

324.42 Devices.

(A) Receptacles. All receptacles, receptacle housings, and self-contained devices used with the FCC system shall be identified for this use and shall be connected to the Type FCC cable and metal shields. Connection from any grounding conductor of the Type FCC cable shall be made to the shield system at each receptacle.

(B) Receptacles and Housings. Receptacle housings and self-contained devices designed either for floor mounting or for in-wall or on-wall mounting shall be permitted for use with the FCC system. Receptacle housings and self-contained devices shall incorporate means for facilitating entry and termination of Type FCC cable and for electrically connecting the housing or device with the metal shield. Receptacles and self-contained devices shall comply with 406.4. Power and communications outlets installed together in common housing shall be permitted in accordance with 800.133(A)(1)(d), Exception No. 2.

324.56 Splices and Taps.

(A) FCC Systems Alterations. Alterations to FCC systems shall be permitted. New cable connectors shall be used at new connection points to make alterations. It shall be permitted to leave unused cable runs and associated cable connectors in place and energized. All cable ends shall be covered with insulating ends.

(B) Transition Assemblies. All transition assemblies shall be identified for their use. Each assembly shall incorporate means for facilitating entry of the Type FCC cable into the assembly, for connecting the Type FCC cable to grounded conductors, and for electrically connecting the assembly to the metal cable shields and to equipment grounding conductors.

324.60 Grounding. All metal shields, boxes, receptacle housings, and self-contained devices shall be electrically continuous to the equipment grounding conductor of the supplying branch circuit. All such electrical connections shall be made with connectors identified for this use. The electrical resistivity of such shield system shall not be more than that of one conductor of the Type FCC cable used in the installation.

Part III. Construction Specifications

324.100 Construction.

(A) Type FCC Cable. Type FCC cable shall be listed for use with the FCC system and shall consist of three, four, or five flat copper conductors, one of which shall be an equipment grounding conductor.

(B) Shields.

(1) Materials and Dimensions. All top and bottom shields shall be of designs and materials identified for their use. Top shields shall be metal. Both metallic and nonmetallic materials shall be permitted for bottom shields.

(2) Resistivity. Metal shields shall have cross-sectional areas that provide for electrical resistivity of not more than that of one conductor of the Type FCC cable used in the installation.

324.101 Corrosion Resistance. Metal components of the system shall be either corrosion resistant, coated with corrosion-resistant materials, or insulated from contact with corrosive substances.

324.112 Insulation. The insulating material of the cable shall be moisture resistant and flame retardant. All insulating materials in the FCC systems shall be identified for their use.

324.120 Markings.

(A) Cable Marking. Type FCC cable shall be clearly and durably marked on both sides at intervals of not more than 610 mm (24 in.) with the information required by 310.120(A) and with the following additional information:

(1) Material of conductors
(2) Maximum temperature rating
(3) Ampacity

(B) Conductor Identification. Conductors shall be clearly and durably identified on both sides throughout their length as specified in 310.110.

ARTICLE 326
Integrated Gas Spacer Cable: Type IGS

Part I. General

326.1 Scope. This article covers the use, installation, and construction specifications for integrated gas spacer cable, Type IGS.

326.2 Definition.

Integrated Gas Spacer Cable, Type IGS. A factory assembly of one or more conductors, each individually insulated and enclosed in a loose fit, nonmetallic flexible conduit as an integrated gas spacer cable rated 0 through 600 volts.

Part II. Installation

326.10 Uses Permitted. Type IGS cable shall be permitted for use underground, including direct burial in the earth, as the following:

(1) Service-entrance conductors
(2) Feeder or branch-circuit conductors
(3) Service conductors, underground

326.12 Uses Not Permitted. Type IGS cable shall not be used as interior wiring or be exposed in contact with buildings.

326.24 Bending Radius. Where the coilable nonmetallic conduit and cable are bent for installation purposes or are flexed or bent during shipment or installation, the radius of the curve of the inner edge measured to the inside of the bend shall not be less than specified in Table 326.24.

326.26 Bends. A run of Type IGS cable between pull boxes or terminations shall not contain more than the equivalent of four quarter bends (360 degrees total), including those bends located immediately at the pull box or terminations.

326.40 Fittings. Terminations and splices for Type IGS cable shall be identified as a type that is suitable for maintaining the gas pressure within the conduit. A valve and cap shall be provided for each length of the cable and conduit to check the gas pressure or to inject gas into the conduit.

326.80 Ampacity. The ampacity of Type IGS cable shall not exceed the values shown in Table 326.80.

Part III. Construction Specifications

326.104 Conductors. The conductors shall be solid aluminum rods, laid parallel, consisting of one to nineteen 12.7 mm (½ in.) diameter rods. The minimum conductor size shall be 250 kcmil, and the maximum size shall be 4750 kcmil.

Table 326.24 Minimum Radii of Bends

Conduit Size		Minimum Radii	
Metric Designator	Trade Size	mm	in.
53	2	600	24
78	3	900	35
103	4	1150	45

Table 326.80 Ampacity of Type IGS Cable

Size (kcmil)	Amperes	Size (kcmil)	Amperes
250	119	2500	376
500	168	3000	412
750	206	3250	429
1000	238	3500	445
1250	266	3750	461
1500	292	4000	476
1750	315	4250	491
2000	336	4500	505
2250	357	4750	519

326.112 Insulation. The insulation shall be dry kraft paper tapes and a pressurized sulfur hexafluoride gas (SF_6), both approved for electrical use. The nominal gas pressure shall be 138 kPa gauge (20 lb/in.² gauge). The thickness of the paper spacer shall be as specified in Table 326.112.

326.116 Conduit. The conduit shall be a medium density polyethylene identified as suitable for use with natural gas rated pipe in metric designator 53, 78, or 103 (trade size 2, 3, or 4). The percent fill dimensions for the conduit are shown in Table 326.116.

The size of the conduit permitted for each conductor size shall be calculated for a percent fill not to exceed those found in Table 1, Chapter 9.

326.120 Marking. The cable shall be marked in accordance with 310.120(A), 310.120(B)(1), and 310.120(D).

Table 326.112 Paper Spacer Thickness

| Size (kcmil) | Thickness | |
	mm	in.
250–1000	1.02	0.040
1250–4750	1.52	0.060

Table 326.116 Conduit Dimensions

| Conduit Size | | Actual Outside Diameter | | Actual Inside Diameter | |
Metric Designator	Trade Size	mm	in.	mm	in.
53	2	60	2.375	49.46	1.947
78	3	89	3.500	73.30	2.886
103	4	114	4.500	94.23	3.710

ARTICLE 328
Medium Voltage Cable: Type MV

Part I. General

328.1 Scope. This article covers the use, installation, and construction specifications for medium voltage cable, Type MV.

328.2 Definition.

Medium Voltage Cable, Type MV. A single or multiconductor solid dielectric insulated cable rated 2001 volts or higher.

N **328.6 Listing Requirements.** Type MV cables and associated fittings shall be listed.

Part II. Installation

328.10 Uses Permitted. Type MV cable shall be permitted for use on power systems rated up to and including 35,000 volts, nominal, as follows:

(1) In wet or dry locations.
(2) In raceways.
(3) In cable trays, where identified for the use, in accordance with 392.10, 392.20(B), (C), and (D), 392.22(C), 392.30(B)(1), 392.46, 392.56, and 392.60. Type MV cable that has an overall metallic sheath or armor, complies with the requirements for Type MC cable, and is identified as "MV or MC" shall be permitted to be installed in cable trays in accordance with 392.10(B)(2).
(4) Direct buried in accordance with 300.50.
(5) In messenger-supported wiring in accordance with Part II of Article 396.
(6) As exposed runs in accordance with 300.37. Type MV cable that has an overall metallic sheath or armor, complies with the requirements for Type MC cable, and is identified as "MV or MC" shall be permitted to be installed as exposed runs of metal-clad cable in accordance with 300.37.

Informational Note: The "Uses Permitted" is not an all-inclusive list.

328.12 Uses Not Permitted. Type MV cable shall not be used where exposed to direct sunlight, unless identified for the use.

328.14 Installation. Type MV cable shall be installed, terminated, and tested by qualified persons.

Informational Note: Information about accepted industry practices and installation procedures for medium-voltage cable are described in ANSI/NECA/NCSCB 600-2014, *Standard for Installing and Maintaining Medium-Voltage Cable* and in IEEE 576-2000, *Recommended Practice for Installation, Termination, and Testing of Insulated Power Cables as Used in Industrial and Commercial Applications*.

N **328.30 Support.** Type MV cable terminated in equipment or installed in pull boxes or vaults shall be secured and supported by metallic or nonmetallic supports suitable to withstand the weight by cable ties listed and identified for securement and support, or other approved means, at intervals not exceeding 1.5 m (5 ft) from terminations or a maximum of 1.8 m (6 ft) between supports.

328.80 Ampacity. The ampacity of Type MV cable shall be determined in accordance with 310.60. The ampacity of Type MV cable installed in cable tray shall be determined in accordance with 392.80(B).

Part III. Construction Specifications

328.100 Construction. Type MV cables shall have copper, aluminum, or copper-clad aluminum conductors and shall comply with Table 310.104(C) and Table 310.104(D) or Table 310.104(E).

328.120 Marking. Medium voltage cable shall be marked as required by 310.120.

ARTICLE 330
Metal-Clad Cable: Type MC

Part I. General

330.1 Scope. This article covers the use, installation, and construction specifications of metal-clad cable, Type MC.

330.2 Definition.

Metal Clad Cable, Type MC. A factory assembly of one or more insulated circuit conductors with or without optical fiber members enclosed in an armor of interlocking metal tape, or a smooth or corrugated metallic sheath.

_N_ **330.6 Listing Requirements.** Type MC cable shall be listed. Fittings used for connecting Type MC cable to boxes, cabinets, or other equipment shall be listed and identified for such use.

Part II. Installation

330.10 Uses Permitted.

(A) General Uses. Type MC cable shall be permitted as follows:

(1) For services, feeders, and branch circuits.
(2) For power, lighting, control, and signal circuits.
(3) Indoors or outdoors.
(4) Exposed or concealed.
(5) To be direct buried where identified for such use.
(6) In cable tray where identified for such use.
(7) In any raceway.
(8) As aerial cable on a messenger.
(9) In hazardous (classified) locations where specifically permitted by other articles in this _Code_.
(10) In dry locations and embedded in plaster finish on brick or other masonry except in damp or wet locations.
(11) In wet locations where a corrosion-resistant jacket is provided over the metallic covering and any of the following conditions are met:

   a. The metallic covering is impervious to moisture.
   b. A jacket resistant to moisture is provided under the metal covering.
   c. The insulated conductors under the metallic covering are listed for use in wet locations.

(12) Where single-conductor cables are used, all phase conductors and, where used, the grounded conductor shall be grouped together to minimize induced voltage on the sheath.

(B) Specific Uses. Type MC cable shall be permitted to be installed in compliance with Parts II and III of Article 725 and 770.133 as applicable and in accordance with 330.10(B)(1) through (B)(4).

Informational Note: The "Uses Permitted" is not an all-inclusive list.

(1) Cable Tray. Type MC cable installed in cable tray shall comply with 392.10, 392.12, 392.18, 392.20, 392.22, 392.30, 392.46, 392.56, 392.60(C), and 392.80.

(2) Direct Buried. Direct-buried cable shall comply with 300.5 or 300.50, as appropriate.

(3) Installed as Service-Entrance Cable. Type MC cable installed as service-entrance cable shall be permitted in accordance with 230.43.

(4) Installed Outside of Buildings or Structures or as Aerial Cable. Type MC cable installed outside of buildings or structures or as aerial cable shall comply with 225.10, 396.10, and 396.12.

330.12 Uses Not Permitted. Type MC cable shall not be used under either of the following conditions:

(1) Where subject to physical damage
(2) Where exposed to any of the destructive corrosive conditions in (a) or (b), unless the metallic sheath or armor is resistant to the conditions or is protected by material resistant to the conditions:

   a. Direct buried in the earth or embedded in concrete unless identified for direct burial
   b. Exposed to cinder fills, strong chlorides, caustic alkalis, or vapors of chlorine or of hydrochloric acids

_N_ **330.15 Exposed Work.** Exposed runs of cable, except as provided in 300.11(A), shall closely follow the surface of the building finish or of running boards. Exposed runs shall also be permitted to be installed on the underside of joists where supported at each joist and located so as not to be subject to physical damage.

330.17 Through or Parallel to Framing Members. Type MC cable shall be protected in accordance with 300.4(A), (C), and (D) where installed through or parallel to framing members.

330.23 In Accessible Attics. The installation of Type MC cable in accessible attics or roof spaces shall also comply with 320.23.

330.24 Bending Radius. Bends in Type MC cable shall be so made that the cable will not be damaged. The radius of the curve of the inner edge of any bend shall not be less than required in 330.24(A) through (C).

(A) Smooth Sheath.

(1) Ten times the external diameter of the metallic sheath for cable not more than 19 mm (¾ in.) in external diameter
(2) Twelve times the external diameter of the metallic sheath for cable more than 19 mm (¾ in.) but not more than 38 mm (1½ in.) in external diameter
(3) Fifteen times the external diameter of the metallic sheath for cable more than 38 mm (1½ in.) in external diameter

(B) Interlocked-Type Armor or Corrugated Sheath. Seven times the external diameter of the metallic sheath.

(C) Shielded Conductors. Twelve times the overall diameter of one of the individual conductors or seven times the overall diameter of the multiconductor cable, whichever is greater.

330.30 Securing and Supporting.

(A) General. Type MC cable shall be supported and secured by staples; cable ties listed and identified for securement and support; straps, hangers, or similar fittings; or other approved means designed and installed so as not to damage the cable.

(B) Securing. Unless otherwise provided, cables shall be secured at intervals not exceeding 1.8 m (6 ft). Cables containing four or fewer conductors sized no larger than 10 AWG shall be secured within 300 mm (12 in.) of every box, cabinet, fitting, or other cable termination. In vertical installations, listed cables with ungrounded conductors 250 kcmil and larger shall be permitted to be secured at intervals not exceeding 3 m (10 ft).

(C) Supporting. Unless otherwise provided, cables shall be supported at intervals not exceeding 1.8 m (6 ft).

Horizontal runs of Type MC cable installed in wooden or metal framing members or similar supporting means shall be considered supported and secured where such support does not exceed 1.8-m (6-ft) intervals.

(D) Unsupported Cables. Type MC cable shall be permitted to be unsupported and unsecured where the cable complies with any of the following:

(1) Is fished between access points through concealed spaces in finished buildings or structures and supporting is impractical.

(2) Is not more than 1.8 m (6 ft) in length from the last point of cable support to the point of connection to luminaires or other electrical equipment and the cable and point of connection are within an accessible ceiling.

(3) Is Type MC of the interlocked armor type in lengths not exceeding 900 mm (3 ft) from the last point where it is securely fastened and is used to connect equipment where flexibility is necessary to minimize the transmission of vibration from equipment or to provide flexibility for equipment that requires movement after installation.

For the purpose of this section, Type MC cable fittings shall be permitted as a means of cable support.

330.31 Single Conductors.
Where single-conductor cables with a nonferrous armor or sheath are used, the installation shall comply with 300.20.

330.80 Ampacity.
The ampacity of Type MC cable shall be determined in accordance with 310.15 or 310.60 for 14 AWG and larger conductors and in accordance with Table 402.5 for 18 AWG and 16 AWG conductors. The installation shall not exceed the temperature ratings of terminations and equipment.

(A) Type MC Cable Installed in Cable Tray. The ampacities for Type MC cable installed in cable tray shall be determined in accordance with 392.80.

(B) Single Type MC Conductors Grouped Together. Where single Type MC conductors are grouped together in a triangular or square configuration and installed on a messenger or exposed with a maintained free airspace of not less than 2.15 times one conductor diameter (2.15 × O.D.) of the largest conductor contained within the configuration and adjacent conductor configurations or cables, the ampacity of the conductors shall not exceed the allowable ampacities in the following tables:

(1) Table 310.15(B)(20) for conductors rated 0 through 2000 volts

(2) Table 310.60(C)(67) and Table 310.60(C)(68) for conductors rated over 2000 volts

Part III. Construction Specifications

330.104 Conductors.
Conductors shall be of copper, aluminum, copper-clad aluminum, nickel or nickel-coated copper, solid or stranded. The minimum conductor size shall be 18 AWG copper, nickel or nickel-coated copper, or 12 AWG aluminum or copper-clad aluminum.

330.108 Equipment Grounding Conductor.
Where Type MC cable is used to provide an equipment grounding conductor, it shall comply with 250.118(10) and 250.122.

330.112 Insulation.
Insulated conductors shall comply with 330.112(A) or (B).

(A) 1000 Volts or Less. Insulated conductors in sizes 18 AWG and 16 AWG shall be of a type listed in Table 402.3, with a maximum operating temperature not less than 90°C (194°F) and as permitted by 725.49. Conductors larger than 16 AWG shall be of a type listed in Table 310.104(A) or of a type identified for use in Type MC cable.

(B) Over 1000 Volts. Insulated conductors shall be of a type listed in Table 310.104(B) and Table 310.104(C).

330.116 Sheath.
Metallic covering shall be one of the following types: smooth metallic sheath, corrugated metallic sheath, or interlocking metal tape armor. The metallic sheath shall be continuous and close fitting. A nonmagnetic sheath or armor shall be used on single conductor Type MC. Supplemental protection of an outer covering of corrosion-resistant material shall be permitted and shall be required where such protection is needed. The sheath shall not be used as a current-carrying conductor.

Informational Note: See 300.6 for protection against corrosion.

ARTICLE 332
Mineral-Insulated, Metal-Sheathed Cable: Type MI

Part I. General

332.1 Scope.
This article covers the use, installation, and construction specifications for mineral-insulated, metal-sheathed cable, Type MI.

332.2 Definition.

Mineral-Insulated, Metal-Sheathed Cable, Type MI. A factory assembly of one or more conductors insulated with a highly compressed refractory mineral insulation and enclosed in a liquidtight and gastight continuous copper or alloy steel sheath.

N **332.6 Listing Requirements.** Type MI cable and associated fittings shall be listed.

Part II. Installation

332.10 Uses Permitted. Type MI cable shall be permitted as follows:

(1) For services, feeders, and branch circuits
(2) For power, lighting, control, and signal circuits
(3) In dry, wet, or continuously moist locations
(4) Indoors or outdoors
(5) Where exposed or concealed
(6) Where embedded in plaster, concrete, fill, or other masonry, whether above or below grade
(7) In hazardous (classified) locations where specifically permitted by other articles in this *Code*
(8) Where exposed to oil and gasoline
(9) Where exposed to corrosive conditions not deteriorating to its sheath
(10) In underground runs where suitably protected against physical damage and corrosive conditions
(11) In or attached to cable tray

Informational Note: The "Uses Permitted" is not an all-inclusive list.

332.12 Uses Not Permitted. Type MI cable shall not be used under the following conditions or in the following locations:

(1) In underground runs unless protected from physical damage, where necessary
(2) Where exposed to conditions that are destructive and corrosive to the metallic sheath, unless additional protection is provided

332.17 Through or Parallel to Framing Members. Type MI cable shall be protected in accordance with 300.4 where installed through or parallel to framing members.

332.24 Bending Radius. Bends in Type MI cable shall be so made that the cable will not be damaged. The radius of the inner edge of any bend shall not be less than required as follows:

(1) Five times the external diameter of the metallic sheath for cable not more than 19 mm (¾ in.) in external diameter
(2) Ten times the external diameter of the metallic sheath for cable greater than 19 mm (¾ in.) but not more than 25 mm (1 in.) in external diameter

332.30 Securing and Supporting. Type MI cable shall be supported and secured by staples, straps, hangers, or similar fittings, designed and installed so as not to damage the cable, at intervals not exceeding 1.8 m (6 ft).

(A) Horizontal Runs Through Holes and Notches. In other than vertical runs, cables installed in accordance with 300.4 shall be considered supported and secured where such support does not exceed 1.8 m (6 ft) intervals.

(B) Unsupported Cable. Type MI cable shall be permitted to be unsupported where the cable is fished between access points

through concealed spaces in finished buildings or structures and supporting is impracticable.

(C) Cable Trays. All MI cable installed in cable trays shall comply with 392.30(A).

332.31 Single Conductors. Where single-conductor cables are used, all phase conductors and, where used, the neutral conductor shall be grouped together to minimize induced voltage on the sheath.

332.40 Boxes and Fittings.

(A) Fittings. Fittings used for connecting Type MI cable to boxes, cabinets, or other equipment shall be identified for such use.

(B) Terminal Seals. Where Type MI cable terminates, an end seal fitting shall be installed immediately after stripping to prevent the entrance of moisture into the insulation. The conductors extending beyond the sheath shall be individually provided with an insulating material.

332.80 Ampacity. The ampacity of Type MI cable shall be determined in accordance with 310.15. The conductor temperature at the end seal fitting shall not exceed the temperature rating of the listed end seal fitting, and the installation shall not exceed the temperature ratings of terminations or equipment.

(A) Type MI Cable Installed in Cable Tray. The ampacities for Type MI cable installed in cable tray shall be determined in accordance with 392.80(A).

(B) Single Type MI Conductors Grouped Together. Where single Type MI conductors are grouped together in a triangular or square configuration, as required by 332.31, and installed on a messenger or exposed with a maintained free air space of not less than 2.15 times one conductor diameter (2.15 × O.D.) of the largest conductor contained within the configuration and adjacent conductor configurations or cables, the ampacity of the conductors shall not exceed the allowable ampacities of Table 310.15(B)(17).

Part III. Construction Specifications

332.104 Conductors. Type MI cable conductors shall be of solid copper, nickel, or nickel-coated copper with a resistance corresponding to standard AWG and kcmil sizes.

332.108 Equipment Grounding Conductor. Where the outer sheath is made of copper, it shall provide an adequate path to serve as an equipment grounding conductor. Where the outer sheath is made of steel, a separate equipment grounding conductor shall be provided.

332.112 Insulation. The conductor insulation in Type MI cable shall be a highly compressed refractory mineral that provides proper spacing for all conductors.

332.116 Sheath. The outer sheath shall be of a continuous construction to provide mechanical protection and moisture seal.

ARTICLE 334
Nonmetallic-Sheathed Cable: Types NM, NMC, and NMS

Part I. General

334.1 Scope. This article covers the use, installation, and construction specifications of nonmetallic-sheathed cable.

334.2 Definitions.

Nonmetallic-Sheathed Cable. A factory assembly of two or more insulated conductors enclosed within an overall nonmetallic jacket.

Type NM. Insulated conductors enclosed within an overall nonmetallic jacket.

Type NMC. Insulated conductors enclosed within an overall, corrosion resistant, nonmetallic jacket.

Type NMS. Insulated power or control conductors with signaling, data, and communications conductors within an overall nonmetallic jacket.

334.6 Listing Requirements. Type NM, Type NMC, and Type NMS cables and associated fittings shall be listed.

Part II. Installation

334.10 Uses Permitted. Type NM, Type NMC, and Type NMS cables shall be permitted to be used in the following, except as prohibited in 334.12:

(1) One- and two-family dwellings and their attached or detached garages, and their storage buildings.
(2) Multi-family dwellings permitted to be of Types III, IV, and V construction.
(3) Other structures permitted to be of Types III, IV, and V construction. Cables shall be concealed within walls, floors, or ceilings that provide a thermal barrier of material that has at least a 15-minute finish rating as identified in listings of fire-rated assemblies.

Informational Note No. 1: Types of building construction and occupancy classifications are defined in NFPA 220 -2015, *Standard on Types of Building Construction*, or the applicable building code, or both.

Informational Note No. 2: See Informative Annex E for determination of building types [NFPA 220, Table 3-1].

(4) Cable trays in structures permitted to be Types III, IV, or V where the cables are identified for the use.

Informational Note: See 310.15(A)(3) for temperature limitation of conductors.

(5) Types I and II construction where installed within raceways permitted to be installed in Types I and II construction.

(A) Type NM. Type NM cable shall be permitted as follows:

(1) For both exposed and concealed work in normally dry locations except as prohibited in 334.10(3)
(2) To be installed or fished in air voids in masonry block or tile walls

(B) Type NMC. Type NMC cable shall be permitted as follows:

(1) For both exposed and concealed work in dry, moist, damp, or corrosive locations, except as prohibited by 334.10(3)
(2) In outside and inside walls of masonry block or tile
(3) In a shallow chase in masonry, concrete, or adobe protected against nails or screws by a steel plate at least 1.59 mm (1/16 in.) thick and covered with plaster, adobe, or similar finish

(C) Type NMS. Type NMS cable shall be permitted as follows:

(1) For both exposed and concealed work in normally dry locations except as prohibited by 334.10(3)
(2) To be installed or fished in air voids in masonry block or tile walls

334.12 Uses Not Permitted.

(A) Types NM, NMC, and NMS. Types NM, NMC, and NMS cables shall not be permitted as follows:

(1) In any dwelling or structure not specifically permitted in 334.10(1), (2), (3), and (5)
(2) Exposed within a dropped or suspended ceiling cavity in other than one- and two-family and multifamily dwellings
(3) As service-entrance cable
(4) In commercial garages having hazardous (classified) locations as defined in 511.3
(5) In theaters and similar locations, except where permitted in 518.4(B)
(6) In motion picture studios
(7) In storage battery rooms
(8) In hoistways or on elevators or escalators
(9) Embedded in poured cement, concrete, or aggregate
(10) In hazardous (classified) locations, except where specifically permitted by other articles in this *Code*

(B) Types NM and NMS. Types NM and NMS cables shall not be used under the following conditions or in the following locations:

(1) Where exposed to corrosive fumes or vapors
(2) Where embedded in masonry, concrete, adobe, fill, or plaster
(3) In a shallow chase in masonry, concrete, or adobe and covered with plaster, adobe, or similar finish
(4) In wet or damp locations

334.15 Exposed Work. In exposed work, except as provided in 300.11(A), cable shall be installed as specified in 334.15(A) through (C).

(A) To Follow Surface. Cable shall closely follow the surface of the building finish or of running boards.

(B) Protection from Physical Damage. Cable shall be protected from physical damage where necessary by rigid metal conduit, intermediate metal conduit, electrical metallic tubing, Schedule 80 PVC conduit, Type RTRC marked with the suffix -XW, or other approved means. Where passing through a floor, the cable shall be enclosed in rigid metal conduit, intermediate metal conduit, electrical metallic tubing, Schedule 80 PVC conduit, Type RTRC marked with the suffix -XW, or other approved means extending at least 150 mm (6 in.) above the floor.

Type NMC cable installed in shallow chases or grooves in masonry, concrete, or adobe shall be protected in accordance with the requirements in 300.4(F) and covered with plaster, adobe, or similar finish.

(C) In Unfinished Basements and Crawl Spaces. Where cable is run at angles with joists in unfinished basements and crawl spaces, it shall be permissible to secure cables not smaller than two 6 AWG or three 8 AWG conductors directly to the lower edges of the joists. Smaller cables shall be run either through bored holes in joists or on running boards. Nonmetallic-sheathed cable installed on the wall of an unfinished basement shall be permitted to be installed in a listed conduit or tubing or shall be protected in accordance with 300.4. Conduit or tubing shall be provided with a suitable insulating bushing or adapter at the point the cable enters the raceway. The sheath of the nonmetallic-sheathed cable shall extend through the conduit or tubing and into the outlet or device box not less than 6 mm (¼ in.). The cable shall be secured within 300 mm (12 in.) of the point where the cable enters the conduit or tubing. Metal conduit, tubing, and metal outlet boxes shall be connected to an equipment grounding conductor complying with the provisions of 250.86 and 250.148.

334.17 Through or Parallel to Framing Members. Types NM, NMC, or NMS cable shall be protected in accordance with 300.4 where installed through or parallel to framing members. Grommets used as required in 300.4(B)(1) shall remain in place and be listed for the purpose of cable protection.

334.23 In Accessible Attics. The installation of cable in accessible attics or roof spaces shall also comply with 320.23.

334.24 Bending Radius. Bends in Types NM, NMC, and NMS cable shall be so made that the cable will not be damaged. The radius of the curve of the inner edge of any bend during or after installation shall not be less than five times the diameter of the cable.

334.30 Securing and Supporting. Nonmetallic-sheathed cable shall be supported and secured by staples; cable ties listed and identified for securement and support; or straps, hangers, or similar fittings designed and installed so as not to damage the cable, at intervals not exceeding 1.4 m (4½ ft) and within 300 mm (12 in.) of every cable entry into enclosures such as outlet boxes, junction boxes, cabinets, or fittings. Flat cables shall not be stapled on edge.

Sections of cable protected from physical damage by raceway shall not be required to be secured within the raceway.

(A) Horizontal Runs Through Holes and Notches. In other than vertical runs, cables installed in accordance with 300.4 shall be considered to be supported and secured where such support does not exceed 1.4-m (4½-ft) intervals and the nonmetallic-sheathed cable is securely fastened in place by an approved means within 300 mm (12 in.) of each box, cabinet, conduit body, or other nonmetallic-sheathed cable termination.

Informational Note: See 314.17(C) for support where nonmetallic boxes are used.

(B) Unsupported Cables. Nonmetallic-sheathed cable shall be permitted to be unsupported where the cable:

(1) Is fished between access points through concealed spaces in finished buildings or structures and supporting is impractical.

(2) Is not more than 1.4 m (4½ ft) from the last point of cable support to the point of connection to a luminaire or other piece of electrical equipment and the cable and point of connection are within an accessible ceiling in one-, two-, or multifamily dwellings.

(C) Wiring Device Without a Separate Outlet Box. A wiring device identified for the use, without a separate outlet box, and incorporating an integral cable clamp shall be permitted where the cable is secured in place at intervals not exceeding 1.4 m (4½ ft) and within 300 mm (12 in.) from the wiring device wall opening, and there shall be at least a 300 mm (12 in.) loop of unbroken cable or 150 mm (6 in.) of a cable end available on the interior side of the finished wall to permit replacement.

334.40 Boxes and Fittings.

(A) Boxes of Insulating Material. Nonmetallic outlet boxes shall be permitted as provided by 314.3.

(B) Devices of Insulating Material. Self-contained switches, self-contained receptacles, and nonmetallic-sheathed cable interconnector devices of insulating material that are listed shall be permitted to be used without boxes in exposed cable wiring and for repair wiring in existing buildings where the cable is concealed. Openings in such devices shall form a close fit around the outer covering of the cable, and the device shall fully enclose the part of the cable from which any part of the covering has been removed. Where connections to conductors are by binding-screw terminals, there shall be available as many terminals as conductors.

(C) Devices with Integral Enclosures. Wiring devices with integral enclosures identified for such use shall be permitted as provided by 300.15(E).

334.80 Ampacity. The ampacity of Types NM, NMC, and NMS cable shall be determined in accordance with 310.15. The allowable ampacity shall not exceed that of a 60°C (140°F) rated conductor. The 90°C (194°F) rating shall be permitted to be used for ampacity adjustment and correction calculations, provided the final calculated ampacity does not exceed that of a 60°C (140°F) rated conductor. The ampacity of Types NM, NMC, and NMS cable installed in cable trays shall be determined in accordance with 392.80(A).

Where more than two NM cables containing two or more current-carrying conductors are installed, without maintaining spacing between the cables, through the same opening in wood framing that is to be sealed with thermal insulation, caulk, or sealing foam, the allowable ampacity of each conductor shall be adjusted in accordance with Table 310.15(B)(3)(a) and the provisions of 310.15(A)(2), Exception, shall not apply.

Where more than two NM cables containing two or more current-carrying conductors are installed in contact with thermal insulation without maintaining spacing between cables, the allowable ampacity of each conductor shall be adjusted in accordance with Table 310.15(B)(3)(a).

Part III. Construction Specifications

334.100 Construction. The outer cable sheath of nonmetallic-sheathed cable shall be a nonmetallic material.

334.104 Conductors. The 600-volt insulated conductors shall be sizes 14 AWG through 2 AWG copper conductors or sizes 12 AWG through 2 AWG aluminum or copper-clad aluminum conductors. The communications conductors shall comply with Part V of Article 800.

334.108 Equipment Grounding Conductor. In addition to the insulated conductors, the cable shall have an insulated, covered, or bare equipment grounding conductor.

334.112 Insulation. The insulated power conductors shall be one of the types listed in Table 310.104(A) that are suitable for branch-circuit wiring or one that is identified for use in these cables. Conductor insulation shall be rated at 90°C (194°F).

Informational Note: Types NM, NMC, and NMS cable identified by the markings NM-B, NMC-B, and NMS-B meet this requirement.

334.116 Sheath. The outer sheath of nonmetallic-sheathed cable shall comply with 334.116(A), (B), and (C).

(A) Type NM. The overall covering shall be flame retardant and moisture resistant.

(B) Type NMC. The overall covering shall be flame retardant, moisture resistant, fungus resistant, and corrosion resistant.

(C) Type NMS. The overall covering shall be flame retardant and moisture resistant. The sheath shall be applied so as to separate the power conductors from the communications conductors.

ARTICLE 336
Power and Control Tray Cable: Type TC

Part I. General

336.1 Scope. This article covers the use, installation, and construction specifications for power and control tray cable, Type TC.

336.2 Definition.

Power and Control Tray Cable, Type TC. A factory assembly of two or more insulated conductors, with or without associated bare or covered grounding conductors, under a nonmetallic jacket.

336.6 Listing Requirements. Type TC cables and associated fittings shall be listed.

Part II. Installation

336.10 Uses Permitted. Type TC cable shall be permitted to be used as follows:

(1) For power, lighting, control, and signal circuits.
(2) In cable trays, including those with mechanically discontinuous segments up to 300 mm (1 ft).
(3) In raceways.
(4) In outdoor locations supported by a messenger wire.
(5) For Class 1 circuits as permitted in Parts II and III of Article 725.
(6) For non–power-limited fire alarm circuits if conductors comply with the requirements of 760.49.

(7) Between a cable tray and the utilization equipment or device(s), provided all of the following apply:
  a. The cable is Type TC-ER.
  b. The cable is installed in industrial establishments where the conditions of maintenance and supervision ensure that only qualified persons service the installation.
  c. The cable is continuously supported and protected against physical damage using mechanical protection such as struts, angles, or channels.
  d. The cable that complies with the crush and impact requirements of Type MC cable and is identified with the marking "TC–ER."
  e. The cable is secured at intervals not exceeding 1.8 m (6 ft).
  f. Equipment grounding for the utilization equipment is provided by an equipment grounding conductor within the cable. In cables containing conductors sized 6 AWG or smaller, the equipment grounding conductor must be provided within the cable or, at the time of installation, one or more insulated conductors must be permanently identified as an equipment grounding conductor in accordance with 250.119(B).

Exception to (7): Where not subject to physical damage, Type TC-ER shall be permitted to transition between cable trays and between cable trays and utilization equipment or devices for a distance not to exceed 1.8 m (6 ft) without continuous support. The cable shall be mechanically supported where exiting the cable tray to ensure that the minimum bending radius is not exceeded.

(8) Where installed in wet locations, Type TC cable shall also be resistant to moisture and corrosive agents.
(9) In one- and two-family dwelling units, Type TC-ER cable containing both power and control conductors that is identified for pulling through structural members shall be permitted. Type TC-ER cable used as interior wiring shall be installed per the requirements of Part II of Article 334.

Exception: Where used to connect a generator and associated equipment having terminals rated 75°C (140°F) or higher, the cable shall not be limited in ampacity by 334.80 or 340.80.

Informational Note No. 1: TC-ER cable that is suitable for pulling through structural members is marked "JP."

Informational Note No. 2: See 725.136 for limitations on Class 2 or 3 circuits contained within the same cable with conductors of electric light, power, or Class 1 circuits.

(10) Direct buried, where identified for such use

Informational Note: See 310.15(A)(3) for temperature limitation of conductors.

336.12 Uses Not Permitted. Type TC tray cable shall not be installed or used as follows:

(1) Installed where it will be exposed to physical damage

(2) Installed outside a raceway or cable tray system, except as permitted in 336.10(4), 336.10(7), 336.10(9), and 336.10(10)

(3) Used where exposed to direct rays of the sun, unless identified as sunlight resistant

336.24 Bending Radius. Bends in Type TC cable shall be made so as not to damage the cable. For Type TC cable without metal shielding, the minimum bending radius shall be as follows:

(1) Four times the overall diameter for cables 25 mm (1 in.) or less in diameter

(2) Five times the overall diameter for cables larger than 25 mm (1 in.) but not more than 50 mm (2 in.) in diameter

(3) Six times the overall diameter for cables larger than 50 mm (2 in.) in diameter

Type TC cables with metallic shielding shall have a minimum bending radius of not less than 12 times the cable overall diameter.

336.80 Ampacity. The ampacity of Type TC tray cable shall be determined in accordance with 392.80(A) for 14 AWG and larger conductors, in accordance with 402.5 for 18 AWG through 16 AWG conductors where installed in cable tray, and in accordance with 310.15 where installed in a raceway or as messenger-supported wiring.

Part III. Construction Specifications

336.100 Construction. A metallic sheath or armor as defined in 330.116 shall not be permitted either under or over the nonmetallic jacket. Metallic shield(s) shall be permitted over groups of conductors, under the outer jacket, or both.

336.104 Conductors. The insulated conductors of Type TC cables shall be in sizes 18 AWG to 1000 kcmil copper, nickel, or nickel-coated copper, and sizes 12 AWG through 1000 kcmil aluminum or copper-clad aluminum. Insulated conductors of sizes 14 AWG, and larger copper, nickel, or nickel-coated copper, and sizes 12 AWG through 1000 kcmil aluminum or copper-clad aluminum shall be one of the types listed in Table 310.104(A) or Table 310.104(B) that is suitable for branch circuit and feeder circuits or one that is identified for such use.

(A) Fire Alarm Systems. Where used for fire alarm systems, conductors shall also be in accordance with 760.49.

(B) Thermocouple Circuits. Conductors in Type TC cable used for thermocouple circuits in accordance with Part III of Article 725 shall also be permitted to be any of the materials used for thermocouple extension wire.

(C) Class 1 Circuit Conductors. Insulated conductors of 18 AWG and 16 AWG copper shall also be in accordance with 725.49.

336.116 Jacket. The outer jacket shall be a flame-retardant, nonmetallic material.

336.120 Marking. There shall be no voltage marking on a Type TC cable employing thermocouple extension wire.

ARTICLE 338
Service-Entrance Cable: Types SE and USE

Part I. General

338.1 Scope. This article covers the use, installation, and construction specifications of service-entrance cable.

338.2 Definitions.

Service-Entrance Cable. A single conductor or multiconductor assembly provided with or without an overall covering, primarily used for services, and of the following types:

Type SE. Service-entrance cable having a flame-retardant moisture-resistant covering.

Type USE. Service-entrance cable, identified for underground use, having a moisture-resistant covering, but not required to have a flame-retardant covering.

N 338.6 Listing Requirements. Type SE and USE cables and associated fittings shall be listed.

Part II. Installation

338.10 Uses Permitted.

(A) Service-Entrance Conductors. Service-entrance cable shall be permitted to be used as service-entrance conductors and shall be installed in accordance with 230.6, 230.7, and Parts II, III, and IV of Article 230.

(B) Branch Circuits or Feeders.

(1) Grounded Conductor Insulated. Type SE service-entrance cables shall be permitted in wiring systems where all of the circuit conductors of the cable are of the thermoset or thermoplastic type.

(2) Use of Uninsulated Conductor. Type SE service-entrance cable shall be permitted for use where the insulated conductors are used for circuit wiring and the uninsulated conductor is used only for equipment grounding purposes.

Exception: In existing installations, uninsulated conductors shall be permitted as a grounded conductor in accordance with 250.32 and 250.140, where the uninsulated grounded conductor of the cable originates in service equipment, and with 225.30 through 225.40.

(3) Temperature Limitations. Type SE service-entrance cable used to supply appliances shall not be subject to conductor temperatures in excess of the temperature specified for the type of insulation involved.

(4) Installation Methods for Branch Circuits and Feeders.

(a)   *Interior Installations.* In addition to the provisions of this article, Type SE service-entrance cable used for interior wiring shall comply with the installation requirements of Part II of Article 334, excluding 334.80.

For Type SE cable with ungrounded conductor sizes 10 AWG and smaller, where installed in thermal insulation, the ampacity shall be in accordance with 60°C (140°F) conductor temperature rating. The maximum conductor temperature rating shall be permitted to be used for ampacity adjustment and correction purposes, if the final derated ampacity does not exceed that for a 60°C (140°F) rated conductor.

Informational Note No. 1: See 310.15(A)(3) for temperature limitation of conductors.

Informational Note No. 2: For the installation of main power feeder conductors in dwelling units refer to 310.15(B)(7).

(b)   *Exterior Installations.* In addition to the provisions of this article, service-entrance cable used for feeders or branch circuits, where installed as exterior wiring, shall be installed in accordance with Part I of Article 225. The cable shall be supported in accordance with 334.30. Type USE cable installed as underground feeder and branch circuit cable shall comply with Part II of Article 340.

Exception: Single-conductor Type USE and multi-rated USE conductors shall not be subject to the ampacity limitations of Part II of Article 340.

338.12 Uses Not Permitted.

(A) Service-Entrance Cable.   Service-entrance cable (SE) shall not be used under the following conditions or in the following locations:

(1)   Where subject to physical damage unless protected in accordance with 230.50(B)
(2)   Underground with or without a raceway
(3)   For exterior branch circuits and feeder wiring unless the installation complies with the provisions of Part I of Article 225 and is supported in accordance with 334.30 or is used as messenger-supported wiring as permitted in Part II of Article 396

(B) Underground Service-Entrance Cable.   Underground service-entrance cable (USE) shall not be used under the following conditions or in the following locations:

(1)   For interior wiring
(2)   For aboveground installations except where USE cable emerges from the ground and is terminated in an enclosure at an outdoor location and the cable is protected in accordance with 300.5(D)
(3)   As aerial cable unless it is a multiconductor cable identified for use aboveground and installed as messenger-supported wiring in accordance with 225.10 and Part II of Article 396

338.24 Bending Radius.   Bends in Types USE and SE cable shall be so made that the cable will not be damaged. The radius of the curve of the inner edge of any bend, during or after installation, shall not be less than five times the diameter of the cable.

Part III. Construction Specifications

338.100 Construction.   Cabled, single-conductor, Type USE constructions recognized for underground use shall be permitted to have a bare copper conductor cabled with the assembly. Type USE single, parallel, or cabled conductor assemblies recognized for underground use shall be permitted to have a bare copper concentric conductor applied. These constructions shall not require an outer overall covering.

Informational Note: See 230.41, Exception, item (2), for directly buried, uninsulated service-entrance conductors.

Type SE or USE cable containing two or more conductors shall be permitted to have one conductor uninsulated.

338.120 Marking.   Service-entrance cable shall be marked as required in 310.120. Cable with the neutral conductor smaller than the ungrounded conductors shall be so marked.

ARTICLE 340
Underground Feeder and Branch-Circuit Cable: Type UF

Part I.   General

340.1 Scope.   This article covers the use, installation, and construction specifications for underground feeder and branch-circuit cable, Type UF.

340.2 Definition.

Underground Feeder and Branch-Circuit Cable, Type UF.   A factory assembly of one or more insulated conductors with an integral or an overall covering of nonmetallic material suitable for direct burial in the earth.

N **340.6 Listing Requirements.**   Type UF cable and associated fittings shall be listed.

Part II.   Installation

340.10 Uses Permitted.   Type UF cable shall be permitted as follows:

(1)   For use underground, including direct burial in the earth. For underground requirements, see 300.5.
(2)   As single-conductor cables. Where installed as single-conductor cables, all conductors of the feeder grounded conductor or branch circuit, including the grounded conductor and equipment grounding conductor, if any, shall be installed in accordance with 300.3.
(3)   For wiring in wet, dry, or corrosive locations under the recognized wiring methods of this *Code.*
(4)   Installed as nonmetallic-sheathed cable. Where so installed, the installation and conductor requirements shall comply with Parts II and III of Article 334 and shall be of the multiconductor type.
(5)   For solar photovoltaic systems in accordance with 690.31.
(6)   As single-conductor cables as the nonheating leads for heating cables as provided in 424.43.

(7) Supported by cable trays. Type UF cable supported by cable trays shall be of the multiconductor type.

Informational Note: See 310.15(A)(3) for temperature limitation of conductors.

340.12 Uses Not Permitted. Type UF cable shall not be used as follows:

(1) As service-entrance cable
(2) In commercial garages
(3) In theaters and similar locations
(4) In motion picture studios
(5) In storage battery rooms
(6) In hoistways or on elevators or escalators
(7) In hazardous (classified) locations, except as specifically permitted by other articles in this *Code*
(8) Embedded in poured cement, concrete, or aggregate, except where embedded in plaster as nonheating leads where permitted in 424.43
(9) Where exposed to direct rays of the sun, unless identified as sunlight resistant
(10) Where subject to physical damage
(11) As overhead cable, except where installed as messenger-supported wiring in accordance with Part II of Article 396

340.24 Bending Radius. Bends in Type UF cable shall be so made that the cable is not damaged. The radius of the curve of the inner edge of any bend shall not be less than five times the diameter of the cable.

340.80 Ampacity. The ampacity of Type UF cable shall be that of 60°C (140°F) conductors in accordance with 310.15.

Part III. Construction Specifications

340.104 Conductors. The conductors shall be sizes 14 AWG copper or 12 AWG aluminum or copper-clad aluminum through 4/0 AWG.

340.108 Equipment Grounding Conductor. In addition to the insulated conductors, the cable shall be permitted to have an insulated or bare equipment grounding conductor.

340.112 Insulation. The conductors of Type UF shall be one of the moisture-resistant types listed in Table 310.104(A) that is suitable for branch-circuit wiring or one that is identified for such use. Where installed as a substitute wiring method for NM cable, the conductor insulation shall be rated 90°C (194°F).

340.116 Sheath. The overall covering shall be flame retardant; moisture, fungus, and corrosion resistant; and suitable for direct burial in the earth.

ARTICLE 342
Intermediate Metal Conduit: Type IMC

Part I. General

342.1 Scope. This article covers the use, installation, and construction specifications for intermediate metal conduit (IMC) and associated fittings.

342.2 Definition.

Intermediate Metal Conduit (IMC). A steel threadable raceway of circular cross section designed for the physical protection and routing of conductors and cables and for use as an equipment grounding conductor when installed with its integral or associated coupling and appropriate fittings.

342.6 Listing Requirements. IMC, factory elbows and couplings, and associated fittings shall be listed.

Part II. Installation

342.10 Uses Permitted.

(A) All Atmospheric Conditions and Occupancies. Use of IMC shall be permitted under all atmospheric conditions and occupancies.

(B) Corrosion Environments. IMC, elbows, couplings, and fittings shall be permitted to be installed in concrete, in direct contact with the earth, or in areas subject to severe corrosive influences where protected by corrosion protection approved for the condition.

(C) Cinder Fill. IMC shall be permitted to be installed in or under cinder fill where subject to permanent moisture where protected on all sides by a layer of noncinder concrete not less than 50 mm (2 in.) thick; where the conduit is not less than 450 mm (18 in.) under the fill; or where protected by corrosion protection approved for the condition.

(D) Wet Locations. All supports, bolts, straps, screws, and so forth, shall be of corrosion-resistant materials or protected against corrosion by corrosion-resistant materials.

Informational Note: See 300.6 for protection against corrosion.

342.14 Dissimilar Metals. Where practicable, dissimilar metals in contact anywhere in the system shall be avoided to eliminate the possibility of galvanic action.

Aluminum fittings and enclosures shall be permitted to be used with galvanized steel IMC where not subject to severe corrosive influences. Stainless steel IMC shall only be used with stainless steel fittings and approved accessories, outlet boxes, and enclosures.

342.20 Size.

(A) Minimum. IMC smaller than metric designator 16 (trade size ½) shall not be used.

(B) Maximum. IMC larger than metric designator 103 (trade size 4) shall not be used.

Informational Note: See 300.1(C) for the metric designators and trade sizes. These are for identification purposes only and do not relate to actual dimensions.

342.22 Number of Conductors. The number of conductors shall not exceed that permitted by the percentage fill specified in Table 1, Chapter 9.

Cables shall be permitted to be installed where such use is not prohibited by the respective cable articles. The number of cables shall not exceed the allowable percentage fill specified in Table 1, Chapter 9.

342.24 Bends — How Made. Bends of IMC shall be so made that the conduit will not be damaged and the internal diameter of the conduit will not be effectively reduced. The radius of the curve of any field bend to the centerline of the conduit shall not be less than indicated in Table 2, Chapter 9.

342.26 Bends — Number in One Run. There shall not be more than the equivalent of four quarter bends (360 degrees total) between pull points, for example, conduit bodies and boxes.

342.28 Reaming and Threading. All cut ends shall be reamed or otherwise finished to remove rough edges. Where conduit is threaded in the field, a standard cutting die with a taper of 1 in 16 (¾ in. taper per foot) shall be used.

Informational Note: See ANSI/ASME B.1.20.1-1983, *Standard for Pipe Threads, General Purpose (Inch)*.

342.30 Securing and Supporting. IMC shall be installed as a complete system in accordance with 300.18 and shall be securely fastened in place and supported in accordance with 342.30(A) and (B).

(A) Securely Fastened. IMC shall be secured in accordance with one of the following:

(1) IMC shall be securely fastened within 900 mm (3 ft) of each outlet box, junction box, device box, cabinet, conduit body, or other conduit termination.
(2) Where structural members do not readily permit fastening within 900 mm (3 ft), fastening shall be permitted to be increased to a distance of 1.5 m (5 ft).
(3) Where approved, conduit shall not be required to be securely fastened within 900 mm (3 ft) of the service head for above-the-roof termination of a mast.

(B) Supports. IMC shall be supported in accordance with one of the following:

(1) Conduit shall be supported at intervals not exceeding 3 m (10 ft).
(2) The distance between supports for straight runs of conduit shall be permitted in accordance with Table 344.30(B)(2), provided the conduit is made up with threaded couplings and supports that prevent transmission of stresses to termination where conduit is deflected between supports.
(3) Exposed vertical risers from industrial machinery or fixed equipment shall be permitted to be supported at intervals not exceeding 6 m (20 ft) if the conduit is made up with threaded couplings, the conduit is supported and securely fastened at the top and bottom of the riser, and no other means of intermediate support is readily available.
(4) Horizontal runs of IMC supported by openings through framing members at intervals not exceeding 3 m (10 ft) and securely fastened within 900 mm (3 ft) of termination points shall be permitted.

342.42 Couplings and Connectors.

(A) Threadless. Threadless couplings and connectors used with conduit shall be made tight. Where buried in masonry or concrete, they shall be the concretetight type. Where installed in wet locations, they shall comply with 314.15. Threadless couplings and connectors shall not be used on threaded conduit ends unless listed for the purpose.

(B) Running Threads. Running threads shall not be used on conduit for connection at couplings.

342.46 Bushings. Where a conduit enters a box, fitting, or other enclosure, a bushing shall be provided to protect the wires from abrasion unless the box, fitting, or enclosure is designed to provide such protection.

Informational Note: See 300.4(G) for the protection of conductors 4 AWG and larger at bushings.

342.56 Splices and Taps. Splices and taps shall be made in accordance with 300.15.

342.60 Grounding. IMC shall be permitted as an equipment grounding conductor.

N 342.100 Construction. IMC shall be made of one of the following:

(1) Steel, with protective coatings
(2) Stainless steel

Part III. Construction Specifications

342.120 Marking. Each length shall be clearly and durably marked at least every 1.5 m (5 ft) with the letters IMC. Each length shall be marked as required in the first sentence of 110.21(A).

342.130 Standard Lengths. The standard length of IMC shall be 3.05 m (10 ft), including an attached coupling, and each end shall be threaded. Longer or shorter lengths with or without coupling and threaded or unthreaded shall be permitted.

ARTICLE 344
Rigid Metal Conduit: Type RMC

Part I. General

344.1 Scope. This article covers the use, installation, and construction specifications for rigid metal conduit (RMC) and associated fittings.

344.2 Definition.

Rigid Metal Conduit (RMC). A threadable raceway of circular cross section designed for the physical protection and routing of conductors and cables and for use as an equipment grounding conductor when installed with its integral or associated coupling and appropriate fittings.

344.6 Listing Requirements. RMC, factory elbows and couplings, and associated fittings shall be listed.

Part II. Installation

344.10 Uses Permitted.

(A) Atmospheric Conditions and Occupancies.

(1) Galvanized Steel and Stainless Steel RMC. Galvanized steel and stainless steel RMC shall be permitted under all atmospheric conditions and occupancies.

(2) Red Brass RMC. Red brass RMC shall be permitted to be installed for direct burial and swimming pool applications.

(3) Aluminum RMC. Aluminum RMC shall be permitted to be installed where approved for the environment. Rigid aluminum conduit encased in concrete or in direct contact with the earth shall be provided with approved supplementary corrosion protection.

(4) Ferrous Raceways and Fittings. Ferrous raceways and fittings protected from corrosion solely by enamel shall be permitted only indoors and in occupancies not subject to severe corrosive influences.

(B) Corrosive Environments.

(1) Galvanized Steel, Stainless Steel, and Red Brass RMC, Elbows, Couplings, and Fittings. Galvanized steel, stainless steel, and red brass RMC elbows, couplings, and fittings shall be permitted to be installed in concrete, in direct contact with the earth, or in areas subject to severe corrosive influences where protected by corrosion protection approved for the condition.

(2) Supplementary Protection of Aluminum RMC. Aluminum RMC shall be provided with approved supplementary corrosion protection where encased in concrete or in direct contact with the earth.

(C) Cinder Fill. Galvanized steel, stainless steel, and red brass RMC shall be permitted to be installed in or under cinder fill where subject to permanent moisture where protected on all sides by a layer of noncinder concrete not less than 50 mm (2 in.) thick; where the conduit is not less than 450 mm (18 in.) under the fill; or where protected by corrosion protection approved for the condition.

(D) Wet Locations. All supports, bolts, straps, screws, and so forth, shall be of corrosion-resistant materials or protected against corrosion by corrosion-resistant materials.

> Informational Note: See 300.6 for protection against corrosion.

344.14 Dissimilar Metals. Where practicable, dissimilar metals in contact anywhere in the system shall be avoided to eliminate the possibility of galvanic action. Aluminum fittings and enclosures shall be permitted to be used with galvanized steel RMC, and galvanized steel fittings and enclosures shall be permitted to be used with aluminum RMC where not subject to severe corrosive influences. Stainless steel RMC shall only be used with stainless steel fittings and approved accessories, outlet boxes, and enclosures.

344.20 Size.

(A) Minimum. RMC smaller than metric designator 16 (trade size ½) shall not be used.

Exception: For enclosing the leads of motors as permitted in 430.245(B).

(B) Maximum. RMC larger than metric designator 155 (trade size 6) shall not be used.

> Informational Note: See 300.1(C) for the metric designators and trade sizes. These are for identification purposes only and do not relate to actual dimensions.

344.22 Number of Conductors. The number of conductors shall not exceed that permitted by the percentage fill specified in Table 1, Chapter 9.

Cables shall be permitted to be installed where such use is not prohibited by the respective cable articles. The number of cables shall not exceed the allowable percentage fill specified in Table 1, Chapter 9.

344.24 Bends — How Made. Bends of RMC shall be so made that the conduit will not be damaged and so that the internal diameter of the conduit will not be effectively reduced. The radius of the curve of any field bend to the centerline of the conduit shall not be less than indicated in Table 2, Chapter 9.

344.26 Bends — Number in One Run. There shall not be more than the equivalent of four quarter bends (360 degrees total) between pull points, for example, conduit bodies and boxes.

344.28 Reaming and Threading. All cut ends shall be reamed or otherwise finished to remove rough edges. Where conduit is threaded in the field, a standard cutting die with a 1 in 16 taper (¾ in. taper per foot) shall be used.

> Informational Note: See ANSI/ASME B.1.20.1-1983, *Standard for Pipe Threads, General Purpose (Inch).*

344.30 Securing and Supporting. RMC shall be installed as a complete system in accordance with 300.18 and shall be securely fastened in place and supported in accordance with 344.30(A) and (B).

(A) Securely Fastened. RMC shall be secured in accordance with one of the following:

(1) RMC shall be securely fastened within 900 mm (3 ft) of each outlet box, junction box, device box, cabinet, conduit body, or other conduit termination.
(2) Fastening shall be permitted to be increased to a distance of 1.5 m (5 ft) where structural members do not readily permit fastening within 900 mm (3 ft).
(3) Where approved, conduit shall not be required to be securely fastened within 900 mm (3 ft) of the service head for above-the-roof termination of a mast.

(B) Supports. RMC shall be supported in accordance with one of the following:

(1) Conduit shall be supported at intervals not exceeding 3 m (10 ft).
(2) The distance between supports for straight runs of conduit shall be permitted in accordance with Table 344.30(B)(2), provided the conduit is made up with threaded couplings and supports that prevent transmission of stresses to termination where conduit is deflected between supports.
(3) Exposed vertical risers from industrial machinery or fixed equipment shall be permitted to be supported at intervals not exceeding 6 m (20 ft) if the conduit is made up with threaded couplings, the conduit is supported and securely fastened at the top and bottom of the riser, and no other means of intermediate support is readily available.
(4) Horizontal runs of RMC supported by openings through framing members at intervals not exceeding 3 m (10 ft) and securely fastened within 900 mm (3 ft) of termination points shall be permitted.

344.42 Couplings and Connectors.

(A) Threadless. Threadless couplings and connectors used with conduit shall be made tight. Where buried in masonry or concrete, they shall be the concrete tight type. Where installed in wet locations, they shall comply with 314.15. Threadless

Table 344.30(B)(2) Supports for Rigid Metal Conduit

Conduit Size		Maximum Distance Between Rigid Metal Conduit Supports	
Metric Designator	Trade Size	m	ft
16–21	½–¾	3.0	10
27	1	3.7	12
35–41	1¼–1½	4.3	14
53–63	2–2½	4.9	16
78 and larger	3 and larger	6.1	20

couplings and connectors shall not be used on threaded conduit ends unless listed for the purpose.

(B) Running Threads. Running threads shall not be used on conduit for connection at couplings.

344.46 Bushings. Where a conduit enters a box, fitting, or other enclosure, a bushing shall be provided to protect the wires from abrasion unless the box, fitting, or enclosure is designed to provide such protection.

> Informational Note: See 300.4(G) for the protection of conductors sizes 4 AWG and larger at bushings.

344.56 Splices and Taps. Splices and taps shall be made in accordance with 300.15.

344.60 Grounding. RMC shall be permitted as an equipment grounding conductor.

Part III. Construction Specifications

344.100 Construction. RMC shall be made of one of the following:

(1) Steel with protective coatings
(2) Aluminum
(3) Red brass
(4) Stainless steel

344.120 Marking. Each length shall be clearly and durably identified in every 3 m (10 ft) as required in the first sentence of 110.21(A). Nonferrous conduit of corrosion-resistant material shall have suitable markings.

344.130 Standard Lengths. The standard length of RMC shall be 3.05 m (10 ft), including an attached coupling, and each end shall be threaded. Longer or shorter lengths with or without coupling and threaded or unthreaded shall be permitted.

ARTICLE 348
Flexible Metal Conduit: Type FMC

Part I. General

348.1 Scope. This article covers the use, installation, and construction specifications for flexible metal conduit (FMC) and associated fittings.

348.2 Definition.

Flexible Metal Conduit (FMC). A raceway of circular cross section made of helically wound, formed, interlocked metal strip.

348.6 Listing Requirements. FMC and associated fittings shall be listed.

Part II. Installation

348.10 Uses Permitted. FMC shall be permitted to be used in exposed and concealed locations.

348.12 Uses Not Permitted. FMC shall not be used in the following:

(1) In wet locations
(2) In hoistways, other than as permitted in 620.21(A)(1)
(3) In storage battery rooms
(4) In any hazardous (classified) location except as permitted by other articles in this *Code*
(5) Where exposed to materials having a deteriorating effect on the installed conductors, such as oil or gasoline
(6) Underground or embedded in poured concrete or aggregate
(7) Where subject to physical damage

348.20 Size.

(A) Minimum. FMC less than metric designator 16 (trade size ½) shall not be used unless permitted in 348.20(A)(1) through (A)(5) for metric designator 12 (trade size ⅜).

(1) For enclosing the leads of motors as permitted in 430.245(B)
(2) In lengths not in excess of 1.8 m (6 ft) for any of the following uses:
   a. For utilization equipment
   b. As part of a listed assembly
   c. For tap connections to luminaires as permitted in 410.117(C)
(3) For manufactured wiring systems as permitted in 604.100(A)
(4) In hoistways as permitted in 620.21(A)(1)
(5) As part of a listed assembly to connect wired luminaire sections as permitted in 410.137(C)

(B) Maximum. FMC larger than metric designator 103 (trade size 4) shall not be used.

> Informational Note: See 300.1(C) for the metric designators and trade sizes. These are for identification purposes only and do not relate to actual dimensions.

348.22 Number of Conductors. The number of conductors shall not exceed that permitted by the percentage fill specified in Table 1, Chapter 9, or as permitted in Table 348.22, or for metric designator 12 (trade size ⅜).

Cables shall be permitted to be installed where such use is not prohibited by the respective cable articles. The number of cables shall not exceed the allowable percentage fill specified in Table 1, Chapter 9.

348.24 Bends — How Made. Bends in conduit shall be made so that the conduit is not damaged and the internal diameter of the conduit is not effectively reduced. Bends shall be permitted to be made manually without auxiliary equipment. The

Table 348.22 Maximum Number of Insulated Conductors in Metric Designator 12 (Trade Size ⅜) Flexible Metal Conduit (FMC)*

Size (AWG)	Types RFH-2, SF-2		Types TF, XHHW, TW		Types TFN, THHN, THWN		Types FEP, FEBP, PF, PGF	
	Fittings Inside Conduit	Fittings Outside Conduit	Fittings Inside Conduit	Fittings Outside Conduit	Fittings Inside Conduit	Fittings Outside Conduit	Fittings Inside Conduit	Fittings Outside Conduit
18	2	3	3	5	5	8	5	8
16	1	2	3	4	4	6	4	6
14	1	2	2	3	3	4	3	4
12	—	—	1	2	2	3	2	3
10	—	—	1	1	1	1	1	2

*In addition, one insulated, covered, or bare equipment grounding conductor of the same size shall be permitted.

radius of the curve to the centerline of any bend shall not be less than shown in Table 2, Chapter 9 using the column "Other Bends."

348.26 Bends — Number in One Run. There shall not be more than the equivalent of four quarter bends (360 degrees total) between pull points, for example, conduit bodies and boxes.

348.28 Trimming. All cut ends shall be trimmed or otherwise finished to remove rough edges, except where fittings that thread into the convolutions are used.

348.30 Securing and Supporting. FMC shall be securely fastened in place and supported in accordance with 348.30(A) and (B).

(A) Securely Fastened. FMC shall be securely fastened in place by an approved means within 300 mm (12 in.) of each box, cabinet, conduit body, or other conduit termination and shall be supported and secured at intervals not to exceed 1.4 m (4½ ft). Where used, cable ties shall be listed and be identified for securement and support.

Exception No. 1: Where FMC is fished between access points through concealed spaces in finished buildings or structures and supporting is impracticable.

Exception No. 2: Where flexibility is necessary after installation, lengths from the last point where the raceway is securely fastened shall not exceed the following:

(1) 900 mm (3 ft) for metric designators 16 through 35 (trade sizes ½ through 1¼)

(2) 1200 mm (4 ft) for metric designators 41 through 53 (trade sizes 1½ through 2)

(3) 1500 mm (5 ft) for metric designators 63 (trade size 2½) and larger

Exception No. 3: Lengths not exceeding 1.8 m (6 ft) from a luminaire terminal connection for tap connections to luminaires as permitted in 410.117(C).

Exception No. 4: Lengths not exceeding 1.8 m (6 ft) from the last point where the raceway is securely fastened for connections within an accessible ceiling to a luminaire(s) or other equipment. For the purposes of this exception, listed flexible metal conduit fittings shall be permitted as a means of securement and support.

(B) Supports. Horizontal runs of FMC supported by openings through framing members at intervals not greater than 1.4 m (4½ ft) and securely fastened within 300 mm (12 in.) of termination points shall be permitted.

348.42 Couplings and Connectors. Angle connectors shall not be concealed.

348.56 Splices and Taps. Splices and taps shall be made in accordance with 300.15.

348.60 Grounding and Bonding. If used to connect equipment where flexibility is necessary to minimize the transmission of vibration from equipment or to provide flexibility for equipment that requires movement after installation, an equipment grounding conductor shall be installed.

Where flexibility is not required after installation, FMC shall be permitted to be used as an equipment grounding conductor when installed in accordance with 250.118(5).

Where required or installed, equipment grounding conductors shall be installed in accordance with 250.134(B).

Where required or installed, equipment bonding jumpers shall be installed in accordance with 250.102.

ARTICLE 350
Liquidtight Flexible Metal Conduit: Type LFMC

Part I. General

350.1 Scope. This article covers the use, installation, and construction specifications for liquidtight flexible metal conduit (LFMC) and associated fittings.

350.2 Definition.

Liquidtight Flexible Metal Conduit (LFMC). A raceway of circular cross section having an outer liquidtight, nonmetallic, sunlight-resistant jacket over an inner flexible metal core with associated couplings, connectors, and fittings for the installation of electric conductors.

350.6 Listing Requirements. LFMC and associated fittings shall be listed.

Part II. Installation

350.10 Uses Permitted. LFMC shall be permitted to be used in exposed or concealed locations as follows:

(1) Where conditions of installation, operation, or maintenance require flexibility or protection from liquids, vapors, or solids

(2) In hazardous (classified) locations where specifically permitted by Chapter 5

(3) For direct burial where listed and marked for the purpose

350.12 Uses Not Permitted. LFMC shall not be used as follows:

(1) Where subject to physical damage

(2) Where any combination of ambient and conductor temperature produces an operating temperature in excess of that for which the material is approved

350.20 Size.

(A) Minimum. LFMC smaller than metric designator 16 (trade size ½) shall not be used.

Exception: LFMC of metric designator 12 (trade size ⅜) shall be permitted as covered in 348.20(A).

(B) Maximum. The maximum size of LFMC shall be metric designator 103 (trade size 4).

> Informational Note: See 300.1(C) for the metric designators and trade sizes. These are for identification purposes only and do not relate to actual dimensions.

350.22 Number of Conductors or Cables.

(A) Metric Designators 16 through 103 (Trade Sizes ½ through 4). The number of conductors shall not exceed that permitted by the percentage fill specified in Table 1, Chapter 9.

Cables shall be permitted to be installed where such use is not prohibited by the respective cable articles. The number of cables shall not exceed the allowable percentage fill specified in Table 1, Chapter 9.

(B) Metric Designator 12 (Trade Size ⅜). The number of conductors shall not exceed that permitted in Table 348.22, "Fittings Outside Conduit" columns.

350.24 Bends — How Made. Bends in conduit shall be so made that the conduit will not be damaged and the internal diameter of the conduit will not be effectively reduced. Bends shall be permitted to be made manually without auxiliary equipment. The radius of the curve to the centerline of any bend shall be not less than required in Table 2, Chapter 9 using the column "Other Bends."

350.26 Bends — Number in One Run. There shall not be more than the equivalent of four quarter bends (360 degrees total) between pull points, for example, conduit bodies and boxes.

350.28 Trimming. All cut ends of conduit shall be trimmed inside and outside to remove rough edges.

350.30 Securing and Supporting. LFMC shall be securely fastened in place and supported in accordance with 350.30(A) and (B).

(A) Securely Fastened. LFMC shall be securely fastened in place by an approved means within 300 mm (12 in.) of each box, cabinet, conduit body, or other conduit termination and shall be supported and secured at intervals not to exceed 1.4 m (4½ ft). Where used, cable ties shall be listed and be identified for securement and support.

Exception No. 1: Where LFMC is fished between access points through concealed spaces in finished buildings or structures and supporting is impractical.

Exception No. 2: Where flexibility is necessary after installation, lengths from the last point where the raceway is securely fastened shall not exceed the following:

(1) 900 mm (3 ft) for metric designators 16 through 35 (trade sizes ½ through 1¼)

(2) 1200 mm (4 ft) for metric designators 41 through 53 (trade sizes 1½ through 2)

(3) 1500 mm (5 ft) for metric designators 63 (trade size 2½) and larger

Exception No. 3: Lengths not exceeding 1.8 m (6 ft) from a luminaire terminal connection for tap conductors to luminaires, as permitted in 410.117(C).

Exception No. 4: Lengths not exceeding 1.8 m (6 ft) from the last point where the raceway is securely fastened for connections within an accessible ceiling to luminaire(s) or other equipment. For the purposes of 350.30, listed LFMC fittings shall be permitted as a means of securement and support.

(B) Supports. Horizontal runs of LFMC supported by openings through framing members at intervals not greater than 1.4 m (4½ ft) and securely fastened within 300 mm (12 in.) of termination points shall be permitted.

350.42 Couplings and Connectors. Only fittings listed for use with LFMC shall be used. Angle connectors shall not be concealed. Straight LFMC fittings shall be permitted for direct burial where marked.

350.56 Splices and Taps. Splices and taps shall be made in accordance with 300.15.

350.60 Grounding and Bonding. If used to connect equipment where flexibility is necessary to minimize the transmission of vibration from equipment or to provide flexibility for equipment that requires movement after installation, an equipment grounding conductor shall be installed.

Where flexibility is not required after installation, LFMC shall be permitted to be used as an equipment grounding conductor when installed in accordance with 250.118(6).

Where required or installed, equipment grounding conductors shall be installed in accordance with 250.134(B).

Where required or installed, equipment bonding jumpers shall be installed in accordance with 250.102.

> Informational Note: See 501.30(B), 502.30(B), 503.30(B), 505.25(B), and 506.25(B) for types of equipment grounding conductors.

Part III. Construction Specifications

350.120 Marking. LFMC shall be marked according to 110.21. The trade size and other information required by the listing shall also be marked on the conduit. Conduit suitable for direct burial shall be so marked.

ARTICLE 352
Rigid Polyvinyl Chloride Conduit: Type PVC

Part I. General

352.1 Scope. This article covers the use, installation, and construction specifications for rigid polyvinyl chloride conduit (PVC) and associated fittings.

> Informational Note: Refer to Article 353 for High Density Polyethylene Conduit: Type HDPE, and Article 355 for Reinforced Thermosetting Resin Conduit: Type RTRC.

352.2 Definition.

Rigid Polyvinyl Chloride Conduit (PVC). A rigid nonmetallic raceway of circular cross section, with integral or associated couplings, connectors, and fittings for the installation of electrical conductors and cables.

352.6 Listing Requirements. PVC conduit, factory elbows, and associated fittings shall be listed.

Part II. Installation

352.10 Uses Permitted. The use of PVC conduit shall be permitted in accordance with 352.10(A) through (I).

> Informational Note: Extreme cold may cause some nonmetallic conduits to become brittle and, therefore, more susceptible to damage from physical contact.

(A) Concealed. PVC conduit shall be permitted in walls, floors, and ceilings.

(B) Corrosive Influences. PVC conduit shall be permitted in locations subject to severe corrosive influences as covered in 300.6 and where subject to chemicals for which the materials are specifically approved.

(C) Cinders. PVC conduit shall be permitted in cinder fill.

(D) Wet Locations. PVC conduit shall be permitted in portions of dairies, laundries, canneries, or other wet locations, and in locations where walls are frequently washed, the entire conduit system, including boxes and fittings used therewith, shall be installed and equipped so as to prevent water from entering the conduit. All supports, bolts, straps, screws, and so forth, shall be of corrosion-resistant materials or be protected against corrosion by approved corrosion-resistant materials.

(E) Dry and Damp Locations. PVC conduit shall be permitted for use in dry and damp locations not prohibited by 352.12.

(F) Exposed. PVC conduit shall be permitted for exposed work. PVC conduit used exposed in areas of physical damage shall be identified for the use.

> Informational Note: PVC Conduit, Type Schedule 80, is identified for areas of physical damage.

(G) Underground Installations. For underground installations, PVC shall be permitted for direct burial and underground encased in concrete. See 300.5 and 300.50.

(H) Support of Conduit Bodies. PVC conduit shall be permitted to support nonmetallic conduit bodies not larger than the largest trade size of an entering raceway. These conduit bodies shall not support luminaires or other equipment and shall not contain devices other than splicing devices as permitted by 110.14(B) and 314.16(C)(2).

(I) Insulation Temperature Limitations. Conductors or cables rated at a temperature higher than the listed temperature rating of PVC conduit shall be permitted to be installed in PVC conduit, provided the conductors or cables are not operated at a temperature higher than the listed temperature rating of the PVC conduit.

352.12 Uses Not Permitted. PVC conduit shall not be used under the conditions specified in 352.12(A) through (E).

(A) Hazardous (Classified) Locations. In any hazardous (classified) location, except as permitted by other articles of this *Code.*

(B) Support of Luminaires. For the support of luminaires or other equipment not described in 352.10(H).

(C) Physical Damage. Where subject to physical damage unless identified for such use.

(D) Ambient Temperatures. Where subject to ambient temperatures in excess of 50°C (122°F) unless listed otherwise.

(E) Theaters and Similar Locations. In theaters and similar locations, except as provided in 518.4 and 520.5.

352.20 Size.

(A) Minimum. PVC conduit smaller than metric designator 16 (trade size ½) shall not be used.

(B) Maximum. PVC conduit larger than metric designator 155 (trade size 6) shall not be used.

> Informational Note: The trade sizes and metric designators are for identification purposes only and do not relate to actual dimensions. See 300.1(C).

352.22 Number of Conductors. The number of conductors shall not exceed that permitted by the percentage fill specified in Table 1, Chapter 9.

Cables shall be permitted to be installed where such use is not prohibited by the respective cable articles. The number of cables shall not exceed the allowable percentage fill specified in Table 1, Chapter 9.

352.24 Bends — How Made. Bends shall be so made that the conduit will not be damaged and the internal diameter of the conduit will not be effectively reduced. Field bends shall be made only with identified bending equipment. The radius of the curve to the centerline of such bends shall not be less than shown in Table 2, Chapter 9.

352.26 Bends — Number in One Run. There shall not be more than the equivalent of four quarter bends (360 degrees total) between pull points, for example, conduit bodies and boxes.

352.28 Trimming. All cut ends shall be trimmed inside and outside to remove rough edges.

352.30 Securing and Supporting. PVC conduit shall be installed as a complete system as provided in 300.18 and shall be fastened so that movement from thermal expansion or contraction is permitted. PVC conduit shall be securely fastened and supported in accordance with 352.30(A) and (B).

(A) Securely Fastened. PVC conduit shall be securely fastened within 900 mm (3 ft) of each outlet box, junction box, device box, conduit body, or other conduit termination. Conduit listed for securing at other than 900 mm (3 ft) shall be permitted to be installed in accordance with the listing.

(B) Supports. PVC conduit shall be supported as required in Table 352.30. Conduit listed for support at spacings other than as shown in Table 352.30 shall be permitted to be installed in accordance with the listing. Horizontal runs of PVC conduit supported by openings through framing members at intervals not exceeding those in Table 352.30 and securely fastened within 900 mm (3 ft) of termination points shall be permitted.

352.44 Expansion Fittings. Expansion fittings for PVC conduit shall be provided to compensate for thermal expansion and contraction where the length change, in accordance with Table 352.44, is expected to be 6 mm (¼ in.) or greater in a straight run between securely mounted items such as boxes, cabinets, elbows, or other conduit terminations.

352.46 Bushings. Where a conduit enters a box, fitting, or other enclosure, a bushing or adapter shall be provided to protect the wire from abrasion unless the box, fitting, or enclosure design provides equivalent protection.

Informational Note: See 300.4(G) for the protection of conductors 4 AWG and larger at bushings.

352.48 Joints. All joints between lengths of conduit, and between conduit and couplings, fittings, and boxes, shall be made by an approved method.

352.56 Splices and Taps. Splices and taps shall be made in accordance with 300.15.

352.60 Grounding. Where equipment grounding is required, a separate equipment grounding conductor shall be installed in the conduit.

Exception No. 1: As permitted in 250.134(B), Exception No. 2, for dc circuits and 250.134(B), Exception No. 1, for separately run equipment grounding conductors.

Exception No. 2: Where the grounded conductor is used to ground equipment as permitted in 250.142.

Part III. Construction Specifications

352.100 Construction. PVC conduit shall be made of rigid (nonplasticized) polyvinyl chloride (PVC). PVC conduit and fittings shall be composed of suitable nonmetallic material that is resistant to moisture and chemical atmospheres. For use aboveground, it shall also be flame retardant, resistant to impact and crushing, resistant to distortion from heat under conditions likely to be encountered in service, and resistant to low temperature and sunlight effects. For use underground, the material shall be acceptably resistant to moisture and corro-

Table 352.30 Support of Rigid Polyvinyl Chloride Conduit (PVC)

Conduit Size		Maximum Spacing Between Supports	
Metric Designator	Trade Size	mm or m	ft
16–27	½–1	900 mm	3
35–53	1¼–2	1.5 m	5
63–78	2½–3	1.8 m	6
91–129	3½–5	2.1 m	7
155	6	2.5 m	8

Table 352.44 Expansion Characteristics of PVC Rigid Nonmetallic Conduit Coefficient of Thermal Expansion = 6.084 × 10⁻⁵ mm/mm/°C (3.38 × 10⁻⁵ in./in./°F)

Temperature Change (°C)	Length Change of PVC Conduit (mm/m)	Temperature Change (°F)	Length Change of PVC Conduit (in./100 ft)	Temperature Change (°F)	Length Change of PVC Conduit (in./100 ft)
5	0.30	5	0.20	105	4.26
10	0.61	10	0.41	110	4.46
15	0.91	15	0.61	115	4.66
20	1.22	20	0.81	120	4.87
25	1.52	25	1.01	125	5.07
30	1.83	30	1.22	130	5.27
35	2.13	35	1.42	135	5.48
40	2.43	40	1.62	140	5.68
45	2.74	45	1.83	145	5.88
50	3.04	50	2.03	150	6.08
55	3.35	55	2.23	155	6.29
60	3.65	60	2.43	160	6.49
65	3.95	65	2.64	165	6.69
70	4.26	70	2.84	170	6.90
75	4.56	75	3.04	175	7.10
80	4.87	80	3.24	180	7.30
85	5.17	85	3.45	185	7.50
90	5.48	90	3.65	190	7.71
95	5.78	95	3.85	195	7.91
100	6.08	100	4.06	200	8.11

sive agents and shall be of sufficient strength to withstand abuse, such as by impact and crushing, in handling and during installation. Where intended for direct burial, without encasement in concrete, the material shall also be capable of withstanding continued loading that is likely to be encountered after installation.

352.120 Marking. Each length of PVC conduit shall be clearly and durably marked at least every 3 m (10 ft) as required in the first sentence of 110.21(A). The type of material shall also be included in the marking unless it is visually identifiable. For conduit recognized for use aboveground, these markings shall be permanent. For conduit limited to underground use only, these markings shall be sufficiently durable to remain legible until the material is installed. Conduit shall be permitted to be surface marked to indicate special characteristics of the material.

Informational Note: Examples of these markings include but are not limited to "limited smoke" and "sunlight resistant."

ARTICLE 353
High Density Polyethylene Conduit: Type HDPE Conduit

Part I. General

353.1 Scope. This article covers the use, installation, and construction specifications for high density polyethylene (HDPE) conduit and associated fittings.

Informational Note: Refer to Article 352 for Rigid Polyvinyl Chloride Conduit: Type PVC and Article 355 for Reinforced Thermosetting Resin Conduit: Type RTRC.

353.2 Definition.

High Density Polyethylene (HDPE) Conduit. A nonmetallic raceway of circular cross section, with associated couplings, connectors, and fittings for the installation of electrical conductors.

353.6 Listing Requirements. HDPE conduit and associated fittings shall be listed.

Part II. Installation

353.10 Uses Permitted. The use of HDPE conduit shall be permitted under the following conditions:

(1) In discrete lengths or in continuous lengths from a reel
(2) In locations subject to severe corrosive influences as covered in 300.6 and where subject to chemicals for which the conduit is listed
(3) In cinder fill
(4) In direct burial installations in earth or concrete

Informational Note to (4): Refer to 300.5 and 300.50 for underground installations.

(5) Above ground, except as prohibited in 353.12, where encased in not less than 50 mm (2 in.) of concrete.
(6) Conductors or cables rated at a temperature higher than the listed temperature rating of HDPE conduit shall be permitted to be installed in HDPE conduit, provided the conductors or cables are not operated at a temperature higher than the listed temperature rating of the HDPE conduit.

353.12 Uses Not Permitted. HDPE conduit shall not be used under the following conditions:

(1) Where exposed
(2) Within a building
(3) In any hazardous (classified) location, except as permitted by other articles in this *Code*
(4) Where subject to ambient temperatures in excess of 50°C (122°F) unless listed otherwise

353.20 Size.

(A) Minimum. HDPE conduit smaller than metric designator 16 (trade size ½) shall not be used.

(B) Maximum. HDPE conduit larger than metric designator 155 (trade size 6) shall not be used.

Informational Note: The trade sizes and metric designators are for identification purposes only and do not relate to actual dimensions. See 300.1(C).

353.22 Number of Conductors. The number of conductors shall not exceed that permitted by the percentage fill specified in Table 1, Chapter 9.

Cables shall be permitted to be installed where such use is not prohibited by the respective cable articles. The number of cables shall not exceed the allowable percentage fill specified in Table 1, Chapter 9.

353.24 Bends — How Made. Bends shall be so made that the conduit will not be damaged and the internal diameter of the conduit will not be effectively reduced. Bends shall be permitted to be made manually without auxiliary equipment, and the radius of the curve to the centerline of such bends shall not be less than shown in Table 354.24. For conduits of metric designators 129 and 155 (trade sizes 5 and 6) the allowable radii of bends shall be in accordance with specifications provided by the manufacturer.

353.26 Bends — Number in One Run. There shall not be more than the equivalent of four quarter bends (360 degrees total) between pull points, for example, conduit bodies and boxes.

353.28 Trimming. All cut ends shall be trimmed inside and outside to remove rough edges.

353.46 Bushings. Where a conduit enters a box, fitting, or other enclosure, a bushing or adapter shall be provided to protect the wire from abrasion unless the box, fitting, or enclosure design provides equivalent protection.

Informational Note: See 300.4(G) for the protection of conductors 4 AWG and larger at bushings.

353.48 Joints. All joints between lengths of conduit, and between conduit and couplings, fittings, and boxes, shall be made by an approved method.

Informational Note: HDPE conduit can be joined using either heat fusion, electrofusion, or mechanical fittings.

353.56 Splices and Taps. Splices and taps shall be made in accordance with 300.15.

353.60 Grounding. Where equipment grounding is required, a separate equipment grounding conductor shall be installed in the conduit.

Exception No. 1: The equipment grounding conductor shall be permitted to be run separately from the conduit where used for grounding dc circuits as permitted in 250.134, Exception No. 2.

Exception No. 2: The equipment grounding conductor shall not be required where the grounded conductor is used to ground equipment as permitted in 250.142.

Part III. Construction Specifications

353.100 Construction. HDPE conduit shall be composed of high density polyethylene that is resistant to moisture and chemical atmospheres. The material shall be resistant to moisture and corrosive agents and shall be of sufficient strength to withstand abuse, such as by impact and crushing, in handling and during installation. Where intended for direct burial, without encasement in concrete, the material shall also be capable of withstanding continued loading that is likely to be encountered after installation.

353.120 Marking. Each length of HDPE shall be clearly and durably marked at least every 3 m (10 ft) as required in 110.21. The type of material shall also be included in the marking.

ARTICLE 354
Nonmetallic Underground Conduit with Conductors: Type NUCC

Part I. General

354.1 Scope. This article covers the use, installation, and construction specifications for nonmetallic underground conduit with conductors (NUCC).

354.2 Definition.

Nonmetallic Underground Conduit with Conductors (NUCC). A factory assembly of conductors or cables inside a nonmetallic, smooth wall raceway with a circular cross section.

354.6 Listing Requirements. NUCC and associated fittings shall be listed.

Part II. Installation

354.10 Uses Permitted. The use of NUCC and fittings shall be permitted in the following:

(1) For direct burial underground installation (For minimum cover requirements, see Table 300.5 and Table 300.50 under Rigid Nonmetallic Conduit.)
(2) Encased or embedded in concrete
(3) In cinder fill
(4) In underground locations subject to severe corrosive influences as covered in 300.6 and where subject to chemicals for which the assembly is specifically approved

(5) Aboveground, except as prohibited in 354.12, where encased in not less than 50 mm (2 in.) of concrete

354.12 Uses Not Permitted. NUCC shall not be used in the following:

(1) In exposed locations
(2) Inside buildings

Exception: The conductor or the cable portion of the assembly, where suitable, shall be permitted to extend within the building for termination purposes in accordance with 300.3.

(3) In any hazardous (classified) location, except as permitted by other articles of this *Code*

354.20 Size.

(A) Minimum. NUCC smaller than metric designator 16 (trade size ½) shall not be used.

(B) Maximum. NUCC larger than metric designator 103 (trade size 4) shall not be used.

> Informational Note: See 300.1(C) for the metric designators and trade sizes. These are for identification purposes only and do not relate to actual dimensions.

354.22 Number of Conductors. The number of conductors or cables shall not exceed that permitted by the percentage fill in Table 1, Chapter 9.

354.24 Bends — How Made. Bends shall be manually made so that the conduit will not be damaged and the internal diameter of the conduit will not be effectively reduced. The radius of the curve of the centerline of such bends shall not be less than shown in Table 354.24.

354.26 Bends — Number in One Run. There shall not be more than the equivalent of four quarter bends (360 degrees total) between termination points.

354.28 Trimming. For termination, the conduit shall be trimmed away from the conductors or cables using an approved method that will not damage the conductor or cable insulation or jacket. All conduit ends shall be trimmed inside and out to remove rough edges.

Table 354.24 Minimum Bending Radius for Nonmetallic Underground Conduit with Conductors (NUCC)

\multicolumn{2}{c}{Conduit Size}		\multicolumn{2}{c}{Minimum Bending Radius}	
Metric Designator	Trade Size	mm	in.
16	½	250	10
21	¾	300	12
27	1	350	14
35	1¼	450	18
41	1½	500	20
53	2	650	26
63	2½	900	36
78	3	1200	48
103	4	1500	60

354.46 Bushings. Where the NUCC enters a box, fitting, or other enclosure, a bushing or adapter shall be provided to protect the conductor or cable from abrasion unless the design of the box, fitting, or enclosure provides equivalent protection.

Informational Note: See 300.4(G) for the protection of conductors size 4 AWG or larger.

354.48 Joints. All joints between conduit, fittings, and boxes shall be made by an approved method.

354.50 Conductor Terminations. All terminations between the conductors or cables and equipment shall be made by an approved method for that type of conductor or cable.

354.56 Splices and Taps. Splices and taps shall be made in junction boxes or other enclosures.

354.60 Grounding. Where equipment grounding is required, an assembly containing a separate equipment grounding conductor shall be used.

Part III. Construction Specifications

354.100 Construction.

(A) General. NUCC is an assembly that is provided in continuous lengths shipped in a coil, reel, or carton.

(B) Nonmetallic Underground Conduit. The nonmetallic underground conduit shall be listed and composed of a material that is resistant to moisture and corrosive agents. It shall also be capable of being supplied on reels without damage or distortion and shall be of sufficient strength to withstand abuse, such as impact or crushing, in handling and during installation without damage to conduit or conductors.

(C) Conductors and Cables. Conductors and cables used in NUCC shall be listed and shall comply with 310.10(C). Conductors of different systems shall be installed in accordance with 300.3(C).

(D) Conductor Fill. The maximum number of conductors or cables in NUCC shall not exceed that permitted by the percentage fill in Table 1, Chapter 9.

354.120 Marking. NUCC shall be clearly and durably marked at least every 3.05 m (10 ft) as required by 110.21. The type of conduit material shall also be included in the marking.

Identification of conductors or cables used in the assembly shall be provided on a tag attached to each end of the assembly or to the side of a reel. Enclosed conductors or cables shall be marked in accordance with 310.120.

ARTICLE 355
Reinforced Thermosetting Resin Conduit: Type RTRC

Part I. General

355.1 Scope. This article covers the use, installation, and construction specification for reinforced thermosetting resin conduit (RTRC) and associated fittings.

Informational Note: Refer to Article 352 for Rigid Polyvinyl Chloride Conduit: Type PVC, and Article 353 for High Density Polyethylene Conduit: Type HDPE.

355.2 Definition.

Reinforced Thermosetting Resin Conduit (RTRC). A rigid nonmetallic raceway of circular cross section, with integral or associated couplings, connectors, and fittings for the installation of electrical conductors and cables.

355.6 Listing Requirements. RTRC, factory elbows, and associated fittings shall be listed.

Part II. Installation

355.10 Uses Permitted. The use of RTRC shall be permitted in accordance with 355.10(A) through (I).

(A) Concealed. RTRC shall be permitted in walls, floors, and ceilings.

(B) Corrosive Influences. RTRC shall be permitted in locations subject to severe corrosive influences as covered in 300.6 and where subject to chemicals for which the materials are specifically approved.

(C) Cinders. RTRC shall be permitted in cinder fill.

(D) Wet Locations. RTRC shall be permitted in portions of dairies, laundries, canneries, or other wet locations, and in locations where walls are frequently washed, the entire conduit system, including boxes and fittings used therewith, shall be installed and equipped so as to prevent water from entering the conduit. All supports, bolts, straps, screws, and so forth, shall be of corrosion-resistant materials or be protected against corrosion by approved corrosion-resistant materials.

(E) Dry and Damp Locations. RTRC shall be permitted for use in dry and damp locations not prohibited by 355.12.

(F) Exposed. RTRC shall be permitted for exposed work if identified for such use.

Informational Note: RTRC, Type XW, is identified for areas of physical damage.

(G) Underground Installations. For underground installations, see 300.5 and 300.50.

(H) Support of Conduit Bodies. RTRC shall be permitted to support nonmetallic conduit bodies not larger than the largest trade size of an entering raceway. These conduit bodies shall not support luminaires or other equipment and shall not contain devices other than splicing devices as permitted by 110.14(B) and 314.16(C)(2).

(I) Insulation Temperature Limitations. Conductors or cables rated at a temperature higher than the listed temperature rating of RTRC conduit shall be permitted to be installed in RTRC conduit, if the conductors or cables are not operated at a temperature higher than the listed temperature rating of the RTRC conduit.

355.12 Uses Not Permitted. RTRC shall not be used under the following conditions.

(A) Hazardous (Classified) Locations.

(1)   In any hazardous (classified) location, except as permitted by other articles in this *Code*

(2) In Class I, Division 2 locations, except as permitted in 501.10(B)(1)(6)

(B) Support of Luminaires. For the support of luminaires or other equipment not described in 355.10(H).

(C) Physical Damage. Where subject to physical damage unless identified for such use.

(D) Ambient Temperatures. Where subject to ambient temperatures in excess of 50°C (122°F) unless listed otherwise.

(E) Theaters and Similar Locations. In theaters and similar locations, except as provided in 518.4 and 520.5.

355.20 Size.

(A) Minimum. RTRC smaller than metric designator 16 (trade size ½) shall not be used.

(B) Maximum. RTRC larger than metric designator 155 (trade size 6) shall not be used.

> Informational Note: The trade sizes and metric designators are for identification purposes only and do not relate to actual dimensions. See 300.1(C).

355.22 Number of Conductors. The number of conductors shall not exceed that permitted by the percentage fill specified in Table 1, Chapter 9. Cables shall be permitted to be installed where such use is not prohibited by the respective cable articles. The number of cables shall not exceed the allowable percentage fill specified in Table 1, Chapter 9.

355.24 Bends — How Made. Bends shall be so made that the conduit will not be damaged and the internal diameter of the conduit will not be effectively reduced. Field bends shall be made only with identified bending equipment. The radius of the curve to the centerline of such bends shall not be less than shown in Table 2, Chapter 9.

355.26 Bends — Number in One Run. There shall not be more than the equivalent of four quarter bends (360 degrees total) between pull points, for example, conduit bodies and boxes.

355.28 Trimming. All cut ends shall be trimmed inside and outside to remove rough edges.

355.30 Securing and Supporting. RTRC shall be installed as a complete system in accordance with 300.18 and shall be securely fastened in place and supported in accordance with 355.30(A) and (B).

(A) Securely Fastened. RTRC shall be securely fastened within 900 mm (3 ft) of each outlet box, junction box, device box, conduit body, or other conduit termination. Conduit listed for securing at other than 900 mm (3 ft) shall be permitted to be installed in accordance with the listing.

(B) Supports. RTRC shall be supported as required in Table 355.30. Conduit listed for support at spacing other than as shown in Table 355.30 shall be permitted to be installed in accordance with the listing. Horizontal runs of RTRC supported by openings through framing members at intervals not exceeding those in Table 355.30 and securely fastened within 900 mm (3 ft) of termination points shall be permitted.

355.44 Expansion Fittings. Expansion fittings for RTRC shall be provided to compensate for thermal expansion and contraction where the length change, in accordance with Table 355.44,

Table 355.30 Support of Reinforced Thermosetting Resin Conduit (RTRC)

Conduit Size		Maximum Spacing Between Supports	
Metric Designator	Trade Size	mm or m	ft
16–27	½–1	900 mm	3
35–53	1¼–2	1.5 m	5
63–78	2½–3	1.8 m	6
91–129	3½–5	2.1 m	7
155	6	2.5 m	8

is expected to be 6 mm (¼ in.) or greater in a straight run between securely mounted items such as boxes, cabinets, elbows, or other conduit terminations.

355.46 Bushings. Where a conduit enters a box, fitting, or other enclosure, a bushing or adapter shall be provided to protect the wire from abrasion unless the box, fitting, or enclosure design provides equivalent protection.

> Informational Note: See 300.4(G) for the protection of conductors 4 AWG and larger at bushings.

355.48 Joints. All joints between lengths of conduit, and between conduit and couplings, fitting, and boxes, shall be made by an approved method.

355.56 Splices and Taps. Splices and taps shall be made in accordance with 300.15.

355.60 Grounding. Where equipment grounding is required, a separate equipment grounding conductor shall be installed in the conduit.

Exception No. 1: As permitted in 250.134(B), Exception No. 2, for dc circuits and 250.134(B), Exception No. 1, for separately run equipment grounding conductors.

Exception No. 2: Where the grounded conductor is used to ground equipment as permitted in 250.142.

Part III. Construction Specifications

355.100 Construction. RTRC and fittings shall be composed of suitable nonmetallic material that is resistant to moisture and chemical atmospheres. For use aboveground, it shall also be flame retardant, resistant to impact and crushing, resistant to distortion from heat under conditions likely to be encountered in service, and resistant to low temperature and sunlight effects. For use underground, the material shall be acceptably resistant to moisture and corrosive agents and shall be of sufficient strength to withstand abuse, such as by impact and crushing, in handling and during installation. Where intended for direct burial, without encasement in concrete, the material shall also be capable of withstanding continued loading that is likely to be encountered after installation.

Table 355.44 Expansion Characteristics of Reinforced Thermosetting Resin Conduit (RTRC) Coefficient of Thermal Expansion = 2.7×10^{-5} mm/mm/°C (1.5×10^{-5} in./in./°F)

Temperature Change (°C)	Length Change of RTRC Conduit (mm/m)	Temperature Change (°F)	Length Change of RTRC Conduit (in./100 ft)	Temperature Change (°F)	Length Change of RTRC Conduit (in./100 ft)
5	0.14	5	0.09	105	1.89
10	0.27	10	0.18	110	1.98
15	0.41	15	0.27	115	2.07
20	0.54	20	0.36	120	2.16
25	0.68	25	0.45	125	2.25
30	0.81	30	0.54	130	2.34
35	0.95	35	0.63	135	2.43
40	1.08	40	0.72	140	2.52
45	1.22	45	0.81	145	2.61
50	1.35	50	0.90	150	2.70
55	1.49	55	0.99	155	2.79
60	1.62	60	1.08	160	2.88
65	1.76	65	1.17	165	2.97
70	1.89	70	1.26	170	3.06
75	2.03	75	1.35	175	3.15
80	2.16	80	1.44	180	3.24
85	2.30	85	1.53	185	3.33
90	2.43	90	1.62	190	3.42
95	2.57	95	1.71	195	3.51
100	2.70	100	1.80	200	3.60

355.120 Marking. Each length of RTRC shall be clearly and durably marked at least every 3 m (10 ft) as required in the first sentence of 110.21(A). The type of material shall also be included in the marking unless it is visually identifiable. For conduit recognized for use aboveground, these markings shall be permanent. For conduit limited to underground use only, these markings shall be sufficiently durable to remain legible until the material is installed. Conduit shall be permitted to be surface marked to indicate special characteristics of the material.

Informational Note: Examples of these markings include but are not limited to "limited smoke" and "sunlight resistant."

ARTICLE 356
Liquidtight Flexible Nonmetallic Conduit: Type LFNC

Part I. General

356.1 Scope. This article covers the use, installation, and construction specifications for liquidtight flexible nonmetallic conduit (LFNC) and associated fittings.

356.2 Definition.

Liquidtight Flexible Nonmetallic Conduit (LFNC). A raceway of circular cross section of various types as follows:

(1) A smooth seamless inner core and cover bonded together and having one or more reinforcement layers between the core and covers, designated as Type LFNC-A
(2) A smooth inner surface with integral reinforcement within the raceway wall, designated as Type LFNC-B
(3) A corrugated internal and external surface without integral reinforcement within the raceway wall, designated as LFNC-C

Informational Note: FNMC is an alternative designation for LFNC.

356.6 Listing Requirements. LFNC and associated fittings shall be listed.

Part II. Installation

356.10 Uses Permitted. LFNC shall be permitted to be used in exposed or concealed locations for the following purposes:

Informational Note: Extreme cold can cause some types of nonmetallic conduits to become brittle and therefore more susceptible to damage from physical contact.

(1) Where flexibility is required for installation, operation, or maintenance.
(2) Where protection of the contained conductors is required from vapors, liquids, or solids.
(3) For outdoor locations where listed and marked as suitable for the purpose.
(4) For direct burial where listed and marked for the purpose.
(5) Type LFNC shall be permitted to be installed in lengths longer than 1.8 m (6 ft) where secured in accordance with 356.30.

(6) Type LFNC-B as a listed manufactured prewired assembly, metric designator 16 through 27 (trade size ½ through 1) conduit.

(7) For encasement in concrete where listed for direct burial and installed in compliance with 356.42.

356.12 Uses Not Permitted. LFNC shall not be used as follows:

(1) Where subject to physical damage
(2) Where any combination of ambient and conductor temperatures is in excess of that for which it is listed
(3) In lengths longer than 1.8 m (6 ft), except as permitted by 356.10(5) or where a longer length is approved as essential for a required degree of flexibility
(4) In any hazardous (classified) location, except as permitted by other articles in this *Code*

356.20 Size.

(A) Minimum. LFNC smaller than metric designator 16 (trade size ½) shall not be used unless permitted in 356.20(A)(1) or (A)(2) for metric designator 12 (trade size ⅜).

(1) For enclosing the leads of motors as permitted in 430.245(B)
(2) In lengths not exceeding 1.8 m (6 ft) as part of a listed assembly for tap connections to luminaires as required in 410.117(C), or for utilization equipment

(B) Maximum. LFNC larger than metric designator 103 (trade size 4) shall not be used.

> Informational Note: See 300.1(C) for the metric designators and trade sizes. These are for identification purposes only and do not relate to actual dimensions.

356.22 Number of Conductors. The number of conductors shall not exceed that permitted by the percentage fill specified in Table 1, Chapter 9.

Cables shall be permitted to be installed where such use is not prohibited by the respective cable articles. The number of cables shall not exceed the allowable percentage fill specified in Table 1, Chapter 9.

356.24 Bends — How Made. Bends in conduit shall be so made that the conduit is not damaged and the internal diameter of the conduit is not effectively reduced. Bends shall be permitted to be made manually without auxiliary equipment. The radius of the curve to the centerline of any bend shall not be less than shown in Table 2, Chapter 9 using the column "Other Bends."

356.26 Bends — Number in One Run. There shall not be more than the equivalent of four quarter bends (360 degrees total) between pull points, for example, conduit bodies and boxes.

356.28 Trimming. All cut ends of conduit shall be trimmed inside and outside to remove rough edges.

356.30 Securing and Supporting. Type LFNC shall be securely fastened and supported in accordance with one of the following:

(1) Where installed in lengths exceeding 1.8 m (6 ft), the conduit shall be securely fastened at intervals not exceeding 900 mm (3 ft) and within 300 mm (12 in.) on each side of every outlet box, junction box, cabinet, or fitting.

Where used, cable ties shall be listed as suitable for the application and for securing and supporting.

(2) Securing or supporting of the conduit shall not be required where it is fished, installed in lengths not exceeding 900 mm (3 ft) at terminals where flexibility is required, or installed in lengths not exceeding 1.8 m (6 ft) from a luminaire terminal connection for tap conductors to luminaires permitted in 410.117(C).

(3) Horizontal runs of LFNC supported by openings through framing members at intervals not exceeding 900 mm (3 ft) and securely fastened within 300 mm (12 in.) of termination points shall be permitted.

(4) Securing or supporting of LFNC shall not be required where installed in lengths not exceeding 1.8 m (6 ft) from the last point where the raceway is securely fastened for connections within an accessible ceiling to a luminaire(s) or other equipment. For the purpose of 356.30, listed liquidtight flexible nonmetallic conduit fittings shall be permitted as a means of support.

356.42 Couplings and Connectors. Only fittings listed for use with LFNC shall be used. Angle connectors shall not be used for concealed raceway installations. Straight LFNC fittings are permitted for direct burial or encasement in concrete.

356.56 Splices and Taps. Splices and taps shall be made in accordance with 300.15.

356.60 Grounding. Where equipment grounding is required, a separate equipment grounding conductor shall be installed in the conduit.

Exception No. 1: As permitted in 250.134(B), Exception No. 2, for dc circuits and 250.134(B), Exception No. 1, for separately run equipment grounding conductors.

Exception No. 2: Where the grounded conductor is used to ground equipment as permitted in 250.142.

Part III. Construction Specifications

356.100 Construction. LFNC-B as a prewired manufactured assembly shall be provided in continuous lengths capable of being shipped in a coil, reel, or carton without damage.

356.120 Marking. LFNC shall be marked at least every 600 mm (2 ft) in accordance with 110.21. The marking shall include a type designation in accordance with 356.2 and the trade size. Conduit that is intended for outdoor use or direct burial shall be marked.

The type, size, and quantity of conductors used in prewired manufactured assemblies shall be identified by means of a printed tag or label attached to each end of the manufactured assembly and either the carton, coil, or reel. The enclosed conductors shall be marked in accordance with 310.120.

ARTICLE 358
Electrical Metallic Tubing: Type EMT

Part I. General

358.1 Scope. This article covers the use, installation, and construction specifications for electrical metallic tubing (EMT) and associated fittings.

358.2 Definition.

Electrical Metallic Tubing (EMT). An unthreaded thinwall raceway of circular cross section designed for the physical protection and routing of conductors and cables and for use as an equipment grounding conductor when installed utilizing appropriate fittings.

358.6 Listing Requirements. EMT, factory elbows, and associated fittings shall be listed.

Part II. Installation

358.10 Uses Permitted.

(A) Exposed and Concealed. The use of EMT shall be permitted for both exposed and concealed work for the following:

(1) In concrete, in direct contact with the earth or in areas subject to severe corrosive influences where installed in accordance with 358.10(B)
(2) In dry, damp, and wet locations
(3) In any hazardous (classified) location as permitted by other articles in this *Code*

(B) Corrosive Environments.

N **(1) Galvanized Steel and Stainless Steel EMT, Elbows, and Fittings.** Galvanized steel and stainless steel EMT, elbows, and fittings shall be permitted to be installed in concrete, in direct contact with the earth, or in areas subject to severe corrosive influences where protected by corrosion protection and approved as suitable for the condition.

N **(2) Supplementary Protection of Aluminum EMT.** Aluminum EMT shall be provided with approved supplementary corrosion protection where encased in concrete or in direct contact with the earth.

N **(C) Cinder Fill.** Galvanized steel and stainless steel EMT shall be permitted to be installed in cinder concrete or cinder fill where subject to permanent moisture when protected on all sides by a layer of noncinder concrete at least 50 mm (2 in.) thick or when the tubing is installed at least 450 mm (18 in.) under the fill.

(D) Wet Locations. All supports, bolts, straps, screws, and so forth shall be of corrosion-resistant materials or protected against corrosion by corrosion-resistant materials.

Informational Note: See 300.6 for protection against corrosion.

358.12 Uses Not Permitted. EMT shall not be used under the following conditions:

(1) Where subject to severe physical damage
(2) For the support of luminaires or other equipment except conduit bodies no larger than the largest trade size of the tubing

N **358.14 Dissimilar Metals.** Where practicable, dissimilar metals in contact anywhere in the system shall be avoided to eliminate the possibility of galvanic action. Aluminum fittings and enclosures shall be permitted to be used with galvanized steel EMT, and galvanized steel fittings and enclosures shall be permitted to be used with aluminum EMT where not subject to severe corrosive influences. Stainless steel EMT shall only be used with stainless steel fittings and approved accessories, outlet boxes, and enclosures.

358.20 Size.

(A) Minimum. EMT smaller than metric designator 16 (trade size ½) shall not be used.

Exception: For enclosing the leads of motors as permitted in 430.245(B).

(B) Maximum. The maximum size of EMT shall be metric designator 103 (trade size 4).

Informational Note: See 300.1(C) for the metric designators and trade sizes. These are for identification purposes only and do not relate to actual dimensions.

358.22 Number of Conductors. The number of conductors shall not exceed that permitted by the percentage fill specified in Table 1, Chapter 9.

Cables shall be permitted to be installed where such use is not prohibited by the respective cable articles. The number of cables shall not exceed the allowable percentage fill specified in Table 1, Chapter 9.

358.24 Bends — How Made. Bends shall be made so that the tubing is not damaged and the internal diameter of the tubing is not effectively reduced. The radius of the curve of any field bend to the centerline of the tubing shall not be less than shown in Table 2, Chapter 9 for one-shot and full shoe benders.

358.26 Bends — Number in One Run. There shall not be more than the equivalent of four quarter bends (360 degrees total) between pull points, for example, conduit bodies and boxes.

358.28 Reaming and Threading.

(A) Reaming. All cut ends of EMT shall be reamed or otherwise finished to remove rough edges.

(B) Threading. EMT shall not be threaded.

Exception: EMT with factory threaded integral couplings complying with 358.100.

358.30 Securing and Supporting. EMT shall be installed as a complete system in accordance with 300.18 and shall be securely fastened in place and supported in accordance with 358.30(A) and (B).

(A) Securely Fastened. EMT shall be securely fastened in place at intervals not to exceed 3 m (10 ft). In addition, each EMT run between termination points shall be securely fastened within 900 mm (3 ft) of each outlet box, junction box, device box, cabinet, conduit body, or other tubing termination.

Exception No. 1: Fastening of unbroken lengths shall be permitted to be increased to a distance of 1.5 m (5 ft) where structural members do not readily permit fastening within 900 mm (3 ft).

Exception No. 2: For concealed work in finished buildings or prefinished wall panels where such securing is impracticable, unbroken lengths (without coupling) of EMT shall be permitted to be fished.

(B) Supports. Horizontal runs of EMT supported by openings through framing members at intervals not greater than 3 m (10 ft) and securely fastened within 900 mm (3 ft) of termination points shall be permitted.

358.42 Couplings and Connectors. Couplings and connectors used with EMT shall be made up tight. Where buried in masonry or concrete, they shall be concretetight type. Where installed in wet locations, they shall comply with 314.15.

358.56 Splices and Taps. Splices and taps shall be made in accordance with 300.15.

358.60 Grounding. EMT shall be permitted as an equipment grounding conductor.

Part III. Construction Specifications

358.100 Construction. EMT shall be made of one of the following:

(1) Steel with protective coatings
(2) Aluminum
(3) Stainless steel

358.120 Marking. EMT shall be clearly and durably marked at least every 3 m (10 ft) as required in the first sentence of 110.21(A).

ARTICLE 360
Flexible Metallic Tubing: Type FMT

Part I. General

360.1 Scope. This article covers the use, installation, and construction specifications for flexible metallic tubing (FMT) and associated fittings.

360.2 Definition.

Flexible Metallic Tubing (FMT). A metal raceway that is circular in cross section, flexible, and liquidtight without a nonmetallic jacket.

360.6 Listing Requirements. FMT and associated fittings shall be listed.

Part II. Installation

360.10 Uses Permitted. FMT shall be permitted to be used for branch circuits as follows:

(1) In dry locations
(2) Where concealed
(3) In accessible locations
(4) For system voltages of 1000 volts maximum

360.12 Uses Not Permitted. FMT shall not be used as follows:

(1) In hoistways
(2) In storage battery rooms
(3) In hazardous (classified) locations unless otherwise permitted under other articles in this *Code*
(4) Underground for direct earth burial, or embedded in poured concrete or aggregate
(5) Where subject to physical damage
(6) In lengths over 1.8 m (6 ft)

360.20 Size.

(A) Minimum. FMT smaller than metric designator 16 (trade size ½) shall not be used.

Exception No. 1: FMT of metric designator 12 (trade size ⅜) shall be permitted to be installed in accordance with 300.22(B) and (C).

Exception No. 2: FMT of metric designator 12 (trade size ⅜) shall be permitted in lengths not in excess of 1.8 m (6 ft) as part of a listed assembly or for luminaires. See 410.117(C).

(B) Maximum. The maximum size of FMT shall be metric designator 21 (trade size ¾).

> Informational Note: See 300.1(C) for the metric designators and trade sizes. These are for identification purposes only and do not relate to actual dimensions.

360.22 Number of Conductors.

(A) FMT — Metric Designators 16 and 21 (Trade Sizes ½ and ¾). The number of conductors in metric designators 16 (trade size ½) and 21 (trade size ¾) shall not exceed that permitted by the percentage fill specified in Table 1, Chapter 9.

Cables shall be permitted to be installed where such use is not prohibited by the respective cable articles. The number of cables shall not exceed the allowable percentage fill specified in Table 1, Chapter 9.

(B) FMT — Metric Designator 12 (Trade Size ⅜). The number of conductors in metric designator 12 (trade size ⅜) shall not exceed that permitted in Table 348.22.

360.24 Bends.

(A) Infrequent Flexing Use. When FMT is infrequently flexed in service after installation, the radii of bends measured to the inside of the bend shall not be less than specified in Table 360.24(A).

(B) Fixed Bends. Where FMT is bent for installation purposes and is not flexed or bent as required by use after installation, the radii of bends measured to the inside of the bend shall not be less than specified in Table 360.24(B).

360.56 Splices and Taps. Splices and taps shall be made in accordance with 300.15.

Table 360.24(A) Minimum Radii for Flexing Use

Metric Designator	Trade Size	Minimum Radii for Flexing Use	
		mm	in.
12	⅜	254.0	10
16	½	317.5	12½
21	¾	444.5	17½

Table 360.24(B) Minimum Radii for Fixed Bends

Metric Designator	Trade Size	Minimum Radii for Fixed Bends	
		mm	in.
12	⅜	88.9	3½
16	½	101.6	4
21	¾	127.0	5

360.60 Grounding. FMT shall be permitted as an equipment grounding conductor where installed in accordance with 250.118(7).

Part III. Construction Specifications

360.120 Marking. FMT shall be marked according to 110.21.

ARTICLE 362
Electrical Nonmetallic Tubing: Type ENT

Part I. General

362.1 Scope. This article covers the use, installation, and construction specifications for electrical nonmetallic tubing (ENT) and associated fittings.

362.2 Definition.

Electrical Nonmetallic Tubing (ENT). A nonmetallic, pliable, corrugated raceway of circular cross section with integral or associated couplings, connectors, and fittings for the installation of electrical conductors. ENT is composed of a material that is resistant to moisture and chemical atmospheres and is flame retardant.

A pliable raceway is a raceway that can be bent by hand with a reasonable force but without other assistance.

362.6 Listing Requirements. ENT and associated fittings shall be listed.

Part II. Installation

362.10 Uses Permitted. For the purpose of this article, the first floor of a building shall be that floor that has 50 percent or more of the exterior wall surface area level with or above finished grade. One additional level that is the first level and not designed for human habitation and used only for vehicle parking, storage, or similar use shall be permitted. The use of ENT and fittings shall be permitted in the following:

(1) In any building not exceeding three floors above grade as follows:
   a. For exposed work, where not prohibited by 362.12
   b. Concealed within walls, floors, and ceilings
(2) In any building exceeding three floors above grade, ENT shall be concealed within walls, floors, and ceilings where the walls, floors, and ceilings provide a thermal barrier of material that has at least a 15-minute finish rating as iden-

tified in listings of fire-rated assemblies. The 15-minute-finish-rated thermal barrier shall be permitted to be used for combustible or noncombustible walls, floors, and ceilings.

Exception to (2): Where a fire sprinkler system(s) is installed in accordance with NFPA 13-2013, Standard for the Installation of Sprinkler Systems, on all floors, ENT shall be permitted to be used within walls, floors, and ceilings, exposed or concealed, in buildings exceeding three floors abovegrade.

Informational Note: A finish rating is established for assemblies containing combustible (wood) supports. The finish rating is defined as the time at which the wood stud or wood joist reaches an average temperature rise of 121°C (250°F) or an individual temperature of 163°C (325°F) as measured on the plane of the wood nearest the fire. A finish rating is not intended to represent a rating for a membrane ceiling.

(3) In locations subject to severe corrosive influences as covered in 300.6 and where subject to chemicals for which the materials are specifically approved.
(4) In concealed, dry, and damp locations not prohibited by 362.12.
(5) Above suspended ceilings where the suspended ceilings provide a thermal barrier of material that has at least a 15-minute finish rating as identified in listings of fire-rated assemblies, except as permitted in 362.10(1)a.

Exception to (5): ENT shall be permitted to be used above suspended ceilings in buildings exceeding three floors above grade where the building is protected throughout by a fire sprinkler system installed in accordance with NFPA 13-2013, Standard for the Installation of Sprinkler Systems.

(6) Encased in poured concrete, or embedded in a concrete slab on grade where ENT is placed on sand or approved screenings, provided fittings identified for this purpose are used for connections.
(7) For wet locations indoors as permitted in this section or in a concrete slab on or belowgrade, with fittings listed for the purpose.
(8) Metric designator 16 through 27 (trade size ½ through 1) as listed manufactured prewired assembly.

Informational Note: Extreme cold may cause some types of nonmetallic conduits to become brittle and therefore more susceptible to damage from physical contact.

(9) Conductors or cables rated at a temperature higher than the listed temperature rating of ENT shall be permitted to be installed in ENT, if the conductors or cables are not operated at a temperature higher than the listed temperature rating of the ENT.

362.12 Uses Not Permitted. ENT shall not be used in the following:

(1) In any hazardous (classified) location, except as permitted by other articles in this *Code*
(2) For the support of luminaires and other equipment
(3) Where subject to ambient temperatures in excess of 50°C (122°F) unless listed otherwise
(4) For direct earth burial
(5) In exposed locations, except as permitted by 362.10(1), 362.10(5), and 362.10(7)
(6) In theaters and similar locations, except as provided in 518.4 and 520.5

(7) Where exposed to the direct rays of the sun, unless identified as sunlight resistant

(8) Where subject to physical damage

362.20 Size.

(A) Minimum. ENT smaller than metric designator 16 (trade size ½) shall not be used.

(B) Maximum. ENT larger than metric designator 63 (trade size 2½) shall not be used.

> Informational Note: See 300.1(C) for the metric designators and trade sizes. These are for identification purposes only and do not relate to actual dimensions.

362.22 Number of Conductors.
The number of conductors shall not exceed that permitted by the percentage fill in Table 1, Chapter 9.

Cables shall be permitted to be installed where such use is not prohibited by the respective cable articles. The number of cables shall not exceed the allowable percentage fill specified in Table 1, Chapter 9.

362.24 Bends — How Made.
Bends shall be so made that the tubing will not be damaged and the internal diameter of the tubing will not be effectively reduced. Bends shall be permitted to be made manually without auxiliary equipment, and the radius of the curve to the centerline of such bends shall not be less than shown in Table 2, Chapter 9 using the column "Other Bends."

362.26 Bends — Number in One Run.
There shall not be more than the equivalent of four quarter bends (360 degrees total) between pull points, for example, conduit bodies and boxes.

362.28 Trimming.
All cut ends shall be trimmed inside and outside to remove rough edges.

362.30 Securing and Supporting.
ENT shall be installed as a complete system in accordance with 300.18 and shall be securely fastened in place by an approved means and supported in accordance with 362.30(A) and (B).

(A) Securely Fastened. ENT shall be securely fastened at intervals not exceeding 900 mm (3 ft). In addition, ENT shall be securely fastened in place within 900 mm (3 ft) of each outlet box, device box, junction box, cabinet, or fitting where it terminates. Where used, cable ties shall be listed as suitable for the application and for securing and supporting.

Exception No. 1: Lengths not exceeding a distance of 1.8 m (6 ft) from a luminaire terminal connection for tap connections to lighting luminaires shall be permitted without being secured.

Exception No. 2: Lengths not exceeding 1.8 m (6 ft) from the last point where the raceway is securely fastened for connections within an accessible ceiling to luminaire(s) or other equipment.

Exception No. 3: For concealed work in finished buildings or prefinished wall panels where such securing is impracticable, unbroken lengths (without coupling) of ENT shall be permitted to be fished.

(B) Supports. Horizontal runs of ENT supported by openings in framing members at intervals not exceeding 900 mm (3 ft) and securely fastened within 900 mm (3 ft) of termination points shall be permitted.

362.46 Bushings.
Where a tubing enters a box, fitting, or other enclosure, a bushing or adapter shall be provided to protect the wire from abrasion unless the box, fitting, or enclosure design provides equivalent protection.

> Informational Note: See 300.4(G) for the protection of conductors size 4 AWG or larger.

362.48 Joints.
All joints between lengths of tubing and between tubing and couplings, fittings, and boxes shall be by an approved method.

362.56 Splices and Taps.
Splices and taps shall be made only in accordance with 300.15.

> Informational Note: See Article 314 for rules on the installation and use of boxes and conduit bodies.

362.60 Grounding.
Where equipment grounding is required, a separate equipment grounding conductor shall be installed in the raceway in compliance with Article 250, Part VI.

Part III. Construction Specifications

362.100 Construction.
ENT shall be made of material that does not exceed the ignitibility, flammability, smoke generation, and toxicity characteristics of rigid (nonplasticized) polyvinyl chloride.

ENT, as a prewired manufactured assembly, shall be provided in continuous lengths capable of being shipped in a coil, reel, or carton without damage.

362.120 Marking.
ENT shall be clearly and durably marked at least every 3 m (10 ft) as required in the first sentence of 110.21(A). The type of material shall also be included in the marking. Marking for limited smoke shall be permitted on the tubing that has limited smoke-producing characteristics.

The type, size, and quantity of conductors used in prewired manufactured assemblies shall be identified by means of a printed tag or label attached to each end of the manufactured assembly and either the carton, coil, or reel. The enclosed conductors shall be marked in accordance with 310.120.

ARTICLE 366
Auxiliary Gutters

Part I. General

366.1 Scope.
This article covers the use, installation, and construction requirements of metal auxiliary gutters and nonmetallic auxiliary gutters and associated fittings.

366.2 Definitions.

Metal Auxiliary Gutter. A sheet metal enclosure used to supplement wiring spaces at meter centers, distribution centers, switchgear, switchboards, and similar points of wiring systems. The enclosure has hinged or removable covers for housing and protecting electrical wires, cable, and busbars. The enclosure is designed for conductors to be laid or set in place after the enclosures have been installed as a complete system.

Nonmetallic Auxiliary Gutter. A flame-retardant, nonmetallic enclosure used to supplement wiring spaces at meter centers, distribution centers, switchgear, switchboards, and similar points of wiring systems. The enclosure has hinged or removable covers for housing and protecting electrical wires, cable, and busbars. The enclosure is designed for conductors to be laid or set in place after the enclosures have been installed as a complete system.

366.6 Listing Requirements.

(A) Outdoors. Nonmetallic auxiliary gutters installed outdoors shall be listed for all of the following conditions:

(1) Exposure to sunlight
(2) Use in wet locations
(3) Maximum ambient temperature of the installation

(B) Indoors. Nonmetallic auxiliary gutters installed indoors shall be listed for the maximum ambient temperature of the installation.

Part II. Installation

366.10 Uses Permitted.

(A) Sheet Metal Auxiliary Gutters.

(1) Indoor and Outdoor Use. Sheet metal auxiliary gutters shall be permitted for indoor and outdoor use.

(2) Wet Locations. Sheet metal auxiliary gutters installed in wet locations shall be suitable for such locations.

(B) Nonmetallic Auxiliary Gutters. Nonmetallic auxiliary gutters shall be listed for the maximum ambient temperature of the installation and marked for the installed conductor insulation temperature rating.

> Informational Note: Extreme cold may cause nonmetallic auxiliary gutters to become brittle and therefore more susceptible to damage from physical contact.

(1) Outdoors. Nonmetallic auxiliary gutters shall be permitted to be installed outdoors where listed and marked as suitable for the purpose.

(2) Indoors. Nonmetallic auxiliary gutters shall be permitted to be installed indoors.

366.12 Uses Not Permitted. Auxiliary gutters shall not be used:

(1) To enclose switches, overcurrent devices, appliances, or other similar equipment
(2) To extend a greater distance than 9 m (30 ft) beyond the equipment that it supplements

Exception: As permitted in 620.35 for elevators, an auxiliary gutter shall be permitted to extend a distance greater than 9 m (30 ft) beyond the equipment it supplements.

> Informational Note: For wireways, see Articles 376 and 378. For busways, see Article 368.

N **366.20 Conductors Connected in Parallel.** Where single conductor cables comprising each phase, neutral, or grounded conductor of an alternating-current circuit are connected in parallel as permitted in 310.10(H), the conductors shall be installed in groups consisting of not more than one conductor per phase, neutral, or grounded conductor to prevent current imbalance in the paralleled conductors due to inductive reactance.

366.22 Number of Conductors.

(A) Sheet Metal Auxiliary Gutters. The sum of the cross-sectional areas of all contained conductors and cables at any cross section of a sheet metal auxiliary gutter shall not exceed 20 percent of the interior cross-sectional area of the sheet metal auxiliary gutter. The adjustment factors in 310.15(B)(3)(a) shall be applied only where the number of current-carrying conductors, including neutral conductors classified as current-carrying under the provisions of 310.15(B)(5), exceeds 30. Conductors for signaling circuits or controller conductors between a motor and its starter and used only for starting duty shall not be considered as current-carrying conductors.

(B) Nonmetallic Auxiliary Gutters. The sum of cross-sectional areas of all contained conductors and cables at any cross section of the nonmetallic auxiliary gutter shall not exceed 20 percent of the interior cross-sectional area of the nonmetallic auxiliary gutter.

366.23 Ampacity of Conductors.

(A) Sheet Metal Auxiliary Gutters. Where the number of current-carrying conductors contained in the sheet metal auxiliary gutter is 30 or less, the adjustment factors specified in 310.15(B)(3)(a) shall not apply. The current carried continuously in bare copper bars in sheet metal auxiliary gutters shall not exceed 1.55 amperes/mm^2 (1000 amperes/in.^{2}) of cross section of the conductor. For aluminum bars, the current carried continuously shall not exceed 1.09 amperes/mm^2 (700 amperes/in.^{2}) of cross section of the conductor.

(B) Nonmetallic Auxiliary Gutters. The adjustment factors specified in 310.15(B)(3)(a) shall be applicable to the current-carrying conductors in the nonmetallic auxiliary gutter.

366.30 Securing and Supporting.

(A) Sheet Metal Auxiliary Gutters. Sheet metal auxiliary gutters shall be supported and secured throughout their entire length at intervals not exceeding 1.5 m (5 ft).

(B) Nonmetallic Auxiliary Gutters. Nonmetallic auxiliary gutters shall be supported and secured at intervals not to exceed 900 mm (3 ft) and at each end or joint, unless listed for other support intervals. In no case shall the distance between supports exceed 3 m (10 ft).

366.44 Expansion Fittings. Expansion fittings shall be installed where expected length change, due to expansion and contraction due to temperature change, is more than 6 mm (0.25 in.).

366.56 Splices and Taps. Splices and taps shall comply with 366.56(A) through (D).

(A) Within Gutters. Splices or taps shall be permitted within gutters where they are accessible by means of removable covers or doors. The conductors, including splices and taps, shall not fill the gutter to more than 75 percent of its area.

(B) Bare Conductors. Taps from bare conductors shall leave the gutter opposite their terminal connections, and conductors shall not be brought in contact with uninsulated current-carrying parts of different voltages.

(C) Suitably Identified. All taps shall be suitably identified at the gutter as to the circuit or equipment that they supply.

(D) Overcurrent Protection. Tap connections from conductors in auxiliary gutters shall be provided with overcurrent protection as required in 240.21.

366.58 Insulated Conductors.

(A) Deflected Insulated Conductors. Where insulated conductors are deflected within an auxiliary gutter, either at the ends or where conduits, fittings, or other raceways or cables enter or leave the gutter, or where the direction of the gutter is deflected greater than 30 degrees, dimensions corresponding to one wire per terminal in Table 312.6(A) shall apply.

(B) Auxiliary Gutters Used as Pull Boxes. Where insulated conductors 4 AWG or larger are pulled through an auxiliary gutter, the distance between raceway and cable entries enclosing the same conductor shall not be less than that required in 314.28(A)(1) for straight pulls and 314.28(A)(2) for angle pulls.

366.60 Grounding. Metal auxiliary gutters shall be connected to an equipment grounding conductor(s), to an equipment bonding jumper, or to the grounded conductor where permitted or required by 250.92(B)(1) or 250.142.

Part III. Construction Specifications

366.100 Construction.

(A) Electrical and Mechanical Continuity. Gutters shall be constructed and installed so that adequate electrical and mechanical continuity of the complete system is secured.

(B) Substantial Construction. Gutters shall be of substantial construction and shall provide a complete enclosure for the contained conductors. All surfaces, both interior and exterior, shall be suitably protected from corrosion. Corner joints shall be made tight, and where the assembly is held together by rivets, bolts, or screws, such fasteners shall be spaced not more than 300 mm (12 in.) apart.

(C) Smooth Rounded Edges. Suitable bushings, shields, or fittings having smooth, rounded edges shall be provided where conductors pass between gutters, through partitions, around bends, between gutters and cabinets or junction boxes, and at other locations where necessary to prevent abrasion of the insulation of the conductors.

(D) Covers. Covers shall be securely fastened to the gutter.

(E) Clearance of Bare Live Parts. Bare conductors shall be securely and rigidly supported so that the minimum clearance between bare current-carrying metal parts of different voltages mounted on the same surface will not be less than 50 mm (2 in.), nor less than 25 mm (1 in.) for parts that are held free in the air. A clearance not less than 25 mm (1 in.) shall be secured between bare current-carrying metal parts and any metal surface. Adequate provisions shall be made for the expansion and contraction of busbars.

366.120 Marking.

(A) Outdoors. Nonmetallic auxiliary gutters installed outdoors shall have the following markings:

(1) Suitable for exposure to sunlight

(2) Suitable for use in wet locations
(3) Installed conductor insulation temperature rating

(B) Indoors. Nonmetallic auxiliary gutters installed indoors shall be marked with the installed conductor insulation temperature rating.

ARTICLE 368
Busways

Part I. General Requirements

368.1 Scope. This article covers service-entrance, feeder, and branch-circuit busways and associated fittings.

368.2 Definition.

Busway. A raceway consisting of a metal enclosure containing factory-mounted, bare or insulated conductors, which are usually copper or aluminum bars, rods, or tubes.

Informational Note: For cablebus, refer to Article 370.

Part II. Installation

368.10 Uses Permitted. Busways shall be permitted to be installed where they are located in accordance with 368.10(A) through (C).

Informational Note: See 300.21 for information concerning the spread of fire or products of combustion.

(A) Exposed. Busways shall be permitted to be located in the open where visible, except as permitted in 368.10(C).

(B) Behind Access Panels. Busways shall be permitted to be installed behind access panels, provided the busways are totally enclosed, of nonventilating-type construction, and installed so that the joints between sections and at fittings are accessible for maintenance purposes. Where installed behind access panels, means of access shall be provided, and either of the following conditions shall be met:

(1) The space behind the access panels shall not be used for air-handling purposes.
(2) Where the space behind the access panels is used for environmental air, other than ducts and plenums, there shall be no provisions for plug-in connections, and the conductors shall be insulated.

(C) Through Walls and Floors. Busways shall be permitted to be installed through walls or floors in accordance with (C)(1) and (C)(2).

(1) Walls. Unbroken lengths of busway shall be permitted to be extended through dry walls.

(2) Floors. Floor penetrations shall comply with (a) and (b):
    (a) Busways shall be permitted to be extended vertically through dry floors if totally enclosed (unventilated) where passing through and for a minimum distance of 1.8 m (6 ft) above the floor to provide adequate protection from physical damage.

(b) In other than industrial establishments, where a vertical riser penetrates two or more dry floors, a minimum 100-mm (4-in.) high curb shall be installed around all floor openings for riser busways to prevent liquids from entering the opening. The curb shall be installed within 300 mm (12 in.) of the floor opening. Electrical equipment shall be located so that it will not be damaged by liquids that are retained by the curb.

368.12 Uses Not Permitted.

(A) Physical Damage. Busways shall not be installed where subject to severe physical damage or corrosive vapors.

(B) Hoistways. Busways shall not be installed in hoistways.

(C) Hazardous Locations. Busways shall not be installed in any hazardous (classified) location, unless specifically approved for such use.

Informational Note: See 501.10(B).

(D) Wet Locations. Busways shall not be installed outdoors or in wet or damp locations unless identified for such use.

(E) Working Platform. Lighting busway and trolley busway shall not be installed less than 2.5 m (8 ft) above the floor or working platform unless provided with an identified cover.

368.17 Overcurrent Protection. Overcurrent protection shall be provided in accordance with 368.17(A) through (D).

(A) Rating of Overcurrent Protection — Feeders. A busway shall be protected against overcurrent in accordance with the allowable current rating of the busway.

Exception No. 1: The applicable provisions of 240.4 shall be permitted.

Exception No. 2: Where used as transformer secondary ties, the provisions of 450.6(A)(3) shall be permitted.

(B) Reduction in Ampacity Size of Busway. Overcurrent protection shall be required where busways are reduced in ampacity.

Exception: For industrial establishments only, omission of overcurrent protection shall be permitted at points where busways are reduced in ampacity, provided that the length of the busway having the smaller ampacity does not exceed 15 m (50 ft) and has an ampacity at least equal to one-third the rating or setting of the overcurrent device next back on the line, and provided that such busway is free from contact with combustible material.

(C) Feeder or Branch Circuits. Where a busway is used as a feeder, devices or plug-in connections for tapping off feeder or branch circuits from the busway shall contain the overcurrent devices required for the protection of the feeder or branch circuits. The plug-in device shall consist of an externally operable circuit breaker or an externally operable fusible switch. Where such devices are mounted out of reach and contain disconnecting means, suitable means such as ropes, chains, or sticks shall be provided for operating the disconnecting means from the floor.

Exception No. 1: As permitted in 240.21.

Exception No. 2: For fixed or semifixed luminaires, where the branch-circuit overcurrent device is part of the luminaire cord plug on cord-connected luminaires.

Exception No. 3: Where luminaires without cords are plugged directly into the busway and the overcurrent device is mounted on the luminaire.

Exception No. 4: Where the branch-circuit overcurrent plug-in device is directly supplying a readily accessible disconnect, a method of floor operation shall not be required.

(D) Rating of Overcurrent Protection — Branch Circuits. A busway used as a branch circuit shall be protected against overcurrent in accordance with 210.20.

368.30 Support. Busways shall be securely supported at intervals not exceeding 1.5 m (5 ft) unless otherwise designed and marked.

368.56 Branches from Busways. Branches from busways shall be permitted to be made in accordance with 368.56(A), (B), and (C).

(A) General. Branches from busways shall be permitted to use any of the following wiring methods:

(1) Type AC armored cable
(2) Type MC metal-clad cable
(3) Type MI mineral-insulated, metal-sheathed cable
(4) Type IMC intermediate metal conduit
(5) Type RMC rigid metal conduit
(6) Type FMC flexible metal conduit
(7) Type LFMC liquidtight flexible metal conduit
(8) Type PVC rigid polyvinyl chloride conduit
(9) Type RTRC reinforced thermosetting resin conduit
(10) Type LFNC liquidtight flexible nonmetallic conduit
(11) Type EMT electrical metallic tubing
(12) Type ENT electrical nonmetallic tubing
(13) Busways
(14) Strut-type channel raceway
(15) Surface metal raceway
(16) Surface nonmetallic raceway

Where a separate equipment grounding conductor is used, connection of the equipment grounding conductor to the busway shall comply with 250.8 and 250.12.

(B) Cord and Cable Assemblies. Suitable cord and cable assemblies approved for extra-hard usage or hard usage and listed bus drop cable shall be permitted as branches from busways for the connection of portable equipment or the connection of stationary equipment to facilitate their interchange in accordance with 400.10 and 400.12 and the following conditions:

(1) The cord or cable shall be attached to the building by an approved means.
(2) The length of the cord or cable from a busway plug-in device to a suitable tension take-up support device shall not exceed 1.8 m (6 ft).
(3) The cord and cable shall be installed as a vertical riser from the tension take-up support device to the equipment served.
(4) Strain relief cable grips shall be provided for the cord or cable at the busway plug-in device and equipment terminations.

Exception to (B)(2): In industrial establishments only, where the conditions of maintenance and supervision ensure that only qualified persons service the installation, lengths exceeding 1.8 m (6 ft) shall be permitted between the busway plug-in device and the tension take-up support device where the cord or cable is supported at intervals not exceeding 2.5 m (8 ft).

(C) Branches from Trolley-Type Busways. Suitable cord and cable assemblies approved for extra-hard usage or hard usage and listed bus drop cable shall be permitted as branches from trolley-type busways for the connection of movable equipment in accordance with 400.10 and 400.12.

368.58 Dead Ends. A dead end of a busway shall be closed.

368.60 Grounding. Busway shall be connected to an equipment grounding conductor(s), to an equipment bonding jumper, or to the grounded conductor where permitted or required by 250.92(B)(1) or 250.142.

Part III. Construction

368.120 Marking. Busways shall be marked with the voltage and current rating for which they are designed, and with the manufacturer's name or trademark in such a manner as to be visible after installation.

Part IV. Requirements for Over 1000 Volts, Nominal

368.214 Adjacent and Supporting Structures. Metal-enclosed busways shall be installed so that temperature rise from induced circulating currents in adjacent ferrous metal parts will not be hazardous to personnel or constitute a fire hazard.

368.234 Barriers and Seals.

(A) Vapor Seals. Busway runs that have sections located both inside and outside of buildings shall have a vapor seal at the building wall to prevent interchange of air between indoor and outdoor sections.

Exception: Vapor seals shall not be required in forced-cooled bus.

(B) Fire Barriers. Fire barriers shall be provided where fire walls, floors, or ceilings are penetrated.

Informational Note: See 300.21 for information concerning the spread of fire or products of combustion.

368.236 Drain Facilities. Drain plugs, filter drains, or similar methods shall be provided to remove condensed moisture from low points in busway run.

368.237 Ventilated Bus Enclosures. Ventilated busway enclosures shall be installed in accordance with Article 110, Part III, and 490.24.

368.238 Terminations and Connections. Where bus enclosures terminate at machines cooled by flammable gas, seal-off bushings, baffles, or other means shall be provided to prevent accumulation of flammable gas in the busway enclosures.

All conductor termination and connection hardware shall be accessible for installation, connection, and maintenance.

368.239 Switches. Switching devices or disconnecting links provided in the busway run shall have the same momentary rating as the busway. Disconnecting links shall be plainly marked to be removable only when bus is de-energized. Switching devices that are not load-break shall be interlocked to prevent operation under load, and disconnecting link enclosures shall be interlocked to prevent access to energized parts.

368.240 Wiring 1000 Volts or Less, Nominal. Secondary control devices and wiring that are provided as part of the metal-enclosed bus run shall be insulated by fire-retardant barriers from all primary circuit elements with the exception of

short lengths of wire, such as at instrument transformer terminals.

368.244 Expansion Fittings. Flexible or expansion connections shall be provided in long, straight runs of bus to allow for temperature expansion or contraction, or where the busway run crosses building vibration insulation joints.

368.258 Neutral Conductor. Neutral bus, where required, shall be sized to carry all neutral load current, including harmonic currents, and shall have adequate momentary and short-circuit rating consistent with system requirements.

368.260 Grounding. Metal-enclosed busway shall be grounded.

368.320 Marking. Each busway run shall be provided with a permanent nameplate on which the following information shall be provided:

(1) Rated voltage.
(2) Rated continuous current; if bus is forced-cooled, both the normal forced-cooled rating and the self-cooled (not forced-cooled) rating for the same temperature rise shall be given.
(3) Rated frequency.
(4) Rated impulse withstand voltage.
(5) Rated 60-Hz withstand voltage (dry).
(6) Rated momentary current.
(7) Manufacturer's name or trademark

Informational Note: See ANSI C37.23-1987 (R1991), *Guide for Metal-Enclosed Bus and Calculating Losses in Isolated-Phase Bus,* for construction and testing requirements for metal-enclosed buses.

ARTICLE 370
Cablebus

Part I. General

370.1 Scope. This article covers the use and installation requirements of cablebus and associated fittings.

370.2 Definition.

Cablebus. An assembly of units or sections with insulated conductors having associated fittings forming a structural system used to securely fasten or support conductors and conductor terminations in a completely enclosed, ventilated, protective metal housing. This assembly is designed to carry fault current and to withstand the magnetic forces of such current.

Informational Note: Cablebus is ordinarily assembled at the point of installation from the components furnished or specified by the manufacturer in accordance with instructions for the specific job.

Part II. Installation

370.10 Uses Permitted. Approved cablebus shall be permitted:

(1) At any voltage or current for which spaced conductors are rated and where installed only for exposed work, except as permitted in 370.18

(2) For branch circuits, feeders, and services

(3) To be installed outdoors or in corrosive, wet, or damp locations where identified for the use

370.12 Uses Not Permitted. Cablebus shall not be permitted to be installed in the following:

(1) Hoistways

(2) Hazardous (classified) locations, unless specifically permitted in Chapter 5

370.18 Cablebus Installation.

(A) Transversely Routed. Cablebus shall be permitted to extend transversely through partitions or walls, other than fire walls, provided that the section within the wall is continuous, protected against physical damage, and unventilated.

(B) Through Dry Floors and Platforms. Except where fire-stops are required, cablebus shall be permitted to extend vertically through dry floors and platforms, provided that the cablebus is totally enclosed at the point where it passes through the floor or platform and for a distance of 1.8 m (6 ft) above the floor or platform.

(C) Through Floors and Platforms in Wet Locations. Except where firestops are required, cablebus shall be permitted to extend vertically through floors and platforms in wet locations where:

(1) There are curbs or other suitable means to prevent water-flow through the floor or platform opening, and

(2) Where the cablebus is totally enclosed at the point where it passes through the floor or platform and for a distance of 1.8 m (6 ft) above the floor or platform.

370.20 Conductor Size and Termination.

(A) Conductors. The current-carrying conductors in cablebus shall:

(1) Have an insulation rating of 75°C (167°F) or higher and be of an approved type suitable for the application.

(2) Be sized in accordance with the design of the cablebus but in no case be smaller than 1/0.

(B) Termination. Approved terminating means shall be used for connections to cablebus conductors.

370.22 Number of Conductors. The number of conductors shall be that for which the cablebus is designed.

370.23 Overcurrent Protection. Cablebus shall be protected against overcurrent in accordance with the allowable ampacity of the cablebus conductors in accordance with 240.4.

Exception: Overcurrent protection shall be permitted in accordance with 240.100 and 240.101.

370.30 Securing and Supporting.

(A) Cablebus Supports. Cablebus shall be securely supported at intervals not exceeding 3.7 m (12 ft). Where spans longer than 3.7 m (12 ft) are required, the structure shall be specifically designed for the required span length.

(B) Conductor Supports. The insulated conductors shall be supported on blocks or other identified mounting means.

The individual conductors in a cablebus shall be supported at intervals not greater than 900 mm (3 ft) for horizontal runs and 450 mm (1½ ft) for vertical runs. Vertical and horizontal spacing between supported conductors shall be not less than one conductor diameter at the points of support.

370.42 Fittings. A cablebus system shall include approved fittings for the following:

(1) Changes in horizontal or vertical direction of the run

(2) Dead ends

(3) Terminations in or on connected apparatus or equipment or the enclosures for such equipment

(4) Additional physical protection where required, such as guards where subject to severe physical damage

370.60 Grounding. A cablebus system shall be grounded and/or bonded as applicable:

(1) Cablebus framework, where bonded, shall be permitted to be used as the equipment grounding conductor for branch circuits and feeders.

(2) A cablebus installation shall be grounded and bonded in accordance with Article 250, excluding 250.86, Exception No. 2.

370.80 Ampacity of Conductors. The ampacity of conductors in cablebus shall be in accordance with Table 310.15(B)(17) and Table 310.15(B)(19) for installations up to and including 2000 volts, or with Table 310.60(C)(69) and Table 310.60(C)(70) for installations 2001 to 35,000 volts.

Part III. Construction Specifications

370.120 Marking. Each section of cablebus shall be marked with the manufacturer's name or trade designation and the maximum diameter, number, voltage rating, and ampacity of the conductors to be installed. Markings shall be located so as to be visible after installation.

ARTICLE 372
Cellular Concrete Floor Raceways

Part I. General

372.1 Scope. This article covers cellular concrete floor raceways, the hollow spaces in floors constructed of precast cellular concrete slabs, together with suitable metal fittings designed to provide access to the floor cells.

372.2 Definitions.

Cell. A single, enclosed tubular space in a floor made of precast cellular concrete slabs, the direction of the cell being parallel to the direction of the floor member.

Header. Transverse metal raceways for electrical conductors, providing access to predetermined cells of a precast cellular concrete floor, thereby permitting the installation of electrical conductors from a distribution center to the floor cells.

Part II. Installations

372.12 Uses Not Permitted. Conductors shall not be installed in precast cellular concrete floor raceways as follows:

(1) Where subject to corrosive vapor
(2) In any hazardous (classified) location, except as permitted by other articles in this *Code*
(3) In commercial garages, other than for supplying ceiling outlets or extensions to the area below the floor but not above

Informational Note: See 300.8 for installation of conductors with other systems.

372.18 Cellular Concrete Floor Raceways Installation. Installation of cellular concrete floor raceways shall comply with 372.18(A) through 372.18(E).

(A) Header. The header shall be installed in a straight line at right angles to the cells. The header shall be mechanically secured to the top of the precast cellular concrete floor. The end joints shall be closed by a metal closure fitting and sealed against the entrance of concrete. The header shall be electrically continuous throughout its entire length and shall be electrically bonded to the enclosure of the distribution center.

(B) Connection to Cabinets and Other Enclosures. Connections from headers to cabinets and other enclosures shall be made by means of listed metal raceways and listed fittings.

(C) Junction Boxes. Junction boxes shall be leveled to the floor grade and sealed against the free entrance of water or concrete. Junction boxes shall be of metal and shall be mechanically and electrically continuous with the header.

(D) Inserts. Inserts shall be leveled and sealed against the entrance of concrete. Inserts shall be of metal and shall be fitted with grounded-type receptacles. A grounding conductor shall connect the insert receptacles to a positive ground connection provided on the header. Where cutting through the cell wall for setting inserts or other purposes (such as providing access openings between header and cells), chips and other dirt shall not be allowed to remain in the raceway, and the tool used shall be designed so as to prevent the tool from entering the cell and damaging the conductors.

(E) Markers. A suitable number of markers shall be installed for the future location of cells.

372.20 Size of Conductors. No conductor larger than 1/0 AWG shall be installed, except by special permission.

372.22 Maximum Number of Conductors. The combined cross-sectional area of all conductors or cables shall not exceed 40 percent of the cross-sectional area of the cell or header.

372.23 Ampacity of Conductors. The ampacity adjustment factors as provided in 310.15(B)(3) shall apply to conductors installed in cellular concrete floor raceways.

372.56 Splices and Taps. Splices and taps shall be made only in header access units or junction boxes. A continuous unbroken conductor connecting the individual outlets is not a splice or tap.

372.58 Discontinued Outlets. When an outlet is abandoned, discontinued, or removed, the sections of circuit conductors supplying the outlet shall be removed from the raceway. No splices or reinsulated conductors, such as would be the case of abandoned outlets on loop wiring, shall be allowed in raceways.

ARTICLE 374
Cellular Metal Floor Raceways

Part I. General

374.1 Scope. This article covers the use and installation requirements for cellular metal floor raceways.

374.2 Definitions.

Cellular Metal Floor Raceway. The hollow spaces of cellular metal floors, together with suitable fittings, that may be approved as enclosed channel for electrical conductors.

Cell. A single enclosed tubular space in a cellular metal floor member, the axis of the cell being parallel to the axis of the metal floor member.

Header. A transverse raceway for electrical conductors, providing access to predetermined cells of a cellular metal floor, thereby permitting the installation of electrical conductors from a distribution center to the cells.

Part II. Installation

374.12 Uses Not Permitted. Conductors shall not be installed in cellular metal floor raceways as follows:

(1) Where subject to corrosive vapor
(2) In any hazardous (classified) location, except as permitted by other articles in this *Code*
(3) In commercial garages, other than for supplying ceiling outlets or extensions to the area below the floor but not above

Informational Note: See 300.8 for installation of conductors with other systems.

374.18 Cellular Metal Floor Raceways Installations. Installation of cellular metal floor raceways shall comply with 374.18(A) through 374.18(D).

(A) Connection to Cabinets and Extensions from Cells. Connections between raceways and distribution centers and wall outlets shall be made by means of liquidtight flexible metal conduit, flexible metal conduit where not installed in concrete, rigid metal conduit, intermediate metal conduit, electrical metallic tubing, or approved fittings. Where there are provisions for the termination of an equipment grounding conductor, rigid polyvinyl chloride conduit, reinforced thermosetting resin conduit, electrical nonmetallic tubing, or liquidtight flexible nonmetallic conduit shall be permitted. Where installed in concrete, liquidtight flexible metal conduit and liquidtight flexible nonmetallic conduit shall be listed and marked for direct burial.

(B) Junction Boxes. Junction boxes shall be leveled to the floor grade and sealed against the free entrance of water or concrete. Junction boxes used with these raceways shall be of metal and shall be electrically continuous with the raceway.

(C) Inserts. Inserts shall be leveled to the floor grade and sealed against the entrance of concrete. Inserts shall be of metal and shall be electrically continuous with the raceway. In cutting through the cell wall and setting inserts, chips and other dirt shall not be allowed to remain in the raceway, and tools shall be used that are designed to prevent the tool from entering the cell and damaging the conductors.

(D) Markers. A suitable number of markers shall be installed for locating cells in the future.

374.20 Size of Conductors. No conductor larger than 1/0 AWG shall be installed, except by special permission.

374.22 Maximum Number of Conductors in Raceway. The combined cross-sectional area of all conductors or cables shall not exceed 40 percent of the interior cross-sectional area of the cell or header.

374.23 Ampacity of Conductors. The ampacity adjustment factors in 310.15(B)(3) shall apply to conductors installed in cellular metal floor raceways.

374.56 Splices and Taps. Splices and taps shall be made only in header access units or junction boxes.

For the purposes of this section, so-called loop wiring (continuous unbroken conductor connecting the individual outlets) shall not be considered to be a splice or tap.

374.58 Discontinued Outlets. When an outlet is abandoned, discontinued, or removed, the sections of circuit conductors supplying the outlet shall be removed from the raceway. No splices or reinsulated conductors, such as would be the case with abandoned outlets on loop wiring, shall be allowed in raceways.

Part III. Construction Specifications

374.100 General. Cellular metal floor raceways shall be constructed so that adequate electrical and mechanical continuity of the complete system will be secured. They shall provide a complete enclosure for the conductors. The interior surfaces shall be free from burrs and sharp edges, and surfaces over which conductors are drawn shall be smooth. Suitable bushings or fittings having smooth rounded edges shall be provided where conductors pass.

ARTICLE 376
Metal Wireways

Part I. General

376.1 Scope. This article covers the use, installation, and construction specifications for metal wireways and associated fittings.

376.2 Definition.

Metal Wireways. Sheet metal troughs with hinged or removable covers for housing and protecting electrical wires and cable and in which conductors are laid in place after the raceway has been installed as a complete system.

Part II. Installation

376.10 Uses Permitted. The use of metal wireways shall be permitted as follows:

(1) For exposed work.
(2) In any hazardous (classified) location, as permitted by other articles in this *Code*.
(3) In wet locations where wireways are listed for the purpose.
(4) In concealed spaces as an extension that passes transversely through walls, if the length passing through the wall is unbroken. Access to the conductors shall be maintained on both sides of the wall.

376.12 Uses Not Permitted. Metal wireways shall not be used in the following:

(1) Where subject to severe physical damage
(2) Where subject to severe corrosive environments

N **376.20 Conductors Connected in Parallel.** Where single conductor cables comprising each phase, neutral, or grounded conductor of an alternating-current circuit are connected in parallel as permitted in 310.10(H), the conductors shall be installed in groups consisting of not more than one conductor per phase, neutral, or grounded conductor to prevent current imbalance in the paralleled conductors due to inductive reactance.

376.21 Size of Conductors. No conductor larger than that for which the wireway is designed shall be installed in any wireway.

376.22 Number of Conductors and Ampacity. The number of conductors or cables and their ampacity shall comply with 376.22(A) and (B).

(A) Cross-Sectional Areas of Wireway. The sum of the cross-sectional areas of all contained conductors and cables at any cross section of a wireway shall not exceed 20 percent of the interior cross-sectional area of the wireway.

(B) Adjustment Factors. The adjustment factors in 310.15(B)(3)(a) shall be applied only where the number of current-carrying conductors, including neutral conductors classified as current-carrying under the provisions of 310.15(B)(5), exceeds 30 at any cross section of the wireway. Conductors for signaling circuits or controller conductors between a motor and its starter and used only for starting duty shall not be considered as current-carrying conductors.

376.23 Insulated Conductors. Insulated conductors installed in a metal wireway shall comply with 376.23(A) and (B).

(A) Deflected Insulated Conductors. Where insulated conductors are deflected within a metal wireway, either at the ends or where conduits, fittings, or other raceways or cables enter or leave the metal wireway, or where the direction of the metal wireway is deflected greater than 30 degrees, dimensions corresponding to one wire per terminal in Table 312.6(A) shall apply.

(B) Metal Wireways Used as Pull Boxes. Where insulated conductors 4 AWG or larger are pulled through a wireway, the distance between raceway and cable entries enclosing the same conductor shall not be less than that required by 314.28(A)(1) for straight pulls and 314.28(A)(2) for angle pulls. When transposing cable size into raceway size, the minimum metric designator (trade size) raceway required for the number and size of conductors in the cable shall be used.

376.30 Securing and Supporting. Metal wireways shall be supported in accordance with 376.30(A) and (B).

(A) Horizontal Support. Wireways shall be supported where run horizontally at each end and at intervals not to exceed 1.5 m (5 ft) or for individual lengths longer than 1.5 m (5 ft) at each end or joint, unless listed for other support intervals. The distance between supports shall not exceed 3 m (10 ft).

(B) Vertical Support. Vertical runs of wireways shall be securely supported at intervals not exceeding 4.5 m (15 ft) and shall not have more than one joint between supports. Adjoining wireway sections shall be securely fastened together to provide a rigid joint.

376.56 Splices, Taps, and Power Distribution Blocks.

(A) Splices and Taps. Splices and taps shall be permitted within a wireway, provided they are accessible. The conductors, including splices and taps, shall not fill the wireway to more than 75 percent of its area at that point.

(B) Power Distribution Blocks.

(1) Installation. Power distribution blocks installed in metal wireways shall be listed. Power distribution blocks installed on the line side of the service equipment shall be marked "suitable for use on the line side of service equipment" or equivalent.

(2) Size of Enclosure. In addition to the wiring space requirement in 376.56(A), the power distribution block shall be installed in a wireway with dimensions not smaller than specified in the installation instructions of the power distribution block.

(3) Wire Bending Space. Wire bending space at the terminals of power distribution blocks shall comply with 312.6(B).

(4) Live Parts. Power distribution blocks shall not have uninsulated live parts exposed within a wireway, whether or not the wireway cover is installed.

(5) Conductors. Conductors shall be arranged so the power distribution block terminals are unobstructed following installation.

376.58 Dead Ends. Dead ends of metal wireways shall be closed.

376.70 Extensions from Metal Wireways. Extensions from wireways shall be made with cord pendants installed in accordance with 400.14 or with any wiring method in Chapter 3 that includes a means for equipment grounding. Where a separate equipment grounding conductor is employed, connection of the equipment grounding conductors in the wiring method to the wireway shall comply with 250.8 and 250.12.

Part III. Construction Specifications

376.100 Construction.

(A) Electrical and Mechanical Continuity. Wireways shall be constructed and installed so that electrical and mechanical continuity of the complete system are assured.

(B) Substantial Construction. Wireways shall be of substantial construction and shall provide a complete enclosure for the contained conductors. All surfaces, both interior and exterior, shall be suitably protected from corrosion. Corner joints shall be made tight, and where the assembly is held together by rivets, bolts, or screws, such fasteners shall be spaced not more than 300 mm (12 in.) apart.

(C) Smooth Rounded Edges. Suitable bushings, shields, or fittings having smooth, rounded edges shall be provided where conductors pass between wireways, through partitions, around bends, between wireways and cabinets or junction boxes, and at other locations where necessary to prevent abrasion of the insulation of the conductors.

(D) Covers. Covers shall be securely fastened to the wireway.

376.120 Marking. Metal wireways shall be so marked that their manufacturer's name or trademark will be visible after installation.

ARTICLE 378
Nonmetallic Wireways

Part I. General

378.1 Scope. This article covers the use, installation, and construction specifications for nonmetallic wireways and associated fittings.

378.2 Definition.

Nonmetallic Wireways. Flame-retardant, nonmetallic troughs with removable covers for housing and protecting electrical wires and cables in which conductors are laid in place after the raceway has been installed as a complete system.

378.6 Listing Requirements. Nonmetallic wireways and associated fittings shall be listed.

Part II. Installation

378.10 Uses Permitted. The use of nonmetallic wireways shall be permitted in the following:

(1) Only for exposed work, except as permitted in 378.10(4).
(2) Where subject to corrosive environments where identified for the use.
(3) In wet locations where listed for the purpose.

   Informational Note: Extreme cold may cause nonmetallic wireways to become brittle and therefore more susceptible to damage from physical contact.

(4) As extensions to pass transversely through walls if the length passing through the wall is unbroken. Access to the conductors shall be maintained on both sides of the wall.

378.12 Uses Not Permitted. Nonmetallic wireways shall not be used in the following:

(1) Where subject to physical damage
(2) In any hazardous (classified) location, except as permitted by other articles in this *Code*
(3) Where exposed to sunlight unless listed and marked as suitable for the purpose
(4) Where subject to ambient temperatures other than those for which nonmetallic wireway is listed

(5) For conductors whose insulation temperature limitations would exceed those for which the nonmetallic wireway is listed

N 378.20 Conductors Connected in Parallel. Where single conductor cables comprising each phase, neutral, or grounded conductor of an alternating-current circuit are connected in parallel as permitted in 310.10(H), the conductors shall be installed in groups consisting of not more than one conductor per phase, neutral, or grounded conductor to prevent current imbalance in the paralleled conductors due to inductive reactance.

378.21 Size of Conductors. No conductor larger than that for which the nonmetallic wireway is designed shall be installed in any nonmetallic wireway.

378.22 Number of Conductors. The sum of cross-sectional areas of all contained conductors or cables at any cross section of the nonmetallic wireway shall not exceed 20 percent of the interior cross-sectional area of the nonmetallic wireway. Conductors for signaling circuits or controller conductors between a motor and its starter and used only for starting duty shall not be considered as current-carrying conductors.

The adjustment factors specified in 310.15(B)(3)(a) shall be applicable to the current-carrying conductors up to and including the 20 percent fill specified in the first paragraph of this section.

378.23 Insulated Conductors. Insulated conductors installed in a nonmetallic wireway shall comply with 378.23(A) and (B).

(A) Deflected Insulated Conductors. Where insulated conductors are deflected within a nonmetallic wireway, either at the ends or where conduits, fittings, or other raceways or cables enter or leave the nonmetallic wireway, or where the direction of the nonmetallic wireway is deflected greater than 30 degrees, dimensions corresponding to one wire per terminal in Table 312.6(A) shall apply.

(B) Nonmetallic Wireways Used as Pull Boxes. Where insulated conductors 4 AWG or larger are pulled through a wireway, the distance between raceway and cable entries enclosing the same conductor shall not be less than that required in 314.28(A)(1) for straight pulls and in 314.28(A)(2) for angle pulls. When transposing cable size into raceway size, the minimum metric designator (trade size) raceway required for the number and size of conductors in the cable shall be used.

378.30 Securing and Supporting. Nonmetallic wireway shall be supported in accordance with 378.30(A) and (B).

(A) Horizontal Support. Nonmetallic wireways shall be supported where run horizontally at intervals not to exceed 900 mm (3 ft), and at each end or joint, unless listed for other support intervals. In no case shall the distance between supports exceed 3 m (10 ft).

(B) Vertical Support. Vertical runs of nonmetallic wireway shall be securely supported at intervals not exceeding 1.2 m (4 ft), unless listed for other support intervals, and shall not have more than one joint between supports. Adjoining nonmetallic wireway sections shall be securely fastened together to provide a rigid joint.

378.44 Expansion Fittings. Expansion fittings for nonmetallic wireway shall be provided to compensate for thermal expansion and contraction where the length change is expected to be 6 mm (0.25 in.) or greater in a straight run.

Informational Note: See Table 352.44 for expansion characteristics of PVC conduit. The expansion characteristics of PVC nonmetallic wireway are identical.

378.56 Splices and Taps. Splices and taps shall be permitted within a nonmetallic wireway, provided they are accessible. The conductors, including splices and taps, shall not fill the nonmetallic wireway to more than 75 percent of its area at that point.

378.58 Dead Ends. Dead ends of nonmetallic wireway shall be closed using listed fittings.

378.60 Grounding. Where equipment grounding is required, a separate equipment grounding conductor shall be installed in the nonmetallic wireway. A separate equipment grounding conductor shall not be required where the grounded conductor is used to ground equipment as permitted in 250.142.

378.70 Extensions from Nonmetallic Wireways. Extensions from nonmetallic wireway shall be made with cord pendants or any wiring method of Chapter 3. A separate equipment grounding conductor shall be installed in, or an equipment grounding connection shall be made to, any of the wiring methods used for the extension.

Part III. Construction Specifications

378.120 Marking. Nonmetallic wireways shall be marked so that the manufacturer's name or trademark and interior cross-sectional area in square inches shall be visible after installation. Marking for limited smoke shall be permitted on the nonmetallic wireways that have limited smoke-producing characteristics.

ARTICLE 380
Multioutlet Assembly

Part I. General

380.1 Scope. This article covers the use and installation requirements for multioutlet assemblies.

Informational Note: See the definition of multioutlet assembly in Article 100.

Part II. Installation

380.10 Uses Permitted. The use of a multioutlet assembly shall be permitted in dry locations.

380.12 Uses Not Permitted. A multioutlet assembly shall not be installed as follows:

(1) Where concealed, except that it shall be permissible to surround the back and sides of a metal multioutlet assembly by the building finish or recess a nonmetallic multioutlet assembly in a baseboard
(2) Where subject to severe physical damage
(3) Where the voltage is 300 volts or more between conductors unless the assembly is of metal having a thickness of not less than 1.02 mm (0.040 in.)

(4) Where subject to corrosive vapors
(5) In hoistways
(6) In any hazardous (classified) location, except as permitted by other articles in this *Code*

380.23 Insulated Conductors. For field-assembled multioutlet assemblies, insulated conductors shall comply with 380.23(A) and (B), as applicable.

(A) Deflected Insulated Conductors. Where insulated conductors are deflected within a multioutlet assembly, either at the ends or where conduits, fittings, or other raceways or cables enter or leave the multioutlet assembly, or where the direction of the multioutlet assembly is deflected greater than 30 degrees, dimensions corresponding to one wire per terminal in Table 312.6(A) shall apply.

(B) Multioutlet Assemblies Used as Pull Boxes. Where insulated conductors 4 AWG or larger are pulled through a multioutlet assembly, the distance between raceway and cable entries enclosing the same conductor shall not be less than that required by 314.28(A)(1) for straight pulls and 314.28(A)(2) for angle pulls. When transposing cable size into raceway size, the minimum metric designator (trade size) raceway required for the number and size of conductors in the cable shall be used.

380.76 Metal Multioutlet Assembly Through Dry Partitions. It shall be permissible to extend a metal multioutlet assembly through (not run within) dry partitions if arrangements are made for removing the cap or cover on all exposed portions and no outlet is located within the partitions.

ARTICLE 382
Nonmetallic Extensions

Part I. General

382.1 Scope. This article covers the use, installation, and construction specifications for nonmetallic extensions.

382.2 Definitions.

Concealable Nonmetallic Extension. A listed assembly of two, three, or four insulated circuit conductors within a nonmetallic jacket, an extruded thermoplastic covering, or a sealed nonmetallic covering. The classification includes surface extensions intended for mounting directly on the surface of walls or ceilings, and concealed with paint, texture, joint compound, plaster, wallpaper, tile, wall paneling, or other similar materials.

Nonmetallic Extension. An assembly of two insulated conductors within a nonmetallic jacket or an extruded thermoplastic covering. The classification includes surface extensions intended for mounting directly on the surface of walls or ceilings.

382.6 Listing Requirements. Concealable nonmetallic extensions and associated fittings and devices shall be listed. The starting/source tap device for the extension shall contain and provide the following protection for all load-side extensions and devices:

(1) Supplementary overcurrent protection
(2) Level of protection equivalent to a Class A GFCI

(3) Level of protection equivalent to a portable GFCI
(4) Line and load-side miswire protection
(5) Provide protection from the effects of arc faults

Part II. Installation

382.10 Uses Permitted. Nonmetallic extensions shall be permitted only in accordance with 382.10(A), (B), and (C).

(A) From an Existing Outlet. The extension shall be from an existing outlet on a 15- or 20-ampere branch circuit. Where a concealable nonmetallic extension originates from a non–grounding-type receptacle, the installation shall comply with 250.130(C), 406.4(D)(2)(b), or 406.4(D)(2)(c).

(B) Exposed and in a Dry Location. The extension shall be run exposed, or concealed as permitted in 382.15, and in a dry location.

(C) Residential or Offices. For nonmetallic surface extensions mounted directly on the surface of walls or ceilings, the building shall be occupied for residential or office purposes and shall not exceed three floors abovegrade. Where identified for the use, concealable nonmetallic extensions shall be permitted more than three floors abovegrade.

Informational Note No. 1: See 310.15(A)(3) for temperature limitation of conductors.

Informational Note No. 2: See 362.10 for definition of *First Floor*.

382.12 Uses Not Permitted. Nonmetallic extensions shall not be used as follows:

(1) In unfinished basements, attics, or roof spaces
(2) Where the voltage between conductors exceeds 150 volts for nonmetallic surface extensions and 300 volts for aerial cable
(3) Where subject to corrosive vapors
(4) Where run through a floor or partition, or outside the room in which it originates

382.15 Exposed.

(A) Nonmetallic Extensions. One or more extensions shall be permitted to be run in any direction from an existing outlet, but not on the floor or within 50 mm (2 in.) from the floor.

(B) Concealable Nonmetallic Extensions. Where identified for the use, nonmetallic extensions shall be permitted to be concealed with paint, texture, concealing compound, plaster, wallpaper, tile, wall paneling, or other similar materials and installed in accordance with 382.15(A).

382.26 Bends.

(A) Nonmetallic Extensions. A bend that reduces the normal spacing between the conductors shall be covered with a cap to protect the assembly from physical damage.

(B) Concealable Nonmetallic Extensions. Concealable extensions shall be permitted to be folded back over themselves and flattened as required for installation.

382.30 Securing and Supporting.

(A) Nonmetallic Extensions. Nonmetallic surface extensions shall be secured in place by approved means at intervals not exceeding 200 mm (8 in.), with an allowance for 300 mm (12 in.) to the first fastening where the connection to the supplying outlet is by means of an attachment plug. There shall be at least one fastening between each two adjacent outlets supplied. An extension shall be attached to only woodwork or plaster finish and shall not be in contact with any metal work or other conductive material other than with metal plates on receptacles.

(B) Concealable Nonmetallic Extensions. All surface-mounted concealable nonmetallic extension components shall be firmly anchored to the wall or ceiling using an adhesive or mechanical anchoring system identified for this use.

382.40 Boxes and Fittings. Each run shall terminate in a fitting, connector, or box that covers the end of the assembly. All fittings, connectors, and devices shall be of a type identified for the use.

382.42 Devices.

(A) Receptacles. All receptacles, receptacle housings, and self-contained devices used with concealable nonmetallic extensions shall be identified for this use.

(B) Receptacles and Housings. Receptacle housings and self-contained devices designed either for surface or for recessed mounting shall be permitted for use with concealable nonmetallic extensions. Receptacle housings and self-contained devices shall incorporate means for facilitating entry and termination of concealable nonmetallic extensions and for electrically connecting the housing or device. Receptacle and self-contained devices shall comply with 406.4. Power and communications outlets installed together in common housing shall be permitted in accordance with 800.133(A)(1)(d), Exception No. 2.

382.56 Splices and Taps. Extensions shall consist of a continuous unbroken length of the assembly, without splices, and without exposed conductors between fittings, connectors, or devices. Taps shall be permitted where approved fittings completely covering the tap connections are used. Aerial cable and its tap connectors shall be provided with an approved means for polarization. Receptacle-type tap connectors shall be of the locking type.

Part III. Construction Specifications (Concealable Nonmetallic Extensions Only)

382.100 Construction. Concealable nonmetallic extensions shall be a multilayer flat conductor design consisting of a center ungrounded conductor enclosed by a sectioned grounded conductor, and an overall sectioned grounding conductor.

382.104 Flat Conductors. Concealable nonmetallic extensions shall be constructed, using flat copper conductors equivalent to 14 AWG or 12 AWG conductor sizes, and constructed per 382.104(A), (B), and (C).

(A) Ungrounded Conductor (Center Layer). The ungrounded conductor shall consist of one or more ungrounded flat conductor(s) enclosed in accordance with 382.104(B) and (C) and identified in accordance with 310.110(C).

(B) Grounded Conductor (Inner Sectioned Layers). The grounded conductor shall consist of two sectioned inner flat conductors that enclose the center ungrounded conductor(s). The sectioned grounded conductor shall be enclosed by the sectioned grounding conductor and identified in accordance with 200.6.

(C) Grounding Conductor (Outer Sectioned Layers). The grounding conductor shall consist of two overall sectioned conductors that enclose the grounded conductor and ungrounded conductor(s) and shall comply with 250.4(A)(5). The grounding conductor layers shall be identified by any one of the following methods:

(1) As permitted in 250.119
(2) A clear covering
(3) One or more continuous green stripes or hash marks
(4) The term "Equipment Ground" printed at regular intervals throughout the cable

382.112 Insulation. The ungrounded and grounded flat conductor layers shall be individually insulated and comply with 310.15(A)(3). The grounding conductor shall be covered or insulated.

382.120 Marking.

(A) Cable. Concealable nonmetallic extensions shall be clearly and durably marked on both sides at intervals of not more than 610 mm (24 in.) with the information required by 310.120(A) and with the following additional information:

(1) Material of conductors
(2) Maximum temperature rating
(3) Ampacity

(B) Conductor Identification. Conductors shall be clearly and durably identified on both sides throughout their length as specified in 382.104.

ARTICLE 384
Strut-Type Channel Raceway

Part I. General

384.1 Scope. This article covers the use, installation, and construction specifications of strut-type channel raceway.

384.2 Definition.

Strut-Type Channel Raceway. A metal raceway that is intended to be mounted to the surface of or suspended from a structure, with associated accessories for the installation of electrical conductors and cables.

384.6 Listing Requirements. Strut-type channel raceways and accessories shall be listed and identified for such use.

Part II. Installation

384.10 Uses Permitted. The use of strut-type channel raceways shall be permitted in the following:

(1) Where exposed.
(2) In dry locations.
(3) In locations subject to corrosive vapors where protected by finishes approved for the condition.
(4) As power poles.
(5) In hazardous (classified) locations as permitted in Chapter 5.

(6) As extensions of unbroken lengths through walls, partitions, and floors where closure strips are removable from either side and the portion within the wall, partition, or floor remains covered.

(7) Ferrous channel raceways and fittings protected from corrosion solely by enamel shall be permitted only indoors.

384.12 Uses Not Permitted. Strut-type channel raceways shall not be used as follows:

(1) Where concealed.
(2) Ferrous channel raceways and fittings protected from corrosion solely by enamel shall not be permitted where subject to severe corrosive influences.

384.21 Size of Conductors. No conductor larger than that for which the raceway is listed shall be installed in strut-type channel raceways.

384.22 Number of Conductors. The number of conductors or cables permitted in strut-type channel raceways shall not exceed the percentage fill using Table 384.22 and applicable cross-sectional area of specific types and sizes of wire given in the tables in Chapter 9.

The adjustment factors of 310.15(B)(3)(a) shall not apply to conductors installed in strut-type channel raceways where all of the following conditions are met:

(1) The cross-sectional area of the raceway exceeds 2500 mm² (4 in.²).
(2) The current-carrying conductors do not exceed 30 in number.
(3) The sum of the cross-sectional areas of all contained conductors does not exceed 20 percent of the interior cross-sectional area of the strut-type channel raceways.

384.30 Securing and Supporting.

(A) Surface Mount. A surface mount strut-type channel raceway shall be secured to the mounting surface with retention straps external to the channel at intervals not exceeding 3 m (10 ft) and within 900 mm (3 ft) of each outlet box, cabinet, junction box, or other channel raceway termination.

(B) Suspension Mount. Strut-type channel raceways shall be permitted to be suspension mounted in air with identified methods at intervals not to exceed 3 m (10 ft) and within 900 mm (3 ft) of channel raceway terminations and ends.

384.56 Splices and Taps. Splices and taps shall be permitted in raceways that are accessible after installation by having a removable cover. The conductors, including splices and taps, shall not fill the raceway to more than 75 percent of its area at that point. All splices and taps shall be made by approved methods.

384.60 Grounding. Strut-type channel raceway enclosures providing a transition to or from other wiring methods shall have a means for connecting an equipment grounding conductor. Strut-type channel raceways shall be permitted as an equipment grounding conductor in accordance with 250.118(13). Where a snap-fit metal cover for strut-type channel raceways is used to achieve electrical continuity in accordance with the listing, this cover shall not be permitted as the means for providing electrical continuity for a receptacle mounted in the cover.

Part III. Construction Specifications

384.100 Construction. Strut-type channel raceways and their accessories shall be of a construction that distinguishes them from other raceways. Raceways and their elbows, couplings, and other fittings shall be designed such that the sections can be electrically and mechanically coupled together and installed without subjecting the wires to abrasion. They shall comply with 384.100(A), (B), and (C).

(A) Material. Raceways and accessories shall be formed of steel, stainless steel, or aluminum.

(B) Corrosion Protection. Steel raceways and accessories shall be protected against corrosion by galvanizing or by an organic coating.

> Informational Note: Enamel and PVC coatings are examples of organic coatings that provide corrosion protection.

(C) Cover. Covers of strut-type channel raceways shall be either metal or nonmetallic.

384.120 Marking. Each length of strut-type channel raceway shall be clearly and durably identified as required in the first sentence of 110.21(A).

Table 384.22 Channel Size and Inside Cross-Sectional Area

Size Channel	Area		40% Area*		25% Area†	
	in.²	mm²	in.²	mm²	in.²	mm²
1⅝ × ¹⁵⁄₁₆	0.887	572	0.355	229	0.222	143
1⅝ × 1	1.151	743	0.460	297	0.288	186
1⅝ × 1⅜	1.677	1076	0.671	433	0.419	270
1⅝ × 1⅝	2.028	1308	0.811	523	0.507	327
1⅝ × 2⁷⁄₁₆	3.169	2045	1.267	817	0.792	511
1⅝ × 3¼	4.308	2780	1.723	1112	1.077	695
1½ × ¾	0.849	548	0.340	219	0.212	137
1½ × 1½	1.828	1179	0.731	472	0.457	295
1½ × 1⅞	2.301	1485	0.920	594	0.575	371
1½ × 3	3.854	2487	1.542	995	0.964	622

*Raceways with external joiners shall use a 40 percent wire fill calculation to determine the number of conductors permitted.
†Raceways with internal joiners shall use a 25 percent wire fill calculation to determine the number of conductors permitted.

ARTICLE 386
Surface Metal Raceways

Part I. General

386.1 Scope. This article covers the use, installation, and construction specifications for surface metal raceways and associated fittings.

386.2 Definition.

Surface Metal Raceway. A metal raceway that is intended to be mounted to the surface of a structure, with associated couplings, connectors, boxes, and fittings for the installation of electrical conductors.

386.6 Listing Requirements. Surface metal raceway and associated fittings shall be listed.

Part II. Installation

386.10 Uses Permitted. The use of surface metal raceways shall be permitted in the following:

(1) In dry locations.
(2) In Class I, Division 2 hazardous (classified) locations as permitted in 501.10(B)(3).
(3) Under raised floors, as permitted in 645.5(E)(2).
(4) Extension through walls and floors. Surface metal raceway shall be permitted to pass transversely through dry walls, dry partitions, and dry floors if the length passing through is unbroken. Access to the conductors shall be maintained on both sides of the wall, partition, or floor.

386.12 Uses Not Permitted. Surface metal raceways shall not be used in the following:

(1) Where subject to severe physical damage, unless otherwise approved
(2) Where the voltage is 300 volts or more between conductors, unless the metal has a thickness of not less than 1.02 mm (0.040 in.) nominal
(3) Where subject to corrosive vapors
(4) In hoistways
(5) Where concealed, except as permitted in 386.10

386.21 Size of Conductors. No conductor larger than that for which the raceway is designed shall be installed in surface metal raceway.

386.22 Number of Conductors or Cables. The number of conductors or cables installed in surface metal raceway shall not be greater than the number for which the raceway is designed. Cables shall be permitted to be installed where such use is not prohibited by the respective cable articles.

The adjustment factors of 310.15(B)(3)(a) shall not apply to conductors installed in surface metal raceways where all of the following conditions are met:

(1) The cross-sectional area of the raceway exceeds 2500 mm² (4 in.²).
(2) The current-carrying conductors do not exceed 30 in number.
(3) The sum of the cross-sectional areas of all contained conductors does not exceed 20 percent of the interior cross-sectional area of the surface metal raceway.

386.30 Securing and Supporting. Surface metal raceways and associated fittings shall be supported in accordance with the manufacturer's installation instructions.

386.56 Splices and Taps. Splices and taps shall be permitted in surface metal raceways having a removable cover that is accessible after installation. The conductors, including splices and taps, shall not fill the raceway to more than 75 percent of its area at that point. Splices and taps in surface metal raceways without removable covers shall be made only in boxes. All splices and taps shall be made by approved methods.

Taps of Type FC cable installed in surface metal raceway shall be made in accordance with 322.56(B).

386.60 Grounding. Surface metal raceway enclosures providing a transition from other wiring methods shall have a means for connecting an equipment grounding conductor.

386.70 Combination Raceways. When combination surface metal raceways are used for both signaling and for lighting and power circuits, the different systems shall be run in separate compartments identified by stamping, imprinting, or color coding of the interior finish.

Part III. Construction Specifications

386.100 Construction. Surface metal raceways shall be of such construction as will distinguish them from other raceways. Surface metal raceways and their elbows, couplings, and similar fittings shall be designed so that the sections can be electrically and mechanically coupled together and installed without subjecting the wires to abrasion.

Where covers and accessories of nonmetallic materials are used on surface metal raceways, they shall be identified for such use.

386.120 Marking. Each length of surface metal raceway shall be clearly and durably identified as required in the first sentence of 110.21(A).

ARTICLE 388
Surface Nonmetallic Raceways

Part I. General

388.1 Scope. This article covers the use, installation, and construction specifications for surface nonmetallic raceways and associated fittings.

388.2 Definition.

Surface Nonmetallic Raceway. A nonmetallic raceway that is intended to be mounted to the surface of a structure, with associated couplings, connectors, boxes, and fittings for the installation of electrical conductors.

388.6 Listing Requirements. Surface nonmetallic raceway and associated fittings shall be listed.

Part II. Installation

388.10 Uses Permitted. Surface nonmetallic raceways shall be permitted as follows:

(1) The use of surface nonmetallic raceways shall be permitted in dry locations.
(2) Extension through walls and floors shall be permitted. Surface nonmetallic raceway shall be permitted to pass transversely through dry walls, dry partitions, and dry floors if the length passing through is unbroken. Access to the conductors shall be maintained on both sides of the wall, partition, or floor.

388.12 Uses Not Permitted. Surface nonmetallic raceways shall not be used in the following:

(1) Where concealed, except as permitted in 388.10(2)

(2) Where subject to severe physical damage

(3) Where the voltage is 300 volts or more between conductors, unless listed for higher voltage

(4) In hoistways

(5) In any hazardous (classified) location, except as permitted by other articles in this *Code*

(6) Where subject to ambient temperatures exceeding those for which the nonmetallic raceway is listed

(7) For conductors whose insulation temperature limitations would exceed those for which the nonmetallic raceway is listed

388.21 Size of Conductors. No conductor larger than that for which the raceway is designed shall be installed in surface nonmetallic raceway.

388.22 Number of Conductors or Cables. The number of conductors or cables installed in surface nonmetallic raceway shall not be greater than the number for which the raceway is designed. Cables shall be permitted to be installed where such use is not prohibited by the respective cable articles.

388.30 Securing and Supporting. Surface nonmetallic raceways and associated fittings shall be supported in accordance with the manufacturer's installation instructions.

388.56 Splices and Taps. Splices and taps shall be permitted in surface nonmetallic raceways having a cover capable of being opened in place that is accessible after installation. The conductors, including splices and taps, shall not fill the raceway to more than 75 percent of its area at that point. Splices and taps in surface nonmetallic raceways without covers capable of being opened in place shall be made only in boxes. All splices and taps shall be made by approved methods.

388.60 Grounding. Where equipment grounding is required, a separate equipment grounding conductor shall be installed in the raceway.

388.70 Combination Raceways. When combination surface nonmetallic raceways are used both for signaling and for lighting and power circuits, the different systems shall be run in separate compartments identified by stamping, imprinting, or color coding of the interior finish.

Part III. Construction Specifications

388.100 Construction. Surface nonmetallic raceways shall be of such construction as will distinguish them from other raceways. Surface nonmetallic raceways and their elbows, couplings, and similar fittings shall be designed so that the sections can be mechanically coupled together and installed without subjecting the wires to abrasion.

Surface nonmetallic raceways and fittings are made of suitable nonmetallic material that is resistant to moisture and chemical atmospheres. It shall also be flame retardant, resistant to impact and crushing, resistant to distortion from heat under conditions likely to be encountered in service, and resistant to low-temperature effects.

388.120 Marking. Surface nonmetallic raceways that have limited smoke-producing characteristics shall be permitted to be so identified. Each length of surface nonmetallic raceway shall be clearly and durably identified as required in the first sentence of 110.21(A).

ARTICLE 390
Underfloor Raceways

390.1 Scope. This article covers the use and installation requirements for underfloor raceways.

390.2 Definition.

Underfloor Raceway. A raceway and associated components designed and intended for installation beneath or flush with the surface of a floor for the installation of cables and electrical conductors.

390.3 Use.

(A) Permitted. The installation of underfloor raceways shall be permitted beneath the surface of concrete or other flooring material or in office occupancies where laid flush with the concrete floor and covered with linoleum or equivalent floor covering.

(B) Not Permitted. Underfloor raceways shall not be installed (1) where subject to corrosive vapors or (2) in any hazardous (classified) locations, except as permitted by 504.20 and in Class I, Division 2 locations as permitted in 501.10(B)(3). Unless made of a material approved for the condition or unless corrosion protection approved for the condition is provided, metal underfloor raceways, junction boxes, and fittings shall not be installed in concrete or in areas subject to severe corrosive influences.

390.4 Covering. Raceway coverings shall comply with 390.4(A) through (D).

(A) Raceways Not over 100 mm (4 in.) Wide. Half-round and flat-top raceways not over 100 mm (4 in.) in width shall have not less than 20 mm (¾ in.) of concrete or wood above the raceway.

Exception: As permitted in 390.4(C) and (D) for flat-top raceways.

(B) Raceways over 100 mm (4 in.) Wide But Not over 200 mm (8 in.) Wide. Flat-top raceways over 100 mm (4 in.) but not over 200 mm (8 in.) wide with a minimum of 25 mm (1 in.) spacing between raceways shall be covered with concrete to a depth of not less than 25 mm (1 in.). Raceways spaced less than 25 mm (1 in.) apart shall be covered with concrete to a depth of 38 mm (1½ in.).

(C) Trench-Type Raceways Flush with Concrete. Trench-type flush raceways with removable covers shall be permitted to be laid flush with the floor surface. Such approved raceways shall be designed so that the cover plates provide adequate mechanical protection and rigidity equivalent to junction box covers.

(D) Other Raceways Flush with Concrete. In office occupancies, approved metal flat-top raceways, if not over 100 mm (4 in.) in width, shall be permitted to be laid flush with the concrete floor surface, provided they are covered with substantial linoleum that is not less than 1.6 mm (¹⁄₁₆ in.) thick or with equivalent floor covering. Where more than one and not more than three single raceways are each installed flush with the concrete, they shall be contiguous with each other and joined to form a rigid assembly.

390.5 Size of Conductors. No conductor larger than that for which the raceway is designed shall be installed in underfloor raceways.

390.6 Maximum Number of Conductors in Raceway. The combined cross-sectional area of all conductors or cables shall not exceed 40 percent of the interior cross-sectional area of the raceway.

390.7 Splices and Taps. Splices and taps shall be made only in junction boxes.

For the purposes of this section, so-called loop wiring (continuous, unbroken conductor connecting the individual outlets) shall not be considered to be a splice or tap.

Exception: Splices and taps shall be permitted in trench-type flush raceway having a removable cover that is accessible after installation. The conductors, including splices and taps, shall not fill more than 75 percent of the raceway area at that point.

390.8 Discontinued Outlets. When an outlet is abandoned, discontinued, or removed, the sections of circuit conductors supplying the outlet shall be removed from the raceway. No splices or reinsulated conductors, such as would be the case with abandoned outlets on loop wiring, shall be allowed in raceways.

390.9 Laid in Straight Lines. Underfloor raceways shall be laid so that a straight line from the center of one junction box to the center of the next junction box coincides with the centerline of the raceway system. Raceways shall be firmly held in place to prevent disturbing this alignment during construction.

390.10 Markers at Ends. A suitable marker shall be installed at or near each end of each straight run of raceways to locate the last insert.

390.11 Dead Ends. Dead ends of raceways shall be closed.

390.13 Junction Boxes. Junction boxes shall be leveled to the floor grade and sealed to prevent the free entrance of water or concrete. Junction boxes used with metal raceways shall be metal and shall be electrically continuous with the raceways.

390.14 Inserts. Inserts shall be leveled and sealed to prevent the entrance of concrete. Inserts used with metal raceways shall be metal and shall be electrically continuous with the raceway. Inserts set in or on fiber raceways before the floor is laid shall be mechanically secured to the raceway. Inserts set in fiber raceways after the floor is laid shall be screwed into the raceway. When cutting through the raceway wall and setting inserts, chips and other dirt shall not be allowed to remain in the raceway, and tools shall be used that are designed so as to prevent the tool from entering the raceway and damaging conductors that may be in place.

390.15 Connections to Cabinets and Wall Outlets. Connections from underfloor raceways to distribution centers and wall outlets shall be made by approved fittings or by any of the wiring methods in Chapter 3, where installed in accordance with the provisions of the respective articles.

390.17 Ampacity of Conductors. The ampacity adjustment factors, in 310.15(B)(3), shall apply to conductors installed in underfloor raceways.

ARTICLE 392
Cable Trays

Part I. General

392.1 Scope. This article covers cable tray systems, including ladder, ventilated trough, ventilated channel, solid bottom, and other similar structures.

Informational Note: For further information on cable trays, see ANSI/NEMA–VE 1-2009, *Metal Cable Tray Systems*; NECA/NEMA 105-2015, *Standard for Installing Metal Cable Tray Systems*; and NEMA–FG 1-1993, *Nonmetallic Cable Tray Systems*.

392.2 Definition.

Cable Tray System. A unit or assembly of units or sections and associated fittings forming a structural system used to securely fasten or support cables and raceways.

Part II. Installation

392.10 Uses Permitted. Cable tray shall be permitted to be used as a support system for service conductors, feeders, branch circuits, communications circuits, control circuits, and signaling circuits. Cable tray installations shall not be limited to industrial establishments. Where exposed to direct rays of the sun, insulated conductors and jacketed cables shall be identified as being sunlight resistant. Cable trays and their associated fittings shall be identified for the intended use.

(A) Wiring Methods. The wiring methods in Table 392.10(A) shall be permitted to be installed in cable tray systems under the conditions described in their respective articles and sections.

(B) In Industrial Establishments. The wiring methods in Table 392.10(A) shall be permitted to be used in any industrial establishment under the conditions described in their respective articles. In industrial establishments only, where conditions of maintenance and supervision ensure that only qualified persons service the installed cable tray system, any of the cables in 392.10(B)(1) and (B)(2) shall be permitted to be installed in ladder, ventilated trough, solid bottom, or ventilated channel cable trays.

(1) Single-conductor cables shall be permitted to be installed in accordance with (B)(1)(a) through (B)(1)(c).

(a) Single-conductor cable shall be 1/0 AWG or larger and shall be of a type listed and marked on the surface for use in cable trays. Where 1/0 AWG through 4/0 AWG single-conductor cables are installed in ladder cable tray, the maximum allowable rung spacing for the ladder cable tray shall be 225 mm (9 in.).

(b) Welding cables shall comply with the provisions of Article 630, Part IV.

(c) Single conductors used as equipment grounding conductors shall be insulated, covered, or bare, and they shall be 4 AWG or larger.

(2) Single- and multiconductor medium voltage cables shall be Type MV cable. Single conductors shall be installed in accordance with 392.10(B)(1).

Table 392.10(A) Wiring Methods

Wiring Method	Article
Armored cable: Type AC	320
CATV cables	820
Class 2 and Class 3 cables	725
Communications cables	800
Communications raceways	725, 770, and 800
Electrical metallic tubing: Type EMT	358
Electrical nonmetallic tubing: Type ENT	362
Fire alarm cables	760
Flexible metal conduit: Type FMC	348
Flexible metallic tubing: Type FMT	360
Instrumentation tray cable: Type ITC	727
Intermediate metal conduit: Type IMC	342
Liquidtight flexible metal conduit: Type LFMC	350
Liquidtight flexible nonmetallic conduit: Type LFNC	356
Metal-clad cable: Type MC	330
Mineral-insulated, metal-sheathed cable: Type MI	332
Network-powered broadband communications cables	830
Nonmetallic-sheathed cable: Types NM, NMC, and NMS	334
Non–power-limited fire alarm cable	760
Optical fiber cables	770
Other factory-assembled, multiconductor control, signal, or power cables that are specifically approved for installation in cable trays	
Power and control tray cable: Type TC	336
Power-limited fire alarm cable	760
Power-limited tray cable	725
Rigid metal conduit: Type RMC	344
Rigid polyvinyl chloride conduit: Type PVC	352
Reinforced thermosetting resin conduit: Type RTRC	355
Service-entrance cable: Types SE and USE	338
Underground feeder and branch-circuit cable: Type UF	340

(C) Hazardous (Classified) Locations. Cable trays in hazardous (classified) locations shall contain only the cable types and raceways permitted by other articles in this *Code*.

(D) Nonmetallic Cable Tray. In addition to the uses permitted elsewhere in 392.10, nonmetallic cable tray shall be permitted in corrosive areas and in areas requiring voltage isolation.

(E) Airfield Lighting Cable Tray. In airports where maintenance and supervision conditions ensure that only qualified persons can access, install, or service the cable, airfield lighting cable used in series circuits that are rated up to 5000 volts and are powered by constant current regulators shall be permitted to be installed in cable trays.

Informational Note: Federal Aviation Administration (FAA) Advisory Circulars (ACs) provide additional practices and methods for airport lighting.

392.12 Uses Not Permitted. Cable tray systems shall not be used in hoistways or where subject to severe physical damage.

392.18 Cable Tray Installation.

(A) Complete System. Cable trays shall be installed as a complete system. Field bends or modifications shall be so made that the electrical continuity of the cable tray system and support for the cables is maintained. Cable tray systems shall be permitted to have mechanically discontinuous segments between cable tray runs or between cable tray runs and equipment.

(B) Completed Before Installation. Each run of cable tray shall be completed before the installation of cables.

(C) Covers. In portions of runs where additional protection is required, covers or enclosures providing the required protection shall be of a material that is compatible with the cable tray.

(D) Through Partitions and Walls. Cable trays shall be permitted to extend transversely through partitions and walls or vertically through platforms and floors in wet or dry locations where the installations, complete with installed cables, are made in accordance with the requirements of 300.21.

(E) Exposed and Accessible. Cable trays shall be exposed and accessible, except as permitted by 392.18(D).

(F) Adequate Access. Sufficient space shall be provided and maintained about cable trays to permit adequate access for installing and maintaining the cables.

(G) Raceways, Cables, Boxes, and Conduit Bodies Supported from Cable Tray Systems. In industrial facilities where conditions of maintenance and supervision ensure that only qualified persons service the installation and where the cable tray systems are designed and installed to support the load, such systems shall be permitted to support raceways and cables, and boxes and conduit bodies covered in 314.1. For raceways terminating at the tray, a listed cable tray clamp or adapter shall be used to securely fasten the raceway to the cable tray system. Additional supporting and securing of the raceway shall be in accordance with the requirements of the appropriate raceway article. For raceways or cables running parallel to and attached to the bottom or side of a cable tray system, fastening and supporting shall be in accordance with the requirements of the appropriate raceway or cable article.

For boxes and conduit bodies attached to the bottom or side of a cable tray system, fastening and supporting shall be in accordance with the requirements of 314.23.

(H) Marking. Cable trays containing conductors rated over 600 volts shall have a permanent, legible warning notice carrying the wording "DANGER — HIGH VOLTAGE — KEEP AWAY" placed in a readily visible position on all cable trays, with the spacing of warning notices not to exceed 3 m (10 ft). The danger marking(s) or labels shall comply with 110.21(B).

Exception: Where not accessible (as applied to equipment), in industrial establishments where the conditions of maintenance and supervision ensure that only qualified persons service the installation, cable tray system warning notices shall be located where necessary for the installation to ensure safe maintenance and operation.

392.20 Cable and Conductor Installation.

(A) Multiconductor Cables Operating at 1000 Volts or Less. Multiconductor cables operating at 1000 volts or less shall be permitted to be installed in the same tray.

(B) Cables Operating at Over 1000 Volts. Cables operating at over 1000 volts and those operating at 1000 volts or less installed in the same cable tray shall comply with either of the following:

(1) The cables operating at over 1000 volts are Type MC.
(2) The cables operating at over 1000 volts are separated from the cables operating at 1000 volts or less by a solid fixed barrier of a material compatible with the cable tray.

(C) Connected in Parallel. Where single conductor cables comprising each phase, neutral, or grounded conductor of an alternating-current circuit are connected in parallel as permitted in 310.10(H), the conductors shall be installed in groups consisting of not more than one conductor per phase, neutral, or grounded conductor to prevent current imbalance in the paralleled conductors due to inductive reactance.

Single conductors shall be securely bound in circuit groups to prevent excessive movement due to fault-current magnetic forces unless single conductors are cabled together, such as triplexed assemblies.

(D) Single Conductors. Where any of the single conductors installed in ladder or ventilated trough cable trays are 1/0 through 4/0 AWG, all single conductors shall be installed in a single layer. Conductors that are bound together to comprise each circuit group shall be permitted to be installed in other than a single layer.

392.22 Number of Conductors or Cables.

(A) Number of Multiconductor Cables, Rated 2000 Volts or Less, in Cable Trays. The number of multiconductor cables, rated 2000 volts or less, permitted in a single cable tray shall not exceed the requirements of this section. The conductor sizes shall apply to both aluminum and copper conductors. Where dividers are used, fill calculations shall apply to each divided section of the cable tray.

(1) Ladder or Ventilated Trough Cable Trays Containing Any Mixture of Cables. Where ladder or ventilated trough cable trays contain multiconductor power or lighting cables, or any mixture of multiconductor power, lighting, control, and signal cables, the maximum number of cables shall conform to the following:

(a)   Where all of the cables are 4/0 AWG or larger, the sum of the diameters of all cables shall not exceed the cable tray width, and the cables shall be installed in a single layer. Where the cable ampacity is determined according to 392.80(A)(1)(c), the cable tray width shall not be less than the sum of the diameters of the cables and the sum of the required spacing widths between the cables.

(b)   Where all of the cables are smaller than 4/0 AWG, the sum of the cross-sectional areas of all cables shall not exceed the maximum allowable cable fill area in Column 1 of Table 392.22(A) for the appropriate cable tray width.

(c)   Where 4/0 AWG or larger cables are installed in the same cable tray with cables smaller than 4/0 AWG, the sum of the cross-sectional areas of all cables smaller than 4/0 AWG shall not exceed the maximum allowable fill area resulting from the calculation in Column 2 of Table 392.22(A) for the appropriate cable tray width. The 4/0 AWG and larger cables shall be installed in a single layer, and no other cables shall be placed on them.

(2) Ladder or Ventilated Trough Cable Trays Containing Multiconductor Control and/or Signal Cables Only. Where a ladder or ventilated trough cable tray having a usable inside depth of 150 mm (6 in.) or less contains multiconductor control and/or signal cables only, the sum of the cross-sectional areas of all cables at any cross section shall not exceed 50 percent of the interior cross-sectional area of the cable tray. A depth of 150 mm (6 in.) shall be used to calculate the allowable interior cross-sectional area of any cable tray that has a usable inside depth of more than 150 mm (6 in.).

(3) Solid Bottom Cable Trays Containing Any Mixture of Cables. Where solid bottom cable trays contain multiconductor power or lighting cables, or any mixture of multiconductor power, lighting, control, and signal cables, the maximum number of cables shall conform to the following:

(a)   Where all of the cables are 4/0 AWG or larger, the sum of the diameters of all cables shall not exceed 90 percent of the cable tray width, and the cables shall be installed in a single layer.

(b)   Where all of the cables are smaller than 4/0 AWG, the sum of the cross-sectional areas of all cables shall not exceed the maximum allowable cable fill area in Column 3 of Table 392.22(A) for the appropriate cable tray width.

(c)   Where 4/0 AWG or larger cables are installed in the same cable tray with cables smaller than 4/0 AWG, the sum of the cross-sectional areas of all cables smaller than 4/0 AWG shall not exceed the maximum allowable fill area resulting from the computation in Column 4 of Table 392.22(A) for the appropriate cable tray width. The 4/0 AWG and larger cables shall be installed in a single layer, and no other cables shall be placed on them.

(4) Solid Bottom Cable Tray Containing Multiconductor Control and/or Signal Cables Only. Where a solid bottom cable tray having a usable inside depth of 150 mm (6 in.) or less contains multiconductor control and/or signal cables only, the sum of the cross sectional areas of all cables at any cross section shall not exceed 40 percent of the interior cross-sectional area of the cable tray. A depth of 150 mm (6 in.) shall be used to calculate the allowable interior cross-sectional area of any cable tray that has a usable inside depth of more than 150 mm (6 in.).

(5) Ventilated Channel Cable Trays Containing Multiconductor Cables of Any Type. Where ventilated channel cable trays contain multiconductor cables of any type, the following shall apply:

(a)   Where only one multiconductor cable is installed, the cross-sectional area shall not exceed the value specified in Column 1 of Table 392.22(A)(5).

(b)   Where more than one multiconductor cable is installed, the sum of the cross-sectional area of all cables shall not exceed the value specified in Column 2 of Table 392.22(A)(5).

(6) Solid Channel Cable Trays Containing Multiconductor Cables of Any Type. Where solid channel cable trays contain multiconductor cables of any type, the following shall apply:

(a)   Where only one multiconductor cable is installed, the cross-sectional area of the cable shall not exceed the value specified in Column 1 of Table 392.22(A)(6).

(b)   Where more than one multiconductor cable is installed, the sum of the cross-sectional area of all cable shall not exceed the value specified in Column 2 of Table 392.22(A)(6).

(B) Number of Single-Conductor Cables, Rated 2000 Volts or Less, in Cable Trays. The number of single conductor cables, rated 2000 volts or less, permitted in a single cable tray section shall not exceed the requirements of this section. The single conductors, or conductor assemblies, shall be evenly distributed across the cable tray. The conductor sizes shall apply to both aluminum and copper conductors.

(1) Ladder or Ventilated Trough Cable Trays. Where ladder or ventilated trough cable trays contain single-conductor cables, the maximum number of single conductors shall conform to the following:

(a) Where all of the cables are 1000 kcmil or larger, the sum of the diameters of all single-conductor cables shall not exceed the cable tray width, and the cables shall be installed in a single layer. Conductors that are bound together to comprise each circuit group shall be permitted to be installed in other than a single layer.

(b) Where all of the cables are from 250 kcmil through 900 kcmil, the sum of the cross-sectional areas of all single-conductor cables shall not exceed the maximum allowable cable fill area in Column 1 of Table 392.22(B)(1) for the appropriate cable tray width.

(c) Where 1000 kcmil or larger single-conductor cables are installed in the same cable tray with single-conductor cables smaller than 1000 kcmil, the sum of the cross sectional areas of all cables smaller than 1000 kcmil shall not exceed the maximum allowable fill area resulting from the computation in Column 2 of Table 392.22(B)(1) for the appropriate cable tray width.

(d) Where any of the single conductor cables are 1/0 through 4/0 AWG, the sum of the diameters of all single conductor cables shall not exceed the cable tray width.

Table 392.22(A)(5) Allowable Cable Fill Area for Multiconductor Cables in Ventilated Channel Cable Trays for Cables Rated 2000 Volts or Less

Inside Width of Cable Tray		Maximum Allowable Fill Area for Multiconductor Cables			
		Column 1 One Cable		Column 2 More Than One Cable	
mm	in.	mm²	in.²	mm²	in.²
75	3	1500	2.3	850	1.3
100	4	2900	4.5	1600	2.5
150	6	4500	7.0	2450	3.8

Table 392.22(A)(6) Allowable Cable Fill Area for Multiconductor Cables in Solid Channel Cable Trays for Cables Rated 2000 Volts or Less

Inside Width of Cable Tray		Column 1 One Cable		Column 2 More Than One Cable	
mm	in.	mm²	in.²	mm²	in.²
50	2	850	1.3	500	0.8
75	3	1300	2.0	700	1.1
100	4	2400	3.7	1400	2.1
150	6	3600	5.5	2100	3.2

Table 392.22(A) Allowable Cable Fill Area for Multiconductor Cables in Ladder, Ventilated Trough, or Solid Bottom Cable Trays for Cables Rated 2000 Volts or Less

Inside Width of Cable Tray		Maximum Allowable Fill Area for Multiconductor Cables							
		Ladder or Ventilated Trough or Wire Mesh Cable Trays, 392.22(A)(1)				Solid Bottom Cable Trays, 392.22(A)(3)			
		Column 1 Applicable for 392.22(A)(1)(b) Only		Column 2[a] Applicable for 392.22(A)(1)(c) Only		Column 3 Applicable for 392.22(A)(3)(b) Only		Column 4[a] Applicable for 392.22(A)(3)(c) Only	
mm	in.	mm²	in.²	mm²	in.²	mm²	in.²	mm²	in.²
50	2.0	1,500	2.5	1,500 − (30 Sd)[b]	2.5 − (1.2 Sd)[b]	1,200	2.0	1,200 − (25 Sd)[b]	2.0 − Sd[b]
100	4.0	3,000	4.5	3,000 − (30 Sd)[b]	4.5 − (1.2 Sd)	2,300	3.5	2,300 − (25 Sd)	3.5 − Sd
150	6.0	4,500	7.0	4,500 − (30 Sd)[b]	7 − (1.2 Sd)	3,500	5.5	3,500 − (25 Sd)[b]	5.5–Sd
200	8.0	6,000	9.5	6,000 − (30 Sd)[b]	9.5 − (1.2 Sd)	4,500	7.0	4,500 − (25 Sd)	7.0 − Sd
225	9.0	6,800	10.5	6,800 − (30 Sd)	10.5 − (1.2 Sd)	5,100	8.0	5,100 − (25 Sd)	8.0 − Sd
300	12.0	9,000	14.0	9,000 − (30 Sd)	14 − (1.2 Sd)	7,100	11.0	7,100 − (25 Sd)	11.0 − Sd
400	16.0	12,000	18.5	12,000 − (30 Sd)	18.5 − (1.2 Sd)	9,400	14.5	9,400 − (25 Sd)	14.5 − Sd
450	18.0	13,500	21.0	13,500 − (30 Sd)	21 − (1.2 Sd)	10,600	16.5	10,600 − (25 Sd)	16.5 − Sd
500	20.0	15,000	23.5	15,000 − (30 Sd)	23.5 − (1.2 Sd)	11,800	18.5	11,800 − (25 Sd)	18.5 − Sd
600	24.0	18,000	28.0	18,000 − (30 Sd)	28 − (1.2 Sd)	14,200	22.0	14,200 − (25 Sd)	22.0 − Sd
750	30.0	22,500	35.0	22,500 − (30 Sd)	35 − (1.2 Sd)	17,700	27.5	17,700 − (25 Sd)	27.5 − Sd
900	36.0	27,000	42.0	27,000 − (30 Sd)	42 − (1.2 Sd)	21,300	33.0	21,300 − (25 Sd)	33.0 − Sd

[a]The maximum allowable fill areas in Columns 2 and 4 shall be calculated. For example, the maximum allowable fill in mm² for a 150-mm wide cable tray in Column 2 shall be 4500 minus (30 multiplied by Sd) [the maximum allowable fill, in square inches, for a 6-in. wide cable tray in Column 2 shall be 7 minus (1.2 multiplied by Sd)].

[b]The term Sd in Columns 2 and 4 is equal to the sum of the diameters, in mm, of all cables 107.2 mm (in inches, of all 4/0 AWG) and larger multiconductor cables in the same cable tray with smaller cables.

(2) Ventilated Channel Cable Trays. Where 50 mm (2 in.), 75 mm (3 in.), 100 mm (4 in.), or 150 mm (6 in.) wide ventilated channel cable trays contain single-conductor cables, the sum of the diameters of all single conductors shall not exceed the inside width of the channel.

(C) Number of Type MV and Type MC Cables (2001 Volts or Over) in Cable Trays. The number of cables rated 2001 volts or over permitted in a single cable tray shall not exceed the requirements of this section.

The sum of the diameters of single-conductor and multiconductor cables shall not exceed the cable tray width, and the cables shall be installed in a single layer. Where single conductor cables are triplexed, quadruplexed, or bound together in circuit groups, the sum of the diameters of the single conductors shall not exceed the cable tray width, and these groups shall be installed in single layer arrangement.

392.30 Securing and Supporting.

(A) Cable Trays. Cable trays shall be supported at intervals in accordance with the installation instructions.

(B) Cables and Conductors. Cables and conductors shall be secured to and supported by the cable tray system in accordance with (1), (2) and (3) as applicable:

(1) In other than horizontal runs, the cables shall be fastened securely to transverse members of the cable runs.

(2) Supports shall be provided to prevent stress on cables where they enter raceways from cable tray systems.

(3) The system shall provide for the support of cables and raceway wiring methods in accordance with their corresponding articles. Where cable trays support individual conductors and where the conductors pass from one cable tray to another, or from a cable tray to raceway(s) or from a cable tray to equipment where the conductors are terminated, the distance between the cable trays or between the cable tray and the raceway(s) or the equipment shall not exceed 1.8 m (6 ft). The conductors shall be secured to the cable tray(s) at the transition, and they shall be protected, by guarding or by location, from physical damage.

392.46 Bushed Conduit and Tubing. A box shall not be required where cables or conductors are installed in bushed conduit and tubing used for support or for protection against physical damage.

392.56 Cable Splices. Cable splices made and insulated by approved methods shall be permitted to be located within a cable tray, provided they are accessible. Splices shall be permitted to project above the side rails where not subject to physical damage.

392.60 Grounding and Bonding.

(A) Metal Cable Trays. Metal cable trays shall be permitted to be used as equipment grounding conductors where continuous maintenance and supervision ensure that qualified persons service the installed cable tray system and the cable tray complies with provisions of this section. Metal cable trays that support electrical conductors shall be grounded as required for conductor enclosures in accordance with 250.96 and Part IV of Article 250. Metal cable trays containing only non-power conductors shall be electrically continuous through approved connections or the use of a bonding jumper.

Informational Note: Examples of non-power conductors include nonconductive optical fiber cables and Class 2 and Class 3 Remote Control Signaling and Power Limiting Circuits.

Table 392.22(B)(1) Allowable Cable Fill Area for Single-Conductor Cables in Ladder, Ventilated Trough, or Wire Mesh Cable Trays for Cables Rated 2000 Volts or Less

		Maximum Allowable Fill Area for Single-Conductor Cables in Ladder, Ventilated Trough, or Wire Mesh Cable Trays			
Inside Width of Cable Tray		Column 1 Applicable for 392.22(B)(1)(b) Only		Column 2[a] Applicable for 392.22(B)(1)(c) Only	
mm	in.	mm²	in.²	mm²	in.²
50	2	1,400	2.0	1,400 – (28 Sd)[b]	2.0 – (1.1 Sd)[b]
100	4	2,800	4.5	2,800 – (28 Sd)	4.5 – (1.1 Sd)
150	6	4,200	6.5	4,200 – (28 Sd)[b]	6.5 – (1.1 Sd)[b]
200	8	5,600	8.5	5,600 – (28 Sd)	8.5 – (1.1 Sd)
225	9	6,100	9.5	6,100 – (28 Sd)	9.5 – (1.1 Sd)
300	12	8,400	13.0	8,400 – (28 Sd)	13.0 – (1.1 Sd)
400	16	11,200	17.5	11,200 – (28 Sd)	17.5 – (1.1 Sd)
450	18	12,600	19.5	12,600 – (28 Sd)	19.5 – (1.1 Sd)
500	20	14,000	21.5	14,000 – (28 Sd)	21.5 – (1.1 Sd)
600	24	16,800	26.0	16,800 – (28 Sd)	26.0 – (1.1 Sd)
750	30	21,000	32.5	21,000 – (28 Sd)	32.5 – (1.1 Sd)
900	36	25,200	39.0	25,200 – (28 Sd)	39.0 – (1.1 Sd)

[a]The maximum allowable fill areas in Column 2 shall be calculated. For example, the maximum allowable fill, in mm², for a 150-mm wide cable tray in Column 2 shall be 4200 minus (28 multiplied by Sd) [the maximum allowable fill, in square inches, for a 6-in. wide cable tray in Column 2 shall be 6.5 minus (1.1 multiplied by Sd)].

[b]The term *Sd* in Column 2 is equal to the sum of the diameters, in mm, of all cables 507 mm² (in inches, of all 1000 kcmil) and larger single-conductor cables in the same cable tray with small cables.

Table 392.60(A) Metal Area Requirements for Cable Trays Used as Equipment Grounding Conductor

Maximum Fuse Ampere Rating, Circuit Breaker Ampere Trip Setting, or Circuit Breaker Protective Relay Ampere Trip Setting for Ground-Fault Protection of Any Cable Circuit in the Cable Tray System	Minimum Cross-Sectional Area of Metal[a]			
	Steel Cable Trays		Aluminum Cable Trays	
	mm²	in.²	mm²	in.²
60	129	0.20	129	0.20
100	258	0.40	129	0.20
200	451.5	0.70	129	0.20
400	645	1.00	258	0.40
600	967.5	1.50[b]	258	0.40
1000	—	—	387	0.60
1200	—	—	645	1.00
1600	—	—	967.5	1.50
2000	—	—	1290	2.00[b]

[a]Total cross-sectional area of both side rails for ladder or trough cable trays; or the minimum cross-sectional area of metal in channel cable trays or cable trays of one-piece construction.

[b]Steel cable trays shall not be used as equipment grounding conductors for circuits with ground-fault protection above 600 amperes. Aluminum cable trays shall not be used as equipment grounding conductors for circuits with ground-fault protection above 2000 amperes.

(B) Steel or Aluminum Cable Tray Systems. Steel or aluminum cable tray systems shall be permitted to be used as equipment grounding conductors, provided all the following requirements are met:

(1) The cable tray sections and fittings are identified as an equipment grounding conductor.

(2) The minimum cross-sectional area of cable trays conform to the requirements in Table 392.60(A).

(3) All cable tray sections and fittings are legibly and durably marked to show the cross-sectional area of metal in channel cable trays, or cable trays of one-piece construction, and the total cross-sectional area of both side rails for ladder or trough cable trays.

(4) Cable tray sections, fittings, and connected raceways are bonded in accordance with 250.96, using bolted mechanical connectors or bonding jumpers sized and installed in accordance with 250.102.

(C) Transitions. Where metal cable tray systems are mechanically discontinuous, as permitted in 392.18(A), a bonding jumper sized in accordance with 250.102 shall connect the two sections of the cable tray, or the cable tray and the raceway or equipment. Bonding shall be in accordance with 250.96.

392.80 Ampacity of Conductors.

(A) Ampacity of Cables, Rated 2000 Volts or Less, in Cable Trays.

Informational Note: See 110.14(C) for conductor temperature limitations due to termination provisions.

(1) Multiconductor Cables. The allowable ampacity of multiconductor cables, nominally rated 2000 volts or less, installed according to the requirements of 392.22(A) shall be as given in Table 310.15(B)(16) and Table 310.15(B)(18), subject to the provisions of (A)(1)(a), (b), (c), and 310.15(A)(2).

(a) The adjustment factors of 310.15(A)(3)(a) shall apply only to multiconductor cables with more than three current-carrying conductors. Adjustment factors shall be limited to the number of current-carrying conductors in the cable and not to the number of conductors in the cable tray.

(b) Where cable trays are continuously covered for more than 1.8 m (6 ft) with solid unventilated covers, not over 95 percent of the allowable ampacities of Table 310.15(B)(16) and Table 310.15(B)(18) shall be permitted for multiconductor cables.

(c) Where multiconductor cables are installed in a single layer in uncovered trays, with a maintained spacing of not less than one cable diameter between cables, the ampacity shall not exceed the allowable ambient temperature-corrected ampacities of multiconductor cables, with not more than three insulated conductors rated 0 through 2000 volts in free air, in accordance with 310.15(C).

Informational Note: See Table B.310.15(B)(2)(3).

(2) Single-Conductor Cables. The allowable ampacity of single-conductor cables shall be as permitted by 310.15(A)(2). The adjustment factors of 310.15(B)(3)(a) shall not apply to the ampacity of cables in cable trays. The ampacity of single-conductor cables, or single conductors cabled together (triplexed, quadruplexed, etc.), nominally rated 2000 volts or less, shall comply with the following:

(a) Where installed according to the requirements of 392.22(B), the ampacities for 600 kcmil and larger single-conductor cables in uncovered cable trays shall not exceed 75 percent of the allowable ampacities in Table 310.15(B)(17) and Table 310.15(B)(19). Where cable trays are continuously covered for more than 1.8 m (6 ft) with solid unventilated covers, the ampacities for 600 kcmil and larger cables shall not exceed 70 percent of the allowable ampacities in Table 310.15(B)(17) and Table 310.15(B)(19).

(b) Where installed according to the requirements of 392.22(B), the ampacities for 1/0 AWG through 500 kcmil single-conductor cables in uncovered cable trays shall not exceed 65 percent of the allowable ampacities in Table 310.15(B)(17) and Table 310.15(B)(19). Where cable trays are continuously covered for more than 1.8 m (6 ft) with solid unventilated covers, the ampacities for 1/0 AWG through 500 kcmil cables shall not exceed 60 percent of the allowable ampacities in Table 310.15(B)(17) and Table 310.15(B)(19).

(c) Where single conductors are installed in a single layer in uncovered cable trays, with a maintained space of not less than one cable diameter between individual conductors, the ampacity of 1/0 AWG and larger cables shall not exceed the allowable ampacities in Table 310.15(B)(17) and Table 310.15(B)(19).

Exception to (2)(3)(c): For solid bottom cable trays the ampacity of single conductor cables shall be determined by 310.15(C).

(d) Where single conductors are installed in a triangular or square configuration in uncovered cable trays, with a maintained free airspace of not less than 2.15 times one conductor diameter (2.15 × O.D.) of the largest conductor contained within the configuration and adjacent conductor configurations or cables, the ampacity of 1/0 AWG and larger cables shall not exceed the allowable ampacities of two or three single insulated conductors rated 0 through 2000 volts supported on a messenger in accordance with 310.15(B).

Informational Note: See Table 310.15(B)(20).

(3) Combinations of Multiconductor and Single-Conductor Cables. Where a cable tray contains a combination of multiconductor and single-conductor cables, the allowable ampacities shall be as given in 392.80(A)(1) for multiconductor cables and 392.80(A)(2) for single-conductor cables, provided that the following conditions apply:

(1) The sum of the multiconductor cable fill area as a percentage of the allowable fill area for the tray calculated in accordance with 392.22(A), and the single-conductor cable fill area as a percentage of the allowable fill area for the tray calculated in accordance with 392.22(B), totals not more than 100 percent.

(2) Multiconductor cables are installed according to 392.22(A), and single-conductor cables are installed according to 392.22(B) and 392.22(C).

(B) Ampacity of Type MV and Type MC Cables (2001 Volts or Over) in Cable Trays. The ampacity of cables, rated 2001 volts, nominal, or over, installed according to 392.22(C) shall not exceed the requirements of this section.

Informational Note: See 110.40 for conductor temperature limitations due to termination provisions.

(1) Multiconductor Cables (2001 Volts or Over). The allowable ampacity of multiconductor cables shall be as given in Table 310.60(C)(75) and Table 310.60(C)(76), subject to the following provisions:

(1) Where cable trays are continuously covered for more than 1.8 m (6 ft) with solid unventilated covers, not more than 95 percent of the allowable ampacities of Table 310.60(C)(75) and Table 310.60(C)(76) shall be permitted for multiconductor cables.

(2) Where multiconductor cables are installed in a single layer in uncovered cable trays, with maintained spacing of not less than one cable diameter between cables, the ampacity shall not exceed the allowable ampacities of Table 310.60(C)(71) and Table 310.60(C)(72).

(2) Single-Conductor Cables (2001 Volts or Over). The ampacity of single-conductor cables, or single conductors cabled together (triplexed, quadruplexed, etc.), shall comply with the following:

(1) The ampacities for 1/0 AWG and larger single-conductor cables in uncovered cable trays shall not exceed 75 percent of the allowable ampacities in Table 310.60(C)(69) and Table 310.60(C)(70). Where the cable trays are covered for more than 1.8 m (6 ft) with solid unventilated covers, the ampacities for 1/0 AWG and larger single-conductor cables shall not exceed 70 percent of the allowable ampacities in Table 310.60(C)(69) and Table 310.60(C)(70).

(2) Where single-conductor cables are installed in a single layer in uncovered cable trays, with a maintained space of not less than one cable diameter between individual conductors, the ampacity of 1/0 AWG and larger cables shall not exceed the allowable ampacities in Table 310.60(C)(69) and Table 310.60(C)(70).

(3) Where single conductors are installed in a triangular or square configuration in uncovered cable trays, with a maintained free air space of not less than 2.15 times the diameter (2.15 × O.D.) of the largest conductor contained within the configuration and adjacent conductor configurations or cables, the ampacity of 1/0 AWG

and larger cables shall not exceed the allowable ampacities in Table 310.60(C)(67) and Table 310.60(C)(68).

Part III. Construction Specifications

392.100 Construction.

(A) Strength and Rigidity. Cable trays shall have suitable strength and rigidity to provide adequate support for all contained wiring.

(B) Smooth Edges. Cable trays shall not have sharp edges, burrs, or projections that could damage the insulation or jackets of the wiring.

(C) Corrosion Protection. Cable tray systems shall be corrosion resistant. If made of ferrous material, the system shall be protected from corrosion as required by 300.6.

(D) Side Rails. Cable trays shall have side rails or equivalent structural members.

(E) Fittings. Cable trays shall include fittings or other suitable means for changes in direction and elevation of runs.

(F) Nonmetallic Cable Tray. Nonmetallic cable trays shall be made of flame-retardant material.

ARTICLE 393
Low-Voltage Suspended Ceiling Power Distribution Systems

Part I. General

393.1 Scope. This article covers the installation of low-voltage suspended ceiling power distribution systems.

393.2 Definitions.

Busbar. A noninsulated conductor electrically connected to the source of supply and physically supported on an insulator providing a power rail for connection to utilization equipment, such as sensors, actuators, A/V devices, low-voltage luminaire assemblies, and similar electrical equipment.

Busbar Support. An insulator that runs the length of a section of suspended ceiling bus rail that serves to support and isolate the busbars from the suspended grid rail.

Connector. A term used to refer to an electromechanical fitting.

Connector, Load. An electromechanical connector used for power from the busbar to utilization equipment.

Connector, Pendant. An electromechanical or mechanical connector used to suspend low-voltage luminaire or utilization equipment below the grid rail and to supply power to connect from the busbar to utilization equipment.

Connector, Power Feed. An electromechanical connector used to connect the power supply to a power distribution cable, to connect directly to the busbar, or to connect from a power distribution cable to the busbar.

Connector, Rail to Rail. An electromechanical connector used to interconnect busbars from one ceiling grid rail to another grid rail.

Grid Bus Rail. A combination of the busbar, the busbar support, and the structural suspended ceiling grid system.

Low-Voltage Suspended Ceiling Power Distribution System. A system that serves as a support for a finished ceiling surface and consists of a busbar and busbar support system to distribute power to utilization equipment supplied by a Class 2 power supply.

Power Supply. A Class 2 power supply connected between the branch-circuit power distribution system and the busbar low-voltage suspended ceiling power distribution system.

Rail. The structural support for the suspended ceiling system typically forming the ceiling grid supporting the ceiling tile and listed utilization equipment, such as sensors, actuators, A/V devices, and low-voltage luminaires and similar electrical equipment.

Reverse Polarity Protection (Backfeed Protection). A system that prevents two interconnected power supplies, connected positive to negative, from passing current from one power source into a second power source.

Suspended Ceiling Grid. A system that serves as a support for a finished ceiling surface and other utilization equipment.

393.6 Listing Requirements. Suspended ceiling power distribution systems and associated fittings shall be listed as in 393.6(A) or (B).

(A) Listed System. Low-voltage suspended ceiling distribution systems operating at 30 volts ac or less or 60 volts dc or less shall be listed as a complete system, with the utilization equipment, power supply, and fittings as part of the same identified system.

(B) Assembly of Listed Parts. A low-voltage suspended ceiling power distribution system assembled from the following parts, listed according to the appropriate function, shall be permitted:

(1) Listed low-voltage utilization equipment
(2) Listed Class 2 power supply
(3) Listed or identified fittings, including connectors and grid rails with bare conductors
(4) Listed low-voltage cables in accordance with 725.179, conductors in raceways, or other fixed wiring methods for the secondary circuit

Part II. Installation

393.10 Uses Permitted. Low-voltage suspended ceiling power distribution systems shall be permanently connected and shall be permitted as follows:

(1) For listed utilization equipment capable of operation at a maximum of 30 volts ac (42.4 volts peak) or 60 volts dc (24.8 volts peak for dc interrupted at a rate of 10 Hz to 200 Hz) and limited to Class 2 power levels in Chapter 9, Table 11(A) and Table 11(B) for lighting, control, and signaling circuits.

(2) In indoor dry locations.
(3) For residential, commercial, and industrial installations.
(4) In other spaces used for environmental air in accordance with 300.22(C), electrical equipment having a metal enclosure, or with a nonmetallic enclosure and fittings, shall be listed for use within an air-handling space and shall have adequate fire-resistant and low-smoke-producing characteristics and associated wiring material suitable for the ambient temperature.

Informational Note: One method of defining adequate fire-resistant and low-smoke producing characteristics for electrical equipment with a nonmetallic enclosure is in ANSI/UL 2043-2008, *Fire Test for Heat and Visible Smoke Release for Discrete Products and Their Accessories Installed in Air-Handling Spaces.*

393.12 Uses Not Permitted. Suspended ceiling power distribution systems shall not be installed in the following:

(1) In damp or wet locations
(2) Where subject to corrosive fumes or vapors, such as storage battery rooms
(3) Where subject to physical damage
(4) In concealed locations
(5) In hazardous (classified) locations
(6) As part of a fire-rated floor-ceiling or roof-ceiling assembly, unless specifically listed as part of the assembly
(7) For lighting in general or critical patient care areas

393.14 Installation.

(A) General Requirements. Support wiring shall be installed in a neat and workmanlike manner. Cables and conductors installed exposed on the surface of ceilings and sidewalls shall be supported by the building structure in such a manner that the cable is not damaged by normal building use. Such cables shall be supported by straps, staples, hangers, cable ties, or similar fittings designed and installed so as not to damage the cable.

Informational Note: Suspended ceiling low-voltage power grid distribution systems should be installed by qualified persons in accordance with the manufacturer's installation instructions.

(B) Insulated Conductors. Exposed insulated secondary circuit conductors shall be listed, of the type, and installed as described as follows:

(1) Class 2 cable supplied by a listed Class 2 power source and installed in accordance with Parts I and III of Article 725
(2) Wiring methods described in Chapter 3

393.21 Disconnecting Means.

(A) Location. A disconnecting means for the Class 2 supply to the power grid system shall be located so as to be accessible and within sight of the Class 2 power source for servicing or maintenance of the grid system.

(B) Multiwire Branch Circuits. Where connected to a multiwire branch circuit, the disconnecting means shall simultaneously disconnect all the supply conductors, including the grounded conductors.

393.30 Securing and Supporting.

(A) Attached to Building Structure. A suspended ceiling low-voltage power distribution system shall be secured to the mounting surface of the building structure by hanging wires, screws, or bolts in accordance with the installation and operation instructions. Mounting hardware, such as screws or bolts, shall be either packaged with the suspended ceiling low-voltage lighting power distribution system, or the installation instructions shall specify the types of mounting fasteners to be used.

(B) Attachment of Power Grid Rails. The individual power grid rails shall be mechanically secured to the overall ceiling grid assembly.

393.40 Connectors and Enclosures.

(A) Connectors. Connections to busbar grid rails, cables, and conductors shall be made with listed insulating devices, and these connections shall be accessible after installation. A soldered connection shall be made mechanically secure before being soldered. Other means of securing leads, such as push-on terminals and spade-type connectors, shall provide a secure mechanical connection. The following connectors shall be permitted to be used as connection or interconnection devices:

(1) Load connectors shall be used for power from the busbar to listed utilization equipment.

(2) A pendant connector shall be permitted to suspend low-voltage luminaires or utilization equipment below the grid rail and to supply power from the busbar to the utilization equipment.

(3) A power feed connector shall be permitted to connect the power supply directly to a power distribution cable and to the busbar.

(4) Rail-to-rail connectors shall be permitted to interconnect busbars from one ceiling grid rail to another grid rail.

Informational Note: For quick-connect terminals, see UL 310, *Standard for Electrical Quick-Connect*, and for mechanical splicing devices, see UL 486A, *Standard for Wire Connectors and Soldering Lugs for Use with Copper Conductors*, and 486B, *Standard for Wire Connectors*.

(B) Enclosures. Where made in a wall, connections shall be installed in an enclosure in accordance with Parts I, II, and III of Article 314.

393.45 Overcurrent and Reverse Polarity (Backfeed) Protection.

(A) Overcurrent Protection. The listed Class 2 power supply or transformer primary shall be protected at not greater than 20 amperes.

(B) Interconnection of Power Sources. Listed Class 2 sources shall not have the output connections paralleled or otherwise interconnected unless listed for such interconnection.

(C) Reverse Polarity (Backfeed) Protection of Direct-Current Systems. A suspended ceiling low-voltage power distribution system shall be permitted to have reverse polarity (backfeed) protection of dc circuits by one of the following means:

(1) If the power supply is provided as part of the system, the power supply is provided with reverse polarity (backfeed) protection; or

(2) If the power supply is not provided as part of the system, reverse polarity or backfeed protection can be provided as part of the grid rail busbar or as a part of the power feed connector.

393.56 Splices. A busbar splice shall be provided with insulation and mechanical protection equivalent to that of the grid rail busbars involved.

393.57 Connections. Connections in busbar grid rails, cables, and conductors shall be made with listed insulating devices and

be accessible after installation. Where made in a wall, connections shall be installed in an enclosure in accordance with Parts I, II, and III of Article 314, as applicable.

393.60 Grounding.

(A) Grounding of Supply Side of Class 2 Power Source. The supply side of the Class 2 power source shall be connected to an equipment grounding conductor in accordance with the applicable requirements in Part IV of Article 250.

(B) Grounding of Load Side of Class 2 Power Source. Class 2 load side circuits for suspended ceiling low-voltage power grid distribution systems shall not be grounded.

Part III. Construction Specifications

393.104 Sizes and Types of Conductors.

(A) Load Side Utilization Conductor Size. Current-carrying conductors for load side utilization equipment shall be copper and shall be 18 AWG minimum.

Exception: Conductors of a size smaller than 18 AWG, but not smaller than 24 AWG, shall be permitted to be used for Class 2 circuits. Where used, these conductors shall be installed using a Chapter 3 wiring method, shall be totally enclosed, shall not be subject to movement or strain, and shall comply with the ampacity requirements in Table 522.22.

(B) Power Feed Bus Rail Conductor Size. The power feed bus rail shall be 16 AWG minimum or equivalent. For a busbar with a circular cross section, the diameter shall be 1.29 mm (0.051 in.) minimum, and, for other than circular busbars, the area shall be 1.32 mm^2 (0.002 in.^{2}) minimum.

ARTICLE 394
Concealed Knob-and-Tube Wiring

Part I. General

394.1 Scope. This article covers the use, installation, and construction specifications of concealed knob-and-tube wiring.

394.2 Definition.

Concealed Knob-and-Tube Wiring. A wiring method using knobs, tubes, and flexible nonmetallic tubing for the protection and support of single insulated conductors.

Part II. Installation

394.10 Uses Permitted. Concealed knob-and-tube wiring shall be permitted to be installed in the hollow spaces of walls and ceilings, or in unfinished attics and roof spaces as provided by 394.23, only as follows:

(1) For extensions of existing installations
(2) Elsewhere by special permission

394.12 Uses Not Permitted. Concealed knob-and-tube wiring shall not be used in the following:

(1) Commercial garages

(2)   Theaters and similar locations

(3)   Motion picture studios

(4)   Hazardous (classified) locations

(5)   Hollow spaces of walls, ceilings, and attics where such spaces are insulated by loose, rolled, or foamed-in-place insulating material that envelops the conductors

394.17 Through or Parallel to Framing Members. Conductors shall comply with 398.17 where passing through holes in structural members. Where passing through wood cross members in plastered partitions, conductors shall be protected by noncombustible, nonabsorbent, insulating tubes extending not less than 75 mm (3 in.) beyond the wood member.

394.19 Clearances.

(A) General. A clearance of not less than 75 mm (3 in.) shall be maintained between conductors and a clearance of not less than 25 mm (1 in.) between the conductor and the surface over which it passes.

(B) Limited Conductor Space. Where space is too limited to provide these minimum clearances, such as at meters, panelboards, outlets, and switch points, the individual conductors shall be enclosed in flexible nonmetallic tubing, which shall be continuous in length between the last support and the enclosure or terminal point.

(C) Clearance from Piping, Exposed Conductors, and So Forth. Conductors shall comply with 398.19 for clearances from other exposed conductors, piping, and so forth.

394.23 In Accessible Attics. Conductors in unfinished attics and roof spaces shall comply with 394.23(A) or (B).

> Informational Note: See 310.15(A)(3) for temperature limitation of conductors.

(A) Accessible by Stairway or Permanent Ladder. Conductors shall be installed along the side of or through bored holes in floor joists, studs, or rafters. Where run through bored holes, conductors in the joists and in studs or rafters to a height of not less than 2.1 m (7 ft) above the floor or floor joists shall be protected by substantial running boards extending not less than 25 mm (1 in.) on each side of the conductors. Running boards shall be securely fastened in place. Running boards and guard strips shall not be required where conductors are installed along the sides of joists, studs, or rafters.

(B) Not Accessible by Stairway or Permanent Ladder. Conductors shall be installed along the sides of or through bored holes in floor joists, studs, or rafters.

Exception: In buildings completed before the wiring is installed, attic and roof spaces that are not accessible by stairway or permanent ladder and have headroom at all points less than 900 mm (3 ft), the wiring shall be permitted to be installed on the edges of rafters or joists facing the attic or roof space.

394.30 Securing and Supporting.

(A) Supporting. Conductors shall be rigidly supported on noncombustible, nonabsorbent insulating materials and shall not contact any other objects. Supports shall be installed as follows:

(1)   Within 150 mm (6 in.) of each side of each tap or splice, and

(2)   At intervals not exceeding 1.4 m (4½ ft).

Where it is impracticable to provide supports, conductors shall be permitted to be fished through hollow spaces in dry locations, provided each conductor is individually enclosed in flexible nonmetallic tubing that is in continuous lengths between supports, between boxes, or between a support and a box.

(B) Securing. Where solid knobs are used, conductors shall be securely tied thereto by tie wires having insulation equivalent to that of the conductor.

394.42 Devices. Switches shall comply with 404.4 and 404.10(B).

394.56 Splices and Taps. Splices shall be soldered unless approved splicing devices are used. In-line or strain splices shall not be used.

Part III. Construction Specifications

394.104 Conductors. Conductors shall be of a type specified by Article 310.

ARTICLE 396
Messenger-Supported Wiring

Part I. General

396.1 Scope. This article covers the use, installation, and construction specifications for messenger-supported wiring.

396.2 Definitions.

N **Insulated Conductor.** For the purposes of this article, an insulated conductor includes the following:

(1)   Conductor types described in 310.104, and

(2)   Overhead service conductors encased in a polymeric material that has been evaluated for the applied nominal voltage.

> Informational Note: Evidence of evaluation for the applied nominal voltage can be given by certification that the conductors have met the requirements of ICEA S-76-474-2011, *Standard for Neutral Supported Power Cable Assemblies with Weather Resistant Extruded Insulation Rated 600 Volts.*

Messenger-Supported Wiring. An exposed wiring support system using a messenger wire to support insulated conductors by any one of the following:

(1)   A messenger with rings and saddles for conductor support

(2)   A messenger with a field-installed lashing material for conductor support

(3)   Factory-assembled aerial cable

(4)   Multiplex cables utilizing a bare conductor, factory assembled and twisted with one or more insulated conductors, such as duplex, triplex, or quadruplex type of construction

Part II. Installation

396.10 Uses Permitted.

(A) Cable Types. The cable types in Table 396.10(A) shall be permitted to be installed in messenger-supported wiring under the conditions described in the article or section referenced for each.

(B) In Industrial Establishments. In industrial establishments only, where conditions of maintenance and supervision ensure that only qualified persons service the installed messenger-supported wiring, the following shall be permitted:

(1) Any of the conductor types shown in Table 310.104(A) or Table 310.104(B)
(2) MV cable

Where exposed to weather, conductors shall be listed for use in wet locations. Where exposed to direct rays of the sun, conductors or cables shall be sunlight resistant.

(C) Hazardous (Classified) Locations. Messenger-supported wiring shall be permitted to be used in hazardous (classified) locations where the contained cables and messenger-supported wiring are specifically permitted by other articles in this *Code*.

396.12 Uses Not Permitted. Messenger-supported wiring shall not be used in hoistways or where subject to physical damage.

396.30 Messenger.

(A) Support. The messenger shall be supported at dead ends and at intermediate locations so as to eliminate tension on the conductors. The conductors shall not be permitted to come into contact with the messenger supports or any structural members, walls, or pipes.

(B) Neutral Conductor. Where the messenger is used as a neutral conductor, it shall comply with the requirements of 225.4, 250.184(A), 250.184(B)(7), and 250.187(B).

(C) Equipment Grounding Conductor. Where the messenger is used as an equipment grounding conductor, it shall comply

with the requirements of 250.32(B), 250.118, 250.184(B)(8), and 250.187(D).

396.56 Conductor Splices and Taps. Conductor splices and taps made and insulated by approved methods shall be permitted in messenger-supported wiring.

396.60 Grounding. The messenger shall be grounded as required by 250.80 and 250.86 for enclosure grounding.

ARTICLE 398
Open Wiring on Insulators

Part I. General

398.1 Scope. This article covers the use, installation, and construction specifications of open wiring on insulators.

398.2 Definition.

Open Wiring on Insulators. An exposed wiring method using cleats, knobs, tubes, and flexible tubing for the protection and support of single insulated conductors run in or on buildings.

Part II. Installation

398.10 Uses Permitted. Open wiring on insulators shall be permitted only for industrial or agricultural establishments on systems of 1000 volts, nominal, or less, as follows:

(1) Indoors or outdoors
(2) In wet or dry locations
(3) Where subject to corrosive vapors
(4) For services

398.12 Uses Not Permitted. Open wiring on insulators shall not be installed where concealed by the building structure.

398.15 Exposed Work.

(A) Dry Locations. In dry locations, where not exposed to physical damage, conductors shall be permitted to be separately enclosed in flexible nonmetallic tubing. The tubing shall be in continuous lengths not exceeding 4.5 m (15 ft) and secured to the surface by straps at intervals not exceeding 1.4 m (4½ ft).

(B) Entering Spaces Subject to Dampness, Wetness, or Corrosive Vapors. Conductors entering or leaving locations subject to dampness, wetness, or corrosive vapors shall have drip loops formed on them and shall then pass upward and inward from the outside of the buildings, or from the damp, wet, or corrosive location, through noncombustible, nonabsorbent insulating tubes.

> Informational Note: See 230.52 for individual conductors entering buildings or other structures.

(C) Exposed to Physical Damage. Conductors within 2.1 m (7 ft) from the floor shall be considered exposed to physical damage. Where open conductors cross ceiling joists and wall studs and are exposed to physical damage, they shall be protected by one of the following methods:

Table 396.10(A) Cable Types

Cable Type	Section	Article
Medium-voltage cable		328
Metal-clad cable		330
Mineral-insulated, metal-sheathed cable		332
Multiconductor service-entrance cable		338
Multiconductor underground feeder and branch-circuit cable		340
Other factory-assembled, multiconductor control, signal, or power cables that are identified for the use		
Power and control tray cable		336
Power-limited tray cable	Table 725.154, 725.135(J), and 725.179(E)	

(1) Guard strips not less than 25 mm (1 in.) nominal in thickness and at least as high as the insulating supports, placed on each side of and close to the wiring.

(2) A substantial running board at least 13 mm (½ in.) thick in back of the conductors with side protections. Running boards shall extend at least 25 mm (1 in.) outside the conductors, but not more than 50 mm (2 in.), and the protecting sides shall be at least 50 mm (2 in.) high and at least 25 mm (1 in.), nominal, in thickness.

(3) Boxing made in accordance with 398.15(C)(1) or (C)(2) and furnished with a cover kept at least 25 mm (1 in.) away from the conductors within. Where protecting vertical conductors on side walls, the boxing shall be closed at the top and the holes through which the conductors pass shall be bushed.

(4) Rigid metal conduit, intermediate metal conduit, rigid nonmetallic conduit, or electrical metallic tubing. When installed in metal piping, the conductors shall be encased in continuous lengths of approved flexible tubing.

398.17 Through or Parallel to Framing Members. Open conductors shall be separated from contact with walls, floors, wood cross members, or partitions through which they pass by tubes or bushings of noncombustible, nonabsorbent insulating material. Where the bushing is shorter than the hole, a waterproof sleeve of noninductive material shall be inserted in the hole and an insulating bushing slipped into the sleeve at each end in such a manner as to keep the conductors absolutely out of contact with the sleeve. Each conductor shall be carried through a separate tube or sleeve.

Informational Note: See 310.15(A)(3) for temperature limitation of conductors.

398.19 Clearances. Open conductors shall be separated at least 50 mm (2 in.) from metal raceways, piping, or other conducting material, and from any exposed lighting, power, or signaling conductor, or shall be separated therefrom by a continuous and firmly fixed nonconductor in addition to the insulation of the conductor. Where any insulating tube is used, it shall be secured at the ends. Where practicable, conductors shall pass over rather than under any piping subject to leakage or accumulations of moisture.

398.23 In Accessible Attics. Conductors in unfinished attics and roof spaces shall comply with 398.23(A) or (B).

(A) Accessible by Stairway or Permanent Ladder. Conductors shall be installed along the side of or through bored holes in floor joists, studs, or rafters. Where run through bored holes, conductors in the joists and in studs or rafters to a height of not less than 2.1 m (7 ft) above the floor or floor joists shall be protected by substantial running boards extending not less than 25 mm (1 in.) on each side of the conductors. Running boards shall be securely fastened in place. Running boards and guard strips shall not be required for conductors installed along the sides of joists, studs, or rafters.

(B) Not Accessible by Stairway or Permanent Ladder. Conductors shall be installed along the sides of or through bored holes in floor joists, studs, or rafters.

Exception: In buildings completed before the wiring is installed, in attic and roof spaces that are not accessible by stairway or permanent ladder and have headroom at all points less than 900 mm (3 ft), the wiring shall be permitted to be installed on the edges of rafters or joists facing the attic or roof space.

398.30 Securing and Supporting.

(A) Conductor Sizes Smaller Than 8 AWG. Conductors smaller than 8 AWG shall be rigidly supported on noncombustible, nonabsorbent insulating materials and shall not contact any other objects. Supports shall be installed as follows:

(1) Within 150 mm (6 in.) from a tap or splice
(2) Within 300 mm (12 in.) of a dead-end connection to a lampholder or receptacle
(3) At intervals not exceeding 1.4 m (4½ ft) and at closer intervals sufficient to provide adequate support where likely to be disturbed

(B) Conductor Sizes 8 AWG and Larger. Supports for conductors 8 AWG or larger installed across open spaces shall be permitted up to 4.5 m (15 ft) apart if noncombustible, nonabsorbent insulating spacers are used at least every 1.4 m (4½ ft) to maintain at least 65 mm (2½ in.) between conductors.

Where not likely to be disturbed in buildings of mill construction, 8 AWG and larger conductors shall be permitted to be run across open spaces if supported from each wood cross member on approved insulators maintaining 150 mm (6 in.) between conductors.

(C) Industrial Establishments. In industrial establishments only, where conditions of maintenance and supervision ensure that only qualified persons service the system, conductors of sizes 250 kcmil and larger shall be permitted to be run across open spaces where supported at intervals up to 9.0 m (30 ft) apart.

(D) Mounting of Conductor Supports. Where nails are used to mount knobs, they shall not be smaller than tenpenny. Where screws are used to mount knobs, or where nails or screws are used to mount cleats, they shall be of a length sufficient to penetrate the wood to a depth equal to at least one-half the height of the knob and the full thickness of the cleat. Cushion washers shall be used with nails.

(E) Tie Wires. Conductors 8 AWG or larger and supported on solid knobs shall be securely tied thereto by tie wires having an insulation equivalent to that of the conductor.

398.42 Devices. Surface-type snap switches shall be mounted in accordance with 404.10(A), and boxes shall not be required. Other type switches shall be installed in accordance with 404.4.

Part III. Construction Specifications

398.104 Conductors. Conductors shall be of a type specified by Article 310.

ARTICLE 399
Outdoor Overhead Conductors over 1000 Volts

399.1 Scope. This article covers the use and installation for outdoor overhead conductors over 1000 volts, nominal.

399.2 Definition.

Outdoor Overhead Conductors. Single conductors, insulated, covered, or bare, installed outdoors on support structures in free air.

399.10 Uses Permitted. Outdoor overhead conductors over 1000 volts, nominal, shall be permitted only for systems rated over 1000 volts, nominal, as follows:

(1) Outdoors in free air
(2) For service conductors, feeders, or branch circuits

> Informational Note: For additional information on outdoor overhead conductors over 1000 volts, see IEEE C2-2012, *National Electrical Safety Code*, and ANSI/IEEE 3001.2, *Recommended Practice for Evaluating the Electrical Service Requirements of Industrial and Commercial Power Systems.*

399.30 Support.

(A) Conductors. Documentation of the engineered design by a licensed professional engineer engaged primarily in the design of such systems for the spacing between conductors shall be available upon request of the authority having jurisdiction and shall include consideration of the following:

(1) Applied voltage
(2) Conductor size
(3) Distance between support structures
(4) Type of structure
(5) Wind/ice loading
(6) Surge protection

(B) Structures. Structures of wood, metal, concrete, or combinations of those materials, shall be provided for support of overhead conductors over 1000 volts, nominal. Documentation of the engineered design by a licensed professional engineer engaged primarily in the design of such systems and the installation of each support structure shall be available upon request of the authority having jurisdiction and shall include consideration of the following:

(1) Soil conditions
(2) Foundations and structure settings
(3) Weight of all supported conductors and equipment
(4) Weather loading and other conditions such as, but not limited to, ice, wind, temperature, and lightning
(5) Angle where change of direction occurs
(6) Spans between adjacent structures
(7) Effect of dead-end structures
(8) Strength of guys and guy anchors
(9) Structure size and material(s)
(10) Hardware

(C) Insulators. Insulators used to support conductors shall be rated for all of the following:

(1) Applied phase-to-phase voltage
(2) Mechanical strength required for each individual installation
(3) Impulse withstand BIL in accordance with Table 490.24

> Informational Note: 399.30(A), (B), and (C) are not all-inclusive lists.

Chapter 4 Equipment for General Use

ARTICLE 400
Flexible Cords and Flexible Cables

Part I. General

400.1 Scope. This article covers general requirements, applications, and construction specifications for flexible cords and flexible cables.

> Informational Note: UL 817, *Cord Sets and Power-Supply Cords*, allows the use of flexible cords manufactured in accordance with UL 62, *Flexible Cords and Cables*. See 400.10 and 400.12 for flexible cords that are part of a listed cord set or power-supply cord.

400.2 Other Articles. Flexible cords and flexible cables shall comply with this article and with the applicable provisions of other articles of this *Code*.

400.3 Suitability. Flexible cords and flexible cables and their associated fittings shall be suitable for the conditions of use and location.

400.4 Types. Flexible cords and flexible cables shall conform to the description in Table 400.4. The use of flexible cords and flexible cables other than those in Table 400.4 shall require permission by the authority having jurisdiction.

400.5 Ampacities for Flexible Cords and Flexible Cables.

(A) Ampacity Tables. Table 400.5(A)(1) provides the allowable ampacities, and Table 400.5(A)(2) provides the ampacities for flexible cords and flexible cables with not more than three current-carrying conductors. These tables shall be used in conjunction with applicable end-use product standards to ensure selection of the proper size and type. Where cords and cables are used in ambient temperatures other than 30°C (86°F), the temperature correction factors from Table 310.15(B)(2)(a) that correspond to the temperature rating of the cord or cable shall be applied to the ampacity in Table 400.5(A)(1) and Table 400.5(A)(2). Cords and cables rated 105°C shall use correction factors in the 90°C column of Table 310.15(B)(2)(a) for temperature correction. Where the number of current-carrying conductors exceeds three, the allowable ampacity or the ampacity of each conductor shall be reduced from the three-conductor rating as shown in Table 400.5(A)(3).

> Informational Note: See Informative Annex B, Table B.310.15(B)(2)(11), for adjustment factors for more than three current-carrying conductors in a raceway or cable with load diversity.

A neutral conductor that carries only the unbalanced current from other conductors of the same circuit shall not be required to meet the requirements of a current-carrying conductor.

In a 3-wire circuit consisting of two phase conductors and the neutral conductor of a 4-wire, 3-phase, wye-connected system, a common conductor carries approximately the same current as the line-to-neutral currents of the other conductors and shall be considered to be a current-carrying conductor.

On a 4-wire, 3-phase, wye circuit where more than 50 percent of the load consists of nonlinear loads, there are harmonic currents present in the neutral conductor and the neutral conductor shall be considered to be a current-carrying conductor.

An equipment grounding conductor shall not be considered a current-carrying conductor.

Where a single conductor is used for both equipment grounding and to carry unbalanced current from other conductors, as provided for in 250.140 for electric ranges and electric clothes dryers, it shall not be considered as a current-carrying conductor.

(B) Ultimate Insulation Temperature. In no case shall conductors be associated together in such a way with respect to the kind of circuit, the wiring method used, or the number of conductors such that the limiting temperature of the conductors is exceeded.

(C) Engineering Supervision. Under engineering supervision, conductor ampacities shall be permitted to be calculated in accordance with 310.15(C).

400.6 Markings.

(A) Standard Markings. Flexible cords and flexible cables shall be marked by means of a printed tag attached to the coil reel or carton. The tag shall contain the information required in 310.120(A). Types S, SC, SCE, SCT, SE, SEO, SEOO, SJ, SJE, SJEO, SJEOO, SJO, SJT, SJTO, SJTOO, SO, SOO, ST, STO, STOO, SEW, SEOW, SEOOW, SJEW, SJEOW, SJEOOW, SJOW, SJTW, SJTOW, SJTOOW, SOW, SOOW, STW, STOW, and STOOW flexible cords and G, G-GC, PPE, and W flexible cables shall be durably marked on the surface at intervals not exceeding 610 mm (24 in.) with the type designation, size, and number of conductors. Required markings on tags, cords, and cables shall also include the maximum operating temperature of the flexible cord or flexible cable.

(B) Optional Markings. Flexible cords and cable types listed in Table 400.4 shall be permitted to be surface marked to indicate special characteristics of the cable materials. These markings include, but are not limited to, markings for limited smoke, sunlight resistance, and so forth.

Table 400.4 Flexible Cords and Flexible Cables

Trade Name	Type Letter	Voltage	AWG or kcmil	Number of Conductors	Insulation	AWG or kcmil	Nominal Insulation Thickness mm	mils	Braid on Each Conductor	Outer Covering	Use		
Lamp cord	C	300 600	18–16 15–10	2 or more	Thermoset or thermoplastic	18–16 15–10	0.76 1.14	30 45	Cotton	None	Pendant or portable	Dry locations	Not hard usage
Elevator cable	E [1,2,3,4]	300 or 600	20–2	2 or more	Thermoset	20–16 15–12 12–10 8–2	0.51 0.76 1.14 1.52	20 30 45 60	Cotton	Three cotton; outer one flame-retardant and moisture-resistant	Elevator lighting and control	Unclassified locations	
						20–16 15–12 12–10 8–2	0.51 0.76 1.14 1.52	20 30 45 60	Flexible nylon jacket				
Elevator cable	EO [1,2,4]	300 or 600	20–2	2 or more	Thermoset	20–16 15–12 12–10 8–2	0.51 0.76 1.14 1.52	20 30 45 60	Cotton	Three cotton; outer one flame-retardant and moisture-resistant	Elevator lighting and control	Unclassified locations	
										One cotton and a neoprene jacket		Hazardous (classified) locations	
Elevator cable	ETP [2,4]	300 or 600							Rayon	Thermoplastic	Hazardous (classified) locations		
	ETT [2,4]	300 or 600							None	One cotton or equivalent and a thermoplastic jacket			
Electric vehicle cable	EV [5,6]	600	18–500	2 or more plus grounding conductor(s), plus optional hybrid data, signal communications, and optical fiber cables	Thermoset with optional nylon	18–15	0.76 (0.51)	30 (20)	Optional	Oil-resistant thermoset	Electric vehicle charging	Wet locations	Extra-hard usage
						14–10	1.14 (0.76)	45 (30)					
						8–2	1.52 (1.14)	60 (45)					
						1–4/0	2.03 (1.52)	80 (60)					
						250–500	2.41 (1.90)	95 (75)					
	EVJ [5,6]	300	18–12			18–12	0.76 (0.51)	30 (20)					Hard usage
	EVE [5,6]	600	18–500	2 or more plus grounding conductor(s), plus optional hybrid data, signal communications, and optical fiber cables	Thermoplastic elastomer with optional nylon	18–15	0.76 (0.51)	30 (20)		Oil-resistant thermoplastic elastomer			Extra-hard usage
						14–10	1.14 (0.76)	45 (30)					
						8–2	1.52 (1.14)	60 (45)					
						1–4/0	2.03 (1.52)	80 (60)					
						250–500	2.41 (1.90)	95 (75)					
	EVJE [5,6]	300	18–12			18–12	0.76 (0.51)	30 (20)					Hard usage

(continues)

Table 400.4 *Continued*

Trade Name	Type Letter	Voltage	AWG or kcmil	Number of Conductors	Insulation	AWG or kcmil	Nominal Insulation Thickness mm	mils	Braid on Each Conductor	Outer Covering	Use		
	EVT [5,6]	600	18–500	2 or more plus grounding conductor(s), plus optional hybrid data, signal communications, and optical fiber cables	Thermoplastic with optional nylon	18–15	0.76 (0.51)	30 (20)	Optional	Oil-resistant thermoplastic	Electric vehicle charging	Wet Locations	Extra-hard usage
						14–10	1.14 (0.76)	45 (30)					
						8–2	1.52 (1.14)	60 (45)					
						1–4/0	2.03 (1.52)	80 (60)					
						250–500	2.41 (1.90)	95 (75)					
	EVJT [5,6]	300	18–12			18–12	0.76 (0.51)	30 (20)					Hard usage
Portable power cable	G	2000	12–500	2–6 plus grounding conductor(s)	Thermoset	12–2 1–4/0 250–500	1.52 2.03 2.41	60 80 95		Oil-resistant thermoset	Portable and extra-hard usage		
	G-GC [7]	2000	12–500	3–6 plus grounding conductors and 1 ground check conductor	Thermoset	12–2 1–4/0 250–500	1.52 2.03 2.41	60 80 95		Oil-resistant thermoset			
Heater cord	HPD	300	18–12	2, 3, or 4	Thermoset	18–16 15–12	0.38 0.76	15 30	None	Cotton or rayon	Portable heaters	Dry locations	Not hard usage
Parallel heater cord	HPN [8]	300	18–12	2 or 3	Oil-resistant thermoset	18–16 15 14 12	1.14 1.52 2.41	45 60 95	None	Oil-resistant thermoset	Portable	Damp locations	Not hard usage
Thermoset jacketed heater cords	HSJ	300	18–12	2, 3, or 4	Thermoset	18–16 15–12	0.76 1.14	30 45	None	Cotton and thermoset	Portable or portable heater	Damp locations	Hard usage
	HSJO	300	18–12							Cotton and oil-resistant thermoset		Damp and wet locations	
	HSJOW [9]	300	18–12									Damp locations	
	HSJOO	300	18–12		Oil-resistant thermoset								
	HSJOOW [9]	300	18–12									Damp and wet locations	
Non-integral parallel cords	NISP-1	300	20–18	2 or 3	Thermoset	20–18	0.38	15	None	Thermoset	Pendant or portable	Damp locations	Not hard usage
	NISP-2	300	18–16			18–16	0.76	30					
	NISPE-1 [8]	300	20–18		Thermoplastic elastomer	20–18	0.38	15		Thermoplastic elastomer			
	NISPE-2 [8]	300	18–16			18–16	0.76	30					
	NISPT-1 [8]	300	20–18		Thermoplastic	20–18	0.38	15		Thermoplastic			
	NISPT-2 [8]	300	18–16			18–16	0.76	30					
Twisted portable cord	PD	300 600	18–16 14–10	2 or more	Thermoset or thermoplastic	18–16 15–10	0.76 1.14	30 45	Cotton	Cotton or rayon	Pendant or portable	Dry locations	Not hard usage
Portable power cable	PPE [7]	2000	12–500	1–6 plus optional grounding conductor(s)	Thermoplastic elastomer	12–2 1–4/0 250–500	1.52 2.03 2.41	60 80 95		Oil-resistant thermoplastic elastomer	Portable, extra-hard usage		
Hard service cord	S [7]	600	18–2	2 or more	Thermoset	18–15 14–10 8–2	0.76 1.14 1.52	30 45 60	None	Thermoset	Pendant or portable	Damp locations	Extra-hard usage

(continues)

Table 400.4 *Continued*

Trade Name	Type Letter	Voltage	AWG or kcmil	Number of Conductors	Insulation	AWG or kcmil	mm	mils	Braid on Each Conductor	Outer Covering	Use		
							Nominal Insulation Thickness						
Flexible stage and lighting power cable	SC[7,10]	600	8–250	1 or more	Thermoset	8–2 / 1–4/0 / 250	1.52 / 2.03 / 2.41	60 / 80 / 95		Thermoset	Portable, extra-hard usage		
	SCE[7,10]	600			Thermoplastic elastomer					Thermoplastic elastomer			
	SCT[7,10]	600			Thermoplastic					Thermoplastic			
Hard service cord	SE[7]	600	18–2	2 or more	Thermoplastic elastomer	18–15 / 14–9 / 8–2	0.76 / 1.14 / 1.52	30 / 45 / 60	None	Thermoplastic elastomer	Pendant or portable	Damp locations	Extra-hard usage
	SEW[7,9]	600										Damp and wet locations	
	SEO[7]	600								Oil-resistant thermoplastic elastomer		Damp locations	
	SEOW[7,9]	600										Damp and wet locations	
	SEOO[7]	600			Oil-resistant thermoplastic elastomer							Damp locations	
	SEOOW[7,9]	600										Damp and wet locations	
Junior hard service cord	SJ	300	18–10	2–6	Thermoset	18–11 / 10	0.76 / 1.14	30 / 45	None	Thermoset	Pendant or portable	Damp locations	Hard usage
	SJE	300			Thermoplastic elastomer					Thermoplastic elastomer			
	SJEW[9]	300										Damp and wet locations	
	SJEO	300								Oil-resistant thermoplastic elastomer		Damp locations	
	SJEOW[9]	300										Damp and wet locations	
	SJEOO	300			Oil-resistant thermoplastic elastomer							Damp locations	
	SJEOOW[9]	300										Damp and wet locations	
	SJO	300			Thermoset					Oil-resistant thermoset		Damp locations	
	SJOW[9]	300										Damp and wet locations	
	SJOO	300			Oil-resistant thermoset							Damp locations	
	SJOOW[9]	300										Damp and wet locations	

(continues)

Table 400.4   *Continued*

Trade Name	Type Letter	Voltage	AWG or kcmil	Number of Conductors	Insulation	AWG or kcmil	Nominal Insulation Thickness		Braid on Each Conductor	Outer Covering		Use	
							mm	mils					
	SJT	300			Thermoplastic					Thermoplastic		Damp locations	
	SJTW[9]	300										Damp and wet locations	
	SJTO	300				18–12 10	0.76 1.14	30 45		Oil-resistant thermolastic		Damp locations	
	SJTOW[9]	300										Damp and wet locations	
	SJTOO	300			Oil-resistant thermoplastic							Damp locations	
	SJTOOW[9]	300										Damp and wet locations	
Hard service cord	SO[7]	600	18–2	2 or more	Thermoset	18–15	0.76	30	None	Oil-resistant thermoset	Pendant or portable	Damp locations	Extra-hard usage
	SOW[7,9]	600										Damp and wet locations	
	SOO[7]	600			Oil-resistant thermoset	14–9 8–2	1.14 1.52	45 60				Damp locations	
	SOOW[7,9]	600										Damp and wet locations	
All thermoset parallel cord	SP-1	300	20–18	2 or 3	Thermoset	20–18	0.76	30	None	None	Pendant or portable	Damp locations	Not hard usage
	SP-2	300	18–16			18–16	1.14	45					
	SP-3	300	18–10			18–16 15, 14 12 10	1.52 2.03 2.41 2.80	60 80 95 110			Refrigerators, room air conditioners, and as permitted in 422.16(B)		
All elastomer (thermoplastic) parallel cord	SPE-1[8]	300	20–18	2 or 3	Thermoplastic elastomer	20–18	0.76	30	None	None	Pendant or portable	Damp locations	Not hard usage
	SPE-2[8]	300	18–16			18–16	1.14	45					
	SPE-3[8]	300	18–10			18–16 15 14 12 10	1.52 2.03 2.41 2.80	60 80 95 110			Refrigerators, room air conditioners, and as permitted in 422.16(B)		
All thermoplastic parallel cord	SPT-1	300	20–18	2 or 3	Thermoplastic	20–18	0.76	30	None	None	Pendant or portable	Damp locations	Not hard usage
	SPT-1W[9]	300		2								Damp and wet locations	

(continues)

Table 400.4 *Continued*

Trade Name	Type Letter	Voltage	AWG or kcmil	Number of Conductors	Insulation	AWG or kcmil	Nominal Insulation Thickness		Braid on Each Conductor	Outer Covering	Use		
							mm	mils					
	SPT-2	300	18–16	2 or 3		18–16	1.14	45				Damp locations	
	SPT-2W[9]	300		2								Damp and wet locations	
	SPT-3	300	18–10	2 or 3		18–16 15 14 12 10	1.52 2.03 2.41 2.80	60 80 95 110			Refrigerators, room air conditioners, and as permitted in 422.16(B)	Damp locations	Not hard usage
Range, dryer cable	SRD	300	10–4	3 or 4	Thermoset	10–4	1.14	45	None	Thermoset	Portable	Damp locations	Ranges, dryers
	SRDE	300	10–4	3 or 4	Thermoplastic elastomer				None	Thermoplastic elastomer			
	SRDT	300	10–4	3 or 4	Thermoplastic				None	Thermoplastic			
Hard service cord	ST[7]	600	18–2	2 or more	Thermoplastic	18–15 14–9 8–2	0.76 1.14 1.52	30 45 60	None	Thermoplastic	Pendant or portable	Damp locations	Extra-hard usage
	STW[7,9]	600										Damp and wet locations	
	STO[7]	600								Oil-resistant thermolastic		Damp locations	
	STOW[7,9]	600										Damp and wet locations	
	STOO[7]	600			Oil-resistant thermoplastic							Damp locations	
	STOOW[7]	600										Damp and wet locations	
Vacuum cleaner cord	SV	300	18–16	2 or 3	Thermoset	18–16	0.38	15	None	Thermoset	Pendant or portable	Damp locations	Not hard usage
	SVE	300			Thermoplastic elastomer					Thermoplastic elastomer			
	SVEO	300								Oil-resistant thermoplastic elastomer			
	SVEOO	300			Oil-resistant thermoplastic elastomer								
	SVO	300			Thermoset					Oil-resistant thermoset			
	SVOO	300			Oil-resistant thermoset					Oil-resistant thermoset			
	SVT	300			Thermoplastic					Thermoplastic			
	SVTO	300			Thermoplastic					Oil-resistant thermoplastic			
	SVTOO	300			Oil-resistant thermoplastic								

(continues)

Table 400.4 *Continued*

Trade Name	Type Letter	Voltage	AWG or kcmil	Number of Conductors	Insulation	AWG or kcmil	Nominal Insulation Thickness		Braid on Each Conductor	Outer Covering	Use		
							mm	mils					
Parallel tinsel cord	TPT[11]	300	27	2	Thermoplastic	27	0.76	30	None	Thermoplastic	Attached to an appliance	Damp locations	Not hard usage
Jacketed tinsel cord	TST[11]	300	27	2	Thermoplastic	27	0.38	15	None	Thermoplastic	Attached to an appliance	Damp locations	Not hard usage
Portable power cable	W[7]	2000	12–500 501–1000	1–6 1	Thermoset	12–2 1–4/0 250–500 501–1000	1.52 2.03 2.41 2.80	60 80 95 110		Oil-resistant thermoset	Portable, extra-hard usage		

Notes:

All types listed in Table 400.4 shall have individual conductors twisted together, except for Types HPN, SP-1, SP-2, SP-3, SPE-1, SPE-2, SPE-3, SPT-1, SPT-2, SPT-3, SPT-1W, SPT-2W, TPT, NISP-1, NISP-2, NISPT-1, NISPT-2, NISPE-1, NISPE-2, and three-conductor parallel versions of SRD, SRDE, and SRDT.

The individual conductors of all cords, except those of heat-resistant cords, shall have a thermoset or thermoplastic insulation, except that the equipment grounding conductor, where used, shall be in accordance with 400.23(B).

Rubber-filled or varnished cambric tapes shall be permitted as a substitute for the inner braids.

Elevator traveling cables for operating control and signal circuits shall contain nonmetallic fillers as necessary to maintain concentricity. Cables shall have steel supporting members as required for suspension by 620.41. In locations subject to excessive moisture or corrosive vapors or gases, supporting members of other materials shall be permitted. Where steel supporting members are used, they shall run straight through the center of the cable assembly and shall not be cabled with the copper strands of any conductor.

In addition to conductors used for control and signaling circuits, Types E, EO, ETP, and ETT elevator cables shall be permitted to incorporate in the construction one or more 20 AWG telephone conductor pairs, one or more coaxial cables, or one or more optical fibers. The 20 AWG conductor pairs shall be permitted to be covered with suitable shielding for telephone, audio, or higher frequency communications circuits; the coaxial cables shall consist of a center conductor, insulation, and a shield for use in video or other radio frequency communications circuits. The optical fiber shall be suitably covered with flame-retardant thermoplastic. The insulation of the conductors shall be rubber or thermoplastic of a thickness not less than specified for the other conductors of the particular type of cable. Metallic shields shall have their own protective covering. Where used, these components shall be permitted to be incorporated in any layer of the cable assembly but shall not run straight through the center.

Insulations and outer coverings that meet the requirements as flame retardant, limited smoke, and are so listed, shall be permitted to be marked for limited smoke after the *Code* type designation.

Elevator cables in sizes 20 AWG through 14 AWG are rated 300 volts, and sizes 10 AWG through 2 AWG are rated 600 volts. 12 AWG is rated 300 volts with a 0.76 mm (30 mil) insulation thickness and 600 volts with a 1.14 mm (45 mil) insulation thickness.

Conductor size for Types EV, EVJ, EVE, EVJE, EVT, and EVJT cables apply to nonpower-limited circuits only. Conductors of power-limited (data, signal, or communications) circuits may extend beyond the stated AWG size range. All conductors shall be insulated for the same cable voltage rating. Insulation thickness for Types EV, EVJ, EVEJE, EVT, and EVJT cables of nylon construction is indicated in parentheses.

Types G, G-GC, S, SC, SCE, SCT, SE, SEO, SEOO, SEW, SEOW, SEOOW, SO, SOO, SOW, SOOW, ST, STO, STOO, STW, STOW, STOOW, PPE, and W shall be permitted for use on theater stages, in garages, and elsewhere where flexible cords are permitted by this *Code*.

The third conductor in Type HPN shall be used as an equipment grounding conductor only. The insulation of the equipment grounding conductor for Types SPE-1, SPE-2, SPE-3, SPT-1, SPT-2, SPT-3, NISPT-1, NISPT-2, NISPE-1, and NISPE-2 shall be permitted to be thermoset polymer.

Cords that comply with the requirements for outdoor cords and are so listed shall be permitted to be designated as weather and water resistant with the suffix "W" after the *Code* type designation. Cords with the "W" suffix are suitable for use in wet locations and are sunlight resistant.

The required outer covering on some single-conductor cables may be integral with the insulation.

Types TPT and TST shall be permitted in lengths not exceeding 2.5 m (8 ft) where attached directly, or by means of a special type of plug, to a portable appliance rated at 50 watts or less and of such nature that extreme flexibility of the cord is essential.

Table 400.5(A)(1) Allowable Ampacity for Flexible Cords and Flexible Cables [Based on Ambient Temperature of 30°C (86°F). See 400.13 and Table 400.4.]

Copper Conductor Size (AWG)	Thermoplastic Types TPT, TST	Thermoset Types C, E, EO, PD, S, SJ, SJO, SJOW, SJOO, SJOOW, SO, SOW, SOO, SOOW, SP-1, SP-2, SP-3, SRD, SV, SVO, SVOO, NISP-1, NISP-2 / Thermoplastic Types ETP, ETT, NISPE-1, NISPE-2, NISPT-1, NISPT-2, SE, SEW, SEO, SEOO, SEOW, SEOOW, SJE, SJEW, SJEO, SJEOO, SJEOW, SJEOOW, SJT, SJTW, SJTO, SJTOW, SJTOO, SJTOOW, SPE-1, SPE-2, SPE-3, SPT-1, SPT-1W, SPT-2, SPT-2W, SPT-3, ST, STW, SRDE, SRDT, STO, STOW, STOO, STOOW, SVE, SVEO, SVEOO, SVT, SVTO, SVTOO		Types HPD, HPN, HSJ, HSJO, HSJOW, HSJOO, HSJOOW
		Column A[a]	Column B[b]	
27[c]	0.5	—	—	—
20	—	5[d]	[e]	—
18	—	7	10	10
17	—	9	12	13
16	—	10	13	15
15	—	12	16	17
14	—	15	18	20
13	—	17	21	—
12	—	20	25	30
11	—	23	27	—
10	—	25	30	35
9	—	29	34	—
8	—	35	40	—
7		40	47	
6	—	45	55	—
5		52	62	
4	—	60	70	—
3		70	82	
2	—	80	95	

[a]The allowable currents under Column A apply to three-conductor cords and other multiconductor cords connected to utilization equipment so that only three-conductors are current-carrying.

[b]The allowable currents under Column B apply to two-conductor cords and other multiconductor cords connected to utilization equipment so that only two conductors are current-carrying.

[c]Tinsel cord.

[d]Elevator cables only.

[e]7 amperes for elevator cables only; 2 amperes for other types.

Table 400.5(A)(2) Ampacity of Cable Types SC, SCE, SCT, PPE, G, G-GC, and W [Based on Ambient Temperature of 30°C (86°F). See Table 400.4.]

Copper Conductor Size (AWG or kcmil)	Temperature Rating of Cable								
	60°C (140°F)			75°C (167°F)			90°C (194°F)		
	D[1]	E[2]	F[3]	D[1]	E[2]	F[3]	D[1]	E[2]	F[3]
12	—	31	26	—	37	31	—	42	35
10	—	44	37	—	52	43	—	59	49
8	60	55	48	70	65	57	80	74	65
6	80	72	63	95	88	77	105	99	87
4	105	96	84	125	115	101	140	130	114
3	120	113	99	145	135	118	165	152	133
2	140	128	112	170	152	133	190	174	152
1	165	150	131	195	178	156	220	202	177
1/0	195	173	151	230	207	181	260	234	205
2/0	225	199	174	265	238	208	300	271	237
3/0	260	230	201	310	275	241	350	313	274
4/0	300	265	232	360	317	277	405	361	316
250	340	296	259	405	354	310	455	402	352
300	375	330	289	445	395	346	505	449	393
350	420	363	318	505	435	381	570	495	433
400	455	392	343	545	469	410	615	535	468
500	515	448	392	620	537	470	700	613	536
600	575	—	—	690	—	—	780	—	—
700	630	—	—	755	—	—	855	—	—
750	655	—	—	785	—	—	885	—	—
800	680	—	—	815	—	—	920	—	—
900	730	—	—	870	—	—	985	—	—
1000	780	—	—	935	—	—	1055	—	—

[1]The ampacities under subheading D shall be permitted for single-conductor Types SC, SCE, SCT, PPE, and W cable only where the individual conductors are not installed in raceways and are not in physical contact with each other except in lengths not to exceed 600 mm (24 in.) where passing through the wall of an enclosure.

[2]The ampacities under subheading E apply to two-conductor cables and other multiconductor cables connected to utilization equipment so that only two conductors are current-carrying.

[3]The ampacities under subheading F apply to three-conductor cables and other multiconductor cables connected to utilization equipment so that only three conductors are current-carrying.

Table 400.5(A)(3) Adjustment Factors for More Than Three Current-Carrying Conductors in a Flexible Cord or Flexible Cable

Number of Conductors	Percent of Value in Table 400.5(A)(1) and Table 400.5(A)(2)
4–6	80
7–9	70
10–20	50
21–30	45
31–40	40
41 and above	35

400.10 Uses Permitted.

(A) Uses. Flexible cords and flexible cables shall be used only for the following:

(1) Pendants.
(2) Wiring of luminaires.
(3) Connection of portable luminaires, portable and mobile signs, or appliances.
(4) Elevator cables.
(5) Wiring of cranes and hoists.
(6) Connection of utilization equipment to facilitate frequent interchange.
(7) Prevention of the transmission of noise or vibration.
(8) Appliances where the fastening means and mechanical connections are specifically designed to permit ready removal for maintenance and repair, and the appliance is intended or identified for flexible cord connection.
(9) Connection of moving parts.
(10) Where specifically permitted elsewhere in this *Code*.
(11) Between an existing receptacle outlet and an inlet, where the inlet provides power to an additional single receptacle outlet. The wiring interconnecting the inlet to the single receptacle outlet shall be a Chapter 3 wiring method. The inlet, receptacle outlet, and Chapter 3 wiring method, including the flexible cord and fittings, shall be a listed assembly specific for this application.

(B) Attachment Plugs. Where used as permitted in 400.10(A)(3), (A)(6), and (A)(8), each flexible cord shall be equipped with an attachment plug and shall be energized from a receptacle outlet or cord connector body.

Exception: As permitted in 368.56.

400.12 Uses Not Permitted. Unless specifically permitted in 400.10, flexible cables, flexible cord sets, and power supply cords shall not be used for the following:

(1) As a substitute for the fixed wiring of a structure
(2) Where run through holes in walls, structural ceilings, suspended ceilings, dropped ceilings, or floors
(3) Where run through doorways, windows, or similar openings
(4) Where attached to building surfaces

Exception to (4): Flexible cord and flexible cable shall be permitted to be attached to building surfaces in accordance with 368.56(B).

(5) Where concealed by walls, floors, or ceilings or located above suspended or dropped ceilings

Exception to (5): Flexible cord and flexible cable shall be permitted if contained within an enclosure for use in Other Spaces Used for Environmental Air as permitted by 300.22(C)(3).

(6) Where installed in raceways, except as otherwise permitted in this *Code*
(7) Where subject to physical damage

400.13 Splices. Flexible cord shall be used only in continuous lengths without splice or tap where initially installed in applications permitted by 400.10(A). The repair of hard-service cord and junior hard-service cord (see Trade Name column in Table 400.4) 14 AWG and larger shall be permitted if conductors are spliced in accordance with 110.14(B) and the completed splice retains the insulation, outer sheath properties, and usage characteristics of the cord being spliced.

400.14 Pull at Joints and Terminals. Flexible cords and flexible cables shall be connected to devices and to fittings so that tension is not transmitted to joints or terminals.

Exception: Listed portable single-pole devices that are intended to accommodate such tension at their terminals shall be permitted to be used with single-conductor flexible cable.

Informational Note: Some methods of preventing pull on a cord from being transmitted to joints or terminals include knotting the cord, winding with tape, and using support or strain-relief fittings.

400.15 In Show Windows and Showcases. Flexible cords used in show windows and showcases shall be Types S, SE, SEO, SEOO, SJ, SJE, SJEO, SJEOO, SJO, SJOO, SJT, SJTO, SJTOO, SO, SOO, ST, STO, STOO, SEW, SEOW, SEOOW, SJEW, SJEOW, SJEOOW, SJOW, SJOOW, SJTW, SJTOW, SJTOOW, SOW, SOOW, STW, STOW, or STOOW.

Exception No. 1: For the wiring of chain-supported luminaires.

Exception No. 2: As supply cords for portable luminaires and other merchandise being displayed or exhibited.

400.16 Overcurrent Protection. Flexible cords not smaller than 18 AWG, and tinsel cords or cords having equivalent characteristics of smaller size approved for use with specific appliances, shall be considered as protected against overcurrent in accordance with 240.5.

400.17 Protection from Damage. Flexible cords and flexible cables shall be protected by bushings or fittings where passing through holes in covers, outlet boxes, or similar enclosures.

In industrial establishments where the conditions of maintenance and supervision ensure that only qualified persons service the installation, flexible cords and flexible cables shall be permitted to be installed in aboveground raceways that are no longer than 15 m (50 ft) to protect the flexible cord or flexible cable from physical damage. Where more than three current-carrying conductors are installed within the raceway, the allowable ampacity shall be reduced in accordance with Table 400.5(A)(3).

Part II. Construction Specifications

400.20 Labels. Flexible cords shall be examined and tested at the factory and labeled before shipment.

400.21 Construction.

(A) Conductors. The individual conductors of a flexible cord or flexible cable shall have copper flexible stranding and shall not be smaller than the sizes specified in Table 400.4.

(B) Nominal Insulation Thickness. The nominal thickness of insulation for conductors of flexible cords and flexible cables shall not be less than specified in Table 400.4.

400.22 Grounded-Conductor Identification. One conductor of flexible cords that is intended to be used as a grounded circuit conductor shall have a continuous marker that readily distinguishes it from the other conductor or conductors. The identification shall consist of one of the methods indicated in 400.22(A) through (F).

(A) Colored Braid. A braid finished to show a white or gray color and the braid on the other conductor or conductors finished to show a readily distinguishable solid color or colors.

(B) Tracer in Braid. A tracer in a braid of any color contrasting with that of the braid and no tracer in the braid of the other conductor or conductors. No tracer shall be used in the braid of any conductor of a flexible cord that contains a conductor having a braid finished to show white or gray.

Exception: In the case of Types C and PD and cords having the braids on the individual conductors finished to show white or gray. In such cords, the identifying marker shall be permitted to consist of the solid white or gray finish on one conductor, provided there is a colored tracer in the braid of each other conductor.

(C) Colored Insulation. A white or gray insulation on one conductor and insulation of a readily distinguishable color or colors on the other conductor or conductors for cords having no braids on the individual conductors.

For jacketed cords furnished with appliances, one conductor having its insulation colored light blue, with the other conductors having their insulation of a readily distinguishable color other than white or gray.

Exception: Cords that have insulation on the individual conductors integral with the jacket.

The insulation shall be permitted to be covered with an outer finish to provide the desired color.

(D) Colored Separator. A white or gray separator on one conductor and a separator of a readily distinguishable solid color on the other conductor or conductors of cords having insulation on the individual conductors integral with the jacket.

(E) Tinned Conductors. One conductor having the individual strands tinned and the other conductor or conductors having the individual strands untinned for cords having insulation on the individual conductors integral with the jacket.

(F) Surface Marking. One or more ridges, grooves, or white stripes located on the exterior of the cord so as to identify one conductor for cords having insulation on the individual conductors integral with the jacket.

400.23 Equipment Grounding Conductor Identification. A conductor intended to be used as an equipment grounding conductor shall have a continuous identifying marker readily distinguishing it from the other conductor or conductors. Conductors having a continuous green color or a continuous green color with one or more yellow stripes shall not be used for other than equipment grounding conductors. Cords or cables consisting of integral insulation and a jacket without a nonintegral grounding conductor shall be permitted to be green. The identifying marker shall consist of one of the methods in 400.23(A) or (B).

(A) Colored Braid. A braid finished to show a continuous green color or a continuous green color with one or more yellow stripes.

(B) Colored Insulation or Covering. For cords having no braids on the individual conductors, an insulation of a continuous green color or a continuous green color with one or more yellow stripes.

400.24 Attachment Plugs. Where a flexible cord is provided with an equipment grounding conductor and equipped with an attachment plug, the attachment plug shall comply with 250.138(A) and (B).

Part III. Portable Cables Over 600 Volts, Nominal

400.30 Scope. Part III applies to single and multiconductor portable cables used to connect mobile equipment and machinery.

400.31 Construction.

(A) Conductors. The conductors shall be 12 AWG copper or larger and shall employ flexible stranding.

(B) Equipment Grounding Conductor(s). An equipment grounding conductor(s) shall be provided in cables with three or more conductors. The total area shall not be less than that of the size of the equipment grounding conductor required in 250.122.

400.32 Shielding. All shields shall be connected to an equipment grounding conductor.

400.33 Equipment Grounding Conductors. Equipment grounding conductors shall be connected in accordance with Parts VI and VII of Article 250.

400.34 Minimum Bending Radii. The minimum bending radii for portable cables during installation and handling in service shall be adequate to prevent damage to the cable.

400.35 Fittings. Connectors used to connect lengths of cable in a run shall be of a type that locks firmly together. Provisions shall be made to prevent opening or closing these connectors while energized. Suitable means shall be used to eliminate tension at connectors and terminations.

400.36 Splices and Terminations. Portable cables shall not contain splices unless the splices are of the permanent molded, vulcanized types in accordance with 110.14(B). Terminations on portable cables rated over 600 volts, nominal, shall be accessible only to authorized and qualified personnel.

ARTICLE 402
Fixture Wires

402.1 Scope. This article covers general requirements and construction specifications for fixture wires.

402.2 Other Articles. Fixture wires shall comply with this article and also with the applicable provisions of other articles of this *Code*.

Informational Note: For application in luminaires, see Article 410.

402.3 Types. Fixture wires shall be of a type listed in Table 402.3, and they shall comply with all requirements of that table. The fixture wires listed in Table 402.3 are all suitable for service at 600 volts, nominal, unless otherwise specified.

Informational Note: Thermoplastic insulation may stiffen at temperatures lower than −10°C (+14°F). Thermoplastic insulation may also be deformed at normal temperatures where subjected to pressure, such as at points of support.

Table 402.3 Fixture Wires

Name	Type Letter	Insulation	AWG	Thickness of Insulation		Outer Covering	Maximum Operating Temperature	Application Provisions
				mm	mils			
Heat-resistant rubber-covered fixture wire — flexible stranding	FFH-2	Heat-resistant rubber	18–16	0.76	30	Nonmetallic covering	75°C	Fixture wiring
		Cross-linked synthetic polymer	18–16	0.76	30		(167°F)	
ECTFE — solid or 7-strand	HF	Ethylene chloro-trifluoroethylene	18–14	0.38	15	None	150°C (302°F)	Fixture wiring
ECTFE — flexible stranding	HFF	Ethylene chlorotrifluo-roethylene	18–14	0.38	15	None	150°C (302°F)	Fixture wiring
Tape insulated fixture wire — solid or 7-strand	KF-1	Aromatic polyimide tape	18–10	0.14	5.5	None	200°C (392°F)	Fixture wiring — limited to 300 volts
	KF-2	Aromatic polyimide tape	18–10	0.21	8.4	None	200°C (392°F)	Fixture wiring
Tape insulated fixture wire — flexible stranding	KFF-1	Aromatic polyimide tape	18–10	0.14	5.5	None	200°C (392°F)	Fixture wiring — limited to 300 volts
	KFF-2	Aromatic polyimide tape	18–10	0.21	8.4	None	200°C (392°F)	Fixture wiring
Perfluoro-alkoxy — solid or 7-strand (nickel or nickel-coated copper)	PAF	Perfluoro-alkoxy	18–14	0.51	20	None	250°C (482°F)	Fixture wiring (nickel or nickel-coated copper)
Perfluoro-alkoxy — flexible stranding	PAFF	Perfluoro-alkoxy	18–14	0.51	20	None	150°C (302°F)	Fixture wiring
Fluorinated ethylene propylene fixture wire — solid or 7-strand	PF	Fluorinated ethylene propylene	18–14	0.51	20	None	200°C (392°F)	Fixture wiring
Fluorinated ethylene propylene fixture wire — flexible stranding	PFF	Fluorinated ethylene propylene	18–14	0.51	20	None	150°C (302°F)	Fixture wiring
Fluorinated ethylene propylene fixture wire — solid or 7-strand	PGF	Fluorinated ethylene propylene	18–14	0.36	14	Glass braid	200°C (392°F)	Fixture wiring
Fluorinated ethylene propylene fixture wire — flexible stranding	PGFF	Fluorinated ethylene propylene	18–14	0.36	14	Glass braid	150°C (302°F)	Fixture wiring
Extruded polytetrafluoroethylene — solid or 7-strand (nickel or nickel-coated copper)	PTF	Extruded polytetrafluo-roethylene	18–14	0.51	20	None	250°C (482°F)	Fixture wiring (nickel or nickel-coated copper)
Extruded polytetrafluoroethylene — flexible stranding 26-36 (AWG silver or nickel-coated copper)	PTFF	Extruded polytetrafluo-roethylene	18–14	0.51	20	None	150°C (302°F)	Fixture wiring (silver or nickel-coated copper)

(continues)

Table 402.3 *Continued*

Name	Type Letter	Insulation	AWG	Thickness of Insulation		Outer Covering	Maximum Operating Temperature	Application Provisions
				mm	mils			
Heat-resistant rubber-covered fixture wire — solid or 7-strand	RFH-1	Heat-resistant rubber	18	0.38	15	Nonmetallic covering	75°C (167°F)	Fixture wiring — limited to 300 volts
	RFH-2	Heat-resistant rubber Cross-linked synthetic polymer	18–16	0.76	30	None or non-metallic covering	75°C (167°F)	Fixture wiring
Heat-resistant cross-linked synthetic polymer-insulated fixture wire — solid or 7-strand	RFHH-2* RFHH-3*	Cross-linked synthetic polymer	18–16 18–16	0.76 1.14	30 45	None or non-metallic covering	90°C (194°F)	Fixture wiring
Silicone insulated fixture wire — solid or 7-strand	SF-1	Silicone rubber	18	0.38	15	Nonmetallic covering	200°C (392°F)	Fixture wiring — limited to 300 volts
	SF-2	Silicone rubber	18–12 10	0.76 1.14	30 45	Nonmetallic covering	200°C (392°F)	Fixture wiring
Silicone insulated fixture wire — flexible stranding	SFF-1	Silicone rubber	18	0.38	15	Nonmetallic covering	150°C (302°F)	Fixture wiring — limited to 300 volts
	SFF-2	Silicone rubber	18–12 10	0.76 1.14	30 45	Nonmetallic covering	150°C (302°F)	Fixture wiring
Thermoplastic covered fixture wire — solid or 7-strand	TF*	Thermoplastic	18–16	0.76	30	None	60°C (140°F)	Fixture wiring
Thermoplastic covered fixture wire — flexible stranding	TFF*	Thermoplastic	18–16	0.76	30	None	60°C (140°F)	Fixture wiring
Heat-resistant thermoplastic covered fixture wire — solid or 7 strand	TFN*	Thermoplastic	18–16	0.38	15	Nylon-jacket-ed or equivalent	90°C (194°F)	Fixture wiring
Heat-resistant thermoplastic covered fixture wire — flexible stranded	TFFN*	Thermoplastic	18–16	0.38	15	Nylon-jacket-ed or equivalent	90°C (194°F)	Fixture wiring
Cross-linked polyolefin insulated fixture wire — solid or 7-strand	XF*	Cross-linked polyolefin	18–14 12–10	0.76 1.14	30 45	None	150°C (302°F)	Fixture wiring — limited to 300 volts
Cross-linked polyolefin insulated fixture wire — flexible stranded	XFF*	Cross-linked polyolefin	18–14 12–10	0.76 1.14	30 45	None	150°C (302°F)	Fixture wiring — limited to 300 volts
Modified ETFE — solid or 7-strand	ZF	Modified ethylene tetrafluoro-ethylene	18–14	0.38	15	None	150°C (302°F)	Fixture wiring
Flexible stranding	ZFF	Modified ethylene tetrafluoro-ethylene	18–14	0.38	15	None	150°C (302°F)	Fixture wiring
High temp. modified ETFE— solid or 7-strand	ZHF	Modified ethylene tetrafluoro-ethylene	18–14	0.38	15	None	200°C (392°F)	Fixture wiring

*Insulations and outer coverings that meet the requirements of flame retardant, limited smoke, and are so listed, shall be permitted to be marked for limited smoke after the *Code* type designation.

402.5 Allowable Ampacities for Fixture Wires. The allowable ampacity of fixture wire shall be as specified in Table 402.5.

No conductor shall be used under such conditions that its operating temperature exceeds the temperature specified in Table 402.3 for the type of insulation involved.

Informational Note: See 310.15(A)(3) for temperature limitation of conductors.

402.6 Minimum Size. Fixture wires shall not be smaller than 18 AWG.

402.7 Number of Conductors in Conduit or Tubing. The number of fixture wires permitted in a single conduit or tubing shall not exceed the percentage fill specified in Table 1, Chapter 9.

402.8 Grounded Conductor Identification. Fixture wires that are intended to be used as grounded conductors shall be identified by one or more continuous white stripes on other than green insulation or by the means described in 400.22(A) through (E).

402.9 Marking.

(A) Method of Marking. Thermoplastic insulated fixture wire shall be durably marked on the surface at intervals not exceeding 610 mm (24 in.). All other fixture wire shall be marked by means of a printed tag attached to the coil, reel, or carton.

(B) Optional Marking. Fixture wire types listed in Table 402.3 shall be permitted to be surface marked to indicate special characteristics of the cable materials. These markings include, but are not limited to, markings for limited smoke, sunlight resistance, and so forth.

402.10 Uses Permitted. Fixture wires shall be permitted (1) for installation in luminaires and in similar equipment where enclosed or protected and not subject to bending or twisting in use, or (2) for connecting luminaires to the branch-circuit conductors supplying the luminaires.

402.12 Uses Not Permitted. Fixture wires shall not be used as branch-circuit conductors except as permitted elsewhere in this *Code*.

402.14 Overcurrent Protection. Overcurrent protection for fixture wires shall be as specified in 240.5.

Table 402.5 Allowable Ampacity for Fixture Wires

Size (AWG)	Allowable Ampacity
18	6
16	8
14	17
12	23
10	28

ARTICLE 404
Switches

Part I. Installation

404.1 Scope. The provisions of this article apply to all switches, switching devices, and circuit breakers used as switches operating at 1000 volts and below, unless specifically referenced elsewhere in this *Code* for higher voltages.

404.2 Switch Connections.

(A) Three-Way and Four-Way Switches. Three-way and four-way switches shall be wired so that all switching is done only in the ungrounded circuit conductor. Where in metal raceways or metal-armored cables, wiring between switches and outlets shall be in accordance with 300.20(A).

Exception: Switch loops shall not require a grounded conductor.

(B) Grounded Conductors. Switches or circuit breakers shall not disconnect the grounded conductor of a circuit.

Exception: A switch or circuit breaker shall be permitted to disconnect a grounded circuit conductor where all circuit conductors are disconnected simultaneously, or where the device is arranged so that the grounded conductor cannot be disconnected until all the ungrounded conductors of the circuit have been disconnected.

(C) Switches Controlling Lighting Loads. The grounded circuit conductor for the controlled lighting circuit shall be installed at the location where switches control lighting loads that are supplied by a grounded general-purpose branch circuit serving bathrooms, hallways, stairways, or rooms suitable for human habitation or occupancy as defined in the applicable building code. Where multiple switch locations control the same lighting load such that the entire floor area of the room or space is visible from the single or combined switch locations, the grounded circuit conductor shall only be required at one location. A grounded conductor shall not be required to be installed at lighting switch locations under any of the following conditions:

(1) Where conductors enter the box enclosing the switch through a raceway, provided that the raceway is large enough for all contained conductors, including a grounded conductor

(2) Where the box enclosing the switch is accessible for the installation of an additional or replacement cable without removing finish materials

(3) Where snap switches with integral enclosures comply with 300.15(E)

(4) Where lighting in the area is controlled by automatic means

(5) Where a switch controls a receptacle load

The grounded conductor shall be extended to any switch location as necessary and shall be connected to switching devices that require line-to-neutral voltage to operate the electronics of the switch in the standby mode and shall meet the requirements of 404.22.

Exception: The connection requirement shall become effective on January 1, 2020. It shall not apply to replacement or retrofit switches installed in locations prior to local adoption of 404.2(C) and where the grounded conductor cannot be extended without removing finish materials. The number of electronic lighting control switches on a branch circuit shall not exceed five, and the number connected to any feeder on the load side of a system or main bonding jumper shall not exceed 25. For the purpose of this exception, a neutral busbar, in compliance with 200.2(B) and to which a main or system bonding jumper is connected shall not be limited as to the number of electronic lighting control switches connected.

Informational Note: The provision for a (future) grounded conductor is to complete a circuit path for electronic lighting control devices.

404.3 Enclosure.

(A) General. Switches and circuit breakers shall be of the externally operable type mounted in an enclosure listed for the intended use. The minimum wire-bending space at terminals and minimum gutter space provided in switch enclosures shall be as required in 312.6.

Exception No. 1: Pendant- and surface-type snap switches and knife switches mounted on an open-face switchboard or panelboard shall be permitted without enclosures.

Exception No. 2: Switches and circuit breakers installed in accordance with 110.27(A)(1), (A)(2), (A)(3), or (A)(4) shall be permitted without enclosures.

(B) Used as a Raceway. Enclosures shall not be used as junction boxes, auxiliary gutters, or raceways for conductors feeding through or tapping off to other switches or overcurrent devices, unless the enclosure complies with 312.8.

404.4 Damp or Wet Locations.

(A) Surface-Mounted Switch or Circuit Breaker. A surface-mounted switch or circuit breaker shall be enclosed in a weatherproof enclosure or cabinet that complies with 312.2.

(B) Flush-Mounted Switch or Circuit Breaker. A flush-mounted switch or circuit breaker shall be equipped with a weatherproof cover.

(C) Switches in Tub or Shower Spaces. Switches shall not be installed within tubs or shower spaces unless installed as part of a listed tub or shower assembly.

404.5 Time Switches, Flashers, and Similar Devices.
Time switches, flashers, and similar devices shall be of the enclosed type or shall be mounted in cabinets or boxes or equipment enclosures. Energized parts shall be barriered to prevent operator exposure when making manual adjustments or switching.

Exception: Devices mounted so they are accessible only to qualified persons shall be permitted without barriers, provided they are located within an enclosure such that any energized parts within 152 mm (6.0 in.) of the manual adjustment or switch are covered by suitable barriers.

404.6 Position and Connection of Switches.

(A) Single-Throw Knife Switches. Single-throw knife switches shall be placed so that gravity will not tend to close them. Single-throw knife switches, approved for use in the inverted position, shall be provided with an integral mechanical means that ensures that the blades remain in the open position when so set.

(B) Double-Throw Knife Switches. Double-throw knife switches shall be permitted to be mounted so that the throw is either vertical or horizontal. Where the throw is vertical, integral mechanical means shall be provided to hold the blades in the open position when so set.

(C) Connection of Switches. Single-throw knife switches and switches with butt contacts shall be connected such that their blades are de-energized when the switch is in the open position. Bolted pressure contact switches shall have barriers that prevent inadvertent contact with energized blades. Single-throw knife switches, bolted pressure contact switches, molded case switches, switches with butt contacts, and circuit breakers used as switches shall be connected so that the terminals supplying the load are de-energized when the switch is in the open position.

Exception: The blades and terminals supplying the load of a switch shall be permitted to be energized when the switch is in the open position where the switch is connected to circuits or equipment inherently capable of providing a backfeed source of power. For such installations, a permanent sign shall be installed on the switch enclosure or immediately adjacent to open switches with the following words or equivalent: WARNING — LOAD SIDE TERMINALS MAY BE ENERGIZED BY BACKFEED. The warning sign or label shall comply with 110.21(B).

404.7 Indicating.
General-use and motor-circuit switches, circuit breakers, and molded case switches, where mounted in an enclosure as described in 404.3, shall clearly indicate whether they are in the open (off) or closed (on) position.

Where these switch or circuit breaker handles are operated vertically rather than rotationally or horizontally, the up position of the handle shall be the closed (on) position.

Exception No. 1: Vertically operated double-throw switches shall be permitted to be in the closed (on) position with the handle in either the up or down position.

Exception No. 2: On busway installations, tap switches employing a center-pivoting handle shall be permitted to be open or closed with either end of the handle in the up or down position. The switch position shall be clearly indicating and shall be visible from the floor or from the usual point of operation.

404.8 Accessibility and Grouping.

(A) Location. All switches and circuit breakers used as switches shall be located so that they may be operated from a readily accessible place. They shall be installed such that the center of the grip of the operating handle of the switch or circuit breaker, when in its highest position, is not more than 2.0 m (6 ft 7 in.) above the floor or working platform.

Exception No. 1: On busway installations, fused switches and circuit breakers shall be permitted to be located at the same level as the busway. Suitable means shall be provided to operate the handle of the device from the floor.

Exception No. 2: Switches and circuit breakers installed adjacent to motors, appliances, or other equipment that they supply shall be permitted to be located higher than 2.0 m (6 ft 7 in.) and to be accessible by portable means.

Exception No. 3: Hookstick operable isolating switches shall be permitted at greater heights.

(B) Voltage Between Adjacent Devices. A snap switch shall not be grouped or ganged in enclosures with other snap switches, receptacles, or similar devices, unless they are arranged so that the voltage between adjacent devices does not exceed 300 volts, or unless they are installed in enclosures equipped with identified, securely installed barriers between adjacent devices.

(C) Multipole Snap Switches. A multipole, general-use snap switch shall not be permitted to be fed from more than a single circuit unless it is listed and marked as a two-circuit or three-circuit switch.

> Informational Note: See 210.7 for disconnect requirements where more than one circuit supplies a switch.

404.9 Provisions for General-Use Snap Switches.

(A) Faceplates. Faceplates provided for snap switches mounted in boxes and other enclosures shall be installed so as to completely cover the opening and, where the switch is flush mounted, seat against the finished surface.

(B) Grounding. Snap switches, including dimmer and similar control switches, shall be connected to an equipment grounding conductor and shall provide a means to connect metal faceplates to the equipment grounding conductor, whether or not a metal faceplate is installed. Metal faceplates shall be grounded. Snap switches shall be considered to be part of an effective ground-fault current path if either of the following conditions is met:

(1) The switch is mounted with metal screws to a metal box or metal cover that is connected to an equipment grounding conductor or to a nonmetallic box with integral means for connecting to an equipment grounding conductor.

(2) An equipment grounding conductor or equipment bonding jumper is connected to an equipment grounding termination of the snap switch.

Exception No. 1 to (B): Where no means exists within the snap-switch enclosure for connecting to the equipment grounding conductor, or where the wiring method does not include or provide an equipment grounding conductor, a snap switch without a connection to an equipment grounding conductor shall be permitted for replacement purposes only. A snap switch wired under the provisions of this exception and located within 2.5 m (8 ft) vertically, or 1.5 m (5 ft) horizontally, of ground or exposed grounded metal objects shall be provided with a faceplate of nonconducting noncombustible material with nonmetallic attachment screws, unless the switch mounting strap or yoke is nonmetallic or the circuit is protected by a ground-fault circuit interrupter.

Exception No. 2 to (B): Listed kits or listed assemblies shall not be required to be connected to an equipment grounding conductor if all of the following conditions are met:

(1) The device is provided with a nonmetallic faceplate that cannot be installed on any other type of device,

(2) The device does not have mounting means to accept other configurations of faceplates,

(3) The device is equipped with a nonmetallic yoke, and

(4) All parts of the device that are accessible after installation of the faceplate are manufactured of nonmetallic materials.

Exception No. 3 to (B): A snap switch with integral nonmetallic enclosure complying with 300.15(E) shall be permitted without a connection to an equipment grounding conductor.

(C) Construction. Metal faceplates shall be of ferrous metal not less than 0.76 mm (0.030 in.) in thickness or of nonferrous metal not less than 1.02 mm (0.040 in.) in thickness. Faceplates of insulating material shall be noncombustible and not less than 2.54 mm (0.100 in.) in thickness, but they shall be permitted to be less than 2.54 mm (0.100 in.) in thickness if formed or reinforced to provide adequate mechanical strength.

404.10 Mounting of Snap Switches.

(A) Surface Type. Snap switches used with open wiring on insulators shall be mounted on insulating material that separates the conductors at least 13 mm (½ in.) from the surface wired over.

(B) Box Mounted. Flush-type snap switches mounted in boxes that are set back of the finished surface as permitted in 314.20 shall be installed so that the extension plaster ears are seated against the surface. Flush-type snap switches mounted in boxes that are flush with the finished surface or project from it shall be installed so that the mounting yoke or strap of the switch is seated against the box. Screws used for the purpose of attaching a snap switch to a box shall be of the type provided with a listed snap switch, or shall be machine screws having 32 threads per inch or part of listed assemblies or systems, in accordance with the manufacturer's instructions.

404.11 Circuit Breakers as Switches. A hand-operable circuit breaker equipped with a lever or handle, or a power-operated circuit breaker capable of being opened by hand in the event of a power failure, shall be permitted to serve as a switch if it has the required number of poles.

> Informational Note: See the provisions contained in 240.81 and 240.83.

404.12 Grounding of Enclosures. Metal enclosures for switches or circuit breakers shall be connected to an equipment grounding conductor as specified in Part IV of Article 250. Metal enclosures for switches or circuit breakers used as service equipment shall comply with the provisions of Part V of Article 250. Where nonmetallic enclosures are used with metal raceways or metal-armored cables, provision shall be made for connecting the equipment grounding conductor(s).

Except as covered in 404.9(B), Exception No. 1, nonmetallic boxes for switches shall be installed with a wiring method that provides or includes an equipment grounding conductor.

404.13 Knife Switches.

(A) Isolating Switches. Knife switches rated at over 1200 amperes at 250 volts or less, and at over 1000 amperes at 251 to 1000 volts, shall be used only as isolating switches and shall not be opened under load.

(B) To Interrupt Currents. To interrupt currents over 1200 amperes at 250 volts, nominal, or less, or over 600 amperes at 251 to 1000 volts, nominal, a circuit breaker or a switch listed for such purpose shall be used.

(C) General-Use Switches. Knife switches of ratings less than specified in 404.13(A) and (B) shall be considered general-use switches.

> Informational Note: See the definition of *General-Use Switch* in Article 100.

(D) Motor-Circuit Switches. Motor-circuit switches shall be permitted to be of the knife-switch type.

> Informational Note: See the definition of a *Motor-Circuit Switch* in Article 100.

404.14 Rating and Use of Switches. Switches shall be used within their ratings and as indicated in 404.14(A) through (F).

> Informational Note No. 1: For switches on signs and outline lighting, see 600.6.

> Informational Note No. 2: For switches controlling motors, see 430.83, 430.109, and 430.110.

(A) Alternating-Current General-Use Snap Switch. A form of general-use snap switch suitable only for use on ac circuits for controlling the following:

(1) Resistive and inductive loads not exceeding the ampere rating of the switch at the voltage applied

(2) Tungsten-filament lamp loads not exceeding the ampere rating of the switch at 120 volts

(3) Motor loads not exceeding 80 percent of the ampere rating of the switch at its rated voltage

(B) Alternating-Current or Direct-Current General-Use Snap Switch. A form of general-use snap switch suitable for use on either ac or dc circuits for controlling the following:

(1) Resistive loads not exceeding the ampere rating of the switch at the voltage applied.

(2) Inductive loads not exceeding 50 percent of the ampere rating of the switch at the applied voltage. Switches rated in horsepower are suitable for controlling motor loads within their rating at the voltage applied.

(3) Tungsten-filament lamp loads not exceeding the ampere rating of the switch at the applied voltage if T-rated.

(C) CO/ALR Snap Switches. Snap switches rated 20 amperes or less directly connected to aluminum conductors shall be listed and marked CO/ALR.

(D) Alternating-Current Specific-Use Snap Switches Rated for 347 Volts. Snap switches rated 347 volts ac shall be listed and shall be used only for controlling the loads permitted by (D)(1) and (D)(2).

(1) Noninductive Loads. Noninductive loads other than tungsten-filament lamps not exceeding the ampere and voltage ratings of the switch.

(2) Inductive Loads. Inductive loads not exceeding the ampere and voltage ratings of the switch. Where particular load characteristics or limitations are specified as a condition of the listing, those restrictions shall be observed regardless of the ampere rating of the load.

The ampere rating of the switch shall not be less than 15 amperes at a voltage rating of 347 volts ac. Flush-type snap switches rated 347 volts ac shall not be readily interchangeable in box mounting with switches identified in 404.14(A) and (B).

(E) Dimmer Switches. General-use dimmer switches shall be used only to control permanently installed incandescent luminaires unless listed for the control of other loads and installed accordingly.

(F) Cord- and Plug-Connected Loads. Where a snap switch or control device is used to control cord- and plug-connected equipment on a general-purpose branch circuit, each snap switch or control device controlling receptacle outlets or cord connectors that are supplied by permanently connected cord pendants shall be rated at not less than the rating of the maximum permitted ampere rating or setting of the overcurrent device protecting the receptacles or cord connectors, as provided in 210.21(B).

> Informational Note: See 210.50(A) and 400.10(A)(1) for equivalency to a receptacle outlet of a cord connector that is supplied by a permanently connected cord pendant.

Exception: Where a snap switch or control device is used to control not more than one receptacle on a branch circuit, the switch or control device shall be permitted to be rated at not less than the rating of the receptacle.

Part II. Construction Specifications

404.20 Marking.

(A) Ratings. Switches shall be marked with the current, voltage, and, if horsepower rated, the maximum rating for which they are designed.

(B) Off Indication. Where in the off position, a switching device with a marked OFF position shall completely disconnect all ungrounded conductors to the load it controls.

404.22 Electronic Lighting Control Switches. Electronic lighting control switches shall be listed. Electronic lighting control switches shall not introduce current on the equipment grounding conductor during normal operation. The requirement to not introduce current on the equipment grounding conductor shall take effect on January 1, 2020.

Exception: Electronic lighting control switches that introduce current on the equipment grounding conductor shall be permitted for applications covered by 404.2(C), Exception. Electronic lighting control switches that introduce current on the equipment grounding conductor shall be listed and marked for use in replacement or retrofit applications only.

404.26 Knife Switches Rated 600 to 1000 Volts. Auxiliary contacts of a renewable or quick-break type or the equivalent shall be provided on all knife switches rated 600 to 1000 volts and designed for use in breaking current over 200 amperes.

404.27 Fused Switches. A fused switch shall not have fuses in parallel except as permitted in 240.8.

404.28 Wire-Bending Space. The wire-bending space required by 404.3 shall meet Table 312.6(B) spacings to the enclosure wall opposite the line and load terminals.

ARTICLE 406
Receptacles, Cord Connectors, and Attachment Plugs (Caps)

406.1 Scope. This article covers the rating, type, and installation of receptacles, cord connectors, and attachment plugs (cord caps).

406.2 Definitions.

Child Care Facility. A building or structure, or portion thereof, for educational, supervisory, or personal care services for more than four children 7 years old or less.

N **Outlet Box Hood.** A housing shield intended to fit over a faceplate for flush-mounted wiring devices, or an integral component of an outlet box or of a faceplate for flush-mounted wiring devices. The hood does not serve to complete the electrical enclosure; it reduces the risk of water coming in contact with electrical components within the hood, such as attachment plugs, current taps, surge protective devices, direct plug-in transformer units, or wiring devices.

406.3 Receptacle Rating and Type.

(A) Receptacles. Receptacles shall be listed and marked with the manufacturer's name or identification and voltage and ampere ratings.

(B) Rating. Receptacles and cord connectors shall be rated not less than 15 amperes, 125 volts, or 15 amperes, 250 volts, and shall be of a type not suitable for use as lampholders.

> Informational Note: See 210.21(B) for receptacle ratings where installed on branch circuits.

(C) Receptacles for Aluminum Conductors. Receptacles rated 20 amperes or less and designed for the direct connection of aluminum conductors shall be marked CO/ALR.

(D) Isolated Ground Receptacles. Receptacles incorporating an isolated grounding conductor connection intended for the reduction of electrical noise (electromagnetic interference) as permitted in 250.146(D) shall be identified by an orange triangle located on the face of the receptacle.

(1) Isolated Equipment Grounding Conductor Required. Receptacles so identified shall be used only with equipment grounding conductors that are isolated in accordance with 250.146(D).

(2) Installation in Nonmetallic Boxes. Isolated ground receptacles installed in nonmetallic boxes shall be covered with a nonmetallic faceplate.

Exception: Where an isolated ground receptacle is installed in a nonmetallic box, a metal faceplate shall be permitted if the box contains a feature or accessory that permits the effective grounding of the faceplate.

(E) Controlled Receptacle Marking. All nonlocking-type, 125-volt, 15- and 20-ampere receptacles that are controlled by an automatic control device, or that incorporate control features that remove power from the receptacle for the purpose of energy management or building automation, shall be permanently marked with the symbol shown in Figure 406.3(E) and the word "controlled."

For receptacles controlled by an automatic control device, the marking shall be located on the receptacle face and visible after installation.

In both cases where a multiple receptacle device is used, the required marking of the word "controlled" and symbol shall denote which contact device(s) are controlled.

Exception: The marking shall not be required for receptacles controlled by a wall switch that provide the required room lighting outlets as permitted by 210.70.

(F) Receptacle with USB Charger. A 125-volt 15- or 20-ampere receptacle that additionally provides Class 2 power shall be listed and constructed such that the Class 2 circuitry is integral with the receptacle.

406.4 General Installation Requirements. Receptacle outlets shall be located in branch circuits in accordance with Part III of Article 210. General installation requirements shall be in accordance with 406.4(A) through (F).

(A) Grounding Type. Except as provided in 406.4(D), receptacles installed on 15- and 20-ampere branch circuits shall be of the grounding type. Grounding-type receptacles shall be installed only on circuits of the voltage class and current for which they are rated, except as provided in Table 210.21(B)(2) and Table 210.21(B)(3).

(B) To Be Grounded. Receptacles and cord connectors that have equipment grounding conductor contacts shall have those contacts connected to an equipment grounding conductor.

Exception No. 1: Receptacles mounted on portable and vehicle-mounted generator sets and generators in accordance with 250.34.

Exception No. 2: Replacement receptacles as permitted by 406.4(D).

(C) Methods of Grounding. The equipment grounding conductor contacts of receptacles and cord connectors shall be grounded by connection to the equipment grounding conductor of the circuit supplying the receptacle or cord connector.

> Informational Note: For installation requirements for the reduction of electrical noise, see 250.146(D).

The branch-circuit wiring method shall include or provide an equipment grounding conductor to which the equipment grounding conductor contacts of the receptacle or cord connector are connected.

> Informational Note No. 1: See 250.118 for acceptable grounding means.

> Informational Note No. 2: For extensions of existing branch circuits, see 250.130.

(D) Replacements. Replacement of receptacles shall comply with 406.4(D)(1) through (D)(6), as applicable. Arc-fault circuit-interrupter type and ground-fault circuit-interrupter type receptacles shall be installed in a readily accessible location.

Controlled

FIGURE 406.3(E)   Controlled Receptacle Marking Symbol.

(1) Grounding-Type Receptacles. Where a grounding means exists in the receptacle enclosure or an equipment grounding conductor is installed in accordance with 250.130(C), grounding-type receptacles shall be used and shall be connected to the equipment grounding conductor in accordance with 406.4(C) or 250.130(C).

(2) Non–Grounding-Type Receptacles. Where attachment to an equipment grounding conductor does not exist in the receptacle enclosure, the installation shall comply with (D)(2)(a), (D)(2)(b), or (D)(2)(c).

(a) A non–grounding-type receptacle(s) shall be permitted to be replaced with another non–grounding-type receptacle(s).

(b) A non–grounding-type receptacle(s) shall be permitted to be replaced with a ground-fault circuit interrupter-type of receptacle(s). These receptacles or their cover plates shall be marked "No Equipment Ground." An equipment grounding conductor shall not be connected from the ground-fault circuit-interrupter-type receptacle to any outlet supplied from the ground-fault circuit-interrupter receptacle.

(c) A non–grounding-type receptacle(s) shall be permitted to be replaced with a grounding-type receptacle(s) where supplied through a ground-fault circuit interrupter. Where grounding-type receptacles are supplied through the ground-fault circuit interrupter, grounding-type receptacles or their cover plates shall be marked "GFCI Protected" and "No Equipment Ground," visible after installation. An equipment grounding conductor shall not be connected between the grounding-type receptacles.

> Informational Note No. 1: Some equipment or appliance manufacturers require that the branch circuit to the equipment or appliance includes an equipment grounding conductor.

> Informational Note No. 2: See 250.114 for a list of a cord-and-plug-connected equipment or appliances that require an equipment grounding conductor.

(3) Ground-Fault Circuit Interrupters. Ground-fault circuit-interrupter protected receptacles shall be provided where replacements are made at receptacle outlets that are required to be so protected elsewhere in this *Code.*

Exception: Where replacement of the receptacle type is impracticable, such as where the outlet box size will not permit the installation of the GFCI receptacle, the receptacle shall be permitted to be replaced with a new receptacle of the existing type, where GFCI protection is provided and the receptacle is marked "GFCI Protected" and "No Equipment Ground," in accordance with 406.4(D)(2)(a), (b), or (c), as applicable.

(4) Arc-Fault Circuit-Interrupter Protection. Where a receptacle outlet is located in any areas specified in 210.12(A) or (B), a replacement receptacle at this outlet shall be one of the following:

(1) A listed outlet branch-circuit type arc-fault circuit-interrupter receptacle

(2) A receptacle protected by a listed outlet branch-circuit type arc-fault circuit-interrupter type receptacle

(3) A receptacle protected by a listed combination type arc-fault circuit-interrupter type circuit breaker

Exception No. 1: Arc-fault circuit-interrupter protection shall not be required where all of the following apply:

(1)   The replacement complies with 406.4(D)(2)(b).

(2)   It is impracticable to provide an equipment grounding conductor as provided by 250.130(C).

(3)   A listed combination type arc-fault circuit-interrupter circuit breaker is not commercially available.

(4)   GFCI/AFCI dual function receptacles are not commercially available.

Exception No. 2: Section 210.12(B), Exception shall not apply to replacement of receptacles.

(5) Tamper-Resistant Receptacles. Listed tamper-resistant receptacles shall be provided where replacements are made at receptacle outlets that are required to be tamper-resistant elsewhere in this *Code,* except where a non-grounding receptacle is replaced with another non-grounding receptacle.

(6) Weather-Resistant Receptacles. Weather-resistant receptacles shall be provided where replacements are made at receptacle outlets that are required to be so protected elsewhere in this *Code.*

(E) Cord- and Plug-Connected Equipment. The installation of grounding-type receptacles shall not be used as a requirement that all cord-and plug-connected equipment be of the grounded type.

> Informational Note: See 250.114 for types of cord-and plug-connected equipment to be grounded.

(F) Noninterchangeable Types. Receptacles connected to circuits that have different voltages, frequencies, or types of current (ac or dc) on the same premises shall be of such design that the attachment plugs used on these circuits are not interchangeable.

406.5 Receptacle Mounting. Receptacles shall be mounted in identified boxes or assemblies. The boxes or assemblies shall be securely fastened in place unless otherwise permitted elsewhere in this *Code.* Screws used for the purpose of attaching receptacles to a box shall be of the type provided with a listed receptacle, or shall be machine screws having 32 threads per inch or part of listed assemblies or systems, in accordance with the manufacturer's instructions.

(A) Boxes That Are Set Back. Receptacles mounted in boxes that are set back from the finished surface as permitted in 314.20 shall be installed such that the mounting yoke or strap of the receptacle is held rigidly at the finished surface.

(B) Boxes That Are Flush. Receptacles mounted in boxes that are flush with the finished surface or project therefrom shall be installed such that the mounting yoke or strap of the receptacle is held rigidly against the box or box cover.

(C) Receptacles Mounted on Covers. Receptacles mounted to and supported by a cover shall be held rigidly against the cover by more than one screw or shall be a device assembly or box cover listed and identified for securing by a single screw.

(D) Position of Receptacle Faces. After installation, receptacle faces shall be flush with or project from faceplates of insulating material and shall project a minimum of 0.4 mm (0.015 in.) from metal faceplates.

Exception: Listed kits or assemblies encompassing receptacles and nonmetallic faceplates that cover the receptacle face, where the plate cannot be installed on any other receptacle, shall be permitted.

(E) Receptacles in Countertops. Receptacle assemblies for installation in countertop surfaces shall be listed for countertop applications. Where receptacle assemblies for countertop applications are required to provide ground-fault circuit-interrupter protection for personnel in accordance with 210.8, such assemblies shall be permitted to be listed as GFCI receptacle assemblies for countertop applications.

N (F) Receptacles in Work Surfaces. Receptacle assemblies and GFCI receptacle assemblies listed for work surface or countertop applications shall be permitted to be installed in work surfaces.

N (G) Receptacle Orientation. Receptacles shall not be installed in a face-up position in or on countertop surfaces or work surfaces unless listed for countertop or work surface applications.

(H) Receptacles in Seating Areas and Other Similar Surfaces. In seating areas or similar surfaces, receptacles shall not be installed in a face-up position unless the receptacle is any of the following:

(1) Part of an assembly listed as a furniture power distribution unit

(2) Part of an assembly listed either as household furnishings or as commercial furnishings

(3) Listed either as a receptacle assembly for countertop applications or as a GFCI receptacle assembly for countertop applications

(4) Installed in a listed floor box

(I) Exposed Terminals. Receptacles shall be enclosed so that live wiring terminals are not exposed to contact.

(J) Voltage Between Adjacent Devices. A receptacle shall not be grouped or ganged in enclosures with other receptacles, snap switches, or similar devices, unless they are arranged so that the voltage between adjacent devices does not exceed 300 volts, or unless they are installed in enclosures equipped with identified, securely installed barriers between adjacent devices.

406.6 Receptacle Faceplates (Cover Plates). Receptacle faceplates shall be installed so as to completely cover the opening and seat against the mounting surface.

Receptacle faceplates mounted inside a box having a recessmounted receptacle shall effectively close the opening and seat against the mounting surface.

(A) Thickness of Metal Faceplates. Metal faceplates shall be of ferrous metal not less than 0.76 mm (0.030 in.) in thickness or of nonferrous metal not less than 1.02 mm (0.040 in.) in thickness.

(B) Grounding. Metal faceplates shall be grounded.

(C) Faceplates of Insulating Material. Faceplates of insulating material shall be noncombustible and not less than 2.54 mm (0.10 in.) in thickness but shall be permitted to be less than 2.54 mm (0.10 in.) in thickness if formed or reinforced to provide adequate mechanical strength.

N (D) Receptacle Faceplate (Cover Plates) with Integral Night Light and/or USB Charger. A flush device cover plate that additionally provides a night light and/or Class 2 output connector(s) shall be listed and constructed such that the night light and/or Class 2 circuitry is integral with the flush device cover plate.

406.7 Attachment Plugs, Cord Connectors, and Flanged Surface Devices. All attachment plugs, cord connectors, and flanged surface devices (inlets and outlets) shall be listed and marked with the manufacturer's name or identification a voltage and ampere ratings.

(A) Construction of Attachment Plugs and Cord Connecto Attachment plugs and cord connectors shall be constructed that there are no exposed current-carrying parts except t prongs, blades, or pins. The cover for wire terminations sh be a part that is essential for the operation of an attachme plug or connector (dead-front construction).

(B) Connection of Attachment Plugs. Attachment plugs sh be installed so that their prongs, blades, or pins are not en gized unless inserted into an energized receptacle or co connectors. No receptacle shall be installed so as to require t insertion of an energized attachment plug as its source supply.

(C) Attachment Plug Ejector Mechanisms. Attachment pl ejector mechanisms shall not adversely affect engagement the blades of the attachment plug with the contacts of t receptacle.

(D) Flanged Surface Inlet. A flanged surface inlet shall installed such that the prongs, blades, or pins are not energiz unless an energized cord connector is inserted into it.

406.8 Noninterchangeability. Receptacles, cord connecto and attachment plugs shall be constructed such that recepta or cord connectors do not accept an attachment plug with different voltage or current rating from that for which t device is intended. However, a 20-ampere T-slot receptacle cord connector shall be permitted to accept a 15-amp attachment plug of the same voltage rating. Non–groundi type receptacles and connectors shall not accept groundi type attachment plugs.

406.9 Receptacles in Damp or Wet Locations.

(A) Damp Locations. A receptacle installed outdoors ir location protected from the weather or in other damp lo tions shall have an enclosure for the receptacle that is weath proof when the receptacle is covered (attachment plug cap inserted and receptacle covers closed).

An installation suitable for wet locations shall also be con ered suitable for damp locations.

A receptacle shall be considered to be in a location pro ted from the weather where located under roofed op porches, canopies, marquees, and the like, and will not subjected to a beating rain or water runoff. All 15- a 20-ampere, 125- and 250-volt nonlocking receptacles shall b listed weather-resistant type.

Informational Note: The types of receptacles covered by this requirement are identified as 5-15, 5-20, 6-15, and 6-20 in ANSI/ NEMA WD 6–2012, *Wiring Devices — Dimensional Specifications.*

(B) Wet Locations.

(1) Receptacles of 15 and 20 Amperes in a Wet Locati Receptacles of 15 and 20 amperes, 125 and 250 volts insta in a wet location shall have an enclosure that is weatherpr whether or not the attachment plug cap is inserted. An ou box hood installed for this purpose shall be listed and shal identified as "extra-duty." Other listed products, enclosures assemblies providing weatherproof protection that do utilize an outlet box hood need not be marked "extra duty."

Informational Note No. 1: Requirements for extra-duty outlet box hoods are found in ANSI/UL 514D–2013, *Cover Plates for Flush-Mounted Wiring Devices.* "Extra duty" identification and requirements are not applicable to listed receptacles, faceplates, outlet boxes, enclosures, or assemblies that are identified as either being suitable for wet locations or rated as one of the outdoor enclosure–type numbers of Table 110.28 that does not utilize an outlet box hood.

Exception: 15- and 20-ampere, 125- through 250-volt receptacles installed in a wet location and subject to routine high-pressure spray washing shall be permitted to have an enclosure that is weatherproof when the attachment plug is removed.

All 15- and 20-ampere, 125- and 250-volt nonlocking-type receptacles shall be listed and so identified as the weather-resistant type.

Informational Note No. 2: The configuration of weather-resistant receptacles covered by this requirement are identified as 5-15, 5-20, 6-15, and 6-20 in ANSI/NEMA WD 6–2012, *Wiring Devices — Dimensional Specifications.*

(3) Other Receptacles. All other receptacles installed in a wet location shall comply with (B)(2)(a) or (B)(2)(b).

(a) A receptacle installed in a wet location, where the product intended to be plugged into it is not attended while in use, shall have an enclosure that is weatherproof with the attachment plug cap inserted or removed.

(b) A receptacle installed in a wet location where the product intended to be plugged into it will be attended while in use (e.g., portable tools) shall have an enclosure that is weatherproof when the attachment plug is removed.

(C) Bathtub and Shower Space. Receptacles shall not be installed within or directly over a bathtub or shower stall.

(D) Protection for Floor Receptacles. Standpipes of floor receptacles shall allow floor-cleaning equipment to be operated without damage to receptacles.

(E) Flush Mounting with Faceplate. The enclosure for a receptacle installed in an outlet box flush-mounted in a finished surface shall be made weatherproof by means of a weatherproof faceplate assembly that provides a watertight connection between the plate and the finished surface.

406.10 Grounding-Type Receptacles, Adapters, Cord Connectors, and Attachment Plugs.

(A) Grounding Poles. Grounding-type receptacles, cord connectors, and attachment plugs shall be provided with one fixed grounding pole in addition to the circuit poles. The grounding contacting pole of grounding-type plug-in ground-fault circuit interrupters shall be permitted to be of the movable, self-restoring type on circuits operating at not over 150 volts between any two conductors or any conductor and ground.

(B) Grounding-Pole Identification. Grounding-type receptacles, adapters, cord connections, and attachment plugs shall have a means for connection of an equipment grounding conductor to the grounding pole.

(1) A terminal for connection to the grounding pole shall be designated by one of the following:

(a) A green-colored hexagonal-headed or -shaped terminal screw or nut, not readily removable.

(b) A green-colored pressure wire connector body (a wire barrel).

(3) A similar green-colored connection device, in the case of adapters. The grounding terminal of a grounding adapter shall be a green-colored rigid ear, lug, or similar device. The equipment grounding connection shall be so designed that it cannot make contact with current-carrying parts of the receptacle, adapter, or attachment plug. The adapter shall be polarized.

(4) If the terminal for the equipment grounding conductor is not visible, the conductor entrance hole shall be marked with the word *green* or *ground*, the letters *G* or *GR*, a grounding symbol, or otherwise identified by a distinctive green color. If the terminal for the equipment grounding conductor is readily removable, the area adjacent to the terminal shall be similarly marked.

Informational Note: See Informational Note Figure 406.10(B)(4).

(C) Grounding Terminal Use. A grounding terminal shall not be used for purposes other than grounding.

(D) Grounding-Pole Requirements. Grounding-type attachment plugs and mating cord connectors and receptacles shall be designed such that the equipment grounding connection is made before the current-carrying connections. Grounding-type devices shall be so designed that grounding poles of attachment plugs cannot be brought into contact with current-carrying parts of receptacles or cord connectors.

(E) Use. Grounding-type attachment plugs shall be used only with a cord having an equipment grounding conductor.

Informational Note: See 250.126 for identification of grounding conductor terminals.

406.11 Connecting Receptacle Grounding Terminal to Box. The connection of the receptacle grounding terminal shall comply with 250.146.

406.12 Tamper-Resistant Receptacles. All 15- and 20-ampere, 125- and 250-volt nonlocking type receptacles in the areas specified in 406.12(1) through (7) shall be listed tamper-resistant receptacles.

(1) Dwelling units in all areas specified in 210.52 and 550.13
(2) Guest rooms and guest suites of hotels and motels
(3) Child care facilities
(4) Preschools and elementary education facilities
(5) Business offices, corridors, waiting rooms and the like in clinics, medical and dental offices and outpatient facilities
(6) Subset of assembly occupancies described in 518.2 to include places of waiting transportation, gymnasiums, skating rinks, and auditoriums
(7) Dormitories

Informational Note: This requirement would include receptacles identified as 5-15, 5-20, 6-15, and 6-20 in ANSI/NEMA WD 6-2016, *Wiring Devices — Dimensional Specifications.*

Exception to (1), (2), (3), (4), (5), (6), and (7): Receptacles in the following locations shall not be required to be tamper resistant:

(1) Receptacles located more than 1.7 m (5 ½ ft) above the floor

Informational Note Figure 406.10(B)(4)   One Example of a Symbol Used to Identify the Termination Point for an Equipment Grounding Conductor.

(2)   *Receptacles that are part of a luminaire or appliance*

(3)   *A single receptacle or a duplex receptacle for two appliances located within the dedicated space for each appliance that, in normal use, is not easily moved from one place to another and that is cord-and-plug-connected in accordance with 400.10(A)(6), (A)(7), or (A)(8)*

(4)   *Nongrounding receptacles used for replacements as permitted in 406.4(D)(2)(a)*

ARTICLE 408
Switchboards, Switchgear, and Panelboards

Part I. General

408.1 Scope. This article covers switchboards, switchgear, and panelboards. It does not apply to equipment operating at over 1000 volts, except as specifically referenced elsewhere in the *Code*.

408.2 Other Articles. Switches, circuit breakers, and overcurrent devices used on switchboards, switchgear, and panelboards and their enclosures shall comply with this article and also with the requirements of Articles 240, 250, 312, 404, and other articles that apply. Switchboards, switchgear, and panelboards in hazardous (classified) locations shall comply with the applicable provisions of Articles 500 through 517.

408.3 Support and Arrangement of Busbars and Conductors.

(A) Conductors and Busbars on a Switchboard, Switchgear, or Panelboard. Conductors and busbars on a switchboard, switchgear, or panelboard shall comply with 408.3(A)(1), (A)(2), and (A)(3) as applicable.

(1) Location. Conductors and busbars shall be located so as to be free from physical damage and shall be held firmly in place.

(2) Service Panelboards, Switchboards, and Switchgear. Barriers shall be placed in all service panelboards, switchboards, and switchgear such that no uninsulated, ungrounded service busbar or service terminal is exposed to inadvertent contact by persons or maintenance equipment while servicing load terminations.

Exception: This requirement shall not apply to service panelboards with provisions for more than one service disconnect within a single enclosure as permitted in 408.36, Exceptions 1, 2, and 3.

(3) Same Vertical Section. Other than the required interconnections and control wiring, only those conductors that are intended for termination in a vertical section of a switchboard or switchgear shall be located in that section.

Exception: Conductors shall be permitted to travel horizontally through vertical sections of switchboards and switchgear where such conductors are isolated from busbars by a barrier.

(B) Overheating and Inductive Effects. The arrangement of busbars and conductors shall be such as to avoid overheating due to inductive effects.

(C) Used as Service Equipment. Each switchboard, switchgear, or panelboard, if used as service equipment, shall be provided with a main bonding jumper sized in accordance with 250.28(D) or the equivalent placed within the panelboard or one of the sections of the switchboard or switchgear for connecting the grounded service conductor on its supply side to the switchboard, switchgear, or panelboard frame. All sections of a switchboard or switchgear shall be bonded together using an equipment-bonding jumper or a supply-side bonding jumper sized in accordance with 250.122 or 250.102(C)(1) as applicable.

Exception: Switchboards, switchgear, and panelboards used as service equipment on high-impedance grounded neutral systems in accordance with 250.36 shall not be required to be provided with a main bonding jumper.

(D) Terminals. In switchboards, switchgear, and panelboards, load terminals for field wiring, including grounded circuit conductor load terminals and connections to the equipment grounding conductor bus for load equipment grounding conductors, shall be so located that it is not necessary to reach across or beyond an uninsulated ungrounded bus in order to make connections.

(E) Bus Arrangement.

(1) AC Phase Arrangement. Alternating-current phase arrangement on 3-phase buses shall be A, B, C from front to back, top to bottom, or left to right, as viewed from the front of the switchboard, switchgear, or panelboard. The B phase shall be that phase having the higher voltage to ground on 3-phase, 4-wire, delta-connected systems. Other busbar arrangements shall be permitted for additions to existing installations and shall be marked.

Exception: Equipment within the same single section or multisection switchboard, switchgear, or panelboard as the meter on 3-phase, 4-wire, delta-connected systems shall be permitted to have the same phase configuration as the metering equipment.

> Informational Note: See 110.15 for requirements on marking the busbar or phase conductor having the higher voltage to ground where supplied from a 4-wire, delta-connected system.

(2) DC Bus Arrangement. Direct-current ungrounded buses shall be permitted to be in any order. Arrangement of dc buses shall be field marked as to polarity, grounding system, and nominal voltage.

(F) Switchboard, Switchgear, or Panelboard Identification. A caution sign(s) or a label(s) provided in accordance with 408.3(F)(1) through (F)(5) shall comply with 110.21(B).

(1) High-Leg Identification. A switchboard, switchgear, or panelboard containing a 4-wire, delta-connected system where the midpoint of one phase winding is grounded shall be legibly and permanently field marked as follows:

"Caution _____ Phase Has _____ Volts to Ground"

(2) Ungrounded AC Systems. A switchboard, switchgear, or panelboard containing an ungrounded ac electrical system as permitted in 250.21 shall be legibly and permanently field marked as follows:

"Caution Ungrounded System Operating — _____ Volts Between Conductors"

(3) High-Impedance Grounded Neutral AC System. A switchboard, switchgear, or panelboard containing a high-impedance grounded neutral ac system in accordance with 250.36 shall be legibly and permanently field marked as follows:

CAUTION: HIGH-IMPEDANCE GROUNDED NEUTRAL AC SYSTEM OPERATING — _____ VOLTS BETWEEN CONDUCTORS AND MAY OPERATE — _____ VOLTS TO GROUND FOR INDEFINITE PERIODS UNDER FAULT CONDITIONS

(4) Ungrounded DC Systems. A switchboard, switchgear, or panelboard containing an ungrounded dc electrical system in accordance with 250.169 shall be legibly and permanently field marked as follows:

CAUTION: UNGROUNDED DC SYSTEM OPERATING — _____ VOLTS BETWEEN CONDUCTORS

(5) Resistively Grounded DC Systems. A switchboard, switchgear, or panelboard containing a resistive connection between current-carrying conductors and the grounding system to stabilize voltage to ground shall be legibly and permanently field marked as follows:

CAUTION: DC SYSTEM OPERATING — _____ VOLTS BETWEEN CONDUCTORS AND
MAY OPERATE — _____ VOLTS TO GROUND FOR INDEFINITE PERIODS UNDER FAULT CONDITIONS

(G) Minimum Wire-Bending Space. The minimum wire-bending space at terminals and minimum gutter space provided in switchboards, switchgear, and panelboards shall be as required in 312.6.

408.4 Field Identification Required.

(A) Circuit Directory or Circuit Identification. Every circuit and circuit modification shall be legibly identified as to its clear, evident, and specific purpose or use. The identification shall include an approved degree of detail that allows each circuit to be distinguished from all others. Spare positions that contain unused overcurrent devices or switches shall be described accordingly. The identification shall be included in a circuit directory that is located on the face or inside of the panel door in the case of a panelboard and at each switch or circuit breaker in a switchboard or switchgear. No circuit shall be described in a manner that depends on transient conditions of occupancy.

(B) Source of Supply. All switchboards, switchgear, and panelboards supplied by a feeder(s) in other than one- or two-family dwellings shall be permanently marked to indicate each device or equipment where the power originates. The label shall be permanently affixed, of sufficient durability to withstand the environment involved, and not handwritten.

408.5 Clearance for Conductor Entering Bus Enclosures. Where conduits or other raceways enter a switchboard, switchgear, floor-standing panelboard, or similar enclosure at the bottom, approved space shall be provided to permit installation of conductors in the enclosure. The wiring space shall not be less than shown in Table 408.5 where the conduit or raceways enter or leave the enclosure below the busbars, their supports, or other obstructions. The conduit or raceways, including their end fittings, shall not rise more than 75 mm (3 in.) above the bottom of the enclosure.

Table 408.5 Clearance for Conductors Entering Bus Enclosures

Conductor	Minimum Spacing Between Bottom of Enclosure and Busbars, Their Supports, or Other Obstructions	
	mm	in.
Insulated busbars, their supports, or other obstructions	200	8
Noninsulated busbars	250	10

408.7 Unused Openings. Unused openings for circuit breakers and switches shall be closed using identified closures, or other approved means that provide protection substantially equivalent to the wall of the enclosure.

Part II. Switchboards and Switchgear

408.16 Switchboards and Switchgear in Damp or Wet Locations. Switchboards and switchgear in damp or wet locations shall be installed in accordance with 312.2.

408.17 Location Relative to Easily Ignitible Material. Switchboards and switchgear shall be placed so as to reduce to a minimum the probability of communicating fire to adjacent combustible materials. Where installed over a combustible floor, suitable protection thereto shall be provided.

408.18 Clearances.

(A) From Ceiling. For other than a totally enclosed switchboard or switchgear, a space not less than 900 mm (3 ft) shall be provided between the top of the switchboard or switchgear and any combustible ceiling, unless a noncombustible shield is provided between the switchboard or switchgear and the ceiling.

(B) Around Switchboards and Switchgear. Clearances around switchboards and switchgear shall comply with the provisions of 110.26.

408.19 Conductor Insulation. An insulated conductor used within a switchboard or switchgear shall be listed, shall be flame retardant, and shall be rated not less than the voltage applied to it and not less than the voltage applied to other conductors or busbars with which it may come into contact.

408.20 Location of Switchboards and Switchgear. Switchboards and switchgear that have any exposed live parts shall be located in permanently dry locations and then only where under competent supervision and accessible only to qualified persons. Switchboards and switchgear shall be located such that the probability of damage from equipment or processes is reduced to a minimum.

408.22 Grounding of Instruments, Relays, Meters, and Instrument Transformers on Switchboards and Switchgear. Instruments, relays, meters, and instrument transformers located on switchboards and switchgear shall be grounded as specified in 250.170 through 250.178.

Part III. Panelboards

408.30 General. All panelboards shall have a rating not less than the minimum feeder capacity required for the load calculated in accordance with Part III, IV, or V of Article 220, as applicable.

408.36 Overcurrent Protection. In addition to the requirement of 408.30, a panelboard shall be protected by an overcurrent protective device having a rating not greater than that of the panelboard. This overcurrent protective device shall be located within or at any point on the supply side of the panelboard.

Exception No. 1: Individual protection shall not be required for a panelboard used as service equipment with multiple disconnecting means in accordance with 230.71. In panelboards protected by three or more main circuit breakers or sets of fuses, the circuit breakers or sets of fuses shall not supply a second bus structure within the same panelboard assembly.

Exception No. 2: Individual protection shall not be required for a panelboard protected on its supply side by two main circuit breakers or two sets of fuses having a combined rating not greater than that of the panelboard. A panelboard constructed or wired under this exception shall not contain more than 42 overcurrent devices. For the purposes of determining the maximum of 42 overcurrent devices, a 2-pole or a 3-pole circuit breaker shall be considered as two or three overcurrent devices, respectively.

Exception No. 3: For existing panelboards, individual protection shall not be required for a panelboard used as service equipment for an individual residential occupancy.

(A) Snap Switches Rated at 30 Amperes or Less. Panelboards equipped with snap switches rated at 30 amperes or less shall have overcurrent protection of 200 amperes or less.

(B) Supplied Through a Transformer. Where a panelboard is supplied through a transformer, the overcurrent protection required by 408.36 shall be located on the secondary side of the transformer.

Exception: A panelboard supplied by the secondary side of a transformer shall be considered as protected by the overcurrent protection provided on the primary side of the transformer where that protection is in accordance with 240.21(C)(1).

(C) Delta Breakers. A 3-phase disconnect or overcurrent device shall not be connected to the bus of any panelboard that has less than 3-phase buses. Delta breakers shall not be installed in panelboards.

(D) Back-Fed Devices. Plug-in-type overcurrent protection devices or plug-in type main lug assemblies that are backfed and used to terminate field-installed ungrounded supply conductors shall be secured in place by an additional fastener that requires other than a pull to release the device from the mounting means on the panel.

408.37 Panelboards in Damp or Wet Locations. Panelboards in damp or wet locations shall be installed to comply with 312.2.

408.38 Enclosure. Panelboards shall be mounted in cabinets, cutout boxes, or identified enclosures and shall be dead-front.

Exception: Panelboards other than of the dead-front, externally operable type shall be permitted where accessible only to qualified persons.

408.39 Relative Arrangement of Switches and Fuses. In panelboards, fuses of any type shall be installed on the load side of any switches.

Exception: Fuses installed as part of service equipment in accordance with the provisions of 230.94 shall be permitted on the line side of the service switch.

408.40 Grounding of Panelboards. Panelboard cabinets and panelboard frames, if of metal, shall be in physical contact with each other and shall be connected to an equipment grounding conductor. Where the panelboard is used with nonmetallic raceway or cable or where separate equipment grounding conductors are provided, a terminal bar for the equipment grounding conductors shall be secured inside the cabinet. The terminal bar shall be bonded to the cabinet and panelboard frame, if of metal; otherwise it shall be connected to the equipment grounding conductor that is run with the conductors feeding the panelboard.

Exception: Where an isolated equipment grounding conductor is provided as permitted by 250.146(D), the insulated equipment grounding conductor that is run with the circuit conductors shall be permitted to pass through the panelboard without being connected to the panelboard's equipment grounding terminal bar.

Equipment grounding conductors shall not be connected to a terminal bar provided for grounded conductors or neutral conductors unless the bar is identified for the purpose and is located where interconnection between equipment grounding conductors and grounded circuit conductors is permitted or required by Article 250.

408.41 Grounded Conductor Terminations. Each grounded conductor shall terminate within the panelboard in an individual terminal that is not also used for another conductor.

Exception: Grounded conductors of circuits with parallel conductors shall be permitted to terminate in a single terminal if the terminal is identified for connection of more than one conductor.

Part IV. Construction Specifications

408.50 Panels. The panels of switchboards and switchgear shall be made of moisture-resistant, noncombustible material.

408.51 Busbars. Insulated or bare busbars shall be rigidly mounted.

408.52 Protection of Instrument Circuits. Instruments, pilot lights, voltage (potential) transformers, and other switchboard or switchgear devices with potential coils shall be supplied by a circuit that is protected by standard overcurrent devices rated 15 amperes or less.

Exception No. 1: Overcurrent devices rated more than 15 amperes shall be permitted where the interruption of the circuit could create a hazard. Short-circuit protection shall be provided.

Exception No. 2: For ratings of 2 amperes or less, special types of enclosed fuses shall be permitted.

408.53 Component Parts. Switches, fuses, and fuseholders used on panelboards shall comply with the applicable requirements of Articles 240 and 404.

408.54 Maximum Number of Overcurrent Devices. A panelboard shall be provided with physical means to prevent the stallation of more overcurrent devices than that number for nich the panelboard was designed, rated, and listed.

For the purposes of this section, a 2-pole circuit breaker or sible switch shall be considered two overcurrent devices; a 3-ole circuit breaker or fusible switch shall be considered three ercurrent devices.

8.55 Wire-Bending Space Within an Enclosure Containing a nelboard.

.) Top and Bottom Wire-Bending Space. The enclosure for a nelboard shall have the top and bottom wire-bending space zed in accordance with Table 312.6(B) for the largest conduc-r entering or leaving the enclosure.

ception No. 1: Either the top or bottom wire-bending space shall be rmitted to be sized in accordance with Table 312.6(A) for a panel-ard rated 225 amperes or less and designed to contain not over 2 overcurrent devices. For the purposes of this exception, a 2-pole or a pole circuit breaker shall be considered as two or three overcurrent vices, respectively.

ception No. 2: Either the top or bottom wire-bending space for any nelboard shall be permitted to be sized in accordance with ble 312.6(A) where at least one side wire-bending space is sized in cordance with Table 312.6(B) for the largest conductor to be termina-d in any side wire-bending space.

ception No. 3: The top and bottom wire-bending space shall be rmitted to be sized in accordance with Table 312.6(A) spacings if the nelboard is designed and constructed for wiring using only a single)-degree bend for each conductor, including the grounded circuit nductor, and the wiring diagram shows and specifies the method of iring that shall be used.

ception No. 4: Either the top or the bottom wire-bending space, but ot both, shall be permitted to be sized in accordance with ble 312.6(A) where there are no conductors terminated in that space.

) Side Wire-Bending Space. Side wire-bending space shall in accordance with Table 312.6(A) for the largest conductor be terminated in that space.

) Back Wire-Bending Space. Where a raceway or cable try is in the wall of the enclosure opposite a removable cover, e distance from that wall to the cover shall be permitted to mply with the distance required for one wire per terminal in ble 312.6(A). The distance between the center of the rear try and the nearest termination for the entering conductors all not be less than the distance given in Table 312.6(B).

8.56 Minimum Spacings. The distance between bare metal rts, busbars, and so forth shall not be less than specified in ble 408.56.

Where close proximity does not cause excessive heating, rts of the same polarity at switches, enclosed fuses, and so rth shall be permitted to be placed as close together as nvenience in handling will allow.

ception: The distance shall be permitted to be less than that specified Table 408.56 at circuit breakers and switches and in listed compo-nts installed in switchboards, switchgear, and panelboards.

Table 408.56 Minimum Spacings Between Bare Metal Parts

AC or DC Voltage	Opposite Polarity Where Mounted on the Same Surface		Opposite Polarity Where Held Free in Air		Live Parts to Ground*	
	mm	in.	mm	in.	mm	in.
Not over 125 volts, nominal	19.1	¾	12.7	½	12.7	½
Not over 250 volts, nominal	31.8	1¼	19.1	¾	12.7	½
Not over 1000 volts, nominal	50.8	2	25.4	1	25.4	1

*For spacing between live parts and doors of cabinets, see 312.11(A) (1), (2), and (3).

408.58 Panelboard Marking. Panelboards shall be durably marked by the manufacturer with the voltage and the current rating and the number of ac phases or dc buses for which they are designed and with the manufacturer's name or trademark in such a manner so as to be visible after installation, without disturbing the interior parts or wiring.

ARTICLE 409
Industrial Control Panels

Part I. General

409.1 Scope. This article covers industrial control panels intended for general use and operating at 1000 volts or less.

> Informational Note: ANSI/UL 508A, *Standard for Industrial Control Panels*, is a safety standard for industrial control panels.

409.3 Other Articles. In addition to the requirements of Arti-cle 409, industrial control panels that contain branch circuits for specific loads or components, or are for control of specific types of equipment addressed in other articles of this *Code*, shall be constructed and installed in accordance with the appli-cable requirements from the specific articles in Table 409.3.

Part II. Installation

409.20 Conductor — Minimum Size and Ampacity. The size of the industrial control panel supply conductor shall have an ampacity not less than 125 percent of the full-load current rating of all heating loads plus 125 percent of the full-load current rating of the highest rated motor plus the sum of the full-load current ratings of all other connected motors and apparatus based on their duty cycle that may be in operation at the same time.

409.21 Overcurrent Protection.

(A) General. Industrial control panels shall be provided with overcurrent protection in accordance with Parts I, II, and IX of Article 240.

Table 409.3 Other Articles

Equipment/Occupancy	Article	Section
Branch circuits	210	
Luminaires	410	
Motors, motor circuits, and controllers	430	
Air-conditioning and refrigerating equipment	440	
Capacitors		460.8, 460.9
Hazardous (classified) locations	500, 501, 502, 503, 504, 505	
Commercial garages; aircraft hangars; motor fuel dispensing facilities; bulk storage plants; spray application, dipping, and coating processes; and inhalation anesthetizing locations	511, 513, 514, 515, 516, and 517 Part IV	
Cranes and hoists	610	
Electrically driven or controlled irrigation machines	675	
Elevators, dumbwaiters, escalators, moving walks, wheelchair lifts, and stairway chair lifts	620	
Industrial machinery	670	
Resistors and reactors	470	
Transformers	450	
Class 1, Class 2, and Class 3 remote-control, signaling, and power-limited circuits	725	

(B) Location. This protection shall be provided for each incoming supply circuit by either of the following:

(1) An overcurrent protective device located ahead of the industrial control panel.

(2) A single main overcurrent protective device located within the industrial control panel. Where overcurrent protection is provided as part of the industrial control panel, the supply conductors shall be considered as either feeders or taps as covered by 240.21.

(C) Rating. The rating or setting of the overcurrent protective device for the circuit supplying the industrial control panel shall not be greater than the sum of the largest rating or setting of the branch-circuit short-circuit and ground-fault protective device provided with the industrial control panel, plus 125 percent of the full-load current rating of all resistance heating loads, plus the sum of the full-load currents of all other motors and apparatus that could be in operation at the same time.

Exception: Where one or more instantaneous trip circuit breakers or motor short-circuit protectors are used for motor branch-circuit short-circuit and ground-fault protection as permitted by 430.52(C), the procedure specified above for determining the maximum rating of the protective device for the circuit supplying the industrial control panel shall apply with the following provision: For the purpose of the calculation, each instantaneous trip circuit breaker or motor short-circuit protector shall be assumed to have a rating not exceeding the maximum percentage of motor full-load current permitted by Table 430.52 for the type of control panel supply circuit protective device employed.

Where no branch-circuit short-circuit and ground-fault protective device is provided with the industrial control panel for motor or combination of motor and non-motor loads, the rating or setting of the overcurrent protective device shall be based on 430.52 and 430.53, as applicable.

409.22 Short-Circuit Current Rating.

(A) Installation. An industrial control panel shall not be installed where the available short-circuit current exceeds its short-circuit current rating as marked in accordance with 409.110(4).

(B) Documentation. If an industrial control panel is required to be marked with a short-circuit current rating in accordance with 409.110(4), the available short-circuit current at the industrial control panel and the date the short-circuit current calculation was performed shall be documented and made available to those authorized to inspect the installation.

409.30 Disconnecting Means. Disconnecting means that supply motor loads shall comply with Part IX of Article 430.

409.60 Grounding. Multisection industrial control panels shall be bonded together with an equipment grounding conductor or an equivalent equipment grounding bus sized in accordance with Table 250.122. Equipment grounding conductors shall be connected to this equipment grounding bus or to an equipment grounding termination point provided in a single-section industrial control panel.

Part III. Construction Specifications

409.100 Enclosures. Table 110.28 shall be used as the basis for selecting industrial control panel enclosures for use in specific locations other than hazardous (classified) locations. The enclosures are not intended to protect against conditions such as condensation, icing, corrosion, or contamination that may occur within the enclosure or enter via the conduit or unsealed openings.

409.102 Busbars and Conductors. Industrial control panels utilizing busbars shall comply with 409.102(A) and (B).

(A) Support and Arrangement. Busbars shall be protected from physical damage and be held firmly in place.

(B) Phase Arrangement. The phase arrangement on 3-phase horizontal common power and vertical buses shall be A, B, C from front to back, top to bottom, or left to right, as viewed from the front of the industrial control panel. The B phase shall be that phase having the higher voltage to ground on 3-phase, 4-wire, delta-connected systems. Other busbar arrangements shall be permitted for additions to existing installations, and the phases shall be permanently marked.

409.104 Wiring Space.

(A) General. Industrial control panel enclosures shall not be used as junction boxes, auxiliary gutters, or raceways for conductors feeding through or tapping off to other switches or overcurrent devices or other equipment, unless the conductors fill less than 40 percent of the cross-sectional area of the wiring space. In addition, the conductors, splices, and taps shall not fill the wiring space at any cross section to more than 75 percent of the cross-sectional area of that space.

(B) Wire Bending Space. Wire bending space within industrial control panels for field wiring terminals shall be in accordance with the requirements in 430.10(B).

409.106 Spacings. Spacings in feeder circuits between uninsulated live parts of adjacent components, between uninsulated live parts of components and grounded or accessible non–current-carrying metal parts, between uninsulated live parts of components and the enclosure, and at field wiring terminals shall be as shown in Table 430.97(D).

Exception: Spacings shall be permitted to be less than those specified in Table 430.97(D) at circuit breakers and switches and in listed components installed in industrial control panels.

409.108 Service Equipment. Where used as service equipment, each industrial control panel shall be of the type that is suitable for use as service equipment.

Where a grounded conductor is provided, the industrial control panel shall be provided with a main bonding jumper, sized in accordance with 250.28(D), for connecting the grounded conductor, on its supply side, to the industrial control panel equipment ground bus or equipment ground terminal.

409.110 Marking. An industrial control panel shall be marked with the following information that is plainly visible after installation:

(1) Manufacturer's name, trademark, or other descriptive marking by which the organization responsible for the product can be identified.
(2) Supply voltage, number of phases, frequency, and full-load current for each incoming supply circuit.
(3) Industrial control panels supplied by more than one electrical source where more than one disconnecting means is required to disconnect all circuits 50-volts or more within the control panel shall be marked to indicate that more than one disconnecting means is required to de-energize the equipment. The location of the means necessary to disconnect all circuits 50-volts or more shall be documented and available.
(4) Short-circuit current rating of the industrial control panel based on one of the following:

a. Short-circuit current rating of a listed and labeled assembly
b. Short-circuit current rating established utilizing an approved method

Informational Note: ANSI/UL 508A, *Standard for Industrial Control Panels*, Supplement SB, is an example of an approved method.

Exception to (4): Short-circuit current rating markings are not required for industrial control panels containing only control circuit components.

(5) If the industrial control panel is intended as service equipment, it shall be marked to identify it as being suitable for use as service equipment.
(6) Electrical wiring diagram or the identification number of a separate electrical wiring diagram or a designation referenced in a separate wiring diagram.
(7) An enclosure type number shall be marked on the industrial control panel enclosure.

ARTICLE 410
Luminaires, Lampholders, and Lamps

Part I. General

410.1 Scope. This article covers luminaires, portable luminaires, lampholders, pendants, incandescent filament lamps, arc lamps, electric-discharge lamps, decorative lighting products, lighting accessories for temporary seasonal and holiday use, portable flexible lighting products, and the wiring and equipment forming part of such products and lighting installations.

410.2 Definition.

Closet Storage Space. The volume bounded by the sides and back closet walls and planes extending from the closet floor vertically to a height of 1.8 m (6 ft) or to the highest clothes-hanging rod and parallel to the walls at a horizontal distance of 600 mm (24 in.) from the sides and back of the closet walls, respectively, and continuing vertically to the closet ceiling parallel to the walls at a horizontal distance of 300 mm (12 in.) or the width of the shelf, whichever is greater; for a closet that permits access to both sides of a hanging rod, this space includes the volume below the highest rod extending 300 mm (12 in.) on either side of the rod on a plane horizontal to the floor extending the entire length of the rod. See Figure 410.2.

410.5 Live Parts. Luminaires, portable luminaires, lampholders, and lamps shall have no live parts normally exposed to contact. Exposed accessible terminals in lampholders and switches shall not be installed in metal luminaire canopies or in open bases of portable table or floor luminaires.

Exception: Cleat-type lampholders located at least 2.5 m (8 ft) above the floor shall be permitted to have exposed terminals.

410.6 Listing Required. All luminaires, lampholders, and retrofit kits shall be listed.

410.8 Inspection. Luminaires shall be installed such that the connections between the luminaire conductors and the circuit conductors can be inspected without requiring the disconnection of any part of the wiring unless the luminaires are connected by attachment plugs and receptacles.

Part II. Luminaire Locations

410.10 Luminaires in Specific Locations.

(A) Wet and Damp Locations. Luminaires installed in wet or damp locations shall be installed such that water cannot enter or accumulate in wiring compartments, lampholders, or other electrical parts. All luminaires installed in wet locations shall be marked, "Suitable for Wet Locations." All luminaires installed in damp locations shall be marked "Suitable for Wet Locations" or "Suitable for Damp Locations."

(B) Corrosive Locations. Luminaires installed in corrosive locations shall be of a type suitable for such locations.

(C) In Ducts or Hoods. Luminaires shall be permitted to be installed in commercial cooking hoods where all of the following conditions are met:

(1) The luminaire shall be identified for use within commercial cooking hoods and installed such that the temperature limits of the materials used are not exceeded.

(2) The luminaire shall be constructed so that all exhaust vapors, grease, oil, or cooking vapors are excluded from the lamp and wiring compartment. Diffusers shall be resistant to thermal shock.

(3) Parts of the luminaire exposed within the hood shall be corrosion resistant or protected against corrosion, and the surface shall be smooth so as not to collect deposits and to facilitate cleaning.

(4) Wiring methods and materials supplying the luminaire(s) shall not be exposed within the cooking hood.

Informational Note: See 110.11 for conductors and equipment exposed to deteriorating agents.

(D) Bathtub and Shower Areas. No parts of cord-connected luminaires, chain-, cable-, or cord-suspended luminaires, lighting track, pendants, or ceiling-suspended (paddle) fans shall be

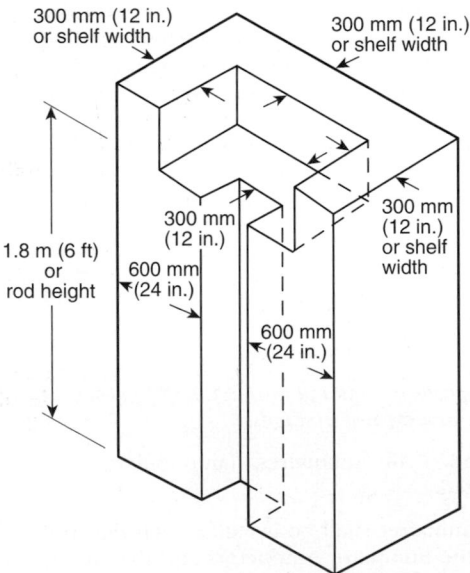

FIGURE 410.2   Closet Storage Space.

located within a zone measured 900 mm (3 ft) horizontally and 2.5 m (8 ft) vertically from the top of the bathtub rim or shower stall threshold. This zone is all encompassing and includes the space directly over the tub or shower stall. Luminaires located within the actual outside dimension of the bathtub or shower to a height of 2.5 m (8 ft) vertically from the top of the bathtub rim or shower threshold shall be marked for damp locations, or marked for wet locations where subject to shower spray.

(E) Luminaires in Indoor Sports, Mixed-Use, and All-Purpose Facilities. Luminaires subject to physical damage, using a mercury vapor or metal halide lamp, installed in playing and spectator seating areas of indoor sports, mixed-use, or all-purpose facilities shall be of the type that protects the lamp with a glass or plastic lens. Such luminaires shall be permitted to have an additional guard.

(F) Luminaires Installed in or Under Roof Decking. Luminaires installed in exposed or concealed locations under metal-corrugated sheet roof decking shall be installed and supported so there is not less than 38 mm (1½ in.) measured from the lowest surface of the roof decking to the top of the luminaire.

410.11 Luminaires Near Combustible Material. Luminaires shall be constructed, installed, or equipped with shades or guards so that combustible material is not subjected to temperatures in excess of 90°C (194°F).

410.12 Luminaires over Combustible Material. Lampholders installed over highly combustible material shall be of the unswitched type. Unless an individual switch is provided for each luminaire, lampholders shall be located at least 2.5 m (8 ft) above the floor or shall be located or guarded so that the lamps cannot be readily removed or damaged.

410.14 Luminaires in Show Windows. Chain-supported luminaires used in a show window shall be permitted to be externally wired. No other externally wired luminaires shall be used.

410.16 Luminaires in Clothes Closets.

(A) Luminaire Types Permitted. Only luminaires of the following types shall be permitted in a closet:

(1) Surface-mounted or recessed incandescent or LED luminaires with completely enclosed light sources

(2) Surface-mounted or recessed fluorescent luminaires

(3) Surface-mounted fluorescent or LED luminaires identified as suitable for installation within the closet storage space

(B) Luminaire Types Not Permitted. Incandescent luminaires with open or partially enclosed lamps and pendant luminaires or lampholders shall not be permitted.

(C) Location. The minimum clearance between luminaires installed in clothes closets and the nearest point of a closet storage space shall be as follows:

(1) 300 mm (12 in.) for surface-mounted incandescent or LED luminaires with a completely enclosed light source installed on the wall above the door or on the ceiling.

(2) 150 mm (6 in.) for surface-mounted fluorescent luminaires installed on the wall above the door or on the ceiling.

(3) 150 mm (6 in.) for recessed incandescent or LED luminaires with a completely enclosed light source installed in the wall or the ceiling.

(4) 150 mm (6 in.) for recessed fluorescent luminaires installed in the wall or the ceiling.

(5) Surface-mounted fluorescent or LED luminaires shall be permitted to be installed within the closet storage space where identified for this use.

410.18 Space for Cove Lighting. Coves shall have adequate space and shall be located so that lamps and equipment can be properly installed and maintained.

Part III. Provisions at Luminaire Outlet Boxes, Canopies, and Pans

410.20 Space for Conductors. Canopies and outlet boxes taken together shall provide sufficient space so that luminaire conductors and their connecting devices are capable of being installed in accordance with 314.16.

410.21 Temperature Limit of Conductors in Outlet Boxes. Luminaires shall be of such construction or installed so that the conductors in outlet boxes shall not be subjected to temperatures greater than that for which the conductors are rated.

Branch-circuit wiring, other than 2-wire or multiwire branch circuits supplying power to luminaires connected together, shall not be passed through an outlet box that is an integral part of a luminaire unless the luminaire is identified for through-wiring.

> Informational Note: See 410.64(C) for wiring supplying power to luminaires connected together.

410.22 Outlet Boxes to Be Covered. In a completed installation, each outlet box shall be provided with a cover unless covered by means of a luminaire canopy, lampholder, receptacle, or similar device.

410.23 Covering of Combustible Material at Outlet Boxes. Any combustible wall or ceiling finish exposed between the edge of a luminaire canopy or pan and an outlet box having a surface area of 1160 mm² (180 in.²) or more shall be covered with noncombustible material.

410.24 Connection of Electric-Discharge and LED Luminaires.

(A) Independent of the Outlet Box. Electric-discharge and LED luminaires supported independently of the outlet box shall be connected to the branch circuit through metal raceway, nonmetallic raceway, Type MC cable, Type AC cable, Type MI cable, nonmetallic sheathed cable, or by flexible cord as permitted in 410.62(B) or 410.62(C).

(B) Access to Boxes. Electric-discharge and LED luminaires surface mounted over concealed outlet, pull, or junction boxes and designed not to be supported solely by the outlet box shall be provided with suitable openings in the back of the luminaire to provide access to the wiring in the box.

Part IV. Luminaire Supports

410.30 Supports.

(A) General. Luminaires and lampholders shall be securely supported. A luminaire that weighs more than 3 kg (6 lb) or exceeds 400 mm (16 in.) in any dimension shall not be supported by the screw shell of a lampholder.

(B) Metal or Nonmetallic Poles Supporting Luminaires. Metal or nonmetallic poles shall be permitted to be used to support luminaires and as a raceway to enclose supply conductors, provided the following conditions are met:

(1) A pole shall have a handhole not less than 50 mm × 100 mm (2 in. × 4 in.) with a cover suitable for use in wet locations to provide access to the supply terminations within the pole or pole base.

Exception No. 1: No handhole shall be required in a pole 2.5 m (8 ft) or less in height abovegrade where the supply wiring method continues without splice or pull point, and where the interior of the pole and any splices are accessible by removing the luminaire.

Exception No. 2: No handhole shall be required in a pole 6.0 m (20 ft) or less in height abovegrade that is provided with a hinged base.

(2) Where raceway risers or cable is not installed within the pole, a threaded fitting or nipple shall be brazed, welded, or attached to the pole opposite the handhole for the supply connection.

(3) A metal pole shall be provided with an equipment grounding terminal as follows:

a. A pole with a handhole shall have the equipment grounding terminal accessible from the handhole.
b. A pole with a hinged base shall have the equipment grounding terminal accessible within the base.

Exception to (3): No grounding terminal shall be required in a pole 2.5 m (8 ft) or less in height abovegrade where the supply wiring method continues without splice or pull, and where the interior of the pole and any splices are accessible by removing the luminaire.

(4) A metal pole with a hinged base shall have the hinged base and pole bonded together.

(5) Metal raceways or other equipment grounding conductors shall be bonded to the metal pole with an equipment grounding conductor recognized by 250.118 and sized in accordance with 250.122.

(6) Conductors in vertical poles used as raceway shall be supported as provided in 300.19.

410.36 Means of Support.

(A) Outlet Boxes. Outlet boxes or fittings installed as required by 314.23 and complying with the provisions of 314.27(A)(1) and 314.27(A)(2) shall be permitted to support luminaires.

(B) Suspended Ceilings. Framing members of suspended ceiling systems used to support luminaires shall be securely fastened to each other and shall be securely attached to the building structure at appropriate intervals. Luminaires shall be securely fastened to the ceiling framing member by mechanical means such as bolts, screws, or rivets. Listed clips identified for use with the type of ceiling framing member(s) and luminaire(s) shall also be permitted.

(C) Luminaire Studs. Luminaire studs that are not a part of outlet boxes, hickeys, tripods, and crowfeet shall be made of steel, malleable iron, or other material suitable for the application.

(D) Insulating Joints. Insulating joints that are not designed to be mounted with screws or bolts shall have an exterior metal casing, insulated from both screw connections.

(E) Raceway Fittings. Raceway fittings used to support a luminaire(s) shall be capable of supporting the weight of the complete fixture assembly and lamp(s).

(F) Busways. Luminaires shall be permitted to be connected to busways in accordance with 368.17(C).

(G) Trees. Outdoor luminaires and associated equipment shall be permitted to be supported by trees.

Informational Note No. 1: See 225.26 for restrictions for support of overhead conductors.

Informational Note No. 2: See 300.5(D) for protection of conductors.

Part V. Grounding

410.40 General. Luminaires and lighting equipment shall be grounded as required in Article 250 and Part V of this article.

410.42 Luminaire(s) with Exposed Conductive Parts. Exposed metal parts shall be connected to an equipment grounding conductor or insulated from the equipment grounding conductor and other conducting surfaces or be inaccessible to unqualified personnel. Lamp tie wires, mounting screws, clips, and decorative bands on glass spaced at least 38 mm (1½ in.) from lamp terminals shall not be required to be grounded.

410.44 Methods of Grounding. Luminaires and equipment shall be mechanically connected to an equipment grounding conductor as specified in 250.118 and sized in accordance with 250.122.

Exception No. 1: Luminaires made of insulating material that is directly wired or attached to outlets supplied by a wiring method that does not provide a ready means for grounding attachment to an equipment grounding conductor shall be made of insulating material and shall have no exposed conductive parts.

Exception No. 2: Replacement luminaires shall be permitted to connect an equipment grounding conductor from the outlet in compliance with 250.130(C). The luminaire shall then comply with 410.42.

Exception No. 3: Where no equipment grounding conductor exists at the outlet, replacement luminaires that are GFCI protected shall not be required to be connected to an equipment grounding conductor.

410.46 Equipment Grounding Conductor Attachment. Luminaires with exposed metal parts shall be provided with a means for connecting an equipment grounding conductor for such luminaires.

Part VI. Wiring of Luminaires

410.48 Luminaire Wiring — General. Wiring on or within luminaires shall be neatly arranged and shall not be exposed to physical damage. Excess wiring shall be avoided. Conductors shall be arranged so that they are not subjected to temperatures above those for which they are rated.

410.50 Polarization of Luminaires. Luminaires shall be wired so that the screw shells of lampholders are connected to the same luminaire or circuit conductor or terminal. The grounded conductor, where connected to a screw shell lampholder, shall be connected to the screw shell.

410.52 Conductor Insulation. Luminaires shall be wired with conductors having insulation suitable for the environmental conditions, current, voltage, and temperature to which the conductors will be subjected.

Informational Note: For ampacity of fixture wire, maximum operating temperature, voltage limitations, minimum wire size, and other information, see Article 402.

410.54 Pendant Conductors for Incandescent Filament Lamps.

(A) Support. Pendant lampholders with permanently attached leads, where used for other than festoon wiring, shall be hung from separate stranded rubber-covered conductors that are soldered directly to the circuit conductors but supported independently thereof.

(B) Size. Unless part of listed decorative lighting assemblies, pendant conductors shall not be smaller than 14 AWG for mogul-base or medium-base screw shell lampholders or smaller than 18 AWG for intermediate or candelabra-base lampholders.

(C) Twisted or Cabled. Pendant conductors longer than 900 mm (3 ft) shall be twisted together where not cabled in a listed assembly.

410.56 Protection of Conductors and Insulation.

(A) Properly Secured. Conductors shall be secured in a manner that does not tend to cut or abrade the insulation.

(B) Protection Through Metal. Conductor insulation shall be protected from abrasion where it passes through metal.

(C) Luminaire Stems. Splices and taps shall not be located within luminaire arms or stems.

(D) Splices and Taps. No unnecessary splices or taps shall be made within or on a luminaire.

Informational Note: For approved means of making connections, see 110.14.

(E) Stranding. Stranded conductors shall be used for wiring on luminaire chains and on other movable or flexible parts.

(F) Tension. Conductors shall be arranged so that the weight of the luminaire or movable parts does not put tension on the conductors.

410.59 Cord-Connected Showcases. Individual showcases, other than fixed, shall be permitted to be connected by flexible cord to permanently installed receptacles, and groups of not more than six such showcases shall be permitted to be coupled together by flexible cord and separable locking-type connectors with one of the group connected by flexible cord to a permanently installed receptacle.

The installation shall comply with 410.59(A) through (E).

(A) Cord Requirements. Flexible cord shall be of the hard-service type, having conductors not smaller than the branch-circuit conductors, having ampacity at least equal to the branch-circuit overcurrent device, and having an equipment grounding conductor.

Informational Note: See Table 250.122 for size of equipment grounding conductor.

(B) Receptacles, Connectors, and Attachment Plugs. Receptacles, connectors, and attachment plugs shall be of a listed grounding type rated 15 or 20 amperes.

(C) Support. Flexible cords shall be secured to the undersides of showcases such that all of the following conditions are ensured:

(1) The wiring is not exposed to physical damage.

(2) The separation between cases is not in excess of 50 mm (2 in.), or more than 300 mm (12 in.) between the first case and the supply receptacle.

(3) The free lead at the end of a group of showcases has a female fitting not extending beyond the case.

(D) No Other Equipment. Equipment other than showcases shall not be electrically connected to showcases.

(E) Secondary Circuit(s). Where showcases are cord-connected, the secondary circuit(s) of each electric-discharge lighting ballast shall be limited to one showcase.

410.62 Cord-Connected Lampholders and Luminaires.

(A) Lampholders. Where a metal lampholder is attached to a flexible cord, the inlet shall be equipped with an insulating bushing that, if threaded, is not smaller than metric designator 12 (trade size ⅜) pipe size. The cord hole shall be of a size appropriate for the cord, and all burrs and fins shall be removed in order to provide a smooth bearing surface for the cord.

Bushing having holes 7 mm (%₃₂ in.) in diameter shall be permitted for use with plain pendant cord and holes 11 mm (¹³⁄₃₂ in.) in diameter with reinforced cord.

(B) Adjustable Luminaires. Luminaires that require adjusting or aiming after installation shall not be required to be equipped with an attachment plug or cord connector, provided the exposed cord is suitable for hard-usage or extra-hard-usage and is not longer than that required for maximum adjustment. The cord shall not be subject to strain or physical damage.

Informational Note: For application provisions, see Table 400.4, "Use" column.

(C) Electric-Discharge and LED Luminaires. Electric-discharge and LED luminaires shall comply with (1), (2), and (3) as applicable.

(1) Cord-Connected Installation. A luminaire or a listed assembly in compliance with any of the conditions in (a) through (c) shall be permitted to be cord connected provided the luminaire is located directly below the outlet or busway, the cord is not subject to strain or physical damage, and the cord is visible over its entire length except at terminations.

(a) A luminaire shall be permitted to be connected with a cord terminating in a grounding-type attachment plug or busway plug.

(b) A luminaire assembly equipped with a strain relief and canopy shall be permitted to use a cord connection between the luminaire assembly and the canopy. The canopy shall be permitted to include a section of raceway not over 150 mm (6 in.) in length and intended to facilitate the connection to an outlet box mounted above a suspended ceiling.

(c) Listed luminaires connected using listed assemblies that incorporate manufactured wiring system connectors in accordance with 604.100(C) shall be permitted to be cord connected.

(2) Provided with Mogul-Base, Screw Shell Lampholders. Electric-discharge luminaires provided with mogul-base, screw shell lampholders shall be permitted to be connected to branch circuits of 50 amperes or less by cords complying with 240.5. Receptacles and attachment plugs shall be permitted to be of a lower ampere rating than the branch circuit but not less than 125 percent of the luminaire full-load current.

(3) Equipped with Flanged Surface Inlet. Electric-discharge luminaires equipped with a flanged surface inlet shall be permitted to be supplied by cord pendants equipped with cord connectors. Inlets and connectors shall be permitted to be of a lower ampere rating than the branch circuit but not less than 125 percent of the luminaire load current.

410.64 Luminaires as Raceways.
Luminaires shall not be used as a raceway for circuit conductors unless they comply with 410.64(A), (B), or (C).

(A) Listed. Luminaires listed and marked for use as a raceway shall be permitted to be used as a raceway.

(B) Through-Wiring. Luminaires identified for through-wiring, as permitted by 410.21, shall be permitted to be used as a raceway.

(C) Luminaires Connected Together. Luminaires designed for end-to-end connection to form a continuous assembly, or luminaires connected together by recognized wiring methods, shall be permitted to contain the conductors of a 2-wire branch circuit, or one multiwire branch circuit, supplying the connected luminaires and shall not be required to be listed as a raceway. One additional 2-wire branch circuit separately supplying one or more of the connected luminaires shall also be permitted.

Informational Note: See Article 100 for the definition of *Multiwire Branch Circuit*.

410.68 Feeder and Branch-Circuit Conductors and Ballasts.
Feeder and branch-circuit conductors within 75 mm (3 in.) of a ballast, LED driver, power supply, or transformer shall have an insulation temperature rating not lower than 90°C (194°F), unless supplying a luminaire marked as suitable for a different insulation temperature.

Part VII. Construction of Luminaires

410.70 Combustible Shades and Enclosures.
Adequate airspace shall be provided between lamps and shades or other enclosures of combustible material.

410.74 Luminaire Rating.

(A) Marking. All luminaires shall be marked with the maximum lamp wattage or electrical rating, manufacturer's name, trademark, or other suitable means of identification. A luminaire requiring supply wire rated higher than 60°C (140°F) shall be marked with the minimum supply wire temperature rating on the luminaire and shipping carton or equivalent.

(B) Electrical Rating. The electrical rating shall include the voltage and frequency and shall indicate the current rating of the unit, including the ballast, transformer, LED driver, power supply, or autotransformer.

410.82 Portable Luminaires.

(A) General. Portable luminaires shall be wired with flexible cord recognized by 400.4 and an attachment plug of the polarized or grounding type. Where used with Edison-base lampholders, the grounded conductor shall be identified and attached to the screw shell and the identified blade of the attachment plug.

(B) Portable Handlamps. In addition to the provisions of 410.82(A), portable handlamps shall comply with the following:

(1) Metal shell, paper-lined lampholders shall not be used.
(2) Handlamps shall be equipped with a handle of molded composition or other insulating material.
(3) Handlamps shall be equipped with a substantial guard attached to the lampholder or handle.
(4) Metallic guards shall be grounded by means of an equipment grounding conductor run with circuit conductors within the power-supply cord.
(5) Portable handlamps shall not be required to be grounded where supplied through an isolating transformer with an ungrounded secondary of not over 50 volts.

410.84 Cord Bushings. A bushing or the equivalent shall be provided where flexible cord enters the base or stem of a portable luminaire. The bushing shall be of insulating material unless a jacketed type of cord is used.

Part VIII. Installation of Lampholders

410.90 Screw Shell Type. Lampholders of the screw shell type shall be installed for use as lampholders only. Where supplied by a circuit having a grounded conductor, the grounded conductor shall be connected to the screw shell.

410.93 Double-Pole Switched Lampholders. Where supplied by the ungrounded conductors of a circuit, the switching device of lampholders of the switched type shall simultaneously disconnect both conductors of the circuit.

410.96 Lampholders in Wet or Damp Locations. Lampholders installed in wet locations shall be listed for use in wet locations. Lampholders installed in damp locations shall be listed for damp locations or shall be listed for wet locations.

410.97 Lampholders Near Combustible Material. Lampholders shall be constructed, installed, or equipped with shades or guards so that combustible material is not subjected to temperatures in excess of 90°C (194°F).

Part IX. Lamps and Auxiliary Equipment

410.103 Bases, Incandescent Lamps. An incandescent lamp for general use on lighting branch circuits shall not be equipped with a medium base if rated over 300 watts, or with a mogul base if rated over 1500 watts. Special bases or other devices shall be used for over 1500 watts.

410.104 Electric-Discharge Lamp Auxiliary Equipment.

(A) Enclosures. Auxiliary equipment for electric-discharge lamps shall be enclosed in noncombustible cases and treated as sources of heat.

(B) Switching. Where supplied by the ungrounded conductors of a circuit, the switching device of auxiliary equipment shall simultaneously disconnect all conductors.

Part X. Special Provisions for Flush and Recessed Luminaire

410.110 General. Luminaires installed in recessed cavities walls or ceilings, including suspended ceilings, shall com with 410.115 through 410.122.

410.115 Temperature.

(A) Combustible Material. Luminaires shall be installed that adjacent combustible material will not be subjected temperatures in excess of 90°C (194°F).

(B) Fire-Resistant Construction. Where a luminaire recessed in fire-resistant material in a building of fire-resista construction, a temperature higher than 90°C (194°F) but n higher than 150°C (302°F) shall be considered acceptable the luminaire is plainly marked for that service.

(C) Recessed Incandescent Luminaires. Incandescent lum aires shall have thermal protection and shall be identified thermally protected.

Exception No. 1: Thermal protection shall not be required in a reces luminaire identified for use and installed in poured concrete.

Exception No. 2: Thermal protection shall not be required in a reces luminaire whose design, construction, and thermal performance cl acteristics are equivalent to a thermally protected luminaire and identified as inherently protected.

410.116 Clearance and Installation.

(A) Clearance.

(1) Non-Type IC. A recessed luminaire that is not identif for contact with insulation shall have all recessed parts spa not less than 13 mm (½ in.) from combustible materials. T points of support and the trim finishing off the openings in ceiling, wall, or other finished surface shall be permitted to in contact with combustible materials.

(2) Type IC. A recessed luminaire that is identified contact with insulation, Type IC, shall be permitted to be contact with combustible materials at recessed parts, points support, and portions passing through or finishing off opening in the building structure.

(B) Installation. Thermal insulation shall not be instal above a recessed luminaire or within 75 mm (3 in.) of recessed luminaire's enclosure, wiring compartment, ball transformer, LED driver, or power supply unless the lumin is identified as Type IC for insulation contact.

410.117 Wiring.

(A) General. Conductors that have insulation suitable for temperature encountered shall be used.

(B) Circuit Conductors. Branch-circuit conductors that h an insulation suitable for the temperature encountered shal permitted to terminate in the luminaire.

(C) Tap Conductors. Tap conductors of a type suitable for temperature encountered shall be permitted to run from luminaire terminal connection to an outlet box placed at l 300 mm (1 ft) from the luminaire. Such tap conductors s be in suitable raceway or Type AC or MC cable of at l 450 mm (18 in.) but not more than 1.8 m (6 ft) in length.

Part XI. Construction of Flush and Recessed Luminaires

410.118 Temperature. Luminaires shall be constructed such that adjacent combustible material is not subject to temperatures in excess of 90°C (194°F).

410.120 Lamp Wattage Marking. Incandescent lamp luminaires shall be marked to indicate the maximum allowable wattage of lamps. The markings shall be permanently installed, in letters at least 6 mm (¼ in.) high, and shall be located where visible during relamping.

410.121 Solder Prohibited. No solder shall be used in the construction of a luminaire recessed housing.

410.122 Lampholders. Lampholders of the screw shell type shall be of porcelain or other suitable insulating materials.

Part XII. Special Provisions for Electric-Discharge Lighting Systems of 1000 Volts or Less

410.130 General.

(A) Open-Circuit Voltage of 1000 Volts or Less. Equipment for use with electric-discharge lighting systems and designed for an open-circuit voltage of 1000 volts or less shall be of a type identified for such service.

(B) Considered as Energized. The terminals of an electric-discharge lamp shall be considered as energized where any lamp terminal is connected to a circuit of over 300 volts.

(C) Transformers of the Oil-Filled Type. Transformers of the oil-filled type shall not be used.

(D) Additional Requirements. In addition to complying with the general requirements for luminaires, such equipment shall comply with Part XII of this article.

(E) Thermal Protection — Fluorescent Luminaires.

(1) Integral Thermal Protection. The ballast of a fluorescent luminaire installed indoors shall have integral thermal protection. Replacement ballasts shall also have thermal protection integral with the ballast.

(2) Simple Reactance Ballasts. A simple reactance ballast in a fluorescent luminaire with straight tubular lamps shall not be required to be thermally protected.

(3) Exit Luminaires. A ballast in a fluorescent exit luminaire shall not have thermal protection.

(4) Egress Luminaires. A ballast in a fluorescent luminaire that is used for egress lighting and energized only during a failure of the normal supply shall not have thermal protection.

(F) High-Intensity Discharge Luminaires.

(1) Recessed. Recessed high-intensity luminaires designed to be installed in wall or ceiling cavities shall have thermal protection and be identified as thermally protected.

(2) Inherently Protected. Thermal protection shall not be required in a recessed high-intensity luminaire whose design, construction, and thermal performance characteristics are equivalent to a thermally protected luminaire and are identified as inherently protected.

(3) Installed in Poured Concrete. Thermal protection shall not be required in a recessed high-intensity discharge luminaire identified for use and installed in poured concrete.

(4) Recessed Remote Ballasts. A recessed remote ballast for a high-intensity discharge luminaire shall have thermal protection that is integral with the ballast and shall be identified as thermally protected.

(5) Metal Halide Lamp Containment. Luminaires that use a metal halide lamp other than a thick-glass parabolic reflector lamp (PAR) shall be provided with a containment barrier that encloses the lamp, or shall be provided with a physical means that only allows the use of a lamp that is Type O.

Informational Note: See ANSI Standard C78.389, *American National Standard for Electric Lamps — High Intensity Discharge, Methods of Measuring Characteristics.*

(G) Disconnecting Means.

(1) General. In indoor locations other than dwellings and associated accessory structures, fluorescent luminaires that utilize double-ended lamps and contain ballast(s) that can be serviced in place shall have a disconnecting means either internal or external to each luminaire. For existing installed luminaires without disconnecting means, at the time a ballast is replaced, a disconnecting means shall be installed. The line side terminals of the disconnecting means shall be guarded.

Exception No. 1: A disconnecting means shall not be required for luminaires installed in hazardous (classified) location(s).

Exception No. 2: A disconnecting means shall not be required for luminaires that provide emergency illumination required in 700.16.

Exception No. 3: For cord-and-plug-connected luminaires, an accessible separable connector or an accessible plug and receptacle shall be permitted to serve as the disconnecting means.

Exception No. 4: Where more than one luminaire is installed and supplied by other than a multiwire branch circuit, a disconnecting means shall not be required for every luminaire when the design of the installation includes disconnecting means, such that the illuminated space cannot be left in total darkness.

(2) Multiwire Branch Circuits. When connected to multiwire branch circuits, the disconnecting means shall simultaneously break all the supply conductors to the ballast, including the grounded conductor.

(3) Location. The disconnecting means shall be located so as to be accessible to qualified persons before servicing or maintaining the ballast. Where the disconnecting means is external to the luminaire, it shall be a single device, and shall be attached to the luminaire or the luminaire shall be located within sight of the disconnecting means.

410.134 Direct-Current Equipment. Luminaires installed on dc circuits shall be equipped with auxiliary equipment and resistors designed for dc operation. The luminaires shall be marked for dc operation.

410.135 Open-Circuit Voltage Exceeding 300 Volts. Equipment having an open-circuit voltage exceeding 300 volts shall not be installed in dwelling occupancies unless such equipment is designed so that there will be no exposed live parts when lamps are being inserted, are in place, or are being removed.

410.136 Luminaire Mounting.

(A) Exposed Components. Luminaires that have exposed ballasts, transformers, LED drivers, or power supplies shall be installed such that ballasts, transformers, LED drivers, or power supplies shall not be in contact with combustible material unless listed for such condition.

(B) Combustible Low-Density Cellulose Fiberboard. Where a surface-mounted luminaire containing a ballast, transformer, LED driver, or power supply is to be installed on combustible low-density cellulose fiberboard, it shall be marked for this condition or shall be spaced not less than 38 mm (1½ in.) from the surface of the fiberboard. Where such luminaires are partially or wholly recessed, the provisions of 410.110 through 410.122 shall apply.

Informational Note: Combustible low-density cellulose fiberboard includes sheets, panels, and tiles that have a density of 320 kg/m^3 (20 lb/ft^3) or less and that are formed of bonded plant fiber material but does not include solid or laminated wood or fiberboard that has a density in excess of 320 kg/m^3 (20 lb/ft^3) or is a material that has been integrally treated with fire-retarding chemicals to the degree that the flame spread index in any plane of the material will not exceed 25, determined in accordance with tests for surface burning characteristics of building materials. See ANSI/ASTM E84–2015a, *Standard Test Method for Surface Burning Characteristics of Building Materials* or ANSI/UL 723–2013, *Standard for Test for Surface Burning Characteristics of Building Materials.*

410.137 Equipment Not Integral with Luminaire.

(A) Metal Cabinets. Auxiliary equipment, including reactors, capacitors, resistors, and similar equipment, where not installed as part of a luminaire assembly, shall be enclosed in accessible, permanently installed metal cabinets.

(B) Separate Mounting. Separately mounted ballasts, transformers, LED drivers, or power supplies that are listed for direct connection to a wiring system shall not be required to be additionally enclosed.

(C) Wired Luminaire Sections. Wired luminaire sections are paired, with a ballast(s) supplying a lamp or lamps in both. For interconnection between paired units, it shall be permissible to use metric designator 12 (trade size ⅜) flexible metal conduit in lengths not exceeding 7.5 m (25 ft), in conformance with Article 348. Luminaire wire operating at line voltage, supplying only the ballast(s) of one of the paired luminaires shall be permitted in the same raceway as the lamp supply wires of the paired luminaires.

410.138 Autotransformers. An autotransformer that is used to raise the voltage to more than 300 volts, as part of a ballast for supplying lighting units, shall be supplied only by a grounded system.

410.139 Switches. Snap switches shall comply with 404.14.

Part XIII. Special Provisions for Electric-Discharge Lighting Systems of More Than 1000 Volts

410.140 General.

(A) Listing. Electric-discharge lighting systems with an open-circuit voltage exceeding 1000 volts shall be listed and installed in conformance with that listing.

(B) Dwelling Occupancies. Equipment that has an open-circuit voltage exceeding 1000 volts shall not be installed in or on dwelling occupancies.

(C) Live Parts. The terminal of an electric-discharge lamp shall be considered as a live part.

(D) Additional Requirements. In addition to complying with the general requirements for luminaires, such equipment shall comply with Part XIII of this article.

Informational Note: For signs and outline lighting, see Article 600.

410.141 Control.

(A) Disconnection. Luminaires or lamp installation shall be controlled either singly or in groups by an externally operable switch or circuit breaker that opens all ungrounded primary conductors.

(B) Within Sight or Locked Type. The switch or circuit breaker shall be located within sight from the luminaires or lamps, or it shall be permitted to be located elsewhere if it is lockable in accordance with 110.25.

410.142 Lamp Terminals and Lampholders. Parts that must be removed for lamp replacement shall be hinged or held captive. Lamps or lampholders shall be designed so that there are no exposed live parts when lamps are being inserted or removed.

410.143 Transformers.

(A) Type. Transformers shall be enclosed, identified for the use, and listed.

(B) Voltage. The secondary circuit voltage shall not exceed 15,000 volts, nominal, under any load condition. The voltage to ground of any output terminals of the secondary circuit shall not exceed 7500 volts under any load conditions.

(C) Rating. Transformers shall have a secondary short-circuit current rating of not more than 150 mA if the open-circuit voltage is over 7500 volts, and not more than 300 mA if the open-circuit voltage rating is 7500 volts or less.

(D) Secondary Connections. Secondary circuit outputs shall not be connected in parallel or in series.

410.144 Transformer Locations.

(A) Accessible. Transformers shall be accessible after installation.

(B) Secondary Conductors. Transformers shall be installed as near to the lamps as practicable to keep the secondary conductors as short as possible.

(C) Adjacent to Combustible Materials. Transformers shall be located so that adjacent combustible materials are not subjected to temperatures in excess of 90°C (194°F).

410.145 Exposure to Damage. Lamps shall not be located where normally exposed to physical damage.

410.146 Marking. Each luminaire or each secondary circuit of tubing having an open-circuit voltage of over 1000 volts shall have a clearly legible marking in letters not less than 6 mm (¼ in.) high reading "Caution ____ volts." The voltage indicated shall be the rated open-circuit voltage. The caution sign(s) or label(s) shall comply with 110.21(B).

Part XIV. Lighting Track

410.151 Installation.

(A) Lighting Track. Lighting track shall be permanently installed and permanently connected to a branch circuit. Only lighting track fittings shall be installed on lighting track. Lighting track fittings shall not be equipped with general-purpose receptacles.

(B) Connected Load. The connected load on lighting track shall not exceed the rating of the track. Lighting track shall be supplied by a branch circuit having a rating not more than that of the track. The load calculation in 220.43(B) shall not be required to limit the length of track on a single branch circuit, and it shall not be required to limit the number of luminaires on a single track.

(C) Locations Not Permitted. Lighting track shall not be installed in the following locations:

(1) Where likely to be subjected to physical damage
(2) In wet or damp locations
(3) Where subject to corrosive vapors
(4) In storage battery rooms
(5) In hazardous (classified) locations
(6) Where concealed
(7) Where extended through walls or partitions
(8) Less than 1.5 m (5 ft) above the finished floor except where protected from physical damage or track operating at less than 30 volts rms open-circuit voltage
(9) Where prohibited by 410.10(D)

(D) Support. Fittings identified for use on lighting track shall be designed specifically for the track on which they are to be installed. They shall be securely fastened to the track, shall maintain polarization and connections to the equipment grounding conductor, and shall be designed to be suspended directly from the track.

410.153 Heavy-Duty Lighting Track. Heavy-duty lighting track is lighting track identified for use exceeding 20 amperes. Each fitting attached to a heavy-duty lighting track shall have individual overcurrent protection.

410.154 Fastening. Lighting track shall be securely mounted so that each fastening is suitable for supporting the maximum weight of luminaires that can be installed. Unless identified for supports at greater intervals, a single section 1.2 m (4 ft) or shorter in length shall have two supports, and, where installed in a continuous row, each individual section of not more than 1.2 m (4 ft) in length shall have one additional support.

410.155 Construction Requirements.

(A) Construction. The housing for the lighting track system shall be of substantial construction to maintain rigidity. The conductors shall be installed within the track housing, permitting insertion of a luminaire, and designed to prevent tampering and accidental contact with live parts. Components of lighting track systems of different voltages shall not be interchangeable. The track conductors shall be a minimum 12 AWG or equal and shall be copper. The track system ends shall be insulated and capped.

(B) Grounding. Lighting track shall be grounded in accordance with Article 250, and the track sections shall be securely coupled to maintain continuity of the circuitry, polarization, and grounding throughout.

Part XV. Decorative Lighting and Similar Accessories

410.160 Listing of Decorative Lighting. Decorative lighting and similar accessories used for holiday lighting and similar purposes, in accordance with 590.3(B), shall be listed.

ARTICLE 411
Low-Voltage Lighting

411.1 Scope. This article covers lighting systems and their associated components operating at no more than 30 volts ac or 60 volts dc. Where wet contact is likely to occur, the limits are 15 volts ac or 30 volts dc.

> Informational Note: Refer to Article 680 for applications involving immersion.

411.3 Low-Voltage Lighting Systems. Low voltage lighting systems shall consist of an isolating power supply, low-voltage luminaires, and associated equipment that are all identified for the use. The output circuits of the power supply shall be rated for 25 amperes maximum under all load conditions.

411.4 Listing Required. Low-voltage lighting systems shall comply with 411.4(A) or 411.4(B).

(A) Listed System. The luminaires, power supply, and luminaire fittings (including the exposed bare conductors) of an exposed bare conductor lighting system shall be listed for the use as part of the same identified lighting system.

(B) Assembly of Listed Parts. A lighting system assembled from the following listed parts shall be permitted:

(1) Low-voltage luminaires
(2) Power supply
(3) Low-voltage luminaire fittings
(4) Suitably rated cord, cable, conductors in conduit, or other fixed Chapter 3 wiring method for the secondary circuit

411.5 Specific Location Requirements.

(A) Walls, Floors, and Ceilings. Conductors concealed or extended through a wall, floor, or ceiling shall be in accordance with (1) or (2):

(1) Installed using any of the wiring methods specified in Chapter 3
(2) Installed using wiring supplied by a listed Class 2 power source and installed in accordance with 725.130

(B) Pools, Spas, Fountains, and Similar Locations. Lighting systems shall be installed not less than 3 m (10 ft) horizontally from the nearest edge of the water, unless permitted by Article 680.

411.6 Secondary Circuits.

(A) Grounding. Secondary circuits shall not be grounded.

(B) Isolation. The secondary circuit shall be insulated from the branch circuit by an isolating transformer.

(C) Bare Conductors. Exposed bare conductors and current-carrying parts shall be permitted for indoor installations only.

Bare conductors shall not be installed less than 2.1 m (7 ft) above the finished floor, unless specifically listed for a lower installation height.

(D) Insulated Conductors. Insulated secondary circuit conductors shall be of the type, and installed as, described in (1), (2), or (3):

(1) Class 2 cable supplied by a Class 2 power source and installed in accordance with Parts I and III of Article 725.

(2) Conductors, cord, or cable of the listed system and installed not less than 2.1 m (7 ft) above the finished floor unless the system is specifically listed for a lower installation height.

(3) Wiring methods described in Chapter 3.

411.7 Branch Circuit. Lighting systems covered by this article shall be supplied from a maximum 20-ampere branch circuit.

411.8 Hazardous (Classified) Locations. Where installed in hazardous (classified) locations, these systems shall conform with Articles 500 through 517 in addition to this article.

ARTICLE 422
Appliances

Part I. General

422.1 Scope. This article covers electrical appliances used in any occupancy.

422.3 Other Articles. The requirements of Article 430 shall apply to the installation of motor-operated appliances, and the requirements of Article 440 shall apply to the installation of appliances containing a hermetic refrigerant motor-compressor(s), except as specifically amended in this article.

422.4 Live Parts. Appliances shall have no live parts normally exposed to contact other than those parts functioning as open-resistance heating elements, such as the heating element of a toaster, which are necessarily exposed.

422.5 Ground-Fault Circuit-Interrupter (GFCI) Protection for Personnel.

(A) General. Appliances identified in 422.5(A)(1) through (5) rated 250 volts or less and 60 amperes or less, single- or 3-phase, shall be provided with GFCI protection for personnel. Multiple GFCI protective devices shall be permitted but shall not be required.

(1) Automotive vacuum machines provided for public use
(2) Drinking water coolers
(3) High-pressure spray washing machines — cord-and-plug-connected
(4) Tire inflation machines provided for public use
(5) Vending machines

N **(B) Type.** The GFCI shall be readily accessible, listed, and located in one or more of the following locations:

(1) Within the branch circuit overcurrent device
(2) A device or outlet within the supply circuit
(3) An integral part of the attachment plug

(4) Within the supply cord not more than 300 mm (12 in.) from the attachment plug
(5) Factory installed within the appliance

N **422.6 Listing Required.** All appliances operating at 50 volts or more shall be listed.

Part II. Installation

422.10 Branch-Circuit Rating. This section specifies the ratings of branch circuits capable of carrying appliance current without overheating under the conditions specified.

(A) Individual Circuits. The rating of an individual branch circuit shall not be less than the marked rating of the appliance or the marked rating of an appliance having combined loads as provided in 422.62.

The rating of an individual branch circuit for motor-operated appliances not having a marked rating shall be in accordance with Part II of Article 430.

The branch-circuit rating for an appliance that is a continuous load, other than a motor-operated appliance, shall not be less than 125 percent of the marked rating, or not less than 100 percent of the marked rating if the branch-circuit device and its assembly are listed for continuous loading at 100 percent of its rating.

Branch circuits and branch-circuit conductors for household ranges and cooking appliances shall be permitted to be in accordance with Table 220.55 and shall be sized in accordance with 210.19(A)(3).

(B) Circuits Supplying Two or More Loads. For branch circuits supplying appliance and other loads, the rating shall be determined in accordance with 210.23.

422.11 Overcurrent Protection. Appliances shall be protected against overcurrent in accordance with 422.11(A) through (G) and 422.10.

(A) Branch-Circuit Overcurrent Protection. Branch circuits shall be protected in accordance with 240.4.

If a protective device rating is marked on an appliance, the branch-circuit overcurrent device rating shall not exceed the protective device rating marked on the appliance.

(B) Household-Type Appliances with Surface Heating Elements. Household-type appliances with surface heating elements having a maximum demand of more than 60 amperes calculated in accordance with Table 220.55 shall have their power supply subdivided into two or more circuits, each of which shall be provided with overcurrent protection rated at not over 50 amperes.

(C) Infrared Lamp Commercial and Industrial Heating Appliances. Infrared lamp commercial and industrial heating appliances shall have overcurrent protection not exceeding 50 amperes.

(D) Open-Coil or Exposed Sheathed-Coil Types of Surface Heating Elements in Commercial-Type Heating Appliances. Open-coil or exposed sheathed-coil types of surface heating elements in commercial-type heating appliances shall be protected by overcurrent protective devices rated at not over 50 amperes.

(E) Single Non–Motor-Operated Appliance. If the branch circuit supplies a single non–motor-operated appliance, the rating of overcurrent protection shall comply with the following:

(1) Not exceed that marked on the appliance.
(2) Not exceed 20 amperes if the overcurrent protection rating is not marked and the appliance is rated 13.3 amperes or less; or
(3) Not exceed 150 percent of the appliance rated current if the overcurrent protection rating is not marked and the appliance is rated over 13.3 amperes. Where 150 percent of the appliance rating does not correspond to a standard overcurrent device ampere rating, the next higher standard rating shall be permitted.

(F) Electric Heating Appliances Employing Resistance-Type Heating Elements Rated More Than 48 Amperes.

(1) Electric Heating Appliances. Electric heating appliances employing resistance-type heating elements rated more than 48 amperes, other than household appliances with surface heating elements covered by 422.11(B), and commercial-type heating appliances covered by 422.11(D), shall have the heating elements subdivided. Each subdivided load shall not exceed 48 amperes and shall be protected at not more than 60 amperes.

These supplementary overcurrent protective devices shall be (1) factory-installed within or on the heater enclosure or provided as a separate assembly by the heater manufacturer; (2) accessible; and (3) suitable for branch-circuit protection.

The main conductors supplying these overcurrent protective devices shall be considered branch-circuit conductors.

(2) Commercial Kitchen and Cooking Appliances. Commercial kitchen and cooking appliances using sheathed-type heating elements not covered in 422.11(D) shall be permitted to be subdivided into circuits not exceeding 120 amperes and protected at not more than 150 amperes where one of the following is met:

(1) Elements are integral with and enclosed within a cooking surface.
(2) Elements are completely contained within an enclosure identified as suitable for this use.
(3) Elements are contained within an ASME-rated and stamped vessel.

(3) Water Heaters and Steam Boilers. Resistance-type immersion electric heating elements shall be permitted to be subdivided into circuits not exceeding 120 amperes and protected at not more than 150 amperes as follows:

(1) Where contained in ASME-rated and stamped vessels
(2) Where included in listed instantaneous water heaters
(3) Where installed in low-pressure water heater tanks or open-outlet water heater vessels

Informational Note: Low-pressure and open-outlet heaters are atmospheric pressure water heaters as defined in IEC 60335-2-21, *Household and similar electrical appliances — Safety — Particular requirements for storage water heaters.*

(G) Motor-Operated Appliances. Motors of motor-operated appliances shall be provided with overload protection in accordance with Part III of Article 430. Hermetic refrigerant motor-compressors in air-conditioning or refrigerating equipment shall be provided with overload protection in accordance with Part VI of Article 440. Where appliance overcurrent protective devices that are separate from the appliance are required, data for selection of these devices shall be marked on the appliance. The minimum marking shall be that specified in 430.7 and 440.4.

422.12 Central Heating Equipment. Central heating equipment other than fixed electric space-heating equipment shall be supplied by an individual branch circuit.

Exception No. 1: Auxiliary equipment, such as a pump, valve, humidifier, or electrostatic air cleaner directly associated with the heating equipment, shall be permitted to be connected to the same branch circuit.

Exception No. 2: Permanently connected air-conditioning equipment shall be permitted to be connected to the same branch circuit.

422.13 Storage-Type Water Heaters. A fixed storage-type water heater that has a capacity of 450 L (120 gal) or less shall be considered a continuous load for the purposes of sizing branch circuits.

Informational Note: For branch-circuit rating, see 422.10.

422.15 Central Vacuum Outlet Assemblies.

(A) Listed central vacuum outlet assemblies shall be permitted to be connected to a branch circuit in accordance with 210.23(A).

(B) The ampacity of the connecting conductors shall not be less than the ampacity of the branch circuit conductors to which they are connected.

(C) Accessible non–current-carrying metal parts of the central vacuum outlet assembly likely to become energized shall be connected to an equipment grounding conductor in accordance with 250.110. Incidental metal parts such as screws or rivets installed into or on insulating material shall not be considered likely to become energized.

422.16 Flexible Cords.

(A) General. Flexible cord shall be permitted (1) for the connection of appliances to facilitate their frequent interchange or to prevent the transmission of noise or vibration or (2) to facilitate the removal or disconnection of appliances that are fastened in place, where the fastening means and mechanical connections are specifically designed to permit ready removal for maintenance or repair and the appliance is intended or identified for flexible cord connection.

(B) Specific Appliances.

(1) Electrically Operated In-Sink Waste Disposers. Electrically operated in-sink waste disposers shall be permitted to be cord-and-plug-connected with a flexible cord identified as suitable in the installation instructions of the appliance manufacturer where all of the following conditions are met:

(1) The flexible cord shall be terminated with a grounding-type attachment plug.

Exception: A listed in-sink waste disposer distinctly marked to identify it as protected by a system of double insulation shall not be required to be terminated with a grounding-type attachment plug.

(2) The length of the cord shall not be less than 450 mm (18 in.) and not over 900 mm (36 in.).

(3) Receptacles shall be located to protect against physical damage to the flexible cord.

(4) The receptacle shall be accessible.

(2) Built-in Dishwashers and Trash Compactors. Built-in dishwashers and trash compactors shall be permitted to be cord-and-plug-connected with a flexible cord identified as suitable for the purpose in the installation instructions of the appliance manufacturer where all of the following conditions are met:

(1) The flexible cord shall be terminated with a grounding-type attachment plug.

Exception: A listed dishwasher or trash compactor distinctly marked to identify it as protected by a system of double insulation shall not be required to be terminated with a grounding-type attachment plug.

(2) For a trash compactor, the length of the cord shall be 0.9 m to 1.2 m (3 ft to 4 ft) measured from the face of the attachment plug to the plane of the rear of the appliance.

(3) For a built-in dishwasher, the length of the cord shall be 0.9 m to 2.0 m (3 ft to 6.5 ft) measured from the face of the attachment plug to the plane of the rear of the appliance.

(4) Receptacles shall be located to protect against physical damage to the flexible cord.

(5) The receptacle for a trash compactor shall be located in the space occupied by the appliance or adjacent thereto.

(6) The receptacle for a built-in dishwasher shall be located in the space adjacent to the space occupied by the dishwasher.

(7) The receptacle shall be accessible.

(3) Wall-Mounted Ovens and Counter-Mounted Cooking Units. Wall-mounted ovens and counter-mounted cooking units complete with provisions for mounting and for making electrical connections shall be permitted to be permanently connected or, only for ease in servicing or for installation, cord-and-plug-connected.

A separable connector or a plug and receptacle combination in the supply line to an oven or cooking unit shall be approved for the temperature of the space in which it is located.

(4) Range Hoods. Range hoods shall be permitted to be cord-and-plug-connected with a flexible cord identified as suitable for use on range hoods in the installation instructions of the appliance manufacturer, where all of the following conditions are met:

(1) The flexible cord is terminated with a grounding-type attachment plug.

Exception: A listed range hood distinctly marked to identify it as protected by a system of double insulation shall not be required to be terminated with a grounding-type attachment plug.

(2) The length of the cord is not less than 450 mm (18 in.) and not over 1.2 m (4 ft).

(3) Receptacles are located to protect against physical damage to the flexible cord.

(4) The receptacle is accessible.

(5) The receptacle is supplied by an individual branch circuit.

422.17 Protection of Combustible Material. Each electrically heated appliance that is intended by size, weight, and service to be located in a fixed position shall be placed so as to provide ample protection between the appliance and adjacent combustible material.

422.18 Support of Ceiling-Suspended (Paddle) Fans. Ceiling-suspended (paddle) fans shall be supported independently of an outlet box or by one of the following:

(1) A listed outlet box or listed outlet box system identified for the use and installed in accordance with 314.27(C)

(2) A listed outlet box system, a listed locking support and mounting receptacle, and a compatible factory installed attachment fitting designed for support, identified for the use and installed in accordance with 314.27(E)

422.19 Space for Conductors. Canopies of ceiling-suspended (paddle) fans and outlet boxes taken together shall provide sufficient space so that conductors and their connecting devices are capable of being installed in accordance with 314.16.

422.20 Outlet Boxes to Be Covered. In a completed installation, each outlet box shall be provided with a cover unless covered by means of a ceiling-suspended (paddle) fan canopy.

422.21 Covering of Combustible Material at Outlet Boxes. Any combustible ceiling finish that is exposed between the edge of a ceiling-suspended (paddle) fan canopy or pan and an outlet box and that has a surface area of 1160 mm² (180 in.²) or more shall be covered with noncombustible material.

422.22 Other Installation Methods. Appliances employing methods of installation other than covered by this article shall be permitted to be used only by special permission.

Part III. Disconnecting Means

422.30 General. A means shall be provided to simultaneously disconnect each appliance from all ungrounded conductors in accordance with the following sections of Part III. If an appliance is supplied by more than one branch circuit or feeder, these disconnecting means shall be grouped and identified as being the multiple disconnecting means for the appliance. Each disconnecting means shall simultaneously disconnect all ungrounded conductors that it controls.

422.31 Disconnection of Permanently Connected Appliances.

(A) Rated at Not over 300 Volt-Amperes or ⅛ Horsepower. For permanently connected appliances rated at not over 300 volt-amperes or ⅛ hp, the branch-circuit overcurrent device shall be permitted to serve as the disconnecting means where the switch or circuit breaker is within sight from the appliance or is lockable in accordance with 110.25.

(B) Appliances Rated over 300 Volt-Amperes. For permanently connected appliances rated over 300 volt-amperes, the branch-circuit switch or circuit breaker shall be permitted to serve as the disconnecting means where the switch or circuit breaker is within sight from the appliance or is lockable in accordance with 110.25.

Informational Note: For appliances employing unit switches, see 422.34.

(C) Motor-Operated Appliances Rated over ⅛ Horsepower. The disconnecting means shall comply with 430.109 and 430.110. For permanently connected motor-operated appliances with motors rated over ⅛ hp, the disconnecting means shall be within sight from the appliance or be capable of being locked in the open position in compliance with 110.25.

Exception: If an appliance of more than ⅛ hp is provided with a unit switch that complies with 422.34(A), (B), (C), or (D), the switch or circuit breaker serving as the other disconnecting means shall be permitted to be out of sight from the appliance.

422.33 Disconnection of Cord-and-Plug-Connected or Attachment Fitting–Connected Appliances.

(A) Separable Connector or an Attachment Plug (or Attachment Fitting) and Receptacle. For cord-and-plug-(or attachment fitting–) connected appliances, an accessible separable connector or an accessible plug (or attachment fitting) and receptacle combination shall be permitted to serve as the disconnecting means. The attachment fitting shall be a factory installed part of the appliance and suitable for disconnection of the appliance. Where the separable connector or plug (or attachment fitting) and receptacle combination are not accessible, cord-and-plug-connected or attachment fitting-and-plug-connected appliances shall be provided with disconnecting means in accordance with 422.31.

(B) Connection at the Rear Base of a Range. For cord-and-plug-connected household electric ranges, an attachment plug and receptacle connection at the rear base of a range, accessible from the front by removal of a drawer, shall meet the intent of 422.33(A).

(C) Rating. The rating of a receptacle or of a separable connector shall not be less than the rating of any appliance connected thereto.

Exception: Demand factors authorized elsewhere in this Code shall be permitted to be applied to the rating of a receptacle or of a separable connector.

422.34 Unit Switch(es) as Disconnecting Means. A unit switch(es) with a marked-off position that is a part of an appliance and disconnects all ungrounded conductors shall be permitted as the disconnecting means required by this article where other means for disconnection are provided in occupancies specified in 422.34(A) through (D).

(A) Multifamily Dwellings. In multifamily dwellings, the other disconnecting means shall be within the dwelling unit, or on the same floor as the dwelling unit in which the appliance is installed, and shall be permitted to control lamps and other appliances.

(B) Two-Family Dwellings. In two-family dwellings, the other disconnecting means shall be permitted either inside or outside of the dwelling unit in which the appliance is installed. In this case, an individual switch or circuit breaker for the dwelling unit shall be permitted and shall also be permitted to control lamps and other appliances.

(C) One-Family Dwellings. In one-family dwellings, the service disconnecting means shall be permitted to be the other disconnecting means.

(D) Other Occupancies. In other occupancies, the branch-circuit switch or circuit breaker, where readily accessible for servicing of the appliance, shall be permitted as the other disconnecting means.

422.35 Switch and Circuit Breaker to Be Indicating. Switches and circuit breakers used as disconnecting means shall be of the indicating type.

Part IV. Construction

422.40 Polarity in Cord-and Plug-Connected Appliances. If the appliance is provided with a manually operated, line-connected, single-pole switch for appliance on–off operation, an Edison-base lampholder, or a 15- or 20-ampere receptacle, the attachment plug shall be of the polarized or grounding type.

A 2-wire, nonpolarized attachment plug shall be permitted to be used on a listed double-insulated shaver.

Informational Note: For polarity of Edison-base lampholders, see 410.82(A).

422.41 Cord-and Plug-Connected Appliances Subject to Immersion. Cord-and plug-connected portable, freestanding hydromassage units and hand-held hair dryers shall be constructed to provide protection for personnel against electrocution when immersed while in the "on" or "off" position.

422.42 Signals for Heated Appliances. In other than dwelling-type occupancies, each electrically heated appliance or group of appliances intended to be applied to combustible material shall be provided with a signal or an integral temperature-limiting device.

422.43 Flexible Cords.

(A) Heater Cords. All cord-and plug-connected smoothing irons and electrically heated appliances that are rated at more than 50 watts and produce temperatures in excess of 121°C (250°F) on surfaces with which the cord is likely to be in contact shall be provided with one of the types of approved heater cords listed in Table 400.4.

(B) Other Heating Appliances. All other cord-and plug-connected electrically heated appliances shall be connected with one of the approved types of cord listed in Table 400.4, selected in accordance with the usage specified in that table.

422.44 Cord-and Plug-Connected Immersion Heaters. Electric heaters of the cord-and plug-connected immersion type shall be constructed and installed so that current-carrying parts are effectively insulated from electrical contact with the substance in which they are immersed.

422.45 Stands for Cord-and Plug-Connected Appliances. Each smoothing iron and other cord-and plug-connected electrically heated appliance intended to be applied to combustible material shall be equipped with an approved stand, which shall be permitted to be a separate piece of equipment or a part of the appliance.

422.46 Flatirons. Electrically heated smoothing irons shall be equipped with an identified temperature-limiting means.

422.47 Water Heater Controls. All storage or instantaneous-type water heaters shall be equipped with a temperature-limiting means in addition to its control thermostat to disconnect all ungrounded conductors. Such means shall comply with both of the following:

(1) Installed to sense maximum water temperature.
(2) Be either a trip-free, manually reset type or a type having a replacement element. Such water heaters shall be marked to require the installation of a temperature and pressure relief valve.

Exception No. 1: Storage water heaters that are identified as being suitable for use with a supply water temperature of 82°C (180°F) or above and a capacity of 60 kW or above.

Exception No. 2: Instantaneous-type water heaters that are identified as being suitable for such use, with a capacity of 4 L (1 gal) or less.

Informational Note: See ANSI Z21.22-1999/CSA 4.4-M99, *Relief Valves for Hot Water Supply Systems.*

422.48 Infrared Lamp Industrial Heating Appliances.

(A) 300 Watts or Less. Infrared heating lamps rated at 300 watts or less shall be permitted with lampholders of the medium-base, unswitched porcelain type or other types identified as suitable for use with infrared heating lamps rated 300 watts or less.

(B) Over 300 Watts. Screw shell lampholders shall not be used with infrared lamps rated over 300 watts, unless the lampholders are identified as being suitable for use with infrared heating lamps rated over 300 watts.

422.50 Cord-and-Plug-Connected Pipe Heating Assemblies.
Cord-and-plug-connected pipe heating assemblies intended to prevent freezing of piping shall be listed.

Part V. Marking

422.60 Nameplate.

(A) Nameplate Marking. Each electrical appliance shall be provided with a nameplate giving the identifying name and the rating in volts and amperes, or in volts and watts. If the appliance is to be used on a specific frequency or frequencies, it shall be so marked.

Where motor overload protection external to the appliance is required, the appliance shall be so marked.

Informational Note: See 422.11 for overcurrent protection requirements.

(B) To Be Visible. Marking shall be located so as to be visible or easily accessible after installation.

422.61 Marking of Heating Elements. All heating elements that are rated over one ampere, replaceable in the field, and a part of an appliance shall be legibly marked with the ratings in volts and amperes, or in volts and watts, or with the manufacturer's part number.

422.62 Appliances Consisting of Motors and Other Loads.

(A) Nameplate Horsepower Markings. Where a motor-operated appliance nameplate includes a horsepower rating, that rating shall not be less than the horsepower rating on the motor nameplate. Where an appliance consists of multiple motors, or one or more motors and other loads, the nameplate value shall not be less than the equivalent horsepower of the combined loads, calculated in accordance with 430.110(C)(1).

(B) Additional Nameplate Markings. Appliances, other than those factory-equipped with cords and attachment plugs and with nameplates in compliance with 422.60, shall be marked in accordance with 422.62(B)(1) or (B)(2).

(1) Marking. In addition to the marking required in 422.60, the marking on an appliance consisting of a motor with other load(s) or motors with or without other load(s) shall specify the minimum supply circuit conductor ampacity and the maximum rating of the circuit overcurrent protective device. This requirement shall not apply to an appliance with a nameplate in compliance with 422.60 where both the minimum supply circuit conductor ampacity and maximum rating of the circuit overcurrent protective device are not more than 15 amperes.

(2) Alternate Marking Method. An alternative marking method shall be permitted to specify the rating of the largest motor in volts and amperes, and the additional load(s) in volts and amperes, or volts and watts in addition to the marking required in 422.60. The ampere rating of a motor ⅛ horsepower or less or a nonmotor load 1 ampere or less shall be permitted to be omitted unless such loads constitute the principal load.

ARTICLE 424
Fixed Electric Space-Heating Equipment

Part I. General

424.1 Scope. This article covers fixed electric equipment used for space heating. For the purpose of this article, heating equipment shall include heating cable, unit heaters, boilers, central systems, or other approved fixed electric space-heating equipment. This article shall not apply to process heating and room air conditioning.

424.2 Other Articles. Fixed electric space-heating equipment incorporating a hermetic refrigerant motor-compressor shall also comply with Article 440.

424.3 Branch Circuits.

(A) Branch-Circuit Requirements. Individual branch circuits shall be permitted to supply any volt-ampere or wattage rating of fixed electric space-heating equipment for which they are rated.

Branch circuits supplying two or more outlets for fixed electric space-heating equipment shall be rated not over 30 amperes. In other than a dwelling unit, fixed infrared heating equipment shall be permitted to be supplied from branch circuits rated not over 50 amperes.

(B) Branch-Circuit Sizing. Fixed electric space-heating equipment and motors shall be considered continuous load.

424.6 Listed Equipment. Electric baseboard heaters, heating cables, duct heaters, and radiant heating systems shall be listed and labeled.

Part II. Installation

424.9 General. Permanently installed electric baseboard heaters equipped with factory-installed receptacle outlets, or outlets provided as a separate listed assembly, shall be permitted in lieu of a receptacle outlet(s) that is required by 210.50(B). Such receptacle outlets shall not be connected to the heater circuits.

> Informational Note: Listed baseboard heaters include instructions that may not permit their installation below receptacle outlets.

424.10 Special Permission. Fixed electric space-heating equipment and systems installed by methods other than covered by this article shall be permitted only by special permission.

424.11 Supply Conductors. Fixed electric space-heating equipment requiring supply conductors with over 60°C insulation shall be clearly and permanently marked. This marking shall be plainly visible after installation and shall be permitted to be adjacent to the field connection box.

424.12 Locations.

(A) Exposed to Physical Damage. Where subject to physical damage, fixed electric space-heating equipment shall be protected in an approved manner.

(B) Damp or Wet Locations. Heaters and related equipment installed in damp or wet locations shall be listed for such locations and shall be constructed and installed so that water or other liquids cannot enter or accumulate in or on wired sections, electrical components, or ductwork.

> Informational Note No. 1: See 110.11 for equipment exposed to deteriorating agents.

> Informational Note No. 2: See 680.27(C) for pool deck areas.

424.13 Spacing from Combustible Materials. Fixed electric space-heating equipment shall be installed to provide the required spacing between the equipment and adjacent combustible material, unless it is listed to be installed in direct contact with combustible material.

Part III. Control and Protection of Fixed Electric Space-Heating Equipment

424.19 Disconnecting Means. Means shall be provided to simultaneously disconnect the heater, motor controller(s), and supplementary overcurrent protective device(s) of all fixed electric space-heating equipment from all ungrounded conductors. Where heating equipment is supplied by more than one source, feeder, or branch circuit, the disconnecting means shall be grouped and identified as having multiple disconnecting means. Each disconnecting means shall simultaneously disconnect all ungrounded conductors that it controls. The disconnecting means specified in 424.19(A) and (B) shall have an ampere rating not less than 125 percent of the total load of the motors and the heaters and shall be lockable in accordance with 110.25.

(A) Heating Equipment with Supplementary Overcurrent Protection. The disconnecting means for fixed electric space-heating equipment with supplementary overcurrent protection shall be within sight from the supplementary overcurrent protective device(s), on the supply side of these devices, if fuses, and, in addition, shall comply with either 424.19(A)(1) or (A)(2).

(1) Heater Containing No Motor Rated over ⅛ Horsepower. The disconnecting means specified in 424.19 or unit switches complying with 424.19(C) shall be permitted to serve as the required disconnecting means for both the motor controller(s) and heater under either of the following conditions:

(1) The disconnecting means provided is also within sight from the motor controller(s) and the heater.
(2) The disconnecting means is lockable in accordance with 110.25.

(2) Heater Containing a Motor(s) Rated over ⅛ Horsepower. The above disconnecting means shall be permitted to serve as the required disconnecting means for both the motor controller(s) and heater under either of the following conditions:

(1) Where the disconnecting means is in sight from the motor controller(s) and the heater and complies with Part IX of Article 430.
(2) Where a motor(s) of more than ⅛ hp and the heater are provided with a single unit switch that complies with 422.34(A), (B), (C), or (D), the disconnecting means shall be permitted to be out of sight from the motor controller.

(B) Heating Equipment Without Supplementary Overcurrent Protection.

(1) Without Motor or with Motor Not over ⅛ Horsepower. For fixed electric space-heating equipment without a motor rated over ⅛ hp, the branch-circuit switch or circuit breaker shall be permitted to serve as the disconnecting means where the switch or circuit breaker is within sight from the heater or is lockable in accordance with 110.25.

(2) Over ⅛ Horsepower. For motor-driven electric space-heating equipment with a motor rated over ⅛ hp, a disconnecting means shall be located within sight from the motor controller or shall be permitted to comply with the requirements in 424.19(A)(2).

(C) Unit Switch(es) as Disconnecting Means. A unit switch(es) with a marked "off" position that is part of a fixed heater and disconnects all ungrounded conductors shall be permitted as the disconnecting means required by this article where other means for disconnection are provided in the types of occupancies in 424.19(C)(1) through (C)(4).

(1) Multifamily Dwellings. In multifamily dwellings, the other disconnecting means shall be within the dwelling unit, or on the same floor as the dwelling unit in which the fixed heater is installed, and shall also be permitted to control lamps and appliances.

(2) Two-Family Dwellings. In two-family dwellings, the other disconnecting means shall be permitted either inside or outside of the dwelling unit in which the fixed heater is installed. In this case, an individual switch or circuit breaker for the dwelling unit shall be permitted and shall also be permitted to control lamps and appliances.

(3) One-Family Dwellings. In one-family dwellings, the service disconnecting means shall be permitted to be the other disconnecting means.

(4) Other Occupancies. In other occupancies, the branch-circuit switch or circuit breaker, where readily accessible for servicing of the fixed heater, shall be permitted as the other disconnecting means.

424.20 Thermostatically Controlled Switching Devices.

(A) Serving as Both Controllers and Disconnecting Means. Thermostatically controlled switching devices and combination thermostats and manually controlled switches shall be permitted to serve as both controllers and disconnecting means, provided they meet all of the following conditions:

(1) Provided with a marked "off" position
(2) Directly open all ungrounded conductors when manually placed in the "off" position
(3) Designed so that the circuit cannot be energized automatically after the device has been manually placed in the "off" position
(4) Located as specified in 424.19

(B) Thermostats That Do Not Directly Interrupt All Ungrounded Conductors. Thermostats that do not directly interrupt all ungrounded conductors and thermostats that operate remote-control circuits shall not be required to meet the requirements of 424.20(A). These devices shall not be permitted as the disconnecting means.

424.21 Switch and Circuit Breaker to Be Indicating.
Switches and circuit breakers used as disconnecting means shall be of the indicating type.

424.22 Overcurrent Protection.

(A) Branch-Circuit Devices. Electric space-heating equipment, other than such motor-operated equipment as required by Articles 430 and 440 to have additional overcurrent protection, shall be permitted to be protected against overcurrent where supplied by one of the branch circuits in Article 210.

(B) Resistance Elements. Resistance-type heating elements in electric space-heating equipment shall be protected at not more than 60 amperes. Equipment rated more than 48 amperes and employing such elements shall have the heating elements subdivided, and each subdivided load shall not exceed 48 amperes. Where a subdivided load is less than 48 amperes, the rating of the supplementary overcurrent protective device shall comply with 424.3(B). A boiler employing resistance-type immersion heating elements contained in an ASME-rated and stamped vessel shall be permitted to comply with 424.72(A).

(C) Overcurrent Protective Devices. The supplementary overcurrent protective devices for the subdivided loads specified in 424.22(B) shall be (1) factory-installed within or on the heater enclosure or supplied for use with the heater as a separate assembly by the heater manufacturer; (2) accessible, but shall not be required to be readily accessible; and (3) suitable for branch-circuit protection.

> Informational Note: See 240.10.

Where cartridge fuses are used to provide this overcurrent protection, a single disconnecting means shall be permitted to be used for the several subdivided loads.

> Informational Note No. 1: For supplementary overcurrent protection, see 240.10.

> Informational Note No. 2: For disconnecting means for cartridge fuses in circuits of any voltage, see 240.40.

(D) Branch-Circuit Conductors. The conductors supplying the supplementary overcurrent protective devices shall be considered branch-circuit conductors.

Where the heaters are rated 50 kW or more, the conductors supplying the supplementary overcurrent protective devices specified in 424.22(C) shall be permitted to be sized at not less than 100 percent of the nameplate rating of the heater, provided all of the following conditions are met:

(1) The heater is marked with a minimum conductor size.
(2) The conductors are not smaller than the marked minimum size.
(3) A temperature-actuated device controls the cyclic operation of the equipment.

(E) Conductors for Subdivided Loads. Field-wired conductors between the heater and the supplementary overcurrent protective devices shall be sized at not less than 125 percent of the load served. The supplementary overcurrent protective devices specified in 424.22(C) shall protect these conductors in accordance with 240.4.

Where the heaters are rated 50 kW or more, the ampacity of field-wired conductors between the heater and the supplementary overcurrent protective devices shall be permitted to be not less than 100 percent of the load of their respective subdivided circuits, provided all of the following conditions are met:

(1) The heater is marked with a minimum conductor size.
(2) The conductors are not smaller than the marked minimum size.
(3) A temperature-activated device controls the cyclic operation of the equipment.

Part IV. Marking of Heating Equipment

424.28 Nameplate.

(A) Marking Required. Each unit of fixed electric space-heating equipment shall be provided with a nameplate giving the identifying name and the normal rating in volts and watts or in volts and amperes.

Electric space-heating equipment intended for use on alternating current only, direct current only, or both shall be marked to so indicate. The marking of equipment consisting of motors over ⅛ hp and other loads shall specify the rating of the motor in volts, amperes, and frequency, and the heating load in volts and watts or in volts and amperes.

(B) Location. This nameplate shall be located so as to be visible or easily accessible after installation.

424.29 Marking of Heating Elements.
All heating elements that are replaceable in the field and are part of an electric heater shall be legibly marked with the ratings in volts and watts or in volts and amperes.

Part V. Electric Space-Heating Cables

424.34 Heating Cable Construction.
Factory-assembled nonheating leads of heating cables, if any, shall be at least 2.1 m (7 ft) in length.

424.35 Marking of Heating Cables.
Each unit shall be marked with the identifying name or identification symbol, catalog number, and ratings in volts and watts or in volts and amperes.

424.36 Clearances of Wiring in Ceilings. Wiring located above heated ceilings shall be spaced not less than 50 mm (2 in.) above the heated ceiling. The ampacity of conductors shall be calculated on the basis of an assumed ambient temperature of 50°C (122°F), applying the correction factors shown in the 0–2000 volt ampacity tables of Article 310. If this wiring is located above thermal insulation having a minimum thickness of 50 mm (2 in.), the wiring shall not require correction for temperature.

424.38 Area Restrictions.

(A) Extending Beyond the Room or Area. Heating cables shall be permitted to extend beyond the room or area in which they originate.

(B) Uses Not Permitted. Heating cables shall not be installed as follows:

(1) In closets, other than as noted in 424.38(C)
(2) Over the top of walls where the wall intersects the ceiling
(3) Over partitions that extend to the ceiling, unless they are isolated single runs of embedded cable
(4) Under or through walls
(5) Over cabinets whose clearance from the ceiling is less than the minimum horizontal dimension of the cabinet to the nearest cabinet edge that is open to the room or area
(6) In tub and shower walls
(7) Under cabinets or similar built-ins having no clearance to the floor

(C) In Closet Ceilings as Low-Temperature Heat Sources to Control Relative Humidity. The provisions of 424.38(B) shall not prevent the use of cable in closet ceilings as low-temperature heat sources to control relative humidity, provided they are used only in those portions of the ceiling that are unobstructed to the floor by shelves or other permanent luminaires.

424.39 Clearance from Other Objects and Openings. Heating elements of cables installed in ceilings shall be separated at least 200 mm (8 in.) from the edge of outlet boxes and junction boxes that are to be used for mounting surface luminaires. A clearance of not less than 50 mm (2 in.) shall be provided from recessed luminaires and their trims, ventilating openings, and other such openings in room surfaces. No heating cable shall be covered by any surface-mounted equipment.

424.40 Splices. The length of heating cable shall only be altered using splices identified in the manufacturer's instructions.

424.41 Ceiling Installation of Heating Cables on Dry Board, in Plaster, and on Concrete.

(A) In Walls. Cables shall not be installed in walls unless it is necessary for an isolated single run of cable to be installed down a vertical surface to reach a dropped ceiling.

(B) Adjacent Runs. Adjacent runs of heating cable shall be installed in accordance with the manufacturer's instructions.

(C) Surfaces to Be Applied. Heating cables shall be applied only to gypsum board, plaster lath, or other fire-resistant material. With metal lath or other electrically conductive surfaces, a coat of plaster or other means employed in accordance with the heating cable manufacturer's instructions shall be applied

to completely separate the metal lath or conductive surface from the cable.

Informational Note: See also 424.41(F).

(D) Splices. All heating cables, the splice between the heating cable and nonheating leads, and 75-mm (3-in.) minimum of the nonheating lead at the splice shall be embedded in plaster or dry board in the same manner as the heating cable.

(E) Ceiling Surface. The entire ceiling surface shall have a finish of thermally noninsulating sand plaster that has a nominal thickness of 13 mm (½ in.), or other noninsulating material identified as suitable for this use and applied according to specified thickness and directions.

(F) Secured. Cables shall be secured by means of approved stapling, tape, plaster, nonmetallic spreaders, or other approved means either at intervals not exceeding 400 mm (16 in.) or at intervals not exceeding 1.8 m (6 ft) for cables identified for such use. Staples or metal fasteners that straddle the cable shall not be used with metal lath or other electrically conductive surfaces.

(G) Dry Board Installations. In dry board installations, the entire ceiling below the heating cable shall be covered with gypsum board not exceeding 13 mm (½ in.) thickness. The void between the upper layer of gypsum board, plaster lath, or other fire-resistant material and the surface layer of gypsum board shall be completely filled with thermally conductive, nonshrinking plaster or other approved material or equivalent thermal conductivity.

(H) Free from Contact with Conductive Surfaces. Cables shall be kept free from contact with metal or other electrically conductive surfaces.

(I) Joists. In dry board applications, cable shall be installed parallel to the joist, leaving a clear space centered under the joist of 65 mm (2½ in.) (width) between centers of adjacent runs of cable. A surface layer of gypsum board shall be mounted so that the nails or other fasteners do not pierce the heating cable.

(J) Crossing Joists. Cables shall cross joists only at the ends of the room unless the cable is required to cross joists elsewhere in order to satisfy the manufacturer's instructions that the installer avoid placing the cable too close to ceiling penetrations and luminaires.

424.42 Finished Ceilings. Finished ceilings shall not be covered with decorative panels or beams constructed of materials that have thermal insulating properties, such as wood, fiber, or plastic. Finished ceilings shall be permitted to be covered with paint, wallpaper, or other approved surface finishes.

424.43 Installation of Nonheating Leads of Cables.

(A) Free Nonheating Leads. Free nonheating leads of cables shall be installed in accordance with approved wiring methods from the junction box to a location within the ceiling. Such installations shall be permitted to be single conductors in approved raceways, single or multiconductor Type UF, Type NMC, Type MI, or other approved conductors.

(B) Leads in Junction Box. Not less than 150 mm (6 in.) of free nonheating lead shall be within the junction box. The marking of the leads shall be visible in the junction box.

(C) Excess Leads. Excess leads of heating cables shall not be cut but shall be secured to the underside of the ceiling and embedded in plaster or other approved material, leaving only a length sufficient to reach the junction box with not less than 150 mm (6 in.) of free lead within the box.

424.44 Installation of Cables in Concrete or Poured Masonry Floors.

(A) Adjacent Runs. Adjacent runs of heating cable shall be installed in accordance with the manufacturer's instructions.

(B) Secured in Place. Cables shall be secured in place by nonmetallic frames or spreaders or other approved means while the concrete or other finish is applied.

(C) Leads Protected. Leads shall be protected where they leave the floor by rigid metal conduit, intermediate metal conduit, rigid nonmetallic conduit, electrical metallic tubing, or by other approved means.

(D) Bushings or Approved Fittings. Bushings or approved fittings shall be used where the leads emerge within the floor slab.

(E) Ground-Fault Circuit-Interrupter Protection. Ground-fault circuit-interrupter protection for personnel shall be provided for cables installed in electrically heated floors of bathrooms and kitchens and in hydromassage bathtub locations.

N **424.45 Installation of Cables Under Floor Coverings.**

(A) Identification. Heating cables for installation under floor covering shall be identified as suitable for installation under floor covering.

(B) Expansion Joints. Heating cables shall not be installed where they bridge expansion joints unless provided with expansion and contraction fittings applicable to the manufacture of the cable.

(C) Connection to Conductors. Heating cables shall be connected to branch-circuit and supply wiring by wiring methods described in the installation instructions or as recognized in Chapter 3.

(D) Anchoring. Heating cables shall be positioned or secured in place under the floor covering, per the manufacturer's instructions.

(E) Ground-Fault Circuit-Interrupter Protection. Ground-fault circuit-interrupter protection for personnel shall be provided.

(F) Grounding Braid or Sheath. Grounding means, such as copper braid, metal sheath, or other approved means, shall be provided as part of the heated length.

424.46 Inspection and Tests. Cable installations shall be made with due care to prevent damage to the cable assembly and shall be inspected and approved before cables are covered or concealed.

N **424.47 Label Provided by Manufacturer.** The manufacturers of electric space-heating cables shall provide marking labels that indicate that the space-heating installation incorporates electric space-heating cables and instructions that the labels shall be affixed to the panelboards to identify which branch circuits supply the circuits to those space-heating installations. If the electric space-heating cable installations are visible and distinguishable after installation, the labels shall not be required to be provided and affixed to the panelboards.

Part VI. Duct Heaters

424.57 General. Part VI shall apply to any heater mounted in the airstream of a forced-air system where the air-moving unit is not provided as an integral part of the equipment.

424.58 Identification. Heaters installed in an air duct shall be identified as suitable for the installation.

424.59 Airflow. Means shall be provided to ensure uniform airflow over the face of the heater in accordance with the manufacturer's instructions.

> Informational Note: Heaters installed within 1.2 m (4 ft) of the outlet of an air-moving device, heat pump, air conditioner, elbows, baffle plates, or other obstructions in ductwork may require turning vanes, pressure plates, or other devices on the inlet side of the duct heater to ensure an even distribution of air over the face of the heater.

424.60 Elevated Inlet Temperature. Duct heaters intended for use with elevated inlet air temperature shall be identified as suitable for use at the elevated temperatures.

424.61 Installation of Duct Heaters with Heat Pumps and Air Conditioners. Heat pumps and air conditioners having duct heaters closer than 1.2 m (4 ft) to the heat pump or air conditioner shall have both the duct heater and heat pump or air conditioner identified as suitable for such installation and so marked.

424.62 Condensation. Duct heaters used with air conditioners or other air-cooling equipment that could result in condensation of moisture shall be identified as suitable for use with air conditioners.

424.63 Fan Circuit Interlock. Means shall be provided to ensure that the fan circuit is energized when any heater circuit is energized. However, time- or temperature-controlled delay in energizing the fan motor shall be permitted.

424.64 Limit Controls. Each duct heater shall be provided with an approved, integral, automatic-reset temperature-limiting control or controllers to de-energize the circuit or circuits.

In addition, an integral independent supplementary control or controllers shall be provided in each duct heater that disconnects a sufficient number of conductors to interrupt current flow. This device shall be manually resettable or replaceable.

424.65 Location of Disconnecting Means. Duct heater controller equipment shall be either accessible with the disconnecting means installed at or within sight from the controller or as permitted by 424.19(A).

424.66 Installation. Duct heaters shall be installed in accordance with the manufacturer's instructions in such a manner that operation does not create a hazard to persons or property. Furthermore, duct heaters shall be located with respect to building construction and other equipment so as to permit access to the heater. Sufficient clearance shall be maintained to permit replacement of controls and heating elements and for adjusting and cleaning of controls and other parts requiring such attention. See 110.26.

Informational Note: For additional installation information, see NFPA 90A-2015, *Standard for the Installation of Air-Conditioning and Ventilating Systems*, and NFPA 90B-2015, *Standard for the Installation of Warm Air Heating and Air-Conditioning Systems*.

Part VII. Resistance-Type Boilers

424.70 Scope. The provisions in Part VII of this article shall apply to boilers employing resistance-type heating elements. See Part VIII of this article for electrode-type boilers.

424.71 Identification. Resistance-type boilers shall be identified as suitable for the installation.

424.72 Overcurrent Protection.

(A) Boiler Employing Resistance-Type Immersion Heating Elements in an ASME-Rated and Stamped Vessel. A boiler employing resistance-type immersion heating elements contained in an ASME-rated and stamped vessel shall have the heating elements protected at not more than 150 amperes. Such a boiler rated more than 120 amperes shall have the heating elements subdivided into loads not exceeding 120 amperes.

Where a subdivided load is less than 120 amperes, the rating of the overcurrent protective device shall comply with 424.3(B).

(B) Boiler Employing Resistance-Type Heating Elements Rated More Than 48 Amperes and Not Contained in an ASME-Rated and Stamped Vessel. A boiler employing resistance-type heating elements not contained in an ASME-rated and stamped vessel shall have the heating elements protected at not more than 60 amperes. Such a boiler rated more than 48 amperes shall have the heating elements subdivided into loads not exceeding 48 amperes.

Where a subdivided load is less than 48 amperes, the rating of the overcurrent protective device shall comply with 424.3(B).

(C) Supplementary Overcurrent Protective Devices. The supplementary overcurrent protective devices for the subdivided loads as required by 424.72(A) and (B) shall be as follows:

(1) Factory-installed within or on the boiler enclosure or provided as a separate assembly by the boiler manufacturer
(2) Accessible, but need not be readily accessible
(3) Suitable for branch-circuit protection

Where cartridge fuses are used to provide this overcurrent protection, a single disconnecting means shall be permitted for the several subdivided circuits. See 240.40.

(D) Conductors Supplying Supplementary Overcurrent Protective Devices. The conductors supplying these supplementary overcurrent protective devices shall be considered branch-circuit conductors.

Where the heaters are rated 50 kW or more, the conductors supplying the overcurrent protective device specified in 424.72(C) shall be permitted to be sized at not less than 100 percent of the nameplate rating of the heater, provided all of the following conditions are met:

(1) The heater is marked with a minimum conductor size.
(2) The conductors are not smaller than the marked minimum size.

(3) A temperature- or pressure-actuated device controls the cyclic operation of the equipment.

(E) Conductors for Subdivided Loads. Field-wired conductors between the heater and the supplementary overcurrent protective devices shall be sized at not less than 125 percent of the load served. The supplementary overcurrent protective devices specified in 424.72(C) shall protect these conductors in accordance with 240.4.

Where the heaters are rated 50 kW or more, the ampacity of field-wired conductors between the heater and the supplementary overcurrent protective devices shall be permitted to be not less than 100 percent of the load of their respective subdivided circuits, provided all of the following conditions are met:

(1) The heater is marked with a minimum conductor size.
(2) The conductors are not smaller than the marked minimum size.
(3) A temperature-activated device controls the cyclic operation of the equipment.

424.73 Overtemperature Limit Control. Each boiler designed so that in normal operation there is no change in state of the heat transfer medium shall be equipped with a temperature-sensitive limiting means. It shall be installed to limit maximum liquid temperature and shall directly or indirectly disconnect all ungrounded conductors to the heating elements. Such means shall be in addition to a temperature-regulating system and other devices protecting the tank against excessive pressure.

424.74 Overpressure Limit Control. Each boiler designed so that in normal operation there is a change in state of the heat transfer medium from liquid to vapor shall be equipped with a pressure-sensitive limiting means. It shall be installed to limit maximum pressure and shall directly or indirectly disconnect all ungrounded conductors to the heating elements. Such means shall be in addition to a pressure-regulating system and other devices protecting the tank against excessive pressure.

Part VIII. Electrode-Type Boilers

424.80 Scope. The provisions in Part VIII of this article shall apply to boilers for operation at 600 volts, nominal, or less, in which heat is generated by the passage of current between electrodes through the liquid being heated.

Informational Note: For over 600 volts, see Part V of Article 490.

424.81 Identification. Electrode-type boilers shall be identified as suitable for the installation.

424.82 Branch-Circuit Requirements. The size of branch-circuit conductors and overcurrent protective devices shall be calculated on the basis of 125 percent of the total load (motors not included). A contactor, relay, or other device, approved for continuous operation at 100 percent of its rating, shall be permitted to supply its full-rated load. See 210.19(A), Exception. The provisions of this section shall not apply to conductors that form an integral part of an approved boiler.

Where an electrode boiler is rated 50 kW or more, the conductors supplying the boiler electrode(s) shall be permitted to be sized at not less than 100 percent of the nameplate rating of the electrode boiler, provided all the following conditions are met:

(1) The electrode boiler is marked with a minimum conductor size.

(2) The conductors are not smaller than the marked minimum size.

(3) A temperature- or pressure-actuated device controls the cyclic operation of the equipment.

424.83 Overtemperature Limit Control. Each boiler, designed so that in normal operation there is no change in state of the heat transfer medium, shall be equipped with a temperature-sensitive limiting means. It shall be installed to limit maximum liquid temperature and shall directly or indirectly interrupt all current flow through the electrodes. Such means shall be in addition to the temperature-regulating system and other devices protecting the tank against excessive pressure.

424.84 Overpressure Limit Control. Each boiler, designed so that in normal operation there is a change in state of the heat transfer medium from liquid to vapor, shall be equipped with a pressure-sensitive limiting means. It shall be installed to limit maximum pressure and shall directly or indirectly interrupt all current flow through the electrodes. Such means shall be in addition to a pressure-regulating system and other devices protecting the tank against excessive pressure.

424.85 Grounding. For those boilers designed such that fault currents do not pass through the pressure vessel, and the pressure vessel is electrically isolated from the electrodes, all exposed non–current-carrying metal parts, including the pressure vessel, supply, and return connecting piping, shall be grounded.

For all other designs, the pressure vessel containing the electrodes shall be isolated and electrically insulated from ground.

424.86 Markings. All electrode-type boilers shall be marked to show the following:

(1) The manufacturer's name.

(2) The normal rating in volts, amperes, and kilowatts.

(3) The electrical supply required specifying frequency, number of phases, and number of wires.

(4) The marking "Electrode-Type Boiler."

(5) A warning marking, "All Power Supplies Shall Be Disconnected Before Servicing, Including Servicing the Pressure Vessel." A field-applied warning marking or label shall comply with 110.21(B).

The nameplate shall be located so as to be visible after installation.

Part IX. Electric Radiant Heating Panels and Heating Panel Sets

424.90 Scope. The provisions of Part IX of this article shall apply to radiant heating panels and heating panel sets.

424.91 Definitions.

Heating Panel. A complete assembly provided with a junction box or a length of flexible conduit for connection to a branch circuit.

Heating Panel Set. A rigid or nonrigid assembly provided with nonheating leads or a terminal junction assembly identified as being suitable for connection to a wiring system.

424.92 Markings.

(A) Location. Markings shall be permanent and in a location that is visible prior to application of panel finish.

(B) Identified as Suitable. Each unit shall be identified as suitable for the installation.

(C) Required Markings. Each unit shall be marked with the identifying name or identification symbol, catalog number, and rating in volts and watts or in volts and amperes.

424.93 Installation.

(A) General.

(1) Manufacturer's Instructions. Heating panels and heating panel sets shall be installed in accordance with the manufacturer's instructions.

(2) Locations Not Permitted. The heating portion shall not be installed as follows:

(1) In or behind surfaces where subject to physical damage

(2) Run through or above walls, partitions, cupboards, or similar portions of structures that extend to the ceiling

(3) Run in or through thermal insulation, but shall be permitted to be in contact with the surface of thermal insulation

(3) Separation from Outlets for Luminaires. Edges of panels and panel sets shall be separated by not less than 200 mm (8 in.) from the edges of any outlet boxes and junction boxes that are to be used for mounting surface luminaires. A clearance of not less than 50 mm (2 in.) shall be provided from recessed luminaires and their trims, ventilating openings, and other such openings in room surfaces, unless the heating panels and panel sets are listed and marked for lesser clearances, in which case they shall be permitted to be installed at the marked clearances. Sufficient area shall be provided to ensure that no heating panel or heating panel set is to be covered by any surface-mounted units.

(4) Surfaces Covering Heating Panels. After the heating panels or heating panel sets are installed and inspected, it shall be permitted to install a surface that has been identified by the manufacturer's instructions as being suitable for the installation. The surface shall be secured so that the nails or other fastenings do not pierce the heating panels or heating panel sets.

(5) Surface Coverings. Surfaces permitted by 424.93(A)(4) shall be permitted to be covered with paint, wallpaper, or other approved surfaces identified in the manufacturer's instructions as being suitable.

(B) Heating Panel Sets.

(1) Mounting Location. Heating panel sets shall be permitted to be secured to the lower face of joists or mounted in between joists, headers, or nailing strips.

(2) Parallel to Joists or Nailing Strips. Heating panel sets shall be installed parallel to joists or nailing strips.

(3) Installation of Nails, Staples, or Other Fasteners. Nailing or stapling of heating panel sets shall be done only through the unheated portions provided for this purpose. Heating panel sets shall not be cut through or nailed through any point closer than 6 mm (1/4 in.) to the element. Nails, staples, or other fasteners shall not be used where they penetrate current-carrying parts.

(4) Installed as Complete Unit. Heating panel sets shall be installed as complete units unless identified as suitable for field cutting in an approved manner.

424.94 Clearances of Wiring in Ceilings. Wiring located above heated ceilings shall be spaced not less than 50 mm (2 in.) above the heated ceiling. The ampacity shall be calculated on the basis of an assumed ambient temperature of 50°C (122°F), applying the correction factors given in the 0–2000 volt ampacity tables of Article 310. If this wiring is located above thermal insulations having a minimum thickness of 50 mm (2 in.), the wiring shall not require correction for temperature.

424.95 Location of Branch-Circuit and Feeder Wiring in Walls.

(A) Exterior Walls. Wiring methods shall comply with Article 300 and 310.15(A)(3).

(B) Interior Walls. The ampacity of any wiring behind heating panels or heating panel sets located in interior walls or partitions shall be calculated on the basis of an assumed ambient temperature of 40°C (104°F), applying correction factors given in the 0–2000 volt ampacity tables of Article 310.

424.96 Connection to Branch-Circuit Conductors.

(A) General. Heating panels or heating panel sets assembled together in the field to form a heating installation in one room or area shall be connected in accordance with the manufacturer's instructions.

(B) Heating Panels. Heating panels shall be connected to branch-circuit wiring by an approved wiring method.

(C) Heating Panel Sets.

(1) Connection to Branch-Circuit Wiring. Heating panel sets shall be connected to branch-circuit wiring by a method identified as being suitable for the purpose.

(2) Panel Sets with Terminal Junction Assembly. A heating panel set provided with terminal junction assembly shall be permitted to have the nonheating leads attached at the time of installation in accordance with the manufacturer's instructions.

424.97 Nonheating Leads. Excess nonheating leads of heating panels or heating panel sets shall be permitted to be cut to the required length as indicated in the manufacturer's installation instructions. Nonheating leads that are an integral part of a heating panel and a heating panel set, either attached or provided by the manufacturer as part of a terminal junction assembly, shall not be subjected to the ampacity requirements of 424.3(B) for branch circuits.

424.98 Installation in Concrete or Poured Masonry.

(A) Secured in Place and Identified as Suitable. Heating panels or heating panel sets shall be secured in place by means specified in the manufacturer's instructions and identified as suitable for the installation.

(B) Expansion Joints. Heating panels or heating panel sets shall not be installed where they bridge expansion joints unless provision is made for expansion and contraction.

(C) Spacings. Spacings shall be maintained between heating panels or heating panel sets and metal embedded in the floor.

Grounded metal-clad heating panels shall be permitted to be in contact with metal embedded in the floor.

(D) Protection of Leads. Leads shall be protected where they leave the floor by rigid metal conduit, intermediate metal conduit, rigid nonmetallic conduit, or electrical metallic tubing, or by other approved means.

(E) Bushings or Fittings Required. Bushings or approved fittings shall be used where the leads emerge within the floor slabs.

424.99 Installation Under Floor Covering.

(A) Identification. Heating panels or heating panel sets for installation under floor covering shall be identified as suitable for installation under floor covering.

(B) Installation. Listed heating panels or panel sets, if installed under floor covering, shall be installed on floor surfaces that are smooth and flat in accordance with the manufacturer's instructions and shall also comply with 424.99(B)(1) through (C)(5).

(1) Expansion Joints. Heating panels or heating panel sets shall not be installed where they bridge expansion joints unless protected from expansion and contraction.

(2) Connection to Conductors. Heating panels and heating panel sets shall be connected to branch-circuit and supply wiring by wiring methods recognized in Chapter 3.

(3) Anchoring. Heating panels and heating panel sets shall be firmly anchored to the floor using an adhesive or anchoring system identified for this use.

(4) Coverings. After heating panels or heating panel sets are installed and inspected, they shall be permitted to be covered by a floor covering that has been identified by the manufacturer as being suitable for the installation.

(5) GFCI Protection. Branch circuits supplying the heating panel or heating panel sets shall have ground-fault circuit-interrupter protection for personnel.

N **(6) Grounding Braid or Sheath.** Excluding nonheating leads, grounding means, such as copper braid, metal sheath, or other approved means, shall be provided with or as an integral part of the heating panel or heating panel set.

N **Part X. Low-Voltage Fixed Electric Space-Heating Equipment**

424.100 Scope. Low-voltage fixed electric space-heating equipment shall consist of an isolating power supply, low-voltage heaters, and associated equipment that are all identified for use in dry locations.

424.101 Energy Source.

(A) Power Unit. The power unit shall be an isolating type with a rated output not exceeding 25 amperes, 30 volts (42.4 volts peak) ac, or 60 volts dc under all load conditions.

(B) Alternate Energy Sources. Listed low-voltage fixed electric space-heating equipment shall be permitted to be supplied directly from an alternate energy source such as solar photovoltaic (PV) or wind power. When supplied from such a source, the source and any power conversion equipment between the source and the heating equipment and its supply shall be listed and comply with the applicable section of the *NEC* for the source used. The output of the source shall meet the limits of 424.101(A).

424.102 Listed Equipment.   Low-voltage fixed electric space-heating equipment shall be listed as a complete system.

424.103 Installation.

(A) General.   Equipment shall be installed per the manufacturer's installation instructions.

(B) Ground.   Secondary circuits shall not be grounded.

(C) Ground-Fault Protection.   Ground-fault protection shall not be required.

424.104 Branch Circuit.

(A)   Equipment shall be permitted to be supplied from branch circuits rated not over 30 amperes.

(B)   The equipment shall be considered a continuous duty load.

ARTICLE 425
Fixed Resistance and Electrode Industrial Process Heating Equipment

Part I. General

425.1 Scope.   This article covers fixed industrial process heating employing electric resistance or electrode heating technology. For the purpose of this article, heating equipment shall include boilers, electrode boilers, duct heaters, strip heaters, immersion heaters, process air heaters, or other approved fixed electric equipment used for industrial process heating. This article shall not apply to heating and room air conditioning for personnel spaces covered by Article 424, fixed heating equipment for pipelines and vessels covered by Article 427, induction and dielectric heating equipment covered by Article 665, and industrial furnaces incorporating silicon carbide, molybdenum, or graphite process heating elements.

425.2 Other Articles.   Fixed industrial process heating equipment incorporating a hermetic refrigerant motor-compressor shall also comply with Article 440.

425.3 Branch Circuits.

(A) Branch-Circuit Requirements.   Individual branch circuits shall be permitted to supply any volt-ampere or wattage rating of fixed industrial process heating equipment for which they are rated.

(B) Branch-Circuit Sizing.   Fixed industrial process heating equipment and motors shall be considered continuous loads.

425.6 Listed Equipment.   Fixed industrial process heating equipment shall be listed.

Part II. Installation

425.8 General.

(A) Location.   Fixed industrial process heating equipment shall be located with respect to building construction and other equipment so as to permit access to the equipment. Sufficient clearance shall be maintained to permit replacement of controls and heating elements and for adjusting and cleaning of controls and other parts requiring such attention.

(B) Working Space.   Working space about electrical enclosures for fixed industrial process heating equipment that require examination, adjustment, servicing, or maintenance while energized shall be accessible, and the work space for personnel shall comply with 110.26 and 110.34, based upon the utilization voltage to ground.

Exception: With special permission, in industrial establishments only, where conditions of maintenance and supervision ensure that only qualified persons will service the installation, working space less than that required in 110.26 or 110.34 shall be permitted.

(C) Above Grade Level, Floor, or Work Platform.   Where the enclosure is located above grade, the floor, or a work platform, all of the following shall apply:

(1)   The enclosure shall be accessible.
(2)   The width of the working space shall be the width of the enclosure or a minimum of 762 mm (30 in.), whichever is greater.
(3)   The depth of the workspace shall comply with 110.26(A) or 110.34 based upon the voltage to ground.
(4)   All doors or hinged panels shall open to at least 90 degrees.

425.9 Approval.   All fixed industrial process heating equipment shall be installed in an approved manner.

425.10 Special Permission.   Fixed industrial process heating equipment and systems installed by methods other than covered by this article shall be permitted only by special permission.

425.11 Supply Conductors.   Fixed industrial process heating equipment requiring supply conductors with over 60°C insulation shall be clearly and permanently marked. This marking shall be plainly visible after installation and shall be permitted to be adjacent to the field connection box.

425.12 Locations.

(A) Exposed to Physical Damage.   Where subject to physical damage, fixed industrial process heating equipment shall be protected in an approved manner.

(B) Damp or Wet Locations.   Fixed industrial process heating equipment installed in damp or wet locations shall be listed for such locations and shall be constructed and installed so that water or other liquids cannot enter or accumulate in or on wired sections, electrical components, or ductwork.

   Informational Note: See 110.11 for equipment exposed to deteriorating agents.

425.13 Spacing from Combustible Materials.   Fixed industrial process heating equipment shall be installed to provide the required spacing between the equipment and adjacent combustible material, unless it is listed to be installed in direct contact with combustible material.

425.14 Infrared Lamp Industrial Heating Equipment. In industrial occupancies, infrared industrial process heating equipment lampholders shall be permitted to be operated in series on circuits of over 150 volts to ground, provided the voltage rating of the lampholders is not less than the circuit voltage.

Each section, panel, or strip carrying a number of infrared lampholders, including the terminal wiring of such section, panel, or strip, shall be considered as infrared industrial heating equipment. The terminal connection block of each assembly shall be considered an individual outlet.

Part III. Control and Protection of Fixed Industrial Process Heating Equipment

425.19 Disconnecting Means. Means shall be provided to simultaneously disconnect the heater, motor controller(s), and supplementary overcurrent protective device(s) of all fixed industrial process heating equipment from all ungrounded conductors. Where heating equipment is supplied by more than one source, feeder, or branch circuit, the disconnecting means shall be grouped and identified as having multiple disconnecting means. Each disconnecting means shall simultaneously disconnect all ungrounded conductors that it controls. The disconnecting means specified in 425.19(A) and (B) shall have an ampere rating not less than 125 percent of the total load of the motors and the heaters and shall be lockable in accordance with 110.25

(A) Heating Equipment with Supplementary Overcurrent Protection. The disconnecting means for fixed industrial process heating equipment with supplementary overcurrent protection shall be within sight from the supplementary overcurrent protective device(s), on the supply side of these devices, if fuses, and, in addition, shall comply with either 425.19(A)(1) or (A)(2).

(1) Heater Containing No Motor Rated over ⅛ Horsepower. The disconnecting means specified in 425.19 or unit switches complying with 425.19(C) shall be permitted to serve as the required disconnecting means for both the motor controller(s) and heater under either of the following conditions:

(1) The disconnecting means provided is also within sight from the motor controller(s) and the heater.
(2) The disconnecting means is lockable in accordance with 110.25.

(2) Heater Containing a Motor(s) Rated over ⅛ Horsepower. The above disconnecting means shall be permitted to serve as the required disconnecting means for both the motor controller(s) and heater under either of the following conditions:

(1) The disconnecting means is in sight from the motor controller(s) and the heater and complies with Part IX of Article 430.
(2) Motor(s) of more than ⅛ hp and the heater are provided with a single unit switch that complies with 422.34(A), (B), (C), or (D), the disconnecting means shall be permitted to be out of sight from the motor controller.

(B) Heating Equipment Without Supplementary Overcurrent Protection.

(1) Without Motor or with Motor Not over ⅛ Horsepower. For fixed industrial process heating equipment without a motor rated over ⅛ hp, the branch-circuit switch or circuit breaker shall be permitted to serve as the disconnecting means where the switch or circuit breaker is within sight from the heater or is lockable in accordance with 110.25.

(2) Over ⅛ Horsepower. For motor-driven fixed industrial process heating equipment with a motor rated over ⅛ hp, a disconnecting means shall be located within sight from the motor controller or shall be permitted to comply with the requirements in 425.19(A)(2).

(C) Unit Switch(es) as Disconnecting Means. A unit switch(es) with a marked "off" position that is part of a fixed heater and disconnects all ungrounded conductors shall be permitted as the disconnecting means required by this article. The branch circuit switch or circuit breaker, where readily accessible for servicing of the fixed heater, shall be permitted as the other disconnecting means.

425.21 Switch and Circuit Breaker to Be Indicating. Switches and circuit breakers used as disconnecting means shall be of the indicating type.

425.22

(A) Branch-Circuit Devices. Fixed industrial process heating equipment other than such motor-operated equipment as required by Articles 430 and 440 to have additional overcurrent protection shall be permitted to be protected against overcurrent where supplied by one of the branch circuits in Article 210.

(B) Resistance Elements. Resistance-type heating elements in fixed industrial process heating equipment shall be protected at not more than 60 amperes. Equipment rated more than 48 amperes and employing such elements shall have the heating elements subdivided, and each subdivided load shall not exceed 48 amperes. Where a subdivided load is less than 48 amperes, the rating of the supplementary overcurrent protective device shall comply with 425.3(B). A boiler employing resistance type immersion heating elements contained in an ASME-rated and stamped vessel shall be permitted to comply with 425.72(A).

(C) Overcurrent Protective Devices. The supplementary overcurrent protective devices for the subdivided loads specified in 425.22(B) shall be (1) factory installed within or on the heater enclosure or supplied for use with the heater as a separate assembly by the heater manufacturer; (2) accessible, but shall not be required to be readily accessible; and (3) suitable for branch-circuit protection.

Informational Note No. 1: See 240.10. Where cartridge fuses are used to provide this overcurrent protection, a single disconnecting means shall be permitted to be used for the several subdivided loads.

Informational Note No. 2: For supplementary overcurrent protection, see 240.10.

Informational Note No. 3: disconnecting means for cartridge fuses in circuits of any voltage, see 240.40.

(D) Branch-Circuit Conductors. The conductors supplying the supplementary overcurrent protective devices shall be considered branch-circuit conductors.

Where the heaters are rated 50 kW or more, the conductors supplying the supplementary overcurrent protective devices specified in 425.22(C) shall be permitted to be sized at not less than 100 percent of the nameplate rating of the heater, provided all of the following conditions are met:

(1)   The heater is marked with a minimum conductor size.
(2)   The conductors are not smaller than the marked minimum size.
(3)   A temperature-actuated device controls the cyclic operation of the equipment.

(E) Conductors for Subdivided Loads.   Field-wired conductors between the heater and the supplementary overcurrent protective devices for fixed industrial process heating equipment shall be sized at not less than 125 percent of the load served. The supplementary overcurrent protective devices specified in 425.22(C) shall protect these conductors in accordance with 240.4. Where the heaters are rated 50 kW or more, the ampacity of field-wired conductors between the heater and the supplementary overcurrent protective devices shall be permitted to be not less than 100 percent of the load of their respective subdivided circuits, provided all of the following conditions are met:

(1)   The heater is marked with a minimum conductor size.
(2)   The conductors are not smaller than the marked minimum size.
(3)   A temperature-activated device controls the cyclic operation of the equipment.

Part IV.   Marking of Heating Equipment

425.28 Nameplate.

(A) Marking Required.   Fixed industrial process heating equipment shall be provided with a nameplate giving the identifying name and the normal rating in volts and watts or in volts and amperes.

Fixed industrial process heating equipment intended for use on alternating current only, direct current only, or both shall be marked to so indicate. The marking of equipment consisting of motors over $\frac{1}{8}$ hp and other loads shall specify the rating of the motor in volts, amperes, and frequency and the heating load in volts and watts or in volts and amperes.

(B) Location.   This nameplate shall be located so as to be visible or easily accessible after installation.

425.29 Marking of Heating Elements.   All heating elements that are replaceable in the field and are part of industrial process heating equipment shall be legibly marked with the ratings in volts or watts or in volts or amperes.

425.45 Concealed Fixed Industrial Heating Equipment — Inspection and Tests.   Concealed fixed industrial heating equipment installations shall be made with due care to prevent damage to the heating equipment and shall be inspected and approved before heating equipment is covered or concealed.

Part V.   Fixed Industrial Process Duct Heaters

425.57 General.   Part V shall apply to any heater mounted in the airstream of a forced-air system where the air-moving unit is not provided as an integral part of the equipment.

425.58 Identification.   Heaters installed in an air duct shall be identified as suitable for the installation.

425.59 Airflow.   Means shall be provided to ensure uniform airflow over the face of the heater in accordance with the manufacturer's instructions.

> Informational Note: Some heaters installed within 1.2 m (4 ft) of the outlet of an air-moving device, elbows, baffle plates, or other obstructions in ductwork use turning vanes, pressure plates, or other devices on the inlet side of the duct heater to ensure an even distribution of air over the face of the heater.

425.60 Elevated Inlet Temperature.   Duct heaters intended for use with elevated inlet air temperature shall be identified as suitable for use at the elevated temperatures.

425.63 Fan Circuit Interlock.   Means shall be provided to ensure that the fan circuit, where present, is energized when any heater circuit is energized. However, time- or temperature-controlled delay in energizing the fan motor shall be permitted.

425.64 Limit Controls.   Each duct heater shall be provided with an approved, integral, automatic-reset temperature limiting control or controllers to de-energize the circuit or circuits. In addition, an integral independent supplementary control or controllers shall be provided in each duct heater that disconnects a sufficient number of conductors to interrupt current flow. This device shall be manually resettable or replaceable.

425.65 Location of Disconnecting Means.   Duct heater controller equipment shall be either accessible with the disconnecting means installed at or within sight from the controller or as permitted by 425.19(A).

Part VI.   Fixed Industrial Process Resistance-Type Boilers

425.70 Scope.   The provisions in Part VI of this article shall apply to boilers employing resistance-type heating elements. Electrode-type boilers shall not be considered as employing resistance-type heating elements. See Part VII of this article.

425.71 Identification.   Resistance-type boilers shall be identified as suitable for the installation.

425.72 Overcurrent Protection.

(A) Boiler Employing Resistance-Type Immersion Heating Elements in an ASME-Rated and Stamped Vessel.   A boiler employing resistance-type immersion heating elements contained in an ASME-rated and stamped vessel shall have the heating elements protected at not more than 150 amperes. Such a boiler rated more than 120 amperes shall have the heating elements subdivided into loads not exceeding 120 amperes. Where a subdivided load is less than 120 amperes, the rating of the overcurrent protective device shall comply with 425.3(B).

(B) Boiler Employing Resistance-Type Heating Elements Rated More Than 48 Amperes and Not Contained in an ASME-Rated and Stamped Vessel.   A boiler employing resistance-type heating elements not contained in an ASME-rated and stamped vessel shall have the heating elements protected at not more than 60 amperes. Such a boiler rated more than 48 amperes shall have the heating elements subdivided into loads not exceeding 48 amperes. Where a subdivided load is less than 48 amperes, the rating of the overcurrent protective device shall comply with 425.3(B).

(C) Supplementary Overcurrent Protective Devices.   The supplementary overcurrent protective devices for the subdivided loads as required by 425.72(A) and (B) shall be as follows:

(1) Factory-installed within or on the boiler enclosure or provided as a separate assembly by the boiler manufacturer.

(2) Accessible, but need not be readily accessible.

(D) Suitable for Branch-Circuit Protection. Where cartridge fuses are used to provide this overcurrent protection, a single disconnecting means shall be permitted for the several subdivided circuits. See 240.40.

(E) Conductors Supplying Supplementary Overcurrent Protective Devices. The conductors supplying these supplementary overcurrent protective devices shall be considered branch-circuit conductors. Where the heaters are rated 50 kW or more, the conductors supplying the overcurrent protective device specified in 424.72(C) shall be permitted to be sized at not less than 100 percent of the nameplate rating of the heater, provided all of the following conditions are met:

(1) The heater is marked with a minimum conductor size.
(2) The conductors are not smaller than the marked minimum size.
(3) A temperature- or pressure-actuated device controls the cyclic operation of the equipment.

(F) Conductors for Subdivided Loads. Field-wired conductors between the heater and the supplementary overcurrent protective devices shall be sized at not less than 125 percent of the load served. The supplementary overcurrent protective devices specified in 425.72(C) shall protect these conductors in accordance with 240.4. Where the heaters are rated 50 kW or more, the ampacity of field-wired conductors between the heater and the supplementary overcurrent protective devices shall be permitted to be not less than 100 percent of the load of their respective subdivided circuits, provided all of the following conditions are met:

(1) The heater is marked with a minimum conductor size.
(2) The conductors are not smaller than the marked minimum size.
(3) A temperature-activated device controls the cyclic operation of the equipment.

425.73 Overtemperature Limit Control. Each boiler designed so that in normal operation there is no change in state of the heat transfer medium shall be equipped with a temperature-sensitive limiting means. It shall be installed to limit maximum liquid temperature and shall directly or indirectly disconnect all ungrounded conductors to the heating elements. Such means shall be in addition to a temperature-regulating system and other devices protecting the tank against excessive pressure.

425.74 Overpressure Limit Control. Each boiler designed so that in normal operation there is a change in state of the heat transfer medium from liquid to vapor shall be equipped with a pressure-sensitive limiting means. It shall be installed to limit maximum pressure and shall directly or indirectly disconnect all ungrounded conductors to the heating elements. Such means shall be in addition to a pressure-regulating system and other devices protecting the tank against excessive pressure.

Part VII. Fixed Industrial Process Electrode-Type Boilers

425.80 Scope. The provisions in Part VII of this article shall apply to boilers for operation at 600 volts, nominal, or less, in

which heat is generated by the passage of current between electrodes through the liquid being heated.

425.81 Identification. Electrode-type boilers shall be identified as suitable for the installation.

425.82 Branch-Circuit Requirements. The size of branch-circuit conductors and overcurrent protective devices shall be calculated on the basis of 125 percent of the total load (motors not included). A contactor, relay, or other device, approved for continuous operation at 100 percent of its rating, shall be permitted to supply its full-rated load. See 210.19(A).

Exception: The provisions of this section shall not apply to conductors that form an integral part of an approved boiler. Where an electrode boiler is rated 50 kW or more, the conductors supplying the boiler electrode(s) shall be permitted to be sized at not less than 100 percent of the nameplate rating of the electrode boiler, provided all the following conditions are met:

(1) The electrode boiler is marked with a minimum conductor size.
(2) The conductors are not smaller than the marked minimum size.
(3) A temperature- or pressure-actuated device controls the cyclic operation of the equipment.

425.83 Overtemperature Limit Control. Each boiler, designed so that in normal operation there is no change in state of the heat transfer medium, shall be equipped with a temperature-sensitive limiting means. It shall be installed to limit maximum liquid temperature and shall directly or indirectly interrupt all current flow through the electrodes. Such means shall be in addition to the temperature regulating system and other devices protecting the tank against excessive pressure.

425.84 Overpressure Limit Control. Each boiler, designed so that in normal operation there is a change in state of the heat transfer medium from liquid to vapor, shall be equipped with a pressure-sensitive limiting means. It shall be installed to limit maximum pressure and shall directly or indirectly interrupt all current flow through the electrodes. Such means shall be in addition to a pressure-regulating system and other devices protecting the tank against excessive pressure.

425.85 Grounding. For those boilers designed such that fault currents do not pass through the pressure vessel, and the pressure vessel is electrically isolated from the electrodes, all exposed non–current-carrying metal parts, including the pressure vessel, supply, and return connecting piping, shall be grounded. For all other designs, the pressure vessel containing the electrodes shall be isolated and electrically insulated from ground.

425.86 Markings. All electrode-type boilers shall be marked to show the following:

(1) The manufacturer's name.
(2) The normal rating in volts, amperes, and kilowatts.
(3) The electrical supply required specifying frequency, number of phases, and number of wires.
(4) The marking "Electrode-Type Process Heating Boiler."
(5) A warning marking, "All Power Supplies Shall Be Disconnected Before Servicing, Including Servicing the Pressure Vessel." A field-applied warning marking or label shall comply with 110.21(B).

The nameplate shall be located so as to be visible after installation.

ARTICLE 426
Fixed Outdoor Electric Deicing and Snow-Melting Equipment

Part I. General

426.1 Scope. The requirements of this article shall apply to electrically energized heating systems and the installation of these systems.

(A) Embedded. Embedded in driveways, walks, steps, and other areas.

(B) Exposed. Exposed on drainage systems, bridge structures, roofs, and other structures.

> Informational Note: For further information, see ANSI/IEEE 515.1-2012, *Standard for the Testing, Design, Installation and Maintenance of Electrical Resistance Trace Heating for Commercial Applications.*

426.2 Definitions.

Heating System. A complete system consisting of components such as heating elements, fastening devices, nonheating circuit wiring, leads, temperature controllers, safety signs, junction boxes, raceways, and fittings.

Impedance Heating System. A system in which heat is generated in a pipe or rod, or combination of pipes and rods, by causing current to flow through the pipe or rod by direct connection to an ac voltage source from an isolating transformer. The pipe or rod shall be permitted to be embedded in the surface to be heated, or constitute the exposed components to be heated.

Resistance Heating Element. A specific separate element to generate heat that is embedded in or fastened to the surface to be heated.

> Informational Note: Tubular heaters, strip heaters, heating cable, heating tape, and heating panels are examples of resistance heaters.

Skin-Effect Heating System. A system in which heat is generated on the inner surface of a ferromagnetic envelope embedded in or fastened to the surface to be heated.

> Informational Note: Typically, an electrically insulated conductor is routed through and connected to the envelope at the other end. The envelope and the electrically insulated conductor are connected to an ac voltage source from an isolating transformer.

426.3 Application of Other Articles. Cord-and-plug-connected fixed outdoor electric deicing and snow-melting equipment intended for specific use and identified as suitable for this use shall be installed according to Article 422.

426.4 Continuous Load. Fixed outdoor electric deicing and snow-melting equipment shall be considered a continuous load.

Part II. Installation

426.10 General. Equipment for outdoor electric deicing and snow melting shall be identified as being suitable for the following:

(1) The chemical, thermal, and physical environment
(2) Installation in accordance with the manufacturer's drawings and instructions

426.11 Use. Electric heating equipment shall be installed in such a manner as to be afforded protection from physical damage.

426.12 Thermal Protection. External surfaces of outdoor electric deicing and snow-melting equipment that operate at temperatures exceeding 60°C (140°F) shall be physically guarded, isolated, or thermally insulated to protect against contact by personnel in the area.

426.13 Identification. The presence of outdoor electric deicing and snow-melting equipment shall be evident by the posting of appropriate caution signs or markings where clearly visible.

426.14 Special Permission. Fixed outdoor deicing and snow-melting equipment employing methods of construction or installation other than covered by this article shall be permitted only by special permission.

Part III. Resistance Heating Elements

426.20 Embedded Deicing and Snow-Melting Equipment.

(A) Watt Density. Panels or units shall not exceed 1300 watts/m^2 (120 watts/ft^2) of heated area.

(B) Spacing. The spacing between adjacent cable runs is dependent upon the rating of the cable and shall be not less than 25 mm (1 in.) on centers.

(C) Cover. Units, panels, or cables shall be installed as follows:

(1) On a substantial asphalt or masonry base at least 50 mm (2 in.) thick and have at least 38 mm (1½ in.) of asphalt or masonry applied over the units, panels, or cables; or
(2) They shall be permitted to be installed over other approved bases and embedded within 90 mm (3½ in.) of masonry or asphalt but not less than 38 mm (1½ in.) from the top surface; or
(3) Equipment that has been listed for other forms of installation shall be installed only in the manner for which it has been identified.

(D) Secured. Cables, units, and panels shall be secured in place by frames or spreaders or other approved means while the masonry or asphalt finish is applied.

(E) Expansion and Contraction. Cables, units, and panels shall not be installed where they bridge expansion joints unless provision is made for expansion and contraction.

426.21 Exposed Deicing and Snow-Melting Equipment.

(A) Secured. Heating element assemblies shall be secured to the surface being heated by approved means.

(B) Overtemperature. Where the heating element is not in direct contact with the surface being heated, the design of the heater assembly shall be such that its temperature limitations shall not be exceeded.

(C) Expansion and Contraction. Heating elements and assemblies shall not be installed where they bridge expansion joints unless provision is made for expansion and contraction.

(D) Flexural Capability. Where installed on flexible structures, the heating elements and assemblies shall have a flexural capability that is compatible with the structure.

426.22 Installation of Nonheating Leads for Embedded Equipment.

(A) Grounding Sheath or Braid. Nonheating leads having a grounding sheath or braid shall be permitted to be embedded in the masonry or asphalt in the same manner as the heating cable without additional physical protection.

(B) Raceways. All but 25 mm to 150 mm (1 in. to 6 in.) of nonheating leads not having a grounding sheath shall be enclosed in a rigid metal conduit, electrical metallic tubing, intermediate metal conduit, or other raceways within asphalt or masonry. The distance from the factory splice to raceway shall not be less than 25 mm (1 in.) or more than 150 mm (6 in.).

(C) Bushings. Insulating bushings shall be used in the asphalt or masonry where leads enter conduit or tubing.

(D) Expansion and Contraction. Leads shall be protected in expansion joints and where they emerge from masonry or asphalt by rigid conduit, electrical metallic tubing, intermediate metal conduit, other raceways, or other approved means.

(E) Leads in Junction Boxes. Not less than 150 mm (6 in.) of free nonheating lead shall be within the junction box.

426.23 Installation of Nonheating Leads for Exposed Equipment.

(A) Nonheating Leads. Power supply nonheating leads (cold leads) for resistance elements shall be identified for the temperature encountered. Not less than 150 mm (6 in.) of nonheating leads shall be provided within the junction box. Preassembled factory-supplied and field-assembled nonheating leads on approved heaters shall be permitted to be shortened if the markings specified in 426.25 are retained.

(B) Protection. Nonheating power supply leads shall be enclosed in a rigid conduit, intermediate metal conduit, electrical metallic tubing, or other approved means.

426.24 Electrical Connection.

(A) Heating Element Connections. Electrical connections, other than factory connections of heating elements to nonheating elements embedded in masonry or asphalt or on exposed surfaces, shall be made with insulated connectors identified for the use.

(B) Circuit Connections. Splices and terminations at the end of the nonheating leads, other than the heating element end, shall be installed in a box or fitting in accordance with 110.14 and 300.15.

426.25 Marking.
Each factory-assembled heating unit shall be legibly marked within 75 mm (3 in.) of each end of the nonheating leads with the permanent identification symbol, catalog number, and ratings in volts and watts or in volts and amperes.

426.26 Corrosion Protection.
Ferrous and nonferrous metal raceways, cable armor, cable sheaths, boxes, fittings, supports, and support hardware shall be permitted to be installed in concrete or in direct contact with the earth, or in areas subject to severe corrosive influences, where made of material suitable for the condition, or where provided with corrosion protection identified as suitable for the condition.

426.27 Grounding Braid or Sheath.
Grounding means, such as copper braid, metal sheath, or other approved means, shall be provided as part of the heated section of the cable, panel, or unit.

426.28 Ground-Fault Protection of Equipment.
Ground-fault protection of equipment shall be provided for fixed outdoor electric deicing and snow-melting equipment.

Part IV. Impedance Heating

426.30 Personnel Protection.
Exposed elements of impedance heating systems shall be physically guarded, isolated, or thermally insulated with a weatherproof jacket to protect against contact by personnel in the area.

426.31 Isolation Transformer.
An isolation transformer with a grounded shield between the primary and secondary windings shall be used to isolate the distribution system from the heating system.

426.32 Voltage Limitations.
The secondary winding of the isolation transformer connected to the impedance heating elements shall not have an output voltage greater than 30 volts ac.

426.33 Induced Currents.
All current-carrying components shall be installed in accordance with 300.20.

426.34 Grounding.
An impedance heating system that is operating at a voltage greater than 30 but not more than 80 shall be grounded at a designated point(s).

Part V. Skin-Effect Heating

426.40 Conductor Ampacity.
The current through the electrically insulated conductor inside the ferromagnetic envelope shall be permitted to exceed the ampacity values shown in Article 310, provided it is identified as suitable for this use.

426.41 Pull Boxes.
Where pull boxes are used, they shall be accessible without excavation by location in suitable vaults or abovegrade. Outdoor pull boxes shall be of watertight construction.

426.42 Single Conductor in Enclosure.
The provisions of 300.20 shall not apply to the installation of a single conductor in a ferromagnetic envelope (metal enclosure).

426.43 Corrosion Protection.
Ferromagnetic envelopes, ferrous or nonferrous metal raceways, boxes, fittings, supports, and support hardware shall be permitted to be installed in concrete or in direct contact with the earth, or in areas subjec-

ted to severe corrosive influences, where made of material suitable for the condition, or where provided with corrosion protection identified as suitable for the condition. Corrosion protection shall maintain the original wall thickness of the ferromagnetic envelope.

426.44 Grounding. The ferromagnetic envelope shall be connected to an equipment grounding conductor at both ends; and, in addition, it shall be permitted to be connected to an equipment grounding conductor at intermediate points as required by its design.

The provisions of 250.30 shall not apply to the installation of skin-effect heating systems.

Informational Note: For grounding methods, see Article 250.

Part VI. Control and Protection

426.50 Disconnecting Means.

(A) Disconnection. All fixed outdoor deicing and snow-melting equipment shall be provided with a means for simultaneous disconnection from all ungrounded conductors. Where readily accessible to the user of the equipment, the branch-circuit switch or circuit breaker shall be permitted to serve as the disconnecting means. The disconnecting means shall be of the indicating type and be capable of being locked in the open (off) position.

(B) Cord-and-Plug-Connected Equipment. The factory-installed attachment plug of cord-and-plug-connected equipment rated 20 amperes or less and 150 volts or less to ground shall be permitted to be the disconnecting means.

426.51 Controllers.

(A) Temperature Controller with "Off" Position. Temperature controlled switching devices that indicate an "off" position and that interrupt line current shall open all ungrounded conductors when the control device is in the "off" position. These devices shall not be permitted to serve as the disconnecting means unless they are lockable in accordance with 110.25.

(B) Temperature Controller Without "Off" Position. Temperature controlled switching devices that do not have an "off" position shall not be required to open all ungrounded conductors and shall not be permitted to serve as the disconnecting means.

(C) Remote Temperature Controller. Remote controlled temperature-actuated devices shall not be required to meet the requirements of 426.51(A). These devices shall not be permitted to serve as the disconnecting means.

(D) Combined Switching Devices. Switching devices consisting of combined temperature-actuated devices and manually controlled switches that serve both as the controller and the disconnecting means shall comply with all of the following conditions:

(1) Open all ungrounded conductors when manually placed in the "off" position
(2) Be so designed that the circuit cannot be energized automatically if the device has been manually placed in the "off" position
(3) Be lockable in accordance with 110.25

426.54 Cord-and-Plug-Connected Deicing and Snow-Melting Equipment. Cord-and-plug-connected deicing and snow-melting equipment shall be listed.

ARTICLE 427
Fixed Electric Heating Equipment for Pipelines and Vessels

Part I. General

427.1 Scope. The requirements of this article shall apply to electrically energized heating systems and the installation of these systems used with pipelines or vessels or both.

Informational Note: For further information, see ANSI/IEEE 515-2011, *Standard for the Testing, Design, Installation and Maintenance of Electrical Resistance Trace Heating for Industrial Applications*; ANSI/IEEE 844-2000, *Recommended Practice for Electrical Impedance, Induction, and Skin Effect Heating of Pipelines and Vessels*; and ANSI/NECA 202-2013, *Standard for Installing and Maintaining Industrial Heat Tracing Systems*.

427.2 Definitions.

Impedance Heating System. A system in which heat is generated in a pipeline or vessel wall by causing current to flow through the pipeline or vessel wall by direct connection to an ac voltage source from a dual-winding transformer.

Induction Heating System. A system in which heat is generated in a pipeline or vessel wall by inducing current and hysteresis effect in the pipeline or vessel wall from an external isolated ac field source.

Pipeline. A length of pipe including pumps, valves, flanges, control devices, strainers, and/or similar equipment for conveying fluids.

Resistance Heating Element. A specific separate element to generate heat that is applied to the pipeline or vessel externally or internally.

Informational Note: Tubular heaters, strip heaters, heating cable, heating tape, heating blankets, and immersion heaters are examples of resistance heaters.

Skin-Effect Heating System. A system in which heat is generated on the inner surface of a ferromagnetic envelope attached to a pipeline or vessel, or both.

Informational Note: Typically, an electrically insulated conductor is routed through and connected to the envelope at the other end. The envelope and the electrically insulated conductor are connected to an ac voltage source from a dual-winding transformer.

Vessel. A container such as a barrel, drum, or tank for holding fluids or other material.

427.3 Application of Other Articles. Cord-connected pipe heating assemblies intended for specific use and identified as suitable for this use shall be installed according to Article 422.

427.4 Continuous Load. Fixed electric heating equipment for pipelines and vessels shall be considered continuous load.

Part II. Installation

427.10 General. Equipment for pipeline and vessel electric heating shall be identified as being suitable for (1) the chemical, thermal, and physical environment and (2) installation in accordance with the manufacturer's drawings and instructions.

427.11 Use. Electric heating equipment shall be installed in such a manner as to be afforded protection from physical damage.

427.12 Thermal Protection. External surfaces of pipeline and vessel heating equipment that operate at temperatures exceeding 60°C (140°F) shall be physically guarded, isolated, or thermally insulated to protect against contact by personnel in the area.

427.13 Identification. The presence of electrically heated pipelines, vessels, or both, shall be evident by the posting of appropriate caution signs or markings at intervals not exceeding 6 m (20 ft) along the pipeline or vessel and on or adjacent to equipment in the piping system that requires periodic servicing.

Part III. Resistance Heating Elements

427.14 Secured. Heating element assemblies shall be secured to the surface being heated by means other than the thermal insulation.

427.15 Not in Direct Contact. Where the heating element is not in direct contact with the pipeline or vessel being heated, means shall be provided to prevent overtemperature of the heating element unless the design of the heater assembly is such that its temperature limitations will not be exceeded.

427.16 Expansion and Contraction. Heating elements and assemblies shall not be installed where they bridge expansion joints unless provisions are made for expansion and contraction.

427.17 Flexural Capability. Where installed on flexible pipelines, the heating elements and assemblies shall have a flexural capability that is compatible with the pipeline.

427.18 Power Supply Leads.

(A) Nonheating Leads. Power supply nonheating leads (cold leads) for resistance elements shall be suitable for the temperature encountered. Not less than 150 mm (6 in.) of nonheating leads shall be provided within the junction box. Preassembled factory-supplied and field-assembled nonheating leads on approved heaters shall be permitted to be shortened if the markings specified in 427.20 are retained.

(B) Power Supply Leads Protection. Nonheating power supply leads shall be protected where they emerge from electrically heated pipeline or vessel heating units by rigid metal conduit, intermediate metal conduit, electrical metallic tubing, or other raceways identified as suitable for the application.

(C) Interconnecting Leads. Interconnecting nonheating leads connecting portions of the heating system shall be permitted to be covered by thermal insulation in the same manner as the heaters.

427.19 Electrical Connections.

(A) Nonheating Interconnections. Nonheating interconnections, where required under thermal insulation, shall be made with insulated connectors identified as suitable for this use.

(B) Circuit Connections. Splices and terminations outside the thermal insulation shall be installed in a box or fitting in accordance with 110.14 and 300.15.

427.20 Marking. Each factory-assembled heating unit shall be legibly marked within 75 mm (3 in.) of an end of the nonheating leads with the permanent identification symbol, catalog number, and ratings in volts and watts or in volts and amperes.

427.22 Ground-Fault Protection of Equipment. Ground-fault protection of equipment shall be provided for electric heat tracing and heating panels. This requirement shall not apply in industrial establishments where there is alarm indication of ground faults and the following conditions apply:

(1) Conditions of maintenance and supervision ensure that only qualified persons service the installed systems.
(2) Continued circuit operation is necessary for safe operation of equipment or processes.

427.23 Grounded Conductive Covering. Electric heating equipment shall be listed and have a grounded conductive covering in accordance with 427.23(A) or (B). The conductive covering shall provide an effective ground path for equipment protection.

(A) Heating Wires or Cables. Heating wires or cables shall have a grounded conductive covering that surrounds the heating element and bus wires, if any, and their electrical insulation.

(B) Heating Panels. Heating panels shall have a grounded conductive covering over the heating element and its electrical insulation on the side opposite the side attached to the surface to be heated.

Part IV. Impedance Heating

427.25 Personnel Protection. All accessible external surfaces of the pipeline, vessel, or both, being heated shall be physically guarded, isolated, or thermally insulated (with a weatherproof jacket for outside installations) to protect against contact by personnel in the area.

427.26 Isolation Transformer. A dual-winding transformer with a grounded shield between the primary and secondary windings shall be used to isolate the distribution system from the heating system.

427.27 Voltage Limitations. The secondary winding of the isolation transformer connected to the pipeline or vessel being heated shall not have an output voltage greater than 30 volts ac.

Exception No. 1: In industrial establishments, the isolation transformer connected to the pipeline or vessel being heated shall be permitted to have an output voltage greater than 30 but not more than 80 volts ac to ground where all of the following conditions apply:

(1)   Conditions of guarding, maintenance, and supervision ensure that only qualified persons have access to the installed systems.
(2)   Ground-fault protection of equipment is provided.

Exception No. 2: In industrial establishments, the isolation transformer connected to the pipeline or vessel being heated shall be permitted to have an output voltage not greater than 132 volts ac to ground where all of the following conditions apply:

(1) Conditions of guarding, maintenance, and supervision ensure that only qualified persons service the installed systems.
(2) Ground-fault protection of equipment is provided.
(3) The pipeline or vessel being heated is completely enclosed in a grounded metal enclosure.
(4) The transformer secondary connections to the pipeline or vessel being heated are completely enclosed in a grounded metal mesh or metal enclosure.

427.28 Induced Currents. All current-carrying components shall be installed in accordance with 300.20.

427.29 Grounding. The pipeline, vessel, or both, that is being heated and operating at a voltage greater than 30 but not more than 80 shall be grounded at designated points.

427.30 Secondary Conductor Sizing. The ampacity of the conductors connected to the secondary of the transformer shall be at least 100 percent of the total load of the heater.

Part V. Induction Heating

427.35 Scope. This part covers the installation of line frequency induction heating equipment and accessories for pipelines and vessels.

Informational Note: See Article 665 for other applications.

427.36 Personnel Protection. Induction coils that operate or may operate at a voltage greater than 30 volts ac shall be enclosed in a nonmetallic or split metallic enclosure, isolated, or made inaccessible by location to protect personnel in the area.

427.37 Induced Current. Induction coils shall be prevented from inducing circulating currents in surrounding metallic equipment, supports, or structures by shielding, isolation, or insulation of the current paths. Stray current paths shall be bonded to prevent arcing.

Part VI. Skin-Effect Heating

427.45 Conductor Ampacity. The ampacity of the electrically insulated conductor inside the ferromagnetic envelope shall be permitted to exceed the values given in Article 310, provided it is identified as suitable for this use.

427.46 Pull Boxes. Pull boxes for pulling the electrically insulated conductor in the ferromagnetic envelope shall be permitted to be buried under the thermal insulation, provided their locations are indicated by permanent markings on the insulation jacket surface and on drawings. For outdoor installations, pull boxes shall be of watertight construction.

427.47 Single Conductor in Enclosure. The provisions of 300.20 shall not apply to the installation of a single conductor in a ferromagnetic envelope (metal enclosure).

427.48 Grounding. The ferromagnetic envelope shall be grounded at both ends, and, in addition, it shall be permitted to be grounded at intermediate points as required by its design. The ferromagnetic envelope shall be bonded at all joints to ensure electrical continuity.

The provisions of 250.30 shall not apply to the installation of skin-effect heating systems.

Informational Note: See Article 250 for grounding methods.

Part VII. Control and Protection

427.55 Disconnecting Means.

(A) Switch or Circuit Breaker. Means shall be provided to simultaneously disconnect all fixed electric pipeline or vessel heating equipment from all ungrounded conductors. The branch-circuit switch or circuit breaker, where readily accessible to the user of the equipment, shall be permitted to serve as the disconnecting means. The disconnecting means shall be of the indicating type and shall be capable of being locked in the open (off) position. The disconnecting means shall be installed in accordance with 110.25.

(B) Cord-and-Plug-Connected Equipment. The factory-installed attachment plug of cord-and-plug-connected equipment rated 20 amperes or less and 150 volts or less to ground shall be permitted to be the disconnecting means.

427.56 Controls.

(A) Temperature Control with "Off" Position. Temperature-controlled switching devices that indicate an "off" position and that interrupt line current shall open all ungrounded conductors when the control device is in this "off" position. These devices shall not be permitted to serve as the disconnecting means unless capable of being locked in the open position.

(B) Temperature Control Without "Off" Position. Temperature controlled switching devices that do not have an "off" position shall not be required to open all ungrounded conductors and shall not be permitted to serve as the disconnecting means.

(C) Remote Temperature Controller. Remote controlled temperature-actuated devices shall not be required to meet the requirements of 427.56(A) and (B). These devices shall not be permitted to serve as the disconnecting means.

(D) Combined Switching Devices. Switching devices consisting of combined temperature-actuated devices and manually controlled switches that serve both as the controllers and the disconnecting means shall comply with all the following conditions:

(1) Open all ungrounded conductors when manually placed in the "off" position
(2) Be designed so that the circuit cannot be energized automatically if the device has been manually placed in the "off" position
(3) Be capable of being locked in the open position

427.57 Overcurrent Protection. Heating equipment shall be considered protected against overcurrent where supplied by a branch circuit as specified in 210.18 and 210.23.

ARTICLE 430
Motors, Motor Circuits, and Controllers

Part I. General

430.1 Scope. This article covers motors, motor branch-circuit and feeder conductors and their protection, motor overload protection, motor control circuits, motor controllers, and motor control centers.

> Informational Note No. 1: Installation requirements for motor control centers are covered in 110.26(E). Air-conditioning and refrigerating equipment are covered in Article 440.

> Informational Note No. 2: Figure 430.1 is for information only.

430.2 Definitions.

Controller. For the purpose of this article, a controller is any switch or device that is normally used to start and stop a motor by making and breaking the motor circuit current.

To Supply	Part II 430.24, 430.25, 430.26
Motor feeder	
Motor feeder short-circuit and ground-fault protection	Part V
Motor disconnecting means	Part IX
Motor branch-circuit short-circuit and ground-fault protection	Part IV
Motor circuit conductor	Part II
Motor controller	Part VII
Motor control circuits	Part VI
Motor overload protection	Part III
Motor	Part I
Thermal protection	Part III
Secondary controller Secondary conductors	Part II 430.23
Secondary resistor	Part II 430.23 and Article 470

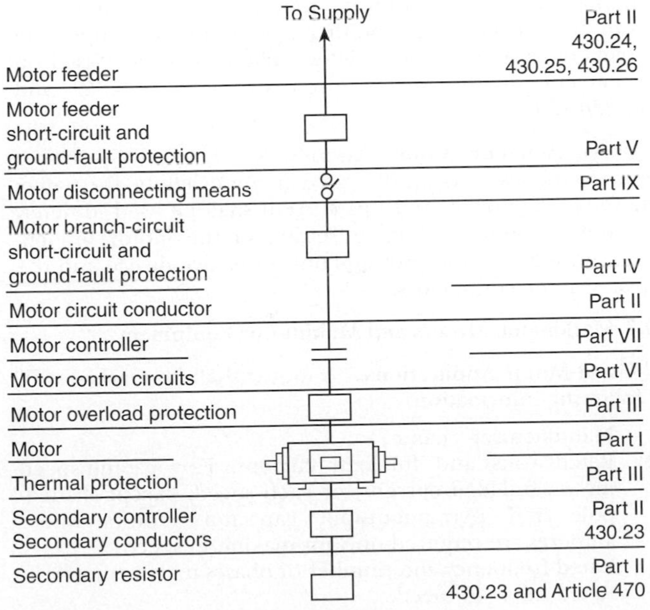

FIGURE 430.1 Article 430 Contents.

Part-Winding Motors. A part-winding start induction or synchronous motor is one that is arranged for starting by first energizing part of its primary (armature) winding and, subsequently, energizing the remainder of this winding in one or more steps. A standard part-winding start induction motor is arranged so that one-half of its primary winding can be energized initially, and, subsequently, the remaining half can be energized, both halves then carrying equal current. A hermetic refrigerant compressor motor shall not be considered a standard part-winding start induction motor.

System Isolation Equipment. A redundantly monitored, remotely operated contactor-isolating system, packaged to provide the disconnection/isolation function, capable of verifiable operation from multiple remote locations by means of lockout switches, each having the capability of being padlocked in the "off" (open) position.

Valve Actuator Motor (VAM) Assemblies. A manufactured assembly, used to operate a valve, consisting of an actuator motor and other components such as controllers, torque switches, limit switches, and overload protection.

> Informational Note: VAMs typically have short-time duty and high-torque characteristics.

430.4 Part-Winding Motors. Where separate overload devices are used with a standard part-winding start induction motor, each half of the motor winding shall be individually protected in accordance with 430.32 and 430.37 with a trip current one-half that specified.

Each motor-winding connection shall have branch-circuit short-circuit and ground-fault protection rated at not more than one-half that specified by 430.52.

Exception: A short-circuit and ground-fault protective device shall be permitted for both windings if the device will allow the motor to start. Where time delay (dual-element) fuses are used, they shall be permitted to have a rating not exceeding 150 percent of the motor full-load current.

430.5 Other Articles. Motors and controllers shall also comply with the applicable provisions of Table 430.5.

430.6 Ampacity and Motor Rating Determination. The size of conductors supplying equipment covered by Article 430 shall be selected from the allowable ampacity tables in accordance with 310.15(B) or shall be calculated in accordance with 310.15(C). Where flexible cord is used, the size of the conductor shall be selected in accordance with 400.5. The required ampacity and motor ratings shall be determined as specified in 430.6(A), (B), (C), and (D).

(A) General Motor Applications. For general motor applications, current ratings shall be determined based on (A)(1) and (A)(2).

(1) Table Values. Other than for motors built for low speeds (less than 1200 RPM) or high torques, and for multispeed motors, the values given in Table 430.247, Table 430.248, Table 430.249, and Table 430.250 shall be used to determine the ampacity of conductors or ampere ratings of switches, branch-circuit short-circuit and ground-fault protection, instead of the actual current rating marked on the motor nameplate. Where a motor is marked in amperes, but not horsepower, the horsepower rating shall be assumed to be that corresponding to the value given in Table 430.247, Table 430.248, Table 430.249, and Table 430.250, interpolated if

Table 430.5 Other Articles

Equipment/Occupancy	Article	Section
Air-conditioning and refrigerating equipment	440	
Capacitors		460.8, 460.9
Commercial garages; aircraft hangars; motor fuel dispensing facilities; bulk storage plants; spray application, dipping, and coating processes; and inhalation anesthetizing locations	511, 513, 514, 515, 516, and 517 Part IV	
Cranes and hoists	610	
Electrically driven or controlled irrigation machines	675	
Elevators, dumbwaiters, escalators, moving walks, wheelchair lifts, and stairway chair lifts	620	
Fire pumps	695	
Hazardous (classified) locations	500–503, 505, and 506	
Industrial machinery	670	
Motion picture projectors		540.11 and 540.20
Motion picture and television studios and similar locations	530	
Resistors and reactors	470	
Theaters, audience areas of motion picture and television studios, and similar locations		520.48
Transformers and transformer vaults	450	

necessary. Motors built for low speeds (less than 1200 RPM) or high torques may have higher full-load currents, and multispeed motors will have full-load current varying with speed, in which case the nameplate current ratings shall be used.

Exception No. 1: Multispeed motors shall be in accordance with 430.22(B) and 430.52.

Exception No. 2: For equipment that employs a shaded-pole or permanent-split capacitor-type fan or blower motor that is marked with the motor type, the full load current for such motor marked on the nameplate of the equipment in which the fan or blower motor is employed shall be used instead of the horsepower rating to determine the ampacity or rating of the disconnecting means, the branch-circuit conductors, the controller, the branch-circuit short-circuit and ground-fault protection, and the separate overload protection. This marking on the equipment nameplate shall not be less than the current marked on the fan or blower motor nameplate.

Exception No. 3: For a listed motor-operated appliance that is marked with both motor horsepower and full-load current, the motor full-load current marked on the nameplate of the appliance shall be used instead of the horsepower rating on the appliance nameplate to determine the ampacity or rating of the disconnecting means, the branch-circuit conductors, the controller, the branch-circuit short-circuit and ground-fault protection, and any separate overload protection.

(2) Nameplate Values. Separate motor overload protection shall be based on the motor nameplate current rating.

(B) Torque Motors. For torque motors, the rated current shall be locked-rotor current, and this nameplate current shall be used to determine the ampacity of the branch-circuit conductors covered in 430.22 and 430.24, the ampere rating of the motor overload protection, and the ampere rating of motor branch-circuit short-circuit and ground-fault protection in accordance with 430.52(B).

Informational Note: For motor controllers and disconnecting means, see 430.83(D) and 430.110.

(C) Alternating-Current Adjustable Voltage Motors. For motors used in alternating-current, adjustable voltage, variable torque drive systems, the ampacity of conductors, or ampere ratings of switches, branch-circuit short-circuit and ground-fault protection, and so forth, shall be based on the maximum operating current marked on the motor or control nameplate, or both. If the maximum operating current does not appear on the nameplate, the ampacity determination shall be based on 150 percent of the values given in Table 430.249 and Table 430.250.

(D) Valve Actuator Motor Assemblies. For valve actuator motor assemblies (VAMs), the rated current shall be the nameplate full-load current, and this current shall be used to determine the maximum rating or setting of the motor branch-circuit short-circuit and ground-fault protective device and the ampacity of the conductors.

430.7 Marking on Motors and Multimotor Equipment.

(A) Usual Motor Applications. A motor shall be marked with the following information:

(1) Manufacturer's name.
(2) Rated volts and full-load current. For a multispeed motor, full-load current for each speed, except shaded-pole and permanent-split capacitor motors where amperes are required only for maximum speed.
(3) Rated frequency and number of phases if an ac motor.
(4) Rated full-load speed.

(5) Rated temperature rise or the insulation system class and rated ambient temperature.

(6) Time rating. The time rating shall be 5, 15, 30, or 60 minutes, or continuous.

(7) Rated horsepower if ⅛ hp or more. For a multispeed motor ⅛ hp or more, rated horsepower for each speed, except shaded-pole and permanent-split capacitor motors ⅛ hp or more where rated horsepower is required only for maximum speed. Motors of arc welders are not required to be marked with the horsepower rating.

(8) Code letter or locked-rotor amperes if an alternating-current motor rated ½ hp or more. On polyphase wound-rotor motors, the code letter shall be omitted.

Informational Note: See 430.7(B).

(9) Design letter for design B, C, or D motors.

Informational Note: Motor design letter definitions are found in ANSI/NEMA MG 1-1993, *Motors and Generators, Part 1, Definitions*, and in IEEE 100-1996, *Standard Dictionary of Electrical and Electronic Terms*.

(10) Secondary volts and full-load current if a wound-rotor induction motor.

(11) Field current and voltage for dc excited synchronous motors.

(12) Winding — straight shunt, stabilized shunt, compound, or series, if a dc motor. Fractional horsepower dc motors 175 mm (7 in.) or less in diameter shall not be required to be marked.

(13) A motor provided with a thermal protector complying with 430.32(A)(2) or (B)(2) shall be marked "Thermally Protected." Thermally protected motors rated 100 watts or less and complying with 430.32(B)(2) shall be permitted to use the abbreviated marking "T.P."

(14) A motor complying with 430.32(B)(4) shall be marked "Impedance Protected." Impedance-protected motors rated 100 watts or less and complying with 430.32(B)(4) shall be permitted to use the abbreviated marking "Z.P."

(15) Motors equipped with electrically powered condensation prevention heaters shall be marked with the rated heater voltage, number of phases, and the rated power in watts.

(B) Locked-Rotor Indicating Code Letters. Code letters marked on motor nameplates to show motor input with locked rotor shall be in accordance with Table 430.7(B).

The code letter indicating motor input with locked rotor shall be in an individual block on the nameplate, properly designated.

(1) Multispeed Motors. Multispeed motors shall be marked with the code letter designating the locked-rotor kilovolt-ampere (kVA) per horsepower (hp) for the highest speed at which the motor can be started.

Exception: Constant horsepower multispeed motors shall be marked with the code letter giving the highest locked-rotor kilovolt-ampere (kVA) per horsepower (hp).

(2) Single-Speed Motors. Single-speed motors starting on wye connection and running on delta connections shall be marked with a code letter corresponding to the locked-rotor kilovolt-ampere (kVA) per horsepower (hp) for the wye connection.

(3) Dual-Voltage Motors. Dual-voltage motors that have a different locked-rotor kilovolt-ampere (kVA) per horsepower

Table 430.7(B) Locked-Rotor Indicating Code Letters

Code Letter	Kilovolt-Amperes per Horsepower with Locked Rotor
A	0–3.14
B	3.15–3.54
C	3.55–3.99
D	4.0–4.49
E	4.5–4.99
F	5.0–5.59
G	5.6–6.29
H	6.3–7.09
J	7.1–7.99
K	8.0–8.99
L	9.0–9.99
M	10.0–11.19
N	11.2–12.49
P	12.5–13.99
R	14.0–15.99
S	16.0–17.99
T	18.0–19.99
U	20.0–22.39
V	22.4 and up

(hp) on the two voltages shall be marked with the code letter for the voltage giving the highest locked-rotor kilovolt-ampere (kVA) per horsepower (hp).

(4) 50/60 Hz Motors. Motors with 50- and 60-Hz ratings shall be marked with a code letter designating the locked-rotor kilovolt-ampere (kVA) per horsepower (hp) on 60 Hz.

(5) Part-Winding Motors. Part-winding start motors shall be marked with a code letter designating the locked-rotor kilovolt-ampere (kVA) per horsepower (hp) that is based on the locked-rotor current for the full winding of the motor.

(C) Torque Motors. Torque motors are rated for operation at standstill and shall be marked in accordance with 430.7(A), except that locked-rotor torque shall replace horsepower.

(D) Multimotor and Combination-Load Equipment.

(1) Factory-Wired. Multimotor and combination-load equipment shall be provided with a visible nameplate marked with the manufacturer's name, the rating in volts, frequency, number of phases, minimum supply circuit conductor ampacity, and the maximum ampere rating of the circuit short-circuit and ground-fault protective device. The conductor ampacity shall be calculated in accordance with 430.24 and counting all of the motors and other loads that will be operated at the same time. The short-circuit and ground-fault protective device rating shall not exceed the value calculated in accordance with 430.53. Multimotor equipment for use on two or more circuits shall be marked with the preceding information for each circuit.

(2) Not Factory-Wired. Where the equipment is not factory-wired and the individual nameplates of motors and other loads are visible after assembly of the equipment, the individual nameplates shall be permitted to serve as the required marking.

430.8 Marking on Controllers. A controller shall be marked with the manufacturer's name or identification, the voltage, the current or horsepower rating, the short-circuit current rating, and other necessary data to properly indicate the applications for which it is suitable.

Exception No. 1: The short-circuit current rating is not required for controllers applied in accordance with 430.81(A) or (B).

Exception No. 2: The short-circuit rating is not required to be marked on the controller when the short-circuit current rating of the controller is marked elsewhere on the assembly.

Exception No. 3: The short-circuit rating is not required to be marked on the controller when the assembly into which it is installed has a marked short-circuit current rating.

Exception No. 4: Short-circuit ratings are not required for controllers rated less than 2 hp at 300 V or less and listed exclusively for general-purpose branch circuits.

A controller that includes motor overload protection suitable for group motor application shall be marked with the motor overload protection and the maximum branch-circuit short-circuit and ground-fault protection for such applications.

Combination controllers that employ adjustable instantaneous trip circuit breakers shall be clearly marked to indicate the ampere settings of the adjustable trip element.

Where a controller is built in as an integral part of a motor or of a motor-generator set, individual marking of the controller shall not be required if the necessary data are on the nameplate. For controllers that are an integral part of equipment approved as a unit, the above marking shall be permitted on the equipment nameplate.

Informational Note: See 110.10 for information on circuit impedance and other characteristics.

430.9 Terminals.

(A) Markings. Terminals of motors and controllers shall be suitably marked or colored where necessary to indicate the proper connections.

(B) Conductors. Motor controllers and terminals of control circuit devices shall be connected with copper conductors unless identified for use with a different conductor.

(C) Torque Requirements. Control circuit devices with screw-type pressure terminals used with 14 AWG or smaller copper conductors shall be torqued to a minimum of 0.8 N·m (7 lb-in.) unless identified for a different torque value.

430.10 Wiring Space in Enclosures.

(A) General. Enclosures for motor controllers and disconnecting means shall not be used as junction boxes, auxiliary gutters, or raceways for conductors feeding through or tapping off to the other apparatus unless designs are employed that provide adequate space for this purpose.

Informational Note: See 312.8 for switch and overcurrent-device enclosures.

(B) Wire-Bending Space in Enclosures. Minimum wire-bending space within the enclosures for motor controllers shall be in accordance with Table 430.10(B) where measured in a straight line from the end of the lug or wire connector (in the direction the wire leaves the terminal) to the wall or barrier.

Where alternate wire termination means are substituted for that supplied by the manufacturer of the controller, they shall be of a type identified by the manufacturer for use with the controller and shall not reduce the minimum wire-bending space.

Table 430.10(B) Minimum Wire-Bending Space at the Terminals of Enclosed Motor Controllers

| Size of Wire (AWG or kcmil) | Wires per Terminal[*] | | | |
| | 1 | | 2 | |
	mm	in.	mm	in.
10 and smaller	Not specified		—	—
8–6	38	1½	—	—
4–3	50	2	—	—
2	65	2½	—	—
1	75	3	—	—
1/0	125	5	125	5
2/0	150	6	150	6
3/0–4/0	175	7	175	7
250	200	8	200	8
300	250	10	250	10
350–500	300	12	300	12
600–700	350	14	400	16
750–900	450	18	475	19

[*]Where provision for three or more wires per terminal exists, the minimum wire-bending space shall be in accordance with the requirements of Article 312.

430.11 Protection Against Liquids. Suitable guards or enclosures shall be provided to protect exposed current-carrying parts of motors and the insulation of motor leads where installed directly under equipment, or in other locations where dripping or spraying oil, water, or other liquid is capable of occurring, unless the motor is designed for the existing conditions.

430.12 Motor Terminal Housings.

(A) Material. Where motors are provided with terminal housings, the housings shall be of metal and of substantial construction.

Exception: In other than hazardous (classified) locations, substantial, nonmetallic, noncombustible housings shall be permitted, provided an internal grounding means between the motor frame and the equipment grounding connection is incorporated within the housing.

(B) Dimensions and Space — Wire-to-Wire Connections. Where these terminal housings enclose wire-to-wire connections, they shall have minimum dimensions and usable volumes in accordance with Table 430.12(B).

Table 430.12(B) Terminal Housings — Wire-to-Wire Connections

Motors 275 mm (11 in.) in Diameter or Less

Horsepower	Cover Opening Minimum Dimension		Usable Volume Minimum	
	mm	in.	cm³	in.³
1 and smaller[a]	41	1⅝	170	10.5
1½, 2, and 3[b]	45	1¾	275	16.8
5 and 7½	50	2	365	22.4
10 and 15	65	2½	595	36.4

Motors Over 275 mm (11 in.) in Diameter — Alternating-Current Motors

Maximum Full Load Current for 3-Phase Motors with Maximum of 12 Leads (Amperes)	Terminal Box Cover Opening Minimum Dimension		Usable Volume Minimum		Typical Maximum Horsepower 3-Phase	
					230 Volt	460 Volt
	mm	in.	cm³	in.³		
45	65	2.5	595	36.4	15	30
70	84	3.3	1,265	77	25	50
110	100	4.0	2,295	140	40	75
160	125	5.0	4,135	252	60	125
250	150	6.0	7,380	450	100	200
400	175	7.0	13,775	840	150	300
600	200	8.0	25,255	1540	250	500

Direct-Current Motors

Maximum Full-Load Current for Motors with Maximum of 6 Leads (Amperes)	Terminal Box Minimum Dimensions		Usable Volume Minimum	
	mm	in.	cm³	in.³
68	65	2.5	425	26
105	84	3.3	900	55
165	100	4.0	1,640	100
240	125	5.0	2,950	180
375	150	6.0	5,410	330
600	175	7.0	9,840	600
900	200	8.0	18,040	1,100

Note: Auxiliary leads for such items as brakes, thermostats, space heaters, and exciting fields shall be permitted to be neglected if their current-carrying area does not exceed 25 percent of the current-carrying area of the machine power leads.

[a]For motors rated 1 hp and smaller, and with the terminal housing partially or wholly integral with the frame or end shield, the volume of the terminal housing shall not be less than 18.0 cm³ (1.1 in.³) per wire-to-wire connection. The minimum cover opening dimension is not specified.

[b]For motors rated 1½, 2, and 3 hp, and with the terminal housing partially or wholly integral with the frame or end shield, the volume of the terminal housing shall not be less than 23.0 cm³ (1.4 in.³) per wire-to-wire connection. The minimum cover opening dimension is not specified.

(C) Dimensions and Space — Fixed Terminal Connections. Where these terminal housings enclose rigidly mounted motor terminals, the terminal housing shall be of sufficient size to provide minimum terminal spacings and usable volumes in accordance with Table 430.12(C)(1) and Table 430.12(C)(2).

(D) Large Wire or Factory Connections. For motors with larger ratings, greater number of leads, or larger wire sizes, or where motors are installed as a part of factory-wired equipment, without additional connection being required at the motor terminal housing during equipment installation, the terminal housing shall be of ample size to make connections, but the foregoing provisions for the volumes of terminal housings shall not be considered applicable.

(E) Equipment Grounding Connections. A means for attachment of an equipment grounding conductor termination in accordance with 250.8 shall be provided at motor terminal housings for wire-to-wire connections or fixed terminal connections. The means for such connections shall be permitted to be located either inside or outside the motor terminal housing.

Exception: Where a motor is installed as a part of factory-wired equipment that is required to be grounded and without additional connection being required at the motor terminal housing during equipment installation, a separate means for motor grounding at the motor terminal housing shall not be required.

430.13 Bushing. Where wires pass through an opening in an enclosure, conduit box, or barrier, a bushing shall be used to protect the conductors from the edges of openings having sharp edges. The bushing shall have smooth, well-rounded surfaces where it may be in contact with the conductors. If used where oils, greases, or other contaminants may be present, the bushing shall be made of material not deleteriously affected.

Informational Note: For conductors exposed to deteriorating agents, see 310.10(G).

430.14 Location of Motors.

(A) Ventilation and Maintenance. Motors shall be located so that adequate ventilation is provided and so that maintenance,

Table 430.12(C)(1) Terminal Spacings — Fixed Terminals

Nominal Volts	Minimum Spacing			
	Between Line Terminals		Between Line Terminals and Other Uninsulated Metal Parts	
	mm	in.	mm	in.
250 or less	6	¼	6	¼
Over 250 – 1000	10	⅜	10	⅜

Table 430.12(C)(2) Usable Volumes — Fixed Terminals

Power-Supply Conductor Size (AWG)	Minimum Usable Volume per Power-Supply Conductor	
	cm³	in.³
14	16	1
12 and 10	20	1¼
8 and 6	37	2¼

such as lubrication of bearings and replacing of brushes, can be readily accomplished.

Exception: Ventilation shall not be required for submersible types of motors.

(B) Open Motors. Open motors that have commutators or collector rings shall be located or protected so that sparks cannot reach adjacent combustible material.

Exception: Installation of these motors on wooden floors or supports shall be permitted.

430.16 Exposure to Dust Accumulations. In locations where dust or flying material collects on or in motors in such quantities as to seriously interfere with the ventilation or cooling of motors and thereby cause dangerous temperatures, suitable types of enclosed motors that do not overheat under the prevailing conditions shall be used.

> Informational Note: Especially severe conditions may require the use of enclosed pipe-ventilated motors, or enclosure in separate dusttight rooms, properly ventilated from a source of clean air.

430.17 Highest Rated or Smallest Rated Motor. In determining compliance with 430.24, 430.53(B), and 430.53(C), the highest rated or smallest rated motor shall be based on the rated full-load current as selected from Table 430.247, Table 430.248, Table 430.249, and Table 430.250.

430.18 Nominal Voltage of Rectifier Systems. The nominal value of the ac voltage being rectified shall be used to determine the voltage of a rectifier derived system.

Exception: The nominal dc voltage of the rectifier shall be used if it exceeds the peak value of the ac voltage being rectified.

Part II. Motor Circuit Conductors

430.21 General. Part II specifies ampacities of conductors that are capable of carrying the motor current without overheating under the conditions specified.

The provisions of Part II shall not apply to motor circuits rated over 1000 volts, nominal.

> Informational Note: For over 1000 volts, nominal, see Part XI.

The provisions of Articles 250, 300, and 310 shall not apply to conductors that form an integral part of equipment, such as motors, motor controllers, motor control centers, or other factory-assembled control equipment.

> Informational Note: See 110.14(C) and 430.9(B) for equipment device terminal requirements.

430.22 Single Motor. Conductors that supply a single motor used in a continuous duty application shall have an ampacity of not less than 125 percent of the motor full-load current rating, as determined by 430.6(A)(1), or not less than specified in 430.22(A) through (G).

(A) Direct-Current Motor-Rectifier Supplied. For dc motors operating from a rectified power supply, the conductor ampacity on the input of the rectifier shall not be less than 125 percent of the rated input current to the rectifier. For dc motors operating from a rectified single-phase power supply, the conductors between the field wiring output terminals of the rectifier and the motor shall have an ampacity of not less

than the following percentages of the motor full-load current rating:

(1)   Where a rectifier bridge of the single-phase, half-wave type is used, 190 percent.
(2)   Where a rectifier bridge of the single-phase, full-wave type is used, 150 percent.

(B) Multispeed Motor. For a multispeed motor, the selection of branch-circuit conductors on the line side of the controller shall be based on the highest of the full-load current ratings shown on the motor nameplate. The ampacity of the branch-circuit conductors between the controller and the motor shall not be less than 125 percent of the current rating of the winding(s) that the conductors energize.

(C) Wye-Start, Delta-Run Motor. For a wye-start, delta-run connected motor, the ampacity of the branch-circuit conductors on the line side of the controller shall not be less than 125 percent of the motor full-load current as determined by 430.6(A)(1). The ampacity of the conductors between the controller and the motor shall not be less than 72 percent of the motor full-load current rating as determined by 430.6(A)(1).

> Informational Note: The individual motor circuit conductors of a wye-start, delta-run connected motor carry 58 percent of the rated load current. The multiplier of 72 percent is obtained by multiplying 58 percent by 1.25.

(D) Part-Winding Motor. For a part-winding connected motor, the ampacity of the branch-circuit conductors on the line side of the controller shall not be less than 125 percent of the motor full-load current as determined by 430.6(A)(1). The ampacity of the conductors between the controller and the motor shall not be less than 62.5 percent of the motor full-load current rating as determined by 430.6(A)(1).

> Informational Note: The multiplier of 62.5 percent is obtained by multiplying 50 percent by 1.25.

(E) Other Than Continuous Duty. Conductors for a motor used in a short-time, intermittent, periodic, or varying duty application shall have an ampacity of not less than the percentage of the motor nameplate current rating shown in Table 430.22(E), unless the authority having jurisdiction grants special permission for conductors of lower ampacity.

(F) Separate Terminal Enclosure. The conductors between a stationary motor rated 1 hp or less and the separate terminal enclosure permitted in 430.245(B) shall be permitted to be smaller than 14 AWG but not smaller than 18 AWG, provided they have an ampacity as specified in 430.22 .

(G) Conductors for Small Motors. Conductors for small motors shall not be smaller than 14 AWG unless otherwise permitted in 430.22(G)(1) or (G)(2).

(1) 18 AWG Copper. 18 AWG individual copper conductors installed in a cabinet or enclosure, copper conductors that are part of a jacketed multiconductor cable assembly, or copper conductors in a flexible cord shall be permitted, under either of the following sets of conditions:

(1)   The circuit supplies a motor with a full-load current rating, as determined by 430.6(A)(1), of greater than 3.5 amperes, and less than or equal to 5 amperes, and all the following conditions are met:

   a.   The circuit is protected in accordance with 430.52.

b. The circuit is provided with maximum Class 10 or Class 10A overload protection in accordance with 430.32.

c. Overcurrent protection is provided in accordance with 240.4(D)(1)(2).

(2) The circuit supplies a motor with a full-load current rating, as determined by 430.6(A)(1), of 3.5 amperes or less, and all the following conditions are met:

a. The circuit is protected in accordance with 430.52.

b. The circuit is provided with maximum Class 20 overload protection in accordance with 430.32.

c. Overcurrent protection is provided in accordance with 240.4(D)(1)(2).

(2) 16 AWG Copper. 16 AWG individual copper conductors installed in a cabinet or enclosure, copper conductors that are part of a jacketed multiconductor cable assembly, or copper conductors in a flexible cord shall be permitted under either of the following sets of conditions:

(1) The circuit supplies a motor with a full-load current rating, as determined by 430.6(A)(1), of greater than 5.5 amperes, and less than or equal to 8 amperes, and all the following conditions are met:

a. The circuit is protected in accordance with 430.52.

b. The circuit is provided with maximum Class 10 or Class 10A overload protection in accordance with 430.32.

c. Overcurrent protection is provided in accordance with 240.4(D)(2)(2).

Table 430.22(E) Duty-Cycle Service

Classification of Service	Nameplate Current Rating Percentages			
	5-Minute Rated Motor	15-Minute Rated Motor	30- & 60-Minute Rated Motor	Continuous Rated Motor
Short-time duty operating valves, raising or lowering rolls, etc.	110	120	150	—
Intermittent duty freight and passenger elevators, tool heads, pumps, drawbridges, turntables, etc. (for arc welders, see 630.11)	85	85	90	140
Periodic duty rolls, ore- and coal-handling machines, etc.	85	90	95	140
Varying duty	110	120	150	200

Note: Any motor application shall be considered as continuous duty unless the nature of the apparatus it drives is such that the motor will not operate continuously with load under any condition of use.

(2) The circuit supplies a motor with a full-load current rating, as determined by 430.6(A)(1), of 5.5 amperes or less, and all the following conditions are met:

a. The circuit is protected in accordance with 430.52.

b. The circuit is provided with maximum Class 20 overload protection in accordance with 430.32.

c. Overcurrent protection is provided in accordance with 240.4(D)(2)(2).

430.23 Wound-Rotor Secondary.

(A) Continuous Duty. For continuous duty, the conductors connecting the secondary of a wound-rotor ac motor to its controller shall have an ampacity not less than 125 percent of the full-load secondary current of the motor.

(B) Other Than Continuous Duty. For other than continuous duty, these conductors shall have an ampacity, in percent of full-load secondary current, not less than that specified in Table 430.22(E).

(C) Resistor Separate from Controller. Where the secondary resistor is separate from the controller, the ampacity of the conductors between controller and resistor shall not be less than that shown in Table 430.23(C).

430.24 Several Motors or a Motor(s) and Other Load(s). Conductors supplying several motors, or a motor(s) and other load(s), shall have an ampacity not less than the sum of each of the following:

(1) 125 percent of the full-load current rating of the highest rated motor, as determined by 430.6(A)

(2) Sum of the full-load current ratings of all the other motors in the group, as determined by 430.6(A)

(3) 100 percent of the noncontinuous non-motor load

(4) 125 percent of the continuous non-motor load.

Informational Note: See Informative Annex D, Example No. D8.

Exception No. 1: Where one or more of the motors of the group are used for short time, intermittent, periodic, or varying duty, the ampere rating of such motors to be used in the summation shall be determined in accordance with 430.22(E). For the highest rated motor, the greater of either the ampere rating from 430.22(E) or the largest continuous duty motor full-load current multiplied by 1.25 shall be used in the summation.

Exception No. 2: The ampacity of conductors supplying motor-operated fixed electric space-heating equipment shall comply with 424.3(B).

Exception No. 3: Where the circuitry is interlocked so as to prevent simultaneous operation of selected motors or other loads, the conductor ampacity shall be permitted to be based on the summation of the currents of the motors and other loads to be operated simultaneously that results in the highest total current.

430.25 Multimotor and Combination-Load Equipment. The ampacity of the conductors supplying multimotor and combination-load equipment shall not be less than the minimum circuit ampacity marked on the equipment in accordance with 430.7(D). Where the equipment is not factory-wired and the individual nameplates are visible in accordance with 430.7(D)(2), the conductor ampacity shall be determined in accordance with 430.24.

430.26 Feeder Demand Factor. Where reduced heating of the conductors results from motors operating on duty-cycle, intermittently, or from all motors not operating at one time, the authority having jurisdiction may grant permission for feeder conductors to have an ampacity less than specified in 430.24, provided the conductors have sufficient ampacity for the maximum load determined in accordance with the sizes and number of motors supplied and the character of their loads and duties.

Informational Note: Demand factors determined in the design of new facilities can often be validated against actual historical experience from similar installations. Refer to ANSI/IEEE Std. 141, *IEEE Recommended Practice for Electric Power Distribution for Industrial Plants,* and ANSI/IEEE Std. 241, *Recommended Practice for Electric Power Systems in Commercial Buildings,* for information on the calculation of loads and demand factor.

430.27 Capacitors with Motors. Where capacitors are installed in motor circuits, conductors shall comply with 460.8 and 460.9.

430.28 Feeder Taps. Feeder tap conductors shall have an ampacity not less than that required by Part II, shall terminate in a branch-circuit protective device, and, in addition, shall meet one of the following requirements:

(1) Be enclosed either by an enclosed controller or by a raceway, be not more than 3.0 m (10 ft) in length, and, for field installation, be protected by an overcurrent device on the line side of the tap conductor, the rating or setting of which shall not exceed 1000 percent of the tap conductor ampacity

(2) Have an ampacity of at least one-third that of the feeder conductors, be suitably protected from physical damage or enclosed in a raceway, and be not more than 7.5 m (25 ft) in length

(3) Have an ampacity not less than the feeder conductors

Exception: Feeder taps over 7.5 m (25 ft) long. In high-bay manufacturing buildings [over 11 m (35 ft) high at walls], where conditions of maintenance and supervision ensure that only qualified persons service the systems, conductors tapped to a feeder shall be permitted to be not over 7.5 m (25 ft) long horizontally and not over 30.0 m (100 ft) in total length where all of the following conditions are met:

(1) The ampacity of the tap conductors is not less than one-third that of the feeder conductors.

(2) The tap conductors terminate with a single circuit breaker or a single set of fuses complying with (1) Part IV, where the load-side conductors are a branch circuit, or (2) Part V, where the load-side conductors are a feeder.

(3) The tap conductors are suitably protected from physical damage and are installed in raceways.

(4) The tap conductors are continuous from end-to-end and contain no splices.

(5) The tap conductors shall be 6 AWG copper or 4 AWG aluminum or larger.

(6) The tap conductors shall not penetrate walls, floors, or ceilings.

(7) The tap shall not be made less than 9.0 m (30 ft) from the floor.

430.29 Constant Voltage Direct-Current Motors — Power Resistors. Conductors connecting the motor controller to separately mounted power accelerating and dynamic braking resistors in the armature circuit shall have an ampacity not less than the value calculated from Table 430.29 using motor full-load current. If an armature shunt resistor is used, the power accelerating resistor conductor ampacity shall be calculated using the total of motor full-load current and armature shunt resistor current.

Armature shunt resistor conductors shall have an ampacity of not less than that calculated from Table 430.29 using rated shunt resistor current as full-load current.

Part III. Motor and Branch-Circuit Overload Protection

430.31 General. Part III specifies overload devices intended to protect motors, motor-control apparatus, and motor branch-circuit conductors against excessive heating due to motor overloads and failure to start.

Informational Note No. 1: See Informative Annex D, Example No. D8.

Informational Note No. 2: See the definition of *Overload* in Article 100.

These provisions shall not require overload protection where a power loss would cause a hazard, such as in the case of fire pumps.

Informational Note: For protection of fire pump supply conductors, see 695.7.

The provisions of Part III shall not apply to motor circuits rated over 1000 volts, nominal.

Informational Note: For over 1000 volts, nominal, see Part XI.

430.32 Continuous-Duty Motors.

(A) More Than 1 Horsepower. Each motor used in a continuous duty application and rated more than 1 hp shall be protected against overload by one of the means in 430.32(A)(1) through (A)(4).

(1) Separate Overload Device. A separate overload device that is responsive to motor current. This device shall be selec-

Table 430.23(C) Secondary Conductor

Resistor Duty Classification	Ampacity of Conductor in Percent of Full-Load Secondary Current
Light starting duty	35
Heavy starting duty	45
Extra-heavy starting duty	55
Light intermittent duty	65
Medium intermittent duty	75
Heavy intermittent duty	85
Continuous duty	110

Table 430.29 Conductor Rating Factors for Power Resistors

Time in Seconds		Ampacity of Conductor in Percent of Full-Load Current
On	Off	
5	75	35
10	70	45
15	75	55
15	45	65
15	30	75
15	15	85
Continuous Duty		110

ted to trip or shall be rated at no more than the following percent of the motor nameplate full-load current rating:

Motors with a marked service factor 1.15 or greater	125%
Motors with a marked temperature rise 40°C or less	125%
All other motors	115%

Modification of this value shall be permitted as provided in 430.32(C). For a multispeed motor, each winding connection shall be considered separately.

Where a separate motor overload device is connected so that it does not carry the total current designated on the motor nameplate, such as for wye-delta starting, the proper percentage of nameplate current applying to the selection or setting of the overload device shall be clearly designated on the equipment, or the manufacturer's selection table shall take this into account.

Informational Note: Where power factor correction capacitors are installed on the load side of the motor overload device, see 460.9.

(2) Thermal Protector. A thermal protector integral with the motor, approved for use with the motor it protects on the basis that it will prevent dangerous overheating of the motor due to overload and failure to start. The ultimate trip current of a thermally protected motor shall not exceed the following percentage of motor full-load current given in Table 430.248, Table 430.249, and Table 430.250:

Motor full-load current 9 amperes or less	170%
Motor full-load current from 9.1 to, and including, 20 amperes	156%
Motor full-load current greater than 20 amperes	140%

If the motor current-interrupting device is separate from the motor and its control circuit is operated by a protective device integral with the motor, it shall be arranged so that the opening of the control circuit will result in interruption of current to the motor.

(3) Integral with Motor. A protective device integral with a motor that will protect the motor against damage due to failure to start shall be permitted if the motor is part of an approved assembly that does not normally subject the motor to overloads.

(4) Larger Than 1500 Horsepower. For motors larger than 1500 hp, a protective device having embedded temperature detectors that cause current to the motor to be interrupted when the motor attains a temperature rise greater than marked on the nameplate in an ambient temperature of 40°C.

(B) One Horsepower or Less, Automatically Started. Any motor of 1 hp or less that is started automatically shall be protected against overload by one of the following means.

(1) Separate Overload Device. By a separate overload device following the requirements of 430.32(A)(1).

For a multispeed motor, each winding connection shall be considered separately. Modification of this value shall be permitted as provided in 430.32(C).

(2) Thermal Protector. A thermal protector integral with the motor, approved for use with the motor that it protects on the basis that it will prevent dangerous overheating of the motor due to overload and failure to start. Where the motor current-interrupting device is separate from the motor and its control circuit is operated by a protective device integral with the motor, it shall be arranged so that the opening of the control circuit results in interruption of current to the motor.

(3) Integral with Motor. A protective device integral with a motor that protects the motor against damage due to failure to start shall be permitted (1) if the motor is part of an approved assembly that does not subject the motor to overloads, or (2) if the assembly is also equipped with other safety controls (such as the safety combustion controls on a domestic oil burner) that protect the motor against damage due to failure to start. Where the assembly has safety controls that protect the motor, it shall be so indicated on the nameplate of the assembly where it will be visible after installation.

(4) Impedance-Protected. If the impedance of the motor windings is sufficient to prevent overheating due to failure to start, the motor shall be permitted to be protected as specified in 430.32(D)(2)(a) for manually started motors if the motor is part of an approved assembly in which the motor will limit itself so that it will not be dangerously overheated.

Informational Note: Many ac motors of less than 1/20 hp, such as clock motors, series motors, and so forth, and also some larger motors such as torque motors, come within this classification. It does not include split-phase motors having automatic switches that disconnect the starting windings.

(C) Selection of Overload Device. Where the sensing element or setting or sizing of the overload device selected in accordance with 430.32(A)(1) and 430.32(B)(1) is not sufficient to start the motor or to carry the load, higher size sensing elements or incremental settings or sizing shall be permitted to be used, provided the trip current of the overload device does not exceed the following percentage of motor nameplate full-load current rating:

Motors with marked service factor 1.15 or greater	140%
Motors with a marked temperature rise 40°C or less	140%
All other motors	130%

If not shunted during the starting period of the motor as provided in 430.35, the overload device shall have sufficient time delay to permit the motor to start and accelerate its load.

Informational Note: A Class 20 overload relay will provide a longer motor acceleration time than a Class 10 or Class 10A overload relay. A Class 30 overload relay will provide a longer motor acceleration time than a Class 20 overload relay. Use of a higher class overload relay may preclude the need for selection of a higher trip current.

(D) One Horsepower or Less, Nonautomatically Started.

(1) Permanently Installed. Overload protection shall be in accordance with 430.32(B).

(2) Not Permanently Installed.

(a) *Within Sight from Controller.* Overload protection shall be permitted to be furnished by the branch-circuit short-circuit and ground-fault protective device; such device, however, shall not be larger than that specified in Part IV of Article 430.

Exception: Any such motor shall be permitted on a nominal 120-volt branch circuit protected at not over 20 amperes.

(b) *Not Within Sight from Controller.* Overload protection shall be in accordance with 430.32(B).

(E) Wound-Rotor Secondaries. The secondary circuits of wound-rotor ac motors, including conductors, controllers, resistors, and so forth, shall be permitted to be protected against overload by the motor-overload device.

430.33 Intermittent and Similar Duty. A motor used for a condition of service that is inherently short-time, intermittent, periodic, or varying duty, as illustrated by Table 430.22(E), shall be permitted to be protected against overload by the branch-circuit short-circuit and ground-fault protective device, provided the protective device rating or setting does not exceed that specified in Table 430.52.

Any motor application shall be considered to be for continuous duty unless the nature of the apparatus it drives is such that the motor cannot operate continuously with load under any condition of use.

430.35 Shunting During Starting Period.

(A) Nonautomatically Started. For a nonautomatically started motor, the overload protection shall be permitted to be shunted or cut out of the circuit during the starting period of the motor if the device by which the overload protection is shunted or cut out cannot be left in the starting position and if fuses or inverse time circuit breakers rated or set at not over 400 percent of the full-load current of the motor are located in the circuit so as to be operative during the starting period of the motor.

(B) Automatically Started. The motor overload protection shall not be shunted or cut out during the starting period if the motor is automatically started.

Exception: The motor overload protection shall be permitted to be shunted or cut out during the starting period on an automatically started motor where the following apply:

(a) The motor starting period exceeds the time delay of available motor overload protective devices, and

(b) Listed means are provided to perform the following:

(1) Sense motor rotation and automatically prevent the shunting or cutout in the event that the motor fails to start, and

(2) Limit the time of overload protection shunting or cutout to less than the locked rotor time rating of the protected motor, and

(3) Provide for shutdown and manual restart if motor running condition is not reached.

430.36 Fuses — In Which Conductor. Where fuses are used for motor overload protection, a fuse shall be inserted in each ungrounded conductor and also in the grounded conductor if the supply system is 3-wire, 3-phase ac with one conductor grounded.

430.37 Devices Other Than Fuses — In Which Conductor. Where devices other than fuses are used for motor overload protection, Table 430.37 shall govern the minimum allowable number and location of overload units such as trip coils or relays.

430.38 Number of Conductors Opened by Overload Device. Motor overload devices, other than fuses or thermal protectors, shall simultaneously open a sufficient number of ungrounded conductors to interrupt current flow to the motor.

Table 430.37 Overload Units

Kind of Motor	Supply System	Number and Location of Overload Units, Such as Trip Coils or Relays
1-phase ac or dc	2-wire, 1-phase ac or dc ungrounded	1 in either conductor
1-phase ac or dc	2-wire, 1-phase ac or dc, one conductor grounded	1 in ungrounded conductor
1-phase ac or dc	3-wire, 1-phase ac or dc, grounded neutral conductor	1 in either ungrounded conductor
1-phase ac	Any 3-phase	1 in ungrounded conductor
2-phase ac	3-wire, 2-phase ac, ungrounded	2, one in each phase
2-phase ac	3-wire, 2-phase ac, one conductor grounded	2 in ungrounded conductors
2-phase ac	4-wire, 2-phase ac, grounded or ungrounded	2, one for each phase in ungrounded conductors
2-phase ac	Grounded neutral or 5-wire, 2-phase ac, ungrounded	2, one for each phase in any ungrounded phase wire
3-phase ac	Any 3-phase	3, one in each phase*

*Exception: An overload unit in each phase shall not be required where overload protection is provided by other approved means.

430.39 Motor Controller as Overload Protection. A motor controller shall also be permitted to serve as an overload device if the number of overload units complies with Table 430.37 and if these units are operative in both the starting and running position in the case of a dc motor, and in the running position in the case of an ac motor.

430.40 Overload Relays. Overload relays and other devices for motor overload protection that are not capable of opening short circuits or ground faults shall be protected by fuses or circuit breakers with ratings or settings in accordance with 430.52 or by a motor short-circuit protector in accordance with 430.52.

Exception: Where approved for group installation and marked to indicate the maximum size of fuse or inverse time circuit breaker by which they must be protected, the overload devices shall be protected in accordance with this marking.

430.42 Motors on General-Purpose Branch Circuits. Overload protection for motors used on general-purpose branch circuits as permitted in Article 210 shall be provided as specified in 430.42(A), (B), (C), or (D).

(A) Not over 1 Horsepower. One or more motors without individual overload protection shall be permitted to be connected to a general-purpose branch circuit only where the installation complies with the limiting conditions specified in 430.32(B) and 430.32(D) and 430.53(A)(1) and (A)(2).

(B) Over 1 Horsepower. Motors of ratings larger than specified in 430.53(A) shall be permitted to be connected to general-purpose branch circuits only where each motor is protected by overload protection selected to protect the motor as specified in 430.32. Both the controller and the motor overload device shall be approved for group installation with the short-circuit and ground-fault protective device selected in accordance with 430.53.

(C) Cord-and Plug-Connected. Where a motor is connected to a branch circuit by means of an attachment plug and a receptacle or a cord connector, and individual overload protection is omitted as provided in 430.42(A), the rating of the attachment plug and receptacle or cord connector shall not exceed 15 amperes at 125 volts or 250 volts. Where individual overload protection is required as provided in 430.42(B) for a motor or motor-operated appliance that is attached to the branch circuit through an attachment plug and a receptacle or a cord connector, the overload device shall be an integral part of the motor or of the appliance. The rating of the attachment plug and receptacle or the cord connector shall determine the rating of the circuit to which the motor may be connected, as provided in 210.21(B).

(D) Time Delay. The branch-circuit short-circuit and ground-fault protective device protecting a circuit to which a motor or motor-operated appliance is connected shall have sufficient time delay to permit the motor to start and accelerate its load.

430.43 Automatic Restarting. A motor overload device that can restart a motor automatically after overload tripping shall not be installed unless approved for use with the motor it protects. A motor overload device that can restart a motor automatically after overload tripping shall not be installed if automatic restarting of the motor can result in injury to persons.

430.44 Orderly Shutdown. If immediate automatic shutdown of a motor by a motor overload protective device(s) would introduce additional or increased hazard(s) to a person(s) and continued motor operation is necessary for safe shutdown of equipment or process, a motor overload sensing device(s) complying with the provisions of Part III of this article shall be permitted to be connected to a supervised alarm instead of causing immediate interruption of the motor circuit, so that corrective action or an orderly shutdown can be initiated.

Part IV. Motor Branch-Circuit Short-Circuit and Ground-Fault Protection

430.51 General. Part IV specifies devices intended to protect the motor branch-circuit conductors, the motor control apparatus, and the motors against overcurrent due to short circuits or ground faults. These rules add to or amend the provisions of Article 240. The devices specified in Part IV do not include the types of devices required by 210.8, 230.95, and 590.6.

Informational Note: See Informative Annex D, Example D8.

The provisions of Part IV shall not apply to motor circuits rated over 1000 volts, nominal.

Informational Note: For over 1000 volts, nominal, see Part XI.

430.52 Rating or Setting for Individual Motor Circuit.

(A) General. The motor branch-circuit short-circuit and ground-fault protective device shall comply with 430.52(B) and either 430.52(C) or (D), as applicable.

(B) All Motors. The motor branch-circuit short-circuit and ground-fault protective device shall be capable of carrying the starting current of the motor.

(C) Rating or Setting.

(1) In Accordance with Table 430.52. A protective device that has a rating or setting not exceeding the value calculated according to the values given in Table 430.52 shall be used.

Exception No. 1: Where the values for branch-circuit short-circuit and ground-fault protective devices determined by Table 430.52 do not correspond to the standard sizes or ratings of fuses, nonadjustable circuit breakers, thermal protective devices, or possible settings of adjustable circuit breakers, a higher size, rating, or possible setting that does not exceed the next higher standard ampere rating shall be permitted.

Exception No. 2: Where the rating specified in Table 430.52, or the rating modified by Exception No. 1, is not sufficient for the starting current of the motor:

(a)   The rating of a nontime-delay fuse not exceeding 600 amperes or a time-delay Class CC fuse shall be permitted to be increased but shall in no case exceed 400 percent of the full-load current.

(b)   The rating of a time-delay (dual-element) fuse shall be permitted to be increased but shall in no case exceed 225 percent of the full-load current.

(c)   The rating of an inverse time circuit breaker shall be permitted to be increased but shall in no case exceed 400 percent for full-load currents of 100 amperes or less or 300 percent for full-load currents greater than 100 amperes.

(d)   The rating of a fuse of 601–6000 ampere classification shall be permitted to be increased but shall in no case exceed 300 percent of the full-load current.

Informational Note: See Informative Annex D, Example D8, and Figure 430.1.

(2) Overload Relay Table. Where maximum branch-circuit short-circuit and ground-fault protective device ratings are shown in the manufacturer's overload relay table for use with a motor controller or are otherwise marked on the equipment, they shall not be exceeded even if higher values are allowed as shown above.

(3) Instantaneous Trip Circuit Breaker. An instantaneous trip circuit breaker shall be used only if adjustable and if part of a listed combination motor controller having coordinated motor overload and short-circuit and ground-fault protection in each conductor, and the setting is adjusted to no more than the value specified in Table 430.52.

Informational Note No. 1: Instantaneous trip circuit breakers are also known as motor-circuit protectors (MCPs).

Informational Note No. 2: For the purpose of this article, instantaneous trip circuit breakers may include a damping means to accommodate a transient motor inrush current without nuisance tripping of the circuit breaker.

Exception No. 1: Where the setting specified in Table 430.52 is not sufficient for the starting current of the motor, the setting of an instantaneous trip circuit breaker shall be permitted to be increased but shall in no case exceed 1300 percent of the motor full-load current for other

Table 430.52 Maximum Rating or Setting of Motor Branch-Circuit Short-Circuit and Ground-Fault Protective Devices

Type of Motor	Percentage of Full-Load Current			
	Nontime Delay Fuse[1]	Dual Element (Time-Delay) Fuse[1]	Instantaneous Trip Breaker	Inverse Time Breaker[2]
Single-phase motors	300	175	800	250
AC polyphase motors other than wound-rotor	300	175	800	250
Squirrel cage — other than Design B energy-efficient	300	175	800	250
Design B energy-efficient	300	175	1100	250
Synchronous[3]	300	175	800	250
Wound-rotor	150	150	800	150
DC (constant voltage)	150	150	250	150

Note: For certain exceptions to the values specified, see 430.54.

[1]The values in the Nontime Delay Fuse column apply to time-delay Class CC fuses.

[2]The values given in the last column also cover the ratings of nonadjustable inverse time types of circuit breakers that may be modified as in 430.52(C)(1), Exceptions No. 1 and No. 2.

[3]Synchronous motors of the low-torque, low-speed type (usually 450 rpm or lower), such as are used to drive reciprocating compressors, pumps, and so forth, that start unloaded, do not require a fuse rating or circuit-breaker setting in excess of 200 percent of full-load current.

than Design B energy-efficient motors and no more than 1700 percent of full-load motor current for Design B energy-efficient motors. Trip settings above 800 percent for other than Design B energy-efficient motors and above 1100 percent for Design B energy-efficient motors shall be permitted where the need has been demonstrated by engineering evaluation. In such cases, it shall not be necessary to first apply an instantaneous-trip circuit breaker at 800 percent or 1100 percent.

Informational Note: For additional information on the requirements for a motor to be classified "energy efficient," see NEMA Standards Publication No. MG1-1993, Revision, *Motors and Generators*, Part 12.59.

Exception No. 2: Where the motor full-load current is 8 amperes or less, the setting of the instantaneous-trip circuit breaker with a continuous current rating of 15 amperes or less in a listed combination motor controller that provides coordinated motor branch-circuit overload and short-circuit and ground-fault protection shall be permitted to be increased to the value marked on the controller.

(4) Multispeed Motor. For a multispeed motor, a single short-circuit and ground-fault protective device shall be permitted for two or more windings of the motor, provided the rating of the protective device does not exceed the above applicable percentage of the nameplate rating of the smallest winding protected.

Exception: For a multispeed motor, a single short-circuit and ground-fault protective device shall be permitted to be used and sized according to the full-load current of the highest current winding, where all of the following conditions are met:

(a) Each winding is equipped with individual overload protection sized according to its full-load current.

(b) The branch-circuit conductors supplying each winding are sized according to the full-load current of the highest full-load current winding.

(c) The controller for each winding has a horsepower rating not less than that required for the winding having the highest horsepower rating.

(5) Power Electronic Devices. Semiconductor fuses intended for the protection of electronic devices shall be permitted in lieu of devices listed in Table 430.52 for power electronic devices, associated electromechanical devices (such as bypass contactors and isolation contactors), and conductors in a solid-state motor controller system, provided that the marking for replacement fuses is provided adjacent to the fuses.

(6) Self-Protected Combination Controller. A listed self-protected combination controller shall be permitted in lieu of the devices specified in Table 430.52. Adjustable instantaneous-trip settings shall not exceed 1300 percent of full-load motor current for other than Design B energy-efficient motors and not more than 1700 percent of full-load motor current for Design B energy-efficient motors.

Informational Note: Proper application of self-protected combination controllers on 3-phase systems, other than solidly grounded wye, particularly on corner grounded delta systems, considers the self-protected combination controllers' individual pole-interrupting capability.

(7) Motor Short-Circuit Protector. A motor short-circuit protector shall be permitted in lieu of devices listed in Table 430.52 if the motor short-circuit protector is part of a listed combination motor controller having coordinated motor overload protection and short-circuit and ground-fault protection in each conductor and it will open the circuit at currents exceeding 1300 percent of motor full-load current for other than Design B energy-efficient motors and 1700 percent of motor full-load motor current for Design B energy-efficient motors.

Informational Note: A motor short-circuit protector, as used in this section, is a fused device and is not an instantaneous trip circuit breaker.

(D) Torque Motors. Torque motor branch circuits shall be protected at the motor nameplate current rating in accordance with 240.4(B).

430.53 Several Motors or Loads on One Branch Circuit. Two or more motors or one or more motors and other loads shall be permitted to be connected to the same branch circuit under conditions specified in 430.53(D) and in 430.53(A), (B), or (C). The branch-circuit protective device shall be fuses or inverse time circuit breakers.

(A) Not Over 1 Horsepower. Several motors, each not exceeding 1 hp in rating, shall be permitted on a nominal 120-volt branch circuit protected at not over 20 amperes or a branch circuit of 1000 volts, nominal, or less, protected at not over 15 amperes, if all of the following conditions are met:

(1) The full-load rating of each motor does not exceed 6 amperes.

(2) The rating of the branch-circuit short-circuit and ground-fault protective device marked on any of the controllers is not exceeded.

(3) Individual overload protection conforms to 430.32.

(B) If Smallest Rated Motor Protected. If the branch-circuit short-circuit and ground-fault protective device is selected not to exceed that allowed by 430.52 for the smallest rated motor, two or more motors or one or more motors and other load(s), with each motor having individual overload protection, shall be permitted to be connected to a branch circuit where it can be determined that the branch-circuit short-circuit and ground-fault protective device will not open under the most severe normal conditions of service that might be encountered.

(C) Other Group Installations. Two or more motors of any rating or one or more motors and other load(s), with each motor having individual overload protection, shall be permitted to be connected to one branch circuit where the motor controller(s) and overload device(s) are (1) installed as a listed factory assembly and the motor branch-circuit short-circuit and ground-fault protective device either is provided as part of the assembly or is specified by a marking on the assembly, or (2) the motor branch-circuit short-circuit and ground-fault protective device, the motor controller(s), and overload device(s) are field-installed as separate assemblies listed for such use and provided with manufacturers' instructions for use with each other, and (3) all of the following conditions are complied with:

(1) Each motor overload device is either (a) listed for group installation with a specified maximum rating of fuse, inverse time circuit breaker, or both, or (b) selected such that the ampere rating of the motor-branch short-circuit and ground-fault protective device does not exceed that permitted by 430.52 for that individual motor overload device and corresponding motor load.

(2) Each motor controller is either (a) listed for group installation with a specified maximum rating of fuse, circuit breaker, or both, or (b) selected such that the ampere

rating of the motor-branch short-circuit and ground-fault protective device does not exceed that permitted by 430.52 for that individual controller and corresponding motor load.

(3) Each circuit breaker is listed and is of the inverse time type.

(4) The branch circuit shall be protected by fuses or inverse time circuit breakers having a rating not exceeding that specified in 430.52 for the highest rated motor connected to the branch circuit plus an amount equal to the sum of the full-load current ratings of all other motors and the ratings of other loads connected to the circuit. Where this calculation results in a rating less than the ampacity of the branch-circuit conductors, it shall be permitted to increase the maximum rating of the fuses or circuit breaker to a value not exceeding that permitted by 240.4(B).

(5) The branch-circuit fuses or inverse time circuit breakers are not larger than allowed by 430.40 for the overload relay protecting the smallest rated motor of the group.

(6) Overcurrent protection for loads other than motor loads shall be in accordance with Parts I through VII of Article 240.

Informational Note: See 110.10 for circuit impedance and other characteristics.

(D) Single Motor Taps. For group installations described above, the conductors of any tap supplying a single motor shall not be required to have an individual branch-circuit short-circuit and ground-fault protective device, provided they comply with one of the following:

(1) No conductor to the motor shall have an ampacity less than that of the branch-circuit conductors.

(2) No conductor to the motor shall have an ampacity less than one-third that of the branch-circuit conductors, with a minimum in accordance with 430.22. The conductors from the point of the tap to the motor overload device shall be not more than 7.5 m (25 ft) long and be protected from physical damage by being enclosed in an approved raceway or by use of other approved means.

(3) Conductors from the point of the tap from the branch circuit to a listed manual motor controller additionally marked "Suitable for Tap Conductor Protection in Group Installations," or to a branch-circuit protective device, shall be permitted to have an ampacity not less than one-tenth the rating or setting of the branch-circuit short-circuit and ground-fault protective device. The conductors from the controller to the motor shall have an ampacity in accordance with 430.22. The conductors from the point of the tap to the controller(s) shall (1) be suitably protected from physical damage and enclosed either by an enclosed controller or by a raceway and be not more than 3 m (10 ft) long or (2) have an ampacity not less than that of the branch-circuit conductors.

(4) Conductors from the point of the tap from the branch circuit to a listed manual motor controller additionally marked "Suitable for Tap Conductor Protection in Group Installations," or to a branch-circuit protective device, shall be permitted to have an ampacity not less than one-third that of the branch-circuit conductors. The conductors from the controller to the motor shall have an ampacity in accordance with 430.22. The conductors from the point of the tap to the controller(s) shall (1) be suitably protected from physical damage and enclosed

either by an enclosed controller or by a raceway and be not more than 7.5 m (25 ft) long or (2) have an ampacity not less than that of the branch-circuit conductors.

430.54 Multimotor and Combination-Load Equipment. The rating of the branch-circuit short-circuit and ground-fault protective device for multimotor and combination-load equipment shall not exceed the rating marked on the equipment in accordance with 430.7(D).

430.55 Combined Overcurrent Protection. Motor branch-circuit short-circuit and ground-fault protection and motor overload protection shall be permitted to be combined in a single protective device where the rating or setting of the device provides the overload protection specified in 430.32.

430.56 Branch-Circuit Protective Devices — In Which Conductor. Branch-circuit protective devices shall comply with the provisions of 240.15.

430.57 Size of Fuseholder. Where fuses are used for motor branch-circuit short-circuit and ground-fault protection, the fuseholders shall not be of a smaller size than required to accommodate the fuses specified by Table 430.52.

Exception: Where fuses having time delay appropriate for the starting characteristics of the motor are used, it shall be permitted to use fuseholders sized to fit the fuses that are used.

430.58 Rating of Circuit Breaker. A circuit breaker for motor branch-circuit short-circuit and ground-fault protection shall have a current rating in accordance with 430.52 and 430.110.

Part V. Motor Feeder Short-Circuit and Ground-Fault Protection

430.61 General. Part V specifies protective devices intended to protect feeder conductors supplying motors against overcurrents due to short circuits or grounds.

Informational Note: See Informative Annex D, Example D8.

430.62 Rating or Setting — Motor Load.

(A) Specific Load. A feeder supplying a specific fixed motor load(s) and consisting of conductor sizes based on 430.24 shall be provided with a protective device having a rating or setting not greater than the largest rating or setting of the branch-circuit short-circuit and ground-fault protective device for any motor supplied by the feeder [based on the maximum permitted value for the specific type of a protective device in accordance with 430.52, or 440.22(A) for hermetic refrigerant motor-compressors], plus the sum of the full-load currents of the other motors of the group.

Where the same rating or setting of the branch-circuit short-circuit and ground-fault protective device is used on two or more of the branch circuits supplied by the feeder, one of the protective devices shall be considered the largest for the above calculations.

Exception No. 1: Where one or more instantaneous trip circuit breakers or motor short-circuit protectors are used for motor branch-circuit short-circuit and ground-fault protection as permitted in 430.52(C), the procedure provided above for determining the maximum rating of the feeder protective device shall apply with the following provision: For the purpose of the calculation, each instantaneous trip circuit breaker or motor short-circuit protector shall be assumed to have a rating not

exceeding the maximum percentage of motor full-load current permitted by Table 430.52 for the type of feeder protective device employed.

Exception No. 2: Where the feeder overcurrent protective device also provides overcurrent protection for a motor control center, the provisions of 430.94 shall apply.

Informational Note: See Informative Annex D, Example D8.

(B) Other Installations. Where feeder conductors have an ampacity greater than required by 430.24, the rating or setting of the feeder overcurrent protective device shall be permitted to be based on the ampacity of the feeder conductors.

430.63 Rating or Setting — Motor Load and Other Load(s). Where a feeder supplies a motor load and other load(s), the feeder protective device shall have a rating not less than that required for the sum of the other load(s) plus the following:

(1) For a single motor, the rating permitted by 430.52
(2) For a single hermetic refrigerant motor-compressor, the rating permitted by 440.22
(3) For two or more motors, the rating permitted by 430.62

Exception: Where the feeder overcurrent device provides the overcurrent protection for a motor control center, the provisions of 430.94 shall apply.

Part VI. Motor Control Circuits

430.71 General. Part VI contains modifications of the general requirements and applies to the particular conditions of motor control circuits.

430.72 Overcurrent Protection.

(A) General. A motor control circuit tapped from the load side of a motor branch-circuit short-circuit and ground-fault protective device(s) and functioning to control the motor(s) connected to that branch circuit shall be protected against overcurrent in accordance with 430.72. Such a tapped control circuit shall not be considered to be a branch circuit and shall be permitted to be protected by either a supplementary or branch-circuit overcurrent protective device(s). A motor control circuit other than such a tapped control circuit shall be protected against overcurrent in accordance with 725.43 or the notes to Table 11(A) and Table 11(B) in Chapter 9, as applicable.

(B) Conductor Protection. The overcurrent protection for conductors shall be provided as specified in 430.72(B)(1) or (B)(2).

Exception No. 1: Where the opening of the control circuit would create a hazard as, for example, the control circuit of a fire pump motor, and the like, conductors of control circuits shall require only short-circuit and ground-fault protection and shall be permitted to be protected by the motor branch-circuit short-circuit and ground-fault protective device(s).

Exception No. 2: Conductors supplied by the secondary side of a single-phase transformer having only a two-wire (single-voltage) secondary shall be permitted to be protected by overcurrent protection provided on the primary (supply) side of the transformer, provided this protection does not exceed the value determined by multiplying the appropriate maximum rating of the overcurrent device for the secondary conductor from Table 430.72(B) by the secondary-to-primary voltage ratio. Transformer secondary conductors (other than two-wire) shall not be considered to be protected by the primary overcurrent protection.

(1) Separate Overcurrent Protection. Where the motor branch-circuit short-circuit and ground-fault protective device does not provide protection in accordance with 430.72(B)(2), separate overcurrent protection shall be provided. The overcurrent protection shall not exceed the values specified in Column A of Table 430.72(B).

(2) Branch-Circuit Overcurrent Protective Device. Conductors shall be permitted to be protected by the motor branch-circuit short-circuit and ground-fault protective device and shall require only short-circuit and ground-fault protection. Where the conductors do not extend beyond the motor control equipment enclosure, the rating of the protective device(s) shall not exceed the value specified in Column B of Table 430.72(B). Where the conductors extend beyond the motor control equipment enclosure, the rating of the protective device(s) shall not exceed the value specified in Column C of Table 430.72(B).

(C) Control Circuit Transformer. Where a motor control circuit transformer is provided, the transformer shall be protected in accordance with 430.72(C)(1), (C)(2), (C)(3), (C)(4), or (C)(5).

Exception: Overcurrent protection shall be omitted where the opening of the control circuit would create a hazard as, for example, the control circuit of a fire pump motor and the like.

(1) Compliance with Article 725. Where the transformer supplies a Class 1 power-limited circuit, Class 2, or Class 3 remote-control circuit complying with the requirements of Article 725, protection shall comply with Article 725.

(2) Compliance with Article 450. Protection shall be permitted to be provided in accordance with 450.3.

(3) Less Than 50 Volt-Amperes. Control circuit transformers rated less than 50 volt-amperes (VA) and that are an integral part of the motor controller and located within the motor controller enclosure shall be permitted to be protected by primary overcurrent devices, impedance limiting means, or other inherent protective means.

(4) Primary Less Than 2 Amperes. Where the control circuit transformer rated primary current is less than 2 amperes, an overcurrent device rated or set at not more than 500 percent of the rated primary current shall be permitted in the primary circuit.

(5) Other Means. Protection shall be permitted to be provided by other approved means.

430.73 Protection of Conductors from Physical Damage. Where damage to a motor control circuit would constitute a hazard, all conductors of such a remote motor control circuit that are outside the control device itself shall be installed in a raceway or be otherwise protected from physical damage.

430.74 Electrical Arrangement of Control Circuits. Where one conductor of the motor control circuit is grounded, the motor control circuit shall be arranged so that a ground fault in the control circuit remote from the motor controller will (1) not start the motor and (2) not bypass manually operated shutdown devices or automatic safety shutdown devices.

430.75 Disconnection.

(A) General. Motor control circuits shall be arranged so that they will be disconnected from all sources of supply when the disconnecting means is in the open position. The disconnecting means shall be permitted to consist of two or more separate devices, one of which disconnects the motor and the controller from the source(s) of power supply for the motor, and the other(s), the motor control circuit(s) from its power supply. Where separate devices are used, they shall be located immediately adjacent to each other.

Exception No. 1: Where more than 12 motor control circuit conductors are required to be disconnected, the disconnecting means shall be permitted to be located other than immediately adjacent to each other where all of the following conditions are complied with:

(a) Access to energized parts is limited to qualified persons in accordance with Part XII of this article.

(b) A warning sign is permanently located on the outside of each equipment enclosure door or cover permitting access to the live parts in the motor control circuit(s), warning that motor control circuit disconnecting means are remotely located and specifying the location and identification of each disconnect. Where energized parts are not in an equipment enclosure as permitted by 430.232 and 430.233, an addi-

Table 430.72(B) Maximum Rating of Overcurrent Protective Device in Amperes

Control Circuit Conductor Size (AWG)	Column A Separate Protection Provided		Protection Provided by Motor Branch-Circuit Protective Device(s)			
			Column B Conductors Within Enclosure		Column C Conductors Extend Beyond Enclosure	
	Copper	Aluminum or Copper-Clad Aluminum	Copper	Aluminum or Copper-Clad Aluminum	Copper	Aluminum or Copper-Clad Aluminum
18	7	—	25	—	7	—
16	10	—	40	—	10	—
14	(Note 1)	—	100	—	45	—
12	(Note 1)	(Note 1)	120	100	60	45
10	(Note 1)	(Note 1)	160	140	90	75
Larger than 10	(Note 1)	(Note 1)	(Note 2)	(Note 2)	(Note 3)	(Note 3)

Notes:
1. Value specified in 310.15 as applicable.
2. 400 percent of value specified in Table 310.15(B)(17) for 60°C conductors.
3. 300 percent of value specified in Table 310.15(B)(16) for 60°C conductors.

tional warning sign(s) shall be located where visible to persons who may be working in the area of the energized parts.

Exception No. 2: The motor control circuit disconnecting means shall be permitted to be remote from the motor controller power supply disconnecting means where the opening of one or more motor control circuit disconnecting means is capable of resulting in potentially unsafe conditions for personnel or property and the conditions of items (a) and (b) of Exception No. 1 are complied with.

(B) Control Transformer in Controller Enclosure. Where a transformer or other device is used to obtain a reduced voltage for the motor control circuit and is located in the controller enclosure, such transformer or other device shall be connected to the load side of the disconnecting means for the motor control circuit.

Part VII. Motor Controllers

430.81 General. Part VII is intended to require suitable controllers for all motors.

(A) Stationary Motor of $\frac{1}{8}$ Horsepower or Less. For a stationary motor rated at $\frac{1}{8}$ hp or less that is normally left running and is constructed so that it cannot be damaged by overload or failure to start, such as clock motors and the like, the branch-circuit disconnecting means shall be permitted to serve as the controller.

(B) Portable Motor of $\frac{1}{3}$ Horsepower or Less. For a portable motor rated at $\frac{1}{3}$ hp or less, the controller shall be permitted to be an attachment plug and receptacle or cord connector.

430.82 Controller Design.

(A) Starting and Stopping. Each controller shall be capable of starting and stopping the motor it controls and shall be capable of interrupting the locked-rotor current of the motor.

(B) Autotransformer. An autotransformer starter shall provide an "off" position, a running position, and at least one starting position. It shall be designed so that it cannot rest in the starting position or in any position that will render the overload device in the circuit inoperative.

(C) Rheostats. Rheostats shall be in compliance with the following:

(1) Motor-starting rheostats shall be designed so that the contact arm cannot be left on intermediate segments. The point or plate on which the arm rests when in the starting position shall have no electrical connection with the resistor.
(2) Motor-starting rheostats for dc motors operated from a constant voltage supply shall be equipped with automatic devices that will interrupt the supply before the speed of the motor has fallen to less than one-third its normal rate.

430.83 Ratings. The controller shall have a rating as specified in 430.83(A), unless otherwise permitted in 430.83(B) or (C), or as specified in (D), under the conditions specified.

(A) General.

(1) Horsepower Ratings. Controllers, other than inverse time circuit breakers and molded case switches, shall have horsepower ratings at the application voltage not lower than the horsepower rating of the motor.

(2) Circuit Breaker. A branch-circuit inverse time circuit breaker rated in amperes shall be permitted as a controller for all motors. Where this circuit breaker is also used for overload protection, it shall conform to the appropriate provisions of this article governing overload protection.

(3) Molded Case Switch. A molded case switch rated in amperes shall be permitted as a controller for all motors.

(B) Small Motors. Devices as specified in 430.81(A) and (B) shall be permitted as a controller.

(C) Stationary Motors of 2 Horsepower or Less. For stationary motors rated at 2 hp or less and 300 volts or less, the controller shall be permitted to be either of the following:

(1) A general-use switch having an ampere rating not less than twice the full-load current rating of the motor
(2) On ac circuits, a general-use snap switch suitable only for use on ac (not general-use ac–dc snap switches) where the motor full-load current rating is not more than 80 percent of the ampere rating of the switch

(D) Torque Motors. For torque motors, the controller shall have a continuous-duty, full-load current rating not less than the nameplate current rating of the motor. For a motor controller rated in horsepower but not marked with the foregoing current rating, the equivalent current rating shall be determined from the horsepower rating by using Table 430.247, Table 430.248, Table 430.249, or Table 430.250.

(E) Voltage Rating. A controller with a straight voltage rating, for example, 240 volts or 480 volts, shall be permitted to be applied in a circuit in which the nominal voltage between any two conductors does not exceed the controller's voltage rating. A controller with a slash rating, for example, 120/240 volts or 480Y/277 volts, shall only be applied in a solidly grounded circuit in which the nominal voltage to ground from any conductor does not exceed the lower of the two values of the controller's voltage rating and the nominal voltage between any two conductors does not exceed the higher value of the controller's voltage rating.

430.84 Need Not Open All Conductors. The controller shall not be required to open all conductors to the motor.

Exception: Where the controller serves also as a disconnecting means, it shall open all ungrounded conductors to the motor as provided in 430.111.

430.85 In Grounded Conductors. One pole of the controller shall be permitted to be placed in a permanently grounded conductor, provided the controller is designed so that the pole in the grounded conductor cannot be opened without simultaneously opening all conductors of the circuit.

430.87 Number of Motors Served by Each Controller. Each motor shall be provided with an individual controller.

Exception No. 1: For motors rated 1000 volts or less, a single controller rated at not less than the equivalent horsepower, as determined in accordance with 430.110(C)(1), of all the motors in the group shall be permitted to serve the group under any of the following conditions:

(a) Where a number of motors drive several parts of a single machine or piece of apparatus, such as metal and woodworking machines, cranes, hoists, and similar apparatus

(b) Where a group of motors is under the protection of one overcurrent device as permitted in 430.53(A)

(c)   Where a group of motors is located in a single room within sight from the controller location

Exception No. 2: A branch-circuit disconnecting means serving as the controller as allowed in 430.81(A) shall be permitted to serve more than one motor.

430.88 Adjustable-Speed Motors.   Adjustable-speed motors that are controlled by means of field regulation shall be equipped and connected so that they cannot be started under a weakened field.

Exception: Starting under a weakened field shall be permitted where the motor is designed for such starting.

430.89 Speed Limitation.   Machines of the following types shall be provided with speed-limiting devices or other speed-limiting means:

(1)   Separately excited dc motors
(2)   Series motors
(3)   Motor-generators and converters that can be driven at excessive speed from the dc end, as by a reversal of current or decrease in load

Exception: Separate speed-limiting devices or means shall not be required under either of the following conditions:

(1)   Where the inherent characteristics of the machines, the system, or the load and the mechanical connection thereto are such as to safely limit the speed
(2)   Where the machine is always under the manual control of a qualified operator

430.90 Combination Fuseholder and Switch as Controller. The rating of a combination fuseholder and switch used as a motor controller shall be such that the fuseholder will accommodate the size of the fuse specified in Part III of this article for motor overload protection.

Exception: Where fuses having time delay appropriate for the starting characteristics of the motor are used, fuseholders of smaller size than specified in Part III of this article shall be permitted.

Part VIII.   Motor Control Centers

430.92 General.   Part VIII covers motor control centers installed for the control of motors, lighting, and power circuits.

430.94 Overcurrent Protection.   Motor control centers shall be provided with overcurrent protection in accordance with Parts I, II, and VIII of Article 240. The ampere rating or setting of the overcurrent protective device shall not exceed the rating of the common power bus. This protection shall be provided by (1) an overcurrent protective device located ahead of the motor control center or (2) a main overcurrent protective device located within the motor control center.

430.95 Service Equipment.   Where used as service equipment, each motor control center shall be provided with a single main disconnecting means to disconnect all ungrounded service conductors.

Exception: A second service disconnect shall be permitted to supply additional equipment.

Where a grounded conductor is provided, the motor control center shall be provided with a main bonding jumper, sized in accordance with 250.28(D), within one of the sections for connecting the grounded conductor, on its supply side, to the motor control center equipment ground bus.

Exception: High-impedance grounded neutral systems shall be permitted to be connected as provided in 250.36.

430.96 Grounding.   Multisection motor control centers shall be connected together with an equipment grounding conductor or an equivalent equipment grounding bus sized in accordance with Table 250.122. Equipment grounding conductors shall be connected to this equipment grounding bus or to a grounding termination point provided in a single-section motor control center.

430.97 Busbars and Conductors.

(A) Support and Arrangement.   Busbars shall be protected from physical damage and be held firmly in place. Other than for required interconnections and control wiring, only those conductors that are intended for termination in a vertical section shall be located in that section.

Exception: Conductors shall be permitted to travel horizontally through vertical sections where such conductors are isolated from the busbars by a barrier.

(B) Phase Arrangement.   The phase arrangement on 3-phase horizontal common power and vertical buses shall be A, B, C from front to back, top to bottom, or left to right, as viewed from the front of the motor control center. The B phase shall be that phase having the higher voltage to ground on 3-phase, 4-wire, delta-connected systems. Other busbar arrangements shall be permitted for additions to existing installations and shall be marked.

Exception: Rear-mounted units connected to a vertical bus that is common to front-mounted units shall be permitted to have a C, B, A phase arrangement where properly identified.

(C) Minimum Wire-Bending Space.   The minimum wire-bending space at the motor control center terminals and minimum gutter space shall be as required in Article 312(D).

(D) Spacings.   Spacings between motor control center bus terminals and other bare metal parts shall not be less than specified in Table 430.97(D).

(E) Barriers.   Barriers shall be placed in all service-entrance motor control centers to isolate service busbars and terminals from the remainder of the motor control center.

430.98 Marking.

(A) Motor Control Centers.   Motor control centers shall be marked according to 110.21, and the marking shall be plainly visible after installation. Marking shall also include common power bus current rating and motor control center short-circuit rating.

(B) Motor Control Units.   Motor control units in a motor control center shall comply with 430.8.

N **430.99 Available Fault Current.**   The available short circuit current at the motor control center and the date the short circuit current calculation was performed shall be documented and made available to those authorized to inspect the installation.

Table 430.97(D) Minimum Spacing Between Bare Metal Parts

Nominal Voltage	Opposite Polarity Where Mounted on the Same Surface		Opposite Polarity Where Held Free in Air		Live Parts to Ground	
	mm	in.	mm	in.	mm	in.
Not over 125 volts, nominal	19.1	¾	12.7	½	12.7	½
Not over 250 volts, nominal	31.8	1¼	19.1	¾	12.7	½
Not over 600 volts, nominal	50.8	2	25.4	1	25.4	1

Part IX. Disconnecting Means

430.101 General. Part IX is intended to require disconnecting means capable of disconnecting motors and controllers from the circuit.

430.102 Location.

(A) Controller. An individual disconnecting means shall be provided for each controller and shall disconnect the controller. The disconnecting means shall be located in sight from the controller location.

Exception No. 1: For motor circuits over 1000 volts, nominal, a controller disconnecting means lockable in accordance with 110.25 shall be permitted to be out of sight of the controller, provided that the controller is marked with a warning label giving the location of the disconnecting means.

Exception No. 2: A single disconnecting means shall be permitted for a group of coordinated controllers that drive several parts of a single machine or piece of apparatus. The disconnecting means shall be located in sight from the controllers, and both the disconnecting means and the controllers shall be located in sight from the machine or apparatus.

Exception No. 3: The disconnecting means shall not be required to be in sight from valve actuator motor (VAM) assemblies containing the controller where such a location introduces additional or increased hazards to persons or property and conditions (a) and (b) are met.
 (a) The valve actuator motor assembly is marked with a warning label giving the location of the disconnecting means.
 (b) The disconnecting means is lockable in accordance with 110.25.

(B) Motor. A disconnecting means shall be provided for a motor in accordance with (B)(1) or (B)(2).

(1) Separate Motor Disconnect. A disconnecting means for the motor shall be located in sight from the motor location and the driven machinery location.

(2) Controller Disconnect. The controller disconnecting means required in accordance with 430.102(A) shall be permitted to serve as the disconnecting means for the motor if it is in sight from the motor location and the driven machinery location.

Exception to (1) and (2): The disconnecting means for the motor shall not be required under either condition (a) or condition (b), which follow, provided that the controller disconnecting means required in 430.102(A) is lockable in accordance with 110.25.
 (a) Where such a location of the disconnecting means for the motor is impracticable or introduces additional or increased hazards to persons or property

Informational Note: Some examples of increased or additional hazards include, but are not limited to, motors rated in excess of 100 hp, multimotor equipment, submersible motors, motors associated with adjustable speed drives, and motors located in hazardous (classified) locations.

 (b) In industrial installations, with written safety procedures, where conditions of maintenance and supervision ensure that only qualified persons service the equipment

Informational Note: For information on lockout/tagout procedures, see *NFPA 70E-2015, Standard for Electrical Safety in the Workplace.*

430.103 Operation. The disconnecting means shall open all ungrounded supply conductors and shall be designed so that no pole can be operated independently. The disconnecting means shall be permitted in the same enclosure with the controller. The disconnecting means shall be designed so that it cannot be closed automatically.

Informational Note: See 430.113 for equipment receiving energy from more than one source.

430.104 To Be Indicating. The disconnecting means shall plainly indicate whether it is in the open (off) or closed (on) position.

430.105 Grounded Conductors. One pole of the disconnecting means shall be permitted to disconnect a permanently grounded conductor, provided the disconnecting means is designed so that the pole in the grounded conductor cannot be opened without simultaneously disconnecting all conductors of the circuit.

430.107 Readily Accessible. At least one of the disconnecting means shall be readily accessible.

430.108 Every Disconnecting Means. Every disconnecting means in the motor circuit between the point of attachment to the feeder or branch circuit and the point of connection to the motor shall comply with the requirements of 430.109 and 430.110.

430.109 Type. The disconnecting means shall be a type specified in 430.109(A), unless otherwise permitted in 430.109(B) through (G), under the conditions specified.

(A) General.

(1) Motor Circuit Switch. A listed motor-circuit switch rated in horsepower.

(2) Molded Case Circuit Breaker. A listed molded case circuit breaker.

(3) Molded Case Switch. A listed molded case switch.

(4) Instantaneous Trip Circuit Breaker. An instantaneous trip circuit breaker that is part of a listed combination motor controller.

(5) Self-Protected Combination Controller. Listed self-protected combination controller.

(6) Manual Motor Controller. Listed manual motor controllers additionally marked "Suitable as Motor Disconnect" shall be permitted as a disconnecting means where installed between the final motor branch-circuit short-circuit protective device and the motor. Listed manual motor controllers additionally marked "Suitable as Motor Disconnect" shall be permitted as disconnecting means on the line side of the fuses permitted in 430.52(C)(5). In this case, the fuses permitted in 430.52(C)(5) shall be considered supplementary fuses, and suitable branch-circuit short-circuit and ground-fault protective devices shall be installed on the line side of the manual motor controller additionally marked "Suitable as Motor Disconnect."

(7) System Isolation Equipment. System isolation equipment shall be listed for disconnection purposes. System isolation equipment shall be installed on the load side of the overcurrent protection and its disconnecting means. The disconnecting means shall be one of the types permitted by 430.109(A)(1) through (A)(3).

(B) Stationary Motors of ⅛ Horsepower or Less. For stationary motors of ⅛ hp or less, the branch-circuit overcurrent device shall be permitted to serve as the disconnecting means.

(C) Stationary Motors of 2 Horsepower or Less. For stationary motors rated at 2 hp or less and 300 volts or less, the disconnecting means shall be permitted to be one of the devices specified in (1), (2), or (3):

(1) A general-use switch having an ampere rating not less than twice the full load current rating of the motor
(2) On ac circuits, a general-use snap switch suitable only for use on ac (not general-use ac–dc snap switches) where the motor full-load current rating is not more than 80 percent of the ampere rating of the switch
(3) A listed manual motor controller having a horsepower rating not less than the rating of the motor and marked "Suitable as Motor Disconnect"

(D) Autotransformer-Type Controlled Motors. For motors of over 2 hp to and including 100 hp, the separate disconnecting means required for a motor with an autotransformer-type controller shall be permitted to be a general-use switch where all of the following provisions are met:

(1) The motor drives a generator that is provided with overload protection.
(2) The controller is capable of interrupting the locked-rotor current of the motors, is provided with a no voltage release, and is provided with running overload protection not exceeding 125 percent of the motor full-load current rating.
(3) Separate fuses or an inverse time circuit breaker rated or set at not more than 150 percent of the motor full-load current is provided in the motor branch circuit.

(E) Isolating Switches. For stationary motors rated at more than 40 hp dc or 100 hp ac, the disconnecting means shall be permitted to be a general-use or isolating switch where plainly marked "Do not operate under load."

(F) Cord-and-Plug-Connected Motors. For a cord-and-plug-connected motor, a horsepower-rated attachment plug and receptacle, flanged surface inlet and cord connector, or attachment plug and cord connector having ratings no less than the motor ratings shall be permitted to serve as the disconnecting means. Horsepower-rated attachment plugs, flanged surface inlets, receptacles, or cord connectors shall not be required for cord-and-plug-connected appliances in accordance with 422.33, room air conditioners in accordance with 440.63, or portable motors rated ⅓ hp or less.

(G) Torque Motors. For torque motors, the disconnecting means shall be permitted to be a general-use switch.

430.110 Ampere Rating and Interrupting Capacity.

(A) General. The disconnecting means for motor circuits rated 1000 volts, nominal, or less shall have an ampere rating not less than 115 percent of the full-load current rating of the motor.

Exception: A listed unfused motor-circuit switch having a horsepower rating not less than the motor horsepower shall be permitted to have an ampere rating less than 115 percent of the full-load current rating of the motor.

(B) For Torque Motors. Disconnecting means for a torque motor shall have an ampere rating of at least 115 percent of the motor nameplate current.

(C) For Combination Loads. Where two or more motors are used together or where one or more motors are used in combination with other loads, such as resistance heaters, and where the combined load may be simultaneous on a single disconnecting means, the ampere and horsepower ratings of the combined load shall be determined as follows.

(1) Horsepower Rating. The rating of the disconnecting means shall be determined from the sum of all currents, including resistance loads, at the full-load condition and also at the locked-rotor condition. The combined full-load current and the combined locked-rotor current so obtained shall be considered as a single motor for the purpose of this requirement as follows.

The full-load current equivalent to the horsepower rating of each motor shall be selected from Table 430.247, Table 430.248, Table 430.249, or Table 430.250. These full-load currents shall be added to the rating in amperes of other loads to obtain an equivalent full-load current for the combined load.

The locked-rotor current equivalent to the horsepower rating of each motor shall be selected from Table 430.251(A) or Table 430.251(B). The locked-rotor currents shall be added to the rating in amperes of other loads to obtain an equivalent locked-rotor current for the combined load. Where two or more motors or other loads cannot be started simultaneously, the largest sum of locked-rotor currents of a motor or group of motors that can be started simultaneously and the full-load currents of other concurrent loads shall be permitted to be used to determine the equivalent locked-rotor current for the simultaneous combined loads. In cases where different current ratings are obtained when applying these tables, the largest value obtained shall be used.

Exception: Where part of the concurrent load is resistance load, and where the disconnecting means is a switch rated in horsepower and amperes, the switch used shall be permitted to have a horsepower rating that is not less than the combined load of the motor(s), if the ampere rating of the switch is not less than the locked-rotor current of the motor(s) plus the resistance load.

(2) Ampere Rating. The ampere rating of the disconnecting means shall not be less than 115 percent of the sum of all currents at the full-load condition determined in accordance with 430.110(C)(1).

Exception: A listed nonfused motor-circuit switch having a horsepower rating equal to or greater than the equivalent horsepower of the combined loads, determined in accordance with 430.110(C)(1), shall be permitted to have an ampere rating less than 115 percent of the sum of all currents at the full-load condition.

(3) Small Motors. For small motors not covered by Table 430.247, Table 430.248, Table 430.249, or Table 430.250, the locked-rotor current shall be assumed to be six times the full-load current.

430.111 Switch or Circuit Breaker as Both Controller and Disconnecting Means. A switch or circuit breaker shall be permitted to be used as both the controller and disconnecting means if it complies with 430.111(A) and is one of the types specified in 430.111(B).

(A) General. The switch or circuit breaker complies with the requirements for controllers specified in 430.83, opens all ungrounded conductors to the motor, and is protected by an overcurrent device in each ungrounded conductor (which shall be permitted to be the branch-circuit fuses). The overcurrent device protecting the controller shall be permitted to be part of the controller assembly or shall be permitted to be separate. An autotransformer-type controller shall be provided with a separate disconnecting means.

(B) Type. The device shall be one of the types specified in 430.111(B)(1), (B)(2), or (B)(3).

(1) Air-Break Switch. An air-break switch, operable directly by applying the hand to a lever or handle.

(2) Inverse Time Circuit Breaker. An inverse time circuit breaker operable directly by applying the hand to a lever or handle. The circuit breaker shall be permitted to be both power and manually operable.

(3) Oil Switch. An oil switch used on a circuit whose rating does not exceed 1000 volts or 100 amperes, or by special permission on a circuit exceeding this capacity where under expert supervision. The oil switch shall be permitted to be both power and manually operable.

430.112 Motors Served by Single Disconnecting Means. Each motor shall be provided with an individual disconnecting means.

Exception: A single disconnecting means shall be permitted to serve a group of motors under any one of the conditions of (a), (b), and (c). The single disconnecting means shall be rated in accordance with 430.110(C).

(a) Where a number of motors drive several parts of a single machine or piece of apparatus, such as metal- and woodworking machines, cranes, and hoists.

(b) Where a group of motors is under the protection of one set of branch-circuit protective devices as permitted by 430.53(A).

(c) Where a group of motors is in a single room within sight from the location of the disconnecting means.

430.113 Energy from More Than One Source. Motor and motor-operated equipment receiving electric energy from more than one source shall be provided with disconnecting means from each source of electric energy immediately adjacent to the equipment served. Each source shall be permitted to have a separate disconnecting means. Where multiple disconnecting means are provided, a permanent warning sign shall be provided on or adjacent to each disconnecting means.

Exception No. 1: Where a motor receives electric energy from more than one source, the disconnecting means for the main power supply to the motor shall not be required to be immediately adjacent to the motor, provided that the controller disconnecting means is lockable in accordance with 110.25.

Exception No. 2: A separate disconnecting means shall not be required for a Class 2 remote-control circuit conforming with Article 725, rated not more than 30 volts, and isolated and ungrounded.

Part X. Adjustable-Speed Drive Systems

430.120 General. The installation provisions of Part I through Part IX are applicable unless modified or supplemented by Part X.

430.122 Conductors — Minimum Size and Ampacity.

(A) Branch/Feeder Circuit Conductors. Circuit conductors supplying power conversion equipment included as part of an adjustable-speed drive system shall have an ampacity not less than 125 percent of the rated input current to the power conversion equipment.

> Informational Note: Power conversion equipment can have multiple power ratings and corresponding input currents.

(B) Bypass Device. For an adjustable-speed drive system that utilizes a bypass device, the conductor ampacity shall not be less than required by 430.6. The ampacity of circuit conductors supplying power conversion equipment included as part of an adjustable-speed drive system that utilizes a bypass device shall be the larger of either of the following:

(1) 125 percent of the rated input current to the power conversion equipment
(2) 125 percent of the motor full-load current rating as determined by 430.6

430.124 Overload Protection. Overload protection of the motor shall be provided.

(A) Included in Power Conversion Equipment. Where the power conversion equipment is marked to indicate that motor overload protection is included, additional overload protection shall not be required.

(B) Bypass Circuits. For adjustable-speed drive systems that utilize a bypass device to allow motor operation at rated full-load speed, motor overload protection as described in Article 430, Part III, shall be provided in the bypass circuit.

(C) Multiple Motor Applications. For multiple motor application, individual motor overload protection shall be provided in accordance with Article 430, Part III.

430.126 Motor Overtemperature Protection.

(A) General. Adjustable-speed drive systems shall protect against motor overtemperature conditions where the motor is not rated to operate at the nameplate rated current over the speed range required by the application. This protection shall be provided in addition to the conductor protection required in 430.32. Protection shall be provided by one of the following means.

(1) Motor thermal protector in accordance with 430.32
(2) Adjustable-speed drive system with load and speed-sensitive overload protection and thermal memory retention upon shutdown or power loss

Exception to (2): Thermal memory retention upon shutdown or power loss is not required for continuous duty loads.

(3) Overtemperature protection relay utilizing thermal sensors embedded in the motor and meeting the requirements of 430.126(A)(2) or (B)(2)
(4) Thermal sensor embedded in the motor whose communications are received and acted upon by an adjustable-speed drive system

Informational Note: The relationship between motor current and motor temperature changes when the motor is operated by an adjustable-speed drive. In certain applications, overheating of motors can occur when operated at reduced speed, even at current levels less than a motor's rated full-load current. The overheating can be the result of reduced motor cooling when its shaft-mounted fan is operating less than rated nameplate RPM. As part of the analysis to determine whether overheating will occur, it is necessary to consider the continuous torque capability curves for the motor given the application requirements. This will assist in determining whether the motor overload protection will be able, on its own, to provide protection against overheating. These overheating protection requirements are only intended to apply to applications where an adjustable-speed drive, as defined in Article 100, is used.

For motors that utilize external forced air or liquid cooling systems, overtemperature can occur if the cooling system is not operating. Although this issue is not unique to adjustable speed applications, externally cooled motors are most often encountered with such applications. In these instances, overtemperature protection using direct temperature sensing is recommended [i.e., 430.126(A)(1), (A)(3), or (A)(4)], or additional means should be provided to ensure that the cooling system is operating (flow or pressure sensing, interlocking of adjustable-speed drive system and cooling system, etc.).

(B) Multiple Motor Applications. For multiple motor applications, individual motor overtemperature protection shall be provided as required in 430.126(A).

(C) Automatic Restarting and Orderly Shutdown. The provisions of 430.43 and 430.44 shall apply to the motor overtemperature protection means.

430.128 Disconnecting Means. The disconnecting means shall be permitted to be in the incoming line to the conversion equipment and shall have a rating not less than 115 percent of the rated input current of the conversion unit.

430.130 Branch-Circuit Short-Circuit and Ground-Fault Protection for Single Motor Circuits Containing Power Conversion Equipment.

(A) Circuits Containing Power Conversion Equipment. Circuits containing power conversion equipment shall be protected by a branch-circuit short-circuit and ground-fault protective device in accordance with the following:

(1) The rating and type of protection shall be determined by 430.52(C)(1), (C)(3), (C)(5), or (C)(6), using the full-load current rating of the motor load as determined by 430.6.
(2) Where maximum branch-circuit short-circuit and ground-fault protective ratings are stipulated for specific device types in the manufacturer's instructions for the power conversion equipment or are otherwise marked on the equipment, they shall not be exceeded even if higher values are permitted by 430.130(A)(1).
(3) A self-protected combination controller shall only be permitted where specifically identified in the manufacturer's instructions for the power conversion equipment or if otherwise marked on the equipment.

Informational Note: The type of protective device, its rating, and its setting are often marked on or provided with the power conversion equipment.

(4) Where an instantaneous trip circuit breaker or semiconductor fuses are permitted in accordance with the drive manufacturer's instructions for use as the branch-circuit short-circuit and ground-fault protective device for listed power conversion equipment, they shall be provided as an integral part of a single listed assembly incorporating both the protective device and power conversion equipment.

(B) Bypass Circuit/Device. Branch-circuit short-circuit and ground-fault protection shall also be provided for a bypass circuit/device(s). Where a single branch-circuit short-circuit and ground-fault protective device is provided for circuits containing both power conversion equipment and a bypass circuit, the branch-circuit protective device type and its rating or setting shall be in accordance with those determined for the power conversion equipment and for the bypass circuit/device(s) equipment.

430.131 Several Motors or Loads on One Branch Circuit Including Power Conversion Equipment. For installations meeting all the requirements of 430.53 that include one or more power converters, the branch-circuit short-circuit and ground-fault protective fuses or inverse time circuit breakers shall be of a type and rating or setting permitted for use with the power conversion equipment using the full-load current rating of the connected motor load in accordance with 430.53. For the purposes of 430.53 and 430.131, power conversion equipment shall be considered to be a motor controller.

Part XI. Over 1000 Volts, Nominal

430.221 General. Part XI recognizes the additional hazard due to the use of higher voltages. It adds to or amends the other provisions of this article.

430.222 Marking on Controllers. In addition to the marking required by 430.8, a controller shall be marked with the control voltage.

430.223 Raceway Connection to Motors. Flexible metal conduit or liquidtight flexible metal conduit not exceeding 1.8 m (6 ft) in length shall be permitted to be employed for raceway connection to a motor terminal enclosure.

430.224 Size of Conductors. Conductors supplying motors shall have an ampacity not less than the current at which the motor overload protective device(s) is selected to trip.

430.225 Motor-Circuit Overcurrent Protection.

(A) General. Each motor circuit shall include coordinated protection to automatically interrupt overload and fault currents in the motor, the motor-circuit conductors, and the motor control apparatus.

Exception: Where a motor is critical to an operation and the motor should operate to failure if necessary to prevent a greater hazard to persons, the sensing device(s) shall be permitted to be connected to a supervised annunciator or alarm instead of interrupting the motor circuit.

(B) Overload Protection.

(1) Type of Overload Device. Each motor shall be protected against dangerous heating due to motor overloads and failure to start by a thermal protector integral with the motor or external current-sensing devices, or both. Protective device settings for each motor circuit shall be determined under engineering supervision.

(2) Wound-Rotor Alternating-Current Motors. The secondary circuits of wound-rotor ac motors, including conductors, controllers, and resistors rated for the application, shall be considered as protected against overcurrent by the motor overload protection means.

(3) Operation. Operation of the overload interrupting device shall simultaneously disconnect all ungrounded conductors.

(4) Automatic Reset. Overload sensing devices shall not automatically reset after trip unless resetting of the overload sensing device does not cause automatic restarting of the motor or there is no hazard to persons created by automatic restarting of the motor and its connected machinery.

(C) Fault-Current Protection.

(1) Type of Protection. Fault-current protection shall be provided in each motor circuit as specified by either (1)(a) or (1)(b).

(a) A circuit breaker of suitable type and rating arranged so that it can be serviced without hazard. The circuit breaker shall simultaneously disconnect all ungrounded conductors. The circuit breaker shall be permitted to sense the fault current by means of integral or external sensing elements.

(b) Fuses of a suitable type and rating placed in each ungrounded conductor. Fuses shall be used with suitable disconnecting means, or they shall be of a type that can also serve as the disconnecting means. They shall be arranged so that they cannot be serviced while they are energized.

(2) Reclosing. Fault-current interrupting devices shall not automatically reclose the circuit.

Exception: Automatic reclosing of a circuit shall be permitted where the circuit is exposed to transient faults and where such automatic reclosing does not create a hazard to persons.

(3) Combination Protection. Overload protection and fault-current protection shall be permitted to be provided by the same device.

430.226 Rating of Motor Control Apparatus. The ultimate trip current of overcurrent (overload) relays or other motor-protective devices used shall not exceed 115 percent of the controller's continuous current rating. Where the motor branch-circuit disconnecting means is separate from the controller, the disconnecting means current rating shall not be less than the ultimate trip setting of the overcurrent relays in the circuit.

430.227 Disconnecting Means. The controller disconnecting means shall be lockable in accordance with 110.25.

Part XII. Protection of Live Parts — All Voltages

430.231 General. Part XII specifies that live parts shall be protected in an approved manner for the hazard involved.

430.232 Where Required. Exposed live parts of motors and controllers operating at 50 volts or more between terminals shall be guarded against accidental contact by enclosure or by location as follows:

(1) By installation in a room or enclosure that is accessible only to qualified persons
(2) By installation on a suitable balcony, gallery, or platform, elevated and arranged so as to exclude unqualified persons
(3) By elevation 2.5 m (8 ft) or more above the floor

Exception: Live parts of motors operating at more than 50 volts between terminals shall not require additional guarding for stationary motors that have commutators, collectors, and brush rigging located inside of motor-end brackets and not conductively connected to supply circuits operating at more than 150 volts to ground.

430.233 Guards for Attendants. Where live parts of motors or controllers operating at over 50 volts to ground are guarded against accidental contact only by location as specified in 430.232, and where adjustment or other attendance may be necessary during the operation of the apparatus, suitable insulating mats or platforms shall be provided so that the attendant cannot readily touch live parts unless standing on the mats or platforms.

Informational Note: For working space, see 110.26 and 110.34.

Part XIII. Grounding — All Voltages

430.241 General. Part XIII specifies the grounding of exposed non–current-carrying metal parts, likely to become energized, of motor and controller frames to prevent a voltage aboveground in the event of accidental contact between energized parts and frames. Insulation, isolation, or guarding are suitable alternatives to grounding of motors under certain conditions.

430.242 Stationary Motors. The frames of stationary motors shall be grounded under any of the following conditions:

(1) Where supplied by metal-enclosed wiring
(2) Where in a wet location and not isolated or guarded
(3) If in a hazardous (classified) location
(4) If the motor operates with any terminal at over 150 volts to ground

Where the frame of the motor is not grounded, it shall be permanently and effectively insulated from the ground.

430.243 Portable Motors. The frames of portable motors that operate over 150 volts to ground shall be guarded or grounded.

Informational Note No. 1: See 250.114(4) for grounding of portable appliances in other than residential occupancies.

Informational Note No. 2: See 250.119(C) for color of equipment grounding conductor.

Exception No. 1: Listed motor-operated tools, listed motor-operated appliances, and listed motor-operated equipment shall not be required to be grounded where protected by a system of double insulation or its equivalent. Double-insulated equipment shall be distinctively marked.

Exception No. 2: Listed motor-operated tools, listed motor-operated appliances, and listed motor-operated equipment connected by a cord and attachment plug other than those required to be grounded in accordance with 250.114.

430.244 Controllers. Controller enclosures shall be connected to the equipment grounding conductor regardless of voltage. Controller enclosures shall have means for attachment of an equipment grounding conductor termination in accordance with 250.8.

Exception: Enclosures attached to ungrounded portable equipment shall not be required to be grounded.

430.245 Method of Grounding. Connection to the equipment grounding conductor shall be done in the manner specified in Part VI of Article 250.

(A) Grounding Through Terminal Housings. Where the wiring to motors is metal-enclosed cable or in metal raceways, junction boxes to house motor terminals shall be provided, and the armor of the cable or the metal raceways shall be connected to them in the manner specified in 250.96(A) and 250.97.

(B) Separation of Junction Box from Motor. The junction box required by 430.245(A) shall be permitted to be separated from the motor by not more than 1.8 m (6 ft), provided the leads to the motor are stranded conductors within Type AC cable, interlocked metal tape Type MC cable where listed and identified in accordance with 250.118(10)(a), or armored cord or are stranded leads enclosed in liquidtight flexible metal conduit, flexible metal conduit, intermediate metal conduit, rigid metal conduit, or electrical metallic tubing not smaller than metric designator 12 (trade size ⅜), the armor or raceway being connected both to the motor and to the box.

Liquidtight flexible nonmetallic conduit and rigid nonmetallic conduit shall be permitted to enclose the leads to the motor, provided the leads are stranded and the required equipment grounding conductor is connected to both the motor and to the box.

Where stranded leads are used, protected as specified above, each strand within the conductor shall be not larger than 10 AWG and shall comply with other requirements of this *Code* for conductors to be used in raceways.

(C) Grounding of Controller-Mounted Devices. Instrument transformer secondaries and exposed non–current-carrying metal or other conductive parts or cases of instrument transformers, meters, instruments, and relays shall be grounded as specified in 250.170 through 250.178.

Part XIV. Tables

Table 430.247 Full-Load Current in Amperes, Direct-Current Motors

The following values of full-load currents* are for motors running at base speed.

| Horsepower | Armature Voltage Rating* | | | | | |
	90 Volts	120 Volts	180 Volts	240 Volts	500 Volts	550 Volts
¼	4.0	3.1	2.0	1.6	—	—
⅓	5.2	4.1	2.6	2.0	—	—
½	6.8	5.4	3.4	2.7	—	—
¾	9.6	7.6	4.8	3.8	—	—
1	12.2	9.5	6.1	4.7	—	—
1½	—	13.2	8.3	6.6	—	—
2	—	17	10.8	8.5	—	—
3	—	25	16	12.2	—	—
5	—	40	27	20	—	—
7½	—	58	—	29	13.6	12.2
10	—	76	—	38	18	16
15	—	—	—	55	27	24
20	—	—	—	72	34	31
25	—	—	—	89	43	38
30	—	—	—	106	51	46
40	—	—	—	140	67	61
50	—	—	—	173	83	75
60	—	—	—	206	99	90
75	—	—	—	255	123	111
100	—	—	—	341	164	148
125	—	—	—	425	205	185
150	—	—	—	506	246	222
200	—	—	—	675	330	294

*These are average dc quantities.

Table 430.248 Full-Load Currents in Amperes, Single-Phase Alternating-Current Motors

The following values of full-load currents are for motors running at usual speeds and motors with normal torque characteristics. The voltages listed are rated motor voltages. The currents listed shall be permitted for system voltage ranges of 110 to 120 and 220 to 240 volts.

Horsepower	115 Volts	200 Volts	208 Volts	230 Volts
$\frac{1}{6}$	4.4	2.5	2.4	2.2
$\frac{1}{4}$	5.8	3.3	3.2	2.9
$\frac{1}{3}$	7.2	4.1	4.0	3.6
$\frac{1}{2}$	9.8	5.6	5.4	4.9
$\frac{3}{4}$	13.8	7.9	7.6	6.9
1	16	9.2	8.8	8.0
$1\frac{1}{2}$	20	11.5	11.0	10
2	24	13.8	13.2	12
3	34	19.6	18.7	17
5	56	32.2	30.8	28
$7\frac{1}{2}$	80	46.0	44.0	40
10	100	57.5	55.0	50

Table 430.249 Full-Load Current, Two-Phase Alternating-Current Motors (4-Wire)

The following values of full-load current are for motors running at speeds usual for belted motors and motors with normal torque characteristics. Current in the common conductor of a 2-phase, 3-wire system will be 1.41 times the value given. The voltages listed are rated motor voltages. The currents listed shall be permitted for system voltage ranges of 110 to 120, 220 to 240, 440 to 480, and 550 to 600 volts.

Horsepower	Induction-Type Squirrel Cage and Wound Rotor (Amperes)				
	115 Volts	230 Volts	460 Volts	575 Volts	2300 Volts
$\frac{1}{2}$	4.0	2.0	1.0	0.8	—
$\frac{3}{4}$	4.8	2.4	1.2	1.0	—
1	6.4	3.2	1.6	1.3	—
$1\frac{1}{2}$	9.0	4.5	2.3	1.8	—
2	11.8	5.9	3.0	2.4	—
3	—	8.3	4.2	3.3	—
5	—	13.2	6.6	5.3	—
$7\frac{1}{2}$	—	19	9.0	8.0	—
10	—	24	12	10	—
15	—	36	18	14	—
20	—	47	23	19	—
25	—	59	29	24	—
30	—	69	35	28	—
40	—	90	45	36	—
50	—	113	56	45	—
60	—	133	67	53	14
75	—	166	83	66	18
100	—	218	109	87	23
125	—	270	135	108	28
150	—	312	156	125	32
200	—	416	208	167	43

Table 430.250 Full-Load Current, Three-Phase Alternating-Current Motors
The following values of full-load currents are typical for motors running at speeds usual for belted motors and motors with normal torque characteristics. The voltages listed are rated motor voltages. The currents listed shall be permitted for system voltage ranges of 110 to 120, 220 to 240, 440 to 480, and 550 to 600 volts.

| Horsepower | Induction-Type Squirrel Cage and Wound Rotor (Amperes) | | | | | | | Synchronous-Type Unity Power Factor* (Amperes) | | | |
	115 Volts	200 Volts	208 Volts	230 Volts	460 Volts	575 Volts	2300 Volts	230 Volts	460 Volts	575 Volts	2300 Volts
½	4.4	2.5	2.4	2.2	1.1	0.9	—	—	—	—	—
¾	6.4	3.7	3.5	3.2	1.6	1.3	—	—	—	—	—
1	8.4	4.8	4.6	4.2	2.1	1.7	—	—	—	—	—
1½	12.0	6.9	6.6	6.0	3.0	2.4	—	—	—	—	—
2	13.6	7.8	7.5	6.8	3.4	2.7	—	—	—	—	—
3	—	11.0	10.6	9.6	4.8	3.9	—	—	—	—	—
5	—	17.5	16.7	15.2	7.6	6.1	—	—	—	—	—
7½	—	25.3	24.2	22	11	9	—	—	—	—	—
10	—	32.2	30.8	28	14	11	—	—	—	—	—
15	—	48.3	46.2	42	21	17	—	—	—	—	—
20	—	62.1	59.4	54	27	22	—	—	—	—	—
25	—	78.2	74.8	68	34	27	—	53	26	21	—
30	—	92	88	80	40	32	—	63	32	26	—
40	—	120	114	104	52	41	—	83	41	33	—
50	—	150	143	130	65	52	—	104	52	42	—
60	—	177	169	154	77	62	16	123	61	49	12
75	—	221	211	192	96	77	20	155	78	62	15
100	—	285	273	248	124	99	26	202	101	81	20
125	—	359	343	312	156	125	31	253	126	101	25
150	—	414	396	360	180	144	37	302	151	121	30
200		552	528	480	240	192	49	400	201	161	40
250	—	—	—	—	302	242	60	—	—	—	—
300	—	—	—	—	361	289	72	—	—	—	—
350	—	—	—	—	414	336	83	—	—	—	—
400	—	—	—	—	477	382	95	—	—	—	—
450	—	—	—	—	515	412	103	—	—	—	—
500	—	—	—	—	590	472	118	—	—	—	—

*For 90 and 80 percent power factor, the figures shall be multiplied by 1.1 and 1.25, respectively.

Table 430.251(A) Conversion Table of Single-Phase Locked-Rotor Currents for Selection of Disconnecting Means and Controllers as Determined from Horsepower and Voltage Rating
For use only with 430.110, 440.12, 440.41, and 455.8(C).

| Rated Horsepower | Maximum Locked-Rotor Current in Amperes, Single Phase | | |
	115 Volts	208 Volts	230 Volts
½	58.8	32.5	29.4
¾	82.8	45.8	41.4
1	96	53	48
1½	120	66	60
2	144	80	72
3	204	113	102
5	336	186	168
7½	480	265	240
10	1000	332	300

Table 430.251(B) Conversion Table of Polyphase Design B, C, and D Maximum Locked-Rotor Currents for Selection of Disconnecting Means and Controllers as Determined from Horsepower and Voltage Rating and Design Letter
For use only with 430.110, 440.12, 440.41, and 455.8(C).

Rated Horsepower	Maximum Motor Locked-Rotor Current in Amperes, Two- and Three-Phase, Design B, C, and D*					
	115 Volts	200 Volts	208 Volts	230 Volts	460 Volts	575 Volts
	B, C, D	B, C, D	B, C, D	B, C, D	B, C, D	B, C, D
½	40	23	22.1	20	10	8
¾	50	28.8	27.6	25	12.5	10
1	60	34.5	33	30	15	12
1½	80	46	44	40	20	16
2	100	57.5	55	50	25	20
3	—	73.6	71	64	32	25.6
5	—	105.8	102	92	46	36.8
7½	—	146	140	127	63.5	50.8
10	—	186.3	179	162	81	64.8
15	—	267	257	232	116	93
20	—	334	321	290	145	116
25	—	420	404	365	183	146
30	—	500	481	435	218	174
40	—	667	641	580	290	232
50	—	834	802	725	363	290
60	—	1001	962	870	435	348
75	—	1248	1200	1085	543	434
100	—	1668	1603	1450	725	580
125	—	2087	2007	1815	908	726
150	—	2496	2400	2170	1085	868
200	—	3335	3207	2900	1450	1160
250	—	—	—	—	1825	1460
300	—	—	—	—	2200	1760
350	—	—	—	—	2550	2040
400	—	—	—	—	2900	2320
450	—	—	—	—	3250	2600
500	—	—	—	—	3625	2900

*Design A motors are not limited to a maximum starting current or locked rotor current.

ARTICLE 440
Air-Conditioning and Refrigerating Equipment

Part I. General

440.1 Scope. The provisions of this article apply to electric motor-driven air-conditioning and refrigerating equipment and to the branch circuits and controllers for such equipment. It provides for the special considerations necessary for circuits supplying hermetic refrigerant motor-compressors and for any air-conditioning or refrigerating equipment that is supplied from a branch circuit that supplies a hermetic refrigerant motor-compressor.

440.2 Definitions.

Branch-Circuit Selection Current. The value in amperes to be used instead of the rated-load current in determining the ratings of motor branch-circuit conductors, disconnecting means, controllers, and branch-circuit short-circuit and ground-fault protective devices wherever the running overload protective device permits a sustained current greater than the specified percentage of the rated-load current. The value of branch-circuit selection current will always be equal to or greater than the marked rated-load current.

Leakage-Current Detector-Interrupter (LCDI). A device provided in a power supply cord or cord set that senses leakage current flowing between or from the cord conductors and interrupts the circuit at a predetermined level of leakage current.

Rated-Load Current. The current of a hermetic refrigerant motor-compressor resulting when it is operated at the rated load, rated voltage, and rated frequency of the equipment it serves.

440.3 Other Articles.

(A) Article 430. These provisions are in addition to, or amendatory of, the provisions of Article 430 and other articles in this *Code*, which apply except as modified in this article.

(B) Articles 422, 424, or 430. The rules of Articles 422, 424, or 430, as applicable, shall apply to air-conditioning and refrigerating equipment that does not incorporate a hermetic refrigerant motor-compressor. This equipment includes devices that employ refrigeration compressors driven by conventional motors, furnaces with air-conditioning evaporator coils installed, fan-coil units, remote forced air-cooled condensers, remote commercial refrigerators, and so forth.

(C) Article 422. Equipment such as room air conditioners, household refrigerators and freezers, drinking water coolers, and beverage dispensers shall be considered appliances, and the provisions of Article 422 shall also apply.

(D) Other Applicable Articles. Hermetic refrigerant motor-compressors, circuits, controllers, and equipment shall also comply with the applicable provisions of Table 440.3(D).

Table 440.3(D) Other Articles

Equipment/Occupancy	Article	Section
Capacitors		460.9
Commercial garages, aircraft hangars, motor fuel dispensing facilities, bulk storage plants, spray application, dipping, and coating processes, and inhalation anesthetizing locations	511, 513, 514, 515, 516, and 517 Part IV	
Hazardous (classified) locations	500–503, 505, and 506	
Motion picture and television studios and similar locations	530	
Resistors and reactors	470	

440.4 Marking on Hermetic Refrigerant Motor-Compressors and Equipment.

(A) Hermetic Refrigerant Motor-Compressor Nameplate. A hermetic refrigerant motor-compressor shall be provided with a nameplate that shall indicate the manufacturer's name, trademark, or symbol; identifying designation; phase; voltage; and frequency. The rated-load current in amperes of the motor-compressor shall be marked by the equipment manufacturer on either or both the motor-compressor nameplate and the nameplate of the equipment in which the motor-compressor is used. The locked-rotor current of each single-phase motor-compressor having a rated-load current of more than 9 amperes at 115 volts, or more than 4.5 amperes at 230 volts, and each polyphase motor-compressor shall be marked on the motor-compressor nameplate. Where a thermal protector complying with 440.52(A)(2) and (B)(2) is used, the motor-compressor nameplate or the equipment nameplate shall be marked with the words "thermally protected." Where a protective system complying with 440.52(A)(4) and (B)(4) is used and is furnished with the equipment, the equipment nameplate shall be marked with the words, "thermally protected system." Where a protective system complying with 440.52(A)(4) and (B)(4) is specified, the equipment nameplate shall be appropriately marked.

(B) Multimotor and Combination-Load Equipment. Multimotor and combination-load equipment shall be provided with a visible nameplate marked with the maker's name, the rating in volts, frequency and number of phases, minimum supply circuit conductor ampacity, the maximum rating of the branch-circuit short-circuit and ground-fault protective device, and the short-circuit current rating of the motor controllers or industrial control panel. The ampacity shall be calculated by using Part IV and counting all the motors and other loads that will be operated at the same time. The branch-circuit short-circuit and ground-fault protective device rating shall not exceed the value calculated by using Part III. Multimotor or combination-load equipment for use on two or more circuits shall be marked with the above information for each circuit.

Exception No. 1: Multimotor and combination-load equipment that is suitable under the provisions of this article for connection to a single 15- or 20-ampere, 120-volt, or a 15-ampere, 208- or 240-volt, single-phase branch circuit shall be permitted to be marked as a single load.

Exception No. 2: The minimum supply circuit conductor ampacity and the maximum rating of the branch-circuit short-circuit and ground-fault protective device shall not be required to be marked on a room air conditioner complying with 440.62(A).

Exception No. 3: Multimotor and combination-load equipment used in one- and two-family dwellings or cord-and-attachment-plug-connected equipment shall not be required to be marked with a short-circuit current rating.

(C) Branch-Circuit Selection Current. A hermetic refrigerant motor-compressor, or equipment containing such a compressor, having a protection system that is approved for use with the motor-compressor that it protects and that permits continuous current in excess of the specified percentage of nameplate rated-load current given in 440.52(B)(2) or (B)(4) shall also be marked with a branch-circuit selection current that complies with 440.52(B)(2) or (B)(4). This marking shall be provided by the equipment manufacturer and shall be on the nameplate(s) where the rated-load current(s) appears.

440.5 Marking on Controllers. A controller shall be marked with the manufacturer's name, trademark, or symbol; identifying designation; voltage; phase; full-load and locked-rotor current (or horsepower) rating; and other data as may be needed to properly indicate the motor-compressor for which it is suitable.

440.6 Ampacity and Rating. The size of conductors for equipment covered by this article shall be selected from Table 310.15(B)(16) through Table 310.15(B)(19) or calculated in accordance with 310.15 as applicable. The required ampacity of conductors and rating of equipment shall be determined according to 440.6(A) and 440.6(B).

(A) Hermetic Refrigerant Motor-Compressor. For a hermetic refrigerant motor-compressor, the rated-load current marked on the nameplate of the equipment in which the motor-compressor is employed shall be used in determining the rating or ampacity of the disconnecting means, the branch-circuit conductors, the controller, the branch-circuit short-circuit and ground-fault protection, and the separate motor overload protection. Where no rated-load current is shown on the equipment nameplate, the rated-load current shown on the compressor nameplate shall be used.

Exception No. 1: Where so marked, the branch-circuit selection current shall be used instead of the rated-load current to determine the rating or ampacity of the disconnecting means, the branch-circuit conductors, the controller, and the branch-circuit short-circuit and ground-fault protection.

Exception No. 2: For cord-and-plug-connected equipment, the nameplate marking shall be used in accordance with 440.22(B), Exception No. 2.

(B) Multimotor Equipment. For multimotor equipment employing a shaded-pole or permanent split-capacitor-type fan or blower motor, the full-load current for such motor marked on the nameplate of the equipment in which the fan or blower motor is employed shall be used instead of the horsepower rating to determine the ampacity or rating of the disconnecting means, the branch-circuit conductors, the controller, the branch-circuit short-circuit and ground-fault protection, and the separate overload protection. This marking on the equipment nameplate shall not be less than the current marked on the fan or blower motor nameplate.

440.7 Highest Rated (Largest) Motor. In determining compliance with this article and with 430.24, 430.53(B) and 430.53(C), and 430.62(A), the highest rated (largest) motor shall be considered to be the motor that has the highest rated-load current. Where two or more motors have the same highest rated-load current, only one of them shall be considered as the highest rated (largest) motor. For other than hermetic refrigerant motor-compressors, and fan or blower motors as covered in 440.6(B), the full-load current used to determine the highest rated motor shall be the equivalent value corresponding to the motor horsepower rating selected from Table 430.248, Table 430.249, or Table 430.250.

Exception: Where so marked, the branch-circuit selection current shall be used instead of the rated-load current in determining the highest rated (largest) motor-compressor.

440.8 Single Machine. An air-conditioning or refrigerating system shall be considered to be a single machine under the provisions of 430.87, Exception No. 1, and 430.112, Exception. The motors shall be permitted to be located remotely from each other.

N **440.9 Grounding and Bonding.** Where multimotor and combination-load equipment is installed outdoors on a roof, an equipment grounding conductor of the wire type shall be installed in outdoor portions of metallic raceway systems that use non-threaded fittings.

N **440.10 Short-Circuit Current Rating.**

(A) Installation. Motor controllers of multimotor and combination-load equipment shall not be installed where the available short-circuit current exceeds its short-circuit current rating as marked in accordance with 440.4(B).

(B) Documentation. When motor controllers or industrial control panels of multimotor and combination load equipment are required to be marked with a short circuit current rating, the available short circuit current and the date the short circuit current calculation was performed shall be documented and made available to those authorized to inspect the installation.

Part II. Disconnecting Means

440.11 General. The provisions of Part II are intended to require disconnecting means capable of disconnecting air-conditioning and refrigerating equipment, including motor-compressors and controllers from the circuit conductors.

440.12 Rating and Interrupting Capacity.

(A) Hermetic Refrigerant Motor-Compressor. A disconnecting means serving a hermetic refrigerant motor-compressor shall be selected on the basis of the nameplate rated-load current or branch-circuit selection current, whichever is greater, and locked-rotor current, respectively, of the motor-compressor as follows.

(1) Ampere Rating. The ampere rating shall be at least 115 percent of the nameplate rated-load current or branch-circuit selection current, whichever is greater.

Exception: A listed unfused motor circuit switch, without fuseholders, having a horsepower rating not less than the equivalent horsepower determined in accordance with 440.12(A)(2) shall be permitted to have an ampere rating less than 115 percent of the specified current.

(2) Equivalent Horsepower. To determine the equivalent horsepower in complying with the requirements of 430.109, the horsepower rating shall be selected from Table 430.248, Table 430.249, or Table 430.250 corresponding to the rated-load current or branch-circuit selection current, whichever is greater, and also the horsepower rating from Table 430.251(A) or Table 430.251(B) corresponding to the locked-rotor current. In case the nameplate rated-load current or branch-circuit selection current and locked-rotor current do not correspond to the currents shown in Table 430.248, Table 430.249, Table 430.250, Table 430.251(A), or Table 430.251(B), the horsepower rating corresponding to the next higher value shall be selected. In case different horsepower ratings are obtained when applying these tables, a horsepower rating at least equal to the larger of the values obtained shall be selected.

(B) Combination Loads. Where the combined load of two or more hermetic refrigerant motor-compressors or one or more hermetic refrigerant motor-compressor with other motors or loads may be simultaneous on a single disconnecting means, the rating for the disconnecting means shall be determined in accordance with 440.12(B)(1) and (B)(2).

(1) Horsepower Rating. The horsepower rating of the disconnecting means shall be determined from the sum of all currents, including resistance loads, at the rated-load condition and also at the locked-rotor condition. The combined rated-load current and the combined locked-rotor current so obtained shall be considered as a single motor for the purpose of this requirement as required by (1)(a) and (1)(b).

(a) The full-load current equivalent to the horsepower rating of each motor, other than a hermetic refrigerant motor-compressor, and fan or blower motors as covered in 440.6(B) shall be selected from Table 430.248, Table 430.249, or Table 430.250. These full-load currents shall be added to the motor-compressor rated-load current(s) or branch-circuit selection current(s), whichever is greater, and to the rating in amperes of other loads to obtain an equivalent full-load current for the combined load.

(b) The locked-rotor current equivalent to the horsepower rating of each motor, other than a hermetic refrigerant motor-compressor, shall be selected from Table 430.251(A) or Table 430.251(B), and, for fan and blower motors of the shaded-pole or permanent split-capacitor type marked with the locked-rotor current, the marked value shall be used. The locked-rotor currents shall be added to the motor-compressor locked-rotor current(s) and to the rating in amperes of other loads to obtain an equivalent locked-rotor current for the combined load. Where two or more motors or other loads such as resistance heaters, or both, cannot be started simultaneously, appropriate combinations of locked-rotor and rated-load current or branch-circuit selection current, whichever is greater, shall be an acceptable means of determining the equivalent locked-rotor current for the simultaneous combined load.

Exception: Where part of the concurrent load is a resistance load and the disconnecting means is a switch rated in horsepower and amperes, the switch used shall be permitted to have a horsepower rating not less than the combined load to the motor-compressor(s) and other motor(s) at the locked-rotor condition, if the ampere rating of the switch is not less than this locked-rotor load plus the resistance load.

(2) Full-Load Current Equivalent. The ampere rating of the disconnecting means shall be at least 115 percent of the sum of all currents at the rated-load condition determined in accordance with 440.12(B)(1).

Exception: A listed unfused motor circuit switch, without fuseholders, having a horsepower rating not less than the equivalent horsepower determined by 440.12(B)(1) shall be permitted to have an ampere rating less than 115 percent of the sum of all currents.

(C) Small Motor-Compressors. For small motor-compressors not having the locked-rotor current marked on the nameplate, or for small motors not covered by Table 430.247, Table 430.248, Table 430.249, or Table 430.250, the locked-rotor current shall be assumed to be six times the rated-load current.

(D) Disconnecting Means. Every disconnecting means in the refrigerant motor-compressor circuit between the point of attachment to the feeder and the point of connection to the refrigerant motor-compressor shall comply with the requirements of 440.12.

(E) Disconnecting Means Rated in Excess of 100 Horsepower. Where the rated-load or locked-rotor current as determined above would indicate a disconnecting means rated in excess of 100 hp, the provisions of 430.109(E) shall apply.

440.13 Cord-Connected Equipment. For cord-connected equipment such as room air conditioners, household refrigerators and freezers, drinking water coolers, and beverage dispensers, a separable connector or an attachment plug and receptacle shall be permitted to serve as the disconnecting means.

Informational Note: For room air conditioners, see 440.63.

440.14 Location. Disconnecting means shall be located within sight from, and readily accessible from the air-conditioning or refrigerating equipment. The disconnecting means shall be permitted to be installed on or within the air-conditioning or refrigerating equipment.

The disconnecting means shall not be located on panels that are designed to allow access to the air-conditioning or refrigeration equipment or to obscure the equipment nameplate(s).

Exception No. 1: Where the disconnecting means provided in accordance with 430.102(A) is lockable in accordance with 110.25 and the refrigerating or air-conditioning equipment is essential to an industrial process in a facility with written safety procedures, and where the conditions of maintenance and supervision ensure that only qualified persons service the equipment, a disconnecting means within sight from the equipment shall not be required.

Exception No. 2: Where an attachment plug and receptacle serve as the disconnecting means in accordance with 440.13, their location shall be accessible but shall not be required to be readily accessible.

Informational Note No. 1: See Parts VII and IX of Article 430 for additional requirements.

Informational Note No. 2: See 110.26.

Part III. Branch-Circuit Short-Circuit and Ground-Fault Protection

440.21 General. The provisions of Part III specify devices intended to protect the branch-circuit conductors, control apparatus, and motors in circuits supplying hermetic refrigerant motor-compressors against overcurrent due to short circuits and ground faults. They are in addition to or amendatory of the provisions of Article 240.

440.22 Application and Selection.

(A) Rating or Setting for Individual Motor-Compressor. The motor-compressor branch-circuit short-circuit and ground-fault protective device shall be capable of carrying the starting current of the motor. A protective device having a rating or setting not exceeding 175 percent of the motor-compressor rated-load current or branch-circuit selection current, whichever is greater, shall be permitted, provided that, where the protection specified is not sufficient for the starting current of the motor, the rating or setting shall be permitted to be increased but shall not exceed 225 percent of the motor rated-load current or branch-circuit selection current, whichever is greater.

Exception: The rating of the branch-circuit short-circuit and ground-fault protective device shall not be required to be less than 15 amperes.

(B) Rating or Setting for Equipment. The equipment branch-circuit short-circuit and ground-fault protective device shall be capable of carrying the starting current of the equipment. Where the hermetic refrigerant motor-compressor is the only load on the circuit, the protection shall comply with 440.22(A). Where the equipment incorporates more than one hermetic refrigerant motor-compressor or a hermetic refrigerant motor-compressor and other motors or other loads, the equipment short-circuit and ground-fault protection shall comply with 430.53 and 440.22(B)(1) and (B)(2).

(1) Motor-Compressor Largest Load. Where a hermetic refrigerant motor-compressor is the largest load connected to the circuit, the rating or setting of the branch-circuit short-circuit and ground-fault protective device shall not exceed the value specified in 440.22(A) for the largest motor-compressor plus the sum of the rated-load current or branch-circuit selection current, whichever is greater, of the other motor-compressor(s) and the ratings of the other loads supplied.

(2) Motor-Compressor Not Largest Load. Where a hermetic refrigerant motor-compressor is not the largest load connected to the circuit, the rating or setting of the branch-circuit short-circuit and ground-fault protective device shall not exceed a value equal to the sum of the rated-load current or branch-circuit selection current, whichever is greater, rating(s) for the motor-compressor(s) plus the value specified in 430.53(C)(4) where other motor loads are supplied, or the value specified in 240.4 where only nonmotor loads are supplied in addition to the motor-compressor(s).

Exception No. 1: Equipment that starts and operates on a 15- or 20-ampere 120-volt, or 15-ampere 208- or 240-volt single-phase branch circuit, shall be permitted to be protected by the 15- or 20-ampere overcurrent device protecting the branch circuit, but if the maximum branch-circuit short-circuit and ground-fault protective device rating marked on the equipment is less than these values, the circuit protective device shall not exceed the value marked on the equipment nameplate.

Exception No. 2: The nameplate marking of cord-and-plug-connected equipment rated not greater than 250 volts, single-phase, such as household refrigerators and freezers, drinking water coolers, and beverage dispensers, shall be used in determining the branch-circuit requirements, and each unit shall be considered as a single motor unless the nameplate is marked otherwise.

(C) Protective Device Rating Not to Exceed the Manufacturer's Values. Where maximum protective device ratings shown on a manufacturer's overload relay table for use with a motor controller are less than the rating or setting selected in accordance with 440.22(A) and (B), the protective device rating shall not exceed the manufacturer's values marked on the equipment.

Part IV. Branch-Circuit Conductors

440.31 General. The provisions of Part IV and Article 310 specify ampacities of conductors required to carry the motor current without overheating under the conditions specified, except as modified in 440.6(A), Exception No. 1.

The provisions of these articles shall not apply to integral conductors of motors, to motor controllers and the like, or to conductors that form an integral part of approved equipment.

440.32 Single Motor-Compressor. Branch-circuit conductors supplying a single motor-compressor shall have an ampacity not less than 125 percent of either the motor-compressor rated-load current or the branch-circuit selection current, whichever is greater.

For a wye-start, delta-run connected motor-compressor, the selection of branch-circuit conductors between the controller and the motor-compressor shall be permitted to be based on 72 percent of either the motor-compressor rated-load current or the branch-circuit selection current, whichever is greater.

> Informational Note: The individual motor circuit conductors of wye-start, delta-run connected motor-compressors carry 58 percent of the rated load current. The multiplier of 72 percent is obtained by multiplying 58 percent by 1.25.

440.33 Motor-Compressor(s) With or Without Additional Motor Loads. Conductors supplying one or more motor-compressor(s) with or without an additional motor load(s) shall have an ampacity not less than the sum of each of the following:

(1) The sum of the rated-load or branch-circuit selection current, whichever is greater, of all motor-compressor(s)
(2) The sum of the full-load current rating of all other motors
(3) 25 percent of the highest motor-compressor or motor full load current in the group

Exception No. 1: Where the circuitry is interlocked so as to prevent the starting and running of a second motor-compressor or group of motor-compressors, the conductor size shall be determined from the largest motor-compressor or group of motor-compressors that is to be operated at a given time.

Exception No. 2: The branch-circuit conductors for room air conditioners shall be in accordance with Part VII of Article 440.

440.34 Combination Load. Conductors supplying a motor-compressor load in addition to other load(s) as calculated from Article 220 and other applicable articles shall have an ampacity sufficient for the other load(s) plus the required ampacity for the motor-compressor load determined in accordance with 440.33 or, for a single motor-compressor, in accordance with 440.32.

Exception: Where the circuitry is interlocked so as to prevent simultaneous operation of the motor-compressor(s) and all other loads connected, the conductor size shall be determined from the largest size required for the motor-compressor(s) and other loads to be operated at a given time.

440.35 Multimotor and Combination-Load Equipment. The ampacity of the conductors supplying multimotor and combination-load equipment shall not be less than the minimum circuit ampacity marked on the equipment in accordance with 440.4(B).

Part V. Controllers for Motor-Compressors

440.41 Rating.

(A) Motor-Compressor Controller. A motor-compressor controller shall have both a continuous-duty full-load current rating and a locked-rotor current rating not less than the nameplate rated-load current or branch-circuit selection current, whichever is greater, and locked-rotor current, respectively, of the compressor. In case the motor controller is rated in horsepower but is without one or both of the foregoing current ratings, equivalent currents shall be determined from the ratings as follows. Table 430.248, Table 430.249, and Table 430.250 shall be used to determine the equivalent full-load current rating. Table 430.251(A) and Table 430.251(B) shall be used to determine the equivalent locked-rotor current ratings.

(B) Controller Serving More Than One Load. A controller serving more than one motor-compressor or a motor-compressor and other loads shall have a continuous-duty full-load current rating and a locked-rotor current rating not less than the combined load as determined in accordance with 440.12(B).

Part VI. Motor-Compressor and Branch-Circuit Overload Protection

440.51 General. The provisions of Part VI specify devices intended to protect the motor-compressor, the motor-control apparatus, and the branch-circuit conductors against excessive heating due to motor overload and failure to start.

> Informational Note: See 240.4(G) for application of Parts III and VI of Article 440.

440.52 Application and Selection.

(A) Protection of Motor-Compressor. Each motor-compressor shall be protected against overload and failure to start by one of the following means:

(1) A separate overload relay that is responsive to motor-compressor current. This device shall be selected to trip at not more than 140 percent of the motor-compressor rated-load current.

(2) A thermal protector integral with the motor-compressor, approved for use with the motor-compressor that it protects on the basis that it will prevent dangerous overheating of the motor-compressor due to overload and failure to start. If the current-interrupting device is separate from the motor-compressor and its control circuit is operated by a protective device integral with the motor-compressor, it shall be arranged so that the opening of the control circuit will result in interruption of current to the motor-compressor.

(3) A fuse or inverse time circuit breaker responsive to motor current, which shall also be permitted to serve as the branch-circuit short-circuit and ground-fault protective device. This device shall be rated at not more than 125 percent of the motor-compressor rated-load current.

It shall have sufficient time delay to permit the motor-compressor to start and accelerate its load. The equipment or the motor-compressor shall be marked with this maximum branch-circuit fuse or inverse time circuit breaker rating.

(4) A protective system, furnished or specified and approved for use with the motor-compressor that it protects on the basis that it will prevent dangerous overheating of the motor-compressor due to overload and failure to start. If the current-interrupting device is separate from the motor-compressor and its control circuit is operated by a protective device that is not integral with the current-interrupting device, it shall be arranged so that the opening of the control circuit will result in interruption of current to the motor-compressor.

(B) Protection of Motor-Compressor Control Apparatus and Branch-Circuit Conductors. The motor-compressor controller(s), the disconnecting means, and the branch-circuit conductors shall be protected against overcurrent due to motor overload and failure to start by one of the following means, which shall be permitted to be the same device or system protecting the motor-compressor in accordance with 440.52(A):

Exception: Overload protection of motor-compressors and equipment on 15- and 20-ampere, single-phase, branch circuits shall be permitted to be in accordance with 440.54 and 440.55.

(1) An overload relay selected in accordance with 440.52(A)(1)

(2) A thermal protector applied in accordance with 440.52(A)(2), that will not permit a continuous current in excess of 156 percent of the marked rated-load current or branch-circuit selection current

(3) A fuse or inverse time circuit breaker selected in accordance with 440.52(A)(3)

(4) A protective system, in accordance with 440.52(A)(4), that will not permit a continuous current in excess of 156 percent of the marked rated-load current or branch-circuit selection current

440.53 Overload Relays. Overload relays and other devices for motor overload protection that are not capable of opening short circuits shall be protected by fuses or inverse time circuit breakers with ratings or settings in accordance with Part III unless identified for group installation or for part-winding motors and marked to indicate the maximum size of fuse or inverse time circuit breaker by which they shall be protected.

Exception: The fuse or inverse time circuit breaker size marking shall be permitted on the nameplate of the equipment in which the overload relay or other overload device is used.

440.54 Motor-Compressors and Equipment on 15- or 20-Ampere Branch Circuits — Not Cord- and Attachment-Plug-Connected. Overload protection for motor-compressors and equipment used on 15- or 20-ampere 120-volt, or 15-ampere 208- or 240-volt single-phase branch circuits as permitted in Article 210 shall be permitted as indicated in 440.54(A) and 440.54(B).

(A) Overload Protection. The motor-compressor shall be provided with overload protection selected as specified in 440.52(A). Both the controller and motor overload protective device shall be identified for installation with the short-circuit and ground-fault protective device for the branch circuit to which the equipment is connected.

(B) Time Delay. The short-circuit and ground-fault protective device protecting the branch circuit shall have sufficient time delay to permit the motor-compressor and other motors to start and accelerate their loads.

440.55 Cord- and Attachment-Plug-Connected Motor-Compressors and Equipment on 15- or 20-Ampere Branch Circuits. Overload protection for motor-compressors and equipment that are cord- and attachment-plug-connected and used on 15- or 20-ampere 120-volt, or 15-ampere 208- or 240-volt, single-phase branch circuits as permitted in Article 210 shall be permitted as indicated in 440.55(A), (B), and (C).

(A) Overload Protection. The motor-compressor shall be provided with overload protection as specified in 440.52(A). Both the controller and the motor overload protective device shall be identified for installation with the short-circuit and ground-fault protective device for the branch circuit to which the equipment is connected.

(B) Attachment Plug and Receptacle or Cord Connector Rating. The rating of the attachment plug and receptacle or cord connector shall not exceed 20 amperes at 125 volts or 15 amperes at 250 volts.

(C) Time Delay. The short-circuit and ground-fault protective device protecting the branch circuit shall have sufficient time delay to permit the motor-compressor and other motors to start and accelerate their loads.

Part VII. Provisions for Room Air Conditioners

440.60 General. The provisions of Part VII shall apply to electrically energized room air conditioners that control temperature and humidity. For the purpose of Part VII, a room air conditioner (with or without provisions for heating) shall be considered as an ac appliance of the air-cooled window, console, or in-wall type that is installed in the conditioned room and that incorporates a hermetic refrigerant motor-compressor(s). The provisions of Part VII cover equipment rated not over 250 volts, single phase, and the equipment shall be permitted to be cord- and attachment-plug-connected.

A room air conditioner that is rated 3-phase or rated over 250 volts shall be directly connected to a wiring method recognized in Chapter 3, and provisions of Part VII shall not apply.

440.61 Grounding. The enclosures of room air conditioners shall be connected to the equipment grounding conductor in accordance with 250.110, 250.112, and 250.114.

440.62 Branch-Circuit Requirements.

(A) Room Air Conditioner as a Single Motor Unit. A room air conditioner shall be considered as a single motor unit in determining its branch-circuit requirements where all the following conditions are met:

(1) It is cord- and attachment-plug-connected.
(2) Its rating is not more than 40 amperes and 250 volts, single phase.
(3) Total rated-load current is shown on the room air-conditioner nameplate rather than individual motor currents.

(4) The rating of the branch-circuit short-circuit and ground-fault protective device does not exceed the ampacity of the branch-circuit conductors or the rating of the receptacle, whichever is less.

(B) Where No Other Loads Are Supplied. The total marked rating of a cord- and attachment-plug-connected room air conditioner shall not exceed 80 percent of the rating of a branch circuit where no other loads are supplied.

(C) Where Lighting Units or Other Appliances Are Also Supplied. The total marked rating of a cord- and attachment-plug-connected room air conditioner shall not exceed 50 percent of the rating of a branch circuit where lighting outlets, other appliances, or general-use receptacles are also supplied. Where the circuitry is interlocked to prevent simultaneous operation of the room air conditioner and energization of other outlets on the same branch circuit, a cord- and attachment-plug-connected room air conditioner shall not exceed 80 percent of the branch-circuit rating.

440.63 Disconnecting Means. An attachment plug and receptacle or cord connector shall be permitted to serve as the disconnecting means for a single-phase room air conditioner rated 250 volts or less if (1) the manual controls on the room air conditioner are readily accessible and located within 1.8 m (6 ft) of the floor, or (2) an approved manually operable disconnecting means is installed in a readily accessible location within sight from the room air conditioner.

440.64 Supply Cords. Where a flexible cord is used to supply a room air conditioner, the length of such cord shall not exceed 3.0 m (10 ft) for a nominal, 120-volt rating or 1.8 m (6 ft) for a nominal, 208- or 240-volt rating.

440.65 Protection Devices. Single-phase cord- and plug-connected room air conditioners shall be provided with one of the following factory-installed devices:

(1) Leakage-current detector-interruptor (LCDI)
(2) Arc-fault circuit interrupter (AFCI)
(3) Heat detecting circuit interrupter (HDCI)

The protection device shall be an integral part of the attachment plug or be located in the power supply cord within 300 mm (12 in.) of the attachment plug.

ARTICLE 445
Generators

445.1 Scope. This article contains installation and other requirements for generators.

445.10 Location. Generators shall be of a type suitable for the locations in which they are installed. They shall also meet the requirements for motors in 430.14.

Informational Note: See NFPA 37, *Standard for the Installation and Use of Stationary Combustion Engines and Gas Turbines*, for information on the location of generator exhaust.

445.11 Marking. Each generator shall be provided with a nameplate giving the manufacturer's name, the rated frequency, the number of phases if ac, the rating in kilowatts or kilovolt-amperes, the power factor, the normal volts and amperes corresponding to the rating, the rated ambient temperature, and the rated temperature rise.

Nameplates or manufacturer's instructions shall provide the following information for all stationary generators and portable generators rated more than 15 kW:

(1) Subtransient, transient, synchronous, and zero sequence reactances
(2) Power rating category
(3) Insulation system class
(4) Indication if the generator is protected against overload by inherent design, an overcurrent protective relay, circuit breaker, or fuse
(5) Maximum short-circuit current for inverter-based generators, in lieu of the synchronous, subtransient, and transient reactances

Marking shall be provided by the manufacturer to indicate whether or not the generator neutral is bonded to its frame. Where the bonding is modified in the field, additional marking shall be required to indicate whether the neutral is bonded to the frame.

445.12 Overcurrent Protection.

(A) Constant-Voltage Generators. Constant-voltage generators, except ac generator exciters, shall be protected from overload by inherent design, circuit breakers, fuses, protective relays, or other identified overcurrent protective means suitable for the conditions of use.

(B) Two-Wire Generators. Two-wire, dc generators shall be permitted to have overcurrent protection in one conductor only if the overcurrent device is actuated by the entire current generated other than the current in the shunt field. The overcurrent device shall not open the shunt field.

(C) 65 Volts or Less. Generators operating at 65 volts or less and driven by individual motors shall be considered as protected by the overcurrent device protecting the motor if these devices will operate when the generators are delivering not more than 150 percent of their full-load rated current.

(D) Balancer Sets. Two-wire, dc generators used in conjunction with balancer sets to obtain neutral points for 3-wire systems shall be equipped with overcurrent devices that disconnect the 3-wire system in case of excessive unbalancing of voltages or currents.

(E) Three-Wire, Direct-Current Generators. Three-wire, dc generators, whether compound or shunt wound, shall be equipped with overcurrent devices, one in each armature lead, and connected so as to be actuated by the entire current from the armature. Such overcurrent devices shall consist either of a double-pole, double-coil circuit breaker or of a 4-pole circuit breaker connected in the main and equalizer leads and tripped by two overcurrent devices, one in each armature lead. Such protective devices shall be interlocked so that no one pole can be opened without simultaneously disconnecting both leads of the armature from the system.

Exception to (A) through (E): Where deemed by the authority having jurisdiction that a generator is vital to the operation of an electrical system and the generator should operate to failure to prevent a greater

hazard to persons, the overload sensing device(s) shall be permitted to be connected to an annunciator or alarm supervised by authorized personnel instead of interrupting the generator circuit.

445.13 Ampacity of Conductors.

(A) General. The ampacity of the conductors from the generator output terminals to the first distribution device(s) containing overcurrent protection shall not be less than 115 percent of the nameplate current rating of the generator. It shall be permitted to size the neutral conductors in accordance with 220.61. Conductors that must carry ground-fault currents shall not be smaller than required by 250.30(A). Neutral conductors of dc generators that must carry ground-fault currents shall not be smaller than the minimum required size of the largest conductor.

Exception: Where the design and operation of the generator prevent overloading, the ampacity of the conductors shall not be less than 100 percent of the nameplate current rating of the generator.

N **(B) Overcurrent Protection Provided.** Where the generator set is equipped with a listed overcurrent protective device or a combination of a current transformer and overcurrent relay, conductors shall be permitted to be tapped from the load side of the protected terminals in accordance with 240.21(B).

Tapped conductors shall not be permitted for portable generators rated 15 kW or less where field wiring connection terminals are not accessible.

445.14 Protection of Live Parts. Live parts of generators operated at more than 50 volts ac or 60 volts dc to ground shall not be exposed to accidental contact where accessible to unqualified persons.

445.15 Guards for Attendants. Where necessary for the safety of attendants, the requirements of 430.233 shall apply.

445.16 Bushings. Where field-installed wiring passes through an opening in an enclosure, a conduit box, or a barrier, a bushing shall be used to protect the conductors from the edges of an opening having sharp edges. The bushing shall have smooth, well-rounded surfaces where it may be in contact with the conductors. If used where oils, grease, or other contaminants may be present, the bushing shall be made of a material not deleteriously affected.

445.17 Generator Terminal Housings. Generator terminal housings shall comply with 430.12. Where a horsepower rating is required to determine the required minimum size of the generator terminal housing, the full-load current of the generator shall be compared with comparable motors in Table 430.247 through Table 430.250. The higher horsepower rating of Table 430.247 and Table 430.250 shall be used whenever the generator selection is between two ratings.

Exception: This section shall not apply to generators rated over 600 volts.

445.18 Disconnecting Means and Shutdown of Prime Mover.

N **(A) Disconnecting Means.** Generators other than cord-and-plug-connected portable shall have one or more disconnecting means. Each disconnecting means shall simultaneously open all associated ungrounded conductors. Each disconnecting means shall be lockable in the open position in accordance with 110.25.

(B) Shutdown of Prime Mover. Generators shall have provisions to shut down the prime mover. The means of shutdown shall comply with all of the following:

(1) Be equipped with provisions to disable all prime mover start control circuits to render the prime mover incapable of starting

(2) Initiate a shutdown mechanism that requires a mechanical reset

The provisions to shut down the prime mover shall be permitted to satisfy the requirements of 445.18(A) where it is capable of being locked in the open position in accordance with 110.25.

Generators with greater than 15 kW rating shall be provided with an additional requirement to shut down the prime mover. This additional shutdown means shall be located outside the equipment room or generator enclosure and shall also meet the requirements of 445.18(B)(1) and (B)(2).

(C) Generators Installed in Parallel. Where a generator is installed in parallel with other generators, the provisions of 445.18(A) shall be capable of isolating the generator output terminals from the paralleling equipment. The disconnecting means shall not be required to be located at the generator.

445.20 Ground-Fault Circuit-Interrupter Protection for Receptacles on 15-kW or Smaller Portable Generators. Receptacle outlets that are a part of a 15-kW or smaller portable generator shall have listed ground-fault circuit-interrupter protection (GFCI) for personnel integral to the generator or receptacle as indicated in either (A) or (B):

(A) Unbonded (Floating Neutral) Generators. Unbonded generators with both 125-volt and 125/250-volt receptacle outlets shall have listed GFCI protection for personnel integral to the generator or receptacle on all 125-volt, 15- and 20-ampere receptacle outlets.

Exception: GFCI protection shall not be required where the 125-volt receptacle outlets(s) is interlocked such that it is not available for use when any 125/250-volt receptacle(s) is in use.

(B) Bonded Neutral Generators. Bonded generators shall be provided with GFCI protection on all 125-volt, 15- and 20-ampere receptacle outlets.

Informational Note: Refer to 590.6(A)(3) for GFCI requirements for 15-kW or smaller portable generators used for temporary electric power and lighting.

Exception to (A) and (B): If the generator was manufactured or remanufactured prior to January 1, 2015, listed cord sets or devices incorporating listed GFCI protection for personnel identified for portable use shall be permitted.

ARTICLE 450
Transformers and Transformer Vaults (Including Secondary Ties)

Part I. General Provisions

450.1 Scope. This article covers the installation of all transformers.

Exception No. 1: Current transformers.

Exception No. 2: Dry-type transformers that constitute a component part of other apparatus and comply with the requirements for such apparatus.

Exception No. 3: Transformers that are an integral part of an X-ray, high-frequency, or electrostatic-coating apparatus.

Exception No. 4: Transformers used with Class 2 and Class 3 circuits that comply with Article 725.

Exception No. 5: Transformers for sign and outline lighting that comply with Article 600.

Exception No. 6: Transformers for electric-discharge lighting that comply with Article 410.

Exception No. 7: Transformers used for power-limited fire alarm circuits that comply with Part III of Article 760.

Exception No. 8: Transformers used for research, development, or testing, where effective arrangements are provided to safeguard persons from contacting energized parts.

This article covers the installation of transformers dedicated to supplying power to a fire pump installation as modified by Article 695.

This article also covers the installation of transformers in hazardous (classified) locations as modified by Articles 501 through 504.

450.2 Definition. For the purpose of this article, the following definition shall apply.

Transformer. An individual transformer, single- or polyphase, identified by a single nameplate, unless otherwise indicated in this article.

450.3 Overcurrent Protection. Overcurrent protection of transformers shall comply with 450.3(A), (B), or (C). As used in this section, the word *transformer* shall mean a transformer or polyphase bank of two or more single-phase transformers operating as a unit.

Informational Note No. 1: See 240.4, 240.21, 240.100, and 240.101 for overcurrent protection of conductors.

Informational Note No. 2: Nonlinear loads can increase heat in a transformer without operating its overcurrent protective device.

(A) Transformers Over 1000 Volts, Nominal. Overcurrent protection shall be provided in accordance with Table 450.3(A).

(B) Transformers 1000 Volts, Nominal, or Less. Overcurrent protection shall be provided in accordance with Table 450.3(B).

Exception: Where the transformer is installed as a motor control circuit transformer in accordance with 430.72(C)(1) through (C)(5).

(C) Voltage (Potential) Transformers. Voltage (potential) transformers installed indoors or enclosed shall be protected with primary fuses.

> Informational Note: For protection of instrument circuits including voltage transformers, see 408.52.

450.4 Autotransformers 1000 Volts, Nominal, or Less.

(A) Overcurrent Protection. Each autotransformer 1000 volts, nominal, or less shall be protected by an individual overcurrent device installed in series with each ungrounded input conductor. Such overcurrent device shall be rated or set at not more than 125 percent of the rated full-load input current of the autotransformer. Where this calculation does not correspond to a standard rating of a fuse or nonadjustable circuit breaker and the rated input current is 9 amperes or more, the next higher standard rating described in 240.6 shall be permitted.

An overcurrent device shall not be installed in series with the shunt winding (the winding common to both the input and the output circuits) of the autotransformer between Points A and B as shown in Figure 450.4(A).

Exception: Where the rated input current of the autotransformer is less than 9 amperes, an overcurrent device rated or set at not more than 167 percent of the input current shall be permitted.

(B) Transformer Field-Connected as an Autotransformer. A transformer field-connected as an autotransformer shall be identified for use at elevated voltage.

> Informational Note: For information on permitted uses of autotransformers, see 210.9 and 215.11.

450.5 Grounding Autotransformers. Grounding autotransformers covered in this section are zigzag or T-connected transformers connected to 3-phase, 3-wire ungrounded systems for the purpose of creating a 3-phase, 4-wire distribution system or providing a neutral point for grounding purposes. Such transformers shall have a continuous per-phase current rating and a continuous neutral current rating. Zigzag-connected transformers shall not be installed on the load side of any system grounding connection, including those made in accordance with 250.24(B), 250.30(A)(1), or 250.32(B), Exception No. 1.

> Informational Note: The phase current in a grounding autotransformer is one-third the neutral current.

Table 450.3(A) Maximum Rating or Setting of Overcurrent Protection for Transformers Over 1000 Volts (as a Percentage of Transformer-Rated Current)

Location Limitations	Transformer Rated Impedance	Primary Protection over 1000 Volts		Secondary Protection (See Note 2.)		
				Over 1000 Volts		1000 Volts or Less
		Circuit Breaker (See Note 4.)	Fuse Rating	Circuit Breaker (See Note 4.)	Fuse Rating	Circuit Breaker or Fuse Rating
Any location	Not more than 6%	600% (See Note 1.)	300% (See Note 1.)	300% (See Note 1.)	250% (See Note 1.)	125% (See Note 1.)
	More than 6% and not more than 10%	400% (See Note 1.)	300% (See Note 1.)	250% (See Note 1.)	225% (See Note 1.)	125% (See Note 1.)
Supervised locations only (See Note 3.)	Any	300% (See Note 1.)	250% (See Note 1.)	Not required	Not required	Not required
	Not more than 6%	600%	300%	300% (See Note 5.)	250% (See Note 5.)	250% (See Note 5.)
	More than 6% and not more than 10%	400%	300%	250% (See Note 5.)	225% (See Note 5.)	250% (See Note 5.)

Notes:

1. Where the required fuse rating or circuit breaker setting does not correspond to a standard rating or setting, a higher rating or setting that does not exceed the following shall be permitted:
a. The next higher standard rating or setting for fuses and circuit breakers 1000 volts and below, or
b. The next higher commercially available rating or setting for fuses and circuit breakers above 1000 volts.

2. Where secondary overcurrent protection is required, the secondary overcurrent device shall be permitted to consist of not more than six circuit breakers or six sets of fuses grouped in one location. Where multiple overcurrent devices are utilized, the total of all the device ratings shall not exceed the allowed value of a single overcurrent device. If both circuit breakers and fuses are used as the overcurrent device, the total of the device ratings shall not exceed that allowed for fuses.

3. A supervised location is a location where conditions of maintenance and supervision ensure that only qualified persons monitor and service the transformer installation.

4. Electronically actuated fuses that may be set to open at a specific current shall be set in accordance with settings for circuit breakers.

5. A transformer equipped with a coordinated thermal overload protection by the manufacturer shall be permitted to have separate secondary protection omitted.

Table 450.3(B) Maximum Rating or Setting of Overcurrent Protection for Transformers 1000 Volts and Less (as a Percentage of Transformer-Rated Current)

Protection Method	Primary Protection			Secondary Protection (See Note 2.)	
	Currents of 9 Amperes or More	Currents Less Than 9 Amperes	Currents Less Than 2 Amperes	Currents of 9 Amperes or More	Currents Less Than 9 Amperes
Primary only protection	125% (See Note 1.)	167%	300%	Not required	Not required
Primary and secondary protection	250% (See Note 3.)	250% (See Note 3.)	250% (See Note 3.)	125% (See Note 1.)	167%

Notes:

1. Where 125 percent of this current does not correspond to a standard rating of a fuse or nonadjustable circuit breaker, a higher rating that does not exceed the next higher standard rating shall be permitted.

2. Where secondary overcurrent protection is required, the secondary overcurrent device shall be permitted to consist of not more than six circuit breakers or six sets of fuses grouped in one location. Where multiple overcurrent devices are utilized, the total of all the device ratings shall not exceed the allowed value of a single overcurrent device.

3. A transformer equipped with coordinated thermal overload protection by the manufacturer and arranged to interrupt the primary current shall be permitted to have primary overcurrent protection rated or set at a current value that is not more than six times the rated current of the transformer for transformers having not more than 6 percent impedance and not more than four times the rated current of the transformer for transformers having more than 6 percent but not more than 10 percent impedance.

(A) Three-Phase, 4-Wire System. A grounding autotransformer used to create a 3-phase, 4-wire distribution system from a 3-phase, 3-wire ungrounded system shall conform to 450.5(A)(1) through (A)(4).

(1) Connections. The transformer shall be directly connected to the ungrounded phase conductors and shall not be switched or provided with overcurrent protection that is independent of the main switch and common-trip overcurrent protection for the 3-phase, 4-wire system.

(2) Overcurrent Protection. An overcurrent sensing device shall be provided that will cause the main switch or common-trip overcurrent protection referred to in 450.5(A)(1) to open if the load on the autotransformer reaches or exceeds 125 percent of its continuous current per-phase or neutral rating. Delayed tripping for temporary overcurrents sensed at the autotransformer overcurrent device shall be permitted for the purpose of allowing proper operation of branch or feeder protective devices on the 4-wire system.

(3) Transformer Fault Sensing. A fault-sensing system that causes the opening of a main switch or common-trip overcurrent device for the 3-phase, 4-wire system shall be provided to guard against single-phasing or internal faults.

> Informational Note: This can be accomplished by the use of two subtractive-connected donut-type current transformers installed to sense and signal when an unbalance occurs in the line current to the autotransformer of 50 percent or more of rated current.

(4) Rating. The autotransformer shall have a continuous neutral-current rating that is not less than the maximum possible neutral unbalanced load current of the 4-wire system.

(B) Ground Reference for Fault Protection Devices. A grounding autotransformer used to make available a specified magnitude of ground-fault current for operation of a ground-responsive protective device on a 3-phase, 3-wire ungrounded system shall conform to 450.5(B)(1) and (B)(2).

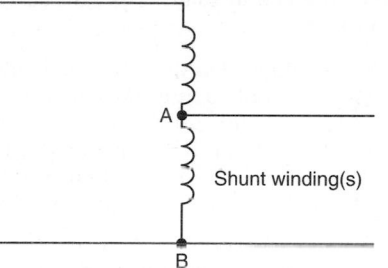

A
Shunt winding(s)
B

FIGURE 450.4(A)   Autotransformer.

(1) Rating. The autotransformer shall have a continuous neutral current rating not less than the specified ground-fault current.

(2) Overcurrent Protection. Overcurrent protection shall comply with (a) and (b).

(a)   *Operation and Interrupting Rating.* An overcurrent protective device having an interrupting rating in compliance with 110.9 and that will open simultaneously all ungrounded conductors when it operates shall be applied in the grounding autotransformer branch circuit.

(b)   *Ampere Rating.* The overcurrent protection shall be rated or set at a current not exceeding 125 percent of the auto-transformer continuous per-phase current rating or 42 percent of the continuous-current rating of any series-connected devices in the autotransformer neutral connection. Delayed tripping for temporary overcurrents to permit the proper operation of ground-responsive tripping devices on the main system shall be permitted but shall not exceed values that would be more than the short-time current rating of the grounding autotransformer or any series connected devices in the neutral connection thereto.

Exception: For high-impedance grounded systems covered in 250.36, where the maximum ground-fault current is designed to be not more than 10 amperes, and where the grounding autotransformer and the grounding impedance are rated for continuous duty, an overcurrent device rated not more than 20 amperes that will simultaneously open all ungrounded conductors shall be permitted to be installed on the line side of the grounding autotransformer.

(C) Ground Reference for Damping Transitory Overvoltages. A grounding autotransformer used to limit transitory overvoltages shall be of suitable rating and connected in accordance with 450.5(A)(1).

450.6 Secondary Ties. As used in this article, a secondary tie is a circuit operating at 1000 volts, nominal, or less between phases that connects two power sources or power supply points, such as the secondaries of two transformers. The tie shall be permitted to consist of one or more conductors per phase or neutral. Conductors connecting the secondaries of transformers in accordance with 450.7 shall not be considered secondary ties.

As used in this section, the word *transformer* means a transformer or a bank of transformers operating as a unit.

(A) Tie Circuits. Tie circuits shall be provided with overcurrent protection at each end as required in Parts I, II, and VIII of Article 240.

Under the conditions described in 450.6(A)(1) and 450.6(A)(2), the overcurrent protection shall be permitted to be in accordance with 450.6(A)(3).

(1) Loads at Transformer Supply Points Only. Where all loads are connected at the transformer supply points at each end of the tie and overcurrent protection is not provided in accordance with Parts I, II, and VIII of Article 240, the rated ampacity of the tie shall not be less than 67 percent of the rated secondary current of the highest rated transformer supplying the secondary tie system.

(2) Loads Connected Between Transformer Supply Points. Where load is connected to the tie at any point between transformer supply points and overcurrent protection is not provided in accordance with Parts I, II, and VIII of Article 240, the rated ampacity of the tie shall not be less than 100 percent of the rated secondary current of the highest rated transformer supplying the secondary tie system.

Exception: Tie circuits comprised of multiple conductors per phase shall be permitted to be sized and protected in accordance with 450.6(A)(4).

(3) Tie Circuit Protection. Under the conditions described in 450.6(A)(1) and (A)(2), both supply ends of each ungrounded tie conductor shall be equipped with a protective device that opens at a predetermined temperature of the tie conductor under short-circuit conditions. This protection shall consist of one of the following: (1) a fusible link cable connector, terminal, or lug, commonly known as a limiter, each being of a size corresponding with that of the conductor and of construction and characteristics according to the operating voltage and the type of insulation on the tie conductors or (2) automatic circuit breakers actuated by devices having comparable time–current characteristics.

(4) Interconnection of Phase Conductors Between Transformer Supply Points. Where the tie consists of more than one conductor per phase or neutral, the conductors of each phase or neutral shall comply with one of the following provisions.

(a) *Interconnected.* The conductors shall be interconnected in order to establish a load supply point, and the protective device specified in 450.6(A)(3) shall be provided in each ungrounded tie conductor at this point on both sides of the interconnection. The means of interconnection shall have an ampacity not less than the load to be served.

(b) *Not Interconnected.* The loads shall be connected to one or more individual conductors of a paralleled conductor tie without interconnecting the conductors of each phase or neutral and without the protection specified in 450.6(A)(3) at load connection points. Where this is done, the tie conductors of each phase or neutral shall have a combined capacity ampacity of not less than 133 percent of the rated secondary current of the highest rated transformer supplying the secondary tie system, the total load of such taps shall not exceed the rated secondary current of the highest rated transformer, and the loads shall be equally divided on each phase and on the individual conductors of each phase as far as practicable.

(5) Tie Circuit Control. Where the operating voltage exceeds 150 volts to ground, secondary ties provided with limiters shall have a switch at each end that, when open, de-energizes the associated tie conductors and limiters. The current rating of the switch shall not be less than the rated current ampacity of the conductors connected to the switch. It shall be capable of interrupting its rated current, and it shall be constructed so that it will not open under the magnetic forces resulting from short-circuit current.

(B) Overcurrent Protection for Secondary Connections. Where secondary ties are used, an overcurrent device rated or set at not more than 250 percent of the rated secondary current of the transformers shall be provided in the secondary connections of each transformer supplying the tie system. In addition, an automatic circuit breaker actuated by a reverse-current relay set to open the circuit at not more than the rated secondary current of the transformer shall be provided in the secondary connection of each transformer.

(C) Grounding. Where the secondary tie system is grounded, each transformer secondary supplying the tie system shall be grounded in accordance with the requirements of 250.30 for separately derived systems.

450.7 Parallel Operation. Transformers shall be permitted to be operated in parallel and switched as a unit, provided the overcurrent protection for each transformer meets the requirements of 450.3(A) for primary and secondary protective devices over 1000 volts, or 450.3(B) for primary and secondary protective devices 1000 volts or less.

450.8 Guarding. Transformers shall be guarded as specified in 450.8(A) through (D).

(A) Mechanical Protection. Appropriate provisions shall be made to minimize the possibility of damage to transformers from external causes where the transformers are exposed to physical damage.

(B) Case or Enclosure. Dry-type transformers shall be provided with a noncombustible moisture-resistant case or enclosure that provides protection against the accidental insertion of foreign objects.

(C) Exposed Energized Parts. Switches or other equipment operating at 1000 volts, nominal, or less and serving only equipment within a transformer enclosure shall be permitted to be installed in the transformer enclosure if accessible to qualified persons only. All energized parts shall be guarded in accordance with 110.27 and 110.34.

(D) Voltage Warning. The operating voltage of exposed live parts of transformer installations shall be indicated by signs or visible markings on the equipment or structures.

450.9 Ventilation. The ventilation shall dispose of the transformer full-load heat losses without creating a temperature rise that is in excess of the transformer rating.

> Informational Note No. 1: See ANSI/IEEE C57.12.00-1993, *General Requirements for Liquid-Immersed Distribution, Power, and Regulating Transformers,* and ANSI/IEEE C57.12.01-1989, *General Requirements for Dry-Type Distribution and Power Transformers.*

> Informational Note No. 2: Additional losses may occur in some transformers where nonsinusoidal currents are present, resulting in increased heat in the transformer above its rating. See ANSI/IEEE C57.110-1993, *Recommended Practice for Establishing Transformer Capability When Supplying Nonsinusoidal Load Currents,* where transformers are utilized with nonlinear loads.

Transformers with ventilating openings shall be installed so that the ventilating openings are not blocked by walls or other obstructions. The required clearances shall be clearly marked on the transformer.

450.10 Grounding.

(A) Dry-Type Transformer Enclosures. Where separate equipment grounding conductors and supply-side bonding jumpers are installed, a terminal bar for all grounding and bonding conductor connections shall be secured inside the transformer enclosure. The terminal bar shall be bonded to the enclosure in accordance with 250.12 and shall not be installed on or over any vented portion of the enclosure.

Exception: Where a dry-type transformer is equipped with wire-type connections (leads), the grounding and bonding connections shall be permitted to be connected together using any of the methods in 250.8 and shall be bonded to the enclosure if of metal.

(B) Other Metal Parts. Where grounded, exposed non–current-carrying metal parts of transformer installations, including fences, guards, and so forth, shall be grounded and bonded under the conditions and in the manner specified for electrical equipment and other exposed metal parts in Parts V, VI, and VII of Article 250.

450.11 Marking.

(A) General. Each transformer shall be provided with a nameplate giving the following information:

(1) Name of manufacturer
(2) Rated kilovolt-amperes
(3) Frequency
(4) Primary and secondary voltage
(5) Impedance of transformers 25 kVA and larger
(6) Required clearances for transformers with ventilating openings
(7) Amount and kind of insulating liquid where used
(8) For dry-type transformers, temperature class for the insulation system

(B) Source Marking. A transformer shall be permitted to be supplied at the marked secondary voltage, provided that the installation is in accordance with the manufacturer's instructions.

450.12 Terminal Wiring Space. The minimum wire-bending space at fixed, 1000-volt and below terminals of transformer line and load connections shall be as required in 312.6. Wiring space for pigtail connections shall conform to Table 314.16(B).

450.13 Accessibility. All transformers and transformer vaults shall be readily accessible to qualified personnel for inspection and maintenance or shall meet the requirements of 450.13(A) or 450.13(B).

(A) Open Installations. Dry-type transformers 1000 volts, nominal, or less, located in the open on walls, columns, or structures, shall not be required to be readily accessible.

(B) Hollow Space Installations. Dry-type transformers 1000 volts, nominal, or less and not exceeding 50 kVA shall be permitted in hollow spaces of buildings not permanently closed in by structure, provided they meet the ventilation requirements of 450.9 and separation from combustible materials requirements of 450.21(A). Transformers so installed shall not be required to be readily accessible.

450.14 Disconnecting Means. Transformers, other than Class 2 or Class 3 transformers, shall have a disconnecting means located either in sight of the transformer or in a remote location. Where located in a remote location, the disconnecting means shall be lockable in accordance with 110.25, and its location shall be field marked on the transformer.

Part II. Specific Provisions Applicable to Different Types of Transformers

450.21 Dry-Type Transformers Installed Indoors.

(A) Not Over 112½ kVA. Dry-type transformers installed indoors and rated 112½ kVA or less shall have a separation of at least 300 mm (12 in.) from combustible material unless separated from the combustible material by a fire-resistant, heat-insulated barrier.

Exception: This rule shall not apply to transformers rated for 1000 volts, nominal, or less that are completely enclosed, except for ventilating openings.

(B) Over 112½ kVA. Individual dry-type transformers of more than 112½ kVA rating shall be installed in a transformer room of fire-resistant construction. Unless specified otherwise in this article, the term *fire resistant* means a construction having a minimum fire rating of 1 hour.

Exception No. 1: Transformers with Class 155 or higher insulation systems and separated from combustible material by a fire-resistant, heat-insulating barrier or by not less than 1.83 m (6 ft) horizontally and 3.7 m (12 ft) vertically.

Exception No. 2: Transformers with Class 155 or higher insulation systems and completely enclosed except for ventilating openings.

> Informational Note: See ANSI/ASTM E119-15, *Method for Fire Tests of Building Construction and Materials.*

(C) Over 35,000 Volts. Dry-type transformers rated over 35,000 volts shall be installed in a vault complying with Part III of this article.

450.22 Dry-Type Transformers Installed Outdoors. Dry-type transformers installed outdoors shall have a weatherproof enclosure.

Transformers exceeding 112½ kVA shall not be located within 300 mm (12 in.) of combustible materials of buildings unless the transformer has Class 155 insulation systems or higher and is completely enclosed except for ventilating openings.

450.23 Less-Flammable Liquid-Insulated Transformers. Transformers insulated with listed less-flammable liquids that have a fire point of not less than 300°C shall be permitted to be installed in accordance with 450.23(A) or 450.23(B).

(A) Indoor Installations. Indoor installations shall be permitted in accordance with one of the following:

(1) In Type I or Type II buildings, in areas where all of the following requirements are met:

  a. The transformer is rated 35,000 volts or less.
  b. No combustible materials are stored.
  c. A liquid confinement area is provided.
  d. The installation complies with all the restrictions provided for in the listing of the liquid.

  > Informational Note: Such restrictions may include, but are not limited to: maximum pressure of the tank, use of a pressure relief valve, appropriate fuse types and proper sizing of overcurrent protection.

  e. With an automatic fire extinguishing system and a liquid confinement area, provided the transformer is rated 35,000 volts or less
  f. In accordance with 450.26

(B) Outdoor Installations. Less-flammable liquid-filled transformers shall be permitted to be installed outdoors, attached to, adjacent to, or on the roof of buildings, where installed in accordance with (1) or (2).

(1) For Type I and Type II buildings, the installation shall comply with all the restrictions provided for in the listing of the liquid.

  > Informational Note No. 1: Installations adjacent to combustible material, fire escapes, or door and window openings may require additional safeguards such as those listed in 450.27.

  > Informational Note No. 2: Such restrictions may include, but are not limited to: maximum pressure of the tank, use of a pressure relief valve, appropriate fuse types, and proper sizing of overcurrent protection.

(2) In accordance with 450.27.

  > Informational Note No. 1: As used in this section, *Type I and Type II buildings* refers to Type I and Type II building construction as defined in NFPA 220-2015, *Standard on Types of Building Construction. Combustible materials* refers to those materials not classified as noncombustible or limited-combustible as defined in NFPA 220-2015, *Standard on Types of Building Construction*.

  > Informational Note No. 2: See definition of *Listed* in Article 100.

450.24 Nonflammable Fluid-Insulated Transformers. Transformers insulated with a dielectric fluid identified as nonflammable shall be permitted to be installed indoors or outdoors. Such transformers installed indoors and rated over 35,000 volts shall be installed in a vault. Such transformers installed indoors shall be furnished with a liquid confinement area and a pressure-relief vent. The transformers shall be furnished with a means for absorbing any gases generated by arcing inside the tank, or the pressure-relief vent shall be connected to a chimney or flue that will carry such gases to an environmentally safe area.

> Informational Note: Safety may be increased if fire hazard analyses are performed for such transformer installations.

For the purposes of this section, a nonflammable dielectric fluid is one that does not have a flash point or fire point and is not flammable in air.

450.25 Askarel-Insulated Transformers Installed Indoors. Askarel-insulated transformers installed indoors and rated over 25 kVA shall be furnished with a pressure-relief vent. Where installed in a poorly ventilated place, they shall be furnished with a means for absorbing any gases generated by arcing inside the case, or the pressure-relief vent shall be connected to a chimney or flue that carries such gases outside the building. Askarel-insulated transformers rated over 35,000 volts shall be installed in a vault.

450.26 Oil-Insulated Transformers Installed Indoors. Oil-insulated transformers installed indoors shall be installed in a vault constructed as specified in Part III of this article.

Exception No. 1: Where the total capacity does not exceed 112½ kVA, the vault specified in Part III of this article shall be permitted to be constructed of reinforced concrete that is not less than 100 mm (4 in.) thick.

Exception No. 2: Where the nominal voltage does not exceed 1000, a vault shall not be required if suitable arrangements are made to prevent a transformer oil fire from igniting other materials and the total capacity in one location does not exceed 10 kVA in a section of the building classified as combustible or 75 kVA where the surrounding structure is classified as fire-resistant construction.

Exception No. 3: Electric furnace transformers that have a total rating not exceeding 75 kVA shall be permitted to be installed without a vault in a building or room of fire-resistant construction, provided suitable arrangements are made to prevent a transformer oil fire from spreading to other combustible material.

Exception No. 4: A transformer that has a total rating not exceeding 75 kVA and a supply voltage of 1000 volts or less that is an integral part of charged-particle-accelerating equipment shall be permitted to be installed without a vault in a building or room of noncombustible or fire-resistant construction, provided suitable arrangements are made to prevent a transformer oil fire from spreading to other combustible material.

Exception No. 5: Transformers shall be permitted to be installed in a detached building that does not comply with Part III of this article if neither the building nor its contents present a fire hazard to any other building or property, and if the building is used only in supplying electric service and the interior is accessible only to qualified persons.

Exception No. 6: Oil-insulated transformers shall be permitted to be used without a vault in portable and mobile surface mining equipment (such as electric excavators) if each of the following conditions is met:

(1) Provision is made for draining leaking fluid to the ground.
(2) Safe egress is provided for personnel.
(3) A minimum 6-mm (¼-in.) steel barrier is provided for personnel protection.

450.27 Oil-Insulated Transformers Installed Outdoors. Combustible material, combustible buildings, and parts of buildings, fire escapes, and door and window openings shall be safeguarded from fires originating in oil-insulated transformers installed on roofs, attached to or adjacent to a building or combustible material.

In cases where the transformer installation presents a fire hazard, one or more of the following safeguards shall be applied according to the degree of hazard involved:

(1) Space separations
(2) Fire-resistant barriers
(3) Automatic fire suppression systems
(4) Enclosures that confine the oil of a ruptured transformer tank

Oil enclosures shall be permitted to consist of fire-resistant dikes, curbed areas or basins, or trenches filled with coarse, crushed stone. Oil enclosures shall be provided with trapped drains where the exposure and the quantity of oil involved are such that removal of oil is important.

Informational Note: For additional information on transformers installed on poles or structures or under ground, see ANSI C2-2007, *National Electrical Safety Code.*

450.28 Modification of Transformers. When modifications are made to a transformer in an existing installation that change the type of the transformer with respect to Part II of this article, such transformer shall be marked to show the type of insulating liquid installed, and the modified transformer installation shall comply with the applicable requirements for that type of transformer.

Part III. Transformer Vaults

450.41 Location. Vaults shall be located where they can be ventilated to the outside air without using flues or ducts wherever such an arrangement is practicable.

450.42 Walls, Roofs, and Floors. The walls and roofs of vaults shall be constructed of materials that have approved structural strength for the conditions with a minimum fire resistance of 3 hours. The floors of vaults in contact with the earth shall be of concrete that is not less than 100 mm (4 in.) thick, but, where the vault is constructed with a vacant space or other stories below it, the floor shall have approved structural strength for the load imposed thereon and a minimum fire resistance of 3 hours. For the purposes of this section, studs and wallboard construction shall not be permitted.

Exception: Where transformers are protected with automatic sprinkler, water spray, carbon dioxide, or halon, construction of 1-hour rating shall be permitted.

Informational Note No. 1: For additional information, see ANSI/ASTM E119-15, *Method for Fire Tests of Building Construction and Materials.*

Informational Note No. 2: A typical 3-hour construction is 150 mm (6 in.) thick reinforced concrete.

450.43 Doorways. Vault doorways shall be protected in accordance with 450.43(A), (B), and (C).

(A) Type of Door. Each doorway leading into a vault from the building interior shall be provided with a tight-fitting door that has a minimum fire rating of 3 hours. The authority having jurisdiction shall be permitted to require such a door for an exterior wall opening where conditions warrant.

Exception: Where transformers are protected with automatic sprinkler, water spray, carbon dioxide, or halon, construction of 1-hour rating shall be permitted.

Informational Note: For additional information, see NFPA 80-2013, *Standard for Fire Doors and Other Opening Protectives.*

(B) Sills. A door sill or curb that is of an approved height that will confine the oil from the largest transformer within the vault shall be provided, and in no case shall the height be less than 100 mm (4 in.).

(C) Locks. Doors shall be equipped with locks, and doors shall be kept locked, access being allowed only to qualified persons. Personnel doors shall open in the direction of egress and be equipped with listed panic hardware.

450.45 Ventilation Openings. Where required by 450.9, openings for ventilation shall be provided in accordance with 450.45(A) through (F).

(A) Location. Ventilation openings shall be located as far as possible from doors, windows, fire escapes, and combustible material.

(B) Arrangement. A vault ventilated by natural circulation of air shall be permitted to have roughly half of the total area of openings required for ventilation in one or more openings near the floor and the remainder in one or more openings in the roof or in the sidewalls near the roof, or all of the area required for ventilation shall be permitted in one or more openings in or near the roof.

(C) Size. For a vault ventilated by natural circulation of air to an outdoor area, the combined net area of all ventilating openings, after deducting the area occupied by screens, gratings, or louvers, shall not be less than 1900 mm^2 (3 in.^{2}) per kVA of transformer capacity in service, and in no case shall the net area be less than 0.1 m^2 (1 ft^2) for any capacity under 50 kVA.

(D) Covering. Ventilation openings shall be covered with durable gratings, screens, or louvers, according to the treatment required in order to avoid unsafe conditions.

(E) Dampers. All ventilation openings to the indoors shall be provided with automatic closing fire dampers that operate in response to a vault fire. Such dampers shall possess a standard fire rating of not less than 1½ hours.

Informational Note: See ANSI/UL 555-2011, *Standard for Fire Dampers.*

(F) Ducts. Ventilating ducts shall be constructed of fire-resistant material.

450.46 Drainage. Where practicable, vaults containing more than 100 kVA transformer capacity shall be provided with a drain or other means that will carry off any accumulation of oil or water in the vault unless local conditions make this impracticable. The floor shall be pitched to the drain where provided.

450.47 Water Pipes and Accessories. Any pipe or duct system foreign to the electrical installation shall not enter or pass through a transformer vault. Piping or other facilities provided for vault fire protection, or for transformer cooling, shall not be considered foreign to the electrical installation.

450.48 Storage in Vaults. Materials shall not be stored in transformer vaults.

ARTICLE 455
Phase Converters

Part I. General

455.1 Scope. This article covers the installation and use of phase converters.

455.2 Definitions.

Manufactured Phase. The manufactured or derived phase originates at the phase converter and is not solidly connected to either of the single-phase input conductors.

Phase Converter. An electrical device that converts single-phase power to 3-phase electric power.

Informational Note: Phase converters have characteristics that modify the starting torque and locked-rotor current of motors served, and consideration is required in selecting a phase converter for a specific load.

Rotary-Phase Converter. A device that consists of a rotary transformer and capacitor panel(s) that permits the operation of 3-phase loads from a single-phase supply.

Static-Phase Converter. A device without rotating parts, sized for a given 3-phase load to permit operation from a single-phase supply.

455.3 Other Articles. Phase converters shall comply with this article and with the applicable provisions of other articles of this *Code*.

455.4 Marking. Each phase converter shall be provided with a permanent nameplate indicating the following:

(1) Manufacturer's name
(2) Rated input and output voltages
(3) Frequency
(4) Rated single-phase input full-load amperes
(5) Rated minimum and maximum single load in kilovolt-amperes (kVA) or horsepower
(6) Maximum total load in kilovolt-amperes (kVA) or horsepower
(7) For a rotary-phase converter, 3-phase amperes at full load

455.5 Equipment Grounding Connection. A means for attachment of an equipment grounding conductor termination in accordance with 250.8 shall be provided.

455.6 Conductors.

(A) Ampacity. The ampacity of the single-phase supply conductors shall be determined by 455.6(A)(1) or (A)(2).

Informational Note: Single-phase conductors sized to prevent a voltage drop not exceeding 3 percent from the source of supply to the phase converter may help ensure proper starting and operation of motor loads.

(1) Variable Loads. Where the loads to be supplied are variable, the conductor ampacity shall not be less than 125 percent of the phase converter nameplate single-phase input full-load amperes.

(2) Fixed Loads. Where the phase converter supplies specific fixed loads, and the conductor ampacity is less than 125 percent of the phase converter nameplate single-phase

input full-load amperes, the conductors shall have an ampacity not less than 250 percent of the sum of the full-load, 3-phase current rating of the motors and other loads served where the input and output voltages of the phase converter are identical. Where the input and output voltages of the phase converter are different, the current as determined by this section shall be multiplied by the ratio of output to input voltage.

(B) Manufactured Phase Marking. The manufactured phase conductors shall be identified in all accessible locations with a distinctive marking. The marking shall be consistent throughout the system and premises.

455.7 Overcurrent Protection. The single-phase supply conductors and phase converter shall be protected from overcurrent by 455.7(A) or (B). Where the required fuse or nonadjustable circuit breaker rating or settings of adjustable circuit breakers do not correspond to a standard rating or setting, a higher rating or setting that does not exceed the next higher standard rating shall be permitted.

(A) Variable Loads. Where the loads to be supplied are variable, overcurrent protection shall be set at not more than 125 percent of the phase converter nameplate single-phase input full-load amperes.

(B) Fixed Loads. Where the phase converter supplies specific fixed loads and the conductors are sized in accordance with 455.6(A)(2), the conductors shall be protected in accordance with their ampacity. The overcurrent protection determined from this section shall not exceed 125 percent of the phase converter nameplate single-phase input amperes.

455.8 Disconnecting Means. Means shall be provided to disconnect simultaneously all ungrounded single-phase supply conductors to the phase converter.

(A) Location. The disconnecting means shall be readily accessible and located in sight from the phase converter.

(B) Type. The disconnecting means shall be a switch rated in horsepower, a circuit breaker, or a molded-case switch. Where only nonmotor loads are served, an ampere-rated switch shall be permitted.

(C) Rating. The ampere rating of the disconnecting means shall not be less than 115 percent of the rated maximum single-phase input full-load amperes or, for specific fixed loads, shall be permitted to be selected from 455.8(C)(1) or (C)(2).

(1) Current Rated Disconnect. The disconnecting means shall be a circuit breaker or molded-case switch with an ampere rating not less than 250 percent of the sum of the following:

(1) Full-load, 3-phase current ratings of the motors
(2) Other loads served

(2) Horsepower Rated Disconnect. The disconnecting means shall be a switch with a horsepower rating. The equivalent locked rotor current of the horsepower rating of the switch shall not be less than 200 percent of the sum of the following:

(1) Nonmotor loads
(2) The 3-phase, locked-rotor current of the largest motor as determined from Table 430.251(B)
(3) The full-load current of all other 3-phase motors operating at the same time

(D) Voltage Ratios. The calculations in 455.8(C) shall apply directly where the input and output voltages of the phase converter are identical. Where the input and output voltages of the phase converter are different, the current shall be multiplied by the ratio of the output to input voltage.

455.9 Connection of Single-Phase Loads. Where single-phase loads are connected on the load side of a phase converter, they shall not be connected to the manufactured phase.

455.10 Terminal Housings. A terminal housing in accordance with the provisions of 430.12 shall be provided on a phase converter.

Part II. Specific Provisions Applicable to Different Types of Phase Converters

455.20 Disconnecting Means. The single-phase disconnecting means for the input of a static phase converter shall be permitted to serve as the disconnecting means for the phase converter and a single load if the load is within sight of the disconnecting means.

455.21 Start-Up. Power to the utilization equipment shall not be supplied until the rotary-phase converter has been started.

455.22 Power Interruption. Utilization equipment supplied by a rotary-phase converter shall be controlled in such a manner that power to the equipment will be disconnected in the event of a power interruption.

Informational Note: Magnetic motor starters, magnetic contactors, and similar devices, with manual or time delay restarting for the load, provide restarting after power interruption.

455.23 Capacitors. Capacitors that are not an integral part of the rotary-phase conversion system but are installed for a motor load shall be connected to the line side of that motor overload protective device.

ARTICLE 460
Capacitors

460.1 Scope. This article covers the installation of capacitors on electrical circuits.

Surge capacitors or capacitors included as a component part of other apparatus and conforming with the requirements of such apparatus are excluded from these requirements.

This article also covers the installation of capacitors in hazardous (classified) locations as modified by Articles 501 through 503.

460.2 Enclosing and Guarding.

(A) Containing More Than 11 L (3 gal) of Flammable Liquid. Capacitors containing more than 11 L (3 gal) of flammable liquid shall be enclosed in vaults or outdoor fenced enclosures complying with Article 110, Part III. This limit shall apply to any single unit in an installation of capacitors.

(B) Accidental Contact. Where capacitors are accessible to unauthorized and unqualified persons, they shall be enclosed, located, or guarded so that persons cannot come into accidental contact or bring conducting materials into accidental

contact with exposed energized parts, terminals, or buses associated with them. However, no additional guarding is required for enclosures accessible only to authorized and qualified persons.

Part I. 1000 Volts, Nominal, and Under

460.6 Discharge of Stored Energy. Capacitors shall be provided with a means of discharging stored energy.

(A) Time of Discharge. The residual voltage of a capacitor shall be reduced to 50 volts, nominal, or less within 1 minute after the capacitor is disconnected from the source of supply.

(B) Means of Discharge. The discharge circuit shall be either permanently connected to the terminals of the capacitor or capacitor bank or provided with automatic means of connecting it to the terminals of the capacitor bank on removal of voltage from the line. Manual means of switching or connecting the discharge circuit shall not be used.

460.8 Conductors.

(A) Ampacity. The ampacity of capacitor circuit conductors shall not be less than 135 percent of the rated current of the capacitor. The ampacity of conductors that connect a capacitor to the terminals of a motor or to motor circuit conductors shall not be less than one-third the ampacity of the motor circuit conductors and in no case less than 135 percent of the rated current of the capacitor.

(B) Overcurrent Protection. An overcurrent device shall be provided in each ungrounded conductor for each capacitor bank. The rating or setting of the overcurrent device shall be as low as practicable.

Exception: A separate overcurrent device shall not be required for a capacitor connected on the load side of a motor overload protective device.

(C) Disconnecting Means. A disconnecting means shall be provided in each ungrounded conductor for each capacitor bank and shall meet the following requirements:

(1) The disconnecting means shall open all ungrounded conductors simultaneously.
(2) The disconnecting means shall be permitted to disconnect the capacitor from the line as a regular operating procedure.
(3) The rating of the disconnecting means shall not be less than 135 percent of the rated current of the capacitor.

Exception: A separate disconnecting means shall not be required where a capacitor is connected on the load side of a motor controller.

460.9 Rating or Setting of Motor Overload Device. Where a motor installation includes a capacitor connected on the load side of the motor overload device, the rating or setting of the motor overload device shall be based on the improved power factor of the motor circuit.

The effect of the capacitor shall be disregarded in determining the motor circuit conductor rating in accordance with 430.22.

460.10 Grounding. Capacitor cases shall be connected to the equipment grounding conductor.

Exception: Capacitor cases shall not be connected to the equipment grounding conductor where the capacitor units are supported on a structure designed to operate at other than ground potential.

460.12 Marking. Each capacitor shall be provided with a nameplate giving the name of the manufacturer, rated voltage, frequency, kilovar or amperes, number of phases, and, if filled with a combustible liquid, the volume of liquid. Where filled with a nonflammable liquid, the nameplate shall so state. The nameplate shall also indicate whether a capacitor has a discharge device inside the case.

Part II. Over 1000 Volts, Nominal

460.24 Switching.

(A) Load Current. Group-operated switches shall be used for capacitor switching and shall be capable of the following:

(1) Carrying continuously not less than 135 percent of the rated current of the capacitor installation
(2) Interrupting the maximum continuous load current of each capacitor, capacitor bank, or capacitor installation that will be switched as a unit
(3) Withstanding the maximum inrush current, including contributions from adjacent capacitor installations
(4) Carrying currents due to faults on capacitor side of switch

(B) Isolation.

(1) General. A means shall be installed to isolate from all sources of voltage each capacitor, capacitor bank, or capacitor installation that will be removed from service as a unit. The isolating means shall provide a visible gap in the electrical circuit adequate for the operating voltage.

(2) Isolating or Disconnecting Switches with No Interrupting Rating. Isolating or disconnecting switches (with no interrupting rating) shall be interlocked with the load-interrupting device or shall be provided with prominently displayed caution signs in accordance with 490.22 to prevent switching load current.

(C) Additional Requirements for Series Capacitors. The proper switching sequence shall be ensured by use of one of the following:

(1) Mechanically sequenced isolating and bypass switches
(2) Interlocks
(3) Switching procedure prominently displayed at the switching location

460.25 Overcurrent Protection.

(A) Provided to Detect and Interrupt Fault Current. A means shall be provided to detect and interrupt fault current likely to cause dangerous pressure within an individual capacitor.

(B) Single Pole or Multipole Devices. Single-pole or multipole devices shall be permitted for this purpose.

(C) Protected Individually or in Groups. Capacitors shall be permitted to be protected individually or in groups.

(D) Protective Devices Rated or Adjusted. Protective devices for capacitors or capacitor equipment shall be rated or adjusted to operate within the limits of the safe zone for individual capacitors. If the protective devices are rated or adjusted to operate within the limits for Zone 1 or Zone 2, the capacitors shall be enclosed or isolated.

In no event shall the rating or adjustment of the protective devices exceed the maximum limit of Zone 2.

Informational Note: For definitions of *Safe Zone*, *Zone 1*, and *Zone 2*, see ANSI/IEEE 18-1992, *Shunt Power Capacitors*.

460.26 Identification. Each capacitor shall be provided with a permanent nameplate giving the manufacturer's name, rated voltage, frequency, kilovar or amperes, number of phases, and the volume of liquid identified as flammable, if such is the case.

460.27 Grounding. Capacitor cases shall be connected to the equipment grounding conductor. If the capacitor neutral point is connected to a grounding electrode conductor, the connection shall be made in accordance with Part III of Article 250.

Exception: Capacitor cases shall not be connected to the equipment grounding conductor where the capacitor units are supported on a structure designed to operate at other than ground potential.

460.28 Means for Discharge.

(A) Means to Reduce the Residual Voltage. A means shall be provided to reduce the residual voltage of a capacitor to 50 volts or less within 5 minutes after the capacitor is disconnected from the source of supply.

(B) Connection to Terminals. A discharge circuit shall be either permanently connected to the terminals of the capacitor or provided with automatic means of connecting it to the terminals of the capacitor bank after disconnection of the capacitor from the source of supply. The windings of motors, transformers, or other equipment directly connected to capacitors without a switch or overcurrent device interposed shall meet the requirements of 460.28(A).

ARTICLE 470
Resistors and Reactors

Part I. 1000 Volts, Nominal, and Under

470.1 Scope. This article covers the installation of separate resistors and reactors on electrical circuits.

Exception: Resistors and reactors that are component parts of other apparatus.

This article also covers the installation of resistors and reactors in hazardous (classified) locations as modified by Articles 501 through 504.

470.2 Location. Resistors and reactors shall not be placed where exposed to physical damage.

470.3 Space Separation. A thermal barrier shall be required if the space between the resistors and reactors and any combustible material is less than 305 mm (12 in.).

470.4 Conductor Insulation. Insulated conductors used for connections between resistance elements and controllers shall be suitable for an operating temperature of not less than 90°C (194°F).

Exception: Other conductor insulations shall be permitted for motor starting service.

Part II. Over 1000 Volts, Nominal

470.18 General.

(A) Protected Against Physical Damage. Resistors and reactors shall be protected against physical damage.

(B) Isolated by Enclosure or Elevation. Resistors and reactors shall be isolated by enclosure or elevation to protect personnel from accidental contact with energized parts.

(C) Combustible Materials. Resistors and reactors shall not be installed in close enough proximity to combustible materials to constitute a fire hazard and shall have a clearance of not less than 305 mm (12 in.) from combustible materials.

(D) Clearances. Clearances from resistors and reactors to grounded surfaces shall be adequate for the voltage involved.

(E) Temperature Rise from Induced Circulating Currents. Metallic enclosures of reactors and adjacent metal parts shall be installed so that the temperature rise from induced circulating currents is not hazardous to personnel or does not constitute a fire hazard.

470.19 Grounding. Resistor and reactor cases or enclosures shall be connected to the equipment grounding conductor.

Exception: Resistor or reactor cases or enclosures supported on a structure designed to operate at other than ground potential shall not be connected to the equipment grounding conductor.

470.20 Oil-Filled Reactors. Installation of oil-filled reactors, in addition to the above requirements, shall comply with applicable requirements of Article 450.

ARTICLE 480
Storage Batteries

480.1 Scope. This article applies to all stationary installations of storage batteries.

Informational Note: The following standards are frequently referenced for the installation of stationary batteries:

(1) IEEE 484, *Recommended Practice for Installation Design and Installation of Vented Lead-Acid Batteries for Stationary Applications*

(2) IEEE 485, *Recommended Practice for Sizing Vented Lead-Acid Storage Batteries for Stationary Applications*

(3) IEEE 1145, *Recommended Practice for Installation and Maintenance of Nickel-Cadmium Batteries for Photovoltaic (PV) Systems*

(4) IEEE 1187, *Recommended Practice for Installation Design, and Installation of Valve-Regulated Lead-Acid Batteries for Stationary Applications*

(5) IEEE 1375, *IEEE Guide for the Protection of Stationary Battery Systems*

(6) IEEE 1578, *Recommended Practice for Stationary Battery Electrolyte Spill Containment and Management*

(7) IEEE 1635/ASHRAE 21, *Guide for the Ventilation and Thermal Management of Batteries for Stationary Applications*

(8) UL 1973, *Standard for Batteries for Use in Light Electric Rail (LER) Applications and Stationary Applications*

(9) UL Subject 2436, *Outline of Investigation for Spill Containment for Stationary Lead Acid Battery Systems*

(10) UL 1989, *Standard for Standby Batteries*

480.2 Definitions.

Cell. The basic electrochemical unit, characterized by an anode and a cathode, used to receive, store, and deliver electrical energy.

Container. A vessel that holds the plates, electrolyte, and other elements of a single unit in a battery.

Informational Note: A container may be single-cell or multi-cell and is sometimes referred to in the industry as a "jar."

Electrolyte. The medium that provides the ion transport mechanism between the positive and negative electrodes of a cell.

Intercell Connector. An electrically conductive bar or cable used to connect adjacent cells.

Intertier Connector. An electrical conductor used to connect two cells on different tiers of the same rack or different shelves of the same rack.

Nominal Voltage (Battery or Cell). The value assigned to a cell or battery of a given voltage class for the purpose of convenient designation. The operating voltage of the cell or battery may vary above or below this value.

Informational Note: The most common nominal cell voltages are 2 volts per cell for the lead-acid systems, 1.2 volts per cell for alkali systems, and 3.6 to 3.8 volts per cell for Li-ion systems. Nominal voltages might vary with different chemistries.

Sealed Cell or Battery. A cell or battery that has no provision for the routine addition of water or electrolyte or for external measurement of electrolyte specific gravity and might contain pressure relief venting.

Storage Battery. A battery comprised of one or more rechargeable cells of the lead-acid, nickel-cadmium, or other rechargeable electrochemical types.

Terminal. That part of a cell, container, or battery to which an external connection is made (commonly identified as post, pillar, pole, or terminal post).

N **480.3 Equipment.** Storage batteries and battery management equipment shall be listed. This requirement shall not apply to lead-acid batteries.

480.4 Battery and Cell Terminations.

(A) Corrosion Prevention. Where mating dissimilar metals, antioxidant material suitable for the battery connection shall be used where recommended by the battery manufacturer.

Informational Note: The battery manufacturer's installation and instruction manual can be used for guidance for acceptable materials.

(B) Intercell and Intertier Conductors and Connections. The ampacity of field-assembled intercell and intertier connectors and conductors shall be of such cross-sectional area that the temperature rise under maximum load conditions and at maximum ambient temperature shall not exceed the safe operating temperature of the conductor insulation or of the material of the conductor supports.

Informational Note: Conductors sized to prevent a voltage drop exceeding 3 percent of maximum anticipated load, and where the maximum total voltage drop to the furthest point of connection does not exceed 5 percent, may not be appropriate for all battery applications. IEEE 1375-2003, *Guide for the Protection of*

Stationary Battery Systems, provides guidance for overcurrent protection and associated cable sizing.

(C) Battery Terminals. Electrical connections to the battery, and the cable(s) between cells on separate levels or racks, shall not put mechanical strain on the battery terminals. Terminal plates shall be used where practicable.

> Informational Note: Conductors are commonly pre-formed to eliminate stress on battery terminations. Fine stranded cables may also eliminate the stress on battery terminations. See the manufacturer's instructions for guidance.

480.5 Wiring and Equipment Supplied from Batteries. Wiring and equipment supplied from storage batteries shall be subject to the applicable provisions of this *Code* applying to wiring and equipment operating at the same voltage, unless otherwise permitted by 480.6.

480.6 Overcurrent Protection for Prime Movers. Overcurrent protection shall not be required for conductors from a battery with a voltage of 60 volts dc or less if the battery provides power for starting, ignition, or control of prime movers. Section 300.3 shall not apply to these conductors.

480.7 DC Disconnect Methods.

(A) Disconnecting Means. A disconnecting means shall be provided for all ungrounded conductors derived from a stationary battery system with a voltage over 60 volts dc. A disconnecting means shall be readily accessible and located within sight of the battery system.

> Informational Note: See 240.21(H) for information on the location of the overcurrent device for battery conductors.

(B) Remote Actuation. Where a disconnecting means, located in accordance with 480.7(A), is provided with remote controls to activate the disconnecting means and the controls for the disconnecting means are not located within sight of the stationary battery system, the disconnecting means shall be capable of being locked in the open position, in accordance with 110.25, and the location of the controls shall be field marked on the disconnecting means.

(C) Busway. Where a DC busway system is installed, the disconnecting means shall be permitted to be incorporated into the busway.

(D) Notification. The disconnecting means shall be legibly marked in the field. A label with the marking shall be placed in a conspicuous location near the battery if a disconnecting means is not provided. The marking shall be of sufficient durability to withstand the environment involved and shall include the following:

(1) Nominal battery voltage
(2) Maximum available short-circuit current derived from the stationary battery system
(3) Date the short-circuit current calculation was performed
(4) The battery disconnecting means shall be marked in accordance with 110.16.

> Informational Note No. 1: Battery equipment suppliers can provide information about short-circuit current on any particular battery model.

> Informational Note No. 2: The available short-circuit current marking(s) addressed in 480.7(D)(2) is related to required short-circuit current ratings of equipment. *NFPA 70E*-2015, *Standard for Electrical Safety in the Workplace,* provides assistance in determining the severity of potential exposure, planning safe work practices, and selecting personal protective equipment.

480.8 Insulation of Batteries. Batteries constructed of an electrically conductive container shall have insulating support if a voltage is present between the container and ground.

480.9 Battery Support Systems. For battery chemistries with corrosive electrolyte, the structure that supports the battery shall be resistant to deteriorating action by the electrolyte. Metallic structures shall be provided with nonconducting support members for the cells, or shall be constructed with a continuous insulating material. Paint alone shall not be considered as an insulating material.

The terminals of all cells or multi-cell units shall be readily accessible for readings, inspection, and cleaning where required by the equipment design. One side of transparent battery containers shall be readily accessible for inspection of the internal components.

480.10 Battery Locations. Battery locations shall conform to 480.10(A), (B), and (C).

(A) Ventilation. Provisions appropriate to the battery technology shall be made for sufficient diffusion and ventilation of gases from the battery, if present, to prevent the accumulation of an explosive mixture.

> Informational Note No. 1: See NFPA 1-2015, *Fire Code,* Chapter 52, for ventilation considerations for specific battery chemistries.

> Informational Note No. 2: Some battery technologies do not require ventilation.

> Informational Note No. 3: For additional information on the ventilation of stationary battery systems, see *IEEE Std 1635-2012/ ASHRAE Guideline 21-2012 Guide for the Ventilation and Thermal Management of Batteries for Stationary Applications.*

(B) Live Parts. Guarding of live parts shall comply with 110.27.

(C) Spaces About Battery Systems. Spaces about battery systems shall comply with 110.26. Working space shall be measured from the edge of the battery cabinet, racks, or trays.

For battery racks, there shall be a minimum clearance of 25 mm (1 in.) between a cell container and any wall or structure on the side not requiring access for maintenance. Battery stands shall be permitted to contact adjacent walls or structures, provided that the battery shelf has a free air space for not less than 90 percent of its length.

> Informational Note: Additional space is often needed to accommodate battery hoisting equipment, tray removal, or spill containment.

(D) Top Terminal Batteries. Where top terminal batteries are installed on tiered racks or on shelves of battery cabinets, working space in accordance with the battery manufacturer's instructions shall be provided between the highest point on a cell and the row, shelf, or ceiling above that point.

> Informational Note: IEEE 1187-2013, *IEEE Recommended Practice for Installation Design and Installation of Valve-Regulated Lead-Acid Batteries for Stationary Applications,* provides guidance for top clearance of valve-regulated lead-acid batteries, which are commonly used in battery cabinets.

(E) Egress. A personnel door(s) intended for entrance to, and egress from, rooms designated as battery rooms shall open in the direction of egress and shall be equipped with listed panic hardware.

(F) Piping in Battery Rooms. Gas piping shall not be permitted in dedicated battery rooms.

(G) Illumination. Illumination shall be provided for working spaces containing battery systems. The lighting outlets shall not be controlled by automatic means only. Additional lighting outlets shall not be required where the work space is illuminated by an adjacent light source. The location of luminaires shall not:

(1) Expose personnel to energized battery components while performing maintenance on the luminaires in the battery space; or

(2) Create a hazard to the battery upon failure of the luminaire.

480.11 Vents.

(A) Vented Cells. Each vented cell shall be equipped with a flame arrester.

Informational Note: A flame arrester prevents destruction of the cell due to ignition of gases within the cell by an external spark or flame.

(B) Sealed Cells. Where the battery is constructed such that an excessive accumulation of pressure could occur within the cell during operation, a pressure-release vent shall be provided.

ARTICLE 490
Equipment Over 1000 Volts, Nominal

Part I. General

490.1 Scope. This article covers the general requirements for equipment operating at more than 1000 volts, nominal.

Informational Note No. 1: See *NFPA 70E*-2015, *Standard for Electrical Safety in the Workplace*, for electrical safety requirements for employee workplaces.

Informational Note No. 2: For further information on hazard signs and labels, see ANSI Z535.4-2011, *Product Signs and Safety Labels*.

490.2 Definition.

High Voltage. For the purposes of this article, more than 1000 volts, nominal.

490.3 Other Articles.

(A) Oil-Filled Equipment. Installation of electrical equipment, other than transformers covered in Article 450, containing more than 38 L (10 gal) of flammable oil per unit shall meet the requirements of Parts II and III of Article 450.

(B) Enclosures in Damp or Wet Locations. Enclosures in damp or wet locations shall meet the requirements of 312.2.

Part II. Equipment — Specific Provisions

490.21 Circuit-Interrupting Devices.

(A) Circuit Breakers.

(1) Location.

(a) Circuit breakers installed indoors shall be mounted either in metal-enclosed units or fire-resistant cell-mounted units, or they shall be permitted to be open-mounted in locations accessible to qualified persons only.

(b) Circuit breakers used to control oil-filled transformers in a vault shall either be located outside the transformer vault or be capable of operation from outside the vault.

(c) Oil circuit breakers shall be arranged or located so that adjacent readily combustible structures or materials are safeguarded in an approved manner.

(2) Operating Characteristics. Circuit breakers shall have the following equipment or operating characteristics:

(1) An accessible mechanical or other identified means for manual tripping, independent of control power

(2) Be release free (trip free)

(3) If capable of being opened or closed manually while energized, main contacts that operate independently of the speed of the manual operation

(4) A mechanical position indicator at the circuit breaker to show the open or closed position of the main contacts

(5) A means of indicating the open and closed position of the breaker at the point(s) from which they may be operated

(3) Nameplate. A circuit breaker shall have a permanent and legible nameplate showing manufacturer's name or trademark, manufacturer's type or identification number, continuous current rating, interrupting rating in megavolt-amperes (MVA) or amperes, and maximum voltage rating. Modification of a circuit breaker affecting its rating(s) shall be accompanied by an appropriate change of nameplate information.

(4) Rating. Circuit breakers shall have the following ratings:

(1) The continuous current rating of a circuit breaker shall not be less than the maximum continuous current through the circuit breaker.

(2) The interrupting rating of a circuit breaker shall not be less than the maximum fault current the circuit breaker will be required to interrupt, including contributions from all connected sources of energy.

(3) The closing rating of a circuit breaker shall not be less than the maximum asymmetrical fault current into which the circuit breaker can be closed.

(4) The momentary rating of a circuit breaker shall not be less than the maximum asymmetrical fault current at the point of installation.

(5) The rated maximum voltage of a circuit breaker shall not be less than the maximum circuit voltage.

(B) Power Fuses and Fuseholders.

(1) Use. Where fuses are used to protect conductors and equipment, a fuse shall be placed in each ungrounded conductor. Two power fuses shall be permitted to be used in parallel to protect the same load if both fuses have identical ratings and both fuses are installed in an identified common mounting with electrical connections that divide the current equally. Power fuses of the vented type shall not be used indoors, underground, or in metal enclosures unless identified for the use.

(2) Interrupting Rating. The interrupting rating of power fuses shall not be less than the maximum fault current the fuse is required to interrupt, including contributions from all connected sources of energy.

(3) Voltage Rating. The maximum voltage rating of power fuses shall not be less than the maximum circuit voltage. Fuses having a minimum recommended operating voltage shall not be applied below this voltage.

(4) Identification of Fuse Mountings and Fuse Units. Fuse mountings and fuse units shall have permanent and legible nameplates showing the manufacturer's type or designation, continuous current rating, interrupting current rating, and maximum voltage rating.

(5) Fuses. Fuses that expel flame in opening the circuit shall be designed or arranged so that they function properly without hazard to persons or property.

(6) Fuseholders. Fuseholders shall be designed or installed so that they are de-energized while a fuse is being replaced. A field-applied permanent and legible sign, in accordance with 110.21(B), shall be installed immediately adjacent to the fuseholders and shall be worded as follows:

DANGER — DISCONNECT CIRCUIT BEFORE REPLACING FUSES.

Exception: Fuses and fuseholders designed to permit fuse replacement by qualified persons using identified equipment without de-energizing the fuseholder shall be permitted.

(7) High-Voltage Fuses. Switchgear and substations that utilize high-voltage fuses shall be provided with a gang-operated disconnecting switch. Isolation of the fuses from the circuit shall be provided by either connecting a switch between the source and the fuses or providing roll-out switch and fuse-type construction. The switch shall be of the load-interrupter type, unless mechanically or electrically interlocked with a load-interrupting device arranged to reduce the load to the interrupting capability of the switch.

Exception: More than one switch shall be permitted as the disconnecting means for one set of fuses where the switches are installed to provide connection to more than one set of supply conductors. The switches shall be mechanically or electrically interlocked to permit access to the fuses only when all switches are open. A conspicuous sign shall be placed at the fuses identifying the presence of more than one source.

(C) Distribution Cutouts and Fuse Links — Expulsion Type.

(1) Installation. Cutouts shall be located so that they may be readily and safely operated and re-fused, and so that the exhaust of the fuses does not endanger persons. Distribution cutouts shall not be used indoors, underground, or in metal enclosures.

(2) Operation. Where fused cutouts are not suitable to interrupt the circuit manually while carrying full load, an approved means shall be installed to interrupt the entire load. Unless the fused cutouts are interlocked with the switch to prevent opening of the cutouts under load, a conspicuous sign shall be placed at such cutouts identifying that they shall not be operated under load.

(3) Interrupting Rating. The interrupting rating of distribution cutouts shall not be less than the maximum fault current the cutout is required to interrupt, including contributions from all connected sources of energy.

(4) Voltage Rating. The maximum voltage rating of cutouts shall not be less than the maximum circuit voltage.

(5) Identification. Distribution cutouts shall have on their body, door, or fuse tube a permanent and legible nameplate or identification showing the manufacturer's type or designation, continuous current rating, maximum voltage rating, and interrupting rating.

(6) Fuse Links. Fuse links shall have a permanent and legible identification showing continuous current rating and type.

(7) Structure Mounted Outdoors. The height of cutouts mounted outdoors on structures shall provide safe clearance between lowest energized parts (open or closed position) and standing surfaces, in accordance with 110.34(E).

(D) Oil-Filled Cutouts.

(1) Continuous Current Rating. The continuous current rating of oil-filled cutouts shall not be less than the maximum continuous current through the cutout.

(2) Interrupting Rating. The interrupting rating of oil-filled cutouts shall not be less than the maximum fault current the oil-filled cutout is required to interrupt, including contributions from all connected sources of energy.

(3) Voltage Rating. The maximum voltage rating of oil-filled cutouts shall not be less than the maximum circuit voltage.

(4) Fault Closing Rating. Oil-filled cutouts shall have a fault closing rating not less than the maximum asymmetrical fault current that can occur at the cutout location, unless suitable interlocks or operating procedures preclude the possibility of closing into a fault.

(5) Identification. Oil-filled cutouts shall have a permanent and legible nameplate showing the rated continuous current, rated maximum voltage, and rated interrupting current.

(6) Fuse Links. Fuse links shall have a permanent and legible identification showing the rated continuous current.

(7) Location. Cutouts shall be located so that they are readily and safely accessible for re-fusing, with the top of the cutout not over 1.5 m (5 ft) above the floor or platform.

(8) Enclosure. Suitable barriers or enclosures shall be provided to prevent contact with nonshielded cables or energized parts of oil-filled cutouts.

(E) Load Interrupters. Load-interrupter switches shall be permitted if suitable fuses or circuit breakers are used in conjunction with these devices to interrupt fault currents. Where these devices are used in combination, they shall be coordinated electrically so that they will safely withstand the effects of closing, carrying, or interrupting all possible currents up to the assigned maximum short-circuit rating.

Where more than one switch is installed with interconnected load terminals to provide for alternate connection to different supply conductors, each switch shall be provided with a conspicuous sign identifying this hazard.

(1) Continuous Current Rating. The continuous current rating of interrupter switches shall equal or exceed the maximum continuous current at the point of installation.

(2) Voltage Rating. The maximum voltage rating of interrupter switches shall equal or exceed the maximum circuit voltage.

(3) Identification. Interrupter switches shall have a permanent and legible nameplate including the following information: manufacturer's type or designation, continuous current rating, interrupting current rating, fault closing rating, maximum voltage rating.

(4) Switching of Conductors. The switching mechanism shall be arranged to be operated from a location where the operator is not exposed to energized parts and shall be arranged to open all ungrounded conductors of the circuit simultaneously with one operation. Switches shall be arranged to be locked in the open position. Metal-enclosed switches shall be operable from outside the enclosure.

(5) Stored Energy for Opening. The stored-energy operator shall be permitted to be left in the uncharged position after the switch has been closed if a single movement of the operating handle charges the operator and opens the switch.

(6) Supply Terminals. The supply terminals of fused interrupter switches shall be installed at the top of the switch enclosure, or, if the terminals are located elsewhere, the equipment shall have barriers installed so as to prevent persons from accidentally contacting energized parts or dropping tools or fuses into energized parts.

490.22 Isolating Means. Means shall be provided to completely isolate an item of equipment from all ungrounded conductors. The use of isolating switches shall not be required where there are other ways of de-energizing the equipment for inspection and repairs, such as draw-out-type switchgear units and removable truck panels.

Isolating switches not interlocked with an approved circuit-interrupting device shall be provided with a sign warning against opening them under load. The warning sign(s) or label(s) shall comply with 110.21(B).

An identified fuseholder and fuse shall be permitted as an isolating switch.

490.23 Voltage Regulators. Proper switching sequence for regulators shall be ensured by use of one of the following:

(1) Mechanically sequenced regulator bypass switch(es)
(2) Mechanical interlocks
(3) Switching procedure prominently displayed at the switching location

490.24 Minimum Space Separation. In field-fabricated installations, the minimum air separation between bare live conductors and between such conductors and adjacent grounded surfaces shall not be less than the values given in Table 490.24. These values shall not apply to interior portions or exterior terminals of equipment designed, manufactured, and tested in accordance with accepted national standards.

490.25 Backfeed. Installations where the possibility of backfeed exists shall comply with (a) and (b), which follow.

(a)   A permanent sign in accordance with 110.21(B) shall be installed on the disconnecting means enclosure or immediately adjacent to open disconnecting means with the following words or equivalent: DANGER — CONTACTS ON EITHER SIDE OF THIS DEVICE MAY BE ENERGIZED BY BACKFEED.

(b)   A permanent and legible single-line diagram of the local switching arrangement, clearly identifying each point of connection to the high-voltage section, shall be provided within sight of each point of connection.

Part III. Equipment — Switchgear and Industrial Control Assemblies

490.30 General. Part III covers assemblies of switchgear and industrial control equipment including, but not limited to, switches and interrupting devices and their control, metering, protection, and regulating equipment where they are an integral part of the assembly, with associated interconnections and supporting structures.

490.31 Arrangement of Devices in Assemblies. Arrangement of devices in assemblies shall be such that individual components can safely perform their intended function without adversely affecting the safe operation of other components in the assembly.

490.32 Guarding of High-Voltage Energized Parts Within a Compartment. Where access for other than visual inspection is required to a compartment that contains energized high-voltage parts, barriers shall be provided to prevent accidental contact by persons, tools, or other equipment with energized parts. Exposed live parts shall only be permitted in compartments accessible to qualified persons. Fuses and fuseholders designed to enable future replacement without de-energizing the fuseholder shall only be permitted for use by qualified persons.

490.33 Guarding of Energized Parts Operating at 1000 Volts, Nominal, or Less Within Compartments. Energized bare parts mounted on doors shall be guarded where the door must be opened for maintenance of equipment or removal of draw-out equipment.

490.34 Clearance for Cable Conductors Entering Enclosure. The unobstructed space opposite terminals or opposite raceways or cables entering a switchgear or control assembly shall be approved for the type of conductor and method of termination.

490.35 Accessibility of Energized Parts.

(A) High-Voltage Equipment. Doors that would provide unqualified persons access to high-voltage energized parts shall be locked. Permanent signs in accordance with 110.21(B) shall be installed on panels or doors that provide access to live parts over 1000 volts and shall read DANGER — HIGH VOLTAGE — KEEP OUT.

(B) Control Equipment. Where operating at 1000 volts, nominal, or less, control equipment, relays, motors, and the like shall not be installed in compartments with exposed high-voltage energized parts or high-voltage wiring, unless either of the following conditions is met:

(1) The access means is interlocked with the high-voltage switch or disconnecting means to prevent the access means from being opened or removed.
(2) The high-voltage switch or disconnecting means is in the isolating position.

(C) High-Voltage Instruments or Control Transformers and Space Heaters. High-voltage instrument or control transformers and space heaters shall be permitted to be installed in the high-voltage compartment without access restrictions beyond those that apply to the high-voltage compartment generally.

490.36 Grounding. Frames of switchgear and control assemblies shall be connected to an equipment grounding conductor or, where permitted, the grounded conductor.

490.37 Grounding of Devices. The metal cases or frames, or both, such as those of instruments, relays, meters, and instrument and control transformers, located in or on switchgear or control assemblies, shall be connected to an equipment grounding conductor or, where permitted, the grounded conductor.

490.38 Door Stops and Cover Plates. External hinged doors or covers shall be provided with stops to hold them in the open position. Cover plates intended to be removed for inspection of energized parts or wiring shall be equipped with lifting handles and shall not exceed 1.1 m² (12 ft²) in area or 27 kg (60 lb) in weight, unless they are hinged and bolted or locked.

490.39 Gas Discharge from Interrupting Devices. Gas discharged during operating of interrupting devices shall be directed so as not to endanger personnel.

490.40 Visual Inspection Windows. Windows intended for visual inspection of disconnecting switches or other devices shall be of suitable transparent material.

490.41 Location of Industrial Control Equipment. Routinely operated industrial control equipment shall meet the requirements of (A) unless infrequently operated, as covered in 490.41(B).

(A) Control and Instrument Transfer Switch Handles or Push Buttons. Control and instrument transfer switch handles or push buttons shall be in a readily accessible location at an elevation of not over 2.0 m (6 ft 7 in.).

Exception: Operating handles requiring more than 23 kg (50 lb) of force shall be located no higher than 1.7 m (66 in.) in either the open or closed position.

(B) Infrequently Operated Devices. Where operating handles for such devices as draw-out fuses, fused potential or control transformers and their primary disconnects, and bus transfer and isolating switches are only operated infrequently, the handles shall be permitted to be located where they are safely operable and serviceable from a portable platform.

490.42 Interlocks — Interrupter Switches. Interrupter switches equipped with stored energy mechanisms shall have mechanical interlocks to prevent access to the switch compartment unless the stored energy mechanism is in the discharged or blocked position.

490.43 Stored Energy for Opening. The stored energy operator shall be permitted to be left in the uncharged position after the switch has been closed if a single movement of the operating handle charges the operator and opens the switch.

490.44 Fused Interrupter Switches.

(A) Supply Terminals. The supply terminals of fused interrupter switches shall be installed at the top of the switch enclosure or, if the terminals are located elsewhere, the equipment shall have barriers installed so as to prevent persons from accidentally contacting energized parts or dropping tools or fuses into energized parts.

Table 490.24 Minimum Clearance of Live Parts

Nominal Voltage Rating (kV)	Impulse Withstand, Basic Impulse Level B.I.L (kV)		Minimum Clearance of Live Parts							
			Phase-to-Phase				Phase-to-Ground			
			Indoors		Outdoors		Indoors		Outdoors	
	Indoors	Outdoors	mm	in.	mm	in.	mm	in.	mm	in.
2.4–4.16	60	95	115	4.5	180	7	80	3.0	155	6
7.2	75	95	140	5.5	180	7	105	4.0	155	6
13.8	95	110	195	7.5	305	12	130	5.0	180	7
14.4	110	110	230	9.0	305	12	170	6.5	180	7
23	125	150	270	10.5	385	15	190	7.5	255	10
34.5	150	150	320	12.5	385	15	245	9.5	255	10
	200	200	460	18.0	460	18	335	13.0	335	13
46	—	200	—	—	460	18	—	—	335	13
	—	250	—	—	535	21	—	—	435	17
69	—	250	—	—	535	21	—	—	435	17
	—	350	—	—	790	31	—	—	635	25
115	—	550	—	—	1350	53	—	—	1070	42
138	—	550	—	—	1350	53	—	—	1070	42
	—	650	—	—	1605	63	—	—	1270	50
161	—	650	—	—	1605	63	—	—	1270	50
	—	750	—	—	1830	72	—	—	1475	58
230	—	750	—	—	1830	72	—	—	1475	58
	—	900	—	—	2265	89	—	—	1805	71
	—	1050	—	—	2670	105	—	—	2110	83

Note: The values given are the minimum clearance for rigid parts and bare conductors under favorable service conditions. They shall be increased for conductor movement or under unfavorable service conditions or wherever space limitations permit. The selection of the associated impulse withstand voltage for a particular system voltage is determined by the characteristics of the surge protective equipment.

(B) Backfeed. Where fuses can be energized by backfeed, a sign shall be placed on the enclosure door identifying this hazard.

(C) Switching Mechanism. The switching mechanism shall be arranged to be operated from a location outside the enclosure where the operator is not exposed to energized parts and shall be arranged to open all ungrounded conductors of the circuit simultaneously with one operation. Switches shall be lockable in accordance with 110.25.

490.45 Circuit Breakers — Interlocks.

(A) Circuit Breakers. Circuit breakers equipped with stored energy mechanisms shall be designed to prevent the release of the stored energy unless the mechanism has been fully charged.

(B) Mechanical Interlocks. Mechanical interlocks shall be provided in the housing to prevent the complete withdrawal of the circuit breaker from the housing when the stored energy mechanism is in the fully charged position, unless a suitable device is provided to block the closing function of the circuit breaker before complete withdrawal.

490.46 Circuit Breaker Locking.
Circuit breakers shall be capable of being locked in the open position or, if they are installed in a drawout mechanism, that mechanism shall be capable of being locked in such a position that the mechanism cannot be moved into the connected position. In either case, the provision for locking shall be lockable in accordance with 110.25.

490.47 Switchgear Used as Service Equipment.
Switchgear installed as high-voltage service equipment shall include a ground bus for the connection of service cable shields and to facilitate the attachment of safety grounds for personnel protection. This bus shall be extended into the compartment where the service conductors are terminated. Where the compartment door or panel provides access to parts that can only be de-energized and visibly isolated by the serving utility, the warning sign required by 490.35(A) shall include a notice that access is limited to the serving utility or is permitted only following an authorization of the serving utility.

490.48 Substation Design, Documentation, and Required Diagram.

(A) Design and Documentation. Substations shall be designed by a qualified licensed professional engineer. Where components or the entirety of the substation are listed by a qualified electrical testing laboratory, documentation of internal design features subject to the listing investigation shall not be required. The design shall address but not be limited to the following topics, and the documentation of this design shall be made available to the authority having jurisdiction.

(1)   Clearances and exits
(2)   Electrical enclosures
(3)   Securing and support of electrical equipment
(4)   Fire protection
(5)   Safety ground connection provisions
(6)   Guarding live parts
(7)   Transformers and voltage regulation equipment
(8)   Conductor insulation, electrical and mechanical protection, isolation, and terminations
(9)   Application, arrangement, and disconnection of circuit breakers, switches, and fuses
(10) Provisions for oil filled equipment
(11) Switchgear
(12) Surge arresters

(B) Diagram. A permanent, single-line diagram of the switchgear shall be provided in a readily visible location within the same room or enclosed area with the switchgear, and this diagram shall clearly identify interlocks, isolation means, and all possible sources of voltage to the installation under normal or emergency conditions and the marking on the switchgear shall cross-reference the diagram.

Exception: Where the equipment consists solely of a single cubicle or metal-enclosed unit substation containing only one set of high-voltage switching devices, diagrams shall not be required.

Part IV. Mobile and Portable Equipment

490.51 General.

(A) Covered. The provisions of this part shall apply to installations and use of high-voltage power distribution and utilization equipment that is portable, mobile, or both, such as substations and switch houses mounted on skids, trailers, or cars; mobile shovels; draglines; cranes; hoists; drills; dredges; compressors; pumps; conveyors; underground excavators; and the like.

(B) Other Requirements. The requirements of this part shall be additional to, or amendatory of, those prescribed in Articles 100 through 725 of this *Code*. Special attention shall be paid to Article 250.

(C) Protection. Approved enclosures or guarding, or both, shall be provided to protect portable and mobile equipment from physical damage.

(D) Disconnecting Means. Disconnecting means shall be installed for mobile and portable high-voltage equipment according to the requirements of Part VIII of Article 230 and shall disconnect all ungrounded conductors.

490.52 Overcurrent Protection.
Motors driving single or multiple dc generators supplying a system operating on a cyclic load basis do not require overload protection, provided that the thermal rating of the ac drive motor cannot be exceeded under any operating condition. The branch-circuit protective device(s) shall provide short-circuit and locked-rotor protection and shall be permitted to be external to the equipment.

490.53 Enclosures.
All energized switching and control parts shall be enclosed in grounded metal cabinets or enclosures. These cabinets or enclosures shall be marked DANGER — HIGH VOLTAGE — KEEP OUT and shall be locked so that only authorized and qualified persons can enter. The danger marking(s) or label(s) shall comply with 110.21(B). Circuit breakers and protective equipment shall have the operating means projecting through the metal cabinet or enclosure so these units can be reset without opening locked doors. With doors closed, safe access for normal operation of these units shall be provided.

490.54 Collector Rings.
The collector ring assemblies on revolving-type machines (shovels, draglines, etc.) shall be guarded to prevent accidental contact with energized parts by personnel on or off the machine.

490.55 Power Cable Connections to Mobile Machines.
A metallic enclosure shall be provided on the mobile machine

for enclosing the terminals of the power cable. The enclosure shall include terminal connections to the machine frame for the equipment grounding conductor. Ungrounded conductors shall be attached to insulators or be terminated in approved high-voltage cable couplers (which include equipment grounding conductor connectors) of proper voltage and ampere rating. The method of cable termination used shall prevent any strain or pull on the cable from stressing the electrical connections. The enclosure shall have provision for locking so that only authorized and qualified persons may open it and shall be marked as follows:

DANGER — HIGH VOLTAGE — KEEP OUT.

The danger marking(s) or label(s) shall comply with 110.21(B).

490.56 High-Voltage Portable Cable for Main Power Supply. Flexible high-voltage cable supplying power to portable or mobile equipment shall comply with Article 250 and Article 400, Part III.

Part V. Electrode-Type Boilers

490.70 General. The provisions of Part V shall apply to boilers operating over 1000 volts, nominal, in which heat is generated by the passage of current between electrodes through the liquid being heated.

490.71 Electrical Supply System. Electrode-type boilers shall be supplied only from a 3-phase, 4-wire solidly grounded wye system, or from isolating transformers arranged to provide such a system. Control circuit voltages shall not exceed 150 volts, shall be supplied from a grounded system, and shall have the controls in the ungrounded conductor.

490.72 Branch-Circuit Requirements.

(A) Rating. Each boiler shall be supplied from an individual branch circuit rated not less than 100 percent of the total load.

(B) Common-Trip Fault-Interrupting Device. The circuit shall be protected by a 3-phase, common-trip fault-interrupting device, which shall be permitted to automatically reclose the circuit upon removal of an overload condition but shall not reclose after a fault condition.

(C) Phase-Fault Protection. Phase-fault protection shall be provided in each phase, consisting of a separate phase-overcurrent relay connected to a separate current transformer in the phase.

(D) Ground Current Detection. Means shall be provided for detection of the sum of the neutral conductor and equipment grounding conductor currents and shall trip the circuit-interrupting device if the sum of those currents exceeds the greater of 5 amperes or 7½ percent of the boiler full-load current for 10 seconds or exceeds an instantaneous value of 25 percent of the boiler full-load current.

(E) Grounded Neutral Conductor. The grounded neutral conductor shall be as follows:

(1) Connected to the pressure vessel containing the electrodes
(2) Insulated for not less than 1000 volts
(3) Have not less than the ampacity of the largest ungrounded branch-circuit conductor
(4) Installed with the ungrounded conductors in the same raceway, cable, or cable tray, or, where installed as open conductors, in close proximity to the ungrounded conductors
(5) Not used for any other circuit

490.73 Pressure and Temperature Limit Control. Each boiler shall be equipped with a means to limit the maximum temperature, pressure, or both, by directly or indirectly interrupting all current flow through the electrodes. Such means shall be in addition to the temperature, pressure, or both, regulating systems and pressure relief or safety valves.

490.74 Bonding. All exposed non–current-carrying metal parts of the boiler and associated exposed metal structures or equipment shall be bonded to the pressure vessel or to the neutral conductor to which the vessel is connected in accordance with 250.102, except the ampacity of the bonding jumper shall not be less than the ampacity of the neutral conductor.

Chapter 5 Special Occupancies

ARTICLE 500
Hazardous (Classified) Locations,
Classes I, II, and III, Divisions 1 and 2

Informational Note: Text that is followed by a reference in brackets has been extracted from NFPA 497-2012, *Recommended Practice for the Classification of Flammable Liquids, Gases, or Vapors and of Hazardous (Classified) Locations for Electrical Installations in Chemical Process Areas*, and NFPA 499-2013, *Recommended Practice for the Classification of Combustible Dusts and of Hazardous (Classified) Locations for Electrical Installation in Chemical Process Areas*. Only editorial changes were made to the extracted text to make it consistent with this *Code*.

500.1 Scope — Articles 500 Through 504. Articles 500 through 504 cover the requirements for electrical and electronic equipment and wiring for all voltages in Class I, Divisions 1 and 2; Class II, Divisions 1 and 2, and Class III, Divisions 1 and 2 locations where fire or explosion hazards may exist due to flammable gases, flammable liquid–produced vapors, combustible liquid–produced vapors, combustible dusts, or ignitible fibers/flyings.

Informational Note No. 1: The unique hazards associated with explosives, pyrotechnics, and blasting agents are not addressed in this article.

Informational Note No. 2: For the requirements for electrical and electronic equipment and wiring for all voltages in Zone 0, Zone 1, and Zone 2 hazardous (classified) locations where fire or explosion hazards may exist due to flammable gases or vapors or flammable liquids, refer to Article 505.

Informational Note No. 3: For the requirements for electrical and electronic equipment and wiring for all voltages in Zone 20, Zone 21, and Zone 22 hazardous (classified) locations where fire or explosion hazards may exist due to combustible dusts or ignitible fibers/flyings, refer to Article 506.

500.3 Other Articles. Except as modified in Articles 500 through 504, all other applicable rules contained in this *Code* shall apply to electrical equipment and wiring installed in hazardous (classified) locations.

500.4 General.

(A) Documentation. All areas designated as hazardous (classified) locations shall be properly documented. This documentation shall be available to those authorized to design, install, inspect, maintain, or operate electrical equipment at the location.

(B) Reference Standards. Important information relating to topics covered in Chapter 5 may be found in other publications.

Informational Note No. 1: Familiarity with the standards of the National Fire Protection Association (NFPA), the American Petroleum Institute (API), and the International Society of Automation (ISA), as well as relevant industrial experience, may be of use in the classification of various locations, the determination of adequate ventilation, and the protection against static electricity and lightning hazards.

Informational Note No. 2: For further information on the classification of locations, see NFPA 30-2015, *Flammable and Combustible Liquids Code*; NFPA 32-2011, *Standard for Drycleaning Plants*; NFPA 33-2015, *Standard for Spray Application Using Flammable or Combustible Materials*; NFPA 34-2015, *Standard for Dipping and Coating Processes Using Flammable or Combustible Liquids*; NFPA 35-2011, *Standard for the Manufacture of Organic Coatings*; NFPA 36-2013, *Standard for Solvent Extraction Plants*; NFPA 45-2015, *Standard on Fire Protection for Laboratories Using Chemicals*; NFPA 55-2013, *Compressed Gases and Cryogenic Fluids Code*; NFPA 58-2014, *Liquefied Petroleum Gas Code*; NFPA 59-2015, *Utility LP-Gas Plant Code*; NFPA 497-2012, *Recommended Practice for the Classification of Flammable Liquids, Gases, or Vapors and of Hazardous (Classified) Locations for Electrical Installations in Chemical Process Areas*; NFPA 499-2013, *Recommended Practice for the Classification of Combustible Dusts and of Hazardous (Classified) Locations for Electrical Installations in Chemical Process Areas*; NFPA 820-2012, *Standard for Fire Protection in Wastewater Treatment and Collection Facilities*; ANSI/API RP 500-2012, *Recommended Practice for Classification of Locations of Electrical Installations at Petroleum Facilities Classified as Class I, Division 1 and Division 2*; ISA-12.10-1988, *Area Classification in Hazardous (Classified) Dust Locations*.

Informational Note No. 3: For further information on protection against static electricity and lightning hazards in hazardous (classified) locations, see NFPA 77-2014, *Recommended Practice on Static Electricity*; NFPA 780-2014, *Standard for the Installation of Lightning Protection Systems*; and API RP 2003-2008, *Protection Against Ignitions Arising Out of Static Lightning and Stray Currents*.

Informational Note No. 4: For further information on ventilation, see NFPA 30-2015, *Flammable and Combustible Liquids Code*, and ANSI/API RP 500-2012, *Recommended Practice for Classification of Locations for Electrical Installations at Petroleum Facilities Classified as Class I, Division 1 and Division 2*.

Informational Note No. 5: For further information on electrical systems for hazardous (classified) locations on offshore oil- and gas-producing platforms, see ANSI/API RP 14F-2013, *Recommended Practice for Design and Installation of Electrical Systems for Fixed and Floating Offshore Petroleum Facilities for Unclassified and Class I, Division 1 and Division 2 Locations*.

Informational Note No. 6: Portable or transportable equipment having self-contained power supplies, such as battery-operated equipment, could potentially become an ignition source in hazardous (classified) locations. See ANSI/ISA-12.12.03-2011, *Standard for Portable Electronic Products Suitable for Use in Class I and II, Division 2, Class I Zone 2 and Class III, Division 1 and 2 Hazardous (Classified) Locations*.

500.5 Classifications of Locations.

(A) General. Locations shall be classified depending on the properties of the flammable gas, flammable liquid–produced vapor, combustible liquid–produced vapors, combustible dusts, or fibers/flyings that could be present, and the likelihood that a flammable or combustible concentration or quantity is present. Each room, section, or area shall be considered individually in determining its classification. Where pyrophoric materials are the only materials used or handled, these locations are outside the scope of this article.

Informational Note No. 1: Through the exercise of ingenuity in the layout of electrical installations for hazardous (classified) locations, it is frequently possible to locate much of the equipment in a reduced level of classification or in an unclassified location and, thus, to reduce the amount of special equipment required.

Refrigerant machinery rooms that contain ammonia refrigeration systems and are equipped with adequate mechanical ventilation that operates continuously or is initiated by a detection system at a concentration not exceeding 150 ppm shall be permitted to be classified as "unclassified" locations.

Informational Note No. 2: For further information regarding classification and ventilation of areas involving closed-circuit ammonia refrigeration systems, see ANSI/ASHRAE 15-2013, *Safety Standard for Refrigeration Systems,* and ANSI/IIAR 2–2014, *Standard for Safe Design of Closed-Circuit Ammonia Refrigeration Systems.*

(B) Class I Locations. Class I locations are those in which flammable gases, flammable liquid–produced vapors, or combustible liquid–produced vapors are or may be present in the air in quantities sufficient to produce explosive or ignitible mixtures. Class I locations shall include those specified in 500.5(B)(1) and (B)(2).

(1) Class I, Division 1. A Class I, Division 1 location is a location:

(1) In which ignitible concentrations of flammable gases, flammable liquid–produced vapors, or combustible liquid–produced vapors can exist under normal operating conditions, or

(2) In which ignitible concentrations of such flammable gases, flammable liquid–produced vapors, or combustible liquids above their flash points may exist frequently because of repair or maintenance operations or because of leakage, or

(3) In which breakdown or faulty operation of equipment or processes might release ignitible concentrations of flammable gases, flammable liquid–produced vapors, or combustible liquid–produced vapors and might also cause simultaneous failure of electrical equipment in such a way as to directly cause the electrical equipment to become a source of ignition

Informational Note No. 1: This classification usually includes the following locations:

(1) Where volatile flammable liquids or liquefied flammable gases are transferred from one container to another

(2) Interiors of spray booths and areas in the vicinity of spraying and painting operations where volatile flammable solvents are used

(3) Locations containing open tanks or vats of volatile flammable liquids

(4) Drying rooms or compartments for the evaporation of flammable solvents

(5) Locations containing fat- and oil-extraction equipment using volatile flammable solvents

(6) Portions of cleaning and dyeing plants where flammable liquids are used

(7) Gas generator rooms and other portions of gas manufacturing plants where flammable gas may escape

(8) Inadequately ventilated pump rooms for flammable gas or for volatile flammable liquids

(9) The interiors of refrigerators and freezers in which volatile flammable materials are stored in open, lightly stoppered, or easily ruptured containers

(10) All other locations where ignitible concentrations of flammable vapors or gases are likely to occur in the course of normal operations

Informational Note No. 2: In some Division 1 locations, ignitible concentrations of flammable gases or vapors may be present continuously or for long periods of time. Examples include the following:

(1) The inside of inadequately vented enclosures containing instruments normally venting flammable gases or vapors to the interior of the enclosure

(2) The inside of vented tanks containing volatile flammable liquids

(3) The area between the inner and outer roof sections of a floating roof tank containing volatile flammable fluids

(4) Inadequately ventilated areas within spraying or coating operations using volatile flammable fluids

(5) The interior of an exhaust duct that is used to vent ignitible concentrations of gases or vapors

Experience has demonstrated the prudence of avoiding the installation of instrumentation or other electrical equipment in these particular areas altogether or where it cannot be avoided because it is essential to the process and other locations are not feasible [see 500.5(A), Informational Note] using electrical equipment or instrumentation approved for the specific application or consisting of intrinsically safe systems as described in Article 504.

(2) Class I, Division 2. A Class I, Division 2 location is a location:

(1) In which volatile flammable gases, flammable liquid–produced vapors, or combustible liquid–produced vapors are handled, processed, or used, but in which the liquids, vapors, or gases will normally be confined within closed containers or closed systems from which they can escape only in case of accidental rupture or breakdown of such containers or systems or in case of abnormal operation of equipment, or

(2) In which ignitible concentrations of flammable gases, flammable liquid–produced vapors, or combustible liquid–produced vapors are normally prevented by positive mechanical ventilation and which might become hazardous through failure or abnormal operation of the ventilating equipment, or

(3) That is adjacent to a Class I, Division 1 location, and to which ignitible concentrations of flammable gases, flammable liquid–produced vapors, or combustible liquid–produced vapors above their flash points might occasionally be communicated unless such communication is prevented by adequate positive-pressure ventilation from a source of clean air and effective safeguards against ventilation failure are provided.

Informational Note No. 1: This classification usually includes locations where volatile flammable liquids or flammable gases or vapors are used but that, in the judgment of the authority having jurisdiction, would become hazardous only in case of an accident or of some unusual operating condition. The quantity of flammable material that might escape in case of accident, the adequacy of ventilating equipment, the total area involved, and the record of the industry or business with respect to explosions or fires are all factors that merit consideration in determining the classification and extent of each location.

Informational Note No. 2: Piping without valves, checks, meters, and similar devices would not ordinarily introduce a hazardous condition even though used for flammable liquids or gases. Depending on factors such as the quantity and size of the containers and ventilation, locations used for the storage of flammable liquids or liquefied or compressed gases in sealed containers may be considered either hazardous (classified) or unclassified locations. See NFPA 30-2015, *Flammable and Combustible Liquids Code,* and NFPA 58-2014, *Liquefied Petroleum Gas Code.*

(C) Class II Locations. Class II locations are those that are hazardous because of the presence of combustible dust. Class II locations shall include those specified in 500.5(C)(1) and (C)(2).

(1) Class II, Division 1. A Class II, Division 1 location is a location:

(1) In which combustible dust is in the air under normal operating conditions in quantities sufficient to produce explosive or ignitible mixtures, or

(2) Where mechanical failure or abnormal operation of machinery or equipment might cause such explosive or ignitible mixtures to be produced, and might also provide a source of ignition through simultaneous failure of electrical equipment, through operation of protection devices, or from other causes, or

(3) In which Group E combustible dusts may be present in quantities sufficient to be hazardous.

Informational Note: Dusts containing magnesium or aluminum are particularly hazardous, and the use of extreme precaution is necessary to avoid ignition and explosion.

(2) Class II, Division 2. A Class II, Division 2 location is a location:

(1) In which combustible dust due to abnormal operations may be present in the air in quantities sufficient to produce explosive or ignitible mixtures; or

(2) Where combustible dust accumulations are present but are normally insufficient to interfere with the normal operation of electrical equipment or other apparatus, but could as a result of infrequent malfunctioning of handling or processing equipment become suspended in the air; or

(3) In which combustible dust accumulations on, in, or in the vicinity of the electrical equipment could be sufficient to interfere with the safe dissipation of heat from electrical equipment, or could be ignitible by abnormal operation or failure of electrical equipment.

Informational Note No. 1: The quantity of combustible dust that may be present and the adequacy of dust removal systems are factors that merit consideration in determining the classification and may result in an unclassified area.

Informational Note No. 2: Where products such as seed are handled in a manner that produces low quantities of dust, the amount of dust deposited may not warrant classification.

(D) Class III Locations. Class III locations are those that are hazardous because of the presence of easily ignitible fibers or where materials producing combustible flyings are handled, manufactured, or used, but in which such fibers/flyings are not likely to be in suspension in the air in quantities sufficient to produce ignitible mixtures. Class III locations shall include those specified in 500.5(D)(1) and (D)(2).

(1) Class III, Division 1. A Class III, Division 1 location is a location in which easily ignitible fibers/flyings are handled, manufactured, or used.

Informational Note No. 1: Such locations usually include some parts of rayon, cotton, and other textile mills; combustible fibers/flyings manufacturing and processing plants; cotton gins and cotton-seed mills; flax-processing plants; clothing manufacturing plants; woodworking plants; and establishments and industries involving similar hazardous processes or conditions.

Informational Note No. 2: Easily ignitible fibers/flyings include rayon, cotton (including cotton linters and cotton waste), sisal or henequen, istle, jute, hemp, tow, cocoa fiber, oakum, baled waste kapok, Spanish moss, excelsior, and other materials of similar nature.

(2) Class III, Division 2. A Class III, Division 2 location is a location in which easily ignitible fibers/flyings are stored or handled other than in the process of manufacture.

500.6 Material Groups. For purposes of testing, approval, and area classification, various air mixtures (not oxygen-enriched) shall be grouped in accordance with 500.6(A) and (B).

Exception: Equipment identified for a specific gas, vapor, dust, or fiber/flying.

Informational Note: This grouping is based on the characteristics of the materials. Facilities are available for testing and identifying equipment for use in the various atmospheric groups.

(A) Class I Group Classifications. Class I groups shall be according to 500.6(A)(1) through (A)(4).

Informational Note No. 1: Informational Note Nos. 2 and 3 apply to 500.6(A).

Informational Note No. 2: The explosion characteristics of air mixtures of gases or vapors vary with the specific material involved. For Class I locations, Groups A, B, C, and D, the classification involves determinations of maximum explosion pressure and maximum safe clearance between parts of a clamped joint in an enclosure. It is necessary, therefore, that equipment be identified not only for class but also for the specific group of the gas or vapor that will be present.

Informational Note No. 3: Certain chemical atmospheres may have characteristics that require safeguards beyond those required for any of the Class I groups. Carbon disulfide is one of these chemicals because of its low autoignition temperature (90°C) and the small joint clearance permitted to arrest its flame.

(1) Group A. Acetylene. [497:3.3.5.1.1]

(2) Group B. Flammable gas, flammable liquid–produced vapor, or combustible liquid–produced vapor mixed with air that may burn or explode, having either a maximum experimental safe gap (MESG) value less than or equal to 0.45 mm or a minimum igniting current ratio (MIC ratio) less than or equal to 0.40. [497:3.3.5.1.2]

Informational Note: A typical Class I, Group B material is hydrogen.

Exception No. 1: Group D equipment shall be permitted to be used for atmospheres containing butadiene, provided all conduit runs into explosionproof equipment are provided with explosionproof seals installed within 450 mm (18 in.) of the enclosure.

Exception No. 2: Group C equipment shall be permitted to be used for atmospheres containing allyl glycidyl ether, n-butyl glycidyl ether, ethylene oxide, propylene oxide, and acrolein, provided all conduit runs into explosionproof equipment are provided with explosionproof seals installed within 450 mm (18 in.) of the enclosure.

(3) Group C. Flammable gas, flammable liquid–produced vapor, or combustible liquid–produced vapor mixed with air that may burn or explode, having either a maximum experimental safe gap (MESG) value greater than 0.45 mm and less than or equal to 0.75 mm, or a minimum igniting current ratio (MIC ratio) greater than 0.40 and less than or equal to 0.80. [**497**:3.3.5.1.3]

Informational Note: A typical Class I, Group C material is ethylene.

(4) Group D. Flammable gas, flammable liquid–produced vapor, or combustible liquid–produced vapor mixed with air that may burn or explode, having either a maximum experimental safe gap (MESG) value greater than 0.75 mm or a minimum igniting current (MIC) ratio greater than 0.80. [**497**:3.3.5.1.4]

Informational Note No. 1: A typical Class I, Group D material is propane. [**497**:3.3.5.1.4]

Informational Note No. 2: For classification of areas involving ammonia atmospheres, see ANSI/ASHRAE 15-2013, *Safety Standard for Refrigeration Systems.*

(B) Class II Group Classifications. Class II groups shall be in accordance with 500.6(B)(1) through (B)(3).

(1) Group E. Atmospheres containing combustible metal dusts, including aluminum, magnesium, and their commercial alloys, or other combustible dusts whose particle size, abrasiveness, and conductivity present similar hazards in the use of electrical equipment. [**499**:3.3.4.1]

Informational Note: Certain metal dusts may have characteristics that require safeguards beyond those required for atmospheres containing the dusts of aluminum, magnesium, and their commercial alloys. For example, zirconium, thorium, and uranium dusts have extremely low ignition temperatures [as low as 20°C (68°F)] and minimum ignition energies lower than any material classified in any of the Class I or Class II groups.

(2) Group F. Atmospheres containing combustible carbonaceous dusts that have more than 8 percent total entrapped volatiles (see ASTM D3175-11, *Standard Test Method for Volatile Matter in the Analysis Sample for Coal and Coke,* for coal and coke dusts) or that have been sensitized by other materials so that they present an explosion hazard. [**499**:3.3.4.2] Coal, carbon black, charcoal, and coke dusts are examples of carbonaceous dusts. [**499**:A.3.3.4.2]

Informational Note: Testing of specific dust samples, following established ASTM testing procedures, is a method used to identify the combustibility of a specific dust and the need to classify those locations containing that material as Group F.

(3) Group G. Atmospheres containing combustible dusts not included in Group E or Group F, including flour, grain, wood, plastic, and chemicals. [**499**:3.3.4.3]

Informational Note No. 1: For additional information on group classification of Class II materials, see NFPA 499-2013, *Recommended Practice for the Classification of Combustible Dusts and of Hazardous (Classified) Locations for Electrical Installations in Chemical Process Areas.*

Informational Note No. 2: The explosion characteristics of air mixtures of dust vary with the materials involved. For Class II locations, Groups E, F, and G, the classification involves the tightness of the joints of assembly and shaft openings to prevent the entrance of dust in the dust-ignitionproof enclosure, the blanketing effect of layers of dust on the equipment that may cause overheating, and the ignition temperature of the dust. It is necessary, therefore, that equipment be identified not only for the class but also for the specific group of dust that will be present.

Informational Note No. 3: Certain dusts may require additional precautions due to chemical phenomena that can result in the generation of ignitible gases. See ANSI/IEEE C2-2012, *National Electrical Safety Code,* Section 127A, Coal Handling Areas.

500.7 Protection Techniques. Section 500.7(A) through (L) shall be acceptable protection techniques for electrical and electronic equipment in hazardous (classified) locations.

(A) Explosionproof Equipment. This protection technique shall be permitted for equipment in Class I, Division 1 or 2 locations.

(B) Dust Ignitionproof. This protection technique shall be permitted for equipment in Class II, Division 1 or 2 locations.

(C) Dusttight. This protection technique shall be permitted for equipment in Class II, Division 2 or Class III, Division 1 or 2 locations.

(D) Purged and Pressurized. This protection technique shall be permitted for equipment in any hazardous (classified) location for which it is identified.

(E) Intrinsic Safety. This protection technique shall be permitted for equipment in Class I, Division 1 or 2; or Class II, Division 1 or 2; or Class III, Division 1 or 2 locations. The provisions of Articles 501 through 503 and Articles 510 through 516 shall not be considered applicable to such installations, except as required by Article 504, and installation of intrinsically safe apparatus and wiring shall be in accordance with the requirements of Article 504.

(F) Nonincendive Circuit. This protection technique shall be permitted for equipment in Class I, Division 2; Class II, Division 2; or Class III, Division 1 or 2 locations.

(G) Nonincendive Equipment. This protection technique shall be permitted for equipment in Class I, Division 2; Class II, Division 2; or Class III, Division 1 or 2 locations.

(H) Nonincendive Component. This protection technique shall be permitted for equipment in Class I, Division 2; Class II, Division 2; or Class III, Division 1 or 2 locations.

(I) Oil Immersion. This protection technique shall be permitted for current-interrupting contacts in Class I, Division 2 locations as described in 501.115(B)(1)(2).

(J) Hermetically Sealed. This protection technique shall be permitted for equipment in Class I, Division 2; Class II, Division 2; or Class III, Division 1 or 2 locations.

(K) Combustible Gas Detection System. A combustible gas detection system shall be permitted as a means of protection in industrial establishments with restricted public access and where the conditions of maintenance and supervision ensure that only qualified persons service the installation. Where such a system is installed, equipment specified in 500.7(K)(1), (K)(2), or (K)(3) shall be permitted.

The type of detection equipment, its listing, installation location(s), alarm and shutdown criteria, and calibration frequency shall be documented where combustible gas detectors are used as a protection technique.

Informational Note No. 1: For further information, see ANSI/ ISA-60079-29-1 (12.13.01)-2013, *Explosive Atmospheres — Part 29-1: Gas detectors — Performance requirements of detectors for flammable gases.*

Informational Note No. 2: For further information, see ANSI/API RP 500–2012, *Recommended Practice for Classification of Locations for Electrical Installations at Petroleum Facilities Classified as Class I, Division I or Division 2.*

Informational Note No. 3: For further information, see ANSI/ ISA-60079-29-2 (12.13.02)-2012, *Explosive Atmospheres — Part 29-2: Gas detectors — Selection, installation, use and maintenance of detectors for flammable gases and oxygen.*

Informational Note No. 4: For further information, see ISA-TR12.13.03-2009, *Guide for Combustible Gas Detection as a Method of Protection.*

(1) Inadequate Ventilation. In a Class I, Division 1 location that is so classified due to inadequate ventilation, electrical equipment suitable for Class I, Division 2 locations shall be permitted. Combustible gas detection equipment shall be listed for Class I, Division 1, for the appropriate material group, and for the detection of the specific gas or vapor to be encountered.

(2) Interior of a Building. In a building located in, or with an opening into, a Class I, Division 2 location where the interior does not contain a source of flammable gas or vapor, electrical equipment for unclassified locations shall be permitted. Combustible gas detection equipment shall be listed for Class I, Division 1 or Class I, Division 2, for the appropriate material group, and for the detection of the specific gas or vapor to be encountered.

(3) Interior of a Control Panel. In the interior of a control panel containing instrumentation utilizing or measuring flammable liquids, gases, or vapors, electrical equipment suitable for Class I, Division 2 locations shall be permitted. Combustible gas detection equipment shall be listed for Class I, Division 1, for the appropriate material group, and for the detection of the specific gas or vapor to be encountered.

(L) Other Protection Techniques. Other protection techniques used in equipment identified for use in hazardous (classified) locations.

500.8 Equipment. Articles 500 through 504 require equipment construction and installation that ensure safe performance under conditions of proper use and maintenance.

Informational Note No. 1: It is important that inspection authorities and users exercise more than ordinary care with regard to installation and maintenance.

Informational Note No. 2: Since there is no consistent relationship between explosion properties and ignition temperature, the two are independent requirements.

Informational Note No. 3: Low ambient conditions require special consideration. Explosionproof or dust-ignitionproof equipment may not be suitable for use at temperatures lower than -25°C (-13°F) unless they are identified for low-temperature service. However, at low ambient temperatures, flammable concentrations of vapors may not exist in a location classified as Class I, Division 1 at normal ambient temperature.

(A) Suitability. Suitability of identified equipment shall be determined by one of the following:

(1) Equipment listing or labeling

(2) Evidence of equipment evaluation from a qualified testing laboratory or inspection agency concerned with product evaluation

(3) Evidence acceptable to the authority having jurisdiction such as a manufacturer's self-evaluation or an owner's engineering judgment

Informational Note: Additional documentation for equipment may include certificates demonstrating compliance with applicable equipment standards, indicating special conditions of use, and other pertinent information. Guidelines for certificates may be found in ANSI/UL 120002:2014, *Certificate Standard for AEx Equipment for Hazardous (Classified) Locations.*

(B) Approval for Class and Properties.

(1) Equipment shall be identified not only for the class of location but also for the explosive, combustible, or ignitible properties of the specific gas, vapor, dust, or fibers/flyings that will be present. In addition, Class I equipment shall not have any exposed surface that operates at a temperature in excess of the autoignition temperature of the specific gas or vapor. Class II equipment shall not have an external temperature higher than that specified in 500.8(D)(2). Class III equipment shall not exceed the maximum surface temperatures specified in 503.5.

Informational Note: Luminaires and other heat-producing apparatus, switches, circuit breakers, and plugs and receptacles are potential sources of ignition and are investigated for suitability in classified locations. Such types of equipment, as well as cable terminations for entry into explosionproof enclosures, are available as listed for Class I, Division 2 locations. Fixed wiring, however, may utilize wiring methods that are not evaluated with respect to classified locations. Wiring products such as cable, raceways, boxes, and fittings, therefore, are not marked as being suitable for Class I, Division 2 locations. Also see 500.8(C)(6)(a).

(2) Equipment that has been identified for a Division 1 location shall be permitted in a Division 2 location of the same class, group, and temperature class and shall comply with (a) or (b) as applicable.

(a) Intrinsically safe apparatus having a control drawing requiring the installation of associated apparatus for a Division 1 installation shall be permitted to be installed in a Division 2 location if the same associated apparatus is used for the Division 2 installation.

(b) Equipment that is required to be explosionproof shall incorporate seals in accordance with 501.15(A) or (D) when the wiring methods of 501.10(B) are employed.

(3) Where specifically permitted in Articles 501 through 503, general-purpose equipment or equipment in general-purpose enclosures shall be permitted to be installed in Division 2 locations if the equipment does not constitute a source of ignition under normal operating conditions.

(4) Equipment that depends on a single compression seal, diaphragm, or tube to prevent flammable or combustible fluids from entering the equipment shall be identified for a Class I, Division 2 location even if installed in an unclassified location. Equipment installed in a Class I, Division 1 location shall be identified for the Class I, Division 1 location.

Informational Note: Equipment used for flow measurement is an example of equipment having a single compression seal, diaphragm, or tube.

(5) Unless otherwise specified, normal operating conditions for motors shall be assumed to be rated full-load steady conditions.

(6) Where flammable gases, flammable liquid–produced vapors, combustible liquid–produced vapors, or combustible dusts are or may be present at the same time, the simultaneous presence of both shall be considered when determining the safe operating temperature of the electrical equipment.

Informational Note: The characteristics of various atmospheric mixtures of gases, vapors, and dusts depend on the specific material involved.

(C) Marking. Equipment shall be marked to show the environment for which it has been evaluated. Unless otherwise specified or allowed in (C)(6), the marking shall include the information specified in (C)(1) through (C)(5).

(1) Class. The marking shall specify the class(es) for which the equipment is suitable.

(2) Division. The marking shall specify the division if the equipment is suitable for Division 2 only. Equipment suitable for Division 1 shall be permitted to omit the division marking.

Informational Note: Equipment not marked to indicate a division, or marked "Division 1" or "Div. 1," is suitable for both Division 1 and 2 locations; see 500.8(B)(2). Equipment marked "Division 2" or "Div. 2" is suitable for Division 2 locations only.

(3) Material Classification Group. The marking shall specify the applicable material classification group(s) or specific gas, vapor, dust, or fiber/flying in accordance with 500.6.

Exception: Fixed luminaires marked for use only in Class I, Division 2 or Class II, Division 2 locations shall not be required to indicate the group.

Informational Note: A specific gas, vapor, dust, or fiber/flying is typically identified by the generic name, chemical formula, CAS number, or combination thereof.

(4) Equipment Temperature. The marking shall specify the temperature class or operating temperature at a 40°C ambient temperature, or at the higher ambient temperature if the equipment is rated and marked for an ambient temperature of greater than 40°C. For equipment installed in a Class II, Division 1 location, the temperature class or operating temperature shall be based on operation of the equipment when blanketed with the maximum amount of dust that can accumulate on the equipment. The temperature class, if provided, shall be indicated using the temperature class (T codes) shown in Table 500.8(C). Equipment for Class I and Class II shall be marked with the maximum safe operating temperature, as determined by simultaneous exposure to the combinations of Class I and Class II conditions.

Exception: Equipment of the non–heat-producing type, such as junction boxes, conduit, and fittings, and equipment of the heat-producing type having a maximum temperature not more than 100°C shall not be required to have a marked operating temperature or temperature class.

Informational Note: More than one marked temperature class or operating temperature, for gases and vapors, dusts, and different ambient temperatures, may appear.

(5) Ambient Temperature Range. Electrical equipment designed for use in the ambient temperature range between –25°C to +40°C shall require no ambient temperature marking. For equipment rated for a temperature range other than –25°C

Table 500.8(C) Classification of Maximum Surface Temperature

Maximum Temperature		Temperature Class (T Code)
°C	°F	
450	842	T1
300	572	T2
280	536	T2A
260	500	T2B
230	446	T2C
215	419	T2D
200	392	T3
180	356	T3A
165	329	T3B
160	320	T3C
135	275	T4
120	248	T4A
100	212	T5
85	185	T6

to +40°C, the marking shall specify the special range of ambient temperatures in degrees Celsius. The marking shall include either the symbol "Ta" or "Tamb."

Informational Note: As an example, such a marking might be "–30°C ≤ Ta ≤ +40°C."

(6) Special Allowances.

(a)   *General-Purpose Equipment.* Fixed general-purpose equipment in Class I locations, other than fixed luminaires, that is acceptable for use in Class I, Division 2 locations shall not be required to be marked with the class, division, group, temperature class, or ambient temperature range.

(b)   *Dusttight Equipment.* Fixed dusttight equipment, other than fixed luminaires, that is acceptable for use in Class II, Division 2 and Class III locations shall not be required to be marked with the class, division, group, temperature class, or ambient temperature range.

(c)   *Associated Apparatus.* Associated intrinsically safe apparatus and associated nonincendive field wiring apparatus that are not protected by an alternative type of protection shall not be marked with the class, division, group, or temperature class. Associated intrinsically safe apparatus and associated nonincendive field wiring apparatus shall be marked with the class, division, and group of the apparatus to which it is to be connected.

(d)   *Simple Apparatus.* "Simple apparatus" as defined in Article 504, shall not be required to be marked with class, division, group, temperature class, or ambient temperature range.

(D) Temperature.

(1) Class I Temperature. The temperature marking specified in 500.8(C) shall not exceed the autoignition temperature of the specific gas or vapor to be encountered.

Informational Note: For information regarding autoignition temperatures of gases and vapors, see NFPA 497-2013, *Recommended Practice for the Classification of Flammable Liquids, Gases, or Vapors, and of Hazardous (Classified) Locations for Electrical Installations in Chemical Process Areas.*

(2) Class II Temperature. The temperature marking specified in 500.8(C) shall be less than the ignition temperature of the specific dust to be encountered. For organic dusts that may dehydrate or carbonize, the temperature marking shall no exceed the lower of either the ignition temperature or 165°C (329°F).

Informational Note: See NFPA 499-2013, *Recommended Practice for the Classification of Combustible Dusts and of Hazardous (Classified) Locations for Electrical Installations in Chemical Process Areas,* for minimum ignition temperatures of specific dusts.

(E) Threading. The supply connection entry thread form shall be NPT or metric. Conduit and fittings shall be made wrenchtight to prevent sparking when fault current flows through the conduit system, and to ensure the explosionproof integrity of the conduit system where applicable. Equipment provided with threaded entries for field wiring connections shall be installed in accordance with 500.8(E)(1) or (E)(2) and with (E)(3).

(1) Equipment Provided with Threaded Entries for NPT-Threaded Conduit or Fittings. For equipment provided with threaded entries for NPT-threaded conduit or fittings, listed conduit, listed conduit fittings, or listed cable fittings shall be used. All NPT-threaded conduit and fittings shall be threaded with a National (American) Standard Pipe Taper (NPT) thread.

NPT-threaded entries into explosionproof equipment shall be made up with at least five threads fully engaged.

Exception: For listed explosionproof equipment, joints with factory-threaded NPT entries shall be made up with at least four and one-half threads fully engaged.

Informational Note No. 1: Thread specifications for male NPT threads are located in ASME B1.20.1-2013, *Pipe Threads, General Purpose (Inch).*

Informational Note No. 2: Female NPT-threaded entries use a modified National Standard Pipe Taper (NPT) thread with thread form per ASME B1.20.1-2013, *Pipe Threads, General Purpose (Inch).* See ANSI/UL 1203-2009, *Explosionproof and Dust-Ignition-Proof Electrical Equipment for Use in Hazardous (Classified) Locations.*

(2) Equipment Provided with Threaded Entries for Metric-Threaded Fittings. For equipment with metric-threaded entries, listed conduit fittings or listed cable fittings shall be used. Such entries shall be identified as being metric, or listed adapters to permit connection to conduit or NPT-threaded fittings shall be provided with the equipment and shall be used for connection to conduit or NPT-threaded fittings.

Metric-threaded fittings installed into explosionproof equipment shall have a class of fit of at least 6g/6H and shall be made up with at least five threads fully engaged.

Informational Note: Threading specifications for metric-threaded entries are located in ISO 965-1-2013, *ISO general purpose metric screw threads — Tolerances — Part 1: Principles and basic data,* and ISO 965-3-1998, *ISO general purpose metric screw threads — Tolerances — Part 3: Deviations for constructional screw threads.*

(3) Unused Openings. All unused openings shall be closed with listed metal close-up plugs. The plug engagement shall comply with 500.8(E)(1) or (E)(2).

(F) Optical Fiber Cables. An optical fiber cable, with or without current-carrying conductors (composite optical fiber cable), shall be installed to address the associated fire hazard and sealed to address the associated explosion hazard in accordance with the requirements of Article 500, 501, 502, or 503, as applicable.

500.9 Specific Occupancies. Articles 510 through 517 cover garages, aircraft hangars, motor fuel dispensing facilities, bulk storage plants, spray application, dipping and coating processes, and health care facilities.

ARTICLE 501
Class I Locations

Part I. General

501.1 Scope. Article 501 covers the requirements for electrical and electronic equipment and wiring for all voltages in Class I, Division 1 and 2 locations where fire or explosion hazards may exist due to flammable gases or vapors or flammable liquids.

Informational Note: For the requirements for electrical and electronic equipment and wiring for all voltages in Zone 0, Zone 1, or Zone 2 hazardous (classified) locations where fire or explosion hazards may exist due to flammable gases or vapors or flammable liquids, refer to Article 505.

501.5 Zone Equipment. Equipment listed and marked in accordance with 505.9(C)(2) for use in Zone 0, 1, or 2 locations shall be permitted in Class I, Division 2 locations for the same gas and with a suitable temperature class. Equipment listed and marked in accordance with 505.9(C)(2) for use in Zone 0 locations shall be permitted in Class I, Division 1 or Division 2 locations for the same gas and with a suitable temperature class.

Part II. Wiring

501.10 Wiring Methods. Wiring methods shall comply with 501.10(A) or (B).

(A) Class I, Division 1.

(1) General. In Class I, Division 1 locations, the wiring methods in (a) through (f) shall be permitted.

(a) Threaded rigid metal conduit or threaded steel intermediate metal conduit.

Exception: Type PVC conduit, Type RTRC conduit, and Type HDPE conduit shall be permitted where encased in a concrete envelope a minimum of 50 mm (2 in.) thick and provided with not less than 600 mm (24 in.) of cover measured from the top of the conduit to grade. The concrete encasement shall be permitted to be omitted where subject to the provisions of 514.8, Exception No. 2, or 515.8(A). Threaded rigid metal conduit or threaded steel intermediate metal conduit shall be used for the last 600 mm (24 in.) of the underground run to emergence or to the point of connection to the aboveground raceway. An equipment grounding conductor shall be included to provide for electrical continuity of the raceway system and for grounding of non–current-carrying metal parts.

(b) Type MI cable terminated with fittings listed for the location. Type MI cable shall be installed and supported in a manner to avoid tensile stress at the termination fittings.

(c) In industrial establishments with restricted public access, where the conditions of maintenance and supervision ensure that only qualified persons service the installation, Type MC-HL cable listed for use in Class I, Zone 1 or Division 1 locations, with a gas/vaportight continuous corrugated metallic sheath, an overall jacket of suitable polymeric material, and a separate equipment grounding conductor(s) in accordance with 250.122, and terminated with fittings listed for the application.

Type MC-HL cable shall be installed in accordance with the provisions of Article 330, Part II.

(d) In industrial establishments with restricted public access, where the conditions of maintenance and supervision ensure that only qualified persons service the installation, Type ITC-HL cable listed for use in Class I, Zone 1 or Division 1 locations, with a gas/vaportight continuous corrugated metallic sheath and an overall jacket of suitable polymeric material, and terminated with fittings listed for the application, and installed in accordance with the provisions of Article 727.

(e) Optical fiber cable Types OFNP, OFCP, OFNR, OFCR, OFNG, OFCG, OFN, and OFC shall be permitted to be installed in raceways in accordance with 501.10(A). These optical fiber cables shall be sealed in accordance with 501.15.

(2) Flexible Connections. Where necessary to employ flexible connections, as at motor terminals, one of the following shall be permitted:

(1) Flexible fittings listed for the location
(2) Flexible cord in accordance with the provisions of 501.140, terminated with cord connectors listed for the location
(3) In industrial establishments with restricted public access, where the conditions of maintenance and supervision ensure that only qualified persons service the installation, for applications limited to 600 volts, nominal, or less, and where protected from damage by location or a suitable guard, listed Type TC-ER-HL cable with an overall jacket and a separate equipment grounding conductor(s) in accordance with 250.122 that is terminated with fittings listed for the location

(3) Boxes and Fittings. All boxes and fittings shall be approved for Class I, Division 1.

Informational Note: For entry into enclosures required to be explosionproof, see the information on construction, testing, and marking of cables, explosionproof cable fittings, and explosionproof cord connectors in ANSI/UL 2225-2011, *Cables and Cable-Fittings for Use in Hazardous (Classified) Locations.*

(B) Class I, Division 2.

(1) General. In Class I, Division 2 locations, all wiring methods permitted in 501.10(A) and the following wiring methods shall be permitted:

(1) Rigid metal conduit (RMC) and intermediate metal conduit (IMC) with listed threadless fittings.
(2) Enclosed gasketed busways and enclosed gasketed wireways.
(3) Type PLTC and Type PLTC-ER cable in accordance with the provisions of Article 725, including installation in cable tray systems. The cable shall be terminated with listed fittings.
(4) Type ITC and Type ITC-ER cable as permitted in 727.4 and terminated with listed fittings.

(5) Type MC, MV, TC, or TC-ER cable, including installation in cable tray systems. The cable shall be terminated with listed fittings.
(6) In industrial establishments with restricted public access, where the conditions of maintenance and supervision ensure that only qualified persons service the installation and where metallic conduit does not provide sufficient corrosion resistance, listed reinforced thermosetting resin conduit (RTRC), factory elbows, and associated fittings, all marked with the suffix -XW, and Schedule 80 PVC conduit, factory elbows, and associated fittings shall be permitted.
(7) Optical fiber cable Types OFNP, OFCP, OFNR, OFCR, OFNG, OFCG, OFN, and OFC shall be permitted to be installed in cable trays or any other raceway in accordance with 501.10(B). Optical fiber cables shall be sealed in accordance with 501.15.
(8) Cablebus.

Where seals are required for boundary conditions as defined in 501.15(A)(4), the Division 1 wiring method shall extend into the Division 2 area to the seal, which shall be located on the Division 2 side of the Division 1–Division 2 boundary.

(2) Flexible Connections. Where provision must be made for flexibility, one or more of the following shall be permitted:

(1) Listed flexible metal fittings.
(2) Flexible metal conduit with listed fittings.
(3) Interlocked armor Type MC cable with listed fittings.
(4) Liquidtight flexible metal conduit with listed fittings.
(5) Liquidtight flexible nonmetallic conduit with listed fittings.
(6) Flexible cord listed for extra-hard usage and terminated with listed fittings. A conductor for use as an equipment grounding conductor shall be included in the flexible cord.
(7) For elevator use, an identified elevator cable of Type EO, ETP, or ETT, shown under the "use" column in Table 400.4 for "hazardous (classified) locations" and terminated with listed fittings.

Informational Note: See 501.30(B) for grounding requirements where flexible conduit is used.

(3) Nonincendive Field Wiring. Nonincendive field wiring shall be permitted using any of the wiring methods permitted for unclassified locations. Nonincendive field wiring systems shall be installed in accordance with the control drawing(s). Simple apparatus, not shown on the control drawing, shall be permitted in a nonincendive field wiring circuit, provided the simple apparatus does not interconnect the nonincendive field wiring circuit to any other circuit.

Informational Note: Simple apparatus is defined in 504.2.

Separate nonincendive field wiring circuits shall be installed in accordance with one of the following:

(1) In separate cables
(2) In multiconductor cables where the conductors of each circuit are within a grounded metal shield
(3) In multiconductor cables or in raceways, where the conductors of each circuit have insulation with a minimum thickness of 0.25 mm (0.01 in.)

(4) Boxes and Fittings. Boxes and fittings shall not be required to be explosionproof except as required by 501.105(B)(2), 501.115(B)(1), and 501.150(B)(1).

Informational Note: For entry into enclosures required to be explosionproof, see the information on construction, testing, and marking of cables, explosionproof cable fittings, and explosionproof cord connectors in ANSI/UL 2225-2011, *Cables and Cable-Fittings for Use in Hazardous (Classified) Locations.*

501.15 Sealing and Drainage. Seals in conduit and cable systems shall comply with 501.15(A) through (F). Sealing compound shall be used in Type MI cable termination fittings to exclude moisture and other fluids from the cable insulation.

Informational Note No. 1: Seals are provided in conduit and cable systems to minimize the passage of gases and vapors and prevent the passage of flames from one portion of the electrical installation to another through the conduit. Such communication through Type MI cable is inherently prevented by construction of the cable. Unless specifically designed and tested for the purpose, conduit and cable seals are not intended to prevent the passage of liquids, gases, or vapors at a continuous pressure differential across the seal. Even at differences in pressure across the seal equivalent to a few inches of water, there may be a slow passage of gas or vapor through a seal and through conductors passing through the seal. Temperature extremes and highly corrosive liquids and vapors can affect the ability of seals to perform their intended function.

Informational Note No. 2: Gas or vapor leakage and propagation of flames may occur through the interstices between the strands of standard stranded conductors larger than 2 AWG. Special conductor constructions, such as compacted strands or sealing of the individual strands, are means of reducing leakage and preventing the propagation of flames.

(A) Conduit Seals, Class I, Division 1. In Class I, Division 1 locations, conduit seals shall be located in accordance with 501.15(A)(1) through (A)(4).

(1) Entering Enclosures. Each conduit entry into an explosionproof enclosure shall have a conduit seal where either of the following conditions apply:

(1) The enclosure contains apparatus, such as switches, circuit breakers, fuses, relays, or resistors that may produce arcs, sparks, or temperatures that exceed 80 percent of the autoignition temperature, in degrees Celsius, of the gas or vapor involved in normal operation.

Exception: Seals shall not be required for conduit entering an enclosure under any one of the following conditions:

a. *The switch, circuit breaker, fuse, relay, or resistor is enclosed within a chamber hermetically sealed against the entrance of gases or vapors.*

b. *The switch, circuit breaker, fuse, relay, or resistor is immersed in oil in accordance with 501.115(B)(1)(2).*

c. *The switch, circuit breaker, fuse, relay, or resistor is enclosed within an enclosure, identified for the location, and marked "Leads Factory Sealed," or "Factory Sealed," "Seal not Required," or equivalent.*

d. *The switch, circuit breaker, fuse, relay, or resistor is part of a nonincendive circuit.*

(2) The entry is metric designator 53 (trade size 2) or larger, and the enclosure contains terminals, splices, or taps.

An enclosure, identified for the location, and marked "Leads Factory Sealed", or "Factory Sealed," or "Seal not Required," or equivalent shall not be considered to serve as a seal for another adjacent enclosure that is required to have a conduit seal.

Conduit seals shall be installed within 450 mm (18 in.) from the enclosure or as required by the enclosure marking. Only explosionproof unions, couplings, reducers, elbows, and capped elbows that are not larger than the trade size of the conduit shall be permitted between the sealing fitting and the explosionproof enclosure.

(2) Pressurized Enclosures. Conduit seals shall be installed within 450 mm (18 in.) of the enclosure in each conduit entry into a pressurized enclosure where the conduit is not pressurized as part of the protection system.

Informational Note No. 1: Installing the seal as close as possible to the enclosure will reduce problems with purging the dead airspace in the pressurized conduit.

Informational Note No. 2: For further information, see NFPA 496-2013, *Standard for Purged and Pressurized Enclosures for Electrical Equipment.*

(3) Two or More Explosionproof Enclosures. Where two or more explosionproof enclosures that require conduit seals are connected by nipples or runs of conduit not more than 900 mm (36 in.) long, a single conduit seal in each such nipple connection or run of conduit shall be considered sufficient if the seal is located not more than 450 mm (18 in.) from either enclosure.

(4) Class I, Division 1 Boundary. A conduit seal shall be required in each conduit run leaving a Division 1 location. The sealing fitting shall be permitted to be installed on either side of the boundary within 3.05 m (10 ft) of the boundary, and it shall be designed and installed to minimize the amount of gas or vapor within the portion of the conduit installed in the Division 1 location that can be communicated beyond the seal. The conduit run between the conduit seal and the point at which the conduit leaves the Division 1 location shall contain no union, coupling, box, or other fitting except for a listed explosionproof reducer installed at the conduit seal.

Exception No. 1: Metal conduit that contains no unions, couplings, boxes, or fittings, that passes completely through a Division 1 location with no fittings installed within 300 mm (12 in.) of either side of the boundary, shall not require a conduit seal if the termination points of the unbroken conduit are located in unclassified locations.

Exception No. 2: For underground conduit installed in accordance with 300.5 where the boundary is below grade, the sealing fitting shall be permitted to be installed after the conduit emerges from below grade, but there shall be no union, coupling, box, or fitting, other than listed explosionproof reducers at the sealing fitting, in the conduit between the sealing fitting and the point at which the conduit emerges from below grade.

(B) Conduit Seals, Class I, Division 2. In Class I, Division 2 locations, conduit seals shall be located in accordance with 501.15(B)(1) and (B)(2).

(1) Entering Enclosures. For connections to enclosures that are required to be explosionproof, a conduit seal shall be provided in accordance with 501.15(A)(1)(1) and (A)(3). All portions of the conduit run or nipple between the seal and enclosure shall comply with 501.10(A).

(2) Class I, Division 2 Boundary. A conduit seal shall be required in each conduit run leaving a Class I, Division 2 location. The sealing fitting shall be permitted to be installed on either side of the boundary within 3.05 m (10 ft) of the boundary and it shall be designed and installed to minimize the amount of gas or vapor within the portion of the conduit installed in the Division 2 location that can be communicated beyond the seal. Rigid metal conduit or threaded steel intermediate metal conduit shall be used between the sealing fitting and the point at which the conduit leaves the Division 2 location, and a threaded connection shall be used at the sealing fitting. The conduit run between the conduit seal and the point at which the conduit leaves the Division 2 location shall contain no union, coupling, box, or other fitting except for a listed explosionproof reducer installed at the conduit seal. Such seals shall not be required to be explosionproof but shall be identified for the purpose of minimizing the passage of gases permitted under normal operating conditions and shall be accessible.

Informational Note: For further information, refer to ANSI/UL 514B-2012, *Conduit, Tubing, and Cable Fittings.*

Exception No. 1: Metal conduit that contains no unions, couplings, boxes, or fittings, that passes completely through a Division 2 location with no fittings installed within 300 mm (12 in.) of either side of the boundary, shall not be required to be sealed if the termination points of the unbroken conduit are located in unclassified locations.

Exception No. 2: Conduit systems terminating in an unclassified location where the metal conduit transitions to cable tray, cablebus, ventilated busway, or Type MI cable, or to cable not installed in any cable tray or raceway system, shall not be required to be sealed where passing from the Division 2 location into the unclassified location under the following conditions:

(1)   *The unclassified location is outdoors, or the unclassified location is indoors and the conduit system is entirely in one room.*

(2)   *The conduits shall not terminate at an enclosure containing an ignition source in normal operation.*

Exception No. 3: Conduit systems passing from an enclosure or a room that is unclassified, as a result of pressurization, into a Division 2 location shall not require a seal at the boundary.

Informational Note: For further information, refer to NFPA 496-2013, *Standard for Purged and Pressurized Enclosures for Electrical Equipment.*

Exception No. 4: Segments of aboveground conduit systems shall not be required to be sealed where passing from a Division 2 location into an unclassified location if all of the following conditions are met:

(1)   *No part of the conduit system segment passes through a Division 1 location where the conduit segment contains unions, couplings, boxes, or fittings that are located within 300 mm (12 in.) of the Division 1 location.*

(2)   *The conduit system segment is located entirely in outdoor locations.*

(3)   *The conduit system segment is not directly connected to canned pumps, process or service connections for flow, pressure, or analysis measurement, and so forth, that depend on a single compression seal, diaphragm, or tube to prevent flammable or combustible fluids from entering the conduit system.*

(4)   *The conduit system segment contains only threaded metal conduit, unions, couplings, conduit bodies, and fittings in the unclassified location.*

(5)   *The conduit system segment is sealed at its entry to each enclosure or fitting located in the Division 2 location that contains terminals, splices, or taps.*

(C) Class I, Divisions 1 and 2. Seals installed in Class I, Division 1 and Division 2 locations shall comply with 501.15(C)(1) through (C)(6).

Exception: Seals that are not required to be explosionproof by 501.15(B)(2) or 504.70 shall not be required to comply with 501.15(C).

(1) Fittings. Enclosures that contain connections or equipment shall be provided with an integral sealing means, or sealing fittings listed for the location shall be used. Sealing fittings shall be listed for use with one or more specific compounds and shall be accessible.

(2) Compound. The compound shall provide a seal to minimize the passage of gas and/or vapors through the sealing fitting and shall not be affected by the surrounding atmosphere or liquids. The melting point of the compound shall not be less than 93°C (200°F).

(3) Thickness of Compounds. The thickness of the sealing compound installed in completed seals, other than listed cable sealing fittings, shall not be less than the metric designator (trade size) of the sealing fitting expressed in the units of measurement employed; however, in no case shall the thickness of the compound be less than 16 mm (⅝ in.).

(4) Splices and Taps. Splices and taps shall not be made in fittings intended only for sealing with compound; nor shall other fittings in which splices or taps are made be filled with compound.

(5) Assemblies. An entire assembly shall be identified for the location where the equipment that may produce arcs, sparks, or high temperatures is located in a compartment that is separate from the compartment containing splices or taps, and an integral seal is provided where conductors pass from one compartment to the other. In Division 1 locations, seals shall be provided in conduit connecting to the compartment containing splices or taps where required by 501.15(A)(1)(2).

(6) Conductor or Optical Fiber Fill. The cross-sectional area of the conductors or optical fiber tubes (metallic or nonmetallic) permitted in a seal shall not exceed 25 percent of the cross-sectional area of a rigid metal conduit of the same trade size unless the seal is specifically identified for a higher percentage of fill.

(D) Cable Seals, Class I, Division 1. In Division 1 locations, cable seals shall be located according to 501.15(D)(1) through (D)(3).

(1) At Terminations. Cables shall be sealed with sealing fittings that comply with 501.15(C) at all terminations. Type MC-HL cables with a gas/vaportight continuous corrugated metallic sheath and an overall jacket of suitable polymeric material shall be sealed with a listed fitting after the jacket and any other covering have been removed so that the sealing compound can surround each individual insulated conductor in such a manner as to minimize the passage of gases and vapors.

Seals for cables entering enclosures shall be installed within 450 mm (18 in.) of the enclosure or as required by the enclosure marking. Only explosionproof unions, couplings, reducers, elbows, and capped elbows that are not larger than the trade size of the enclosure entry shall be permitted between the sealing fitting and the enclosure.

Exception: Shielded cables and twisted pair cables shall not require the removal of the shielding material or separation of the twisted pairs, provided the termination is sealed by an approved means to minimize the entrance of gases or vapors and prevent propagation of flame into the cable core.

(2) Cables Capable of Transmitting Gases or Vapors. Cables with a gas/vaportight continuous sheath capable of transmitting gases or vapors through the cable core, installed in conduit, shall be sealed in the Class 1, Division 1 location after the jacket and any other coverings have been removed so that the sealing compound can surround each individual insulated conductor or optical fiber tube and the outer jacket.

Exception: Multiconductor cables with a gas/vaportight continuous sheath capable of transmitting gases or vapors through the cable core shall be permitted to be considered as a single conductor by sealing the cable in the conduit within 450 mm (18 in.) of the enclosure and the cable end within the enclosure by an approved means to minimize the entrance of gases or vapors and prevent the propagation of flame into the cable core, or by other approved methods. It shall not be required to remove the shielding material or separate the twisted pairs of shielded cables and twisted pair cables.

(3) Cables Incapable of Transmitting Gases or Vapors. Each multiconductor cable installed in conduit shall be considered as a single conductor if the cable is incapable of transmitting gases or vapors through the cable core. These cables shall be sealed in accordance with 501.15(A).

(E) Cable Seals, Class I, Division 2. In Division 2 locations, cable seals shall be located in accordance with 501.15(E)(1) through (E)(4).

Exception: Cables with an unbroken gas/vaportight continuous sheath shall be permitted to pass through a Division 2 location without seals.

(1) Terminations. Cables entering enclosures that are required to be explosionproof shall be sealed at the point of entrance. The sealing fitting shall comply with 501.15(B)(1) or be explosionproof. Multiconductor or optical multifiber cables with a gas/vaportight continuous sheath capable of transmitting gases or vapors through the cable core that are installed in a Division 2 location shall be sealed with a listed fitting after the jacket and any other coverings have been removed, so that the sealing compound can surround each individual insulated conductor or optical fiber tube in such a manner as to minimize the passage of gases and vapors. Multiconductor or optical multifiber cables installed in conduit shall be sealed as described in 501.15(D).

Exception No. 1: Cables leaving an enclosure or room that is unclassified as a result of Type Z pressurization and entering into a Division 2 location shall not require a seal at the boundary.

Exception No. 2: Shielded cables and twisted pair cables shall not require the removal of the shielding material or separation of the twisted pairs, provided the termination is by an approved means to minimize the entrance of gases or vapors and prevent propagation of flame into the cable core.

(2) Cables That Do Not Transmit Gases or Vapors. Cables that have a gas/vaportight continuous sheath and do not transmit gases or vapors through the cable core in excess of the quantity permitted for seal fittings shall not be required to be sealed except as required in 501.15(E)(1). The minimum length of such a cable run shall not be less than the length needed to limit gas or vapor flow through the cable core, excluding the interstices of the conductor strands, to the rate permitted for seal fittings [200 cm^3/hr (0.007 ft^3/hr) of air at a pressure of 1500 pascals (6 in. of water)].

(3) Cables Capable of Transmitting Gases or Vapors. Cables with a gas/vaportight continuous sheath capable of transmitting gases or vapors through the cable core shall not be required to be sealed except as required in 501.15(E)(1), unless the cable is attached to process equipment or devices that may cause a pressure in excess of 1500 pascals (6 in. of water) to be exerted at a cable end, in which case a seal, a barrier, or other means shall be provided to prevent migration of flammables into an unclassified location.

(4) Cables Without Gas/Vaportight Sheath. Cables that do not have a gas/vaportight continuous sheath shall be sealed at the boundary of the Division 2 and unclassified location in such a manner as to minimize the passage of gases or vapors into an unclassified location.

(F) Drainage.

(1) Control Equipment. Where there is a probability that liquid or other condensed vapor may be trapped within enclosures for control equipment or at any point in the raceway system, approved means shall be provided to prevent accumulation or to permit periodic draining of such liquid or condensed vapor.

(2) Motors and Generators. Where liquid or condensed vapor may accumulate within motors or generators, joints and conduit systems shall be arranged to minimize the entrance of liquid. If means to prevent accumulation or to permit periodic draining are necessary, such means shall be provided at the time of manufacture and shall be considered an integral part of the machine.

501.17 Process Sealing. This section shall apply to process-connected equipment, which includes, but is not limited to, canned pumps, submersible pumps, flow, pressure, temperature, or analysis measurement instruments. A process seal is a device to prevent the migration of process fluids from the designed containment into the external electrical system. Process-connected electrical equipment that incorporates a single process seal, such as a single compression seal, diaphragm, or tube to prevent flammable or combustible fluids from entering a conduit or cable system capable of transmitting fluids, shall be provided with an additional means to mitigate a single process seal failure. The additional means may include, but is not limited to, the following:

(1) A suitable barrier meeting the process temperature and pressure conditions that the barrier will be subjected to upon failure of the single process seal. There shall be a vent or drain between the single process seal and the suitable barrier. Indication of the single process seal failure shall be provided by visible leakage, an audible whistle, or other means of monitoring.

(2) A listed Type MI cable assembly, rated at not less than 125 percent of the process pressure and not less than 125 percent of the maximum process temperature (in degrees Celsius), installed between the cable or conduit and the single process seal.

(3) A drain or vent located between the single process seal and a conduit or cable seal. The drain or vent shall be sufficiently sized to prevent overpressuring the conduit or cable seal above 6 in. water column (1493 Pa). Indication of the single process seal failure shall be provided by visible leakage, an audible whistle, or other means of monitoring.

(4) An add-on secondary seal marked "secondary seal" and rated for the pressure and temperature conditions to which it will be subjected upon failure of the single process seal.

Process-connected electrical equipment that does not rely on a single process seal or is listed and marked "single seal" or "dual seal" shall not be required to be provided with an additional means of sealing.

Informational Note: For construction and testing requirements for process sealing for listed and marked single seal, dual seal, or secondary seal equipment, refer to ANSI/ISA-12.27.01-2011, *Requirements for Process Sealing Between Electrical Systems and Flammable or Combustible Process Fluids.*

501.20 Conductor Insulation, Class I, Divisions 1 and 2. Where condensed vapors or liquids may collect on, or come in contact with, the insulation on conductors, such insulation shall be of a type identified for use under such conditions; or the insulation shall be protected by a sheath of lead or by other approved means.

501.25 Uninsulated Exposed Parts, Class I, Divisions 1 and 2. There shall be no uninsulated exposed parts, such as electrical conductors, buses, terminals, or components, that operate at more than 30 volts (15 volts in wet locations). These parts shall additionally be protected by a protection technique according to 500.7(E), (F), or (G) that is suitable for the location.

501.30 Grounding and Bonding, Class I, Divisions 1 and 2. Regardless of the voltage of the electrical system, wiring and equipment in Class I, Division 1 and 2 locations shall be grounded as specified in Article 250 and in accordance with the requirements of 501.30(A) and (B).

(A) Bonding. The locknut-bushing and double-locknut types of contacts shall not be depended on for bonding purposes, but bonding jumpers with proper fittings or other approved means of bonding shall be used. Such means of bonding shall apply to all intervening raceways, fittings, boxes, enclosures, and so forth between Class I locations and the point of grounding for service equipment or point of grounding of a separately derived system.

Exception: The specific bonding means shall be required only to the nearest point where the grounded circuit conductor and the grounding electrode are connected together on the line side of the building or structure disconnecting means as specified in 250.32(B), provided the branch-circuit overcurrent protection is located on the load side of the disconnecting means.

(B) Types of Equipment Grounding Conductors. Flexible metal conduit and liquidtight flexible metal conduit shall include an equipment bonding jumper of the wire type in compliance with 250.102.

Exception: In Class I, Division 2 locations, the bonding jumper shall be permitted to be deleted where all of the following conditions are met:

(1) Listed liquidtight flexible metal conduit 1.8 m (6 ft) or less in length, with fittings listed for grounding, is used.
(2) Overcurrent protection in the circuit is limited to 10 amperes or less.
(3) The load is not a power utilization load.

501.35 Surge Protection.

(A) Class I, Division 1. Surge arresters, surge-protective devices, and capacitors shall be installed in enclosures identified for Class I, Division 1 locations. Surge-protective capacitors shall be of a type designed for specific duty.

(B) Class I, Division 2. Surge arresters and surge-protective devices shall be nonarcing, such as metal-oxide varistor (MOV) sealed type, and surge-protective capacitors shall be of a type designed for specific duty. Enclosures shall be permitted to be of the general-purpose type. Surge protection of types other than described in this paragraph shall be installed in enclosures identified for Class I, Division 1 locations.

Part III. Equipment

501.100 Transformers and Capacitors.

(A) Class I, Division 1. In Class I, Division 1 locations, transformers and capacitors shall comply with 501.100(A)(1) and (A)(2).

(1) Containing Liquid That Will Burn. Transformers and capacitors containing a liquid that will burn shall be installed only in vaults that comply with 450.41 through 450.48 and with (1) through (4) as follows:

(1) There shall be no door or other communicating opening between the vault and the Division 1 location.
(2) Ample ventilation shall be provided for the continuous removal of flammable gases or vapors.
(3) Vent openings or ducts shall lead to a safe location outside of buildings.
(4) Vent ducts and openings shall be of sufficient area to relieve explosion pressures within the vault, and all portions of vent ducts within the buildings shall be of reinforced concrete construction.

(2) Not Containing Liquid That Will Burn. Transformers and capacitors that do not contain a liquid that will burn shall be installed in vaults complying with 501.100(A)(1) or be identified for Class I locations.

(B) Class I, Division 2. In Class I, Division 2 locations, transformers shall comply with 450.21 through 450.27, and capacitors shall comply with 460.2 through 460.28.

501.105 Meters, Instruments, and Relays.

(A) Class I, Division 1. In Class I, Division 1 locations, meters, instruments, and relays, including kilowatt-hour meters, instrument transformers, resistors, rectifiers, and thermionic tubes, shall be provided with enclosures identified for Class I, Division 1 locations. Enclosures for Class I, Division 1 locations include explosionproof enclosures and purged and pressurized enclosures.

Informational Note: See NFPA 496-2013, *Standard for Purged and Pressurized Enclosures for Electrical Equipment.*

(B) Class I, Division 2. In Class I, Division 2 locations, meters, instruments, and relays shall comply with 501.105(B)(2) through (B)(6).

(1) General-Purpose Assemblies. Where an assembly is made up of components for which general-purpose enclosures are acceptable as provided in 501.105(B)(1), (B)(2), and (B)(3), a single general-purpose enclosure shall be acceptable for the assembly. Where such an assembly includes any of the equipment described in 501.105(B)(1), 501.105(B)(2), and 501.105(B)(3), the maximum obtainable surface temperature of any component of the assembly that exceeds 100°C shall be clearly and permanently indicated on the outside of the enclosure. Alternatively, equipment shall be permitted to be marked to indicate the temperature class for which it is suitable, using the temperature class (T Code) of Table 500.8(C).

(2) Contacts. Switches, circuit breakers, and make-and-break contacts of pushbuttons, relays, alarm bells, and horns shall have enclosures identified for Class I, Division 1 locations in accordance with 501.105(A).

Exception: General-purpose enclosures shall be permitted if current-interrupting contacts comply with one of the following:

(1) Are immersed in oil
(2) Are enclosed within a chamber that is hermetically sealed against the entrance of gases or vapors
(3) Are in nonincendive circuits
(4) Are listed for Division 2

(3) Resistors and Similar Equipment. Resistors, resistance devices, thermionic tubes, rectifiers, and similar equipment that are used in or in connection with meters, instruments, and relays shall comply with 501.105(A).

Exception: General-purpose-type enclosures shall be permitted if such equipment is without make-and-break or sliding contacts [other than as provided in 501.105(B)(2)] and if the marked maximum operating temperature of any exposed surface will not exceed 80 percent of the autoignition temperature in degrees Celsius of the gas or vapor involved or has been tested and found incapable of igniting the gas or vapor. This exception shall not apply to thermionic tubes.

(4) Without Make-or-Break Contacts. Transformer windings, impedance coils, solenoids, and other windings that do not incorporate sliding or make-or-break contacts shall be provided with enclosures. General-purpose-type enclosures shall be permitted.

(5) Fuses. Where general-purpose enclosures are permitted in 501.105(B)(2) through (B)(4), fuses for overcurrent protection of instrument circuits not subject to overloading in normal use shall be permitted to be mounted in general-purpose enclosures if each such fuse is preceded by a switch complying with 501.105(B)(2).

(6) Connections. To facilitate replacements, process control instruments shall be permitted to be connected through flexible cord by means of attachment plug and receptacle, provided that all of the following conditions apply:

(1) The attachment plug and receptacle are listed for use in Class I, Division 2 locations and for use with flexible cords and shall be of the locking and grounding type.

Exception: A Class I, Division 2 listing shall not be required if the circuit is nonincendive field wiring.

(2) Unless the attachment plug and receptacle are interlocked mechanically or electrically, or otherwise designed so that they cannot be separated when the contacts are energized and the contacts cannot be energized when the plug and socket outlet are separated, a switch complying with 501.105(B)(2) is provided so that the attachment plug or receptacle is not depended on to interrupt current.

Exception: The switch shall not be required if the circuit is nonincendive field wiring.

(3) The flexible cord does not exceed 900 mm (3 ft) and is of a type listed for extra-hard usage or for hard usage if protected by location, if applicable.
(4) Only necessary receptacles are provided.
(5) The circuit has a maximum current of 3 amps.

501.115 Switches, Circuit Breakers, Motor Controllers, and Fuses.

(A) Class I, Division 1. In Class I, Division 1 locations, switches, circuit breakers, motor controllers, and fuses, including pushbuttons, relays, and similar devices, shall be provided with enclosures, and the enclosure in each case, together with the enclosed apparatus, shall be identified as a complete assembly for use in Class I locations.

(B) Class I, Division 2. Switches, circuit breakers, motor controllers, and fuses in Class I, Division 2 locations shall comply with 501.115(B)(1) through (B)(4).

(1) Type Required. Circuit breakers, motor controllers, and switches intended to interrupt current in the normal performance of the function for which they are installed shall be provided with enclosures identified for Class I, Division 1 locations in accordance with 501.105(A), unless general-purpose enclosures are provided and any of the following apply:

(1) The interruption of current occurs within a chamber hermetically sealed against the entrance of gases and vapors.
(2) The current make-and-break contacts are oil-immersed and of the general-purpose type having a 50-mm (2-in.) minimum immersion for power contacts and a 25-mm (1-in.) minimum immersion for control contacts.
(3) The interruption of current occurs within an enclosure, identified for the location, and marked "Leads Factory Sealed", or "Factory Sealed", or "Seal not Required", or equivalent.
(4) The device is a solid state, switching control without contacts, where the surface temperature does not exceed 80 percent of the autoignition temperature in degrees Celsius of the gas or vapor involved.

(2) Isolating Switches. Fused or unfused disconnect and isolating switches for transformers or capacitor banks that are not intended to interrupt current in the normal performance of the function for which they are installed shall be permitted to be installed in general-purpose enclosures.

(3) Fuses. For the protection of motors, appliances, and lamps, other than as provided in 501.115(B)(4), standard plug or cartridge fuses shall be permitted, provided they are placed within enclosures identified for the location; or fuses shall be permitted if they are within general-purpose enclosures, and if they are of a type in which the operating element is immersed in oil or other approved liquid, or the operating element is enclosed within a chamber hermetically sealed against the entrance of gases and vapors, or the fuse is a nonindicating, filled, current-limiting type.

(4) Fuses Internal to Luminaires. Listed cartridge fuses shall be permitted as supplementary protection within luminaires.

501.120 Control Transformers and Resistors. Transformers, impedance coils, and resistors used as, or in conjunction with, control equipment for motors, generators, and appliances shall comply with 501.120(A) and (B).

(A) Class I, Division 1. In Class I, Division 1 locations, transformers, impedance coils, and resistors, together with any switching mechanism associated with them, shall be provided with enclosures identified for Class I, Division 1 locations in accordance with 501.105(A).

(B) Class I, Division 2. In Class I, Division 2 locations, control transformers and resistors shall comply with 501.120(B)(1) through (B)(3).

(1) Switching Mechanisms. Switching mechanisms used in conjunction with transformers, impedance coils, and resistors shall comply with 501.115(B).

(2) Coils and Windings. Enclosures for windings of transformers, solenoids, or impedance coils shall be permitted to be of the general-purpose type.

(3) Resistors. Resistors shall be provided with enclosures; and the assembly shall be identified for Class I locations, unless resistance is nonvariable and maximum operating temperature, in degrees Celsius, will not exceed 80 percent of the autoignition temperature of the gas or vapor involved or the resistor has been tested and found incapable of igniting the gas or vapor.

501.125 Motors and Generators.

(A) Class I, Division 1. In Class I, Division 1 locations, motors, generators, and other rotating electrical machinery shall be one of the following:

(1) Identified for Class I, Division 1 locations
(2) Of the totally enclosed type supplied with positive-pressure ventilation from a source of clean air with discharge to a safe area, so arranged to prevent energizing of the machine until ventilation has been established and the enclosure has been purged with at least 10 volumes of air, and also arranged to automatically de-energize the equipment when the air supply fails
(3) Of the totally enclosed inert gas–filled type supplied with a suitable reliable source of inert gas for pressurizing the enclosure, with devices provided to ensure a positive pressure in the enclosure and arranged to automatically de-energize the equipment when the gas supply fails
(4) For machines that are for use only in industrial establishments with restricted public access, where the conditions of maintenance and supervision ensure that only qualified persons service the installation, the machine is permitted to be of a type designed to be submerged in a liquid that is flammable only when vaporized and mixed with air, or in a gas or vapor at a pressure greater than atmospheric and that is flammable only when mixed with air; and the machine is so arranged to prevent energizing it until it has been purged with the liquid or gas to exclude air, and also arranged to automatically de-energize the equipment when the supply of liquid or gas or vapor fails or the pressure is reduced to atmospheric

Totally enclosed motors of the types specified in 501.125(A)(2) or (A)(3) shall have no external surface with an operating temperature in degrees Celsius in excess of 80 percent of the autoignition temperature of the gas or vapor involved. Appropriate devices shall be provided to detect and automatically de-energize the motor or provide an adequate alarm if there is any increase in temperature of the motor beyond designed limits. Auxiliary equipment shall be of a type identified for the location in which it is installed.

(B) Class I, Division 2. In Class I, Division 2 locations, motors, generators, and other rotating electrical machinery shall comply with (1), (2), or (3). They shall also comply with (4) and (5), if applicable.

(1) Be identified for Class I, Division 2 locations, or
(2) Be identified for Class I, Division 1 locations where sliding contacts, centrifugal or other types of switching mechanism (including motor overcurrent, overloading, and overtemperature devices), or integral resistance devices, either while starting or while running, are employed, or
(3) Be open or nonexplosionproof enclosed motors, such as squirrel-cage induction motors without brushes, switching mechanisms, or similar arc-producing devices that are not identified for use in a Class I, Division 2 location.
(4) The exposed surface of space heaters used to prevent condensation of moisture during shutdown periods shall not exceed 80 percent of the autoignition temperature in degrees Celsius of the gas or vapor involved when operated at rated voltage, and the maximum space heater surface temperature [based on a 40°C or higher marked ambient] shall be permanently marked on a visible nameplate mounted on the motor. Otherwise, space heaters shall be identified for Class I, Division 2 locations.
(5) A sliding contact shaft bonding device used for the purpose of maintaining the rotor at ground potential, shall be permitted where the potential discharge energy is determined to be nonincendive for the application. The shaft bonding device shall be permitted to be installed on the inside or the outside of the motor.

Informational Note No. 1: It is important to consider the temperature of internal and external surfaces that may be exposed to the flammable atmosphere.

Informational Note No. 2: It is important to consider the risk of ignition due to currents arcing across discontinuities and overheating of parts in multisection enclosures of large motors and generators. Such motors and generators may need equipotential bonding jumpers across joints in the enclosure and from enclosure to ground. Where the presence of ignitible gases or vapors is suspected, clean-air purging may be needed immediately prior to and during start-up periods.

Informational Note No. 3: For further information on the application of electric motors in Class I, Division 2 hazardous (classified) locations, see IEEE 1349-2011, *IEEE Guide for the Application of Electric Motors in Class I, Division 2 and Class I, Zone 2 Hazardous (Classified) Locations.*

Informational Note No. 4: Reciprocating engine–driven generators, compressors, and other equipment installed in Class I, Division 2 locations may present a risk of ignition of flammable materials associated with fuel, starting, compression, and so forth, due to inadvertent release or equipment malfunction by the engine ignition system and controls. For further information on the requirements for ignition systems for reciprocating engines installed in Class I, Division 2 hazardous (classified) locations, see ANSI/UL 122001-2014, *General Requirements for Electrical Ignition Systems for Internal Combustion Engines in Class I, Division 2 or Zone 2, Hazardous (Classified) Locations.*

Informational Note No. 5: For details of the evaluation process to determine incendivity, refer to Annex A and Figure A1 of UL 1836–2014, *Outline of Investigation for Electric Motors and Generators for Use in Class I, Division 2, Class I, Zone 2, Class II, Division 2 and Zone 22 Hazardous (Classified) Locations.*

501.130 Luminaires. Luminaires shall comply with 501.130(A) or (B).

(A) Class I, Division 1. In Class I, Division 1 locations, luminaires shall comply with 501.130(A)(1) through (A)(4).

(1) Luminaires. Each luminaire shall be identified as a complete assembly for the Class I, Division 1 location and shall be clearly marked to indicate the maximum wattage of lamps for which it is identified. Luminaires intended for portable use shall be specifically listed as a complete assembly for that use.

(2) Physical Damage. Each luminaire shall be protected against physical damage by a suitable guard or by location.

(3) Pendant Luminaires. Pendant luminaires shall be suspended by and supplied through threaded rigid metal conduit stems or threaded steel intermediate conduit stems, and threaded joints shall be provided with set-screws or other effective means to prevent loosening. For stems longer than 300 mm (12 in.), permanent and effective bracing against lateral displacement shall be provided at a level not more than 300 mm (12 in.) above the lower end of the stem, or flexibility in the form of a fitting or flexible connector identified for the Class I, Division 1 location shall be provided not more than 300 mm (12 in.) from the point of attachment to the supporting box or fitting.

(4) Supports. Boxes, box assemblies, or fittings used for the support of luminaires shall be identified for Class I locations.

(B) Class I, Division 2. In Class I, Division 2 locations, luminaires shall comply with 501.130(B)(1) through (B)(6).

(1) Luminaires. Where lamps are of a size or type that may, under normal operating conditions, reach surface temperatures exceeding 80 percent of the autoignition temperature in degrees Celsius of the gas or vapor involved, luminaires shall comply with 501.130(A)(1) or shall be of a type that has been tested in order to determine the marked operating temperature or temperature class (T code).

(2) Physical Damage. Luminaires shall be protected from physical damage by suitable guards or by location. Where there is danger that falling sparks or hot metal from lamps or luminaires might ignite localized concentrations of flammable vapors or gases, suitable enclosures or other effective protective means shall be provided.

(3) Pendant Luminaires. Pendant luminaires shall be suspended by threaded rigid metal conduit stems, threaded steel intermediate metal conduit stems, or other approved means. For rigid stems longer than 300 mm (12 in.), permanent and effective bracing against lateral displacement shall be provided at a level not more than 300 mm (12 in.) above the lower end of the stem, or flexibility in the form of an identified fitting or flexible connector shall be provided not more than 300 mm (12 in.) from the point of attachment to the supporting box or fitting.

(4) Portable Lighting Equipment. Portable lighting equipment shall comply with 501.130(B)(1).

Exception: Where portable lighting equipment is mounted on movable stands and is connected by flexible cords, as covered in 501.140, it shall be permitted to comply with 501.130(B)(1), where mounted in any position, provided that it also complies with 501.130(B)(2).

(5) Switches. Switches that are a part of an assembled fixture or of an individual lampholder shall comply with 501.115(B)(1).

(6) Starting Equipment. Starting and control equipment for electric-discharge lamps shall comply with 501.120(B).

Exception: A thermal protector potted into a thermally protected fluorescent lamp ballast if the luminaire is identified for the location.

501.135 Utilization Equipment.

(A) Class I, Division 1. In Class I, Division 1 locations, all utilization equipment shall be identified for Class I, Division 1 locations.

(B) Class I, Division 2. In Class I, Division 2 locations, all utilization equipment shall comply with 501.135(B)(1) through (B)(3).

(1) Heaters. Electrically heated utilization equipment shall conform with either item (1) or item (2):

(1)   The heater shall not exceed 80 percent of the autoignition temperature in degrees Celsius of the gas or vapor involved on any surface that is exposed to the gas or vapor when continuously energized at the maximum rated ambient temperature. If a temperature controller is not provided, these conditions shall apply when the heater is operated at 120 percent of rated voltage.

Exception No. 1: For motor-mounted anticondensation space heaters, see 501.125.

Exception No. 2: Where a current-limiting device is applied to the circuit serving the heater to limit the current in the heater to a value less than that required to raise the heater surface temperature to 80 percent of the autoignition temperature.

(2)   The heater shall be identified for Class I, Division 1 locations.

Exception to (2): Electrical resistance heat tracing identified for Class I, Division 2 locations.

(2) Motors. Motors of motor-driven utilization equipment shall comply with 501.125(B).

(3) Switches, Circuit Breakers, and Fuses. Switches, circuit breakers, and fuses shall comply with 501.115(B).

501.140 Flexible Cords, Class I, Divisions 1 and 2.

(A) Permitted Uses. Flexible cord shall be permitted:

(1)   For connection between portable lighting equipment or other portable utilization equipment and the fixed portion of their supply circuit. The flexible cord shall be attached to the utilization equipment with a cord connector listed for the protection technique of the equipment wiring compartment. An attachment plug in accordance with 501.140(B)(4) shall be employed.

(2)   For that portion of the circuit where the fixed wiring methods of 501.10(A) cannot provide the necessary degree of movement for fixed and mobile electrical utilization equipment, and the flexible cord is protected by location or by a suitable guard from damage and only in an industrial establishment where conditions of maintenance and engineering supervision ensure that only qualified persons install and service the installation.

(3)   For electric submersible pumps with means for removal without entering the wet-pit. The extension of the flexible cord within a suitable raceway between the wet-pit and the power source shall be permitted.

(4) For electric mixers intended for travel into and out of open-type mixing tanks or vats.

(5) For temporary portable assemblies consisting of receptacles, switches, and other devices that are not considered portable utilization equipment but are individually listed for the location.

(B) Installation. Where flexible cords are used, the cords shall comply with all of the following:

(1) Be of a type listed for extra-hard usage

(2) Contain, in addition to the conductors of the circuit, an equipment grounding conductor complying with 400.23

(3) Be supported by clamps or by other suitable means in such a manner that there is no tension on the terminal connections

(4) In Division 1 locations or in Division 2 locations where the boxes, fittings, or enclosures are required to be explosionproof, the cord shall be terminated with a cord connector or attachment plug listed for the location or a listed cord connector installed with a seal listed for the location. In Division 2 locations where explosionproof equipment is not required, the cord shall be terminated with a listed cord connector or listed attachment plug.

(5) Be of continuous length. Where 501.140(A)(5) is applied, cords shall be of continuous length from the power source to the temporary portable assembly and from the temporary portable assembly to the utilization equipment.

Informational Note: See 501.20 for flexible cords exposed to liquids having a deleterious effect on the conductor insulation.

501.145 Receptacles and Attachment Plugs, Class I, Divisions 1 and 2.

(A) Receptacles. Receptacles shall be part of the premises wiring, except as permitted by 501.140(A).

(B) Attachment Plugs. Attachment plugs shall be of the type providing for connection to the equipment grounding conductor of a permitted flexible cord and shall be identified for the location.

501.150 Signaling, Alarm, Remote-Control, and Communications Systems.

(A) Class I, Division 1. In Class I, Division 1 locations, all apparatus and equipment of signaling, alarm, remote-control, and communications systems, regardless of voltage, shall be identified for Class I, Division 1 locations, and all wiring shall comply with 501.10(A), 501.15(A), and 501.15(C).

(B) Class I, Division 2. In Class I, Division 2 locations, signaling, alarm, remote-control, and communications systems shall comply with 501.150(B)(1) through (B)(4).

(1) Contacts. Switches, circuit breakers, and make-and-break contacts of pushbuttons, relays, alarm bells, and horns shall have enclosures identified for Class I, Division 1 locations in accordance with 501.105(A).

Exception: General-purpose enclosures shall be permitted if current-interrupting contacts are one of the following:

(1) Immersed in oil

(2) Enclosed within a chamber hermetically sealed against the entrance of gases or vapors

(3) In nonincendive circuits

(4) Part of a listed nonincendive component

(2) Resistors and Similar Equipment. Resistors, resistance devices, thermionic tubes, rectifiers, and similar equipment shall comply with 501.105(B)(3) .

(3) Protectors. Enclosures shall be provided for lightning protective devices and for fuses. Such enclosures shall be permitted to be of the general-purpose type.

(4) Wiring and Sealing. All wiring shall comply with 501.10(B), 501.15(B), and 501.15(C).

ARTICLE 502
Class II Locations

Part I. General

502.1 Scope. Article 502 covers the requirements for electrical and electronic equipment and wiring for all voltages in Class II, Division 1 and 2 locations where fire or explosion hazards may exist due to combustible dust.

502.5 Explosionproof Equipment. Explosionproof equipment and wiring shall not be required and shall not be acceptable in Class II locations unless also identified for such locations.

502.6 Zone Equipment. Equipment listed and marked in accordance with 506.9(C)(2) for Zone 20 locations shall be permitted in Class II, Division 1 locations for the same dust atmosphere; and with a suitable temperature class.

Equipment listed and marked in accordance with 506.9(C)(2) for Zone 20, 21, or 22 locations shall be permitted in Class II, Division 2 locations for the same dust atmosphere and with a suitable temperature class.

Part II. Wiring

502.10 Wiring Methods. Wiring methods shall comply with 502.10(A) or (B).

(A) Class II, Division 1.

(1) General. In Class II, Division 1 locations, the wiring methods in (1) through (5) shall be permitted:

(1) Threaded rigid metal conduit, or threaded steel intermediate metal conduit.

(2) Type MI cable with termination fittings listed for the location. Type MI cable shall be installed and supported in a manner to avoid tensile stress at the termination fittings.

(3) In industrial establishments with limited public access, where the conditions of maintenance and supervision ensure that only qualified persons service the installation, Type MC-HL cable, listed for use in Class II, Division 1 locations, with a gas/vaportight continuous corrugated metallic sheath, an overall jacket of suitable polymeric material, a separate equipment grounding conductor(s) in accordance with 250.122, and provided with termination fittings listed for the location, shall be permitted.

(4) Optical fiber cable Types OFNP, OFCP, OFNR, OFCR, OFNG, OFCG, OFN, and OFC shall be permitted to be installed in raceways in accordance with 502.10(A). Optical fiber cables shall be sealed in accordance with 502.15.

(5) In industrial establishments with restricted public access, where the conditions of maintenance and supervision ensure that only qualified persons service the installation, listed Type ITC-HL cable with a gas/vaportight continuous corrugated metallic sheath and an overall jacket of suitable polymeric material, and terminated with fittings listed for the application, and installed in accordance with the provisions of Article 727.

(2) Flexible Connections. Where necessary to employ flexible connections, one or more of the following shall also be permitted:

(1) Dusttight flexible connectors.
(2) Liquidtight flexible metal conduit with listed fittings.
(3) Liquidtight flexible nonmetallic conduit with listed fittings.
(4) Interlocked armor Type MC cable having an overall jacket of suitable polymeric material and provided with termination fittings listed for Class II, Division 1 locations.
(5) Flexible cord listed for extra-hard usage and terminated with listed dusttight cord connectors. Where flexible cords are used, they shall comply with 502.140.
(6) For elevator use, an identified elevator cable of Type EO, ETP, or ETT, shown under the "use" column in Table 400.4 for "hazardous (classified) locations" and terminated with listed dusttight fittings.

Informational Note: See 502.30(B) for grounding requirements where flexible conduit is used.

(3) Boxes and Fittings. Boxes and fittings shall be provided with threaded bosses for connection to conduit or cable terminations and shall be dusttight. Boxes and fittings in which taps, joints, or terminal connections are made, or that are used in Group E locations, shall be identified for Class II locations.

Informational Note: For entry into enclosures required to be dust-ignitionproof, see the information on construction, testing, and marking of cables, dust-ignitionproof cable fittings, and dust-ignitionproof cord connectors in ANSI/UL 2225-2011, *Cables and Cable-Fittings for Use in Hazardous (Classified) Locations.*

(B) Class II, Division 2.

(1) General. In Class II, Division 2 locations, the following wiring methods shall be permitted:

(1) All wiring methods permitted in 502.10(A).
(2) Rigid metal conduit, intermediate metal conduit, electrical metallic tubing, dusttight wireways.
(3) Type MC or MI cable with listed termination fittings.
(4) Type PLTC and Type PLTC-ER cable in accordance with the provisions of Article 725, including installation in cable tray systems. The cable shall be terminated with listed fittings.
(5) Type ITC and Type ITC-ER cable as permitted in 727.4 and terminated with listed fittings.
(6) Type MC, MI, MV, TC, or TC-ER cable installed in ladder, ventilated trough, or ventilated channel cable trays in a single layer, with a space not less than the larger cable diameter between the two adjacent cables, shall be the wiring method employed.

Exception to (6): Type MC cable listed for use in Class II, Division 1 locations shall be permitted to be installed without the spacings required by (6).

(7) In industrial establishments with restricted public access, where the conditions of maintenance and supervision ensure that only qualified persons service the installation and where metallic conduit does not provide sufficient corrosion resistance, reinforced thermosetting resin conduit (RTRC) factory elbows, and associated fittings, all marked with suffix -XW, and Schedule 80 PVC conduit, factory elbows and associated fittings shall be permitted.

(8) Optical fiber cable Types OFNP, OFCP, OFNR, OFCR, OFNG, OFCG, OFN, and OFC shall be permitted to be installed in cable trays or any other raceway in accordance with 502.10(B). Optical fiber cables shall be sealed in accordance with 502.15.

(9) Cablebus.

(2) Flexible Connections. Where provision must be made for flexibility, 502.10(A)(2) shall apply.

(3) Nonincendive Field Wiring. Nonincendive field wiring shall be permitted using any of the wiring methods permitted for unclassified locations. Nonincendive field wiring systems shall be installed in accordance with the control drawing(s). Simple apparatus, not shown on the control drawing, shall be permitted in a nonincendive field wiring circuit, provided the simple apparatus does not interconnect the nonincendive field wiring circuit to any other circuit.

Informational Note: Simple apparatus is defined in 504.2.

Separate nonincendive field wiring circuits shall be installed in accordance with one of the following:

(1) In separate cables
(2) In multiconductor cables where the conductors of each circuit are within a grounded metal shield
(3) In multiconductor cables or in raceways where the conductors of each circuit have insulation with a minimum thickness of 0.25 mm (0.01 in.)

(4) Boxes and Fittings. All boxes and fittings shall be dusttight.

502.15 Sealing, Class II, Divisions 1 and 2. Where a raceway provides communication between an enclosure that is required to be dust-ignitionproof and one that is not, suitable means shall be provided to prevent the entrance of dust into the dust-ignitionproof enclosure through the raceway. One of the following means shall be permitted:

(1) A permanent and effective seal
(2) A horizontal raceway not less than 3.05 m (10 ft) long
(3) A vertical raceway not less than 1.5 m (5 ft) long and extending downward from the dust-ignitionproof enclosure
(4) A raceway installed in a manner equivalent to (2) or (3) that extends only horizontally and downward from the dust-ignition proof enclosures

Where a raceway provides communication between an enclosure that is required to be dust-ignitionproof and an enclosure in an unclassified location, seals shall not be required.

Sealing fittings shall be accessible.

Seals shall not be required to be explosionproof.

Informational Note: Electrical sealing putty is a method of sealing.

502.25 Uninsulated Exposed Parts, Class II, Divisions 1 and 2. There shall be no uninsulated exposed parts, such as electrical conductors, buses, terminals, or components, that operate at more than 30 volts (15 volts in wet locations). These parts shall additionally be protected by a protection technique according to 500.7(E), (F), or (G) that is suitable for the location.

502.30 Grounding and Bonding, Class II, Divisions 1 and 2. Regardless of the voltage of the electrical system, wiring and equipment in Class II, Division 1 and 2 locations shall be grounded as specified in Article 250 and in accordance with the requirements of 502.30(A) and (B).

(A) Bonding. The locknut-bushing and double-locknut types of contact shall not be depended on for bonding purposes, but bonding jumpers with proper fittings or other approved means of bonding shall be used. Such means of bonding shall apply to all intervening raceways, fittings, boxes, enclosures, and so forth, between Class II locations and the point of grounding for service equipment or point of grounding of a separately derived system.

Exception: The specific bonding means shall only be required to the nearest point where the grounded circuit conductor and the grounding electrode conductor are connected together on the line side of the building or structure disconnecting means as specified in 250.32(B) if the branch-circuit overcurrent protection is located on the load side of the disconnecting means.

(B) Types of Equipment Grounding Conductors. Liquidtight flexible metal conduit shall include an equipment bonding jumper of the wire type in compliance with 250.102.

Exception: In Class II, Division 2 locations, the bonding jumper shall be permitted to be deleted where all of the following conditions are met:

(1) *Listed liquidtight flexible metal conduit 1.8 m (6 ft) or less in length, with fittings listed for grounding, is used.*
(2) *Overcurrent protection in the circuit is limited to 10 amperes or less.*
(3) *The load is not a power utilization load.*

502.35 Surge Protection — Class II, Divisions 1 and 2. Surge arresters and surge-protective devices installed in a Class II, Division 1 location shall be in suitable enclosures. Surge-protective capacitors shall be of a type designed for specific duty.

Part III. Equipment

502.100 Transformers and Capacitors.

(A) Class II, Division 1. In Class II, Division 1 locations, transformers and capacitors shall comply with 502.100(A)(1) through (A)(3).

(1) Containing Liquid That Will Burn. Transformers and capacitors containing a liquid that will burn shall be installed only in vaults complying with 450.41 through 450.48, and, in addition, (1), (2), and (3) shall apply.

(1) Doors or other openings communicating with the Division 1 location shall have self-closing fire doors on both sides of the wall, and the doors shall be carefully fitted and provided with suitable seals (such as weather stripping) to minimize the entrance of dust into the vault.
(2) Vent openings and ducts shall communicate only with the outside air.
(3) Suitable pressure-relief openings communicating with the outside air shall be provided.

(2) Not Containing Liquid That Will Burn. Transformers and capacitors that do not contain a liquid that will burn shall be

installed in vaults complying with 450.41 through 450.48 or be identified as a complete assembly, including terminal connections.

(3) Group E. No transformer or capacitor shall be installed in a Class II, Division 1, Group E location.

(B) Class II, Division 2. In Class II, Division 2 locations, transformers and capacitors shall comply with 502.100(B)(1) through (B)(3).

(1) Containing Liquid That Will Burn. Transformers and capacitors containing a liquid that will burn shall be installed in vaults that comply with 450.41 through 450.48.

(2) Containing Askarel. Transformers containing askarel and rated in excess of 25 kVA shall be as follows:

(1) Provided with pressure-relief vents
(2) Provided with a means for absorbing any gases generated by arcing inside the case, or the pressure-relief vents shall be connected to a chimney or flue that will carry such gases outside the building
(3) Have an airspace of not less than 150 mm (6 in.) between the transformer cases and any adjacent combustible material

(3) Dry-Type Transformers. Dry-type transformers shall be installed in vaults or shall have their windings and terminal connections enclosed in tight metal housings without ventilating or other openings and shall operate at not over 600 volts, nominal.

502.115 Switches, Circuit Breakers, Motor Controllers, and Fuses.

(A) Class II, Division 1. In Class II, Division 1 locations, switches, circuit breakers, motor controllers, fuses, pushbuttons, relays, and similar devices shall be provided with enclosures identified for the location.

(B) Class II, Division 2. In Class II, Division 2 locations, enclosures for fuses, switches, circuit breakers, and motor controllers, including push buttons, relays, and similar devices, shall be dusttight or otherwise identified for the location.

502.120 Control Transformers and Resistors.

(A) Class II, Division 1. In Class II, Division 1 locations, control transformers, solenoids, impedance coils, resistors, and any overcurrent devices or switching mechanisms associated with them shall be provided with enclosures identified for the location.

(B) Class II, Division 2. In Class II, Division 2 locations, transformers and resistors shall comply with 502.120(B)(1) through (B)(3).

(1) Switching Mechanisms. Switching mechanisms (including overcurrent devices) associated with control transformers, solenoids, impedance coils, and resistors shall be provided with enclosures that are dusttight or otherwise identified for the location.

(2) Coils and Windings. Where not located in the same enclosure with switching mechanisms, control transformers, solenoids, and impedance coils shall be provided with enclosures that are dusttight or otherwise identified for the location.

(3) Resistors. Resistors and resistance devices shall have dust-ignitionproof enclosures that are dusttight or otherwise identified for the location.

502.125 Motors and Generators.

(A) Class II, Division 1. In Class II, Division 1 locations, motors, generators, and other rotating electrical machinery shall be in conformance with either of the following:

(1) Identified for the location
(2) Totally enclosed pipe-ventilated

(B) Class II, Division 2. In Class II, Division 2 locations, motors, generators, and other rotating electrical equipment shall be totally enclosed nonventilated, totally enclosed pipe-ventilated, totally enclosed water-air-cooled, totally enclosed fan-cooled, or dust-ignitionproof, for which maximum full-load external temperature shall be in accordance with 500.8(D)(2) for normal operation when operating in free air (not dust blanketed) and shall have no external openings.

Exception: If the authority having jurisdiction believes accumulations of nonconductive, nonabrasive dust will be moderate and if machines can be easily reached for routine cleaning and maintenance, the following shall be permitted to be installed:

(1) Standard open-type machines without sliding contacts, centrifugal or other types of switching mechanism (including motor overcurrent, overloading, and overtemperature devices), or integral resistance devices
(2) Standard open-type machines with such contacts, switching mechanisms, or resistance devices enclosed within dusttight housings without ventilating or other openings
(3) Self-cleaning textile motors of the squirrel-cage type
(4) Machines with sealed bearings, bearing isolators, and seals

502.128 Ventilating Piping.

Ventilating pipes for motors, generators, or other rotating electrical machinery, or for enclosures for electrical equipment, shall be of metal not less than 0.53 mm (0.021 in.) in thickness or of equally substantial noncombustible material and shall comply with all of the following:

(1) Lead directly to a source of clean air outside of buildings
(2) Be screened at the outer ends to prevent the entrance of small animals or birds
(3) Be protected against physical damage and against rusting or other corrosive influences

Ventilating pipes shall also comply with 502.128(A) and (B).

(A) Class II, Division 1. In Class II, Division 1 locations, ventilating pipes, including their connections to motors or to the dust-ignitionproof enclosures for other equipment, shall be dusttight throughout their length. For metal pipes, seams and joints shall comply with one of the following:

(1) Be riveted and soldered
(2) Be bolted and soldered
(3) Be welded
(4) Be rendered dusttight by some other equally effective means

(B) Class II, Division 2. In Class II, Division 2 locations, ventilating pipes and their connections shall be sufficiently tight to prevent the entrance of appreciable quantities of dust into the ventilated equipment or enclosure and to prevent the escape of sparks, flame, or burning material that might ignite dust accumulations or combustible material in the vicinity. For metal pipes, lock seams and riveted or welded joints shall be permitted; and tight-fitting slip joints shall be permitted where some flexibility is necessary, as at connections to motors.

502.130 Luminaires.

(A) Class II, Division 1. In Class II, Division 1 locations, luminaires for fixed and portable lighting shall comply with 502.130(A)(1) through (A)(4).

(1) Marking. Each luminaire shall be identified for the location and shall be clearly marked to indicate the type and maximum wattage of the lamp for which it is designed.

(2) Physical Damage. Each luminaire shall be protected against physical damage by a suitable guard or by location.

(3) Pendant Luminaires. Pendant luminaires shall be suspended by threaded rigid metal conduit stems, by threaded steel intermediate metal conduit stems, by chains with approved fittings, or by other approved means. For rigid stems longer than 300 mm (12 in.), permanent and effective bracing against lateral displacement shall be provided at a level not more than 300 mm (12 in.) above the lower end of the stem, or flexibility in the form of a fitting or a flexible connector listed for the location shall be provided not more than 300 mm (12 in.) from the point of attachment to the supporting box or fitting. Threaded joints shall be provided with set screws or other effective means to prevent loosening. Where wiring between an outlet box or fitting and a pendant luminaire is not enclosed in conduit, flexible cord listed for hard usage shall be permitted to be used in accordance with 502.10(A)(2)(5). Flexible cord shall not serve as the supporting means for a luminaire.

(4) Supports. Boxes, box assemblies, or fittings used for the support of luminaires shall be identified for Class II locations.

(B) Class II, Division 2. In Class II, Division 2 locations, luminaires shall comply with 502.130(B)(1) through (B)(5).

(1) Portable Lighting Equipment. Portable lighting equipment shall be identified for the location. They shall be clearly marked to indicate the maximum wattage of lamps for which they are designed.

(2) Fixed Lighting. Luminaires for fixed lighting shall be provided with enclosures that are dusttight or otherwise identified for the location. Each luminaire shall be clearly marked to indicate the maximum wattage of the lamp that shall be permitted without exceeding an exposed surface temperature in accordance with 500.8(D)(2) under normal conditions of use.

(3) Physical Damage. Luminaires for fixed lighting shall be protected from physical damage by suitable guards or by location.

(4) Pendant Luminaires. Pendant luminaires shall be suspended by threaded rigid metal conduit stems, by threaded steel intermediate metal conduit stems, by chains with approved fittings, or by other approved means. For rigid stems longer than 300 mm (12 in.), permanent and effective bracing against lateral displacement shall be provided at a level not more than 300 mm (12 in.) above the lower end of the stem, or flexibility in the form of an identified fitting or a flexible connector shall be provided not more than 300 mm (12 in.) from the point of attachment to the supporting box or fitting. Where wiring between an outlet box or fitting and a pendant luminaire is not enclosed in conduit, flexible cord listed for hard usage shall be permitted if terminated with a listed cord connector that maintains the protection technique. Flexible cord shall not serve as the supporting means for a luminaire.

(5) Electric-Discharge Lamps. Starting and control equipment for electric-discharge lamps shall comply with the requirements of 502.120(B).

502.135 Utilization Equipment.

(A) Class II, Division 1. In Class II, Division 1 locations, all utilization equipment shall be identified for the location.

(B) Class II, Division 2. In Class II, Division 2 locations, all utilization equipment shall comply with 502.135(B)(1) through (B)(4).

(1) Heaters. Electrically heated utilization equipment shall be identified for the location.

Exception: Metal-enclosed radiant heating panel equipment shall be permitted to be dusttight and marked in accordance with 500.8(C).

(2) Motors. Motors of motor-driven utilization equipment shall comply with 502.125(B).

(3) Switches, Circuit Breakers, and Fuses. Enclosures for switches, circuit breakers, and fuses shall comply with 502.115(B).

(4) Transformers, Solenoids, Impedance Coils, and Resistors. Transformers, solenoids, impedance coils, and resistors shall comply with 502.120(B).

502.140 Flexible Cords — Class II, Divisions 1 and 2.

(A) Permitted Uses. Flexible cords used in Class II locations shall comply with all of the following:

(1) For connection between portable lighting equipment or other portable utilization equipment and the fixed portion of its supply circuit. The flexible cord shall be attached to the utilization equipment with a cord connector listed for the protection technique of the equipment wiring compartment. An attachment plug in accordance with 502.145 shall be employed.

(2) Where flexible cord is permitted by 502.10(A)(2) for fixed and mobile electrical utilization equipment; where the flexible cord is protected by location or by a suitable guard from damage; and only in an industrial establishment where conditions of maintenance and engineering supervision ensure that only qualified persons install and service the installation.

(3) For electric submersible pumps with means for removal without entering the wet-pit. The extension of the flexible cord within a suitable raceway between the wet-pit and the power source shall be permitted.

(4) For electric mixers intended for travel into and out of open-type mixing tanks or vats.

(5) For temporary portable assemblies consisting of receptacles, switches, and other devices that are not considered portable utilization equipment but are individually listed for the location.

(B) Installation. Where flexible cords are used, the cords shall comply with all of the following:

(1) Be of a type listed for extra-hard usage.

Exception: Flexible cord listed for hard usage as permitted by 502.130(A)(3) and (B)(4).

(2) Contain, in addition to the conductors of the circuit, an equipment grounding conductor complying with 400.23.

(3) Be supported by clamps or by other suitable means in such a manner that there will be no tension on the terminal connections.

(4) In Division 1 locations, the cord shall be terminated with a cord connector listed for the location or a listed cord connector installed with a seal listed for the location. In Division 2 locations, the cord shall be terminated with a listed dusttight cord connector.

(5) Be of continuous length. Where 502.140(A)(5) is applied, cords shall be of continuous length from the power source to the temporary portable assembly and from the temporary portable assembly to the utilization equipment.

502.145 Receptacles and Attachment Plugs. Receptacles and attachment plugs shall be identified for the location.

(A) Class II, Division 1.

(1) Receptacles. In Class II, Division 1 locations, receptacles shall be part of the premises wiring.

(2) Attachment Plugs. Attachment plugs shall be of the type that provides for connection to the equipment grounding conductor of the flexible cord.

(B) Class II, Division 2.

(1) Receptacles. In Class II, Division 2 locations, receptacles shall be part of the premises wiring.

(2) Attachment Plugs. Attachment plugs shall be of the type that provides for connection to the equipment grounding conductor of the flexible cord.

502.150 Signaling, Alarm, Remote-Control, and Communications Systems; and Meters, Instruments, and Relays.

Informational Note: See Article 800 for rules governing the installation of communications circuits.

(A) Class II, Division 1. In Class II, Division 1 locations, signaling, alarm, remote-control, and communications systems; and meters, instruments, and relays shall comply with 502.150(A)(1) through (A)(3).

(1) Contacts. Switches, circuit breakers, relays, contactors, fuses and current-breaking contacts for bells, horns, howlers, sirens, and other devices in which sparks or arcs may be produced shall be provided with enclosures identified for the location.

Exception: Where current-breaking contacts are immersed in oil or where the interruption of current occurs within a chamber sealed against the entrance of dust, enclosures shall be permitted to be of the general-purpose type.

(2) Resistors and Similar Equipment. Resistors, transformers, choke coils, rectifiers, thermionic tubes, and other heat-generating equipment shall be provided with enclosures identified for the location.

Exception: Where resistors or similar equipment are immersed in oil or enclosed in a chamber sealed against the entrance of dust, enclosures shall be permitted to be of the general-purpose type.

(3) Rotating Machinery. Motors, generators, and other rotating electrical machinery shall comply with 502.125(A).

(B) Class II, Division 2. In Class II, Division 2 locations, signaling, alarm, remote-control, and communications systems; and meters, instruments, and relays shall comply with 502.150(B)(1) through (B)(4).

(1) Contacts. Contacts shall comply with 502.150(A)(1) or shall be installed in enclosures that are dusttight or otherwise identified for the location.

Exception: In nonincendive circuits, enclosures shall be permitted to be of the general-purpose type.

(2) Transformers and Similar Equipment. The windings and terminal connections of transformers, choke coils, and similar equipment shall comply with 502.120(B)(2).

(3) Resistors and Similar Equipment. Resistors, resistance devices, thermionic tubes, rectifiers, and similar equipment shall comply with 502.120(B)(3).

(4) Rotating Machinery. Motors, generators, and other rotating electrical machinery shall comply with 502.125(B).

ARTICLE 503
Class III Locations

Part I. General

503.1 Scope. Article 503 covers the requirements for electrical and electronic equipment and wiring for all voltages in Class III, Division 1 and 2 locations where fire or explosion hazards may exist due to ignitible fibers/flyings.

503.5 General. Equipment installed in Class III locations shall be able to function at full rating without developing surface temperatures high enough to cause excessive dehydration or gradual carbonization of accumulated fibers/flyings. Organic material that is carbonized or excessively dry is highly susceptible to spontaneous ignition. The maximum surface temperatures under operating conditions shall not exceed 165°C (329°F) for equipment that is not subject to overloading, and 120°C (248°F) for equipment (such as motors or power transformers) that may be overloaded. In a Class III, Division 1 location, the operating temperature shall be the temperature of the equipment when blanketed with the maximum amount of dust (simulating fibers/flyings) that can accumulate on the equipment.

> Informational Note: For electric trucks, see NFPA 505-2013, *Fire Safety Standard for Powered Industrial Trucks Including Type Designations, Areas of Use, Conversions, Maintenance, and Operation.*

503.6 Zone Equipment. Equipment listed and marked in accordance with 506.9(C)(2) for Zone 20 locations and with a temperature class of not greater than T120°C (for equipment that may be overloaded) or not greater than T165°C (for equipment not subject to overloading) shall be permitted in Class III, Division 1 locations.

Equipment listed and marked in accordance with 506.9(C)(2) for Zone 20, 21, or 22 locations and with a temperature class of not greater than T120°C (for equipment that may be overloaded) or not greater than T165°C (for equipment not subject to overloading) shall be permitted in Class III, Division 2 locations.

Part II. Wiring

503.10 Wiring Methods. Wiring methods shall comply with 503.10(A) or (B).

(A) Class III, Division 1.

(1) General. In Class III, Division 1 locations, the wiring method shall be in accordance with (1) through (5):

(1) Rigid metal conduit, Type PVC conduit, Type RTRC conduit, intermediate metal conduit, electrical metallic tubing, dusttight wireways, or Type MC or MI cable with listed termination fittings.

(2) Type PLTC and Type PLTC-ER cable in accordance with the provisions of Article 725, including installation in cable tray systems. The cable shall be terminated with listed fittings.

(3) Type ITC and Type ITC-ER cable as permitted in 727.4 and terminated with listed fittings.

(4) Type MC, MI, MV, TC, or TC-ER cable installed in ladder, ventilated trough, or ventilated channel cable trays in a single layer, with a space not less than the larger cable diameter between the two adjacent cables, shall be the wiring method employed. The cable shall be terminated with listed fittings.

Exception to (4): Type MC cable listed for use in Class II, Division 1 locations shall be permitted to be installed without the spacings required by 503.10(A)(1)(4).

(5) Cablebus.

(2) Boxes and Fittings. All boxes and fittings shall be dusttight.

(3) Flexible Connections. Where necessary to employ flexible connections, one or more of the following shall be permitted:

(1) Dusttight flexible connectors
(2) Liquidtight flexible metal conduit with listed fittings
(3) Liquidtight flexible nonmetallic conduit with listed fittings
(4) Interlocked armor Type MC cable having an overall jacket of suitable polymeric material and installed with listed dusttight termination fittings
(5) Flexible cord in compliance with 503.140

> Informational Note: See 503.30(B) for grounding requirements where flexible conduit is used.

(6) For elevator use, an identified elevator cable of Type EO, ETP, or ETT, shown under the "use" column in Table 400.4 for "hazardous (classified) locations" and terminated with listed dusttight fittings

(4) Nonincendive Field Wiring. Nonincendive field wiring shall be permitted using any of the wiring methods permitted for unclassified locations. Nonincendive field wiring systems shall be installed in accordance with the control drawing(s). Simple apparatus, not shown on the control drawing, shall be permitted in a nonincendive field wiring circuit, provided the simple apparatus does not interconnect the nonincendive field wiring circuit to any other circuit.

Informational Note: Simple apparatus is defined in 504.2.

Separate nonincendive field wiring circuits shall be installed in accordance with one of the following:

(1) In separate cables
(2) In multiconductor cables where the conductors of each circuit are within a grounded metal shield
(3) In multiconductor cables where the conductors of each circuit have insulation with a minimum thickness of 0.25 mm (0.01 in.)

(B) Class III, Division 2. In Class III, Division 2 locations, the wiring method shall comply with 503.10(A).

Exception: In sections, compartments, or areas used solely for storage and containing no machinery, open wiring on insulators shall be permitted where installed in accordance with Article 398, but only on condition that protection as required by 398.15(C) be provided where conductors are not run in roof spaces and are well out of reach of sources of physical damage.

503.25 Uninsulated Exposed Parts, Class III, Divisions 1 and 2. There shall be no uninsulated exposed parts, such as electrical conductors, buses, terminals, or components, that operate at more than 30 volts (15 volts in wet locations). These parts shall additionally be protected by a protection technique according to 500.7(E), (F), or (G) that is suitable for the location.

Exception: As provided in 503.155.

503.30 Grounding and Bonding — Class III, Divisions 1 and 2. Regardless of the voltage of the electrical system, wiring and equipment in Class III, Division 1 and 2 locations shall be grounded as specified in Article 250 and with the following additional requirements in 503.30(A) and (B).

(A) Bonding. The locknut-bushing and double-locknut types of contacts shall not be depended on for bonding purposes, but bonding jumpers with proper fittings or other approved means of bonding shall be used. Such means of bonding shall apply to all intervening raceways, fittings, boxes, enclosures, and so forth, between Class III locations and the point of grounding for service equipment or point of grounding of a separately derived system.

Exception: The specific bonding means shall only be required to the nearest point where the grounded circuit conductor and the grounding electrode conductor are connected together on the line side of the building or structure disconnecting means as specified in 250.32(B) if the branch-circuit overcurrent protection is located on the load side of the disconnecting means.

(B) Types of Equipment Bonding Conductors. Liquidtight flexible metal conduit shall include an equipment bonding jumper of the wire type in compliance with 250.102.

Exception: In Class III, Division 1 and 2 locations, the bonding jumper shall be permitted to be deleted where all of the following conditions are met:

(1) Listed liquidtight flexible metal conduit 1.8 m (6 ft) or less in length, with fittings listed for grounding, is used.
(2) Overcurrent protection in the circuit is limited to 10 amperes or less.
(3) The load is not a power utilization load.

Part III. Equipment

503.100 Transformers and Capacitors — Class III, Divisions 1 and 2. Transformers and capacitors shall comply with 502.100(B).

503.115 Switches, Circuit Breakers, Motor Controllers, and Fuses — Class III, Divisions 1 and 2. Switches, circuit breakers, motor controllers, and fuses, including pushbuttons, relays, and similar devices, shall be provided with dusttight enclosures.

503.120 Control Transformers and Resistors — Class III, Divisions 1 and 2. Transformers, impedance coils, and resistors used as, or in conjunction with, control equipment for motors, generators, and appliances shall be provided with dusttight enclosures complying with the temperature limitations in 503.5.

503.125 Motors and Generators — Class III, Divisions 1 and 2. In Class III, Divisions 1 and 2 locations, motors, generators, and other rotating machinery shall be totally enclosed nonventilated, totally enclosed pipe ventilated, or totally enclosed fan cooled.

Exception: In locations where, in the judgment of the authority having jurisdiction, only moderate accumulations of lint or flyings are likely to collect on, in, or in the vicinity of a rotating electrical machine and where such machine is readily accessible for routine cleaning and maintenance, one of the following shall be permitted:

(1) Self-cleaning textile motors of the squirrel-cage type
(2) Standard open-type machines without sliding contacts, centrifugal or other types of switching mechanisms, including motor overload devices
(3) Standard open-type machines having such contacts, switching mechanisms, or resistance devices enclosed within tight housings without ventilating or other openings

503.128 Ventilating Piping — Class III, Divisions 1 and 2. Ventilating pipes for motors, generators, or other rotating electrical machinery, or for enclosures for electric equipment, shall be of metal not less than 0.53 mm (0.021 in.) in thickness, or of equally substantial noncombustible material, and shall comply with the following:

(1) Lead directly to a source of clean air outside of buildings
(2) Be screened at the outer ends to prevent the entrance of small animals or birds
(3) Be protected against physical damage and against rusting or other corrosive influences

Ventilating pipes shall be sufficiently tight, including their connections, to prevent the entrance of appreciable quantities of fibers/flyings into the ventilated equipment or enclosure and to prevent the escape of sparks, flame, or burning material that might ignite accumulations of fibers/flyings or combustible material in the vicinity. For metal pipes, lock seams and riveted or welded joints shall be permitted; and tight-fitting slip joints shall be permitted where some flexibility is necessary, as at connections to motors.

503.130 Luminaires — Class III, Divisions 1 and 2.

(A) Fixed Lighting. Luminaires for fixed lighting shall provide enclosures for lamps and lampholders that are designed to minimize entrance of fibers/flyings and to prevent the escape of sparks, burning material, or hot metal. Each luminaire shall be clearly marked to show the maximum wattage of the lamps that shall be permitted without exceeding an exposed surface temperature of 165°C (329°F) under normal conditions of use.

(B) Physical Damage. A luminaire that may be exposed to physical damage shall be protected by a suitable guard.

(C) Pendant Luminaires. Pendant luminaires shall be suspended by stems of threaded rigid metal conduit, threaded intermediate metal conduit, threaded metal tubing of equivalent thickness, or by chains with approved fittings. For stems longer than 300 mm (12 in.), permanent and effective bracing against lateral displacement shall be provided at a level not more than 300 mm (12 in.) above the lower end of the stem, or flexibility in the form of an identified fitting or a flexible connector shall be provided not more than 300 mm (12 in.) from the point of attachment to the supporting box or fitting.

(D) Portable Lighting Equipment. Portable lighting equipment shall be equipped with handles and protected with substantial guards. Lampholders shall be of the unswitched type with no provision for receiving attachment plugs. There shall be no exposed current-carrying metal parts, and all exposed non–current-carrying metal parts shall be grounded. In all other respects, portable lighting equipment shall comply with 503.130(A).

503.135 Utilization Equipment — Class III, Divisions 1 and 2.

(A) Heaters. Electrically heated utilization equipment shall be identified for Class III locations.

(B) Motors. Motors of motor-driven utilization equipment shall comply with 503.125.

(C) Switches, Circuit Breakers, Motor Controllers, and Fuses. Switches, circuit breakers, motor controllers, and fuses shall comply with 503.115.

503.140 Flexible Cords — Class III, Divisions 1 and 2. Flexible cords shall comply with the following:

(1) Be of a type listed for extra-hard usage
(2) Contain, in addition to the conductors of the circuit, an equipment grounding conductor complying with 400.23
(3) Be supported by clamps or other suitable means in such a manner that there will be no tension on the terminal connections
(4) Be terminated with a listed dusttight cord connector.

503.145 Receptacles and Attachment Plugs — Class III, Divisions 1 and 2. Receptacles and attachment plugs shall be of the grounding type, shall be designed so as to minimize the accumulation or the entry of fibers/flyings, and shall prevent the escape of sparks or molten particles.

Exception: In locations where, in the judgment of the authority having jurisdiction, only moderate accumulations of lint or flyings are likely to collect in the vicinity of a receptacle, and where such receptacle is readily accessible for routine cleaning, general-purpose grounding-type receptacles mounted so as to minimize the entry of fibers/flyings shall be permitted.

503.150 Signaling, Alarm, Remote-Control, and Local Loudspeaker Intercommunications Systems — Class III, Divisions 1 and 2. Signaling, alarm, remote-control, and local loudspeaker intercommunications systems shall comply with the requirements of Article 503 regarding wiring methods, switches, transformers, resistors, motors, luminaires, and related components.

503.155 Electric Cranes, Hoists, and Similar Equipment — Class III, Divisions 1 and 2. Where installed for operation over combustible fibers or accumulations of flyings, traveling cranes and hoists for material handling, traveling cleaners for textile machinery, and similar equipment shall comply with 503.155(A) through (D).

(A) Power Supply. The power supply to contact conductors shall be electrically isolated from all other systems, ungrounded, and shall be equipped with an acceptable ground detector that gives an alarm and automatically de-energizes the contact conductors in case of a fault to ground or gives a visual and audible alarm as long as power is supplied to the contact conductors and the ground fault remains.

(B) Contact Conductors. Contact conductors shall be located or guarded so as to be inaccessible to other than authorized persons and shall be protected against accidental contact with foreign objects.

(C) Current Collectors. Current collectors shall be arranged or guarded so as to confine normal sparking and prevent escape of sparks or hot particles. To reduce sparking, two or more separate surfaces of contact shall be provided for each contact conductor. Reliable means shall be provided to keep contact conductors and current collectors free of accumulations of lint or flyings.

(D) Control Equipment. Control equipment shall comply with 503.115 and 503.120.

503.160 Storage Battery Charging Equipment — Class III, Divisions 1 and 2. Storage battery charging equipment shall be located in separate rooms built or lined with substantial noncombustible materials. The rooms shall be constructed to prevent the entrance of ignitible amounts of flyings or lint and shall be well ventilated.

ARTICLE 504
Intrinsically Safe Systems

504.1 Scope. This article covers the installation of intrinsically safe (I.S.) apparatus, wiring, and systems for Articles 500 through 516.

Informational Note: For further information, see ANSI/ISA-RP 12.06.01-2003, *Recommended Practice for Wiring Methods for Hazardous (Classified) Locations Instrumentation — Part 1: Intrinsic Safety.*

504.2 Definitions.

Different Intrinsically Safe Circuits. Intrinsically safe circuits in which the possible interconnections have not been evaluated and identified as intrinsically safe.

Intrinsically Safe Circuit [as applied to Hazardous (Classified) Locations]. A circuit in which any spark or thermal effect is incapable of causing ignition of a mixture of flammable or combustible material in air under prescribed test conditions.

Informational Note: Test conditions are described in ANSI/UL 913-2006, *Standard for Safety, Intrinsically Safe Apparatus and Associated Apparatus for Use in Class I, II, and III, Division 1, Hazardous (Classified) Locations.*

504.3 Application of Other Articles. Except as modified by this article, all applicable articles of this *Code* shall apply.

504.4 Equipment. All intrinsically safe apparatus and associated apparatus shall be listed.

Exception: Simple apparatus, as described on the control drawing, shall not be required to be listed.

504.10 Equipment Installation.

(A) Control Drawing. Intrinsically safe apparatus, associated apparatus, and other equipment shall be installed in accordance with the control drawing(s).

A simple apparatus, whether or not shown on the control drawing(s), shall be permitted to be installed provided the simple apparatus does not interconnect intrinsically safe circuits.

Informational Note No. 1: The control drawing identification is marked on the apparatus.

Informational Note No. 2: Associated apparatus with a marked Um of less than 250 V may require additional overvoltage protection at the inputs to limit any possible fault voltages to less than the Um marked on the product.

(B) Location. Intrinsically safe apparatus shall be permitted to be installed in any hazardous (classified) location for which it has been identified.

Associated apparatus shall be permitted to be installed in any hazardous (classified) location for which it has been identified.

Simple apparatus shall be permitted to be installed in any hazardous (classified) location in accordance with 504.10(D).

(C) Enclosures. General-purpose enclosures shall be permitted for intrinsically safe apparatus and associated apparatus unless otherwise specified in the manufacturer's documentation.

(D) Simple Apparatus. Simple apparatus shall be permitted to be installed in any hazardous (classified) location in which the maximum surface temperature of the simple apparatus does not exceed the ignition temperature of the flammable gases or vapors, flammable liquids, combustible dusts, or ignitible fibers/flyings present. The maximum surface temperature can be determined from the values of the output power from the associated apparatus or apparatus to which it is connected to obtain the temperature class. The temperature class can be determined by:

(1) Reference to Table 504.10(D)
(2) Calculation using the following equation:

[504.10(D)]

$$T = P_o R_{th} + T_{amb}$$

where:
T = surface temperature
P_o = output power marked on the associated apparatus or intrinsically safe apparatus
R_{th} = thermal resistance of the simple apparatus
T_{amb} = ambient temperature (normally 40°C) and reference Table 500.8(C)

In addition, components with a surface area smaller than 10 cm² (excluding lead wires) may be classified as T5 if their surface temperature does not exceed 150°C.

504.20 Wiring Methods. Any of the wiring methods suitable for unclassified locations, including those covered by Chapter 7 and Chapter 8, shall be permitted for installing intrinsically safe apparatus. Sealing shall be as provided in 504.70, and separation shall be as provided in 504.30.

504.30 Separation of Intrinsically Safe Conductors.

(A) From Nonintrinsically Safe Circuit Conductors.

(1) In Raceways, Cable Trays, and Cables. Conductors of intrinsically safe circuits shall not be placed in any raceway, cable tray, or cable with conductors of any nonintrinsically safe circuit.

Exception No. 1: Where conductors of intrinsically safe circuits are separated from conductors of nonintrinsically safe circuits by a distance of at least 50 mm (2 in.) and secured, or by a grounded metal partition or an approved insulating partition.

Informational Note: No. 20 gauge sheet metal partitions 0.91 mm (0.0359 in.) or thicker are generally considered acceptable.

Exception No. 2: Where either (1) all of the intrinsically safe circuit conductors or (2) all of the nonintrinsically safe circuit conductors are in grounded metal-sheathed or metal-clad cables where the sheathing or cladding is capable of carrying fault current to ground.

Informational Note: Cables meeting the requirements of Articles 330 and 332 are typical of those considered acceptable.

Exception No. 3: Intrinsically safe circuits in a Division 2 or Zone 2 location shall be permitted to be installed in a raceway, cable tray, or cable along with nonincendive field wiring circuits when installed in accordance with 504.30(B).

Exception No. 4: Intrinsically safe circuits passing through a Division 2 or Zone 2 location to supply apparatus that is located in a Division 1, Zone 0 or Zone 1 location shall be permitted to be installed in a raceway, cable tray, or cable along with nonincendive field wiring circuits when installed in accordance with 504.30(B).

Informational Note: Nonincendive field wiring circuits are described in 501.10(B)(3), 502.10(B)(3), and 503.10(A)(4).

(2) Within Enclosures. Conductors of intrinsically safe circuits shall be secured so that any conductor that might come loose from a terminal is unlikely to come into contact with another terminal. The conductors shall be separated from conductors of nonintrinsically safe circuits by one of the methods in (1) through (4).

(1) Separation by at least 50 mm (2 in.) from conductors of any nonintrinsically safe circuits

Table 504.10(D) Assessment for T4 Classification According to Component Size and Temperature

Total Surface Area Excluding Lead Wires	Requirement for T4 Classification
<20 mm²	Surface temperature ≤275°C
≥20 mm² ≤10 cm²	Surface temperature ≤200°C
≥20 mm²	Power not exceeding 1.3 W*

*Based on 40°C ambient temperature. Reduce to 1.2 W with an ambient of 60°C or 1.0 W with 80°C ambient temperature.

(2) Separation from conductors of nonintrinsically safe circuits by use of a grounded metal partition 0.91 mm (0.0359 in.) or thicker

(3) Separation from conductors of nonintrinsically safe circuits by use of an approved insulating partition that extends to within 1.5 mm (0.0625 in.) of the enclosure walls

(4) Where either (1) all of the intrinsically safe circuit conductors or (2) all of the nonintrinsically safe circuit conductors are in grounded metal-sheathed or metal-clad cables where the sheathing or cladding is capable of carrying fault current to ground

Informational Note No. 1: Cables meeting the requirements of Articles 330 and 332 are typical of those considered acceptable.

Informational Note No. 2: The use of separate wiring compartments for the intrinsically safe and nonintrinsically safe terminals is a typical method of complying with this requirement.

Informational Note No. 3: Physical barriers such as grounded metal partitions or approved insulating partitions or approved restricted access wiring ducts separated from other such ducts by at least 19 mm (¾ in.) can be used to help ensure the required separation of the wiring.

(3) Other (Not in Raceway or Cable Tray Systems). Conductors and cables of intrinsically safe circuits run in other than raceway or cable tray systems shall be separated by at least 50 mm (2 in.) and secured from conductors and cables of any nonintrinsically safe circuits.

Exception: Where either (1) all of the intrinsically safe circuit conductors are in Type MI or MC cables or (2) all of the nonintrinsically safe circuit conductors are in raceways or Type MI or MC cables where the sheathing or cladding is capable of carrying fault current to ground.

(B) From Different Intrinsically Safe Circuit Conductors. The clearance between two terminals for connection of field wiring of different intrinsically safe circuits shall be at least 6 mm (0.25 in.), unless this clearance is permitted to be reduced by the control drawing. Different intrinsically safe circuits shall be separated from each other by one of the following means:

(1) The conductors of each circuit are within a grounded metal shield.

(2) The conductors of each circuit have insulation with a minimum thickness of 0.25 mm (0.01 in.).

Exception: Unless otherwise identified.

(C) From Grounded Metal. The clearance between the uninsulated parts of field wiring conductors connected to terminals and grounded metal or other conducting parts shall be at least 3 mm (0.125 in.).

504.50 Grounding.

(A) Intrinsically Safe Apparatus, Enclosures, and Raceways. Intrinsically safe apparatus, enclosures, and raceways, if of metal, shall be connected to the equipment grounding conductor.

Informational Note: In addition to an equipment grounding conductor connection, a connection to a grounding electrode may be needed for some associated apparatus; for example, zener diode barriers, if specified in the control drawing. See ANSI/ISA-RP 12.06.01-2003, *Recommended Practice for Wiring Methods for Hazardous (Classified) Locations Instrumentation — Part 1: Intrinsic Safety.*

(B) Associated Apparatus and Cable Shields. Associated apparatus and cable shields shall be grounded in accordance with the required control drawing. See 504.10(A).

Informational Note: Supplementary connection(s) to the grounding electrode may be needed for some associated apparatus; for example, zener diode barriers, if specified in the control drawing. See ANSI/ISA RP 12.06.01-2003, *Recommended Practice for Wiring Methods for Hazardous (Classified) Locations Instrumentation — Part 1: Intrinsic Safety.*

(C) Connection to Grounding Electrodes. Where connection to a grounding electrode is required, the grounding electrode shall be as specified in 250.52(A)(1), (A)(2), (A)(3), and (A)(4) and shall comply with 250.30(A)(4). Sections 250.52(A)(5), (A)(7), and (A)(8) shall not be used if any of the electrodes specified in 250.52(A)(1), (A)(2), (A)(3), or (A)(4) are present.

504.60 Bonding.

(A) Intrinsically Safe Apparatus. Intrinsically safe apparatus, if of metal, shall be bonded in the hazardous (classified) location in accordance with 501.30(A), 502.30(A), 503.30(A), 505.25, or 506.25, as applicable.

(B) Metal Raceways. Where metal raceways are used for intrinsically safe system wiring, bonding at all ends of the raceway, regardless of the location, shall be in accordance with 501.30(A), 502.30(A), 503.30(A), 505.25, or 506.25, as applicable.

504.70 Sealing. Conduits and cables that are required to be sealed by 501.15, 502.15, 505.16, and 506.16 shall be sealed to minimize the passage of gases, vapors, or dusts. Such seals shall not be required to be explosionproof or flameproof but shall be identified for the purpose of minimizing passage of gases, vapors, or dusts under normal operating conditions and shall be accessible.

Exception: Seals shall not be required for enclosures that contain only intrinsically safe apparatus, except as required by 501.17.

504.80 Identification. Labels required by this section shall be suitable for the environment where they are installed, with consideration given to exposure to chemicals and sunlight.

(A) Terminals. Intrinsically safe circuits shall be identified at terminal and junction locations in a manner that is intended to prevent unintentional interference with the circuits during testing and servicing.

(B) Wiring. Raceways, cable trays, and other wiring methods for intrinsically safe system wiring shall be identified with permanently affixed labels with the wording "Intrinsic Safety Wiring" or equivalent. The labels shall be located so as to be visible after installation and placed so that they may be readily traced through the entire length of the installation. Intrinsic safety circuit labels shall appear in every section of the wiring system that is separated by enclosures, walls, partitions, or floors. Spacing between labels shall not be more than 7.5 m (25 ft).

Exception: Circuits run underground shall be permitted to be identified where they become accessible after emergence from the ground.

Informational Note No. 1: Wiring methods permitted in unclassified locations may be used for intrinsically safe systems in hazardous (classified) locations. Without labels to identify the application of the wiring, enforcement authorities cannot determine that an installation is in compliance with this *Code.*

Informational Note No. 2: In unclassified locations, identification is necessary to ensure that nonintrinsically safe wire will not be inadvertently added to existing raceways at a later date.

(C) Color Coding. Color coding shall be permitted to identify intrinsically safe conductors where they are colored light blue and where no other conductors colored light blue are used. Likewise, color coding shall be permitted to identify raceways, cable trays, and junction boxes where they are colored light blue and contain only intrinsically safe wiring.

ARTICLE 505
Zone 0, 1, and 2 Locations

Informational Note: Text that is followed by a reference in brackets has been extracted from NFPA 497-2012, *Recommended Practice for the Classification of Flammable Liquids, Gases, or Vapors and of Hazardous (Classified) Locations for Electrical Installations in Chemical Process Areas.* Only editorial changes were made to the extracted text to make it consistent with this *Code.*

505.1 Scope. This article covers the requirements for the zone classification system as an alternative to the division classification system covered in Article 500 for electrical and electronic equipment and wiring for all voltages in Class I, Zone 0, Zone 1, and Zone 2 hazardous (classified) locations where fire or explosion hazards may exist due to flammable gases, vapors, or liquids.

Informational Note: For the requirements for electrical and electronic equipment and wiring for all voltages in Class I, Division 1 or Division 2; Class II, Division 1 or Division 2; and Class III, Division 1 or Division 2 hazardous (classified) locations where fire or explosion hazards may exist due to flammable gases or vapors, flammable liquids, or combustible dusts or fibers, refer to Articles 500 through 504.

505.2 Definitions.

Encapsulation "m". Type of protection where electrical parts that could ignite an explosive atmosphere by either sparking or heating are enclosed in a compound in such a way that this explosive atmosphere cannot be ignited.

Informational Note No. 1: See ISA-60079-18-2012, *Explosive atmospheres — Part 18: Equipment protection by encapsulation "m";* and ANSI/UL 60079-18-2009, *Explosive atmospheres — Part 18: Equipment protection by encapsulation "m."*

Informational Note No. 2: Encapsulation is designated type of protection "ma" for use in Zone 0 locations. Encapsulation is designated type of protection "m" or "mb" for use in Zone 1 locations. Encapsulation is designated type of protection "mc" for use in Zone 2 locations.

Flameproof "d". Type of protection where the enclosure will withstand an internal explosion of a flammable mixture that has penetrated into the interior, without suffering damage and without causing ignition, through any joints or structural openings in the enclosure of an external explosive gas atmosphere consisting of one or more of the gases or vapors for which it is designed.

Informational Note: See ISA-60079-1-2009, *Explosive Atmospheres, Part 1: Equipment protection by flameproof enclosures "d";* and ANSI/

UL 60079-1-2009, *Electrical Apparatus for Explosive Gas Atmospheres — Part 1: Flameproof Enclosures "d."*

Increased Safety "e". Type of protection applied to electrical equipment that does not produce arcs or sparks in normal service and under specified abnormal conditions, in which additional measures are applied so as to give increased security against the possibility of excessive temperatures and of the occurrence of arcs and sparks.

Informational Note: See ISA-60079-7-2013, *Explosive Atmospheres — Part 7: Equipment protection by increased safety "e";* and ANSI/UL 60079-7–2008, *Electrical Apparatus for Explosive Gas Atmospheres — Part 7: Increased Safety "e."*

Intrinsic Safety "i". Type of protection where any spark or thermal effect is incapable of causing ignition of a mixture of flammable or combustible material in air under prescribed test conditions.

Informational Note No. 1: See UL 913-2015, *Intrinsically Safe Apparatus and Associated Apparatus for Use in Class I, II, and III, Division 1 Hazardous (Classified) Locations;* ISA-60079-11 (12.02.01)-2011, *Explosive Atmospheres — Part 11: Equipment protection by intrinsic safety "i";* and ANSI/UL 60079-11-2011, *Explosive Atmospheres — Part 11: Equipment protection by intrinsic safety "i."*

Informational Note No. 2: Intrinsic safety is designated type of protection "ia" for use in Zone 0 locations. Intrinsic safety is designated type of protection "ib" for use in Zone 1 locations. Intrinsic safety is designated type of protection "ic" for use in Zone 2 locations.

Informational Note No. 3: Intrinsically safe associated apparatus, designated by [ia], [ib], or [ic], is connected to intrinsically safe apparatus ("ia," "ib," or "ic," respectively) but is located outside the hazardous (classified) location unless also protected by another type of protection (such as flameproof).

Oil Immersion "o". Type of protection where electrical equipment is immersed in a protective liquid in such a way that an explosive atmosphere that may be above the liquid or outside the enclosure cannot be ignited.

Informational Note: See ANSI/ISA-60079-6-2009, *Explosive Atmospheres — Part 6: Equipment protection by oil immersion "o";* and ANSI/UL 60079-6-2009, *Electrical Apparatus for Explosive Gas Atmospheres — Part 6: Oil-Immersion "o."*

Powder Filling "q". Type of protection where electrical parts capable of igniting an explosive atmosphere are fixed in position and completely surrounded by filling material (glass or quartz powder) to prevent the ignition of an external explosive atmosphere.

Informational Note: See ANSI/ISA-60079-5-2009, *Explosive Atmospheres — Part 5: Equipment protection by powder filling "q";* and ANSI/UL 60079-5-2009, *Electrical Apparatus for Explosive Gas Atmospheres — Part 5: Powder Filling "q."*

Pressurization "p". Type of protection for electrical equipment that uses the technique of guarding against the ingress of the external atmosphere, which may be explosive, into an enclosure by maintaining a protective gas therein at a pressure above that of the external atmosphere.

Informational Note: See ANSI/UL-60079-2-2015, *Explosive Atmospheres — Part 2: Equipment protection by pressurized enclosures "p";* and IEC 60079-13-2010, *Electrical apparatus for explosive gas atmospheres — Part 13: Construction and use of rooms or buildings protected by pressurization.*

Type of Protection "n". Type of protection where electrical equipment, in normal operation, is not capable of igniting a surrounding explosive gas atmosphere and a fault capable of causing ignition is not likely to occur.

Informational Note: See ANSI/UL 60079-15-2009, *Electrical Apparatus for Explosive Gas Atmospheres — Part 15: Type of Protection "n"*; and ANSI/ISA-60079-15-2012, *Explosive Atmospheres — Part 15: Equipment protection by type of protection "n."*

505.3 Other Articles. All other applicable rules contained in this *Code* shall apply to electrical equipment and wiring installed in hazardous (classified) locations.

Exception: As modified by Article 504 and this article.

505.4 General.

(A) Documentation for Industrial Occupancies. All areas in industrial occupancies designated as hazardous (classified) locations shall be properly documented. This documentation shall be available to those authorized to design, install, inspect, maintain, or operate electrical equipment at the location.

Informational Note No. 1: For examples of area classification drawings, see ANSI/API RP 505-1997, *Recommended Practice for Classification of Locations for Electrical Installations at Petroleum Facilities Classified as Class I, Zone 0, Zone 1, or Zone 2*; ANSI/ISA-60079-10-1 (12.24.01)-2014 *Explosive Atmospheres — Part 10-1: Classification of Areas — Explosive gas atmospheres*; and *Model Code of Safe Practice in the Petroleum Industry, Part 15: Area Classification Code for Installations Handling Flammable Fluids, EI 15:2005*, Energy Institute, London.

Informational Note No. 2: Where gas detection equipment is used as a means of protection in accordance with 505.8(I)(1), (I)(2), or (I)(3), the documentation typically includes the type of detection equipment, its listing, installation location(s), alarm and shutdown criteria, and calibration frequency.

(B) Reference Standards. Important information relating to topics covered in Chapter 5 may be found in other publications.

Informational Note No. 1: It is important that the authority having jurisdiction be familiar with recorded industrial experience as well as with standards of the National Fire Protection Association (NFPA), the American Petroleum Institute (API), the International Society of Automation (ISA), and the International Electrotechnical Commission (IEC) that may be of use in the classification of various locations, the determination of adequate ventilation, and the protection against static electricity and lightning hazards.

Informational Note No. 2: For further information on the classification of locations, see NFPA 497-2012, *Recommended Practice for the Classification of Flammable Liquids, Gases, or Vapors and of Hazardous (Classified) Locations for Electrical Installations in Chemical Process Areas*; ANSI/API RP 505-1997, *Recommended Practice for Classification of Locations for Electrical Installations at Petroleum Facilities Classified as Class I, Zone 0, Zone 1, or Zone 2*; ANSI/ISA-60079-10-1 (12.24.01)-2014, *Explosive Atmospheres — Part 10-1: Classification of Areas — Explosive gas atmospheres*; and *Model Code of Safe Practice in the Petroleum Industry, Part 15: Area Classification Code for Installations Handling Flammable Fluids, EI 15:2005*, Energy Institute, London.

Informational Note No. 3: For further information on protection against static electricity and lightning hazards in hazardous (classified) locations, see NFPA 77-2014, *Recommended Practice on Static Electricity*; NFPA 780-2014, *Standard for the Installation of Lightning Protection Systems*; and API RP 2003-2008, *Protection Against Ignitions Arising Out of Static Lightning and Stray Currents.*

Informational Note No. 4: For further information on ventilation, see NFPA 30-2015, *Flammable and Combustible Liquids Code*, and ANSI/API RP 505-1997, *Recommended Practice for Classification of Locations for Electrical Installations at Petroleum Facilities Classified as Class I, Zone 0, Zone 1, or Zone 2.*

Informational Note No. 5: For further information on electrical systems for hazardous (classified) locations on offshore oil and gas producing platforms, see ANSI/API RP 14FZ-2013, *Recommended Practice for Design and Installation of Electrical Systems for Fixed and Floating Offshore Petroleum Facilities for Unclassified and Class I, Zone 0, Zone 1, and Zone 2 Locations.*

Informational Note No. 6: For further information on the installation of electrical equipment in hazardous (classified) locations in general, see IEC 60079-14-2013, *Electrical apparatus for explosive gas atmospheres — Part 14: Electrical installations in explosive gas atmospheres (other than mines)*, and IEC 60079-16-1990, *Electrical apparatus for explosive gas atmospheres — Part 16: Artificial ventilation for the protection of analyzer(s) houses.*

Informational Note No. 7: For further information on application of electrical equipment in hazardous (classified) locations in general, see ANSI/ISA-60079-0 (12.00.01)-2013, *Explosive Atmospheres — Part 0: Equipment — General Requirements*; ANSI/ISA-12.01.01-2013, *Definitions and Information Pertaining to Electrical Apparatus in Hazardous (Classified) Locations*; and ANSI/UL 60079-0:2013, *Electrical Apparatus for Explosive Gas Atmospheres — Part 0: General Requirements.*

Informational Note No. 8: Portable or transportable equipment having self-contained power supplies, such as battery-operated equipment, could potentially become an ignition source in hazardous (classified) locations. See ANSI/ISA-12.12.03-2011, *Standard for Portable Electronic Products Suitable for Use in Class I and II, Division 2, Class I Zone 2 and Class III, Division 1 and 2 Hazardous (Classified) Locations.*

Informational Note No. 9: For additional information concerning the installation of equipment utilizing optical emissions technology (such as laser equipment) that could potentially become an ignition source in hazardous (classified) locations, see ANSI/ISA-60079-28 (12.21.02)-2013, *Explosive Atmospheres — Part 28: Protection of equipment and transmission systems using optical radiation.*

505.5 Classifications of Locations.

(A) General. Locations shall be classified depending on the properties of the flammable gases, flammable liquid–produced vapors, combustible liquid–produced vapors, combustible dusts, or fibers/flyings that could be present and the likelihood that a flammable or combustible concentration or quantity is present. Each room, section, or area shall be considered individually in determining its classification. Where pyrophoric materials are the only materials used or handled, these locations are outside the scope of this article.

Informational Note No. 1: See 505.7 for restrictions on area classification.

Informational Note No. 2: Through the exercise of ingenuity in the layout of electrical installations for hazardous (classified) locations, it is frequently possible to locate much of the equipment in reduced level of classification or in an unclassified location and, thus, to reduce the amount of special equipment required.

Refrigerant machinery rooms that contain ammonia refrigeration systems and are equipped with adequate mechanical ventilation that operates continuously or is initiated by a detection system at a concentration not exceeding 150 ppm shall be permitted to be classified as "unclassified" locations.

Informational Note: For further information regarding classification and ventilation of areas involving closed-circuit ammonia refrigeration systems, see ANSI/ASHRAE 15-2013, *Safety Standard for Refrigeration Systems*, and ANSI/IIAR 2-2014, *Standard for Safe Design of Closed-Circuit Ammonia Refrigeration Systems*.

(B) Class I, Zone 0, 1, and 2 Locations. Class I, Zone 0, 1, and 2 locations are those in which flammable gases or vapors are or may be present in the air in quantities sufficient to produce explosive or ignitible mixtures. Class I, Zone 0, 1, and 2 locations shall include those specified in 505.5(B)(1), (B)(2), and (B)(3).

(1) Class I, Zone 0. A Class I, Zone 0 location is a location in which one of the following conditions exists:

(1) Ignitible concentrations of flammable gases or vapors are present continuously

(2) Ignitible concentrations of flammable gases or vapors are present for long periods of time

Informational Note No. 1: As a guide in determining when flammable gases or vapors are present continuously or for long periods of time, refer to ANSI/API RP 505-1997, *Recommended Practice for Classification of Locations for Electrical Installations of Petroleum Facilities Classified as Class I, Zone 0, Zone 1 or Zone 2;* ANSI/ISA 60079-10-1-2014, *Explosive Atmospheres — Part 10-1: Classification of Areas — explosive gas atmospheres.*

Informational Note No. 2: This classification includes locations inside vented tanks or vessels that contain volatile flammable liquids; inside inadequately vented spraying or coating enclosures where volatile flammable solvents are used; between the inner and outer roof sections of a floating roof tank containing volatile flammable liquids; inside open vessels, tanks, and pits containing volatile flammable liquids; the interior of an exhaust duct that is used to vent ignitible concentrations of gases or vapors; and inside inadequately ventilated enclosures that contain normally venting instruments utilizing or analyzing flammable fluids and venting to the inside of the enclosures.

(2) Class I, Zone 1. A Class I, Zone 1 location is a location

(1) In which ignitible concentrations of flammable gases or vapors are likely to exist under normal operating conditions; or

(2) In which ignitible concentrations of flammable gases or vapors may exist frequently because of repair or maintenance operations or because of leakage; or

(3) In which equipment is operated or processes are carried on, of such a nature that equipment breakdown or faulty operations could result in the release of ignitible concentrations of flammable gases or vapors and also cause simultaneous failure of electrical equipment in a mode to cause the electrical equipment to become a source of ignition; or

(4) That is adjacent to a Class I, Zone 0 location from which ignitible concentrations of vapors could be communicated, unless communication is prevented by adequate positive pressure ventilation from a source of clean air and effective safeguards against ventilation failure are provided.

Informational Note No. 1: Normal operation is considered the situation when plant equipment is operating within its design parameters. Minor releases of flammable material may be part of normal operations. Minor releases include the releases from mechanical packings on pumps. Failures that involve repair or shutdown (such as the breakdown of pump seals and flange gaskets, and spillage caused by accidents) are not considered normal operation.

Informational Note No. 2: This classification usually includes locations where volatile flammable liquids or liquefied flammable gases are transferred from one container to another. In areas in the vicinity of spraying and painting operations where flammable solvents are used; adequately ventilated drying rooms or compartments for evaporation of flammable solvents; adequately ventilated locations containing fat and oil extraction equipment using volatile flammable solvents; portions of cleaning and dyeing plants where volatile flammable liquids are used; adequately ventilated gas generator rooms and other portions of gas manufacturing plants where flammable gas may escape; inadequately ventilated pump rooms for flammable gas or for volatile flammable liquids; the interiors of refrigerators and freezers in which volatile flammable materials are stored in the open, lightly stoppered, or in easily ruptured containers; and other locations where ignitible concentrations of flammable vapors or gases are likely to occur in the course of normal operation but not classified Zone 0.

(3) Class I, Zone 2. A Class I, Zone 2 location is a location

(1) In which ignitible concentrations of flammable gases or vapors are not likely to occur in normal operation and, if they do occur, will exist only for a short period; or

(2) In which volatile flammable liquids, flammable gases, or flammable vapors are handled, processed, or used but in which the liquids, gases, or vapors normally are confined within closed containers of closed systems from which they can escape, only as a result of accidental rupture or breakdown of the containers or system, or as a result of the abnormal operation of the equipment with which the liquids or gases are handled, processed, or used; or

(3) In which ignitible concentrations of flammable gases or vapors normally are prevented by positive mechanical ventilation but which may become hazardous as a result of failure or abnormal operation of the ventilation equipment; or

(4) That is adjacent to a Class I, Zone 1 location, from which ignitible concentrations of flammable gases or vapors could be communicated, unless such communication is prevented by adequate positive-pressure ventilation from a source of clean air and effective safeguards against ventilation failure are provided.

Informational Note: The Zone 2 classification usually includes locations where volatile flammable liquids or flammable gases or vapors are used but which would become hazardous only in case of an accident or of some unusual operating condition.

505.6 Material Groups. For purposes of testing, approval, and area classification, various air mixtures (not oxygen enriched) shall be grouped as required in 505.6(A), (B), and (C).

Informational Note No. 1: : Group I is intended for use in describing atmospheres that contain firedamp (a mixture of gases, composed mostly of methane, found underground, usually in mines). This *Code* does not apply to installations underground in mines. See 90.2(B).

Informational Note No. 2: The gas and vapor subdivision as described above is based on the maximum experimental safe gap (MESG), minimum igniting current (MIC), or both. Test equipment for determining the MESG is described in IEC 60079-1A-1975, Amendment No. 1 (1993), *Construction and verification tests of flameproof enclosures of electrical apparatus;* and *UL Technical Report No. 58* (1993). The test equipment for determining MIC is described in IEC 60079-11-1999, *Electrical apparatus for explosive gas atmospheres — Part 11: Intrinsic safety "i."* The classification of gases or vapors according to their maximum experimental safe gaps and minimum igniting currents is described in IEC 60079-12-1978, *Classification of mixtures of gases or vapours with air according to their maximum experimental safe gaps and minimum igniting currents.*

Informational Note No. 3: Group II is currently subdivided into Group IIA, Group IIB, and Group IIC. Prior marking requirements permitted some types of protection to be marked without a subdivision, showing only Group II.

Informational Note No. 4: It is necessary that the meanings of the different equipment markings and Group II classifications be carefully observed to avoid confusion with Class I, Divisions 1 and 2, Groups A, B, C, and D.

Class I, Zone 0, 1, and 2, groups shall be as follows:

(A) Group IIC. Atmospheres containing acetylene, hydrogen, or flammable gas, flammable liquid–produced vapor, or combustible liquid–produced vapor mixed with air that may burn or explode, having either a maximum experimental safe gap (MESG) value less than or equal to 0.50 mm or minimum igniting current (MIC) ratio less than or equal to 0.45. [**497:** 3.3.5.2.3]

Informational Note: Group IIC is equivalent to a combination of Class I, Group A, and Class I, Group B, as described in 500.6(A)(1) and (A)(2).

(B) Group IIB. Atmospheres containing acetaldehyde, ethylene, or flammable gas, flammable liquid–produced vapor, or combustible liquid–produced vapor mixed with air that may burn or explode, having either maximum experimental safe gap (MESG) values greater than 0.50 mm and less than or equal to 0.90 mm or minimum igniting current ratio (MIC ratio) greater than 0.45 and less than or equal to 0.80. [**497:**3.3.5.2.2]

Informational Note: Group IIB is equivalent to Class I, Group C, as described in 500.6(A)(3).

(C) Group IIA. Atmospheres containing acetone, ammonia, ethyl alcohol, gasoline, methane, propane, or flammable gas, flammable liquid–produced vapor, or combustible liquid–produced vapor mixed with air that may burn or explode, having either a maximum experimental safe gap (MESG) value greater than 0.90 mm or minimum igniting current (MIC) ratio greater than 0.80. [**497:** 3.3.5.2.1]

Informational Note: Group IIA is equivalent to Class I, Group D as described in 500.6(A)(4).

505.7 Special Precaution. Article 505 requires equipment construction and installation that ensures safe performance under conditions of proper use and maintenance.

Informational Note No. 1: It is important that inspection authorities and users exercise more than ordinary care with regard to the installation and maintenance of electrical equipment in hazardous (classified) locations.

Informational Note No. 2: Low ambient conditions require special consideration. Electrical equipment depending on the protection techniques described by 505.8(A) may not be suitable for use at temperatures lower than -20°C (-4°F) unless they are identified for use at lower temperatures. However, at low ambient temperatures, flammable concentrations of vapors may not exist in a location classified Class I, Zones 0, 1, or 2 at normal ambient temperature.

(A) Implementation of Zone Classification System. Classification of areas, engineering and design, selection of equipment and wiring methods, installation, and inspection shall be performed by qualified persons.

(B) Dual Classification. In instances of areas within the same facility classified separately, Class I, Zone 2 locations shall be permitted to abut, but not overlap, Class I, Division 2 locations. Class I, Zone 0 or Zone 1 locations shall not abut Class I, Division 1 or Division 2 locations. [**33:**6.2.4]

(C) Reclassification Permitted. A Class I, Division 1 or Division 2 location shall be permitted to be reclassified as a Class I, Zone 0, Zone 1, or Zone 2 location, provided all of the space that is classified because of a single flammable gas or vapor source is reclassified under the requirements of this article.

(D) Solid Obstacles. Flameproof equipment with flanged joints shall not be installed such that the flange openings are closer than the distances shown in Table 505.7(D) to any solid obstacle that is not a part of the equipment (such as steelworks, walls, weather guards, mounting brackets, pipes, or other electrical equipment) unless the equipment is listed for a smaller distance of separation.

(E) Simultaneous Presence of Flammable Gases and Combustible Dusts or Fibers/Flyings. Where flammable gases, combustible dusts, or fibers/flyings are or may be present at the same time, the simultaneous presence shall be considered during the selection and installation of the electrical equipment and the wiring methods, including the determination of the safe operating temperature of the electrical equipment.

(F) Available Short-Circuit Current for Type of Protection "e". Unless listed and marked for connection to circuits with higher available short-circuit current, the available short-circuit current for electrical equipment using type of protection "e" for the field wiring connections in Zone 1 locations shall be limited to 10,000 rms symmetrical amperes to reduce the likelihood of ignition of a flammable atmosphere by an arc during a short-circuit event.

Informational Note: Limitation of the available short-circuit current to this level may require the application of current-limiting fuses or current-limiting circuit breakers.

505.8 Protection Techniques. Acceptable protection techniques for electrical and electronic equipment in hazardous (classified) locations shall be as described in 505.8(A) through (I).

Informational Note: For additional information, see ANSI/ISA-60079-0 (12.00.01)-2009, *Explosive Atmospheres — Part 0: Equipment — General Requirements*; ANSI/ISA-12.01.01-1999, *Definitions and Information Pertaining to Electrical Apparatus in Hazardous (Classified) Locations*; and ANSI/UL 60079-0, *Electrical Apparatus for Explosive Gas Atmospheres — Part 0: General Requirements*.

(A) Flameproof "d". This protection technique shall be permitted for equipment in Class I, Zone 1 or Zone 2 locations.

Table 505.7(D) Minimum Distance of Obstructions from Flameproof "d" Flange Openings

Gas Group	Minimum Distance	
	mm	in.
IIC	40	$1\frac{37}{64}$
IIB	30	$1\frac{3}{16}$
IIA	10	$\frac{25}{64}$

(B) Pressurization "p". This protection technique shall be permitted for equipment in those Class I, Zone 1 or Zone 2 locations for which it is identified.

(C) Intrinsic Safety "i". This protection technique shall be permitted for apparatus and associated apparatus in Class I, Zone 0, Zone 1, or Zone 2 locations for which it is listed.

(D) Type of Protection "n". This protection technique shall be permitted for equipment in Class I, Zone 2 locations. Type of protection "n" is further subdivided into nA, nC, and nR.

Informational Note: See Table 505.9(C)(2)(4) for the descriptions of subdivisions for type of protection "n".

(E) Oil Immersion "o". This protection technique shall be permitted for equipment in Class I, Zone 1 or Zone 2 locations.

(F) Increased Safety "e". This protection technique shall be permitted for equipment in Class I, Zone 1 or Zone 2 locations.

(G) Encapsulation "m". This protection technique shall be permitted for equipment in Class I, Zone 0, Zone 1, or Zone 2 locations for which it is identified.

Informational Note: See Table 505.9(C)(2)(4) for the descriptions of subdivisions for encapsulation.

(H) Powder Filling "q". This protection technique shall be permitted for equipment in Class I, Zone 1 or Zone 2 locations.

(I) Combustible Gas Detection System. A combustible gas detection system shall be permitted as a means of protection in industrial establishments with restricted public access and where the conditions of maintenance and supervision ensure that only qualified persons service the installation. Where such a system is installed, equipment specified in 505.8(I)(1), (I)(2), or (I)(3) shall be permitted. The type of detection equipment, its listing, installation location(s), alarm and shutdown criteria, and calibration frequency shall be documented when combustible gas detectors are used as a protection technique.

Informational Note No. 1: For further information, see ANSI/API RP 505-1997, *Recommended Practice for Classification of Locations for Electrical Installations at Petroleum Facilities Classified as Class I, Zone 0, Zone 1, and Zone 2.*

Informational Note No. 2: For further information, see ANSI/ISA-60079-29-2, *Explosive Atmospheres — Part 29-2: Gas detectors — Selection, installation, use and maintenance of detectors for flammable gases and oxygen.*

Informational Note No. 3: For further information, see ANSI/ISA-TR12.13.03-2009, *Guide for Combustible Gas Detection as a Method of Protection.*

(1) Inadequate Ventilation. In a Class I, Zone 1 location that is so classified due to inadequate ventilation, electrical equipment suitable for Class I, Zone 2 locations shall be permitted. Combustible gas detection equipment shall be listed for Class I, Zone 1, for the appropriate material group, and for the detection of the specific gas or vapor to be encountered.

(2) Interior of a Building. In a building located in, or with an opening into, a Class I, Zone 2 location where the interior does not contain a source of flammable gas or vapor, electrical equipment for unclassified locations shall be permitted. Combustible gas detection equipment shall be listed for Class I, Zone 1 or Class I, Zone 2, for the appropriate material group, and for the detection of the specific gas or vapor to be encountered.

(3) Interior of a Control Panel. In the interior of a control panel containing instrumentation utilizing or measuring flammable liquids, gases, or vapors, electrical equipment suitable for Class I, Zone 2 locations shall be permitted. Combustible gas detection equipment shall be listed for Class I, Zone 1, for the appropriate material group, and for the detection of the specific gas or vapor to be encountered.

505.9 Equipment.

(A) Suitability. Suitability of identified equipment shall be determined by one of the following:

(1) Equipment listing or labeling
(2) Evidence of equipment evaluation from a qualified testing laboratory or inspection agency concerned with product evaluation
(3) Evidence acceptable to the authority having jurisdiction such as a manufacturer's self-evaluation or an owner's engineering judgment

Informational Note: Additional documentation for equipment may include certificates demonstrating compliance with applicable equipment standards, indicating special conditions of use, and other pertinent information.

(B) Listing.

(1) Equipment that is listed for a Zone 0 location shall be permitted in a Zone 1 or Zone 2 location of the same gas or vapor, provided that it is installed in accordance with the requirements for the marked type of protection. Equipment that is listed for a Zone 1 location shall be permitted in a Zone 2 location of the same gas or vapor, provided that it is installed in accordance with the requirements for the marked type of protection.
(2) Equipment shall be permitted to be listed for a specific gas or vapor, specific mixtures of gases or vapors, or any specific combination of gases or vapors.

Informational Note: One common example is equipment marked for "IIB. + H2."

(C) Marking. Equipment shall be marked in accordance with 505.9(C)(1) or (C)(2).

(1) Division Equipment. Equipment identified for Class I, Division 1 or Class I, Division 2 shall, in addition to being marked in accordance with 500.8(C), be permitted to be marked with all of the following:

(1) Class I, Zone 1 or Class I, Zone 2 (as applicable)
(2) Applicable gas classification group(s) in accordance with Table 505.9(C)(1)(2)
(3) Temperature classification in accordance with 505.9(D)(1)

(2) Zone Equipment. Equipment meeting one or more of the protection techniques described in 505.8 shall be marked with all of the following in the order shown:

(1) Class
(2) Zone

Table 505.9(C)(1)(2) Material Groups

Material Group	Comment
IIC	See 505.6(A)
IIB	See 505.6(B)
IIA	See 505.6(C)

(3) Symbol "AEx"

(4) Protection technique(s) in accordance with Table 505.9(C)(2)(4)

(5) Applicable material group in accordance with Table 505.9(C)(1)(2) or a specific gas or vapor

(6) Temperature classification in accordance with 505.9(D)(1)

Exception No. 1: Associated apparatus NOT suitable for installation in a hazardous (classified) location shall be required to be marked only with (3), (4), and (5), but BOTH the symbol AEx (3) and the symbol for the type of protection (4) shall be enclosed within the same square brackets, for example, [AEx ia] IIC.

Exception No. 2: Simple apparatus as defined in 504.2 shall not be required to have a marked operating temperature or temperature class.

Exception No. 3: Fittings for the termination of cables shall not be required to have a marked operating temperature or temperature class.

Informational Note No. 1: An example of the required marking for intrinsically safe apparatus for installation in Class I, Zone 0 is "Class I, Zone 0, AEx ia IIC T6." An explanation of the marking that is required is shown in Informational Note Figure 505.9(C)(2), No. 1.

Informational Note No. 2: An example of the required marking for intrinsically safe associated apparatus mounted in a flameproof enclosure for installation in Class I, Zone 1 is "Class I, Zone 1 AEx d[ia] IIC T4."

Informational Note No. 3: An example of the required marking for intrinsically safe associated apparatus NOT for installation in a hazardous (classified) location is "[AEx ia] IIC."

Informational Note No. 4: The EPL (or equipment protection level) may appear in the product marking. EPLs are designated as G for gas, D for dust, or M for mining and are then followed by a letter (a, b, or c) to give the user a better understanding as to whether the equipment provides either (a) a "very high," (b) a "high," or (c) an "enhanced" level of protection against ignition of an explosive atmosphere. For example, a Class I, Zone 1, AEx d IIC T4 motor (which is suitable by protection concept for application in Zone 1) may additionally be marked with an EPL of "Gb" to indicate that it was provided with a high level of protection, such as Class I, Zone 1 AEx d IIC T4 Gb.

Informational Note No. 5: Equipment installed outside a Zone 0 location, electrically connected to equipment located inside a Zone 0 location, may be marked Class I, Zone 0/1. The "/" indicates that equipment contains a separation element and can be installed at the boundary between a Zone 0 and a Zone 1 location. See ANSI/ISA-60079-26, *Electrical Apparatus for Use in Class I, Zone 0 Hazardous (Classified) Locations.*

(D) Class I Temperature. The temperature marking specified in 505.9(D)(1) shall not exceed the autoignition temperature of the specific gas or vapor to be encountered.

Informational Note: For information regarding autoignition temperatures of gases and vapors, see NFPA 497-2012, *Recommended Practice for the Classification of Flammable Liquids, Gases, or Vapors and of Hazardous (Classified) Locations for Electrical Installations in Chemical Process Areas*; and IEC 60079-20-1996, *Electrical Apparatus for Explosive Gas Atmospheres, Data for Flammable Gases and Vapours, Relating to the Use of Electrical Apparatus.*

(1) Temperature Classifications. Equipment shall be marked to show the operating temperature or temperature class referenced to a 40°C ambient, or at the higher ambient temperature if the equipment is rated and marked for an ambient temperature of greater than 40°C. The temperature class, if

Table 505.9(C)(2)(4) Types of Protection Designation

Designation	Technique	Zone*
d	Flameproof enclosure	1
db	Flameproof enclosure	1
e	Increased safety	1
eb	Increased safety	1
ia	Intrinsic safety	0
ib	Intrinsic safety	1
ic	Intrinsic safety	2
[ia]	Associated apparatus	Unclassified**
[ib]	Associated apparatus	Unclassified**
[ic]	Associated apparatus	Unclassified**
ma	Encapsulation	0
m	Encapsulation	1
mb	Encapsulation	1
mc	Encapsulation	2
nA	Nonsparking equipment	2
nAc	Nonsparking equipment	2
nC	Sparking equipment in which the contacts are suitably protected other than by restricted breathing enclosure	2
nCc	Sparking equipment in which the contacts are suitably protected other than by restricted breathing enclosure	2
nR	Restricted breathing enclosure	2
nRc	Restricted breathing enclosure	2
o	Oil immersion	1
ob	Oil immersion	1
px	Pressurization	1
pxb	Pressurization	1
py	Pressurization	1
pyb	Pressurization	1
pz	Pressurization	2
pzc	Pressurization	2
q	Powder filled	1
qb	Powder filled	1

*Does not address use where a combination of techniques is used.
**Associated apparatus is permitted to be installed in a hazardous (classified) location if suitably protected using another type of protection.

provided, shall be indicated using the temperature class (T code) shown in Table 505.9(D)(1).

Electrical equipment designed for use in the ambient temperature range between -20°C and +40°C shall require no ambient temperature marking.

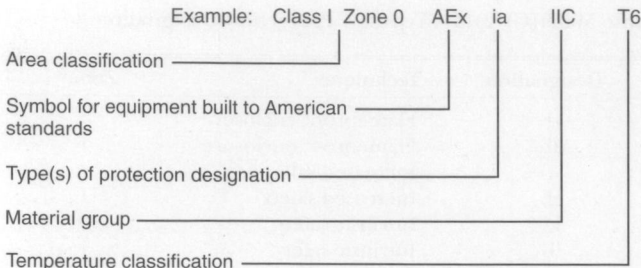

Informational Note Figure 505.9(C)(2), No. 1, Zone Equipment Marking.

Electrical equipment that is designed for use in a range of ambient temperatures other than -20°C to +40°C is considered to be special; and the ambient temperature range shall then be marked on the equipment, including either the symbol "Ta" or "Tamb" together with the special range of ambient temperatures, in degrees Celsius.

> Informational Note: As an example, such a marking might be "-30°C to +40°C."

Exception No. 1: Equipment of the non–heat-producing type, such as conduit fittings, and equipment of the heat-producing type having a maximum temperature of not more than 100°C (212°F) shall not be required to have a marked operating temperature or temperature class.

Exception No. 2: Equipment identified for Class I, Division 1 or Division 2 locations as permitted by 505.20(A), (B), and (C) shall be permitted to be marked in accordance with 505.8(C) and Table 500.8(C).

(E) Threading. The supply connection entry thread form shall be NPT or metric. Conduit and fittings shall be made wrenchtight to prevent sparking when fault current flows through the conduit system, and to ensure the explosionproof or flameproof integrity of the conduit system where applicable. Equipment provided with threaded entries for field wiring connections shall be installed in accordance with 505.9(E)(1) or (E)(2) and with (E)(3).

(1) Equipment Provided with Threaded Entries for NPT Threaded Conduit or Fittings. For equipment provided with threaded entries for NPT threaded conduit or fittings, listed conduit, listed conduit fittings, or listed cable fittings shall be used.

All NPT threaded conduit and fittings shall be threaded with a National (American) Standard Pipe Taper (NPT) thread.

Table 505.9(D)(1) Classification of Maximum Surface Temperature for Group II Electrical Equipment

Temperature Class (T Code)	Maximum Surface Temperature (°C)
T1	≤450
T2	≤300
T3	≤200
T4	≤135
T5	≤100
T6	≤85

NPT threaded entries into explosionproof or flameproof equipment shall be made up with at least five threads fully engaged.

Exception: For listed explosionproof or flameproof equipment, factory-threaded NPT entries shall be made up with at least 4½ threads fully engaged.

> Informational Note No. 1: Thread specifications for male NPT threads are located in ASME B1.20.1-2013, *Pipe Threads, General Purpose (Inch)*.

> Informational Note No. 2: Female NPT threaded entries use a modified National Standard Pipe Taper (NPT) thread with thread form per ASME B1.20.1-2013, *Pipe Threads, General Purpose (Inch)*. See ANSI/UL 60079-1:2013, *Explosive Atmospheres — Part 1: Equipment Protection by Flameproof Enclosures "d"*; and ANSI/ISA 60079-1:2013, *Explosive Atmospheres — Part 1: Equipment Protection by Flameproof Enclosures "d"*.

(2) Equipment Provided with Threaded Entries for Metric Threaded Conduit or Fittings. For equipment with metric threaded entries, listed conduit fittings or listed cable fittings shall be used. Such entries shall be identified as being metric, or listed adapters to permit connection to conduit or NPT threaded fittings shall be provided with the equipment and shall be used for connection to conduit or NPT threaded fittings.

Metric threaded fittings installed into explosionproof or flameproof equipment entries shall have a class of fit of at least 6g/6H and be made up with at least five threads fully engaged for Groups C, D, IIB, or IIA and not less than eight threads fully engaged and wrenchtight.

> Informational Note: Threading specifications for metric threaded entries are located in ISO 965-1-2013, *ISO general purpose metric screw threads — Tolerances — Part 1: Principles and basic data*; and ISO 965-3-1998, *ISO general purpose metric screw threads — Tolerances — Part 3: Deviations for constructional screw threads*.

(3) Unused Openings. All unused openings shall be closed with close-up plugs listed for the location and shall maintain the type of protection. The plug engagement shall comply with 505.9(E)(1) or 505.9(E)(2).

(F) Optical Fiber Cables. An optical fiber cable, with or without current-carrying current (composite optical fiber cable), shall be installed to address the associated fire hazard and sealed to address the associated explosion hazard in accordance with the requirements of 505.15 and 505.16.

505.15 Wiring Methods. Wiring methods shall maintain the integrity of protection techniques and shall comply with 505.15(A) through (C).

(A) Class I, Zone 0. In Class I, Zone 0 locations, equipment protected by intrinsic safety "ia" and equipment protected by encapsulation "ma" shall be connected using intrinsically safe "ia" circuits with wiring methods in accordance with Article 504.

(B) Class I, Zone 1.

(1) General. In Class I, Zone 1 locations, the wiring methods in 505.15(B)(1)(a) through (B)(1)(i) shall be permitted.
    (a) All wiring methods permitted by 505.15(A).
    (b) In industrial establishments with restricted public access, where the conditions of maintenance and supervision ensure that only qualified persons service the installation, and where the cable is not subject to physical damage, Type MC-HL

cable listed for use in Class I, Zone 1 or Division 1 locations, with a gas/vaportight continuous corrugated metallic sheath, an overall jacket of suitable polymeric material, and a separate equipment grounding conductor(s) in accordance with 250.122, and terminated with fittings listed for the application. Type MC-HL cable shall be installed in accordance with the provisions of Article 330, Part II.

(c) In industrial establishments with restricted public access, where the conditions of maintenance and supervision ensure that only qualified persons service the installation, and where the cable is not subject to physical damage, Type ITC-HL cable listed for use in Class I, Zone 1 or Division 1 locations, with a gas/vaportight continuous corrugated metallic sheath and an overall jacket of suitable polymeric material, and terminated with fittings listed for the application. Type ITC-HL cable shall be installed in accordance with the provisions of Article 727.

Informational Note: See 727.4 and 727.5 for restrictions on use of Type ITC cable.

(d) Type MI cable terminated with fittings listed for Class I, Zone 1 or Division 1 locations. Type MI cable shall be installed and supported in a manner to avoid tensile stress at the termination fittings.

(e) Threaded rigid metal conduit, or threaded steel intermediate metal conduit.

(f) Type PVC conduit and Type RTRC conduit shall be permitted where encased in a concrete envelope a minimum of 50 mm (2 in.) thick and provided with not less than 600 mm (24 in.) of cover measured from the top of the conduit to grade. Threaded rigid metal conduit or threaded steel intermediate metal conduit shall be used for the last 600 mm (24 in.) of the underground run to emergence or to the point of connection to the aboveground raceway. An equipment grounding conductor shall be included to provide for electrical continuity of the raceway system and for grounding of non–current-carrying metal parts.

(g) Intrinsic safety type of protection "ib" shall be permitted using the wiring methods specified in Article 504.

Informational Note: For entry into enclosures required to be flameproof, explosionproof, or of increased safety, see the information on construction, testing, and marking of cables; flameproof and increased safety cable fittings; and flameproof and increased safety cord connectors in ANSI/UL 2225-2013, *Cables and Cable-Fittings for Use in Hazardous (Classified) Locations.*

(h) Optical fiber cable Types OFNP, OFCP, OFNR, OFCR, OFNG, OFCG, OFN, and OFC shall be permitted to be installed in raceways in accordance with 505.15(B). Optical fiber cable shall be sealed in accordance with 505.16.

Informational Note: For entry into enclosures required to be flameproof, explosionproof, or of increased safety, see the information on construction, testing, and marking of cables; flameproof and increased safety cable fittings; and flameproof and increased safety cord connectors in ANSI/UL 2225-2013, *Cables and Cable-Fittings for Use in Hazardous (Classified) Locations.*

(i) In industrial establishments with restricted public access, where the conditions of maintenance and supervision ensure that only qualified persons service the installation, for applications limited to 600 volts nominal or less, for cable diameters 25 mm (1 in.) or less, and where the cable is not subject to physical damage, Type TC-ER-HL cable listed for use in Class I, Zone 1 locations, with an overall jacket and a separate equipment grounding conductor(s) in accordance with

250.122, and terminated with fittings listed for the location, Type TC-ER-HL cable shall be installed in accordance with the provisions of Article 336, including the restrictions of 336.10(7).

(2) Flexible Connections. Where necessary to employ flexible connections, flexible fittings listed for Class I, Zone 1 or Division 1 locations, or flexible cord in accordance with the provisions of 505.17(A) terminated with a listed cord connector that maintains the type of protection of the terminal compartment, shall be permitted.

(C) Class I, Zone 2.

(1) General. In Class I, Zone 2 locations, the following wiring methods shall be permitted:

(1) All wiring methods permitted by 505.15(B).
(2) Types MC, MV, TC, or TC-ER cable, including installation in cable tray systems. The cable shall be terminated with listed fittings. Single conductor Type MV cables shall be shielded or metallic-armored.
(3) Type ITC and Type ITC-ER cable as permitted in 727.4 and terminated with listed fittings.
(4) Type PLTC and Type PLTC-ER cable in accordance with the provisions of Article 725, including installation in cable tray systems. The cable shall be terminated with listed fittings.
(5) Enclosed gasketed busways, enclosed gasketed wireways.
(6) In industrial establishments with restricted public access, where the conditions of maintenance and supervision ensure that only qualified persons service the installation, and where metallic conduit does not provide sufficient corrosion resistance, listed reinforced thermosetting resin conduit (RTRC), factory elbows, and associated fittings, all marked with the suffix -XW, and Schedule 80 PVC conduit, factory elbows, and associated fittings shall be permitted. Where seals are required for boundary conditions as defined in 505.16(C)(1)(b), the Zone 1 wiring method shall extend into the Zone 2 area to the seal, which shall be located on the Zone 2 side of the Zone 1/Zone 2 boundary.
(7) Intrinsic safety type of protection "ic" shall be permitted using any of the wiring methods permitted for unclassified locations. Intrinsic safety type of protection "ic" systems shall be installed in accordance with the control drawing(s). Simple apparatus, not shown on the control drawing, shall be permitted in an intrinsic safety type of protection "ic" circuit, provided that the simple apparatus does not interconnect the intrinsic safety type of protection "ic" systems to any other circuit.

Informational Note: Simple apparatus is defined in 504.2.

(8) Optical fiber cable of Types OFNP, OFCP, OFNR, OFCR, OFNG, OFCG, OFN, and OFC shall be permitted to be installed in cable trays or any other raceway in accordance with 505.15(C). Optical fiber cable shall be sealed in accordance with 505.16.
(9) Cablebus.

Separate intrinsic safety type of protection "ic" systems shall be installed in accordance with one of the following:

(1) In separate cables
(2) In multiconductor cables where the conductors of each circuit are within a grounded metal shield

(3) In multiconductor cables where the conductors of each circuit have insulation with a minimum thickness of 0.25 mm (0.01 in.)

(2) Flexible Connections. Where provision must be made for flexibility, flexible metal fittings, flexible metal conduit with listed fittings, liquidtight flexible metal conduit with listed fittings, liquidtight flexible nonmetallic conduit with listed fittings, or flexible cord in accordance with the provisions of 505.17 terminated with a listed cord connector that maintains the type of protection of the terminal compartment shall be permitted.

> Informational Note: See 505.25(B) for grounding requirements where flexible conduit is used.

Exception: For elevator use, an identified elevator cable of Type EO, ETP, or ETT, shown under the "use" column in Table 400.4 for "hazardous (classified) locations," that is terminated with listed connectors that maintain the type of protection of the terminal compartment, shall be permitted.

505.16 Sealing and Drainage. Seals in conduit and cable systems shall comply with 505.16(A) through (E). Sealing compound shall be used in Type MI cable termination fittings to exclude moisture and other fluids from the cable insulation.

> Informational Note No. 1: Seals are provided in conduit and cable systems to minimize the passage of gases and vapors and prevent the passage of flames from one portion of the electrical installation to another through the conduit. Such communication through Type MI cable is inherently prevented by construction of the cable. Unless specifically designed and tested for the purpose, conduit and cable seals are not intended to prevent the passage of liquids, gases, or vapors at a continuous pressure differential across the seal. Even at differences in pressure across the seal equivalent to a few inches of water, there may be a slow passage of gas or vapor through a seal and through conductors passing through the seal. See 505.16(C)(2)(b). Temperature extremes and highly corrosive liquids and vapors can affect the ability of seals to perform their intended function. See 505.16(D)(2).

> Informational Note No. 2: Gas or vapor leakage and propagation of flames may occur through the interstices between the strands of standard stranded conductors larger than 2 AWG. Special conductor constructions, for example, compacted strands or sealing of the individual strands, are means of reducing leakage and preventing the propagation of flames.

(A) Zone 0. In Class I, Zone 0 locations, seals shall be located according to 505.16(A)(1), (A)(2), and (A)(3).

(1) Conduit Seals. Seals shall be provided within 3.05 m (10 ft) of where a conduit leaves a Zone 0 location. There shall be no unions, couplings, boxes, or fittings, except listed reducers at the seal, in the conduit run between the seal and the point at which the conduit leaves the location.

Exception: A rigid unbroken conduit that passes completely through the Zone 0 location with no fittings less than 300 mm (12 in.) beyond each boundary shall not be required to be sealed if the termination points of the unbroken conduit are in unclassified locations.

(2) Cable Seals. Seals shall be provided on cables at the first point of termination after entry into the Zone 0 location.

(3) Not Required to Be Explosionproof or Flameproof. Seals shall not be required to be explosionproof or flameproof.

(B) Zone 1. In Class I, Zone 1 locations, seals shall be located in accordance with 505.16(B)(1) through (B)(8).

(1) Type of Protection "d" or "e" Enclosures. Conduit seals shall be provided within 50 mm (2 in.) for each conduit entering enclosures having type of protection "d" or "e."

Exception No. 1: Where the enclosure having type of protection "d" is marked to indicate that a seal is not required.

Exception No. 2: For type of protection "e," conduit and fittings employing only NPT to NPT raceway joints or fittings listed for type of protection "e" shall be permitted between the enclosure and the seal, and the seal shall not be required to be within 50 mm (2 in.) of the entry.

> Informational Note: Examples of fittings employing other than NPT threads include conduit couplings, capped elbows, unions, and breather drains.

Exception No. 3: For conduit installed between type of protection "e" enclosures employing only NPT to NPT raceway joints or conduit fittings listed for type of protection "e," a seal shall not be required.

(2) Explosionproof Equipment. Conduit seals shall be provided for each conduit entering explosionproof equipment according to 505.16(B)(2)(a), (B)(2)(b), and (B)(2)(c).

(a) In each conduit entry into an explosionproof enclosure where either of the following conditions apply:

(1) The enclosure contains apparatus, such as switches, circuit breakers, fuses, relays, or resistors that may produce arcs, sparks, or high temperatures that are considered to be an ignition source in normal operation. For the purposes of this section, high temperatures shall be considered to be any temperatures exceeding 80 percent of the autoignition temperature in degrees Celsius of the gas or vapor involved.

Exception: Seals shall not be required for conduit entering an enclosure where such switches, circuit breakers, fuses, relays, or resistors comply with one of the following:
(a) Are enclosed within a chamber hermetically sealed against the entrance of gases or vapors.
(b) Are immersed in oil.
(c) Are enclosed within an enclosure, identified for the location, and marked "Leads Factory Sealed," "Factory Sealed," "Seal not Required," or equivalent.

(2) The entry is metric designator 53 (trade size 2) or larger and the enclosure contains terminals, splices, or taps.

> An enclosure, identified for the location, and marked "Leads Factory Sealed," or "Factory Sealed," Seal not Required," or equivalent shall not be considered to serve as a seal for another adjacent explosionproof enclosure that is required to have a conduit seal.

(b) Conduit seals shall be installed within 450 mm (18 in.) from the enclosure. Only explosionproof unions, couplings, reducers, elbows, capped elbows, and conduit bodies similar to L, T, and cross types that are not larger than the trade size of the conduit shall be permitted between the sealing fitting and the explosionproof enclosure.

(c) Where two or more explosionproof enclosures for which conduit seals are required under 505.16(B)(2) are connected by nipples or by runs of conduit not more than 900 mm (36 in.) long, a single conduit seal in each such nipple connection or run of conduit shall be considered sufficient if located not more than 450 mm (18 in.) from either enclosure.

(3) Pressurized Enclosures. Conduit seals shall be provided in each conduit entry into a pressurized enclosure where the conduit is not pressurized as part of the protection system.

Conduit seals shall be installed within 450 mm (18 in.) from the pressurized enclosure.

> Informational Note No. 1: Installing the seal as close as possible to the enclosure reduces problems with purging the dead airspace in the pressurized conduit.

> Informational Note No. 2: For further information, see NFPA 496-2013, *Standard for Purged and Pressurized Enclosures for Electrical Equipment.*

(4) Class I, Zone 1 Boundary. Conduit seals shall be provided in each conduit run leaving a Class I, Zone 1 location. The sealing fitting shall be permitted on either side of the boundary of such location within 3.05 m (10 ft) of the boundary and shall be designed and installed so as to minimize the amount of gas or vapor within the Zone 1 portion of the conduit from being communicated to the conduit beyond the seal. Except for listed explosionproof reducers at the conduit seal, there shall be no union, coupling, box, or fitting between the conduit seal and the point at which the conduit leaves the Zone 1 location.

Exception: Metal conduit containing no unions, couplings, boxes, or fittings and passing completely through a Class I, Zone 1 location with no fittings less than 300 mm (12 in.) beyond each boundary shall not require a conduit seal if the termination points of the unbroken conduit are in unclassified locations.

(5) Cables Capable of Transmitting Gases or Vapors. Conduits containing cables with a gas/vaportight continuous sheath capable of transmitting gases or vapors through the cable core shall be sealed in the Zone 1 location after removing the jacket and any other coverings so that the sealing compound surrounds each individual insulated conductor or optical fiber tube and the outer jacket.

Exception: Multiconductor cables with a gas/vaportight continuous sheath capable of transmitting gases or vapors through the cable core shall be permitted to be considered as a single conductor by sealing the cable in the conduit within 450 mm (18 in.) of the enclosure and the cable end within the enclosure by an approved means to minimize the entrance of gases or vapors and prevent the propagation of flame into the cable core, or by other approved methods. For shielded cables and twisted pair cables, it shall not be required to remove the shielding material or separate the twisted pair.

(6) Cables Incapable of Transmitting Gases or Vapors. Each multiconductor or optical multifiber cable in conduit shall be considered as a single conductor or single optical fiber tube if the cable is incapable of transmitting gases or vapors through the cable core. These cables shall be sealed in accordance with 505.16(D).

(7) Cables Entering Enclosures. Cable seals shall be provided for each cable entering flameproof or explosionproof enclosures. The seal shall comply with 505.16(D).

(8) Class I, Zone 1 Boundary. Cables shall be sealed at the point at which they leave the Zone 1 location.

Exception: Where cable is sealed at the termination point.

(C) Zone 2. In Class I, Zone 2 locations, seals shall be located in accordance with 505.16(C)(1) and (C)(2).

(1) Conduit Seals. Conduit seals shall be located in accordance with (C)(1)(a) and (C)(1)(b).

(a)   For connections to enclosures that are required to be flameproof or explosionproof, a conduit seal shall be provided in accordance with 505.16(B)(1) and (B)(2). All portions of the conduit run or nipple between the seal and enclosure shall comply with 505.16(B).

(b)   In each conduit run passing from a Class I, Zone 2 location into an unclassified location. The sealing fitting shall be permitted on either side of the boundary of such location within 3.05 m (10 ft) of the boundary and shall be designed and installed so as to minimize the amount of gas or vapor within the Zone 2 portion of the conduit from being communicated to the conduit beyond the seal. Rigid metal conduit or threaded steel intermediate metal conduit shall be used between the sealing fitting and the point at which the conduit leaves the Zone 2 location, and a threaded connection shall be used at the sealing fitting. Except for listed explosionproof reducers at the conduit seal, there shall be no union, coupling, box, or fitting between the conduit seal and the point at which the conduit leaves the Zone 2 location. Conduits shall be sealed to minimize the amount of gas or vapor within the Class I, Zone 2 portion of the conduit from being communicated to the conduit beyond the seal. Such seals shall not be required to be flameproof or explosionproof but shall be identified for the purpose of minimizing passage of gases under normal operating conditions and shall be accessible.

Exception No. 1: Metal conduit containing no unions, couplings, boxes, or fittings and passing completely through a Class I, Zone 2 location with no fittings less than 300 mm (12 in.) beyond each boundary shall not be required to be sealed if the termination points of the unbroken conduit are in unclassified locations.

Exception No. 2: Conduit systems terminating at an unclassified location where a wiring method transition is made to cable tray, cablebus, ventilated busway, Type MI cable, or cable that is not installed in a raceway or cable tray system shall not be required to be sealed where passing from the Class I, Zone 2 location into the unclassified location. The unclassified location shall be outdoors or, if the conduit system is all in one room, it shall be permitted to be indoors. The conduits shall not terminate at an enclosure containing an ignition source in normal operation.

Exception No. 3: Conduit systems passing from an enclosure or room that is unclassified as a result of pressurization into a Class I, Zone 2 location shall not require a seal at the boundary.

> Informational Note: For further information, refer to NFPA 496-2013, *Standard for Purged and Pressurized Enclosures for Electrical Equipment.*

Exception No. 4: Segments of aboveground conduit systems shall not be required to be sealed where passing from a Class I, Zone 2 location into an unclassified location if all the following conditions are met:

(1)   No part of the conduit system segment passes through a Zone 0 or Zone 1 location where the conduit contains unions, couplings, boxes, or fittings within 300 mm (12 in.) of the Zone 0 or Zone 1 location.

(2)   The conduit system segment is located entirely in outdoor locations.

(3)   The conduit system segment is not directly connected to canned pumps, process or service connections for flow, pressure, or analysis measurement, and so forth, that depend on a single compression seal, diaphragm, or tube to prevent flammable or combustible fluids from entering the conduit system.

(4)   The conduit system segment contains only threaded metal conduit, unions, couplings, conduit bodies, and fittings in the unclassified location.

(5)   The conduit system segment is sealed at its entry to each enclosure or fitting housing terminals, splices, or taps in Zone 2 locations.

(2) Cable Seals. Cable seals shall be located in accordance with (C)(2)(a), (C)(2)(b), and (C)(2)(c).

(a) *Explosionproof and Flameproof Enclosures.* Cables entering enclosures required to be flameproof or explosionproof shall be sealed at the point of entrance. The seal shall comply with 505.16(D). Multiconductor or optical multifiber cables with a gas/vaportight continuous sheath capable of transmitting gases or vapors through the cable core shall be sealed in the Zone 2 location after removing the jacket and any other coverings so that the sealing compound surrounds each individual insulated conductor or optical fiber tube in such a manner as to minimize the passage of gases and vapors. Multiconductor or optical multifiber cables in conduit shall be sealed as described in 505.16(B)(4).

Exception No. 1: Cables passing from an enclosure or room that is unclassified as a result of Type Z pressurization into a Zone 2 location shall not require a seal at the boundary.

Exception No. 2: Shielded cables and twisted pair cables shall not require the removal of the shielding material or separation of the twisted pairs, provided the termination is by an approved means to minimize the entrance of gases or vapors and prevent propagation of flame into the cable core.

(b) *Cables That Will Not Transmit Gases or Vapors.* Cables with a gas/vaportight continuous sheath and that will not transmit gases or vapors through the cable core in excess of the quantity permitted for seal fittings shall not be required to be sealed except as required in 505.16(C)(2)(b). The minimum length of such cable run shall not be less than the length that limits gas or vapor flow through the cable core to the rate permitted for seal fittings [200 cm³/hr (0.007 ft³/hr) of air at a pressure of 1500 pascals (6 in. of water)].

> Informational Note No. 1: For further information on construction, testing, and marking of cables, cable fittings, and cord connectors, see ANSI/UL 2225-2011, *Cables and Cable-Fittings for Use in Hazardous (Classified) Locations.*

> Informational Note No. 2: The cable core does not include the interstices of the conductor strands.

(c) *Cables Capable of Transmitting Gases or Vapors.* Cables with a gas/vaportight continuous sheath capable of transmitting gases or vapors through the cable core shall not be required to be sealed except as required in 505.16(C)(2)(b), unless the cable is attached to process equipment or devices that may cause a pressure in excess of 1500 pascals (6 in. of water) to be exerted at a cable end, in which case a seal, barrier, or other means shall be provided to prevent migration of flammables into an unclassified area.

Exception: Cables with an unbroken gas/vaportight continuous sheath shall be permitted to pass through a Class I, Zone 2 location without seals.

(d) *Cables Without Gas/Vaportight Continuous Sheath.* Cables that do not have gas/vaportight continuous sheath shall be sealed at the boundary of the Zone 2 and unclassified location in such a manner as to minimize the passage of gases or vapors into an unclassified location.

> Informational Note: The cable sheath may be either metal or a nonmetallic material.

(D) Class I, Zones 0, 1, and 2. Where required, seals in Class I, Zones 0, 1, and 2 locations shall comply with 505.16(D)(1) through (D)(5).

(1) Fittings. Enclosures for connections or equipment shall be provided with an integral means for sealing, or sealing fittings listed for the location shall be used. Sealing fittings shall be listed for use with one or more specific compounds and shall be accessible.

(2) Compound. The compound shall provide a seal against passage of gas or vapors through the seal fitting, shall not be affected by the surrounding atmosphere or liquids, and shall not have a melting point less than 93°C (200°F).

(3) Thickness of Compounds. In a completed seal, the minimum thickness of the sealing compound shall not be less than the trade size of the sealing fitting and, in no case, less than 16 mm (⅝ in.).

Exception: Listed cable sealing fittings shall not be required to have a minimum thickness equal to the trade size of the fitting.

(4) Splices and Taps. Splices and taps shall not be made in fittings intended only for sealing with compound, nor shall other fittings in which splices or taps are made be filled with compound.

(5) Conductor or Optical Fiber Fill. The cross-sectional area of the conductors or optical fiber tubes (metallic or nonmetallic) permitted in a seal shall not exceed 25 percent of the cross-sectional area of a rigid metal conduit of the same trade size unless it is specifically listed for a higher percentage of fill.

(E) Drainage.

(1) Control Equipment. Where there is a probability that liquid or other condensed vapor may be trapped within enclosures for control equipment or at any point in the raceway system, approved means shall be provided to prevent accumulation or to permit periodic draining of such liquid or condensed vapor.

(2) Motors and Generators. Where liquid or condensed vapor may accumulate within motors or generators, joints and conduit systems shall be arranged to minimize entrance of liquid. If means to prevent accumulation or to permit periodic draining are necessary, such means shall be provided at the time of manufacture and shall be considered an integral part of the machine.

505.17 Flexible Cables, Cords and Connections.

(A) Flexible Cords, Class I, Zones 1 and 2. A flexible cord shall be permitted for connection between portable lighting equipment or other portable utilization equipment and the fixed portion of their supply circuit. Flexible cord shall also be permitted for that portion of the circuit where the fixed wiring methods of 505.15(B) and (C) cannot provide the necessary degree of movement for fixed and mobile electrical utilization equipment in an industrial establishment where conditions of maintenance and engineering supervision ensure that only qualified persons install and service the installation, and where the flexible cord is protected by location or by a suitable guard from damage. The length of the flexible cord shall be continuous. Where flexible cords are used, the cords shall comply with the following:

(1) Be of a type listed for extra-hard usage.
(2) Contain, in addition to the conductors of the circuit, an equipment grounding conductor complying with 400.23.
(3) Be connected to terminals or to supply conductors in an approved manner.

(4) Be supported by clamps or by other suitable means in such a manner that there will be no tension on the terminal connections.

(5) Be terminated with a listed cord connector that maintains the type of protection where the flexible cord enters boxes, fittings, or enclosures that are required to be explosionproof or flameproof.

(6) Cord entering an increased safety "e" enclosure shall be terminated with a listed increased safety "e" cord connector.

Informational Note: See 400.10 for permitted uses of flexible cords.

Electric submersible pumps with means for removal without entering the wet-pit shall be considered portable utilization equipment. The extension of the flexible cord within a suitable raceway between the wet-pit and the power source shall be permitted.

Electric mixers intended for travel into and out of open-type mixing tanks or vats shall be considered portable utilization equipment.

Informational Note: See 505.18 for flexible cords exposed to liquids having a deleterious effect on the conductor insulation.

(B) Instrumentation Connections for Zone 2. To facilitate replacements, process control instruments shall be permitted to be connected through flexible cords, attachment plugs, and receptacles, provided that all of the following conditions apply:

(1) A switch listed for Zone 2 is provided so that the attachment plug is not depended on to interrupt current, unless the circuit is type "ia," "ib," or "ic" protection, in which case the switch is not required.

(2) The current does not exceed 3 amperes at 120 volts, nominal.

(3) The power-supply cord does not exceed 900 mm (3 ft), is of a type listed for extra-hard usage or for hard usage if protected by location, and is supplied through an attachment plug and receptacle of the locking and grounding type.

(4) Only necessary receptacles are provided.

(5) The receptacle carries a label warning against unplugging under load.

505.18 Conductors and Conductor Insulation.

(A) Conductors. For type of protection "e," field wiring conductors shall be copper. Every conductor (including spares) that enters Type "e" equipment shall be terminated at a Type "e" terminal.

(B) Conductor Insulation. Where condensed vapors or liquids may collect on, or come in contact with, the insulation on conductors, such insulation shall be of a type identified for use under such conditions, or the insulation shall be protected by a sheath of lead or by other approved means.

505.19 Uninsulated Exposed Parts. There shall be no uninsulated exposed parts, such as electrical conductors, buses, terminals, or components that operate at more than 30 volts (15 volts in wet locations). These parts shall additionally be protected by type of protection "ia," "ib," or "nA" that is suitable for the location.

505.20 Equipment Requirements.

(A) Zone 0. In Class I, Zone 0 locations, only equipment specifically listed and marked as suitable for the location shall be permitted.

Exception: Intrinsically safe apparatus listed for use in Class I, Division 1 locations for the same gas, or as permitted by 505.9(B)(2), and with a suitable temperature class shall be permitted.

(B) Zone 1. In Class I, Zone 1 locations, only equipment specifically listed and marked as suitable for the location shall be permitted.

Exception No. 1: Equipment identified for use in Class I, Division 1 or listed for use in Zone 0 locations for the same gas, or as permitted by 505.9(B)(2), and with a suitable temperature class shall be permitted.

Exception No. 2: Equipment identified for Class I, Zone 1 or Zone 2 type of protection "p" shall be permitted.

(C) Zone 2. In Class I, Zone 2 locations, only equipment specifically listed and marked as suitable for the location shall be permitted.

Exception No. 1: Equipment listed for use in Zone 0 or Zone 1 locations for the same gas, or as permitted by 505.9(B)(2), and with a suitable temperature class, shall be permitted.

Exception No. 2: Equipment identified for Class I, Zone 1 or Zone 2 type of protection "p" shall be permitted.

Exception No. 3: Equipment identified for use in Class I, Division 1 or Division 2 locations for the same gas, or as permitted by 505.9(B)(2), and with a suitable temperature class shall be permitted.

Exception No. 4: In Class I, Zone 2 locations, the installation of open or nonexplosionproof or nonflameproof enclosed motors, such as squirrel-cage induction motors without brushes, switching mechanisms, or similar arc-producing devices that are not identified for use in a Class I, Zone 2 location shall be permitted.

Informational Note No. 1: It is important to consider the temperature of internal and external surfaces that may be exposed to the flammable atmosphere.

Informational Note No. 2: It is important to consider the risk of ignition due to currents arcing across discontinuities and overheating of parts in multisection enclosures of large motors and generators. Such motors and generators may need equipotential bonding jumpers across joints in the enclosure and from enclosure to ground. Where the presence of ignitible gases or vapors is suspected, clean air purging may be needed immediately prior to and during start-up periods.

Informational Note No. 3: For further information on the application of electric motors in Class I, Zone 2 hazardous (classified) locations, see IEEE 1349-2011, *IEEE Guide for the Application of Electric Motors in Class I, Division 2 and Class I, Zone 2 Hazardous (Classified) Locations.*

(D) Materials. Equipment marked Group IIC shall be permitted for applications requiring Group IIA or Group IIB equipment. Similarly, equipment marked Group IIB shall be permitted for applications requiring Group IIA equipment.

Equipment marked for a specific gas or vapor shall be permitted for applications where the specific gas or vapor may be encountered.

Informational Note: One common example combines these markings with equipment marked IIB +H2. This equipment is suitable for applications requiring Group IIA equipment, Group IIB equipment, or equipment for hydrogen atmospheres.

(E) Manufacturer's Instructions. Electrical equipment installed in hazardous (classified) locations shall be installed in accordance with the instructions (if any) provided by the manufacturer.

505.22 Increased Safety "e" Motors and Generators. In Class I, Zone 1 locations, increased safety "e" motors and generators of all voltage ratings shall be listed for Zone 1 locations, and shall comply with all of the following:

(1) Motors shall be marked with the current ratio, I_A / I_N, and time, t_E.

(2) Motors shall have controllers marked with the model or identification number, output rating (horsepower or kilowatt), full-load amperes, starting current ratio (I_A / I_N), and time (t_E) of the motors that they are intended to protect; the controller marking shall also include the specific overload protection type (and setting, if applicable) that is listed with the motor or generator.

(3) Connections shall be made with the specific terminals listed with the motor or generator.

(4) Terminal housings shall be permitted to be of substantial, nonmetallic, nonburning material, provided an internal grounding means between the motor frame and the equipment grounding connection is incorporated within the housing.

(5) The provisions of Part III of Article 430 shall apply regardless of the voltage rating of the motor.

(6) The motors shall be protected against overload by a separate overload device that is responsive to motor current. This device shall be selected to trip or shall be rated in accordance with the listing of the motor and its overload protection.

(7) Sections 430.32(C) and 430.44 shall not apply to such motors.

(8) The motor overload protection shall not be shunted or cut out during the starting period.

Informational Note: Reciprocating engine-driven generators, compressors, and other equipment installed in Class I, Zone 2 locations may present a risk of ignition of flammable materials associated with fuel, starting, compression, and so forth, due to inadvertent release or equipment malfunction by the engine ignition system and controls. For further information on the requirements for ignition systems for reciprocating engines installed in Class I, Zone 2 hazardous (classified) locations, see ANSI/UL 122001:2014, *General Requirements for Electrical Ignition Systems for Internal Combustion Engines in Class I, Division 2 or Zone 2, Hazardous (Classified) Locations.*

505.25 Grounding and Bonding. Regardless of the voltage of the electrical system, grounding and bonding shall comply with Article 250 and the requirements in 505.25(A) and (B).

(A) Bonding. The locknut-bushing and double-locknut types of contacts shall not be depended on for bonding purposes, but bonding jumpers with proper fittings or other approved means of bonding shall be used. Such means of bonding shall apply to all intervening raceways, fittings, boxes, enclosures, and so forth, between Class I locations and the point of grounding for service equipment or point of grounding of a separately derived system.

Exception: The specific bonding means shall be required only to the nearest point where the grounded circuit conductor and the grounding electrode are connected together on the line side of the building or struc-

ture disconnecting means as specified in 250.32(B), provided the branch-circuit overcurrent protection is located on the load side of the disconnecting means.

(B) Types of Equipment Grounding Conductors. Flexible metal conduit and liquidtight flexible metal conduit shall include an equipment bonding jumper of the wire type in compliance with 250.102.

Exception: In Class I, Zone 2 locations, the bonding jumper shall be permitted to be deleted where all of the following conditions are met:

　(a)　Listed liquidtight flexible metal conduit 1.8 m (6 ft) or less in length, with fittings listed for grounding, is used.

　(b)　Overcurrent protection in the circuit is limited to 10 amperes or less.

　(c)　The load is not a power utilization load.

505.26 Process Sealing. This section shall apply to process-connected equipment, which includes, but is not limited to, canned pumps, submersible pumps, flow, pressure, temperature, or analysis measurement instruments. A process seal is a device to prevent the migration of process fluids from the designed containment into the external electrical system. Process connected electrical equipment that incorporates a single process seal, such as a single compression seal, diaphragm, or tube to prevent flammable or combustible fluids from entering a conduit or cable system capable of transmitting fluids, shall be provided with an additional means to mitigate a single process seal failure. The additional means may include, but is not limited to, the following:

(1) A suitable barrier meeting the process temperature and pressure conditions that the barrier is subjected to upon failure of the single process seal. There shall be a vent or drain between the single process seal and the suitable barrier. Indication of the single process seal failure shall be provided by visible leakage, an audible whistle, or other means of monitoring.

(2) A listed Type MI cable assembly, rated at not less than 125 percent of the process pressure and not less than 125 percent of the maximum process temperature (in degrees Celsius), installed between the cable or conduit and the single process seal.

(3) A drain or vent located between the single process seal and a conduit or cable seal. The drain or vent shall be sufficiently sized to prevent overpressuring the conduit or cable seal above 6 in. water column (1493 Pa). Indication of the single process seal failure shall be provided by visible leakage, an audible whistle, or other means of monitoring.

(4) An add-on secondary seal marked "secondary seal" and rated for the pressure and temperature conditions to which it will be subjected upon failure of the single process seal.

Process-connected electrical equipment that does not rely on a single process seal or is listed and marked "single seal" or "dual seal" shall not be required to be provided with an additional means of sealing.

Informational Note: For construction and testing requirements for process sealing for listed and marked single seal, dual seal, or secondary seal equipment, refer to ANSI/ISA-12.27.01-2011, *Requirements for Process Sealing Between Electrical Systems and Flammable or Combustible Process Fluids.*

ARTICLE 506
Zone 20, 21, and 22 Locations for Combustible Dusts or Ignitible Fibers/Flyings

Informational Note: Text that is followed by a reference in brackets has been extracted from NFPA 499-2013, *Recommended Practice for the Classification of Combustible Dusts and of Hazardous (Classified) Locations for Electrical Installation in Chemical Process Areas.* Only editorial changes were made to the extracted text to make it consistent with this *Code.*

506.1 Scope. This article covers the requirements for the zone classification system as an alternative to the division classification system covered in Article 500, Article 502, and Article 503 for electrical and electronic equipment and wiring for all voltages in Zone 20, Zone 21, and Zone 22 hazardous (classified) locations where fire and explosion hazards may exist due to combustible dusts or ignitible fibers/flyings.

Informational Note No. 1: For the requirements for electrical and electronic equipment and wiring for all voltages in Class I, Division 1 or Division 2; Class II, Division 1 or Division 2; Class III, Division 1 or Division 2; and Class I, Zone 0 or Zone 1 or Zone 2 hazardous (classified) locations where fire or explosion hazards may exist due to flammable gases or vapors, flammable liquids, or combustible dusts or fibers, refer to Articles 500 through 505.

Informational Note No. 2: Zone 20, Zone 21, and Zone 22 area classifications are based on the modified IEC area classification system as defined in ANSI/ISA 60079– 10–2 (12.10.05)–2013, *Explosive Atmospheres — Part 10–2: Classification of Areas — Combustible Dust Atmospheres.*

Informational Note No. 3: The unique hazards associated with explosives, pyrotechnics, and blasting agents are not addressed in this article.

506.2 Definitions. For purposes of this article, the following definitions apply.

Protection by Encapsulation "m". Type of protection where electrical parts that could cause ignition of a mixture of combustible dust or fibers/flyings in air are protected by enclosing them in a compound in such a way that the explosive atmosphere cannot be ignited.

Informational Note No. 1: For additional information, see ANSI/ISA-60079-18 (12.23.01)-2012, *Explosive atmospheres — Part 18: Equipment protection by encapsulation "m"*; ANSI/ UL 60079-18-2009, *Explosive atmospheres — Part 18: Equipment protection by encapsulation "m"*; and ANSI/ISA-61241-18 (12.10.07)-2011, *Electrical Apparatus for Use in Zone 20, Zone 21 and Zone 22 Hazardous (Classified) Locations — Protection by Encapsulation "m."*

Informational Note No. 2: Encapsulation is designated level of protection "maD" or "ma" for use in Zone 20 locations. Encapsulation is designated level of protection "mbD" or "mb" for use in Zone 21 locations. Encapsulation is designated type of protection "mc" for use in Zone 22 locations.

Protection by Enclosure "t". Type of protection for explosive dust atmospheres where electrical apparatus is provided with an enclosure providing dust ingress protection and a means to limit surface temperatures.

Informational Note No. 1: For additional information, see ANSI/UL 60079-31-2015, *Explosive Atmospheres — Part 31: Equipment Dust Ignition Protection by Enclosure "t"*; and ANSI/ ISA-61241-1 (12.10.03)-2006, *Electrical Apparatus for Use in Zone 21 and Zone 22 Hazardous (Classified) Locations — Protection by Enclosure "tD".*

Informational Note No. 2: Protection by enclosure is designated level of protection "ta" for use in Zone 20 locations. Protection by enclosure is designated level of protection "tb" or "tD" for use in Zone 21 locations. Protection by enclosure is designated level of protection "tc" or "tD" for use in Zone 22 locations.

Protection by Intrinsic Safety "i". Type of protection where any spark or thermal effect is incapable of causing ignition of a mixture of combustible dust, fibers, or flyings in air under prescribed test conditions.

Informational Note No. 1: For additional information, see ANSI/ISA-60079-11 (12.01.01)-2014, *Electrical Apparatus for Explosive Gas Atmospheres — Part 11: intrinsic safety "i"*; ANSI/ UL 60079-11-2013, *Electrical Apparatus for Explosive Gas Atmospheres — Part 11: Intrinsic safety "i"*; and ANSI/ISA- 61241-11 (12.10.04)-2006, *Electrical Apparatus for Use in Zone 20, Zone 21 and Zone 22 Hazardous (Classified) Locations — Protection by Intrinsic Safety "iD."*

Informational Note No. 2: Intrinsic safety is designated level of protection "iaD" or "ia" for use in Zone 20 locations. Intrinsic safety is designated level of protection "ibD" or "ib" for use in Zone 21 locations. Intrinsic safety is designated type of protection "ic" for use in Zone 22 locations.

Protection by Pressurization "p". Type of protection that guards against the ingress of a mixture of combustible dust or fibers/flyings in air into an enclosure containing electrical equipment by providing and maintaining a protective gas atmosphere inside the enclosure at a pressure above that of the external atmosphere.

Informational Note: For additional information, see ANSI/ ISA 61241 2 (12.10.06)-2006, *Electrical Apparatus for Use in Zone 21 and Zone 22 Hazardous (Classified) Locations — Protection by Pressurization "pD."*

506.3 Other Articles. All other applicable rules contained in this *Code* shall apply to electrical equipment and wiring installed in hazardous (classified) locations.

Exception: As modified by Article 504 and this article.

506.4 General.

(A) Documentation for Industrial Occupancies. Areas designated as hazardous (classified) locations shall be properly documented. This documentation shall be available to those authorized to design, install, inspect, maintain, or operate electrical equipment.

(B) Reference Standards. Important information relating to topics covered in Chapter 5 can be found in other publications.

Informational Note No. 1: It is important that the authority having jurisdiction be familiar with the recorded industrial experience as well as with standards of the National Fire Protection Association (NFPA), the International Society of Automation (ISA), and the International Electrotechnical Commission (IEC) that may be of use in the classification of various locations, the determination of adequate ventilation, and the protection against static electricity and lightning hazards.

Informational Note No. 2: For additional information concerning the installation of equipment utilizing optical emissions technology (such as laser equipment) that could potentially become an ignition source in hazardous (classified) locations, see ANSI/ISA-60079-28 (12.21.02)-2013, *Explosive Atmospheres — Part 28: Protection of equipment and transmission systems using optical radiation.*

506.5 Classification of Locations.

(A) Classifications of Locations.   Locations shall be classified on the basis of the properties of the combustible dust or ignitible fibers/flyings that may be present, and the likelihood that a combustible or combustible concentration or quantity is present. Each room, section, or area shall be considered individually in determining its classification. Where pyrophoric materials are the only materials used or handled, these locations are outside of the scope of this article.

(B) Zone 20, Zone 21, and Zone 22 Locations.   Zone 20, Zone 21, and Zone 22 locations are those in which combustible dust or ignitible fibers/flyings are or may be present in the air or in layers, in quantities sufficient to produce explosive or ignitible mixtures. Zone 20, Zone 21, and Zone 22 locations shall include those specified in 506.5(B)(1), (B)(2), and (B)(3).

> Informational Note: Through the exercise of ingenuity in the layout of electrical installations for hazardous (classified) locations, it is frequently possible to locate much of the equipment in a reduced level of classification and, thus, to reduce the amount of special equipment required.

(1) Zone 20.   A Zone 20 location is a location in which either of the following occur:

(1)   Ignitible concentrations of combustible dust or ignitible fibers/flyings are present continuously.

(2)   Ignitible concentrations of combustible dust or ignitible fibers/flyings are present for long periods of time.

> Informational Note No. 1: As a guide to classification of Zone 20 locations, refer to ANSI/ISA-60079-10-2 (12.10.05)-2013, *Explosive Atmospheres — Part 10-2: Classification of areas — Combustible dust atmospheres.*

> Informational Note No. 2: Zone 20 classification includes locations inside dust containment systems; hoppers, silos, etc., cyclones and filters, dust transport systems, except some parts of belt and chain conveyors, etc.; blenders, mills, dryers, bagging equipment, etc.

(2) Zone 21.   A Zone 21 location is a location where one of the following apply:

(1)   Ignitible concentrations of combustible dust or ignitible fibers/flyings are likely to exist occasionally under normal operating conditions; or

(2)   Ignitible concentrations of combustible dust or ignitible fibers/flyings may exist frequently because of repair or maintenance operations or because of leakage; or

(3)   Equipment is operated or processes are carried on, of such a nature that equipment breakdown or faulty operations could result in the release of ignitible concentrations of combustible dust or ignitible fibers/flyings and also cause simultaneous failure of electrical equipment in a mode to cause the electrical equipment to become a source of ignition; or

(4)   The location is adjacent to a Zone 20 location from which ignitible concentrations of dust or ignitible fibers/flyings could be communicated.

Exception: When communication from an adjacent Zone 20 location is minimized by adequate positive pressure ventilation from a source of clean air, and effective safeguards against ventilation failure are provided.

> Informational Note No. 1: As a guide to classification of Zone 21 locations, refer to ANSI/ISA-60079-10-2 (12.10.05)-2013, *Explosive Atmospheres — Part 10-2: Classification of areas — Combustible dust atmospheres.*

> Informational Note No. 2: This classification usually includes locations outside dust containment and in the immediate vicinity of access doors subject to frequent removal or opening for operation purposes when internal combustible mixtures are present; locations outside dust containment in the proximity of filling and emptying points, feed belts, sampling points, truck dump stations, belt dump over points, etc., where no measures are employed to prevent the formation of combustible mixtures; locations outside dust containment where dust accumulates and where due to process operations the dust layer is likely to be disturbed and form combustible mixtures; locations inside dust containment where explosive dust clouds are likely to occur (but neither continuously, nor for long periods, nor frequently) as, for example, silos (if filled and/or emptied only occasionally) and the dirty side of filters if large self-cleaning intervals are occurring.

(3) Zone 22.   A Zone 22 location is a location where one of the following apply:

(1)   Ignitible concentrations of combustible dust or ignitible fibers/flyings are not likely to occur in normal operation and, if they do occur, will only persist for a short period; or

(2)   Combustible dust or fibers/flyings are handled, processed, or used but in which the dust or fibers/flyings are normally confined within closed containers of closed systems from which they can escape only as a result of the abnormal operation of the equipment with which the dust or fibers/flyings are handled, processed, or used; or

(3)   The location is adjacent to a Zone 21 location, from which ignitible concentrations of dust or fibers/flyings could be communicated.

Exception: When communication from an adjacent Zone 21 location is minimized by adequate positive pressure ventilation from a source of clean air, and effective safeguards against ventilation failure are provided.

> Informational Note No. 1: As a guide to classification of Zone 22 locations, refer to ANSI/ISA-60079-10-2 (12.10.05)-2013, *Explosive Atmospheres — Part 10-2: Classification of areas — Combustible dust atmospheres.*

> Informational Note No. 2: Zone 22 locations usually include outlets from bag filter vents, because in the event of a malfunction there can be emission of combustible mixtures; locations near equipment that has to be opened at infrequent intervals or equipment that from experience can easily form leaks where, due to pressure above atmospheric, dust will blow out; pneumatic equipment, flexible connections that can become damaged, etc.; storage locations for bags containing dusty product, since failure of bags can occur during handling, causing dust leakage; and locations where controllable dust layers are formed that are likely to be raised into explosive dust–air mixtures. Only if the layer is removed by cleaning before hazardous dust–air mixtures can be formed is the area designated unclassified.

> Informational Note No. 3: Locations that normally are classified as Zone 21 can fall into Zone 22 when measures are employed to prevent the formation of explosive dust–air mixtures. Such measures include exhaust ventilation. The measures should be used in the vicinity of (bag) filling and emptying points, feed

belts, sampling points, truck dump stations, belt dump over points, etc.

506.6 Material Groups. For the purposes of testing, approval, and area classification, various air mixtures (not oxygen enriched) shall be grouped as required in 506.6(A), (B), and (C).

(A) Group IIIC. Combustible metal dust. Group IIIC shall be considered to be equivalent to Class II, Group E.

(B) Group IIIB. Combustible dust other than combustible metal dust. Group IIIB shall be considered to be equivalent to Class II, Groups F and G.

(C) Group IIIA. Solid particles, including fibers, greater than 500 μm in nominal size, which could be suspended in air and could settle out of the atmosphere under their own weight. Group IIIA shall be considered to be equivalent to Class III.

Informational Note: Examples of flyings include rayon, cotton (including cotton linters and cotton waste), sisal, jute, hemp, cocoa fiber, oakum, and baled waste kapok.

506.7 Special Precaution. Article 506 requires equipment construction and installation that ensures safe performance under conditions of proper use and maintenance.

Informational Note: It is important that inspection authorities and users exercise more than ordinary care with regard to the installation and maintenance of electrical equipment in hazardous (classified) locations.

(A) Implementation of Zone Classification System. Classification of areas, engineering and design, selection of equipment and wiring methods, installation, and inspection shall be performed by qualified persons.

(B) Dual Classification. In instances of areas within the same facility classified separately, Zone 22 locations shall be permitted to abut, but not overlap, Class II or Class III, Division 2 locations. Zone 20 or Zone 21 locations shall not abut Class II or Class III, Division 1 or Division 2 locations.

(C) Reclassification Permitted. A Class II or Class III, Division 1 or Division 2 location shall be permitted to be reclassified as a Zone 20, Zone 21, or Zone 22 location, provided that all of the space that is classified because of a single combustible dust or ignitible fiber/flying source is reclassified under the requirements of this article.

(D) Simultaneous Presence of Flammable Gases and Combustible Dusts or Fibers/Flyings. Where flammable gases, combustible dusts, or fibers/flyings are or may be present at the same time, the simultaneous presence shall be considered during the selection and installation of the electrical equipment and the wiring methods, including the determination of the safe operating temperature of the electrical equipment.

506.8 Protection Techniques. Acceptable protection techniques for electrical and electronic equipment in hazardous (classified) locations shall be as described in 506.8(A) through (I).

(A) Dust Ignitionproof. This protection technique shall be permitted for equipment in Zone 20, Zone 21, and Zone 22 locations for which it is identified.

(B) Pressurized. This protection technique shall be permitted for equipment in Zone 21 and Zone 22 locations for which it is identified.

(C) Intrinsic Safety. This protection technique shall be permitted for equipment in Zone 20, Zone 21, and Zone 22 locations for which it is identified.

(D) Dusttight. This protection technique shall be permitted for equipment in Zone 22 locations for which it is identified.

(E) Protection by Encapsulation "m". This protection technique shall be permitted for equipment in Zone 20, Zone 21, and Zone 22 locations for which it is identified.

Informational Note: See Table 506.9(C)(2)(3) for the descriptions of subdivisions for encapsulation.

(F) Nonincendive Equipment. This protection technique shall be permitted for equipment in Zone 22 locations for which it is identified.

(G) Protection by Enclosure "t". This protection technique shall be permitted for equipment in Zone 20, Zone 21, and Zone 22 locations for which it is identified.

Informational Note: See Table 506.9(C)(2)(3) for the descriptions of subdivisions for protection by enclosure "t."

(H) Protection by Pressurization "pD". This protection technique shall be permitted for equipment in Zone 21 and Zone 22 locations for which it is identified.

(I) Protection by Intrinsic Safety "iD". This protection technique shall be permitted for equipment in Zone 20, Zone 21, and Zone 22 locations for which it is listed.

506.9 Equipment Requirements.

(A) Suitability. Suitability of identified equipment shall be determined by one of the following:

(1) Equipment listing or labeling
(2) Evidence of equipment evaluation from a qualified testing laboratory or inspection agency concerned with product evaluation
(3) Evidence acceptable to the authority having jurisdiction such as a manufacturer's self-evaluation or an owner's engineering judgment

Informational Note: Additional documentation for equipment may include certificates demonstrating compliance with applicable equipment standards, indicating special conditions of use, and other pertinent information.

(B) Listing. Equipment that is listed for Zone 20 shall be permitted in a Zone 21 or Zone 22 location of the same dust or ignitible fiber/flying. Equipment that is listed for Zone 21 may be used in a Zone 22 location of the same dust or ignitible fiber/flying.

(C) Marking.

(1) Division Equipment. Equipment identified for Class II, Division 1 or Class II, Division 2 shall, in addition to being marked in accordance with 500.8(C), be permitted to be marked with all of the following:

(1) Zone 20, 21, or 22 (as applicable)
(2) Material group in accordance with 506.6
(3) Maximum surface temperature in accordance with 506.9(D), marked as a temperature value in degrees C, preceded by "T" and followed by the symbol ""°C"

(2) Zone Equipment. Equipment meeting one or more of the protection techniques described in 506.8 shall be marked with the following in the order shown:

(1) Zone
(2) Symbol "AEx"
(3) Protection technique(s) in accordance with Table 506.9(C)(2)(3)
(4) Material group in accordance with 506.6
(5) Maximum surface temperature in accordance with 506.9(D), marked as a temperature value in degrees C, preceded by "T" and followed by the symbol "°C"
(6) Ambient temperature marking in accordance with 506.9(D)

Informational Note: The EPL (or equipment protection level) can appear in the product marking. EPLs are designated as G for gas, D for dust, or M for mining, and are then followed by a letter (a, b, or c) to give the user a better understanding as to whether the equipment provides (a) a "very high," (b) a "high," or (c) an "enhanced" level of protection against ignition of an explosive atmosphere. For example, a Zone 21 AEx pb IIIB T165°C motor can additionally be marked with an EPL of "Db", Zone 21 AEx p IIIB T165°C Db.

Exception: Associated apparatus NOT suitable for installation in a hazardous (classified) location shall be required to be marked only with 506.9(C)(2)(2) and (3), and where applicable (4), but BOTH the symbol AEx in 506.9(C)(2)(2) and the symbol for the type of protection in 506.9(C)(2)(3) shall be enclosed within the same square brackets; for example, [AEx iaD] or [AEx ia] IIIC.

Table 506.9(C)(2)(3) Types of Protection Designation

Designation	Technique	Zone*
iaD	Protection by intrinsic safety	20
ia	Protection by intrinsic safety	20
ibD	Protection by intrinsic safety	21
ib	Protection by intrinsic safety	21
ic	Protection by intrinsic safety	22
[iaD]	Associated apparatus	Unclassified**
[ia]	Associated apparatus	Unclassified**
[ibD]	Associated apparatus	Unclassified**
[ib]	Associated apparatus	Unclassified**
[ic]	Associated apparatus	Unclassified**
maD	Protection by encapsulation	20
ma	Protection by encapsulation	20
mbD	Protection by encapsulation	21
mb	Protection by encapsulation	21
mc	Protection by encapsulation	22
pD	Protection by pressurization	21
p	Protection by pressurization	21
pb	Protection by pressurization	21
tD	Protection by enclosures	21
ta	Protection by enclosures	20
tb	Protection by enclosures	21
tc	Protection by enclosures	22

*Does not address use where a combination of techniques is used.
**Associated apparatus is permitted to be installed in a hazardous (classified) location if suitably protected using another type of protection.

Informational Note: The "D" suffix on the type of protection designation was employed prior to the introduction of Group IIIA, IIIB, and IIIC; which is now used to distinguish between the type of protection employed for Group II (Gases) or Group III (Dusts).

(D) Temperature Classifications. Equipment shall be marked to show the maximum surface temperature referenced to a 40°C ambient, or at the higher marked ambient temperature if the equipment is rated and marked for an ambient temperature of greater than 40°C. For equipment installed in a Zone 20 or Zone 21 location, the operating temperature shall be based on operation of the equipment when blanketed with the maximum amount of dust (or with dust-simulating fibers/flyings) that can accumulate on the equipment. Electrical equipment designed for use in the ambient temperature range between -20°C and +40°C shall require no additional ambient temperature marking. Electrical equipment that is designed for use in a range of ambient temperatures other than -20°C and +40°C is considered to be special; and the ambient temperature range shall then be marked on the equipment, including either the symbol "Ta" or "Tamb" together with the special range of ambient temperatures.

Informational Note: As an example, such a marking might be "–30°C ≤ Ta ≤ +40°C."

Exception No. 1: Equipment of the non–heat-producing type, such as conduit fittings, shall not be required to have a marked operating temperature.

Exception No. 2: Equipment identified for Class II, Division 1 or Class II, Division 2 locations as permitted by 506.20(B) and (C) shall be permitted to be marked in accordance with 500.8(C) and Table 500.8(C).

(E) Threading. The supply connection entry thread form shall be NPT or metric. Conduit and fittings shall be made wrenchtight to prevent sparking when the fault current flows through the conduit system and to ensure the integrity of the conduit system. Equipment provided with threaded entries for field wiring connections shall be installed in accordance with 506.9(E)(1) or (E)(2) and with (E)(3).

(1) Equipment Provided with Threaded Entries for NPT-Threaded Conduit or Fittings. For equipment provided with threaded entries for NPT-threaded conduit or fittings, listed conduit fittings or listed cable fittings shall be used. All NPT-threaded conduit and fittings shall be threaded with a National (American) Standard Pipe Taper (NPT) thread.

Informational Note: Thread specifications for NPT threads are located in ASME B1.20.1-2013, *Pipe Threads, General Purpose (Inch)*.

(2) Equipment Provided with Threaded Entries for Metric-Threaded Fittings. For equipment with metric-threaded entries, listed conduit fittings or listed cable fittings shall be used. Such entries shall be identified as being metric, or listed adapters to permit connection to conduit or NPT-threaded fittings shall be provided with the equipment and shall be used for connection to conduit or NPT-threaded fittings. Metric-threaded fittings installed into equipment entries shall be made up with at least five threads fully engaged.

(3) Unused Openings. All unused openings shall be closed with listed metal close-up plugs. The plug engagement shall comply with 506.9(E)(1) or (E)(2).

(F) Optical Fiber Cables. An optical fiber cable, with or without current-carrying conductors (composite optical fiber cable), shall be installed to address the associated fire hazard and sealed to address the associated explosion hazard in accordance with the requirements of 506.15 and 506.16.

506.15 Wiring Methods. Wiring methods shall maintain the integrity of the protection techniques and shall comply with 506.15(A), (B), or (C).

(A) Zone 20. In Zone 20 locations, the following wiring methods shall be permitted:

(1) Threaded rigid metal conduit or threaded steel intermediate metal conduit.

(2) Type MI cable terminated with fittings listed for the location. Type MI cable shall be installed and supported in a manner to avoid tensile stress at the termination fittings.

Exception No. 1: MI cable and fittings listed for Class II, Division 1 locations shall be permitted to be used.

Exception No. 2: Equipment identified as intrinsically safe "iaD" or "ia" shall be permitted to be connected using the wiring methods identified in 504.20.

(3) In industrial establishments with limited public access, where the conditions of maintenance and supervision ensure that only qualified persons service the installation, Type MC-HL cable listed for use in Zone 20 locations, with a continuous corrugated metallic sheath, an overall jacket of suitable polymeric material, and a separate equipment grounding conductor(s) in accordance with 250.122, and terminated with fittings listed for the application, shall be permitted. Type MC-HL cable shall be installed in accordance with the provisions of Article 330, Part II.

Exception: Type MC-HL cable and fittings listed for Class II, Division 1 locations shall be permitted to be used.

(4) In industrial establishments with restricted public access, where the conditions of maintenance and supervision ensure that only qualified persons service the installation, and where the cable is not subject to physical damage, Type ITC-HL cable listed for use in Zone 1 or Class I, Division 1 locations, with a gas/vaportight continuous corrugated metallic sheath and an overall jacket of suitable polymeric material, and terminated with fittings listed for the application. Type ITC-HL cable shall be installed in accordance with the provisions of Article 727.

(5) Fittings and boxes shall be identified for use in Zone 20 locations.

Exception: Boxes and fittings listed for Class II, Division 1 locations shall be permitted to be used.

(6) Where necessary to employ flexible connections, liquidtight flexible metal conduit with listed fittings, liquidtight flexible nonmetallic conduit with listed fittings, or flexible cord listed for extra-hard usage and provided with listed fittings shall be used. Where flexible cords are used, they shall also comply with 506.17 and shall be terminated with a listed cord connector that maintains the type of protection of the terminal compartment. Where flexible connections are subject to oil or other corrosive conditions, the insulation of the conductors shall be of a type listed for the condition or shall be protected by means of a suitable sheath.

Exception No. 1: Flexible conduit and flexible conduit and cord fittings listed for Class II, Division 1 locations shall be permitted to be used.

Exception No. 2: For elevator use, an identified elevator cable of Type EO, ETP, or ETT, shown under the "use" column in Table 400.4 for "hazardous (classified) locations," and terminated with listed connectors that maintain the type of protection of the terminal compartment shall be permitted.

Informational Note No. 1: See 506.25 for grounding requirements where flexible conduit is used.

Informational Note No. 2: For further information on construction, testing, and marking of cables, cable fittings, and cord connectors, see ANSI/UL 2225-2011, *Cables and Cable-Fittings for Use in Hazardous (Classified) Locations.*

(7) Optical fiber cable Types OFNP, OFCP, OFNR, OFCR, OFNG, OFCG, OFN, and OFC shall be permitted to be installed in raceways in accordance with 506.15(A). Optical fiber cables shall be sealed in accordance with 506.16.

(B) Zone 21. In Zone 21 locations, the wiring methods in (B)(1) and (B)(2) shall be permitted:

(1) All wiring methods permitted in 506.15(A).

(2) Fittings and boxes that are dusttight, provided with threaded bosses for connection to conduit, in which taps, joints, or terminal connections are not made, and are not used in locations where metal dust is present, may be used.

Informational Note: For further information on construction, testing, and marking of cables, cable fittings, and cord connectors, see ANSI/UL 2225-2011, *Cables and Cable-Fittings for Use in Hazardous (Classified) Locations.*

Exception: Equipment identified as intrinsically safe "ibD" or "ib" shall be permitted to be connected using the wiring methods identified in 504.20.

(C) Zone 22. In Zone 22 locations, the following wiring methods shall be permitted:

(1) All wiring methods permitted in 506.15(B).

(2) Rigid metal conduit, intermediate metal conduit, electrical metallic tubing, dusttight wireways.

(3) Type MC or MI cable with listed termination fittings.

(4) Type PLTC and Type PLTC-ER cable in accordance with the provisions of Article 725, including installation in cable tray systems. The cable shall be terminated with listed fittings.

(5) Type ITC and Type ITC-ER cable as permitted in 727.4 and terminated with listed fittings.

(6) Type MC, MI, MV, TC, or TC-ER cable installed in ladder, ventilated trough, or ventilated channel cable trays in a single layer, with a space not less than the larger cable diameter between two adjacent cables, shall be the wiring method employed. Single-conductor Type MV cables shall be shielded or metallic armored. The cable shall be terminated with listed fittings.

(7) Intrinsic safety type of protection "ic" shall be permitted using any of the wiring methods permitted for unclassified locations. Intrinsic safety type of protection "ic" systems shall be installed in accordance with the control drawing(s). Simple apparatus, not shown on the control drawing, shall be permitted in a circuit of intrinsic safety type of protection "ic", provided that the simple apparatus does not interconnect the intrinsic safety type of protection "ic" circuit to any other circuit.

Informational Note: The term *Simple Apparatus* is defined in 504.2.

Separation of circuits of intrinsic safety type of protection "ic" shall be in accordance with one of the following:

a. Be in separate cables

b. Be in multiconductor cables where the conductors of each circuit are within a grounded metal shield

c. Be in multiconductor cables where the conductors have insulation with a minimum thickness of 0.25 mm (0.01 in.)

(8) Boxes and fittings shall be dusttight.

(9) Optical fiber cable Types OFNP, OFCP, OFNR, OFCR, OFNG, OFCG, OFN, and OFC shall be permitted to be installed in cable trays or any raceway in accordance with 506.15(C). Optical fiber cables shall be sealed in accordance with 506.16.

(10) Cablebus.

506.16 Sealing. Where necessary to protect the ingress of combustible dust or ignitible fibers/flyings, or to maintain the type of protection, seals shall be provided. The seal shall be identified as capable of preventing the ingress of combustible dust or ignitible fibers/flyings and maintaining the type of protection but need not be explosionproof or flameproof.

506.17 Flexible Cords. Flexible cords used in Zone 20, Zone 21, and Zone 22 locations shall comply with all of the following:

(1) Be of a type listed for extra-hard usage

(2) Contain, in addition to the conductors of the circuit, an equipment grounding conductor complying with 400.23

(3) Be connected to terminals or to supply conductors in an approved manner

(4) Be supported by clamps or by other suitable means in such a manner to minimize tension on the terminal connections

(5) Be terminated with a listed cord connector that maintains the protection technique of the terminal compartment

Informational Note: For further information on construction, testing, and marking of cables, cable fittings, and cord connectors, see ANSI/UL 2225-2011, *Cables and Cable-Fittings for Use in Hazardous (Classified) Locations.*

506.20 Equipment Installation.

(A) Zone 20. In Zone 20 locations, only equipment listed and marked as suitable for the location shall be permitted.

Exception: Equipment listed for use in Class II, Division 1 locations with a suitable temperature class shall be permitted.

(B) Zone 21. In Zone 21 locations, only equipment listed and marked as suitable for the location shall be permitted.

Exception No. 1: Apparatus listed for use in Class II, Division 1 locations with a suitable temperature class shall be permitted.

Exception No. 2: Pressurized equipment identified for Class II, Division 1 shall be permitted.

(C) Zone 22. In Zone 22 locations, only equipment listed and marked as suitable for the location shall be permitted.

Exception No. 1: Apparatus listed for use in Class II, Division 1 or Class II, Division 2 locations with a suitable temperature class shall be permitted.

Exception No. 2: Pressurized equipment identified for Class II, Division 1 or Division 2 shall be permitted.

(D) Material Group. Equipment marked Group IIIC shall be permitted for applications requiring IIIA or IIIB equipment. Similarly, equipment marked Group IIIB shall be permitted for applications requiring IIIA equipment.

(E) Manufacturer's Instructions. Electrical equipment installed in hazardous (classified) locations shall be installed in accordance with the instructions (if any) provided by the manufacturer.

(F) Temperature. The temperature marking specified in 506.9(C)(2)(5) shall comply with (F)(1) or (F)(2):

(1) For combustible dusts, less than the lower of either the layer or cloud ignition temperature of the specific combustible dust. For organic dusts that may dehydrate or carbonize, the temperature marking shall not exceed the lower of either the ignition temperature or 165°C (329°F).

(2) For ignitible fibers/flyings, less than 165°C (329°F) for equipment that is not subject to overloading, or 120°C (248°F) for equipment (such as motors or power transformers) that may be overloaded.

Informational Note: See NFPA 499-2013, *Recommended Practice for the Classification of Combustible Dusts and of Hazardous (Classified) Locations for Electrical Installations in Chemical Processing Areas,* for minimum ignition temperatures of specific dusts.

506.25 Grounding and Bonding. Regardless of the voltage of the electrical system, grounding and bonding shall comply with Article 250 and the requirements in 506.25(A) and (B).

(A) Bonding. The locknut-bushing and double-locknut types of contacts shall not be depended on for bonding purposes, but bonding jumpers with proper fittings or other approved means of bonding shall be used. Such means of bonding shall apply to all intervening raceways, fittings, boxes, enclosures, and so forth, between Zone 20, Zone 21, and Zone 22 locations and the point of grounding for service equipment or point of grounding of a separately derived system.

Exception: The specific bonding means shall be required only to the nearest point where the grounded circuit conductor and the grounding electrode conductor are connected together on the line side of the building or structure disconnecting means as specified in 250.32(B) if the branch side overcurrent protection is located on the load side of the disconnecting means.

(B) Types of Equipment Grounding Conductors. Liquidtight flexible metal conduit shall include an equipment bonding jumper of the wire type in compliance with 250.102.

Exception: In Zone 22 locations, the bonding jumper shall be permitted to be deleted where all of the following conditions are met:

(1) *Listed liquidtight flexible metal conduit 1.8 m (6 ft) or less in length, with fittings listed for grounding, is used.*

(2) *Overcurrent protection in the circuit is limited to 10 amperes or less.*

(3) *The load is not a power utilization load.*

ARTICLE 510
Hazardous (Classified) Locations — Specific

510.1 Scope. Articles 511 through 517 cover occupancies or parts of occupancies that are or may be hazardous because of atmospheric concentrations of flammable liquids, gases, or vapors, or because of deposits or accumulations of materials that may be readily ignitible.

510.2 General. The general rules of this *Code* and the provisions of Articles 500 through 504 shall apply to electrical wiring and equipment in occupancies within the scope of Articles 511 through 517, except as such rules are modified in Articles 511 through 517. Where unusual conditions exist in a specific occupancy, the authority having jurisdiction shall judge with respect to the application of specific rules.

ARTICLE 511
Commercial Garages, Repair and Storage

Informational Note: Text that is followed by a reference in brackets has been extracted from NFPA 30A-2015, *Code for Motor Fuel Dispensing Facilities and Repair Garages*. Only editorial changes were made to the extracted text to make it consistent with this *Code*.

511.1 Scope. These occupancies shall include locations used for service and repair operations in connection with self-propelled vehicles (including, but not limited to, passenger automobiles, buses, trucks, and tractors) in which volatile flammable liquids or flammable gases are used for fuel or power.

511.2 Definitions.

Major Repair Garage. A building or portions of a building where major repairs, such as engine overhauls, painting, body and fender work, and repairs that require draining of the motor vehicle fuel tank are performed on motor vehicles, including associated floor space used for offices, parking, or showrooms. [**30A**:3.3.12.1]

Minor Repair Garage. A building or portions of a building used for lubrication, inspection, and minor automotive maintenance work, such as engine tune-ups, replacement of parts, fluid changes (e.g., oil, antifreeze, transmission fluid, brake fluid, air-conditioning refrigerants), brake system repairs, tire rotation, and similar routine maintenance work, including associated floor space used for offices, parking, or showrooms. [**30A**:3.3.12.2]

511.3 Area Classification, General. Where Class I liquids or gaseous fuels are stored, handled, or transferred, electrical wiring and electrical utilization equipment shall be designed in accordance with the requirements for Class I, Division 1 or 2 hazardous (classified) locations as classified in accordance with 500.5 and 500.6, and this article. A Class I location shall not extend beyond an unpierced wall, roof, or other solid partition that has no openings. [**30A**: 8.3.1, 8.3.3]

(A) Parking Garages. Parking garages used for parking or storage shall be permitted to be unclassified.

Informational Note: For further information, see NFPA 88A-2015, *Standard for Parking Structures*, and NFPA 30A-2015, *Code for Motor Fuel Dispensing Facilities and Repair Garages.*

(B) Repair Garages, with Dispensing. Major and minor repair garages that dispense motor fuels into the fuel tanks of vehicles, including flammable liquids having a flash point below 38°C (100°F) such as gasoline, or gaseous fuels such as natural

gas, hydrogen, or LPG, shall have the dispensing functions and components classified in accordance with Table 514.3(B)(1) in addition to any classification required by this section. Where Class I liquids, other than fuels, are dispensed, the area within 900 mm (3 ft) of any fill or dispensing point, extending in all directions, shall be a Class I, Division 2 location.

(C) Repair Garages, Major and Minor. Where vehicles using Class I liquids or heavier-than-air gaseous fuels (such as LPG) are repaired, hazardous area classification guidance is found in Table 511.3(C).

Informational Note: For additional information, see NFPA 30A-2015, *Code for Motor Fuel Dispensing Facilities and Repair Garages*, Table 8.3.2.

(D) Repair Garages, Major. Where vehicles using lighter-than-air gaseous fuels (such as hydrogen and natural gas) are repaired or stored, hazardous area classification guidance is found in Table 511.3(D).

Informational Note: For additional information see NFPA 30A-2015, *Code for Motor Fuel Dispensing Facilities and Repair Garages*, Table 8.3.2.

(E) Modifications to Classification.

(1) Specific Areas Adjacent to Classified Locations. Areas adjacent to classified locations in which flammable vapors are not likely to be released, such as stock rooms, switchboard rooms, and other similar locations, shall be unclassified where mechanically ventilated at a rate of four or more air changes per hour, or designed with positive air pressure, or where effectively cut off by walls or partitions.

(2) Alcohol-Based Windshield Washer Fluid. The area used for storage, handling, or dispensing into motor vehicles of alcohol-based windshield washer fluid in repair garages shall be unclassified unless otherwise classified by a provision of 511.3. [**30A**: 8.3.1, Exception]

511.4 Wiring and Equipment in Class I Locations.

(A) Wiring Located in Class I Locations. Within Class I locations as classified in 511.3, wiring shall conform to applicable provisions of Article 501.

(B) Equipment Located in Class I Locations. Within Class I locations as defined in 511.3, equipment shall conform to applicable provisions of Article 501.

(1) Fuel-Dispensing Units. Where fuel-dispensing units (other than liquid petroleum gas, which is prohibited) are located within buildings, the requirements of Article 514 shall govern.

Where mechanical ventilation is provided in the dispensing area, the control shall be interlocked so that the dispenser cannot operate without ventilation, as prescribed in 500.5(B)(2).

(2) Portable Lighting Equipment. Portable lighting equipment shall be equipped with handle, lampholder, hook, and substantial guard attached to the lampholder or handle. All exterior surfaces that might come in contact with battery terminals, wiring terminals, or other objects shall be of nonconducting material or shall be effectively protected with insulation. Lampholders shall be of an unswitched type and shall not provide means for plug-in of attachment plugs. The outer shell shall be of molded composition or other suitable material. Unless the lamp and its cord are supported or arranged in such

N **Table 511.3(C) Extent of Classified Locations for Major and Minor Repair Garages with Heavier-Than-Air Fuel**

Location	Class I Division (Group D)	Class I Zone (Group IIA)	Extent of Classified Location
Repair garage, major (where Class I liquids or gaseous fuels are transferred or dispensed*)	1	1	Entire space within any pit, belowgrade work area, or subfloor work area that is not ventilated
	2	2	Entire space within any pit, belowgrade work area, or subfloor work area that is provided with ventilation of at least 0.3 m³/min/m² (1 ft³/min/ft²) of floor area, with suction taken from a point within 300 mm (12 in.) of floor level
	2	2	Up to 450 mm (18 in.) above floor level of the room, except as noted below, for entire floor area
	Unclassified	Unclassified	Up to 450 mm (18 in.) above floor level of the room where room is provided with ventilation of at least 0.3 m³/min/m² (1 ft³/min/ft²) of floor area, with suction taken from a point within 300 mm (12 in.) of floor level
	2	2	Within 0.9 m (3 ft) of any fill or dispensing point, extending in all directions
Specific areas adjacent to classified locations	Unclassified	Unclassified	Areas adjacent to classified locations where flammable vapors are not likely to be released, such as stock rooms, switchboard rooms, and other similar locations, where mechanically ventilated at a rate of four or more air changes per hour or designed with positive air pressure or where effectively cut off by walls or partitions
Repair garage, minor (where Class I liquids or gaseous fuels are not transferred or dispensed*)	2	2	Entire space within any pit, belowgrade work area, or subfloor work area that is not ventilated
	2	2	Up to 450 mm (18 in.) above floor level, extending 0.9 m (3 ft) horizontally in all directions from opening to any pit, belowgrade work area, or subfloor work area that is not ventilated
	Unclassified	Unclassified	Entire space within any pit, belowgrade work area, or subfloor work area that is provided with ventilation of at least 0.3 m³/min/m² (1 ft³/min/ft²) of floor area, with suction taken from a point within 300 mm (12 in.) of floor level
Specific areas adjacent to classified locations	Unclassified	Unclassified	Areas adjacent to classified locations where flammable vapors are not likely to be released, such as stock rooms, switchboard rooms, and other similar locations, where mechanically ventilated at a rate of four or more air changes per hour or designed with positive air pressure, or where effectively cut off by walls or partitions

*Includes draining of Class I liquids from vehicles.

a manner that they cannot be used in the locations classified in 511.3, they shall be of a type identified for Class I, Division 1 locations.

511.7 Wiring and Equipment Installed Above Class I Locations.

(A) Wiring in Spaces Above Class I Locations.

(1) Fixed Wiring Above Class I Locations. All fixed wiring above Class I locations shall be in metal raceways, rigid nonmetallic conduit, electrical nonmetallic tubing, flexible metal conduit, liquidtight flexible metal conduit, or liquidtight flexible nonmetallic conduit, or shall be Type MC, AC, MI, manufactured wiring systems, or PLTC cable in accordance with Article 725, or Type TC cable or Type ITC cable in accordance with Article 727. Cellular metal floor raceways or cellular concrete floor raceways shall be permitted to be used only for supplying ceiling outlets or extensions to the area below the floor, but such raceways shall have no connections leading into or through any Class I location above the floor.

(2) Pendant. For pendants, flexible cord suitable for the type of service and listed for hard usage shall be used.

Table 511.3(D) Extent of Classified Locations for Major Repair Garages with Lighter-than-Air Fuel

| Location | Class I | | Extent of Classified Location |
	Division[2]	Zone[3]	
Repair garage, major (where lighter-than-air gaseous fueled[1] vehicles are repaired or stored)	2	2	Within 450 mm (18 in.) of ceiling, except as noted below
	Unclassified	Unclassified	Within 450 mm (18 in.) of ceiling where ventilation of at least 0.3 m^3/min/m^2 (1 ft^3/min/ft^2) of floor area, with suction taken from a point within 450 mm (18 in.) of the highest point in the ceiling
Specific areas adjacent to classified locations	Unclassified	Unclassified	Areas adjacent to classified locations where flammable vapors are not likely to be released, such as stock rooms, switchboard rooms, and other similar locations, where mechanically ventilated at a rate of four or more air changes per hour or designed with positive air pressure, or where effectively cut off by walls or partitions

[1]Includes fuels such as hydrogen and natural gas, but not LPG.
[2]For hydrogen (lighter than air) Group B, or natural gas Group D.
[3]For hydrogen (lighter than air) Group IIC or IIB+H2, or natural gas Group IIA.

(B) Electrical Equipment Installed Above Class I Locations.

(1) Fixed Electrical Equipment. Electrical equipment in a fixed position shall be located above the level of any defined Class I location or shall be identified for the location.

(a) *Arcing Equipment.* Equipment that is less than 3.7 m (12 ft) above the floor level and that may produce arcs, sparks, or particles of hot metal, such as cutouts, switches, charging panels, generators, motors, or other equipment (excluding receptacles, lamps, and lampholders) having make-and-break or sliding contacts, shall be of the totally enclosed type or constructed so as to prevent the escape of sparks or hot metal particles.

(b) *Fixed Lighting.* Lamps and lampholders for fixed lighting that is located over lanes through which vehicles are commonly driven or that may otherwise be exposed to physical damage shall be located not less than 3.7 m (12 ft) above floor level, unless of the totally enclosed type or constructed so as to prevent escape of sparks or hot metal particles.

511.8 Underground Wiring. Underground wiring shall be installed in threaded rigid metal conduit or intermediate metal conduit.

Exception: Type PVC conduit, Type RTRC conduit, and Type HDPE conduit shall be permitted where buried under not less than 600 mm (2 ft) of cover. Where Type PVC conduit, Type RTRC conduit, or Type HDPE conduit is used, threaded rigid metal conduit or threaded steel intermediate metal conduit shall be used for the last 600 mm (2 ft) of the underground run to emergence or to the point of connection to the aboveground raceway, and an equipment grounding conductor shall be included to provide electrical continuity of the raceway system and for grounding of non–current-carrying metal parts.

511.9 Sealing. Seals complying with the requirements of 501.15 and 501.15(B)(2) shall be provided and shall apply to horizontal as well as vertical boundaries of the defined Class I locations.

511.10 Special Equipment.

(A) Battery Charging Equipment. Battery chargers and their control equipment, and batteries being charged, shall not be located within locations classified in 511.3.

(B) Electric Vehicle Charging Equipment.

(1) General. All electrical equipment and wiring shall be installed in accordance with Article 625, except as noted in 511.10(B)(2) and (B)(3). Flexible cords shall be of a type identified for extra-hard usage.

(2) Connector Location. No connector shall be located within a Class I location as defined in 511.3.

(3) Plug Connections to Vehicles. Where the cord is suspended from overhead, it shall be arranged so that the lowest point of sag is at least 150 mm (6 in.) above the floor. Where an automatic arrangement is provided to pull both cord and plug beyond the range of physical damage, no additional connector shall be required in the cable or at the outlet.

511.12 Ground-Fault Circuit-Interrupter Protection for Personnel. All 125-volt, single-phase, 15- and 20-ampere receptacles installed in areas where electrical diagnostic equipment, electrical hand tools, or portable lighting equipment are to be used shall have ground-fault circuit-interrupter protection for personnel.

511.16 Grounding and Bonding Requirements.

(A) General Grounding Requirements. All metal raceways, the metal armor or metallic sheath on cables, and all non–current-carrying metal parts of fixed or portable electrical equipment, regardless of voltage, shall be grounded.

(B) Supplying Circuits with Grounded and Grounding Conductors in Class I Locations. Grounding in Class I locations shall comply with 501.30.

(1) Circuits Supplying Portable Equipment or Pendants. Where a circuit supplies portables or pendants and includes a grounded conductor as provided in Article 200, receptacles, attachment plugs, connectors, and similar devices shall be of the grounding type, and the grounded conductor of the flexible cord shall be connected to the screw shell of any lampholder or to the grounded terminal of any utilization equipment supplied.

(2) Approved Means. Approved means shall be provided for maintaining continuity of the equipment grounding conductor between the fixed wiring system and the non–current-carrying metal portions of pendant luminaires, portable luminaires, and portable utilization equipment.

ARTICLE 513
Aircraft Hangars

513.1 Scope. This article shall apply to buildings or structures in any part of which aircraft containing Class I (flammable) liquids or Class II (combustible) liquids whose temperatures are above their flash points are housed or stored and in which aircraft might undergo service, repairs, or alterations. It shall not apply to locations used exclusively for aircraft that have never contained fuel or unfueled aircraft.

Informational Note No. 1: For definitions of aircraft hangar and unfueled aircraft, see NFPA 409-2011, *Standard on Aircraft Hangars.*

Informational Note No. 2: For further information on fuel classification see NFPA 30 -2015, *Flammable and Combustible Liquids Code.*

513.2 Definitions. For the purpose of this article, the following definitions shall apply.

Aircraft Painting Hangar. An aircraft hangar constructed for the express purpose of spray/coating/dipping applications and provided with dedicated ventilation supply and exhaust.

513.3 Classification of Locations.

(A) Below Floor Level. Any pit or depression below the level of the hangar floor shall be classified as a Class I, Division 1 or Zone 1 location that shall extend up to said floor level.

(B) Areas Not Cut Off or Ventilated. The entire area of the hangar, including any adjacent and communicating areas not suitably cut off from the hangar, shall be classified as a Class I, Division 2 or Zone 2 location up to a level 450 mm (18 in.) above the floor.

(C) Vicinity of Aircraft.

(1) Aircraft Maintenance and Storage Hangars. The area within 1.5 m (5 ft) horizontally from aircraft power plants or aircraft fuel tanks shall be classified as a Class I, Division 2 or Zone 2 location that shall extend upward from the floor to a level 1.5 m (5 ft) above the upper surface of wings and of engine enclosures.

(2) Aircraft Painting Hangars. The area within 3 m (10 ft) horizontally from aircraft surfaces from the floor to 3 m (10 ft) above the aircraft shall be classified as Class I, Division 1 or

Class I, Zone 1. The area horizontally from aircraft surfaces between 3.0 m (10 ft) and 9.0 m (30 ft) from the floor to 9.0 m (30 ft) above the aircraft surface shall be classified as Class I, Division 2 or Class I, Zone 2.

Informational Note: See NFPA 33-2015, *Standard for Spray Application Using Flammable or Combustible Materials,* for information on ventilation and grounding for static protection in spray painting areas.

(D) Areas Suitably Cut Off and Ventilated. Adjacent areas in which flammable liquids or vapors are not likely to be released, such as stock rooms, electrical control rooms, and other similar locations, shall be unclassified where adequately ventilated and where effectively cut off from the hangar itself by walls or partitions.

513.4 Wiring and Equipment in Class I Locations.

(A) General. All wiring and equipment that is or may be installed or operated within any of the Class I locations defined in 513.3 shall comply with the applicable provisions of Article 501 or Article 505 for the division or zone in which they are used.

Attachment plugs and receptacles in Class I locations shall be identified for Class I locations or shall be designed such that they cannot be energized while the connections are being made or broken.

(B) Stanchions, Rostrums, and Docks. Electrical wiring, outlets, and equipment (including lamps) on or attached to stanchions, rostrums, or docks that are located or likely to be located in a Class I location, as defined in 513.3(C), shall comply with the applicable provisions of Article 501 or Article 505 for the division or zone in which they are used.

513.7 Wiring and Equipment Not Installed in Class I Locations.

(A) Fixed Wiring. All fixed wiring in a hangar but not installed in a Class I location as classified in 513.3 shall be installed in metal raceways or shall be Type MI, TC, or MC cable.

Exception: Wiring in unclassified locations, as described in 513.3(D), shall be permitted to be any suitable type wiring method recognized in Chapter 3.

(B) Pendants. For pendants, flexible cord suitable for the type of service and identified for hard usage or extra-hard usage shall be used. Each such cord shall include a separate equipment grounding conductor.

(C) Arcing Equipment. In locations above those described in 513.3, equipment that is less than 3.0 m (10 ft) above wings and engine enclosures of aircraft and that may produce arcs, sparks, or particles of hot metal, such as lamps and lampholders for fixed lighting, cutouts, switches, receptacles, charging panels, generators, motors, or other equipment having make-and-break or sliding contacts, shall be of the totally enclosed type or constructed so as to prevent the escape of sparks or hot metal particles.

Exception: Equipment in areas described in 513.3(D) shall be permitted to be of the general-purpose type.

(D) Lampholders. Lampholders of metal-shell, fiber-lined types shall not be used for fixed incandescent lighting.

(E) Stanchions, Rostrums, or Docks. Where stanchions, rostrums, or docks are not located or likely to be located in a Class I location, as defined in 513.3(C), wiring and equipment shall comply with 513.7, except that such wiring and equipment not more than 457 mm (18 in.) above the floor in any position shall comply with 513.4(B). Receptacles and attachment plugs shall be of a locking type that will not readily disconnect.

(F) Mobile Stanchions. Mobile stanchions with electrical equipment complying with 513.7(E) shall carry at least one permanently affixed warning sign with the following words or equivalent:

> **WARNING**
> **KEEP 5 FT CLEAR OF AIRCRAFT ENGINES AND FUEL TANK AREAS**

or

> **WARNING**
> **KEEP 1.5 METERS CLEAR OF AIRCRAFT ENGINES AND FUEL TANK AREAS**

513.8 Underground Wiring.

(A) Wiring and Equipment Embedded, Under Slab, or Underground. All wiring installed in or under the hangar floor shall comply with the requirements for Class I, Division 1 locations. Where such wiring is located in vaults, pits, or ducts, adequate drainage shall be provided.

(B) Uninterrupted Raceways, Embedded, Under Slab, or Underground. Uninterrupted raceways that are embedded in a hangar floor or buried beneath the hangar floor shall be considered to be within the Class I location above the floor, regardless of the point at which the raceway descends below or rises above the floor.

513.9 Sealing. Seals shall be provided in accordance with 501.15 or 505.16, as applicable. Sealing requirements specified shall apply to horizontal as well as to vertical boundaries of the defined Class I locations.

513.10 Special Equipment.

(A) Aircraft Electrical Systems.

(1) De-energizing Aircraft Electrical Systems. Aircraft electrical systems shall be de-energized when the aircraft is stored in a hangar and, whenever possible, while the aircraft is undergoing maintenance.

(2) Aircraft Batteries. Aircraft batteries shall not be charged where installed in an aircraft located inside or partially inside a hangar.

(B) Aircraft Battery Charging and Equipment. Battery chargers and their control equipment shall not be located or operated within any of the Class I locations defined in 513.3 and shall preferably be located in a separate building or in an area such as defined in 513.3(D). Mobile chargers shall carry at least one permanently affixed warning sign with the following words or equivalent:

> **WARNING**
> **KEEP 5 FT CLEAR OF AIRCRAFT ENGINES AND FUEL TANK AREAS**

or

> **WARNING**
> **KEEP 1.5 METERS CLEAR OF AIRCRAFT ENGINES AND FUEL TANK AREAS**

Tables, racks, trays, and wiring shall not be located within a Class I location and, in addition, shall comply with Article 480.

(C) External Power Sources for Energizing Aircraft.

(1) Not Less Than 450 mm (18 in.) Above Floor. Aircraft energizers shall be designed and mounted such that all electrical equipment and fixed wiring will be at least 450 mm (18 in.) above floor level and shall not be operated in a Class I location as defined in 513.3(C).

(2) Marking for Mobile Units. Mobile energizers shall carry at least one permanently affixed warning sign with the following words or equivalent:

> **WARNING**
> **KEEP 5 FT CLEAR OF AIRCRAFT ENGINES AND FUEL TANK AREAS**

or

> **WARNING**
> **KEEP 1.5 METERS CLEAR OF AIRCRAFT ENGINES AND FUEL TANK AREAS**

(3) Cords. Flexible cords for aircraft energizers and ground support equipment shall be identified for the type of service and extra-hard usage and shall include an equipment grounding conductor.

(D) Mobile Servicing Equipment with Electrical Components.

(1) General. Mobile servicing equipment (such as vacuum cleaners, air compressors, air movers) having electrical wiring and equipment not suitable for Class I, Division 2 or Zone 2 locations shall be so designed and mounted that all such fixed wiring and equipment will be at least 450 mm (18 in.) above the floor. Such mobile equipment shall not be operated within the Class I location defined in 513.3(C) and shall carry at least one permanently affixed warning sign with the following words or equivalent:

> **WARNING**
> **KEEP 5 FT CLEAR OF AIRCRAFT ENGINES AND FUEL TANK AREAS**

or

> **WARNING**
> **KEEP 1.5 METERS CLEAR OF AIRCRAFT ENGINES AND FUEL TANK AREAS**

(2) Cords and Connectors. Flexible cords for mobile equipment shall be suitable for the type of service and identified for extra-hard usage and shall include an equipment grounding conductor. Attachment plugs and receptacles shall be identified for the location in which they are installed and shall provide for connection of the equipment grounding conductor.

(3) Restricted Use. Equipment that is not identified as suitable for Class I, Division 2 locations shall not be operated in locations where maintenance operations likely to release flammable liquids or vapors are in progress.

(E) Portable Equipment.

(1) Portable Lighting Equipment. Portable lighting equipment that is used within a hangar shall be identified for the location in which they are used. For portable luminaires, flexible cord suitable for the type of service and identified for extra-hard usage shall be used. Each such cord shall include a separate equipment grounding conductor.

(2) Portable Utilization Equipment. Portable utilization equipment that is or may be used within a hangar shall be of a type suitable for use in Class I, Division 2 or Zone 2 locations. For portable utilization equipment, flexible cord suitable for the type of service and approved for extra-hard usage shall be used. Each such cord shall include a separate equipment grounding conductor.

513.12 Ground-Fault Circuit-Interrupter Protection for Personnel. All 125-volt, 50/60-Hz, single-phase, 15– and 20-ampere receptacles installed in areas where electrical diagnostic equipment, electrical hand tools, or portable lighting equipment are to be used shall have ground-fault circuit-interrupter protection for personnel.

513.16 Grounding and Bonding Requirements.

(A) General Grounding Requirements. All metal raceways, the metal armor or metallic sheath on cables, and all non–current-carrying metal parts of fixed or portable electrical equipment, regardless of voltage, shall be grounded. Grounding in Class I locations shall comply with 501.30 for Class I, Division 1 and 2 locations and 505.25 for Class I, Zone 0, 1, and 2 locations.

(B) Supplying Circuits with Grounded and Grounding Conductors in Class I Locations.

(1) Circuits Supplying Portable Equipment or Pendants. Where a circuit supplies portables or pendants and includes a grounded conductor as provided in Article 200, receptacles, attachment plugs, connectors, and similar devices shall be of the grounding type, and the grounded conductor of the flexible cord shall be connected to the screw shell of any lampholder or to the grounded terminal of any utilization equipment supplied.

(2) Approved Means. Approved means shall be provided for maintaining continuity of the grounding conductor between the fixed wiring system and the non–current-carrying metal portions of pendant luminaires, portable luminaires, and portable utilization equipment.

ARTICLE 514
Motor Fuel Dispensing Facilities

Informational Note: Text that is followed by a reference in brackets has been extracted from NFPA 30A-2015, *Code for Motor Fuel Dispensing Facilities and Repair Garages.* Only editorial changes were made to the extracted text to make it consistent with this *Code.*

514.1 Scope. This article shall apply to motor fuel dispensing facilities, marine/motor fuel dispensing facilities, motor fuel dispensing facilities located inside buildings, and fleet vehicle motor fuel dispensing facilities.

Informational Note: For further information regarding safeguards for motor fuel dispensing facilities, see NFPA 30A-2015, *Code for Motor Fuel Dispensing Facilities and Repair Garages.*

514.2 Definition.

Motor Fuel Dispensing Facility. That portion of a property where motor fuels are stored and dispensed from fixed equipment into the fuel tanks of motor vehicles or marine craft or into approved containers, including all equipment used in connection therewith. [**30A:**3.3.11]

Informational Note: Refer to Articles 510 and 511 with respect to electrical wiring and equipment for other areas used as lubritoriums, service rooms, repair rooms, offices, salesrooms, compressor rooms, and similar locations.

514.3 Classification of Locations. [See Figure 514.3.]

(A) Unclassified Locations. Where the authority having jurisdiction can satisfactorily determine that flammable liquids having a flash point below 38°C (100°F), such as gasoline, will not be handled, such location shall not be required to be classified.

(B) Classified Locations. [See Figure 514.3(B).]

(1) Class I Locations. Table 514.3(B)(1) shall be applied where Class I liquids are stored, handled, or dispensed and shall be used to delineate and classify motor fuel dispensing facilities and commercial garages as defined in Article 511. Table 515.3 shall be used for the purpose of delineating and

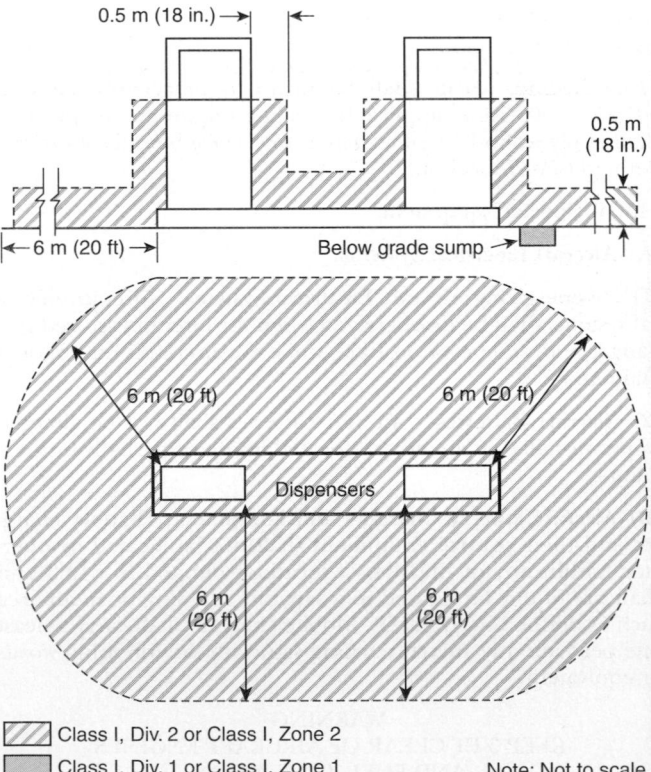

FIGURE 514.3   Classified Areas Adjacent to Dispensers. [30A: Figure 8.3.2(a)]

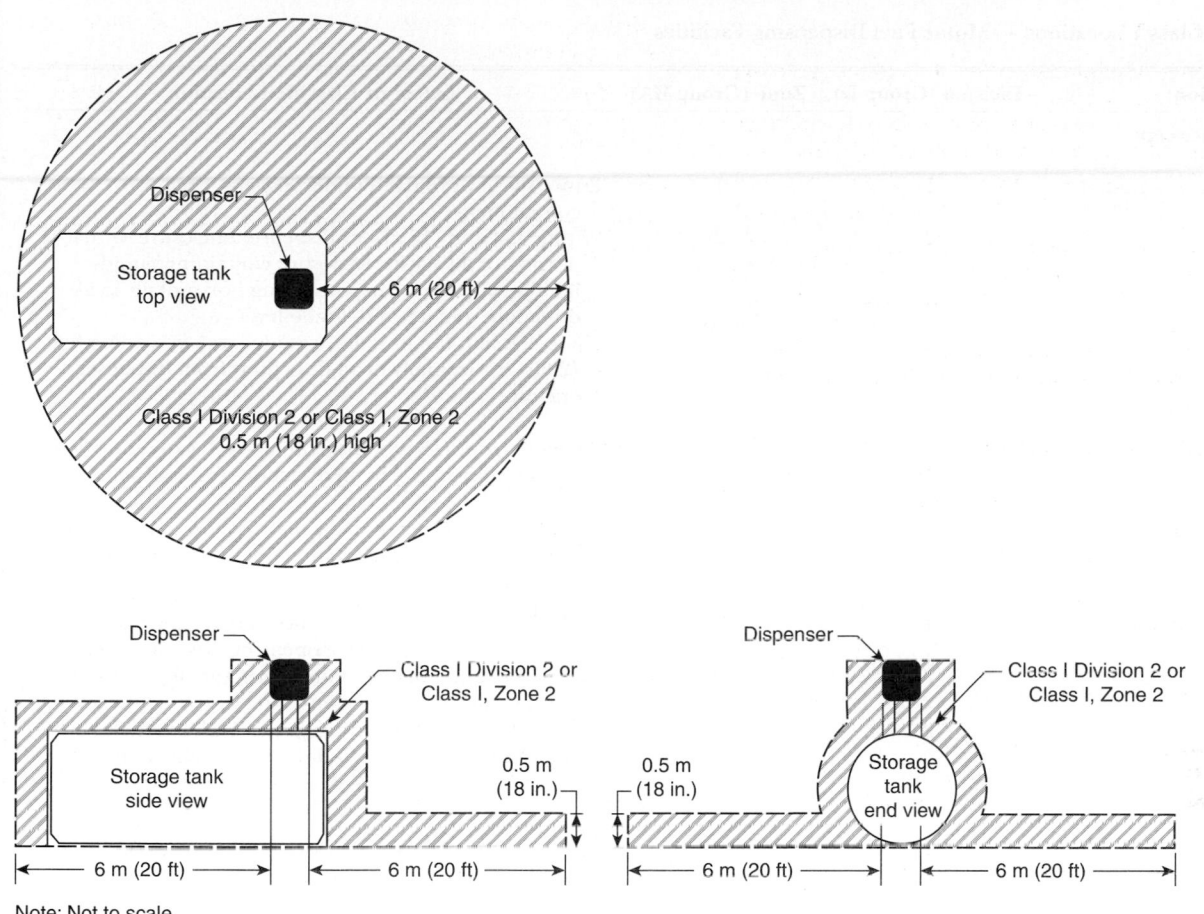

Note: Not to scale.

FIGURE 514.3(B)   Classified Areas Adjacent to Dispenser Mounted on Aboveground Storage Tank. [30A: Figure 8.3.2(b)]

classifying aboveground tanks. A Class I location shall not extend beyond an unpierced wall, roof, or other solid partition. [**30A:**8.1, 8.2, 8.3]

(2) Compressed Natural Gas, Liquefied Natural Gas, and Liquefied Petroleum Gas Areas.   Table 514.3(B)(2) shall be used to delineate and classify areas where CNG, LNG, compressed or liquefied hydrogen, LP-Gas, or combinations of these, are dispensed as motor vehicle fuels along with Class I or Class II liquids that are also dispensed as motor vehicle fuels. [**30A:**12.1]

Where CNG or LNG dispensers are installed beneath a canopy or enclosure, either the canopy or enclosure shall be designed to prevent accumulation or entrapment of ignitible vapors or all electrical equipment installed beneath the canopy or enclosure shall be suitable for Class I, Division 2 hazardous (classified) locations. [**30A:**12.4]

Dispensing devices for LP-Gas shall be located as follows:

(1)   At least 3 m (10 ft) from any dispensing device for Class I liquids

(2)   At least 1.5 m (5 ft) from any dispensing device for Class I liquids where the following conditions exist:

a.   The LP-Gas deliver nozzle and filler valve release no more than 4 cm³ (0.1 oz) of liquid upon disconnection.

b.   The fixed maximum liquid level gauge remains closed during the entire refueling process.

[**30A:**12.5.2]

Informational Note No. 1: Applicable requirements for dispensing devices for LP-Gas is found in NFPA 58-2014, *Liquefied Petroleum Gas Code*.

Informational Note No. 2: For information on classified areas pertaining to LP-Gas systems other than residential or commercial, see NFPA 58-2014, *Liquefied Petroleum Gas Code*, and NFPA 59-2012, *Utility LP-Gas Plant Code*.

Informational Note No. 3: See 514.3(C) for motor fuel dispensing stations in marinas and boatyards.

Table 514.3(B)(1) Class I Locations — Motor Fuel Dispensing Facilities

Location	Division (Group D)	Zone (Group IIA)	Extent of Classified Location[1]
Dispensing Device (except Overhead Type) [2, 3]			
Under dispenser containment	1	1	Entire space within and under dispenser pit or containment
Dispenser	2	2	Within 450 mm (18 in.) of dispenser enclosure or that portion of dispenser enclosure containing liquid-handling components, extending horizontally in all directions and down to grade level
Outdoor	2	2	Up to 450 mm (18 in.) above grade level, extending 6 m (20 ft) horizontally in all directions from dispenser enclosure
Indoor			
- with mechanical ventilation	2	2	Up to 450 mm (18 in.) above floor level, extending 6 m (20 ft) horizontally in all directions from dispenser enclosure
- with gravity ventilation	2	2	Up to 450 mm (18 in.) above floor level, extending 7.5 m (25 ft) horizontally in all directions from dispenser enclosure
Dispensing Device — Overhead Type[4]	1	1	Space within dispenser enclosure and all electrical equipment integral with dispensing hose or nozzle
	2	2	Within 450 mm (18 in.) of dispenser enclosure, extending horizontally in all directions and down to grade level
	2	2	Up to 450 mm (18 in.) above grade level, extending 6 m (20 ft) horizontally in all directions from a point vertically below edge of dispenser enclosure
Remote Pump —			
Outdoor	1	1	Entire space within any pit or box below grade level, any part of which is within 3 m (10 ft) horizontally from any edge of pump
	2	2	Within 900 mm (3 ft) of any edge of pump, extending horizontally in all directions
	2	2	Up to 450 mm (18 in.) above grade level, extending 3 m (10 ft) horizontally in all directions from any edge of pump
Indoor	1	1	Entire space within any pit
	2	2	Within 1.5 m (5 ft) of any edge of pump, extending in all directions
	2	2	Up to 900 mm (3 ft) above floor level, extending 7.5 m (25 ft) horizontally in all directions from any edge of pump
Sales, Storage, Rest Rooms	unclassified	unclassified	Except as noted below
including structures (such as the attendant's kiosk) on or adjacent to dispensers	1	1	Entire volume, if there is any opening to room within the extent of a Division 1 or Zone 1 location
	2	2	Entire volume, if there is any opening to room within the extent of a Division 2 or Zone 2 location
Tank, Aboveground			
Inside tank	1	0	Entire inside volume
Shell, ends, roof, dike area	1	1	Entire space within dike, where dike height exceeds distance from tank shell to inside of dike wall for more than 50 percent of tank circumference
	2	2	Entire space within dike, where dike height does not exceed distance from tank shell to inside of dike wall for more than 50 percent of tank circumference
Vent	2	2	Within 3 m (10 ft) of shell, ends, or roof of tank
	1	1	Within 1.5 m (5 ft) of open end of vent, extending in all directions
	2	2	Between 1.5 m and 3 m (5 ft and 10 ft) from open end of vent, extending in all directions

(continues)

Table 514.3(B)(1) *Continued*

Location	Division (Group D)	Zone (Group IIA)	Extent of Classified Location[1]
Tank, Underground			
Inside tank	1	0	Entire inside volume
Fill Opening	1	1	Entire space within any pit or box below grade level, any part of which is within a Division 1 or Division 2 classified location or within a Zone 1 or Zone 2 classified location
	2	2	Up to 450 mm (18 in.) above grade level, extending 1.5 m (5 ft) horizontally in all directions from any tight-fill connection and extending 3 m (10 ft) horizontally in all directions from any loose-fill connection
Vent	1	1	Within 1.5 m (5 ft) of open end of vent, extending in all directions
	2	2	Between 1.5 m and 3 m (5 ft and 10 ft) from open end of vent, extending in all directions
Vapor Processing System			
Pits	1	1	Entire space within any pit or box below grade level, any part of which: (1) is within a Division 1 or Division 2 classified location; (2) is within a Zone 1 or Zone 2 classified location; (3) houses any equipment used to transfer or process vapors
Equipment in protective enclosures	2	2	Entire space within enclosure
Equipment *not* within protective enclosure	2	2	Within 450 mm (18 in.) of equipment containing flammable vapors or liquid, extending horizontally in all directions and down to grade level
	2	2	Up to 450 mm (18 in.) above grade level within 3 m (10 ft) horizontally of the vapor processing equipment
- Equipment enclosure	1	1	Entire space within enclosure, if flammable vapor or liquid is present under normal operating conditions
	2	2	Entire space within enclosure, if flammable vapor or liquid is not present under normal operating conditions
- Vacuum assist blower	2	2	Within 450 mm (18 in.) of blower, extending horizontally in all directions and down to grade level
	2	2	Up to 450 mm (18 in.) above grade level, extending 3 m (10 ft) horizontally in all directions
Vault	1	1	Entire interior space, if Class I liquids are stored within

[1]For marine application, *grade level* means the surface of a pier, extending down to water level.
[2]Refer to Figure 514.3(a) and Figure 514.3(b) for an illustration of classified location around dispensing devices.
[3]Area classification inside the dispenser enclosure is covered in UL 87, *Standard for Power-Operated Dispensing Devices for Petroleum Products.*
[4]Ceiling-mounted hose reel. [**30A:**Table 8.3.1]

Table 514.3(B)(2) Electrical Equipment Classified Areas for Dispensing Devices

Dispensing Device	Extent of Classified Area	
	Class I, Division 1	Class I, Division 2
Compressed natural gas (CNG)	Entire space within the dispenser enclosure	1.5 m (5 ft) in all directions from dispenser enclosure
Liquefied natural gas (LNG)	Entire space within the dispenser enclosure	3 m (10 ft) in all directions from the dispenser enclosure
Liquefied petroleum gas (LP-Gas)	Entire space within the dispenser enclosure; 450 mm (18 in.) from the exterior surface of the dispenser enclosure to an elevation of 1.22 m (4 ft) above the base of the dispenser; the entire pit or open space beneath the dispenser and within 6 m (20 ft) horizontally from any edge of the dispenser when the pit or trench is not mechanically ventilated	Up to 450 mm (18 in.) above ground and within 6 m (20 ft) horizontally from any edge of the dispenser enclosure, including pits or trenches within this area when provided with adequate mechanical ventilation

[**30A:**Table 12.6.2]

N **(3) Fuel Storage.**

(a) Aboveground tanks storing CNG or LNG shall be separated from any adjacent property line that is or can be built upon, any public way, and the nearest important building on the same property. [**30A:**12.3.1]

Informational Note: The relevant distances are given in Section 8.4 of NFPA 52-2013, *Vehicular Gaseous Fuel Systems Code.*

(b) Aboveground tanks storing hydrogen shall be separated from any adjacent property line that is or can be built upon, any public way, and the nearest important building on the same property. [**30A:**12.3.2]

Informational Note: The relevant distances given in NFPA 2-2011, *Hydrogen Technologies Code.*

(c) Aboveground tanks storing LP-Gas shall be separated from any adjacent property line that is or can be built upon, any public way, and the nearest important building on the same property. [**30A:**12.3.3]

Informational Note: The relevant distances are given in Section 6.3 of NFPA 58-2014, *Liquefied Petroleum Gas Code.*

(d) Aboveground tanks storing CNG, LNG, or LP-Gas shall be separated from each other by at least 6 m (20 ft) and from dispensing devices that dispense liquid or gaseous motor vehicle fuels by at least 6 m (20 ft). [**30A:**12.3.3]

Exception No. 1: The required separation shall not apply to tanks or dispensers storing or handling fuels of the same chemical composition.

Exception No. 2: The required separation shall not apply when both the gaseous fuel storage and dispensing equipment are at least 15 m (50 ft) from any other aboveground motor fuel storage or dispensing equipment.

Informational Note: For further information, see NFPA 52-2013, *Vehicular Gaseous Fuel Systems Code,* or NFPA 58-2014, *Liquefied Petroleum Gas Code,* as applicable.

(e) *Dispenser Installations Beneath Canopies.* Where CNG or LNG dispensers are installed beneath a canopy or enclosure, either the canopy or enclosure shall be designed to prevent accumulation or entrapment of ignitible vapors or all electrical equipment installed beneath the canopy or enclosure shall be suitable for Class I, Division 2 hazardous (classified) locations. [**30A:**12.4]

(f) *Specific Requirements for LP-Gas Dispensing Devices.* [**30A:**12.5] Dispensing devices for LP-Gas shall be located as follows:

(1) At least 3 m (10 ft) from any dispensing device for Class I liquids

(2) At least 1.5 m (5 ft) from any dispensing device for Class I liquids where the following conditions exist:

a. The LP-Gas deliver nozzle and filler valve release no more than 4 cm^3 (0.1 oz) of liquid upon disconnection.

b. The fixed maximum liquid level gauge remains closed during the entire refueling process. [**30A:**12.5.2]

Table 514.3(B)(2) shall be used to delineate and classify areas for the purpose of installation of electrical wiring and electrical utilization equipment.

(C) Motor Fuel Dispensing Stations in Boatyards and Marinas.

Informational Note: For additional information, see NFPA 303-2011, *Fire Protection Standard for Marinas and Boatyards,* and NFPA 30A-2015, *Motor Fuel Dispensing Facilities and Repair Garages.*

(1) General. Electrical wiring and equipment located at or serving motor fuel dispensing locations shall be installed on the side of the wharf, pier, or dock opposite from the liquid piping system.

(2) Classification of Class I, Division 1 and 2 Areas. The following criteria shall be used for the purposes of applying Table 514.3(B)(1) and Table 514.3(B)(2) to motor fuel dispensing equipment on floating or fixed piers, wharfs, or docks.

(D) Closed Construction. Where the construction of floating docks, piers, or wharfs is closed so that there is no space between the bottom of the dock, pier, or wharf and the water, as in the case of concrete-enclosed expanded foam or similar construction, and the construction includes integral service boxes with supply chases, the following shall apply:

(1) The space above the surface of the floating dock, pier, or wharf shall be a Class I, Division 2 location with distances as specified in Table 514.3(B)(1) for dispenser and outdoor locations.

(2) Spaces below the surface of the floating dock, pier, or wharf that have areas or enclosures, such as tubs, voids, pits, vaults, boxes, depressions, fuel piping chases, or similar spaces, where flammable liquid or vapor can accumulate shall be a Class I, Division 1 location.

Exception No. 1: Dock, pier, or wharf sections that do not support fuel dispensers and abut, but are located 6.0 m (20 ft) or more from, dock sections that support a fuel dispenser(s) shall be permitted to be Class I, Division 2 locations where documented air space is provided between dock sections to allow flammable liquids or vapors to dissipate without traveling to such dock sections. The documentation shall comply with 500.4(A).

Exception No. 2: Dock, pier, or wharf sections that do not support fuel dispensers and do not directly abut sections that support fuel dispensers shall be permitted to be unclassified where documented air space is provided and where flammable liquids or vapors cannot travel to such dock sections. The documentation shall comply with 500.4(A).

(E) Open Construction. Where the construction of piers, wharfs, or docks is open, as in the case of decks built on stringers supported by pilings, floats, pontoons, or similar construction, the following shall apply:

(1) The area 450 mm (18 in.) above the surface of the dock, pier, or wharf and extending 6.0 m (20 ft) horizontally in all directions from the outside edge of the dispenser and down to the water level shall be a Class 1, Division 2 location.

(2) Enclosures such as tubs, voids, pits, vaults, boxes, depressions, piping chases, or similar spaces where flammable liquids or vapors can accumulate within 6.0 m (20 ft) of the dispenser shall be a Class I, Division 1 location.

514.4 Wiring and Equipment Installed in Class I Locations. All electrical equipment and wiring installed in Class I locations as classified in 514.3 shall comply with the applicable provisions of Article 501.

Exception: As permitted in 514.8.

Informational Note: For special requirements for conductor insulation, see 501.20.

514.7 Wiring and Equipment Above Class I Locations. Wiring and equipment above the Class I locations as classified in 514.3 shall comply with 511.7.

514.8 Underground Wiring. Underground wiring shall be installed in threaded rigid metal conduit or threaded steel intermediate metal conduit. Any portion of electrical wiring that is below the surface of a Class I, Division 1 or a Class I, Division 2 location [as classified in Table 514.3(B)(1) and Table 514.3(B)(2)] shall be sealed within 3.05 m (10 ft) of the point of emergence above grade. Except for listed explosion-proof reducers at the conduit seal, there shall be no union, coupling, box, or fitting between the conduit seal and the point of emergence above grade. Refer to Table 300.5.

Exception No. 1: Type MI cable shall be permitted where it is installed in accordance with Article 332.

Exception No. 2: Type PVC conduit, Type RTRC conduit, and Type HDPE conduit shall be permitted where buried under not less than 600 mm (2 ft) of cover. Where Type PVC conduit, Type RTRC conduit, or Type HDPE conduit is used, threaded rigid metal conduit or threaded steel intermediate metal conduit shall be used for the last 600 mm (2 ft) of the underground run to emergence or to the point of connection to the aboveground raceway, and an equipment grounding conductor shall be included to provide electrical continuity of the raceway system and for grounding of non–current-carrying metal parts.

514.9 Sealing.

(A) At Dispenser. A listed seal shall be provided in each conduit run entering or leaving a dispenser or any cavities or enclosures in direct communication therewith. The sealing fitting or listed explosionproof reducer at the seal shall be the first fitting after the conduit emerges from the earth or concrete.

(B) At Boundary. Additional seals shall be provided in accordance with 501.15. Sections 501.15(A)(4) and (B)(2) shall apply to horizontal as well as to vertical boundaries of the defined Class I locations.

514.11 Circuit Disconnects.

(A) Emergency Electrical Disconnects. Fuel dispensing systems shall be provided with one or more clearly identified emergency shutoff devices or electrical disconnects. Such devices or disconnects shall be installed in approved locations but not less than 6 m (20 ft) or more than 30 m (100 ft) from the fuel dispensing devices that they serve. Emergency shutoff devices or electrical disconnects shall disconnect power to all dispensing devices; to all remote pumps serving the dispensing devices; to all associated power, control, and signal circuits; and to all other electrical equipment in the hazardous (classified) locations surrounding the fuel dispensing devices. When more than one emergency shutoff device or electrical disconnect is provided, all devices shall be interconnected. Resetting from an emergency shutoff condition shall require manual intervention and the manner of resetting shall be approved by the authority having jurisdiction. [30A:6.7]

Exception: Intrinsically safe electrical equipment need not meet this requirement. [30A:6.7]

(B) Attended Self-Service Motor Fuel Dispensing Facilities. At attended motor fuel dispensing facilities, the devices or disconnects shall be readily accessible to the attendant. [30A:6.7.1]

(C) Unattended Self-Service Motor Fuel Dispensing Facilities. At unattended motor fuel dispensing facilities, the devices or disconnects shall be readily accessible to patrons and at least one additional device or disconnect shall be readily accessible to each group of dispensing devices on an individual island. [30A:6.7.2]

514.13 Provisions for Maintenance and Service of Dispensing Equipment. Each dispensing device shall be provided with a means to remove all external voltage sources, including power, communications, data, and video circuits and including feedback, during periods of maintenance and service of the dispensing equipment. The location of this means shall be permitted to be other than inside or adjacent to the dispensing device. The means shall be capable of being locked in the open position in accordance with 110.25.

514.16 Grounding and Bonding. All metal raceways, the metal armor or metallic sheath on cables, and all non–current-carrying metal parts of fixed and portable electrical equipment, regardless of voltage, shall be grounded and bonded. Grounding and bonding in Class I locations shall comply with 501.30.

ARTICLE 515
Bulk Storage Plants

Informational Note: Text that is followed by a reference in brackets has been extracted from NFPA 30-2015, *Flammable and Combustible Liquids Code*. Only editorial changes were made to the extracted text to make it consistent with this *Code*.

515.1 Scope. This article covers a property or portion of a property where flammable liquids are received by tank vessel, pipelines, tank car, or tank vehicle and are stored or blended in bulk for the purpose of distributing such liquids by tank vessel, pipeline, tank car, tank vehicle, portable tank, or container.

515.3 Class I Locations. Table 515.3 shall be applied where Class I liquids are stored, handled, or dispensed and shall be used to delineate and classify bulk storage plants. The class location shall not extend beyond a floor, wall, roof, or other solid partition that has no communicating openings. [**30:** 7.3, 7.4]

> Informational Note No. 1: The area classifications listed in Table 515.3 are based on the premise that the installation meets the applicable requirements of NFPA 30 -2015, *Flammable and Combustible Liquids Code*, Chapter 5, in all respects. Should this not be the case, the authority having jurisdiction has the authority to classify the extent of the classified space.
>
> Informational Note No. 2: See 514.3(C) through (E) for gasoline dispensing stations in marinas and boatyards.

515.4 Wiring and Equipment Located in Class I Locations. All electrical wiring and equipment within the Class I locations defined in 515.3 shall comply with the applicable provisions of Article 501 or Article 505 for the division or zone in which they are used.

Exception: As permitted in 515.8.

Table 515.3 Electrical Area Classifications

Location	Division	Zone	Extent of Classified Area
Indoor equipment installed where flammable vapor–air mixtures can exist under normal operation (see Informational Note)	1	0	The entire area associated with such equipment where flammable gases or vapors are present continuously or for long periods of time
	1	1	Area within 1.5 m (5 ft) of any edge of such equipment, extending in all directions
	2	2	Area between 1.5 m and 2.5 m (5 ft and 8 ft) of any edge of such equipment, extending in all directions; also, space up to 900 mm (3 ft) above floor or grade level within 1.5 m to 7.5 m (5 ft to 25 ft) horizontally from any edge of such equipment[1]
Outdoor equipment installed where flammable vapor–air mixtures can exist under normal operation	1	0	The entire area associated with such equipment where flammable gases or vapors are present continuously or for long periods of time
	1	1	Area within 900 mm (3 ft) of any edge of such equipment, extending in all directions
	2	2	Area between 900 mm (3 ft) and 2.5 m (8 ft) of any edge of such equipment, extending in all directions; also, space up to 900 mm (3 ft) above floor or grade level within 900 mm to 3.0 m (3 ft to 10 ft) horizontally from any edge of such equipment
Tank storage installations inside buildings	1	1	All equipment located below grade level
	2	2	Any equipment located at or above grade level
Tank — aboveground, fixed roof	1	0	Inside fixed roof tank
	1	1	Area inside dike where dike height is greater than the distance from the tank to the dike for more than 50 percent of the tank circumference
	2	2	Within 3.0 m (10 ft) from shell, ends, or roof of tank; also, area inside dike to level of top of dike wall
	1	0	Area inside of vent piping or opening
	1	1	Within 1.5 m (5 ft) of open end of vent, extending in all directions
	2	2	Area between 1.5 m and 3.0 m (5 ft and 10 ft) from open end of vent, extending in all directions
Tank — aboveground, floating roof			
With fixed outer roof	1	0	Area between the floating and fixed roof sections and within the shell
With no fixed outer roof	1	1	Area above the floating roof and within the shell
Tank vault — interior	1	1	Entire interior volume, if Class I liquids are stored within
Underground tank fill opening	1	1	Any pit, box, or space below grade level, if any part is within a Division 1 or 2, or Zone 1 or 2 classified location
	2	2	Up to 450 mm (18 in.) above grade level within a horizontal radius of 3.0 m (10 ft) from a loose fill connection, and within a horizontal radius of 1.5 m (5 ft) from a tight fill connection
Vent — discharging upward	1	0	Area inside of vent piping or opening
	1	1	Within 900 mm (3 ft) of open end of vent, extending in all directions
	2	2	Area between 900 mm and 1.5 m (3 ft and 5 ft) of open end of vent, extending in all directions
Drum and container filling — outdoors or indoors	1	0	Area inside the drum or container
	1	1	Within 900 mm (3 ft) of vent and fill openings, extending in all directions
	2	2	Area between 900 mm and 1.5 m (3 ft and 5 ft) from vent or fill opening, extending in all directions; also, up to 450 mm (18 in.) above floor or grade level within a horizontal radius of 3.0 m (10 ft) from vent or fill opening
Pumps, bleeders, withdrawal fittings			
Indoor	2	2	Within 1.5 m (5 ft) of any edge of such devices, extending in all directions; also, up to 900 mm (3 ft) above floor or grade level within 7.5 m (25 ft) horizontally from any edge of such devices

(continues)

Table 515.3 *Continued*

Location	Division	Zone	Extent of Classified Area
Outdoor	2	2	Within 900 mm (3 ft) of any edge of such devices, extending in all directions. Also, up to 450 mm (18 in.) above grade level within 3.0 m (10 ft) horizontally from any edge of such devices
Pits and sumps			
Without mechanical ventilation	1	1	Entire area within a pit or sump if any part is within a Division 1 or 2 or Zone 1 or 2 classified location
With adequate mechanical ventilation	2	2	Entire area within a pit or sump if any part is within a Division 1 or 2 or Zone 1 or 2 classified location
Containing valves, fittings, or piping, and not within a Division 1 or 2 or Zone 1 or 2 classified location	2	2	Entire pit or sump
Drainage ditches, separators, impounding basins			
Outdoor	2	2	Area up to 450 mm (18 in.) above ditch, separator, or basin; also, area up to 450 mm (18 in.) above grade within 4.5 m (15 ft) horizontally from any edge
Indoor			Same as pits and sumps
Tank vehicle and tank car[2]			
Loading through open dome	1	0	Area inside of the tank
	1	1	Within 900 mm (3 ft) of edge of dome, extending in all directions
	2	2	Area between 900 mm and 4.5 m (3 ft and 15 ft) from edge of dome, extending in all directions
Loading through bottom connections with atmospheric venting	1	0	Area inside of the tank
	1	1	Within 900 mm (3 ft) of point of venting to atmosphere, extending in all directions
	2	2	Area between 900 mm and 4.5 m (3 ft and 15 ft) from point of venting to atmosphere, extending in all directions; also, up to 450 mm (18 in.) above grade within a horizontal radius of 3.0 m (10 ft) from point of loading connection
Loading through closed dome with atmospheric venting	1	1	Within 900 mm (3 ft) of open end of vent, extending in all directions
	2	2	Area between 900 mm and 4.5 m (3 ft and 15 ft) from open end of vent, extending in all directions; also, within 900 mm (3 ft) of edge of dome, extending in all directions
Loading through closed dome with vapor control	2	2	Within 900 mm (3 ft) of point of connection of both fill and vapor lines extending in all directions
Bottom loading with vapor control or any bottom unloading	2	2	Within 900 mm (3 ft) of point of connections, extending in all directions; also up to 450 mm (18 in.) above grade within a horizontal radius of 3.0 m (10 ft) from point of connections
Storage and repair garage for tank vehicles	1	1	All pits or spaces below floor level
	2	2	Area up to 450 mm (18 in.) above floor or grade level for entire storage or repair garage
Garages for other than tank vehicles	Unclassified		If there is any opening to these rooms within the extent of an outdoor classified location, the entire room shall be classified the same as the area classification at the point of the opening.
Outdoor drum storage	Unclassified		
Inside rooms or storage lockers used for the storage of Class I liquids	2	2	Entire room or locker

(continues)

Table 515.3   *Continued*

Location	Division	Zone	Extent of Classified Area
Indoor warehousing where there is no flammable liquid transfer	Unclassified		If there is any opening to these rooms within the extent of an indoor classified location, the classified location shall extend through the opening to the same extent as if the wall, curb, or partition did not exist.
Office and rest rooms	Unclassified		If there is any opening to these rooms within the extent of an indoor classified location, the room shall be classified the same as if the wall, curb, or partition did not exist.
Piers and wharves			See Figure 515.3.

[1]The release of Class I liquids can generate vapors to the extent that the entire building, and possibly an area surrounding it, should be considered a Class I, Division 2 or Zone 2 location.

[2]When classifying extent of area, consideration shall be given to the fact that tank cars or tank vehicles can be spotted at varying points. Therefore, the extremities of the loading or unloading positions shall be used. [**30:**Table 7.3.3]

Informational Note: See Section 7.3 of NFPA 30-2015, *Flammable and Combustible Liquids Code*, for additional information.

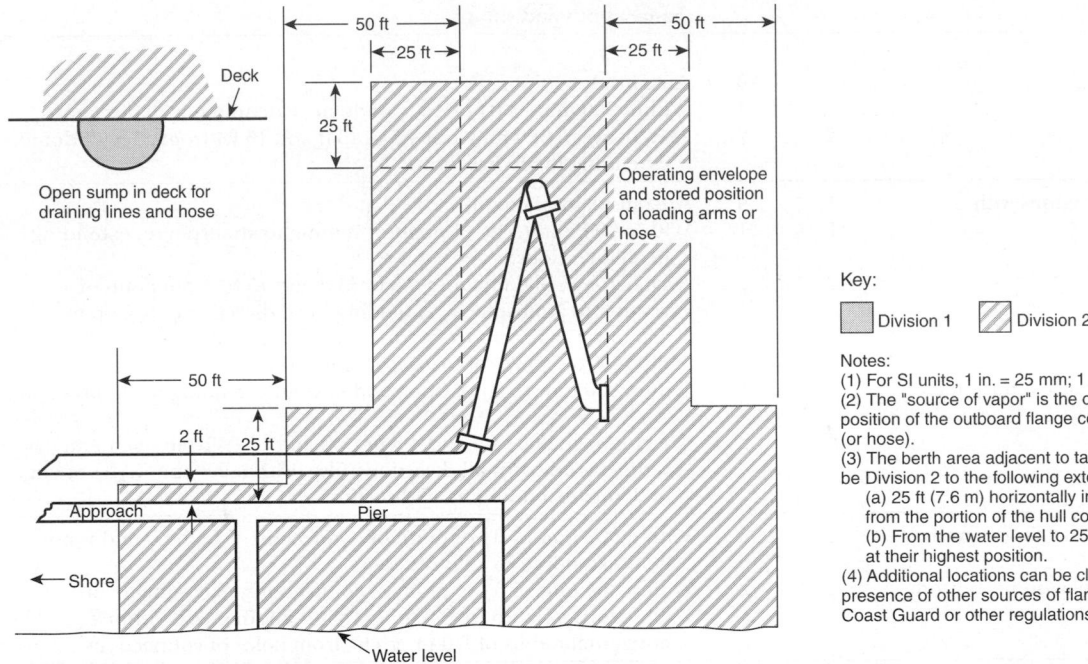

FIGURE 515.3   **Area Classification for a Marine Terminal Handling Flammable Liquids. [30:Figure 29.3.22]**

515.7 Wiring and Equipment Above Class I Locations.

(A) Fixed Wiring.   All fixed wiring above Class I locations shall be in metal raceways, Schedule 80 PVC conduit, Type RTRC marked with the suffix -XW, or Type MI, Type TC, or Type MC cable, or Type PLTC and Type PLTC-ER cable in accordance with the provisions of Article 725, including installation in cable tray systems or Type ITC and Type ITC-ER cable as permitted in 727.4. The cable shall be terminated with listed fittings.

(B) Fixed Equipment.   Fixed equipment that may produce arcs, sparks, or particles of hot metal, such as lamps and lampholders for fixed lighting, cutouts, switches, receptacles, motors, or other equipment having make-and-break or sliding contacts, shall be of the totally enclosed type or be constructed so as to prevent the escape of sparks or hot metal particles.

(C) Portable Luminaires or Other Utilization Equipment.   Portable luminaires or other utilization equipment and their flexible cords shall comply with the provisions of Article 501 or Article 505 for the class of location above which they are connected or used.

515.8 Underground Wiring.

(A) Wiring Method.   Underground wiring shall be installed in threaded rigid metal conduit or threaded steel intermediate metal conduit or, where buried under not less than 600 mm (2 ft) of cover, shall be permitted in Type PVC conduit, Type RTRC conduit, or a listed cable. Where Type PVC conduit or Type RTRC conduit is used, threaded rigid metal conduit or threaded steel intermediate metal conduit shall be used for no less than the last 600 mm (2 ft) of the conduit run to the conduit point of emergence from the underground location or to the point of connection to an aboveground raceway. When cable is used, it shall be enclosed in threaded rigid metal conduit or threaded steel intermediate metal conduit from the point of lowest buried cable level to the point of connection to the aboveground raceway.

(B) Insulation. Conductor insulation shall comply with 501.20.

(C) Nonmetallic Wiring. Where Type PVC conduit, Type RTRC conduit, or cable with a nonmetallic sheath is used, an equipment grounding conductor shall be included to provide for electrical continuity of the raceway system and for grounding of non–current-carrying metal parts.

515.9 Sealing. Sealing requirements shall apply to horizontal as well as to vertical boundaries of the defined Class I locations. Buried raceways and cables under defined Class I locations shall be considered to be within a Class I, Division 1 or Zone 1 location.

515.10 Special Equipment — Gasoline Dispensers. Where gasoline or other volatile flammable liquids or liquefied flammable gases are dispensed at bulk stations, the applicable provisions of Article 514 shall apply.

515.16 Grounding and Bonding. All metal raceways, the metal armor or metallic sheath on cables, and all non–current carrying metal parts of fixed or portable electrical equipment, regardless of voltage, shall be grounded and bonded as provided in Article 250.

Grounding and bonding in Class I locations shall comply with 501.30 for Class I, Division 1 and 2 locations and 505.25 for Class I, Zone 0, 1, and 2 locations.

Informational Note: For information on grounding for static protection, see 4.5.3.4 and 4.5.3.5 of NFPA 30-2015, *Flammable and Combustible Liquids Code.*

ARTICLE 516
Spray Application, Dipping, Coating, and Printing Processes Using Flammable or Combustible Materials

Part I. General

516.1 Scope. This article covers the regular or frequent application of flammable liquids, combustible liquids, and combustible powders by spray operations and the application of flammable liquids or combustible liquids at temperatures above their flashpoint by spraying, dipping, coating, printing, or other means.

Informational Note No. 1: For further information regarding safeguards for these processes, such as fire protection, posting of warning signs, and maintenance, see NFPA 33-2016, *Standard for Spray Application Using Flammable and Combustible Materials,* and NFPA 34-2015, *Standard for Dipping, Coating, and Printing Processes Using Flammable or Combustible Liquids.* For additional information regarding ventilation, see NFPA 91 -2015, *Standard for Exhaust Systems for Air Conveying of Vapors, Gases, Mists, and Particulate Solids.*

Informational Note No. 2: Text that is followed by a reference in brackets has been extracted from NFPA 33-2016, *Standard for Spray Application Using Flammable and Combustible Materials,* or NFPA 34-2015, *Standard for Dipping, Coating, and Printing Processes Using Flammable or Combustible Liquids.* Only editorial changes were made to the extracted text to make it consistent with this *Code.*

516.2 Definitions. For the purpose of this article, the following definitions shall apply.

Limited Finishing Workstation. An apparatus that is capable of confining the vapors, mists, residues, dusts, or deposits that are generated by a spray application process but does not meet the requirements of a spray booth or spray room, as herein defined. [**33:**3.3.18.1]

Informational Note: See Section 14.3 of NFPA 33, *Standard for Spray Application Using Flammable or Combustible Materials,* for limited finishing workstations.

N **Membrane Enclosure.** A temporary enclosure used for the spraying of workpieces that cannot be moved into a spray booth where open spraying is not practical due to the proximity to other operations, finish quality, or concerns such as the collection of overspray.

Informational Note: See Chapter 18 of NFPA 33 2016, *Standard for Spray Application Using Flammable or Combustible Materials,* for information on the construction and use of membrane enclosures.

N **Outdoor Spray Area.** A spray area that is outside the confines of a building or that has a canopy or roof that does not limit the dissipation of the heat of a fire or dispersion of flammable vapors and does not restrict fire-fighting access and control. For the purpose of this standard, an outdoor spray area can be treated as an unenclosed spray area. [**33:**3.3.2.3.1]

Spray Area. Any fully enclosed, partly enclosed, or unenclosed area in which dangerous quantities of flammable or combustible vapors, mists, residues, dusts, or deposits are present due to the operation of spray processes, including (1) any area in the direct path of a spray application process; (2) the interior of a spray booth, spray room, or limited finishing workstation, as herein defined; (3) the interior of any exhaust plenum, eliminator section, or scrubber section; (4) the interior of any exhaust duct or exhaust stack leading from a spray application process; (5) the interior of any air recirculation path up to and including recirculation particulate filters; (6) any solvent concentrator (pollution abatement) unit or solvent recovery (distillation) unit; and (7) the inside of a membrane enclosure. The following are not part of the spray area: (1) fresh air make-up units; (2) air supply ducts and air supply plenums; (3) recirculation air supply ducts downstream of recirculation particulate filters; and (4) exhaust ducts from solvent concentrator (pollution abatement) units. [**33:**3.3.2.3]

Informational Note: Unenclosed spray areas are locations outside of buildings or are localized operations within a larger room or space. Such are normally provided with some local vapor extraction/ventilation system. In automated operations, the area limits are the maximum area in the direct path of spray operations. In manual operations, the area limits are the maximum area of spray when aimed at 90 degrees to the application surface.

Spray Booth. A power-ventilated enclosure for a spray application operation or process that confines and limits the escape of the material being sprayed, including vapors, mists, dusts, and residues that are produced by the spraying operation and conducts or directs these materials to an exhaust system. [**33**:3.3.15]

Informational Note: A spray booth is an enclosure or insert within a larger room used for spray/coating/dipping applications. A spray booth can be fully enclosed or have open front or face and can include separate conveyor entrance and exit. The spray booth is provided with a dedicated ventilation exhaust with supply air from the larger room or from a dedicated air supply.

Spray Room. A power-ventilated fully enclosed room used exclusively for open spraying of flammable or combustible materials. [**33**:3.3.16]

Unenclosed Spray Area. Any spray area that is not confined by a limited finishing workstation, spray booth, or spray room, as herein defined. [**33**:3.3.2.3.2]

N Part II. Open Containers

516.4 Area Classification. For open containers, supply containers, waste containers, spray gun cleaners, and solvent distillation units that contain Class I liquids that are located in ventilated areas, area classification shall be in accordance with the following:

(1) The area within 915 mm (3 ft) in all directions from any such container or equipment and extending to the floor or grade level shall be classified as Class I, Division 1 or Class I, Zone 1, whichever is applicable. [**33**:6.5.5.1]

(2) The area extending 610 mm (2 ft) beyond the Division 1 or Zone 1 location shall be classified as Class I, Division 2 or Class I, Zone 2, whichever is applicable. [**33**:6.5.5.1]

(3) The area extending 1525 mm (5 ft) horizontally beyond the area described in 516.4(2) up to a height of 460 mm (18 in.) above the floor or grade level shall be classified as Class I, Division 2 or Class I, Zone 2, whichever is applicable. [**33**:6.5.5.1]

(4) The area inside any tank or container shall be classified as Class I, Division 1 or Class I, Zone 0, whichever is applicable. [**33**:6.5.5.1]

(5) Sumps, pits, or below grade channels within 3.5 m (10 ft) horizontally of a vapor source shall be classified as Class I, Division 1 or Zone 1. If the sump, pit, or channel extends beyond 3.5 m (10 ft) from the vapor source, it shall be provided with a vapor stop or it shall be classified as Class I, Division 1 or Zone 1 for its entire length.

For the purposes of electrical area classification, the Division system and the Zone system shall not be intermixed for any given source of release. [**33**:6.2.3]

Electrical wiring and utilization equipment installed in these areas shall be suitable for the location, as shown in Figure 516.4. [**33**:6.5.5.2]

Part III. Spray Application Processes

516.5 Area Classification. For spray application processes, the area classification is based on quantities of flammable vapors, combustible mists, residues, dusts, or deposits that are present or might be present in quantities sufficient to produce ignitable or explosive mixtures with air.

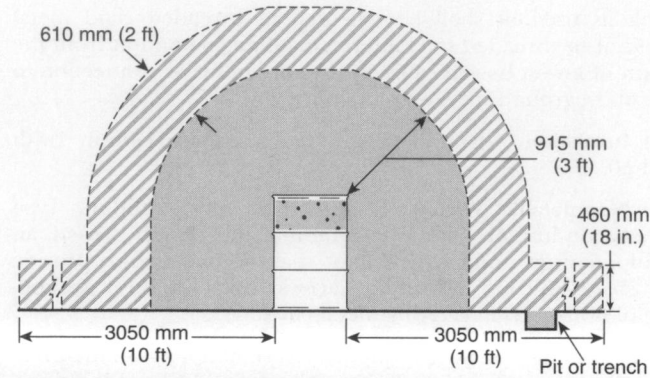

810 mm (2 ft)
915 mm (3 ft)
460 mm (18 in.)
3050 mm (10 ft)
3050 mm (10 ft)
Pit or trench

Class I, Division 1 or Zone 0 (e.g., vapor space in container)

Class I, Division 1 or Zone 1

Class I, Division 2 or Zone 2

N **FIGURE 516.4 Electrical Area Classification for Class I Liquid Operations Around Open Containers, Supply Containers, Waste Containers, Spray Gun Cleaners, and Solvent Distillation Units. [33:Figure 6.5.5.2]**

(A) Zone Classification of Locations.

(1) Classification of Locations. The Zone system of electrical area classification shall be applied as follows:

(1) The inside of closed containers or vessels shall be considered a Class I, Zone 0 location.

(2) A Class I, Division 1 location shall be permitted to be alternatively classified as a Class I, Zone 1 location.

(3) A Class I, Division 2 location shall be permitted to be alternatively classified as a Class I, Zone 2 location.

(4) A Class II, Division 1 location shall be permitted to be alternatively classified as a Zone 21 location.

(5) A Class II, Division 2 location shall be permitted to be alternatively classified as a Zone 22 location. [**33**:6.2.2]

(2) Classification Systems. For the purposes of electrical area classification, the Division system and the Zone system shall not be intermixed for any given source of release. [**33**:6.2.3]

In instances of areas within the same facility classified separately, Class I, Zone 2 locations shall be permitted to abut, but not overlap, Class I, Division 2 locations. Class I, Zone 0 or Zone 1 locations shall not abut Class I, Division 1 or Division 2 locations. [**33**:6.2.4]

(3) Equipment. Open flames, spark-producing equipment or processes, and equipment whose exposed surfaces exceed the autoignition temperature of the material being sprayed shall not be located in a spray area or in any surrounding area that is classified as Division 2, Zone 2, or Zone 22. [**33**:6.2.5]

Exception: This requirement shall not apply to drying, curing, or fusing apparatus.

Any utilization equipment or apparatus that is capable of producing sparks or particles of hot metal and that is located above or adjacent to either the spray area or the surrounding Division 2, Zone 2, or Zone 22 areas shall be of the totally enclosed type or shall be constructed to prevent the escape of sparks or particles of hot metal. [**33**:6.2.6]

(B) Class I, Division 1 or Class I, Zone 0 Locations. The interior of any open or closed container or vessel of a flammable liquid shall be considered Class I, Division 1, or Class I, Zone 0, as applicable:

Informational Note: For additional guidance, see Chapter 6 of NFPA 33-2016, *Standard for Spray Application Using Flammable or Combustible Materials.*

(C) Class I, Division 1; Class I, Zone 1; Class II, Division 1; or Zone 21 Locations. The following spaces shall be considered Class I, Division 1; Class I, Zone 1; Class II, Division 1; or Zone 21 locations, as applicable:

(1) The interior of spray booths and rooms except as specifically provided in 516.5(D).

(2) The interior of exhaust ducts.

(3) Any area in the direct path of spray operations.

(4) Sumps, pits, or below grade channels within 7620 mm (25 ft) horizontally of a vapor source. If the sump, pit, or channel extends beyond 7620 mm (25 ft) from the vapor source, it shall be provided with a vapor stop or it shall be classified as Class I, Division 1 for its entire length. [34:6.4.1]

(5) All space in all directions outside of but within 900 mm (3 ft) of open containers, supply containers, spray gun cleaners, and solvent distillation units containing flammable liquids.

(6) For limited finishing workstations, the area inside the curtains or partitions. *[See Figure 516.5(D)(5).]*

(D) Class I, Division 2; Class I, Zone 2; Class II, Division 2; or Zone 22 Locations. The spaces listed in 516.5(D)(1) through (D)(5) shall be considered Class I, Division 2; Class I, Zone 2; Class II, Division 2; or Zone 22 as applicable.

(1) Unenclosed Spray Processes. Electrical wiring and utilization equipment located outside but within 6100 mm (20 ft) horizontally and 3050 mm (10 ft) vertically of an enclosed spray area and not separated from the spray area by partitions extending to the boundaries of the area designated as Division 2, Zone 2 or Zone 22 in Figure 516.5(D)(1) shall be suitable for Class I, Division 2; Class I, Zone 2; Class II, Division 2; or Zone 22 locations, whichever is applicable. [33:6.5.1] *[See Figure 516.5(D)(1).]*

(2) Closed-Top, Open-Face, and Open-Front Spray Booths and Spray Rooms. If spray application operations are conducted within a closed-top, open-face, or open-front booth or room, as shown in Figure 516.5(D)(2), any electrical wiring or utilization equipment located outside of the booth or room but within 915 mm (3 ft) of any opening shall be suitable for Class I, Division 2; Class I, Zone 2; Class II, Division 2; or Zone 22 locations, whichever is applicable. The Class I, Division 2; Class I, Zone 2; Class II, Division 2; or Zone 22 locations shown in Figure 516.5(D)(2) shall extend from the edges of the open face or open front of the booth or room.

(3) Open-Top Spray Booths. For spraying operations conducted within an open top spray booth, the space 915 mm (3 ft) vertically above the booth and within 915 mm (3 ft) of other booth openings shall be considered Class I, Division 2; Class I, Zone 2; Class II, Division 2; or Zone 22 whichever is applicable. [33:6.5.3]

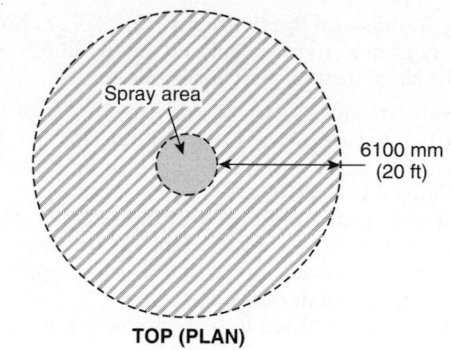

TOP (PLAN)

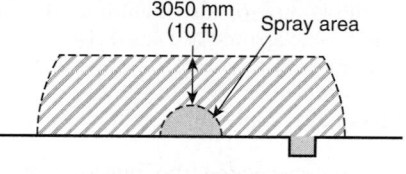

FRONT (ELEVATION)

Class I, Division 1; Class I, Zone 1; Class II, Division 1; or Zone 21

Class I, Division 2; Class I, Zone 2; Class II, Division 2; or Zone 22

N **FIGURE 516.5(D)(1)   Electrical Area Classification for Unenclosed Spray Areas. [33:Figure 6.5.1]**

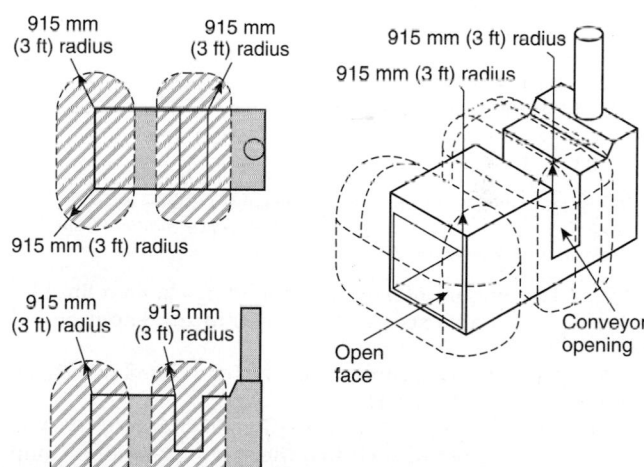

N **FIGURE 516.5(D)(2)   Class I, Division 2; Class I, Zone 2; Class II, Division 2; or Zone 22 Locations Adjacent to a Closed Top, Open Face, or Open Front Spray Booth or Room. [33:Figure 6.5.2]**

(4) Enclosed Spray Booths and Spray Rooms. For spray application operations confined to an enclosed spray booth or room, electrical area classification shall be as follows:

(1) The area within 915 mm (3 ft) of any opening shall be classified as Class I, Division 2; Class I, Zone 2; Class II, Division 2; or Zone 22 locations, whichever is applicable, as shown in Figure 516.5(D)(4).

(2) Where automated spray application equipment is used, the area outside the access doors shall be unclassified provided the door interlock prevents the spray application operations when the door is open.

(3) Where exhaust air is permitted to be recirculated, both of the following shall apply:

  a. The interior of any recirculation path from the secondary particulate filters up to and including the air supply plenum shall be classified as Class I, Division 2; Class I, Zone 2; Class II, Division 2; or Zone 22 locations, whichever is applicable.

  b. The interior of fresh air supply ducts shall be unclassified.

(4) Where exhaust air is not recirculated, the interior of fresh air supply ducts and fresh air supply plenums shall be unclassified.

[**33**:6.5.4]

(5) Limited Finishing Workstations.

  (a) For limited finishing workstations, the area inside the 915 mm (3 ft) space horizontally and vertically beyond the volume enclosed by the outside surface of the curtains or partitions shall be classified as Class I, Division 2; Class I, Zone 2; Class II, Division 2; or Zone 22, as shown in Figure 516.5(D)(5).

  (b) A limited finishing workstation shall be designed and constructed to have all of the following:

(1) A dedicated make-up air supply
(2) Curtains or partitions that are noncombustible or limited combustible
(3) A dedicated mechanical exhaust and filtration system
(4) An approved automatic extinguishing system
[**33**:14.3.1]

Informational Note: For limited combustible curtains or partitions see NFPA 701-2015, *Standard Methods of Fire Tests for Flame Propagation of Textiles and Films.*

  (c) The amount of material sprayed in a limited finishing workstation shall not exceed 3.8 L (1 gal) in any 8-hour period. [**33**:14.3.2]

  (d) Curtains or partitions shall be fully closed during any spray operations. [**33**:14.3.4]

  (e) The equipment within the limited finishing workstation shall be interlocked such that the spray application equipment cannot be operated unless the exhaust ventilation system is operating and functioning properly and spray application is automatically stopped if the exhaust ventilation system fails.

  (f) Any limited finishing workstation used for spray application operations shall not be used for any operation that is capable of producing sparks or particles of hot metal or for operations that involve open flames or electrical utilization equipment capable of producing sparks or particles of hot metal. [**33**:14.3.6]

  (g) Where industrial air heaters are used to elevate the air temperature for drying, curing, or fusing operations, a high limit switch shall be provided to automatically shut off the drying apparatus if the air temperature in the limited finishing workstation exceeds the maximum discharge-air temperature allowed by the standard that the heater is listed to or 93°C (200°F), whichever is less. [**33**:14.3.7.1]

  (h) A means shall be provided to show that the limited finishing workstation is in the drying or curing mode of operation and that the limited finishing workstation is to be unoccupied. [**33**:14.3.7.2]

  (i) Any containers of flammable or combustible liquids shall be removed from the limited finishing workstation before the drying apparatus is energized. [**33**:14.3.7.3]

  (j) Portable spot-drying, curing, or fusion apparatus shall be permitted to be used in a limited finishing workstation, provided that it is not located within the hazardous (classified) location defined in 14.3.5 of NFPA 33 when spray application operations are being conducted. [**33**:14.3.8]

  (k) Recirculation of exhaust air shall be permitted when the provisions of 516.5(D)(4)(3) are both met. [**33**:14.3.9]

516.6 Wiring and Equipment in Class I Locations.

(A) Wiring and Equipment — Vapors. All electrical wiring and equipment within the Class I location (containing vapor only — not residues) defined in 516.5 shall comply with the applicable provisions of Article 501 or Article 505, as applicable.

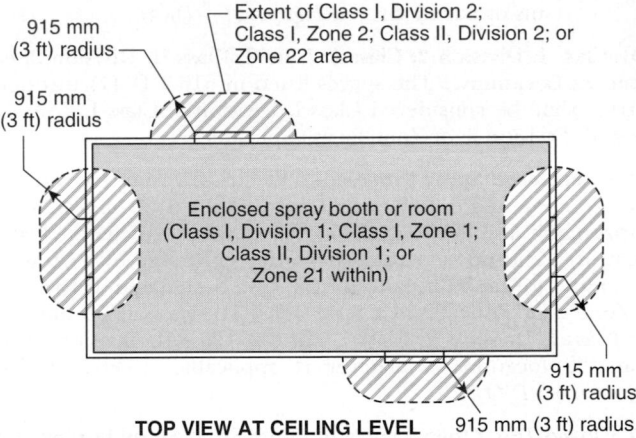

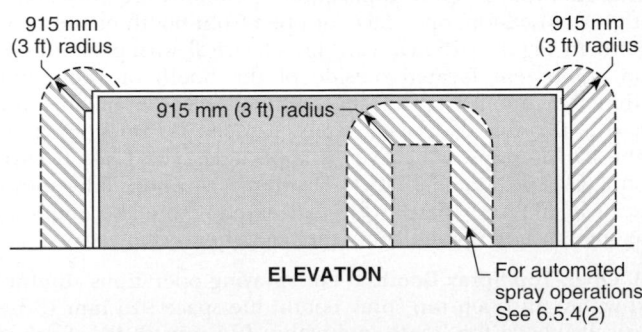

N **FIGURE 516.5(D)(4)   Class I, Division 2; Class I, Zone 2; Class II, Division 2; or Zone 22 Locations Adjacent to an Enclosed Spray Booth or Spray Room. [33:Figure 6.5.4]**

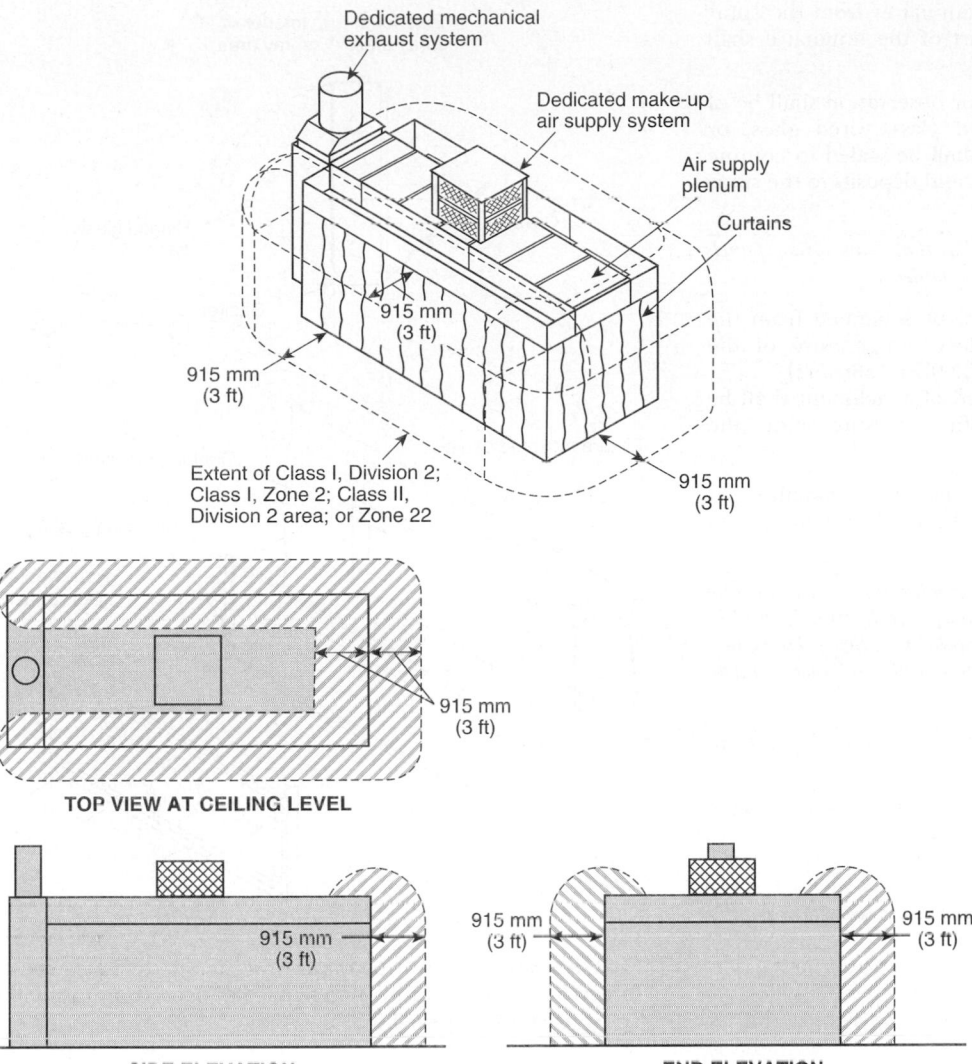

Dedicated mechanical exhaust system

Dedicated make-up air supply system

Air supply plenum

Curtains

915 mm (3 ft)

915 mm (3 ft)

Extent of Class I, Division 2; Class I, Zone 2; Class II, Division 2 area; or Zone 22

915 mm (3 ft)

915 mm (3 ft)

TOP VIEW AT CEILING LEVEL

915 mm (3 ft)

915 mm (3 ft)

915 mm (3 ft)

SIDE ELEVATION

END ELEVATION

V **FIGURE 516.5(D)(5)** **Class I, Division 2; Class I, Zone 2; Class II, Division 2; or Zone 22 Locations Adjacent to a Limited Finishing Workstation. [33:Figure 14.3.5.1]**

(B) Wiring and Equipment — Vapors and Residues. Unless specifically listed for locations containing deposits of dangerous quantities of flammable or combustible vapors, mists, residues, dusts, or deposits (as applicable), there shall be no electrical equipment in any spray area as herein defined whereon deposits of combustible residue could readily accumulate, except wiring in rigid metal conduit, intermediate metal conduit, Type MI cable, or in metal boxes or fittings containing no taps, splices, or terminal connections. [**33:**6.4.2]

(C) Illumination. Luminaires shall be permitted to be installed as follows:

(1) Luminaires, like that shown in Figure 516.6(C)(a), that are attached to the walls or ceiling of a spray area but that are outside any classified area and are separated from the spray area by glass panels shall be suitable for use in unclassified locations. Such fixtures shall be serviced from outside the spray area. [**33:**6.6.1]

(2) Luminaires, like that shown in Figure 516.6(C)(a), that are attached to the walls or ceiling of a spray area; that are separated from the spray area by glass panels and that are located within a Class I, Division 2; a Class I, Zone 2; a Class II, Division 2; or a Zone 22 location shall be suitable for such location. Such fixtures shall be serviced from outside the spray area. [**33:**6.6.2]

(3) Luminaires, like that shown in Figure 516.6(C)(b), that are an integral part of the walls or ceiling of a spray area shall be permitted to be separated from the spray area by glass panels that are an integral part of the fixture. Such fixtures shall be listed for use in Class I, Division 2; Class I, Zone 2; Class II, Division 2; or Zone 22 locations, whichever is applicable, and also shall be listed for accumulations of deposits of combustible residues. Such fixtures shall be permitted to be serviced from inside the spray area. [**33:**6.6.3]

(4) Glass panels used to separate luminaires from the spray area or that are an integral part of the luminaire shall meet the following requirements.

  a. Panels for light fixtures or for observation shall be of heat-treated glass, laminated glass, wired glass, or hammered-wired glass and shall be sealed to confine vapors, mists, residues, dusts, and deposits to the spray area. [**33**:5.5.1]

  Exception: Listed spray booth assemblies that have vision panels constructed of other materials shall be permitted.

  b. Panels for light fixtures shall be separated from the fixture to prevent the surface temperature of the panel from exceeding 93°C (200°F). [**33**:5.5.2]

  c. The panel frame and method of attachment shall be designed to not fail under fire exposure before the vision panel fails. [**33**:5.5.3]

(D) Portable Equipment. Portable electric luminaires or other utilization equipment shall not be used in a spray area during spray operations.

Exception No. 1: Where portable electric luminaires are required for operations in spaces not readily illuminated by fixed lighting within the spraying area, they shall be of the type identified for Class I, Division 1 or Class 1, Zone 1 locations where readily ignitible residues could be present. [33:6.9 Exception]

Exception No. 2: Where portable electric drying apparatus is used in spray booths and the following requirements are met:

(1) The apparatus and its electrical connections are not located within the spray enclosure during spray operations.

(2) Electrical equipment within 450 mm (18 in.) of the floor is identified for Class I, Division 2 or Class I, Zone 2 locations.

(3) All metallic parts of the drying apparatus are electrically bonded and grounded.

(4) Interlocks are provided to prevent the operation of spray equipment while drying apparatus is within the spray enclosure, to allow for a 3-minute purge of the enclosure before energizing the drying apparatus and to shut off drying apparatus on failure of ventilation system.

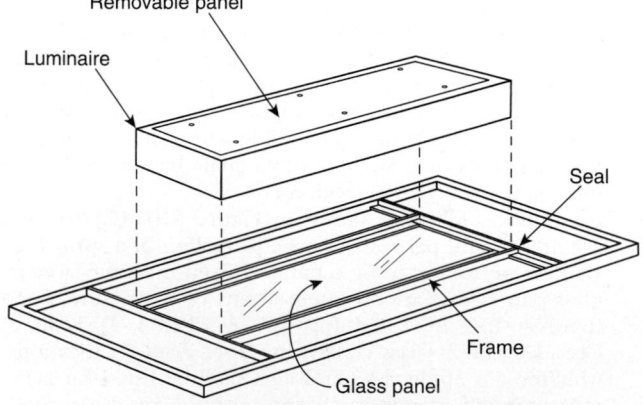

N **FIGURE 516.6(C)(a)   Example of a Luminaire that is Mounted Outside of the Spray Area and is Serviced from Outside the Spray Area. [33:Figure 6.6.1]**

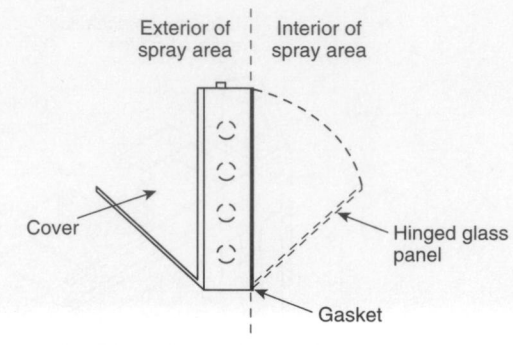

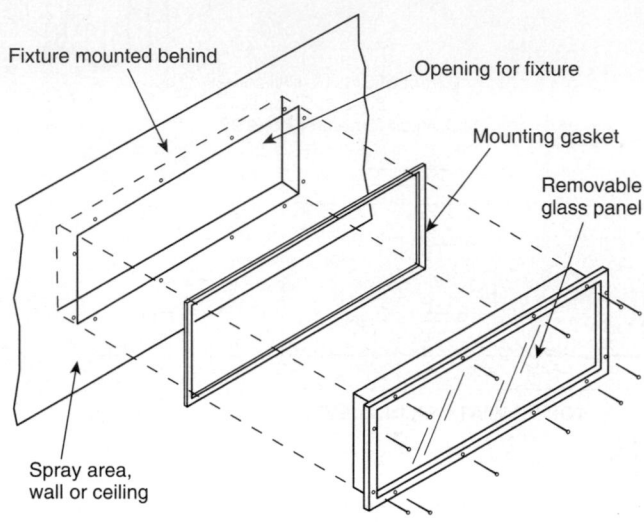

N **FIGURE 516.6(C)(b)   Example of a Luminaire that is an Integral Part of the Spray Area and is Serviced from Inside the Spray Area. [33:Figure 6.6.3]**

(E) Electrostatic Equipment. Electrostatic spraying or detearing equipment shall be installed and used only as provided in 516.10.

  Informational Note: For further information, see NFPA 33-2016, *Standard for Spray Application Using Flammable or Combustible Materials.*

(F) Static Electric Discharges. All persons and all electrically conductive objects, including any metal parts of the process equipment or apparatus, containers of material, exhaust ducts, and piping systems that convey flammable or combustible liquids, shall be electrically grounded. [**34**:6.8.1]

516.7 Wiring and Equipment Not Within Classified Locations.

(A) Wiring. All fixed wiring above the Class I and II locations shall be in metal raceways, Type PVC conduit, Type RTRC conduit, or electrical nonmetallic tubing; where cables are used, they shall be Type MI, Type TC, or Type MC cable. Cellular metal floor raceways shall only be permitted to supply ceiling outlets or as extensions to the area below the floor of a Class I or II location. Where cellular metal raceways are used, they shall not have connections leading into or passing through the Class I or II location unless suitable seals are provided.

(B) Equipment. Equipment that could produce arcs, sparks, or particles of hot metal, such as lamps and lampholders for fixed lighting, cutouts, switches, receptacles, motors, or other equipment having make-and-break or sliding contacts, where installed above a classified location or above a location where freshly finished goods are handled, shall be of the totally enclosed type or be constructed so as to prevent the escape of sparks or hot metal particles.

516.10 Special Equipment.

(A) Fixed Electrostatic Equipment. This section shall apply to any equipment using electrostatically charged elements for the atomization, charging, and/or precipitation of hazardous materials for coatings on articles or for other similar purposes in which the charging or atomizing device is attached to a mechanical support or manipulator. This shall include robotic devices. This section shall not apply to devices that are held or manipulated by hand. Where robot or programming procedures involve manual manipulation of the robot arm while spraying with the high voltage on, the provisions of 516.10(B) shall apply. The installation of electrostatic spraying equipment shall comply with 516.10(A)(1) through (A)(10). Spray equipment shall be listed. All automatic electrostatic equipment systems shall comply with 516.6(A) through (E).

(1) Power and Control Equipment. Transformers, high-voltage supplies, control apparatus, and all other electrical portions of the equipment shall be installed outside of the Class I location or be of a type identified for the location.

Exception: High-voltage grids, electrodes, electrostatic atomizing heads, and their connections shall be permitted within the Class I location.

(2) Electrostatic Equipment. Electrodes and electrostatic atomizing heads shall be adequately supported in permanent locations and shall be effectively insulated from ground. Electrodes and electrostatic atomizing heads that are permanently attached to their bases, supports, reciprocators, or robots shall be deemed to comply with this section.

(3) High-Voltage Leads. High-voltage leads shall be properly insulated and protected from mechanical damage or exposure to destructive chemicals. Any exposed element at high voltage shall be effectively and permanently supported on suitable insulators and shall be effectively guarded against accidental contact or grounding.

(4) Support of Goods. Goods being coated using this process shall be supported on conveyors or hangers. The conveyors or hangers shall be arranged (1) to ensure that the parts being coated are electrically connected to ground with a resistance of 1 megohm or less and (2) to prevent parts from swinging.

(5) Automatic Controls. Electrostatic apparatus shall be equipped with automatic means that will rapidly de-energize the high-voltage elements under any of the following conditions:

(1) Stoppage of ventilating fans or failure of ventilating equipment from any cause
(2) Stoppage of the conveyor carrying goods through the high-voltage field unless stoppage is required by the spray process
(3) Occurrence of excessive current leakage at any point in the high-voltage system
(4) De-energizing the primary voltage input to the power supply

(6) Grounding. All electrically conductive objects in the spray area, except those objects required by the process to be at high voltage, shall be adequately grounded. This requirement shall apply to paint containers, wash cans, guards, hose connectors, brackets, and any other electrically conductive objects or devices in the area.

Informational Note: For more information on grounding and bonding for static electricity purposes, see NFPA 33-2016, *Standard for Spray Application Using Flammable or Combustible Materials*; NFPA 34-2015, *Standard for Dipping, Coating, and Printing Processes Using Flammable or Combustible Liquids*; and NFPA 77-2014, *Recommended Practice on Static Electricity*.

(7) Isolation. Safeguards such as adequate booths, fencing, railings, interlocks, or other means shall be placed about the equipment or incorporated therein so that they, either by their location, character, or both, ensure that a safe separation of the process is maintained.

(8) Signs. Signs shall be conspicuously posted to convey the following:

(1) Designate the process zone as dangerous with regard to fire and accident
(2) Identify the grounding requirements for all electrically conductive objects in the spray area
(3) Restrict access to qualified personnel only

(9) Insulators. All insulators shall be kept clean and dry.

(10) Other Than Nonincendive Equipment. Spray equipment that cannot be classified as nonincendive shall comply with 516.10(A)(10)(a) and (A)(10)(b).

(a) Conveyors, hangers, and application equipment shall be arranged so that a minimum separation of at least twice the sparking distance is maintained between the workpiece or material being sprayed and electrodes, electrostatic atomizing heads, or charged conductors. Warnings defining this safe distance shall be posted. [**33**:11.4.1]

(b) The equipment shall provide an automatic means of rapidly de-energizing the high-voltage elements in the event the distance between the goods being painted and the electrodes or electrostatic atomizing heads falls below that specified in 516.10(A)(10)(a). [**33**:11.3.8]

(B) Hand-Spraying Electrostatic Equipment. This section shall apply to any equipment using electrostatically charged elements for the atomization, charging, or precipitation of flammable and combustible materials for coatings on articles, or for other similar purposes in which the charging or atomizing device is hand-held and manipulated during the spraying operation. Electrostatic hand-spraying equipment and devices used in connection with paint-spraying operations shall be of listed types and shall comply with 516.10(B)(1) through (B)(5).

(1) General. The high-voltage circuits shall be designed so as not to produce a spark of sufficient intensity to ignite the most readily ignitible of those vapor–air mixtures likely to be encountered or result in appreciable shock hazard upon coming in contact with a grounded object under all normal operating conditions. The electrostatically charged exposed elements of the handgun shall be capable of being energized only by an actuator that also controls the coating material supply.

(2) Power Equipment. Transformers, power packs, control apparatus, and all other electrical portions of the equipment shall be located outside of the Class I location or be identified for the location.

Exception: The handgun itself and its connections to the power supply shall be permitted within the Class I location.

(3) Handle. The handle of the spraying gun shall be electrically connected to ground by a conductive material and be constructed so that the operator in normal operating position is in electrical contact with the grounded handle with a resistance of not more than 1 megohm to prevent buildup of a static charge on the operator's body. Signs indicating the necessity for grounding other persons entering the spray area shall be conspicuously posted.

(4) Electrostatic Equipment. All electrically conductive objects in the spraying area, except those objects required by the process to be at high voltage shall be electrically connected to ground with a resistance of not more than 1 megohm. This requirement shall apply to paint containers, wash cans, and any other electrical conductive objects or devices in the area. The equipment shall carry a prominent, permanently installed warning regarding the necessity for this grounding feature.

> Informational Note: For more information on grounding and bonding for static electricity purposes, see NFPA 33-2016, *Standard for Spray Application Using Flammable or Combustible Materials*; NFPA 34-2015, *Standard for Dipping, Coating, and Printing Processes Using Flammable or Combustible Liquids*; and NFPA 77-2007, *Recommended Practice on Static Electricity*.

(5) Support of Objects. Objects being painted shall be maintained in electrical contact with the conveyor or other grounded support. Hooks shall be regularly cleaned to ensure adequate grounding of 1 megohm or less. Areas of contact shall be sharp points or knife edges where possible. Points of support of the object shall be concealed from random spray where feasible, and, where the objects being sprayed are supported from a conveyor, the point of attachment to the conveyor shall be located so as to not collect spray material during normal operation.

(C) Powder Coating. This section shall apply to processes in which combustible dry powders are applied. The hazards associated with combustible dusts are present in such a process to a degree, depending on the chemical composition of the material, particle size, shape, and distribution.

(1) Electrical Equipment and Sources of Ignition. Electrical equipment and other sources of ignition shall comply with the requirements of Article 502. Portable electric luminaires and other utilization equipment shall not be used within a Class II location during operation of the finishing processes. Where such luminaires or utilization equipment are used during cleaning or repairing operations, they shall be of a type identified for Class II, Division 1 locations, and all exposed metal parts shall be connected to an equipment grounding conductor.

Exception: Where portable electric luminaires are required for operations in spaces not readily illuminated by fixed lighting within the spraying area, they shall be of the type listed for Class II, Division 1 locations where readily ignitible residues may be present.

(2) Fixed Electrostatic Spraying Equipment. The provisions of 516.10(A) and 516.10(C)(1) shall apply to fixed electrostatic spraying equipment.

(3) Electrostatic Hand-Spraying Equipment. The provisions of 516.10(B) and 516.10(C)(1) shall apply to electrostatic hand-spraying equipment.

(4) Electrostatic Fluidized Beds. Electrostatic fluidized beds and associated equipment shall be of identified types. The high-voltage circuits shall be designed such that any discharge produced when the charging electrodes of the bed are approached or contacted by a grounded object shall not be of sufficient intensity to ignite any powder–air mixture likely to be encountered or to result in an appreciable shock hazard.

(a) Transformers, power packs, control apparatus, and all other electrical portions of the equipment shall be located outside the powder-coating area or shall otherwise comply with the requirements of 516.10(C)(1).

Exception: The charging electrodes and their connections to the power supply shall be permitted within the powder-coating area.

(b) All electrically conductive objects within the powder-coating area shall be adequately grounded. The powder-coating equipment shall carry a prominent, permanently installed warning regarding the necessity for grounding these objects.

> Informational Note: For more information on grounding and bonding for static electricity purposes, see NFPA 33-2016, *Standard for Spray Application Using Flammable or Combustible Materials*; NFPA 34-2015, *Standard for Dipping, Coating, and Printing Processes Using Flammable or Combustible Liquids*; and NFPA 77-2014, *Recommended Practice on Static Electricity*.

(c) Objects being coated shall be maintained in electrical contact (less than 1 megohm) with the conveyor or other support in order to ensure proper grounding. Hangers shall be regularly cleaned to ensure effective electrical contact. Areas of electrical contact shall be sharp points or knife edges where possible.

(d) The electrical equipment and compressed air supplies shall be interlocked with a ventilation system so that the equipment cannot be operated unless the ventilating fans are in operation. [**33**:Chapter 15]

516.16 Grounding. All metal raceways, the metal armors or metallic sheath on cables, and all non–current-carrying metal parts of fixed or portable electrical equipment, regardless of voltage, shall be grounded and bonded. Grounding and bonding shall comply with 501.30, 502.30, or 505.25, as applicable.

Part IV. Spray Application Operations in Membrane Enclosures

516.18 Area Classification for Temporary Membrane Enclosures. Electrical area classification shall be as follows:

(1) The area within the membrane enclosure shall be considered a Class I, Division 1 area, as shown in Figure 516.18.
(2) A 1.5 m (5 ft) zone outside of the membrane enclosure shall be considered Class I, Division 2, as shown in Figure 516.18.

> Informational Note No. 1: The risks to people and property are unique when spray painting within the confined spaces of temporary membrane enclosures. See NFPA 33-2016, *Standard for Spray Application Using Flammable or Combustible Materials*, for information on occupancy, ventilation, fire protection, and permitting for spray application operations in membrane enclosures. NFPA 33-2016 limits spray application operations within

both outdoor and indoor temporary membrane enclosures, as well as use and time constraints.

Informational Note No. 2: Section 18.6 of NFPA 33-2016, *Standard for Spray Application Using Flammable or Combustible Materials,* limits material used in a vertical plane for membrane enclosures. See also NFPA 701-2015, *Standard Methods of Fire Tests for Flame Propagation of Textiles and Films,* Test Method 2 for construction information.

Informational Note No. 3: See 18.3.2.1.1 of NFPA 33-2016, *Standard for Spray Application Using Flammable or Combustible Materials,* for membrane installation beneath sprinklers. See also 8.15.15 of NFPA 13-2014, *Standard for the Installation of Sprinkler Systems,* for protection of membrane structures.

516.23 Electrical and Other Sources of Ignition. Electrical wiring and utilization equipment used within the classified areas inside and outside of membrane enclosures during spray painting shall be suitable for the location and shall comply with all of the following:

(1)   All power to the workpiece shall be removed during spray painting.
(2)   Workpieces shall be grounded.

(3)   Spray paint equipment shall be grounded.
(4)   Scaffolding shall be bonded to the workpiece and grounded by an approved method.

Part V. Printing, Dipping, and Coating Processes

516.29 Classification of Locations.   Classification is based on quantities of flammable vapors, combustible mists, residues, dusts, or deposits that are present or might be present in quantities sufficient to produce ignitable or explosive mixtures with air. Electrical wiring and electrical utilization equipment located adjacent to open processes shall comply with the requirements as follows. Examples of these requirements are illustrated in Figure 516.29(a), Figure 516.29(b), Figure 516.29(c), and Figure 516.29(d).

Informational Note: For additional guidance, see Chapter 6 of NFPA 33-2016, *Standard for Spray Application Using Flammable or Combustible Materials,* and Chapter 6 of NFPA 34-2015, *Standard for Dipping, Coating, and Printing Processes Using Flammable or Combustible Liquids.*

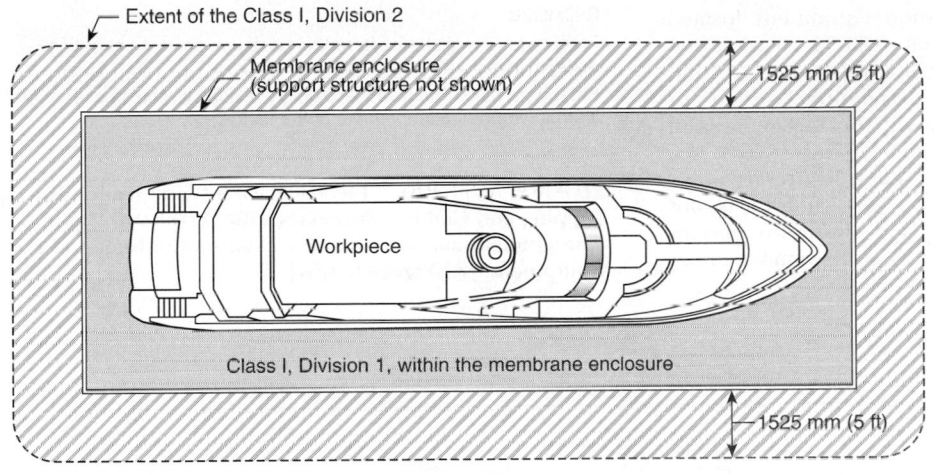

PLAN VIEW

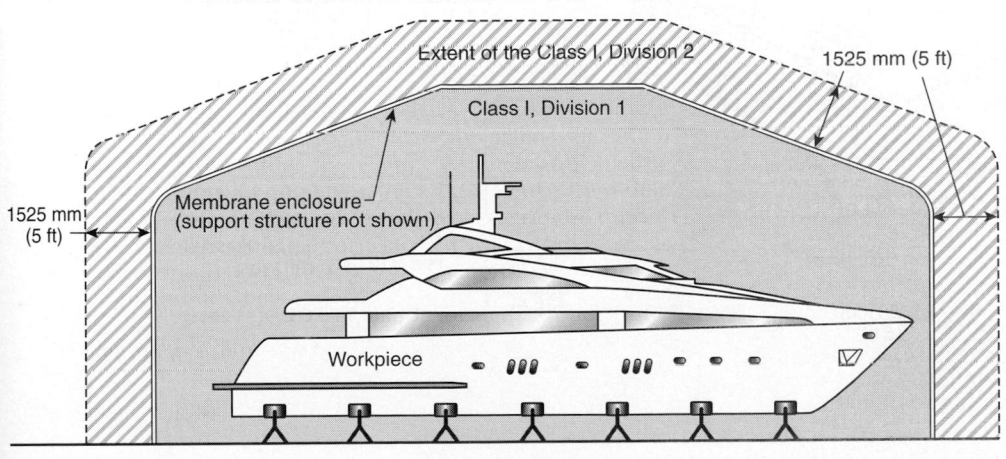

ELEVATION

FIGURE 516.18   **Electrical Classifications for Temporary Outdoor Membrane Enclosures [33:Figure 18.7.1.1]**

(1) Electrical wiring and electrical utilization equipment located in any sump, pit, or below grade channel that is within 7620 mm (25 ft) horizontally of a vapor source, as defined by this standard, shall be suitable for Class I, Division 1 or Class I, Zone 1 locations. If the sump, pit, or channel extends beyond 7620 mm (25 ft) of the vapor source, it shall be provided with a vapor stop, or it shall be classified as Class I, Division 1 or Class I, Zone 1 for its entire length. [34:6.4.1]

(2) Electrical wiring and electrical utilization equipment located within 1525 mm (5 ft) of a vapor source shall be suitable for Class I, Division 1 or Class I, Zone 1 locations. The space inside a dip tank, ink fountain, ink reservoir, or ink tank shall be classified as Class I, Division 1 or Class I, Zone 0, whichever is applicable.

(3) Electrical wiring and electrical utilization equipment located within 915 mm (3 ft) of the Class I, Division 1 or Class I, Zone 1 location shall be suitable for Class I, Division 2 or Class I, Zone 2 locations, whichever is applicable.

(4) The space 915 mm (3 ft) above the floor and extending 6100 mm (20 ft) horizontally in all directions from the Class I, Division 1 or Class I, Zone 1 location shall be classified as Class I, Division 2 or Class I, Zone 2, and electrical wiring and electrical utilization equipment located within this space shall be suitable for Class I, Division 2 or Class I, Zone 2 locations, whichever is applicable.

(5) This space shall be permitted to be nonclassified for purposes of electrical installations if the surface area of the vapor source does not exceed 0.5 m² (5 ft²), the contents of the dip tank, ink fountain, ink reservoir, or ink tank do not exceed 19 L (5 gal), and the vapor concentration during operating and shutdown periods does not exceed 25 percent of the lower flammable limit.

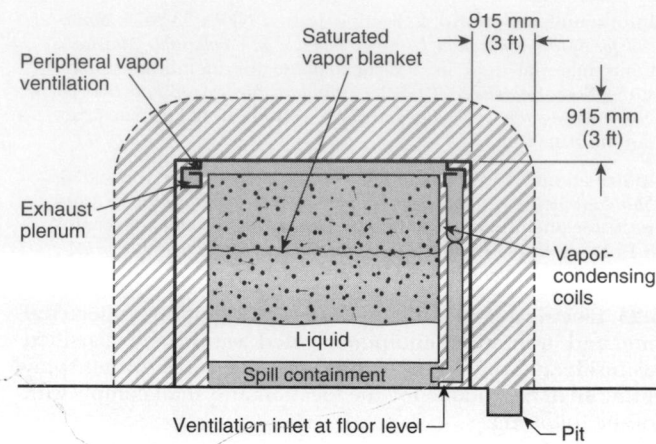

☐ Class I, Division 1 or Zone 0
☐ Class I, Division 1 or Zone 1
☐ Class I, Division 2 or Zone 2

Definitions

Freeboard: The distance from the maximum solvent or coating material level to the top of the tank

Freeboard ratio: The freeboard height divided by the smaller of the interior length or interior width of the tank

N **FIGURE 516.29(b)   Electrical Area Classification for Open Dipping and Coating Processes with Peripheral Vapor Containment and Ventilation — Vapors Confined to Process Equipment. [34:Figure 6.4(b)]**

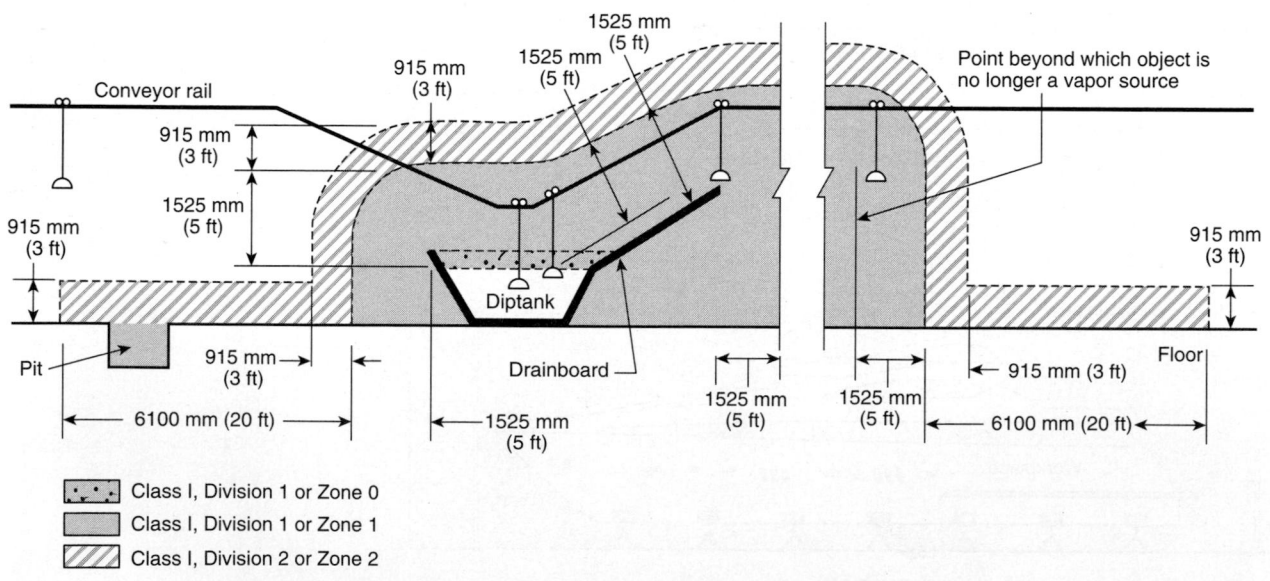

☐ Class I, Division 1 or Zone 0
☐ Class I, Division 1 or Zone 1
☐ Class I, Division 2 or Zone 2

N **FIGURE 516.29(a)   Electrical Area Classification for Open Dipping and Coating Processes Without Vapor Containment or Ventilation. [34:Figure 6.4(a)]**

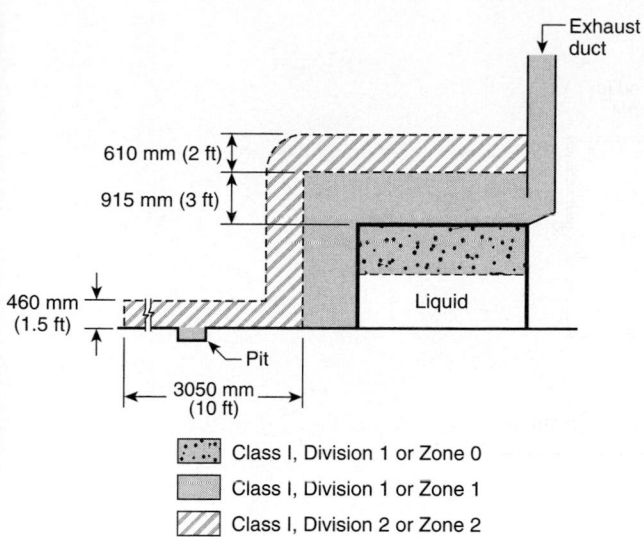

FIGURE 516.29(c)   Electrical Area Classification for Open Dipping and Coating Processes with Partial Peripheral Vapor Containment and Ventilation — Vapors NOT Confined to Process Equipment. [34:Figure 6.4(c)]

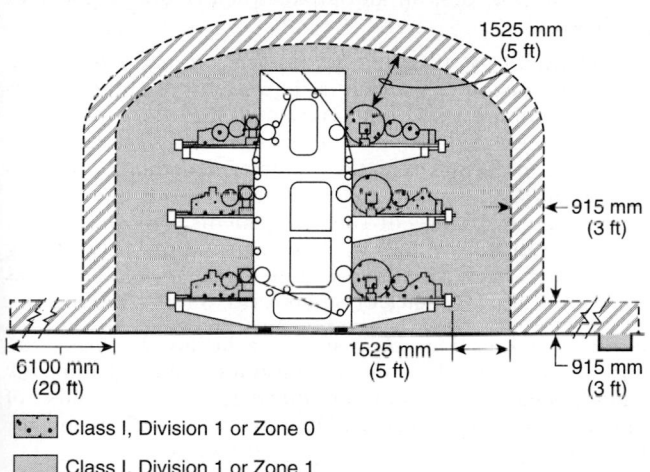

FIGURE 516.29(d)   Electrical Area Classification for a Typical Printing Process. [34:Figure 6.4(d)]

516.35 Areas Adjacent to Enclosed Dipping and Coating Processes. Areas adjacent to enclosed dipping and coating processes are illustrated by Figure 516.35 and shall be classified as follows:

(1) The interior of any enclosed dipping or coating process or apparatus shall be a Class I, Division 1 or Class I, Zone 1 location, and electrical wiring and electrical utilization equipment located within this space shall be suitable for Class I, Division 1 or Class I, Zone 1 locations, whichever is applicable. The area inside the dip tank shall be classified as Class I, Division 1 or Class I, Zone 0, whichever is applicable.

(2) The space within 915 mm (3 ft) in all directions from any opening in the enclosure and extending to the floor or grade level shall be classified as Class I, Division 2 or Class I, Zone 2, and electrical wiring and electrical utilization equipment located within this space shall be suitable for Class I, Division 2 locations or Class I, Zone 2 locations, whichever is applicable.

(3) All other spaces adjacent to an enclosed dipping or coating process or apparatus shall be classified as nonhazardous for purposes of electrical installations.

516.36 Equipment and Containers in Ventilated Areas. Open containers, supply containers, waste containers, and solvent distillation units that contain Class I liquids shall be located in areas ventilated in accordance with 516.4.

516.37 Luminaires. For printing, coating, and dipping equipment where the process area is enclosed by glass panels that are sealed to confine vapors and mists to the inside of the enclosure, luminaires that are attached to the walls or ceilings of a process enclosure and that are located outside of any classified area shall be permitted to be of general purpose construction. Such luminaires shall be serviced from outside the enclosure.

Luminaires that are attached to the walls or ceilings of a process enclosure, are located within the Class I, Division 2 or Class I, Zone 2 location, and are separated from the process area by glass panels that are sealed to confine vapors and mists shall be suitable for use in that location. Such fixtures shall be serviced from outside the enclosure.

516.38 Wiring and Equipment Not Within Classified Locations.

(A) Wiring. All fixed wiring above the Class I and II locations shall be in metal raceways, Type PVC conduit, Type RTRC conduit, or electrical nonmetallic tubing; where cables are used, they shall be Type MI, Type TC, or Type MC cable. Cellular metal floor raceways shall only be permitted to supply ceiling outlets or as extensions to the area below the floor of a Class I or II location. Where cellular metal raceways are used, they shall not have connections leading into or passing through the Class I or II location unless suitable seals are provided.

(B) Equipment. Equipment that is capable of producing arcs, sparks, or particles of hot metal, such as lamps and lampholders for fixed lighting, cutouts, switches, receptacles, motors, or other equipment having make-and-break or sliding contacts, where installed above a classified location or above a location where freshly finished goods are handled, shall be of the totally enclosed type or be constructed so as to prevent the escape of sparks or hot metal particles.

516.40 Static Electric Discharges. All persons and all electrically conductive objects, including any metal parts of the process equipment or apparatus, containers of material, exhaust ducts, and piping systems that convey flammable or combustible liquids, shall be electrically grounded.

Provision shall be made to dissipate static electric charges from all nonconductive substrates in printing processes.

Informational Note: For additional guidance on reducing the risk of ignition from electrostatic discharges, see NFPA 77-2014, *Recommended Practice on Static Electricity.*

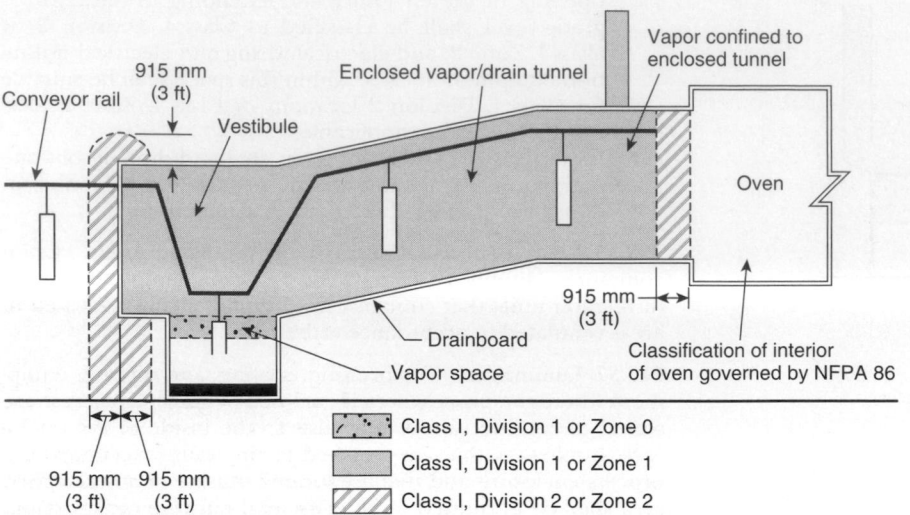

FIGURE 516.35   Electrical Area Classification Around Enclosed Dipping and Coating Processes. [34:Figure 6.5]

ARTICLE 517
Health Care Facilities

Informational Note: Text that is followed by a reference in brackets has been extracted from NFPA 99-2015, *Health Care Facilities Code*, and NFPA 101-2015, *Life Safety Code*. Only editorial changes were made to the extracted text to make it consistent with this *Code*.

Part I. General

517.1 Scope. The provisions of this article shall apply to electrical construction and installation criteria in health care facilities that provide services to human beings.

The requirements in Parts II and III not only apply to single-function buildings but are also intended to be individually applied to their respective forms of occupancy within a multi-function building (e.g., a doctor's examining room located within a limited care facility would be required to meet the provisions of 517.10).

Informational Note: For information concerning performance, maintenance, and testing criteria, refer to the appropriate health care facilities documents.

517.2 Definitions.

Alternate Power Source. One or more generator sets, or battery systems where permitted, intended to provide power during the interruption of the normal electrical service; or the public utility electrical service intended to provide power during interruption of service normally provided by the generating facilities on the premises. [99:3.3.4]

Ambulatory Health Care Occupancy. An occupancy used to provide services or treatment simultaneously to four or more patients that provides, on an outpatient basis, one or more of the following:

(1) Treatment for patients that renders the patients incapable of taking action for self-preservation under emergency conditions without assistance of others.
(2) Anesthesia that renders the patients incapable of taking action for self-preservation under emergency conditions without the assistance of others.
(3) Emergency or urgent care for patients who, due to the nature of their injury or illness, are incapable of taking action for self-preservation under emergency conditions without the assistance of others. [*101:* 3.3.188.1]

Anesthetizing Location. Any area of a facility that has been designated to be used for the administration of any flammable or nonflammable inhalation anesthetic agent in the course of examination or treatment, including the use of such agents for relative analgesia.

Battery-Powered Lighting Units. Individual unit equipment for backup illumination consisting of the following:

(1) Rechargeable battery
(2) Battery-charging means
(3) Provisions for one or more lamps mounted on the equipment, or with terminals for remote lamps, or both
(4) Relaying device arranged to energize the lamps automatically upon failure of the supply to the unit equipment

Critical Branch. A system of feeders and branch circuits supplying power for task illumination, fixed equipment, select receptacles, and select power circuits serving areas and functions related to patient care that are automatically connected to alternate power sources by one or more transfer switches during interruption of normal power source. [99:3.3.27]

Electrical Life-Support Equipment. Electrically powered equipment whose continuous operation is necessary to maintain a patient's life. [99:3.3.39]

Equipment Branch. A system of feeders and branch circuits arranged for delayed, automatic, or manual connection to the alternate power source and that serves primarily 3-phase power equipment. [**99:**3.3.43].

Essential Electrical System. A system comprised of alternate sources of power and all connected distribution systems and ancillary equipment, designed to ensure continuity of electrical power to designated areas and functions of a health care facility during disruption of normal power sources, and also to minimize disruption within the internal wiring system. [**99:**3.3.45]

Exposed Conductive Surfaces. Those surfaces that are capable of carrying electric current and that are unprotected, uninsulated, unenclosed, or unguarded, permitting personal contact. [**99:**3.3.47]

> Informational Note: Paint, anodizing, and similar coatings are not considered suitable insulation, unless they are listed for such use.

Fault Hazard Current. See *Hazard Current.*

Flammable Anesthetics. Gases or vapors, such as fluroxene, cyclopropane, divinyl ether, ethyl chloride, ethyl ether, and ethylene, which may form flammable or explosive mixtures with air, oxygen, or reducing gases such as nitrous oxide.

Flammable Anesthetizing Location. Any area of the facility that has been designated to be used for the administration of any flammable inhalation anesthetic agents in the normal course of examination or treatment.

N Governing Body. The person or persons who have the overall legal responsibility for the operation of a health care facility. [**99:**3.3.62]

Hazard Current. For a given set of connections in an isolated power system, the total current that would flow through a low impedance if it were connected between either isolated conductor and ground.

Fault Hazard Current. The hazard current of a given isolated system with all devices connected except the line isolation monitor.

Monitor Hazard Current. The hazard current of the line isolation monitor alone.

Total Hazard Current. The hazard current of a given isolated system with all devices, including the line isolation monitor, connected.

Health Care Facilities. Buildings, portions of buildings, or mobile enclosures in which human medical, dental, psychiatric, nursing, obstetrical, or surgical care are provided. [**99:**3.3.67]

> Informational Note: Examples of health care facilities include, but are not limited to, hospitals, nursing homes, limited care facilities, clinics, medical and dental offices, and ambulatory care centers, whether permanent or movable.

Hospital. A building or portion thereof used on a 24-hour basis for the medical, psychiatric, obstetrical, or surgical care of four or more inpatients. [*101:* 3.3.142]

Isolated Power System. A system comprising an isolating transformer or its equivalent, a line isolation monitor, and its ungrounded circuit conductors. [**99:**3.3.83]

Isolation Transformer. A transformer of the multiple-winding type, with the primary and secondary windings physically separated, that inductively couples its ungrounded secondary winding(s) to the grounded feeder system that energizes its primary winding(s). [**99:**3.3.84]

N Invasive Procedure. Any procedure that penetrates the protective surfaces of a patient's body (i.e., skin, mucous membrane, cornea) and that is performed with an aseptic field (procedural site). Not included in this category are placement of peripheral intravenous needles or catheters used to administer fluids and/or medications, gastrointestinal endoscopies (i.e., sigmoidoscopies), insertion of urethral catheters, and other similar procedures. [**99:**3.3.81]

Life Safety Branch. A system of feeders and branch circuits supplying power for lighting, receptacles, and equipment essential for life safety that is automatically connected to alternate power sources by one or more transfer switches during interruption of the normal power source. [**99:**3.3.87]

Limited Care Facility. A building or portion thereof used on a 24-hour basis for the housing of four or more persons who are incapable of self-preservation because of age; physical limitation due to accident or illness; or limitations such as mental retardation/developmental disability, mental illness, or chemical dependency.

Line Isolation Monitor. A test instrument designed to continually check the balanced and unbalanced impedance from each line of an isolated circuit to ground and equipped with a built-in test circuit to exercise the alarm without adding to the leakage current hazard. [**99:**3.3.89]

N Medical Office (Dental Office). A building or part thereof in which the following occur: (1) examinations and minor treatments or procedures are performed under the continuous supervision of a medical or dental professional; (2) only sedation or local anesthesia is involved and treatment or procedures do not render the patient incapable of self preservation under emergency conditions; and (3) overnight stays for patients or 24-hour operation are not provided. [**99:**3.3.98]

Monitor Hazard Current. See *Hazard Current.*

Nurses' Stations. Areas intended to provide a center of nursing activity for a group of nurses serving bed patients, where the patient calls are received, nurses are dispatched, nurses' notes written, inpatient charts prepared, and medications prepared for distribution to patients. Where such activities are carried on in more than one location within a nursing unit, all such separate areas are considered a part of the nurses' station.

Nursing Home. A building or portion of a building used on a 24-hour basis for the housing and nursing care of four or more persons who, because of mental or physical incapacity, might be unable to provide for their own needs and safety without the assistance of another person. [*101:* 3.3.142.2]

Patient Bed Location. The location of a patient sleeping bed, or the bed or procedure table of a critical care space. [**99:**3.3.125]

Patient Care Space. Any space of a health care facility wherein patients are intended to be examined or treated. [**99:**3.3.127]

> Informational Note No. 1: The governing body of the facility designates patient care space in accordance with the type of patient care anticipated. [**99:**1.3.4.1]

Informational Note No. 2: Business offices, corridors, lounges, day rooms, dining rooms, or similar areas typically are not classified as patient care spaces. [**99**:A.3.3.127]

Basic Care (Category 3) Space. Space in which failure of equipment or a system is not likely to cause injury to the patients, staff, or visitors but can cause patient discomfort. [**99**:3.3.127.3]

Informational Note: [Category 3] spaces, formerly known as basic care rooms [(spaces)], are typically where basic medical or dental care, treatment, or examinations are performed. Examples include, but are not limited to, examination or treatment rooms in clinics, medical and dental offices, nursing homes, and limited care facilities. [**99**:A.3.3.127.3]

General Care (Category 2) Space. Space in which failure of equipment or a system is likely to cause minor injury to patients, staff, or visitors. [**99**:3.3.127.2]

Informational Note: [Category 2] spaces were formerly known as general care rooms [(spaces)]. Examples include, but are not limited to, inpatient bedrooms, dialysis rooms, in vitro fertilization rooms, procedural rooms, and similar rooms. [**99**:A.3.3.127.2]

Critical Care (Category 1) Space. Space in which failure of equipment or a system is likely to cause major injury or death of patients, staff, or visitors. [**99**:3.3.127.1]

Informational Note: [Category 1] spaces, formerly known as critical care rooms [(spaces)], are typically where patients are intended to be subjected to invasive procedures and connected to line-operated, patient care–related appliances. Examples include, but are not limited to, special care patient rooms used for critical care, intensive care, and special care treatment rooms such as angiography laboratories, cardiac catheterization laboratories, delivery rooms, operating rooms, post-anesthesia care units, trauma rooms, and other similar rooms. [**99**:A.3.3.127.1]

N *Support (Category 4) Space.* Space in which failure of equipment or a system is not likely to have a physical impact on patient care. [**99**:3.3.127.4]

Informational Note: [Category 4] spaces were formerly known as support rooms [(spaces)]. Examples of support spaces include, but are not limited to, anesthesia work rooms, sterile supply, laboratories, morgues, waiting rooms, utility rooms, and lounges. [**99**:A.3.3.127.4]

Patient Care Vicinity. A space, within a location intended for the examination and treatment of patients, extending 1.8 m (6 ft) beyond the normal location of the patient bed, chair, table, treadmill, or other device that supports the patient during examination and treatment and extending vertically to 2.3 m (7 ft 6 in.) above the floor. [**99**:3.3.128]

Patient Equipment Grounding Point. A jack or terminal that serves as the collection point for redundant grounding of electrical appliances serving a patient care vicinity or for grounding other items in order to eliminate electromagnetic interference problems. [**99**:3.3.129]

Psychiatric Hospital. A building used exclusively for the psychiatric care, on a 24-hour basis, of four or more inpatients.

Reference Grounding Point. The ground bus of the panelboard or isolated power system panel supplying the patient care room. [**99**:3.3.143]

Relative Analgesia. A state of sedation and partial block of pain perception produced in a patient by the inhalation of concentrations of nitrous oxide insufficient to produce loss of consciousness (conscious sedation).

Selected Receptacles. A minimum number of receptacles selected by the governing body of a facility as necessary to provide essential patient care and facility services during loss of normal power. [**99**:3.3.148]

Task Illumination. Provisions for the minimum lighting required to carry out necessary tasks in the described areas, including safe access to supplies and equipment and access to exits. [**99**:63.3.161]

Total Hazard Current. The hazard current of a given isolated system with all devices, including the line isolation monitor, connected. [**99**:3.3.66.3]

Wet Procedure Location. The area in a patient care space where a procedure is performed that is normally subject to wet conditions while patients are present, including standing fluids on the floor or drenching of the work area, either of which condition is intimate to the patient or staff. [**99**:3.3.171]

Informational Note: Routine housekeeping procedures and incidental spillage of liquids do not define a wet procedure location. [**99**:A.3.3.171]

X-Ray Installations, Long-Time Rating. A rating based on an operating interval of 5 minutes or longer.

X-Ray Installations, Mobile. X-ray equipment mounted on a permanent base with wheels, casters, or a combination of both to facilitate moving the equipment while completely assembled.

X-Ray Installations, Momentary Rating. A rating based on an operating interval that does not exceed 5 seconds.

X-Ray Installations, Portable. X-ray equipment designed to be hand carried.

X-Ray Installations, Transportable. X-ray equipment to be conveyed by a vehicle or that is readily disassembled for transport by a vehicle.

Part II. Wiring and Protection

517.10 Applicability.

(A) Applicability. Part II shall apply to patient care space of all health care facilities.

(B) Not Covered. Part II shall not apply to the following:

(1) Business offices, corridors, waiting rooms, and the like in clinics, medical and dental offices, and outpatient facilities

(2) Areas of nursing homes and limited care facilities wired in accordance with Chapters 1 through 4 of this *Code* where these areas are used exclusively as patient sleeping rooms

Informational Note: See NFPA *101*-2015, *Life Safety Code*®.

517.11 General Installation — Construction Criteria. The purpose of this article is to specify the installation criteria and wiring methods that minimize electrical hazards by the maintenance of adequately low potential differences only between exposed conductive surfaces that are likely to become energized and could be contacted by a patient.

Informational Note: In a health care facility, it is difficult to prevent the occurrence of a conductive or capacitive path from the patient's body to some grounded object, because that path may be established accidentally or through instrumentation directly connected to the patient. Other electrically conductive surfaces that may make an additional contact with the patient, or instruments that may be connected to the patient, then become possible sources of electric currents that can traverse the patient's body. The hazard is increased as more apparatus is associated with the patient, and, therefore, more intensive precautions are needed. Control of electric shock hazard requires the limitation of electric current that might flow in an electrical circuit involving the patient's body by raising the resistance of the conductive circuit that includes the patient, or by insulating exposed surfaces that might become energized, in addition to reducing the potential difference that can appear between exposed conductive surfaces in the patient care vicinity, or by combinations of these methods. A special problem is presented by the patient with an externalized direct conductive path to the heart muscle. The patient may be electrocuted at current levels so low that additional protection in the design of appliances, insulation of the catheter, and control of medical practice is required.

517.12 Wiring Methods. Except as modified in this article, wiring methods shall comply with the applicable provisions of Chapters 1 through 4 of this *Code*.

517.13 Grounding of Receptacles and Fixed Electrical Equipment in Patient Care Spaces. Wiring in patient care spaces shall comply with 517.13(A) and (B).

(A) Wiring Methods. All branch circuits serving patient care spaces shall be provided with an effective ground-fault current path by installation in a metal raceway system or a cable having a metallic armor or sheath assembly. The metal raceway system, metallic cable armor, or sheath assembly shall itself qualify as an equipment grounding conductor in accordance with 250.118.

(B) Insulated Equipment Grounding Conductors and Insulated Equipment Bonding Jumpers.

(1) General. The following shall be directly connected to an insulated copper equipment grounding conductor that is clearly identified along its entire length by green insulation and installed with the branch circuit conductors in the wiring methods as provided in 517.13(A):

(1) The grounding terminals of all receptacles other than isolated ground receptacles

(2) Metal outlet boxes, metal device boxes, or metal enclosures

(3) All non–current-carrying conductive surfaces of fixed electrical equipment likely to become energized that are subject to personal contact, operating at over 100 volts

Exception No. 1: For other than isolated ground receptacles, an insulated equipment bonding jumper that directly connects to the equipment grounding conductor is permitted to connect the box and receptacle(s) to the equipment grounding conductor. Isolated ground receptacles shall be connected in accordance with 517.16.

Exception No. 2: Metal faceplates shall be permitted to be connected to the equipment grounding conductor by means of a metal mounting screw(s) securing the faceplate to a grounded outlet box or grounded wiring device.

Exception No. 3: Luminaires more than 2.3 m (7½ ft) above the floor and switches located outside of the patient care vicinity shall be permitted to be connected to an equipment grounding return path complying with 517.13(A) or (B).

(2) Sizing. Equipment grounding conductors and equipment bonding jumpers shall be sized in accordance with 250.122.

517.14 Panelboard Bonding. The equipment grounding terminal buses of the normal and essential branch-circuit panelboards serving the same individual patient care vicinity shall be connected together with an insulated continuous copper conductor not smaller than 10 AWG. Where two or more panelboards serving the same individual patient care vicinity are served from separate transfer switches on the essential electrical system, the equipment grounding terminal buses of those panelboards shall be connected together with an insulated continuous copper conductor not smaller than 10 AWG. This conductor shall be permitted to be broken in order to terminate on the equipment grounding terminal bus in each panelboard.

517.16 Use of Isolated Ground Receptacles.

(A) Inside of a Patient Care Vicinity. An isolated grounding receptacle shall not be installed within a patient care vicinity. [**99:**6.3.2.2.7.1(B)]

(B) Outside of a Patient Care Vicinity. Isolated ground receptacle(s) installed in patient care spaces outside of a patient care vicinity(s) shall comply with 517.16(B)(1) and (2).

N **(1)** The grounding terminals of isolated ground receptacles installed in branch circuits for patient care spaces shall be connected to an insulated equipment grounding conductor in accordance with 250.146(D) in addition to the equipment grounding conductor path required in 517.13(A).

The equipment grounding conductor connected to the grounding terminals of isolated ground receptacles in patient care spaces shall be clearly identified along the equipment grounding conductor's entire length by green insulation with one or more yellow stripes.

N **(2)** The insulated grounding conductor required in 517.13(B)(1) shall be clearly identified along its entire length by green insulation, with no yellow stripes, and shall not be connected to the grounding terminals of isolated ground receptacles but shall be connected to the box or enclosure indicated in 517.13(B)(1)(2) and to non–current-carrying conductive surfaces of fixed electrical equipment indicated in 517.13(B)(1)(3).

Informational Note No. 1: This type of installation is typically used where a reduction of electrical noise (electromagnetic interference) is necessary, and parallel grounding paths are to be avoided.

Informational Note No. 2: Care should be taken in specifying a system containing isolated ground receptacles, because the grounding impedance is controlled only by the grounding wires and does not benefit from any conduit or building structure in parallel with the grounding path. [**99:**A.6.3.2.2.7.1]

517.17 Ground-Fault Protection.

(A) Applicability. The requirements of 517.17 shall apply to hospitals, and other buildings (including multiple-occupancy buildings) with critical care (Category 1) spaces or utilizing electrical life-support equipment, and buildings that provide the required essential utilities or services for the operation of critical care (Category 1) spaces or electrical life-support equipment.

(B) Feeders. Where ground-fault protection is provided for operation of the service disconnecting means or feeder disconnecting means as specified by 230.95 or 215.10, an additional step of ground-fault protection shall be provided in all next level feeder disconnecting means downstream toward the load. Such protection shall consist of overcurrent devices and current transformers or other equivalent protective equipment that shall cause the feeder disconnecting means to open.

The additional levels of ground-fault protection shall not be installed on the load side of an essential electrical system transfer switch.

(C) Selectivity. Ground-fault protection for operation of the service and feeder disconnecting means shall be fully selective such that the feeder device, but not the service device, shall open on ground faults on the load side of the feeder device. Separation of ground-fault protection time-current characteristics shall conform to manufacturer's recommendations and shall consider all required tolerances and disconnect operating time to achieve 100 percent selectivity.

Informational Note: See 230.95, informational note, for transfer of alternate source where ground-fault protection is applied.

(D) Testing. When equipment ground-fault protection is first installed, each level shall be performance tested to ensure compliance with 517.17(C).

517.18 General Care (Category 2) Spaces.

(A) Patient Bed Location. Each patient bed location shall be supplied by at least two branch circuits, one from the critical branch and one from the normal system. All branch circuits from the normal system shall originate in the same panelboard. The electrical receptacles or the cover plate for the electrical receptacles supplied from the critical branch shall have a distinctive color or marking so as to be readily identifiable and shall also indicate the panelboard and branch-circuit number supplying them.

Branch circuits serving patient bed locations shall not be part of a multiwire branch circuit.

Exception No. 1: Branch circuits serving only special purpose outlets or receptacles, such as portable X-ray outlets, shall not be required to be served from the same distribution panel or panels.

Exception No. 2: The requirements of 517.18(A) shall not apply to patient bed locations in clinics, medical and dental offices, and outpatient facilities; psychiatric, substance abuse, and rehabilitation hospitals; sleeping rooms of nursing homes; and limited care facilities meeting the requirements of 517.10(B)(2).

Exception No. 3: A general care (Category 2) patient bed location served from two separate transfer switches on the critical branch shall not be required to have circuits from the normal system.

(B) Patient Bed Location Receptacles. Each patient bed location shall be provided with a minimum of eight receptacles. They shall be permitted to be of the single, duplex, or quadruplex type or any combination of the three. All receptacles shall be listed "hospital grade" and shall be so identified. The grounding terminal of each receptacle shall be connected to an insulated copper equipment grounding conductor sized in accordance with Table 250.122.

Exception No. 1: The requirements of 517.18(B) shall not apply to psychiatric, substance abuse, and rehabilitation hospitals meeting the requirements of 517.10(B)(2).

Exception No. 2: Psychiatric security rooms shall not be required to have receptacle outlets installed in the room.

Informational Note: It is not intended that there be a total, immediate replacement of existing non–hospital grade receptacles. It is intended, however, that non–hospital grade receptacles be replaced with hospital grade receptacles upon modification of use, renovation, or as existing receptacles need replacement.

(C) Designated General Care (Category 2) Pediatric Locations. Receptacles that are located within the patient rooms, bathrooms, playrooms, and activity rooms of pediatric units or spaces with similar risk as determined by the governing body, other than nurseries, shall be listed tamper-resistant or shall employ a listed tamper-resistant cover. [**99**:6.3.2.2.6.2(F)]

517.19 Critical Care (Category 1) Spaces.

(A) Patient Bed Location Branch Circuits. Each patient bed location shall be supplied by at least two branch circuits, one or more from the critical branch and one or more circuits from the normal system. At least one branch circuit from the critical branch shall supply an outlet(s) only at that bed location.

The electrical receptacles or the cover plates for the electrical receptacles supplied from the life safety and critical branches shall have a distinctive color or marking so as to be readily identifiable. [**99**:6.4.2.2.6.2(C)]

All branch circuits from the normal system shall be from a single panelboard. Critical branch receptacles shall be identified and shall also indicate the panelboard and circuit number supplying them.

The branch circuit serving patient bed locations shall not be part of a multiwire branch circuit.

Exception No. 1: Branch circuits serving only special-purpose receptacles or equipment in critical care (Category 1) spaces shall be permitted to be served by other panelboards.

Exception No. 2: Critical care (Category 1) spaces served from two separate critical branch transfer switches shall not be required to have circuits from the normal system.

(B) Patient Bed Location Receptacles.

(1) Minimum Number and Supply. Each patient bed location shall be provided with a minimum of 14 receptacles, at least one of which shall be connected to either of the following:

(1)   The normal system branch circuit required in 517.19(A)
(2)   A critical branch circuit supplied by a different transfer switch than the other receptacles at the same patient bed location

(2) Receptacle Requirements. The receptacles required in 517.19(B)(1) shall be permitted to be single, duplex, or quadruplex type or any combination thereof. All receptacles shall be listed "hospital grade" and shall be so identified. The grounding terminal of each receptacle shall be connected to the reference grounding point by means of an insulated copper equipment grounding conductor.

(C) Operating Room Receptacles.

(1) Minimum Number and Supply. Each operating room shall be provided with a minimum of 36 receptacles divided between at least two branch circuits. At least 12 receptacles, but no more than 24, shall be connected to either of the following:

(1)   The normal system branch circuit required in 517.19(A)
(2)   A critical branch circuit supplied by a different transfer switch than the other receptacles at the same location

(2) Receptacle Requirements. The receptacles shall be permitted to be of the locking or nonlocking type, single, duplex, or quadruplex types or any combination of the three.

All nonlocking-type receptacles shall be listed hospital grade and so identified. The grounding terminal of each receptacle shall be connected to the reference grounding point by means of an insulated copper equipment grounding conductor.

(D) Patient Care Vicinity Grounding and Bonding (Optional). A patient care vicinity shall be permitted to have a patient equipment grounding point. The patient equipment grounding point, where supplied, shall be permitted to contain one or more listed grounding and bonding jacks. An equipment bonding jumper not smaller than 10 AWG shall be used to connect the grounding terminal of all grounding-type receptacles to the patient equipment grounding point. The bonding conductor shall be permitted to be arranged centrically or looped as convenient.

> Informational Note: Where there is no patient equipment grounding point, it is important that the distance between the reference grounding point and the patient care vicinity be as short as possible to minimize any potential differences.

(E) Equipment Grounding and Bonding. Where a grounded electrical distribution system is used and metal feeder raceway or Type MC or MI cable that qualifies as an equipment grounding conductor in accordance with 250.118 is installed, grounding of enclosures and equipment, such as panelboards, switchboards, and switchgear, shall be ensured by one of the following bonding means at each termination or junction point of the metal raceway or Type MC or MI cable:

(1) A grounding bushing and a continuous copper bonding jumper, sized in accordance with 250.122, with the bonding jumper connected to the junction enclosure or the ground bus of the panel

(2) Connection of feeder raceways or Type MC or MI cable to threaded hubs or bosses on terminating enclosures

(3) Other approved devices such as bonding-type locknuts or bushings. Standard locknuts shall not be used for bonding.

(F) Additional Protective Techniques in Critical Care (Category 1) Spaces (Optional). Isolated power systems shall be permitted to be used for critical care (Category 1) spaces, and, if used, the isolated power system equipment shall be listed as isolated power equipment. The isolated power system shall be designed and installed in accordance with 517.160.

Exception: The audible and visual indicators of the line isolation monitor shall be permitted to be located at the nursing station for the area being served.

(G) Isolated Power System Equipment Grounding. Where an isolated ungrounded power source is used and limits the first-fault current to a low magnitude, the equipment grounding conductor associated with the secondary circuit shall be permitted to be run outside of the enclosure of the power conductors in the same circuit.

> Informational Note: Although it is permitted to run the grounding conductor outside of the conduit, it is safer to run it with the power conductors to provide better protection in case of a second ground fault.

(H) Special-Purpose Receptacle Grounding. The equipment grounding conductor for special-purpose receptacles, such as the operation of mobile X-ray equipment, shall be extended to the reference grounding points of branch circuits for all locations likely to be served from such receptacles. Where such a circuit is served from an isolated ungrounded system, the grounding conductor shall not be required to be run with the power conductors; however, the equipment grounding terminal of the special-purpose receptacle shall be connected to the reference grounding point.

517.20 Wet Procedure Locations.

(A) Receptacles and Fixed Equipment. Wet procedure locations shall be provided with special protection against electric shock by one of the following means:

(1) Power distribution system that inherently limits the possible ground-fault current due to a first fault to a low value, without interrupting the power supply

(2) Power distribution system in which the power supply is interrupted if the ground-fault current does, in fact, exceed a value of 6 mA

Exception: Branch circuits supplying only listed, fixed, therapeutic and diagnostic equipment shall be permitted to be supplied from a grounded service, single- or 3-phase system, provided that
(a) Wiring for grounded and isolated circuits does not occupy the same raceway, and
(b) All conductive surfaces of the equipment are connected to an insulated copper equipment grounding conductor.

(B) Isolated Power Systems. Where an isolated power system is utilized, the isolated power equipment shall be listed as isolated power equipment, and the isolated power system shall be designed and installed in accordance with 517.160.

> Informational Note: For requirements for installation of therapeutic pools and tubs, see Part VI of Article 680.

517.21 Ground-Fault Circuit-Interrupter Protection for Personnel. Ground-fault circuit-interrupter protection for personnel shall not be required for receptacles installed in those critical care (Category 1) spaces where the toilet and basin are installed within the patient room.

Part III. Essential Electrical System

517.25 Scope. The essential electrical system for these facilities shall comprise a system capable of supplying a limited amount of lighting and power service, which is considered essential for life safety and orderly cessation of procedures during the time normal electrical service is interrupted for any reason. This includes clinics, medical and dental offices, outpatient facilities, nursing homes, limited care facilities, hospitals, and other health care facilities serving patients.

> Informational Note: For information on the need for an essential electrical system, see NFPA 99-2015, *Health Care Facilities Code.*

517.26 Application of Other Articles. The life safety branch of the essential electrical system shall meet the requirements of Article 700, except as amended by Article 517.

> Informational Note No. 1: For additional information, see NFPA 110-2013, *Standard for Emergency and Standby Power Systems.*

> Informational Note No. 2: For additional information, see 517.29 and NFPA 99-2015, *Health Care Facilities Code.*

517.29 Essential Electrical Systems for Hospitals and Other Health Care Facilities.

(A) Applicability. The requirements of Part III, 517.29 through 517.30, shall apply to critical care (Category 1) and general care (Category 2) hospitals and other health care facilities using Type 1 essential electrical systems where patients are sustained by electrical life-support equipment.

> Informational Note No. 1: For performance, maintenance, and testing requirements of essential electrical systems in hospitals, see NFPA 99-2015, *Health Care Facilities Code*. For installation of centrifugal fire pumps, see NFPA 20-2013, *Standard for the Installation of Stationary Pumps for Fire Protection*.

> Informational Note No. 2: For additional information on Type 1 and Type 2 essential electrical systems, see NFPA 99-2015, *Health Care Facilities Code*.

N **(B)** Critical care (Category 1) spaces shall be served only by a Type 1 essential electrical system. [**99:**6.3.2.2.10.1]

517.30 Sources of Power.

(A) Two Independent Power Sources. Essential electrical systems shall have a minimum of the following two independent sources of power: a normal source generally supplying the entire electrical system and one or more alternate source(s) for use when the normal source is interrupted. [**99:**6.4.1.1.4]

(B) Types of Power Sources.

(1) Generating Units. Where the normal source consists of generating units on the premises, the alternate source shall be either another generating set or an external utility service. [**99:**6.4.1.1.5]

N **(2) Fuel Cell Systems.** Fuel cell systems shall be permitted to serve as the alternate source for all or part of an essential electrical system, provided the following conditions apply:

(1) Installation of fuel cells shall comply with the requirements in Parts I through VII of Article 692 for 1000 volts or less and Part VIII for over 1000 volts.

> Informational Note: For information on installation of stationary fuel cells, see NFPA 853-2015, *Standard for Installation of Stationary Fuel Cell Power Systems*. [**99:**6.4.1.1.7]

(2) N + 1 units shall be provided where N units have sufficient capacity to supply the demand loads of the portion of the system served. [**99:**6.4.1.7.2]
(3) System shall be able to assume loads within 10 seconds of loss of normal power source.
(4) System shall have a continuing source of fuel supply, together with sufficient on-site fuel storage for the essential system type.
(5) A connection shall be provided for a portable diesel generator to supply life safety and critical portions of the distribution system. [**99:**6.4.1.1.7.5(1) through (5)]
(6) Fuel cell systems shall be listed for emergency system use.

(C) Location of Essential Electrical System Components. Essential electrical system components shall be located to minimize interruptions caused by natural forces common to the area (e.g., storms, floods, earthquakes, or hazards created by adjoining structures or activities). Installations of electrical services shall be located to reduce possible interruption of normal electrical services resulting from similar causes as well as possible disruption of normal electrical service due to internal wiring and equipment failures. Feeders shall be located to provide physical separation of the feeders of the alternate source and from the feeders of the normal electrical source to prevent possible simultaneous interruption.

> Informational Note: Facilities in which the normal source of power is supplied by two or more separate central station-fed services experience greater than normal electrical service reliability than those with only a single feed. Such a dual source of normal power consists of two or more electrical services fed from separate generator sets or a utility distribution network that has multiple power input sources and is arranged to provide mechanical and electrical separation so that a fault between the facility and the generating sources is not likely to cause an interruption of more than one of the facility service feeders.

517.31 Requirements for the Essential Electrical System.

(A) Separate Branches. Essential electrical systems for hospitals shall be comprised of three separate branches capable of supplying a limited amount of lighting and power service that is considered essential for life safety and effective hospital operation during the time the normal electrical service is interrupted for any reason. The three branches are life safety, critical, and equipment.

The division between the branches shall occur at transfer switches where more than one transfer switch is required [**99:**6.4.2.2.1.2]

(B) Transfer Switches. The number of transfer switches to be used shall be based on reliability and design. Each branch of the essential electrical system shall have one or more transfer switches.

One transfer switch and downstream distribution system shall be permitted to serve one or more branches in a facility with a maximum demand on the essential electrical system of 150 kVA.

> Informational Note No. 1: See NFPA 99-2015, *Health Care Facilities Code*, 6.4.3.2, Transfer Switches; 6.4.2.1.5, Automatic Transfer Switch Features; 6.4.2.1.5.15, Nonautomatic Transfer Switch Features; and 6.4.2.1.7, Nonautomatic Transfer Device Features.

> Informational Note No. 2: See Informational Note Figure 517.31(a).

> Informational Note No. 3: See Informational Note Figure 517.31(b).

(1) Optional Loads. Loads served by the generating equipment not specifically named in Article 517 shall be served by their own transfer switches such that the following conditions apply:

(1) These loads shall not be transferred if the transfer will overload the generating equipment.
(2) These loads shall be automatically shed upon generating equipment overloading.

(2) Contiguous Facilities. Hospital power sources and alternate power sources shall be permitted to serve the essential electrical systems of contiguous or same site facilities.

(C) Wiring Requirements.

(1) Separation from Other Circuits. The life safety branch and critical branch of the essential electrical system shall be kept entirely independent of all other wiring and equipment and shall not enter the same raceways, boxes, or cabinets with each other or other wiring.

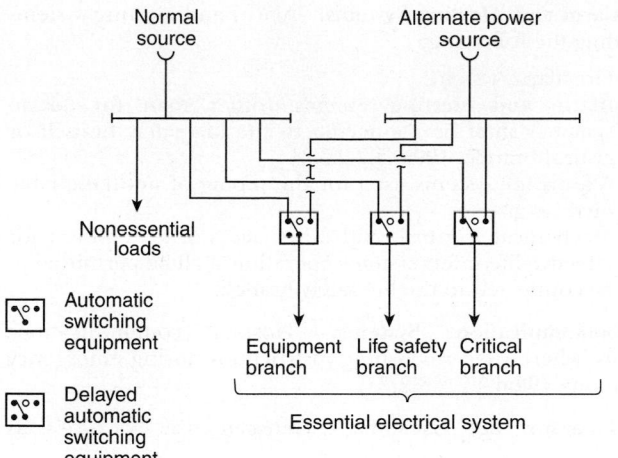

Informational Note Figure 517.31(a)   Hospital — Minimum Requirement (greater than 150 kVA) for Transfer Switch Arrangement.

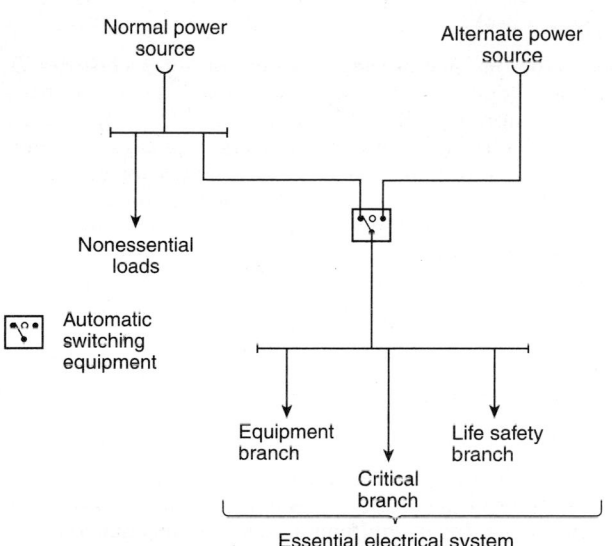

Informational Note Figure 517.31(b)   Hospital — Minimum Requirement (150 kVA or less) for Transfer Switch Arrangement.

Where general care locations are served from two separate transfer switches on the essential electrical system in accordance with 517.18(A), Exception No. 3, the general care circuits from the two separate systems shall be kept independent of each other.

Where critical care locations are served from two separate transfer switches on the essential electrical system in accordance with 517.19(A), Exception No. 2, the critical care circuits from the two separate systems shall be kept independent of each other.

Wiring of the life safety branch and the critical branch shall be permitted to occupy the same raceways, boxes, or cabinets of other circuits not part of the branch where such wiring complies with one of the following:

(1)   Is in transfer equipment enclosures
(2)   Is in exit or emergency luminaires supplied from two sources
(3)   Is in a common junction box attached to exit or emergency luminaires supplied from two sources
(4)   Is for two or more circuits supplied from the same branch and same transfer switch

The wiring of the equipment branch shall be permitted to occupy the same raceways, boxes, or cabinets of other circuits that are not part of the essential electrical system.

(2) Isolated Power Systems. Where isolated power systems are installed in any of the areas in 517.34(A)(1) and (A)(2), each system shall be supplied by an individual circuit serving no other load.

(3) Mechanical Protection of the Essential Electrical System. The wiring of the life safety and critical branches shall be mechanically protected. Where installed as branch circuits in patient care spaces, the installation shall comply with the requirements of 517.13(A) and (B). Only the following wiring methods shall be permitted:

(1)   Nonflexible metal raceways, Type MI cable, Type RTRC marked with the suffix –XW, or Schedule 80 PVC conduit. Nonmetallic raceways shall not be used for branch circuits that supply patient care areas.
(2)   Where encased in not less than 50 mm (2 in.) of concrete, Schedule 40 PVC conduit, flexible nonmetallic or jacketed metallic raceways, or jacketed metallic cable assemblies listed for installation in concrete. Nonmetallic raceways shall not be used for branch circuits that supply patient care areas.
(3)   Listed flexible metal raceways and listed metal sheathed cable assemblies in any of the following:
    a.   Where used in listed prefabricated medical headwalls
    b.   In listed office furnishings
    c.   Where fished into existing walls or ceilings, not otherwise accessible and not subject to physical damage
    d.   Where necessary for flexible connection to equipment
    e.   For equipment that requires a flexible connection due to movement, vibration, or operation
    f.   Luminaires installed in rigid ceiling structures where there is no access above the ceiling space after the luminaire is installed
(4)   Flexible power cords of appliances or other utilization equipment connected to the emergency system.
(5)   Cables for Class 2 or Class 3 systems permitted by Part VI of this Article, with or without raceways.

Informational Note: See 517.13 for additional grounding requirements in patient care areas.

(D) Capacity of Systems. The essential electrical system shall have the capacity and rating to meet the maximum actual demand likely to be produced by the connected load.

Feeders shall be sized in accordance with 215.2 and Part III of Article 220. The generator set(s) shall have the capacity and rating to meet the demand produced by the load at any given time.

Demand calculations for sizing of the generator set(s) shall be based on any of the following:

(1)   Prudent demand factors and historical data

(2)   Connected load
(3)   Feeder calculation procedures described in Article 220
(4)   Any combination of the above

The sizing requirements in 700.4 and 701.4 shall not apply to hospital generator set(s).

(E) Receptacle Identification. The cover plates for the electrical receptacles or the electrical receptacles themselves supplied from the essential electrical system shall have a distinctive color or marking so as to be readily identifiable. [**99**:6.4.2.2.6.2(C)]

(F) Feeders from Alternate Power Source. A single feeder supplied by a local or remote alternate source shall be permitted to supply the essential electrical system to the point at which the life safety, critical, and equipment branches are separated. Installation of the transfer equipment shall be permitted at other than the location of the alternate power source.

(G) Coordination. Overcurrent protective devices serving the essential electrical system shall be coordinated for the period of time that a fault's duration extends beyond 0.1 second.

Exception No. 1: Between transformer primary and secondary overcurrent protective devices, where only one overcurrent protective device or set of overcurrent protective devices exists on the transformer secondary.

Exception No. 2: Between overcurrent protective devices of the same size (ampere rating) in series.

> Informational Note: The terms *coordination* and *coordinated* as used in this section do not cover the full range of overcurrent conditions.

517.32 Branches Requiring Automatic Connection.

(A) Those functions of patient care depending on lighting or appliances that are connected to the essential electrical system shall be divided into the life safety branch and the critical branch, as described in 517.33 and 517.34 .

(B) The life safety and critical branches shall be installed and connected to the alternate power source specified in 517.30(A) and (B) so that all functions specified herein for the life safety and critical branches are automatically restored to operation within 10 seconds after interruption of the normal source. [**99**:6.4.3.1]

517.33 Life Safety Branch.
No functions other than those listed in 517.33(A) through (H) shall be connected to the life safety branch. The life safety branch of the essential electrical system shall supply power for the following lighting, receptacles, and equipment.

(A) Illumination of Means of Egress. Illumination of means of egress, such as lighting required for corridors, passageways, stairways, and landings at exit doors, and all necessary ways of approach to exits. Switching arrangements to transfer patient corridor lighting in hospitals from general illumination circuits to night illumination circuits shall be permitted, provided only one of two circuits can be selected and both circuits cannot be extinguished at the same time.

> Informational Note: See NFPA *101* -2015, *Life Safety Code*, Sections 7.8 and 7.9.

(B) Exit Signs. Exit signs and exit directional signs.

> Informational Note: See NFPA *101* -2012, *Life Safety Code*, Section 7.10.

(C) Alarm and Alerting Systems. Alarm and alerting systems including the following:

(1)   Fire alarm systems
(2)   Alarm and alerting systems (other than fire alarm systems) shall be connected to the life safety branch or critical branch. [**99**:6.4.2.2.3.3]
(3)   Alarms for systems used for the piping of nonflammable medical gases
(4)   Mechanical, control, and other accessories required for effective life safety systems operation shall be permitted to be connected to the life safety branch.

(D) Communications Systems. Hospital communications systems, where used for issuing instructions during emergency conditions. [**99**:6.4.2.2.3.2(3)]

(E) Generator Set Locations. Generator set locations as follows:

(1)   Task illumination
(2)   Battery charger for emergency battery-powered lighting unit(s)
(3)   Select receptacles at the generator set location and essential electrical system transfer switch locations. [**99**: 6.4.2.2.3.2(4)]

(F) Generator Set Accessories. Generator set accessories as required for generator performance. Loads dedicated to a specific generator, including the fuel transfer pump(s), ventilation fans, electrically operated louvers, controls, cooling system, and other generator accessories essential for generator operation, shall be connected to the life safety branch or to the output terminals of the generator with overcurrent protective devices. [**99**:6.4.2.2.3.4]

(G) Elevators. Elevator cab lighting, control, communications, and signal systems. [**99**:6.4.2.2.3.2(5)]

(H) Automatic Doors. Electrically powered doors used for building egress. [**99**:6.4.2.2.3.2(6)]

517.34 Critical Branch.

(A) Task Illumination and Selected Receptacles. The critical branch of the essential electrical system shall supply power for task illumination, fixed equipment, selected receptacles, and special power circuits serving the following areas and functions related to patient care:

(1)   Critical care (Category 1) spaces that utilize anesthetizing gases, task illumination, selected receptacles, and fixed equipment
(2)   The isolated power systems in special environments
(3)   Patient care spaces, task illumination, and selected receptacles in the following:

    a.   Infant nurseries
    b.   Medication preparation areas
    c.   Pharmacy dispensing areas
    d.   Selected acute nursing areas
    e.   Psychiatric bed areas (omit receptacles)
    f.   Ward treatment rooms
    g.   Nurses' stations (unless adequately lighted by corridor luminaires)

(4)   Additional specialized patient care task illumination and receptacles, where needed
(5)   Nurse call systems
(6)   Blood, bone, and tissue banks
(7)   Telephone and data equipment rooms and closets

(8) Task illumination, selected receptacles, and selected power circuits for the following:

   a. General care (Category 2) beds (at least one duplex receptacle in each patient bedroom)
   b. Angiographic labs
   c. Cardiac catheterization labs
   d. Coronary care units
   e. Hemodialysis rooms or areas
   f. Emergency room treatment areas (selected)
   g. Human physiology labs
   h. Intensive care units
   i. Postoperative recovery rooms (selected)

(9) Additional task illumination, receptacles, and selected power circuits needed for effective facility operation, including single-phase fractional horsepower motors, shall be permitted to be connected to the critical branch. [**99:**6.4.2.2.4.2(9)]

N **(B) Switching.** It shall be permitted to control task illumination on the critical branch.

(C) Subdivision of the Critical Branch. It shall be permitted to subdivide the critical branch into two or more branches.

> Informational Note: It is important to analyze the consequences of supplying an area with only critical care branch power when failure occurs between the area and the transfer switch. Some proportion of normal and critical power or critical power from separate transfer switches may be appropriate.

517.35 Equipment Branch Connection to Alternate Power Source. The equipment branch shall be installed and connected to the alternate power source such that the equipment described in 517.35(A) is automatically restored to operation at appropriate time-lag intervals following the energizing of the essential electrical system. Its arrangement shall also provide for the subsequent connection of equipment described in 517.35(B). [**99:**6.4.2.2.5.2]

Exception: For essential electrical systems under 150 kVA, deletion of the time-lag intervals feature for delayed automatic connection to the equipment system shall be permitted.

(A) Equipment for Delayed Automatic Connection. The following equipment shall be permitted to be arranged for delayed automatic connection to the alternate power source:

(1) Central suction systems serving medical and surgical functions, including controls. Such suction systems shall be permitted on the critical branch.
(2) Sump pumps and other equipment required to operate for the safety of major apparatus, including associated control systems and alarms.
(3) Compressed air systems serving medical and surgical functions, including controls. Such air systems shall be permitted on the critical branch.
(4) Smoke control and stair pressurization systems, or both.
(5) Kitchen hood supply or exhaust systems, or both, if required to operate during a fire in or under the hood.
(6) Supply, return, and exhaust ventilating systems for airborne infectious/isolation rooms, protective environment rooms, exhaust fans for laboratory fume hoods, nuclear medicine areas where radioactive material is used, ethylene oxide evacuation, and anesthesia evacuation. Where delayed automatic connection is not appropriate, such ventilation systems shall be permitted to be placed on the critical branch. [**99:**6.4.2.2.5.3(A)(6) and (B)]

(7) Supply, return, and exhaust ventilating systems for operating and delivery rooms.
(8) Supply, return, exhaust ventilating systems and/or air-conditioning systems serving telephone equipment rooms and closets and data equipment rooms and closets.

Exception: Sequential delayed automatic connection to the alternate power source to prevent overloading the generator shall be permitted where engineering studies indicate it is necessary.

(B) Equipment for Delayed Automatic or Manual Connection. The following equipment shall be permitted to be arranged for either delayed automatic or manual connection to the alternate power source:

(1) Heating equipment to provide heating for operating, delivery, labor, recovery, intensive care, coronary care, nurseries, infection/isolation rooms, emergency treatment spaces, and general patient rooms and pressure maintenance (jockey or make-up) pump(s) for water-based fire protection systems

Exception: Heating of general patient rooms and infection/isolation rooms during disruption of the normal source shall not be required under any of the following conditions:

   (a) The outside design temperature is higher than −6.7°C (20°F).

   (b) The outside design temperature is lower than −6.7°C (20°F), and where a selected room(s) is provided for the needs of all confined patients, only such room(s) need be heated.

   (c) The facility is served by a dual source of normal power.

> Informational Note No. 1: The design temperature is based on the 97.5 percent design value as shown in Chapter 24 of the ASHRAE *Handbook of Fundamentals* (2013).

> Informational Note No. 2: For a description of a dual source of normal power, see 517.30(C).

(2) An elevator(s) selected to provide service to patient, surgical, obstetrical, and ground floors during interruption of normal power. In instances where interruption of normal power would result in other elevators stopping between floors, throw-over facilities shall be provided to allow the temporary operation of any elevator for the release of patients or other persons who may be confined between floors.
(3) Hyperbaric facilities.
(4) Hypobaric facilities.
(5) Automatically operated doors.
(6) Minimal electrically heated autoclaving equipment shall be permitted to be arranged for either automatic or manual connection to the alternate source.
(7) Controls for equipment listed in 517.35.
(8) Other selected equipment shall be permitted to be served by the equipment system. [**99:**6.4.2.2.5.4(9)]

(C) AC Equipment for Nondelayed Automatic Connection. Generator accessories, including but not limited to, the transfer fuel pump, electrically operated louvers, and other generator accessories essential for generator operation shall be arranged for automatic connection to the alternate power source. [**99:**6.5.2.2.3.2]

517.40 Type 2 Essential Electrical Systems for Nursing Homes and Limited Care Facilities.

Informational Note: Nursing homes and other limited care facilities can be classified as critical care (Category 1) or general care (Category 2) patient care space depending on the design and type of care administered in the facility. For small, less complex facilities, only minimal alternate lighting and alarm service may be required. At nursing homes and other limited care facilities where patients are not sustained by electrical life-support equipment or inpatient hospital care the requirements of 517.40 through 517.41 apply. If the level of care is comparable to that provided in a hospital, see the essential electrical system requirements of 517.29 through 517.30.

(A) Applicability. The requirements of Part III, 517.40(C) through 517.41, shall apply to nursing homes and limited care facilities.

Exception: The requirements of Part III, 517.40(C) through 517.41, shall not apply to freestanding buildings used as nursing homes and limited care facilities, provided that the following apply:

(1) Admitting and discharge policies are maintained that preclude the provision of care for any patient or resident who may need to be sustained by electrical life-support equipment.

(2) No surgical treatment requiring general anesthesia is offered.

(3) An automatic battery-operated system(s) or equipment shall be effective for at least 1½ hours and is otherwise in accordance with 700.12 and that shall be capable of supplying lighting for exit lights, exit corridors, stairways, nursing stations, medical preparation areas, boiler rooms, and communications areas. This system shall also supply power to operate all alarm systems.

Informational Note: See NFPA *101*-2015, *Life Safety Code.*

(B) Inpatient Hospital Care Facilities. For those nursing homes and limited care facilities that admit patients who need to be sustained by electrical life support equipment, the essential electrical system from the source to the portion of the facility where such patients are treated shall comply with the requirements of Part III, 517.29 through 517.30.

(C) Facilities Contiguous or Located on the Same Site with Hospitals. Nursing homes and limited care facilities that are contiguous or located on the same site with a hospital shall be permitted to have their essential electrical systems supplied by the hospital.

Informational Note No. 1: For performance, maintenance, and testing requirements of essential electrical systems in nursing homes and limited care facilities, see NFPA 99-2015, *Health Care Facilities Code.*

Informational Note No. 2: Where optional loads include contiguous or same-site facilities not covered in this *Code,* see the requirements of Article 700 of this *Code;* NFPA *101*-2015, *Life Safety Code;* and other applicable NFPA requirements for emergency egress under loadshed conditions.

517.41 Required Power Sources.

(A) Two Independent Power Sources. Essential electrical systems shall have a minimum of the following two independent sources of power: a normal source generally supplying the entire electrical system and one or more alternate sources for use when the normal source is interrupted. [**99**:6.4.1.1.4]

(B) Types of Power Sources. Where the normal source consists of generating units on the premises, the alternate source shall be either another generating set or an external utility service. [**99**:6.4.1.1.5]

(C) Location of Essential Electrical System Components. Essential electrical systems shall be located to minimize inter-

ruptions caused by natural forces common to the area (e.g., storms, floods, earthquakes, or hazards created by adjoining structures or activities). Installations of electrical services shall be located to reduce possible interruption of normal electrical services resulting from similar causes as well as possible disruption of normal electrical service due to internal wiring and equipment failures. Feeders shall be located to give physical separation of the feeders of the alternate source and from the feeders of the normal electrical source to prevent possible simultaneous interruption.

517.42 Essential Electrical Systems.

(A) General. Essential electrical systems for nursing homes and limited care facilities shall be divided into the following two branches, the life safety branch and the equipment branch. [**99**:6.5.2.2.1.2]

The division between the branches shall occur at transfer switches where more than one transfer switch is required.

Informational Note No. 1: Essential electrical systems are comprised of two separate branches capable of supplying a limited amount of lighting and power service, which is considered essential for the protection of life and safety and effective operation of the institution during the time normal electrical service is interrupted for any reason.

Informational Note No. 2: For more information see NFPA 99-2015, *Health Care Facilities Code.*

(B) Transfer Switches. The number of transfer switches to be used shall be based on reliability, design, and load considerations. [**99**:6.5.2.2.1.4]

(1) Each branch of the essential electrical system shall have one or more transfer switches. [**99**:6.5.2.2.1.4(A)]
(2) One transfer switch shall be permitted to serve one or more branches or systems in a facility with a continuous load on the switch of 150 kVA (120 kW) or less. [**99**: 6.5.2.2.1.4(B)]

Informational Note No. 1: See NFPA 99-2015, *Health Care Facilities Code,* 6.5.3.2, Transfer Switch Operation Type II; 6.4.2.1.5, Automatic Transfer Switch Features; and 6.4.2.1.7, Nonautomatic Transfer Device Features.

Informational Note No. 2: See Informational Note Figure 517.42(a).

Informational Note No. 3: See Informational Note Figure 517.42(b).

(C) Capacity of System. The essential electrical system shall have adequate capacity to meet the demand for the operation of all functions and equipment to be served by each branch at one time.

(D) Separation from Other Circuits. The life safety branch and equipment branch shall be kept entirely independent of all other wiring and equipment. [**99**:6.5.2.2.4.1]

These circuits shall not enter the same raceways, boxes, or cabinets with other wiring except as follows:

(1) In transfer switches
(2) In exit or emergency luminaires supplied from two sources
(3) In a common junction box attached to exit or emergency luminaires supplied from two sources

Informational Note: For further information see NFPA 99-2015 *Health Care Facilities Code,* A.6.5.2.2.4.1.

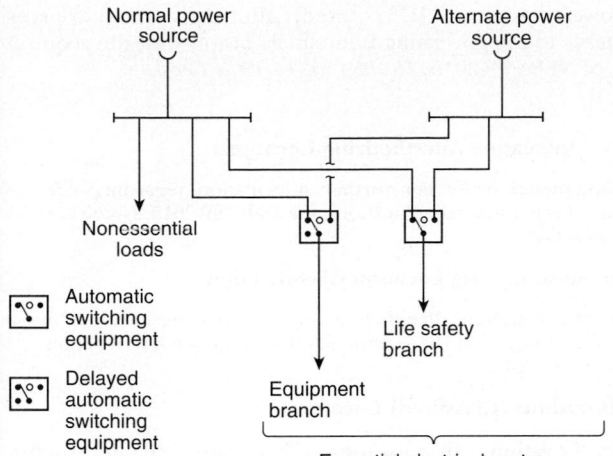

Informational Note Figure 517.42(a) Nursing Home and Limited Health Care Facilities — Minimum Requirement (greater than 150 kVA) for Transfer Switch Arrangement.

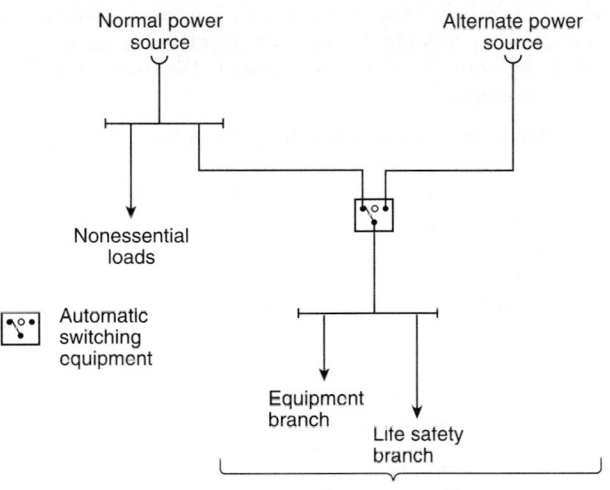

Informational Note Figure 517.42(b) Nursing Home and Limited Health Care Facilities — Minimum Requirement (150 kVA or less) for Transfer Switch Arrangement.

(E) Receptacle Identification. The electrical receptacles or the cover plates for the electrical receptacles supplied from the life safety or equipment branches shall have a distinctive color or marking to be readily identifiable. [99:6.5.2.2.4.2]

Informational Note: If color is used to identify these receptacles, the same color should be used throughout the facility. [99:A.6.5.2.2.4.2]

517.43 Automatic Connection to Life Safety Branch. The life safety branch shall be installed and connected to the alternate source of power so that all functions specified herein shall be automatically restored to operation within 10 seconds after the interruption of the normal source. No functions other than those listed in 517.43(A) through (G) shall be connected to the life safety branch. The life safety branch shall supply power for the following lighting, receptacles, and equipment.

(A) Illumination of Means of Egress. Illumination of means of egress as is necessary for corridors, passageways, stairways, landings, and exit doors and all ways of approach to exits. Switching arrangement to transfer patient corridor lighting from general illumination circuits shall be permitted, providing only one of two circuits can be selected and both circuits cannot be extinguished at the same time.

Informational Note: See NFPA *101* -2015, *Life Safety Code*, Sections 7.8 and 7.9.

(B) Exit Signs. Exit signs and exit directional signs.

Informational Note: See NFPA *101* -2015, *Life Safety Code*, Section 7.10.

(C) Alarm and Alerting Systems. Alarm and alerting systems, including the following:

(1)   Fire alarms

Informational Note: See NFPA *101* -2015, *Life Safety Code*, Sections 9.6 and 18.3.4.

(2)   Alarms required for systems used for the piping of nonflammable medical gases

Informational Note: See NFPA 99 -2015, *Health Care Facilities Code*, 6.5.2.2.2.1(3).

(D) Communications Systems. Communications systems, where used for issuing instructions during emergency conditions. [**99:**6.5.2.2.2.1(4)]

(E) Dining and Recreation Areas. Sufficient lighting in dining and recreation areas to provide illumination to exit ways at a minimum of 5 ft-candles. [**99:**6.5.2.2.2.1.(5)]

(F) Generator Set Location. Task illumination and selected receptacles in the generator set location. [**99:** 6.5.2.2.2.1(6)]

(G) Elevators. Elevator cab lighting, control, communications, and signal systems. [**99:** 6.5.2.2.2.1(7)]

517.44 Connection to Equipment Branch. The equipment branch shall be installed and connected to the alternate power source so that the equipment listed in 517.44(A) shall be automatically restored to operation at appropriate time-lag intervals following the restoration of the life safety branch to operation. [**99:**6.5.2.2.3.1(A)]

The equipment branch arrangement shall also provide for the additional connection of equipment listed in 517.44(B). [**99:**6.5.2.2.3.1]

Exception: For essential electrical systems under 150 kVA, deletion of the time-lag intervals feature for delayed automatic connection to the equipment branch shall be permitted.

(A) Delayed Automatic Connections to Equipment Branch. The following equipment shall be permitted to be connected to the equipment branch and shall be arranged for delayed automatic connection to the alternate power source:

(1)   Task illumination and selected receptacles in the following:
    (a)   Patient care spaces
    (b)   Medication preparation spaces
    (c)   Pharmacy dispensing areas
    (d)   Nurses' stations (unless adequately lighted by corridor luminaires)
(2)   Supply, return, and exhaust ventilating systems for airborne infectious isolation rooms

(3) Sump pumps and other equipment required to operate for the safety of major apparatus and associated control systems and alarms

(4) Smoke control and stair pressurization systems

(5) Kitchen hood supply or exhaust systems, or both, if required to operate during a fire in or under the hood

(6) Nurse call systems

[**99:**6.5.2.2.3.3]

(B) Delayed Automatic or Manual Connection to the Equipment Branch. The following equipment shall be permitted to be connected to the critical equipment branch and shall be arranged for either delayed automatic or manual connection to the alternate power source:

(1) Heating equipment to provide heating for patient rooms.

Exception: Heating of general patient rooms during disruption of the normal source shall not be required under any of the following conditions:

(1) The outside design temperature is higher than –6.7°C (20°F).

(2) The outside design temperature is lower than –6.7°C (20°F) and where a selected room(s) is provided for the needs of all confined patients, only such room(s) need be heated.

(3) The facility is served by a dual source of normal power as described in 517.41(C), Informational Note.

Informational Note: The outside design temperature is based on the 97.5 percent design values, as shown in Chapter 24 of the ASHRAE *Handbook of Fundamentals* (2013).

(4) Elevator service — in instances where disruption of power would result in elevators stopping between floors, throw-over facilities shall be provided to allow the temporary operation of any elevator for the release of passengers. For elevator cab lighting, control, and signal system requirements, see 517.43(G).

(5) Additional illumination, receptacles, and equipment shall be permitted to be connected only to the critical branch.

[**99:**6.5.2.2.3.4(A), (B), and (C)]

517.45 Essential Electrical Systems for Other Health Care Facilities.

(A) Essential Electrical Distribution. If required by the governing body, the essential electrical distribution system for basic care (Category 3) patient care spaces shall be comprised of an alternate power system capable of supplying a limited amount of lighting and power service for the orderly cessation of procedures during a time normal electrical service is interrupted.

Informational Note: See NFPA 99-2015, *Health Care Facilities Code.*

(B) Electrical Life Support Equipment. Where electrical life support equipment is required, the essential electrical distribution system shall be as described in 517.29 through 517.30.

(C) Critical Care (Category 1) Patient Care Spaces. Where critical patient care (Category 1) spaces are present, the essential electrical distribution system shall be as described in 517.29 through 517.30.

N **(D) General Care (Category 2) Patent Care Spaces.** Where general care (Category 2) patent care spaces are present, the essential electrical distribution system shall be as described in 517.40 through 517.45.

(E) Power Systems. If required, alternate power sources acceptable to the governing body shall comply with the requirements of NFPA 99-2015, *Health Care Facilities Code.*

Part IV. Inhalation Anesthetizing Locations

Informational Note: For further information regarding safeguards for anesthetizing locations, see NFPA 99-2015, *Health Care Facilities Code.*

517.60 Anesthetizing Location Classification.

Informational Note: If either of the anesthetizing locations in 517.60(A) or 517.60(B) is designated a wet procedure location, refer to 517.20.

(A) Hazardous (Classified) Location.

(1) Use Location. In a location where flammable anesthetics are employed, the entire area shall be considered to be a Class I, Division 1 location that extends upward to a level 1.52 m (5 ft) above the floor. The remaining volume up to the structural ceiling is considered to be above a hazardous (classified) location. [**99:** Annex E, E.1, and E.2]

(2) Storage Location. Any room or location in which flammable anesthetics or volatile flammable disinfecting agents are stored shall be considered to be a Class I, Division 1 location from floor to ceiling.

(B) Other-Than-Hazardous (Classified) Location. Any inhalation anesthetizing location designated for the exclusive use of nonflammable anesthetizing agents shall be considered to be an other-than-hazardous (classified) location.

517.61 Wiring and Equipment.

(A) Within Hazardous (Classified) Anesthetizing Locations.

(1) Isolation. Except as permitted in 517.160, each power circuit within, or partially within, a flammable anesthetizing location as referred to in 517.60 shall be isolated from any distribution system by the use of an isolated power system.

(2) Design and Installation. Where an isolated power system is utilized, the isolated power equipment shall be listed as isolated power equipment, and the isolated power system shall be designed and installed in accordance with 517.160.

(3) Equipment Operating at More Than 10 Volts. In hazardous (classified) locations referred to in 517.60, all fixed wiring and equipment and all portable equipment, including lamps and other utilization equipment, operating at more than 10 volts between conductors shall comply with the requirements of 501.1 through 501.25, and 501.100 through 501.150, and 501.30(A) and 501.30(B) for Class I, Division 1 locations. All such equipment shall be specifically approved for the hazardous atmospheres involved.

(4) Extent of Location. Where a box, fitting, or enclosure is partially, but not entirely, within a hazardous (classified) location(s), the hazardous (classified) location(s) shall be considered to be extended to include the entire box, fitting, or enclosure.

(5) Receptacles and Attachment Plugs. Receptacles and attachment plugs in a hazardous (classified) location(s) shall be listed for use in Class I, Group C hazardous (classified) locations and shall have provision for the connection of a grounding conductor.

(6) Flexible Cord Type. Flexible cords used in hazardous (classified) locations for connection to portable utilization equipment, including lamps operating at more than 8 volts between conductors, shall be of a type approved for extra-hard usage in accordance with Table 400.4 and shall include an additional conductor for grounding.

(7) Flexible Cord Storage. A storage device for the flexible cord shall be provided and shall not subject the cord to bending at a radius of less than 75 mm (3 in.).

(B) Above Hazardous (Classified) Anesthetizing Locations.

(1) Wiring Methods. Wiring above a hazardous (classified) location referred to in 517.60 shall be installed in rigid metal conduit, electrical metallic tubing, intermediate metal conduit, Type MI cable, or Type MC cable that employs a continuous, gas/vaportight metal sheath.

(2) Equipment Enclosure. Installed equipment that may produce arcs, sparks, or particles of hot metal, such as lamps and lampholders for fixed lighting, cutouts, switches, generators, motors, or other equipment having make-and-break or sliding contacts, shall be of the totally enclosed type or be constructed so as to prevent escape of sparks or hot metal particles.

Exception: Wall-mounted receptacles installed above the hazardous (classified) location in flammable anesthetizing locations shall not be required to be totally enclosed or have openings guarded or screened to prevent dispersion of particles.

(3) Luminaires. Surgical and other luminaires shall conform to 501.130(B).

Exception No. 1: The surface temperature limitations set forth in 501.130(B)(1) shall not apply.

Exception No. 2: Integral or pendant switches that are located above and cannot be lowered into the hazardous (classified) location(s) shall not be required to be explosionproof.

(4) Seals. Listed seals shall be provided in conformance with 501.15, and 501.15(A)(4) shall apply to horizontal as well as to vertical boundaries of the defined hazardous (classified) locations.

(5) Receptacles and Attachment Plugs. Receptacles and attachment plugs located above hazardous (classified) anesthetizing locations shall be listed for hospital use for services of prescribed voltage, frequency, rating, and number of conductors with provision for the connection of the grounding conductor. This requirement shall apply to attachment plugs and receptacles of the 2-pole, 3-wire grounding type for single-phase, 120-volt, nominal, ac service.

(6) 250-Volt Receptacles and Attachment Plugs Rated 50 and 60 Amperes. Receptacles and attachment plugs rated 250 volts, for connection of 50-ampere and 60-ampere ac medical equipment for use above hazardous (classified) locations, shall be arranged so that the 60-ampere receptacle will accept either the 50-ampere or the 60-ampere plug. Fifty-ampere receptacles shall be designed so as not to accept the 60-ampere attachment plug. The attachment plugs shall be of the 2-pole, 3-wire design with a third contact connecting to the insulated (green or green with yellow stripe) equipment grounding conductor of the electrical system.

(C) Other-Than-Hazardous (Classified) Anesthetizing Locations.

(1) Wiring Methods. Wiring serving other-than-hazardous (classified) locations, as defined in 517.60, shall be installed in a metal raceway system or cable assembly. The metal raceway system or cable armor or sheath assembly shall qualify as an equipment grounding conductor in accordance with 250.118. Type MC and Type MI cable shall have an outer metal armor, sheath, or sheath assembly that is identified as an acceptable equipment grounding conductor.

Exception: Pendant receptacle installations that employ listed Type SJO or equivalent hard usage or extra-hard usage, flexible cords suspended not less than 1.8 m (6 ft) from the floor shall not be required to be installed in a metal raceway or cable assembly.

(2) Receptacles and Attachment Plugs. Receptacles and attachment plugs installed and used in other-than-hazardous (classified) locations shall be listed "hospital grade" for services of prescribed voltage, frequency, rating, and number of conductors with provision for connection of the grounding conductor. This requirement shall apply to 2-pole, 3-wire grounding type for single-phase, 120-, 208-, or 240-volt, nominal, ac service.

(3) 250-Volt Receptacles and Attachment Plugs Rated 50 Amperes and 60 Amperes. Receptacles and attachment plugs rated 250 volts, for connection of 50-ampere and 60-ampere ac medical equipment for use in other-than-hazardous (classified) locations, shall be arranged so that the 60-ampere receptacle will accept either the 50-ampere or the 60 ampere plug. Fifty-ampere receptacles shall be designed so as not to accept the 60-ampere attachment plug. The attachment plugs shall be of the 2-pole, 3-wire design with a third contact connecting to the insulated (green or green with yellow stripe) equipment grounding conductor of the electrical system.

517.62 Grounding. In any anesthetizing area, all metal raceways and metal-sheathed cables and all normally non–current-carrying conductive portions of fixed electrical equipment shall be connected to an equipment grounding conductor. Grounding and bonding in Class I locations shall comply with 501.30.

Exception: Equipment operating at not more than 10 volts between conductors shall not be required to be connected to an equipment grounding conductor.

517.63 Grounded Power Systems in Anesthetizing Locations.

(A) Battery-Powered Lighting Units. One or more battery-powered lighting units shall be provided and shall be permitted to be wired to the critical lighting circuit in the area and connected ahead of any local switches.

(B) Branch-Circuit Wiring. Branch circuits supplying only listed, fixed, therapeutic and diagnostic equipment, permanently installed above the hazardous (classified) location and in other-than-hazardous (classified) locations, shall be permitted to be supplied from a normal grounded service, single- or three-phase system, provided the following apply:

(1) Wiring for grounded and isolated circuits does not occupy the same raceway or cable.
(2) All conductive surfaces of the equipment are connected to an equipment grounding conductor.

(3) Equipment (except enclosed X-ray tubes and the leads to the tubes) is located at least 2.5 m (8 ft) above the floor or outside the anesthetizing location.

(4) Switches for the grounded branch circuit are located outside the hazardous (classified) location.

Exception: Sections 517.63(B)(3) and (B)(4) shall not apply in other-than-hazardous (classified) locations.

(C) Fixed Lighting Branch Circuits. Branch circuits supplying only fixed lighting shall be permitted to be supplied by a normal grounded service, provided the following apply:

(1) Such luminaires are located at least 2.5 m (8 ft) above the floor.

(2) All conductive surfaces of luminaires are connected to an equipment grounding conductor.

(3) Wiring for circuits supplying power to luminaires does not occupy the same raceway or cable for circuits supplying isolated power.

(4) Switches are wall-mounted and located above hazardous (classified) locations.

Exception: Sections 517.63(C)(1) and (C)(4) shall not apply in other-than-hazardous (classified) locations.

(D) Remote-Control Stations. Wall-mounted remote-control stations for remote-control switches operating at 24 volts or less shall be permitted to be installed in any anesthetizing location.

(E) Location of Isolated Power Systems. Where an isolated power system is utilized, the isolated power equipment shall be listed as isolated power equipment. Isolated power system equipment and its supply circuit shall be permitted to be located in an anesthetizing location, provided it is installed above a hazardous (classified) location or in an other-than-hazardous (classified) location.

(F) Circuits in Anesthetizing Locations. Except as permitted above, each power circuit within, or partially within, a flammable anesthetizing location as referred to in 517.60 shall be isolated from any distribution system supplying other-than-anesthetizing locations.

517.64 Low-Voltage Equipment and Instruments.

(A) Equipment Requirements. Low-voltage equipment that is frequently in contact with the bodies of persons or has exposed current-carrying elements shall comply with one of the following:

(1) Operate on an electrical potential of 10 volts or less

(2) Be approved as intrinsically safe or double-insulated equipment

(3) Be moisture resistant

(B) Power Supplies. Power shall be supplied to low-voltage equipment from one of the following:

(1) An individual portable isolating transformer (autotransformers shall not be used) connected to an isolated power circuit receptacle by means of an appropriate cord and attachment plug

(2) A common low-voltage isolating transformer installed in an other-than-hazardous (classified) location

(3) Individual dry-cell batteries

(4) Common batteries made up of storage cells located in an other-than-hazardous (classified) location

(C) Isolated Circuits. Isolating-type transformers for supplying low-voltage circuits shall have both of the following:

(1) Approved means for insulating the secondary circuit from the primary circuit

(2) The core and case connected to an equipment grounding conductor

(D) Controls. Resistance or impedance devices shall be permitted to control low-voltage equipment but shall not be used to limit the maximum available voltage to the equipment.

(E) Battery-Powered Appliances. Battery-powered appliances shall not be capable of being charged while in operation unless their charging circuitry incorporates an integral isolating-type transformer.

(F) Receptacles or Attachment Plugs. Any receptacle or attachment plug used on low-voltage circuits shall be of a type that does not permit interchangeable connection with circuits of higher voltage.

> Informational Note: Any interruption of the circuit, even circuits as low as 10 volts, either by any switch or loose or defective connections anywhere in the circuit, may produce a spark that is sufficient to ignite flammable anesthetic agents.

Part V. X-Ray Installations

517.70 Applicability. Nothing in this part shall be construed as specifying safeguards against the useful beam or stray X-ray radiation.

> Informational Note No. 1: Radiation safety and performance requirements of several classes of X-ray equipment are regulated under Public Law 90-602 and are enforced by the Department of Health and Human Services.

> Informational Note No. 2: In addition, information on radiation protection by the National Council on Radiation Protection and Measurements is published as *Reports of the National Council on Radiation Protection and Measurement.* These reports are obtainable from NCRP Publications, P.O. Box 30175, Washington, DC 20014.

517.71 Connection to Supply Circuit.

(A) Fixed and Stationary Equipment. Fixed and stationary X-ray equipment shall be connected to the power supply by means of a wiring method complying with applicable requirements of Chapters 1 through 4 of this *Code,* as modified by this article.

Exception: Equipment properly supplied by a branch circuit rated at not over 30 amperes shall be permitted to be supplied through a suitable attachment plug and hard-service cable or cord.

(B) Portable, Mobile, and Transportable Equipment. Individual branch circuits shall not be required for portable, mobile, and transportable medical X-ray equipment requiring a capacity of not over 60 amperes.

(C) Over 1000-Volt Supply. Circuits and equipment operated on a supply circuit of over 1000 volts shall comply with Article 490.

517.72 Disconnecting Means.

(A) Capacity. A disconnecting means of adequate capacity for at least 50 percent of the input required for the momentary rating or 100 percent of the input required for the long-time rating of the X-ray equipment, whichever is greater, shall be provided in the supply circuit.

(B) Location. The disconnecting means shall be operable from a location readily accessible from the X-ray control.

(C) Portable Equipment. For equipment connected to a 120-volt branch circuit of 30 amperes or less, a grounding-type attachment plug and receptacle of proper rating shall be permitted to serve as a disconnecting means.

517.73 Rating of Supply Conductors and Overcurrent Protection.

(A) Diagnostic Equipment.

(1) Branch Circuits. The ampacity of supply branch-circuit conductors and the current rating of overcurrent protective devices shall not be less than 50 percent of the momentary rating or 100 percent of the long-time rating, whichever is greater.

(2) Feeders. The ampacity of supply feeders and the current rating of overcurrent protective devices supplying two or more branch circuits supplying X-ray units shall not be less than 50 percent of the momentary demand rating of the largest unit plus 25 percent of the momentary demand rating of the next largest unit plus 10 percent of the momentary demand rating of each additional unit. Where simultaneous biplane examinations are undertaken with the X-ray units, the supply conductors and overcurrent protective devices shall be 100 percent of the momentary demand rating of each X-ray unit.

> Informational Note: The minimum conductor size for branch and feeder circuits is also governed by voltage regulation requirements. For a specific installation, the manufacturer usually specifies minimum distribution transformer and conductor sizes, rating of disconnecting means, and overcurrent protection.

(B) Therapeutic Equipment. The ampacity of conductors and rating of overcurrent protective devices shall not be less than 100 percent of the current rating of medical X-ray therapy equipment.

> Informational Note: The ampacity of the branch-circuit conductors and the ratings of disconnecting means and overcurrent protection for X-ray equipment are usually designated by the manufacturer for the specific installation.

517.74 Control Circuit Conductors.

(A) Number of Conductors in Raceway. The number of control circuit conductors installed in a raceway shall be determined in accordance with 300.17.

(B) Minimum Size of Conductors. Size 18 AWG or 16 AWG fixture wires as specified in 725.49 and flexible cords shall be permitted for the control and operating circuits of X-ray and auxiliary equipment where protected by not larger than 20-ampere overcurrent devices.

517.75 Equipment Installations. All equipment for new X-ray installations and all used or reconditioned X-ray equipment moved to and reinstalled at a new location shall be of an approved type.

517.76 Transformers and Capacitors. Transformers and capacitors that are part of X-ray equipment shall not be required to comply with Articles 450 and 460.

Capacitors shall be mounted within enclosures of insulating material or grounded metal.

517.77 Installation of High-Tension X-Ray Cables. Cables with grounded shields connecting X-ray tubes and image intensifiers shall be permitted to be installed in cable trays or cable troughs along with X-ray equipment control and power supply conductors without the need for barriers to separate the wiring.

517.78 Guarding and Grounding.

(A) High-Voltage Parts. All high-voltage parts, including X-ray tubes, shall be mounted within grounded enclosures. Air, oil, gas, or other suitable insulating media shall be used to insulate the high-voltage from the grounded enclosure. The connection from the high-voltage equipment to X-ray tubes and other high-voltage components shall be made with high-voltage shielded cables.

(B) Low-Voltage Cables. Low-voltage cables connecting to oil-filled units that are not completely sealed, such as transformers, condensers, oil coolers, and high-voltage switches, shall have insulation of the oil-resistant type.

(C) Non–Current-Carrying Metal Parts. Non–current-carrying metal parts of X-ray and associated equipment (controls, tables, X-ray tube supports, transformer tanks, shielded cables, X-ray tube heads, etc.) shall be connected to an equipment grounding conductor in the manner specified in Part VII of Article 250, as modified by 517.13(A) and (B).

Part VI. Communications, Signaling Systems, Data Systems, Fire Alarm Systems, and Systems Less Than 120 Volts, Nominal

517.80 Patient Care Spaces. Equivalent insulation and isolation to that required for the electrical distribution systems in patient care areas shall be provided for communications, signaling systems, data system circuits, fire alarm systems, and systems less than 120 volts, nominal.

Class 2 and Class 3 signaling and communications systems and power-limited fire alarm systems shall not be required to comply with the grounding requirements of 517.13, to comply with the mechanical protection requirements of 517.31(C)(3)(5), or to be enclosed in raceways, unless otherwise specified by Chapter 7 or 8.

Secondary circuits of transformer-powered communications or signaling systems shall not be required to be enclosed in raceways unless otherwise specified by Chapter 7 or 8. [**99:**6.4.2.2.6.6]

517.81 Other-Than-Patient-Care Areas. In other-than-patient-care areas, installations shall be in accordance with the applicable provisions of other parts of this *Code*.

517.82 Signal Transmission Between Appliances.

(A) General. Permanently installed signal cabling from an appliance in a patient location to remote appliances shall employ a signal transmission system that prevents hazardous grounding interconnection of the appliances.

> Informational Note: See 517.13(A) for additional grounding requirements in patient care areas.

(B) Common Signal Grounding Wire. Common signal grounding wires (i.e., the chassis ground for single-ended transmission) shall be permitted to be used between appliances all located within the patient care vicinity, provided the appliances are served from the same reference grounding point.

Part VII. Isolated Power Systems

517.160 Isolated Power Systems.

(A) Installations.

(1) Isolated Power Circuits. Each isolated power circuit shall be controlled by a switch or circuit breaker that has a disconnecting pole in each isolated circuit conductor to simultaneously disconnect all power. Such isolation shall be accomplished by means of one or more isolation transformers, by means of generator sets, or by means of electrically isolated batteries. Conductors of isolated power circuits shall not be installed in cables, raceways, or other enclosures containing conductors of another system.

(2) Circuit Characteristics. Circuits supplying primaries of isolating transformers shall operate at not more than 600 volts between conductors and shall be provided with proper overcurrent protection. The secondary voltage of such transformers shall not exceed 600 volts between conductors of each circuit. All circuits supplied from such secondaries shall be ungrounded and shall have an approved overcurrent device of proper ratings in each conductor. Circuits supplied directly from batteries or from motor generator sets shall be ungrounded and shall be protected against overcurrent in the same manner as transformer-fed secondary circuits. If an electrostatic shield is present, it shall be connected to the reference grounding point. [**99:**6.3.2.6.1]

(3) Equipment Location. The isolating transformers, motor generator sets, batteries and battery chargers, and associated primary or secondary overcurrent devices shall not be installed in hazardous (classified) locations. The isolated secondary circuit wiring extending into a hazardous anesthetizing location shall be installed in accordance with 501.10.

(4) Isolation Transformers. An isolation transformer shall not serve more than one operating room except as covered in (A)(4)(a) and (A)(4)(b).

For purposes of this section, anesthetic induction rooms are considered part of the operating room or rooms served by the induction rooms.

   (a)   *Induction Rooms.* Where an induction room serves more than one operating room, the isolated circuits of the induction room shall be permitted to be supplied from the isolation transformer of any one of the operating rooms served by that induction room.

   (b)   *Higher Voltages.* Isolation transformers shall be permitted to serve single receptacles in several patient areas where the following apply:

(1)   The receptacles are reserved for supplying power to equipment requiring 150 volts or higher, such as portable X-ray units.

(2)   The receptacles and mating plugs are not interchangeable with the receptacles on the local isolated power system.
     [**99:**13.4.1.2.6.6]

(5) Conductor Identification. The isolated circuit conductors shall be identified as follows:

(1)   Isolated Conductor No. 1 — Orange with at least one distinctive colored stripe other than white, green, or gray along the entire length of the conductor

(2)   Isolated Conductor No. 2 — Brown with at least one distinctive colored stripe other than white, green, or gray along the entire length of the conductor

For 3-phase systems, the third conductor shall be identified as yellow with at least one distinctive colored stripe other than white, green, or gray along the entire length of the conductor. Where isolated circuit conductors supply 125-volt, single-phase, 15- and 20-ampere receptacles, the striped orange conductor(s) shall be connected to the terminal(s) on the receptacles that are identified in accordance with 200.10(B) for connection to the grounded circuit conductor.

(6) Wire-Pulling Compounds. Wire-pulling compounds that increase the dielectric constant shall not be used on the secondary conductors of the isolated power supply.

> Informational Note No. 1: It is desirable to limit the size of the isolation transformer to 10 kVA or less and to use conductor insulation with low leakage to meet impedance requirements.

> Informational Note No. 2: Minimizing the length of branch-circuit conductors and using conductor insulations with a dielectric constant less than 3.5 and insulation resistance constant greater than 6100 megohm-meters (20,000 megohm-feet) at 16°C (60°F) reduces leakage from line to ground, reducing the hazard current.

(B) Line Isolation Monitor.

(1) Characteristics. In addition to the usual control and overcurrent protective devices, each isolated power system shall be provided with a listed continually operating line isolation monitor that indicates total hazard current. The monitor shall be designed such that a green signal lamp, conspicuously visible to persons in each area served by the isolated power system, remains lighted when the system is adequately isolated from ground. An adjacent red signal lamp and an audible warning signal (remote if desired) shall be energized when the total hazard current (consisting of possible resistive and capacitive leakage currents) from either isolated conductor to ground reaches a threshold value of 5 mA under nominal line voltage conditions. The line monitor shall not alarm for a fault hazard of less than 3.7 mA or for a total hazard current of less than 5 mA.

Exception: A system shall be permitted to be designed to operate at a lower threshold value of total hazard current. A line isolation monitor for such a system shall be permitted to be approved, with the provision that the fault hazard current shall be permitted to be reduced but not to less than 35 percent of the corresponding threshold value of the total hazard current, and the monitor hazard current is to be correspondingly reduced to not more than 50 percent of the alarm threshold value of the total hazard current.

(2) Impedance. The line isolation monitor shall be designed to have sufficient internal impedance such that, when properly connected to the isolated system, the maximum internal current that can flow through the line isolation monitor, when any point of the isolated system is grounded, shall be 1 mA.

Exception: The line isolation monitor shall be permitted to be of the low impedance type such that the current through the line isolation monitor when any point of the isolated system is grounded, will not exceed twice the alarm threshold value for a period not exceeding 5 milliseconds.

Informational Note: Reduction of the monitor hazard current, provided this reduction results in an increased "not alarm" threshold value for the fault hazard current, will increase circuit capacity.

(3) Ammeter. An ammeter calibrated in the total hazard current of the system (contribution of the fault hazard current plus monitor hazard current) shall be mounted in a plainly visible place on the line isolation monitor with the "alarm on" zone at approximately the center of the scale.

Exception: The line isolation monitor shall be permitted to be a composite unit, with a sensing section cabled to a separate display panel section on which the alarm or test functions are located.

Informational Note: It is desirable to locate the ammeter so that it is conspicuously visible to persons in the anesthetizing location.

ARTICLE 518
Assembly Occupancies

518.1 Scope. Except for the assembly occupancies explicitly covered by 520.1, this article covers all buildings or portions of buildings or structures designed or intended for the gathering together of 100 or more persons for such purposes as deliberation, worship, entertainment, eating, drinking, amusement, awaiting transportation, or similar purposes.

518.2 General Classification.

(A) Examples. Assembly occupancies shall include, but not be limited to, the following:

Armories	Exhibition halls
Assembly halls	Gymnasiums
Auditoriums	Mortuary chapels
Bowling lanes	Multipurpose rooms
Club rooms	Museums
Conference rooms	Places of awaiting transportation
Courtrooms	Places of religious worship
Dance halls	Pool rooms
Dining and drinking facilities	Restaurants
	Skating rinks

(B) Multiple Occupancies. Where an assembly occupancy forms a portion of a building containing other occupancies, Article 518 applies only to that portion of the building considered an assembly occupancy. Occupancy of any room or space for assembly purposes by less than 100 persons in a building of other occupancy, and incidental to such other occupancy, shall be classified as part of the other occupancy and subject to the provisions applicable thereto.

(C) Theatrical Areas. Where any such building structure, or portion thereof, contains a projection booth or stage platform or area for the presentation of theatrical or musical productions, either fixed or portable, the wiring for that area, including associated audience seating areas, and all equipment that is

used in the referenced area, and portable equipment and wiring for use in the production that will not be connected to permanently installed wiring, shall comply with Article 520.

Informational Note: For methods of determining population capacity, see local building code or, in its absence, NFPA *101 -2015, Life Safety Code.*

518.3 Other Articles.

(A) Hazardous (Classified) Areas. Electrical installations in hazardous (classified) areas located in assembly occupancies shall comply with Article 500.

(B) Temporary Wiring. In exhibition halls used for display booths, as in trade shows, the temporary wiring shall be permitted to be installed in accordance with Article 590. Flexible cables and cords approved for hard or extra-hard usage shall be permitted to be laid on floors where protected from contact by the general public. The ground-fault circuit-interrupter requirements of 590.6 shall not apply. All other ground-fault circuit-interrupter requirements of this *Code* shall apply.

Where ground-fault circuit interrupter protection for personnel is supplied by plug-and-cord-connection to the branch circuit or to the feeder, the ground fault circuit interrupter protection shall be listed as portable ground fault circuit interrupter protection or provide a level of protection equivalent to a portable ground fault circuit interrupter, whether assembled in the field or at the factory.

Exception: Where conditions of supervision and maintenance ensure that only qualified persons will service the installation, flexible cords or cables identified in Table 400.1 for hard usage or extra-hard usage shall be permitted in cable trays used only for temporary wiring. All cords or cables shall be installed in a single layer. A permanent sign shall be attached to the cable tray at intervals not to exceed 7.5 m (25 ft). The sign shall read

CABLE TRAY FOR TEMPORARY WIRING ONLY

(C) Emergency Systems. Control of emergency systems shall comply with Article 700.

518.4 Wiring Methods.

(A) General. The fixed wiring methods shall be metal raceways, flexible metal raceways, nonmetallic raceways encased in not less than 50 mm (2 in.) of concrete, Type MI, MC, or AC cable. The wiring method shall itself qualify as an equipment grounding conductor according to 250.118 or shall contain an insulated equipment grounding conductor sized in accordance with Table 250.122.

Exception: Fixed wiring methods shall be as provided in
    (a) Audio signal processing, amplification, and reproduction equipment — Article 640
    (b) Communications circuits — Article 800
    (c) Class 2 and Class 3 remote-control and signaling circuits — Article 725
    (d) Fire alarm circuits — Article 760

(B) Nonrated Construction. In addition to the wiring methods of 518.4(A), nonmetallic-sheathed cable, Type AC cable, electrical nonmetallic tubing, and rigid nonmetallic conduit shall be permitted to be installed in those buildings or portions thereof that are not required to be of fire-rated construction by the applicable building code.

Informational Note: Fire-rated construction is the fire-resistive classification used in building codes.

(C) Spaces with Finish Rating. Electrical nonmetallic tubing and rigid nonmetallic conduit shall be permitted to be installed in club rooms, conference and meeting rooms in hotels or motels, courtrooms, dining facilities, restaurants, mortuary chapels, museums, libraries, and places of religious worship where the following apply:

(1) The electrical nonmetallic tubing or rigid nonmetallic conduit is installed concealed within walls, floors, and ceilings where the walls, floors, and ceilings provide a thermal barrier of material that has at least a 15-minute finish rating as identified in listings of fire-rated assemblies.

(2) The electrical nonmetallic tubing or rigid nonmetallic conduit is installed above suspended ceilings where the suspended ceilings provide a thermal barrier of material that has at least a 15-minute finish rating as identified in listings of fire-rated assemblies.

Electrical nonmetallic tubing and rigid nonmetallic conduit are not recognized for use in other space used for environmental air in accordance with 300.22(C).

Informational Note: A finish rating is established for assemblies containing combustible (wood) supports. The finish rating is defined as the time at which the wood stud or wood joist reaches an average temperature rise of 121°C (250°F) or an individual temperature rise of 163°C (325°F) as measured on the plane of the wood nearest the fire. A finish rating is not intended to represent a rating for a membrane ceiling.

518.5 Supply. Portable switchboards and portable power distribution equipment shall be supplied only from listed power outlets of sufficient voltage and ampere rating. Such power outlets shall be protected by overcurrent devices. Such overcurrent devices and power outlets shall not be accessible to the general public. Provisions for connection of an equipment grounding conductor shall be provided. The neutral conductor of feeders supplying solid-state phase control, 3-phase, 4-wire dimmer systems shall be considered a current-carrying conductor for purposes of ampacity adjustment. The neutral conductor of feeders supplying solid-state sine wave, 3-phase, 4-wire dimming systems shall not be considered a current-carrying conductor for purposes of ampacity adjustment.

Exception: The neutral conductor of feeders supplying systems that use or may use both phase-control and sine-wave dimmers shall be considered as current-carrying for purposes of ampacity adjustment.

Informational Note: For definitions of solid-state dimmer types, see 520.2.

ARTICLE 520
Theaters, Audience Areas of Motion Picture and Television Studios, Performance Areas, and Similar Locations

Part I. General

520.1 Scope. This article covers all buildings or that part of a building or structure, indoor or outdoor, designed or used for presentation, dramatic, musical, motion picture projection, or similar purposes and to specific audience seating areas within motion picture or television studios.

520.2 Definitions.

N **Adapter.** A device used to adapt a circuit from one configuration of an attachment plug or receptacle to another configuration with the same current rating.

Border Light. A permanently installed overhead strip light.

Breakout Assembly. An adapter used to connect a multipole connector containing two or more branch circuits to multiple individual branch-circuit connectors.

Bundled. Cables or conductors that are tied, wrapped, taped, or otherwise periodically bound together.

Connector Strip. A metal wireway containing pendant or flush receptacles.

Drop Box. A box containing pendant- or flush-mounted receptacles attached to a multiconductor cable via strain relief or a multipole connector.

Footlight. A border light installed on or in the stage.

Grouped. Cables or conductors positioned adjacent to one another but not in continuous contact with each other.

Performance Area. The stage and audience seating area associated with a temporary stage structure, whether indoors or outdoors, constructed of scaffolding, truss, platforms, or similar devices, that is used for the presentation of theatrical or musical productions or for public presentations.

Portable Equipment. Equipment fed with portable cords or cables intended to be moved from one place to another.

Portable Power Distribution Unit. A power distribution box containing receptacles and overcurrent devices.

Proscenium. The wall and arch that separates the stage from the auditorium (house).

Solid-State Phase-Control Dimmer. A solid-state dimmer where the wave shape of the steady-state current does not follow the wave shape of the applied voltage, such that the wave shape is nonlinear.

Solid-State Sine Wave Dimmer. A solid-state dimmer where the wave shape of the steady-state current follows the wave shape of the applied voltage such that the wave shape is linear.

Stage Equipment. Equipment at any location on the premises integral to the stage production including, but not limited to, equipment for lighting, audio, special effects, rigging, motion control, projection, or video.

Stage Lighting Hoist. A motorized lifting device that contains a mounting position for one or more luminaires, with wiring devices for connection of luminaires to branch circuits, and integral flexible cables to allow the luminaires to travel over the lifting range of the hoist while energized.

Stage Switchboard. A permanently installed switchboard, panelboard, or rack containing dimmers or relays with associated overcurrent protective devices, or overcurrent protective devices alone, used primarily to feed stage equipment.

N **Stage Switchboard, Portable.** A portable rack or pack containing dimmers or relays with associated overcurrent protective devices, or overcurrent protective devices alone that are used to feed stage equipment.

Stand Lamp (Work Light). A portable stand that contains a general-purpose luminaire or lampholder with guard for the purpose of providing general illumination on the stage or in the auditorium.

Strip Light. A luminaire with multiple lamps arranged in a row.

Two-Fer. An assembly containing one male plug and two female cord connectors used to connect two loads to one branch circuit.

520.3 Motion Picture Projectors. Motion picture equipment and its installation and use shall comply with Article 540.

520.4 Audio Signal Processing, Amplification, and Reproduction Equipment. Audio signal processing, amplification, and reproduction equipment and its installation shall comply with Article 640.

520.5 Wiring Methods.

(A) General. The fixed wiring method shall be metal raceways, nonmetallic raceways encased in at least 50 mm (2 in.) of concrete, Type MI cable, MC cable, or AC cable containing an insulated equipment grounding conductor sized in accordance with Table 250.122.

Exception: Fixed wiring methods shall be as provided in Article 640 for audio signal processing, amplification, and reproduction equipment, in Article 800 for communications circuits, in Article 725 for Class 2 and Class 3 remote-control and signaling circuits, and in Article 760 for fire alarm circuits.

(B) Portable Equipment. The wiring for portable switchboards, stage set lighting, stage effects, and other wiring not fixed as to location shall be permitted with approved flexible cords and cables as provided elsewhere in Article 520. Fastening such cables and cords by uninsulated staples or nailing shall not be permitted.

(C) Nonrated Construction. Nonmetallic-sheathed cable, Type AC cable, electrical nonmetallic tubing, and rigid nonmetallic conduit shall be permitted to be installed in those buildings or portions thereof that are not required to be of fire-rated construction by the applicable building code.

520.6 Number of Conductors in Raceway. The number of conductors permitted in any metal conduit, rigid nonmetallic conduit as permitted in this article, or electrical metallic tubing for circuits or for remote-control conductors shall not exceed the percentage fill shown in Table 1 of Chapter 9. Where contained within an auxiliary gutter or a wireway, the sum of the cross-sectional areas of all contained conductors at any cross section shall not exceed 20 percent of the interior cross-sectional area of the auxiliary gutter or wireway. The 30-conductor limitation of 366.22 and 376.22 shall not apply.

520.7 Enclosing and Guarding Live Parts. Live parts shall be enclosed or guarded to prevent accidental contact by persons and objects. All switches shall be of the externally operable type. Dimmers, including rheostats, shall be placed in cases or cabinets that enclose all live parts.

520.8 Emergency Systems. Control of emergency systems shall comply with Article 700.

520.9 Branch Circuits. A branch circuit of any size supplying one or more receptacles shall be permitted to supply stage set lighting. The voltage rating of the receptacles shall be not less than the circuit voltage. Receptacle ampere ratings and branch-circuit conductor ampacity shall be not less than the branch-circuit overcurrent device ampere rating. Table 210.21(B)(2) and 210.23 shall not apply. The application of 210.8(B)(4) shall not be required.

520.10 Portable Equipment Used Outdoors. Portable stage and studio lighting equipment and portable power distribution equipment not identified for outdoor use shall be permitted for temporary use outdoors, provided the equipment is supervised by qualified personnel while energized and barriered from the general public.

Part II. Fixed Stage Switchboards

520.21 General. Fixed stage switchboards shall comply with 520.21(1) through (4):

(1) Fixed stage switchboards shall be listed.
(2) Fixed stage switchboards shall be readily accessible but shall not be required to be located on or adjacent to the stage. Multiple fixed stage switchboards shall be permitted at different locations.
(3) A fixed stage switchboard shall contain overcurrent protective devices for all branch circuits supplied by that switchboard.
(4) A fixed stage switchboard shall be permitted to supply both stage and non-stage equipment.

520.25 Dimmers. Dimmers shall comply with 520.25(A) through (D).

(A) Disconnection and Overcurrent Protection. Where dimmers are installed in ungrounded conductors, each dimmer shall have overcurrent protection not greater than 125 percent of the dimmer rating and shall be disconnected from all ungrounded conductors when the master or individual switch or circuit breaker supplying such dimmer is in the open position.

(B) Resistance- or Reactor-Type Dimmers. Resistance- or series reactor-type dimmers shall be permitted to be placed in either the grounded or the ungrounded conductor of the circuit. Where designed to open either the supply circuit to the dimmer or the circuit controlled by it, the dimmer shall then comply with 404.2(B). Resistance- or reactor-type dimmers placed in the grounded neutral conductor of the circuit shall not open the circuit.

(C) Autotransformer-Type Dimmers. The circuit supplying an autotransformer-type dimmer shall not exceed 150 volts between conductors. The grounded conductor shall be common to the input and output circuits.

Informational Note: See 210.9 for circuits derived from auto-transformers.

(D) Solid-State-Type Dimmers. The circuit supplying a solid-state dimmer shall not exceed 150 volts between conductors unless the dimmer is listed specifically for higher voltage operation. Where a grounded conductor supplies a dimmer, it shall be common to the input and output circuits. Dimmer chassis shall be connected to the equipment grounding conductor.

520.26 Type of Switchboard. A stage switchboard shall be either one or a combination of the types specified in 520.26(A), (B), (C), and (D).

(A) Manual. Dimmers and switches are operated by handles mechanically linked to the control devices.

(B) Remotely Controlled. Devices are operated electrically from a pilot-type control console or panel. Pilot control panels either shall be part of the switchboard or shall be permitted to be at another location.

(C) Intermediate. A stage switchboard with circuit interconnections is a secondary switchboard (patch panel) or panelboard remote to the primary stage switchboard. It shall contain overcurrent protection. Where the required branch-circuit overcurrent protection is provided in the dimmer panel, it shall be permitted to be omitted from the intermediate switchboard.

(D) Constant Power. A stage switchboard containing only overcurrent protective devices and no control elements.

520.27 Stage Switchboard Feeders.

(A) Type of Feeder. Feeders supplying stage switchboards shall be one of the types in 520.27(A)(1) through (A)(3).

(1) Single Feeder. A single feeder disconnected by a single disconnect device.

(2) Multiple Feeders to Intermediate Stage Switchboard (Patch Panel). Multiple feeders of unlimited quantity shall be permitted, provided that all multiple feeders are part of a single system. Where combined, neutral conductors in a given raceway shall be of sufficient ampacity to carry the maximum unbalanced current supplied by multiple feeder conductors in the same raceway, but they need not be greater than the ampacity of the neutral conductor supplying the primary stage switchboard. Parallel neutral conductors shall comply with 310.10(H).

(3) Separate Feeders to Single Primary Stage Switchboard (Dimmer Bank). Installations with separate feeders to a single primary stage switchboard shall have a disconnecting means for each feeder. The primary stage switchboard shall have a permanent and obvious label stating the number and location of disconnecting means. If the disconnecting means are located in more than one distribution switchboard, the primary stage switchboard shall be provided with barriers to correspond with these multiple locations.

(B) Neutral Conductor. For the purpose of ampacity adjustment, the following shall apply:

(1) The neutral conductor of feeders supplying solid-state, phase-control 3-phase, 4-wire dimming systems shall be considered a current-carrying conductor.

(2) The neutral conductor of feeders supplying solid-state, sine wave 3-phase, 4-wire dimming systems shall not be considered a current-carrying conductor.

(3) The neutral conductor of feeders supplying systems that use or may use both phase-control and sine wave dimmers shall be considered as current-carrying.

(C) Supply Capacity. For the purposes of calculating supply capacity to switchboards, it shall be permissible to consider the maximum load that the switchboard is intended to control in a given installation, provided that the following apply:

(1) All feeders supplying the switchboard shall be protected by an overcurrent device with a rating not greater than the ampacity of the feeder.

(2) The opening of the overcurrent device shall not affect the proper operation of the egress or emergency lighting systems.

Informational Note: For calculation of stage switchboard feeder loads, see 220.40.

Part III. Fixed Stage Equipment Other Than Switchboards

520.40 Stage Lighting Hoists. Where a stage lighting hoist is listed as a complete assembly and contains an integral cable-handling system and cable to connect a moving wiring device to a fixed junction box for connection to permanent wiring, the extra-hard usage requirement of 520.44(C)(1) shall not apply.

520.41 Circuit Loads.

(A) Circuits Rated 20 Amperes or Less. Footlights, border lights, and proscenium sidelights shall be arranged so that no branch circuit supplying such equipment carries a load exceeding 20 amperes.

(B) Circuits Rated Greater Than 20 Amperes. Where only heavy-duty lampholders are used, such circuits shall be permitted to comply with Article 210 for circuits supplying heavy-duty lampholders.

520.42 Conductor Insulation. Foot, border, proscenium, or portable strip lights and connector strips shall be wired with conductors that have insulation suitable for the temperature at which the conductors are operated, but not less than 125°C (257°F). The ampacity of the 125°C (257°F) conductors shall be that of 60°C (140°F) conductors. All drops from connector strips shall be 90°C (194°F) wire sized to the ampacity of 60°C (140°F) cords and cables with no more than 150 mm (6 in.) of conductor extending into the connector strip. Section 310.15(B)(3)(a) shall not apply.

Informational Note: See Table 310.104(A) for conductor types.

520.43 Footlights.

(A) Metal Trough Construction. Where metal trough construction is employed for footlights, the trough containing the circuit conductors shall be made of sheet metal not lighter than 0.81 mm (0.032 in.) and treated to prevent oxidation. Lampholder terminals shall be kept at least 13 mm (½ in.) from the metal of the trough. The circuit conductors shall be soldered to the lampholder terminals.

(B) Other-Than-Metal Trough Construction. Where the metal trough construction specified in 520.43(A) is not used, footlights shall consist of individual outlets with lampholders wired with rigid metal conduit, intermediate metal conduit, or flexible metal conduit, Type MC cable, or mineral-insulated, metal-sheathed cable. The circuit conductors shall be soldered to the lampholder terminals.

(C) Disappearing Footlights. Disappearing footlights shall be arranged so that the current supply is automatically disconnected when the footlights are replaced in the storage recesses designed for them.

520.44 Borders, Proscenium Sidelights, Drop Boxes, and Connector Strips.

(A) General. Borders and proscenium sidelights shall be as follows:

(1) Constructed as specified in 520.43
(2) Suitably stayed and supported
(3) Designed so that the flanges of the reflectors or other adequate guards protect the lamps from mechanical damage and from accidental contact with scenery or other combustible material

(B) Connector Strips and Drop Boxes. Connector strips and drop boxes shall be as follows:

(1) Suitably stayed and supported
(2) Listed as stage and studio wiring devices

(C) Cords and Cables for Border Lights, Drop Boxes, and Connector Strips.

(1) General. Cords and cables for supply to border lights, drop boxes, and connector strips shall be listed for extra-hard usage. The cords and cables shall be suitably supported. Such cords and cables shall be employed only where flexible conductors are necessary. Ampacity of the conductors shall be as provided in 400.5.

(2) Cords and Cables Not in Contact with Heat-Producing Equipment. Listed multiconductor extra-hard-usage-type cords and cables not in direct contact with equipment containing heat-producing elements shall be permitted to have their ampacity determined by Table 520.44(C)(3). Maximum load current in any conductor with an ampacity determined by Table 520.44(C)(3) shall not exceed the values in Table 520.44(C)(3).

(3) Identification of Conductors in Multiconductor Extra-Hard-Usage Cords and Cables. Grounded (neutral) conductors shall be white without stripe or shall be identified by a distinctive white marking at their terminations. Grounding conductors shall be green with or without yellow stripe or shall be identified by a distinctive green marking at their terminations.

Table 520.44(C)(3) Ampacity of Listed Extra-Hard-Usage Cords and Cables with Temperature Ratings of 75°C (167°F) and 90°C (194°F)* [Based on Ambient Temperature of 30°C (86°F)]

Size (AWG)	Temperature Rating of Cords and Cables		Maximum Rating of Overcurrent Device
	75°C (167°F)	90°C (194°F)	
14	24	28	15
12	32	35	20
10	41	47	25
8	57	65	35
6	77	87	45
4	101	114	60
2	133	152	80

*Ampacity shown is the ampacity for multiconductor cords and cables where only three copper conductors are current-carrying as described in 400.5. If the number of current-carrying conductors in a cord or cable exceeds three and the load diversity is 50 percent or less, the ampacity of each conductor shall be reduced as shown in the following table:

Table 520.44(C)(3)(a) Ampacity Adjustment Factors for More Than Three Current-Carrying Conductors in a Cord or Cable Where Load Diversity Is 50% or Less

Number of Conductors	Percent of Ampacity Value in Table 520.44(C)(3)
4–6	80
7–24	70
25–42	60
43 and above	50

Note: Ultimate insulation temperature. In no case shall conductors be associated together in such a way with respect to the kind of circuit, the wiring method used, or the number of conductors such that the temperature limit of the conductors is exceeded.

A neutral conductor that carries only the unbalanced current from other conductors of the same circuit need not be considered as a current-carrying conductor.

In a 3-wire circuit consisting of two-phase conductors and the neutral conductor of a 4-wire, 3-phase, wye-connected system, the neutral conductor carries approximately the same current as the line-to-neutral currents of the other conductors and shall be considered to be a current-carrying conductor.

On a 4-wire, 3-phase wye circuit where the major portion of the load consists of nonlinear loads, there are harmonic currents in the neutral conductor. Therefore, the neutral conductor shall be considered to be a current-carrying conductor.

> Informational Note: For the purposes of Table 520.44(C)(3)(a), load diversity is the percentage of the total current of all simultaneously energized circuits fed by the cable to the sum of the ampacity ratings of all circuits in that cable.

520.45 Receptacles. Receptacles for electrical equipment on stages shall be rated in amperes. Conductors supplying receptacles shall be in accordance with Articles 310 and 400.

520.46 Connector Strips, Drop Boxes, Floor Pockets, and Other Outlet Enclosures. Receptacles for the connection of portable stage-lighting equipment shall be pendant or mounted in suitable pockets or enclosures and shall comply with 520.45. Supply cables for connector strips and drop boxes shall be as specified in 520.44(C).

520.47 Backstage Lamps (Bare Bulbs). Lamps (bare bulbs) installed in backstage and ancillary areas where they can come in contact with scenery shall be located and guarded so as to be free from physical damage and shall provide an air space of not less than 50 mm (2 in.) between such lamps and any combustible material.

Exception: Decorative lamps installed in scenery shall not be considered to be backstage lamps for the purpose of this section.

520.48 Curtain Machines. Curtain machines shall be listed.

520.49 Smoke Ventilator Control. Where stage smoke ventilators are released by an electrical device, the circuit operating the device shall be normally closed and shall be controlled by at least two externally operable switches, one switch being placed at a readily accessible location on stage and the other where designated by the authority having jurisdiction. The device shall be designed for the full voltage of the circuit to which it is connected, no resistance being inserted. The device shall be enclosed in a suitable metal box having a door that shall remain closed except during service to the equipment.

Part IV. Portable Switchboards on Stage

520.50 Road Show Connection Panel (A Type of Patch Panel). A panel designed to allow for road show connection of portable stage switchboards to fixed lighting outlets by means of permanently installed supplementary circuits. The panel, supplementary circuits, and outlets shall comply with 520.50(A) through (D).

(A) Load Circuits. Circuits shall originate from grounding-type polarized inlets of current and voltage rating that match the fixed-load receptacle.

(B) Circuit Transfer. Circuits that are transferred between fixed and portable switchboards shall have all circuit conductors transferred simultaneously.

(C) Overcurrent Protection. The supply devices of these supplementary circuits shall be protected by branch-circuit overcurrent protective devices. Each supplementary circuit, within the road show connection panel and theater, shall be protected by branch-circuit overcurrent protective devices installed within the road show connection panel.

(D) Enclosure. Panel construction shall be in accordance with Article 408.

520.51 Supply. Portable switchboards shall be supplied only from power outlets of sufficient voltage and ampere rating. Such power outlets shall include only externally operable, enclosed fused switches or circuit breakers mounted on stage or at the permanent switchboard in locations readily accessible from the stage floor. Provisions for connection of an equipment grounding conductor shall be provided. For the purposes of ampacity adjustment, the requirements of 520.27(B) shall apply.

520.52 Overcurrent Protection for Branch Circuits. Portable switchboards shall contain overcurrent protection for branch circuits. The requirements of 210.23 shall not apply.

520.53 Construction. Portable stage switchboards shall be listed and shall comply with 520.53(A) through (E).

(A) Pilot Light. A pilot light shall be provided for each ungrounded conductor feeding the switchboard. The pilot light(s) shall be connected to the incoming feeder so that operation of the main overcurrent protective device or master switch shall not affect the operation of the pilot light(s).

N **(B) Neutral Terminal.** In portable switchboard equipment designed for use with 3-phase, 4-wire with ground supply, the current rating of the supply neutral terminal, and the ampacity of its associated busbar or equivalent wiring, or both, shall have an ampacity equal to at least twice the ampacity of the largest ungrounded supply terminal.

Exception: Where portable switchboard equipment is specifically constructed and identified to be internally converted in the field, in an approved manner, from use with a balanced 3-phase, 4-wire with ground supply to a balanced single-phase, 3-wire with ground supply, the supply neutral terminal and its associated busbar, equivalent wiring, or both, shall have an ampacity equal to at least that of the largest ungrounded single-phase supply terminal.

N **(C) Single-Pole Separable Connectors.** Where single-pole portable cable connectors are used on a portable stage switchboard, they shall be listed and of the locking type. Sections 406.7 and 406.8 shall not apply to listed single-pole separable connectors and single-conductor cable assemblies utilizing listed single-pole separable connectors. Where paralleled sets of current-carrying, single-pole separable connectors are provided as input devices, they shall be prominently labeled with a warning indicating the presence of internal parallel connections. The use of single-pole separable connectors shall comply with at least one of the following conditions:

(1) Connection and disconnection of connectors are possible only where the supply connectors are interlocked to the source, and it is not possible to connect or disconnect connectors when the supply is energized.

(2) Line connectors are of the listed sequential-interlocking type so that load connectors shall be connected in the following sequence:

    a. Equipment grounding conductor connection
    b. Grounded circuit conductor connection, if provided
    c. Ungrounded conductor connection, and that disconnection shall be in the reverse order

(3) A caution notice shall be provided adjacent to the line connectors indicating that plug connection shall be in the following order:

    a. Equipment grounding conductor connectors
    b. Grounded circuit conductor connectors, if provided
    c. Ungrounded conductor connectors, and that disconnection shall be in the reverse order

The warning sign(s) or label(s) shall comply with 110.21(B).

N **(D) Supply Feed-Through.** Where a portable stage switchboard contains a feed-through outlet of the same rating as its supply inlet, the feed-through outlet shall not require overcurrent protection in the switchboard.

(E) Interior Conductors. All conductors other than busbars within the switchboard enclosure shall be stranded.

520.54 Supply Conductors.

(A) General. The supply to a portable stage switchboard shall be by means of listed extra-hard usage cords or cables. The supply cords or cables shall terminate within the switchboard enclosure in an externally operable fused master switch or circuit breaker or in an identified connector assembly. The supply cords or cable (and connector assembly) shall have current ratings not less than the total load connected to the switchboard and shall be protected by overcurrent devices.

N **(B) Conductor Sizing.** The power supply conductors for portable stage switchboards utilizing solid-state phase control dimmers shall be sized considering the neutral conductor as a current-carrying conductor for ampacity adjustment purposes. The power supply conductors for portable stage switchboards utilizing only solid-state sine wave dimmers shall be sized considering the neutral conductor as a non–current carrying conductor for ampacity adjustment purposes.

(C) Single-Conductor Cables. Single-conductor portable supply cable sets shall be not smaller than 2 AWG conductors. The equipment grounding conductor shall not be smaller than 6 AWG conductor. Single-conductor grounded neutral cables for a supply shall be sized in accordance with 520.54(J). Where single conductors are paralleled for increased ampacity, the paralleled conductors shall be of the same length and size. Single-conductor supply cables shall be grouped together but not bundled. The equipment grounding conductor shall be permitted to be of a different type, provided it meets the other requirements of this section, and it shall be permitted to be reduced in size as permitted by 250.122. Grounded (neutral) and equipment grounding conductors shall be identified in accordance with 200.6, 250.119, and 310.110. Grounded conductors shall be permitted to be identified by marking at least the first 150 mm (6 in.) from both ends of each length of conductor with white or gray. Equipment grounding conductors shall be permitted to be identified by marking at least the first 150 mm (6 in.) from both ends of each length of conductor with green or green with yellow stripes. Where more than one nominal voltage exists within the same premises, each ungrounded conductor shall be identified by system.

(D) Supply Conductors Not Over 3 m (10 ft) Long. Where supply conductors do not exceed 3 m (10 ft) in length between supply and switchboard or supply and a subsequent overcurrent device, the supply conductors shall be permitted to be reduced in size where all of the following conditions are met:

(1) The ampacity of the supply conductors shall be at least one-quarter of the current rating of the supply overcurrent protective device.

(2) The supply conductors shall terminate in a single overcurrent protective device that will limit the load to the ampacity of the supply conductors. This single overcurrent device shall be permitted to supply additional overcurrent devices on its load side.

(3) The supply conductors shall not penetrate walls, floors, or ceilings or be run through doors or traffic areas. The supply conductors shall be adequately protected from physical damage.

(4) The supply conductors shall be suitably terminated in an approved manner.

(5) Conductors shall be continuous without splices or connectors.

(6) Conductors shall not be bundled.

(7) Conductors shall be supported above the floor in an approved manner.

(E) Supply Conductors Not Over 6 m (20 ft) Long. Where supply conductors do not exceed 6 m (20 ft) in length between supply and switchboard or supply and a subsequent overcurrent protection device, the supply conductors shall be permitted to be reduced in size where all of the following conditions are met:

(1) The ampacity of the supply conductors shall be at least one-half of the current rating of the supply overcurrent protective device.

(2) The supply conductors shall terminate in a single overcurrent protective device that limits the load to the ampacity of the supply conductors. This single overcurrent device shall be permitted to supply additional overcurrent devices on its load side.

(3) The supply conductors shall not penetrate walls, floors, or ceilings or be run through doors or traffic areas. The supply conductors shall be adequately protected from physical damage.

(4) The supply conductors shall be suitably terminated in an approved manner.

(5) The supply conductors shall be supported in an approved manner at least 2.1 m (7 ft) above the floor except at terminations.

(6) The supply conductors shall not be bundled.

(7) Tap conductors shall be in unbroken lengths.

(F) Supply Conductors Not Reduced in Size. Supply conductors not reduced in size under provisions of 520.54(D) or (E) shall be permitted to pass through holes in walls specifically designed for the purpose. If penetration is through the fire-resistant–rated wall, it shall be in accordance with 300.21.

N **(G) Protection of Supply Conductors and Connectors.** All supply conductors and connectors shall be protected against physical damage by an approved means. This protection shall not be required to be raceways.

(H) Number of Supply Interconnections. Where connectors are used in a supply conductor, there shall be a maximum number of three interconnections (mated connector pairs) where the total length from supply to switchboard does not exceed 30 m (100 ft). In cases where the total length from supply to switchboard exceeds 30 m (100 ft), one additional interconnection shall be permitted for each additional 30 m (100 ft) of supply conductor.

(I) Single-Pole Separable Connectors. Where single-pole portable cable connectors are used, they shall be listed and of the locking type. Sections 406.7 and 406.8 shall not apply to listed single-pole separable connectors and single-conductor cable assemblies utilizing listed single-pole separable connectors.

(J) Supply Neutral Conductor. Supply neutral conductors shall comply with (1) and (2) below:

(1) Marking. Grounded neutral conductors shall be permitted to be identified by marking at least the first 150 mm (6 in.) from both ends of each length of conductor with white or gray.

(2) Conductor Sizing. Where single-conductor feeder cables not installed in raceways are used on multiphase circuits feeding portable stage switchboards containing solid-state phase-control dimmers, the grounded neutral conductor shall have an ampacity of at least 130 percent of the ungrounded circuit conductors feeding the portable stage switchboard. Where such feeders are supplying only solid-state sine wave dimmers, the grounded neutral conductor shall have an ampacity of at least 100 percent of the ungrounded circuit conductors feeding the portable stage switchboard.

(K) Qualified Personnel.   The routing of portable supply conductors, the making and breaking of supply connectors and other supply connections, and the energization and de-energization of supply services shall be performed by qualified personnel, and portable switchboards shall be so marked, indicating this requirement in a permanent and conspicuous manner.

Exception: A portable switchboard shall be permitted to be connected to a permanently installed supply receptacle by other than qualified personnel provided that the supply receptacle is protected for its current rating by an overcurrent device of not greater than 150 amperes, and where the receptacle, interconnection, and switchboard comply with all of the following:

(a) They employ listed multipole connectors suitable for the purpose for every supply interconnection.

(b) They prevent access to all supply connections by the general public.

(c) They employ listed extra-hard usage multiconductor cords or cables with an ampacity not less than the load and not less than the ampere rating of the connectors.

Part V. Portable Stage Equipment Other Than Switchboards

520.61 Arc Lamps.   Arc lamps, including enclosed arc lamps and associated ballasts, shall be listed. Interconnecting cord sets and interconnecting cords and cables shall be extra-hard usage type and listed.

520.62 Portable Power Distribution Units.   Portable power distribution units shall comply with 520.62(A) through (F).

(A) Enclosure.   The construction shall be such that no current-carrying part will be exposed.

(B) Receptacles and Overcurrent Protection.   Receptacles shall comply with 520.45 and shall have branch-circuit overcurrent protection in the box. Fuses and circuit breakers shall be protected against physical damage. Flexible cords or cables supplying pendant receptacles or cord connectors shall be listed for extra-hard usage.

(C) Busbars and Terminals.   Busbars shall have an ampacity equal to the sum of the ampere ratings of all the circuits connected to the busbar. Lugs shall be provided for the connection of the master cable.

(D) Flanged Surface Inlets.   Flanged surface inlets (recessed plugs) that are used to accept the power shall be rated in amperes.

(E) Cable Arrangement.   Cables shall be adequately protected where they pass through enclosures and be arranged so that tension on the cable is not transmitted to the terminations.

N **(F) Single-Conductor Feeders.**   Portable power distribution equipment fed by single-conductor feeder systems shall comply with 520.53(C) and (D) and 520.54.

520.63 Bracket Fixture Wiring.

(A) Bracket Wiring.   Brackets for use on scenery shall be wired internally, and the fixture stem shall be carried through to the back of the scenery where a bushing shall be placed on the end of the stem. Externally wired brackets or other fixtures shall be permitted where wired with cords designed for hard usage that extend through scenery and without joint or splice in canopy of fixture back and terminate in an approved-type stage connector located, where practical, within 450 mm (18 in.) of the fixture.

(B) Mounting.   Fixtures shall be securely fastened in place.

520.64 Portable Strips.   Portable strips shall be constructed in accordance with the requirements for border lights and proscenium sidelights in 520.44(A). The supply cable shall be protected by bushings where it passes through metal and shall be arranged so that tension on the cable will not be transmitted to the connections.

> Informational Note No. 1: See 520.42 for wiring of portable strips.

> Informational Note No. 2: See 520.68(A)(3) for insulation types required on single conductors.

520.65 Festoons.   Joints in festoon wiring shall be staggered. Where such lampholders have terminals of a type that puncture the insulation and make contact with the conductors, they shall be attached only to conductors of the stranded type. Lamps enclosed in lanterns or similar devices of combustible material shall be equipped with guards.

520.66 Special Effects.   Electrical devices used for simulating lightning, waterfalls, and the like shall be constructed and located so that flames, sparks, or hot particles cannot come in contact with combustible material.

520.67 Multipole Branch-Circuit Cable Connectors.   Multipole branch-circuit cable connectors, male and female, for flexible conductors shall be constructed so that tension on the cord or cable is not transmitted to the connections. The female half shall be attached to the load end of the power supply cord or cable. The connector shall be rated in amperes and designed so that differently rated devices cannot be connected together; however, a 20-ampere T-slot receptacle shall be permitted to accept a 15-ampere attachment plug of the same voltage rating. Alternating-current multipole connectors shall be polarized and comply with 406.7 and 406.10.

> Informational Note: See 400.14 for pull at terminals.

520.68 Conductors for Portables.

(A) Conductor Type.

(1) General.   Flexible conductors, including cable extensions, used to supply portable stage equipment shall be listed extra-hard usage cords or cables.

(2) Stand Lamps.   Listed, hard usage cord shall be permitted to supply stand lamps where the cord is not subject to physical damage and is protected by an overcurrent device rated at not over 20 amperes.

(3) Luminaire Supply Cords. Listed hard usage supply cords shall be permitted to supply luminaires when all of the following conditions are met:

(1) The supply cord is not longer than 2.0 m (6.6 ft).
(2) The supply cord is attached at one end to the luminaire or a luminaire-specific listed connector that mates with a panel-mounted inlet on the body of the luminaire.
(3) The supply cord is protected by an overcurrent protective device of not more than 20 amperes.
(4) The luminaire is listed.
(5) The supply cord is not subject to physical damage.

(4) High-Temperature Applications. A special assembly of conductors in sleeving not longer than 1.0 m (3.3 ft) shall be permitted to be employed in lieu of flexible cord if the individual wires are stranded and rated not less than 125°C (257°F) and the outer sleeve is glass fiber with a wall thickness of at least 0.635 mm (0.025 in.).

Portable stage equipment requiring flexible supply conductors with a higher temperature rating where one end is permanently attached to the equipment shall be permitted to employ alternate, suitable conductors as determined by a qualified testing laboratory and recognized test standards.

(5) Breakouts. Listed, hard usage (junior hard service) cords shall be permitted in breakout assemblies where all of the following conditions are met:

(1) The cords are utilized to connect between a single multipole connector containing two or more branch circuits and multiple 2-pole, 3-wire connectors.
(2) The longest cord in the breakout assembly does not exceed 6.0 m (20 ft).
(3) The breakout assembly is protected from physical damage by attachment over its entire length to a pipe, truss, tower, scaffold, or other substantial support structure.
(4) All branch circuits feeding the breakout assembly are protected by overcurrent devices rated at not over 20 amperes.

(B) Conductor Ampacity. The ampacity of conductors shall be as given in 400.5, except multiconductor, listed, extra-hard usage portable cords that are not in direct contact with equipment containing heat-producing elements shall be permitted to have their ampacity determined by Table 520.44(C)(3). Maximum load current in any conductor with an ampacity determined by Table 520.44(C)(3) shall not exceed the values in Table 520.44(C)(3). Where the ampacity adjustment factors of Table 520.44(C)(3)(a) are applied for more than three current-carrying conductors in a portable cord, the load diversity shall be 50 percent or less.

Exception: Where alternate conductors are allowed in 520.68(A)(4), their ampacity shall be as given in the appropriate table in this Code for the types of conductors employed.

(C) Overcurrent Protection. Overcurrent protection of conductors for portables shall comply with 240.5.

520.69 Adapters. Adapters, two-fers, and other single- and multiple-circuit outlet devices shall comply with 520.69(A), (B), and (C).

(A) No Reduction in Current Rating. Each receptacle and its corresponding cable shall have the same current and voltage rating as the plug supplying it. It shall not be utilized in a stage circuit with a greater current rating.

(B) Connectors. All connectors shall be wired in accordance with 520.67.

(C) Conductor Type. Conductors for adapters and two-fers shall be listed extra-hard usage or listed hard usage (junior hard service) cord. Hard usage (junior hard service) cord shall be restricted in overall length to 2.0 m (6.6 ft).

Part VI. Dressing Rooms, Dressing Areas, and Makeup Areas.

520.71 Pendant Lampholders. Pendant lampholders shall not be installed in dressing or makeup rooms.

520.72 Lamp Guards. All exposed lamps in dressing or makeup areas including rooms where they are less than 2.5 m (8 ft) from the floor shall be equipped with open-end guards riveted to the outlet box cover or otherwise sealed or locked in place. Recessed lamps shall not be required to be equipped with guards.

520.73 Switches Required. All luminaires, lampholders, and any receptacles adjacent to the mirror(s) and above the dressing or makeup counter(s) installed in dressing or makeup rooms shall be controlled by wall switches installed in the dressing or makeup room(s). Other outlets installed in the dressing or makeup rooms shall not be required to be switched.

N **520.74 Pilot Lights Required.** Each switch required in 520.73 shall be provided with a pilot light located outside of and adjacent to the door of the room being controlled to indicate when the circuit is energized. Each pilot light shall be permanently identified indicating a description of the circuit controlled. Pilot lights shall be neon, LED, or other extended-life lamp. Pilot lights shall be recessed or provided with a mechanical guard.

Part VII. Grounding

520.81 Grounding. All metal raceways and metal-sheathed cables shall be connected to an equipment grounding conductor. The metal frames and enclosures of all equipment, including border lights and portable luminaires, shall be connected to an equipment grounding conductor.

ARTICLE 522
Control Systems for Permanent Amusement Attractions

Part I. General

522.1 Scope. This article covers the installation of control circuit power sources and control circuit conductors for electrical equipment, including associated control wiring in or on all structures, that are an integral part of a permanent amusement attraction.

522.2 Definitions.

Entertainment Device. A mechanical or electromechanical device that provides an entertainment experience.

Informational Note: These devices may include animated props, show action equipment, animated figures, and special effects, coordinated with audio and lighting to provide an entertainment experience.

Permanent Amusement Attraction. Ride devices, entertainment devices, or combination thereof, that are installed so that portability or relocation is impracticable.

Ride Device. A device or combination of devices that carry, convey, or direct a person(s) over or through a fixed or restricted course within a defined area for the primary purpose of amusement or entertainment.

522.5 Voltage Limitations. Control voltage shall be a maximum of 150 volts, nominal, ac to ground or 300 volts dc to ground.

522.7 Maintenance. The conditions of maintenance and supervision shall ensure that only qualified persons service the permanent amusement attraction.

Part II. Control Circuits

522.10 Power Sources for Control Circuits.

(A) Power-Limited Control Circuits. Power-limited control circuits shall be supplied from a source that has a rated output of not more than 30 volts and 1000 volt-amperes.

(1) Control Transformers. Transformers used to supply power-limited control circuits shall comply with the applicable sections within Parts I and II of Article 450.

(2) Other Power-Limited Control Power Sources. Power-limited control power sources, other than transformers, shall be protected by overcurrent devices rated at not more than 167 percent of the volt-ampere rating of the source divided by the rated voltage. The fusible overcurrent devices shall not be interchangeable with fusible overcurrent devices of higher ratings. The overcurrent device shall be permitted to be an integral part of the power source.

To comply with the 1000 volt-ampere limitation of 522.10(A), the maximum output of power sources, other than transformers, shall be limited to 2500 volt-amperes, and the product of the maximum current and maximum voltage shall not exceed 10,000 volt-amperes. These ratings shall be determined with any overcurrent-protective device bypassed.

(B) Non–Power-Limited Control Circuits. Non–power-limited control circuits shall not exceed 300 volts. The power output of the source shall not be required to be limited.

(1) Control Transformers. Transformers used to supply non–power-limited control circuits shall comply with the applicable sections within Parts I and II of Article 450.

(2) Other Non–Power-Limited Control Power Sources. Non–power-limited control power sources, other than transformers, shall be protected by overcurrent devices rated at not more than 125 percent of the volt-ampere rating of the source divided by the rated voltage. The fusible overcurrent devices shall not be interchangeable with fusible overcurrent devices of higher ratings. The overcurrent device shall be permitted to be an integral part of the power source.

Part III. Control Circuit Wiring Methods

522.20 Conductors, Busbars, and Slip Rings. Insulated control circuit conductors shall be copper and shall be permitted to be stranded or solid. Listed multiconductor cable assemblies shall be permitted.

Exception No. 1: Busbars and slip rings shall be permitted to be materials other than copper.

Exception No. 2: Conductors used as specific-purpose devices, such as thermocouples and resistive thermal devices, shall be permitted to be materials other than copper.

522.21 Conductor Sizing.

(A) Conductors Within a Listed Component or Assembly. Conductors of size 30 AWG or larger shall be permitted within a listed component or as part of the wiring of a listed assembly.

(B) Conductors Within an Enclosure or Operator Station. Conductors of size 30 AWG or larger shall be permitted in a listed and jacketed multiconductor cable within an enclosure or operator station. Conductors in a non-jacketed multiconductor cable, such as ribbon cable, shall not be smaller than 26 AWG. Single conductors shall not be smaller than 24 AWG.

Exception: Single conductors 30 AWG or larger shall be permitted for jumpers and special wiring applications.

(C) Conductors Outside of an Enclosure or Operator Station. The size of conductors in a listed and jacketed, multiconductor cable shall not be smaller than 26 AWG. Single conductors shall not be smaller than 18 AWG and shall be installed only where part of a recognized wiring method of Chapter 3.

522.22 Conductor Ampacity. Conductors sized 16 AWG and smaller shall not exceed the continuous current values provided in Table 522.22.

522.23 Overcurrent Protection for Conductors. Conductors 30 AWG through 16 AWG shall have overcurrent protection in accordance with the appropriate conductor ampacity in Table 522.22. Conductors larger than 16 AWG shall have overcurrent protection in accordance with the appropriate conductor ampacity in Table 310.15(B)(16).

Table 522.22 Conductor Ampacity Based on Copper Conductors with 60°C and 75°C Insulation in an Ambient Temperature of 30°C

Conductor Size (AWG)	Ampacity	
	60°C	75°C
30	–	0.5
28	–	0.8
26	–	1
24	2	2
22	3	3
20	5	5
18	7	7
16	10	10

Notes:
1. For ambient temperatures other than 30°C, use Table 310.15(B)(2)(a) temperature correction factors.
2. Ampacity adjustment for conductors with 90°C or greater insulation shall be based on ampacities in the 75°C column.

522.24 Conductors of Different Circuits in the Same Cable, Cable Tray, Enclosure, or Raceway. Control circuits shall be permitted to be installed with other circuits as specified in 522.24(A) and (B).

(A) Two or More Control Circuits. Control circuits shall be permitted to occupy the same cable, cable tray, enclosure, or raceway without regard to whether the individual circuits are alternating current or direct current, provided all conductors are insulated for the maximum voltage of any conductor in the cable, cable tray, enclosure, or raceway.

(B) Control Circuits with Power Circuits. Control circuits shall be permitted to be installed with power conductors as specified in 522.24(B)(1) through (B)(3).

(1) In a Cable, Enclosure, or Raceway. Control circuits and power circuits shall be permitted to occupy the same cable, enclosure, or raceway only where the equipment powered is functionally associated.

(2) In Factory- or Field-Assembled Control Centers. Control circuits and power circuits shall be permitted to be installed in factory- or field-assembled control centers.

(3) In a Manhole. Control circuits and power circuits shall be permitted to be installed as underground conductors in a manhole in accordance with one of the following:

(1) The power or control circuit conductors are in a metal-enclosed cable or Type UF cable.
(2) The conductors are permanently separated from the power conductors by a continuous firmly fixed nonconductor, such as flexible tubing, in addition to the insulation on the wire.
(3) The conductors are permanently and effectively separated from the power conductors and securely fastened to racks, insulators, or other approved supports.
(4) In cable trays, where the control circuit conductors and power conductors not functionally associated with them are separated by a solid fixed barrier of a material compatible with the cable tray, or where the power or control circuit conductors are in a metal-enclosed cable.

522.25 Ungrounded Control Circuits. Separately derived ac circuits and systems 50 volts or greater and 2-wire dc circuits and systems 60 volts or greater shall be permitted to be ungrounded, provided that all the following conditions are met:

(1) Continuity of control power is required for orderly shutdown.
(2) Ground detectors are installed on the control system.

522.28 Control Circuits in Wet Locations. Where wet contact is likely to occur, ungrounded 2-wire direct-current control circuits shall be limited to 30 volts maximum for continuous dc or 12.4 volts peak for direct current that is interrupted at a rate of 10 to 200 Hz.

ARTICLE 525
Carnivals, Circuses, Fairs, and Similar Events

Part I. General Requirements

525.1 Scope. This article covers the installation of portable wiring and equipment for carnivals, circuses, fairs, and similar functions, including wiring in or on all structures.

525.2 Definitions.

Operator. The individual responsible for starting, stopping, and controlling an amusement ride or supervising a concession.

Portable Structures. Units designed to be moved including, but not limited to, amusement rides, attractions, concessions, tents, trailers, trucks, and similar units.

525.3 Other Articles.

(A) Portable Wiring and Equipment. Wherever the requirements of other articles of this *Code* and Article 525 differ, the requirements of Article 525 shall apply to the portable wiring and equipment.

(B) Permanent Structures. Articles 518 and 520 shall apply to wiring in permanent structures.

(C) Audio Signal Processing, Amplification, and Reproduction Equipment. Article 640 shall apply to the wiring and installation of audio signal processing, amplification, and reproduction equipment.

(D) Attractions Utilizing Pools, Fountains, and Similar Installations with Contained Volumes of Water. This equipment shall be installed to comply with the applicable requirements of Article 680.

525.5 Overhead Conductor Clearances.

(A) Vertical Clearances. Conductors shall have a vertical clearance to ground in accordance with 225.18. These clearances shall apply only to wiring installed outside of tents and concessions.

(B) Clearance to Portable Structures.

(1) 600 Volts (or Less). Portable structures shall be maintained not less than 4.5 m (15 ft) in any direction from overhead conductors operating at 600 volts or less, except for the conductors supplying the portable structure. Portable structures included in 525.3(D) shall comply with Table 680.9(A).

(2) Over 600 Volts. Portable structures shall not be located under or within a space that is located 4.5 m (15 ft) horizontally and extending vertically to grade of conductors operating in excess of 600 volts.

525.6 Protection of Electrical Equipment. Electrical equipment and wiring methods in or on portable structures shall be provided with mechanical protection where such equipment or wiring methods are subject to physical damage.

Part II. Power Sources

525.10 Services.   Services shall comply with 525.10(A) and (B).

(A) Guarding.   Service equipment shall not be installed in a location that is accessible to unqualified persons, unless the equipment is lockable.

(B) Mounting and Location.   Service equipment shall be securely fastened to a solid backing and be installed so as to be protected from the weather, unless of weatherproof construction.

525.11 Multiple Sources of Supply.   Where multiple services or separately derived systems, or both, supply portable structures, the equipment grounding conductors of all the sources of supply that serve such structures separated by less than 3.7 m (12 ft) shall be bonded together at the portable structures. The bonding conductor shall be copper and sized in accordance with Table 250.122 based on the largest overcurrent device supplying the portable structures, but not smaller than 6 AWG.

Part III. Wiring Methods

525.20 Wiring Methods.

(A) Type.   Where flexible cords or cables are used, they shall be listed for extra-hard usage. Where flexible cords or cables are used and are not subject to physical damage, they shall be permitted to be listed for hard usage. Where used outdoors, flexible cords and cables shall also be listed for wet locations and shall be sunlight resistant. Extra-hard usage flexible cords or cables shall be permitted for use as permanent wiring on portable amusement rides and attractions where not subject to physical damage.

(B) Single-Conductor.   Single-conductor cable shall be permitted only in sizes 2 AWG or larger.

(C) Open Conductors.   Open conductors shall be prohibited except as part of a listed assembly or festoon lighting installed in accordance with Article 225.

(D) Splices.   Flexible cords or cables shall be continuous without splice or tap between boxes or fittings.

(E) Cord Connectors.   Cord connectors shall not be laid on the ground unless listed for wet locations. Connectors and cable connections shall not be placed in audience traffic paths or within areas accessible to the public unless guarded.

(F) Support.   Wiring for an amusement ride, attraction, tent, or similar structure shall not be supported by any other ride or structure unless specifically designed for the purpose.

(G) Protection.   Flexible cords or cables accessible to the public shall be arranged to minimize the tripping hazard and shall be permitted to be covered with nonconductive matting, provided that the matting does not constitute a greater tripping hazard than the uncovered cables. It shall be permitted to bury cables. The requirements of 300.5 shall not apply.

(H) Boxes and Fittings.   A box or fitting shall be installed at each connection point, outlet, switchpoint, or junction point.

525.21 Rides, Tents, and Concessions.

(A) Disconnecting Means.   A means to disconnect each portable structure from all ungrounded conductors shall be provi-
ded. The disconnecting means shall be located within sight of and within 1.8 m (6 ft) of the operator's station. The disconnecting means shall be readily accessible to the operator, including when the ride is in operation. Where accessible to unqualified persons, the disconnecting means shall be lockable. A shunt trip device that opens the fused disconnect or circuit breaker when a switch located in the ride operator's console is closed shall be a permissible method of opening the circuit.

(B) Portable Wiring Inside Tents and Concessions.   Electrical wiring for lighting, where installed inside of tents and concessions, shall be securely installed and, where subject to physical damage, shall be provided with mechanical protection. All lamps for general illumination shall be protected from accidental breakage by a suitable luminaire or lampholder with a guard.

525.22 Portable Distribution or Termination Boxes.   Portable distribution or termination boxes shall comply with 525.22(A) through (D).

(A) Construction.   Boxes shall be designed so that no live parts are exposed except when necessary for examination, adjustment, servicing, or maintenance by qualified persons. Where installed outdoors, the box shall be of weatherproof construction and mounted so that the bottom of the enclosure is not less than 150 mm (6 in.) above the ground.

(B) Busbars and Terminals.   Busbars shall have an ampere rating not less than the overcurrent device supplying the feeder supplying the box. Where conductors terminate directly on busbars, busbar connectors shall be provided.

(C) Receptacles and Overcurrent Protection.   Receptacles shall have overcurrent protection installed within the box. The overcurrent protection shall not exceed the ampere rating of the receptacle, except as permitted in Article 430 for motor loads.

(D) Single-Pole Connectors.   Where single-pole connectors are used, they shall comply with 530.22.

525.23 Ground-Fault Circuit-Interrupter (GFCI) Protection.

(A) Where GFCI Protection Is Required.   GFCI protection for personnel shall be provided for the following:

(1)   All 125-volt, single-phase, 15- and 20-ampere non-locking-type receptacles used for disassembly and reassembly or readily accessible to the general public
(2)   Equipment that is readily accessible to the general public and supplied from a 125-volt, single-phase, 15- or 20-ampere branch circuit

The GFCI shall be permitted to be an integral part of the attachment plug or located in the power-supply cord within 300 mm (12 in.) of the attachment plug. Listed cord sets incorporating GFCI for personnel shall be permitted.

(B) Where GFCI Protection Is Not Required.   Receptacles that are not accessible from grade level and that only facilitate quick disconnecting and reconnecting of electrical equipment shall not be required to be provided with GFCI protection. These receptacles shall be of the locking type.

(C) Where GFCI Protection Is Not Permitted.   Egress lighting shall not be protected by a GFCI.

N **(D) Receptacles Supplied by Portable Cords.** Where GFCI protection is provided through the use of GFCI receptacles, and the branch circuits supplying receptacles utilize flexible cord, the GFCI protection shall be listed, labeled, and identified for portable use.

Part IV. Grounding and Bonding

525.30 Equipment Bonding. The following equipment connected to the same source shall be bonded:

(1) Metal raceways and metal-sheathed cable
(2) Metal enclosures of electrical equipment
(3) Metal frames and metal parts of portable structures, trailers, trucks, or other equipment that contain or support electrical equipment

The equipment grounding conductor of the circuit supplying the equipment in items (1), (2) or (3) that is likely to energize the metal frame or part shall be permitted to serve as the bonding means.

525.31 Equipment Grounding. All equipment to be grounded shall be connected to an equipment grounding conductor of a type recognized by 250.118 and installed in accordance with Parts VI and VII of Article 250. The equipment grounding conductor shall be connected to the system grounded conductor at the service disconnecting means or, in the case of a separately derived system such as a generator, at the generator or first disconnecting means supplied by the generator. The grounded circuit conductor shall not be connected to the equipment grounding conductor on the load side of the service disconnecting means or on the load side of a separately derived system disconnecting means.

525.32 Equipment Grounding Conductor Continuity Assurance. The continuity of the equipment grounding conductors shall be verified each time that portable electrical equipment is connected.

ARTICLE 530
Motion Picture and Television Studios and Similar Locations

Part I. General

530.1 Scope. The requirements of this article shall apply to television studios and motion picture studios using either film or electronic cameras, except as provided in 520.1, and exchanges, factories, laboratories, stages, or a portion of the building in which film or tape more than 22 mm (⅞ in.) in width is exposed, developed, printed, cut, edited, rewound, repaired, or stored.

> Informational Note: For methods of protecting against cellulose nitrate film hazards, see NFPA 40-2011, *Standard for the Storage and Handling of Cellulose Nitrate Film.*

530.2 Definitions.

Alternating-Current Power Distribution Box (Alternating-Current Plugging Box, Scatter Box). An ac distribution center or box that contains one or more grounding-type polarized receptacles that may contain overcurrent protective devices.

Bull Switch. An externally operated wall-mounted safety switch that may or may not contain overcurrent protection and is designed for the connection of portable cables and cords.

Location (Shooting Location). A place outside a motion picture studio where a production or part of it is filmed or recorded.

Location Board (Deuce Board). Portable equipment containing a lighting contactor or contactors and overcurrent protection designed for remote control of stage lighting.

Motion Picture Studio (Lot). A building or group of buildings and other structures designed, constructed, or permanently altered for use by the entertainment industry for the purpose of motion picture or television production.

Plugging Box. A dc device consisting of one or more 2-pole, 2-wire, nonpolarized, nongrounding-type receptacles intended to be used on dc circuits only.

Portable Equipment. Equipment intended to be moved from one place to another.

Single-Pole Separable Connector. A device that is installed at the ends of portable, flexible, single-conductor cable that is used to establish connection or disconnection between two cables or one cable and a single-pole, panel-mounted separable connector.

Spider (Cable Splicing Block). A device that contains busbars that are insulated from each other for the purpose of splicing or distributing power to portable cables and cords that are terminated with single-pole busbar connectors.

Stage Effect (Special Effect). An electrical or electromechanical piece of equipment used to simulate a distinctive visual or audible effect such as wind machines, lightning simulators, sunset projectors, and the like.

Stage Property. An article or object used as a visual element in a motion picture or television production, except painted backgrounds (scenery) and costumes.

Stage Set. A specific area set up with temporary scenery and properties designed and arranged for a particular scene in a motion picture or television production.

Stand Lamp (Work Light). A portable stand that contains a general-purpose luminaire or lampholder with guard for the purpose of providing general illumination in the studio or stage.

Television Studio or Motion Picture Stage (Sound Stage). A building or portion of a building usually insulated from the outside noise and natural light for use by the entertainment industry for the purpose of motion picture, television, or commercial production.

530.6 Portable Equipment. Portable stage and studio lighting equipment and portable power distribution equipment shall be permitted for temporary use outdoors if the equipment is supervised by qualified personnel while energized and barriered from the general public.

Part II. Stage or Set

530.11 Permanent Wiring. The permanent wiring shall be Type MC cable, Type AC cable containing an insulated equipment grounding conductor sized in accordance with Table 250.122, Type MI cable, or in approved raceways.

Exception: Communications circuits; audio signal processing, amplification, and reproduction circuits; Class 1, Class 2, and Class 3 remote-control or signaling circuits and power-limited fire alarm circuits shall be permitted to be wired in accordance with Articles 640, 725, 760, and 800.

530.12 Portable Wiring.

(A) Stage Set Wiring. The wiring for stage set lighting and other supply wiring not fixed as to location shall be done with listed hard usage flexible cords and cables. Where subject to physical damage, such wiring shall be listed extra-hard usage flexible cords and cables. Splices or taps in cables shall be permitted if the total connected load does not exceed the maximum ampacity of the cable.

(B) Stage Effects and Electrical Equipment Used as Stage Properties. The wiring for stage effects and electrical equipment used as stage properties shall be permitted to be wired with single- or multiconductor listed flexible cords or cables if the conductors are protected from physical damage and secured to the scenery by approved cable ties or by insulated staples. Splices or taps shall be permitted where such are made with listed devices and the circuit is protected at not more than 20 amperes.

(C) Other Electrical Equipment. Cords and cables other than extra-hard usage, where supplied as a part of a listed assembly, shall be permitted.

530.13 Stage Lighting and Effects Control. Switches used for studio stage set lighting and effects (on the stages and lots and on location) shall be of the externally operable type. Where contactors are used as the disconnecting means for fuses, an individual externally operable switch, suitably rated, for the control of each contactor shall be located at a distance of not more than 1.8 m (6 ft) from the contactor, in addition to remote-control switches. A single externally operable switch shall be permitted to simultaneously disconnect all the contactors on any one location board, where located at a distance of not more than 1.8 m (6 ft) from the location board.

530.14 Plugging Boxes. Each receptacle of dc plugging boxes shall be rated at not less than 30 amperes.

530.15 Enclosing and Guarding Live Parts.

(A) Live Parts. Live parts shall be enclosed or guarded to prevent accidental contact by persons and objects.

(B) Switches. All switches shall be of the externally operable type.

(C) Rheostats. Rheostats shall be placed in approved cases or cabinets that enclose all live parts, having only the operating handles exposed.

(D) Current-Carrying Parts. Current-carrying parts of bull switches, location boards, spiders, and plugging boxes shall be enclosed, guarded, or located so that persons cannot accidentally come into contact with them or bring conductive material into contact with them.

530.16 Portable Luminaires. Portable luminaires and work lights shall be equipped with flexible cords, composition or metal-sheathed porcelain sockets, and substantial guards.

Exception: Portable luminaires used as properties in a motion picture set or television stage set, on a studio stage or lot, or on location shall not be considered to be portable luminaires for the purpose of this section.

530.17 Portable Arc Lamps.

(A) Portable Carbon Arc Lamps. Portable carbon arc lamps shall be substantially constructed. The arc shall be provided with an enclosure designed to retain sparks and carbons and to prevent persons or materials from coming into contact with the arc or bare live parts. The enclosures shall be ventilated. All switches shall be of the externally operable type.

(B) Portable Noncarbon Arc Electric-Discharge Lamps. Portable noncarbon arc lamps, including enclosed arc lamps, and associated ballasts shall be listed. Interconnecting cord sets and interconnecting cords and cables shall be extra-hard usage type and listed.

530.18 Overcurrent Protection — General. Automatic overcurrent protective devices (circuit breakers or fuses) for motion picture studio stage set lighting and the stage cables for such stage set lighting shall be as given in 530.18(A) through (G). The maximum ampacity allowed on a given conductor, cable, or cord size shall be as given in the applicable tables of Articles 310 and 400.

(A) Stage Cables. Stage cables for stage set lighting shall be protected by means of overcurrent devices set at not more than 400 percent of the ampacity given in the applicable tables of Articles 310 and 400.

(B) Feeders. In buildings used primarily for motion picture production, the feeders from the substations to the stages shall be protected by means of overcurrent devices (generally located in the substation) having a suitable ampere rating. The overcurrent devices shall be permitted to be multipole or single-pole gang operated. No pole shall be required in the neutral conductor. The overcurrent device setting for each feeder shall not exceed 400 percent of the ampacity of the feeder, as given in the applicable tables of Article 310.

(C) Cable Protection. Cables shall be protected by bushings where they pass through enclosures and shall be arranged so that tension on the cable is not transmitted to the connections. Where power conductors pass through metal, the requirements of 300.20 shall apply.

Portable feeder cables shall be permitted to temporarily penetrate fire-rated walls, floors, or ceilings provided that all of the following apply:

(1) The opening is of noncombustible material.
(2) When in use, the penetration is sealed with a temporary seal of a listed firestop material.
(3) When not in use, the opening shall be capped with a material of equivalent fire rating.

(D) Location Boards. Overcurrent protection (fuses or circuit breakers) shall be provided at the location boards. Fuses in the location boards shall have an ampere rating of not over 400 percent of the ampacity of the cables between the location boards and the plugging boxes.

(E) Plugging Boxes. Cables and cords supplied through plugging boxes shall be of copper. Cables and cords smaller than 8 AWG shall be attached to the plugging box by means of a plug containing two cartridge fuses or a 2-pole circuit breaker. The rating of the fuses or the setting of the circuit breaker shall not be over 400 percent of the rated ampacity of the cables or cords as given in the applicable tables of Articles 310 and 400. Plugging boxes shall not be permitted on ac systems.

(F) Alternating-Current Power Distribution Boxes. Alternating-current power distribution boxes used on sound stages and shooting locations shall contain connection receptacles of a polarized, grounding type.

(G) Lighting. Work lights, stand lamps, and luminaires rated 1000 watts or less and connected to dc plugging boxes shall be by means of plugs containing two cartridge fuses not larger than 20 amperes, or they shall be permitted to be connected to special outlets on circuits protected by fuses or circuit breakers rated at not over 20 amperes. Plug fuses shall not be used unless they are on the load side of the fuse or circuit breakers on the location boards.

530.19 Sizing of Feeder Conductors for Television Studio Sets.

(A) General. It shall be permissible to apply the demand factors listed in Table 530.19(A) to that portion of the maximum possible connected load for studio or stage set lighting for all permanently installed feeders between substations and stages and to all permanently installed feeders between the main stage switchboard and stage distribution centers or location boards.

(B) Portable Feeders. A demand factor of 50 percent of maximum possible connected load shall be permitted for all portable feeders.

530.20 Grounding. Type MC cable, Type MI cable, Type AC cable containing an insulated equipment grounding conductor, metal raceways, and all non–current-carrying metal parts of appliances, devices, and equipment shall be connected to an equipment grounding conductor. This shall not apply to pendant and portable lamps, to portable stage lighting and stage sound equipment, or to other portable and special stage equipment operating at not over 150 volts dc to ground.

530.21 Plugs and Receptacles.

(A) Rating. Plugs and receptacles, including cord connectors and flanged surface devices, shall be rated in amperes. The voltage rating of the plugs and receptacles shall not be less than the nominal circuit voltage. Plug and receptacle ampere ratings for ac circuits shall not be less than the feeder or branch-circuit overcurrent device ampere rating. Table 210.21(B)(2) shall not apply.

Table 530.19(A) Demand Factors for Stage Set Lighting

Portion of Stage Set Lighting Load to Which Demand Factor Applied (volt-amperes)	Feeder Demand Factor (%)
First 50,000 or less at	100
From 50,001 to 100,000 at	75
From 100,001 to 200,000 at	60
Remaining over 200,000 at	50

(B) Interchangeability. Plugs and receptacles used in portable professional motion picture and television equipment shall be permitted to be interchangeable for ac or dc use on the same premises, provided they are listed for ac/dc use and marked in a suitable manner to identify the system to which they are connected.

530.22 Single-Pole Separable Connectors.

(A) General. Where ac single-pole portable cable connectors are used, they shall be listed and of the locking type. Sections 400.14, 406.7, and 406.8 shall not apply to listed single-pole separable connections and single-conductor cable assemblies utilizing listed single-pole separable connectors. Where paralleled sets of current-carrying single-pole separable connectors are provided as input devices, they shall be prominently labeled with a warning indicating the presence of internal parallel connections. The use of single-pole separable connectors shall comply with at least one of the following conditions:

(1) Connection and disconnection of connectors are only possible where the supply connectors are interlocked to the source and it is not possible to connect or disconnect connectors when the supply is energized.

(2) Line connectors are of the listed sequential-interlocking type so that load connectors shall be connected in the following sequence:

   a. Equipment grounding conductor connection

   b. Grounded circuit conductor connection, if provided

   c. Ungrounded conductor connection, and that disconnection shall be in the reverse order

(3) A caution notice shall be provided adjacent to the line connectors, indicating that plug connection shall be in the following order:

   a. Equipment grounding conductor connectors

   b. Grounded circuit-conductor connectors, if provided

   c. Ungrounded conductor connectors, and that disconnection shall be in the reverse order

The warning sign(s) or label(s) shall comply with 110.21(B).

(B) Interchangeability. Single-pole separable connectors used in portable professional motion picture and television equipment shall be permitted to be interchangeable for ac or dc use or for different current ratings on the same premises, provided they are listed for ac/dc use and marked in a suitable manner to identify the system to which they are connected.

530.23 Branch Circuits. A branch circuit of any size supplying one or more receptacles shall be permitted to supply stage set lighting loads.

The application of 210.8(B)(4) shall not be required.

Part III. Dressing Rooms

530.31 Dressing Rooms. Fixed wiring in dressing rooms shall be installed in accordance with the wiring methods covered in Chapter 3. Wiring for portable dressing rooms shall be approved.

Part IV. Viewing, Cutting, and Patching Tables

530.41 Lamps at Tables. Only composition or metal-sheathed, porcelain, keyless lampholders equipped with suitable means to guard lamps from physical damage and from film and film scrap shall be used at patching, viewing, and cutting tables.

Part V. Cellulose Nitrate Film Storage Vaults

530.51 Lamps in Cellulose Nitrate Film Storage Vaults. Lamps in cellulose nitrate film storage vaults shall be installed in rigid luminaires of the glass-enclosed and gasketed type. Lamps shall be controlled by a switch having a pole in each ungrounded conductor. This switch shall be located outside of the vault and provided with a pilot light to indicate whether the switch is on or off. This switch shall disconnect from all sources of supply all ungrounded conductors terminating in any outlet in the vault.

530.52 Electrical Equipment in Cellulose Nitrate Film Storage Vaults. Except as permitted in 530.51, no receptacles, outlets, heaters, portable lights, or other portable electrical equipment shall be located in cellulose nitrate film storage vaults. Electric motors shall be permitted, provided they are listed for the application and comply with Article 500, Class I, Division 2.

Part VI. Substations

530.61 Substations. Wiring and equipment of over 1000 volts, nominal, shall comply with Article 490.

530.62 Portable Substations. Wiring and equipment in portable substations shall conform to the sections applying to installations in permanently fixed substations, but, due to the limited space available, the working spaces shall be permitted to be reduced, provided that the equipment shall be arranged so that the operator can work safely and so that other persons in the vicinity cannot accidentally come into contact with current-carrying parts or bring conducting objects into contact with them while they are energized.

530.63 Overcurrent Protection of Direct-Current Generators. Three-wire generators shall have overcurrent protection in accordance with 445.12(E).

530.64 Direct-Current Switchboards.

(A) General. Switchboards of not over 250 volts dc between conductors, where located in substations or switchboard rooms accessible to qualified persons only, shall not be required to be dead-front.

(B) Circuit Breaker Frames. Frames of dc circuit breakers installed on switchboards shall not be required to be connected to an equipment grounding conductor.

ARTICLE 540
Motion Picture Projection Rooms

Part I. General

540.1 Scope. The provisions of this article apply to motion picture projection rooms, motion picture projectors, and associated equipment of the professional and nonprofessional types using incandescent, carbon arc, xenon, or other light source equipment that develops hazardous gases, dust, or radiation.

> Informational Note: For further information, see NFPA 40-2011, *Standard for the Storage and Handling of Cellulose Nitrate Film.*

540.2 Definitions.

Nonprofessional Projector. Nonprofessional projectors are those types of projectors that do not comply with the definition of *Professional-Type Projector.*

Professional-Type Projector. A type of projector using 35- or 70-mm film that has a minimum width of 35 mm (1⅜ in.) and has on each edge 212 perforations per meter (5.4 perforations per inch), or a type using carbon arc, xenon, or other light source equipment that develops hazardous gases, dust, or radiation.

Part II. Equipment and Projectors of the Professional Type

540.10 Motion Picture Projection Room Required. Every professional-type projector shall be located within a projection room. Every projection room shall be of permanent construction, approved for the type of building in which the projection room is located. All projection ports, spotlight ports, viewing ports, and similar openings shall be provided with glass or other approved material so as to completely close the opening. Such rooms shall not be considered as hazardous (classified) locations as defined in Article 500.

> Informational Note: For further information on protecting openings in projection rooms handling cellulose nitrate motion picture film, see NFPA 101-2015, *Life Safety Code.*

540.11 Location of Associated Electrical Equipment.

(A) Motor Generator Sets, Transformers, Rectifiers, Rheostats, and Similar Equipment. Motor-generator sets, transformers, rectifiers, rheostats, and similar equipment for the supply or control of current to projection or spotlight equipment shall, where nitrate film is used, be located in a separate room. Where placed in the projection room, they shall be located or guarded so that arcs or sparks cannot come in contact with film, and the commutator end or ends of motor-generator sets shall comply with one of the conditions in 540.11(A)(1) through (A)(6).

(1) Types. Be of the totally enclosed, enclosed fan-cooled, or enclosed pipe-ventilated type.

(2) Separate Rooms or Housings. Be enclosed in separate rooms or housings built of noncombustible material constructed so as to exclude flyings or lint with approved ventilation from a source of clean air.

(3) Solid Metal Covers. Have the brush or sliding-contact end of motor-generator enclosed by solid metal covers.

(4) Tight Metal Housings. Have brushes or sliding contacts enclosed in substantial, tight metal housings.

(5) Upper and Lower Half Enclosures. Have the upper half of the brush or sliding-contact end of the motor-generator enclosed by a wire screen or perforated metal and the lower half enclosed by solid metal covers.

(6) Wire Screens or Perforated Metal. Have wire screens or perforated metal placed at the commutator of brush ends. No dimension of any opening in the wire screen or perforated metal shall exceed 1.27 mm (0.05 in.), regardless of the shape of the opening and of the material used.

(B) Switches, Overcurrent Devices, or Other Equipment. Switches, overcurrent devices, or other equipment not normally required or used for projectors, sound reproduction, flood or other special effect lamps, or other equipment shall not be installed in projection rooms.

Exception No. 1: In projection rooms approved for use only with cellulose acetate (safety) film, the installation of appurtenant electrical equipment used in conjunction with the operation of the projection equipment and the control of lights, curtains, and audio equipment, and so forth, shall be permitted. In such projection rooms, a sign reading "Safety Film Only Permitted in This Room" shall be posted on the outside of each projection room door and within the projection room itself in a conspicuous location.

Exception No. 2: Remote-control switches for the control of auditorium lights or switches for the control of motors operating curtains and masking of the motion picture screen shall be permitted to be installed in projection rooms.

(C) Emergency Systems. Control of emergency systems shall comply with Article 700.

540.12 Work Space. Each motion picture projector, floodlight, spotlight, or similar equipment shall have clear working space not less than 750 mm (30 in.) wide on each side and at the rear thereof.

Exception: One such space shall be permitted between adjacent pieces of equipment.

540.13 Conductor Size. Conductors supplying outlets for arc and xenon projectors of the professional type shall not be smaller than 8 AWG and shall have an ampacity not less than the projector current rating. Conductors for incandescent-type projectors shall conform to normal wiring standards as provided in 210.24.

540.14 Conductors on Lamps and Hot Equipment. Insulated conductors having a rated operating temperature of not less than 200°C (392°F) shall be used on all lamps or other equipment where the ambient temperature at the conductors as installed will exceed 50°C (122°F).

540.15 Flexible Cords. Cords approved for hard usage, as provided in Table 400.4, shall be used on portable equipment.

540.20 Listing Requirements. Projectors and enclosures for arc, xenon, and incandescent lamps and rectifiers, transformers, rheostats, and similar equipment shall be listed.

540.21 Marking. Projectors and other equipment shall be marked with the manufacturer's name or trademark and with the voltage and current for which they are designed in accordance with 110.21.

Part III. Nonprofessional Projectors

540.31 Motion Picture Projection Room Not Required. Projectors of the nonprofessional or miniature type, where employing cellulose acetate (safety) film, shall be permitted to be operated without a projection room.

540.32 Listing Requirements. Projection equipment shall be listed.

Part IV. Audio Signal Processing, Amplification, and Reproduction Equipment

540.50 Audio Signal Processing, Amplification, and Reproduction Equipment. Audio signal processing, amplification, and reproduction equipment shall be installed as provided in Article 640.

ARTICLE 545
Manufactured Buildings

545.1 Scope. This article covers requirements for a manufactured building and building components as herein defined.

545.2 Definitions.

Building Component. Any subsystem, subassembly, or other system designed for use in or integral with or as part of a structure, which can include structural, electrical, mechanical, plumbing, and fire protection systems, and other systems affecting health and safety.

Building System. Plans, specifications, and documentation for a system of manufactured building or for a type or a system of building components, which can include structural, electrical, mechanical, plumbing, and fire protection systems, and other systems affecting health and safety, and including such variations thereof as are specifically permitted by regulation, and which variations are submitted as part of the building system or amendment thereto.

Closed Construction. Any building, building component, assembly, or system manufactured in such a manner that all concealed parts of processes of manufacture cannot be inspected after installation at the building site without disassembly, damage, or destruction.

Manufactured Building. Any building that is of closed construction and is made or assembled in manufacturing facilities on or off the building site for installation, or for assembly and installation on the building site, other than manufactured homes, mobile homes, park trailers, or recreational vehicles.

545.4 Wiring Methods.

(A) Methods Permitted. All raceway and cable wiring methods included in this *Code* and other wiring systems specifically intended and listed for use in manufactured buildings shall be permitted with listed fittings and with fittings listed and identified for manufactured buildings.

(B) Securing Cables. In closed construction, cables shall be permitted to be secured only at cabinets, boxes, or fittings where 10 AWG or smaller conductors are used and protection against physical damage is provided.

545.5 Supply Conductors. Provisions shall be made to route the service-entrance conductors, underground service conductors, service-lateral, feeder, or branch-circuit supply to the service or building disconnecting means conductors.

545.6 Installation of Service-Entrance Conductors. Service-entrance conductors shall be installed after erection at the building site.

Exception: Where point of attachment is known prior to manufacture.

545.7 Service Equipment. Service equipment shall be installed in accordance with 230.70.

545.8 Protection of Conductors and Equipment. Protection shall be provided for exposed conductors and equipment during processes of manufacturing, packaging, in transit, and erection at the building site.

545.9 Boxes.

(A) Other Dimensions. Boxes of dimensions other than those required in Table 314.16(A) shall be permitted to be installed where tested, identified, and listed to applicable standards.

(B) Not Over 1650 cm³ (100 in.³). Any box not over 1650 cm³ (100 in.³) in size, intended for mounting in closed construction, shall be affixed with anchors or clamps so as to provide a rigid and secure installation.

545.10 Receptacle or Switch with Integral Enclosure. A receptacle or switch with integral enclosure and mounting means, where tested, identified, and listed to applicable standards, shall be permitted to be installed.

545.11 Bonding and Grounding. Prewired panels and building components shall provide for the bonding, or bonding and grounding, of all exposed metals likely to become energized, in accordance with Article 250, Parts V, VI, and VII.

545.12 Grounding Electrode Conductor. Provisions shall be made to route a grounding electrode conductor from the service, feeder, or branch-circuit supply to the point of attachment to the grounding electrode.

545.13 Component Interconnections. Fittings and connectors that are intended to be concealed at the time of on-site assembly, where tested, identified, and listed to applicable standards, shall be permitted for on-site interconnection of modules or other building components. Such fittings and connectors shall be equal to the wiring method employed in insulation, temperature rise, and fault-current withstand and shall be capable of enduring the vibration and minor relative motions occurring in the components of manufactured buildings.

ARTICLE 547
Agricultural Buildings

547.1 Scope. The provisions of this article shall apply to the following agricultural buildings or that part of a building or adjacent areas of similar or like nature as specified in 547.1(A) or (B).

(A) Excessive Dust and Dust with Water. Agricultural buildings where excessive dust and dust with water may accumulate, including all areas of poultry, livestock, and fish confinement systems, where litter dust or feed dust, including mineral feed particles, may accumulate.

(B) Corrosive Atmosphere. Agricultural buildings where a corrosive atmosphere exists. Such buildings include areas where the following conditions exist:

(1) Poultry and animal excrement may cause corrosive vapors.
(2) Corrosive particles may combine with water.
(3) The area is damp and wet by reason of periodic washing for cleaning and sanitizing with water and cleansing agents.
(4) Similar conditions exist.

547.2 Definitions.

Distribution Point. An electrical supply point from which service drops, service conductors, feeders, or branch circuits to buildings or structures utilized under single management are supplied.

> Informational Note No. 1: Distribution points are also known as the center yard pole, meterpole, or the common distribution point.

> Informational Note No. 2: The service point as defined in Article 100 is typically at the distribution point.

Equipotential Plane. An area where wire mesh or other conductive elements are embedded in or placed under concrete, bonded to all metal structures and fixed nonelectrical equipment that may become energized, and connected to the electrical grounding system to minimize voltage differences within the plane and between the planes, the grounded equipment, and the earth.

Site-Isolating Device. A disconnecting means installed at the distribution point for the purposes of isolation, system maintenance, emergency disconnection, or connection of optional standby systems.

547.3 Other Articles. For buildings and structures not having conditions as specified in 547.1, the electrical installations shall be made in accordance with the applicable articles in this *Code*.

547.4 Surface Temperatures. Electrical equipment or devices installed in accordance with the provisions of this article shall be installed in a manner such that they will function at full rating without developing surface temperatures in excess of the specified normal safe operating range of the equipment or device.

547.5 Wiring Methods.

(A) Wiring Systems. Types UF, NMC, copper SE cables, jacketed Type MC cable, rigid nonmetallic conduit, liquidtight flexible nonmetallic conduit, or other cables or raceways suitable for the location, with approved termination fittings, shall be the wiring methods employed. The wiring methods of Article 502, Part II, shall be permitted for areas described in 547.1(A).

> Informational Note: See 300.7, 352.44, and 355.44 for installation of raceway systems exposed to widely different temperatures.

(B) Mounting. All cables shall be secured within 200 mm (8 in.) of each cabinet, box, or fitting. Nonmetallic boxes, fittings, conduit, and cables shall be permitted to be mounted directly to any building surface covered by this article without maintaining the 6 mm (¼ in.) airspace in accordance with 300.6(D).

(C) Equipment Enclosures, Boxes, Conduit Bodies, and Fittings.

(1) Excessive Dust. Equipment enclosures, boxes, conduit bodies, and fittings installed in areas of buildings where excessive dust may be present shall be designed to minimize the entrance of dust and shall have no openings (such as holes for attachment screws) through which dust could enter the enclosure.

(2) Damp or Wet Locations. In damp or wet locations, equipment enclosures, boxes, conduit bodies, and fittings shall be placed or equipped so as to prevent moisture from entering or accumulating within the enclosure, box, conduit body, or fitting. In wet locations, including normally dry or damp locations where surfaces are periodically washed or sprayed with water, boxes, conduit bodies, and fittings shall be listed for use in wet locations and equipment enclosures shall be weatherproof.

(3) Corrosive Atmosphere. Where wet dust, excessive moisture, corrosive gases or vapors, or other corrosive conditions may be present, equipment enclosures, boxes, conduit bodies, and fittings shall have corrosion resistance properties suitable for the conditions.

> Informational Note No. 1: See Table 110.28 for appropriate enclosure type designations.

> Informational Note No. 2: Aluminum and magnetic ferrous materials may corrode in agricultural environments.

(D) Flexible Connections. Where necessary to employ flexible connections, dusttight flexible connectors, liquidtight flexible metal conduit, liquidtight flexible nonmetallic conduit, or flexible cord listed and identified for hard usage shall be used.

(E) Physical Protection. All electrical wiring and equipment subject to physical damage shall be protected.

(F) Separate Equipment Grounding Conductor. Where an equipment grounding conductor is installed underground within a location falling under the scope of Article 547, it shall be insulated.

> Informational Note: For further information on aluminum and copper-clad aluminum conductors, see 250.120(B).

(G) Receptacles. All 125-volt, single-phase, 15- and 20-ampere receptacles installed in the locations listed in (1) through (4) shall have ground-fault circuit-interrupter protection:

(1) Areas having an equipotential plane
(2) Outdoors
(3) Damp or wet locations
(4) Dirt confinement areas for livestock

547.6 Switches, Receptacles, Circuit Breakers, Controllers, and Fuses. Switches, including pushbuttons, relays, and similar devices, receptacles, circuit breakers, controllers, and fuses, shall be provided with enclosures as specified in 547.5(C).

547.7 Motors. Motors and other rotating electrical machinery shall be totally enclosed or designed so as to minimize the entrance of dust, moisture, or corrosive particles.

547.8 Luminaires. Luminaires shall comply with 547.8(A) through (C).

(A) Minimize the Entrance of Dust. Luminaires shall be installed to minimize the entrance of dust, foreign matter, moisture, and corrosive material.

(B) Exposed to Physical Damage. Luminaires exposed to physical damage shall be protected by a suitable guard.

(C) Exposed to Water. Luminaires exposed to water from condensation, building cleansing water, or solution shall be listed for use in wet locations.

547.9 Electrical Supply to Building(s) or Structure(s) from a Distribution Point. A distribution point shall be permitted to supply any building or structure located on the same premises. The overhead electrical supply shall comply with 547.9(A) and (B), or with 547.9(C). The underground electrical supply shall comply with 547.9(C).

(A) Site-Isolating Device. Site-isolating devices shall comply with 547.9(A)(1) through (A)(10).

(1) Where Required. A site-isolating device shall be installed at the distribution point where two or more buildings or structures are supplied from the distribution point.

(2) Location. The site-isolating device shall be pole-mounted and be not less than the height above grade required by 230.24 for the conductors it supplies.

(3) Operation. The site-isolating device shall simultaneously disconnect all ungrounded service conductors from the premises wiring.

(4) Bonding Provisions. The site-isolating device enclosure shall be connected to the grounded circuit conductor and the grounding electrode system.

(5) Grounding. At the site-isolating device, the system grounded conductor shall be connected to a grounding electrode system via a grounding electrode conductor.

(6) Rating. The site-isolating device shall be rated for the calculated load as determined by Part V of Article 220.

(7) Overcurrent Protection. The site-isolating device shall not be required to provide overcurrent protection.

(8) Accessibility. The site-isolating device shall be capable of being remotely operated by an operating handle installed at a readily accessible location. The operating handle of the site-isolating device, when in its highest position, shall not be more than 2.0 m (6 ft 7 in.) above grade or a working platform.

(9) Series Devices. An additional site-isolating device for the premises wiring system shall not be required where a site-isolating device meeting all applicable requirements of this section is provided by the serving utility as part of their service requirements.

(10) Marking. A site-isolating device shall be permanently marked to identify it as a site-isolating device. This marking shall be located on the operating handle or immediately adjacent thereto.

(B) Service Disconnecting Means and Overcurrent Protection at the Building(s) or Structure(s). Where the service disconnecting means and overcurrent protection are located at the building(s) or structure(s), the requirements of 547.9(B)(1) through (B)(3) shall apply.

(1) Conductor Sizing. The supply conductors shall be sized in accordance with Part V of Article 220.

(2) Conductor Installation. The supply conductors shall be installed in accordance with the requirements of Part II of Article 225.

(3) Grounding and Bonding. For each building or structure, grounding and bonding of the supply conductors shall be in accordance with the requirements of 250.32, and the following conditions shall be met:

(1)   The equipment grounding conductor is not smaller than the largest supply conductor if of the same material, or is adjusted in size in accordance with the equivalent size columns of Table 250.122 if of different materials.

(2)   The equipment grounding conductor is connected to the grounded circuit conductor and the site-isolating device enclosure at the distribution point.

(C) Service Disconnecting Means and Overcurrent Protection at the Distribution Point. Where the service disconnecting means and overcurrent protection for each set of feeders or branch circuits are located at the distribution point, the feeders or branch circuits to buildings or structures shall comply with the provisions of 250.32 and Article 225, Parts I and II.

> Informational Note: Methods to reduce neutral-to-earth voltages in livestock facilities include supplying buildings or structures with 4-wire single-phase services, sizing 3-wire single-phase service and feeder conductors to limit voltage drop to 2 percent, and connecting loads line-to-line.

(D) Identification. Where a site is supplied by more than one distribution point, a permanent plaque or directory shall be installed at each of these distribution points denoting the location of each of the other distribution points and the buildings or structures served by each.

547.10 Equipotential Planes and Bonding of Equipotential Planes. The installation and bonding of equipotential planes shall comply with 547.10(A) and (B). For the purposes of this section, the term *livestock* shall not include poultry.

(A) Where Required. Equipotential planes shall be installed where required in (A)(1) and (A)(2).

(1) Indoors. Equipotential planes shall be installed in confinement areas with concrete floors where metallic equipment is located that may become energized and is accessible to livestock.

(2) Outdoors. Equipotential planes shall be installed in concrete slabs where metallic equipment is located that may become energized and is accessible to livestock.

The equipotential plane shall encompass the area where the livestock stands while accessing metallic equipment that may become energized.

(B) Bonding. Equipotential planes shall be connected to the electrical grounding system. The bonding conductor shall be solid copper, insulated, covered or bare, and not smaller than 8 AWG. The means of bonding to wire mesh or conductive elements shall be by pressure connectors or clamps of brass, copper, copper alloy, or an equally substantial approved means. Slatted floors that are supported by structures that are a part of an equipotential plane shall not require bonding.

> Informational Note No. 1: Methods to establish equipotential planes are described in American Society of Agricultural and

Biological Engineers (ASABE) EP473.2-2001, *Equipotential Planes in Animal Containment Areas.*

Informational Note No. 2: Methods for safe installation of livestock waterers are described in American Society of Agricultural and Biological Engineers (ASABE) EP342.3-2010, *Safety for Electrically Heated Livestock Waterers.*

Informational Note No. 3: Low grounding electrode system resistances may reduce voltage differences in livestock facilities.

ARTICLE 550
Mobile Homes, Manufactured Homes, and Mobile Home Parks

Part I. General

550.1 Scope. The provisions of this article cover the electrical conductors and equipment installed within or on mobile and manufactured homes, the conductors that connect mobile and manufactured homes to a supply of electricity, and the installation of electrical wiring, luminaires, equipment, and appurtenances related to electrical installations within a mobile home park up to the mobile home service-entrance conductors or, if none, the mobile home service equipment.

> Informational Note: For additional information on manufactured housing see NFPA 501-2013, *Standard on Manufactured Housing,* and Part 3280, *Manufactured Home Construction and Safety Standards,* of the Federal Department of Housing and Urban Development.

550.2 Definitions.

Appliance, Fixed. An appliance that is fastened or otherwise secured at a specific location.

Appliance, Portable. An appliance that is actually moved or can easily be moved from one place to another in normal use.

> Informational Note: For the purpose of this article, the following major appliances, other than built-in, are considered portable if cord connected: refrigerators, range equipment, clothes washers, dishwashers without booster heaters, or other similar appliances.

Feeder Assembly. The overhead or under-chassis feeder conductors, including the grounding conductor, together with the necessary fittings and equipment or a power-supply cord listed for mobile home use, identified for the delivery of energy from the source of electrical supply to the panelboard within the mobile home.

Laundry Area. An area containing or designed to contain a laundry tray, clothes washer, or a clothes dryer.

Manufactured Home. A structure, transportable in one or more sections, which in the traveling mode is 2.4 m (8 ft) or more in width or 12.2 m (40 ft) or more in length, or when erected on site is 29.77 m² (320 ft²) or more is built on a permanent chassis and is designed to be used as a dwelling with or without a permanent foundation, whether or not connected to the utilities, and includes plumbing, heating, air conditioning, and electrical systems contained therein. The term *manufactured home* includes any structure that meets all the requirements of this paragraph except the size requirements and with respect to which the manufacturer voluntarily files a certification required by the regulatory agency. Calculations used to determine the number of square meters (square feet) in a structure are based on the structure's exterior dimensions and include all expandable rooms, cabinets, and other projections containing interior space, but do not include bay windows [501:1.2.14]. For the purpose of this *Code* and unless otherwise indicated, the term *mobile home* includes manufactured homes and excludes park trailers defined in Article 552.4.

> Informational Note No. 1: See the applicable building code for definition of the term *permanent foundation*.
>
> Informational Note No. 2: See 24 CFR Part 3280, *Manufactured Home Construction and Safety Standards, of the Federal Department of Housing and Urban Development,* for additional information on the definition.

Mobile Home. A factory-assembled structure or structures transportable in one or more sections that are built on a permanent chassis and designed to be used as a dwelling without a permanent foundation where connected to the required utilities and that include the plumbing, heating, air-conditioning, and electrical systems contained therein.

For the purpose of this *Code* and unless otherwise indicated, the term *mobile home* includes manufactured homes.

Mobile Home Accessory Building or Structure. Any awning, cabana, ramada, storage cabinet, carport, fence, windbreak, or porch established for the use of the occupant of the mobile home on a mobile home lot.

Mobile Home Lot. A designated portion of a mobile home park designed for the accommodation of one mobile home and its accessory buildings or structures for the exclusive use of its occupants.

Mobile Home Park. A contiguous parcel of land that is used for the accommodation of occupied mobile homes.

Mobile Home Service Equipment. The equipment containing the disconnecting means, overcurrent protective devices, and receptacles or other means for connecting a mobile home feeder assembly.

Park Electrical Wiring Systems. All of the electrical wiring, luminaires, equipment, and appurtenances related to electrical installations within a mobile home park, including the mobile home service equipment.

550.4 General Requirements.

(A) Mobile Home Not Intended as a Dwelling Unit. A mobile home not intended as a dwelling unit — for example, those equipped for sleeping purposes only, contractor's on-site offices, construction job dormitories, mobile studio dressing rooms, banks, clinics, mobile stores, or intended for the display or demonstration of merchandise or machinery — shall not be required to meet the provisions of this article pertaining to the number or capacity of circuits required. It shall, however, meet all other applicable requirements of this article if provided with an electrical installation intended to be energized from a 120-volt or 120/240-volt ac power supply system. Where different voltage is required by either design or available power supply system, adjustment shall be made in accordance with other articles and sections for the voltage used.

(B) In Other Than Mobile Home Parks. Mobile homes installed in other than mobile home parks shall comply with the provisions of this article.

(C) Connection to Wiring System. The provisions of this article shall apply to mobile homes intended for connection to a wiring system rated 120/240 volts, nominal, 3-wire ac, with a grounded neutral conductor.

(D) Listed and Labeled. All electrical materials, devices, appliances, fittings, and other equipment shall be listed and labeled by a qualified testing agency and shall be connected in an approved manner when installed.

Part II. Mobile and Manufactured Homes

550.10 Power Supply.

(A) Feeder. The power supply to the mobile home shall be a feeder assembly consisting of not more than one listed 50-ampere mobile home power-supply cord or a permanently installed feeder.

Exception No. 1: A mobile home that is factory equipped with gas or oil-fired central heating equipment and cooking appliances shall be permitted to be provided with a listed mobile home power-supply cord rated 40 amperes.

Exception No. 2: A feeder assembly shall not be required for manufactured homes constructed in accordance with 550.32(B).

(B) Power-Supply Cord. If the mobile home has a power-supply cord, it shall be permanently attached to the panelboard, or to a junction box permanently connected to the panelboard, with the free end terminating in an attachment plug cap.

Cords with adapters and pigtail ends, extension cords, and similar items shall not be attached to, or shipped with, a mobile home.

A suitable clamp or the equivalent shall be provided at the panelboard knockout to afford strain relief for the cord to prevent strain from being transmitted to the terminals when the power-supply cord is handled in its intended manner.

The cord shall be a listed type with four conductors, one of which shall be identified by a continuous green color or a continuous green color with one or more yellow stripes for use as the grounding conductor.

(C) Attachment Plug Cap. The attachment plug cap shall be a 3-pole, 4-wire, grounding type, rated 50 amperes, 125/250 volts with a configuration as shown in Figure 550.10(C) and intended for use with the 50-ampere, 125/250-volt receptacle configuration shown in Figure 550.10(C). It shall be listed, by itself or as part of a power-supply cord assembly, for the purpose and shall be molded to or installed on the flexible cord so that it is secured tightly to the cord at the point where the cord enters the attachment plug cap. If a right-angle cap is used, the configuration shall be oriented so that the grounding member is farthest from the cord.

> Informational Note: Complete details of the 50-ampere plug and receptacle configuration can be found in ANSI/NEMA WD 6-2002 (R2008), *Standard for Dimensions of Attachment Plugs and Receptacles,* Figure 14-50.

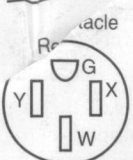
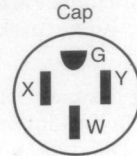

125/250-V, 50-A, 3-pole, 4-wire, grounding type

FIGURE 550.10(C)   **50-Ampere, 125/250-Volt Receptacle and Attachment Plug Cap Configurations, 3-Pole, 4-Wire, Grounding-Types, Used for Mobile Home Supply Cords and Mobile Home Parks.**

(D) Overall Length of a Power-Supply Cord. The overall length of a power-supply cord, measured from the end of the cord, including bared leads, to the face of the attachment plug cap shall not be less than 6.4 m (21 ft) and shall not exceed 11 m (36½ ft). The length of the cord from the face of the attachment plug cap to the point where the cord enters the mobile home shall not be less than 6.0 m (20 ft).

(E) Marking. The power-supply cord shall bear the following marking:

> FOR USE WITH MOBILE HOMES — 40 AMPERES

or

> FOR USE WITH MOBILE HOMES — 50 AMPERES

(F) Point of Entrance. The point of entrance of the feeder assembly to the mobile home shall be in the exterior wall, floor, or roof.

(G) Protected. Where the cord passes through walls or floors, it shall be protected by means of conduits and bushings or equivalent. The cord shall be permitted to be installed within the mobile home walls, provided a continuous raceway having a maximum size of 32 mm (1¼ in.) is installed from the branch-circuit panelboard to the underside of the mobile home floor.

(H) Protection Against Corrosion and Mechanical Damage. Permanent provisions shall be made for the protection of the attachment plug cap of the power-supply cord and any connector cord assembly or receptacle against corrosion and mechanical damage if such devices are in an exterior location while the mobile home is in transit.

(I) Mast Weatherhead or Raceway. Where the calculated load exceeds 50 amperes or where a permanent feeder is used, the supply shall be by means of either of the following:

(1) One mast weatherhead installation, installed in accordance with Article 230, containing four continuous, insulated, color-coded feeder conductors, one of which shall be an equipment grounding conductor

(2) A metal raceway or rigid nonmetallic conduit from the disconnecting means in the mobile home to the underside of the mobile home, with provisions for the attachment to a suitable junction box or fitting to the raceway on the underside of the mobile home [with or without conductors as in 550.10(I)(1)]. The manufacturer shall provide written installation instructions stating the proper feeder conductor sizes for the raceway and the size of the junction box to be used.

550.11 Disconnecting Means and Branch-Circuit Protective Equipment. The branch-circuit equipment shall be permitted to be combined with the disconnecting means as a single assembly. Such a combination shall be permitted to be designated as a panelboard. If a fused panelboard is used, the maximum fuse size for the mains shall be plainly marked with lettering at least 6 mm (¼ in.) high and visible when fuses are changed.

Where plug fuses and fuseholders are used, they shall be tamper-resistant Type S, enclosed in dead-front fuse panelboards. Electrical panelboards containing circuit breakers shall also be dead-front type.

> Informational Note: See 110.22 concerning identification of each disconnecting means and each service, feeder, or branch circuit at the point where it originated and the type marking needed.

(A) Disconnecting Means. A single disconnecting means shall be provided in each mobile home consisting of a circuit breaker, or a switch and fuses and its accessories installed in a readily accessible location near the point of entrance of the supply cord or conductors into the mobile home. The main circuit breakers or fuses shall be plainly marked "Main." This equipment shall contain a solderless type of grounding connector or bar for the purposes of grounding, with sufficient terminals for all grounding conductors. The terminations of the grounded circuit conductors shall be insulated in accordance with 550.16(A). The disconnecting equipment shall have a rating not less than the calculated load. The distribution equipment, either circuit breaker or fused type, shall be located a minimum of 600 mm (24 in.) from the bottom of such equipment to the floor level of the mobile home.

> Informational Note: See 550.20(B) for information on disconnecting means for branch circuits designed to energize heating or air-conditioning equipment, or both, located outside the mobile home, other than room air conditioners.

A panelboard shall be rated not less than 50 amperes and employ a 2-pole circuit breaker rated 40 amperes for a 40-ampere supply cord, or 50 amperes for a 50-ampere supply cord. A panelboard employing a disconnect switch and fuses shall be rated 60 amperes and shall employ a single 2-pole, 60-ampere fuseholder with 40- or 50-ampere main fuses for 40- or 50-ampere supply cords, respectively. The outside of the panelboard shall be plainly marked with the fuse size.

The panelboard shall be located in an accessible location but shall not be located in a bathroom or a clothes closet. A clear working space at least 750 mm (30 in.) wide and 750 mm (30 in.) in front of the panelboard shall be provided. This space shall extend from the floor to the top of the panelboard.

(B) Branch-Circuit Protective Equipment. Branch-circuit distribution equipment shall be installed in each mobile home and shall include overcurrent protection for each branch circuit consisting of either circuit breakers or fuses.

The branch-circuit overcurrent devices shall be rated as follows:

(1) Not more than the circuit conductors; and

(2) Not more than 150 percent of the rating of a single appliance rated 13.3 amperes or more that is supplied by an individual branch circuit; but

(3) Not more than the overcurrent protection size and of the type marked on the air conditioner or other motor-operated appliance.

(C) Two-Pole Circuit Breakers. Where circuit breakers are provided for branch-circuit protection, 240-volt circuits shall be protected by a 2-pole common or companion trip, or by circuit breakers with identified handle ties.

(D) Electrical Nameplates. A metal nameplate on the outside adjacent to the feeder assembly entrance shall read as follows:

THIS CONNECTION FOR 120/240-VOLT,
3-POLE, 4 WIRE, 60 HERTZ,
_____ AMPERE SUPPLY

The correct ampere rating shall be marked in the blank space.

Exception: For manufactured homes, the manufacturer shall provide in its written installation instructions or in the data plate the minimum ampere rating of the feeder assembly or, where provided, the service-entrance conductors intended for connection to the manufactured home. The rating provided shall not be less than the minimum load calculated in accordance with 550.18.

550.12 Branch Circuits. The number of branch circuits required shall be determined in accordance with 550.12(A) through (E).

(A) Lighting. The number of branch circuits shall be based on 33 volt-amperes/m² (3 VA/ft²) times outside dimensions of the mobile home (coupler excluded) divided by 120 volts to determine the number of 15- or 20-ampere lighting area circuits, for example,

[550.12(A)]

$$\frac{3 \times length \times width}{120 \times 15 \ (or \ 20)}$$
$$= \text{No. of 15- (or 20-) ampere circuits}$$

(B) Small Appliances. In kitchens, pantries, dining rooms, and breakfast rooms, two or more 20-ampere small-appliance circuits, in addition to the number of circuits required elsewhere in this section, shall be provided for all receptacle outlets required by 550.13(D) in these rooms. Such circuits shall have no other outlets.

Exception No. 1: Receptacle outlets installed solely for the electrical supply and support of an electric clock in any the rooms specified in 550.12(B) shall be permitted.

Exception No. 2: Receptacle outlets installed to provide power for supplemental equipment and lighting on gas-fired ranges, ovens, or counter-mounted cooking units shall be permitted.

Exception No. 3: A single receptacle for refrigeration equipment shall be permitted to be supplied from an individual branch circuit rated 15 amperes or greater.

Countertop receptacle outlets installed in the kitchen shall be supplied by not less than two small-appliance circuit branch circuits, either or both of which shall be permitted to supply receptacle outlets in the kitchen and other locations specified in 550.12(B).

(C) Laundry Area. Where a laundry area is provided, a 20-ampere branch circuit shall be provided to supply the laundry receptacle outlet(s). This circuit shall have no other outlets.

(D) General Appliances. (Including furnace, water heater, range, and central or room air conditioner, etc.). There shall be one or more circuits of adequate rating in accordance with the following:

Informational Note: For central air conditioning, see Article 440.

(1) The ampere rating of fixed appliances shall be not over 50 percent of the circuit rating if lighting outlets (receptacles, other than kitchen, dining area, and laundry, considered as lighting outlets) are on the same circuit.
(2) For fixed appliances on a circuit without lighting outlets, the sum of rated amperes shall not exceed the branch-circuit rating. Motor loads or continuous loads shall not exceed 80 percent of the branch-circuit rating.
(3) The rating of a single cord-and-plug-connected appliance on a circuit having no other outlets shall not exceed 80 percent of the circuit rating.
(4) The rating of a range branch circuit shall be based on the range demand as specified for ranges in 550.18(B)(5).

(E) Bathrooms. Bathroom receptacle outlets shall be supplied by at least one 20-ampere branch circuit. Such circuits shall have no outlets other than as provided for in 550.13(E)(2).

550.13 Receptacle Outlets.

(A) Grounding-Type Receptacle Outlets. All receptacle outlets shall comply with the following:

(1) Be of grounding type
(2) Be installed according to 406.4
(3) Except where supplying specific appliances, be 15- or 20-ampere, 125-volt, either single or multiple type, and accept parallel-blade attachment plugs

(B) Ground-Fault Circuit Interrupters (GFCI). All 125-volt, single-phase, 15- and 20-ampere receptacle outlets installed in the locations specified in 550.13(B)(1) through (5) shall have GFCI protection for personnel.

(1) Outdoors, including compartments accessible from outside the unit
(2) Bathrooms, including receptacles in luminaires
(3) Kitchens, where receptacles are installed to serve counter-top surfaces
(4) Sinks, where receptacles are installed within 1.8 m (6 ft) of the outer edge of the sink
(5) Dishwashers

Informational Note: For information on protection of dishwashers, see 422.5.

(C) Cord-Connected Fixed Appliance. A grounding-type receptacle outlet shall be provided for each cord-connected fixed appliance installed.

(D) Receptacle Outlets Required. Except in the bath, closet, and hallway areas, receptacle outlets shall be installed at wall spaces 600 mm (2 ft) wide or more so that no point along the floor line is more than 1.8 m (6 ft) measured horizontally from an outlet in that space. In addition, a receptacle outlet shall be installed in the following locations:

(1) Over or adjacent to countertops in the kitchen [at least one on each side of the sink if countertops are on each side and are 300 mm (12 in.) or over in width].
(2) Adjacent to the refrigerator and freestanding gas-range space. A multiple-type receptacle shall be permitted to serve as the outlet for a countertop and a refrigerator.

(3) At countertop spaces for built-in vanities.

(4) At countertop spaces under wall-mounted cabinets.

(5) In the wall at the nearest point to where a bar-type counter attaches to the wall.

(6) In the wall at the nearest point to where a fixed room divider attaches to the wall.

(7) In laundry areas within 1.8 m (6 ft) of the intended location of the laundry appliance(s).

(8) At least one receptacle outlet located outdoors and accessible at grade level and not more than 2.0 m (6½ ft) above grade. A receptacle outlet located in a compartment accessible from the outside of the unit shall be considered an outdoor receptacle.

(9) At least one receptacle outlet shall be installed in bathrooms within 900 mm (36 in.) of the outside edge of each basin. The receptacle outlet shall be located above or adjacent to the basin location. This receptacle shall be in addition to any receptacle that is a part of a luminaire or appliance. The receptacle shall not be enclosed within a bathroom cabinet or vanity.

(E) Pipe Heating Cable(s) Outlet. For the connection of pipe heating cable(s), a receptacle outlet shall be located on the underside of the unit as follows:

(1) Within 600 mm (2 ft) of the cold water inlet.

(2) Connected to an interior branch circuit, other than a small-appliance branch circuit. It shall be permitted to use a bathroom receptacle circuit for this purpose.

(3) On a circuit where all of the outlets are on the load side of the ground-fault circuit-interrupter.

(4) This outlet shall not be considered as the receptacle required by 550.13(D)(8).

(F) Receptacle Outlets Not Permitted. Receptacle outlets shall not be permitted in the following locations:

(1) Receptacle outlets shall not be installed within or directly over a bathtub or shower space.

(2) A receptacle shall not be installed in a face-up position in any countertop.

(3) Receptacle outlets shall not be installed above electric baseboard heaters, unless provided for in the listing or manufacturer's instructions.

(G) Receptacle Outlets Not Required. Receptacle outlets shall not be required in the following locations:

(1) In the wall space occupied by built-in kitchen or wardrobe cabinets

(2) In the wall space behind doors that can be opened fully against a wall surface

(3) In room dividers of the lattice type that are less than 2.5 m (8 ft) long, not solid, and within 150 mm (6 in.) of the floor

(4) In the wall space afforded by bar-type counters

550.14 Luminaires and Appliances.

(A) Fasten Appliances in Transit. Means shall be provided to securely fasten appliances when the mobile home is in transit. (See 550.16 for provisions on grounding.)

(B) Accessibility. Every appliance shall be accessible for inspection, service, repair, or replacement without removal of permanent construction.

(C) Pendants. Listed pendant-type luminaires or pendant cords shall be permitted.

(D) Bathtub and Shower Luminaires. Where a luminaire is installed over a bathtub or in a shower stall, it shall be of the enclosed and gasketed type listed for wet locations.

550.15 Wiring Methods and Materials. Except as specifically limited in this section, the wiring methods and materials included in this *Code* shall be used in mobile homes. Aluminum conductors, aluminum alloy conductors, and aluminum core conductors such as copper-clad aluminum shall not be acceptable for use as branch-circuit wiring.

(A) Nonmetallic Boxes. Nonmetallic boxes shall be permitted only with nonmetallic cable or nonmetallic raceways.

(B) Nonmetallic Cable Protection. Nonmetallic cable located 380 mm (15 in.) or less above the floor, if exposed, shall be protected from physical damage by covering boards, guard strips, or raceways. Cable likely to be damaged by stowage shall be so protected in all cases.

(C) Metal-Covered and Nonmetallic Cable Protection. Metal-covered and nonmetallic cables shall be permitted to pass through the centers of the wide side of 2 by 4 studs. However, they shall be protected where they pass through 2 by 2 studs or at other studs or frames where the cable or armor would be less than 32 mm (1¼ in.) from the inside or outside surface of the studs where the wall covering materials are in contact with the studs. Steel plates on each side of the cable, or a tube, with not less than 1.35 mm (0.053 in.) wall thickness shall be required to protect the cable. These plates or tubes shall be securely held in place.

(D) Metal Faceplates. Where metal faceplates are used, they shall be grounded.

(E) Installation Requirements. Where a range, clothes dryer, or other appliance is connected by metal-covered cable or flexible metal conduit, a length of not less than 900 mm (3 ft) of unsupported cable or conduit shall be provided to service the appliance. The cable or flexible metal conduit shall be secured to the wall. Type NM or Type SE cable shall not be used to connect a range or dryer. This shall not prohibit the use of Type NM or Type SE cable between the branch-circuit overcurrent protective device and a junction box or range or dryer receptacle.

(F) Raceways. Where rigid metal conduit or intermediate metal conduit is terminated at an enclosure with a locknut and bushing connection, two locknuts shall be provided, one inside and one outside of the enclosure. Rigid nonmetallic conduit, electrical nonmetallic tubing, or surface raceway shall be permitted. All cut ends of conduit and tubing shall be reamed or otherwise finished to remove rough edges.

(G) Switches. Switches shall be rated as follows:

(1) For lighting circuits, switches shall be rated not less than 10 amperes, 120 to 125 volts, and in no case less than the connected load.

(2) Switches for motor or other loads shall comply with the provisions of 404.14.

(H) Under-Chassis Wiring (Exposed to Weather).

(1) Where outdoor or under-chassis line-voltage (120 volts, nominal, or higher) wiring is exposed, it shall be protected by a conduit or raceway identified for use in wet locations. The conductors shall be listed for use in wet locations.

(2) Where wiring is exposed to physical damage, it shall be protected by a raceway, conduit, or other means.

(I) Boxes, Fittings, and Cabinets. Boxes, fittings, and cabinets shall be securely fastened in place and shall be supported from a structural member of the home, either directly or by using a substantial brace.

Exception: Snap-in-type boxes. Boxes provided with special wall or ceiling brackets and wiring devices with integral enclosures that securely fasten to walls or ceilings and are identified for the use shall be permitted without support from a structural member or brace. The testing and approval shall include the wall and ceiling construction systems for which the boxes and devices are intended to be used.

(J) Appliance Terminal Connections. Appliances having branch-circuit terminal connections that operate at temperatures higher than 60°C (140°F) shall have circuit conductors as described in the following:

(1) Branch-circuit conductors having an insulation suitable for the temperature encountered shall be permitted to be run directly to the appliance.

(2) Conductors having an insulation suitable for the temperature encountered shall be run from the appliance terminal connection to a readily accessible outlet box placed at least 300 mm (1 ft) from the appliance. These conductors shall be in a suitable raceway or Type AC or MC cable of at least 450 mm (18 in.) but not more than 1.8 m (6 ft) in length.

(K) Component Interconnections. Fittings and connectors that are intended to be concealed at the time of assembly shall be listed and identified for the interconnection of building components. Such fittings and connectors shall be equal to the wiring method employed in insulation, temperature rise, and fault-current withstanding and shall be capable of enduring the vibration and shock occurring in mobile home transportation.

Informational Note: See 550.19 for interconnection of multiple section units.

550.16 Grounding. Grounding of both electrical and nonelectrical metal parts in a mobile home shall be through connection to a grounding bus in the mobile home panelboard and shall be connected through the green-colored insulated conductor in the supply cord or the feeder wiring to the grounding bus in the service-entrance equipment located adjacent to the mobile home location. Neither the frame of the mobile home nor the frame of any appliance shall be connected to the grounded circuit conductor in the mobile home. Where the panelboard is the service equipment as permitted by 550.32(B), the neutral conductors and the equipment grounding bus shall be connected.

(A) Grounded Conductor.

(1) Insulated. The grounded circuit conductor shall be insulated from the grounding conductors and from equipment enclosures and other grounded parts. The grounded circuit conductor terminals in the panelboard and in ranges, clothes dryers, counter-mounted cooking units, and wall-mounted ovens shall be insulated from the equipment enclosure. Bonding screws, straps, or buses in the panelboard or in appliances shall be removed and discarded. Where the panelboard is the service equipment as permitted by 550.32(B), the neutral conductors and the equipment grounding bus shall be connected.

(2) **Connections of Ranges and Clothes Dryers.** Connections of ranges and clothes dryers with 120/240-volt, 3-wire ratings shall be made with 4-conductor cord and 3-pole, 4-wire, grounding-type plugs or by Type AC cable, Type MC cable, or conductors enclosed in flexible metal conduit.

(B) Equipment Grounding Means.

(1) Supply Cord or Permanent Feeder. The green-colored insulated grounding wire in the supply cord or permanent feeder wiring shall be connected to the grounding bus in the panelboard or disconnecting means.

(2) Electrical System. In the electrical system, all exposed metal parts, enclosures, frames, luminaire canopies, and so forth, shall be effectively bonded to the grounding terminal or enclosure of the panelboard.

(3) Cord-Connected Appliances. Cord-connected appliances, such as washing machines, clothes dryers, and refrigerators, and the electrical system of gas ranges and so forth, shall be grounded by means of a cord with an equipment grounding conductor and grounding-type attachment plug.

(C) Bonding of Non–Current-Carrying Metal Parts.

(1) Exposed Non–Current-Carrying Metal Parts. All exposed non–current-carrying metal parts that are likely to become energized shall be effectively bonded to the grounding terminal or enclosure of the panelboard. A bonding conductor shall be connected between the panelboard and an accessible terminal on the chassis.

(2) Grounding Terminals. Grounding terminals shall be of the solderless type and listed as pressure-terminal connectors recognized for the wire size used. The bonding conductor shall be solid or stranded, insulated or bare, and shall be 8 AWG copper minimum, or equivalent. The bonding conductor shall be routed so as not to be exposed to physical damage.

(3) Metallic Piping and Ducts. Metallic gas, water, and waste pipes and metallic air-circulating ducts shall be considered bonded if they are connected to the terminal on the chassis [see 550.16(C)(1)] by clamps, solderless connectors, or by suitable grounding-type straps.

(4) Metallic Roof and Exterior Coverings. Any metallic roof and exterior covering shall be considered bonded if the following conditions are met:

(1) The metal panels overlap one another and are securely attached to the wood or metal frame parts by metallic fasteners.

(2) The lower panel of the metallic exterior covering is secured by metallic fasteners at a cross member of the chassis by two metal straps per mobile home unit or section at opposite ends.

The bonding strap material shall be a minimum of 100 mm (4 in.) in width of material equivalent to the skin or a material of equal or better electrical conductivity. The straps shall be fastened with paint-penetrating fittings such as screws and star-washers or equivalent.

550.17 Testing.

(A) Dielectric Strength Test. The wiring of each mobile home shall be subjected to a 1-minute, 900-volt, dielectric strength test (with all switches closed) between live parts (including neutral conductor) and the mobile home ground. Alternatively, the test shall be permitted to be performed at 1080 volts for 1 second. This test shall be performed after branch circuits are complete and after luminaires or appliances are installed.

Exception: Listed luminaires or appliances shall not be required to withstand the dielectric strength test.

(B) Continuity and Operational Tests and Polarity Checks. Each mobile home shall be subjected to all of the following:

(1) An electrical continuity test to ensure that all exposed electrically conductive parts are properly bonded

(2) An electrical operational test to demonstrate that all equipment, except water heaters and electric furnaces, is connected and in working order

(3) Electrical polarity checks of permanently wired equipment and receptacle outlets to determine that connections have been properly made

550.18 Calculations. The following method shall be employed in calculating the supply-cord and distribution-panelboard load for each feeder assembly for each mobile home in lieu of the procedure shown in Article 220 and shall be based on a 3-wire, 120/240-volt supply with 120-volt loads balanced between the two ungrounded conductors of the 3-wire system.

(A) Lighting, Small-Appliance, and Laundry Load.

(1) Lighting Volt-Amperes. Length times width of mobile home floor (outside dimensions) times 33 volt-amperes/m² (3 VA/ft²), for example, length × width × 3 = lighting volt-amperes.

(2) Small-Appliance Volt-Amperes. Number of circuits times 1500 volt-amperes for each 20-ampere appliance receptacle circuit (see definition of *Appliance, Portable*, with a fine print note in 550.2), for example, number of circuits × 1500 = small-appliance volt-amperes.

(3) Laundry Area Circuit Volt-Amperes. 1500 volt-amperes.

(4) Total Volt-Amperes. Lighting volt-amperes plus small-appliance volt-amperes plus laundry area volt-amperes equals total volt-amperes.

(5) Net Volt-Amperes. First 3000 total volt-amperes at 100 percent plus remainder at 35 percent equals volt-amperes to be divided by 240 volts to obtain current (amperes) per leg.

(B) Total Load for Determining Power Supply. Total load for determining power supply is the sum of the following:

(1) Lighting and small-appliance load as calculated in 550.18(A)(5)

(2) Nameplate amperes for motors and heater loads (exhaust fans, air conditioners, electric, gas, or oil heating). Omit smaller of the heating and cooling loads, except include blower motor if used as air-conditioner evaporator motor. Where an air conditioner is not installed and a 40-ampere power-supply cord is provided, allow 15 amperes per leg for air conditioning.

(3) Twenty-five percent of current of largest motor in item (2).

(4) Total of nameplate amperes for waste disposer, dishwasher, water heater, clothes dryer, wall-mounted oven, cooking units. Where the number of these appliances exceeds three, use 75 percent of total.

(5) Derive amperes for freestanding range (as distinguished from separate ovens and cooking units) by dividing the following values by 240 volts as shown in the table below:

(6) If outlets or circuits are provided for other than factory-installed appliances, include the anticipated load.

Informational Note: Refer to Informative Annex D, Example D11, for an illustration of the application of this calculation.

Nameplate Rating (watts)	Use (volt-amperes)
0–10,000	80 percent of rating
Over 10,000–12,500	8,000
Over 12,500–13,500	8,400
Over 13,500–14,500	8,800
Over 14,500–15,500	9,200
Over 15,500–16,500	9,600
Over 16,500–17,500	10,000

(C) Optional Method of Calculation for Lighting and Appliance Load. The optional method for calculating lighting and appliance load shown in 220.82 shall be permitted.

550.19 Interconnection of Multiple-Section Mobile or Manufactured Home Units.

(A) Wiring Methods. Approved and listed fixed-type wiring methods shall be used to join portions of a circuit that must be electrically joined and are located in adjacent sections after the home is installed on its support foundation. The circuit's junction shall be accessible for disassembly when the home is prepared for relocation.

Informational Note: See 550.15(K) for component interconnections.

(B) Disconnecting Means. Expandable or multiunit manufactured homes, not having permanently installed feeders, that are to be moved from one location to another shall be permitted to have disconnecting means with branch-circuit protective equipment in each unit when so located that after assembly or joining together of units, the requirements of 550.10 will be met.

550.20 Outdoor Outlets, Luminaires, Air-Cooling Equipment, and So Forth.

(A) Listed for Outdoor Use. Outdoor luminaires and equipment shall be listed for wet locations or outdoor use. Outdoor receptacles shall comply with 406.9. Where located on the underside of the home or located under roof extensions or similarly protected locations, outdoor luminaires and equipment shall be listed for use in damp locations.

(B) Outside Heating Equipment, Air-Conditioning Equipment, or Both. A mobile home provided with a branch circuit designed to energize outside heating equipment, air-conditioning equipment, or both, located outside the mobile home, other than room air conditioners, shall have such branch-circuit conductors terminate in a listed outlet box, or disconnecting means, located on the outside of the mobile home. A label shall be permanently affixed adjacent to the outlet box and shall contain the following information:

THIS CONNECTION IS FOR HEATING
AND/OR AIR-CONDITIONING EQUIPMENT.
THE BRANCH CIRCUIT IS RATED AT NOT MORE THAN

_____ AMPERES, AT _____ VOLTS, 60 HERTZ,
_____ CONDUCTOR AMPACITY.
A DISCONNECTING MEANS SHALL BE LOCATED
WITHIN SIGHT OF THE EQUIPMENT.

The correct voltage and ampere rating shall be given. The tag shall be not less than 0.51 mm (0.020 in.) thick etched brass, stainless steel, anodized or alclad aluminum, or equivalent. The tag shall not be less than 75 mm by 45 mm (3 in. by 1¾ in.) minimum size.

550.25 Arc-Fault Circuit-Interrupter Protection.

(A) Definition. Arc-fault circuit interrupters are defined in Article 100.

(B) Mobile Homes and Manufactured Homes. All 120-volt branch circuits that supply 15- and 20-ampere outlets shall comply with 210.12.

Part III. Services and Feeders

550.30 Distribution System. The mobile home park secondary electrical distribution system to mobile home lots shall be single-phase, 120/240 volts, nominal. For the purpose of Part III, where the park service exceeds 240 volts, nominal, transformers and secondary panelboards shall be treated as services.

550.31 Allowable Demand Factors. Park electrical wiring systems shall be calculated (at 120/240 volts) on the larger of the following:

(1)   16,000 volt-amperes for each mobile home lot
(2)   The load calculated in accordance with 550.18 for the largest typical mobile home that each lot will accept

It shall be permissible to calculate the feeder or service load in accordance with Table 550.31. No demand factor shall be allowed for any other load, except as provided in this *Code*.

550.32 Service Equipment.

(A) Mobile Home Service Equipment. The mobile home service equipment shall be located adjacent to the mobile home and not mounted in or on the mobile home. The service equipment shall be located in sight from and not more than 9.0 m (30 ft) from the exterior wall of the mobile home it serves. The service equipment shall be permitted to be located

Table 550.31 Demand Factors for Services and Feeders

Number of Mobile Homes	Demand Factor (%)
1	100
2	55
3	44
4	39
5	33
6	29
7–9	28
10–12	27
13–15	26
16–21	25
22–40	24
41–60	23
61 and over	22

elsewhere on the premises, if a disconnecting means suitable for use as service equipment is located within sight from and not more than 9.0 m (30 ft) from the exterior wall of the mobile home it serves and is rated not less than that required for service equipment in accordance with 550.32(C). Grounding at the disconnecting means shall be in accordance with 250.32.

(B) Manufactured Home Service Equipment. The manufactured home service equipment shall be permitted to be installed in or on a manufactured home, provided that all of the following conditions are met:

(1)   The manufacturer shall include in its written installation instructions information indicating that the home shall be secured in place by an anchoring system or installed on and secured to a permanent foundation.
(2)   The installation of the service shall comply with Part I through Part VII of Article 230.
(3)   Means shall be provided for the connection of a grounding electrode conductor to the service equipment and routing it outside the structure.
(4)   Bonding and grounding of the service shall be in accordance with Part I through Part V of Article 250.
(5)   The manufacturer shall include in its written installation instructions one method of grounding the service equipment at the installation site. The instructions shall clearly state that other methods of grounding are found in Article 250.
(6)   The minimum size grounding electrode conductor shall be specified in the instructions.
(7)   A red warning label shall be mounted on or adjacent to the service equipment. The label shall state the following:

WARNING
DO NOT PROVIDE ELECTRICAL POWER
UNTIL THE GROUNDING ELECTRODE(S)
IS INSTALLED AND CONNECTED
(SEE INSTALLATION INSTRUCTIONS).

Where the service equipment is not installed in or on the unit, the installation shall comply with the other provisions of this section.

(C) Rating. Mobile home service equipment shall be rated at not less than 100 amperes at 120/240 volts, and provisions shall be made for connecting a mobile home feeder assembly by a permanent wiring method. Power outlets used as mobile home service equipment shall also be permitted to contain receptacles rated up to 50 amperes with appropriate overcurrent protection. Fifty-ampere receptacles shall conform to the configuration shown in Figure 550.10(C).

Informational Note: Complete details of the 50-ampere plug and receptacle configuration can be found in ANSI/NEMA WD 6-2002 (Rev. 2008), *Standard for Wiring Devices — Dimensional Requirements*, Figure 14-50.

(D) Additional Outside Electrical Equipment. Means for connecting a mobile home accessory building or structure or additional electrical equipment located outside a mobile home by a fixed wiring method shall be provided in either the mobile home service equipment or the local external disconnecting means permitted in 550.32(A).

(E) Additional Receptacles. Additional receptacles shall be permitted for connection of electrical equipment located outside the mobile home, and all such 125-volt, single-phase, 15- and 20-ampere receptacles shall be protected by a listed ground-fault circuit interrupter.

(F) Mounting Height. Outdoor mobile home disconnecting means shall be installed so the bottom of the enclosure containing the disconnecting means is not less than 600 mm (2 ft) above finished grade or working platform. The disconnecting means shall be installed so that the center of the grip of the operating handle, when in the highest position, is not more than 2.0 m (6 ft 7 in.) above the finished grade or working platform.

(G) Marking. Where a 125/250-volt receptacle is used in mobile home service equipment, the service equipment shall be marked as follows:

TURN DISCONNECTING SWITCH OR
CIRCUIT BREAKER OFF BEFORE INSERTING
OR REMOVING PLUG. PLUG MUST BE FULLY
INSERTED OR REMOVED.

The marking shall be located on the service equipment adjacent to the receptacle outlet.

550.33 Feeder.

(A) Feeder Conductors. Feeder conductors shall comply with the following:

(1) Feeder conductors shall consist of either a listed cord, factory installed in accordance with 550.10(B), or a permanently installed feeder consisting of four insulated, color-coded conductors that shall be identified by the factory or field marking of the conductors in compliance with 310.110. Equipment grounding conductors shall not be identified by stripping the insulation.
(2) Feeder conductors shall be installed in compliance with 250.32(B).

Exception: For an existing feeder that is installed between the service equipment and a disconnecting means as covered in 550.32(A), it shall be permitted to omit the equipment grounding conductor where the grounded circuit conductor is grounded at the disconnecting means in accordance with 250.32(B) Exception.

(B) Feeder Capacity. Mobile home and manufactured home lot feeder circuit conductors shall have a capacity not less than the loads supplied, shall be rated at not less than 100 amperes, and shall be permitted to be sized in accordance with 310.15(B)(7).

ARTICLE 551
Recreational Vehicles and Recreational Vehicle Parks

Part I. General

551.1 Scope. The provisions of this article cover the electrical conductors and equipment other than low-voltage and automotive vehicle circuits or extensions thereof, installed within or on recreational vehicles, the conductors that connect recreational vehicles to a supply of electricity, and the installation of equipment and devices related to electrical installations within a recreational vehicle park.

Informational Note: For information on low-voltage systems, refer to NFPA 1192-2015, *Standard on Recreational Vehicles,* and ANSI/RVIA LV-2014, *Standard for Low Voltage Systems in Conversion and Recreational Vehicles.*

551.2 Definitions. (See Article 100 for additional definitions.)

Air-Conditioning or Comfort-Cooling Equipment. All of that equipment intended or installed for the purpose of processing the treatment of air so as to control simultaneously or individually its temperature, humidity, cleanliness, and distribution to meet the requirements of the conditioned space.

Appliance, Fixed. An appliance that is fastened or otherwise secured at a specific location.

Camping Trailer. A vehicular portable unit mounted on wheels and constructed with collapsible partial side walls that fold for towing by another vehicle and unfold at the campsite to provide temporary living quarters for recreational, camping, or travel use. *(See Recreational Vehicle.)*

Converter. A device that changes electrical energy from one form to another, as from alternating current to direct current.

Dead Front (as applied to switches, circuit breakers, switchboards, and panelboards). Designed, constructed, and installed so that no current-carrying parts are normally exposed on the front.

Disconnecting Means. The necessary equipment usually consisting of a circuit breaker or switch and fuses, and their accessories, located near the point of entrance of supply conductors in a recreational vehicle and intended to constitute the means of cutoff for the supply to that recreational vehicle.

Frame. Chassis rail and any welded addition thereto of metal thickness of 1.35 mm (0.053 in.) or greater.

Low Voltage. An electromotive force rated 24 volts, nominal, or less.

Motor Home. A vehicular unit designed to provide temporary living quarters for recreational, camping, or travel use built on or permanently attached to a self-propelled motor vehicle chassis or on a chassis cab or van that is an integral part of the completed vehicle. *(See Recreational Vehicle.)*

Power-Supply Assembly. The conductors, including ungrounded, grounded, and equipment grounding conductors, the connectors, attachment plug caps, and all other fittings, grommets, or devices installed for the purpose of delivering energy from the source of electrical supply to the distribution panel within the recreational vehicle.

Recreational Vehicle. A vehicular-type unit primarily designed as temporary living quarters for recreational, camping, or travel use, which either has its own motive power or is mounted on or drawn by another vehicle.

Informational Note: The basic entities are travel trailer, camping trailer, truck camper, and motor home as referenced in NFPA 1192-2015, *Standard on Recreational Vehicles.* See 3.3.46, *Recreational Vehicle,* and A.3.3.46 of NFPA 1192.

Recreational Vehicle Park. Any parcel or tract of land under the control of any person, organization, or governmental entity wherein two or more recreational vehicle, recreational park trailer, and/or other camping sites are offered for use by the public or members of an organization for overnight stays.

Recreational Vehicle Site. A specific area within a recreational vehicle park or campground that is set aside for use by a camping unit.

Recreational Vehicle Site Feeder Circuit Conductors. The conductors from the park service equipment to the recreational vehicle site supply equipment.

Recreational Vehicle Site Supply Equipment. The necessary equipment, usually a power outlet, consisting of a circuit breaker or switch and fuse and their accessories, located near the point of entrance of supply conductors to a recreational vehicle site and intended to constitute the disconnecting means for the supply to that site.

Recreational Vehicle Stand. That area of a recreational vehicle site intended for the placement of a recreational vehicle.

Travel Trailer. A vehicular unit, mounted on wheels, designed to provide temporary living quarters for recreational, camping, or travel use, of such size or weight as not to require special highway movement permits when towed by a motorized vehicle, and of gross trailer area less than 30 m^2 (320 ft^2). *(See Recreational Vehicle.)*

Truck Camper. A portable unit constructed to provide temporary living quarters for recreational, travel, or camping use, consisting of a roof, floor, and sides, designed to be loaded onto and unloaded from the bed of a pickup truck. *(See Recreational Vehicle.)*

551.4 General Requirements.

(A) Not Covered. A recreational vehicle not used for the purposes as defined in 551.2 shall not be required to meet the provisions of Part IV pertaining to the number or capacity of circuits required. It shall, however, meet all other applicable requirements of this article if the recreational vehicle is provided with an electrical installation intended to be energized from a 120-volt, 208Y/120-volt, or 120/240-volt, nominal, ac power-supply system.

(B) Systems. This article covers combination electrical systems, generator installations, and 120-volt, 208Y/120-volt, or 120/240-volt, nominal, systems.

> Informational Note: For information on low-voltage systems, refer to NFPA 1192-2015, *Standard on Recreational Vehicles*, and ANSI/RVIA 12V-2011, *Standard for Low Voltage Systems in Conversion and Recreational Vehicles.*

(C) Labels. Labels required by Article 551 shall be made of etched, metal-stamped, or embossed brass; stainless steel; plastic laminates not less than 0.13 mm (0.005 in.) thick; or anodized or alclad aluminum not less than 0.5 mm (0.020 in.) thick or the equivalent.

> Informational Note: For guidance on other label criteria used in the recreational vehicle industry, refer to ANSI Z535.4-2011, *Product Safety Signs and Labels.*

Part II. Combination Electrical Systems

551.20 Combination Electrical Systems.

(A) General. Vehicle wiring suitable for connection to a battery or dc supply source shall be permitted to be connected to a 120-volt source, provided the entire wiring system and equipment are rated and installed in full conformity with Parts I, II, III, IV, and V requirements of this article covering 120-volt electrical systems. Circuits fed from ac transformers shall not supply dc appliances.

(B) Voltage Converters (120-Volt Alternating Current to Low-Voltage Direct Current). The 120-volt ac side of the voltage converter shall be wired in full conformity with the requirements of Parts I, II, and IV of this article for 120-volt electrical systems.

Exception: Converters supplied as an integral part of a listed appliance shall not be subject to 551.20(B).

All converters and transformers shall be listed for use in recreational vehicles and designed or equipped to provide overtemperature protection. To determine the converter rating, the following percentages shall be applied to the total connected load, including average battery-charging rate, of all 12-volt equipment:

The first 20 amperes of load at 100 percent plus

The second 20 amperes of load at 50 percent plus

All load above 40 amperes at 25 percent

Exception: A low-voltage appliance that is controlled by a momentary switch (normally open) that has no means for holding in the closed position or refrigerators with a 120-volt function shall not be considered as a connected load when determining the required converter rating. Momentarily energized appliances shall be limited to those used to prepare the vehicle for occupancy or travel.

(C) Bonding Voltage Converter Enclosures. The non–current-carrying metal enclosure of the voltage converter shall be connected to the frame of the vehicle with a minimum 8 AWG copper conductor. The voltage converter shall be provided with a separate chassis bonding conductor that shall not be used as a current-carrying conductor.

(D) Dual-Voltage Fixtures, Including Luminaires or Appliances. Fixtures, including luminaires, or appliances having both 120-volt and low-voltage connections shall be listed for dual voltage.

(E) Autotransformers. Autotransformers shall not be used.

(F) Receptacles and Plug Caps. Where a recreational vehicle is equipped with an ac system, a low-voltage system, or both, receptacles and plug caps of the low-voltage system shall differ in configuration from those of the ac system. Where a vehicle equipped with a battery or other low-voltage system has an external connection for low-voltage power, the connector shall have a configuration that will not accept ac power.

Part III. Other Power Sources

551.30 Generator Installations.

(A) Mounting. Generators shall be mounted in such a manner as to be effectively bonded to the recreational vehicle chassis.

(B) Generator Protection. Equipment shall be installed to ensure that the current-carrying conductors from the engine generator and from an outside source are not connected to a vehicle circuit at the same time. Automatic transfer switches in such applications shall be listed for use in one of the following:

(1) Emergency systems
(2) Optional standby systems

Receptacles used as disconnecting means shall be accessible (as applied to wiring methods) and capable of interrupting their rated current without hazard to the operator.

(C) Installation of Storage Batteries and Generators. Storage batteries and internal-combustion-driven generator units (subject to the provisions of this *Code*) shall be secured in place to avoid displacement from vibration and road shock.

(D) Ventilation of Generator Compartments. Compartments accommodating internal-combustion-driven generator units shall be provided with ventilation in accordance with instructions provided by the manufacturer of the generator unit.

> Informational Note: For generator compartment construction requirements, see NFPA 1192 -2015, *Standard on Recreational Vehicles.*

(E) Supply Conductors. The supply conductors from the engine generator to the first termination on the vehicle shall be of the stranded type and be installed in listed flexible conduit or listed liquidtight flexible conduit. The point of first termination shall be in one of the following:

(1) Panelboard
(2) Junction box with a blank cover
(3) Junction box with a receptacle
(4) Enclosed transfer switch
(5) Receptacle assembly listed in conjunction with the generator

The panelboard, enclosed transfer switch, or junction box with a receptacle shall be installed within 450 mm (18 in.) of the point of entry of the supply conductors into the vehicle. A junction box with a blank cover shall be mounted on the compartment wall inside or outside the compartment; to any part of the generator-supporting structure (but not to the generator); to the vehicle floor on the outside of the vehicle; or within 450 mm (18 in.) of the point of entry of the supply conductors into the vehicle. A receptacle assembly listed in conjunction with the generator shall be mounted in accordance with its listing.

551.31 Multiple Supply Source.

(A) Multiple Supply Sources. Where a multiple supply system consisting of an alternate power source and a power-supply cord is installed, the feeder from the alternate power source shall be protected by an overcurrent protective device. Installation shall be in accordance with 551.30(A), 551.30(B), and 551.40.

(B) Multiple Supply Sources Capacity. The multiple supply sources shall not be required to be of the same capacity.

(C) Alternate Power Sources Exceeding 30 Amperes. If an alternate power source exceeds 30 amperes, 120 volts, nominal, it shall be permissible to wire it as a 120-volt, nominal, system, a 208Y/120-volt, nominal, system, or a 120/240-volt, nominal, system, provided an overcurrent protective device of the proper rating is installed in the feeder.

(D) Power-Supply Assembly Not Less Than 30 Amperes. The external power-supply assembly shall be permitted to be less than the calculated load but not less than 30 amperes and shall have overcurrent protection not greater than the capacity of the external power-supply assembly.

551.32 Other Sources. Other sources of ac power, such as inverters, motor generators, or engine generators, shall be listed for use in recreational vehicles and shall be installed in accordance with the terms of the listing. Other sources of ac power shall be wired in full conformity with the requirements in Parts I, II, III, IV, and V of this article covering 120-volt electrical systems.

551.33 Alternate Source Restrictions. Transfer equipment, if not integral with the listed power source, shall be installed to ensure that the current-carrying conductors from other sources of ac power and from an outside source are not connected to the vehicle circuit at the same time. Automatic transfer switches in such applications shall be listed for use in one of the following:

(1) Emergency systems
(2) Optional standby systems

Part IV. Nominal 120-Volt or 120/240-Volt Systems

551.40 120-Volt or 120/240-Volt, Nominal, Systems.

(A) General Requirements. The electrical equipment and material of recreational vehicles indicated for connection to a wiring system rated 120 volts, nominal, 2-wire with equipment grounding conductor, or a wiring system rated 120/240 volts, nominal, 3-wire with equipment grounding conductor, shall be listed and installed in accordance with the requirements of Parts I, II, III, IV, and V of this article. Electrical equipment connected line-to-line shall have a voltage rating of 208–230 volts.

(B) Materials and Equipment. Electrical materials, devices, appliances, fittings, and other equipment installed in, intended for use in, or attached to the recreational vehicle shall be listed. All products shall be used only in the manner in which they have been tested and found suitable for the intended use.

(C) Ground-Fault Circuit-Interrupter Protection. The internal wiring of a recreational vehicle having only one 15- or 20-ampere branch circuit as permitted in 551.42(A) and (B) shall have ground-fault circuit-interrupter protection for personnel. The ground-fault circuit interrupter shall be installed at the point where the power supply assembly terminates within the recreational vehicle. Where a separable cord set is not employed, the ground-fault circuit interrupter shall be permitted to be an integral part of the attachment plug of the power supply assembly. The ground-fault circuit interrupter shall provide protection also under the conditions of an open grounded circuit conductor, interchanged circuit conductors, or both.

551.41 Receptacle Outlets Required.

(A) Spacing. Receptacle outlets shall be installed at wall spaces 600 mm (2 ft) wide or more so that no point along the floor line is more than 1.8 m (6 ft), measured horizontally, from an outlet in that space.

Exception No. 1: Bath and hallway areas.

Exception No. 2: Wall spaces occupied by kitchen cabinets, wardrobe cabinets, built-in furniture, behind doors that may open fully against a wall surface, or similar facilities.

(B) Location. Receptacle outlets shall be installed as follows:

(1) Adjacent to countertops in the kitchen [at least one on each side of the sink if countertops are on each side and are 300 mm (12 in.) or over in width and depth].

(2) Adjacent to the refrigerator and gas range space, except where a gas-fired refrigerator or cooking appliance, requiring no external electrical connection, is factory installed.

(3) Adjacent to countertop spaces of 300 mm (12 in.) or more in width and depth that cannot be reached from a receptacle required in 551.41(B)(1) by a cord of 1.8 m (6 ft) without crossing a traffic area, cooking appliance, or sink.

(4) Rooftop decks that are accessible from inside the RV shall have at least one receptacle installed within the perimeter of the rooftop deck. The receptacle shall not be located more than 1.2 m (4 ft) above the balcony, deck, or porch surface. The receptacle shall comply with the requirements of 406.9(B) for wet locations.

(C) Ground-Fault Circuit-Interrupter Protection. Where provided, each 125-volt, single-phase, 15- or 20-ampere receptacle outlet shall have ground-fault circuit-interrupter protection for personnel in the following locations:

(1) Adjacent to a bathroom lavatory

(2) Where the receptacles are installed to serve the counter-top surfaces and are within 1.8 m (6 ft) of any lavatory or sink

Exception No. 1: Receptacles installed for appliances in dedicated spaces, such as for dishwashers, disposals, refrigerators, freezers, and laundry equipment.

Exception No. 2: Single receptacles for interior connections of expandable room sections.

Exception No. 3: De-energized receptacles that are within 1.8 m (6 ft) of any sink or lavatory due to the retraction of the expandable room section.

(3) In the area occupied by a toilet, shower, tub, or any combination thereof

(4) On the exterior of the vehicle

Exception: Receptacles that are located inside of an access panel that is installed on the exterior of the vehicle to supply power for an installed appliance shall not be required to have ground-fault circuit-interrupter protection.

The receptacle outlet shall be permitted in a listed luminaire. A receptacle outlet shall not be installed in a tub or combination tub–shower compartment.

(D) Face-Up Position. A receptacle shall not be installed in a face-up position in any countertop or similar horizontal surface.

551.42 Branch Circuits Required. Each recreational vehicle containing an ac electrical system shall contain one of the circuit arrangements in 551.42(A) through (D).

(A) One 15-Ampere Circuit. One 15-ampere circuit to supply lights, receptacle outlets, and fixed appliances. Such recreational vehicles shall be equipped with one 15-ampere switch and fuse or one 15-ampere circuit breaker.

(B) One 20-Ampere Circuit. One 20-ampere circuit to supply lights, receptacle outlets, and fixed appliances. Such recreational vehicles shall be equipped with one 20-ampere switch and fuse or one 20-ampere circuit breaker.

(C) Two to Five 15- or 20-Ampere Circuits. Two to five 15- or 20-ampere circuits to supply lights, receptacle outlets, and fixed appliances shall be permitted. Such recreational vehicles shall be permitted to be equipped with panelboards rated 120 volt maximum or 120/240 volt maximum and listed for 30-ampere application supplied by the appropriate power-supply assemblies. Not more than two 120-volt thermostatically controlled appliances shall be installed in such systems unless appliance isolation switching, energy management systems, or similar methods are used.

Exception No. 1: Additional 15- or 20-ampere circuits shall be permitted where a listed energy management system rated at 30-ampere maximum is employed within the system.

Exception No. 2: Six 15- or 20-ampere circuits shall be permitted without employing an energy management system, provided that the added sixth circuit serves only the power converter, and the combined load of all six circuits does not exceed the allowable load that was designed for use by the original five circuits.

Informational Note: See 210.23(A) for permissible loads. See 551.45(C) for main disconnect and overcurrent protection requirements.

(D) More Than Five Circuits Without a Listed Energy Management System. A 50-ampere, 120/208–240-volt power-supply assembly and a minimum 50-ampere-rated panelboard shall be used where six or more circuits are employed. The load distribution shall ensure a reasonable current balance between phases.

551.43 Branch-Circuit Protection.

(A) Rating. The branch-circuit overcurrent devices shall be rated as follows:

(1) Not more than the circuit conductors, and

(2) Not more than 150 percent of the rating of a single appliance rated 13.3 amperes or more and supplied by an individual branch circuit, but

(3) Not more than the overcurrent protection size marked on an air conditioner or other motor-operated appliances

(B) Protection for Smaller Conductors. A 20-ampere fuse or circuit breaker shall be permitted for protection for fixtures, including luminaires, leads, cords, or small appliances, and 14 AWG tap conductors, not over 1.8 m (6 ft) long for recessed luminaires.

(C) Fifteen-Ampere Receptacles Considered Protected by 20 Amperes. If more than one receptacle or load is on a branch circuit, 15-ampere receptacles shall be permitted to be protected by a 20-ampere fuse or circuit breaker.

551.44 Power-Supply Assembly. Each recreational vehicle shall have only one of the main power-supply assemblies covered in 551.44(A) through (D).

(A) Fifteen-Ampere Main Power-Supply Assembly. Recreational vehicles wired in accordance with 551.42(A) shall use a listed 15-ampere or larger main power-supply assembly.

(B) Twenty-Ampere Main Power-Supply Assembly. Recreational vehicles wired in accordance with 551.42(B) shall use a listed 20-ampere or larger main power-supply assembly.

(C) Thirty-Ampere Main Power-Supply Assembly. Recreational vehicles wired in accordance with 551.42(C) shall use a listed 30-ampere or larger main power-supply assembly.

(D) Fifty-Ampere Power-Supply Assembly. Recreational vehicles wired in accordance with 551.42(D) shall use a listed 50-ampere, 120/208–240-volt main power-supply assembly.

551.45 Panelboard.

(A) Listed and Appropriately Rated. A listed and appropriately rated panelboard or other equipment specifically listed for this purpose shall be used. The grounded conductor termination bar shall be insulated from the enclosure as provided in 551.54(C). An equipment grounding terminal bar shall be attached inside the enclosure of the panelboard.

(B) Location. The panelboard shall be installed in a readily accessible location with the RV in the setup mode. Working clearance for the panelboard with the RV in the setup mode shall be not less than 600 mm (24 in.) wide and 750 mm (30 in.) deep.

Exception No. 1: Where the panelboard cover is exposed to the inside aisle space, one of the working clearance dimensions shall be permitted to be reduced to a minimum of 550 mm (22 in.). A panelboard is considered exposed where the panelboard cover is within 50 mm (2 in.) of the aisle's finished surface or not more than 25 mm (1 in.) from the backside of doors that enclose the space.

Exception No. 2: Compartment doors used for access to a generator shall be permitted to be equipped with a locking system.

(C) Dead-Front Type. The panelboard shall be of the dead-front type and shall consist of one or more circuit breakers or Type S fuseholders. A main disconnecting means shall be provided where fuses are used or where more than two circuit breakers are employed. A main overcurrent protective device not exceeding the power-supply assembly rating shall be provided where more than two branch circuits are employed.

551.46 Means for Connecting to Power Supply.

(A) Assembly. The power-supply assembly or assemblies shall be factory supplied or factory installed and be of one of the types specified herein.

(1) Separable. Where a separable power-supply assembly consisting of a cord with a female connector and molded attachment plug cap is provided, the vehicle shall be equipped with a permanently mounted, flanged surface inlet (male, recessed-type motor-base attachment plug) wired directly to the panelboard by an approved wiring method. The attachment plug cap shall be of a listed type.

(2) Permanently Connected. Each power-supply assembly shall be connected directly to the terminals of the panelboard or conductors within a junction box and provided with means to prevent strain from being transmitted to the terminals. The ampacity of the conductors between each junction box and the terminals of each panelboard shall be at least equal to the ampacity of the power-supply cord. The supply end of the assembly shall be equipped with an attachment plug of the type described in 551.46(C). Where the cord passes through the walls or floors, it shall be protected by means of conduit and

bushings or equivalent. The cord assembly shall have permanent provisions for protection against corrosion and mechanical damage while the vehicle is in transit and while the cord assembly is being stored or removed for use.

(B) Cord. The cord exposed usable length shall be measured from the point of entrance to the recreational vehicle or the face of the flanged surface inlet (motor-base attachment plug) to the face of the attachment plug at the supply end.

The cord exposed usable length, measured to the point of entry on the vehicle exterior, shall be a minimum of 7.5 m (25 ft) where the point of entrance is at the side of the vehicle or shall be a minimum 9.0 m (30 ft) where the point of entrance is at the rear of the vehicle.

Where the cord entrance into the vehicle is more than 900 mm (3 ft) above the ground, the minimum cord lengths above shall be increased by the vertical distance of the cord entrance heights above 900 mm (3 ft).

Informational Note: See 551.46(E) for location of point of entrance of a power-supply assembly on the recreational vehicle exterior.

(C) Attachment Plugs.

(1) Units with One 15-Ampere Branch Circuit. Recreational vehicles having only one 15-ampere branch circuit as permitted by 551.42(A) shall have an attachment plug that shall be 2-pole, 3-wire grounding type, rated 15 amperes, 125 volts, conforming to the configuration shown in Figure 551.46(C)(1).

Informational Note: Complete details of this configuration can be found in ANSI/NEMA WD 6-2002, *Standard for Dimensions of Attachment Plugs and Receptacle*, Figure 5.15.

(2) Units with One 20-Ampere Branch Circuit. Recreational vehicles having only one 20-ampere branch circuit as permitted in 551.42(B) shall have an attachment plug that shall be 2-pole,

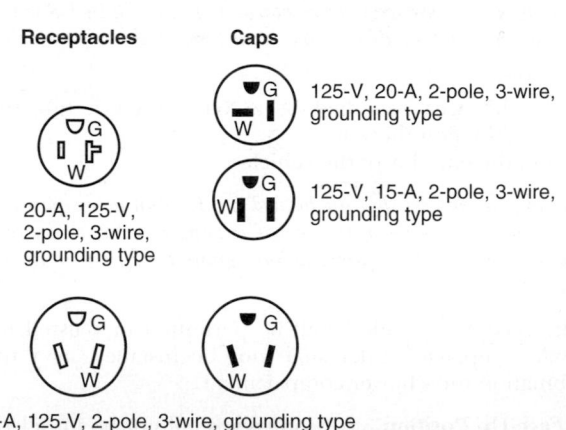

Receptacles   Caps

125-V, 20-A, 2-pole, 3-wire, grounding type

125-V, 15-A, 2-pole, 3-wire, grounding type

20-A, 125-V, 2-pole, 3-wire, grounding type

30-A, 125-V, 2-pole, 3-wire, grounding type

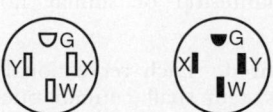

50-A, 125/250-V, 3-pole, 4-wire, grounding type

FIGURE 551.46(C)(1)   Configurations for Grounding-Type Receptacles and Attachment Plug Caps Used for Recreational Vehicle Supply Cords and Recreational Vehicle Lots.

3-wire grounding type, rated 20 amperes, 125 volts, conforming to the configuration shown in Figure 551.46(C)(1).

> Informational Note: Complete details of this configuration can be found in ANSI/NEMA WD 6-2002, National Electrical Manufacturers Association's *Standard for Dimensions of Attachment Plugs and Receptacles*, Figure 5.20.

(3) Units with Two to Five 15- or 20-Ampere Branch Circuits. Recreational vehicles wired in accordance with 551.42(C) shall have an attachment plug that shall be 2-pole, 3-wire grounding type, rated 30 amperes, 125 volts, conforming to the configuration shown in Figure 551.46(C)(1) intended for use with units rated at 30 amperes, 125 volts.

> Informational Note: Complete details of this configuration can be found in ANSI/NEMA WD 6-2002, National Electrical Manufacturers Association's *Standard for Dimensions of Attachment Plugs and Receptacles*, Figure TT.

(4) Units with 50-Ampere Power-Supply Assembly. Recreational vehicles having a power-supply assembly rated 50 amperes as permitted by 551.42(D) shall have a 3-pole, 4-wire grounding-type attachment plug rated 50 amperes, 125/250 volts, conforming to the configuration shown in Figure 551.46(C)(1).

> Informational Note: Complete details of this configuration can be found in ANSI/NEMA WD 6-2002, *Standard for Dimensions of Attachment Plugs and Receptacles*, Figure 14.50.

(D) Labeling at Electrical Entrance. Each recreational vehicle shall have a safety label with the signal word WARNING in minimum 6-mm (¼-in.) high letters and body text in minimum 3-mm (⅛-in.) high letters on a contrasting background. The safety label shall be affixed to the exterior skin, at or near the point of entrance of the power-supply cord(s), and shall read, using one of the following warnings, as appropriate:

WARNING

THIS CONNECTION IS FOR 110–125-VOLT AC,
60 HZ, _____ AMPERE SUPPLY.

or

THIS CONNECTION IS FOR 208Y/120-VOLT or 120/240-
VOLT AC, 3-POLE, 4-WIRE,
60 HZ, _____ AMPERE SUPPLY.

DO NOT EXCEED CIRCUIT RATING.
EXCEEDING THE CIRCUIT RATING MAY CAUSE A
FIRE AND RESULT IN DEATH OR SERIOUS INJURY.

The correct ampere rating shall be marked in the blank space.

(E) Location. The point of entrance of a power-supply assembly shall be located within 4.5 m (15 ft) of the rear, on the left (road) side or at the rear, left of the longitudinal center of the vehicle, within 450 mm (18 in.) of the outside wall.

Exception No. 1: A recreational vehicle equipped with only a listed flexible drain system or a side-vent drain system shall be permitted to have the electrical point of entrance located on either side, provided the drain(s) for the plumbing system is (are) located on the same side.

Exception No. 2: A recreational vehicle shall be permitted to have the electrical point of entrance located more than 4.5 m (15 ft) from the rear. Where this occurs, the distance beyond the 4.5-m (15-ft) dimension shall be added to the cord's minimum length as specified in 551.46(B).

Exception No. 3: Recreational vehicles designed for transporting livestock shall be permitted to have the electrical point of entrance located on either side or the front.

551.47 Wiring Methods.

(A) Wiring Systems. Cables and raceways installed in accordance with Articles 320, 322, 330 through 340, 342 through 362, 386, and 388 shall be permitted in accordance with their applicable article, except as otherwise specified in this article. An equipment grounding means shall be provided in accordance with 250.118.

(B) Conduit and Tubing. Where rigid metal conduit or intermediate metal conduit is terminated at an enclosure with a locknut and bushing connection, two locknuts shall be provided, one inside and one outside of the enclosure. All cut ends of conduit and tubing shall be reamed or otherwise finished to remove rough edges.

(C) Nonmetallic Boxes. Nonmetallic boxes shall be acceptable only with nonmetallic-sheathed cable or nonmetallic raceways.

(D) Boxes. In walls and ceilings constructed of wood or other combustible material, boxes and fittings shall be flush with the finished surface or project therefrom.

(E) Mounting. Wall and ceiling boxes shall be mounted in accordance with Article 314.

Exception No. 1: Snap-in-type boxes or boxes provided with special wall or ceiling brackets that securely fasten boxes in walls or ceilings shall be permitted.

Exception No. 2: A wooden plate providing a 38-mm (1½-in.) minimum width backing around the box and of a thickness of 13 mm (½ in.) or greater (actual) attached directly to the wall panel shall be considered as approved means for mounting outlet boxes.

(F) Raceway and Cable Continuity. Raceways and cable sheaths shall be continuous between boxes and other enclosures.

(G) Protected. Metal-clad, Type AC, or nonmetallic-sheathed cables and electrical nonmetallic tubing shall be permitted to pass through the centers of the wide side of 2 by 4 wood studs. However, they shall be protected where they pass through 2 by 2 wood studs or at other wood studs or frames where the cable or tubing would be less than 32 mm (1¼ in.) from the inside or outside surface. Steel plates on each side of the cable or tubing or a steel tube, with not less than 1.35 mm (0.053 in.) wall thickness, shall be installed to protect the cable or tubing. These plates or tubes shall be securely held in place. Where nonmetallic-sheathed cables pass through punched, cut, or drilled slots or holes in metal members, the cable shall be protected by bushings or grommets securely fastened in the opening prior to installation of the cable.

(H) Bends. No bend shall have a radius of less than five times the cable diameter.

(I) Cable Supports. Where connected with cable connectors or clamps, cables shall be secured and supported within 300 mm (12 in.) of outlet boxes, panelboards, and splice boxes on appliances. Supports and securing shall be provided at intervals not exceeding 1.4 m (4½ ft) at other places.

(J) Nonmetallic Box Without Cable Clamps. Nonmetallic-sheathed cables shall be secured and supported within 200 mm

(8 in.) of a nonmetallic outlet box without cable clamps. Where wiring devices with integral enclosures are employed with a loop of extra cable to permit future replacement of the device, the cable loop shall be considered as an integral portion of the device.

(K) Physical Damage. Where subject to physical damage, exposed nonmetallic cable shall be protected by covering boards, guard strips, raceways, or other means.

(L) Receptacle Faceplates. Metal faceplates shall comply with Section 406.5(A). Nonmetallic faceplates shall comply with Section 406.5(C).

(M) Metal Faceplates Grounded. Where metal faceplates are used, they shall be grounded.

(N) Moisture or Physical Damage. Where outdoor or under-chassis wiring is 120 volts, nominal, or over and is exposed to moisture or physical damage, the wiring shall be protected by rigid metal conduit, by intermediate metal conduit, or by electrical metallic tubing, rigid nonmetallic conduit, or Type MI cable, that is closely routed against frames and equipment enclosures or other raceway or cable identified for the application.

(O) Component Interconnections. Fittings and connectors that are intended to be concealed at the time of assembly shall be listed and identified for the interconnection of building components. Such fittings and connectors shall be equal to the wiring method employed in insulation, temperature rise, and fault-current withstanding and shall be capable of enduring the vibration and shock occurring in recreational vehicles.

(P) Method of Connecting Expandable Units. The method of connecting expandable units to the main body of the vehicle shall comply with 551.47(P)(1) or (P)(2):

(1) Cord-and-Plug-Connected. Cord-and-plug connections shall comply with (a) through (d).

(a) That portion of a branch circuit that is installed in an expandable unit shall be permitted to be connected to the portion of the branch circuit in the main body of the vehicle by means of an attachment plug and cord listed for hard usage. The cord and its connections shall comply with all provisions of Article 400 and shall be considered as a permitted use under 400.10. Where the attachment plug and cord are located within the vehicle's interior, use of plastic thermoset or elastomer parallel cord Type SPT-3, SP-3, or SPE shall be permitted.

(b) Where the receptacle provided for connection of the cord to the main circuit is located on the outside of the vehicle, it shall be protected with a ground-fault circuit interrupter for personnel and be listed for wet locations. A cord located on the outside of a vehicle shall be identified for outdoor use.

(c) Unless removable or stored within the vehicle interior, the cord assembly shall have permanent provisions for protection against corrosion and mechanical damage while the vehicle is in transit.

(d) The attachment plug and cord shall be installed so as not to permit exposed live attachment plug pins.

(2) Direct Wired. That portion of a branch circuit that is installed in an expandable unit shall be permitted to be connected to the portion of the branch circuit in the main body of the vehicle by means of flexible cord installed in accordance with 551.47(P)(2)(a) through (P)(2)(e) or other approved wiring method.

(a) The flexible cord shall be listed for hard usage and for use in wet locations.

(b) The flexible cord shall be permitted to be exposed on the underside of the vehicle.

(c) The flexible cord shall be permitted to pass through the interior of a wall or floor assembly or both a maximum concealed length of 600 mm (24 in.) before terminating at an outlet or junction box.

(d) Where concealed, the flexible cord shall be installed in nonflexible conduit or tubing that is continuous from the outlet or junction box inside the recreational vehicle to a weatherproof outlet box, junction box, or strain relief fitting listed for use in wet locations that is located on the underside of the recreational vehicle. The outer jacket of the flexible cord shall be continuous into the outlet or junction box.

(e) Where the flexible cord passes through the floor to an exposed area inside of the recreational vehicle, it shall be protected by means of conduit and bushings or equivalent.

Where subject to physical damage, the flexible cord shall be protected with RMC, IMC, Schedule 80 PVC, reinforced thermosetting resin conduit (RTRC) listed for exposure to physical damage, or other approved means and shall extend at least 150 mm (6 in.) above the floor. A means shall be provided to secure the flexible cord where it enters the recreational vehicle.

(Q) Prewiring for Air-Conditioning Installation. Prewiring installed for the purpose of facilitating future air-conditioning installation shall comply with the applicable portions of this article and the following:

(1) An overcurrent protective device with a rating compatible with the circuit conductors shall be installed in the panelboard and wiring connections completed.

(2) The load end of the circuit shall terminate in a junction box with a blank cover or other listed enclosure. Where a junction box with a blank cover is used, the free ends of the conductors shall be adequately capped or taped.

(3) A safety label with the signal word WARNING in minimum 6-mm (¼-in.) high letters and body text in minimum 3-mm (⅛-in.) high letters on a contrasting background shall be affixed on or adjacent to the junction box and shall read as follows:

> WARNING
> AIR-CONDITIONING CIRCUIT.
> THIS CONNECTION IS FOR AIR CONDITIONERS
> RATED 110–125-VOLT AC, 60 HZ,
> ___ AMPERES MAXIMUM.
> DO NOT EXCEED CIRCUIT RATING.
> EXCEEDING THE CIRCUIT RATING MAY
> CAUSE A FIRE AND RESULT IN DEATH
> OR SERIOUS INJURY.

An ampere rating, not to exceed 80 percent of the circuit rating, shall be legibly marked in the blank space.

(4) The circuit shall serve no other purpose.

(R) Prewiring for Generator Installation. Prewiring installed for the purpose of facilitating future generator installation shall comply with the other applicable portions of this article and the following:

(1) Circuit conductors shall be appropriately sized in relation to the anticipated load as stated on the label required in (R)(4).

(2) Where junction boxes are utilized at either of the circuit originating or terminus points, free ends of the conductors shall be adequately capped or taped.

(3) Where devices such as receptacle outlet, transfer switch, and so forth, are installed, the installation shall be complete, including circuit conductor connections.

(4) A safety label with the signal word WARNING in minimum 6-mm (¼-in.) high letters and body text in minimum 3-mm (⅛-in.) high letters on a contrasting background shall be affixed on the cover of each junction box containing incomplete circuitry and shall read, using one of the following warnings, as appropriate:

> WARNING
> GENERATOR
> ONLY INSTALL A GENERATOR LISTED
> SPECIFICALLY FOR RV USE
> HAVING OVERCURRENT PROTECTION
> RATED 110–125-VOLT AC,
> 60 HZ, _____ AMPERES MAXIMUM.

or

> GENERATOR ONLY INSTALL A GENERATOR LISTED
> SPECIFICALLY FOR RV USE
> HAVING OVERCURRENT PROTECTION RATED 120–
> 240-VOLT AC,
> 60 HZ, _____ AMPERES MAXIMUM.

The correct ampere rating shall be legibly marked in the blank space.

(S) Prewiring for Other Circuits. Prewiring installed for the purpose of installing other appliances or devices shall comply with the applicable portions of this article and the following:

(1) An overcurrent protection device with a rating compatible with the circuit conductors shall be installed in the panelboard with wiring connections completed.

(2) The load end of the circuit shall terminate in a junction box with a blank cover or a device listed for the purpose. Where a junction box with blank cover is used, the free ends of the conductors shall be adequately capped or taped.

(3) A safety label with the signal word WARNING in minimum 6-mm (¼-in.) high letters and body text in minimum 3-mm (⅛-in.) high letters on a contrasting background shall be affixed on or adjacent to the junction box or device listed for the purpose and shall read as follows:

> WARNING
> THIS CONNECTION IS FOR _____ RATED _____ VOLT
> AC, 60 HZ, _____ AMPERES MAXIMUM. DO NOT
> EXCEED CIRCUIT RATING.
>
> EXCEEDING THE CIRCUIT RATING MAY CAUSE A
> FIRE AND RESULT IN DEATH OR
> SERIOUS INJURY.

An ampere rating not to exceed 80 percent of the circuit rating shall be legibly marked in the blank space.

551.48 Conductors and Boxes. The maximum number of conductors permitted in boxes shall be in accordance with 314.16.

551.49 Grounded Conductors. The identification of grounded conductors shall be in accordance with 200.6.

551.50 Connection of Terminals and Splices. Conductor splices and connections at terminals shall be in accordance with 110.14.

551.51 Switches.

(A) Rating. Switches shall be rated in accordance with 551.51(A)(1) and (A)(2).

(1) Lighting Circuits. For lighting circuits, switches shall be rated not less than 10 amperes, 120–125 volts and in no case less than the connected load.

(2) Motors or Other Loads. Switches for motor or other loads shall comply with the provisions of 404.14.

(B) Location. Switches shall not be installed within wet locations in tub or shower spaces unless installed as part of a listed tub or shower assembly.

551.52 Receptacles. All receptacle outlets shall be of the grounding type and installed in accordance with 406.4 and 210.21.

551.53 Luminaires and Other Equipment.

(A) General. Any combustible wall or ceiling finish exposed between the edge of a canopy or pan of a luminaire or ceiling-suspended (paddle) fan and the outlet box shall be covered with noncombustible material.

(B) Shower Luminaires. If a luminaire is provided over a bathtub or in a shower stall, it shall be of the enclosed and gasketed type and listed for the type of installation, and it shall be ground-fault circuit-interrupter protected.

(C) Outdoor Outlets, Luminaires, Air-Cooling Equipment, and So On. Outdoor luminaires and other equipment shall be listed for outdoor use.

551.54 Grounding. (See also 551.56 on bonding of non–current-carrying metal parts.)

(A) Power-Supply Grounding. The grounding conductor in the supply cord or feeder shall be connected to the grounding bus or other approved grounding means in the panelboard.

(B) Panelboard. The panelboard shall have a grounding bus with terminals for all grounding conductors or other approved grounding means.

(C) Insulated Grounded Conductor (Neutral Conductor). The grounded circuit conductor (neutral conductor) shall be insulated from the equipment grounding conductors and from equipment enclosures and other grounded parts. The grounded circuit conductor (neutral conductor) terminals in the panelboard and in ranges, clothes dryers, counter-mounted cooking units, and wall-mounted ovens shall be insulated from the equipment enclosure. Bonding screws, straps, or buses in the panelboard or in appliances shall be removed and discarded. Connection of electric ranges and electric clothes dryers utilizing a grounded conductor, if cord-connected, shall be made with 4-conductor cord and 3-pole, 4-wire grounding-type plug caps and receptacles.

551.55 Interior Equipment Grounding.

(A) Exposed Metal Parts. In the electrical system, all exposed metal parts, enclosures, frames, luminaire canopies, and so forth, shall be effectively bonded to the grounding terminals or enclosure of the panelboard.

(B) Equipment Grounding and Bonding Conductors. Bare wires, insulated wire with an outer finish that is green or green with one or more yellow stripes, shall be used for equipment grounding or bonding conductors only.

(C) Grounding of Electrical Equipment. Grounding of electrical equipment shall be accomplished by one or more of the following methods:

(1) Connection of metal raceway, the sheath of Type MC and Type MI cable where the sheath is identified for grounding, or the armor of Type AC cable to metal enclosures.

(2) A connection between the one or more equipment grounding conductors and a metal enclosure by means of a grounding screw, which shall be used for no other purpose, or a listed grounding device.

(3) The equipment grounding conductor in nonmetallic-sheathed cable shall be permitted to be secured under a screw threaded into the luminaire canopy other than a mounting screw or cover screw, or attached to a listed grounding means (plate) in a nonmetallic outlet box for luminaire mounting. [Grounding means shall also be permitted for luminaire attachment screws.]

(D) Grounding Connection in Nonmetallic Box. A connection between the one or more equipment grounding conductors brought into a nonmetallic outlet box shall be so arranged that a connection of the equipment grounding conductor can be made to any fitting or device in that box that requires grounding.

(E) Grounding Continuity. Where more than one equipment grounding or bonding conductor of a branch circuit enters a box, all such conductors shall be in good electrical contact with each other, and the arrangement shall be such that the disconnection or removal of a receptacle, luminaire, or other device fed from the box will not interfere with or interrupt the grounding continuity.

(F) Cord-Connected Appliances. Cord-connected appliances, such as washing machines, clothes dryers, refrigerators, and the electrical system of gas ranges, and so forth, shall be grounded by means of an approved cord with equipment grounding conductor and grounding-type attachment plug.

551.56 Bonding of Non–Current-Carrying Metal Parts.

(A) Required Bonding. All exposed non–current-carrying metal parts that are likely to become energized shall be effectively bonded to the grounding terminal or enclosure of the panelboard.

(B) Bonding Chassis. A bonding conductor shall be connected between any panelboard and an accessible terminal on the chassis. Aluminum or copper-clad aluminum conductors shall not be used for bonding if such conductors or their terminals are exposed to corrosive elements.

Exception: Any recreational vehicle that employs a unitized metal chassis-frame construction to which the panelboard is securely fastened with a bolt(s) and nut(s) or by welding or riveting shall be considered to be bonded.

(C) Bonding Conductor Requirements. Grounding terminals shall be of the solderless type and listed as pressure terminal connectors recognized for the wire size used. The bonding conductor shall be solid or stranded, insulated or bare, and shall be 8 AWG copper minimum, or equal.

(D) Metallic Roof and Exterior Bonding. The metal roof and exterior covering shall be considered bonded where both of the following conditions apply:

(1) The metal panels overlap one another and are securely attached to the wood or metal frame parts by metal fasteners.

(2) The lower panel of the metal exterior covering is secured by metal fasteners at each cross member of the chassis, or the lower panel is connected to the chassis by a metal strap.

(E) Gas, Water, and Waste Pipe Bonding. The gas, water, and waste pipes shall be considered grounded if they are bonded to the chassis.

(F) Furnace and Metal Air Duct Bonding. Furnace and metal circulating air ducts shall be bonded.

551.57 Appliance Accessibility and Fastening. Every appliance shall be accessible for inspection, service, repair, and replacement without removal of permanent construction. Means shall be provided to securely fasten appliances in place when the recreational vehicle is in transit.

Part V. Factory Tests

551.60 Factory Tests (Electrical). Each recreational vehicle designed with a 120-volt or a 120/240-volt electrical system shall withstand the applied voltage without electrical breakdown of a 1-minute, 900-volt ac or 1280-volt dc dielectric strength test, or a 1-second, 1080-volt ac or 1530-volt dc dielectric strength test, with all switches closed, between ungrounded and grounded conductors and the recreational vehicle ground. During the test, all switches and other controls shall be in the "on" position. Fixtures, including luminaires and permanently installed appliances, shall not be required to withstand this test. The test shall be performed after branch circuits are complete prior to energizing the system and again after all outer coverings and cabinetry have been secured. The dielectric test shall be performed in accordance with the test equipment manufacturer's written instructions.

Each recreational vehicle shall be subjected to all of the following:

(1) A continuity test to ensure that all metal parts are properly bonded

(2) Operational tests to demonstrate that all equipment is properly connected and in working order

(3) Polarity checks to determine that connections have been properly made

(4) GFCI test to demonstrate that the ground fault protection device(s) installed on the recreational vehicle are operating properly

Part VI. Recreational Vehicle Parks

551.71 Type Receptacles Provided.

(A) 20-Ampere. Every recreational vehicle site with electrical supply shall be equipped with recreational vehicle site supply equipment with at least one 20-ampere, 125-volt receptacle.

(B) 30-Ampere. A minimum of 70 percent of all recreational vehicle sites with electrical supply shall each be equipped with a 30-ampere, 125-volt receptacle conforming to Figure

551.46(C)(1). This supply shall be permitted to include additional receptacle configurations conforming to 551.81. The remainder of all recreational vehicle sites with electrical supply shall be equipped with one or more of the receptacle configurations conforming to 551.81.

(C) 50-Ampere. A minimum of 20 percent of existing and 40 percent of all new recreational vehicle sites, with electrical supply, shall each be equipped with a 50-ampere, 125/250-volt receptacle conforming to the configuration as identified in Figure 551.46(C)(1). Every recreational vehicle site equipped with a 50-ampere receptacle shall also be equipped with a 30-ampere, 125-volt receptacle conforming to Figure 551.46(C)(1). These electrical supplies shall be permitted to include additional receptacles that have configurations in accordance with 551.81.

(D) Tent Sites. Dedicated tent sites with a 15- or 20-ampere electrical supply shall be permitted to be excluded when determining the percentage of recreational vehicle sites with 30- or 50-ampere receptacles.

(E) Additional Receptacles. Additional receptacles shall be permitted for the connection of electrical equipment outside the recreational vehicle within the recreational vehicle park.

(F) GFCI Protection. All 125-volt, single-phase, 15- and 20-ampere receptacles shall have listed ground-fault circuit-interrupter protection for personnel. The GFCI devices used in RV site electrical equipment shall not be required to be weather or tamper resistant in accordance with 406.9 and 406.12.

> Informational Note: The percentage of 50 ampere sites required by 551.71 could be inadequate for seasonal recreational vehicle sites serving a higher percentage of recreational vehicles with 50 ampere electrical systems. In that type of recreational vehicle park, the percentage of 50 ampere sites could approach 100 percent.

551.72 Distribution System.

(A) Systems. Distribution systems shall provide the voltage and have a capacity for the receptacles provided in the recreational vehicle (RV) site supply equipment as calculated according to 551.73 and shall have an ampacity not less than 30 amperes. Systems permitted include 120 volts, 1-phase; 120/240 volts, 1-phase; and 120/208 volts, 1-phase.

(B) Three-Phase Systems. Feeders from 208Y/120-volt, 3-phase systems shall be permitted to include two ungrounded conductors and shall include one grounded conductor and one equipment grounding conductor. So far as practicable, the loads shall be equally distributed on the 3-phase system.

(C) Receptacles. Receptacles rated at 50 amperes shall be supplied from a branch circuit of the voltage class and rating of the receptacle. Other recreational vehicle sites with 125-volt, 20- and 30-ampere receptacles shall be permitted to be derived from any grounded distribution system that supplies 120-volt, single-phase power. The neutral conductors shall not be reduced in size below the size of the ungrounded conductors for the site distribution.

(D) Neutral Conductors. Neutral conductors shall be permitted to be reduced in size below the minimum required size of the ungrounded conductors for 240-volt, line-to-line, permanently connected loads only.

Informational Note: Due to the long circuit lengths typical in most recreational vehicle parks, feeder conductor sizes found in the ampacity tables of Article 310 may be inadequate to maintain the voltage regulation suggested in 215.2(A)(1) Informational Note No. 2. Total circuit voltage drop is a sum of the voltage drops of each serial circuit segment, where the load for each segment is calculated using the load that segment sees and the demand factors shown in Table 551.73(A).

551.73 Calculated Load.

(A) Basis of Calculations. Electrical services and feeders shall be calculated on the basis of not less than 12,000 volt-amperes per site equipped with 50-ampere, 208Y/120 or 120/240-volt supply facilities; 3600 volt-amperes per site equipped with both 20-ampere and 30-ampere supply facilities; 2400 volt-amperes per site equipped with only 20-ampere supply facilities; and 600 volt-amperes per site equipped with only 20-ampere supply facilities that are dedicated to tent sites. The demand factors set forth in Table 551.73(A) shall be the minimum allowable demand factors that shall be permitted in calculating load for service and feeders. Where the electrical supply for a recreational vehicle site has more than one receptacle, the calculated load shall be calculated only for the highest rated receptacle.

Where the electrical supply is in a location that serves two recreational vehicles, the equipment for both sites shall comply with 551.77, and the calculated load shall only be calculated for the two receptacles with the highest rating.

(B) Demand Factors. The demand factor for a given number of sites shall apply to all sites indicated. For example, 20 sites calculated at 45 percent of 3600 volt-amperes results in a permissible demand of 1620 volt-amperes per site or a total of 32,400 volt-amperes for 20 sites.

> Informational Note: These demand factors may be inadequate in areas of extreme hot or cold temperature with loaded circuits for heating or air conditioning.

Loads for other amenities such as, but not limited to, service buildings, recreational buildings, and swimming pools shall be calculated separately and then be added to the value calculated for the recreational vehicle sites where they are all supplied by a common service.

551.74 Overcurrent Protection. Overcurrent protection shall be provided in accordance with Article 240.

Table 551.73(A) Demand Factors for Site Feeders and Service-Entrance Conductors for Park Sites

Number of Recreational Vehicle Sites	Demand Factor (%)
1	100
2	90
3	80
4	75
5	65
6	60
7–9	55
10–12	50
13–15	48
16–18	47
19–21	45
22–24	43
25–35	42
36 plus	41

551.75 Grounding.

(A) General. All electrical equipment and installations in recreational vehicle parks shall be grounded as required by Article 250.

N **(B) Grounding Electrode.** Power outlets or recreational vehicle site supply equipment, other than those used as service equipment, shall not be required to have a grounding electrode. An auxiliary grounding electrode(s) in accordance with 250.54 shall be permitted to be installed.

551.76 Grounding — Recreational Vehicle Site Supply Equipment.

(A) Exposed Non–Current-Carrying Metal Parts. Exposed non–current-carrying metal parts of fixed equipment, metal boxes, cabinets, and fittings that are not electrically connected to grounded equipment shall be grounded by an equipment grounding conductor run with the circuit conductors from the service equipment or from the transformer of a secondary distribution system. Equipment grounding conductors shall be sized in accordance with 250.122 and shall be permitted to be spliced by listed means.

The arrangement of equipment grounding connections shall be such that the disconnection or removal of a receptacle or other device will not interfere with, or interrupt, the grounding continuity.

(B) Secondary Distribution System. Each secondary distribution system shall be grounded at the transformer.

(C) Grounded Conductor Not to Be Used as an Equipment Ground. The grounded conductor shall not be used as an equipment grounding conductor for recreational vehicles or equipment within the recreational vehicle park.

(D) No Connection on the Load Side. No connection to a grounding electrode shall be made to the grounded conductor on the load side of the service disconnecting means except as covered in 250.30(A) for separately derived systems, and 250.32(B) Exception No. 1 for separate buildings.

551.77 Recreational Vehicle Site Supply Equipment.

(A) Location. Where provided on back-in sites, the recreational vehicle site electrical supply equipment shall be located on the left (road) side of the parked vehicle, on a line that is 1.5 m to 2.1 m (5 ft to 7 ft) from the left edge (driver's side of the parked RV) of the stand and shall be located at any point on this line from the rear of the stand to 4.5 m (15 ft) forward of the rear of the stand.

For pull-through sites, the electrical supply equipment shall be permitted to be located at any point along the line that is 1.5 m to 2.1 m (5 ft to 7 ft) from the left edge (driver's side of the parked RV) from 4.9 m (16 ft) forward of the rear of the stand to the center point between the two roads that gives access to and egress from the pull-through sites.

The left edge (driver's side of the parked RV) of the stand shall be marked.

(B) Disconnecting Means. A disconnecting switch or circuit breaker shall be provided in the site supply equipment for disconnecting the power supply to the recreational vehicle.

(C) Access. All site supply equipment shall be accessible by an unobstructed entrance or passageway not less than 600 mm (2 ft) wide and 2.0 m (6 ft 6 in.) high.

(D) Mounting Height. Site supply equipment shall be located not less than 600 mm (2 ft) or more than 2.0 m (6 ft 6 in.) above the ground.

(E) Working Space. Sufficient space shall be provided and maintained about all electrical equipment to permit ready and safe operation, in accordance with 110.26.

(F) Marking. Where the site supply equipment contains a 125/250-volt receptacle, the equipment shall be marked as follows: "Turn disconnecting switch or circuit breaker off before inserting or removing plug. Plug must be fully inserted or removed." The marking shall be located on the equipment adjacent to the receptacle outlet.

551.78 Protection of Outdoor Equipment.

(A) Wet Locations. All switches, circuit breakers, receptacles, control equipment, and metering devices located in wet locations shall be weatherproof.

(B) Meters. If secondary meters are installed, meter sockets without meters installed shall be blanked off with an approved blanking plate.

551.79 Clearance for Overhead Conductors. Open conductors of not over 1000 volts, nominal, shall have a vertical clearance of not less than 5.5 m (18 ft) and a horizontal clearance of not less than 900 mm (3 ft) in all areas subject to recreational vehicle movement. In all other areas, clearances shall conform to 225.18 and 225.19.

Informational Note: For clearances of conductors over 600 volts, nominal, see 225.60 and 225.61.

551.80 Underground Service, Feeder, Branch-Circuit, and Recreational Vehicle Site Feeder-Circuit Conductors.

(A) General. All direct-burial conductors, including the equipment grounding conductor if of aluminum, shall be insulated and identified for the use. All conductors shall be continuous from equipment to equipment. All splices and taps shall be made in approved junction boxes or by use of listed material.

(B) Protection Against Physical Damage. Direct-buried conductors and cables entering or leaving a trench shall be protected by rigid metal conduit, intermediate metal conduit, electrical metallic tubing with supplementary corrosion protection, rigid polyvinyl chloride conduit (PVC), nonmetallic underground conduit with conductors (NUCC), high density polyethylene conduit (HDPE), reinforced thermosetting resin conduit (RTRC), liquidtight flexible nonmetallic conduit, liquidtight flexible metal conduit, or other approved raceways or enclosures. Where subject to physical damage, the conductors or cables shall be protected by rigid metal conduit, intermediate metal conduit, Schedule 80 PVC conduit, or RTRC listed for exposure to physical damage. All such protection shall extend at least 450 mm (18 in.) into the trench from finished grade.

Informational Note: See 300.5 and Article 340 for conductors or Type UF cable used underground or in direct burial in earth.

551.81 Receptacles. A receptacle to supply electric power to a recreational vehicle shall be one of the configurations shown in Figure 551.46(C)(1) in the following ratings:

(1) 50-ampere — 125/250-volt, 50-ampere, 3-pole, 4-wire grounding type for 120/240-volt systems

(2) 30-ampere — 125-volt, 30-ampere, 2-pole, 3-wire grounding type for 120-volt systems

(3) 20-ampere — 125-volt, 20-ampere, 2-pole, 3-wire grounding type for 120-volt systems

Informational Note: Complete details of these configurations can be found in ANSI/NEMA WD 6-2002, National Electrical Manufacturers Association's *Standard for Dimensions of Attachment Plugs and Receptacles*, Figures 14-50, TT, and 5-20.

ARTICLE 552
Park Trailers

Part I. General

552.1 Scope. The provisions of this article cover the electrical conductors and equipment installed within or on park trailers not covered fully under Articles 550 and 551.

552.2 Definition. (See Articles 100, 550, and 551 for additional definitions.)

Park Trailer. A unit that is built on a single chassis mounted on wheels and has a gross trailer area not exceeding 37 m² (400 ft²) in the set-up mode.

552.4 General Requirements. A park trailer as specified in 552.2 is intended for seasonal use. It is not intended as a permanent dwelling unit or for commercial uses such as banks, clinics, offices, or similar.

N **552.5 Labels.** Labels required by Article 552 shall be made of etched, metal-stamped, or embossed brass or stainless steel; plastic laminates not less than 0.13 mm (0.005 in.) thick; or anodized or alclad aluminum not less than 0.5 mm (0.020 in.) thick or the equivalent.

Informational Note: For guidance on other label criteria used in the park trailer industry, refer to ANSI Z535.4-2011, *Product Safety Signs and Labels.*

Part II. Low-Voltage Systems

552.10 Low-Voltage Systems.

(A) Low-Voltage Circuits. Low-voltage circuits furnished and installed by the park trailer manufacturer, other than those related to braking, shall be subject to this *Code*. Circuits supplying lights subject to federal or state regulations shall comply with applicable government regulations and this *Code*.

(B) Low-Voltage Wiring.

(1) Material. Copper conductors shall be used for low-voltage circuits.

Exception: A metal chassis or frame shall be permitted as the return path to the source of supply.

(2) Conductor Types. Conductors shall conform to the requirements for Type GXL, HDT, SGT, SGR, or Type SXL or shall have insulation in accordance with Table 310.104(A) or the equivalent. Conductor sizes 6 AWG through 18 AWG or SAE shall be listed. Single-wire, low-voltage conductors shall be of the stranded type.

Informational Note: See SAE J1128-2011, *Low Tension Primary Cable,* for Types GXL, HDT, and SXL, and SAE J1127-2010, *Battery Cable,* for Types SGT and SGR.

(3) Marking. All insulated low-voltage conductors shall be surface marked at intervals not greater than 1.2 m (4 ft) as follows:

(1) Listed conductors shall be marked as required by the listing agency.

(2) SAE conductors shall be marked with the name or logo of the manufacturer, specification designation, and wire gauge.

(3) Other conductors shall be marked with the name or logo of the manufacturer, temperature rating, wire gauge, conductor material, and insulation thickness.

(C) Low-Voltage Wiring Methods.

(1) Physical Protection. Conductors shall be protected against physical damage and shall be secured. Where insulated conductors are clamped to the structure, the conductor insulation shall be supplemented by an additional wrap or layer of equivalent material, except that jacketed cables shall not be required to be so protected. Wiring shall be routed away from sharp edges, moving parts, or heat sources.

(2) Splices. Conductors shall be spliced or joined with splicing devices that provide a secure connection or by brazing, welding, or soldering with a fusible metal or alloy. Soldered splices shall first be spliced or joined to be mechanically and electrically secure without solder, and then soldered. All splices, joints, and free ends of conductors shall be covered with an insulation equivalent to that on the conductors.

(3) Separation. Battery and other low-voltage circuits shall be physically separated by at least a 13-mm (½-in.) gap or other approved means from circuits of a different power source. Acceptable methods shall be by clamping, routing, or equivalent means that ensure permanent total separation. Where circuits of different power sources cross, the external jacket of the nonmetallic-sheathed cables shall be deemed adequate separation.

(4) Ground Connections. Ground connections to the chassis or frame shall be made in an accessible location and shall be mechanically secure. Ground connections shall be by means of copper conductors and copper or copper-alloy terminals of the solderless type identified for the size of wire used. The surface on which ground terminals make contact shall be cleaned and be free from oxide or paint or shall be electrically connected through the use of a cadmium, tin, or zinc-plated internal/external-toothed lockwasher or locking terminals. Ground terminal attaching screws, rivets or bolts, nuts, and lockwashers shall be cadmium, tin, or zinc-plated except rivets shall be permitted to be unanodized aluminum where attaching to aluminum structures.

The chassis-grounding terminal of the battery shall be connected to the unit chassis with a minimum 8 AWG copper conductor. In the event the unbonded lead from the battery exceeds 8 AWG, the bonding conductor size shall be not less than that of the unbonded lead.

(D) Battery Installations. Storage batteries subject to the provisions of this *Code* shall be securely attached to the unit and installed in an area vaportight to the interior and ventilated directly to the exterior of the unit. Where batteries are installed in a compartment, the compartment shall be ventilated with openings having a minimum area of 1100 mm² (1.7 in.²) at both the top and at the bottom. Where compartment doors are equipped for ventilation, the openings shall be within 50 mm (2 in.) of the top and bottom. Batteries shall not be installed in a compartment containing spark- or flame-producing equipment.

(E) Overcurrent Protection.

(1) Rating. Low-voltage circuit wiring shall be protected by overcurrent protective devices rated not in excess of the ampacity of copper conductors, in accordance with Table 552.10(E)(1).

(2) Type. Circuit breakers or fuses shall be of an approved type, including automotive types. Fuseholders shall be clearly marked with maximum fuse size and shall be protected against shorting and physical damage by a cover or equivalent means.

> Informational Note: For further information, see ANSI/SAE J554-1987, *Standard for Electric Fuses (Cartridge Type)*; SAE J1284-1988, *Standard for Blade Type Electric Fuses*; and UL 275-2005, *Standard for Automotive Glass Tube Fuses*.

(3) Appliances. Appliances such as pumps, compressors, heater blowers, and similar motor-driven appliances shall be installed in accordance with the manufacturer's instructions.

Motors that are controlled by automatic switching or by latching-type manual switches shall be protected in accordance with 430.32(B).

(4) Location. The overcurrent protective device shall be installed in an accessible location on the unit within 450 mm (18 in.) of the point where the power supply connects to the unit circuits. If located outside the park trailer, the device shall be protected against weather and physical damage.

Exception: External low-voltage supply shall be permitted to have the overcurrent protective device within 450 mm (18 in.) after entering the unit or after leaving a metal raceway.

(F) Switches. Switches shall have a dc rating not less than the connected load.

(G) Luminaires. All low-voltage interior luminaires rated more than 4 watts, employing lamps rated more than 1.2 watts, shall be listed.

Table 552.10(E)(1) Low-Voltage Overcurrent Protection

Wire Size (AWG)	Ampacity	Wire Type
18	6	Stranded only
16	8	Stranded only
14	15	Stranded or solid
12	20	Stranded or solid
10	30	Stranded or solid

Part III. Combination Electrical Systems

552.20 Combination Electrical Systems.

(A) General. Unit wiring suitable for connection to a battery or other low-voltage supply source shall be permitted to be connected to a 120-volt source, provided that the entire wiring system and equipment are rated and installed in full conformity with Parts I, III, IV, and V requirements of this article covering 120-volt electrical systems. Circuits fed from ac transformers shall not supply dc appliances.

(B) Voltage Converters (120-Volt Alternating Current to Low-Voltage Direct Current). The 120-volt ac side of the voltage converter shall be wired in full conformity with the requirements of Parts I and IV of this article for 120-volt electrical systems.

Exception: Converters supplied as an integral part of a listed appliance shall not be subject to 552.20(B).

All converters and transformers shall be listed for use in recreation units and designed or equipped to provide over-temperature protection. To determine the converter rating, the following percentages shall be applied to the total connected load, including average battery-charging rate, of all 12-volt equipment:

The first 20 amperes of load at 100 percent plus

The second 20 amperes of load at 50 percent plus

All load above 40 amperes at 25 percent

Exception: A low-voltage appliance that is controlled by a momentary switch (normally open) that has no means for holding in the closed position shall not be considered as a connected load when determining the required converter rating. Momentarily energized appliances shall be limited to those used to prepare the unit for occupancy or travel.

(C) Bonding Voltage Converter Enclosures. The non–current-carrying metal enclosure of the voltage converter shall be connected to the frame of the unit with an 8 AWG copper conductor minimum. The grounding conductor for the battery and the metal enclosure shall be permitted to be the same conductor.

(D) Dual-Voltage Fixtures Including Luminaires or Appliances. Fixtures, including luminaires, or appliances having both 120-volt and low-voltage connections shall be listed for dual voltage.

(E) Autotransformers. Autotransformers shall not be used.

(F) Receptacles and Plug Caps. Where a park trailer is equipped with a 120-volt or 120/240-volt ac system, a low-voltage system, or both, receptacles and plug caps of the low-voltage system shall differ in configuration from those of the 120-volt or 120/240-volt system. Where a unit equipped with a battery or dc system has an external connection for low-voltage power, the connector shall have a configuration that will not accept 120-volt power.

Part IV. Nominal 120-Volt or 120/240-Volt Systems

552.40 120-Volt or 120/240-Volt, Nominal, Systems.

(A) General Requirements. The electrical equipment and material of park trailers indicated for connection to a wiring system rated 120 volts, nominal, 2-wire with an equipment grounding conductor, or a wiring system rated 120/240 volts, nominal, 3-wire with an equipment grounding conductor, shall be listed and installed in accordance with the requirements of Parts I, III, IV, and V of this article.

(B) Materials and Equipment. Electrical materials, devices, appliances, fittings, and other equipment installed, intended for use in, or attached to the park trailer shall be listed. All products shall be used only in the manner in which they have been tested and found suitable for the intended use.

552.41 Receptacle Outlets Required.

(A) Spacing. Receptacle outlets shall be installed at wall spaces 600 mm (2 ft) wide or more so that no point along the floor line is more than 1.8 m (6 ft), measured horizontally, from an outlet in that space.

Exception No. 1: Bath and hallway areas.

Exception No. 2: Wall spaces occupied by kitchen cabinets, wardrobe cabinets, built-in furniture; behind doors that may open fully against a wall surface; or similar facilities.

(B) Location. Receptacle outlets shall be installed as follows:

(1) Adjacent to countertops in the kitchen [at least one on each side of the sink if countertops are on each side and are 300 mm (12 in.) or over in width and depth]

(2) Adjacent to the refrigerator and gas range space, except where a gas-fired refrigerator or cooking appliance, requiring no external electrical connection, is factory-installed

(3) Adjacent to countertop spaces of 300 mm (12 in.) or more in width and depth that cannot be reached from a receptacle required in 552.41(B)(1) by a cord of 1.8 m (6 ft) without crossing a traffic area, cooking appliance, or sink

(C) Ground-Fault Circuit-Interrupter Protection. Each 125-volt, single-phase, 15- or 20-ampere receptacle shall have ground-fault circuit-interrupter protection for personnel in the following locations:

(1) Where the receptacles are installed to serve kitchen countertop surfaces

(2) Within 1.8 m (6 ft) of any lavatory or sink

Exception: Receptacles installed for appliances in dedicated spaces, such as for dishwashers, disposals, refrigerators, freezers, and laundry equipment.

(3) In the area occupied by a toilet, shower, tub, or any combination thereof

(4) On the exterior of the unit

Exception: Receptacles that are located inside of an access panel that is installed on the exterior of the unit to supply power for an installed appliance shall not be required to have ground-fault circuit-interrupter protection.

The receptacle outlet shall be permitted in a listed luminaire. A receptacle outlet shall not be installed in a tub or combination tub–shower compartment.

(D) Pipe Heating Cable Outlet. Where a pipe heating cable outlet is installed, the outlet shall be as follows:

(1) Located within 600 mm (2 ft) of the cold water inlet

(2) Connected to an interior branch circuit, other than a small-appliance branch circuit

(3) On a circuit where all of the outlets are on the load side of the ground-fault circuit-interrupter protection for personnel

(4) Mounted on the underside of the park trailer and shall not be considered to be the outdoor receptacle outlet required in 552.41(E)

(E) Outdoor Receptacle Outlets. At least one receptacle outlet shall be installed outdoors. A receptacle outlet located in a compartment accessible from the outside of the park trailer shall be considered an outdoor receptacle. Outdoor receptacle outlets shall be protected as required in 552.41(C)(4).

(F) Receptacle Outlets Not Permitted.

(1) Shower or Bathtub Space. Receptacle outlets shall not be installed in or within reach [750 mm (30 in.)] of a shower or bathtub space.

(2) Face-Up Position. A receptacle shall not be installed in a face-up position in any countertop or other similar horizontal surface.

N 552.42 Branch-Circuit Protection.

(A) Rating. The branch-circuit overcurrent devices shall be rated as follows:

(1) Not more than the circuit conductors

(2) Not more than 150 percent of the rating of a single appliance rated 13.3 amperes or more and supplied by an individual branch circuit

(3) Not more than the overcurrent protection size marked on an air conditioner or other motor-operated appliances.

(B) Protection for Smaller Conductors. A 20-ampere fuse or circuit breaker shall be permitted for protection for fixtures, including luminaires, leads, cords, or small appliances, and 14 AWG tap conductors, not over 1.8 m (6 ft) long for recessed luminaires.

(C) Fifteen-Ampere Receptacle Considered Protected by 20 Amperes. If more than one receptacle or load is on a branch circuit, 15-ampere receptacles shall be permitted to be protected by a 20-ampere fuse or circuit breaker.

552.43 Power Supply.

(A) Feeder. The power supply to the park trailer shall be a feeder assembly consisting of not more than one listed 30-ampere or 50-ampere park trailer power-supply cord, with an integrally molded or securely attached cap, or a permanently installed feeder.

(B) Power-Supply Cord. If the park trailer has a power-supply cord, it shall be permanently attached to the panelboard, or to a junction box permanently connected to the panelboard, with the free end terminating in a molded-on attachment plug cap.

Cords with adapters and pigtail ends, extension cords, and similar items shall not be attached to, or shipped with, a park trailer.

A suitable clamp or the equivalent shall be provided at the panelboard knockout to afford strain relief for the cord to prevent strain from being transmitted to the terminals when the power-supply cord is handled in its intended manner.

The cord shall be a listed type with 3-wire, 120-volt or 4-wire, 120/240-volt conductors, one of which shall be identified by a continuous green color or a continuous green color with one or more yellow stripes for use as the grounding conductor.

(C) Mast Weatherhead or Raceway. Where the calculated load exceeds 50 amperes or where a permanent feeder is used, the supply shall be by means of one of the following:

(1) One mast weatherhead installation, installed in accordance with Article 230, containing four continuous, insulated, color-coded feeder conductors, one of which shall be an equipment grounding conductor

(2) A metal raceway, rigid nonmetallic conduit, or liquidtight flexible conduit from the disconnecting means in the park trailer to the underside of the park trailer

552.44 Cord.

(A) Permanently Connected. Each power-supply assembly shall be factory supplied or factory installed and connected directly to the terminals of the panelboard or conductors within a junction box and provided with means to prevent strain from being transmitted to the terminals. The ampacity of the conductors between each junction box and the terminals of each panelboard shall be at least equal to the ampacity of the power-supply cord. The supply end of the assembly shall be equipped with an attachment plug of the type described in 552.44(C). Where the cord passes through the walls or floors, it shall be protected by means of conduit and bushings or equivalent. The cord assembly shall have permanent provisions for protection against corrosion and mechanical damage while the unit is in transit.

(B) Cord Length. The cord-exposed usable length shall be measured from the point of entrance to the park trailer or the face of the flanged surface inlet (motor-base attachment plug) to the face of the attachment plug at the supply end.

The cord-exposed usable length, measured to the point of entry on the unit exterior, shall be a minimum of 7.0 m (23 ft) where the point of entrance is at the side of the unit, or shall be a minimum 8.5 m (28 ft) where the point of entrance is at the rear of the unit. The maximum length shall not exceed 11 m (36½ ft).

Where the cord entrance into the unit is more than 900 mm (3 ft) above the ground, the minimum cord lengths above shall be increased by the vertical distance of the cord entrance heights above 900 mm (3 ft).

(C) Attachment Plugs.

(1) Units with Two to Five 15- or 20-Ampere Branch Circuits. Park trailers wired in accordance with 552.46(A) shall have an attachment plug that shall be 2-pole, 3-wire grounding type, rated 30 amperes, 125 volts, conforming to the configuration shown in Figure 552.44(C)(1) intended for use with units rated at 30 amperes, 125 volts.

Informational Note: Complete details of this configuration can be found in ANSI/NEMA WD 6-2002 (Rev. 2008), *Standard for Dimensions of Attachment Plugs and Receptacles*, Figure TT.

(2) Units with 50-Ampere Power Supply Assembly. Park trailers having a power-supply assembly rated 50 amperes as permitted by 552.43(B) shall have a 3-pole, 4-wire grounding-type attachment plug rated 50 amperes, 125/250 volts, conforming to the configuration shown in Figure 552.44(C)(1).

Informational Note: Complete details of this configuration can be found in ANSI/NEMA WD 6-2002 (Rev. 2008), *Standard for Dimensions of Attachment Plugs and Receptacles*, Figure 14-50.

(D) Labeling at Electrical Entrance. Each park trailer shall have a safety label with the signal word WARNING in minimum 6 mm (¼ in.) high letters and body text in minimum 3 mm (⅛ in.) high letters on a contrasting background. The safety label shall be affixed to the exterior skin, at or near the point of entrance of the power-supply assembly and shall read, as appropriate:

THIS CONNECTION IS FOR 110–125-VOLT AC,
60 HZ, 30 AMPERE SUPPLY

or

THIS CONNECTION IS FOR 208Y/120-VOLT OR 120/240-VOLT AC, 3-POLE, 4-WIRE, 60 HZ, _____ AMPERE SUPPLY.

followed by

DO NOT EXCEED THE CIRCUIT RATING. EXCEEDING
THE CIRCUIT RATING MAY CAUSE A FIRE AND RESULT IN
DEATH OR SERIOUS INJURY.

The correct ampere rating shall be marked in the blank space.

(E) Location. The point of entrance of a power-supply assembly shall be located on either side or the rear, within 450 mm (18 in.), of an outside wall.

552.45 Panelboard.

(A) Listed and Appropriately Rated. A listed and appropriately rated panelboard shall be used. The grounded conductor termination bar shall be insulated from the enclosure as provided in 552.55(C). An equipment grounding terminal bar shall be attached inside the metal enclosure of the panelboard.

(B) Location. The panelboard shall be installed in a readily accessible location. Working clearance for the panelboard shall be not less than 600 mm (24 in.) wide and 750 mm (30 in.) deep.

Exception: Where the panelboard cover is exposed to the inside aisle space, one of the working clearance dimensions shall be permitted to be reduced to a minimum of 550 mm (22 in.). A panelboard shall be considered exposed where the panelboard cover is within 50 mm (2 in.) of the aisle's finished surface or not more than 25 mm (1 in.) from the backside of doors that enclose the space.

Receptacles	Caps

30-A,125-V, 2-pole, 3-wire, grounding type

50-A,125/250-V, 3-pole, 4-wire, grounding type

FIGURE 552.44(C)(1)   Attachment Cap and Receptacle Configurations.

(C) Dead-Front Type. The panelboard shall be of the dead-front type. A main disconnecting means shall be provided where fuses are used or where more than two circuit breakers are employed. A main overcurrent protective device not exceeding the power-supply assembly rating shall be provided where more than two branch circuits are employed.

552.46 Branch Circuits. Branch circuits shall be determined in accordance with 552.46(A) and (B).

(A) Two to Five 15- or 20-Ampere Circuits. A maximum of five 15- or 20-ampere circuits to supply lights, receptacle outlets, and fixed appliances shall be permitted. Such park trailers shall be permitted to be equipped with panelboards rated at 120 volt maximum or 120/240 volt maximum and listed for a 30-ampere-rated main power supply assembly. Not more than two 120-volt thermostatically controlled appliances shall be installed in such systems unless appliance isolation switching, energy management systems, or similar methods are used.

Exception No. 1: Additional 15- or 20-ampere circuits shall be permitted where a listed energy management system rated at 30 amperes maximum is employed within the system.

Exception No. 2: Six 15- or 20-ampere circuits shall be permitted without employing an energy management system, provided that the added sixth circuit serves only the power converter, and the combined load of all six circuits does not exceed the allowable load that was designed for use by the original five circuits.

Informational Note: See 210.23(A) for permissible loads. See 552.45(C) for main disconnect and overcurrent protection requirements.

(B) More Than Five Circuits. Where more than five circuits are needed, they shall be determined in accordance with 552.46(B)(1), (B)(2), and (B)(3).

(1) Lighting. Based on 33 volt-amperes/m² (3 VA/ft²) multiplied by the outside dimensions of the park trailer (coupler excluded) divided by 120 volts to determine the number of 15- or 20-ampere lighting area circuits, for example,

$$[552.46(B)(1)]$$

$$\frac{3 \times \text{length} \times \text{width}}{120 \times 15 \ (\text{or } 20)}$$

$$= \text{No. of 15- (or 20-) ampere circuits}$$

The lighting circuits shall be permitted to serve listed cord-connected kitchen waste disposers and to provide power for supplemental equipment and lighting on gas-fired ranges, ovens, or counter-mounted cooking units.

(2) Small Appliances. Small-appliance branch circuits shall be installed in accordance with 210.11(C)(1).

(3) General Appliances. (including furnace, water heater, space heater, range, and central or room air conditioner, etc.) An individual branch circuit shall be permitted to supply any load for which it is rated. There shall be one or more circuits of adequate rating in accordance with (a) through (d).

Informational Note No. 1: For the laundry branch circuit, see 210.11(C)(2).

Informational Note No. 2: For central air conditioning, see Article 440.

(a) The total rating of fixed appliances shall not exceed 50 percent of the circuit rating if lighting outlets, general-use receptacles, or both are also supplied.

(b) For fixed appliances with a motor(s) larger than ⅛ horsepower, the total calculated load shall be based on 125 percent of the largest motor plus the sum of the other loads. Where a branch circuit supplies continuous load(s) or any combination of continuous and noncontinuous loads, the branch-circuit conductor size shall be in accordance with 210.19(A).

(c) The rating of a single cord-and-plug-connected appliance supplied by other than an individual branch circuit shall not exceed 80 percent of the circuit rating.

(d) The rating of a range branch circuit shall be based on the range demand as specified for ranges in 552.47(B)(5).

552.47 Calculations. The following method shall be employed in computing the supply-cord and distribution-panelboard load for each feeder assembly for each park trailer in lieu of the procedure shown in Article 220 and shall be based on a 3-wire, 208Y/120-volt or 120/240-volt supply with 120-volt loads balanced between the two phases of the 3-wire system.

(A) Lighting and Small-Appliance Load. Lighting Volt-Amperes: Length times width of park trailer floor (outside dimensions) times 33 volt-amperes/m² (3 VA/ft²). For example,

$$\text{Length} \times \text{width} \times 3 = \text{lighting volt-amperes}$$

Small-Appliance Volt-Amperes: Number of circuits times 1500 volt-amperes for each 20-ampere appliance receptacle circuit (see definition of *Appliance, Portable* with fine print note) including 1500 volt-amperes for laundry circuit. For example,

$$\text{No. of circuits} \times 1500 = \text{small appliance volt-amperes}$$

Total. Lighting volt-amperes plus small-appliance volt-amperes = total volt-amperes

First 3000 total volt-amperes at 100 percent plus remainder at 35 percent = volt-amperes to be divided by 240 volts to obtain current (amperes) per leg.

(B) Total Load for Determining Power Supply. Total load for determining power supply is the sum of the following:

(1) Lighting and small-appliance load as calculated in 552.47(A).

(2) Nameplate amperes for motors and heater loads (exhaust fans, air conditioners, electric, gas, or oil heating). Omit smaller of the heating and cooling loads, except include blower motor if used as air-conditioner evaporator motor. Where an air conditioner is not installed and a 50-ampere power-supply cord is provided, allow 15 amperes per phase for air conditioning.

(3) Twenty-five percent of current of largest motor in (B)(2).

(4) Total of nameplate amperes for disposal, dishwasher, water heater, clothes dryer, wall-mounted oven, cooking units. Where the number of these appliances exceeds three, use 75 percent of total.

(5) Derive amperes for freestanding range (as distinguished from separate ovens and cooking units) by dividing the following values by 240 volts as shown in the table below:

(6) If outlets or circuits are provided for other than factory-installed appliances, include the anticipated load.

Informational Note: Refer to Informative Annex D, Example D12, for an illustration of the application of this calculation.

Nameplate Rating (watts)	Use (volt-amperes)
0–10,000	80 percent of rating
Over 10,000–12,500	8,000
Over 12,500–13,500	8,400
Over 13,500–14,500	8,800
Over 14,500–15,500	9,200
Over 15,500–16,500	9,600
Over 16,500–17,500	10,000

(C) Optional Method of Calculation for Lighting and Appliance Load. For park trailers, the optional method for calculating lighting and appliance load shown in 220.82 shall be permitted.

552.48 Wiring Methods.

(A) Wiring Systems. Cables and raceways installed in accordance with Articles 320, 322, 330 through 340, 342 through 362, 386, and 388 shall be permitted in accordance with their applicable article, except as otherwise specified in this article. An equipment grounding means shall be provided in accordance with 250.118.

(B) Conduit and Tubing. Where rigid metal conduit or intermediate metal conduit is terminated at an enclosure with a locknut and bushing connection, two locknuts shall be provided, one inside and one outside of the enclosure. All cut ends of conduit and tubing shall be reamed or otherwise finished to remove rough edges.

(C) Nonmetallic Boxes. Nonmetallic boxes shall be acceptable only with nonmetallic-sheathed cable or nonmetallic raceways.

(D) Boxes. In walls and ceilings constructed of wood or other combustible material, boxes and fittings shall be flush with the finished surface or project therefrom.

(E) Mounting. Wall and ceiling boxes shall be mounted in accordance with Article 314.

Exception No. 1: Snap-in-type boxes or boxes provided with special wall or ceiling brackets that securely fasten boxes in walls or ceilings shall be permitted.

Exception No. 2: A wooden plate providing a 38-mm (1½-in.) minimum width backing around the box and of a thickness of 13 mm (½ in.) or greater (actual) attached directly to the wall panel shall be considered as approved means for mounting outlet boxes.

(F) Cable Sheath. The sheath of nonmetallic-sheathed cable, and the armor of metal-clad cable and Type AC cable, shall be continuous between outlet boxes and other enclosures.

(G) Protected. Metal-clad, Type AC, or nonmetallic-sheathed cables and electrical nonmetallic tubing shall be permitted to pass through the centers of the wide side of 2 by 4 wood studs. However, they shall be protected where they pass through 2 by 2 wood studs or at other wood studs or frames where the cable or tubing would be less than 32 mm (1¼ in.) from the inside or outside surface. Steel plates on each side of the cable or tubing, or a steel tube, with not less than 1.35 mm (0.053 in.) wall thickness, shall be installed to protect the cable or tubing. These plates or tubes shall be securely held in place. Where nonmetallic-sheathed cables pass through punched, cut, or drilled slots or holes in metal members, the cable shall be protected by bushings or grommets securely fastened in the opening prior to installation of the cable.

(H) Cable Supports. Where connected with cable connectors or clamps, cables shall be secured and supported within 300 mm (12 in.) of outlet boxes, panelboards, and splice boxes on appliances. Supports and securing shall be provided at intervals not exceeding 1.4 m (4½ ft) at other places.

(I) Nonmetallic Box Without Cable Clamps. Nonmetallic-sheathed cables shall be secured and supported within 200 mm (8 in.) of a nonmetallic outlet box without cable clamps. Where wiring devices with integral enclosures are employed with a loop of extra cable to permit future replacement of the device, the cable loop shall be considered as an integral portion of the device.

(J) Physical Damage. Where subject to physical damage, exposed nonmetallic cable shall be protected by covering boards, guard strips, raceways, or other means.

(K) Receptacle Faceplates. Metal faceplates shall comply with 406.5(A). Nonmetallic faceplates shall comply with 406.5(C).

(L) Metal Faceplates Grounded. Where metal faceplates are used, they shall be grounded.

(M) Moisture or Physical Damage. Where outdoor or under-chassis wiring is 120 volts, nominal, or over and is exposed to moisture or physical damage, the wiring shall be protected by rigid metal conduit, by intermediate metal conduit, by electrical metallic tubing, by rigid nonmetallic conduit, or by Type MI cable that is closely routed against frames and equipment enclosures or other raceway or cable identified for the application.

(N) Component Interconnections. Fittings and connectors that are intended to be concealed at the time of assembly shall be listed and identified for the interconnection of building components. Such fittings and connectors shall be equal to the wiring method employed in insulation, temperature rise, and fault-current withstanding, and shall be capable of enduring the vibration and shock occurring in park trailers.

(O) Method of Connecting Expandable Units. The method of connecting expandable units to the main body of the park trailer shall comply with 552.48(O)(1) and 552.48(O)(2) as applicable.

(1) Cord-and-Plug Connected. Cord-and-plug connections shall comply with 552.48(O)(1)(a) through (O)(1)(d).

(a) The portion of a branch circuit that is installed in an expandable unit shall be permitted to be connected to the portion of the branch circuit in the main body of the vehicle by means of an attachment plug and cord listed for hard usage. The cord and its connections shall comply with all provisions of Article 400 and shall be considered as a permitted use under 400.10. Where the attachment plug and cord are located within the park trailer's interior, use of plastic thermoset or elastomer parallel cord Type SPT-3, SP-3, or SPE shall be permitted.

(b) Where the receptacle provided for connection of the cord to the main circuit is located on the outside of the park trailer, it shall be protected with a ground-fault circuit interrupter for personnel and be listed for wet locations. A cord located on the outside of a park trailer shall be identified for outdoor use.

(c) Unless removable or stored within the park trailer interior, the cord assembly shall have permanent provisions for protection against corrosion and mechanical damage while the park trailer is in transit.

(d) The attachment plug and cord shall be installed so as not to permit exposed live attachment plug pins.

N **(2) Direct Wires Connected.** That portion of a branch circuit that is installed in an expandable unit shall be permitted to be connected to the portion of the branch circuit in the main body of the park trailer by means of flexible cord installed in accordance with 552.48(O)(2)(a) through (O)(2)(f) or other approved wiring method.

(a) The flexible cord shall be listed for hard usage and for use in wet locations.

(b) The flexible cord shall be permitted to be exposed on the underside of the vehicle.

(c) The flexible cord shall be permitted to pass through the interior of a wall or floor assembly or both a maximum concealed length of 600 mm (24 in.) before terminating at an outlet or junction box.

(d) Where concealed, the flexible cord shall be installed in nonflexible conduit or tubing that is continuous from the outlet or junction box inside the park trailer to a weatherproof outlet box, junction box, or strain relief fitting listed for use in wet locations that is located on the underside of the park trailer. The outer jacket of flexible cord shall be continuous into the outlet or junction box.

(e) Where the flexible cord passes through the floor to an exposed area inside of the park trailer, it shall be protected by means of conduit and bushings or equivalent.

(f) Where subject to physical damage, the flexible cord shall be protected with RMC, IMC, Schedule 80 PVC, reinforced thermosetting resin conduit (RTRC) listed for exposure to physical damage, or other approved means and shall extend at least 150 mm (6 in.) above the floor. A means shall be provided to secure the flexible cord where it enters the park trailer.

(P) Prewiring for Air-Conditioning Installation. Prewiring installed for the purpose of facilitating future air-conditioning installation shall comply with the applicable portions of this article and the following:

(1) An overcurrent protective device with a rating compatible with the circuit conductors shall be installed in the panelboard and wiring connections completed.

(2) The load end of the circuit shall terminate in a junction box with a blank cover or other listed enclosure. Where a junction box with a blank cover is used, the free ends of the conductors shall be adequately capped or taped.

(3) A safety label with the word WARNING in minimum 6 mm (¼ in.) high letters and body text in minimum 3 mm (⅛ in.) high letters on a contrasting background shall be affixed on or adjacent to the junction box and shall read as follows:

WARNING
AIR-CONDITIONING CIRCUIT.
THIS CONNECTION IS FOR AIR CONDITIONERS

RATED 110–125-VOLT AC, 60 HZ, ____ AMPERES
MAXIMUM.
DO NOT EXCEED CIRCUIT RATING.

EXCEEDING THE CIRCUIT RATING MAY
CAUSE A FIRE AND RESULT IN
DEATH OR SERIOUS INJURY

An ampere rating not to exceed 80 percent of the circuit rating shall be legibly marked in the blank space.

(4) The circuit shall serve no other purpose.

N **(Q) Prewiring for Other Circuits.** Prewiring installed for the purpose of installing other appliances or devices shall comply with the applicable portions of this article and the following:

(1) An overcurrent protection device with a rating compatible with the circuit conductors shall be installed in the panelboard with wiring connections completed.

(2) The load end of the circuit shall terminate in a junction box with a blank cover or a device listed for the purpose. Where a junction box with blank cover is used, the free ends of the conductors shall be adequately capped or taped.

(3) A safety label with the signal word WARNING in minimum 6 mm (¼ in.) high letters and body text in minimum 3 mm (⅛ in.) high letters on a contrasting background shall be affixed on or adjacent to the junction box or device listed for the purpose and shall read as follows:

WARNING
THIS CONNECTION IS FOR ____ RATED ____ VOLT AC,
60 HZ, ____ AMPERES MAXIMUM. DO NOT EXCEED
CIRCUIT RATING. EXCEEDING THE CIRCUIT RATING
MAY CAUSE A FIRE AND RESULT IN DEATH OR SERIOUS
INJURY.

An ampere rating not to exceed 80 percent of the circuit rating shall be legibly marked in the blank space.

552.49 Maximum Number of Conductors in Boxes. The maximum number of conductors permitted in boxes shall be in accordance with 314.16.

552.50 Grounded Conductors. The identification of grounded conductors shall be in accordance with 200.6.

552.51 Connection of Terminals and Splices. Conductor splices and connections at terminals shall be in accordance with 110.14.

552.52 Switches. Switches shall be rated as required by 552.52(A) and (B).

(A) Lighting Circuits. For lighting circuits, switches shall be rated not less than 10 amperes, 120/125 volts, and in no case less than the connected load.

(B) Motors or Other Loads. For motors or other loads, switches shall have ampere or horsepower ratings, or both, adequate for loads controlled. (An ac general-use snap switch shall be permitted to control a motor 2 hp or less with full-load current not over 80 percent of the switch ampere rating.)

N **(C) Location.** Switches shall not be installed within wet locations in tub or shower spaces unless installed as part of a listed tub or shower assembly.

552.53 Receptacles. All receptacle outlets shall be of the grounding type and installed in accordance with 210.21 and 406.4.

552.54 Luminaires.

(A) General. Any combustible wall or ceiling finish exposed between the edge of a canopy or pan of a luminaire or ceiling suspended (paddle) fan and the outlet box shall be covered with noncombustible material or a material identified for the purpose.

(B) Shower Luminaires. If a luminaire is provided over a bathtub or in a shower stall, it shall be of the enclosed and gasketed type and listed for the type of installation, and it shall be ground-fault circuit-interrupter protected.

(C) Outdoor Outlets, Luminaires, Air-Cooling Equipment, and So On. Outdoor luminaires and other equipment shall be listed for outdoor use or wet locations.

552.55 Grounding. (See also 552.57 on bonding of non–current-carrying metal parts.)

(A) Power-Supply Grounding. The grounding conductor in the supply cord or feeder shall be connected to the grounding bus or other approved grounding means in the panelboard.

(B) Panelboard. The panelboard shall have a grounding bus with sufficient terminals for all grounding conductors or other approved grounding means.

(C) Insulated Grounded Conductor. The grounded circuit conductor shall be insulated from the equipment grounding conductors and from equipment enclosures and other grounded parts. The grounded circuit conductor terminals in the panelboard and in ranges, clothes dryers, counter-mounted cooking units, and wall-mounted ovens shall be insulated from the equipment enclosure. Bonding screws, straps, or buses in the panelboard or in appliances shall be removed and discarded. Connection of electric ranges and electric clothes dryers utilizing a grounded conductor, if cord-connected, shall be made with 4-conductor cord and 3-pole, 4-wire, grounding-type plug caps and receptacles.

552.56 Interior Equipment Grounding.

(A) Exposed Metal Parts. In the electrical system, all exposed metal parts, enclosures, frames, luminaire canopies, and so forth, shall be effectively bonded to the grounding terminals or enclosure of the panelboard.

(B) Equipment Grounding Conductors. Bare conductors or conductors with insulation or individual covering that is green or green with one or more yellow stripes shall be used for equipment grounding conductors only.

(C) Grounding of Electrical Equipment. Where grounding of electrical equipment is specified, it shall be permitted as follows:

(1) Connection of metal raceway (conduit or electrical metallic tubing), the sheath of Type MC and Type MI cable where the sheath is identified for grounding, or the armor of Type AC cable to metal enclosures.

(2) A connection between the one or more equipment grounding conductors and a metal box by means of a grounding screw, which shall be used for no other purpose, or a listed grounding device.

(3) The equipment grounding conductor in nonmetallic-sheathed cable shall be permitted to be secured under a screw threaded into the luminaire canopy other than a mounting screw or cover screw or attached to a listed grounding means (plate) in a nonmetallic outlet box for luminaire mounting (grounding means shall also be permitted for luminaire attachment screws).

(D) Grounding Connection in Nonmetallic Box. A connection between the one or more grounding conductors brought into a nonmetallic outlet box shall be arranged so that a connection can be made to any fitting or device in that box that requires grounding.

(E) Grounding Continuity. Where more than one equipment grounding conductor of a branch circuit enters a box, all such conductors shall be in good electrical contact with each other, and the arrangement shall be such that the disconnection or removal of a receptacle, fixture, including a luminaire, or other device fed from the box will not interfere with or interrupt the grounding continuity.

(F) Cord-Connected Appliances. Cord-connected appliances, such as washing machines, clothes dryers, refrigerators, and the electrical system of gas ranges, and so on, shall be grounded by means of an approved cord with equipment grounding conductor and grounding-type attachment plug.

552.57 Bonding of Non–Current-Carrying Metal Parts.

(A) Required Bonding. All exposed non–current-carrying metal parts that are likely to become energized shall be effectively bonded to the grounding terminal or enclosure of the panelboard.

(B) Bonding Chassis. A bonding conductor shall be connected between any panelboard and an accessible terminal on the chassis. Aluminum or copper-clad aluminum conductors shall not be used for bonding if such conductors or their terminals are exposed to corrosive elements.

Exception: Any park trailer that employs a unitized metal chassis-frame construction to which the panelboard is securely fastened with a bolt(s) and nut(s) or by welding or riveting shall be considered to be bonded.

(C) Bonding Conductor Requirements. Grounding terminals shall be of the solderless type and listed as pressure terminal connectors recognized for the wire size used. The bonding conductor shall be solid or stranded, insulated or bare, and shall be 8 AWG copper minimum or equivalent.

(D) Metallic Roof and Exterior Bonding. The metal roof and exterior covering shall be considered bonded where both of the following conditions apply:

(1) The metal panels overlap one another and are securely attached to the wood or metal frame parts by metal fasteners.

(2) The lower panel of the metal exterior covering is secured by metal fasteners at each cross member of the chassis, or the lower panel is connected to the chassis by a metal strap.

(E) Gas, Water, and Waste Pipe Bonding. The gas, water, and waste pipes shall be considered grounded if they are bonded to the chassis.

(F) Furnace and Metal Air Duct Bonding. Furnace and metal circulating air ducts shall be bonded.

552.58 Appliance Accessibility and Fastening. Every appliance shall be accessible for inspection, service, repair, and replacement without removal of permanent construction. Means shall be provided to securely fasten appliances in place when the park trailer is in transit.

552.59 Outdoor Outlets, Fixtures, Including Luminaires, Air-Cooling Equipment, and So On.

(A) Listed for Outdoor Use. Outdoor fixtures, including luminaires, and equipment shall be listed for outdoor use. Outdoor receptacle outlets shall be in accordance with 406.9(A) and (B). Switches and circuit breakers installed outdoors shall comply with 404.4.

(B) Outside Heating Equipment, Air-Conditioning Equipment, or Both. A park trailer provided with a branch circuit designed to energize outside heating equipment or air-conditioning equipment, or both, located outside the park trailer, other than room air conditioners, shall have such branch-circuit conductors terminate in a listed outlet box or disconnecting means located on the outside of the park trailer. A safety label with the word WARNING in minimum 6 mm (¼ in.) high letters and body text in minimum 3 mm (⅛ in.) high letters on a contrasting background shall be affixed within 150 mm (6 in.) from the listed box or disconnecting means and shall read as follows:

> WARNING
> THIS CONNECTION IS FOR HEATING
> AND/OR AIR-CONDITIONING EQUIPMENT.
> THE BRANCH CIRCUIT IS RATED AT NOT MORE THAN
> _____ AMPERES, AT _____ VOLTS, 60 HZ, _____
> CONDUCTOR AMPACITY.
> A DISCONNECTING MEANS SHALL BE
> LOCATED WITHIN SIGHT OF THE EQUIPMENT.
> EXCEEDING THE CIRCUIT RATING MAY CAUSE A FIRE
> AND RESULT IN DEATH OR SERIOUS INJURY.

The correct voltage and ampere rating shall be given.

Part V. Factory Tests

552.60 Factory Tests (Electrical). Each park trailer shall be subjected to the tests required by 552.60(A) and (B).

(A) Circuits of 120 Volts or 120/240 Volts. Each park trailer designed with a 120-volt or a 120/240-volt electrical system shall withstand the applied voltage without electrical breakdown of a 1 minute, 900-volt dielectric strength test, or a 1 second, 1080-volt dielectric strength test, with all switches closed, between ungrounded and grounded conductors and the park trailer ground. During the test, all switches and other controls shall be in the "on" position. Fixtures, including luminaires, and permanently installed appliances shall not be required to withstand this test.

Each park trailer shall be subjected to the following:

(1) A continuity test to ensure that all metal parts are properly bonded
(2) Operational tests to demonstrate that all equipment is properly connected and in working order
(3) Polarity checks to determine that connections have been properly made
(4) Receptacles requiring GFCI protection shall be tested for correct function by the use of a GFCI testing device

(B) Low-Voltage Circuits. An operational test of low-voltage circuits shall be conducted to demonstrate that all equipment is connected and in electrical working order. This test shall be performed in the final stages of production after all outer coverings and cabinetry have been secured.

ARTICLE 553
Floating Buildings

Part I. General

553.1 Scope. This article covers wiring, services, feeders, and grounding for floating buildings.

553.2 Definition.

Floating Building. A building unit, as defined in Article 100, that floats on water, is moored in a permanent location, and has a premises wiring system served through connection by permanent wiring to an electrical supply system not located on the premises.

Part II. Services and Feeders

553.4 Location of Service Equipment. The service equipment for a floating building shall be located adjacent to, but not in or on, the building or any floating structure. The main overcurrent protective device that feeds the floating structure shall have ground fault protection not exceeding 100 mA. Ground fault protection of each individual branch or feeder circuit shall be permitted as a suitable alternative.

553.5 Service Conductors. One set of service conductors shall be permitted to serve more than one set of service equipment.

553.6 Feeder Conductors. Each floating building shall be supplied by a single set of feeder conductors from its service equipment.

Exception: Where the floating building has multiple occupancy, each occupant shall be permitted to be supplied by a single set of feeder conductors extended from the occupant's service equipment to the occupant's panelboard.

553.7 Installation of Services and Feeders.

(A) Flexibility. Flexibility of the wiring system shall be maintained between floating buildings and the supply conductors. All wiring shall be installed so that motion of the water surface and changes in the water level will not result in unsafe conditions.

(B) Wiring Methods. Liquidtight flexible metal conduit or liquidtight flexible nonmetallic conduit with approved fittings shall be permitted for feeders and where flexible connections are required for services. Extra-hard usage portable power cable listed for both wet locations and sunlight resistance shall be permitted for a feeder to a floating building where flexibility is required. Other raceways suitable for the location shall be permitted to be installed where flexibility is not required.

Part III. Grounding

553.8 General Requirements. Grounding at floating buildings shall comply with 553.8(A) through (D).

(A) Grounding of Electrical and Nonelectrical Parts. Grounding of both electrical and nonelectrical parts in a floating building shall be through connection to a grounding bus in the building panelboard.

(B) Installation and Connection of Equipment Grounding Conductor. The equipment grounding conductor shall be installed with the feeder conductors and connected to a grounding terminal in the service equipment.

(C) Identification of Equipment Grounding Conductor. The equipment grounding conductor shall be an insulated copper conductor with a continuous outer finish that is either green or green with one or more yellow stripes. For conductors larger than 6 AWG, or where multiconductor cables are used, re-identification of conductors allowed in 250.119(A)(2)b. and (A)(2)c. shall be permitted.

(D) Grounding Electrode Conductor Connection. The grounding terminal in the service equipment shall be grounded by connection through an insulated grounding electrode conductor to a grounding electrode on shore.

553.9 Insulated Neutral. The grounded circuit conductor (neutral) shall be an insulated conductor identified in compliance with 200.6. The neutral conductor shall be connected to the equipment grounding terminal in the service equipment, and, except for that connection, it shall be insulated from the equipment grounding conductors, equipment enclosures, and all other grounded parts. The neutral conductor terminals in the panelboard and in ranges, clothes dryers, counter-mounted cooking units, and the like shall be insulated from the enclosures.

553.10 Equipment Grounding.

(A) Electrical Systems. All enclosures and exposed metal parts of electrical systems shall be connected to the grounding bus.

(B) Cord-Connected Appliances. Where required to be grounded, cord-connected appliances shall be grounded by means of an equipment grounding conductor in the cord and a grounding-type attachment plug.

553.11 Bonding of Non–Current-Carrying Metal Parts. All metal parts in contact with the water, all metal piping, and all non–current-carrying metal parts that are likely to become energized shall be connected to the grounding bus in the panelboard.

ARTICLE 555
Marinas, Boatyards, and Commercial and Noncommercial Docking Facilities

555.1 Scope. This article covers the installation of wiring and equipment in the areas comprising fixed or floating piers, wharves, docks, and other areas in marinas, boatyards, boat basins, boathouses, yacht clubs, boat condominiums, docking facilities associated with one-family dwellings, two-family dwellings, multifamily dwellings, and residential condominiums; any multiple docking facility or similar occupancies; and facilities that are used, or intended for use, for the purpose of repair, berthing, launching, storage, or fueling of small craft and the moorage of floating buildings.

Informational Note: See NFPA 303-2011, *Fire Protection Standard for Marinas and Boatyards,* for additional information.

555.2 Definitions.

Electrical Datum Plane. The electrical datum plane is defined as follows:

(1) In land areas subject to tidal fluctuation, the electrical datum plane is a horizontal plane 606 mm (2 ft) above the highest tide level for the area occurring under normal circumstances, that is, highest high tide.

(2) In land areas not subject to tidal fluctuation, the electrical datum plane is a horizontal plane 606 mm (2 ft) above the highest water level for the area occurring under normal circumstances.

(3) The electrical datum plane for floating piers and landing stages that are (a) installed to permit rise and fall response to water level, without lateral movement, and (b) that are so equipped that they can rise to the datum plane established for (1) or (2), is a horizontal plane 762 mm (30 in.) above the water level at the floating pier or landing stage and a minimum of 305 mm (12 in.) above the level of the deck.

Marine Power Outlet. An enclosed assembly that can include equipment such as receptacles, circuit breakers, fused switches, fuses, a watt-hour meter(s), panelboards, and monitoring means approved for marine use.

555.3 Ground-Fault Protection. The overcurrent protective devices that supply the marina, boatyards, and commercial and noncommercial docking facilities shall have ground-fault protection not exceeding 30 mA.

555.4 Distribution System. Yard and pier distribution systems shall not exceed 1000 volts phase to phase.

555.5 Transformers. Transformers and enclosures shall be specifically approved for the intended location. The bottom of enclosures for transformers shall not be located below the electrical datum plane.

555.7 Location of Service Equipment. The service equipment for floating docks or marinas shall be located adjacent to, but not on or in, the floating structure.

555.9 Electrical Connections. Electrical connections shall be located at least 305 mm (12 in.) above the deck of a floating pier. Conductor splices, within approved junction boxes, utilizing sealed wire connector systems listed and identified for submersion shall be permitted where located above the waterline but below the electrical datum plane for floating piers.

All electrical connections shall be located at least 305 mm (12 in.) above the deck of a fixed pier but not below the electrical datum plane.

555.10 Electrical Equipment Enclosures.

(A) Securing and Supporting. Electrical equipment enclosures installed on piers above deck level shall be securely and substantially supported by structural members, independent of any conduit connected to them. If enclosures are not attached to mounting surfaces by means of external ears or lugs, the internal screw heads shall be sealed to prevent seepage of water through mounting holes.

(B) Location. Electrical equipment enclosures on piers shall be located so as not to interfere with mooring lines.

555.11 Circuit Breakers, Switches, Panelboards, and Marine Power Outlets. Circuit breakers and switches installed in gasketed enclosures shall be arranged to permit required manual operation without exposing the interior of the enclosure. All such enclosures shall be arranged with a weep hole to discharge condensation.

555.12 Load Calculations for Service and Feeder Conductors. General lighting and other loads shall be calculated in accordance with Part III of Article 220, and, in addition, the demand factors set forth in Table 555.12 shall be permitted for each service and/or feeder circuit supplying receptacles that provide shore power for boats. These calculations shall be permitted to be modified as indicated in notes (1) and (2) to Table 555.12. Where demand factors of Table 555.12 are applied, the demand factor specified in 220.61(B) shall not be permitted.

> Informational Note: These demand factors may be inadequate in areas of extreme hot or cold temperatures with loaded circuits for heating, air-conditioning, or refrigerating equipment.

555.13 Wiring Methods and Installation.

(A) Wiring Methods.

(1) General. Wiring methods of Chapter 3 shall be permitted where identified for use in wet locations.

(2) Portable Power Cables. Extra-hard usage portable power cables rated not less than 75°C (167°F), 600 volts; listed for both wet locations and sunlight resistance; and having an outer jacket rated to be resistant to temperature extremes, oil, gasoline, ozone, abrasion, acids, and chemicals shall be permitted as follows:

Table 555.12 Demand Factors

Number of Shore Power Receptacles	Sum of the Rating of the Receptacles (%)
1–4	100
5–8	90
9–14	80
15–30	70
31–40	60
41–50	50
51–70	40
≥71	30

Notes:
1. Where shore power accommodations provide two receptacles specifically for an individual boat slip and these receptacles have different voltages (for example, one 30 ampere, 125 volt and one 50 ampere, 125/250 volt), only the receptacle with the larger kilowatt demand shall be required to be calculated.
2. If the facility being installed includes individual kilowatt-hour submeters for each slip and is being calculated using the criteria listed in Table 555.12, the total demand amperes may be multiplied by 0.9 to achieve the final demand amperes.

(1) As permanent wiring on the underside of piers (floating or fixed)
(2) Where flexibility is necessary as on piers composed of floating sections

(3) Temporary Wiring. Temporary wiring, except as permitted by Article 590, shall not be used to supply power to boats.

(B) Installation.

(1) Overhead Wiring. Overhead wiring shall be installed to avoid possible contact with masts and other parts of boats being moved in the yard.

Conductors and cables shall be routed to avoid wiring closer than 6.0 m (20 ft) from the outer edge or any portion of the yard that can be used for moving vessels or stepping or unstepping masts.

(2) Outside Branch Circuits and Feeders. Outside branch circuits and feeders shall comply with Article 225 except that clearances for overhead wiring in portions of the yard other than those described in 555.13(B)(1) shall not be less than 5.49 m (18 ft) abovegrade.

(3) Wiring Over and Under Navigable Water. Wiring over and under navigable water shall be subject to approval by the authority having jurisdiction.

> Informational Note: See NFPA 303-2011, *Fire Protection Standard for Marinas and Boatyards*, for warning sign requirements.

(4) Portable Power Cables.
   (a) Where portable power cables are permitted by 555.13(A)(2), the installation shall comply with the following:

(1) Cables shall be properly supported
(2) Cables shall be located on the underside of the pier.
(3) Cables shall be securely fastened by nonmetallic clips to structural members other than the deck planking.
(4) Cables shall not be installed where subject to physical damage.
(5) Where cables pass through structural members, they shall be protected against chafing by a permanently installed oversized sleeve of nonmetallic material.

   (b) Where portable power cables are used as permitted in 555.13(A)(2)(2), there shall be an approved junction box of corrosion-resistant construction with permanently installed terminal blocks on each pier section to which the feeder and feeder extensions are to be connected. A listed marine power outlet employing terminal blocks/bars shall be permitted in lieu of a junction box. Metal junction boxes and their covers, and metal screws and parts that are exposed externally to the boxes, shall be of corrosion-resistant materials or protected by material resistant to corrosion.

(5) Protection. Rigid metal conduit, reinforced thermosetting resin conduit (RTRC) listed for aboveground use, or rigid polyvinyl chloride (PVC) conduit suitable for the location, shall be installed to protect wiring above decks of piers and landing stages and below the enclosure that it serves. The conduit shall be connected to the enclosure by full standard threads or fittings listed for use in damp or wet locations, as applicable.

555.15 Grounding. Wiring and equipment within the scope of this article shall be grounded as specified in Article 250 and as required by 555.15(A) through (E).

(A) Equipment to Be Grounded. The following items shall be connected to an equipment grounding conductor run with the circuit conductors in the same raceway, cable, or trench:

(1) Metal boxes, metal cabinets, and all other metal enclosures
(2) Metal frames of utilization equipment
(3) Grounding terminals of grounding-type receptacles

(B) Type of Equipment Grounding Conductor. The equipment grounding conductor shall be an insulated conductor with a continuous outer finish that is either green or green with one or more yellow stripes. The equipment grounding conductor of Type MI cable shall be permitted to be identified at terminations. For conductors larger than 6 AWG, or where multiconductor cables are used, re-identification of conductors allowed in 250.119(A)(2)b. and (A)(2)c. or 250.119(B)(2) and (B)(3) shall be permitted.

(C) Size of Equipment Grounding Conductor. The insulated equipment grounding conductor shall be sized in accordance with 250.122 but not smaller than 12 AWG.

(D) Branch-Circuit Equipment Grounding Conductor. The insulated equipment grounding conductor for branch circuits shall terminate at a grounding terminal in a remote panelboard or the grounding terminal in the main service equipment.

(E) Feeder Equipment Grounding Conductors. Where a feeder supplies a remote panelboard, an insulated equipment grounding conductor shall extend from a grounding terminal in the service equipment to a grounding terminal in the remote panelboard.

555.17 Disconnecting Means for Shore Power Connection(s). Disconnecting means shall be provided to isolate each boat from its supply connection(s).

(A) Type. The disconnecting means shall consist of a circuit breaker, switch, or both, and shall be properly identified as to which receptacle it controls.

(B) Location. The disconnecting means shall be readily accessible, located not more than 762 mm (30 in.) from the receptacle it controls, and shall be located in the supply circuit ahead of the receptacle. Circuit breakers or switches located in marine power outlets complying with this section shall be permitted as the disconnecting means.

555.19 Receptacles. Receptacles shall be mounted not less than 305 mm (12 in.) above the deck surface of the pier and not below the electrical datum plane on a fixed pier.

(A) Shore Power Receptacles.

(1) Enclosures. Receptacles intended to supply shore power to boats shall be housed in marine power outlets listed as marina power outlets or listed for set locations, or shall be installed in listed enclosures protected from the weather or in listed weatherproof enclosures. The integrity of the assembly shall not be affected when the receptacles are in use with any type of booted or nonbooted attachment plug/cap inserted.

(2) Strain Relief. Means shall be provided where necessary to reduce the strain on the plug and receptacle caused by the weight and catenary angle of the shore power cord.

(3) Branch Circuits. Each single receptacle that supplies shore power to boats shall be supplied from a marine power outlet or panelboard by an individual branch circuit of the voltage class and rating corresponding to the rating of the receptacle.

Informational Note: Supplying receptacles at voltages other than the voltages marked on the receptacle may cause overheating or malfunctioning of connected equipment, for example, supplying single-phase, 120/240-volt, 3-wire loads from a 208Y/120-volt, 3-wire source.

(4) Ratings. Shore power for boats shall be provided by single receptacles rated not less than 30 amperes.

Informational Note: For locking- and grounding-type receptacles for auxiliary power to boats, see NFPA 303-2011, *Fire Protection Standard for Marinas and Boatyards.*

(a) Receptacles rated 30 amperes and 50 amperes shall be of the locking and grounding type.

Informational Note: For various configurations and ratings of locking- and grounding-type receptacles and caps, see ANSI/NEMA WD 6-2002 (Rev. 2008), *Standard for Dimensions of Attachment Plugs and Receptacles.*

(b) Receptacles rated 60 amperes or higher shall be of the pin and sleeve type.

Informational Note: For various configurations and ratings of pin and sleeve receptacles, see ANSI/UL 1686, *UL Standard for Safety Pin and Sleeve Configurations.*

(B) Other Than Shore Power.

(1) Ground-Fault Circuit-Interrupter (GFCI) Protection for Personnel. Fifteen- and 20-ampere, single-phase, 125-volt receptacles installed outdoors, in boathouses, in buildings or structures used for storage, maintenance, or repair shall be provided with GFCI protection for personnel. Receptacles in other locations shall be protected in accordance with 210.8(B).

(2) Marking. Receptacles other than those supplying shore power to boats shall be permitted to be housed in marine power outlets with the receptacles that provide shore power to boats, provided they are marked to clearly indicate that they are not to be used to supply power to boats.

555.21 Motor Fuel Dispensing Stations — Hazardous (Classified) Locations. Electrical wiring and equipment located at or serving motor fuel dispensing locations shall comply with Article 514 in addition to the requirements of this article.

555.22 Repair Facilities — Hazardous (Classified) Locations. Electrical wiring and equipment located at facilities for the repair of marine craft containing flammable or combustible liquids or gases shall comply with Article 511 in addition to the requirements of this article.

555.23 Marine Hoists, Railways, Cranes, and Monorails. Motors and controls for marine hoists, railways, cranes, and monorails shall not be located below the electrical datum plane. Where it is necessary to provide electric power to a mobile crane or hoist in the yard and a trailing cable is utilized, it shall be a listed portable power cable rated for the conditions of use and be provided with an outer jacket of distinctive color for safety.

N **555.24 Signage.** Permanent safety signs shall be installed to give notice of electrical shock hazard risks to persons using or swimming near a boat dock or marina and shall comply with all of the following:

(1) The signage shall comply with 110.21(B)(1) and be of sufficient durability to withstand the environment.

(2) The signs shall be clearly visible from all approaches to a marina or boatyard facility.

(3) The signs shall state "WARNING — POTENTIAL SHOCK HAZARD — ELECTRICAL CURRENTS MAY BE PRESENT IN THE WATER."

ARTICLE 590
Temporary Installations

590.1 Scope. The provisions of this article apply to temporary electric power and lighting installations.

590.2 All Wiring Installations.

(A) Other Articles. Except as specifically modified in this article, all other requirements of this *Code* for permanent wiring shall apply to temporary wiring installations.

(B) Approval. Temporary wiring methods shall be acceptable only if approved based on the conditions of use and any special requirements of the temporary installation.

590.3 Time Constraints.

(A) During the Period of Construction. Temporary electric power and lighting installations shall be permitted during the period of construction, remodeling, maintenance, repair, or demolition of buildings, structures, equipment, or similar activities.

(B) 90 Days. Temporary electric power and lighting installations shall be permitted for a period not to exceed 90 days for holiday decorative lighting and similar purposes.

(C) Emergencies and Tests. Temporary electric power and lighting installations shall be permitted during emergencies and for tests, experiments, and developmental work.

(D) Removal. Temporary wiring shall be removed immediately upon completion of construction or purpose for which the wiring was installed.

590.4 General.

(A) Services. Services shall be installed in conformance with Parts I through VIII of Article 230, as applicable.

(B) Feeders. Overcurrent protection shall be provided in accordance with 240.4, 240.5, 240.100, and 240.101. Conductors shall be permitted within cable assemblies or within multiconductor cords or cables of a type identified in Table 400.4 for hard usage or extra-hard usage. For the purpose of this section, the following wiring methods shall be permitted:

(1) Type NM, Type NMC, and Type SE cables shall be permitted to be used in any dwelling, building, or structure without any height limitation or limitation by building construction type and without concealment within walls, floors, or ceilings.

(2) Type SE cable shall be permitted to be installed in a raceway in an underground installation.

Exception: Single insulated conductors shall be permitted where installed for the purpose(s) specified in 590.3(C), where accessible only to qualified persons.

(C) Branch Circuits. All branch circuits shall originate in an approved power outlet, switchgear, switchboard or panelboard, motor control center, or fused switch enclosure. Conductors shall be permitted within cable assemblies or within multiconductor cord or cable of a type identified in Table 400.4 for hard usage or extra-hard usage. Conductors shall be protected from overcurrent as provided in 240.4, 240.5, and 240.100. For the purposes of this section, the following wiring methods shall be permitted:

(1) Type NM, Type NMC, and Type SE cables shall be permitted to be used in any dwelling, building, or structure without any height limitation or limitation by building construction type and without concealment within walls, floors, or ceilings.

(2) Type SE cable shall be permitted to be installed in a raceway in an underground installation.

Exception: Branch circuits installed for the purposes specified in 590.3(B) or 590.3(C) shall be permitted to be run as single insulated conductors. Where the wiring is installed in accordance with 590.3(B), the voltage to ground shall not exceed 150 volts, the wiring shall not be subject to physical damage, and the conductors shall be supported on insulators at intervals of not more than 3.0 m (10 ft); or, for festoon lighting, the conductors shall be so arranged that excessive strain is not transmitted to the lampholders.

(D) Receptacles.

(1) All Receptacles. All receptacles shall be of the grounding type. Unless installed in a continuous metal raceway that qualifies as an equipment grounding conductor in accordance with 250.118 or a continuous metal-covered cable that qualifies as an equipment grounding conductor in accordance with 250.118, all branch circuits shall include a separate equipment grounding conductor, and all receptacles shall be electrically connected to the equipment grounding conductor(s). Receptacles on construction sites shall not be installed on any branch circuit that supplies temporary lighting.

(2) Receptacles in Wet Locations. All 15- and 20-ampere, 125- and 250-volt receptacles installed in a wet location shall comply with 406.9(B)(1).

(E) Disconnecting Means. Suitable disconnecting switches or plug connectors shall be installed to permit the disconnection of all ungrounded conductors of each temporary circuit. Multiwire branch circuits shall be provided with a means to disconnect simultaneously all ungrounded conductors at the power outlet or panelboard where the branch circuit originated. Identified handle ties shall be permitted.

(F) Lamp Protection. All lamps for general illumination shall be protected from accidental contact or breakage by a suitable luminaire or lampholder with a guard.

Brass shell, paper-lined sockets, or other metal-cased sockets shall not be used unless the shell is grounded.

(G) Splices. A box, conduit body, or other enclosure, with a cover installed, shall be required for all splices except where:

(1) The circuit conductors being spliced are all from nonmetallic multiconductor cord or cable assemblies, provided that the equipment grounding continuity is maintained with or without the box.

(2) The circuit conductors being spliced are all from metal sheathed cable assemblies terminated in listed fittings that mechanically secure the cable sheath to maintain effective electrical continuity.

(H) Protection from Accidental Damage. Flexible cords and cables shall be protected from accidental damage. Sharp corners and projections shall be avoided. Where passing through doorways or other pinch points, protection shall be provided to avoid damage.

(I) Termination(s) at Devices. Flexible cords and cables entering enclosures containing devices requiring termination shall be secured to the box with fittings listed for connecting flexible cords and cables to boxes designed for the purpose.

(J) Support. Cable assemblies and flexible cords and cables shall be supported in place at intervals that ensure that they will be protected from physical damage. Support shall be in the form of staples, cable ties, straps, or similar type fittings installed so as not to cause damage. Cable assemblies and flexible cords and cables installed as branch circuits or feeders shall not be installed on the floor or on the ground. Extension cords shall not be required to comply with 590.4(J). Vegetation shall not be used for support of overhead spans of branch circuits or feeders.

Exception: For holiday lighting in accordance with 590.3(B), where the conductors or cables are arranged with strain relief devices, tension take-up devices, or other approved means to avoid damage from the movement of the live vegetation, trees shall be permitted to be used for support of overhead spans of branch-circuit conductors or cables.

590.5 Listing of Decorative Lighting. Decorative lighting used for holiday lighting and similar purposes, in accordance with 590.3(B), shall be listed and shall be labeled on the product.

590.6 Ground-Fault Protection for Personnel. Ground-fault protection for personnel for all temporary wiring installations shall be provided to comply with 590.6(A) and (B). This section shall apply only to temporary wiring installations used to supply temporary power to equipment used by personnel during construction, remodeling, maintenance, repair, or demolition of buildings, structures, equipment, or similar activities. This section shall apply to power derived from an electric utility company or from an on-site-generated power source.

(A) Receptacle Outlets. Temporary receptacle installations used to supply temporary power to equipment used by personnel during construction, remodeling, maintenance, repair, or demolition of buildings, structures, equipment, or similar activities shall comply with the requirements of 590.6(A)(1) through (A)(3), as applicable.

Exception: In industrial establishments only, where conditions of maintenance and supervision ensure that only qualified personnel are involved, an assured equipment grounding conductor program as specified in 590.6(B)(3) shall be permitted for only those receptacle outlets used to supply equipment that would create a greater hazard if power were interrupted or having a design that is not compatible with GFCI protection.

(1) Receptacle Outlets Not Part of Permanent Wiring. All 125-volt, single-phase, 15-, 20-, and 30-ampere receptacle outlets that are not a part of the permanent wiring of the building or structure and that are in use by personnel shall have ground-fault circuit-interrupter protection for personnel. In addition to this required ground-fault circuit-interrupter protection for personnel, listed cord sets or devices incorporating listed ground-fault circuit-interrupter protection for personnel identified for portable use shall be permitted.

(2) Receptacle Outlets Existing or Installed as Permanent Wiring. Ground-fault circuit-interrupter protection for personnel shall be provided for all 125-volt, single-phase, 15-, 20-, and 30-ampere receptacle outlets installed or existing as part of the permanent wiring of the building or structure and used for temporary electric power. Listed cord sets or devices incorporating listed ground-fault circuit-interrupter protection for personnel identified for portable use shall be permitted.

(3) Receptacles on 15-kW or less Portable Generators. All 125-volt and 125/250-volt, single-phase, 15-, 20-, and 30-ampere receptacle outlets that are a part of a 15-kW or smaller portable generator shall have listed ground-fault circuit-interrupter protection for personnel. All 15- and 20-ampere, 125- and 250-volt receptacles, including those that are part of a portable generator, used in a damp or wet location shall comply with 406.9(A) and (B). Listed cord sets or devices incorporating listed ground-fault circuit-interrupter protection for personnel identified for portable use shall be permitted for use with 15-kW or less portable generators manufactured or remanufactured prior to January 1, 2011.

(B) Use of Other Outlets. For temporary wiring installations, receptacles, other than those covered by 590.6(A)(1) through (A)(3) used to supply temporary power to equipment used by personnel during construction, remodeling, maintenance, repair, or demolition of buildings, structures, or equipment, or similar activities, shall have protection in accordance with (B)(1), (B)(2), or the assured equipment grounding conductor program in accordance with (B)(3).

(1) GFCI Protection. Ground-fault circuit-interrupter protection for personnel.

N **(2) SPGFCI Protection.** Special purpose ground-fault circuit-interrupter protection for personnel.

(3) Assured Equipment Grounding Conductor Program. A written assured equipment grounding conductor program continuously enforced at the site by one or more designated persons to ensure that equipment grounding conductors for all cord sets, receptacles that are not a part of the permanent wiring of the building or structure, and equipment connected by cord and plug are installed and maintained in accordance with the applicable requirements of 250.114, 250.138, 406.4(C), and 590.4(D).

(a) The following tests shall be performed on all cord sets, receptacles that are not part of the permanent wiring of the building or structure, and cord-and-plug-connected equipment required to be connected to an equipment grounding conductor:

(1) All equipment grounding conductors shall be tested for continuity and shall be electrically continuous.

(2) Each receptacle and attachment plug shall be tested for correct attachment of the equipment grounding conductor. The equipment grounding conductor shall be connected to its proper terminal.

(3) All required tests shall be performed as follows:

    a. Before first use on site

    b. When there is evidence of damage

    c. Before equipment is returned to service following any repairs

    d. At intervals not exceeding 3 months

(b) The tests required in item (3)(a) shall be recorded and made available to the authority having jurisdiction.

590.7 Guarding. For wiring over 600 volts, nominal, suitable fencing, barriers, or other effective means shall be provided to limit access only to authorized and qualified personnel.

Chapter 6   Special Equipment

ARTICLE 600
Electric Signs and Outline Lighting

Part I. General

600.1 Scope. This article covers the installation of conductors, equipment, and field wiring for electric signs, retrofit kits, and outline lighting, regardless of voltage. All installations and equipment using neon tubing, such as signs, decorative elements, skeleton tubing, or art forms, are covered by this article.

> Informational Note: Sign and outline lighting illumination systems include, but are not limited to, cold cathode neon tubing, high-intensity discharge lamps (HID), fluorescent or incandescent lamps, light-emitting diodes (LEDs), and electroluminescent and inductance lighting.

600.2 Definitions.

LED Sign Illumination System. A complete lighting system for use in signs and outline lighting consisting of light-emitting diode (LED) light sources, power supplies, wire, and connectors to complete the installation.

Neon Tubing. Electric-discharge luminous tubing, including cold cathode luminous tubing, that is manufactured into shapes to illuminate signs, form letters, parts of letters, skeleton tubing, outline lighting, other decorative elements, or art forms and filled with various inert gases.

N **Photovoltaic (PV) Powered Sign.** A complete sign powered by solar energy consisting of all components and subassemblies for installation either as an off-grid stand-alone, on-grid interactive, or non-grid interactive system.

Section Sign. A sign or outline lighting system, shipped as subassemblies, that requires field-installed wiring between the subassemblies to complete the overall sign. The subassemblies are either physically joined to form a single sign unit or are installed as separate remote parts of an overall sign.

Sign Body. A portion of a sign that may provide protection from the weather but is not an electrical enclosure.

Skeleton Tubing. Neon tubing that is itself the sign or outline lighting and is not attached to an enclosure or sign body.

600.3 Listing. Fixed, mobile, or portable electric signs, section signs, outline lighting, photovoltaic (PV) powered signs, and retrofit kits, regardless of voltage, shall be listed, provided with installation instructions, and installed in conformance with that listing, unless otherwise approved by special permission.

(A) Field-Installed Skeleton Tubing. Field-installed skeleton tubing shall not be required to be listed where installed in conformance with this *Code*.

(B) Outline Lighting. Outline lighting shall not be required to be listed as a system when it consists of listed luminaires wired in accordance with Chapter 3.

600.4 Markings.

(A) Signs and Outline Lighting Systems. Signs and outline lighting systems shall be listed; marked with the manufacturer's name, trademark, or other means of identification; and input voltage and current rating.

N **(B) Signs with a Retrofitted Illumination System.**

(1) The retrofitted sign shall be marked that the illumination system has been replaced.
(2) The marking shall include the kit providers and installer's name, logo, or unique identifier.
(3) Signs equipped with tubular light-emitting diode lamps powered by the existing sign sockets shall include a label alerting the service personnel that the sign has been modified. The label shall meet the requirements of 110.21(B). The label shall also include a warning not to install fluorescent lamps and shall also be visible during relamping.

(C) Signs with Lampholders for Incandescent Lamps. Signs and outline lighting systems with lampholders for incandescent lamps shall be marked to indicate the maximum allowable lamp wattage per lampholder. The markings shall be permanently installed, in letters at least 6 mm ($\frac{1}{4}$ in.) high, and shall be located where visible during relamping.

(D) Visibility. The markings required in 600.4(A) and listing labels shall not be required to be visible after installation but shall be permanently applied in a location visible during servicing.

(E) Durability. Marking labels shall be permanent, durable and, when in wet locations, shall be weatherproof.

(F) Installation Instructions. All signs, outline lighting, skeleton tubing systems, and retrofit kits shall be marked to indicate that field wiring and installation instructions are required.

Exception: Portable, cord-connected signs are not required to be marked.

600.5 Branch Circuits.

(A) Required Branch Circuit. Each commercial building and each commercial occupancy accessible to pedestrians shall be provided with at least one outlet in an accessible location at each entrance to each tenant space for sign or outline lighting system use. The outlet(s) shall be supplied by a branch circuit rated at least 20 amperes that supplies no other load. Service hallways or corridors shall not be considered accessible to pedestrians.

(B) Rating. Branch circuits that supply signs shall be rated in accordance with 600.5(B)(1) or (B)(2) and shall be considered to be continuous loads for the purposes of calculations.

(1) Neon Signs. Branch circuits that supply neon tubing installations shall not be rated in excess of 30 amperes.

(2) All Other Signs. Branch circuits that supply all other signs and outline lighting systems shall be rated not to exceed 20 amperes.

(C) Wiring Methods. Wiring methods used to supply signs shall comply with 600.5(C)(1), (C)(2), and (C)(3).

(1) Supply. The wiring method used to supply signs and outline lighting systems shall terminate within a sign, an outline lighting system enclosure, a suitable box, or a conduit body.

(2) Enclosures as Pull Boxes. Signs and transformer enclosures shall be permitted to be used as pull or junction boxes for conductors supplying other adjacent signs, outline lighting systems, or floodlights that are part of a sign and shall be permitted to contain both branch and secondary circuit conductors.

(3) Metal or Nonmetallic Poles. Metal or nonmetallic poles used to support signs shall be permitted to enclose supply conductors, provided the poles and conductors are installed in accordance with 410.30(B).

600.6 Disconnects. Each sign and outline lighting system, feeder conductor(s), or branch circuit(s) supplying a sign, outline lighting system, or skeleton tubing shall be controlled by an externally operable switch or circuit breaker that opens all ungrounded conductors and controls no other load. The switch or circuit breaker shall open all ungrounded conductors simultaneously on multi-wire branch circuits in accordance with 210.4(B). Signs and outline lighting systems located within fountains shall have the disconnect located in accordance with 680.13.

Exception No. 1: A disconnecting means shall not be required for an exit directional sign located within a building.

Exception No. 2: A disconnecting means shall not be required for cord-connected signs with an attachment plug.

Informational Note: The location of the disconnect is intended to allow service or maintenance personnel complete and local control of the disconnecting means.

(A) Location. The disconnecting means shall be permitted to be located in accordance with 600.6(A)(1), (A)(2), and (A)(3):

(1) At Point of Entry to a Sign. The disconnect shall be located at the point the feeder circuit or branch circuit(s) supplying a sign or outline lighting system enters a sign enclosure, a sign body, or a pole in accordance with 600.5(C)(3) . The disconnect shall open all ungrounded conductors where it enters the enclosure of the sign or pole.

Exception No. 1: A disconnect shall not be required for branch circuit(s) or feeder conductor(s) passing through the sign where enclosed in a Chapter 3 listed raceway or metal-jacketed cable identified for the location.

Exception No. 2: A disconnect shall not be required at the point of entry to a sign enclosure or sign body for branch circuit(s) or feeder conductor(s) that supply an internal panelboard(s) in a sign enclosure or sign body. The conductors shall be enclosed in a Chapter 3 listed raceway or metal-jacketed cable identified for the location. A field-applied permanent warning label that is visible during servicing shall be applied to the raceway at or near the point of entry into the sign enclosure or sign body. The warning label shall comply with 110.21(B) and state the following: "Danger. This raceway contains energized conductors." The marking shall include the location of the disconnecting means for the energized conductor(s). The disconnecting means shall be capable of being locked in the open position in accordance with 110.25.

(2) Within Sight of the Sign. The disconnecting means shall be within sight of the sign or outline lighting system that it controls. Where the disconnecting means is out of the line of sight from any section that is able to be energized, the disconnecting means shall be lockable in accordance with 110.25. A permanent field-applied marking identifying the location of the disconnecting means shall be applied to the sign in a location visible during servicing. The warning label shall comply with 110.21(B).

(3) Within Sight of the Controller. The following shall apply for signs or outline lighting systems operated by electronic or electromechanical controllers located external to the sign or outline lighting system:

(1) The disconnecting means shall be located within sight of the controller or in the same enclosure with the controller.

(2) The disconnecting means shall disconnect the sign or outline lighting system and the controller from all ungrounded supply conductors.

(3) The disconnecting means shall be designed such that no pole can be operated independently and shall be lockable in accordance with 110.25.

Exception: Where the disconnecting means is not located within sight of the controller, a permanent field-applied marking identifying the location of the disconnecting means shall be applied to the controller in a location visible during servicing. The warning label shall comply with 110.21(B).

(B) Control Switch Rating. Switches, flashers, and similar devices controlling transformers and electronic power supplies shall be rated for controlling inductive loads or have a current rating not less than twice the current rating of the transformer or the electronic power supply.

600.7 Grounding and Bonding.

(A) Grounding.

(1) Equipment Grounding. Metal equipment of signs, outline lighting, and skeleton tubing systems shall be grounded by connection to the equipment grounding conductor of the supply branch circuit(s) or feeder using the types of equipment grounding conductors specified in 250.118.

Exception: Portable cord-connected signs shall not be required to be connected to the equipment grounding conductor where protected by a system of double insulation or its equivalent. Double insulated equipment shall be distinctively marked.

(2) Size of Equipment Grounding Conductor. The equipment grounding conductor size shall be in accordance with 250.122 based on the rating of the overcurrent device protecting the branch circuit or feeder conductors supplying the sign or equipment.

(3) Connections. Equipment grounding conductor connections shall be made in accordance with 250.130 and in a method specified in 250.8.

(4) Auxiliary Grounding Electrode. Auxiliary grounding electrode(s) shall be permitted for electric signs and outline lighting systems covered by this article and shall meet the requirements of 250.54.

(5) Metal Building Parts. Metal parts of a building shall not be permitted as a secondary return conductor or an equipment grounding conductor.

OUTLINE LIGHTING

(B) Bonding.

(1) Bonding of Metal Parts. Metal parts and equipment of signs and outline lighting systems shall be bonded together and to the associated transformer or power-supply equipment grounding conductor of the branch circuit or feeder supplying the sign or outline lighting system and shall meet the requirements of 250.90.

Exception: Remote metal parts of a section sign or outline lighting system only supplied by a remote Class 2 power supply shall not be required to be bonded to an equipment grounding conductor.

(2) Bonding Connections. Bonding connections shall be made in accordance with 250.8.

(3) Metal Building Parts. Metal parts of a building shall not be permitted to be used as a means for bonding metal parts and equipment of signs or outline lighting systems together or to the transformer or power-supply equipment grounding conductor of the supply circuit.

(4) Flexible Metal Conduit Length. Listed flexible metal conduit or listed liquidtight flexible metal conduit that encloses the secondary circuit conductor from a transformer or power supply for use with neon tubing shall be permitted as a bonding means if the total accumulative length of the conduit in the secondary circuit does not exceed 30 m (100 ft).

(5) Small Metal Parts. Small metal parts not exceeding 50 mm (2 in.) in any dimension, not likely to be energized, and spaced at least 19 mm ($\frac{3}{4}$ in.) from neon tubing shall not require bonding.

(6) Nonmetallic Conduit. Where listed nonmetallic conduit is used to enclose the secondary circuit conductor from a transformer or power supply and a bonding conductor is required, the bonding conductor shall be installed separate and remote from the nonmetallic conduit and be spaced at least 38 mm ($1\frac{1}{2}$ in.) from the conduit when the circuit is operated at 100 Hz or less or 45 mm ($1\frac{3}{4}$ in.) when the circuit is operated at over 100 Hz.

(7) Bonding Conductors. Bonding conductors shall comply with (1) and (2).

(1) Bonding conductors shall be copper and not smaller than 14 AWG.
(2) Bonding conductors installed externally of a sign or raceway shall be protected from physical damage.

(8) Signs in Fountains. Signs or outline lighting installed inside a fountain shall have all metal parts bonded to the equipment grounding conductor of the branch circuit for the fountain recirculating system. The bonding connection shall be as near as practicable to the fountain and shall be permitted to be made to metal piping systems that are bonded in accordance with 680.53.

Informational Note: Refer to 600.32(J) for restrictions on length of high-voltage secondary conductors.

600.8 Enclosures. Live parts, other than lamps, and neon tubing shall be enclosed. Transformers and power supplies provided with an integral enclosure, including a primary and secondary circuit splice enclosure, shall not require an additional enclosure.

(A) Strength. Enclosures shall have ample structural strength and rigidity.

(B) Material. Sign and outline lighting system enclosures shall be constructed of metal or shall be listed.

(C) Minimum Thickness of Enclosure Metal. Sheet copper or aluminum shall be at least 0.51 mm (0.020 in.) thick. Sheet steel shall be at least 0.41 mm (0.016 in.) thick.

(D) Protection of Metal. Metal parts of equipment shall be protected from corrosion.

600.9 Location.

(A) Vehicles. Sign or outline lighting system equipment shall be at least 4.3 m (14 ft) above areas accessible to vehicles unless protected from physical damage.

(B) Pedestrians. Neon tubing, other than listed, dry-location, portable signs, readily accessible to pedestrians shall be protected from physical damage.

Informational Note: See 600.41(D) for additional requirements.

(C) Adjacent to Combustible Materials. Signs and outline lighting systems shall be installed so that adjacent combustible materials are not subjected to temperatures in excess of 90°C (194°F).

The spacing between wood or other combustible materials and an incandescent or HID lamp or lampholder shall not be less than 50 mm (2 in.).

(D) Wet Location. Signs and outline lighting system equipment for wet location use, other than listed watertight type, shall be weatherproof and have drain holes, as necessary, in accordance with the following:

(1) Drain holes shall not be larger than 13 mm ($\frac{1}{2}$ in.) or smaller than 6 mm ($\frac{1}{4}$ in.).
(2) Every low point or isolated section of the equipment shall have at least one drain hole.
(3) Drain holes shall be positioned such that there will be no external obstructions.

600.10 Portable or Mobile Signs.

(A) Support. Portable or mobile signs shall be adequately supported and readily movable without the use of tools.

(B) Attachment Plug. An attachment plug shall be provided for each portable or mobile sign.

(C) Wet or Damp Location. Portable or mobile signs in wet or damp locations shall comply with 600.10(C)(1) and (C)(2).

(1) Cords. All cords shall be junior hard-service or hard-service types as designated in Table 400.4 and have an equipment grounding conductor.

(2) Ground-Fault Circuit Interrupter. The manufacturer of portable or mobile signs shall provide listed ground-fault circuit-interrupter protection for personnel. The ground-fault circuit interrupter shall be an integral part of the attachment plug or shall be located in the power-supply cord within 300 mm (12 in.) of the attachment plug.

(D) Dry Location. Portable or mobile signs in dry locations shall meet the following:

(1) Cords shall be SP-2, SPE-2, SPT-2, or heavier, as designated in Table 400.4.
(2) The cord shall not exceed 4.5 m (15 ft) in length.

600.12 Field-Installed Secondary Wiring. Field-installed secondary circuit wiring for electric signs, retrofit kits, outline lighting systems, skeleton tubing, and photovoltaic (PV) powered sign systems shall be in accordance with their installation instructions and 600.12(A), (B), or (C).

(A) 1000 Volts or Less. Neon and secondary circuit wiring of 1000 volts or less shall comply with 600.31.

(B) Over 1000 Volts. Neon secondary circuit wiring of over 1000 volts shall comply with 600.32.

(C) Class 2. Where the installation complies with 600.33 and the power source provides a Class 2 output that complies with 600.24, either of the following wiring methods shall be permitted as determined by the installation instructions and conditions.

(1) Wiring methods identified in Chapter 3
(2) Class 2 cables complying with Table 600.33(A)(1) and Table 600.33(A)(2)

600.21 Ballasts, Transformers, Electronic Power Supplies, and Class 2 Power Sources. Ballasts, transformers, electronic power supplies, and Class 2 power sources shall be of the self-contained type or be enclosed by placement in a listed sign body or listed separate enclosure.

(A) Accessibility. Ballasts, transformers, electronic power supplies, and Class 2 power sources shall be located where accessible and shall be securely fastened in place.

(B) Location. Ballasts, transformers, electronic power supplies, and Class 2 power sources shall be installed as near to the lamps or neon tubing as practicable to keep the secondary conductors as short as possible.

(C) Wet Location. Ballasts, transformers, electronic power supplies, and Class 2 power sources used in wet locations shall be of the weatherproof type or be of the outdoor type and protected from the weather by placement in a sign body or separate enclosure.

(D) Working Space. A working space at least 900 mm (3 ft) high × 900 mm (3 ft) wide × 900 mm (3 ft) deep shall be provided at each ballast, transformer, electronic power supply, and Class 2 power source or at its enclosure where not installed in a sign.

(E) Attic and Soffit Locations. Ballasts, transformers, electronic power supplies, and Class 2 power sources shall be permitted to be located in attics and soffits, provided there is an access door at least 900 mm × 562.5 mm (36 in. × 22½ in.) and a passageway of at least 900 mm (3 ft) high × 600 mm (2 ft) wide with a suitable permanent walkway at least 300 mm (12 in.) wide extending from the point of entry to each component. At least one lighting outlet containing a switch or controlled by a wall switch shall be installed in such spaces. At least one point of control shall be at the usual point of entry to these spaces. The lighting outlet shall be provided at or near the equipment requiring servicing.

(F) Suspended Ceilings. Ballasts, transformers, electronic power supplies, and Class 2 power sources shall be permitted to be located above suspended ceilings, provided that their enclosures are securely fastened in place and not dependent on the suspended-ceiling grid for support. Ballasts, transformers, and electronic power supplies installed in suspended ceilings shall not be connected to the branch circuit by flexible cord.

600.22 Ballasts.

(A) Type. Ballasts shall be identified for the use and shall be listed.

(B) Thermal Protection. Ballasts shall be thermally protected.

600.23 Transformers and Electronic Power Supplies.

(A) Type. Transformers and electronic power supplies shall be identified for the use and shall be listed.

(B) Secondary-Circuit Ground-Fault Protection. Transformers and electronic power supplies other than the following shall have secondary-circuit ground-fault protection:

(1) Transformers with isolated ungrounded secondaries and with a maximum open circuit voltage of 7500 volts or less
(2) Transformers with integral porcelain or glass secondary housing for the neon tubing and requiring no field wiring of the secondary circuit

(C) Voltage. Secondary-circuit voltage shall not exceed 15,000 volts, nominal, under any load condition. The voltage to ground of any output terminals of the secondary circuit shall not exceed 7500 volts, under any load condition.

(D) Rating. Transformers and electronic power supplies shall have a secondary-circuit current rating of not more than 300 mA.

(E) Secondary Connections. Secondary circuit outputs shall not be connected in parallel or in series.

(F) Marking. Transformers and electronic power supplies that are equipped with secondary-circuit ground-fault protection shall be so marked.

600.24 Class 2 Power Sources. Class 2 transformers, power supplies, and power sources shall comply with the requirements of Class 2 circuits and 600.24(A), (B), (C), and (D).

(A) Listing. Class 2 power supplies and power sources shall be listed for use with electric signs and outline lighting systems or shall be a component in a listed electric sign.

(B) Grounding. Metal parts of Class 2 power supplies and power sources shall be grounded by connecting to the equipment grounding conductor.

(C) Wiring Methods on the Supply Side of the Class 2 Power Supply. Conductors and equipment on the supply side of the power source shall be installed in accordance with the appropriate requirements of Chapter 3.

(D) Secondary Wiring. Secondary wiring on the load side of a Class 2 power source shall comply with 600.12(C) and 600.33.

Part II. Field-Installed Skeleton Tubing, Outline Lighting, and Secondary Wiring

600.30 Applicability. Part II of this article shall apply to all of the following:

(1) Field-installed skeleton tubing
(2) Field-installed secondary circuits
(3) Outline lighting

These requirements shall be in addition to the requirements of Part I.

600.31 Neon Secondary-Circuit Wiring, 1000 Volts or Less, Nominal.

(A) Wiring Method. Conductors shall be installed using any wiring method included in Chapter 3 suitable for the conditions.

(B) Insulation and Size. Conductors shall be listed, insulated, and not smaller than 18 AWG.

(C) Number of Conductors in Raceway. The number of conductors in a raceway shall be in accordance with Table 1 of Chapter 9.

(D) Installation. Conductors shall be installed so they are not subject to physical damage.

(E) Protection of Leads. Bushings shall be used to protect wires passing through an opening in metal.

600.32 Neon Secondary-Circuit Wiring, over 1000 Volts, Nominal.

(A) Wiring Methods.

(1) Installation. Conductors shall be installed in rigid metal conduit, intermediate metal conduit, liquidtight flexible nonmetallic conduit, flexible metal conduit, liquidtight flexible metal conduit, electrical metallic tubing, metal enclosures; on insulators in metal raceways; or in other equipment listed for use with neon secondary circuits over 1000 volts.

(2) Number of Conductors. Conduit or tubing shall contain only one conductor.

(3) Size. Conduit or tubing shall be a minimum of metric designator 16 (trade size ½).

(4) Spacing from Grounded Parts. Other than at the location of connection to a metal enclosure or sign body, nonmetallic conduit or flexible nonmetallic conduit shall be spaced no less than 38 mm (1½ in.) from grounded or bonded parts when the conduit contains a conductor operating at 100 Hz or less, and shall be spaced no less than 45 mm (1¾ in.) from grounded or bonded parts when the conduit contains a conductor operating at more than 100 Hz.

(5) Metal Building Parts. Metal parts of a building shall not be permitted as a secondary return conductor or an equipment grounding conductor.

(B) Insulation and Size. Conductors shall be insulated, listed as gas tube sign and ignition cable type GTO, rated for 5, 10, or 15 kV, not smaller than 18 AWG, and have a minimum temperature rating of 105°C (221°F).

(C) Installation. Conductors shall be so installed that they are not subject to physical damage.

(D) Bends in Conductors. Sharp bends in insulated conductors shall be avoided.

(E) Spacing. Secondary conductors shall be separated from each other and from all objects other than insulators or neon tubing by a spacing of not less than 38 mm (1½ in.). GTO cable installed in metal conduit or tubing shall not require spacing between the cable insulation and the conduit or tubing.

(F) Insulators and Bushings. Insulators and bushings for conductors shall be listed for use with neon secondary circuits over 1000 volts.

(G) Conductors in Raceways. The insulation on all conductors shall extend not less than 65 mm (2½ in.) beyond the metal conduit or tubing.

(H) Between Neon Tubing and Midpoint Return. Conductors shall be permitted to run between the ends of neon tubing or to the secondary circuit midpoint return of listed transformers or listed electronic power supplies and provided with terminals or leads at the midpoint.

(I) Dwelling Occupancies. Equipment having an open circuit voltage exceeding 1000 volts shall not be installed in or on dwelling occupancies.

(J) Length of Secondary Circuit Conductors.

(1) Secondary Conductor to the First Electrode. The length of secondary circuit conductors from a high-voltage terminal or lead of a transformer or electronic power supply to the first neon tube electrode shall not exceed the following:

(1) 6 m (20 ft) where installed in metal conduit or tubing
(2) 15 m (50 ft) where installed in nonmetallic conduit

(2) Other Secondary Circuit Conductors. All other sections of secondary circuit conductor in a neon tube circuit shall be as short as practicable.

(K) Splices. Splices in high-voltage secondary circuit conductors shall be made in listed enclosures rated over 1000 volts. Splice enclosures shall be accessible after installation and listed for the location where they are installed.

600.33 Class 2 Sign Illumination Systems, Secondary Wiring. The wiring methods and materials used shall be in accordance with the sign manufacturer's installation instructions using any applicable wiring methods from Chapter 3, Wiring Methods, and the requirements for Class 2 circuits contained in 600.12(C), 600.24, and 600.33(A), (B), (C), and (D).

(A) Insulation and Sizing of Class 2 Conductors. Class 2 cable listed for the application that complies with Table 600.33(A)(1) or Table 600.33(A)(2) for substitutions shall be installed on the load side of the Class 2 power source. The conductors shall have an ampacity not less than the load to be supplied and shall not be sized smaller than 18 AWG.

N **(1) General Use.** CL2 or CL3, PLTC, or any listed applicable cable for general use shall be installed within and on buildings or structures.

N **(2) Other Building Locations.** In other locations, any listed applicable cable permitted in 600.33(A)(1), (A)(2), (A)(3), and (A)(4) and Table 600.33(A)(1) and (A)(2) shall be permitted to be used as follows:

(1) CL2P or CL3P — Ducts, plenums, or other spaces used for environmental air
(2) CL2R or CL3R — Vertical shafts and risers
(3) Substitutions from Table 600.33(A)(2)

(3) Wet Locations. Class 2 cable used in a wet location shall be listed and marked suitable for use in a wet location.

(4) Other Locations. Class 2 cable exposed to sunlight shall be listed and marked sunlight resistant suitable for outdoor use.

(B) Installation. Secondary wiring shall be installed in accordance with (B)(1) and (B)(2).

Table 600.33(A)(1) Applications of Power Limited Cable in Signs and Outline Lighting

Location	CL2	CL3	CL2R	CL3R	CL2P	CL3P	PLTC
Non-concealed spaces inside buildings	Y	Y	Y	Y	Y	Y	Y
Concealed spaces inside buildings that are not used as plenums or risers	Y	Y	Y	Y	Y	Y	Y
Environmental air spaces plenums- or risers	N	N	N	N	Y	Y	N
Wet locations	N	N	N	N	N	N	Y

Y = Permitted. N = Not Permitted.

(1) Wiring shall be installed and supported in a neat and workmanlike manner. Cables and conductors installed exposed on the surface of ceilings and sidewalls shall be supported by the building structure in such a manner that the cable is not damaged by normal building use. The cable shall be supported and secured at intervals not exceeding 1.8 m (6 ft). Such cables shall be supported by straps, staples, hangers, cable ties, or similar fittings designed and installed so as not to damage the cable. The installation shall also comply with 300.4(D).

(2) Connections in cable and conductors shall be made with listed insulating devices and be accessible after installation. Where made in a wall, connections shall be enclosed in a listed box.

(C) Protection Against Physical Damage. Where subject to physical damage, the conductors shall be protected and installed in accordance with 300.4.

(D) Grounding and Bonding. Grounding and bonding shall be in accordance with 600.7.

N 600.34 Photovoltaic (PV) Powered Sign. All field wiring of components and subassemblies for an off-grid stand-alone, on-grid interactive, or non-grid interactive PV installation shall be installed in accordance with Article 690, as applicable, 600.34, and the PV powered sign installation instructions.

(A) Equipment. Inverters, motor generators, PV modules, PV panels, ac PV modules, dc combiners, dc-ac converters, and charge controllers intended for use in PV powered sign systems shall be listed for PV application.

(B) Wiring. Wiring from a photovoltaic panel or wiring external to the PV sign body shall be:

(1) Listed, labeled, and suitable for photovoltaic applications
(2) Routed to closely follow the sign body or enclosure
(3) As short as possible and secured at intervals not exceeding 0.91 m (3 ft)
(4) Protected where subject to physical damage

(C) Flexible Cords and Cables. Flexible cords and cables shall comply with Article 400 and be identified as extra hard usage, rated for outdoor use, and water and sunlight resistant.

(D) Grounding. Grounding a PV powered sign shall comply with Article 690, Part V and 600.7.

(E) Disconnecting Means. The disconnecting means for a PV powered sign shall comply with Article 690, Part III and 600.6.

N Table 600.33(A)(2) Class 2 Cable Substitutions

Cable Type	Permitted Substitutions
CL3P	CMP
CL2P	CMP, CL3P
CL3R	CMP, CL3P, CMR
CL2R	CMP, CL3P, CL2P, CMR, CL3R
CL3	CMP, CL3P, CMR, CL3R, CMG, CM, PLTC
CL2	CMP, CL3P, CL2P, CMR, CL3R, CL2R, CMG, CM, PLTC, CL3
CL3X	CMP, CL3P, CMR, CL3R, CMG, CM, PLTC, CL3, CMX
CL2X	CMP, CL3P, CL2P, CMR, CL3R, CL2R, CMG, CM, PLTC, CL3, CL2, CMX, CL3X

(F) Battery Compartments. Battery compartments shall require a tool to open.

600.41 Neon Tubing.

(A) Design. The length and design of the tubing shall not cause a continuous overcurrent beyond the design loading of the transformer or electronic power supply.

(B) Support. Tubing shall be supported by listed tube supports. The neon tubing shall be supported within 150 mm (6 in.) from the electrode connection.

(C) Spacing. A spacing of not less than 6 mm (¼ in.) shall be maintained between the tubing and the nearest surface, other than its support.

(D) Protection. Field-installed skeleton tubing shall not be subject to physical damage. Where the tubing is readily accessible to other than qualified persons, field-installed skeleton tubing shall be provided with suitable guards or protected by other approved means.

600.42 Electrode Connections.

(A) Points of Transition. Where the high-voltage secondary circuit conductors emerge from the wiring methods specified in 600.32(A), they shall be enclosed in a listed assembly.

(B) Accessibility. Terminals of the electrode shall not be accessible to unqualified persons.

(C) Electrode Connections. Connections shall be made by use of a connection device, twisting of the wires together, or use of an electrode receptacle. Connections shall be electrically and mechanically secure and shall be in an enclosure listed for the purpose.

(D) Support. Neon secondary conductor(s) shall be supported not more than 150 mm (6 in.) from the electrode connection to the tubing.

(E) Receptacles. Electrode receptacles shall be listed.

(F) Bushings. Where electrodes penetrate an enclosure, bushings listed for the purpose shall be used unless receptacles are provided.

(G) Wet Locations. A listed cap shall be used to close the opening between neon tubing and a receptacle where the receptacle penetrates a building. Where a bushing or neon tubing penetrates a building, the opening between neon tubing and the bushing shall be sealed.

(H) Electrode Enclosures. Electrode enclosures shall be listed.

(1) Dry Locations. Electrode enclosures that are listed, labeled, and identified for use in dry, damp, or wet locations shall be permitted to be installed and used in such locations.

(2) Damp and Wet Locations. Electrode enclosures installed in damp and wet locations shall be specifically listed, labeled, and identified for use in such locations.

> Informational Note: See 110.3(B) covering installation and use of electrical equipment.

ARTICLE 604
Manufactured Wiring Systems

604.1 Scope. The provisions of this article apply to field-installed wiring using off-site manufactured subassemblies for branch circuits, remote-control circuits, signaling circuits, and communications circuits in accessible areas.

604.2 Definition.

Manufactured Wiring System. A system containing component parts that are assembled in the process of manufacture and cannot be inspected at the building site without damage or destruction to the assembly and used for the connection of luminaires, utilization equipment, continuous plug-in type busways, and other devices.

N **604.6 Listing Requirements.** Manufactured wiring systems and associated components shall be listed.

> Informational Note: ANSI/UL 183, *Standard for Manufacturing Wiring Systems*, is a safety standard for manufactured wiring systems.

604.7 Installation. Manufactured wiring systems shall be secured and supported in accordance with the applicable cable or conduit article for the cable or conduit type employed.

604.10 Uses Permitted. Manufactured wiring systems shall be permitted in accessible and dry locations and in ducts, plenums, and other air-handling spaces where listed for this application and installed in accordance with 300.22.

Exception No. 1: In concealed spaces, one end of tapped cable shall be permitted to extend into hollow walls for direct termination at switch and outlet points.

Exception No. 2: Manufactured wiring system assemblies installed outdoors shall be listed for use in outdoor locations.

604.12 Uses Not Permitted. Manufactured wiring system types shall not be permitted where limited by the applicable article in Chapter 3 for the wiring method used in its construction.

604.100 Construction.

(A) Cable or Conduit Types.

(1) Cables. Cable shall be one of the following:

(1) Listed Type AC cable containing nominal 600-volt, 8 to 12 AWG insulated copper conductors with a bare or insulated copper equipment grounding conductor equivalent in size to the ungrounded conductor.
(2) Listed Type MC cable containing nominal 600-volt, 8 to 12 AWG insulated copper conductors with a bare or insulated copper equipment grounding conductor equivalent in size to the ungrounded conductor.
(3) Listed Type MC cable containing nominal 600-volt, 8 to 12 AWG insulated copper conductors with a grounding conductor and armor assembly listed and identified for grounding in accordance with 250.118(10). The combined metallic sheath and grounding conductor shall have a current-carrying capacity equivalent to that of the ungrounded copper conductor.

Other cables as listed in 725.154, 800.113, 820.113, and 830.179 shall be permitted in manufactured wiring systems for wiring of equipment within the scope of their respective articles.

(2) Conduits. Conduit shall be listed flexible metal conduit or listed liquidtight flexible conduit containing nominal 600-volt, 8 to 12 AWG insulated copper conductors with a bare or insulated copper equipment grounding conductor equivalent in size to the ungrounded conductor.

Exception No. 1 to (1) and (2): A luminaire tap, no longer than 1.8 m (6 ft) and intended for connection to a single luminaire, shall be permitted to contain conductors smaller than 12 AWG but not smaller than 18 AWG.

Exception No. 2 to (1) and (2): Listed manufactured wiring assemblies containing conductors smaller than 12 AWG shall be permitted for remote-control, signaling, or communication circuits.

Exception No. 3 to (2): Listed manufactured wiring systems containing unlisted flexible metal conduit of noncircular cross section or trade sizes smaller than permitted by 348.20(A), or both, shall be permitted where the wiring systems are supplied with fittings and conductors at the time of manufacture.

(3) Flexible Cord. Flexible cord suitable for hard usage, with minimum 12 AWG conductors, shall be permitted as part of a listed factory-made assembly not exceeding 1.8 m (6 ft) in length when making a transition between components of a manufactured wiring system and utilization equipment not permanently secured to the building structure. The cord shall be visible for the entire length, shall not be subject to physical damage, and shall be provided with identified strain relief.

Exception: Listed electric-discharge luminaires that comply with 410.62(C) shall be permitted with conductors smaller than 12 AWG.

(4) Busways. Busways shall be listed continuous plug-in type containing factory-mounted, bare or insulated conductors, which shall be copper or aluminum bars, rods, or tubes. The busway shall be provided with an equipment ground. The busway shall be rated nominal 600 volts, 20, 30, or 40 amperes. Busways shall be installed in accordance with 368.12, 368.17(D), and 368.30.

(5) Raceway. Prewired, modular, surface-mounted raceways shall be listed for the use, rated nominal 600 volts, 20 amperes, and installed in accordance with 386.12, 386.30, 386.60, and 386.100.

(B) Marking. Each section shall be marked to identify the type of cable, flexible cord, or conduit.

(C) Receptacles and Connectors. Receptacles and connectors shall be of the locking type, uniquely polarized and identified for the purpose, and shall be part of a listed assembly for the appropriate system. All connector openings shall be designed to prevent inadvertent contact with live parts or capped to effectively close the connector openings.

(D) Other Component Parts. Other component parts shall be listed for the appropriate system.

ARTICLE 605
Office Furnishings

605.1 Scope. This article covers electrical equipment, lighting accessories, and wiring systems used to connect, contained within, or installed on office furnishings.

605.2 Definition.

Office Furnishing. Cubicle panels, partitions, study carrels, workstations, desks, shelving systems, and storage units that may be mechanically and electrically interconnected to form an office furnishing system.

605.3 General. Wiring systems shall be identified as suitable for providing power for lighting accessories and utilization equipment used within office furnishings. A wired partition shall not extend from floor to ceiling.

Exception: Where permitted by the authority having jurisdiction, these relocatable wired partitions shall be permitted to extend to, but shall not penetrate, the ceiling.

(A) Use. These assemblies shall be installed and used only as provided for by this article.

(B) Hazardous (Classified) Locations. Where used in hazardous (classified) locations, these assemblies shall comply with Articles 500 through 517 in addition to this article.

605.4 Wireways. All conductors and connections shall be contained within wiring channels of metal or other material identified as suitable for the conditions of use. Wiring channels shall be free of projections or other conditions that might damage conductor insulation.

605.5 Office Furnishing Interconnections. The electrical connection between office furnishings shall be a flexible assembly identified for use with office furnishings or shall be permitted to be installed using flexible cord, provided that all the following conditions are met:

(1) The cord is extra-hard usage type with 12 AWG or larger conductors, with an insulated equipment grounding conductor.
(2) The office furnishings are mechanically contiguous.
(3) The cord is not longer than necessary for maximum positioning of the office furnishing but is in no case to exceed 600 mm (2 ft).
(4) The cord is terminated at an attachment plug-and-cord connector with strain relief.

605.6 Lighting Accessories. Lighting equipment shall be listed, labeled, and identified for use with office furnishings and shall comply with 605.6(A), (B), and (C).

(A) Support. A means for secure attachment or support shall be provided.

(B) Connection. Where cord and plug connection is provided, it shall comply with all of the following:

(1) The cord length shall be suitable for the intended application but shall not exceed 2.7 m (9 ft) in length.
(2) The cord shall not be smaller than 18 AWG.
(3) The cord shall contain an equipment grounding conductor, except as specified in 605.6(B)(4).
(4) Cords on the load side of a listed Class 2 power source shall not be required to contain an equipment grounding conductor.
(5) The cord shall be of the hard usage type, except as specified in 605.6(B)(6).
(6) A cord provided on a listed Class 2 power source shall be of the type provided with the listed luminaire assembly or of the type specified in 725.130 and 725.127.
(7) Connection by other means shall be identified as suitable for the conditions of use.

(C) Receptacle Outlet. Receptacles shall not be permitted in lighting accessories.

605.7 Fixed-Type Office Furnishings. Office furnishings that are fixed (secured to building surfaces) shall be permanently connected to the building electrical system by one of the wiring methods of Chapter 3.

605.8 Freestanding-Type Office Furnishings. Office furnishings of the freestanding type (not fixed) shall be permitted to be connected to the building electrical system by one of the wiring methods of Chapter 3.

605.9 Freestanding-Type Office Furnishings, Cord- and Plug-Connected. Individual office furnishings of the freestanding type, or groups of individual office furnishings that are electrically connected, are mechanically contiguous, and do not exceed 9.0 m (30 ft) when assembled, shall be permitted to be connected to the building electrical system by a single flexible cord and plug, provided that all of the conditions of 605.9(A) through (D) are met.

(A) Flexible Power-Supply Cord. The flexible power supply cord shall be extra-hard usage type with 12 AWG or larger conductors, with an insulated equipment grounding conductor, and shall not exceed 600 mm (2 ft) in length.

(B) Receptacle Supplying Power. The receptacle(s) supplying power shall be on a separate circuit serving only the office furnishing and no other loads and shall be located not more than 300 mm (12 in.) from the office furnishing that is connected to it.

(C) Receptacle, Maximum. An individual office furnishing or groups of interconnected individual office furnishings shall not contain more than 13 15-ampere, 125-volt receptacles. For purposes of this requirement, a receptacle is considered (1) up to two (simplex) receptacles provided within a single enclosure and that are within 0.3 m (1 ft) of each other or (2) one duplex receptacle.

(D) Multiwire Circuits, Not Permitted. An individual office furnishing or groups of interconnected office furnishings shall not contain multiwire circuits.

> Informational Note: See 210.4 for circuits supplying office furnishings in 605.7 and 605.8.

ARTICLE 610
Cranes and Hoists

Part I. General

610.1 Scope. This article covers the installation of electrical equipment and wiring used in connection with cranes, monorail hoists, hoists, and all runways.

> Informational Note: For further information, see ASME B30, *Safety Standards for Cableways, Cranes, Derricks, Hoists, Hooks, Jacks, and Slings.*

610.2 Definition.

Festoon Cable. Single- and multiple-conductor cable intended for use and installation in accordance with Article 610 where flexibility is required.

610.3 Special Requirements for Particular Locations.

(A) Hazardous (Classified) Locations. All equipment that operates in a hazardous (classified) location shall conform to Article 500.

(1) Class I Locations. Equipment used in locations that are hazardous because of the presence of flammable gases or vapors shall conform to Article 501.

(2) Class II Locations. Equipment used in locations that are hazardous because of combustible dust shall conform to Article 502.

(3) Class III Locations. Equipment used in locations that are hazardous because of the presence of easily ignitible fibers or flyings shall conform to Article 503.

(B) Combustible Materials. Where a crane, hoist, or monorail hoist operates over readily combustible material, the resistors shall be located as permitted in the following:

(1) A well ventilated cabinet composed of noncombustible material constructed so that it does not emit flames or molten metal

(2) A cage or cab constructed of noncombustible material that encloses the sides of the cage or cab from the floor to a point at least 150 mm (6 in.) above the top of the resistors

(C) Electrolytic Cell Lines. See 668.32.

Part II. Wiring

610.11 Wiring Method. Conductors shall be enclosed in raceways or be Type AC cable with insulated grounding conductor, Type MC cable, or Type MI cable unless otherwise permitted or required in 610.11(A) through (E).

(A) Contact Conductor. Contact conductors shall not be required to be enclosed in raceways.

(B) Exposed Conductors. Short lengths of exposed conductors at resistors, collectors, and other equipment shall not be required to be enclosed in raceways.

(C) Flexible Connections to Motors and Similar Equipment. Where flexible connections are necessary, flexible stranded conductors shall be used. Conductors shall be in flexible metal conduit, liquidtight flexible metal conduit, liquidtight flexible nonmetallic conduit, multiconductor cable, or an approved nonmetallic flexible raceway.

(D) Pushbutton Station Multiconductor Cable. Where multiconductor cable is used with a suspended pushbutton station, the station shall be supported in some satisfactory manner that protects the electrical conductors against strain.

(E) Flexibility to Moving Parts. Where flexibility is required for power or control to moving parts, listed festoon cable or a cord suitable for the purpose shall be permitted, provided the following apply:

(1) Suitable strain relief and protection from physical damage is provided.
(2) In Class I, Division 2 locations, the cord is approved for extra-hard usage.

610.12 Raceway or Cable Terminal Fittings. Conductors leaving raceways or cables shall comply with either 610.12(A) or (B).

(A) Separately Bushed Hole. A box or terminal fitting that has a separately bushed hole for each conductor shall be used wherever a change is made from a raceway or cable to exposed wiring. A fitting used for this purpose shall not contain taps or splices and shall not be used at luminaire outlets.

(B) Bushing in Lieu of a Box. A bushing shall be permitted to be used in lieu of a box at the end of a rigid metal conduit, intermediate metal conduit, or electrical metallic tubing where the raceway terminates at unenclosed controls or similar equipment, including contact conductors, collectors, resistors, brakes, power-circuit limit switches, and dc split-frame motors.

610.13 Types of Conductors. Conductors shall comply with Table 310.104(A) unless otherwise permitted in 610.13(A) through (D).

(A) Exposed to External Heat or Connected to Resistors. A conductor(s) exposed to external heat or connected to resistors shall have a flame-resistant outer covering or be covered with flame-resistant tape individually or as a group.

(B) Contact Conductors. Contact conductors along runways, crane bridges, and monorails shall be permitted to be bare and shall be copper, aluminum, steel, or other alloys or combinations thereof in the form of hard-drawn wire, tees, angles, tee rails, or other stiff shapes.

(C) Flexibility. Where flexibility is required, listed flexible cord or cable, or listed festoon cable, shall be permitted to be used and, where necessary, cable reels or take-up devices shall be used.

(D) Class 1, Class 2, and Class 3 Circuits. Conductors for Class 1, Class 2, and Class 3 remote-control, signaling, and power-limited circuits, installed in accordance with Article 725, shall be permitted.

610.14 Rating and Size of Conductors.

(A) Ampacity. The allowable ampacities of conductors shall be as shown in Table 610.14(A).

Informational Note: For the ampacities of conductors between controllers and resistors, see 430.23.

Table 610.14(A) Ampacities of Insulated Copper Conductors Used with Short-Time Rated Crane and Hoist Motors. Based on Ambient Temperature of 30°C (86°F).

Maximum Operating Temperature	Up to Four Simultaneously Energized Conductors in Raceway or Cable[1]				Up to Three ac[2] or Four dc[1] Simultaneously Energized Conductors in Raceway or Cable		Maximum Operating Temperature
	75°C (167°F)		90°C (194°F)		125°C (257°F)		
Size (AWG or kcmil)	Types MTW, RHW, THW, THWN, XHHW, USE, ZW		Types TA, TBS, SA, SIS, PFA, FEP, FEPB, RHH, THHN, XHHW, Z, ZW		Types FEP, FEPB, PFA, PFAH, SA, TFE, Z, ZW		Size (AWG or kcmil)
	60 Min	30 Min	60 Min	30 Min	60 Min	30 Min	
16	10	12	—	—	—	—	16
14	25	26	31	32	38	40	14
12	30	33	36	40	45	50	12
10	40	43	49	52	60	65	10
8	55	60	63	69	73	80	8
6	76	86	83	94	101	119	6
5	85	95	95	106	115	134	5
4	100	117	111	130	133	157	4
3	120	141	131	153	153	183	3
2	137	160	148	173	178	214	2
1	143	175	158	192	210	253	1
1/0	190	233	211	259	253	304	1/0
2/0	222	267	245	294	303	369	2/0
3/0	280	341	305	372	370	452	3/0
4/0	300	369	319	399	451	555	4/0
250	364	420	400	461	510	635	250
300	455	582	497	636	587	737	300
350	486	646	542	716	663	837	350
400	538	688	593	760	742	941	400
450	600	765	660	836	818	1042	450
500	660	847	726	914	896	1143	500

AMPACITY CORRECTION FACTORS

Ambient Temperature (°C)	For ambient temperatures other than 30°C (86°F), multiply the ampacities shown above by the appropriate factor shown below.						Ambient Temperature (°F)
21–25	1.05	1.05	1.04	1.04	1.02	1.02	70–77
26–30	1.00	1.00	1.00	1.00	1.00	1.00	79–86
31–35	0.94	0.94	0.96	0.96	0.97	0.97	88–95
36–40	0.88	0.88	0.91	0.91	0.95	0.95	97–104
41–45	0.82	0.82	0.87	0.87	0.92	0.92	106–113
46–50	0.75	0.75	0.82	0.82	0.89	0.89	115–122
51–55	0.67	0.67	0.76	0.76	0.86	0.86	124–131

(continues)

AMPACITY CORRECTION FACTORS

Ambient Temperature (°C)	For ambient temperatures other than 30°C (86°F), multiply the ampacities shown above by the appropriate factor shown below.						Ambient Temperature (°F)
56–60	0.58	0.58	0.71	0.71	0.83	0.83	133–140
61–70	0.33	0.33	0.58	0.58	0.76	0.76	142–158
71–80	—	—	0.41	0.41	0.69	0.69	160–176
81–90	—	—	—	—	0.61	0.61	177–194
91–100	—	—	—	—	0.51	0.51	195–212
101–120	—	—	—	—	0.40	0.40	213–248

Note: Other insulations shown in Table 310.104(A) and approved for the temperature and location shall be permitted to be substituted for those shown in Table 610.14(A). The allowable ampacities of conductors used with 15-minute motors shall be the 30-minute ratings increased by 12 percent.

[1] For 5 to 8 simultaneously energized power conductors in raceway or cable, the ampacity of each power conductor shall be reduced to a value of 80 percent of that shown in this table.

[2] For 4 to 6 simultaneously energized 125°C (257°F) ac power conductors in raceway or cable, the ampacity of each power conductor shall be reduced to a value of 80 percent of that shown in this table.

(B) Secondary Resistor Conductors. Where the secondary resistor is separate from the controller, the minimum size of the conductors between controller and resistor shall be calculated by multiplying the motor secondary current by the appropriate factor from Table 610.14(B) and selecting a wire from Table 610.14(A).

(C) Minimum Size. Conductors external to motors and controls shall be not smaller than 16 AWG unless otherwise permitted in (1) or (2):

(1) 18 AWG wire in multiconductor cord shall be permitted for control circuits not exceeding 7 amperes.

(2) Wires not smaller than 20 AWG shall be permitted for electronic circuits.

(D) Contact Conductors. Contact wires shall have an ampacity not less than that required by Table 610.14(A) for 75°C (167°F) wire, and in no case shall they be smaller than as shown in Table 610.14(D).

Table 610.14(B) Secondary Conductor Rating Factors

Time in Seconds		Ampacity of Wire in Percent of Full-Load Secondary Current
On	Off	
5	75	35
10	70	45
15	75	55
15	45	65
15	30	75
15	15	85
Continuous Duty		110

Table 610.14(D) Minimum Contact Conductor Size Based on Distance Between Supports

Minimum Size of Wire (AWG)	Maximum Distance Between End Strain Insulators or Clamp-Type Intermediate Supports
6	9.0 m (30 ft) or less
4	18 m (60 ft) or less
2	Over 18 m (60 ft)

(E) Calculation of Motor Load.

(1) Single Motor. For one motor, 100 percent of motor nameplate full-load ampere rating shall be used.

(2) Multiple Motors on Single Crane or Hoist. For multiple motors on a single crane or hoist, the minimum ampacity of the power supply conductors shall be the nameplate full-load ampere rating of the largest motor or group of motors for any single crane motion, plus 50 percent of the nameplate full-load ampere rating of the next largest motor or group of motors, using that column of Table 610.14(A) that applies to the longest time-rated motor.

(3) Multiple Cranes or Hoists on a Common Conductor System. For multiple cranes, hoists, or both, supplied by a common conductor system, calculate the motor minimum ampacity shall be calculated for each crane as defined in 610.14(E), added them together, and the sum multiplied by the appropriate demand factor from Table 610.14(E).

(F) Other Loads. Additional loads, such as heating, lighting, and air conditioning, shall be provided for by application of the appropriate sections of this *Code*.

(G) Nameplate. Each crane, monorail, or hoist shall be provided with a visible nameplate marked with the manufacturer's name, rating in volts, frequency, number of phases, and circuit amperes as calculated in 610.14(E) and (F).

610.15 Common Return. Where a crane or hoist is operated by more than one motor, a common-return conductor of proper ampacity shall be permitted.

Table 610.14(E) Demand Factors

Number of Cranes or Hoists	Demand Factor
2	0.95
3	0.91
4	0.87
5	0.84
6	0.81
7	0.78

Part III. Contact Conductors

610.21 Installation of Contact Conductors. Contact conductors shall comply with 610.21(A) through (H).

(A) Locating or Guarding Contact Conductors. Runway contact conductors shall be guarded, and bridge contact conductors shall be located or guarded in such a manner that persons cannot inadvertently touch energized current-carrying parts.

(B) Contact Wires. Wires that are used as contact conductors shall be secured at the ends by means of approved strain insulators and shall be mounted on approved insulators so that the extreme limit of displacement of the wire does not bring the latter within less than 38 mm (1½ in.) from the surface wired over.

(C) Supports Along Runways. Main contact conductors carried along runways shall be supported on insulating supports placed at intervals not exceeding 6.0 m (20 ft) unless otherwise permitted in 610.21(F).

Such conductors shall be separated at not less than 150 mm (6 in.), other than for monorail hoists where a spacing of not less than 75 mm (3 in.) shall be permitted. Where necessary, intervals between insulating supports shall be permitted to be increased up to 12 m (40 ft), the separation between conductors being increased proportionately.

(D) Supports on Bridges. Bridge wire contact conductors shall be kept at least 65 mm (2½ in.) apart, and, where the span exceeds 25 m (80 ft), insulating saddles shall be placed at intervals not exceeding 15 m (50 ft).

(E) Supports for Rigid Conductors. Conductors along runways and crane bridges, that are of the rigid type specified in 610.13(B) and not contained within an approved enclosed assembly, shall be carried on insulating supports spaced at intervals of not more than 80 times the vertical dimension of the conductor, but in no case greater than 4.5 m (15 ft), and spaced apart sufficiently to give a clear electrical separation of conductors or adjacent collectors of not less than 25 mm (1 in.).

(F) Track as Circuit Conductor. Monorail, tram rail, or crane runway tracks shall be permitted as a conductor of current for one phase of a 3-phase, ac system furnishing power to the carrier, crane, or trolley, provided all of the following conditions are met:

(1) The conductors supplying the other two phases of the power supply are insulated.
(2) The power for all phases is obtained from an insulating transformer.
(3) The voltage does not exceed 300 volts.
(4) The rail serving as a conductor shall be bonded to the equipment grounding conductor at the transformer and also shall be permitted to be grounded by the fittings used for the suspension or attachment of the rail to a building or structure.

(G) Electrical Continuity of Contact Conductors. All sections of contact conductors shall be mechanically joined to provide a continuous electrical connection.

(H) Not to Supply Other Equipment. Contact conductors shall not be used as feeders for any equipment other than the crane(s) or hoist(s) that they are primarily designed to serve.

610.22 Collectors. Collectors shall be designed so as to reduce to a minimum sparking between them and the contact conductor; and, where operated in rooms used for the storage of easily ignitible combustible fibers and materials, they shall comply with 503.155.

Part IV. Disconnecting Means

610.31 Runway Conductor Disconnecting Means. A disconnecting means that has a continuous ampere rating not less than that calculated in 610.14(E) and (F) shall be provided between the runway contact conductors and the power supply. The disconnecting means shall comply with 430.109. This disconnecting means shall be as follows:

(1) Readily accessible and operable from the ground or floor level
(2) Lockable open in accordance with 110.25
(3) Open all ungrounded conductors simultaneously
(4) Placed within view of the runway contact conductors

Exception: The runway conductor disconnecting means for electrolytic cell lines shall be permitted to be placed out of view of the runway contact conductors where either of the following conditions are met:

(1) Where a location in view of the contact conductors is impracticable or introduces additional or increased hazards to persons or property
(2) In industrial installations, with written safety procedures, where conditions of maintenance and supervision ensure that only qualified persons service the equipment

610.32 Disconnecting Means for Cranes and Monorail Hoists. A disconnecting means in compliance with 430.109 shall be provided in the leads from the runway contact conductors or other power supply on all cranes and monorail hoists. The disconnecting means shall be lockable open in accordance with 110.25.

Where a monorail hoist or hand-propelled crane bridge installation meets all of the following, the disconnecting means shall be permitted to be omitted:

(1) The unit is controlled from the ground or floor level.
(2) The unit is within view of the power supply disconnecting means.
(3) No fixed work platform has been provided for servicing the unit.

Means shall be provided at the operating station to open the power circuit to all motors of the crane or monorail hoist.

610.33 Rating of Disconnecting Means. The continuous ampere rating of the switch or circuit breaker required by 610.32 shall not be less than 50 percent of the combined short-time ampere rating of the motors or less than 75 percent of the sum of the short-time ampere rating of the motors required for any single motion.

Part V. Overcurrent Protection

610.41 Feeders, Runway Conductors.

(A) Single Feeder. The runway supply conductors and main contact conductors of a crane or monorail shall be protected by an overcurrent device(s) that shall not be greater than the largest rating or setting of any branch-circuit protective device plus the sum of the nameplate ratings of all the other loads with application of the demand factors from Table 610.14(E).

(B) More Than One Feeder Circuit. Where more than one feeder circuit is installed to supply runway conductors, each feeder circuit shall be sized and protected in compliance with 610.41(A).

610.42 Branch-Circuit Short-Circuit and Ground-Fault Protection. Branch circuits shall be protected in accordance with 610.42(A). Branch-circuit taps, where made, shall comply with 610.42(B).

(A) Fuse or Circuit Breaker Rating. Crane, hoist, and monorail hoist motor branch circuits shall be protected by fuses or inverse-time circuit breakers that have a rating in accordance with Table 430.52. Where two or more motors operate a single motion, the sum of their nameplate current ratings shall be considered as that of a single motor.

(B) Taps.

(1) Multiple Motors. Where two or more motors are connected to the same branch circuit, each tap conductor to an individual motor shall have an ampacity not less than one-third that of the branch circuit. Each motor shall be protected from overload according to 610.43.

(2) Control Circuits. Where taps to control circuits originate on the load side of a branch-circuit protective device, each tap and piece of equipment shall be protected in accordance with 430.72.

610.43 Overload Protection.

(A) Motor and Branch-Circuit Overload Protection. Each motor, motor controller, and branch-circuit conductor shall be protected from overload by one of the following means:

(1) A single motor shall be considered as protected where the branch-circuit overcurrent device meets the rating requirements of 610.42.
(2) Overload relay elements in each ungrounded circuit conductor, with all relay elements protected from short circuit by the branch-circuit protection.
(3) Thermal sensing devices, sensitive to motor temperature or to temperature and current, that are thermally in contact with the motor winding(s). Hoist functions shall be considered to be protected if the sensing device limits the hoist to lowering only during an overload condition. Traverse functions shall be considered to be protected if the sensing device limits the travel in both directions for the affected function during an overload condition of either motor.

(B) Manually Controlled Motor. If the motor is manually controlled, with spring return controls, the overload protective device shall not be required to protect the motor against stalled rotor conditions.

(C) Multimotor. Where two or more motors drive a single trolley, truck, or bridge and are controlled as a unit and protected by a single set of overload devices with a rating equal to the sum of their rated full-load currents, a hoist or trolley shall be considered to be protected if the sensing device is connected in the hoist's upper limit switch circuit so as to prevent further hoisting during an overtemperature condition of either motor.

(D) Hoists and Monorail Hoists. Hoists and monorail hoists and their trolleys that are not used as part of an overhead traveling crane shall not require individual motor overload protection, provided the largest motor does not exceed 7½ hp and all motors are under manual control of the operator.

Part VI. Control

610.51 Separate Controllers. Each motor shall be provided with an individual controller unless otherwise permitted in 610.51(A) or (B).

(A) Motions with More Than One Motor. Where two or more motors drive a single hoist, carriage, truck, or bridge, they shall be permitted to be controlled by a single controller.

(B) Multiple Motion Controller. One controller shall be permitted to be switched between motors, under the following conditions:

(1) The controller has a horsepower rating that is not lower than the horsepower rating of the largest motor.
(2) Only one motor is operated at one time.

610.53 Overcurrent Protection. Conductors of control circuits shall be protected against overcurrent. Control circuits shall be considered as protected by overcurrent devices that are rated or set at not more than 300 percent of the ampacity of the control conductors, unless otherwise permitted in 610.53(A) or (B).

(A) Taps to Control Transformers. Taps to control transformers shall be considered as protected where the secondary circuit is protected by a device rated or set at not more than 200 percent of the rated secondary current of the transformer and not more than 200 percent of the ampacity of the control circuit conductors.

(B) Continuity of Power. Where the opening of the control circuit would create a hazard, as for example, the control circuit of a hot metal crane, the control circuit conductors shall be considered as being properly protected by the branch-circuit overcurrent devices.

610.57 Clearance. The dimension of the working space in the direction of access to live parts that are likely to require examination, adjustment, servicing, or maintenance while energized shall be a minimum of 750 mm (2½ ft). Where controls are enclosed in cabinets, the door(s) shall either open at least 90 degrees or be removable.

Part VII. Grounding

610.61 Grounding. All exposed non–current-carrying metal parts of cranes, monorail hoists, hoists, and accessories, including pendant controls, shall be bonded either by mechanical connections or bonding jumpers, where applicable, so that the entire crane or hoist is a ground-fault current path as required or permitted by Article 250, Parts V and VII.

Moving parts, other than removable accessories, or attachments that have metal-to-metal bearing surfaces, shall be considered to be electrically bonded to each other through bearing surfaces for grounding purposes. The trolley frame and bridge frame shall not be considered as electrically grounded through the bridge and trolley wheels and its respective tracks. A separate bonding conductor shall be provided.

ARTICLE 620
Elevators, Dumbwaiters, Escalators, Moving Walks, Platform Lifts, and Stairway Chairlifts

Part I. General

620.1 Scope. This article covers the installation of electrical equipment and wiring used in connection with elevators, dumbwaiters, escalators, moving walks, platform lifts, and stairway chairlifts.

> Informational Note No. 1: For further information, see ASME A17.1-2013/CSA B44-13, *Safety Code for Elevators and Escalators.*

> Informational Note No. 2: For further information, see CSA B44.1-11/ASME-A17.5-2014, *Elevator and Escalator Electrical Equipment*.

> Informational Note No. 3: The term *wheelchair lift* has been changed to *platform lift*. For further information, see ASME A18.1-2014, *Safety Standard for Platform Lifts and Stairway Chairlifts*.

620.2 Definitions.

> Informational Note No. 1: The motor controller, motion controller, and operation controller are located in a single enclosure or a combination of enclosures.

> Informational Note No. 2: Informational Note Figure 620.2, No. 2 is for information only.

Control Room (for Elevator, Dumbwaiter). An enclosed control space outside the hoistway, intended for full bodily entry, that contains the elevator motor controller. The room could also contain electrical and/or mechanical equipment used directly in connection with the elevator or dumbwaiter but not the electric driving machine or the hydraulic machine.

Control Space (for Elevator, Dumbwaiter). A space inside or outside the hoistway, intended to be accessed with or without full bodily entry, that contains the elevator motor controller. This space could also contain electrical and/or mechanical equipment used directly in connection with the elevator or dumbwaiter but not the electrical driving machine or the hydraulic machine.

Control System. The overall system governing the starting, stopping, direction of motion, acceleration, speed, and retardation of the moving member.

Controller, Motion. The electrical device(s) for that part of the control system that governs the acceleration, speed, retardation, and stopping of the moving member.

Controller, Motor. The operative units of the control system comprised of the starter device(s) and power conversion equipment used to drive an electric motor, or the pumping unit used to power hydraulic control equipment.

Controller, Operation. The electrical device(s) for that part of the control system that initiates the starting, stopping, and direction of motion in response to a signal from an operating device.

Machine Room (for Elevator, Dumbwaiter). An enclosed machinery space outside the hoistway, intended for full bodily entry, that contains the electrical driving machine or the hydraulic machine. The room could also contain electrical and/or mechanical equipment used directly in connection with the elevator or dumbwaiter.

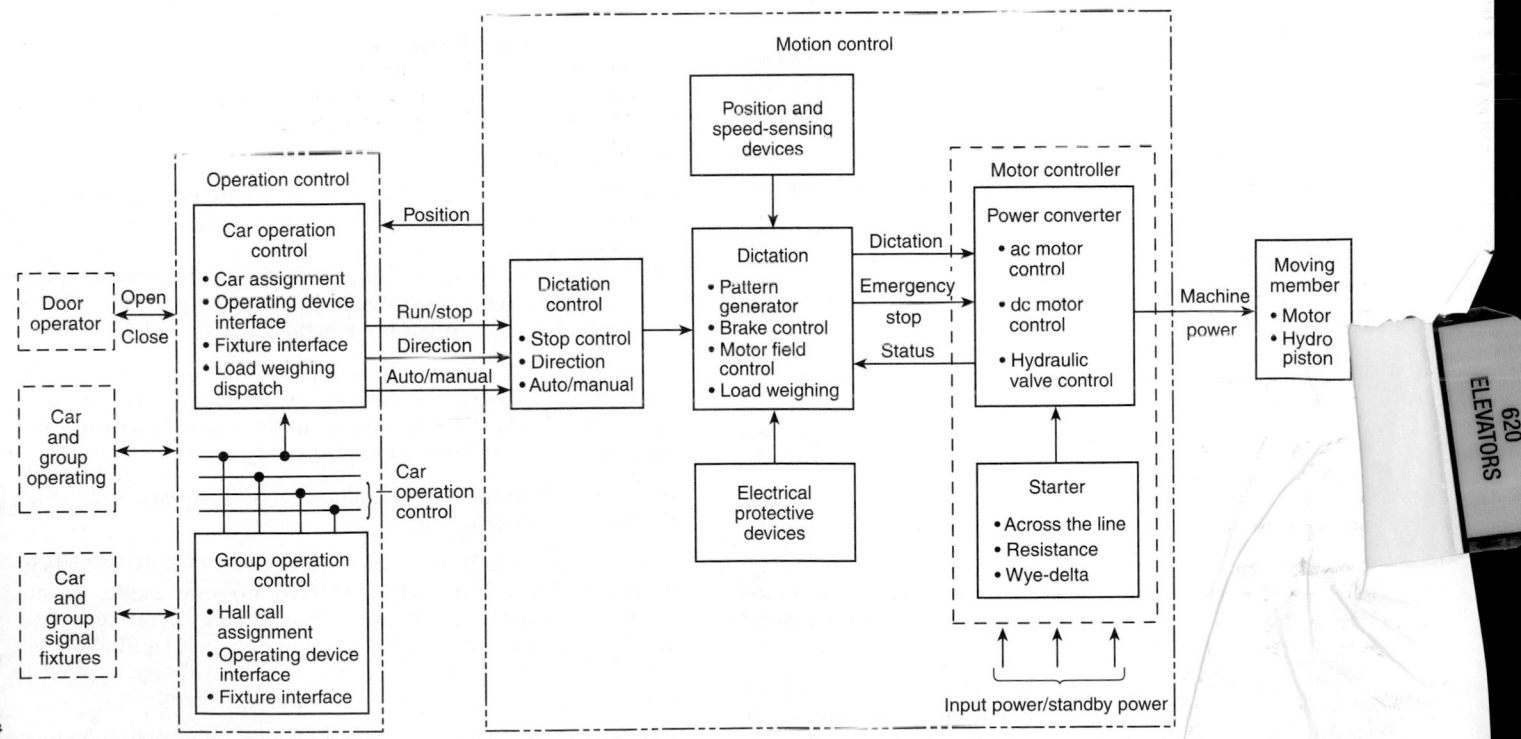

Informational Note Figure 620.2, No. 2   Control System.

Machinery Space (for Elevator, Dumbwaiter). A space inside or outside the hoistway, intended to be accessed with or without full bodily entry, that contains elevator or dumbwaiter mechanical equipment, and could also contain electrical equipment used directly in connection with the elevator or dumbwaiter. This space could also contain the electrical driving machine or the hydraulic machine.

Operating Device. The car switch, pushbuttons, key or toggle switch(s), or other devices used to activate the operation controller.

Remote Machine Room and Control Room (for Elevator, Dumbwaiter). A machine room or control room that is not attached to the outside perimeter or surface of the walls, ceiling, or floor of the hoistway.

Remote Machinery Space and Control Space (for Elevator, Dumbwaiter). A machinery space or control space that is not within the hoistway, machine room, or control room and that is not attached to the outside perimeter or surface of the walls, ceiling, or floor of the hoistway.

Signal Equipment. Includes audible and visual equipment such as chimes, gongs, lights, and displays that convey information to the user.

620.3 Voltage Limitations. The supply voltage shall not exceed 300 volts between conductors unless otherwise permitted in 620.3(A) through (C).

(A) Power Circuits. Branch circuits to door operator controllers and door motors and branch circuits and feeders to motor controllers, driving machine motors, machine brakes, and motor-generator sets shall not have a circuit voltage in excess of 1000 volts. Internal voltages of power conversion equipment and functionally associated equipment, and the operating voltages of wiring interconnecting the equipment, shall be permitted to be higher, provided that all such equipment and wiring shall be listed for the higher voltages. Where the voltage exceeds 600 volts, warning labels or signs that read "DANGER — HIGH VOLTAGE" shall be attached to the equipment and shall be plainly visible. The danger sign(s) or label(s) shall comply with 110.21(B).

(B) Lighting Circuits. Lighting circuits shall comply with the requirements of Article 410.

(C) Heating and Air-Conditioning Circuits. Branch circuits for heating and air-conditioning equipment located on the elevator car shall not have a circuit voltage in excess of 1000 volts.

620.4 Live Parts Enclosed. All live parts of electrical apparatus in the hoistways, at the landings, in or on the cars of elevators and dumbwaiters, in the wellways or the landings of escalators or moving walks, or in the runways and machinery spaces of platform lifts and stairway chairlifts shall be enclosed to protect against accidental contact.

Informational Note: See 110.27 for guarding of live parts (1000 volts, nominal, or less).

620.5 Working Clearances. Working space shall be provided about controllers, disconnecting means, and other electrical equipment in accordance with 110.26(A).

Where conditions of maintenance and supervision ensure that only qualified persons examine, adjust, service, and maintain the equipment the clearance requirements of 110.26(A)

shall not be required where any of the conditions in 620.5(A) through (D) are met.

(A) Flexible Connections to Equipment. Electrical equipment in (A)(1) through (A)(4) is provided with flexible leads to all external connections so that it can be repositioned to meet the clear working space requirements of 110.26:

(1) Controllers and disconnecting means for dumbwaiters, escalators, moving walks, platform lifts, and stairway chairlifts installed in the same space with the driving machine

(2) Controllers and disconnecting means for elevators installed in the hoistway or on the car

(3) Controllers for door operators

(4) Other electrical equipment installed in the hoistway or on the car

(B) Guards. Live parts of the electrical equipment are suitably guarded, isolated, or insulated to reduce the likelihood of inadvertent contact with live parts operating at voltages greater than 30 volts ac rms, 42 volts ac peak, or 60 volts dc, and the equipment can be examined, adjusted, serviced, or maintained while energized without removal of this protection.

(C) Examination, Adjusting, and Servicing. Electrical equipment is not required to be examined, adjusted, serviced, or maintained while energized.

(D) Low Voltage. Uninsulated parts are at a voltage not greater than 30 volts rms, 42 volts peak, or 60 volts dc.

Part II. Conductors

620.11 Insulation of Conductors. The insulation of conductors shall comply with 620.11(A) through (D).

Informational Note: One method of determining that the insulation of conductors is flame retardant is by testing the conductors or cables to the VW-1 (Vertical-Wire) Flame Test in ANSI/UL 1581-2011, *Reference Standard for Electrical Wires, Cables, and Flexible Cords.*

(A) Hoistway Door Interlock Wiring. The conductors to the hoistway door interlocks from the hoistway riser shall be one of the following:

(1) Flame retardant and suitable for a temperature of not less than 200°C (392°F). Conductors shall be Type SF or equivalent.

(2) Physically protected using an approved method, such that the conductor assembly is flame retardant and suitable for a temperature of not less than 200°C (392°F).

(B) Traveling Cables. Traveling cables used as flexible connections between the elevator or dumbwaiter car or counterweight and the raceway shall be of the types of elevator cable listed in Table 400.4 or other approved types.

(C) Other Wiring. All conductors in raceways shall have flame-retardant insulation.

Conductors shall be Type MTW, TF, TFF, TFN, TFFN, THHN, THW, THWN, TW, XHHW, hoistway cable, or any other conductor with insulation designated as flame retardant. Shielded conductors shall be permitted if such conductors are insulated for the maximum nominal circuit voltage applied to any conductor within the cable or raceway system.

(D) Insulation. All conductors shall have an insulation voltage rating equal to at least the maximum nominal circuit voltage applied to any conductor within the enclosure, cable, or raceway. Insulations and outer coverings that are marked for limited smoke and are so listed shall be permitted.

620.12 Minimum Size of Conductors. The minimum size of conductors, other than conductors that form an integral part of control equipment, shall be in accordance with 620.12(A) and (B).

(A) Traveling Cables.

(1) Lighting Circuits. For lighting circuits, 14 AWG copper, 20 AWG copper or larger conductors shall be permitted in parallel, provided the ampacity is equivalent to at least that of 14 AWG copper.

(2) Other Circuits. For other circuits, 20 AWG copper.

(B) Other Wiring. 24 AWG copper. Smaller size listed conductors shall be permitted.

620.13 Feeder and Branch-Circuit Conductors. Conductors shall have an ampacity in accordance with 620.13(A) through (D). With generator field control, the conductor ampacity shall be based on the nameplate current rating of the driving motor of the motor-generator set that supplies power to the elevator motor.

> Informational Note No. 1: The heating of conductors depends on root-mean-square current values, which, with generator field control, are reflected by the nameplate current rating of the motor-generator driving motor rather than by the rating of the elevator motor, which represents actual but short-time and intermittent full-load current values.

> Informational Note No. 2: See Informational Note, Figure 620.13, No. 2.

(A) Conductors Supplying Single Motor. Conductors supplying a single motor shall have an ampacity not less than the percentage of motor nameplate current determined from 430.22(A) and (E).

> Informational Note: Some elevator motor currents, or those motor currents of similar function, exceed the motor nameplate value. Heating of the motor and conductors is dependent on the root-mean square (rms) current value and the length of operation time. Because this motor application is inherently intermittent duty, conductors are sized for duty cycle service as shown in Table 430.22(E).

(B) Conductors Supplying a Single Motor Controller. Conductors supplying a single motor controller shall have an ampacity not less than the motor controller nameplate current rating, plus all other connected loads. Motor controller nameplate current ratings shall be permitted to be derived based on the rms value of the motor current using an intermittent duty cycle and other control system loads, if present.

(C) Conductors Supplying a Single Power Transformer. Conductors supplying a single power transformer shall have an ampacity not less than the nameplate current rating of the power transformer plus all other connected loads.

> Informational Note No. 1: The nameplate current rating of a power transformer supplying a motor controller reflects the nameplate current rating of the motor controller at line voltage (transformer primary).

> Informational Note No. 2: See Informative Annex D, Example No. D10.

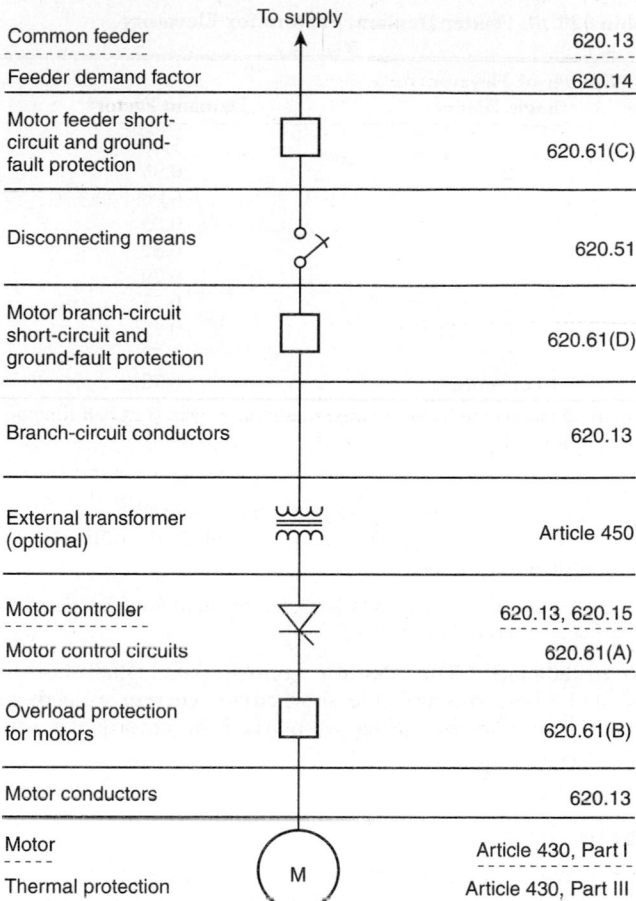

FIGURE 620.13   Informational Note Single-Line Diagram, No. 2.

(D) Conductors Supplying More Than One Motor, Motor Controller, or Power Transformer. Conductors supplying more than one motor, motor controller, or power transformer shall have an ampacity not less than the sum of the nameplate current ratings of the equipment plus all other connected loads. The ampere ratings of motors to be used in the summation shall be determined from Table 430.22(E), 430.24, and 430.24, Exception No. 1.

> Informational Note: See Informative Annex D, Example Nos. D9 and D10.

620.14 Feeder Demand Factor. Feeder conductors of less ampacity than required by 620.13 shall be permitted, subject to the requirements of Table 620.14.

620.15 Motor Controller Rating. The motor controller rating shall comply with 430.83. The rating shall be permitted to be less than the nominal rating of the elevator motor, when the controller inherently limits the available power to the motor and is marked as power limited.

> Informational Note: For controller markings, see 430.8.

N 620.16 Short-Circuit Current Rating.

(A) Marking. Where an elevator control panel is installed, it shall be marked with its short-circuit current rating, based on one of the following:

Table 620.14 Feeder Demand Factors for Elevators

Number of Elevators on a Single Feeder	Demand Factor*
1	1.00
2	0.95
3	0.90
4	0.85
5	0.82
6	0.79
7	0.77
8	0.75
9	0.73
10 or more	0.72

* Demand factors are based on 50 percent duty cycle (i.e., half time on and half time off).

(1)   Short-circuit current rating of a listed assembly
(2)   Short-circuit current rating established utilizing an approved method

Informational Note: UL 508A-2013, Supplement SB, is an example of an approved method.

(B) Installation.   The elevator control panel shall not be installed where the available short-circuit current exceeds its short-circuit current rating, as marked in accordance with 620.16(A).

Part III. Wiring

620.21 Wiring Methods.   Conductors and optical fibers located in hoistways, in escalator and moving walk wellways, in platform lifts, stairway chairlift runways, machinery spaces, control spaces, in or on cars, in machine rooms and control rooms, not including the traveling cables connecting the car or counterweight and hoistway wiring, shall be installed in rigid metal conduit, intermediate metal conduit, electrical metallic tubing, rigid nonmetallic conduit, or wireways, or shall be Type MC, MI, or AC cable unless otherwise permitted in 620.21(A) through (C).

Exception: Cords and cables of listed cord- and plug-connected equipment shall not be required to be installed in a raceway.

(A) Elevators.

(1) Hoistways and Pits.
(a)   Cables used in Class 2 power-limited circuits shall be permitted, provided the cables are supported and protected from physical damage and are of a jacketed and flame-retardant type.
(b)   Flexible cords and cables that are components of listed equipment and used in circuits operating at 30 volts rms or less or 42 volts dc or less shall be permitted, provided the cords and cables are supported and protected from physical damage and are of a jacketed and flame-retardant type.
(c)   The following wiring methods shall be permitted in the hoistway in lengths not to exceed 1.8 m (6 ft):

(1)   Flexible metal conduit
(2)   Liquidtight flexible metal conduit
(3)   Liquidtight flexible nonmetallic conduit
(4)   Flexible cords and cables, or conductors grouped together and taped or corded, shall be permitted to be installed without a raceway. They shall be located to be protected from physical damage and shall be of a flame-retardant type and shall be part of the following:

a.   Listed equipment
b.   A driving machine, or
c.   A driving machine brake

Exception 620.21(A)(1)(c)(1), (2), and (3): The conduit length shall not be required to be limited between risers and limit switches, interlocks, operating buttons, and similar devices.

(d)   A sump pump or oil recovery pump located in the pit shall be permitted to be cord connected. The cord shall be a hard usage oil-resistant type, of a length not to exceed 1.8 m (6 ft), and shall be located to be protected from physical damage.

(2) Cars.
(a)   Flexible metal conduit, liquidtight flexible metal conduit, or liquidtight flexible nonmetallic conduit of metric designator 12 (trade size 3⁄8), or larger, not exceeding 1.8 m (6 ft) in length, shall be permitted on cars where so located as to be free from oil and if securely fastened in place.

Exception: Liquidtight flexible nonmetallic conduit of metric designator 12 (trade size 3⁄8), or larger, as defined by 356.2(2), shall be permitted in lengths in excess of 1.8 m (6 ft).

(b)   Hard-service cords and junior hard-service cords that conform to the requirements of Article 400 (Table 400.4) shall be permitted as flexible connections between the fixed wiring on the car and devices on the car doors or gates. Hard-service cords only shall be permitted as flexible connections for the top-of-car operating device or the car-top work light. Devices or luminaires shall be grounded by means of an equipment grounding conductor run with the circuit conductors. Cables with smaller conductors and other types and thicknesses of insulation and jackets shall be permitted as flexible connections between the fixed wiring on the car and devices on the car doors or gates, if listed for this use.
(c)   Flexible cords and cables that are components of listed equipment and used in circuits operating at 30 volts rms or less or 42 volts dc or less shall be permitted, provided the cords and cables are supported and protected from physical damage and are of a jacketed and flame-retardant type.
(d)   The following wiring methods shall be permitted on the car assembly in lengths not to exceed 1.8 m (6 ft):

(1)   Flexible metal conduit
(2)   Liquidtight flexible metal conduit
(3)   Liquidtight flexible nonmetallic conduit
(4)   Flexible cords and cables, or conductors grouped together and taped or corded, shall be permitted to be installed without a raceway. They shall be located to be protected from physical damage and shall be of a flame-retardant type and shall be part of the following:

a.   Listed equipment
b.   A driving machine, or
c.   A driving machine brake

(3) Within Machine Rooms, Control Rooms, and Machinery Spaces and Control Spaces.
(a)   Flexible metal conduit, liquidtight flexible metal conduit, or liquidtight flexible nonmetallic conduit of metric designator 12 (trade size 3⁄8), or larger, not exceeding 1.8 m (6 ft) in length, shall be permitted between control panels and

machine motors, machine brakes, motor-generator sets, disconnecting means, and pumping unit motors and valves.

Exception: Liquidtight flexible nonmetallic conduit metric designator 12 (trade size ⅜) or larger, as defined in 356.2(2), shall be permitted to be installed in lengths in excess of 1.8 m (6 ft).

(b) Where motor-generators, machine motors, or pumping unit motors and valves are located adjacent to or underneath control equipment and are provided with extra-length terminal leads not exceeding 1.8 m (6 ft) in length, such leads shall be permitted to be extended to connect directly to controller terminal studs without regard to the carrying-capacity requirements of Articles 430 and 445. Auxiliary gutters shall be permitted in machine and control rooms between controllers, starters, and similar apparatus.

(c) Flexible cords and cables that are components of listed equipment and used in circuits operating at 30 volts rms or less or 42 volts dc or less shall be permitted, provided the cords and cables are supported and protected from physical damage and are of a jacketed and flame-retardant type.

(d) On existing or listed equipment, conductors shall also be permitted to be grouped together and taped or corded without being installed in a raceway. Such cable groups shall be supported at intervals not over 900 mm (3 ft) and located so as to be protected from physical damage.

(e) Flexible cords and cables in lengths not to exceed 1.8 m (6 ft) that are of a flame-retardant type and located to be protected from physical damage shall be permitted in these rooms and spaces without being installed in a raceway. They shall be part of the following:

(1) Listed equipment
(2) A driving machine, or
(3) A driving machine brake

(4) Counterweight. The following wiring methods shall be permitted on the counterweight assembly in lengths not to exceed 1.8 m (6 ft):

(1) Flexible metal conduit
(2) Liquidtight flexible metal conduit
(3) Liquidtight flexible nonmetallic conduit
(4) Flexible cords and cables, or conductors grouped together and taped or corded, shall be permitted to be installed without a raceway. They shall be located to be protected from physical damage, shall be of a flame-retardant type, and shall be part of the following:

   a. Listed equipment
   b. A driving machine, or
   c. A driving machine brake

(B) Escalators.

(1) Wiring Methods. Flexible metal conduit, liquidtight flexible metal conduit, or liquidtight flexible nonmetallic conduit shall be permitted in escalator and moving walk wellways. Flexible metal conduit or liquidtight flexible conduit of metric designator 12 (trade size ⅜) shall be permitted in lengths not in excess of 1.8 m (6 ft).

Exception: Metric designator 12 (trade size ⅜), nominal, or larger liquidtight flexible nonmetallic conduit, as defined in 356.2(2), shall be permitted to be installed in lengths in excess of 1.8 m (6 ft).

(2) Class 2 Circuit Cables. Cables used in Class 2 power-limited circuits shall be permitted to be installed within escalators and moving walkways, provided the cables are supported

and protected from physical damage and are of a jacketed and flame-retardant type.

(3) Flexible Cords. Hard-service cords that conform to the requirements of Article 400 (Table 400.4) shall be permitted as flexible connections on escalators and moving walk control panels and disconnecting means where the entire control panel and disconnecting means are arranged for removal from machine spaces as permitted in 620.5.

(C) Platform Lifts and Stairway Chairlift Raceways.

(1) Wiring Methods. Flexible metal conduit or liquidtight flexible metal conduit shall be permitted in platform lifts and stairway chairlift runways and machinery spaces. Flexible metal conduit or liquidtight flexible conduit of metric designator 12 (trade size ⅜) shall be permitted in lengths not in excess of 1.8 m (6 ft).

Exception: Metric designator 12 (trade size ⅜) or larger liquidtight flexible nonmetallic conduit, as defined in 356.2(2), shall be permitted to be installed in lengths in excess of 1.8 m (6 ft).

(2) Class 2 Circuit Cables. Cables used in Class 2 power-limited circuits shall be permitted to be installed within platform lifts and stairway chairlift runways and machinery spaces, provided the cables are supported and protected from physical damage and are of a jacketed and flame-retardant type.

(3) Flexible Cords and Cables. Flexible cords and cables that are components of listed equipment and used in circuits operating at 30 volts rms or less or 42 volts dc or less shall be permitted in lengths not to exceed 1.8 m (6 ft), provided the cords and cables are supported and protected from physical damage and are of a jacketed and flame-retardant type.

620.22 Branch Circuits for Car Lighting, Receptacle(s), Ventilation, Heating, and Air-Conditioning.

(A) Car Light Source. A separate branch circuit shall supply the car lights, receptacle(s), auxiliary lighting power source, and ventilation on each elevator car. The overcurrent device protecting the branch circuit shall be located in the elevator machine room or control room/machinery space or control space.

Required lighting shall not be connected to the load side of a ground-fault circuit interrupter.

(B) Air-Conditioning and Heating Source. A separate branch circuit shall supply the air-conditioning and heating units on each elevator car. The overcurrent device protecting the branch circuit shall be located in the elevator machine room or control room/machinery space or control space.

620.23 Branch Circuits for Machine Room or Control Room/ Machinery Space or Control Space Lighting and Receptacle(s).

(A) Separate Branch Circuits. The branch circuit(s) supplying the lighting for machine rooms, control rooms, machinery spaces, or control spaces shall be separate from the branch circuit(s) supplying the receptacle(s) in those places. These circuits shall supply no other loads.

Required lighting shall not be connected to the load side of a ground-fault circuit interrupter.

(B) Lighting Switch. The machine room or control room/ machinery space or control space lighting switch shall be located at the point of entry.

(C) Duplex Receptacle. At least one 125-volt, single-phase, 15- or 20-ampere duplex receptacle shall be provided in each machine room or control room and machinery space or control space.

Informational Note: See ASME A17.1-2013/CSA B44-13, *Safety Code for Elevators and Escalators,* for illumination levels.

620.24 Branch Circuit for Hoistway Pit Lighting and Receptacles.

(A) Separate Branch Circuits. Separate branch circuits shall supply the hoistway pit lighting and receptacles.

Required lighting shall not be connected to the load side of a ground-fault circuit interrupter.

(B) Lighting Switch. The lighting switch shall be so located as to be readily accessible from the pit access door.

(C) Duplex Receptacle. At least one 125-volt, single-phase, 15- or 20-ampere duplex receptacle shall be provided in the hoistway pit.

Informational Note No. 1: See ASME A17.1-2013/CSA B44-13, *Safety Code for Elevators and Escalators,* for illumination levels.

Informational Note No. 2: See 620.85 for ground-fault circuit-interrupter requirements.

620.25 Branch Circuits for Other Utilization Equipment.

(A) Additional Branch Circuits. Additional branch circuit(s) shall supply utilization equipment not identified in 620.22, 620.23, and 620.24. Other utilization equipment shall be restricted to that equipment identified in 620.1.

(B) Overcurrent Devices. The overcurrent devices protecting the branch circuit(s) shall be located in the elevator machinery room or control room/machinery space or control space.

Part IV. Installation of Conductors

620.32 Metal Wireways and Nonmetallic Wireways. The sum of the cross-sectional area of the individual conductors in a wireway shall not be more than 50 percent of the interior cross-sectional area of the wireway.

Vertical runs of wireways shall be securely supported at intervals not exceeding 4.5 m (15 ft) and shall have not more than one joint between supports. Adjoining wireway sections shall be securely fastened together to provide a rigid joint.

620.33 Number of Conductors in Raceways. The sum of the cross-sectional area of the individual conductors in raceways shall not exceed 40 percent of the interior cross-sectional area of the raceway, except as permitted in 620.32 for wireways.

620.34 Supports. Supports for cables or raceways in a hoistway or in an escalator or moving walk wellway or platform lift and stairway chairlift runway shall be securely fastened to the guide rail; escalator or moving walk truss; or to the hoistway, wellway, or runway construction.

620.35 Auxiliary Gutters. Auxiliary gutters shall not be subject to the restrictions of 366.12(2) covering length or of 366.22 covering number of conductors.

620.36 Different Systems in One Raceway or Traveling Cable. Optical fiber cables and conductors for operating devices, operation and motion control, power, signaling, fire alarm, lighting, heating, and air-conditioning circuits of 1000 volts or less shall be permitted to be run in the same traveling cable or raceway system if all conductors are insulated for the maximum voltage applied to any conductor within the cables or raceway system and if all live parts of the equipment are insulated from ground for this maximum voltage. Such a traveling cable or raceway shall also be permitted to include shielded conductors and/or one or more coaxial cables if such conductors are insulated for the maximum voltage applied to any conductor within the cable or raceway system. Conductors shall be permitted to be covered with suitable shielding for telephone, audio, video, or higher frequency communications circuits.

620.37 Wiring in Hoistways, Machine Rooms, Control Rooms, Machinery Spaces, and Control Spaces.

(A) Uses Permitted. Only such electrical wiring, raceways, and cables used directly in connection with the elevator or dumbwaiter, including wiring for signals, for communication with the car, for lighting, heating, air conditioning, and ventilating the elevator car, for fire detecting systems, for pit sump pumps, and for heating, lighting, and ventilating the hoistway, shall be permitted inside the hoistway, machine rooms, control rooms, machinery spaces, and control spaces.

(B) Lightning Protection. Bonding of elevator rails (car and/or counterweight) to a lightning protection system down conductor(s) shall be permitted. The lightning protection system down conductor(s) shall not be located within the hoistway. Elevator rails or other hoistway equipment shall not be used as the down conductor for lightning protection systems.

Informational Note: See 250.106 for bonding requirements. For further information, see NFPA 780-2014, *Standard for the Installation of Lightning Protection Systems.*

(C) Main Feeders. Main feeders for supplying power to elevators and dumbwaiters shall be installed outside the hoistway unless as follows:

(1) By special permission, feeders for elevators shall be permitted within an existing hoistway if no conductors are spliced within the hoistway.

(2) Feeders shall be permitted inside the hoistway for elevators with driving machine motors located in the hoistway or on the car or counterweight.

620.38 Electrical Equipment in Garages and Similar Occupancies. Electrical equipment and wiring used for elevators, dumbwaiters, escalators, moving walks, and platform lifts and stairway chairlifts in garages shall comply with the requirements of Article 511.

Informational Note: Garages used for parking or storage and where no repair work is done in accordance with 511.3(A) are not classified.

Part V. Traveling Cables

620.41 Suspension of Traveling Cables. Traveling cables shall be suspended at the car and hoistways' ends, or counterweight end where applicable, so as to reduce the strain on the individual copper conductors to a minimum.

Traveling cables shall be supported by one of the following means:

(1) By their steel supporting member(s)

(2) By looping the cables around supports for unsupported lengths less than 30 m (100 ft)

(3) By suspending from the supports by a means that automatically tightens around the cable when tension is increased for unsupported lengths up to 60 m (200 ft)

Unsupported length for the hoistway suspension means shall be that length of cable measured from the point of suspension in the hoistway to the bottom of the loop, with the elevator car located at the bottom landing. Unsupported length for the car suspension means shall be that length of cable measured from the point of suspension on the car to the bottom of the loop, with the elevator car located at the top landing.

620.42 Hazardous (Classified) Locations. In hazardous (classified) locations, traveling cables shall be of a type approved for hazardous (classified) locations as permitted in 501.10(B)(2)(7), 502.10(B)(2)(6), 503.10(A)(3)(6), 505.15(C)(2), and 506.15(A)(6).

620.43 Location of and Protection for Cables. Traveling cable supports shall be located so as to reduce to a minimum the possibility of damage due to the cables coming in contact with the hoistway construction or equipment in the hoistway. Where necessary, suitable guards shall be provided to protect the cables against damage.

620.44 Installation of Traveling Cables. Traveling cables that are suitably supported and protected from physical damage shall be permitted to be run without the use of a raceway in either or both of the following:

(1) When used inside the hoistway, on the elevator car, hoistway wall, counterweight, or controllers and machinery that are located inside the hoistway, provided the cables are in the original sheath.

(2) From inside the hoistway, to elevator controller enclosures and to elevator car and machine room, control room, machinery space, and control space connections that are located outside the hoistway for a distance not exceeding 1.8 m (6 ft) in length as measured from the first point of support on the elevator car or hoistway wall, or counterweight where applicable, provided the conductors are grouped together and taped or corded, or in the original sheath. These traveling cables shall be permitted to be continued to this equipment.

Part VI. Disconnecting Means and Control

620.51 Disconnecting Means. A single means for disconnecting all ungrounded main power supply conductors for each elevator, dumbwaiter, escalator, moving walk, platform lift, or stairway chairlift shall be provided and be designed so that no pole can be operated independently. Where multiple driving machines are connected to a single elevator, escalator, moving walk, or pumping unit, there shall be one disconnecting means to disconnect the motor(s) and control valve operating magnets.

The disconnecting means for the main power supply conductors shall not disconnect the branch circuit required in 620.22, 620.23, and 620.24.

(A) Type. The disconnecting means shall be an enclosed externally operable fused motor circuit switch or circuit breaker that is lockable open in accordance with 110.25.

The disconnecting means shall be a listed device.

Informational Note: For additional information, see ASME A17.1-2013/CSA B44-13, *Safety Code for Elevators and Escalators.*

Exception No. 1: Where an individual branch circuit supplies a platform lift, the disconnecting means required by 620.51(C)(4) shall be permitted to comply with 430.109(C). This disconnecting means shall be listed and shall be lockable open in accordance with 110.25.

Exception No. 2: Where an individual branch circuit supplies a stairway chairlift, the stairway chairlift shall be permitted to be cord-and-plug-connected, provided it complies with 422.16(A) and the cord does not exceed 1.8 m (6 ft) in length.

(B) Operation. No provision shall be made to open or close this disconnecting means from any other part of the premises. If sprinklers are installed in hoistways, machine rooms, control rooms, machinery spaces, or control spaces, the disconnecting means shall be permitted to automatically open the power supply to the affected elevator(s) prior to the application of water. No provision shall be made to automatically close this disconnecting means. Power shall only be restored by manual means.

Informational Note: To reduce hazards associated with water on live elevator electrical equipment.

(C) Location. The disconnecting means shall be located where it is readily accessible to qualified persons.

(1) On Elevators Without Generator Field Control. On elevators without generator field control, the disconnecting means shall be located within sight of the motor controller. Where the motor controller is located in the elevator hoistway, the disconnecting means required by 620.51(A) shall be located outside the hoistway and accessible to qualified persons only. An additional fused or non-fused, enclosed, externally operable motor-circuit switch that is lockable open in accordance with 110.25 to disconnect all ungrounded main power-supply conductors shall be located within sight of the motor controller. The additional switch shall be a listed device and shall comply with 620.91(C).

Driving machines or motion and operation controllers not within sight of the disconnecting means shall be provided with a manually operated switch installed in the control circuit to prevent starting. The manually operated switch(es) shall be installed adjacent to this equipment.

Where the driving machine of an electric elevator or the hydraulic machine of a hydraulic elevator is located in a remote machine room or remote machinery space, a single means for disconnecting all ungrounded main power-supply conductors shall be provided and be lockable open in accordance with 110.25.

(2) On Elevators with Generator Field Control. On elevators with generator field control, the disconnecting means shall be located within sight of the motor controller for the driving motor of the motor-generator set. Driving machines, motor-generator sets, or motion and operation controllers not within sight of the disconnecting means shall be provided with a manually operated switch installed in the control circuit to prevent starting. The manually operated switch(es) shall be installed adjacent to this equipment.

Where the driving machine or the motor-generator set is located in a remote machine room or remote machinery space, a single means for disconnecting all ungrounded main power-supply conductors shall be provided and be lockable open in accordance with 110.25.

(3) On Escalators and Moving Walks. On escalators and moving walks, the disconnecting means shall be installed in the space where the controller is located.

(4) On Platform Lifts and Stairway Chairlifts. On platform lifts and stairway chairlifts, the disconnecting means shall be located within sight of the motor controller.

(D) Identification and Signs.

(1) More than One Driving Machine. Where there is more than one driving machine in a machine room, the disconnecting means shall be numbered to correspond to the identifying number of the driving machine that they control.

The disconnecting means shall be provided with a sign to identify the location of the supply side overcurrent protective device.

N **(2) Available Short-Circuit Current Field Marking.** Where an elevator control panel is used, it shall be legibly marked in the field with the maximum available short-circuit current at its line terminals. The field marking(s) shall include the date the short-circuit current calculation was performed and be of sufficient durability to withstand the environment involved.

When modifications to the electrical installation occur that affect the maximum available short-circuit current at the elevator control panel, the maximum available short-circuit current shall be verified or recalculated as necessary to ensure the elevator control panel's short-circuit current rating is sufficient for the maximum available short-circuit current at the line terminals of the equipment. The required field marking(s) shall be adjusted to reflect the new level of maximum available short-circuit current.

N **(E) Surge Protection.** Where any of the disconnecting means in 620.51 has been designated as supplying an emergency system load, surge protection shall be provided.

620.52 Power from More Than One Source.

(A) Single-Car and Multicar Installations. On single-car and multicar installations, equipment receiving electrical power from more than one source shall be provided with a disconnecting means for each source of electrical power. The disconnecting means shall be within sight of the equipment served.

(B) Warning Sign for Multiple Disconnecting Means. Where multiple disconnecting means are used and parts of the controllers remain energized from a source other than the one disconnected, a warning sign shall be mounted on or next to the disconnecting means. The sign shall be clearly legible and shall read as follows:

WARNING
PARTS OF THE CONTROLLER ARE NOT
DE-ENERGIZED BY THIS SWITCH.

The warning sign(s) or label(s) shall comply with 110.21(B).

(C) Interconnection Multicar Controllers. Where interconnections between controllers are necessary for the operation of the system on multicar installations that remain energized from a source other than the one disconnected, a warning sign in accordance with 620.52(B) shall be mounted on or next to the disconnecting means.

620.53 Car Light, Receptacle(s), and Ventilation Disconnecting Means. Elevators shall have a single means for disconnecting all ungrounded car light, receptacle(s), and ventilation power-supply conductors for that elevator car.

The disconnecting means shall be an enclosed, externally operable, fused motor-circuit switch or circuit breaker that is lockable open in accordance with 110.25 and shall be located in the machine room or control room for that elevator car. Where there is no machine room or control room, the disconnecting means shall be located in a machinery space or control space outside the hoistway that is readily accessible to only qualified persons.

Disconnecting means shall be numbered to correspond to the identifying number of the elevator car whose light source they control.

The disconnecting means shall be provided with a sign to identify the location of the supply side overcurrent protective device.

Exception: Where a separate branch circuit supplies car lighting, a receptacle(s), and a ventilation motor not exceeding 2 hp, the disconnecting means required by 620.53 shall be permitted to comply with 430.109(C). This disconnecting means shall be listed and shall be lockable open in accordance with 110.25.

620.54 Heating and Air-Conditioning Disconnecting Means. Elevators shall have a single means for disconnecting all ungrounded car heating and air-conditioning power-supply conductors for that elevator car.

The disconnecting means shall be an enclosed, externally operable, fused motor-circuit switch or circuit breaker that is lockable open in accordance with 110.25 and shall be located in the machine room or control room for that elevator car. Where there is no machine room or control room, the disconnecting means shall be located in a machinery space or control space outside the hoistway that is readily accessible to only qualified persons.

Where there is equipment for more than one elevator car in the machine room, the disconnecting means shall be numbered to correspond to the identifying number of the elevator car whose heating and air-conditioning source they control.

The disconnecting means shall be provided with a sign to identify the location of the supply side overcurrent protective device.

620.55 Utilization Equipment Disconnecting Means. Each branch circuit for other utilization equipment shall have a single means for disconnecting all ungrounded conductors. The disconnecting means shall be lockable open in accordance with 110.25.

Where there is more than one branch circuit for other utilization equipment, the disconnecting means shall be numbered to correspond to the identifying number of the equipment served. The disconnecting means shall be provided with a sign to identify the location of the supply side overcurrent protective device.

Part VII. Overcurrent Protection

620.61 Overcurrent Protection. Overcurrent protection shall be provided in accordance with 620.61(A) through (D)

(A) Operating Devices and Control and Signaling Circuits. Operating devices and control and signaling circuits shall be protected against overcurrent in accordance with the requirements of 725.43 and 725.45.

Class 2 power-limited circuits shall be protected against overcurrent in accordance with the requirements of Chapter 9, Notes to Tables 11(A) and 11(B).

(B) Overload Protection for Motors. Motor and branch-circuit overload protection shall conform to Article 430, Part III, and (B)(1) through (B)(4).

(1) Duty Rating on Elevator, Dumbwaiter, and Motor-Generator Sets Driving Motors. Duty on elevator and dumbwaiter driving machine motors and driving motors of motor-generators used with generator field control shall be rated as intermittent. Such motors shall be permitted to be protected against overload in accordance with 430.33.

(2) Duty Rating on Escalator Motors. Duty on escalator and moving walk driving machine motors shall be rated as continuous. Such motors shall be protected against overload in accordance with 430.32.

(3) Overload Protection. Escalator and moving walk driving machine motors and driving motors of motor-generator sets shall be protected against running overload as provided in Table 430.37.

(4) Duty Rating and Overload Protection on Platform Lift and Stairway Chairlift Motors. Duty on platform lift and stairway chairlift driving machine motors shall be rated as intermittent. Such motors shall be permitted to be protected against overload in accordance with 430.33.

> Informational Note. For further information, see 430.44 for orderly shutdown.

(C) Motor Feeder Short-Circuit and Ground-Fault Protection. Motor feeder short-circuit and ground-fault protection shall be as required in Article 430, Part V.

(D) Motor Branch-Circuit Short-Circuit and Ground-Fault Protection. Motor branch-circuit short-circuit and ground-fault protection shall be as required in Article 430, Part IV.

620.62 Selective Coordination. Where more than one driving machine disconnecting means is supplied by a single feeder, the overcurrent protective devices in each disconnecting means shall be selectively coordinated with any other supply side overcurrent protective devices.

Selective coordination shall be selected by a licensed professional engineer or other qualified person engaged primarily in the design, installation, or maintenance of electrical systems. The selection shall be documented and made available to those authorized to design, install, inspect, maintain, and operate the system.

Part VIII. Machine Rooms, Control Rooms, Machinery Spaces, and Control Spaces

620.71 Guarding Equipment. Elevator, dumbwaiter, escalator, and moving walk driving machines; motor-generator sets; motor controllers; and disconnecting means shall be installed in a room or space set aside for that purpose unless otherwise permitted in 620.71(A) or (B). The room or space shall be secured against unauthorized access.

(A) Motor Controllers. Motor controllers shall be permitted outside the spaces herein specified, provided they are in enclosures with doors or removable panels that are capable of being locked in the closed position and the disconnecting means is located adjacent to or is an integral part of the motor controller. Motor controller enclosures for escalator or moving walks shall be permitted in the balustrade on the side located away from the moving steps or moving treadway. If the disconnecting means is an integral part of the motor controller, it shall be operable without opening the enclosure.

(B) Driving Machines. Elevators with driving machines located on the car, on the counterweight, or in the hoistway, and driving machines for dumbwaiters, platform lifts, and stairway lifts, shall be permitted outside the spaces herein specified.

Part IX. Grounding

620.81 Metal Raceways Attached to Cars. Metal raceways, Type MC cable, Type MI cable, or Type AC cable attached to elevator cars shall be bonded to metal parts of the car that are bonded to the equipment grounding conductor.

620.82 Electric Elevators. For electric elevators, the frames of all motors, elevator machines, controllers, and the metal enclosures for all electrical equipment in or on the car or in the hoistway shall be bonded in accordance with Article 250, Parts V and VII.

620.83 Nonelectric Elevators. For elevators other than electric having any electrical conductors attached to the car, the metal frame of the car, where normally accessible to persons, shall be bonded in accordance with Article 250, Parts V and VII.

620.84 Escalators, Moving Walks, Platform Lifts, and Stairway Chairlifts. Escalators, moving walks, platform lifts, and stairway chairlifts shall comply with Article 250.

620.85 Ground-Fault Circuit-Interrupter Protection for Personnel. Each 125-volt, single-phase, 15- and 20-ampere receptacle installed in pits, in hoistways, on the cars of elevators and dumbwaiters associated with wind turbine tower elevators, on the platforms or in the runways and machinery spaces of platform lifts and stairway chairlifts, and in escalator and moving walk wellways shall be of the ground-fault circuit-interrupter type.

All 125-volt, single-phase, 15- and 20-ampere receptacles installed in machine rooms, control spaces, and control rooms shall have ground-fault circuit-interrupter protection for personnel.

A single receptacle supplying a permanently installed sump pump shall not require ground-fault circuit-interrupter protection.

Part X. Emergency and Standby Power Systems

620.91 Emergency and Standby Power Systems. An elevator(s) shall be permitted to be powered by an emergency or standby power system.

> Informational Note: See ASME A17.1-2013/CSA B44-13, *Safety Code for Elevators and Escalators,* 2.27.2, for additional information.

(A) Regenerative Power. For elevator systems that regenerate power back into the power source that is unable to absorb the regenerative power under overhauling elevator load conditions, a means shall be provided to absorb this power.

(B) Other Building Loads. Other building loads, such as power and lighting, shall be permitted as the energy absorption means required in 620.91(A), provided that such loads are automatically connected to the emergency or standby power system operating the elevators and are large enough to absorb the elevator regenerative power.

(C) Disconnecting Means. The disconnecting means required by 620.51 shall disconnect the elevator from both the emergency or standby power system and the normal power system.

Where an additional power source is connected to the load side of the disconnecting means, which allows automatic movement of the car to permit evacuation of passengers, the disconnecting means required in 620.51 shall be provided with an auxiliary contact that is positively opened mechanically, and the opening shall not be solely dependent on springs. This contact shall cause the additional power source to be disconnected from its load when the disconnecting means is in the open position.

ARTICLE 625
Electric Vehicle Charging System

Part I. General

625.1 Scope. This article covers the electrical conductors and equipment external to an electric vehicle that connect an electric vehicle to a supply of electricity by conductive, inductive, or wireless power transfer (contactless inductive charging) means, and the installation of equipment and devices related to electric vehicle charging.

> Informational Note No. 1: For industrial trucks, see NFPA 505 -2013, *Fire Safety Standard for Powered Industrial Trucks Including Type Designations, Areas of Use, Conversions, Maintenance, and Operation.*

> Informational Note No. 2: UL 2594-2013, *Standard for Electric Vehicle Supply Equipment,* is a safety standard for conductive electric vehicle supply equipment. UL 2202-2009, *Standard for Electric Vehicle Charging System Equipment,* is a safety standard for conductive electric vehicle charging equipment.

625.2 Definitions.

Cable Management System. An apparatus designed to control and organize the output cable to the electric vehicle or to the primary pad.

N **Charger Power Converter.** The device used to convert energy from the power grid to a high-frequency output for wireless power transfer.

Electric Vehicle. An automotive-type vehicle for on-road use, such as passenger automobiles, buses, trucks, vans, neighborhood electric vehicles, electric motorcycles, and the like, primarily powered by an electric motor that draws current from a rechargeable storage battery, fuel cell, photovoltaic array, or other source of electric current. Plug-in hybrid electric vehicles (PHEV) are considered electric vehicles. For the purpose of this article, off-road, self-propelled electric vehicles, such as industrial trucks, hoists, lifts, transports, golf carts, airline ground support equipment, tractors, boats, and the like, are not included.

Electric Vehicle Connector. A device that, when electrically coupled (conductive or inductive) to an electric vehicle inlet, establishes an electrical connection to the electric vehicle for the purpose of power transfer and information exchange. This device is part of the electric vehicle coupler.

> Informational Note: For further information, see 625.48 for interactive systems.

Electric Vehicle Coupler. A mating electric vehicle inlet and electric vehicle connector set.

Electric Vehicle Inlet. The device on the electric vehicle into which the electric vehicle connector is electrically coupled (conductive or inductive) for power transfer and information exchange. This device is part of the electric vehicle coupler. For the purposes of this *Code,* the electric vehicle inlet is considered to be part of the electric vehicle and not part of the electric vehicle supply equipment.

> Informational Note: For further information, see 625.48 for interactive systems.

Electric Vehicle Storage Battery. A battery, comprised of one or more rechargeable electrochemical cells, that has no provision for the release of excessive gas pressure during normal charging and operation, or for the addition of water or electrolyte for external measurements of electrolyte-specific gravity.

Electric Vehicle Supply Equipment. The conductors, including the ungrounded, grounded, and equipment grounding conductors, and the electric vehicle connectors, attachment plugs, and all other fittings, devices, power outlets, or apparatus installed specifically for the purpose of transferring energy between the premises wiring and the electric vehicle.

> Informational Note No. 1: For further information, see 625.48 for interactive systems.

> Informational Note No. 2: Within this article, the terms *electric vehicle supply equipment* and *electric vehicle charging system equipment* are considered to be equivalent.

N **Fastened in Place.** Mounting means of an EVSE in which the fastening means are specifically designed to permit periodic removal for relocation, interchangeability, maintenance, or repair without the use of a tool.

N **Fixed in Place.** Mounting means of an EVSE attached to a wall or surface with fasteners that require a tool to be removed.

Output Cable to the Electric Vehicle. An assembly consisting of a length of flexible EV cable and an electric vehicle connector (supplying power to the electric vehicle).

N **Output Cable to the Primary Pad.** A multi-conductor, shielded cable assembly consisting of conductors to carry the high-frequency energy and any status signals between the charger power converter and the primary pad.

Personnel Protection System. A system of personnel protection devices and constructional features that when used together provide protection against electric shock of personnel.

Plug-In Hybrid Electric Vehicle (PHEV). A type of electric vehicle intended for on-road use with the ability to store and use off-vehicle electrical energy in the rechargeable energy storage system, and having a second source of motive power.

N **Portable (as applied to EVSE).** A device intended for indoor or outdoor use that can be carried from charging location to charging location and is designed to be transported in the vehicle when not in use.

Power-Supply Cord. An assembly consisting of an attachment plug and length of flexible cord that connects equipment to a receptacle.

N **Primary Pad.** A device external to the EV that provides power via the contactless coupling and may include the charger power converter.

Rechargeable Energy Storage System. Any power source that has the capability to be charged and discharged.

> Informational Note: Batteries, capacitors, and electromechanical flywheels are examples of rechargeable energy storage systems.

N **Wireless Power Transfer (WPT).** The transfer of electrical energy from a power source to an electrical load via electric and magnetic fields or waves by a contactless inductive means between a primary and a secondary device.

N **Wireless Power Transfer Equipment (WPTE).** Equipment consisting of a charger power converter and a primary pad. The two devices are either separate units or contained within one enclosure.

625.4 Voltages. Unless other voltages are specified, the nominal ac system voltages of 120, 120/240, 208Y/120, 240, 480Y/277, 480, 600Y/347, 600, and 1000 volts and dc system voltages of up to 1000 volts shall be used to supply equipment covered by this article.

625.5 Listed. EVSE or WPTE shall be listed.

Part II. Equipment Construction

625.10 Electric Vehicle Coupler. The electric vehicle coupler shall comply with 625.10(A) through (D).

(A) Construction and Installation. The electric vehicle coupler shall be constructed and installed so as to guard against inadvertent contact by persons with parts made live from the electric vehicle supply equipment or the electric vehicle battery.

(B) Unintentional Disconnection. The electric vehicle coupler shall be provided with a positive means to prevent unintentional disconnection.

(C) Grounding Pole. The electric vehicle coupler shall be provided with a grounding pole, unless provided as part of a listed isolated electric vehicle supply equipment system.

(D) Grounding Pole Requirements. If a grounding pole is provided, the electric vehicle coupler shall be so designed that the grounding pole connection is the first to make and the last to break contact.

625.15 Markings. The equipment shall comply with 625.15(A) through (C).

(A) General. All equipment shall be marked by the manufacturer as follows:

FOR USE WITH ELECTRIC VEHICLES

(B) Ventilation Not Required. Where marking is required by 625.52(A), the equipment shall be clearly marked by the manufacturer as follows:

VENTILATION NOT REQUIRED

The marking shall be located so as to be clearly visible after installation.

(C) Ventilation Required. Where marking is required by 625.52(B), the equipment shall be clearly marked by the manufacturer, "Ventilation Required." The marking shall be located so as to be clearly visible after installation.

625.16 Means of Coupling. The means of coupling to the electric vehicle shall be conductive, inductive, or wireless power transfer. Attachment plugs, electric vehicle connectors, and electric vehicle inlets shall be listed or labeled for the purpose.

625.17 Cords and Cables.

(A) Power-Supply Cord. The cable for cord-connected equipment shall comply with all of the following:

(1) Be any of the types specified in 625.17(B) or hard service cord, junior hard service cord, or portable power cable types in accordance with Table 400.4. Hard service cord, junior hard service cord, or portable power cable types shall be listed, as applicable, for exposure to oil and damp and wet locations.

(2) Have an ampacity as specified in Table 400.5(A)(1) or, for 8 AWG and larger, in the 60°C columns of Table 400.5(A)(2).

(3) Have an overall length as specified in 625.17(A)(3)a. or b. as follows:

   a. When the interrupting device of the personnel protection system specified in 625.22 is located within the enclosure of the supply equipment or charging system, the power-supply cord shall be not more than 300 mm (12 in.) long,

   b. When the interrupting device of the personnel protection system specified in 625.22 is located at the attachment plug, or within the first 300 mm (12 in.) of the power-supply cord, the overall cord length shall be a minimum of 1.8 m (6 ft) and shall be not greater than 4.6 m (15 ft).

(B) Output Cable to the Electric Vehicle. The output cable to the electric vehicle shall be Type EV, EVJ, EVE, EVJE, EVT, or EVJT flexible cable as specified in Table 400.4.

> Informational Note: Listed electric vehicle supply equipment may incorporate output cables having ampacities greater than 60°C based on the permissible temperature limits for the components and the cable.

(C) Overall Cord and Cable Length. The overall usable length shall not exceed 7.5 m (25 ft) unless equipped with a cable management system that is part of the listed electric vehicle supply equipment.

(1) Not Fastened in Place. Where the electric vehicle supply equipment or charging system is not fastened in place, the cord-exposed usable length shall be measured from the face of the attachment plug to the face of the electric vehicle connector.

(2) Fastened in Place. Where the electric vehicle supply equipment or charging system is fastened in place, the usable length of the output cable shall be measured from the cable exit of the electric vehicle supply equipment or charging system to the face of the electric vehicle connector.

625.18 Interlock. Electric vehicle supply equipment shall be provided with an interlock that de-energizes the electric vehicle connector whenever the electrical connector is uncoupled from the electric vehicle. An interlock shall not be required for portable cord-and-plug-connected electric vehicle supply equipment intended for connection to receptacle outlets rated at 125 volts, single phase, 15 and 20 amperes. An interlock shall not be required for dc supplies less than 60 volts dc.

625.19 Automatic De-Energization of Cable. The electric vehicle supply equipment or the cable-connector combination of the equipment shall be provided with an automatic means to de-energize the cable conductors and electric vehicle connector upon exposure to strain that could result in either cable rupture or separation of the cable from the electric connector and exposure of live parts. Automatic means to de-energize the cable conductors and electric vehicle connector shall not be required for portable electric vehicle supply equipment constructed in accordance with 625.44(A).

625.22 Personnel Protection System. The equipment shall have a listed system of protection against electric shock of personnel. Where cord-and-plug-connected equipment is used, the interrupting device of a listed personnel protection system shall be provided and shall be an integral part of the attachment plug or shall be located in the power-supply cord not more than 300 mm (12 in.) from the attachment plug. A personnel protection system shall not be required for supplies less than 60 volts dc.

Part III. Installation

N **625.40 Electric Vehicle Branch Circuit.** Each outlet installed for the purpose of charging electric vehicles shall be supplied by an individual branch circuit. Each circuit shall have no other outlets.

625.41 Overcurrent Protection. Overcurrent protection for feeders and branch circuits supplying equipment shall be sized for continuous duty and shall have a rating of not less than 125 percent of the maximum load of the equipment. Where noncontinuous loads are supplied from the same feeder, the overcurrent device shall have a rating of not less than the sum of the noncontinuous loads plus 125 percent of the continuous loads.

625.42 Rating. The equipment shall have sufficient rating to supply the load served. Electric vehicle charging loads shall be considered to be continuous loads for the purposes of this article. Where an automatic load management system is used, the maximum equipment load on a service and feeder shall be the maximum load permitted by the automatic load management system.

625.43 Disconnecting Means. For equipment rated more than 60 amperes or more than 150 volts to ground, the disconnecting means shall be provided and installed in a readily accessible location. The disconnecting means shall be lockable open in accordance with 110.25.

625.44 Equipment Connection. Equipment shall be connected to the premises wiring system in accordance with one of the following:

(A) Portable Equipment. Portable equipment shall be connected to the premises wiring systems by one of the following methods:

(1)   A nonlocking, 2-pole, 3-wire grounding-type receptacle outlet rated at 125 volt, single phase, 15 or 20 amperes

(2)   A nonlocking, 2-pole, 3-wire grounding-type receptacle outlet rated 60 volt dc maximum, 15 or 20 amperes

The length of the power supply cord, if provided, between the receptacle outlet and the equipment shall be in accordance with 625.17(A)(3).

(B) Stationary Equipment. Stationary equipment intended to be fastened in place in such a way as to permit ready removal for interchange, facilitation of maintenance or repair, or repositioning shall be connected to the premises wiring system by one of the following methods:

(1)   A nonlocking, 2-pole, 3-wire grounding-type receptacle outlet rated 125 volt or 250 volt, single phase, up to 50 amperes

(2)   A nonlocking, 3-pole, 4-wire grounding-type receptacle outlet rated 250 volt, three phase, up to 50 amperes

(3)   Any of the receptacle outlets in 625.44(A)(1) or (2)

The length of the power supply cord, if provided, between the receptacle outlet and the equipment shall be in accordance with 625.17(A)(3).

N **(C) Fixed Equipment.** All other equipment shall be permanently wired and fixed in place to the supporting surface.

625.46 Loss of Primary Source. Means shall be provided such that, upon loss of voltage from the utility or other electrical system(s), energy cannot be back fed through the electric vehicle and the supply equipment to the premises wiring system unless permitted by 625.48.

625.47 Multiple Feeder or Branch Circuits. Where equipment is identified for the application, more than one feeder or branch circuit shall be permitted to supply equipment.

625.48 Interactive Systems. Electric vehicle supply equipment that is part of an interactive system that serves as an optional standby system, an electric power production source, or a bidirectional power feed shall be listed, evaluated for use with the specific electric vehicles, and marked as suitable for that purpose. When used as an optional standby system, the requirements of Article 702 shall apply; when used as an electric power production source, the requirements of Article 705 shall apply.

> Informational Note: For further information on supply equipment, see ANSI/UL 1741, *Standard for Inverters, Converters, Controllers and Interconnection System Equipment for Use with Distributed Energy Resources*, and ANSI/UL 9741, *Bidirectional Electric Vehicle (EV) Charging System Equipment*; for vehicle interactive systems, see SAE J3072, *Standard for Interconnection Requirements for Onboard, Utility-Interactive Inverter Systems*.

625.50 Location. The electric vehicle supply equipment shall be located for direct electrical coupling of the EV connector (conductive or inductive) to the electric vehicle. Unless specifically listed and marked for the location, the coupling means of the electric vehicle supply equipment shall be stored or located at a height of not less than 450 mm (18 in.) above the floor level for indoor locations or 600 mm (24 in.) above the grade level for outdoor locations. This requirement does not apply to portable electric vehicle supply equipment constructed in accordance with 625.44(A).

625.52 Ventilation. The ventilation requirement for charging an electric vehicle in an indoor enclosed space shall be determined by 625.52(A) or (B).

(A) Ventilation Not Required. Where electric vehicle storage batteries are used or where the equipment is listed for charging electric vehicles indoors without ventilation and marked in accordance with 625.15(B), mechanical ventilation shall not be required.

(B) Ventilation Required. Where the equipment is listed for charging electric vehicles that require ventilation for indoor charging, and is marked in accordance with 625.15(C), mechanical ventilation, such as a fan, shall be provided. The ventilation shall include both supply and exhaust equipment and shall be permanently installed and located to intake from, and vent directly to, the outdoors. Positive-pressure ventilation systems shall be permitted only in vehicle charging buildings or areas that have been specifically designed and approved for that application. Mechanical ventilation requirements shall be determined by one of the methods specified in 625.52(B)(1) through (B)(4).

(1) Table Values. For supply voltages and currents specified in Table 625.52(B)(1)(a) or Table 625.52(B)(1)(b), the minimum ventilation requirements shall be as specified in Table 625.52(B)(1)(a) or Table 625.52(B)(1)(b) for each of the total number of electric vehicles that can be charged at one time.

(2) Other Values. For supply voltages and currents other than specified in Table 625.52(B)(1)(a) or Table 625.52(B)(1)(b), the minimum ventilation requirements shall be calculated by means of the following general formulas, as applicable:

(1) Single-phase ac or dc:

Ventilation$_{\text{single-phase ac or dc}}$ in cubic meters per minute (m^3/min) =

[625.52(B)(2)a]

$$\frac{(\text{volts})(\text{amperes})}{1718}$$

Ventilation$_{\text{single-phase ac or dc}}$ in cubic feet per minute (cfm) =

[625.52(B)(2)b]

$$\frac{(\text{volts})(\text{amperes})}{48.7}$$

Table 625.52(B)(1)(a) Minimum Ventilation Required in Cubic Meters per Minute (m^3/min) for Each of the Total Number of Electric Vehicles That Can Be Charged at One Time

| | Branch-Circuit Voltage | | | | | | | |
| | Single Phase | | | | 3 Phase | | | |
Branch- Circuit Ampere Rating	DC ≥ 50 V	120 V	208 V	240 V or 120/240 V	208 V or 208Y/120 V	240 V	480 V or 480Y/ 277 V	600 V or 600Y/ 347 V
15	0.5	1.1	1.8	2.1	—	—	—	—
20	0.6	1.4	2.4	2.8	4.2	4.8	9.7	12
30	0.9	2.1	3.6	4.2	6.3	7.2	15	18
40	1.2	2.8	4.8	5.6	8.4	9.7	19	24
50	1.5	3.5	6.1	7.0	10	12	24	30
60	1.8	4.2	7.3	8.4	13	15	29	36
100	2.9	7.0	12	14	21	24	48	60
150	—	—	—	—	31	36	73	91
200	—	—	—	—	42	48	97	120
250	—	—	—	—	52	60	120	150
300	—	—	—	—	63	73	145	180
350	—	—	—	—	73	85	170	210
400	—	—	—	—	84	97	195	240

Table 625.52(B)(1)(b) Minimum Ventilation Required in Cubic Feet per Minute (cfm) for Each of the Total Number of Electric Vehicles That Can Be Charged at One Time

Branch-Circuit Ampere Rating	Branch-Circuit Voltage							
	Single Phase				3 Phase			
	DC ≥ 50V	120 V	208 V	240 V or 120/240 V	208 V or 208Y/ 120 V	240 V	480 V or 480Y/ 277 V	600 V or 600Y/ 347 V
15	15.4	37	64	74	—	—	—	—
20	20.4	49	85	99	148	171	342	427
30	30.8	74	128	148	222	256	512	641
40	41.3	99	171	197	296	342	683	854
50	51.3	123	214	246	370	427	854	1066
60	61.7	148	256	296	444	512	1025	1281
100	102.5	246	427	493	740	854	1708	2135
150	—	—	—	—	1110	1281	2562	3203
200	—	—	—	—	1480	1708	3416	4270
250	—	—	—	—	1850	2135	4270	5338
300	—	—	—	—	2221	2562	5125	6406
350	—	—	—	—	2591	2989	5979	7473
400	—	—	—	—	2961	3416	6832	8541

(2) Three-phase ac:

Ventilation$_{3\text{-phase}}$ in cubic meters per minute (m³/min) =

[625.52(B)(2)c]

$$\frac{1.732(\text{volts})(\text{amperes})}{1718}$$

Ventilation$_{3\text{-phase}}$ in cubic feet per minute (cfm) =

[625.52(B)(2)d]

$$\frac{1.732(\text{volts})(\text{amperes})}{48.7}$$

(3) Engineered Systems. For an equipment ventilation system designed by a person qualified to perform such calculations as an integral part of a building's total ventilation system, the minimum ventilation requirements shall be permitted to be determined in accordance with calculations specified in the engineering study.

(4) Supply Circuits. The supply circuit to the mechanical ventilation equipment shall be electrically interlocked with the equipment and shall remain energized during the entire electric vehicle charging cycle. Equipment shall be marked in accordance with 625.15. Equipment receptacles rated at 125 volts, single phase, 15 and 20 amperes shall be marked in accordance with 625.15 and shall be switched, and the mechanical ventilation system shall be electrically interlocked through the switch supply power to the receptacle. Equipment supplied from less than 50 volts dc shall be marked in accordance with 625.15(C) and shall be switched, and the mechanical ventilation system shall be electrically interlocked through the switch supply power to the equipment.

N **Part IV. Wireless Power Transfer Equipment**

625.101 Grounding. The primary pad base plate shall be of a non-ferrous metal and shall be grounded unless the listed WPTE employs a double-insulation system. The base plate shall be sized to match the size of the primary pad enclosure.

625.102 Construction.

(A) Type. The charger power converter, where integral to the primary pad, shall comply with 625.102(C). The charger power converter, if not integral to the primary pad, shall be provided with a minimum Type 3R enclosure rating.

(B) Installation. If the charger power converter is not integral to the primary pad, it shall be mounted at a height of not less than 450 mm (18 in.) above the floor level for indoor locations or 600 mm (24 in.) above grade level for outdoor locations. The charger power converter shall be mounted in one of the following forms:

(1) Pedestal
(2) Wall or pole
(3) Building or structure
(4) Raised concrete pad

(C) Primary Pad. The primary pad shall be installed on the surface, embedded in the surface of the floor with its top flush with the surface, or embedded in the surface of the floor with its top below the surface. This includes primary pad constructions with the charger power converter located in the primary pad enclosure.

(1) If the primary pad is located in an area requiring snow removal, it shall not be located on or above the surface.

Exception: Where installed on private property where snow removal is done manually, the primary pad shall be permitted to be located on or above the surface.

(2) The enclosure shall be provided with a suitable enclosure rating minimum Type 3. If the primary pad is located in an area subject to severe climatic conditions (e.g., flood-

ing), it shall be suitably rated for those conditions or be provided with a suitably rated enclosure.

(D) Protection of the Output Cable. The output cable to the primary pad shall be secured in place over its entire length for the purpose of restricting its movement and to prevent strain at the connection points. If installed in conditions where drive-over could occur, the cable shall be provided with supplemental protection. Where the charger power converter is a part of the primary pad assembly, the power supply cord to the primary pad shall also be protected.

(E) Other Wiring Systems. Other wiring systems and fittings specifically listed for use on the WPTE shall be permitted.

ARTICLE 626
Electrified Truck Parking Spaces

Part I. General

626.1 Scope. The provisions of this article cover the electrical conductors and equipment external to the truck or transport refrigerated unit that connect trucks or transport refrigerated units to a supply of electricity, and the installation of equipment and devices related to electrical installations within an electrified truck parking space.

626.2 Definitions.

Cable Management System (Electrified Truck Parking Spaces). An apparatus designed to control and organize unused lengths of cable or cord at electrified truck parking spaces.

Cord Connector. A device that, by inserting it into a truck flanged surface inlet, establishes an electrical connection to the truck for the purpose of providing power for the on-board electric loads and may provide a means for information exchange. This device is part of the truck coupler.

Disconnecting Means, Parking Space. The necessary equipment usually consisting of a circuit breaker or switch and fuses, and their accessories, located near the point of entrance of supply conductors in an electrified truck parking space and intended to constitute the means of cutoff for the supply to that truck.

Electrified Truck Parking Space. A truck parking space that has been provided with an electrical system that allows truck operators to connect their vehicles while stopped and to use off-board power sources in order to operate on-board systems such as air conditioning, heating, and appliances, without any engine idling.

Informational Note: An electrified truck parking space also includes dedicated parking areas for heavy-duty trucks at travel plazas, warehouses, shipper and consignee yards, depot facilities, and border crossings. It does not include areas such as the shoulders of highway ramps and access roads, camping and recreational vehicle sites, residential and commercial parking areas used for automotive parking or other areas where ac power is provided solely for the purpose of connecting automotive and other light electrical loads, such as engine block heaters, and at private residences.

Electrified Truck Parking Space Wiring Systems. All of the electrical wiring, equipment, and appurtenances related to electrical installations within an electrified truck parking space, including the electrified parking space supply equipment.

Overhead Gantry. A structure consisting of horizontal framework, supported by vertical columns spanning above electrified truck parking spaces, that supports equipment, appliances, raceway, and other necessary components for the purpose of supplying electrical, HVAC, internet, communications, and other services to the spaces.

Separable Power Supply Cable Assembly. A flexible cord or cable, including ungrounded, grounded, and equipment grounding conductors, provided with a cord connector, an attachment plug, and all other fittings, grommets, or devices installed for the purpose of delivering energy from the source of electrical supply to the truck or TRU flanged surface inlet.

Transport Refrigerated Unit (TRU). A trailer or container, with integrated cooling or heating, or both, used for the purpose of maintaining the desired environment of temperature-sensitive goods or products.

Truck. A motor vehicle designed for the transportation of goods, services, and equipment.

Truck Coupler. A truck flanged surface inlet and mating cord connector.

Truck Flanged Surface Inlet. The device(s) on the truck into which the connector(s) is inserted to provide electric energy and other services. This device is part of the truck coupler. For the purposes of this article, the truck flanged surface inlet is considered to be part of the truck and not part of the electrified truck parking space supply equipment.

626.3 Other Articles. Wherever the requirements of other articles of this *Code* and Article 626 differ, the requirements of Article 626 shall apply. Unless electrified truck parking space wiring systems are supported or arranged in such a manner that they cannot be used in or above locations classified in 511.3 or 514.3, or both, they shall comply with 626.3(A) and (B) in addition to the requirements of this article.

(A) Vehicle Repair and Storage Facilities. Electrified truck parking space electrical wiring systems located at facilities for the repair or storage of self-propelled vehicles that use volatile flammable liquids or flammable gases for fuel or power shall comply with Article 511.

(B) Motor Fuel Dispensing Stations. Electrified truck parking space electrical wiring systems located at or serving motor fuel dispensing stations shall comply with Article 514.

Informational Note: For additional information, see NFPA 88A -2015, *Standard for Parking Structures*, and NFPA 30A -2015, *Code for Motor Fuel Dispensing Facilities and Repair Garages.*

626.4 General Requirements.

(A) Not Covered. The provisions of this article shall not apply to that portion of other equipment in residential, commercial, or industrial facilities that requires electric power used to load and unload cargo, operate conveyors, and for other equipment used on the site or truck.

(B) Distribution System Voltages. Unless other voltages are specified, the nominal ac system voltages of 120, 120/240,

208Y/120, 240, or 480Y/277 shall be used to supply equipment covered by this article.

(C) Connection to Wiring System. The provisions of this article shall apply to the electrified truck parking space supply equipment intended for connection to a wiring system as defined in 626.4(B).

Part II. Electrified Truck Parking Space Electrical Wiring Systems

626.10 Branch Circuits. Electrified truck parking space single-phase branch circuits shall be derived from a 208Y/120-volt, 3-phase, 4-wire system or a 120/240-volt, single-phase, 3-wire system.

Exception: A 120-volt distribution system shall be permitted to supply existing electrified truck parking spaces.

626.11 Feeder and Service Load Calculations.

(A) Parking Space Load. The calculated load of a feeder or service shall be not less than the sum of the loads on the branch circuits. Electrical service and feeders shall be calculated on the basis of not less than 11 kVA per electrified truck parking space.

(B) Demand Factors. Electrified truck parking space electrical wiring system demand factors shall be based upon the climatic temperature zone in which the equipment is installed. The demand factors set forth in Table 626.11(B) shall be the minimum allowable demand factors that shall be permitted for calculating load for service and feeders. No demand factor shall be allowed for any other load, except as provided in this article.

> Informational Note: The U.S. Department of Agriculture (USDA) has developed a commonly used "Plant Hardiness Zone" map that is publicly available. The map provides guidance for determining the Climatic Temperature Zone. Data indicate that the HVAC has the highest power requirement in cold climates, with the heating demand representing the greatest load, which in turn is dependent on outside temperature. In very warm climates, where no heating load is necessary, the cooling load increases as the outdoor temperature rises.

(C) Two or More Electrified Truck Parking Spaces. Where the electrified truck parking space wiring system is in a location that serves two or more electrified truck parking spaces, the equipment for each space shall comply with 626.11(A), and the calculated load shall be calculated on the basis of each parking space.

(D) Conductor Rating. Truck space branch-circuit supplied loads shall be considered to be continuous.

Part III. Electrified Truck Parking Space Supply Equipment

626.22 Wiring Methods and Materials.

(A) Electrified Truck Parking Space Supply Equipment Type. The electrified truck parking space supply equipment shall be provided in one of the following forms:

(1) Pedestal
(2) Overhead gantry
(3) Raised concrete pad

Table 626.11(B) Demand Factors for Services and Feeders

Climatic Temperature Zone (USDA Hardiness Zone) See Note	Demand Factor (%)
1	70%
2a	67%
2b	62%
3a	59%
3b	57%
4a	55%
4b	51%
5a	47%
5b	43%
6a	39%
6b	34%
7a	29%
7b	24%
8a	21%
8b	20%
9a	20%
9b	20%
10a	21%
10b	23%
11	24%

Note: The climatic temperature zones shown in Table 626.11(B) correlate with those found on the "USDA Plant Hardiness Zone Map," and the climatic temperature zone selected for use with the table shall be determined through the use of this map based on the installation location.

(B) Mounting Height. Post, pedestal, and raised concrete pad types of electrified truck parking space supply equipment shall be not less than 600 mm (2 ft) aboveground or above the point identified as the prevailing highest water level mark or an equivalent benchmark based on seasonal or storm-driven flooding from the authority having jurisdiction.

(C) Access to Working Space. All electrified truck parking space supply equipment shall be accessible by an unobstructed entrance or passageway not less than 600 mm (2 ft) wide and not less than 2.0 m (6 ft 6 in.) high.

(D) Disconnecting Means. A disconnecting switch or circuit breaker shall be provided to disconnect one or more electrified truck parking space supply equipment sites from a remote location. The disconnecting means shall be provided and installed in a readily accessible location and shall be lockable open in accordance with 110.25.

626.23 Overhead Gantry or Cable Management System.

(A) Cable Management. Electrified truck parking space equipment provided from either overhead gantry or cable management systems shall utilize a permanently attached power supply cable in electrified truck parking space supply equipment. Other cable types and assemblies listed as being suitable for the purpose, including optional hybrid communications, signal, and composite optical fiber cables, shall be permitted.

(B) Strain Relief. Means to prevent strain from being transmitted to the wiring terminals shall be provided. Permanently attached power supply cable(s) shall be provided with a means to de-energize the cable conductors and power service delivery device upon exposure to strain that could result in either cable damage or separation from the power service delivery device and exposure of live parts.

626.24 Electrified Truck Parking Space Supply Equipment Connection Means.

(A) General. Each truck shall be supplied from electrified truck parking space supply equipment through suitable extra-hard service cables or cords. Each connection to the equipment shall be by a single separable power supply cable assembly.

(B) Receptacle. All receptacles shall be listed and of the grounding type. Every truck parking space with electrical supply shall be equipped with (B)(1) and (B)(2).

(1)   A maximum of three receptacles, each 2-pole, 3-wire grounding type and rated 20 amperes, 125 volts, and two of the three connected to two separate branch circuits.

Informational Note: For the nonlocking-type and grounding-type 20-ampere receptacle configuration, see ANSI/NEMA WD6-2012, *Wiring Devices — Dimensional Specifications*, Figure 5-20.

(2)   One single receptacle, 3-pole, 4-wire grounding type, single phase rated either 30 amperes 208Y/120 volts or 125/250 volts. The 125/250-volt receptacle shall be permitted to be used on a 208Y/120-volt, single-phase circuit.

Informational Note: For various configurations of 30-ampere pin and sleeve receptacles, see ANSI/UL1686, *Standard for Pin and Sleeve Configurations*, Figure C2.9 or Part C3.

Exception: Where electrified truck parking space supply equipment provides the heating, air-conditioning, and comfort-cooling function without requiring a direct electrical connection at the truck, only two receptacles identified in 626.24(B)(1) shall be required.

(C) Disconnecting Means, Parking Space. The electrified truck parking space supply equipment shall be provided with a switch or circuit breaker for disconnecting the power supply to the electrified truck parking space. A disconnecting means shall be provided and installed in a readily accessible location and shall be lockable open in accordance with 110.25.

(D) Ground-Fault Circuit-Interrupter Protection for Personnel. The electrified truck parking space equipment shall be designed and constructed such that all receptacle outlets in 626.24 are provided with ground-fault circuit-interrupter protection for personnel.

626.25 Separable Power-Supply Cable Assembly. A separable power-supply cable assembly, consisting of a power-supply cord, a cord connector, and an attachment plug intended for connection with a truck flanged surface inlet, shall be of a listed type. The power-supply cable assembly or assemblies shall be identified and be one of the types and ratings specified in 626.25(A) and (B). Cords with adapters and pigtail ends, extension cords, and similar items shall not be used.

(A) Rating(s).

(1) Twenty-Ampere Power-Supply Cable Assembly. Equipment with a 20-ampere, 125-volt receptacle, in accordance with 626.24(B)(1), shall use a listed 20-ampere power-supply cable assembly.

Exception: It shall be permitted to use a listed separable power-supply cable assembly, either hard service or extra-hard service and rated 15 amperes, 125 volts, for connection to an engine block heater for legacy vehicles.

(2) Thirty-Ampere Power-Supply Cable Assembly. Equipment with a 30-ampere, 208Y/120-volt or 125/250-volt receptacle, in accordance with 626.24(B)(2), shall use a listed 30-ampere main power-supply cable assembly.

(B) Power-Supply Cord.

(1) Conductors. The cord shall be a listed type with three or four conductors, for single-phase connection, one conductor of which shall be identified in accordance with 400.23.

Exception: It shall be permitted to use a separate listed three-conductor separable power-supply cable assembly, one conductor of which shall be identified in accordance with 400.23 and rated 15 amperes, 125 volts for connection to an engine block heater for existing vehicles.

(2) Cord. Extra-hard usage flexible cords and cables rated not less than 90°C (194°F), 600 volts; listed for both wet locations and sunlight resistance; and having an outer jacket rated to be resistant to temperature extremes, oil, gasoline, ozone, abrasion, acids, and chemicals shall be permitted where flexibility is necessary between the electrified truck parking space supply equipment, the panel board, and flanged surface inlet(s) on the truck.

Exception: Cords for the separable power supply cable assembly for 15- and 20-ampere connections shall be permitted to be a hard service type.

(3) Cord Overall Length. The exposed cord length shall be measured from the face of the attachment plug to the point of entrance to the truck or the face of the flanged surface inlet or to the point where the cord enters the truck. The overall length of the cable shall not exceed 7.5 m (25 ft) unless equipped with a cable management system that is listed as suitable for the purpose.

(4) Attachment Plug. The attachment plug(s) shall be listed, by itself or as part of a cord set, for the purpose and shall be molded to or installed on the flexible cord so that it is secured tightly to the cord at the point where the cord enters the attachment plug. If a right-angle cap is used, the configuration shall be oriented so that the grounding member is farthest from the cord. Where a flexible cord is provided, the attachment plug shall comply with 250.138(A).

(a)   *Connection to a 20-Ampere Receptacle.* A separable power-supply cable assembly for connection to a truck flanged surface inlet, rated at 20 amperes, shall have a nonlocking-type attachment plug that shall be 2-pole, 3-wire grounding type rated 20 amperes, 125 volts and intended for use with the 20-ampere, 125-volt receptacle.

Exception: A separable power-supply cable assembly, rated 15 amperes, provided for the connection of an engine block heater, only, shall have an attachment plug that shall be 2-pole, 3-wire grounding type rated 15 amperes, 125 volts.

Informational Note: For nonlocking- and grounding-type 15- or 20-ampere plug and receptacle configurations, see ANSI/NEMA WD6-2002, *Standard for Dimensions of Attachment Plugs and Receptacles*, Figure 5-15 or Figure 5-20.

(b)   *Connection to a 30-Ampere Receptacle.* A separable power-supply cable assembly for connection to a truck flanged surface inlet, rated at 30 amperes, shall have an attachment plug that shall be 3-pole, 4-wire grounding type rated 30-amperes, 208Y/120 volts or 125/250 volts, and intended for use with the receptacle in accordance with 626.24(B)(2). The 125/250-volt attachment plug shall be permitted to be used on a 208Y/120-volt, single-phase circuit.

> Informational Note: For various configurations of 30-ampere pin and sleeve plugs, see ANSI/UL 1686-2012, *Standard for Pin and Sleeve Configurations*, Figure C2.10 or Part C3.

(5) Cord Connector. The cord connector for a separable power-supply cable assembly, as specified in 626.25(A)(1), shall be a 2-pole, 3-wire grounding type rated 20 amperes, 125 volts. The cord connector for a separable power-supply cable assembly, as specified in 626.25(A)(2), shall be a 3-pole, 4-wire grounding type rated 30 amperes, 208Y/120 volts or 125/250 volts. The 125/250-volt cord connector shall be permitted to be used on a 208Y/120-volt, single-phase circuit.

Exception: The cord connector for a separable power supply cable assembly, rated 15 amperes, provided for the connection of an engine block heater for existing vehicles, shall have an attachment plug that shall be 2-pole, 3-wire grounding type rated 15 amperes, 125 volts.

> Informational Note: For various configurations of 30-ampere cord connectors, see ANSI/UL 1686-2012, *Standard for Pin and Sleeve Configurations*, Figure C2.9 or Part C3.

626.26 Loss of Primary Power. Means shall be provided such that, upon loss of voltage from the utility or other electric supply system(s), energy cannot be back-fed through the truck and the truck supply equipment to the electrified truck parking space wiring system unless permitted by 626.27.

626.27 Interactive Systems. Electrified truck parking space supply equipment and other parts of a system, either on-board or off-board the vehicle, that are identified for and intended to be interconnected to a vehicle and also serve as an optional standby system or an electric power production source or provide for bi-directional power feed shall be listed as suitable for that purpose. When used as an optional standby system, the requirements of Article 702 shall apply, and when used as an electric power production source, the requirements of Article 705 shall apply.

Part IV. Transport Refrigerated Units (TRUs)

626.30 Transport Refrigerated Units. Electrified truck parking spaces intended to supply transport refrigerated units (TRUs) shall include an individual branch circuit and receptacle for operation of the refrigeration/heating units. The receptacle associated with the TRUs shall be provided in addition to the receptacles required in 626.24(B).

(A) Branch Circuits. TRU spaces shall be supplied from 208-volt, 3-phase or 480-volt, 3-phase branch circuits and with an equipment grounding conductor.

(B) Electrified Truck Parking Space Supply Equipment. The electrified truck parking space supply equipment, or portion thereof, providing electric power for the operation of TRUs shall be independent of the loads in Part III of Article 626.

626.31 Disconnecting Means and Receptacles.

(A) Disconnecting Means. Disconnecting means shall be provided to isolate each refrigerated unit from its supply connection. A disconnecting means shall be provided and installed in a readily accessible location and shall be lockable open in accordance with 110.25.

(B) Location. The disconnecting means shall be readily accessible, located not more than 750 mm (30 in.) from the receptacle it controls, and located in the supply circuit ahead of the receptacle. Circuit breakers or switches located in power outlets complying with this section shall be permitted as the disconnecting means.

(C) Receptacles. All receptacles shall be listed and of the grounding type. Every electrified truck parking space intended to provide an electrical supply for transport refrigerated units shall be equipped with one or more of the following:

(1) A 30-ampere, 480-volt, 3-phase, 3-pole, 4-wire receptacle
(2) A 60-ampere, 208-volt, 3-phase, 3-pole, 4-wire receptacle
(3) A 20-ampere, 1000-volt, 3-phase, 3-pole, 4-wire receptacle, pin and sleeve type

> Informational Note: Complete details of the 30-ampere pin and sleeve receptacle configuration for refrigerated containers (transport refrigerated units) can be found in ANSI/UL 1686-2012, *Standard for Pin and Sleeve Configurations*, Figure C2.11. For various configurations of 60-ampere pin and sleeve receptacles, see ANSI/UL1686.

626.32 Separable Power Supply Cable Assembly. A separable power supply cable assembly, consisting of a cord with an attachment plug and cord connector, shall be one of the types and ratings specified in 626.32(A), (B), and (C). Cords with adapters and pigtail ends, extension cords, and similar items shall not be used.

(A) Rating(s). The power supply cable assembly shall be listed and be rated in accordance with one of the following:

(1)   30 ampere, 480-volt, 3-phase
(2)   60 ampere, 208-volt, 3-phase
(3)   A 20-ampere, 1000-volt, 3-phase

(B) Cord Assemblies. The cord shall be a listed type with four conductors, for 3-phase connection, one of which shall be identified in accordance with 400.23 for use as the equipment grounding conductor. Extra-hard usage cables rated not less than 90°C (194°F), 600 volts, listed for both wet locations and sunlight resistance, and having an outer jacket rated to be resistant to temperature extremes, oil, gasoline, ozone, abrasion, acids, and chemicals, shall be permitted where flexibility is necessary between the electrified truck parking space supply equipment and the inlet(s) on the TRU.

(C) Attachment Plug(s) and Cord Connector(s). Where a flexible cord is provided with an attachment plug and cord connector, they shall comply with 250.138(A). The attachment plug(s) and cord connector(s) shall be listed, by itself or as part of the power-supply cable assembly, for the purpose and shall be molded to or installed on the flexible cord so that it is secured tightly to the cord at the point where the cord enters the attachment plug or cord connector. If a right-angle cap is used, the configuration shall be oriented so that the grounding member is farthest from the cord. An attachment plug and cord connector for the connection of a truck or trailer shall be rated in accordance with one of the following:

(1)   30-ampere, 480-volt, 3-phase, 3-pole, 4-wire and intended for use with a 30-ampere 480-volt, 3-phase, 3-pole, 4-wire receptacles and inlets, respectively

(2) 60-ampere, 208-volt, 3-phase, 3-pole, 4-wire and intended for use with a 60-ampere, 208-volt, 3-phase, 3-pole, 4-wire receptacles and inlets, respectively, or

(3) 20-ampere, 1000-volt, 3-phase, 3-pole, 4-wire and intended for use with a 20-ampere, 1000-volt, 3-phase, 3-pole, 4-wire receptacles and inlets, respectively.

Informational Note: Complete details of the 30-ampere pin and sleeve attachment plug and cord connector configurations for refrigerated containers (transport refrigerated units) can be found in ANSI/UL 1686-2012, *Standard for Pin and Sleeve Configurations*, Figures C2.12 and C2.11. For various configurations of 60-ampere pin and sleeve attachment plugs and cord connectors, see ANSI/UL1686.

ARTICLE 630
Electric Welders

Part I. General

630.1 Scope. This article covers apparatus for electric arc welding, resistance welding, plasma cutting, and other similar welding and cutting process equipment that is connected to an electrical supply system.

N **630.6 Listing.** All welding and cutting power equipment under the scope of this article shall be listed.

Part II. Arc Welders

630.11 Ampacity of Supply Conductors. The ampacity of conductors for arc welders shall be in accordance with 630.11(A) and (B).

(A) Individual Welders. The ampacity of the supply conductors shall be not less than the $I_{1\mathrm{eff}}$ value on the rating plate. Alternatively, if the $I_{1\mathrm{eff}}$ is not given, the ampacity of the supply conductors shall not be less than the current value determined by multiplying the rated primary current in amperes given on the welder rating plate by the factor shown in Table 630.11(A) based on the duty cycle of the welder.

(B) Group of Welders. Minimum conductor ampacity shall be based on the individual currents determined in 630.11(A) as the sum of 100 percent of the two largest welders, plus

Table 630.11(A) Duty Cycle Multiplication Factors for Arc Welders

Duty Cycle	Multiplier for Arc Welders	
	Nonmotor Generator	Motor Generator
100	1.00	1.00
90	0.95	0.96
80	0.89	0.91
70	0.84	0.86
60	0.78	0.81
50	0.71	0.75
40	0.63	0.69
30	0.55	0.62
20 or less	0.45	0.55

85 percent of the third largest welder, plus 70 percent of the fourth largest welder, plus 60 percent of all remaining welders.

Exception: Percentage values lower than those given in 630.11(B) shall be permitted in cases where the work is such that a high-operating duty cycle for individual welders is impossible.

Informational Note: Duty cycle considers welder loading based on the use to be made of each welder and the number of welders supplied by the conductors that will be in use at the same time. The load value used for each welder considers both the magnitude and the duration of the load while the welder is in use.

630.12 Overcurrent Protection. Overcurrent protection for arc welders shall be as provided in 630.12(A) and (B). Where the values as determined by this section do not correspond to the standard ampere ratings provided in 240.6 or where the rating or setting specified results in unnecessary opening of the overcurrent device, the next higher standard rating or setting shall be permitted.

(A) For Welders. Each welder shall have overcurrent protection rated or set at not more than 200 percent of $I_{1\mathrm{max}}$. Alternatively, if the $I_{1\mathrm{max}}$ is not given, the overcurrent protection shall be rated or set at not more than 200 percent of the rated primary current of the welder.

An overcurrent device shall not be required for a welder that has supply conductors protected by an overcurrent device rated or set at not more than 200 percent of $I_{1\mathrm{max}}$ or at the rated primary current of the welder.

If the supply conductors for a welder are protected by an overcurrent device rated or set at not more than 200 percent of $I_{1\mathrm{max}}$ or at the rated primary current of the welder, a separate overcurrent device shall not be required.

(B) For Conductors. Conductors that supply one or more welders shall be protected by an overcurrent device rated or set at not more than 200 percent of the conductor ampacity.

Informational Note: $I_{1\mathrm{max}}$ is the maximum value of the rated supply current at maximum rated output. $I_{1\mathrm{eff}}$ is the maximum value of the effective supply current, calculated from the rated supply current (I_1), the corresponding duty cycle (duty factor) (X), and the supply current at no-load (I_0) by the following equation:

[630.12(B)]

$$I_{1\mathrm{eff}} = \sqrt{I_1^2 X + I_0^2 (1 - X)}$$

630.13 Disconnecting Means. A disconnecting means shall be provided in the supply circuit for each arc welder that is not equipped with a disconnect mounted as an integral part of the welder. The disconnecting means identity shall be marked in accordance with 110.22(A).

The disconnecting means shall be a switch or circuit breaker, and its rating shall be not less than that necessary to accommodate overcurrent protection as specified under 630.12.

630.14 Marking. A rating plate shall be provided for arc welders giving the following information:

(1) Name of manufacturer
(2) Frequency
(3) Number of phases
(4) Primary voltage

(5) I_{1max} and I_{1eff}, or rated primary current
(6) Maximum open-circuit voltage
(7) Rated secondary current
(8) Basis of rating, such as the duty cycle

630.15 Grounding of Welder Secondary Circuit. The secondary circuit conductors of an arc welder, consisting of the electrode conductor and the work conductor, shall not be considered as premises wiring for the purpose of applying Article 250.

> Informational Note: Connecting welder secondary circuits to grounded objects can create parallel paths and can cause objectionable current over equipment grounding conductors.

Part III. Resistance Welders

630.31 Ampacity of Supply Conductors. The ampacity of the supply conductors for resistance welders shall be in accordance with 630.31(A) and (B).

> Informational Note: The ampacity of the supply conductors for resistance welders necessary to limit the voltage drop to a value permissible for the satisfactory performance of the welder is usually greater than that required to prevent overheating.

(A) Individual Welders. The rated ampacity for conductors for individual welders shall comply with the following:

(1) The ampacity of the supply conductors for a welder that may be operated at different times at different values of primary current or duty cycle shall not be less than 70 percent of the rated primary current for seam and automatically fed welders, and 50 percent of the rated primary current for manually operated nonautomatic welders.
(2) The ampacity of the supply conductors for a welder wired for a specific operation for which the actual primary current and duty cycle are known and remain unchanged shall not be less than the product of the actual primary current and the multiplier specified in Table 630.31(A)(2) for the duty cycle at which the welder will be operated.

Table 630.31(A)(2) Duty Cycle Multiplication Factors for Resistance Welders

Duty Cycle (%)	Multiplier
50	0.71
40	0.63
30	0.55
25	0.50
20	0.45
15	0.39
10	0.32
7.5	0.27
5 or less	0.22

(B) Groups of Welders. The ampacity of conductors that supply two or more welders shall not be less than the sum of the value obtained in accordance with 630.31(A) for the largest welder supplied and 60 percent of the values obtained for all the other welders supplied.

> Informational Note: **Explanation of Terms**
>
> (1) The *rated primary current* is the rated kilovolt-amperes (kVA) multiplied by 1000 and divided by the rated primary voltage, using values given on the nameplate.
> (2) The *actual primary current* is the current drawn from the supply circuit during each welder operation at the particular heat tap and control setting used.
> (3) The *duty cycle* is the percentage of the time during which the welder is loaded. For instance, a spot welder supplied by a 60-Hz system (216,000 cycles per hour) and making 400 15-cycle welds per hour would have a duty cycle of 2.8 percent (400 multiplied by 15, divided by 216,000, multiplied by 100). A seam welder operating 2 cycles "on" and 2 cycles "off" would have a duty cycle of 50 percent.

630.32 Overcurrent Protection. Overcurrent protection for resistance welders shall be as provided in 630.32(A) and (B). Where the values as determined by this section do not correspond with the standard ampere ratings provided in 240.6 or where the rating or setting specified results in unnecessary opening of the overcurrent device, a higher rating or setting that does not exceed the next higher standard ampere rating shall be permitted.

(A) For Welders. Each welder shall have an overcurrent device rated or set at not more than 300 percent of the rated primary current of the welder. If the supply conductors for a welder are protected by an overcurrent device rated or set at not more than 200 percent of the rated primary current of the welder, a separate overcurrent device shall not be required.

(B) For Conductors. Conductors that supply one or more welders shall be protected by an overcurrent device rated or set at not more than 300 percent of the conductor ampacity.

630.33 Disconnecting Means. A switch or circuit breaker shall be provided by which each resistance welder and its control equipment can be disconnected from the supply circuit. The ampere rating of this disconnecting means shall not be less than the supply conductor ampacity determined in accordance with 630.31. The supply circuit switch shall be permitted as the welder disconnecting means where the circuit supplies only one welder.

630.34 Marking. A nameplate shall be provided for each resistance welder, giving the following information:

(1) Name of manufacturer
(2) Frequency
(3) Primary voltage
(4) Rated kilovolt-amperes (kVA) at 50 percent duty cycle
(5) Maximum and minimum open-circuit secondary voltage
(6) Short-circuit secondary current at maximum secondary voltage
(7) Specified throat and gap setting

Part IV. Welding Cable

630.41 Conductors. Insulation of conductors intended for use in the secondary circuit of electric welders shall be flame retardant.

630.42 Installation. Cables shall be permitted to be installed in a dedicated cable tray as provided in 630.42(A), (B), and (C).

(A) Cable Support. The cable tray shall provide support at not greater than 150-mm (6-in.) intervals.

(B) Spread of Fire and Products of Combustion. The installation shall comply with 300.21.

(C) Signs. A permanent sign shall be attached to the cable tray at intervals not greater than 6.0 m (20 ft). The sign shall read as follows:

CABLE TRAY FOR WELDING CABLES ONLY

ARTICLE 640
Audio Signal Processing, Amplification, and Reproduction Equipment

Part I. General

640.1 Scope.

(A) Covered. This article covers equipment and wiring for audio signal generation, recording, processing, amplification, and reproduction; distribution of sound; public address; speech input systems; temporary audio system installations; and electronic organs or other electronic musical instruments. This also includes audio systems subject to Article 517, Part VI, and Articles 518, 520, 525, and 530.

> Informational Note: Examples of permanently installed distributed audio system locations include, but are not limited to, restaurant, hotel, business office, commercial and retail sales environments, churches, and schools. Both portable and permanently installed equipment locations include, but are not limited to, residences, auditoriums, theaters, stadiums, and movie and television studios. Temporary installations include, but are not limited to, auditoriums, theaters, stadiums (which use both temporary and permanently installed systems), and outdoor events such as fairs, festivals, circuses, public events, and concerts.

(B) Not Covered. This article does not cover the installation and wiring of fire and burglary alarm signaling devices.

640.2 Definitions. For purposes of this article, the following definitions apply.

Abandoned Audio Distribution Cable. Installed audio distribution cable that is not terminated at equipment and not identified for future use with a tag.

Audio Amplifier or Pre-Amplifier. Electronic equipment that increases the current or voltage, or both, of an audio signal intended for use by another piece of audio equipment. *Amplifier* is the term used within this article to denote an audio amplifier.

Audio Autotransformer. A transformer with a single winding and multiple taps intended for use with an amplifier loudspeaker signal output.

Audio Signal Processing Equipment. Electrically operated equipment that produces, processes, or both, electronic signals that, when appropriately amplified and reproduced by a loudspeaker, produce an acoustic signal within the range of normal human hearing (typically 20–20 kHz). Within this article, the terms *equipment* and *audio equipment* are assumed to be equivalent to audio signal processing equipment.

> Informational Note: This equipment includes, but is not limited to, loudspeakers; headphones; pre-amplifiers; microphones and their power supplies; mixers; MIDI (musical instrument digital interface) equipment or other digital control systems; equalizers, compressors, and other audio signal processing equipment; and audio media recording and playback equipment, including turntables, tape decks and disk players (audio and multimedia), synthesizers, tone generators, and electronic organs. Electronic organs and synthesizers may have integral or separate amplification and loudspeakers. With the exception of amplifier outputs, virtually all such equipment is used to process signals (utilizing analog or digital techniques) that have nonhazardous levels of voltage or current.

Audio System. Within this article, the totality of all equipment and interconnecting wiring used to fabricate a fully functional audio signal processing, amplification, and reproduction system.

Audio Transformer. A transformer with two or more electrically isolated windings and multiple taps intended for use with an amplifier loudspeaker signal output.

Equipment Rack. A framework for the support, enclosure, or both, of equipment; can be portable or stationary.

> Informational Note: See EIA/ECIA 310-E-2005, *Cabinets, Racks, Panels and Associated Equipment.*

Loudspeaker. Equipment that converts an ac electric signal into an acoustic signal. The term *speaker* is commonly used to mean *loudspeaker*.

Maximum Output Power. The maximum power delivered by an amplifier into its rated load as determined under specified test conditions.

> Informational Note: The maximum output power can exceed the manufacturer's rated output power for the same amplifier.

Mixer. Equipment used to combine and level match a multiplicity of electronic signals, such as from microphones, electronic instruments, and recorded audio.

Portable Equipment. Equipment fed with portable cords or cables intended to be moved from one place to another.

Rated Output Power. The amplifier manufacturer's stated or marked output power capability into its rated load.

Technical Power System. An electrical distribution system with grounding in accordance with 250.146(D), where the equipment grounding conductor is isolated from the premises grounded conductor and the premises equipment grounding conductor except at a single grounded termination point within a branch-circuit panelboard, at the originating (main breaker) branch-circuit panelboard, or at the premises grounding electrode.

Temporary Equipment. Portable wiring and equipment intended for use with events of a transient or temporary nature where all equipment is presumed to be removed at the conclusion of the event.

640.3 Locations and Other Articles. Circuits and equipment shall comply with 640.3(A) through (M), as applicable.

(A) Spread of Fire or Products of Combustion. Section 300.21 shall apply.

(B) Ducts, Plenums, and Other Air-Handling Spaces. Section 300.22(B) shall apply to circuits and equipment installed in ducts specifically fabricated for environmental air. Section 300.22(C) shall apply to circuits and equipment installed in other spaces used for environmental air (plenums).

Exception No. 1: Class 2 and Class 3 cables installed in accordance with 725.135(B) and Table 725.154 shall be permitted to be installed in ducts specifically fabricated for environmental air.

Exception No. 2: Class 2 and Class 3 cables installed in accordance with 725.135(C) and Table 725.154 shall be permitted to be installed in other spaces used for environmental air (plenums).

Informational Note: NFPA 90A -2015, *Standard for the Installation of Air-Conditioning and Ventilating Systems,* 4.3.10.2.6.5, permits loudspeakers, loudspeaker assemblies, and their accessories listed in accordance with UL 2043-2013, *Fire Test for Heat and Visible Smoke Release for Discrete Products and Their Accessories Installed in Air-Handling Spaces,* to be installed in other spaces used for environmental air (ceiling cavity plenums).

(C) Cable Trays. Cable trays and cable tray systems shall be installed in accordance with Article 392.

Informational Note: See 725.135(H), 725.136(G), and Table 725.154 for the use of Class 2, Class 3, and Type PLTC cable in cable trays.

(D) Hazardous (Classified) Locations. Equipment used in hazardous (classified) locations shall comply with the applicable requirements of Chapter 5.

(E) Assembly Occupancies. Equipment used in assembly occupancies shall comply with Article 518.

(F) Theaters, Audience Areas of Motion Picture and Television Studios, and Similar Locations. Equipment used in theaters, audience areas of motion picture and television studios, and similar locations shall comply with Article 520.

(G) Carnivals, Circuses, Fairs, and Similar Events. Equipment used in carnivals, circuses, fairs, and similar events shall comply with Article 525.

(H) Motion Picture and Television Studios. Equipment used in motion picture and television studios shall comply with Article 530.

(I) Swimming Pools, Fountains, and Similar Locations. Audio equipment used in or near swimming pools, fountains, and similar locations shall comply with Article 680.

(J) Combination Systems. Where the authority having jurisdiction permits audio systems for paging or music, or both, to be combined with fire alarm systems, the wiring shall comply with Article 760.

Informational Note: For installation requirements for such combination systems, refer to *NFPA 72 -2013, National Fire Alarm and Signaling Code,* and NFPA 101-2015, *Life Safety Code.*

(K) Antennas. Equipment used in audio systems that contain an audio or video tuner and an antenna input shall comply with Article 810. Wiring other than antenna wiring that connects such equipment to other audio equipment shall comply with this article.

(L) Generators. Generators shall be installed in accordance with 445.10 through 445.12, 445.14 through 445.16, and 445.18. Grounding of portable and vehicle-mounted generators shall be in accordance with 250.34.

(M) Organ Pipes. Additions of pipe organ pipes to an electronic organ shall be in accordance with 650.4 through 650.9 .

640.4 Protection of Electrical Equipment. Amplifiers, loudspeakers, and other equipment shall be so located or protected as to guard against environmental exposure or physical damage, such as might result in fire, shock, or personal hazard.

640.5 Access to Electrical Equipment Behind Panels Designed to Allow Access. Access to equipment shall not be denied by an accumulation of wires and cables that prevents removal of panels, including suspended ceiling panels.

640.6 Mechanical Execution of Work.

(A) Installation of Audio Distribution Cables. Cables installed exposed on the surface of ceilings and sidewalls shall be supported in such a manner that the audio distribution cables will not be damaged by normal building use. Such cables shall be secured by straps, staples, cable ties, hangers, or similar fittings designed and installed so as not to damage the cable. The installation shall conform to 300.4 and 300.11(A).

(B) Abandoned Audio Distribution Cables. The accessible portion of abandoned audio distribution cables shall be removed.

(C) Installed Audio Distribution Cable Identified for Future Use.

(1) Cables identified for future use shall be marked with a tag of sufficient durability to withstand the environment involved.

(2) Cable tags shall have the following information:
(1) Date cable was identified for future use
(2) Date of intended use
(3) Information related to the intended future use of cable

640.7 Grounding.

(A) General. Wireways and auxiliary gutters shall be connected to an equipment grounding conductor(s), to an equipment bonding jumper, or to the grounded conductor where permitted or required by 250.92(B)(1) or 250.142. Where the wireway or auxiliary gutter does not contain power-supply wires, the equipment grounding conductor shall not be required to be larger than 14 AWG copper or its equivalent. Where the wireway or auxiliary gutter contains power-supply wires, the equipment grounding conductor shall not be smaller than specified in 250.122.

(B) Separately Derived Systems with 60 Volts to Ground. Grounding of separately derived systems with 60 volts to ground shall be in accordance with 647.6.

(C) Isolated Ground Receptacles. Isolated grounding-type receptacles shall be permitted as described in 250.146(D), and for the implementation of other technical power systems in compliance with Article 250. For separately derived systems with 60 volts to ground, the branch-circuit equipment grounding conductor shall be terminated as required in 647.6(B).

Informational Note: See 406.3(D) for grounding-type receptacles and required identification.

640.8 Grouping of Conductors. Insulated conductors of different systems grouped or bundled so as to be in close physical contact with each other in the same raceway or other enclosure, or in portable cords or cables, shall comply with 300.3(C)(1).

640.9 Wiring Methods.

(A) Wiring to and Between Audio Equipment.

(1) Power Wiring. Wiring and equipment from source of power to and between devices connected to the premises wiring systems shall comply with the requirements of Chapters 1 through 4, except as modified by this article.

(2) Separately Derived Power Systems. Separately derived systems shall comply with the applicable articles of this *Code*, except as modified by this article. Separately derived systems with 60 volts to ground shall be permitted for use in audio system installations as specified in Article 647.

(3) Other Wiring. All wiring not connected to the premises wiring system or to a wiring system separately derived from the premises wiring system shall comply with Article 725.

(B) Auxiliary Power Supply Wiring. Equipment that has a separate input for an auxiliary power supply shall be wired in compliance with Article 725. Battery installation shall be in accordance with Article 480. This section shall not apply to the use of uninterruptible power supply (UPS) equipment, or other sources of supply, that are intended to act as a direct replacement for the primary circuit power source and are connected to the primary circuit input.

Informational Note: Refer to *NFPA 72-2013, National Fire Alarm and Signaling Code*, where equipment is used for a fire alarm system.

(C) Output Wiring and Listing of Amplifiers. Amplifiers with output circuits carrying audio program signals shall be permitted to employ Class 1, Class 2, or Class 3 wiring where the amplifier is listed and marked for use with the specific class of wiring method. Such listing shall ensure the energy output is equivalent to the shock and fire risk of the same class as stated in Article 725. Overcurrent protection shall be provided and shall be permitted to be inherent in the amplifier.

Audio amplifier output circuits wired using Class 1 wiring methods shall be considered equivalent to Class 1 circuits and shall be installed in accordance with 725.46, where applicable.

Audio amplifier output circuits wired using Class 2 or Class 3 wiring methods shall be considered equivalent to Class 2 or Class 3 circuits, respectively. They shall use conductors insulated at not less than the requirements of 725.179 and shall be installed in accordance with 725.133 and 725.154.

Informational Note No. 1: ANSI/UL 1711-2006, *Amplifiers for Fire Protective Signaling Systems*, contains requirements for the listing of amplifiers used for fire alarm systems in compliance with *NFPA 72-2013, National Fire Alarm and Signaling Code*.

Informational Note No. 2: Examples of requirements for listing amplifiers used in residential, commercial, and professional use are found in ANSI/UL 813-1996, *Commercial Audio Equipment*; ANSI/UL 1419-2011, *Professional Video and Audio Equipment*; ANSI/UL 1492-2010, *Audio-Video Products and Accessories*; ANSI/UL 6500-2006, *Audio/Video and Musical Instrument Apparatus for Household, Commercial, and Similar Use*; and UL 62368-1-2012, *Audio/Video, Information and Communication Technology Equipment — Part 1: Safety Requirements*.

(D) Use of Audio Transformers and Autotransformers. Audio transformers and autotransformers shall be used only for audio signals in a manner so as not to exceed the manufacturer's stated input or output voltage, impedance, or power limitations. The input or output wires of an audio transformer or autotransformer shall be allowed to connect directly to the amplifier or loudspeaker terminals. No electrical terminal or lead shall be required to be grounded or bonded.

640.10 Audio Systems Near Bodies of Water. Audio systems near bodies of water, either natural or artificial, shall be subject to the restrictions specified in 640.10(A) and (B).

Exception: This section does not include audio systems intended for use on boats, yachts, or other forms of land or water transportation used near bodies of water, whether or not supplied by branch-circuit power.

Informational Note: See 680.27(A) for installation of underwater audio equipment.

(A) Equipment Supplied by Branch-Circuit Power. Audio system equipment supplied by branch-circuit power shall not be placed horizontally within 1.5 m (5 ft) of the inside wall of a pool, spa, hot tub, or fountain, or within 1.5 m (5 ft) of the prevailing or tidal high water mark. The equipment shall be provided with branch-circuit power protected by a ground-fault circuit interrupter where required by other articles.

(B) Equipment Not Supplied by Branch-Circuit Power. Audio system equipment powered by a listed Class 2 power supply or by the output of an amplifier listed as permitting the use of Class 2 wiring shall be restricted in placement only by the manufacturer's recommendations.

Informational Note: See 640.10(A) for placement of the power supply or amplifier if supplied by branch-circuit power.

Part II. Permanent Audio System Installations

640.21 Use of Flexible Cords and Cables.

(A) Between Equipment and Branch-Circuit Power. Power supply cords for audio equipment shall be suitable for the use and shall be permitted to be used where the interchange, maintenance, or repair of such equipment is facilitated through the use of a power-supply cord.

(B) Between Loudspeakers and Amplifiers or Between Loudspeakers. Cables used to connect loudspeakers to each other or to an amplifier shall comply with Article 725. Other listed cable types and assemblies, including optional hybrid communications, signal, and composite optical fiber cables, shall be permitted.

(C) Between Equipment. Cables used for the distribution of audio signals between equipment shall comply with Article 725. Other listed cable types and assemblies, including optional hybrid communications, signal, and composite optical fiber cables, shall be permitted. Other cable types and assemblies specified by the equipment manufacturer as acceptable for the use shall be permitted in accordance with 110.3(B).

Informational Note: See 770.3 for the classification of composite optical fiber cables.

(D) Between Equipment and Power Supplies Other Than Branch-Circuit Power. The following power supplies, other than branch-circuit power supplies, shall be installed and wired between equipment in accordance with the requirements of this *Code* for the voltage and power delivered:

(1)   Storage batteries
(2)   Transformers
(3)   Transformer rectifiers
(4)   Other ac or dc power supplies

> Informational Note: For some equipment, these sources such as in items (1) and (2) serve as the only source of power. These could, in turn, be supplied with intermittent or continuous branch-circuit power.

(E) Between Equipment Racks and Premises Wiring System. Flexible cords and cables shall be permitted for the electrical connection of permanently installed equipment racks to the premises wiring system to facilitate access to equipment or for the purpose of isolating the technical power system of the rack from the premises ground. Connection shall be made either by using approved plugs and receptacles or by direct connection within an approved enclosure. Flexible cords and cables shall not be subjected to physical manipulation or abuse while the rack is in use.

640.22 Wiring of Equipment Racks and Enclosures. Metal equipment racks and enclosures shall be bonded and grounded. Bonding shall not be required if the rack is connected to a technical power ground.

Wires, cables, structural components, or other equipment shall not be placed in such a manner as to prevent reasonable access to equipment power switches and resettable or replaceable circuit overcurrent protection devices.

Supply cords or cables, if used, shall terminate within the equipment rack enclosure in an identified connector assembly. The supply cords or cable (and connector assembly if used) shall have sufficient ampacity to carry the total load connected to the equipment rack and shall be protected by overcurrent devices.

640.23 Conduit or Tubing.

(A) Number of Conductors. The number of conductors permitted in a single conduit or tubing shall not exceed the percentage fill specified in Table 1, Chapter 9.

(B) Nonmetallic Conduit or Tubing and Insulating Bushings. The use of nonmetallic conduit or tubing and insulating bushings shall be permitted where a technical power system is employed and shall comply with applicable articles.

640.24 Wireways, Gutters, and Auxiliary Gutters. The use of metallic and nonmetallic wireways, gutters, and auxiliary gutters shall be permitted for use with audio signal conductors and shall comply with applicable articles with respect to permitted locations, construction, and fill.

640.25 Loudspeaker Installation in Fire Resistance–Rated Partitions, Walls, and Ceilings. Loudspeakers installed in a fire resistance–rated partition, wall, or ceiling shall be listed and labeled, or identified as speaker assemblies for fire resistance, or installed in an enclosure or recess that maintains the fire resistance rating.

> Informational Note: Fire-rated construction is the fire-resistive classification used in building codes.

Part III.   Portable and Temporary Audio System Installations

640.41 Multipole Branch-Circuit Cable Connectors. Multipole branch-circuit cable connectors, male and female, for power-supply cords and cables shall be so constructed that tension on the cord or cable is not transmitted to the connections. The female half shall be attached to the load end of the power supply cord or cable. The connector shall be rated in amperes and designed so that differently rated devices cannot be connected together. Alternating-current multipole connectors shall be polarized and comply with 406.7(A) and (B) and 406.10. Alternating-current or direct-current multipole connectors utilized for connection between loudspeakers and amplifiers, or between loudspeakers, shall not be compatible with nonlocking 15- or 20-ampere rated connectors intended for branch-circuit power or with connectors rated 250 volts or greater and of either the locking or nonlocking type. Signal cabling not intended for such loudspeaker and amplifier interconnection shall not be permitted to be compatible with multipole branch-circuit cable connectors of any accepted configuration.

> Informational Note: See 400.14 for pull at terminals.

640.42 Use of Flexible Cords and Cables.

(A) Between Equipment and Branch-Circuit Power. Power supply cords for audio equipment shall be listed and shall be permitted to be used where the interchange, maintenance, or repair of such equipment is facilitated through the use of a power-supply cord.

(B) Between Loudspeakers and Amplifiers, or Between Loudspeakers. Installation of flexible cords and cables used to connect loudspeakers to each other or to an amplifier shall comply with Part I of Article 400 and Parts I, II, III, and IV of Article 725, respectively. Cords and cables listed for portable use, either hard or extra-hard usage as defined by Article 400, shall also be permitted. Other listed cable types and assemblies, including optional hybrid communications, signal, and composite optical fiber cables, shall be permitted.

(C) Between Equipment and/or Between Equipment Racks. Installation of flexible cords and cables used for the distribution of audio signals between equipment shall comply with Parts I and II of Article 400 and Parts I, II, and III of Article 725, respectively. Cords and cables listed for portable use, either hard or extra-hard service as defined by Article 400, shall also be permitted. Other listed cable types and assemblies, including optional hybrid communications, signal, and composite optical fiber cables, shall be permitted.

(D) Between Equipment, Equipment Racks, and Power Supplies Other Than Branch-Circuit Power. Wiring between the following power supplies, other than branch-circuit power supplies, shall be installed, connected, or wired in accordance with the requirements of this *Code* for the voltage and power required:

(1)   Storage batteries
(2)   Transformers
(3)   Transformer rectifiers
(4)   Other ac or dc power supplies

(E) Between Equipment Racks and Branch-Circuit Power. The supply to a portable equipment rack shall be by means of listed extra-hard usage cords or cables, as defined in Table 400.4. For outdoor portable or temporary use, the cords or cables shall be further listed as being suitable for wet locations and sunlight resistant. Sections 520.5, 520.10, and 525.3 shall apply as appropriate when the following conditions exist:

(1) Where equipment racks include audio and lighting and/or power equipment
(2) When using or constructing cable extensions, adapters, and breakout assemblies

640.43 Wiring of Equipment Racks. Equipment racks fabricated of metal shall be bonded and grounded. Nonmetallic racks with covers (if provided) removed shall not allow access to Class 1, Class 3, or primary circuit power without the removal of covers over terminals or the use of tools.

Wires, cables, structural components, or other equipment shall not be placed in such a manner as to prevent reasonable access to equipment power switches and resettable or replaceable circuit overcurrent protection devices.

Wiring that exits the equipment rack for connection to other equipment or to a power supply shall be relieved of strain or otherwise suitably terminated such that a pull on the flexible cord or cable will not increase the risk of damage to the cable or connected equipment such as to cause an unreasonable risk of fire or electric shock.

640.44 Environmental Protection of Equipment. Portable equipment not listed for outdoor use shall be permitted only where appropriate protection of such equipment from adverse weather conditions is provided to prevent risk of fire or electric shock. Where the system is intended to remain operable during adverse weather, arrangements shall be made for maintaining operation and ventilation of heat-dissipating equipment.

640.45 Protection of Wiring. Where accessible to the public, flexible cords and cables laid or run on the ground or on the floor shall be covered with approved nonconductive mats. Cables and mats shall be arranged so as not to present a tripping hazard. The cover requirements of 300.5 shall not apply to wiring protected by burial.

640.46 Equipment Access. Equipment likely to present a risk of fire, electric shock, or physical injury to the public shall be protected by barriers or supervised by qualified personnel so as to prevent public access.

ARTICLE 645
Information Technology Equipment

Informational Note: Text that is followed by a reference in brackets has been extracted from NFPA 75-2013, *Standard for the Fire Protection of Information Technology Equipment.* Only editorial changes were made to the extracted text to make it consistent with this Code.

645.1 Scope. This article covers equipment, power-supply wiring, equipment interconnecting wiring, and grounding of information technology equipment and systems in an information technology equipment room.

Informational Note: For further information, see NFPA 75 -2017, *Standard for the Fire Protection of Information Technology Equipment,* which covers the requirements for the protection of information technology equipment and information technology equipment areas.

645.2 Definitions.

Abandoned Supply Circuits and Interconnecting Cables. Installed supply circuits and interconnecting cables that are not terminated at equipment and not identified for future use with a tag.

Critical Operations Data System. An information technology equipment system that requires continuous operation for reasons of public safety, emergency management, national security, or business continuity.

Information Technology Equipment Room. A room within the information technology equipment area that contains the information technology equipment. [**75**:3.3.9]

Remote Disconnect Control. An electric device and circuit that controls a disconnecting means through a relay or equivalent device.

Zone. A physically identifiable area (such as barriers or separation by distance) within an information technology equipment room, with dedicated power and cooling systems for the information technology equipment or systems.

645.3 Other Articles. Circuits and equipment shall comply with 645.3(A) through (I), as applicable.

(A) Spread of Fire or Products of Combustion. Sections 300.21, 770.26, 800.26, and 820.26 shall apply to penetrations of the fire-resistant room boundary.

(B) Wiring and Cabling in Other Spaces Used for Environmental Air (Plenums). The following sections and tables shall apply to wiring and cabling in other spaces used for environmental air (plenums) above an information technology equipment room:

(1) Wiring methods: 300.22(C)(1)
(2) Class 2, Class 3, and PLTC cables: 725.135(C) and Table 725.154
(3) Fire alarm systems: 760.53(B)(2), 760.135(C), and Table 760.154
(4) Optical fiber cables: 770.113(C), and Table 770.154(a)
(5) Communications circuits: 800.113(C) and Table 800.154(a), (b), and (c)
(6) CATV and radio distribution systems: 820.113(C) and Table 820.154(a)

(C) Bonding and Grounding. The non-current-carrying conductive members of optical fiber cables in an information technology equipment room shall be bonded and grounded in accordance with 770.114.

(D) Electrical Classification of Data Circuits. Section 725.121(A)(4) shall apply to the electrical classification of listed information technology equipment signaling circuits. Sections 725.139(D)(1) and 800.133(A)(1) (c) shall apply to the electrical classification of Class 2 and Class 3 circuits in the same cable with communications circuits.

(E) Fire Alarm Cables and Equipment. Parts I, II, and III of Article 760 shall apply to fire alarm systems cables and equipment installed in an information technology equipment room. Only fire alarm cables listed in accordance with Part IV of Article 760 and listed fire alarm equipment shall be permitted to be installed in an information technology equipment room.

(F) Cable Routing Assemblies, Communications Wires, Cables, Raceways, and Equipment. Parts I, II, III, IV, and V of Article 800 shall apply to cable routing assemblies, communications wires, cables, raceways, and equipment installed in an information technology equipment room. Only communications wires and cables listed in accordance with 800.179, cable routing assemblies and communications raceways listed in accordance with 800.182, and communications equipment listed in accordance with 800.170 shall be permitted to be installed in an information technology equipment room. Article 645 shall apply to the powering of communications equipment in an information technology equipment room.

Informational Note: See Part I of Article 100, Definitions, for a definition of communications equipment.

(G) Community Antenna Television and Radio Distribution Systems Cables and Equipment. Parts I, II, III, IV, and V of Article 820 shall apply to community antenna television and radio distribution systems cables and equipment installed in an information technology equipment room. Only community antenna television and radio distribution cables listed in accordance with 820.179 and listed CATV equipment shall be permitted to be installed in an information technology equipment room. Article 645 shall apply to the powering of community antenna television and radio distribution systems equipment installed in an information technology equipment room.

N **(H) Optical Fiber Cables.** Only optical fiber cables listed in accordance with 770.179 shall be permitted to be installed in an information technology equipment room.

(I) Cables Not in Information Technology Equipment Room. Cables extending beyond the information technology equipment room shall be subject to the applicable requirements of this *Code.*

645.4 Special Requirements for Information Technology Equipment Room. The alternative wiring methods to Chapter 3 and Parts I and III of Article 725 for signaling wiring and Parts I and V of Article 770 for optical fiber cabling shall be permitted where all of the following conditions are met:

(1) Disconnecting means complying with 645.10 are provided.
(2) A heating/ventilating/air-conditioning (HVAC) system is provided in one of the methods identified in 645.4(2) a or b.

   a. A separate HVAC system that is dedicated for information technology equipment use and is separated from other areas of occupancy; or
   b. An HVAC system that serves other occupancies and meets all of the following:

     1. Also serves the information technology equipment room
     2. Provides fire/smoke dampers at the point of penetration of the room boundary
     3. Activates the damper operation upon initiation by smoke detector alarms, by operation of the disconnecting means required by 645.10, or by both

Informational Note: For further information, see NFPA 75 -2017, *Standard for the Fire Protection of Information Technology Equipment,* Chapter 10, 10.1, 10.1.1, 10.1.2, and 10.1.3.

(3) All information technology and communications equipment installed in the room is listed.
(4) The room is occupied by, and accessible to, only those personnel needed for the maintenance and functional operation of the installed information technology equipment.
(5) The room is separated from other occupancies by fire-resistant-rated walls, floors, and ceilings with protected openings.

Informational Note: For further information on room construction requirements, see NFPA 75 -2017, *Standard for the Fire Protection of Information Technology Equipment,* Chapter 5.

(6) Only electrical equipment and wiring associated with the operation of the information technology room is installed in the room.

Informational Note: HVAC systems, communications systems, and monitoring systems such as telephone, fire alarm systems, security systems, water detection systems, and other related protective equipment are examples of equipment associated with the operation of the information technology room.

645.5 Supply Circuits and Interconnecting Cables.

(A) Branch-Circuit Conductors. The branch-circuit conductors supplying one or more units of information technology equipment shall have an ampacity not less than 125 percent of the total connected load.

(B) Power-Supply Cords. Information technology equipment shall be permitted to be connected to a branch circuit by a power-supply cord.

(1) Power-supply cords shall not exceed 4.5 m (15 ft).
(2) Power cords shall be listed and a type permitted for use on listed information technology equipment or shall be constructed of listed flexible cord and listed attachment plugs and cord connectors of a type permitted for information technology equipment.

Informational Note: One method of determining if cords are of a type permitted for the purpose is found in UL 60950-1-2007, *Safety of Information Technology Equipment — Safety — Part 1: General Requirements;* or UL 62368-1-2012, *Audio/Video, Information and Communication Technology Equipment — Part 1: Safety Requirements.*

(C) Interconnecting Cables. Separate information technology equipment units shall be permitted to be interconnected by means of listed cables and cable assemblies. The 4.5 m (15 ft) limitation in 645.5(B)(1) shall not apply to interconnecting cables.

(D) Physical Protection. Where exposed to physical damage, supply circuits and interconnecting cables shall be protected.

(E) Under Raised Floors. Where the area under the floor is accessible and openings minimize the entrance of debris beneath the floor, power cables, communication cables, connecting cables, interconnecting cables, cord-and-plug connections, and receptacles associated with the information technology equipment shall be permitted under a raised floor of approved construction. The installation requirement shall comply with 645.5(E)(1) through (3).

N **(1) Installation Requirements for Branch Circuit Supply Conductors Under a Raised Floor.**

(a)   The supply conductors shall be installed in accordance with the requirements of 300.11.

(b)   In addition to the wiring methods of 300.22(C), the following wiring methods shall also be permitted:

(1) Rigid metal conduit
(2) Rigid nonmetallic conduit
(3) Intermediate metal conduit
(4) Electrical metallic tubing
(5) Electrical nonmetallic tubing
(6) Metal wireway
(7) Nonmetallic wireway
(8) Surface metal raceway with metal cover
(9) Surface nonmetallic raceway
(10) Flexible metal conduit
(11) Liquidtight flexible metal conduit
(12) Liquidtight flexible nonmetallic conduit
(13) Type MI cable
(14) Type MC cable
(15) Type AC cable
(16) Associated metallic and nonmetallic boxes or enclosures
(17) Type TC power and control tray cable

N **(2) Installation Requirements for Electrical Supply Cords, Data Cables, Interconnecting Cables, and Grounding Conductors Under a Raised Floor.** The following cords, cables, and conductors shall be permitted to be installed under a raised floor:

(1) Supply cords of listed information technology equipment in accordance with 645.5(B)
(2) Interconnecting cables enclosed in a raceway
(3) Equipment grounding conductors
(4) In addition to wiring installed in compliance with 725.135(C), Types CL2R, CL3R, CL2, and CL3 and substitute cables including CMP, CMR, CM, and CMG installed in accordance with 725.154(A), shall be permitted under raised floors.

Informational Note: Figure 725.154(A) illustrates the cable substitution hierarchy for Class 2 and Class 3 cables.

(5) Listed Type DP cable having adequate fire-resistant characteristics suitable for use under raised floors of an information technology equipment room

Informational Note: One method of defining *fire resistance* is by establishing that the cables do not spread fire to the top of the tray in the "UL Flame Exposure, Vertical Tray Flame Test" in UL 1685-2011, *Standard for Safety for Vertical Tray Fire-Propagation and Smoke-Release Test for Electrical and Optical-Fiber Cables.* The smoke measurements in the test method are not applicable.

Another method of defining *fire resistance* is for the damage (char length) not to exceed 1.5 m (4 ft 11 in.) when performing the CSA "Vertical Flame Test — Cables in Cable Trays," as described in CSA C22.2 No. 0.3-09, *Test Methods for Electrical Wires and Cables.*

N **(3) Installation Requirements for Optical Fiber Cables Under a Raised Floor.** In addition to optical fiber cables installed in accordance with 770.113(C), Types OFNR, OFCR, OFN, and OFC shall be permitted under raised floors.

(F) Securing in Place. Power cables; communications cables, connecting cables, interconnecting cables, and associated boxes, connectors, plugs, and receptacles that are listed as part of, or for, information technology equipment shall not be required to be secured in place where installed under raised floors.

Informational Note: Securement requirements for raceways and cables not listed as part of, or for, information technology equipment are found in 300.11.

(G) Abandoned Supply Circuits and Interconnecting Cables. The accessible portion of abandoned supply circuits and interconnecting cables shall be removed unless contained in a raceway.

(H) Installed Supply Circuits and Interconnecting Cables Identified for Future Use.

(1) Supply circuits and interconnecting cables identified for future use shall be marked with a tag of sufficient durability to withstand the environment involved.

(2) Supply circuit tags and interconnecting cable tags shall have the following information:

(1) Date identified for future use
(2) Date of intended use
(3) Information relating to the intended future use

645.10 Disconnecting Means. An approved means shall be provided to disconnect power to all electronic equipment in the information technology equipment room or in designated zones within the room. There shall also be a similar approved means to disconnect the power to all dedicated HVAC systems serving the room or designated zones and to cause all required fire/smoke dampers to close. The disconnecting means shall comply with either 645.10(A) or (B).

Exception: Installations qualifying under the provisions of Article 685.

(A) Remote Disconnect Controls.

(1) Remote disconnect controls shall be located at approved locations readily accessible in case of fire to authorized personnel and emergency responders.

(2) The remote disconnect means for the control of electronic equipment power and HVAC systems shall be grouped and identified. A single means to control both systems shall be permitted.

(3) Where multiple zones are created, each zone shall have an approved means to confine fire or products of combustion to within the zone.

(4) Additional means to prevent unintentional operation of remote disconnect controls shall be permitted.

Informational Note: For further information, see NFPA 75 -2017, *Standard for the Fire Protection of Information Technology Equipment.*

(B) Critical Operations Data Systems. Remote disconnecting controls shall not be required for critical operations data systems when all of the following conditions are met:

(1) An approved procedure has been established and maintained for removing power and air movement within the room or zone.
(2) Qualified personnel are continuously available to advise emergency responders and to instruct them of disconnecting methods.
(3) A smoke-sensing fire detection system is in place.

Informational Note: For further information, see *NFPA 72* -2016, *National Fire Alarm and Signaling Code.*

(4) An approved fire suppression system suitable for the application is in place.

(5) Cables installed under a raised floor, other than branch-circuit wiring, and power cords are installed in compliance with 645.5(E)(2) or (E)(3), or in compliance with Table 645.10(B)(5).

645.11 Uninterruptible Power Supplies (UPSs). Except for installations and constructions covered in 645.11(1)or (2), UPS systems installed within the information technology equipment room, and their supply and output circuits, shall comply with 645.10. The disconnecting means shall also disconnect the battery from its load.

(1) Installations qualifying under the provisions of Article 685

(2) Power sources limited to 750 volt-amperes or less derived either from UPS equipment or from battery circuits integral to electronic equipment

645.14 System Grounding. Separately derived power systems shall be installed in accordance with the provisions of Parts I and II of Article 250. Power systems derived within listed information technology equipment that supply information technology systems through receptacles or cable assemblies supplied as part of this equipment shall not be considered separately derived for the purpose of applying 250.30.

645.15 Equipment Grounding and Bonding. All exposed non–current-carrying metal parts of an information technology system shall be bonded to the equipment grounding conductor in accordance with Parts I, V, VI, VII, and VIII of Article 250 or shall be double insulated. Where signal reference structures are installed, they shall be bonded to the equipment grounding conductor provided for the information technology equipment. Any auxiliary grounding electrode(s) installed for information technology equipment shall be installed in accordance with 250.54.

> Informational Note No. 1: The bonding requirements in the product standards governing this listed equipment ensure that it complies with Article 250.
>
> Informational Note No. 2: Where isolated grounding-type receptacles are used, see 250.146(D) and 406.3(D).

645.16 Marking. Each unit of an information technology system supplied by a branch circuit shall be provided with a manufacturer's nameplate, which shall also include the input power requirements for voltage, frequency, and maximum rated load in amperes.

645.17 Power Distribution Units. Power distribution units that are used for information technology equipment shall be permitted to have multiple panelboards within a single cabinet if the power distribution unit is utilization equipment listed for information technology application.

N 645.18 Surge Protection for Critical Operations Data Systems. Surge protection shall be provided for critical operations data systems.

645.25 Engineering Supervision. As an alternative to the feeder and service load calculations required by Parts III and IV of Article 220, feeder and service load calculations for new or existing loads shall be permitted to be used if provided by qualified persons under engineering supervision.

645.27 Selective Coordination. Critical operations data system(s) overcurrent protective devices shall be selectively coordinated with all supply-side overcurrent protective devices.

ARTICLE 646
Modular Data Centers

Part I. General

646.1 Scope. This article covers modular data centers.

> Informational Note No. 1: Modular data centers include the installed information technology equipment (ITE) and support equipment, electrical supply and distribution, wiring and protection, working space, grounding, HVAC, and the like, that are located in an equipment enclosure.
>
> Informational Note No. 2: For further information, see NFPA 75-2017, *Standard for the Protection of Information Technology Equipment*, which covers the requirements for the protection of information technology equipment and systems in an information technology equipment room.

N Table 645.10(B)(5) Cables Installed Under Raised Floors

Cable Type	Applicable Sections
Branch circuits under raised floors	645.5(E)(1)
Supply cords of listed information technology equipment	645.5(E)(2)(a), 300.22(C)
Class 2 and Class 3 remote control and PLTC cables in other spaces used for environmental air (plenums)	725.135(C) and Table 725.154
Optical fiber cable in other spaces used for environmental air (plenums)	770.113(C) and Table 770.154(a)
Communications wire and cable, cable routing assemblies, and communications raceways in other spaces used for environmental air (plenums)	800.113(C) and Tables 800.154(a), (b), & (c)
Coaxial CATV and radio distribution cables in other spaces used for environmental air (plenums)	820.113(C) and Table 820.154(a)

646.2 Definitions. The definitions in 645.2 shall apply. For the purposes of this article, the following additional definition applies.

Modular Data Center (MDC). Prefabricated units, rated 1000 volts or less, consisting of an outer enclosure housing multiple racks or cabinets of information technology equipment (ITE) (e.g., servers) and various support equipment, such as electrical service and distribution equipment, HVAC systems, and the like.

> Informational Note No. 1: A typical construction may use a standard ISO shipping container or other structure as the outer enclosure, racks or cabinets of ITE, service-entrance equipment and power distribution components, power storage such as a UPS, and an air or liquid cooling system. Modular data centers are intended for fixed installation, either indoors or outdoors, based on their construction and resistance to environmental conditions. MDCs can be configured as an all-in-one system housed in a single equipment enclosure or as a system with the support equipment housed in separate equipment enclosures.

> Informational Note No. 2: For information on listing requirements for both information technology equipment and communications equipment contained within a modular data center, see UL 60950-1-2014, *Information Technology Equipment — Safety — Part 1: General Requirements,* and UL 62368-1-2012, *Audio/Video, Information and Communication Technology Equipment — Part 1: Safety Requirements.*

> Informational Note No. 3: *Modular data centers* as defined in this article are sometimes referred to as containerized data centers.

> Informational Note No. 4: Equipment enclosures housing only support equipment (e.g., HVAC or power distribution equipment) that are not part of a specific modular data center are not considered a modular data center as defined in this article.

646.3 Other Articles. Circuits and equipment shall comply with 646.3(A) through (N) as applicable. Wherever the requirements of other articles of this *Code* and Article 646 differ, the requirements of Article 646 shall apply.

(A) Spread of Fire or Products of Combustion. Sections 300.21, 770.26, 800.26, and 820.26 shall apply to penetrations of a fire-resistant room boundary, if provided.

(B) Wiring and Cabling in Other Spaces Used for Environmental Air (Plenums). The following sections and tables shall apply to wiring and cabling in other spaces used for environmental air (plenums) within a modular data center space:

(1) Wiring methods: 300.22(C)(1)
(2) Class 2, Class 3, and PLTC cables: 725.135(C) and Table 725.154
(3) Fire alarm systems: 760.53(B)(2), 760.135(C) and Table 760.154
(4) Optical fiber cables: 770.113(C) and Table 770.154(a)
(5) Communications circuits: 800.113(C) and Table 800.154(a), (b), and (c)
(6) CATV and radio distribution systems: 820.113(C) and Table 820.154(a)

> Informational Note: Environmentally controlled working spaces, aisles, and equipment areas in an MDC are not considered a plenum.

(C) Grounding. Grounding and bonding of an MDC shall comply with Article 250. The non–current-carrying conductive members of optical fiber cables in an MDC shall be grounded in accordance with 770.114. Grounding and bonding of communications protectors, cable shields, and non–current-carrying metallic members of cable shall comply with Part IV of Article 800.

(D) Electrical Classification of Data Circuits. Section 725.121(A)(4) shall apply to the electrical classification of listed information technology equipment signaling circuits. Sections 725.139(D)(1) and 800.133(A)(1)(c) shall apply to the electrical classification of Class 2 and Class 3 circuits in the same cable with communications circuits.

(E) Fire Alarm Equipment. Parts I, II, and III of Article 760 shall apply to fire alarm systems, cables, and equipment installed in an MDC, where provided. Only fire alarm cables listed in accordance with Part IV of Article 760 and listed fire alarm equipment shall be permitted to be installed in an MDC.

(F) Cable Routing Assemblies and Communications Wires, Cables, Raceways, and Equipment. Parts I, II, III, IV, and V of Article 800 shall apply to cable routing assemblies, communications wires, cables, raceways, and equipment installed in an MDC. Only communications wires and cables listed in accordance with 800.179, cable routing assemblies and communications raceways listed in accordance with 800.182, and communications equipment listed in accordance with 800.170 shall be permitted to be installed in an MDC.

> Informational Note: See Part I of Article 100 for a definition of *communications equipment.*

(G) Community Antenna Television and Radio Distribution Systems Cables and Equipment. Parts I, II, III, IV, and V of Article 820 shall apply to community antenna television and radio distribution systems equipment installed in an MDC. Only community antenna television and radio distribution cables listed in accordance with 820.179 and listed CATV equipment shall be permitted to be installed in an MDC.

(H) Storage Batteries. Installation of storage batteries shall comply with Article 480.

Exception: Batteries that are part of listed and labeled equipment and installed in accordance with the listing requirements.

(I) Surge-Protective Devices (SPDs). Where provided, surge-protective devices shall be listed and labeled and installed in accordance with Article 285.

(J) Lighting. Lighting shall be installed in accordance with Article 410.

(K) Power Distribution Wiring and Wiring Protection. Power distribution wiring and wiring protection within an MDC shall comply with Article 210 for branch circuits.

(L) Wiring Methods and Materials.

(1) Unless modified elsewhere in this article, wiring methods and materials for power distribution shall comply with Chapter 3. Wiring shall be suitable for its use and installation and shall be listed and labeled.

Exception: This requirement shall not apply to wiring that is part of listed and labeled equipment.

(2) The following wiring methods shall not be permitted:

    a. Integrated gas spacer cable: Type IGS (Article 326)
    b. Concealed knob-and-tube wiring (Article 394)
    c. Messenger-supported wiring (Article 396)
    d. Open wiring on insulators (Article 398)

e. Outdoor overhead conductors over 600 volts (Article 399)

(3) Wiring in areas under a raised floor that are constructed and used for ventilation as described in 645.5(E) shall be permitted to use the wiring methods described in 645.5(E) if the conditions of 645.4 are met.

(4) Installation of wiring for remote-control, signaling, and power-limited circuits shall comply with Part III of Article 725.

(5) Installation of optical fiber cables shall comply with Part V of Article 770.

(6) Alternate wiring methods as permitted by Article 645 shall be permitted for MDCs, provided that all of the conditions of 645.4 are met.

(M) Service Equipment. For an MDC that is designed such that it may be powered from a separate electrical service, the service equipment for control and protection of services and their installation shall comply with Article 230. The service equipment and their arrangement and installation shall permit the installation of the service-entrance conductors in accordance with Article 230. Service equipment shall be listed and labeled and marked as being suitable for use as service equipment.

(N) Disconnecting Means. An approved means shall be provided to disconnect power to all electronic equipment in the MDC in accordance with 645.10. There shall also be a similar approved means to disconnect the power to all dedicated HVAC systems serving the MDC that shall cause all required fire/smoke dampers to close.

646.4 Applicable Requirements. All MDCs shall:

(1) Be listed and labeled and comply with 646.3(N) and 646.5 through 646.9, or

Informational Note: For information on listing requirements for modular data centers, see UL Subject 2755, *Outline of Investigation for Modular Data Centers.*

(2) Comply with the provisions of this article.

646.5 Nameplate Data. A permanent nameplate shall be attached to each equipment enclosure of an MDC and shall be plainly visible after installation. The nameplate shall include the information in 646.5(1) through (6), as applicable:

(1) Supply voltage, number of phases, frequency, and full-load current. The full-load current shown on the nameplate shall not be less than the sum of the full-load currents required for all motors and other equipment that may be in operation at the same time under normal conditions of use. Where unusual type loads, duty cycles, and so forth, require oversized conductors or permit reduced-size conductors, the required capacity shall be included in the marked full-load current. Where more than one incoming supply circuit is to be provided, the nameplate shall state the preceding information for each circuit.

Informational Note No. 1: See 430.22(E) and 430.26 for duty cycle requirements.

Informational Note No. 2: For listed equipment, the full-load current shown on the nameplate may be the maximum, measured, 15-minute, average full-load current.

(2) For MDCs powered by a separate service, the short-circuit current rating of the service equipment provided as part of the MDC.

Informational Note: This rating may be part of the service equipment marking.

(3) For MDCs powered by a separate service, if the required service as determined by Parts III and IV of Article 220 is less than the rating of the service panel used, the required service shall be included on the nameplate.

Informational Note: Branch circuits supplying ITE loads are assumed to be loaded not less than 80 percent of the branch-circuit rating with a 100 percent duty cycle. As an alternative to the feeder and service load calculations required by Parts III and IV of Article 220, feeder and service load calculations for new, future, or existing loads may be permitted to be used if performed by qualified persons under engineering supervision.

(4) Electrical diagram number(s) or the number of the index to the electrical drawings.

(5) For MDC equipment enclosures that are not powered by a separate service, feeder, or branch circuit, a reference to the powering equipment.

(6) Manufacturer's name or trademark.

646.6 Supply Conductors and Overcurrent Protection.

(A) Size. The size of the supply conductor shall be such as to have an ampacity not less than 125 percent of the full-load current rating.

Informational Note No. 1: See the 0–2000-volt ampacity tables of Article 310 for ampacity of conductors rated 600 V and below.

Informational Note No. 2: See 430.22(E) and 430.26 for duty cycle requirements.

(B) Overcurrent Protection. Where overcurrent protection for supply conductors is furnished as part of the MDC, overcurrent protection for each supply circuit shall comply with 646.6(B)(1) through (B)(2):

(1) Service Equipment — Overcurrent Protection. Service conductors shall be provided with overcurrent protection in accordance with 230.90 through 230.95.

(2) Taps and Feeders. Where overcurrent protection for supply conductors is furnished as part of the MDC as permitted by 240.21, the overcurrent protection shall comply with the following:

(1) The overcurrent protection shall consist of a single circuit breaker or set of fuses.

(2) The MDC shall be marked "OVERCURRENT PROTECTION PROVIDED AT MDC SUPPLY TERMINALS."

(3) The supply conductors shall be considered either as feeders or as taps and be provided with overcurrent protection complying with 240.21.

646.7 Short-Circuit Current Rating.

(A) Service Equipment. The service equipment of an MDC that connects directly to a service shall have a short-circuit current rating not less than the available fault current of the service.

(B) MDCs Connected to Branch Circuits and Feeders. Modular data centers that connect to a branch circuit or a feeder circuit shall have a short-circuit current rating not less than the available fault current of the branch circuit or feeder. The short-circuit current rating of the MDC shall be based on the short-circuit current rating of a listed and labeled MDC or the short-circuit current rating established utilizing an approved method.

Exception: This requirement shall not apply to listed and labeled equipment connected to branch circuits located inside of the MDC equipment enclosure.

Informational Note: UL 508A-2013, *Standard for Industrial Control Panels, Supplement SB*, is an example of an approved method.

(C) MDCs Powered from Separate MDC System Enclosures. Modular data center equipment enclosures, powered from a separate MDC system enclosure that is part of the specific MDC system, shall have a short-circuit current rating coordinated with the powering module in accordance with 110.10.

Informational Note: UL 508A-2013, *Standard for Industrial Control Panels, Supplement SB*, is an example of an approved method for determining short-circuit current ratings.

646.8 Field-Wiring Compartments. A field-wiring compartment in which service or feeder connections are to be made shall be readily accessible and comply with 646.8(1) through (3) as follows:

(1) Permit the connection of the supply wires after the MDC is installed

(2) Permit the connection to be introduced and readily connected

(3) Be located so that the connections may be readily inspected after the MDC is installed

646.9 Flexible Power Cords and Cables for Connecting Equipment Enclosures of an MDC System.

(A) Uses Permitted. Flexible power cords and cables shall be permitted to be used for connections between equipment enclosures of an MDC system where not subject to physical damage.

Informational Note: One example of flexible power cord usage for connections between equipment enclosures of an MDC system is between an MDC enclosure containing only servers and one containing power distribution equipment.

(B) Uses Not Permitted. Flexible power cords and cables shall not be used for connection to external sources of power.

Informational Note: Examples of external sources of power are electrical services, feeders, and premises branch circuits.

(C) Listing. Where flexible power cords or cables are used, they shall be listed as suitable for extra-hard usage. Where used outdoors, flexible power cords and cables shall also be listed as suitable for wet locations and shall be sunlight resistant.

(D) Single-Conductor Cable. Single-conductor power cable shall be permitted to be used only in sizes 2 AWG or larger.

Part II. Equipment

646.10 Electrical Supply and Distribution. Equipment used for electrical supply and distribution in an MDC, including fittings, devices, luminaires, apparatus, machinery, and the like, shall comply with Parts I and II of Article 110.

646.11 Distribution Transformers.

(A) Utility-Owned Transformers. Utility-owned distribution transformers shall not be permitted in an MDC.

(B) Non-Utility-Owned Premises Transformers. Non-utility-owned premises distribution transformers installed in the vicinity of an MDC shall be of the dry type or the type filled with a noncombustible dielectric medium. Such transformers shall be installed in accordance with the requirements of Article 450. Non-utility-owned premises distribution transformers shall not be permitted in an MDC.

(C) Power Transformers. Power transformers that supply power only to the MDC shall be permitted to be installed in the MDC equipment enclosure. Only dry-type transformers shall be permitted to be installed in the MDC equipment enclosure. Such transformers shall be installed in accordance with the requirements of Article 450.

646.12 Receptacles. At least one 125-volt ac, 15- or 20-ampere-rated duplex convenience outlet shall be provided in each work area of the MDC to facilitate the powering of test and measurement equipment that may be required during routine maintenance and servicing, without having to route flexible power cords through or across doorways or around line-ups of equipment, or the like.

646.13 Other Electrical Equipment. Electrical equipment that is an integral part of the MDC, including information technology equipment, lighting, control, power, HVAC (heating, ventilation, and air-conditioning), emergency lighting, alarm circuits, and so forth, shall comply with the requirements for its use and installation and shall be listed and labeled.

646.14 Installation and Use. Listed and labeled equipment shall be installed and used in accordance with any instructions or limitations included in the listing.

Part III. Lighting

646.15 General Illumination. Illumination shall be provided for all workspaces and areas that are used for exit access and exit discharge. The illumination shall be arranged so that the failure of any single lighting unit does not result in a complete loss of illumination.

Informational Note: See NFPA *101*® -2015, *Life Safety Code*, Section 7.8, for information on illumination of means of egress.

646.16 Emergency Lighting. Areas that are used for exit access and exit discharge shall be provided with emergency lighting. Emergency lighting systems shall be listed and labeled equipment installed in accordance with the manufacturer's instructions.

Informational Note: See NFPA *101*® -2015, *Life Safety Code*, Section 7.9, for information on emergency lighting.

646.17 Emergency Lighting Circuits. No appliances or lamps, other than those specified as required for emergency use, shall be supplied by emergency lighting circuits. Branch circuits supplying emergency lighting shall be installed to provide service from storage batteries, generator sets, UPS, separate service, fuel cells, or unit equipment. No other equipment shall be connected to these circuits unless the emergency lighting system includes a backup system where only the lighting is supplied by battery circuits under power failure conditions. All boxes and enclosures (including transfer switches, generators, and power panels) for emergency circuits shall be marked to identify them as components of an emergency circuit or system.

Part IV. Workspace

646.18 General. Space about electrical equipment shall comply with 110.26.

646.19 Entrance to and Egress from Working Space. For equipment over 1.8 m (6 ft) wide or deep, there shall be one entrance to and egress from the required working space not less than 610 mm (24 in.) wide and 2.0 m (6 ½ ft) high at each end of the working space. The door(s) shall open in the direction of egress and be equipped with panic bars, pressure plates, or other devices that are normally latched but open under simple pressure. A single entrance to and egress from the required working space shall be permitted where either of the conditions in 646.19(A) or (B) is met.

(A) Unobstructed Egress. Where the location permits a continuous and unobstructed way of egress travel, a single entrance to the working space shall be permitted.

(B) Extra Working Space. Where the depth of the working space is twice that required by 110.26(A)(1), a single entrance shall be permitted. It shall be located such that the distance from the equipment to the nearest edge of the entrance is not less than the minimum clear distance specified in Table 110.26(A)(1) for equipment operating at that voltage and in that condition.

646.20 Working Space for ITE.

(A) Low-Voltage Circuits. The working space about ITE where any live parts that may be exposed during routine servicing operate at not greater than 30 volts rms, 42 volts peak, or 60 volts dc shall not be required to comply with the workspace requirements of 646.19.

(B) Other Circuits. Any areas of ITE that require servicing of parts that are greater than 30 volts rms, 42 volts peak, or 60 volts dc shall comply with the workspace requirements of 646.19.

> Informational Note No. 1: For example, field-wiring compartments for ac mains connections, power distribution units, and so forth.

> Informational Note No. 2: It is assumed that ITE operates at voltages not exceeding 1000 volts.

646.21 Work Areas and Working Space About Batteries. Working space about a battery system shall comply with 110.26. Working space shall be measured from the edges of the battery racks, cabinets, or trays.

646.22 Workspace for Routine Service and Maintenance. Workspace shall be provided to facilitate routine servicing and maintenance (those tasks involving operations that can be accomplished by employees and where extensive disassembly of equipment is not required). Routine servicing and maintenance shall be able to be performed without exposing the worker to a risk of electric shock or personal injury.

> Informational Note: An example of such routine maintenance is cleaning or replacing an air filter.

ARTICLE 647
Sensitive Electronic Equipment

647.1 Scope. This article covers the installation and wiring of separately derived systems operating at 120 volts line-to-line and 60 volts to ground for sensitive electronic equipment.

647.3 General. Use of a separately derived 120-volt single-phase 3-wire system with 60 volts on each of two ungrounded conductors to an equipment grounding conductor shall be permitted for the purpose of reducing objectionable noise in sensitive electronic equipment locations, provided the following conditions apply:

(1)   The system is installed only in commercial or industrial occupancies.
(2)   The system's use is restricted to areas under close supervision by qualified personnel.
(3)   All of the requirements in 647.4 through 647.8 are met.

647.4 Wiring Methods.

(A) Panelboards and Overcurrent Protection. Use of standard single-phase panelboards and distribution equipment with a higher voltage rating shall be permitted. The system shall be clearly marked on the face of the panel or on the inside of the panel doors. Common trip two-pole circuit breakers or a combination two-pole fused disconnecting means that are identified for use at the system voltage shall be provided for both ungrounded conductors in all feeders and branch circuits. Branch circuits and feeders shall be provided with a means to simultaneously disconnect all ungrounded conductors.

(B) Junction Boxes. All junction box covers shall be clearly marked to indicate the distribution panel and the system voltage.

(C) Conductor Identification. All feeders and branch-circuit conductors installed under this section shall be identified as to system at all splices and terminations by color, marking, tagging, or equally effective means. The means of identification shall be posted at each branch-circuit panelboard and at the disconnecting means for the building.

(D) Voltage Drop. The voltage drop on any branch circuit shall not exceed 1.5 percent. The combined voltage drop of feeder and branch-circuit conductors shall not exceed 2.5 percent.

(1) Fixed Equipment. The voltage drop on branch circuits supplying equipment connected using wiring methods in Chapter 3 shall not exceed 1.5 percent. The combined voltage drop of feeder and branch-circuit conductors shall not exceed 2.5 percent.

(2) Cord-Connected Equipment. The voltage drop on branch circuits supplying receptacles shall not exceed 1 percent. For the purposes of making this calculation, the load connected to the receptacle outlet shall be considered to be 50 percent of the branch-circuit rating. The combined voltage drop of feeder and branch-circuit conductors shall not exceed 2.0 percent.

> Informational Note: The purpose of this provision is to limit voltage drop to 1.5 percent where portable cords may be used as a means of connecting equipment.

647.5 Three-Phase Systems. Where 3-phase power is supplied, a separately derived 6-phase "wye" system with 60 volts to ground installed under this article shall be configured as three separately derived 120-volt single-phase systems having a combined total of no more than six disconnects.

647.6 Grounding.

(A) General. The transformer secondary center tap of the 60/120-volt, 3-wire system shall be grounded as provided in 250.30.

(B) Grounding Conductors Required. Permanently wired utilization equipment and receptacles shall be grounded by means of an equipment grounding conductor run with the circuit conductors to an equipment grounding bus prominently marked "Technical Equipment Ground" in the originating branch-circuit panelboard. The grounding bus shall be connected to the grounded conductor on the line side of the separately derived system's disconnecting means. The grounding conductor shall not be smaller than that specified in Table 250.122 and run with the feeder conductors. The technical equipment grounding bus need not be bonded to the panelboard enclosure. Other grounding methods authorized elsewhere in this *Code* shall be permitted where the impedance of the grounding return path does not exceed the impedance of equipment grounding conductors sized and installed in accordance with this article.

> Informational Note No. 1: See 250.122 for equipment grounding conductor sizing requirements where circuit conductors are adjusted in size to compensate for voltage drop.

> Informational Note No. 2: These requirements limit the impedance of the ground fault path where only 60 volts apply to a fault condition instead of the usual 120 volts.

647.7 Receptacles.

(A) General. Where receptacles are used as a means of connecting equipment, the following conditions shall be met:

(1) All 15- and 20-ampere receptacles shall be GFCI protected.

(2) All receptacle outlet strips, adapters, receptacle covers, and faceplates shall be marked with the following words or equivalent:

WARNING — TECHNICAL POWER

Do not connect to lighting equipment.

For electronic equipment use only.

60/120 V. 1ϕac

GFCI protected

The warning sign(s) or label(s) shall comply with 110.21(B).

(3) A 125-volt, single-phase, 15- or 20-ampere-rated receptacle having one of its current-carrying poles connected to a grounded circuit conductor shall be located within 1.8 m (6 ft) of all permanently installed 15- or 20-ampere-rated 60/120-volt technical power-system receptacles.

(4) All 125-volt receptacles used for 60/120-volt technical power shall have a unique configuration and be identified for use with this class of system.

Exception: Receptacles and attachment plugs rated 125-volt, single-phase, 15- or 20-amperes, and that are identified for use with grounded circuit conductors, shall be permitted in machine rooms, control rooms, equipment rooms, equipment racks, and other similar locations that are restricted to use by qualified personnel.

(B) Isolated Ground Receptacles. Isolated ground receptacles shall be permitted as described in 250.146(D); however, the branch-circuit equipment grounding conductor shall be terminated as required in 647.6(B).

647.8 Lighting Equipment. Lighting equipment installed under this article for the purpose of reducing electrical noise originating from lighting equipment shall meet the conditions of 647.8(A) through (C).

(A) Disconnecting Means. All luminaires connected to separately derived systems operating at 60 volts to ground, and associated control equipment if provided, shall have a disconnecting means that simultaneously opens all ungrounded conductors. The disconnecting means shall be located within sight of the luminaire or be lockable open in accordance with 110.25.

(B) Luminaires. All luminaires shall be permanently installed and listed for connection to a separately derived system at 120 volts line-to-line and 60 volts to ground.

(C) Screw Shell. Luminaires installed under this section shall not have an exposed lamp screw shell.

ARTICLE 650
Pipe Organs

650.1 Scope. This article covers those electrical circuits and parts of electrically operated pipe organs that are employed for the control of the keyboards and of the sounding apparatus, typically organ pipes.

> Informational Note: The typical pipe organ is a very large musical instrument that is built as part of a building or structure.

N 650.2 Definitions.

Electronic Organ. A musical instrument that imitates the sound of a pipe organ by producing sound electronically.

> Informational Note: Most new electronic organs produce sound digitally and are called digital organs.

Pipe Organ. A musical instrument that produces sound by driving pressurized air (called wind) through pipes selected via a keyboard.

Sounding Apparatus. The sound-producing part of a pipe organ, including, but not limited to, pipes, chimes, bells, the pressurized air (wind)-producing equipment (blower), associated controls, and power equipment.

> Informational Note: The sounding apparatus is also referred to as the "pipe organ chamber."

650.3 Other Articles. Installations of circuits and equipment shall comply with 650.3(A) and (B) as applicable. Wherever the requirements of other articles in Chapters 1 through 7 of this *Code* and Article 650 differ, the requirements of Article 650 shall apply.

(A) Electronic Organ Equipment. Installations of digital/analog–sampled sound production technology and associated audio signal processing, amplification, reproduction equipment, and wiring installed as part of a pipe organ shall be in accordance with Article 640.

(B) Optical Fiber Cable. Installations of optical fiber cables shall be in accordance with Parts I and V of Article 770.

650.4 Source of Energy. DC power shall be supplied by a listed dc power supply with a maximum output of 30 volts.

> Informational Note: Class 1 power-limited power supplies are often utilized in pipe organ applications.

650.5 Grounding or Double Insulation of the DC Power Supply. The installation of the dc power supply shall comply with either of the following:

(1) The dc power supply shall be double insulated.
(2) The metallic case of the dc power supply shall be bonded to the input equipment grounding conductor.

650.6 Conductors. Conductors shall comply with 650.6(A) through (D).

(A) Size. The minimum conductor size shall be not less than 28 AWG for electronic signal circuits and not less than 26 AWG for electromagnetic valve supply and the like. The minimum conductor size of a main common-return conductor in the electromagnetic supply shall not be less than 14 AWG.

(B) Insulation. Conductors shall have thermoplastic or thermosetting insulation.

(C) Conductors to Be Cabled. Except for the common-return conductor and conductors inside the organ proper, the organ sections and the organ console conductors shall be cabled. The common-return conductors shall be permitted under an additional covering enclosing both cable and return conductor, or they shall be permitted as a separate conductor and shall be permitted to be in contact with the cable.

(D) Cable Covering. Each cable shall be provided with an outer covering, either overall or on each of any subassemblies of grouped conductors. Tape shall be permitted in place of a covering. Where not installed in metal raceway, the covering shall be resistant to flame spread, or the cable or each cable subassembly shall be covered with a closely wound listed fireproof tape.

> Informational Note: One method of determining that cable is resistant to flame spread is by testing the cable to the VW-1 (vertical-wire) flame test in ANSI/UL 1581-2011, *Reference Standard for Electrical Wires, Cables and Flexible Cords.*

650.7 Installation of Conductors. Cables shall be securely fastened in place and shall be permitted to be attached directly to the organ structure without insulating supports. Splices shall not be required to be enclosed in boxes or other enclosures. Control equipment and busbars connecting common-return conductors shall be permitted to be attached directly to the organ structure without insulation supports. Abandoned cables that are not terminated at equipment shall be identified with a tag of sufficient durability to withstand the environment involved.

650.8 Overcurrent Protection. Circuits shall be so arranged that 20 AWG through 28 AWG conductors shall be protected by an overcurrent device rated at not more than 6 amperes. Other conductor sizes shall be protected in accordance with their ampacity. A common return conductor shall not require overcurrent protection.

N 650.9 Protection from Accidental Contact. The wiring of the sounding apparatus shall be within the lockable enclosure (organ chamber) where the exterior pipes shall be permitted to form part of the enclosure.

> Informational Note: Access to the sounding apparatus and the associated circuitry is restricted by an enclosure. In most pipe organ installations, exterior pipes form part of the enclosure. In other installations, the pipes are covered by millwork that permits the passage of sound.

ARTICLE 660
X-Ray Equipment

Part I. General

660.1 Scope. This article covers all X-ray equipment operating at any frequency or voltage for industrial or other nonmedical or nondental use.

> Informational Note: See Article 517, Part V, for X-ray installations in health care facilities.

Nothing in this article shall be construed as specifying safeguards against the useful beam or stray X-ray radiation.

> Informational Note No. 1: Radiation safety and performance requirements of several classes of X-ray equipment are regulated under Public Law 90-602 and are enforced by the Department of Health and Human Services.

> Informational Note No. 2: In addition, information on radiation protection by the National Council on Radiation Protection and Measurements is published as *Reports of the National Council on Radiation Protection and Measurement.* These reports can be obtained from NCRP Publications, 7910 Woodmont Ave., Suite 1016, Bethesda, MD 20814.

660.2 Definitions.

Long-Time Rating. A rating based on an operating interval of 5 minutes or longer.

Mobile. X-ray equipment mounted on a permanent base with wheels and/or casters for moving while completely assembled.

Momentary Rating. A rating based on an operating interval that does not exceed 5 seconds.

Portable. X-ray equipment designed to be hand-carried.

Transportable. X-ray equipment that is to be installed in a vehicle or that may be readily disassembled for transport in a vehicle.

660.3 Hazardous (Classified) Locations. Unless identified for the location, X-ray and related equipment shall not be installed or operated in hazardous (classified) locations.

> Informational Note: See Article 517, Part IV.

660.4 Connection to Supply Circuit.

(A) Fixed and Stationary Equipment. Fixed and stationary X-ray equipment shall be connected to the power supply by means of a wiring method meeting the general requirements of this *Code*. Equipment properly supplied by a branch circuit rated at not over 30 amperes shall be permitted to be supplied through a suitable attachment plug cap and hard-service cable or cord.

(B) Portable, Mobile, and Transportable Equipment. Individual branch circuits shall not be required for portable, mobile, and transportable X-ray equipment requiring a capacity of not over 60 amperes. Portable and mobile types of X-ray equipment of any capacity shall be supplied through a suitable hard-service cable or cord. Transportable X-ray equipment of any capacity shall be permitted to be connected to its power supply by suitable connections and hard-service cable or cord.

(C) Over 1000 Volts, Nominal. Circuits and equipment operated at more than 1000 volts, nominal, shall comply with Article 490.

660.5 Disconnecting Means. A disconnecting means of adequate capacity for at least 50 percent of the input required for the momentary rating, or 100 percent of the input required for the long-time rating, of the X-ray equipment, whichever is greater, shall be provided in the supply circuit. The disconnecting means shall be located within sight from the X-ray control and readily accessible.

Exception: The disconnecting means for the X-ray equipment shall not be required under either of the following conditions, provided that the controller disconnecting means is lockable in accordance with 110.25:

(1) Where such a location of the disconnecting means for the X-ray equipment is impracticable or introduces additional or increased hazards to persons or property

(2) In industrial installations, with written safety procedures, where conditions of maintenance and supervision ensure that only qualified persons service the equipment

660.6 Rating of Supply Conductors and Overcurrent Protection.

(A) Branch-Circuit Conductors. The ampacity of supply branch-circuit conductors and the overcurrent protective devices shall not be less than 50 percent of the momentary rating or 100 percent of the long-time rating, whichever is greater.

(B) Feeder Conductors. The rated ampacity of conductors and overcurrent devices of a feeder for two or more branch circuits supplying X-ray units shall not be less than 100 percent of the momentary demand rating [as determined by 660.6(A)] of the two largest X-ray apparatus plus 20 percent of the momentary ratings of other X-ray apparatus.

> Informational Note: The minimum conductor size for branch and feeder circuits is also governed by voltage regulation requirements. For a specific installation, the manufacturer usually specifies minimum distribution transformer and conductor sizes, rating of disconnect means, and overcurrent protection.

660.7 Wiring Terminals. X-ray equipment not provided with a permanently attached cord or cord set shall be provided with suitable wiring terminals or leads for the connection of power-supply conductors of the size required by the rating of the branch circuit for the equipment.

660.9 Minimum Size of Conductors. Size 18 AWG or 16 AWG fixture wires, as specified in 725.49, and flexible cords shall be permitted for the control and operating circuits of X-ray and auxiliary equipment where protected by not larger than 20-ampere overcurrent devices.

660.10 Equipment Installations. All equipment for new X-ray installations and all used or reconditioned X-ray equipment moved to and reinstalled at a new location shall be of an approved type.

Part II. Control

660.20 Fixed and Stationary Equipment.

(A) Separate Control Device. A separate control device, in addition to the disconnecting means, shall be incorporated in the X-ray control supply or in the primary circuit to the high-voltage transformer. This device shall be a part of the X-ray equipment but shall be permitted in a separate enclosure immediately adjacent to the X-ray control unit.

(B) Protective Device. A protective device, which shall be permitted to be incorporated into the separate control device, shall be provided to control the load resulting from failures in the high-voltage circuit.

660.21 Portable and Mobile Equipment. Portable and mobile equipment shall comply with 660.20, but the manually controlled device shall be located in or on the equipment.

660.23 Industrial and Commercial Laboratory Equipment.

(A) Radiographic and Fluoroscopic Types. All radiographic- and fluoroscopic-type equipment shall be effectively enclosed or shall have interlocks that de-energize the equipment automatically to prevent ready access to live current-carrying parts.

(B) Diffraction and Irradiation Types. Diffraction- and irradiation-type equipment or installations not effectively enclosed or not provided with interlocks to prevent access to uninsulated live parts during operation shall be provided with a positive means to indicate when they are energized. The indicator shall be a pilot light, readable meter deflection, or equivalent means.

660.24 Independent Control. Where more than one piece of equipment is operated from the same high-voltage circuit, each piece or each group of equipment as a unit shall be provided with a high-voltage switch or equivalent disconnecting means. This disconnecting means shall be constructed, enclosed, or located so as to avoid contact by persons with its live parts.

Part III. Transformers and Capacitors

660.35 General. Transformers and capacitors that are part of an X-ray equipment shall not be required to comply with Articles 450 and 460.

660.36 Capacitors. Capacitors shall be mounted within enclosures of insulating material or grounded metal.

Part IV. Guarding and Grounding

660.47 General.

(A) High-Voltage Parts. All high-voltage parts, including X-ray tubes, shall be mounted within grounded enclosures. Air, oil, gas, or other suitable insulating media shall be used to insulate the high voltage from the grounded enclosure. The connection from the high-voltage equipment to X-ray tubes and other high-voltage components shall be made with high-voltage shielded cables.

(B) Low-Voltage Cables. Low-voltage cables connecting to oil-filled units that are not completely sealed, such as transformers, condensers, oil coolers, and high-voltage switches, shall have insulation of the oil-resistant type.

660.48 Grounding. Non–current-carrying metal parts of X-ray and associated equipment (controls, tables, X-ray tube supports, transformer tanks, shielded cables, X-ray tube heads, and so forth) shall be grounded in the manner specified in Article 250. Portable and mobile equipment shall be provided with an approved grounding-type attachment plug cap.

Exception: Battery-operated equipment.

ARTICLE 665
Induction and Dielectric Heating Equipment

Part I. General

665.1 Scope. This article covers the construction and installation of dielectric heating, induction heating, induction melting, and induction welding equipment and accessories for industrial and scientific applications. Medical or dental applications, appliances, or line frequency pipeline and vessel heating are not covered in this article.

Informational Note: See Article 427, Part V, for line frequency induction heating of pipelines and vessels.

665.2 Definitions.

Applicator. The device used to transfer energy between the output circuit and the object or mass to be heated.

Converting Device. That part of the heating equipment that converts input mechanical or electrical energy to the voltage, current, and frequency used for the heating applicator. A converting device consists of equipment using line frequency, all static multipliers, oscillator-type units using vacuum tubes, inverters using solid-state devices, or motor-generator equipment.

Dielectric Heating. Heating of a nominally insulating material due to its own dielectric losses when the material is placed in a varying electric field.

Heating Equipment. As used in this article, any equipment that is used for heating purposes and whose heat is generated by induction or dielectric methods.

Induction Heating, Melting, and Welding. The heating, melting, or welding of a nominally conductive material due to its own I^2R losses when the material is placed in a varying electromagnetic field.

665.4 Hazardous (Classified) Locations. Heating equipment shall not be installed in hazardous (classified) locations as defined in Article 500 unless the equipment and wiring are designed and approved for the hazardous (classified) locations.

665.5 Output Circuit. The output circuit shall include all output components external to the converting device, including contactors, switches, busbars, and other conductors. The current flow from the output circuit to ground under operating and ground-fault conditions shall be limited to a value that does not cause 50 volts or more to ground to appear on any accessible part of the heating equipment and its load. The output circuit shall be permitted to be isolated from ground.

665.7 Remote Control.

(A) Multiple Control Points. Where multiple control points are used for applicator energization, a means shall be provided and interlocked so that the applicator can be energized from only one control point at a time. A means for de-energizing the applicator shall be provided at each control point.

(B) Foot Switches. Switches operated by foot pressure shall be provided with a shield over the contact button to avoid accidental closing of a foot switch.

665.10 Ampacity of Supply Conductors. The ampacity of supply conductors shall be determined by 665.10(A) or (B).

(A) Nameplate Rating. The ampacity of conductors supplying one or more pieces of equipment shall be not less than the sum of the nameplate ratings for the largest group of machines capable of simultaneous operation, plus 100 percent of the standby currents of the remaining machines. Where standby currents are not given on the nameplate, the nameplate rating shall be used as the standby current.

(B) Motor-Generator Equipment. The ampacity of supply conductors for motor-generator equipment shall be determined in accordance with Article 430, Part II.

665.11 Overcurrent Protection. Overcurrent protection for the heating equipment shall be provided as specified in Article 240. This overcurrent protection shall be permitted to be provided separately or as a part of the equipment.

665.12 Disconnecting Means. A readily accessible disconnecting means shall be provided to disconnect each heating equipment from its supply circuit. The disconnecting means shall be located within sight from the controller or be lockable open in accordance with 110.25.

The rating of this disconnecting means shall not be less than the nameplate rating of the heating equipment. Motor-generator equipment shall comply with Article 430, Part IX. The supply circuit disconnecting means shall be permitted to serve as the heating equipment disconnecting means where only one heating equipment is supplied.

Part II. Guarding, Grounding, and Labeling

665.19 Component Interconnection. The interconnection components required for a complete heating equipment installation shall be guarded.

665.20 Enclosures. The converting device (excluding the component interconnections) shall be completely contained within an enclosure(s) of noncombustible material.

665.21 Control Panels. All control panels shall be of dead-front construction.

665.22 Access to Internal Equipment. Access doors or detachable access panels shall be employed for internal access to heating equipment. Access doors to internal compartments containing equipment employing voltages from 150 volts to 1000 volts ac or dc shall be capable of being locked closed or shall be interlocked to prevent the supply circuit from being energized while the door(s) is open. The provision for locking or adding a lock to the access doors shall be installed on or at the access door and shall remain in place with or without the lock installed.

Access doors to internal compartments containing equipment employing voltages exceeding 1000 volts ac or dc shall be provided with a disconnecting means equipped with mechanical lockouts to prevent access while the heating equipment is energized, or the access doors shall be capable of being locked closed and interlocked to prevent the supply circuit from being energized while the door(s) is open. Detachable panels not normally used for access to such parts shall be fastened in a manner that makes them inconvenient to remove.

665.23 Warning Labels or Signs. Warning labels or signs that read "DANGER — HIGH VOLTAGE — KEEP OUT" shall be attached to the equipment and shall be plainly visible where persons might come in contact with energized parts when doors are open or closed or when panels are removed from compartments containing over 150 volts ac or dc. The warning sign(s) or label(s) shall comply with 110.21(B).

665.24 Capacitors. The time and means of discharge shall be in accordance with 460.6 for capacitors rated 600 volts, nominal, and under. The time and means of discharge shall be in accordance with 460.28 for capacitors rated over 600 volts, nominal. Capacitor internal pressure switches connected to a circuit-interrupter device shall be permitted for capacitor overcurrent protection.

665.25 Dielectric Heating Applicator Shielding. Protective cages or adequate shielding shall be used to guard dielectric heating applicators. Interlock switches shall be used on all hinged access doors, sliding panels, or other easy means of access to the applicator. All interlock switches shall be connected in such a manner as to remove all power from the applicator when any one of the access doors or panels is open.

665.26 Grounding and Bonding. Bonding to the equipment grounding conductor or inter-unit bonding, or both, shall be used wherever required for circuit operation, and for limiting to a safe value radio frequency voltages between all exposed non–current-carrying parts of the equipment and earth ground, between all equipment parts and surrounding objects, and between such objects and earth ground. Such connection to the equipment grounding conductor and bonding shall be installed in accordance with Article 250, Parts II and V.

Informational Note: Under certain conditions, contact between the object being heated and the applicator results in an unsafe condition, such as eruption of heated materials. Grounding of the object being heated and ground detection can be used to prevent this unsafe condition.

665.27 Marking. Each heating equipment shall be provided with a nameplate giving the manufacturer's name and model identification and the following input data: line volts, frequency, number of phases, maximum current, full-load kilovolt-amperes (kVA), and full-load power factor. Additional data shall be permitted.

ARTICLE 668
Electrolytic Cells

668.1 Scope. This article applies to the installation of the electrical components and accessory equipment of electrolytic cells, electrolytic cell lines, and process power supply for the production of aluminum, cadmium, chlorine, copper, fluorine, hydrogen peroxide, magnesium, sodium, sodium chlorate, and zinc.

Not covered by this article are cells used as a source of electric energy and for electroplating processes and cells used for the production of hydrogen.

Informational Note No. 1: In general, any cell line or group of cell lines operated as a unit for the production of a particular metal, gas, or chemical compound may differ from any other cell lines producing the same product because of variations in the particular raw materials used, output capacity, use of proprietary methods or process practices, or other modifying factors to the extent that detailed *Code* requirements become overly restrictive and do not accomplish the stated purpose of this *Code*.

Informational Note No. 2: For further information, see IEEE 463-2013, *Standard for Electrical Safety Practices in Electrolytic Cell Line Working Zones.*

668.2 Definitions.

Cell Line. An assembly of electrically interconnected electrolytic cells supplied by a source of direct-current power.

Cell Line Attachments and Auxiliary Equipment. As applied to this article, a term that includes, but is not limited to, auxiliary tanks; process piping; ductwork; structural supports; exposed cell line conductors; conduits and other raceways; pumps, positioning equipment, and cell cutout or bypass electrical devices. Auxiliary equipment includes tools, welding machines, crucibles, and other portable equipment used for operation and maintenance within the electrolytic cell line working zone.

In the cell line working zone, auxiliary equipment includes the exposed conductive surfaces of ungrounded cranes and crane-mounted cell-servicing equipment.

Electrically Connected. A connection capable of carrying current as distinguished from connection through electromagnetic induction.

Electrolytic Cell. A tank or vat in which electrochemical reactions are caused by applying electric energy for the purpose of refining or producing usable materials.

Electrolytic Cell Line Working Zone. The space envelope wherein operation or maintenance is normally performed on or in the vicinity of exposed energized surfaces of electrolytic cell lines or their attachments.

668.3 Other Articles.

(A) Lighting, Ventilating, Material Handling. Chapters 1 through 4 shall apply to services, feeders, branch circuits, and apparatus for supplying lighting, ventilating, material handling, and the like that are outside the electrolytic cell line working zone.

(B) Systems Not Electrically Connected. Those elements of a cell line power-supply system that are not electrically connected to the cell supply system, such as the primary winding of a two-winding transformer, the motor of a motor-generator set, feeders, branch circuits, disconnecting means, motor controllers, and overload protective equipment, shall be required to comply with all applicable provisions of this *Code*.

(C) Electrolytic Cell Lines. Electrolytic cell lines shall comply with the provisions of Chapters 1 through 4 except as amended in 668.3(C)(1) through (C)(4).

(1) Conductors. The electrolytic cell line conductors shall not be required to comply with the provisions of Articles 110, 210, 215, 220, and 225. See 668.11.

(2) Overcurrent Protection. Overcurrent protection of electrolytic cell dc process power circuits shall not be required to comply with the requirements of Article 240.

(3) Grounding. Equipment located or used within the electrolytic cell line working zone or associated with the cell line direct-current power circuits shall not be required to comply with the provisions of Article 250.

(4) Working Zone. The electrolytic cells, cell line attachments, and the wiring of auxiliary equipment and devices within the cell line working zone shall not be required to comply with the provisions of Articles 110, 210, 215, 220, and 225. See 668.30.

> Informational Note: See 668.15 for equipment, apparatus, and structural component grounding.

668.10 Cell Line Working Zone.

(A) Area Covered. The space envelope of the cell line working zone shall encompass spaces that meet any of the following conditions:

(1)   Is within 2.5 m (96 in.) above energized surfaces of electrolytic cell lines or their energized attachments
(2)   Is below energized surfaces of electrolytic cell lines or their energized attachments, provided the headroom in the space beneath is less than 2.5 m (96 in.)
(3)   Is within 1.0 m (42 in.) horizontally from energized surfaces of electrolytic cell lines or their energized attachments or from the space envelope described in 668.10(A) (1) or (A)(2)

(B) Area Not Covered. The cell line working zone shall not be required to extend through or beyond walls, floors, roofs, partitions, barriers, or the like.

668.11 Direct-Current Cell Line Process Power Supply.

(A) Not Grounded. The direct-current cell line process power-supply conductors shall not be required to be grounded.

(B) Metal Enclosures Grounded. All metal enclosures of power-supply apparatus for the direct-current cell line process operating with a power supply over 50 volts shall be grounded by either of the following means:

(1)   Through protective relaying equipment
(2)   By a minimum 2/0 AWG copper grounding conductor or a conductor of equal or greater conductance

(C) Grounding Requirements. The grounding connections required by 668.11(B) shall be installed in accordance with 250.8, 250.10, 250.12, 250.68, and 250.70.

668.12 Cell Line Conductors.

(A) Insulation and Material. Cell line conductors shall be either bare, covered, or insulated and of copper, aluminum, copper-clad aluminum, steel, or other suitable material.

(B) Size. Cell line conductors shall be of such cross-sectional area that the temperature rise under maximum load conditions and at maximum ambient shall not exceed the safe operating temperature of the conductor insulation or the material of the conductor supports.

(C) Connections. Cell line conductors shall be joined by bolted, welded, clamped, or compression connectors.

668.13 Disconnecting Means.

(A) More Than One Process Power Supply. Where more than one direct-current cell line process power supply serves the same cell line, a disconnecting means shall be provided on the cell line circuit side of each power supply to disconnect it from the cell line circuit.

(B) Removable Links or Conductors. Removable links or removable conductors shall be permitted to be used as the disconnecting means.

668.14 Shunting Means.

(A) Partial or Total Shunting. Partial or total shunting of cell line circuit current around one or more cells shall be permitted.

(B) Shunting One or More Cells. The conductors, switches, or combination of conductors and switches used for shunting one or more cells shall comply with the applicable requirements of 668.12.

668.15 Grounding. For equipment, apparatus, and structural components that are required to be grounded by provisions of Article 668, the provisions of Article 250 shall apply, except a water pipe electrode shall not be required to be used. Any electrode or combination of electrodes described in 250.52 shall be permitted.

668.20 Portable Electrical Equipment.

(A) Portable Electrical Equipment Not to Be Grounded. The frames and enclosures of portable electrical equipment used within the cell line working zone shall not be grounded.

Exception No. 1: Where the cell line voltage does not exceed 200 volts dc, these frames and enclosures shall be permitted to be grounded.

Exception No. 2: These frames and enclosures shall be permitted to be grounded where guarded.

(B) Isolating Transformers. Electrically powered, hand-held, cord-connected portable equipment with ungrounded frames or enclosures used within the cell line working zone shall be connected to receptacle circuits that have only ungrounded conductors such as a branch circuit supplied by an isolating transformer with an ungrounded secondary.

(C) Marking. Ungrounded portable electrical equipment shall be distinctively marked and shall employ plugs and receptacles of a configuration that prevents connection of this equipment to grounding receptacles and that prevents inadvertent interchange of ungrounded and grounded portable electrical equipments.

668.21 Power-Supply Circuits and Receptacles for Portable Electrical Equipment.

(A) Isolated Circuits. Circuits supplying power to ungrounded receptacles for hand-held, cord-connected equipment shall be electrically isolated from any distribution system supplying areas other than the cell line working zone and shall be ungrounded. Power for these circuits shall be supplied through isolating transformers. Primaries of such transformers shall operate at not more than 1000 volts between conductors and shall be provided with proper overcurrent protection. The secondary voltage of such transformers shall not exceed 300 volts between conductors, and all circuits supplied from such secondaries shall be ungrounded and shall have an approved overcurrent device of proper rating in each conductor.

(B) Noninterchangeability. Receptacles and their mating plugs for ungrounded equipment shall not have provision for a grounding conductor and shall be of a configuration that prevents their use for equipment required to be grounded.

(C) Marking. Receptacles on circuits supplied by an isolating transformer with an ungrounded secondary shall be a distinctive configuration, shall be distinctively marked, and shall not be used in any other location in the plant.

668.30 Fixed and Portable Electrical Equipment.

(A) Electrical Equipment Not Required to Be Grounded. Alternating-current systems supplying fixed and portable electrical equipment within the cell line working zone shall not be required to be grounded.

(B) Exposed Conductive Surfaces Not Required to Be Grounded. Exposed conductive surfaces, such as electrical equipment housings, cabinets, boxes, motors, raceways, and the like, that are within the cell line working zone shall not be required to be grounded.

(C) Wiring Methods. Auxiliary electrical equipment such as motors, transducers, sensors, control devices, and alarms, mounted on an electrolytic cell or other energized surface, shall be connected to premises wiring systems by any of the following means:

(1) Multiconductor hard usage cord.
(2) Wire or cable in suitable raceways or metal or nonmetallic cable trays. If metal conduit, cable tray, armored cable, or similar metallic systems are used, they shall be installed with insulating breaks such that they do not cause a potentially hazardous electrical condition.

(D) Circuit Protection. Circuit protection shall not be required for control and instrumentation that are totally within the cell line working zone.

(E) Bonding. Bonding of fixed electrical equipment to the energized conductive surfaces of the cell line, its attachments, or auxiliaries shall be permitted. Where fixed electrical equipment is mounted on an energized conductive surface, it shall be bonded to that surface.

668.31 Auxiliary Nonelectrical Connections. Auxiliary nonelectrical connections, such as air hoses, water hoses, and the like, to an electrolytic cell, its attachments, or auxiliary equipment shall not have continuous conductive reinforcing wire, armor, braids, and the like. Hoses shall be of a nonconductive material.

668.32 Cranes and Hoists.

(A) Conductive Surfaces to Be Insulated from Ground. The conductive surfaces of cranes and hoists that enter the cell line working zone shall not be required to be grounded. The portion of an overhead crane or hoist that contacts an energized electrolytic cell or energized attachments shall be insulated from ground.

(B) Hazardous Electrical Conditions. Remote crane or hoist controls that could introduce hazardous electrical conditions into the cell line working zone shall employ one or more of the following systems:

(1) Isolated and ungrounded control circuit in accordance with 668.21(A)
(2) Nonconductive rope operator
(3) Pendant pushbutton with nonconductive supporting means and having nonconductive surfaces or ungrounded exposed conductive surfaces
(4) Radio

668.40 Enclosures. General-purpose electrical equipment enclosures shall be permitted where a natural draft ventilation system prevents the accumulation of gases.

ARTICLE 669
Electroplating

669.1 Scope. The provisions of this article apply to the installation of the electrical components and accessory equipment that supply the power and controls for electroplating, anodizing, electropolishing, and electrostripping. For purposes of this article, the term *electroplating* shall be used to identify any or all of these processes.

669.3 General. Equipment for use in electroplating processes shall be identified for such service.

669.5 Branch-Circuit Conductors. Branch-circuit conductors supplying one or more units of equipment shall have an ampacity of not less than 125 percent of the total connected load. The ampacities for busbars shall be in accordance with 366.23.

669.6 Wiring Methods. Conductors connecting the electrolyte tank equipment to the conversion equipment shall be in accordance with 669.6(A) and (B).

(A) Systems Not Exceeding 60 Volts Direct Current. Insulated conductors shall be permitted to be run without insulated support, provided they are protected from physical damage. Bare copper or aluminum conductors shall be permitted where supported on insulators.

(B) Systems Exceeding 60 Volts Direct Current. Insulated conductors shall be permitted to be run on insulated supports, provided they are protected from physical damage. Bare copper or aluminum conductors shall be permitted where supported on insulators and guarded against accidental contact up to the point of termination in accordance with 110.27.

669.7 Warning Signs. Warning signs shall be posted to indicate the presence of bare conductors. The warning sign(s) or label(s) shall comply with 110.21(B).

669.8 Disconnecting Means.

(A) More Than One Power Supply. Where more than one power supply serves the same dc system, a disconnecting means shall be provided on the dc side of each power supply.

(B) Removable Links or Conductors. Removable links or removable conductors shall be permitted to be used as the disconnecting means.

669.9 Overcurrent Protection. Direct-current conductors shall be protected from overcurrent by one or more of the following:

(1) Fuses or circuit breakers
(2) A current-sensing device that operates a disconnecting means
(3) Other approved means

ARTICLE 670
Industrial Machinery

670.1 Scope. This article covers the definition of, the nameplate data for, and the size and overcurrent protection of supply conductors to industrial machinery.

Informational Note No. 1: For further information, see NFPA 79-2015, *Electrical Standard for Industrial Machinery.*

Informational Note No. 2: For information on the workspace requirements for equipment containing supply conductor terminals, see 110.26. For information on the workspace requirements for machine power and control equipment, see NFPA 79-2015, *Electrical Standard for Industrial Machinery.*

670.2 Definition.

Industrial Machinery (Machine). A power-driven machine (or a group of machines working together in a coordinated manner), not portable by hand while working, that is used to process material by cutting; forming; pressure; electrical, thermal, or optical techniques; lamination; or a combination of these processes. It can include associated equipment used to transfer material or tooling, including fixtures, to assemble/disassemble, to inspect or test, or to package. [The associated electrical equipment, including the logic controller(s) and associated software or logic together with the machine actuators and sensors, are considered as part of the industrial machine.]

670.3 Machine Nameplate Data.

Informational Note: See 430.22(E) and 430.26 for duty cycle requirements.

(A) Permanent Nameplate. A permanent nameplate shall be attached to the control equipment enclosure or machine and shall be plainly visible after installation. The nameplate shall include the following information:

(1) Supply voltage, number of phases, frequency, and full-load current

(2) Maximum ampere rating of the short-circuit and ground-fault protective device
(3) Ampere rating of largest motor, from the motor nameplate, or load
(4) Short-circuit current rating of the machine industrial control panel based on one of the following:

    a. Short-circuit current rating of a listed and labeled machine control enclosure or assembly
    b. Short-circuit current rating established utilizing an approved method

    Informational Note: UL 508A-2001, Supplement SB, is an example of an approved method.

(5) Electrical diagram number(s) or the number of the index to the electrical drawings

The full-load current shown on the nameplate shall not be less than the sum of the full-load currents required for all motors and other equipment that may be in operation at the same time under normal conditions of use. Where unusual type loads, duty cycles, and so forth require oversized conductors or permit reduced-size conductors, the required capacity shall be included in the marked "full-load current." Where more than one incoming supply circuit is to be provided, the nameplate shall state the preceding information for each circuit.

(B) Overcurrent Protection. Where overcurrent protection is provided in accordance with 670.4(C), the machine shall be marked "overcurrent protection provided at machine supply terminals."

670.4 Supply Conductors and Overcurrent Protection.

(A) Size. The size of the supply conductor shall be such as to have an ampacity not less than 125 percent of the full-load current rating of all resistance heating loads plus 125 percent of the full-load current rating of the highest rated motor plus the sum of the full-load current ratings of all other connected motors and apparatus, based on their duty cycle, that may be in operation at the same time.

Informational Note No. 1: See Table 310.15(B)(16) through Table 310.15(B)(20) for ampacity of conductors rated 2000 volts and below.

Informational Note No. 2: See 430.22(E) and 430.26 for duty cycle requirements.

(B) Disconnecting Means. A machine shall be considered as an individual unit and therefore shall be provided with disconnecting means. The disconnecting means shall be permitted to be supplied by branch circuits protected by either fuses or circuit breakers. The disconnecting means shall not be required to incorporate overcurrent protection.

(C) Overcurrent Protection. Where furnished as part of the machine, overcurrent protection for each supply circuit shall consist of a single circuit breaker or set of fuses, the machine shall bear the marking required in 670.3, and the supply conductors shall be considered either as feeders or as taps as covered by 240.21.

The rating or setting of the overcurrent protective device for the circuit supplying the machine shall not be greater than the sum of the largest rating or setting of the branch-circuit short-circuit and ground-fault protective device provided with the machine, plus 125 percent of the full-load current rating of all resistance heating loads, plus the sum of the full-load currents of all other motors and apparatus that could be in operation at the same time.

Exception: Where one or more instantaneous trip circuit breakers or motor short-circuit protectors are used for motor branch-circuit short-circuit and ground-fault protection as permitted by 430.52(C), the procedure specified in 670.4(C) for determining the maximum rating of the protective device for the circuit supplying the machine shall apply with the following provision: For the purpose of the calculation, each instantaneous trip circuit breaker or motor short-circuit protector shall be assumed to have a rating not exceeding the maximum percentage of motor full-load current permitted by Table 430.52 for the type of machine supply circuit protective device employed.

Where no branch-circuit short-circuit and ground-fault protective device is provided with the machine, the rating or setting of the overcurrent protective device shall be based on 430.52 and 430.53, as applicable.

670.5 Short-Circuit Current Rating.

(1)   Industrial machinery shall not be installed where the available short-circuit current exceeds its short-circuit current rating as marked in accordance with 670.3(A)(4).

N (2)   Industrial machinery shall be legibly marked in the field with the maximum available short-circuit current. The field marking(s) shall include the date the short-circuit current calculation was performed and be of sufficient durability to withstand the environment involved.

N **670.6 Surge Protection.**   Industrial machinery with safety interlock circuits shall have surge protection installed.

ARTICLE 675
Electrically Driven or Controlled Irrigation Machines

Part I.   General

675.1 Scope.   The provisions of this article apply to electrically driven or controlled irrigation machines, and to the branch circuits and controllers for such equipment.

675.2 Definitions.

Center Pivot Irrigation Machine.   A multimotored irrigation machine that revolves around a central pivot and employs alignment switches or similar devices to control individual motors.

Collector Rings.   An assembly of slip rings for transferring electric energy from a stationary to a rotating member.

Irrigation Machine.   An electrically driven or controlled machine, with one or more motors, not hand-portable, and used primarily to transport and distribute water for agricultural purposes.

675.4 Irrigation Cable.

(A) Construction.   The cable used to interconnect enclosures on the structure of an irrigation machine shall be an assembly of stranded, insulated conductors with nonhygroscopic and nonwicking filler in a core of moisture- and flame-resistant nonmetallic material overlaid with a metallic covering and jacketed with a moisture-, corrosion-, and sunlight-resistant nonmetallic material.

The conductor insulation shall be of a type listed in Table 310.104(A) for an operating temperature of 75°C (167°F) and for use in wet locations. The core insulating material thickness shall not be less than 0.76 mm (30 mils), and the metallic overlay thickness shall be not less than 0.20 mm (8 mils). The jacketing material thickness shall be not less than 1.27 mm (50 mils).

A composite of power, control, and grounding conductors in the cable shall be permitted.

(B) Alternate Wiring Methods.   Installation of other listed cables complying with the construction requirements of 675.4(A) shall be permitted.

(C) Supports.   Irrigation cable shall be secured by straps, hangers, or similar fittings identified for the purpose and so installed as not to damage the cable. Cable shall be supported at intervals not exceeding 1.2 m (4 ft).

(D) Fittings.   Fittings shall be used at all points where irrigation cable terminates. The fittings shall be designed for use with the cable and shall be suitable for the conditions of service.

675.5 More Than Three Conductors in a Raceway or Cable.   The signal and control conductors of a raceway or cable shall not be counted for the purpose of ampacity adjustment as required in 310.15(B)(3)(a).

675.6 Marking on Main Control Panel.   The main control panel shall be provided with a nameplate giving the following information:

(1)   The manufacturer's name, the rated voltage, the phase, and the frequency
(2)   The current rating of the machine
(3)   The rating of the main disconnecting means and size of overcurrent protection required

675.7 Equivalent Current Ratings.   Where intermittent duty is not involved, the provisions of Article 430 shall be used for determining ratings for controllers, disconnecting means, conductors, and the like. Where irrigation machines have inherent intermittent duty, the determinations of equivalent current ratings in 675.7(A) and (B) shall be used.

(A) Continuous-Current Rating.   The equivalent continuous-current rating for the selection of branch-circuit conductors and overcurrent protection shall be equal to 125 percent of the motor nameplate full-load current rating of the largest motor, plus a quantity equal to the sum of each of the motor nameplate full-load current ratings of all remaining motors on the circuit, multiplied by the maximum percent duty cycle at which they can continuously operate.

(B) Locked-Rotor Current.   The equivalent locked-rotor current rating shall be equal to the numerical sum of the locked-rotor current of the two largest motors plus 100 percent of the sum of the motor nameplate full-load current ratings of all the remaining motors on the circuit.

675.8 Disconnecting Means.

(A) Main Controller. A controller that is used to start and stop the complete machine shall meet all of the following requirements:

(1) An equivalent continuous current rating not less than specified in 675.7(A) or 675.22(A)
(2) A horsepower rating not less than the value from Table 430.251(A) and Table 430.251(B), based on the equivalent locked-rotor current specified in 675.7(B) or 675.22(B)

Exception: A listed molded case switch shall not require a horsepower rating.

(B) Main Disconnecting Means. The main disconnecting means for the machine shall provide overcurrent protection, shall be at the point of connection of electric power to the machine, or shall be in sight from the machine, and it shall be readily accessible and lockable in accordance with 110.25. This disconnecting means shall have a horsepower and current rating not less than required for the main controller.

Exception No. 1: Circuit breakers without marked horsepower ratings shall be permitted in accordance with 430.109.

Exception No. 2: A listed molded case switch without marked horsepower ratings shall be permitted.

(C) Disconnecting Means for Individual Motors and Controllers. A disconnecting means shall be provided to simultaneously disconnect all ungrounded conductors for each motor and controller and shall be located as required by Article 430, Part IX. The disconnecting means shall not be required to be readily accessible.

675.9 Branch-Circuit Conductors. The branch-circuit conductors shall have an ampacity not less than specified in 675.7(A) or 675.22(A).

675.10 Several Motors on One Branch Circuit.

(A) Protection Required. Several motors, each not exceeding 2 hp rating, shall be permitted to be used on an irrigation machine circuit protected at not more than 30 amperes at 1000 volts, nominal, or less, provided all of the following conditions are met:

(1) The full-load rating of any motor in the circuit shall not exceed 6 amperes.
(2) Each motor in the circuit shall have individual overload protection in accordance with 430.32.
(3) Taps to individual motors shall not be smaller than 14 AWG copper and not more than 7.5 m (25 ft) in length.

(B) Individual Protection Not Required. Individual branch-circuit short-circuit protection for motors and motor controllers shall not be required where the requirements of 675.10(A) are met.

675.11 Collector Rings.

(A) Transmitting Current for Power Purposes. Collector rings shall have a current rating not less than 125 percent of the full-load current of the largest device served plus the full-load current of all other devices served, or as determined from 675.7(A) or 675.22(A).

(B) Control and Signal Purposes. Collector rings for control and signal purposes shall have a current rating not less than 125 percent of the full-load current of the largest device served plus the full-load current of all other devices served.

(C) Grounding. The collector ring used for grounding shall have a current rating not less than that sized in accordance with 675.11(A).

(D) Protection. Collector rings shall be protected from the expected environment and from accidental contact by means of a suitable enclosure.

675.12 Grounding. The following equipment shall be grounded:

(1) All electrical equipment on the irrigation machine
(2) All electrical equipment associated with the irrigation machine
(3) Metal junction boxes and enclosures
(4) Control panels or control equipment that supplies or controls electrical equipment to the irrigation machine

Exception: Grounding shall not be required on machines where all of the following provisions are met:
   (a)   The machine is electrically controlled but not electrically driven.
   (b)   The control voltage is 30 volts or less.
   (c)   The control or signal circuits are current limited as specified in Chapter 9, Tables 11(A) and 11(B).

675.13 Methods of Grounding. Machines that require grounding shall have a non–current-carrying equipment grounding conductor provided as an integral part of each cord, cable, or raceway. This grounding conductor shall be sized not less than the largest supply conductor in each cord, cable, or raceway. Feeder circuits supplying power to irrigation machines shall have an equipment grounding conductor sized according to Table 250.122.

675.14 Bonding. Where electrical grounding is required on an irrigation machine, the metallic structure of the machine, metallic conduit, or metallic sheath of cable shall be connected to the grounding conductor. Metal-to-metal contact with a part that is connected to the grounding conductor and the non–current-carrying parts of the machine shall be considered as an acceptable bonding path.

675.15 Lightning Protection. If an irrigation machine has a stationary point, a grounding electrode system in accordance with Article 250, Part III, shall be connected to the machine at the stationary point for lightning protection.

675.16 Energy from More Than One Source. Equipment within an enclosure receiving electric energy from more than one source shall not be required to have a disconnecting means for the additional source if its voltage is 30 volts or less and it meets the requirements of Part III of Article 725.

675.17 Connectors. External plugs and connectors on the equipment shall be of the weatherproof type.

Unless provided solely for the connection of circuits meeting the requirements of Part III of Article 725, external plugs and connectors shall be constructed as specified in 250.124(A).

Part II. Center Pivot Irrigation Machines

675.21 General. The provisions of Part II are intended to cover additional special requirements that are peculiar to center pivot irrigation machines. See 675.2 for the definition of *Center Pivot Irrigation Machine.*

675.22 Equivalent Current Ratings. To establish ratings of controllers, disconnecting means, conductors, and the like, for the inherent intermittent duty of center pivot irrigation machines, the determinations in 675.22(A) and (B) shall be used.

(A) Continuous-Current Rating. The equivalent continuous-current rating for the selection of branch-circuit conductors and branch-circuit devices shall be equal to 125 percent of the motor nameplate full-load current rating of the largest motor plus 60 percent of the sum of the motor nameplate full-load current ratings of all remaining motors on the circuit.

(B) Locked-Rotor Current. The equivalent locked-rotor current rating shall be equal to the numerical sum of two times the locked-rotor current of the largest motor plus 80 percent of the sum of the motor nameplate full-load current ratings of all the remaining motors on the circuit.

ARTICLE 680
Swimming Pools, Fountains, and Similar Installations

Part I. General

680.1 Scope. The provisions of this article apply to the construction and installation of electrical wiring for, and equipment in or adjacent to, all swimming, wading, therapeutic, and decorative pools; fountains; hot tubs; spas; and hydromassage bathtubs, whether permanently installed or storable, and to metallic auxiliary equipment, such as pumps, filters, and similar equipment. The term *body of water* used throughout Part I applies to all bodies of water covered in this scope unless otherwise amended.

680.2 Definitions.

Cord-and-Plug-Connected Lighting Assembly. A lighting assembly consisting of a luminaire intended for installation in the wall of a spa, hot tub, or storable pool, and a cord-and-plug-connected transformer.

Dry-Niche Luminaire. A luminaire intended for installation in the floor or wall of a pool, spa, or fountain in a niche that is sealed against the entry of water.

N **Electrically Powered Pool Lift.** An electrically powered lift that provides accessibility to and from a pool or spa for people with disabilities.

Fixed (as applied to equipment). Equipment that is fastened or otherwise secured at a specific location.

Forming Shell. A structure designed to support a wet-niche luminaire assembly and intended for mounting in a pool or fountain structure.

Fountain. Fountains, ornamental pools, display pools, and reflection pools. The definition does not include drinking fountains.

Hydromassage Bathtub. A permanently installed bathtub equipped with a recirculating piping system, pump, and associated equipment. It is designed so it can accept, circulate, and discharge water upon each use.

Low Voltage Contact Limit. A voltage not exceeding the following values:

(1) 15 volts (RMS) for sinusoidal ac
(2) 21.2 volts peak for nonsinusoidal ac
(3) 30 volts for continuous dc
(4) 12.4 volts peak for dc that is interrupted at a rate of 10 to 200 Hz

Maximum Water Level. The highest level that water can reach before it spills out.

No-Niche Luminaire. A luminaire intended for installation above or below the water without a niche.

Packaged Spa or Hot Tub Equipment Assembly. A factory-fabricated unit consisting of water-circulating, heating, and control equipment mounted on a common base, intended to operate a spa or hot tub. Equipment can include pumps, air blowers, heaters, lights, controls, sanitizer generators, and so forth.

Packaged Therapeutic Tub or Hydrotherapeutic Tank Equipment Assembly. A factory-fabricated unit consisting of water-circulating, heating, and control equipment mounted on a common base, intended to operate a therapeutic tub or hydrotherapeutic tank. Equipment can include pumps, air blowers, heaters, lights, controls, sanitizer generators, and so forth.

Permanently Installed Decorative Fountains and Reflection Pools. Those that are constructed in the ground, on the ground, or in a building in such a manner that the fountain cannot be readily disassembled for storage, whether or not served by electrical circuits of any nature. These units are primarily constructed for their aesthetic value and are not intended for swimming or wading.

Permanently Installed Swimming, Wading, Immersion, and Therapeutic Pools. Those that are constructed in the ground or partially in the ground, and all others capable of holding water in a depth greater than 1.0 m (42 in.), and all pools installed inside of a building, regardless of water depth, whether or not served by electrical circuits of any nature.

Pool. Manufactured or field-constructed equipment designed to contain water on a permanent or semipermanent basis and used for swimming, wading, immersion, or therapeutic purposes.

Pool Cover, Electrically Operated. Motor-driven equipment designed to cover and uncover the water surface of a pool by means of a flexible sheet or rigid frame.

Portable (as applied to equipment). Equipment that is actually moved or can easily be moved from one place to another in normal use.

Self-Contained Spa or Hot Tub. Factory-fabricated unit consisting of a spa or hot tub vessel with all water-circulating, heating, and control equipment integral to the unit. Equipment can include pumps, air blowers, heaters, lights, controls, sanitizer generators, and so forth.

Self-Contained Therapeutic Tubs or Hydrotherapeutic Tanks. A factory-fabricated unit consisting of a therapeutic tub or hydrotherapeutic tank with all water-circulating, heating, and control equipment integral to the unit. Equipment may include pumps, air blowers, heaters, light controls, sanitizer generators, and so forth.

Spa or Hot Tub. A hydromassage pool, or tub for recreational or therapeutic use, not located in health care facilities, designed for immersion of users, and usually having a filter, heater, and motor-driven blower. It may be installed indoors or outdoors, on the ground or supporting structure, or in the ground or supporting structure. Generally, a spa or hot tub is not designed or intended to have its contents drained or discharged after each use.

Stationary (as applied to equipment). Equipment that is not moved from one place to another in normal use.

Storable Swimming, Wading, or Immersion Pools; or Storable/Portable Spas and Hot Tubs. Swimming, wading, or immersion pools that are intended to be stored when not in use, constructed on or above the ground and are capable of holding water to a maximum depth of 1.0 m (42 in.), or a pool, spa, or hot tub constructed on or above the ground, with nonmetallic, molded polymeric walls or inflatable fabric walls regardless of dimension.

Through-Wall Lighting Assembly. A lighting assembly intended for installation above grade, on or through the wall of a pool, consisting of two interconnected groups of components separated by the pool wall.

Wet-Niche Luminaire. A luminaire intended for installation in a forming shell mounted in a pool or fountain structure where the luminaire will be completely surrounded by water.

680.4 Approval of Equipment. All electrical equipment installed in the water, walls, or decks of pools, fountains, and similar installations shall comply with the provisions of this article. Equipment and products shall be listed.

680.5 Ground-Fault Circuit Interrupters. Ground-fault circuit interrupters (GFCIs) shall be self-contained units, circuit-breaker or receptacle types, or other listed types.

680.6 Grounding. Electrical equipment shall be grounded in accordance with Parts V, VI, and VII of Article 250 and connected by wiring methods of Chapter 3, except as modified by this article. The following equipment shall be grounded:

(1) Through-wall lighting assemblies and underwater luminaires, other than those low-voltage lighting products listed for the application without a grounding conductor
(2) All electrical equipment located within 1.5 m (5 ft) of the inside wall of the specified body of water
(3) All electrical equipment associated with the recirculating system of the specified body of water
(4) Junction boxes
(5) Transformer and power supply enclosures
(6) Ground-fault circuit interrupters
(7) Panelboards that are not part of the service equipment and that supply any electrical equipment associated with the specified body of water

N 680.7 Grounding and Bonding Terminals. Grounding and bonding terminals shall be identified for use in wet and corrosive environments. Field-installed grounding and bonding connections in a damp, wet, or corrosive environment shall be composed of copper, copper alloy, or stainless steel. They shall be listed for direct burial use.

680.8 Cord-and-Plug-Connected Equipment. Fixed or stationary equipment, other than underwater luminaires, for a permanently installed pool shall be permitted to be connected with a flexible cord and plug to facilitate the removal or disconnection for maintenance or repair.

(A) Length. For other than storable pools, the flexible cord shall not exceed 900 mm (3 ft) in length.

(B) Equipment Grounding. The flexible cord shall have a copper equipment grounding conductor sized in accordance with 250.122 but not smaller than 12 AWG. The cord shall terminate in a grounding-type attachment plug.

(C) Construction. The equipment grounding conductors shall be connected to a fixed metal part of the assembly. The removable part shall be mounted on or bonded to the fixed metal part.

680.9 Overhead Conductor Clearances. Overhead conductors shall meet the clearance requirements in this section. Where a minimum clearance from the water level is given, the measurement shall be taken from the maximum water level of the specified body of water.

(A) Power. With respect to service-drop conductors, overhead service conductors, and open overhead wiring, swimming pool and similar installations shall comply with the minimum clearances given in Table 680.9(A) and illustrated in Figure 680.9(A).

> Informational Note: Open overhead wiring as used in this article typically refers to conductor(s) not in an enclosed raceway.

(B) Communications Systems. Communications, radio, and television coaxial cables within the scope of Articles 800 through 820 shall be permitted at a height of not less than 3.0 m (10 ft) above swimming and wading pools, diving structures, and observation stands, towers, or platforms.

(C) Network-Powered Broadband Communications Systems. The minimum clearances for overhead network-powered broadband communications systems conductors from pools or fountains shall comply with the provisions in Table 680.9(A) for conductors operating at 0 to 750 volts to ground.

680.10 Electric Pool Water Heaters. All electric pool water heaters shall have the heating elements subdivided into loads not exceeding 48 amperes and protected at not over 60 amperes. The ampacity of the branch-circuit conductors and the rating or setting of overcurrent protective devices shall not be less than 125 percent of the total nameplate-rated load.

680.11 Underground Wiring Location. Underground wiring shall be permitted where installed in rigid metal conduit, intermediate metal conduit, rigid polyvinyl chloride conduit, reinforced thermosetting resin conduit, or Type MC cable, suitable for the conditions subject to that location. Underground wiring shall not be permitted under the pool unless this wiring is necessary to supply pool equipment permitted by this article. Minimum cover depths shall be as given in Table 300.5.

Table 680.9(A) Overhead Conductor Clearances

		Insulated Cables, 0–750 Volts to Ground, Supported on and Cabled Together with a Solidly Grounded Bare Messenger or Solidly Grounded Neutral Conductor		All Other Conductors Voltage to Ground			
				0 through 15 kV		Over 15 through 50 kV	
	Clearance Parameters	m	ft	m	ft	m	ft
A.	Clearance in any direction to the water level, edge of water surface, base of diving platform, or permanently anchored raft	6.9	22.5	7.5	25	8.0	27
B.	Clearance in any direction to the observation stand, tower, or diving platform	4.4	14.5	5.2	17	5.5	18
C.	Horizontal limit of clearance measured from inside wall of the pool	This limit shall extend to the outer edge of the structures listed in A and B of this table but not less than 3 m (10 ft).					

680.12 Equipment Rooms and Pits. Electrical equipment shall not be installed in rooms or pits that do not have drainage that prevents water accumulation during normal operation or filter maintenance. Equipment shall be suitable for the environment in accordance with 300.6.

> Informational Note: Chemicals such as chlorine cause severe corrosive and deteriorating effects on electrical connections, equipment, and enclosures when stored and kept in the same vicinity. Adequate ventilation of indoor spaces such as equipment and storage rooms is addressed by ANSI/APSP-11, *Standard for Water Quality in Public Pools and Spas*, and can reduce the likelihood of the accumulation of corrosive vapors.

680.13 Maintenance Disconnecting Means. One or more means to simultaneously disconnect all ungrounded conductors shall be provided for all utilization equipment other than lighting. Each means shall be readily accessible and within sight from its equipment and shall be located at least 1.5 m (5 ft) horizontally from the inside walls of a pool, spa, fountain, or hot tub unless separated from the open water by a permanently installed barrier that provides a 1.5 m (5 ft) reach path or greater. This horizontal distance shall be measured from the water's edge along the shortest path required to reach the disconnect.

N 680.14 Corrosive Environment.

(A) General. Areas where pool sanitation chemicals are stored, as well as areas with circulation pumps, automatic chlorinators, filters, open areas under decks adjacent to or abutting the pool structure, and similar locations shall be considered to be a corrosive environment. The air in such areas shall be considered to be laden with acid, chlorine, and bromine vapors, or any combination of acid, chlorine, or bromine vapors, and any liquids or condensation in those areas shall be considered to be laden with acids, chlorine, and bromine vapors, or any combination of acid, chlorine, or bromine vapors.

(B) Wiring Methods. Wiring methods in the areas described in 680.14(A) shall be listed and identified for use in such areas. Rigid metal conduit, intermediate metal conduit, rigid polyvinyl chloride conduit, and reinforced thermosetting resin conduit shall be considered to be resistant to the corrosive environment specified in 680.14(A).

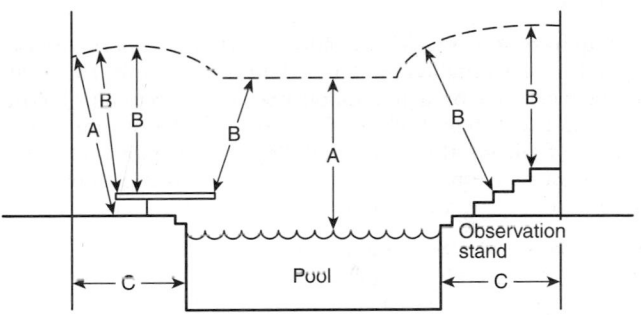

FIGURE 680.9(A) Clearances from Pool Structures.

Part II. Permanently Installed Pools

680.20 General. Electrical installations at permanently installed pools shall comply with the provisions of Part I and Part II of this article.

680.21 Motors.

(A) Wiring Methods. The wiring to a pool motor shall comply with (A)(1) unless modified for specific circumstances by (A)(2), (A)(3), (A)(4), or (A)(5).

(1) General. Wiring methods installed in the corrosive environment described in 680.14 shall comply with 680.14(B) or shall be type MC cable listed for that location. Wiring methods installed in these locations shall contain an insulated copper equipment grounding conductor sized in accordance with Table 250.122 but not smaller than 12 AWG.

Where installed in noncorrosive environments, branch circuits shall comply with the general requirements in Chapter 3.

(2) Flexible Connections. Where necessary to employ flexible connections at or adjacent to the motor, liquidtight flexible metal or liquidtight flexible nonmetallic conduit with listed fittings shall be permitted.

(3) Cord-and-Plug Connections. Pool-associated motors shall be permitted to employ cord-and-plug connections. The flexible cord shall not exceed 900 mm (3 ft) in length. The flexible cord shall include a copper equipment grounding conductor sized in accordance with 250.122 but not smaller than 12 AWG. The cord shall terminate in a grounding-type attachment plug.

(B) Double Insulated Pool Pumps. A listed cord-and-plug-connected pool pump incorporating an approved system of double insulation that provides a means for grounding only the internal and nonaccessible, non–current-carrying metal parts of the pump shall be connected to any wiring method recognized in Chapter 3 that is suitable for the location. Where the bonding grid is connected to the equipment grounding conductor of the motor circuit in accordance with the second sentence of 680.26(B)(6)(a), the branch-circuit wiring shall comply with 680.21(A).

(C) GFCI Protection. Outlets supplying pool pump motors connected to single-phase, 120-volt through 240-volt branch circuits, whether by receptacle or by direct connection, shall be provided with ground-fault circuit-interrupter protection for personnel.

680.22 Lighting, Receptacles, and Equipment.

(A) Receptacles.

(1) Required Receptacle, Location. Where a permanently installed pool is installed, no fewer than one 125-volt, 15- or 20-ampere receptacle on a general-purpose branch circuit shall be located not less than 1.83 m (6 ft) from, and not more than 6.0 m (20 ft) from, the inside wall of the pool. This receptacle shall be located not more than 2.0 m (6 ft 6 in.) above the floor, platform, or grade level serving the pool.

(2) Circulation and Sanitation System, Location. Receptacles that provide power for water-pump motors or for other loads directly related to the circulation and sanitation system shall be located at least 1.83 m (6 ft) from the inside walls of the pool. These receptacles shall have GFCI protection and be of the grounding type.

(3) Other Receptacles, Location. Other receptacles shall be not less than 1.83 m (6 ft) from the inside walls of a pool.

(4) GFCI Protection. All 15- and 20-ampere, single-phase, 125-volt receptacles located within 6.0 m (20 ft) of the inside walls of a pool shall be protected by a ground-fault circuit interrupter.

(5) Measurements. In determining the dimensions in this section addressing receptacle spacings, the distance to be measured shall be the shortest path the supply cord of an appliance connected to the receptacle would follow without piercing a floor, wall, ceiling, doorway with hinged or sliding door, window opening, or other effective permanent barrier.

(B) Luminaires, Lighting Outlets, and Ceiling-Suspended (Paddle) Fans.

(1) New Outdoor Installation Clearances. In outdoor pool areas, luminaires, lighting outlets, and ceiling-suspended (paddle) fans installed above the pool or the area extending 1.5 m (5 ft) horizontally from the inside walls of the pool shall be installed at a height not less than 3.7 m (12 ft) above the maximum water level of the pool.

(2) Indoor Clearances. For installations in indoor pool areas, the clearances shall be the same as for outdoor areas unless modified as provided in this paragraph. If the branch circuit supplying the equipment is protected by a ground-fault circuit interrupter, the following equipment shall be permitted at a

height not less than 2.3 m (7 ft 6 in.) above the maximum pool water level:

(1) Totally enclosed luminaires
(2) Ceiling-suspended (paddle) fans identified for use beneath ceiling structures such as provided on porches or patios

(3) Existing Installations. Existing luminaires and lighting outlets located less than 1.5 m (5 ft) measured horizontally from the inside walls of a pool shall be not less than 1.5 m (5 ft) above the surface of the maximum water level, shall be rigidly attached to the existing structure, and shall be protected by a ground-fault circuit interrupter.

(4) GFCI Protection in Adjacent Areas. Luminaires, lighting outlets, and ceiling-suspended (paddle) fans installed in the area extending between 1.5 m (5 ft) and 3.0 m (10 ft) horizontally from the inside walls of a pool shall be protected by a ground-fault circuit interrupter unless installed not less than 1.5 m (5 ft) above the maximum water level and rigidly attached to the structure adjacent to or enclosing the pool.

(5) Cord-and-Plug-Connected Luminaires. Cord-and-plug-connected luminaires shall comply with the requirements of 680.8 where installed within 4.9 m (16 ft) of any point on the water surface, measured radially.

(6) Low-Voltage Luminaires. Listed low-voltage luminaires not requiring grounding, not exceeding the low-voltage contact limit, and supplied by listed transformers or power supplies that comply with 680.23(A)(2) shall be permitted to be located less than 1.5 m (5 ft) from the inside walls of the pool.

N **(7) Low-Voltage Gas-Fired Luminaires, Decorative Fireplaces, Fire Pits, and Similar Equipment.** Listed low-voltage gas-fired-luminaires, decorative fireplaces, fire pits, and similar equipment using low-voltage ignitors that do not require grounding, and are supplied by listed transformers or power supplies that comply with 680.23(A)(2) with outputs that do not exceed the low-voltage contact limit shall be permitted to be located less than 1.5 m (5 ft) from the inside walls of the pool. Metallic equipment shall be bonded in accordance with the requirements in 680.26(B). Transformers or power supplies supplying this type of equipment shall be installed in accordance with the requirements in 680.24. Metallic gas piping shall be bonded in accordance with the requirements in 250.104(B) and 680.26(B)(7).

(C) Switching Devices. Switching devices shall be located at least 1.5 m (5 ft) horizontally from the inside walls of a pool unless separated from the pool by a solid fence, wall, or other permanent barrier. Alternatively, a switch that is listed as being acceptable for use within 1.5 m (5 ft) shall be permitted.

(D) Other Outlets. Other outlets shall be not less than 3.0 m (10 ft) from the inside walls of the pool. Measurements shall be determined in accordance with 680.22(A)(5).

> Informational Note: Other outlets may include, but are not limited to, remote-control, signaling, fire alarm, and communications circuits.

680.23 Underwater Luminaires. This section covers all luminaires installed below the maximum water level of the pool.

(A) General.

(1) Luminaire Design, Normal Operation. The design of an underwater luminaire supplied from a branch circuit either directly or by way of a transformer or power supply meeting the requirements of this section shall be such that, where the luminaire is properly installed without a ground-fault circuit interrupter, there is no shock hazard with any likely combination of fault conditions during normal use (not relamping).

(2) Transformers and Power Supplies. Transformers and power supplies used for the supply of underwater luminaires, together with the transformer or power supply enclosure, shall be listed, labeled, and identified for swimming pool and spa use. The transformer or power supply shall incorporate either a transformer of the isolated winding type, with an ungrounded secondary that has a grounded metal barrier between the primary and secondary windings, or one that incorporates an approved system of double insulation between the primary and secondary windings.

(3) GFCI Protection, Lamping, Relamping, and Servicing. Ground-fault circuit-interrupter protection for personnel shall be installed in the branch circuit supplying luminaires operating at voltages greater than the low-voltage contact limit.

(4) Voltage Limitation. No luminaires shall be installed for operation on supply circuits over 150 volts between conductors.

(5) Location, Wall-Mounted Luminaires. Luminaires mounted in walls shall be installed with the top of the luminaire lens not less than 450 mm (18 in.) below the normal water level of the pool, unless the luminaire is listed and identified for use at lesser depths. No luminaire shall be installed less than 100 mm (4 in.) below the normal water level of the pool.

(6) Bottom-Mounted Luminaires. A luminaire facing upward shall comply with either (1) or (2):

(1) Have the lens guarded to prevent contact by any person
(2) Be listed for use without a guard

(7) Dependence on Submersion. Luminaires that depend on submersion for safe operation shall be inherently protected against the hazards of overheating when not submerged.

(8) Compliance. Compliance with these requirements shall be obtained by the use of a listed underwater luminaire and by installation of a listed ground-fault circuit interrupter in the branch circuit or a listed transformer or power supply for luminaires operating at not more than the low voltage contact limit.

(B) Wet-Niche Luminaires.

(1) Forming Shells. Forming shells shall be installed for the mounting of all wet-niche underwater luminaires and shall be equipped with provisions for conduit entries. Metal parts of the luminaire and forming shell in contact with the pool water shall be of brass or other approved corrosion-resistant metal. All forming shells used with nonmetallic conduit systems, other than those that are part of a listed low-voltage lighting system not requiring grounding, shall include provisions for terminating an 8 AWG copper conductor.

(2) Wiring Extending Directly to the Forming Shell. Conduit shall be installed from the forming shell to a junction box or other enclosure conforming to the requirements in 680.24. Conduit shall be rigid metal, intermediate metal, liquidtight flexible nonmetallic, or rigid nonmetallic.

(a) *Metal Conduit.* Metal conduit shall be approved and shall be of brass or other approved corrosion-resistant metal.

(b) *Nonmetallic Conduit.* Where a nonmetallic conduit is used, an 8 AWG insulated solid or stranded copper bonding jumper shall be installed in this conduit unless a listed low-voltage lighting system not requiring grounding is used. The bonding jumper shall be terminated in the forming shell, junction box or transformer enclosure, or ground-fault circuit-interrupter enclosure. The termination of the 8 AWG bonding jumper in the forming shell shall be covered with, or encapsulated in, a listed potting compound to protect the connection from the possible deteriorating effect of pool water.

(3) Equipment Grounding Provisions for Cords. Other than listed low-voltages lighting systems not requiring grounding wet-niche luminaires that are supplied by a flexible cord or cable shall have all exposed non–current-carrying metal parts grounded by an insulated copper equipment grounding conductor that is an integral part of the cord or cable. This grounding conductor shall be connected to a grounding terminal in the supply junction box, transformer enclosure, or other enclosure. The grounding conductor shall not be smaller than the supply conductors and not smaller than 16 AWG.

(4) Luminaire Grounding Terminations. The end of the flexible-cord jacket and the flexible-cord conductor terminations within a luminaire shall be covered with, or encapsulated in, a suitable potting compound to prevent the entry of water into the luminaire through the cord or its conductors. If present, the grounding connection within a luminaire shall be similarly treated to protect such connection from the deteriorating effect of pool water in the event of water entry into the luminaire.

(5) Luminaire Bonding. The luminaire shall be bonded to, and secured to, the forming shell by a positive locking device that ensures a low-resistance contact and requires a tool to remove the luminaire from the forming shell. Bonding shall not be required for luminaires that are listed for the application and have no non–current-carrying metal parts.

(6) Servicing. All wet-niche luminaires shall be removable from the water for inspection, relamping, or other maintenance. The forming shell location and length of cord in the forming shell shall permit personnel to place the removed luminaire on the deck or other dry location for such maintenance. The luminaire maintenance location shall be accessible without entering or going in the pool water.

(C) Dry-Niche Luminaires.

(1) Construction. A dry-niche luminaire shall have provision for drainage of water. Other than listed low voltage luminaires not requiring grounding, a dry-niche luminaire shall have means for accommodating one equipment grounding conductor for each conduit entry.

(2) Junction Box. A junction box shall not be required but, if used, shall not be required to be elevated or located as specified in 680.24(A)(2) if the luminaire is specifically identified for the purpose.

(D) No-Niche Luminaires. A no-niche luminaire shall meet the construction requirements of 680.23(B)(3) and be installed in accordance with the requirements of 680.23(B). Where connection to a forming shell is specified, the connection shall be to the mounting bracket.

(E) Through-Wall Lighting Assembly. A through-wall lighting assembly shall be equipped with a threaded entry or hub, or a nonmetallic hub, for the purpose of accommodating the termination of the supply conduit. A through-wall lighting assembly shall meet the construction requirements of 680.23(B)(3) and be installed in accordance with the requirements of 680.23. Where connection to a forming shell is specified, the connection shall be to the conduit termination point.

(F) Branch-Circuit Wiring.

(1) Wiring Methods. Where branch-circuit wiring on the supply side of enclosures and junction boxes connected to conduits run to underwater luminaires are installed in corrosive environments as described in 680.14, the wiring method of that portion of the branch circuit shall be as required in 680.14(B) or shall be liquidtight flexible nonmetallic conduit. Wiring methods installed in corrosive environments as described in 680.14 shall contain an insulated copper equipment grounding conductor sized in accordance with Table 250.122, but not smaller than 12 AWG.

Where installed in noncorrosive environments, branch circuits shall comply with the general requirements in Chapter 3.

Exception: Where connecting to transformers or power supplies for pool lights, liquidtight flexible metal conduit shall be permitted. The length shall not exceed 1.8 m (6 ft) for any one length or exceed 3.0 m (10 ft) in total length used.

(2) Equipment Grounding. Other than listed low-voltage luminaires not requiring grounding, all through-wall lighting assemblies, wet-niche, dry-niche, or no-niche luminaires shall be connected to an insulated copper equipment grounding conductor installed with the circuit conductors. The equipment grounding conductor shall be installed without joint or splice except as permitted in (F)(2)(a) and (F)(2)(b). The equipment grounding conductor shall be sized in accordance with Table 250.122 but shall not be smaller than 12 AWG.

Exception: An equipment grounding conductor between the wiring chamber of the secondary winding of a transformer and a junction box shall be sized in accordance with the overcurrent device in this circuit.

(a)   If more than one underwater luminaire is supplied by the same branch circuit, the equipment grounding conductor, installed between the junction boxes, transformer enclosures, or other enclosures in the supply circuit to wet-niche luminaires, or between the field-wiring compartments of dry-niche luminaires, shall be permitted to be terminated on grounding terminals.

(b)   If the underwater luminaire is supplied from a transformer, ground-fault circuit interrupter, clock-operated switch, or a manual snap switch that is located between the panelboard and a junction box connected to the conduit that extends directly to the underwater luminaire, the equipment grounding conductor shall be permitted to terminate on grounding terminals on the transformer, ground-fault circuit interrupter, clock-operated switch enclosure, or an outlet box used to enclose a snap switch.

(3) Conductors. Conductors on the load side of a ground-fault circuit interrupter or of a transformer, used to comply with the provisions of 680.23(A)(8), shall not occupy raceways, boxes, or enclosures containing other conductors unless one of the following conditions applies:

(1)   The other conductors are protected by ground-fault circuit interrupters.

(2)   The other conductors are equipment grounding conductors and bonding jumpers as required per 680.23(B)(2)(b).

(3)   The other conductors are supply conductors to a feed-through-type ground-fault circuit interrupter.

(4)   Ground-fault circuit interrupters shall be permitted in a panelboard that contains circuits protected by other than ground-fault circuit interrupters.

680.24 Junction Boxes and Electrical Enclosures for Transformers or Ground-Fault Circuit Interrupters.

(A) Junction Boxes. A junction box connected to a conduit that extends directly to a forming shell or mounting bracket of a no-niche luminaire shall meet the requirements of this section.

(1) Construction. The junction box shall be listed, labeled, and identified as a swimming pool junction box and shall comply with the following conditions:

(1)   Be equipped with threaded entries or hubs or a nonmetallic hub
(2)   Be comprised of copper, brass, suitable plastic, or other approved corrosion-resistant material
(3)   Be provided with electrical continuity between every connected metal conduit and the grounding terminals by means of copper, brass, or other approved corrosion-resistant metal that is integral with the box

(2) Installation. Where the luminaire operates over the low voltage contact limit, the junction box location shall comply with (A)(2)(a) and (A)(2)(b). Where the luminaire operates at the low voltage contact limit or less, the junction box location shall be permitted to comply with (A)(2)(c).

(a)   *Vertical Spacing.* The junction box shall be located not less than 100 mm (4 in.), measured from the inside of the bottom of the box, above the ground level, or pool deck, or not less than 200 mm (8 in.) above the maximum pool water level, whichever provides the greater elevation.

(b)   *Horizontal Spacing.* The junction box shall be located not less than 1.2 m (4 ft) from the inside wall of the pool, unless separated from the pool by a solid fence, wall, or other permanent barrier.

(c)   *Flush Deck Box.* If used on a lighting system operating at the low voltage contact limit or less, a flush deck box shall be permitted if both of the following conditions are met:

(1)   An approved potting compound is used to fill the box to prevent the entrance of moisture.
(2)   The flush deck box is located not less than 1.2 m (4 ft) from the inside wall of the pool.

(B) Other Enclosures. An enclosure for a transformer, ground-fault circuit interrupter, or a similar device connected to a conduit that extends directly to a forming shell or mounting bracket of a no-niche luminaire shall meet the requirements of this section.

(1) Construction. The enclosure shall be listed and labeled for the purpose and meet the following requirements:

(1)   Equipped with threaded entries or hubs or a nonmetallic hub
(2)   Comprised of copper, brass, suitable plastic, or other approved corrosion-resistant material
(3)   Provided with an approved seal, such as duct seal at the conduit connection, that prevents circulation of air between the conduit and the enclosures

(4) Provided with electrical continuity between every connected metal conduit and the grounding terminals by means of copper, brass, or other approved corrosion-resistant metal that is integral with the box

(2) Installation.

(a) *Vertical Spacing.* The enclosure shall be located not less than 100 mm (4 in.), measured from the inside of the bottom of the box, above the ground level, or pool deck, or not less than 200 mm (8 in.) above the maximum pool water level, whichever provides the greater elevation.

(b) *Horizontal Spacing.* The enclosure shall be located not less than 1.2 m (4 ft) from the inside wall of the pool, unless separated from the pool by a solid fence, wall, or other permanent barrier.

(C) Protection. Junction boxes and enclosures mounted above the grade of the finished walkway around the pool shall not be located in the walkway unless afforded additional protection, such as by location under diving boards, adjacent to fixed structures, and the like.

(D) Grounding Terminals. Junction boxes, transformer and power-supply enclosures, and ground-fault circuit-interrupter enclosures connected to a conduit that extends directly to a forming shell or mounting bracket of a no-niche luminaire shall be provided with a number of grounding terminals that shall be no fewer than one more than the number of conduit entries.

(E) Strain Relief. The termination of a flexible cord of an underwater luminaire within a junction box, transformer or power-supply enclosure, ground-fault circuit interrupter, or other enclosure shall be provided with a strain relief.

(F) Grounding. The equipment grounding conductor terminals of a junction box, transformer enclosure, or other enclosure in the supply circuit to a wet-niche or no-niche luminaire and the field-wiring chamber of a dry-niche luminaire shall be connected to the equipment grounding terminal of the panelboard. This terminal shall be directly connected to the panelboard enclosure.

680.25 Feeders. These provisions shall apply to any feeder on the supply side of panelboards supplying branch circuits for pool equipment covered in Part II of this article and on the load side of the service equipment or the source of a separately derived system.

(A) Feeders. Where feeders are installed in corrosive environments as described in 680.14, the wiring method of that portion of the feeder shall be as required in 680.14(B) or shall be liquidtight flexible nonmetallic conduit. Wiring methods installed in corrosive environments as described in 680.14 shall contain an insulated copper equipment grounding conductor sized in accordance with Table 250.122, but not smaller than 12 AWG.

Where installed in noncorrosive environments, feeders shall comply with the general requirements in Chapter 3.

(B) Aluminum Conduit. Aluminum conduit shall not be permitted in the pool area where subject to corrosion.

680.26 Equipotential Bonding.

(A) Performance. The equipotential bonding required by this section shall be installed to reduce voltage gradients in the pool area.

(B) Bonded Parts. The parts specified in 680.26(B)(1) through (B)(7) shall be bonded together using solid copper conductors, insulated covered, or bare, not smaller than 8 AWG or with rigid metal conduit of brass or other identified corrosion-resistant metal. Connections to bonded parts shall be made in accordance with 250.8. An 8 AWG or larger solid copper bonding conductor provided to reduce voltage gradients in the pool area shall not be required to be extended or attached to remote panelboards, service equipment, or electrodes.

(1) Conductive Pool Shells. Bonding to conductive pool shells shall be provided as specified in 680.26(B)(1)(a) or (B)(1)(b). Poured concrete, pneumatically applied or sprayed concrete, and concrete block with painted or plastered coatings shall all be considered conductive materials due to water permeability and porosity. Vinyl liners and fiberglass composite shells shall be considered to be nonconductive materials.

(a) *Structural Reinforcing Steel.* Unencapsulated structural reinforcing steel shall be bonded together by steel tie wires or the equivalent. Where structural reinforcing steel is encapsulated in a nonconductive compound, a copper conductor grid shall be installed in accordance with 680.26(B)(1)(b).

(b) *Copper Conductor Grid.* A copper conductor grid shall be provided and shall comply with (b)(1) through (b)(4).

(1) Be constructed of minimum 8 AWG bare solid copper conductors bonded to each other at all points of crossing. The bonding shall be in accordance with 250.8 or other approved means.
(2) Conform to the contour of the pool
(3) Be arranged in a 300-mm (12-in.) by 300-mm (12-in.) network of conductors in a uniformly spaced perpendicular grid pattern with a tolerance of 100 mm (4 in.)
(4) Be secured within or under the pool no more than 150 mm (6 in.) from the outer contour of the pool shell

(2) Perimeter Surfaces. The perimeter surface to be bonded shall be considered to extend for 1 m (3 ft) horizontally beyond the inside walls of the pool and shall include unpaved surfaces and other types of paving. Perimeter surfaces separated from the pool by a permanent wall or building 1.5 m (5 ft) in height or more shall require equipotential bonding only on the pool side of the permanent wall or building. Bonding to perimeter surfaces shall be provided as specified in 680.26(B)(2)(a) or (2)(b) and shall be attached to the pool reinforcing steel or copper conductor grid at a minimum of four (4) points uniformly spaced around the perimeter of the pool. For nonconductive pool shells, bonding at four points shall not be required.

(a) *Structural Reinforcing Steel.* Structural reinforcing steel shall be bonded in accordance with 680.26(B)(1)(a).

(b) *Alternate Means.* Where structural reinforcing steel is not available or is encapsulated in a nonconductive compound, a copper conductor(s) shall be utilized where the following requirements are met:

(1) At least one minimum 8 AWG bare solid copper conductor shall be provided.
(2) The conductors shall follow the contour of the perimeter surface.
(3) Only listed splices shall be permitted.
(4) The required conductor shall be 450 mm to 600 mm (18 in. to 24 in.) from the inside walls of the pool.
(5) The required conductor shall be secured within or under the perimeter surface 100 mm to 150 mm (4 in. to 6 in.) below the subgrade.

(3) Metallic Components. All metallic parts of the pool structure, including reinforcing metal not addressed in 680.26(B)(1)(a), shall be bonded. Where reinforcing steel is encapsulated with a nonconductive compound, the reinforcing steel shall not be required to be bonded.

(4) Underwater Lighting. All metal forming shells and mounting brackets of no-niche luminaires shall be bonded.

Exception: Listed low-voltage lighting systems with nonmetallic forming shells shall not require bonding.

(5) Metal Fittings. All metal fittings within or attached to the pool structure shall be bonded. Isolated parts that are not over 100 mm (4 in.) in any dimension and do not penetrate into the pool structure more than 25 mm (1 in.) shall not require bonding.

(6) Electrical Equipment. Metal parts of electrical equipment associated with the pool water circulating system, including pump motors and metal parts of equipment associated with pool covers, including electric motors, shall be bonded.

Exception: Metal parts of listed equipment incorporating an approved system of double insulation shall not be bonded.
    (a)   *Double-Insulated Water Pump Motors.* Where a double-insulated water pump motor is installed under the provisions of this rule, a solid 8 AWG copper conductor of sufficient length to make a bonding connection to a replacement motor shall be extended from the bonding grid to an accessible point in the vicinity of the pool pump motor. Where there is no connection between the swimming pool bonding grid and the equipment grounding system for the premises, this bonding conductor shall be connected to the equipment grounding conductor of the motor circuit.
    (b)   *Pool Water Heaters.* For pool water heaters rated at more than 50 amperes and having specific instructions regarding bonding and grounding, only those parts designated to be bonded shall be bonded and only those parts designated to be grounded shall be grounded.

(7) Fixed Metal Parts. All fixed metal parts shall be bonded including, but not limited to, metal-sheathed cables and raceways, metal piping, metal awnings, metal fences, and metal door and window frames.

Exception No. 1: Those separated from the pool by a permanent barrier that prevents contact by a person shall not be required to be bonded.

Exception No. 2: Those greater than 1.5 m (5 ft) horizontally from the inside walls of the pool shall not be required to be bonded.

Exception No. 3: Those greater than 3.7 m (12 ft) measured vertically above the maximum water level of the pool, or as measured vertically above any observation stands, towers, or platforms, or any diving structures, shall not be required to be bonded.

(C) Pool Water. Where none of the bonded parts is in direct connection with the pool water, the pool water shall be in direct contact with an approved corrosion-resistant conductive surface that exposes not less than 5800 mm² (9 in.²) of surface area to the pool water at all times. The conductive surface shall be located where it is not exposed to physical damage or dislodgement during usual pool activities, and it shall be bonded in accordance with 680.26(B).

680.27 Specialized Pool Equipment.

(A) Underwater Audio Equipment. All underwater audio equipment shall be identified.

(1) Speakers. Each speaker shall be mounted in an approved metal forming shell, the front of which is enclosed by a captive metal screen, or equivalent, that is bonded to, and secured to, the forming shell by a positive locking device that ensures a low-resistance contact and requires a tool to open for installation or servicing of the speaker. The forming shell shall be installed in a recess in the wall or floor of the pool.

(2) Wiring Methods. Rigid metal conduit of brass or other identified corrosion-resistant metal, liquidtight flexible nonmetallic conduit (LFNC), rigid polyvinyl chloride conduit, or reinforced thermosetting resin conduit shall extend from the forming shell to a listed junction box or other enclosure as provided in 680.24. Where rigid polyvinyl chloride conduit, reinforced thermosetting resin conduit, or liquidtight flexible nonmetallic conduit is used, an 8 AWG insulated solid or stranded copper bonding jumper shall be installed in this conduit. The bonding jumper shall be terminated in the forming shell and the junction box. The termination of the 8 AWG bonding jumper in the forming shell shall be covered with, or encapsulated in, a listed potting compound to protect such connection from the possible deteriorating effect of pool water.

(3) Forming Shell and Metal Screen. The forming shell and metal screen shall be of brass or other approved corrosion-resistant metal. All forming shells shall include provisions for terminating an 8 AWG copper conductor.

(B) Electrically Operated Pool Covers.

(1) Motors and Controllers. The electric motors, controllers, and wiring shall be located not less than 1.5 m (5 ft) from the inside wall of the pool unless separated from the pool by a wall, cover, or other permanent barrier. Electric motors installed below grade level shall be of the totally enclosed type. The device that controls the operation of the motor for an electrically operated pool cover shall be located such that the operator has full view of the pool.

Exception: Motors that are part of listed systems with ratings not exceeding the low-voltage contact limit that are supplied by listed transformers or power supplies that comply with 680.23(A)(2) shall be permitted to be located less than 1.5 m (5 ft) from the inside walls of the pool.

(2) Protection. The electric motor and controller shall be connected to a branch circuit protected by a ground-fault circuit interrupter.

Exception: Motors that are part of listed systems with ratings not exceeding the low-voltage contact limit that are supplied by listed transformers or power supplies that comply with 680.23(A)(2).

(C) Deck Area Heating. The provisions of this section shall apply to all pool deck areas, including a covered pool, where electrically operated comfort heating units are installed within 6.0 m (20 ft) of the inside wall of the pool.

(1) Unit Heaters. Unit heaters shall be rigidly mounted to the structure and shall be of the totally enclosed or guarded type. Unit heaters shall not be mounted over the pool or within the area extending 1.5 m (5 ft) horizontally from the inside walls of a pool.

(2) Permanently Wired Radiant Heaters. Radiant electric heaters shall be suitably guarded and securely fastened to their mounting device(s). Heaters shall not be installed over a pool or within the area extending 1.5 m (5 ft) horizontally from the inside walls of the pool and shall be mounted at least 3.7 m (12 ft) vertically above the pool deck unless otherwise approved.

(3) Radiant Heating Cables Not Permitted. Radiant heating cables embedded in or below the deck shall not be permitted.

N **680.28 Gas-Fired Water Heater.** Circuits serving gas-fired swimming pool and spa water heaters operating at voltages above the low-voltage contact limit shall be provided with ground-fault circuit-interrupter protection for personnel.

Part III. Storable Pools, Storable Spas, and Storable Hot Tubs

680.30 General. Electrical installations at storable pools, storable spas, or storable hot tubs shall comply with the provisions of Part I and Part III of this article.

680.31 Pumps. A cord-connected pool filter pump shall incorporate an approved system of double insulation or its equivalent and shall be provided with means for grounding only the internal and nonaccessible non–current-carrying metal parts of the appliance.

The means for grounding shall be an equipment grounding conductor run with the power-supply conductors in the flexible cord that is properly terminated in a grounding-type attachment plug having a fixed grounding contact member.

Cord-connected pool filter pumps shall be provided with a ground-fault circuit interrupter that is an integral part of the attachment plug or located in the power supply cord within 300 mm (12 in.) of the attachment plug.

680.32 Ground-Fault Circuit Interrupters Required. All electrical equipment, including power-supply cords, used with storable pools shall be protected by ground-fault circuit interrupters.

All 125-volt, 15- and 20-ampere receptacles located within 6.0 m (20 ft) of the inside walls of a storable pool, storable spa, or storable hot tub shall be protected by a ground-fault circuit interrupter. In determining these dimensions, the distance to be measured shall be the shortest path the supply cord of an appliance connected to the receptacle would follow without piercing a floor, wall, ceiling, doorway with hinged or sliding door, window opening, or other effective permanent barrier.

Informational Note: For flexible cord usage, see 400.4.

680.33 Luminaires. An underwater luminaire, if installed, shall be installed in or on the wall of the storable pool, storable spa, or storable hot tub. It shall comply with either 680.33(A) or (B).

(A) Within the Low Voltage Contact Limit. A luminaire shall be part of a cord-and plug connected lighting assembly. This assembly shall be listed as an assembly for the purpose and have the following construction features:

(1) No exposed metal parts
(2) A luminaire lamp that is suitable for use at the supplied voltage
(3) An impact-resistant polymeric lens, luminaire body, and transformer enclosure

(4) A transformer or power supply meeting the requirements of 680.23(A)(2) with a primary rating not over 150 V

(B) Over the Low Voltage Contact Limit But Not over 150 Volts. A lighting assembly without a transformer or power supply and with the luminaire lamp(s) operating at not over 150 volts shall be permitted to be cord-and-plug-connected where the assembly is listed as an assembly for the purpose. The installation shall comply with 680.23(A)(5), and the assembly shall have the following construction features:

(1) No exposed metal parts
(2) An impact-resistant polymeric lens and luminaire body
(3) A ground-fault circuit interrupter with open neutral conductor protection as an integral part of the assembly
(4) The luminaire lamp permanently connected to the ground-fault circuit interrupter with open-neutral protection
(5) Compliance with the requirements of 680.23(A)

680.34 Receptacle Locations. Receptacles shall not be located less than 1.83 m (6 ft) from the inside walls of a storable pool, storable spa, or storable hot tub. In determining these dimensions, the distance to be measured shall be the shortest path the supply cord of an appliance connected to the receptacle would follow without piercing a floor, wall, ceiling, doorway with hinged or sliding door, window opening, or other effective permanent barrier.

Part IV. Spas and Hot Tubs

680.40 General. Electrical installations at spas and hot tubs shall comply with the provisions of Part I and Part IV of this article.

680.41 Emergency Switch for Spas and Hot Tubs. A clearly labeled emergency shutoff or control switch for the purpose of stopping the motor(s) that provides power to the recirculation system and jet system shall be installed at a point readily accessible to the users and not less than 1.5 m (5 ft) away, adjacent to, and within sight of the spa or hot tub. This requirement shall not apply to one-family dwellings.

680.42 Outdoor Installations. A spa or hot tub installed outdoors shall comply with the provisions of Parts I and II of this article, except as permitted in 680.42(A) and (B), that would otherwise apply to pools installed outdoors.

(A) Flexible Connections. Listed packaged spa or hot tub equipment assemblies or self-contained spas or hot tubs utilizing a factory-installed or assembled control panel or panelboard shall be permitted to use flexible connections as covered in 680.42(A)(1) and (A)(2).

(1) Flexible Conduit. Liquidtight flexible metal conduit or liquidtight flexible nonmetallic conduit shall be permitted.

(2) Cord-and-Plug Connections. Cord-and-plug connections with a cord not longer than 4.6 m (15 ft) shall be permitted where protected by a ground-fault circuit interrupter.

(B) Bonding. Bonding by metal-to-metal mounting on a common frame or base shall be permitted. The metal bands or hoops used to secure wooden staves shall not be required to be bonded as required in 680.26.

Equipotential bonding of perimeter surfaces in accordance with 680.26(B)(2) shall not be required to be provided for spas and hot tubs where all of the following conditions apply:

(1) The spa or hot tub shall be listed, labeled, and identified as a self-contained spa for aboveground use.

(2) The spa or hot tub shall not be identified as suitable only for indoor use.

(3) The installation shall be in accordance with the manufacturer's instructions and shall be located on or above grade.

(4) The top rim of the spa or hot tub shall be at least 710 mm (28 in.) above all perimeter surfaces that are within 760 mm (30 in.), measured horizontally from the spa or hot tub. The height of nonconductive external steps for entry to or exit from the self-contained spa shall not be used to reduce or increase this rim height measurement.

Informational Note: For information regarding listing requirements for self-contained spas and hot tubs, see ANSI/UL 1563–2010, *Standard for Electric Spas, Equipment Assemblies, and Associated Equipment.*

(C) Interior Wiring to Outdoor Installations. In the interior of a dwelling unit or in the interior of another building or structure associated with a dwelling unit, any of the wiring methods recognized or permitted in Chapter 3 of this *Code* shall be permitted to be used for the connection to motor disconnecting means and the motor, heating, and control loads that are part of a self-contained spa or hot tub or a packaged spa or hot tub equipment assembly. Wiring to an underwater luminaire shall comply with 680.23 or 680.33.

Informational Note: See 680.25 for feeders.

680.43 Indoor Installations. A spa or hot tub installed indoors shall comply with the provisions of Parts I and II of this article except as modified by this section and shall be connected by the wiring methods of Chapter 3.

Exception No. 1: Listed spa and hot tub packaged units rated 20 amperes or less shall be permitted to be cord-and-plug-connected to facilitate the removal or disconnection of the unit for maintenance and repair.

Exception No. 2: The equipotential bonding requirements for perimeter surfaces in 680.26(B)(2) shall not apply to a listed self-contained spa or hot tub installed above a finished floor.

Exception No. 3: For a dwelling unit(s) only, where a listed spa or hot tub is installed indoors, the wiring method requirements of 680.42(C) shall also apply.

(A) Receptacles. At least one 125-volt, 15- or 20-ampere receptacle on a general-purpose branch circuit shall be located not less than 1.83 m (6 ft) from, and not exceeding 3.0 m (10 ft) from, the inside wall of the spa or hot tub.

(1) Location. Receptacles shall be located at least 1.83 m (6 ft) measured horizontally from the inside walls of the spa or hot tub.

(2) Protection, General. Receptacles rated 125 volts and 30 amperes or less and located within 3.0 m (10 ft) of the inside walls of a spa or hot tub shall be protected by a ground-fault circuit interrupter.

(3) Protection, Spa or Hot Tub Supply Receptacle. Receptacles that provide power for a spa or hot tub shall be ground-fault circuit-interrupter protected.

(4) Measurements. In determining the dimensions in this section addressing receptacle spacings, the distance to be measured shall be the shortest path the supply cord of an appliance connected to the receptacle would follow without piercing a floor, wall, ceiling, doorway with hinged or sliding door, window opening, or other effective permanent barrier.

(B) Installation of Luminaires, Lighting Outlets, and Ceiling-Suspended (Paddle) Fans.

(1) Elevation. Luminaires, except as covered in 680.43(B)(2), lighting outlets, and ceiling-suspended (paddle) fans located over the spa or hot tub or within 1.5 m (5 ft) from the inside walls of the spa or hot tub shall comply with the clearances specified in (B)(1)(a), (B)(1)(b), and (B)(1)(c) above the maximum water level.

(a) *Without GFCI.* Where no GFCI protection is provided, the mounting height shall be not less than 3.7 m (12 ft).

(b) *With GFCI.* Where GFCI protection is provided, the mounting height shall be permitted to be not less than 2.3 m (7 ft 6 in.).

(c) *Below 2.3 m (7 ft 6 in.).* Luminaires meeting the requirements of item (1) or (2) and protected by a ground-fault circuit interrupter shall be permitted to be installed less than 2.3 m (7 ft 6 in.) over a spa or hot tub:

(1) Recessed luminaires with a glass or plastic lens, nonmetallic or electrically isolated metal trim, and suitable for use in damp locations

(2) Surface-mounted luminaires with a glass or plastic globe, a nonmetallic body, or a metallic body isolated from contact, and suitable for use in damp locations

(2) Underwater Applications. Underwater luminaires shall comply with the provisions of 680.23 or 680.33.

(C) Switches. Switches shall be located at least 1.5 m (5 ft), measured horizontally, from the inside walls of the spa or hot tub.

(D) Bonding. The following parts shall be bonded together:

(1) All metal fittings within or attached to the spa or hot tub structure.

(2) Metal parts of electrical equipment associated with the spa or hot tub water circulating system, including pump motors, unless part of a listed, labeled, and identified self-contained spa or hot tub.

(3) Metal raceway and metal piping that are within 1.5 m (5 ft) of the inside walls of the spa or hot tub and that are not separated from the spa or hot tub by a permanent barrier.

(4) All metal surfaces that are within 1.5 m (5 ft) of the inside walls of the spa or hot tub and that are not separated from the spa or hot tub area by a permanent barrier.

Exception: Small conductive surfaces not likely to become energized, such as air and water jets and drain fittings, where not connected to metallic piping, towel bars, mirror frames, and similar nonelectrical equipment, shall not be required to be bonded.

(5) Electrical devices and controls that are not associated with the spas or hot tubs and that are located less than 1.5 m (5 ft) from such units; otherwise, they shall be bonded to the spa or hot tub system.

(E) Methods of Bonding. All metal parts associated with the spa or hot tub shall be bonded by any of the following methods:

(1) The interconnection of threaded metal piping and fittings
(2) Metal-to-metal mounting on a common frame or base
(3) The provisions of a solid copper bonding jumper, insulated, covered, or bare, not smaller than 8 AWG

(F) Grounding. The following equipment shall be grounded:

(1) All electrical equipment located within 1.5 m (5 ft) of the inside wall of the spa or hot tub
(2) All electrical equipment associated with the circulating system of the spa or hot tub

(G) Underwater Audio Equipment. Underwater audio equipment shall comply with the provisions of Part II of this article.

680.44 Protection. Except as otherwise provided in this section, the outlet(s) that supplies a self-contained spa or hot tub, a packaged spa or hot tub equipment assembly, or a field-assembled spa or hot tub shall be protected by a ground-fault circuit interrupter.

(A) Listed Units. If so marked, a listed, labeled, and identified self-contained unit or a listed, labeled, and identified packaged equipment assembly that includes integral ground-fault circuit-interrupter protection for all electrical parts within the unit or assembly (pumps, air blowers, heaters, lights, controls, sanitizer generators, wiring, and so forth) shall be permitted without additional GFCI protection.

(B) Other Units. A field-assembled spa or hot tub rated 3 phase or rated over 250 volts or with a heater load of more than 50 amperes shall not require the supply to be protected by a ground-fault circuit interrupter.

Informational Note: See 680.2 for definitions of *self-contained spa or hot tub* and for *packaged spa or hot tub equipment assembly.*

Part V. Fountains

680.50 General. The provisions of Part I and Part V of this article shall apply to all permanently installed fountains as defined in 680.2. Fountains that have water common to a pool shall additionally comply with the requirements in Part II of this article. Part V does not cover self-contained, portable fountains. Portable fountains shall comply with Parts II and III of Article 422.

680.51 Luminaires, Submersible Pumps, and Other Submersible Equipment.

(A) Ground-Fault Circuit Interrupter. Luminaires, submersible pumps, and other submersible equipment, unless listed for operation at low voltage contact limit or less and supplied by a transformer or power supply that complies with 680.23(A)(2), shall be protected by a ground-fault circuit interrupter.

(B) Operating Voltage. No luminaires shall be installed for operation on supply circuits over 150 volts between conductors. Submersible pumps and other submersible equipment shall operate at 300 volts or less between conductors.

(C) Luminaire Lenses. Luminaires shall be installed with the top of the luminaire lens below the normal water level of the fountain unless listed for above-water locations. A luminaire facing upward shall comply with either (1) or (2):

(1) Have the lens guarded to prevent contact by any person
(2) Be listed for use without a guard

(D) Overheating Protection. Electrical equipment that depends on submersion for safe operation shall be protected against overheating by a low-water cutoff or other approved means when not submerged.

(E) Wiring. Equipment shall be equipped with provisions for threaded conduit entries or be provided with a suitable flexible cord. The maximum length of each exposed cord in the fountain shall be limited to 3.0 m (10 ft). Cords extending beyond the fountain perimeter shall be enclosed in approved wiring enclosures. Metal parts of equipment in contact with water shall be of brass or other approved corrosion-resistant metal.

(F) Servicing. All equipment shall be removable from the water for relamping or normal maintenance. Luminaires shall not be permanently embedded into the fountain structure such that the water level must be reduced or the fountain drained for relamping, maintenance, or inspection.

(G) Stability. Equipment shall be inherently stable or be securely fastened in place.

680.52 Junction Boxes and Other Enclosures.

(A) General. Junction boxes and other enclosures used for other than underwater installation shall comply with 680.24.

(B) Underwater Junction Boxes and Other Underwater Enclosures. Junction boxes and other underwater enclosures shall meet the requirements of 680.52(B)(1) and (B)(2).

(1) Construction.
　(a) Underwater enclosures shall be equipped with provisions for threaded conduit entries or compression glands or seals for cord entry.
　(b) Underwater enclosures shall be submersible and made of copper, brass, or other approved corrosion-resistant material.

(2) Installation. Underwater enclosure installations shall comply with (a) and (b).
　(a) Underwater enclosures shall be filled with an approved potting compound to prevent the entry of moisture.
　(b) Underwater enclosures shall be firmly attached to the supports or directly to the fountain surface and bonded as required. Where the junction box is supported only by conduits in accordance with 314.23(E) and (F), the conduits shall be of copper, brass, stainless steel, or other approved corrosion-resistant metal. Where the box is fed by nonmetallic conduit, it shall have additional supports and fasteners of copper, brass, or other approved corrosion-resistant material.

680.53 Bonding. All metal piping systems associated with the fountain shall be bonded to the equipment grounding conductor of the branch circuit supplying the fountain.

Informational Note: See 250.122 for sizing of these conductors.

680.54 Grounding. The following equipment shall be grounded:

(1) Other than listed low-voltage luminaires not requiring grounding, all electrical equipment located within the fountain or within 1.5 m (5 ft) of the inside wall of the fountain
(2) All electrical equipment associated with the recirculating system of the fountain
(3) Panelboards that are not part of the service equipment and that supply any electrical equipment associated with the fountain

680.55 Methods of Grounding.

(A) Applied Provisions. The provisions of 680.21(A), 680.23(B)(3), 680.23(F)(1) and (F)(2), 680.24(F), and 680.25 shall apply.

(B) Supplied by a Flexible Cord. Electrical equipment that is supplied by a flexible cord shall have all exposed non–current-carrying metal parts grounded by an insulated copper equipment grounding conductor that is an integral part of this cord. The equipment grounding conductor shall be connected to an equipment grounding terminal in the supply junction box, transformer enclosure, power supply enclosure, or other enclosure.

680.56 Cord-and-Plug-Connected Equipment.

(A) Ground-Fault Circuit Interrupter. All electrical equipment, including power-supply cords, shall be protected by ground-fault circuit interrupters.

(B) Cord Type. Flexible cord immersed in or exposed to water shall be of a type for extra-hard usage, as designated in Table 400.4, and shall be a listed type with a "W" suffix.

(C) Sealing. The end of the flexible cord jacket and the flexible cord conductor termination within equipment shall be covered with, or encapsulated in, a suitable potting compound to prevent the entry of water into the equipment through the cord or its conductors. In addition, the ground connection within equipment shall be similarly treated to protect such connections from the deteriorating effect of water that may enter into the equipment.

(D) Terminations. Connections with flexible cord shall be permanent, except that grounding-type attachment plugs and receptacles shall be permitted to facilitate removal or disconnection for maintenance, repair, or storage of fixed or stationary equipment not located in any water-containing part of a fountain.

680.57 Signs.

(A) General. This section covers electric signs installed within a fountain or within 3.0 m (10 ft) of the fountain edge.

(B) Ground-Fault Circuit-Interrupter Protection for Personnel. Branch circuits or feeders supplying the sign shall have ground-fault circuit-interrupter protection for personnel.

(C) Location.

(1) Fixed or Stationary. A fixed or stationary electric sign installed within a fountain shall be not less than 1.5 m (5 ft) inside the fountain measured from the outside edges of the fountain.

(2) Portable. A portable electric sign shall not be placed within a pool or fountain or within 1.5 m (5 ft) measured horizontally from the inside walls of the fountain.

(D) Disconnect. A sign shall have a local disconnecting means in accordance with 600.6 and 680.13.

(E) Bonding and Grounding. A sign shall be grounded and bonded in accordance with 600.7.

680.58 GFCI Protection for Adjacent Receptacle Outlets. All 15- or 20-ampere, single-phase 125-volt through 250-volt receptacles located within 6.0 m (20 ft) of a fountain edge shall be provided with GFCI protection.

Part VI. Pools and Tubs for Therapeutic Use

680.60 General. The provisions of Part I and Part VI of this article shall apply to pools and tubs for therapeutic use in health care facilities, gymnasiums, athletic training rooms, and similar areas. Portable therapeutic appliances shall comply with Parts II and III of Article 422.

Informational Note: See 517.2 for definition of health care facilities.

680.61 Permanently Installed Therapeutic Pools. Therapeutic pools that are constructed in the ground, on the ground, or in a building in such a manner that the pool cannot be readily disassembled shall comply with Parts I and II of this article.

Exception: The limitations of 680.22(B)(1) through (C)(4) shall not apply where all luminaires are of the totally enclosed type.

680.62 Therapeutic Tubs (Hydrotherapeutic Tanks). Therapeutic tubs, used for the submersion and treatment of patients, that are not easily moved from one place to another in normal use or that are fastened or otherwise secured at a specific location, including associated piping systems, shall comply with Part VI.

(A) Protection. Except as otherwise provided in this section, the outlet(s) that supplies a self-contained therapeutic tub or hydrotherapeutic tank, a packaged therapeutic tub or hydrotherapeutic tank, or a field-assembled therapeutic tub or hydrotherapeutic tank shall be protected by a ground-fault circuit interrupter.

(1) Listed Units. If so marked, a listed, labeled, and identified self-contained unit or a listed, labeled, and identified packaged equipment assembly that includes integral ground-fault circuit-interrupter protection for all electrical parts within the unit or assembly (pumps, air blowers, heaters, lights, controls, sanitizer generators, wiring, and so forth) shall be permitted without additional GFCI protection.

(2) Other Units. A therapeutic tub or hydrotherapeutic tank rated 3 phase or rated over 250 volts or with a heater load of more than 50 amperes shall not require the supply to be protected by a ground-fault circuit interrupter.

(B) Bonding. The following parts shall be bonded together:

(1) All metal fittings within or attached to the tub structure
(2) Metal parts of electrical equipment associated with the tub water circulating system, including pump motors
(3) Metal-sheathed cables and raceways and metal piping that are within 1.5 m (5 ft) of the inside walls of the tub and not separated from the tub by a permanent barrier
(4) All metal surfaces that are within 1.5 m (5 ft) of the inside walls of the tub and not separated from the tub area by a permanent barrier
(5) Electrical devices and controls that are not associated with the therapeutic tubs and located within 1.5 m (5 ft) from such units

Exception: Small conductive surfaces not likely to become energized, such as air and water jets and drain fittings not connected to metallic piping, and towel bars, mirror frames, and similar nonelectrical equipment not connected to metal framing, shall not be required to be bonded.

(C) Methods of Bonding. All metal parts required to be bonded by this section shall be bonded by any of the following methods:

(1) The interconnection of threaded metal piping and fittings
(2) Metal-to-metal mounting on a common frame or base
(3) Connections by suitable metal clamps
(4) By the provisions of a solid copper bonding jumper, insulated, covered, or bare, not smaller than 8 AWG

(D) Grounding.

(1) Fixed or Stationary Equipment. The equipment specified in (a) and (b) shall be connected to the equipment grounding conductor.

　(a) *Location.* All electrical equipment located within 1.5 m (5 ft) of the inside wall of the tub shall be connected to the equipment grounding conductor.

　(b) *Circulation System.* All electrical equipment associated with the circulating system of the tub shall be connected to the equipment grounding conductor.

(2) Portable Equipment. Portable therapeutic appliances shall meet the grounding requirements in 250.114.

(E) Receptacles. All receptacles within 1.83 m (6 ft) of a therapeutic tub shall be protected by a ground-fault circuit interrupter.

(F) Luminaires. All luminaires used in therapeutic tub areas shall be of the totally enclosed type.

Part VII. Hydromassage Bathtubs

680.70 General. Hydromassage bathtubs as defined in 680.2 shall comply with Part VII of this article. They shall not be required to comply with other parts of this article.

680.71 Protection. Hydromassage bathtubs and their associated electrical components shall be on an individual branch circuit(s) and protected by a readily accessible ground-fault circuit interrupter. All 125-volt, single-phase receptacles not exceeding 30 amperes and located within 1.83 m (6 ft) measured horizontally of the inside walls of a hydromassage tub shall be protected by a ground-fault circuit interrupter.

680.72 Other Electrical Equipment. Luminaires, switches, receptacles, and other electrical equipment located in the same room, and not directly associated with a hydromassage bathtub, shall be installed in accordance with the requirements of Chapters 1 through 4 in this *Code* covering the installation of that equipment in bathrooms.

680.73 Accessibility. Hydromassage bathtub electrical equipment shall be accessible without damaging the building structure or building finish. Where the hydromassage bathtub is cord- and plug-connected with the supply receptacle accessible only through a service access opening, the receptacle shall be installed so that its face is within direct view and not more than 300 mm (1 ft) of the opening.

680.74 Bonding.

N **(A) General.** The following parts shall be bonded together:

(1) All metal fittings within or attached to the tub structure that are in contact with the circulating water
(2) Metal parts of electrical equipment associated with the tub water circulating system, including pump and blower motors

(3) Metal-sheathed cables and raceways and metal piping that are within 1.5 m (5 ft) of the inside walls of the tub and not separated from the tub by a permanent barrier
(4) All exposed metal surfaces that are within 1.5 m (5 ft) of the inside walls of the tub and not separated from the tub area by a permanent barrier
(5) Electrical devices and controls that are not associated with the hydromassage tubs and that are located within 1.5 m (5 ft) from such units

Exception No. 1: Small conductive surfaces not likely to become energized, such as air and water jets, supply valve assemblies, and drain fittings not connected to metallic piping, and towel bars, mirror frames, and similar nonelectrical equipment not connected to metal framing shall not be required to be bonded.

Exception No. 2: Double-insulated motors and blowers shall not be bonded.

(B) All metal parts required to be bonded by this section shall be bonded together using a solid copper bonding jumper, insulated, covered, or bare, not smaller than 8 AWG. The bonding jumper(s) shall be required for equipotential bonding in the area of the hydromassage bathtub and shall not be required to be extended or attached to any remote panelboard, service equipment, or any electrode. In all installations a bonding jumper long enough to terminate on a replacement non-double-insulated pump or blower motor shall be provided and shall be terminated to the equipment grounding conductor of the branch circuit of the motor when a double-insulated circulating pump or blower motor is used.

N Part VIII. Electrically Powered Pool Lifts

680.80 General. Electrically powered pool lifts as defined in 680.2 shall comply with Part VIII of this article. They shall not be required to comply with other parts of this article.

680.81 Equipment Approval. Lifts shall be listed, labeled, and identified for swimming pool and spa use.

Exception No. 1: Lifts where the battery is removed for charging at another location and the battery is rated less than or equal to the low-voltage contact limit shall not be required to be listed or labeled.

Exception No. 2: Solar-operated or solar-recharged lifts where the solar panel is attached to the lift and the battery is rated less than or equal to 24 volts shall not be required to be listed or labeled.

Exception No. 3: Lifts that are supplied from a source not exceeding the low-voltage contact limit and supplied by listed transformers or power supplies that comply with 680.23(A)(2) shall not be required to be listed or labeled.

680.82 Protection. Pool lifts connected to premises wiring and operated above the low-voltage contact limit shall be provided with GFCI protection for personnel.

680.83 Bonding. Lifts shall be bonded in accordance with 680.26(B)(5) and (B)(7).

680.84 Switching Devices. Switches and switching devices that are operated above the low-voltage contact limit shall comply with 680.22(C).

680.85 Nameplate Marking. Electrically powered pool lifts shall be provided with a nameplate giving the identifying name and model and rating in volts and amperes, or in volts and watts. If the lift is to be used on a specific frequency or frequencies, it shall be so marked. Battery-powered pool lifts shall indicate the type reference of the battery or battery pack to be used. Batteries and battery packs shall be provided with a battery type reference and voltage rating.

Exception: Nameplate ratings for battery-powered pool lifts shall only need to provide a rating in volts in addition to the identifying name and model.

ARTICLE 682
Natural and Artificially Made Bodies of Water

Part I. General

682.1 Scope. This article applies to the installation of electrical wiring for, and equipment in and adjacent to, natural or artificially made bodies of water not covered by other articles in this *Code*, such as but not limited to aeration ponds, fish farm ponds, storm retention basins, treatment ponds, and irrigation (channels) facilities.

682.2 Definitions.

Artificially Made Bodies of Water. Bodies of water that have been constructed or modified to fit some decorative or commercial purpose such as, but not limited to, aeration ponds, fish farm ponds, storm retention basins, treatment ponds, and irrigation (channel) facilities. Water depths may vary seasonally or be controlled.

Electrical Datum Plane. The electrical datum plane as used in this article is defined as follows:

(1) In land areas subject to tidal fluctuation, the electrical datum plane is a horizontal plane 600 mm (2 ft) above the highest tide level for the area occurring under normal circumstances, that is, highest high tide.

(2) In land areas not subject to tidal fluctuation, the electrical datum plane is a horizontal plane 600 mm (2 ft) above the highest water level for the area occurring under normal circumstances.

(3) In land areas subject to flooding, the electrical datum plane based on (1) or (2) above is a horizontal plane 600 mm (2 ft) above the point identified as the prevailing high water mark or an equivalent benchmark based on seasonal or storm-driven flooding from the authority having jurisdiction.

(4) The electrical datum plane for floating structures and landing stages that are (a) installed to permit rise and fall response to water level, without lateral movement, and (b) that are so equipped that they can rise to the datum plane established for (1) or (2) above, is a horizontal plane 750 mm (30 in.) above the water level at the floating structure or landing stage and a minimum of 300 mm (12 in.) above the level of the deck.

Equipotential Plane. An area where wire mesh or other conductive elements are on, embedded in, or placed under the walk surface within 75 mm (3 in.), bonded to all metal structures and fixed nonelectrical equipment that may become energized, and connected to the electrical grounding system to prevent a difference in voltage from developing within the plane.

Natural Bodies of Water. Bodies of water such as lakes, streams, ponds, rivers, and other naturally occurring bodies of water, which may vary in depth throughout the year.

Shoreline. The farthest extent of standing water under the applicable conditions that determine the electrical datum plane for the specified body of water.

682.3 Other Articles. If the water is subject to boat traffic, the wiring shall comply with 555.13(B).

Part II. Installation

682.10 Electrical Equipment and Transformers. Electrical equipment and transformers, including their enclosures, shall be specifically approved for the intended location. No portion of an enclosure for electrical equipment not identified for operation while submerged shall be located below the electrical datum plane.

682.11 Location of Service Equipment. On land, the service equipment for floating structures and submersible electrical equipment shall be located no closer than 1.5 m (5 ft) horizontally from the shoreline and live parts shall be elevated a minimum of 300 mm (12 in.) above the electrical datum plane. Service equipment shall disconnect when the water level reaches the height of the established electrical datum plane.

682.12 Electrical Connections. All electrical connections not intended for operation while submerged shall be located at least 300 mm (12 in.) above the deck of a floating or fixed structure, but not below the electrical datum plane.

682.13 Wiring Methods and Installation. Liquidtight flexible metal conduit or liquidtight flexible nonmetallic conduit with approved fittings shall be permitted for feeders and where flexible connections are required for services. Extra-hard usage portable power cable listed for both wet locations and sunlight resistance shall be permitted for a feeder or a branch circuit where flexibility is required. Other wiring methods suitable for the location shall be permitted to be installed where flexibility is not required. Temporary wiring in accordance with 590.4 shall be permitted.

682.14 Submersible or Floating Equipment Power Connection(s). Submersible or floating equipment shall be cord- and plug-connected, using extra-hard usage cord, as designated in Table 400.4, and listed with a "W" suffix. The plug and receptacle combination shall be arranged to be suitable for the location while in use. Disconnecting means shall be provided to isolate each submersible or floating electrical equipment from its supply connection(s) without requiring the plug to be removed from the receptacle.

Exception: Equipment listed for direct connection and equipment anchored in place and incapable of routine movement caused by water currents or wind shall be permitted to be connected using wiring methods covered in 682.13.

(A) Type and Marking. The disconnecting means shall consist of a circuit breaker, a switch, or both, or a molded case switch, and shall be specifically marked to designate which receptacle or other outlet it controls.

(B) Location. The disconnecting means shall be readily accessible on land, located not more than 750 mm (30 in.) from the receptacle it controls, and shall be located in the supply circuit ahead of the receptacle. The disconnecting means shall be located within sight of but not closer than 1.5 m (5 ft) from the shoreline and shall be elevated not less than 300 mm (12 in.) above the datum plane.

682.15 Ground-Fault Circuit-Interrupter (GFCI) Protection. Fifteen- and 20-ampere single-phase, 125-volt through 250-volt receptacles installed outdoors and in or on floating buildings or structures within the electrical datum plane area shall be provided with GFCI protection for personnel. The GFCI protection device shall be located not less than 300 mm (12 in.) above the established electrical datum plane.

Part III. Grounding and Bonding

682.30 Grounding. Wiring and equipment within the scope of this article shall be grounded as specified in Part III of 553, 555.15, and with the requirements in Part III of this article.

682.31 Equipment Grounding Conductors.

(A) Type. Equipment grounding conductors shall be insulated copper conductors sized in accordance with 250.122 but not smaller than 12 AWG.

(B) Feeders. Where a feeder supplies a remote panelboard or other distribution equipment, an insulated equipment grounding conductor shall extend from a grounding terminal in the service to a grounding terminal and busbar in the remote panelboard or other distribution equipment.

(C) Branch Circuits. The insulated equipment grounding conductor for branch circuits shall terminate at a grounding terminal in a remote panelboard or other distribution equipment or the grounding terminal in the main service equipment.

(D) Cord-and-Plug-Connected Appliances. Where grounded, cord-and-plug-connected appliances shall be grounded by means of an equipment grounding conductor in the cord and a grounding-type attachment plug.

682.32 Bonding of Non–Current-Carrying Metal Parts. All metal parts in contact with the water, all metal piping, tanks, and all non–current-carrying metal parts that are likely to become energized shall be bonded to the grounding terminal in the distribution equipment.

682.33 Equipotential Planes and Bonding of Equipotential Planes. An equipotential plane shall be installed where required in this section to mitigate step and touch voltages at electrical equipment.

(A) Areas Requiring Equipotential Planes. Equipotential planes shall be installed adjacent to all outdoor service equipment or disconnecting means that control equipment in or on water, that have a metallic enclosure and controls accessible to personnel, and that are likely to become energized. The equipotential plane shall encompass the area around the equipment and shall extend from the area directly below the equipment out not less than 900 mm (36 in.) in all directions from which a person would be able to stand and come in contact with the equipment.

(B) Areas Not Requiring Equipotential Planes. Equipotential planes shall not be required for the controlled equipment supplied by the service equipment or disconnecting means. All circuits rated not more than 60 amperes at 120 through 250 volts, single phase, shall have GFCI protection.

(C) Bonding. Equipotential planes shall be bonded to the electrical grounding system. The bonding conductor shall be solid copper, insulated, covered or bare, and not smaller than 8 AWG. Connections shall be made by exothermic welding or by listed pressure connectors or clamps that are labeled as being suitable for the purpose and are of stainless steel, brass, copper, or copper alloy.

ARTICLE 685
Integrated Electrical Systems

Part I. General

685.1 Scope. This article covers integrated electrical systems, other than unit equipment, in which orderly shutdown is necessary to ensure safe operation. An *integrated electrical system* as used in this article is a unitized segment of an industrial wiring system where all of the following conditions are met:

(1) An orderly shutdown is required to minimize personnel hazard and equipment damage.
(2) The conditions of maintenance and supervision ensure that qualified persons service the system. The name(s) of the qualified person(s) shall be kept in a permanent record at the office of the establishment in charge of the completed installation.
    A person designated as a qualified person shall possess the skills and knowledge related to the construction and operation of the electrical equipment and installation and shall have received documented safety training on the hazards involved. Documentation of their qualifications shall be on file with the office of the establishment in charge of the completed installation.
(3) Effective safeguards acceptable to the authority having jurisdiction are established and maintained.

685.3 Application of Other Articles. The articles/sections in Table 685.3 apply to particular cases of installation of conductors and equipment, where there are orderly shutdown requirements that are in addition to those of this article or are modifications of them.

Part II. Orderly Shutdown

685.10 Location of Overcurrent Devices in or on Premises. Location of overcurrent devices that are critical to integrated electrical systems shall be permitted to be accessible, with mounting heights permitted to ensure security from operation by unqualified personnel.

685.12 Direct-Current System Grounding. Two-wire dc circuits shall be permitted to be ungrounded.

Table 685.3 Application of Other Articles

Conductor/Equipment	Section
More than one building or other structure	225, Part II
Ground-fault protection of equipment	230.95, Exception
Protection of conductors	240.4
Electrical system coordination	240.12
Ground-fault protection of equipment	240.13(1)
Grounding ac systems of 50 volts to less than 1000 volts	250.21
Equipment protection	427.22
Orderly shutdown	430.44
Disconnection	430.75, Exception Nos. 1 and 2
Disconnecting means in sight from controller	430.102(A), Exception No. 2
Energy from more than one source	430.113, Exception Nos. 1 and 2
Disconnecting means	645.10, Exception
Uninterruptible power supplies (UPS)	645.11(1)
Point of connection	705.12

685.14 Ungrounded Control Circuits. Where operational continuity is required, control circuits of 150 volts or less from separately derived systems shall be permitted to be ungrounded.

ARTICLE 690
Solar Photovoltaic (PV) Systems

Part I. General

690.1 Scope. This article applies to solar PV systems, other than those covered by Article 691, including the array circuit(s), inverter(s), and controller(s) for such systems. *[See Figure 690.1(a) and Figure 690.1(b).]* The systems covered by this article may be interactive with other electrical power production sources or stand-alone or both, and may or may not be connected to energy storage systems such as batteries. These PV systems may have ac or dc output for utilization.

> Informational Note: Article 691 covers the installation of large-scale PV electric supply stations.

690.2 Definitions.

Alternating-Current (ac) Module (Alternating-Current Photovoltaic Module). A complete, environmentally protected unit consisting of solar cells, optics, inverter, and other components, exclusive of tracker, designed to generate ac power when exposed to sunlight.

Array. A mechanically integrated assembly of module(s) or panel(s) with a support structure and foundation, tracker, and other components, as required, to form a dc or ac power-producing unit.

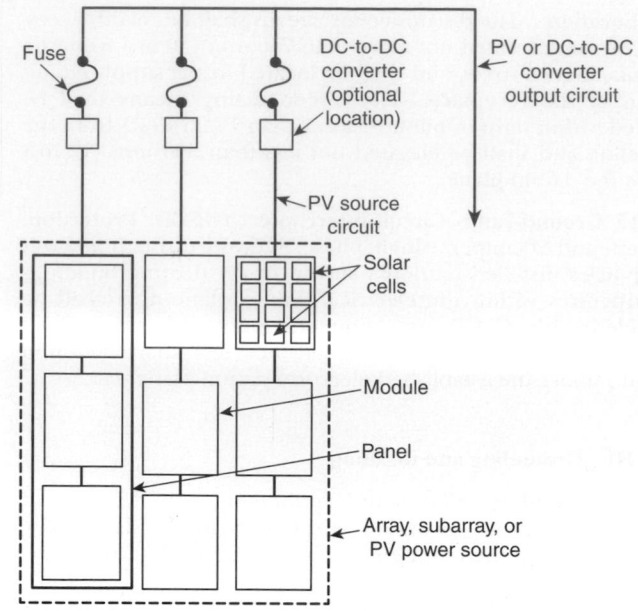

Notes:
(1) These diagrams are intended to be a means of identification for PV power source components, circuits, and connections that make up the PV power source.
(2) Custom PV power source designs occur, and some components are optional.

FIGURE 690.1(a)   Identification of PV Power Source Components.

Bipolar Photovoltaic Array. A dc PV array that has two outputs, each having opposite polarity to a common reference point or center tap.

DC-to-DC Converter. A device installed in the PV source circuit or PV output circuit that can provide an output dc voltage and current at a higher or lower value than the input dc voltage and current.

N **DC-to-DC Converter Output Circuit.** Circuit conductors between the dc-to-dc converter source circuit(s) and the inverter or dc utilization equipment.

N **DC-to-DC Converter Source Circuit.** Circuits between dc-to-dc converters and from dc-to-dc converters to the common connection point(s) of the dc system.

Direct-Current (dc) Combiner. A device used in the PV source and PV output circuits to combine two or more dc circuit inputs and provide one dc circuit output.

Diversion Charge Controller. Equipment that regulates the charging process of a battery by diverting power from energy storage to direct-current or alternating-current loads or to an interconnected utility service.

Electrical Production and Distribution Network. A power production, distribution, and utilization system, such as a utility system and connected loads, that is external to and not controlled by the PV power system.

Functional Grounded PV System. A PV system that has an electrical reference to ground that is not solidly grounded.

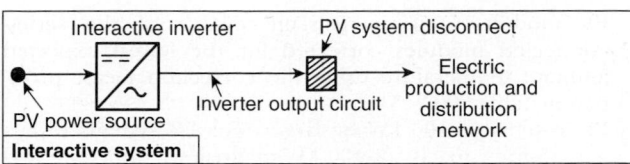

Interactive system

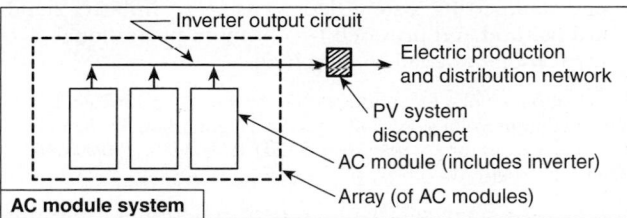

AC module system

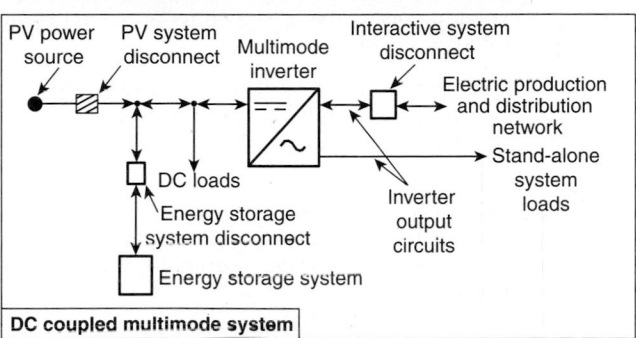

DC coupled multimode system

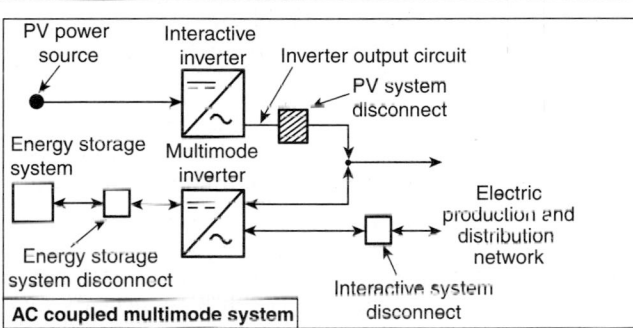

AC coupled multimode system

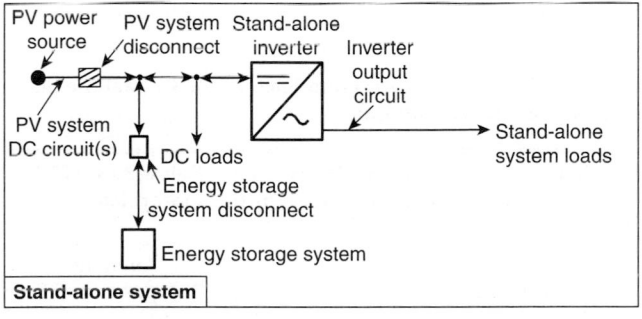

Stand-alone system

Notes:
(1) These diagrams are intended to be a means of identification for PV system components, circuits, and connections.
(2) The PV system disconnect in these diagrams separates the PV system from all other systems.
(3) Not all disconnecting means required by Article 690, Part III are shown.
(4) System grounding and equipment grounding are not shown. See Article 690, Part V.
(5) Custom designs occur in each configuration, and some components are optional.

FIGURE 690.1(b)   Identification of PV System Components in Common Configurations.

Informational Note: A functional grounded PV system is often connected to ground through a fuse, circuit breaker, resistance device, non-isolated grounded ac circuit, or electronic means that is part of a listed ground-fault protection system. Conductors in these systems that are normally at ground potential may have voltage to ground during fault conditions.

N **Generating Capacity.** The sum of parallel-connected inverter maximum continuous output power at 40°C in kilowatts.

Interactive System. A PV system that operates in parallel with and may deliver power to an electrical production and distribution network.

N **Interactive Inverter Output Circuit.** The conductors between the interactive inverter and the service equipment or another electrical power production and distribution network.

Inverter. Equipment that is used to change voltage level or waveform, or both, of electrical energy. Commonly, an inverter [also known as a power conditioning unit (PCU) or power conversion system (PCS)] is a device that changes dc input to an ac output. Inverters may also function as battery chargers that use alternating current from another source and convert it into direct current for charging batteries.

Inverter Input Circuit. Conductors connected to the dc input of an inverter.

Inverter Output Circuit. Conductors connected to the ac output of an inverter.

Module. A complete, environmentally protected unit consisting of solar cells, optics, and other components, exclusive of tracker, designed to generate dc power when exposed to sunlight.

Monopole Subarray. A PV subarray that has two conductors in the output circuit, one positive (+) and one negative (−). Two monopole PV subarrays are used to form a bipolar PV array.

Multimode Inverter. Equipment having the capabilities of both the interactive inverter and the stand-alone inverter.

Panel. A collection of modules mechanically fastened together, wired, and designed to provide a field-installable unit.

Photovoltaic Output Circuit. Circuit conductors between the PV source circuit(s) and the inverter or dc utilization equipment.

Photovoltaic Power Source. An array or aggregate of arrays that generates dc power at system voltage and current.

Photovoltaic Source Circuit. Circuits between modules and from modules to the common connection point(s) of the dc system.

N **Photovoltaic System DC Circuit.** Any dc conductor supplied by a PV power source, including PV source circuits, PV output circuits, dc-to-dc converter source circuits, or dc-to-dc converter output circuits.

Solar Cell. The basic PV device that generates electricity when exposed to light.

Stand-Alone System. A solar PV system that supplies power independently of an electrical production and distribution network.

Subarray. An electrical subset of a PV array.

690.4 General Requirements.

(A) Photovoltaic Systems. Photovoltaic systems shall be permitted to supply a building or other structure in addition to any other electrical supply system(s).

(B) Equipment. Inverters, motor generators, PV modules, PV panels, ac modules, dc combiners, dc-to-dc converters, and charge controllers intended for use in PV systems shall be listed or field labeled for the PV application.

(C) Qualified Personnel. The installation of equipment and all associated wiring and interconnections shall be performed only by qualified persons.

Informational Note: See Article 100 for the definition of *qualified person*.

(D) Multiple PV Systems. Multiple PV systems shall be permitted to be installed in or on a single building or structure. Where the PV systems are remotely located from each other, a directory in accordance with 705.10 shall be provided at each PV system disconnecting means.

N **(E) Locations Not Permitted.** PV system equipment and disconnecting means shall not be installed in bathrooms.

690.6 Alternating-Current (ac) Modules.

(A) Photovoltaic Source Circuits. The requirements of Article 690 pertaining to PV source circuits shall not apply to ac modules. The PV source circuit, conductors, and inverters shall be considered as internal wiring of an ac module.

(B) Inverter Output Circuit. The output of an ac module shall be considered an inverter output circuit.

Part II. Circuit Requirements

690.7 Maximum Voltage. The maximum voltage of PV system dc circuits shall be the highest voltage between any two circuit conductors or any conductor and ground. PV system dc circuits on or in one- and two-family dwellings shall be permitted to have a maximum voltage of 600 volts or less. PV system dc circuits on or in other types of buildings shall be permitted to have a maximum voltage of 1000 volts or less. Where not located on or in buildings, listed dc PV equipment, rated at a maximum voltage of 1500 volts or less, shall not be required to comply with Parts II and III of Article 490.

(A) Photovoltaic Source and Output Circuits. In a dc PV source circuit or output circuit, the maximum PV system voltage for that circuit shall be calculated in accordance with one of the following methods:

Informational Note: One source for lowest-expected, ambient temperature design data for various locations is the chapter titled Extreme Annual Mean Minimum Design Dry Bulb Temperature found in the *ASHRAE Handbook — Fundamentals*, 2013. These temperature data can be used to calculate maximum voltage.

(1) Instructions in listing or labeling of the module: The sum of the PV module–rated open-circuit voltage of the series-connected modules corrected for the lowest expected ambient temperature using the open-circuit voltage temperature coefficients in accordance with the instructions included in the listing or labeling of the module

(2) Crystalline and multicrystalline modules: For crystalline and multicrystalline silicon modules, the sum of the

PV module–rated open-circuit voltage of the series-connected modules corrected for the lowest expected ambient temperature using the correction factor provided in Table 690.7(A)

(3) PV systems of 100 kW or larger: For PV systems with a generating capacity of 100 kW or greater, a documented and stamped PV system design, using an industry standard method and provided by a licensed professional electrical engineer, shall be permitted.

Informational Note: One industry standard method for calculating maximum voltage of a PV system is published by Sandia National Laboratories, reference SAND 2004-3535, *Photovoltaic Array Performance Model.*

The maximum voltage shall be used to determine the voltage rating of conductors, cables, disconnects, overcurrent devices, and other equipment.

(B) DC-to-DC Converter Source and Output Circuits. In a dc-to-dc converter source and output circuit, the maximum voltage shall be calculated in accordance with 690.7(B)(1) or (B)(2).

(1) Single DC-to-DC Converter. For circuits connected to the output of a single dc-to-dc converter, the maximum voltage shall be the maximum rated voltage output of the dc-to-dc converter.

(2) Two or More Series Connected DC-to-DC Converters. For circuits connected to the output of two or more series-connected dc-to-dc converters, the maximum voltage shall be determined in accordance with the instructions included in the listing or labeling of the dc-to-dc converter. If these instructions do not state the rated voltage of series-connected dc-to-dc converters, the maximum voltage shall be the sum of the maximum rated voltage output of the dc-to-dc converters in series.

(C) Bipolar Source and Output Circuits. For 2-wire dc circuits connected to bipolar PV arrays, the maximum voltage shall be the highest voltage between the 2-wire circuit conductors where one conductor of the 2-wire circuit is connected to the functional ground reference (center tap). To prevent overvoltage in the event of a ground-fault or arc-fault, the array shall be

Table 690.7(A) Voltage Correction Factors for Crystalline and Multicrystalline Silicon Modules

Correction Factors for Ambient Temperatures Below 25°C (77°F). (Multiply the rated open-circuit voltage by the appropriate correction factor shown below.)

Ambient Temperature (°C)	Factor	Ambient Temperature (°F)
24 to 20	1.02	76 to 68
19 to 15	1.04	67 to 59
14 to 10	1.06	58 to 50
9 to 5	1.08	49 to 41
4 to 0	1.10	40 to 32
−1 to −5	1.12	31 to 23
−6 to −10	1.14	22 to 14
−11 to −15	1.16	13 to 5
−16 to −20	1.18	4 to −4
−21 to −25	1.20	−5 to −13
−26 to −30	1.21	−14 to −22
−31 to −35	1.23	−23 to −31
−36 to −40	1.25	−32 to −40

isolated from the ground reference and isolated into two 2-wire circuits.

690.8 Circuit Sizing and Current.

(A) Calculation of Maximum Circuit Current. The maximum current for the specific circuit shall be calculated in accordance with 690.8(A)(1) through (A)(6).

Informational Note: Where the requirements of 690.8(A)(1) and (B)(1) are both applied, the resulting multiplication factor is 156 percent.

(1) Photovoltaic Source Circuit Currents. The maximum current shall be calculated by one of the following methods:

(1) The sum of parallel-connected PV module–rated short-circuit currents multiplied by 125 percent

(2) For PV systems with a generating capacity of 100 kW or greater, a documented and stamped PV system design, using an industry standard method and provided by a licensed professional electrical engineer, shall be permitted. The calculated maximum current value shall be based on the highest 3-hour current average resulting from the simulated local irradiance on the PV array accounting for elevation and orientation. The current value used by this method shall not be less than 70 percent of the value calculated using 690.8(A)(1)(1).

Informational Note: One industry standard method for calculating maximum current of a PV system is available from Sandia National Laboratories, reference SAND 2004-3535, *Photovoltaic Array Performance Model*. This model is used by the System Advisor Model simulation program provided by the National Renewable Energy Laboratory.

(2) Photovoltaic Output Circuit Currents. The maximum current shall be the sum of parallel source circuit maximum currents as calculated in 690.8(A)(1).

(3) Inverter Output Circuit Current. The maximum current shall be the inverter continuous output current rating.

(4) Stand-Alone Inverter Input Circuit Current. The maximum current shall be the stand-alone continuous inverter input current rating when the inverter is producing rated power at the lowest input voltage.

(5) DC-to-DC Converter Source Circuit Current. The maximum current shall be the dc-to-dc converter continuous output current rating.

N **(6) DC-to-DC Converter Output Circuit Current.** The maximum current shall be the sum of parallel connected dc-to-dc converter source circuit currents as calculated in 690.8(A)(5).

(B) Conductor Ampacity. PV system currents shall be considered to be continuous. Circuit conductors shall be sized to carry not less than the larger of 690.8(B)(1) or (B)(2) or where protected by a listed adjustable electronic overcurrent protective device in accordance 690.9(B)(3), not less than the current in 690.8(B)(3).

(1) Before Application of Adjustment and Correction Factors. One hundred twenty-five percent of the maximum currents calculated in 690.8(A) before the application of adjustment and correction factors.

Exception: Circuits containing an assembly, together with its overcurrent device(s), that is listed for continuous operation at 100 percent of its rating shall be permitted to be used at 100 percent of its rating.

(2) After Application of Adjustment and Correction Factors. The maximum currents calculated in 690.8(A) after the application of adjustment and correction factors.

N **(3) Adjustable Electronic Overcurrent Protective Device.** The rating or setting of an adjustable electronic overcurrent protective device installed in accordance with 240.6.

(C) Systems with Multiple Direct-Current Voltages. For a PV power source that has multiple output circuit voltages and employs a common-return conductor, the ampacity of the common-return conductor shall not be less than the sum of the ampere ratings of the overcurrent devices of the individual output circuits.

(D) Sizing of Module Interconnection Conductors. Where a single overcurrent device is used to protect a set of two or more parallel-connected module circuits, the ampacity of each of the module interconnection conductors shall not be less than the sum of the rating of the single overcurrent device plus 125 percent of the short-circuit current from the other parallel-connected modules.

690.9 Overcurrent Protection.

(A) Circuits and Equipment. PV system dc circuit and inverter output conductors and equipment shall be protected against overcurrent. Overcurrent protective devices shall not be required for circuits with sufficient ampacity for the highest available current. Circuits connected to current limited supplies (e.g., PV modules, dc-to-dc converters, interactive inverter output circuits) and also connected to sources having higher current availability (e.g., parallel strings of modules, utility power) shall be protected at the higher current source connection.

Exception: An overcurrent device shall not be required for PV modules or PV source circuit or dc-to-dc converters source circuit conductors sized in accordance with 690.8(B) where one of the following applies:

(1)   There are no external sources such as parallel-connected source circuits, batteries, or backfeed from inverters.

(2)   The short-circuit currents from all sources do not exceed the ampacity of the conductors and the maximum overcurrent protective device size rating specified for the PV module or dc-to-dc converter.

Informational Note: Photovoltaic system dc circuits are current limited circuits that only need overcurrent protection when connected in parallel to higher current sources. The overcurrent device is often installed at the higher current source end of the circuit.

(B) Overcurrent Device Ratings. Overcurrent devices used in PV system dc circuits shall be listed for use in PV systems. Overcurrent devices, where required, shall be rated in accordance with one of the following:

(1) Not less than 125 percent of the maximum currents calculated in 690.8(A).

(2) An assembly, together with its overcurrent device(s), that is listed for continuous operation at 100 percent of its rating shall be permitted to be used at 100 percent of its rating.

(3) Adjustable electronic overcurrent protective devices rated or set in accordance with 240.6.

Informational Note: Some electronic overcurrent protective devices prevent backfeed current.

(C) Photovoltaic Source and Output Circuits. A single overcurrent protective device, where required, shall be permitted to protect the PV modules and conductors of each source circuit or the conductors of each output circuit. Where single overcurrent protection devices are used to protect PV source or output circuits, all overcurrent devices shall be placed in the same polarity for all circuits within a PV system. The overcurrent devices shall be accessible but shall not be required to be readily accessible.

Informational Note: Due to improved ground-fault protection required in PV systems by 690.41(B), a single overcurrent protective device in either the positive or negative conductors of a PV system in combination with this ground-fault protection provides adequate overcurrent protection.

(D) Power Transformers. Overcurrent protection for a transformer with a source(s) on each side shall be provided in accordance with 450.3 by considering first one side of the transformer, then the other side of the transformer, as the primary.

Exception: A power transformer with a current rating on the side connected toward the interactive inverter output, not less than the rated continuous output current of the inverter, shall be permitted without overcurrent protection from the inverter.

690.10 Stand-Alone Systems. The wiring system connected to a stand-alone system shall be installed in accordance with 710.15.

690.11 Arc-Fault Circuit Protection (Direct Current). Photovoltaic systems operating at 80 volts dc or greater between any two conductors shall be protected by a listed PV arc-fault circuit interrupter or other system components listed to provide equivalent protection. The system shall detect and interrupt arcing faults resulting from a failure in the intended continuity of a conductor, connection, module, or other system component in the PV system dc circuits.

Informational Note: Annex A includes the reference for the Photovoltaic DC Arc-Fault Circuit Protection product standard.

Exception: For PV systems not installed on or in buildings, PV output circuits and dc-to-dc converter output circuits that are direct buried, installed in metallic raceways, or installed in enclosed metallic cable trays are permitted without arc-fault circuit protection. Detached structures whose sole purpose is to house PV system equipment shall not be considered buildings according to this exception.

690.12 Rapid Shutdown of PV Systems on Buildings. PV system circuits installed on or in buildings shall include a rapid shutdown function to reduce shock hazard for emergency responders in accordance with 690.12(A) through (D).

Exception: Ground mounted PV system circuits that enter buildings, of which the sole purpose is to house PV system equipment, shall not be required to comply with 690.12.

N **(A) Controlled Conductors.** Requirements for controlled conductors shall apply to PV circuits supplied by the PV system.

N **(B) Controlled Limits.** The use of the term *array boundary* in this section is defined as 305 mm (1 ft) from the array in all directions. Controlled conductors outside the array boundary shall comply with 690.12(B)(1) and inside the array boundary shall comply with 690.12(B)(2).

(1) Outside the Array Boundary. Controlled conductors located outside the boundary or more than 1 m (3 ft) from the point of entry inside a building shall be limited to not more

than 30 volts within 30 seconds of rapid shutdown initiation. Voltage shall be measured between any two conductors and between any conductor and ground.

(2) Inside the Array Boundary. The PV system shall comply with one of the following:

(1) The PV array shall be listed or field labeled as a rapid shutdown PV array. Such a PV array shall be installed and used in accordance with the instructions included with the rapid shutdown PV array listing or field labeling.

Informational Note: A listed or field labeled rapid shutdown PV array is evaluated as an assembly or system as defined in the installation instructions to reduce but not eliminate risk of electric shock hazard within a damaged PV array during fire-fighting procedures. These rapid shutdown PV arrays are designed to reduce shock hazards by methods such as limiting access to energized components, reducing the voltage difference between energized components, limiting the electric current that might flow in an electrical circuit involving personnel with increased resistance of the conductive circuit, or by a combination of such methods.

(2) Controlled conductors located inside the boundary or not more than 1 m (3 ft) from the point of penetration of the surface of the building shall be limited to not more than 80 volts within 30 seconds of rapid shutdown initiation. Voltage shall be measured between any two conductors and between any conductor and ground.

(3) PV arrays with no exposed wiring methods, no exposed conductive parts, and installed more than 2.5 m (8 ft) from exposed grounded conductive parts or ground shall not be required to comply with 690.12(B)(2).

The requirement of 690.12(B)(2) shall become effective January 1, 2019.

N **(C) Initiation Device.** The initiation device(s) shall initiate the rapid shutdown function of the PV system. The device "off" position shall indicate that the rapid shutdown function has been initiated for all PV systems connected to that device. For one-family and two-family dwellings, an initiation device(s) shall be located at a readily accessible location outside the building.

The rapid shutdown initiation device(s) shall consist of at least one of the following:

(1) Service disconnecting means
(2) PV system disconnecting means
(3) Readily accessible switch that plainly indicates whether it is in the "off" or "on" position

Informational Note: One example of why an initiation device that complies with 690.12(C)(3) would be used is where a PV system is connected to an optional standby system that remains energized upon loss of utility voltage.

Where multiple PV systems are installed with rapid shutdown functions on a single service, the initiation device(s) shall consist of not more than six switches or six sets of circuit breakers, or a combination of not more than six switches and sets of circuit breakers, mounted in a single enclosure, or in a group of separate enclosures. These initiation device(s) shall initiate the rapid shutdown of all PV systems with rapid shutdown functions on that service. Where auxiliary initiation devices are installed, these auxiliary devices shall control all PV systems with rapid shutdown functions on that service.

N **(D) Equipment.** Equipment that performs the rapid shutdown functions, other than initiation devices such as listed disconnect switches, circuit breakers, or control switches, shall be listed for providing rapid shutdown protection.

> Informational Note: Inverter input circuit conductors often remain energized for up to 5 minutes with inverters not listed for rapid shutdown.

Part III. Disconnecting Means

690.13 Photovoltaic System Disconnecting Means. Means shall be provided to disconnect the PV system from all wiring systems including power systems, energy storage systems, and utilization equipment and its associated premises wiring.

(A) Location. The PV system disconnecting means shall be installed at a readily accessible location.

> Informational Note: PV systems installed in accordance with 690.12 address the concerns related to energized conductors entering a building.

(B) Marking. Each PV system disconnecting means shall plainly indicate whether in the open (off) or closed (on) position and be permanently marked "PV SYSTEM DISCONNECT" or equivalent. Additional markings shall be permitted based upon the specific system configuration. For PV system disconnecting means where the line and load terminals may be energized in the open position, the device shall be marked with the following words or equivalent:

WARNING

ELECTRIC SHOCK HAZARD
TERMINALS ON THE LINE AND LOAD
SIDES MAY BE
ENERGIZED IN THE OPEN POSITION

The warning sign(s) or label(s) shall comply with 110.21(B).

(C) Suitable for Use. If the PV system is connected to the supply side of the service disconnecting means as permitted in 230.82(6), the PV system disconnecting means shall be listed as suitable for use as service equipment.

(D) Maximum Number of Disconnects. Each PV system disconnecting means shall consist of not more than six switches or six sets of circuit breakers, or a combination of not more than six switches and sets of circuit breakers, mounted in a single enclosure, or in a group of separate enclosures. A single PV system disconnecting means shall be permitted for the combined ac output of one or more inverters or ac modules in an interactive system.

> Informational Note: This requirement does not limit the number of PV systems connected to a service as permitted in 690.4(D). This requirement allows up to six disconnecting means to disconnect a single PV system. For PV systems where all power is converted through interactive inverters, a dedicated circuit breaker, in 705.12(B)(1), is an example of a single PV system disconnecting means.

(E) Ratings. The PV system disconnecting means shall have ratings sufficient for the maximum circuit current available short-circuit current, and voltage that is available at the terminals of the PV system disconnect.

N **(F) Type of Disconnect.**

(1) Simultaneous Disconnection. The PV system disconnecting means shall simultaneously disconnect the PV system conductors of the circuit from all conductors of other wiring systems. The PV system disconnecting means shall be an externally operable general-use switch or circuit breaker, or other approved means. A dc PV system disconnecting means shall be marked for use in PV systems or be suitable for backfeed operation.

(2) Devices Marked "Line" and "Load." Devices marked with "line" and "load" shall not be permitted for backfeed or reverse current.

(3) DC-Rated Enclosed Switches, Open-Type Switches, and Low-Voltage Power Circuit Breakers. DC-rated, enclosed switches, open-type switches, and low-voltage power circuit breakers shall be permitted for backfeed operation.

690.15 Disconnection of Photovoltaic Equipment. Isolating devices shall be provided to isolate PV modules, ac PV modules, fuses, dc-to-dc converters inverters, and charge controllers from all conductors that are not solidly grounded. An equipment disconnecting means or a PV system disconnecting means shall be permitted in place of an isolating device. Where the maximum circuit current is greater than 30 amperes for the output circuit of a dc combiner or the input circuit of a charge controller or inverter, an equipment disconnecting means shall be provided for isolation. Where a charge controller or inverter has multiple input circuits, a single equipment disconnecting means shall be permitted to isolate the equipment from the input circuits.

> Informational Note: The purpose of these isolating devices are for the safe and convenient replacement or service of specific PV system equipment without exposure to energized conductors.

(A) Location. Isolating devices or equipment disconnecting means shall be installed in circuits connected to equipment at a location within the equipment, or within sight and within 3 m (10 ft) of the equipment. An equipment disconnecting means shall be permitted to be remote from the equipment where the equipment disconnecting means can be remotely operated from within 3 m (10 ft) of the equipment.

N **(B) Interrupting Rating.** An equipment disconnecting means shall have an interrupting rating sufficient for the maximum short-circuit current and voltage that is available at the terminals of the equipment. An isolating device shall not be required to have an interrupting rating.

N **(C) Isolating Device.** An isolating device shall not be required to simultaneously disconnect all current-carrying conductors of a circuit. The isolating device shall be one of the following:

(1) A connector meeting the requirements of 690.33 and listed and identified for use with specific equipment
(2) A finger safe fuse holder
(3) An isolating switch that requires a tool to open
(4) An isolating device listed for the intended application

An isolating device shall be rated to open the maximum circuit current under load or be marked "Do Not Disconnect Under Load" or "Not for Current Interrupting."

N **(D) Equipment Disconnecting Means.** An equipment disconnecting means shall simultaneously disconnect all current-carrying conductors that are not solidly grounded of the circuit to which it is connected. An equipment disconnecting means shall be externally operable without exposing the operator to contact with energized parts, shall indicate whether in the open (off) or closed (on) position, and shall be lockable in accordance with 110.25. An equipment disconnecting means shall be one of the following devices:

(1) A manually operable switch or circuit breaker
(2) A connector meeting the requirements of 690.33(E)(1)
(3) A load break fused pull out switch
(4) A remote-controlled circuit breaker that is operable locally and opens automatically when control power is interrupted

For equipment disconnecting means, other than those complying with 690.33, where the line and load terminals can be energized in the open position, the device shall be marked in accordance with the warning in 690.13(B).

Part IV. Wiring Methods

690.31 Methods Permitted.

(A) Wiring Systems. All raceway and cable wiring methods included in this *Code*, other wiring systems and fittings specifically listed for use on PV arrays, and wiring as part of a listed system shall be permitted. Where wiring devices with integral enclosures are used, sufficient length of cable shall be provided to facilitate replacement.

Where PV source and output circuits operating at voltages greater than 30 volts are installed in readily accessible locations, circuit conductors shall be guarded or installed in Type MC cable or in raceway. For ambient temperatures exceeding 30°C (86°F), conductor ampacities shall be corrected in accordance with Table 690.31(A).

(B) Identification and Grouping. PV source circuits and PV output circuits shall not be contained in the same raceway, cable tray, cable, outlet box, junction box, or similar fitting as conductors, feeders, branch circuits of other non-PV systems, or inverter output circuits, unless the conductors of the different systems are separated by a partition. PV system circuit conductors shall be identified and grouped as required by 690.31(B)(1) through (2). The means of identification shall be permitted by separate color coding, marking tape, tagging, or other approved means.

(1) Identification. PV system circuit conductors shall be identified at all accessible points of termination, connection, and splices.

The means of identification shall be permitted by separate color coding, marking tape, tagging, or other approved means. Only solidly grounded PV system circuit conductors, in accordance with 690.41(A)(5), shall be marked in accordance with 200.6.

Exception: Where the identification of the conductors is evident by spacing or arrangement, further identification shall not be required.

(2) Grouping. Where the conductors of more than one PV system occupy the same junction box or raceway with a removable cover(s), the ac and dc conductors of each system shall be grouped separately by cable ties or similar means at least once and shall then be grouped at intervals not to exceed 1.8 m (6 ft).

Exception: The requirement for grouping shall not apply if the circuit enters from a cable or raceway unique to the circuit that makes the grouping obvious.

(C) Single-Conductor Cable.

(1) **General.** Single-conductor cable Type USE-2 and single-conductor cable listed and identified as photovoltaic (PV) wire shall be permitted in exposed outdoor locations in PV source circuits within the PV array. PV wire shall be installed in accordance with 338.10(B)(4)(b) and 334.30.

(2) **Cable Tray.** PV source circuits and PV output circuits using single-conductor cable listed and identified as photovoltaic (PV) wire of all sizes, with or without a cable tray marking/rating, shall be permitted in cable trays installed in outdoor locations, provided that the cables are supported at intervals not to exceed 300 mm (12 in.) and secured at intervals not to exceed 1.4 m (4½ ft).

Informational Note: Photovoltaic wire and PV cable have a nonstandard outer diameter. Table 1 of Chapter 9 contains the allowable percent of cross section of conduit and tubing for conductors and cables.

(D) Multiconductor Cable. Jacketed multiconductor cable assemblies listed and identified for the application shall be permitted in outdoor locations. The cable shall be secured at intervals not exceeding 1.8 m (6 ft).

N **Table 690.31(A) Correction Factors**

Ambient Temperature (°C)	Temperature Rating of Conductor				Ambient Temperature (°F)
	60°C (140°F)	75°C (167°F)	90°C (194°F)	105°C (221°F)	
30	1.00	1.00	1.00	1.00	86
31–35	0.91	0.94	0.96	0.97	87–95
36–40	0.82	0.88	0.91	0.93	96–104
41–45	0.71	0.82	0.87	0.89	105–113
46–50	0.58	0.75	0.82	0.86	114–122
51–55	0.41	0.67	0.76	0.82	123–131
56–60	—	0.58	0.71	0.77	132–140
61–70	—	0.33	0.58	0.68	141–158
71–80	—	—	0.41	0.58	159–176

(E) Flexible Cords and Cables Connected to Tracking PV Arrays. Flexible cords and flexible cables, where connected to moving parts of tracking PV arrays, shall comply with Article 400 and shall be of a type identified as a hard service cord or portable power cable; they shall be suitable for extra-hard usage, listed for outdoor use, water resistant, and sunlight resistant. Allowable ampacities shall be in accordance with 400.5. Stranded copper PV wire shall be permitted to be connected to moving parts of tracking PV arrays in accordance with the minimum number of strands specified in Table 690.31(E).

(F) Small-Conductor Cables. Single-conductor cables listed for outdoor use that are sunlight resistant and moisture resistant in sizes 16 AWG and 18 AWG shall be permitted for module interconnections where such cables meet the ampacity requirements of 400.5. Section 310.15 shall be used to determine the cable ampacity adjustment and correction factors.

(G) Photovoltaic System Direct Current Circuits on or in a Building. Where PV system dc circuits run inside a building, they shall be contained in metal raceways, Type MC metal-clad cable that complies with 250.118(10), or metal enclosures from the point of penetration of the surface of the building to the first readily accessible disconnecting means. The disconnecting means shall comply with 690.13(B) and (C) and 690.15(A) and (B). The wiring methods shall comply with the additional installation requirements in 690.31(G)(1) through (4).

(1) Embedded in Building Surfaces. Where circuits are embedded in built-up, laminate, or membrane roofing materials in roof areas not covered by PV modules and associated equipment, the location of circuits shall be clearly marked using a marking protocol that is approved as being suitable for continuous exposure to sunlight and weather.

(2) Flexible Wiring Methods. Where flexible metal conduit (FMC) smaller than metric designator 21 (trade size ¾) or Type MC cable smaller than 25 mm (1 in.) in diameter containing PV power circuit conductors is installed across ceilings or floor joists, the raceway or cable shall be protected by substantial guard strips that are at least as high as the raceway or cable. Where run exposed, other than within 1.8 m (6 ft) of their connection to equipment, these wiring methods shall closely follow the building surface or be protected from physical damage by an approved means.

(3) Marking and Labeling Required. The following wiring methods and enclosures that contain PV system dc circuit conductors shall be marked with the wording WARNING: PHOTOVOLTAIC POWER SOURCE by means of permanently affixed labels or other approved permanent marking:

(1) Exposed raceways, cable trays, and other wiring methods
(2) Covers or enclosures of pull boxes and junction boxes

Table 690.31(E) Minimum PV Wire Strands

PV Wire AWG	Minimum Strands
18	17
16–10	19
8–4	49
2	130
1 AWG–1000 MCM	259

(3) Conduit bodies in which any of the available conduit openings are unused

(4) Marking and Labeling Methods and Locations. The labels or markings shall be visible after installation. The labels shall be reflective, and all letters shall be capitalized and shall be a minimum height of 9.5 mm (⅜ in.) in white on a red background. PV system dc circuit labels shall appear on every section of the wiring system that is separated by enclosures, walls, partitions, ceilings, or floors. Spacing between labels or markings, or between a label and a marking, shall not be more than 3 m (10 ft). Labels required by this section shall be suitable for the environment where they are installed.

(H) Flexible, Fine-Stranded Cables. Flexible, fine-stranded cables shall be terminated only with terminals, lugs, devices, or connectors in accordance with 110.14.

(I) Bipolar Photovoltaic Systems. Where the sum, without consideration of polarity, of the voltages of the two monopole subarrays exceeds the rating of the conductors and connected equipment, monopole subarrays in a bipolar PV system shall be physically separated, and the electrical output circuits from each monopole subarray shall be installed in separate raceways until connected to the inverter. The disconnecting means and overcurrent protective devices for each monopole subarray output shall be in separate enclosures. All conductors from each separate monopole subarray shall be routed in the same raceway. Solidly grounded bipolar PV systems shall be clearly marked with a permanent, legible warning notice indicating that the disconnection of the grounded conductor(s) may result in overvoltage on the equipment.

Exception: Listed switchgear rated for the maximum voltage between circuits and containing a physical barrier separating the disconnecting means for each monopole subarray shall be permitted to be used instead of disconnecting means in separate enclosures.

690.32 Component Interconnections. Fittings and connectors that are intended to be concealed at the time of on-site assembly, where listed for such use, shall be permitted for on-site interconnection of modules or other array components. Such fittings and connectors shall be equal to the wiring method employed in insulation, temperature rise, and fault-current withstand, and shall be capable of resisting the effects of the environment in which they are used.

690.33 Connectors. Connectors, other than those covered by 690.32, shall comply with 690.33(A) through (E).

(A) Configuration. The connectors shall be polarized and shall have a configuration that is noninterchangeable with receptacles in other electrical systems on the premises.

(B) Guarding. The connectors shall be constructed and installed so as to guard against inadvertent contact with live parts by persons.

(C) Type. The connectors shall be of the latching or locking type. Connectors that are readily accessible and that are used in circuits operating at over 30 volts dc or 15 volts ac shall require a tool for opening.

(D) Grounding Member. The grounding member shall be the first to make and the last to break contact with the mating connector.

(E) Interruption of Circuit. Connectors shall be either (1) or (2):

(1) Be rated for interrupting current without hazard to the operator.

(2) Be a type that requires the use of a tool to open and marked "Do Not Disconnect Under Load" or "Not for Current Interrupting."

690.34 Access to Boxes. Junction, pull, and outlet boxes located behind modules or panels shall be so installed that the wiring contained in them can be rendered accessible directly or by displacement of a module(s) or panel(s) secured by removable fasteners and connected by a flexible wiring system.

Part V. Grounding and Bonding

690.41 System Grounding.

(A) PV System Grounding Configurations. One or more of the following system grounding configurations shall be employed:

(1) 2-wire PV arrays with one functional grounded conductor
(2) Bipolar PV arrays according to 690.7(C) with a functional ground reference (center tap)
(3) PV arrays not isolated from the grounded inverter output circuit
(4) Ungrounded PV arrays
(5) Solidly grounded PV arrays as permitted in 690.41(B) Exception
(6) PV systems that use other methods that accomplish equivalent system protection in accordance with 250.4(A) with equipment listed and identified for the use

(B) Ground-Fault Protection. DC PV arrays shall be provided with dc ground-fault protection meeting the requirements of 690.41(B)(1) and (2) to reduce fire hazards.

Exception: PV arrays with not more than two PV source circuits and with all PV system dc circuits not on or in buildings shall be permitted without ground-fault protection where solidly grounded.

(1) Ground-Fault Detection. The ground fault protective device or system shall detect ground fault(s) in the PV array dc current–carrying conductors and components, including any functional grounded conductors, and be listed for providing PV ground-fault protection.

(2) Isolating Faulted Circuits. The faulted circuits shall be isolated by one of the following methods:

(1) The current-carrying conductors of the faulted circuit shall be automatically disconnected.
(2) The inverter or charge controller fed by the faulted circuit shall automatically cease to supply power to output circuits and isolate the PV system dc circuits from the ground reference in a functional grounded system.

690.42 Point of System Grounding Connection. Systems with a ground-fault protective device in accordance with 690.41(B) shall have any current-carrying conductor-to-ground connection made by the ground-fault protective device. For solidly grounded PV systems, the dc circuit grounding connection shall be made at any single point on the PV output circuit.

690.43 Equipment Grounding and Bonding. Exposed non–current-carrying metal parts of PV module frames, electrical equipment, and conductor enclosures of PV systems shall be grounded in accordance with 250.134 or 250.136(A), regard-less of voltage. Equipment grounding conductors and devices shall comply with 690.43(A) through (C).

(A) Photovoltaic Module Mounting Systems and Devices. Devices and systems used for mounting PV modules that are also used for bonding module frames shall be listed, labeled, and identified for bonding PV modules. Devices that mount adjacent PV modules shall be permitted to bond adjacent PV modules.

(B) Equipment Secured to Grounded Metal Supports. Devices listed, labeled, and identified for bonding and grounding the metal parts of PV systems shall be permitted to bond the equipment to grounded metal supports. Metallic support structures shall have identified bonding jumpers connected between separate metallic sections or shall be identified for equipment bonding and shall be connected to the equipment grounding conductor.

(C) With Circuit Conductors. Equipment grounding conductors for the PV array and support structure (where installed) shall be contained within the same raceway, cable, or otherwise run with the PV array circuit conductors when those circuit conductors leave the vicinity of the PV array.

690.45 Size of Equipment Grounding Conductors. Equipment grounding conductors for PV source and PV output circuits shall be sized in accordance with 250.122. Where no overcurrent protective device is used in the circuit, an assumed overcurrent device rated in accordance with 690.9(B) shall be used when applying Table 250.122. Increases in equipment grounding conductor size to address voltage drop considerations shall not be required. An equipment grounding conductor shall not be smaller than 14 AWG.

690.46 Array Equipment Grounding Conductors. For PV modules, equipment grounding conductors smaller than 6 AWG shall comply with 250.120(C).

690.47 Grounding Electrode System.

(A) Buildings or Structures Supporting a PV Array. A building or structure supporting a PV array shall have a grounding electrode system installed in accordance with Part III of Article 250.

PV array equipment grounding conductors shall be connected to the grounding electrode system of the building or structure supporting the PV array in accordance with Part VII of Article 250. This connection shall be in addition to any other equipment grounding conductor requirements in 690.43(C). The PV array equipment grounding conductors shall be sized in accordance with 690.45.

For PV systems that are not solidly grounded, the equipment grounding conductor for the output of the PV system, connected to associated distribution equipment, shall be permitted to be the connection to ground for ground-fault protection and equipment grounding of the PV array.

For solidly grounded PV systems, as permitted in 690.41(A)(5), the grounded conductor shall be connected to a grounding electrode system by means of a grounding electrode conductor sized in accordance with 250.166.

Informational Note: Most PV systems installed in the past decade are actually functional grounded systems rather than solidly grounded systems as defined in this *Code*. For functional grounded PV systems with an interactive inverter output, the ac equipment grounding conductor is connected to associated grounded ac distribution equipment. This connection is often the connection to ground for ground-fault protection and equipment grounding of the PV array.

N **(B) Additional Auxiliary Electrodes for Array Grounding.** Grounding electrodes shall be permitted to be installed in accordance with 250.52 and 250.54 at the location of ground- and roof-mounted PV arrays. The electrodes shall be permitted to be connected directly to the array frame(s) or structure. The grounding electrode conductor shall be sized according to 250.66. The structure of a ground-mounted PV array shall be permitted to be considered a grounding electrode if it meets the requirements of 250.52. Roof mounted PV arrays shall be permitted to use the metal frame of a building or structure if the requirements of 250.52(A)(2) are met.

690.50 Equipment Bonding Jumpers. Equipment bonding jumpers, if used, shall comply with 250.120(C).

Part VI. Marking

690.51 Modules. Modules shall be marked with identification of terminals or leads as to polarity, maximum overcurrent device rating for module protection, and with the following ratings:

(1) Open-circuit voltage
(2) Operating voltage
(3) Maximum permissible system voltage
(4) Operating current
(5) Short-circuit current
(6) Maximum power

690.52 Alternating-Current Photovoltaic Modules. Alternating-current modules shall be marked with identification of terminals or leads and with identification of the following ratings:

(1) Nominal operating ac voltage
(2) Nominal operating ac frequency
(3) Maximum ac power
(4) Maximum ac current
(5) Maximum overcurrent device rating for ac module protection

690.53 Direct-Current Photovoltaic Power Source. A permanent label for the dc PV power source indicating the information specified in (1) through (3) shall be provided by the installer at dc PV system disconnecting means and at each dc equipment disconnecting means required by 690.15. Where a disconnecting means has more than one dc PV power source, the values in 690.53(1) through (3) shall be specified for each source.

(1) Maximum voltage

Informational Note to (1): See 690.7 for voltage.

(2) Maximum circuit current

Informational Note to (2): See 690.8(A) for calculation of maximum circuit current.

(3) Maximum rated output current of the charge controller or dc-to-dc converter (if installed)

690.54 Interactive System Point of Interconnection. All interactive system(s) points of interconnection with other sources shall be marked at an accessible location at the disconnecting means as a power source and with the rated ac output current and the nominal operating ac voltage.

690.55 Photovoltaic Systems Connected to Energy Storage Systems. The PV system output circuit conductors shall be marked to indicate the polarity where connected to energy storage systems.

690.56 Identification of Power Sources.

(A) Facilities with Stand-Alone Systems. Any structure or building with a PV power system that is not connected to a utility service source and is a stand-alone system shall have a permanent plaque or directory installed on the exterior of the building or structure at a readily visible location. The plaque or directory shall indicate the location of system disconnecting means and that the structure contains a stand-alone electrical power system.

(B) Facilities with Utility Services and Photovoltaic Systems. Plaques or directories shall be installed in accordance with 705.10.

(C) Buildings with Rapid Shutdown. Buildings with PV systems shall have permanent labels as described in 690.56(C)(1) through (C)(3).

N **(1) Rapid Shutdown Type.** The type of PV system rapid shutdown shall be labeled as described in 690.56(C)(1)(a) or (1)(b):

(a) For PV systems that shut down the array and conductors leaving the array:

SOLAR PV SYSTEM IS EQUIPPED WITH RAPID SHUTDOWN.
TURN RAPID SHUTDOWN SWITCH TO THE "OFF" POSITION TO SHUT DOWN PV SYSTEM AND REDUCE SHOCK HAZARD IN ARRAY.

The title "SOLAR PV SYSTEM IS EQUIPPED WITH RAPID SHUTDOWN" shall utilize capitalized characters with a minimum height of 9.5 mm (⅜ in.) in black on yellow background, and the remaining characters shall be capitalized with a minimum height of 4.8 mm (³⁄₁₆ in.) in black on white background. [See Figure 690.56(C)(1)(a).]

(b) For PV systems that only shut down conductors leaving the array:

SOLAR PV SYSTEM IS EQUIPPED WITH RAPID SHUTDOWN
TURN RAPID SHUTDOWN SWITCH TO THE "OFF" POSITION TO SHUT DOWN
CONDUCTORS OUTSIDE THE ARRAY. CONDUCTORS IN ARRAY REMAIN
ENERGIZED IN SUNLIGHT.

The title "SOLAR PV SYSTEM IS EQUIPPED WITH RAPID SHUTDOWN" shall utilize capitalized characters with a minimum height of 9.5 mm (⅜ in.) in white on red background, and the remaining characters shall be capitalized with a minimum height of 4.8 mm (³⁄₁₆ in.) in black on white background. [See Figure 690.56(C)(1)(b).]

The labels in 690.56(C)(1)(a) and (b) shall include a simple diagram of a building with a roof. The diagram shall have sections in red to signify sections of the PV system that are not shut down when the rapid shutdown switch is operated.

The rapid shutdown label in 690.56(C)(1) shall be located on or no more than 1 m (3 ft) from the service disconnecting means to which the PV systems are connected and shall indi-

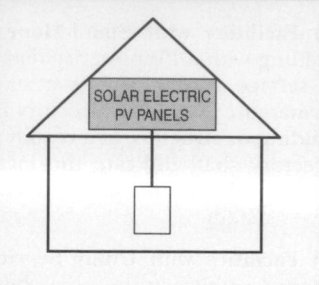

N **FIGURE 690.56(C)(1)(a)   Label for PV Systems that Shut Down the Array and the Conductors Leaving the Array.**

cate the location of all identified rapid shutdown switches if not at the same location.

N **(2) Buildings with More Than One Rapid Shutdown Type.** For buildings that have PV systems with both rapid shutdown types or a PV system with a rapid shutdown type and a PV system with no rapid shutdown, a detailed plan view diagram of the roof shall be provided showing each different PV system and a dotted line around areas that remain energized after the rapid shutdown switch is operated.

N **(3) Rapid Shutdown Switch.** A rapid shutdown switch shall have a label located on or no more than 1 m (3 ft) from the switch that includes the following wording:

RAPID SHUTDOWN SWITCH FOR SOLAR PV SYSTEM

The label shall be reflective, with all letters capitalized and having a minimum height of 9.5 mm (⅜ in.), in white on red background.

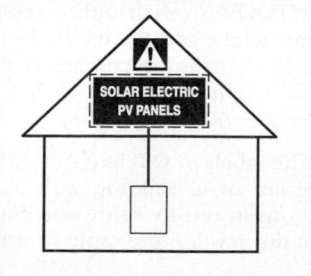

N **FIGURE 690.56(C)(1)(b)   Label for PV Systems that Shut Down the Conductors Leaving the Array Only.**

Part VII.   Connection to Other Sources

N **690.59 Connection to Other Sources.** PV systems connected to other sources shall be installed in accordance with Parts I and II of Article 705.

Part VIII.   Energy Storage Systems

690.71 General. An energy storage system connected to a PV system shall be installed in accordance with Article 706.

690.72 Self-Regulated PV Charge Control. The PV source circuit shall be considered to comply with the requirements of 706.23 if:

(1) The PV source circuit is matched to the voltage rating and charge current requirements of the interconnected battery cells and,

(2) The maximum charging current multiplied by 1 hour is less than 3 percent of the rated battery capacity expressed in ampere-hours or as recommended by the battery manufacturer

N
ARTICLE 691
Large-Scale Photovoltaic (PV) Electric Power Production Facility

691.1 Scope. This article covers the installation of large-scale PV electric power production facilities with a generating capacity of no less than 5000 kW, and not under exclusive utility control.

> Informational Note No. 1: Facilities covered by this article have specific design and safety features unique to large-scale PV facilities and are operated for the sole purpose of providing electric supply to a system operated by a regulated utility for the transfer of electric energy.

> Informational Note No. 2: Section 90.2(B)(5) includes information about utility-owned properties not covered under this *Code*. For additional information on electric supply stations, see ANSI/IEEE C2-2012, *National Electrical Safety Code*.

691.2 Definitions.

Electric Supply Stations. Locations containing the generating stations and substations, including their associated generator, storage battery, transformer, and switchgear areas.

Generating Capacity. The sum of the parallel-connected inverter rated maximum continuous output power at 40°C in kilowatts (kW).

Generating Station. A plant wherein electric energy is produced by conversion from some other form of energy (e.g., chemical, nuclear, solar, wind, mechanical, or hydraulic) by means of suitable apparatus.

691.4 Special Requirements for Large-Scale PV Electric Supply Stations. Large-scale PV electric supply stations shall be accessible only to authorized personnel and comply with the following:

(1) Electrical circuits and equipment shall be maintained and operated only by qualified personnel.

Informational Note: Refer to NFPA 70E-2015, *Standard for Electrical Safety in the Workplace,* for electrical safety requirements.

(2) Access to PV electric supply stations shall be restricted by fencing or other adequate means in accordance with 110.31. Field-applied hazard markings shall be applied in accordance with 110.21(B).
(3) The connection between the PV electric supply station and the system operated by a utility for the transfer of electrical energy shall be through medium- or high-voltage switch gear, substation, switch yard, or similar methods whose sole purpose shall be to safely and effectively interconnect the two systems.
(4) The electrical loads within the PV electric supply station shall only be used to power auxiliary equipment for the generation of the PV power.
(5) Large-scale PV electric supply stations shall not be installed on buildings.

691.5 Equipment Approval. All electrical equipment shall be approved for installation by one of the following:

(1) Listing and labeling
(2) Field labeling
(3) Where products complying with 691.5(1) or (2) are not available, by engineering review validating that the electrical equipment is tested to relevant standards or industry practice

691.6 Engineered Design. Documentation of the electrical portion of the engineered design of the electric supply station shall be stamped and provided upon request of the AHJ. Additional stamped independent engineering reports detailing compliance of the design with applicable electrical standards and industry practice shall be provided upon request of the AHJ. The independent engineer shall be a licensed professional electrical engineer retained by the system owner or installer. This documentation shall include details of conformance of the design with Article 690, and any alternative methods to Article 690, or other articles of this *Code.*

691.7 Conformance of Construction to Engineered Design. Documentation that the construction of the electric supply station conforms to the electrical engineered design shall be provided upon request of the AHJ. Additional stamped independent engineering reports detailing the construction conforms with this *Code,* applicable standards and industry practice shall be provided upon request of the AHJ. The independent engineer shall be a licensed professional electrical engineer retained by the system owner or installer. This documentation, where requested, shall be available prior to commercial operation of the station.

691.8 Direct Current Operating Voltage. For large-scale PV electric supply stations, calculations shall be included in the documentation required in 691.6.

691.9 Disconnection of Photovoltaic Equipment. Isolating devices shall be permitted to be more than 1.8 m (6 ft) from the equipment where written safety procedures and conditions of maintenance and supervision ensure that only qualified persons service the equipment.

Informational Note: For information on lockout/tagout procedures, see NFPA 70E-2015, *Standard for Electrical Safety in the Workplace.*

Buildings whose sole purpose is to house and protect supply station equipment shall not be required to comply with 690.12. Written standard operating procedures shall be available at the site detailing necessary shutdown procedures in the event of an emergency.

691.10 Arc-Fault Mitigation. PV systems that do not comply with the requirements of 690.11 shall include details of fire mitigation plans to address dc arc-faults in the documentation required in 691.6.

691.11 Fence Grounding. Fence grounding requirements and details shall be included in the documentation required in 691.6.

ARTICLE 692
Fuel Cell Systems

Part I. General

692.1 Scope. This article applies to the installation of fuel cell systems.

Informational Note: Some fuel cell systems can be interactive with other electrical power production sources, are stand-alone, or both. Some fuel cell systems are connected to electric energy storage systems such as batteries. Fuel cell systems can have ac output(s), dc output(s), or both for utilization.

692.2 Definitions.

Fuel Cell. An electrochemical system that consumes fuel to produce an electric current. In such cells, the main chemical reaction used for producing electric power is not combustion. However, there may be sources of combustion used within the overall cell system, such as reformers/fuel processors.

Fuel Cell System. The complete aggregate of equipment used to convert chemical fuel into usable electricity and typically consisting of a reformer, stack, power inverter, and auxiliary equipment.

Interactive System. A fuel cell system that operates in parallel with and may deliver power to an electrical production and distribution network. For the purpose of this definition, an energy storage subsystem of a fuel cell system, such as a battery, is not another electrical production source.

Maximum System Voltage. The highest fuel cell inverter output voltage between any ungrounded conductors present at accessible output terminals.

Output Circuit. The conductors used to connect the fuel cell system to its electrical point of delivery.

Informational Note: In the case of sites that have series- or parallel-connected multiple units, the term *output circuit* also refers to the conductors used to electrically interconnect the fuel cell system(s).

Point of Common Coupling. The point at which the power production and distribution network and the customer interface occurs in an interactive system. Typically, this is the load side of the power network meter.

Stand-Alone System. A fuel cell system that supplies power independently of an electrical production and distribution network.

692.4 Installation.

(A) Fuel Cell System. A fuel cell system shall be permitted to supply a building or other structure in addition to any service(s) of another electricity supply system(s).

(B) Identification. A permanent plaque or directory, denoting all electric power sources on or in the premises, shall be installed at each service equipment location.

(C) System Installation. Fuel cell systems including all associated wiring and interconnections shall be installed by only qualified persons.

Informational Note: See Article 100 for the definition of *qualified person.*

692.6 Listing Requirement. The fuel cell system shall be listed or field labeled for its intended application.

Part II. Circuit Requirements

692.8 Circuit Sizing and Current.

(A) Nameplate Rated Circuit Current. The nameplate(s) rated circuit current shall be the rated current indicated on the fuel cell nameplate(s).

(B) Conductor Ampacity and Overcurrent Device Ratings. The ampacity of the feeder circuit conductors from the fuel cell system(s) to the premises wiring system shall not be less than the greater of (1) nameplate(s) rated circuit current or (2) the rating of the fuel cell system(s) overcurrent protective device(s).

(C) Ampacity of Grounded or Neutral Conductor. If an interactive single-phase, 2-wire fuel cell output(s) is connected to the grounded or neutral conductor and a single ungrounded conductor of a 3-wire system or of a 3-phase, 4-wire, wye-connected system, the maximum unbalanced neutral load current plus the fuel cell system(s) output rating shall not exceed the ampacity of the grounded or neutral conductor.

692.9 Overcurrent Protection.

(A) Circuits and Equipment. If the fuel cell system is provided with overcurrent protection sufficient to protect the circuit conductors that supply the load, additional circuit overcurrent devices shall not be required. Equipment and conductors connected to more than one electrical source shall be protected.

(B) Accessibility. Overcurrent devices shall be readily accessible.

692.10 Stand-Alone Systems. The premises wiring system shall meet the requirements of this *Code* except as modified by 692.10(A), (B), and (C).

(A) Fuel Cell System Output. The fuel cell system output from a stand-alone system shall be permitted to supply ac power to the building or structure disconnecting means at current levels below the rating of that disconnecting means.

(B) Sizing and Protection. The circuit conductors between the fuel cell system(s) output and the building or structure disconnecting means shall be sized based on the output rating of the fuel cell system(s). These conductors shall be protected from overcurrents in accordance with 240.4. The overcurrent protection shall be located at the output of the fuel cell system(s).

(C) Single 120-Volt Nominal Supply. The inverter output of a stand-alone fuel cell system shall be permitted to supply 120 volts, nominal, to single-phase, 3-wire 120/240-volt service equipment or distribution panels where there are no 240-volt loads and where there are no multiwire branch circuits. In all installations, the rating of the overcurrent device connected to the output of the fuel cell system(s) shall be less than the rating of the service equipment. This equipment shall be marked as follows:

WARNING

SINGLE 120-VOLT SUPPLY.
DO NOT CONNECT MULTIWIRE
BRANCH CIRCUITS!

The warning sign(s) or label(s) shall comply with 110.21(B).

Part III. Disconnecting Means

692.13 All Conductors. Means shall be provided to disconnect all current-carrying conductors of a fuel cell system power source from all other conductors in a building or other structure.

692.17 Switch or Circuit Breaker. The disconnecting means for ungrounded conductors shall consist of readily accessible, manually operable switch(es) or circuit breaker(s).

Where all terminals of the disconnecting means may be energized in the open position, a warning sign shall be mounted on or adjacent to the disconnecting means. The sign shall be clearly legible and shall have the following words or equivalent:

DANGER
ELECTRIC SHOCK HAZARD. DO NOT TOUCH TERMINALS.
TERMINALS ON BOTH THE LINE AND
LOAD SIDES MAY BE ENERGIZED
IN THE OPEN POSITION.

The danger sign(s) or label(s) shall comply with 110.21(B).

Part IV. Wiring Methods

692.31 Wiring Systems. All raceway and cable wiring methods included in Chapter 3 of this *Code* and other wiring systems and fittings specifically intended and identified for use with fuel cell systems shall be permitted. Where wiring devices with integral enclosures are used, sufficient length of cable shall be provided to facilitate replacement.

Part V. Grounding

692.41 System Grounding.

(A) AC Systems. Grounding of ac systems shall be in accordance with 250.20, and with 250.30 for stand-alone systems.

(B) DC Systems. Grounding of dc systems shall be in accordance with 250.160.

(C) Systems with Alternating-Current and Direct-Current Grounding Requirements. When fuel cell power systems have both alternating-current (ac) and direct-current (dc) grounding requirements, the dc grounding system shall be bonded to the ac grounding system. The bonding conductor shall be sized according to 692.45. A single common grounding electrode and grounding bar may be used for both systems, in which case the common grounding electrode conductor shall be sized to meet the requirements of both 250.66 (ac) and 250.166 (dc).

692.44 Equipment Grounding Conductor. A separate equipment grounding conductor shall be installed.

692.45 Size of Equipment Grounding Conductor. The equipment grounding conductor shall be sized in accordance with 250.122.

692.47 Grounding Electrode System. Any auxiliary grounding electrode(s) required by the manufacturer shall be connected to the equipment grounding conductor specified in 250.118.

Part VI. Marking

692.53 Fuel Cell Power Sources. A marking specifying the fuel cell system, output voltage, output power rating, and continuous output current rating shall be provided at the disconnecting means for the fuel cell power source at an accessible location on the site.

692.54 Fuel Shut-Off. The location of the manual fuel shut-off valve shall be marked at the location of the primary disconnecting means of the building or circuits supplied.

692.56 Stored Energy. A fuel cell system that stores electrical energy shall require the following warning sign, or equivalent, at the location of the service disconnecting means of the premises:

WARNING

FUEL CELL POWER SYSTEM CONTAINS
ELECTRICAL ENERGY STORAGE DEVICES.

The warning sign(s) or label(s) shall comply with 110.21(B).

Part VII. Connection to Other Circuits

692.59 Transfer Switch. A transfer switch shall be required in non–grid-interactive systems that use utility grid backup. The transfer switch shall maintain isolation between the electrical production and distribution network and the fuel cell system. The transfer switch shall be permitted to be located externally or internally to the fuel cell system unit. Where the utility service conductors of the structure are connected to the transfer switch, the switch shall comply with Article 230, Part V.

692.60 Identified Interactive Equipment. Only fuel cell systems listed and marked as interactive shall be permitted in interactive systems.

692.61 Output Characteristics. Output characteristics shall be in accordance with 705.14.

692.62 Loss of Interactive System Power. The fuel cell system shall be provided with a means of detecting when the electrical production and distribution network has become de-energized and shall not feed the electrical production and distribution network side of the point of common coupling during this condition. The fuel cell system shall remain in that state until the electrical production and distribution network voltage has been restored.

A normally interactive fuel cell system shall be permitted to operate as a stand-alone system to supply loads that have been disconnected from electrical production and distribution network sources.

692.64 Unbalanced Interconnections. Unbalanced interconnections shall be in accordance with 705.100.

692.65 Utility-Interactive Point of Connection. Point of connection shall be in accordance with 705.12.

ARTICLE 694
Wind Electric Systems

Part I. General

694.1 Scope. This article applies to wind (turbine) electric systems that consist of one or more wind electric generators and their related alternators, generators, inverters, controllers, and associated equipment.

> Informational Note: Some wind electric systems are interactive with other electric power sources *[see Figure 694.1(a)]* and some are stand-alone systems *[see Figure 694.1(b)]*. Some systems have ac output and some have dc output. Some systems contain electrical energy storage, such as batteries.

694.2 Definitions.

Diversion Charge Controller. Equipment that regulates the charging process of a battery or other energy storage device by

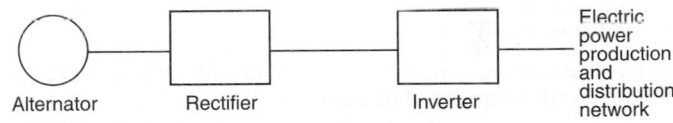

FIGURE 694.1(a) **Identification of Wind Electric System Components — Interactive System.**

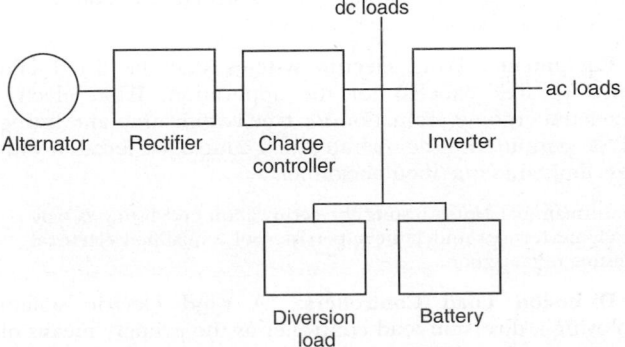

FIGURE 694.1(b) **Identification of Wind Electric System Components — Stand-Alone System.**

diverting power from energy storage to dc or ac loads, or to an interconnected utility service.

Diversion Load. A load connected to a diversion charge controller or diversion load controller, also known as a dump load.

Diversion Load Controller. Equipment that regulates the output of a wind generator by diverting power from the generator to dc or ac loads or to an interconnected utility service.

Inverter Output Circuit. The conductors between an inverter and an ac panelboard for stand-alone systems, or the conductors between an inverter and service equipment or another electric power production source, such as a utility, for an electrical production and distribution network.

Maximum Output Power. The maximum 1 minute average power output a wind turbine produces in normal steady-state operation (instantaneous power output can be higher).

Maximum Voltage. The maximum voltage the wind turbine produces in operation including open circuit conditions.

Nacelle. An enclosure housing the alternator and other parts of a wind turbine.

Rated Power. The output power of a wind turbine at its rated wind speed.

> Informational Note: The method for measuring wind turbine power output is specified in IEC 61400-12-1, *Power Performance Measurements of Electricity Producing Wind Turbines.*

Tower (as applied to wind electric systems). A pole or other structure that supports a wind turbine.

Wind Turbine. A mechanical device that converts wind energy to electrical energy.

Wind Turbine Output Circuit. The circuit conductors between the internal components of a wind turbine (which might include an alternator, integrated rectifier, controller, and/or inverter) and other equipment.

> Informational Note: See also definitions for interconnected systems in Article 705.

694.7 Installation. Systems covered by this article shall be installed only by qualified persons.

> Informational Note: See Article 100 for the definition of *Qualified Person.*

(A) Wind Electric Systems. A wind electric system(s) shall be permitted to supply a building or other structure in addition to other sources of supply.

(B) Equipment. Wind electric systems shall be listed and labeled or field labeled for the application. Wind electric systems undergoing evaluation for type certification and listing shall be permitted to be operated in a controlled location with access limited to qualified personnel.

> Informational Note: Testing for certification and listing is typically performed under the supervision of a qualified electrical testing organization.

(C) Diversion Load Controllers. A wind electric system employing a diversion load controller as the primary means of regulating the speed of a wind turbine rotor shall be equipped with an additional, independent, reliable means to prevent

over-speed operation. An interconnected utility service shall not be considered to be a reliable diversion load.

(D) Surge Protective Devices (SPD). A surge protective device shall be installed between a wind electric system and any loads served by the premises electrical system. The surge protective device shall be permitted to be a Type 3 SPD on the circuit serving a wind electric system or a Type 2 SPD located anywhere on the load side of the service disconnect. Surge protective devices shall be installed in accordance with Part II of Article 285.

(E) Receptacles. A receptacle shall be permitted to be supplied by a wind electric system branch or feeder circuit for maintenance or data acquisition use. Receptacles shall be protected with an overcurrent device with a rating not to exceed the current rating of the receptacle. All 125-volt, single-phase, 15- and 20-ampere receptacles installed for maintenance of the wind turbine shall have ground-fault circuit-interrupter protection for personnel.

(F) Poles or Towers Supporting Wind Turbines Used as a Raceway. A pole or tower shall be permitted to be used as a raceway if evaluated as part of the listing for the wind turbine or otherwise shall be listed or field labeled for the purpose.

(G) Working Clearances. Working space shall be provided for electrical cabinets and other electrical equipment in accordance with 110.26(A).

For large wind turbines where service personnel enter the equipment, where conditions of maintenance and supervision ensure that only qualified persons perform the work, working clearances shall be permitted to comply with Table 694.7 for systems up to 1000 V nominal.

Part II. Circuit Requirements

694.10 Maximum Voltage.

(A) Wind Turbine Output Circuits. For wind turbines connected to one- and two-family dwellings, turbine output circuits shall be permitted to have a maximum voltage up to 600 volts.

(B) Direct-Current Utilization Circuits. The voltage of dc utilization circuits shall comply with 210.6.

(C) Circuits over 150 Volts to Ground. In one- and two-family dwellings, live parts in circuits over 150 volts to ground shall not be accessible to other than qualified persons while energized.

> Informational Note: See 110.27 for guarding of live parts and 210.6 for branch circuit voltage limitations.

N **Table 694.7 Working Spaces**

Nominal Voltage to Ground	Condition 1	Condition 2	Condition 3
0–150	900 mm (3 ft)	900 mm (3 ft)	900 mm (3 ft)
151–1000	900 mm (3 ft)	1.0 m (3 ft 6 in.)	1.2 m (4 ft)

694.12 Circuit Sizing and Current.

(A) Calculation of Maximum Circuit Current. The maximum current for a circuit shall be calculated in accordance with 694.12(A)(1) through (A)(3).

(1) Turbine Output Circuit Currents. The maximum current shall be based on the circuit current of the wind turbine operating at maximum output power.

(2) Inverter Output Circuit Current. The maximum output current shall be the inverter continuous output current rating.

(3) Stand-Alone Inverter Input Circuit Current. The maximum input current shall be the stand-alone continuous inverter input current rating of the inverter producing rated power at the lowest input voltage.

(B) Ampacity and Overcurrent Device Ratings.

(1) Continuous Current. Wind turbine electric system currents shall be considered to be continuous.

(2) Sizing of Conductors and Overcurrent Devices. Circuit conductors and overcurrent devices shall be sized to carry not less than 125 percent of the maximum current as calculated in 694.12(A). The rating or setting of overcurrent devices shall be permitted in accordance with 240.4(B) and (C).

Exception: Circuits containing an assembly, together with its overcurrent devices, listed for continuous operation at 100 percent of its rating shall be permitted to be used at 100 percent of its rating.

694.15 Overcurrent Protection.

(A) Circuits and Equipment. Turbine output circuits, inverter output circuits, and storage battery circuit conductors and equipment shall be protected in accordance with the requirements of Article 240. Circuits connected to more than one electrical source shall have overcurrent devices located so as to provide overcurrent protection from all sources.

Exception: An overcurrent device shall not be required for circuit conductors sized in accordance with 694.12(B) where the maximum current from all sources does not exceed the ampacity of the conductors.

Informational Note: Possible backfeed of current from any source of supply, including a supply through an inverter to the wind turbine output circuit, is a consideration in determining whether overcurrent protection from all sources is provided. Some wind electric systems rely on the turbine output circuit to regulate turbine speed. Inverters may also operate in reverse for turbine startup or speed control.

(B) Power Transformers. Overcurrent protection for a transformer with sources on each side shall be provided in accordance with 450.3 by considering first one side of the transformer, then the other side of the transformer, as the primary.

Exception: A power transformer with a current rating on the side connected to the inverter output, which is not less than the rated continuous output current rating of the inverter, shall not be required to have overcurrent protection at the inverter.

(C) Direct-Current Rating. Overcurrent devices, either fuses or circuit breakers, used in any dc portion of a wind electric system shall be listed for use in dc circuits and shall have appropriate voltage, current, and interrupting ratings.

Part III. Disconnecting Means

694.20 All Conductors. Means shall be provided to disconnect all current-carrying conductors of a wind electric power source from all other conductors in a building or other structure. A switch, circuit breaker, or other device, either ac or dc, shall not be installed in a grounded conductor if operation of that switch, circuit breaker, or other device leaves the marked, grounded conductor in an ungrounded and energized state.

Exception: A wind turbine that uses the turbine output circuit for regulating turbine speed shall not require a turbine output circuit disconnecting means.

694.22 Additional Provisions. Disconnecting means shall comply with 694.22(A) through (D).

(A) Disconnecting Means. The disconnecting means shall not be required to be suitable for use as service equipment. The disconnecting means for ungrounded conductors shall consist of manually operable switches or circuit breakers complying with all of the following requirements:

(1) They shall be located where readily accessible.
(2) They shall be externally operable without exposing the operator to contact with live parts.
(3) They shall plainly indicate whether in the open or closed position.
(4) They shall have an interrupting rating sufficient for the nominal circuit voltage and the current that is available at the line terminals of the equipment.

Where all terminals of the disconnecting means are capable of being energized in the open position, a warning sign shall be mounted on or adjacent to the disconnecting means. The sign shall be clearly legible and shall have the following words or equivalent.

WARNING.

ELECTRIC SHOCK HAZARD.

DO NOT TOUCH TERMINALS.

TERMINALS ON BOTH THE LINE

AND LOAD SIDES MAY BE

ENERGIZED IN THE OPEN POSITION.

The warning sign(s) or label(s) shall comply with 110.21(B).

(B) Equipment. Equipment such as rectifiers, controllers, output circuit isolating and shorting switches, and overcurrent devices shall be permitted on the wind turbine side of the disconnecting means.

(C) Requirements for Disconnecting Means.

(1) Location. The wind electric system disconnecting means shall be installed at a readily accessible location either on or adjacent to the turbine tower, on the outside of a building or structure or inside, at the point of entrance of the wind system conductors.

Exception: Installations that comply with 694.30(C) shall be permitted to have the disconnecting means located remotely from the point of entry of the wind system conductors.

A wind turbine disconnecting means shall not be required to be located at the nacelle or tower.

The disconnecting means shall not be installed in bathrooms.

(2) Marking. Each turbine system disconnecting means shall be permanently marked to identify it as a wind electric system disconnect. A plaque shall be installed in accordance with 705.10.

(3) Suitable for Use. Turbine system disconnecting means shall be suitable for the prevailing conditions.

(4) Maximum Number of Disconnects. The turbine disconnecting means shall consist of not more than six switches or six circuit breakers mounted in a single enclosure, in a group of separate enclosures, or in or on a switchgear.

(D) Equipment That Is Not Readily Accessible. Rectifiers, controllers, and inverters shall be permitted to be mounted in nacelles or other exterior areas that are not readily accessible.

694.23 Turbine Shutdown.

(A) Manual Shutdown. Wind turbines shall be required to have a readily accessible manual shutdown button or switch. Operation of the button or switch shall result in a parked turbine state that shall either stop the turbine rotor or allow limited rotor speed combined with a means to de-energize the turbine output circuit.

Exception: Turbines with a swept area of less than 50 m² (538 ft²) shall not be required to have a manual shutdown button or switch.

(B) Shutdown Procedure. The shutdown procedure for a wind turbine shall be defined and permanently posted at the location of a shutdown means and at the location of the turbine controller or disconnect, if the location is different.

694.24 Disconnection of Wind Electric System Equipment. Means shall be provided to disconnect equipment, such as inverters, batteries, and charge controllers, from all ungrounded conductors of all sources. If the equipment is energized from more than one source, the disconnecting means shall be grouped and identified.

A single disconnecting means in accordance with 694.22 shall be permitted for the combined ac output of one or more inverters in an interactive system.

A shorting switch or plug shall be permitted to be used as an alternative to a disconnect in systems that regulate turbine speed using the turbine output circuit.

Exception: Equipment housed in a turbine nacelle shall not be required to have a disconnecting means.

694.26 Fuses. Means shall be provided to disconnect a fuse from all sources of supply where the fuse is energized from both directions and is accessible to other than qualified persons. Switches, pullouts, or similar devices that are rated for the application shall be permitted to serve as a means to disconnect fuses from all sources of supply.

694.28 Installation and Service of a Wind Turbine. Open circuiting, short circuiting, or mechanical brakes shall be used to disable a turbine for installation and service.

> Informational Note: Some wind turbines rely on the connection from the alternator to a remote controller for speed regulation. Opening turbine output circuit conductors may cause mechanical damage to a turbine and create excessive voltages that could damage equipment or expose persons to electric shock.

Part IV. Wiring Methods

694.30 Permitted Methods.

(A) Wiring Systems. All raceway and cable wiring methods included in this *Code*, and other wiring systems and fittings specifically intended for use on wind turbines, shall be permitted. In readily accessible locations, turbine output circuits that operate at voltages greater than 30 volts shall be installed in raceways.

(B) Flexible Cords and Cables. Flexible cords and cables, where used to connect the moving parts of turbines or where used for ready removal for maintenance and repair, shall comply with Article 400 and shall be of a type identified as hard service cord or portable power cable, shall be suitable for extra-hard usage, shall be listed for outdoor use, and shall be water resistant. Cables exposed to sunlight shall be sunlight resistant. Flexible, fine-stranded cables shall be terminated only with terminals, lugs, devices, or connectors in accordance with 110.14(A).

(C) Direct-Current Turbine Output Circuits Inside a Building. Direct-current turbine output circuits installed inside a building or structure shall be enclosed in metal raceways or installed in metal enclosures, or run in Type MC metal-clad cable that complies with 250.118(10), from the point of penetration of the surface of the building or structure to the first readily accessible disconnecting means.

Part V. Grounding and Bonding

694.40 Equipment Grounding and Bonding.

(A) General. Exposed non–current-carrying metal parts of towers, turbine nacelles, other equipment, and conductor enclosures shall be grounded and bonded to the premises grounding and bonding system. Attached metal parts, such as turbine blades and tails that are not likely to become energized, shall not be required to be grounded or bonded.

(B) Tower Grounding and Bonding.

(1) Grounding Electrodes and Grounding Electrode Conductors. A wind turbine tower shall be connected to a grounding electrode system. Where installed in close proximity to galvanized foundation or tower anchor components, galvanized grounding electrodes shall be used.

> Informational Note: Copper and copper-clad grounding electrodes, where used in highly conductive soils, can cause electrolytic corrosion of galvanized foundation and tower anchor components.

(2) Bonding Conductor. Equipment grounding conductors or supply-side bonding jumpers, as applicable, shall be required between turbines, towers, and the premises grounding system.

(3) Tower Connections. Equipment grounding, bonding, and grounding electrode conductors, where used, shall be connected to metallic towers using listed means. All mechanical elements used to terminate these conductors shall be accessible.

(4) Guy Wires. Guy wires used to support turbine towers shall not be required to be connected to an equipment grounding conductor or to comply with the requirements of 250.110.

Informational Note: Guy wires supporting grounded towers are unlikely to become energized under normal conditions, but partial lightning currents could flow through guy wires when exposed to a lightning environment. Grounding of metallic guy wires may be required by lightning standards. For information on lightning protection systems, see NFPA 780-2014, *Standard for the Installation of Lightning Protection Systems.*

Part VI. Marking

694.50 Interactive System Point of Interconnection. All interactive system points of interconnection with other sources shall be marked at an accessible location at the disconnecting means and with the rated ac output current and the nominal operating ac voltage.

694.52 Power Systems Employing Energy Storage. Wind electric systems employing energy storage shall be marked with the maximum operating voltage, any equalization voltage, and the polarity of the grounded circuit conductor.

694.54 Identification of Power Sources.

(A) Facilities with Stand-Alone Systems. Any structure or building with a stand-alone system and not connected to a utility service source shall have a permanent plaque or directory installed on the exterior of the building or structure at a readily visible location. The plaque or directory shall indicate the location of system disconnecting means and shall indicate that the structure contains a stand-alone electrical power system.

(B) Facilities with Utility Services and Wind Electric Systems. Buildings or structures with both utility service and wind electric systems shall have a permanent plaque or directory providing the location of the service disconnecting means and the wind electric system disconnecting means.

694.56 Instructions for Disabling Turbine. A plaque shall be installed at or adjacent to the turbine location providing basic instructions for disabling the turbine.

Part VII. Connection to Other Sources

694.60 Identified Interactive Equipment. Only inverters that are listed, labeled, and identified as interactive shall be permitted in interactive systems.

694.62 Installation. Wind electric systems, where connected to utility electric sources, shall comply with the requirements of Article 705.

694.66 Operating Voltage Range. Wind electric systems connected to dedicated branch or feeder circuits shall be permitted to exceed normal voltage operating ranges on these circuits, provided that the voltage at any distribution equipment supplying other loads remains within normal ranges.

Informational Note: Wind turbines might use the electric grid to dump energy from short-term wind gusts. Normal operating voltages are defined in ANSI C84.1-2006, *Voltage Ratings for Electric Power Systems and Equipment (60 Hz).*

694.68 Point of Connection. Points of connection to interconnected electric power sources shall comply with 705.12.

ARTICLE 695
Fire Pumps

695.1 Scope.

Informational Note: Text that is followed by a reference in brackets has been extracted from NFPA 20-2013, *Standard for the Installation of Stationary Pumps for Fire Protection.* Only editorial changes were made to the extracted text to make it consistent with this *Code.*

(A) Covered. This article covers the installation of the following:

(1) Electric power sources and interconnecting circuits
(2) Switching and control equipment dedicated to fire pump drivers

(B) Not Covered. This article does not cover the following:

(1) The performance, maintenance, and acceptance testing of the fire pump system, and the internal wiring of the components of the system
(2) The installation of pressure maintenance (jockey or makeup) pumps

Informational Note: For the installation of pressure maintenance (jockey or makeup) pumps supplied by the fire pump circuit or another source, see Article 430.

(3) Transfer equipment upstream of the fire pump transfer switch(es)

Informational Note: See NFPA 20-2013, *Standard for the Installation of Stationary Pumps for Fire Protection*, for further information.

695.2 Definitions.

Fault-Tolerant External Control Circuits. Those control circuits either entering or leaving the fire pump controller enclosure, which if broken, disconnected, or shorted will not prevent the controller from starting the fire pump from all other internal or external means and may cause the controller to start the pump under these conditions.

On-Site Power Production Facility. The normal supply of electric power for the site that is expected to be constantly producing power.

On-Site Standby Generator. A facility producing electric power on site as the alternate supply of electric power. It differs from an on-site power production facility, in that it is not constantly producing power.

695.3 Power Source(s) for Electric Motor-Driven Fire Pumps. Electric motor-driven fire pumps shall have a reliable source of power.

Informational Note: See Sections 9.3.2 and A.9.3.2 from NFPA 20-2013, *Standard for the Installation of Stationary Pumps for Fire Protection*, for guidance on the determination of power source reliability.

(A) Individual Sources. Where reliable, and where capable of carrying indefinitely the sum of the locked-rotor current of the fire pump motor(s) and the pressure maintenance pump motor(s) and the full-load current of the associated fire pump accessory equipment when connected to this power supply, the power source for an electric motor driven fire pump shall be one or more of the following.

(1) Electric Utility Service Connection. A fire pump shall be permitted to be supplied by a separate service, or from a connection located ahead of and not within the same cabinet, enclosure, vertical switchgear section, or vertical switchboard section as the service disconnecting means. The connection shall be located and arranged so as to minimize the possibility of damage by fire from within the premises and from exposing hazards. A tap ahead of the service disconnecting means shall comply with 230.82(5). The service equipment shall comply with the labeling requirements in 230.2 and the location requirements in 230.72(B). [**20**:9.2.2(1)]

(2) On-Site Power Production Facility. A fire pump shall be permitted to be supplied by an on-site power production facility. The source facility shall be located and protected to minimize the possibility of damage by fire. [**20**:9.2.2(3)]

(3) Dedicated Feeder. A dedicated feeder shall be permitted where it is derived from a service connection as described in 695.3(A)(1). [**20**:9.2.2(3)]

(B) Multiple Sources. If reliable power cannot be obtained from a source described in 695.3(A), power shall be supplied by one of the following: [**20**:9.3.2]

(1) Individual Sources. An approved combination of two or more of the sources from 695.3(A).

(2) Individual Source and On-site Standby Generator. An approved combination of one or more of the sources in 695.3(A) and an on-site standby generator complying with 695.3(D). [**20**:9.3.4]

Exception to (B)(1) and (B)(2): An alternate source of power shall not be required where a back-up engine-driven or back-up steam turbine-driven fire pump is installed. [20:9.3.3]

(C) Multibuilding Campus-Style Complexes. If the sources in 695.3(A) are not practicable and the installation is part of a multibuilding campus-style complex, feeder sources shall be permitted if approved by the authority having jurisdiction and installed in accordance with either (C)(1) and (C)(3) or (C)(2) and (C)(3).

(1) Feeder Sources. Two or more feeders shall be permitted as more than one power source if such feeders are connected to, or derived from, separate utility services. The connection(s), overcurrent protective device(s), and disconnecting means for such feeders shall meet the requirements of 695.4(B)(1) (b).

(2) Feeder and Alternate Source. A feeder shall be permitted as a normal source of power if an alternate source of power independent from the feeder is provided. The connection(s), overcurrent protective device(s), and disconnecting means for such feeders shall meet the requirements of 695.4(B)(1) (b).

(3) Selective Coordination. The overcurrent protective device(s) in each disconnecting means shall be selectively coordinated with any other supply-side overcurrent protective device(s).

(D) On-Site Standby Generator as Alternate Source. An on-site standby generator(s) used as an alternate source of power shall comply with (D)(1) through (D)(3). [**20**:9.6.2.1]

(1) Capacity. The generator shall have sufficient capacity to allow normal starting and running of the motor(s) driving the fire pump(s) while supplying all other simultaneously operated load(s). [**20**:9.6.1.1]

Automatic shedding of one or more optional standby loads in order to comply with this capacity requirement shall be permitted.

(2) Connection. A tap ahead of the generator disconnecting means shall not be required. [**20**:9.6.1.2]

(3) Adjacent Disconnects. The requirements of 430.113 shall not apply.

(E) Arrangement. All power supplies shall be located and arranged to protect against damage by fire from within the premises and exposing hazards. [**20**:9.1.4]

Multiple power sources shall be arranged so that a fire at one source does not cause an interruption at the other source.

(F) Transfer of Power. Transfer of power to the fire pump controller between the individual source and one alternate source shall take place within the pump room. [**20**:9.6.4]

(G) Power Source Selection. Selection of power source shall be performed by a transfer switch listed for fire pump service. [**20**:10.8.1.3.1]

(H) Overcurrent Device Selection. An instantaneous trip circuit breaker shall be permitted in lieu of the overcurrent devices specified in 695.4(B)(2)(a)(1), provided that it is part of a transfer switch assembly listed for fire pump service that complies with 695.4(B)(2)(a)(2).

(I) Phase Converters. Phase converters shall not be permitted to be used for fire pump service. [**20**:9.1.7]

695.4 Continuity of Power. Circuits that supply electric motor–driven fire pumps shall be supervised from inadvertent disconnection as covered in 695.4(A) or (B).

(A) Direct Connection. The supply conductors shall directly connect the power source to a listed fire pump controller, a listed combination fire pump controller and power transfer switch, or a listed fire pump power transfer switch.

(B) Connection Through Disconnecting Means and Overcurrent Device.

(1) Number of Disconnecting Means.

    (a)   *General.* A single disconnecting means and associated overcurrent protective device(s) shall be permitted to be installed between the fire pump power source(s) and one of the following: [**20**:9.1.2]

(1)   A listed fire pump controller
(2)   A listed fire pump power transfer switch
(3)   A listed combination fire pump controller and power transfer switch

    (b)   *Feeder Sources.* For systems installed under the provisions of 695.3(C) only, additional disconnecting means and the associated overcurrent protective device(s) shall be permitted.

    (c)   *On-Site Standby Generator.* Where an on-site standby generator is used to supply a fire pump, an additional disconnecting means and an associated overcurrent protective device(s) shall be permitted.

(2) Overcurrent Device Selection. Overcurrent devices shall comply with 695.4(B)(2)(a) or (b).

    (a)   *Individual Sources.* Overcurrent protection for individual sources shall comply with 695.4(B)(2)(a)(1) or (2).

(1) Overcurrent protective device(s) shall be rated to carry indefinitely the sum of the locked-rotor current of the largest fire pump motor and the pressure maintenance pump motor(s) and the full-load current of all of the other pump motors and associated fire pump accessory equipment when connected to this power supply. Where the locked-rotor current value does not correspond to a standard overcurrent device size, the next standard overcurrent device size shall be used in accordance with 240.6. The requirement to carry the locked-rotor currents indefinitely shall not apply to conductors or devices other than overcurrent devices in the fire pump motor circuit(s). The requirement to carry the locked rotor currents indefinitely shall not apply to feeder overcurrent protective devices installed in accordance with 695.3(C). [20:9.2.3.4]

(2) Overcurrent protection shall be provided by an assembly listed for fire pump service and complying with the following:

  a. The overcurrent protective device shall not open within 2 minutes at 600 percent of the full-load current of the fire pump motor(s).

  b. The overcurrent protective device shall not open with a re-start transient of 24 times the full-load current of the fire pump motor(s).

  c. The overcurrent protective device shall not open within 10 minutes at 300 percent of the full-load current of the fire pump motor(s).

  d. The trip point for circuit breakers shall not be field adjustable. [20:9.2.3.4.1]

  (b) *On-Site Standby Generators.* Overcurrent protective devices between an on-site standby generator and a fire pump controller shall be selected and sized to allow for instantaneous pickup of the full pump room load, but shall not be larger than the value selected to comply with 430.62 to provide short-circuit protection only. [20:9.6.1.1]

(3) **Disconnecting Means.** All disconnecting devices that are unique to the fire pump loads shall comply with items (a) through (e).

  (a) *Features and Location — Normal Power Source.* The disconnecting means for the normal power source shall comply with all of the following: [20:9.2.3.1]

(1) Be identified as suitable for use as service equipment.

(2) Be lockable in the closed position. The provision for locking or adding a lock to the disconnecting means shall be installed on or at the switch or circuit breaker used as the disconnecting means and shall remain in place with or without the lock installed.

(3) Not be located within the same enclosure, panelboard, switchboard, switchgear, or motor control center, with or without common bus, that supplies loads other than the fire pump.

(4) Be located sufficiently remote from other building or other fire pump source disconnecting means such that inadvertent operation at the same time would be unlikely.

Exception to 695.4(B)(3)(a): For a multibuilding campus-style complex(s) installed under the provisions of 695.3(C), only the requirements in 695.4(B)(3)(a)(2) shall apply for normal power source disconnects.

  (b) *Features and Location — On-Site Standby Generator.* The disconnecting means for an on-site standby generator(s) used as the alternate power source shall be installed in accordance

with 700.10(B)(5) for emergency circuits and shall be lockable in the closed position. The provision for locking or adding a lock to the disconnecting means shall be installed on or at the switch or circuit breaker used as the disconnecting means and shall remain in place with or without the lock installed.

  (c) *Disconnect Marking.* The disconnecting means shall be marked "Fire Pump Disconnecting Means." The letters shall be at least 25 mm (1 in.) in height, and they shall be visible without opening enclosure doors or covers. [20:9.2.3.1(5)]

  (d) *Controller Marking.* A placard shall be placed adjacent to the fire pump controller, stating the location of this disconnecting means and the location of the key (if the disconnecting means is locked). [20:9.2.3.2]

  (e) *Supervision.* The disconnecting means shall be supervised in the closed position by one of the following methods:

(1) Central station, proprietary, or remote station signal device

(2) Local signaling service that causes the sounding of an audible signal at a constantly attended point

(3) Locking the disconnecting means in the closed position

  (f) Sealing of disconnecting means and approved weekly recorded inspections when the disconnecting means are located within fenced enclosures or in buildings under the control of the owner [20:9.2.3.3]

695.5 Transformers. Where the service or system voltage is different from the utilization voltage of the fire pump motor, transformer(s) protected by disconnecting means and overcurrent protective devices shall be permitted to be installed between the system supply and the fire pump controller in accordance with 695.5(A) and (B), or with (C). Only transformers covered in 695.5(C) shall be permitted to supply loads not directly associated with the fire pump system.

(A) Size. Where a transformer supplies an electric motor driven fire pump, it shall be rated at a minimum of 125 percent of the sum of the fire pump motor(s) and pressure maintenance pump(s) motor loads, and 100 percent of the associated fire pump accessory equipment supplied by the transformer.

(B) Overcurrent Protection. The primary overcurrent protective device(s) shall be selected or set to carry indefinitely the sum of the locked-rotor current of the fire pump motor(s) and the pressure maintenance pump motor(s) and the full-load current of the associated fire pump accessory equipment when connected to this power supply. Secondary overcurrent protection shall not be permitted. The requirement to carry the locked-rotor currents indefinitely shall not apply to conductors or devices other than overcurrent devices in the fire pump motor circuit(s).

(C) Feeder Source. Where a feeder source is provided in accordance with 695.3(C), transformers supplying the fire pump system shall be permitted to supply other loads. All other loads shall be calculated in accordance with Article 220, including demand factors as applicable.

(1) Size. Transformers shall be rated at a minimum of 125 percent of the sum of the fire pump motor(s) and pressure maintenance pump(s) motor loads, and 100 percent of the remaining load supplied by the transformer.

(2) Overcurrent Protection. The transformer size, the feeder size, and the overcurrent protective device(s) shall be coordinated such that overcurrent protection is provided for the transformer in accordance with 450.3 and for the feeder in accordance with 215.3, and such that the overcurrent protective device(s) is selected or set to carry indefinitely the sum of the locked-rotor current of the fire pump motor(s), the pressure maintenance pump motor(s), the full-load current of the associated fire pump accessory equipment, and 100 percent of the remaining loads supplied by the transformer. The requirement to carry the locked-rotor currents indefinitely shall not apply to conductors or devices other than overcurrent devices in the fire pump motor circuit(s).

695.6 Power Wiring. Power circuits and wiring methods shall comply with the requirements in 695.6(A) through (J), and as permitted in 230.90(A), Exception No. 4; 230.94, Exception No. 4; 240.13; 230.208; 240.4(A); and 430.31.

(A) Supply Conductors.

(1) Services and On-Site Power Production Facilities. Service conductors and conductors supplied by on-site power production facilities shall be physically routed outside a building(s) and shall be installed as service-entrance conductors in accordance with 230.6, 230.9, and Parts III and IV of Article 230. Where supply conductors cannot be physically routed outside of buildings, the conductors shall be permitted to be routed through the building(s) where installed in accordance with 230.6(1) or (2).

(2) Feeders. Fire pump supply conductors on the load side of the final disconnecting means and overcurrent device(s) permitted by 695.4(B), or conductors that connect directly to an on-site standby generator, shall comply with all of the following:

(a)   *Independent Routing.* The conductors shall be kept entirely independent of all other wiring.

(b)   *Associated Fire Pump Loads.* The conductors shall supply only loads that are directly associated with the fire pump system.

(c)   *Protection from Potential Damage.* The conductors shall be protected from potential damage by fire, structural failure, or operational accident.

(d)   *Inside of a Building.* Where routed through a building, the conductors shall be protected from fire for 2 hours using one of the following methods:

(1)   The cable or raceway is encased in a minimum 50 mm (2 in.) of concrete.
(2)   The cable or raceway is a listed fire-resistive cable system.

Informational Note 1: Fire-resistive cables are tested to ANSI/UL 2196, *Tests for Fire Resistive Cables.*

Informational Note 2: The listing organization provides information for fire-resistive cable systems on proper installation requirements to maintain the fire rating.

(3)   The cable or raceway is a listed electrical circuit protective system.

Informational Note 1: Electrical circuit protective systems could include, but are not limited to, thermal barriers or a protective shaft and are tested in accordance with UL 1724, *Fire Tests for Electrical Circuit Protection Systems.*

Informational Note 2: The listing organization provides information for electrical circuit protective systems on proper installation requirements to maintain the fire rating.

Exception to (A)(2)(d): The supply conductors located in the electrical equipment room where they originate and in the fire pump room shall not be required to have the minimum 2-hour fire separation or fire-resistance rating, unless otherwise required by 700.10(D) of this Code.

(B) Conductor Size.

(1) Fire Pump Motors and Other Equipment. Conductors supplying a fire pump motor(s), pressure maintenance pumps, and associated fire pump accessory equipment shall have a rating not less than 125 percent of the sum of the fire pump motor(s) and pressure maintenance motor(s) full-load current(s), and 100 percent of the associated fire pump accessory equipment.

(2) Fire Pump Motors Only. Conductors supplying only a fire pump motor shall have a minimum ampacity in accordance with 430.22 and shall comply with the voltage drop requirements in 695.7.

(C) Overload Protection. Power circuits shall not have automatic protection against overloads. Except for protection of transformer primaries provided in 695.5(C)(2), branch-circuit and feeder conductors shall be protected against short circuit only. Where a tap is made to supply a fire pump, the wiring shall be treated as service conductors in accordance with 230.6. The applicable distance and size restrictions in 240.21 shall not apply.

Exception No. 1: Conductors between storage batteries and the engine shall not require overcurrent protection or disconnecting means.

Exception No. 2: For an on-site standby generator(s) rated to produce continuous current in excess of 225 percent of the full-load amperes of the fire pump motor, the conductors between the on-site generator(s) and the combination fire pump transfer switch controller or separately mounted transfer switch shall be installed in accordance with 695.6(A)(2).

The protection provided shall be in accordance with the short-circuit current rating of the combination fire pump transfer switch controller or separately mounted transfer switch.

(D) Pump Wiring. All wiring from the controllers to the pump motors shall be in rigid metal conduit, intermediate metal conduit, electrical metallic tubing, liquidtight flexible metal conduit, or liquidtight flexible nonmetallic conduit, listed Type MC cable with an impervious covering, or Type MI cable. Electrical connections at motor terminal boxes shall be made with a listed means of connection. Twist-on, insulation-piercing–type, and soldered wire connectors shall not be permitted to be used for this purpose.

(E) Loads Supplied by Controllers and Transfer Switches. A fire pump controller and fire pump power transfer switch, if provided, shall not serve any load other than the fire pump for which it is intended.

(F) Mechanical Protection. All wiring from engine controllers and batteries shall be protected against physical damage and shall be installed in accordance with the controller and engine manufacturer's instructions.

(G) Ground-Fault Protection of Equipment. Ground-fault protection of equipment shall not be installed in any fire pump power circuit. [20:9.1.8.1]

(H) Listed Electrical Circuit Protective System to Controller Wiring. Electrical circuit protective system installation shall comply with any restrictions provided in the listing of the electrical circuit protective system used, and the following also shall apply:

(1)   A junction box shall be installed ahead of the fire pump controller a minimum of 300 mm (12 in.) beyond the fire-rated wall or floor bounding the fire zone.

(2) Where required by the manufacturer of a listed electrical circuit protective system or by the listing, or as required elsewhere in this *Code*, the raceway between a junction box and the fire pump controller shall be sealed at the junction box end as required and in accordance with the instructions of the manufacturer. [**20**:9.8.2]

(3) Standard wiring between the junction box and the controller shall be permitted. [**20**:9.8.3]

(I) Junction Boxes. Where fire pump wiring to or from a fire pump controller is routed through a junction box, the following requirements shall be met:

(1) The junction box shall be securely mounted. [**20**:9.7(1)]

(2) Mounting and installing of a junction box shall not violate the enclosure type rating of the fire pump controller(s). [**20**:9.7(2)]

(3) Mounting and installing of a junction box shall not violate the integrity of the fire pump controller(s) and shall not affect the short-circuit rating of the controller(s). [**20**:9.7(3)]

(4) As a minimum, a Type 2, drip-proof enclosure (junction box) shall be used where installed in the fire pump room. The enclosure shall be listed to match the fire pump controller enclosure type rating. [**20**:9.7(4)]

(5) Terminals, junction blocks, wire connectors, and splices, where used, shall be listed. [**20**:9.7(5)]

(6) A fire pump controller or fire pump power transfer switch, where provided, shall not be used as a junction box to supply other equipment, including a pressure maintenance (jockey) pump(s).

(J) Raceway Terminations. Where raceways are terminated at a fire pump controller, the following requirements shall be met: [**20**:9.9]

(1) Listed conduit hubs shall be used. [**20**:9.9.1]

(2) The type rating of the conduit hub(s) shall be at least equal to that of the fire pump controller. [**20**:9.9.2]

(3) The installation instructions of the manufacturer of the fire pump controller shall be followed. [**20**:9.9.3]

(4) Alterations to the fire pump controller, other than conduit entry as allowed elsewhere in this *Code*, shall be approved by the authority having jurisdiction. [**20**:9.9.4]

695.7 Voltage Drop.

(A) Starting. The voltage at the fire pump controller line terminals shall not drop more than 15 percent below normal (controller-rated voltage) under motor starting conditions.

Exception: This limitation shall not apply for emergency run mechanical starting. [20: 9.4.2]

(B) Running. The voltage at the load terminals of the fire pump controller shall not drop more than 5 percent below the voltage rating of the motor connected to those terminals when the motor is operating at 115 percent of the full-load current rating of the motor.

695.10 Listed Equipment. Diesel engine fire pump controllers, electric fire pump controllers, electric motors, fire pump power transfer switches, foam pump controllers, and limited service controllers shall be listed for fire pump service. [**20**:9.5.1.1, 10.1.2.1, 12.1.3.1]

695.12 Equipment Location.

(A) Controllers and Transfer Switches. Electric motor-driven fire pump controllers and power transfer switches shall be located as close as practicable to, and within sight of, the motors that they control.

(B) Engine-Drive Controllers. Engine-drive fire pump controllers shall be located as close as is practical to, and within sight of, the engines that they control.

(C) Storage Batteries. Storage batteries for fire pump engine drives shall be supported above the floor, secured against displacement, and located where they are not subject to physical damage, flooding with water, excessive temperature, or excessive vibration.

(D) Energized Equipment. All energized equipment parts shall be located at least 300 mm (12 in.) above the floor level.

(E) Protection Against Pump Water. Fire pump controller and power transfer switches shall be located or protected so that they are not damaged by water escaping from pumps or pump connections.

(F) Mounting. All fire pump control equipment shall be mounted in a substantial manner on noncombustible supporting structures.

695.14 Control Wiring.

(A) Control Circuit Failures. External control circuits that extend outside the fire pump room shall be arranged so that failure of any external circuit (open or short circuit) shall not prevent the operation of a pump(s) from all other internal or external means. Breakage, disconnecting, shorting of the wires, or loss of power to these circuits could cause continuous running of the fire pump but shall not prevent the controller(s) from starting the fire pump(s) due to causes other than these external control circuits. All control conductors within the fire pump room that are not fault tolerant shall be protected against physical damage. [**20**:10.5.2.6, 12.5.2.5]

(B) Sensor Functioning. No undervoltage, phase-loss, frequency-sensitive, or other sensor(s) shall be installed that automatically or manually prohibits actuation of the motor contactor. [**20**:10.4.5.6]

Exception: A phase-loss sensor(s) shall be permitted only as a part of a listed fire pump controller.

(C) Remote Device(s). No remote device(s) shall be installed that will prevent automatic operation of the transfer switch. [**20**:10.8.1.3]

(D) Engine-Drive Control Wiring. All wiring between the controller and the diesel engine shall be stranded and sized to continuously carry the charging or control currents as required by the controller manufacturer. Such wiring shall be protected against physical damage. Controller manufacturer's specifications for distance and wire size shall be followed. [**20**:12.3.5.1]

(E) Electric Fire Pump Control Wiring Methods. All electric motor–driven fire pump control wiring shall be in rigid metal conduit, intermediate metal conduit, liquidtight flexible metal conduit, electrical metallic tubing, liquidtight flexible nonmetallic conduit, listed Type MC cable with an impervious covering, or Type MI cable.

(F) Generator Control Wiring Methods. Control conductors installed between the fire pump power transfer switch and the standby generator supplying the fire pump during normal power loss shall be kept entirely independent of all other wiring. The integrity of the generator control wiring shall be continuously monitored. Loss of integrity of the remote start circuit(s) shall initiate visual and audible annunciation of generator malfunction at the generator local and remote annunciator(s) and start the generator(s).

Informational Note: See NFPA 20-2013, *Standard for the Installation of Stationary Pumps for Fire Protection*, Section 3.3.7.2, for more information on fault-tolerant external control circuits.

The control conductors shall be protected to resist potential damage by fire or structural failure. They shall be permitted to be routed through a building(s) using one of the following methods:

(1) Be encased in a minimum 50 mm (2 in.) of concrete.

(2) Be protected by a fire-rated assembly listed to achieve a minimum fire rating of 2 hours and dedicated to the fire pump circuits.

(3) Be a listed electrical circuit protective system with a minimum 2-hour fire rating. The installation shall comply with any restrictions provided in the listing of the electrical circuit protective system used.

Informational Note: The listing organization provides information for electrical circuit protective systems on proper installation requirements to maintain the fire rating.

N 695.15 Surge Protection. A listed surge protection device shall be installed in or on the fire pump controller.

Chapter 7 Special Conditions

ARTICLE 700
Emergency Systems

Part I. General

700.1 Scope. This article applies to the electrical safety of the installation, operation, and maintenance of emergency systems consisting of circuits and equipment intended to supply, distribute, and control electricity for illumination, power, or both, to required facilities when the normal electrical supply or system is interrupted.

> Informational Note No. 1: For further information regarding wiring and installation of emergency systems in health care facilities, see Article 517.

> Informational Note No. 2: For further information regarding performance and maintenance of emergency systems in health care facilities, see NFPA 99-2015, *Health Care Facilities Code.*

> Informational Note No. 3: For specification of locations where emergency lighting is considered essential to life safety, see NFPA 101-2015, *Life Safety Code.*

> Informational Note No. 4: For further information regarding performance of emergency and standby power systems, see NFPA 110-2013, *Standard for Emergency and Standby Power Systems.*

700.2 Definitions.

N **Branch Circuit Emergency Lighting Transfer Switch.** A device connected on the load side of a branch circuit overcurrent protective device that transfers only emergency lighting loads from the normal supply to an emergency supply.

> Informational Note: See ANSI/UL 1008, *Transfer Switch Equipment,* for information covering branch circuit emergency lighting transfer switches.

Emergency Systems. Those systems legally required and classed as emergency by municipal, state, federal, or other codes, or by any governmental agency having jurisdiction. These systems are intended to automatically supply illumination, power, or both, to designated areas and equipment in the event of failure of the normal supply or in the event of accident to elements of a system intended to supply, distribute, and control power and illumination essential for safety to human life.

> Informational Note: Emergency systems are generally installed in places of assembly where artificial illumination is required for safe exiting and for panic control in buildings subject to occupancy by large numbers of persons, such as hotels, theaters, sports arenas, health care facilities, and similar institutions. Emergency systems may also provide power for such functions as ventilation where essential to maintain life, fire detection and alarm systems, elevators, fire pumps, public safety communications systems, industrial processes where current interruption would produce serious life safety or health hazards, and similar functions.

N **Luminaire, Directly Controlled.** An emergency luminaire that has a control input for an integral dimming or switching function that drives the luminaire to full illumination upon loss of normal power.

N > Informational Note: See ANSI/UL 924, *Emergency Lighting and Power Equipment,* for information covering directly controlled luminaires.

Relay, Automatic Load Control. A device used to set normally dimmed or normally-off switched emergency lighting equipment to full power illumination levels in the event of a loss of the normal supply by bypassing the dimming/switching controls, and to return the emergency lighting equipment to normal status when the device senses the normal supply has been restored.

> Informational Note: See ANSI/UL 924, *Emergency Lighting and Power Equipment,* for the requirements covering automatic load control relays.

700.3 Tests and Maintenance.

(A) Conduct or Witness Test. The authority having jurisdiction shall conduct or witness a test of the complete system upon installation and periodically afterward.

(B) Tested Periodically. Systems shall be tested periodically on a schedule acceptable to the authority having jurisdiction to ensure the systems are maintained in proper operating condition.

(C) Maintenance. Emergency system equipment shall be maintained in accordance with manufacturer instructions and industry standards.

(D) Written Record. A written record shall be kept of such tests and maintenance.

(E) Testing Under Load. Means for testing all emergency lighting and power systems during maximum anticipated load conditions shall be provided.

> Informational Note: For information on testing and maintenance of emergency power supply systems (EPSSs), see NFPA 110-2013, *Standard for Emergency and Standby Power Systems.*

N **(F) Temporary Source of Power for Maintenance or Repair of the Alternate Source of Power.** If the emergency system relies on a single alternate source of power, which will be disabled for maintenance or repair, the emergency system shall include permanent switching means to connect a portable or temporary alternate source of power, which shall be available for the duration of the maintenance or repair. The permanent switching means to connect a portable or temporary alternate source of power shall comply with the following:

(1) Connection to the portable or temporary alternate source of power shall not require modification of the permanent system wiring.

(2) Transfer of power between the normal power source and the emergency power source shall be in accordance with 700.12.

(3) The connection point for the portable or temporary alternate source shall be marked with the phase rotation and system bonding requirements.

(4) Mechanical or electrical interlocking shall prevent inadvertent interconnection of power sources.

(5) The switching means shall include a contact point that shall annunciate at a location remote from the generator or at another facility monitoring system to indicate that the permanent emergency source is disconnected from the emergency system.

It shall be permissible to utilize manual switching to switch from the permanent source of power to the portable or temporary alternate source of power and to utilize the switching means for connection of a load bank.

Informational Note: There are many possible methods to achieve the requirements of 700.3(F). See Figure 700.3(F) for one example.

Exception: The permanent switching means to connect a portable or temporary alternate source of power, for the duration of the maintenance or repair, shall not be required where any of the following conditions exists:

(1) *All processes that rely on the emergency system source are capable of being disabled during maintenance or repair of the emergency source of power.*
(2) *The building or structure is unoccupied and fire suppression systems are fully functional and do not require an alternate power source.*
(3) *Other temporary means can be substituted for the emergency system.*
(4) *A permanent alternate emergency source, such as, but not limited to, a second on-site standby generator or separate electric utility service connection, capable of supporting the emergency system, exists.*

700.4 Capacity.

(A) Capacity and Rating. An emergency system shall have adequate capacity and rating for all loads to be operated simultaneously. The emergency system equipment shall be suitable for the maximum available fault current at its terminals.

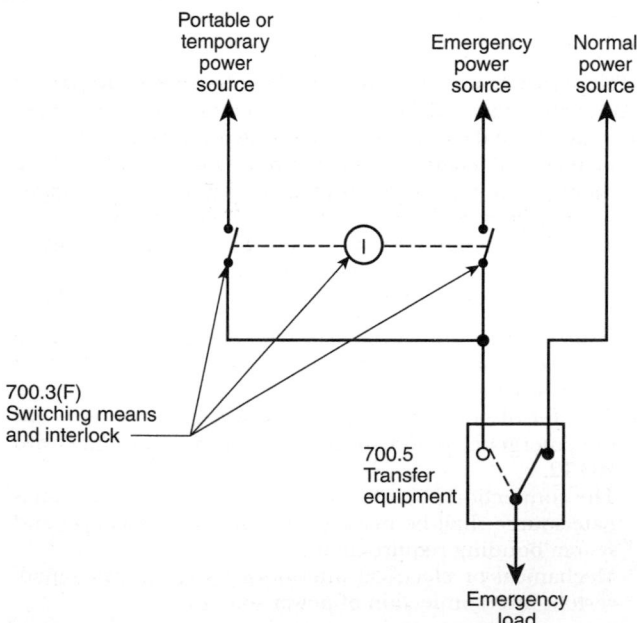

700.3(F)
Switching means
and interlock

Portable or
temporary
power
source

Emergency
power
source

Normal
power
source

700.5
Transfer
equipment

Emergency
load

N FIGURE 700.3(F)

(B) Selective Load Pickup, Load Shedding, and Peak Load Shaving. The alternate power source shall be permitted to supply emergency, legally required standby, and optional standby system loads where the source has adequate capacity or where automatic selective load pickup and load shedding is provided as needed to ensure adequate power to (1) the emergency circuits, (2) the legally required standby circuits, and (3) the optional standby circuits, in that order of priority. The alternate power source shall be permitted to be used for peak load shaving, provided these conditions are met.

Peak load shaving operation shall be permitted for satisfying the test requirement of 700.3(B), provided all other conditions of 700.3 are met.

700.5 Transfer Equipment.

(A) General. Transfer equipment, including automatic transfer switches, shall be automatic, identified for emergency use, and approved by the authority having jurisdiction. Transfer equipment shall be designed and installed to prevent the inadvertent interconnection of normal and emergency sources of supply in any operation of the transfer equipment. Transfer equipment and electric power production systems installed to permit operation in parallel with the normal source shall meet the requirements of Article 705.

(B) Bypass Isolation Switches. Means shall be permitted to bypass and isolate the transfer equipment. Where bypass isolation switches are used, inadvertent parallel operation shall be avoided.

(C) Automatic Transfer Switches. Automatic transfer switches shall be electrically operated and mechanically held. Automatic transfer switches shall be listed for emergency system use.

(D) Use. Transfer equipment shall supply only emergency loads.

N (E) Documentation. The short-circuit current rating of the transfer equipment, based on the specific overcurrent protective device type and settings protecting the transfer equipment, shall be field marked on the exterior of the transfer equipment.

700.6 Signals. Audible and visual signal devices shall be provided, where practicable, for the purpose described in 700.6(A) through (D).

(A) Malfunction. To indicate malfunction of the emergency source.

(B) Carrying Load. To indicate that the battery is carrying load.

(C) Not Functioning. To indicate that the battery charger is not functioning.

(D) Ground Fault. To indicate a ground fault in solidly grounded wye emergency systems of more than 150 volts to ground and circuit-protective devices rated 1000 amperes or more. The sensor for the ground-fault signal devices shall be located at, or ahead of, the main system disconnecting means for the emergency source, and the maximum setting of the signal devices shall be for a ground-fault current of 1200 amperes. Instructions on the course of action to be taken in event of indicated ground fault shall be located at or near the sensor location.

For systems with multiple emergency sources connected to a paralleling bus, the ground fault sensor shall be permitted to be at an alternative location.

700.7 Signs.

(A) Emergency Sources. A sign shall be placed at the service-entrance equipment, indicating type and location of each on-site emergency power source.

Exception: A sign shall not be required for individual unit equipment as specified in 700.12(F).

(B) Grounding. Where removal of a grounding or bonding connection in normal power source equipment interrupts the grounding electrode conductor connection to the alternate power source(s) grounded conductor, a warning sign shall be installed at the normal power source equipment stating:

WARNING
SHOCK HAZARD EXISTS IF GROUNDING
ELECTRODE CONDUCTOR OR BONDING JUMPER
CONNECTION IN THIS EQUIPMENT IS REMOVED
WHILE ALTERNATE SOURCE(S) IS ENERGIZED.

The warning sign(s) or label(s) shall comply with 110.21(B).

700.8 Surge Protection. A listed SPD shall be installed in or on all emergency systems switchboards and panelboards.

Part II. Circuit Wiring

700.10 Wiring, Emergency System.

(A) Identification. Emergency circuits shall be permanently marked so they will be readily identified as a component of an emergency circuit or system by the following methods:

(1) All boxes and enclosures (including transfer switches, generators, and power panels) for emergency circuits shall be permanently marked as a component of an emergency circuit or system.

(2) Where boxes or enclosures are not encountered, exposed cable or raceway systems shall be permanently marked to be identified as a component of an emergency circuit or system, at intervals not to exceed 7.6 m (25 ft).

Receptacles supplied from the emergency system shall have a distinctive color or marking on the receptacle cover plates or the receptacles.

(B) Wiring. Wiring of two or more emergency circuits supplied from the same source shall be permitted in the same raceway, cable, box, or cabinet. Wiring from an emergency source or emergency source distribution overcurrent protection to emergency loads shall be kept entirely independent of all other wiring and equipment, unless otherwise permitted in 700.10(B)(1) through (5):

(1) Wiring from the normal power source located in transfer equipment enclosures

(2) Wiring supplied from two sources in exit or emergency luminaires

(3) Wiring from two sources in a listed load control relay supplying exit or emergency luminaires, or in a common junction box, attached to exit or emergency luminaires

(4) Wiring within a common junction box attached to unit equipment, containing only the branch circuit supplying the unit equipment and the emergency circuit supplied by the unit equipment

(5) Wiring from an emergency source to supply emergency and other (nonemergency) loads in accordance with 700.10(B)(5)a., b., c., and d. as follows:

a. Separate vertical switchgear sections or separate vertical switchboard sections, with or without a common bus, or individual disconnects mounted in separate enclosures shall be used to separate emergency loads from all other loads.

b. The common bus of separate sections of the switchgear, separate sections of the switchboard, or the individual enclosures shall be either of the following:

(i) Supplied by single or multiple feeders without overcurrent protection at the source

(ii) Supplied by single or multiple feeders with overcurrent protection, provided that the overcurrent protection that is common to an emergency system and any non-emergency system(s) is selectively coordinated with the next downstream overcurrent protective device in the nonemergency system(s)

Informational Note: For further information, see Informational Note Figure 700.10(B)(5)(b)(1) and Informational Note Figure 700.10(B)(5)(b)(2).

c. Emergency circuits shall not originate from the same vertical switchgear section, vertical switchboard section, panelboard enclosure, or individual disconnect enclosure as other circuits.

d. It shall be permissible to utilize single or multiple feeders to supply distribution equipment between an emergency source and the point where the emergency loads are separated from all other loads.

(C) Wiring Design and Location. Emergency wiring circuits shall be designed and located so as to minimize the hazards that might cause failure due to flooding, fire, icing, vandalism, and other adverse conditions.

(D) Fire Protection. Emergency systems shall meet the additional requirements in (D)(1) through (D)(3) in the following occupancies:

(1) Assembly occupancies for not less than 1000 persons

(2) Buildings above 23 m (75 ft) in height

(3) Health care occupancies where persons are not capable of self preservation

(4) Educational occupancies with more than 300 occupants

(1) Feeder-Circuit Wiring. Feeder-circuit wiring shall meet one of the following conditions:

(1) The cable or raceway is installed in spaces or areas that are fully protected by an approved automatic fire suppression system.

(2) The cable or raceway is protected by a listed electrical circuit protective system with a minimum 2-hour fire rating.

Informational Note No. 1: Electrical circuit protective systems could include but not be limited to thermal barriers or a protective shaft and are tested to UL 1724, *Fire Tests for Electrical Circuit Protection Systems.*

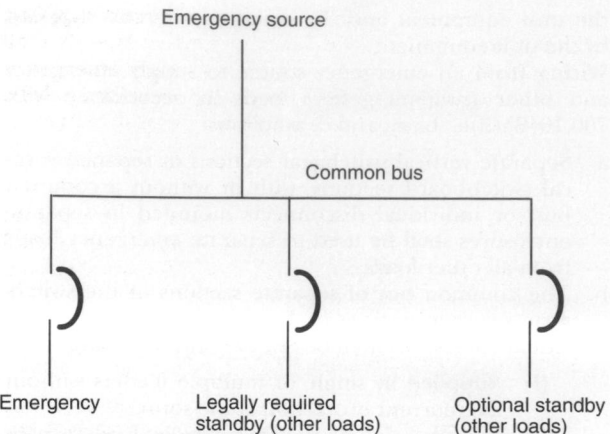

Informational Note Figure 700.10(B)(5)(b)(1)   Single or Multiple Feeders without Overcurrent Protection

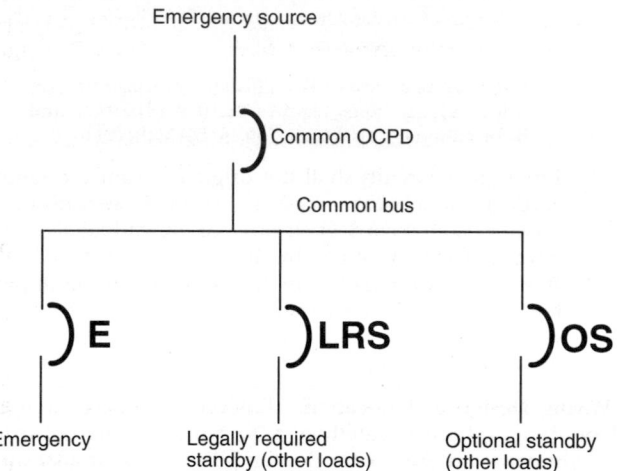

Informational Note Figure 700.10(B)(5)(b)(2)   Single or Multiple Feeders with Overcurrent Protection

> Informational Note No. 2: The listing organization provides information for electrical circuit protective systems on proper installation requirements to maintain the fire rating.

(3)   The cable or raceway is a listed fire-resistive cable system.

> Informational Note No. 1: Fire-resistive cables are tested to ANSI/UL 2196, *Tests for Fire Resistive Cables*.

> Informational Note No. 2: The listing organization provides information for fire-resistive cable systems on proper installation requirements to maintain the fire rating.

(4)   The cable or raceway is protected by a listed fire-rated assembly that has a minimum fire rating of 2 hours and contains only emergency circuits.

(5)   The cable or raceway is encased in a minimum of 50 mm (2 in.) of concrete.

(2) Feeder-Circuit Equipment. Equipment for feeder circuits (including transfer switches, transformers, and panelboards) shall be located either in spaces fully protected by approved automatic fire suppression systems (including sprinklers, carbon dioxide systems) or in spaces with a 2-hour fire resistance rating.

(3) Generator Control Wiring. Control conductors installed between the transfer equipment and the emergency generator shall be kept entirely independent of all other wiring and shall meet the conditions of 700.10(D)(1). The integrity of the generator control wiring shall be continuously monitored. Loss of integrity of the remote start circuit(s) shall initiate visual and audible annunciation of generator malfunction at the generator local and remote annunciator(s) and start the generator(s).

Part III. Sources of Power

700.12 General Requirements. Current supply shall be such that, in the event of failure of the normal supply to, or within, the building or group of buildings concerned, emergency lighting, emergency power, or both shall be available within the time required for the application but not to exceed 10 seconds. The supply system for emergency purposes, in addition to the normal services to the building and meeting the general requirements of this section, shall be one or more of the types of systems described in 700.12(A) through (E). Unit equipment in accordance with 700.12(F) shall satisfy the applicable requirements of this article.

In selecting an emergency source of power, consideration shall be given to the occupancy and the type of service to be rendered, whether of minimum duration, as for evacuation of a theater, or longer duration, as for supplying emergency power and lighting due to an indefinite period of current failure from trouble either inside or outside the building.

Equipment shall be designed and located so as to minimize the hazards that might cause complete failure due to flooding, fires, icing, and vandalism.

Equipment for sources of power as described in 700.12(A) through (E) shall be installed either in spaces fully protected by approved automatic fire suppression systems (sprinklers, carbon dioxide systems, and so forth) or in spaces with a 1-hour fire rating where located within the following:

(1)   Assembly occupancies for more than 1000 persons
(2)   Buildings above 23 m (75 ft) in height with any of the following occupancy classes — assembly, educational, residential, detention and correctional, business, and mercantile
(3)   Health care occupancies where persons are not capable of self-preservation
(4)   Educational occupancies with more than 300 occupants

> Informational Note No. 1: For the definition of *Occupancy Classification*, see Section 6.1 of NFPA *101*-2015, *Life Safety Code*.

> Informational Note No. 2: For further information, see ANSI/IEEE 493-2007, *Recommended Practice for the Design of Reliable Industrial and Commercial Power Systems*.

(A) Storage Battery. Storage batteries shall be of suitable rating and capacity to supply and maintain the total load for a minimum period of 1½ hours, without the voltage applied to the load falling below 87½ percent of normal. Automotive-type batteries shall not be used.

An automatic battery charging means shall be provided.

(B) Generator Set.

(1) Prime Mover-Driven. For a generator set driven by a prime mover acceptable to the authority having jurisdiction and sized in accordance with 700.4, means shall be provided for automatically starting the prime mover on failure of the normal service and for automatic transfer and operation of all required electrical circuits. A time-delay feature permitting a 15-minute setting shall be provided to avoid retransfer in case of short-time reestablishment of the normal source.

(2) Internal Combustion Engines as Prime Movers. Where internal combustion engines are used as the prime mover, an on-site fuel supply shall be provided with an on-premises fuel supply sufficient for not less than 2 hours' full-demand operation of the system. Where power is needed for the operation of the fuel transfer pumps to deliver fuel to a generator set day tank, this pump shall be connected to the emergency power system.

(3) Dual Supplies. Prime movers shall not be solely dependent on a public utility gas system for their fuel supply or municipal water supply for their cooling systems. Means shall be provided for automatically transferring from one fuel supply to another where dual fuel supplies are used.

Exception: Where acceptable to the authority having jurisdiction, the use of other than on-site fuels shall be permitted where there is a low probability of a simultaneous failure of both the off-site fuel delivery system and power from the outside electrical utility company.

(4) Battery Power and Dampers. Where a storage battery is used for control or signal power or as the means of starting the prime mover, it shall be suitable for the purpose and shall be equipped with an automatic charging means independent of the generator set. Where the battery charger is required for the operation of the generator set, it shall be connected to the emergency system. Where power is required for the operation of dampers used to ventilate the generator set, the dampers shall be connected to the emergency system.

(5) Auxiliary Power Supply. Generator sets that require more than 10 seconds to develop power shall be permitted if an auxiliary power supply energizes the emergency system until the generator can pick up the load.

(6) Outdoor Generator Sets. Where an outdoor housed generator set is equipped with a readily accessible disconnecting means in accordance with 445.18, and the disconnecting means is located within sight of the building or structure supplied, an additional disconnecting means shall not be required where ungrounded conductors serve or pass through the building or structure. Where the generator supply conductors terminate at a disconnecting means in or on a building or structure, the disconnecting means shall meet the requirements of 225.36.

Exception: For installations under single management, where conditions of maintenance and supervision ensure that only qualified persons will monitor and service the installation and where documented safe switching procedures are established and maintained for disconnection, the generator set disconnecting means shall not be required to be located within sight of the building or structure served.

(C) Uninterruptible Power Supplies. Uninterruptible power supplies used to provide power for emergency systems shall comply with the applicable provisions of 700.12(A) and (B).

(D) Separate Service. Where approved by the authority having jurisdiction as suitable for use as an emergency source of power, an additional service shall be permitted. This service shall be in accordance with the applicable provisions of Article 230 and the following additional requirements:

(1) Separate overhead service conductors, service drops, underground service conductors, or service laterals shall be installed.
(2) The service conductors for the separate service shall be installed sufficiently remote electrically and physically from any other service conductors to minimize the possibility of simultaneous interruption of supply.

(E) Fuel Cell System. Fuel cell systems used as a source of power for emergency systems shall be of suitable rating and capacity to supply and maintain the total load for not less than 2 hours of full-demand operation.

Installation of a fuel cell system shall meet the requirements of Parts II through VIII of Article 692.

Where a single fuel cell system serves as the normal supply for the building or group of buildings concerned, it shall not serve as the sole source of power for the emergency standby system.

(F) Unit Equipment.

(1) Components of Unit Equipment. Individual unit equipment for emergency illumination shall consist of the following:

(1) A rechargeable battery
(2) A battery charging means
(3) Provisions for one or more lamps mounted on the equipment, or shall be permitted to have terminals for remote lamps, or both
(4) A relaying device arranged to energize the lamps automatically upon failure of the supply to the unit equipment

(2) Installation of Unit Equipment. Unit equipment shall be installed in accordance with 700.12(F)(2)(1) through (6).

(1) The batteries shall be of suitable rating and capacity to supply and maintain the total lamp load associated with the unit in accordance with (a) or (b):
    (a) For a period of at least 1½ hours without the voltage falling below 87½ percent of normal battery voltage
    (b) The unit equipment shall supply and maintain not less than 60 percent of the initial emergency illumination for a period of at least 1½ hours
(2) Unit equipment shall be permanently fixed (i.e., not portable) in place and shall have all wiring to each unit installed in accordance with the requirements of any of the wiring methods in Chapter 3. Flexible cord-and-plug connection shall be permitted, provided that the cord does not exceed 900 mm (3 ft) in length.
(3) The branch circuit feeding the unit equipment shall be the same branch circuit as that serving the normal lighting in the area and connected ahead of any local switches.

Exception: In a separate and uninterrupted area supplied by a minimum of three normal lighting circuits that are not part of a multiwire branch circuit, a separate branch circuit for unit equipment shall be permitted if it originates from the same panelboard as that of the normal lighting circuits and is provided with a lock-on feature.

(4) The branch circuit that feeds unit equipment shall be clearly identified at the distribution panel.
(5) Emergency luminaires that obtain power from a unit equipment and are not part of the unit equipment shall be wired to the unit equipment as required by 700.10 and by one of the wiring methods of Chapter 3.

(6) Remote heads providing lighting for the exterior of an exit door shall be permitted to be supplied by the unit equipment serving the area immediately inside the exit door.

Part IV. Emergency System Circuits for Lighting and Power

700.15 Loads on Emergency Branch Circuits. No appliances and no lamps, other than those specified as required for emergency use, shall be supplied by emergency lighting circuits.

700.16 Emergency Illumination. Emergency illumination shall include means of egress lighting, illuminated exit signs, and all other luminaires specified as necessary to provide required illumination.

Emergency lighting systems shall be designed and installed so that the failure of any individual lighting element, such as the burning out of a lamp, cannot leave in total darkness any space that requires emergency illumination.

Where high-intensity discharge lighting such as high- and low-pressure sodium, mercury vapor, and metal halide is used as the sole source of normal illumination, the emergency lighting system shall be required to operate until normal illumination has been restored.

Where an emergency system is installed, emergency illumination shall be provided in the area of the disconnecting means required by 225.31 and 230.70, as applicable, where the disconnecting means are installed indoors.

Exception: Alternative means that ensure that the emergency lighting illumination level is maintained shall be permitted.

700.17 Branch Circuits for Emergency Lighting. Branch circuits that supply emergency lighting shall be installed to provide service from a source complying with 700.12 when the normal supply for lighting is interrupted. Such installations shall provide either of the following:

(1) An emergency lighting supply, independent of the normal lighting supply, with provisions for automatically transferring the emergency lights upon the event of failure of the normal lighting branch circuit

(2) Two or more branch circuits supplied from separate and complete systems with independent power sources. One of the two power sources and systems shall be part of the emergency system, and the other shall be permitted to be part of the normal power source and system. Each system shall provide sufficient power for emergency lighting purposes.

Unless both systems are used for regular lighting purposes and are both kept lighted, means shall be provided for automatically energizing either system upon failure of the other. Either or both systems shall be permitted to be a part of the general lighting of the protected occupancy if circuits supplying lights for emergency illumination are installed in accordance with other sections of this article.

700.18 Circuits for Emergency Power. For branch circuits that supply equipment classed as emergency, there shall be an emergency supply source to which the load will be transferred automatically upon the failure of the normal supply.

700.19 Multiwire Branch Circuits. The branch circuit serving emergency lighting and power circuits shall not be part of a multiwire branch circuit.

Part V. Control — Emergency Lighting Circuits

700.20 Switch Requirements. The switch or switches installed in emergency lighting circuits shall be arranged so that only authorized persons have control of emergency lighting.

Exception No. 1: Where two or more single-throw switches are connected in parallel to control a single circuit, at least one of these switches shall be accessible only to authorized persons.

Exception No. 2: Additional switches that act only to put emergency lights into operation but not disconnect them shall be permissible.

Switches connected in series or 3- and 4-way switches shall not be used.

700.21 Switch Location. All manual switches for controlling emergency circuits shall be in locations convenient to authorized persons responsible for their actuation. In facilities covered by Articles 518 and 520, a switch for controlling emergency lighting systems shall be located in the lobby or at a place conveniently accessible thereto.

In no case shall a control switch for emergency lighting be placed in a motion-picture projection booth or on a stage or platform.

Exception: Where multiple switches are provided, one such switch shall be permitted in such locations where arranged so that it can only energize the circuit but cannot de-energize the circuit.

700.22 Exterior Lights. Those lights on the exterior of a building that are not required for illumination when there is sufficient daylight shall be permitted to be controlled by an automatic light-actuated device.

700.23 Dimmer and Relay Systems. A dimmer or relay system containing more than one dimmer or relay and listed for use in emergency systems shall be permitted to be used as a control device for energizing emergency lighting circuits. Upon failure of normal power, the dimmer or relay system shall be permitted to selectively energize only those branch circuits required to provide minimum emergency illumination. All branch circuits supplied by the dimmer or relay system cabinet shall comply with the wiring methods of Article 700.

700.24 Directly Controlled Luminaires. Where emergency illumination is provided by one or more directly controlled luminaires that respond to an external control input to bypass normal control upon loss of normal power, such luminaires and external bypass controls shall be individually listed for use in emergency systems.

N **700.25 Branch Circuit Emergency Lighting Transfer Switch.** Emergency lighting loads supplied by branch circuits rated at not greater than 20 amperes shall be permitted to be transferred from the normal branch circuit to an emergency branch circuit using a listed branch circuit emergency lighting transfer switch. The mechanically held requirement of 700.5(C) shall not apply to listed branch circuit emergency lighting transfer switches.

700.26 Automatic Load Control Relay. If an emergency lighting load is automatically energized upon loss of the normal supply, a listed automatic load control relay shall be permitted to energize the load. The load control relay shall not be used as transfer equipment.

Part VI. Overcurrent Protection

700.30 Accessibility. The branch-circuit overcurrent devices in emergency circuits shall be accessible to authorized persons only.

700.31 Ground-Fault Protection of Equipment. The alternate source for emergency systems shall not be required to provide ground-fault protection of equipment with automatic disconnecting means. Ground-fault indication at the emergency source shall be provided in accordance with 700.6(D) if ground-fault protection of equipment with automatic disconnecting means is not provided.

700.32 Selective Coordination. Emergency system(s) overcurrent devices shall be selectively coordinated with all supply-side overcurrent protective devices.

Selective coordination shall be selected by a licensed professional engineer or other qualified persons engaged primarily in the design, installation, or maintenance of electrical systems. The selection shall be documented and made available to those authorized to design, install, inspect, maintain, and operate the system.

Exception: Selective coordination shall not be required between two overcurrent devices located in series if no loads are connected in parallel with the downstream device.

ARTICLE 701
Legally Required Standby Systems

Part I. General

701.1 Scope. The provisions of this article apply to the electrical safety of the installation, operation, and maintenance of legally required standby systems consisting of circuits and equipment intended to supply, distribute, and control electricity to required facilities for illumination or power, or both, when the normal electrical supply or system is interrupted.

The systems covered by this article consist only of those that are permanently installed in their entirety, including the power source.

Informational Note No. 1: For further information, see NFPA 99-2015, *Health Care Facilities Code.*

Informational Note No. 2: For further information regarding performance of emergency and standby power systems, see NFPA 110-2013, *Standard for Emergency and Standby Power Systems.*

Informational Note No. 3: For further information, see ANSI/IEEE 446-1995, *Recommended Practice for Emergency and Standby Power Systems for Industrial and Commercial Applications.*

701.2 Definition.

Legally Required Standby Systems. Those systems required and so classed as legally required standby by municipal, state, federal, or other codes or by any governmental agency having jurisdiction. These systems are intended to automatically supply power to selected loads (other than those classed as emergency systems) in the event of failure of the normal source.

Informational Note: Legally required standby systems are typically installed to serve loads, such as heating and refrigeration systems, communications systems, ventilation and smoke removal systems, sewage disposal, lighting systems, and industrial processes, that, when stopped during any interruption of the normal electrical supply, could create hazards or hamper rescue or fire-fighting operations.

701.3 Tests and Maintenance.

(A) Conduct or Witness Test. The authority having jurisdiction shall conduct or witness a test of the complete system upon installation.

(B) Tested Periodically. Systems shall be tested periodically on a schedule and in a manner acceptable to the authority having jurisdiction to ensure the systems are maintained in proper operating condition.

(C) Maintenance. Legally required standby system equipment shall be maintained in accordance with manufacturer instructions and industry standards.

(D) Written Record. A written record shall be kept on such tests and maintenance.

(E) Testing Under Load. Means for testing legally required standby systems under load shall be provided.

Informational Note: For information on testing and maintenance of emergency power supply systems (EPSSs), see NFPA 110-2013, *Standard for Emergency and Standby Power Systems.*

701.4 Capacity and Rating. A legally required standby system shall have adequate capacity and rating for the supply of all equipment intended to be operated at one time. Legally required standby system equipment shall be suitable for the maximum available fault current at its terminals.

The legally required standby alternate power source shall be permitted to supply both legally required standby and optional standby system loads under either of the following conditions:

(1) Where the alternate source has adequate capacity to handle all connected loads
(2) Where automatic selective load pickup and load shedding is provided that will ensure adequate power to the legally required standby circuits

701.5 Transfer Equipment.

(A) General. Transfer equipment, including automatic transfer switches, shall be automatic and identified for standby use and approved by the authority having jurisdiction. Transfer equipment shall be designed and installed to prevent the inadvertent interconnection of normal and alternate sources of supply in any operation of the transfer equipment. Transfer equipment and electric power production systems installed to permit operation in parallel with the normal source shall meet the requirements of Article 705.

(B) Bypass Isolation Switches. Means to bypass and isolate the transfer switch equipment shall be permitted. Where bypass isolation switches are used, inadvertent parallel operation shall be avoided.

(C) Automatic Transfer Switches. Automatic transfer switches shall be electrically operated and mechanically held. Automatic transfer switches shall be listed for emergency use.

N **(D) Documentation.** The short-circuit current rating of the transfer equipment, based on the specific overcurrent protective device type and settings protecting the transfer equipment, shall be field marked on the exterior of the transfer equipment.

701.6 Signals. Audible and visual signal devices shall be provided, where practicable, for the purposes described in 701.6(A), (B), (C), and (D).

(A) Malfunction. To indicate malfunction of the standby source.

(B) Carrying Load. To indicate that the standby source is carrying load.

(C) Not Functioning. To indicate that the battery charger is not functioning.

Informational Note: For signals for generator sets, see NFPA 110-2013, *Standard for Emergency and Standby Power Systems.*

(D) Ground Fault. To indicate a ground fault in solidly grounded wye, legally required standby systems of more than 150 volts to ground and circuit-protective devices rated 1000 amperes or more. The sensor for the ground-fault signal devices shall be located at, or ahead of, the main system disconnecting means for the legally required standby source, and the maximum setting of the signal devices shall be for a ground-fault current of 1200 amperes. Instructions on the course of action to be taken in event of indicated ground fault shall be located at or near the sensor location.

For systems with multiple emergency sources connected to a paralleling bus, the ground fault sensor shall be permitted at an alternate location.

Informational Note: For signals for generator sets, see NFPA 110-2013, *Standard for Emergency and Standby Power Systems.*

701.7 Signs.

(A) Mandated Standby. A sign shall be placed at the service entrance indicating type and location of each on-site legally required standby power source.

Exception: A sign shall not be required for individual unit equipment as specified in 701.12(G).

(B) Grounding. Where removal of a grounding or bonding connection in normal power source equipment interrupts the grounding electrode conductor connection to the alternate power source(s) grounded conductor, a warning sign shall be installed at the normal power source equipment stating:

WARNING
SHOCK HAZARD EXISTS IF GROUNDING
ELECTRODE CONDUCTOR OR BONDING JUMPER
CONNECTION IN THIS EQUIPMENT IS REMOVED
WHILE ALTERNATE SOURCE(S) IS ENERGIZED.

The warning sign(s) or label(s) shall comply with 110.21(B).

Part II. Circuit Wiring

701.10 Wiring Legally Required Standby Systems. The legally required standby system wiring shall be permitted to occupy the same raceways, cables, boxes, and cabinets with other general wiring.

Part III. Sources of Power

701.12 General Requirements. Current supply shall be such that, in the event of failure of the normal supply to, or within, the building or group of buildings concerned, legally required standby power will be available within the time required for the application but not to exceed 60 seconds. The supply system for legally required standby purposes, in addition to the normal services to the building, shall be permitted to comprise one or more of the types of systems described in 701.12(A) through (F). Unit equipment in accordance with 701.12(G) shall satisfy the applicable requirements of this article.

In selecting a legally required standby source of power, consideration shall be given to the type of service to be rendered, whether of short-time duration or long duration.

Consideration shall be given to the location or design, or both, of all equipment to minimize the hazards that might cause complete failure due to floods, fires, icing, and vandalism.

Informational Note: For further information, see ANSI/IEEE 493-2007, *Recommended Practice for the Design of Reliable Industrial and Commercial Power Systems.*

(A) Storage Battery. Storage batteries shall be of suitable rating and capacity to supply and maintain the total load for a minimum period of 1½ hours without the voltage applied to the load falling below 87½ percent of normal. Automotive-type batteries shall not be used.

An automatic battery charging means shall be provided.

(B) Generator Set.

(1) Prime Mover-Driven. For a generator set driven by a prime mover acceptable to the authority having jurisdiction and sized in accordance with 701.4, means shall be provided for automatically starting the prime mover upon failure of the normal service and for automatic transfer and operation of all required electrical circuits. A time-delay feature permitting a 15-minute setting shall be provided to avoid retransfer in case of short-time re-establishment of the normal source.

(2) Internal Combustion Engines as Prime Mover. Where internal combustion engines are used as the prime mover, an on-site fuel supply shall be provided with an on-premises fuel supply sufficient for not less than 2 hours of full-demand operation of the system. Where power is needed for the operation of the fuel transfer pumps to deliver fuel to a generator set day tank, the pumps shall be connected to the legally required standby power system.

(3) Dual Supplies. Prime movers shall not be solely dependent on a public utility gas system for their fuel supply or on a municipal water supply for their cooling systems. Means shall be provided for automatically transferring one fuel supply to another where dual fuel supplies are used.

Exception: Where acceptable to the authority having jurisdiction, the use of other than on-site fuels shall be permitted where there is a low probability of a simultaneous failure of both the off-site fuel delivery system and power from the outside electrical utility company.

(4) Battery Power. Where a storage battery is used for control or signal power or as the means of starting the prime mover, it shall be suitable for the purpose and shall be equipped with an automatic charging means independent of the generator set.

(5) Outdoor Generator Sets. Where an outdoor housed generator set is equipped with a readily accessible disconnecting means in accordance with 445.18, and the disconnecting means is located within sight of the building or structure supplied, an additional disconnecting means shall not be required where ungrounded conductors serve or pass through the building or structure. Where the generator supply conductors terminate at a disconnecting means in or on a building or structure, the disconnecting means shall meet the requirements of 225.36.

(C) Uninterruptible Power Supplies. Uninterruptible power supplies used to provide power for legally required standby systems shall comply with the applicable provisions of 701.12(A) and (B).

(D) Separate Service. Where approved, a separate service shall be permitted as a legally required source of standby power. This service shall be in accordance with the applicable provisions of Article 230, with a separate service drop or lateral or a separate set of overhead or underground service conductors sufficiently remote electrically and physically from any other service to minimize the possibility of simultaneous interruption of supply from an occurrence in another service.

(E) Connection Ahead of Service Disconnecting Means. Where acceptable to the authority having jurisdiction, connections located ahead of and not within the same cabinet, enclosure, vertical switchgear section, or vertical switchboard section as the service disconnecting means shall be permitted. The legally required standby service shall be sufficiently separated from the normal main service disconnecting means to minimize simultaneous interruption of supply through an occurrence within the building or groups of buildings served.

> Informational Note: See 230.82 for equipment permitted on the supply side of a service disconnecting means.

(F) Fuel Cell System. Fuel cell systems used as a source of power for legally required standby systems shall be of suitable rating and capacity to supply and maintain the total load for not less than 2 hours of full-demand operation.

Installation of a fuel cell system shall meet the requirements of Parts II through VIII of Article 692.

Where a single fuel cell system serves as the normal supply for the building or group of buildings concerned, it shall not serve as the sole source of power for the legally required standby system.

(G) Unit Equipment. Individual unit equipment for legally required standby illumination shall consist of the following:

(1) A rechargeable battery
(2) A battery charging means
(3) Provisions for one or more lamps mounted on the equipment and shall be permitted to have terminals for remote lamps
(4) A relaying device arranged to energize the lamps automatically upon failure of the supply to the unit equipment

The batteries shall be of suitable rating and capacity to supply and maintain the total lamp load associated with the unit for not less than (a) or (b):

(a) For a period of 1½ hours, without the voltage falling below 87½ percent of normal voltage

(b) The unit equipment shall supply and maintain not less than 60 percent of the initial emergency illumination for a period of at least 1½ hours.

Unit equipment shall be permanently fixed in place (i.e., not portable) and shall have all wiring to each unit installed in accordance with the requirements of any of the wiring methods in Chapter 3. Flexible cord-and-plug connection shall be permitted, provided that the cord does not exceed 900 mm (3 ft) in length. The branch circuit feeding the unit equipment shall be the same branch circuit as that serving the normal lighting in the area and connected ahead of any local switches. Legally required standby luminaires that obtain power from a unit equipment and are not part of the unit equipment shall be wired to the unit equipment by one of the wiring methods of Chapter 3.

Exception: In a separate and uninterrupted area supplied by a minimum of three normal lighting circuits, a separate branch circuit for unit equipment shall be permitted if it originates from the same panelboard as that of the normal lighting circuits and is provided with a lock-on feature.

Part IV. Overcurrent Protection

701.25 Accessibility. The branch-circuit overcurrent devices in legally required standby circuits shall be accessible to authorized persons only.

701.26 Ground-Fault Protection of Equipment. The alternate source for legally required standby systems shall not be required to provide ground-fault protection of equipment with automatic disconnecting means. Ground-fault indication at the legally required standby source shall be provided in accordance with 701.6(D) if ground-fault protection of equipment with automatic disconnecting means is not provided.

701.27 Selective Coordination. Legally required standby system(s) overcurrent devices shall be selectively coordinated with all supply-side overcurrent protective devices.

Selective coordination shall be selected by a licensed professional engineer or other qualified persons engaged primarily in the design, installation, or maintenance of electrical systems. The selection shall be documented and made available to those authorized to design, install, inspect, maintain, and operate the system.

Exception: Selective coordination shall not be required between two overcurrent devices located in series if no loads are connected in parallel with the downstream device.

ARTICLE 702
Optional Standby Systems

Part I. General

702.1 Scope. The provisions of this article apply to the installation and operation of optional standby systems.

The systems covered by this article consist of those that are permanently installed in their entirety, including prime movers, and those that are arranged for a connection to a premises wiring system from a portable alternate power supply.

702.2 Definition.

Optional Standby Systems. Those systems intended to supply power to public or private facilities or property where life safety does not depend on the performance of the system. These systems are intended to supply on-site generated power to selected loads either automatically or manually.

> Informational Note: Optional standby systems are typically installed to provide an alternate source of electric power for such facilities as industrial and commercial buildings, farms, and residences and to serve loads such as heating and refrigeration systems, data processing and communications systems, and industrial processes that, when stopped during any power outage, could cause discomfort, serious interruption of the process, damage to the product or process, or the like.

702.4 Capacity and Rating.

(A) Available Short-Circuit Current. Optional standby system equipment shall be suitable for the maximum available short-circuit current at its terminals.

(B) System Capacity. The calculations of load on the standby source shall be made in accordance with Article 220 or by another approved method.

(1) Manual Transfer Equipment. Where manual transfer equipment is used, an optional standby system shall have adequate capacity and rating for the supply of all equipment intended to be operated at one time. The user of the optional standby system shall be permitted to select the load connected to the system.

(2) Automatic Transfer Equipment. Where automatic transfer equipment is used, an optional standby system shall comply with (2)(a) or (2)(b).

(a) *Full Load.* The standby source shall be capable of supplying the full load that is transferred by the automatic transfer equipment.

(b) *Load Management.* Where a system is employed that will automatically manage the connected load, the standby source shall have a capacity sufficient to supply the maximum load that will be connected by the load management system.

702.5 Transfer Equipment.
Transfer equipment shall be suitable for the intended use and designed and installed so as to prevent the inadvertent interconnection of normal and alternate sources of supply in any operation of the transfer equipment. Transfer equipment and electric power production systems installed to permit operation in parallel with the normal source shall meet the requirements of Article 705.

Transfer equipment, located on the load side of branch circuit protection, shall be permitted to contain supplemental overcurrent protection having an interrupting rating sufficient for the available fault current that the generator can deliver. The supplementary overcurrent protection devices shall be part of a listed transfer equipment.

Transfer equipment shall be required for all standby systems subject to the provisions of this article and for which an electric utility supply is either the normal or standby source.

Exception: Temporary connection of a portable generator without transfer equipment shall be permitted where conditions of maintenance and supervision ensure that only qualified persons service the installation

and where the normal supply is physically isolated by a lockable disconnecting means or by disconnection of the normal supply conductors.

The short-circuit current rating of the transfer equipment, based on the specific overcurrent protective device type and settings protecting the transfer equipment, shall be field marked on the exterior of the transfer equipment.

702.6 Signals. Audible and visual signal devices shall be provided, where practicable, for the following purposes specified in 702.6(A) and (B).

(A) Malfunction. To indicate malfunction of the optional standby source.

(B) Carrying Load. To indicate that the optional standby source is carrying load.

Exception: Signals shall not be required for portable standby power sources.

702.7 Signs.

(A) Standby. A sign shall be placed at the service-entrance equipment that indicates the type and location of each on-site optional standby power source. A sign shall not be required for individual unit equipment for standby illumination.

(B) Grounding. Where removal of a grounding or bonding connection in normal power source equipment interrupts the grounding electrode conductor connection to the alternate power source(s) grounded conductor, a warning sign shall be installed at the normal power source equipment stating:

> WARNING
> SHOCK HAZARD EXISTS IF GROUNDING
> ELECTRODE CONDUCTOR OR BONDING JUMPER
> CONNECTION IN THIS EQUIPMENT IS REMOVED
> WHILE ALTERNATE SOURCE(S) IS ENERGIZED.

The warning sign(s) or label(s) shall comply with 110.21(B).

(C) Power Inlet. Where a power inlet is used for a temporary connection to a portable generator, a warning sign shall be placed near the inlet to indicate the type of derived system that the system is capable of based on the wiring of the transfer equipment. The sign shall display one of the following warnings:

> WARNING:
> FOR CONNECTION OF A SEPARATELY DERIVED
> (BONDED NEUTRAL) SYSTEM ONLY
> or
> WARNING:
> FOR CONNECTION OF A NONSEPARATELY DERIVED
> (FLOATING NEUTRAL) SYSTEM ONLY

Part II. Wiring

702.10 Wiring Optional Standby Systems. The optional standby system wiring shall be permitted to occupy the same raceways, cables, boxes, and cabinets with other general wiring.

702.11 Portable Generator Grounding.

(A) Separately Derived System. Where a portable optional standby source is used as a separately derived system, it shall be grounded to a grounding electrode in accordance with 250.30.

(B) Nonseparately Derived System. Where a portable optional standby source is used as a nonseparately derived system, the equipment grounding conductor shall be bonded to the system grounding electrode.

702.12 Outdoor Generator Sets.

(A) Portable Generators Greater Than 15 kW and Permanently Installed Generators. Where an outdoor housed generator set is equipped with a readily accessible disconnecting means in accordance with 445.18, and the disconnecting means is located within sight of the building or structure supplied, an additional disconnecting means shall not be required where ungrounded conductors serve or pass through the building or structure. Where the generator supply conductors terminate at a disconnecting means in or on a building or structure, the disconnecting means shall meet the requirements of 225.36.

(B) Portable Generators 15 kW or Less. Where a portable generator, rated 15 kW or less, is installed using a flanged inlet or other cord- and plug-type connection, a disconnecting means shall not be required where ungrounded conductors serve or pass through a building or structure.

N (C) Power Inlets Rated at 100 Amperes or Greater, for Portable Generators. Equipment containing power inlets for the connection of a generator source shall be listed for the intended use. Systems with power inlets shall be equipped with an interlocked disconnecting means.

Exception No. 1: If the inlet device is rated as a disconnecting means

Exception No. 2: Supervised industrial installations where permanent space is identified for the portable generator located within line of sight of the power inlets shall not be required to have interlocked disconnecting means nor inlets rated as disconnects.

ARTICLE 705
Interconnected Electric Power Production Sources

Part I. General

705.1 Scope. This article covers installation of one or more electric power production sources operating in parallel with a primary source(s) of electricity.

Informational Note: Examples of the types of primary sources include a utility supply or an on-site electric power source(s).

705.2 Definitions.

Interactive Inverter Out⌐ ⌐ircuit. The conductors between the interactive inverter ⌐ the service equipment or another electric power produ ⌐urce, such as a utility, for electrical production and dis ⌐ n network.

Microgrid Interc ⌐ Device (MID). A device that allows a microgrid syste⌐ ⌐arate from and reconnect to a primary power source.

**Microgrid ⌐ ⌐ premises wiring system that has genera-tion, ener ⌐ , and load(s), or any combination thereof,

that includes the ability to disconnect from and parallel with the primary source.

Informational Note: The application of Article 705 to microgrid systems is limited by the exclusions in 90.2(B)(5) related to electric utilities.

Multimode Inverter. Equipment having the capabilities of both the interactive inverter and the stand-alone inverter.

Power Production Equipment. The generating source, and all distribution equipment associated with it, that generates electricity from a source other than a utility supplied service.

Informational Note: Examples of power production equipment include such items as generators, solar photovoltaic systems, and fuel cell systems.

705.3 Other Articles. Interconnected electric power production sources shall comply with this article and also with the applicable requirements of the articles in Table 705.3.

705.6 Equipment Approval. All equipment shall be approved for the intended use. Interactive inverters for interconnection to systems interactive equipment intended to operate in parallel with the electric power system including, but not limited to, interactive inverters, engine generators, energy storage equipment, and wind turbines shall be listed and or field labeled for the intended use of interconnection service.

705.8 System Installation. Installation of one or more electrical power production sources operating in parallel with a primary source(s) of electricity shall be performed only by qualified persons.

Informational Note: See Article 100 for the definition of *Qualified Person.*

705.10 Directory. A permanent plaque or directory denoting the location of all electric power source disconnecting means on or in the premises shall be installed at each service equipment location and at the location(s) of the system disconnect(s) for all electric power production sources capable of being interconnected. The marking shall comply with 110.21(B).

Exception: Installations with large numbers of power production sources shall be permitted to be designated by groups.

705.12 Point of Connection. The output of an interconnected electric power source shall be connected as specified in 705.12(A) or (B).

Table 705.3 Other Articles

Equipment/System	Article
Generators	445
Solar photovoltaic systems	690
Fuel cell systems	692
Wind electric systems	694
Emergency systems	700
Legally required standby systems	701
Optional standby systems	702
Energy storage systems	706
Stand-alone systems	710
DC microgrids	712

(A) Supply Side. An electric power production source shall be permitted to be connected to the supply side of the service disconnecting means as permitted in 230.82(6). The sum of the ratings of all overcurrent devices connected to power production sources shall not exceed the rating of the service.

(B) Load Side. The output of an interconnected electric power source shall be permitted to be connected to the load side of the service disconnecting means of the other source(s) at any distribution equipment on the premises. Where distribution equipment, including switchgear, switchboards, or panelboards, is fed simultaneously by a primary source(s) of electricity and one or more other power source(s), and where this distribution equipment is capable of supplying multiple branch circuits or feeders, or both, the interconnecting provisions for other power sources shall comply with 705.12(B)(1) through (B)(5).

(1) Dedicated Overcurrent and Disconnect. Each source interconnection of one or more power sources installed in one system shall be made at a dedicated circuit breaker or fusible disconnecting means.

(2) Bus or Conductor Ampere Rating. One hundred twenty-five percent of the power source output circuit current shall be used in ampacity calculations for the following:

(1) *Feeders.* Where the power source output connection is made to a feeder at a location other than the opposite end of the feeder from the primary source overcurrent device, that portion of the feeder on the load side of the power source output connection shall be protected by one of the following:

    a. The feeder ampacity shall be not less than the sum of the primary source overcurrent device and 125 percent of the power source output circuit current.

    b. An overcurrent device on the load side of the power source connection shall be rated not greater than the ampacity of the feeder.

(2) *Taps.* In systems where power source output connections are made at feeders, any taps shall be sized based on the sum of 125 percent of the power source(s) output circuit current and the rating of the overcurrent device protecting the feeder conductors as calculated in 240.21(B).

(3) *Busbars.* One of the methods that follows shall be used to determine the ratings of busbars in panelboards.

    (a) The sum of 125 percent of the power source(s) output circuit current and the rating of the overcurrent device protecting the busbar shall not exceed the ampacity of the busbar.

> Informational Note: This general rule assumes no limitation in the number of the loads or sources applied to busbars or their locations.

    (b) Where two sources, one a primary power source and the other another power source, are located at opposite ends of a busbar that contains loads, the sum of 125 percent of the power source(s) output circuit current and the rating of the overcurrent device protecting the busbar shall not exceed 120 percent of the ampacity of the busbar. The busbar shall be sized for the loads connected in accordance with Article 220. A permanent warning label shall be applied to the distribution equipment adjacent to the back-fed breaker from the power source that displays the following or equivalent wording:

> WARNING:
> POWER SOURCE OUTPUT CONNECTION —
> DO NOT RELOCATE THIS OVERCURRENT DEVICE.

The warning sign(s) or label(s) shall comply with 110.21(B).

    (c) The sum of the ampere ratings of all overcurrent devices on panelboards, both load and supply devices, excluding the rating of the overcurrent device protecting the busbar, shall not exceed the ampacity of the busbar. The rating of the overcurrent device protecting the busbar shall not exceed the rating of the busbar. Permanent warning labels shall be applied to distribution equipment displaying the following or equivalent wording:

> WARNING:
> THIS EQUIPMENT FED BY MULTIPLE SOURCES.
> TOTAL RATING OF ALL OVERCURRENT DEVICES
> EXCLUDING MAIN SUPPLY OVERCURRENT DEVICE
> SHALL NOT EXCEED AMPACITY OF BUSBAR.

The warning sign(s) or label(s) shall comply with 110.21(B).

    (d) A connection at either end, but not both ends, of a center-fed panelboard in dwellings shall be permitted where the sum of 125 percent of the power source(s) output circuit current and the rating of the overcurrent device protecting the busbar does not exceed 120 percent of the current rating of the busbar.

    (e) Connections shall be permitted on multiple-ampacity busbars where designed under engineering supervision that includes available fault current and busbar load calculations.

(3) Marking. Equipment containing overcurrent devices in circuits supplying power to a busbar or conductor supplied from multiple sources shall be marked to indicate the presence of all sources.

(4) Suitable for Backfeed. Circuit breakers, if backfed, shall be suitable for such operation.

> Informational Note: Fused disconnects, unless otherwise marked, are suitable for backfeeding.

(5) Fastening. Listed plug-in-type circuit breakers backfed from electric power sources that are listed and identified as interactive shall be permitted to omit the additional fastener normally required by 408.36(D) for such applications.

705.14 Output Characteristics. The output of a generator or other electric power production source operating in parallel with an electrical supply system shall be compatible with the voltage, wave shape, and frequency of the system to which it is connected.

> Informational Note: The term *compatible* does not necessarily mean matching the primary source wave shape.

705.16 Interrupting and Short-Circuit Current Rating. Consideration shall be given to the contribution of fault currents from all interconnected power sources for the interrupting and short-circuit current ratings of equipment on interactive systems.

705.20 Disconnecting Means, Sources. Means shall be provided to disconnect all ungrounded conductors of an electric power production source(s) from all other conductors.

705.21 Disconnecting Means, Equipment. Means shall be provided to disconnect power production equipment, such as interactive inverters or transformers associated with a power production source, from all ungrounded conductors of all sources of supply. Equipment intended to be operated and maintained as an integral part of a power production source exceeding 1000 volts shall not be required to have a disconnecting means.

705.22 Disconnect Device. The disconnecting means for ungrounded conductors shall consist of a manual or power operated switch(es) or circuit breaker(s) that complies with the following:

(1) Located where readily accessible
(2) Externally operable without exposing the operator to contact with live parts and, if power operated, of a type that is opened by hand in the event of a power-supply failure
(3) Plainly indicate whether in the open (off) or closed (on) position
(4) Have ratings sufficient for the maximum circuit current, available short-circuit current, and voltage that is available at the terminals
(5) Where the line and load terminals are capable of being energized in the open position, marked in accordance with the warning in 690.13(B)

Informational Note : In parallel generation systems, some equipment, including knife blade switches and fuses, is likely to be energized from both directions. See 240.40.

(6) Simultaneously disconnect all ungrounded conductors of the circuit
(7) Be lockable in the open (off) position in accordance with 110.25

705.23 Interactive System Disconnecting Means. A readily accessible means shall be provided to disconnect the interactive system from all wiring systems including power systems, energy storage systems, and utilization equipment and its associated premises wiring.

705.30 Overcurrent Protection. Conductors shall be protected in accordance with Article 240. Equipment and conductors connected to more than one electrical source shall have a sufficient number of overcurrent devices located so as to provide protection from all sources.

(A) Solar Photovoltaic Systems. Solar photovoltaic systems shall be protected in accordance with Article 690.

(B) Transformers. Overcurrent protection for a transformer with a source(s) on each side shall be provided in accordance with 450.3 by considering first one side of the transformer, then the other side of the transformer, as the primary.

(C) Fuel Cell Systems. Fuel cell systems shall be protected in accordance with Article 692.

(D) Interactive Inverters. Interactive inverters shall be protected in accordance with 705.65.

(E) Generators. Generators shall be protected in accordance with 705.130.

705.31 Location of Overcurrent Protection. Overcurrent protection for electric power production source conductors, connected to the supply side of the service disconnecting means in accordance with 705.12(A), shall be located within 3 m (10 ft) of the point where the electric power production source conductors are connected to the service.

Informational Note: This overcurrent protection protects against short-circuit current supplied from the primary source(s) of electricity.

Exception: Where the overcurrent protection for the power production source is located more than 3 m (10 ft) from the point of connection for the electric power production source to the service, cable limiters or current-limited circuit breakers for each ungrounded conductor shall be installed at the point where the electric power production conductors are connected to the service.

705.32 Ground-Fault Protection. Where ground-fault protection is used, the output of an interactive system shall be connected to the supply side of the ground-fault protection.

Exception: Connection shall be permitted to be made to the load side of ground-fault protection, if there is ground-fault protection for equipment from all ground-fault current sources.

705.40 Loss of Primary Source. Upon loss of primary source, an electric power production source shall be automatically disconnected from all ungrounded conductors of the primary source and shall not be reconnected until the primary source is restored.

Exception: A listed interactive inverter shall be permitted to automatically cease exporting power upon loss of primary source and shall not be required to automatically disconnect all ungrounded conductors from the primary source. A listed interactive inverter shall be permitted to automatically or manually resume exporting power to the utility once the primary source is restored.

Informational Note No. 1: Risks to personnel and equipment associated with the primary source could occur if an utility interactive electric power production source can operate as an intentional island. Special detection methods are required to determine that a primary source supply system outage has occurred and whether there should be automatic disconnection. When the primary source supply system is restored, special detection methods can be required to limit exposure of power production sources to out-of-phase reconnection.

Informational Note No. 2: Induction-generating equipment on systems with significant capacitance can become self-excited upon loss of the primary source and experience severe overvoltage as a result.

An interactive inverter shall be permitted to operate as a stand-alone system to supply loads that have been disconnected from electrical production and distribution network sources.

705.42 Loss of 3-Phase Primary Source. A 3-phase electric power production source shall be automatically disconnected from all ungrounded conductors of the interconnected systems when one of the phases of that source opens. This requirement shall not be applicable to an electric power production source providing power for an emergency or legally required standby system.

Exception: A listed interactive inverter shall be permitted to automatically cease exporting power when one of the phases of the source opens and shall not be required to automatically disconnect all ungrounded conductors from the primary source. A listed interactive inverter shall be permitted to automatically or manually resume exporting power to the utility once all phases of the source are restored.

705.50 Grounding. Interconnected electric power production sources shall be grounded in accordance with Article 250.

Exception: For direct-current systems connected through an inverter directly to a grounded service, other methods that accomplish equivalent system protection and that utilize equipment listed and identified for the use shall be permitted.

Part II. Interactive Inverters

705.60 Circuit Sizing and Current.

(A) Calculation of Maximum Circuit Current. The maximum current for the specific circuit shall be calculated in accordance with 705.60(A)(1) and (A)(2).

(1) Inverter Input Circuit Currents. The maximum current shall be the maximum rated input current of the inverter.

(2) Inverter Output Circuit Current. The maximum current shall be the inverter continuous output current rating.

(B) Ampacity and Overcurrent Device Ratings. Inverter system currents shall be considered to be continuous. The circuit conductors and overcurrent devices shall be sized to carry not less than 125 percent of the maximum currents as calculated in 705.60(A). The rating or setting of overcurrent devices shall be permitted in accordance with 240.4(B) and (C).

Exception: Circuits containing an assembly together with its overcurrent device(s) that is listed for continuous operation at 100 percent of its rating shall be permitted to be utilized at 100 percent of its rating.

705.65 Overcurrent Protection.

(A) Circuits and Equipment. Inverter input circuits, inverter output circuits, and storage battery circuit conductors and equipment shall be protected in accordance with the requirements of Article 240. Circuits connected to more than one electrical source shall have overcurrent devices located so as to provide overcurrent protection from all sources.

Exception: An overcurrent device shall not be required for circuit conductors sized in accordance with 705.60(B) and located where one of the following applies:

(1) There are no external sources such as parallel-connected source circuits, batteries, or backfeed from inverters.

(2) The short-circuit currents from all sources do not exceed the ampacity of the conductors.

> Informational Note: Possible backfeed of current from any source of supply, including a supply through an inverter into the inverter output circuit and inverter source circuits, is a consideration in determining whether adequate overcurrent protection from all sources is provided for conductors and modules.

(B) Power Transformers. Overcurrent protection for a transformer with a source(s) on each side shall be provided in accordance with 450.3 by considering first one side of the transformer, then the other side of the transformer, as the primary.

Exception: A power transformer with a current rating on the side connected toward the interactive inverter output that is not less than the rated continuous output current of the inverter shall be permitted without overcurrent protection from that source.

N **(C) Conductor Ampacity.** Power source output circuit conductors that are connected to a feeder, if smaller than the feeder conductors, shall be sized to carry not less than the larger of the current as calculated in 705.60(B) or as calculated in accordance with 240.21(B) based on the over-current device protecting the feeder.

705.70 Interactive Inverters Mounted in Not Readily Accessible Locations. Interactive inverters shall be permitted to be mounted on roofs or other exterior areas that are not readily accessible. These installations shall comply with (1) through (4):

(1) A dc disconnecting means shall be mounted within sight of or in the inverter.

(2) An ac disconnecting means shall be mounted within sight of or in the inverter.

(3) An additional ac disconnecting means for the inverter shall comply with 705.22.

(4) A plaque shall be installed in accordance with 705.10.

705.80 Utility-Interactive Power Systems Employing Energy Storage. Utility-interactive power systems employing energy storage shall also be marked with the maximum operating voltage, including any equalization voltage, and the polarity of the grounded circuit conductor.

705.82 Hybrid Systems. Hybrid systems shall be permitted to be interconnected with interactive inverters.

705.95 Ampacity of Neutral Conductor. The ampacity of the neutral conductors shall comply with either (A) or (B).

(A) Neutral Conductor for Single Phase, 2-Wire Inverter Output. If a single-phase, 2-wire inverter output is connected to the neutral and one ungrounded conductor (only) of a 3-wire system or of a 3-phase, 4-wire, wye-connected system, the maximum load connected between the neutral and any one ungrounded conductor plus the inverter output rating shall not exceed the ampacity of the neutral conductor.

(B) Neutral Conductor for Instrumentation, Voltage, Detection or Phase Detection. A conductor used solely for instrumentation, voltage detection, or phase detection and connected to a single-phase or 3-phase interactive inverter, shall be permitted to be sized at less than the ampacity of the other current-carrying conductors and shall be sized equal to or larger than the equipment grounding conductor.

705.100 Unbalanced Interconnections.

(A) Single Phase. Single-phase inverters for hybrid systems and ac modules in interactive hybrid systems shall be connected to 3-phase power systems in order to limit unbalanced voltages to not more than 3 percent.

> Informational Note: For interactive single-phase inverters, unbalanced voltages can be minimized by the same methods that are used for single-phase loads on a 3-phase power system. See ANSI/C84.1-2011, *Electric Power Systems and Equipment — Voltage Ratings (60 Hertz).*

(B) Three Phase. Three-phase inverters and 3-phase ac modules in interactive systems shall have all phases automatically de-energized upon loss of, or unbalanced, voltage in one or more phases unless the interconnected system is designed so that significant unbalanced voltages will not result.

Part III. Generators

705.130 Overcurrent Protection. Conductors shall be protected in accordance with Article 240. Equipment and conductors connected to more than one electrical source shall have overcurrent devices located so as to provide protection from all sources. Generators shall be protected in accordance with 445.12.

705.143 Synchronous Generators. Synchronous generators in a parallel system shall be provided with the necessary equipment to establish and maintain a synchronous condition.

N **Part IV. Microgrid Systems**

705.150 System Operation. Microgrid systems shall be permitted to disconnect from the primary source of power or other interconnected electric power production sources and operate as a separate microgrid system.

705.160 Primary Power Source Connection. Connections to primary power sources that are external to the microgrid system shall comply with the requirements of 705.12.

705.165 Reconnection to Primary Power Source. Microgrid systems that reconnect to primary power sources shall be provided with the necessary equipment to establish a synchronous transition.

705.170 Microgrid Interconnect Devices (MID). Microgrid interconnect devices shall comply with the following:

(1) Be required for any connection between a microgrid system and a primary power source
(2) Be listed or field labeled for the application
(3) Have sufficient number of overcurrent devices located to provide overcurrent protection from all sources

> Informational Note: MID functionality is often incorporated in an interactive or multimode inverter, energy storage system, or similar device identified for interactive operation.

ARTICLE 706
Energy Storage Systems

N

Part I. General

706.1 Scope. This article applies to all permanently installed energy storage systems (ESS) operating at over 50 volts ac or 60 volts dc that may be stand-alone or interactive with other electric power production sources.

> Informational Note: The following standards are frequently referenced for the installation of energy storage systems:
>
> (1) NFPA 111-2013, *Standard on Stored Electrical Energy Emergency and Standby Systems*
> (2) IEEE 484-2008, *Recommended Practice for Installation Design and Installation of Vented Lead-Acid Batteries for Stationary Applications*
> (3) IEEE 485-1997, *Recommended Practice for Sizing Vented Lead-Acid Storage Batteries for Stationary Applications*
> (4) IEEE 1145-2007, *Recommended Practice for Installation and Maintenance of Nickel-Cadmium Batteries for Photovoltaic (PV) Systems*

> (5) IEEE 1187-2002, *Recommended Practice for Installation Design, and Installation of Valve-Regulated Lead-Acid Batteries for Stationary Applications*
> (6) IEEE 1578-2007, *Recommended Practice for Stationary Battery Electrolyte Spill Containment and Management*
> (7) IEEE 1635/ASHRAE 21-2012, *Guide for the Ventilation and Thermal Management of Batteries for Stationary Applications*
> (8) UL 810A, *Electrochemical Capacitors*
> (9) UL 1973, *Batteries for Use in Light Electric Rail (LER) Applications and Stationary Applications*
> (10) UL 1989, *Standard for Standby Batteries*
> (11) UL Subject 2436, *Spill Containment For Stationary Lead Acid Battery Systems*
> (12) UL Subject 9540, *Safety of Energy Storage Systems and Equipment*

706.2 Definitions.

Battery. Two or more cells connected together electrically in series, in parallel, or a combination of both to provide the required operating voltage and current levels.

Cell. The basic electrochemical unit, characterized by an anode and a cathode, used to receive, store, and deliver electrical energy.

Container. A vessel that holds the plates, electrolyte, and other elements of a single unit, comprised of one or more cells, in a battery. It can be referred to as a jar or case.

Diversion Charge Controller. Equipment that regulates the charging process of an ESS by diverting power from energy storage to direct-current or alternating-current loads or to an interconnected utility service.

Electrolyte. The medium that provides the ion transport mechanism between the positive and negative electrodes of a cell.

Energy Storage System (ESS). One or more components assembled together capable of storing energy for use at a future time. ESS(s) can include but is not limited to batteries, capacitors, and kinetic energy devices (e.g., flywheels and compressed air). These systems can have ac or dc output for utilization and can include inverters and converters to change stored energy into electrical energy.

Energy Storage System, Self-Contained. Energy storage systems where the components such as cells, batteries, or modules and any necessary controls, ventilation, illumination, fire suppression, or alarm systems are assembled, installed, and packaged into a singular energy storage container or unit.

> Informational Note: Self-contained systems will generally be manufactured by a single entity, tested and listed to safety standards relevant to the system, and readily connected on site to the electrical system and in the case of multiple systems to each other.

Energy Storage System, Pre-Engineered of Matched Components. Energy storage systems that are not self-contained systems but instead are pre-engineered and field-assembled using separate components supplied as a system by a singular entity that are matched and intended to be assembled as an energy storage system at the system installation site.

> Informational Note: Pre-engineered systems of matched components for field assembly as a system will generally be designed by a single entity and comprised of components that are tested and listed separately or as an assembly.

Energy Storage System, Other. Energy storage systems that are not self-contained or pre-engineered systems of matched components but instead are composed of individual components assembled as a system.

Informational Note: Other systems will generally be comprised of different components combined on site to create an ESS. Those components would generally be tested and listed to safety standards relevant to the application.

Flow Battery. An energy storage component similar to a fuel cell that stores its active materials in the form of two electrolytes external to the reactor interface. When in use, the electrolytes are transferred between reactor and storage tanks.

Informational Note: Two commercially available flow battery technologies are zinc bromine and vanadium redox, sometimes referred to as pumped electrolyte ESS.

Intercell Connector. An electrically conductive bar or cable used to connect adjacent cells.

Intertier Connector. In a battery system, an electrical conductor used to connect two cells on different tiers of the same rack or different shelves of the same rack.

Inverter Input Circuit. Conductors between the inverter and the ESS in stand-alone and multimode inverter systems.

Inverter Output Circuit. Conductors between the inverter and another electric power production source, such as a utility for an electrical production and distribution network.

Inverter Utilization Output Circuit. Conductors between the multimode or standalone inverter and utilization equipment.

Nominal Voltage (Battery or Cell). The value assigned to a cell or battery of a given voltage class for the purpose of convenient designation. The operating voltage of the cell or battery may vary above or below this value.

Sealed Cell or Battery. A cell or battery that has no provision for the routine addition of water or electrolyte or for external measurement of electrolyte specific gravity.

Informational Note: Some cells that are considered to be sealed under conditions of normal use, such as valve-regulated lead-acid or some lithium cells, contain pressure relief valves.

Terminal. That part of a cell, container, or battery to which an external connection is made (commonly identified as a post, pillar, pole, or terminal post).

706.3 Other Articles. Wherever the requirements of other articles of this *Code* and Article 706 differ, the requirements of Article 706 shall apply. If the ESS is capable of being operated in parallel with a primary source(s) of electricity, the requirements in 705.6, 705.12, 705.14, 705.16, 705.32, 705.40, 705.100, 705.143, and Part IV of Article 705 shall apply.

706.4 System Classification. ESS shall be classified as one of the types described as follows:

(1) ESS, self-contained

Informational Note: Some self-contained systems may be listed.

(2) ESS, pre-engineered of matched components
(3) ESS, other

706.5 Equipment. Monitors, controls, switches, fuses, circuit breakers, power conversion systems, inverters and transformers, energy storage components, and other components of the energy storage system other than lead-acid batteries, shall be

listed. Alternatively, self-contained ESS shall be listed as a complete energy storage system.

706.6 Multiple Systems. Multiple ESSs shall be permitted to be installed in or on a single building or structure.

706.7 Disconnecting Means.

(A) ESS Disconnecting Means. A disconnecting means shall be provided for all ungrounded conductors derived from an ESS. A disconnecting means shall be readily accessible and located within sight of the ESS.

Informational Note: See 240.21(H) for information on the location of the overcurrent device for conductors.

(B) Remote Actuation. Where controls to activate the disconnecting means of an ESS are not located within sight of the system, the disconnecting means shall be capable of being locked in the open position, in accordance with 110.25, and the location of the controls shall be field marked on the disconnecting means.

(C) Busway. Where a dc busway system is installed, the disconnecting means shall be permitted to be incorporated into the busway.

(D) Notification. The disconnecting means shall be legibly marked in the field. The marking shall meet the requirements of 110.21(B) and shall include the following:

(1) Nominal ESS voltage
(2) Maximum available short-circuit current derived from the ESS
(3) The associated clearing time or arc duration based on the available short-circuit current from the ESS and associated overcurrent protective devices if applicable
(4) Date the calculation was performed

Exception: The labeling in 706.7(D)(1) through (D)(4) shall not be required if an arc flash label is applied in accordance with acceptable industry practice.

Informational Note No. 1: Industry practices for equipment labeling are described in NFPA 70E-2015, *Standard for Electrical Safety in the Workplace.* This standard provides specific criteria for developing arc-flash labels for equipment that provides nominal system voltage, incident energy levels, arc-flash boundaries, minimum required levels of personal protective equipment, and so forth.

Informational Note No. 2: Battery equipment suppliers can provide information about short-circuit current on any particular battery model.

(E) Partitions and Distance. Where energy storage system input and output terminals are more than 1.5 m (5 ft) from connected equipment, or where the circuits from these terminals pass through a wall or partition, the installation shall comply with the following:

(1) A disconnecting means shall be provided at the energy storage system end of the circuit. Fused disconnecting means or circuit breakers shall be permitted to be used.
(2) A second disconnecting means located at the connected equipment shall be installed where the disconnecting means required by 706.7(E)(1) is not within sight of the connected equipment.

Informational Note No. 1: For remote disconnect controls in information technology equipment rooms, see 645.10.

Informational Note No. 2: For overcurrent protection of batteries, see 240.21(H).

(3) Where fused disconnecting means are used, the line terminals of the disconnecting means shall be connected toward the energy storage system terminals.

(4) Disconnecting means shall be permitted to be installed in energy storage system enclosures where explosive atmospheres can exist if listed for hazardous locations.

(5) Where the disconnecting means in (1) is not within sight of the disconnecting means in (2), placards or directories shall be installed at the locations of all disconnecting means indicating the location of all other disconnecting means.

706.8 Connection to Other Energy Sources. Connection to other energy sources shall comply with the requirements of 705.12.

(A) Load Disconnect. A load disconnect that has multiple sources of power shall disconnect all energy sources when in the off position.

(B) Identified Interactive Equipment. Only inverters and ac modules listed and identified as interactive shall be permitted on interactive systems.

(C) Loss of Interactive System Power. Upon loss of primary source, an ESS with a utility interactive inverter shall comply with the requirements of 705.40.

(D) Unbalanced Interconnections. Unbalanced connections between an energy storage system and electric power production sources shall be in accordance with 705.100.

(E) Point of Connection. The point of connection between an energy storage system and electric power production sources shall be in accordance with 705.12.

706.10 Energy Storage System Locations. Battery locations shall conform to 706.10(A), (B), and (C).

(A) Ventilation. Provisions appropriate to the energy storage technology shall be made for sufficient diffusion and ventilation of any possible gases from the storage device, if present, to prevent the accumulation of an explosive mixture. A pre-engineered or self-contained ESS shall be permitted to provide ventilation in accordance with the manufacturer's recommendations and listing for the system.

Informational Note No. 1: See NFPA 1-2015, *Fire Code*, Chapter 52, for ventilation considerations for specific battery chemistries.

Informational Note No. 2: Some storage technologies do not require ventilation.

Informational Note No. 3: A source for design of ventilation of battery systems is IEEE 1635-2012/ASHRAE Guideline 21-2012 *Guide for the Ventilation and Thermal Management of Batteries for Stationary Applications*, and the UBC.

Informational Note No. 4: Fire protection considerations are addressed in NFPA 1-2015, *Fire Code*.

(B) Guarding of Live Parts. Guarding of live parts shall comply with 110.27.

(C) Spaces About ESS Components. Spaces about the ESS shall comply with 110.26. Working space shall be measured from the edge of the ESS modules, battery cabinets, racks, or trays. For battery racks, there shall be a minimum clearance of 25 mm (1 in.) between a cell container and any wall or structure on the side not requiring access for maintenance. ESS modules, battery cabinets, racks, or trays shall be permitted to contact adjacent walls or structures, provided that the battery shelf has a free air space for not less than 90 percent of its length. Pre-engineered and self-contained ESSs shall be permitted to have working space between components within the system in accordance with the manufacturer's recommendations and listing of the system.

Informational Note: Additional space is often needed to accommodate ESS equipment hoisting equipment, tray removal, or spill containment.

(D) Egress. A personnel door(s) intended for entrance to and egress from rooms designated as ESS rooms shall open in the direction of egress and shall be equipped with listed panic hardware.

(E) Illumination. Illumination shall be provided for working spaces associated with ESS and their equipment and components. Luminaires shall not be controlled by automatic means only. Additional luminaires shall not be required where the work space is illuminated by an adjacent light source. The location of luminaires shall not do either of the following:

(1) Expose personnel to energized system components while performing maintenance on the luminaires in the system space

(2) Create a hazard to the system or system components upon failure of the luminaire

706.11 Directory. ESS shall be indicated by 706.11(A) and (B). The markings or labels shall be in accordance with 110.21(B).

(A) Directory. A permanent plaque or directory denoting all electric power sources on or in the premises shall be installed at each service equipment location and at locations of all electric power production sources capable of being interconnected.

Exception: Installations with large numbers of power production sources shall be permitted to be designated by groups.

(B) Facilities with Stand-Alone Systems. Any structure or building with an ESS that is not connected to a utility service source and is a stand-alone system shall have a permanent plaque or directory installed on the exterior of the building or structure at a readily visible location acceptable to the authority having jurisdiction. The plaque or directory shall indicate the location of system disconnecting means and that the structure contains a stand-alone electrical power system.

Part II. Circuit Requirements

706.20 Circuit Sizing and Current.

(A) Maximum Rated Current for a Specific Circuit. The maximum current for the specific circuit shall be calculated in accordance with 706.20(A)(1) through (A)(5).

(1) Nameplate-Rated Circuit Current. The nameplate(s)-rated circuit current shall be the rated current indicated on the ESS nameplate(s) or system listing for pre-engineered or self-contained systems of matched components intended for field assembly as a system.

(2) Inverter Output Circuit Current. The maximum current shall be the inverter continuous output current rating.

(3) Inverter Input Circuit Current. The maximum current shall be the continuous inverter input current rating when the inverter is producing rated power at the lowest input voltage.

(4) Inverter Utilization Output Circuit Current. The maximum current shall be the continuous inverter output current rating when the inverter is producing rated power at the lowest input voltage.

(5) DC to DC Converter Output Current. The maximum current shall be the dc-to-dc converter continuous output current rating.

(B) Conductor Ampacity and Overcurrent Device Ratings. The ampacity of the feeder circuit conductors from the ESS(s) to the wiring system serving the loads to be serviced by the system shall not be less than the greater of the (1) nameplate(s) rated circuit current as determined in accordance with 706.20(A) or (2) the rating of the ESS(s) overcurrent protective device(s).

(C) Ampacity of Grounded or Neutral Conductor. If the output of a single-phase, 2-wire ESS output(s) is connected to the grounded or neutral conductor and a single ungrounded conductor of a 3-wire system or of a 3-phase, 4-wire, wye-connected system, the maximum unbalanced neutral load current plus the ESS(s) output rating shall not exceed the ampacity of the grounded or neutral conductor.

706.21 Overcurrent Protection.

(A) Circuits and Equipment. ESS circuit conductors shall be protected in accordance with the requirements of Article 240. Protection devices for ESS circuits shall be in accordance with the requirements of 706.21(B) through (F). Circuits shall be protected at the source from overcurrent.

(B) Overcurrent Device Ampere Ratings. Overcurrent protective devices, where required, shall be rated in accordance with Article 240 and the rating provided on systems serving the ESS and shall be not less than 125 percent of the maximum currents calculated in 706.20(A).

(C) Direct Current Rating. Overcurrent protective devices, either fuses or circuit breakers, used in any dc portion of an ESS shall be listed and for dc and shall have the appropriate voltage, current, and interrupting ratings for the application.

(D) Current Limiting. A listed and labeled current-limiting overcurrent protective device shall be installed adjacent to the ESS for each dc output circuit.

Exception: Where current-limiting overcurrent protection is provided for the dc output circuits of a listed ESS, additional current-limiting overcurrent devices shall not be required.

(E) Fuses. Means shall be provided to disconnect any fuses associated with ESS equipment and components when the fuse is energized from both directions and is accessible to other than qualified persons. Switches, pullouts, or similar devices that are rated for the application shall be permitted to serve as a means to disconnect fuses from all sources of supply.

(F) Location. Where ESS input and output terminals are more than 1.5 m (5 ft) from connected equipment, or where the circuits from these terminals pass through a wall or partition, overcurrent protection shall be provided at the ESS.

706.23 Charge Control.

(A) General. Provisions shall be provided to control the charging process of the ESS. All adjustable means for control of the charging process shall be accessible only to qualified persons.

Informational Note: Certain types of energy storage equipment such as valve-regulated lead acid or nickel cadmium can experience thermal failure when overcharged.

(B) Diversion Charge Controller.

(1) Sole Means of Regulating Charging. An ESS employing a diversion charge controller as the sole means of regulating charging shall be equipped with a second independent means to prevent overcharging of the storage device.

(2) Circuits with Diversion Charge Controller and Diversion Load. Circuits containing a diversion charge controller and a diversion load shall comply with the following:

(1) The current rating of the diversion load shall be less than or equal to the current rating of the diversion load charge controller. The voltage rating of the diversion load shall be greater than the maximum ESS voltage. The power rating of the diversion load shall be at least 150 percent of the power rating of the charging source.

(2) The conductor ampacity and the rating of the overcurrent device for this circuit shall be at least 150 percent of the maximum current rating of the diversion charge controller.

(3) Energy Storage Systems Using Utility-Interactive Inverters. Systems using utility-interactive inverters to control energy storage state-of-charge by diverting excess power into the utility system shall comply with 706.23(B)(3)(a) and (B)(3)(b).

(a) These systems shall not be required to comply with 706.23(B)(2).

(b) These systems shall have a second, independent means of controlling the ESS charging process for use when the utility is not present or when the primary charge controller fails or is disabled.

(C) Charge Controllers and DC-to-DC Converters. Where charge controllers and other DC-to-DC power converters that increase or decrease the output current or output voltage with respect to the input current or input voltage are installed, all of the following shall apply:

(1) The ampacity of the conductors in output circuits shall be based on the maximum rated continuous output current of the charge controller or converter for the selected output voltage range.

(2) The voltage rating of the output circuits shall be based on the maximum voltage output of the charge controller or converter for the selected output voltage range.

Part III. Electrochemical Energy Storage Systems
Part III of this article applies to ESSs that are comprised of sealed and non-sealed cells or batteries or system modules that are comprised of multiple sealed cells or batteries that are not components within a listed product.

Informational Note: An energy storage component, such as batteries, that are integrated into a larger piece of listed equipment, such as an uninterruptible power supply (UPS), are examples of components within a listed product.

706.30 Installation of Batteries.

(A) Dwelling Units. An ESS for dwelling units shall not exceed 100 volts between conductors or to ground.

Exception: Where live parts are not accessible during routine ESS maintenance, an ESS voltage exceeding 100 volts shall be permitted.

(B) Disconnection of Series Battery Circuits. Battery circuits subject to field servicing, where exceeding 240 volts nominal between conductors or to ground, shall have provisions to disconnect the series-connected strings into segments not exceeding 240 volts nominal for maintenance by qualified persons. Non–load-break bolted or plug-in disconnects shall be permitted.

(C) Storage System Maintenance Disconnecting Means. ESS exceeding 100 volts between conductors or to ground shall have a disconnecting means, accessible only to qualified persons, that disconnects ungrounded and grounded circuit conductor(s) in the electrical storage system for maintenance. This disconnecting means shall not disconnect the grounded circuit conductor(s) for the remainder of any other electrical system. A non–load-break-rated switch shall be permitted to be used as a disconnecting means.

(D) Storage Systems of More Than 100 Volts. On ESS exceeding 100 volts between the conductors or to ground, the battery circuits shall be permitted to operate with ungrounded conductors, provided a ground-fault detector and indicator is installed to monitor for ground faults within the storage system.

706.31 Battery and Cell Terminations.

(A) Corrosion Prevention. Antioxidant material suitable for the battery connection shall be used when recommended by the battery or cell manufacturer.

Informational Note: The battery manufacturer's installation and instruction manual can be used for guidance for acceptable materials.

(B) Intercell and Intertier Conductors and Connections. The ampacity of field-assembled intercell and intertier connectors and conductors shall be of such cross-sectional area that the temperature rise under maximum load conditions and at maximum ambient temperature shall not exceed the safe operating temperature of the conductor insulation or of the material of the conductor supports.

Informational Note: Conductors sized to prevent a voltage drop exceeding 3 percent of maximum anticipated load, and where the maximum total voltage drop to the furthest point of connection does not exceed 5 percent, may not be appropriate for all battery applications. IEEE 1375-2003, *Guide for the Protection of Stationary Battery Systems*, provides guidance for overcurrent protection and associated cable sizing.

(C) Battery Terminals. Electrical connections to the battery and the cable(s) between cells on separate levels or racks shall not put mechanical strain on the battery terminals. Terminal plates shall be used where practicable.

706.32 Battery Interconnections. Flexible cables, as identified in Article 400, in sizes 2/0 AWG and larger shall be permitted within the battery enclosure from battery terminals to a nearby junction box where they shall be connected to an approved wiring method. Flexible battery cables shall also be permitted between batteries and cells within the battery enclosure. Such cables shall be listed and identified as moisture resistant. Flexible, fine-stranded cables shall only be used with terminals, lugs, devices, or connectors in accordance with 110.14.

706.33 Accessibility. The terminals of all cells or multicell units shall be readily accessible for readings, inspection, and cleaning where required by the equipment design. One side of transparent battery containers shall be readily accessible for inspection of the internal components.

706.34 Battery Locations. Battery locations shall conform to 706.34(A), (B), and (C).

(A) Live Parts. Guarding of live parts shall comply with 110.27.

(B) Top Terminal Batteries. Where top terminal batteries are installed on tiered racks or on shelves of battery cabinets, working space in accordance with the storage equipment manufacturer's instructions shall be provided between the highest point on a storage system component and the row, shelf, or ceiling above that point.

Informational Note: IEEE 1187 provides guidance for top clearance of VRLA batteries, which are the most commonly used battery in cabinets.

(C) Gas Piping. Gas piping shall not be permitted in dedicated battery rooms.

Part IV. Flow Battery Energy Storage Systems
Part IV applies to ESSs composed of or containing flow batteries.

706.40 General. All electrical connections to and from the system and system components shall be in accordance with the applicable provisions of Article 692. The system and system components shall also meet the provisions of Parts I and II of this article. Unless otherwise directed by this article, flow battery ESS shall comply with the applicable provisions of Article 692.

706.41 Electrolyte Classification. The electrolyte(s) that are acceptable for use in the batteries associated with the ESS shall be identified by name and chemical composition. Such identification shall be provided by readily discernable signage adjacent to every location in the system where the electrolyte can be put into or taken out of the system.

706.42 Electrolyte Containment. Flow battery systems shall be provided with a means for electrolyte containment to prevent spills of electrolyte from the system. An alarm system shall be provided to signal an electrolyte leak from the system. Electrical wiring and connections shall be located and routed in a manner that mitigates the potential for exposure to electrolytes.

706.43 Flow Controls. Controls shall be provided to safely shut down the system in the event of electrolyte blockage.

706.44 Pumps and Other Fluid Handling Equipment. Pumps and other fluid handling equipment are to be rated/specified suitable for exposure to the electrolytes.

Part V. Other Energy Storage Technologies
The provisions of Part V apply to ESSs using other technologies intended to store energy and when there is a demand for elec-

trical power to use the stored energy to generate the needed power.

706.50 General. All electrical connections to and from the system and system components shall be in accordance with the applicable provisions of this *Code*. Unless otherwise directed by this article, other energy storage technologies shall comply with the applicable provisions of Part III of Article 705.

ARTICLE 708
Critical Operations Power Systems (COPS)

Informational Note: Text that is followed by a reference in brackets has been extracted from *NFPA 1600* -2013, *Standard on Disaster/Emergency Management and Business Continuity Programs*. Only editorial changes were made to the extracted text to make it consistent with this *Code*.

Part I. General

708.1 Scope. The provisions of this article apply to the installation, operation, monitoring, control, and maintenance of the portions of the premises wiring system intended to supply, distribute, and control electricity to designated critical operations areas (DCOA) in the event of disruption to elements of the normal system.

Critical operations power systems are those systems so classed by municipal, state, federal, or other codes by any governmental agency having jurisdiction or by facility engineering documentation establishing the necessity for such a system. These systems include but are not limited to power systems, HVAC, fire alarm, security, communications, and signaling for designated critical operations areas.

Informational Note No. 1: Critical operations power systems are generally installed in vital infrastructure facilities that, if destroyed or incapacitated, would disrupt national security, the economy, public health or safety; and where enhanced electrical infrastructure for continuity of operation has been deemed necessary by governmental authority.

Informational Note No. 2: For further information on disaster and emergency management, see *NFPA 1600* -2013, *Standard on Disaster/Emergency Management and Business Continuity Programs*.

Informational Note No. 3: For further information regarding performance of emergency and standby power systems, see NFPA 110-2013, *Standard for Emergency and Standby Power Systems*.

Informational Note No. 4: For further information regarding performance and maintenance of emergency systems in health care facilities, see NFPA 99-2015, *Health Care Facilities Code*.

Informational Note No. 5: For specification of locations where emergency lighting is considered essential to life safety, see NFPA 101-2015, *Life Safety Code*, or the applicable building code.

Informational Note No. 6: For further information regarding physical security, see NFPA 730-2014, *Guide for Premises Security*.

Informational Note No. 7: Threats to facilities that may require transfer of operation to the critical systems include both naturally occurring hazards and human-caused events. See also

A.5.3.2 of *NFPA 1600* -2013, *Standard on Disaster/Emergency Management and Business Continuity Programs*.

Informational Note No. 8: See Informative Annex F, Availability and Reliability for Critical Operations Power Systems; and Development and Implementation of Functional Performance Tests (FPTs) for Critical Operations Power Systems.

Informational Note No. 9: See Informative Annex G, Supervisory Control and Data Acquisition (SCADA).

708.2 Definitions.

Commissioning. The acceptance testing, integrated system testing, operational tune-up, and start-up testing is the process by which baseline test results verify the proper operation and sequence of operation of electrical equipment, in addition to developing baseline criteria by which future trend analysis can identify equipment deterioration.

Critical Operations Power Systems (COPS). Power systems for facilities or parts of facilities that require continuous operation for the reasons of public safety, emergency management, national security, or business continuity.

Designated Critical Operations Areas (DCOA). Areas within a facility or site designated as requiring critical operations power.

Supervisory Control and Data Acquisition (SCADA). An electronic system that provides monitoring and controls for the operation of the critical operations power system. This can include the fire alarm system, security system, control of the HVAC, the start/stop/monitoring of the power supplies and electrical distribution system, annunciation and communications equipment to emergency personnel, facility occupants, and remote operators.

708.4 Risk Assessment. Risk assessment for critical operations power systems shall be documented and shall be conducted in accordance with 708.4(A) through (C).

Informational Note: Chapter 5 of *NFPA 1600*-2013, *Standard on Disaster/Emergency Management and Business Continuity Programs*, provides additional guidance concerning risk assessment and hazard analysis.

(A) Conducting Risk Assessment. In critical operations power systems, risk assessment shall be performed to identify hazards, the likelihood of their occurrence, and the vulnerability of the electrical system to those hazards.

(B) Identification of Hazards. Hazards to be considered at a minimum shall include, but shall not be limited to, the following:

(1) Naturally occurring hazards (geological, meteorological, and biological)
(2) Human-caused events (accidental and intentional) [**1600**:5.3.2]

(C) Developing Mitigation Strategy. Based on the results of the risk assessment, a strategy shall be developed and implemented to mitigate the hazards that have not been sufficiently mitigated by the prescriptive requirements of this *Code*.

708.5 Physical Security. Physical security shall be provided for critical operations power systems in accordance with 708.5(A) and (B).

(A) Risk Assessment. Based on the results of the risk assessment, a strategy for providing physical security for critical operations power systems shall be developed, documented, and implemented.

(B) Restricted Access. Electrical circuits and equipment for critical operations power systems shall be accessible to qualified personnel only.

708.6 Testing and Maintenance.

(A) Conduct or Witness Test. The authority having jurisdiction shall conduct or witness a test of the complete system upon installation and periodically afterward.

(B) Tested Periodically. Systems shall be tested periodically on a schedule acceptable to the authority having jurisdiction to ensure the systems are maintained in proper operating condition.

(C) Maintenance. The authority having jurisdiction shall require a documented preventive maintenance program for critical operations power systems.

> Informational Note: For information concerning maintenance, see NFPA 70B-2013, *Recommended Practice for Electrical Equipment Maintenance.*

(D) Written Record. A written record shall be kept of such tests and maintenance.

(E) Testing Under Load. Means for testing all critical power systems during maximum anticipated load conditions shall be provided.

> Informational Note: For information concerning testing and maintenance of emergency power supply systems (EPSSs) that are also applicable to COPS, see NFPA 110-2013, *Standard for Emergency and Standby Power Systems.*

708.8 Commissioning.

(A) Commissioning Plan. A commissioning plan shall be developed and documented.

> Informational Note: For further information on developing a commissioning program see NFPA 70B-2013, *Recommended Practice for Electrical Equipment Maintenance.*

(B) Component and System Tests. The installation of the equipment shall undergo component and system tests to ensure that, when energized, the system will function properly.

(C) Baseline Test Results. A set of baseline test results shall be documented for comparison with future periodic maintenance testing to identify equipment deterioration.

(D) Functional Performance Tests. A functional performance test program shall be established, documented, and executed upon complete installation of the critical system in order to establish a baseline reference for future performance requirements.

> Informational Note: See Informative Annex F for more information on developing and implementing a functional performance test program.

Part II. Circuit Wiring and Equipment

708.10 Feeder and Branch Circuit Wiring.

(A) Identification.

(1) Boxes and Enclosures. In a building or at a structure where a critical operations power system and any other type of power system are present, all boxes and enclosures (including transfer switches, generators, and power panels) for critical operations power system circuits shall be permanently marked so they will be readily identified as a component of the critical operations power system.

(2) Receptacle Identification. In a building in which COPS are present with other types of power systems described in other sections in this article, the cover plates for the receptacles or the receptacles themselves supplied from the COPS shall have a distinctive color or marking so as to be readily identifiable. Nonlocking-type, 125-volt, 15- and 20-ampere receptacles supplied from the COPS shall have an illuminated face or an indicator light to indicate that there is power to the receptacle.

> *Exception: If the COPS supplies power to a DCOA that is a stand-alone building, receptacle cover plates or the receptacles themselves shall not be required to have distinctive marking.*

(B) Wiring. Wiring of two or more COPS circuits supplied from the same source shall be permitted in the same raceway, cable, box, or cabinet. Wiring from a COPS source or COPS source distribution overcurrent protection to critical loads shall be kept entirely independent of all other wiring and equipment.

> *Exception: Where the COPS feeder is installed in transfer equipment enclosures.*

(C) COPS Feeder Wiring Requirements. COPS feeders shall comply with 708.10(C)(1) through (C)(3).

(1) Protection Against Physical Damage. The wiring of the COPS system shall be protected against physical damage. Only the following wiring methods shall be permitted:

(1) Rigid metal conduit, intermediate metal conduit, or Type MI cable.

(2) Where encased in not less than 50 mm (2 in.) of concrete, any of the following wiring methods shall be permitted:

    a. Schedule 40 or Schedule 80 rigid polyvinyl chloride conduit (Type PVC)
    b. Reinforced thermosetting resin conduit (Type RTRC)
    c. Electrical metallic tubing (Type EMT)
    d. Flexible nonmetallic or jacketed metallic raceways
    e. Jacketed metallic cable assemblies listed for installation in concrete

(3) Where provisions must be made for flexibility at equipment connection, one or more of the following shall also be permitted:

    a. Flexible metal fittings
    b. Flexible metal conduit with listed fittings
    c. Liquidtight flexible metal conduit with listed fittings

(2) Fire Protection for Feeders. Feeders shall meet one of the following conditions:

(1) The cable or raceway is protected by a listed electrical circuit protective system with a minimum 2-hour fire rating.

> Informational Note: The listing organization provides information for electrical circuit protection systems on proper installation requirements to maintain the fire rating.

(2) The cable or raceway is a listed fire-resistive cable system with a minimum 2-hour fire rating.

Informational Note No. 1: Fire-resistive cables are tested to ANSI/UL 2196, *Tests for Fire Resistive Cables*.

Informational Note No. 2: The listing organization provides information for fire-resistive cable systems on proper installation requirements to maintain the fire rating.

(3) The cable or raceway is protected by a listed fire-rated assembly that has a minimum fire rating of 2 hours.

(4) The cable or raceway is encased in a minimum of 50 mm (2 in.) of concrete.

(3) Floodplain Protection. Where COPS feeders are installed below the level of the 100-year floodplain, the insulated circuit conductors shall be listed for use in a wet location and be installed in a wiring method that is permitted for use in wet locations.

(D) COPS Branch Circuit Wiring.

(1) *Outside the DCOA.* COPS branch circuits installed outside the DCOA shall comply with the physical and fire protection requirements of 708.10(C)(1) through (C)(3).

(2) *Within the DCOA.* Any of the wiring methods recognized in Chapter 3 of this *Code* shall be permitted within the DCOA.

708.11 Branch Circuit and Feeder Distribution Equipment.

(A) Branch Circuit Distribution Equipment. COPS branch circuit distribution equipment shall be located within the same DCOA as the branch circuits it supplies.

(B) Feeder Distribution Equipment. Equipment for COPS feeder circuits (including transfer equipment, transformers, and panelboards) shall comply with (1) and (2):

(1) Be located in spaces with a 2-hour fire resistance rating

(2) Be located above the 100-year floodplain

708.12 Feeders and Branch Circuits Supplied by COPS. Feeders and branch circuits supplied by the COPS shall supply only equipment specified as required for critical operations use.

708.14 Wiring of HVAC, Fire Alarm, Security, Emergency Communications, and Signaling Systems. All conductors or cables shall be installed using any of the metal wiring methods permitted by 708.10(C)(1) and, in addition, shall comply with 708.14(1) through (8), as applicable.

(1) All cables for fire alarm, security, signaling systems, and emergency communications shall be shielded twisted pair cables or installed to comply with the performance requirements of the system.

(2) Shields of cables for fire alarm, security, signaling systems, and emergency communications shall be arranged in accordance with the manufacturer's published installation instructions.

(3) Optical fiber cables shall be used for connections between two or more buildings on the property and under single management.

(4) A listed primary protector shall be provided on all communications circuits. Listed secondary protectors shall be provided at the terminals of the communications circuits.

(5) Conductors for all control circuits rated above 50 volts shall be rated not less than 600 volts.

(6) Communications, fire alarm, and signaling circuits shall use relays with contact ratings that exceed circuit voltage and current ratings in the controlled circuit.

(7) All cables for fire alarm, security, and signaling systems shall be riser-rated and shall be a listed 2-hour electrical circuit protective system. Emergency communication cables shall be Type CMR-CI or shall be riser-rated and shall be a listed 2-hour electrical circuit protective system.

(8) Control, monitoring, and power wiring to HVAC systems shall be a listed 2-hour electrical circuit protective system.

Part III. Power Sources and Connection

708.20 Sources of Power.

(A) General Requirements. Current supply shall be such that, in the event of failure of the normal supply to the DCOA, critical operations power shall be available within the time required for the application. The supply system for critical operations power, in addition to the normal services to the building and meeting the general requirements of this section, shall be one or more of the types of systems described in 708.20(E) through (H).

Informational Note: Assignment of degree of reliability of the recognized critical operations power system depends on the careful evaluation in accordance with the risk assessment.

(B) Fire Protection. Where located within a building, equipment for sources of power as described in 708.20(E) through (H) shall be installed either in spaces fully protected by approved automatic fire suppression systems (sprinklers, carbon dioxide systems, and so forth) or in spaces with a 2-hour fire rating.

(C) Grounding. All sources of power shall be grounded as a separately derived source in accordance with 250.30.

Exception: Where the equipment containing the main bonding jumper or system bonding jumper for the normal source and the feeder wiring to the transfer equipment are installed in accordance with 708.10(C) and 708.11(B).

(D) Surge Protection Devices. Surge protection devices shall be provided at all facility distribution voltage levels.

(E) Storage Battery. An automatic battery charging means shall be provided. Batteries shall be compatible with the charger for that particular installation. Automotive-type batteries shall not be used.

(F) Generator Set.

(1) Prime Mover-Driven. Generator sets driven by a prime mover shall be provided with means for automatically starting the prime mover on failure of the normal service. A time-delay feature permitting a minimum 15-minute setting shall be provided to avoid retransfer in case of short-time reestablishment of the normal source.

(2) Power for fuel transfer pumps. Where power is needed for the operation of the fuel transfer pumps to deliver fuel to a generator set day tank, this pump shall be connected to the COPS.

(3) Dual Supplies. Prime movers shall not be solely dependent on a public utility gas system for their fuel supply or municipal water supply for their cooling systems. Means shall be provided for automatically transferring from one fuel supply to another where dual fuel supplies are used.

(4) Battery Power and Dampers. Where a storage battery is used for control or signal power or as the means of starting the prime mover, it shall be suitable for the purpose and shall be equipped with an automatic charging means independent of the generator set. Where the battery charger is required for the operation of the generator set, it shall be connected to the COPS. Where power is required for the operation of dampers used to ventilate the generator set, the dampers shall be connected to the COPS.

(5) Outdoor Generator Sets.

(a) *Permanently Installed Generators and Portable Generators Greater Than 15 kW.* Where an outdoor housed generator set is equipped with a readily accessible disconnecting means in accordance with 445.18, and the disconnecting means is located within sight of the building or structure supplied, an additional disconnecting means shall not be required where ungrounded conductors serve or pass through the building or structure. Where the generator supply conductors terminate at a disconnecting means in or on a building or structure, the disconnecting means shall meet the requirements of 225.36.

(b) *Portable Generators 15 kW or Less.* Where a portable generator, rated 15 kW or less, is installed using a flanged inlet or other cord-and plug-type connection, a disconnecting means shall not be required where ungrounded conductors serve or pass through a building or structure.

(6) Means for Connecting Portable or Vehicle-Mounted Generator. Where the COPS is supplied by a single generator, a means to connect a portable or vehicle-mounted generator shall be provided.

(7) On-Site Fuel Supply. Where internal combustion engines are used as the prime mover, an on-site fuel supply shall be provided. The on-site fuel supply shall be secured and protected in accordance with the risk assessment.

(G) Uninterruptible Power Supplies. Uninterruptible power supplies used as the sole source of power for COPS shall comply with the applicable provisions of 708.20(E) and (F).

(H) Fuel Cell System. Installation of a fuel cell system shall meet the requirements of Parts II through VIII of Article 692.

708.21 Ventilation. Adequate ventilation shall be provided for the alternate power source for continued operation under maximum anticipated ambient temperatures.

Informational Note: NFPA 110-2013, *Standard for Emergency and Standby Power Systems*, and NFPA 111-2013, *Standard on Stored Energy Emergency and Standby Power Systems*, include additional information on ventilation air for combustion and cooling.

708.22 Capacity of Power Sources.

(A) Capacity and Rating. A COPS shall have capacity and rating for all loads to be operated simultaneously for continuous operation with variable load for an unlimited number of hours, except for required maintenance of the power source. A portable, temporary, or redundant alternate power source shall be available for use whenever the COPS power source is out of service for maintenance or repair.

(B) Selective Load Pickup, Load Shedding, and Peak Load Shaving. The alternate power source shall be permitted to supply COPS emergency, legally required standby, and optional loads where the source has adequate capacity or where automatic selective load pickup and load shedding is provided as needed to ensure adequate power to (1) the COPS and emergency circuits, (2) the legally required standby circuits, and (3) the optional standby circuits, in that order of priority. The alternate power source shall be permitted to be used for peak load shaving, provided these conditions are met.

Peak load-shaving operation shall be permitted for satisfying the test requirement of 708.6(B), provided all other conditions of 708.6 are met.

(C) Duration of COPS Operation. The alternate power source shall be capable of operating the COPS for a minimum of 72 hours at full load of DCOA with a steady-state voltage within ±10 percent of nominal utilization voltage.

708.24 Transfer Equipment.

(A) General. Transfer equipment, including automatic transfer switches, shall be automatic and identified for emergency use. Transfer equipment shall be designed and installed to prevent the inadvertent interconnection of normal and critical operations sources of supply in any operation of the transfer equipment. Transfer equipment and electric power production systems installed to permit operation in parallel with the normal source shall meet the requirements of Article 705.

(B) Bypass Isolation Switches. Means shall be permitted to bypass and isolate the transfer equipment. Where bypass isolation switches are used, inadvertent parallel operation shall be avoided.

(C) Automatic Transfer Switches. Where used with sources that are not inherently synchronized, automatic transfer switches shall comply with (C)(1) and (C)(2).

(1) Automatic transfer switches shall be listed for emergency use.
(2) Automatic transfer switches shall be electrically operated and mechanically held.

(D) Use. Transfer equipment shall supply only COPS loads.

N **(E) Documentation.** The short-circuit current rating of the transfer equipment, based on the specific overcurrent protective device type and settings protecting the transfer equipment, shall be field marked on the exterior of the transfer equipment.

708.30 Branch Circuits Supplied by COPS. Branch circuits supplied by the COPS shall only supply equipment specified as required for critical operations use.

Part IV. Overcurrent Protection

708.50 Accessibility. The feeder- and branch-circuit overcurrent devices shall be accessible to authorized persons only.

708.52 Ground-Fault Protection of Equipment.

(A) Applicability. The requirements of 708.52 shall apply to critical operations (including multiple occupancy buildings) with critical operation areas.

(B) Feeders. Where ground-fault protection is provided for operation of the service disconnecting means or feeder disconnecting means as specified by 230.95 or 215.10, an additional step of ground-fault protection shall be provided in all next level feeder disconnecting means downstream toward the load. Such protection shall consist of overcurrent devices and current transformers or other equivalent protective equipment that causes the feeder disconnecting means to open.

(C) Testing. When equipment ground-fault protection is first installed, each level shall be tested to ensure that ground-fault protection is operational.

> Informational Note: Testing is intended to verify the ground-fault function is operational. The performance test is not intended to verify selectivity in 708.52(D), as this is often coordinated similarly to circuit breakers by reviewing time and current curves and properly setting the equipment. (Selectivity of fuses and circuit breakers is not performance tested for overload and short circuit.)

(D) Selectivity. Ground-fault protection for operation of the service and feeder disconnecting means shall be fully selective such that the feeder device, but not the service device, shall open on ground faults on the load side of the feeder device. Separation of ground-fault protection time-current characteristics shall conform to the manufacturer's recommendations and shall consider all required tolerances and disconnect operating time to achieve 100 percent selectivity.

> Informational Note: See 230.95, Informational Note No. 4, for transfer of alternate source where ground-fault protection is applied.

708.54 Selective Coordination. Critical operations power system(s) overcurrent devices shall be selectively coordinated with all supply-side overcurrent protective devices.

Selective coordination shall be selected by a licensed professional engineer or other qualified persons engaged primarily in the design, installation, or maintenance of electrical systems. The selection shall be documented and made available to those authorized to design, install, inspect, maintain, and operate the system.

Exception: Selective coordination shall not be required between two overcurrent devices located in series if no loads are connected in parallel with the downstream device.

Part V. System Performance and Analysis

708.64 Emergency Operations Plan. A facility with a COPS shall have documented an emergency operations plan. The plan shall consider emergency operations and response, recovery, and continuity of operations.

> Informational Note: *NFPA 1600*-2013, *Standard on Disaster/Emergency Management and Business Continuity Programs*, Section 5.7, provides guidance for the development and implementation of emergency plans.

ARTICLE 710
Stand-Alone Systems

710.1 Scope. This article covers electric power production sources operating in stand-alone mode.

N 710.6 Equipment Approval. All equipment shall be listed or field labeled for the intended use.

710.15 General. Premises wiring systems shall be adequate to meet the requirements of this *Code* for similar installations supplied by a feeder or service. The wiring on the supply side of the building or structure disconnecting means shall comply with the requirements of this *Code*, except as modified by 710.15(A) through (F).

(A) Supply Output. Power supply to premises wiring systems shall be permitted to have less capacity than the calculated load. The capacity of the stand-alone supply shall be equal to or greater than the load posed by the largest single utilization equipment connected to the system. Calculated general lighting loads shall not be considered as a single load.

(B) Sizing and Protection. The circuit conductors between a stand-alone source and a building or structure disconnecting means shall be sized based on the sum of the output ratings of the stand-alone sources.

(C) Single 120-Volt Supply. Stand-alone systems shall be permitted to supply 120 volts to single-phase, 3-wire, 120/240-volt service equipment or distribution panels where there are no 240-volt outlets and where there are no multiwire branch circuits. In all installations, the sum of the ratings of the power sources shall be less than the rating of the neutral bus in the service equipment. This equipment shall be marked with the following words or equivalent:

WARNING:
SINGLE 120-VOLT SUPPLY. DO NOT CONNECT MULTI-
WIRE BRANCH CIRCUITS!

The warning sign(s) or label(s) shall comply with 110.21(B).

(D) Energy Storage or Backup Power System Requirements. Energy storage or backup power supplies are not required.

(E) Back-Fed Circuit Breakers. Plug-in type back-fed circuit breakers connected to an interconnected supply shall be secured in accordance with 408.36(D). Circuit breakers marked "line" and "load" shall not be back-fed.

(F) Voltage and Frequency Control. The stand-alone supply shall be controlled so that voltage and frequency remain within suitable limits for the connected loads.

N

ARTICLE 712
Direct Current Microgrids

Part I. General

712.1 Scope. This article applies to direct current microgrids.

712.2 Definitions.

Direct Current Microgrid (DC Microgrid). A direct current microgrid is a power distribution system consisting of more than one interconnected dc power source, supplying dc-dc converter(s), dc load(s), and/or ac load(s) powered by dc-ac inverter(s). A dc microgrid is typically not directly connected to an ac primary source of electricity, but some dc microgrids interconnect via one or more dc-ac bidirectional converters or dc–ac inverters.

Informational Note: Direct current power sources include ac-dc converters (rectifiers), bidirectional dc-ac inverters/converters, photovoltaic systems, wind generators, energy storage systems (including batteries), and fuel cells.

Grounded Two-Wire DC System A system that has a solid connection or reference-ground between one of the current carrying conductors and the equipment grounding system.

Grounded Three-Wire DC System. A system with a solid connection or reference-ground between the center point of a bipolar dc power source and the equipment grounding system.

Nominal Voltage. A value assigned to a circuit or system for the purpose of conveniently designating its dc voltage class.

Informational Note: The actual voltage at which a circuit operates can vary from the nominal voltage within a range that permits satisfactory operation of equipment.

Reference-Grounded DC System. A system that is not solidly grounded but has a low-resistance electrical reference that maintains voltage to ground in normal operation.

Resistively Grounded. A system with a high-resistance connection between the current carrying conductors and the equipment grounding system.

Primary DC Source. A source that supplies the majority of the dc load in a dc microgrid.

Ungrounded DC System. A system that has no direct or resistive connection between the current carrying conductors and the equipment grounding system.

712.3 Other Articles. Wherever the requirements of other articles of this *Code* and Article 712 differ, the requirements of Article 712 shall apply. DC microgrids interconnected through an inverter or bi-directional converter with ac electric power production sources shall comply with Article 705.

712.4 Listing and Labeling. Any equipment used in the dc circuits of a direct-current micro grid shall be listed and labeled for dc use.

712.10 Directory. A permanent directory denoting all dc electric power sources operating to supply the dc microgrid shall be installed at each source location capable of acting as the primary dc source.

Part II. Circuit Requirements

712.25 Identification of Circuit Conductors

(A) Ungrounded circuit conductors in dc microgrids shall be identified according to the requirements of 210.5(C)(2) for branch circuits and 215.12(C)(2) for feeders.

(B) Ungrounded conductors of 6 AWG or smaller shall be permitted to be identified by polarity at all termination, connection, and splice points by marking tape, tagging, or other approved means.

712.30 System Voltage. The system voltage of a dc microgrid shall be determined by one of the following methods:

(1) The nominal voltage to ground for solidly grounded systems
(2) The nominal voltage to ground for reference-grounded systems

(3) The highest nominal voltage between conductors for resistively grounded dc systems and ungrounded dc systems.

Informational Note: Examples of nominal dc system voltages include but are not limited to 24, 48, 125, 190/380, or 380 volts.

Part III. Disconnecting Means

712.34 DC Source Disconnecting Means. The output of each dc source shall have a readily accessible, disconnecting means that is lockable in the open position and adjacent to the source.

712.35 Disconnection of Ungrounded Conductors. In solidly grounded two- and three-wire systems, the disconnecting means shall simultaneously open all ungrounded conductors. In ungrounded, resistively grounded and reference-grounded systems, such devices shall open all current-carrying conductors.

712.37 Directional Current Devices. Disconnecting means shall be listed, be marked for use in a single current direction, and only be used in the designated current direction.

Informational Note: Examples of directional current devices are magnetically quenched contactors and semiconductor switches in overcurrent devices.

Part IV. Wiring Methods

712.52 System Grounding.

(A) General. Direct-current microgrids shall be grounded in accordance with 250.162.

(B) Over 300 Volts. DC microgrids operating at voltages greater than 300 volts dc shall be reference-grounded dc systems or resistively grounded dc systems.

712.55 Ground Fault Detection Equipment. Ungrounded, reference grounded, or resistively grounded dc microgrids operating at greater than 60 volts dc shall have ground fault detection that indicates that a fault has occurred. The ground fault equipment shall be marked in accordance with 250.167(C).

712.57 Arc Fault Protection. Where required elsewhere in this *Code*, specific systems within the DC microgrid shall have arc fault protection. The arc fault protection equipment shall be listed.

Informational Note: Section 90.4 applies when suitable equipment for arc fault protection is not available.

Part V. Marking

712.62 Distribution Equipment and Conductors. Distribution equipment and conductors shall be marked as required elsewhere in this *Code*.

712.65 Available DC Short-Circuit Current.

(A) Field Marking. The maximum available dc short-circuit current on the dc microgrid shall be field marked at the dc source(s). The field marking(s) shall include the date the short-circuit current calculation was performed and be of sufficient durability to withstand the environment involved.

(B) Modifications. When modifications to the electrical installation occur that affect the maximum available short-circuit current at the dc source, the maximum available short-circuit current shall be verified or recalculated as necessary to ensure the equipment ratings are sufficient for the maximum available short-circuit current at the line terminals of the equipment. The required field marking(s) in 712.65(A) shall indicate the new maximum available short-circuit current and date.

Part VI. Protection

712.70 Overcurrent Protection. Equipment and conductors connected to more than one electrical source shall have overcurrent protective devices to provide protection from all sources.

712.72 Interrupting and Short-Circuit Current Ratings. Consideration shall be given to the contribution of short-circuit currents from all interconnected power sources for the interrupting ratings and short-circuit current ratings of equipment in the dc microgrid system(s). Overcurrent protective devices and equipment used within a dc microgrid shall have an interrupting rating at nominal circuit voltage or a short-circuit current rating sufficient for the available short-circuit current at the line terminals of the equipment.

Part VII. Systems over 1000 Volts

712.80 General. Systems with a maximum voltage between conductors of over 1000 volts dc shall comply with Article 490 and other requirements in this *Code* applicable to installations rated over 1000 volts.

ARTICLE 720
Circuits and Equipment Operating at Less Than 50 Volts

720.1 Scope. This article covers installations operating at less than 50 volts, direct current or alternating current.

720.2 Other Articles. Direct current or alternating-current installations operating at less than 50 volts, as covered in 411.1 through 411.8; Part VI of Article 517; Part II of Article 551; Parts II and III and 552.60(B) of Article 552; 650.1 through 650.8; 669.1 through 669.9; Parts I and VIII of Article 690; Parts I and III of Article 725; or Parts I and III of Article 760 shall not be required to comply with this article.

720.3 Hazardous (Classified) Locations. Installations within the scope of this article and installed in hazardous (classified) locations shall also comply with the appropriate provisions for hazardous (classified) locations in other applicable articles of this *Code*.

720.4 Conductors. Conductors shall not be smaller than 12 AWG copper or equivalent. Conductors for appliance branch circuits supplying more than one appliance or appliance receptacle shall not be smaller than 10 AWG copper or equivalent.

720.5 Lampholders. Standard lampholders that have a rating of not less than 660 watts shall be used.

720.6 Receptacle Rating. Receptacles shall have a rating of not less than 15 amperes.

720.7 Receptacles Required. Receptacles of not less than 20-ampere rating shall be provided in kitchens, laundries, and other locations where portable appliances are likely to be used.

720.9 Batteries. Installations of storage batteries shall comply with 480.1 through 480.6 and 480.9 through 480.11.

720.11 Mechanical Execution of Work. Circuits operating at less than 50 volts shall be installed in a neat and workmanlike manner. Cables shall be supported by the building structure in such a manner that the cable will not be damaged by normal building use.

ARTICLE 725
Class 1, Class 2, and Class 3 Remote-Control, Signaling, and Power-Limited Circuits

Part I. General

725.1 Scope. This article covers remote-control, signaling, and power-limited circuits that are not an integral part of a device or of utilization equipment.

> Informational Note: The circuits described herein are characterized by usage and electrical power limitations that differentiate them from electric light and power circuits; therefore, alternative requirements to those of Chapters 1 through 4 are given with regard to minimum wire sizes, ampacity adjustment and correction factors, overcurrent protection, insulation requirements, and wiring methods and materials.

725.2 Definitions.

Abandoned Class 2, Class 3, and PLTC Cable. Installed Class 2, Class 3, and PLTC cable that is not terminated at equipment and not identified for future use with a tag.

Circuit Integrity (CI) Cable. Cable(s) used for remote-control, signaling, or power-limited systems that supply critical circuits to ensure survivability for continued circuit operation for a specified time under fire conditions.

Class 1 Circuit. The portion of the wiring system between the load side of the overcurrent device or power-limited supply and the connected equipment.

> Informational Note: See 725.41 for voltage and power limitations of Class 1 circuits.

Class 2 Circuit. The portion of the wiring system between the load side of a Class 2 power source and the connected equipment. Due to its power limitations, a Class 2 circuit considers safety from a fire initiation standpoint and provides acceptable protection from electric shock.

Class 3 Circuit. The portion of the wiring system between the load side of a Class 3 power source and the connected equipment. Due to its power limitations, a Class 3 circuit considers safety from a fire initiation standpoint. Since higher levels of voltage and current than for Class 2 are permitted, additional safeguards are specified to provide protection from an electric shock hazard that could be encountered.

Power-Limited Tray Cable (PLTC).   A factory assembly of two or more insulated conductors rated at 300 V, with or without associated bare or insulated equipment grounding conductors, under a nonmetallic jacket.

725.3 Other Articles.   Circuits and equipment shall comply with the articles or sections listed in 725.3(A) through (N). Only those sections of Article 300 referenced in this article shall apply to Class 1, Class 2, and Class 3 circuits.

(A) Number and Size of Conductors in Raceway.   Section 300.17.

(B) Spread of Fire or Products of Combustion.   Installation of Class 1, Class 2, and Class 3 circuits shall comply with 300.21.

(C) Ducts, Plenums, and Other Air-Handling Spaces.   Class 1, Class 2, and Class 3 circuits installed in ducts, plenums, or other space used for environmental air shall comply with 300.22.

Exception No. 1:   Class 2 and Class 3 cables selected in accordance with Table 725.154 and installed in accordance with 725.135(B) and 300.22(B), Exception shall be permitted to be installed in ducts specifically fabricated for environmental air.

Exception No. 2: Class 2 and Class 3 cables selected in accordance with Table 725.154 and installed in accordance with 725.135(C) shall be permitted to be installed in other spaces used for environmental air (plenums).

(D) Hazardous (Classified) Locations.   Articles 500 through 516 and Article 517, Part IV, where installed in hazardous (classified) locations.

(E) Cable Trays.   Article 392, where installed in cable tray.

(F) Motor Control Circuits.   Article 430, Part VI, where tapped from the load side of the motor branch-circuit protective device(s) as specified in 430.72(A).

(G) Instrumentation Tray Cable.   See Article 727.

(H) Raceways Exposed to Different Temperatures.   Installations shall comply with 300.7(A).

(I) Vertical Support for Fire-Rated Cables and Conductors.   Vertical installations of circuit integrity (CI) cables and conductors installed in a raceway or conductors and cables of electrical circuit protective systems shall be installed in accordance with 300.19.

(J) Bushing.   A bushing shall be installed where cables emerge from raceway used for mechanical support or protection in accordance with 300.15(C).

(K) Installation of Conductors with Other Systems.   Installations shall comply with 300.8.

(L) Corrosive, Damp, or Wet Locations.   Class 2 and Class 3 cables installed in corrosive, damp, or wet locations shall comply with the applicable requirements in 110.11, 300.5(B), 300.6, 300.9, and 310.10(G).

N **(M) Cable Routing Assemblies.**   Class 2, Class 3, and Type PLTC cables shall be permitted to be installed in plenum cable routing assemblies, riser cable routing assemblies, and general-purpose cable routing assemblies selected in accordance with Table 800.154(c), listed in accordance with the provisions of 800.182, and installed in accordance with 800.110(C) and 800.113.

N **(N) Communications Raceways.**   Class 2, Class 3, and Type PLTC cables shall be permitted to be installed in plenum communications raceways, riser communications raceways, and general-purpose communications raceways selected in accordance with the provisions of Table 800.154(b), listed in accordance with 800.182, and installed in accordance with 800.113 and 362.24 through 362.56, where the requirements applicable to electrical nonmetallic tubing (ENT) apply.

725.21 Access to Electrical Equipment Behind Panels Designed to Allow Access.   Access to electrical equipment shall not be denied by an accumulation of wires and cables that prevents removal of panels, including suspended ceiling panels.

725.24 Mechanical Execution of Work.   Class 1, Class 2, and Class 3 circuits shall be installed in a neat and workmanlike manner. Cables and conductors installed exposed on the surface of ceilings and sidewalls shall be supported by the building structure in such a manner that the cable will not be damaged by normal building use. Such cables shall be supported by straps, staples, hangers, cable ties, or similar fittings designed and installed so as not to damage the cable. The installation shall also comply with 300.4(D).

725.25 Abandoned Cables.   The accessible portion of abandoned Class 2, Class 3, and PLTC cables shall be removed. Where cables are identified for future use with a tag, the tag shall be of sufficient durability to withstand the environment involved.

725.30 Class 1, Class 2, and Class 3 Circuit Identification.   Class 1, Class 2, and Class 3 circuits shall be identified at terminal and junction locations in a manner that prevents unintentional interference with other circuits during testing and servicing.

725.31 Safety-Control Equipment.

(A) Remote-Control Circuits.   Remote-control circuits for safety-control equipment shall be classified as Class 1 if the failure of the equipment to operate introduces a direct fire or life hazard. Room thermostats, water temperature regulating devices, and similar controls used in conjunction with electrically controlled household heating and air conditioning shall not be considered safety-control equipment.

(B) Physical Protection.   Where damage to remote-control circuits of safety-control equipment would introduce a hazard, as covered in 725.31(A), all conductors of such remote-control circuits shall be installed in rigid metal conduit, intermediate metal conduit, rigid nonmetallic conduit, electrical metallic tubing, Type MI cable, or Type MC cable, or be otherwise suitably protected from physical damage.

725.35 Class 1, Class 2, and Class 3 Circuit Requirements.   A remote-control, signaling, or power-limited circuit shall comply with the following parts of this article:

(1)   Class 1 Circuits: Parts I and II
(2)   Class 2 and Class 3 Circuits: Parts I and III

Part II. Class 1 Circuits

725.41 Class 1 Circuit Classifications and Power Source Requirements. Class 1 circuits shall be classified as either Class 1 power-limited circuits where they comply with the power limitations of 725.41(A) or as Class 1 remote-control and signaling circuits where they are used for remote-control or signaling purposes and comply with the power limitations of 725.41(B).

(A) Class 1 Power-Limited Circuits. These circuits shall be supplied from a source that has a rated output of not more than 30 volts and 1000 volt-amperes.

(1) Class 1 Transformers. Transformers used to supply power-limited Class 1 circuits shall comply with the applicable sections within Parts I and II of Article 450.

(2) Other Class 1 Power Sources. Power sources other than transformers shall be protected by overcurrent devices rated at not more than 167 percent of the volt-ampere rating of the source divided by the rated voltage. The overcurrent devices shall not be interchangeable with overcurrent devices of higher ratings. The overcurrent device shall be permitted to be an integral part of the power supply.

To comply with the 1000 volt-ampere limitation of 725.41(A), the maximum output (VA_{max}) of power sources other than transformers shall be limited to 2500 volt-amperes, and the product of the maximum current (I_{max}) and maximum voltage (V_{max}) shall not exceed 10,000 volt-amperes. These ratings shall be determined with any overcurrent-protective device bypassed.

VA_{max} is the maximum volt-ampere output after one minute of operation regardless of load and with overcurrent protection bypassed, if used. Current-limiting impedance shall not be bypassed when determining VA_{max}.

I_{max} is the maximum output current under any noncapacitive load, including short circuit, and with overcurrent protection bypassed, if used. Current-limiting impedance should not be bypassed when determining I_{max}. Where a current-limiting impedance, listed for the purpose or as part of a listed product, is used in combination with a stored energy source, for example, storage battery, to limit the output current, I_{max} limits apply after 5 seconds.

V_{max} is the maximum output voltage regardless of load with rated input applied.

(B) Class 1 Remote-Control and Signaling Circuits. These circuits shall not exceed 600 volts. The power output of the source shall not be required to be limited.

725.43 Class 1 Circuit Overcurrent Protection. Overcurrent protection for conductors 14 AWG and larger shall be provided in accordance with the conductor ampacity, without applying the ampacity adjustment and correction factors of 310.15 to the ampacity calculation. Overcurrent protection shall not exceed 7 amperes for 18 AWG conductors and 10 amperes for 16 AWG.

Exception: Where other articles of this Code permit or require other overcurrent protection.

> Informational Note: For example, see 430.72 for motors, 610.53 for cranes and hoists, and 517.74(B) and 660.9 for X-ray equipment.

725.45 Class 1 Circuit Overcurrent Device Location. Overcurrent devices shall be located as specified in 725.45(A), (B), (C), (D), or (E).

(A) Point of Supply. Overcurrent devices shall be located at the point where the conductor to be protected receives its supply.

(B) Feeder Taps. Class 1 circuit conductors shall be permitted to be tapped, without overcurrent protection at the tap, where the overcurrent device protecting the circuit conductor is sized to protect the tap conductor.

(C) Branch-Circuit Taps. Class 1 circuit conductors 14 AWG and larger that are tapped from the load side of the overcurrent protective device(s) of a controlled light and power circuit shall require only short-circuit and ground-fault protection and shall be permitted to be protected by the branch-circuit overcurrent protective device(s) where the rating of the protective device(s) is not more than 300 percent of the ampacity of the Class 1 circuit conductor.

(D) Primary Side of Transformer. Class 1 circuit conductors supplied by the secondary of a single-phase transformer having only a 2-wire (single-voltage) secondary shall be permitted to be protected by overcurrent protection provided on the primary side of the transformer, provided this protection is in accordance with 450.3 and does not exceed the value determined by multiplying the secondary conductor ampacity by the secondary-to-primary transformer voltage ratio. Transformer secondary conductors other than 2-wire shall not be considered to be protected by the primary overcurrent protection.

(E) Input Side of Electronic Power Source. Class 1 circuit conductors supplied by the output of a single-phase, listed electronic power source, other than a transformer, having only a 2-wire (single-voltage) output for connection to Class 1 circuits shall be permitted to be protected by overcurrent protection provided on the input side of the electronic power source, provided this protection does not exceed the value determined by multiplying the Class 1 circuit conductor ampacity by the output-to-input voltage ratio. Electronic power source outputs, other than 2-wire (single voltage), shall not be considered to be protected by the primary overcurrent protection.

725.46 Class 1 Circuit Wiring Methods. Class 1 circuits shall be installed in accordance with Part I of Article 300 and with the wiring methods from the appropriate articles in Chapter 3.

Exception No. 1: The provisions of 725.48 through 725.51 shall be permitted to apply in installations of Class 1 circuits.

Exception No. 2: Methods permitted or required by other articles of this Code shall apply to installations of Class 1 circuits.

725.48 Conductors of Different Circuits in the Same Cable, Cable Tray, Enclosure, or Raceway. Class 1 circuits shall be permitted to be installed with other circuits as specified in 725.48(A) and (B).

(A) Two or More Class 1 Circuits. Class 1 circuits shall be permitted to occupy the same cable, cable tray, enclosure, or raceway without regard to whether the individual circuits are alternating current or direct current, provided all conductors are insulated for the maximum voltage of any conductor in the cable, cable tray, enclosure, or raceway.

(B) Class 1 Circuits with Power-Supply Circuits. Class 1 circuits shall be permitted to be installed with power-supply conductors as specified in 725.48(B)(1) through (B)(4).

(1) In a Cable, Enclosure, or Raceway. Class 1 circuits and power-supply circuits shall be permitted to occupy the same cable, enclosure, or raceway only where the equipment powered is functionally associated.

(2) In Factory- or Field-Assembled Control Centers. Class 1 circuits and power-supply circuits shall be permitted to be installed in factory- or field-assembled control centers.

(3) In a Manhole. Class 1 circuits and power-supply circuits shall be permitted to be installed as underground conductors in a manhole in accordance with one of the following:

(1) The power-supply or Class 1 circuit conductors are in a metal-enclosed cable or Type UF cable.
(2) The conductors are permanently separated from the power-supply conductors by a continuous firmly fixed nonconductor, such as flexible tubing, in addition to the insulation on the wire.
(3) The conductors are permanently and effectively separated from the power supply conductors and securely fastened to racks, insulators, or other approved supports.

(4) In Cable Trays. Installations in cable trays shall comply with 725.48(B)(4)(1) or (B)(4)(2).

(1) Class 1 circuit conductors and power-supply conductors not functionally associated with the Class 1 circuit conductors shall be separated by a solid fixed barrier of a material compatible with the cable tray.
(2) Class 1 circuit conductors and power-supply conductors not functionally associated with the Class 1 circuit conductors shall be permitted to be installed in a cable tray without barriers where all of the conductors are installed with separate multiconductor Type AC, Type MC, Type MI, or Type TC cables and all the conductors in the cables are insulated at 600 volts or greater.

725.49 Class 1 Circuit Conductors.

(A) Sizes and Use. Conductors of sizes 18 AWG and 16 AWG shall be permitted to be used, provided they supply loads that do not exceed the ampacities given in 402.5 and are installed in a raceway, an approved enclosure, or a listed cable. Conductors larger than 16 AWG shall not supply loads greater than the ampacities given in 310.15. Flexible cords shall comply with Article 400.

(B) Insulation. Insulation on conductors shall be rated for the system voltage and not less than 600 volts. Conductors larger than 16 AWG shall comply with Article 310. Conductors in sizes 18 AWG and 16 AWG shall be Type FFH-2, KF-2, KFF-2, PAF, PAFF, PF, PFF, PGF, PGFF, PTF, PTFF, RFH-2, RFHH-2, RFHH-3, SF-2, SFF-2, TF, TFF, TFFN, TFN, ZF, or ZFF. Conductors with other types and thicknesses of insulation shall be permitted if listed for Class 1 circuit use.

725.51 Number of Conductors in Cable Trays and Raceway, and Ampacity Adjustment.

(A) Class 1 Circuit Conductors. Where only Class 1 circuit conductors are in a raceway, the number of conductors shall be determined in accordance with 300.17. The ampacity adjustment factors given in 310.15(B)(3)(a) shall apply only if such conductors carry continuous loads in excess of 10 percent of the ampacity of each conductor.

(B) Power-Supply Conductors and Class 1 Circuit Conductors. Where power-supply conductors and Class 1 circuit conductors are permitted in a raceway in accordance with 725.48, the number of conductors shall be determined in accordance with 300.17. The ampacity adjustment factors given in 310.15(B)(3)(a) shall apply as follows:

(1) To all conductors where the Class 1 circuit conductors carry continuous loads in excess of 10 percent of the ampacity of each conductor and where the total number of conductors is more than three
(2) To the power-supply conductors only, where the Class 1 circuit conductors do not carry continuous loads in excess of 10 percent of the ampacity of each conductor and where the number of power-supply conductors is more than three

(C) Class 1 Circuit Conductors in Cable Trays. Where Class 1 circuit conductors are installed in cable trays, they shall comply with the provisions of 392.22 and 392.80(A).

725.52 Circuits Extending Beyond One Building. Class 1 circuits that extend aerially beyond one building shall also meet the requirements of Article 225.

Part III. Class 2 and Class 3 Circuits

725.121 Power Sources for Class 2 and Class 3 Circuits.

(A) Power Source. The power source for a Class 2 or a Class 3 circuit shall be as specified in 725.121(A)(1), (A)(2), (A)(3), (A)(4), or (A)(5):

Informational Note No. 1: Informational Note Figure 725.121, No. 1 illustrates the relationships between Class 2 or Class 3 power sources, their supply, and the Class 2 or Class 3 circuits.

Informational Note No. 2: Table 11(A) and Table 11(B) in Chapter 9 provide the requirements for listed Class 2 and Class 3 power sources.

(1) A listed Class 2 or Class 3 transformer
(2) A listed Class 2 or Class 3 power supply
(3) Other listed equipment marked to identify the Class 2 or Class 3 power source

Exception No. 1 to (3): Thermocouples shall not require listing as a Class 2 power source.

Exception No. 2 to (3): Limited power circuits of listed equipment where these circuits have energy levels rated at or below the limits established in Chapter 9, Table 11(A) and Table 11(B).

Informational Note: Examples of other listed equipment are as follows:

(1) A circuit card listed for use as a Class 2 or Class 3 power source where used as part of a listed assembly
(2) A current-limiting impedance, listed for the purpose, or part of a listed product, used in conjunction with a non–power-limited transformer or a stored energy source, for example, storage battery, to limit the output current
(3) A thermocouple
(4) Limited voltage/current or limited impedance secondary communications circuits of listed industrial control equipment

(4) Listed audio/video information technology (computer), communications, and industrial equipment limited-power circuits.

> Informational Note: One way to determine applicable requirements for listing of information technology (computer) equipment is to refer to UL 60950-1-2011, *Standard for Safety of Information Technology Equipment.* Another way to determine applicable requirements for listing of audio/video, information and communication technology equipment is to refer to UL 62368-1-2014, *Safety of audio/video, information and communication technology equipment.* Typically such circuits are used to interconnect data circuits for the purpose of exchanging information data. One way to determine applicable requirements for listing of industrial equipment is to refer to UL 61010-2-201, *Safety requirements for electrical equipment for measurement, control, and laboratory use –Part 2-201: Particular requirements for control equipment,* and/or UL 61800-5-1, *Adjustable speed electrical power drive systems – Part 5-1: Safety requirements –Electrical, thermal and energy.*

(5) A dry cell battery shall be considered an inherently limited Class 2 power source, provided the voltage is 30 volts or less and the capacity is equal to or less than that available from series connected No. 6 carbon zinc cells.

(B) Interconnection of Power Sources. Class 2 or Class 3 power sources shall not have the output connections paralleled or otherwise interconnected unless listed for such interconnection.

N (C) Marking. The power sources for limited power circuits in 725.121(A)(3) and limited power circuits for listed audio/video information technology (equipment) and listed industrial equipment in 725.121(A)(4) shall have a label indicating the maximum voltage and current output for each connection point. The effective date shall be January 1, 2018.

725.124 Circuit Marking. The equipment supplying the circuits shall be durably marked where plainly visible to indicate each circuit that is a Class 2 or Class 3 circuit.

725.127 Wiring Methods on Supply Side of the Class 2 or Class 3 Power Source. Conductors and equipment on the supply side of the power source shall be installed in accordance with the appropriate requirements of Chapters 1 through 4. Transformers or other devices supplied from electric light or power circuits shall be protected by an overcurrent device rated not over 20 amperes.

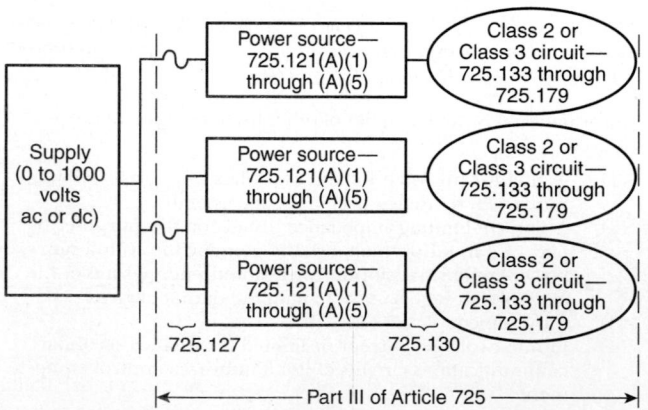

Informational Note Figure 725.121 No. 1 Class 2 and Class 3 Circuits.

Exception: The input leads of a transformer or other power source supplying Class 2 and Class 3 circuits shall be permitted to be smaller than 14 AWG, but not smaller than 18 AWG if they are not over 305 mm (12 in.) long and if they have insulation that complies with 725.49(B).

725.130 Wiring Methods and Materials on Load Side of the Class 2 or Class 3 Power Source. Class 2 and Class 3 circuits on the load side of the power source shall be permitted to be installed using wiring methods and materials in accordance with either 725.130(A) or (B).

(A) Class 1 Wiring Methods and Materials. Installation shall be in accordance with 725.46.

Exception No. 1: The ampacity adjustment factors given in 310.15(B)(3)(a) shall not apply.

Exception No. 2: Class 2 and Class 3 circuits shall be permitted to be reclassified and installed as Class 1 circuits if the Class 2 and Class 3 markings required in 725.124 are eliminated and the entire circuit is installed using the wiring methods and materials in accordance with Part II, Class 1 circuits.

> Informational Note: Class 2 and Class 3 circuits reclassified and installed as Class 1 circuits are no longer Class 2 or Class 3 circuits, regardless of the continued connection to a Class 2 or Class 3 power source.

(B) Class 2 and Class 3 Wiring Methods. Conductors on the load side of the power source shall be insulated at not less than the requirements of 725.179 and shall be installed in accordance with 725.133 and 725.154.

Exception No. 1: As provided for in 620.21 for elevators and similar equipment.

Exception No. 2: Other wiring methods and materials installed in accordance with the requirements of 725.3 shall be permitted to extend or replace the conductors and cables described in 725.179 and permitted by 725.130(B).

Exception No. 3: Bare Class 2 conductors shall be permitted as part of a listed intrusion protection system where installed in accordance with the listing instructions for the system.

725.133 Installation of Conductors and Equipment in Cables, Compartments, Cable Trays, Enclosures, Manholes, Outlet Boxes, Device Boxes, Raceways, and Cable Routing Assemblies for Class 2 and Class 3 Circuits. Conductors and equipment for Class 2 and Class 3 circuits shall be installed in accordance with 725.135 through 725.144.

725.135 Installation of Class 2, Class 3, and PLTC Cables. Installation of Class 2, Class 3, and PLTC cables shall comply with 725.135(A) through (M).

(A) Listing. Class 2, Class 3, and PLTC cables installed in buildings shall be listed.

(B) Ducts Specifically Fabricated for Environmental Air. The following wires and cables shall be permitted in ducts specifically fabricated for environmental air as described in 300.22(B) if directly associated with the air distribution system:

(1) Types CL2P and CL3P cables in lengths as short as practicable to perform the required function
(2) Types CL2P, CL3P, CL2R, CL3R, CL2, CL3, CL2X, CL3X, and PLTC cables installed in raceways that are installed in compliance with 300.22(B)

Informational Note: For information on fire protection of wiring installed in fabricated ducts, see 4.3.4.1 and 4.3.11.3.3 of NFPA 90A-2015, *Standard for the Installation of Air-Conditioning and Ventilating Systems.*

(C) Other Spaces Used for Environmental Air (Plenums). The following cables shall be permitted in other spaces used for environmental air as described in 300.22(C):

(1) Types CL2P and CL3P cables
(2) Types CL2P and CL3P cables installed in plenum communications raceways
(3) Types CL2P and CL3P cables installed in plenum cable routing assemblies
(4) Types CL2P and CL3P cables and plenum communications raceways supported by open metallic cable trays or cable tray systems
(5) Types CL2P, CL3P, CL2R, CL3R, CL2, CL3, CL2X, CL3X, and PLTC cables installed in raceways that are installed in compliance with 300.22(C)
(6) Types CL2P, CL3P, CL2R, CL3R, CL2, CL3, CL2X, CL3X, and PLTC cables supported by solid bottom metal cable trays with solid metal covers in other spaces used for environmental air (plenums) as described in 300.22(C)
(7) Types CL2P, CL3P, CL2R, CL3R, CL2, CL3, CL2X, CL3X, and PLTC cables installed in plenum communications raceways, riser communications raceways, and general-purpose communications raceways supported by solid bottom metal cable trays with solid metal covers in other spaces used for environmental air (plenums) as described in 300.22(C)

(D) Risers — Cables in Vertical Runs. The following cables shall be permitted in vertical runs penetrating one or more floors and in vertical runs in a shaft:

(1) Types CL2P, CL3P, CL2R, and CL3R cables
(2) Types CL2P, CL3P, CL2R, and CL3R cables installed in the following:

 a. Plenum communications raceways
 b. Plenum cable routing assemblies
 c. Riser communications raceways
 d. Riser cable routing assemblies

Informational Note: See 300.21 for firestop requirements for floor penetrations.

(E) Risers — Cables in Metal Raceways. The following cables shall be permitted in metal raceways in a riser having firestops at each floor:

(1) Types CL2P, CL3P, CL2R, CL3R, CL2, CL3, CL2X, CL3X, and PLTC cables
(2) Types CL2P, CL3P, CL2R, CL3R, CL2, CL3, CL2X, CL3X, and PLTC cables installed in the following:

 a. Plenum communications raceways
 b. Riser communications raceways
 c. General-purpose communications raceways

Informational Note: See 300.21 for firestop requirements for floor penetrations.

(F) Risers — Cables in Fireproof Shafts. The following shall be permitted to be installed in fireproof riser shafts having firestops at each floor:

(1) Types CL2P, CL3P, CL2R, CL3R, CL2, CL3, CL2X, CL3X, and PLTC cables

(2) Types CL2P, CL3P, CL2R, CL3R, CL2, CL3, and PLTC cables installed in the following:

 a. Plenum communications raceways
 b. Plenum cable routing assemblies
 c. Riser communications raceways
 d. Riser cable routing assemblies
 e. General-purpose communications raceways
 f. General-purpose cable routing assemblies

Informational Note: See 300.21 for firestop requirements for floor penetrations.

(G) Risers — One- and Two-Family Dwellings. The following cables shall be permitted in one- and two-family dwellings:

(1) Types CL2P, CL3P, CL2R, CL3R, CL2, CL3, and PLTC cables
(2) Types CL2X and CL3X cables less than 6 mm (0.25 in.) in diameter
(3) Types CL2P, CL3P, CL2R, CL3R, CL2, CL3, and PLTC cables installed in the following:

 a. Plenum communications raceways
 b. Plenum cable routing assemblies
 c. Riser communications raceways
 d. Riser cable routing assemblies
 e. General-purpose communications raceways
 f. General-purpose cable routing assemblies

(H) Cable Trays. Cables installed in cable trays outdoors shall be Type PLTC. The following cables shall be permitted to be supported by cable trays in buildings:

(1) Types CM CL2P, CL3P, CL2R, CL3R, CL2, CL3, and PLTC cables
(2) Types CL2P, CL3P, CL2R, CL3R, CL2, CL3, and PLTC cables installed in the following:

 a. Plenum communications raceways
 b. Riser communications raceways
 c. General-purpose communications raceways

(I) Cross-Connect Arrays. The following cables shall be permitted to be installed in cross-connect arrays:

(1) Types CL2P, CL3P, CL2R, CL3R, CL2, CL3, and PLTC cables
(2) Types CL2P, CL3P, CL2R, CL3R, CL2, CL3, and PLTC cables installed in the following:

 a. Plenum communications raceways
 b. Plenum cable routing assemblies
 c. Riser communications raceways
 d. Riser cable routing assemblies
 e. General-purpose communications raceways
 f. General-purpose cable routing assemblies

(J) Industrial Establishments. In industrial establishments where the conditions of maintenance and supervision ensure that only qualified persons service the installation, Type PLTC cable shall be permitted in accordance with either (1) or (2) as follows:

(1) Where the cable is not subject to physical damage, Type PLTC cable that complies with the crush and impact requirements of Type MC cable and is identified as PLTC-ER for such use shall be permitted to be exposed between the cable tray and the utilization equipment or device. The cable shall be continuously supported and protected against physical damage using mechanical protection such as dedicated struts, angles, or channels. The cable

shall be supported and secured at intervals not exceeding 1.8 m (6 ft). Where not subject to physical damage, Type PLTC-ER cable shall be permitted to transition between cable trays and between cable trays and utilization equipment or devices for a distance not to exceed 1.8 m (6 ft) without continuous support. The cable shall be mechanically supported where exiting the cable tray to ensure that the minimum bending radius is not exceeded.

(2) Type PLTC cable, with a metallic sheath or armor in accordance with 725.179(E), shall be permitted to be installed exposed. The cable shall be continuously supported and protected against physical damage using mechanical protection such as dedicated struts, angles, or channels. The cable shall be secured at intervals not exceeding 1.8 m (6 ft).

(K) Other Building Locations. The following wires and cables shall be permitted to be installed in building locations other than the locations covered in 725.135(B) through (I):

(1) Types CL2P, CL3P, CL2R, CL3R, CL2, CL3, and PLTC cables

(2) A maximum of 3 m (10 ft) of exposed Type CL2X wires and cables in nonconcealed spaces

(3) A maximum of 3 m (10 ft) of exposed Type CL3X wires and cables in nonconcealed spaces

(4) Types CL2P, CL3P, CL2R, CL3R, CL2, CL3, and PLTC cables installed in the following:

    a. Plenum communications raceways
    b. Plenum cable routing assemblies
    c. Riser communications raceways
    d. Riser cable routing assemblies
    e. General-purpose communications raceways
    f. General-purpose cable routing assemblies

(5) Types CL2P, CL3P, CL2R, CL3R, CL2, CL3, CL2X, CL3X, and PLTC cables installed in raceways recognized in Chapter 3

(6) Type CMUC undercarpet communications wires and cables installed under carpet, modular flooring, and planks

(L) Multifamily Dwellings. The following wires and cables shall be permitted to be installed in multifamily dwellings in locations other than the locations covered in 725.135(B) through (I):

(1) Types CL2P, CL3P, CL2R, CL3R, CL2, CL3, and PLTC wires and cables

(2) Type CL2X wires and cables less than 6 mm ($\frac{1}{4}$ in.) in diameter in nonconcealed spaces

(3) Type CL3X wires and cables less than 6 mm ($\frac{1}{4}$ in.) in diameter in nonconcealed spaces

(4) Types CL2P, CL3P, CL2R, CL3R, CL2, CL3, and PLTC wires and cables installed in the following:

    a. Plenum communications raceways
    b. Plenum cable routing assemblies
    c. Riser communications raceways
    d. Riser cable routing assemblies
    e. General-purpose communications raceways
    f. General-purpose cable routing assemblies

(5) Types CL2P, CL3P, CL2R, CL3R, CL2, CL3, CL2X, CL3X, and PLTC wires and cables installed in raceways recognized in Chapter 3

(6) Type CMUC undercarpet communications wires and cables installed under carpet, modular flooring, and planks

(M) One- and Two-Family Dwellings. The following wires and cables shall be permitted to be installed in one- and two-family dwellings in locations other than the locations covered in 725.135(B) through (I):

(1) Types CL2P, CL3P, CL2R, CL3R, CL2, CL3, and PLTC wires and cables

(2) Type CL2X wires and cables less than 6 mm ($\frac{1}{4}$ in.) in diameter

(3) Type CL3X wires and cables less than 6 mm ($\frac{1}{4}$ in.) in diameter

(4) Communications wires and Types CL2P, CL3P, CL2R, CL3R, CL2, CL3, and PLTC cables installed in the following:

    a. Plenum communications raceways
    b. Plenum cable routing assemblies
    c. Riser communications raceways
    d. Riser cable routing assemblies
    e. General-purpose communications raceways
    f. General-purpose cable routing assemblies

(5) Types CL2P, CL3P, CL2R, CL3R, CL2, CL3, CL2X, CL3X, and PLTC wires and cables installed in raceways recognized in Chapter 3

(6) Type CMUC undercarpet communications wires and cables installed under carpet, modular flooring, and planks

725.136 Separation from Electric Light, Power, Class 1, Non–Power-Limited Fire Alarm Circuit Conductors, and Medium-Power Network-Powered Broadband Communications Cables.

(A) General. Cables and conductors of Class 2 and Class 3 circuits shall not be placed in any cable, cable tray, compartment, enclosure, manhole, outlet box, device box, raceway, or similar fitting with conductors of electric light, power, Class 1, non–power-limited fire alarm circuits, and medium-power network-powered broadband communications circuits unless permitted by 725.136(B) through (I).

(B) Separated by Barriers. Class 2 and Class 3 circuits shall be permitted to be installed together with the conductors of electric light, power, Class 1, non–power-limited fire alarm and medium power network-powered broadband communications circuits where they are separated by a barrier.

(C) Raceways Within Enclosures. In enclosures, Class 2 and Class 3 circuits shall be permitted to be installed in a raceway to separate them from Class 1, non–power-limited fire alarm and medium-power network-powered broadband communications circuits.

(D) Associated Systems Within Enclosures. Class 2 and Class 3 circuit conductors in compartments, enclosures, device boxes, outlet boxes, or similar fittings shall be permitted to be installed with electric light, power, Class 1, non–power-limited fire alarm, and medium-power network-powered broadband communications circuits where they are introduced solely to connect the equipment connected to Class 2 and Class 3 circuits, and where (1) or (2) applies:

(1) The electric light, power, Class 1, non–power-limited fire alarm, and medium-power network-powered broadband communications circuit conductors are routed to main-

tain a minimum of 6 mm (0.25 in.) separation from the conductors and cables of Class 2 and Class 3 circuits.

(2) The circuit conductors operate at 150 volts or less to ground and also comply with one of the following:

    a. The Class 2 and Class 3 circuits are installed using Type CL3, CL3R, or CL3P or permitted substitute cables, provided these Class 3 cable conductors extending beyond the jacket are separated by a minimum of 6 mm (0.25 in.) or by a nonconductive sleeve or nonconductive barrier from all other conductors.

    b. The Class 2 and Class 3 circuit conductors are installed as a Class 1 circuit in accordance with 725.41.

(E) Enclosures with Single Opening. Class 2 and Class 3 circuit conductors entering compartments, enclosures, device boxes, outlet boxes, or similar fittings shall be permitted to be installed with Class 1, non–power-limited fire alarm and medium-power network-powered broadband communications circuits where they are introduced solely to connect the equipment connected to Class 2 and Class 3 circuits. Where Class 2 and Class 3 circuit conductors must enter an enclosure that is provided with a single opening, they shall be permitted to enter through a single fitting (such as a tee), provided the conductors are separated from the conductors of the other circuits by a continuous and firmly fixed nonconductor, such as flexible tubing.

(F) Manholes. Underground Class 2 and Class 3 circuit conductors in a manhole shall be permitted to be installed with Class 1, non–power-limited fire alarm and medium-power network-powered broadband communications circuits where one of the following conditions is met:

(1) The electric light, power, Class 1, non–power-limited fire alarm and medium-power network-powered broadband communications circuit conductors are in a metal-enclosed cable or Type UF cable.

(2) The Class 2 and Class 3 circuit conductors are permanently and effectively separated from the conductors of other circuits by a continuous and firmly fixed nonconductor, such as flexible tubing, in addition to the insulation or covering on the wire.

(3) The Class 2 and Class 3 circuit conductors are permanently and effectively separated from conductors of the other circuits and securely fastened to racks, insulators, or other approved supports.

(G) Cable Trays. Class 2 and Class 3 circuit conductors shall be permitted to be installed in cable trays, where the conductors of the electric light, Class 1, and non–power-limited fire alarm circuits are separated by a solid fixed barrier of a material compatible with the cable tray or where the Class 2 or Class 3 circuits are installed in Type MC cable.

(H) In Hoistways. In hoistways, Class 2 or Class 3 circuit conductors shall be installed in rigid metal conduit, rigid nonmetallic conduit, intermediate metal conduit, liquidtight flexible nonmetallic conduit, or electrical metallic tubing. For elevators or similar equipment, these conductors shall be permitted to be installed as provided in 620.21.

(I) Other Applications. For other applications, conductors of Class 2 and Class 3 circuits shall be separated by at least 50 mm (2 in.) from conductors of any electric light, power, Class 1 non–power-limited fire alarm or medium power network-powered broadband communications circuits unless one of the following conditions is met:

(1) Either (a) all of the electric light, power, Class 1, non–power-limited fire alarm and medium-power network-powered broadband communications circuit conductors or (b) all of the Class 2 and Class 3 circuit conductors are in a raceway or in metal-sheathed, metal-clad, non-metallic-sheathed, or Type UF cables.

(2) All of the electric light, power, Class 1 non–power-limited fire alarm, and medium-power network-powered broadband communications circuit conductors are permanently separated from all of the Class 2 and Class 3 circuit conductors by a continuous and firmly fixed nonconductor, such as porcelain tubes or flexible tubing, in addition to the insulation on the conductors.

725.139 Installation of Conductors of Different Circuits in the Same Cable, Enclosure, Cable Tray, Raceway, or Cable Routing Assembly.

(A) Two or More Class 2 Circuits. Conductors of two or more Class 2 circuits shall be permitted within the same cable, enclosure, raceway, or cable routing assembly.

(B) Two or More Class 3 Circuits. Conductors of two or more Class 3 circuits shall be permitted within the same cable, enclosure, raceway, or cable routing assembly.

(C) Class 2 Circuits with Class 3 Circuits. Conductors of one or more Class 2 circuits shall be permitted within the same cable, enclosure, raceway, or cable routing assembly with conductors of Class 3 circuits, provided that the insulation of the Class 2 circuit conductors in the cable, enclosure, raceway, or cable routing assembly is at least that required for Class 3 circuits.

(D) Class 2 and Class 3 Circuits with Communications Circuits.

(1) Classified as Communications Circuits. Class 2 and Class 3 circuit conductors shall be permitted in the same cable with communications circuits, in which case the Class 2 and Class 3 circuits shall be classified as communications circuits and shall be installed in accordance with the requirements of Article 800. The cables shall be listed as communications cables.

(2) Composite Cables. Cables constructed of individually listed Class 2, Class 3, and communications cables under a common jacket shall be permitted to be classified as communications cables. The fire resistance rating of the composite cable shall be determined by the performance of the composite cable.

(E) Class 2 or Class 3 Cables with Other Circuit Cables. Jacketed cables of Class 2 or Class 3 circuits shall be permitted in the same enclosure, cable tray, raceway, or cable routing assembly with jacketed cables of any of the following:

(1) Power-limited fire alarm systems in compliance with Parts I and III of Article 760

(2) Nonconductive and conductive optical fiber cables in compliance with Parts I and IV of Article 770

(3) Communications circuits in compliance with Parts I and IV of Article 800

(4) Community antenna television and radio distribution systems in compliance with Parts I and IV of Article 820

(5) Low-power, network-powered broadband communications in compliance with Parts I and IV of Article 830

(F) Class 2 or Class 3 Conductors or Cables and Audio System Circuits. Audio system circuits described in 640.9(C), and installed using Class 2 or Class 3 wiring methods in compliance with 725.133 and 725.154, shall not be permitted to be installed in the same cable, raceway, or cable routing assembly with Class 2 or Class 3 conductors or cables.

725.141 Installation of Circuit Conductors Extending Beyond One Building. Where Class 2 or Class 3 circuit conductors extend beyond one building and are run so as to be subject to accidental contact with electric light or power conductors operating over 300 volts to ground, or are exposed to lightning on interbuilding circuits on the same premises, the requirements of the following shall also apply:

(1) Sections 800.44, 800.50, 800.53, 800.93, 800.100, 800.170(A), and 800.170(B) for other than coaxial conductors

(2) Sections 820.44, 820.93, and 820.100 for coaxial conductors

725.143 Support of Conductors. Class 2 or Class 3 circuit conductors shall not be strapped, taped, or attached by any means to the exterior of any conduit or other raceway as a means of support. These conductors shall be permitted to be installed as permitted by 300.11(C)(2).

N **725.144 Transmission of Power and Data.** The requirements of 725.144(A) and (B) shall apply to Class 2 and Class 3 circuits that transmit power and data to a powered device. The requirements of Parts I and III of Article 725 and 300.11 shall apply to Class 2 and Class 3 circuits that transmit power and data. The conductors that carry power for the data circuits shall be copper. The current in the power circuit shall not exceed the current limitation of the connectors.

Informational No. 1: One example of the use of cables that transmit power and data is the connection of closed-circuit TV cameras (CCTV).

Informational Note No. 2: The 8P8C connector is in widespread use with powered communications systems. These connectors are typically rated at 1.3 amperes maximum.

(A) Use of Class 2 or Class 3 Cables to Transmit Power and Data. Where Types CL3P, CL2P, CL3R, CL2R, CL3, or CL2 transmit power and data, the following shall apply, as applicable:

(1) The ampacity ratings in Table 725.144 shall apply at an ambient temperature of 30°C (86°F).

(2) For ambient temperatures above 30°C (86°F), the correction factors of 310.15(B)(2) shall apply.

Informational Note: One example of the use of Class 2 cables is a network of closed-circuit TV cameras using 24 AWG, 60°C rated, Type CL2R, Category 5e local area network (LAN) cables.

(B) Use of Class 2-LP or Class 3-LP Cables to Transmit Power and Data. Types CL3P-LP, CL2P-LP, CL3R-LP, CL2R-LP, CL3-LP, or CL2-LP shall be permitted to supply power to equipment at a current level up to the marked ampere limit located immediately following the suffix LP and shall be permitted to transmit data to the equipment. The Class 2-LP and Class 3-LP cables shall comply with the following, as applicable:

Informational Note 1: The "(xxA)" following the suffix -LP indicates the ampacity of each conductor in a cable.

Informational Note 2: An example of a limited power (LP) cable is a cable marked Type CL2-LP(0.5A), 23 AWG. A Type CL2-LP(0.5), 23 AWG could be used in any location where a Type CL2 could be used; however, the LP cable would be suitable for carrying up to 0.5 A per conductor, regardless of the number of cables in a bundle. If used in a 7-cable bundle, the same cable could carry up to 1.2 amperes per conductor.

(1) Cables with the suffix "-LP" shall be permitted to be installed in bundles, raceways, cable trays, communications raceways, and cable routing assemblies.

(2) Cables with the suffix "-LP" and a marked ampere level shall follow the substitution hierarchy of Table 725.154 and Figure 725.154(A) for the cable type without the suffix "LP" and without the marked ampere level.

(3) System design shall be permitted by qualified persons under engineering supervision.

725.154 Applications of Listed Class 2, Class 3, and PLTC Cables. Class 2, Class 3, and PLTC cables shall comply with any of the requirements described in 725.154(A) through (C) and as indicated in Table 725.154.

Table 725.144 Ampacities of Each Conductor in Amperes in 4-Pair Class 2 or Class 3 Data Cables Based on Copper Conductors at an Ambient Temperature of 30°C (86°F) with All Conductors in All Cables Carrying Current, 60°C (140°F), 75°C (167°F), and 90°C (194°F) Rated Cables

AWG	Number of 4-Pair Cables in a Bundle																				
	1			2–7			8–19			20–37			38–61			62–91			92–192		
	Temperature Rating			Temperature Rating			Temperature Rating			Temperature Rating			Temperature Rating			Temperature Rating			Temperature Rating		
	60°C	75°C	90°C	60°C	75°C	90°C	60°C	75°C	90°C	60°C	75°C	90°C	60°C	75°C	90°C	60°C	75°C	90°C	60°C	75°C	90°C
26	1	1	1	1	1	1	0.7	0.8	1	0.5	0.6	0.7	0.4	0.5	0.6	0.4	0.5	0.6	NA	NA	NA
24	2	2	2	1	1.4	1.6	0.8	1	1.1	0.6	0.7	0.9	0.5	0.6	0.7	0.4	0.5	0.6	0.3	0.4	0.5
23	2.5	2.5	2.5	1.2	1.5	1.7	0.8	1.1	1.2	0.6	0.8	0.9	0.5	0.7	0.8	0.5	0.7	0.8	0.4	0.5	0.6
22	3	3	3	1.4	1.8	2.1	1	1.2	1.4	0.7	0.9	1.1	0.6	0.8	0.9	0.6	0.8	0.9	0.5	0.6	0.7

Note 1: For bundle sizes over 192 cables, or for conductor sizes smaller than 26 AWG, ampacities shall be permitted to be determined by qualified personnel under engineering supervision.
Note 2: Where only half of the conductors in each cable are carrying current, the values in the table shall be permitted to be increased by a factor of 1.4.
Informational Note: The conductor sizes in data cables in wide-spread use are typically 22–26 AWG.

Table 725.154 Applications of Listed Class 2, Class 3, CMUC, and PLTC Cables in Buildings

	Applications	Cable Type					
		CL2P & CL3P	CL2R & CL3R	CL2 & CL3	CL2X & CL3X	CMUC	PLTC
In fabricated ducts as described in 300.22(B)	In fabricated ducts	Y*	N	N	N	N	N
	In metal raceway that complies with 300.22(B)	Y*	Y*	Y*	Y*	N	Y*
In other spaces used for environmental air as described in 300.22(C)	In other spaces used for environmental air	Y*	N	N	N	N	N
	In metal raceway that complies with 300.22(C)	Y*	Y*	Y*	Y*	N	Y*
	In plenum communications raceways	Y*	N	N	N	N	N
	In plenum cable routing assemblies	Y*	N	N	N	N	N
	Supported by open metal cable trays	Y*	N	N	N	N	N
	Supported by solid bottom metal cable trays with solid metal covers	Y*	Y*	Y*	Y*	N	N
In risers	In vertical runs	Y*	Y*	N	N	N	N
	In metal raceways	Y*	Y*	Y*	Y*	N	Y*
	In fireproof shafts	Y*	Y*	Y*	Y*	N	Y*
	In plenum communications raceways	Y*	Y*	N	N	N	N
	In plenum cable routing assemblies	Y*	Y*	N	N	N	N
	In riser communications raceways	Y*	Y*	N	N	N	N
	In riser cable routing assemblies	Y*	Y*	N	N	N	N
	In one- and two-family dwellings	Y*	Y*	Y*	Y*	N	Y*
Within buildings in other than air-handling spaces and risers	General	Y*	Y*	Y*	Y*	N	Y*
	In one- and two-family dwellings	Y*	Y*	Y*	Y*	Y*	Y*
	In multifamily dwellings	Y*	Y*	Y*	Y*	Y*	Y*
	In nonconcealed spaces	Y*	Y*	Y*	Y*	Y*	Y*
	Supported by cable trays	Y*	Y*	Y*	N	N	Y*
	Under carpet	N	N	N	N	Y*	N
	In cross-connect arrays	Y*	Y*	Y*	N	N	Y*
	In any raceway recognized in Chapter 3	Y*	Y*	Y*	Y*	N	Y*
	In plenum communications raceways	Y*	Y*	Y*	N	N	Y*
	In plenum cable routing assemblies	Y*	Y*	Y*	N	N	Y*
	In riser communications raceways	Y*	Y*	Y*	N	N	Y*
	In riser cable routing assemblies	Y*	Y*	Y*	N	N	Y*
	In general-purpose communications raceways	Y*	Y*	Y*	N	N	Y*
	In general-purpose cable routing assemblies	Y*	Y*	Y*	N	N	Y*

Note: "N" indicates that the cable type shall not be permitted to be installed in the application.
"Y*" indicates that the cable type shall be permitted to be installed in the application, subject to the limitations described in 725.130 through 725.143.

(A) Class 2 and Class 3 Cable Substitutions. The substitutions for Class 2 and Class 3 cables listed in Table 725.154(A) and illustrated in Figure 725.154(A) shall be permitted. Where substitute cables are installed, the wiring requirements of Article 725, Parts I and III, shall apply.

Informational Note: For information on Types CMP, CMR, CM, and CMX, see 800.179.

(B) Class 2, Class 3, PLTC Circuit Integrity (CI) Cable or Electrical Circuit Protective System. Circuit integrity (CI) cable or a listed electrical circuit protective system shall be permitted for use in remote control, signaling, or power-limited systems that supply critical circuits to ensure survivability for continued circuit operation for a specified time under fire conditions.

(C) Thermocouple Circuits. Conductors in Type PLTC cables used for Class 2 thermocouple circuits shall be permitted to be any of the materials used for thermocouple extension wire.

Table 725.154(A) Cable Substitutions

Cable Type	Permitted Substitutions
CL3P	CMP
CL2P	CMP, CL3P
CL3R	CMP, CL3P, CMR
CL2R	CMP, CL3P, CL2P, CMR, CL3R
PLTC	
CL3	CMP, CL3P, CMR, CL3R, CMG, CM, PLTC
CL2	CMP, CL3P, CL2P, CMR, CL3R, CL2R, CMG, CM, PLTC, CL3
CL3X	CMP, CL3P, CMR, CL3R, CMG, CM, PLTC, CL3, CMX
CL2X	CMP, CL3P, CL2P, CMR, CL3R, CL2R, CMG, CM, PLTC, CL3, CL2, CMX, CL3X

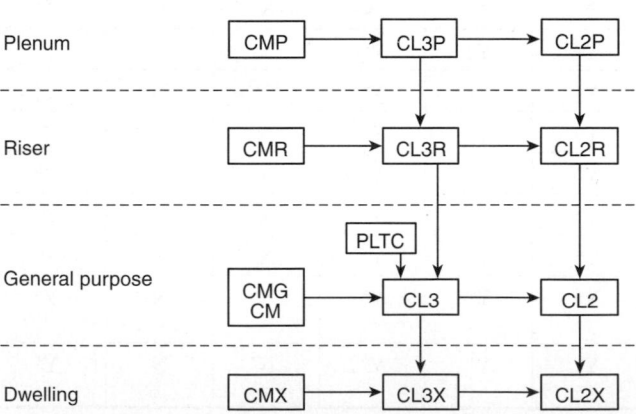

Type CM—Communications wires and cables
Type CL2 and CL3—Class 2 and Class 3 remote-control, signaling, and power-limited cables
Type PLTC—Power-limited tray cable

[A]→[B]   Cable A shall be permitted to be used in place of cable B.

FIGURE 725.154(A)   Cable Substitution Hierarchy.

Part IV. Listing Requirements

N **725.170 Listing and Marking of Equipment for Power and Data Transmission.** The listed power source for circuits intended to provide power and data over Class 2 cables to remote equipment shall be as specified in 725.121(A)(1), (A)(2), (A)(3), or (A)(4). In accordance with 725.121(B), the power sources shall not have the output connections paralleled or otherwise interconnected, unless listed for such interconnection. Powered devices connected to a circuit supplying data and power shall be listed. Marking of equipment output connections shall be in accordance with 725.121(C).

725.179 Listing and Marking of Class 2, Class 3, and Type PLTC Cables. Class 2, Class 3, and Type PLTC cables, installed as wiring methods within buildings, shall be listed as resistant to the spread of fire and other criteria in accordance with 725.179(A) through (I) and shall be marked in accordance with 725.179(J).

(A) Types CL2P and CL3P. Types CL2P and CL3P plenum cable shall be listed as suitable for use in ducts, plenums, and other space for environmental air and shall be listed as having adequate fire-resistant and low-smoke producing characteristics.

Informational Note: One method of defining a cable that is low-smoke producing and fire resistant is that the cable exhibits a maximum peak optical density of 0.50 or less, an average optical density of 0.15 or less, and a maximum flame spread distance of 1.52 m (5 ft) or less when tested in accordance with NFPA 262 -2015, *Standard Method of Test for Flame Travel and Smoke of Wires and Cables for Use in Air-Handling Spaces.*

(B) Types CL2R and CL3R. Types CL2R and CL3R riser cables shall be marked as Type CL2R or CL3R, respectively, and be listed as suitable for use in a vertical run in a shaft or from floor to floor and shall be listed as having fire-resistant characteristics capable of preventing the carrying of fire from floor to floor.

Informational Note: One method of defining fire-resistant characteristics capable of preventing the carrying of fire from floor to floor is that the cables pass the requirements of ANSI/ UL 1666-2012, *Test for Flame Propagation Height of Electrical and Optical-Fiber Cable Installed Vertically in Shafts.*

(C) Types CL2 and CL3. Types CL2 and CL3 cables shall be marked as Type CL2 or CL3, respectively, and be listed as suitable for general-purpose use, with the exception of risers, ducts, plenums, and other space used for environmental air, and shall be listed as resistant to the spread of fire.

Informational Note: One method of defining *resistant to the spread of fire* is that the cables do not spread fire to the top of the tray in the UL flame exposure, vertical tray flame test in ANSI/ UL 1685-2010, *Standard for Safety for Vertical-Tray Fire-Propagation and Smoke-Release Test for Electrical and Optical-Fiber Cables.* The smoke measurements in the test method are not applicable.

Another method of defining *resistant to the spread of fire* is for the damage (char length) not to exceed 1.5 m (4 ft 11 in.) when performing the CSA vertical flame test for— cables in cable trays, as described in CSA C22.2 No. 0.3-M-2001, *Test Methods for Electrical Wires and Cables.*

(D) Types CL2X and CL3X. Types CL2X and CL3X limited-use cables shall be marked as Type CL2X or CL3X, and be listed as suitable for use in dwellings and raceways and shall be listed as resistant to flame spread.

Informational Note: One method of determining that cable is resistant to flame spread is by testing the cable to the VW-1 (vertical wire) flame test in ANSI/UL 1581-2011, *Reference Standard for Electrical Wires, Cables and Flexible Cords.*

(E) Type PLTC. Type PLTC nonmetallic-sheathed, power-limited tray cable shall be listed as being suitable for cable trays and shall consist of a factory assembly of two or more insulated conductors under a nonmetallic jacket. The insulated conductors shall be 22 AWG through 12 AWG. The conductor material shall be copper (solid or stranded). Insulation on conductors shall be rated for 300 volts. The cable core shall be two or more parallel conductors, one or more group assemblies of twisted or parallel conductors, or a combination thereof. A metallic shield or a metallized foil shield with drain wire(s) shall be permitted to be applied over the cable core, over groups of conductors, or both. The cable shall be listed as resistant to the spread of fire. The outer jacket shall be a sunlight- and moisture-resistant nonmetallic material. Type PLTC cable used in a wet location shall be listed for use in wet locations or have a moisture-impervious metal sheath.

Exception No. 1: Where a smooth metallic sheath, continuous corrugated metallic sheath, or interlocking tape armor is applied over the nonmetallic jacket, an overall nonmetallic jacket shall not be required. On metallic-sheathed cable without an overall nonmetallic jacket, the information required in 310.120 shall be located on the nonmetallic jacket under the sheath.

Exception No. 2: Conductors in PLTC cables used for Class 2 thermocouple circuits shall be permitted to be any of the materials used for thermocouple extension wire.

Informational Note: One method of defining *resistant to the spread of fire* is that the cables do not spread fire to the top of the tray in the UL flame exposure, vertical tray flame test in ANSI/UL 1685-2010, *Standard for Safety for Vertical-Tray Fire-Propagation and Smoke-Release Test for Electrical and Optical-Fiber Cables.* The smoke measurements in the test method are not applicable.

Another method of defining *resistant to the spread of fire* is for the damage (char length) not to exceed 1.5 m (4 ft 11 in.) when performing the CSA vertical tray flame test for cables in cable trays, as described in CSA C22.2 No. 0.3-M-2001, *Test Methods for Electrical Wires and Cables.*

(F) Circuit Integrity (CI) Cable or Electrical Circuit Protective System. Cables that are used for survivability of critical circuits under fire conditions shall meet either 725.179(F)(1) or (F)(2) as follows:

(1) Circuit Integrity (CI) Cables. Circuit Integrity (CI) cables, specified in 725.179(A), (B), (C), and (E), and used for survivability of critical circuits, shall have the additional classification using the suffix "CI." Circuit integrity (CI) cables shall only be permitted to be installed in a raceway where specifically listed and marked as part of an electrical circuit protective system as covered in 725.179(F)(2).

(2) Electrical Circuit Protective System. Cables specified in 725.179(A), (B), (C), (E), and (F)(1) that are part of an electrical circuit protective system shall be identified with the protective system number and hourly rating printed on the outer jacket of the cable and installed in accordance with the listing of the protective system.

Informational Note No. 1: One method of defining circuit integrity (CI) cable or an electrical circuit protective system is by establishing a minimum 2-hour fire-resistive rating when tested in accordance with UL 2196-2012, *Standard for Tests of Fire Resistive Cables.*

Informational Note No. 2: UL guide information for electrical circuit protective systems (FHIT) contains information on proper installation requirements to maintain the fire rating.

(G) Class 2 and Class 3 Cable Voltage Ratings. Class 2 cables shall have a voltage rating of not less than 150 volts. Class 3 cables shall have a voltage rating of not less than 300 volts. Class 2 and Class 3 cables shall have a temperature rating of not less than 60°C (140°F).

(H) Class 3 Single Conductors. Class 3 single conductors used as other wiring within buildings shall not be smaller than 18 AWG and shall be Type CL3. Conductor types described in 725.49(B) that are also listed as Type CL3 shall be permitted.

Informational Note: One method of defining *resistant to the spread of fire* is that the cables do not spread fire to the top of the tray in the UL flame exposure, vertical tray flame test in ANSI/UL 1685-2010, *Standard for Safety for Vertical-Tray Fire-Propagation and Smoke-Release Test for Electrical and Optical-Fiber Cables.* The smoke measurements in the test method are not applicable.

Another method of defining *resistant to the spread of fire* is for the damage (char length) not to exceed 1.5 m (4 ft 11 in.) when performing the CSA vertical tray flame test for cables in cable trays, as described in CSA C22.2 No. 0.3-M-2001, *Test Methods for Electrical Wires and Cables.*

N (I) Limited Power (LP) Cables. Limited power (LP) cables shall be listed as suitable for carrying power and data circuits up to a specified current limit for each conductor without exceeding the temperature rating of the cable where the cable is installed in cable bundles in free air or installed within a raceway, cable tray, or cable routing assembly. The cables shall be marked with the suffix "-LP" with the ampere limit located immediately following the suffix LP, where the current limit is in amperes per conductor.

Informational Note: The ampere limit located immediately following the suffix LP is the ampacity of each conductor in a cable. For example, 1 ampere Class 2 limited-power cables would be marked CL2-LP (1.0A), CL2R-LP (1.0A), or CL2-LP (1.0A).

(J) Marking. Cables shall be marked in accordance with 310.120(A)(2), (A)(3), (A)(4), (A)(5), and Table 725.179(J). Voltage ratings shall not be marked on the cables.

Informational Note: Voltage markings on cables may be misinterpreted to suggest that the cables may be suitable for Class 1 electric light and power applications.

Exception: Voltage markings shall be permitted where the cable has multiple listings and a voltage marking is required for one or more of the listings.

Temperature rating shall be marked on the jacket of Class 2 and Class 3 cables that have a temperature rating exceeding 60°C (140°F).

Table 725.179(J) Cable Marking

Cable Marking	Type
CL3P	Class 3 plenum cable
CL2P	Class 2 plenum cable
CL3R	Class 3 riser cable
CL2R	Class 2 riser cable
PLTC	Power-limited tray cable
CL3	Class 3 cable
CL2	Class 2 cable
CL3X	Class 3 cable, limited use
CL2X	Class 2 cable, limited use

Informational Note: Class 2 and Class 3 cable types are listed in descending order of fire resistance rating, and Class 3 cables are listed above Class 2 cables because Class 3 cables can substitute for Class 2 cables.

ARTICLE 727
Instrumentation Tray Cable: Type ITC

727.1 Scope. This article covers the use, installation, and construction specifications of instrumentation tray cable for application to instrumentation and control circuits operating at 150 volts or less and 5 amperes or less.

727.2 Definition.

Type ITC Instrumentation Tray Cable. A factory assembly of two or more insulated conductors, with or without a grounding conductor(s), enclosed in a nonmetallic sheath.

727.3 Other Articles. In addition to the provisions of this article, installation of Type ITC cable shall comply with other applicable articles of this *Code.*

727.4 Uses Permitted. Type ITC cable shall be permitted to be used as follows in industrial establishments where the conditions of maintenance and supervision ensure that only qualified persons service the installation:

(1) In cable trays.
(2) In raceways.
(3) In hazardous locations as permitted in 501.10, 502.10, 503.10, 504.20, 504.30, 504.80, and 505.15.
(4) Enclosed in a smooth metallic sheath, continuous corrugated metallic sheath, or interlocking tape armor applied over the nonmetallic sheath in accordance with 727.6. The cable shall be supported and secured at intervals not exceeding 1.8 m (6 ft).
(5) Cable, without a metallic sheath or armor, that complies with the crush and impact requirements of Type MC cable and is identified for such use with the marking *ITC-ER* shall be permitted to be installed exposed. The cable shall be continuously supported and protected against physical damage using mechanical protection such as dedicated struts, angles, or channels. The cable shall be secured at intervals not exceeding 1.8 m (6 ft).

Exception to (5): Where not subject to physical damage, Type ITC-ER shall be permitted to transition between cable trays and between cable trays and utilization equipment or devices for a distance not to exceed 1.8 m (6 ft) without continuous support. The cable shall be mechanically supported where exiting the cable tray to ensure that the minimum bending radius is not exceeded.

(6) As aerial cable on a messenger.
(7) Direct buried where identified for the use.
(8) Under raised floors in rooms containing industrial process control equipment and rack rooms where arranged to prevent damage to the cable.
(9) Under raised floors in information technology equipment rooms in accordance with 645.5(E)(5)(b).

727.5 Uses Not Permitted. Type ITC cable shall not be installed on circuits operating at more than 150 volts or more than 5 amperes.

Installation of Type ITC cable with other cables shall be subject to the stated provisions of the specific articles for the other cables. Where the governing articles do not contain stated provisions for installation with Type ITC cable, the installation of Type ITC cable with the other cables shall not be permitted.

Type ITC cable shall not be installed with power, lighting, Class 1 circuits that are not power limited, or non–power-limited circuits.

Exception No. 1: Where terminated within equipment or junction boxes and separations are maintained by insulating barriers or other means.

Exception No. 2: Where a metallic sheath or armor is applied over the nonmetallic sheath of the Type ITC cable.

727.6 Construction. The insulated conductors of Type ITC cable shall be in sizes 22 AWG through 12 AWG. The conductor material shall be copper or thermocouple alloy. Insulation on the conductors shall be rated for 300 volts. Shielding shall be permitted.

The cable shall be listed as being resistant to the spread of fire. The outer jacket shall be sunlight and moisture resistant.

Where a smooth metallic sheath, continuous corrugated metallic sheath, or interlocking tape armor is applied over the nonmetallic sheath, an overall nonmetallic jacket shall not be required.

Informational Note: One method of defining *resistant to the spread of fire* is that the cables do not spread fire to the top of the tray in the UL flame exposure, vertical tray flame test in ANSI/UL 1685-2010, *Standard for Safety for Vertical-Tray Fire-Propagation and Smoke-Release Test for Electrical and Optical-Fiber Cables.* The smoke measurements in the test method are not applicable.

Another method of defining *resistant to the spread of fire* is for the damage (char length) not to exceed 1.5 m (4 ft 11 in.) when performing the CSA vertical flame test — cables in cable trays, as described in CSA C22.2 No. 0.3-M-2001, *Test Methods for Electrical Wires and Cables.*

727.7 Marking. The cable shall be marked in accordance with 310.120(A)(2), (A)(3), (A)(4), and (A)(5). Voltage ratings shall not be marked on the cable.

727.8 Allowable Ampacity. The allowable ampacity of the conductors shall be 5 amperes, except for 22 AWG conductors, which shall have an allowable ampacity of 3 amperes.

727.9 Overcurrent Protection. Overcurrent protection shall not exceed 5 amperes for 20 AWG and larger conductors, and 3 amperes for 22 AWG conductors.

727.10 Bends. Bends in Type ITC cables shall be made so as not to damage the cable.

ARTICLE 728
Fire-Resistive Cable Systems

728.1 Scope. This article covers the installation of fire-resistive cables, fire-resistive conductors, and other system components used for survivability of critical circuits to ensure continued operation during a specified time under fire conditions as required in this *Code*.

728.2 Definition.

Fire-Resistive Cable System. A cable and components used to ensure survivability of critical circuits for a specified time under fire conditions.

728.3 Other Articles. Wherever the requirements of other articles of this *Code* and Article 728 differ, the requirements of Article 728 shall apply.

728.4 General. Fire-resistive cables, fire-resistive conductors, and components shall be tested and listed as a complete system, shall be designated for use in a specific fire-rated system, and shall not be interchangeable between systems.

> Informational Note No. 1: One method of defining the fire rating is by testing the system in accordance with UL 2196-2012, *Standard for Tests of Fire Resistive Cables.*

> Informational Note No. 2: Fire-resistive cable systems are considered part of an electrical circuit protective system.

728.5 Installations. Fire-resistive cable systems installed outside the fire-rated rooms that they serve, such as the electrical room or the fire pump room, shall comply with the requirements of 728.5(A) through (H) and all other installation instructions provided in the listing.

(A) Mounting. The fire-resistive cable system shall be secured to the building structure in accordance with the listing and the manufacturer's installation instructions.

(B) Supports. The fire-resistive system shall be supported in accordance with the listing and the manufacturer's installation instructions.

> Informational Note: The supports are critical for survivability of the system. Each system has its specific support requirements.

(C) Raceways and Couplings. Where the fire-resistive system is listed to be installed in a raceway, the raceways enclosing the system, any couplings, and connectors shall be listed as part of the fire-rated system.

The raceway fill for each system shall comply with the listing requirements for the system and shall not be greater than the fill permitted in Table 1, Chapter 9.

> Informational Note: Raceway fill may not be the same for all listed fire-resistive systems.

(D) Cable Trays. Cable trays used as part of a fire-resistive system shall be listed as part of the fire-resistive system.

(E) Boxes. Boxes or enclosures used as part of a fire-resistive system shall be listed as part of the fire-resistive system and shall be secured to the building structure independently of the raceways or cables listed in the system.

(F) Pulling Lubricants. Fire-resistive cable systems installed in a raceway shall only use pulling lubricants listed as part of the fire-resistive cable system.

(G) Vertical Supports. Cables and conductors installed in vertical raceways shall be supported in accordance with the listing of the fire-resistive cable system.

(H) Splices. Only splices that are part of the listing for the fire-resistive cable system shall be used. Splices shall have manufacturer's installation instructions.

728.60 Grounding. Fire-resistive systems installed in a raceway requiring an equipment grounding conductor shall use the same fire-rated cable described in the system, unless alternative equipment grounding conductors are listed with the system. Any alternative equipment grounding conductor shall be marked with the system number. The system shall specify a permissible equipment grounding conductor. If not specified, the equipment grounding conductor shall be the same as the fire-rated cable described in the system.

728.120 Marking. In addition to the marking required in 310.120, system cables and conductors shall be surface marked with the suffix "FRR" (fire-resistive rating), along with the circuit integrity duration in hours, and with the system identifier.

ARTICLE 750
Energy Management Systems

750.1 Scope. This article applies to the installation and operation of energy management systems.

> Informational Note: Performance provisions in other codes establish prescriptive requirements that may further restrict the requirements contained in this article.

750.2 Definitions. For the purpose of this article, the following definitions shall apply.

Control. The predetermined process of connecting, disconnecting, increasing, or reducing electric power.

Energy Management System. A system consisting of any of the following: a monitor(s), communications equipment, a controller(s), a timer(s), or other device(s) that monitors and /or controls an electrical load or a power production or storage source.

Monitor. An electrical or electronic means to observe, record, or detect the operation or condition of the electric power system or apparatus.

750.20 Alternate Power Sources. An energy management system shall not override any control necessary to ensure continuity of an alternate power source for the following:

(1) Fire pumps
(2) Health care facilities

(3)   Emergency systems
(4)   Legally required standby systems
(5)   Critical operations power systems

750.30 Load Management. Energy management systems shall be permitted to monitor and control electrical loads unless restricted in accordance with 750.30(A) through (C).

(A) Load Shedding Controls. An energy management system shall not override the load shedding controls put in place to ensure the minimum electrical capacity for the following:

(1)   Fire pumps
(2)   Emergency systems
(3)   Legally required standby systems
(4)   Critical operations power systems

(B) Disconnection of Power. An energy management system shall not be permitted to cause disconnection of power to the following:

(1)   Elevators, escalators, moving walks, or stairway lift chairs
(2)   Positive mechanical ventilation for hazardous (classified) locations
(3)   Ventilation used to exhaust hazardous gas or reclassify an area
(4)   Circuits supplying emergency lighting
(5)   Essential electrical systems in health care facilities

(C) Capacity of Branch Circuit, Feeder, or Service. An energy management system shall not cause a branch circuit, feeder, or service to be overloaded at any time.

750.50 Field Markings. Where an energy management system is employed to control electrical power through the use of a remote means, a directory identifying the controlled device(s) and circuit(s) shall be posted on the enclosure of the controller, disconnect, or branch-circuit overcurrent device.

> Informational Note: The use of the term *remote* is intended to convey that a controller can be operated via another means or location through communications without a direct operator interface with the controlled device.

ARTICLE 760
Fire Alarm Systems

Part I. General

760.1 Scope. This article covers the installation of wiring and equipment of fire alarm systems, including all circuits controlled and powered by the fire alarm system.

> Informational Note No. 1: Fire alarm systems include fire detection and alarm notification, guard's tour, sprinkler waterflow, and sprinkler supervisory systems. Circuits controlled and powered by the fire alarm system include circuits for the control of building systems safety functions, elevator capture, elevator shutdown, door release, smoke doors and damper control, fire doors and damper control and fan shutdown, but only where these circuits are powered by and controlled by the fire alarm system. For further information on the installation and monitoring for integrity requirements for fire alarm systems, refer to the *NFPA 72-2013, National Fire Alarm and Signaling Code.*

> Informational Note No. 2: Class 1, 2, and 3 circuits are defined in Article 725.

760.2 Definitions.

Abandoned Fire Alarm Cable. Installed fire alarm cable that is not terminated at equipment other than a connector and not identified for future use with a tag.

Fire Alarm Circuit. The portion of the wiring system between the load side of the overcurrent device or the power-limited supply and the connected equipment of all circuits powered and controlled by the fire alarm system. Fire alarm circuits are classified as either non–power-limited or power-limited.

Fire Alarm Circuit Integrity (CI) Cable. Cable used in fire alarm systems to ensure continued operation of critical circuits during a specified time under fire conditions.

Non–Power-Limited Fire Alarm Circuit (NPLFA). A fire alarm circuit powered by a source that complies with 760.41 and 760.43.

Power-Limited Fire Alarm Circuit (PLFA). A fire alarm circuit powered by a source that complies with 760.121.

760.3 Other Articles. Circuits and equipment shall comply with 760.3(A) through (M). Only those sections of Article 300 referenced in this article shall apply to fire alarm systems.

(A) Spread of Fire or Products of Combustion. See 300.21.

(B) Ducts, Plenums, and Other Air-Handling Spaces. Power-limited and non-power-limited fire alarm cables installed in ducts, plenums, or other spaces used for environmental air shall comply with 300.22.

Exception No. 1: Power-limited fire alarm cables selected in accordance with Table 760.154 and installed in accordance with 760.135(B) and 300.22(B), Exception shall be permitted to be installed in ducts specifically fabricated for environmental air.

Exception No. 2: Power-limited fire alarm cables selected in accordance with Table 760.154 and installed in accordance with 760.135(C) shall be permitted to be installed in other spaces used for environmental air (plenums).

(C) Hazardous (Classified) Locations. Articles 500 through 516 and Article 517, Part IV, where installed in hazardous (classified) locations.

(D) Corrosive, Damp, or Wet Locations. Sections 110.11, 300.5(B), 300.6, 300.9, and 310.10(G), where installed in corrosive, damp, or wet locations.

(E) Building Control Circuits. Article 725, where building control circuits (e.g., elevator capture, fan shutdown) are associated with the fire alarm system.

(F) Optical Fiber Cables. Where optical fiber cables are utilized for fire alarm circuits, the cables shall be installed in accordance with Article 770.

(G) Installation of Conductors with Other Systems. Installations shall comply with 300.8.

(H) Raceways or Sleeves Exposed to Different Temperatures. Installations shall comply with 300.7(A).

(I) Vertical Support for Fire Rated Cables and Conductors. Vertical installations of circuit integrity (CI) cables and conductors installed in a raceway or conductors and cables of electrical circuit protective systems shall be installed in accordance with 300.19.

(J) Number and Size of Cables and Conductors in Raceway. Installations shall comply with 300.17.

(K) Bushing. A bushing shall be installed where cables emerge from raceway used for mechanical support or protection in accordance with 300.15(C).

N **(L) Cable Routing Assemblies.** Power-limited fire alarm cables shall be permitted to be installed in plenum cable routing assemblies, riser cable routing assemblies, and general-purpose cable routing assemblies selected in accordance with Table 800.154(c), listed in accordance with the provisions of 800.182, and installed in accordance with 800.110(C) and 800.113.

N **(M) Communications Raceways.** Power-limited fire alarm cables shall be permitted to be installed in plenum communications raceways, riser communications raceways, and general-purpose communications raceways selected in accordance with Table 800.154(b), listed in accordance with the provisions of 800.182, and installed in accordance with 800.113 and 362.24 through 362.56, where the requirements applicable to electrical nonmetallic tubing apply.

760.21 Access to Electrical Equipment Behind Panels Designed to Allow Access. Access to electrical equipment shall not be denied by an accumulation of conductors and cables that prevents removal of panels, including suspended ceiling panels.

760.24 Mechanical Execution of Work.

(A) General. Fire alarm circuits shall be installed in a neat workmanlike manner. Cables and conductors installed exposed on the surface of ceilings and sidewalls shall be supported by the building structure in such a manner that the cable will not be damaged by normal building use. Such cables shall be supported by straps, staples, cable ties, hangers, or similar fittings designed and installed so as not to damage the cable. The installation shall also comply with 300.4(D).

(B) Circuit Integrity (CI) Cable. Circuit integrity (CI) cables shall be supported at a distance not exceeding 610 mm (24 in.). Where located within 2.1 m (7 ft) of the floor, as covered in 760.53(A)(1) and 760.130(1), as applicable, the cable shall be fastened in an approved manner at intervals of not more than 450 mm (18 in.). Cable supports and fasteners shall be steel.

760.25 Abandoned Cables. The accessible portion of abandoned fire alarm cables shall be removed. Where cables are identified for future use with a tag, the tag shall be of sufficient durability to withstand the environment involved.

760.30 Fire Alarm Circuit Identification. Fire alarm circuits shall be identified at terminal and junction locations in a manner that helps to prevent unintentional signals on fire alarm system circuit(s) during testing and servicing of other systems.

760.32 Fire Alarm Circuits Extending Beyond One Building. Non–power-limited fire alarm circuits and power-limited fire alarm circuits that extend beyond one building and run outdoors shall meet the installation requirements of Parts II, III, and IV of Article 800 and shall meet the installation requirements of Part I of Article 300.

Informational Note: An example of a protective device suitable to provide protection is a device tested to the requirements of ANSI/UL 497B, *Protectors for Data Communications.*

760.35 Fire Alarm Circuit Requirements. Fire alarm circuits shall comply with 760.35(A) and (B).

(A) Non–Power-Limited Fire Alarm (NPLFA) Circuits. See Parts I and II.

(B) Power-Limited Fire Alarm (PLFA) Circuits. See Parts I and III.

Part II. Non–Power-Limited Fire Alarm (NPLFA) Circuits

760.41 NPLFA Circuit Power Source Requirements.

(A) Power Source. The power source of non–power-limited fire alarm circuits shall comply with Chapters 1 through 4, and the output voltage shall be not more than 600 volts, nominal. The fire alarm circuit disconnect shall be permitted to be secured in the "on" position.

(B) Branch Circuit. The branch circuit supplying the fire alarm equipment(s) shall supply no other loads. The location of the branch-circuit overcurrent protective device shall be permanently identified at the fire alarm control unit. The circuit disconnecting means shall have red identification, shall be accessible only to qualified personnel, and shall be identified as "FIRE ALARM CIRCUIT." The red identification shall not damage the overcurrent protective devices or obscure the manufacturer's markings. This branch circuit shall not be supplied through ground-fault circuit interrupters or arc-fault circuit-interrupters.

Informational Note: See 210.8(A)(5), Exception, for receptacles in dwelling-unit unfinished basements that supply power for fire alarm systems.

760.43 NPLFA Circuit Overcurrent Protection. Overcurrent protection for conductors 14 AWG and larger shall be provided in accordance with the conductor ampacity without applying the ampacity adjustment and correction factors of 310.15 to the ampacity calculation. Overcurrent protection shall not exceed 7 amperes for 18 AWG conductors and 10 amperes for 16 AWG conductors.

Exception: Where other articles of this Code permit or require other overcurrent protection.

760.45 NPLFA Circuit Overcurrent Device Location. Overcurrent devices shall be located at the point where the conductor to be protected receives its supply.

Exception No. 1: Where the overcurrent device protecting the larger conductor also protects the smaller conductor.

Exception No. 2: Transformer secondary conductors. Non–power-limited fire alarm circuit conductors supplied by the secondary of a single-phase transformer that has only a 2-wire (single-voltage) secondary shall be permitted to be protected by overcurrent protection provided by the primary (supply) side of the transformer, provided the protection is in accordance with 450.3 and does not exceed the value determined by multiplying the secondary conductor ampacity by the secondary-to-primary transformer voltage ratio. Transformer secondary conductors other than 2-wire shall not be considered to be protected by the primary overcurrent protection.

Exception No. 3: Electronic power source output conductors. Non–power-limited circuit conductors supplied by the output of a single-phase, listed electronic power source, other than a transformer, having only a 2-wire (single-voltage) output for connection to non–power-limited circuits shall be permitted to be protected by overcurrent protection provided on the input side of the electronic power source, provided this protection does not exceed the value determined by multiplying the non–power-limited circuit conductor ampacity by the output-to-input voltage ratio. Electronic power source outputs, other than 2-wire (single voltage), connected to non–power-limited circuits shall not be considered to be protected by overcurrent protection on the input of the electronic power source.

Informational Note: A single-phase, listed electronic power supply whose output supplies a 2-wire (single-voltage) circuit is an example of a non–power-limited power source that meets the requirements of 760.41.

760.46 NPLFA Circuit Wiring. Installation of non–power-limited fire alarm circuits shall be in accordance with 110.3(B), 300.7, 300.11, 300.15, 300.17, 300.19(B), and other appropriate articles of Chapter 3.

Exception No. 1: As provided in 760.48 through 760.53.

Exception No. 2: Where other articles of this Code require other methods.

760.48 Conductors of Different Circuits in Same Cable, Enclosure, or Raceway.

(A) Class 1 with NPLFA Circuits. Class 1 and non–power-limited fire alarm circuits shall be permitted to occupy the same cable, enclosure, or raceway without regard to whether the individual circuits are alternating current or direct current, provided all conductors are insulated for the maximum voltage of any conductor in the enclosure or raceway.

(B) Fire Alarm with Power-Supply Circuits. Power-supply and fire alarm circuit conductors shall be permitted in the same cable, enclosure, or raceway only where connected to the same equipment.

760.49 NPLFA Circuit Conductors.

(A) Sizes and Use. Only copper conductors shall be permitted to be used for fire alarm systems. Size 18 AWG and 16 AWG conductors shall be permitted to be used, provided they supply loads that do not exceed the ampacities given in Table 402.5 and are installed in a raceway, an approved enclosure, or a listed cable. Conductors larger than 16 AWG shall not supply loads greater than the ampacities given in 310.15, as applicable.

(B) Insulation. Insulation on conductors shall be rated for the system voltage and not less than 600 volts. Conductors larger than 16 AWG shall comply with Article 310. Conductors 18 AWG and 16 AWG shall be Type KF-2, KFF-2, PAFF, PTFF, PF, PFF, PGF, PGFF, RFH-2, RFHH-2, RFHH-3, SF-2, SFF-2, TF, TFF, TFN, TFFN, ZF, or ZFF. Conductors with other types and thickness of insulation shall be permitted if listed for non–power-limited fire alarm circuit use.

Informational Note: For application provisions, see Table 402.3.

(C) Conductor Materials. Conductors shall be solid or stranded copper.

Exception to (B) and (C): Wire Types PAF and PTF shall be permitted only for high-temperature applications between 90°C (194°F) and 250°C (482°F).

760.51 Number of Conductors in Cable Trays and Raceways, and Ampacity Adjustment Factors.

(A) NPLFA Circuits and Class 1 Circuits. Where only non–power-limited fire alarm circuit and Class 1 circuit conductors are in a raceway, the number of conductors shall be determined in accordance with 300.17. The ampacity adjustment factors given in 310.15(B)(3)(a) shall apply if such conductors carry continuous load in excess of 10 percent of the ampacity of each conductor.

(B) Power-Supply Conductors and NPLFA Circuit Conductors. Where power-supply conductors and non–power-limited fire alarm circuit conductors are permitted in a raceway in accordance with 760.48, the number of conductors shall be determined in accordance with 300.17. The ampacity adjustment factors given in 310.15(B)(3)(a) shall apply as follows:

(1) To all conductors where the fire alarm circuit conductors carry continuous loads in excess of 10 percent of the ampacity of each conductor and where the total number of conductors is more than three
(2) To the power-supply conductors only, where the fire alarm circuit conductors do not carry continuous loads in excess of 10 percent of the ampacity of each conductor and where the number of power-supply conductors is more than three

(C) Cable Trays. Where fire alarm circuit conductors are installed in cable trays, they shall comply with 392.22 and 392.80(A).

760.53 Multiconductor NPLFA Cables. Multiconductor non–power-limited fire alarm cables that meet the requirements of 760.176 shall be permitted to be used on fire alarm circuits operating at 150 volts or less and shall be installed in accordance with 760.53(A) and (B).

(A) NPLFA Wiring Method. Multiconductor non–power-limited fire alarm circuit cables shall be installed in accordance with 760.53(A)(1)(1), (A)(2), and (A)(3).

(1) In Raceways, Exposed on Ceilings or Sidewalls, or Fished in Concealed Spaces. Cable splices or terminations shall be made in listed fittings, boxes, enclosures, fire alarm devices, or utilization equipment. Where installed exposed, cables shall be adequately supported and installed in such a way that maximum protection against physical damage is afforded by building construction such as baseboards, door frames, ledges, and so forth. Where located within 2.1 m (7 ft) of the floor, cables shall be securely fastened in an approved manner at intervals of not more than 450 mm (18 in.).

(2) Passing Through a Floor or Wall. Cables shall be installed in metal raceway or rigid nonmetallic conduit where passing through a floor or wall to a height of 2.1 m (7 ft) above the floor, unless adequate protection can be afforded by building construction such as detailed in 760.53(A)(1), or unless an equivalent solid guard is provided.

(3) In Hoistways. Cables shall be installed in rigid metal conduit, rigid nonmetallic conduit, intermediate metal conduit, liquidtight flexible nonmetallic conduit, or electrical metallic tubing where installed in hoistways.

Exception: As provided for in 620.21 for elevators and similar equipment.

(B) Applications of Listed NPLFA Cables. The use of non–power-limited fire alarm circuit cables shall comply with 760.53(B)(1) through (B)(4).

(1) Ducts Specifically Fabricated for Environmental Air. Multiconductor non–power-limited fire alarm circuit cables, Types NPLFP, NPLFR, and NPLF, shall not be installed exposed in ducts specifically fabricated for environmental air.

Informational Note: See 300.22(B).

(2) Other Spaces Used for Environmental Air (Plenums). Cables installed in other spaces used for environmental air shall be Type NPLFP.

Exception No. 1: Types NPLFR and NPLF cables installed in compliance with 300.22(C).

Exception No. 2: Other wiring methods in accordance with 300.22(C) and conductors in compliance with 760.49(C).

Exception No. 3: Type NPLFP-CI cable shall be permitted to be installed to provide a 2-hour circuit integrity rated cable.

(3) Riser. Cables installed in vertical runs and penetrating one or more floors, or cables installed in vertical runs in a shaft, shall be Type NPLFR. Floor penetrations requiring Type NPLFR shall contain only cables suitable for riser or plenum use.

Exception No. 1: Type NPLF or other cables that are specified in Chapter 3 and are in compliance with 760.49(C) and encased in metal raceway.

Exception No. 2: Type NPLF cables located in a fireproof shaft having firestops at each floor.

Informational Note: See 300.21 for firestop requirements for floor penetrations.

Exception No. 3: Type NPLF-CI cable shall be permitted to be installed to provide a 2-hour circuit integrity rated cable.

(4) Other Wiring Within Buildings. Cables installed in building locations other than the locations covered in 760.53(B)(1), (B)(2), and (B)(3) shall be Type NPLF.

Exception No. 1: Chapter 3 wiring methods with conductors in compliance with 760.49(C).

Exception No. 2: Type NPLFP or Type NPLFR cables shall be permitted.

Exception No. 3: Type NPLFR-CI cable shall be permitted to be installed to provide a 2-hour circuit integrity rated cable.

Part III. Power-Limited Fire Alarm (PLFA) Circuits

760.121 Power Sources for PLFA Circuits.

(A) Power Source. The power source for a power-limited fire alarm circuit shall be as specified in 760.121(A)(1), (A)(2), or (A)(3).

Informational Note No. 1: Tables 12(A) and 12(B) in Chapter 9 provide the listing requirements for power-limited fire alarm circuit sources.

Informational Note No. 2: See 210.8(A)(5), Exception, for receptacles in dwelling-unit unfinished basements that supply power for fire alarm systems.

(1)   A listed PLFA or Class 3 transformer

(2)   A listed PLFA or Class 3 power supply

(3)   Listed equipment marked to identify the PLFA power source

Informational Note: Examples of listed equipment are a fire alarm control panel with integral power source; a circuit card listed for use as a PLFA source, where used as part of a listed assembly; a current-limiting impedance, listed for the purpose or part of a listed product, used in conjunction with a non–power-limited transformer or a stored energy source, for example, storage battery, to limit the output current.

(B) Branch Circuit. The branch circuit supplying the fire alarm equipment(s) shall supply no other loads. The location of the branch-circuit overcurrent protective device shall be permanently identified at the fire alarm control unit. The circuit disconnecting means shall have red identification, shall be accessible only to qualified personnel, and shall be identified as "FIRE ALARM CIRCUIT." The red identification shall not damage the overcurrent protective devices or obscure the manufacturer's markings. This branch circuit shall not be supplied through ground-fault circuit interrupters or arc-fault circuit interrupters.

760.124 Circuit Marking. The equipment supplying PLFA circuits shall be durably marked where plainly visible to indicate each circuit that is a power-limited fire alarm circuit.

Informational Note: See 760.130(A), Exception No. 3, where a power-limited circuit is to be reclassified as a non–power-limited circuit.

760.127 Wiring Methods on Supply Side of the PLFA Power Source. Conductors and equipment on the supply side of the power source shall be installed in accordance with the appropriate requirements of Part II and Chapters 1 through 4. Transformers or other devices supplied from power-supply conductors shall be protected by an overcurrent device rated not over 20 amperes.

Exception: The input leads of a transformer or other power source supplying power-limited fire alarm circuits shall be permitted to be smaller than 14 AWG, but not smaller than 18 AWG, if they are not over 300 mm (12 in.) long and if they have insulation that complies with 760.49(B).

760.130 Wiring Methods and Materials on Load Side of the PLFA Power Source. Fire alarm circuits on the load side of the power source shall be permitted to be installed using wiring methods and materials in accordance with 760.130(A), (B), or a combination of (A) and (B).

(A) NPLFA Wiring Methods and Materials. Installation shall be in accordance with 760.46, and conductors shall be solid or stranded copper.

Exception No. 1: The ampacity adjustment factors given in 310.15(B)(3)(a) shall not apply.

Exception No. 2: Conductors and multiconductor cables described in and installed in accordance with 760.49 and 760.53 shall be permitted.

Exception No. 3: Power-limited circuits shall be permitted to be reclassified and installed as non–power-limited circuits if the power-limited fire alarm circuit markings required by 760.124 are eliminated and the entire circuit is installed using the wiring methods and materials in accordance with Part II, Non–Power-Limited Fire Alarm Circuits.

Informational Note: Power-limited circuits reclassified and installed as non–power-limited circuits are no longer power-limited circuits, regardless of the continued connection to a power-limited source.

(B) PLFA Wiring Methods and Materials. Power-limited fire alarm conductors and cables described in 760.179 shall be installed as detailed in 760.130(B)(1), (B)(2), or (B)(3) of this section and 300.7. Devices shall be installed in accordance with 110.3(B), 300.11(A), and 300.15.

(1) In Raceways, Exposed on Ceilings or Sidewalls, or Fished in Concealed Spaces. Cable splices or terminations shall be made in listed fittings, boxes, enclosures, fire alarm devices, or utilization equipment. Where installed exposed, cables shall be adequately supported and installed in such a way that maximum protection against physical damage is afforded by building construction such as baseboards, door frames, ledges, and so forth. Where located within 2.1 m (7 ft) of the floor, cables shall be securely fastened in an approved manner at intervals of not more than 450 mm (18 in.).

(2) Passing Through a Floor or Wall. Cables shall be installed in metal raceways or rigid nonmetallic conduit where passing through a floor or wall to a height of 2.1 m (7 ft) above the floor, unless adequate protection can be afforded by building construction such as detailed in 760.130(B)(1), or unless an equivalent solid guard is provided.

(3) In Hoistways. Cables shall be installed in rigid metal conduit, rigid nonmetallic conduit, intermediate metal conduit, or electrical metallic tubing where installed in hoistways.

Exception: As provided for in 620.21 for elevators and similar equipment.

760.133 Installation of Conductors and Equipment in Cables, Compartments, Cable Trays, Enclosures, Manholes, Outlet Boxes, Device Boxes, Raceways, and Cable Routing Assemblies for Power-Limited Fire Alarm Circuits. Conductors and equipment for power-limited fire alarm circuits shall be installed in accordance with 760.135 through 760.143.

760.135 Installation of PLFA Cables in Buildings. Installation of power-limited fire alarm cables in buildings shall comply with 760.135(A) through (J).

(A) Listing. PLFA cables installed in buildings shall be listed.

(B) Ducts Specifically Fabricated for Environmental Air. The following cables shall be permitted in ducts specifically fabricated for environmental air as described in 300.22(B), if they are directly associated with the air distribution system:

(1) Types FPLP and FPLP-CI cables in lengths as short as practicable to perform the required function
(2) Types FPLP, FPLP-CI, FPLR, FPLR-CI, FPL, and FPL-CI cables installed in raceways that are installed in compliance with 300.22(B)

Informational Note: For information on fire protection of wiring installed in fabricated ducts, see 4.3.4.1 and 4.3.11.3.3 of NFPA 90A-2015, *Standard for the Installation of Air-Conditioning and Ventilating Systems.*

(C) Other Spaces Used For Environmental Air (Plenums). The following cables shall be permitted in other spaces used for environmental air as described in 300.22(C):

(1) Type FPLP cables
(2) Type FPLP cables installed in plenum communications raceways

(3) Type FPLP cables installed in plenum routing assemblies
(4) Types FPLP and FPLP-CI cables supported by open metallic cable trays or cable tray systems
(5) Types FPLP, FPLR, and FPL cables installed in raceways that are installed in compliance with 300.22(C)
(6) Types FPLP, FPLR, and FPL cables supported by solid bottom metal cable trays with solid metal covers in other spaces used for environmental air (plenums) as described in 300.22(C)
(7) Types FPLP, FPLR, and FPL cables installed in plenum communications raceways, riser communications raceways, or general-purpose communications raceways supported by solid bottom metal cable trays with solid metal covers in other spaces used for environmental air (plenums) as described in 300.22(C)

(D) Risers — Cables in Vertical Runs. The following cables shall be permitted in vertical runs penetrating one or more floors and in vertical runs in a shaft:

(1) Types FPLP and FPLR cables
(2) Types FPLP and FPLR cables installed in the following:

    a. Plenum communications raceways
    b. Plenum cable routing assemblies
    c. Riser communications raceways
    d. Riser cable routing assemblies

Informational Note: See 300.21 for firestop requirements for floor penetrations.

(E) Risers — Cables in Metal Raceways. The following cables shall be permitted in metal raceways in a riser having firestops at each floor:

(1) Types FPLP, FPLR, and FPL cables
(2) Types FPLP, FPLR, and FPL cables installed in the following:

    a. Plenum communications raceways
    b. Riser communications raceways
    c. General-purpose communications raceways

Informational Note: See 300.21 for firestop requirements for floor penetrations.

(F) Risers — Cables in Fireproof Shafts. The following cables shall be permitted to be installed in fireproof riser shafts having firestops at each floor:

(1) Types FPLP, FPLR, and FPL cables
(2) Types FPLP, FPLR, and FPL cables installed in the following:

    a. Plenum communications raceways
    b. Plenum cable routing assemblies
    c. Riser communications raceways
    d. Riser cable routing assemblies
    e. General-purpose communications raceways
    f. General-purpose cable routing assemblies

Informational Note: See 300.21 for firestop requirements for floor penetrations.

(G) Risers — One- and Two-Family Dwellings. The following cables shall be permitted in one- and two-family dwellings:

(1) Types FPLP, FPLR, and FPL cables
(2) Types FPLP, FPLR, and FPL cables installed in the following:

    a. Plenum communications raceways
    b. Plenum cable routing assemblies

   c. Riser communications raceways
   d. Riser cable routing assemblies
   e. General-purpose communications raceways
   f. General-purpose cable routing assemblies

(H) Other Building Locations. The following cables shall be permitted to be installed in building locations other than the locations covered in 770.113(B) through (H):

(1) Types FPLP, FPLR, and FPL cables
(2) Types FPLP, FPLR, and FPL cables installed in the following:

   a. Plenum communications raceways
   b. Plenum cable routing assemblies
   c. Riser communications raceways
   d. Riser cable routing assemblies
   e. General-purpose communications raceways
   f. General-purpose cable routing assemblies

(3) Types FPLP, FPLR, and FPL cables installed in a raceway of a type recognized in Chapter 3

(I) Nonconcealed Spaces. Cables specified in Chapter 3 and meeting the requirements of 760.179(A) and (B) shall be permitted to be installed in nonconcealed spaces where the exposed length of cable does not exceed 3 m (10 ft).

(J) Portable Fire Alarm System. A portable fire alarm system provided to protect a stage or set when not in use shall be permitted to use wiring methods in accordance with 530.12.

760.136 Separation from Electric Light, Power, Class 1, NPLFA, and Medium-Power Network-Powered Broadband Communications Circuit Conductors.

(A) General. Power-limited fire alarm circuit cables and conductors shall not be placed in any cable, cable tray, compartment, enclosure, manhole, outlet box, device box, raceway, or similar fitting with conductors of electric light, power, Class 1, non–power-limited fire alarm circuits, and medium-power network-powered broadband communications circuits unless permitted by 760.136(B) through (G).

(B) Separated by Barriers. Power-limited fire alarm circuit cables shall be permitted to be installed together with Class 1, non–power-limited fire alarm, and medium-power network-powered broadband communications circuits where they are separated by a barrier.

(C) Raceways Within Enclosures. In enclosures, power-limited fire alarm circuits shall be permitted to be installed in a raceway within the enclosure to separate them from Class 1, non–power-limited fire alarm, and medium-power network-powered broadband communications circuits.

(D) Associated Systems Within Enclosures. Power-limited fire alarm conductors in compartments, enclosures, device boxes, outlet boxes, or similar fittings shall be permitted to be installed with electric light, power, Class 1, non–power-limited fire alarm, and medium power network-powered broadband communications circuits where they are introduced solely to connect the equipment connected to power-limited fire alarm circuits, and comply with either of the following conditions:

(1) The electric light, power, Class 1, non–power-limited fire alarm, and medium-power network-powered broadband communications circuit conductors are routed to maintain a minimum of 6 mm (0.25 in.) separation from the conductors and cables of power-limited fire alarm circuits.

(2) The circuit conductors operate at 150 volts or less to ground and also comply with one of the following:

   a. The fire alarm power-limited circuits are installed using Type FPL, FPLR, FPLP, or permitted substitute cables, provided these power-limited cable conductors extending beyond the jacket are separated by a minimum of 6 mm (0.25 in.) or by a nonconductive sleeve or nonconductive barrier from all other conductors.

   b. The power-limited fire alarm circuit conductors are installed as non–power-limited circuits in accordance with 760.46.

(E) Enclosures with Single Opening. Power-limited fire alarm circuit conductors entering compartments, enclosures, device boxes, outlet boxes, or similar fittings shall be permitted to be installed with electric light, power, Class 1, non–power-limited fire alarm, and medium-power network-powered broadband communications circuits where they are introduced solely to connect the equipment connected to power-limited fire alarm circuits or to other circuits controlled by the fire alarm system to which the other conductors in the enclosure are connected. Where power-limited fire alarm circuit conductors must enter an enclosure that is provided with a single opening, they shall be permitted to enter through a single fitting (such as a tee), provided the conductors are separated from the conductors of the other circuits by a continuous and firmly fixed nonconductor, such as flexible tubing.

(F) In Hoistways. In hoistways, power-limited fire alarm circuit conductors shall be installed in rigid metal conduit, rigid nonmetallic conduit, intermediate metal conduit, liquid-tight flexible nonmetallic conduit, or electrical metallic tubing. For elevators or similar equipment, these conductors shall be permitted to be installed as provided in 620.21.

(G) Other Applications. For other applications, power-limited fire alarm circuit conductors shall be separated by at least 50 mm (2 in.) from conductors of any electric light, power, Class 1, non–power-limited fire alarm, or medium-power network-powered broadband communications circuits unless one of the following conditions is met:

(1) Either (a) all of the electric light, power, Class 1, non–power-limited fire alarm, and medium-power network-powered broadband communications circuit conductors or (b) all of the power-limited fire alarm circuit conductors are in a raceway or in metal-sheathed, metal-clad, nonmetallic-sheathed, or Type UF cables.

(2) All of the electric light, power, Class 1, non–power-limited fire alarm, and medium-power network-powered broadband communications circuit conductors are permanently separated from all of the power-limited fire alarm circuit conductors by a continuous and firmly fixed nonconductor, such as porcelain tubes or flexible tubing, in addition to the insulation on the conductors.

760.139 Installation of Conductors of Different PLFA Circuits, Class 2, Class 3, and Communications Circuits in the Same Cable, Enclosure, Cable Tray, Raceway, or Cable Routing Assembly.

(A) Two or More PLFA Circuits. Cable and conductors of two or more power-limited fire alarm circuits, communications circuits, or Class 3 circuits shall be permitted within the same cable, enclosure, cable tray, raceway, or cable routing assembly.

(B) Class 2 Circuits with PLFA Circuits. Conductors of one or more Class 2 circuits shall be permitted within the same cable, enclosure, cable tray, raceway, or cable routing assembly with conductors of power-limited fire alarm circuits, provided that the insulation of the Class 2 circuit conductors in the cable, enclosure, raceway, or cable routing assembly is at least that required by the power-limited fire alarm circuits.

(C) Low-Power Network-Powered Broadband Communications Cables and PLFA Cables. Low-power network-powered broadband communications circuits shall be permitted in the same enclosure, cable tray, raceway, or cable routing assembly with PLFA cables.

(D) Audio System Circuits and PLFA Circuits. Audio system circuits described in 640.9(C) and installed using Class 2 or Class 3 wiring methods in compliance with 725.133 and 725.154 shall not be permitted to be installed in the same cable, cable tray, raceway, or cable routing assembly with power-limited conductors or cables.

760.142 Conductor Size. Conductors of 26 AWG shall be permitted only where spliced with a connector listed as suitable for 26 AWG to 24 AWG or larger conductors that are terminated on equipment or where the 26 AWG conductors are terminated on equipment listed as suitable for 26 AWG conductors. Single conductors shall not be smaller than 18 AWG.

760.143 Support of Conductors. Power-limited fire alarm circuit conductors shall not be strapped, taped, or attached by any means to the exterior of any conduit or other raceway as a means of support.

760.145 Current-Carrying Continuous Line-Type Fire Detectors.

(A) Application. Listed continuous line-type fire detectors, including insulated copper tubing of pneumatically operated detectors, employed for both detection and carrying signaling currents shall be permitted to be used in power-limited circuits.

(B) Installation. Continuous line-type fire detectors shall be installed in accordance with 760.124 through 760.130 and 760.133.

760.154 Applications of Listed PLFA Cables. PLFA cables shall comply with the requirements described in Table 760.154 or where cable substitutions are made as shown in 760.154(A). Where substitute cables are installed, the wiring requirements of Article 760, Parts I and III, shall apply. Types FPLP-CI, FPLR-CI, and FPL-CI cables shall be permitted to be installed to provide 2-hour circuit integrity rated cables.

(A) Fire Alarm Cable Substitutions. The substitutions for fire alarm cables listed in Table 760.154(A) and illustrated in Figure 760.154(A) shall be permitted. Where substitute cables are installed, the wiring requirements of Article 760, Parts I and III, shall apply.

Informational Note: For information on communications cables (CMP, CMR, CMG, CM), see 800.179.

Part IV. Listing Requirements

760.176 Listing and Marking of NPLFA Cables. Non-power-limited fire alarm cables installed as wiring within buildings shall be listed in accordance with 760.176(A) and (B) and as being resistant to the spread of fire in accordance with

760.176(C) through (F), and shall be marked in accordance with 760.176(G). Cable used in a wet location shall be listed for use in wet locations or have a moisture-impervious metal sheath. Non-power-limited fire alarm cables shall have a temperature rating of not less than 60°C (140°F).

(A) NPLFA Conductor Materials. Conductors shall be 18 AWG or larger solid or stranded copper.

(B) Insulated Conductors. Insulation on conductors shall be rated for the system voltage and not less than 600 V. Insulated conductors 14 AWG and larger shall be one of the types listed in Table 310.104(A) or one that is identified for this use. Insulated conductors 18 AWG and 16 AWG shall be in accordance with 760.49.

(C) Type NPLFP. Type NPLFP non–power-limited fire alarm cable for use in other space used for environmental air shall be listed as being suitable for use in other space used for environmental air as described in 300.22(C) and shall also be listed as having adequate fire-resistant and low smoke–producing characteristics.

Informational Note: One method of defining a cable that is low-smoke producing cable and fire-resistant cable is that the cable exhibits a maximum peak optical density of 0.50 or less, an average optical density of 0.15 or less, and a maximum flame spread distance of 1.52 m (5 ft) or less when tested in accordance with NFPA 262-2015, *Standard Method of Test for Flame Travel and Smoke of Wires and Cables for Use in Air-Handling Spaces.*

(D) Type NPLFR. Type NPLFR non–power-limited fire alarm riser cable shall be listed as being suitable for use in a vertical run in a shaft or from floor to floor and shall also be listed as having fire-resistant characteristics capable of preventing the carrying of fire from floor to floor.

Informational Note: One method of defining fire-resistant characteristics capable of preventing the carrying of fire from floor to floor is that the cables pass ANSI/UL 1666-2012, *Test for Flame Propagation Height of Electrical and Optical-Fiber Cables Installed Vertically in Shafts.*

(E) Type NPLF. Type NPLF non–power-limited fire alarm cable shall be listed as being suitable for general-purpose fire alarm use, with the exception of risers, ducts, plenums, and other space used for environmental air, and shall also be listed as being resistant to the spread of fire.

Informational Note: One method of defining *resistant to the spread of fire* is that the cables do not spread fire to the top of the tray in the "UL Flame Exposure, Vertical Tray Flame Test" in ANSI/UL 1685-2010, *Standard for Safety for Vertical-Tray Fire-Propagation and Smoke-Release Test for Electrical and Optical-Fiber Cables.* The smoke measurements in the test method are not applicable.

Another method of defining *resistant to the spread of fire* is for the damage (char length) not to exceed 1.5 m (4 ft 11 in.) when performing the CSA "Vertical Flame Test — Cables in Cable Trays," as described in CSA C22.2 No. 0.3-M-2001, *Test Methods for Electrical Wires and Cables.*

Table 760.154 Applications of Listed PLFA Cables in Buildings

Applications		Cable Type		
		FPLP & FPLP-CI	FPLR & FPLR-CI	FPL & FPL-CI
In fabricated ducts as described in 300.22(B)	In fabricated ducts	Y*	N	N
	In metal raceway that complies with 300.22(B)	Y*	Y*	Y*
In other spaces used for environmental air as described in 300.22(C)	In other spaces used for environmental air	Y*	N	N
	In metal raceway that complies with 300.22(C)	Y*	Y*	Y*
	In plenum communications raceways	Y*	N	N
	In plenum cable routing assemblies	Y*	N	N
	Supported by open metal cable trays	Y*	N	N
	Supported by solid bottom metal cable trays with solid metal covers	Y*	Y*	Y*
In risers	In vertical runs	Y*	Y*	N
	In metal raceways	Y*	Y*	Y*
	In fireproof shafts	Y*	Y*	Y*
	In plenum communications raceways	Y*	Y*	N
	In plenum cable routing assemblies	Y*	Y*	N
	In riser communications raceways	Y*	Y*	N
	In riser cable routing assemblies	Y*	Y*	N
	In one- and two-family dwellings	Y*	Y*	Y*
Within buildings in other than air-handling spaces and risers	General	Y*	Y*	Y*
	Supported by cable trays	Y*	Y*	Y*
	In any raceway recognized in Chapter 3	Y*	Y*	Y*
	In plenum communications raceway	Y*	Y*	Y*
	In plenum cable routing assemblies	Y*	Y*	Y*
	In riser communications raceways	Y*	Y*	Y*
	In riser cable routing assemblies	Y*	Y*	Y*
	In general-purpose communications raceways	Y*	Y*	Y*
	In general-purpose cable routing assemblies	Y*	Y*	Y*

Note:
"N" indicates that the cable type shall not be permitted to be installed in the application.
"Y*" indicates that the cable type shall be permitted to be installed in the application subject to the limitations described in 760.130 through 760.145.

Table 760.154(A) Cable Substitutions

Cable Type	Permitted Substitutions
FPLP	CMP
FPLR	CMP, FPLP, CMR
FPL	CMP, FPLP, CMR, FPLR, CMG, CM

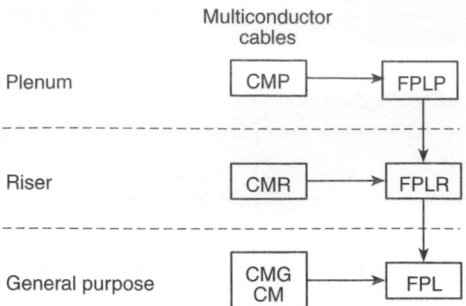

Type CM—Communications wires and cables
Type FPL—Power-limited fire alarm cables

[A] ▸ [B]   Cable A shall be permitted to be used in place of cable B, 26 AWG minimum

FIGURE 760.154(A)   Cable Substitution Hierarchy.

(F) Fire Alarm Circuit Integrity (CI) Cable or Electrical Circuit Protective System. Cables that are used for survivability of critical circuits under fire conditions shall meet either 760.176(F)(1) or (F)(2) as follows:

> Informational Note No. 1: Fire alarm circuit integrity (CI) cable and electrical circuit protective systems may be used for fire alarm circuits to comply with the survivability requirements of *NFPA 72 -2013, National Fire Alarm and Signaling Code*, 12.4.3 and 12.4.4, that the circuit maintain its electrical function during fire conditions for a defined period of time.

> Informational Note No. 2: One method of defining circuit integrity (CI) cable or an electrical circuit protective system is by establishing a minimum 2-hour fire-resistive rating for the cable when tested in accordance with UL 2196-2012, *Standard for Tests of Fire Resistive Cables*.

> Informational Note No. 3: UL guide information for electrical circuit protective systems (FHIT) contains information on proper installation requirements for maintaining the fire rating.

(1) Circuit Integrity (CI) Cables. Circuit integrity (CI) cables, specified in 760.176(C), (D), and (E), and used for survivability of critical circuits, shall have an additional classification using the suffix "CI." Circuit integrity (CI) cables shall only be permitted to be installed in a raceway where specifically listed and marked as part of an electrical circuit protective system as covered in 760.176(F)(2).

(2) Electrical Circuit Protective System. Cables specified in 760.176(C), (D), (E), and (F)(1), that are part of an electrical circuit protective system, shall be identified with the protective system number and hourly rating printed on the outer jacket of the cable and installed in accordance with the listing of the protective system.

(G) NPLFA Cable Markings. Multiconductor non–power-limited fire alarm cables shall be marked in accordance with Table 760.176(G). Non–power-limited fire alarm circuit cables shall be permitted to be marked with a maximum usage voltage rating of 150 volts. Cables that are listed for circuit integrity shall be identified with the suffix "CI" as defined in 760.176(F). Temperature rating shall be marked on the jacket of NPLFA cables that have a temperature rating exceeding 60°C (140°F). The jacket of NPLFA cables shall be marked with the conductor size.

> Informational Note: Cable types are listed in descending order of fire resistance rating.

760.179 Listing and Marking of PLFA Cables and Insulated Continuous Line-Type Fire Detectors. PLFA cables installed as wiring within buildings shall be listed as being resistant to the spread of fire and other criteria in accordance with 760.179(A) through (H) and shall be marked in accordance with 760.179(I). Insulated continuous line-type fire detectors shall be listed in accordance with 760.179(J). Cable used in a wet location shall be listed for use in wet locations or have a moisture-impervious metal sheath.

(A) Conductor Materials. Conductors shall be solid or stranded copper.

(B) Conductor Size. The size of conductors in a multiconductor cable shall not be smaller than 26 AWG. Single conductors shall not be smaller than 18 AWG.

(C) Ratings. The cable shall have a voltage rating of not less than 300 volts. The cable shall have a temperature rating of not less than 60°C (140°F).

(D) Type FPLP. Type FPLP power-limited fire alarm plenum cable shall be listed as being suitable for use in ducts, plenums, and other space used for environmental air and shall also be listed as having adequate fire-resistant and low smoke-producing characteristics.

> Informational Note: One method of defining a cable that is low-smoke producing cable and fire-resistant cable is that the cable exhibits a maximum peak optical density of 0.50 or less, an average optical density of 0.15 or less, and a maximum flame spread distance of 1.52 m (5 ft) or less when tested in accordance with NFPA 262-2015, *Standard Method of Test for Flame Travel and Smoke of Wires and Cables for Use in Air-Handling Spaces*.

(E) Type FPLR. Type FPLR power-limited fire alarm riser cable shall be listed as being suitable for use in a vertical run in

Table 760.176(G) NPLFA Cable Markings

Cable Marking	Type	Reference
NPLFP	Non–power-limited fire alarm circuit cable for use in "other space used for environmental air"	760.176(C) and (G)
NPLFR	Non–power-limited fire alarm circuit riser cable	760.176(D) and (G)
NPLF	Non–power-limited fire alarm circuit cable	760.176(E) and (G)

Note: Cables identified in 760.176(C), (D), and (E) and meeting the requirements for circuit integrity shall have the additional classification using the suffix "CI" (for example, NPLFP-CI, NPLFR-CI, and NPLF-CI).

a shaft or from floor to floor and shall also be listed as having fire-resistant characteristics capable of preventing the carrying of fire from floor to floor.

Informational Note: One method of defining fire-resistant characteristics capable of preventing the carrying of fire from floor to floor is that the cables pass the requirements of ANSI/UL 1666-2012, *Standard Test for Flame Propagation Height of Electrical and Optical-Fiber Cable Installed Vertically in Shafts.*

(F) Type FPL. Type FPL power-limited fire alarm cable shall be listed as being suitable for general-purpose fire alarm use, with the exception of risers, ducts, plenums, and other spaces used for environmental air, and shall also be listed as being resistant to the spread of fire.

Informational Note: One method of defining *resistant to the spread of fire* is that the cables do not spread fire to the top of the tray in the "UL Flame Exposure, Vertical Tray Flame Test" in ANSI/UL 1685-2012, *Standard for Safety for Vertical-Tray Fire-Propagation and Smoke-Release Test for Electrical and Optical-Fiber Cables.* The smoke measurements in the test method are not applicable.

Another method of defining *resistant to the spread of fire* is for the damage (char length) not to exceed 1.5 m (4 ft 11 in.) when performing the CSA "Vertical Flame Test — Cables in Cable Trays," as described in CSA C22.2 No. 0.3-M-2001, *Test Methods for Electrical Wires and Cables.*

(G) Fire Alarm Circuit Integrity (CI) Cable or Electrical Circuit Protective System. Cables that are used for survivability of critical circuits under fire conditions shall meet either 760.179(G)(1) or (G)(2) as follows:

Informational Note No. 1: Fire alarm circuit integrity (CI) cable and electrical circuit protective systems may be used for fire alarm circuits to comply with the survivability requirements of *NFPA 72-2013, National Fire Alarm and Signaling Code,* 12.4.3 and 12.4.4, that the circuit maintain its electrical function during fire conditions for a defined period of time.

Informational Note No. 2: One method of defining circuit integrity (CI) cable or an electrical circuit protective system is by establishing a minimum 2-hour fire-resistive rating for the cable when tested in accordance with UL 2196-2012, *Standard for Tests of Fire Resistive Cables.*

Informational Note No. 3: UL guide information for electrical circuit protective systems (FHIT) contains information on proper installation requirements for maintaining the fire rating.

(1) Circuit Integrity (CI) Cables. Circuit integrity (CI) cables specified in 760.179(D), (E), (F), and (H), and used for survivability of critical circuits, shall have an additional classification using the suffix "CI." Circuit integrity (CI) cables shall only be permitted to be installed in a raceway where specifically listed and marked as part of an electrical circuit protective system as covered in 760.179(G)(2).

(2) Electrical Circuit Protective System. Cables specified in 760.179(D), (E), (F), (H), and (G)(1), that are part of an electrical circuit protective system, shall be identified with the protective system number and hourly rating printed on the outer jacket of the cable and installed in accordance with the listing of the protective system.

(H) Coaxial Cables. Coaxial cables shall be permitted to use 30 percent conductivity copper-covered steel center conductor wire and shall be listed as Type FPLP, FPLR, or FPL cable.

(I) Cable Marking. The cable shall be marked in accordance with Table 760.179(I). The voltage rating shall not be marked

on the cable. Cables that are listed for circuit integrity shall be identified with the suffix CI as defined in 760.179(G). Temperature rating shall be marked on the jacket of PLFA cables that have a temperature rating exceeding 60°C (140°F). The jacket of PLFA cables shall be marked with the conductor size.

Informational Note: Voltage ratings on cables may be misinterpreted to suggest that the cables may be suitable for Class 1, electric light, and power applications.

Exception: Voltage markings shall be permitted where the cable has multiple listings and voltage marking is required for one or more of the listings.

Table 760.179(I) Cable Markings

Cable Marking	Type
FPLP	Power-limited fire alarm plenum cable
FPLR	Power-limited fire alarm riser cable
FPL	Power-limited fire alarm cable

Note: Cables identified in 760.179(D), (E), and (F) as meeting the requirements for circuit integrity shall have the additional classification using the suffix "CI" (for example, FPLP-CI, FPLR-CI, and FPL-CI).

Informational Note: Cable types are listed in descending order of fire resistance rating.

(J) Insulated Continuous Line-Type Fire Detectors. Insulated continuous line-type fire detectors shall be rated in accordance with 760.179(C), listed as being resistant to the spread of fire in accordance with 760.179(D) through (F), and marked in accordance with 760.179(I), and the jacket compound shall have a high degree of abrasion resistance.

ARTICLE 770
Optical Fiber Cables

Informational Note: See Informational Note Figure 800(a) and Informational Note Figure 800(b) for illustrative application of a bonding conductor or grounding electrode conductor.

Part I. General

770.1 Scope. This article covers the installation of optical fiber cables. This article does not cover the construction of optical fiber cables.

770.2 Definitions. See Part I of Article 100. For purposes of this article, the following additional definitions apply.

Abandoned Optical Fiber Cable. Installed optical fiber cable that is not terminated at equipment other than a connector and not identified for future use with a tag.

Cable Sheath. A covering over the optical fiber assembly that includes one or more jackets and may include one or more metallic members or strength members.

Exposed (to Accidental Contact). A conductive optical fiber cable in such a position that, in case of failure of supports or insulation, contact between the cable's non–current-carrying conductive members and an electrical circuit might result.

> Informational Note: See Part I of Article 100 for two other definitions of Exposed: Exposed (as applied to live parts) and Exposed (as applied to wiring methods).

Point of Entrance. The point within a building at which the optical fiber cable emerges from an external wall or from a concrete floor slab.

770.3 Other Articles. Installations of optical fiber cables shall comply with 770.3(A) and (B). Only those sections of Chapter 2 and Article 300 referenced in this article shall apply to optical fiber cables.

(A) Hazardous (Classified) Locations. Listed optical fiber cables shall be permitted to be installed in hazardous (classified) locations. The cables shall be sealed in accordance with the requirements of 501.15, 502.15, 505.16, or 506.16, as applicable.

(B) Cables in Ducts for Dust, Loose Stock, or Vapor Removal. The requirements of 300.22(A) for wiring systems shall apply to conductive optical fiber cables.

(C) Composite Cables. Composite optical fiber cables shall be classified as electrical cables in accordance with the type of electrical conductors. They shall be constructed, listed, and marked in accordance with the appropriate article for each type of electrical cable.

770.21 Access to Electrical Equipment Behind Panels Designed to Allow Access. Access to electrical equipment shall not be denied by an accumulation of optical fiber cables that prevents removal of panels, including suspended ceiling panels.

770.24 Mechanical Execution of Work. Optical fiber cables shall be installed in a neat and workmanlike manner. Cables installed exposed on the surface of ceilings and sidewalls shall be supported by the building structure in such a manner that the cable will not be damaged by normal building use. Such cables shall be secured by hardware including straps, staples, cable ties, hangers, or similar fittings designed and installed so as not to damage the cable. The installation shall also conform with 300.4(D) through (G) and 300.11. Nonmetallic cable ties and other nonmetallic cable accessories used to secure and support cables in other spaces used for environmental air (plenums) shall be listed as having low smoke and heat release properties.

> Informational Note No. 1: Accepted industry practices are described in ANSI/NECA/BICSI 568-2006, *Standard for Installing Commercial Building Telecommunications Cabling*; ANSI/NECA/FOA 301-2009, *Standard for Installing and Testing Fiber Optic Cables*; and other ANSI-approved installation standards.

> Informational Note No. 2: See 4.3.11.2.6.5 and 4.3.11.5.5.6 of NFPA 90A-2012, *Standard for the Installation of Air-Conditioning and Ventilating Systems*, for discrete combustible components installed in accordance with 300.22(C).

> Informational Note No. 3: Paint, plaster, cleaners, abrasives, corrosive residues, or other contaminants may result in an undetermined alteration of optical fiber cable properties.

770.25 Abandoned Cables. The accessible portion of abandoned optical fiber cables shall be removed. Where cables are

identified for future use with a tag, the tag shall be of sufficient durability to withstand the environment involved.

770.26 Spread of Fire or Products of Combustion. Installations of optical fiber cables and communications raceways in hollow spaces, vertical shafts, and ventilation or air-handling ducts shall be made so that the possible spread of fire or products of combustion will not be substantially increased. Openings around penetrations of optical fiber cables and communications raceways through fire-resistant–rated walls, partitions, floors, or ceilings shall be firestopped using approved methods to maintain the fire resistance rating.

> Informational Note: Directories of electrical construction materials published by qualified testing laboratories contain many listing installation restrictions necessary to maintain the fire-resistive rating of assemblies where penetrations or openings are made. Building codes also contain restrictions on membrane penetrations on opposite sides of a fire resistance–rated wall assembly. An example is the 600-mm (24-in.) minimum horizontal separation that usually applies between boxes installed on opposite sides of the wall. Assistance in complying with 770.26 can be found in building codes, fire resistance directories, and product listings.

Part II. Cables Outside and Entering Buildings

N **770.44 Overhead (Aerial) Optical Fiber Cables.** Overhead optical fiber cables containing a non–current-carrying metallic member entering buildings shall comply with 840.44(A) and (B).

(A) On Poles and In-Span. Where outside plant optical fiber cables and electric light or power conductors are supported by the same pole or are run parallel to each other in-span, the conditions described in 770.44(A)(1) through (A)(4) shall be met.

(1) Relative Location. Where practicable, the outside plant optical fiber cables shall be located below the electric light or power conductors.

(2) Attachment to Cross-Arms. Attachment of outside plant optical fiber cables to a cross-arm that carries electric light or power conductors shall not be permitted.

(3) Climbing Space. The climbing space through outside plant optical fiber cables shall comply with the requirements of 225.14(D).

(4) Clearance. Supply service drops and sets of overhead service conductors of 0 to 750 volts running above and parallel to optical fiber cable service drops shall have a minimum separation of 300 mm (12 in.) at any point in the span, including the point of their attachment to the building. Clearance of not less than 1.0 m (40 in.) shall be maintained between the two services at the pole.

(B) Above Roofs. Outside plant optical fiber cables shall have a vertical clearance of not less than 2.5 m (8 ft) from all points of roofs above which they pass.

Exception No. 1: The requirement of 770.44(B) shall not apply to auxiliary buildings such as garages and the like.

Exception No. 2: A reduction in clearance above only the overhanging portion of the roof to not less than 450 mm (18 in.) shall be permitted if (a) not more than 1.2 m (4 ft) of optical fiber cable service drop cable passes above the roof overhang, and (b) the cable is terminated at a through- or above-the-roof raceway or approved support.

Exception No. 3: Where the roof has a slope of not less than 100 mm in 300 mm (4 in. in 12 in.), a reduction in clearance to not less than 900 mm (3 ft) shall be permitted.

> Informational Note: For additional information regarding overhead wires and cables, see ANSI/IEEE C2-2012, *National Electric Safety Code, Part 2, Safety Rules for Overhead Lines.*

770.47 Underground Optical Fiber Cables Entering Buildings.
Underground optical fiber cables entering buildings shall comply with 770.47(A) and (B).

(A) Underground Systems with Electric Light, Power, Class 1, or Non–Power-Limited Fire Alarm Circuit Conductors. Underground conductive optical fiber cables entering buildings with electric light, power, Class 1, or non–power-limited fire alarm circuit conductors in a raceway, handhole enclosure, or manhole shall be located in a section separated from such conductors by means of brick, concrete, or tile partitions or by means of a suitable barrier.

(B) Direct-Buried Cables and Raceways. Direct-buried conductive optical fiber cables shall be separated by at least 300 mm (12 in.) from conductors of any electric light, power, non–power-limited fire alarm circuit conductors, or Class 1 circuit.

Exception No. 1: Direct-buried conductive optical fiber cables shall not be required to be separated by at least 300 mm (12 in.) from electric service conductors where electric service conductors are installed in raceways or have metal cable armor.

Exception No. 2: Direct-buried conductive optical fiber cables shall not be required to be separated by at least 300 mm (12 in.) from electric light or power branch circuit or feeder conductors, non–power-limited fire alarm circuit conductors, or Class 1 circuit conductors where electric light or power branch circuit or feeder conductors, non–power-limited fire alarm circuit conductors, or Class 1 circuit conductors are installed in a raceway or in metal-sheathed, metal-clad, or Type UF or Type USE cables.

770.48 Unlisted Cables Entering Buildings.

(A) Conductive and Nonconductive Cables. Unlisted conductive and nonconductive outside plant optical fiber cables shall be permitted to be installed in building spaces, other than risers, ducts used for environmental air, plenums used for environmental air, and other spaces used for environmental air, where the length of the cable within the building, measured from its point of entrance, does not exceed 15 m (50 ft) and the cable enters the building from the outside and is terminated in an enclosure.

The point of entrance shall be permitted to be extended from the penetration of the external wall or floor slab by continuously enclosing the entrance optical fiber cables in rigid metal conduit (RMC) or intermediate metal conduit (IMC) to the point of emergence.

> Informational Note: Splice cases or terminal boxes, both metallic and plastic types, typically are used as enclosures for splicing or terminating optical fiber cables.

(B) Nonconductive Cables in Raceway. Unlisted nonconductive outside plant optical fiber cables shall be permitted to enter the building from the outside and shall be permitted to be installed in any of the following raceways:

(1) Intermediate metal conduit (IMC)
(2) Rigid metal conduit (RMC)
(3) Rigid polyvinyl chloride conduit (PVC)
(4) Electrical metallic tubing (EMT)

Unlisted nonconductive outside plant cables installed in rigid polyvinyl chloride conduit (PVC) or electrical metallic tubing (EMT) shall not be permitted to be installed in risers, ducts used for environmental air, plenums used for environmental air, and other spaces used for environmental air.

770.49 Metallic Entrance Conduit Grounding.
Metallic conduit containing optical fiber entrance cable shall be connected by a bonding conductor or grounding electrode conductor to a grounding electrode in accordance with 770.100(B).

Part III. Protection

770.93 Grounding or Interruption of Non–Current-Carrying Metallic Members of Optical Fiber Cables.
Optical fiber cables entering the building or terminating on the outside of the building shall comply with 770.93(A) or (B).

(A) Entering Buildings. In installations where an optical fiber cable is exposed to contact with electric light or power conductors and the cable enters the building, the non–current-carrying metallic members shall be either grounded as specified in 770.100, or interrupted by an insulating joint or equivalent device. The grounding or interruption shall be as close as practicable to the point of entrance.

(B) Terminating on the Outside of Buildings. In installations where an optical fiber cable is exposed to contact with electric light or power conductors and the cable is terminated on the outside of the building, the non–current-carrying metallic members shall be either grounded as specified in 770.100, or interrupted by an insulating joint or equivalent device. The grounding or interruption shall be as close as practicable to the point of termination of the cable.

Part IV. Grounding Methods

770.100 Entrance Cable Bonding and Grounding.
Where required, the non–current-carrying metallic members of optical fiber cables entering buildings shall be bonded or grounded as specified in 770.100(A) through (D).

(A) Bonding Conductor or Grounding Electrode Conductor.

(1) Insulation. The bonding conductor or grounding electrode conductor shall be listed and shall be permitted to be insulated, covered, or bare.

(2) Material. The bonding conductor or grounding electrode conductor shall be copper or other corrosion-resistant conductive material, stranded or solid.

(3) Size. The bonding conductor or grounding electrode conductor shall not be smaller than 14 AWG. It shall have a current-carrying capacity not less than that of the grounded metallic member(s). The bonding conductor or grounding electrode conductor shall not be required to exceed 6 AWG.

(4) Length. The bonding conductor or grounding electrode conductor shall be as short as practicable. In one- and two-family dwellings, the bonding conductor or grounding electrode conductor shall be as short as practicable not to exceed 6.0 m (20 ft) in length.

Informational Note: Similar bonding conductor or grounding electrode conductor length limitations applied at apartment buildings and commercial buildings help to reduce voltages that may develop between the building's power and communications systems during lightning events.

Exception: In one- and two-family dwellings where it is not practicable to achieve an overall maximum bonding conductor or grounding electrode conductor length of 6.0 m (20 ft), a separate ground rod meeting the minimum dimensional criteria of 770.100(B)(3)(2) shall be driven, the grounding electrode conductor shall be connected to the separate ground rod in accordance with 770.100(C), and the separate ground rod shall be bonded to the power grounding electrode system in accordance with 770.100(D).

(5) Run in Straight Line. The bonding conductor or grounding electrode conductor shall be run in as straight a line as practicable.

(6) Physical Protection. Bonding conductors and grounding electrode conductors shall be protected where exposed to physical damage. Where the bonding conductor or grounding electrode conductor is installed in a metal raceway, both ends of the raceway shall be bonded to the contained conductor or to the same terminal or electrode to which the bonding conductor or grounding electrode conductor is connected.

(B) Electrode. The bonding conductor and grounding electrode conductor shall be connected in accordance with 770.100(B)(1), (B)(2), or (B)(3).

(1) In Buildings or Structures with an Intersystem Bonding Termination. If the building or structure served has an intersystem bonding termination as required by 250.94, the bonding conductor shall be connected to the intersystem bonding termination.

(2) In Buildings or Structures with Grounding Means. If an intersystem bonding termination is established, 250.94(A) shall apply.

If the building or structure served has no intersystem bonding termination, the bonding conductor or grounding electrode conductor shall be connected to the nearest accessible location on one of the following:

(1) The building or structure grounding electrode system as covered in 250.50
(2) The grounded interior metal water piping system, within 1.5 m (5 ft) from its point of entrance to the building, as covered in 250.52
(3) The power service accessible means external to enclosures using the options identified in 250.94(A), Exception
(4) The nonflexible metallic power service raceway
(5) The service equipment enclosure
(6) The grounding electrode conductor or the grounding electrode conductor metal enclosure of the power service
(7) The grounding electrode conductor or the grounding electrode of a building or structure disconnecting means that is grounded to an electrode as covered in 250.32

(3) In Buildings or Structures Without Intersystem Bonding Termination or Grounding Means. If the building or structure served has no intersystem bonding termination or grounding

means, as described in 770.100(B)(2), the grounding electrode conductor shall be connected to either of the following:

(1) To any one of the individual grounding electrodes described in 250.52(A)(1), (A)(2), (A)(3), or (A)(4).
(2) If the building or structure served has no grounding means, as described in 770.100(B)(2) or (B)(3)(1), to any one of the individual grounding electrodes described in 250.52(A)(7) and (A)(8) or to a ground rod or pipe not less than 1.5 m (5 ft) in length and 12.7 mm (½ in.) in diameter, driven, where practicable, into permanently damp earth and separated from lightning protection system conductors as covered in 800.53 and at least 1.8 m (6 ft) from electrodes of other systems. Steam, hot water pipes, or lightning protection system conductors shall not be employed as electrodes for non–current-carrying metallic members.

(C) Electrode Connection. Connections to grounding electrodes shall comply with 250.70.

(D) Bonding of Electrodes. A bonding jumper not smaller than 6 AWG copper or equivalent shall be connected between the grounding electrode and power grounding electrode system at the building or structure served where separate electrodes are used.

Exception: At mobile homes as covered in 770.106.

Informational Note No. 1: See 250.60 for connection to a lightning protection system.

Informational Note No. 2: Bonding together of all separate electrodes limits potential differences between them and between their associated wiring systems.

770.106 Grounding and Bonding of Entrance Cables at Mobile Homes.

(A) Grounding. Grounding shall comply with 770.106(A)(1) and (A)(2).

(1) Where there is no mobile home service equipment located within 9.0 m (30 ft) of the exterior wall of the mobile home it serves, the non–current-carrying metallic members of optical fiber cables entering the mobile home shall be grounded in accordance with 770.100(B)(3).
(2) Where there is no mobile home disconnecting means grounded in accordance with 250.32 and located within 9.0 m (30 ft) of the exterior wall of the mobile home it serves, the non–current-carrying metallic members of optical fiber cables entering the mobile home shall be grounded in accordance with 770.100(B)(3).

(B) Bonding. The grounding electrode shall be bonded to the metal frame or available grounding terminal of the mobile home with a copper conductor not smaller than 12 AWG under either of the following conditions:

(1) Where there is no mobile home service equipment or disconnecting means as in 770.106(A)
(2) Where the mobile home is supplied by cord and plug

Part V. Installation Methods Within Buildings

770.110 Raceways and Cable Routing Assemblies for Optical Fiber Cables.

(A) Types of Raceways. Optical fiber cables shall be permitted to be installed in any raceway that complies with either 770.110(A)(1) or (A)(2) and in cable routing assemblies installed in compliance with 770.110(C).

(1) Raceways Recognized in Chapter 3. Optical fiber cables shall be permitted to be installed in any raceway included in Chapter 3. The raceways shall be installed in accordance with the requirements of Chapter 3.

(2) Communications Raceways. Optical fiber cables shall be permitted to be installed in plenum communications raceways, riser communications raceways, and general-purpose communications raceways selected in accordance with Table 800.154(b), listed in accordance with 800.113, and installed in accordance with 362.24 through 362.56, where the requirements applicable to electrical nonmetallic tubing (ENT) apply.

(3) Innerduct for Optical Fiber Cables. Listed plenum communications raceway, listed riser communications raceway, and listed general-purpose communications raceway selected in accordance with the provisions of Table 800.154(b) shall be permitted to be installed as innerduct in any type of listed raceway permitted in Chapter 3.

(B) Raceway Fill for Optical Fiber Cables. Raceway fill for optical fiber cables shall comply with either 770.110(B)(1) or (B)(2).

(1) Without Electric Light or Power Conductors. Where optical fiber cables are installed in raceway without electric light or power conductors, the raceway fill requirements of Chapters 3 and 9 shall not apply.

(2) Nonconductive Optical Fiber Cables with Electric Light or Power Conductors. Where nonconductive optical fiber cables are installed with electric light or power conductors in a raceway, the raceway fill requirements of Chapters 3 and 9 shall apply.

(C) Cable Routing Assemblies. Optical fiber cables shall be permitted to be installed in plenum cable routing assemblies, riser cable routing assemblies, and general-purpose cable routing assemblies selected in accordance with Table 800.154(c), listed in accordance with 800.182, and installed in accordance with 800.110(C) and 800.113.

770.113 Installation of Optical Fiber Cables.
Installation of optical fiber cables shall comply with 770.113(A) through (J). Installation of raceways and cable routing assemblies shall comply with 770.110.

(A) Listing. Optical fiber cables installed in buildings shall be listed in accordance with 770.179.

Exception: Optical fiber cables that are installed in compliance with 770.48 shall not be required to be listed.

(B) Ducts Specifically Fabricated for Environmental Air. The following cables shall be permitted in ducts specifically fabricated for environmental air as described in 300.22(B) if they are directly associated with the air distribution system:

(1)   Up to 1.22 m (4 ft) of Types OFNP and OFCP

(2)   Types OFNP, OFCP, OFNR, OFCR, OFNG, OFCG, OFN, and OFC installed in raceways that are installed in compliance with 300.22(B)

Informational Note: For information on fire protection of wiring installed in fabricated ducts, see 4.3.4.1 and 4.3.11.3.3 of NFPA 90A-2015, *Standard for the Installation of Air-Conditioning and Ventilating Systems.*

(C) Other Spaces Used for Environmental Air (Plenums). The following cables shall be permitted in other spaces used for environmental air as described in 300.22(C):

(1)   Types OFNP and OFCP
(2)   Types OFNP and OFCP installed in plenum communications raceways listed in accordance with 800.182
(3)   Types OFNP and OFCP installed in plenum cable routing assemblies listed in accordance with 800.182
(4)   Types OFNP and OFCP supported by open metallic cable trays or cable tray systems
(5)   Types OFNP, OFCP, OFNR, OFCR, OFNG, OFCG, OFN, and OFC installed in raceways that are installed in compliance with 300.22(C)
(6)   Types OFNP, OFCP, OFNR, OFCR, OFNG, OFCG, OFN, and OFC supported by solid bottom metal cable trays with solid metal covers in other spaces used for environmental air (plenums), as described in 300.22(C)
(7)   Types OFNP, OFCP, OFNR, OFCR, OFNG, OFCG, OFN, and OFC installed in plenum communications raceways, riser communications raceways, or general-purpose communications raceways listed in accordance with 800.182 or supported by solid bottom metal cable trays with solid metal covers in other spaces used for environmental air (plenums), as described in 300.22(C)

Informational Note: For information on fire protection of wiring installed in other spaces used for environmental air, see 4.3.11.2, 4.3.11.4, and 4.3.11.5 of NFPA 90A-2015, *Standard for the Installation of Air-Conditioning and Ventilating Systems.*

(D) Risers — Cables in Vertical Runs. The following cables shall be permitted in vertical runs penetrating one or more floors and in vertical runs in a shaft:

(1)   Types OFNP, OFCP, OFNR, and OFCR
(2)   Types OFNP, OFCP, OFNR, and OFCR installed in:

   a.   Plenum communications raceways
   b.   Plenum cable routing assemblies
   c.   Riser communications raceways
   d.   Riser cable routing assemblies

Informational Note: See 770.26 for firestop requirements for floor penetrations.

(E) Risers — Cables and Innerducts in Metal Raceways. The following cables and innerducts shall be permitted in metal raceways in a riser having firestops at each floor:

(1)   Types OFNP, OFCP, OFNR, OFCR, OFNG, OFCG, OFN, and OFC
(2)   Types OFNP, OFCP, OFNR, OFCR, OFNG, OFCG, OFN, and OFC installed in:

   a.   Plenum communications raceways (innerduct)
   b.   Riser communications raceways (innerduct)
   c.   General-purpose communications raceways (innerduct)

Informational Note: See 770.26 for firestop requirements for floor penetrations.

(F) Risers — Cables in Fireproof Shafts. The following cables shall be permitted to be installed in fireproof riser shafts having firestops at each floor:

(1) Types OFNP, OFCP, OFNR, OFCR, OFNG, OFCG, OFN, and OFC

(2) Types OFNP, OFCP, OFNR, OFCR, OFNG, OFCG, OFN, and OFC installed in:

    a. Plenum communications raceways
    b. Plenum cable routing assemblies
    c. Riser communications raceways
    d. Riser cable routing assemblies
    e. General-purpose communications raceways
    f. General-purpose cable routing assemblies

Informational Note: See 770.26 for firestop requirements for floor penetrations.

(G) Risers — One- and Two-Family Dwellings. The following cables shall be permitted in one- and two-family dwellings:

(1) Types OFNP, OFCP, OFNR, OFCR, OFNG, OFCG, OFN, and OFC

(2) Types OFNP, OFCP, OFNR, OFCR, OFNG, OFCG, OFN, and OFC installed in:

    a. Plenum communications raceways
    b. Plenum cable routing assemblies
    c. Riser communications raceways
    d. Riser cable routing assemblies
    e. General-purpose communications raceways
    f. General-purpose cable routing assemblies

(H) Cable Trays. The following cables shall be permitted to be supported by cable trays:

(1) Types OFNP, OFCP, OFNR, OFCR, OFNG, OFCG, OFN, and OFC

(2) Types OFNP, OFCP, OFNR, OFCR, OFNG, OFCG, OFN, and OFC installed in:

    a. Plenum communications raceways
    b. Riser communications raceways
    c. General-purpose communications raceways

(I) Distributing Frames and Cross-Connect Arrays. The following cables shall be permitted to be installed in distributing frames and cross-connect arrays:

(1) Types OFNP, OFCP, OFNR, OFCR, OFNG, OFCG, OFN, and OFC

(2) Types OFNP, OFCP, OFNR, OFCR, OFNG, OFCG, OFN, and OFC installed in:

    a. Plenum communications raceways
    b. Plenum cable routing assemblies
    c. Riser communications raceways
    d. Riser cable routing assemblies
    e. General-purpose communications raceways
    f. General-purpose cable routing assemblies

(J) Other Building Locations. The following cables shall be permitted to be installed in building locations other than the locations covered in 770.113(B) through (I):

(1) Types OFNP, OFCP, OFNR, OFCR, OFNG, OFCG, OFN, and OFC

(2) Types OFNP, OFCP, OFNR, OFCR, OFNG, OFCG, OFN, and OFC installed in:

    a. Plenum communications raceways
    b. Plenum cable routing assemblies

    c. Riser communications raceways
    d. Riser cable routing assemblies
    e. General-purpose communications raceways
    f. General-purpose cable routing assemblies

(3) Types OFNP, OFCP, OFNR, OFCR, OFNG, OFCG, OFN, and OFC installed in a raceway of a type recognized in Chapter 3

770.114 Grounding. Non–current-carrying conductive members of optical fiber cables shall be bonded to a grounded equipment rack or enclosure, or grounded in accordance with the grounding methods specified by 770.110(B)(2).

770.133 Installation of Optical Fibers and Electrical Conductors.

(A) With Conductors for Electric Light, Power, Class 1, Non–Power-Limited Fire Alarm, or Medium Power Network-Powered Broadband Communications Circuits. When optical fibers are within the same composite cable for electric light, power, Class 1, non–power-limited fire alarm, or medium-power network-powered broadband communications circuits operating at 1000 volts or less, they shall be permitted to be installed only where the functions of the optical fibers and the electrical conductors are associated.

Nonconductive optical fiber cables shall be permitted to occupy the same cable tray or raceway with conductors for electric light, power, Class 1, non–power-limited fire alarm, Type ITC, or medium-power network-powered broadband communications circuits operating at 1000 volts or less. Conductive optical fiber cables shall not be permitted to occupy the same cable tray or raceway with conductors for electric light, power, Class 1, non–power-limited fire alarm, Type ITC, or medium-power network-powered broadband communications circuits.

Optical fibers in composite optical fiber cables containing only current-carrying conductors for electric light, power, or Class 1 circuits rated 1000 volts or less shall be permitted to occupy the same cabinet, cable tray, outlet box, panel, raceway, or other termination enclosure with conductors for electric light, power, or Class 1 circuits operating at 1000 volts or less.

Nonconductive optical fiber cables shall not be permitted to occupy the same cabinet, outlet box, panel, or similar enclosure housing the electrical terminations of an electric light, power, Class 1, non–power-limited fire alarm, or medium-power network-powered broadband communications circuit.

Exception No. 1: Occupancy of the same cabinet, outlet box, panel, or similar enclosure shall be permitted where nonconductive optical fiber cable is functionally associated with the electric light, power, Class 1, non–power-limited fire alarm, or medium-power network-powered broadband communications circuit.

Exception No. 2: Occupancy of the same cabinet, outlet box, panel, or similar enclosure shall be permitted where nonconductive optical fiber cables are installed in factory- or field-assembled control centers.

Exception No. 3: In industrial establishments only, where conditions of maintenance and supervision ensure that only qualified persons service the installation, nonconductive optical fiber cables shall be permitted with circuits exceeding 1000 volts.

Exception No. 4: In industrial establishments only, where conditions of maintenance and supervision ensure that only qualified persons service the installation, optical fibers in composite optical fiber cables contain-

ing current-carrying conductors operating over 1000 volts shall be permitted to be installed.

Exception No. 5: Where all of the conductors of electric light, power, Class 1, nonpower-limited fire alarm, and medium-power network-powered broadband communications circuits are separated from all of the optical fiber cables by a permanent barrier or listed divider.

(B) With Other Circuits. Optical fibers shall be permitted in the same cable, and conductive and nonconductive optical fiber cables shall be permitted in the same raceway, cable tray, box, enclosure, or cable routing assembly, with conductors of any of the following:

(1) Class 2 and Class 3 remote-control, signaling, and power-limited circuits in compliance with Article 645 or Parts I and III of Article 725

(2) Power-limited fire alarm systems in compliance with Parts I and III of Article 760

(3) Communications circuits in compliance with Parts I and V of Article 800

(4) Community antenna television and radio distribution systems in compliance with Parts I and V of Article 820

(5) Low-power network-powered broadband communications circuits in compliance with Parts I and V of Article 830

(C) Support of Optical Fiber Cables. Raceways shall be used for their intended purpose. Optical fiber cables shall not be strapped, taped, or attached by any means to the exterior of any conduit or raceway as a means of support.

Exception: Overhead (aerial) spans of optical fiber cables shall be permitted to be attached to the exterior of a raceway-type mast intended for the attachment and support of such cables.

770.154 Applications of Listed Optical Fiber Cables. Permitted and nonpermitted applications of listed optical fiber cables shall be as indicated in Table 770.154(a). The permitted applications shall be subject to the installation requirements of 770.110 and 770.113. The substitutions for optical fiber cables in Table 770.154(b) and illustrated in Figure 770.154 shall be permitted.

Part VI. Listing Requirements

770.179 Optical Fiber Cables. Optical fiber cables shall be listed and identified in accordance with 770.179(A) through (F) and shall be marked in accordance with Table 770.179. Optical fiber cables shall have a temperature rating of not less than 60°C (140°F). Temperature rating shall be marked on the jacket of optical fiber cables that have a temperature rating exceeding 60°C (140°F).

(A) Types OFNP and OFCP. Types OFNP and OFCP nonconductive and conductive optical fiber plenum cables shall be suitable for use in ducts, plenums, and other space used for environmental air and shall also have adequate fire-resistant and low smoke producing characteristics.

Informational Note: One method of defining a cable that has adequate fire-resistant and low-smoke producing characteristics is that the cable exhibits a maximum peak optical density of 0.50 or less, an average optical density of 0.15 or less, and a maximum flame spread distance of 1.52 m (5 ft) or less when tested in accordance with NFPA 262-2015, *Standard Method of Test for Flame Travel and Smoke of Wires and Cables for Use in Air-Handling Spaces.*

(B) Types OFNR and OFCR. Types OFNR and OFCR nonconductive and conductive optical fiber riser cables shall be suitable for use in a vertical run in a shaft or from floor to floor and shall also have the fire-resistant characteristics capable of preventing the carrying of fire from floor to floor.

Informational Note: One method of defining fire-resistant characteristics capable of preventing the carrying of fire from floor to floor is that the cables pass the requirements of ANSI/UL 1666-2011, *Standard Test for Flame Propagation Height of Electrical and Optical-Fiber Cable Installed Vertically in Shafts.*

(C) Types OFNG and OFCG. Types OFNG and OFCG nonconductive and conductive general-purpose optical fiber cables shall be suitable for general-purpose use, with the exception of risers and plenums, and shall also be resistant to the spread of fire.

Informational Note: One method of defining *resistant to the spread of fire* is for the damage (char length) not to exceed 1.5 m (4 ft 11 in.) when performing the CSA vertical flame test — cables in cable trays, as described in CSA C22.2 No. 0.3-M-2001, *Test Methods for Electrical Wires and Cables.*

(D) Types OFN and OFC. Types OFN and OFC nonconductive and conductive optical fiber cables shall be suitable for general-purpose use, with the exception of risers, plenums, and other spaces used for environmental air, and shall also be resistant to the spread of fire.

Informational Note No. 1: One method of defining *resistant to the spread of fire* is that the cables do not spread fire to the top of the tray in the UL flame exposure, vertical tray flame test in ANSI/UL 1685-2010, *Standard for Safety for Vertical-Tray Fire-Propagation and Smoke-Release Test for Electrical and Optical-Fiber Cables.* The smoke measurements in the test method are not applicable.

Another method of defining *resistant to the spread of fire* is for the damage (char length) not to exceed 1.5 m (4 ft 11 in.) when performing the CSA vertical flame test — cables in cable trays, as described in CSA C22.2 No. 0.3-M-2001, *Test Methods for Electrical Wires and Cables.*

Informational Note No. 2: Cable types are listed in descending order of fire resistance rating. Within each fire resistance rating, nonconductive cable is listed first because it is often substituted for conductive cable.

(E) Circuit Integrity (CI) Cable or Electrical Circuit Protective System. Cables that are used for survivability of critical circuits under fire conditions shall meet either 770.179(E)(1) or (E)(2).

Informational Note: The listing organization provides information for circuit integrity (CI) cable and electrical circuit protective systems, including installation requirements necessary to maintain the fire rating.

(1) Circuit Integrity (CI) Cables. Circuit integrity (CI) cables specified in 770.179(A) through (D), and used for survivability of critical circuits, shall have an additional classification using the suffix "CI." In order to maintain its listed fire rating, circuit integrity (CI) cable shall only be installed in free air.

Informational Note: One method of defining circuit integrity (CI) cable is by establishing a minimum 2-hour fire resistance rating for the cable when tested in accordance with ANSI/UL 2196-2006, *Standard for Tests of Fire-Resistive Cable.*

Table 770.154(a) Applications of Listed Optical Fiber Cables in Buildings

Applications		Listed Optical Fiber Cable Type		
		OFNP, OFCP	OFNR, OFCR	OFNG, OFCG, OFN, OFC
In ducts specifically fabricated for environmental air as described in 300.22(B)	In fabricated ducts	Y*	N	N
	In metal raceway that complies with 300.22(B)	Y*	Y*	Y*
In other spaces used for environmental air (plenums) as described in 300.22(C)	In other spaces used for environmental air	Y*	N	N
	In metal raceway that complies with 300.22(C)	Y*	Y*	Y*
	In plenum communications raceways	Y*	N	N
	In plenum cable routing assemblies	Y*	N	N
	Supported by open metal cable trays	Y*	N	N
	Supported by solid bottom metal cable trays with solid metal covers	Y*	Y*	Y*
In risers	In vertical runs	Y*	Y*	N
	In metal raceways	Y*	Y*	Y*
	In fireproof shafts	Y*	Y*	Y*
	In plenum communications raceways	Y*	Y*	N
	In plenum cable routing assemblies	Y*	Y*	N
	In riser communications raceways	Y*	Y*	N
	In riser cable routing assemblies	Y*	Y*	N
	In one- and two-family dwellings	Y*	Y*	Y*
Within buildings in other than air-handling spaces and risers	General	Y*	Y*	Y*
	Supported by cable trays	Y*	Y*	Y*
	In distributing frames and cross-connect arrays	Y*	Y*	Y*
	In any raceway recognized in Chapter 3	Y*	Y*	Y*
	In plenum communications raceways	Y*	Y*	Y*
	In plenum cable routing assemblies	Y*	Y*	Y*
	In riser communications raceways	Y*	Y*	Y*
	In riser cable routing assemblies	Y*	Y*	Y*
	In general-purpose communications raceways	Y*	Y*	Y*
	In general-purpose cable routing assemblies	Y*	Y*	Y*

Note: "N" indicates that the cable type shall not be permitted to be installed in the application.
"Y*" indicates that the cable type shall be permitted to be installed in the application subject to the limitations described in 770.110 and 770.113.
Informational Note No. 1: Part V of Article 770 covers installation methods within buildings. This table covers the applications of listed optical fiber cables in buildings. The definition of *Point of Entrance* is in 770.2.
Informational Note No. 2: For information on the restrictions to the installation of optical fiber cables in ducts specifically fabricated for environmental air, see 770.113(B).

Table 770.154(b) Cable Substitutions

Cable Type	Permitted Substitutions
OFNP	None
OFCP	OFNP
OFNR	OFNP
OFCR	OFNP, OFCP, OFNR
OFNG, OFN	OFNP, OFNR
OFCG, OFC	OFNP, OFCP, OFNR, OFCR, OFNG, OFN

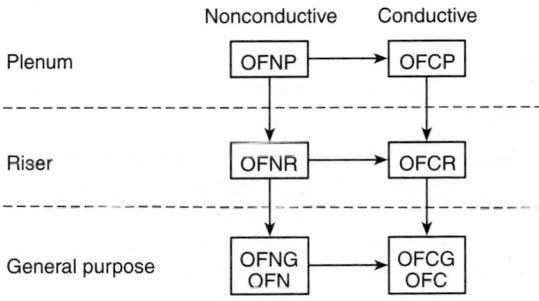

[A] ⟶ [B] Cable A shall be permitted to be used in place of cable B.

FIGURE 770.154   Cable Substitution Hierarchy.

Table 770.179 Cable Markings

Cable Marking	Type
OFNP	Nonconductive optical fiber plenum cable
OFCP	Conductive optical fiber plenum cable
OFNR	Nonconductive optical fiber riser cable
OFCR	Conductive optical fiber riser cable
OFNG	Nonconductive optical fiber general-purpose cable
OFCG	Conductive optical fiber general-purpose cable
OFN	Nonconductive optical fiber general-purpose cable
OFC	Conductive optical fiber general-purpose cable

(2) Fire-Resistive Cables. Cables specified in 770.179(A) through (D) and 770.179(E)(1) that are part of an electrical circuit protective system shall be fire-resistive cable and identified with the protective system number on the product or on the smallest unit container in which the product is packaged and installed in accordance with the listing of the protective system.

> Informational Note No. 1: One method of defining an electrical circuit protective system is by establishing a minimum 2-hour fire resistance rating for the system when tested in accordance with UL Subject 1724, *Outline of Investigation for Fire Tests for Electrical Circuit Protective Systems.*

> Informational Note No. 2: The listing organization provides information for electrical circuit protective systems (FHIT), including installation requirements for maintaining the fire rating.

(F) Field-Assembled Optical Fiber Cables. Field-assembled optical fiber cable shall comply with 770.179(F)(1) through (4).

(1) The specific combination of jacket and optical fibers intended to be installed as a field-assembled optical fiber cable shall be one of the types in 770.179(A), (B), or (D) and shall be marked in accordance with Table 770.179.

(2) The jacket of a field-assembled optical fiber cable shall have a surface marking indicating the specific optical fibers with which it is identified for use.

(3) The optical fibers shall have a permanent marking, such as a marker tape, indicating the jacket with which they are identified for use.

(4) The jacket without fibers shall meet the listing requirements for communications raceways in 800.182(A), (B), or (C) in accordance with the cable marking.

770.180 Grounding Devices. Where bonding or grounding is required, devices used to connect a shield, a sheath, or non–current-carrying metallic members of a cable to a bonding conductor or grounding electrode conductor shall be listed or be part of listed equipment.

Chapter 8   Communications Systems

ARTICLE 800
Communications Circuits

Informational Note: Informational Note Figure 800(a) and Informational Note Figure 800(b) illustrate the application of bonding and grounding electrode conductors in communications installations.

Part I. General

800.1 Scope. This article covers communications circuits and equipment.

Informational Note No. 1: See 90.2(B)(4) for installations of communications circuits and equipment that are not covered.

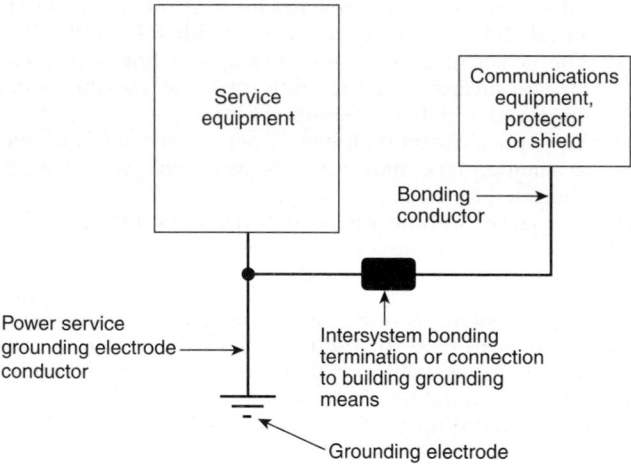

Informational Note Figure 800(a)   Illustration of a Bonding Conductor in a Communications Installation.

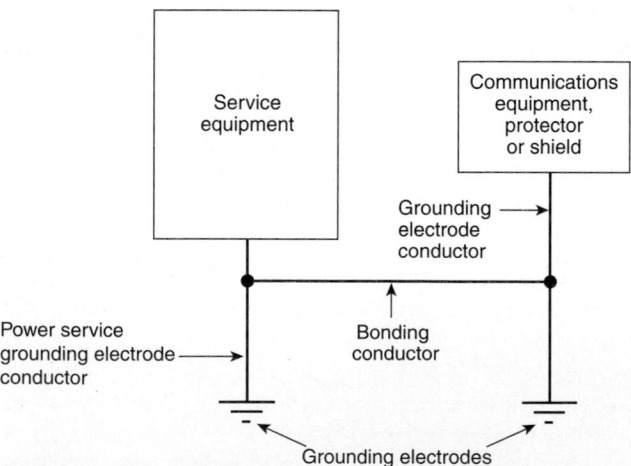

Informational Note Figure 800(b)   Illustration of a Grounding Electrode Conductor in a Communications Installation.

Informational Note No. 2: For further information for remote-control, signaling, and power-limited circuits, see Article 725.

Informational Note No. 3: For further information for fire alarm systems, see Article 760.

800.2 Definitions. See Part I of Article 100. For the purposes of this article, the following additional definitions apply.

Abandoned Communications Cable. Installed communications cable that is not terminated at both ends at a connector or other equipment and not identified for future use with a tag.

Block. A square or portion of a city, town, or village enclosed by streets and including the alleys so enclosed, but not any street.

Cable. A factory assembly of two or more conductors having an overall covering.

Cable Sheath. A covering over the conductor assembly that may include one or more metallic members, strength members, or jackets.

Communications Circuit. The circuit that extends voice, audio, video, data, interactive services, telegraph (except radio), outside wiring for fire alarm and burglar alarm from the communications utility to the customer's communications equipment up to and including terminal equipment such as a telephone, fax machine, or answering machine.

Communications Circuit Integrity (CI) Cable. Cable used in communications systems to ensure continued operation of critical circuits during a specified time under fire conditions.

Exposed (to Accidental Contact). A circuit that is in such a position that, in case of failure of supports or insulation, contact with another circuit may result.

Informational Note: See Part I of Article 100 for two other definitions of *Exposed*.

Point of Entrance. The point within a building at which the communications wire or cable emerges from an external wall or from a concrete floor slab.

Premises. The land and buildings of a user located on the user side of the utility-user network point of demarcation.

Wire. A factory assembly of one or more insulated conductors without an overall covering.

800.3 Other Articles.

(A) Hazardous (Classified) Locations. Communications circuits and equipment installed in a location that is classified in accordance with 500.5 and 505.5 shall comply with the applicable requirements of Chapter 5.

(B) Wiring in Ducts for Dust, Loose Stock, or Vapor Removal. The requirements of 300.22(A) shall apply.

(C) Equipment in Other Space Used for Environmental Air. The requirements of 300.22(C)(3) shall apply.

(D) Installation and Use. The requirements of 110.3(B) shall apply.

(E) Network-Powered Broadband Communications Systems. Article 830 shall apply to network-powered broadband communications systems.

(F) Premises-Powered Broadband Communications Systems. Article 840 shall apply to premises-powered broadband communications systems.

(G) Optical Fiber Cable. Where optical fiber cable is used, either in whole or in part, to provide a communications circuit within a building, Article 770 shall apply to the installation of the optical fiber portion of the communications circuit.

N (H) Temperature Limitation of Conductors. Section 310.15(A)(3) shall apply.

800.18 Installation of Equipment. Equipment electrically connected to a communications network shall be listed in accordance with 800.170.

Exception: This listing requirement shall not apply to test equipment that is intended for temporary connection to a telecommunications network by qualified persons during the course of installation, maintenance, or repair of telecommunications equipment or systems.

800.21 Access to Electrical Equipment Behind Panels Designed to Allow Access. Access to electrical equipment shall not be denied by an accumulation of communications wires and cables that prevents removal of panels, including suspended ceiling panels.

800.24 Mechanical Execution of Work. Communications circuits and equipment shall be installed in a neat and workmanlike manner. Cables installed exposed on the surface of ceilings and sidewalls shall be supported by the building structure in such a manner that the cable will not be damaged by normal building use. Such cables shall be secured by hardware, including straps, staples, cable ties, hangers, or similar fittings, designed and installed so as not to damage the cable. The installation shall also conform to 300.4(D) and 300.11. Nonmetallic cable ties and other nonmetallic cable accessories used to secure and support cables in other spaces used for environmental air (plenums) shall be listed as having low smoke and heat release properties in accordance with 800.170(C).

Informational Note No. 1: Accepted industry practices are described in ANSI/NECA/BICSI 568-2006, *Standard for Installing Commercial Building Telecommunications Cabling*; ANSI/TIA-568.1-D-2015, *Commercial Building Telecommunications Infrastructure Standard*; ANSI/TIA-569-D-2015, *Telecommunications Pathways and Spaces*; ANSI/TIA-570-C-2012, *Residential Telecommunications Infrastructure Standard* ; ANSI/TIA-1005-A-2012, *Telecommunications Infrastructure Standard for Industrial Premises*; ANSI/TIA-1179 2010, *Healthcare Facility Telecommunications Infrastructure Standard*; ANSI/TIA-4966-2014, *Telecommunications Infrastructure Standard for Educational Facilities*; and other ANSI-approved installation standards.

Informational Note No. 2: See 4.3.11.2.6.5 and 4.3.11.5.5.6 of NFPA 90A-2015, *Standard for the Installation of Air-Conditioning and Ventilating Systems*, for discrete combustible components installed in accordance with 300.22(C).

Informational Note No. 3: Paint, plaster, cleaners, abrasives, corrosive residues, or other contaminants may result in an undetermined alteration of communications wire and cable properties.

800.25 Abandoned Cables. The accessible portion of abandoned communications cables shall be removed. Where cables are identified for future use with a tag, the tag shall be of sufficient durability to withstand the environment involved.

800.26 Spread of Fire or Products of Combustion. Installations of communications cables, communications raceways, cable routing assemblies in hollow spaces, vertical shafts, and ventilation or air-handling ducts shall be made so that the possible spread of fire or products of combustion will not be substantially increased. Openings around penetrations of communications cables, communications raceways, and cable routing assemblies through fire-resistant-rated walls, partitions, floors, or ceilings shall be firestopped using approved methods to maintain the fire resistance rating.

Informational Note: Directories of electrical construction materials published by qualified testing laboratories contain many listing installation restrictions necessary to maintain the fire-resistive rating of assemblies where penetrations or openings are made. Building codes also contain restrictions on membrane penetrations on opposite sides of a fire resistance–rated wall assembly. An example is the 600 mm (24 in.) minimum horizontal separation that usually applies between boxes installed on opposite sides of the wall. Assistance in complying with 800.26 can be found in building codes, fire resistance directories, and product listings.

Part II. Wires and Cables Outside and Entering Buildings

800.44 Overhead (Aerial) Communications Wires and Cables. Overhead (aerial) communications wires and cables entering buildings shall comply with 800.44(A) and (B).

(A) On Poles and In-Span. Where communications wires and cables and electric light or power conductors are supported by the same pole or are run parallel to each other in-span, the conditions described in 800.44(A)(1) through 800.44(A)(4) shall be met.

(1) Relative Location. Where practicable, the communications wires and cables shall be located below the electric light or power conductors.

(2) Attachment to Cross-Arms. Communications wires and cables shall not be attached to a cross-arm that carries electric light or power conductors.

(3) Climbing Space. The climbing space through communications wires and cables shall comply with the requirements of 225.14(D).

(4) Clearance. Supply service drops and sets of overhead service conductors of 0 to 750 volts running above and parallel to communications service drops shall have a minimum separation of 300 mm (12 in.) at any point in the span, including the point of and at their attachment to the building, provided that the ungrounded conductors are insulated and that a clearance of not less than 1.0 m (40 in.) is maintained between the two services at the pole.

(B) Above Roofs. Communications wires and cables shall have a vertical clearance of not less than 2.5 m (8 ft) from all points of roofs above which they pass.

Exception No. 1: Communications wires and cables shall not be required to have a vertical clearance of not less than 2.5 m (8 ft) above auxiliary buildings, such as garages and the like.

Exception No. 2: A reduction in clearance above only the overhanging portion of the roof to not less than 450 mm (18 in.) shall be permitted if (a) not more than 1.2 m (4 ft) of communications service-drop conductors pass above the roof overhang and (b) they are terminated at a through- or above-the-roof raceway or approved support.

Exception No. 3: Where the roof has a slope of not less than 100 mm in 300 mm (4 in. in 12 in.), a reduction in clearance to not less than 900 mm (3 ft) shall be permitted.

> Informational Note: For additional information regarding overhead (aerial) wires and cables, see ANSI/IEEE C2-2012, *National Electrical Safety Code, Part 2, Safety Rules for Overhead Lines.*

800.47 Underground Communications Wires and Cables Entering Buildings. Underground communications wires and cables entering buildings shall comply with 800.47(A) and (B). The requirements of 310.10(C) shall not apply to communications wires and cables.

(A) Underground Systems with Electric Light, Power, Class 1, or Non–Power-Limited Fire Alarm Circuit Conductors. Underground communications wires and cables in a raceway, handhole enclosure, or manhole containing electric light, power, Class 1, or non–power-limited fire alarm circuit conductors shall be in a section separated from such conductors by means of brick, concrete, or tile partitions or by means of a suitable barrier.

(B) Underground Block Distribution. Where the entire street circuit is run underground and the circuit within the block is placed so as to be free from the likelihood of accidental contact with electric light or power circuits of over 300 volts to ground, the insulation requirements of 800.50(A) and 800.50(C) shall not apply, insulating supports shall not be required for the conductors, and bushings shall not be required where the conductors enter the building.

800.48 Unlisted Cables Entering Buildings. Unlisted outside plant communications cables shall be permitted to be installed in building spaces other than risers, ducts used for environmental air, plenums used for environmental air, and other spaces used for environmental air, where the length of the cable within the building, measured from its point of entrance, does not exceed 15 m (50 ft) and the cable enters the building from the outside and is terminated in an enclosure or on a listed primary protector. The point of entrance shall be permitted to be extended from the penetration of the external wall or floor slab by continuously enclosing the entrance cables in rigid metal conduit (RMC) or intermediate metal conduit (IMC) to the point of emergence.

> Informational Note No. 1: Splice cases or terminal boxes, both metallic and plastic types, are typically used as enclosures for splicing or terminating telephone cables.

> Informational Note No. 2: This section limits the length of unlisted outside plant cable to 15 m (50 ft), while 800.90(B) requires that the primary protector be located as close as practicable to the point at which the cable enters the building. Therefore, in installations requiring a primary protector, the outside plant cable may not be permitted to extend 15 m (50 ft) into the building if it is practicable to place the primary protector closer than 15 m (50 ft) to the point of entrance.

800.49 Metallic Entrance Conduit Grounding. Metallic conduit containing communications entrance wire or cable shall be connected by a bonding conductor or grounding electrode conductor to a grounding electrode in accordance with 800.100(B).

800.50 Circuits Requiring Primary Protectors. Circuits that require primary protectors as provided in 800.90 shall comply with 800.50(A), 800.50(B), and 800.50(C).

(A) Insulation, Wires, and Cables. Communications wires and cables without a metallic shield, running from the last outdoor support to the primary protector, shall be listed in accordance with 800.173.

(B) On Buildings. Communications wires and cables in accordance with 800.50(A) shall be separated at least 100 mm (4 in.) from electric light or power conductors not in a raceway or cable or be permanently separated from conductors of the other systems by a continuous and firmly fixed nonconductor in addition to the insulation on the wires, such as porcelain tubes or flexible tubing. Communications wires and cables in accordance with 800.50(A) exposed to accidental contact with electric light and power conductors operating at over 300 volts to ground and attached to buildings shall be separated from woodwork by being supported on glass, porcelain, or other insulating material.

Exception: Separation from woodwork shall not be required where fuses are omitted as provided for in 800.90(A)(1), or where conductors are used to extend circuits to a building from a cable having a grounded metal sheath.

(C) Entering Buildings. Where a primary protector is installed inside the building, the communications wires and cables shall enter the building either through a noncombustible, nonabsorbent insulating bushing or through a metal raceway. The insulating bushing shall not be required where the entering communications wires and cables (1) are in metal-sheathed cable, (2) pass through masonry, (3) meet the requirements of 800.50(A) and fuses are omitted as provided in 800.90(A)(1), or (4) meet the requirements of 800.50(A) and are used to extend circuits to a building from a cable having a grounded metallic sheath. Raceways or bushings shall slope upward from the outside or, where this cannot be done, drip loops shall be formed in the communications wires and cables immediately before they enter the building.

Raceways shall be equipped with an approved service head. More than one communications wire and cable shall be permitted to enter through a single raceway or bushing. Conduits or other metal raceways located ahead of the primary protector shall be grounded.

800.53 Lightning Conductors. Where practicable, a separation of at least 1.8 m (6 ft) shall be maintained between communications wires and cables on buildings and lightning conductors.

> Informational Note: Specific separation distances may be calculated from the sideflash equation in NFPA 780-2014, *Standard for the Installation of Lightning Protection Systems,* 4.16.2.

Part III. Protection

800.90 Protective Devices.

(A) Application. A listed primary protector shall be provided on each circuit run partly or entirely in aerial wire or aerial cable not confined within a block. Also, a listed primary protector shall be provided on each circuit, aerial or underground, located within the block containing the building served so as to be exposed to accidental contact with electric light or power conductors operating at over 300 volts to ground. In addition, where there exists a lightning exposure, each interbuilding circuit on a premises shall be protected by a listed primary protector at each end of the interbuilding circuit. Installation of primary protectors shall also comply with 110.3(B).

Informational Note No. 1: On a circuit not exposed to accidental contact with power conductors, providing a listed primary protector in accordance with this article helps protect against other hazards, such as lightning and above-normal voltages induced by fault currents on power circuits in proximity to the communications circuit.

Informational Note No. 2: Interbuilding circuits are considered to have a lightning exposure unless one or more of the following conditions exist:

(1) Circuits in large metropolitan areas where buildings are close together and sufficiently high to intercept lightning.
(2) Interbuilding cable runs of 42 m (140 ft) or less, directly buried or in underground conduit, where a continuous metallic cable shield or a continuous metallic conduit containing the cable is connected to each building grounding electrode system.
(3) Areas having an average of five or fewer thunderstorm days per year and earth resistivity of less than 100 ohm-meters. Such areas are found along the Pacific coast.

Informational Note: For information on lightning protection systems, see NFPA 780- 2014, *Standard for the Installation of Lightning Protection Systems.*

(1) Fuseless Primary Protectors. Fuseless-type primary protectors shall be permitted under any of the conditions given in (A)(1)(a) through (A)(1)(e).

(a) Where conductors enter a building through a cable with grounded metallic sheath member(s) and where the conductors in the cable safely fuse on all currents greater than the current-carrying capacity of the primary protector and of the primary protector bonding conductor or grounding electrode conductor

(b) Where insulated conductors in accordance with 800.50(A) are used to extend circuits to a building from a cable with an effectively grounded metallic sheath member(s) and where the conductors in the cable or cable stub, or the connections between the insulated conductors and the plant exposed to accidental contact with electric light or power conductors operating at greater than 300 volts to ground, safely fuse on all currents greater than the current-carrying capacity of the primary protector, or the associated insulated conductors and of the primary protector bonding conductor or grounding electrode conductor

(c) Where insulated conductors in accordance with 800.50(A) or (B) are used to extend circuits to a building from other than a cable with metallic sheath member(s), where (1) the primary protector is listed as being suitable for this purpose for application with circuits extending from other than a cable with metallic sheath members, and (2) the connections of the insulated conductors to the plant exposed to accidental contact with electric light or power conductors operating at greater than 300 volts to ground or the conductors of the plant exposed to accidental contact with electric light or power conductors operating at greater than 300 volts to ground safely fuse on all currents greater than the current-carrying capacity of the primary protector, or associated insulated conductors and of the primary protector bonding conductor or grounding electrode conductor

(d) Where insulated conductors in accordance with 800.50(A) are used to extend circuits aerially to a building from a buried or underground circuit that is unexposed to accidental contact with electric light or power conductors operating at greater than 300 volts to ground

(e) Where insulated conductors in accordance with 800.50(A) are used to extend circuits to a building from cable with an effectively grounded metallic sheath member(s), and where (1) the combination of the primary protector and insulated conductors is listed as being suitable for this purpose for application with circuits extending from a cable with an effectively grounded metallic sheath member(s), and (2) the insulated conductors safely fuse on all currents greater than the current-carrying capacity of the primary protector and of the primary protector bonding conductor or grounding electrode conductor

Informational Note: Section 9 of ANSI/IEEE C2-2012, *National Electrical Safety Code,* provides an example of methods of protective grounding that can achieve effective grounding of communications cable sheaths for cables from which communications circuits are extended.

(2) Fused Primary Protectors. Where the requirements listed under 800.90(A)(1)(a) through (A)(1)(e) are not met, fused-type primary protectors shall be used. Fused-type primary protectors shall consist of an arrester connected between each line conductor and ground, a fuse in series with each line conductor, and an appropriate mounting arrangement. Primary protector terminals shall be marked to indicate line, instrument, and ground, as applicable.

(B) Location. The primary protector shall be located in, on, or immediately adjacent to the structure or building served and as close as practicable to the point of entrance.

For purposes of this section, primary protectors located at mobile home service equipment within 9.0 m (30 ft) of the exterior wall of the mobile home it serves, or at a mobile home disconnecting means connected to an electrode by a grounding electrode conductor in accordance with 250.32 and located within 9.0 m (30 ft) of the exterior wall of the mobile home it serves, shall be considered to meet the requirements of this section.

Informational Note: Selecting a primary protector location to achieve the shortest practicable primary protector bonding conductor or grounding electrode conductor helps limit potential differences between communications circuits and other metallic systems.

(C) Hazardous (Classified) Locations. The primary protector shall not be located in any hazardous (classified) locations, as defined in 500.5 and 505.5, or in the vicinity of easily ignitible material.

Exception: As permitted in 501.150, 502.150, and 503.150.

(D) Secondary Protectors. Where a secondary protector is installed in series with the indoor communications wire and cable between the primary protector and the equipment, it shall be listed for the purpose in accordance with 800.170(B).

Informational Note: Secondary protectors on circuits exposed to accidental contact with electric light or power conductors operating at greater than 300 volts to ground are not intended for use without primary protectors.

800.93 Grounding or Interruption of Non–Current-Carrying Metallic Sheath Members of Communications Cables. Communications cables entering the building or terminating on the outside of the building shall comply with 800.93(A) or 800.93(B).

(A) Entering Buildings. In installations where the communications cable enters a building, the metallic sheath members of the cable shall be either grounded as specified in 800.100 or interrupted by an insulating joint or equivalent device. The grounding or interruption shall be as close as practicable to the point of entrance.

(B) Terminating on the Outside of Buildings. In installations where the communications cable is terminated on the outside of the building, the metallic sheath members of the cable shall be either grounded as specified in 800.100 or interrupted by an insulating joint or equivalent device. The grounding or interruption shall be as close as practicable to the point of termination of the cable.

Part IV. Grounding Methods

800.100 Cable and Primary Protector Bonding and Grounding. The primary protector and the metallic member(s) of the cable sheath shall be bonded or grounded as specified in 800.100(A) through 800.100(D).

(A) Bonding Conductor or Grounding Electrode Conductor.

(1) Insulation. The bonding conductor or grounding electrode conductor shall be listed and shall be permitted to be insulated, covered, or bare.

(2) Material. The bonding conductor or grounding electrode conductor shall be copper or other corrosion-resistant conductive material, stranded or solid.

(3) Size. The bonding conductor or grounding electrode conductor shall not be smaller than 14 AWG. It shall have a current-carrying capacity not less than the grounded metallic sheath member(s) and protected conductor(s) of the communications cable. The bonding conductor or grounding electrode conductor shall not be required to exceed 6 AWG.

(4) Length. The primary protector bonding conductor or grounding electrode conductor shall be as short as practicable. In one- and two-family dwellings, the primary protector bonding conductor or grounding electrode conductor shall be as short as practicable, not to exceed 6.0 m (20 ft) in length.

> Informational Note: Similar bonding conductor or grounding electrode conductor length limitations applied at apartment buildings and commercial buildings help to reduce voltages that may be developed between the building's power and communications systems during lightning events.

Exception: In one- and two-family dwellings where it is not practicable to achieve an overall maximum primary protector bonding conductor or grounding electrode conductor length of 6.0 m (20 ft), a separate communications ground rod meeting the minimum dimensional criteria of 800.100(B)(3)(2) shall be driven, the primary protector shall be connected to the communications ground rod in accordance with 800.100(C), and the communications ground rod shall be connected to the power grounding electrode system in accordance with 800.100(D).

(5) Run in Straight Line. The bonding conductor or grounding electrode conductor shall be run in as straight a line as practicable.

(6) Physical Protection. Bonding conductors and grounding electrode conductors shall be protected where exposed to physical damage. Where the bonding conductor or grounding electrode conductor is installed in a metal raceway, both ends of the raceway shall be bonded to the contained conductor or to the same terminal or electrode to which the bonding conductor or grounding electrode conductor is connected.

(B) Electrode. The bonding conductor or grounding electrode conductor shall be connected in accordance with 800.100(B)(1), 800.100(B)(2), or 800.100(B)(3).

(1) In Buildings or Structures with an Intersystem Bonding Termination. If the building or structure served has an intersystem bonding termination as required by 250.94, the bonding conductor shall be connected to the intersystem bonding termination.

(2) In Buildings or Structures with Grounding Means. If an intersystem bonding termination is established, 250.94(A) shall apply.

If the building or structure served has no intersystem bonding termination, the bonding conductor or grounding electrode conductor shall be connected to the nearest accessible location on one of the following:

(1)　The building or structure grounding electrode system as covered in 250.50

(2)　The grounded interior metal water piping system, within 1.5 m (5 ft) from its point of entrance to the building, as covered in 250.52

(3)　The power service accessible means external to enclosures using the options identified in 250.94(A), Exception

(4)　The nonflexible metallic power service raceway

(5)　The service equipment enclosure

(6)　The grounding electrode conductor or the grounding electrode conductor metal enclosure of the power service

(7)　The grounding electrode conductor or the grounding electrode of a building or structure disconnecting means that is grounded to an electrode as covered in 250.32

A bonding device intended to provide a termination point for the bonding conductor (intersystem bonding) shall not interfere with the opening of an equipment enclosure. A bonding device shall be mounted on nonremovable parts. A bonding device shall not be mounted on a door or cover even if the door or cover is nonremovable.

For purposes of this section, the mobile home service equipment or the mobile home disconnecting means, as described in 800.90(B), shall be considered accessible.

(3) In Buildings or Structures Without an Intersystem Bonding Termination or Grounding Means. If the building or structure served has no intersystem bonding termination or grounding means, as described in 800.100(B)(2), the grounding electrode conductor shall be connected to either of the following:

(1)　To any one of the individual grounding electrodes described in 250.52(A)(1), (A)(2), (A)(3), or (A)(4).

(2)　If the building or structure served has no intersystem bonding termination or has no grounding means, as described in 800.100(B)(2) or (B)(3)(1), to any one of the individual grounding electrodes described in 250.52(A)(7) and (A)(8) or to a ground rod or pipe not less than 1.5 m (5 ft) in length and 12.7 mm (½ in.) in diameter, driven, where practicable, into permanently damp earth and separated from lightning protection system conductors as covered in 800.53 and at least 1.8 m (6 ft) from electrodes of other systems. Steam, hot water pipes, or lightning protection system conductors shall not be employed as electrodes for protectors and grounded metallic members.

(C) Electrode Connection. Connections to grounding electrodes shall comply with 250.70.

(D) Bonding of Electrodes. A bonding jumper not smaller than 6 AWG copper or equivalent shall be connected between the communications grounding electrode and power grounding electrode system at the building or structure served where separate electrodes are used.

Exception: At mobile homes as covered in 800.106.

Informational Note No. 1: See 250.60 for connection to a lightning protection system.

Informational Note No. 2: Bonding together of all separate electrodes limits potential differences between them and between their associated wiring systems.

800.106 Primary Protector Grounding and Bonding at Mobile Homes.

(A) Grounding. Grounding shall comply with 800.106(A)(1) and 800.106(A)(2).

(1) Where there is no mobile home service equipment located within 9.0 m (30 ft) of the exterior wall of the mobile home it serves, the primary protector grounding terminal shall be connected to a grounding electrode conductor or grounding electrode in accordance with 800.100(B)(3).

(2) Where there is no mobile home disconnecting means grounded in accordance with 250.32 and located within 9.0 m (30 ft) of the exterior wall of the mobile home it serves, the primary protector grounding terminal shall be connected to a grounding electrode in accordance with 800.100(B)(3).

(B) Bonding. The primary protector grounding terminal or grounding electrode shall be connected to the metal frame or available grounding terminal of the mobile home with a copper conductor not smaller than 12 AWG under either of the following conditions:

(1) Where there is no mobile home service equipment or disconnecting means as in 800.106(A)

(2) Where the mobile home is supplied by cord and plug

Part V. Installation Methods Within Buildings

800.110 Raceways and Cable Routing Assemblies for Communications Wires and Cables.

(A) Types of Raceways. Communications wires and cables shall be permitted to be installed in any raceway that complies with either (A)(1) or (A)(2) and in cable routing assemblies installed in compliance with 800.110(C).

(1) Raceways Recognized in Chapter 3. Communications wires and cables shall be permitted to be installed in any raceway included in Chapter 3. The raceways shall be installed in accordance with the requirements of Chapter 3.

(2) Communications Raceways. Communications wires and cables shall be permitted to be installed in plenum communications raceways, riser communications raceways, and general-purpose communications raceways selected in accordance with Table 800.154(b), listed in accordance with 800.182, and installed in accordance with 800.113 and 362.24 through 362.56, where the requirements applicable to electrical nonmetallic tubing (ENT) apply.

(3) Innerduct for Communications Wires and Cables. Listed plenum communications raceway, listed riser communications raceway, and listed general-purpose communications raceway selected in accordance with Table 800.154(b) shall be permitted to be installed as innerduct in any type of listed raceway permitted in Chapter 3.

(B) Raceway Fill for Communications Wires and Cables. The raceway fill requirements of Chapters 3 and 9 shall not apply to communications wires and cables.

(C) Cable Routing Assemblies. Communications wires and cables shall be permitted to be installed in plenum cable routing assemblies, riser cable routing assemblies, and general-purpose cable routing assemblies selected in accordance with Table 800.154(c), listed in accordance with 800.182, and installed in accordance with 800.110(C)(1) and (C)(2) and 800.113.

(1) Horizontal Support. Cable routing assemblies shall be supported where run horizontally at intervals not to exceed 900 mm (3 ft), and at each end or joint, unless listed for other support intervals. In no case shall the distance between supports exceed 3 m (10 ft).

(2) Vertical Support. Vertical runs of cable routing assemblies shall be supported at intervals not exceeding 1.2 m (4 ft), unless listed for other support intervals, and shall not have more than one joint between supports.

800.113 Installation of Communications Wires, Cables and Raceways, and Cable Routing Assemblies.
Installation of communications wires, cables and raceways, and cable routing assemblies shall comply with 800.113(A) through (L). Installation of raceways and cable routing assemblies shall also comply with 800.110.

(A) Listing. Communications wires, communications cables, communications raceways, and cable routing assemblies installed in buildings shall be listed.

Exception: Communications cables that are installed in compliance with 800.48 shall not be required to be listed.

(B) Ducts Specifically Fabricated for Environmental Air. The following wires and cables shall be permitted in ducts used for environmental air as described in 300.22(B) if they are directly associated with the air distribution system:

(1) Up to 1.22 m (4 ft) of Type CMP cable
(2) Types CMP, CMR, CMG, CM, and CMX cables and communications wires installed in raceways that are installed in compliance with 300.22(B)

Informational Note: For information on fire protection of wiring installed in fabricated ducts see 4.3.4.1 and 4.3.11.3.3 of NFPA 90A-2015, *Standard for the Installation of Air-Conditioning and Ventilating Systems.*

(C) Other Spaces Used for Environmental Air (Plenums). The following wires, cables, raceways, and cable routing assemblies shall be permitted in other spaces used for environmental air as described in 300.22(C):

(1) Type CMP cables
(2) Plenum communications raceways
(3) Plenum cable routing assemblies
(4) Type CMP cables installed in plenum communications raceways
(5) Type CMP cables installed in plenum cable routing assemblies
(6) Type CMP cables and plenum communications raceways supported by open metallic cable trays or cable tray systems
(7) Types CMP, CMR, CMG, CM, and CMX cables and communications wires installed in raceways that are installed in compliance with 300.22(C)
(8) Types CMP, CMR, CMG, CM, and CMX cables, plenum communications raceways, riser communications raceways, and general-purpose communications raceways supported by solid bottom metal cable trays with solid metal covers in other spaces used for environmental air (plenums) as described in 300.22(C)
(9) Types CMP, CMR, CMG, CM, and CMX cables installed in plenum communications raceways, riser communications raceways, and general-purpose communications raceways supported by solid bottom metal cable trays with solid metal covers in other spaces used for environmental air (plenums) as described in 300.22(C)

Informational Note: For information on fire protection of wiring installed in other spaces used for environmental air, see 4.3.11.2, 4.3.11.4, and 4.3.11.5 of NFPA 90A-2015, *Standard for the Installation of Air-Conditioning and Ventilating Systems.*

(D) Risers — Cables, Raceways, and Cable Routing Assemblies in Vertical Runs. The following cables, raceways, and cable routing assemblies shall be permitted in vertical runs penetrating one or more floors and in vertical runs in a shaft:

(1) Types CMP and CMR cables
(2) Plenum and riser communications raceways
(3) Plenum and riser cable routing assemblies
(4) Types CMP and CMR cables installed in:

   a. Plenum communications raceways
   b. Riser communications raceways
   c. Plenum cable routing assemblies
   d. Riser cable routing assemblies

Informational Note: See 800.26 for firestop requirements for floor penetrations.

(E) Risers — Cables and Innerducts in Metal Raceways. The following cables and innerducts shall be permitted in metal raceways in a riser having firestops at each floor:

(1) Types CMP, CMR, CMG, CM, and CMX cables
(2) Plenum, riser, and general-purpose communications raceways
(3) Types CMP, CMR, CMG, CM, and CMX cables installed in:

   a. Plenum communications raceways (innerduct)
   b. Riser communications raceways (innerduct)
   c. General-purpose communications raceways (innerduct)

Informational Note: See 800.26 for firestop requirements for floor penetrations.

(F) Risers — Cables, Raceways, and Cable Routing Assemblies in Fireproof Shafts. The following cables, raceways, and cable routing assemblies shall be permitted to be installed in fireproof riser shafts having firestops at each floor:

(1) Types CMP, CMR, CMG, CM, and CMX cables
(2) Plenum, riser, and general-purpose communications raceways
(3) Plenum, riser, and general-purpose cable routing assemblies
(4) Types CMP, CMR, CMG, and CM cables installed in:

   a. Plenum communications raceways
   b. Riser communications raceways
   c. General-purpose communications raceways
   d. Plenum cable routing assemblies
   e. Riser cable routing assemblies
   f. General-purpose cable routing assemblies

Informational Note: See 800.26 for firestop requirements for floor penetrations.

(G) Risers — One- and Two-Family Dwellings. The following cables, raceways, and cable routing assemblies shall be permitted in one- and two-family dwellings:

(1) Types CMP, CMR, CMG, and CM cables
(2) Type CMX cables less than 6 mm (0.25 in.) in diameter
(3) Plenum, riser, and general-purpose communications raceways
(4) Plenum, riser, and general-purpose cable routing assemblies
(5) Types CMP, CMR, CMG, and CM cables installed in:

   a. Plenum communications raceways
   b. Riser communications raceways
   c. General-purpose communications raceways
   d. Plenum cable routing assemblies
   e. Riser cable routing assemblies
   f. General-purpose cable routing assemblies

(H) Cable Trays. The following wires, cables, and raceways shall be permitted to be supported by cable trays:

(1) Types CMP, CMR, CMG, and CM cables
(2) Plenum, riser, and general-purpose communications raceways
(3) Communications wires and Types CMP, CMR, CMG, and CM cables installed in:

   a. Plenum communications raceways
   b. Riser communications raceways
   c. General-purpose communications raceways

(I) Distributing Frames and Cross-Connect Arrays. The following wires, cables, raceways, and cable routing assemblies shall be permitted to be installed in distributing frames and cross-connect arrays:

(1) Types CMP, CMR, CMG, and CM cables and communications wires
(2) Plenum, riser, and general-purpose communications raceways
(3) Plenum, riser, and general-purpose cable routing assemblies
(4) Communications wires and Types CMP, CMR, CMG, and CM cables installed in:

   a. Plenum communications raceways
   b. Riser communications raceways
   c. General-purpose communications raceways

d. Plenum cable routing assemblies

e. Riser cable routing assemblies

f. General-purpose cable routing assemblies

(J) Other Building Locations. The following wires, cables, raceways, and cable routing assemblies shall be permitted to be installed in building locations other than the locations covered in 800.113(B) through (I):

(1) Types CMP, CMR, CMG, and CM cables

(2) A maximum of 3 m (10 ft) of exposed Type CMX in nonconcealed spaces

(3) Plenum, riser, and general-purpose communications raceways

(4) Plenum, riser, and general-purpose cable routing assemblies

(5) Communications wires and Types CMP, CMR, CMG, and CM cables installed in:

a. Plenum communications raceways

b. Riser communications raceways

c. General-purpose communications raceways

(6) Types CMP, CMR, CMG, and CM cables installed in:

a. Plenum cable routing assemblies

b. Riser cable routing assemblies

c. General-purpose cable routing assemblies

(7) Communications wires and Types CMP, CMR, CMG, CM, and CMX cables installed in raceways recognized in Chapter 3

(8) Type CMUC under-carpet communications wires and cables installed under carpet, modular tiles, and planks

(K) Multifamily Dwellings. The following cables, raceways, and cable routing assemblies shall be permitted to be installed in multifamily dwellings in locations other than the locations covered in 800.113(B) through (G):

(1) Types CMP, CMR, CMG, and CM cables

(2) Type CMX cables less than 6 mm (0.25 in.) in diameter in nonconcealed spaces

(3) Plenum, riser, and general-purpose communications raceways

(4) Plenum, riser, and general-purpose cable routing assemblies

(5) Communications wires and Types CMP, CMR, CMG, and CM cables installed in:

a. Plenum communications raceways

b. Riser communications raceways

c. General-purpose communications raceways

(6) Types CMP, CMR, CMG, and CM cables installed in:

a. Plenum cable routing assemblies

b. Riser cable routing assemblies

c. General-purpose cable routing assemblies

(7) Communications wires and Types CMP, CMR, CMG, CM, and CMX cables installed in raceways recognized in Chapter 3

(8) Type CMUC under-carpet communications wires and cables installed under carpet, modular tiles, and planks

(L) One- and Two-Family Dwellings. The following cables, raceways, and cable routing assemblies shall be permitted to be installed in one- and two-family dwellings in locations other than the locations covered in 800.113(B) through 800.113(F):

(1) Types CMP, CMR, CMG, and CM cables

(2) Type CMX cables less than 6 mm (0.25 in.) in diameter

(3) Plenum, riser, and general-purpose communications raceways

(4) Plenum, riser, and general-purpose cable routing assemblies

(5) Communications wires and Types CMP, CMR, CMG, and CM cables installed in:

a. Plenum communications raceways

b. Riser communications raceways

c. General-purpose communications raceways

(6) Types CMP, CMR, CMG, and CM cables installed in:

a. Plenum cable routing assemblies

b. Riser cable routing assemblies

c. General-purpose cable routing assemblies

(7) Communications wires and Types CMP, CMR, CMG, CM, and CMX cables installed in raceways recognized in Chapter 3

(8) Type CMUC under-carpet communications wires and cables installed under carpet, modular tiles, and planks

(9) Hybrid power and communications cable listed in accordance with 800.179(I)

800.133 Installation of Communications Wires, Cables, and Equipment. Communications wires and cables from the protector to the equipment or, where no protector is required, communications wires and cables attached to the outside or inside of the building shall comply with 800.133(A) and 800.133(B).

(A) Separation from Other Conductors.

(1) In Raceways, Cable Trays, Boxes, Cables, Enclosures, and Cable Routing Assemblies.

(a) *Other Circuits.* Communications cables shall be permitted in the same raceway, cable tray, box, enclosure, or cable routing assembly with cables of any of the following:

(1) Class 2 and Class 3 remote-control, signaling, and power-limited circuits in compliance with Article 645 or Parts I and III of Article 725

(2) Power-limited fire alarm systems in compliance with Parts I and III of Article 760

(3) Nonconductive and conductive optical fiber cables in compliance with Parts I and V of Article 770

(4) Community antenna television and radio distribution systems in compliance with Parts I and V of Article 820

(5) Low-power network-powered broadband communications circuits in compliance with Parts I and V of Article 830

(b) *Class 2 and Class 3 Circuits.* Class 1 circuits shall not be run in the same cable with communications circuits. Class 2 and Class 3 circuit conductors shall be permitted in the same cable with communications circuits, in which case the Class 2 and Class 3 circuits shall be classified as communications circuits and shall meet the requirements of this article. The cables shall be listed as communications cables.

Exception: Cables constructed of individually listed Class 2, Class 3, and communications cables under a common jacket shall not be required to be classified as communications cable. The fire-resistance rating of the composite cable shall be determined by the performance of the composite cable.

(c) *Electric Light, Power, Class 1, Non–Power-Limited Fire Alarm, and Medium-Power Network-Powered Broadband Communications Circuits in Raceways, Compartments, and Boxes.* Communications conductors shall not be placed in any raceway, compartment, outlet box, junction box, or similar fitting with conductors of electric light, power, Class 1, non–power-limited fire alarm, or medium-power network-powered broadband communications circuits.

Exception No. 1: Section 800.133(A)(1)(c) shall not apply if all of the conductors of electric light, power, Class 1, non–power-limited fire alarm, and medium-power network-powered broadband communications circuits are separated from all of the conductors of communications circuits by a permanent barrier or listed divider.

Exception No. 2: Power conductors in outlet boxes, junction boxes, or similar fittings or compartments where such conductors are introduced solely for power supply to communications equipment. The power circuit conductors shall be routed within the enclosure to maintain a minimum of 6 mm (1/4 in.) separation from the communications circuit conductors.

Exception No. 3: As permitted by 620.36.

(2) Other Applications. Communications wires and cables shall be separated at least 50 mm (2 in.) from conductors of any electric light, power, Class 1, non–power-limited fire alarm, or medium-power network-powered broadband communications circuits.

Exception No. 1: Section 800.133(A)(2) shall not apply where either (1) all of the conductors of the electric light, power, Class 1, non–power-limited fire alarm, and medium-power network-powered broadband communications circuits are in a raceway or in metal-sheathed, metal-clad, nonmetallic-sheathed, Type AC, or Type UF cables, or (2) all of the conductors of communications circuits are encased in raceway.

Exception No. 2: Section 800.133(A)(2) shall not apply where the communications wires and cables are permanently separated from the conductors of electric light, power, Class 1, non–power-limited fire alarm, and medium-power network-powered broadband communications circuits by a continuous and firmly fixed nonconductor, such as porcelain tubes or flexible tubing, in addition to the insulation on the wire.

(B) Support of Communications Wires and Cables. Raceways shall be used for their intended purpose. Communications wires and cables shall not be strapped, taped, or attached by any means to the exterior of any raceway as a means of support.

Exception: Overhead (aerial) spans of communications wires and cables shall be permitted to be attached to the exterior of a raceway-type mast intended for the attachment and support of such wires and cables.

800.154 Applications of Listed Communications Wires, Cables, and Raceways, and Listed Cable Routing Assemblies. Permitted and nonpermitted applications of listed communications wires, cables, and raceways, and listed cable routing assemblies, shall be in accordance with one of the following:

(1) Listed communications wires and cables as indicated in Table 800.154(a)
(2) Listed communications raceways as indicated in Table 800.154(b)
(3) Listed cable routing assemblies as indicated in Table 800.154(c)

The permitted applications shall be subject to the installation requirements of 800.110 and 800.113. The substitutions for communications cables listed in Table 800.154(d) and illustrated in Figure 800.154 shall be permitted.

Table 800.154(d) Cable Substitutions

Cable Type	Permitted Substitutions
CMR	CMP
CMG, CM	CMP, CMR
CMX	CMP, CMR, CMG, CM

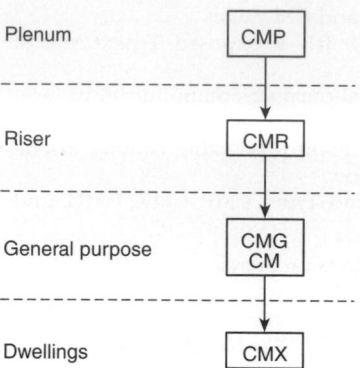

Type CM—Communications cables

[A] Cable A shall be permitted to be used in place of cable B.

[B]

FIGURE 800.154   Cable Substitution Hierarchy.

800.156 Dwelling Unit Communications Outlet. For new construction, a minimum of one communications outlet shall be installed within the dwelling in a readily accessible area and cabled to the service provider demarcation point.

Part VI. Listing Requirements

800.170 Equipment. Communications equipment shall be listed as being suitable for electrical connection to a communications network.

Informational Note: One way to determine applicable requirements is to refer to ANSI/UL 60950-1-2014, *Standard for Safety of Information Technology Equipment;* UL 1459-1998, *Standard for Safety Telephone Equipment;* ANSI/UL 1863-2012, *Standard for Safety Communications Circuit Accessories;* or ANSI/UL 62368-1-2014, *Audio/Video, Information and Communication Technology Equipment – Part 1: Safety Requirements.*

(A) Primary Protectors. The primary protector shall consist of an arrester connected between each line conductor and ground in an appropriate mounting. Primary protector terminals shall be marked to indicate line and ground as applicable.

Informational Note: One way to determine applicable requirements for a listed primary protector is to refer to ANSI/UL 497-2013, *Standard for Protectors for Paired Conductor Communications Circuits.*

Table 800.154(a) Applications of Listed Communications Wires and Cables in Buildings

Applications		CMP	CMR	CMG CM	CMX	CMUC	Hybrid power and Communications cables	Communications wires
In ducts specifically fabricated for environmental air as described in 300.22(B)	In fabricated ducts	Y*	N	N	N	N	N	N
	In metal raceway that complies with 300.22(B)	Y*	Y*	Y*	Y*	N	N	Y*
In other spaces used for environmental air as (plenums) described in 300.22(C)	In other spaces used for environmental air	Y*	N	N	N	N	N	N
	In metal raceway that complies with 300.22(C)	Y*	Y*	Y*	Y*	N	N	Y*
	In plenum communications raceways	Y*	N	N	N	N	N	N
	In plenum cable routing assemblies	Y*	N	N	N	N	N	N
	Supported by open metal cable trays	Y*	N	N	N	N	N	N
	Supported by solid bottom metal cable trays with solid metal covers	Y*	Y*	Y*	Y*	N	N	N
In risers	In vertical runs	Y*	Y*	N	N	N	N	N
	In metal raceways	Y*	Y*	Y*	Y*	N	N	N
	In fireproof shafts	Y*	Y*	Y*	Y*	N	N	N
	In plenum communications raceways	Y*	Y*	N	N	N	N	N
	In plenum cable routing assemblies	Y*	Y*	N	N	N	N	N
	In riser communications raceways	Y*	Y*	N	N	N	N	N
	In riser cable routing assemblies	Y*	Y*	N	N	N	N	N
	In one- and two-family dwellings	Y*	Y*	Y*	Y*	N	Y*	N
Within buildings in other than air-handling spaces and risers	General	Y*	Y*	Y*	Y*	N	N	N
	In one- and two-family dwellings	Y*	Y*	Y*	Y*	Y*	Y*	N
	In multifamily dwellings	Y*	Y*	Y*	Y*	Y*	N	N
	In nonconcealed spaces	Y*	Y*	Y*	Y*	Y*	N	N
	Supported by cable trays	Y*	Y*	Y*	N	N	N	N
	Under carpet or under floor covering, modular tiles, and planks	N	N	N	N	Y*	N	N
	In distributing frames and cross-connect arrays	Y*	Y*	Y*	N	N	N	Y*
	In any raceway recognized in Chapter 3	Y*	Y*	Y*	Y*	N	N	Y*

(continues)

Table 800.154(a) *Continued*

Applications	Wire and Cable Type						
	CMP	CMR	CMG CM	CMX	CMUC	Hybrid power and Communications cables	Communications wires
In plenum communications raceways	Y*	Y*	Y*	N	N	N	Y*
In plenum cable routing assemblies	Y*	Y*	Y*	N	N	N	Y*
In riser communications raceways	Y*	Y*	Y*	N	N	N	Y*
In riser cable routing assemblies	Y*	Y*	Y*	N	N	N	Y*
In general-purpose communications raceways	Y*	Y*	Y*	N	N	N	Y*
In general-purpose cable routing assemblies	Y*	Y*	Y*	N	N	N	Y*

Note: An "N" in the table indicates that the cable type is not permitted to be installed in the application. A "Y*" indicates that the cable type is permitted to be installed in the application subject to the limitations described in 800.113.
Informational Note No. 1: Part V of Article 800 covers installation methods within buildings. This table covers the applications of listed communications wires, cables, and raceways in buildings. See the definition of Point of Entrance in 800.2.
Informational Note No. 2: For information on the restrictions to the installation of communications cables in fabricated ducts, see 800.113(B).

(B) Secondary Protectors. The secondary protector shall be listed as suitable to provide means to safely limit currents to less than the current-carrying capacity of listed indoor communications wire and cable, listed telephone set line cords, and listed communications terminal equipment having ports for external wire line communications circuits. Any overvoltage protection, arresters, or grounding connection shall be connected on the equipment terminals side of the secondary protector current-limiting means.

> Informational Note: One way to determine applicable requirements for a listed secondary protector is to refer to ANSI/UL 497A-2012, *Standard for Secondary Protectors for Communications Circuits.*

(C) Plenum Grade Cable Ties. Cable ties intended for use in other space used for environmental air (plenums) shall be listed as having low smoke and heat release properties.

> Informational Note: See NFPA 90A-2015, *Standard for the Installation of Air-Conditioning and Ventilating Systems,* and ANSI/UL 2043-2013, *Standard for Safety Fire Test for Heat and Visible Smoke Release for Discrete Products and Their Accessories Installed in Air-Handling Spaces,* for information on listing discrete products as having low smoke and heat release properties.

800.173 Drop Wire and Cable. Communications wires and cables without a metallic shield, running from the last outdoor support to the primary protector, shall be listed as being suitable for the purpose and shall have current-carrying capacity as specified in 800.90(A)(1)(b) or (A)(1)(c).

800.179 Communications Wires and Cables. Communications wires and cables shall be listed in accordance with 800.179(A) through (I) and marked in accordance with Table 800.179. Conductors in communications cables, other than in a coaxial cable, shall be copper.

Communications wires and cables shall have a voltage rating of not less than 300 volts. The insulation for the individual conductors, other than the outer conductor of a coaxial cable, shall be rated for 300 volts minimum. The cable voltage rating shall not be marked on the cable or on the undercarpet communications wire. Communications wires and cables shall have a temperature rating of not less than 60°C (140°F). The temperature rating shall be marked on the jacket of communications cables that have a temperature rating exceeding 60°C (140°F).

Exception: Voltage markings shall be permitted where the cable has multiple listings and voltage marking is required for one or more of the listings.

> Informational Note: Voltage markings on cables may be misinterpreted to suggest that the cables may be suitable for Class 1, electric light, and power applications.

(A) Type CMP. Type CMP communications plenum cables shall be listed as being suitable for use in ducts, plenums, and other spaces used for environmental air and shall also be listed as having adequate fire-resistant and low smoke-producing characteristics.

> Informational Note: One method of defining a cable that is low-smoke producing cable and fire-resistant cable is that the cable exhibits a maximum peak optical density of 0.50 or less, an average optical density of 0.15 or less, and a maximum flame spread distance of 1.52 m (5 ft) or less when tested in accordance with NFPA 262-2015, *Standard Method of Test for Flame Travel and Smoke of Wires and Cables for Use in Air-Handling Spaces.*

Table 800.154(b) Applications of Listed Communications Raceways in Buildings

Applications		Listed Communications Raceway Type		
		Plenum	Riser	General-Purpose
In ducts specifically fabricated for environmental air as described in 300.22(B)	In fabricated ducts	N	N	N
	In metal raceway that complies with 300.22(B)	N	N	N
In other spaces used for environmental air (plenums) as described in 300.22(C)	In other spaces used for environmental air	Y*	N	N
	In metal raceway that complies with 300.22(C)	Y*	Y*	Y*
	In plenum cable routing assemblies	N	N	N
	Supported by open metal cable trays	Y*	N	N
	Supported by solid bottom metal cable trays with solid metal covers	Y*	Y*	Y*
In risers	In vertical runs	Y*	Y*	N
	In metal raceways	Y*	Y*	Y*
	In fireproof shafts	Y*	Y*	Y*
	In plenum cable routing assemblies	N	N	N
	In riser cable routing assemblies	N	N	N
	In one- and two-family dwellings	Y*	Y*	Y*
Within buildings in other than air-handling spaces and risers	General	Y*	Y*	Y*
	In one- and two-family dwellings	Y*	Y*	Y*
	In multifamily dwellings	Y*	Y*	Y*
	In nonconcealed spaces	Y*	Y*	Y*
	Supported by cable trays	Y*	Y*	Y*
	Under carpet or under floor covering, modular tiles, and planks	N	N	N
	In distributing frames and cross-connect arrays	Y*	Y*	Y*
	In any raceway recognized in Chapter 3	Y*	Y*	Y*
	In plenum cable routing assemblies	N	N	N
	In riser cable routing assemblies	N	N	N
	In general-purpose cable routing assemblies	N	N	N

Note: An "N" in the table indicates that the communications raceway type shall not be permitted to be installed in the application. A "Y*" indicates that the communications raceway type shall be permitted to be installed in the application, subject to the limitations described in 800.110 and 800.113.

(B) Type CMR. Type CMR communications riser cables shall be listed as being suitable for use in a vertical run in a shaft or from floor to floor and shall also be listed as having fire-resistant characteristics capable of preventing the carrying of fire from floor to floor.

Informational Note: One method of defining fire-resistant characteristics capable of preventing the carrying of fire from floor to floor is that the cables pass the requirements of ANSI/UL 1666-2011, *Standard Test for Flame Propagation Height of Electrical and Optical-Fiber Cable Installed Vertically in Shafts.*

(C) Type CMG. Type CMG general-purpose communications cables shall be listed as being suitable for general-purpose communications use, with the exception of risers and plenums, and shall also be listed as being resistant to the spread of fire.

Informational Note: One method of defining *resistant to the spread of fire* is for the damage (char length) not to exceed 1.5 m (4 ft 11 in.) when performing the CSA "Vertical Flame Test — Cables in Cable Trays," as described in CSA C22.2 No. 0.3-09, *Test Methods for Electrical Wires and Cables.*

Table 800.154(c) Applications of Listed Cable Routing Assemblies in Buildings

Applications		Listed Cable Routing Assembly Type		
		Plenum	Riser	General-Purpose
In ducts specifically fabricated for environmental air as described in 300.22(B)	In fabricated ducts	N	N	N
	In metal raceway that complies with 300.22(B)	N	N	N
In other spaces used for environmental air (plenums) as described in 300.22(C)	In other spaces used for environmental air	Y*	N	N
	In metal raceway that complies with 300.22(C)	N	N	N
	In plenum communications raceways	N	N	N
	Supported by open metal cable trays	N	N	N
	Supported by solid bottom metal cable trays with solid metal covers	N	N	N
In risers	In vertical runs	Y*	Y*	N
	In metal raceways	N	N	N
	In fireproof shafts	Y*	Y*	Y*
	In plenum communications raceways	N	N	N
	In riser communications raceways	N	N	N
	In one- and two-family dwellings	Y*	Y*	Y*
Within buildings in other than air-handling spaces and risers	General	Y*	Y*	Y*
	In one- and two-family dwellings	Y*	Y*	Y*
	In multifamily dwellings	Y*	Y*	Y*
	In nonconcealed spaces	Y*	Y*	Y*
	Supported by cable trays	N	N	N
	Under carpet or under floor covering, modular tiles, and planks	N	N	N
	In distributing frames and cross-connect arrays	Y*	Y*	Y*
	In any raceway recognized in Chapter 3	N	N	N
	In plenum communications raceways	N	N	N
	In riser communications raceways	N	N	N
	In general-purpose communications raceways	N	N	N

Note: An "N" in the table indicates that the cable routing assembly type shall not be permitted to be installed in the application. A "Y*" indicates that the cable routing assembly type shall be permitted to be installed in the application subject to the limitations described in 800.113.

(D) Type CM. Type CM communications cables shall be listed as being suitable for general-purpose communications use, with the exception of risers and plenums, and shall also be listed as being resistant to the spread of fire.

Informational Note: One method of defining *resistant to the spread of fire* is that the cables do not spread fire to the top of the tray in the "UL Flame Exposure, Vertical Flame Tray Test" in ANSI/UL 1685-2010, *Standard for Safety for Vertical-Tray Fire-Propagation and Smoke-Release Test for Electrical and Optical-Fiber Cables*. The smoke measurements in the test method are not applicable.

Another method of defining *resistant to the spread of fire* is for the damage (char length) not to exceed 1.5 m (4 ft 11 in.) when performing the CSA "Vertical Flame Test — Cables in Cable Trays," as described in CSA C22.2 No. 0.3-09, *Test Methods for Electrical Wires and Cables*.

Table 800.179 Cable Markings

Cable Marking	Type
CMP	Communications plenum cable
CMR	Communications riser cable
CMG	Communications general-purpose cable
CM	Communications general-purpose cable
CMX	Communications cable, limited use
CMUC	Under-carpet communications wire and cable

Informational Note: Cable types are listed in descending order of fire resistance rating.

(E) Type CMX. Type CMX limited-use communications cables shall be listed as being suitable for use in dwellings and for use in raceway and shall also be listed as being resistant to flame spread.

Informational Note: One method of determining that cable is resistant to flame spread is by testing the cable to the VW-1 (vertical-wire) flame test in ANSI/UL 1581-2011, *Reference Standard for Electrical Wires, Cables and Flexible Cords.*

(F) Type CMUC Undercarpet Wires and Cables. Type CMUC undercarpet communications wires and cables shall be listed as being suitable for undercarpet use and shall also be listed as being resistant to flame spread.

Informational Note: One method of determining that cable is resistant to flame spread is by testing the cable to the VW-1 (vertical-wire) flame test in ANSI/UL 1581-2011, *Reference Standard for Electrical Wires, Cables and Flexible Cords.*

(G) Circuit Integrity (CI) Cable or Electrical Circuit Protective System. Cables that are used for survivability of critical circuits under fire conditions shall be listed and meet either 800.179(G)(1) or 800.179(G)(2) as follows:

Informational Note: The listing organization provides information for circuit integrity (CI) cable and electrical circuit protective systems, including installation requirements required to maintain the fire rating.

(1) Circuit Integrity (CI) Cables. Circuit integrity (CI) cables specified in 800.179(A) through (D), and used for survivability of critical circuits, shall have an additional classification using the suffix "CI." In order to maintain its listed fire rating, circuit integrity (CI) cable shall only be installed in free air.

Informational Note: One method of defining circuit integrity (CI) cable is by establishing a minimum 2-hour fire resistance rating for the cable when tested in accordance with ANSI/UL 2196-2006, *Standard for Tests of Fire-Resistive Cable.*

(2) Fire-Resistive Cables. Cables specified in 800.179(A) through (D) and 800.179(G)(1), that are part of an electrical circuit protective system, shall be fire-resistive cable identified with the protective system number on the product, or on the smallest unit container in which the product is packaged, and shall be installed in accordance with the listing of the protective system.

Informational Note No. 1: One method of defining an electrical circuit protective system is by establishing a minimum 2-hour fire resistance rating for the system when tested in accordance with UL Subject 1724, *Outline of Investigation for Fire Tests for Electrical Circuit Protective Systems.*

Informational Note No. 2: The listing organization provides information for electrical circuit protective systems (FHIT), including installation requirements for maintaining the fire rating.

(H) Communications Wires. Communications wires, such as distributing frame wire and jumper wire, shall be listed as being resistant to the spread of fire.

Informational Note: One method of defining *resistant to the spread of fire* is that the cables do not spread fire to the top of the tray in the "UL Flame Exposure, Vertical Flame Tray Test" in ANSI/UL 1685-2010, *Standard for Safety for Vertical-Tray Fire-Propagation and Smoke-Release Test for Electrical and Optical-Fiber Cables.* The smoke measurements in the test method are not applicable.
Another method of defining *resistant to the spread of fire* is for the damage (char length) not to exceed 1.5 m (4 ft 11 in.) when

performing the CSA "Vertical Flame Test— Cables in Cable Trays," as described in CSA C22.2 No. 0.3-09, *Test Methods for Electrical Wires and Cables.*

(I) Hybrid Power and Communications Cables. Listed hybrid power and communications cables shall be permitted where the power cable is a listed Type NM or NM-B, conforming to Part III of Article 334, and the communications cable is a listed Type CM, the jackets on the listed NM or NM-B, and listed CM cables are rated for 600 volts minimum, and the hybrid cable is listed as being resistant to the spread of fire.

Informational Note: One method of defining *resistant to the spread of fire* is that the cables do not spread fire to the top of the tray in the "UL Flame Exposure, Vertical Flame Tray Test" in ANSI/UL 1685-2010, *Standard for Safety for Vertical-Tray Fire-Propagation and Smoke-Release Test for Electrical and Optical-Fiber Cables.* The smoke measurements in the test method are not applicable.
Another method of defining *resistant to the spread of fire* is for the damage (char length) not to exceed 1.5 m (4 ft 11 in.) when performing the CSA "Vertical Flame Test — Cables in Cable Trays," as described in CSA C22.2 No. 0.3-09, *Test Methods for Electrical Wires and Cables.*

800.180 Grounding Devices. Where bonding or grounding is required, devices used to connect a shield, a sheath, or non–current-carrying metallic members of a cable to a bonding conductor or grounding electrode conductor shall be listed or be part of listed equipment.

800.182 Cable Routing Assemblies and Communications Raceways. Cable routing assemblies and communications raceways shall be listed in accordance with 800.182(A) through (C). Cable routing assemblies shall be marked in accordance with Table 800.182(a). Communications raceways shall be marked in accordance with Table 800.182(b).

Informational Note: For information on listing requirements for both communications raceways and cable routing assemblies, see ANSI/UL 2024-5-2015, *Cable Routing Assemblies and Communications Raceways.*

(A) Plenum Cable Routing Assemblies and Plenum Communications Raceways. Plenum cable routing assemblies and plenum communications raceways shall be listed as having adequate fire-resistant and low-smoke-producing characteristics.

Table 800.182(a) Cable Routing Assembly Markings

Type	Marking
Plenum Cable Routing Assembly	Plenum Cable Routing Assembly
Riser Cable Routing Assembly	Riser Cable Routing Assembly
General-Purpose Cable Routing Assembly	General-Purpose Cable Routing Assembly

Table 800.182(b) Communications Raceway Markings

Type	Marking
Plenum Communications Raceway	Plenum Communications Raceway
Riser Communications Raceway	Riser Communications Raceway
General-Purpose Communications Raceway	General-Purpose Communications Raceway

Informational Note No. 1: One method of defining cable routing assemblies and communications raceways that have adequate fire-resistant and low-smoke-producing characteristics is that they exhibit a maximum flame spread index of 25 and a maximum smoke developed index of 50 when tested in accordance with ASTM E84-15a, *Standard Test Method for Surface Burning Characteristics of Building Materials,* or ANSI/UL 723-2013, *Standard Test Method for Surface Burning Characteristics of Building Materials.*

Informational Note No. 2: Another method of defining communications raceways that have adequate fire-resistant and low-smoke-producing characteristics is that they exhibit a maximum peak optical density of 0.50 or less, an average optical density of 0.15 or less, and a maximum flame spread distance of 1.52 m (5 ft) or less when tested in accordance with NFPA 262-2015, *Standard Method of Test for Flame Travel and Smoke of Wires and Cables for Use in Air-Handling Spaces.*

Informational Note No. 3: See 4.3.11.2.6 or 4.3.11.5.5 of NFPA 90A-2015, *Standard for the Installation of Air-Conditioning and Ventilating Systems,* for information on materials exposed to the airflow in ceiling cavity and raised floor plenums.

(B) Riser Cable Routing Assemblies and Riser Communications Raceways. Riser cable routing assemblies and riser communications raceways shall be listed as having adequate fire-resistant characteristics capable of preventing the carrying of fire from floor to floor.

Informational Note: One method of defining fire-resistant characteristics capable of preventing the carrying of fire from floor to floor is that the cable routing assemblies and communications raceways pass the requirements of ANSI/UL 1666-2011, *Standard Test for Flame Propagation Height of Electrical and Optical-Fiber Cable Installed Vertically in Shafts.*

(C) General-Purpose Cable Routing Assemblies and General-Purpose Communication Raceways. General-purpose cable routing assemblies and general-purpose communications raceways shall be listed as being resistant to the spread of fire.

Informational Note: One method of defining *resistant to the spread of fire* is that the cable routing assemblies and communications raceways do not spread fire to the top of the tray in the "UL Flame Exposure, Vertical Flame Tray Test" in ANSI/UL 1685-2011, *Standard for Safety for Vertical-Tray Fire-Propagation and Smoke-Release Test for Electrical and Optical-Fiber Cables.*

ARTICLE 810
Radio and Television Equipment

Informational Note: See Informational Note Figure 800(a) and Informational Note Figure 800(b) for an illustrative application of a bonding conductor or grounding electrode conductor.

Part I. General

810.1 Scope. This article covers antenna systems for radio and television receiving equipment, amateur and citizen band radio transmitting and receiving equipment, and certain features of transmitter safety. This article covers antennas such as wire-strung type, multi-element, vertical rod, flat, or parabolic and also covers the wiring and cabling that connect them to equipment. This article does not cover equipment and antennas used for coupling carrier current to power line conductors.

810.2 Definitions. For definitions applicable to this article, see Part I of Article 100.

810.3 Other Articles. Wiring from the source of power to and between devices connected to the interior wiring system shall comply with Chapters 1 through 4 other than as modified by Parts I and II of Article 640. Wiring for audio signal processing, amplification, and reproduction equipment shall comply with Article 640. Coaxial cables that connect antennas to equipment shall comply with Article 820.

810.4 Community Television Antenna. The antenna shall comply with this article. The distribution system shall comply with Article 820.

810.5 Radio Noise Suppressors. Radio interference eliminators, interference capacitors, or noise suppressors connected to power-supply leads shall be of a listed type. They shall not be exposed to physical damage.

810.6 Antenna Lead-In Protectors. Where an antenna lead-in surge protector is installed, it shall be listed as being suitable for limiting surges on the cable that connects the antenna to the receiver/transmitter electronics and shall be connected between the conductors and the grounded shield or other ground connection. The antenna lead-in protector shall be grounded using a bonding conductor or grounding electrode conductor installed in accordance with 810.21(F).

Informational Note: For requirements covering protectors for antenna lead-in conductors, refer to UL 497E, *Outline of Investigation for Protectors for Antenna Lead-In Conductors.*

810.7 Grounding Devices. Where bonding or grounding is required, devices used to connect a shield, a sheath, non–current-carrying metallic members of a cable, or metal parts of equipment or antennas to a bonding conductor or grounding electrode conductor shall be listed or be part of listed equipment.

Part II. Receiving Equipment — Antenna Systems

810.11 Material. Antennas and lead-in conductors shall be of hard-drawn copper, bronze, aluminum alloy, copper-clad steel, or other high-strength, corrosion-resistant material.

Exception: Soft-drawn or medium-drawn copper shall be permitted for lead-in conductors where the maximum span between points of support is less than 11 m (35 ft).

810.12 Supports. Outdoor antennas and lead-in conductors shall be securely supported. The antennas or lead-in conductors shall not be attached to the electric service mast. They shall not be attached to poles or similar structures carrying open electric light or power wires or trolley wires of over 250 volts between conductors. Insulators supporting the antenna conductors shall have sufficient mechanical strength to safely support the conductors. Lead-in conductors shall be securely attached to the antennas.

810.13 Avoidance of Contacts with Conductors of Other Systems. Outdoor antennas and lead-in conductors from an antenna to a building shall not cross over open conductors of electric light or power circuits and shall be kept well away from all such circuits so as to avoid the possibility of acc...

contact. Where proximity to open electric light or power service conductors of less than 250 volts between conductors cannot be avoided, the installation shall be such as to provide a clearance of at least 600 mm (2 ft).

Where practicable, antenna conductors shall be installed so as not to cross under open electric light or power conductors.

810.14 Splices. Splices and joints in antenna spans shall be made mechanically secure with approved splicing devices or by such other means as will not appreciably weaken the conductors.

810.15 Grounding. Masts and metal structures supporting antennas shall be grounded in accordance with 810.21, unless the antenna and its related supporting mast or structure are within a zone of protection defined by a 46 m (150 ft) radius rolling sphere.

> Informational Note: See 4.8.3.1 of NFPA 780-2014, *Standard for the Installation of Lightning Protection Systems,* for the application of the term *rolling sphere.*

810.16 Size of Wire-Strung Antenna — Receiving Station.

(A) Size of Antenna Conductors. Outdoor antenna conductors for receiving stations shall be of a size not less than given in Table 810.16(A).

(B) Self-Supporting Antennas. Outdoor antennas, such as vertical rods and flat, parabolic, or dipole structures, shall be of corrosion-resistant materials and of strength suitable to withstand ice and wind loading conditions and shall be located well away from overhead conductors of electric light and power circuits of over 150 volts to ground, so as to avoid the possibility of the antenna or structure falling into or making accidental contact with such circuits.

810.17 Size of Lead-in — Receiving Station. Lead-in conductors from outside antennas for receiving stations shall, for various maximum open span lengths, be of such size as to have a tensile strength at least as great as that of the conductors for antennas as specified in 810.16. Where the lead-in consists of two or more conductors that are twisted together, are enclosed in the same covering, or are concentric, the conductor size shall, for various maximum open span lengths, be such that the tensile strength of the combination is at least as great as that of the conductors for antennas as specified in 810.16.

Table 810.16(A) Size of Receiving Station Outdoor Antenna Conductors

Material	Minimum Size of Conductors (AWG) Where Maximum Open Span Length Is		
	Less Than 11 m (35 ft)	11 m to 45 m (35 ft to 150 ft)	Over 45 m (150 ft)
Aluminum alloy, hard-drawn copper	19	14	12
Copper-clad steel, bronze, or other high-strength material	20	17	14

810.18 Clearances — Receiving Stations.

(A) Outside of Buildings. Lead-in conductors attached to buildings shall be installed so that they cannot swing closer than 600 mm (2 ft) to the conductors of circuits of 250 volts or less between conductors, or 3.0 m (10 ft) to the conductors of circuits of over 250 volts between conductors, except that in the case of circuits not over 150 volts between conductors, where all conductors involved are supported so as to ensure permanent separation, the clearance shall be permitted to be reduced but shall not be less than 100 mm (4 in.). The clearance between lead-in conductors and any conductor forming a part of a lightning protection system shall not be less than 1.8 m (6 ft). Underground conductors shall be separated at least 300 mm (12 in.) from conductors of any light or power circuits or Class 1 circuits.

Exception: Where the electric light or power conductors, Class 1 conductors, or lead-in conductors are installed in raceways or metal cable armor.

> Informational Note No. 1: See 250.60 for grounding associated with lightning protection components — strike termination devices. For further information, see NFPA 780-2014, *Standard for the Installation of Lightning Protection Systems,* which contains detailed information on grounding, bonding, and spacing from lightning protection systems, and the calculation of specific separation distances using the sideflash equation in Section 4.6.

> Informational Note No. 2: Metal raceways, enclosures, frames, and other non–current-carrying metal parts of electrical equipment installed on a building equipped with a lightning protection system may require bonding or spacing from the lightning protection conductors in accordance with NFPA 780-2014, *Standard for the Installation of Lightning Protection Systems.* Separation from lightning protection conductors is typically 1.8 m (6 ft) through air or 900 mm (3 ft) through dense materials such as concrete, brick, or wood.

(B) Antennas and Lead-ins — Indoors. Indoor antennas and indoor lead-ins shall not be run nearer than 50 mm (2 in.) to conductors of other wiring systems in the premises.

Exception No. 1: Where such other conductors are in metal raceways or cable armor.

Exception No. 2: Where permanently separated from such other conductors by a continuous and firmly fixed nonconductor, such as porcelain tubes or flexible tubing.

(C) In Boxes or Other Enclosures. Indoor antennas and indoor lead-ins shall be permitted to occupy the same box or enclosure with conductors of other wiring systems where separated from such other conductors by an effective permanently installed barrier.

810.19 Electrical Supply Circuits Used in Lieu of Antenna — Receiving Stations. Where an electrical supply circuit is used in lieu of an antenna, the device by which the radio receiving set is connected to the supply circuit shall be listed.

810.20 Antenna Discharge Units — Receiving Stations.

(A) Where Required. Each conductor of a lead-in from an outdoor antenna shall be provided with a listed antenna discharge unit.

Exception: Where the lead-in conductors are enclosed in a continuous metallic shield that either is grounded with a conductor in accordance with 810.21 or is protected by an antenna discharge unit.

(B) Location. Antenna discharge units shall be located outside the building or inside the building between the point of entrance of the lead-in and the radio set or transformers and as near as practicable to the entrance of the conductors to the building. The antenna discharge unit shall not be located near combustible material or in a hazardous (classified) location as defined in Article 500.

(C) Grounding. The antenna discharge unit shall be grounded in accordance with 810.21.

810.21 Bonding Conductors and Grounding Electrode Conductors — Receiving Stations. Bonding conductors and grounding electrode conductors shall comply with 810.21(A) through 810.21(K).

(A) Material. The bonding conductor or grounding electrode conductor shall be of copper, aluminum, copper-clad steel, bronze, or similar corrosion-resistant material. Aluminum or copper-clad aluminum bonding conductors or grounding electrode conductors shall not be used where in direct contact with masonry or the earth or where subject to corrosive conditions. Where used outside, aluminum or copper-clad aluminum conductors shall not be installed within 450 mm (18 in.) of the earth.

(B) Insulation. Insulation on bonding conductors or grounding electrode conductors shall not be required.

(C) Supports. The bonding conductor or grounding electrode conductor shall be securely fastened in place and shall be permitted to be directly attached to the surface wired over without the use of insulating supports.

Exception: Where proper support cannot be provided, the size of the bonding conductors or grounding electrode conductors shall be increased proportionately.

(D) Physical Protection. Bonding conductors and grounding electrode conductors shall be protected where exposed to physical damage. Where the bonding conductor or grounding electrode conductor is installed in a metal raceway, both ends of the raceway shall be bonded to the contained conductor or to the same terminal or electrode to which the bonding conductor or grounding electrode conductor is connected.

(E) Run in Straight Line. The bonding conductor or grounding electrode conductor for an antenna mast or antenna discharge unit shall be run in as straight a line as practicable.

(F) Electrode. The bonding conductor or grounding electrode conductor shall be connected as required in 810.21(F)(1) through 810.21(F)(3).

(1) In Buildings or Structures with an Intersystem Bonding Termination. If the building or structure served has an intersystem bonding termination as required by 250.94, the bonding conductor shall be connected to the intersystem bonding termination.

(2) In Buildings or Structures with Grounding Means. If the building or structure served has no intersystem bonding termination, the bonding conductor or grounding electrode conductor shall be connected to the nearest accessible location on the following:

(1) The building or structure grounding electrode system as covered in 250.50

(2) The grounded interior metal water piping systems, within 1.52 m (5 ft) from its point of entrance to the building, as covered in 250.52

(3) The power service accessible means external to the building, as covered in 250.94

(4) The nonflexible metallic power service raceway

(5) The service equipment enclosure, or

(6) The grounding electrode conductor or the grounding electrode conductor metal enclosures of the power service

A bonding device intended to provide a termination point for the bonding conductor (intersystem bonding) shall not interfere with the opening of an equipment enclosure. A bonding device shall be mounted on nonremovable parts. A bonding device shall not be mounted on a door or cover even if the door or cover is nonremovable.

(3) In Buildings or Structures Without an Intersystem Bonding Termination or Grounding Means. If the building or structure served has no intersystem bonding termination or grounding means as described in 810.21(F)(2), the grounding electrode conductor shall be connected to a grounding electrode as described in 250.52.

(G) Inside or Outside Building. The bonding conductor or grounding electrode conductor shall be permitted to be run either inside or outside the building.

(H) Size. The bonding conductor or grounding electrode conductor shall not be smaller than 10 AWG copper, 8 AWG aluminum, or 17 AWG copper-clad steel or bronze.

(I) Common Ground. A single bonding conductor or grounding electrode conductor shall be permitted for both protective and operating purposes.

(J) Bonding of Electrodes. A bonding jumper not smaller than 6 AWG copper or equivalent shall be connected between the radio and television equipment grounding electrode and the power grounding electrode system at the building or structure served where separate electrodes are used.

(K) Electrode Connection. Connections to grounding electrodes shall comply with 250.70.

Part III. Amateur and Citizen Band Transmitting and Receiving Stations — Antenna Systems

810.51 Other Sections. In addition to complying with Part III, antenna systems for amateur and citizen band transmitting and receiving stations shall also comply with 810.11 through 810.15.

810.52 Size of Antenna. Antenna conductors for transmitting and receiving stations shall be of a size not less than given in Table 810.52.

810.53 Size of Lead-in Conductors. Lead-in conductors for transmitting stations shall, for various maximum span lengths, be of a size at least as great as that of conductors for antennas as specified in 810.52.

810.54 Clearance on Building. Antenna conductors for transmitting stations, attached to buildings, shall be firmly mounted at least 75 mm (3 in.) clear of the surface of the building on nonabsorbent insulating supports, such as treated pins or brackets equipped with insulators having not less than 75-mm (3-in.) creepage and airgap distances. Lead-in conductors

Table 810.52 Size of Outdoor Antenna Conductors

Material	Minimum Size of Conductors (AWG) Where Maximum Open Span Length Is	
	Less Than 45 m (150 ft)	Over 45 m (150 ft)
Hard-drawn copper	14	10
Copper-clad steel, bronze, or other high-strength material	14	12

attached to buildings shall also comply with these requirements.

Exception: Where the lead-in conductors are enclosed in a continuous metallic shield that is grounded with a conductor in accordance with 810.58, they shall not be required to comply with these requirements. Where grounded, the metallic shield shall also be permitted to be used as a conductor.

810.55 Entrance to Building. Except where protected with a continuous metallic shield that is grounded with a conductor in accordance with 810.58, lead-in conductors for transmitting stations shall enter buildings by one of the following methods:

(1) Through a rigid, noncombustible, nonabsorbent insulating tube or bushing

(2) Through an opening provided for the purpose in which the entrance conductors are firmly secured so as to provide a clearance of at least 50 mm (2 in.)

(3) Through a drilled window pane

810.56 Protection Against Accidental Contact. Lead-in conductors to radio transmitters shall be located or installed so as to make accidental contact with them difficult.

810.57 Antenna Discharge Units — Transmitting Stations. Each conductor of a lead-in for outdoor antennas shall be provided with an antenna discharge unit or other suitable means that drain static charges from the antenna system.

Exception No. 1: Where the lead-in is protected by a continuous metallic shield that is grounded with a conductor in accordance with 810.58, an antenna discharge unit or other suitable means shall not be required.

Exception No. 2: Where the antenna is grounded with a conductor in accordance with 810.58, an antenna discharge unit or other suitable means shall not be required.

810.58 Bonding Conductors and Grounding Electrode Conductors — Amateur and Citizen Band Transmitting and Receiving Stations. Bonding conductors and grounding electrode conductors shall comply with 810.58(A) through 810.58(C).

(A) Other Sections. All bonding conductors and grounding electrode conductors for amateur and citizen band transmitting and receiving stations shall comply with 810.21(A) through 810.21(C).

(B) Size of Protective Bonding Conductor or Grounding Electrode Conductor. The protective bonding conductor or grounding electrode conductor for transmitting stations shall be as large as the lead-in but not smaller than 10 AWG copper, bronze, or copper-clad steel.

(C) Size of Operating Bonding Conductor or Grounding Electrode Conductor. The operating bonding conductor or grounding electrode conductor for transmitting stations shall not be less than 14 AWG copper or its equivalent.

Part IV. Interior Installation — Transmitting Stations

810.70 Clearance from Other Conductors. All conductors inside the building shall be separated at least 100 mm (4 in.) from the conductors of any electric light, power, or signaling circuit.

Exception No. 1: As provided in Article 640.

Exception No. 2: Where separated from other conductors by raceway or some firmly fixed nonconductor, such as porcelain tubes or flexible tubing.

810.71 General. Transmitters shall comply with 810.71(A) through (C).

(A) Enclosing. The transmitter shall be enclosed in a metal frame or grille or separated from the operating space by a barrier or other equivalent means, all metallic parts of which are effectively connected to a bonding conductor or grounding electrode conductor.

(B) Grounding of Controls. All external metal handles and controls accessible to the operating personnel shall be effectively connected to an equipment grounding conductor if the transmitter is powered by the premises wiring system or grounded with a conductor in accordance with 810.21.

(C) Interlocks on Doors. All access doors shall be provided with interlocks that disconnect all voltages of over 350 volts between conductors when any access door is opened.

ARTICLE 820
Community Antenna Television and Radio Distribution Systems

Informational Note: See Informational Note Figure 800(a) and Informational Note Figure 800(b) for an illustrative application of a bonding conductor or grounding electrode conductor.

Part I. General

820.1 Scope. This article covers coaxial cable distribution of radio frequency signals typically employed in community antenna television (CATV) systems.

Informational Note: See 90.2(B)(4) for installations of CATV and radio distribution systems that are not covered.

820.2 Definitions. See Part I of Article 100. For the purposes of this article, the following additional definitions apply.

Abandoned Coaxial Cable. Installed coaxial cable that is not terminated at equipment other than a coaxial connector and not identified for future use with a tag.

Exposed (to Accidental Contact). A circuit in such a position that, in case of failure of supports and or insulation, contact with another circuit may result.

Informational Note: See Part I of Article 100 for two other definitions of *Exposed*.

Point of Entrance. The point within a building at which the coaxial cable emerges from an external wall or from a concrete floor slab.

Premises. The land and buildings of a user located on the user side of utility-user network point of demarcation.

820.3 Other Articles. Circuits and equipment shall comply with 820.3(A) through 820.3(I).

(A) Hazardous (Classified) Locations. CATV equipment installed in a location that is classified in accordance with 500.5 and 505.5 shall comply with the applicable requirements of Chapter 5.

(B) Wiring in Ducts for Dust, Loose Stock, or Vapor Removal. The requirements of 300.22(A) shall apply.

(C) Equipment in Other Space Used for Environmental Air. The requirements of 300.22(C)(3) shall apply.

(D) Installation and Use. The requirements of 110.3(B) shall apply.

(E) Installations of Conductive and Nonconductive Optical Fiber Cables. The requirements of Article 770 shall apply.

(F) Communications Circuits. The requirements of Article 800 shall apply.

(G) Network-Powered Broadband Communications Systems. The requirements of Article 830 shall apply.

(H) Premises-Powered Broadband Communications Systems. The requirements of Article 840 shall apply.

(I) Alternate Wiring Methods. The wiring methods of Article 830 shall be permitted to substitute for the wiring methods of Article 820.

Informational Note: Use of Article 830 wiring methods will facilitate the upgrading of Article 820 installations to network-powered broadband applications.

820.15 Power Limitations. Coaxial cable shall be permitted to deliver power to equipment that is directly associated with the radio frequency distribution system if the voltage is not over 60 volts and if the current is supplied by a transformer or other device that has power-limiting characteristics.

Power shall be blocked from premises devices on the network that are not intended to be powered via the coaxial cable.

820.21 Access to Electrical Equipment Behind Panels Designed to Allow Access. Access to electrical equipment shall not be denied by an accumulation of coaxial cables that prevents removal of panels, including suspended ceiling panels.

820.24 Mechanical Execution of Work. Community television and radio distribution systems shall be installed in a neat and workmanlike manner. Coaxial cables installed exposed on the surface of ceiling and sidewalls shall be supported by the building structure in such a manner that the cables will not be damaged by normal building use. Such cables shall be secured by hardware including straps, staples, cable ties, hangers, or similar fittings designed and installed so as not to damage the cable. The installation shall also conform to 300.4(D) and 300.11. Nonmetallic cable ties and other nonmetallic cable accessories used to secure and support cables in other spaces used for environmental air (plenums) shall be listed as having low smoke and heat release properties in accordance with 800.170(C).

Informational Note No. 1: Accepted industry practices are described in ANSI/NECA/BICSI 568–2006, *Standard for Installing Commercial Building Telecommunications Cabling*; ANSI/TIA-568.1-D-2015, *Commercial Building Telecommunications Infrastructure Standard*; ANSI/TIA-569-D-2015, *Telecommunications Pathways and Spaces*; ANSI/TIA-570-C-2012, *Residential Telecommunications Infrastructure Standard*; ANSI/TIA-1005-A-2012, *Telecommunications Infrastructure Standard for Industrial Premises*; ANSI/TIA-1179-2010, *Healthcare Facility Telecommunications Infrastructure Standard*; ANSI/TIA-4966-2014, *Telecommunications Infrastructure Standard for Educational Facilities*; and other ANSI-approved installation standards.

Informational Note No. 2: See 4.3.11.2.6.5 and 4.3.11.5.5.6 of NFPA 90A-2015, *Standard for the Installation of Air-Conditioning and Ventilating Systems*, for discrete combustible components installed in accordance with 300.22(C).

Informational Note No. 3: Paint, plaster, cleaners, abrasives, corrosive residues, or other contaminants may result in an undetermined alteration of coaxial cable properties.

820.25 Abandoned Cables. The accessible portion of abandoned coaxial cables shall be removed. Where cables are identified for future use with a tag, the tag shall be of sufficient durability to withstand the environment involved.

820.26 Spread of Fire or Products of Combustion. Installations of coaxial cables and communications raceways in hollow spaces, vertical shafts, and ventilation or air-handling ducts shall be made so that the possible spread of fire or products of combustion will not be substantially increased. Openings around penetrations of coaxial cables and communications raceways through fire-resistant-rated walls, partitions, floors, or ceilings shall be firestopped using approved methods to maintain the fire resistance rating.

Informational Note: Directories of electrical construction materials published by qualified testing laboratories contain many listing installation restrictions necessary to maintain the fire-resistive rating of assemblies where penetrations or openings are made. Building codes also contain restrictions on membrane penetrations on opposite sides of a fire resistance–rated wall assembly. An example is the 600-mm (24-in.) minimum horizontal separation that usually applies between boxes installed on opposite sides of the wall. Assistance in complying with 820.26 can be found in building codes, fire resistance directories, and product listings.

Part II. Coaxial Cables Outside and Entering Buildings

820.44 Overhead (Aerial) Coaxial Cables. Overhead (aerial) coaxial cables, prior to the point of grounding, as specified in 820.93, shall comply with 820.44(A) through 820.44(E).

(A) On Poles and In-Span. Where coaxial cables and electric light or power conductors are supported by the same pole or are run parallel to each other in-span, the conditions described in 820.44(A)(1) through 820.44(A)(4) shall be met.

(1) Relative Location. Where practicable, the coaxial cables shall be located below the electric light or power conductors.

(2) Attachment to Cross-Arms. Coaxial cables shall not be attached to cross-arm that carries electric light or power conductors.

(3) Climbing Space. The climbing space through coaxial cables shall comply with the requirements of 225.14(D).

(4) Clearance. Lead-in or overhead (aerial) -drop coaxial cables from a pole or other support, including the point of initial attachment to a building or structure, shall be kept away from electric light, power, Class 1, or non–power-limited fire alarm circuit conductors so as to avoid the possibility of accidental contact.

Exception: Where proximity to electric light, power, Class 1, or non–power-limited fire alarm circuit conductors cannot be avoided, the installation shall provide clearances of not less than 300 mm (12 in.) from electric light, power, Class 1, or non–power-limited fire alarm circuit conductors. The clearance requirement shall apply at all points along the drop, and it shall increase to 1.0 m (40 in.) at the pole.

(B) Above Roofs. Coaxial cables shall have a vertical clearance of not less than 2.5 m (8 ft) from all points of roofs above which they pass.

Exception No. 1: Vertical clearance requirements shall not apply to auxiliary *buildings such as garages and the like.*

Exception No. 2: A reduction in clearance above only the overhanging portion of the roof to not less than 450 mm (18 in.) shall be permitted if (1) not more than 1.2 m (4 ft) of communications service drop conductors pass above the roof overhang, and (2) they are terminated at a raceway mast or other approved support.

Exception No. 3: Where the roof has a slope of not less than 100 mm in 300 mm (4 in. in 12 in.), a reduction in clearance to not less than 900 mm (3 ft) shall be permitted.

(C) On Masts. Overhead (aerial) coaxial cables shall be permitted to be attached to an above-the-roof raceway mast that does not enclose or support conductors of electric light or power circuits.

(D) Between Buildings. Coaxial cables extending between buildings or structures, and also the supports or attachment fixtures, shall be identified and shall have sufficient strength to withstand the loads to which they might be subjected.

Exception: Where a coaxial cable does not have sufficient strength to be self-supporting, it shall be attached to a supporting messenger cable that, together with the attachment fixtures or supports, shall be acceptable for the purpose and shall have sufficient strength to withstand the loads to which they may be subjected.

(E) On Buildings. Where attached to buildings, coaxial cables shall be securely fastened in such a manner that they will be separated from other conductors in accordance with 820.44(E)(1), 820.44(E)(2), and 820.44(E)(3).

(1) Electric Light or Power. The coaxial cable shall have a separation of at least 100 mm (4 in.) from electric light, power, Class 1, or non–power-limited fire alarm circuit conductors not in raceway or cable, or shall be permanently separated from conductors of the other system by a continuous and firmly fixed nonconductor in addition to the insulation on the wires.

(2) Other Communications Systems. Coaxial cable shall be installed so that there will be no unnecessary interference in the maintenance of the separate systems. In no case shall the conductors, cables, messenger strand, or equipment of one system cause abrasion to the conductors, cable, messenger strand, or equipment of any other system.

(3) Lightning Conductors. Where practicable, a separation of at least 1.8 m (6 ft) shall be maintained between any coaxial cable and lightning conductors.

> Informational Note No. 1: For additional information regarding overhead (aerial) wires and cables, see ANSI C2-2012, *National Electrical Safety Code*, Part 2, Safety Rules for Overhead Lines.

> Informational Note No. 2: See Section 4.6 of NFPA 780-2014, *Standard for the Installation of Lightning Protection Systems*, for the calculation of sideflash distance.

820.47 Underground Coaxial Cables Entering Buildings. Underground coaxial cables entering buildings shall comply with 820.47(A) and 820.47(B).

(A) Underground Systems with Electric Light, Power, Class 1, or Non–Power-Limited Fire Alarm Circuit Conductors. Underground coaxial cables in a duct, pedestal, handhole enclosure, or manhole that contains electric light, power, or Class 1 or non–power-limited fire alarm circuit conductors shall be in a section permanently separated from such conductors by means of a suitable barrier.

(B) Direct-Buried Cables and Raceways. Direct-buried coaxial cable shall be separated at least 300 mm (12 in.) from conductors of any light or power, non–power-limited fire alarm circuit conductors, or Class 1 circuit.

Exception No. 1: Separation shall not be required where *electric service conductors or coaxial cables are installed in raceways or have metal cable armor.*

Exception No. 2: Separation shall not be required where *electric light or power branch-circuit or feeder conductors or Class 1 circuit conductors are installed in a raceway or in metal-sheathed, metal-clad, or Type UF or Type USE cables; or the coaxial cables have metal cable armor or are installed in a raceway.*

820.48 Unlisted Cables Entering Buildings. Unlisted outside plant coaxial cables shall be permitted to be installed in building spaces other than risers, ducts used for environmental air, plenums used for environmental air, and other spaces used for environmental air, where the length of the cable within the building, measured from its point of entrance, does not exceed 15 m (50 ft) and the cable enters the building from the outside and is terminated at a grounding block. The point of entrance shall be permitted to be extended from the penetration of the external wall or floor slab by continuously enclosing the entrance cables in rigid metal conduit (RMC) or intermediate metal conduit (IMC) to the point of emergence.

820.49 Metallic Entrance Conduit Grounding. Metallic conduit containing entrance coaxial cable shall be connected by a bonding conductor or grounding electrode conductor to a grounding electrode in accordance with 820.100(B).

Part III. Protection

820.93 Grounding of the Outer Conductive Shield of Coaxial Cables. Coaxial cables entering buildings or attached to buildings shall comply with 820.93(A) or (B). Where the outer conductive shield of a coaxial cable is grounded, no other protective devices shall be required. For purposes of this section, grounding located at mobile home service equipment located within 9.0 m (30 ft) of the exterior wall of the mobile home it serves, or at a mobile home disconnecting means grounded in accordance with 250.32 and located within 9.0 m (30 ft) of the exterior wall of the mobile home it serves, shall be considered to meet the requirements of this section.

Informational Note: Selecting a grounding block location to achieve the shortest practicable bonding conductor or grounding electrode conductor helps limit potential differences between CATV and other metallic systems.

(A) Entering Buildings.   In installations where the coaxial cable enters the building, the outer conductive shield shall be grounded in accordance with 820.100. The grounding shall be as close as practicable to the point of entrance.

(B) Terminating Outside of the Building.   In installations where the coaxial cable is terminated outside of the building, the outer conductive shield shall be grounded in accordance with 820.100. The grounding shall be as close as practicable to the point of attachment or termination.

(C) Location.   Where installed, a listed primary protector shall be applied on each community antenna and radio distribution (CATV) cable external to the premises. The listed primary protector shall be located as close as practicable to the entrance point of the cable on either side or integral to the ground block.

(D) Hazardous (Classified) Locations.   Where a primary protector or equipment providing the primary protection function is used, it shall not be located in any hazardous (classified) location as defined in 500.5 and 505.5 or in the vicinity of easily ignitible material.

Exception: As permitted in 501.150, 502.150, and 503.150.

Part IV. Grounding Methods

820.100 Cable Bonding and Grounding.   The shield of the coaxial cable shall be bonded or grounded as specified in 820.100(A) through (E).

Exception: For communications systems using coaxial cable completely contained within the building (i.e., they do not exit the building) or the exterior zone of protection defined by a 46 m (150 ft) radius rolling sphere and isolated from outside cable plant, the shield shall be permitted to be grounded by a connection to an equipment grounding conductor as described in 250.118. Connecting to an equipment grounding conductor through a grounded receptacle using a dedicated bonding jumper and a permanently connected listed device shall be permitted. Use of a cord and plug for the connection to an equipment grounding conductor shall not be permitted.

Informational Note: See 4.8.3.1 of NFPA 780-2014, *Standard for the Installation of Lightning Protection Systems,* for the theory of the term *rolling sphere.*

(A) Bonding Conductor or Grounding Electrode Conductor.

(1) Insulation.   The bonding conductor or grounding electrode conductor shall be listed and shall be permitted to be insulated, covered, or bare.

(2) Material.   The bonding conductor or grounding electrode conductor shall be copper or other corrosion-resistant conductive material, stranded or solid.

(3) Size.   The bonding conductor or grounding electrode conductor shall not be smaller than 14 AWG. It shall have a current-carrying capacity not less than the outer sheath of the coaxial cable. The bonding conductor or grounding electrode conductor shall not be required to exceed 6 AWG.

(4) Length.   The bonding conductor or grounding electrode conductor shall be as short as practicable. In one- and two-family dwellings, the bonding conductor or grounding electrode conductor shall be as short as practicable, not to exceed 6.0 m (20 ft) in length.

Informational Note: Similar bonding conductor or grounding electrode conductor length limitations applied at apartment buildings and commercial buildings help to reduce voltages that may be developed between the building's power and communications systems during lightning events.

Exception: In one- and two-family dwellings where it is not practicable to achieve an overall maximum bonding conductor or grounding electrode conductor length of 6.0 m (20 ft), a separate grounding electrode as specified in 250.52(A)(5), (A)(6), or (A)(7) shall be used, the grounding electrode conductor shall be connected to the separate grounding electrode in accordance with 250.70, and the separate grounding electrode shall be connected to the power grounding electrode system in accordance with 820.100(D).

(5) Run in Straight Line.   The bonding conductor or grounding electrode conductor shall be run in as straight a line as practicable.

(6) Physical Protection.   Bonding conductors and grounding electrode conductors shall be protected where exposed to physical damage. Where the bonding conductor or grounding electrode conductor is installed in a metal raceway, both ends of the raceway shall be bonded to the contained conductor or to the same terminal or electrode to which the bonding conductor or grounding electrode conductor is connected.

(B) Electrode.   The bonding conductor or grounding electrode conductor shall be connected in accordance with 820.100(B)(1), 820.100(B)(2), or 820.100(B)(3).

(1) In Buildings or Structures with an Intersystem Bonding Termination.   If the building or structure served has an intersystem bonding termination as required by 250.94, the bonding conductor shall be connected to the intersystem bonding termination.

(2) In Buildings or Structures with Grounding Means.   If an intersystem bonding termination is established, 250.94(A) shall apply.

If the building or structure served has no intersystem bonding termination, the bonding conductor or grounding electrode conductor shall be connected to the nearest accessible location on one of the following:

(1)   The building or structure grounding electrode system as covered in 250.50
(2)   The grounded interior metal water piping system, within 1.5 m (5 ft) from its point of entrance to the building, as covered in 250.52
(3)   The power service accessible means external to enclosures using the options identified in 250.94(A), Exception

(4) The nonflexible metallic power service raceway

(5) The service equipment enclosure

(6) The grounding electrode conductor or the grounding electrode conductor metal enclosure of the power service

(7) The grounding electrode conductor or the grounding electrode of a building or structure disconnecting means that is connected to an electrode as covered in 250.32

A bonding device intended to provide a termination point for the bonding conductor (intersystem bonding) shall not interfere with the opening of an equipment enclosure. A bonding device shall be mounted on nonremovable parts. A bonding device shall not be mounted on a door or cover even if the door or cover is nonremovable.

For purposes of this section, the mobile home service equipment or the mobile home disconnecting means, as described in 820.93, shall be considered accessible.

(3) In Buildings or Structures Without an Intersystem Bonding Termination or Grounding Means. If the building or structure served has no intersystem bonding termination or grounding means, as described in 820.100(B)(2), the grounding electrode conductor shall be connected to either of the following:

(1) To any one of the individual grounding electrodes described in 250.52(A)(1), (A)(2), (A)(3), or (A)(4).

(2) If the building or structure served has no intersystem bonding termination or grounding means, as described in 820.100(B)(2) or (B)(3)(1), to any one of the individual grounding electrodes described in 250.52(A)(5), (A)(7), and (A)(8). Steam, hot water pipes, or lightning protection system conductors shall not be employed as grounding electrodes for bonding conductors or grounding electrode conductors.

(C) Electrode Connection. Connections to grounding electrodes shall comply with 250.70.

(D) Bonding of Electrodes. A bonding jumper not smaller than 6 AWG copper or equivalent shall be connected between the community antenna television system's grounding electrode and the power grounding electrode system at the building or structure served where separate electrodes are used.

Exception: At mobile homes as covered in 820.106.

Informational Note No. 1: See 250.60 for connection to a lightning protection system.

Informational Note No. 2: Bonding together of all separate electrodes limits potential differences between them and between their associated wiring systems.

(E) Shield Protection Devices. Grounding of a coaxial drop cable shield by means of a protective device that does not interrupt the grounding system within the premises shall be permitted.

820.103 Equipment Grounding. Unpowered equipment and enclosures or equipment powered by the coaxial cable shall be considered grounded where connected to the metallic cable shield.

820.106 Grounding and Bonding at Mobile Homes.

(A) Grounding. Grounding shall comply with 820.106(A)(1) and (A)(2).

(1) Where there is no mobile home service equipment located within 9.0 m (30 ft) of the exterior wall of the mobile home it serves, the coaxial cable shield ground, or surge arrester grounding terminal, shall be connected to a grounding electrode conductor or grounding electrode in accordance with 820.100(B)(3).

(2) Where there is no mobile home disconnecting means grounded in accordance with 250.32 and located within 9.0 m (30 ft) of the exterior wall of the mobile home it serves, the coaxial cable shield ground, or surge arrester grounding terminal, shall be connected to a grounding electrode in accordance with 820.100(B)(3).

(B) Bonding. The coaxial cable shield grounding terminal, surge arrester grounding terminal, or grounding electrode shall be connected to the metal frame or available grounding terminal of the mobile home with a copper conductor not smaller than 12 AWG under any of the following conditions:

(1) Where there is no mobile home service equipment or disconnecting means as in 820.106(A)

(2) Where the mobile home is supplied by cord and plug

Part V. Installation Methods Within Buildings

820.110 Raceways and Cable Routing Assemblies for Coaxial Cables.

(A) Types of Raceways. Coaxial cables shall be permitted to be installed in any raceway that complies with either (A)(1) or (A)(2) and in cable routing assemblies installed in compliance with 820.110(C).

(1) Raceways Recognized in Chapter 3. Coaxial cables shall be permitted to be installed in any raceway included in Chapter 3. The raceways shall be installed in accordance with the requirements of Chapter 3.

(2) Communications Raceways. Coaxial cables shall be permitted to be installed in plenum communications raceways, riser communications raceways, and general-purpose communications raceways, selected in accordance with Table 800.154(b), listed in accordance with 800.182, and installed in accordance with 800.113 and 362.24 through 362.56, where the requirements applicable to electrical nonmetallic tubing (ENT) apply.

N **(3) Innerduct for Coaxial Cables.** Listed plenum communications raceways, listed riser communications raceways, and listed general-purpose communications raceways selected in accordance with Table 800.154(b) shall be permitted to be installed as innerduct in any type of listed raceway permitted in Chapter 3.

(B) Raceway Fill for Coaxial Cables. The raceway fill requirements of Chapters 3 and 9 shall not apply to coaxial cables.

(C) Cable Routing Assemblies. Coaxial cables shall be permitted to be installed in plenum cable routing assemblies, riser cable routing assemblies, and general-purpose cable routing assemblies selected in accordance with Table 800.154(c), listed in accordance with 800.182, and installed in accordance with 800.110(C) and 800.113.

820.113 Installation of Coaxial Cables. Installation of coaxial cables shall comply with 820.113(A) through (K). Installation of raceways and cable routing assemblies shall comply with 820.110.

(A) Listing. Coaxial cables installed in buildings shall be listed.

Exception: Coaxial cables that are installed in compliance with 820.48 shall not be required to be listed.

(B) Ducts Specifically Fabricated for Environmental Air. The following cables shall be permitted in ducts specifically fabricated for environmental air as described in 300.22(B) if they are directly associated with the air distribution system:

(1)   Up to 1.22 m (4 ft) of Type CATVP
(2)   Types CATVP, CATVR, CATV, and CATVX installed in raceways that are installed in compliance with 300.22(B)

Informational Note: For information on fire protection of wiring installed in fabricated ducts see 4.3.4.1 and 4.3.11.3.3 of NFPA 90A-2015, Standard for the Installation of Air-Conditioning and Ventilating Systems.

(C) Other Spaces Used For Environmental Air (Plenums). The following cables shall be permitted in other spaces used for environmental air as described in 300.22(C):

(1)   Type CATVP
(2)   Type CATVP installed in plenum communications raceways
(3)   Type CATVP installed in plenum cable routing assemblies
(4)   Type CATVP supported by open metallic cable trays or cable tray systems
(5)   Types CATVP, CATVR, CATV, and CATVX installed in raceways that are installed in compliance with 300.22(C)
(6)   Types CATVP, CATVR, CATV, and CATVX supported by solid-bottom metal cable trays with solid metal covers in other spaces used for environmental air (plenums) as described in 300.22(C)
(7)   Types CATVP, CATVR, CATV, and CATVX installed in plenum communications raceways, riser communications raceways, or general-purpose communications raceways supported by solid-bottom metal cable trays with solid metal covers in other spaces used for environmental air (plenums) as described in 300.22(C)

Informational Note: For information on fire protection of wiring installed in other spaces used for environmental air, see 4.3.11.2, 4.3.11.4, and 4.3.11.5 of NFPA 90A-2015, Standard for the Installation of Air-Conditioning and Ventilating Systems.

(D) Risers — Cables in Vertical Runs. The following cables shall be permitted in vertical runs penetrating one or more floors and in vertical runs in a shaft:

(1)   Types CATVP and CATVR
(2)   Types CATVP and CATVR installed in the following:

   a.   Plenum communications raceways
   b.   Plenum cable routing assemblies
   c.   Riser communications raceways
   d.   Riser cable routing assemblies

Informational Note: See 820.26 for firestop requirements for floor penetrations.

(E) Risers — Cables and Innerducts in Metal Raceways. The following cables and innerducts shall be permitted in metal raceways in a riser having firestops at each floor:

(1)   Types CATVP, CATVR, CATV, and CATVX
(2)   Types CATVP, CATVR, CATV, and CATVX installed in the following:

   a.   Plenum communications raceways (innerduct)
   b.   Riser communications raceways (innerduct)
   c.   General-purpose communications raceways (innerduct)

Informational Note: See 820.26 for firestop requirements for floor penetrations.

(F) Risers — Cables in Fireproof Shafts. The following cables shall be permitted to be installed in fireproof riser shafts with firestops at each floor:

(1)   Types CATVP, CATVR, CATV, and CATVX
(2)   Types CATVP, CATVR, and CATV installed in the following:

   a.   Plenum communications raceways
   b.   Plenum cable routing assemblies
   c.   Riser communications raceways
   d.   Riser cable routing assemblies
   e.   General-purpose communications raceways
   f.   General-purpose cable routing assemblies

Informational Note: See 820.26 for firestop requirements for floor penetrations.

(G) Risers — One- and Two-Family Dwellings. The following cables shall be permitted in one- and two-family dwellings:

(1)   Types CATVP, CATVR, and CATV
(2)   Type CATVX less than 10 mm (⅜ in.) in diameter
(3)   Types CATVP, CATVR, and CATV installed in the following:

   a.   Plenum communications raceways
   b.   Plenum cable routing assemblies
   c.   Riser communications raceways
   d.   Riser cable routing assemblies
   e.   General-purpose communications raceways
   f.   General-purpose cable routing assemblies

Informational Note: See 820.26 for firestop requirements for floor penetrations.

(H) Cable Trays. The following cables shall be permitted to be supported by cable trays:

(1)   Types CATVP, CATVR, and CATV
(2)   Types CATVP, CATVR, and CATV installed in the following:

   a.   Plenum communications raceways
   b.   Riser communications raceways
   c.   General-purpose communications raceways

(I) Distributing Frames and Cross-Connect Arrays. The following cables shall be permitted to be installed in distributing frames and cross-connect arrays:

(1)   Types CATVP, CATVR, and CATV
(2)   Types CATVP, CATVR, and CATV installed in the following:

   a.   Plenum communications raceways
   b.   Plenum cable routing assemblies
   c.   Riser communications raceways
   d.   Riser cable routing assemblies
   e.   General-purpose communications raceways
   f.   General-purpose cable routing assemblies

(J) Other Building Locations. The following cables shall be permitted to be installed in building locations other than the locations covered in 820.113(B) through (I):

(1)   Types CATVP, CATVR, and CATV
(2)   A maximum of 3 m (10 ft) of exposed Type CATVX in nonconcealed spaces

(3) Types CATVP, CATVR, and CATV installed in the following:

   a. Plenum communications raceways
   b. Plenum cable routing assemblies
   c. Riser communications raceways
   d. Riser cable routing assemblies
   e. General-purpose communications raceways
   f. General-purpose cable routing assemblies

(4) Types CATVP, CATVR, CATV, and CATVX installed in a raceway of a type recognized in Chapter 3

(K) One- and Two-Family and Multifamily Dwellings. The following cables shall be permitted to be installed in one- and two-family and multifamily dwellings in locations other than those locations covered in 820.113(B) through (I):

(1) Types CATVP, CATVR, and CATV
(2) Type CATVX less than 10 mm (⅜ in.) in diameter
(3) Types CATVP, CATVR, and CATV installed in the following:

   a. Plenum communications raceways
   b. Plenum cable routing assemblies
   c. Riser communications raceways
   d. Riser cable routing assemblies
   e. General-purpose communications raceways
   f. General-purpose cable routing assemblies

(4) Types CATVP, CATVR, CATV, and CATVX installed in a raceway of a type recognized in Chapter 3

820.133 Installation of Coaxial Cables and Equipment. Beyond the point of grounding, as defined in 820.93, the coaxial cable installation shall comply with 820.133(A) and (B).

(A) Separation from Other Conductors.

(1) In Raceways, Cable Trays, Boxes, Enclosures, and Cable Routing Assemblies.

   (a) *Other Circuits.* Coaxial cables shall be permitted in the same raceway, cable tray, box, enclosure, or cable routing assembly with jacketed cables of any of the following:

(1) Class 2 and Class 3 remote control, signaling, and power-limited circuits in compliance with Article 645 or Parts I and III of Article 725
(2) Power-limited fire alarm systems in compliance with Parts I and III of Article 760
(3) Nonconductive and conductive optical fiber cables in compliance with Parts I and V of Article 770
(4) Communications circuits in compliance with Parts I and V of Article 800
(5) Low-power network-powered broadband communications circuits in compliance with Parts I and V of Article 830

   (b) *Electric Light, Power, Class 1, Non–Power-Limited Fire Alarm, and Medium-Power Network-Powered Broadband Communications Circuits.* Coaxial cable shall not be placed in any raceway, compartment, outlet box, junction box, or other enclosures with conductors of electric light, power, Class 1, non–power-limited fire alarm, or medium-power network-powered broadband communications circuits.

Exception No. 1: Coaxial cable shall be permitted to be placed in any raceway, compartment, outlet box, junction box, or other enclosures with conductors of electric light, power, Class 1, non–power-limited fire alarm, or medium-power network-powered broadband communications circuits where all of the conductors of electric light, power, Class 1, non–power-limited fire alarm, and medium-power network-powered broad-band communications circuits are separated from all of the coaxial cables by a permanent barrier or listed divider.

Exception No. 2: Coaxial cable shall be permitted to be placed in outlet boxes, junction boxes, or similar fittings or compartments with power conductors where such conductors are introduced solely for power supply to the coaxial cable system distribution equipment. The power circuit conductors shall be routed within the enclosure to maintain a minimum 6 mm (¼ in.) separation from coaxial cables.

(2) Other Applications. Coaxial cable shall be separated at least 50 mm (2 in.) from conductors of any electric light, power, Class 1, non–power-limited fire alarm, or medium-power network-powered broadband communications circuits.

Exception No. 1: Separation shall not be required where either (1) all of the conductors of electric light, power, Class 1, non–power-limited fire alarm, and medium-power network-powered broadband communications circuits are in a raceway, or in metal-sheathed, metal-clad, nonmetallic-sheathed, Type AC or Type UF cables, or (2) all of the coaxial cables are encased in a raceway.

Exception No. 2: Separation shall not be required where the coaxial cables are permanently separated from the conductors of electric light, power, Class 1, non–power-limited fire alarm, and medium-power network-powered broadband communications circuits by a continuous and firmly fixed nonconductor, such as porcelain tubes or flexible tubing, in addition to the insulation on the wire.

(B) Support of Coaxial Cables. Raceways shall be used for their intended purpose. Coaxial cables shall not be strapped, taped, or attached by any means to the exterior of any conduit or raceway as a means of support.

Exception: Overhead (aerial) spans of coaxial cables shall be permitted to be attached to the exterior of a raceway-type mast intended for the attachment and support of such cables.

820.154 Applications of Listed CATV Cables. Permitted and nonpermitted applications of listed coaxial cables shall be as indicated in Table 820.154(a). The permitted applications shall be subject to the installation requirements of 820.110 and 820.113. The substitutions for coaxial cables in Table 820.154(b) and illustrated in Figure 820.154 shall be permitted.

   Informational Note: The substitute cables in Table 820.154(b) and Figure 820.154 are only coaxial-type cables.

Part VI. Listing Requirements

820.179 Coaxial Cables. Cables shall be listed in accordance with 820.179(A) through (D) and marked in accordance with Table 820.179. The cable voltage rating shall not be marked on the cable. Coaxial cables shall have a temperature rating of not less than 60°C (140°F). The temperature rating shall be marked on the jacket of coaxial cables that have a temperature rating exceeding 60°C (140°F).

   Informational Note: Voltage markings on cables could be misinterpreted to suggest that the cables may be suitable for Class 1, electric light, and power applications.

Exception: Voltage markings shall be permitted where the cable has multiple listings and voltage marking is required for one or more of the listings.

Table 820.154(a) Applications of Listed Coaxial Cables in Buildings

Applications		Listed Coaxial Cable Type			
		CATVP	CATVR	CATV	CATVX
In ducts specifically fabricated for environmental air as described in 300.22(B)	In fabricated ducts as described in 300.22(B)	Y*	N	N	N
	In metal raceway that complies with 300.22(B)	Y*	Y*	Y*	Y*
In other spaces used for environmental air (plenums) as described in 300.22(C)	In other spaces used for environmental air (plenums) as described in 300.22(C)	Y*	N	N	N
	In metal raceway that complies with 300.22(C)	Y*	Y*	Y*	Y*
	In plenum communications raceways	Y*	N	N	N
	In plenum cable routing assemblies	Y*	N	N	N
	Supported by open metal cable trays	Y*	N	N	N
	Supported by solid-bottom metal cable trays with solid metal covers	Y*	Y*	Y*	Y*
In risers	In vertical runs	Y*	Y*	N	N
	In metal raceways	Y*	Y*	Y*	Y*
	In fireproof shafts	Y*	Y*	Y*	Y*
	In plenum communications raceways	Y*	Y*	N	N
	In plenum cable routing assemblies	Y*	Y*	N	N
	In riser communications raceways	Y*	Y*	N	N
	In riser cable routing assemblies	Y*	Y*	N	N
	In one- and two- family dwellings	Y*	Y*	Y*	Y*
Within buildings in other than air-handling spaces and risers	General	Y*	Y*	Y*	Y*
	In one- and two-family dwellings	Y*	Y*	Y*	Y*
	In multifamily dwellings	Y*	Y*	Y*	Y*
	In nonconcealed spaces	Y*	Y*	Y*	Y*
	Supported by cable trays	Y*	Y*	Y*	N
	In distributing frames and cross-connect arrays	Y*	Y*	Y*	N
	In any raceway recognized in Chapter 3	Y*	Y*	Y*	Y*
	In plenum communications raceways	Y*	Y*	Y*	N
	In plenum cable routing assemblies	Y*	Y*	Y*	N
	In riser communications raceways	Y*	Y*	Y*	N
	In riser cable routing assemblies	Y*	Y*	Y*	N
	In general-purpose communications raceways	Y*	Y*	Y*	N
	In general-purpose cable routing assemblies	Y*	Y*	Y*	N

Note: An "N" in the table indicates that the cable type is not permitted to be installed in the application. A "Y*" indicates that the cable type is permitted to be installed in the application, subject to the limitations described in 820.113.

Informational Note No. 1: Part V of Article 820 covers installation methods within buildings. This table covers the applications of listed coaxial cables in buildings. The definition of *Point of Entrance* is in 820.2.

Informational Note No. 2: For information on the restrictions to the installation of communications cables in ducts specifically fabricated for environmental air, see 820.113(B).

Table 820.154(b) Coaxial Cable Uses and Permitted Substitutions

Cable Type	Permitted Substitutions
CATVP	CMP, BLP
CATVR	CATVP, CMP, CMR, BMR, BLP, BLR
CATV	CATVP, CMP, CATVR, CMR, CMG, CM, BMR, BM, BLP, BLR, BL
CATVX	CATVP, CMP, CATVR, CMR, CATV, CMG, CM, BMR, BM, BLP, BLR, BL, BLX

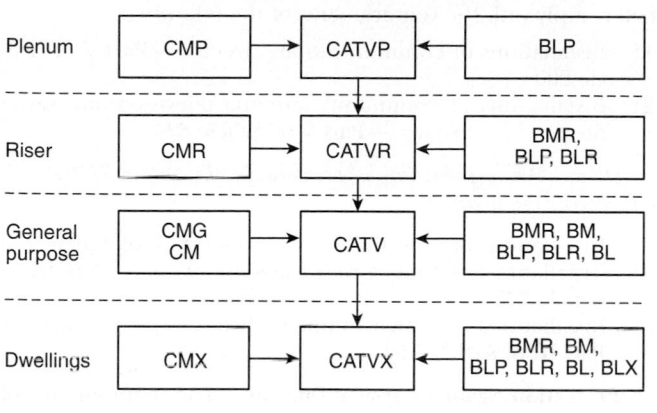

  Coaxial cable A shall be permitted to be used in place of coaxial cable B.

Type BL—Network-powered broadband communications low-power cables

Type BM—Network-powered broadband communications medium-power cables

Type CATV—Community antenna television cables

Type CM—Communications cables

FIGURE 820.154   Cable Substitution Hierarchy.

(A) Type CATVP. Type CATVP community antenna television plenum coaxial cables shall be listed as being suitable for use in ducts, plenums, and other spaces used for environmental air and shall also be listed as having adequate fire-resistant and low-smoke-producing characteristics.

Informational Note: One method of defining a cable that is low-smoke-producing cable and fire-resistant cable is that the cable exhibits a maximum peak optical density of 0.50 or less, an average optical density of 0.15 or less, and a maximum flame spread distance of 1.52 m (5 ft) or less when tested in accordance with NFPA 262 -2015, *Standard Method of Test for Flame Travel and Smoke of Wires and Cables for Use in Air-Handling Spaces.*

(B) Type CATVR. Type CATVR community antenna television riser coaxial cables shall be listed as being suitable for use in a vertical run in a shaft or from floor to floor and shall also be listed as having fire-resistant characteristics capable of preventing the carrying of fire from floor to floor.

Informational Note: One method of defining fire-resistant characteristics capable of preventing the carrying of fire from floor to floor is that the cables pass the requirements of ANSI/UL 1666-2012, *Standard Test for Flame Propagation Height of Electrical and Optical-Fiber Cable Installed Vertically in Shafts.*

(C) Type CATV. Type CATV community antenna television coaxial cables shall be listed as being suitable for general-purpose CATV use, with the exception of risers and plenums, and shall also be listed as being resistant to the spread of fire.

Informational Note: One method of defining *resistant to the spread of fire* is that the cables do not spread fire to the top of the tray in the "UL Flame Exposure, Vertical Tray Flame Test" in ANSI/UL 1685-2010, *Standard for Safety for Vertical-Tray Fire-Propagation and Smoke-Release Test for Electrical and Optical-Fiber Cables.* The smoke measurements in the test method are not applicable.

Another method of defining *resistant to the spread of fire* is for the damage (char length) not to exceed 1.5 m (4 ft 11 in.) when performing the CSA "Vertical Flame Test — Cables in Cable Trays," as described in CSA C22.2 No. 0.3-09, *Test Methods for Electrical Wires and Cables.*

(D) Type CATVX. Type CATVX limited-use community antenna television coaxial cables shall be listed as being suitable for use in dwellings and for use in raceways and shall also be listed as being resistant to flame spread.

Informational Note: One method of determining that cable is resistant to flame spread is by testing the cable to the VW-1 (vertical-wire) flame test in ANSI/UL 1581-2013, *Reference Standard for Electrical Wires, Cables and Flexible Cords.*

Table 820.179 Coaxial Cable Markings

Cable Marking	Type
CATVP	CATV plenum cable
CATVR	CATV riser cable
CATV	CATV cable
CATVX	CATV cable, limited use

Informational Note: Cable types are listed in descending order of fire resistance rating.

820.180 Grounding Devices. Where bonding or grounding is required, devices used to connect a shield, a sheath, or non-current-carrying metallic members of a cable to a bonding conductor, or grounding electrode conductor, shall be listed or be part of listed equipment.

ARTICLE 830
Network-Powered Broadband Communications Systems

Informational Note: See Informational Note Figure 800(a) and Informational Note Figure 800(b) for an illustrative application of a bonding conductor or grounding electrode conductor.

Part I. General

830.1 Scope. This article covers network-powered broadband communications systems that provide any combination of voice, audio, video, data, and interactive services through a network interface unit.

Informational Note No. 1: A typical basic system configuration includes a cable supplying power and broadband signal to a network interface unit that converts the broadband signal to the component signals. Typical cables are coaxial cable with both broadband signal and power on the center conductor, composite metallic cable with a coaxial member(s) or twisted pair members for the broadband signal and twisted pair members for power, and composite optical fiber cable with a pair of conductors for power. Larger systems may also include network components such as amplifiers that require network power.

Informational Note No. 2: See 90.2(B)(4) for installations of broadband communications systems that are not covered.

830.2 Definitions. See Part I of Article 100. For purposes of this article, the following additional definitions apply.

Informational Note: A typical single-family network-powered communications circuit consists of a communications drop or communications service cable and an NIU and includes the communications utility's serving terminal or tap where it is not under the exclusive control of the communications utility.

Abandoned Network-Powered Broadband Communications Cable. Installed network-powered broadband communications cable that is not terminated at equipment other than a connector and not identified for future use with a tag.

Block. A square or portion of a city, town, or village enclosed by streets, including the alleys so enclosed but not any street.

Exposed (to Accidental Contact). A circuit in such a position that, in case of failure of supports or insulation, contact with another circuit may result.

Informational Note: See Part I of Article 100 for two other definitions of *Exposed*: Exposed (as applied to live parts) and Exposed (as applied to wiring methods).

Fault Protection Device. An electronic device that is intended for the protection of personnel and functions under fault conditions, such as network-powered broadband communications cable short or open circuit, to limit the current or voltage, or both, for a low-power network-powered broadband communications circuit and provide acceptable protection from electric shock.

Network Interface Unit (NIU). A device that converts a broadband signal into component voice, audio, video, data, and interactive services signals and provides isolation between the network power and the premises signal circuits. These devices often contain primary and secondary protectors.

Network-Powered Broadband Communications Circuit. The circuit extending from the communications utility's serving terminal or tap up to and including the NIU.

Informational Note: A typical one-family dwelling network-powered communications circuit consists of a communications drop or communications service cable and an NIU and includes the communications utility's serving terminal or tap where it is not under the exclusive control of the communications utility.

Point of Entrance. The point within a building at which the network-powered broadband communications cable emerges from an external wall, from a concrete floor slab, from rigid metal conduit (RMC), or from intermediate metal conduit (IMC).

830.3 Other Articles. Circuits and equipment shall comply with 830.3(A) through (G).

(A) Hazardous (Classified) Locations. Network-powered broadband communications circuits and equipment installed in a location that is classified in accordance with 500.5 and 505.5 shall comply with the applicable requirements of Chapter 5.

(B) Wiring in Ducts for Dust, Loose Stock, or Vapor Removal. The requirements of 300.22(A) shall apply.

(C) Equipment in Other Space Used for Environmental Air. The requirements of 300.22(C)(3) shall apply.

(D) Installation and Use. The requirements of 110.3(B) shall apply.

(E) Output Circuits. As appropriate for the services provided, the output circuits derived from the network interface unit shall comply with the requirements of the following:

(1) Installations of communications circuits — Part V of Article 800

(2) Installations of community antenna television and radio distribution circuits — Part V of Article 820

Exception: Where protection is provided in the output of the NIU 830.90(B)(3) shall apply.

(3) Installations of optical fiber cables — Part V of Article 770

(4) Installations of Class 2 and Class 3 circuits — Part III of Article 725

(5) Installations of power-limited fire alarm circuits — Part III of Article 760

(F) Protection Against Physical Damage. The requirements of 300.4 shall apply.

830.15 Power Limitations. Network-powered broadband communications systems shall be classified as having low- or medium-power sources as specified in 830.15(1) or (2).

(1) Sources shall be classified as defined in Table 830.15.

(2) Direct-current power sources exceeding 150 volts to ground, but no more than 200 volts to ground, with the current to ground limited to 10 mA dc, that meet the current and power limitation for medium-power sources in Table 830.15 shall be classified as medium-power sources.

Informational Note: One way to determine compliance with 830.15(2) is listed information technology equipment intended to supply power via a communications network that complies with the requirements for RFT-V circuits as defined in UL 60950-21-2007, *Standard for Safety for Information Technology Equipment — Safety — Part 21: Remote Power Feeding.*

830.21 Access to Electrical Equipment Behind Panels Designed to Allow Access. Access to electrical equipment shall not be denied by an accumulation of network-powered broadband communications cables that prevents removal of panels, including suspended ceiling panels.

830.24 Mechanical Execution of Work. Network-powered broadband communications circuits and equipment shall be installed in a neat and workmanlike manner. Cables installed exposed on the surface of ceilings and sidewalls shall be supported by the building structure in such a manner that the cable will not be damaged by normal building use. Such cables shall be secured by hardware including straps, staples, cable ties, hangers, or similar fittings designed and installed so as not to damage the cable. The installation shall also conform to 300.4(A), (D), (E), (F), and 300.11. Nonmetallic cable ties and other nonmetallic cable accessories used to secure and support cables in other spaces used for environmental air (plenums) shall be listed as having low smoke and heat release properties in accordance with 800.170(C).

Informational Note No. 1: Accepted industry practices are described in ANSI/NECA/BICSI 568-2006, *Standard for Installing Commercial Building Telecommunications Cabling*; ANSI/TIA-568.1-D-2015, *Commercial Building Telecommunications Infrastructure Standard*; ANSI/TIA-569-D-2015, *Telecommunications Pathways and Spaces*; ANSI/TIA-570-C-2012, *Residential Telecommunications Infrastructure Standard* ; ANSI/TIA-1005-A-2012, *Telecommunications Infrastructure Standard for Industrial Premises*; ANSI/TIA-1179-2010, *Healthcare Facility Telecommunications Infrastructure Standard*; ANSI/TIA-4966-2014, *Telecommunications Infrastructure Standard for Educational Facilities*; and other ANSI-approved installation standards.

Informational Note No. 2: See 4.3.11.2.6.5 and 4.3.11.5.5.6 of NFPA 90A-2015, *Standard for the Installation of Air-Conditioning and Ventilating Systems*, for discrete combustible components installed in accordance with 300.22(C).

Informational Note No. 3: Paint, plaster, cleaners, abrasives, corrosive residues, or other contaminants may result in an undetermined alteration of network-powered broadband cable properties.

830.25 Abandoned Cables. The accessible portion of abandoned network-powered broadband cables shall be removed. Where cables are identified for future use with a tag, the tag shall be of sufficient durability to withstand the environment involved.

830.26 Spread of Fire or Products of Combustion. Installations of network-powered broadband cables in hollow spaces, vertical shafts, and ventilation or air-handling ducts shall be made so that the possible spread of fire or products of combustion will not be substantially increased. Openings around penetrations of network-powered broadband cables through fire-resistant-rated walls, partitions, floors, or ceilings shall be

Table 830.15 Limitations for Network-Powered Broadband Communications Systems

Network Power Source	Low	Medium
Circuit voltage, V_{max} (volts)[1]	0–100	0–150
Power limitation, VA_{max} (volt-amperes)[1]	250	250
Current limitation, I_{max} (amperes)[1]	$1000/V_{max}$	$1000/V_{max}$
Maximum power rating (volt-amperes)	100	100
Maximum voltage rating (volts)	100	150
Maximum overcurrent protection (amperes)[2]	$100/V_{max}$	NA

[1] V_{max}, I_{max}, and VA_{max} are determined with the current-limiting impedance in the circuit (not bypassed) as follows:

V_{max} — Maximum system voltage regardless of load with rated input applied.

I_{max} — Maximum system current under any noncapacitive load, including short circuit, and with overcurrent protection bypassed if used. I_{max} limits apply after 1 minute of operation.

VA_{max} — Maximum volt-ampere output after 1 minute of operation regardless of load and overcurrent protection bypassed if used.

[2]Overcurrent protection is not required where the current-limiting device provides equivalent current limitation and the current-limiting device does not reset until power or the load is removed.

firestopped using approved methods to maintain the fire resistance rating.

Informational Note: Directories of electrical construction materials published by qualified testing laboratories contain many listing installation restrictions necessary to maintain the fire-resistive rating of assemblies where penetrations or openings are made. Building codes also contain restrictions on membrane penetrations on opposite sides of a fire resistance–rated wall assembly. An example is the 600-mm (24-in.) minimum horizontal separation that usually applies between boxes installed on opposite sides of the wall. Assistance in complying with 830.26 can be found in building codes, fire resistance directories, and product listings.

Part II. Cables Outside and Entering Buildings

830.40 Entrance Cables. Network-powered broadband communications cables located outside and entering buildings shall comply with 830.40(A) and (B).

(A) Medium-Power Circuits. Medium-power network-powered broadband communications circuits located outside and entering buildings shall be installed using Type BMU, Type BM, or Type BMR network-powered broadband communications medium-power cables.

(B) Low-Power Circuits. Low-power network-powered broadband communications circuits located outside and entering buildings shall be installed using Type BLU or Type BLX low-power network-powered broadband communications cables. Cables shown in Table 830.154(b) shall be permitted to substitute.

Exception: Outdoor community antenna television and radio distribution system coaxial cables installed prior to January 1, 2000, and installed in accordance with Article 820, shall be permitted for low-power-type, network-powered broadband communications circuits.

830.44 Overhead (Aerial) Cables. Overhead (aerial) network-powered broadband communications cables shall comply with 830.44(A) through (G).

Informational Note: For additional information regarding overhead (aerial) wires and cables, see ANSI C2-2007, *National Electrical Safety Code*, Part 2, Safety Rules for Overhead Lines.

(A) On Poles and In-Span. Where network-powered broadband communications cables and electric light or power conductors are supported by the same pole or are run parallel to each other in-span, the conditions described in 830.44(A) (1) through (A)(4) shall be met.

(1) Relative Location. Where practicable, the network-powered broadband communications cables shall be located below the electric light or power conductors.

(2) Attachment to Cross-Arms. Network-powered broadband communications cables shall not be attached to a cross-arm that carries electric light or power conductors.

(3) Climbing Space. The climbing space through network-powered broadband communications wires and cables shall comply with the requirements of 225.14(D).

(4) Clearance. Lead-in or overhead (aerial)-drop network-powered broadband communications cables from a pole or other support, including the point of initial attachment to a building or structure, shall be kept away from electric light, power, Class 1, or non–power-limited fire alarm circuit conductors so as to avoid the possibility of accidental contact.

Exception: Where proximity to electric light, power, Class 1, or non–power-limited fire alarm circuit conductors cannot be avoided, the installation shall provide clearances of not less than 300 mm (12 in.) from electric light, power, Class 1, or non–power-limited fire alarm circuit conductors. The clearance requirement shall apply to all points along the drop, and it shall increase to 1.02 m (40 in.) at the pole.

(B) Above Roofs. Network-powered broadband communications cables shall have a vertical clearance of not less than 2.5 m (8 ft) from all points of roofs above which they pass.

Exception No. 1: Network-powered broadband communications cables shall not be required to have a vertical clearance of 2.5 m (8 ft) above auxillary buildings such as garages and the like.

Exception No. 2: A reduction in clearance above only the overhanging portion of the roof to not less than 450 mm (18 in.) shall be permitted if (1) not more than 1.2 m (4 ft) of the broadband communications drop cables pass above the roof overhang, and (2) they are terminated at a through-the-roof raceway or support.

Exception No. 3: Where the roof has a slope of not less than 100 mm in 300 mm (4 in. in 12 in.), a reduction in clearance to not less than 900 mm (3 ft) shall be permitted.

(C) Clearance from Ground. Overhead (aerial) spans of network-powered broadband communications cables shall conform to not less than the following:

(1) 2.9 m (9½ ft) — above finished grade, sidewalks, or from any platform or projection from which they might be reached and accessible to pedestrians only

(2) 3.5 m (11½ ft) — over residential property and driveways, and those commercial areas not subject to truck traffic

(3) 4.7 m (15½ ft) — over public streets, alleys, roads, parking areas subject to truck traffic, driveways on other than residential property, and other land traversed by vehicles such as cultivated, grazing, forest, and orchard

Informational Note: These clearances have been specifically chosen to correlate with ANSI/IEEE C2-2012, *National Electrical Safety Code,* Table 232-1, which provides for clearances of wires, conductors, and cables above ground and roadways, rather than using the clearances referenced in 225.18. Because Article 800 and Article 820 have had no required clearances, the communications industry has used the clearances from the NESC for their installed cable plant.

(D) Over Pools. Clearance of network-powered broadband communications cable in any direction from the water level, edge of pool, base of diving platform, or anchored raft shall comply with those clearances in 680.9.

(E) Final Spans. Final spans of network-powered broadband communications cables without an outer jacket shall be permitted to be attached to the building, but they shall be kept not less than 900 mm (3 ft) from windows that are designed to be opened, doors, porches, balconies, ladders, stairs, fire escapes, or similar locations.

Exception: Conductors run above the top level of a window shall be permitted to be less than the 900-mm (3-ft) requirement above.

Overhead (aerial) network-powered broadband communications cables shall not be installed beneath openings through which materials might be moved, such as openings in farm and commercial buildings, and shall not be installed where they obstruct entrance to these building openings.

(F) Between Buildings. Network-powered broadband communications cables extending between buildings or structures, and also the supports or attachment fixtures, shall be identified as suitable for outdoor aerial applications and shall have sufficient strength to withstand the loads to which they may be subjected.

Exception: Where a network-powered broadband communications cable does not have sufficient strength to be self-supporting, it shall be attached to a supporting messenger cable that, together with the attachment fixtures or supports, shall be acceptable for the purpose and shall have sufficient strength to withstand the loads to which they may be subjected.

(G) On Buildings. Where attached to buildings, network-powered broadband communications cables shall be securely fastened in such a manner that they are separated from other conductors in accordance with 830.44(G)(1) through (G)(4).

(1) Electric Light or Power. The network-powered broadband communications cable shall have a separation of at least 100 mm (4 in.) from electric light, power, Class 1, or non–power-limited fire alarm circuit conductors not in raceway or cable, or be permanently separated from conductors of the other system by a continuous and firmly fixed nonconductor in addition to the insulation on the wires.

(2) Other Communications Systems. Network-powered broadband communications cables shall be installed so that there will be no unnecessary interference in the maintenance of the separate systems. In no case shall the conductors, cables, messenger strand, or equipment of one system cause abrasion to the conductors, cables, messenger strand, or equipment of any other system.

(3) Lightning Conductors. Where practicable, a separation of at least 1.8 m (6 ft) shall be maintained between any network-powered broadband communications cable and lightning conductors.

Informational Note: Specific separation distances may be calculated from the sideflash formula found in NFPA 780-2014, *Standard for the Installation of Lightning Protection Systems,* Section 4.6.

(4) Protection from Damage. Network-powered broadband communications cables attached to buildings or structures and located within 2.5 m (8 ft) of finished grade shall be protected by enclosures, raceways, or other approved means.

Exception: A low-power network-powered broadband communications circuit that is equipped with a listed fault protection device, appropriate to the network-powered broadband communications cable used, and located on the network side of the network-powered broadband communications cable shall not be required to be additionally protected by enclosures, raceways, or other approved means.

830.47 Underground Network-Powered Broadband Communications Cables Entering Buildings. Underground network-powered broadband communications cables entering buildings shall comply with 830.47(A) through 830.47(D).

(A) Underground Systems with Electric Light and Power, Class 1, or Non–Power-Limited Fire Alarm Circuit Conductors. Underground network-powered broadband communications cables in a duct, pedestal, handhole enclosure, or manhole that contains electric light, power conductors, non–power-limited fire alarm circuit conductors, or Class 1 circuits shall be in a section permanently separated from such conductors by means of a suitable barrier.

(B) Direct-Buried Cables and Raceways. Direct-buried network-powered broadband communications cables shall be separated by at least 300 mm (12 in.) from conductors of any light, power, non–power-limited fire alarm circuit conductors or Class 1 circuit.

Exception No. 1: Separation shall not be required where electric service conductors or network-powered broadband communications cables are installed in raceways or have metal cable armor.

Exception No. 2: Separation shall not be required where electric light or power branch-circuit or feeder conductors, non–power-limited fire alarm circuit conductors, or Class 1 circuit conductors are installed in a raceway or in metal-sheathed, metal-clad, or Type UF or Type USE cables; or the network-powered broadband communications cables have metal cable armor or are installed in a raceway.

(C) Mechanical Protection. Direct-buried cable, conduit, or other raceways shall be installed to meet the minimum cover requirements of Table 830.47(C). In addition, direct-buried cables emerging from the ground shall be protected by enclosures, raceways, or other approved means extending from the minimum cover distance required by Table 830.47(C) below grade to a point at least 2.5 m (8 ft) above finished grade. In no case shall the protection be required to exceed 450 mm (18 in.) below finished grade. Types BMU and BLU direct-buried cables emerging from the ground shall be installed in rigid metal conduit (RMC), intermediate metal conduit (IMC), rigid nonmetallic conduit, or other approved means extending

from the minimum cover distance required by Table 830.47(C) below grade to the point of entrance.

Exception: A low-power network-powered broadband communications circuit that is equipped with a listed fault protection device, appropriate to the network-powered broadband communications cable used, and located on the network side of the network-powered broadband communications cable being protected.

(D) Pools. Cables located under the pool or within the area extending 1.5 m (5 ft) horizontally from the inside wall of the pool shall meet those clearances and requirements specified in 680.11.

830.49 Metallic Entrance Conduit Grounding. Metallic conduit containing network-powered broadband communications entrance cable shall be connected by a bonding conductor or grounding electrode conductor to a grounding electrode in accordance with 830.100(B).

Part III. Protection

830.90 Primary Electrical Protection.

(A) Application. Primary electrical protection shall be provided on all network-powered broadband communications conductors that are neither grounded nor interrupted and are run partly or entirely in aerial cable not confined within a block. Also, primary electrical protection shall be provided on all aerial or underground network-powered broadband communications conductors that are neither grounded nor interrupted and are located within the block containing the

Table 830.47(C) Network-Powered Broadband Communications Systems Minimum Cover Requirements (*Cover* is the shortest distance measured between a point on the top surface of any direct-buried cable, conduit, or other raceway and the top surface of finished grade, concrete, or similar cover.)

Location of Wiring Method or Circuit	Direct Burial Cables		Rigid Metal Conduit (RMC) or Intermediate Metal Conduit (IMC)		Nonmetallic Raceways Listed for Direct Burial; Without Concrete Encasement or Other Approved Raceways	
	mm	in.	mm	in.	mm	in.
All locations not specified below	450	18	150	6	300	12
In trench below 50-mm (2-in.) thick concrete or equivalent	300	12	150	6	150	6
Under a building (in raceway only)	0	0	0	0	0	0
Under minimum of 100-mm (4 in.) thick concrete exterior slab with no vehicular traffic and the slab extending not less than 150 mm (6 in.) beyond the underground installation	300	12	100	4	100	4
One- and two-family dwelling driveways and outdoor parking areas and used only for dwelling-related purposes	300	12	300	12	300	12

Notes:
1. Raceways approved for burial only where concrete encased shall require a concrete envelope not less than 50 mm (2 in.) thick.
2. Lesser depths shall be permitted where cables rise for terminations or splices or where access is otherwise required.
3. Where solid rock is encountered, all wiring shall be installed in metal or nonmetallic raceway permitted for direct burial. The raceways shall be covered by a minimum of 50 mm (2 in.) of concrete extending down to rock.
4. Low-power network-powered broadband communications circuits using directly buried community antenna television and radio distribution system coaxial cables that were installed outside and entering buildings prior to January 1, 2000, in accordance with Article 820 shall be permitted where buried to a minimum depth of 300 mm (12 in.).

building served so as to be exposed to lightning or accidental contact with electric light or power conductors operating at over 300 volts to ground.

Exception: Primary electrical protection shall not be required on the network-powered broadband communications conductors where electrical protection is provided on the derived circuit(s) (output side of the NIU) in accordance with 830.90(B)(3).

> Informational Note No. 1: On network-powered broadband communications conductors not exposed to lightning or accidental contact with power conductors, providing primary electrical protection in accordance with this article helps protect against other hazards, such as ground potential rise caused by power fault currents, and above-normal voltages induced by fault currents on power circuits in proximity to the network-powered broadband communications conductors.

> Informational Note No. 2: Network-powered broadband communications circuits are considered to have a lightning exposure unless one or more of the following conditions exist:
>
> (1) Circuits in large metropolitan areas where buildings are close together and sufficiently high to intercept lightning.
> (2) Areas having an average of five or fewer thunderstorm days each year and earth resistivity of less than 100 ohm-meters. Such areas are found along the Pacific coast.

> Informational Note No. 3: For information on lightning protection systems, see NFPA 780-2014, *Standard for the Installation of Lightning Protection Systems.*

(1) Fuseless Primary Protectors. Fuseless-type primary protectors shall be permitted where power fault currents on all protected conductors in the cable are safely limited to a value no greater than the current-carrying capacity of the primary protector and of the primary protector bonding conductor or grounding electrode conductor.

(2) Fused Primary Protectors. Where the requirements listed in 830.90(A)(1) are not met, fused-type primary protectors shall be used. Fused-type primary protectors shall consist of an arrester connected between each conductor to be protected and ground, a fuse in series with each conductor to be protected, and an appropriate mounting arrangement. Fused primary protector terminals shall be marked to indicate line, instrument, and ground, as applicable.

(B) Location. The location of the primary protector, where required, shall comply with (B)(1), (B)(2), or (B)(3):

(1) A listed primary protector shall be applied on each network-powered broadband communications cable external to and on the network side of the network interface unit.

(2) The primary protector function shall be an integral part of and contained in the network interface unit. The network interface unit shall be listed as being suitable for application with network-powered broadband communications systems and shall have an external marking indicating that it contains primary electrical protection.

(3) The primary protector(s) shall be provided on the derived circuit(s) (output side of the NIU), and the combination of the NIU and the protector(s) shall be listed as being suitable for application with network-powered broadband communications systems.

A primary protector, whether provided integrally or external to the network interface unit, shall be located as close as practicable to the point of entrance.

For purposes of this section, a network interface unit and any externally provided primary protectors located at mobile home service equipment located in sight from and not more than 9.0 m (30 ft) from the exterior wall of the mobile home it serves, or at a mobile home disconnecting means grounded in accordance with 250.32 and located in sight from and not more than 9.0 m (30 ft) from the exterior wall of the mobile home it serves, shall be considered to meet the requirements of this section.

> Informational Note: Selecting a network interface unit and primary protector location to achieve the shortest practicable primary protector bonding conductor or grounding electrode conductor helps limit potential differences between communications circuits and other metallic systems.

(C) Hazardous (Classified) Locations. The primary protector or equipment providing the primary protection function shall not be located in any hazardous (classified) location as defined in 500.5 and 505.5 or in the vicinity of easily ignitible material.

Exception: As permitted in 501.150, 502.150, and 503.150.

830.93 Grounding or Interruption of Metallic Members of Network-Powered Broadband Communications Cables. Network-powered communications cables entering buildings or attaching to buildings shall comply with 830.93(A) or (B).

For purposes of this section, grounding located at mobile home service equipment located within 9.0 m (30 ft) of the exterior wall of the mobile home it serves, or at a mobile home disconnecting means grounded in accordance with 250.32 and located within 9.0 m (30 ft) of the exterior wall of the mobile home it serves, shall be considered to meet the requirements of this section.

> Informational Note: Selecting a grounding location to achieve the shortest practicable bonding conductor or grounding electrode conductor helps limit potential differences between the network-powered broadband communications circuits and other metallic systems.

(A) Entering Buildings. In installations where the network-powered communications cable enters the building, the shield shall be grounded in accordance with 830.100, and metallic members of the cable not used for communications or powering shall be grounded in accordance with 830.100 or interrupted by an insulating joint or equivalent device. The grounding or interruption shall be as close as practicable to the point of entrance.

(B) Terminating Outside of the Building. In installations where the network-powered communications cable is terminated outside of the building, the shield shall be grounded in accordance with 830.100, and metallic members of the cable not used for communications or powering shall be grounded in accordance with 830.100 or interrupted by an insulating joint or equivalent device. The grounding or interruption shall be as close as practicable to the point of attachment of the NIU.

Part IV. Grounding Methods

830.100 Cable, Network Interface Unit, and Primary Protector Bonding and Grounding. Network interface units containing protectors, NIUs with metallic enclosures, primary protectors, and the metallic members of the network-powered broadband communications cable that are intended to be bonded or grounded shall be connected as specified in 830.100(A) through 830.100(D).

(A) Bonding Conductor or Grounding Electrode Conductor.

(1) Insulation. The bonding conductor or grounding electrode conductor shall be listed and shall be permitted to be insulated, covered, or bare.

(2) Material. The bonding conductor or grounding electrode conductor shall be copper or other corrosion-resistant conductive material, stranded or solid.

(3) Size. The bonding conductor or grounding electrode conductor shall not be smaller than 14 AWG and shall have a current-carrying capacity not less than that of the grounded metallic member(s) and protected conductor(s) of the network-powered broadband communications cable. The bonding conductor or grounding electrode conductor shall not be required to exceed 6 AWG.

(4) Length. The bonding conductor or grounding electrode conductor shall be as short as practicable. In one- and two-family dwellings, the bonding conductor or grounding electrode conductor shall be as short as practicable, not to exceed 6.0 m (20 ft) in length.

> Informational Note: Similar bonding conductor or grounding electrode conductor length limitations applied at apartment buildings and commercial buildings help to reduce voltages that may be developed between the building's power and communications systems during lightning events.

Exception: In one- and two-family dwellings where it is not practicable to achieve an overall maximum bonding conductor or grounding electrode conductor length of 6.0 m (20 ft), a separate communications ground rod meeting the minimum dimensional criteria of 830.100(B)(3)(2) shall be driven, and the grounding electrode conductor shall be connected to the communications ground rod in accordance with 830.100(C). The communications ground rod shall be bonded to the power grounding electrode system in accordance with 830.100(D).

(5) Run in Straight Line. The bonding conductor or grounding electrode conductor shall be run in as straight a line as practicable.

(6) Physical Protection. Bonding conductors and grounding electrode conductors shall be protected where exposed to physical damage. Where the bonding conductor or grounding electrode conductor is installed in a metal raceway, both ends of the raceway shall be bonded to the contained conductor or to the same terminal or electrode to which the bonding conductor or grounding electrode conductor is connected.

(B) Electrode. The bonding conductor or grounding electrode conductor shall be connected in accordance with 830.100(B)(1), 830.100(B)(2), or 830.100(B)(3).

(1) In Buildings or Structures with an Intersystem Bonding Termination. If the building or structure served has an intersystem bonding termination as required by 250.94, the bonding conductor shall be connected to the intersystem bonding termination.

(2) In Buildings or Structures with Grounding Means. If an intersystem bonding termination is established, 250.94(A) shall apply.

If the building or structure served has no intersystem bonding termination, the bonding conductor or grounding electrode conductor shall be connected to the nearest accessible location on one of the following:

(1) The building or structure grounding electrode system as covered in 250.50
(2) The grounded interior metal water piping system, within 1.5 m (5 ft) from its point of entrance to the building, as covered in 250.52
(3) The power service accessible means external to enclosures using the options identified in 250.94(A), Exception
(4) The nonflexible metallic power service raceway
(5) The service equipment enclosure
(6) The grounding electrode conductor or the grounding electrode conductor metal enclosure of the power service
(7) The grounding electrode conductor or the grounding electrode of a building or structure disconnecting means that is connected to an electrode as covered in 250.32

A bonding device intended to provide a termination point for the bonding conductor (intersystem bonding) shall not interfere with the opening of an equipment enclosure. A bonding device shall be mounted on nonremovable parts. A bonding device shall not be mounted on a door or cover even if the door or cover is nonremovable.

For purposes of this section, the mobile home service equipment or the mobile home disconnecting means, as described in 830.93, shall be considered accessible.

(3) In Buildings or Structures Without an Intersystem Bonding Termination or Grounding Means. If the building or structure served has no intersystem bonding termination or grounding means, as described in 830.100(B)(2), the grounding electrode conductor shall be connected to either of the following:

(1) To any one of the individual grounding electrodes described in 250.52(A)(1), (A)(2), (A)(3), or (A)(4).
(2) If the building or structure served has no intersystem bonding termination or has no grounding means, as described in 830.100(B)(2) or (B)(3)(1), to any one of the individual grounding electrodes described in 250.52(A)(7) and (A)(8), or to a ground rod or pipe not less than 1.5 m (5 ft) in length and 12.7 mm ($\frac{1}{2}$ in.) in diameter, driven, where practicable, into permanently damp earth and separated from lightning conductors as covered in 800.53 and at least 1.8 m (6 ft) from electrodes of other systems. Steam, hot water pipes, or lightning-protection system conductors shall not be employed as grounding electrodes for protectors, NIUs with integral protection, grounded metallic members, NIUs with metallic enclosures, and other equipment.

(C) Electrode Connection. Connections to grounding electrodes shall comply with 250.70.

(D) Bonding of Electrodes. A bonding jumper not smaller than 6 AWG copper or equivalent shall be connected between the network-powered broadband communications system grounding electrode and the power grounding electrode system at the building or structure served where separate electrodes are used.

Exception: At mobile homes as covered in 830.106.

> Informational Note No. 1: See 250.60 for use of a connection to a lightning protection system.

Informational Note No. 2: Bonding together of all separate electrodes limits potential differences between them and between their associated wiring systems.

830.106 Grounding and Bonding at Mobile Homes.

(A) Grounding. Grounding shall comply with 830.106(A)(1) or (A)(2).

(1)   Where there is no mobile home service equipment located within 9.0 m (30 ft) of the exterior wall of the mobile home it serves, the network-powered broadband communications cable shield, network-powered broadband communications cable metallic members not used for communications or powering, network interface unit, and primary protector grounding terminal shall be connected to a grounding electrode conductor or grounding electrode in accordance with 830.100(B)(3).

(2)   Where there is no mobile home disconnecting means grounded in accordance with 250.32 and located within 9.0 m (30 ft) of the exterior wall of the mobile home it serves, the network-powered broadband communications cable shield, network-powered broadband communications cable metallic members not used for communications or powering, network interface unit, and primary protector grounding terminal shall be connected to a grounding electrode in accordance with 830.100(B)(3).

(B) Bonding. The network-powered broadband communications cable grounding terminal, network interface unit grounding terminal, if present, and primary protector grounding terminal shall be bonded together with a copper bonding conductor not smaller than 12 AWG. The network-powered broadband communications cable grounding terminal, network interface unit grounding terminal, primary protector grounding terminal, or the grounding electrode shall be bonded to the metal frame or available grounding terminal of the mobile home with a copper bonding conductor not smaller than 12 AWG under any of the following conditions:

(1)   Where there is no mobile home service equipment or disconnecting means as in 830.106(A)
(2)   Where the mobile home is supplied by cord and plug

Part V. Installation Methods Within Buildings

830.110 Raceways and Cable Routing Assemblies for Network-Powered Broadband Communications Cables.

(A) Types of Raceways. Low-power network-powered broadband communications cables shall be permitted to be installed in any raceway that complies with either 830.110(A)(1) or (A)(2) and in cable routing assemblies installed in compliance with 830.110(C). Medium-power network-powered broadband communications cables shall be permitted to be installed in any raceway that complies with 830.110(A)(1).

(1) Raceways Recognized in Chapter 3. Low- and medium-power network-powered broadband communications cables shall be permitted to be installed in any raceway included in Chapter 3. The raceways shall be installed in accordance with the requirements of Chapter 3.

(2) Communications Raceways. Low-power network-powered broadband communications cables shall be permitted to be installed in plenum communications raceways, riser communications raceways, and general-purpose communications raceways, selected in accordance with Table 800.154(b), listed in

accordance with 800.182, and installed in accordance with 800.113 and 362.24 through 362.56, where the requirements applicable to electrical nonmetallic tubing apply.

(3) Innerduct for Low-Power Network-Powered Broadband Communications Cables. Listed plenum communications raceways, listed riser communications raceways, and listed general-purpose communications raceways selected in accordance with Table 800.154(b) shall be permitted to be installed as innerducts in any type of listed raceway permitted in Chapter 3.

(B) Raceway Fill for Network-Powered Broadband Communications Cables. Raceway fill for network-powered broadband communications cables shall comply with either (B)(1) or (B)(2).

(1) Low-Power Network-Powered Broadband Communications Cables. The raceway fill requirements of Chapters 3 and 9 shall not apply to low-power network-powered broadband communications cables.

(2) Medium-Power Network-Powered Broadband Communications Cables. Where medium-power network-powered broadband communications cables are installed in a raceway, the raceway fill requirements of Chapters 3 and 9 shall apply.

(C) Cable Routing Assemblies. Low-power network-powered broadband communications cables shall be permitted to be installed in plenum cable routing assemblies, riser cable routing assemblies, and general-purpose cable routing assemblies selected in accordance with Table 800.154(c), listed in accordance with 800.182, and installed in accordance with 800.110(C) and 800.113.

830.113 Installation of Network-Powered Broadband Communications Cables. Installation of network-powered broadband communications cables shall comply with 830.113(A) through (I). Installation of raceways and cable routing assemblies shall comply with 830.110.

(A) Listing. Network-powered broadband communications cables installed in buildings shall be listed.

(B) Ducts Specifically Fabricated for Environmental Air. The following cables shall be permitted in ducts specifically fabricated for environmental air as described in 300.22(B) if they are directly associated with the air distribution system:

(1)   Up to 1.22 m (4 ft) of Type BLP
(2)   Types BLP, BMR, BLR, BM, BL, and BLX installed in raceways that are installed in compliance with 300.22(B)

Informational Note: For information on fire protection of wiring installed in fabricated ducts, see 4.3.4.1 and 4.3.11.3.3 in NFPA 90A-2015, *Standard for the Installation of Air-Conditioning and Ventilating Systems.*

(C) Other Spaces Used For Environmental Air (Plenums). The following cables shall be permitted in other spaces used for environmental air as described in 300.22(C):

(1)   Type BLP
(2)   Type BLP installed in plenum communications raceways
(3)   Type BLP installed in plenum cable routing assemblies
(4)   Type BLP supported by open metallic cable trays or cable tray systems
(5)   Types BLP, BMR, BLR, BM, BL, and BLX installed in raceways that are installed in compliance with 300.22(C)
(6)   Types BLP, BMR, BLR, BM, BL, and BLX supported by solid-bottom metal cable trays with solid metal covers in

other spaces used for environmental air (plenums) as described in 300.22(C)

(7) Types BLP, BLR, BM, BL, and BLX installed in plenum communications raceways, riser communications raceways, or general-purpose communications raceways supported by solid bottom metal cable trays with solid metal covers in other spaces used for environmental air (plenums) as described in 300.22(C)

Informational Note: For information on fire protection of wiring installed in other spaces used for environmental air, see 4.3.11.2, 4.3.11.4, and 4.3.11.5 of NFPA 90A-2015, *Standard for the Installation of Air-Conditioning and Ventilating Systems.*

(D) Risers — Cables in Vertical Runs. The following cables shall be permitted in vertical runs penetrating one or more floors and in vertical runs in a shaft:

(1) Types BLP, BMR, and BLR
(2) Types BLP and BLR installed in the following:

    a. Plenum communications raceways
    b. Plenum cable routing assemblies
    c. Riser communications raceways
    d. Riser cable routing assemblies

Informational Note: See 830.26 for firestop requirements for floor penetrations.

(E) Risers — Cables and Innerducts in Metal Raceways. The following cables and innerducts shall be permitted in a metal raceway in a riser with firestops at each floor:

(1) Types BLP, BMR, BLR, BM, BL, and BLX
(2) Types BLP, BLR, and BL installed in the following:

    a. Plenum communications raceways (innerduct)
    b. Riser communications raceways (innerduct)
    c. General-purpose communications raceways (innerduct)

Informational Note: See 830.26 for firestop requirements for floor penetrations.

(F) Risers — Cables in Fireproof Shafts. The following cables shall be permitted to be installed in fireproof riser shafts with firestops at each floor:

(1) Types BLP, BMR, BLR, BM, BL, and BLX
(2) Types BLP, BLR, and BL installed in the following:

    a. Plenum communications raceways
    b. Plenum cable routing assemblies
    c. Riser communications raceways
    d. Riser cable routing assemblies
    e. General-purpose communications raceways
    f. General-purpose cable routing assemblies

Informational Note: See 830.26 for firestop requirements for floor penetrations.

(G) Risers — One- and Two-Family Dwellings. The following cables shall be permitted in one- and two-family dwellings:

(1) Types BLP, BMR, BLR, BM, BL, and BLX less than 10 mm (⅜ in.) in diameter
(2) Types BLP, BLR, and BL installed in the following:

    a. Plenum communications raceways
    b. Plenum cable routing assemblies
    c. Riser communications raceways
    d. Riser cable routing assemblies
    e. General-purpose communications raceways
    f. General-purpose cable routing assemblies

Informational Note: See 830.26 for firestop requirements for floor penetrations.

N **(H) Cable Trays.** The following cables shall be permitted to be supported by cable trays:

(1) Types BLP, BMR, BLR, BM, and BL
(2) Types BLP, BLR, and BL installed in the following:

    a. Plenum communications raceways
    b. Riser communications raceways
    c. General-purpose communications raceways

(I) Other Building Locations. The following cables shall be permitted to be installed in building locations other than those covered in 830.113(B) through (H):

(1) Types BLP, BMR, BLR, BM, and BL
(2) Types BLP, BMR, BLR, BM, BL, and BLX installed in raceways recognized in Chapter 3
(3) Types BLP, BLR, and BL installed in the following:

    a. Plenum communications raceways
    b. Plenum cable routing assemblies
    c. Riser communications raceways
    d. Riser cable routing assemblies
    e. General-purpose communications raceways
    f. General-purpose cable routing assemblies

(4) Type BLX less than 10 mm (⅜ in.) in diameter in one- and two-family dwellings
(5) Types BMU and BLU entering the building from outside and run in rigid metal conduit (RMC) or intermediate metal conduit (IMC) where the conduit is connected by a bonding conductor or grounding electrode conductor in accordance with 830.100(B)

    Informational Note: This provision limits the length of Type BLX cable to 15 m (50 ft), while 830.90(B) requires that the primary protector, or NIU with integral protection, be located as close as practicable to the point at which the cable enters the building. Therefore, in installations requiring a primary protector, or NIU with integral protection, Type BLX cable may not be permitted to extend 15 m (50 ft) into the building if it is practicable to place the primary protector closer than 15 m (50 ft) to the entrance point.

(6) A maximum length of 15 m (50 ft), within the building, of Type BLX cable entering the building from outside and terminating at an NIU or a primary protection location

830.133 Installation of Network-Powered Broadband Communications Cables and Equipment. Cable and equipment installations within buildings shall comply with 830.133(A) and (B), as applicable.

(A) Separation of Conductors.

(1) In Raceways, Cable Trays, Boxes, Enclosures, and Cable Routing Assemblies.

    (a) *Low- and Medium-Power Network-Powered Broadband Communications Circuit Cables.* Low- and medium-power network-powered broadband communications cables shall be permitted in the same raceway, cable tray, box, enclosure, or cable routing assembly.

    (b) *Low-Power Network-Powered Broadband Communications Circuit Cables with Other Circuits.* Low-power network-powered broadband communications cables shall be permitted in the same raceway, cable tray, box, enclosure, or cable routing assembly with jacketed cables of any of the following circuits:

(1) Class 2 and Class 3 remote-control, signaling, and power-limited circuits in compliance with Parts I and III of Article 725

(2) Power-limited fire alarm systems in compliance with Parts I and III of Article 760

(3) Communications circuits in compliance with Parts I and V of Article 800

(4) Nonconductive and conductive optical fiber cables in compliance with Parts I and V of Article 770

(5) Community antenna television and radio distribution systems in compliance with Parts I and V of Article 820

(c) *Medium-Power Network-Powered Broadband Communications Circuit Cables with Optical Fiber Cables and Other Communications Cables.* Medium-power network-powered broadband communications cables shall not be permitted in the same raceway, cable tray, box, enclosure, or cable routing assembly with conductors of any of the following circuits:

(1) Communications circuits in compliance with Parts I and V of Article 800

(2) Conductive optical fiber cables in compliance with Parts I and V of Article 770

(3) Community antenna television and radio distribution systems in compliance with Parts I and V of Article 820

(d) *Medium-Power Network-Powered Broadband Communications Circuit Cables with Other Circuits.* Medium-power network-powered broadband communications cables shall not be permitted in the same raceway, cable tray, box, enclosure, or cable routing assembly with conductors of any of the following circuits:

(1) Class 2 and Class 3 remote-control, signaling, and power-limited circuits in compliance with Parts I and III of Article 725

(2) Power-limited fire alarm systems in compliance with Parts I and III of Article 760

(e) *Electric Light, Power, Class 1, Non–Powered Broadband Communications Circuit Cables.* Network-powered broadband communications cable shall not be placed in any raceway, cable tray, compartment, outlet box, junction box, or similar fittings with conductors of electric light, power, Class 1, or non–power-limited fire alarm circuit cables.

Exception No. 1: Where all of the conductors of electric light, power, Class 1, non–power-limited fire alarm circuits are separated from all of the network-powered broadband communications cables by a permanent barrier or listed divider.

Exception No. 2: Power circuit conductors in outlet boxes, junction boxes, or similar fittings or compartments where such conductors are introduced solely for power supply to the network-powered broadband communications system distribution equipment. The power circuit conductors shall be routed within the enclosure to maintain a minimum 6 mm (¼ in.) separation from network-powered broadband communications cables.

(2) Other Applications. Network-powered broadband communications cable shall be separated at least 50 mm (2 in.) from conductors of any electric light, power, Class 1, and non–power-limited fire alarm circuits.

Exception No. 1: Separation shall not be required where: (1) all of the conductors of electric light, power, Class 1, and non–power-limited fire alarm circuits are in a raceway, or in metal-sheathed, metal-clad, nonmetallic-sheathed, Type AC, or Type UF cables, or (2) all of the network-powered broadband communications cables are encased in a raceway.

Exception No. 2: Separation shall not be required where the network-powered broadband communications cables are permanently separated from the conductors of electric light, power, Class 1, and non–power-limited fire alarm circuits by a continuous and firmly fixed nonconductor, such as porcelain tubes or flexible tubing, in addition to the insulation on the wire.

(B) Support of Network-Powered Broadband Communications Cables. Raceways shall be used for their intended purpose. Network-powered broadband communications cables shall not be strapped, taped, or attached by any means to the exterior of any conduit or raceway as a means of support.

830.154 Applications of Network-Powered Broadband Communications System Cables. Permitted and nonpermitted applications of listed network-powered broadband communications system cables shall be as indicated in Table 830.154(a). The permitted applications shall be subject to the installation requirements of 830.40, 830.110, and 830.113. The substitutions for network-powered broadband system cables listed in Table 830.154(b) shall be permitted.

830.160 Bends. Bends in network broadband cable shall be made so as not to damage the cable.

Part VI. Listing Requirements

830.179 Network-Powered Broadband Communications Equipment and Cables. Network-powered broadband communications equipment and cables shall be listed and marked in accordance with 830.179(A) or (B). Network-powered broadband communications cables shall have a temperature rating of not less than 60°C (140°F). Temperature rating shall be marked on the jacket of network-powered broadband communications cables that have a temperature rating exceeding 60°C (140°F).

Exception No. 1: This listing requirement shall not apply to community antenna television and radio distribution system coaxial cables that were installed prior to January 1, 2000, in accordance with Article 820 and are used for low-power network-powered broadband communications circuits.

Exception No. 2: Substitute cables for network-powered broadband communications cables shall be permitted as shown in Table 830.154(b).

(A) Network-Powered Broadband Communications Medium-Power Cables. Network-powered broadband communications medium-power cables shall be factory-assembled cables consisting of a jacketed coaxial cable, a jacketed combination of coaxial cable and multiple individual conductors, or a jacketed combination of an optical fiber cable and multiple individual conductors. The insulation for the individual conductors shall be rated for 300 volts minimum. Cables intended for outdoor use shall be listed as suitable for the application. Cables shall be marked in accordance with 310.120.

(1) Type BMR. Type BMR cables shall be listed as being suitable for use in a vertical run in a shaft or from floor to floor and shall also be listed as having fire-resistant characteristics capable of preventing the carrying of fire from floor to floor.

Informational Note: One method of defining fire-resistant characteristics capable of preventing the carrying of fire from floor to floor is that the cables pass the requirements of ANSI/UL 1666-2011, *Standard Test for Flame Propagation Height of Electrical and Optical-Fiber Cable Installed Vertically in Shafts.*

Table 830.154(a) Applications of Listed Network-Powered Broadband Cables in Buildings

Applications		Listed Network-Powered Broadband Cable Types						
		BLP	BLR	BL	BMR	BM	BLX	BMU, BLU
Inducts specifically fabricated for environmental air as described in 300.22(B)	In fabricated ducts as described in 300.22(B)	Y*	N	N	N	N	N	N
	In metal raceway that complies with 300.22(B)	Y*	Y*	Y*	Y*	Y*	Y*	N
In other spaces used for environmental air (plenums) as described in 300.22(C)	In other spaces used for environmental air as described in 300.22(C)	Y*	N	N	N	N	N	N
	In metal raceway that complies with 300.22(C)	Y*	Y*	Y*	Y*	Y*	Y*	N
	In plenum communications raceways	Y*	N	N	N	N	N	N
	In plenum cable routing assemblies	Y*	N	N	N	N	N	N
	Supported by open metal cable trays	Y*	N	N	N	N	N	N
	Supported by solid-bottom metal cable trays with solid metal covers	Y*	Y*	Y*	Y*	Y*	Y*	N
In risers	In vertical runs	Y*	Y*	N	Y*	N	N	N
	In metal raceways	Y*	Y*	Y*	Y*	Y*	Y*	N
	In fireproof shafts	Y*	Y*	Y*	Y*	Y*	Y*	N
	In plenum communications raceways	Y*	Y*	N	N	N	N	N
	In plenum cable routing assemblies	Y*	Y*	N	N	N	N	N
	In riser communications raceways	Y*	Y*	N	N	N	N	N
	In riser cable routing assemblies	Y*	Y*	N	N	N	N	N
	In one- and two-family dwellings	Y*	Y*	Y*	Y*	Y*	Y*	N
Within buildings in other than air-handling spaces and risers	General	Y*	Y*	Y*	Y*	Y*	Y*	N
	In one- and two-family dwellings	Y*	Y*	Y*	Y*	Y*	Y*	N
	Supported by cable trays	Y*	Y*	Y*	Y*	Y*	N	N
	In rigid metal conduit (RMC) and intermediate metal conduit (IMC)	Y*	Y*	Y*	Y*	Y*	Y*	Y*
	In any raceway recognized in Chapter 3	Y*	Y*	Y*	Y*	Y*	Y*	N
	In plenum communications raceways	Y*	Y*	Y*	N	N	N	N
	In plenum cable routing assemblies	Y*	Y*	Y*	N	N	N	N
	In riser communications raceways	Y*	Y*	Y*	N	N	N	N
	In riser cable routing assemblies	Y*	Y*	Y*	N	N	N	N
	In general-purpose communications raceways	Y*	Y*	Y*	N	N	N	N
	In general-purpose cable routing assemblies	Y*	Y*	Y*	N	N	N	N

Note: An "N" in the table indicates that the cable type shall not be permitted to be installed in the application. A "Y*" indicates that the cable type shall be permitted to be installed in the application subject to the limitations described in 830.113.

Informational Note No. 1: Part V of Article 830 covers installation methods within buildings. This table covers the applications of listed network-powered broadband communications cables in buildings. The definition of Point of Entrance is in 830.2.

Informational Note No. 2: For information on the restrictions to the installation of network-powered broadband communications cables in ducts specifically fabricated for environmental air, see 830.113(B).

Table 830.154(b) Cable Substitutions

Cable Type	Permitted Cable Substitutions
BM	BMR
BLP	CMP, CL3P
BLR	CMP, CL3P, CMR, CL3R, BLP, BMR
BL	CMP, CMR, CM, CMG, CL3P, CL3R, CL3, BMR, BM, BLP, BLR
BLX	CMP, CMR, CM, CMG, CMX, CL3P, CL3R, CL3, CL3X, BMR, BM, BLP, BRP, BL

(2) Type BM. Type BM cables shall be listed as being suitable for general-purpose use, with the exception of risers and plenums, and shall also be listed as being resistant to the spread of fire.

> Informational Note: One method of defining resistant to the spread of fire is that the cables do not spread fire to the top of the tray in the UL Flame Exposure, Vertical Tray Flame Test in ANSI/UL 1685-2010, *Standard for Safety for Vertical-Tray Fire-Propagation and Smoke-Release Test for Electrical and Optical-Fiber Cables.* The smoke measurements in the test method are not applicable.
>
> Another method of defining resistant to the spread of fire is for the damage (char length) not to exceed 1.5 m (4 ft 11 in.) when performing the CSA Vertical Flame Test — Cables in Cable Trays, as described in CSA C22.2 No. 0.3-09, *Test Methods for Electrical Wires and Cables.*

(3) Type BMU. Type BMU cables shall be jacketed and listed as being suitable for outdoor underground use.

(B) Network-Powered Broadband Communication Low-Power Cables. Network-powered broadband communications low-power cables shall be factory-assembled cables consisting of a jacketed coaxial cable, a jacketed combination of coaxial cable and multiple individual conductors, or a jacketed combination of an optical fiber cable and multiple individual conductors. The insulation for the individual conductors shall be rated for 300 volts minimum. Cables intended for outdoor use shall be listed as suitable for the application. Cables shall be marked in accordance with 310.120.

(1) Type BLP. Type BLP cables shall be listed as being suitable for use in ducts, plenums, and other spaces used for environmental air and shall also be listed as having adequate fire-resistant and low-smoke producing characteristics.

> Informational Note: One method of defining a cable that is low-smoke producing cable and fire-resistant cable is that the cable exhibits a maximum peak optical density of 0.50 or less, an average optical density of 0.15 or less, and a maximum flame spread distance of 1.52 m (5'ft) or less when tested in accordance with NFPA 262-2015, *Standard Method of Test for Flame Travel and Smoke of Wires and Cables for Use in Air-Handling Spaces.*

(2) Type BLR. Type BLR cables shall be listed as being suitable for use in a vertical run in a shaft, or from floor to floor, and shall also be listed as having fire-resistant characteristics capable of preventing the carrying of fire from floor to floor.

> Informational Note: One method of defining fire-resistant characteristics capable of preventing the carrying of fire from floor to floor is that the cables pass the requirements of ANSI/UL 1666-2011, *Standard Test for Flame Propagation Height of Electrical and Optical-Fiber Cable Installed Vertically in Shafts.*

(3) Type BL. Type BL cables shall be listed as being suitable for general-purpose use, with the exception of risers and

plenums, and shall also be listed as being resistant to the spread of fire.

> Informational Note: One method of defining resistant to the spread of fire is that the cables do not spread fire to the top of the tray in the UL Flame Exposure, Vertical Tray Flame Test in ANSI/UL 1685-2010, *Standard for Safety for Vertical-Tray Fire-Propagation and Smoke-Release Test for Electrical and Optical-Fiber Cables.* The smoke measurements in the test method are not applicable.
>
> Another method of defining resistant to the spread of fire is for the damage (char length) not to exceed 1.5 m (4 ft 11 in.) when performing the CSA Vertical Flame Test — Cables in Cable Trays, as described in CSA C22.2 No. 0.3-09, *Test Methods for Electrical Wires and Cables.*

(4) Type BLX. Type BLX limited-use cables shall be listed as being suitable for use outside, for use in dwellings, and for use in raceways and shall also be listed as being resistant to flame spread.

> Informational Note: One method of determining that cable is resistant to flame spread is by testing the cable to VW-1 (vertical-wire) flame test in ANSI/UL 1581-2011, *Reference Standard for Electrical Wires, Cables and Flexible Cords.*

(5) Type BLU. Type BLU cables shall be jacketed and listed as being suitable for outdoor underground use.

830.180 Grounding Devices. Where bonding or grounding is required, devices used to connect a shield, a sheath, or non-current-carrying metallic members of a cable to a bonding conductor, or grounding electrode conductor, shall be listed or be part of listed equipment.

ARTICLE 840
Premises-Powered Broadband Communications Systems

Part I. General

840.1 Scope. This article covers premises-powered broadband communications systems.

> Informational Note No. 1: A typical basic system configuration consists of an optical fiber, twisted pair, or coaxial cable to the premises supplying a broadband signal to a network terminal that converts the broadband signal into component electrical signals, such as traditional telephone, video, high-speed Internet, and interactive services. Powering for the network terminal and network devices is typically accomplished through a premises power supply that might be built into the network terminal or provided as a separate unit. In order to provide communications in the event of a power interruption, a battery backup unit or an uninterruptible power supply (UPS) is typically part of the powering system.
>
> Informational Note No. 2: See 90.2(B)(4) for installations of premises-powered broadband communications systems that are not covered in this article.

840.2 Definitions. The definitions in Part I of Article 100 and 645.2, 770.2, 800.2, and 820.2 shall apply. For purposes of this article, the following additional definitions apply.

Network Terminal. A device that converts network-provided signals (optical, electrical, or wireless) into component signals, including voice, audio, video, data, wireless, optical, and interactive services, and is considered a network device on the premises that is connected to a communications service provider and is powered at the premises.

Premises Communications Circuit. The circuit that extends voice, audio, video, data, interactive services, telegraph (except radio), and outside wiring for fire alarm and burglar alarm from the service provider's network terminal to the customer's communications equipment up to and including terminal equipment, such as a telephone, a fax machine, or an answering machine.

Premises Community Antenna Television (CATV) Circuit. The circuit that extends community antenna television (CATV) systems for audio, video, data, and interactive services from the service provider's network terminal to the appropriate customer equipment.

840.3 Other Articles.

(A) Hazardous (Classified) Locations. Premises-powered broadband communications circuits and equipment installed in a location that is classified in accordance with 500.5 and 505.5 shall comply with the applicable requirements of Chapter 5.

(B) Cables in Ducts for Dust, Loose Stock, or Vapor Removal. The requirements of 300.22(A), 770.3(B), 800.3(B), and 820.3(B) shall apply.

(C) Equipment in Other Space Used for Environmental Air. The requirements of 300.22(C)(3) shall apply.

(D) Installation and Use. The requirements of 110.3(B) shall apply.

(E) Output Circuits. As appropriate for the services provided, the output circuits derived from the network terminal shall comply with the requirements of the following:

(1) Installations of communications circuits — Part V of Article 800
(2) Installations of premises (within buildings) community antenna television and radio distribution circuits — Part V of Article 820
(3) Installations of optical fiber cables — Part V of Article 770
(4) Installations of Class 2 and Class 3 circuits — Part III of Article 725

> Informational Note: See 725.121 for information on the classification of information technology equipment circuits.

(5) Installations of power-limited fire alarm circuits — Part III of Article 760

N **(F) Other Communications Systems.** As appropriate for the system involved, traditional communications systems shall comply with the requirements of the following:

(1) Communications Circuits — Article 800
(2) Radio and Television Equipment — Article 810
(3) Community Antenna Television and Radio Distribution Systems — Article 820
(4) Network-Powered Broadband Communications Systems — Article 830

N **(G) Electrical Classification of Data Circuits and Cables.** Sections 725.139(D)(1) and 800.133(A)(1)(c) shall apply to the electrical classification of Class 2 and Class 3 circuits in the same cable with communications circuits.

840.21 Access to Electrical Equipment Behind Panels Designed to Allow Access. Access to electrical equipment shall not be denied by an accumulation of premises-powered broadband cables that prevents removal of panels, including suspended ceiling panels.

840.24 Mechanical Execution of Work. The requirements of 770.24, 800.24, and 820.24 shall apply.

840.25 Abandoned Cables. The requirements of 770.25, 800.25, and 820.25 shall apply.

840.26 Spread of Fire or Products of Combustion. The requirements of 770.26, 800.26, and 820.26 shall apply.

Part II. Cables Outside and Entering Buildings

840.44 Overhead (Aerial) Optical Fiber Cables. Overhead (aerial) optical fiber cables containing a non–current-carrying metallic member entering buildings shall comply with 840.44(A) and (B).

(A) On Poles and In-Span. Where outside plant optical fiber cables and electric light or power conductors are supported by the same pole or are run parallel to each other in-span, the conditions described in 840.44(A)(1) through (A)(4) shall be met.

(1) Relative Location. Where practicable, the outside plant optical fiber cables shall be located below the electric light or power conductors.

(2) Attachment to Cross-Arms. Attachment of outside plant optical fiber cables to a cross-arm that carries electric light or power conductors shall not be permitted.

(3) Climbing Space. The climbing space through outside plant optical fiber cables shall comply with the requirements of 225.14(D).

(4) Clearance. Supply service drops and sets of overhead service conductors of 0 to 750 volts running above and parallel to broadband communications service drops shall have a minimum separation of 300 mm (12 in.) at any point in the span, including the point of and at their attachment to the building. Clearance of not less than 1.0 m (40 in.) shall be maintained between the two services at the pole.

(B) Above Roofs. Outside plant optical fiber cables shall have a vertical clearance of not less than 2.5 m (8 ft) from all points of roofs above which they pass.

Exception No. 1: Vertical clearance requirements shall not apply to auxiliary buildings, such as garages and the like.

Exception No. 2: A reduction in clearance above only the overhanging portion of the roof, to not less than 450 mm (18 in.), shall be permitted if (a) not more than 1.2 m (4 ft) of premises-powered broadband communications service-drop cable passes above the roof overhang, and (b) the cable is terminated at a through- or above-the-roof raceway or approved support.

Exception No. 3: Where the roof has a slope of not less than 100 mm in 300 mm (4 in. in 12 in.), a reduction in clearance to not less than 900 mm (3 ft) shall be permitted.

Informational Note: For additional information regarding overhead wires and cables, see ANSI/IEEE C2-2012, *National Electrical Safety Code, Part 2, Safety Rules for Overhead Lines.*

N 840.45 Overhead (Aerial) Communications Wires and Cables. Section 800.44 shall apply to overhead (aerial) communications wires and multipair communications cables.

N 840.46 Overhead (Aerial) Coaxial Cables. Section 820.44 shall apply to overhead (aerial) coaxial cables.

840.47 Underground Wires and Cables Entering Buildings. Underground wires and cables entering buildings shall comply with 840.47(A) through (C).

(A) Optical Fiber Cables.

(1) Class 1 or Non–Power-Limited Fire Alarm Circuits. Underground conductive optical fiber cables entering buildings with electric light, power, Class 1, or non–power-limited fire alarm circuit conductors in a raceway, handhole enclosure, or manhole shall be located in a section separated from such conductors by means of brick, concrete, or tile partitions or by means of a suitable barrier.

(2) Direct-Buried Cables. Direct-buried conductive optical fiber cables shall be separated by at least 300 mm (12 in.) from conductors of any electric light, power, or non–power-limited fire alarm circuit conductors or Class 1 circuit.

Exception No. 1: Separation shall not be required where the electric service conductors are installed in raceways or have metal cable armor.

Exception No. 2: Separation shall not be required where the electric light or power branch-circuit or feeder conductors, non–power-limited fire alarm circuit conductors, or Class 1 circuit conductors are installed in a raceway or in metal-sheathed, metal-clad, or Type UF or Type USE cables.

(3) Mechanical Protection. Direct-buried cable, conduit, or other raceway shall be installed to have a minimum cover of 150 mm (6 in.).

N (B) Communications Wires and Cables. Installations of communications wires and multipair communications cables shall comply with 800.47.

N (C) Coaxial Cables. Installations of coaxial cables shall comply with 820.47.

840.48 Unlisted Wires and Cables Entering Buildings. Installations of unlisted cables entering buildings shall comply with 840.48(A), (B), or (C), as applicable.

N (A) Optical Fiber Cables. Installations of unlisted optical fiber cables entering buildings shall comply with 770.48.

N (B) Communications Wires and Cables. Installations of unlisted communications wires and unlisted multipair communications cables entering buildings shall comply with 800.48.

N (C) Coaxial Cables. Installations of unlisted coaxial cables entering buildings shall comply with 820.48.

840.49 Metallic Entrance Conduit Grounding. The requirements of 770.49, 800.49, and 820.49 shall apply, as applicable.

Part III. Protection

840.90 Protective Devices. The requirements of 800.90 shall apply.

840.93 Grounding or Interruption. Non–current-carrying metallic members of optical fiber cables, communications cables, or coaxial cables entering buildings or attaching to buildings shall comply with 840.93(A), (B), or (C), respectively.

(A) Non–Current-Carrying Metallic Members of Optical Fiber Cables. Non–current-carrying metallic members of optical fiber cables entering a building or terminating on the outside of a building shall comply with 770.93(A) or (B).

(B) Communications Cables. The grounding or interruption of the metallic sheath of communications cable shall comply with 800.93.

(C) Coaxial Cables. Where the network terminal is installed inside or outside of the building, with coaxial cables terminating at the network terminal, and is either entering, exiting, or attached to the outside of the building, 820.93 shall apply.

Part IV. Grounding Methods

840.100 Network Terminal and Cable Grounding. Grounding required for protection of the network terminal, conductive optical fiber cables, multipair communications cables, antenna lead-in conductors, and coaxial cables shall comply with 770.100, 800.100, 810.21 , or 820.100, as applicable.

840.101 Premises Circuits Not Leaving the Building. Where the network terminal is served by a nonconductive optical fiber cable, or where any non–current-carrying metallic member of a conductive optical fiber cable is interrupted by an insulating joint or equivalent device, and circuits that terminate at the network terminal are completely contained within the building (i.e., they do not exit the building), 840.101(A), (B), or (C) shall apply, as applicable.

(A) Coaxial Cable Shield Grounding. The shield of coaxial cable shall be grounded by one of the following:

(1) Any of the methods described in 820.100 or 820.106
(2) A fixed connection to an equipment grounding conductor as described in 250.118
(3) Connection to the network terminal grounding terminal provided that the terminal is connected to ground by one of the methods described in 820.100 or 820.106, or to an equipment grounding conductor through a listed grounding device that will retain the ground connection if the network terminal is unplugged

(B) Communications Circuit Grounding. Communications circuits shall not be required to be grounded.

(C) Network Terminal Grounding. The network terminal shall not be required to be grounded unless required by its listing. If the coaxial cable shield is separately grounded as described in 840.101(A)(1) or 840.101(A)(2), the use of a cord and plug for the connection to the network terminal grounding connection shall be permitted.

Informational Note: Where required to be grounded, a listed device that extends the equipment grounding conductor from the receptacle to the network terminal equipment grounding terminal is permitted. Sizing of the extended equipment grounding conductor is covered in Table 250.122.

840.106 Grounding and Bonding at Mobile Homes.

(A) Grounding. Grounding shall comply with (1) and (2).

(1) Where there is no mobile home service equipment located within 9.0 m (30 ft) of the exterior wall of the mobile home it serves, the non–current-carrying metallic members of optical fiber cables shall be connected to a grounding electrode in accordance with 770.106(A)(1). The network terminal, if required to be grounded, shall be connected to a grounding electrode in accordance with 800.106(A)(1). Premises CATV circuits shall be grounded in accordance with 820.106(A)(1), unless the network terminal is listed to provide the grounding path for the shield of the coaxial cable. The grounding electrode shall be bonded in accordance with 770.106(B).

(2) Where there is no mobile home disconnecting means grounded in accordance with 250.32 and located within 9.0 m (30 ft) of the exterior wall of the mobile home it serves, the non–current-carrying metallic members of optical fiber cables shall be connected to a grounding electrode in accordance with 770.106(A)(2). The network terminal, if required to be grounded, shall be connected to a grounding electrode in accordance with 800.106(A)(2). Premises CATV circuits shall be grounded in accordance with 820.106(A)(2), unless the network terminal is listed to provide the grounding path for the shield of the coaxial cable. The grounding electrode shall be bonded in accordance with 770.106(B).

(B) Bonding. The network terminal grounding terminal or grounding electrode shall be connected to the metal frame or available grounding terminal of the mobile home with a copper conductor not smaller than 12 AWG under any of the following conditions:

(1) Where there is no mobile home service equipment or disconnecting means as specified in 840.106(A)

(2) Where the mobile home is supplied by cord and plug.

Part V. Installation Methods Within Buildings

840.110 Raceways and Cable Routing Assemblies. Installations of raceways and cable routing assemblies for premises-powered broadband communications cables shall comply with 840.110(A)840.110(A), (B), or (C) as applicable.

N **(A) Optical Fiber Cables.** The requirements of 770.110 shall apply.

N **(B) Multipair Communications Cables.** The requirements of 800.110 shall apply.

N **(C) Coaxial Cables.** The requirements of 820.110 shall apply.

840.113 Installation on the Customer Premises Side of the Network Terminal. Installation of premises communications circuits and premises coaxial circuits shall comply with 840.113(A) or (B) as applicable.

(A) Premises Communications Circuits. Premises communications wires and multipair cables installed in a building from the network terminal shall be listed in accordance with 800.179, and the installation shall comply with 800.113 and 800.133.

(B) Premises Community Antenna Television (CATV) Circuits. Premises CATV coaxial cables installed in a building from the network terminal shall be listed in accordance with 820.179, and the installation shall comply with 820.113 and 820.133.

840.133 Installation of Optical Fibers and Electrical Conductors. The requirements of 770.133 shall apply.

840.154 Applications of Listed Optical Fiber Cables. The requirements of 770.154 shall apply.

N **Part VI. Premises Powering of Communications Equipment over Communications Cables**

840.160 Powering Circuits. Communications cables, in addition to carrying the communications circuit, shall also be permitted to carry circuits for powering communications equipment. Where the power supplied over a communications cable to communications equipment is greater than 60 watts, communication cables and the power circuit shall comply with 725.144 where communications cables are used in place of Class 2 and Class 3 cables.

Part VII. Listing Requirements

840.170 Equipment and Cables. Premises-powered broadband communications systems equipment and cables shall comply with 840.170(A) through (H).

(A) Network Terminal. The network terminal and applicable grounding means shall be listed for application with premises-powered broadband communications systems.

> Informational Note No. 1: One way to determine applicable requirements is to refer to ANSI/UL 60950-1-2014, *Standard for Safety of Information Technology Equipment;* ANSI/UL 498A-2015, *Current Taps and Adapters;* ANSI/UL 467-2013, *Grounding and Bonding Equipment;* or ANSI/UL 62368-1-2014, *Audio/Video, Information and Communication Technology Equipment – Part 1: Safety Requirements.*

> Informational Note No. 2: There are no requirements on the network terminal and its grounding methodologies except for those covered by the listing of the product.

(B) Optical Fiber Cables. Optical fiber cables shall be listed in accordance with 770.179(A) through (D) and shall be marked in accordance with Table 770.179.

(C) Communications Equipment. Communications equipment shall be listed in accordance with 800.170. Premises communications wires and cables connecting to the network terminal shall be listed in accordance with 800.179.

N **(D) Cable Routing Assemblies and Communications Raceways.** Cable routing assemblies and communications raceways shall be listed in accordance with 800.182.

N **(E) Premises Communications Wires and Cables.** Communications wires and cable shall be listed and marked in accordance with 800.179.

(F) Premises Community Antenna Television (CATV) Circuits. Premises community antenna television (CATV) coaxial cables connecting to the network terminal shall be listed in accordance with 820.179. Applicable grounding means shall be listed for application with premises-powered broadband communications systems.

N **(G) Power Source.** The power source for circuits intended to provide power over communications cables to remote equipment shall be limited in accordance with Table 11(B) in Chapter 9 for voltage sources up to 60 V dc and be listed as specified in either of the following:

(1) A power source shall be listed as specified in 725.121(A)(1), (A)(2), (A)(3), or (A)(4). The power sources shall not have the output connections paralleled or otherwise interconnected unless listed for such interconnection.

(2) A power source shall be listed as communications equipment for limited-power circuits.

Informational Note: One way to determine applicable requirements is to refer to ANSI/UL 60950-1-2014, *Standard for Safety of Information Technology Equipment-Safety — Part 1*; or ANSI/UL 62368-1-2014, *Audio/Video, Information and Communication Technology Equipment — Part 1: Safety Requirements*. Typically, such circuits are used to interconnect equipment for the purpose of exchanging information (data).

N **(H) Accessory Equipment.** Communications accessory equipment and/or assemblies shall be listed for application with premises-powered communications systems.

Informational Note: One way to determine applicable requirements is to refer to ANSI/UL 1863-2004, *Communications-Circuit Accessories*.

840.180 Grounding Devices. Where bonding or grounding is required, devices used to connect a shield, a sheath, or non–current-carrying metallic members of a cable to a bonding conductor, or grounding electrode conductor, shall be listed or be part of listed equipment.

Chapter 9 Tables

Table 1 Percent of Cross Section of Conduit and Tubing for Conductors and Cables

Number of Conductors and/or Cables	Cross-Sectional Area (%)
1	53
2	31
Over 2	40

Informational Note No. 1: Table 1 is based on common conditions of proper cabling and alignment of conductors where the length of the pull and the number of bends are within reasonable limits. It should be recognized that, for certain conditions, a larger size conduit or a lesser conduit fill should be considered.

Informational Note No. 2: When pulling three conductors or cables into a raceway, if the ratio of the raceway (inside diameter) to the conductor or cable (outside diameter) is between 2.8 and 3.2, jamming can occur. While jamming can occur when pulling four or more conductors or cables into a raceway, the probability is very low.

Notes to Tables

(1) See Informative Annex C for the maximum number of conductors and fixture wires, all of the same size (total cross-sectional area including insulation) permitted in trade sizes of the applicable conduit or tubing.

(2) Table 1 applies only to complete conduit or tubing systems and is not intended to apply to sections of conduit or tubing used to protect exposed wiring from physical damage.

(3) Equipment grounding or bonding conductors, where installed, shall be included when calculating conduit or tubing fill. The actual dimensions of the equipment grounding or bonding conductor (insulated or bare) shall be used in the calculation.

(4) Where conduit or tubing nipples having a maximum length not to exceed 600 mm (24 in.) are installed between boxes, cabinets, and similar enclosures, the nipples shall be permitted to be filled to 60 percent of

their total cross-sectional area, and 310.15(B)(3)(a) adjustment factors need not apply to this condition.

(5) For conductors not included in Chapter 9, such as multiconductor cables and optical fiber cables, the actual dimensions shall be used.

(6) For combinations of conductors of different sizes, use actual dimensions or Table 5 and Table 5A for dimensions of conductors and Table 4 for the applicable conduit or tubing dimensions.

(7) When calculating the maximum number of conductors or cables permitted in a conduit or tubing, all of the same size (total cross-sectional area including insulation), the next higher whole number shall be used to determine the maximum number of conductors permitted when the calculation results in a decimal greater than or equal to 0.8. When calculating the size for conduit or tubing permitted for a single conductor, one conductor shall be permitted when the calculation results in a decimal greater than or equal to 0.8.

(8) Where bare conductors are permitted by other sections of this Code, the dimensions for bare conductors in Table 8 shall be permitted.

(9) A multiconductor cable, optical fiber cable, or flexible cord of two or more conductors shall be treated as a single conductor for calculating percentage conduit or tubing fill area. For cables that have elliptical cross sections, the cross-sectional area calculation shall be based on using the major diameter of the ellipse as a circle diameter. Assemblies of single insulated conductors without an overall covering shall not be considered a cable when determining conduit or tubing fill area. The conduit or tubing fill for the assemblies shall be calculated based upon the individual conductors.

(10) The values for approximate conductor diameter and area shown in Table 5 are based on worst-case scenario and indicate round concentric-lay-stranded conductors. Solid and round concentric-lay-stranded conductor values are grouped together for the purpose of Table 5. Round compact-stranded conductor values are shown in Table 5A. If the actual values of the conductor diameter and area are known, they shall be permitted to be used.

Table 2 Radius of Conduit and Tubing Bends

Conduit or Tubing Size		One Shot and Full Shoe Benders		Other Bends	
Metric Designator	Trade Size	mm	in.	mm	in.
16	½	101.6	4	101.6	4
21	¾	114.3	4½	127	5
27	1	146.05	5¾	152.4	6
35	1¼	184.15	7¼	203.2	8
41	1½	209.55	8¼	254	10
53	2	241.3	9½	304.8	12
63	2½	266.7	10½	381	15
78	3	330.2	13	457.2	18
91	3½	381	15	533.4	21
103	4	406.4	16	609.6	24
129	5	609.6	24	762	30
155	6	762	30	914.4	36

Table 4 Dimensions and Percent Area of Conduit and Tubing (Areas of Conduit or Tubing for the Combinations of Wires Permitted in Table 1, Chapter 9)

Article 358 — Electrical Metallic Tubing (EMT)

Metric Designator	Trade Size	Over 2 Wires 40%		60%		1 Wire 53%		2 Wires 31%		Nominal Internal Diameter		Total Area 100%	
		mm²	in.²	mm²	in.²	mm²	in.²	mm²	in.²	mm	in.	mm²	in.²
16	½	78	0.122	118	0.182	104	0.161	61	0.094	15.8	0.622	196	0.304
21	¾	137	0.213	206	0.320	182	0.283	106	0.165	20.9	0.824	343	0.533
27	1	222	0.346	333	0.519	295	0.458	172	0.268	26.6	1.049	556	0.864
35	1¼	387	0.598	581	0.897	513	0.793	300	0.464	35.1	1.380	968	1.496
41	1½	526	0.814	788	1.221	696	1.079	407	0.631	40.9	1.610	1314	2.036
53	2	866	1.342	1299	2.013	1147	1.778	671	1.040	52.5	2.067	2165	3.356
63	2½	1513	2.343	2270	3.515	2005	3.105	1173	1.816	69.4	2.731	3783	5.858
78	3	2280	3.538	3421	5.307	3022	4.688	1767	2.742	85.2	3.356	5701	8.846
91	3½	2980	4.618	4471	6.927	3949	6.119	2310	3.579	97.4	3.834	7451	11.545
103	4	3808	5.901	5712	8.852	5046	7.819	2951	4.573	110.1	4.334	9521	14.753

Article 362 — Electrical Nonmetallic Tubing (ENT)

Metric Designator	Trade Size	Over 2 Wires 40%		60%		1 Wire 53%		2 Wires 31%		Nominal Internal Diameter		Total Area 100%	
		mm²	in.²	mm²	in.²	mm²	in.²	mm²	in.²	mm	in.	mm²	in.²
16	½	73	0.114	110	0.171	97	0.151	57	0.088	15.3	0.602	184	0.285
21	¾	131	0.203	197	0.305	174	0.269	102	0.157	20.4	0.804	328	0.508
27	1	215	0.333	322	0.499	284	0.441	166	0.258	26.1	1.029	537	0.832
35	1¼	375	0.581	562	0.872	497	0.770	291	0.450	34.5	1.36	937	1.453
41	1½	512	0.794	769	1.191	679	1.052	397	0.616	40.4	1.59	1281	1.986
53	2	849	1.316	1274	1.975	1125	1.744	658	1.020	52	2.047	2123	3.291
63	2½	—	—	—	—	—	—	—	—	—	—	—	—
78	3	—	—	—	—	—	—	—	—	—	—	—	—
91	3½	—	—	—	—	—	—	—	—	—	—	—	—

Article 348 — Flexible Metal Conduit (FMC)

Metric Designator	Trade Size	Over 2 Wires 40%		60%		1 Wire 53%		2 Wires 31%		Nominal Internal Diameter		Total Area 100%	
		mm²	in.²	mm²	in.²	mm²	in.²	mm²	in.²	mm	in.	mm²	in.²
12	⅜	30	0.046	44	0.069	39	0.061	23	0.036	9.7	0.384	74	0.116
16	½	81	0.127	122	0.190	108	0.168	63	0.098	16.1	0.635	204	0.317
21	¾	137	0.213	206	0.320	182	0.283	106	0.165	20.9	0.824	343	0.533
27	1	211	0.327	316	0.490	279	0.433	163	0.253	25.9	1.020	527	0.817
35	1¼	330	0.511	495	0.766	437	0.677	256	0.396	32.4	1.275	824	1.277
41	1½	480	0.743	720	1.115	636	0.985	372	0.576	39.1	1.538	1201	1.858
53	2	843	1.307	1264	1.961	1117	1.732	653	1.013	51.8	2.040	2107	3.269
63	2½	1267	1.963	1900	2.945	1678	2.602	982	1.522	63.5	2.500	3167	4.909
78	3	1824	2.827	2736	4.241	2417	3.746	1414	2.191	76.2	3.000	4560	7.069
91	3½	2483	3.848	3724	5.773	3290	5.099	1924	2.983	88.9	3.500	6207	9.621
103	4	3243	5.027	4864	7.540	4297	6.660	2513	3.896	101.6	4.000	8107	12.566

Article 342 — Intermediate Metal Conduit (IMC)

Metric Designator	Trade Size	Over 2 Wires 40%		60%		1 Wire 53%		2 Wires 31%		Nominal Internal Diameter		Total Area 100%	
		mm²	in.²	mm²	in.²	mm²	in.²	mm²	in.²	mm	in.	mm²	in.²
12	⅜	—	—	—	—	—	—	—	—	—	—	—	—
16	½	89	0.137	133	0.205	117	0.181	69	0.106	16.8	0.660	222	0.342
21	¾	151	0.235	226	0.352	200	0.311	117	0.182	21.9	0.864	377	0.586

(continues)

Article 342 — Intermediate Metal Conduit (IMC)

Metric Designator	Trade Size	Over 2 Wires 40%		60%		1 Wire 53%		2 Wires 31%		Nominal Internal Diameter		Total Area 100%	
		mm²	in.²	mm²	in.²	mm²	in.²	mm²	in.²	mm	in.	mm²	in.²
27	1	248	0.384	372	0.575	329	0.508	192	0.297	28.1	1.105	620	0.959
35	1¼	425	0.659	638	0.988	564	0.873	330	0.510	36.8	1.448	1064	1.647
41	1½	573	0.890	859	1.335	759	1.179	444	0.690	42.7	1.683	1432	2.225
53	2	937	1.452	1405	2.178	1241	1.924	726	1.125	54.6	2.150	2341	3.630
63	2½	1323	2.054	1985	3.081	1753	2.722	1026	1.592	64.9	2.557	3308	5.135
78	3	2046	3.169	3069	4.753	2711	4.199	1586	2.456	80.7	3.176	5115	7.922
91	3½	2729	4.234	4093	6.351	3616	5.610	2115	3.281	93.2	3.671	6822	10.584
103	4	3490	5.452	5235	8.179	4624	7.224	2705	4.226	105.4	4.166	8725	13.631

Article 356 — Liquidtight Flexible Nonmetallic Conduit (LFNC-A*)

Metric Designator	Trade Size	Over 2 Wires 40%		60%		1 Wire 53%		2 Wires 31%		Nominal Internal Diameter		Total Area 100%	
		mm²	in.²	mm²	in.²	mm²	in.²	mm²	in.²	mm	in.	mm²	in.²
12	⅜	50	0.077	75	0.115	66	0.102	39	0.060	12.6	0.495	125	0.192
16	½	80	0.125	121	0.187	107	0.165	62	0.097	16.0	0.630	201	0.312
21	¾	139	0.214	208	0.321	184	0.283	107	0.166	21.0	0.825	346	0.535
27	1	221	0.342	331	0.513	292	0.453	171	0.265	26.5	1.043	552	0.854
35	1¼	387	0.601	581	0.901	513	0.796	300	0.466	35.1	1.383	968	1.502
41	1½	520	0.807	781	1.211	690	1.070	403	0.626	40.7	1.603	1301	2.018
53	2	863	1.337	1294	2.006	1143	1.772	669	1.036	52.4	2.063	2157	3.343

*Corresponds to 356.2(1).

Article 356 — Liquidtight Flexible Nonmetallic Conduit (LFNC-B*)

Metric Designator	Trade Size	Over 2 Wires 40%		60%		1 Wire 53%		2 Wires 31%		Nominal Internal Diameter		Total Area 100%	
		mm²	in.²	mm²	in.²	mm²	in.²	mm²	in.²	mm	in.	mm²	in.²
12	⅜	49	0.077	74	0.115	65	0.102	38	0.059	12.5	0.494	123	0.192
16	½	81	0.125	122	0.188	108	0.166	63	0.097	16.1	0.632	204	0.314
21	¾	140	0.216	210	0.325	185	0.287	108	0.168	21.1	0.880	350	0.541
27	1	226	0.349	338	0.524	299	0.462	175	0.270	26.8	1.054	564	0.873
35	1¼	394	0.611	591	0.917	522	0.810	305	0.474	35.4	1.395	984	1.528
41	1½	510	0.792	765	1.188	676	1.050	395	0.614	40.3	1.588	1276	1.981
53	2	836	1.298	1255	1.948	1108	1.720	648	1.006	51.6	2.033	2091	3.246

*Corresponds to 356.2(2)

Article 356 — Liquidtight Flexible Nonmetallic Conduit (LFNC-C*)

Metric Designator	Trade Size	Over 2 Wires 40%		60%		1 Wire 53%		2 Wires 31%		Nominal Internal Diameter		Total Area 100%	
		mm²	in.²	mm²	in.²	mm²	in.²	mm²	in.²	mm	in.	mm²	in.²
12	⅜	47.7	0.074	71.5	0.111	63.2	0.098	36.9	0.057	12.3	0.485	119.19	0.185
16	½	77.9	0.121	116.9	0.181	103.2	0.160	60.4	0.094	15.7	0.620	194.778	0.302
21	¾	134.6	0.209	201.9	0.313	178.4	0.276	104.3	0.162	20.7	0.815	336.568	0.522
27	1	215.0	0.333	322.5	0.500	284.9	0.442	166.6	0.258	26.2	1.030	537.566	0.833
35	1¼	380.4	0.590	570.6	0.884	504.1	0.781	294.8	0.457	34.8	1.370	951.039	1.474
41	1½	509.2	0.789	763.8	1.184	674.7	1.046	394.6	0.612	40.3	1.585	1272.963	1.973
53	2	847.6	1.314	1271.4	1.971	1123.1	1.741	656.9	1.018	51.9	2.045	2119.063	3.285

*Corresponds to 356.2(3).

Article 350 — Liquidtight Flexible Metal Conduit (LFMC)

Metric Designator	Trade Size	Over 2 Wires 40%		60%		1 Wire 53%		2 Wires 31%		Nominal Internal Diameter		Total Area 100%	
		mm²	in.²	mm²	in.²	mm²	in.²	mm²	in.²	mm	in.	mm²	in.²
12	⅜	49	0.077	74	0.115	65	0.102	38	0.059	12.5	0.494	123	0.192
16	½	81	0.125	122	0.188	108	0.166	63	0.097	16.1	0.632	204	0.314
21	¾	140	0.216	210	0.325	185	0.287	108	0.168	21.1	0.830	350	0.541
27	1	226	0.349	338	0.524	299	0.462	175	0.270	26.8	1.054	564	0.873
35	1¼	394	0.611	591	0.917	522	0.810	305	0.474	35.4	1.395	984	1.528
41	1½	510	0.792	765	1.188	676	1.050	395	0.614	40.3	1.588	1276	1.981
53	2	836	1.298	1255	1.948	1108	1.720	648	1.006	51.6	2.033	2091	3.246
63	2½	1259	1.953	1888	2.929	1668	2.587	976	1.513	63.3	2.493	3147	4.881
78	3	1931	2.990	2896	4.485	2559	3.962	1497	2.317	78.4	3.085	4827	7.475
91	3½	2511	3.893	3766	5.839	3327	5.158	1946	3.017	89.4	3.520	6277	9.731
103	4	3275	5.077	4912	7.615	4339	6.727	2538	3.935	102.1	4.020	8187	12.692
129	5	—	—	—	—	—	—	—	—	—	—	—	—
155	6	—	—	—	—	—	—	—	—	—	—	—	—

Article 344 — Rigid Metal Conduit (RMC)

Metric Designator	Trade Size	Over 2 Wires 40%		60%		1 Wire 53%		2 Wires 31%		Nominal Internal Diameter		Total Area 100%	
		mm²	in.²	mm²	in.²	mm²	in.²	mm²	in.²	mm	in.	mm²	in.²
12	⅜	—	—	—	—	—	—	—	—	—	—	—	—
16	½	81	0.125	122	0.188	108	0.166	63	0.097	16.1	0.632	204	0.314
21	¾	141	0.220	212	0.329	187	0.291	109	0.170	21.2	0.836	353	0.549
27	1	229	0.355	344	0.532	303	0.470	177	0.275	27.0	1.063	573	0.887
35	1¼	394	0.610	591	0.916	522	0.809	305	0.473	35.4	1.394	984	1.526
41	1½	533	0.829	800	1.243	707	1.098	413	0.642	41.2	1.624	1333	2.071
53	2	879	1.363	1319	2.045	1165	1.806	681	1.056	52.9	2.083	2198	3.408
63	2½	1255	1.946	1882	2.919	1663	2.579	972	1.508	63.2	2.489	3137	4.866
78	3	1936	3.000	2904	4.499	2565	3.974	1500	2.325	78.5	3.090	4840	7.499
91	3½	2584	4.004	3877	6.006	3424	5.305	2003	3.103	90.7	3.570	6461	10.010
103	4	3326	5.153	4990	7.729	4408	6.828	2578	3.994	102.9	4.050	8316	12.882
129	5	5220	8.085	7830	12.127	6916	10.713	4045	6.266	128.9	5.073	13050	20.212
155	6	7528	11.663	11292	17.495	9975	15.454	5834	9.039	154.8	6.093	18821	29.158

Article 352 — Rigid PVC Conduit (PVC), Schedule 80

Metric Designator	Trade Size	Over 2 Wires 40%		60%		1 Wire 53%		2 Wires 31%		Nominal Internal Diameter		Total Area 100%	
		mm²	in.²	mm²	in.²	mm²	in.²	mm²	in.²	mm	in.	mm²	in.²
12	⅜	—	—	—	—	—	—	—	—	—	—	—	—
16	½	56	0.087	85	0.130	75	0.115	44	0.067	13.4	0.526	141	0.217
21	¾	105	0.164	158	0.246	139	0.217	82	0.127	18.3	0.722	263	0.409
27	1	178	0.275	267	0.413	236	0.365	138	0.213	23.8	0.936	445	0.688
35	1¼	320	0.495	480	0.742	424	0.656	248	0.383	31.9	1.255	799	1.237
41	1½	442	0.684	663	1.027	585	0.907	342	0.530	37.5	1.476	1104	1.711
53	2	742	1.150	1113	1.725	983	1.523	575	0.891	48.6	1.913	1855	2.874
63	2½	1064	1.647	1596	2.471	1410	2.183	825	1.277	58.2	2.290	2660	4.119
78	3	1660	2.577	2491	3.865	2200	3.414	1287	1.997	72.7	2.864	4151	6.442
91	3½	2243	3.475	3365	5.213	2972	4.605	1738	2.693	84.5	3.326	5608	8.688
103	4	2907	4.503	4361	6.755	3852	5.967	2253	3.490	96.2	3.786	7268	11.258
129	5	4607	7.142	6911	10.713	6105	9.463	3571	5.535	121.1	4.768	11518	17.855
155	6	6605	10.239	9908	15.359	8752	13.567	5119	7.935	145.0	5.709	16513	25.598

Articles 352 and 353 — Rigid PVC Conduit (PVC), Schedule 40, and HDPE Conduit (HDPE)

Metric Designator	Trade Size	Over 2 Wires 40%		60%		1 Wire 53%		2 Wires 31%		Nominal Internal Diameter		Total Area 100%	
		mm²	in.²	mm²	in.²	mm²	in.²	mm²	in.²	mm	in.	mm²	in.²
12	⅜	—	—	—	—	—	—	—	—	—	—	—	—
16	½	74	0.114	110	0.171	97	0.151	57	0.088	15.3	0.602	184	0.285
21	¾	131	0.203	196	0.305	173	0.269	101	0.157	20.4	0.804	327	0.508
27	1	214	0.333	321	0.499	284	0.441	166	0.258	26.1	1.029	535	0.832
35	1¼	374	0.581	561	0.872	495	0.770	290	0.450	34.5	1.360	935	1.453
41	1½	513	0.794	769	1.191	679	1.052	397	0.616	40.4	1.590	1282	1.986
53	2	849	1.316	1274	1.975	1126	1.744	658	1.020	52.0	2.047	2124	3.291
63	2½	1212	1.878	1817	2.817	1605	2.488	939	1.455	62.1	2.445	3029	4.695
78	3	1877	2.907	2816	4.361	2487	3.852	1455	2.253	77.3	3.042	4693	7.268
91	3½	2511	3.895	3766	5.842	3327	5.161	1946	3.018	89.4	3.521	6277	9.737
103	4	3237	5.022	4855	7.532	4288	6.654	2508	3.892	101.5	3.998	8091	12.554
129	5	5099	7.904	7649	11.856	6756	10.473	3952	6.126	127.4	5.016	12748	19.761
155	6	7373	11.427	11060	17.140	9770	15.141	5714	8.856	153.2	6.031	18433	28.567

Article 352 — Type A, Rigid PVC Conduit (PVC)

Metric Designator	Trade Size	Over 2 Wires 40%		60%		1 Wire 53%		2 Wires 31%		Nominal Internal Diameter		Total Area 100%	
		mm²	in.²	mm²	in.²	mm²	in.²	mm²	in.²	mm	in.	mm²	in.²
16	½	100	0.154	149	0.231	132	0.204	77	0.119	17.8	0.700	249	0.385
21	¾	168	0.260	251	0.390	222	0.345	130	0.202	23.1	0.910	419	0.650
27	1	279	0.434	418	0.651	370	0.575	216	0.336	29.8	1.175	697	1.084
35	1¼	456	0.707	684	1.060	604	0.937	353	0.548	38.1	1.500	1140	1.767
41	1½	600	0.929	900	1.394	795	1.231	465	0.720	43.7	1.720	1500	2.324
53	2	940	1.459	1410	2.188	1245	1.933	728	1.131	54.7	2.155	2350	3.647
63	2½	1406	2.181	2109	3.272	1863	2.890	1090	1.690	66.9	2.635	3515	5.453
78	3	2112	3.278	3169	4.916	2799	4.343	1637	2.540	82.0	3.230	5281	8.194
91	3½	2758	4.278	4137	6.416	3655	5.668	2138	3.315	93.7	3.690	6896	10.694
103	4	3543	5.489	5315	8.234	4695	7.273	2746	4.254	106.2	4.180	8858	13.723
129	5	—	—	—	—	—	—	—	—	—	—	—	—
155	6	—	—	—	—	—	—	—	—	—	—	—	—

Article 352 — Type EB, Rigid PVC Conduit (PVC)

Metric Designator	Trade Size	Over 2 Wires 40%		60%		1 Wire 53%		2 Wires 31%		Nominal Internal Diameter		Total Area 100%	
		mm²	in.²	mm²	in.²	mm²	in.²	mm²	in.²	mm	in.	mm²	in.²
16	½	—	—	—	—	—	—	—	—	—	—	—	—
21	¾	—	—	—	—	—	—	—	—	—	—	—	—
27	1	—	—	—	—	—	—	—	—	—	—	—	—
35	1¼	—	—	—	—	—	—	—	—	—	—	—	—
41	1½	—	—	—	—	—	—	—	—	—	—	—	—
53	2	999	1.550	1499	2.325	1324	2.053	774	1.201	56.4	2.221	2498	3.874
63	2½	—	—	—	—	—	—	—	—	—	—	—	—
78	3	2248	3.484	3373	5.226	2979	4.616	1743	2.700	84.6	3.330	5621	8.709
91	3½	2932	4.546	4397	6.819	3884	6.023	2272	3.523	96.6	3.804	7329	11.365
103	4	3726	5.779	5589	8.669	4937	7.657	2887	4.479	108.9	4.289	9314	14.448
129	5	5726	8.878	8588	13.317	7586	11.763	4437	6.881	135.0	5.316	14314	22.195
155	6	8133	12.612	12200	18.918	10776	16.711	6303	9.774	160.9	6.336	20333	31.530

Table 5 Dimensions of Insulated Conductors and Fixture Wires

Type	Size (AWG or kcmil)	Approximate Area		Approximate Diameter	
		mm²	in.²	mm	in.
Type: FFH-2, RFH-1, RFH-2, RFHH-2, RHH*, RHW* , RHW-2*, RHH, RHW, RHW-2, SF-1, SF-2, SFF-1, SFF-2, TF, TFF, THHW, THW, THW-2, TW, XF, XFF					
RFH-2, FFH-2, RFHH-2	18	9.355	0.0145	3.454	0.136
	16	11.10	0.0172	3.759	0.148
RHH, RHW, RHW-2	14	18.90	0.0293	4.902	0.193
	12	22.77	0.0353	5.385	0.212
	10	28.19	0.0437	5.994	0.236
	8	53.87	0.0835	8.280	0.326
	6	67.16	0.1041	9.246	0.364
	4	86.00	0.1333	10.46	0.412
	3	98.13	0.1521	11.18	0.440
	2	112.9	0.1750	11.99	0.472
	1	171.6	0.2660	14.78	0.582
	1/0	196.1	0.3039	15.80	0.622
	2/0	226.1	0.3505	16.97	0.668
	3/0	262.7	0.4072	18.29	0.720
	4/0	306.7	0.4754	19.76	0.778
	250	405.9	0.6291	22.73	0.895
	300	457.3	0.7088	24.13	0.950
	350	507.7	0.7870	25.43	1.001
	400	556.5	0.8626	26.62	1.048
	500	650.5	1.0082	28.78	1.133
	600	782.9	1.2135	31.57	1.243
	700	874.9	1.3561	33.38	1.314
	750	920.8	1.4272	34.24	1.348
	800	965.0	1.4957	35.05	1.380
	900	1057	1.6377	36.68	1.444
	1000	1143	1.7719	38.15	1.502
	1250	1515	2.3479	43.92	1.729
	1500	1738	2.6938	47.04	1.852
	1750	1959	3.0357	49.94	1.966
	2000	2175	3.3719	52.63	2.072
SF-2, SFF-2	18	7.419	0.0115	3.073	0.121
	16	8.968	0.0139	3.378	0.133
	14	11.10	0.0172	3.759	0.148
SF-1, SFF-1	18	4.194	0.0065	2.311	0.091
RFH-1,TF, TFF, XF, XFF	18	5.161	0.0088	2.692	0.106
TF, TFF, XF, XFF	16	7.032	0.0109	2.997	0.118
TW, XF, XFF, THHW, THW, THW-2	14	8.968	0.0139	3.378	0.133
TW, THHW, THW, THW-2	12	11.68	0.0181	3.861	0.152
	10	15.68	0.0243	4.470	0.176
	8	28.19	0.0437	5.994	0.236
RHH*, RHW*, RHW-2*	14	13.48	0.0209	4.140	0.163
RHH*, RHW*, RHW-2*, XF, XFF	12	16.77	0.0260	4.623	0.182
Type: RHH*, RHW*, RHW-2*, THHN, THHW, THW, THW-2, TFN, TFFN, THWN, THWN-2, XF, XFF					
RHH,* RHW,* RHW-2,* XF, XFF	10	21.48	0.0333	5.232	0.206

(continues)

Table 5 *Continued*

Type	Size (AWG or kcmil)	Approximate Area		Approximate Diameter	
		mm²	in.²	mm	in.
RHH*, RHW*, RHW-2*	8	35.87	0.0556	6.756	0.266
TW, THW, THHW, THW-2, RHH*, RHW*, RHW-2*	6	46.84	0.0726	7.722	0.304
	4	62.77	0.0973	8.941	0.352
	3	73.16	0.1134	9.652	0.380
	2	86.00	0.1333	10.46	0.412
	1	122.6	0.1901	12.50	0.492
	1/0	143.4	0.2223	13.51	0.532
	2/0	169.3	0.2624	14.68	0.578
	3/0	201.1	0.3117	16.00	0.630
	4/0	239.9	0.3718	17.48	0.688
	250	296.5	0.4596	19.43	0.765
	300	340.7	0.5281	20.83	0.820
	350	384.4	0.5958	22.12	0.871
	400	427.0	0.6619	23.32	0.918
	500	509.7	0.7901	25.48	1.003
	600	627.7	0.9729	28.27	1.113
	700	710.3	1.1010	30.07	1.184
	750	751.7	1.1652	30.94	1.218
	800	791.7	1.2272	31.75	1.250
	900	874.9	1.3561	33.38	1.314
	1000	953.8	1.4784	34.85	1.372
	1250	1200	1.8602	39.09	1.539
	1500	1400	2.1695	42.21	1.662
	1750	1598	2.4773	45.11	1.776
	2000	1795	2.7818	47.80	1.882
TFN, TFFN	18	3.548	0.0055	2.134	0.084
	16	4.645	0.0072	2.438	0.096
THHN, THWN, THWN-2	14	6.258	0.0097	2.819	0.111
	12	8.581	0.0133	3.302	0.130
	10	13.61	0.0211	4.166	0.164
	8	23.61	0.0366	5.486	0.216
	6	32.71	0.0507	6.452	0.254
	4	53.16	0.0824	8.230	0.324
	3	62.77	0.0973	8.941	0.352
	2	74.71	0.1158	9.754	0.384
	1	100.8	0.1562	11.33	0.446
	1/0	119.7	0.1855	12.34	0.486
	2/0	143.4	0.2223	13.51	0.532
	3/0	172.8	0.2679	14.83	0.584
	4/0	208.8	0.3237	16.31	0.642
	250	256.1	0.3970	18.06	0.711
	300	297.3	0.4608	19.46	0.766
Type: FEP, FEPB, PAF, PAFF, PF, PFA, PFAH, PFF, PGF, PGFF, PTF, PTFF, TFE, THHN, THWN, THWN-2, Z, ZF, ZFF, ZHF					
THHN, THWN, THWN-2	350	338.2	0.5242	20.75	0.817
	400	378.3	0.5863	21.95	0.864
	500	456.3	0.7073	24.10	0.949
	600	559.7	0.8676	26.70	1.051
	700	637.9	0.9887	28.50	1.122
	750	677.2	1.0496	29.36	1.156
	800	715.2	1.1085	30.18	1.188
	900	794.3	1.2311	31.80	1.252
	1000	869.5	1.3478	33.27	1.310

(continues)

Table 5   *Continued*

Type	Size (AWG or kcmil)	Approximate Area		Approximate Diameter	
		mm^2	in.^{2}	mm	in.
PF, PGFF, PGF, PFF, PTF, PAF, PTFF, PAFF	18	3.742	0.0058	2.184	0.086
	16	4.839	0.0075	2.489	0.098
PF, PGFF, PGF, PFF, PTF, PAF, PTFF, PAFF, TFE, FEP, PFA, FEPB, PFAH	14	6.452	0.0100	2.870	0.113
TFE, FEP, PFA, FEPB, PFAH	12	8.839	0.0137	3.353	0.132
	10	12.32	0.0191	3.962	0.156
	8	21.48	0.0333	5.232	0.206
	6	30.19	0.0468	6.198	0.244
	4	43.23	0.0670	7.417	0.292
	3	51.87	0.0804	8.128	0.320
	2	62.77	0.0973	8.941	0.352
TFE, PFAH, PFA	1	90.26	0.1399	10.72	0.422
TFE, PFA, PFAH, Z	1/0	108.1	0.1676	11.73	0.462
	2/0	130.8	0.2027	12.90	0.508
	3/0	158.9	0.2463	14.22	0.560
	4/0	193.5	0.3000	15.70	0.618
ZF, ZFF, ZHF	18	2.903	0.0045	1.930	0.076
	16	3.935	0.0061	2.235	0.088
Z, ZF, ZFF, ZHF	14	5.355	0.0083	2.616	0.103
Z	12	7.548	0.0117	3.099	0.122
	10	12.32	0.0191	3.962	0.156
	8	19.48	0.0302	4.978	0.196
	6	27.74	0.0430	5.944	0.234
	4	40.32	0.0625	7.163	0.282
	3	55.16	0.0855	8.382	0.330
	2	66.39	0.1029	9.195	0.362
	1	81.87	0.1269	10.21	0.402
Type: KF-1, KF-2, KFF-1, KFF-2, XHH, XHHW, XHHW-2, ZW					
XHHW, ZW, XHHW-2, XHH	14	8.968	0.0139	3.378	0.133
	12	11.68	0.0181	3.861	0.152
	10	15.68	0.0243	4.470	0.176
	8	28.19	0.0437	5.994	0.236
	6	38.06	0.0590	6.960	0.274
	4	52.52	0.0814	8.179	0.322
	3	62.06	0.0962	8.890	0.350
	2	73.94	0.1146	9.703	0.382

(continues)

Table 5 *Continued*

Type	Size (AWG or kcmil)	Approximate Area		Approximate Diameter	
		mm²	in.²	mm	in.
XHHW, XHHW-2, XHH	1	98.97	0.1534	11.23	0.442
	1/0	117.7	0.1825	12.24	0.482
	2/0	141.3	0.2190	13.41	0.528
	3/0	170.5	0.2642	14.73	0.58
	4/0	206.3	0.3197	16.21	0.638
	250	251.9	0.3904	17.91	0.705
	300	292.6	0.4536	19.30	0.76
	350	333.3	0.5166	20.60	0.811
	400	373.0	0.5782	21.79	0.858
	500	450.6	0.6984	23.95	0.943
	600	561.9	0.8709	26.75	1.053
	700	640.2	0.9923	28.55	1.124
	750	679.5	1.0532	29.41	1.158
	800	717.5	1.1122	30.23	1.190
	900	796.8	1.2351	31.85	1.254
	1000	872.2	1.3519	33.32	1.312
	1250	1108	1.7180	37.57	1.479
	1500	1300	2.0156	40.69	1.602
	1750	1492	2.3127	43.59	1.716
	2000	1682	2.6073	46.28	1.822
KF-2, KFF-2	18	2.000	0.003	1.575	0.062
	16	2.839	0.0043	1.88	0.074
	14	4.129	0.0064	2.286	0.090
	12	6.000	0.0092	2.743	0.108
	10	8.968	0.0139	3.378	0.133
KF-1, KFF-1	18	1.677	0.0026	1.448	0.057
	16	2.387	0.0037	1.753	0.069
	14	3.548	0.0055	2.134	0.084
	12	5.355	0.0083	2.616	0.103
	10	8.194	0.0127	3.226	0.127

*Types RHH, RHW, and RHW-2 without outer covering.

Table 5A Compact Copper and Aluminum Building Wire Nominal Dimensions* and Areas

Size (AWG or kcmil)	Bare Conductor Diameter		Types RHH**, RHW**, or USE				Types THW and THHW				Type THHN				Type XHHW				Size (AWG or kcmil)
			Approximate Diameter		Approximate Area		Approximate Diameter		Approximate Area		Approximate Diameter		Approximate Area		Approximate Diameter		Approximate Area		
	mm	in.	mm	in.	mm²	in.²	mm	in.	mm²	in.²	mm	in.	mm²	in.²	mm	in.	mm²	in.²	
8	3.404	0.134	6.604	0.260	34.25	0.0531	6.477	0.255	32.90	0.0510	—	—	—	—	5.690	0.224	25.42	0.0394	8
6	4.293	0.169	7.493	0.295	44.10	0.0683	7.366	0.290	42.58	0.0660	6.096	0.240	29.16	0.0452	6.604	0.260	34.19	0.0530	6
4	5.410	0.213	8.509	0.335	56.84	0.0881	8.509	0.335	56.84	0.0881	7.747	0.305	47.10	0.0730	7.747	0.305	47.10	0.0730	4
2	6.807	0.268	9.906	0.390	77.03	0.1194	9.906	0.390	77.03	0.1194	9.144	0.360	65.61	0.1017	9.144	0.360	65.61	0.1017	2
1	7.595	0.299	11.81	0.465	109.5	0.1698	11.81	0.465	109.5	0.1698	10.54	0.415	87.23	0.1352	10.54	0.415	87.23	0.1352	1
1/0	8.534	0.336	12.70	0.500	126.6	0.1963	12.70	0.500	126.6	0.1963	11.43	0.450	102.6	0.1590	11.43	0.450	102.6	0.1590	1/0
2/0	9.550	0.376	13.72	0.540	147.8	0.2290	13.84	0.545	150.5	0.2332	12.57	0.495	124.1	0.1924	12.45	0.490	121.6	0.1885	2/0
3/0	10.74	0.423	14.99	0.590	176.3	0.2733	14.99	0.590	176.3	0.2733	13.72	0.540	147.7	0.2290	13.72	0.540	147.7	0.2290	3/0
4/0	12.07	0.475	16.26	0.640	207.6	0.3217	16.38	0.645	210.8	0.3267	15.11	0.595	179.4	0.2780	14.99	0.590	176.3	0.2733	4/0
250	13.21	0.520	18.16	0.715	259.0	0.4015	18.42	0.725	266.3	0.4128	17.02	0.670	227.4	0.3525	16.76	0.660	220.7	0.3421	250
300	14.48	0.570	19.43	0.765	296.5	0.4596	19.69	0.775	304.3	0.4717	18.29	0.720	262.6	0.4071	18.16	0.715	259.0	0.4015	300
350	15.65	0.616	20.57	0.810	332.3	0.5153	20.83	0.820	340.7	0.5281	19.56	0.770	300.4	0.4656	19.30	0.760	292.6	0.4536	350
400	16.74	0.659	21.72	0.855	370.5	0.5741	21.97	0.865	379.1	0.5876	20.70	0.815	336.5	0.5216	20.32	0.800	324.3	0.5026	400
500	18.69	0.736	23.62	0.930	438.2	0.6793	23.88	0.940	447.7	0.6939	22.48	0.885	396.8	0.6151	22.35	0.880	392.4	0.6082	500
600	20.65	0.813	26.29	1.035	542.8	0.8413	26.67	1.050	558.6	0.8659	25.02	0.985	491.6	0.7620	24.89	0.980	486.6	0.7542	600
700	22.28	0.877	27.94	1.100	613.1	0.9503	28.19	1.110	624.3	0.9676	26.67	1.050	558.6	0.8659	26.67	1.050	558.6	0.8659	700
750	23.06	0.908	28.83	1.135	652.8	1.0118	29.21	1.150	670.1	1.0386	27.31	1.075	585.5	0.9076	27.69	1.090	602.0	0.9331	750
900	25.37	0.999	31.50	1.240	779.3	1.2076	31.09	1.224	759.1	1.1766	30.33	1.194	722.5	1.1196	29.69	1.169	692.3	1.0733	900
1000	26.92	1.060	32.64	1.285	836.6	1.2968	32.64	1.285	836.6	1.2968	31.88	1.255	798.1	1.2370	31.24	1.230	766.6	1.1882	1000

*Dimensions are from industry sources.

**Types RHH and RHW without outer coverings.

Table 8 Conductor Properties

Size (AWG or kcmil)	Area mm²	Area Circular mils	Stranding Quantity	Stranding Diameter mm	Stranding Diameter in.	Overall Diameter mm	Overall Diameter in.	Overall Area mm²	Overall Area in.²	Copper Uncoated ohm/km	Copper Uncoated ohm/kFT	Copper Coated ohm/km	Copper Coated ohm/kFT	Aluminum ohm/km	Aluminum ohm/kFT
18	0.823	1620	1	—	—	1.02	0.040	0.823	0.001	25.5	7.77	26.5	8.08	42.0	12.8
18	0.823	1620	7	0.39	0.015	1.16	0.046	1.06	0.002	26.1	7.95	27.7	8.45	42.8	13.1
16	1.31	2580	1	—	—	1.29	0.051	1.31	0.002	16.0	4.89	16.7	5.08	26.4	8.05
16	1.31	2580	7	0.49	0.019	1.46	0.058	1.68	0.003	16.4	4.99	17.3	5.29	26.9	8.21
14	2.08	4110	1	—	—	1.63	0.064	2.08	0.003	10.1	3.07	10.4	3.19	16.6	5.06
14	2.08	4110	7	0.62	0.024	1.85	0.073	2.68	0.004	10.3	3.14	10.7	3.26	16.9	5.17
12	3.31	6530	1	—	—	2.05	0.081	3.31	0.005	6.34	1.93	6.57	2.01	10.45	3.18
12	3.31	6530	7	0.78	0.030	2.32	0.092	4.25	0.006	6.50	1.98	6.73	2.05	10.69	3.25
10	5.261	10380	1	—	—	2.588	0.102	5.26	0.008	3.984	1.21	4.148	1.26	6.561	2.00
10	5.261	10380	7	0.98	0.038	2.95	0.116	6.76	0.011	4.070	1.24	4.226	1.29	6.679	2.04
8	8.367	16510	1	—	—	3.264	0.128	8.37	0.013	2.506	0.764	2.579	0.786	4.125	1.26
8	8.367	16510	7	1.23	0.049	3.71	0.146	10.76	0.017	2.551	0.778	2.653	0.809	4.204	1.28
6	13.30	26240	7	1.56	0.061	4.67	0.184	17.09	0.027	1.608	0.491	1.671	0.510	2.652	0.808
4	21.15	41740	7	1.96	0.077	5.89	0.232	27.19	0.042	1.010	0.308	1.053	0.321	1.666	0.508
3	26.67	52620	7	2.20	0.087	6.60	0.260	34.28	0.053	0.802	0.245	0.833	0.254	1.320	0.403
2	33.62	66360	7	2.47	0.097	7.42	0.292	43.23	0.067	0.634	0.194	0.661	0.201	1.045	0.319
1	42.41	83690	19	1.69	0.066	8.43	0.332	55.80	0.087	0.505	0.154	0.524	0.160	0.829	0.253
1/0	53.49	105600	19	1.89	0.074	9.45	0.372	70.41	0.109	0.399	0.122	0.415	0.127	0.660	0.201
2/0	67.43	133100	19	2.13	0.084	10.62	0.418	88.74	0.137	0.3170	0.0967	0.329	0.101	0.523	0.159
3/0	85.01	167800	19	2.39	0.094	11.94	0.470	111.9	0.173	0.2512	0.0766	0.2610	0.0797	0.413	0.126
4/0	107.2	211600	19	2.68	0.106	13.41	0.528	141.1	0.219	0.1996	0.0608	0.2050	0.0626	0.328	0.100
250	127	—	37	2.09	0.082	14.61	0.575	168	0.260	0.1687	0.0515	0.1753	0.0535	0.2778	0.0847
300	152	—	37	2.29	0.090	16.00	0.630	201	0.312	0.1409	0.0429	0.1463	0.0446	0.2318	0.0707
350	177	—	37	2.47	0.097	17.30	0.681	235	0.364	0.1205	0.0367	0.1252	0.0382	0.1984	0.0605
400	203	—	37	2.64	0.104	18.49	0.728	268	0.416	0.1053	0.0321	0.1084	0.0331	0.1737	0.0529
500	253	—	37	2.95	0.116	20.65	0.813	336	0.519	0.0845	0.0258	0.0869	0.0265	0.1391	0.0424
600	304	—	61	2.52	0.099	22.68	0.893	404	0.626	0.0704	0.0214	0.0732	0.0223	0.1159	0.0353
700	355	—	61	2.72	0.107	24.49	0.964	471	0.730	0.0603	0.0184	0.0622	0.0189	0.0994	0.0303
750	380	—	61	2.82	0.111	25.35	0.998	505	0.782	0.0563	0.0171	0.0579	0.0176	0.0927	0.0282
800	405	—	61	2.91	0.114	26.16	1.030	538	0.834	0.0528	0.0161	0.0544	0.0166	0.0868	0.0265
900	456	—	61	3.09	0.122	27.79	1.094	606	0.940	0.0470	0.0143	0.0481	0.0147	0.0770	0.0235
1000	507	—	61	3.25	0.128	29.26	1.152	673	1.042	0.0423	0.0129	0.0434	0.0132	0.0695	0.0212
1250	633	—	91	2.98	0.117	32.74	1.289	842	1.305	0.0338	0.0103	0.0347	0.0106	0.0554	0.0169
1500	760	—	91	3.26	0.128	35.86	1.412	1011	1.566	0.02814	0.00858	0.02814	0.00883	0.0464	0.0141
1750	887	—	127	2.98	0.117	38.76	1.526	1180	1.829	0.02410	0.00735	0.02410	0.00756	0.0397	0.0121
2000	1013	—	127	3.19	0.126	41.45	1.632	1349	2.092	0.02109	0.00643	0.02109	0.00662	0.0348	0.0106

Notes:

1. These resistance values are valid only for the parameters as given. Using conductors having coated strands, different stranding type, and, especially, other temperatures changes the resistance.

2. Equation for temperature change: $R_2 = R_1 [1 + \alpha (T_2 - 75)]$, where $\alpha_{cu} = 0.00323$, $\alpha_{AL} = 0.00330$ at 75°C.

3. Conductors with compact and compressed stranding have about 9 percent and 3 percent, respectively, smaller bare conductor diameters than those shown. See Table 5A for actual compact cable dimensions.

4. The IACS conductivities used: bare copper = 100%, aluminum = 61%.

5. Class B stranding is listed as well as solid for some sizes. Its overall diameter and area are those of its circumscribing circle.

Informational Note: The construction information is in accordance with NEMA WC/70-2009 or ANSI/UL 1581-2011. The resistance is calculated in accordance with National Bureau of Standards Handbook 100, dated 1966, and Handbook 109, dated 1972.

Table 9 Alternating-Current Resistance and Reactance for 600-Volt Cables, 3-Phase, 60 Hz, 75°C (167°F) — Three Single Conductors in Conduit

	Ohms to Neutral per Kilometer														
	Ohms to Neutral per 1000 Feet														
	X_L (Reactance) for All Wires		Alternating-Current Resistance for Uncoated Copper Wires			Alternating-Current Resistance for Aluminum Wires			Effective Z at 0.85 PF for Uncoated Copper Wires			Effective Z at 0.85 PF for Aluminum Wires			
Size (AWG or kcmil)	PVC, Aluminum Conduits	Steel Conduit	PVC Conduit	Aluminum Conduit	Steel Conduit	PVC Conduit	Aluminum Conduit	Steel Conduit	PVC Conduit	Aluminum Conduit	Steel Conduit	PVC Conduit	Aluminum Conduit	Steel Conduit	Size (AWG or kcmil)
14	0.190 0.058	0.240 0.073	10.2 3.1	10.2 3.1	10.2 3.1	— —	— —	— —	8.9 2.7	8.9 2.7	8.9 2.7	— —	— —	— —	14
12	0.177 0.054	0.223 0.068	6.6 2.0	6.6 2.0	6.6 2.0	10.5 3.2	10.5 3.2	10.5 3.2	5.6 1.7	5.6 1.7	5.6 1.7	9.2 2.8	9.2 2.8	9.2 2.8	12
10	0.164 0.050	0.207 0.063	3.9 1.2	3.9 1.2	3.9 1.2	6.6 2.0	6.6 2.0	6.6 2.0	3.6 1.1	3.6 1.1	3.6 1.1	5.9 1.8	5.9 1.8	5.9 1.8	10
8	0.171 0.052	0.213 0.065	2.56 0.78	2.56 0.78	2.56 0.78	4.3 1.3	4.3 1.3	4.3 1.3	2.26 0.69	2.26 0.69	2.30 0.70	3.6 1.1	3.6 1.1	3.6 1.1	8
6	0.167 0.051	0.210 0.064	1.61 0.49	1.61 0.49	1.61 0.49	2.66 0.81	2.66 0.81	2.66 0.81	1.44 0.44	1.48 0.45	1.48 0.45	2.33 0.71	2.36 0.72	2.36 0.72	6
4	0.157 0.048	0.197 0.060	1.02 0.31	1.02 0.31	1.02 0.31	1.67 0.51	1.67 0.51	1.67 0.51	0.95 0.29	0.95 0.29	0.98 0.30	1.51 0.46	1.51 0.46	1.51 0.46	4
3	0.154 0.047	0.194 0.059	0.82 0.25	0.82 0.25	0.82 0.25	1.31 0.40	1.35 0.41	1.31 0.40	0.75 0.23	0.79 0.24	0.79 0.24	1.21 0.37	1.21 0.37	1.21 0.37	3
2	0.148 0.045	0.187 0.057	0.62 0.19	0.66 0.20	0.66 0.20	1.05 0.32	1.05 0.32	1.05 0.32	0.62 0.19	0.62 0.19	0.66 0.20	0.98 0.30	0.98 0.30	0.98 0.30	2
1	0.151 0.046	0.187 0.057	0.49 0.15	0.52 0.16	0.52 0.16	0.82 0.25	0.85 0.26	0.82 0.25	0.52 0.16	0.52 0.16	0.52 0.16	0.79 0.24	0.79 0.24	0.82 0.25	1
1/0	0.144 0.044	0.180 0.055	0.39 0.12	0.43 0.13	0.39 0.12	0.66 0.20	0.69 0.21	0.66 0.20	0.43 0.13	0.43 0.13	0.43 0.13	0.62 0.19	0.66 0.20	0.66 0.20	1/0
2/0	0.141 0.043	0.177 0.054	0.33 0.10	0.33 0.10	0.33 0.10	0.52 0.16	0.52 0.16	0.52 0.16	0.36 0.11	0.36 0.11	0.36 0.11	0.52 0.16	0.52 0.16	0.52 0.16	2/0
3/0	0.138 0.042	0.171 0.052	0.253 0.077	0.269 0.082	0.259 0.079	0.43 0.13	0.43 0.13	0.43 0.13	0.289 0.088	0.302 0.092	0.308 0.094	0.43 0.13	0.43 0.13	0.46 0.14	3/0
4/0	0.135 0.041	0.167 0.051	0.203 0.062	0.220 0.067	0.207 0.063	0.33 0.10	0.36 0.11	0.33 0.10	0.243 0.074	0.256 0.078	0.262 0.080	0.36 0.11	0.36 0.11	0.36 0.11	4/0
250	0.135 0.041	0.171 0.052	0.171 0.052	0.187 0.057	0.177 0.054	0.279 0.085	0.295 0.090	0.282 0.086	0.217 0.066	0.230 0.070	0.240 0.073	0.308 0.094	0.322 0.098	0.33 0.10	250
300	0.135 0.041	0.167 0.051	0.144 0.044	0.161 0.049	0.148 0.045	0.233 0.071	0.249 0.076	0.236 0.072	0.194 0.059	0.207 0.063	0.213 0.065	0.269 0.082	0.282 0.086	0.289 0.088	300
350	0.131 0.040	0.164 0.050	0.125 0.038	0.141 0.043	0.128 0.039	0.200 0.061	0.217 0.066	0.207 0.063	0.174 0.053	0.190 0.058	0.197 0.060	0.240 0.073	0.253 0.077	0.262 0.080	350
400	0.131 0.040	0.161 0.049	0.108 0.033	0.125 0.038	0.115 0.035	0.177 0.054	0.194 0.059	0.180 0.055	0.161 0.049	0.174 0.053	0.184 0.056	0.217 0.066	0.233 0.071	0.240 0.073	400
500	0.128 0.039	0.157 0.048	0.089 0.027	0.105 0.032	0.095 0.029	0.141 0.043	0.157 0.048	0.148 0.045	0.141 0.043	0.157 0.048	0.164 0.050	0.187 0.057	0.200 0.061	0.210 0.064	500
600	0.128 0.039	0.157 0.048	0.075 0.023	0.092 0.028	0.082 0.025	0.118 0.036	0.135 0.041	0.125 0.038	0.131 0.040	0.144 0.044	0.154 0.047	0.167 0.051	0.180 0.055	0.190 0.058	600
750	0.125 0.038	0.157 0.048	0.062 0.019	0.079 0.024	0.069 0.021	0.095 0.029	0.112 0.034	0.102 0.031	0.118 0.036	0.131 0.040	0.141 0.043	0.148 0.045	0.161 0.049	0.171 0.052	750
1000	0.121 0.037	0.151 0.046	0.049 0.015	0.062 0.019	0.059 0.018	0.075 0.023	0.089 0.027	0.082 0.025	0.105 0.032	0.118 0.036	0.131 0.040	0.128 0.039	0.138 0.042	0.151 0.046	1000

Notes:

1. These values are based on the following constants: UL-Type RHH wires with Class B stranding, in cradled configuration. Wire conductivities are 100 percent IACS copper and 61 percent IACS aluminum, and aluminum conduit is 45 percent IACS. Capacitive reactance is ignored, since it is negligible at these voltages. These resistance values are valid only at 75°C (167°F) and for the parameters as given, but are representative for 600-volt wire types operating at 60 Hz.

2. *Effective Z* is defined as $R\cos(\theta) + X\sin(\theta)$, where θ is the power factor angle of the circuit. Multiplying current by effective impedance gives a good approximation for line-to-neutral voltage drop. Effective impedance values shown in this table are valid only at 0.85 power factor. For another circuit power factor (*PF*), effective impedance (*Ze*) can be calculated from R and X_L values given in this table as follows: $Ze = R \times PF + X_L\sin[\arccos(PF)]$.

Table 10 Conductor Stranding

Conductor Size		Number of Strands		
		Copper		Aluminum
AWG or kcmil	mm²	Class B[a]	Class C	Class B[a]
24–30	0.20–0.05	[b]	—	—
22	0.32	7	—	—
20	0.52	10	—	—
18	0.82	16	—	—
16	1.3	26	—	—
14–2	2.1–33.6	7	19	7[c]
1–4/0	42.4–107	19	37	19
250–500	127–253	37	61	37
600–1000	304–508	61	91	61
1250–1500	635–759	91	127	91
1750–2000	886–1016	127	271	127

[a]Conductors with a lesser number of strands shall be permitted based on an evaluation for connectability and bending.
[b]Number of strands vary.
[c]Aluminum 14 AWG (2.1 mm²) is not available.
With the permission of Underwriters Laboratories, Inc., material is reproduced from UL Standard 486A-B, Wire Connectors, which is copyrighted by Underwriters Laboratories, Inc., Northbrook, Illinois. While use of this material has been authorized, UL shall not be responsible for the manner in which the information is presented, nor for any interpretations thereof. For more information on UL or to purchase standards, please visit our Standards website at www.comm-2000.com or call 1-888-853-3503.

Table 11(A) and Table 11(B)

For listing purposes, Table 11(A) and Table 11(B) provide the required power source limitations for Class 2 and Class 3 power sources. Table 11(A) applies for alternating current sources, and Table 11(B) applies for direct-current sources.

The power for Class 2 and Class 3 circuits shall be either (1) inherently limited, requiring no overcurrent protection, or (2) not inherently limited, requiring a combination of power source and overcurrent protection. Power sources designed for interconnection shall be listed for the purpose.

As part of the listing, the Class 2 or Class 3 power source shall be durably marked where plainly visible to indicate the class of supply and its electrical rating. A Class 2 power source not suitable for wet location use shall be so marked.

Exception: Limited power circuits used by listed information technology equipment.

Overcurrent devices, where required, shall be located at the point where the conductor to be protected receives its supply and shall not be interchangeable with devices of higher ratings. The overcurrent device shall be permitted as an integral part of the power source.

Table 11(A) Class 2 and Class 3 Alternating-Current Power Source Limitations

Power Source		Inherently Limited Power Source (Overcurrent Protection Not Required)				Not Inherently Limited Power Source (Overcurrent Protection Required)			
		Class 2			Class 3	Class 2			Class 3
Source voltage V_{max} (volts) (see Note 1)		0 through 20*	Over 20 and through 30*	Over 30 and through 150	Over 30 and through 100	0 through 20*	Over 20 and through 30*	Over 30 and through 100	Over 100 and through 150
Power limitations VA_{max} (volt-amperes) (see Note 1)		—	—	—	—	250 (see Note 3)	250	250	N.A.
Current limitations I_{max} (amperes) (see Note 1)		8.0	8.0	0.005	$150/V_{max}$	$1000/V_{max}$	$1000/V_{max}$	$1000/V_{max}$	1.0
Maximum overcurrent protection (amperes)		—	—	—	—	5.0	$100/V_{max}$	$100/V_{max}$	1.0
Power source maximum nameplate rating	VA (volt-amperes)	$5.0 \times V_{max}$	100	$0.005 \times V_{max}$	100	$5.0 \times V_{max}$	100	100	100
	Current (amperes)	5.0	$100/V_{max}$	0.005	$100/V_{max}$	5.0	$100/V_{max}$	$100/V_{max}$	$100/V_{max}$

Note: Notes for this table can be found following Table 11(B).

*Voltage ranges shown are for sinusoidal ac in indoor locations or where wet contact is not likely to occur.

For nonsinusoidal or wet contact conditions, see Note 2.

Table 11(B) Class 2 and Class 3 Direct-Current Power Source Limitations

Power Source		Inherently Limited Power Source (Overcurrent Protection Not Required)					Not Inherently Limited Power Source (Overcurrent Protection Required)			
		Class 2				Class 3	Class 2			Class 3
Source voltage V_{max} (volts) (see Note 1)		0 through 20*	Over 20 and through 30*	Over 30 and through 60*	Over 60 and through 150	Over 60 and through 100	0 through 20*	Over 20 and through 60*	Over 60 and through 100	Over 100 and through 150
Power limitations VA_{max} (volt-amperes) (see Note 1)		—	—	—	—	—	250 (see Note 3)	250	250	N.A.
Current limitations I_{max} (amperes) (see Note 1)		8.0	8.0	$150/V_{max}$	0.005	$150/V_{max}$	$1000/V_{max}$	$1000/V_{max}$	$1000/V_{max}$	1.0
Maximum overcurrent protection (amperes)		—	—	—	—	—	5.0	$100/V_{max}$	$100/V_{max}$	1.0
Power source maximum nameplate rating	VA (volt-amperes)	$5.0 \times V_{max}$	100	100	$0.005 \times V_{max}$	100	$5.0 \times V_{max}$	100	100	100
	Current (amperes)	5.0	$100/V_{max}$	$100/V_{max}$	0.005	$100/V_{max}$	5.0	$100/V_{max}$	$100/V_{max}$	$100/V_{max}$

*Voltage ranges shown are for continuous dc in indoor locations or where wet contact is not likely to occur.

For interrupted dc or wet contact conditions, see Note 4.

Notes for Table 11(A) and Table 11(B)

1. V_{max}, I_{max}, and VA_{max} are determined with the current-limiting impedance in the circuit (not bypassed) as follows:

V_{max}: Maximum output voltage regardless of load with rated input applied.

I_{max}: Maximum output current under any noncapacitive load, including short circuit, and with overcurrent protection bypassed if used. Where a transformer limits the output current, I_{max} limits apply after 1 minute of operation. Where a current-limiting impedance, listed for the purpose, or as part of a listed product, is used in combination with a nonpower-limited transformer or a stored energy source, e.g., storage battery, to limit the output current, I_{max} limits apply after 5 seconds.

VA_{max}: Maximum volt-ampere output after 1 minute of operation regardless of load and overcurrent protection bypassed if used.

2. For nonsinusoidal ac, V_{max} shall not be greater than 42.4 volts peak. Where wet contact (immersion not included) is likely to occur, Class 3 wiring methods shall be used or V_{max} shall not be greater than 15 volts for sinusoidal ac and 21.2 volts peak for nonsinusoidal ac.

3. If the power source is a transformer, VA_{max} is 350 or less when V_{max} is 15 or less.

4. For dc interrupted at a rate of 10 to 200 Hz, V_{max} shall not be greater than 24.8 volts peak. Where wet contact (immersion not included) is likely to occur, Class 3 wiring methods shall be used, or V_{max} shall not be greater than 30 volts for continuous dc; 12.4 volts peak for dc that is interrupted at a rate of 10 to 200 Hz.

Table 12(A) and Table 12(B)

For listing purposes, Table 12(A) and Table 12(B) provide the required power source limitations for power-limited fire alarm sources. Table 12(A) applies for alternating-current sources, and Table 12(B) applies for direct-current sources. The power for power-limited fire alarm circuits shall be either (1) inherently limited, requiring no overcurrent protection, or (2) not inherently limited, requiring the power to be limited by a combination of power source and overcurrent protection.

As part of the listing, the PLFA power source shall be durably marked where plainly visible to indicate that it is a power-limited fire alarm power source. The overcurrent device, where required, shall be located at the point where the conductor to be protected receives its supply and shall not be interchangeable with devices of higher ratings. The overcurrent device shall be permitted as an integral part of the power source.

Table 12(A) PLFA Alternating-Current Power Source Limitations

Power Source		Inherently Limited Power Source (Overcurrent Protection Not Required)			Not Inherently Limited Power Source (Overcurrent Protection Required)		
Circuit voltage V_{max} (volts) (see Note 1)		0 through 20	Over 20 and through 30	Over 30 and through 100	0 through 20	Over 20 and through 100	Over 100 and through 150
Power limitations VA_{max} (volt-amperes) (see Note 1)		—	—	—	250 (see Note 2)	250	N.A.
Current limitations I_{max} (amperes) (see Note 1)		8.0	8.0	$150/V_{max}$	$1000/V_{max}$	$1000/V_{max}$	1.0
Maximum overcurrent protection (amperes)		—	—	—	5.0	$100/V_{max}$	1.0
Power source maximum nameplate ratings	VA (volt-amperes)	$5.0 \times V_{max}$	100	100	$5.0 \times V_{max}$	100	100
	Current (amperes)	5.0	$100/V_{max}$	$100/V_{max}$	5.0	$100/V_{max}$	$100/V_{max}$

Note: Notes for this table can be found following Table 12(B).

Table 12(B) PLFA Direct-Current Power Source Limitations

Power Source		Inherently Limited Power Source (Overcurrent Protection Not Required)			Not Inherently Limited Power Source (Overcurrent Protection Required)		
Circuit voltage V_{max} (volts) (see Note 1)		0 through 20	Over 20 and through 30	Over 30 and through 100	0 through 20	Over 20 and through 100	Over 100 and through 150
Power limitations VA_{max} (volt-amperes) (see Note 1)		—	—	—	250 (see Note 2)	250	N.A.
Current limitations I_{max} (amperes) (see Note 1)		8.0	8.0	$150/V_{max}$	$1000/V_{max}$	$1000/V_{max}$	1.0
Maximum overcurrent protection (amperes)		—	—	—	5.0	$100/V_{max}$	1.0
Power source maximum nameplate ratings	VA (volt-amperes)	$5.0 \times V_{max}$	100	100	$5.0 \times V_{max}$	100	100
	Current (amperes)	5.0	$100/V_{max}$	$100/V_{max}$	5.0	$100/V_{max}$	$100/V_{max}$

Notes for Table 12(A) and Table 12(B)

1. V_{max}, I_{max}, and VA_{max} are determined as follows:

V_{max}: Maximum output voltage regardless of load with rated input applied.

I_{max}: Maximum output current under any noncapacitive load, including short circuit, and with overcurrent protection bypassed if used. Where a transformer limits the output current, I_{max} limits apply after 1 minute of operation. Where a current-limiting impedance, listed for the purpose, is used in combination with a nonpower-limited transformer or a stored energy source, e.g., storage battery, to limit the output current, I_{max} limits apply after 5 seconds.

VA_{max}: Maximum volt-ampere output after 1 minute of operation regardless of load and overcurrent protection bypassed if used. Current limiting impedance shall not be bypassed when determining I_{max} and VA_{max}.

2. If the power source is a transformer, VA_{max} is 350 or less when V_{max} is 15 or less.

Informative Annex A Product Safety Standards

Informative Annex A is not a part of the requirements of this NFPA document but is included for informational purposes only.

This informative annex provides a list of product safety standards used for product listing where that listing is required by this *Code*. It is recognized that this list is current at the time of publication but that new standards or modifications to existing standards can occur at any time while this edition of the *Code* is in effect.

This informative annex does not form a mandatory part of the requirements of this *Code* but is intended only to provide *Code* users with informational guidance about the product characteristics about which *Code* requirements have been based.

Product Standard Name	Product Standard Number
Aboveground Reinforced Thermosetting Resin Conduit (RTRC) and Fittings	UL 2515
Adjustable Speed Electrical Power Drive Systems — Part 5-1: Safety Requirements — Electrical, Thermal and Energy	UL 61800-5-1
Antenna-Discharge Units	UL 452
Arc-Fault Circuit-Interrupters	UL 1699
Armored Cable	UL 4
Attachment Plugs and Receptacles	UL 498
Audio, Video and Similar Electronic Apparatus — Safety Requirements	UL 60065
Audio/Video, Information and Communication Technology Equipment — Part 1: Safety Requirements	UL 62368-1
Automatic Electrical Controls	UL 60730-1
Batteries for Use in Electric Vehicles	UL 2580
Batteries for Use in Light Electric Rail (LER) Applications and Stationary Applications	UL 1973
Belowground Reinforced Thermosetting Resin Conduit (RTRC) and Fittings	UL 2420
Bidirectional Electric Vehicle (EV) Charging System Equipment	UL Subject 9741
Busways	UL 857
Cables — Thermoplastic-Insulated Underground Feeder and Branch-Circuit Cables	UL 493
Cables — Thermoplastic-Insulated Wires and Cables	UL 83
Cables — Thermoset-Insulated Wires and Cables	UL 44
Cable and Cable Fittings for Use in Hazardous (Classified) Locations	UL 2225
Cable Routing Assemblies and Communications Raceways	UL 2024
Cables for Non–Power-Limited Fire-Alarm Circuits	UL 1425
Cables for Power-Limited Fire-Alarm Circuits	UL 1424
Capacitors	UL 810
Cellular Metal Floor Raceways and Fittings	UL 209
Circuit Breakers for Use in Communication Equipment	UL 489A
Circuit Integrity (CI) Cable — Fire Tests for Electrical Circuit Protective Systems	Subject 1724
Circuit Integrity (CI) Cable — Tests for Fire Resistive Cables	UL 2196
Class 2 Power Units	UL 1310
Communications-Circuit Accessories	UL 1863
Communications Cables	UL 444
Community-Antenna Television Cables	UL 1655
Concentrator Photovoltaic Modules and Assemblies	Subject 8703
Conduit, Tubing, and Cable Fittings	UL 514B
Connectors for Use in Photovoltaic Systems	Subject 6703
Cord Sets and Power-Supply Cords	UL 817
Cover Plates for Flush-Mounted Wiring Devices	UL 514D
Data-Processing Cable	UL 1690
Distributed Generation Wiring Harnesses	Subject 9703
Electric Duct Heaters	UL 1996
Electric Generators	UL 1004-4
Electric Heating Appliances	UL 499
Electric Sign Components	UL 879
Electric Signs	UL 48
Electric Spas, Equipment Assemblies, and Associated Equipment	UL 1563
Electric Vehicle (EV) Charging System Equipment	UL 2202
Electric Vehicle Supply Equipment	UL 2594
Electric Water Heaters for Pools and Tubs	UL 1261
Electrical Apparatus for Explosive Gas Atmospheres — Part 15: Type of Protection "n"	UL 60079-15
Electrical Apparatus for Use in Class I, Zone 1 Hazardous (Classified) Locations Type of Protection — Encapsulation "m"	UL 60079-18
Electrical Apparatus for Use in Zone 20, Zone 21, and Zone 22 Hazardous (Classified) Locations — Protection by Encapsulation "mD"	UL 61241-18
Electrical Apparatus for Use in Zone 21 and Zone 22 Hazardous (Classified) Locations — Protection by Enclosure "tD"	UL 61241-1
Electrical Apparatus for Use in Zone 20, Zone 21, and Zone 22 Hazardous (Classified) Locations — General Requirements	UL 61241-0

(continues)

Product Standard Name	Product Standard Number
Electrical Apparatus for Use in Zone 20, Zone 21, and Zone 22 Hazardous (Classified) Locations — Protection by Intrinsic Safety "iD"	UL 61241-11
Electrical Apparatus for Use in Zone 21 and Zone 22 Hazardous (Classified) Locations — Protection by Pressurization "pD"	UL 61241-2
Electrical Equipment for Measurement, Control, and Laboratory Use — Part 2-201: Particular Requirements for Control Equipment	UL 61010-2-201
Electrical Intermediate Metal Conduit — Steel	UL 1242
Electrical Metallic Tubing — Aluminum and Stainless Steel	UL 797A
Electrical Metallic Tubing — Steel	UL 797
Electrical Nonmetallic Tubing	UL 1653
Electrical Resistance Heat Tracing for Industrial Applications	IEEE 515
Electrical Rigid Metal Conduit — Steel	UL 6
Electric-Battery-Powered Industrial Trucks	UL 583
Electrochemical Capacitors	UL 810A
Emergency Lighting and Power Equipment	UL 924
Enclosed and Dead-Front Switches	UL 98
Enclosed and Dead-Front Switches for Use in Photovoltaic Systems	Subject 98B
Enclosures for Electrical Equipment, Non-Environmental Considerations	UL 50
Enclosures for Electrical Equipment, Environmental Considerations	UL 50E
Energy Management Equipment	UL 916
Energy Storage Systems and Equipment	UL 9540
Explosion-Proof and Dust-Ignition-Proof Electrical Equipment for Use in Hazardous (Classified) Locations	UL 1203
Explosive Gas Atmospheres — Part 0: Equipment- General requirements	UL 60079-0
Explosive Gas Atmospheres — Part 7: Increased safety "e"	UL 60079-7
Explosive Gas Atmospheres — Part 1: Type of protection – Flameproof "d"	UL 60079-1
Explosive Gas Atmospheres — Part 5: Type of protection – Powder filling "q"	UL 60079-5
Explosive Gas Atmospheres — Part 6: Type of protection – Oil immersion "o"	UL 60079-6
Fire Pump Controllers	UL 218
Fire Pump Motors	UL 1004-5
Fire Resistive Cables, Test for	UL 2196
Fixture Wire	UL 66
Flame Propagation Height of Electrical and Optical-Fiber Cables Installed Vertically in Shafts, Test for	UL 1666
Flat-Plate Photovoltaic Modules and Panels	UL 1703
Flexible Cords and Cables	UL 62
Flexible Lighting Products	UL 2388
Flexible Metal Conduit	UL 1
Fluorescent-Lamp Ballasts	UL 935
Gas and Vapor Detectors and Sensors	UL 2075
Gas-Burning Heating Appliances for Manufactured Homes and Recreational Vehicles	UL 307B
Gas-Tube-Sign Cable	UL 814
General-Use Snap Switches	UL 20
Ground-Fault Circuit-Interrupters	UL 943
Ground-Fault Sensing and Relaying Equipment	UL 1053
Grounding and Bonding Equipment	UL 467
Hardware for the Support of Conduit, Tubing and Cable	UL 2239
Heating and Cooling Equipment	UL 1995
High-Intensity-Discharge Lamp Ballasts	UL 1029
Household and Similar Electrical Appliances, Part 2: Particular Requirements for Electrical Heat Pumps, Air Conditioners and Dehumidifiers	UL 60335-2-40
Household and Similar Electrical Appliances, Part 2: Particular Requirements for Refrigerating Appliances, Ice-Cream Appliances, and Ice-makers	UL 60335-2-24
Household Refrigerators and Freezers	UL 250
Impedance Protected Motors	UL 1004-2
Industrial Battery Chargers	UL 1564
Industrial Control Equipment	UL 508
Industrial Control Panels	UL 508A
Information Technology Equipment Safety — Part 1: General Requirements	UL 60950-1
Information Technology Equipment Safety — Part 21: Remote Power Feeding	UL 60950-21
Information Technology Equipment Safety — Part 22: Equipment to be Installed Outdoors	UL 60950-22
Information Technology Equipment Safety — Part 23: Large Data Storage Equipment	UL 60950-23
Instrumentation Tray Cable	UL 2250
Insulated Multi-Pole Splicing Wire Connectors	UL 2459
Inverters, Converters, Controllers and Interconnection System Equipment for Use with Distributed Energy Resources	UL 1741
Isolated Power Systems Equipment	UL 1047
Junction Boxes for Swimming Pool Luminaires	UL 1241
Light Emitting Diode (LED) Equipment for Use in Lighting Products	UL 8750
Line Insolation Monitors	UL 1022

(continues)

Product Standard Name	Product Standard Number
Liquid Fuel-Burning Heating Appliances for Manufactured Homes and Recreational Vehicles	UL 307A
Liquid-Tight Flexible Nonmetallic Conduit	UL 1660
Liquid-Tight Flexible Metal Conduit	UL 360
Lithium Batteries	UL 1642
Low-Voltage Fuses — Fuses for Photovoltaic Systems	Subject 2579
Low-Voltage Fuses — Part 1: General Requirements	UL 248-1
Low-Voltage Fuses — Part 2: Class C Fuses	UL 248-2
Low-Voltage Fuses — Part 3: Class CA and CB Fuses	UL 248-3
Low-Voltage Fuses — Part 4: Class CC Fuses	UL 248-4
Low-Voltage Fuses — Part 5: Class G Fuses	UL 248-5
Low-Voltage Fuses — Part 6: Class H Non-Renewable Fuses	UL 248-6
Low-Voltage Fuses — Part 7: Class H Renewable Fuses	UL 248-7
Low-Voltage Fuses — Part 8: Class J Fuses	UL 248-8
Low-Voltage Fuses — Part 9: Class K Fuses	UL 248-9
Low-Voltage Fuses — Part 10: Class L Fuses	UL 249-10
Low-Voltage Fuses — Part 11: Plug Fuses	UL 248-11
Low-Voltage Fuses — Part 12: Class R Fuses	UL 248-12
Low-Voltage Fuses — Part 13: Semiconductor Fuses	UL 248–13
Low-Voltage Fuses — Part 14: Supplemental Fuses	UL 248–14
Low-Voltage Fuses — Part 15: Class T Fuses	UL 248-15
Low-Voltage Fuses — Part 16: Test Limiters	UL 248-16
Low-Voltage Landscape Lighting Systems	UL 1838
Low-Voltage Lighting Fixtures for Use in Recreational Vehicles	UL 234
Low-Voltage Lighting Systems	UL 2108
Low-Voltage Switchgear and Controlgear — Part 1: General Rules	UL 60947-1
Low-Voltage Switchgear and Controlgear — Part 4-1: Contactors and Motor-Starters — Electromechanical Contactors and Motor-Starters	UL 60947-4-1
Low-Voltage Switchgear and Controlgear — Part 4-2: Contactors and Motor-Starters — AC Semiconductor Motor Controllers and Starters	UL 60947-4-2
Low-Voltage Switchgear and Controlgear — Part 5-1: Control Circuit Devices and Switching Elements — Electromechanical Control Circuit Devices	UL 60947-5-1
Low-Voltage Switchgear and Controlgear — Part 5-2: Control Circuit Devices and Switching Elements — Proximity Switches	UL 60947-5-2
Low-Voltage Switchgear and Controlgear — Part 7-1: Ancillary Equipment — Terminal Blocks for Copper Conductors	UL 60947-7-1
Low-Voltage Switchgear and Controlgear — Part 7-2: Ancillary Equipment — Protective Conductor Terminal Blocks for Copper Conductors	UL 60947-7-2
Low-Voltage Switchgear and Controlgear — Part 7-3: Ancillary Equipment — Safety Requirements for Fuse Terminal Blocks	UL 60947-7-3
Low Voltage Transformers — Part 1: General Requirements	UL 5085-1
Low Voltage Transformers — Part 3: Class 2 and Class 3 Transformers	UL 5085-3
Luminaire Reflector Kits for Installation on Previously Installed Fluorescent Luminaires, Supplemental Requirements	UL 1598B
Luminaires	UL 1598
Machine-Tool Wires and Cables	UL 1063
Manufactured Wiring Systems	UL 183
Medical Electrical Equipment — Part 1: General Requirements for Safety	UL 60601–1
Medium-Voltage AC Contactors, Controllers, and Control Centers	UL 347
Medium-Voltage Power Cables	UL 1072
Metal-Clad Cables	UL 1569
Metallic Outlet Boxes	UL 514A
Mobile Home Pipe Heating Cable	Subject 1462
Modular Data Centers	UL Subject 2755
Molded-Case Circuit Breakers, Molded-Case Switches, and Circuit-Breaker Enclosures	UL 489
Molded-Case Circuit Breakers, Molded-Case Switches, and Circuit-Breaker Enclosures for Use with Photovoltaic (PV) Systems	Subject 489B
Molded-Case Circuit Breakers and Molded-Case Switches for Use with Wind Turbines	Subject 489C
Motor Control Centers	UL 845
Motor-Operated Appliances	UL 73
Multi-Pole Connectors for Use in Photovoltaic Systems	Subject 6703A
Neon Transformers and Power Supplies	UL 2161
Nonincendive Electrical Equipment for Use in Class I and II, Division 2 and Class III, Divisions 1 and 2 Hazardous (Classified) Locations	ANSI/ISA-12.12.01
Nonmetallic Outlet Boxes, Flush-Device Boxes, and Covers	UL 514C
Nonmetallic Surface Raceways and Fittings	UL 5A
Nonmetallic Underground Conduit with Conductors	UL 1990
Office Furnishings	UL 1286
Optical Fiber Cable	UL 1651
Panelboards	UL 67
Performance Requirements of Detectors for Flammable Gases	UL 60079-29-1

(continues)

Product Standard Name	Product Standard Number
Personnel Protection Systems for Electric Vehicle (EV) Supply Circuits: Part 1: General Requirements	UL 2231–1
Personnel Protection Systems for Electric Vehicle (EV) Supply Circuits: Part 2: Particular Requirements for Protection Devices for Use in Charging Systems	UL 2231–2
Photovoltaic DC Arc-Fault Circuit Protection	Subject 1699B
Photovoltaic Junction Boxes	Subject 3730
Photovoltaic Wire	UL 4703
Plugs, Receptacles and Couplers for Electrical Vehicles	UL 2251
Portable Electric Luminaires	UL 153
Portable Power-Distribution Equipment	UL 1640
Potting Compounds for Swimming Pool, Fountain, and Spa Equipment	UL 676A
Power Distribution Blocks	UL Subject 1953
Power Outlets	UL 231
Power Units Other Than Class 2	UL 1012
Power-Limited Circuit Cables	UL 13
Power Ventilators	UL 705
Professional Video and Audio Equipment	UL 1419
Programmable Controllers – Part 2: Equipment Requirements and Tests	UL 61131-2
Protectors for Coaxial Communications Circuits	UL 497C
Protectors for Data Communication and Fire Alarm Circuits	UL 497B
Protectors for Paired-Conductor Communications Circuits	UL 497
Reference Standard for Electrical Wires, Cables, and Flexible Cords	UL 1581
Requirements for Process Sealing Between Electrical Systems and Flammable or Combustible Process Fluids	ANSI/ISA-12.27.01
Residential Pipe Heating Cable	Subject 2049
Roof and Gutter De-Icing Cable Units	Subject 1588
Room Air Conditioners	UL 484
Rotating Electrical Machines — General Requirements	UL 1004-1
Safety of Power Converters for Use in Photovoltaic Power Systems — Part 1: General Requirements	UL 62109-1
Safety of Power Converters for Use in Photovoltaic Power Systems — Part 2: Particular Requirements for Inverters	UL 62109-2
Schedule 40, 80, Type EB and A Rigid PVC Conduit and Fittings	UL 651
Schedule 40 and 80 High Density Polyethylene (HDPE) Conduit	UL 651A
Sealed Wire Connector Systems	UL 486D
Seasonal and Holiday Decorative Products	UL 588
Secondary Protectors for Communications Circuits	UL 497A
Self-Ballasted Lamps and Lamp Adapters	UL 1993
Service Entrance Cables	UL 854
Smoke Detectors for Fire Alarm Signaling Systems	UL 268
Solar Trackers	Subject 3703
Solid State Overcurrent Protectors	UL 2367
Specialty Transformers	UL 506
Splicing Wire Connectors	UL 486C
Stage and Studio Luminaires and Connector Strips	UL 1573
Standby Batteries	UL 1989
Stationary Engine Generator Assemblies	UL 2200
Strut-Type Channel Raceways and Fittings	UL 5B
Supplemental Requirements for Extra-Heavy Wall Reinforced Thermosetting Resin Conduit (RTRC) and Fittings	UL 2515A
Surface Metal Raceways and Fittings	UL 5
Surface Raceways and Fittings for Use with Data, Signal and Control Circuits	UL 5C
Surge Arresters — Gapped Silicon-Carbide Surge Arresters for AC Power Circuits	IEEE C62.1
Surge Arresters — Metal-Oxide Surge Arresters for AC Power Circuits	IEEE C62.11
Surge Protective Devices	UL 1449
Swimming Pool Pumps, Filters, and Chlorinators	UL 1081
Switchboards	UL 891
Thermally Protected Motors	UL 1004-3
Transfer Switch Equipment	UL 1008
Underfloor Raceways and Fittings	UL 884
Underwater Luminaires and Submersible Junction Boxes	UL 676
Uninterruptible Power Systems	UL 1778
Vacuum Cleaners, Blower Cleaners, and Household Floor Finishing Machines	UL 1017
Waste Disposers	UL 430
Wind Turbine Generating Systems	Subject 6140
Wind Turbine Generating Systems — Large	UL 6141
Wind Turbine Generating Systems — Small	UL 6142
Wire Connectors	UL 486A, UL 486B
Wireways, Auxiliary Gutters, and Associated Fittings	UL 870

Informative Annex B   Application Information for Ampacity Calculation

This informative annex is not a part of the requirements of this NFPA document but is included for informational purposes only.

B.1 Equation Application Information. This informative annex provides application information for ampacities calculated under engineering supervision.

B.2 Typical Applications Covered by Tables. Typical ampacities for conductors rated 0 through 2000 volts are shown in Table B.310.15(B)(2)(1) through Table B.310.15(B)(2)(10). Table B.310.15(B)(2)(11) provides the adjustment factors for more than three current-carrying conductors in a raceway or cable with load diversity. Underground electrical duct bank configurations, as detailed in Figure B.310.15(B)(2)(3), Figure B.310.15(B)(2)(4), and Figure B.310.15(B)(2)(5), are utilized for conductors rated 0 through 5000 volts. In Figure B.310.15(B)(2)(2) through Figure B.310.15(B)(2)(5), where adjacent duct banks are used, a separation of 1.5 m (5 ft) between the centerlines of the closest ducts in each bank or 1.2 m (4 ft) between the extremities of the concrete envelopes is sufficient to prevent derating of the conductors due to mutual heating. These ampacities were calculated as detailed in the basic ampacity paper, AIEE Paper 57-660, *The Calculation of the Temperature Rise and Load Capability of Cable Systems*, by J. H. Neher and M. H. McGrath. For additional information concerning the application of these ampacities, see IEEE STD 835-1994, *Standard Power Cable Ampacity Tables*.

Typical values of thermal resistivity (Rho) are as follows:

Average soil (90 percent of USA) = 90

Concrete = 55

Damp soil (coastal areas, high water table) = 60

Paper insulation = 550

Polyethylene (PE) = 450

Polyvinyl chloride (PVC) = 650

Rubber and rubber-like = 500

Very dry soil (rocky or sandy) = 120

Thermal resistivity, as used in this informative annex, refers to the heat transfer capability through a substance by conduction. It is the reciprocal of thermal conductivity and is normally expressed in the units°C-cm/watt. For additional information on determining soil thermal resistivity (Rho), see ANSI/IEEE STD 442-1996, *Guide for Soil Thermal Resistivity Measurements*.

B.3 Criteria Modifications. Where values of load factor and Rho are known for a particular electrical duct bank installation and they are different from those shown in a specific table or figure, the ampacities shown in the table or figure can be modified by the application of factors derived from the use of Figure B.310.15(B)(2)(1).

Where two different ampacities apply to adjacent portions of a circuit, the higher ampacity can be used beyond the point of transition, a distance equal to 3 m (10 ft) or 10 percent of the circuit length calculated at the higher ampacity, whichever is less.

Where the burial depth of direct burial or electrical duct bank circuits are modified from the values shown in a figure or table, ampacities can be modified as shown in (a) and (b) as follows.

(a) Where burial depths are increased in part(s) of an electrical duct run to avoid underground obstructions, no decrease in ampacity of the conductors is needed, provided the total length of parts of the duct run increased in depth to avoid obstructions is less than 25 percent of the total run length.

(b) Where burial depths are deeper than shown in a specific underground ampacity table or figure, an ampacity derating factor of 6 percent per increased 300 mm (foot) of depth for all values of Rho can be utilized. No rating change is needed where the burial depth is decreased.

B.4 Electrical Ducts. The term *electrical duct(s)* is defined in 310.60.

B.5 Table B.310.15(B)(2)(6) and Table B.310.15(B)(2)(7).

(a) To obtain the ampacity of cables installed in two electrical ducts in one horizontal row with 190-mm (7.5-in.) center-to-center spacing between electrical ducts, similar to Figure B.310.15(B)(2)(2), Detail 1, multiply the ampacity shown for one duct in Table B.310.15(B)(2)(6) and Table B.310.15(B)(2)(7) by 0.88.

(b) To obtain the ampacity of cables installed in four electrical ducts in one horizontal row with 190-mm (7.5-in.) center-to-center spacing between electrical ducts, similar to Figure B.310.15(B)(2)(2), Detail 2, multiply the ampacity shown for three electrical ducts in Table B.310.15(B)(2)(6) and Table B.310.15(B)(2)(7) by 0.94.

B.6 Electrical Ducts Used in Figure B.310.15(B)(2)(2). If spacing between electrical ducts, as shown in Figure B.310.15(B)(2)(2), is less than as specified where electrical ducts enter equipment enclosures from underground, the ampacity of conductors contained within such electrical ducts need not be reduced.

B.7 Examples Showing Use of Figure B.310.15(B)(2)(1) for Electrical Duct Bank Ampacity Modifications. Figure B.310.15(B)(2)(1) is used for interpolation or extrapolation for values of Rho and load factor for cables installed in electrical ducts. The upper family of curves shows the variation in ampacity and Rho at unity load factor in terms of I_1, the ampacity for Rho = 60, and 50 percent load factor. Each curve is designated for a particular ratio I_2/I_1, where I_2 is the ampacity at Rho = 120 and 100 percent load factor.

The lower family of curves shows the relationship between Rho and load factor that will give substantially the same ampacity as the indicated value of Rho at 100 percent load factor.

As an example, to find the ampacity of a 500-kcmil copper cable circuit for six electrical ducts as shown in Table B.310.15(B)(2)(5): At the Rho = 60, LF = 50, $I_1 = 583$; for Rho = 120 and LF = 100, $I_2 = 400$. The ratio $I_2/I_1 = 0.686$. Locate Rho = 90 at the bottom of the chart and follow the 90 Rho line to the intersection with 100 percent load factor where the equivalent Rho = 90. Then follow the 90 Rho line to I_2/I_1 ratio of 0.686 where $F = 0.74$. The desired ampacity = 0.74×583 = 431, which agrees with the table for Rho = 90, LF = 100.

To determine the ampacity for the same circuit where Rho = 80 and LF = 75, using Figure B.310.15(B)(2)(1), the equivalent Rho = 43, F = 0.855, and the desired ampacity = 0.855 × 583 = 498 amperes. Values for using Figure B.310.15(B)(2)(1) are found in the electrical duct bank ampacity tables of this informative annex.

Where the load factor is less than 100 percent and can be verified by measurement or calculation, the ampacity of electrical duct bank installations can be modified as shown. Different values of Rho can be accommodated in the same manner.

Informational Note: The ampacity limit for 10 through 85 current-carrying conductors is based on the following equation. For more than 85 conductors, special calculations are required that are beyond the scope of this table.

[B.310.15(B)(7)a]

$$A_2 = \left[\sqrt{\frac{0.5N}{E}} \times (A_1) \right] \text{ or } A_1, \text{ whichever is less}$$

where:

A_1	=	ampacity from Table 310.15(B)(16), Table 310.15(B)(18), Table B.310.15(B)(2)(1), Table B.310.15(B)(2)(6), or Table B.310.15(B)(2)(7) multiplied by the appropriate adjustment factor from Table B.310.15(B)(2)(11).
N	=	total number of conductors used to select adjustment factor from Table B.310.15(B)(2)(11)
E	=	number of conductors carrying current simultaneously in the raceway or cable
A_2	=	ampacity limit for the current-carrying conductors in the raceway or cable

Example 1

Calculate the ampacity limit for twelve 14 AWG THWN current-carrying conductors (75°C) in a raceway that contains 24 conductors that may, at different times, be current-carrying.

[B.310.15(B)(7)b]

$$A_2 = \sqrt{\frac{(0.5)(24)}{12}} \times 20(0.7)$$

$$= 14 \text{ amperes } \left(\text{i.e., 50 percent diversity} \right)$$

Example 2

Calculate the ampacity limit for eighteen 14 AWG THWN current-carrying conductors (75°C) in a raceway that contains 24 conductors that may, at different times, be current-carrying.

[B.310.15(B)(7)c]

$$A_2 = \sqrt{\frac{(0.5)(24)}{18}} \times 20(0.7) = 11.5 \text{ amperes}$$

Table B.310.15(B)(2)(1) Ampacities of Two or Three Insulated Conductors, Rated 0 Through 2000 Volts, Within an Overall Covering (Multiconductor Cable), in Raceway in Free Air Based on Ambient Air Temperature of 30°C (86°F)*

Size (AWG or kcmil)	Temperature Rating of Conductor. [See Table 310.104(A).]						Size (AWG or kcmil)
	60°C (140°F)	75°C (167°F)	90°C (194°F)	60°C (140°F)	75°C (167°F)	90°C (194°F)	
	Types TW, UF	Types RHW, THHW, THW, THWN, XHHW, ZW	Types THHN, THHW, THW-2, THWN-2, RHH, RWH-2, USE-2, XHHW, XHHW-2, ZW-2	Type TW	Types RHW, THHW, THW, THWN, XHHW	Types THHN, THHW, THW-2, THWN-2, RHH, RWH-2, USE-2, XHHW, XHHW-2, ZW-2	
	COPPER			ALUMINUM OR COPPER-CLAD ALUMINUM			
14**	16	18	21	—	—	—	14
12**	20	24	27	16	18	21	12
10**	27	33	36	21	25	28	10
8	36	43	48	28	33	37	8
6	48	58	65	38	45	51	6
4	66	79	89	51	61	69	4
3	76	90	102	59	70	79	3
2	88	105	119	69	83	93	2
1	102	121	137	80	95	106	1
1/0	121	145	163	94	113	127	1/0
2/0	138	166	186	108	129	146	2/0
3/0	158	189	214	124	147	167	3/0
4/0	187	223	253	147	176	197	4/0
250	205	245	276	160	192	217	250
300	234	281	317	185	221	250	300
350	255	305	345	202	242	273	350
400	274	328	371	218	261	295	400
500	315	378	427	254	303	342	500
600	343	413	468	279	335	378	600
700	376	452	514	310	371	420	700
750	387	466	529	321	384	435	750
800	397	479	543	331	397	450	800
900	415	500	570	350	421	477	900
1000	448	542	617	382	460	521	1000

*Refer to 310.15(B)(2) for the ampacity correction factors where the ambient temperature is other than 30°C (86°F).

**Refer to 240.4(D) for conductor overcurrent protection limitations.

Table B.310.15(B)(2)(3) Ampacities of Multiconductor Cables with Not More Than Three Insulated Conductors, Rated 0 Through 2000 Volts, in Free Air Based on Ambient Air Temperature of 40°C (104°F) (for Types TC, MC, MI, UF, and USE Cables)*

Size (AWG or kcmil)	Temperature Rating of Conductor. [See Table 310.104(A).]								Size (AWG or kcmil)
	60°C (140°F)	75°C (167°F)	85°C (185°F)	90°C (194°F)	60°C (140°F)	75°C (167°F)	85°C (185°F)	90°C (194°F)	
	COPPER				ALUMINUM OR COPPER-CLAD ALUMINUM				
18	—	—	—	11	—	—	—	—	18
16	—	—	—	16	—	—	—	—	16
14**	18	21	24	25	—	—	—	—	14
12**	21	28	30	32	18	21	24	25	12
10**	28	36	41	43	21	28	30	32	10
8	39	50	56	59	30	39	44	46	8
6	52	68	75	79	41	53	59	61	6
4	69	89	100	104	54	70	78	81	4
3	81	104	116	121	63	81	91	95	3
2	92	118	132	138	72	92	103	108	2
1	107	138	154	161	84	108	120	126	1
1/0	124	160	178	186	97	125	139	145	1/0
2/0	143	184	206	215	111	144	160	168	2/0
3/0	165	213	238	249	129	166	185	194	3/0
4/0	190	245	274	287	149	192	214	224	4/0
250	212	274	305	320	166	214	239	250	250
300	237	306	341	357	186	240	268	280	300
350	261	337	377	394	205	265	296	309	350
400	281	363	406	425	222	287	317	334	400
500	321	416	465	487	255	330	368	385	500
600	354	459	513	538	284	368	410	429	600
700	387	502	562	589	306	405	462	473	700
750	404	523	586	615	328	424	473	495	750
800	415	539	604	633	339	439	490	513	800
900	438	570	639	670	362	469	514	548	900
1000	461	601	674	707	385	499	558	584	1000

*Refer to 310.15(B)(2) for the ampacity correction factors where the ambient temperature is other than 40°C (104°F).
**Refer to 240.4(D) for conductor overcurrent protection limitations.

Table B.310.15(B)(2)(5) Ampacities of Single Insulated Conductors, Rated 0 Through 2000 Volts, in Nonmagnetic Underground Electrical Ducts (One Conductor per Electrical Duct), Based on Ambient Earth Temperature of 20°C (68°F), Electrical Duct Arrangement in Accordance with Figure B.310.15(B)(2)(2), Conductor Temperature 75°C (167°F)

Size (kcmil)	3 Electrical Ducts [Fig. B.310.15(B)(2)(2), Detail 2] Types RHW, THHW, THW, THWN, XHHW, USE			6 Electrical Ducts [Fig. B.310.15(B)(2)(2), Detail 3] Types RHW, THHW, THW, THWN, XHHW, USE			9 Electrical Ducts [Fig. B.310.15(B)(2)(2), Detail 4] Types RHW, THHW, THW, THWN, XHHW, USE			3 Electrical Ducts [Fig. B.310.15(B)(2)(2), Detail 2] Types RHW, THHW, THW, THWN, XHHW, USE			6 Electrical Ducts [Fig. B.310.15(B)(2)(2), Detail 3] Types RHW, THHW, THW, THWN, XHHW, USE			9 Electrical Ducts [Fig. B.310.15(B)(2)(2), Detail 4] Types RHW, THHW, THW, THWN, XHHW, USE			Size (kcmil)
	COPPER									ALUMINUM OR COPPER-CLAD ALUMINUM									
	RHO 60 LF 50	RHO 90 LF 100	RHO 120 LF 100	RHO 60 LF 50	RHO 90 LF 100	RHO 120 LF 100	RHO 60 LF 50	RHO 90 LF 100	RHO 120 LF 100	RHO 60 LF 50	RHO 90 LF 100	RHO 120 LF 100	RHO 60 LF 50	RHO 90 LF 100	RHO 120 LF 100	RHO 60 LF 50	RHO 90 LF 100	RHO 120 LF 100	
250	410	344	327	386	295	275	369	270	252	320	269	256	302	230	214	288	211	197	250
350	503	418	396	472	355	330	446	322	299	393	327	310	369	277	258	350	252	235	350
500	624	511	484	583	431	400	545	387	360	489	401	379	457	337	313	430	305	284	500
750	794	640	603	736	534	494	674	469	434	626	505	475	581	421	389	538	375	347	750
1000	936	745	700	864	617	570	776	533	493	744	593	557	687	491	453	629	432	399	1000
1250	1055	832	781	970	686	632	854	581	536	848	668	627	779	551	508	703	478	441	1250
1500	1160	907	849	1063	744	685	918	619	571	941	736	689	863	604	556	767	517	477	1500
1750	1250	970	907	1142	793	729	975	651	599	1026	796	745	937	651	598	823	550	507	1750
2000	1332	1027	959	1213	836	768	1030	683	628	1103	850	794	1005	693	636	877	581	535	2000
Ambient Temp. (°C)	Correction Factors																		**Ambient Temp. (°F)**
6–10	1.09			1.09			1.09			1.09			1.09			1.09			43–50
11–15	1.04			1.04			1.04			1.04			1.04			1.04			52–59
16–20	1.00			1.00			1.00			1.00			1.00			1.00			61–68
21–25	0.95			0.95			0.95			0.95			0.95			0.95			70–77
26–30	0.90			0.90			0.90			0.90			0.90			0.90			79–86

Table B.310.15(B)(2)(6) Ampacities of Three Insulated Conductors, Rated 0 Through 2000 Volts, Within an Overall Covering (Three-Conductor Cable) in Underground Electrical Ducts (One Cable per Electrical Duct) Based on Ambient Earth Temperature of 20°C (68°F), Electrical Duct Arrangement in Accordance with Figure B.310.15(B)(2)(2), Conductor Temperature 75°C (167°F)

Size (AWG or kcmil)	1 Electrical Duct [Fig. B.310.15(B)(2)(2), Detail 1] Types RHW, THHW, THW, THWN, XHHW, USE COPPER			3 Electrical Ducts [Fig. B.310.15(B)(2)(2), Detail 2] Types RHW, THHW, THW, THWN, XHHW, USE			6 Electrical Ducts [Fig. B.310.15(B)(2)(2), Detail 3] Types RHW, THHW, THW, THWN, XHHW, USE			1 Electrical Duct [Fig. B.310.15(B)(2)(2), Detail 1] Types RHW, THHW, THW, THWN, XHHW, USE ALUMINUM OR COPPER-CLAD ALUMINUM			3 Electrical Ducts [Fig. B.310.15(B)(2)(2), Detail 2] Types RHW, THHW, THW, THWN, XHHW, USE			6 Electrical Ducts [Fig. B.310.15(B)(2)(2), Detail 3] Types RHW, THHW, THW, THWN, XHHW, USE			Size (AWG or kcmil)
	RHO 60 LF 50	RHO 90 LF 100	RHO 120 LF 100	RHO 60 LF 50	RHO 90 LF 100	RHO 120 LF 100	RHO 60 LF 50	RHO 90 LF 100	RHO 120 LF 100	RHO 60 LF 50	RHO 90 LF 100	RHO 120 LF 100	RHO 60 LF 50	RHO 90 LF 100	RHO 120 LF 100	RHO 60 LF 50	RHO 90 LF 100	RHO 120 LF 100	
8	58	54	53	56	48	46	53	42	39	45	42	41	43	37	36	41	32	30	8
6	77	71	69	74	63	60	70	54	51	60	55	54	57	49	47	54	42	39	6
4	101	93	91	96	81	77	91	69	65	78	72	71	75	63	60	71	54	51	4
2	132	121	118	126	105	100	119	89	83	103	94	92	98	82	78	92	70	65	2
1	154	140	136	146	121	114	137	102	95	120	109	106	114	94	89	107	79	74	1
1/0	177	160	156	168	137	130	157	116	107	138	125	122	131	107	101	122	90	84	1/0
2/0	203	183	178	192	156	147	179	131	121	158	143	139	150	122	115	140	102	95	2/0
3/0	233	210	204	221	178	158	205	148	137	182	164	159	172	139	131	160	116	107	3/0
4/0	268	240	232	253	202	190	234	168	155	209	187	182	198	158	149	183	131	121	4/0
250	297	265	256	280	222	209	258	184	169	233	207	201	219	174	163	202	144	132	250
350	363	321	310	340	267	250	312	219	202	285	252	244	267	209	196	245	172	158	350
500	444	389	375	414	320	299	377	261	240	352	308	297	328	254	237	299	207	190	500
750	552	478	459	511	388	362	462	314	288	446	386	372	413	314	293	374	254	233	750
1000	628	539	518	579	435	405	522	351	321	521	447	430	480	361	336	433	291	266	1000

Ambient Temp. (°C)	Correction Factors						Ambient Temp. (°F)
6–10	1.09	1.09	1.09	1.09	1.09	1.09	43–50
11–15	1.04	1.04	1.04	1.04	1.04	1.04	52–59
16–20	1.00	1.00	1.00	1.00	1.00	1.00	61–68
21–25	0.95	0.95	0.95	0.95	0.95	0.95	70–77
26–30	0.90	0.90	0.90	0.90	0.90	0.90	79–86

Table B.310.15(B)(2)(7) Ampacities of Three Single Insulated Conductors, Rated 0 Through 2000 Volts, in Underground Electrical Ducts (Three Conductors per Electrical Duct) Based on Ambient Earth Temperature of 20°C (68°F), Electrical Duct Arrangement in Accordance with Figure B.310.15(B)(2)(2), Conductor Temperature 75°C (167°F)

Size (AWG or kcmil)	1 Electrical Duct [Fig. B.310.15(B)(2)(2), Detail 1] Types RHW, THHW, THW, THWN, XHHW, USE			3 Electrical Ducts [Fig. B.310.15(B)(2)(2), Detail 2] Types RHW, THHW, THW, THWN, XHHW, USE			6 Electrical Ducts [Fig. B.310.15(B)(2)(2), Detail 3] Types RHW, THHW, THW, THWN, XHHW, USE			1 Electrical Duct [Fig. B.310.15(B)(2)(2), Detail 1] Types RHW, THHW, THW, THWN, XHHW, USE			3 Electrical Ducts [Fig. B.310.15(B)(2)(2), Detail 2] Types RHW, THHW, THW, THWN, XHHW, USE			6 Electrical Ducts [Fig. B.310.15(B)(2)(2), Detail 3] Types RHW, THHW, THW, THWN, XHHW, USE			Size (AWG or kcmil)
	COPPER									ALUMINUM OR COPPER-CLAD ALUMINUM									
	RHO 60 LF 50	RHO 90 LF 100	RHO 120 LF 100	RHO 60 LF 50	RHO 90 LF 100	RHO 120 LF 100	RHO 60 LF 50	RHO 90 LF 100	RHO 120 LF 100	RHO 60 LF 50	RHO 90 LF 100	RHO 120 LF 100	RHO 60 LF 50	RHO 90 LF 100	RHO 120 LF 100	RHO 60 LF 50	RHO 90 LF 100	RHO 120 LF 100	
8	63	58	57	61	51	49	57	44	41	49	45	44	47	40	38	45	34	32	8
6	84	77	75	80	67	63	75	56	53	66	60	58	63	52	49	59	44	41	6
4	111	100	98	105	86	81	98	73	67	86	78	76	79	67	63	77	57	52	4
3	129	116	113	122	99	94	113	83	77	101	91	89	83	77	73	84	65	60	3
2	147	132	128	139	112	106	129	93	86	115	103	100	108	87	82	101	73	67	2
1	171	153	148	161	128	121	149	106	98	133	119	115	126	100	94	116	83	77	1
1/0	197	175	169	185	146	137	170	121	111	153	136	132	144	114	107	133	94	87	1/0
2/0	226	200	193	212	166	156	194	136	126	176	156	151	165	130	121	151	106	98	2/0
3/0	260	228	220	243	189	177	222	154	142	203	178	172	189	147	138	173	121	111	3/0
4/0	301	263	253	280	215	201	255	175	161	235	205	198	219	168	157	199	137	126	4/0
250	334	290	279	310	236	220	281	192	176	261	227	218	242	185	172	220	150	137	250
300	373	321	308	344	260	242	310	210	192	293	252	242	272	204	190	245	165	151	300
350	409	351	337	377	283	264	340	228	209	321	276	265	296	222	207	266	179	164	350
400	442	376	361	394	302	280	368	243	223	349	297	284	321	238	220	288	191	174	400
500	503	427	409	460	341	316	412	273	249	397	338	323	364	270	250	326	216	197	500
600	552	468	447	511	371	343	457	296	270	446	373	356	408	296	274	365	236	215	600
700	602	509	486	553	402	371	492	319	291	488	408	389	443	321	297	394	255	232	700
750	632	529	505	574	417	385	509	330	301	508	425	405	461	334	309	409	265	241	750
800	654	544	520	597	428	395	527	338	308	530	439	418	481	344	318	427	273	247	800
900	692	575	549	628	450	415	554	355	323	563	466	444	510	365	337	450	288	261	900
1000	730	605	576	659	472	435	581	372	338	597	494	471	538	385	355	475	304	276	1000

Ambient Temp. (°C)	Correction Factors																		Ambient Temp. (°F)
6–10	1.09			1.09			1.09			1.09			1.09			1.09			43–50
11–15	1.04			1.04			1.04			1.04			1.04			1.04			52–59
16–20	1.00			1.00			1.00			1.00			1.00			1.00			61–68
21–25	0.95			0.95			0.95			0.95			0.95			0.95			70–77
26–30	0.90			0.90			0.90			0.90			0.90			0.90			79–86

Table B.310.15(B)(2)(8) Ampacities of Two or Three Insulated Conductors, Rated 0 Through 2000 Volts, Cabled Within an Overall (Two- or Three-Conductor) Covering, Directly Buried in Earth, Based on Ambient Earth Temperature of 20°C (68°F), Electrical Duct Arrangement in Accordance with Figure B.310.15(B)(2)(2), 100 Percent Load Factor, Thermal Resistance (Rho) of 90

	1 Cable [Fig. B.310.15(B)(2)(2), Detail 5]		2 Cables [Fig. B.310.15(B)(2)(2), Detail 6]		1 Cable [Fig. B.310.15(B)(2)(2), Detail 5]		2 Cables [Fig. B.310.15(B)(2)(2), Detail 6]		
	60°C (140°F)	75°C (167°F)	60°C (140°F)	75°C (167°F)	60°C (140°F)	75°C (167°F)	60°C (140°F)	75°C (167°F)	
	TYPES				TYPES				
Size (AWG or kcmil)	UF	RHW, THHW, THW, THWN, XHHW, USE	UF	RHW, THHW, THW, THWN, XHHW, USE	UF	RHW, THHW, THW, THWN, XHHW, USE	UF	RHW, THHW, THW, THWN, XHHW, USE	Size (AWG or kcmil)
	COPPER				ALUMINUM OR COPPER-CLAD ALUMINUM				
8	64	75	60	70	51	59	47	55	8
6	85	100	81	95	68	75	60	70	6
4	107	125	100	117	83	97	78	91	4
2	137	161	128	150	107	126	110	117	2
1	155	182	145	170	121	142	113	132	1
1/0	177	208	165	193	138	162	129	151	1/0
2/0	201	236	188	220	157	184	146	171	2/0
3/0	229	269	213	250	179	210	166	195	3/0
4/0	259	304	241	282	203	238	188	220	4/0
250	—	333	—	308	—	261	—	241	250
350	—	401	—	370	—	315	—	290	350
500	—	481	—	442	—	381	—	350	500
750	—	585	—	535	—	473	—	433	750
1000	—	657	—	600	—	545	—	497	1000
Ambient Temp. (°C)	Correction Factors								Ambient Temp. (°F)
6–10	1.12	1.09	1.12	1.09	1.12	1.09	1.12	1.09	43–50
11–15	1.06	1.04	1.06	1.04	1.06	1.04	1.06	1.04	52–59
16–20	1.00	1.00	1.00	1.00	1.00	1.00	1.00	1.00	61–68
21–25	0.94	0.95	0.94	0.95	0.94	0.95	0.94	0.95	70–77
26–30	0.87	0.90	0.87	0.90	0.87	0.90	0.87	0.90	79–86

Note: For ampacities of Type UF cable in underground electrical ducts, multiply the ampacities shown in the table by 0.74.

Table B.310.15(B)(2)(9) Ampacities of Three Triplexed Single Insulated Conductors, Rated 0 Through 2000 Volts, Directly Buried in Earth Based on Ambient Earth Temperature of 20°C (68°F), Electrical Duct Arrangement in Accordance with Figure B.310.15(B)(2) (2), 100 Percent Load Factor, Thermal Resistance (Rho) of 90

	See Fig. B.310.15(B)(2)(2), Detail 7		See Fig. B.310.15(B)(2)(2), Detail 8		See Fig. B.310.15(B)(2)(2), Detail 7		See Fig. B.310.15(B)(2)(2), Detail 8		
	60°C (140°F)	75°C (167°F)	60°C (140°F)	75°C (167°F)	60°C (140°F)	75°C (167°F)	60°C (140°F)	75°C (167°F)	
	TYPES				TYPES				
Size (AWG or kcmil)	UF	USE	UF	USE	UF	USE	UF	USE	Size (AWG or kcmil)
	COPPER				ALUMINUM OR COPPER-CLAD ALUMINUM				
8	72	84	66	77	55	65	51	60	8
6	91	107	84	99	72	84	66	77	6
4	119	139	109	128	92	108	85	100	4
2	153	179	140	164	119	139	109	128	2
1	173	203	159	186	135	158	124	145	1
1/0	197	231	181	212	154	180	141	165	1/0
2/0	223	262	205	240	175	205	159	187	2/0
3/0	254	298	232	272	199	233	181	212	3/0
4/0	289	339	263	308	226	265	206	241	4/0
250	—	370	—	336	—	289	—	263	250
350	—	445	—	403	—	349	—	316	350
500	—	536	—	483	—	424	—	382	500
750	—	654	—	587	—	525	—	471	750
1000	—	744	—	665	—	608	—	544	1000
Ambient Temp. (°C)	Correction Factors								Ambient Temp. (°F)
6–10	1.12	1.09	1.12	1.09	1.12	1.09	1.12	1.09	43–50
11–15	1.06	1.04	1.06	1.04	1.06	1.04	1.06	1.04	52–59
16–20	1.00	1.00	1.00	1.00	1.00	1.00	1.00	1.00	61–68
21–25	0.94	0.95	0.94	0.95	0.94	0.95	0.94	0.95	70–77
26–30	0.87	0.90	0.87	0.90	0.87	0.90	0.87	0.90	79–86

Table B.310.15(B)(2)(10) Ampacities of Three Single Insulated Conductors, Rated 0 Through 2000 Volts, Directly Buried in Earth Based on Ambient Earth Temperature of 20°C (68°F), Electrical Duct Arrangement in Accordance with Figure B.310.15(B)(2)(2), 100 Percent Load Factor, Thermal Resistance (Rho) of 90

Size (AWG or kcmil)	See Fig. B.310.15(B)(2)(2), Detail 9		See Fig. B.310.15(B)(2)(2), Detail 10		See Fig. B.310.15(B)(2)(2), Detail 9		See Fig. B.310.15(B)(2)(2), Detail 10		Size (AWG or kcmil)
	60°C (140°F)	75°C (167°F)	60°C (140°F)	75°C (167°F)	60°C (140°F)	75°C (167°F)	60°C (140°F)	75°C (167°F)	
	TYPES				TYPES				
	UF	USE	UF	USE	UF	USE	UF	USE	
	COPPER				ALUMINUM OR COPPER-CLAD ALUMINUM				
8	84	98	78	92	66	77	61	72	8
6	107	126	101	118	84	98	78	92	6
4	139	163	130	152	108	127	101	118	4
2	178	209	165	194	139	163	129	151	2
1	201	236	187	219	157	184	146	171	1
1/0	230	270	212	249	179	210	165	194	1/0
2/0	261	306	241	283	204	239	188	220	2/0
3/0	297	348	274	321	232	272	213	250	3/0
4/0	336	394	309	362	262	307	241	283	4/0
250	—	429	—	394	—	335	—	308	250
350	—	516	—	474	—	403	—	370	350
500	—	626	—	572	—	490	—	448	500
750	—	767	—	700	—	605	—	552	750
1000	—	887	—	808	—	706	—	642	1000
1250	—	979	—	891	—	787	—	716	1250
1500	—	1063	—	965	—	862	—	783	1500
1750	—	1133	—	1027	—	930	—	843	1750
2000	—	1195	—	1082	—	990	—	897	2000
Ambient Temp. (°C)	Correction Factors								Ambient Temp. (°F)
6–10	1.12	1.09	1.12	1.09	1.12	1.09	1.12	1.09	43–50
11–15	1.06	1.04	1.06	1.04	1.06	1.04	1.06	1.04	52–59
16–20	1.00	1.00	1.00	1.00	1.00	1.00	1.00	1.00	61–68
21–25	0.94	0.95	0.94	0.95	0.94	0.95	0.94	0.95	70–77
26–30	0.87	0.90	0.87	0.90	0.87	0.90	0.87	0.90	79–86

Table B.310.15(B)(2)(11) Adjustment Factors for More Than Three Current-Carrying Conductors in a Raceway or Cable with Load Diversity

Number of Conductors*	Percent of Values in Tables as Adjusted for Ambient Temperature if Necessary
4–6	80
7–9	70
10–24	70**
25–42	60**
43–85	50**

*Number of conductors is the total number of conductors in the raceway or cable adjusted in accordance with 310.15(B)(4) and (5).
**These factors include the effects of a load diversity of 50 percent.

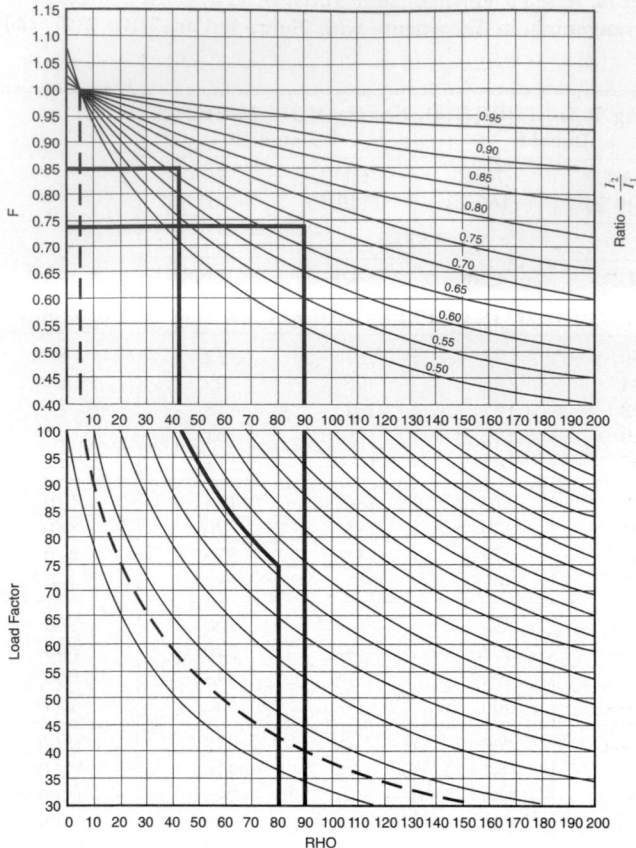

FIGURE B.310.15(B)(2)(1) Interpolation Chart for Cables in a Duct Bank. I_1 = ampacity for Rho = 60, 50 LF; I_2 = ampacity for Rho = 120, 100 LF (load factor); desired ampacity = F × I_1.

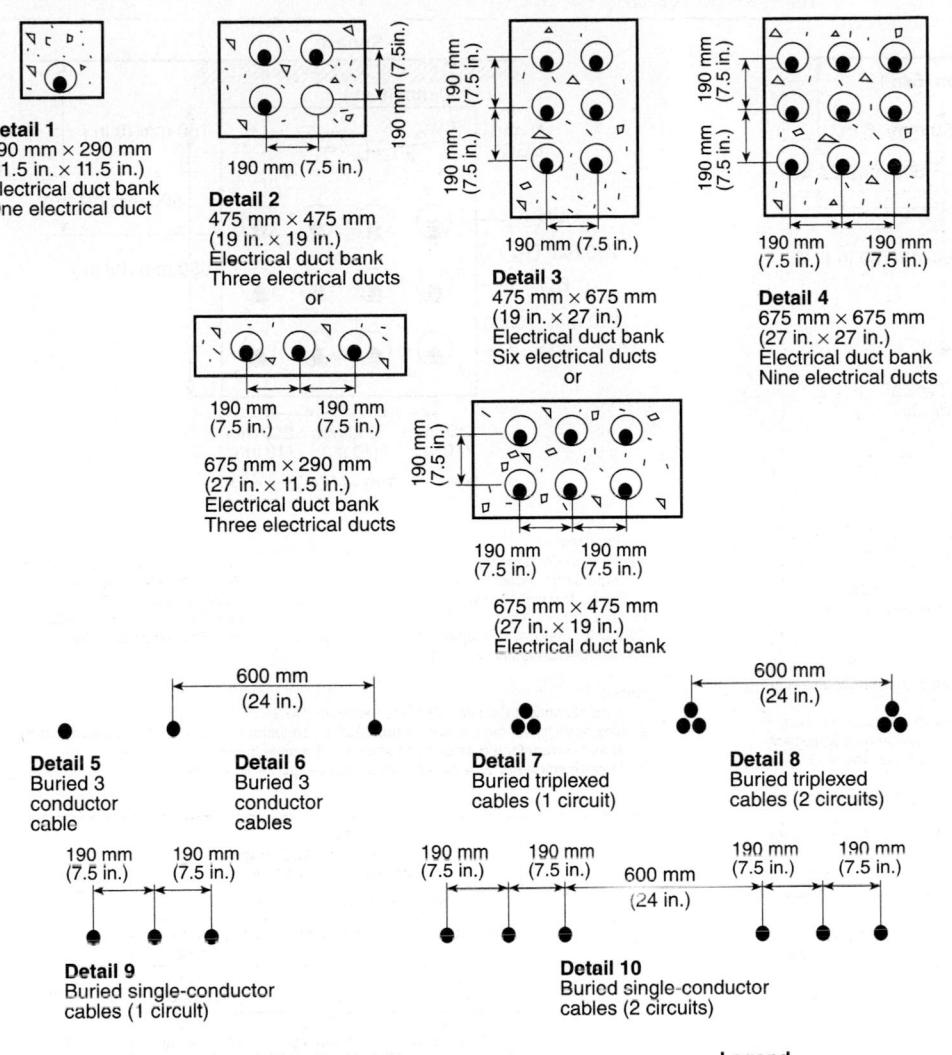

Detail 1
290 mm × 290 mm
(11.5 in. × 11.5 in.)
Electrical duct bank
One electrical duct

Detail 2
475 mm × 475 mm
(19 in. × 19 in.)
Electrical duct bank
Three electrical ducts
or

675 mm × 290 mm
(27 in. × 11.5 in.)
Electrical duct bank
Three electrical ducts

Detail 3
475 mm × 675 mm
(19 in. × 27 in.)
Electrical duct bank
Six electrical ducts
or

675 mm × 475 mm
(27 in. × 19 in.)
Electrical duct bank

Detail 4
675 mm × 675 mm
(27 in. × 27 in.)
Electrical duct bank
Nine electrical ducts

Detail 5
Buried 3
conductor
cable

Detail 6
Buried 3
conductor
cables

Detail 7
Buried triplexed
cables (1 circuit)

Detail 8
Buried triplexed
cables (2 circuits)

Detail 9
Buried single-conductor
cables (1 circuit)

Detail 10
Buried single-conductor
cables (2 circuits)

Note 1: Minimum burial depths to top electrical ducts or cables shall be in accordance with 300.5. Maximum depth to the top of electrical duct banks shall be 750 mm (30 in.) and maximum depth to the top of direct-buried cables shall be 900 mm (36 in.).

Note 2: For two and four electrical duct installations with electrical ducts installed in a single row, see B.310.15(B)(5).

Legend

▨ Backfill (earth or concrete)

◯ Electrical duct

● Cable or cables

FIGURE B.310.15(B)(2)(2)   Cable Installation Dimensions for Use with Table B.310.15(B)(2)(5) Through Table B.310.15(B)(2)(10).

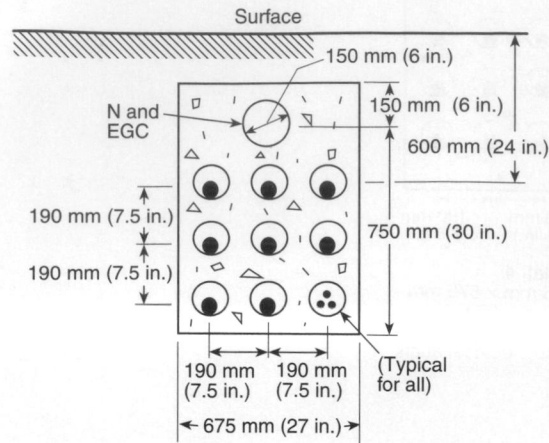

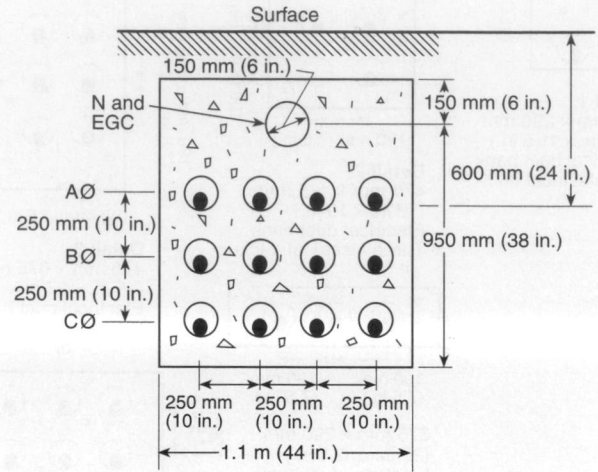

Design Criteria

Neutral and Equipment
  Grounding conductor (EGC)
  Duct = 150 mm (6 in.)
Phase Ducts = 75 to 125 mm (3 to 5 in.)
Conductor Material = Copper
Number of Cables per Duct = 3

Number of Cables per Phase = 9
Rho Concrete = Rho Earth – 5
Rho PVC Duct = 650
Rho Cable Insulation = 500
Rho Cable Jacket = 650

Notes:
1. Neutral configuration per 300.5(I), Exception No. 2, for isolated phase installations in nonmagnetic ducts.
2. Phasing is A, B, C in rows or columns. Where magnetic electrical ducts are used, conductors are installed A, B, C per electrical duct with the neutral and all equipment grounding conductors in the same electrical duct. In this case, the 6-in. trade size neutral duct is eliminated.
3. Maximum harmonic loading on the neutral conductor cannot exceed 50 percent of the phase current for the ampacities shown in the table below.
4. Metallic shields of Type MV-90 cable shall be grounded at one point only where using A, B, C phasing in rows or columns.

Size kcmil	TYPES RHW, THHW, THW, THWN, XHHW, USE, OR MV-90*			Size kcmil
	Total per Phase Ampere Rating			
	RHO EARTH 60 LF 50	RHO EARTH 90 LF 100	RHO EARTH 120 LF 100	
250	2340 (260A/Cable)	1530 (170A/Cable)	1395 (155A/Cable)	250
350	2790 (310A/Cable)	1800 (200A/Cable)	1665 (185A/Cable)	350
500	3375 (375A/Cable)	2160 (240A/Cable)	1980 (220A/Cable)	500

Ambient Temp. (°C)	For ambient temperatures other than 20°C (68°F), multiply the ampacities shown above by the appropriate factor shown below.					Ambient Temp. (°F)
6–10	1.09	1.09	1.09	1.09	1.09	43–50
11–15	1.04	1.04	1.04	1.04	1.04	52–59
16–20	1.00	1.00	1.00	1.00	1.00	61–68
21–25	0.95	0.95	0.95	0.95	0.95	70–77
26–30	0.90	0.90	0.90	0.90	0.90	79–86

*Limited to 75°C conductor temperature.

Informational Note Figure B.310.15(B)(2)(3) Ampacities of Single Insulated Conductors Rated 0 Through 5000 Volts in Underground Electrical Ducts (Three Conductors per Electrical Duct), Nine Single-Conductor Cables per Phase Based on Ambient Earth Temperature of 20°C (68°F), Conductor Temperature 75°C (167°F).

Design Criteria

Neutral and Equipment
  Grounding conductor (EGC)
  Duct = 150 mm (6 in.)
Phase Ducts = 75 mm (3 in.)
Conductor Material = Copper
Number of Cables per Duct = 1

Number of Cables per Phase = 4
Rho Concrete = Rho Earth – 5
Rho PVC Duct = 650
Rho Cable Insulation = 500
Rho Cable Jacket = 650

Notes:
1. Neutral configuration per 300.5(I), Exception No 2.
2. Maximum harmonic loading on the neutral conductor cannot exceed 50 percent of the phase current for the ampacities shown in the table below.
3. Metallic shields of Type MV-90 cable shall be grounded at one point only.

Size kcmil	TYPES RHW,THHW,THW,THWN, XHHW,USE,OR MV-90*			Size kcmil
	Total per Phase Ampere Rating			
	RHO EARTH 60 LF 50	RHO EARTH 90 LF 100	RHO EARTH 120 LF 100	
750	2820 (705A/Cable)	1860 (465A/Cable)	1680 (420A/Cable)	750
1000	3300 (825A/Cable)	2140 (535A/Cable)	1920 (480A/Cable)	1000
1250	3700 (925A/Cable)	2380 (595A/Cable)	2120 (530A/Cable)	1250
1500	4060 (1015A/Cable)	2580 (645A/Cable)	2300 (575A/Cable)	1500
1750	4360 (1090A/Cable)	2740 (685A/Cable)	2460 (615A/Cable)	1750

Ambient Temp. (°C)	For ambient temperatures other than 20°C (68°F), multiply the ampacities shown above by the appropriate factor shown below.					Ambient Temp. (°F)
6–10	1.09	1.09	1.09	1.09	1.09	43–50
11–15	1.04	1.04	1.04	1.04	1.04	52–59
16–20	1.00	1.00	1.00	1.00	1.00	61–68
21–25	0.95	0.95	0.95	0.95	0.95	70–77
26–30	0.90	0.90	0.90	0.90	0.90	79–86

*Limited to 75°C conductor temperature.

Informational Note Figure B.310.15(B)(2)(4) Ampacities of Single Insulated Conductors Rated 0 Through 5000 Volts in Nonmagnetic Underground Electrical Ducts (One Conductor per Electrical Duct), Four Single-Conductor Cables per Phase Based on Ambient Earth Temperature of 20°C (68°F), Conductor Temperature 75°C (167°F).

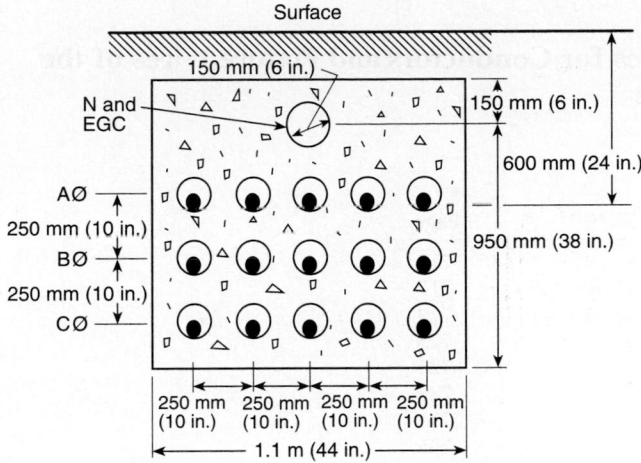

Design Criteria

Neutral and Equipment	Number of Cables per Phase = 5
Grounding conductor (EGC)	Rho Concrete = Rho Earth – 5
Duct = 150 mm (6 in.)	Rho PVC Duct = 650
Phase Ducts = 75 mm (3 in.)	Rho Cable Insulation = 500
Conductor Material = Copper	Rho Cable Jacket = 650
Number of Cables per Duct = 1	

Notes:
1. Neutral configuration per 300.5(I), Exception No. 2.
2. Maximum harmonic loading on the neutral conductor cannot exceed 50 percent of the phase current for the ampacities shown in the table below.
3. Metallic shields of Type MV-90 cable shall be grounded at one point only.

Size kcmil	TYPES RHW, THHW, THW, THWN, XHHW, USE, or MV-90*			Size kcmil
	Total per Phase Ampere Rating			
	RHO EARTH 60 LF 50	RHO EARTH 90 LF 100	RHO EARTH 120 LF 100	
2000	5575 (1115A/Cable)	3375 (675A/Cable)	3000 (600A/Cable)	2000

Ambient Temp. (°C)	For ambient temperatures other than 20°C (68°F), multiply the ampacities shown above by the appropriate factor shown below.					Ambient Temp. (°F)
6–10	1.09	1.09	1.09	1.09	1.09	43–50
11–15	1.04	1.04	1.04	1.04	1.04	52–59
16–20	1.00	1.00	1.00	1.00	1.00	61–68
21–25	0.95	0.95	0.95	0.95	0.95	70–77
26–30	0.90	0.90	0.90	0.90	0.90	79–86

*Limited to 75°C conductor temperature.

Informational Note Figure B.310.15(B)(2)(5)   **Ampacities of Single Insulated Conductors Rated 0 Through 5000 Volts in Nonmagnetic Underground Electrical Ducts (One Conductor per Electrical Duct), Five Single-Conductor Cables per Phase Based on Ambient Earth Temperature of 20°C (68°F), Conductor Temperature 75°C (167°F).**

Informative Annex C   Conduit and Tubing Fill Tables for Conductors and Fixture Wires of the Same Size

This informative annex is not a part of the requirements of this NFPA document but is included for informational purposes only.

*Where this table is used in conjunction with Tables C.1 through C.13, the conductors installed must be of the compact type.

Table C.1 Maximum Number of Conductors or Fixture Wires in Electrical Metallic Tubing (EMT)
(Based on Chapter 9: Table 1, Table 4, and Table 5)

Type	Conductor Size (AWG/kcmil)	Trade Size (Metric Designator)												
		⅜ (12)	½ (16)	¾ (21)	1 (27)	1¼ (35)	1½ (41)	2 (53)	2½ (63)	3 (78)	3½ (91)	4 (103)	5 (129)	6 (155)
CONDUCTORS														
RHH, RHW, RHW-2	14	—	4	7	11	20	27	46	80	120	157	201	—	—
	12	—	3	6	9	17	23	38	66	100	131	167	—	—
	10	—	2	5	8	13	18	30	53	81	105	135	—	—
	8	—	1	2	4	7	9	16	28	42	55	70	—	—
	6	—	1	1	3	5	8	13	22	34	44	56	—	—
	4	—	1	1	2	4	6	10	17	26	34	44	—	—
	3	—	1	1	1	4	5	9	15	23	30	38	—	—
	2	—	1	1	1	3	4	7	13	20	26	33	—	—
	1	—	0	1	1	1	3	5	9	13	17	22	—	—
	1/0	—	0	1	1	1	2	4	7	11	15	19	—	—
	2/0	—	0	1	1	1	2	4	6	10	13	17	—	—
	3/0	—	0	0	1	1	1	3	5	8	11	14	—	—
	4/0	—	0	0	1	1	1	3	5	7	9	12	—	—
	250	—	0	0	0	1	1	1	3	5	7	9	—	—
	300	—	0	0	0	1	1	1	3	5	6	8	—	—
	350	—	0	0	0	1	1	1	3	4	6	7	—	—
	400	—	0	0	0	1	1	1	2	4	5	7	—	—
	500	—	0	0	0	0	1	1	2	3	4	6	—	—
	600	—	0	0	0	0	1	1	1	3	4	5	—	—
	700	—	0	0	0	0	0	1	1	2	3	4	—	—
	750	—	0	0	0	0	0	1	1	2	3	4	—	—
	800	—	0	0	0	0	0	1	1	2	3	4	—	—
	900	—	0	0	0	0	0	1	1	1	3	3	—	—
	1000	—	0	0	0	0	0	1	1	1	2	3	—	—
	1250	—	0	0	0	0	0	0	1	1	1	2	—	—
	1500	—	0	0	0	0	0	0	1	1	1	1	—	—
	1750	—	0	0	0	0	0	0	1	1	1	1	—	—
	2000	—	0	0	0	0	0	0	1	1	1	1	—	—
TW, THHW, THW, THW-2	14	—	8	15	25	43	58	96	168	254	332	424	—	—
	12	—	6	11	19	33	45	74	129	195	255	326	—	—
	10	—	5	8	14	24	33	55	96	145	190	243	—	—
	8	—	2	5	8	13	18	30	53	81	105	135	—	—
RHH*, RHW*, RHW-2*	14	—	6	10	16	28	39	64	112	169	221	282	—	—
	12	—	4	8	13	23	31	51	90	136	177	227	—	—
	10	—	3	6	10	18	24	40	70	106	138	177	—	—
	8	—	1	4	6	10	14	24	42	63	83	106	—	—
TW, THW, THHW, THW-2, RHH*, RHW*, RHW-2*	6	—	1	3	4	8	11	18	32	48	63	81	—	—
	4	—	1	1	3	6	8	13	24	36	47	60	—	—
	3	—	1	1	3	5	7	12	20	31	40	52	—	—
	2	—	1	1	2	4	6	10	17	26	34	44	—	—
	1	—	1	1	1	3	4	7	12	18	24	31	—	—
	1/0	—	0	1	1	2	3	6	10	16	20	26	—	—
	2/0	—	0	1	1	1	3	5	9	13	17	22	—	—
	3/0	—	0	1	1	1	2	4	7	11	15	19	—	—
	4/0	—	0	0	1	1	1	3	6	9	12	16	—	—
	250	—	0	0	1	1	1	3	5	7	10	13	—	—
	300	—	0	0	1	1	1	2	4	6	8	11	—	—
	350	—	0	0	0	1	1	1	4	6	7	10	—	—
	400	—	0	0	0	1	1	1	3	5	7	9	—	—
	500	—	0	0	0	1	1	1	3	4	6	7	—	—
	600	—	0	0	0	1	1	1	2	3	4	6	—	—
	700	—	0	0	0	0	1	1	1	3	4	5	—	—
	750	—	0	0	0	0	1	1	1	3	4	5	—	—

(continues)

Table C.1 *Continued*

Type	Conductor Size (AWG/kcmil)	3/8 (12)	1/2 (16)	3/4 (21)	1 (27)	1¼ (35)	1½ (41)	2 (53)	2½ (63)	3 (78)	3½ (91)	4 (103)	5 (129)	6 (155)
										Trade Size (Metric Designator)				
	800	—	0	0	0	0	1	1	1	3	3	5	—	—
	900	—	0	0	0	0	0	1	1	2	3	4	—	—
	1000	—	0	0	0	0	0	1	1	2	3	4	—	—
	1250	—	0	0	0	0	0	1	1	1	2	3	—	—
	1500	—	0	0	0	0	0	1	1	1	1	2	—	—
	1750	—	0	0	0	0	0	0	1	1	1	2	—	—
	2000	—	0	0	0	0	0	0	1	1	1	1	—	—
THHN, THWN, THWN-2	14	—	12	22	35	61	84	138	241	364	476	608	—	—
	12	—	9	16	26	45	61	101	176	266	347	443	—	—
	10	—	5	10	16	28	38	63	111	167	219	279	—	—
	8	—	3	6	9	16	22	36	64	96	126	161	—	—
	6	—	2	4	7	12	16	26	46	69	91	116	—	—
	4	—	1	2	4	7	10	16	28	43	56	71	—	—
	3	—	1	1	3	6	8	13	24	36	47	60	—	—
	2	—	1	1	3	5	7	11	20	30	40	51	—	—
	1	—	1	1	1	4	5	8	15	22	29	37	—	—
	1/0	—	1	1	1	3	4	7	12	19	25	32	—	—
	2/0	—	0	1	1	2	3	6	10	16	20	26	—	—
	3/0	—	0	1	1	1	3	5	8	13	17	22	—	—
	4/0	—	0	1	1	1	2	4	7	11	14	18	—	—
	250	—	0	0	1	1	1	3	6	9	11	15	—	—
	300	—	0	0	1	1	1	3	5	7	10	13	—	—
	350	—	0	0	1	1	1	2	4	6	9	11	—	—
	400	—	0	0	0	1	1	1	4	6	8	10	—	—
	500	—	0	0	0	1	1	1	3	5	6	8	—	—
	600	—	0	0	0	1	1	1	2	4	5	7	—	—
	700	—	0	0	0	1	1	1	2	3	4	6	—	—
	750	—	0	0	0	0	1	1	1	3	4	5	—	—
	800	—	0	0	0	0	1	1	1	3	4	5	—	—
	900	—	0	0	0	0	1	1	1	3	3	4	—	—
	1000	—	0	0	0	0	1	1	1	2	3	4	—	—
FEP, FEPB, PFA, PFAH, TFE	14	—	12	21	34	60	81	134	234	354	462	590	—	—
	12	—	9	15	25	43	59	98	171	258	337	430	—	—
	10	—	6	11	18	31	42	70	122	185	241	309	—	—
	8	—	3	6	10	18	24	40	70	106	138	177	—	—
	6	—	2	4	7	12	17	28	50	75	98	126	—	—
	4	—	1	3	5	9	12	20	35	53	69	88	—	—
	3	—	1	2	4	7	10	16	29	44	57	73	—	—
	2	—	1	1	3	6	8	13	24	36	47	60	—	—
PFA, PFAH, TFE	1	—	1	1	2	4	6	9	16	25	33	42	—	—
PFA, PFAH, TFE, Z	1/0	—	1	1	1	3	5	8	14	21	27	35	—	—
	2/0	—	0	1	1	3	4	6	11	17	22	29	—	—
	3/0	—	0	1	1	2	3	5	9	14	18	24	—	—
	4/0	—	0	1	1	1	2	4	8	11	15	19	—	—
Z	14	—	14	25	41	72	98	161	282	426	556	711	—	—
	12	—	10	18	29	51	69	114	200	302	394	504	—	—
	10	—	6	11	18	31	42	70	122	185	241	309	—	—
	8	—	4	7	11	20	27	44	77	117	153	195	—	—
	6	—	3	5	8	14	19	31	54	82	107	137	—	—
	4	—	1	3	5	9	13	21	37	56	74	94	—	—
	3	—	1	2	4	7	9	15	27	41	54	69	—	—
	2	—	1	1	3	6	8	13	22	34	45	57	—	—
	1	—	1	1	2	4	6	10	18	28	36	46	—	—

(continues)

Table C.1   *Continued*

Type	Conductor Size (AWG/kcmil)	Trade Size (Metric Designator)												
		⅜ (12)	½ (16)	¾ (21)	1 (27)	1¼ (35)	1½ (41)	2 (53)	2½ (63)	3 (78)	3½ (91)	4 (103)	5 (129)	6 (155)
XHHW, ZW, XHHW-2, XHH	14	—	8	15	25	43	58	96	168	254	332	424	—	—
	12	—	6	11	19	33	45	74	129	195	255	326	—	—
	10	—	5	8	14	24	33	55	96	145	190	243	—	—
	8	—	2	5	8	13	18	30	53	81	105	135	—	—
	6	—	1	3	6	10	14	22	39	60	78	100	—	—
	4	—	1	2	4	7	10	16	28	43	56	72	—	—
	3	—	1	1	3	6	8	14	24	36	48	61	—	—
	2	—	1	1	3	5	7	11	20	31	40	51	—	—
XHHW, XHHW-2, XHH	1	—	1	1	1	4	5	8	15	23	30	38	—	—
	1/0	—	1	1	1	3	4	7	13	19	25	32	—	—
	2/0	—	0	1	1	2	3	6	10	16	21	27	—	—
	3/0	—	0	1	1	1	3	5	9	13	17	22	—	—
	4/0	—	0	1	1	1	2	4	7	11	14	18	—	—
	250	—	0	0	1	1	1	3	6	9	12	15	—	—
	300	—	0	0	1	1	1	3	5	8	10	13	—	—
	350	—	0	0	1	1	1	2	4	7	9	11	—	—
	400	—	0	0	0	1	1	1	4	6	8	10	—	—
	500	—	0	0	0	1	1	1	3	5	6	8	—	—
	600	—	0	0	0	1	1	1	2	4	5	6	—	—
	700	—	0	0	0	0	1	1	2	3	4	6	—	—
	750	—	0	0	0	0	1	1	1	3	4	5	—	—
	800	—	0	0	0	0	1	1	1	3	4	5	—	—
	900	—	0	0	0	0	1	1	1	3	3	4	—	—
	1000	—	0	0	0	0	0	1	1	2	3	4	—	—
	1250	—	0	0	0	0	0	1	1	1	2	3	—	—
	1500	—	0	0	0	0	0	1	1	1	1	3	—	—
	1750	—	0	0	0	0	0	0	1	1	1	2	—	—
	2000	—	0	0	0	0	0	0	1	1	1	1	—	—
FIXTURE WIRES														
RFH-2, FFH-2, RFHH-2	18	—	8	14	24	41	56	92	161	244	318	407	—	—
	16	—	7	12	20	34	47	78	136	205	268	343	—	—
SF-2, SFF-2	18	—	10	18	30	52	71	116	203	307	401	513	—	—
	16	—	8	15	25	43	58	96	168	254	332	424	—	—
	14	—	7	12	20	34	47	78	136	205	268	343	—	—
SF-1, SFF-1	18	—	18	33	53	92	125	206	360	544	710	908	—	—
RFH-1, TF, TFF, XF, XFF	18	—	14	24	39	68	92	152	266	402	524	670	—	—
	16	—	11	19	31	55	74	123	215	324	423	541	—	—
XF, XFF	14	—	8	15	25	43	58	96	168	254	332	424	—	—
TFN, TFFN	18	—	22	38	63	109	148	244	426	643	839	1073	—	—
	16	—	17	29	48	83	113	186	325	491	641	819	—	—
PF, PFF, PGF, PGFF, PAF, PTF, PTFF, PAFF	18	—	21	36	59	103	140	231	404	610	796	1017	—	—
	16	—	16	28	46	79	108	179	312	471	615	787	—	—
	14	—	12	21	34	60	81	134	234	354	462	590	—	—
ZF, ZFF, ZHF	18	—	27	47	77	133	181	298	520	786	1026	1311	—	—
	16	—	20	35	56	98	133	220	384	580	757	967	—	—
	14	—	14	25	41	72	98	161	282	426	556	711	—	—
KF-2, KFF-2	18	—	40	71	115	199	271	447	781	1179	1539	1967	—	—
	16	—	28	49	80	139	189	312	545	823	1074	1372	—	—
	14	—	19	33	54	93	127	209	366	553	721	922	—	—
	12	—	13	23	37	65	88	146	254	384	502	641	—	—
	10	—	8	15	25	43	58	96	168	254	332	424	—	—

(continues)

Table C.1 *Continued*

Type	Conductor Size (AWG/kcmil)	Trade Size (Metric Designator)												
		⅜ (12)	½ (16)	¾ (21)	1 (27)	1¼ (35)	1½ (41)	2 (53)	2½ (63)	3 (78)	3½ (91)	4 (103)	5 (129)	6 (155)
KF-1, KFF-1	18	—	46	82	133	230	313	516	901	1361	1776	2269	—	—
	16	—	33	57	93	161	220	363	633	956	1248	1595	—	—
	14	—	22	38	63	109	148	244	426	643	839	1073	—	—
	12	—	14	25	41	72	98	161	282	426	556	711	—	—
	10	—	9	16	27	47	64	105	184	278	363	464	—	—
XF, XFF	12	—	4	8	13	23	31	51	90	136	177	227	—	—
	10	—	3	6	10	18	24	40	70	106	138	177	—	—

Notes:

1. This table is for concentric stranded conductors only. For compact stranded conductors, Table C.1(A) should be used.

2. Two-hour fire-rated RHH cable has ceramifiable insulation, which has much larger diameters than other RHH wires. Consult manufacturer's conduit fill tables.

*Types RHH, RHW, and RHW-2 without outer covering.

Table C.1(A) Maximum Number of Conductors or Fixture Wires in Electrical Metallic Tubing (EMT) (Based on Chapter 9: Table 1, Table 4, and Table 5A)

Type	Conductor Size (AWG/ kcmil)	Trade Size (Metric Designator)												
		⅜ (12)	½ (16)	¾ (21)	1 (27)	1¼ (35)	1½ (41)	2 (53)	2½ (63)	3 (78)	3½ (91)	4 (103)	5 (129)	6 (155)
COMPACT CONDUCTORS														
THW, THW-2, THHW	8	—	2	4	6	11	16	26	46	69	90	115	—	—
	6	—	1	3	5	9	12	20	35	53	70	89	—	—
	4	—	1	2	4	6	9	15	26	40	52	67	—	—
	2	—	1	1	3	5	7	11	19	29	38	49	—	—
	1	—	1	1	1	3	4	8	13	21	27	34	—	—
	1/0	—	1	1	1	3	4	7	12	18	23	30	—	—
	2/0	—	0	1	1	2	3	5	10	15	20	25	—	—
	3/0	—	0	1	1	1	3	5	8	13	17	21	—	—
	4/0	—	0	1	1	1	2	4	7	11	14	18	—	—
	250	—	0	0	1	1	1	3	5	8	11	14	—	—
	300	—	0	0	1	1	1	3	5	7	9	12	—	—
	350	—	0	0	1	1	1	2	4	6	8	11	—	—
	400	—	0	0	0	1	1	1	4	6	8	10	—	—
	500	—	0	0	0	1	1	1	3	5	6	8	—	—
	600	—	0	0	0	1	1	1	2	4	5	7	—	—
	700	—	0	0	0	1	1	1	2	3	4	6	—	—
	750	—	0	0	0	0	1	1	1	3	4	5	—	—
	900	—	0	0	0	0	1	1	1	3	4	5	—	—
	1000	—	0	0	0	0	1	1	1	2	3	4	—	—
THHN, THWN, THWN-2	8	—	—	—	—	—	—	—	—	—	—	—	—	—
	6	—	2	4	7	13	18	29	52	78	102	130	—	—
	4	—	1	3	4	8	11	18	32	48	63	81	—	—
	2	—	1	1	3	6	8	13	23	34	45	58	—	—
	1	—	1	1	2	4	6	10	17	26	34	43	—	—
	1/0	—	1	1	1	3	5	8	14	22	29	37	—	
	2/0	—	1	1	1	3	4	7	12	18	24	30	—	—
	3/0	—	0	1	1	2	3	6	10	15	20	25	—	—
	4/0	—	0	1	1	1	3	5	8	12	16	21	—	—
	250	—	0	1	1	1	1	4	6	10	13	16	—	—
	300	—	0	0	1	1	1	3	5	8	11	14	—	—
	350	—	0	0	1	1	1	3	5	7	10	12	—	—
	400	—	0	0	1	1	1	2	4	6	9	11	—	—
	500	—	0	0	0	1	1	1	4	5	7	9	—	—
	600	—	0	0	0	1	1	1	3	4	6	7	—	—
	700	—	0	0	0	1	1	1	2	4	5	7	—	—
	750	—	0	0	0	1	1	1	2	4	5	6	—	—
	900	—	0	0	0	0	1	1	1	3	4	5	—	—
	1000	—	0	0	0	0	1	1	1	3	3	4	—	—
XHHW, XHHW-2	8	—	3	5	8	15	20	34	59	90	117	149	—	—
	6	—	1	4	6	11	15	25	44	66	87	111	—	—
	4	—	1	3	4	8	11	18	32	48	63	81	—	—
	2	—	1	1	3	6	8	13	23	34	45	58	—	—
	1	—	1	1	2	4	6	10	17	26	34	43	—	—
	1/0	—	1	1	1	3	5	8	14	22	29	37	—	—
	2/0	—	1	1	1	3	4	7	12	18	24	31	—	—
	3/0	—	0	1	1	2	3	6	10	15	20	25	—	—
	4/0	—	0	1	1	1	3	5	8	13	17	21	—	—
	250	—	0	1	1	1	2	4	7	10	13	17	—	—
	300	—	0	0	1	1	1	3	6	9	11	14	—	—
	350	—	0	0	1	1	1	3	5	8	10	13	—	—
	400	—	0	0	1	1	1	2	4	7	9	11	—	—
	500	—	0	0	0	1	1	1	4	6	7	9	—	—

(continues)

Table C.1(A) *Continued*

Type	Conductor Size (AWG/ kcmil)	Trade Size (Metric Designator)												
		⅜ (12)	½ (16)	¾ (21)	1 (27)	1¼ (35)	1½ (41)	2 (53)	2½ (63)	3 (78)	3½ (91)	4 (103)	5 (129)	6 (155)
	600	—	0	0	0	1	1	1	3	4	6	8	—	—
	700	—	0	0	0	1	1	1	2	4	5	7	—	—
	750	—	0	0	0	1	1	1	2	3	5	6	—	—
	900	—	0	0	0	0	1	1	1	3	4	5	—	—
	1000	—	0	0	0	0	1	1	1	3	4	5	—	—

Definition: *Compact stranding* is the result of a manufacturing process where the stranded conductor is compressed to the extent that the interstices (voids between strand wires) are virtually eliminated.

Table C.2 Maximum Number of Conductors or Fixture Wires in Electrical Nonmetallic Tubing (ENT) (Based on Chapter 9: Table 1, Table 4, and Table 5)

Type	Conductor Size (AWG/ kcmil)	Trade Size (Metric Designator)												
		⅜ (12)	½ (16)	¾ (21)	1 (27)	1¼ (35)	1½ (41)	2 (53)	2½ (63)	3 (78)	3½ (91)	4 (103)	5 (129)	6 (155)
CONDUCTORS														
RHH, RHW, RHW-2	14	—	4	7	11	20	27	45	—	—	—	—	—	—
	12	—	3	5	9	16	22	37	—	—	—	—	—	—
	10	—	2	4	7	13	18	30	—	—	—	—	—	—
	8	—	1	2	4	7	9	15	—	—	—	—	—	—
	6	—	1	1	3	5	7	12	—	—	—	—	—	—
	4	—	1	1	2	4	6	10	—	—	—	—	—	—
	3	—	1	1	1	4	5	8	—	—	—	—	—	—
	2	—	1	1	1	3	4	7	—	—	—	—	—	—
	1	—	0	1	1	1	3	5	—	—	—	—	—	—
	1/0	—	0	1	1	1	2	4	—	—	—	—	—	—
	2/0	—	0	0	1	1	1	3	—	—	—	—	—	—
	3/0	—	0	0	1	1	1	3	—	—	—	—	—	—
	4/0	—	0	0	1	1	1	2	—	—	—	—	—	—
	250	—	0	0	0	1	1	1	—	—	—	—	—	—
	300	—	0	0	0	1	1	1	—	—	—	—	—	—
	350	—	0	0	0	1	1	1	—	—	—	—	—	—
	400	—	0	0	0	1	1	1	—	—	—	—	—	—
	500	—	0	0	0	0	1	1	—	—	—	—	—	—
	600	—	0	0	0	0	1	1	—	—	—	—	—	—
	700	—	0	0	0	0	0	1	—	—	—	—	—	—
	750	—	0	0	0	0	0	1	—	—	—	—	—	—
	800	—	0	0	0	0	0	1	—	—	—	—	—	—
	900	—	0	0	0	0	0	1	—	—	—	—	—	—
	1000	—	0	0	0	0	0	1	—	—	—	—	—	—
	1250	—	0	0	0	0	0	0	—	—	—	—	—	—
	1500	—	0	0	0	0	0	0	—	—	—	—	—	—
	1750	—	0	0	0	0	0	0	—	—	—	—	—	—
	2000	—	0	0	0	0	0	0	—	—	—	—	—	—
TW, THHW, THW, THW-2	14	—	8	14	24	42	57	94	—	—	—	—	—	—
	12	—	6	11	18	32	44	72	—	—	—	—	—	—
	10	—	4	8	13	24	32	54	—	—	—	—	—	—
	8	—	2	4	7	13	18	30	—	—	—	—	—	—
RHH*, RHW*, RHW-2*	14	—	5	9	16	28	38	63	—	—	—	—	—	—
	12	—	4	8	13	22	30	50	—	—	—	—	—	—
	10	—	3	6	10	17	24	39	—	—	—	—	—	—
	8	—	1	3	6	10	14	23	—	—	—	—	—	—
TW, THW, THHW, THW-2, RHH*, RHW*, RHW-2*	6	—	1	2	4	8	11	18	—	—	—	—	—	—
	4	—	1	1	3	6	8	13	—	—	—	—	—	—
	3	—	1	1	3	5	7	11	—	—	—	—	—	—
	2	—	1	1	2	4	6	10	—	—	—	—	—	—
	1	—	0	1	1	3	4	7	—	—	—	—	—	—
	1/0	—	0	1	1	2	3	6	—	—	—	—	—	—
	2/0	—	0	1	1	1	3	5	—	—	—	—	—	—
	3/0	—	0	1	1	1	2	4	—	—	—	—	—	—
	4/0	—	0	0	1	1	1	3	—	—	—	—	—	—
	250	—	0	0	1	1	1	3	—	—	—	—	—	—
	300	—	0	0	1	1	1	2	—	—	—	—	—	—
	350	—	0	0	0	1	1	1	—	—	—	—	—	—
	400	—	0	0	0	1	1	1	—	—	—	—	—	—
	500	—	0	0	0	1	1	1	—	—	—	—	—	—
	600	—	0	0	0	0	1	1	—	—	—	—	—	—
	700	—	0	0	0	0	1	1	—	—	—	—	—	—
	750	—	0	0	0	0	1	1	—	—	—	—	—	—

(continues)

Table C.2   *Continued*

Type	Conductor Size (AWG/ kcmil)	⅜ (12)	½ (16)	¾ (21)	1 (27)	1¼ (35)	1½ (41)	2 (53)	2½ (63)	3 (78)	3½ (91)	4 (103)	5 (129)	6 (155)
						Trade Size (Metric Designator)								
	800	—	0	0	0	0	1	1	—	—	—	—	—	—
	900	—	0	0	0	0	0	1	—	—	—	—	—	—
	1000	—	0	0	0	0	0	1	—	—	—	—	—	—
	1250	—	0	0	0	0	0	1	—	—	—	—	—	—
	1500	—	0	0	0	0	0	1	—	—	—	—	—	—
	1750	—	0	0	0	0	0	0	—	—	—	—	—	—
	2000	—	0	0	0	0	0	0	—	—	—	—	—	—
THHN, THWN, THWN-2	14	—	11	21	34	60	82	135	—	—	—	—	—	—
	12	—	8	15	25	43	59	99	—	—	—	—	—	—
	10	—	5	9	15	27	37	62	—	—	—	—	—	—
	8	—	3	5	9	16	21	36	—	—	—	—	—	—
	6	—	1	4	6	11	15	26	—	—	—	—	—	—
	4	—	1	2	4	7	9	16	—	—	—	—	—	—
	3	—	1	1	3	6	8	13	—	—	—	—	—	—
	2	—	1	1	3	5	7	11	—	—	—	—	—	—
	1	—	1	1	1	3	5	8	—	—	—	—	—	—
	1/0	—	1	1	1	3	4	7	—	—	—	—	—	—
	2/0	—	0	1	1	2	3	6	—	—	—	—	—	—
	3/0	—	0	1	1	1	3	5	—	—	—	—	—	—
	4/0	—	0	1	1	1	2	4	—	—	—	—	—	—
	250	—	0	0	1	1	1	3	—	—	—	—	—	—
	300	—	0	0	1	1	1	3	—	—	—	—	—	—
	350	—	0	0	1	1	1	2	—	—	—	—	—	—
	400	—	0	0	0	1	1	1	—	—	—	—	—	—
	500	—	0	0	0	1	1	1	—	—	—	—	—	—
	600	—	0	0	0	1	1	1	—	—	—	—	—	—
	700	—	0	0	0	0	1	1	—	—	—	—	—	—
	750	—	0	0	0	0	1	1	—	—	—	—	—	—
	800	—	0	0	0	0	1	1	—	—	—	—	—	—
	900	—	0	0	0	0	1	1	—	—	—	—	—	—
	1000	—	0	0	0	0	0	1	—	—	—	—	—	—
FEP, FEPB, PFA, PFAH, TFE	14	—	11	20	33	58	79	131	—	—	—	—	—	—
	12	—	8	15	24	42	58	96	—	—	—	—	—	—
	10	—	6	10	17	30	41	69	—	—	—	—	—	—
	8	—	3	6	10	17	24	39	—	—	—	—	—	—
	6	—	2	4	7	12	17	28	—	—	—	—	—	—
	4	—	1	3	5	8	12	19	—	—	—	—	—	—
	3	—	1	2	4	7	10	16	—	—	—	—	—	—
	2	—	1	1	3	6	8	13	—	—	—	—	—	—
PFA, PFAH, TFE	1	—	1	1	2	4	5	9	—	—	—	—	—	—
PFA, PFAH, TFE, Z	1/0	—	1	1	1	3	4	8	—	—	—	—	—	—
	2/0	—	0	1	1	3	4	6	—	—	—	—	—	—
	3/0	—	0	1	1	2	3	5	—	—	—	—	—	—
	4/0	—	0	1	1	1	2	4	—	—	—	—	—	—
Z	14	—	13	24	40	70	95	158	—	—	—	—	—	—
	12	—	9	17	28	49	68	112	—	—	—	—	—	—
	10	—	6	10	17	30	41	69	—	—	—	—	—	—
	8	—	3	6	11	19	26	43	—	—	—	—	—	—
	6	—	2	4	7	13	18	30	—	—	—	—	—	—
	4	—	1	3	5	9	12	21	—	—	—	—	—	—
	3	—	1	2	4	6	9	15	—	—	—	—	—	—
	2	—	1	1	3	5	7	12	—	—	—	—	—	—
	1	—	1	1	2	4	6	10	—	—	—	—	—	—

(continues)

Table C.2   *Continued*

Type	Conductor Size (AWG/ kcmil)	Trade Size (Metric Designator)												
		⅜ (12)	½ (16)	¾ (21)	1 (27)	1¼ (35)	1½ (41)	2 (53)	2½ (63)	3 (78)	3½ (91)	4 (103)	5 (129)	6 (155)
XHHW, ZW, XHHW-2, XHH	14	—	8	14	24	42	57	94	—	—	—	—	—	—
	12	—	6	11	18	32	44	72	—	—	—	—	—	—
	10	—	4	8	13	24	32	54	—	—	—	—	—	—
	8	—	2	4	7	13	18	30	—	—	—	—	—	—
	6	—	1	3	5	10	13	22	—	—	—	—	—	—
	4	—	1	2	4	7	9	16	—	—	—	—	—	—
	3	—	1	1	3	6	8	13	—	—	—	—	—	—
	2	—	1	1	3	5	7	11	—	—	—	—	—	—
XHHW, XHHW-2, XHH	1	—	1	1	1	3	5	8	—	—	—	—	—	—
	1/0	—	0	1	1	3	4	7	—	—	—	—	—	—
	2/0	—	0	1	1	2	3	6	—	—	—	—	—	—
	3/0	—	0	1	1	1	3	5	—	—	—	—	—	—
	4/0	—	0	1	1	1	2	4	—	—	—	—	—	—
	250	—	0	0	1	1	1	3	—	—	—	—	—	—
	300	—	0	0	1	1	1	3	—	—	—	—	—	—
	350	—	0	0	1	1	1	2	—	—	—	—	—	—
	400	—	0	0	0	1	1	1	—	—	—	—	—	—
	500	—	0	0	0	1	1	1	—	—	—	—	—	—
	600	—	0	0	0	1	1	1	—	—	—	—	—	—
	700	—	0	0	0	0	1	1	—	—	—	—	—	—
	750	—	0	0	0	0	1	1	—	—	—	—	—	—
	800	—	0	0	0	0	1	1	—	—	—	—	—	—
	900	—	0	0	0	0	1	1	—	—	—	—	—	—
	1000	—	0	0	0	0	0	1	—	—	—	—	—	—
	1250	—	0	0	0	0	0	1	—	—	—	—	—	—
	1500	—	0	0	0	0	0	1	—	—	—	—	—	—
	1750	—	0	0	0	0	0	0	—	—	—	—	—	—
	2000	—	0	0	0	0	0	0	—	—	—	—	—	—
FIXTURE WIRES														
RFH-2, FFH-2, RFH1-2	18	—	8	14	23	40	54	90	—	—	—	—	—	—
	16	—	6	12	19	33	46	76	—	—	—	—	—	—
SF-2, SFF-2	18	—	10	17	29	50	69	114	—	—	—	—	—	—
	16	—	8	14	24	42	57	94	—	—	—	—	—	—
	14	—	6	12	19	33	46	76	—	—	—	—	—	—
SF-1, SFF-1	18	—	17	31	51	89	122	202	—	—	—	—	—	—
RFH-1, TF, TFF, XF, XFF	18	—	13	23	38	66	90	149	—	—	—	—	—	—
	16	—	10	18	30	53	73	120	—	—	—	—	—	—
XF, XFF	14	—	8	14	24	42	57	94	—	—	—	—	—	—
TFN, TFFN	18	—	20	37	60	105	144	239	—	—	—	—	—	—
	16	—	16	28	46	80	110	183	—	—	—	—	—	—
PF, PFF, PGF, PGFF, PAF, PTF, PTFF, PAFF	18	—	19	35	57	100	137	227	—	—	—	—	—	—
	16	—	15	27	44	77	106	175	—	—	—	—	—	—
	14	—	11	20	33	58	79	131	—	—	—	—	—	—
ZF, ZFF, ZHF	18	—	25	45	74	129	176	292	—	—	—	—	—	—
	16	—	18	33	54	95	130	216	—	—	—	—	—	—
	14	—	13	24	40	70	95	158	—	—	—	—	—	—
KF-2, KFF-2	18	—	38	67	111	193	265	439	—	—	—	—	—	—
	16	—	26	47	77	135	184	306	—	—	—	—	—	—
	14	—	18	31	52	91	124	205	—	—	—	—	—	—
	12	—	12	22	36	63	86	143	—	—	—	—	—	—
	10	—	8	14	24	42	57	94	—	—	—	—	—	—

(continues)

Table C.2 *Continued*

Type	Conductor Size (AWG/ kcmil)	Trade Size (Metric Designator)												
		3/8 (12)	1/2 (16)	3/4 (21)	1 (27)	1 1/4 (35)	1 1/2 (41)	2 (53)	2 1/2 (63)	3 (78)	3 1/2 (91)	4 (103)	5 (129)	6 (155)
KF-1, KFF-1	18	—	44	78	128	223	305	506	—	—	—	—	—	—
	16	—	31	55	90	157	214	355	—	—	—	—	—	—
	14	—	20	37	60	105	144	239	—	—	—	—	—	—
	12	—	13	24	40	70	95	158	—	—	—	—	—	—
	10	—	9	16	26	45	62	103	—	—	—	—	—	—
XF, XFF	12	—	4	8	13	22	30	50	—	—	—	—	—	—
	10	—	3	6	10	17	24	39	—	—	—	—	—	—

Notes:

1. This table is for concentric stranded conductors only. For compact stranded conductors, Table C.2(A) should be used.

2. Two-hour fire-rated RHH cable has ceramifiable insulation, which has much larger diameters than other RHH wires. Consult manufacturer's conduit fill tables.

*Types RHH, RHW, and RHW-2 without outer covering.

Table C.2(A) Maximum Number of Conductors or Fixture Wires in Electrical Nonmetallic Tubing (ENT) (Based on Chapter 9: Table 1, Table 4, and Table 5A)

Type	Conductor Size (AWG/ kcmil)	Trade Size (Metric Designator)												
		⅜ (12)	½ (16)	¾ (21)	1 (27)	1¼ (35)	1½ (41)	2 (53)	2½ (63)	3 (78)	3½ (91)	4 (103)	5 (129)	6 (155)
COMPACT CONDUCTORS														
THW, THW-2, THHW	8	—	1	4	6	11	15	26	—	—	—	—	—	—
	6	—	1	3	5	9	12	20	—	—	—	—	—	—
	4	—	1	1	3	6	9	15	—	—	—	—	—	—
	2	—	1	1	2	5	6	11	—	—	—	—	—	—
	1	—	1	1	1	3	4	7	—	—	—	—	—	—
	1/0	—	0	1	1	3	4	6	—	—	—	—	—	—
	2/0	—	0	1	1	2	3	5	—	—	—	—	—	—
	3/0	—	0	1	1	1	3	5	—	—	—	—	—	—
	4/0	—	0	1	1	1	2	4	—	—	—	—	—	—
	250	—	0	0	1	1	1	3	—	—	—	—	—	—
	300	—	0	0	1	1	1	2	—	—	—	—	—	—
	350	—	0	0	1	1	1	2	—	—	—	—	—	—
	400	—	0	0	0	1	1	1	—	—	—	—	—	—
	500	—	0	0	0	1	1	1	—	—	—	—	—	—
	600	—	0	0	0	1	1	1	—	—	—	—	—	—
	700	—	0	0	0	0	1	1	—	—	—	—	—	—
	750	—	0	0	0	0	1	1	—	—	—	—	—	—
	900	—	0	0	0	0	1	1	—	—	—	—	—	—
	1000	—	0	0	0	0	1	1	—	—	—	—	—	—
THHN, THWN, THWN-2	8	—	—	—	—	—	—	—	—	—	—	—	—	—
	6	—	2	4	7	13	17	29	—	—	—	—	—	—
	4	—	1	2	4	8	11	18	—	—	—	—	—	—
	2	—	1	1	3	5	8	13	—	—	—	—	—	—
	1	—	1	1	2	4	6	9	—	—	—	—	—	—
	1/0	—	1	1	1	3	5	8	—	—	—	—	—	—
	2/0	—	0	1	1	3	4	7	—	—	—	—	—	—
	3/0	—	0	1	1	2	3	5	—	—	—	—	—	—
	4/0	—	0	1	1	1	3	4	—	—	—	—	—	—
	250	—	0	0	1	1	1	3	—	—	—	—	—	—
	300	—	0	0	1	1	1	3	—	—	—	—	—	—
	350	—	0	0	1	1	1	3	—	—	—	—	—	—
	400	—	0	0	1	1	1	2	—	—	—	—	—	—
	500	—	0	0	0	1	1	1	—	—	—	—	—	—
	600	—	0	0	0	1	1	1	—	—	—	—	—	—
	700	—	0	0	0	1	1	1	—	—	—	—	—	—
	750	—	0	0	0	1	1	1	—	—	—	—	—	—
	900	—	0	0	0	0	1	1	—	—	—	—	—	—
	1000	—	0	0	0	0	1	1	—	—	—	—	—	—
XHHW, XHHW-2	8	—	3	5	8	14	20	33	—	—	—	—	—	—
	6	—	1	4	6	11	15	25	—	—	—	—	—	—
	4	—	1	2	4	8	11	18	—	—	—	—	—	—
	2	—	1	1	3	5	8	13	—	—	—	—	—	—
	1	—	1	1	2	4	6	9	—	—	—	—	—	—
	1/0	—	1	1	1	3	5	8	—	—	—	—	—	—
	2/0	—	1	1	1	3	4	7	—	—	—	—	—	—
	3/0	—	0	1	1	2	3	5	—	—	—	—	—	—
	4/0	—	0	1	1	1	3	5	—	—	—	—	—	—
	250	—	0	0	1	1	1	4	—	—	—	—	—	—
	300	—	0	0	1	1	1	3	—	—	—	—	—	—
	350	—	0	0	1	1	1	3	—	—	—	—	—	—
	400	—	0	0	1	1	1	2	—	—	—	—	—	—
	500	—	0	0	0	1	1	1	—	—	—	—	—	—
	600	—	0	0	0	1	1	1	—	—	—	—	—	—

(continues)

Table C.2(A)   *Continued*

Type	Conductor Size (AWG/ kcmil)	Trade Size (Metric Designator)												
		³⁄₈ (12)	¹⁄₂ (16)	³⁄₄ (21)	1 (27)	1¹⁄₄ (35)	1¹⁄₂ (41)	2 (53)	2¹⁄₂ (63)	3 (78)	3¹⁄₂ (91)	4 (103)	5 (129)	6 (155)
	700	—	0	0	0	1	1	1	—	—	—	—	—	—
	750	—	0	0	0	1	1	1	—	—	—	—	—	—
	900	—	0	0	0	0	1	1	—	—	—	—	—	—
	1000	—	0	0	0	0	1	1	—	—	—	—	—	—

Definition: *Compact stranding* is the result of a manufacturing process where the stranded conductor is compressed to the extent that the interstices (voids between strand wires) are virtually eliminated.

Table C.3 Maximum Number of Conductors or Fixture Wires in Flexible Metal Conduit (FMC)
(Based on Chapter 9: Table 1, Table 4, and Table 5)

Type	Conductor Size (AWG/kcmil)	Trade Size (Metric Designator)												
		3/8 (12)	1/2 (16)	3/4 (21)	1 (27)	1¼ (35)	1½ (41)	2 (53)	2½ (63)	3 (78)	3½ (91)	4 (103)	5 (129)	6 (155)
CONDUCTORS														
RHH, RHW, RHW-2	14	1	4	7	11	17	25	44	67	96	131	171	—	—
	12	1	3	6	9	14	21	37	55	80	109	142	—	—
	10	1	3	5	7	11	17	30	45	64	88	115	—	—
	8	0	1	2	4	6	9	15	23	34	46	60	—	—
	6	0	1	1	3	5	7	12	19	27	37	48	—	—
	4	0	1	1	2	4	5	10	14	21	29	37	—	—
	3	0	1	1	1	3	5	8	13	18	25	33	—	—
	2	0	1	1	1	3	4	7	11	16	22	28	—	—
	1	0	0	1	1	1	2	5	7	10	14	19	—	—
	1/0	0	0	1	1	1	2	4	6	9	12	16	—	—
	2/0	0	0	1	1	1	1	3	5	8	11	14	—	—
	3/0	0	0	0	1	1	1	3	5	7	9	12	—	—
	4/0	0	0	0	1	1	1	2	4	6	8	10	—	—
	250	0	0	0	0	1	1	1	3	4	6	8	—	—
	300	0	0	0	0	1	1	1	2	4	5	7	—	—
	350	0	0	0	0	1	1	1	2	3	5	6	—	—
	400	0	0	0	0	0	1	1	1	3	4	6	—	—
	500	0	0	0	0	0	1	1	1	3	4	5	—	—
	600	0	0	0	0	0	1	1	1	2	3	4	—	—
	700	0	0	0	0	0	0	1	1	1	3	3	—	—
	750	0	0	0	0	0	0	1	1	1	2	3	—	—
	800	0	0	0	0	0	0	1	1	1	2	3	—	—
	900	0	0	0	0	0	0	1	1	1	2	3	—	—
	1000	0	0	0	0	0	0	1	1	1	1	3	—	—
	1250	0	0	0	0	0	0	0	1	1	1	1	—	—
	1500	0	0	0	0	0	0	0	1	1	1	1	—	—
	1750	0	0	0	0	0	0	0	1	1	1	1	—	—
	2000	0	0	0	0	0	0	0	0	1	1	1	—	—
TW, THHW, THW, THW-2	14	3	9	15	23	36	53	94	141	203	277	361	—	—
	12	2	7	11	18	28	41	72	108	156	212	277	—	—
	10	1	5	8	13	21	30	54	81	116	158	207	—	—
	8	1	3	5	7	11	17	30	45	64	88	115	—	—
RHH*, RHW*, RHW-2*	14	1	6	10	15	24	35	62	94	135	184	240	—	—
	12	1	5	8	12	19	28	50	75	108	148	193	—	—
	10	1	4	6	10	15	22	39	59	85	115	151	—	—
	8	1	1	4	6	9	13	23	35	51	69	90	—	—
TW, THW, THHW, THW-2, RHH*, RHW*, RHW-2*	6	1	1	3	4	7	10	18	27	39	53	69	—	—
	4	0	1	1	3	5	7	13	20	29	39	51	—	—
	3	0	1	1	3	4	6	11	17	25	34	44	—	—
	2	0	1	1	2	4	5	10	14	21	29	37	—	—
	1	0	1	1	1	2	4	7	10	15	20	26	—	—
	1/0	0	0	1	1	1	3	6	9	12	17	22	—	—
	2/0	0	0	1	1	1	3	5	7	10	14	19	—	—
	3/0	0	0	1	1	1	2	4	6	9	12	16	—	—
	4/0	0	0	0	1	1	1	3	5	7	10	13	—	—
	250	0	0	0	1	1	1	3	4	6	8	11	—	—
	300	0	0	0	1	1	1	2	3	5	7	9	—	—
	350	0	0	0	0	1	1	1	3	4	6	8	—	—
	400	0	0	0	0	1	1	1	3	4	6	7	—	—
	500	0	0	0	0	1	1	1	2	3	5	6	—	—
	600	0	0	0	0	0	1	1	1	3	4	5	—	—
	700	0	0	0	0	0	1	1	1	2	3	4	—	—
	750	0	0	0	0	0	1	1	1	2	3	4	—	—

(continues)

Table C.3 *Continued*

Type	Conductor Size (AWG/ kcmil)	Trade Size (Metric Designator)												
		⅜ (12)	½ (16)	¾ (21)	1 (27)	1¼ (35)	1½ (41)	2 (53)	2½ (63)	3 (78)	3½ (91)	4 (103)	5 (129)	6 (155)
	800	0	0	0	0	0	1	1	1	1	3	4	—	—
	900	0	0	0	0	0	0	1	1	1	3	3	—	—
	1000	0	0	0	0	0	0	1	1	1	2	3	—	—
	1250	0	0	0	0	0	0	1	1	1	1	2	—	—
	1500	0	0	0	0	0	0	0	1	1	1	1	—	—
	1750	0	0	0	0	0	0	0	1	1	1	1	—	—
	2000	0	0	0	0	0	0	0	1	1	1	1	—	—
THHN, THWN, THWN-2	14	4	13	22	33	52	76	135	202	291	396	518	—	—
	12	3	9	16	24	38	56	98	147	212	289	378	—	—
	10	1	6	10	15	24	35	62	93	134	182	238	—	—
	8	1	3	6	9	14	20	35	53	77	105	137	—	—
	6	1	2	4	6	10	14	25	38	55	76	99	—	—
	4	0	1	2	4	6	9	16	24	34	46	61	—	—
	3	0	1	1	3	5	7	13	20	29	39	51	—	—
	2	0	1	1	3	4	6	11	17	24	33	43	—	—
	1	0	1	1	1	3	4	8	12	18	24	32	—	—
	1/0	0	1	1	1	2	4	7	10	15	20	27	—	—
	2/0	0	0	1	1	1	3	6	9	12	17	22	—	—
	3/0	0	0	1	1	1	2	5	7	10	14	18	—	—
	4/0	0	0	1	1	1	1	4	6	8	12	15	—	—
	250	0	0	0	1	1	1	3	5	7	9	12	—	—
	300	0	0	0	1	1	1	3	4	6	8	11	—	—
	350	0	0	0	1	1	1	2	3	5	7	9	—	—
	400	0	0	0	0	1	1	1	3	5	6	8	—	—
	500	0	0	0	0	1	1	1	2	4	5	7	—	—
	600	0	0	0	0	0	1	1	1	3	4	5	—	—
	700	0	0	0	0	0	1	1	1	3	4	5	—	—
	750	0	0	0	0	0	1	1	1	2	3	4	—	—
	800	0	0	0	0	0	1	1	1	2	3	4	—	—
	900	0	0	0	0	0	0	1	1	1	3	4	—	—
	1000	0	0	0	0	0	0	1	1	1	3	3	—	—
FEP, FEPB, PFA, PFAH, TFE	14	4	12	21	32	51	74	130	196	282	385	502	—	—
	12	3	9	15	24	37	54	95	143	206	281	367	—	—
	10	2	6	11	17	26	39	68	103	148	201	263	—	—
	8	1	4	6	10	15	22	39	59	85	115	151	—	—
	6	1	2	4	7	11	16	28	42	60	82	107	—	—
	4	1	1	3	5	7	11	19	29	42	57	75	—	—
	3	0	1	2	4	6	9	16	24	35	48	62	—	—
	2	0	1	1	3	5	7	13	20	29	39	51	—	—
PFA, PFAH, TFE	1	0	1	1	2	3	5	9	14	20	27	36	—	—
PFA, PFAH, TFE, Z	1/0	0	1	1	1	3	4	8	11	17	23	30	—	—
	2/0	0	1	1	1	2	3	6	9	14	19	24	—	—
	3/0	0	0	1	1	1	3	5	8	11	15	20	—	—
	4/0	0	0	1	1	1	2	4	6	9	13	16	—	—
Z	14	5	15	25	39	61	89	157	236	340	463	605	—	—
	12	4	11	18	28	43	63	111	168	241	329	429	—	—
	10	2	6	11	17	26	39	68	103	148	201	263	—	—
	8	1	4	7	11	17	24	43	65	93	127	166	—	—
	6	1	3	5	7	12	17	30	45	65	89	117	—	—
	4	1	1	3	5	8	12	21	31	45	61	80	—	—
	3	0	1	2	4	6	8	15	23	33	45	58	—	—
	2	0	1	1	3	5	7	12	19	27	37	49	—	—
	1	0	1	1	2	4	6	10	15	22	30	39	—	—

(continues)

Table C.3 *Continued*

Type	Conductor Size (AWG/kcmil)	Trade Size (Metric Designator)												
		⅜ (12)	½ (16)	¾ (21)	1 (27)	1¼ (35)	1½ (41)	2 (53)	2½ (63)	3 (78)	3½ (91)	4 (103)	5 (129)	6 (155)
XHHW, ZW, XHHW-2, XHH	14	3	9	15	23	36	53	94	141	203	277	361	—	—
	12	2	7	11	18	28	41	72	108	156	212	277	—	—
	10	1	5	8	13	21	30	54	81	116	158	207	—	—
	8	1	3	5	7	11	17	30	45	64	88	115	—	—
	6	1	1	3	5	8	12	22	33	48	65	85	—	—
	4	0	1	2	4	6	9	16	24	34	47	61	—	—
	3	0	1	1	3	5	7	13	20	29	40	52	—	—
	2	0	1	1	3	4	6	11	17	24	33	44	—	—
XHH, XHHW, XHHW-2	1	0	1	1	1	3	5	8	13	18	25	32	—	—
	1/0	0	1	1	1	2	4	7	10	15	21	27	—	—
	2/0	0	0	1	1	2	3	6	9	13	17	23	—	—
	3/0	0	0	1	1	1	3	5	7	10	14	19	—	—
	4/0	0	0	1	1	1	2	4	6	9	12	15	—	—
	250	0	0	0	1	1	1	3	5	7	10	13	—	—
	300	0	0	0	1	1	1	3	4	6	8	11	—	—
	350	0	0	0	1	1	1	2	4	5	7	9	—	—
	400	0	0	0	0	1	1	1	3	5	6	8	—	—
	500	0	0	0	0	1	1	1	3	4	5	7	—	—
	600	0	0	0	0	0	1	1	1	3	4	5	—	—
	700	0	0	0	0	0	1	1	1	3	4	5	—	—
	750	0	0	0	0	0	1	1	1	2	3	4	—	—
	800	0	0	0	0	0	1	1	1	2	3	4	—	—
	900	0	0	0	0	0	0	1	1	1	3	4	—	—
	1000	0	0	0	0	0	0	1	1	1	3	3	—	—
	1250	0	0	0	0	0	0	1	1	1	1	3	—	—
	1500	0	0	0	0	0	0	1	1	1	1	2	—	—
	1750	0	0	0	0	0	0	0	1	1	1	1	—	—
	2000	0	0	0	0	0	0	0	1	1	1	1	—	—
FIXTURE WIRES														
RFH-2, FFH-2, RFHH-2	18	3	8	14	22	35	51	90	135	195	265	346	—	—
	16	2	7	12	19	29	43	76	114	164	223	292	—	—
SF-2, SFF-2	18	4	11	18	28	44	64	113	170	246	334	437	—	—
	16	3	9	15	23	36	53	94	141	203	277	361	—	—
	14	2	7	12	19	29	43	76	114	164	223	292	—	—
SF-1, SFF-1	18	7	19	33	50	78	114	201	302	435	592	773	—	—
RFH-1, TF, TFF, XF, XFF	18	5	14	24	37	58	84	148	223	321	437	571	—	—
	16	4	11	19	30	47	68	120	180	259	353	461	—	—
XF, XFF	14	3	9	15	23	36	53	94	141	203	277	361	—	—
TFN, TFFN	18	8	23	38	59	93	135	237	357	514	699	914	—	—
	16	6	17	29	45	71	103	181	272	392	534	698	—	—
PF, PFF, PGF, PGFF, PAF, PTF, PTFF, PAFF	18	8	22	36	56	88	128	225	338	487	663	866	—	—
	16	6	17	28	43	68	99	174	262	377	513	670	—	—
	14	4	12	21	32	51	74	130	196	282	385	502	—	—
ZF, ZFF, ZHF	18	10	28	47	72	113	165	290	436	628	855	1117	—	—
	16	7	20	35	53	83	122	214	322	463	631	824	—	—
	14	5	15	25	39	61	89	157	236	340	463	605	—	—
KF-2, KFF-2	18	15	42	71	109	170	247	436	654	942	1282	1675	—	—
	16	10	29	49	76	118	173	304	456	657	895	1169	—	—
	14	7	20	33	51	80	116	204	307	442	601	785	—	—
	12	5	13	23	35	55	80	142	213	307	418	546	—	—
	10	3	9	15	23	36	53	94	141	203	277	361	—	—
KF-1, KFF-1	18	18	48	82	125	196	286	503	755	1087	1480	1933	—	—

(continues)

Table C.3 *Continued*

Type	Conductor Size (AWG/ kcmil)	Trade Size (Metric Designator)												
		⅜ (12)	½ (16)	¾ (21)	1 (27)	1¼ (35)	1½ (41)	2 (53)	2½ (63)	3 (78)	3½ (91)	4 (103)	5 (129)	6 (155)
	16	12	34	57	88	138	201	353	530	764	1040	1358	—	—
	14	8	23	38	59	93	135	237	357	514	699	914	—	—
	12	5	15	25	39	61	89	157	236	340	463	605	—	—
	10	3	10	16	25	40	58	103	154	222	303	395	—	—
XF, XFF	12	1	5	8	12	19	28	50	75	108	148	193	—	—
	10	1	4	6	10	15	22	39	59	85	115	151	—	—

Notes:

1. This table is for concentric stranded conductors only. For compact stranded conductors, Table C.3(A) should be used.

2. Two-hour fire-rated RHH cable has ceramifiable insulation, which has much larger diameters than other RHH wires. Consult manufacturer's conduit fill tables.

*Types RHH, RHW, and RHW-2 without outer covering.

Table C.3(A) Maximum Number of Conductors or Fixture Wires in Flexible Metal Conduit (FMC)
(Based on Chapter 9: Table 1, Table 4, and Table 5A)

Type	Conductor Size (AWG/ kcmil)	⅜ (12)	½ (16)	¾ (21)	1 (27)	1¼ (35)	1½ (41)	2 (53)	2½ (63)	3 (78)	3½ (91)	4 (103)	5 (129)	6 (155)
						COMPACT CONDUCTORS								
THW, THW-2, THHW	8	1	2	4	6	10	14	25	38	55	75	98	—	—
	6	1	1	3	5	7	11	20	29	43	58	76	—	—
	4	0	1	2	3	5	8	15	22	32	43	57	—	—
	2	0	1	1	2	4	6	11	16	23	32	42	—	—
	1	0	1	1	1	3	4	7	11	16	22	29	—	—
	1/0	0	1	1	1	2	3	6	10	14	19	25	—	—
	2/0	0	0	1	1	1	3	5	8	12	16	21	—	—
	3/0	0	0	1	1	1	2	4	7	10	14	18	—	—
	4/0	0	0	1	1	1	1	4	6	8	11	15	—	—
	250	0	0	0	1	1	1	3	4	7	9	12	—	—
	300	0	0	0	1	1	1	2	4	6	8	10	—	—
	350	0	0	0	1	1	1	2	3	5	7	9	—	—
	400	0	0	0	0	1	1	1	3	5	6	8	—	—
	500	0	0	0	0	1	1	1	3	4	5	7	—	—
	600	0	0	0	0	0	1	1	1	3	4	6	—	—
	700	0	0	0	0	0	1	1	1	3	4	5	—	—
	750	0	0	0	0	0	1	1	1	2	3	5	—	—
	900	0	0	0	0	0	1	1	1	2	3	4	—	—
	1000	0	0	0	0	0	0	1	1	1	3	4	—	—
THHN, THWN, THWN-2	8	—	—	—	—	—	—	—	—	—	—	—	—	—
	6	1	3	4	7	11	16	29	43	62	85	111	—	—
	4	1	1	3	4	7	10	18	27	38	52	69	—	—
	2	0	1	1	3	5	7	13	19	28	38	49	—	—
	1	0	1	1	2	3	5	9	14	21	28	37	—	—
	1/0	0	1	1	1	3	4	8	12	17	24	31	—	—
	2/0	0	1	1	1	2	4	6	10	14	20	26	—	—
	3/0	0	0	1	1	1	3	5	8	12	17	22	—	—
	4/0	0	0	1	1	1	2	4	7	10	14	18	—	—
	250	0	0	1	1	1	1	3	5	8	11	14	—	—
	300	0	0	0	1	1	1	3	5	7	9	12	—	—
	350	0	0	0	1	1	1	3	4	6	8	10	—	—
	400	0	0	0	1	1	1	2	3	5	7	9	—	—
	500	0	0	0	0	1	1	1	3	4	6	8	—	—
	600	0	0	0	0	1	1	1	2	3	5	6	—	—
	700	0	0	0	0	0	1	1	1	3	4	6	—	—
	750	0	0	0	0	0	1	1	1	3	4	5	—	—
	900	0	0	0	0	0	1	1	1	2	3	4	—	—
	1000	0	0	0	0	0	0	1	1	1	3	4	—	—
XHHW, XHHW-2	8	1	3	5	8	13	19	33	50	71	97	127	—	—
	6	1	2	4	6	9	14	24	37	53	72	95	—	—
	4	1	1	3	4	7	10	18	27	38	52	69	—	—
	2	0	1	1	3	5	7	13	19	28	38	49	—	—
	1	0	1	1	2	3	5	9	14	21	28	37	—	—
	1/0	0	1	1	1	3	4	8	12	17	24	31	—	—
	2/0	0	1	1	1	2	4	7	10	15	20	26	—	—
	3/0	0	0	1	1	1	3	5	8	12	17	22	—	—
	4/0	0	0	1	1	1	2	4	7	10	14	18	—	—
	250	0	0	1	1	1	1	4	5	8	11	14	—	—
	300	0	0	0	1	1	1	3	5	7	9	12	—	—
	350	0	0	0	1	1	1	3	4	6	8	11	—	—
	400	0	0	0	1	1	1	2	4	5	7	10	—	—
	500	0	0	0	0	1	1	1	3	4	6	8	—	—

(continues)

Table C.3(A)   *Continued*

Type	Conductor Size (AWG/ kcmil)	Trade Size (Metric Designator)												
		⅜ (12)	½ (16)	¾ (21)	1 (27)	1¼ (35)	1½ (41)	2 (53)	2½ (63)	3 (78)	3½ (91)	4 (103)	5 (129)	6 (155)
	600	0	0	0	0	1	1	1	2	3	5	6	—	—
	700	0	0	0	0	0	1	1	1	3	4	6	—	—
	750	0	0	0	0	0	1	1	1	3	4	5	—	—
	900	0	0	0	0	0	1	1	1	2	3	4	—	—
	1000	0	0	0	0	0	1	1	1	2	3	4	—	—

Definition: *Compact stranding* is the result of a manufacturing process where the stranded conductor is compressed to the extent that the interstices (voids between strand wires) are virtually eliminated.

Table C.4 Maximum Number of Conductors or Fixture Wires in Intermediate Metal Conduit (IMC) (Based on Chapter 9: Table 1, Table 4, and Table 5)

Type	Conductor Size (AWG/kcmil)	Trade Size (Metric Designator)												
		⅜ (12)	½ (16)	¾ (21)	1 (27)	1¼ (35)	1½ (41)	2 (53)	2½ (63)	3 (78)	3½ (91)	4 (103)	5 (129)	6 (155)
CONDUCTORS														
RHH, RHW, RHW-2	14	—	4	8	13	22	30	49	70	108	144	186	—	—
	12	—	4	6	11	18	25	41	58	89	120	154	—	—
	10	—	3	5	8	15	20	33	47	72	97	124	—	—
	8	—	1	3	4	8	10	17	24	38	50	65	—	—
	6	—	1	1	3	6	8	14	19	30	40	52	—	—
	4	—	1	1	3	5	6	11	15	23	31	41	—	—
	3	—	1	1	2	4	6	9	13	21	28	36	—	—
	2	—	1	1	1	3	5	8	11	18	24	31	—	—
	1	—	0	1	1	2	3	5	7	12	16	20	—	—
	1/0	—	0	1	1	1	3	4	6	10	14	18	—	—
	2/0	—	0	1	1	1	2	4	6	9	12	15	—	—
	3/0	—	0	0	1	1	1	3	5	7	10	13	—	—
	4/0	—	0	0	1	1	1	3	4	6	9	11	—	—
	250	—	0	0	1	1	1	1	3	5	6	8	—	—
	300	—	0	0	0	1	1	1	3	4	6	7	—	—
	350	—	0	0	0	1	1	1	2	4	5	7	—	—
	400	—	0	0	0	1	1	1	2	3	5	6	—	—
	500	—	0	0	0	1	1	1	1	3	4	5	—	—
	600	—	0	0	0	0	1	1	1	2	3	4	—	—
	700	—	0	0	0	0	1	1	1	2	3	4	—	—
	750	—	0	0	0	0	1	1	1	1	3	4	—	—
	800	—	0	0	0	0	0	1	1	1	3	3	—	—
	900	—	0	0	0	0	0	1	1	1	2	3	—	—
	1000	—	0	0	0	0	0	1	1	1	2	3	—	—
	1250	—	0	0	0	0	0	1	1	1	1	1	—	—
	1500	—	0	0	0	0	0	0	1	1	1	1	—	—
	1750	—	0	0	0	0	0	0	1	1	1	1	—	—
	2000	—	0	0	0	0	0	0	1	1	1	1	—	—
TW, THHW, THW, THW-2	14	—	10	17	27	47	64	104	147	228	304	392	—	—
	12	—	7	13	21	36	49	80	113	175	234	301	—	—
	10	—	5	9	15	27	36	59	84	130	174	224	—	—
	8	—	3	5	8	15	20	33	47	72	97	124	—	—
RHH*, RHW*, RHW-2*	14	—	6	11	18	31	42	69	98	151	202	261	—	—
	12	—	5	9	14	25	34	56	79	122	163	209	—	—
	10	—	4	7	11	19	26	43	61	95	127	163	—	—
	8	—	2	4	7	12	16	26	37	57	76	98	—	—
TW, THW, THHW, THW-2, RHH*, RHW*, RHW-2*	6	—	1	3	5	9	12	20	28	43	58	75	—	—
	4	—	1	2	4	6	9	15	21	32	43	56	—	—
	3	—	1	1	3	6	8	13	18	28	37	48	—	—
	2	—	1	1	3	5	6	11	15	23	31	41	—	—
	1	—	1	1	1	3	4	7	11	16	22	28	—	—
	1/0	—	1	1	1	3	4	6	9	14	19	24	—	—
	2/0	—	0	1	1	2	3	5	8	12	16	20	—	—
	3/0	—	0	1	1	1	3	4	6	10	13	17	—	—
	4/0	—	0	1	1	1	2	4	5	8	11	14	—	—
	250	—	0	0	1	1	1	3	4	7	9	12	—	—
	300	—	0	0	1	1	1	2	4	6	8	10	—	—
	350	—	0	0	1	1	1	2	3	5	7	9	—	—
	400	—	0	0	0	1	1	1	3	4	6	8	—	—

(continues)

Table C.4 *Continued*

Type	Conductor Size (AWG/ kcmil)	Trade Size (Metric Designator)												
		⅜ (12)	½ (16)	¾ (21)	1 (27)	1¼ (35)	1½ (41)	2 (53)	2½ (63)	3 (78)	3½ (91)	4 (103)	5 (129)	6 (155)
	500	—	0	0	0	1	1	1	2	4	5	7	—	—
	600	—	0	0	0	1	1	1	1	3	4	5	—	—
	700	—	0	0	0	0	1	1	1	3	4	5	—	—
	750	—	0	0	0	0	1	1	1	2	3	4	—	—
	800	—	0	0	0	0	1	1	1	2	3	4	—	—
	900	—	0	0	0	0	1	1	1	2	3	4	—	—
	1000	—	0	0	0	0	0	1	1	1	3	3	—	—
	1250	—	0	0	0	0	0	1	1	1	1	3	—	—
	1500	—	0	0	0	0	0	1	1	1	1	2	—	—
	1750	—	0	0	0	0	0	0	1	1	1	1	—	—
	2000	—	0	0	0	0	0	0	1	1	1	1	—	—
THHN, THWN, THWN-2	14	—	14	24	39	68	91	149	211	326	436	562	—	—
	12	—	10	17	29	49	67	109	154	238	318	410	—	—
	10	—	6	11	18	31	42	69	97	150	200	258	—	—
	8	—	3	6	10	18	24	39	56	86	115	149	—	—
	6	—	2	4	7	13	17	28	40	62	83	107	—	—
	4	—	1	3	4	8	11	17	25	38	51	66	—	—
	3	—	1	2	4	6	9	15	21	32	43	56	—	—
	2	—	1	1	3	5	7	12	17	27	36	47	—	—
	1	—	1	1	2	4	5	9	13	20	27	35	—	—
	1/0	—	1	1	1	3	4	8	11	17	23	29	—	—
	2/0	—	1	1	1	3	4	6	9	14	19	24	—	—
	3/0	—	0	1	1	2	3	5	7	12	16	20	—	—
	4/0	—	0	1	1	1	2	4	6	9	13	17	—	—
	250	—	0	0	1	1	1	3	5	8	10	13	—	—
	300	—	0	0	1	1	1	3	4	7	9	12	—	—
	350	—	0	0	1	1	1	2	4	6	8	10	—	—
	400	—	0	0	1	1	1	2	3	5	7	9	—	—
	500	—	0	0	0	1	1	1	3	4	6	7	—	—
	600	—	0	0	0	1	1	1	2	3	5	6	—	—
	700	—	0	0	0	1	1	1	1	3	4	5	—	—
	750	—	0	0	0	1	1	1	1	3	4	5	—	—
	800	—	0	0	0	0	1	1	1	3	4	5	—	—
	900	—	0	0	0	0	1	1	1	2	3	4	—	—
	1000	—	0	0	0	0	1	1	1	2	3	4	—	—
FEP, FEPB, PFA, PFAH, TFE	14	—	13	23	38	66	89	145	205	317	423	545	—	—
	12	—	10	17	28	48	65	106	150	231	309	398	—	—
	10	—	7	12	20	34	46	76	107	166	221	285	—	—
	8	—	4	7	11	19	26	43	61	95	127	163	—	—
	6	—	3	5	8	14	19	31	44	67	90	116	—	—
	4	—	1	3	5	10	13	21	30	47	63	81	—	—
	3	—	1	3	4	8	11	18	25	39	52	68	—	—
	2	—	1	2	4	6	9	15	21	32	43	56	—	—
PFA, PFAH, TFE	1	—	1	1	2	4	6	10	14	22	30	39	—	—
PFA, PFAH, TFE, Z	1/0	—	1	1	1	4	5	8	12	19	25	32	—	—
	2/0	—	1	1	1	3	4	7	10	15	21	27	—	—
	3/0	—	0	1	1	2	3	6	8	13	17	22	—	—
	4/0	—	0	1	1	1	3	5	7	10	14	18	—	—
Z	14	—	16	28	46	79	107	175	247	381	510	657	—	—
	12	—	11	20	32	56	76	124	175	271	362	466	—	—
	10	—	7	12	20	34	46	76	107	166	221	285	—	—
	8	—	4	7	12	22	29	48	68	105	140	180	—	—

(continues)

Table C.4 *Continued*

Type	Conductor Size (AWG/ kcmil)	³⁄₈ (12)	½ (16)	¾ (21)	1 (27)	1¼ (35)	1½ (41)	2 (53)	2½ (63)	3 (78)	3½ (91)	4 (103)	5 (129)	6 (155)
								Trade Size (Metric Designator)						
	6	—	3	5	9	15	20	33	47	73	98	127	—	—
	4	—	1	3	6	10	14	23	33	50	67	87	—	—
	3	—	1	2	4	7	10	17	24	37	49	63	—	—
	2	—	1	1	3	6	8	14	20	30	41	53	—	—
	1	—	1	1	3	5	7	11	16	25	33	43	—	—
XHHW, ZW, XHHW-2, XHH	14	—	10	17	27	47	64	104	147	228	304	392	—	—
	12	—	7	13	21	36	49	80	113	175	234	301	—	—
	10	—	5	9	15	27	36	59	84	130	174	224	—	—
	8	—	3	5	8	15	20	33	47	72	97	124	—	—
	6	—	1	4	6	11	15	24	35	53	71	92	—	—
	4	—	1	3	4	8	11	18	25	39	52	67	—	—
	3	—	1	2	4	7	9	15	21	33	44	56	—	—
	2	—	1	1	3	5	7	12	18	27	37	47	—	—
XHHW, XHHW-2, XHH	1	—	1	1	2	4	6	9	13	20	27	35	—	—
	1/0	—	1	1	1	3	5	8	11	17	23	30	—	—
	2/0	—	1	1	1	3	4	6	9	14	19	25	—	—
	3/0	—	0	1	1	2	3	5	7	12	16	20	—	—
	4/0	—	0	1	1	1	2	4	6	10	13	17	—	—
	250	—	0	0	1	1	1	3	5	8	11	14	—	—
	300	—	0	0	1	1	1	3	4	7	9	12	—	—
	350	—	0	0	1	1	1	3	4	6	8	10	—	—
	400	—	0	0	1	1	1	2	3	5	7	9	—	—
	500	—	0	0	0	1	1	1	3	4	6	8	—	—
	600	—	0	0	0	1	1	1	2	3	5	6		
	700	—	0	0	0	1	1	1	1	3	4	5	—	—
	750	—	0	0	0	1	1	1	1	3	4	5	—	—
	800	—	0	0	0	0	1	1	1	3	4	5	—	—
	900	—	0	0	0	0	1	1	1	2	3	4	—	—
	1000	—	0	0	0	0	1	1	1	2	3	4	—	—
	1250	—	0	0	0	0	0	1	1	1	2	3	—	—
	1500	—	0	0	0	0	0	1	1	1	1	2	—	—
	1750	—	0	0	0	0	0	1	1	1	1	2	—	—
	2000	—	0	0	0	0	0	0	1	1	1	1	—	—
FIXTURE WIRES														
RFH-2, FFH-2, RFHH-2	18	—	9	16	26	45	61	100	141	218	292	376	—	—
	16	—	8	13	22	38	51	84	119	184	246	317	—	—
SF-2, SFF-2	18	—	12	20	33	57	77	126	178	275	368	474	—	—
	16	—	10	17	27	47	64	104	147	228	304	392	—	—
	14	—	8	13	22	38	51	84	119	184	246	317	—	—
SF-1, SFF-1	18	—	21	36	59	101	137	223	316	487	651	839	—	—
RFH-1, TF, TFF, XF, XFF	18	—	15	26	43	75	101	165	233	360	481	619	—	—
	16	—	12	21	35	60	81	133	188	290	388	500	—	—
XF, XFF	14	—	10	17	27	47	64	104	147	228	304	392	—	—
TFN, TFFN	18	—	25	42	69	119	162	264	373	576	769	991	—	—
	16	—	19	32	53	91	123	201	285	440	588	757	—	—
PF, PFF, PGF, PGFF, PAF, PTF, PTFF, PAFF	18	—	23	40	66	113	153	250	354	546	730	940	—	—
	16	—	18	31	51	88	118	193	274	422	564	727	—	—
	14	—	13	23	38	66	89	145	205	317	423	545	—	—

(continues)

Table C.4 *Continued*

Type	Conductor Size (AWG/ kcmil)	Trade Size (Metric Designator)												
		⅜ (12)	½ (16)	¾ (21)	1 (27)	1¼ (35)	1½ (41)	2 (53)	2½ (63)	3 (78)	3½ (91)	4 (103)	5 (129)	6 (155)
ZF, ZFF, ZHF	18	—	30	52	85	146	197	322	456	704	941	1211	—	—
	16	—	22	38	63	108	146	238	336	519	694	894	—	—
	14	—	16	28	46	79	107	175	247	381	510	657	—	—
KF-2, KFF-2	18	—	45	78	128	219	296	484	684	1056	1411	1817	—	—
	16	—	32	54	89	153	207	337	477	737	984	1268	—	—
	14	—	21	36	60	103	139	227	321	495	661	852	—	—
	12	—	15	25	41	71	96	158	223	344	460	592	—	—
	10	—	10	17	27	47	64	104	147	228	304	392	—	—
KF-1, KFF-1	18	—	52	90	147	253	342	558	790	1218	1628	2097	—	—
	16	—	37	63	103	178	240	392	555	856	1144	1473	—	—
	14	—	25	42	69	119	162	264	373	576	769	991	—	—
	12	—	16	28	46	79	107	175	247	381	510	657	—	—
	10	—	10	18	30	52	70	114	161	249	333	429	—	—
XF, XFF	12	—	5	9	14	25	34	56	79	122	163	209	—	—
	10	—	4	7	11	19	26	43	61	95	127	163	—	—

Notes:

1. This table is for concentric stranded conductors only. For compact stranded conductors, Table C.4(A) should be used.

2. Two-hour fire-rated RHH cable has ceramifiable insulation, which has much larger diameters than other RHH wires. Consult manufacturer's conduit fill tables.

*Types RHH, RHW, and RHW-2 without outer covering.

Table C.4(A) Maximum Number of Conductors or Fixture Wires in Intermediate Metal Conduit (IMC)
(Based on Chapter 9: Table 1, Table 4, and Table 5A)

Type	Conductor Size (AWG/ kcmil)	Trade Size (Metric Designator)												
		⅜ (12)	½ (16)	¾ (21)	1 (27)	1¼ (35)	1½ (41)	2 (53)	2½ (63)	3 (78)	3½ (91)	4 (103)	5 (129)	6 (155)
COMPACT CONDUCTORS														
THW, THW-2, THHW	8	—	2	4	7	13	17	28	40	62	83	107	—	—
	6	—	1	3	6	10	13	22	31	48	64	82	—	—
	4	—	1	2	4	7	10	16	23	36	48	62	—	—
	2	—	1	1	3	5	7	12	17	26	35	45	—	—
	1	—	1	1	1	4	5	8	12	18	25	32	—	—
	1/0	—	1	1	1	3	4	7	10	16	21	27	—	—
	2/0	—	0	1	1	3	4	6	9	13	18	23	—	—
	3/0	—	0	1	1	2	3	5	7	11	15	20	—	—
	4/0	—	0	1	1	1	2	4	6	9	13	16	—	—
	250	—	0	0	1	1	1	3	5	7	10	13	—	—
	300	—	0	0	1	1	1	3	4	6	9	11	—	—
	350	—	0	0	1	1	1	2	4	6	8	10	—	—
	400	—	0	0	1	1	1	2	3	5	7	9	—	—
	500	—	0	0	0	1	1	1	3	4	6	8	—	—
	600	—	0	0	0	1	1	1	2	3	5	6	—	—
	700	—	0	0	0	1	1	1	1	3	4	5	—	—
	750	—	0	0	0	1	1	1	1	3	4	5	—	—
	900	—	0	0	0	0	1	1	1	2	3	4	—	—
	1000	—	0	0	0	0	1	1	1	2	3	4	—	—
THHN, THWN, THWN-2	8	—	—	—	—	—	—	—	—	—	—	—	—	—
	6	—	3	5	8	14	19	32	45	70	93	120	—	—
	4	—	1	3	5	9	12	20	28	43	58	74	—	—
	2	—	1	1	3	6	8	14	20	31	41	53	—	—
	1	—	1	1	3	5	6	10	15	23	31	40	—	—
	1/0	—	1	1	2	4	5	9	13	20	26	34	—	—
	2/0	—	1	1	1	3	4	7	10	16	22	28	—	—
	3/0	—	0	1	1	3	4	6	9	14	18	24	—	—
	4/0	—	0	1	1	2	3	5	7	11	15	19	—	—
	250	—	0	1	1	1	2	4	6	9	12	15	—	—
	300	—	0	0	1	1	1	3	5	7	10	13	—	—
	350	—	0	0	1	1	1	3	4	7	9	11	—	—
	400	—	0	0	1	1	1	2	4	6	8	10	—	—
	500	—	0	0	1	1	1	2	3	5	7	9	—	—
	600	—	0	0	0	1	1	1	2	4	5	7	—	—
	700	—	0	0	0	1	1	1	2	3	5	6	—	—
	750	—	0	0	0	1	1	1	1	3	4	6	—	—
	900	—	0	0	0	0	1	1	1	3	3	5	—	—
	1000	—	0	0	0	0	1	1	1	2	3	4	—	—
XHHW, XHHW-2	8	—	3	6	9	16	22	37	52	80	107	138	—	—
	6	—	2	4	7	12	16	27	38	59	80	103	—	—
	4	—	1	3	5	9	12	20	28	43	58	74	—	—
	2	—	1	1	3	6	8	14	20	31	41	53	—	—
	1	—	1	1	3	5	6	10	15	23	31	40	—	—
	1/0	—	1	1	2	4	5	9	13	20	26	34	—	—
	2/0	—	1	1	1	3	4	7	11	17	22	29	—	—
	3/0	—	0	1	1	3	4	6	9	14	18	24	—	—
	4/0	—	0	1	1	2	3	5	7	11	15	20	—	—
	250	—	0	1	1	1	2	4	6	9	12	16	—	—
	300	—	0	0	1	1	1	3	5	8	10	13	—	—
	350	—	0	0	1	1	1	3	4	7	9	12	—	—
	400	—	0	0	1	1	1	3	4	6	8	11	—	—
	500	—	0	0	1	1	1	2	3	5	7	9	—	—
	600	—	0	0	0	1	1	1	2	4	5	7	—	—

(continues)

Table C.4(A) *Continued*

Type	Conductor Size (AWG/ kcmil)	Trade Size (Metric Designator)												
		⅜ (12)	½ (16)	¾ (21)	1 (27)	1¼ (35)	1½ (41)	2 (53)	2½ (63)	3 (78)	3½ (91)	4 (103)	5 (129)	6 (155)
	700	—	0	0	0	1	1	1	2	3	5	6	—	—
	750	—	0	0	0	1	1	1	1	3	4	6	—	—
	900	—	0	0	0	1	1	1	1	3	4	5	—	—
	1000	—	0	0	0	0	1	1	1	2	3	4	—	—

Definition: *Compact stranding* is the result of a manufacturing process where the stranded conductor is compressed to the extent that interstices (voids between strand wires) are virtually eliminated.

Table C.5 Maximum Number of Conductors or Fixture Wires in Liquidtight Flexible Nonmetallic Conduit (Type LFNC-A) (Based on Chapter 9: Table 1, Table 4, and Table 5)

Type	Conductor Size (AWG/ kcmil)	Trade Size (Metric Designator)												
		⅜ (12)	½ (16)	¾ (21)	1 (27)	1¼ (35)	1½ (41)	2 (53)	2½ (63)	3 (78)	3½ (91)	4 (103)	5 (129)	6 (155)
CONDUCTORS														
RHH, RHW, RHW-2	14	2	4	7	11	20	27	45	—	—	—	—	—	—
	12	1	3	6	9	17	23	38	—	—	—	—	—	—
	10	1	3	5	8	13	18	30	—	—	—	—	—	—
	8	1	1	2	4	7	9	16	—	—	—	—	—	—
	6	1	1	1	3	5	7	13	—	—	—	—	—	—
	4	0	1	1	2	4	6	10	—	—	—	—	—	—
	3	0	1	1	1	4	5	8	—	—	—	—	—	—
	2	0	1	1	1	3	4	7	—	—	—	—	—	—
	1	0	0	1	1	1	3	5	—	—	—	—	—	—
	1/0	0	0	1	1	1	2	4	—	—	—	—	—	—
	2/0	0	0	1	1	1	1	4	—	—	—	—	—	—
	3/0	0	0	0	1	1	1	3	—	—	—	—	—	—
	4/0	0	0	0	1	1	1	3	—	—	—	—	—	—
	250	0	0	0	0	1	1	1	—	—	—	—	—	—
	300	0	0	0	0	1	1	1	—	—	—	—	—	—
	350	0	0	0	0	1	1	1	—	—	—	—	—	—
	400	0	0	0	0	1	1	1	—	—	—	—	—	—
	500	0	0	0	0	0	1	1	—	—	—	—	—	—
	600	0	0	0	0	0	1	1	—	—	—	—	—	—
	700	0	0	0	0	0	0	1	—	—	—	—	—	—
	750	0	0	0	0	0	0	1	—	—	—	—	—	—
	800	0	0	0	0	0	0	1	—	—	—	—	—	—
	900	0	0	0	0	0	0	1	—	—	—	—	—	—
	1000	0	0	0	0	0	0	1	—	—	—	—	—	—
	1250	0	0	0	0	0	0	0	—	—	—	—	—	—
	1500	0	0	0	0	0	0	0	—	—	—	—	—	—
	1750	0	0	0	0	0	0	0	—	—	—	—	—	—
	2000	0	0	0	0	0	0	0	—	—	—	—	—	—
TW, THHW, THW, THW-2	14	5	9	15	24	43	58	96	—	—	—	—	—	—
	12	4	7	12	19	33	44	74	—	—	—	—	—	—
	10	3	5	9	14	24	33	55	—	—	—	—	—	—
	8	1	3	5	8	13	18	30	—	—	—	—	—	—
RHH*, RHW*, RHW-2*	14	3	6	10	16	28	38	64	—	—	—	—	—	—
	12	3	5	8	13	23	31	51	—	—	—	—	—	—
	10	1	3	6	10	18	24	40	—	—	—	—	—	—
	8	1	1	4	6	11	14	24	—	—	—	—	—	—
TW, THW, THHW, THW-2, RHH*, RHW*, RHW-2*	6	1	1	3	4	8	11	18	—	—	—	—	—	—
	4	1	1	1	3	6	8	13	—	—	—	—	—	—
	3	1	1	1	3	5	7	11	—	—	—	—	—	—
	2	0	1	1	2	4	6	10	—	—	—	—	—	—
	1	0	1	1	1	3	4	7	—	—	—	—	—	—
	1/0	0	0	1	1	2	3	6	—	—	—	—	—	—
	2/0	0	0	1	1	1	3	5	—	—	—	—	—	—
	3/0	0	0	1	1	1	2	4	—	—	—	—	—	—
	4/0	0	0	0	1	1	1	3	—	—	—	—	—	—
	250	0	0	0	1	1	1	3	—	—	—	—	—	—
	300	0	0	0	1	1	1	2	—	—	—	—	—	—
	350	0	0	0	0	1	1	1	—	—	—	—	—	—
	400	0	0	0	0	1	1	1	—	—	—	—	—	—
	500	0	0	0	0	1	1	1	—	—	—	—	—	—
	600	0	0	0	0	1	1	1	—	—	—	—	—	—
	700	0	0	0	0	0	1	1	—	—	—	—	—	—
	750	0	0	0	0	0	1	1	—	—	—	—	—	—

(continues)

Table C.5 *Continued*

Type	Conductor Size (AWG/ kcmil)	Trade Size (Metric Designator)												
		⅜ (12)	½ (16)	¾ (21)	1 (27)	1¼ (35)	1½ (41)	2 (53)	2½ (63)	3 (78)	3½ (91)	4 (103)	5 (129)	6 (155)
	800	0	0	0	0	0	1	1	—	—	—	—	—	—
	900	0	0	0	0	0	0	1	—	—	—	—	—	—
	1000	0	0	0	0	0	0	1	—	—	—	—	—	—
	1250	0	0	0	0	0	0	1	—	—	—	—	—	—
	1500	0	0	0	0	0	0	1	—	—	—	—	—	—
	1750	0	0	0	0	0	0	0	—	—	—	—	—	—
	2000	0	0	0	0	0	0	0	—	—	—	—	—	—
THHN, THWN, THWN-2	14	8	13	22	35	62	83	138	—	—	—	—	—	—
	12	5	9	16	25	45	60	100	—	—	—	—	—	—
	10	3	6	10	16	28	38	63	—	—	—	—	—	—
	8	1	3	6	9	16	22	36	—	—	—	—	—	—
	6	1	2	4	6	12	16	26	—	—	—	—	—	—
	4	1	1	2	4	7	9	16	—	—	—	—	—	—
	3	1	1	1	3	6	8	13	—	—	—	—	—	—
	2	1	1	1	3	5	7	11	—	—	—	—	—	—
	1	0	1	1	1	4	5	8	—	—	—	—	—	—
	1/0	0	1	1	1	3	4	7	—	—	—	—	—	—
	2/0	0	0	1	1	2	3	6	—	—	—	—	—	—
	3/0	0	0	1	1	1	3	5	—	—	—	—	—	—
	4/0	0	0	1	1	1	2	4	—	—	—	—	—	—
	250	0	0	0	1	1	1	3	—	—	—	—	—	—
	300	0	0	0	1	1	1	3	—	—	—	—	—	—
	350	0	0	0	1	1	1	2	—	—	—	—	—	—
	400	0	0	0	0	1	1	1	—	—	—	—	—	—
	500	0	0	0	0	1	1	1	—	—	—	—	—	—
	600	0	0	0	0	1	1	1	—	—	—	—	—	—
	700	0	0	0	0	1	1	1	—	—	—	—	—	—
	750	0	0	0	0	0	1	1	—	—	—	—	—	—
	800	0	0	0	0	0	1	1	—	—	—	—	—	—
	900	0	0	0	0	0	1	1	—	—	—	—	—	—
	1000	0	0	0	0	0	0	1	—	—	—	—	—	—
FEP, FEPB, PFA, PFAH, TFE	14	7	12	21	34	60	80	133	—	—	—	—	—	—
	12	5	9	15	25	44	59	97	—	—	—	—	—	—
	10	4	6	11	18	31	42	70	—	—	—	—	—	—
	8	1	3	6	10	18	24	40	—	—	—	—	—	—
	6	1	2	4	7	13	17	28	—	—	—	—	—	—
	4	1	1	3	5	9	12	20	—	—	—	—	—	—
	3	1	1	2	4	7	10	16	—	—	—	—	—	—
	2	1	1	1	3	6	8	13	—	—	—	—	—	—
PFA, PFAH, TFE	1	0	1	1	2	4	5	9	—	—	—	—	—	—
PFA, PFAH, TFE, Z	1/0	0	1	1	1	3	5	8	—	—	—	—	—	—
	2/0	0	1	1	1	3	4	6	—	—	—	—	—	—
	3/0	0	0	1	1	2	3	5	—	—	—	—	—	—
	4/0	0	0	1	1	1	2	4	—	—	—	—	—	—
Z	14	9	15	25	41	72	97	161	—	—	—	—	—	—
	12	6	10	18	29	51	69	114	—	—	—	—	—	—
	10	4	6	11	18	31	42	70	—	—	—	—	—	—
	8	2	4	7	11	20	26	44	—	—	—	—	—	—
	6	1	3	5	8	14	18	31	—	—	—	—	—	—
	4	1	1	3	5	9	13	21	—	—	—	—	—	—
	3	1	1	2	4	7	9	15	—	—	—	—	—	—
	2	1	1	1	3	6	8	13	—	—	—	—	—	—
	1	1	1	1	2	4	6	10	—	—	—	—	—	—

(continues)

Table C.5   *Continued*

Type	Conductor Size (AWG/kcmil)	3/8 (12)	1/2 (16)	3/4 (21)	1 (27)	1¼ (35)	1½ (41)	2 (53)	2½ (63)	3 (78)	3½ (91)	4 (103)	5 (129)	6 (155)
XHHW, ZW, XHHW-2, XHH	14	5	9	15	24	43	58	96	—	—	—	—	—	—
	12	4	7	12	19	33	44	74	—	—	—	—	—	—
	10	3	5	9	14	24	33	55	—	—	—	—	—	—
	8	1	3	5	8	13	18	30	—	—	—	—	—	—
	6	1	1	3	5	10	13	22	—	—	—	—	—	—
	4	1	1	2	4	7	10	16	—	—	—	—	—	—
	3	1	1	1	3	6	8	14	—	—	—	—	—	—
	2	1	1	1	3	5	7	11	—	—	—	—	—	—
XHHW, XHHW-2, XHH	1	0	1	1	1	4	5	8	—	—	—	—	—	—
	1/0	0	1	1	1	3	4	7	—	—	—	—	—	—
	2/0	0	0	1	1	2	3	6	—	—	—	—	—	—
	3/0	0	0	1	1	1	3	5	—	—	—	—	—	—
	4/0	0	0	1	1	1	2	4	—	—	—	—	—	—
	250	0	0	0	1	1	1	3	—	—	—	—	—	—
	300	0	0	0	1	1	1	3	—	—	—	—	—	—
	350	0	0	0	1	1	1	2	—	—	—	—	—	—
	400	0	0	0	0	1	1	1	—	—	—	—	—	—
	500	0	0	0	0	1	1	1	—	—	—	—	—	—
	600	0	0	0	0	1	1	1	—	—	—	—	—	—
	700	0	0	0	0	1	1	1	—	—	—	—	—	—
	750	0	0	0	0	0	1	1	—	—	—	—	—	—
	800	0	0	0	0	0	1	1	—	—	—	—	—	—
	900	0	0	0	0	0	1	1	—	—	—	—	—	—
	1000	0	0	0	0	0	0	1	—	—	—	—	—	—
	1250	0	0	0	0	0	0	1	—	—	—	—	—	—
	1500	0	0	0	0	0	0	1	—	—	—	—	—	—
	1750	0	0	0	0	0	0	0	—	—	—	—	—	—
	2000	0	0	0	0	0	0	0	—	—	—	—	—	—
FIXTURE WIRES														
RFH-2, FFH-2, RFHH-2	18	5	8	14	23	41	55	92	—	—	—	—	—	—
	16	4	7	12	20	35	47	77	—	—	—	—	—	—
SF-2, SFF-2	18	6	11	18	29	52	70	116	—	—	—	—	—	—
	16	5	9	15	24	43	58	96	—	—	—	—	—	—
	14	4	7	12	20	35	47	77	—	—	—	—	—	—
SF-1, SFF-1	18	12	19	33	52	92	124	205	—	—	—	—	—	—
RFH-1, TF, TFF, XF, XFF	18	8	14	24	39	68	91	152	—	—	—	—	—	—
	16	7	11	19	31	55	74	122	—	—	—	—	—	—
XF, XFF	14	5	9	15	24	43	58	96	—	—	—	—	—	—
TFN, TFFN	18	14	22	39	62	109	146	243	—	—	—	—	—	—
	16	10	17	29	47	83	112	185	—	—	—	—	—	—
PF, PFF, PGF, PGFF, PAF, PTF, PTFF, PAFF	18	13	21	37	59	103	139	230	—	—	—	—	—	—
	16	10	16	28	45	80	107	178	—	—	—	—	—	—
	14	7	12	21	34	60	80	133	—	—	—	—	—	—
ZF, ZFF, ZHF	18	17	27	47	76	133	179	297	—	—	—	—	—	—
	16	12	20	35	56	98	132	219	—	—	—	—	—	—
	14	9	15	25	41	72	97	161	—	—	—	—	—	—
KF-2, KFF-2	18	25	41	71	114	200	269	445	—	—	—	—	—	—
	16	18	29	49	79	139	187	311	—	—	—	—	—	—
	14	12	19	33	53	94	126	209	—	—	—	—	—	—
	12	8	13	23	37	65	87	145	—	—	—	—	—	—
	10	5	9	15	24	43	58	96	—	—	—	—	—	—

(continues)

Table C.5   *Continued*

Type	Conductor Size (AWG/ kcmil)	Trade Size (Metric Designator)												
		⅜ (12)	½ (16)	¾ (21)	1 (27)	1¼ (35)	1½ (41)	2 (53)	2½ (63)	3 (78)	3½ (91)	4 (103)	5 (129)	6 (155)
KF-1, KFF-1	18	29	48	82	131	231	310	514	—	—	—	—	—	—
	16	20	33	58	92	162	218	361	—	—	—	—	—	—
	14	14	22	39	62	109	146	243	—	—	—	—	—	—
	12	9	15	25	41	72	97	161	—	—	—	—	—	—
	10	6	10	17	27	47	63	105	—	—	—	—	—	—
XF, XFF	12	3	5	8	13	23	31	51	—	—	—	—	—	—
	10	1	3	6	10	18	24	40	—	—	—	—	—	—

Notes:

1. This table is for concentric stranded conductors only. For compact stranded conductors, Table C.6(A) should be used.

2. Two-hour fire-rated RHH cable has ceramifiable insulation, which has much larger diameters than other RHH wires. Consult manufacturer's conduit fill tables.

*Types RHH, RHW, and RHW-2 without outer covering.

Table C.5(A) Maximum Number of Conductors or Fixture Wires in Liquidtight Flexible Nonmetallic Conduit (Type LFNC-A)
(Based on Chapter 9: Table 1, Table 4, and Table 5A)

Type	Conductor Size (AWG/kcmil)	⅜ (12)	½ (16)	¾ (21)	1 (27)	1¼ (35)	1½ (41)	2 (53)	2½ (63)	3 (78)	3½ (91)	4 (103)	5 (129)	6 (155)
COMPACT CONDUCTORS														
THW, THW-2, THHW	8	1	2	4	6	11	16	—	—	—	—	—	—	—
	6	1	1	3	5	9		26	—	—	—	—	—	—
	4	1	1	2	4	7	12	20	—	—	—	—	—	—
	2	1	1	1	3	5	9	15	—	—	—	—	—	—
	1	0	1	1	1		6	11	—	—	—	—	—	—
	1/0	0	1	1	1	3	4	8	—	—	—	—	—	—
	2/0	0	0	1	1			7	—	—	—	—	—	—
	3/0	0	0	1		3	4	5	—	—	—	—	—	—
	4/0	0	0	1		2	3	5	—	—	—	—	—	—
	250	0	0		1	1	2	4	—	—	—	—	—	—
	300	0	0	0	1	1	1	3	—	—	—	—	—	—
	350	0	0	0	1	1	1	3	—	—	—	—	—	—
	400	0		0	1	1	1	2	—	—	—	—	—	—
	500	0	0	0	0	1	1	1	—	—	—	—	—	—
	600	0	0	0	0	1	1	1	—	—	—	—	—	—
	700		0	0	0	1	1	1	—	—	—	—	—	—
	750	0	0	0	0	1	1	1	—	—	—	—	—	—
	900	0	0	0	0	0	1	1	—	—	—	—	—	—
	1000	0	0	0	0	0	1	1	—	—	—	—	—	—
THHN, THWN, THWN-2	8	—	—	—	—	—	—	—	—	—	—	—	—	—
	6	1	2	4	7	13	18	29	—	—	—	—	—	—
	4	1	1	3	4	8	11	18	—	—	—	—	—	—
	2	1	1	1	3	6	8	13	—	—	—	—	—	—
	1	0	1	1	2	4	6	10	—	—	—	—	—	—
	1/0	0	1	1	1	3	5	8	—	—	—	—	—	—
	2/0	0	1	1	1	3	4	7	—	—	—	—	—	—
	3/0	0	0	1	1	2	3	6	—	—	—	—	—	—
	4/0	0	0	1	1	1	3	5	—	—	—	—	—	—
	250	0	0	1	1	1	1	3	—	—	—	—	—	—
	300	0	0	0	1	1	1	3	—	—	—	—	—	—
	350	0	0	0	1	1	1	3	—	—	—	—	—	—
	400	0	0	0	1	1	1	2	—	—	—	—	—	—
	500	0	0	0	0	1	1	1	—	—	—	—	—	—
	600	0	0	0	0	1	1	1	—	—	—	—	—	—
	700	0	0	0	0	1	1	1	—	—	—	—	—	—
	750	0	0	0	0	1	1	1	—	—	—	—	—	—
	900	0	0	0	0	0	1	1	—	—	—	—	—	—
	1000	0	0	0	0	0	1	1	—	—	—	—	—	—
XHHW, XHHW-2	8	1	3	5	8	15	20	34	—	—	—	—	—	—
	6	1	2	4	6	11	15	25	—	—	—	—	—	—
	4	1	1	3	4	8	11	18	—	—	—	—	—	—
	2	1	1	1	3	6	8	13	—	—	—	—	—	—
	1	0	1	1	2	4	6	10	—	—	—	—	—	—
	1/0	0	1	1	1	3	5	8	—	—	—	—	—	—
	2/0	0	1	1	1	3	4	7	—	—	—	—	—	—
	3/0	0	0	1	1	2	3	6	—	—	—	—	—	—
	4/0	0	0	1	1	1	3	5	—	—	—	—	—	—
	250	0	0	1	1	1	2	4	—	—	—	—	—	—
	300	0	0	0	1	1	1	3	—	—	—	—	—	—
	350	0	0	0	1	1	1	3	—	—	—	—	—	—
	400	0	0	0	1	1	1	2	—	—	—	—	—	—
	500	0	0	0	0	1	1	1	—	—	—	—	—	—
	600	0	0	0	0	1	1	1	—	—	—	—	—	—

(continues)

Table C.5(A) *Continued*

Type	Conductor Size (AWG/ kcmil)	⅜ (12)	½ (16)	¾ (21)	1 (27)	1¼ (35)	1½ (41)	2 (53)	2½ (63)	3 (78)	3½ (91)	4 (103)	5 (129)	6 (155)
						Trade Size (Metric Designator)								
	700	0	0	0	0	1	1	1	—	—	—	—	—	—
	750	0	0	0	0	1	1	1	—	—	—	—	—	—
	900	0	0	0	0	0	1	1	—	—	—	—	—	—
	1000	0	0	0	0	0	1	1	—	—	—	—	—	—

Definition: *Compact stranding* is the result of a manufacturing process where the stranded conductor is compressed to the extent that the interstices (voids between strand wires) are virtually eliminated.

Table C.6 Maximum Number of Conductors or Fixture Wires in Liquidtight Flexible Nonmetallic Conduit (Type LFNC-B*) (Based on Chapter 9: Table 1, Table 4, and Table 5)

Type	Conductor Size (AWG/ kcmil)	⅜ (12)	½ (16)	¾ (21)	1 (27)	1¼ (35)	1½ (41)	2 (53)	2½ (63)	3 (78)	3½ (91)	4 (103)	5 (129)	6 (155)
					CONDUCTORS									
RHH, RHW, RHW-2	14	2	4	7	12	21	27	44	—	—	—	—	—	—
	12	1	3	6	10	17	22	36	—	—	—	—	—	—
	10	1	3	5	8	14	18	29	—	—	—	—	—	—
	8	1	1	2	4	7	9	15	—	—	—	—	—	—
	6	1	1	1	3	6	7	12	—	—	—	—	—	—
	4	0	1	1	2	4	6	9	—	—	—	—	—	—
	3	0	1	1	1	4	5	8	—	—	—	—	—	—
	2	0	1	1	1	3	4	7	—	—	—	—	—	—
	1	0	0	1	1	1	3	5	—	—	—	—	—	—
	1/0	0	0	1	1	1	2	4	—	—	—	—	—	—
	2/0	0	0	1	1	1	1	3	—	—	—	—	—	—
	3/0	0	0	0	1	1	1	3	—	—	—	—	—	—
	4/0	0	0	0	1	1	1	2	—	—	—	—	—	—
	250	0	0	0	0	1	1	1	—	—	—	—	—	—
	300	0	0	0	0	1	1	1	—	—	—	—	—	—
	350	0	0	0	0	1	1	1	—	—	—	—	—	—
	400	0	0	0	0	1	1	1	—	—	—	—	—	—
	500	0	0	0	0	1	1	1	—	—	—	—	—	—
	600	0	0	0	0	0	1	1	—	—	—	—	—	—
	700	0	0	0	0	0	0	1	—	—	—	—	—	—
	750	0	0	0	0	0	0	1	—	—	—	—	—	—
	800	0	0	0	0	0	0	1	—	—	—	—	—	—
	900	0	0	0	0	0	0	1	—	—	—	—	—	—
	1000	0	0	0	0	0	0	1	—	—	—	—	—	—
	1250	0	0	0	0	0	0	0	—	—	—	—	—	—
	1500	0	0	0	0	0	0	0	—	—	—	—	—	—
	1750	0	0	0	0	0	0	0	—	—	—	—	—	—
	2000	0	0	0	0	0	0	0	—	—	—	—	—	—
TW, THHW, THW, THW-2	14	5	9	15	25	44	57	93	—	—	—	—	—	—
	12	4	7	12	19	33	43	71	—	—	—	—	—	—
	10	3	5	9	14	25	32	53	—	—	—	—	—	—
	8	1	3	5	8	14	18	29	—	—	—	—	—	—
RHH*, RHW*, RHW-2*	14	3	6	10	16	29	38	62	—	—	—	—	—	—
	12	3	5	8	13	23	30	50	—	—	—	—	—	—
	10	1	3	6	10	18	23	39	—	—	—	—	—	—
	8	1	1	4	6	11	14	23	—	—	—	—	—	—
TW, THW, THHW, THW-2, RHH*, RHW*, RHW-2*	6	1	1	3	5	8	11	18	—	—	—	—	—	—
	4	1	1	1	3	6	8	13	—	—	—	—	—	—
	3	1	1	1	3	5	7	11	—	—	—	—	—	—
	2	0	1	1	2	4	6	9	—	—	—	—	—	—
	1	0	1	1	1	3	4	7	—	—	—	—	—	—
	1/0	0	0	1	1	2	3	6	—	—	—	—	—	—
	2/0	0	0	1	1	2	3	5	—	—	—	—	—	—
	3/0	0	0	1	1	1	2	4	—	—	—	—	—	—
	4/0	0	0	0	1	1	1	3	—	—	—	—	—	—
	250	0	0	0	1	1	1	3	—	—	—	—	—	—
	300	0	0	0	1	1	1	2	—	—	—	—	—	—
	350	0	0	0	0	1	1	1	—	—	—	—	—	—
	400	0	0	0	0	1	1	1	—	—	—	—	—	—
	500	0	0	0	0	1	1	1	—	—	—	—	—	—
	600	0	0	0	0	1	1	1	—	—	—	—	—	—
	700	0	0	0	0	0	1	1	—	—	—	—	—	—
	750	0	0	0	0	0	1	1	—	—	—	—	—	—

(continues)

Table C.6 *Continued*

Type	Conductor Size (AWG/kcmil)	⅜ (12)	½ (16)	¾ (21)	1 (27)	1¼ (35)	1½ (41)	2 (53)	2½ (63)	3 (78)	3½ (91)	4 (103)	5 (129)	6 (155)
								Trade Size (Metric Designator)						
	800	0	0	0	0	0	1	1	—	—	—	—	—	—
	900	0	0	0	0	0	0	1	—	—	—	—	—	—
	1000	0	0	0	0	0	0	1	—	—	—	—	—	—
	1250	0	0	0	0	0	0	1	—	—	—	—	—	—
	1500	0	0	0	0	0	0	0	—	—	—	—	—	—
	1750	0	0	0	0	0	0	0	—	—	—	—	—	—
	2000	0	0	0	0	0	0	0	—	—	—	—	—	—
THHN, THWN, THWN-2	14	8	13	22	36	63	81	134	—	—	—	—	—	—
	12	5	9	16	26	46	59	97	—	—	—	—	—	—
	10	3	6	10	16	29	37	61	—	—	—	—	—	—
	8	1	3	6	9	16	21	35	—	—	—	—	—	—
	6	1	2	4	7	12	15	25	—	—	—	—	—	—
	4	1	1	2	4	7	9	15	—	—	—	—	—	—
	3	1	1	1	3	6	8	13	—	—	—	—	—	—
	2	1	1	1	3	5	7	11	—	—	—	—	—	—
	1	0	1	1	1	4	5	8	—	—	—	—	—	—
	1/0	0	1	1	1	3	4	7	—	—	—	—	—	—
	2/0	0	0	1	1	2	3	6	—	—	—	—	—	—
	3/0	0	0	1	1	1	3	5	—	—	—	—	—	—
	4/0	0	0	1	1	1	2	4	—	—	—	—	—	—
	250	0	0	0	1	1	1	3	—	—	—	—	—	—
	300	0	0	0	1	1	1	3	—	—	—	—	—	—
	350	0	0	0	1	1	1	2	—	—	—	—	—	—
	400	0	0	0	0	1	1	1	—	—	—	—	—	—
	500	0	0	0	0	1	1	1	—	—	—	—	—	—
	600	0	0	0	0	1	1	1	—	—	—	—	—	—
	700	0	0	0	0	1	1	1	—	—	—	—	—	—
	750	0	0	0	0	0	1	1	—	—	—	—	—	—
	800	0	0	0	0	0	1	1	—	—	—	—	—	—
	900	0	0	0	0	0	1	1	—	—	—	—	—	—
	1000	0	0	0	0	0	0	1	—	—	—	—	—	—
FEP, FEPB, PFA, PFAH, TFE	14	7	12	21	35	61	79	130	—	—	—	—	—	—
	12	5	9	15	25	44	58	94	—	—	—	—	—	—
	10	4	6	11	18	32	41	68	—	—	—	—	—	—
	8	1	3	6	10	18	23	39	—	—	—	—	—	—
	6	1	2	4	7	13	17	27	—	—	—	—	—	—
	4	1	1	3	5	9	12	19	—	—	—	—	—	—
	3	1	1	2	4	7	10	16	—	—	—	—	—	—
	2	1	1	1	3	6	8	13	—	—	—	—	—	—
PFA, PFAH, TFE	1	0	1	1	2	4	5	9	—	—	—	—	—	—
PFA, PFAH, TFE, Z	1/0	0	1	1	1	3	4	7	—	—	—	—	—	—
	2/0	0	1	1	1	3	4	6	—	—	—	—	—	—
	3/0	0	0	1	1	2	3	5	—	—	—	—	—	—
	4/0	0	0	1	1	1	2	4	—	—	—	—	—	—
Z	14	9	15	26	42	73	95	156	—	—	—	—	—	—
	12	6	10	18	30	52	67	111	—	—	—	—	—	—
	10	4	6	11	18	32	41	68	—	—	—	—	—	—
	8	2	4	7	11	20	26	43	—	—	—	—	—	—
	6	1	3	5	8	14	18	30	—	—	—	—	—	—
	4	1	1	3	5	9	12	20	—	—	—	—	—	—
	3	1	1	2	4	7	9	15	—	—	—	—	—	—
	2	1	1	1	3	6	7	12	—	—	—	—	—	—
	1	1	1	1	2	5	6	10	—	—	—	—	—	—

(continues)

Table C.6   *Continued*

Type	Conductor Size (AWG/ kcmil)	³⁄₈ (12)	½ (16)	¾ (21)	1 (27)	1¼ (35)	1½ (41)	2 (53)	2½ (63)	3 (78)	3½ (91)	4 (103)	5 (129)	6 (155)
XHHW, ZW, XHHW-2, XHH	14	5	9	15	25	44	57	93	—	—	—	—	—	—
	12	4	7	12	19	33	43	71	—	—	—	—	—	—
	10	3	5	9	14	25	32	53	—	—	—	—	—	—
	8	1	3	5	8	14	18	29	—	—	—	—	—	—
	6	1	1	3	6	10	13	22	—	—	—	—	—	—
	4	1	1	2	4	7	9	16	—	—	—	—	—	—
	3	1	1	1	3	6	8	13	—	—	—	—	—	—
	2	1	1	1	3	5	7	11	—	—	—	—	—	—
XHHW, XHHW-2, XHH	1	0	1	1	1	4	5	8	—	—	—	—	—	—
	1/0	0	1	1	1	3	4	7	—	—	—	—	—	—
	2/0	0	0	1	1	2	3	6	—	—	—	—	—	—
	3/0	0	0	1	1	1	3	5	—	—	—	—	—	—
	4/0	0	0	1	1	1	2	4	—	—	—	—	—	—
	250	0	0	0	1	1	1	3	—	—	—	—	—	—
	300	0	0	0	1	1	1	3	—	—	—	—	—	—
	350	0	0	0	1	1	1	2	—	—	—	—	—	—
	400	0	0	0	1	1	1	1	—	—	—	—	—	—
	500	0	0	0	0	1	1	1	—	—	—	—	—	—
	600	0	0	0	0	1	1	1	—	—	—	—	—	—
	700	0	0	0	0	1	1	1	—	—	—	—	—	—
	750	0	0	0	0	0	1	1	—	—	—	—	—	—
	800	0	0	0	0	0	1	1	—	—	—	—	—	—
	900	0	0	0	0	0	1	1	—	—	—	—	—	—
	1000	0	0	0	0	0	0	1	—	—	—	—	—	—
	1250	0	0	0	0	0	0	1	—	—	—	—	—	—
	1500	0	0	0	0	0	0	1	—	—	—	—	—	—
	1750	0	0	0	0	0	0	0	—	—	—	—	—	—
	2000	0	0	0	0	0	0	0	—	—	—	—	—	—
FIXTURE WIRES														
RFH-2, FFH-2, RFHH-2	18	5	8	15	24	42	54	89	—	—	—	—	—	—
	16	4	7	12	20	35	46	75	—	—	—	—	—	—
SF-2, SFF-2	18	6	11	19	30	53	69	113	—	—	—	—	—	—
	16	5	9	15	25	44	57	93	—	—	—	—	—	—
	14	4	7	12	20	35	46	75	—	—	—	—	—	—
SF-1, SFF-1	18	12	19	33	53	94	122	199	—	—	—	—	—	—
RFH-1, TF, TFF, XF, XFF	18	8	14	24	39	69	90	147	—	—	—	—	—	—
	16	7	11	20	32	56	72	119	—	—	—	—	—	—
XF, XFF	14	5	9	15	25	44	57	93	—	—	—	—	—	—
TFN, TFFN	18	14	23	39	63	111	144	236	—	—	—	—	—	—
	16	10	17	30	48	85	110	180	—	—	—	—	—	—
PF, PFF, PGF, PGFF, PAF, PTF, PTFF, PAFF	18	13	21	37	60	105	136	224	—	—	—	—	—	—
	16	10	16	29	46	81	105	173	—	—	—	—	—	—
	14	7	12	21	35	61	79	130	—	—	—	—	—	—
ZF, ZFF, ZHF	18	17	28	48	77	136	176	288	—	—	—	—	—	—
	16	12	20	35	57	100	130	213	—	—	—	—	—	—
	14	9	15	26	42	73	95	156	—	—	—	—	—	—
KF-2, KFF-2	18	25	42	72	116	203	264	433	—	—	—	—	—	—
	16	18	29	50	81	142	184	302	—	—	—	—	—	—
	14	12	19	34	54	95	124	203	—	—	—	—	—	—
	12	8	13	23	38	66	86	141	—	—	—	—	—	—
	10	5	9	15	25	44	57	93	—	—	—	—	—	—

(continues)

Table C.6   *Continued*

Type	Conductor Size (AWG/kcmil)	Trade Size (Metric Designator)												
		³⁄₈ (12)	½ (16)	¾ (21)	1 (27)	1¼ (35)	1½ (41)	2 (53)	2½ (63)	3 (78)	3½ (91)	4 (103)	5 (129)	6 (155)
KF-1, KFF-1	18	29	48	83	134	235	304	499	—	—	—	—	—	—
	16	20	34	58	94	165	214	351	—	—	—	—	—	—
	14	14	23	39	63	111	144	236	—	—	—	—	—	—
	12	9	15	26	42	73	95	156	—	—	—	—	—	—
	10	6	10	17	27	48	62	102	—	—	—	—	—	—
XF, XFF	12	3	5	8	13	23	30	50	—	—	—	—	—	—
	10	1	3	6	10	18	23	39	—	—	—	—	—	—

Notes:

1. This table is for concentric stranded conductors only. For compact stranded conductors, Table C.5(A) should be used.

2. Two-hour fire-rated RHH cable has ceramifiable insulation, which has much larger diameters than other RHH wires. Consult manufacturer's conduit fill tables.

*Types RHH, RHW, and RHW-2 without outer covering.

**Table C.6(A) Maximum Number of Conductors or Fixture Wires in Liquidtight Flexible Nonmetallic Conduit (Type LFNC-B)
(Based on Chapter 9: Table 1, Table 4, and Table 5A)**

Type	Conductor Size (AWG/ kcmil)	Trade Size (Metric Designator)												
		⅜ (12)	½ (16)	¾ (21)	1 (27)	1¼ (35)	1½ (41)	2 (53)	2½ (63)	3 (78)	3½ (91)	4 (103)	5 (129)	6 (155)
COMPACT CONDUCTORS														
THW, THW-2, THHW	8	1	2	4	7	12	15	25	—	—	—	—	—	—
	6	1	1	3	5	9	12	19	—	—	—	—	—	—
	4	1	1	2	4	7	9	14	—	—	—	—	—	—
	2	1	1	1	3	5	6	11	—	—	—	—	—	—
	1	0	1	1	1	3	4	7	—	—	—	—	—	—
	1/0	0	1	1	1	3	4	6	—	—	—	—	—	—
	2/0	0	0	1	1	2	3	5	—	—	—	—	—	—
	3/0	0	0	1	1	1	3	4	—	—	—	—	—	—
	4/0	0	0	1	1	1	2	4	—	—	—	—	—	—
	250	0	0	0	1	1	1	3	—	—	—	—	—	—
	300	0	0	0	1	1	1	2	—	—	—	—	—	—
	350	0	0	0	1	1	1	2	—	—	—	—	—	—
	400	0	0	0	0	1	1	1	—	—	—	—	—	—
	500	0	0	0	0	1	1	1	—	—	—	—	—	—
	600	0	0	0	0	1	1	1	—	—	—	—	—	—
	700	0	0	0	0	1	1	1	—	—	—	—	—	—
	750	0	0	0	0	0	1	1	—	—	—	—	—	—
	900	0	0	0	0	0	1	1	—	—	—	—	—	—
	1000	0	0	0	0	0	1	1	—	—	—	—	—	—
THHN, THWN, THWN-2	8	—	—	—	—	—	—	—	—	—	—	—	—	—
	6	1	2	4	7	13	17	28	—	—	—	—	—	—
	4	1	1	3	4	8	11	17	—	—	—	—	—	—
	2	1	1	1	3	6	7	12	—	—	—	—	—	—
	1	0	1	1	2	4	6	9	—	—	—	—	—	—
	1/0	0	1	1	1	4	5	8	—	—	—	—	—	—
	2/0	0	1	1	1	3	4	6	—	—	—	—	—	—
	3/0	0	0	1	1	2	3	5	—	—	—	—	—	—
	4/0	0	0	1	1	1	3	4	—	—	—	—	—	—
	250	0	0	1	1	1	1	3	—	—	—	—	—	—
	300	0	0	0	1	1	1	3	—	—	—	—	—	—
	350	0	0	0	1	1	1	2	—	—	—	—	—	—
	400	0	0	0	1	1	1	2	—	—	—	—	—	—
	500	0	0	0	0	1	1	1	—	—	—	—	—	—
	600	0	0	0	0	1	1	1	—	—	—	—	—	—
	700	0	0	0	0	1	1	1	—	—	—	—	—	—
	750	0	0	0	0	1	1	1	—	—	—	—	—	—
	900	0	0	0	0	0	1	1	—	—	—	—	—	—
	1000	0	0	0	0	0	1	1	—	—	—	—	—	—
XHHW, XHHW-2	8	1	3	5	9	15	20	33	—	—	—	—	—	—
	6	1	2	4	6	11	15	24	—	—	—	—	—	—
	4	1	1	3	4	8	11	17	—	—	—	—	—	—
	2	1	1	1	3	6	7	12	—	—	—	—	—	—
	1	0	1	1	2	4	6	9	—	—	—	—	—	—
	1/0	0	1	1	1	4	5	8	—	—	—	—	—	—
	2/0	0	1	1	1	3	4	7	—	—	—	—	—	—
	3/0	0	0	1	1	2	3	5	—	—	—	—	—	—
	4/0	0	0	1	1	1	3	4	—	—	—	—	—	—
	250	0	0	1	1	1	1	3	—	—	—	—	—	—
	300	0	0	0	1	1	1	3	—	—	—	—	—	—
	350	0	0	0	1	1	1	3	—	—	—	—	—	—
	400	0	0	0	1	1	1	2	—	—	—	—	—	—
	500	0	0	0	0	1	1	1	—	—	—	—	—	—
	600	0	0	0	0	1	1	1	—	—	—	—	—	—

(continues)

Table C.6(A)   *Continued*

Type	Conductor Size (AWG/ kcmil)	⅜ (12)	½ (16)	¾ (21)	1 (27)	1¼ (35)	1½ (41)	2 (53)	2½ (63)	3 (78)	3½ (91)	4 (103)	5 (129)	6 (155)
						Trade Size (Metric Designator)								
	700	0	0	0	0	1	1	1	—	—	—	—	—	—
	750	0	0	0	0	1	1	1	—	—	—	—	—	—
	900	0	0	0	0	0	1	1	—	—	—	—	—	—
	1000	0	0	0	0	0	1	1	—	—	—	—	—	—

Definition: *Compact stranding* is the result of a manufacturing process where the stranded conductor is compressed to the extent that the interstices (voids between strand wires) are virtually eliminated.

N Table C.7 Maximum Number of Conductors of Fixture Wires in Liquidtight Flexible Nonmetallic Conduit (Type LFNC-C) (Based on Chapter 9: Table 1, Table 4, and Table 5)

Type	Conductor Size (AWG/kcmil)	Trade Size (Metric Designator)												
		⅜ (12)	½ (16)	¾ (21)	1 (27)	1¼ (35)	1½ (41)	2 (53)	2½ (63)	3 (78)	3½ (91)	4 (103)	5 (129)	6 (155)
		CONDUCTORS												
RHH, RHW, RHW-2	14	2	4	7	11	20	27	45	—	—	—	—	—	—
	12	1	3	6	9	16	22	37	—	—	—	—	—	—
	10	1	2	4	7	13	18	30	—	—	—	—	—	—
	8	1	1	2	4	7	9	15	—	—	—	—	—	—
	6	0	1	1	3	5	7	12	—	—	—	—	—	—
	4	0	1	1	2	4	6	10	—	—	—	—	—	—
	3	0	1	1	1	4	5	8	—	—	—	—	—	—
	2	0	0	1	1	3	4	7	—	—	—	—	—	—
	1	0	0	1	1	1	3	5	—	—	—	—	—	—
	1/0	0	0	0	1	1	2	4	—	—	—	—	—	—
	2/0	0	0	0	1	1	1	3	—	—	—	—	—	—
	3/0	0	0	0	1	1	1	3	—	—	—	—	—	—
	4/0	0	0	0	0	1	1	2	—	—	—	—	—	—
	250	0	0	0	0	1	1	1	—	—	—	—	—	—
	300	0	0	0	0	1	1	1	—	—	—	—	—	—
	350	0	0	0	0	0	1	1	—	—	—	—	—	—
	400	0	0	0	0	0	1	1	—	—	—	—	—	—
	500	0	0	0	0	0	1	1	—	—	—	—	—	—
	600	0	0	0	0	0	0	1	—	—	—	—	—	—
	700	0	0	0	0	0	0	1	—	—	—	—	—	—
	750	0	0	0	0	0	0	1	—	—	—	—	—	—
	800	0	0	0	0	0	0	1	—	—	—	—	—	—
	900	0	0	0	0	0	0	1	—	—	—	—	—	—
	1000	0	0	0	0	0	0	0	—	—	—	—	—	—
	1250	0	0	0	0	0	0	0	—	—	—	—	—	—
	1500	0	0	0	0	0	0	0	—	—	—	—	—	—
	1750	0	0	0	0	0	0	0	—	—	—	—	—	—
	2000	0	0	0	0	0	0	0	—	—	—	—	—	—
TW, THHW, THW, THW-2	14	5	8	15	24	42	56	94	—	—	—	—	—	—
	12	4	6	11	18	32	43	72	—	—	—	—	—	—
	10	3	5	8	13	24	32	54	—	—	—	—	—	—
	8	1	2	4	7	13	18	30	—	—	—	—	—	—
RHH*, RHW*, RHW-2*	14	2	5	10	16	28	37	63	—	—	—	—	—	—
	12	2	4	8	13	22	30	50	—	—	—	—	—	—
	10	1	3	6	10	17	23	39	—	—	—	—	—	—
	8	1	1	3	6	10	14	23	—	—	—	—	—	—

(continues)

Table C.7 *Continued*

Type	Conductor Size (AWG/kcmil)	Trade Size (Metric Designator)												
		⅜ (12)	½ (16)	¾ (21)	1 (27)	1¼ (35)	1½ (41)	2 (53)	2½ (63)	3 (78)	3½ (91)	4 (103)	5 (129)	6 (155)
TW, THW, THHW, THW-2, RHH*, RHW*, RHW-2	6	1	1	3	4	8	11	18	—	—	—	—	—	—
	4	1	1	1	3	6	8	13	—	—	—	—	—	—
	3	0	1	1	3	5	7	11	—	—	—	—	—	—
	2	0	1	1	2	4	6	10	—	—	—	—	—	—
	1	0	0	1	1	3	4	7	—	—	—	—	—	—
	1/0	0	0	1	1	2	3	6	—	—	—	—	—	—
	2/0	0	0	1	1	1	3	5	—	—	—	—	—	—
	3/0	0	0	0	1	1	2	4	—	—	—	—	—	—
	4/0	0	0	0	1	1	1	3	—	—	—	—	—	—
	250	0	0	0	0	1	1	3	—	—	—	—	—	—
	300	0	0	0	0	1	1	2	—	—	—	—	—	—
	350	0	0	0	0	1	1	1	—	—	—	—	—	—
	400	0	0	0	0	1	1	1	—	—	—	—	—	—
	500	0	0	0	0	0	1	1	—	—	—	—	—	—
	600	0	0	0	0	0	1	1	—	—	—	—	—	—
	700	0	0	0	0	0	0	1	—	—	—	—	—	—
	750	0	0	0	0	0	0	1	—	—	—	—	—	—
	800	0	0	0	0	0	0	1	—	—	—	—	—	—
	900	0	0	0	0	0	0	1	—	—	—	—	—	—
	1000	0	0	0	0	0	0	1	—	—	—	—	—	—
	1250	0	0	0	0	0	0	0	—	—	—	—	—	—
	1500	0	0	0	0	0	0	0	—	—	—	—	—	—
	1750	0	0	0	0	0	0	0	—	—	—	—	—	—
	2000	0	0	0	0	0	0	0	—	—	—	—	—	—
THHW, THWN, THWN-2	14	7	12	21	34	61	81	135	—	—	—	—	—	—
	12	5	9	15	25	44	59	98	—	—	—	—	—	—
	10	3	5	10	15	28	37	62	—	—	—	—	—	—
	8	1	3	5	9	16	21	36	—	—	—	—	—	—
	6	1	2	4	6	11	15	26	—	—	—	—	—	—
	4	1	1	2	4	7	9	16	—	—	—	—	—	—
	3	0	1	1	3	6	8	13	—	—	—	—	—	—
	2	0	1	1	3	5	7	11	—	—	—	—	—	—
	1	0	1	1	1	3	5	8	—	—	—	—	—	—
	1/0	0	1	1	1	3	4	7	—	—	—	—	—	—
	2/0	0	0	1	1	2	3	6	—	—	—	—	—	—
	3/0	0	0	1	1	1	3	5	—	—	—	—	—	—
	4/0	0	0	1	1	1	2	4	—	—	—	—	—	—
	250	0	0	0	1	1	1	3	—	—	—	—	—	—
	300	0	0	0	1	1	1	3	—	—	—	—	—	—
	350	0	0	0	1	1	1	2	—	—	—	—	—	—
	400	0	0	0	0	1	1	1	—	—	—	—	—	—
	500	0	0	0	0	1	1	1	—	—	—	—	—	—
	600	0	0	0	0	1	1	1	—	—	—	—	—	—
	700	0	0	0	0	0	1	1	—	—	—	—	—	—
	750	0	0	0	0	0	1	1	—	—	—	—	—	—
	800	0	0	0	0	0	1	1	—	—	—	—	—	—
	900	0	0	0	0	0	1	1	—	—	—	—	—	—
	1000	0	0	0	0	0	0	1	—	—	—	—	—	—

(continues)

Table C.7 *Continued*

Type	Conductor Size (AWG/kcmil)	³⁄₈ (12)	½ (16)	¾ (21)	1 (27)	1¼ (35)	1½ (41)	2 (53)	2½ (63)	3 (78)	3½ (91)	4 (103)	5 (129)	6 (155)
FEP, FEPB, PFA, PFAH, TFE	14	7	12	21	33	59	79	131	—	—	—	—	—	—
	12	5	9	15	24	43	57	96	—	—	—	—	—	—
	10	4	6	11	17	31	41	68	—	—	—	—	—	—
	8	1	3	6	10	17	23	39	—	—	—	—	—	—
	6	1	2	4	7	12	17	28	—	—	—	—	—	—
	4	1	1	3	5	9	11	19	—	—	—	—	—	—
	3	1	1	2	4	7	10	16	—	—	—	—	—	—
	2	1	1	1	3	6	8	13	—	—	—	—	—	—
PFA, PFAH, TFE	1	0	1	1	2	4	5	9	—	—	—	—	—	—
PFA, PFAH, TFE, Z	1/0	0	1	1	1	3	4	8	—	—	—	—	—	—
	2/0	0	0	1	1	3	4	6	—	—	—	—	—	—
	3/0	0	0	1	1	2	3	5	—	—	—	—	—	—
	4/0	0	0	1	1	1	2	4	—	—	—	—	—	—
Z	14	9	14	25	40	71	95	158	—	—	—	—	—	—
	12	6	10	18	28	50	67	112	—	—	—	—	—	—
	10	4	6	11	17	31	41	68	—	—	—	—	—	—
	8	2	4	7	11	19	26	43	—	—	—	—	—	—
	6	1	3	5	7	13	18	30	—	—	—	—	—	—
	4	1	1	3	5	9	12	21	—	—	—	—	—	—
	3	1	1	2	4	7	9	15	—	—	—	—	—	—
	2	1	1	1	3	5	7	12	—	—	—	—	—	—
	1	0	1	1	2	4	6	10	—	—	—	—	—	—
XHHW, ZW, XHHW-2, XHH	14	5	8	15	24	42	56	94	—	—	—	—	—	—
	12	4	6	11	18	32	43	72	—	—	—	—	—	—
	10	3	5	8	13	24	32	54	—	—	—	—	—	—
	8	1	2	4	7	13	18	30	—	—	—	—	—	—
	6	1	1	3	5	10	13	22	—	—	—	—	—	—
	4	1	1	2	4	7	9	16	—	—	—	—	—	—
	3	1	1	1	3	6	8	13	—	—	—	—	—	—
	2	1	1	1	3	5	7	11	—	—	—	—	—	—
XHHW, XHHW-2, XHH	1	0	1	1	1	4	5	8	—	—	—	—	—	—
	1/0	0	1	1	1	3	4	7	—	—	—	—	—	—
	2/0	0	0	1	1	2	3	6	—	—	—	—	—	—
	3/0	0	0	1	1	2	3	5	—	—	—	—	—	—
	4/0	0	0	1	1	1	2	4	—	—	—	—	—	—
	250	0	0	0	1	1	1	3	—	—	—	—	—	—
	300	0	0	0	1	1	1	3	—	—	—	—	—	—
	350	0	0	0	1	1	1	2	—	—	—	—	—	—
	400	0	0	0	0	1	1	1	—	—	—	—	—	—
	500	0	0	0	0	1	1	1	—	—	—	—	—	—
	600	0	0	0	0	1	1	1	—	—	—	—	—	—
	700	0	0	0	0	0	1	1	—	—	—	—	—	—
	750	0	0	0	0	0	1	1	—	—	—	—	—	—
	800	0	0	0	0	0	1	1	—	—	—	—	—	—
	900	0	0	0	0	0	1	1	—	—	—	—	—	—
	1000	0	0	0	0	0	0	1	—	—	—	—	—	—
	1250	0	0	0	0	0	0	1	—	—	—	—	—	—
	1500	0	0	0	0	0	0	1	—	—	—	—	—	—
	1750	0	0	0	0	0	0	0	—	—	—	—	—	—
	2000	0	0	0	0	0	0	0	—	—	—	—	—	—

(continues)

Table C.7 *Continued*

Type	Conductor Size (AWG/kcmil)	3/8 (12)	1/2 (16)	3/4 (21)	1 (27)	1 1/4 (35)	1 1/2 (41)	2 (53)	2 1/2 (63)	3 (78)	3 1/2 (91)	4 (103)	5 (129)	6 (155)
						Trade Size (Metric Designator)								
						FIXTURE WIRES								
RFH-2, FFH-2, RFHH-2	18	5	8	14	23	40	54	90	—	—	—	—	—	—
	16	4	7	12	19	34	46	76	—	—	—	—	—	—
SF-2, SFF-2	18	6	10	18	29	51	68	114	—	—	—	—	—	—
	16	5	8	15	24	42	56	94	—	—	—	—	—	—
	14	4	7	12	19	34	46	76	—	—	—	—	—	—
SF-1, SFF-1	18	11	18	32	51	90	121	202	—	—	—	—	—	—
RFH-1, TF, TFF, XF, XFF	18	8	13	23	38	67	89	149	—	—	—	—	—	—
	16	6	11	19	30	54	72	120	—	—	—	—	—	—
XF, XFF	14	5	8	15	24	42	56	94	—	—	—	—	—	—
TFN, TFFN	18	13	22	38	60	107	143	239	—	—	—	—	—	—
	16	10	17	29	46	82	109	182	—	—	—	—	—	—
PF, PFF, PGF, PGFF, PAF, PTF, PTFF, PAFF	18	12	21	36	57	101	136	226	—	—	—	—	—	—
	16	10	16	28	44	78	105	175	—	—	—	—	—	—
	14	7	12	21	33	59	79	131	—	—	—	—	—	—
ZF, ZFF, ZHF	18	16	27	46	74	131	175	292	—	—	—	—	—	—
	16	12	20	34	54	96	129	215	—	—	—	—	—	—
	14	9	14	25	40	71	95	131	—	—	—	—	—	—
KF-2, KFF-2	18	24	40	69	111	196	263	438	—	—	—	—	—	—
	16	17	28	48	77	137	183	305	—	—	—	—	—	—
	14	11	19	32	52	92	123	205	—	—	—	—	—	—
	12	8	13	22	36	64	85	142	—	—	—	—	—	—
	10	5	8	15	24	42	56	94	—	—	—	—	—	—
KF-1, KFF-1	18	28	46	80	128	227	303	505	—	—	—	—	—	—
	16	20	32	56	90	159	213	355	—	—	—	—	—	—
	14	13	22	38	60	107	143	239	—	—	—	—	—	—
	12	9	14	25	40	71	95	158	—	—	—	—	—	—
	10	6	9	16	26	46	62	103	—	—	—	—	—	—
XF, XFF	12	3	4	8	13	22	30	50	—	—	—	—	—	—
	10	1	3	6	10	17	23	39	—	—	—	—	—	—

Notes:
1. This table is for concentric stranded conductors only. For compact stranded conductors, Table C.5(A) should be used.
2. Two-hour fire-rated RHH cable has ceramifiable insulation, which has larger diameters than other RHH wires. Consult manufacturer's conduit fill tables.
*Types RHH, RHW, and RHW-2 without outer covering.

Table C.7(A) Maximum Number of Conductors of Fixture Wires in Liquidtight Flexible Nonmetallic Conduit (Type LFNC-C) (Based on Chapter 9: Table 1, Table 4, and Table 5A)

| Type | Conductor Size (AWG/kcmil) | Trade Size (Metric Designator) | | | | | | | | | | | | |
		⅜ (12)	½ (16)	¾ (21)	1 (27)	1¼ (35)	1½ (41)	2 (53)	2½ (63)	3 (78)	3½ (91)	4 (103)	5 (129)	6 (155)
		COMPACT CONDUCTORS												
THW, THW-2, THHW	8	1	2	4	6	11	15	25	—	—	—	—	—	—
	6	1	1	3	5	9	12	20	—	—	—	—	—	—
	4	1	1	2	3	6	9	15	—	—	—	—	—	—
	2	1	1	1	2	5	6	11	—	—	—	—	—	—
	1	0	1	1	1	3	4	7	—	—	—	—	—	—
	1/0	0	1	1	1	3	4	6	—	—	—	—	—	—
	2/0	0	0	1	1	2	3	5	—	—	—	—	—	—
	3/0	0	0	1	1	1	3	5	—	—	—	—	—	—
	4/0	0	0	1	1	1	2	4	—	—	—	—	—	—
	250	0	0	0	1	1	1	3	—	—	—	—	—	—
	300	0	0	0	1	1	1	2	—	—	—	—	—	—
	350	0	0	0	1	1	1	2	—	—	—	—	—	—
	400	0	0	0	0	1	1	1	—	—	—	—	—	—
	500	0	0	0	0	1	1	1	—	—	—	—	—	—
	600	0	0	0	0	1	1	1	—	—	—	—	—	—
	700	0	0	0	0	1	1	1	—	—	—	—	—	—
	750	0	0	0	0	0	1	1	—	—	—	—	—	—
	900	0	0	0	0	0	1	1	—	—	—	—	—	—
	1000	0	0	0	0	0	1	1	—	—	—	—	—	—
THHN, THWN, THWN-2	8	—	—	—	—	—	—	—	—	—	—	—	—	—
	6	1	2	4	7	13	17	29	—	—	—	—	—	—
	4	1	1	3	4	8	11	18	—	—	—	—	—	—
	2	1	1	1	3	6	7	13	—	—	—	—	—	—
	1	0	1	1	2	4	6	9	—	—	—	—	—	—
	1/0	0	1	1	1	3	5	8	—	—	—	—	—	—
	2/0	0	1	1	1	3	4	7	—	—	—	—	—	—
	3/0	0	0	1	1	2	3	5	—	—	—	—	—	—
	4/0	0	0	1	1	1	3	4	—	—	—	—	—	—
	250	0	0	0	1	1	1	3	—	—	—	—	—	—
	300	0	0	0	1	1	1	3	—	—	—	—	—	—
	350	0	0	0	1	1	1	3	—	—	—	—	—	—
	400	0	0	0	1	1	1	2	—	—	—	—	—	—
	500	0	0	0	0	1	1	1	—	—	—	—	—	—
	600	0	0	0	0	1	1	1	—	—	—	—	—	—
	700	0	0	0	0	1	1	1	—	—	—	—	—	—
	750	0	0	0	0	1	1	1	—	—	—	—	—	—
	900	0	0	0	0	0	1	1	—	—	—	—	—	—
	1000	0	0	0	0	0	1	1	—	—	—	—	—	—
XHHW, XHHW-2	8	1	3	5	8	15	20	33	—	—	—	—	—	—
	6	1	1	4	6	11	15	24	—	—	—	—	—	—
	4	1	1	3	4	8	11	18	—	—	—	—	—	—
	2	1	1	1	3	6	7	13	—	—	—	—	—	—
	1	0	1	1	2	4	6	9	—	—	—	—	—	—
	1/0	0	1	1	1	3	5	8	—	—	—	—	—	—
	2/0	0	1	1	1	3	4	7	—	—	—	—	—	—
	3/0	0	0	1	1	2	3	5	—	—	—	—	—	—
	4/0	0	0	1	1	1	3	5	—	—	—	—	—	—
	250	0	0	1	1	1	1	4	—	—	—	—	—	—
	300	0	0	0	1	1	1	3	—	—	—	—	—	—
	350	0	0	0	1	1	1	3	—	—	—	—	—	—
	400	0	0	0	1	1	1	2	—	—	—	—	—	—
	500	0	0	0	0	1	1	1	—	—	—	—	—	—

(continues)

Table C.7(A) *Continued*

Type	Conductor Size (AWG/kcmil)	⅜ (12)	½ (16)	¾ (21)	1 (27)	1¼ (35)	1½ (41)	2 (53)	2½ (63)	3 (78)	3½ (91)	4 (103)	5 (129)	6 (155)
						Trade Size (Metric Designator)								
	600	0	0	0	0	1	1	1	—	—	—	—	—	—
	700	0	0	0	0	1	1	1	—	—	—	—	—	—
	750	0	0	0	0	1	1	1	—	—	—	—	—	—
	900	0	0	0	0	0	1	1	—	—	—	—	—	—
	1000	0	0	0	0	0	1	1	—	—	—	—	—	—

Definition: Compact stranding is the result of a manufacturing process where the stranded conductor is compressed to the extent that the interstices (voids between stranded wires) are virtually eliminated.

Table C.8 Maximum Number of Conductors or Fixture Wires in Liquidtight Flexible Metal Conduit (LFMC)
(Based on Chapter 9: Table 1, Table 4, and Table 5)

Type	Conductor Size (AWG/ kcmil)	Trade Size (Metric Designator)												
		⅜ (12)	½ (16)	¾ (21)	1 (27)	1¼ (35)	1½ (41)	2 (53)	2½ (63)	3 (78)	3½ (91)	4 (103)	5 (129)	6 (155)
		CONDUCTORS												
RHH, RHW, RHW-2	14	2	4	7	12	21	27	44	66	102	133	173	—	—
	12	1	3	6	10	17	22	36	55	84	110	144	—	—
	10	1	3	5	8	14	18	29	44	68	89	116	—	—
	8	1	1	2	4	7	9	15	23	36	46	61	—	—
	6	1	1	1	3	6	7	12	18	28	37	48	—	—
	4	0	1	1	2	4	6	9	14	22	29	38	—	—
	3	0	1	1	1	4	5	8	13	19	25	33	—	—
	2	0	1	1	1	3	4	7	11	17	22	29	—	—
	1	0	0	1	1	1	3	5	7	11	14	19	—	—
	1/0	0	0	1	1	1	2	4	6	10	13	16	—	—
	2/0	0	0	1	1	1	1	3	5	8	11	14	—	—
	3/0	0	0	0	1	1	1	3	4	7	9	12	—	—
	4/0	0	0	0	1	1	1	2	4	6	8	10	—	—
	250	0	0	0	0	1	1	1	3	4	6	8	—	—
	300	0	0	0	0	1	1	1	2	4	5	7	—	—
	350	0	0	0	0	1	1	1	2	3	5	6	—	—
	400	0	0	0	0	1	1	1	1	3	4	6	—	—
	500	0	0	0	0	1	1	1	1	3	4	5	—	—
	600	0	0	0	0	0	1	1	1	2	3	4	—	—
	700	0	0	0	0	0	0	1	1	1	3	3	—	—
	750	0	0	0	0	0	0	1	1	1	2	3	—	—
	800	0	0	0	0	0	0	1	1	1	2	3	—	—
	900	0	0	0	0	0	0	1	1	1	2	3	—	—
	1000	0	0	0	0	0	0	1	1	1	1	3	—	—
	1250	0	0	0	0	0	0	0	1	1	1	1	—	—
	1500	0	0	0	0	0	0	0	1	1	1	1	—	—
	1750	0	0	0	0	0	0	0	1	1	1	1	—	—
	2000	0	0	0	0	0	0	0	0	1	1	1	—	—
TW, THHW, THW, THW-2	14	5	9	15	25	44	57	93	140	215	280	365	—	—
	12	4	7	12	19	33	43	71	108	165	215	280	—	—
	10	3	5	9	14	25	32	53	80	123	160	209	—	—
	8	1	3	5	8	14	18	29	44	68	89	116	—	—
RHH*, RHW*, RHW-2*	14	3	6	10	16	29	38	62	93	143	186	243	—	—
	12	3	5	8	13	23	30	50	75	115	149	195	—	—
	10	1	3	6	10	18	23	39	58	89	117	152	—	—
	8	1	1	4	6	11	14	23	35	53	70	91	—	—
TW, THW, THHW, THW-2, RHH*, RHW*, RHW-2*	6	1	1	3	5	8	11	18	27	41	53	70	—	—
	4	1	1	1	3	6	8	13	20	30	40	52	—	—
	3	1	1	1	3	5	7	11	17	26	34	44	—	—
	2	0	1	1	2	4	6	9	14	22	29	38	—	—
	1	0	1	1	1	3	4	7	10	15	20	26	—	—
	1/0	0	0	1	1	2	3	6	8	13	17	23	—	—
	2/0	0	0	1	1	2	3	5	7	11	15	19	—	—
	3/0	0	0	1	1	1	2	4	6	9	12	16	—	—
	4/0	0	0	0	1	1	1	3	5	8	10	13	—	—
	250	0	0	0	1	1	1	3	4	6	8	11	—	—
	300	0	0	0	1	1	1	2	3	5	7	9	—	—
	350	0	0	0	0	1	1	1	3	5	6	8	—	—
	400	0	0	0	0	1	1	1	3	4	6	7	—	—
	500	0	0	0	0	1	1	1	2	3	5	6	—	—
	600	0	0	0	0	1	1	1	1	3	4	5	—	—
	700	0	0	0	0	0	1	1	1	2	3	4	—	—

(continues)

Table C.8 *Continued*

Type	Conductor Size (AWG/kcmil)	3/8 (12)	1/2 (16)	3/4 (21)	1 (27)	1¼ (35)	1½ (41)	2 (53)	2½ (63)	3 (78)	3½ (91)	4 (103)	5 (129)	6 (155)
							Trade Size (Metric Designator)							
	750	0	0	0	0	0	1	1	1	2	3	4	—	—
	800	0	0	0	0	0	1	1	1	2	3	4	—	—
	900	0	0	0	0	0	0	1	1	1	3	3	—	—
	1000	0	0	0	0	0	0	1	1	1	2	3	—	—
	1250	0	0	0	0	0	0	1	1	1	1	2	—	—
	1500	0	0	0	0	0	0	0	1	1	1	2	—	—
	1750	0	0	0	0	0	0	0	1	1	1	1	—	—
	2000	0	0	0	0	0	0	0	1	1	1	1	—	—
THHN, THWN, THWN-2	14	8	13	22	36	63	81	134	201	308	401	523	—	—
	12	5	9	16	26	46	59	97	146	225	292	381	—	—
	10	3	6	10	16	29	37	61	92	141	184	240	—	—
	8	1	3	6	9	16	21	35	53	81	106	138	—	—
	6	1	2	4	7	12	15	25	38	59	76	100	—	—
	4	1	1	2	4	7	9	15	23	36	47	61	—	—
	3	1	1	1	3	6	8	13	20	30	40	52	—	—
	2	1	1	1	3	5	7	11	17	26	33	44	—	—
	1	0	1	1	1	4	5	8	12	19	25	32	—	—
	1/0	0	1	1	1	3	4	7	10	16	21	27	—	—
	2/0	0	0	1	1	2	3	6	8	13	17	23	—	—
	3/0	0	0	1	1	1	3	5	7	11	14	19	—	—
	4/0	0	0	1	1	1	2	4	6	9	12	15	—	—
	250	0	0	0	1	1	1	3	5	7	10	12	—	—
	300	0	0	0	1	1	1	3	4	6	8	11	—	—
	350	0	0	0	1	1	1	2	3	5	7	9	—	—
	400	0	0	0	0	1	1	1	3	5	6	8	—	—
	500	0	0	0	0	1	1	1	2	4	5	7	—	—
	600	0	0	0	0	1	1	1	1	3	4	6	—	—
	700	0	0	0	0	1	1	1	1	3	4	5	—	—
	750	0	0	0	0	0	1	1	1	3	3	5	—	—
	800	0	0	0	0	0	1	1	1	2	3	4	—	—
	900	0	0	0	0	0	1	1	1	2	3	4	—	—
	1000	0	0	0	0	0	0	1	1	1	3	3	—	—
FEP, FEPB, PFA, PFAH, TFE	14	7	12	21	35	61	79	130	195	299	389	507	—	—
	12	5	9	15	25	44	58	94	142	218	284	370	—	—
	10	4	6	11	18	32	41	68	102	156	203	266	—	—
	8	1	3	6	10	18	23	39	58	89	117	152	—	—
	6	1	2	4	7	13	17	27	41	64	83	108	—	—
	4	1	1	3	5	9	12	19	29	44	58	75	—	—
	3	1	1	2	4	7	10	16	24	37	48	63	—	—
	2	1	1	1	3	6	8	13	20	30	40	52	—	—
PFA, PFAH, TFE	1	0	1	1	2	4	5	9	14	21	28	36	—	—
PFA, PFAH, TFE, Z	1/0	0	1	1	1	3	4	7	11	18	23	30	—	—
	2/0	0	1	1	1	3	4	6	9	14	19	25	—	—
	3/0	0	0	1	1	2	3	5	8	12	16	20	—	—
	4/0	0	0	1	1	1	2	4	6	10	13	17	—	—
Z	14	9	15	26	42	73	95	156	235	360	469	611	—	—
	12	6	10	18	30	52	67	111	167	255	332	434	—	—
	10	4	6	11	18	32	41	68	102	156	203	266	—	—
	8	2	4	7	11	20	26	43	64	99	129	168	—	—
	6	1	3	5	8	14	18	30	45	69	90	118	—	—
	4	1	1	3	5	9	12	20	31	48	62	81	—	—
	3	1	1	2	4	7	9	15	23	35	45	59	—	—
	2	1	1	1	3	6	7	12	19	29	38	49	—	—
	1	1	1	1	2	5	6	10	15	23	30	40	—	—

(continues)

Table C.8   *Continued*

Type	Conductor Size (AWG/kcmil)	Trade Size (Metric Designator)												
		⅜ (12)	½ (16)	¾ (21)	1 (27)	1¼ (35)	1½ (41)	2 (53)	2½ (63)	3 (78)	3½ (91)	4 (103)	5 (129)	6 (155)
XHHW, ZW, XHHW-2, XHH	14	5	9	15	25	44	57	93	140	215	280	365	—	—
	12	4	7	12	19	33	43	71	108	165	215	280	—	—
	10	3	5	9	14	25	32	53	80	123	160	209	—	—
	8	1	3	5	8	14	18	29	44	68	89	116	—	—
	6	1	1	3	6	10	13	22	33	50	66	86	—	—
	4	1	1	2	4	7	9	16	24	36	48	62	—	—
	3	1	1	1	3	6	8	13	20	31	40	52	—	—
	2	1	1	1	3	5	7	11	17	26	34	44	—	—
XHHW, XHHW-2, XHH	1	0	1	1	1	4	5	8	12	19	25	33	—	—
	1/0	0	1	1	1	3	4	7	10	16	21	28	—	—
	2/0	0	0	1	1	2	3	6	9	13	17	23	—	—
	3/0	0	0	1	1	1	3	5	7	11	14	19	—	—
	4/0	0	0	1	1	1	2	4	6	9	12	16	—	—
	250	0	0	0	1	1	1	3	5	7	10	13	—	—
	300	0	0	0	1	1	1	3	4	6	8	11	—	—
	350	0	0	0	1	1	1	2	3	5	7	10	—	—
	400	0	0	0	1	1	1	1	3	5	6	8	—	—
	500	0	0	0	0	1	1	1	2	4	5	7	—	—
	600	0	0	0	0	1	1	1	1	3	4	6	—	—
	700	0	0	0	0	1	1	1	1	3	4	5	—	—
	750	0	0	0	0	0	1	1	1	3	3	5	—	—
	800	0	0	0	0	0	1	1	1	2	3	4	—	—
	900	0	0	0	0	0	1	1	1	2	3	4	—	—
	1000	0	0	0	0	0	0	1	1	1	3	3	—	—
	1250	0	0	0	0	0	0	1	1	1	1	3	—	—
	1500	0	0	0	0	0	0	1	1	1	1	2	—	—
	1750	0	0	0	0	0	0	0	1	1	1	1	—	—
	2000	0	0	0	0	0	0	0	1	1	1	1	—	—

FIXTURE WIRES

Type	Conductor Size (AWG/kcmil)	⅜ (12)	½ (16)	¾ (21)	1 (27)	1¼ (35)	1½ (41)	2 (53)	2½ (63)	3 (78)	3½ (91)	4 (103)	5 (129)	6 (155)
RFH-2, FFH-2, RFHH-2	18	5	8	15	24	42	54	89	134	206	268	350	—	—
	16	4	7	12	20	35	46	75	113	174	226	295	—	—
SF-2, SFF-2	18	6	11	19	30	53	69	113	169	260	338	441	—	—
	16	5	9	15	25	44	57	93	140	215	280	365	—	—
	14	4	7	12	20	35	46	75	113	174	226	295	—	—
SF-1, SFF-1	18	12	19	33	53	94	122	199	300	460	599	781	—	—
RFH-1, TF, TFF, XF, XFF	18	8	14	24	39	69	90	147	222	339	442	577	—	—
	16	7	11	20	32	56	72	119	179	274	357	465	—	—
XF, XFF	14	5	9	15	25	44	57	93	140	215	280	365	—	—
TFN, TFFN	18	14	23	39	63	111	144	236	355	543	707	923	—	—
	16	10	17	30	48	85	110	180	271	415	540	705	—	—
PF, PFF, PGF, PGFF, PAF, PTF, PTFF, PAFF	18	13	21	37	60	105	136	224	336	515	671	875	—	—
	16	10	16	29	46	81	105	173	260	398	519	677	—	—
	14	7	12	21	35	61	79	130	195	299	389	507	—	—
ZF, ZFF, ZHF	18	17	28	48	77	136	176	288	434	664	865	1128	—	—
	16	12	20	35	57	100	130	213	320	490	638	832	—	—
	14	9	15	26	42	73	95	156	235	360	469	611	—	—
KF-2, KFF-2	18	25	42	72	116	203	264	433	651	996	1297	1692	—	—
	16	18	29	50	81	142	184	302	454	695	905	1180	—	—
	14	12	19	34	54	95	124	203	305	467	608	793	—	—
	12	8	13	23	38	66	86	141	212	325	423	552	—	—
	10	5	9	15	25	44	57	93	140	215	280	365	—	—

(continues)

Table C.8 *Continued*

Type	Conductor Size (AWG/ kcmil)	Trade Size (Metric Designator)												
		⅜ (12)	½ (16)	¾ (21)	1 (27)	1¼ (35)	1½ (41)	2 (53)	2½ (63)	3 (78)	3½ (91)	4 (103)	5 (129)	6 (155)
KF-1, KFF-1	18	29	48	83	134	235	304	499	751	1150	1497	1952	—	—
	16	20	34	58	94	165	214	351	527	808	1052	1372	—	—
	14	14	23	39	63	111	144	236	355	543	707	923	—	—
	12	9	15	26	42	73	95	156	235	360	469	611	—	—
	10	6	10	17	27	48	62	102	153	235	306	399	—	—
XF, XFF	12	3	5	8	13	23	30	50	75	115	149	195	—	—
	10	1	3	6	10	18	23	39	58	89	117	152	—	—

Notes:

1. This table is for concentric stranded conductors only. For compact stranded conductors, Table C.7(A) should be used.

2. Two-hour fire-rated RHH cable has ceramifiable insulation, which has much larger diameters than other RHH wires. Consult manufacturer's conduit fill tables.

*Types RHH, RHW, and RHW-2 without outer covering.

Table C.8(A) Maximum Number of Conductors or Fixture Wires in Liquidtight Flexible Metal Conduit (LFMC)
(Based on Chapter 9: Table 1, Table 4, and Table 5A)

Type	Conductor Size (AWG/ kcmil)	Trade Size (Metric Designator)												
		⅜ (12)	½ (16)	¾ (21)	1 (27)	1¼ (35)	1½ (41)	2 (53)	2½ (63)	3 (78)	3½ (91)	4 (103)	5 (129)	6 (155)
		COMPACT CONDUCTORS												
THW, THW-2, THHW	8	1	2	4	7	12	15	25	38	58	76	99	—	—
	6	1	1	3	5	9	12	19	29	45	59	77	—	—
	4	1	1	2	4	7	9	14	22	34	44	57	—	—
	2	1	1	1	3	5	6	11	16	25	32	42	—	—
	1	0	1	1	1	3	4	7	11	17	23	30	—	—
	1/0	0	1	1	1	3	4	6	10	15	20	26	—	—
	2/0	0	0	1	1	2	3	5	8	13	16	21	—	—
	3/0	0	0	1	1	1	3	4	7	11	14	18	—	—
	4/0	0	0	1	1	1	2	4	6	9	12	15	—	—
	250	0	0	0	1	1	1	3	4	7	9	12	—	—
	300	0	0	0	1	1	1	2	4	6	8	10	—	—
	350	0	0	0	1	1	1	2	3	5	7	9	—	—
	400	0	0	0	0	1	1	1	3	5	6	8	—	—
	500	0	0	0	0	1	1	1	3	4	5	7	—	—
	600	0	0	0	0	1	1	1	1	3	4	6	—	—
	700	0	0	0	0	1	1	1	1	3	4	5	—	—
	750	0	0	0	0	0	1	1	1	3	3	5	—	—
	900	0	0	0	0	0	1	1	1	2	3	4	—	—
	1000	0	0	0	0	0	1	1	1	1	3	4	—	—
THHN, THWN, THWN-2	8	—	—	—	—	—	—	—	—	—	—	—	—	—
	6	1	2	4	7	13	17	28	43	66	86	112	—	—
	4	1	1	3	4	8	11	17	26	41	53	69	—	—
	2	1	1	1	3	6	7	12	19	29	38	50	—	—
	1	0	1	1	2	4	6	9	14	22	28	37	—	—
	1/0	0	1	1	1	4	5	8	12	19	24	32	—	—
	2/0	0	1	1	1	3	4	6	10	15	20	26	—	—
	3/0	0	0	1	1	2	3	5	8	13	17	22	—	—
	4/0	0	0	1	1	1	3	4	7	10	14	18	—	—
	250	0	0	1	1	1	1	3	5	8	11	14	—	—
	300	0	0	0	1	1	1	3	4	7	9	12	—	—
	350	0	0	0	1	1	1	2	4	6	8	11	—	—
	400	0	0	0	1	1	1	2	3	5	7	9	—	—
	500	0	0	0	0	1	1	1	3	5	6	8	—	—
	600	0	0	0	0	1	1	1	2	4	5	6	—	—
	700	0	0	0	0	1	1	1	1	3	4	6	—	—
	750	0	0	0	0	1	1	1	1	3	4	5	—	—
	900	0	0	0	0	0	1	1	1	2	3	4	—	—
	1000	0	0	0	0	0	1	1	1	2	3	4	—	—
XHHW, XHHW-2	8	1	3	5	9	15	20	33	49	76	98	129	—	—
	6	1	2	4	6	11	15	24	37	56	73	95	—	—
	4	1	1	3	4	8	11	17	26	41	53	69	—	—
	2	1	1	1	3	6	7	12	19	29	38	50	—	—
	1	0	1	1	2	4	6	9	14	22	28	37	—	—
	1/0	0	1	1	1	4	5	8	12	19	24	32	—	—
	2/0	0	1	1	1	3	4	7	10	16	20	27	—	—
	3/0	0	0	1	1	2	3	5	8	13	17	22	—	—
	4/0	0	0	1	1	1	3	4	7	11	14	18	—	—
	250	0	0	1	1	1	1	3	5	8	11	15	—	—
	300	0	0	0	1	1	1	3	5	7	9	12	—	—
	350	0	0	0	1	1	1	3	4	6	8	11	—	—
	400	0	0	0	1	1	1	2	4	6	7	10	—	—
	500	0	0	0	0	1	1	1	3	5	6	8	—	—
	600	0	0	0	0	1	1	1	2	4	5	6	—	—

(continues)

Table C.8(A) *Continued*

Type	Conductor Size (AWG/ kcmil)	⅜ (12)	½ (16)	¾ (21)	1 (27)	1¼ (35)	1½ (41)	2 (53)	2½ (63)	3 (78)	3½ (91)	4 (103)	5 (129)	6 (155)
								Trade Size (Metric Designator)						
	700	0	0	0	0	1	1	1	1	3	4	6	—	—
	750	0	0	0	0	1	1	1	1	3	4	5	—	—
	900	0	0	0	0	0	1	1	1	2	3	4	—	—
	1000	0	0	0	0	0	1	1	1	2	3	4	—	—

Definition: *Compact stranding* is the result of a manufacturing process where the stranded conductor is compressed to the extent that the interstices (voids between strand wires) are virtually eliminated.

**Table C.9 Maximum Number of Conductors or Fixture Wires in Rigid Metal Conduit (RMC)
(Based on Chapter 9: Table 1, Table 4, and Table 5)**

Type	Conductor Size (AWG/ kcmil)	Trade Size (Metric Designator)												
		⅜ (12)	½ (16)	¾ (21)	1 (27)	1¼ (35)	1½ (41)	2 (53)	2½ (63)	3 (78)	3½ (91)	4 (103)	5 (129)	6 (155)
CONDUCTORS														
RHH, RHW, RHW-2	14	—	4	7	12	21	28	46	66	102	136	176	276	398
	12	—	3	6	10	17	23	38	55	85	113	146	229	330
	10	—	3	5	8	14	19	31	44	68	91	118	185	267
	8	—	1	2	4	7	10	16	23	36	48	61	97	139
	6	—	1	1	3	6	8	13	18	29	38	49	77	112
	4	—	1	1	2	4	6	10	14	22	30	38	60	87
	3	—	1	1	2	4	5	9	12	19	26	34	53	76
	2	—	1	1	1	3	4	7	11	17	23	29	46	66
	1	—	0	1	1	1	3	5	7	11	15	19	30	44
	1/0	—	0	1	1	1	2	4	6	10	13	17	26	38
	2/0	—	0	1	1	1	2	4	5	8	11	14	23	33
	3/0	—	0	0	1	1	1	3	4	7	10	12	20	28
	4/0	—	0	0	1	1	1	3	4	6	8	11	17	24
	250	—	0	0	0	1	1	1	3	4	6	8	13	18
	300	—	0	0	0	1	1	1	2	4	5	7	11	16
	350	—	0	0	0	1	1	1	2	4	5	6	10	15
	400	—	0	0	0	1	1	1	1	3	4	6	9	13
	500	—	0	0	0	1	1	1	1	3	4	5	8	11
	600	—	0	0	0	0	1	1	1	2	3	4	6	9
	700	—	0	0	0	0	1	1	1	1	3	3	6	8
	750	—	0	0	0	0	0	1	1	1	3	3	5	8
	800	—	0	0	0	0	0	1	1	1	2	3	5	7
	900	—	0	0	0	0	0	1	1	1	2	3	5	7
	1000	—	0	0	0	0	0	1	1	1	1	3	4	6
	1250	—	0	0	0	0	0	0	1	1	1	1	3	5
	1500	—	0	0	0	0	0	0	1	1	1	1	3	4
	1750	—	0	0	0	0	0	0	1	1	1	1	2	4
	2000	—	0	0	0	0	0	0	0	1	1	1	2	3
TW, THHW, THW, THW-2	14	—	9	15	25	44	59	98	140	215	288	370	581	839
	12	—	7	12	19	33	45	75	107	165	221	284	446	644
	10	—	5	9	14	25	34	56	80	123	164	212	332	480
	8	—	3	5	8	14	19	31	44	68	91	118	185	267
RHH*, RHW*, RHW-2*	14	—	6	10	17	29	39	65	93	143	191	246	387	558
	12	—	5	8	13	23	32	52	75	115	154	198	311	448
	10	—	3	6	10	18	25	41	58	90	120	154	242	350
	8	—	1	4	6	11	15	24	35	54	72	92	145	209
TW, THW, THHW, THW-2, RHH*, RHW*, RHW-2*	6	—	1	3	5	8	11	18	27	41	55	71	111	160
	4	—	1	1	3	6	8	14	20	31	41	53	83	120
	3	—	1	1	3	5	7	12	17	26	35	45	71	103
	2	—	1	1	2	4	6	10	14	22	30	38	60	87
	1	—	1	1	1	3	4	7	10	15	21	27	42	61
	1/0	—	0	1	1	2	3	6	8	13	18	23	36	52
	2/0	—	0	1	1	2	3	5	7	11	15	19	31	44
	3/0	—	0	1	1	1	2	4	6	9	13	16	26	37
	4/0	—	0	0	1	1	1	3	5	8	10	14	21	31
	250	—	0	0	1	1	1	3	4	6	8	11	17	25
	300	—	0	0	1	1	1	2	3	5	7	9	15	22
	350	—	0	0	0	1	1	1	3	5	6	8	13	19
	400	—	0	0	0	1	1	1	3	4	6	7	12	17
	500	—	0	0	0	1	1	1	2	3	5	6	10	14
	600	—	0	0	0	1	1	1	1	3	4	5	8	12
	700	—	0	0	0	0	1	1	1	2	3	4	7	10

(continues)

Table C.9   *Continued*

Type	Conductor Size (AWG/ kcmil)	⅜ (12)	½ (16)	¾ (21)	1 (27)	1¼ (35)	1½ (41)	2 (53)	2½ (63)	3 (78)	3½ (91)	4 (103)	5 (129)	6 (155)
							Trade Size (Metric Designator)							
	750	—	0	0	0	0	1	1	1	2	3	4	7	10
	800	—	0	0	0	0	1	1	1	2	3	4	6	9
	900	—	0	0	0	0	1	1	1	1	3	3	6	8
	1000	—	0	0	0	0	0	1	1	1	2	3	5	8
	1250	—	0	0	0	0	0	1	1	1	1	2	4	6
	1500	—	0	0	0	0	0	1	1	1	1	2	3	5
	1750	—	0	0	0	0	0	0	1	1	1	1	3	4
	2000	—	0	0	0	0	0	0	1	1	1	1	3	4
THHN, THWN, THWN-2	14	—	13	22	36	63	85	140	200	309	412	531	833	1202
	12	—	9	16	26	46	62	102	146	225	301	387	608	877
	10	—	6	10	17	29	39	64	92	142	189	244	383	552
	8	—	3	6	9	16	22	37	53	82	109	140	221	318
	6	—	2	4	7	12	16	27	38	59	79	101	159	230
	4	—	1	2	4	7	10	16	23	36	48	62	98	141
	3	—	1	1	3	6	8	14	20	31	41	53	83	120
	2	—	1	1	3	5	7	11	17	26	34	44	70	100
	1	—	1	1	1	4	5	8	12	19	25	33	51	74
	1/0	—	1	1	1	3	4	7	10	16	21	27	43	63
	2/0	—	0	1	1	2	3	6	8	13	18	23	36	52
	3/0	—	0	1	1	1	3	5	7	11	15	19	30	43
	4/0	—	0	1	1	1	2	4	6	9	12	16	25	36
	250	—	0	0	1	1	1	3	5	7	10	13	20	29
	300	—	0	0	1	1	1	3	4	6	8	11	17	25
	350	—	0	0	1	1	1	2	3	5	7	10	15	22
	400	—	0	0	1	1	1	2	3	5	7	8	13	20
	500		0	0	0	1	1	1	2	4	5	7	11	16
	600	—	0	0	0	1	1	1	1	3	4	6	9	13
	700	—	0	0	0	1	1	1	1	3	4	5	8	11
	750	—	0	0	0	0	1	1	1	3	4	5	7	11
	800	—	0	0	0	0	1	1	1	2	3	4	7	10
	900	—	0	0	0	0	1	1	1	2	3	4	6	9
	1000	—	0	0	0	0	1	1	1	1	3	4	6	8
FEP, FEPB, PFA, PFAH, TFE	14	—	12	22	35	61	83	136	194	300	400	515	808	1166
	12	—	9	16	26	44	60	99	142	219	292	376	590	851
	10	—	6	11	18	32	43	71	102	157	209	269	423	610
	8	—	3	6	10	18	25	41	58	90	120	154	242	350
	6	—	2	4	7	13	17	29	41	64	85	110	172	249
	4	—	1	3	5	9	12	20	29	44	59	77	120	174
	3	—	1	2	4	7	10	17	24	37	50	64	100	145
	2		1	1	3	6	8	14	20	31	41	53	83	120
PFA, PFAH, TFE	1		1	1	2	4	6	9	14	21	28	37	57	83
PFA, PFAH, TFE, Z	1/0	—	1	1	1	3	5	8	11	18	24	30	48	69
	2/0	—	1	1	1	3	4	6	9	14	19	25	40	57
	3/0	—	0	1	1	2	3	5	8	12	16	21	33	47
	4/0	—	0	1	1	1	2	4	6	10	13	17	27	39
Z	14	—	15	26	42	73	100	164	234	361	482	621	974	1405
	12	—	10	18	30	52	71	116	166	256	342	440	691	997
	10	—	6	11	18	32	43	71	102	157	209	269	423	610
	8	—	4	7	11	20	27	45	64	99	132	170	267	386
	6	—	3	5	8	14	19	31	45	69	93	120	188	271
	4	—	1	3	5	9	13	22	31	48	64	82	129	186
	3	—	1	2	4	7	9	16	22	35	47	60	94	136
	2		1	1	3	6	8	13	19	29	39	50	78	113
	1		1	1	2	5	6	10	15	23	31	40	63	92

(continues)

Table C.9 *Continued*

Type	Conductor Size (AWG/ kcmil)	Trade Size (Metric Designator)												
		⅜ (12)	½ (16)	¾ (21)	1 (27)	1¼ (35)	1½ (41)	2 (53)	2½ (63)	3 (78)	3½ (91)	4 (103)	5 (129)	6 (155)
XHHW, ZW. XHHW-2, XHH	14	—	9	15	25	44	59	98	140	215	288	370	581	839
	12	—	7	12	19	33	45	75	107	165	221	284	446	644
	10	—	5	9	14	25	34	56	80	123	164	212	332	480
	8	—	3	5	8	14	19	31	44	68	91	118	185	267
	6	—	1	3	6	10	14	23	33	51	68	87	137	197
	4	—	1	2	4	7	10	16	24	37	49	63	99	143
	3	—	1	1	3	6	8	14	20	31	41	53	84	121
	2	—	1	1	3	5	7	12	17	26	35	45	70	101
XHHW, XHHW-2, XHH	1	—	1	1	1	4	5	9	12	19	26	33	52	76
	1/0	—	1	1	1	3	4	7	10	16	22	28	44	64
	2/0	—	0	1	1	2	3	6	9	13	18	23	37	53
	3/0	—	0	1	1	1	3	5	7	11	15	19	30	44
	4/0	—	0	1	1	1	2	4	6	9	12	16	25	36
	250	—	0	0	1	1	1	3	5	7	10	13	20	30
	300	—	0	0	1	1	1	3	4	6	9	11	18	25
	350	—	0	0	1	1	1	2	3	6	7	10	15	22
	400	—	0	0	1	1	1	2	3	5	7	9	14	20
	500	—	0	0	0	1	1	1	2	4	5	7	11	16
	600	—	0	0	0	1	1	1	1	3	4	6	9	13
	700	—	0	0	0	1	1	1	1	3	4	5	8	11
	750	—	0	0	0	0	1	1	1	3	4	5	7	11
	800	—	0	0	0	0	1	1	1	2	3	4	7	10
	900	—	0	0	0	0	1	1	1	2	3	4	6	9
	1000	—	0	0	0	0	1	1	1	1	3	4	6	8
	1250	—	0	0	0	0	0	1	1	1	2	3	4	6
	1500	—	0	0	0	0	0	1	1	1	1	2	4	5
	1750	—	0	0	0	0	0	0	1	1	1	1	3	5
	2000	—	0	0	0	0	0	0	1	1	1	1	3	4
FIXTURE WIRES														
RFH-2, FFH-2, RFHH-2	18	—	8	15	24	42	57	94	134	207	276	355	557	804
	16	—	7	12	20	35	48	79	113	174	232	299	470	678
SF-2, SFF-2	18	—	11	19	31	53	72	118	169	261	348	448	703	1014
	16	—	9	15	25	44	59	98	140	215	288	370	581	839
	14	—	7	12	20	35	48	79	113	174	232	299	470	678
SF-1, SFF-1	18	—	19	33	54	94	127	209	299	461	616	792	1244	1794
RFH-1, TF, TFF, XF, XFF	18	—	14	25	40	69	94	155	221	341	455	585	918	1325
	16	—	11	20	32	56	76	125	178	275	367	472	741	1070
XF, XFF	14	—	9	15	25	44	59	98	140	215	288	370	581	839
TFN, TFFN	18	—	23	40	64	111	150	248	354	545	728	937	1470	2120
	16	—	17	30	49	84	115	189	270	416	556	715	1123	1620
PF, PFF, PGF, PGFF, PAF, PTF, PTFF, PAFF	18	—	21	38	61	105	143	235	335	517	690	888	1394	2011
	16	—	16	29	47	81	110	181	259	400	534	687	1078	1555
	14	—	12	22	35	61	83	136	194	300	400	515	808	1166
ZF, ZFF, ZHF	18	—	28	49	79	135	184	303	432	666	889	1145	1796	2592
	16	—	20	36	58	100	136	223	319	491	656	844	1325	1912
	14	—	15	26	42	73	100	164	234	361	482	621	974	1405
KF-2, KFF-2	18	—	42	73	118	203	276	454	648	1000	1334	1717	2695	3887
	16	—	29	51	82	142	192	317	452	697	931	1198	1880	2712
	14	—	19	34	55	95	129	213	304	468	625	805	1263	1822
	12	—	13	24	38	66	90	148	211	326	435	560	878	1267
	10	—	9	15	25	44	59	98	140	215	288	370	581	839

(continues)

Table C.9   *Continued*

Type	Conductor Size (AWG/ kcmil)	Trade Size (Metric Designator)												
		⅜ (12)	½ (16)	¾ (21)	1 (27)	1¼ (35)	1½ (41)	2 (53)	2½ (63)	3 (78)	3½ (91)	4 (103)	5 (129)	6 (155)
KF-1, KFF-1	18	—	48	84	136	234	318	524	748	1153	1540	1982	3109	4486
	16	—	34	59	96	165	224	368	526	810	1082	1392	2185	3152
	14	—	23	40	64	111	150	248	354	545	728	937	1470	2120
	12	—	15	26	42	73	100	164	234	361	482	621	974	1405
	10	—	10	17	28	48	65	107	153	236	315	405	636	918
XF, XFF	12	—	5	8	13	23	32	52	75	115	154	198	311	448
	10	—	3	6	10	18	25	41	58	90	120	154	242	350

Notes:

1. This table is for concentric stranded conductors only. For compact stranded conductors, Table C.8(A) should be used.

2. Two-hour fire-rated RHH cable has ceramifiable insulation, which has much larger diameters than other RHH wires. Consult manufacturer's conduit fill tables.

*Types RHH, RHW, and RHW-2 without outer covering.

Table C.9(A) Maximum Number of Conductors or Fixture Wires in Rigid Metal Conduit (RMC)
(Based on Chapter 9: Table 1, Table 4, and Table 5A)

Type	Conductor Size (AWG/ kcmil)	Trade Size (Metric Designator)												
		³⁄₈ (12)	½ (16)	¾ (21)	1 (27)	1¼ (35)	1½ (41)	2 (53)	2½ (63)	3 (78)	3½ (91)	4 (103)	5 (129)	6 (155)
		COMPACT CONDUCTORS												
THW, THW-2, THHW	8	—	2	4	7	12	16	26	38	59	78	101	158	228
	6	—	1	3	5	9	12	20	29	45	60	78	122	176
	4	—	1	2	4	7	9	15	22	34	45	58	91	132
	2	—	1	1	3	5	7	11	16	25	33	43	67	97
	1	—	1	1	1	3	5	8	11	17	23	30	47	68
	1/0	—	1	1	1	3	4	7	10	15	20	26	41	59
	2/0	—	0	1	1	2	3	6	8	13	17	22	34	50
	3/0	—	0	1	1	1	3	5	7	11	14	19	29	42
	4/0	—	0	1	1	1	2	4	6	9	12	15	24	35
	250	—	0	0	1	1	1	3	4	7	9	12	19	28
	300	—	0	0	1	1	1	3	4	6	8	11	17	24
	350	—	0	0	1	1	1	2	3	5	7	9	15	22
	400	—	0	0	1	1	1	1	3	5	7	8	13	20
	500	—	0	0	0	1	1	1	3	4	5	7	11	17
	600	—	0	0	0	1	1	1	1	3	4	6	9	13
	700	—	0	0	0	1	1	1	1	3	4	5	8	12
	750	—	0	0	0	0	1	1	1	3	4	5	7	11
	900	—	0	0	0	0	1	1	1	2	3	4	7	10
	1000	—	0	0	0	0	1	1	1	1	3	4	6	9
THHN, THWN, THWN-2	8	—	—	—	—	—	—	—	—	—	—	—	—	—
	6	—	2	5	8	13	18	30	43	66	88	114	179	258
	4		1	3	5	8	11	18	26	41	55	70	110	159
	2	—	1	1	3	6	8	13	19	29	39	50	79	114
	1	—	1	1	2	4	6	10	14	22	29	38	59	86
	1/0	—	1	1	1	4	5	8	12	19	25	32	51	73
	2/0	—	1	1	1	3	4	7	10	15	21	26	42	60
	3/0	—	0	1	1	2	3	6	8	13	17	22	35	51
	4/0	—	0	1	1	1	3	5	7	10	14	18	29	42
	250	—	0	1	1	1	2	4	5	8	11	14	23	33
	300	—	0	0	1	1	1	3	4	7	10	12	20	28
	350	—	0	0	1	1	1	3	4	6	8	11	17	25
	400	—	0	0	1	1	1	2	3	5	7	10	15	22
	500	—	0	0	0	1	1	1	3	5	6	8	13	19
	600	—	0	0	0	1	1	1	2	4	5	6	10	15
	700	—	0	0	0	1	1	1	1	3	4	6	9	13
	750	—	0	0	0	1	1	1	1	3	4	5	9	13
	900	—	0	0	0	0	1	1	1	2	3	4	7	10
	1000	—	0	0	0	0	1	1	1	2	3	4	6	9
XHHW, XHHW-2	8	—	3	5	9	15	21	34	49	76	101	130	205	296
	6	—	2	4	6	11	15	25	36	56	75	97	152	220
	4	—	1	3	5	8	11	18	26	41	55	70	110	159
	2	—	1	1	3	6	8	13	19	29	39	50	79	114
	1	—	1	1	2	4	6	10	14	22	29	38	59	86
	1/0	—	1	1	1	4	5	8	12	19	25	32	51	73
	2/0	—	1	1	1	3	4	7	10	16	21	27	43	62
	3/0	—	0	1	1	2	3	6	8	13	17	22	35	51
	4/0	—	0	1	1	1	3	5	7	11	14	19	29	42
	250	—	0	1	1	1	2	4	5	8	11	15	23	34
	300	—	0	0	1	1	1	3	5	7	10	13	20	29
	350	—	0	0	1	1	1	3	4	6	9	11	18	25
	400	—	0	0	1	1	1	2	4	6	8	10	16	23
	500	—	0	0	0	1	1	1	3	5	6	8	13	19
	600	—	0	0	0	1	1	1	2	4	5	7	10	15

(continues)

Table C.9(A) *Continued*

Type	Conductor Size (AWG/ kcmil)	Trade Size (Metric Designator)												
		³⁄₈ (12)	½ (16)	¾ (21)	1 (27)	1¼ (35)	1½ (41)	2 (53)	2½ (63)	3 (78)	3½ (91)	4 (103)	5 (129)	6 (155)
	700	—	0	0	0	1	1	1	1	3	4	6	9	13
	750	—	0	0	0	1	1	1	1	3	4	5	8	12
	900	—	0	0	0	0	1	1	1	2	3	5	7	11
	1000	—	0	0	0	1	1	1	2	3	4	7	10

Definition: *Compact stranding* is the result of a manufacturing process where the stranded conductor is compressed to the extent that the interstices (voids between strand wires) are virtually eliminated.

Table C.10 Maximum Number of Conductors or Fixture Wires in Rigid PVC Conduit, Schedule 80 (Based on Chapter 9: Table 1, Table 4, and Table 5)

Type	Conductor Size (AWG/kcmil)	⅜ (12)	½ (16)	¾ (21)	1 (27)	1¼ (35)	1½ (41)	2 (53)	2½ (63)	3 (78)	3½ (91)	4 (103)	5 (129)	6 (155)
CONDUCTORS														
RHH, RHW, RHW-2	14	—	3	5	9	17	23	39	56	88	118	153	243	349
	12	—	2	4	7	14	19	32	46	73	98	127	202	290
	10	—	1	3	6	11	15	26	37	59	79	103	163	234
	8	—	1	1	3	6	8	13	19	31	41	54	85	122
	6	—	1	1	2	4	6	11	16	24	33	43	68	98
	4	—	1	1	1	3	5	8	12	19	26	33	53	77
	3	—	0	1	1	3	4	7	11	17	23	29	47	67
	2	—	0	1	1	3	4	6	9	14	20	25	41	58
	1	—	0	1	1	1	2	4	6	9	13	17	27	38
	1/0	—	0	0	1	1	1	3	5	8	11	15	23	33
	2/0	—	0	0	1	1	1	3	4	7	10	13	20	29
	3/0	—	0	0	1	1	1	3	4	6	8	11	17	25
	4/0	—	0	0	0	1	1	2	3	5	7	9	15	21
	250	—	0	0	0	1	1	1	2	4	5	7	11	16
	300	—	0	0	0	1	1	1	2	3	5	6	10	14
	350	—	0	0	0	1	1	1	1	3	4	5	9	13
	400	—	0	0	0	0	1	1	1	3	4	5	8	12
	500	—	0	0	0	0	1	1	1	2	3	4	7	10
	600	—	0	0	0	0	0	1	1	1	3	3	6	8
	700	—	0	0	0	0	0	1	1	1	2	3	5	7
	750	—	0	0	0	0	0	1	1	1	2	3	5	7
	800	—	0	0	0	0	0	1	1	1	2	3	4	7
	900	—	0	0	0	0	0	1	1	1	1	2	4	6
	1000	—	0	0	0	0	0	1	1	1	1	2	4	5
	1250	—	0	0	0	0	0	0	1	1	1	1	3	4
	1500	—	0	0	0	0	0	0	1	1	1	1	2	4
	1750	—	0	0	0	0	0	0	0	1	1	1	2	3
	2000	—	0	0	0	0	0	0	0	1	1	1	1	3
TW, THHW, THW, THW-2	14	—	6	11	19	35	49	82	118	185	250	324	514	736
	12	—	4	9	15	27	38	63	91	142	192	248	394	565
	10	—	3	6	11	20	28	47	68	106	143	185	294	421
	8	—	1	3	6	11	15	26	37	59	79	103	163	234
RHH*, RHW*, RHW-2*	14	—	4	8	13	23	32	55	79	123	166	215	341	490
	12	—	3	6	10	19	26	44	63	99	133	173	274	394
	10	—	2	5	8	15	20	34	49	77	104	135	214	307
	8	—	1	3	5	9	12	20	29	46	62	81	128	184
TW, THW, THHW, THW-2, RHH*, RHW*, RHW-2*	6	—	1	1	3	7	9	16	22	35	48	62	98	141
	4	—	1	1	3	5	7	12	17	26	35	46	73	105
	3	—	1	1	2	4	6	10	14	22	30	39	63	90
	2	—	1	1	1	3	5	8	12	19	26	33	53	77
	1	—	0	1	1	2	3	6	8	13	18	23	37	54
	1/0	—	0	1	1	1	3	5	7	11	15	20	32	46
	2/0	—	0	1	1	1	2	4	6	10	13	17	27	39
	3/0	—	0	0	1	1	1	3	5	8	11	14	23	33
	4/0	—	0	0	1	1	1	3	4	7	9	12	19	27
	250	—	0	0	0	1	1	2	3	5	7	9	15	22
	300	—	0	0	0	1	1	1	3	5	6	8	13	19
	350	—	0	0	0	1	1	1	2	4	6	7	12	17
	400	—	0	0	0	1	1	1	2	4	5	7	10	15
	500	—	0	0	0	1	1	1	1	3	4	5	9	13
	600	—	0	0	0	0	1	1	1	2	3	4	7	10
	700	—	0	0	0	0	1	1	1	2	3	4	6	9

(continues)

Table C.10 *Continued*

Type	Conductor Size (AWG/ kcmil)	Trade Size (Metric Designator)												
		⅜ (12)	½ (16)	¾ (21)	1 (27)	1¼ (35)	1½ (41)	2 (53)	2½ (63)	3 (78)	3½ (91)	4 (103)	5 (129)	6 (155)
	750	—	0	0	0	0	0	1	1	1	3	4	6	8
	800	—	0	0	0	0	0	1	1	1	3	3	6	8
	900	—	0	0	0	0	0	1	1	1	2	3	5	7
	1000	—	0	0	0	0	0	1	1	1	2	3	5	7
	1250	—	0	0	0	0	0	1	1	1	1	2	4	5
	1500	—	0	0	0	0	0	0	1	1	1	1	3	4
	1750	—	0	0	0	0	0	0	1	1	1	1	3	4
	2000	—	0	0	0	0	0	0	0	1	1	1	2	3
THHN, THWN, THWN-2	14	—	9	17	28	51	70	118	170	265	358	464	736	1055
	12	—	6	12	20	37	51	86	124	193	261	338	537	770
	10	—	4	7	13	23	32	54	78	122	164	213	338	485
	8	—	2	4	7	13	18	31	45	70	95	123	195	279
	6	—	1	3	5	9	13	22	32	51	68	89	141	202
	4	—	1	1	3	6	8	14	20	31	42	54	86	124
	3	—	1	1	3	5	7	12	17	26	35	46	73	105
	2	—	1	1	2	4	6	10	14	22	30	39	61	88
	1	—	0	1	1	3	4	7	10	16	22	29	45	65
	1/0	—	0	1	1	2	3	6	9	14	18	24	38	55
	2/0	—	0	1	1	1	3	5	7	11	15	20	32	46
	3/0	—	0	1	1	1	2	4	6	9	13	17	26	38
	4/0	—	0	0	1	1	1	3	5	8	10	14	22	31
	250	—	0	0	1	1	1	3	4	6	8	11	18	25
	300	—	0	0	0	1	1	2	3	5	7	9	15	22
	350	—	0	0	0	1	1	1	3	5	6	8	13	19
	400	—	0	0	0	1	1	1	3	4	6	7	12	17
	500	—	0	0	0	1	1	1	2	3	5	6	10	14
	600	—	0	0	0	0	1	1	1	3	4	5	8	12
	700	—	0	0	0	0	1	1	1	2	3	4	7	10
	750	—	0	0	0	0	1	1	1	2	3	4	7	9
	800	—	0	0	0	0	1	1	1	2	3	4	6	9
	900	—	0	0	0	0	0	1	1	1	3	3	6	8
	1000	—	0	0	0	0	0	1	1	1	2	3	5	7
FEP, FEPB, PFA, PFAH, TFE	14	—	8	16	27	49	68	115	164	257	347	450	714	1024
	12	—	6	12	20	36	50	84	120	188	253	328	521	747
	10	—	4	8	14	26	36	60	86	135	182	235	374	536
	8	—	2	5	8	15	20	34	49	77	104	135	214	307
	6	—	1	3	6	10	14	24	35	55	74	96	152	218
	4	—	1	2	4	7	10	17	24	38	52	67	106	153
	3	—	1	1	3	6	8	14	20	32	43	56	89	127
	2	—	1	1	3	5	7	12	17	26	35	46	73	105
PFA, PFAH, TFE	1	—	1	1	1	3	5	8	11	18	25	32	51	73
PFA, PFAH, TFE, Z	1/0	—	0	1	1	3	4	7	10	15	20	27	42	61
	2/0	—	0	1	1	2	3	5	8	12	17	22	35	50
	3/0	—	0	1	1	1	2	4	6	10	14	18	29	41
	4/0	—	0	0	1	1	1	4	5	8	11	15	24	34
Z	14	—	10	19	33	59	82	138	198	310	418	542	860	1233
	12	—	7	14	23	42	58	98	141	220	297	385	610	875
	10	—	4	8	14	26	36	60	86	135	182	235	374	536
	8	—	3	5	9	16	22	38	54	85	115	149	236	339
	6	—	1	4	6	11	16	26	38	60	81	104	166	238
	4	—	1	2	4	8	11	18	26	41	55	72	114	164
	3	—	1	1	3	5	8	13	19	30	40	52	83	119
	2	—	1	1	2	5	6	11	16	25	33	43	69	99
	1	—	1	1	1	4	5	9	13	20	27	35	56	80

(continues)

Table C.10   *Continued*

Type	Conductor Size (AWG/kcmil)	⅜ (12)	½ (16)	¾ (21)	1 (27)	1¼ (35)	1½ (41)	2 (53)	2½ (63)	3 (78)	3½ (91)	4 (103)	5 (129)	6 (155)
								Trade Size (Metric Designator)						
XHHW, ZW, XHHW-2, XHH	14	—	6	11	19	35	49	82	118	185	250	324	514	736
	12	—	4	9	15	27	38	63	91	142	192	248	394	565
	10	—	3	6	11	20	28	47	68	106	143	185	294	421
	8	—	1	3	6	11	15	26	37	59	79	103	163	234
	6	—	1	2	4	8	11	19	28	43	59	76	121	173
	4	—	1	1	3	6	8	14	20	31	42	55	87	125
	3	—	1	1	3	5	7	12	17	26	36	47	74	106
	2	—	1	1	2	4	6	10	14	22	30	39	62	89
XHHW, XHHW-2, XHH	1	—	0	1	1	3	4	7	10	16	22	29	46	66
	1/0	—	0	1	1	2	3	6	9	14	19	24	39	56
	2/0	—	0	1	1	1	3	5	7	11	16	20	32	46
	3/0	—	0	1	1	1	2	4	6	9	13	17	27	38
	4/0	—	0	0	1	1	1	3	5	8	11	14	22	32
	250	—	0	0	1	1	1	3	4	6	9	11	18	26
	300	—	0	0	1	1	1	2	3	5	7	10	15	22
	350	—	0	0	0	1	1	1	3	5	6	8	14	20
	400	—	0	0	0	1	1	1	3	4	6	7	12	17
	500	—	0	0	0	1	1	1	2	3	5	6	10	14
	600	—	0	0	0	0	1	1	1	3	4	5	8	11
	700	—	0	0	0	0	1	1	1	2	3	4	7	10
	750	—	0	0	0	0	1	1	1	2	3	4	6	9
	800	—	0	0	0	0	1	1	1	1	3	4	6	9
	900	—	0	0	0	0	0	1	1	1	3	3	5	8
	1000	—	0	0	0	0	1	1	1	2	3	5	7
	1250	—	0	0	0	0	1	1	1	1	2	4	6
	1500	—	0	0	0	0	0	1	1	1	1	3	5
	1750	—	0	0	0	0	0	1	1	1	1	3	4
	2000	—	0	0	0	0	0	1	1	1	1	2	4

FIXTURE WIRES

Type	Conductor Size (AWG/kcmil)	⅜ (12)	½ (16)	¾ (21)	1 (27)	1¼ (35)	1½ (41)	2 (53)	2½ (63)	3 (78)	3½ (91)	4 (103)	5 (129)	6 (155)
RFH-2, FFH-2, RFHH-2	18	—	6	11	19	34	47	79	113	177	239	310	492	706
	16	—	5	9	16	28	39	67	95	150	202	262	415	595
SF-2, SFF-2	18	—	7	14	24	43	59	100	143	224	302	391	621	890
	16	—	6	11	19	35	49	82	118	185	250	324	514	736
	14	—	5	9	16	28	39	67	95	150	202	262	415	595
SF-1, SFF-1	18	—	13	25	42	76	105	177	253	396	534	692	1098	1575
RFH-1, TF, TFF, XF, XFF	18	—	10	18	31	56	77	130	187	293	395	511	811	1163
	16	—	8	15	25	45	62	105	151	236	319	413	655	939
XF, XFF	14	—	6	11	19	35	49	82	118	185	250	324	514	736
TFN, TFFN	18	—	15	29	50	90	124	209	299	468	632	818	1298	1861
	16	—	12	22	38	68	95	159	229	358	482	625	992	1422
PF, PFF, PGF, PGFF, PAF, PTF, PTFF, PAFF	18	—	15	28	47	85	118	198	284	444	599	776	1231	1765
	16	—	11	22	36	66	91	153	219	343	463	600	952	1365
	14	—	8	16	27	49	68	115	164	257	347	450	714	1024
ZF, ZFF, ZHF	18	—	19	36	61	110	152	255	366	572	772	1000	1587	2275
	16	—	14	27	45	81	112	188	270	422	569	738	1171	1678
	14	—	10	19	33	59	82	138	198	310	418	542	860	1233
KF-2, KFF-2	18	—	29	54	91	165	228	383	549	859	1158	1501	2380	3413
	16	—	20	38	64	115	159	267	383	599	808	1047	1661	2381
	14	—	13	25	43	77	107	179	257	402	543	703	1116	1600
	12	—	9	17	30	53	74	125	179	280	377	489	776	1113
	10	—	6	11	19	35	49	82	118	185	250	324	514	736

(continues)

Table C.10   *Continued*

Type	Conductor Size (AWG/ kcmil)	Trade Size (Metric Designator)												
		⅜ (12)	½ (16)	¾ (21)	1 (27)	1¼ (35)	1½ (41)	2 (53)	2½ (63)	3 (78)	3½ (91)	4 (103)	5 (129)	6 (155)
KF-1, KFF-1	18	—	33	63	106	190	263	442	633	991	1336	1732	2747	3938
	16	—	23	44	74	133	185	310	445	696	939	1217	1930	2767
	14	—	15	29	50	90	124	209	299	468	632	818	1298	1861
	12	—	10	19	33	59	82	138	198	310	418	542	860	1233
	10	—	7	13	21	39	54	90	129	203	273	354	562	806
XF, XFF	12	—	3	6	10	19	26	44	63	99	133	173	274	394
	10	—	2	5	8	15	20	34	49	77	104	135	214	307

Notes:

1. This table is for concentric stranded conductors only. For compact stranded conductors, Table C.9(A) should be used.

2. Two-hour fire-rated RHH cable has ceramifiable insulation, which has much larger diameters than other RHH wires. Consult manufacturer's conduit fill tables.

*Types RHH, RHW, and RHW-2 without outer covering.

Table C.10(A) Maximum Number of Conductors or Fixture Wires in Rigid PVC Conduit, Schedule 80 (Based on Chapter 9: Table 1, Table 4, and Table 5A)

Type	Conductor Size (AWG/ kcmil)	⅜ (12)	½ (16)	¾ (21)	1 (27)	1¼ (35)	1½ (41)	2 (53)	2½ (63)	3 (78)	3½ (91)	4 (103)	5 (129)	6 (155)
						COMPACT CONDUCTORS								
THW, THW-2, THHW	8	—	1	3	5	9	13	22	32	50	68	88	140	200
	6	—	1	2	4	7	10	17	25	39	52	68	108	155
	4	—	1	1	3	5	7	13	18	29	39	51	81	116
	2	—	1	1	1	4	5	9	13	21	29	37	60	85
	1	—	0	1	1	3	4	6	9	15	20	26	42	60
	1/0	—	0	1	1	2	3	6	8	13	17	23	36	52
	2/0	—	0	1	1	1	3	5	7	11	15	19	30	44
	3/0	—	0	0	1	1	2	4	6	9	12	16	26	37
	4/0	—	0	0	1	1	1	3	5	8	10	13	22	31
	250	—	0	0	1	1	1	2	4	6	8	11	17	25
	300	—	0	0	0	1	1	2	3	5	7	9	15	21
	350	—	0	0	0	1	1	1	3	5	6	8	13	19
	400	—	0	0	0	1	1	1	3	4	6	7	12	17
	500	—	0	0	0	1	1	1	2	3	5	6	10	14
	600	—	0	0	0	0	1	1	1	3	4	5	8	12
	700	—	0	0	0	0	1	1	1	2	3	4	7	10
	750	—	0	0	0	0	1	1	1	2	3	4	7	10
	900	—	0	0	0	0	0	1	1	1	3	4	6	8
	1000	—	0	0	0	0	0	1	1	1	2	3	5	8
THHN, THWN, THWN-2	8	—	—	—	—	—	—	—	—	—	—	—	—	—
	6	—	1	3	6	11	15	25	36	57	77	99	158	226
	4	—	1	1	3	6	9	15	22	35	47	61	98	140
	2	—	1	1	2	5	6	11	16	25	34	44	70	100
	1	—	1	1	1	3	5	8	12	19	25	33	53	75
	1/0	—	0	1	1	3	4	7	10	16	22	28	45	64
	2/0	—	0	1	1	2	3	6	8	13	18	23	37	53
	3/0	—	0	1	1	1	3	5	7	11	15	19	31	44
	4/0	—	0	0	1	1	2	4	6	9	12	16	25	37
	250	—	0	0	1	1	1	3	4	7	10	12	20	29
	300	—	0	0	1	1	1	3	4	6	8	11	17	25
	350	—	0	0	0	1	1	2	3	5	7	9	15	22
	400	—	0	0	0	1	1	1	3	5	6	8	13	19
	500	—	0	0	0	1	1	1	2	4	5	7	11	16
	600	—	0	0	0	1	1	1	1	3	4	6	9	13
	700	—	0	0	0	0	1	1	1	3	4	5	8	12
	750	—	0	0	0	0	1	1	1	3	4	5	8	11
	900	—	0	0	0	0	1	1	1	1	3	4	6	9
	1000	—	0	0	0	0	0	1	1	1	3	3	5	8
XHHW, XHHW-2	8	—	1	4	7	12	17	29	42	65	88	114	181	260
	6	—	1	3	5	9	13	21	31	48	65	85	134	193
	4	—	1	1	3	6	9	15	22	35	47	61	98	140
	2	—	1	1	2	5	6	11	16	25	34	44	70	100
	1	—	1	1	1	3	5	8	12	19	25	33	53	75
	1/0	—	0	1	1	3	4	7	10	16	22	28	45	64
	2/0	—	0	1	1	2	3	6	8	13	18	24	38	54
	3/0	—	0	1	1	1	3	5	7	11	15	19	31	44
	4/0	—	0	0	1	1	2	4	6	9	12	16	26	37
	250	—	0	0	1	1	1	3	5	7	10	13	21	30
	300	—	0	0	1	1	1	3	4	6	8	11	17	25
	350	—	0	0	1	1	1	2	3	5	7	10	15	22
	400	—	0	0	0	1	1	1	3	5	7	9	14	20
	500	—	0	0	0	1	1	1	2	4	5	7	11	17
	600	—	0	0	0	1	1	1	1	3	4	6	9	13

(continues)

Table C.10(A)   *Continued*

Type	Conductor Size (AWG/ kcmil)	Trade Size (Metric Designator)												
		⅜ (12)	½ (16)	¾ (21)	1 (27)	1¼ (35)	1½ (41)	2 (53)	2½ (63)	3 (78)	3½ (91)	4 (103)	5 (129)	6 (155)
	700	—	0	0	0	0	1	1	1	3	4	5	8	12
	750	—	0	0	0	0	1	1	1	2	3	5	7	11
	900	—	0	0	0	0	1	1	1	2	3	4	6	9
	1000	—	0	0	0	0	0	1	1	1	3	3	6	8

Definition: *Compact stranding* is the result of a manufacturing process where the stranded conductor is compressed to the extent that the interstices (voids between strand wires) are virtually eliminated.

Table C.11 Maximum Number of Conductors or Fixture Wires in Rigid PVC Conduit, Schedule 40 and HDPE Conduit (Based on Chapter 9: Table 1, Table 4, and Table 5)

Type	Conductor Size (AWG/ kcmil)	⅜ (12)	½ (16)	¾ (21)	1 (27)	1¼ (35)	1½ (41)	2 (53)	2½ (63)	3 (78)	3½ (91)	4 (103)	5 (129)	6 (155)
							CONDUCTORS							
RHH, RHW, RHW-2	14	—	4	7	11	20	27	45	64	99	133	171	269	390
	12	—	3	5	9	16	22	37	53	82	110	142	224	323
	10	—	2	4	7	13	18	30	43	66	89	115	181	261
	8	—	1	2	4	7	9	15	22	35	46	60	94	137
	6	—	1	1	3	5	7	12	18	28	37	48	76	109
	4	—	1	1	2	4	6	10	14	22	29	37	59	85
	3	—	1	1	1	4	5	8	12	19	25	33	52	75
	2	—	1	1	1	3	4	7	10	16	22	28	45	65
	1	—	0	1	1	1	3	5	7	11	14	19	29	43
	1/0	—	0	1	1	1	2	4	6	9	13	16	26	37
	2/0	—	0	0	1	1	1	3	5	8	11	14	22	32
	3/0	—	0	0	1	1	1	3	4	7	9	12	19	28
	4/0	—	0	0	1	1	1	2	4	6	8	10	16	24
	250	—	0	0	0	1	1	1	3	4	6	8	12	18
	300	—	0	0	0	1	1	1	2	4	5	7	11	16
	350	—	0	0	0	1	1	1	2	3	5	6	10	14
	400	—	0	0	0	1	1	1	1	3	4	6	9	13
	500	—	0	0	0	0	1	1	1	3	4	5	8	11
	600	—	0	0	0	0	1	1	1	2	3	4	6	9
	700	—	0	0	0	0	0	1	1	1	3	3	6	8
	750	—	0	0	0	0	0	1	1	1	2	3	5	8
	800	—	0	0	0	0	0	1	1	1	2	3	5	7
	900	—	0	0	0	0	0	1	1	1	2	3	5	7
	1000	—	0	0	0	0	0	1	1	1	1	3	4	6
	1250	—	0	0	0	0	0	0	1	1	1	1	3	5
	1500	—	0	0	0	0	0	0	1	1	1	1	3	4
	1750	—	0	0	0	0	0	0	1	1	1	1	2	3
	2000	—	0	0	0	0	0	0	0	1	1	1	2	3
TW, THHW, THW, THW-2	14	—	8	14	24	42	57	94	135	209	280	361	568	822
	12	—	6	11	18	32	44	72	103	160	215	277	436	631
	10	—	4	8	13	24	32	54	77	119	160	206	325	470
	8	—	2	4	7	13	18	30	43	66	89	115	181	261
RHH*, RHW*, RHW-2*	14	—	5	9	16	28	38	63	90	139	186	240	378	546
	12	—	4	8	13	22	30	50	72	112	150	193	304	439
	10	—	3	6	10	17	24	39	56	87	117	150	237	343
	8	—	1	3	6	10	14	23	33	52	70	90	142	205
TW, THW, THHW, THW-2, RHH*, RHW*, RHW-2*	6	—	1	2	4	8	11	18	26	40	53	69	109	157
	4	—	1	1	3	6	8	13	19	30	40	51	81	117
	3	—	1	1	3	5	7	11	16	25	34	44	69	100
	2	—	1	1	2	4	6	10	14	22	29	37	59	85
	1	—	0	1	1	3	4	7	10	15	20	26	41	60
	1/0	—	0	1	1	2	3	6	8	13	17	22	35	51
	2/0	—	0	1	1	1	3	5	7	11	15	19	30	43
	3/0	—	0	1	1	1	2	4	6	9	12	16	25	36
	4/0	—	0	0	1	1	1	3	5	8	10	13	21	30
	250	—	0	0	1	1	1	3	4	6	8	11	17	25
	300	—	0	0	1	1	1	2	3	5	7	9	15	21
	350	—	0	0	0	1	1	1	3	5	6	8	13	19
	400	—	0	0	0	1	1	1	3	4	6	7	12	17
	500	—	0	0	0	1	1	1	2	3	5	6	10	14
	600	—	0	0	0	0	1	1	1	3	4	5	8	11
	700	—	0	0	0	0	1	1	1	2	3	4	7	10
	750	—	0	0	0	0	1	1	1	2	3	4	6	10

(continues)

Table C.11 *Continued*

Type	Conductor Size (AWG/ kcmil)	⅜ (12)	½ (16)	¾ (21)	1 (27)	1¼ (35)	1½ (41)	2 (53)	2½ (63)	3 (78)	3½ (91)	4 (103)	5 (129)	6 (155)
												Trade Size (Metric Designator)		
	800	—	0	0	0	0	1	1	1	2	3	4	6	9
	900	—	0	0	0	0	0	1	1	1	3	3	6	8
	1000	—	0	0	0	0	0	1	1	1	2	3	5	7
	1250	—	0	0	0	0	0	1	1	1	1	2	4	6
	1500	—	0	0	0	0	0	1	1	1	1	1	3	5
	1750	—	0	0	0	0	0	0	1	1	1	1	3	4
	2000	—	0	0	0	0	0	0	1	1	1	1	3	4
THHN, THWN, THWN-2	14	—	11	21	34	60	82	135	193	299	401	517	815	1178
	12	—	8	15	25	43	59	99	141	218	293	377	594	859
	10	—	5	9	15	27	37	62	89	137	184	238	374	541
	8	—	3	5	9	16	21	36	51	79	106	137	216	312
	6	—	1	4	6	11	15	26	37	57	77	99	156	225
	4	—	1	2	4	7	9	16	22	35	47	61	96	138
	3	—	1	1	3	6	8	13	19	30	40	51	81	117
	2	—	1	1	3	5	7	11	16	25	33	43	68	98
	1	—	1	1	1	3	5	8	12	18	25	32	50	73
	1/0	—	1	1	1	3	4	7	10	15	21	27	42	61
	2/0	—	0	1	1	2	3	6	8	13	17	22	35	51
	3/0	—	0	1	1	1	3	5	7	11	14	18	29	42
	4/0	—	0	1	1	1	2	4	6	9	12	15	24	35
	250	—	0	0	1	1	1	3	4	7	10	12	20	28
	300	—	0	0	1	1	1	3	4	6	8	11	17	24
	350	—	0	0	1	1	1	2	3	5	7	9	15	21
	400	—	0	0	0	1	1	1	3	5	6	8	13	19
	500	—	0	0	0	1	1	1	2	4	5	7	11	16
	600	—	0	0	0	1	1	1	1	3	4	5	9	13
	700	—	0	0	0	0	1	1	1	3	4	5	8	11
	750	—	0	0	0	0	1	1	1	2	3	4	7	11
	800	—	0	0	0	0	1	1	1	2	3	4	7	10
	900	—	0	0	0	0	1	1	1	2	3	4	6	9
	1000	—	0	0	0	0	0	1	1	1	3	3	6	8
FEP, FEPB, PFA, PFAH, TFE	14	—	11	20	33	58	79	131	188	290	389	502	790	1142
	12	—	8	15	24	42	58	96	137	212	284	366	577	834
	10	—	6	10	17	30	41	69	98	152	204	263	414	598
	8	—	3	6	10	17	24	39	56	87	117	150	237	343
	6	—	2	4	7	12	17	28	40	62	83	107	169	244
	4	—	1	3	5	8	12	19	28	43	58	75	118	170
	3	—	1	2	4	7	10	16	23	36	48	62	98	142
	2	—	1	1	3	6	8	13	19	30	40	51	81	117
PFA, PFAH, TFE	1	—	1	1	2	4	5	9	13	20	28	36	56	81
PFA, PFAH, TFE, Z	1/0	—	1	1	1	3	4	8	11	17	23	30	47	68
	2/0	—	0	1	1	3	4	6	9	14	19	24	39	56
	3/0	—	0	1	1	2	3	5	7	12	16	20	32	46
	4/0	—	0	1	1	1	2	4	6	9	13	16	26	38
Z	14	—	13	24	40	70	95	158	226	350	469	605	952	1376
	12	—	9	17	28	49	68	112	160	248	333	429	675	976
	10	—	6	10	17	30	41	69	98	152	204	263	414	598
	8	—	3	6	11	19	26	43	62	96	129	166	261	378
	6	—	2	4	7	13	18	30	43	67	90	116	184	265
	4	—	1	3	5	9	12	21	30	46	62	80	126	183
	3	—	1	2	4	6	9	15	22	34	45	58	92	133
	2	—	1	1	3	5	7	12	18	28	38	49	77	111
	1	—	1	1	2	4	6	10	14	23	30	39	62	90

(continues)

Table C.11   *Continued*

Type	Conductor Size (AWG/ kcmil)	Trade Size (Metric Designator)												
		⅜ (12)	½ (16)	¾ (21)	1 (27)	1¼ (35)	1½ (41)	2 (53)	2½ (63)	3 (78)	3½ (91)	4 (103)	5 (129)	6 (155)
XHHW, ZW, XHHW-2, XHH	14	—	8	14	24	42	57	94	135	209	280	361	568	822
	12	—	6	11	18	32	44	72	103	160	215	277	436	631
	10	—	4	8	13	24	32	54	77	119	160	206	325	470
	8	—	2	4	7	13	18	30	43	66	89	115	181	261
	6	—	1	3	5	10	13	22	32	49	66	85	134	193
	4	—	1	2	4	7	9	16	23	35	48	61	97	140
	3	—	1	1	3	6	8	13	19	30	40	52	82	118
	2	—	1	1	3	5	7	11	16	25	34	44	69	99
XHHW, XHHW-2, XHH	1	—	1	1	1	3	5	8	12	19	25	32	51	74
	1/0	—	1	1	1	3	4	7	10	16	21	27	43	62
	2/0	—	0	1	1	2	3	6	8	13	17	23	36	52
	3/0	—	0	1	1	1	3	5	7	11	14	19	30	43
	4/0	—	0	1	1	1	2	4	6	9	12	15	24	35
	250	—	0	0	1	1	1	3	5	7	10	13	20	29
	300	—	0	0	1	1	1	3	4	6	8	11	17	25
	350	—	0	0	1	1	1	2	3	5	7	9	15	22
	400	—	0	0	0	1	1	1	3	5	6	8	13	19
	500	—	0	0	0	1	1	1	2	4	5	7	11	16
	600	—	0	0	0	1	1	1	1	3	4	5	9	13
	700	—	0	0	0	0	1	1	1	3	4	5	8	11
	750	—	0	0	0	0	1	1	1	2	3	4	7	11
	800	—	0	0	0	0	1	1	1	2	3	4	7	10
	900	—	0	0	0	0	1	1	1	2	3	4	6	9
	1000	—	0	0	0	0	0	1	1	1	3	3	6	8
	1250	—	0	0	0	0	0	1	1	1	1	3	4	6
	1500	—	0	0	0	0	0	1	1	1	1	2	4	5
	1750	—	0	0	0	0	0	0	1	1	1	1	3	5
	2000	—	0	0	0	0	0	0	1	1	1	1	3	4
FIXTURE WIRES														
RFH-2, FFH-2, RFHH-2	18	—	8	14	23	40	54	90	129	200	268	346	545	788
	16	—	6	12	19	33	46	76	109	169	226	292	459	664
SF-2, SFF-2	18	—	10	17	29	50	69	114	163	253	338	436	687	993
	16	—	8	14	24	42	57	94	135	209	280	361	568	822
	14	—	6	12	19	33	46	76	109	169	226	292	459	664
SF-1, SFF-1	18	—	17	31	51	89	122	202	289	447	599	772	1216	1758
RFH-1, TF, TFF, XF, XFF	18	—	13	23	38	66	90	149	213	330	442	570	898	1298
	16	—	10	18	30	53	73	120	172	266	357	460	725	1048
XF, XFF	14	—	8	14	24	42	57	94	135	209	280	361	568	822
TFN, TFFN	18	—	20	37	60	105	144	239	341	528	708	913	1437	2077
	16	—	16	28	46	80	110	183	261	403	541	697	1098	1587
PF, PFF, PGF, PGFF, PAF, PTF, PTFF, PAFF	18	—	19	35	57	100	137	227	323	501	671	865	1363	1970
	16	—	15	27	44	77	106	175	250	387	519	669	1054	1523
	14	—	11	20	33	58	79	131	188	290	389	502	790	1142
ZF, ZFF, ZHF	18	—	25	45	74	129	176	292	417	646	865	1116	1756	2539
	16	—	18	33	54	95	130	216	308	476	638	823	1296	1873
	14	—	13	24	40	70	95	158	226	350	469	605	952	1376
KF-2, KFF-2	18	—	38	67	111	193	265	439	626	969	1298	1674	2634	3809
	16	—	26	47	77	135	184	306	436	676	905	1168	1838	2657
	14	—	18	31	52	91	124	205	293	454	608	784	1235	1785
	12	—	12	22	36	63	86	143	204	316	423	546	859	1242
	10	—	8	14	24	42	57	94	135	209	280	361	568	822
KF-1, KFF-1	18	—	44	78	128	223	305	506	722	1118	1498	1931	3040	4395

(continues)

Table C.11 *Continued*

Type	Conductor Size (AWG/ kcmil)	Trade Size (Metric Designator)												
		³⁄₈ (12)	½ (16)	¾ (21)	1 (27)	1¼ (35)	1½ (41)	2 (53)	2½ (63)	3 (78)	3½ (91)	4 (103)	5 (129)	6 (155)
	16	—	31	55	90	157	214	355	507	785	1052	1357	2136	3088
	14	—	20	37	60	105	144	239	341	528	708	913	1437	2077
	12	—	13	24	40	70	95	158	226	350	469	605	952	1376
	10	—	9	16	26	45	62	103	148	229	306	395	622	899
XF, XFF	12	—	4	8	13	22	30	50	72	112	150	193	304	439
	10	—	3	6	10	17	24	39	56	87	117	150	237	343

Notes:

1. This table is for concentric stranded conductors only. For compact stranded conductors, Table C.10(A) should be used.

2. Two-hour fire-rated RHH cable has ceramifiable insulation, which has much larger diameters than other RHH wires. Consult manufacturer's conduit fill tables.

*Types RHH, RHW, and RHW-2 without outer covering.

Table C.11(A) Maximum Number of Conductors or Fixture Wires in Rigid PVC Conduit, Schedule 40 and HDPE Conduit (Based on Chapter 9: Table 1, Table 4, and Table 5A)

Type	Conductor Size (AWG/kcmil)	Trade Size (Metric Designator)												
		⅜ (12)	½ (16)	¾ (21)	1 (27)	1¼ (35)	1½ (41)	2 (53)	2½ (63)	3 (78)	3½ (91)	4 (103)	5 (129)	6 (155)
COMPACT CONDUCTORS														
THW, THW-2, THHW	8	—	1	4	6	11	15	26	37	57	76	98	155	224
	6	—	1	3	5	9	12	20	28	44	59	76	119	173
	4	—	1	1	3	6	9	15	21	33	44	57	89	129
	2	—	1	1	2	5	6	11	15	24	32	42	66	95
	1	—	1	1	1	3	4	7	11	17	23	29	46	67
	1/0	—	0	1	1	3	4	6	9	15	20	25	40	58
	2/0	—	0	1	1	2	3	5	8	12	16	21	34	49
	3/0	—	0	1	1	1	3	5	7	10	14	18	29	42
	4/0	—	0	1	1	1	2	4	5	9	12	15	24	35
	250	—	0	0	1	1	1	3	4	7	9	12	19	27
	300	—	0	0	1	1	1	2	4	6	8	10	16	24
	350	—	0	0	1	1	1	2	3	5	7	9	15	21
	400	—	0	0	0	1	1	1	3	5	6	8	13	19
	500	—	0	0	0	1	1	1	2	4	5	7	11	16
	600	—	0	0	0	1	1	1	1	3	4	5	9	13
	700	—	0	0	0	0	1	1	1	3	4	5	8	12
	750	—	0	0	0	0	1	1	1	2	3	5	7	11
	900	—	0	0	0	0	1	1	1	2	3	4	6	9
	1000	—	0	0	0	0	1	1	1	1	3	4	6	9
THHN, THWN, THWN-2	8	—	—	—	—	—	—	—	—	—	—	—	—	—
	6	—	2	4	7	13	17	29	41	64	86	111	175	253
	4	—	1	2	4	8	11	18	25	40	53	68	108	156
	2	—	1	1	3	5	8	13	18	28	38	49	77	112
	1	—	1	1	2	4	6	9	14	21	29	37	58	84
	1/0	—	1	1	1	3	5	8	12	18	24	31	49	72
	2/0	—	0	1	1	3	4	7	9	15	20	26	41	59
	3/0	—	0	1	1	2	3	5	8	12	17	22	34	50
	4/0	—	0	1	1	1	3	4	6	10	14	18	28	41
	250	—	0	0	1	1	1	3	5	8	11	14	22	32
	300	—	0	0	1	1	1	3	4	7	9	12	19	28
	350	—	0	0	1	1	1	3	4	6	8	10	17	24
	400	—	0	0	1	1	1	2	3	5	7	9	15	22
	500	—	0	0	0	1	1	1	3	4	6	8	13	18
	600	—	0	0	0	1	1	1	2	4	5	6	10	15
	700	—	0	0	0	1	1	1	1	3	4	5	9	13
	750	—	0	0	0	1	1	1	1	3	4	5	8	12
	900	—	0	0	0	0	1	1	1	2	3	4	7	10
	1000	—	0	0	0	0	1	1	1	2	3	4	6	9
XHHW, XHHW-2	8	—	3	5	8	14	20	33	47	73	99	127	200	290
	6	—	1	4	6	11	15	25	35	55	73	94	149	215
	4	—	1	2	4	8	11	18	25	40	53	68	108	156
	2	—	1	1	3	5	8	13	18	28	38	49	77	112
	1	—	1	1	2	4	6	9	14	21	29	37	58	84
	1/0	—	1	1	1	3	5	8	12	18	24	31	49	72
	2/0	—	1	1	1	3	4	7	10	15	20	26	42	60
	3/0	—	0	1	1	2	3	5	8	12	17	22	34	50
	4/0	—	0	1	1	1	3	5	7	10	14	18	29	42
	250	—	0	0	1	1	1	4	5	8	11	14	23	33
	300	—	0	0	1	1	1	3	4	7	9	12	19	28
	350	—	0	0	1	1	1	3	4	6	8	11	17	25
	400	—	0	0	1	1	1	2	3	5	7	10	15	22
	500	—	0	0	0	1	1	1	3	4	6	8	13	18
	600	—	0	0	0	1	1	1	2	4	5	6	10	15

(continues)

Table C.11(A)   *Continued*

Type	Conductor Size (AWG/ kcmil)	Trade Size (Metric Designator)												
		³⁄₈ (12)	½ (16)	¾ (21)	1 (27)	1¼ (35)	1½ (41)	2 (53)	2½ (63)	3 (78)	3½ (91)	4 (103)	5 (129)	6 (155)
	700	—	0	0	0	1	1	1	1	3	4	5	9	13
	750	—	0	0	0	1	1	1	1	3	4	5	8	12
	900	—	0	0	0	0	1	1	1	2	3	4	7	10
	1000	—	0	0	0	0	1	1	1	2	3	4	6	9

Definition: *Compact stranding* is the result of a manufacturing process where the stranded conductor is compressed to the extent that the interstices (voids between strand wires) are virtually eliminated.

Table C.12 Maximum Number of Conductors or Fixture Wires in Type A, Rigid PVC Conduit (Based on Chapter 9: Table 1, Table 4, and Table 5)

Type	Conductor Size (AWG/ kcmil)	Trade Size (Metric Designator)												
		⅜ (12)	½ (16)	¾ (21)	1 (27)	1¼ (35)	1½ (41)	2 (53)	2½ (63)	3 (78)	3½ (91)	4 (103)	5 (129)	6 (155)
		CONDUCTORS												
RHH, RHW, RHW-2	14	—	5	9	14	24	31	49	74	112	146	187	—	—
	12	—	4	7	12	20	26	41	61	93	121	155	—	—
	10	—	3	6	10	16	21	33	50	75	98	125	—	—
	8	—	1	3	5	8	11	17	26	39	51	65	—	—
	6	—	1	2	4	6	9	14	21	31	41	52	—	—
	4	—	1	1	3	5	7	11	16	24	32	41	—	—
	3	—	1	1	3	4	6	9	14	21	28	36	—	—
	2	—	1	1	2	4	5	8	12	18	24	31	—	—
	1	—	0	1	1	2	3	5	8	12	16	20	—	—
	1/0	—	0	1	1	2	3	5	7	10	14	18	—	—
	2/0	—	0	1	1	1	2	4	6	9	12	15	—	—
	3/0	—	0	1	1	1	1	3	5	8	10	13	—	—
	4/0	—	0	0	1	1	1	3	4	7	9	11	—	—
	250	—	0	0	1	1	1	1	3	5	6	8	—	—
	300	—	0	0	1	1	1	1	3	4	6	7	—	—
	350	—	0	0	0	1	1	1	2	4	5	7	—	—
	400	—	0	0	0	1	1	1	2	3	5	6	—	—
	500	—	0	0	0	1	1	1	1	3	4	5	—	—
	600	—	0	0	0	0	1	1	1	2	3	4	—	—
	700	—	0	0	0	0	1	1	1	2	3	4	—	—
	750	—	0	0	0	0	1	1	1	1	3	4	—	—
	800	—	0	0	0	0	1	1	1	1	3	3	—	—
	900	—	0	0	0	0	0	1	1	1	2	3	—	—
	1000	—	0	0	0	0	0	1	1	1	2	3	—	—
	1250	—	0	0	0	0	0	1	1	1	1	2	—	—
	1500	—	0	0	0	0	0	0	1	1	1	1	—	—
	1750	—	0	0	0	0	0	0	1	1	1	1	—	—
	2000	—	0	0	0	0	0	0	1	1	1	1	—	—
TW, THHW, THW, THW-2	14	—	11	18	31	51	67	105	157	235	307	395	—	—
	12	—	8	14	24	39	51	80	120	181	236	303	—	—
	10	—	6	10	18	29	38	60	89	135	176	226	—	—
	8	—	3	6	10	16	21	33	50	75	98	125	—	—
RHH*, RHW*, RHW-2*	14	—	7	12	20	34	44	69	104	157	204	262	—	—
	12	—	6	10	16	27	35	56	84	126	164	211	—	—
	10	—	4	8	13	21	28	44	65	98	128	165	—	—
	8	—	2	4	7	12	16	26	39	59	77	98	—	—
TW, THW, THHW, THW-2, RHH*, RHW*, RHW-2*	6	—	1	3	6	9	13	20	30	45	59	75	—	—
	4	—	1	2	4	7	9	15	22	33	44	56	—	—
	3	—	1	1	4	6	8	13	19	29	37	48	—	—
	2	—	1	1	3	5	7	11	16	24	32	41	—	—
	1	—	1	1	1	3	5	7	11	17	22	29	—	—
	1/0	—	1	1	1	3	4	6	10	14	19	24	—	—
	2/0	—	0	1	1	2	3	5	8	12	16	21	—	—
	3/0	—	0	1	1	1	3	4	7	10	13	17	—	—
	4/0	—	0	1	1	1	2	4	6	9	11	14	—	—
	250	—	0	0	1	1	1	3	4	7	9	12	—	—
	300	—	0	0	1	1	1	2	4	6	8	10	—	—
	350	—	0	0	1	1	1	2	3	5	7	9	—	—
	400	—	0	0	1	1	1	1	3	5	6	8	—	—
	500	—	0	0	0	1	1	1	2	4	5	7	—	—
	600	—	0	0	0	1	1	1	1	3	4	5	—	—
	700	—	0	0	0	1	1	1	1	3	4	5	—	—
	750	—	0	0	0	1	1	1	1	3	3	4	—	—

(continues)

Table C.12 *Continued*

Type	Conductor Size (AWG/kcmil)	Trade Size (Metric Designator)												
		⅜ (12)	½ (16)	¾ (21)	1 (27)	1¼ (35)	1½ (41)	2 (53)	2½ (63)	3 (78)	3½ (91)	4 (103)	5 (129)	6 (155)
	800	—	0	0	0	0	1	1	1	2	3	4	—	—
	900	—	0	0	0	0	1	1	1	2	3	4	—	—
	1000	—	0	0	0	0	1	1	1	1	3	3	—	—
	1250	—	0	0	0	0	0	1	1	1	1	3	—	—
	1500	—	0	0	0	0	0	1	1	1	1	2	—	—
	1750	—	0	0	0	0	0	0	1	1	1	1	—	—
	2000	—	0	0	0	0	0	0	1	1	1	1	—	—
THHN, THWN, THWN-2	14	—	16	27	44	73	96	150	225	338	441	566	—	—
	12	—	11	19	32	53	70	109	164	246	321	412	—	—
	10	—	7	12	20	33	44	69	103	155	202	260	—	—
	8	—	4	7	12	19	25	40	59	89	117	150	—	—
	6	—	3	5	8	14	18	28	43	64	84	108	—	—
	4	—	1	3	5	8	11	17	26	39	52	66	—	—
	3	—	1	2	4	7	9	15	22	33	44	56	—	—
	2	—	1	1	3	6	8	12	19	28	37	47	—	—
	1	—	1	1	2	4	6	9	14	21	27	35	—	—
	1/0	—	1	1	2	4	5	8	11	17	23	29	—	—
	2/0	—	1	1	1	3	4	6	10	14	19	24	—	—
	3/0	—	0	1	1	2	3	5	8	12	16	20	—	—
	4/0	—	0	1	1	1	3	4	6	10	13	17	—	—
	250	—	0	1	1	1	2	3	5	8	10	14	—	—
	300	—	0	0	1	1	1	3	4	7	9	12	—	—
	350	—	0	0	1	1	1	2	4	6	8	10	—	—
	400	—	0	0	1	1	1	2	3	5	7	9	—	—
	500	—	0	0	1	1	1	1	3	4	6	7	—	—
	600	—	0	0	0	1	1	1	2	3	5	6	—	—
	700	—	0	0	0	1	1	1	1	3	4	5	—	—
	750	—	0	0	0	1	1	1	1	3	4	5	—	—
	800	—	0	0	0	1	1	1	1	3	4	5	—	—
	900	—	0	0	0	0	1	1	1	2	3	4	—	—
	1000	—	0	0	0	0	1	1	1	2	3	4	—	—
FEP, FEPB, PFA, PFAH, TFE	14	—	15	26	43	70	93	146	218	327	427	549	—	—
	12	—	11	19	31	51	68	106	159	239	312	400	—	—
	10	—	8	13	22	37	48	76	114	171	224	287	—	—
	8	—	4	8	13	21	28	44	65	98	128	165	—	—
	6	—	3	5	9	15	20	31	46	70	91	117	—	—
	4	—	1	4	6	10	14	21	32	49	64	82	—	—
	3	—	1	3	5	8	11	18	27	40	53	68	—	—
	2	—	1	2	4	7	9	15	22	33	44	56	—	—
PFA, PFAH, TFE	1	—	1	1	3	5	6	10	15	23	30	39	—	—
PFA, PFAH, TFE, Z	1/0	—	1	1	2	4	5	8	13	19	25	32	—	—
	2/0	—	1	1	1	3	4	7	10	16	21	27	—	—
	3/0	—	1	1	1	3	3	6	9	13	17	22	—	—
	4/0	—	0	1	1	2	3	5	7	11	14	18	—	—
Z	14	—	18	31	52	85	112	175	262	395	515	661	—	—
	12	—	13	22	37	60	79	124	186	280	365	469	—	—
	10	—	8	13	22	37	48	76	114	171	224	287	—	—
	8	—	5	8	14	23	30	48	72	108	141	181	—	—
	6	—	3	6	10	16	21	34	50	76	99	127	—	—
	4	—	2	4	7	11	15	23	35	52	68	88	—	—
	3	—	1	3	5	8	11	17	25	38	50	64	—	—
	2	—	1	2	4	7	9	14	21	32	41	53	—	—
	1	—	1	1	3	5	7	11	17	26	33	43	—	—

(continues)

Table C.12 *Continued*

Type	Conductor Size (AWG/kcmil)	Trade Size (Metric Designator)												
		⅜ (12)	½ (16)	¾ (21)	1 (27)	1¼ (35)	1½ (41)	2 (53)	2½ (63)	3 (78)	3½ (91)	4 (103)	5 (129)	6 (155)
XHHW, ZW, XHHW-2, XHH	14	—	11	18	31	51	67	105	157	235	307	395	—	—
	12	—	8	14	24	39	51	80	120	181	236	303	—	—
	10	—	6	10	18	29	38	60	89	135	176	226	—	—
	8	—	3	6	10	16	21	33	50	75	98	125	—	—
	6	—	2	4	7	12	15	24	37	55	72	93	—	—
	4	—	1	3	5	8	11	18	26	40	52	67	—	—
	3	—	1	2	4	7	9	15	22	34	44	57	—	—
	2	—	1	1	3	6	8	12	19	28	37	48	—	—
XHHW, XHHW-2, XHH	1	—	1	1	3	4	6	9	14	21	28	35	—	—
	1/0	—	1	1	2	4	5	8	12	18	23	30	—	—
	2/0	—	1	1	1	3	4	6	10	15	19	25	—	—
	3/0	—	0	1	1	2	3	5	8	12	16	20	—	—
	4/0	—	0	1	1	1	3	4	7	10	13	17	—	—
	250	—	0	1	1	1	2	3	5	8	11	14	—	—
	300	—	0	0	1	1	1	3	5	7	9	12	—	—
	350	—	0	0	1	1	1	3	4	6	8	10	—	—
	400	—	0	0	1	1	1	2	3	5	7	9	—	—
	500	—	0	0	1	1	1	1	3	4	6	8	—	—
	600	—	0	0	0	1	1	1	2	3	5	6	—	—
	700	—	0	0	0	1	1	1	1	3	4	5	—	—
	750	—	0	0	0	1	1	1	1	3	4	5	—	—
	800	—	0	0	0	1	1	1	1	3	4	5	—	—
	900	—	0	0	0	0	1	1	1	2	3	4	—	—
	1000	—	0	0	0	0	1	1	1	2	3	4	—	—
	1250	—	0	0	0	0	0	1	1	1	2	3	—	—
	1500	—	0	0	0	0	0	1	1	1	1	2	—	—
	1750	—	0	0	0	0	0	1	1	1	1	2	—	—
	2000	—	0	0	0	0	0	0	1	1	1	1	—	—
FIXTURE WIRES														
RFH-2, FFH-2, RFHH-2	18	—	10	18	30	48	64	100	150	226	295	378	—	—
	16	—	9	15	25	41	54	85	127	190	248	319	—	—
SF-2, SFF-2	18	—	13	22	37	61	81	127	189	285	372	477	—	—
	16	—	11	18	31	51	67	105	157	235	307	395	—	—
	14	—	9	15	25	41	54	85	127	190	248	319	—	—
SF-1, SFF-1	18	—	23	40	66	108	143	224	335	504	658	844	—	—
RFH-1, TF, TFF, XF, XFF	18	—	17	29	49	80	105	165	248	372	486	623	—	—
	16	—	14	24	39	65	85	134	200	300	392	503	—	—
XF, XFF	14	—	11	18	31	51	67	105	157	235	307	395	—	—
TFN, TFFN	18	—	28	47	79	128	169	265	396	596	777	998	—	—
	16	—	21	36	60	98	129	202	303	455	594	762	—	—
PF, PFF, PGF, PGFF, PAF, PTF, PTFF, PAFF	18	—	26	45	74	122	160	251	376	565	737	946	—	—
	16	—	20	34	58	94	124	194	291	437	570	732	—	—
	14	—	15	26	43	70	93	146	218	327	427	549	—	—
ZF, ZFF, ZHF	18	—	34	57	96	157	206	324	484	728	950	1220	—	—
	16	—	25	42	71	116	152	239	357	537	701	900	—	—
	14	—	18	31	52	85	112	175	262	395	515	661	—	—
KF-2, KFF-2	18	—	51	86	144	235	310	486	727	1092	1426	1829	—	—
	16	—	36	60	101	164	216	339	507	762	994	1276	—	—
	14	—	24	40	67	110	145	228	341	512	668	857	—	—
	12	—	16	28	47	77	101	158	237	356	465	596	—	—
	10	—	11	18	31	51	67	105	157	235	307	395	—	—

(continues)

Table C.12   *Continued*

Type	Conductor Size (AWG/ kcmil)	Trade Size (Metric Designator)												
		⅜ (12)	½ (16)	¾ (21)	1 (27)	1¼ (35)	1½ (41)	2 (53)	2½ (63)	3 (78)	3½ (91)	4 (103)	5 (129)	6 (155)
KF-1, KFF-1	18	—	59	100	166	272	357	561	839	1260	1645	2111	—	—
	16	—	41	70	117	191	251	394	589	886	1156	1483	—	—
	14	—	28	47	79	128	169	265	396	596	777	998	—	—
	12	—	18	31	52	85	112	175	262	395	515	661	—	—
	10	—	12	20	34	55	73	115	171	258	337	432	—	—
XF, XFF	12	—	6	10	16	27	35	56	84	126	164	211	—	—
	10	—	4	8	13	21	28	44	65	98	128	165	—	—

Notes:

1. This table is for concentric stranded conductors only. For compact stranded conductors, Table C.11(A) should be used.

2. Two-hour fire-rated RHH cable has ceramifiable insulation, which has much larger diameters than other RHH wires. Consult manufacturer's conduit fill tables.

*Types RHH, RHW, and RHW-2 without outer covering.

Table C.12(A) Maximum Number of Conductors or Fixture Wires in Type A, Rigid PVC Conduit (Based on Chapter 9: Table 1, Table 4, and Table 5A)

Type	Conductor Size (AWG/ kcmil)	⅜ (12)	½ (16)	¾ (21)	1 (27)	1¼ (35)	1½ (41)	2 (53)	2½ (63)	3 (78)	3½ (91)	4 (103)	5 (129)	6 (155)
						COMPACT CONDUCTORS								
THW, THW-2, THHW	8	—	3	5	8	14	18	28	42	64	84	107	—	—
	6	—	2	4	6	10	14	22	33	49	65	83	—	—
	4	—	1	3	5	8	10	16	24	37	48	62	—	—
	2	—	1	1	3	6	7	12	18	27	36	46	—	—
	1	—	1	1	2	4	5	8	13	19	25	32	—	—
	1/0	—	1	1	1	3	4	7	11	16	21	28	—	—
	2/0	—	1	1	1	3	4	6	9	14	18	23	—	—
	3/0	—	0	1	1	2	3	5	8	12	15	20	—	—
	4/0	—	0	1	1	1	3	4	6	10	13	17	—	—
	250	—	0	1	1	1	1	3	5	8	10	13	—	—
	300	—	0	0	1	1	1	3	4	7	9	11	—	—
	350	—	0	0	1	1	1	2	4	6	8	10	—	—
	400	—	0	0	1	1	1	2	3	5	7	9	—	—
	500	—	0	0	1	1	1	1	3	4	6	8	—	—
	600	—	0	0	0	1	1	1	2	3	5	6	—	—
	700	—	0	0	0	1	1	1	1	3	4	5	—	—
	750	—	0	0	0	1	1	1	1	3	4	5	—	—
	900	—	0	0	0	0	1	1	1	2	3	4	—	—
	1000	—	0	0	0	0	1	1	1	2	3	4	—	—
THHN, THWN, THWN-2	8	—	—	—	—	—	—	—	—	—	—	—	—	—
	6	—	3	5	9	15	20	32	48	72	94	121	—	—
	4		1	3	6	9	12	20	30	45	58	75	—	—
	2	—	1	2	4	7	9	14	21	32	42	54	—	—
	1	—	1	1	3	5	7	10	16	24	31	40	—	—
	1/0	—	1	1	2	4	6	9	13	20	27	34	—	—
	2/0	—	1	1	1	3	5	7	11	17	22	28	—	—
	3/0	—	1	1	1	3	4	6	9	14	18	24	—	—
	4/0	—	0	1	1	2	3	5	8	11	15	19	—	—
	250	—	0	1	1	1	2	4	6	9	12	15	—	—
	300	—	0	1	1	1	1	3	5	8	10	13	—	—
	350	—	0	0	1	1	1	3	4	7	9	11	—	—
	400	—	0	0	1	1	1	2	4	6	8	10	—	—
	500	—	0	0	1	1	1	2	3	5	7	9	—	—
	600	—	0	0	0	1	1	1	3	4	5	7	—	—
	700	—	0	0	0	1	1	1	2	3	5	6	—	—
	750	—	0	0	0	1	1	1	2	3	4	6	—	—
	900	—	0	0	0	1	1	1	1	3	4	5	—	—
	1000	—	0	0	0	0	1	1	1	2	3	4	—	—
XHHW, XHHW-2	8	—	4	6	11	18	23	37	55	83	108	139	—	—
	6	—	3	5	8	13	17	27	41	62	80	103	—	—
	4	—	1	3	6	9	12	20	30	45	58	75	—	—
	2	—	1	2	4	7	9	14	21	32	42	54	—	—
	1	—	1	1	3	5	7	10	16	24	31	40	—	—
	1/0	—	1	1	2	4	6	9	13	20	27	34	—	—
	2/0	—	1	1	1	3	5	7	11	17	22	29	—	—
	3/0	—	1	1	1	3	4	6	9	14	18	24	—	—
	4/0	—	0	1	1	2	3	5	8	12	15	20	—	—
	250	—	0	1	1	1	2	4	6	9	12	16	—	—
	300	—	0	1	1	1	1	3	5	8	10	13	—	—
	350	—	0	0	1	1	1	3	5	7	9	12	—	—
	400	—	0	0	1	1	1	3	4	6	8	11	—	—
	500	—	0	0	1	1	1	2	3	5	7	9	—	—
	600	—	0	0	0	1	1	1	3	4	5	7	—	—

(continues)

Table C.12(A) *Continued*

Type	Conductor Size (AWG/ kcmil)	Trade Size (Metric Designator)												
		³⁄₈ (12)	¹⁄₂ (16)	³⁄₄ (21)	1 (27)	1¼ (35)	1½ (41)	2 (53)	2½ (63)	3 (78)	3½ (91)	4 (103)	5 (129)	6 (155)
	700	—	0	0	0	1	1	1	2	3	5	6	—	—
	750	—	0	0	0	1	1	1	2	3	4	6	—	—
	900	—	0	0	0	1	1	1	1	3	4	5	—	—
	1000	—	0	0	0	0	1	1	1	2	3	4	—	—

Definition: *Compact stranding* is the result of a manufacturing process where the stranded conductor is compressed to the extent that the interstices (voids between strand wires) are virtually eliminated.

Table C.13 Maximum Number of Conductors or Fixture Wires in Type EB, PVC Conduit (Based on Chapter 9: Table 1, Table 4, and Table 5)

Type	Conductor Size (AWG/ kcmil)	Trade Size (Metric Designator)												
		³⁄₈ (12)	½ (16)	¾ (21)	1 (27)	1¼ (35)	1½ (41)	2 (53)	2½ (63)	3 (78)	3½ (91)	4 (103)	5 (129)	6 (155)
CONDUCTORS														
RHH, RHW, RHW-2	14	—	—	—	—	—	—	53	—	119	155	197	303	430
	12	—	—	—	—	—	—	44	—	98	128	163	251	357
	10	—	—	—	—	—	—	35	—	79	104	132	203	288
	8	—	—	—	—	—	—	18	—	41	54	69	106	151
	6	—	—	—	—	—	—	15	—	33	43	55	85	121
	4	—	—	—	—	—	—	11	—	26	34	43	66	94
	3	—	—	—	—	—	—	10	—	23	30	38	58	83
	2	—	—	—	—	—	—	9	—	20	26	33	50	72
	1	—	—	—	—	—	—	6	—	13	17	21	33	47
	1/0	—	—	—	—	—	—	5	—	11	15	19	29	41
	2/0	—	—	—	—	—	—	4	—	10	13	16	25	36
	3/0	—	—	—	—	—	—	4	—	8	11	14	22	31
	4/0	—	—	—	—	—	—	3	—	7	9	12	18	26
	250	—	—	—	—	—	—	2	—	5	7	9	14	20
	300	—	—	—	—	—	—	1	—	5	6	8	12	17
	350	—	—	—	—	—	—	1	—	4	5	7	11	16
	400	—	—	—	—	—	—	1	—	4	5	6	10	14
	500	—	—	—	—	—	—	1	—	3	4	5	9	12
	600	—	—	—	—	—	—	1	—	3	3	4	7	10
	700	—	—	—	—	—	—	1	—	2	3	4	6	9
	750	—	—	—	—	—	—	1	—	2	3	4	6	9
	800	—	—	—	—	—	—	1	—	2	3	4	6	8
	900	—	—	—	—	—	—	1	—	1	2	3	5	7
	1000	—	—	—	—	—	—	1	—	1	2	3	5	7
	1250	—	—	—	—	—	—	1	—	1	1	2	3	5
	1500	—	—	—	—	—	—	0	—	1	1	1	3	4
	1750	—	—	—	—	—	—	0	—	1	1	1	3	4
	2000	—	—	—	—	—	—	0	—	1	1	1	2	3
TW, THHW, THW, THW-2	14	—	—	—	—	—	—	111	—	250	327	415	638	907
	12	—	—	—	—	—	—	85	—	192	251	319	490	696
	10	—	—	—	—	—	—	63	—	143	187	238	365	519
	8	—	—	—	—	—	—	35	—	79	104	132	203	288
RHH*, RHW*, RHW-2*	14	—	—	—	—	—	—	74	—	166	217	276	424	603
	12	—	—	—	—	—	—	59	—	134	175	222	341	485
	10	—	—	—	—	—	—	46	—	104	136	173	266	378
	8	—	—	—	—	—	—	28	—	62	81	104	159	227
TW, THW, THHW, THW-2, RHH*, RHW*, RHW-2*	6	—	—	—	—	—	—	21	—	48	62	79	122	173
	4	—	—	—	—	—	—	16	—	36	46	59	91	129
	3	—	—	—	—	—	—	13	—	30	40	51	78	111
	2	—	—	—	—	—	—	11	—	26	34	43	66	94
	1	—	—	—	—	—	—	8	—	18	24	30	46	66
	1/0	—	—	—	—	—	—	7	—	15	20	26	40	56
	2/0	—	—	—	—	—	—	6	—	13	17	22	34	48
	3/0	—	—	—	—	—	—	5	—	11	14	18	28	40
	4/0	—	—	—	—	—	—	4	—	9	12	15	24	34
	250	—	—	—	—	—	—	3	—	7	10	12	19	27
	300	—	—	—	—	—	—	3	—	6	8	11	17	24
	350	—	—	—	—	—	—	2	—	6	7	9	15	21
	400	—	—	—	—	—	—	2	—	5	7	8	13	19

(continues)

Table C.13 *Continued*

Type	Conductor Size (AWG/ kcmil)	³⁄₈ (12)	½ (16)	¾ (21)	1 (27)	1¼ (35)	1½ (41)	2 (53)	2½ (63)	3 (78)	3½ (91)	4 (103)	5 (129)	6 (155)
	500	—	—	—	—	—	—	1	—	4	5	7	11	16
	600	—	—	—	—	—	—	1	—	3	4	6	9	13
	700	—	—	—	—	—	—	1	—	3	4	5	8	11
	750	—	—	—	—	—	—	1	—	3	4	5	7	11
	800	—	—	—	—	—	—	1	—	3	3	4	7	10
	900	—	—	—	—	—	—	1	—	2	3	4	6	9
	1000	—	—	—	—	—	—	1	—	2	3	4	6	8
	1250	—	—	—	—	—	—	1	—	1	2	3	4	6
	1500	—	—	—	—	—	—	1	—	1	1	2	4	6
	1750	—	—	—	—	—	—	1	—	1	1	2	3	5
	2000	—	—	—	—	—	—	0	—	1	1	1	3	4
THHN, THWN, THWN-2	14	—	—	—	—	—	—	159	—	359	468	595	915	1300
	12	—	—	—	—	—	—	116	—	262	342	434	667	948
	10	—	—	—	—	—	—	73	—	165	215	274	420	597
	8	—	—	—	—	—	—	42	—	95	124	158	242	344
	6	—	—	—	—	—	—	30	—	68	89	114	175	248
	4	—	—	—	—	—	—	19	—	42	55	70	107	153
	3	—	—	—	—	—	—	16	—	36	46	59	91	129
	2	—	—	—	—	—	—	13	—	30	39	50	76	109
	1	—	—	—	—	—	—	10	—	22	29	37	57	80
	1/0	—	—	—	—	—	—	8	—	18	24	31	48	68
	2/0	—	—	—	—	—	—	7	—	15	20	26	40	56
	3/0	—	—	—	—	—	—	5	—	13	17	21	33	47
	4/0	—	—	—	—	—	—	4	—	10	14	18	27	39
	250	—	—	—	—	—	—	4	—	8	11	14	22	31
	300	—	—	—	—	—	—	3	—	7	10	12	19	27
	350	—	—	—	—	—	—	3	—	6	8	11	17	24
	400	—	—	—	—	—	—	2	—	6	7	10	15	21
	500	—	—	—	—	—	—	1	—	5	6	8	12	18
	600	—	—	—	—	—	—	1	—	4	5	6	10	14
	700	—	—	—	—	—	—	1	—	3	4	6	9	12
	750	—	—	—	—	—	—	1	—	3	4	5	8	12
	800	—	—	—	—	—	—	1	—	3	4	5	8	11
	900	—	—	—	—	—	—	1	—	3	3	4	7	10
	1000	—	—	—	—	—	—	1	—	2	3	4	6	9
FEP, FEPB, PFA, PFAH, TFE	14	—	—	—	—	—	—	155	—	348	454	578	887	1261
	12	—	—	—	—	—	—	113	—	254	332	422	648	920
	10	—	—	—	—	—	—	81	—	182	238	302	465	660
	8	—	—	—	—	—	—	46	—	104	136	173	266	378
	6	—	—	—	—	—	—	33	—	74	97	123	189	269
	4	—	—	—	—	—	—	23	—	52	68	86	132	188
	3	—	—	—	—	—	—	19	—	43	56	72	110	157
	2	—	—	—	—	—	—	16	—	36	46	59	91	129
PFA, PFAH, TFE	1	—	—	—	—	—	—	11	—	25	32	41	63	90
PFA, PFAH, TFE, Z	1/0	—	—	—	—	—	—	9	—	20	27	34	53	75
	2/0	—	—	—	—	—	—	7	—	17	22	28	43	62
	3/0	—	—	—	—	—	—	6	—	14	18	23	36	51
	4/0	—	—	—	—	—	—	5	—	11	15	19	29	42

(continues)

Table C.13   *Continued*

Type	Conductor Size (AWG/kcmil)	3/8 (12)	1/2 (16)	3/4 (21)	1 (27)	1 1/4 (35)	1 1/2 (41)	2 (53)	2 1/2 (63)	3 (78)	3 1/2 (91)	4 (103)	5 (129)	6 (155)
Z	14	—	—	—	—	—	—	186	—	419	547	696	1069	1519
	12	—	—	—	—	—	—	132	—	297	388	494	759	1078
	10	—	—	—	—	—	—	81	—	182	238	302	465	660
	8	—	—	—	—	—	—	51	—	115	150	191	294	417
	6	—	—	—	—	—	—	36	—	81	105	134	206	293
	4	—	—	—	—	—	—	24	—	55	72	92	142	201
	3	—	—	—	—	—	—	18	—	40	53	67	104	147
	2	—	—	—	—	—	—	15	—	34	44	56	86	122
	1	—	—	—	—	—	—	12	—	27	36	45	70	99
XHHW, ZW, XHHW-2, XHH	14	—	—	—	—	—	—	111	—	250	327	415	638	907
	12	—	—	—	—	—	—	85	—	192	251	319	490	696
	10	—	—	—	—	—	—	63	—	143	187	238	365	519
	8	—	—	—	—	—	—	35	—	79	104	132	203	288
	6	—	—	—	—	—	—	26	—	59	77	98	150	213
	4	—	—	—	—	—	—	19	—	42	56	71	109	155
	3	—	—	—	—	—	—	16	—	36	47	60	92	131
	2	—	—	—	—	—	—	13	—	30	39	50	77	110
XHHW, XHHW-2, XHH	1	—	—	—	—	—	—	10	—	22	29	37	58	82
	1/0	—	—	—	—	—	—	8	—	19	25	31	48	69
	2/0	—	—	—	—	—	—	7	—	16	20	26	40	57
	3/0	—	—	—	—	—	—	6	—	13	17	22	33	47
	4/0	—	—	—	—	—	—	5	—	11	14	18	27	39
	250	—	—	—	—	—	—	4	—	9	11	15	22	32
	300	—	—	—	—	—	—	3	—	7	10	12	19	28
	350	—	—	—	—	—	—	3	—	6	8	11	17	24
	400	—	—	—	—	—	—	2	—	6	8	10	15	22
	500	—	—	—	—	—	—	1	—	5	6	8	12	18
	600	—	—	—	—	—	—	1	—	4	5	6	10	14
	700	—	—	—	—	—	—	1	—	3	4	6	9	12
	750	—	—	—	—	—	—	1	—	3	4	5	8	12
	800	—	—	—	—	—	—	1	—	3	4	5	8	11
	900	—	—	—	—	—	—	1	—	3	3	4	7	10
	1000	—	—	—	—	—	—	1	—	2	3	4	6	9
	1250	—	—	—	—	—	—	1	—	1	2	3	5	7
	1500	—	—	—	—	—	—	1	—	1	1	3	4	6
	1750	—	—	—	—	—	—	1	—	1	1	2	4	5
	2000	—	—	—	—	—	—	0	—	1	1	1	3	5

FIXTURE WIRES

Type	Conductor Size (AWG/kcmil)	3/8 (12)	1/2 (16)	3/4 (21)	1 (27)	1 1/4 (35)	1 1/2 (41)	2 (53)	2 1/2 (63)	3 (78)	3 1/2 (91)	4 (103)	5 (129)	6 (155)
RFH-2, FFH-2, RFHH-2	18	—	—	—	—	—	—	107	—	240	313	398	612	869
	16	—	—	—	—	—	—	90	—	202	264	336	516	733
SF-2, SFF-2	18	—	—	—	—	—	—	134	—	303	395	502	772	1096
	16	—	—	—	—	—	—	111	—	250	327	415	638	907
	14	—	—	—	—	—	—	90	—	202	264	336	516	733
SF-1, SFF-1	18	—	—	—	—	—	—	238	—	536	699	889	1366	1940
RFH-1, TF, TFF, XF, XFF	18	—	—	—	—	—	—	176	—	396	516	656	1009	1433
	16	—	—	—	—	—	—	142	—	319	417	530	814	1157
XF, XFF	14	—	—	—	—	—	—	111	—	250	327	415	638	907
TFN, TFFN	18	—	—	—	—	—	—	281	—	633	826	1050	1614	2293
	16	—	—	—	—	—	—	215	—	484	631	802	1233	1751

(continues)

Table C.13 *Continued*

Type	Conductor Size (AWG/kcmil)	⅜ (12)	½ (16)	¾ (21)	1 (27)	1¼ (35)	1½ (41)	2 (53)	2½ (63)	3 (78)	3½ (91)	4 (103)	5 (129)	6 (155)
PF, PFF, PGF, PGFF, PAF, PTF, PTFF, PAFF	18	—	—	—	—	—	—	267	—	600	783	996	1530	2174
	16	—	—	—	—	—	—	206	—	464	606	770	1183	1681
	14	—	—	—	—	—	—	155	—	348	454	578	887	1261
ZF, ZFF, ZHF	18	—	—	—	—	—	—	344	—	774	1010	1284	1973	2802
	16	—	—	—	—	—	—	254	—	571	745	947	1455	2067
	14	—	—	—	—	—	—	186	—	419	547	696	1069	1519
KF-2, KFF-2	18	—	—	—	—	—	—	516	—	1161	1515	1926	2959	4204
	16	—	—	—	—	—	—	360	—	810	1057	1344	2064	2933
	14	—	—	—	—	—	—	242	—	544	710	903	1387	1970
	12	—	—	—	—	—	—	168	—	378	494	628	965	1371
	10	—	—	—	—	—	—	111	—	250	327	415	638	907
KF-1, KFF-1	18	—	—	—	—	—	—	596	—	1340	1748	2222	3414	4850
	16	—	—	—	—	—	—	419	—	941	1228	1562	2399	3408
	14	—	—	—	—	—	—	281	—	633	826	1050	1614	2293
	12	—	—	—	—	—	—	186	—	419	547	696	1069	1519
	10	—	—	—	—	—	—	122	—	274	358	455	699	993
XF, XFF	12	—	—	—	—	—	—	59	—	134	175	222	341	485
	10	—	—	—	—	—	—	46	—	104	136	173	266	378

Notes:

1. This table is for concentric stranded conductors only. For compact stranded conductors, Table C.12(A) should be used.

2. Two-hour fire-rated RHH cable has ceramifiable insulation, which has much larger diameters than other RHH wires. Consult manufacturer's conduit fill tables.

*Types RHH, RHW, and RHW-2 without outer covering.

Table C.13(A) Maximum Number of Conductors or Fixture Wires in Type EB, PVC Conduit (Based on Chapter 9: Table 1, Table 4, and Table 5A)

Type	Conductor Size (AWG/ kcmil)	Trade Size (Metric Designator)												
		⅜ (12)	½ (16)	¾ (21)	1 (27)	1¼ (35)	1½ (41)	2 (53)	2½ (63)	3 (78)	3½ (91)	4 (103)	5 (129)	6 (155)
COMPACT CONDUCTORS														
THW, THW-2, THHW	8	—	—	—	—	—	—	30	—	68	89	113	174	247
	6	—	—	—	—	—	—	23	—	52	69	87	134	191
	4	—	—	—	—	—	—	17	—	39	51	65	100	143
	2	—	—	—	—	—	—	13	—	29	38	48	74	105
	1	—	—	—	—	—	—	9	—	20	26	34	52	74
	1/0	—	—	—	—	—	—	8	—	17	23	29	45	64
	2/0	—	—	—	—	—	—	6	—	15	19	24	38	54
	3/0	—	—	—	—	—	—	5	—	12	16	21	32	46
	4/0	—	—	—	—	—	—	4	—	10	14	17	27	38
	250	—	—	—	—	—	—	3	—	8	11	14	21	30
	300	—	—	—	—	—	—	3	—	7	9	12	19	26
	350	—	—	—	—	—	—	3	—	6	8	11	17	24
	400	—	—	—	—	—	—	2	—	6	7	10	15	21
	500	—	—	—	—	—	—	1	—	5	6	8	12	18
	600	—	—	—	—	—	—	1	—	4	5	6	10	14
	700	—	—	—	—	—	—	1	—	3	4	6	9	13
	750	—	—	—	—	—	—	1	—	3	4	5	8	12
	900	—	—	—	—	—	—	1	—	3	4	5	7	10
	1000	—	—	—	—	—	—	1	—	2	3	4	7	9
THHN, THWN, THWN-2	8	—	—	—	—	—	—	—	—	—	—	—	—	—
	6	—	—	—	—	—	—	34	—	77	100	128	196	279
	4	—	—	—	—	—	—	21	—	47	62	79	121	172
	2	—	—	—	—	—	—	15	—	34	44	57	87	124
	1	—	—	—	—	—	—	11	—	25	33	42	65	93
	1/0	—	—	—	—	—	—	9	—	22	28	36	56	79
	2/0	—	—	—	—	—	—	8	—	18	23	30	46	65
	3/0	—	—	—	—	—	—	6	—	15	20	25	38	55
	4/0	—	—	—	—	—	—	5	—	12	16	20	32	45
	250	—	—	—	—	—	—	4	—	10	13	16	25	35
	300	—	—	—	—	—	—	4	—	8	11	14	22	31
	350	—	—	—	—	—	—	3	—	7	9	12	19	27
	400	—	—	—	—	—	—	3	—	6	8	11	17	24
	500	—	—	—	—	—	—	2	—	5	7	9	14	20
	600	—	—	—	—	—	—	1	—	4	6	7	11	16
	700	—	—	—	—	—	—	1	—	4	5	6	10	14
	750	—	—	—	—	—	—	1	—	4	5	6	9	14
	900	—	—	—	—	—	—	1	—	3	4	5	8	11
	1000	—	—	—	—	—	—	1	—	3	3	4	7	10
XHHW, XHHW-2	8	—	—	—	—	—	—	39	—	88	115	146	225	320
	6	—	—	—	—	—	—	29	—	65	85	109	167	238
	4	—	—	—	—	—	—	21	—	47	62	79	121	172
	2	—	—	—	—	—	—	15	—	34	44	57	87	124
	1	—	—	—	—	—	—	11	—	25	33	42	65	93
	1/0	—	—	—	—	—	—	9	—	22	28	36	56	79
	2/0	—	—	—	—	—	—	8	—	18	24	30	47	67
	3/0	—	—	—	—	—	—	6	—	15	20	25	38	55
	4/0	—	—	—	—	—	—	5	—	12	16	21	32	46
	250	—	—	—	—	—	—	4	—	10	13	17	26	37
	300	—	—	—	—	—	—	4	—	8	11	14	22	31
	350	—	—	—	—	—	—	3	—	7	10	12	19	28
	400	—	—	—	—	—	—	3	—	7	9	11	17	25
	500	—	—	—	—	—	—	2	—	5	7	9	14	20
	600	—	—	—	—	—	—	1	—	4	6	7	11	16

(continues)

Table C.13(A) *Continued*

Type	Conductor Size (AWG/ kcmil)	Trade Size (Metric Designator)												
		³⁄₈ (12)	½ (16)	¾ (21)	1 (27)	1¼ (35)	1½ (41)	2 (53)	2½ (63)	3 (78)	3½ (91)	4 (103)	5 (129)	6 (155)
	700	—	—	—	—	—	—	1	—	4	5	6	10	14
	750	—	—	—	—	—	—	1	—	3	5	6	9	13
	900	—	—	—	—	—	—	1	—	3	4	5	8	11
	1000	—	—	—	—	—	—	1	—	3	4	5	7	10

Definition: *Compact stranding* is the result of a manufacturing process where the stranded conductor is compressed to the extent that the interstices (voids between strand wires) are virtually eliminated.

Informative Annex D   Examples

This informative annex is not a part of the requirements of this NFPA document but is included for informational purposes only.

Selection of Conductors. In the following examples, the results are generally expressed in amperes (A). To select conductor sizes, refer to the 0 through 2000 volt (V) ampacity tables of Article 310 and the rules of 310.15 that pertain to these tables.

Voltage. For uniform application of Articles 210, 215, and 220, a nominal voltage of 120, 120/240, 240, and 208Y/120 V is used in calculating the ampere load on the conductor.

Fractions of an Ampere. Except where the calculations result in a major fraction of an ampere (0.5 or larger), such fractions are permitted to be dropped.

Power Factor. Calculations in the following examples are based, for convenience, on the assumption that all loads have the same power factor (PF).

Ranges. For the calculation of the range loads in these examples, Column C of Table 220.55 has been used. For optional methods, see Columns A and B of Table 220.55. Except where the calculations result in a major fraction of a kilowatt (0.5 or larger), such fractions are permitted to be dropped.

SI Units. For metric conversions, $0.093 \text{ m}^2 = 1 \text{ ft}^2$ and $0.3048 \text{ m} = 1 \text{ ft}$.

Example D1(a) One-Family Dwelling

The dwelling has a floor area of 1500 ft², exclusive of an unfinished cellar not adaptable for future use, unfinished attic, and open porches. Appliances are a 12-kW range and a 5.5-kW, 240-V dryer. Assume range and dryer kW ratings equivalent to kVA ratings in accordance with 220.54 and 220.55.

Calculated Load *(see 220.40)*

General Lighting Load 1500 ft² at 3 VA/ft² = 4500 VA

Minimum Number of Branch Circuits Required *[see 210.11(A)]*

General Lighting Load: 4500 VA ÷ 120 V = 38 A

This requires three 15-A, 2-wire or two 20-A, 2-wire circuits.

Small-Appliance Load: Two 2-wire, 20-A circuits *[see 210.11(C)(1)]*

Laundry Load: One 2-wire, 20-A circuit *[see 210.11(C)(2)]*

Bathroom Branch Circuit: One 2-wire, 20-A circuit (no additional load calculation is required for this circuit) *[see 210.11(C)(3)]*

Minimum Size Feeder Required *[see 220.40]*

General Lighting	4,500 VA
Small Appliance	3,000 VA
Laundry	1,500 VA
Total	9,000 VA
3000 VA at 100%	3,000 VA
9000 VA – 3000 VA = 6000 VA at 35%	2,100 VA
Net Load	5,100 VA
Range *(see Table 220.55)*	8,000 VA
Dryer Load *(see Table 220.54)*	5,500 VA
Net Calculated Load	18,600 VA

Net Calculated Load for 120/240-V, 3-wire, single-phase service or feeder

18,600 VA ÷ 240 V = 78 A

Sections 230.42(B) and 230.79 require service conductors and disconnecting means rated not less than 100 amperes.

Calculation for Neutral for Feeder and Service

Lighting and Small-Appliance Load	5,100 VA
Range: 8000 VA at 70% *(see 220.61)*	5,600 VA
Dryer: 5500 VA at 70% *(see 220.61)*	3,850 VA
Total	14,550 VA

Calculated Load for Neutral

14,550 VA ÷ 240 V = 61 A

Example D1(b) One-Family Dwelling

Assume same conditions as Example No. D1(a), plus addition of one 6-A, 230-V, room air-conditioning unit and one 12-A, 115-V, room air-conditioning unit,* one 8-A, 115-V, rated waste disposer, and one 10-A, 120-V, rated dishwasher. See Article 430 for general motors and Article 440, Part VII, for air-conditioning equipment. Motors have nameplate ratings of 115 V and 230 V for use on 120-V and 240-V nominal voltage systems.

*(For feeder neutral, use larger of the two appliances for unbalance.)

From Example D1(a), feeder current is 78 A (3-wire, 240 V).

	Line A	Neutral	Line B
Amperes from Example D1(a)	78	61	78
One 230-V air conditioner	6	—	6
One 115-V air conditioner and 120-V dishwasher	12	12	10
One 115-V disposer	—	8	8
25% of largest motor (see 430.24)	3	3	2
Total amperes per conductor	99	84	104

Therefore, the service would be rated 110 A.

Example D2(a) Optional Calculation for One-Family Dwelling, Heating Larger Than Air Conditioning

(see 220.82)
The dwelling has a floor area of 1500 ft², exclusive of an unfinished cellar not adaptable for future use, unfinished attic, and open porches. It has a 12-kW range, a 2.5-kW water heater, a 1.2-kW dishwasher, 9 kW of electric space heating installed in five rooms, a 5-kW clothes dryer, and a 6-A, 230-V, room air-conditioning unit. Assume range, water heater, dishwasher, space heating, and clothes dryer kW ratings equivalent to kVA.

Air Conditioner kVA Calculation
6 A × 230 V ÷ 1000 = 1.38 kVA
This 1.38 kVA [item 1 from 220.82(C)] is less than 40% of 9 kVA of separately controlled electric heat [item 6 from 220.82(C)], so the 1.38 kVA need not be included in the service calculation.

General Load

1500 ft² at 3 VA	4,500 VA
Two 20-A appliance outlet circuits at 1500 VA each	3,000 VA
Laundry circuit	1,500 VA
Range (at nameplate rating)	12,000 VA
Water heater	2,500 VA
Dishwasher	1,200 VA
Clothes dryer	5,000 VA
Total	29,700 VA

Application of Demand Factor [see 220.82(B)]

First 10 kVA of general load at 100%	10,000 VA
Remainder of general load at 40% (19.7 kVA × 0.4)	7,880 VA
Total of general load	17,880 VA
9 kVA of heat at 40% (9000 VA × 0.4) =	3,600 VA
Total	21,480 VA

Calculated Load for Service Size

21.48 kVA = 21,480 VA

21,480 VA ÷ 240 V = 90 A

Therefore, the minimum service rating would be 100 A in accordance with 230.42 and 230.79.

Feeder Neutral Load in Accordance with 220.61

1500 ft² at 3 VA	4,500 VA
Three 20-A circuits at 1500 VA	4,500 VA
Total	9,000 VA
3000 VA at 100%	3,000 VA
9000 VA - 3000 VA = 6000 VA at 35%	2,100 VA
Subtotal	5,100 VA
Range: 8 kVA at 70%	5,600 VA
Clothes dryer: 5 kVA at 70%	3,500 VA
Dishwasher	1,200 VA
Total	15,400 VA

Calculated Load for Neutral

15,400 VA ÷ 240 V = 64 A

Example D2(b) Optional Calculation for One-Family Dwelling, Air Conditioning Larger Than Heating

[see 220.82(A) and 220.82(C)]
The dwelling has a floor area of 1500 ft², exclusive of an unfinished cellar not adaptable for future use, unfinished attic, and open porches. It has two 20-A small appliance circuits, one 20-A laundry circuit, two 4-kW wall-mounted ovens, one 5.1-kW counter-mounted cooking unit, a 4.5-kW water heater, a 1.2-kW dishwasher, a 5-kW combination clothes washer and dryer, six 7-A, 230-V room air-conditioning units, and a 1.5-kW permanently installed bathroom space heater. Assume wall-mounted ovens, counter-mounted cooking unit, water heater, dishwasher, and combination clothes washer and dryer kW ratings equivalent to kVA.

Air Conditioning kVA Calculation
Total amperes = 6 units × 7 A = 42 A

42 A × 240 V ÷ 1000 = 10.08 kVA (assume PF = 1.0)

Load Included at 100%
Air Conditioning: Included below [see item 1 in 220.82(C)]

Space Heater: Omit [see item 5 in 220.82(C)]

General Load

1500 ft² at 3 VA	4,500 VA
Two 20-A small-appliance circuits at 1500 VA each	3,000 VA
Laundry circuit	1,500 VA
Two ovens	8,000 VA
One cooking unit	5,100 VA
Water heater	4,500 VA
Dishwasher	1,200 VA
Washer/dryer	5,000 VA
Total general load	32,800 VA
First 10 kVA at 100%	10,000 VA
Remainder at 40% (22.8 kVA × 0.4 × 1000)	9,120 VA
Subtotal general load	19,120 VA
Air conditioning	10,080 VA
Total	29,200 VA

Calculated Load for Service

29,200 VA ÷ 240 V = 122 A (service rating)

Feeder Neutral Load, in accordance with 220.61

Assume that the two 4-kVA wall-mounted ovens are supplied by one branch circuit, the 5.1-kVA counter-mounted cooking unit by a separate circuit.

1500 ft² at 3 VA		4,500 VA
Three 20-A circuits at 1500 VA		4,500 VA
	Subtotal	9,000 VA
3000 VA at 100%		3,000 VA
9000 VA - 3000 VA = 6000 VA at 35%		2,100 VA
	Subtotal	5,100 VA

Two 4-kVA ovens plus one 5.1-kVA cooking unit = 13.1 kVA. Table 220.55 permits 55% demand factor or 13.1 kVA × 0.55 = 7.2 kVA feeder capacity.

	Subtotal from above	5,100 VA
Ovens and cooking unit: 7200 VA × 70% for neutral load		5,040 VA
Clothes washer/dryer: 5 kVA × 70% for neutral load		3,500 VA
Dishwasher		1,200 VA
	Total	14,840 VA

Calculated Load for Neutral

14,840 VA ÷ 240 V = 62

Example D2(c) Optional Calculation for One-Family Dwelling with Heat Pump (Single-Phase, 240/120-Volt Service)

(see 220.82)

The dwelling has a floor area of 2000 ft², exclusive of an unfinished cellar not adaptable for future use, unfinished attic, and open porches. It has a 12-kW range, a 4.5-kW water heater, a 1.2-kW dishwasher, a 5-kW clothes dryer, and a 2½-ton (24-A) heat pump with 15 kW of backup heat.

Heat Pump kVA Calculation

24 A × 240 V ÷ 1000 = 5.76 kVA

This 5.76 kVA is less than 15 kVA of the backup heat; therefore, the heat pump load need not be included in the service calculation *[see 220.82(C)]*.

General Load

2000 ft² at 3 VA	6,000 VA
Two 20-A appliance outlet circuits at 1500 VA each	3,000 VA
Laundry circuit	1,500 VA
Range (at nameplate rating)	12,000 VA
Water heater	4,500 VA
Dishwasher	1,200 VA
Clothes dryer	5,000 VA
Subtotal general load	33,200 VA
First 10 kVA at 100%	10,000 VA

(continues)

Remainder of general load at 40% (23,200 VA× 0.4)	9,280 VA
Total net general load	19,280 VA

Heat Pump and Supplementary Heat*

240 V × 24 A = 5760 VA

15 kW Electric Heat:

5760 VA + (15,000 VA × 65%) = 5.76 kVA + 9.75 kVA = 15.51 kVA

*If supplementary heat is not on at same time as heat pump, heat pump kVA need not be added to total.

Totals

Net general load	19,280 VA
Heat pump and supplementary heat	15,510 VA
Total	34,790 VA

Calculated Load for Service

34.79 kVA × 1000 ÷ 240 V= 145 A

Therefore, this dwelling unit would be permitted to be served by a 150-A service.

Example D3 Store Building

A store 50 ft by 60 ft, or 3000 ft², has 30 ft of show window. There are a total of 80 duplex receptacles. The service is 120/240 V, single phase 3-wire service. Actual connected lighting load is 8500 VA.

Calculated Load *(see 220.40)*

Noncontinuous Loads

Receptacle Load *(see 220.44)*		
80 receptacles at 180 VA		14,400 VA
10,000 VA at 100%		10,000 VA
14,400 VA – 10,000 VA = 4400 at 50%		2,200 VA
	Subtotal	12,200 VA

Continuous Loads

General Lighting*		
3000 ft² at 3 VA/ft²		9,000 VA
Show Window Lighting Load		
30 ft at 200 VA/ft *[see 220.14(G)]*		6,000 VA
Outside Sign Circuit *[see 220.14(F)]*		1,200 VA
	Subtotal	16,200 VA
	Subtotal from noncontinuous	12,200 VA
	Total noncontinuous loads + continuous loads =	28,400 VA

*In the example, the actual connected lighting load (8500 VA) is less than the load from Table 220.12, so the minimum lighting load from Table 220.12 is used in the calculation. Had the actual lighting load been greater than the value calculated from Table 220.12, the actual connected lighting load would have been used.

Minimum Number of Branch Circuits Required

General Lighting: Branch circuits need only be installed to supply the actual connected load *[see 210.11(B)]*.

8500 VA × 1.25 = 10,625 VA

10,625 VA ÷ 240 V = 44 A for 3-wire, 120/240 V

The lighting load would be permitted to be served by 2-wire or 3-wire, 15- or 20-A circuits with combined capacity equal to 44 A or greater for 3-wire circuits or 88 A or greater for 2-wire circuits. The feeder capacity as well as the number of branch-circuit positions available for lighting circuits in the panelboard must reflect the full calculated load of 9000 VA × 1.25 = 11,250 VA.

Show Window

6000 VA × 1.25 = 7500 VA

7500 VA ÷ 240 V = 31 A for 3-wire, 120/240 V

The show window lighting is permitted to be served by 2-wire or 3-wire circuits with a capacity equal to 31 A or greater for 3-wire circuits or 62 A or greater for 2-wire circuits.

Receptacles required by 210.62 are assumed to be included in the receptacle load above if these receptacles do not supply the show window lighting load.

Receptacles

Receptacle Load: 14,400 VA ÷ 240 V = 60 A for 3-wire, 120/240 V

The receptacle load would be permitted to be served by 2-wire or 3-wire circuits with a capacity equal to 60 A or greater for 3-wire circuits or 120 A or greater for 2-wire circuits.

Minimum Size Feeder (or Service) Overcurrent Protection
(see 215.3 or 230.90)

Subtotal noncontinuous loads	12,200 VA
Subtotal continuous load at 125%	20,250 VA
(16,200 VA × 1.25)	
Total	32,450 VA

32,450 VA ÷ 240 V = 135 A

The next higher standard size is 150 A *(see 240.6)*.

Minimum Size Feeders (or Service Conductors) Required *[see 215.2, 230.42(A)]*

For 120/240 V, 3-wire system, 32,450 VA ÷ 240 V = 135 A
Service or feeder conductor is 1/0 Cu in accordance with 215.3 and Table 310.15(B)(16) (with 75°C terminations).

Example D3(a) Industrial Feeders in a Common Raceway

An industrial multi-building facility has its service at the rear of its main building, and then provides 480Y/277-volt feeders to additional buildings behind the main building in order to segregate certain processes. The facility supplies its remote buildings through a partially enclosed access corridor that extends from the main switchboard rearward along a path that provides convenient access to services within 15 m (50 ft) of each additional building supplied. Two building feeders share a common raceway for approximately 45 m (150 ft) and run in the access corridor along with process steam and control and communications cabling. The steam raises the ambient temperature around the power raceway to as much as 35°C. At a tee fitting, the individual building feeders then run to each of the two buildings involved. The feeder neutrals are not connected to the equipment grounding conductors in the remote buildings. All distribution equipment terminations are listed as being suitable for 75°C connections. Each of the two buildings has the following loads:

Lighting, 11,600 VA, comprised of electric-discharge luminaires connected at 277 V

Receptacles, 22 125-volt, 20-ampere receptacles on general-purpose branch circuits, supplied by separately derived systems in each of the buildings

1 Air compressor, 460 volt, three phase, 5 hp

1 Grinder, 460 volt, three phase, 1.5 hp

3 Welders, AC transformer type (nameplate: 23 amperes, 480 volts, 60 percent duty cycle)

3 Industrial Process Dryers, 480 volt, three phase,15 kW each (assume continuous use throughout certain shifts)

Determine the overcurrent protection and conductor size for the feeders in the common raceway, assuming the use of XHHW-2 insulation (90°C):

Calculated Load {Note: For reasonable precision, volt-ampere calculations are carried to three significant figures only; where loads are converted to amperes, the results are rounded to the nearest ampere *[see 220.5(B)]*}.

Noncontinuous Loads
Receptacle Load *(see 220.44)*

22 receptacles at 180 VA	3,960 VA

Welder Load *[see 630.11(A), Table 630.11(A)]*
Each welder: 480V × 23A × 0.78 = 8,610 VA
All 3 welders *[see 630.11(B)]* (demand factors 100%, 100%, 85% respectively)

8,610 VA + 8,610 VA + 7,320 VA =	24,500 VA
Subtotal, Noncontinuous Loads	**28,500 VA**

Motor Loads *(see 430.24, Table 430.250)*

(continues)

Air compressor: 7.6 A × 480 V × √3 =	6,310 VA
Grinder: 3 A × 480 V × √3 =	2,490 VA
Largest motor, additional 25%:	1,580 VA

Subtotal, Motor Loads   **10,400 VA**

By using 430.24, the motor loads and the noncontinuous loads can be combined for the remaining calculation.

Subtotal for load calculations,		
Noncontinuous Loads		**38,900 VA**
Continuous Loads		
General Lighting		11,600 VA
3 Industrial Process Dryers	15 kW each	45,000 VA
Subtotal, Continuous Loads:		**56,600 VA**

Overcurrent protection *(see 215.3)*

The overcurrent protective device must accommodate 125% of the continuous load, plus the noncontinuous load:

Continuous load	56,600 VA
Noncontinuous load	38,900 VA
Subtotal, actual load [actual load in amperes]	**95,500 VA**
[99,000 VA ÷ (480V × √3) = 119 A]	
(25% of 56,600 VA) *(See 215.3)*	**14,200 VA**
Total VA	**109,700 VA**

Conversion to amperes using three significant
   figures: 109,700 VA / (480V × √3) = 132 A
Minimum size overcurrent protective device: 132 A
Minimum standard size overcurrent protective device *(see 240.6)*:
   150 amperes

Where the overcurrent protective device and its assembly are listed for operation at 100 percent of its rating, a 125 ampere overcurrent protective device would be permitted. However, overcurrent protective device assemblies listed for 100 percent of their rating are typically not available at the 125-ampere rating. *(See 215.3 Exception.)*

Ungrounded Feeder Conductors

The conductors must independently meet requirements for (1) terminations, and (2) conditions of use throughout the raceway run.

Minimum size conductor at the overcurrent device termination *[see 110.14(C) and 215.2(A)(1), using 75°C ampacity column in Table 310.15(B)(16)]*: 1/0 AWG.

Minimum size conductors in the raceway based on actual load *[see Article 100, Ampacity, and 310.15(B)(3)(a) and correction factors to Table 310.15(B)(16)]*:

95,500 VA / 0.7 / 0.96 = 142,000 VA
[70% = 310.15(B)(3)(a)] & [0.96 = Correction factors to Table
   310.15(B)(16)]
Conversion to amperes:
142,000 VA / (480V × √3) = 171 A

Note that the neutral conductors are counted as current-carrying conductors *[see 310.15(B)(5)(c)]* in this example because the discharge lighting has substantial nonlinear content. This requires a 2/0 AWG conductor based on the 90°C column of Table 310.15(B)(16). Therefore, the worst case is given by the raceway conditions, and 2/0 AWG conductors

must be used. If the utility corridor were at normal temperatures [(30°C (86°F)], and if the lighting at each building were supplied from the local separately derived system (thus requiring no neutrals in the supply feeders), the raceway result (95,500 VA / 0.8 = 119,000 VA; 119,000 VA / (480V × √3) = 143 A, or a 1 AWG conductor @ 90°C) could not be used, because the termination result (1/0 AWG) based on the 75°C column of Table 310.15(B)(16) would become the worst case, requiring the larger conductor.

In every case, the overcurrent protective device shall provide overcurrent protection for the feeder conductors in accordance with their ampacity as provided by this *Code (see 240.4)*. A 90°C 2/0 AWG conductor has a Table 310.15(B)(16) ampacity of 195 amperes. Adjusting for the conditions of use (35°C ambient temperature, 8 current-carrying conductors in the common raceway),

195 amperes × 0.96 × 0.7 = 131 A

The 150-ampere circuit breaker protects the 2/0 AWG feeder conductors, because 240.4(B) permits the use of the next higher standard size overcurrent protective device. Note that the feeder layout precludes the application of 310.15(A)(2) Exception.

Feeder Neutral Conductor *(see 220.61)*

Because 210.11(B) does not apply to these buildings, the load cannot be assumed to be evenly distributed across phases. Therefore the maximum imbalance must be assumed to be the full lighting load in this case, or 11,600 VA. (11,600 VA / 277V = 42 amperes.) The ability of the neutral to return fault current *[see 250.32(B) Exception(2)]* is not a factor in this calculation.

Because the neutral runs between the main switchboard and the building panelboard, likely terminating on a busbar at both locations, and not on overcurrent devices, the effects of continuous loading can be disregarded in evaluating its terminations *[see 215.2(A)(1) Exception No. 2]*. That calculation is (11,600 VA ÷ 277V) – 42 amperes, to be evaluated under the 75°C column of Table 310.15(B)(16). The minimum size of the neutral might seem to be 8 AWG, but that size would not be sufficient to be depended upon in the event of a line-to-neutral short circuit *[see 215.2(A)(1), second paragraph]*. Therefore, since the minimum size equipment grounding conductor for a 150 ampere circuit, as covered in Table 250.122, is 6 AWG, that is the minimum neutral size required for this feeder.

Example D4(a) Multifamily Dwelling

A multifamily dwelling has 40 dwelling units.
Meters are in two banks of 20 each with individual feeders to each dwelling unit.
One-half of the dwelling units are equipped with electric ranges not exceeding 12 kW each. Assume range kW rating equivalent to kVA rating in accordance with 220.55. Other half of ranges are gas ranges.
Area of each dwelling unit is 840 ft².
Laundry facilities on premises are available to all tenants. Add no circuit to individual dwelling unit.

Calculated Load for Each Dwelling Unit *(see Article 220)*
General Lighting: 840 ft² at 3 VA/ft² = 2520 VA
Special Appliance: Electric range *(see 220.55)* = 8000 VA

Minimum Number of Branch Circuits Required for Each Dwelling Unit *[see 210.11(A)]*
General Lighting Load: 2520 VA ÷ 120 V = 21 A or two 15-A, 2-wire circuits; or two 20-A, 2-wire circuits
Small-Appliance Load: Two 2-wire circuits of 12 AWG wire *[see 210.11(C)(1)]*
Range Circuit: 8000 VA ÷ 240 V = 33 A or a circuit of two 8 AWG conductors and one 10 AWG conductor in accordance with 210.19(A)(3)

Minimum Size Feeder Required for Each Dwelling Unit *(see 215.2)*

Calculated Load *(see Article 220)*:

General Lighting	2,520 VA
Small Appliance (two 20-ampere circuits)	3,000 VA
Subtotal Calculated Load (without ranges)	5,520 VA

Application of Demand Factor *(see Table 220.42)*

First 3000 VA at 100%	3,000 VA
5520 VA - 3000 VA = 2520 VA at 35%	882 VA
Net Calculated Load (without ranges)	3,882 VA
Range Load	8,000 VA
Net Calculated Load (with ranges)	11,882 VA

Size of Each Feeder *(see 215.2)*
For 120/240-V, 3-wire system (without ranges)
Net calculated load of 3882 VA ÷ 240 V = 16 A For 120/240-V, 3-wire system (with ranges)
Net calculated load, 11,882 VA ÷ 240 V = 50 A

Feeder Neutral

Lighting and Small-Appliance Load	3,882 VA
Range Load: 8000 VA at 70% *(see 220.61)*	5,600 VA
(only for apartments with electric range)	5,600 VA
Net Calculated Load (neutral)	9,482 VA

Calculated Load for Neutral

9482 VA ÷ 240 V = 39.5 A

Minimum Size Feeders Required from Service Equipment to Meter Bank (For 20 Dwelling Units — 10 with Ranges)

Total Calculated Load:
Lighting and Small Appliance

20 units × 5520 VA	110,400 VA

Application of Demand Factor

First 3000 VA at 100%	3,000 VA
110,400 VA - 3000 VA = 107,400 VA at 35%	37,590 VA
Net Calculated Load	40,590 VA
Range Load: 10 ranges (not over 12 kVA) *(see Col. C, Table 220.55, 25 kW)*	25,000 VA
Net Calculated Load (with ranges)	65,590 VA

Net calculated load for 120/240-V, 3-wire system,

65,590 VA ÷ 240 V = 273 A

Feeder Neutral

Lighting and Small-Appliance Load	40,590 VA
Range Load: 25,000 VA at 70% *[see 220.61(B)]*	17,500 VA
Calculated Load (neutral)	58,090 VA

Calculated Load for Neutral

58,090 VA ÷ 240 V = 242 A

Further Demand Factor *[220.61(B)]*

200 A at 100%	200 A
242 A - 200 A = 42 A at 70%	29 A
Net Calculated Load (neutral)	229 A

Minimum Size Main Feeders (or Service Conductors) Required (Less House Load) (For 40 Dwelling Units — 20 with Ranges)

Total Calculated Load:
Lighting and Small-Appliance Load

40 units × 5520 VA	220,800 VA

Application of Demand Factor *(from Table 220.42)*

First 3000 VA at 100%	3,000 VA
Next 120,000 VA - 3000 VA = 117,000 VA at 35%	40,950 VA
Remainder 220,800 VA - 120,000 VA = 100,800 VA at 25%	25,200 VA
Net Calculated Load	69,150 VA
Range Load: 20 ranges (less than 12 kVA) *(see Col. C, Table 220.55)*	35,000 VA
Net Calculated Load	104,150 VA

For 120/240-V, 3-wire system

Net calculated load of 104,150 VA ÷ 240 V = 434 A

Feeder Neutral

Lighting and Small-Appliance Load	69,150 VA
Range: 35,000 VA at 70% *[see 220.61(B)]*	24,500 VA
Calculated Load (neutral)	93,650 VA

93,650 VA ÷ 240 V = 390 A

Further Demand Factor *[see 220.61(B)]*

200 A at 100%	200 A
390 A - 200 A = 190 A at 70%	133 A
Net Calculated Load (neutral)	333 A

[See Table 310.15(B)(16) through Table 310.15(B)(21), and 310.15(B)(2), (B)(3), and (B)(5).]

Example D4(b) Optional Calculation for Multifamily Dwelling

A multifamily dwelling equipped with electric cooking and space heating or air conditioning has 40 dwelling units.

Meters are in two banks of 20 each plus house metering and individual feeders to each dwelling unit.

Each dwelling unit is equipped with an electric range of 8-kW nameplate rating, four 1.5-kW separately controlled 240-V electric space heaters, and a 2.5-kW, 240-V electric water heater. Assume range, space heater, and water heater kW ratings equivalent to kVA. Calculate the load for the individual dwelling unit by the standard calculation (Part III of Article 220).

A common laundry facility is available to all tenants [see 210.52(F), Exception No. 1].

Area of each dwelling unit is 840 ft².

Calculated Load for Each Dwelling Unit *(see Part II and Part III of Article 220)*

General Lighting Load:	
840 ft² at 3 VA/ft²	2,520 VA
Electric range	8,000 VA
Electric heat: 6 kVA (or air conditioning if larger)	6,000 VA
Electric water heater	2,500 VA

Minimum Number of Branch Circuits Required for Each Dwelling Unit

General Lighting Load: 2520 VA ÷ 120 V = 21 A or two 15-A, 2-wire circuits, or two 20-A, 2-wire circuits

Small-Appliance Load: Two 2-wire circuits of 12 AWG [see 210.11(C)(1)]

Range Circuit (See Table 220.55, Column B):

8000 VA × 80% ÷ 240 V = 27 A on a circuit of three

10 AWG conductors in accordance with 210.19(A)(3) Space Heating: 6000 VA ÷ 240 V = 25 A
Number of circuits (see 210.11)

Minimum Size Feeder Required for Each Dwelling Unit *(see 215.2)*

Calculated Load (see Article 220):	
General Lighting	2,520 VA
Small Appliance (two 20-A circuits)	3,000 VA
Subtotal Calculated Load (without range and space heating)	5,520 VA

Application of Demand Factor

First 3000 VA at 100%	3,000 VA
5520 VA − 3000 VA = 2520 VA at 35%	882 VA
Net Calculated Load (without range and space heating)	3,882 VA

(continues)

Range	6,400 VA
Space Heating *(see 220.51)*	6,000 VA
Water Heater	2,500 VA
Net Calculated Load (for individual dwelling unit)	18,782 VA

Size of Each Feeder

For 120/240-V, 3-wire system,

Net calculated load of 18,782 VA ÷ 240 V = 78 A

Feeder Neutral *(see 220.61)*

Lighting and Small Appliance	3,882 VA
Range Load: 6400 VA at 70% *[see 220.61(B)]*	4,480 VA
Space and Water Heating (no neutral): 240 V	0 VA
Net Calculated Load (neutral)	8,362 VA

Calculated Load for Neutral

8362 VA ÷ 240 V = 35 A

Minimum Size Feeder Required from Service Equipment to Meter Bank (For 20 Dwelling Units)

Total Calculated Load:	
Lighting and Small-Appliance Load	
20 units × 5520 VA	110,400 VA
Water and Space Heating Load	
20 units × 8500 VA	170,000 VA
Range Load: 20 × 8000 VA	160,000 VA
Net Calculated Load (20 dwelling units)	440,400 VA
Net Calculated Load Using Optional Calculation *(see Table 220.84)*	
440,400 VA × 0.38	167,352 VA

167,352 VA ÷ 240 V = 697 A

Minimum Size Main Feeder Required (Less House Load) (For 40 Dwelling Units)

Calculated Load:	
Lighting and Small-Appliance Load	
40 units × 5520 VA	220,800 VA
Water and Space Heating Load	340,000 VA
40 units × 8500 VA	
Range: 40 ranges × 8000 VA	320,000 VA
Net Calculated Load (40 dwelling units)	880,800 VA

Net Calculated Load Using Optional Calculation *(see Table 220.84)*

880,800 VA × 0.28 = 246,624 VA

246,624 VA ÷ 240 V = 1028 A

Feeder Neutral Load for Feeder from Service Equipment to Meter Bank (For 20 Dwelling Units)

Lighting and Small-Appliance Load	
20 units × 5520 VA	110,400 VA

(continues)

First 3000 VA at 100%   3,000 VA
110,400 VA - 3000 VA = 107,400 VA   37,590 VA
    at 35%

             Net Calculated Load   40,590 VA
20 ranges: 35,000 VA at 70%   24,500 VA
   [see Table 220.55 and 220.61(B)]

                     Total   65,090 VA

65,090 VA ÷ 240 V = 271 A

Further Demand Factor *[see 220.61(B)]*

First 200 A at 100%   200 A
Balance: 271 A - 200 A = 71 A at 70%   50 A

                     Total   250 A

Feeder Neutral Load of Main Feeder (Less House Load) (For 40 Dwelling Units)

Lighting and Small-Appliance Load
40 units × 5520 VA   220,800 VA
First 3000 VA at 100%   3,000 VA
Next 120,000 VA - 3000 VA = 117,000 VA at 35%   40,950 VA
Remainder 220,800 VA - 120,000 VA = 100,800 VA at   25,200 VA
   25%

             Net Calculated Load   69,150 VA
40 ranges: 55,000 VA at 70% *[see Table 220.55 and*   38,500 VA
   220.61(B)]

                     Total 107,650 VA

107,650 VA ÷ 240 V = 449 A

Further Demand Factor *[see 220.61(B)]*

First 200 A at 100%   200 A
Balance: 449 - 200 A = 249 A at 70%   174 A

                     Total   374 A

Example D5(a) Multifamily Dwelling Served at 208Y/120 Volts, Three Phase

All conditions and calculations are the same as for the multifamily dwelling [Example D4(a)] served at 120/240 V, single phase except as follows:

   Service to each dwelling unit would be two phase legs and neutral.

Minimum Number of Branch Circuits Required for Each Dwelling Unit *(see 210.11)*
Range Circuit: 8000 VA ÷ 208 V = 38 A or a circuit of two 8 AWG conductors and one 10 AWG conductor in accordance with 210.19(A)(3)

Minimum Size Feeder Required for Each Dwelling Unit *(see 215.2)*

   For 120/208-V, 3-wire system (without ranges),

Net calculated load of 3882 VA ÷ 2 legs ÷ 120 V/leg = 16 A

For 120/208-V, 3-wire system (with ranges),

Net calculated load (range) of 8000 VA ÷ 208 V = 39 A

Total load (range + lighting) = 39 A + 16 A = 55 A

Reducing the neutral load on the feeder to each dwelling unit is not permitted *[see 220.61(C)(1)]*.

Minimum Size Feeders Required from Service Equipment to Meter Bank (For 20 Dwelling Units — 10 with Ranges)

For 208Y/120-V, 3-phase, 4-wire system,

Ranges: Maximum number between any two phase legs = 4

   2 × 4 = 8.

Table 220.55 demand = 23,000 VA

Per phase demand = 23,000 VA ÷ 2 = 11,500 VA

Equivalent 3-phase load = 34,500 VA

Net Calculated Load (total):

40,590 VA + 34,500 VA = 75,090 VA

75,090 VA ÷ (208 V)(1.732) = 208 A

Feeder Neutral Size

Net Calculated Lighting and Appliance Load & Equivalent Range Load:

40,590 VA + (34,500 VA at 70%) = 64,700 VA

Net Calculated Neutral Load:

64,700 VA ÷ (208 V)(1.732) = 180 A

Minimum Size Main Feeder (Less House Load) (For 40 Dwelling Units — 20 with Ranges)

For 208Y/120-V, 3-phase, 4-wire system,

Ranges:

   Maximum number between any two phase legs = 7

   2 × 7 = 14.

Table 220.55 demand = 29,000 VA

Per phase demand = 29,000 VA ÷ 2 = 14,500 VA

Equivalent 3-phase load = 43,500 VA

Net Calculated Load (total):

   69,150 VA + 43,500 VA = 112,650 VA

   112,650 VA ÷ (208 V)(1.732) = 313 A

Main Feeder Neutral Size:

   69,150 VA + (43,500 VA at 70%) = 99,600 VA

   99,600 VA ÷ (208 V)(1.732) = 277 A

Further Demand Factor *(see 220.61)*

200 A at 100%   200.0 A
277 A - 200 A = 77 A at 70%   54 A

         Net Calculated Load (neutral)   254 A

Example D5(b) Optional Calculation for Multifamily Dwelling Served at 208Y/120 Volts, Three Phase

All conditions and calculations are the same as for Optional Calculation for the Multifamily Dwelling [Example D4(b)] served at 120/240 V, single phase except as follows:

Service to each dwelling unit would be two phase legs and neutral.

Minimum Number of Branch Circuits Required for Each Dwelling Unit (see 210.11)

Range Circuit (see Table 220.55, Column B): 8000 VA at 80% ÷ 208 V = 31 A or a circuit of two 8 AWG conductors and one 10 AWG conductor in accordance with 210.19(A)(3)

Space Heating: 6000 VA ÷ 208 V = 29 A

Two 20-ampere, 2-pole circuits required, 12 AWG conductors

Minimum Size Feeder Required for Each Dwelling Unit

120/208-V, 3-wire circuit

Net calculated load of 18,782 VA ÷ 208 V = 90 A

Net calculated load (lighting line to neutral):

3882 VA ÷ 2 legs ÷ 120 V per leg = 16 amperes

Line to line = 14,900 VA ÷ 208 V = 72 A

Total load = 16.2 A + 71.6 A = 88 A

Minimum Size Feeder Required for Service Equipment to Meter Bank (For 20 Dwelling Units)

Net Calculated Load

167,352 VA ÷ (208 V)(1.732) = 465 A

Feeder Neutral Load

65,080 VA ÷ (208 V)(1.732) = 181 A

Minimum Size Main Feeder Required (Less House Load) (For 40 Dwelling Units)

Net Calculated Load

246,624 VA ÷ (208 V)(1.732) = 685 A

Main Feeder Neutral Load

107,650 VA ÷ (208 V)(1.732) = 299 A

Further Demand Factor [see 220.61(B)]

200 A at 100%	200.0 A
299 A - 200 A = 99 A at 70%	69 A
Net Calculated Load (neutral)	269 A

Example D6 Maximum Demand for Range Loads

Table 220.55, Column C, applies to ranges not over 12 kW. The application of Note 1 to ranges over 12 kW (and not over 27 kW) and Note 2 to ranges over 8¾ kW (and not over 27 kW) is illustrated in the following two examples.

A. Ranges All the Same Rating (see Table 220.55, Note 1)

Assume 24 ranges, each rated 16 kW.

From Table 220.55, Column C, the maximum demand for 24 ranges of 12-kW rating is 39 kW. 16 kW exceeds 12 kW by 4.

5% × 4 = 20% (5% increase for each kW in excess of 12)

39 kW × 20% = 7.8 kW increase

39 + 7.8 = 46.8 kW (value to be used in selection of feeders)

B. Ranges of Unequal Rating (see Table 220.55, Note 2)

Assume 5 ranges, each rated 11 kW; 2 ranges, each rated 12 kW; 20 ranges, each rated 13.5 kW; 3 ranges, each rated 18 kW.

5 ranges	× 12 kW =	60 kW (use 12 kW for range rated less than 12)
2 ranges	× 12 kW =	24 kW
20 ranges	× 13.5 kW =	270 kW
3 ranges	× 18 kW =	54 kW
30 ranges, Total kW =		408 kW

408 ÷ 30 ranges = 13.6 kW (average to be used for calculation)

From Table 220.55, Column C, the demand for 30 ranges of 12-kW rating is 15 kW + 30 (1 kW × 30 ranges) = 45 kW. 13.6 kW exceeds 12 kW by 1.6 kW (use 2 kW).

5% × 2 = 10% (5% increase for each kW in excess of 12 kW)

45 kW × 10% = 4.5 kW increase

45 kW + 4.5 kW = 49.5 kW (value to be used in selection of feeders)

Example D7 Sizing of Service Conductors for Dwelling(s)

Service conductors and feeders for certain dwellings are permitted to be sized in accordance with 310.15(B)(7).

With No Required Adjustment or Correction Factors. If a 175-ampere service rating is selected, a service conductor is then sized as follows:

175 amperes × 0.83 = 145.25 amperes per 310.15(B)(7).

If no other adjustments or corrections are required for the installation, then, in accordance with Table 310.15(B)(16), a 1/0 AWG Cu or a 3/0 AWG Al meets this rating at 75°C (167°F).

N **With Required Temperature Correction Factor.** If a 175-ampere service rating is selected, a service conductor is then

175 amperes × 0.83 = 145.25 amperes per 310.15(B)(7).

If the conductors are installed in an ambient temperature of 40°C (104°F), the conductor ampacity must be multiplied by the appropriate correction factor in Table 310.15(B)(2)(a). In this case, we will use an XHHW-2 conductor, so we use a correction factor of 0.91 to find the minimum conductor ampacity and size:

145.25/.91 = 159.6 amperes

In accordance with Table 310.15(B)(16), a 2/0 AWG Cu or a 4/0 AWG Al would be required.

If no temperature correction or ampacity adjustment factors are required, the following table includes conductor sizes calculated using the requirements in 310.15(B)(7). This table is based on 75°C terminations and without any adjustment or correction factors.

| Service or Feeder Rating (Amperes) | Conductor (AWG or kcmil) | |
	Copper	Aluminum or Copper-Clad Aluminum
100	4	2
110	3	1
125	2	1/0
150	1	2/0
175	1/0	3/0
200	2/0	4/0
225	3/0	250
250	4/0	300
300	250	350
350	350	500
400	400	600

Example D8 Motor Circuit Conductors, Overload Protection, and Short-Circuit and Ground-Fault Protection

(see 240.6, 430.6, 430.22, 430.23, 430.24, 430.32, 430.52, and 430.62, Table 430.52, and Table 430.250)

Determine the minimum required conductor ampacity, the motor overload protection, the branch-circuit short-circuit and ground-fault protection, and the feeder protection, for three induction-type motors on a 480-V, 3-phase feeder, as follows:

(a) One 25-hp, 460-V, 3-phase, squirrel-cage motor, nameplate full-load current 32 A, Design B, Service Factor 1.15

(b) Two 30-hp, 460-V, 3-phase, wound-rotor motors, nameplate primary full-load current 38 A, nameplate secondary full-load current 65 A, 40°C rise.

Conductor Ampacity The full-load current value used to determine the minimum required conductor ampacity is obtained from Table 430.250 *[see 430.6(A)]* for the squirrel-cage motor and the primary of the wound-rotor motors. To obtain the minimum required conductor ampacity, the full-load current is multiplied by 1.25 *[see 430.22 and 430.23(A)]*.

For the 25-hp motor,

34 A × 1.25 = 43 A

For the 30-horsepower motors,

40 A × 1.25 = 50 A

65 A × 1.25 = 81 A

Motor Overload Protection Where protected by a separate overload device, the motors are required to have overload protection rated or set to trip at not more than 125% of the nameplate full-load current *[see 430.6(A) and 430.32(A)(1)]*.

For the 25-hp motor,

32 A × 1.25 = 40.0 A

For the 30-hp motors,

38 A × 1.25 = 48 A

Where the separate overload device is an overload relay (not a fuse or circuit breaker), and the overload device selected at 125% is not sufficient to start the motor or carry the load, the trip setting is permitted to be increased in accordance with 430.32(C).

Branch-Circuit Short-Circuit and Ground-Fault Protection The selection of the rating of the protective device depends on the type of protective device selected, in accordance with 430.52 and Table 430.52. The following is for the 25-hp motor.

(a) Nontime-Delay Fuse: The fuse rating is 300% × 34 A = 102 A. The next larger standard fuse is 110 A *[see 240.6 and 430.52(C)(1), Exception No. 1]*. If the motor will not start with a 110-A nontime-delay fuse, the fuse rating is permitted to be increased to 125 A because this rating does not exceed 400% *[see 430.52(C)(1), Exception No. 2(a)]*.

(b) Time-Delay Fuse: The fuse rating is 175% × 34 A = 59.5 A. The next larger standard fuse is 60 A *[see 240.6 and 430.52(C)(1), Exception No. 1]*. If the motor will not start with a 60-A time-delay fuse, the fuse rating is permitted to be increased to 70 A because this rating does not exceed 225% *[see 430.52(C)(1), Exception No. 2(b)]*.

Feeder Short-Circuit and Ground-Fault Protection (a) Example using nontime delay fuse. The rating of the feeder protective device is based on the sum of the largest branch-circuit protective device for the specific type of device protecting the feeder: 300% × 34 A = 102 A (therefore the next largest standard size, 110 A, would be used) plus the sum of the full-load currents of the other motors, or 110 A + 40 A + 40 A = 190 A. The nearest standard fuse that does not exceed this value is 175 A *[see 240.6 and 430.62(A)]*.

(b) Example using inverse time circuit breaker. The largest branch-circuit protective device for the specific type of device protecting the feeder, 250% × 34 A = 85. The next larger standard size is 90 A, plus the sum of the full-load currents of the other motors, or 90 A + 40 A + 40 A = 170 A. The nearest standard inverse time circuit breaker that does not exceed this value is 150 A *[see 240.6 and 430.62(A)]*.

Example D9 Feeder Ampacity Determination for Generator Field Control

[see 215.2, 430.24, 430.24 Exception No. 1, 620.13, 620.14, 620.61, and Table 430.22(E) and 620.14]

Determine the conductor ampacity for a 460-V 3-phase, 60-Hz ac feeder supplying a group of six elevators. The 460-V ac drive motor nameplate rating of the largest MG set for one elevator is 40 hp and 52 A, and the remaining elevators each have a 30-hp, 40-A, ac drive motor rating for their MG sets. In addition to a motor controller, each elevator has a separate motion/operation controller rated 10 A continuous to operate microprocessors, relays, power supplies, and the elevator car door operator. The MG sets are rated continuous.

Conductor Ampacity Conductor ampacity is determined as follows:

(a) In accordance with 620.13(D) and 620.61(B)(1), use Table 430.22(E), for intermittent duty (elevators). For intermittent duty using a continuous rated motor, the percentage of nameplate current rating to be used is 140%.

(b) For the 30-hp ac drive motor,

140% × 40 A = 56 A

(c) For the 40-hp ac drive motor,

140% × 52 A = 73 A(I)

(d) The total conductor ampacity is the sum of all the motor currents:

(1 motor × 73 A) + (5 motors × 56 A) = 353 A

(e) In accordance with 620.14 and Table 620.14, the conductor (feeder) ampacity would be permitted to be reduced by the use of a demand factor. Constant loads are not included (*see 620.14, Informational Note*). For six elevators, the demand factor is 0.79. The feeder diverse ampacity is, therefore, 0.79 × 353 A = 279 A.

(f) In accordance with 430.24 and 215.3, the controller continuous current is 125% × 10 A = 13 A

(g) The total feeder ampacity is the sum of the diverse current and all the controller continuous current.

I_{total} = 279 A + (6 elevators × 12.5 A) = 354 A

(h) This ampacity would be permitted to be used to select the wire size.

See Figure D9.

Example D10 Feeder Ampacity Determination for Adjustable Speed Drive Control [see 215.2, 430.24, 620.13, 620.14, 620.61, and Table 430.22(E)]

Determine the conductor ampacity for a 460-V, 3-phase, 60-Hz ac feeder supplying a group of six identical elevators. The system is adjustable-speed SCR dc drive. The power transformers are external to the drive (motor controller) cabinet. Each elevator has a separate motion/operation controller connected to the load side of the main line disconnect switch rated 10 A continuous to operate microprocessors, relays, power supplies, and the elevator car door operator. Each transformer is rated 95 kVA with an efficiency of 90%.

Conductor Ampacity

Conductor ampacity is determined as follows:

(a) Calculate the nameplate rating of the transformer:

[D10]

$$I = \frac{95 \text{ kVA} \times 1000}{\sqrt{3} \times 460 \text{ V} \times 0.90_{\text{eff.}}} = 133 \text{ A}$$

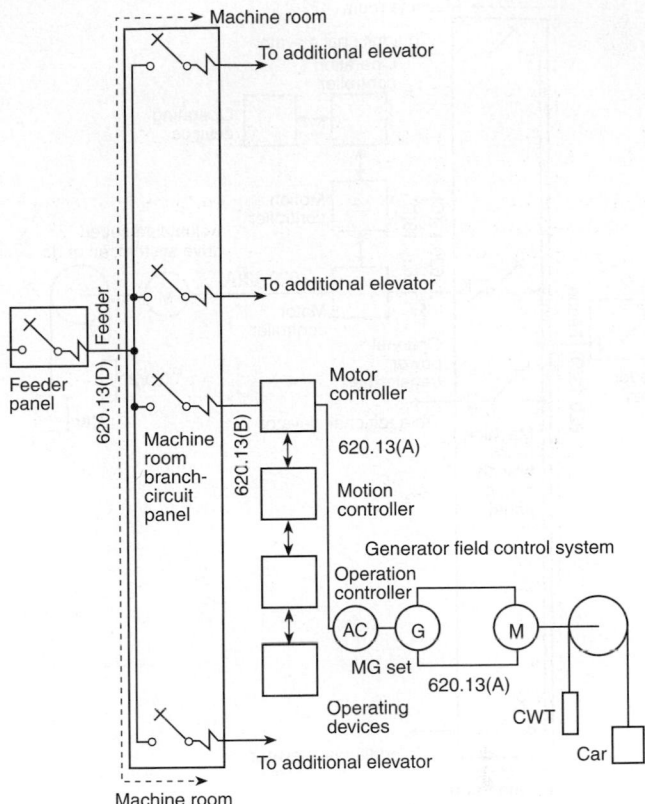

FIGURE D9   Generator Field Control.

(b) In accordance with 620.13(D), for six elevators, the total conductor ampacity is the sum of all the currents.

6 elevators × 133 A = 798 A

(c) In accordance with 620.14 and Table 620.14, the conductor (feeder) ampacity would be permitted to be reduced by the use of a demand factor. Constant loads are not included (*see 620.13, Informational Note No. 2*). For six elevators, the demand factor is 0.79. The feeder diverse ampacity is, therefore, 0.79 × 798 A = 630 A.

(d) In accordance with 430.24 and 215.3, the controller continuous current is 125% × 10 A = 13 A.

(e) The total feeder ampacity is the sum of the diverse current and all the controller constant current.

I_{total} = 630 A + (6 elevators × 12.5 A) = 705 A

(f) This ampacity would be permitted to be used to select the wire size.

See Figure D10.

Example D11 Mobile Home

(*see 550.18*)

A mobile home floor is 70 ft by 10 ft and has two small appliance circuits; a 1000-VA, 240-V heater; a 200-VA, 120-V exhaust fan; a 400-VA, 120-V dishwasher; and a 7000-VA electric range.

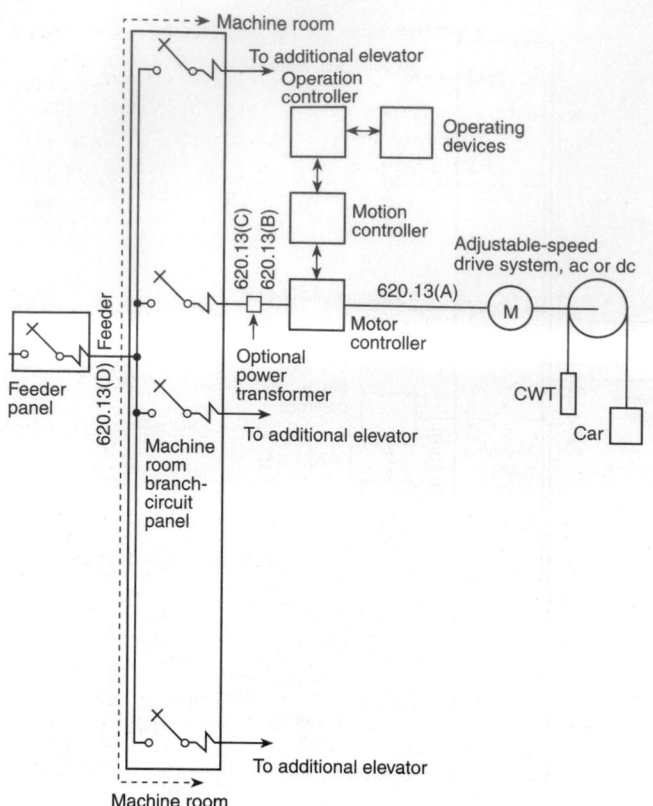

FIGURE D10   Adjustable Speed Drive Control.

Lighting and Small-Appliance Load

Lighting (70 ft × 10 ft × 3 VA per ft²)		2,100 VA
Small-appliance (1500 VA× 2 circuits)		3,000 VA
Laundry (1500 VA × 1 circuit)		1,500 VA
	Subtotal	6,600 VA
First 3000 VA at 100%		3,000 VA
Remainder (6600 VA – 3000 VA = 3600 VA) × 35%		1,260 VA
	Total	4,260 VA

4260 VA ÷ 240 V = 17.75 A per leg

Amperes per Leg	Leg A	Leg B
Lighting and appliances	18	18
Heater (1000 VA ÷ 240 V)	4	4
Fan (200 VA × 125% ÷ 120 V)	2	—
Dishwasher (400 VA ÷ 120 V)	—	3
Range (7000 VA × 0.8 ÷ 240 V)	23	23
Total amperes per leg	47	48

Based on the higher current calculated for either leg, a minimum 50-A supply cord would be required.

For SI units, 0.093 m² = 1 ft² and 0.3048 m = 1 ft.

Example D12 Park Trailer

(see 552.47)

A park trailer floor is 40 ft by 10 ft and has two small appliance circuits, a 1000-VA, 240-V heater, a 200-VA, 120-V exhaust fan, a 400-VA, 120-V dishwasher, and a 7000-VA electric range.

Lighting and Small-Appliance Load

Lighting (40 ft × 10 ft × 3 VA per ft²)		1,200 VA
Small-appliance (1500 VA× 2 circuits)		3,000 VA
Laundry (1500 VA × 1 circuit)		1,500 VA
	Subtotal	5,700 VA
First 3000 VA at 100%		3,000 VA
Remainder (5700 VA – 3000 VA = 2700 VA) × 35%		945 VA
	Total	3,945 VA

3945 VA ÷ 240 V = 16.44 A per leg

Amperes per Leg	Leg A	Leg B
Lighting and appliances	16	16
Heater (1000 VA ÷ 240 V)	4	4
Fan (200 VA × 125% ÷ 120 V)	2	—
Dishwasher (400 VA ÷ 120 V)	—	3
Range (7000 VA × 0.8 ÷ 240 V)	23	23
Totals	45	46

Based on the higher current calculated for either leg, a minimum 50-A supply cord would be required.

For SI units, 0.093 m² = 1 ft² and 0.3048 m = 1 ft.

Example D13 Cable Tray Calculations

(See Article 392)

D13(a) Multiconductor Cables 4/0 AWG and Larger

Use: *NEC* 392.22(A)(1)(a)

Cable tray must have an inside width equal to or greater than the sum of the diameters (Sd) of the cables, which must be installed in a single layer.

Example: Cable tray width is obtained as follows:

Cable Size Being Used	(OD) Cable Outside Diameters (in.)	(N) Number of Cables	SD = (OD) × (N) (Sum of the Cable Diameters) (in.)
3–conductor Type MC cable — 4/0 AWG	1.57	12	18.84

The sum of the diameters (Sd) of all cables = 18.84 in., therefore a cable tray with an inside width of at least 18.84 in. is required.

Note: Cable outside diameter is a nominal diameter from catalog data.

D13(b) Multiconductor Cables Smaller Than 4/0 AWG

Use: *NEC* 392.22(A)(1)(b)

The sum of the cross-sectional areas of all the cables to be installed in the cable tray must be equal to or less than the allowable cable area for the tray width, as indicated in Table 392.22(A), Column 1.

Table D13(b) from Table 392.22(A), Column 1

Inside Width of Cable Tray (in.)	Allowable Cable Area (in.^{2})
6	7.0
9	10.5
12	14.0
18	21.0
24	28.0
30	35.0
36	42.0

Example: Cable tray width is obtained as follows:

Cable Size Being Used	(A) Cable Cross-Sectional Area (in.^{2})	(N) Number of Cables	Multiply (A) × (N) (Which Is a Total Cable Cross-Sectional Area in in.^{2})
4-conductor Type MC cable — 1 AWG	1.1350	9	12.15

The total cable cross-sectional area is 12.15 in.^{2}. Using Table D13(b) above, the next higher allowable cable area must be used, which is 14.0 in.^{2}. The table specifies that the cable tray inside width for an allowable cable area of 14.0 in.^{2} is 12 in.

Note: Cable cross-sectional area is a nominal area from catalog data.

D13(c) Single Conductor Cables 1/0 AWG through 4/0 AWG

Use: *NEC* 392.22(B)(1)(d)

Cable tray must have an inside width equal to or greater than the sum of the diameters (Sd) of the cables. The cables must be evenly distributed across the cable tray.

Example: Cable tray width is obtained as follows:

Single Conductor Cable Size Being Used	(OD) Cable Outside Diameters (in.)	(N) Number of Cables	Sd = (OD) × (N) (Sum of the Cable Diameters) (in.)
THHN — 4/0 AWG	0.642	18	11.556

The sum of the diameters (Sd) of all cables = 11.56 in., therefore, a cable tray with an inside width of at least 11.56 in. is required.

Note: Cable outside diameter from Chapter 9, Table 5.

D.13(d) Single Conductor Cables 250 kcmil through 900 kcmil

Use: *NEC* 392.22(B)(1)(b)

The sum of the cross-sectional areas of all the cables to be installed in the cable tray must be equal to or less than the allowable cable area for the tray width, as indicated in Table 392.22(B)(1), Column 1.

Table D13(d) from Table 392.22(B)(1), Column 1

Inside Width of Cable Tray (in.)	Allowable Cable Area (in.^{2})
6	6.5
9	9.5
12	13.0
18	19.5
24	26.0
30	32.5
36	39.0

Example: Cable tray width is obtained as follows:

Cable Size Being Used	(A) Cable Cross-Sectional Area (in.^{2})	(N) Number of Cables	Multiply (A) × (N) (Which Is a Total Cable Cross-Sectional Area in in.^{2})
THHN — 500 kcmil	0.707	9	6.36

The total cable cross-sectional area is 6.36 in.^{2}. Using Table D13(d), the next higher allowable cable area must be used, which is 6.5 in.^{2}. The table specifies that the cable tray inside width for an allowable cable area of 6.5 in.^{2} is 6 in.

Note: Single-conductor cable cross-sectional area from Chapter 9, Table 5.

Informative Annex E Types of Construction

This informative annex is not a part of the requirements of this NFPA document but is included for informational purposes only.

Table E.1 contains the fire resistance rating, in hours, for Types I through V construction. The five different types of construction can be summarized briefly as follows (see also Table E.2):

Type I is a fire-resistive construction type. All structural elements and most interior elements are required to be noncombustible. Interior, nonbearing partitions are permitted to be 1 or 2 hour rated. For nearly all occupancy types, Type I construction can be of unlimited height.

Type II construction has three categories: fire-resistive, one-hour rated, and non-rated. The number of stories permitted for multifamily dwellings varies from two for non-rated and four for one-hour rated to 12 for fire-resistive construction.

Type III construction has two categories: one-hour rated and non-rated. Both categories require the structural framework and exterior walls to be of noncombustible material. One-hour

rated construction requires all interior partitions to be one-hour rated. Non-rated construction allows nonbearing interior partitions to be of non-rated construction. The maximum permitted number of stories for multifamily dwellings and other structures is two for non-rated and four for one-hour rated.

Type IV is a single construction category that provides for heavy timber construction. Both the structural framework and the exterior walls are required to be noncombustible except that wood members of certain minimum sizes are allowed. This construction type is seldom used for multifamily dwellings but, if used, would be permitted to be four stories high.

Type V construction has two categories: one-hour rated and non-rated. One-hour rated construction requires a minimum of one-hour rated construction throughout the building. Non-rated construction allows non-rated interior partitions with certain restrictions. The maximum permitted number of stories for multifamily dwellings and other structures is two for non-rated and three for one-hour rated.

Table E.1 Fire Resistance Ratings for Type I Through Type V Construction (hr)

	Type I		Type II			Type III		Type IV	Type V	
	442	332	222	111	000	211	200	2HH	111	000
Exterior Bearing Walls [a]										
Supporting more than one floor, columns, or other bearing walls	4	3	2	1	0[b]	2	2	2	1	0[b]
Supporting one floor only	4	3	2	1	0[b]	2	2	2	1	0[b]
Supporting a roof only	4	3	1	1	0[b]	2	2	2	1	0[b]
Interior Bearing Walls										
Supporting more than one floor, columns, or other bearing walls	4	3	2	1	0	1	0	2	1	0
Supporting one floor only	3	2	2	1	0	1	0	1	1	0
Supporting roofs only	3	2	1	1	0	1	0	1	1	0
Columns										
Supporting more than one floor, columns, or other bearing walls	4	3	2	1	0	1	0	H	1	0
Supporting one floor only	3	2	2	1	0	1	0	H	1	0
Supporting roofs only	3	2	1	1	0	1	0	H	1	0
Beams, Girders, Trusses, and Arches										
Supporting more than one floor, columns, or other bearing walls	4	3	2	1	0	1	0	H	1	0
Supporting one floor only	2	2	2	1	0	1	0	H	1	0
Supporting roofs only	2	2	1	1	0	1	0	H	1	0
Floor/Ceiling Assemblies	2	2	2	1	0	1	0	H	1	0
Roof/Ceiling Assemblies	2	1½	1	1	0	1	0	H	1	0
Interior Nonbearing Walls	0	0	0	0	0	0	0	0	0	0
Exterior Nonbearing Walls [c]	0[b]	0[b]	0[b]	0[b]	0[b]	0[b]	0[b]	0[b]	0[b]	0[b]

Source: Table 7.2.1.1 from *NFPA 5000, Building Construction and Safety Code*, 2012 edition.
H: Heavy timber members.
[a]See 7.3.2.1 in *NFPA 5000*.
[b]See Section 7.3 in *NFPA 5000*.
[c]See 7.2.3.2.12, 7.2.4.2.3, and 7.2.5.6.8 in *NFPA 5000*.

Table E.2 Maximum Number of Stories for Types V, IV, and III Construction

Construction Type	Maximum Number of Stories Permitted
V Non-rated	2
V Non-rated, Sprinklered	3
V One-Hour Rated	3
V One-Hour Rated, Sprinklered	4
IV Heavy Timber	4
IV Heavy Timber, Sprinklered	5
III Non-rated	2
III Non-rated, Sprinklered	3
III One-Hour Rated	4
III One-Hour Rated, Sprinklered	5

In Table E.1 the system of designating types of construction also includes a specific breakdown of the types of construction through the use of arabic numbers. These arabic numbers follow the roman numeral notation where identifying a type of construction [for example, Type I(442), Type II(111), Type III(200)] and indicate the fire resistance rating requirements for certain structural elements as follows:

(1) First arabic number — exterior bearing walls
(2) Second arabic number — columns, beams, girders, trusses and arches, supporting bearing walls, columns, or loads from more than one floor
(3) Third arabic number — floor construction

Table E.3 provides a comparison of the types of construction for various model building codes. [*5000* : A.7.2.1.1]

Table E.3 Cross-Reference of Building Construction Types

NFPA 5000	I (442)	I (332)	II (222)	II (111)	II (000)	III (211)	III (200)	IV (2HH)	V (111)	V (000)
UBC	—	I FR	II FR	II 1 hr	II N	III 1 hr	III N	IV HT	V 1 hr	V N
B/NBC	1A	1B	2A	2B	2C	3A	3B	4	5A	5B
SBC	I	II	—	IV 1 hr	IV UNP	V 1 hr	V UNP	III	VI 1 hr	VI UNP
IBC	—	IA	1B	IIA	IIB	IIIA	IIIB	IV	VA	VB

Source: Table A.7.2.1.1 from *NFPA 5000, Building Construction and Safety Code*, 2012 edition.
UBC: *Uniform Building Code.*
FR: Fire rated.
N: Nonsprinklered.
HT: Heavy timber.
B/NBC: *National Building Code.*
SBC: *Standard Building Code.*
UNP: Unprotected.
IBC: *International Building Code.*

Informative Annex F   Availability and Reliability for Critical Operations Power Systems; and Development and Implementation of Functional Performance Tests (FPTs) for Critical Operations Power Systems

This informative annex is not a part of the requirements of this NFPA document but is included for informational purposes only.

I. Availability and Reliability for Critical Operations Power Systems.   Critical operations power systems may support facilities with a variety of objectives that are vital to public safety. Often these objectives are of such critical importance that system downtime is costly in terms of economic losses, loss of security, or loss of mission. For those reasons, the availability of the critical operations power system, the percentage of time that the system is in service, is important to those facilities. Given a specified level of availability, the reliability and maintainability requirements are then derived based on that availability requirement.

Availability. Availability is defined as the percentage of time that a system is available to perform its function(s). Availability is measured in a variety of ways, including the following:

$$\text{Availability} = \frac{MTBF}{MTBF + MTTR}$$

where:
$MTBF$ = mean time between failures
$MTTF$ = mean time to failure
$MTTR$ = mean time to repair

See the following table for an example of how to establish required availability for critical operation power systems:

Availability	Hours of Downtime[*]
0.9	876
0.99	87.6
0.999	8.76
0.9999	0.876
0.99999	0.0876
0.999999	0.00876
0.9999999	0.000876

[*]Based on a year of 8760 hours.

Availability of a system in actual operations is determined by the following:

(1)   The frequency of occurrence of failures. Failures may prevent the system from performing its function or may cause a degraded effect on system operation. Frequency of failures is directly related to the system's level of reliability.

(2)   The time required to restore operations following a system failure or the time required to perform maintenance to prevent a failure. These times are determined in part by the system's level of maintainability.

(3)   The logistics provided to support maintenance of the system. The number and availability of spares, maintenance personnel, and other logistics resources (refueling, etc.) combined with the system's level of maintainability determine the total downtime following a system failure.

Reliability. Reliability is concerned with the probability and frequency of failures (or lack of failures). A commonly used measure of reliability for repairable systems is *MTBF*. The equivalent measure for nonrepairable items is *MTTF*. Reliability is more accurately expressed as a probability over a given duration of time, cycles, or other parameter. For example, the reliability of a power plant might be stated as 95 percent probability of no failure over a 1000-hour operating period while generating a certain level of power. Reliability is usually defined in two ways (the electrical power industry has historically not used these definitions):

(1)   The duration or probability of failure-free performance under stated conditions
(2)   The probability that an item can perform its intended function for a specified interval under stated conditions [For nonredundant items, this is equivalent to the preceding definition (1). For redundant items this is equivalent to the definition of mission reliability.]

Maintainability. Maintainability is a measure of how quickly and economically failures can be prevented through preventive maintenance, or system operation can be restored following failure through corrective maintenance. A commonly used measure of maintainability in terms of corrective maintenance is the mean time to repair (*MTTR*). Maintainability is not the same thing as maintenance. It is a design parameter, while maintenance consists of actions to correct or prevent a failure event.

Improving Availability. The appropriate methods to use for improving availability depend on whether the facility is being designed or is already in use. For both cases, a reliability/availability analysis should be performed to determine the availability of the old system or proposed new system in order to ascertain the hours of downtime (see the preceding table). The AHJ or government agency should dictate how much downtime is acceptable.

Existing facilities: For a facility that is being operated, two basic methods are available for improving availability when the current level of availability is unacceptable: (1) Selectively adding redundant units (e.g., generators, chillers, fuel supply) to eliminate sources of single-point failure, and (2) optimizing maintenance using a reliability-centered maintenance (RCM) approach to minimize downtime. [Refer to NFPA 70B-2010, *Recommended Practice for Electrical Equipment Maintenance.*] A combination of the previous two methods can also be implemented. A third very expensive method is to redesign subsystems or to replace components and subsystems with higher reliability items. [Refer to NFPA 70B.]

New facilities: The opportunity for high availability and reliability is greatest when designing a new facility. By applying an effective reliability strategy, designing for maintainability, and ensuring that manufacturing and commissioning do not negatively affect the inherent levels of reliability and maintainability, a highly available facility will result. The approach should be as follows:

(1) *Develop and determine a reliability strategy* (establish goals, develop a system model, design for reliability, conduct reliability development testing, conduct reliability acceptance testing, design system delivery, maintain design reliability, maintain design reliability in operation).

(2) *Develop a reliability program.* This is the application of the reliability strategy to a specific system, process, or function. Each step in the preceding strategy requires the selection and use of specific methods and tools. For example, various tools can be used to develop requirements or evaluate potential failures. To derive requirements, analytical models can be used, for example, quality function development (a technique for deriving more detailed, lower-level requirements from one level to another, beginning with mission requirements, i.e., customer needs). This model was developed as part of the total quality management movement. Parametric models can also be used to derive design values of reliability from operational values and vice versa. Analytical methods include but are not limited to things such as thermal analysis, durability analysis, and predictions. Finally, one should evaluate possible failures. A failure modes and effects criticality analysis (FMECA) and fault tree analysis (FTA) are two methods for evaluating possible failures. The mission facility engineer should determine which method to use or whether to use both.

(3) *Identify Reliability Requirements.* The entire effort for designing for reliability begins with identifying the mission critical facility's reliability requirements. These requirements are stated in a variety of ways, depending on the customer and the specific system. For a mission-critical facility, it would be the mission success probability.

II. Development and Implementation of Functional Performance Tests (FPTs) for Critical Operations Power Systems Development of FPT

(1) Submit Functional Performance Tests (FPTs). System/component tests or FPTs are developed from submitted drawings, systems operating documents (SODs), and systems operation and maintenance manuals (SOMMs), including large component testing (i.e., transformers, cable, generators, UPS), and how components operate as part of the total system. The commissioning authority develops the test and cannot be the installation contractor (or subcontractor).

As the equipment/components/systems are installed, quality assurance procedures are administered to verify that components are installed in accordance with minimum manufacturers' recommendations, safety codes, and acceptable installation practices. Quality assurance discrepancies are then identified and added to a "commissioning action list" that must be rectified as part of the commissioning program. These items would usually be discussed during commissioning meetings. Discrepancies are usually identified initially by visual inspection.

(2) Review FPTs. The tests must be reviewed by the customer, electrical contractors, quality assurance personnel, maintenance personnel, and other key personnel (the commissioning team). Areas of concern include, among others, all functions of the system being tested, all major components included, whether the tests reflect the system operating documents, and verification that the tests make sense.

(3) Make Changes to FPTs as Required. The commissioning authority then implements the corrections, questions answered, and additions.

(4) FPTs Approval. After the changes are made to the FPTs, they are submitted to the commissioning team. When it is acceptable, the customer or the designated approval authority approves the FPTs. It should be noted that even though the FPT is approved, problems that arise during the test (or areas not covered) must be addressed.

Testing Implementation for FPTs. The final step in the successful commissioning plan is testing and proper execution of system-integrated tests.

(1) Systems Ready to Operate. The FPTs can be implemented as various systems become operative (i.e., test for the generator system) or when the entire system is installed. However, the final "pull the plug" test is performed only after all systems are completely installed. If the electrical contractor (or subcontractor) implements the FPTs, a witness must initial each step of the test. The electrical contractor cannot employ the witness directly or indirectly.

(2) Perform Tests (FPTs). If the system fails the test, the problem must be resolved and the equipment or system retested or the testing requirements re-analyzed until successful tests are witnessed. Once the system or equipment passes testing, it is verified by designated commissioning official.

(3) Customer Receives System. After all tests are completed (including the "pull the plug" test), the system is turned over to the customer.

Informative Annex G Supervisory Control and Data Acquisition (SCADA)

This informative annex is not a part of the requirements of this NFPA document, but is included for informational purposes only.

(A) General. Where provided, the general requirements in (A)(1) through (A)(11) shall apply to SCADA systems.

(1) The SCADA system for the COPS loads shall be separate from the building management SCADA system.

(2) No single point failure shall be able to disable the SCADA system.

(3) The SCADA system shall be permitted to provide control and monitor electrical and mechanical utility systems related to mission critical loads, including, but not limited to, the following:

    a. The fire alarm system
    b. The security system
    c. Power distribution
    d. Power generation
    e. HVAC and ventilation (damper position, airflow speed and direction)
    f. Load shedding
    g. Fuel levels or hours of operation

(4) Before installing or employing a SCADA system, an operations and maintenance analysis and risk assessment shall be performed to provide the maintenance parameter data

(5) A redundant system shall be provided in either warm or hot standby.

(6) The controller shall be a programmable logic controller (PLC).

(7) The SCADA system shall utilize open, not proprietary, protocols.

(8) The SCADA system shall be able to assess the damage and determine system integrity after the "event."

(9) The monitor display shall provide graphical user interface for all major components monitored and controlled by the SCADA system, with color schemes readily recognized by the typical user.

(10) The SCADA system shall have the capability to provide storage of critical system parameters at a 15-minute rate or more often when out-of-limit conditions exist.

(11) The SCADA system shall have a separate data storage facility not located in the same vicinity.

(B) Power Supply. The SCADA system power supply shall comply with (B)(1) through (B)(3):

(1) The power supply shall be provided with a direct-current station battery system, rated between 24 and 125 volts dc, with a 72-hour capacity.

(2) The batteries of the SCADA system shall be separate from the batteries for other electrical systems.

(3) The power supply shall be provided with a properly installed surge-protective device (TVSS) at its terminals with a direct low-impedance path to ground. Protected and unprotected circuits shall be physically separated to prevent coupling.

(C) Security Against Hazards. Security against hazards shall be provided in accordance with (C)(1) through (C)(6):

(1) Controlled physical access by authorized personnel to only the system operational controls and software shall be provided.

(2) The SCADA system shall be protected against dust, dirt, water, and other contaminants by specifying enclosures appropriate for the environment.

(3) Conduit and tubing shall not violate the integrity of the SCADA system enclosure.

(4) The SCADA system shall be located in the same secure locations as the secured systems that they monitor and control.

(5) The SCADA system shall be provided with dry agent fire protection systems or double interlocked preaction sprinkler systems using cross-zoned detection, to minimize the threat of accidental water discharge into unprotected equipment. The fire protection systems shall be monitored by the fire alarm system in accordance with *NFPA 72*-2013, *National Fire Alarm and Signaling Code.*

(6) The SCADA system shall not be connected to other network communications outside the secure locations without encryption or use of fiber optics.

(D) Maintenance and Testing. SCADA systems shall be maintained and tested in accordance with (D)(1) and (D)(2).

(1) Maintenance. The maintenance program for SCADA systems shall consist of the following components:

(1) A documented preventive maintenance program

(2) Concurrent maintenance capabilities, to allow the testing, troubleshooting, repair, and/or replacement of a component or subsystem while redundant component(s) or subsystem(s) are serving the load

(3) Retention of operational data — the deleted material goes well beyond requirements to ensure proper maintenance and operation

(2) Testing. SCADA systems shall be tested periodically under actual or simulated contingency conditions.

> Informational Note No. 1: Periodic system testing procedures can duplicate or be derived from the recommended functional performance testing procedures of individual components, as provided by the manufacturers.

> Informational Note No. 2: For more information on maintenance and testing of SCADA, see NFPA 70B-2013, *Recommended Practice for Electrical Equipment Maintenance.*

Informative Annex H Administration and Enforcement

Informative Annex H is not a part of the requirements of this NFPA document and is included for informational purposes only. This informative annex is informative unless specifically adopted by the local jurisdiction adopting the National Electrical Code®.

80.1 Scope. The following functions are covered:

(1) The inspection of electrical installations as covered by 90.2
(2) The investigation of fires caused by electrical installations
(3) The review of construction plans, drawings, and specifications for electrical systems
(4) The design, alteration, modification, construction, maintenance, and testing of electrical systems and equipment
(5) The regulation and control of electrical installations at special events including but not limited to exhibits, trade shows, amusement parks, and other similar special occupancies

80.2 Definitions.

Authority Having Jurisdiction. The organization, office, or individual responsible for approving equipment, materials, an installation, or a procedure.

Chief Electrical Inspector. An electrical inspector who either is the authority having jurisdiction or is designated by the authority having jurisdiction and is responsible for administering the requirements of this *Code.*

Electrical Inspector. An individual meeting the requirements of 80.27 and authorized to perform electrical inspections.

80.3 Purpose. The purpose of this article shall be to provide requirements for administration and enforcement of the *National Electrical Code.*

80.5 Adoption. Article 80 shall not apply unless specifically adopted by the local jurisdiction adopting the *National Electrical Code.*

80.7 Title. The title of this *Code* shall be *NFPA 70, National Electrical Code* ®, of the National Fire Protection Association. The short title of this *Code* shall be the *NEC* ®.

80.9 Application.

(A) New Installations. This *Code* applies to new installations. Buildings with construction permits dated after adoption of this *Code* shall comply with its requirements.

(B) Existing Installations. Existing electrical installations that do not comply with the provisions of this *Code* shall be permitted to be continued in use unless the authority having jurisdiction determines that the lack of conformity with this *Code* presents an imminent danger to occupants. Where changes are required for correction of hazards, a reasonable amount of time shall be given for compliance, depending on the degree of the hazard.

(C) Additions, Alterations, or Repairs. Additions, alterations, or repairs to any building, structure, or premises shall conform to that required of a new building without requiring the existing building to comply with all the requirements of this *Code.* Additions, alterations, installations, or repairs shall not cause an existing building to become unsafe or to adversely affect the performance of the building as determined by the authority having jurisdiction. Electrical wiring added to an existing service, feeder, or branch circuit shall not result in an installation that violates the provisions of the *Code* in force at the time the additions are made.

80.11 Occupancy of Building or Structure.

(A) New Construction. No newly constructed building shall be occupied in whole or in part in violation of the provisions of this *Code.*

(B) Existing Buildings. Existing buildings that are occupied at the time of adoption of this *Code* shall be permitted to remain in use provided the following conditions apply:

(1) The occupancy classification remains unchanged.
(2) There exists no condition deemed hazardous to life or property that would constitute an imminent danger.

80.13 Authority. Where used in this article, the term *authority having jurisdiction* shall include the chief electrical inspector or other individuals designated by the governing body. This *Code* shall be administered and enforced by the authority having jurisdiction designated by the governing authority as follows.

(1) The authority having jurisdiction shall be permitted to render interpretations of this *Code* in order to provide clarification to its requirements, as permitted by 90.4.
(2) When the use of any electrical equipment or its installations is found to be dangerous to human life or property, the authority having jurisdiction shall be empowered to have the premises disconnected from its source of electric supply, as established by the Board. When such equipment or installation has been so condemned or disconnected, a notice shall be placed thereon listing the causes for the condemnation, the disconnection, or both, and the penalty under 80.23 for the unlawful use thereof. Written notice of such condemnation or disconnection and the causes therefor shall be given within 24 hours to the owners, the occupant, or both, of such building, structure, or premises. It shall be unlawful for any person to remove said notice, to reconnect the electrical equipment to its source of electric supply, or to use or permit to be used electric power in any such electrical equipment until such causes for the condemnation or disconnection have been remedied to the satisfaction of the inspection authorities.
(3) The authority having jurisdiction shall be permitted to delegate to other qualified individuals such powers as necessary for the proper administration and enforcement of this *Code.*
(4) Police, fire, and other enforcement agencies shall have authority to render necessary assistance in the enforcement of this *Code* when requested to do so by the authority having jurisdiction.
(5) The authority having jurisdiction shall be authorized to inspect, at all reasonable times, any building or premises for dangerous or hazardous conditions or equipment as set forth in this *Code.* The authority having jurisdiction shall be permitted to order any person(s) to remove or remedy such dangerous or hazardous condition or

equipment. Any person(s) failing to comply with such order shall be in violation of this *Code*.

(6) Where the authority having jurisdiction deems that conditions hazardous to life and property exist, he or she shall be permitted to require that such hazardous conditions in violation of this *Code* be corrected.

(7) To the full extent permitted by law, any authority having jurisdiction engaged in inspection work shall be authorized at all reasonable times to enter and examine any building, structure, or premises for the purpose of making electrical inspections. Before entering a premises, the authority having jurisdiction shall obtain the consent of the occupant thereof or obtain a court warrant authorizing entry for the purpose of inspection except in those instances where an emergency exists. As used in this section, *emergency* means circumstances that the authority having jurisdiction knows, or has reason to believe, exist and that reasonably can constitute immediate danger to persons or property.

(8) Persons authorized to enter and inspect buildings, structures, and premises as herein set forth shall be identified by proper credentials issued by this governing authority.

(9) Persons shall not interfere with an authority having jurisdiction carrying out any duties or functions prescribed by this *Code*.

(10) Persons shall not use a badge, uniform, or other credentials to impersonate the authority having jurisdiction.

(11) The authority having jurisdiction shall be permitted to investigate the cause, origin, and circumstances of any fire, explosion, or other hazardous condition.

(12) The authority having jurisdiction shall be permitted to require plans and specifications to ensure compliance with this *Code*.

(13) Whenever any installation subject to inspection prior to use is covered or concealed without having first been inspected, the authority having jurisdiction shall be permitted to require that such work be exposed for inspection. The authority having jurisdiction shall be notified when the installation is ready for inspection and shall conduct the inspection within ___ days.

(14) The authority having jurisdiction shall be permitted to order the immediate evacuation of any occupied building deemed unsafe when such building has hazardous conditions that present imminent danger to building occupants.

(15) The authority having jurisdiction shall be permitted to waive specific requirements in this *Code* or permit alternative methods where it is assured that equivalent objectives can be achieved by establishing and maintaining effective safety. Technical documentation shall be submitted to the authority having jurisdiction to demonstrate equivalency and that the system, method, or device is approved for the intended purpose.

(16) Each application for a waiver of a specific electrical requirement shall be filed with the authority having jurisdiction and shall be accompanied by such evidence, letters, statements, results of tests, or other supporting information as required to justify the request. The authority having jurisdiction shall keep a record of actions on such applications, and a signed copy of the authority having jurisdiction's decision shall be provided for the applicant.

80.15 Electrical Board.

(A) Creation of the Electrical Board. There is hereby created the Electrical Board of the _____ of _____, hereinafter designated as the Board.

(B) Appointments. Board members shall be appointed by the Governor with the advice and consent of the Senate (or by the Mayor with the advice and consent of the Council, or the equivalent).

(1) Members of the Board shall be chosen in a manner to reflect a balanced representation of individuals or organizations. The Chair of the Board shall be elected by the Board membership.

(2) The Chief Electrical Inspector in the jurisdiction adopting this Article authorized in (B)(3)(a) shall be the nonvoting secretary of the Board. Where the Chief Electrical Inspector of a local municipality serves a Board at a state level, he or she shall be permitted to serve as a voting member of the Board.

(3) The board shall consist of not fewer than five voting members. Board members shall be selected from the following:

   a. Chief Electrical Inspector from a local government (for State Board only)
   b. An electrical contractor operating in the jurisdiction
   c. A licensed professional engineer engaged primarily in the design or maintenance of electrical installations
   d. A journeyman electrician

(4) Additional membership shall be selected from the following:

   a. A master (supervising) electrician
   b. The Fire Marshal (or Fire Chief)
   c. A representative of the property/casualty insurance industry
   d. A representative of an electric power utility operating in the jurisdiction
   e. A representative of electrical manufacturers primarily and actively engaged in producing materials, fittings, devices, appliances, luminaires, or apparatus used as part of or in connection with electrical installations
   f. A member of the labor organization that represents the primary electrical workforce
   g. A member from the public who is not affiliated with any other designated group
   h. A representative of a telecommunications utility operating in the jurisdiction

(C) Terms. Of the members first appointed, _____ shall be appointed for a term of 1 year, _____ for a term of 2 years, _____ for a term of 3 years, and _____ for a term of 4 years, and thereafter each appointment shall be for a term of 4 years or until a successor is appointed. The Chair of the Board shall be appointed for a term not to exceed ____ years.

(D) Compensation. Each appointed member shall receive the sum of _____dollars ($_____) for each day during which the member attends a meeting of the Board and, in addition thereto, shall be reimbursed for direct lodging, travel, and meal expenses as covered by policies and procedures established by the jurisdiction.

(E) Quorum. A quorum as established by the Board operating procedures shall be required to conduct Board business. The Board shall hold such meetings as necessary to carry out the purposes of Article 80. The Chair or a majority of the members of the Board shall have the authority to call meetings of the Board.

(F) Duties. It shall be the duty of the Board to perform the following:

(1) Adopt the necessary rules and regulations to administer and enforce Article 80.

(2) Establish qualifications of electrical inspectors.

(3) Revoke or suspend the recognition of any inspector's certificate for the jurisdiction.

(4) After advance notice of the public hearings and the execution of such hearings, as established by law, the Board is authorized to establish and update the provisions for the safety of electrical installations to conform to the current edition of the *National Electrical Code* (NFPA 70) and other nationally recognized safety standards for electrical installations.

(5) Establish procedures for recognition of electrical safety standards and acceptance of equipment conforming to these standards.

(G) Appeals.

(1) *Review of Decisions.* Any person, firm, or corporation may register an appeal with the Board for a review of any decision of the Chief Electrical Inspector or of any Electrical Inspector, provided that such appeal is made in writing within fifteen (15) days after such person, firm, or corporation shall have been notified. Upon receipt of such appeal, said Board shall, if requested by the person making the appeal, hold a public hearing and proceed to determine whether the action of the Board, or of the Chief Electrical Inspector, or of the Electrical Inspector complies with this law and, within fifteen (15) days after receipt of the appeal or after holding the hearing, shall make a decision in accordance with its findings.

(2) *Conditions.* Any person shall be permitted to appeal a decision of the authority having jurisdiction to the Board when it is claimed that any one or more of the following conditions exist:

     a. The true intent of the codes or ordinances described in this *Code* has been incorrectly interpreted.

     b. The provisions of the codes or ordinances do not fully apply.

     c. A decision is unreasonable or arbitrary as it applies to alternatives or new materials.

(3) *Submission of Appeals.* A written appeal, outlining the *Code* provision from which relief is sought and the remedy proposed, shall be submitted to the authority having jurisdiction within 15 calendar days of notification of violation.

(H) Meetings and Records. Meetings and records of the Board shall conform to the following:

(1) Meetings of the Board shall be open to the public as required by law.

(2) Records of meetings of the Board shall be available for review during normal business hours, as required by law.

80.17 Records and Reports. The authority having jurisdiction shall retain records in accordance with (A) and (B).

(A) Retention. The authority having jurisdiction shall keep a record of all electrical inspections, including the date of such inspections and a summary of any violations found to exist, the date of the services of notices, and a record of the final disposition of all violations. All required records shall be maintained until their usefulness has been served or as otherwise required by law.

(B) Availability. A record of examinations, approvals, and variances granted shall be maintained by the authority having jurisdiction and shall be available for public review as prescribed by law during normal business hours.

80.19 Permits and Approvals. Permits and approvals shall conform to (A) through (H).

(A) Application.

(1) Activity authorized by a permit issued under this *Code* shall be conducted by the permittee or the permittee's agents or employees in compliance with all requirements of this *Code* applicable thereto and in accordance with the approved plans and specifications. No permit issued under this *Code* shall be interpreted to justify a violation of any provision of this *Code* or any other applicable law or regulation. Any addition or alteration of approved plans or specifications shall be approved in advance by the authority having jurisdiction, as evidenced by the issuance of a new or amended permit.

(2) A copy of the permit shall be posted or otherwise readily accessible at each work site or carried by the permit holder as specified by the authority having jurisdiction.

(B) Content. Permits shall be issued by the authority having jurisdiction and shall bear the name and signature of the authority having jurisdiction or that of the authority having jurisdiction's designated representative. In addition, the permit shall indicate the following:

(1) Operation or activities for which the permit is issued

(2) Address or location where the operation or activity is to be conducted

(3) Name and address of the permittee

(4) Permit number and date of issuance

(5) Period of validity of the permit

(6) Inspection requirements

(C) Issuance of Permits. The authority having jurisdiction shall be authorized to establish and issue permits, certificates, notices, and approvals, or orders pertaining to electrical safety hazards pursuant to 80.23, except that no permit shall be required to execute any of the classes of electrical work specified in the following:

(1) Installation or replacement of equipment such as lamps and of electric utilization equipment approved for connection to suitable permanently installed receptacles

(2) Replacement of flush or snap switches, fuses, lamp sockets, and receptacles, and other minor maintenance and repair work, such as replacing worn cords and tightening connections on a wiring device

(3) The process of manufacturing, testing, servicing, or repairing electrical equipment or apparatus

(D) Annual Permits. In lieu of an individual permit for each installation or alteration, an annual permit shall, upon application, be issued to any person, firm, or corporation regularly employing one or more employees for the installation, alteration, and maintenance of electrical equipment in or on buildings or premises owned or occupied by the applicant for the permit. Upon application, an electrical contractor as agent for the owner or tenant shall be issued an annual permit. The applicant shall keep records of all work done, and the records shall be transmitted periodically to the electrical inspector.

(E) Fees. Any political subdivision that has been provided for electrical inspection in accordance with the provisions of Article 80 may establish fees that shall be paid by the applicant for a permit before the permit is issued.

(F) Inspection and Approvals.

(1) Upon the completion of any installation of electrical equipment that has been made under a permit other than an annual permit, it shall be the duty of the person, firm, or corporation making the installation to notify the Electrical Inspector having jurisdiction, who shall inspect the work within a reasonable time.

(2) Where the Inspector finds the installation to be in conformity with the statutes of all applicable local ordinances and all rules and regulations, the Inspector shall issue to the person, firm, or corporation making the installation a certificate of approval, with duplicate copy for delivery to the owner, authorizing the connection to the supply of electricity and shall send written notice of such authorization to the supplier of electric service. When a certificate of temporary approval is issued authorizing the connection of an installation, such certificates shall be issued to expire at a time to be stated therein and shall be revocable by the Electrical Inspector for cause.

(3) When any portion of the electrical installation within the jurisdiction of an Electrical Inspector is to be hidden from view by the permanent placement of parts of the building, the person, firm, or corporation installing the equipment shall notify the Electrical Inspector, and the equipment shall not be concealed until it has been approved by the Electrical Inspector or until _____ days have elapsed from the time of such notification, provided that on large installations, where the concealment of equipment proceeds continuously, the person, firm, or corporation installing the equipment shall give the Electrical Inspector due notice in advance, and inspections shall be made periodically during the progress of the work.

(4) At regular intervals, the Electrical Inspector having jurisdiction shall visit all buildings and premises where work may be done under annual permits and shall inspect all electrical equipment installed under such permits since the date of the previous inspection. The Electrical Inspector shall issue a certificate of approval for such work as is found to be in conformity with the provisions of Article 80 and all applicable ordinances, orders, rules, and regulations, after payments of all required fees.

(5) If, upon inspection, any installation is found not to be fully in conformity with the provisions of Article 80, and all applicable ordinances, rules, and regulations, the Inspector making the inspection shall at once forward to the person, firm, or corporation making the installation a written notice stating the defects that have been found to exist.

(G) Revocation of Permits. Revocation of permits shall conform to the following:

(1) The authority having jurisdiction shall be permitted to revoke a permit or approval issued if any violation of this *Code* is found upon inspection or in case there have been any false statements or misrepresentations submitted in the application or plans on which the permit or approval was based.

(2) Any attempt to defraud or otherwise deliberately or knowingly design, install, service, maintain, operate, sell, represent for sale, falsify records, reports, or applications, or other related activity in violation of the requirements prescribed by this *Code* shall be a violation of this *Code*. Such violations shall be cause for immediate suspension or revocation of any related licenses, certificates, or permits issued by this jurisdiction. In addition, any such violation shall be subject to any other criminal or civil penalties as available by the laws of this jurisdiction.

(3) Revocation shall be constituted when the permittee is duly notified by the authority having jurisdiction.

(4) Any person who engages in any business, operation, or occupation, or uses any premises, after the permit issued therefor has been suspended or revoked pursuant to the provisions of this *Code*, and before such suspended permit has been reinstated or a new permit issued, shall be in violation of this *Code*.

(5) A permit shall be predicated upon compliance with the requirements of this *Code* and shall constitute written authority issued by the authority having jurisdiction to install electrical equipment. Any permit issued under this *Code* shall not take the place of any other license or permit required by other regulations or laws of this jurisdiction.

(6) The authority having jurisdiction shall be permitted to require an inspection prior to the issuance of a permit.

(7) A permit issued under this *Code* shall continue until revoked or for the period of time designated on the permit. The permit shall be issued to one person or business only and for the location or purpose described in the permit. Any change that affects any of the conditions of the permit shall require a new or amended permit.

(H) Applications and Extensions. Applications and extensions of permits shall conform to the following:

(1) The authority having jurisdiction shall be permitted to grant an extension of the permit time period upon presentation by the permittee of a satisfactory reason for failure to start or complete the work or activity authorized by the permit.

(2) Applications for permits shall be made to the authority having jurisdiction on forms provided by the jurisdiction and shall include the applicant's answers in full to inquiries set forth on such forms. Applications for permits shall be accompanied by such data as required by the authority having jurisdiction, such as plans and specifications, location, and so forth. Fees shall be determined as required by local laws.

(3) The authority having jurisdiction shall review all applications submitted and issue permits as required. If an application for a permit is rejected by the authority having jurisdiction, the applicant shall be advised of the reasons for such rejection. Permits for activities requiring evidence of financial responsibility by the jurisdiction shall not be issued unless proof of required financial responsibility is furnished.

80.21 Plans Review. Review of plans and specifications shall conform to (A) through (C).

(A) Authority. For new construction, modification, or rehabilitation, the authority having jurisdiction shall be permitted to review construction documents and drawings.

(B) Responsibility of the Applicant. It shall be the responsibility of the applicant to ensure the following:

(1) The construction documents include all of the electrical requirements.

(2) The construction documents and drawings are correct and in compliance with the applicable codes and standards.

(C) Responsibility of the Authority Having Jurisdiction. It shall be the responsibility of the authority having jurisdiction to promulgate rules that cover the following:

(1) Review of construction documents and drawings shall be completed within established time frames for the purpose of acceptance or to provide reasons for nonacceptance.

(2) Review and approval by the authority having jurisdiction shall not relieve the applicant of the responsibility of compliance with this *Code*.

(3) Where field conditions necessitate any substantial change from the approved plan, the authority having jurisdiction shall be permitted to require that the corrected plans be submitted for approval.

80.23 Notice of Violations, Penalties. Notice of violations and penalties shall conform to (A) and (B).

(A) Violations.

(1) Whenever the authority having jurisdiction determines that there are violations of this *Code*, a written notice shall be issued to confirm such findings.

(2) Any order or notice issued pursuant to this *Code* shall be served upon the owner, operator, occupant, or other person responsible for the condition or violation, either by personal service or mail or by delivering the same to, and leaving it with, some person of responsibility upon the premises. For unattended or abandoned locations, a copy of such order or notice shall be posted on the premises in a conspicuous place at or near the entrance to such premises and the order or notice shall be mailed by registered or certified mail, with return receipt requested, to the last known address of the owner, occupant, or both.

(B) Penalties.

(1) Any person who fails to comply with the provisions of this *Code* or who fails to carry out an order made pursuant to this *Code* or violates any condition attached to a permit, approval, or certificate shall be subject to the penalties established by this jurisdiction.

(2) Failure to comply with the time limits of an abatement notice or other corrective notice issued by the authority having jurisdiction shall result in each day that such violation continues being regarded as a new and separate offense.

(3) Any person, firm, or corporation who shall willfully violate any of the applicable provisions of this article shall be guilty of a misdemeanor and, upon conviction thereof, shall be punished by a fine of not less than _____ dollars ($_____) or more than _____ dollars ($_____) for each offense, together with the costs of prosecution, imprisonment, or both, for not less than _____ (_____) days or more than _____ (_____) days.

80.25 Connection to Electricity Supply. Connections to the electric supply shall conform to (A) through (E).

(A) Authorization. Except where work is done under an annual permit and except as otherwise provided in 80.25, it shall be unlawful for any person, firm, or corporation to make connection to a supply of electricity or to supply electricity to any electrical equipment installation for which a permit is required or that has been disconnected or ordered to be disconnected.

(B) Special Consideration. By special permission of the authority having jurisdiction, temporary power shall be permitted to be supplied to the premises for specific needs of the construction project. The Board shall determine what needs are permitted under this provision.

(C) Notification. If, within _____ business days after the Electrical Inspector is notified of the completion of an installation of electric equipment, other than a temporary approval installation, the Electrical Inspector has neither authorized connection nor disapproved the installation, the supplier of electricity is authorized to make connections and supply electricity to such installation.

(D) Other Territories. If an installation or electric equipment is located in any territory where an Electrical Inspector has not been authorized or is not required to make inspections, the supplier of electricity is authorized to make connections and supply electricity to such installations.

(E) Disconnection. Where a connection is made to an installation that has not been inspected, as outlined in the preceding paragraphs of this section, the supplier of electricity shall immediately report such connection to the Chief Electrical Inspector. If, upon subsequent inspection, it is found that the installation is not in conformity with the provisions of Article 80, the Chief Electrical Inspector shall notify the person, firm, or corporation making the installation to rectify the defects and, if such work is not completed within fifteen (15) business days or a longer period as may be specified by the Board, the Board shall have the authority to cause the disconnection of that portion of the installation that is not in conformity.

80.27 Inspector's Qualifications.

(A) Certificate. All electrical inspectors shall be certified by a nationally recognized inspector certification program accepted by the Board. The certification program shall specifically qualify the inspector in electrical inspections. No person shall be employed as an Electrical Inspector unless that person is the holder of an Electrical Inspector's certificate of qualification issued by the Board, except that any person who on the date on which this law went into effect was serving as a legally appointed Electrical Inspector of _____ shall, upon application and payment of the prescribed fee and without examination, be issued a special certificate permitting him or her to continue to serve as an Electrical Inspector in the same territory.

(B) Experience. Electrical inspector applicants shall demonstrate the following:

(1) Have a demonstrated knowledge of the standard materials and methods used in the installation of electric equipment

(2) Be well versed in the approved methods of construction for safety to persons and property

(3) Be well versed in the statutes of _____ relating to electrical work and the *National Electrical Code*, as approved by the American National Standards Institute

(4) Have had at least _____ years' experience as an Electrical Inspector or _____ years in the installation of electrical equipment. In lieu of such experience, the applicant shall be a graduate in electrical engineering or of a similar curriculum of a college or university considered by the Board as having suitable requirements for graduation and shall have had two years' practical electrical experience.

(C) Recertification. Electrical inspectors shall be recertified as established by provisions of the applicable certification program.

(D) Revocation and Suspension of Authority. The Board shall have the authority to revoke an inspector's authority to conduct inspections within a jurisdiction.

80.29 Liability for Damages. Article 80 shall not be construed to affect the responsibility or liability of any party owning, designing, operating, controlling, or installing any electrical equipment for damages to persons or property caused by a defect therein, nor shall the _____ or any of its employees be held as assuming any such liability by reason of the inspection, reinspection, or other examination authorized.

80.31 Validity. If any section, subsection, sentence, clause, or phrase of Article 80 is for any reason held to be unconstitutional, such decision shall not affect the validity of the remaining portions of Article 80.

80.33 Repeal of Conflicting Acts. All acts or parts of acts in conflict with the provisions of Article 80 are hereby repealed.

80.35 Effective Date. Article 80 shall take effect _____ (_____) days after its passage and publication.

Informative Annex I Recommended Tightening Torque Tables from UL Standard 486A-B

This informative annex is not a part of the requirements of this NFPA document, but is included for informational purposes only.

In the absence of connector or equipment manufacturer's recommended torque values, Table I.1, Table I.2, and Table I.3 may be used to correctly tighten screw-type connections for power and lighting circuits*. Control and signal circuits may require different torque values, and the manufacturer should be contacted for guidance.

*For proper termination of conductors, it is very important that field connections be properly tightened. In the absence of manufacturer's instructions on the equipment, the torque values given in these tables are recommended. Because it is normal for some relaxation to occur in service, checking torque values sometime after installation is not a reliable means of determining the values of torque applied at installation.

Table I.1 Tightening Torque for Screws

Test Conductor Installed in Connector		Tightening Torque, N-m (lbf-in.)							
		Slotted head No. 10 and larger*							
		Slot width 1.2 mm (0.047 in.) or less and slot length 6.4 mm (¼ in.) or less		Slot width over 1.2 mm (0.047 in.) or slot length over 6.4 mm (¼ in.)		Split-bolt connectors		Other connectors	
AWG or kcmil	mm²								
30–10	0.05–5.3	2.3	(20)	4.0	(35)	9.0	(80)	8.5	(75)
8	8.4	2.8	(25)	4.5	(40)	9.0	(80)	8.5	(75)
6–4	13.2–21.2	4.0	(35)	5.1	(45)	18.5	(165)	12.4	(110)
3	26.7	4.0	(35)	5.6	(50)	31.1	(275)	16.9	(150)
2	33.6	4.5	(40)	5.6	(50)	31.1	(275)	16.9	(150)
1	42.4	—		5.6	(50)	31.1	(275)	16.9	(150)
1/0–2/0	53.5–67.4	—		5.6	(50)	43.5	(385)	20.3	(180)
3/0–4/0	85.0–107.2	—		5.6	(50)	56.5	(500)	28.2	(250)
250–350	127–177	—		5.6	(50)	73.4	(650)	36.7	(325)
400	203	—		5.6	(50)	93.2	(825)	36.7	(325)
500	253	—		5.6	(50)	93.2	(825)	42.4	(375)
600–750	304–380	—		5.6	(50)	113.0	(1000)	42.4	(375)
800–1000	405–508	—		5.6	(50)	124.3	(1100)	56.5	(500)
1250–2000	635–1010	—		—		124.3	(1100)	67.8	(600)

*For values of slot width or length not corresponding to those specified, select the largest torque value associated with the conductor size. Slot width is the nominal design value. Slot length shall be measured at the bottom of the slot.

Table I.2 Tightening Torque for Slotted Head Screws Smaller Than No. 10 Intended for Use with 8 AWG (8.4 mm²) or Smaller Conductors

Slot Length of Screw[a]		Tightening Torque, N-m (lbf-in.)	
mm	in.	Slot width of screw smaller than 1.2 mm (0.047 in.)[b]	Slot width of screw 1.2 mm (0.047 in.) and larger[b]
Less than 4	Less than ⁵⁄₃₂	0.79 (7)	1.0 (9)
4	⁵⁄₃₂	0.79 (7)	1.4 (12)
4.8	³⁄₁₆	0.79 (7)	1.4 (12)
5.5	⁷⁄₃₂	0.79 (7)	1.4 (12)
6.4	¼	1.0 (9)	1.4 (12)
7.1	⁹⁄₃₂		1.7 (15)
Above 7.1	Above ⁹⁄₃₂		2.3 (20)

[a]For slot lengths of intermediate values, select torques pertaining to next shorter slot lengths. Also, see 9.1.9.6 of UL 486A-2003, *Wire Connectors and Soldering Lugs for Use with Copper Conductors,* for screws with multiple tightening means. Slot length shall be measured at the bottom of the slot.
[b]Slot width is the nominal design value.

Table I.3 Tightening Torque for Screws with Recessed Allen or Square Drives

Socket Width Across Flats[a]		Tightening Torque, N-m (lbf-in.)	
mm	in.		
3.2	1/8	5.1	(45)
4.0	5/32	11.3	(100)
4.8	3/16	13.5	(120)
5.5	7/32	16.9	(150)
6.4	1/4	22.5	(200)
7.9	5/16	31.1	(275)
9.5	3/8	42.4	(375)
12.7	1/2	56.5	(500)
14.3	9/16	67.8	(600)

[a]See 9.1.9.6 of UL 486A-2003, *Wire Connectors and Soldering Lugs for Use with Copper Conductors*, for screws with multiple tightening means.

With the permission of Underwriters Laboratories Inc., material is reproduced from UL 486A-486B-2013, *Wire Connectors*, which is copyrighted by Underwriters Laboratories Inc., Northbrook, Illinois. While use of this material has been authorized, UL shall not be responsible for the manner in which the information is presented, nor for any interpretations thereof. For more information on UL, or to purchase standards, please visit their website at www.comm-2000.com or call 1-888-853-3503.

Informative Annex J   ADA Standards for Accessible Design

This informative annex is not a part of the requirements of this NFPA document, but is included for informational purposes only.

The provisions cited in Informative Annex J are intended to assist the users of the *Code* in properly considering the various electrical design constraints of other building systems and are part of the 2010 ADA Standards for Accessible Design. They are the same provisions as those found in ANSI/ICC A117.1-2009, *Accessible and Usable Buildings and Facilities.*

J.1 Protruding Objects. Protruding objects shall comply with Section J.2.

J.2 Protrusion Limits. Objects with leading edges more than 685 mm (27 in.) and not more than 2030 mm (80 in.) above the finish floor or ground shall protrude a maximum of 100 mm (4 in.) horizontally into the circulation path. *(See Figure J.2.)*

Exception: Handrails shall be permitted to protrude 115 mm (4½ in.) maximum.

J.3 Post-Mounted Objects. Freestanding objects mounted on posts or pylons shall overhang circulation paths 305 mm (12 in.) maximum where located 685 mm (27 in.) minimum and 2030 mm (80 in.) maximum above the finish floor or ground. Where a sign or other obstruction is mounted between posts or pylons, and the clear distance between the posts or pylons is greater than 305 mm (12 in.), the lowest edge of such sign or obstruction shall be 685 mm (27 in.) maximum or 2030 mm (80 in.) minimum above the finish floor or ground. *(See Figure J.3.)*

Exception: The sloping portions of handrails serving stairs and ramps shall not be required to comply with Section J.3.

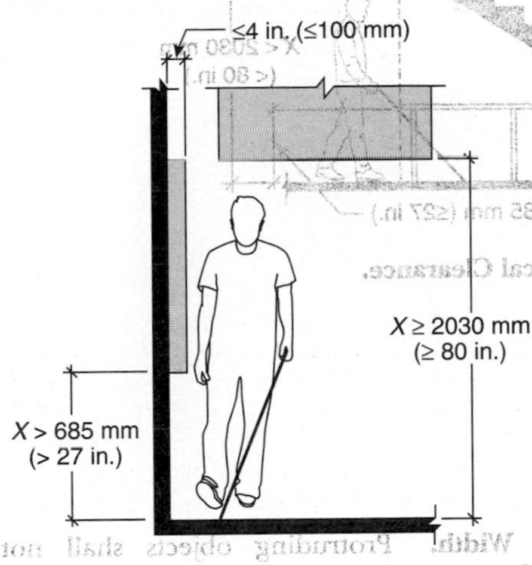

FIGURE J.2   Limits of Protruding Objects.

J.4 Vertical Clearance. Vertical clearance shall be 2030 mm (80 in.) high minimum. Guardrails or other barriers shall be provided where the vertical clearance is less than 2030 mm (80 in.) high. The leading edge of such guardrail or barrier shall be located 685 mm (27 in.) maximum above the finish floor or ground. *(See Figure J.4.)*

Exception: Door closers and door stops shall be permitted to be 1980 mm (78 in.) minimum above the finish floor or ground.

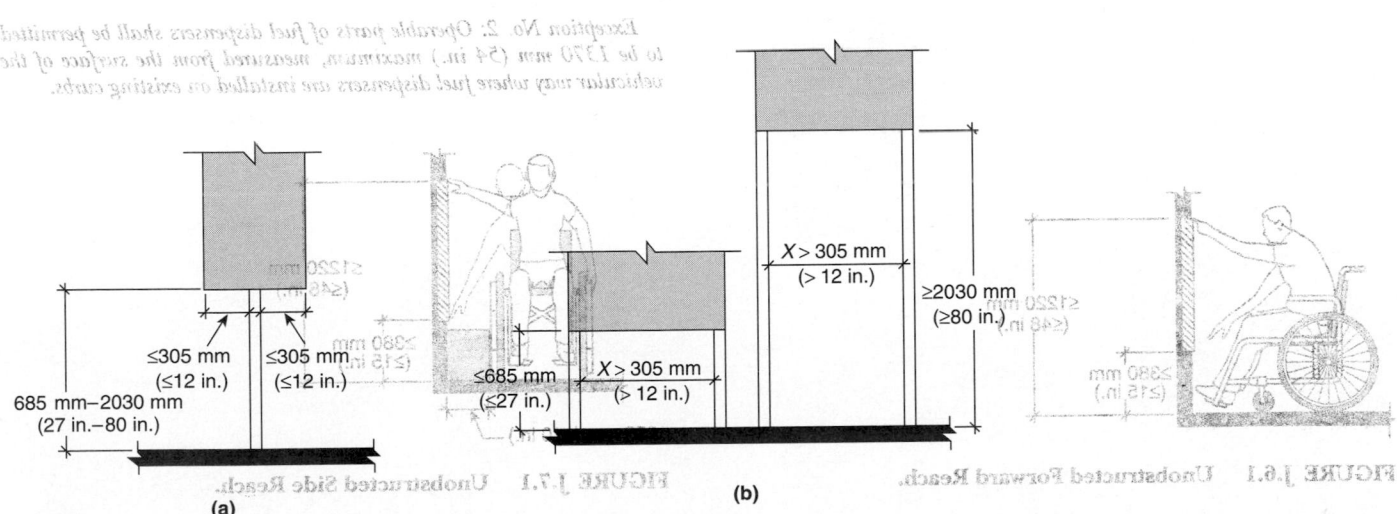

(a)   **(b)**

FIGURE J.3   Post-Mounted Protruding Objects.

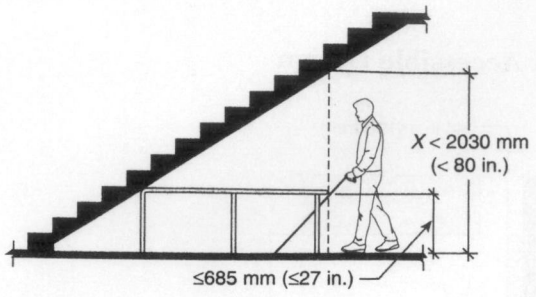

FIGURE J.4   Vertical Clearance.

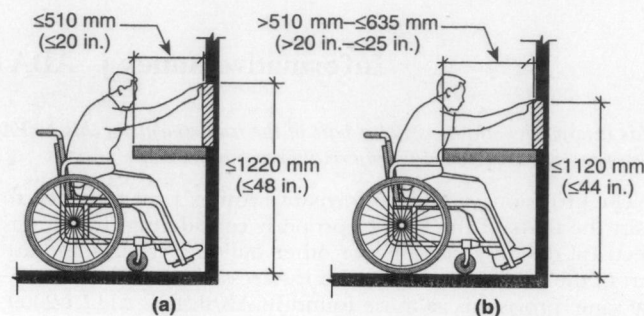

FIGURE J.6.2   Obstructed High Forward Reach.

J.5 Required Clear Width. Protruding objects shall not reduce the clear width required for accessible routes.

J.6 Forward Reach.

J.6.1 Unobstructed. Where a forward reach is unobstructed, the high forward reach shall be 1220 mm (48 in.) maximum, and the low forward reach shall be 380 mm (15 in.) minimum above the finish floor or ground. *(See Figure J.6.1.)*

J.6.2 Obstructed High Reach. Where a high forward reach is over an obstruction, the clear floor space shall extend beneath the element for a distance not less than the required reach depth over the obstruction. The high forward reach shall be 1220 mm (48 in.) maximum where the reach depth is 510 mm (20 in.) maximum. Where the reach depth exceeds 510 mm (20 in.), the high forward reach shall be 1120 mm (44 in.) maximum, and the reach depth shall be 635 mm (25 in.) maximum. *(See Figure J.6.2.)*

J.7 Side Reach.

J.7.1 Unobstructed. Where a clear floor or ground space allows a parallel approach to an element, and the side reach is unobstructed, the high side reach shall be 1220 mm (48 in.) maximum, and the low side reach shall be 380 mm (15 in.) minimum above the finish floor or ground. *(See Figure J.7.1.)*

Exception No. 1: An obstruction shall be permitted between the clear floor or ground space and the element where the depth of the obstruction is 255 mm (10 in.) maximum.

Exception No. 2: Operable parts of fuel dispensers shall be permitted to be 1370 mm (54 in.) maximum, measured from the surface of the vehicular way where fuel dispensers are installed on existing curbs.

J.7.2 Obstructed High Reach. Where a clear floor or ground space allows a parallel approach to an element and the high side reach is over an obstruction, the height of the obstruction shall be 865 mm (34 in.) maximum, and the depth of the obstruction shall be 610 mm (24 in.) maximum. The high side reach shall be 1220 mm (48 in.) maximum for a reach depth of 255 mm (10 in.) maximum. Where the reach depth exceeds 255 mm (10 in.), the high side reach shall be 1170 mm (46 in.) maximum for a reach depth of 610 mm (24 in.) maximum. *(See Figure J.7.2.)*

Exception No. 1: The top of washing machines and clothes dryers shall be permitted to be 915 mm (36 in.) maximum above the finish floor.

Exception No. 2: Operable parts of fuel dispensers shall be permitted to be 1370 mm (54 in.) maximum, measured from the surface of the vehicular way where fuel dispensers are installed on existing curbs.

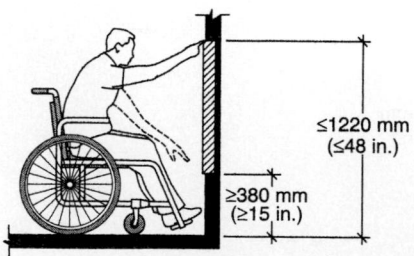

FIGURE J.6.1   Unobstructed Forward Reach.

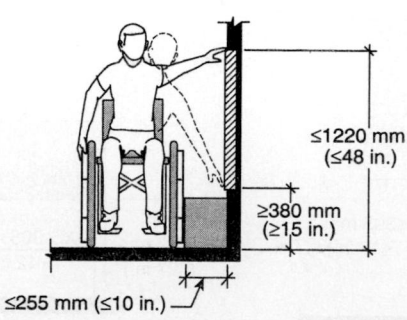

FIGURE J.7.1   Unobstructed Side Reach.

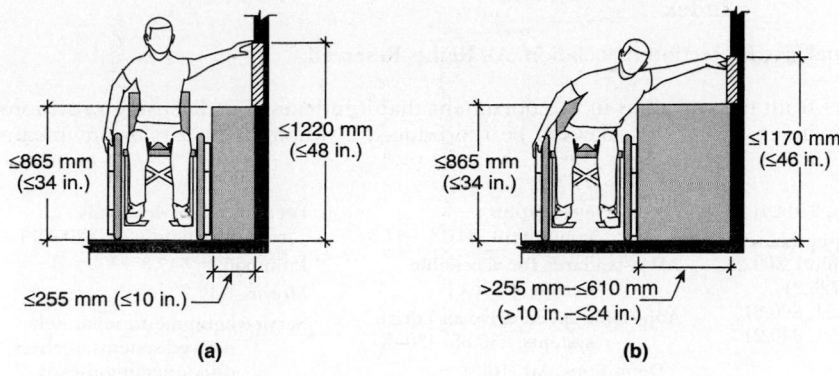

FIGURE J.7.2   Obstructed High Side Reach.

Index

Copyright © 2016 National Fire Protection Association. All Rights Reserved.

The copyright in this index is separate and distinct from the copyright in the document that it indexes. The licensing provisions set forth for the document are not applicable to this index. This index may not be reproduced in whole or in part by any means without the express written permission of NFPA.

Field-installed secondary
wiring, 600.12
Field-installed skeleton
tubing, 600–II
Applicability, neon secondary
circuit
conductors, 600.30
Grounding, 250.112(G),
600.7(A), 600.24(B),
600.33(D)
Listing, 600.3
Location, 600.9, 600.10(D),
600.21, 600.33(A),
Table 600.33(A)(1),
600.42(G), 600.42(H)
Markings, 600.4
Portable or mobile, 600.10
Section signs
Definition, 600.2
Solar photovoltaic powered
sign, 600.34
Definition, 600.2
Exit, health care
facilities, 517.33(B),
517.43(B)
Mandated standby, 701.7(A)
Outline lighting, Art. 600
Standby, 702.7
Warning, *see* Warning signs
(labels), at equipment
Simple apparatus, 504.4 Ex.
Definition, Art.100
Installation, 501.10(B)(3),
502.10(B)(3),
503.10(A)(4), 504.10,
505.15(C)(1)(7),
506.15(C)(7)
Marking, 500.8(C)(6),
505.9(C)(2) Ex. 2
Site-isolating devices, 547.9(A)
Definition, 547.2
Skeleton tubing, 600.4(F), 600.6,
600.7(A)(1), 600–II
Definition, 600.2
Slices and taps
Service-entrance
conductors, 230.46
**Smoke ventilator control,
stage**, 520.49
Snap switches
AC general use, 404.14
Accessibility, grouping, 404.8(B)
Electric-discharge
lighting, 410.138
Enclosures, 404.3(A) Ex. 1
General-use
Definition, Art. 100
Grounding, 404.9(B)
Motors, 430.83(C),
430.109(A)(3),
430.109(C), 430.109(G)

Mounting, 404.10
Multipole, 404.8(C)
Ratings, 404.14
Snow melting, *see* Fixed outdoor
electric deicing and
snow-melting equipment
Solar cell
Definition, 690.2
**Solar photovoltaic powered
signs**, 600.34
Definition, 600.2
Solar photovoltaic systems
AC modules, 690.6, 690.52
Definition, 690.2
Building integrated
photovoltaics
Definition, 690.2
Rapid shutdown, 690.12
Circuit sizing and current, 690.8
Connection to supply side of
service
disconnect, 230.82(6)
DC combiner, 600.34(A),
690.4(B), 690.15
Definition, 690.2
DC-to-DC converter, 690.4(B),
690.7(B)
Definition, 690.2
Definitions, Art. 100, 690.2,
691.2
Disconnecting means, 690–III,
690.56, 691.9
Ground-fault
protection, 690.41(B)
Large-scale, Art. 691
Marking, 690.13(B),
690.31(B)(1), 690.31(G)
Overcurrent protection, 690.9
Stand-alone systems,
690.8(A)(4), 690.10,
690.56(A)
Definition, 690.2
Underground feeder and
branch-circuit
cable, 340.10(5)
Voltage, 690.7
**Solderless (pressure)
connectors**, 250.8(3),
250.70
Definition, Art. 100
Solidly grounded, 430.83(E),
490.71, 712.30
Definition, Art. 100
Photovoltaic systems, 690.31(B),
690.31(I), 690.41,
690.42, 690.47
Solidly, 110.54(A)
Swimming pools, fountains, and
similar installations,
Table 680.9(A)

**Sound recording
equipment**, Art. 640
Audio signal processing,
amplification, and
reproduction
equipment, 540.50
Theaters, 520.4
Space
Cabinets and cutout
boxes, 312.7, 312.9,
312.11
Climbing space, line conductors
on poles, 225.14(D)
Lightning rods, conductor
enclosures,
equipment, 250.60,
250.106 IN No. 2
Outside branch circuits and
feeders, 225.14
Over 1000 volts,
separation, 110.32
through 110.34
Working, *see* Working space
Space heating, fixed, *see* Fixed
electric space-heating
equipment
**Spacing between bare metal
parts**, 408.56, Table
408.56
Spas and hot tubs, 680–IV
Definitions, 680.2
Indoor installations, 680.43
Outdoor installations, 680.42
Packaged equipment assembly
Definition, 680.2
Protection, 680.43, 680.43(A)(2)
Special permission
Definition, Art. 100
Spider (cable splicing block),
530.15(D)
Definition, 530.2
Splices and taps
Antennas, 810.14
Auxiliary gutters, 366.56
Cabinets and cutout boxes, 312.8
Cable trays, in, 392.56
Cellular concrete floor
raceways, 372.56
Cellular metal floor
raceways, 374.56
Concealed knob-and-
tube, 394.56
Conduit bodies, 300.15,
314.16(C)(2)
Construction sites, 590.4(G)
Deicing and snow-
melting, 426.24(B)
Electrical metallic tubing, 358.56
Electrical nonmetallic
tubing, 362.56

Equipment grounding
conductors, 250.122(G)
Fire resistive system, 728.5(H)
Flat cable assemblies, 322.56
Flat conductor cable, 324.56
Flexible cords and
cables, 400.13, 400.36,
530.12(A), 530.12(B),
590.4(G)
Flexible metal conduit, 348.56
Flexible metallic tubing, 360.56
General provisions, 110.14
Hazardous (classified) Class I
locations, 501.15(C)(4)
High density polyethylene
conduit, 353.56
Intermediate metal
conduit, 342.56
Liquidtight flexible metal
conduit, 350.56
Liquidtight flexible nonmetallic
conduit, 356.56
Low-voltage suspended ceiling
power distribution
systems, 393.56
Luminaires, 410.56(C),
410.56(D)
Messenger-supported
wiring, 396.56
Nonmetallic extensions, 382.56
Nonmetallic underground
conduit with
conductors, 354.56
Reinforced thermosetting resin
conduit, 355.56
Rigid metal conduit, 344.56
Rigid polyvinyl chloride
conduit, 352.56
Space-heating cables, 424.40,
424.41(D)
Strut-type channel
raceways, 384.56
Surface raceways, 386.56, 388.56
Underfloor raceways, 390.7
Underground, 300.5(E),
300.50(D)
Wireways, 376.56, 378.56
**Spray application, dipping,
coating, and printing
processes**, Art. 516
Classification of locations, 516.5,
516.18, 516.29
Definitions, 516.2
Equipment, 516.5(A)(3), 516.6,
516.7, 516.10, 516.38(B)
Grounding, 516.16
Membrane enclosures, 516–IV
Definition, 516.2
Open containers, 516–II
Printing, dipping, and coating
processes, 516–V

Electrical nonmetallic, *see*
Electrical nonmetallic
tubing
Flexible metallic, Art. 360 *see*
Flexible metallic tubing
Neon, 600.32, 600.41 *see also*
Skeleton tubing
Definition, 600.2

Tunnels
Access, 110.76
Installations over 1000 volts,
nominal, 110–IV
Ventilation, 110.57, 110.77,
110.78
TV, *see* Radio and television
equipment
Two-fer, 520.69
Definition, 520.2

-U-

Unclassified locations
Definition, Art. 100
Underfloor raceways, Art. 390
Conductors
Ampacity, 390.17
Number in raceway, 390.6
Size of, 390.5
Connections to cabinets, wall
outlets, 390.15
Covering, 390.4
Dead ends, 390.11
Definition, 390.2
Discontinued outlets, 390.8
Inserts, 390.14
Junction boxes, 390.13
Laid in straight lines, 390.9
Markers, 390.10
Splices and taps, junction
boxes, 390.7
Uses not permitted, 390.3(B)
Uses permitted, 390.3(A)
**Underground circuits,
communication**, 800.47
Underground enclosures, 110.59
see also Manholes; Vaults
**Underground feeder and branch-
circuit cable
(Type UF)**, Art. 340
Ampacity, 340.80
Bending radius, 340.24
Conductors, 340.104
Construction
specifications, 340–III
Definition, 340.2
Equipment grounding
conductor, 340.108
Installation, 340–II
Insulation, 340.112
Listing, 340.6
Sheath, 340.116

Support, 340.10(7)
Uses not permitted, 340.12
Uses permitted, 340.10
Underground installations, *see*
Manholes; Tunnels;
Vaults
**Underground service
conductors**, 300.5(D)(3),
Table 310.104(A)
Definition, Art. 100
**Underground service-entrance
cable (USE)**, *see* Service-
entrance cable
(Types SE and USE)
Underground wiring, 300.5 *see also*
Hazardous (classified)
locations
Aircraft hangars, 513.8
Ampacities, 310.60(C)(2), Fig.
310.60(C)(3), Tables
310.60(C)(77) through
310.60(C)(85)
Bulk storage plants, 515.8
Buried conductors, Types USE,
UF, 340.10
CATV coaxial cables, 820.47
Conductor types in
raceways, 310.10(F)
Dry and damp
locations, 310.10(B)
Earth movement and, 300.5(J)
Flexible metallic
tubing, 360.12(4)
High density polyethylene
conduit, 353.10(4)
Intermediate metal
conduit, 342.10(B) and
(C)
Liquidtight flexible metal
conduit, 350.10(3)
Liquidtight flexible nonmetallic
conduit, 356.10(4),
356.10(7)
Minimum cover
requirements, 300.5(A)
Motor fuel dispensing
facilities, 514.8
Optical fiber cables, 770.47
Over 1000 volts, 300.50
Protection of, 300.5(D), 300.5(J)
Raceways, service, 250.84
Recreational vehicle
parks, 551.80
Reinforcing thermosetting resin
conduit, 355.10(G)
Rigid metal conduit, 344.10
Rigid nonmetallic underground
conduit, Art. 354
Rigid polyvinyl chloride
conduit, 352.10(G)
Service cable, 250.84
Services, 230–III

"S" loops, 300.5(J) IN
Splices and taps, 300.5(E),
300.50(D)
Swimming pools, 680.11
Wet locations, 310.10(C)
Ungrounded, *see also* Conductors,
Ungrounded
DC system, 712.35, 712.55
Definition, 712.2
Definition, Art. 100
Separately derived
systems, 250.32(C)(2)
**Uninterruptible power supplies
(UPS)**, 645.11,
700.12(C), 701.12(C),
708.20(G)
Definition, Art. 100
**Unit equipment, emergency and
standby
systems**, 700.12(F),
701.12(G)
Unused openings
Boxes and fittings, 110.12(A)
Utility-interactive inverters, 200.3
Ex., 705.30(D) *see also*
Interactive inverters
Definition, Art. 100
Energy storage
systems, 706.23(B)(3)
Fuel cell systems, 692.65
Hybrid systems, 705.80
Utilization equipment
Aircraft hangars, portable
utilization equipment
in, 513.10(E)(2)
Boxes
Minimum depth of, 314.24
Outlet, 314.24(B), 314.27(D)
Branch circuits, permissible
loads, 210.23
Bulk storage plants, 515.7(C)
Definition, Art. 100

-V-

**Vacuum outlet assemblies,
central**, 422.15
**Valve actuator motor (VAM)
assemblies**, 430.102(A)
Ex. 3
Definition, 430.2
**Vapors, flammable liquid-
produced**, *see*
Hazardous (classified)
locations
Varying duty
Definition, Art. 100
Vaults, 110.71, 110.73, 110–V
Access, 110.76
Capacitors, 460.2(A)
Film storage, 530–V

Service over 1000 volts, 110.31,
230.212
Transformers, 230.6(3), 450–III
Ventilation, 110.77, 110.78
Vehicles, *see* Electric vehicles;
Recreational vehicles
Ventilated
Busway enclosures, 368.238
Cable trays, 392.22
Definition, Art. 100
Ventilating ducts, wiring, 300.21,
300.22
**Ventilating piping for motors,
etc.**, 502.128, 503.128
Ventilation
Aircraft hangars, 513.3(D)
Battery locations, 480.10(A)
Critical operations power
systems, 708.21
Electric vehicle charging system
equipment, 625.52,
Table 625.52(B)(1)(a),
Table 625.52(B)(1)(b)
Energy storage
systems, 706.10(A)
Equipment, general, 110.13(B)
Garages, commercial, 511.3(C)
through (E)
Manholes, tunnels, and
vaults, 110.57, 110.77,
110.78
Motor fuel dispensing facilities,
lubrication and service
rooms — without
dispensing, Table
514.3(B)(1), Table
514.3(B)(2)
Motors, 430.14(A), 430.16
Printing, dipping, and coating
processes, 516.36
Transformers, 450.9, 450.45
Vessels, *see also* Fixed electric
heating equipment for
pipelines and vessels
Definition, 427.2
**Viewing tables, motion
picture**, 530–IV
Volatile flammable liquid
Definition, Art. 100
Voltage and volts
Branch circuits, limits, 210.6
Circuit
Definition, Art. 100
Drop
Branch circuits, 210.19(A) IN
No. 4
Conductors, 310.15(A)(1) IN
No. 1
Sensitive electronic
equipment, 647.4(D)

Sequence of Events for the Standards Development Process

Once the current edition is published, a Standard is opened for Public Input.

Step 1 – Input Stage

- Input accepted from the public or other committees for consideration to develop the First Draft
- Technical Committee holds First Draft Meeting to revise Standard (23 weeks); Technical Committee(s) with Correlating Committee (10 weeks)
- Technical Committee ballots on First Draft (12 weeks); Technical Committee(s) with Correlating Committee (11 weeks)
- Correlating Committee First Draft Meeting (9 weeks)
- Correlating Committee ballots on First Draft (5 weeks)
- First Draft Report posted on the document information page

Step 2 – Comment Stage

- Public Comments accepted on First Draft (10 weeks) following posting of First Draft Report
- If Standard does not receive Public Comments and the Technical Committee chooses not to hold a Second Draft meeting, the Standard becomes a Consent Standard and is sent directly to the Standards Council for issuance (see Step 4) or
- Technical Committee holds Second Draft Meeting (21 weeks); Technical Committee(s) with Correlating Committee (7 weeks)
- Technical Committee ballots on Second Draft (11 weeks); Technical Committee(s) with Correlating Committee (10 weeks)
- Correlating Committee Second Draft Meeting (9 weeks)
- Correlating Committee ballots on Second Draft (8 weeks)
- Second Draft Report posted on the document information page

Step 3 – NFPA Technical Meeting

- Notice of Intent to Make a Motion (NITMAM) accepted (5 weeks) following the posting of Second Draft Report
- NITMAMs are reviewed and valid motions are certified by the Motions Committee for presentation at the NFPA Technical Meeting
- NFPA membership meets each June at the NFPA Technical Meeting to act on Standards with "Certified Amending Motions" (certified NITMAMs)
- Committee(s) vote on any successful amendments to the Technical Committee Reports made by the NFPA membership at the NFPA Technical Meeting

Step 4 – Council Appeals and Issuance of Standard

- Notification of intent to file an appeal to the Standards Council on Technical Meeting action must be filed within 20 days of the NFPA Technical Meeting
- Standards Council decides, based on all evidence, whether to issue the standard or to take other action

Notes:

1. Time periods are approximate; refer to published schedules for actual dates.
2. Annual revision cycle documents with certified amending motions take approximately 101 weeks to complete.
3. Fall revision cycle documents receiving certified amending motions take approximately 141 weeks to complete.

Committee Membership Classifications[1,2,3,4]

The following classifications apply to Committee members and represent their principal interest in the activity of the Committee.

1. M *Manufacturer:* A representative of a maker or marketer of a product, assembly, or system, or portion thereof, that is affected by the standard.
2. U *User:* A representative of an entity that is subject to the provisions of the standard or that voluntarily uses the standard.
3. IM *Installer/Maintainer:* A representative of an entity that is in the business of installing or maintaining a product, assembly, or system affected by the standard.
4. L *Labor:* A labor representative or employee concerned with safety in the workplace.
5. RT *Applied Research/Testing Laboratory:* A representative of an independent testing laboratory or independent applied research organization that promulgates and/or enforces standards.
6. E *Enforcing Authority:* A representative of an agency or an organization that promulgates and/or enforces standards.
7. I *Insurance:* A representative of an insurance company, broker, agent, bureau, or inspection agency.
8. C *Consumer:* A person who is or represents the ultimate purchaser of a product, system, or service affected by the standard, but who is not included in (2).
9. SE *Special Expert:* A person not representing (1) through (8) and who has special expertise in the scope of the standard or portion thereof.

NOTE 1: "Standard" connotes code, standard, recommended practice, or guide.

NOTE 2: A representative includes an employee.

NOTE 3: While these classifications will be used by the Standards Council to achieve a balance for Technical Committees, the Standards Council may determine that new classifications of member or unique interests need representation in order to foster the best possible Committee deliberations on any project. In this connection, the Standards Council may make such appointments as it deems appropriate in the public interest, such as the classification of "Utilities" in the National Electrical Code Committee.

NOTE 4: Representatives of subsidiaries of any group are generally considered to have the same classification as the parent organization.

Submitting Public Input / Public Comment Through the Online Submission System

Soon after the current edition is published, a Standard is open for Public Input.

Before accessing the Online Submission System, you must first sign in at www.nfpa.org. *Note: You will be asked to sign-in or create a free online account with NFPA before using this system:*

a. Click on Sign In at the upper right side of the page.
b. Under the Codes and Standards heading, click on the "List of NFPA Codes & Standards," and then select your document from the list or use one of the search features.

OR

a. Go directly to your specific document information page by typing the convenient shortcut link of www.nfpa.org/document# (Example: NFPA 921 would be www.nfpa.org/921). Sign in at the upper right side of the page.

To begin your Public Input, select the link "The next edition of this standard is now open for Public Input" located on the About tab, Current & Prior Editions tab, and the Next Edition tab. Alternatively, the Next Edition tab includes a link to Submit Public Input online.

At this point, the NFPA Standards Development Site will open showing details for the document you have selected. This "Document Home" page site includes an explanatory introduction, information on the current document phase and closing date, a left-hand navigation panel that includes useful links, a document Table of Contents, and icons at the top you can click for Help when using the site. The Help icons and navigation panel will be visible except when you are actually in the process of creating a Public Input.

Once the First Draft Report becomes available there is a Public Comment period during which anyone may submit a Public Comment on the First Draft. Any objections or further related changes to the content of the First Draft must be submitted at the Comment stage.

To submit a Public Comment you may access the online submission system utilizing the same steps as previously explained for the submission of Public Input.

For further information on submitting public input and public comments, go to: http://www.nfpa.org/publicinput.

Other Resources Available on the Document Information Pages

About tab: View general document and subject-related information.

Current & Prior Editions tab: Research current and previous edition information on a Standard.

Next Edition tab: Follow the committee's progress in the processing of a Standard in its next revision cycle.

Technical Committee tab: View current committee member rosters or apply to a committee.

Technical Questions tab: For members and Public Sector Officials/AHJs to submit questions about codes and standards to NFPA staff. Our Technical Questions Service provides a convenient way to receive timely and consistent technical assistance when you need to know more about NFPA codes and standards relevant to your work. Responses are provided by NFPA staff on an informal basis.

Products & Training tab: List of NFPA's publications and training available for purchase.

Information on the NFPA Standards Development Process

I. Applicable Regulations. The primary rules governing the processing of NFPA standards (codes, standards, recommended practices, and guides) are the NFPA *Regulations Governing the Development of NFPA Standards (Regs)*. Other applicable rules include NFPA *Bylaws*, NFPA *Technical Meeting Convention Rules*, NFPA *Guide for the Conduct of Participants in the NFPA Standards Development Process*, and the NFPA *Regulations Governing Petitions to the Board of Directors from Decisions of the Standards Council*. Most of these rules and regulations are contained in the *NFPA Standards Directory*. For copies of the *Directory*, contact Codes and Standards Administration at NFPA Headquarters; all these documents are also available on the NFPA website at "www.nfpa.org."

The following is general information on the NFPA process. All participants, however, should refer to the actual rules and regulations for a full understanding of this process and for the criteria that govern participation.

II. Technical Committee Report. The Technical Committee Report is defined as "the Report of the responsible Committee(s), in accordance with the Regulations, in preparation of a new or revised NFPA Standard." The Technical Committee Report is in two parts and consists of the First Draft Report and the Second Draft Report. (See *Regs* at Section 1.4.)

III. Step 1: First Draft Report. The First Draft Report is defined as "Part one of the Technical Committee Report, which documents the Input Stage." The First Draft Report consists of the First Draft, Public Input, Committee Input, Committee and Correlating Committee Statements, Correlating Notes, and Ballot Statements. (See *Regs* at 4.2.5.2 and Section 4.3.) Any objection to an action in the First Draft Report must be raised through the filing of an appropriate Comment for consideration in the Second Draft Report or the objection will be considered resolved. [See *Regs* at 4.3.1(b).]

IV. Step 2: Second Draft Report. The Second Draft Report is defined as "Part two of the Technical Committee Report, which documents the Comment Stage." The Second Draft Report consists of the Second Draft, Public Comments with corresponding Committee Actions and Committee Statements, Correlating Notes and their respective Committee Statements, Committee Comments, Correlating Revisions, and Ballot Statements. (See *Regs* at 4.2.5.2 and Section 4.4.) The First Draft Report and the Second Draft Report together constitute the Technical Committee Report. Any outstanding objection following the Second Draft Report must be raised through an appropriate Amending Motion at the NFPA Technical Meeting or the objection will be considered resolved. [See *Regs* at 4.4.1(b).]

V. Step 3a: Action at NFPA Technical Meeting. Following the publication of the Second Draft Report, there is a period during which those wishing to make proper Amending Motions on the Technical Committee Reports must signal their intention by submitting a Notice of Intent to Make a Motion (NITMAM). (See *Regs* at 4.5.2.) Standards that receive notice of proper Amending Motions (Certified Amending Motions) will be presented for action at the annual June NFPA Technical Meeting. At the meeting, the NFPA membership can consider and act on these Certified Amending Motions as well as Follow-up Amending Motions, that is, motions that become necessary as a result of a previous successful Amending Motion. (See 4.5.3.2 through 4.5.3.6 and Table 1, Columns 1-3 of *Regs* for a summary of the available Amending Motions and who may make them.) Any outstanding objection following action at an NFPA Technical Meeting (and any further Technical Committee consideration following successful Amending Motions, see *Regs* at 4.5.3.7 through 4.6.5.3) must be raised through an appeal to the Standards Council or it will be considered to be resolved.

VI. Step 3b: Documents Forwarded Directly to the Council. Where no NITMAM is received and certified in accordance with the Technical Meeting Convention Rules, the standard is forwarded directly to the Standards Council for action on issuance. Objections are deemed to be resolved for these documents. (See *Regs* at 4.5.2.5.)

VII. Step 4a: Council Appeals. Anyone can appeal to the Standards Council concerning procedural or substantive matters related to the development, content, or issuance of any document of the NFPA or on matters within the purview of the authority of the Council, as established by the Bylaws and as determined by the Board of Directors. Such appeals must be in written form and filed with the Secretary of the Standards Council (see *Regs* at Section 1.6). Time constraints for filing an appeal must be in accordance with 1.6.2 of the *Regs*. Objections are deemed to be resolved if not pursued at this level.

VIII. Step 4b: Document Issuance. The Standards Council is the issuer of all documents (see Article 8 of *Bylaws*). The Council acts on the issuance of a document presented for action at an NFPA Technical Meeting within 75 days from the date of the recommendation from the NFPA Technical Meeting, unless this period is extended by the Council (see *Regs* at 4.7.2). For documents forwarded directly to the Standards Council, the Council acts on the issuance of the document at its next scheduled meeting, or at such other meeting as the Council may determine (see *Regs* at 4.5.2.5 and 4.7.4).

IX. Petitions to the Board of Directors. The Standards Council has been delegated the responsibility for the administration of the codes and standards development process and the issuance of documents. However, where extraordinary circumstances requiring the intervention of the Board of Directors exist, the Board of Directors may take any action necessary to fulfill its obligations to preserve the integrity of the codes and standards development process and to protect the interests of the NFPA. The rules for petitioning the Board of Directors can be found in the *Regulations Governing Petitions to the Board of Directors from Decisions of the Standards Council* and in Section 1.7 of the *Regs*.

X. For More Information. The program for the NFPA Technical Meeting (as well as the NFPA website as information becomes available) should be consulted for the date on which each report scheduled for consideration at the meeting will be presented. To view the First Draft Report and Second Draft Report as well as information on NFPA rules and for up-to-date information on schedules and deadlines for processing NFPA documents, check the NFPA website (www.nfpa.org/docinfo) or contact NFPA Codes & Standards Administration at (617) 984-7246.

An online community that connects you with peers worldwide and directly with NFPA staff

Have a question about the code or standard you're reading now? NFPA Xchange™ can help!

NFPA Xchange™ is our free online community, and we want to invite you to join us there! NFPA Xchange lets you connect with professionals worldwide, explore content, share ideas, and ask questions, as the latest way to stay up-to-date on codes- and standards-related information. Plus you can:

- ❯ **ASK** your peers
- ❯ **SEARCH** questions & answers
- ❯ **READ** NFPA blogs

NFPA Members also enjoy exclusive access to the 'Members Only' section on NFPA Xchange, where you can submit technical standards questions* directly to NFPA staff or search questions that have already been submitted by others!

Join NFPA Xchange TODAY – IT'S FREE AND EASY!

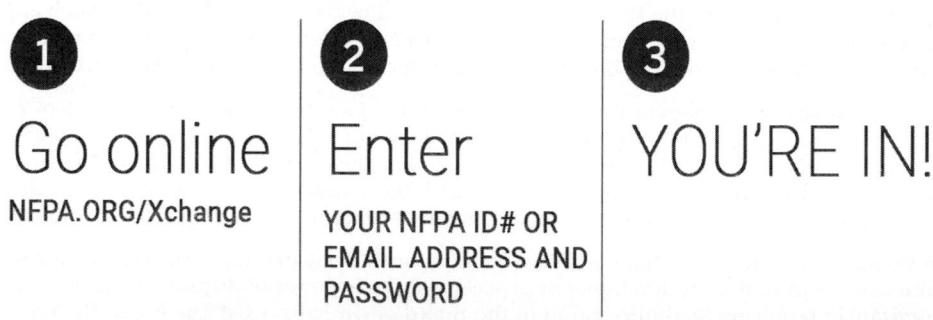

1 Go online
NFPA.ORG/Xchange

2 Enter
YOUR NFPA ID# OR EMAIL ADDRESS AND PASSWORD

3 YOU'RE IN!

Search. Share. Xchange Ideas. Get Involved Today!

*For the full terms of use, please visit nfpa.org/standard_items/terms-of-use#xchange. NFPA® is a registered trademark of the National Fire Protection Association, Quincy, MA 02169.